P9-DVG-186

VOLUME ONE

MEDICAL PHYSIOLOGY

VOLUME ONE

MEDICAL PHYSIOLOGY

Edited by

VERNON B. MOUNTCASTLE, M.D.

Professor and Director, Department of Physiology
The Johns Hopkins University
Baltimore, Maryland

with 2133 illustrations

Thirteenth edition

THE C. V. MOSBY COMPANY

Saint Louis / 1974

THIRTEENTH EDITION

Previous editions copyrighted 1918, 1919, 1920, 1922, 1926, 1930, 1935, 1938, 1941, 1956, 1961, 1968

Printed in the United States of America

Distributed in Great Britain by Henry Kimpton, London

Library of Congress Cataloging in Publication Data

Mountcastle, Vernon B
Medical physiology.

Includes bibliographies.
1. Physiology. 2. Physiology, Pathological.
I. Title. [DNLM: 1. Physiology. QT104 M928m 1974]
QP34.5.M76 1974 612 73-14503
ISBN 0-8016-3550-0

CB/CB/B 9 8 7 6 5 4 3 2

CONTRIBUTORS

G. D. AURBACH
National Institutes of Health
Bethesda, Maryland

ROSS J. BALDESSARINI
Harvard University
Boston, Massachusetts

PHILIP BARD
The Johns Hopkins University
Baltimore, Maryland

LLOYD M. BEIDLER
The Florida State University
Tallahassee, Florida

F. J. BRINLEY, Jr.
The Johns Hopkins University
Baltimore, Maryland

JOHN R. BROBECK
University of Pennsylvania
Philadelphia, Pennsylvania

CHANDLER McC. BROOKS
State University of New York
Brooklyn, New York

KENNETH T. BROWN
University of California
San Francisco, California

C. LOCKARD CONLEY
The Johns Hopkins University
Baltimore, Maryland

ROBERT D. DeVOE
The Johns Hopkins University
Baltimore, Maryland

MOÏSE H. GOLDSTEIN, Jr.
The Johns Hopkins University
Baltimore, Maryland

H. MAURICE GOODMAN
University of Massachusetts
Worcester, Massachusetts

CARL W. GOTTSCHALK
The University of North Carolina
Chapel Hill, North Carolina

JAMES D. HARDY
John B. Pierce Foundation Laboratory
New Haven, Connecticut

THOMAS R. HENDRIX
The Johns Hopkins University
Baltimore, Maryland

ELWOOD HENNEMAN
Harvard University
Boston, Massachusetts

JAMES HOUK
The Johns Hopkins University
Baltimore, Maryland

KIYOMI KOIZUMI
State University of New York
Brooklyn, New York

CHRISTIAN J. LAMBERTSEN
University of Pennsylvania
Philadelphia, Pennsylvania

WILLIAM E. LASSITER
The University of North Carolina
Chapel Hill, North Carolina

JANICE W. MARAN
Stanford University
Stanford, California

THOMAS H. MAREN
University of Florida
Gainesville, Florida

DONALD J. MARSH
University of Southern California
Los Angeles, California

JEAN M. MARSHALL
Brown University
Providence, Rhode Island

WILLIAM R. MILNOR
The Johns Hopkins University
Baltimore, Maryland

VERNON B. MOUNTCASTLE
The Johns Hopkins University
Baltimore, Maryland

WILLIAM L. NASTUK
Columbia University
New York, New York

JAMES M. PHANG
National Institutes of Health
Bethesda, Maryland

GIAN F. POGGIO
The Johns Hopkins University
Baltimore, Maryland

SID ROBINSON
Indiana University
Bloomington, Indiana

J. A. J. STOLWIJK
John B. Pierce Foundation Laboratory
New Haven, Connecticut

KAREL V. S. TOLL
Boston University
Boston, Massachusetts

LESTER VAN MIDDLESWORTH
The University of Tennessee
Memphis, Tennessee

GERHARD WERNER
University of Pittsburgh
Pittsburgh, Pennsylvania

GERALD WESTHEIMER
University of California
Berkeley, California

F. EUGENE YATES
University of Southern California
Los Angeles, California

LAURENCE R. YOUNG
Massachusetts Institute of Technology
Cambridge, Massachusetts

KENNETH L. ZIERLER
The Johns Hopkins University
Baltimore, Maryland

PREFACE

The general principles on which this textbook is organized are those described in the preface to its twelfth edition. An extensive revision of that book has been made for the present edition. Completely new sections appear on endocrinology, renal physiology and the regulation of the internal milieu, gastrointestinal physiology, and the autonomic nervous system. Chapters on new subjects appear in several other sections as well. This edition thus contains twenty-eight chapters that are wholly new. The elimination and combination of others has reduced to seventy-three the eighty chapters of the twelfth edition, printed in approximately the same space. Each of the remaining forty-five chapters has been substantially revised and in many cases rewritten to take into account advances in the relevant fields of physiology since 1968.

This edition has been written by thirty-eight authors, of whom seventeen have joined this endeavor for the first time. Each is busily engaged in the work of research and teaching in physiology; each has taken time from that dedicated life to summarize here the state of knowledge in his particular field of interest. Whatever value the result of our common effort may possess is wholly due to their depth of understanding and skill in exposition and their devotion to the task. For this I am greatly indebted to each.

For them and for myself I wish to thank those authors and publishers who have allowed us to reproduce illustrations previously published elsewhere.

Vernon B. Mountcastle

PREFACE
TO TWELFTH EDITION

The twelfth edition of *Medical Physiology* presents a cross section of knowledge of the physiologic sciences, as viewed by a group of thirty-one individuals, twenty-three of whom are actively engaged in physiologic research and teaching. Each section of the book provides statements of the central core of information in a particular field of physiology, reflecting, by virtue of the daily occupations of its authors, the questioning and explorative attitude of the investigator and indeed some of the excitement of the search. These statements vary along a continuum from those with a high probability for continuing certainty to those that are speculative but, it is hoped, of heuristic value. An attempt has been made to maintain a balanced point of view. I hope this book will convey to the student who reads it the fact that physiology is a living and changing science, continuously perfecting its basic propositions and laws in the light of new discoveries that permit new conceptual advances. The student should retain for himself a questioning attitude toward all, for commonly the most important advances are made when young investigators doubt those statements others have come to regard as absolutely true. This is not a book that sets forth in stately order a series of facts which, if learned, will be considered adequate for success in a course in physiology. Many such "facts" are likely to be obsolete before the student of physiology reaches the research laboratory, or the student of medicine the bedside. Nor is it a book that provides ready-made correlations and integrations of the various fields of physiology necessary for a comprehensive understanding of bodily function. Those integrations are an essential part of scholarly endeavor not readily gained from books alone. It is my hope, however, that study of this book, combined with laboratory experience and scholarly reflection, will provide the student with a method and an attitude that will serve him long after the concepts presented here are replaced by new and more cogent ones.

The title *Medical Physiology* has been retained, for one of the purposes of this edition, in common with earlier ones, is "to present that part of physiology which is of special concern to the medical student, the practitioner of medicine, and the medical scientist in terms of the experimental inquiries that have led to our present state of knowledge." The scope of the book was and is still broader, however, and attempts to present mammalian physiology as an independent biologic discipline as well as a basic medical science. Mammalian physiology has its base in cellular physiology and biophysics, and it is from this point of view that many of the subjects treated here are approached. Above all, mammalian physiology must deal with problems of the interactions between large populations of cells, organs, and organ systems and, finally, the integrated function of an entire animal. Physiology thus must bridge the distance from cellular biology on the one hand to systems analysis and control theory on the other: each is important and any one is incomplete without the others. This approach to the problems of internal homeostasis, of reaction to the environment, and of action upon the environment is evidenced in several sections of this book.

Of the eighty chapters composing this book, twenty-nine are wholly new in this edition; forty-five from the last edition have been extensively revised either by their original authors or by new ones. Six have been allowed to stand substantially as previously written, for these seemed to comprise as balanced and modern a survey as any presently possible. The names and affiliations of my colleagues in this effort have been listed. They have taken time from busy lives to survey their fields of interest; for this I am greatly indebted to each. If this book possesses any worth it is in large part due to their continuing devotion to the task of its preparation.

For them and for myself I wish to thank those authors and publishers who have allowed us to reproduce illustrations previously published elsewhere.

Vernon B. Mountcastle
Baltimore, Maryland

CONTENTS

VOLUME ONE

PART I CELLULAR PHYSIOLOGY

PART II INTERACTIONS BETWEEN EXCITABLE TISSUES

PART III GENERAL PHYSIOLOGY OF THE FOREBRAIN

PART IV CENTRAL NERVOUS MECHANISMS IN SENSATION

PART V NEURAL CONTROL OF MOVEMENT AND POSTURE

PART VI THE AUTONOMIC NERVOUS SYSTEM, HYPOTHALAMUS, AND INTEGRATION OF BODY FUNCTIONS

VOLUME TWO

PART VII THE CIRCULATION

PART XI RESPIRATION

PART XII ENDOCRINE GLANDS

I

CELLULAR PHYSIOLOGY

1

ROBERT D. DeVOE

Principles of cell homeostasis

Cells are semiautonomous units of tissue; isolated cells can survive for long periods of time in tissue culture media that mimic their normal environments. The internal environments of cells are very different from the fluids around them, however, and there are constant exchanges of metabolites, waste products, and other substances between a cell and its environment. For this molecular traffic to be possible, the cell cannot wall itself off altogether. On the other hand, it must have some barrier between its different internal and external environments, simply in order to survive. The problem all cells face, therefore, is how to surround themselves with barriers (cell membranes) that allow desired substances to pass in and out while maintaining their own internal constancies. The maintenance of this constancy is what is meant by cell homeostasis.

The beginning point here is therefore this boundary between a cell and its environment. A "typical" cell is depicted in Fig. 1-1, *A*, but for our purposes it may be simplified to the hollow shell depicted in Fig. 1-1, *B*. This shell consists of a uniform cell membrane that surrounds a fluid of one composition and is itself surrounded by a fluid of a different composition. By way of illustration, the compositions of intra- and extracellular fluids are given for a number of cells in Table 1-1 (with some other quantities that will be explained later). Proceeding from values given in this table, and by reference to Fig. 1-1, *B*, the basic principles of cell homeostasis that will be developed in this chapter are sixfold.

First, water is in general in osmotic equilibrium across cell membranes and easily passes back and forth across them. Second, large internal organic molecules, both charged and uncharged, are retained within the cell. They are designated collectively in Fig. 1-1, *B*, as P^-. Third, the presence of osmotically active organic molecules held within the cell by the membrane must be balanced by the external presence of some substance(s) impeded by the membrane from entering the cell. If this were not the case, the cell could not be in osmotic equilibrium. By and large, this external substance is sodium, shown in Fig. 1-1, *B*, in large concentration outside and in low concentration inside. Fourth, the charge on the internal organic ions is, in the aggregate, negative, and some cation must be present to give electroneutrality within the cell. As sodium is impeded by the membrane from entering but potassium is less impeded, the predominant intracellular cation tends to be potassium. This is shown in Fig. 1-1, *B*, as a large internal potassium concentration

Table 1-1. Some representative values for intracellular and extracellular ionic concentrations (in mM/L cell water or extracellular volume), equilibrium potentials, and resting potentials

	Squid giant axon	Frog sartorius muscle	Human red blood cell
Intracellular concentrations			
Na	65.0	13.0	19.0
K	344.0	138.0	136.0
Mg	10.0	14.0	5.5
Ca	3.5	5.0	0.0
Cl	80.0	3.0	78.0
Extracellular concentrations*			
Na	460.0	110.0	155.0
K	10.0	2.5	5.0
Mg	53.0	1.0	2.2
Ca	10.0	2.0	5.0
Cl	540.0	90.0	112.0
SO_4	25.0	2.0	1.0
Equilibrium potentials			
E_{Na}	+49	+55	+55
E_K	−89	−101	−86
E_{Cl}	−48	−86	−9
Resting potentials	−77	−99	−6 to −10

*Values for squid giant axon are concentrations in seawater; values for frog sartorius muscle and human red blood cell are concentrations in plasma.

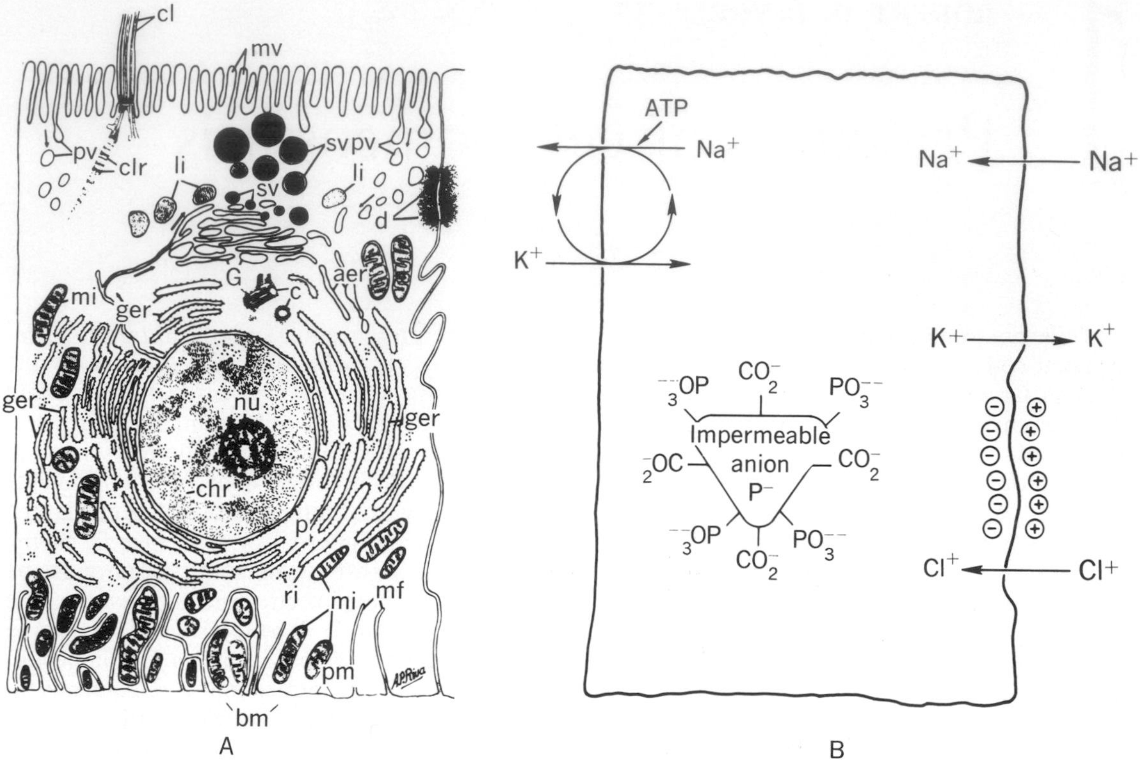

Fig. 1-1. A, Diagram of ultrastructure of ideal animal cell. (See De Robertis et al.[2a] for key to details.) **B,** "Hollow-shell" depiction of plasma membrane around cell, ionic movements and electric potential across membrane, and relative concentrations of substances on either side (shown by relative sizes of lettering.) (**A** From De Robertis et al.[2a])

and a low external concentration. Fifth, there is a negative potential difference between the inside and the outside of the cell. This membrane potential, as it is called, is due to the tendency of a potassium ion, especially, to equalize its internal and external concentrations by diffusing out of the cell, thus upsetting electroneutrality across the membrane. In any event, the effect of the negative membrane potential is to exclude negative ions, particularly chloride, from the cell interior, while accelerating the slow tendency of sodium to equalize *its* internal and external concentrations by diffusing into the cell. Sixth, and finally, in order to prevent even a slow net inward movement of sodium, which would upset the osmotic equilibrium, cells utilize metabolic energy to transport, or pump, the excess sodium out of the cell. In many instances there is a linked inward movement of potassium ions when this is necessary to maintain the internal potassium concentration.

It can thus be seen that there are three points at which cells can control the states they will achieve in homeostasis by means of metabolic, synthetic, or other activity. These are (1) the permeabilities of their membranes to water, ions, and nonelectrolytes, (2) the osmolarities and amounts of charge of internal organic molecules, and (3) the rate of ion transport. Even in a given cell one or more of these may be a variable, depending on physiologic activity. Thus nerve cells, which like other cells have low membrane permeabilities to sodium, transiently increase this permeability during the generation of action potentials. Cells such as those in kidney collecting tubules have water permeabilities that are under endocrine control. Oxidation and reduction of hemoglobin in erythrocytes, which occur as these cells transport oxygen and carbon dioxide to and from tissues, respectively, involve changes in internal anionic charge and hence in internal ionic distributions. Rates of ion transport may depend on internal ion concentrations or, in some epithelial cells, on hormones. It should thus be understood that there may be many different combinations of membrane permeabilities, internal anions, and ion transport mechanisms that cells use both to maintain themselves and to carry out their physiologic functions.

Whatever the particular combination that a given cell has evolved, it will result in homeostasis, a steady-state condition in which there are no *net* molecular movements of consequence into and out of the cell. There will be molecular traffic across the membrane, but the movement of osmotically active particles into the cell must be matched by an equal outward movement if the cell is to stay in osmotic equilibrium. In general, since cell membranes show selectivity toward each chemical species, this means that the inward movement of a given substance must be matched by an equal outward movement, unless it is consumed. Again, since the major osmotically active substances in body fluids are the three principal ions sodium, chloride, and potassium, the matched movements in question will be of these ions. Movement of any substance across a membrane per unit of time is called its *flux,* and the flux per 1 cm^2 unit area has the dimensions of moles per second per square centimeter. For a steady state, then, the outflux of each ion (and water) must be of the same size but opposite in sign (i.e., direction) to the influx; the net flux, their sum, must be zero. This is indicated in Fig. 1-1, *B,* when the sums of arrows showing sizes of inward and outward movements are zero for all substances.

To complete this initial picture of cell homeostasis, recall that the *causes* of inward and outward movements of substances can be both the tendencies of these substances to distribute themselves according to concentration and electrical differences across the cell membrane as well as the use of metabolic energy by the cell to transport substances. The term "passive fluxes" is used to describe movements of substances due to kinetic forces, i.e., concentration gradients, electric potential gradients (in the case of ions), and other gradients such as pressure and temperature. When influxes and outfluxes of a substance are both solely passive and equal to each other, that substance is in equilibrium across the membrane. As stated earlier, water is thought to be in equilibrium across cell membranes,[76] and in resting muscle in situ, potassium and chloride may likewise be in equilibrium.[5] However, there is in addition an *active transport* of ions such as sodium across cell membranes that results from the expenditure of metabolic energy by the cell. Intuitively it can be seen that if sodium ions are transported out of a cell interior the negative organic anions are left behind, and so there will be movements of the other ions to reestablish electroneutrality. Thus active transport of only one ion out of a cell, with given amounts of internal indiffusible anion and with given membrane permeabilities to the various ions, can result in extensive redistributions of other ions as well. In general, the process of this redistribution cannot be observed; what is usually seen is the final result. Nonetheless, cell homeostasis may be approached as the study of what active fluxes there are and hence what the requisite concentration and electric gradients there must be in order to set up, across the membrane in question, the passive fluxes that will just balance the active fluxes. Once this is done, the distributions of ions (and water) can be determined from the necessity for electrical neutrality and for osmotic equilibrium, i.e., for the distribution that will bring these other substances into equilibrium.

To proceed further, it is necessary to consider first the structure of membranes and how, to the extent known, they exert their selective effects upon the movements of water, ions, and nonelectrolytes into and out of cells. Second, considerable emphasis will given to the manner in which ions and water move under the concentration and electric gradients established by ion pumps. Initially, emphasis will be placed on the homeostasis of individual cells—their ionic and water distributions and how these affect membrane potentials and ionic equilibriums. Finally, the properties of the active transport mechanisms will be discussed, since these are the key to understanding the ultimate ionic and water distributions that are seen in cells.

COMPOSITION AND STRUCTURE OF CELL MEMBRANES

Controversy surrounds the exact arrangements of molecules in cell membranes,[4] and cell membranes undeniably differ from one cell type to another. However, it is agreed that the following characteristics apply to all membranes:

1. Cell membranes are thin, on the order of 70 to 100 Å.
2. Cell membranes are electric insulators, or dielectrics, with capacitances of about 1 $\mu F/cm^2$.
3. Cell membranes contain proteins and various mixtures of lipids.

4. Water, small molecules, and lipid-soluble substances pass easily through cell membranes; charged molecules and large molecules do not.
5. Portions of membranes are specialized for transport, ion excitation and conduction, metabolism, etc.

Thickness of membranes

Before the advent of electron microscopy, cell membranes could not be seen, as they are below the limits of resolution of light microscopy. Their presence was inferred from a large variety of osmotic, electrical, and cytosurgical experiments.[52, 79] In particular, electrical measurements of conductivities of cells indicated that cell membranes might be very thin. Beginning in the 1920s with studies of packed suspensions of erythrocytes and continuing in the 1930s with studies of giant axons of squid and with frog muscles, it was found that cells had highly conducting interiors but were bounded by poorly conducting layers with capacitances on the order of 1 μF/cm^2 (Table 1-2).[2] The capacitance (C) of any insulating layer is related to its thickness (d) as follows:

$$C = \frac{\epsilon A}{4\pi d} \quad (1)$$

where A is the area of the layer and ε is its dielectric constant. The problem in using such a relation to determine the thickness of the insulating membrane around cells is identifying the dielectric constant of this layer. By the beginning of this century, it was well known that lipid-soluble substances penetrated easily into cells,[52] and in the 1920s it was argued that this was because the membrane itself was made up of lipids[39] (see below). Thus, if the membrane were taken to have a dielectric constant typical of a lipid, such as 3, the membrane thickness would be estimated at approximately 33 Å. Although such estimates of membrane thicknesses were off by a factor of 2 or 3, they indicated both that the membrane thickness was of the order of the length of a single lipid molecule and that such high capacitances, far more efficient than man-made capacitors per unit area, were indicative of a high-degree of molecular organization.

Table 1-2. Comparison of some properties of artificial lipid bilayer membranes with biologic membranes*

Property	Biologic membranes	Artificial membranes
Electron microscope image	Trilaminar	Trilaminar
Thickness (Å)	60-100	40-90
Capacitance (μF/cm^2)	0.5-1.3	0.38-1.0
Resistance (ohm-cm^2)	10^2-10^5	10^6-10^9
Dielectric breakdown (mv)	100-200	150-200
Surface tension (dynes/cm)	0.03-1.0	0.5-2.0
Water permeability (μ/sec)	0.37-400	31.7

*Adapted from Henn and Thompson.[42]

The advent of electron microscopy in the 1950s first allowed the visualization of cell membranes in stained sections as very thin lines at the outer edges of cells. As Robertson[74] pointed out, it was by no means certain originally which or how many of the electron-dense lines in electron micrographs corresponded to the physiologic membrane. The clue came from studies of myelin formation. Nerve axons, both myelinated and unmyelinated, are nearly surrounded by the cytoplasm of Schwann cells, but in the case of the myelinated axons, the myelin is found to be made up of layer upon layer of Schwann cell membrane wrapped around the axon.[36] Robertson found that the *unit membrane* of which myelin is built up is a trilaminar structure consisting of two dense lines each 25 Å thick and separated by a clear space 25 Å thick. The attachments of the Schwann cell unit membranes to each other in myelin or in their initial infolding, the mesaxon, could be made to come unstuck or be made to adhere more tightly by the use of distilled water or hypertonic solutions, respectively, but the dimensions of the unit membrane remained constant.

Fig. 1-2 illustrates the electron microscopic appearance of a myelinated axon, *D;* the two Schwann cell membranes infolding to form myelin, *E;* and myelin fixed with osmium, *A,* and permanganate, *B* and *C.* One of the chief points is that when the inner surfaces of Schwann cell membranes meet (arrow in *E*), a darker fusion line results than when the outer membrane surfaces meet, as though there were differently staining substances lining the inner as opposed to the outer surfaces of these membranes. This is not unlikely, since many cells have "extraneous coats" on their cell membranes, consisting in the case of the ameba, for example, of mucopolysaccharide.[15] These coats in other cells may be the antigens for immunologic

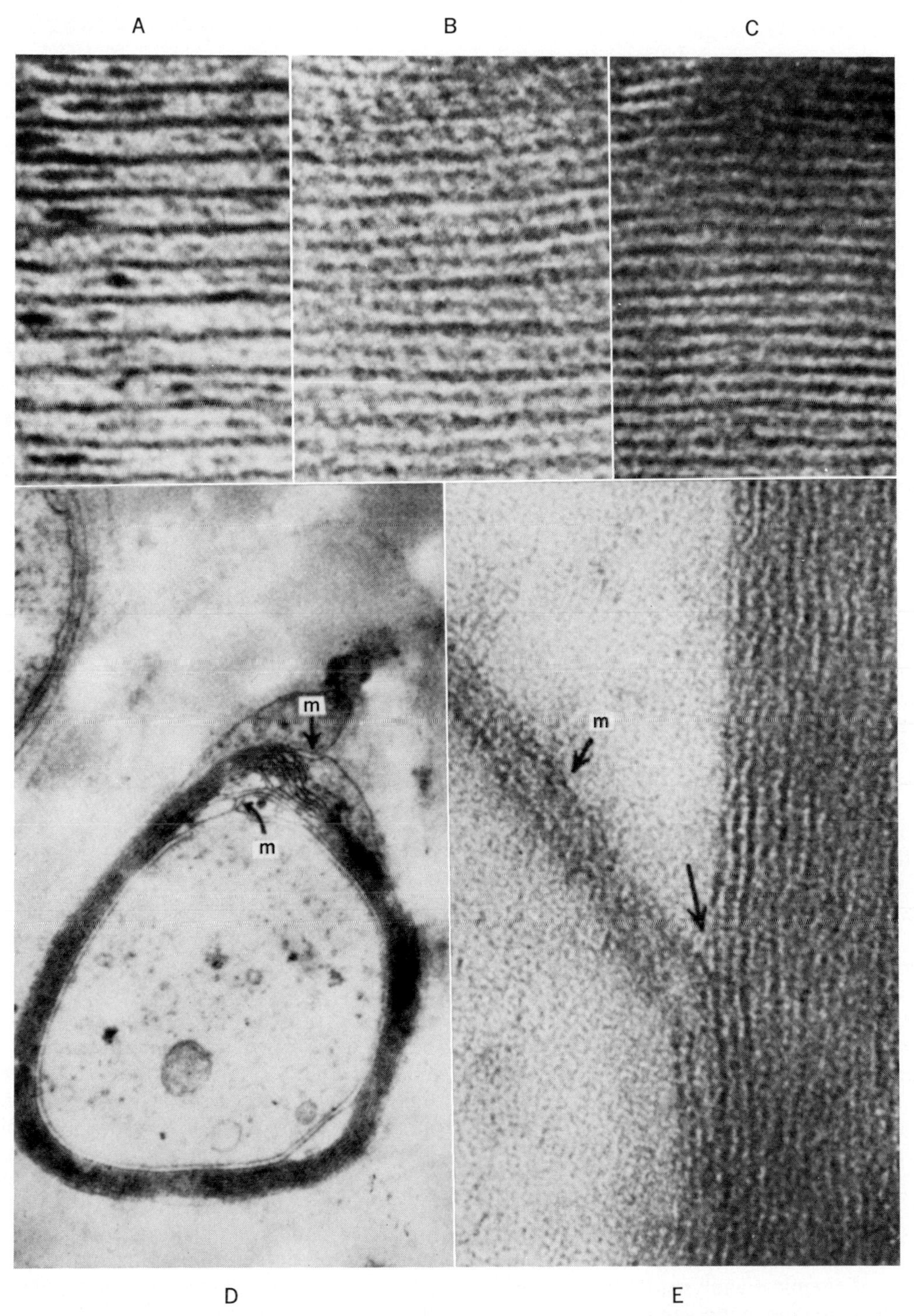

Fig. 1-2. Appearance of myelin in electron micrographs. **A** to **C** show that myelin has major repeat structure of 100 to 120 Å and a minor repeat of 50 to 60 Å. **D** and **E** show formation of mesaxon, *m*, by infolding and joining of plasma membrane. (From Robertson.[74])

reactions.[21] From such considerations, Robertson[74] supposed that in Schwann cell membranes, too, there might be polysaccharide on the outside and probably protein on the inside.

Support for the unit membrane hypothesis has come from low-angle x-ray diffraction experiments,[30] from which it has been found that there is a major repeat structure of 180 Å and a half repeat of 90 Å in fresh myelin. Treatment with the fixatives used for electron microscopy reduced the major repeat to 150 Å, and the major repeat seen in electron micrographs was 120 Å.[74] The major repeats seen in the x-ray diffraction patterns presumably are twice the thickness of the unit membrane and correspond to the thick (or thin) lines seen in Fig. 1-2, *A* to *C*. The half repeat of 90 Å corresponds then to the thickness (in fresh myelin) of the unit membrane itself. Thus the x-ray diffraction data support the observation made in electron micrographs, and both depend fundamentally on the derivation of the myelin from and its similarity to the membranes of the Schwann cells.

However, unit membranes are by no means restricted to Schwann cells; indeed, the term derives from the appearance of most internal and external cell membranes as trilamellar in electron micrographs (some exceptions will be cited later). The molecular organization that might underlie such an appearance had been first proposed in 1925 by Gorter and Grendel,[39] who extracted the lipids from erythrocytes and spread them out on an air-water interface (a Langmuir trough) to determine the area they would occupy as a monomolecular film when compressed. They used erythrocytes from dogs, sheep, goats, rabbits, guinea pigs, and man and in all cells calculated that the surface areas of extracted lipids were twice as great as the surface areas of the cells. They concluded from this that erythrocytes "are covered by a layer of fatty substances that is two molecules thick."[39] Later and quite independently, Danielli and Davson[25] proposed their now classic "paucimolecular" model of the cell membrane from entirely different considerations. They supposed, first, that since lipid-soluble materials passed through membranes easily, the membrane was most likely lipid itself; second, on the basis of chemical analysis, that these lipids had both polar and nonpolar regions[55]; third, on the basis of membrane capacitances, that the lipids might be arranged in as few as one

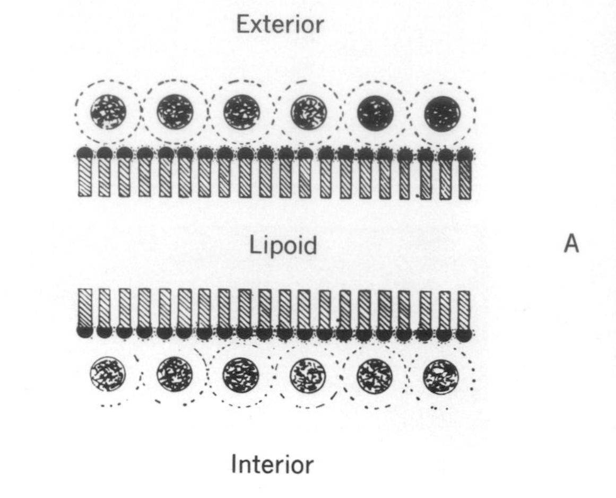

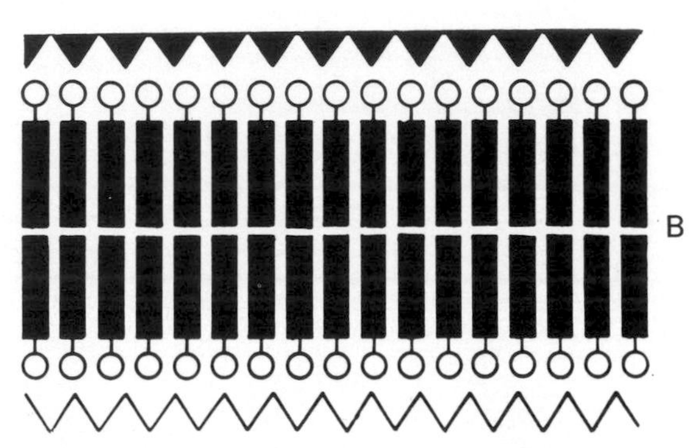

Fig. 1-3. A, Original paucimolar membrane model of Danielli and Davson. **B,** Diagram of Robertson's unit membrane model. (**A** from Danielli and Davson[25]; **B** from Robertson.[75])

to as many as three layers; and finally, because cell surface tensions were as low as 0.03 to 1.0 dyne/cm, compared to surface tensions as great as 9.0 dyne/cm for oil droplets, that the lipid layers were covered with surface-active proteins. Since the proteins, to have been surface active, would have had to be polar, Danielli and Davson suggested that the polar proteins were situated next to the polar ends of the lipid molecules at the outsides of the membrane, while the nonpolar lipid regions were associated together in the interior of what was thus at least a bimolecular lipid leaflet. This model is shown in Fig. 1-3, *A,* while the model in Fig. 1-3, *B,* shows Robertson's modification with different substances on the two membrane faces, in light of the differences in the electron micrographic appearance of Schwann cell membranes when they fuse to form myelin. In this view, then, the trilamellar appearance of membranes in electron micrographs is a direct consequence

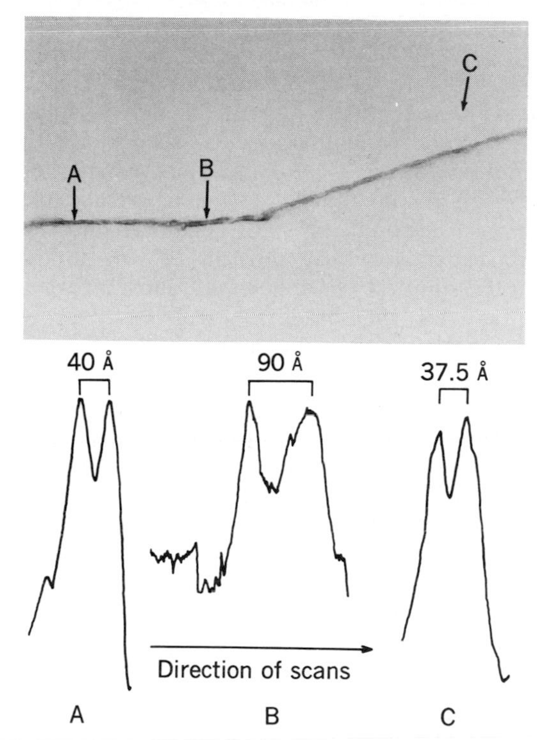

Fig. 1-4. Top: Electron micrograph of transverse section of artificial lipid bilayer membrane. Microdensitometer tracings shown at bottom were taken at points *A, B,* and *C*. Peak-to-peak distances between dark lines are given above each tracing. (From Henn et al.[42a])

of their bimolecular structure: the electron-dense material is deposited on the outside faces, whereas the nonpolar interiors do not stain and are seen as clear layers.

In 1962 Mueller et al.[66] reported the fabrication of artificial lipid bilayers from mixtures of polar lipids and organic solvents, and it became possible to test the properties of such lipid bilayers for comparison with natural cell membranes. The extensive work in this field has been reviewed by Tien and Diana[81] and by Henn and Thompson.[42] In Table 1-2 some of the comparisons between natural and such artificial bilayer membranes are listed. It is possible to show by optical means that these artificial membranes are about 72 Å, or only two molecular layers thick. Under the electron microscope such membranes also have a trilamellar appearance, as shown in Fig. 1-4. From the low surface tensions of these purely lipid membranes, it must be concluded that for energetic stability the polar groups (rather than the nonpolar groups) are in contact with water at the membranes' outer faces.[42] Thus these artificial membranes have the lipid arrangement postulated to underlie the paucimolecular membrane. However, their surface tensions are already as low as those of cell membranes, so that the original assumption of Danielli and Davson[25] that the low surface tensions of cell membranes are due to outer protein layers is unnecessary. Moreover, some of the other properties of naked lipid bilayers, such as their capacitances, dielectric breakdowns, and water permeabilities (to be discussed in the next section), are also very much like those of cell membranes. On the other hand, resistances of artificial membranes are much higher, which is to say that their permeabilities to ions as well as to other polar molecules are much lower than those of biologic membranes. Artificial membranes may be adulterated with various substances that, like valinomycin, selectively increase the permeability of these membranes to potassium ions or that, like cholesterol, appear to diminish their permeability to water.[31] The role of "adulterants" in biologic membranes is undoubtedly played by proteins, among other substances, since there are many enzymes and enzymelike processes, associated with metabolic reactions at, and transport across, membranes.

The disposition of proteins in the cell membrane is an altogether different matter. The original Danielli and Davson concept[25] that layers of protein are spread across both membrane faces is not only unnecessary to account for the low surface tensions of membranes but also may be impossible. Korn[55] has claimed that there is not enough protein to cover both faces of the membrane. He has argued further that the trilamellar appearance of membranes in electron micrographs gives no confirmation of a lipid bilayer structure, since such fixatives as osmium could be expected to react primarily with unsaturated bonds in the nonpolar interior of the membrane. In fact, there is no deposition of the electron-dense osmium in this region at all. As an alternative to the Danielli-Davson model of the membrane, he advocates one in which the membrane consists of a protein matrix upon which lipid is inserted, rather than the other way around. Other alternative models of membranes have also been proposed. Examples are the micellar model (Lucy[63]) and the repeating-unit membrane for mitochondria (Fernandez-Moran[29]). It is uncertain, however, whether any of these alternative membrane models mimics the biologic membranes as effectively as the bimolecular lipid membrane in, for example, capacitance or water permeability. Moreover, that there might be too little protein to fully cover a bimolecular lipid leaflet is of little consequence; as has been pointed out, this part of the paucimolecular membrane model is no longer needed, while Robertson[74] has emphasized that asymmetries in appearances of the membranes in myelin might indicate that there may be different substances on the two sides of the membrane anyway. Another objection has been that there is too little lipid in erythrocyte membranes, for example, to cover them with a bilayer,[72] but redeterminations of the areas covered by lipid in red blood cell membranes[12] show that at physiologic surface tensions the lipid easily spreads out to cover these cells as bilayers.

All in all, the concept that biologic membranes are built on the common substratum of a bimolecular lipid leaflet, with various substances on various sides and with specializations due to proteins at various sites, seems to be the closest that one can come to an understanding of membrane structure at the present time. Indeed, most exceptions to this concept are largely based on physiologic rather than anatomic grounds, and it must be said that just as the original concept of the paucimolecular membrane arose from functional considerations, so too are the various undoubted specializations of the cell membrane defined, in most cases, solely on the basis of physiologic evidence. Some of the various specializations of membranes will be described in the remainder of this chapter, while others will be presented elsewhere in the book.

DIFFUSION AND PERMEATION ACROSS CELL MEMBRANES

The most important physiologic attributes of membranes are the ways in which substances can permeate, or pass through, them. The permeabilities of membranes describe their selectivities toward substances on the insides and on the outsides of cells and thus reflect the degree to which cell integrity depends on keeping these substances in or out or letting them pass back and forth. Metabolites must enter, waste products must leave, but large osmotic shifts of water that could lead to lysis must be prevented.

The starting point for a study of cell permeabilities is the diffusion of substances in free solution. Permeation can be looked on as diffusion up to a membrane in a water phase, diffusion through the lipid matrix of the membrane, and diffusion away from the membrane on the other side. Permeation thus involves more than the mere passage through the membrane, and sometimes this passage is not the limiting factor, as, for example, when there are only convoluted, restricted extracellular spaces around cells through which substances must diffuse to reach the membrane. In the simplest cases, however, substances will cross cell membranes, in the absence of metabolic activity, under the driving force of concentration gradients, or more strictly, under the driving force of gradients in their chemical potentials (Appendix A, p. 28). The physical basis for such movements is the empiric first law of Fick, which relates the number of moles (dS) of a substance that will diffuse through a given area (A) in a given time (dt) under a concentration gradient (dC/dx):

$$\text{Flux} = \frac{dS}{dt} = -DA\frac{dC}{dx} \qquad (2)$$

Fick's law has the same form as Ohm's law, for in both a flow (flux, current) is related to a driving force (concentration gradient, potential difference) through a proportionality factor characteristic of the system: the diffusion constant (D) for diffusion, 1/resistance for Ohm's law. The dimensions of the diffusion constant can be found from the following:

$$\text{Moles/sec} = \frac{\text{cm}^2 \cdot \text{moles/cm}^3 \cdot D}{\text{cm}}$$

or

$$D = \text{moles/sec} \cdot \frac{\text{cm}^3\ \text{cm}}{\text{cm}^2\ \text{moles}} = \text{cm}^2\text{/sec}$$

When it comes to cell membranes, however, neither the thickness (d) nor the concentration gradient (dC/dx) is generally known, although the thickness may be approximated as 75 to 100 Å. On the other hand, the concentrations in the solutions at either side of the membrane can be measured. Since the membrane, of uncertain thickness (d), is the boundary between the two solutions, the entire concentration difference must be developed across this thickness:

$$dS/dt = -DA(C_i - C_o)/d = -(D/d)A(C_i - C_o) \qquad (3)$$

where:

C_o = External concentration

C_i = Intracellular concentration

The ratio of the diffusion constant (D) to the unknown thickness (d) is called the *permeability constant* (P) and has the dimensions of a velocity (centimeters per second). It defines quantitatively the ease with which a given substance can permeate a given cell membrane, and it is a property of the membrane.

The direct application of equation 3 to cell membranes is, however, beset with difficulties. First of all, concentrations of permeating substances at the cell membrane may not be the same as in the applied external solutions or as measured by analyses of total cell contents. Substances inside cells may be bound or compartmentalized; tortuous paths through connective tissue or within greatly restricted extracellular spaces may impede diffusion from the outside to the membrane. Second, some metabolites such as sugars and amino acids may cross membranes by means of specialized carriers or transport sites and exhibit saturating kinetics, unlike the simple permeation described by equation 3. This sort of permeation has been called "facilitated diffusion"[7] and is often linked to the passive inward movements of sodium ions down their concentration gradients (the so called sodium gradient hypothesis).[77] Third, movements of ions are influenced by differences in electric potentials that exist across membranes (discussed later), and the permeabilities of excitable membranes such as those of nerves and muscles depend on these membrane potentials (Chapter 2). Finally, some of the older measurements of permeabilities[26] were made by studying bulk entry of permeants into cells, with accompanying osmotic movements of water. When water enters simultaneously with permeant, it may drag with it more permeant that would move solely as a result of the concentration difference across the membrane. Such *solvent drag* and related interactions between solvent and solute, which are treated by the irreversible thermodynamics,[7] render some of these older permeability measurements invalid. However, it is possible to apply equation 3 directly to inextensible cells (such as those of plants) that do not swell and admit water upon bulk entry of permeant or when isotopic tracers are used.

Valid permeability measurements indicate that the smaller the permeating molecule and the more lipid soluble it is, the more likely it is to cross membranes. Small molecules such as water, methanol, and dissolved gases cross membranes with great ease (the molecular sieve effect), as do lipid-soluble molecules such as ether. Larger size need not result in lower permeability if the larger molecule has more nonpolar groups. In the plant cell *Chara,* ethylene glycol has a lower permeability than the larger propylene glycol, which is less polar because of its extra $-CH_2$ group.[26] Stein[7] has proposed a "lattice" model for permeation in an attempt to take into account a permeant's ability to leave the hydrogen-bonded, latticelike structure of water and enter the lipid matrix of the membrane.

Movement of water across membranes

By assuming a given thickness for membranes (say, 100 Å), it is possible to convert permeabilities to diffusion constants and thus compare how well a substance diffuses in a membrane as compared to free water solution. On this basis, even such highly permeable substances as water and methanol are found to diffuse 3 to 6 orders of magnitude less rapidly than in free solution. (The self-diffusion constant of water in water is about 2.4×10^{-5} cm/sec; the "diffusion constant" in biologic membranes is 4 to $4{,}000 \times 10^{-11}$ cm/sec.) A further complication, however, is that water permeabilities, in particular, are more than twice as great under osmotic gradients as permeabilities measured using isotopic water.[3] Since movement under an osmotic gradient is formally equivalent to movement under a hydrostatic pressure gradient,[3] one explanation for these differences in water permeabilities is that osmotic water flows through pores in the same way that water flows through small tubes (Poiseuille's flow). If water also diffuses through these same pores and not through the bulk of the membrane, then the differences between diffusive and osmotic water flows across membranes can be explained.[7] However, an alternative view is that during diffusive water flows, *unstirred layers* near the membrane effectively impede movements of the isotopes, whereas during osmotic flows the bulk movements of water flush out such unstirred layers. When great care is taken to stir the water next to membranes, these differences in osmotic and diffusive water permeabilities disappear,[24, 31] as they do when diffusional permeabilities of the unstirred layers are measured and taken into account inside perfused cells of the giant marine alga *Valonia*.[40] All this implies that, in membranes, water molecules move independently of each other (as in diffusion) and not through pores. Indeed, it is possible to describe water movements in artificial membranes as proportional to the solubility of water in the lipids of the membrane, the rate of diffusion of water through the lipid, and the mole fractions of water at each interface (i.e., to the water concentration differences).[31] These water permeabilities of artifical membranes appear sufficient to explain the water permeabilities of biologic membranes, since they are both of the same order of magnitude (Table 1-2).

State of intracellular substances

The concentrations of intracellular substances that can be measured by chemical means do not necessarily correspond to the concentrations at the insides of cell membranes, and so it is important to consider in what state these intracellular constituents may be in terms of their osmotic properties, their binding to protoplasmic constituents, if any, and their abilities to move about by diffusion in comparison with free solution. For the study of the osmotic properties of the intracellular constituents, it may be assumed that these will not leave the cell if it shrinks or swells during movements of water. That is, the cell membrane may be considered to be semipermeable and the cell an osmometer. Then the number of internal osmoles (n_o) will remain constant (assuming no changes in dissociation of ions, for example) and equal to the cell's concentration times its volume, for any volume:

$$C_o \cdot V_o = n_o = C_1 \cdot V_1 = C_2 \cdot V_2, \text{ etc.} \qquad (4)$$

Since the osmotic pressure (π) is proportional to the concentration of a (completely dissociated) substance, equation 3 is a form of an osmotic Boyle's law:

$$C_oRT \cdot V_o = \pi_o \cdot V_o = \pi_1 \cdot V_1, \text{ etc.} \qquad (5)$$

However, it appears that equations 4 and 5 are not strictly followed by cells, although the deviations may be as small as 6%.[63] Rather, it appears that the entire volume of the cell is not osmotically active, the remainder being filled with protein, fats, glycogen, etc. Equation 4 may be modified to include an osmotically inactive volume (b):

$$\pi_o(V_o - b) = \pi_1(V_1 - b) = k$$

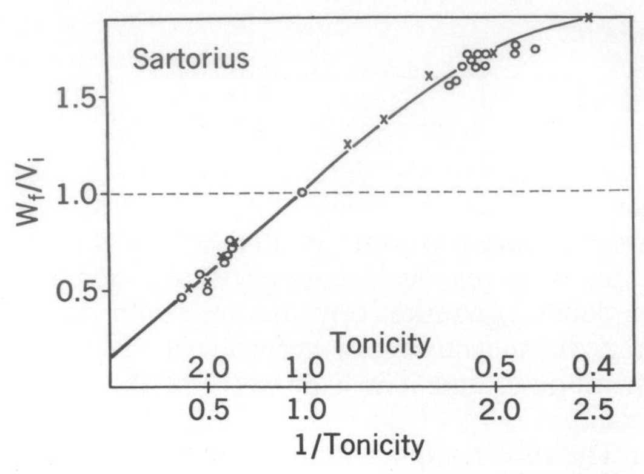

Fig. 1-5. Change in volume of fiber water of frog sartorius muscle with change in concentration of Ringer's solution, expressed as tonicity relative to normal Ringer's solution. Ordinate gives ratio of final volume of fiber water, W_f, to initial volume, V_i. (From Bozler.[14])

where k is a constant, or, from equation 4:

$$c(V - b) = k$$
$$V - b = k/C$$
$$V = k/C + b$$

Hence a plot of cell volume versus the reciprocal of the cell concentration should be a straight line with an intercept, for zero $1/C$, of b. So long as the cell is in osmotic equilibrium its internal concentration will be the same as the external concentration; so the reciprocal of the concentration of the bathing solution may be plotted versus cell volume to give b, the osmotically inactive volume. An example of such an experiment with frog muscle is illustrated in Fig. 1-5. In frog sartorius muscle the inactive volume is about 15%[14]; in red blood cells, 38%[58]; and in eggs of the sea urchin *Arbacia*, 12%.[62] The osmotically inactive volume does not appear to be made up solely of solids, however. In erythrocytes, water associated with hemoglobin in the inactive volume may be something like 7% of the total cell water,[58] while in barnacle muscle the inactive volume of 25% contains 25% of the total cell water.[45] This inactive water appears in nuclear magnetic resonance studies[22] to be highly structured; it may well be "bound" to the highly structured myoplasm and appears unavailable to act as a solvent for intracellular ions.[45]

Not only water but ions, too, may be bound intracellularly. Ling[59] originally proposed that the high intracellular potassium and low intracellular sodium concentrations found in muscle were due to preferential binding of potassium ions to cell proteins, excluding sodium ions by virtue of their larger hydrated radius (as inferred from their lower mobility in solution) and hence smaller charge density. However, it appears more likely that it is sodium that may be "bound." Measurements have been made of intracellular sodium and potassium activities with ion-selective glass microelectrodes in squid axoplasm,[43] barnacle muscle,[44] and frog skeletal muscle.[10] The potassium activities are what would be expected if the activity coefficients were the same intracellularly as in Ringer's solution, but the sodium activities are far lower, a condition that could result if some of the sodium were "bound." Estimates of bound sodium range from 25% in squid axoplasm,[43] 83% in barnacle muscle,[44] to "most" in frog muscle.[10] Similarly, "over half" of the sodium in nerve cell bodies of the puffer fish is estimated to be bound.[53] Binding of potassium is not excluded by any of these studies, but it is probably a small fraction (approximately 10% at most) of the total intracellular potassium.[10, 45] Where the binding takes place is a matter of some question; it may be on proteins in nerve or in the sarcoplasmic reticulum in muscle.

The remaining unbound ions appear to be able to diffuse nearly as readily within cells as in free solution. Potassium in squid axons[49] and water, urea, and glycerol in barnacle muscles[18] have diffusion constants that are the same as in free solution, while diffusion of potassium, sodium, sulfate, and ATP ions as well as of sorbital and sucrose in frog muscles is reduced by about one half.[57] These reduced diffusivities are presumably due to physical (tortuosity) rather than chemical (binding) restraints, since the ions and nonelectrolytes were similarly affected. In sum, therefore, it seems that intracellular substances may not always exist in the concentrations that can be measured by gross chemical means. However, those substances that are dissolved in the "free" cell water appear to move as though they were in free solution under, at most, physical restraints. From this it follows that the chemistry of solutions may indeed be applied to movements of substances within cells and across their membranes, provided only that notice is taken of the "free" concentration of each substance.

Diffusion potentials

Electrolytes diffuse in free solution and across cell membranes just as do nonelectrolytes. However, because of their polar nature, they do not cross lipid membranes with ease. They also move in solution under the influence of electric potentials, and indeed, their very diffusion may set up electric potentials. Such diffusion potentials can arise spontaneously (i.e., due to kinetic forces) whenever one ion species, for example, diffuses at a different rate than its counterion(s). An example of such diffusion would be at the junction between a salt bridge filled with a concentrated monovalent salt solution and a dilute bath of the same salt. The concentrated salt will tend to diffuse down its concentration gradient, according to Fick's law, into the more dilute salt across the liquid junction separating them. In general, one of the two ions (cation or anion) will diffuse faster, leaving its counterion behind. Such a charge separation immediately leads to an electric potential. If the faster ion is the cation, the solution into which it is diffusing becomes positive, while the solution it is leaving becomes negative. This potential decelerates the cation and accelerates the anion; both finally move at the same rate, and equal numbers of both ions cross the liquid junction per unit time. The ultimate liquid junction potential is whatever is required to make the two ions move at equal rates, which is to say that there must be a potential if they are to move at the same rate, and that if two ions are moving at the same rate, this does not mean that there can be no potential. The size of the potential is proportional to the difference between the rates at which each ion alone moves in free solution under a given force (i.e., proportional to the differences in ionic mobilities). The source of energy for the potential is, of course, the concentration gradient. If diffusion proceeds long enough to abolish the concentration gradient, the diffusion potential will likewise be abolished.

Membrane potentials

Concentration gradients of ions likewise exist across cell membranes. In general, none of the major extracellular ions (chloride, sodium, potassium) is at the same concentration on the two sides of cell membranes. As might be expected, therefore, there are diffusion potentials set up across cell membranes. The ultimate source of the concentration gradients, and hence of the membrane diffusion potentials, is the metabolic energy expended in active transport of ions. If this transport is abolished by metabolic poisons or by cold, the cell begins to "run down" as ion concentrations tend to equalize across its membrane. However, if the permeabilities of the membrane to ions are low and especially if the cell is large and contains much intracellular electrolyte, it may take many hours for a cell to run down. During this period of time, cells continue to exhibit membrane potentials, so that active transport per se may not be necessary for the existence of a membrane potential (as opposed to its maintenance). For the moment, therefore, the emphasis will be on those membrane potentials that are created by the diffusion of ions down the concentration gradients that active transport processes have set up. Since these diffusion potentials depend only on kinetic movements of ions, they are called *passive* potentials.

Equilibrium potentials

Unlike a liquid junction, a membrane may act as a physical restraint on the diffusion of ions to the extent that an ion permeating a membrane may not be followed by its counterion. In such a case a diffusion potential will still result, but the net movement of the permeating ion will rapidly stop as the developing potential difference across the membrane opposes further ions crossing the membrane. The situation is analogous to that at the liquid junction, where the diffusion potential arises from the differences in ionic mobilities. The mobilities of various ions in the membrane may, and generally do, differ, and so lead to diffusion potentials too. More strictly, it is the permeabilities of the ions that differ. However, the permeability constant is taken to be the diffusion constant of a substance in a membrane of unknown thickness, and the diffusion constant is proportional to the mobility (Appendix B, p. 28). As an example, large internal organic anions in cells may not be able to diffuse out through the membrane if their counterions, the internal cations, do so, and a membrane potential opposing further net cation movement therefore results. Similarly, sodium permeates most resting cell membranes poorly (there are exceptions, however), so that if chloride in extracellular fluids (which contain much sodium chloride) permeates cells, it leaves sodium behind. The resulting membrane potential opposes further chloride entry.

For any given ion, the membrane potential that just stops net diffusion of this ion across the membrane is called its equilibrium potential. Equilibrium potentials are denoted by E_{ion}. The three main equilibrium potentials of interest for cells are those of sodium (E_{Na}), potassium (E_K), and chloride (E_{Cl}). The equilibrium potential for an ion is found by equating the diffusion force on this ion (proportional to its concentration gradient) to the electric force on this ion (proportional to the electric field). The result (Appendix B, p. 29) is the well-known Nernst potential. As applied to membranes, this potential is taken to be the intracellular potential minus the extracellular potential. With this convention, and at 20° C:

$$E_{Na} = -58 \log \frac{[Na]_i}{[Na]_o} \text{ mv} \quad (6)$$

$$E_K = -58 \log \frac{[K]_i}{[K]_o} \text{ mv} \quad (7)$$

$$E_{Cl} = -58 \log \frac{[Cl]_i}{[Cl]_o} \text{ mv} \quad (8)$$

where the subscripts i and o represent the inside and the outside of the cell, respectively.

If the membrane potential of a cell equals the equilibrium potential of a given ion, there will be no net movement of that ion into or out of the cell. (This does not mean that passive influxes and outfluxes of the ion across the membrane cease. The use of radioisotopic tracers has long since shown that there is a steady traffic of ions both ways across membranes. However, for an ion that is at equilibrium, the influx equals the outflux.) If an ion is not in equilibrium with the resting potential, there will be net passive movements of this ion (its permeability permitting) in the direction that would bring it into equilibrium. If the membrane potential changes during electrical activity, for example, so that it reaches the equilibrium potential of an ion not formerly in equilibrium, then net passive movements of this ion cease. This is an important point, for the species of ions that move during electrical activity can often be identified by the membrane potentials at which their movements (currents) cease. This membrane potential will be their equilibrium potentials, which can be compared with the Nernst potential if the concentrations are known.

In general, all ions do not have the same equilibrium potentials, and therefore they cannot all be in equilibrium at any given cell membrane potential. However, the membrane potential of a "resting" cell (the resting potential) may be very close to the equilibrium potential of one or more ions. Examples of equilibrium potentials and resting potentials from a number of cells are cited in Table 1-1. In nerve and muscle cells the membrane potentials are close to the potassium equilibrium potentials, whereas in erythrocytes they are close to the chloride equilibrium potential. In none of these cells is sodium at or near equilibrium, so there is a steady tendency for sodium to enter down its electric and concentration gradients, a tendency that must be counteracted by active transport.

Ionic basis of membrane potentials

Although ions such as potassium and chloride may sometimes be in equilibrium across cell membranes, they need not be the *cause* of the membrane potential, even though their diffusion would set up potentials. It is conceivable, for example, that some other factor such as an electrogenic sodium pump (discussed later in this chapter) might set up a membrane potential, whereupon all permeable ions would move until they came into equilibrium with this potential. On the other hand, it has long been known that increases in the concentration of extracellular potassium ions can depolarize nerve and muscle, and this has led to the belief that the diffusion of potassium ions results in the membrane potential in the first place. The advent of intracellular recording techniques made it possible to measure membrane potentials directly and test their dependence upon potassium and other ions.

Resting bioelectric potentials were measured earlier as the difference in potential between a "normal" part of a nerve or a muscle and a region that had been crushed, cut, or narcotized so as to make contact with the interior. These were called "demarcation potentials" or "injury potentials" and had their origins in the resting potentials. The injury potential rarely represents the full magnitude of the resting potential, however, since extracellular fluids may provide a short-circuiting path between the cut and normal tissue unless special precautions are taken. One such method for measuring the resting potential of single nerve fibers of frogs, using the demarcation potential, was devised by Huxley and Stämpfli.[51]

Resting potentials of large invertebrate nerve axons are measurable by threading an axial electrode into the axon from the cut end until it rests under normal membrane unaffected by the injury at the cut end.[47] However, it was the development by Ling and Gerard[60] of the glass micropipet electrode, with tips smaller than 0.5 μm, that made possible the measurement of resting and other bioelectric potentials from a wide variety of cells. Such electrodes may be inserted through cell membranes with very little injury to the cell and little or no alteration of the membrane potential; membranes appear to "seal" around the electrode tip. Therefore the amount of short circuiting between the inside and the outside of the cell can be quite small, allowing accurate and reproducible measurements of resting potentials. Many of the measurements cited in this chapter were made using such micropipet electrodes.

If normal resting potentials indeed arise as the diffusion potentials of potassium (and other ions), then alteration of the potassium concentration ratios should produce equal changes in the resting potential and the potassium equilibrium potential. Numerous experiments have been made on a variety of cells to test this hypothesis; several examples follow. Fig. 1-6 shows resting potentials of frog sartorius muscle fibers with varying concentration ratios of potassium across the muscle membrane.[20] Concentration ratios were altered by adding solid KCl to Ringer's fluid and allowing the muscle to reach a balanced state in the cold overnight. The solid line, *A*, represents the magnitude of the potassium equilibrium potential, and it can be seen that the membrane potential is the same as the potassium equilibrium potential only for external concentrations of potassium above 10 mM (vertical dashed line *C*). This is a common finding, for at low external potassium concentrations, membrane potentials of excised tissues may deviate from the theoretical potential. However, the resting potential is

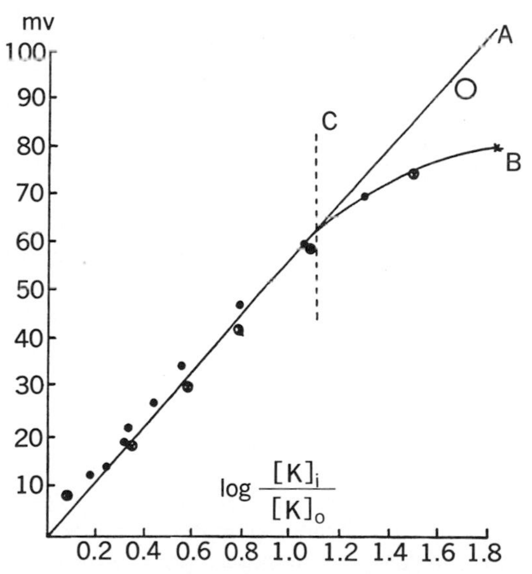

Fig. 1-6. Mean values of resting potentials of fibers in frog sartorius muscles at various ratios of $[K]_i/[K]_o$. Ordinate gives absolute values of measured membrane potentials. Filled circles and circles with crosses represent experimental points, while open circle represents initial membrane potential in 2.5 mM $[K]_o$ immediately after excision of muscle. Line *A* is theoretical line form from equation 7; vertical line *C* denotes 10 mM $[K]_o$. (From Conway.[20])

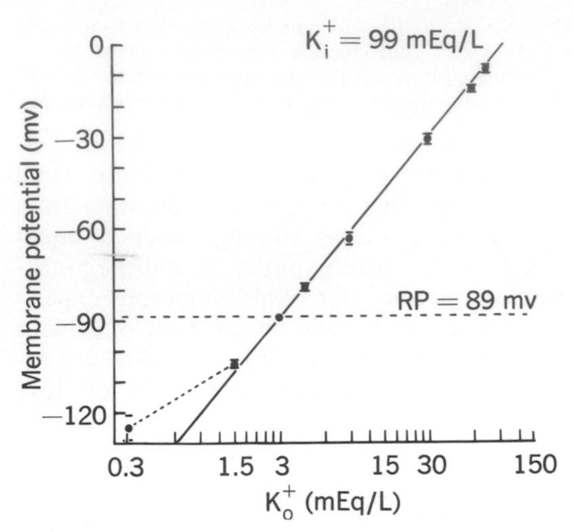

Fig. 1-7. Amphibian neuroglia. Dependence of membrane potentials of mudpuppy glial cells on external potassium concentrations. Solid line is theoretical line form from equation 7. (From Kuffler et al.[56])

close to the potassium equilibrium potential immediately after excision (large circle in Fig. 1-6), and similar results have been found in muscles carefully dissected out in plasma.[5] The trauma of excision of tissue may make the cells more leaky to other ions or may activate electrogenic pumps. Both of these possibilities, which will be considered later, would make the resting potential depart from the equilibrium potential. In another example of potassium determination of membrane potential, Kuffler et al.[56] altered potassium bathing the glial cells of the optic nerve of the mudpuppy *Necturus*. These cells were found to have a normal resting potential of 90 mv. Alterations of extracellular potassium due to activity of the optic nerve occur physiologically[71] (Chapter 6), and the membrane of the glial cells depolarizes as predicted from alterations of the potassium equilibrium potential (Fig. 1-7). In these glial cells, unlike frog muscle cells (discussed later), chloride has no effect on the membrane potential.

Numerous other examples of the dependence of membrane potential on external (and internal) potassium exist (squid,[11, 23] frog muscle,[8, 60, 70] plant cells,[32, 78] and many other cells). In none of these does a change in external sodium concentration result in a significant membrane potential change.[48, 56, 70]

Establishment of the resting potential

In order to get a better physical picture of what ions do at cell membranes to set up membrane potentials, consider what would happen if intra- and extracellular fluids were suddenly placed on either side of a membrane. At the initial instant there would be no membrane potential, and there would be electroneutrality on each side of the membrane. Assume that only one ion species (such as potassium) is permeable, so that it alone can diffuse across the membrane. Thus, since it leaves its counterion behind, the diffusion potential that it generates very rapidly increases up to the equilibrium potential of the permeable ion, and its net diffusion comes to a stop. Then the amount of ion that moved across the membrane, an insulating dielectric as described earlier, is balanced by an equal amount of counterion on the other side of the membrane. The membrane has literally charged up to the equilibrium potential and acts like a capacitor. Since most biologic membranes have capacitances of about 1 $\mu F/cm^2$, the amount of charge on 1 cm^2 of membrane is:

$$Q = CV$$

where:

$$Q = \text{Charge in coulombs}$$

For a typical 100 mv resting potential of a muscle, the charge is:

$$Q = 10^{-6}\ F/cm^2 \cdot 10^{-1}\ V = 10^{-7}\ \text{coulombs}/cm^2$$

Since there are about 10^5 coulombs per mole, the number of moles is:

$$\text{Moles}/cm^2 = 10^{-7}\ \text{coulombs}/cm^2 \cdot$$
$$10^{-5}\ \text{mole/coulombs} = 10^{-12}\ \text{moles}/cm^2$$

This quantity of ions, 10^{-12} moles, reappears often in calculations of amounts of ions involved in bioelectric potentials and is termed a picomole (abbreviated pmole). Such an amount of cations on one side of the membrane, and an equal number of anions on the other side, is that amount by which electroneutrality on each side fails: there are 1 pmole more cations on one side than anions and vice versa on the other side. However, this may be an extremely small proportion of all ions in a cell. For example, in a muscle cell of 20 μm radius, length 1, internal potassium concentration of 140 mM, and with 85% of its volume occupied by fiber water (the rest being inactive volume), the proportion of potassium ions held in the surface ion clouds is as follows:

$$\frac{\text{Amount of ion on surface}}{\text{Amount of ion in cell}} = \frac{(10^{-12}\ \text{mole/cm}^2)\ (2\pi)\ (20 \cdot 10^{-4}\ \text{cm})\ (1)}{(140 \cdot 10^{-3}\ \text{mole/cm}^3)\ (0.85)\ (\pi)\ (20 \cdot 10^{-4}\ \text{cm})^2\ (1)} = \text{about } 10^{-5}$$

Hence only 1/100,000th of this cell's potassium ion needs to diffuse across the cell membrane to set up a membrane potential of 100 mv. It is for this reason that the membrane may be "discharged" again and again, as during an action potential (described in following chapters), and the membrane recharged again and again by the outward diffusion of potassium, without an appreciable change in the internal amounts of potassium.

The Donnan equilibrium

From the preceding, it can be seen that a membrane potential results when the charge on one side of a membrane exceeds that on the other. Clearly, it does matter what the ionic species is that diffuses; it could be one of a number of positive charges diffusing out, one of a number of negative charges diffusing in, or both. In Table 1-1 it can be seen that in cells such as muscle or erythrocytes, several ions are in equilibrium with the membrane potential. This means that a number of ionic species have crossed the membrane to charge up the membrane potential. Since the equilibrium potential of each is equal to the membrane potential, they must be equal to each other. This leads to the result, in the case of frog muscle, for example:

$$E_K = -58 \log [K]_i/[K]_o = -58 \log [Cl]_o/[Cl]_i = E_{Cl}$$

$$[K]_i/[K]_o = [Cl]_o/[Cl]_i = r \qquad (9)$$

or

$$[K]_i[Cl]_i = [K]_o[Cl]_o \qquad (10)$$

General relations of this kind were derived by Donnan[28] in 1924 for equilibriums in inanimate systems in which solutions of nonpermeating organic electrolytes (such as phenol red) were initially separated by colloidin membranes from permeating solutions of salts. By way of illustration, consider such a system where the permeating salt is uni-univalent (such as KCl) and the organic impermeant (P) has a valency of –1. The initial condition can be depicted as follows for KCl.

Initial condition

SIDE 1	SIDE 2
c_1 K^+	K^+ c_2
c_1 P^-	Cl^- c_2

Equilibrium condition

SIDE 1	SIDE 2
$c_1 + x$ K^+	K^+ c_2-x
x Cl^-	Cl^- c_2-x
c_1 P^-	

The c's represent the concentrations. After an amount x of chloride has diffused into side 1, accompanied by a like amount of potassium, the resulting membrane potential stops further ion movements and equilibrium results (neglecting osmotic movements of water for the moment). At such an equilibrium Donnan[28] showed that equations 9 and 10 hold. Inspection of the equilibrium condition shows that an asymmetry of ions across the membranes has been set up solely by the presence of the impermeant anion. Nonetheless, all ions but this impermeant (and water, as explained later) are in equilibrium, so that the maintenance of the ionic asymmetries requires the expenditure of no energy.

It will now be seen to what extent equilibriums of the Donnan type exist across cell membranes. Inside, charged impermeable anions certainly exist; these include amino acids and proteins whose net charge is negative, phosphorylated intermediates of metabolism, etc. If two or more ions are in electrochemical equilibrium across such cell membranes, as Table 1-1 indicates some may be, then these Donnan relations (equations 9 and 10) should apply to their intra- and extracellular concentrations. In 1941 Boyle and Conway[13] used chemical analyses of frog muscle to show that potassium and chloride could move across the muscle membrane and distribute themselves in accordance with the Donnan relations. Table 1-3 gives results from muscles soaked overnight in Ringer's solution, to which solid KCl has been added. Over a 25-fold variation in extracellular potassium concentration, there was a near equality of the product of intracellular and extracellular concentrations.

Further evidence that the Donnan relations

Table 1-3. Donnan relations in frog muscle*

$[K]_o$	$[K]_o\ [Cl]_o \times 10^{-3}$	$[K]_i\ [Cl]_i \times 10^{-3}$	Ratio
12	1.05	1.00	1.05
18	1.69	1.72	0.98
30	3.18	2.99	1.06
60	8.16	8.61	0.94
90	14.90	15.80	0.94
120	23.50	24.20	0.97
150	33.90	34.40	0.99
210	60.00	52.80	1.14
300	112.80	118.70	1.05
			(average 1.01)

*Adapted from Boyle and Conway.[13]

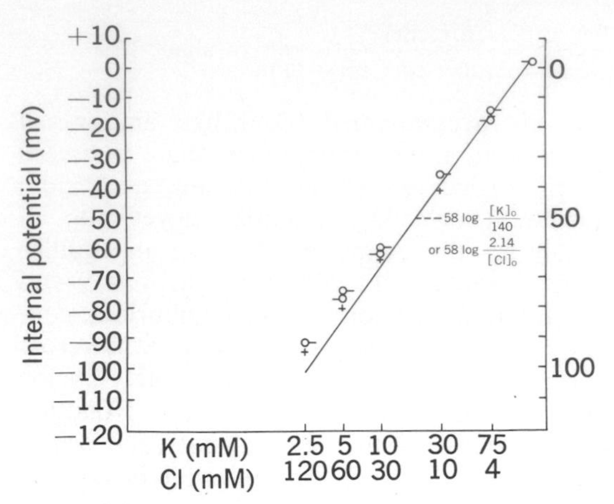

Fig. 1-8. Relation between membrane potential and log $[K]_o$ or −log $[Cl]_o$ when using solutions with $[K]_o[Cl]_o$ = 300 mM^2. (From Hodgkin and Horowicz.[46])

might apply to frog muscle was obtained by Hodgkin and Horowicz,[46] who altered the membrane potentials of these cells with changes in external potassium ion charges while keeping the product $(K)_o$ $(Cl)_o$ constant. In this situation, both potassium and chloride ions should always be in equilibrium, and the membrane potential should be given by both equations 7 and 8. Fig. 1-8 indicates that this is, in fact, the case. The measured resting potentials agree well with those predicted. (At low external potassium concentrations there is a departure from the theoretical line, as there was in Fig. 1-6. Such departures will be considered later when polyionic potentials are discussed.)

Two very important conclusions follow from all this. First, as mentioned earlier, concentration ratios of the permeant ions (potassium and chloride in muscle, anions in erythrocytes) can be maintained with little or no expenditure of energy, since the distribution of these ions is set by the concentration of the internal impermeant anion. However, there is another less permeant ion for most cells, and this is extracellular sodium. Boyle and Conway[13] felt that it was "practically perfectly excluded" in their experiments; i.e., it was also an impermeant. The only effect of impermeant sodium ion on the Donnan equilibrium would be to ensure that there would be more extracellular chloride than potassium ions to balance the sodium ions. However, the use of radioisotopes has shown that membranes are not impermeable to sodium ions, merely less permeable than to potassium and chloride in muscle, for example. Rather, it is the intervention of the sodium pump to eject the sodium that passively enters the cell that makes sodium seem *effectively* impermeable, at least in the resting state of most cells.

The second important conclusion to follow from the applicability of the Donnan systems to cells is that no animal cell can be in osmotic equilibrium without extracellular sodium being effectively impermeant. This is because the osmolarity of the intracellular impermeant(s) must be balanced by some extracellular impermeant in order for an osmotic equilibrium to be present. No simple Donnan system of the type just depicted can reach osmotic equilibrium without the application of hydrostatic pressure.[28] Plant cells, with inextensible cell walls, could exert such hydrostatic pressures, but animal cells could not, for they would swell. This has already been demonstrated in the behavior of animal cells as osmometers. What can now be added is that C in equation 5 is almost entirely represented by extracellular sodium concentrations; in frog muscle, for example, the cell volume is inversely proportional to sodium alone.[13] However, the complete dependence of cell volume on extracellular sodium depends on the constancy of the intracellular impermeant; yet there are times when this undergoes normal physiologic changes. For example, during the oxidation and reduction of hemoglobin in erythrocytes (Chapter 59), the charge on the hemoglobin alters, as do the amounts of associated ions. With changes in the amount of associated ions, there are osmotic changes in cell water and hence in cell volume.

Taken all together, it is remarkable how many aspects of ion distributions and of resting membrane potentials seem to be explained on the simple basis of ions equilibrating across membranes. However, as previously emphasized, not all ions (especially not sodium) are in equilibrium with the resting membrane, and these have a tendency to disrupt the resting conditions by their passive movements. It is to counteract such movements that cells utilize metabolic energy to "pump" ions against their concentration gradients and so maintain constant concentrations. In the following discussion the subject of metabolic pumps will be considered in more detail.

NATURE OF ACTIVE TRANSPORT

The most compelling evidence for active (i.e., nonpassive) movements of ions across membranes is that cells that gain sodium and lose potassium in the cold can expel the added sodium and reaccumulate the lost potassium when returned to room temperature.[28, 80] The sodium is expelled against the electrochemical gradients noted earlier, its expulsion is prevented by metabolic inhibitors or by a return to the cold, and this expulsion may be linked in part to active, inward movements of potassium ions (the potassium effect). The mechanism responsible for this phenomenon is called the "sodium pump." If there is a 1:1 exchange of potassium for sodium, no net charge transport occurs, no electric potential results, and the pump is called *nonelectrogenic.* (A nonelectrogenic pump can also result when a cation and an anion are transported together in the same direction; here, too, there is no net charge transfer.) If the amount of sodium extruded exceeds the amount of potassium taken up, there is a net charge transfer, a potential due to the action of the pump results, and the pump is considered to be *electrogenic.* An electrogenic pump can therefore alter or add to diffusion potentials set up by permeating ions, and because of this it can cause these permeating ions to redistribute themselves in order to return to electrochemical equilibrium. Therefore the activities of metabolic pumps can have profound effects upon ion distributions and membrane potentials, either by acting simply to set up concentration gradients or by altering the membrane potential as well. In the steady state that is homeostasis the net fluxes of all ions must be zero; that is to say, ions must distribute themselves so that their passive fluxes just balance the active fluxes, if any. The emphasis here will be on the means by which measured fluxes across cell membranes may be broken into their active and passive components and how the active fluxes are driven by metabolic energy.

Division of ion fluxes into active and passive components

A number of different methods have been used with varying success to separate active from passive components of ion fluxes. The simplest in principle is to selectively poison the ion pumps with cardiac glycosides, e.g., strophanthidin or ouabain. The principal action of these drugs appears to be on a Na-K–sensitive ATPase that is presumably involved in transport (p. 22). The differences between "uphill" fluxes (e.g., sodium outflux, potassium influx) in normal and poisoned cells is then taken to be a measure of passive uphill flux. However, cardiac glycosides rarely inhibit ion pumps by more than 80%,[16] and they may in addition alter membrane properties so as to affect passive as well as active fluxes.[17, 37]

A second method of estimating passive fluxes is to calculate the ratio of passive fluxes (influx to outflux), given known concentrations and membrane potential, and to compare this result with the measured ratio. The equation used for this purpose is the flux ratio equation, first derived by Ussing.[83] (It comes from the equations used in Appendix B, p. 28, to describe passive ion movements down electric and chemical gradients.)

$$\frac{\text{Flux}_{in}}{\text{Flux}_{out}} = \frac{C_{out}}{C_{in}}\, e^{-\frac{VF}{RT}}$$

where, as before:

$$V = V_{in} - V_{out}$$

This equation can be applied to measurements made in poisoned cells, for example, to see whether all the residual fluxes are passive. Since downhill fluxes may generally be assumed to be passive, it is possible to calculate the passive uphill flux using this equation. The difference between passive and total uphill flux is then the active flux. There are times when all fluxes measured are passive, but the flux ratio equation still does not apply. These cases include the "long pore effect"[7] and exchange diffusion (a 1:1 exchange of two ions of the same species). Exchange diffusion of sodium (the primary ion involved) can usually be abolished (and thus measured) by reducing or removing external sodium.

A third means of measuring active ion fluxes is to abolish the electric and concentration gradients responsible for passive ion movements. This method has been used primarily in studies of ion transport by epithelial tissues (gut, kidney, etc.), on both sides of which the same bathing solutions can be used and across which the potential can be abolished by passing electric current. This *short-circuit current* has been shown to exactly equal the active flux of sodium across the frog skin as measured with isotopes.[84] However, transport across such epithelial cells involves at least two membranes in parallel. At single membranes the short-circuit current of a nonelectrogenic pump would be zero, and even if more sodium is pumped out than potassium taken up, the short-circuit current would still give only the difference between

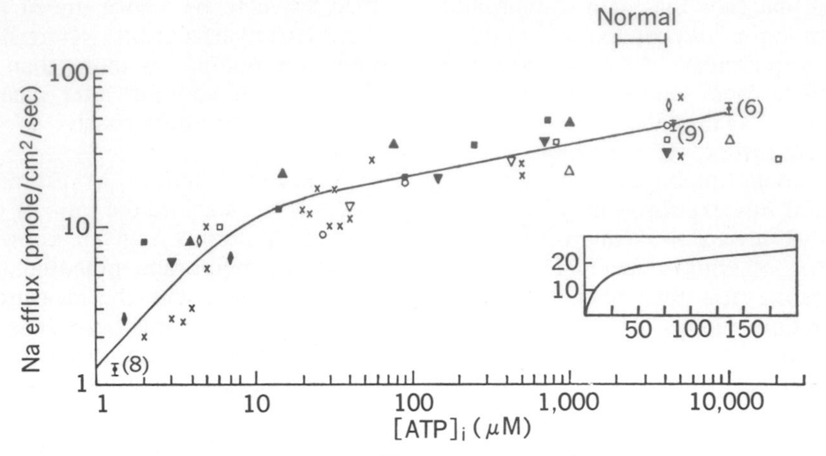

Fig. 1-9. Relation between $[ATP]_i$ and Na efflux in internally dialyzed squid axons. Inset shows data for range 0 to 200 μM ATP plotted on linear scale. (From Brinley and Mullins.[17])

fluxes of the two ions. This objection does not apply to epithelial tissues, however, since there are net active ion movements from one side of the tissue to the other.

Metabolic requirements of ion pumping

In frog muscle, squid giant axon, and erythrocytes about 20% of all metabolic energy is used for active transport of sodium ions.[17]

This figure is derived from the following: The total power (energy per second) available to the cell is calculated from the oxygen consumption or for the erythrocyte from the lactate production. The power (potential × current) used for sodium pumping is the product of active sodium flux (expressed as a current) and the potential against which it must be moved, i.e., the difference between the membrane potential and the sodium equilibrium potential.

In other tissues, such as retina, practically all the available metabolic energy must be used by the sodium pump.[41] The evidence does not support the conclusion of Ling[59] that there is insufficient metabolic energy available to support measured ion fluxes, and it is therefore unnecessary to suppose, as Ling did, that ion binding is required to maintain measured concentration ratios across the cell membrane.

The immediate source of energy for pumping appears to be adenosine triphosphate (ATP). High-energy phosphate bonds were implicated in pumping by the experiments of Caldwell et al.,[19] in which injection of ATP, arginine phosphate, or phosphenolpyruvate into squid axons poisoned with cyanide temporarily restored the sodium pump. Of these three sources of high-energy phosphate, it appears that ATP "fuels" the pump, while the other two serve solely to transphosphorylate adenosine diphosphate (ADP) to make ATP. In squid axons that are internally dialyzed to wash out all ADP, only when ATP is in the dialysis fluid is there any pumping of sodium (as much as 80 to 100% of the normal rate).[17] When the concentration of ATP drops to 1 μM, pumping also ceases (Fig. 1-9). In erythrocytes, too, the fuel for the sodium pump appears to be ATP. This was shown in an ingenious way by Garrahan and Glynn,[35] who ran the pump "backward" and observed incorporation of radioactive phosphate into ATP. In their experiments they filled ghosts (lysed erythrocytes) with solutions that were high in potassium, low in sodium, and that contained ADP and radioactive phosphate. The external solution was high in sodium and contained no potassium. These concentration gradients were too great for the sodium pump to work against; on the contrary, the pump was driven backward. Instead of hydrolyzing ATP, therefore, ATP was formed. As in normal pumping, ouabain poisoned the process. The point is that when the pump is run backward, the high-energy phosphate compound formed is ATP, and so presumably this is the source of energy for driving the pump forward.

Ionic requirements of the sodium pump

In dialyzed squid axons in which the internal ion concentrations are determined by the composition of the dialysis fluid, the rate of sodium pumping is directly proportional to

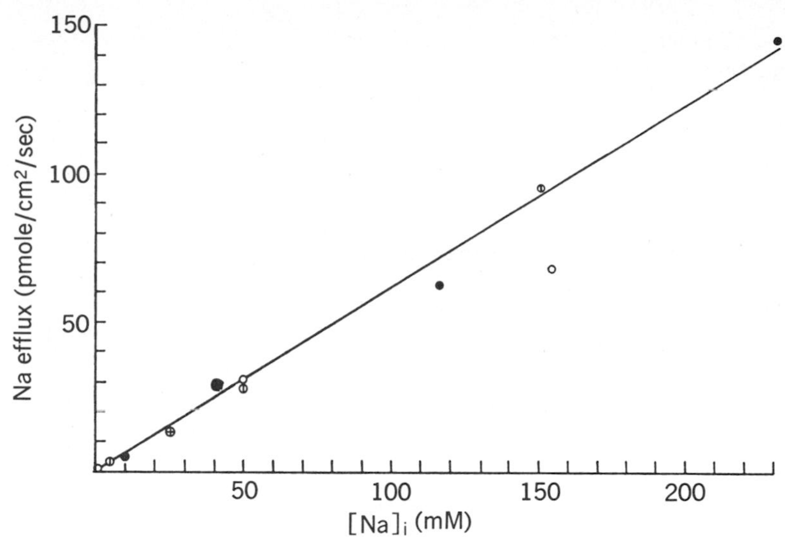

Fig. 1-10. Linear relationship between Na efflux and $[Na]_i$ in internally dialyzed squid axons. (From Brinley and Mullins.[17])

the internal sodium concentrations, as shown in Fig. 1-10.[17] In frog muscle, on the other hand, the rate of pumping appears to be a sigmoidal function of internal sodium concentration.[68] Regardless of the dependence on sodium concentration, in these cells and in the erythrocyte[34] about three sodium ions are pumped for every ATP split (see also Stein[7]). In the erythrocyte the free energy change involved in the sodium transport (plus some from potassium transport) was about 9 kcal/mole, compared with a liberation of about 13 kcal/mole from the hydrolysis of ATP under the given experimental conditions.[85] Thus it appears that there is ample energy available from ATP for the amounts of ion transported, even if it is not all utilized for the osmotic work.

In addition to pumping sodium out, most cells likewise appear to pump some potassium in. In animal cells the two processes seem to be linked, inasmuch as 50 to 70% of the active sodium efflux is inhibited when potassium is removed from outside the cell.[50] The "potassium-free effect" has been interpreted to mean that there is an obligatory coupling between sodium and potassium transport, possibly because they occupied the same "carrier" at different points on its movement back and forth across the membrane. Some credence is given to this point of view by experiments on erythrocytes in potassium-free solution, in which sodium is apparently able to substitute for potassium.[33] When the carrier is moving sodium both ways across the membrane (exchange diffusion), there is no hydrolysis of ATP, but the movements are poisoned by ouabain. Likewise, in squid giant axon, active influx of potassium ion appears to depend in part on the internal sodium concentration.[67]

One model for the active transport of ions, then, is a carrier that transports sodium from the inside to the outside of a cell and then combines with potassium and moves from the outside to the inside. In this model the effect of changing external potassium concentrations on outward sodium transport is that, without external potassium, carriers will "pile up" on the outside of the membrane, and the concentration of carriers inside capable of transporting sodium outward will diminish. Similar arguments relate the effects of internal sodium concentrations on inward potassium transport. However, several lines of evidence indicate that the carrier may be able to move more easily from the outside to the inside without potassium than it can move from the inside to the outside without sodium. First of all, inhibition of sodium transport by removal of external potassium ions is never complete. Second, the coupling ratio between outward sodium transport and inward potassium appears to be variable. In erythrocytes the ratio is always slightly more than 1.[34, 82] In the giant axon of the squid the coupling ratio appears to depend on the internal sodium concentration. This is illustrated in Fig. 1-11. The lower graph shows that the coupling ratio is constant at about 3:1 (sodium:potassium)

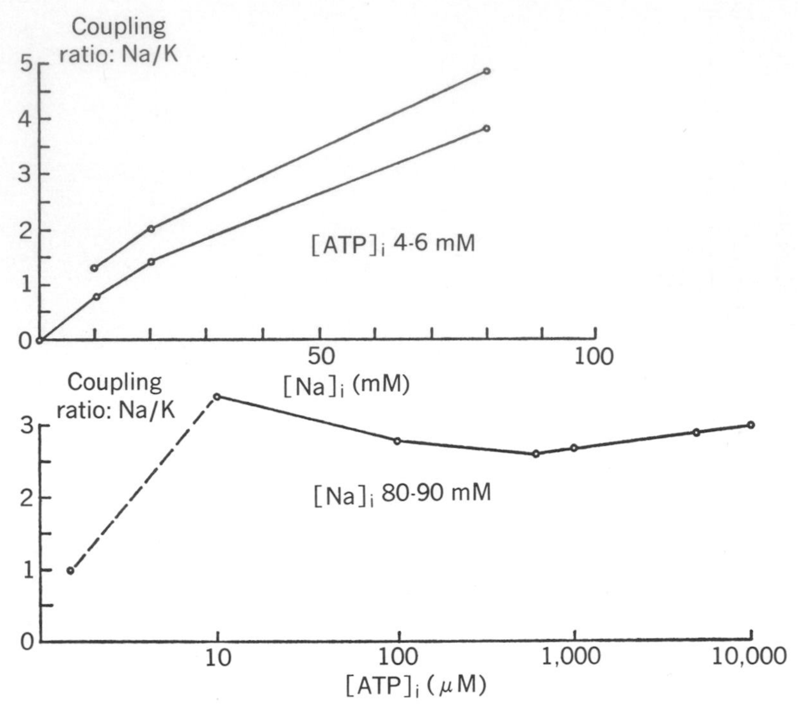

Fig. 1-11. Coupling ratio between Na and K in sodium pump. Dependence of this ratio on $[Na]_i$ is shown at top, and dependence on $[ATP]_i$ is shown at bottom. (From Mullins and Brinley.[67])

for large variations in the concentration of ATP. The upper graph shows that the coupling ratio can approach 1:1 when the internal sodium concentration is low. This is because over the sodium concentration range given in the upper graph the sodium transport rate decreases by about 10 times (Fig. 1-10), while the total potassium transport rate (corresponding to the upper line in the top graph) decreases somewhat more than one-half. Therefore there is clearly not a fixed, obligatory ratio between potassium and sodium ions pumped, even if the same carrier is involved in both. In a cell in which potassium is near equilibrium (e.g., the giant axon of the squid) the active potassium influx needed to maintain the internal potassium concentration constant should be small and nearly constant. The active sodium outflux, however, might need to be quite a bit larger, since sodium is far from being at equilibrium and therefore has a much greater tendency to enter. A variable sodium:potassium coupling ratio is one means by which a cell can maintain a somewhat constant rate of potassium uptake while increasing its rate of sodium extrusion if the internal sodium concentration rises.

Sodium-potassium–activated ATPase

A molecular basis for active transport of ions has long been sought. Of late, ATPases that require sodium and potassium for their enzymatic activity have been isolated from a wide variety of transporting tissues. The relationship of such ATPases to other ATPases and to the actual physical movements of ions across membranes is not clear. However, Skou[6] has claimed that in eight different ways the Na-K–activated ATPases fit the requirements of known sodium and potassium transport systems:

1. They are located in cell membranes.
2. They have an affinity for Na that is higher than for K at sites located on the insides of membranes.
3. They have an affinity for K that is higher than for Na at sites located on the outsides of membranes.
4. They are enzyme systems that can catalyze the hydrolysis of ATP (and convert the energy into the movement of cations).
5. They are capable of hydrolyzing ATP at rates dependent on the concentrations of Na inside the cell and of K outside the cell.

6. They are found in all cells in which active, linked transport of Na and K occurs.
7. There is a close correlation between the effects of cardiac glycosides on cation transport in intact cells and on Na-K–activated ATPases.
8. The enzyme systems have the same quantitative relations to Na and K as do the transport systems of intact cells.

For maximal activity of this enzyme system, both Na and K must be present. However, phosphorylation of an intermediate can occur when only Na is present; release of inorganic phosphate then occurs when K is added.[1] One scheme that would link this phosphorylation and dephosphorylation to ion movements is the following[1]:

1. $E_1 + MgATP \overset{Na}{\rightleftharpoons} E_1 \sim P + MgADP$
2. $E_1 \sim P \rightleftharpoons E_2 - P$
3. $E_2 - P \overset{K}{\longrightarrow} E_2 + P$
4. $E_2 \rightleftharpoons E_1$

Transport could occur if the various forms of E_1 were inwardly oriented sites with high affinities for Na, while the various forms of E_2 were outwardly oriented sites with high affinities for K. Rega et al.[73] have found evidence that the phosphatase aspect of the system (step 3) may be so oriented. This phosphatase will split *p*-nitrophenyl phosphate, too, but only when the substrate is on the inside of erythrocyte ghosts and K is on the outside. Nonetheless, this scheme must be viewed as speculative at present, and it still leaves open the actual physical mechanisms whereby the ions move through the lipid bulk of the membrane from one side to the other.

POLYIONIC MEMBRANE POTENTIALS

So far, membrane potentials have been considered in cells in which one or a few ions are highly permeable, and others are not or are rendered effectively impermeable by ion pumps. In these cells the highly permeable ions attempt to diffuse down their concentration gradients, but after the movements of minute numbers of ions, restraining membrane potentials are set up by the resulting charge separations and net ion movements come to an end. There are cells such as those of the retina, however, where no ion appears to be at or near equilibrium, and where there are therefore net passive fluxes of all ions (to be made up by ion pumping, of course).[41] Similarly, the equilibrium theory of membrane potentials appears to be less satisfactory for excised tissues that are "running down" and whose membrane potentials are different from their in situ values. Examples of the latter are evident in Fig. 1-6, where the membrane potentials of excised muscle cells differ increasingly from the potassium equilibrium potential at external potassium concentrations of less than 10 mM. From the point of view of understanding just how bioelectric potentials are maintained in the normal state, these nonequilibrium states are often valuable for allowing interactions of ionic potentials to be assessed. Sometimes, indeed, cell systems are experimentally thrown out of homeostasis in order to watch the mechanisms that cells use to return to homeostasis, e.g., loading cells with sodium in the cold, which stops the sodium pump, and then watching the cells pump out the sodium at room temperature. When cells are far from homeostasis or no ion is at equilibrium, so that ion pumping is great, the pump itself may contribute to the membrane potential of a cell if it is electrogenic. Thus nonhomeostatic conditions allow the study of a number of ionic parameters whose effect at homeostasis might be small and difficult to measure.

Constant field equation

A common approach to nonequilibrium membrane potentials is to apply the constant field equation.[38, 48] The purpose of the equation, which is given below, is to assess how a number of permeating ions, none at equilibrium, would determine the membrane potential. The approach is to consider the diffusion of a number of different ions, and to predict the membrane potential that would cause no net membrane current. (If there were a 1:1 exchange of sodium and potassium ions, or if equal numbers of sodium and chloride ions, or combinations of these, moved across the membrane, there would be no net current. There would be net fluxes of each of the permeating ions, however.) The basic assumption in solving the diffusion equations is that the membrane potential, measured by determining the difference between the inside and the outside of the membrane, changes uniformly within the membrane (hence the term "constant field"). In essence this means that within the membrane (or within "channels" through which ions could pass) there are charges (dipoles) that orient in the electric field across the membrane and thus result in its changing uniformly from one side to the other. There is evidence that such charged membrane channels may exist; erythrocytes are highly permeable to anions, but hardly at all to cations, as though the membrane were positively charged. Similarly, as mentioned earlier, ions move very poorly through purely lipid, artificial membranes, and the dielectric constant of these lipids is so low (about 3, compared with 80 for water) that few dissociated ions would move through membranes. Since flux studies and those of the action potential (Chapter 2) indicate that ions do move independently of each other through membranes under the influence of an electric field, this must mean that ions do move through membranes in the dissociated state. A reasonable interpretation therefore is that ions move through regions of the membrane with a high dielectric constant (compared to the lipid bulk of the membrane); i.e., through polar, charged regions. Therefore the assumption of a constant electric field within the membrane is reasonable, although it remains to be proved.

Given the assumption of a constant electric field across the membrane and the case of zero net ionic current, the membrane potential based on sodium, chloride, and potassium ions (the three most prevalent ions in biologic fluids) is predicted to be, at 20° C (Appendix C, p. 29):

$$V = -58 \log \frac{P_K[K]_i + P_{Na}[Na]_i + P_{Cl}[Cl]_o}{P_K[K]_o + P_{Na}[Na]_o + P_{Cl}[Cl]_i}$$

The constant field equation has been applied to membrane potentials of squid axons,[48] frog muscles,[46] and molluscan neurons.[64, 65] Several simplifications of the equation are possible. The membrane potentials of molluscan neurons appear to be nearly independent of external chloride ion concentrations, while frog muscles, which are permeable to chloride ions, may be studied in chloride-free solutions. In these instances the chloride terms in the equation may be dropped. Likewise, expressing all permeabilities with reference to the permeability to potassium, the equation may be rewritten as:

$$V = -58 \log \frac{[K]_i + P_{Na}/P_K\ [Na]_i}{[K]_o + P_{Na}/P_K\ [Na]_o}$$

Since in general the resting permeabilities of sodium ions are less than those of potassium, and the internal concentrations of sodium ions are low (especially compared with intracellular concentrations of potassium), the term P_{Na}/P_K $(Na)_i$ may be ignored. This means that if intracellular and extracellular ion concentrations are known, only the ratio of permeabilities (which may not be known) need be estimated. (If the intracellular concentrations are not known but the equation can be fitted to the data, then these concentrations, too, may be esti-

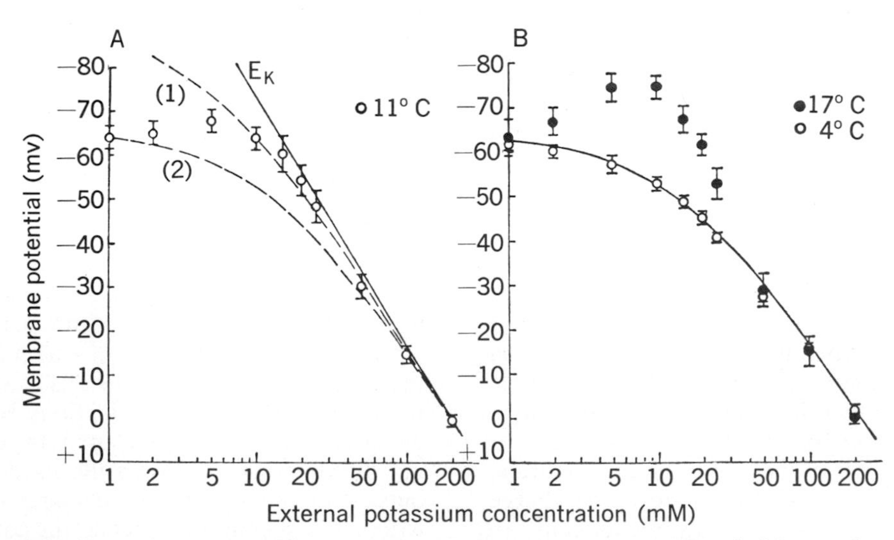

Fig. 1-12. Dependence of resting potential of molluscan neurons on $[K]_o$. **A,** Measured resting potentials at 11° C are not fitted by either potassium equilibrium potential, E_K, except at high $[K]_o$, or by constant field equation, *1* and *2*, at two values of P_{Na}/P_K. **B,** At 4° C, points (open circles) *are* fitted by constant field equation—$P_{Na}/P_K = 0.033$; $[K]_i = 235$ mM for solid curve—but not at 17° C. (From Marmor and Gorman.[64])

mated.)[65] Finally, it can be seen that at high extracellular potassium concentrations the potassium concentration term in the denominator dominates and the equation simplifies to the Nernst potential.

Fig. 1-12 from Marmor and Gorman illustrates the application of the equation to resting potentials of neurons of a mollusc as well as the effects of electrogenic sodium pumps, which will be discussed subsequently. In Fig. 1-12, *A*, with high external potassium concentrations, the resting potentials of these neurons are indeed predicted by the Nernst potential for E_K (solid line). However, at low external potassium concentrations the resting potentials are described by neither the Nernst potential nor the constant field equation (dashed lines, two sets of parameters). Upon cooling (Fig. 1-12, *B*), the resting potential is well described by the constant field equation (solid line). The difference is seen to be due to an electrogenic sodium pump active at 17° and 11° C but not at 4° C; it is also blocked by ouabain (not shown). The membrane potential thus has two components in these experiments: a polyionic diffusion potential and a potential due to an electrogenic pump. In other experiments to which the constant field equation has been applied (see previous discussion), hyperpolarizations due to electrogenic pumps were not considered but might have been present. This is because the constant field equation was applied to cells that were running down, and such cells may well attempt to restore homeostasis by enhanced ion pumping. If the pumping is electrogenic, membrane hyperpolarization over the polyionic diffusion potentials of the constant field type can result. These will be considered now in more detail.

Restoration of cell homeostasis

Excised tissues used for experimentation may well be running down, in part because the trauma of dissection probably lessens the membrane's selectivity toward ions and in part because in physiologic salt solutions (as opposed to plasma with all its proteins) the membranes lose selectivity and become "leaky." On the other hand, there are normal physiologic situations in which cells gain sodium and lose potassium and must restore their original states by enhanced ion pumping. Such situations include normal nerve and muscle activity (Chapters 3 and 4), and following such activity the sodium pumps may hyperpolarize the membrane. For example, large (10 to 20 mv) hyperpolarizations are recorded in cells of some crustacean eyes and stretch receptors after activity.[54, 69] These hyperpolarizations can be differentiated from positive afterpotentials that are due to potassium ions[48] (Chapter 2), since they are abolished by ouabain and are set off by the intracellular injection of sodium ions. Undoubtedly the primary effect of the pumps is to remove sodium and restore potassium ions that have passed through the membrane during nerve activity, and since the movements of these two ions by the pumps are unequal (3:1 ratio of sodium and potassium; see previous discussion), the pumps are electrogenic. A secondary effect for normal nerve is that the hyperpolarization itself brings about a redistribution of ions. The more permeable potassium ions, no longer at equilibrium when the cell is hyperpolarized, are electrically attracted into the cell interior, so that the internal concentration of potassium rises, presumably back to its original level prior to the activity. Similarly, as the internal sodium concentration falls due to pumping, the pump rate decreases and the effect of the pump on the membrane potential falls to imperceptible levels.

It is just because the pump has an imperceptible effect on the normal resting potential that it is necessary to discern its effect upon cell homeostasis by deliberately stopping it for a time, allowing the intracellular concentrations of ions to change, and then observing the manner in which the cell restores its normal internal environment. Such experiments have been performed by loading frog muscle cells with sodium in the cold and observing their recovery at warmer temperatures. Fig. 1-13 illustrates one such experiment. In the cold in potassium-free Ringer's solution for 48 hr, about half the normal potassium was replaced by sodium. Upon warming to 18° C in a Ringer's solution with 10 mM external potassium, the potassium was regained and the sodium extruded in 3 to 4 hr (not shown). Fig. 1-13 shows what the membrane potential was doing during this ion pumping. The cell was first allowed to equilibrate with the 10 mM potassium recovery solution, which depolarized it (as compared to the solution with no external potassium). Then it was warmed to 21° C. Immediately the cell hyperpolarized, but after 3 hr or so, the hyperpolarization decreased and the membrane potential came toward (and eventually to) the potassium equilibrium potential, E_K (solid

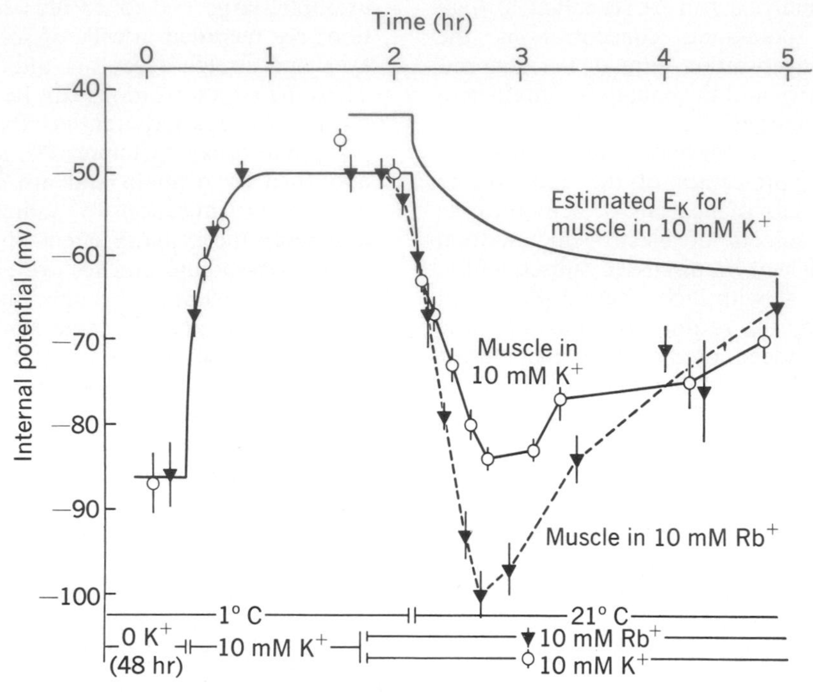

Fig. 1-13. Membrane potentials of sodium-rich fibers of frog sartorius muscles during recovery in solutions containing either potassium or rubidium. (From Adrian and Slayman.[9])

line in Fig. 1-13). Similarly treated muscles that were allowed to recover in solutions with rubidium, which is physiologically similar to potassium, instead of potassium itself, showed even greater amounts of hyperpolarization. The hyperpolarizations were prevented by removal of external potassium (or rubidium) or by ouabain, so they undoubtedly reflect the activities of coupled electrogenic ion pumps.

A detailed analysis of all that is involved as these muscles reestablished homeostasis has not yet been done, but a tentative outline can be sketched. Frog muscle in plasma has a resting potential (−99.2 mv) very near the potassium equilibrium potential (−101.1 mv)[5] but very far from the sodium equilibrium potential (+52 mv).[70] Chloride likewise is at equilibrium across the membrane.[13, 46] Therefore there are no net passive fluxes of either potassium or chloride since these are at electrochemical equilibrium and have already diffused out of or into the cell, respectively, to the point at which the resulting diffusion potential (the resting potential) has stopped any further net movements. There is a net passive inward flux of sodium down both its electric and concentration gradients, but as the permeability of the muscle membrane to sodium is only about 1/100th of its permeability to potassium,[46] it requires only a slow rate of ion pumping to maintain constant the normally low intracellular sodium concentration. Indeed, at these low concentrations the pumping rate itself slows down[68] and is most likely nonelectrogenic or imperceptibly electrogenic. (This could be so if, as in the case of the squid previously discussed, the sodium:potassium coupling ratio were variable and close to 1:1 at low intracellular sodium concentrations.)

In the sodium-loaded muscle, by contrast, both sodium and potassium (and most likely chloride, too) are not in electrochemical equilibrium, there is a large amount of intracellular sodium so that the pump rate is high (as it is in the squid when intracellular sodium is high), and the pump is electrogenic; i.e., the pump expels more sodium ions than it takes up in potassium ions. If this were to continue for a long period of time, and if the only way that potassium could enter the cell were via the pump, the excess of sodium pumped over potassium would soon lead to very large voltages. However, the hyperpolarizations caused by the electrogenic pump accelerate the passive movements of potassium into the cell and chloride out of the cell, and the final amount of hyperpolarization is that at which the number of sodium ions being actively pumped out are exactly balanced by the number of potassium ions actively pumped in, plus the number of potassium ions passively accelerated inward, minus the number of chloride ions passively accelerated outward. In other words, the net membrane current must be zero, for if this were not the case, any net current would continue to charge up the membrane capacitance to some new voltage at which net current *would* be zero. This, then, is why the membrane hyperpolarization during recovery in rubidium solutions is greater than in potassium solutions; rubidium is less permeable than potassium and so it requires a greater membrane hyperpolarization to accelerate it inward.[9] (However, about 90% of the rubidium may move by the pump, whereas it is uncertain how much potassium is moved by the pump and how much passively.)

As the pump restores normal intracellular potassium concentrations to these initially sodium-loaded muscles, the potassium equilibrium potential will become more negative (solid line in Fig. 1-13), while the pump rate will drop as the intracellular sodium concentration drops. Qualitatively it can be seen where this will stop: as the sodium is pumped out and the intracellular concentration falls, the pump will continue to slow down and its active flux will decrease. Simultaneously, however, the sodium concentration gradient increases as the intracellular sodium drops, so the passive influx of sodium increases. Similarly, the rise in intracellular potassium results in increased negativity of the membrane potential (to the extent that the resting potential depends solely on the potassium and chloride equilibrium potential), and this, too, results in increased passive sodium influx. Assuming the potassium and chloride to be at equilibrium in homeostasis, then homeostasis will occur when the potassium and chloride ions have distributed themselves across the membrane so that the membrane potential they set up attracts sodium into the cell; the internal sodium concentration is such that the pump now expels any further amounts of sodium ions that attempt to enter by exactly matching the rate of net passive influx. If potassium (and chloride) are not exactly at equilibrium, they will have net passive fluxes across the membrane, and homeostasis will result at that combination of intracellular potassium, chloride, and sodium concentrations and membrane potential at which the active fluxes of an ion pump will exactly balance all net passive fluxes.

Homeostasis of erythrocytes

Tosteson and Hoffman[82] have considered the requirements for homeostasis in sheep erythrocytes both theoretically and experimentally. The erythrocyte differs fundamentally from muscle and nerve cells considered thus far in that anions, not cations, are in equilibrium across its membrane and determine the low (approximately 9 mv) resting potential.[5] Consequently, both sodium and potassium must be actively transported. However, there are two genetically distinct types of sheep erythrocytes, those with normally high potassium and low sodium intracellularly (HK cells) and those with high sodium and low potassium intracellularly (LK cells). Tosteson and Hoffman attempted to evaluate the intracellular cation contents in terms of selectivity of the membrane to potassium and sodium, the ratio of potassium pumped to that leaked passively through the membrane, and the ratio of sodium ion pumped to potassium ion pumped. For their theoretical calculations, they made five assumptions that can be summarized as follows:

1. Water is in equilibrium across the membrane; i.e., there is no hydrostatic pressure gradient (if there were, these extensible cells would swell or shrink).
2. Anions are in equilibrium across the membrane; their distribution follows the Donnan relations, and the membrane potential is determined by their distribution.
3. There is electroneutrality inside and outside the cell.
4. The passive fluxes of sodium and potassium follow the flux ratio equations (given earlier in this chapter).
5. The ratio of sodium pumped to potassium pumped is a constant.

Proceeding from these assumptions and from the experimental finding that the coupling of the pump for sodium and potassium was slightly greater than 1:1, they made the following correct predictions: (1) in LK cells the membrane would be more selective to potassium than in HK cells but (2) in HK cells there would be far more potassium pumped in relative to the passive leakage than in LK cells, and the pump would work 4 times faster. In other words, LK cells had more sodium than potassium internally, not because they did not discriminate against sodium, but because the potassium pump did not work fast enough to build up the internal potassium concentration over what leaked back out. Conversely, HK cells had high intracellular concentrations of potassium, despite the fact that sodium and potassium were about equally permeable, because the pump rate was high compared to the rate of leakage.

Finally, Tosteson and Hoffman were able to account quantitatively for cell swelling that occurred when the ion pumps were poisoned and internal cation contents changed. Stein,[7] building on this work, argued that in cells such as erythrocytes in which neither sodium nor potassium were in equilibrium, there would be less initial swelling than in cells in which one or more ions were initially in equilibrium across the membrane. In poisoned erythrocytes, for example, sodium would passively leak in and potassium would leak out, but if there were nearly 1:1 exchanges, the changes in net internal cation content and hence in cell water would be slow. Conversely, in cells such as muscle, in which potassium and chloride are in and remain in equilibrium due to their high permeabilities, the principal ion movements will be those of sodium inward, together with accompanying water. The latter cells should therefore swell more rapidly than the former when poisoned.

Appendix A

DIFFUSION EQUATION

The force on a mole of a substance in solution is the negative gradient of the chemical potential (μ):

$$\text{Force} = \frac{-d\mu}{dx} = -RT\frac{d(\ln a)}{dx} \qquad \text{(A-1)}$$

where:

R = Gas constant
T = Absolute temperature
a = Activity of the substance

In most biologic work the activities are not known, so the concentration (C) is used instead:

$$\text{Force} = -RT\frac{d(\ln C)}{C} = \frac{-RT}{C}\frac{dC}{dx} \qquad \text{(A-2)}$$

The velocity imposed upon this substance is the product of the force and of its generalized mobility (u′):

$$\text{Velocity} = \frac{-u'RT}{C}\frac{dC}{dx} \qquad \text{(A-3)}$$

where the units of u′ are (cm/sec)/(dyne/mole).

Finally, the flux $\frac{ds}{dt}$ of the substance through an area of solution (A) is the product of its velocity, its concentration, and the area:

$$\frac{ds}{dt} = -A(C)\frac{u'RT}{C}\frac{dC}{dx} = u'RT\frac{dC}{dx} \qquad \text{(A-4)}$$

This can be compared with equation 2 in the text:

$$\frac{ds}{dt} = -DA\frac{dC}{dx}$$

From this it can be seen that the diffusion coefficient (D) is equal to u′RT.

Appendix B

DIFFUSION OF IONS

Just as the driving force for diffusion of nonelectrolytes was the negative gradient of the chemical potential (u) in Appendix A, so the driving force for diffusion of ions is the negative gradient of the electrochemical potential (μ'):

$$\mu' = \mu + V \qquad \text{(B-1)}$$

$$\text{Force/mole} = \frac{-d\mu'}{dx} = -\left(\frac{d\mu}{dx} + \frac{zFdV}{dx}\right) \qquad \text{(B-2)}$$

$$= -\left(\frac{RT}{C}\frac{dC}{dx} + zF\frac{dV}{dx}\right)$$

where:

F (Faraday) = 10^5 coulomb/mole
z = Ionic valency

The velocity of an ion will be proportional to the force on it and to its mobility u′ (in [cm/sec]/[dyne/mole]):

$$\text{Velocity} = -u'\left(\frac{RT}{C}\frac{dC}{dx} + zF\frac{dV}{dx}\right) \qquad \text{(B-3)}$$

The flux will be proportional to the concentration of ions moving at the given velocity across a given area (A):

$$\text{Flux} = \frac{ds}{dt} = -u'C\ A\left(\frac{RT}{C}\frac{dC}{dx} + zF\frac{dV}{dx}\right) \quad \text{(B-4)}$$

When dealing with ions, it is more convenient to use the electric mobility (u), whose dimensions are (cm/sec)/(volts/cm). The generalized mobility (u′) and the electric mobility u are related by the following:

$$u' = zFu \quad \text{(B-5)}$$

Likewise, ionic fluxes are currents (movements of charge) and can be so written:

$$\text{Flux} = \frac{ds}{dt} = \frac{I'}{zF} \quad \text{(B-6)}$$

Combining B-5 and B-6 with B-4, and considering the current per unit area I (=I′/A), there results:

$$I = -zu\left(RT\frac{dC}{dx} + zCF\frac{dV}{dx}\right) \quad \text{(B-7)}$$

This is the basic equation for an ionic species moving passively under an electrochemical gradient. In a special case of great practical interest there is a certain electrochemical gradient for each ionic species in which there is no net movement at all. That is, the ionic current (I_{ion}) is zero. Such an electrochemical gradient may exist across cell membranes. So long as the mobility u of the ion in the membrane is not zero (i.e., the membrane is permeable to the ion), then no ionic current means that:

$$RT\frac{dC}{dx} = -zFC\frac{dV}{dx}$$

or

$$\frac{RT}{C}\frac{dC}{dx} = -zF\frac{dV}{dx} \quad \text{(B-8)}$$

Such an equation states that the forces due to the gradient of the chemical potential and to the gradient of the electric potential are equal and opposite. That is, there is no net force on the ions of this species.

For a given concentration ratio, $(C)_2/(C)_1$, there may be calculated the potential difference, $V_2 - V_1 = V$, at which there will be this zero net force on these ions. Rearranging equation B-8, then:

$$\frac{-RT}{zF} \cdot \frac{dC}{C} = dV \quad \text{(B-9)}$$

Integrating equation B-9 between the two sides of the membrane:

$$\frac{-RT}{zF}\int_{C_1}^{C_2}\frac{dC}{C} = \int_{V_1}^{V_2} dV$$

There results the well-known Nernst equation:

$$V_2 - V_1 = V = \frac{-RT}{zF}\ \ln\frac{(C)_2}{(C)_1} \quad \text{(B-10)}$$

Several simplifications of this equation are possible. First, the predominant ions in tissue are monovalent (K^+, Na^+, and Cl^-), and so z is ±1. Second, it is more convenient to use common logarithms than natural logarithms, so that equation B-10 becomes:

$$V = \mp 2.303\frac{RT}{F}\log\frac{(C)_2}{(C)_1} \quad \text{(B-11)}$$

Finally, at 20° C, 2.303 RT/F = about 58 mv; at 37° C, 2.303 RT/F = about 61 mv.

Appendix C

CONSTANT FIELD EQUATION[38]

The approach here follows that of Hodgkin and Katz[48] and begins with equation B-7. The assumption here is that the there is a constant electric field across the membrane, so that:

$$\frac{dV}{dx} = \frac{V}{a} \quad \text{(C 1)}$$

where:

a = Thickness of the membrane

Then:

$$I = -zu\left(RT\frac{dC}{dx} + zCF\frac{V}{a}\right) \quad \text{(C-2)}$$

rearranging:

$$I + \frac{uz^2VFC}{a} = -zuRT\frac{dC}{dx}$$

$$dx = -zuRT\left(\frac{dC}{I + \frac{uVFC}{a}}\right)$$

where:

$z^2 = 1$ for univalent ions

This equation may be integrated across the membrane, using the substitution:

$$y = I + \frac{uVFC}{a}$$

$$\frac{dy}{\left(\frac{uVF}{a}\right)} = dC \quad \text{(C-3)}$$

Then:

$$\int_0^a dx = \frac{-uzRT}{\left(\frac{uVF}{a}\right)} \int_{I + \frac{UVF}{a}C_o}^{I + \frac{UVF}{a}C_i} \frac{dy}{y} = \frac{-azRT}{VF} \int_{I + \frac{UVF}{a}C_o}^{I + \frac{UVF}{a}C_i} \frac{dy}{y} \quad \text{(C-4)}$$

The result of integration is:

$$a = \frac{-azRT}{VF} \ln \frac{I + \left(\frac{uVF}{a}\right)C_i}{I + \left(\frac{uVF}{a}\right)C_o} \quad \text{(C-5)}$$

The a's cancel out, and upon rearranging and taking exponentials of both sides:

$$\exp\left(\frac{-VF}{zRT}\right) = \frac{I + \left(\frac{uVF}{a}\right)C_i}{I + \left(\frac{uVF}{a}\right)C_o}$$

This equation may be solved for I:

$$I = \frac{uVF}{a}\left(\frac{C_o\exp\left(\frac{-VF}{zRT}\right) - C_i}{1 - \exp\left(\frac{-VF}{zRT}\right)}\right) \quad \text{(C-6)}$$

Since the diffusion constant D = uRT/F for an ion, and the permeability constant P = D/a = uRT/aF, equation C-6 becomes:

$$I = \left(\frac{F^2V}{RT}\right) \cdot P \cdot \left(\frac{C_o\exp\left(\frac{-VF}{zRT}\right) - C_i}{1 - \exp\left(\frac{-VF}{zRT}\right)}\right) \quad \text{(C-7)}$$

This is the basic constant field equation for passive ionic current across a membrane under concentration and potential gradients. For cations, the equation becomes:

$$I_{cat} = \left(\frac{F^2V}{RT}\right)P_{cat}\left(\frac{(C)_o\exp\left(\frac{-VF}{RT}\right) - (C)_i}{1 - \exp\left(\frac{-VF}{RT}\right)}\right) \quad \text{(C-8)}$$

For an anion:

$$I_{an} = \left(\frac{F^2V}{RT}\right)P_{an}\left(\frac{(A)_i\exp\left(\frac{-VF}{RT}\right) - (A)_o}{1 - \exp\left(-\frac{VF}{RT}\right)}\right) \quad \text{(C-9)}$$

What is often of interest is the potential that will result when there are equal numbers of anions and cations diffusing across a membrane, so that their charges neutralize each other and there is no net current, although there is, of course, net flux. The currents will be assumed to be made up of sodium, potassium, and chloride fluxes, since these are the predominant biologic ions of interest. Then:

$$I_{net} = I_{Na} + I_K + I_{Cl} = 0 \quad \text{(C-10)}$$

Call:

$$w = P_K(K)_o + P_{Na}(Na)_o + P_{Cl}(Cl)_i$$
$$y = P_K(K)_i + P_{Na}(Na)_i + P_{Cl}(Cl)_o$$

The net current I_{net} is thus:

$$I_{net} = \left(\frac{F^2V}{RT}\right)\frac{1}{1 - \exp\left(\frac{-VF}{RT}\right)}\left[P_K\left(K_o\exp\left(\frac{-VF}{RT}\right) - K_i\right) + P_{Na}\left(Na_o\exp\left(\frac{-VF}{RT}\right) - Na_i\right) + P_{Cl}\left(Cl_i\exp\left(\frac{-VF}{RT}\right) - Cl_o\right)\right] = \left(\frac{F^2V}{RT}\right)\frac{1}{1 - \exp\left(\frac{-VF}{RT}\right)}\left(w\exp\left(\frac{-VF}{RT}\right) - y\right) = 0 \quad \text{(C-11)}$$

Thus:

$$\exp\left(\frac{-VF}{RT}\right) = \frac{y}{w} \qquad (C\text{-}12)$$

Upon taking logarithms of both sides:

$$\frac{-VF}{RT} = \ln\left(\frac{y}{w}\right)$$

Finally, upon rearranging terms and writing out y and w, there results what has come to be known as the Goldman-Hodgkin-Katz equation:

$$V = -RT/F \ln \frac{P_K(K)_i + P_{Na}(Na)_i + P_{Cl}(Cl)_o}{P_K(K)_o + P_{Na}(Na)_o + P_{Cl}(Cl)_i} \qquad (C\text{-}13)$$

REFERENCES

General reviews

1. Albers, R. W.: Biochemical aspects of active transport, Ann. Rev. Biochem. **36:**727, 1967.
2. Cole, K. S.: Membranes, ions and impulses, Berkeley, 1968, University of California Press.

2a. DeRobertis, E. D. P., Nowinski, W. W., and Saez, F. A.: Cell biology, Philadelphia, 1965, W. B. Saunders Co.

3. Dick, D. A. T.: Cell water, Washington, D. C., 1966, Butterworth & Co.
4. Hendler, R. W.: Biological membrane ultrastructure, Physiol. Rev. **51:**66, 1971.
5. Kernan, R. P.: Cell K, Washington, D. C., 1965, Butterworth & Co.
6. Skou, J. C.: Enzymatic basis for active transport of Na^+ and K^+ across cell membranes, Physiol. Rev. **45:**596, 1965.
7. Stein, W. D.: The movement of molecules across cell membranes, New York, 1967, Academic Press, Inc.

Original papers

8. Adrian, R. H.: The effect of internal and external potassium concentration on the membrane potential of frog muscle, J. Physiol. **133:**631, 1956.
9. Adrian, R. H., and Slayman, C. L.: Membrane potential and conductance during transport of sodium, potassium, and rubidium in frog muscle, J. Physiol. **184:**970, 1966.
10. Armstrong, W. McD., and Lee, C. O.: Sodium and potassium activities in normal and "sodium-rich" frog skeletal muscle, Science **171:**413, 1971.
11. Baker, P. F., Hodgkin, A. L., and Shaw, T. I.: The effects of changes in internal ionic concentrations on the electrical properties of perfused giant axons, J. Physiol. **164:**355, 1962.
12. Bar, R. S., Deamer, D. W., and Cornwall, D. G.: Surface area of human erythrocyte lipids: reinvestigation of experiments on plasma membrane, Science **153:**1010, 1966.
13. Boyle, P. J., and Conway, E. J.: Potassium accumulation in muscle and associated changes, J. Physiol. **100:**1, 1941.
14. Bozler, E.: Osmotic properties of amphibian muscle, J. Gen. Physiol. **49:**37, 1965.
15. Brandt, P. W.: A consideration of the extraneous coats of the plasma membrane, Circulation **26:**1075, 1962.
16. Brinley, F. J., Jr.: Sodium and potassium fluxes in isolated barnacle muscle fibers, J. Gen. Physiol. **51:**445, 1968.
17. Brinley, F. J., Jr., and Mullins, L. J.: Sodium fluxes in internally dialyzed squid axons, J. Gen. Physiol. **52:** 181, 1968.
18. Bunch, W. H., and Kallsen, G.: Rate of intracellular diffusion as measured in barnacle muscles, Science **164:**1178, 1969.
19. Caldwell, P. C., Hodgkin, A. L., Keynes, R. D., and Shaw, T. I.: Effects of injecting energy-rich phosphate compounds on the active transport of ions in the giant axons of Loligo, J. Physiol. **152:**561, 1960.
20. Conway, E. J.: Nature and significance of concentration relations of potassium and sodium ions in skeletal muscle, Physiol. Rev. **37:**84, 1957.
21. Coombs, R. R. A., and Lachmann, P. J.: Immunological reactions at the cell surface, Br. Med. Bull. **24:**113, 1968.
22. Cope, F. W.: Nuclear magnetic resonance evidence using D_2O for structured water in muscle and brain, Biophys. J. **9:**303, 1969.
23. Curtis, H. J., and Cole, K. S.: Membrane resting and action potentials from the squid giant axon, J. Cell. Comp. Physiol. **19:**135, 1942.
24. Dainty, J., and House, C. R.: An examination of the evidence for membrane pores in frog skin, J. Physiol. **185:**172, 1966.
25. Danielli, J. F., and Davson, H.: A contribution to the theory of permeability of thin films, J. Cell. Comp. Physiol. **5:**495, 1935.
26. Davson, H., and Danielli, J. F.: The permeability of natural membranes, ed. 2, London, 1952, Cambridge University Press.
27. Desmedt, J. E.: Electrical activity and intracellular sodium concentration in frog muscle, J. Physiol. **121:**191, 1953.
28. Donnan, F. G.: The theory of membrane equilibria, Chem. Rev. **1:**73, 1924.
29. Fernandez-Moran, H.: Cell-membrane ultrastructure, Circulation **26:**1039, 1962.
30. Finean, J. B.: The nature and stability of the plasma membrane, Circulation **26:**1151, 1962.
31. Finklestein, A., and Cass, A.: Permeability and electrical properties of thin lipid membranes, J. Gen. Physiol. **52:**1455, 1968.
32. Gaffey, C. T., and Mullins, L. J.: Ion fluxes during the action potential in Chara, J. Physiol. **144:**505, 1958.
33. Garrahan, P. J., and Glynn, I. M.: The behaviour of the sodium pump in red cells in the absence of external potassium, J. Physiol. **192:**159, 1967.
34. Garrahan, P. J., and Glynn, I. M.: The stoichiometry of the sodium pump, J. Physiol. **192:**217, 1967.
35. Garrahan, P. J., and Glynn, I. M.: The incorporation of inorganic phosphate into adenosine triphosphate by reversal of the sodium pump, J. Physiol. **192:**237, 1967.

36. Geren, B. B.: The formation from the Schwann cell surface of myelin in the peripheral nerves of chick embryos, Exp. Cell Res. **7**:558, 1954.
37. Glynn, I. M.: The action of cardiac glycosides on sodium and potassium movements in human red cells, J. Physiol. **136**:148, 1957.
38. Goldman, D. E.: Potential, impedance, and rectification in membranes, J. Gen. Physiol. **27**:37, 1943.
39. Gorter, E., and Grendel, F.: On bimolecular layers of lipoids on the chromatocytes of the blood, J. Exp. Med. **41**:439, 1925.
40. Gutknecht, J.: Membranes of Valonia ventricosa: apparent absence of water-filled pores, Science **158**:787, 1967.
41. Hagins, W. A., Penn, R. D., and Yoshikama, S.: Dark current and photocurrent in retinal rods, Biophys. J. **10**:380, 1970.
42. Henn, F. A., and Thompson, T. E.: Synthetic lipid bilayer membranes, Ann. Rev. Biochem. **38**:241, 1969.
42a. Henn, F. A., Decker, G. L., Greenawalt, J. W., and Thompson, T. E.: Properties of lipid bilayer membranes separating two aqueous phases: electron microscope studies. J. Mol. Biol. **24**:51, 1967.
43. Hinke, J. A. M.: The measurement of sodium and potassium activities in the squid axon by means of cation-selective glass microelectrodes, J. Physiol. **156**:314, 1961.
44. Hinke, J. A. M.: Glass microelectrodes in the study of binding and compartmentalization of intracellular ions. In Lavallee, M., Schanne, O. F., and Hébert, N. C., editors: Glass microelectrodes, New York, 1969, John Wiley & Sons, Inc.
45. Hinke, J. A. M.: Solvent water for electrolytes in the muscle fiber of the giant barnacle, J. Gen. Physiol. **56**:521, 1970.
46. Hodgkin, A. L., and Horowicz, P.: The effect of sudden changes in ionic concentrations on the membrane potential of single muscle fibers, J. Physiol. **148**:127, 1959.
47. Hodgkin, A. L., and Huxley, A. F.: Resting and action potentials in single nerve fibers, J. Physiol. **104**:176, 1945.
48. Hodgkin, A. L., and Katz, B.: The effect of sodium ions on the electrical activity of the giant axon of the squid, J. Physiol. **108**:37, 1949.
49. Hodgkin, A. L., and Keynes, R. D.: The mobility and diffusion coefficient of potassium in giant axons from Sepia, J. Physiol. **119**:513, 1953.
50. Hodgkin, A. L., and Keynes, R. D.: Active transport of cations in giant axons from Sepia and Loligo, J. Physiol. **128**:28, 1955.
51. Huxley, A. F., and Stämpfli, R.: Direct determination of membrane resting potential and action potential in single myelinated nerve fibers, J. Physiol. **112**:476, 1951.
52. Jacobs, M. H.: Early osmotic history of the plasma membrane, Circulation **26**:1013, 1962.
53. Katzman, R., Lehrer, G. M., and Wilson, C. E.: Sodium and potassium distribution in puffer fish supramedullary nerve cell bodies, J. Gen. Physiol. **54**:232, 1969.
54. Koike, H., Brown, H. M., and Hagiwara, S.: Hyperpolarization of a barnacle photoreceptor membrane following illumination, J. Gen. Physiol. **57**:723, 1971.
55. Korn, E. D.: Cell membranes: structure and synthesis, Ann. Rev. Biochem. **38**:263, 1969.
56. Kuffler, S. W., Nicholls, J. G., and Orkand, R. K.: Physiological properties of glial cells in the central nervous system of amphibia, J. Neurophysiol. **29**:768, 1966.
57. Kushmerick, M. J., and Podolsky, R. J.: Ionic mobility in muscle cells, Science **166**:1297, 1969.
58. LeFevre, P. G.: The osmotically functional water content of the human erythrocyte, J. Gen. Physiol. **47**:585, 1964.
59. Ling, G.: The role of phosphorus in the maintenance of the resting potential and selective ionic accumulation in frog muscle cells. In McElroy, W. D., and Glass, H. B., editors: Phosphorus metabolism, Baltimore, 1952, The Johns Hopkins University Press, vol. 2.
60. Ling, G., and Gerard, R. W.: The normal membrane potential of frog sartorius fibers, J. Cell. Comp. Physiol. **34**:383, 1949.
61. Lucké, B., and McCutcheon, M.: The living cell as an osmotic system and its permeability to water, Physiol. Rev. **12**:68, 1932.
62. Lucké, B., Larrabee, M. G., and Hartline, H. K.: Studies on osmotic equilibrium and on the kinetics of osmosis in living cells by a diffraction method, J. Gen. Physiol. **19**:1, 1935.
63. Lucy, J. A.: Ultrastructure of membranes: micellar organization, Br. Med. Bull. **24**:127, 1968.
64. Marmor, M. F., and Gorman, A. L. F.: Membrane potential as the sum of ionic and metabolic components, Science **167**:65, 1970.
65. Moreton, R. B.: An application of the constant-field theory to the behaviour of giant neurones of the snail, Helix aspersa, J. Exp. Biol. **48**:611, 1968.
66. Mueller, P., Rudin, D. O., Tien, H. T., and Wescott, W. E.: Reconstitution of excitable cell membrane structure in vitro, Circulation **26**:1167, 1962.
67. Mullins, L. J., and Brinley, F. J., Jr.: Potassium fluxes in dialyzed squid axons, J. Gen. Physiol. **53**:704, 1969.
68. Mullins, L. J., and Frumento, A. S.: The concentration dependence of sodium efflux from muscle, J. Gen. Physiol. **46**:629, 1963.
69. Nakajima, S., and Onodera, K.: Post-tetanic hyperpolarization and electrogenic Na pump in stretch receptor neurone of crayfish, J. Physiol. **187**:105, 1966.
70. Nastuk, W. L., and Hodgkin, A. L.: The electrical activity of single muscle fibers, J. Cell. Comp. Physiol. **35**:39, 1950.
71. Orkand, R. K., Nicholls, J. G., and Kuffler, S. W.: Effect of nerve impulses on the membrane potential of glial cells in the central nervous system of amphibians, J. Neurophysiol. **29**:788, 1966.
72. Rand, R. P.: The structure of a model membrane in relation to the viscoelastic properties of the red cell membrane, J. Gen. Physiol. **52**:1735, 1968.
73. Rega, A. F., Pouchan, M. I., and Garrahan, P. J.: Potassium ions asymmetrically activate

erythrocyte membrane phosphatase, Science **167:**55, 1970.
74. Robertson, J. D.: The ultrastructure of cell membranes and their derivatives, Biochem. Soc. Symp. **16:**3, 1959.
75. Robertson, J. D.: Unit membranes: a review with recent new studies of experimental alterations and a new subunit structure in synaptic membranes. In Locke, M., editor: Cellular membranes in development, New York, 1964, Academic Press, Inc.
76. Robinson, J. R.: Metabolism of intracellular water, Physiol. Rev. **40:**112, 1960.
77. Schultz, S. G., and Curran, P. F.: Coupled transport of sodium and organic solutes, Physiol. Rev. **50:**637, 1970.
78. Slayman, C. L., and Slayman, C. W.: Measurement of membrane potentials in Neurospora, Science **136:**875, 1962.
79. Smith, H. W.: The plasma membrane, with notes on the history of botany, Circulation **26:** 987, 1962.
80. Steinbach, H. B.: Sodium extrusion from isolated frog muscle, Am. J. Physiol. **167:**284, 1951.
81. Tien, H. T., and Diana, A. L.: Bimolecular lipid membranes: a review and a summary of some recent studies, Chem. Phys. Lipids **2:** 55, 1968.
82. Tosteson, D. C., and Hoffman, J. F.: Regulation of cell volume by active cation transport in high and low potassium sheep red cells, J. Gen. Physiol. **44:**169, 1960.
83. Ussing, H. H.: Distinction by means of tracers between active transport and diffusion. The transfer of iodide across the isolated frog skin, Acta Physiol. Scand. **19:**43, 1949.
84. Ussing, H. H.: Active transport of inorganic ions, Symp. Soc. Exp. Biol. **8:**407, 1954.
85. Whittam, R., and Ager, M. E.: The connexion between active cation transport and metabolism in erythrocytes, Biochem. J. **97:**214, 1965.

2

F. J. BRINLEY, Jr.

Excitation and conduction in nerve fibers

FUNCTION OF THE NERVE IMPULSE IN THE PERIPHERAL NERVOUS SYSTEM

The peripheral nervous system functions as a communications network, permitting transmission of information from one part of the organism to another. The message unit in this information transfer is the propagated or conducted nerve impulse.

Several physical changes occur in the nerve during the passage of an impulse, but the alteration most directly related to propagation occurs in the electrical properties of the membrane. The passage of an impulse along a nerve produces an abrupt, brief decrease in the steady resting potential across the axolemma and an associated flow of transmembrane electric current (in the form of ions). The potential change is commonly referred to either as the "action potential" because it indicates activity in the fiber or as the "spike" because electrical recording of the action potential produces a brief spikelike deflection on an oscilloscope trace. The flow of current is referred to as the "action current." Typical action potentials recorded from nerve fibers and muscle cells are shown in Fig. 2-1. Although the shape of the action potential varies considerably in different cells, all are characterized by a very rapid depolarization (upward deflection) and somewhat slower repolarization to the resting or steady potential. Another characteristic feature is the transient reversal of potential polarity at the peak of the spike, the inside of the cell becoming positive with respect to the extracellular fluid, whereas at rest the inside is negative.

Other physical changes in excitable cells do occur during an action potential, e.g., changes in temperature, birefringence, light scattering, and ionic concentrations. However, the relation of these events, which have been less thoroughly studied, to the transient potential change across the membrane is uncertain at present. Some of the changes in optical properties may reflect changes in ionic concentrations immediately outside the axolemma.

There are several important phenomena observed in excitable cells that can be explained on the basis of certain selective ion permeability changes in the cell membrane and ion fluxes across the membrane. A comprehensive discussion of these phenomena is best deferred until after consideration of the classic experiments that have provided the basis for our modern understanding of the properties of excitable membrane. However, in order to provide the working vocabulary necessary for discussion, it is necessary to define and illustrate certain terms commonly used in discussing electrical activity observed in peripheral nerve.

Excitability

The ability to produce an action potential is termed "excitability," the process of generating an action potential is referred to as "excitation," and the class of cells exhibiting such behavior is called "excitable." Although excitation is frequently regarded as a property solely of surface membranes, the process actually depends not only on events within the membrane but also on transmembrane ionic gradients. Broadly viewed, excitability is a property of the entire cell, including its immediate external environment. In mammals the only normally excitable cells are nerve cells and muscle fibers. In lower organisms, tissues other than nerve or muscle can become excitable. Under certain experimental conditions, for example, frog epithelium and algae can produce regenerative action potentials.

Stimulus

The event or process that elicits an action potential in excitable cells is called a stimulus. One of the most common experimental stimuli

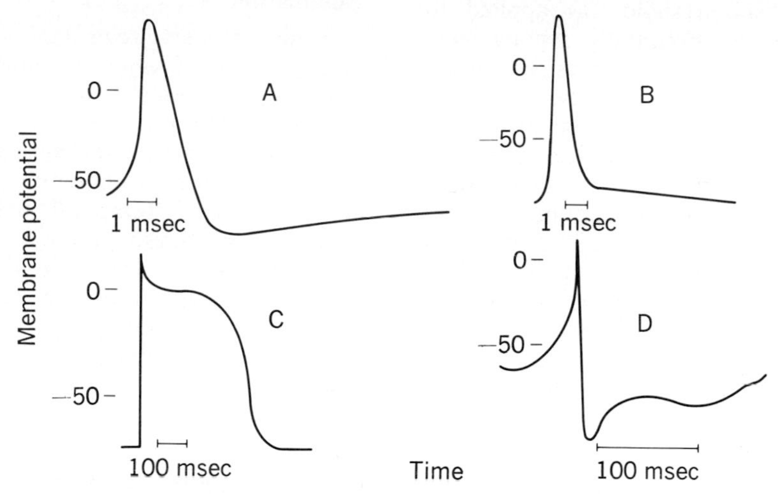

Fig. 2-1. Representative action potentials recorded from excitable cells with intracellular microelectrodes. **A,** Isolated squid axon. **B,** Frog sartorius muscle fiber. **C,** Dog ventricular muscle fiber. **D,** Guinea pig vas deferens. Zero level of potential is indicated on vertical axis. Beginning of trace indicates membrane potential (inside of cell negative) before excitation occurs. Note variation in time scale.

is electricity. By means of an electric shock, current is passed across a membrane to produce a transient depolarization of the resting potential, which, if it is of sufficient duration and magnitude, can initiate the train of events that produces an action potential. Although electric currents are a convenient laboratory means of initiating excitation, they are, strictly speaking, not physiologic stimuli. Examples of physiologic stimuli are as follows: hormonal (acetylcholine acting on the postjunctional membrane of the neuromuscular junction), thermal (skin temperature receptors), mechanical (displacement of outer lamellae of paccinian corpuscle, auditory hair cells), electromagnetic radiation (retinal rods) and chemical (protons and salts acting on gustatory receptors of the tongue).

Threshold

Not all physiologic or experimental stimuli will produce conduced action potentials. Only those stimuli with greater than minimum magnitude (intensity) and duration suffice. The minimum necessary intensity of a stimulus can be referred to as the threshold *stimulus.* (It is equally correct to define a threshold *duration* in a comparable way, but this is not generally done.) A stimulus of less than threshold intensity is referred to as *subthreshold;* one of greater than threshold intensity as *superthreshold.*

The term "threshold" is commonly used, especially by electrophysiologists, to refer either to the absolute magnitude of the cell membrane potential at which an action potential is initiated or to the magnitude of the depolarization from resting potential required to initiate an action potential. The context usually makes the intended usage clear.

The threshold potential for excitation should not be considered a fixed parameter for all cells. The thresholds of different types of cells may vary considerably. Furthermore, the threshold of a single cell can change, either rapidly as after a train of impulses or more slowly in response to metabolic or hormonal influences. The significance of the threshold phenomenon lies in the fact that it allows an excitable cell to function as a signal discriminator. Only those stimuli equal to or greater than the threshold intensity will produce information transfer in the peripheral nervous system.

Local, graded, or subthreshold (subliminal) response

Although a subthreshold stimulus fails to initiate an action potential, it still induces physiologically significant alterations in the membrane potential of a nerve fiber. The time course and magnitude of the stimulus determines, although in a mathematically very complicated manner, the response of the membrane potential. These responses are not propagated or conducted along an excitable cell and are generally manifest only a very short distance from the stimulus point; hence the term "local response." The terms "local," "graded," or "subthreshold" as applied to these responses are essentially synonymous. The exact expression used serves to emphasize a particular aspect of the response. "Local" emphasizes that the response is not propagated, "graded" emphasizes that the configuration of the response is continuously variable or graded with the stimulus, and "subthreshold" indicates that the stimulus cannot initiate an action potential. Strictly speaking, local responses can be observed in any cell, whether excitable or not, since they depend only on the resistivity of the cell membrane and of the internal and external fluids and on the membrane capacitance (see discussion of cable theory, p. 40).

All-or-none response

The expression "all or none" describes the ability of a nerve fiber, once a superthreshold response has been applied to its surface, to initiate an action potential whose configuration is determined solely by the properties of the cell independently of the precise configuration of the exciting stimulus, and to propagate such an action potential a very long distance along the nerve fiber *without variation of wave form* at essentially constant velocity. Although the condition of the fiber may change with time, the action potential configuration and conduction velocity is invariant for a single nerve fiber for at least a short period of time.

The expression "all or none" does not adequately describe the process of initiation of an action potential very close to the stimulus point where variations of configuration or velocity with stimulus may occur, but it does serve to describe a fundamental property of the peripheral nervous system, i.e., *the arrival of an action potential in the central nervous system signals only that a suprathreshold stimulus (with respect to both magnitude and duration) has occurred in the periphery*. An action potential cannot signal the occurrence of a subthreshold stimulus, nor can a single impulse indicate either the magnitude or duration of a suprathreshold stimulus. This latter type of information is coded in the interval between the action potentials. Information transfer within the nervous system, peripheral as well as central, is therefore frequency rather than amplitude modulated. Special cells called receptors, which exist in the periphery of the nervous system, function essentially as amplitude to frequency transducers. (Receptor physiology is discussed in Chapter 9.)

Summation

Under proper circumstances, two or more stimuli, each of which is subthreshold when occurring individually, may combine to cause excitation. This phenomenon is called summation. In the peripheral nervous system the most commonly encountered type of summation is *temporal*. Temporal summation occurs when two subthreshold stimuli are applied in close succession. The local depolarizing response resulting from the second stimulus adds to the residual depolarizing response from the first stimulus, net resultant depolarization of the membrane exceeds threshold, and excitation results. A second type of summation of great significance in the integrative function of the central nervous system is called *spatial* summation. Two subthreshold stimuli occur simultaneously but at different loci on a neuron. A local response is greatest at the point of stimulus application but does produce depolarization in adjacent regions. Therefore subthreshold responses arising from two (or more) loci can sum to produce threshold depolarization at another locus and cause excitation.

Refractory period

During the period in which an excitable membrane is producing an action potential in response to a suprathreshold stimulus, the ability of the membrane to respond to a second stimulus of any sort is markedly altered. During the initial portion of the spike the membrane cannot respond to any stimulus no matter how intense; this interval is called the *absolute* refractory period. Following the absolute refractory period, an action potential can be produced first by very intense stimuli and then gradually by stimuli of progressively lesser magnitude. This interval is called the *relative* refractory period or sometimes the *subnormal* period. This refractory behavior of the excitable membrane is also frequently described as a change in threshold. Initially the threshold is infinite (absolute refractory period); then it declines (relative refractory period) to normal. In some cases following an action potential there are intervals in which small, long-lasting changes in threshold occur; these are designated as subnormal (increased threshold) and supranormal (decreased threshold) periods.

Accommodation

Accommodation refers to the fact that the rate of membrane potential change during the

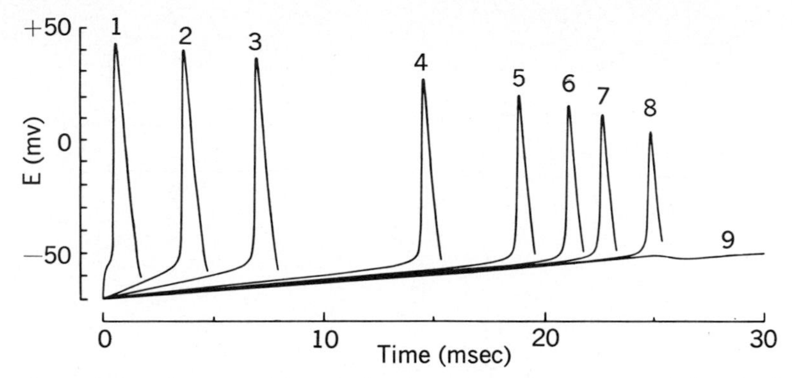

Fig. 2-2. Theoretical action potentials calculated to show effect of rate of rise of stimulus on final threshold for excitation (frog nerve). Traces show action potential produced by linearly varying electrical stimuli with various rates of rise. Threshold depolarization was 21 mv for step pulse (trace *1*) and 28 mv for slowest rising stimulus capable of producing an action potential (trace *8*). In trace *9*, rate of rise was too slow to initiate an action potential and only a subthreshold response resulted. (From Frankenhaeuser and Vallbo.[9])

application of a stimulus can affect the threshold voltage at which excitation finally occurs. The effect is illustrated in Fig. 2-2 for electrical stimuli with various rise times. The more slowly the stimulus depolarizes the membrane, the greater the depolarization required to initiate an action potential, i.e., the lower the absolute potential at which excitation occurs, and the greater the total current required to stimulate. The membrane behaves as if it were becoming less excitable during the period of application of a stimulus and accommodates itself to the presence of the stimulus; hence the term "accommodation."

Electrotonic conduction

The potential changes associated with an action potential in excitable tissues propagate along the cell length with a velocity ranging from centimeters per second to tens of meters per second for different cells with no essential loss of amplitude. In sharp contrast the alterations in potential induced by subthreshold depolarizations attenuate very rapidly with distance from the stimulus point, but the disturbance is manifested (i.e., propagated) extremely rapidly, the delay being due to the distributed membrane capacitance. The terms "electrotonic" conduction and "decremental" conduction are used to describe this disturbance produced in adjacent membrane by a localized subthreshold stimulus.

Stimulus artifact

The technical term "stimulus artifact" refers to any deflection of the recording trace that is produced by the stimulus itself and is not due to any response of the tissue being studied. Various arrangements of stimulating and recording devices are used to control the size of the stimulus artifact and to prevent distortion of the physiologic response. In this chapter we are primarily concerned with the electrical activity of nerve fibers recorded on an oscilliscope trace in response to an electrical stimulus. Under these conditions the stimulus artifact, or "shock artifact" as it is commonly called by electrophysiologists, appears as a large abrupt deflection of the baseline trace.

ELECTRICAL MANIFESTATIONS OF NERVOUS ACTIVITY

Study of the electrical signs of activity in excitable cells is important to a fundamental understanding of nervous phenomena as well as clinical medicine. This section will consider some of the problems of recording and interpreting electrical phenomena in tissues.

Electrical recording of nerve activity

There are several techniques used for recording electrical activity in excitable cells. Since the action potential is manifest as a change in membrane potential, all of these methods require, directly or indirectly, a measurement of the transmembrane potential. Several methods are illustrated schematically in Fig. 2-3. The easiest to understand conceptually is shown in Fig. 2-3, *A*. The transmembrane potential is measured directly as the potential difference between two electrodes, one inside the cell and the other outside. In most cases the membrane resistance

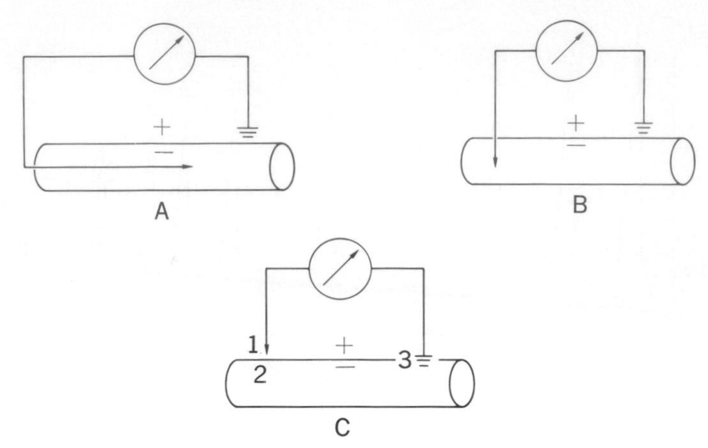

Fig. 2-3. Schematic illustration of three methods of measuring absolute size of resting potential. Potential is measured between tip of recording electrode (arrow) and an indifferent electrode (ground symbol). Membrane polarity inside is negative with respect to outside.

is so large compared to the internal and external resistance that the potential recorded is essentially independent of the position of the recording electrode inside the cell. This method measures not only the action potential produced as the nerve impulse passes along the surface membrane but also the steady resting potential present in the absence of excitation. In fact, the importance of knowing the absolute magnitude of the resting potential was the impetus that led to the development of intracellular recording techniques. The first successful intracellular potential measurements were made as shown in Fig. 2-3, *A*, by inserting a salt-filled glass capillary longitudinally along the axis of a nerve fiber. Given the minimum practical diameter of such long glass capillaries, on the order of 100μ, this approach is feasible with only a few very large cells from marine invertebrates, e.g., the squid giant axon or the barnacle giant muscle fiber. A more common approach (Fig. 2-3, *B*) is to impale the cell transversely with a saline-filled glass microcapillary (called a microelectrode), which terminates in a very short, fine tip (approximately 0.1μ in diameter). In this way the transverse membrane potential of a variety of small structures such as vertebrate muscle fibers, neuromuscular junctions, neurons, and glia has been determined. However, the microelectrode method does considerable damage to the membrane at the impalement site in very small cells and has not generally been satisfactory for use in vertebrate nerve fibers.

Fig. 2-3, *C*, illustrates an approach used to measure action potentials from these fibers. The method involves depolarizing the fiber at one recording site (position *3*, Fig. 2-3, *C*), either by injury or by high concentrations of extracellular potassium, and placing the second recording electrode sufficiently distant so that the electronic depolarization from the cut end is negligible. If the external resistance is made very large, then the transverse membrane potential between points *1* and *2* in Fig. 2-3, *C*, can be recorded, with negligible error, between the external electrodes at points *1* and *3*. This method is generally described as the "air" gap or "sucrose" gap method, depending on the nature of the insulating material used to raise the external resistance.

This description of recording techniques designed to measure the absolute membrane potential belies the extreme technical difficulty of making such measurements. It is far easier and much more common to place a pair of external electrodes outside the cell and record external potential drop between active and inactive regions of the cell surface (Fig. 2-4). The amplitude of these externally recorded action potentials is considerably less than the intracellular action potential. Since the magnitude is a function of the current density at the two recording sites, it depends on the placement of the electrodes with respect to the membrane surface and also on the electrical resistance and geometry of the external medium.

Several arrangements of external recording electrodes are shown in Fig. 2-4. The direction of impulse conduction is from left to right, and the region of axon undergoing depolarization is shown as stippled. *Monophasic* recording is illustrated in Fig. 2-4, *A*. One

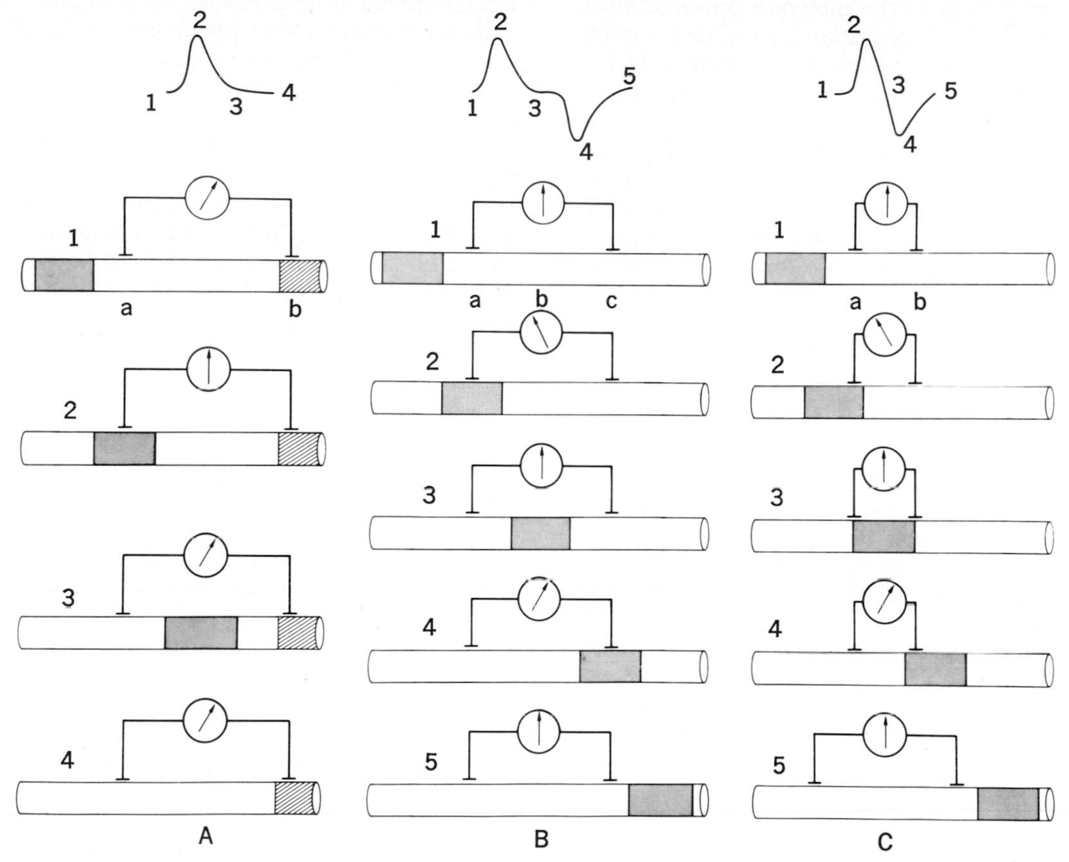

Fig. 2-4. Monophasic and diphasic recording of extracellular action potentials in nerve fibers. Curves at top of figure are actual recorded action potentials. Numbers on curves indicate instantaneous membrane potential when nerve impulse occupies various positions relative to recording electrodes, as shown in diagrams below. Stippled areas represent location of nerve impulse; coarsely hatched regions indicate permanently depolarized nerve. Direction of current flow is indicated by positions of arrows in ammeters.

electrode is placed on healthy tissue and the other is placed on a cut or damaged region. Since the resting potential is intact at position *a* and zero at position *b,* a steady "injury" current flows between the two points, giving rise to a steady "injury" potential measured in the external medium (stage 1). The magnitude of this injury potential depends on the external resistance. It is proportional, but not generally equal, to the resting potential. The direction of current flow in the external medium is indicated by the arrow in the ammeter. As the action potential approaches point *a,* the membrane becomes depolarized and the injury current flow in the external circuit is greatly reduced. Consequently, the potential difference between points *a* and *b* largely disappears, as indicated by the position of the arrow (stage 2). As the wave of excitation passes point *a,* the membrane is repolarized and the injury current is reestablished, with its former magnitude and direction (stages 3 and 4). Since the current flow has not changed direction but only varied from positive to zero and then back to positive, the potential change is *monophasic;* hence the term for this method of recording.

A second type of external recording that does not involve nerve damage is illustrated in Fig. 2-4, *B*. Since both electrodes are placed on intact membrane, the steady current flow between them is zero, as indicated by the arrow (stage 1). As the action potential passes under point *a,* the deflection of the arrow is to the left (stage 2). When the active region is between the electrodes, at point *b,* the current flow from *a* to *b* and *c* to *b* will be opposite in direction and nearly equal in magnitude. Consequently, the net current flow between *a* and *c* and therefore the potential difference is nearly zero (stage 3). When the impulse reaches point *c,* the current flow is in the direction opposite what it was when the excitation passed under point *a*. The resulting potential is therefore in the opposite direction, as indicated by the arrow (stage 4). As the action potential recedes from point *c*, the potential difference between the electrodes returns to zero (stage 5). Thus the complete sequence of potential change consists of two more or less symmetric phases of opposite polarity. This recording situation is therefore called *biphasic*. The exact configuration observed depends, among other variables, on the relative length of the propagating wave and the electrode separation. Fig. 2-4, *C*, illustrates a situation in which the distance between *a* and *b* is less than the wavelength of the nerve impulse (actually the more common experimental circumstance). In such a case there is no time interval during which the propagating wave is entirely contained between the recording electrodes. The isopotential interval of stage 3 in Fig. 2-4, *B*, is therefore reduced to an instant and the biphasic configuration shown in Fig. 2-4, *C*, results. Further shortening of the electrode separation leads to a progressively greater algebraic summation of the separate phases.

Cable theory

Any physical system that consists of a conducting core of material enclosed by a surface layer of relatively high resistance and immersed in a conducting medium evidences *cable properties;* i.e., a displacement of potential across the surface layer at any one point on the cable results in a displacement across adjacent regions of the surface layer. Nervous tissues are, of course, cables in this sense, and the existence of electrotonic potential spread is responsible for such diverse and important nervous phenomena as spatial summation and impulse propagation.

A complete discussion of cable effects in nervous tissue involves a consideration of complicated special cases beyond the scope of this text. Fortunately, an adequate insight into the physiologic significance of cable properties can be gained by examining the distribution of electrotonic potential following a square-wave voltage pulse applied to the axolemma of a hypothetical axon infinitely long, unmyelinated, and of uniform diameter. In order to render the mathematics tractable, a number of simplifying assumptions are made. Although arbitrary, these assumptions appear to be sufficiently realistic so as to introduce no serious error in interpretation. The resistivities of both the axoplasm and external fluid are considered isotropic and constant, although not necessarily equal. The impedance of the membrane is considered as consisting of a capacitance in parallel with a resistance. The electric circuit of a very small segment of the hypothetical axon is shown in Fig. 2-5. A complete circuit for the entire length of the axon would consist of an endless series of such networks connected together.

The basic problem in cable theory is to formulate equations relating potential differences and current flows through an arbitrary segment and then solve the resulting differential equation for various boundary conditions of interest. Since the electrotonic potential is entirely separate from the resting potential and linearly superposable upon it, the resting potential is ignored in cable theory. However, one must remember that the total trans-

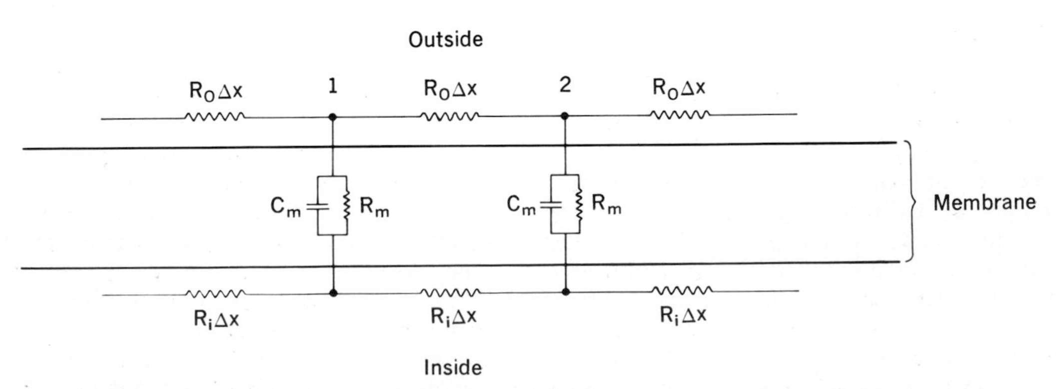

Fig. 2-5. Equivalent electrical circuit of cell. Membrane resistance and capacitance are represented as occurring in discrete packets for convenience; actually they are uniformly distributed along cell surface. Similarly, internal and external resistances are represented as discrete elements, although resistance is distributed throughout intra- and extracellular spaces.

membrane potential will be the algebraic sum of the resting potential and the electrotonic potential.

For convenience, the major assumptions and definitions used in the present derivation of the cable equations are as follows:

Assumptions:

1. Axon represented as infinite circular cylinder
2. Inside of cylinder (cytoplasm) of relatively low electrical resistance
3. Shell of cylinder (membrane) of relatively high electrical resistance
4. Region external to cylinder (extracellular space) of relatively low electrical resistance
5. Surface potential radially symmetric
6. Internal and external potentials independent of radial distance
7. Resistances are ohmic (i.e., voltage and time independent)
8. Electrotonic potential linearly superposable on resting potential

Definitions:

V_i = Inside electrotonic potential (volts)
V_o = Outside electrotonic potential (volts)
R_i = Internal longitudinal resistance (ohm/cm)
R_o = External longitudinal resistance (ohm/cm)
I_m = Radial membrane current (amp/cm)
I_i = Internal longitudinal current (amp)
I_o = External longitudinal current (amp)
V_m = $V_i - V_o$ = Electrotonic potential across membrane (volts)
C_m = Membrane capacitance per unit length of cylinder (farads/cm)
R_m = Membrane resistance per unit length of cylinder (ohm-cm)

The derivation involves repeated use of Ohm's law (current flow is proportional to voltage) and Kirchoff's law (the sum of the currents flowing into a circuit node is zero). In the external fluid (Fig. 2-5), current flow from point *1* to point *2* is I_o. The potential difference is:

$$V_o(1) - V_o(2) = \Delta V_o = I_o R_o \Delta X \qquad (1)$$

$$\left(\text{amp} \times \frac{\text{ohms}}{\text{cm}} \times \text{cm}\right)$$

If this is true for a finite increment (ΔX), it must be true for an infinitesimal increment (dx). Hence:

$$\frac{\partial V_o}{\partial x} = I_o R_o \qquad (2)$$

and similarly:

$$\frac{\partial V_i}{\partial x} = I_i R_i \qquad (3)$$

By subtraction, we get a relation between the electrotonic potential and longitudinal current flow inside and outside the cylinder:

$$\frac{\partial (V_o - V_i)}{\partial x} = \frac{\partial V_m}{\partial x} = I_o R_o - I_i R_i \qquad (4)$$

Next we derive an expression relating change in external longitudinal current (I_o) to the transmembrane current (I_m):

$$\Delta I_o = -\Delta I_i = I_m \Delta X \text{ (amp)} \qquad (5)$$

Hence:

$$\frac{\partial I_o}{\partial x} = \frac{\partial I_i}{\partial x} = I_m \qquad (6)$$

By differentiating equation 4 and substituting into equation 6, we get a fundamental relationship:

$$\frac{\partial^2 V_m}{\partial x^2} = \frac{\partial (I_o R_o - I_i R_i)}{\partial x} = I_m (R_o + R_i) \qquad (7)$$

Now examine the components of I_m, the transmembrane current. It has been shown that the membrane behaves electrically as though it were a leaky condensor, i.e., as though it were a pure capacitance connected in parallel with a pure resistance. The total transmembrane current (I_m) will be the sum of the current through the resistor plus the displacement current across the capacitor (Fig. 2-6). Thus:

$$I_m = I_R + I_C = V_m/R_m + C_m \frac{\partial V_m}{\partial t} \qquad (8)$$

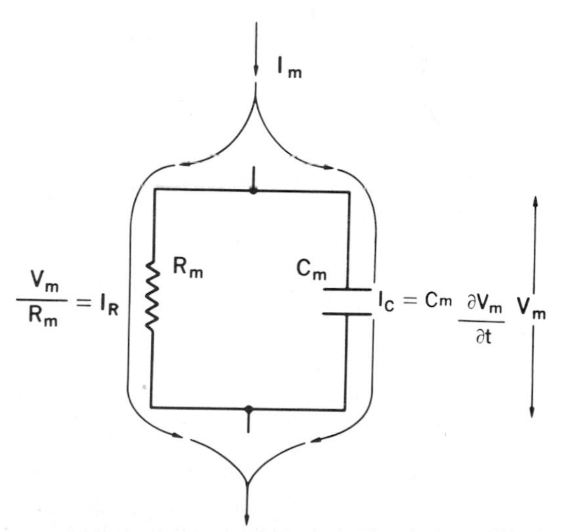

Fig. 2-6. Schematic diagrams showing passage of membrane current, I_m, across small patch of membrane. Part of membrane current passes across resistive element; this is ionic current, I_R. The other component, I_C, results from change in membrane potential across membrane capacitance. This charge, moving in response to potential change, does *not* physically cross capacitor but leaves (or enters) plates.

Combining equations 7 and 8 gives a partial differential equation relating distance along the cylinder, time after initiation of signal, and known membrane parameters:

$$\frac{\partial^2 V_m}{\partial x^2} = \left(\frac{V_m}{R_m} + C_m\frac{\partial V_m}{\partial t}\right) \times (R_o + R_i) \qquad (9)$$

It is convenient to rearrange equation 9 and to define two new variables:

$$-\lambda^2 \frac{\partial^2 V_m}{\partial x^2} + \tau\frac{\partial V_m}{\partial t} + V_m = 0 \qquad (10)$$

where:

$$\tau = R_m C_m$$

$$\lambda^2 = \frac{R_m}{R_o + R_i}$$

Note that the membrane time constant (τ) depends only on the membrane parameters, whereas the length constant (λ) involves not only the membrane resistance but also the internal and external resistance.

The form of solutions of equation 9 depends on boundary conditions. One simple form is the time-independent displacement of potential along an infinite nerve fiber, which results when an initial displacement of potential (V_0) is made at some point on the nerve surface ($x = 0$).

$$V_m = V_o \exp(-x/\lambda) \qquad (11)$$

Recall that total potential across the membrane equals the electrotonic potential with a spatial variation plus the resting potential, which is everywhere constant.

A note on units

R_m, R_i, and R_o have been defined in terms appropriate to a particular fiber. In order to compare data from fibers of different parameters, it is convenient to have formulas that contain dimensions and cell parameters explicitly. Consider a fiber of radius a and length L, and recall the definition of resistance. $R = \frac{L\rho}{A}$, where ρ is the resistivity (ohm-cm) of the interior medium, and A is the cross-sectional area. Therefore the interior resistance per unit length is $\frac{\rho}{A} = \frac{\rho}{\pi a^2}$. In most cases fibers are immersed in large volumes of salt-containing solutions, so that the external resistance per unit length is negligible compared to R_i, even though the external and internal resistivities may be comparable.

If the fiber has a radius of a, then its surface area per unit length is $2\pi a$. If the resistance per unit length is R_m, then another quantity, the specific membrane resistance (r_m) can be defined as $r_m = R_m \times 2\pi a$. The specific membrane resistance is the transverse resistance of 1 cm^2 of membrane.

When λ is expressed in these derived units:

$$\lambda = \left[\frac{\frac{r_m}{2\pi a}}{R_o + \frac{\rho}{\pi a^2}}\right]^{1/2} \qquad (12)$$

If $R_o \ll R_i$, we get:

$$\lambda = \left[\frac{r_m a}{2\rho}\right]^{1/2} \quad \left[\frac{(\text{ohm-cm}^2) \times \text{cm}}{\text{ohm-cm}}\right]^{1/2} \qquad (13)$$

This permits comparison of the space constant in fibers with different dimensions, membrane resistance, and internal resistance. In particular, note that the space constant increases as the square root of membrane resistance and fiber diameter.

As an example of the use of the cable theory, one can calculate the length constant for attentuation of the signal along a C fiber using equation 13 and the following typical values for the various parameters:

Membrane resistance	1,000 ohm-cm^2
Internal resistivity	110 ohm-cm
Fiber diameter	1μ

The length constant is calculated to be about 330μ, which means, from equation 11, that an electrotonic signal is attenuated ε-fold every 330μ along the fiber. Clearly, electrotonic conduction is an ineffective way of propagating information along a nerve fiber, since the signal would be undetectable more than a few hundred microns away from the source.

ACTION POTENTIAL

Between 1950 and 1952, Hodgkin and Huxley conducted experiments, the results of which permitted description of the action potential on the basis of specific sequential changes in the sodium and potassium permeability of the axolemma. They used a technique commonly referred to as a "voltage clamp" because the nerve membrane is held at a fixed arbitrary potential for a period sufficient to permit measurement of the ionic current that flows in response to the imposed potential. The initial experiments were performed on unmyelinated axons isolated from squid because the large diameter of these fibers (300 to 1,000μ) facilitated the placement of electrodes inside the fiber. Subsequently, voltage clamp analyses of action potentials from other excitable tissues using different methods have been performed; the results lead to the conclusion that regenerative (i.e., self-perpetuating) potential changes in excitable cells are generally explicable in terms of sequential changes in specific ion permeabilities, although the charge carriers need not be solely sodium and potassium.

In addition to providing a quantitative description of the nerve impulse, voltage clamp analysis of ion conductances in nerve membrane has also provided a satisfactory explanation for other rapidly occurring, physiologically important phenomena in excitable tissue, e.g., accommodation, refractory period, and pacemaker activity. The analysis does not provide, nor was it intended to, any explanation for other slower phenomena such as adaptation or afterpotentials, which seem to have entirely different explanations.

Before presenting the somewhat complicated analysis necessary to derive the changes in ionic permeabilities from voltage clamp data, we will describe three observations antedating the voltage clamp experiments, which when considered together provide an intuitive basis for focusing on the important variables (sodium and potassium permeability) out of the many possible parameters that could have been investigated as an explanation of the action potential.

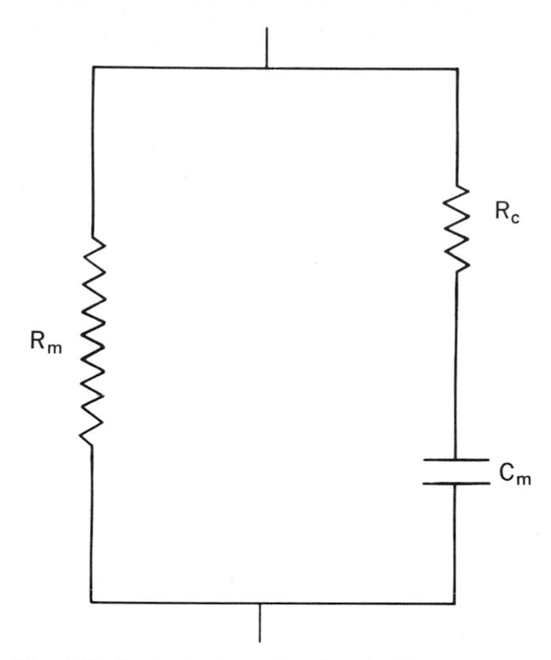

Fig. 2-7. Equivalent circuit of excitable membrane showing, in addition to main membrane resistance, R_m, a numerically smaller resistance, R_c, associated with capacitative elements, C_m.

Membrane impedance changes during action potential

It has already been shown (Chapter 1) that the membrane behaves electrically as if it were a resistance in parallel with a capacitance. A slight refinement, necessary for the present purpose, is the recognition that any capacitance element has a small resistive element in series with it. The circuit for the membrane will therefore be taken as shown in Fig. 2-7.

One standard approach to the analysis of such a reactive circuit is to determine the frequency dependence of the equivalent resistive and reactive elements and then to plot the impedance as a frequency-dependent locus in the R-X plane. For the circuit diagrammed in Fig. 2-7, the impedance locus is a segment of a circle, such as that illustrated in Fig. 2-8, *A*. The absolute magnitude of the impedance vector remains constant due to the fact that the locus is a circle, whereas the impedance of the equivalent resistive and reactive elements vary as shown. The numbers on the arc of the circle refer to frequency in kilohertz and indicate the position of the impedance vector as well as the magnitudes of the equivalent resistive and reactive components at representative frequencies.

The experimental impedance loci for many cells closely approximate the circle loci shown in Fig. 2-8, *A,* which supports the circuit shown in Fig. 2-7 as one possible, though not unique, representation of the electrical behavior of the membrane.

Of greater importance to the present discussion is the fact that when the impedance locus was measured during the action potential another circle diagram, with a different radius and center, was obtained. One result is shown in Fig. 2-8, *A* (open circles) as the impedance locus (determined at various signal frequencies) for the peak of the action potential. Other circle diagrams were obtained for the membrane impedance locus determined at other points on the action potential.

This result indicates that during the action potential there has been a change in one or more of the membrane parameters; R_m, R_c, or C_m. If the impedance at a fixed frequency (e.g., 10 kHz) is determined during the entire course of the action potential, the impedance locus is seen to follow the path shown as the insert in Fig. 2-8. The experimental points on both the rising phase (solid circles) and falling phase (open circles) of the action potential lie on the dotted line, which is the calculated impedance locus for a circuit in which *only* the membrane resistance (R_m) changes. Although not obvious from the figure, calculation shows that the

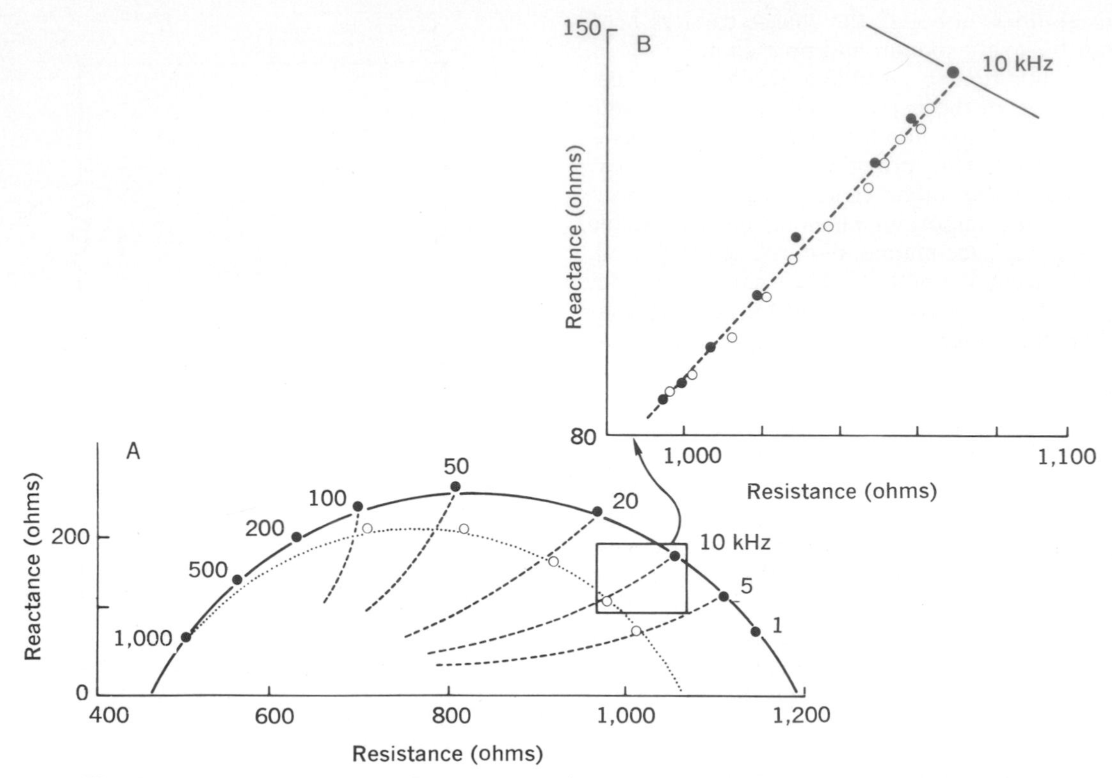

Fig. 2-8. Impedance loci for squid axolemma. **A,** Impedance locus as a function of frequency. Solid circles represent locus of resting membrane; open circles represent locus at peak of action potential. Solid and dotted lines are theoretical loci calculated on the assumption that equivalent circuit of membrane is that shown in Fig. 2-7. Dashed lines are theoretical loci at a fixed frequency drawn on the assumption that only R_m changes during action potential. **B,** Detailed impedance locus at 10 kHz showing correspondence of theoretical curve and experimental points during rise of action potential (closed circles) and fall (open circles). (From Cole.[1])

membrane resistance during activity falls profoundly by a factor of 10 to 40, whereas the capacitance changes by a few percent at most.

The inference to be drawn from these impedance studies, therefore, is that the basic structure of the membrane, as indicated by the capacitance, remains intact during activity, but the ion traffic or concentration of charge carriers crossing the membrane is greatly increased.

Overshoot

In 1902 Bernstein proposed that the resting membrane was permeable to potassium but not to other ions. The action potential was thought to result from a collapse of the membrane with a general increase in electrolyte permeability. At the time the hypothesis was proposed there was no way of directly measuring the intracellular potential. Since the theory provided a qualitatively correct description of the polarity of the resting potential and the monophasic action potential, it was generally accepted until the development of intracellular recording techniques about the time of World War II. The first published records of action potentials from isolated squid axons showed conclusively that the transmembrane potential did not approach zero during the peak of the action potential. Instead the membrane polarity was reversed and the inside became transiently positive. The magnitude of this overshoot, as the reversed potential is called, was much too large to be ascribed either to a liquid junction potential between the recording electrode and axoplasm or to a residual diffusion across a collapsed membrane.

Several representative examples of resting and action potentials recorded intracellularly from several excitable cells are illustrated in Fig. 2-1. Although the exact form of the action potential shows considerable variation, in all cases the peak of the action potential

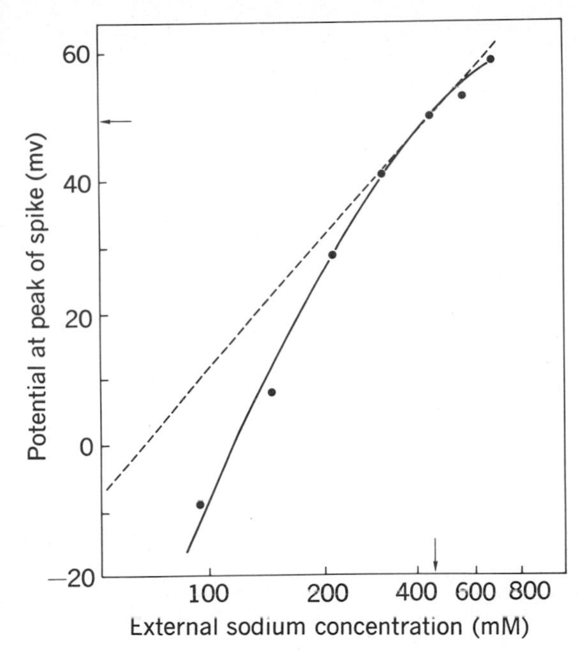

Fig. 2-9. Variation of squid membrane potential at peak of spike with external sodium concentration in artificial seawater bathing axon. Arrows indicate normal spike overshoot for squid axons and standard external sodium concentration of seawater. Dashed line indicates theoretical relation between sodium and action potential for membrane permeable only to sodium. Deviation shown for experimental points at low external sodium indicates that under these conditions ions other than sodium can flow during action potential. (From Hodgkin and Katz.[17])

overshoots the resting potential. These relatively straightforward, although technically difficult, measurements proved conclusively that during the action potential it was Bernstein's hypothesis that collapsed rather than the excitable membrane. The membrane does *not* become a nonspecific sieve during the passage of the nerve impulse.

Reversible block of action potentials by removal of external sodium

A third piece of evidence bearing on the mechanism of the action potential came to light at the end of the 1940s when it was shown that the action potential depended on the presence of external sodium in the bathing medium. Excitation was completely blocked if less than 10% of the normal sodium was present in the outside solution. At intermediate concentrations the amplitude of the overshoot appeared to correlate with the sodium equilibrium potential as calculated from the Nernst relation, implying that at the peak of the spike the membrane was selectively permeable to sodium, in sharp contrast to the resting situation in which the membrane was permeable mainly to potassium. The effect is illustrated in Fig. 2-9. The dotted line is drawn according to the Nernst relation for a purely sodium-permeable membrane, whereas the solid line shows the experimental curves. The deviation, increasing at lower sodium concentrations, is in the direction to be expected if the membrane were not solely sodium permeable, because at low external sodium concentrations other ions would make a significant contribution to the membrane potential.

Although external sodium is a general requirement for excitability in mammalian tissues, there may be exceptions. Some recent work suggests that part of the ionic current flowing during action potentials in smooth muscle cells may be composed of calcium ions. In many invertebrate preparations, calcium, chloride, or other charge carriers are responsible for the initial displacement of the membrane potential from the resting levels.

Summary

Considered together, the existence of an action potential overshoot plus the requirement for external sodium suggest that, at least during the initial phase of the action potential, there is an inrush of sodium that depolarizes the membrane and in fact transiently reverses the polarity. The studies on membrane impedance changes during the spike indicate that the ion flows occur in such restricted regions as to cause no great disorganization of membrane structure. Moreover, the fact that the conductance changes continuously during the entire spike suggests that intermediate potentials during both the rise and the fall of the action potential might also be associated with increased conductance and ion currents through the membrane. Although the ion substitution technique was not suitable for accurate measurements of changes in relative ion permeabilities during the spike, such data were obtained by the voltage clamp technique that is described next.

VOLTAGE CLAMP

In addition to the experimental difficulty of dealing with an electric potential change of very short duration, study of the propagated action potential also poses another

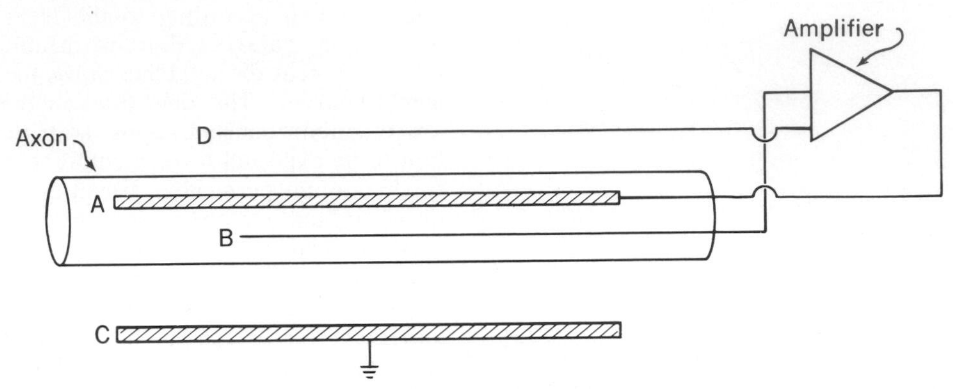

Fig. 2-10. Schematic diagram of voltage clamp technique as applied to squid axons. Membrane potential recorded between electrodes *B* and *C* can be held at any arbitrary level by passing appropriate currents between electrodes *A* and *D*.

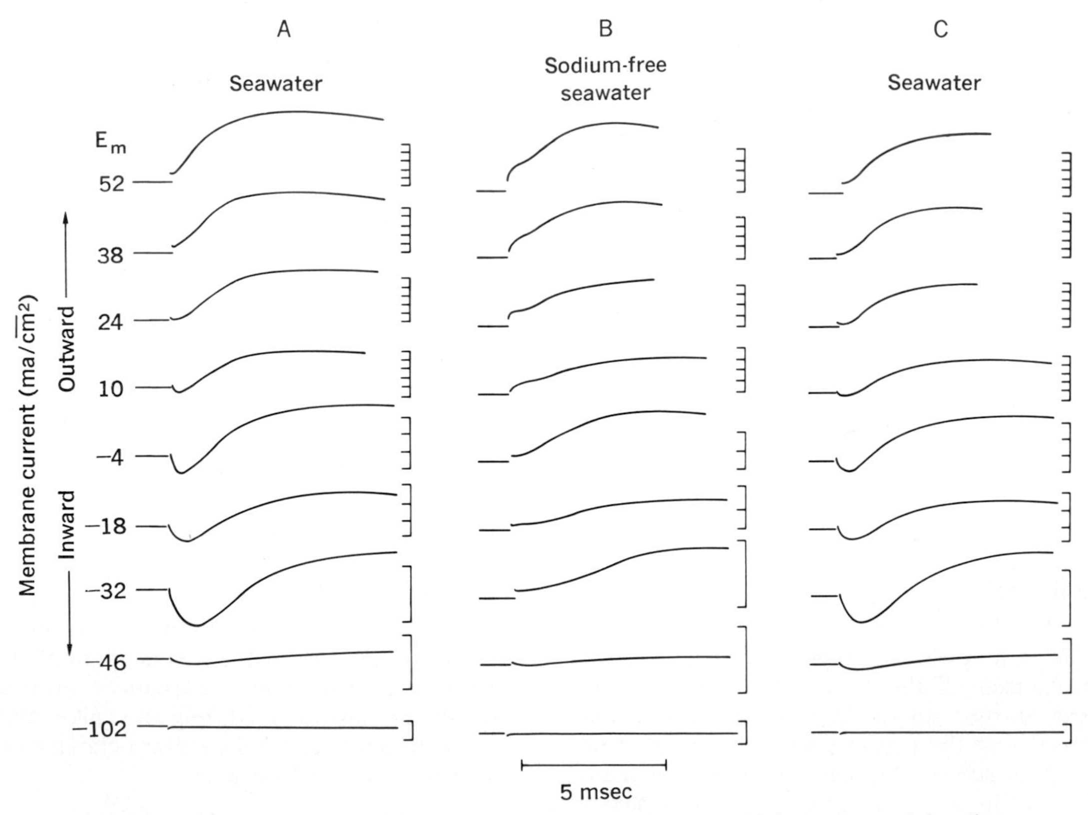

Fig. 2-11. Voltage clamp currents recorded from squid axon. Voltage-clamped membrane potential is indicated opposite each record. Inward current is plotted downward. Each division on current scale to right of each column is 0.5 ma/cm^2 (note scale changes). Temperature of axon is 8.5° C. Records were taken as follows: **A,** Axon in seawater. **B,** Axon in sodium-free seawater (choline replacing sodium). **C,** Axon in seawater as control to show reversibility. (From Hodgkin and Huxley.[13])

problem. Because of the cable properties of nerve membrane and the finite conduction velocity of the nerve impulse, the action potential causes a continuous spatial variation in membrane potential along the nerve. Measurements of membrane conductance or current obtained from a length of membrane would therefore represent values averaged from regions of widely varying potential.

The problem of spatial variation can be solved either by limiting measurements to very short lengths of nerve or by shunting the core resistance with an internal longitudinal wire inside the axon. In the presence of a reasonably low external resistance the latter procedure effectively makes the membrane space constant infinite (i.e., the denominator of equation 13 becomes very small), thus achieving spatial uniformity of membrane potential. Since the technique of inserting an internal electrode is feasible only for very large axons, the first voltage clamp records were obtained from the giant axon of the squid, whose relatively large diameter permits such a procedure.

Description. The problem of temporal variation of potential was solved for squid axon through the use of the apparatus diagrammed schematically in Fig. 2-10. In addition to the shunt wire, *A*, used to achieve spatial uniformity of potential, a second electrode, *B*, is also inserted longitudinally into the axoplasm. The transmembrane potential recorded between this electrode and the reference electrode, *C*, can be held constant at any desired level for several tens of milliseconds (i.e., achieving temporal clamping of the membrane potential) by passing currents of appropriate magnitude and direction between the shunt electrode, *A*, and the external current collecting electrode, *D*. The control and measurement of these currents is accomplished by complicated electronic circuity involving multiple amplifiers but shown only as a single device in the figure.

The basic voltage clamp experiment consists of obtaining a set of membrane currents at constant potential for a series of voltage steps both above (hyperpolarized) and below (depolarized) the resting potential. Representative curves are shown in Fig. 2-11, *A*.

The numbers at left represent the absolute magnitude of the membrane potential. Inward currents are shown as downward on the diagram. The salient features of these curves are as follows:

1. At moderate depolarizations the ionic current is biphasic, initially inward followed by a sustained outward current.
2. At large depolarizations, i.e., positive potentials, there is no inward current. The current is outward at all times.
3. With hyperpolarizing voltage steps, there is little current flow in either direction.

When the experiment is repeated in the absence of external sodium, an additional effect is obvious (Fig. 2-11, *B*). The inward phase of the current disappears, whereas the outward currents are essentially unchanged.

Close inspection of the currents in Fig. 2-11 shows that they initially appear discontinuous; i.e., there is an initial break in the current record lasting about 50 to 100 μsec. This apparent discontinuity results from the time required for the amplifier system to pass sufficient current onto the membrane capacitor to change the charge to an amount appropriate to the new potential. The charge required is $\Delta Q = C\Delta V$. The time required to pass this current from the electrodes depends on such factors as the current capacity of the electrodes and diffusion time from wire to axolemma and may require several tens of microseconds.

Identification of charge carriers comprising the ionic current. The dependence of the initial inward current on external sodium suggests that the charge carrier for this phase of the current is the sodium ion flowing inward across the membrane from a region of high electrochemical potential on the outside to a region of lower potential on the inside. This interpretation is supported by the relation between external sodium concentration and the equilibrium potential for the early current calculated from the Nernst relation. Experimentally the equilibrium potential for the initial current can be found from voltage clamp records by noting the potential at which the current flow initially is zero; i.e., there is no flow of current in either direction. Equilibrium potentials determined in this manner are approximately linear with the logarithm of the external sodium concentration, as required by the theoretical Nernst relation, and also agree reasonably well with the equilibrium potentials calculated from analytic determination of the internal sodium concentration. The equilibrium potential for sodium in squid nerve is about +45 mv and probably somewhat lower for mammalian nerve. The selectivity of the process responsible for the early current is not absolute; other monovalent ions can pass through the early current channels. However, with the exception of lithium, which passes about as readily as sodium, the other monovalent cations are considerably less permeable. Thus, under physiologic circumstances in which sodium is the predominant extracellular cation, the early inward current in a voltage clamp is carried almost exclusively by this ion.

From a consideration of the experiments discussed previously, one might have anticipated that the late outward current in voltage-clamped axons would be carried by potassium ions. Unfortunately the identification of the ionic species carrying the late currents is not entirely satisfactory. The attempt to demonstrate the dependence of outward currents on internal potassium analogous to the dependence of inward currents upon external sodium fails because of the technical difficulties of achieving voltage clamps in the presence of low internal potassium concentrations. Comparison of the equilibrium potential for the late current with the calculated Nernst potential is difficult because hyperpolarizing voltage steps to the vicinity of E_K are too small to permit an accurate measurement of the potential corresponding to zero late current flow. Present opinion is that the process responsible for the late current is primarily potassium selective. However, the relative selectivity of this process for potassium is not nearly as great as the relative selectivity of the early current process for sodium. It is possible that under physiologic conditions a significant portion of the late current is carried by monovalent ions other than potassium; however, for convenience, the late current will be considered the "potassium current."

In addition to the two major ion currents just described, there is a third current, called the *leakage* or *leak* current. This current does not appear to have the marked time dependence or ion specificity characteristic of early and late currents. As its name implies, it may be regarded as a nonspecific leak of ionic charge across the membrane. While not of major theoretical interest, the leak current must be considered in calculations involving voltage clamp parameters.

The relative magnitudes and time courses of the two major components of the voltage clamp current are shown in Fig. 2-12 for a voltage step to approximately zero absolute membrane potential. Since the leak current does not change with time during the voltage clamp and is only a few percent of the total current, it is not distinguishable from the baseline. Fig. 2-12 shows clearly the different responses of the sodium and potassium currents to a voltage clamp. The sodium current begins to flow promptly in response to a voltage change but the response is transient, whereas the potassium current begins more slowly but persists for the duration of the pulse. For depolarizing steps to potentials less positive than the sodium equilibrium potential, the sodium current is directed inward across the membrane, whereas the potassium current is directed outward.

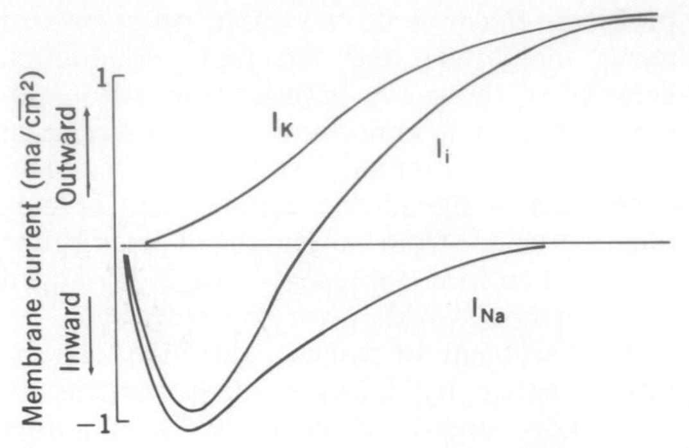

Fig. 2-12. Schematic drawing showing how total ionic current, I_i, is composed of two separate ionic currents, one due to potassium, I_K, and one to sodium, I_{Na}. Time course of third component, the leak current, is not shown. In magnitude it corresponds to residual difference between I_i and I_K at the end of long duration voltage clamp of several milliseconds. (Redrawn from Woodbury.[5])

As has been discussed in the section on cell physiology, ion flow across a membrane depends on both the electric and chemical gradients existing across the membrane. Ion flowing during a voltage clamp is no exception, the magnitude of the driving force on the sodium and potassium ions being due to the difference between the electric and chemical potential calculated from the Nernst relation. The relation between specific ion flow and electrochemical potential gradient for that ion is conveniently expressed as the specific ion conductance, written as:

$$g_{Na} = \frac{I_{Na}}{V - E_{Na}} \tag{14}$$

$$g_K = \frac{I_K}{V - E_K} \tag{15}$$

$$g_L = \frac{I_L}{V - E_L} \tag{16}$$

The leak conductance is commonly formulated in this way simply to give it the same form as the specific ion conductances. E_L is an empirical quantity without theoretical significance.

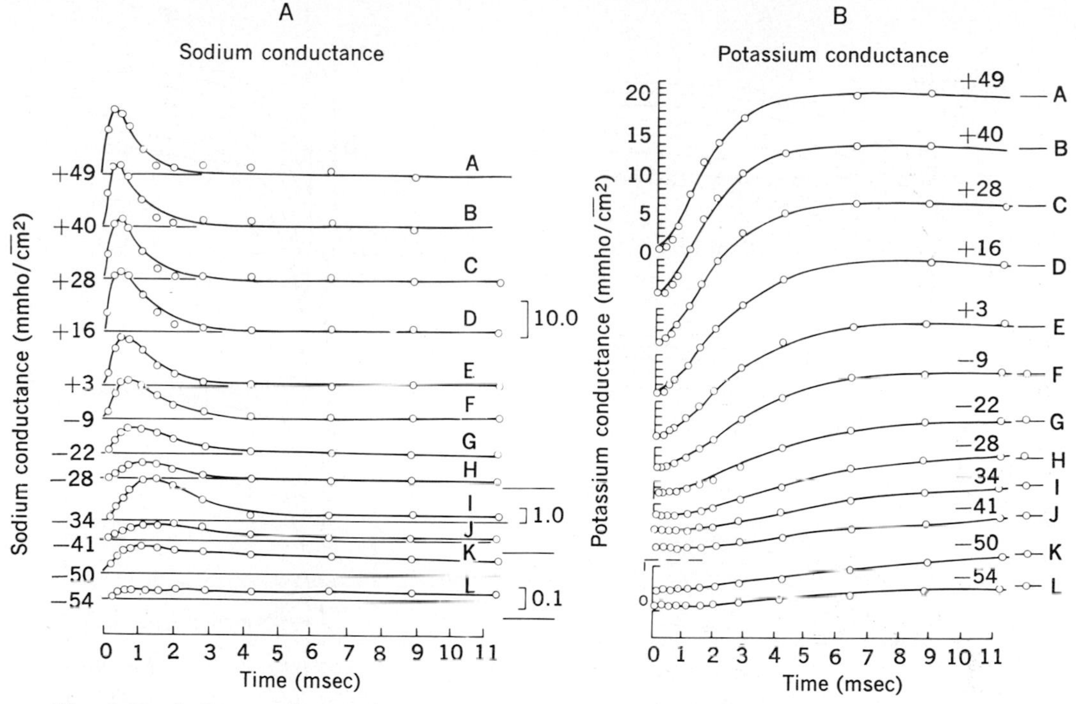

Fig. 2-13. Sodium and potassium conductance curves in voltage-clamped squid axons. **A,** Sodium conductance. **B,** Potassium conductance. Absolute value of clamped membrane potential is given opposite each curve. Note scale changes in conductance curves. Circles represent experimental points; solid curves are drawn according to theoretical equations developed in text. Temperature of axon is 6° to 7° C. (From Hodgkin and Huxley.[16])

The ionic conductances calculated according to equations 14 to 16 for squid axons are shown in Fig. 2-13; those for other excitable tissues are similar. The sodium conductance rises rapidly following a step depolarization, but falls to near zero within 1 to 2 msec. The dependence of the peak conductance upon membrane potential is especially steep for very small depolarization, increasing nearly ε-fold for a depolarization of 4 mv (Fig. 2-13, *A*). In contrast to the behavior of the sodium conductance, the rise in potassium conductance (Fig. 2-13, *B*) following a voltage depolarization is much slower, rising to a plateau only after several milliseconds and remaining high for the duration of the pulse. The delay in rise of potassium conductance is especially noticeable at low depolarizations, although the peak potassium conductance has only a slightly less steep voltage dependence than does the sodium conductance.

Since these conductance data provide a reasonably complete description of nearly all short-term electrical phenomena in nervous tissue and in addition are related to permeability changes within the membrane, it is not surprising that a number of theoretical models have been devised in an attempt to explain these permeability changes on a molecular basis. Given the complexity of the phenomena to be explained and the paucity of information concerning membrane structure, it is equally predictable that such attempts have been largely unsuccessful. The model presented here is that originally devised by Hodgkin and Huxley to explain their voltage clamp data on squid axons. It seems adaptable with minor modification to other excitable tissues and is as physically reasonable as any other.

Before presenting the details of the mathematical analysis that follows from the model, the major underlying assumptions will be considered, since they are similar to those of other models.

1. The voltage-dependent ion flows are controlled by independent permeability mechanisms. This assumption seems intuitively reasonable in view of the marked difference in behavior of the two currents. It is also supported by the fact that certain drugs alter voltage clamp currents in such a way as to suggest that two distinct currents exist.

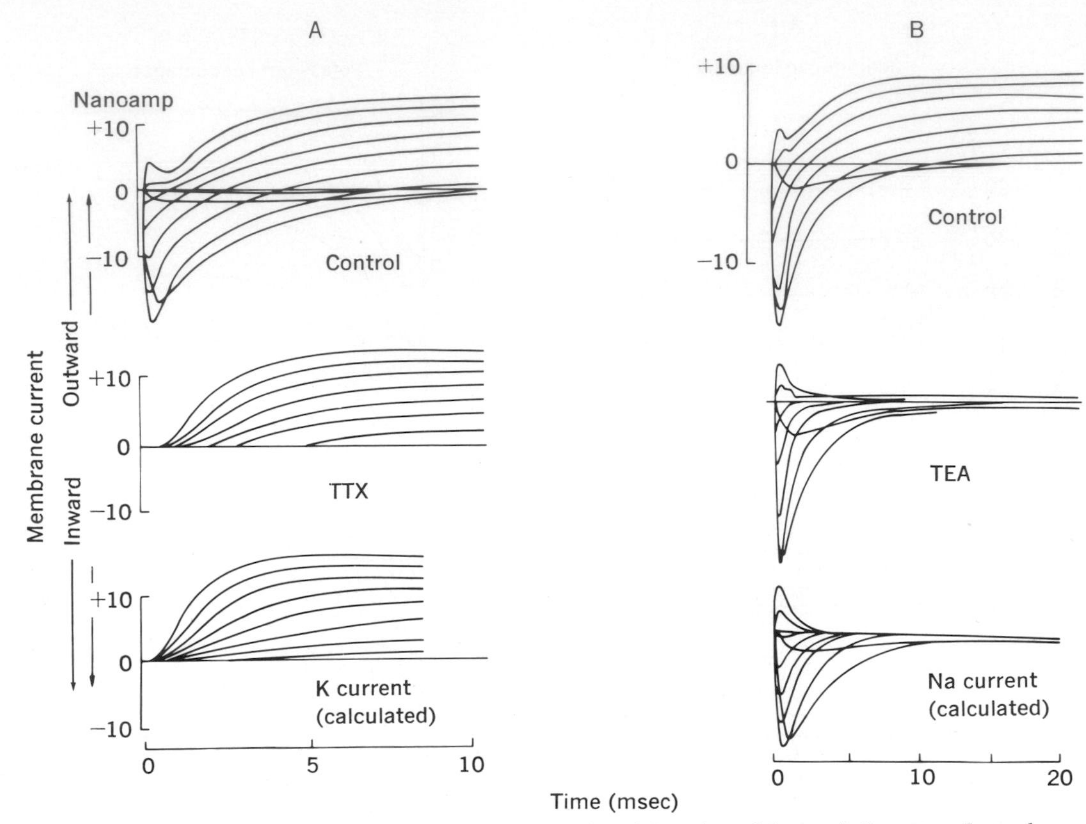

Fig. 2-14. Voltage clamp records from single node of Ranvier of frog sciatic nerve. Inward current is downward. Because of uncertainty about nodal membrane area, current flow is expressed as nanoamperes rather than current density. Note change in time scale between **A** and **B**. **A,** Effects of tetrodotoxin (TTX) on voltage clamp currents. Time course of current flow in presence of TTX is similar to that calculated for potassium current, leading to conclusion that TTX blocks that portion of total membrane current carried by sodium ions but not by potassium. **B,** Effects of tetraethylammonium ion (TEA) on voltage clamp currents. Effects of TEA on voltage clamp current indicate that this agent blocks potassium, but not sodium current. (Modified from Hille.[12])

Fig. 2-14 shows the effects of selected agents on voltage clamp currents recorded from a single node of Ranvier in frog myelinated nerve. The top set of records in Fig. 2-14, *A,* shows a normal family of clamp currents for various voltage pulses. The center set of records in Fig. 2-14, *A,* shows the membrane currents after the application of tetrodotoxin to the node. The lower tracing shows the theoretical time course of potassium currents in response to voltage clamps. The close correspondence between the calculated potassium currents and those actually observed in the presence of tetrodotoxin leads to the inference that the effect of this drug is to block only the early sodium currents, while leaving the late potassium currents unaffected.

In an analogous fashion, the drug tetraethylammonium chloride acts to block the late potassium current, while leaving the early sodium current unaffected. Fig. 2-14, *B,* shows the close correspondence between the calculated sodium currents and the voltage clamp currents observed after treatment of the fiber with tetraethylammonium ions.

Although this type of pharamacologic analysis indicates that the early and late current permeability control mechanisms are independent, it does not necessarily imply that they occupy physically distinct regions of the membrane.

2. The charge is carried by sodium, potassium, and "leakage" ions. The experimental reasons for identifying the charge carriers as sodium, potassium and "leakage" have already been discussed.

3. In the mechanism of permeability control the particles involved in controlling the

permeability changes reorient themselves with finite velocity in response to changes in the transmembrane electric field. This assumption is little more than a restatement of the experimental results that the conductances are strong functions of the membrane potential but do not change instantaneously. No unique mathematical formalism follows from this assumption, which simply requires that the variables introduced have both a time and potential dependence.

Formulation of the voltage clamp equations

The text follows the analysis developed by Hodgkin and Huxley. The equations giving the sodium and potassium conductances are as follows:

$$g_K = \bar{g}_K n^4 \qquad (17a)$$

$$\frac{dn}{dt} = \alpha_n(1 - n) - \beta_n n \qquad (17b)$$

$$g_{Na} = \bar{g}_{Na} m^3 h \qquad (18a)$$

$$\frac{dm}{dt} = \alpha_m(1 - m) - \beta_m m \qquad (18b)$$

$$\frac{dh}{dt} = \alpha_h(1 - h) - \beta_h h$$

In these equations the $\bar{g}_K$ and $\bar{g}_{Na}$ represent the maximum possible values of the potassium and sodium conductances and are essentially scaling factors to allow the variables m, n, and h to range between 0 and 1.

The equations may be given a plausible, although not unique, physical basis by the assumption that passage of sodium or potassium ions across the membrane requires the cooperative interaction of several identical particles. The potassium conductance process is considered first because the mathematical form is simpler. One may assume that potassium ions can cross the membrane only when 4 N particles occupy some critical region of the membrane. The fraction of particles in this location is n, and the fraction elsewhere is 1-n. The potassium conductance is therefore proportional to n^4. The N particles may be thought of as associating together to constitute a carrier for the potassium ion. The α_n and β_n are voltage-dependent parameters that may be thought of as representing position coordinates of polar molecules whose orientation is determined by the electric field across the membrane and that control the movement of the N particles into or out of the potassium-activating region. Although these polar particles reorient themselves instantaneously, or nearly so, in response to a step change in the electric field (i.e., a voltage clamp), the N particles will still require a finite time to move into the proper position for carrying potassium ions. Therefore the conductance will not change instantaneously but will increase gradually with time, even though the rate constants α_n and β_n are not time dependent.

One must realize that the foregoing "explanation" of the potassium conductance process is little more than a verbal description of one set of equations that describe the time and voltage dependence of the potassium conductance. There is no a priori reason for assuming that 4 N particles are required to form a carrier. This number derives from the fact that the original voltage clamp conductance data were reasonably well fitted by an exponent of 4 in equation 17. Actually more recent data obtained with improved techniques are better matched by higher powers of the n parameter, but the improvement is minor.

Fig. 2-15, *A,* is a purely schematic representation of the process controlling conductance. The orientation of the asymmetric "gate" between the two pools of N particles is controlled by the transverse potential gradient; this orientation in turn regulates the magnitude of the rate constants α_n and β_n that describe the velocity of movement of the N particles into and out of the region of the membrane where activation of the potassium conductance can occur. The shape of the gate is entirely arbitrary and serves merely to indicate that it is asymmetric so that it can respond to the electric field and that the potential gradient affects the rate constants unequally.

There is no evidence to indicate the actual dimensions of the conductance sites, but calculations based on the estimated ion conductance of a single site indicate that in aggregate they occupy no more than a very small region of the membrane, possibly on the order of 10^{-5} of the total area. Since there appear to be many potassium channels in the membrane, one supposes that both gates and particle pools are distributed over the entire nerve surface, although the density of sites may be quite low, with individual channels separated by hundreds to thousands of angstroms.

The sodium conductance formula can be given a heuristic basis similar to that provided

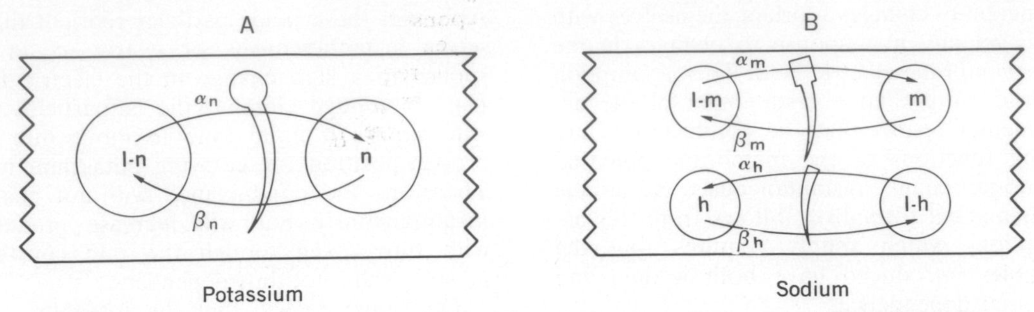

Fig. 2-15. Schematic diagram indicates manner in which macromolecules (gates) whose orientation (or configuration) depends on transverse electric field could control passage of hypothetical m, n, and h particles into regions where they could form channels for passage of sodium and potassium ions. As transmembrane potential varies, gates change position, thus altering rate constants α and β that control movement of particles into or out of active regions. Repositioning of gate after a potential change is assumed to be nearly instantaneous (i.e., time independent); thus rate constants depend only on voltage and not on time. However, movement (diffusion?) of particles into or out of active position is slow compared to gate repositioning and is both time and potential dependent. Therefore conductance, which is proportional to the appropriate power of the fraction of particles in active region, does not change instantaneously but exhibits time as well as potential dependence.

for potassium; however, since the sodium conductance response to a depolarizing voltage step is transient, it is necessary to suppose that two classes of regulating particles exist—those whose association promotes sodium passage, i.e., the M particles, and those whose presence blocks passage, the H particles. In equation 18, m represents the fraction of activating particles present at the activating site for sodium conductance; h refers to the fraction of inactivating particles *not* present at the inactivating site. The defining relation for m is intuitive and is of the same form as for n, the potassium conductance–activating parameter. That for h was chosen so as to preserve the symmetry of the differential equations defining the dimensionless parameters m, n, and h.

Since the m or activation process is more rapid than the h or inactivation process, the initial response to a voltage depolarization is a rise in the sodium conductance. The h or inactivation process develops more slowly but eventually blocks all sodium movement, and the sodium conductance falls to very low levels. A scheme representing both of the processes involved in the regulation of sodium conductance is shown in Fig. 2-15, *B,* where it may be compared with that for potassium conductance.

As previously indicated, the equations for the conductances are essentially empiric. The sodium conductance, for example, could probably be fitted by a single variable obeying a second-order, differential equation describing a highly damped oscillator. However, study of the effect of a number of agents upon voltage clamp currents has revealed a few that selectively affect only one of the voltage clamp parameters.

For example, cesium ions and a toxin isolated from the sea anemone *Condylactis* seem to affect primarily the inactivation process, h, in the sodium conductance, leaving the m and n parameters unaffected. Similarly, procaine, at least in squid axons, seems to reduce $\bar{g}_{Na}$ to a very low level, leaving $\bar{g}_K$ as well as m, n, and h unaffected. In other nerve fibers the effect is less selective. The implication of these facts is that the membrane may indeed contain separate structures that behave kinetically as do the hypothetical M, N, and H particles, thus providing some justification for the form of equations 17 and 18.

Quantitative reconstruction of action potential

The conductance parameters can be used to calculate theoretical action potentials that are remarkably close to those observed experimentally. The method will be illustrated first by discussing a somewhat artificial situation in which the nerve is stimulated simultaneously everywhere along its length, producing what is called a *membrane action potential.* The considerations involved in calculating a more physiologic situation, the propagated action potential, are similar but the equations are more complicated. The total current across the membrane (I_m) is:

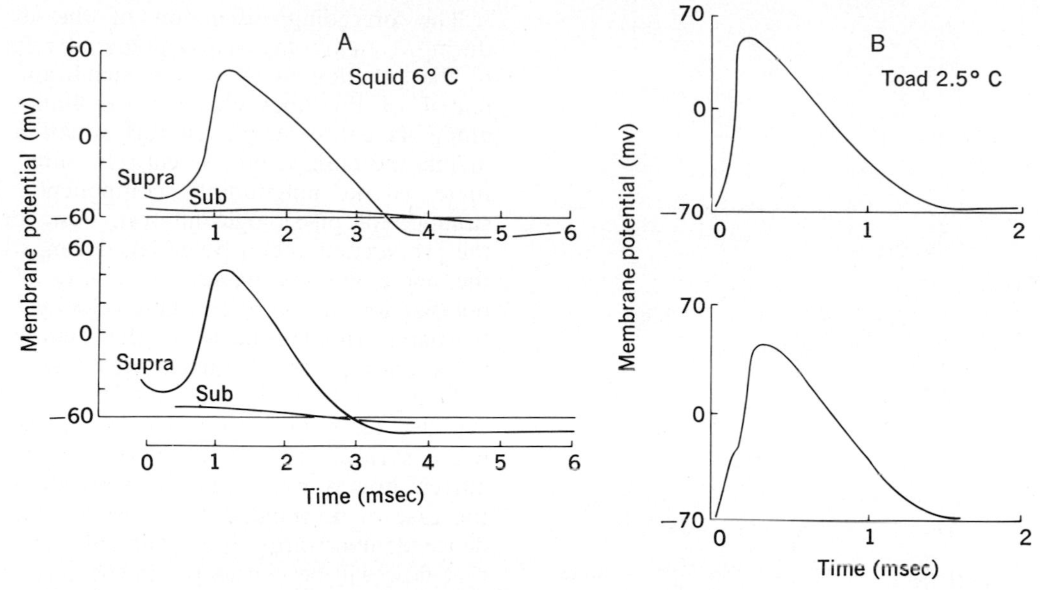

Fig. 2-16. Comparison of calculated (upper curves) and experimental (lower curves) action potentials. **A,** Unmyelinated (squid) nerve fibers. **B,** Myelinated (toad) nerve fibers. Considering the complex shape of the action potential, correspondence between calculated and experimental curves is remarkable. **A** includes both supra- and subthreshold responses for comparison. (**A** From Hodgkin and Huxley[16]; **B** from Frankenhaeuser and Huxley.[8])

$$I_m = I_C + I_i \quad (19)$$

that is, the sum of the capacitance current and the ionic current. The capacitative current is simply:

$$I_C = C_m \frac{\partial V}{\partial t} \quad (20)$$

whereas the ionic current is the sum of the sodium, potassium, and leakage currents. Thus:

$$I_i = I_K + I_{Na} + I_L \quad (21)$$

If we recall the definition of conductance used in the voltage clamp experiments and use the empirical conductance formula, we have for the membrane current:

$$I_m = C_m \frac{\partial V}{\partial t} + \bar{g}_K n^4 (V - E_K) + \bar{g}_{Na} m^3 h (V - E_{Na}) + \bar{g}_L (V - E_L) \quad (22)$$

For a membrane action potential, the net current across the membrane (I_m) must always be zero. With $I_m = 0$, equation 22 becomes an ordinary differential equation capable of numerical solution by standard methods, giving E as E(t).

Calculated membrane action potentials for both unmyelinated (squid) and myelinated (toad) axons are shown in the upper part of Fig. 2-16 and may be compared with experimental curves obtained for similar depolarizations shown in the lower part of Fig. 2-16. The theoretical and experimental curves agree closely for suprathreshold as well as for subthreshold stimuli.

The time course during an action potential of the membrane conductance and the various specific ion conductance components as well as the m, n, and h parameters are shown in Fig. 2-17. During an action potential the transmembrane voltage is not maintained constant as it is during the voltage clamp; therefore the conductance parameters vary not only with time but also in response to the continuing variation of m, n, and h, as determined by the instantaneous voltage. For this reason the time course of the conductances shown in Fig. 2-17 relating to the action potential are quite different from those that relate to voltage clamps, as illustrated in Fig. 2-13.

Fig. 2-17, *B,* shows that during the rising phase of the action potential the major part of the conductance increase is due to sodium. The potassium conductance does not increase appreciably until near the peak of the spike. Thereafter the sodium conductance decreases rapidly and a proportionately greater fraction of the total conductance is potassium. In this particular preparation of squid axon the potassium conductance remains elevated for a

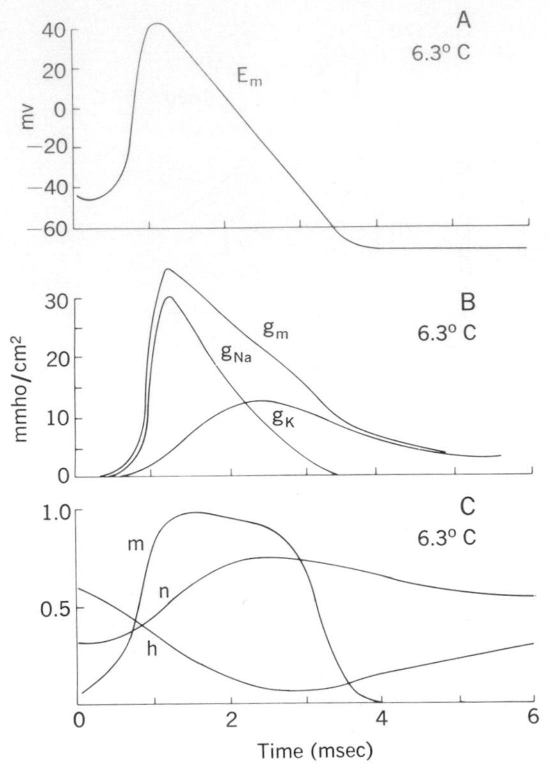

Fig. 2-17. Calculated parameters for squid nerve action potential. **A,** Membrane potential. **B,** Membrane conductance (g_m), sodium conductance (g_{Na}), and potassium conductance (g_K). Note that during rising phase of spike, increase in membrane conductance is due largely to sodium, whereas late on falling phase and during postspike hyperpolarization, increase in membrane conductance is due largely to potassium. **C,** Variation of m, n, and h parameters during spike. Note that whereas m remains high until well on falling phase of spike, h begins to fall immediately after stimulation. Net result of variations is reduction of sodium conductance to low level during falling phase of spike. The n parameter rises slowly but remains elevated until the end of postspike hyperpolarization. (**A** and **C** from Cole[1]; **B** from Michalov et al.[21])

short time after the end of the spike, thus producing the small postspike hyperpolarization of the membrane potential seen in Fig. 2-17, *A*.

The reason for the fall of the sodium conductance (Fig. 2-17, *C*) is a decrease in the h or sodium-inactivation parameter. Actually the m or sodium-activation process that tends to increase sodium conductance remains near its maximum value until well past the peak of the spike. The rather rapid development and persistence of sodium inactivation is important in determining the refractory behavior of nerve and will be considered further in that regard.

Propagated action potentials

The preceding discussion of the events during a membrane action potential referred to the situation in which the membrane potential of the fiber changed simultaneously along its entire length. In this situation the membrane current flow is entirely radial and there is no longitudinal component. The situation of physiologic interest, however, is the propagated action potential, during which the nerve impulse passes as a wave of depolarization at nearly constant velocity along the nerve. In consequence of the longitudinal variation in potential, current must flow longitudinally both inside and outside the nerve cylinder as well as transversely across the nerve surface. The existence of longitudinal current flow is most elegantly demonstrated in the case of myelinated nerve fibers, and evidence demonstrating the reality of such current flow will be considered in the section on saltatory conduction.

Since the wavelength of an action potential, on the order of a few centimeters, is rather short compared to the total length of the nerve, the longitudinal variation of transverse membrane potential and hence the longitudinal current itself is restricted to a correspondingly short segment of nerve. For this reason the currents are commonly referred to as "local currents."

The transverse membrane current flowing in the local circuit can be related to longitudinal potential variation by use of the cable theory (p. 40):

$$I_m = \frac{1}{R_o + R_i} \times \frac{\partial^2 V}{\partial x^2} \qquad (23)$$

In the case of a propagated action potential, unlike the membrane action potential, the membrane current is not zero, but in fact provides the current for the local circuit.

This membrane current flow consists of a capacitative component and a resistive component in parallel.

$$I_m = C_m \frac{\partial V}{\partial t} + \frac{V}{R_m} = C_m \frac{\partial V}{\partial t} + I_i \qquad (24)$$

The resistive component is the sum of the ionic currents that flow across the membrane as the physical consequence of a change in membrane permeability. Their time course is given by the Hodgkin-Huxley equations.

The capacitative current, by contrast, does not result from a change in the physical properties of the membrane but only from the change in potential across the capacitative

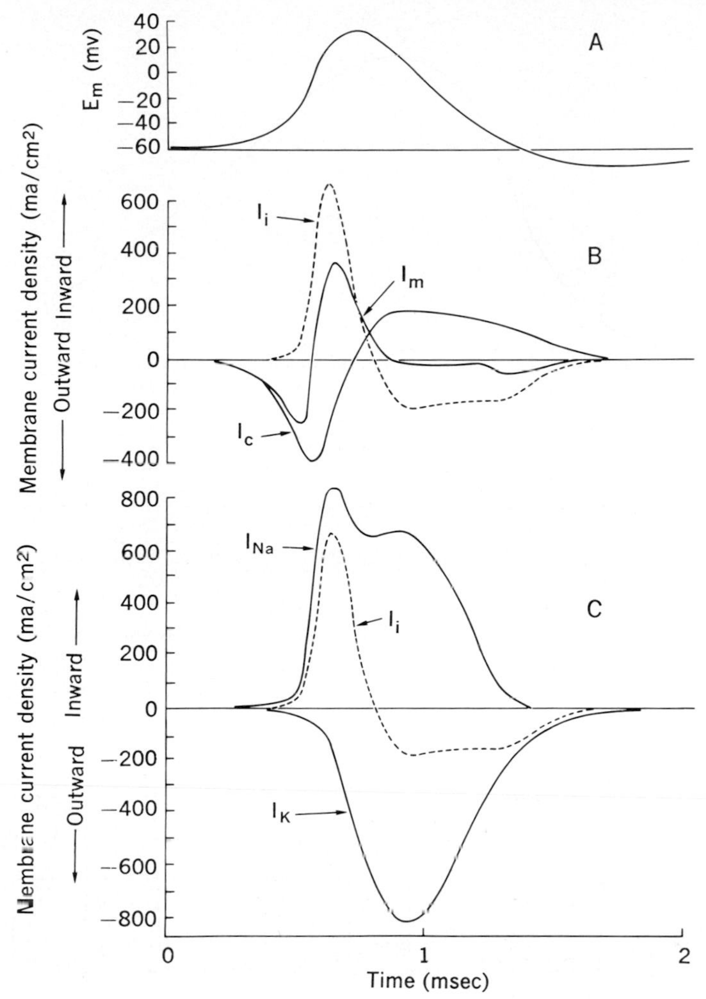

Fig. 2-18. Calculated time course of components of membrane current during propagated action potential in squid axon. Inward current is plotted upward. I_m is the total membrane current, consisting of capacitative component (I_c) and ionic component (I_i). Ionic component is further subdivided into sodium current (I_{Na}) and potassium current (I_K). Note that initial flow of membrane current is *outward* (flowing to active membrane behind wave front) and is due entirely to displacement current leaking from membrane capacitance. This displacement of charge is responsible for initial membrane depolarization of nearly 10 mv, which reduces potential to point where sodium conductance increases (shown as rapid rise of sodium current, I_{Na}). Repolarization is accomplished by outward flow of potassium. During this period, charge gradually reaccumulates on membrane capacitance to return potential to resting level. (From Hodgkin and Huxley.[16])

element. No ions cross the membrane as a result of capacitative current flow; rather, charge is removed from or added to either side of the membrane in response to the potential change. The actual ionic species carrying the charge away from either membrane surface will be determined by the transference numbers and concentrations of ion present in the internal and external media.

The contribution of each of the components of the membrane current at various times during the course of the propagated action potential is shown in Fig. 2-18. In particular, note that the initial flow of membrane current at the beginning of the propagated spike is due entirely to the capacitative current and is actually outward, whereas the ionic current, which is initially inward, does not begin until the threshold has been reached at about 10 mv of depolarization. A second significant point is that during the falling phase of the spike, the net membrane current is rather small, since the ionic current (mainly potassium) and the capacitative current are nearly equal and flow in opposite directions.

The distribution of the ionic and capacita-

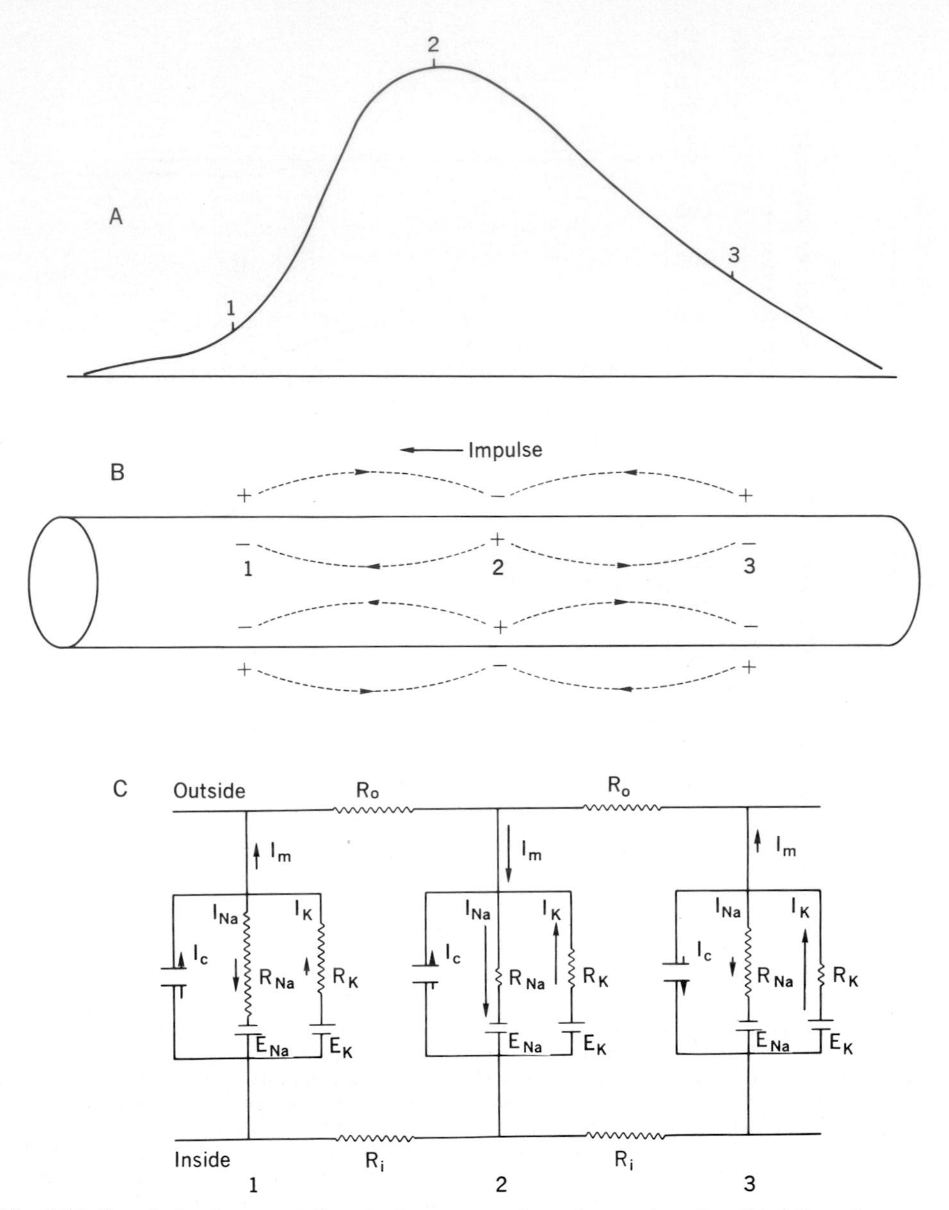

Fig. 2-19. Local circuit current flow during propagation of nerve impulse. Direction of propagation is from right to left. Current flow is illustrated as occurring only at the three indicated regions corresponding to beginning of spike, at peak, and late on repolarizing phase. Circuit diagram in **C** shows directions and approximate magnitude of components of membrane current at the three selected points. In reality, membrane current does not flow only at discrete points but varies smoothly and continuously along entire length of nerve occupied by nerve impulse.

tive currents during impulse conduction along a nerve fiber at three representative stages during the action potential is shown schematically in Fig. 2-19. These three stages are indicated by the numbers on the action potential in Fig. 2-19, *A*. The direction of longitudinal current flow at these three points is presented schematically in Fig. 2-19, *B*. For simplicity, the transmembrane current is illustrated as occurring only at these three points, although actually it is distributed along the entire surface of the fiber, as determined by the cable properties of the fiber.

Position *1* represents an area at the foot of the advancing wave front that is not yet active. In this region the membrane has the usual polarity, inside negative. The active region of the nerve is at position *2*. In this region the membrane is largely permeable to sodium so that ionic current is passing inward across the membrane, making the inside of the nerve positive. In position *3* the peak of the spike has passed and the membrane is repolarizing.

The components of transverse membrane current at these three points are shown schematically in Fig. 2-19, *C*. The length of the resistive elements R_K and R_{Na} indicates qualitatively the magnitude of the specific ionic resistances at various stages of the spike. The direction of the chemical gradients is indicated by the polarity of the batteries labeled E_{Na} and E_K. The arrows show the direction and, qualitatively, the magnitude of the current flow through the elements.

At point *1*, ahead of the active region, charge is being drawn off of the membrane capacitance and is moving as an electric current in the direction indicated by the arrow. As charge is drawn from the capacitor, the transmembrane potential will begin to drop. As the potential drops, the sodium and potassium conductances will begin to increase in accordance with the empirical equations relating conductance to potential. However, the absolute change in ionic conductances is rather small, so that the net ionic current flowing is rather small compared to the capacitative current. If the capacitative current could be suddenly interrupted at this moment, the membrane potential would return to the resting level because the sodium current would become inactivated and be inadequate to maintain the small depolarization initiated by the prior capacitative flow. In this case the response of the membrane would be subthreshold. However, usually the capacitative current is continually drawn out of the region, lowering the membrane potential to the threshold level, which for most fibers is a depolarization on the order of 5 to 15 mv. At this moment the ionic conductances, principally sodium, increase abruptly. A large sodium current begins to flow inward across the membrane, potential polarity reverses, and the active region propagates forward from points *1* to *2*. Capacitative current at point *2* is less than at point *1* because the rate of change of the membrane potential is smaller.

As the peak of the spike passes, the sodium current begins to decline and the potassium current begins to increase. The situation is now as indicated at position *3*. The net ionic current is outward repolarizing the membrane; the capacitative current is, however, in the opposite direction. As the potential difference across the capacitance increases, charges accumulate on either surface of the membrane. Because the ionic and capacitative currents are in opposite directions, the net membrane current (I_m) during the falling phase is rather small, as indicated in Fig. 2-18, *B*.

Fig. 2-19 illustrates why propagation is only in one direction; i.e., the impulse does not spontaneously turn around and propagate toward the direction of origin. At point *3* the sodium conductance is lower than in the resting case because of inactivation. Even if the capacitance charge on the membrane were somehow reduced to the usual threshold, the sodium conductance would be incapable of increasing sufficiently to permit the inrush of sodium current necessary to initiate a spike. Moreover, the increased potassium permeability during this time results in an increased outflow of potassium ions in a hyperpolarizing direction, thus opposing the effect of a small depolarization.

Two factors that affect propagation velocity can be identified from Fig. 2-19. These are membrane capacitance and fiber diameter.

Membrane capacitance. Since the membrane capacitance must be discharged to the threshold level with a time constant approximately equal to $R_m \times C_m$ before excitation occurs, fibers with a larger capacitance per unit area will require longer to depolarize to threshold the patch of membrane immediately ahead of the advancing wave front and will therefore conduct impulses more slowly.

Fiber diameter. The magnitude of the initial local circuit current that flows is limited by the resistance of the internal and external longitudinal current paths. For fibers immersed in large volumes of solution, the external resistance will generally be negligible; however, the internal longitudinal resistance will decrease with increasing diameter. Consequently, the initial local circuit flow will be greater and the speed of propagation faster in larger fibers. A more exact dimensional analysis indicates that the velocity of propagation increases as the square root of the diameter for unmyelinated fibers and approximately linearly with the diameter for myelinated fibers.

USE OF VOLTAGE CLAMP PARAMETERS TO EXPLAIN OTHER PHENOMENA IN EXCITABLE TISSUES

Subthreshold response

Subthreshold responses that closely resemble the experimental response (Fig. 2-16) can be calculated from the empirical equa-

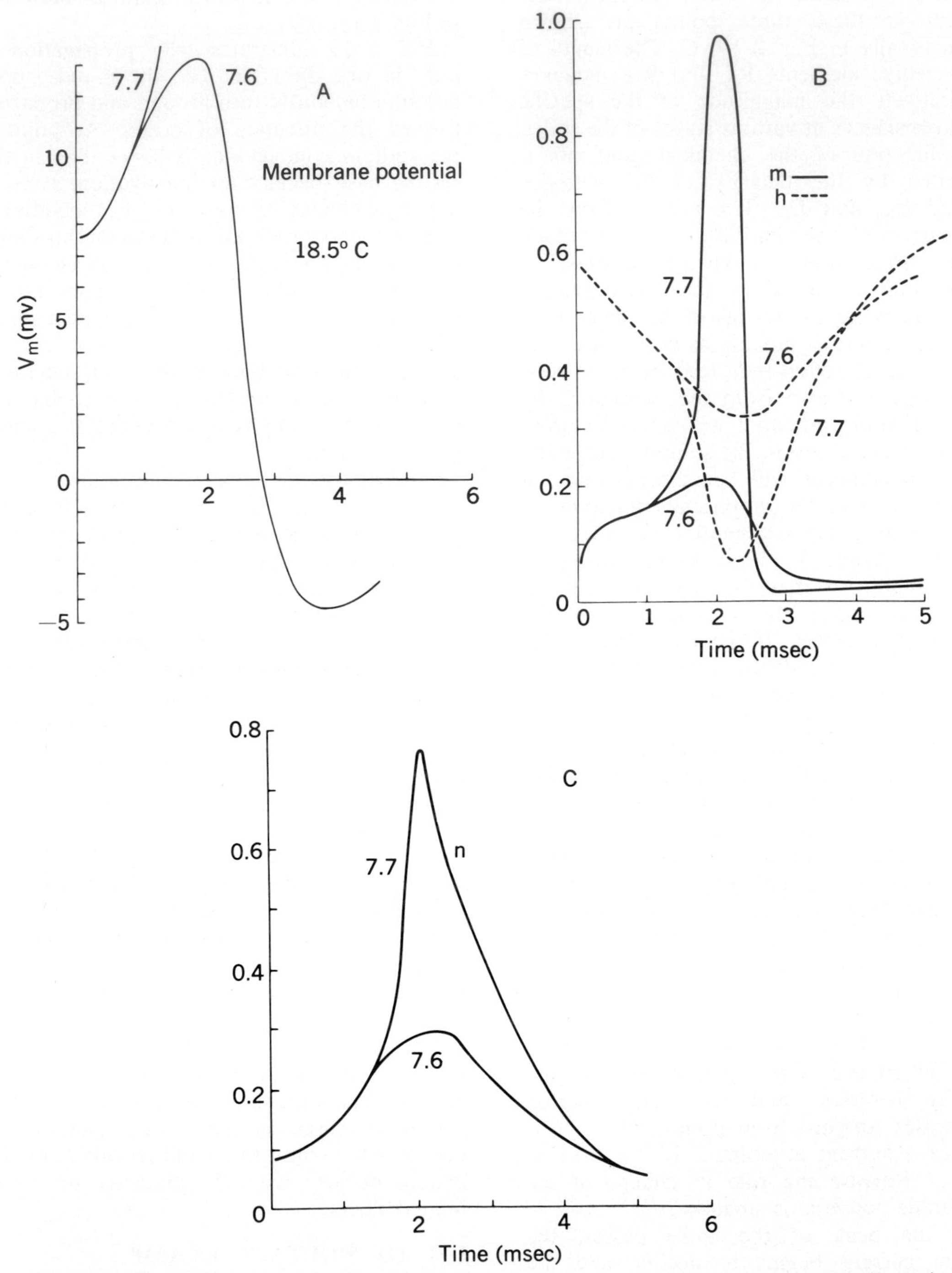

Fig. 2-20. Calculated responses of squid axon membrane to near-threshold electrical depolarization (subthreshold, 7.6 mv; suprathreshold, 7.7 mv). **A,** Membrane potential. Only foot of suprathreshold (7.7 mv) response is shown because complete spike is off scale, which was chosen to show subthreshold response. **B,** Time course of m and h parameters governing sodium conductance. **C,** Time course of n parameter governing potassium conductance. (From Michalov et al.[21])

tions by the same methods used for action potentials. Such responses can be understood by noting that the initial depolarization (i.e., stimulus), if it is to produce an action potential, must increase the sodium conductance sufficiently to cause a net inward current adequate to carry the membrane rapidly to still greater depolarizations. Otherwise the combined effects of sodium inactivation and potassium activation will act to reverse the net membrane current from inward to outward, and the membrane potential will gradually return to normal values. A calculated subthreshold response and the time course of the variables m, n, and h are shown in Fig. 2-20, together with corresponding parameters for a calculated action potential. The time course of the Hodgkin-Huxley parameters, which determine the shape of the potential response, are similar for both subthreshold and suprathreshold stimuli. Thus both action potentials and subthreshold responses are produced by the same specific ionic permeability changes showing the same general time variation, although there are important quantitative differences in magnitude. In particular, note that the m parameter controlling the turning on of sodium conductance (and hence promoting depolarization) is much more sensitive to a small increment in initial stimulus than is the n parameter controlling the turning on of potassium conductance (hence opposing depolarization).

Threshold

The Hodgkin-Huxley equations provide no physical insight into the mechanism of threshold, i.e., how the sodium and potassium conductances are adjusted so as to permit the regenerative response to occur only in response to stimuli exceeding a sharp cut-off value. Calculations from these equations, however, serve better than any experiment to emphasize how sharp the threshold really is. Calculated responses to stimuli of various intensities are shown in Fig. 2-21. The transition from a subthreshold to a threshold response occurs over a stimulus range of less than 1 part in 5,000. Since a suprathreshold stimulus puts the membrane into a state in which the sodium conductance is strongly regenerative, the stronger the suprathreshold stimulus, the sooner the action potential begins, but the configuration of the response is essentially independent of the intensity of the stimulus. Detailed investigation of the Hodgkin-Huxley equations indicates

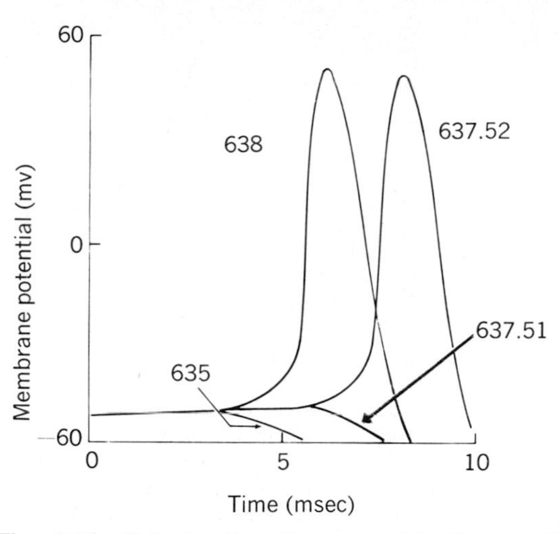

Fig. 2-21. Calculated action potentials for squid axons produced in response to closely graded electrical stimuli. Numbers opposite each curve are proportional to intensity of stimulation. (Redrawn from Cole.[1])

that the membrane's response to a stimulus can be continuously graded from typically subthreshold to typically regenerative if the range of stimulus intensities varies by no more than 1 part in 10^{11}. This calculation demonstrates that the explanation for the marked qualitative differences between subthreshold and suprathreshold responses lies in the quantitative relations between the Hodgkin-Huxley parameters rather than in the existence of separate membrane mechanisms for the two types of responses. However, from a practical point of view, graded action potentials cannot exist because the minute voltage increments required to produce them are a great deal smaller than the gradations of stimulus intensity that occur physiologically.

Refractory period

The refractory period following a single stimulation to a peripheral nerve can be adequately explained by the changes in potassium conductance and the h parameter component of sodium conductance. The time courses of the two variables in relation to the spike are shown in Fig. 2-17. Following a single stimulus, the h parameter falls rapidly toward zero, while g_K begins to rise, although more slowly. Both effects reach their maximum on the falling phase of the spike and tend to reduce the ability of the membrane to respond regeneratively to a second threshold depolarization. The low value of h reduces the level to which the sodium conductance can be raised by a second depolari-

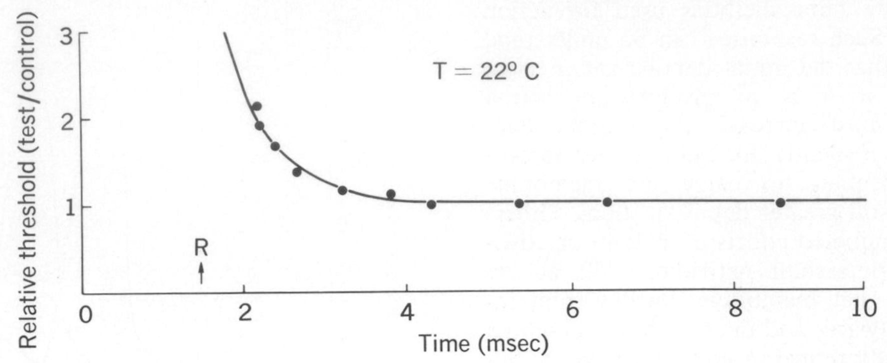

Fig. 2-22. Relative refractory period of frog node of Ranvier. Intensity of threshold test shock is plotted relative to intensity of preceding threshold control shock. End of absolute refractory period is indicated by arrow, *R*. During this period, node is unresponsive regardless of intensity of test shock. As node passes through relative refractory period, threshold for test shock rapidly decreases until, after 3 to 4 msec, fiber has regained normal excitability. (From Tasaki and Takeuchi.[23])

zation, thus reducing the inward current that can flow. The increased potassium conductance allows a larger than usual outward current to flow, thus tending to negate the effect of a transient surge of inward sodium current.

During the earlier part of the falling phase of the spike these effects combine to prevent a regenerative response to any stimulus regardless of intensity; i.e., the threshold for excitation is infinite. During this interval of time the fiber is said to be *absolutely* refractory. After a short interval the fiber becomes able to respond to stimuli of greater than normal intensity; i.e., the threshold for excitation has been increased but is not infinite. During this interval of time the fiber is said to be *relatively* refractory. During the relative refractory period the threshold declines toward the normal level, with a time course determined by many factors, e.g., temperature, nature of the fiber, and extent of previous excitation.

The time course of recovery of excitability for a single node of Ranvier of a frog nerve fiber is shown in Fig. 2-22. The threshold for a response relative to the control threshold is plotted against time after the first threshold stimulus. In this preparation the absolute refractory period, *R,* lasts about 1.5 msec. During the absolute refractory period the nerve is completely unresponsive, no matter how intense the second shock. By the end of the absolute refractory period the node has become slightly responsive. As the nodal membrane passes through the relative refractory period, the threshold for a second response declines, and after a few milliseconds the fiber has regained its normal excitability.

Nerve refractoriness is also expressed in terms of a parameter, the *excitability,* defined as the reciprocal of threshold. The complete time course for alteration of excitability for mammalian A fibers following the single test stimulus is shown in Fig. 2-23. On the time scale used in the figure, the absolute and relative refractory periods are not shown in detail but can be seen to last a few milliseconds. Following the refractory periods the fiber is actually hyperexcitable for some 10 to 20 msec (the supranormal period) and then becomes hypoexcitable for a longer period of some 50 msec (the subnormal period). Fig. 2-23 shows clearly that the refractory periods produce the most profound alterations in nerve excitability but that they are transient phenomena succeeded by smaller, prolonged excitability changes. The explanations for these more slowly developing and longer lasting supranormal and subnormal periods are not related directly to the voltage-dependent conductance but are better considered in relation to the phenomena of afterpotentials discussed on p. 66.

Accommodation

Although accommodation has been studied and defined in many ways (see definition of terms, p. 36), the basic observation is that any subthreshold depolarization of the membrane of arbitrary duration and wave form decreases the membrane response to a second stimulus applied during the duration of the first; i.e., the threshold to subsequent

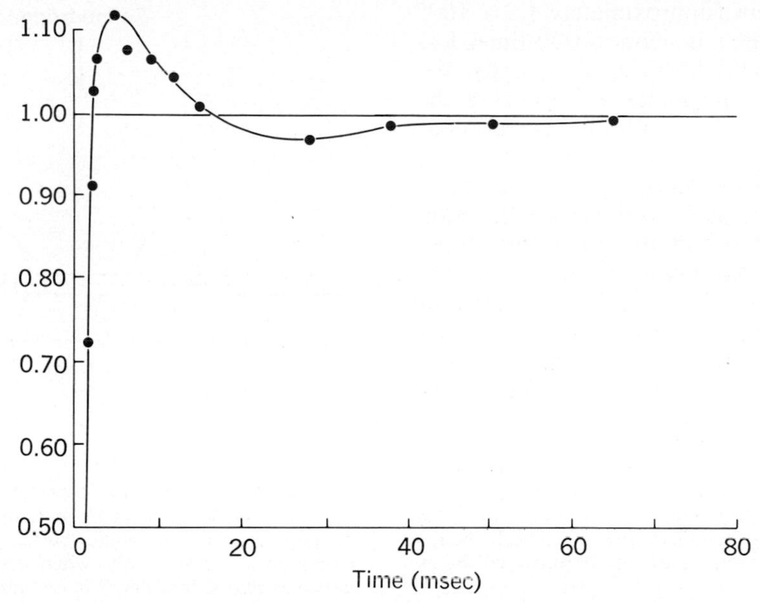

Fig. 2-23. Excitability curves for mammalian A fibers. Excitability is measured as reciprocal of relative threshold and is plotted relative to excitability of fiber at first shock. Absolute refractory period is not shown. Relative refractory period is represented only by first two points. It is followed by intervals of slightly altered excitability related to afterpotentials but not to refractory periods. (Redrawn from Erlanger and Gasser.[2])

stimuli is increased. The phenomenon can be explained in the same general terms as refractoriness. The transient increase in inward sodium current induced by any subthreshold depolarization is ultimately overcompensated by the outward potassium current, which rises more slowly but is of greater magnitude. The net steady-state ionic current flowing across the membrane is therefore outward for all sustained depolarizations, thus tending to counteract the effect of an induced depolarization and raising the threshold for a regenerative response.

One convenient way of demonstrating accommodation is to stimulate a nerve with a linearly increasing current, i.e., a ramp-shaped pulse (Fig. 2-2). As the slope of the ramp is decreased, i.e., a more slowly rising current, the threshold at which excitation occurs increases. Fig. 2-2 shows this effect for frog fibers and also shows that there is a minimum rate of rise below which no excitation will occur. The extent to which a fiber can accommodate and still produce an action potential is therefore somewhat limited. In Fig. 2-2 the threshold increased about 30%.

SALTATORY CONDUCTION

The process of nerve impulse propagation in unmyelinated fibers by a continuous progression of local circuit flow along the length of the fiber has already been described. In myelinated fibers the process by which excitation arises is nearly the same quantitatively as in unmyelinated nerves, but a major difference occurs during propagation of the impulse in that transmembrane ionic current does not flow across the myelin sheath but is constrained to flow across the axolemma only at the nodes of Ranvier. Myelinization has several consequences of tremendous significance for the development of the vertebrate nervous system.

Size. Myelinization permits many fibers of high conduction velocity to be contained in a relatively small volume of nerve trunk. Since conduction velocity in unmyelinated fibers is proportional to the square root of the diameter, an unmyelinated nerve, i.e., a C fiber, would need a diameter on the order of 4 mm if it were to conduct with the same velocity as the fastest myelinated A fibers, i.e., around 120 m/sec.

Energy conservation. Since the ionic membrane currents only flow at the nodes, the quantity of sodium gain or potassium loss per centimeter of nerve per impulse is much less for myelinated fibers. For example, the conduction velocities of squid axons and frog fibers are comparable; however, the sodium gain in frog fibers (approximately 1.3×10^{-16} moles/cm/impulse) is about 5,000 times less than that for squid fibers. Consequently, the work done by ion pumps in restoring the concentration gradients will be correspondingly less.

Rapid repetitive firing. The large internodal regions act as a reservoir for diffusion of ions into and out of the node, thus effectively increasing the nodal volume somewhat and reducing the effect, on internal concentration, of the large surges of ionic current that occur during the spike and that might otherwise render the fiber inexcitable after relatively few impulses.

This effect can be significant, as shown by a simple calculation of the effect of excitation on internal sodium concentration. Assuming the axis cylinder is roughly cylindric, the fractional change in concentration produced by one impulse will be as follows:

$$\frac{\Delta C}{C} = \frac{M}{C} \cdot \frac{A}{V} = \frac{M}{C} \cdot \frac{2}{r}$$

where M is the net flux per impulse, r is the radius of the axis cylinder, and C is the internal concentration. Use of reasonable parameters for frog myelinated fibers yields the following:

$$M = 10 \times 10^{-12} \text{ moles/cm}^2\text{/impulse}$$
$$r = 3.5 \times 10^{-4} \text{ cm}$$
$$C = 50 \times 10^{-6} \text{ moles/ml}$$

From these data the calculated fractional increase in sodium concentration will be about 0.1%/impulse at the node. Clearly, the node will become inexcitable after a few hundred impulses unless the sodium concentration in the node is kept at a relatively low level. Ultimately, of course, the extra sodium that enters during activity will have to be pumped out, but for short intervals the nodal concentration of sodium can be somewhat reduced by diffusion into the internodes. Comparable arguments apply to the internal potassium concentration but with less force, since the fractional change per impulse is much smaller.

In the section that follows we will consider, first, the evidence that electric current actually flows in the external medium during propagation; second, the evidence that excitation occurs only at the nodes; and finally, a few details of the excitation process, as revealed by voltage clamp analysis, that differ from unmyelinated invertebrate axons.

Evidence for local circuit flow

If local electric currents flow during nerve impulse propagation, then an external pathway for current flow must exist. In the absence of an external pathway the local circuit will be broken, no current can flow, and propagation cannot occur. The validity of this reasoning is demonstrated by a simple experiment illustrated in Fig. 2-24. A length of single nerve fiber is placed in two pools of Ringer's solution that are separated by a short air gap. In such a circumstance, conduction is blocked because local currents cannot flow. However, conduction can be restored reversibly simply by providing a conducting bridge of solution across the air gap. The fact that conduction is restored as soon as contact is made precludes diffusion of some activator substance in the external pathway as an explanation for the restoration of conduction. The design of the experiment with an internode in the air gap also demonstrates the absence of significant local current flow *within* the myelin sheath, for if there were such a flow, removal of the saline bridge could not block propagation.

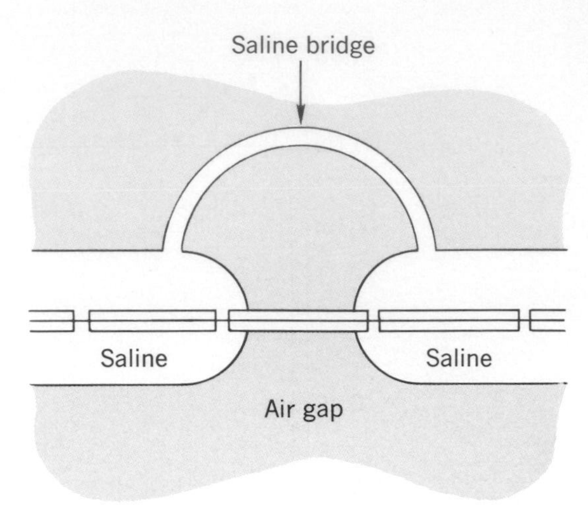

Fig. 2-24. Diagram of experiment demonstrating local circuit flow in single myelinated fibers. Short air gap separates two nodes bathed in saline solution. Conduction occurs only when electrical continuity between two saline pools is completed with external saline bridge.

Evidence for localization of membrane current flow to nodes

It will be recalled from the discussion on nerve impulse propagation that when a regenerative potential change occurs in a patch of nerve membrane, causing it to become active, the net membrane current becomes inward, whereas when the active region is elsewhere, the net membrane current through the patch is outward (Figs. 2-18 and 2-19). Applying these concepts to myelinated fibers allows identification of the active regions along a length of myelinated nerve simply by locating those areas in which inward membrane current flows occur during propagation. The re-

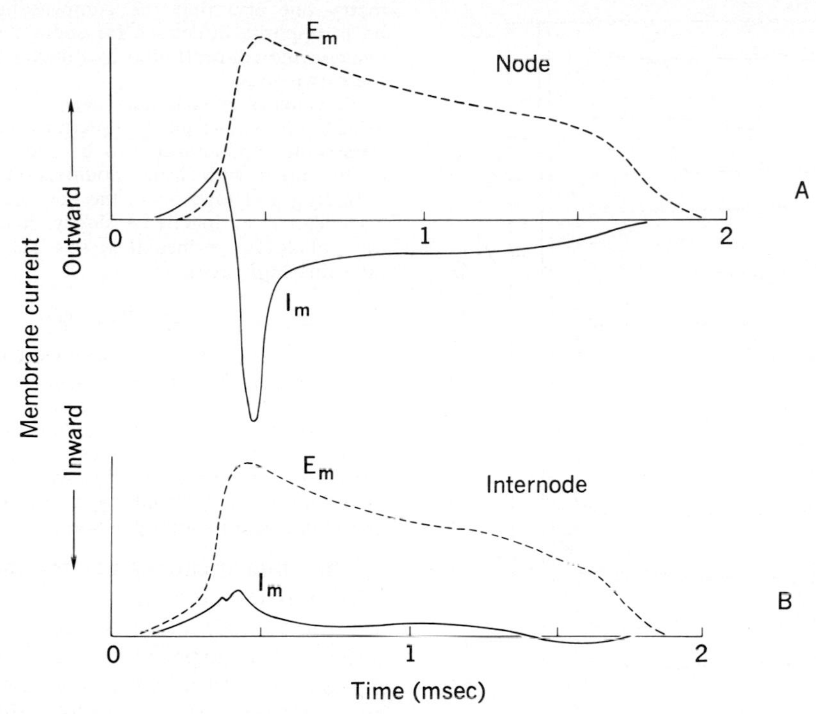

Fig. 2-25. Action potentials and membrane currents of frog sciatic nerve fiber. **A,** Recorded at node. **B,** Recorded at internode. Note that large inward component of current occurs near peak of spike at node, implying ionic component to membrane current. In internodal region, net current is outward except during end of falling phase of spike, implying that internodal region supplies membrane current in the form of displacement charge leak flowing from membrane capacitance, but does not pass inward ionic current. Compare with I_e and I_i in Fig. 2-18. (Redrawn from Huxley and Stämpfli.[18])

sults of such a survey are shown for a frog node (Fig. 2-25, *A*) and internode (Fig. 2-25, *B*). Only in the nodal regions does the membrane current have a large inward component, and it occurs during the rising phase of the action potential. The membrane current in the internodal region is not only smaller in absolute magnitude but also flows essentially only in an outward direction.

The progress of the action current over a distance of several nodes is shown in Fig. 2-26, in which the inward component of membrane current, plotted downward, is seen only when records are taken opposite the nodes. One therefore concludes that excitation must leap in a spatially discontinuous manner from node to node, thereby justifying the designation of the conduction process as saltatory.

Voltage clamp analysis of the action potential in myelinated fibers

Application of the voltage clamp technique to a single vertebrate node is less direct and technically much more difficult than its application to squid axons because the extremely small size of the node precludes direct use of internal electrodes or even direct micropuncture. Nevertheless, a complete voltage clamp analysis of toad and frog nodes using less direct techniques has been successfully accomplished by Frankenhauser and his colleagues. The computed action potentials agree closely with those found experimentally; as in the case of squid axons, the voltage clamp parameters serve to describe other short-term electrical phenomena in myelinated nerve, i.e., subthreshold responses, the refractory period, etc. Although the results have largely confirmed the squid data, the experiments have been extremely important in demonstrating that the occurrence of selective sequential sodium-potassium permeability changes in the axolemma can be a general explanation of nervous activity in vertebrates and is not a unique specialization in an unusually large invertebrate nerve fiber.

Separation of ionic currents. The identification of the late voltage clamp currents as being due to potassium is even less satisfactory than it is for the squid axon. In toad nerve there is a component of late current that is apparently carried by sodium,

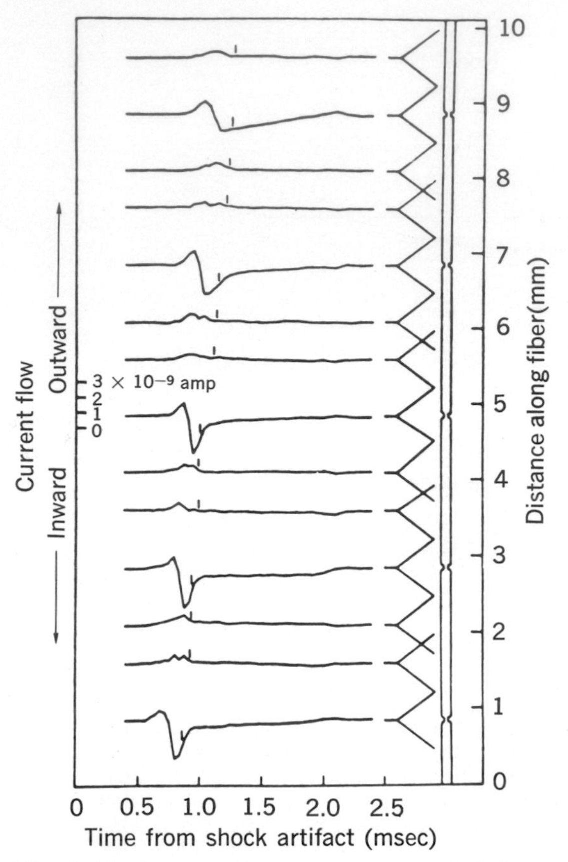

Fig. 2-26. Transmembrane current recorded from nodes and internodes along single myelinated fiber isolated from sciatic nerve of frog. Large inward current occurs opposite node but not internode (compare with Fig. 2-25). Vertical bars on current curves mark occurrence of peak of action potential. (From Huxley and Stämpfli.[18])

but it has a different time course than either the early sodium current or the late potassium current.

Expression of current voltage data as conductances. In squid axons the relations between current and voltage could be expressed satisfactorily as a conductance without explicitly involving the concentration. Such is not the case in amphibian myelinated fibers, in which the specific ion currents must be explicit expressions involving the ionic concentration on either side of the membrane. A suitable expression that fits the data is that for the ionic current flowing across a constant field type of membrane.

$$I_{Na} = -P_{Na} \times \frac{VF^2}{RT} \times \frac{Na_o - Na_i \exp(-VF/RT)}{1 - \exp(-VF/RT)}$$

with a similar expression for I_K.

Although more complex mathematically, such an equation is actually easier to understand theoretically because it is what one would expect from the unequal concentrations on either side of the membrane, whereas the apparently simple result from squid axons seems to require not only some compensating factor in the membrane such as fixed-charge asymmetry but also that the compensating asymmetry be in opposite directions for sodium and potassium because the concentration asymmetry is opposite for these two ions.

Formulation of empirical equations. The formalism of the Hodgkin-Huxley equations to sodium and potassium conductance can be successfully applied to the node with little modification. However, in both frog and toad fibers the potassium current has been shown to inactivate slowly, and the equation for potassium permeability has been modified to take this into account:

$$P_K = \overline{P}_K \times n^2 k$$

where k is the potassium inactivation factor analogous to the h factor for sodium inactivation.

Potassium inactivation has also been found in other systems under special circumstances, including squid axons, so that it can be considered a general physiologic phenomenon. However, it is not a major factor normally contributing to membrane repolarization during an action potential, even in toad fibers.

The numerical value of the Hodgkin-Huxley parameters m, n, and h for frog nodes are remarkably close to those obtained for squid axons, especially considering that the nerve membranes in the two cases are bathed by solutions of markedly different ionic strengths. The steady-state values of the parameters for the two species (i.e., those obtained after long voltage clamps) are shown as a function of membrane voltage in Fig. 2-27. The major difference is that the node shows considerably greater sodium inactivation, i.e., lower value of h, at a given potential than does the squid axon, thus tending to reduce somewhat the sodium currents during an action potential.

Determinants of conduction velocity in myelinated fibers

Although the presence of a myelin sheath increases the velocity of conduction of a nerve impulse by confining the local circuits to the node, it does not follow that the node spacing is the sole or even the most important determinant of conduction velocity. There are several other factors to be considered.

For example, some local circuit current must flow in the internodal region to discharge the internodal capacitance, even though this region is passive. (Note the small outward current in the internodal region during the rising phase of the action potential in Fig. 2-25, *B*.) Increasing the distance between nodes will therefore have two opposing effects: increase the velocity by increasing the "jump" distance for excitation and reduce the velocity by increasing the fraction of local circuit current flowing through the internode,

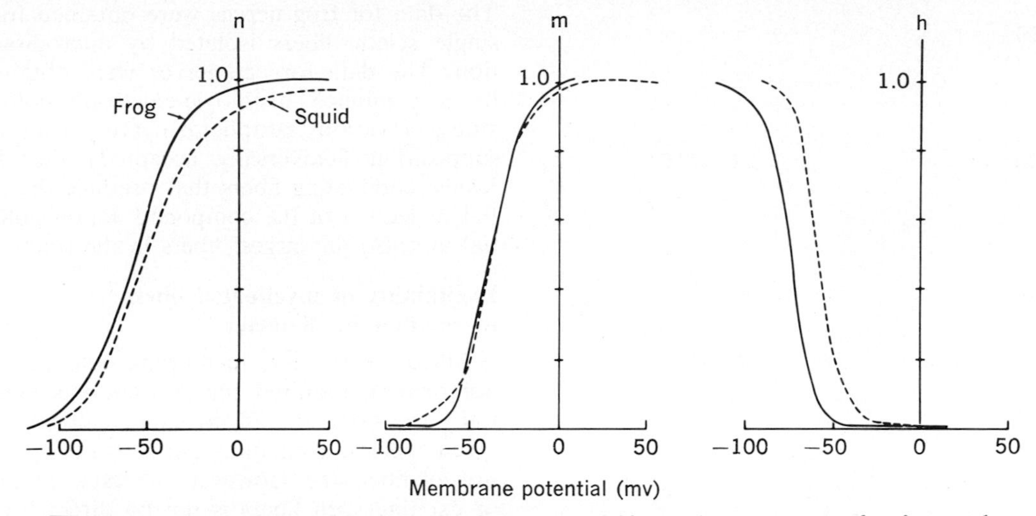

Fig. 2-27. Comparison of steady-state values of Hodgkin-Huxley parameters for frog nodes and squid axons. Solid lines, frog node; dashed lines, squid axon. Steady-state values are those obtained after long-duration voltage clamps at indicated membrane potentials. (Redrawn from Cole.[1])

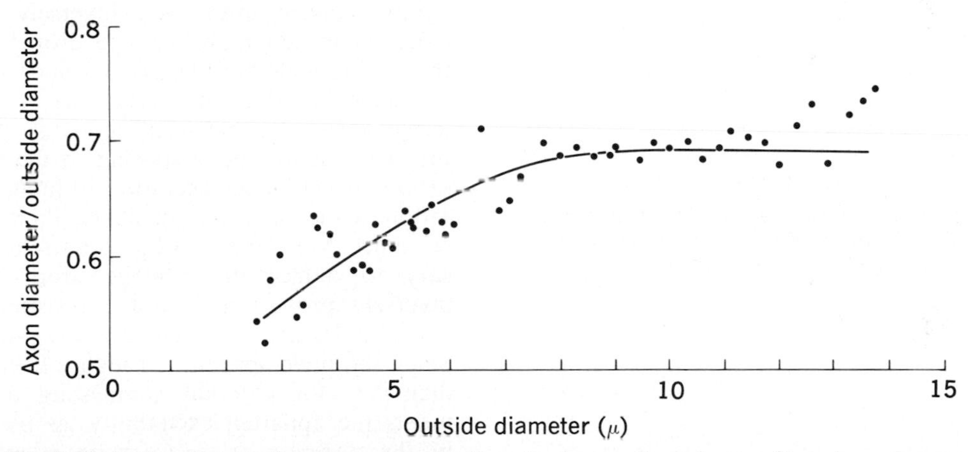

Fig. 2-28. Ratio of axon diameter to outside diameter for myelinated nerve fibers in saphenous nerve of cat. (From Gasser and Grundfest.[10])

thereby reducing the current available to produce depolarization and excitation at the node. To complicate matters further, a simple increase in myelin thickness, to reduce the myelin capacitance and hence the internode capacitance loss, would require reduction of the axis cylinder diameter and hence reduce local circuit flow because of a higher core resistance in the interior of the axis cylinder. A more quantitative analysis of the problem indicates that there is a relation between the internode spacing, the fiber diameter, and the myelin thickness such that the fastest conduction velocity for a given external diameter occurs when the ratio of the axis cylinder diameter to the outside diameter is about 0.7, and that conduction velocity and internode spacing should be proportional to diameter.

These theoretical conclusions are generally substantiated by experiments. Fig. 2-28 shows that the observed ratio of axon diameter to fiber diameter is close to the theoretical optimum (0.7), at least for large myelinated fibers, although there appears to be a relative thinning of the myelin sheath in smaller fibers. Furthermore, in accordance with theory, the conduction velocity for myelinated fibers is roughly linear with diameter. Fig. 2-29 shows this relationship for both frog and cat nerves.

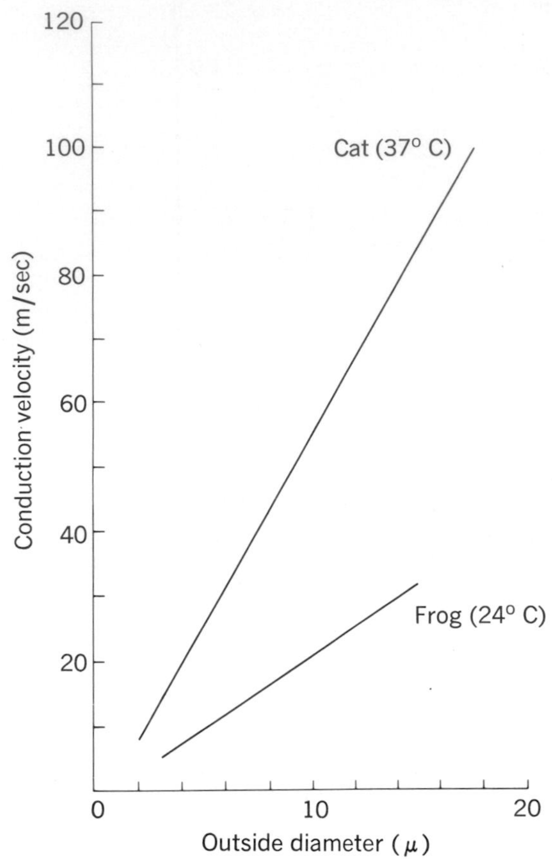

Fig. 2-29. Relation between outside diameter and conduction velocity at physiologic temperatures for myelinated nerve fibers of cat and frog.

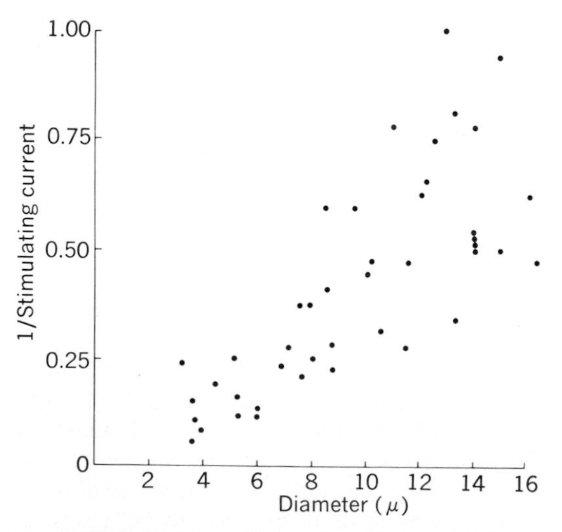

Fig. 2-30. Relation between stimulating current and outside diameter of single myelinated fibers isolated from sciatic nerve of frog. Stimulating current has been normalized relative to maximum current used. Reciprocal of stimulating current is proportional to excitability. (From Tasaki.[1])

The data for frog nerves were obtained from single sciatic fibers isolated by microdissection. The data for cat nerves were obtained by a combined histologic-electrophysiologic study of various peripheral nerves, using the supposition (universally accepted) that the fastest conducting fibers that produce the initial deflection of the compound action potential are also the largest fibers in the tract.

Excitability of myelinated fibers in relation to diameter

Since the absolute membrane potential depolarization required for excitation is essentially the same for all myelinated fibers of a given species, excitability per se is not a function of fiber size. However, the usual method of exciting such fibers is not by direct transmembrane depolarization but rather by passing currents in the extracellular media between electrodes separated by distances much greater than the node spacing. Under these circumstances the magnitude of the stimulating current must vary inversely as the node separation if the voltage drop between the resting node and the excited node is to be the same for fibers of varying size. (The external resistance between nodes is proportional to the internode spacing. If the voltage drop between nodes necessary to induce excitation is the same for all fibers, then, in accordance with Ohm's law, the current necessary to induce this voltage drop will be inversely proportional to the resistance between nodes and hence the internode spacing.) Since the node spacing is roughly linear with diameter, for external stimulating arrangements the apparent excitability, as measured by the current required to excite, should vary inversely with the diameter. This relation is seen to be approximately true for frog fibers (Fig. 2-30) with a rather wide scatter of data. In mammalian nerve fibers it is generally stated that the ratio of conduction velocity to fiber diameter is 6 to 9 m/sec/μ.

AFTERPOTENTIALS

The term "afterpotential" refers to one of several small-amplitude, long-duration potential changes occurring subsequent to the action potential spike. The afterpotentials of a large mammalian myelinated fiber, together with the spike potential drawn to scale, is illustrated in Fig. 2-31.

The first afterpotential is a residual depolarization of the fibeı persisting after the spike. It is called the negative afterpotential,

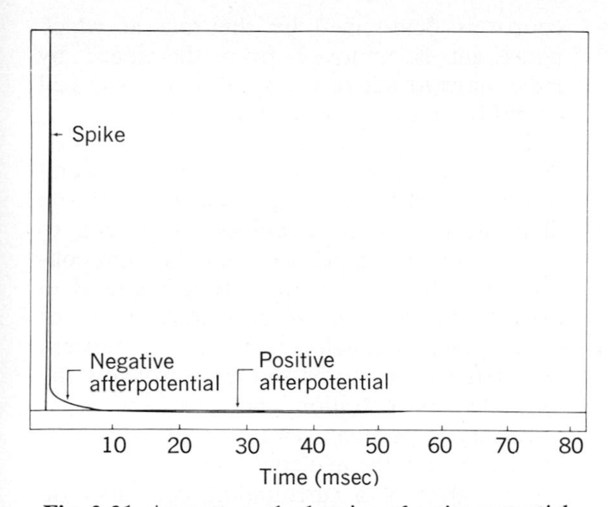

Fig. 2-31. Accurate scale drawing of action potential and related afterpotentials recorded extracellularly from saphenous nerve of cat. Temperature about 25° C.

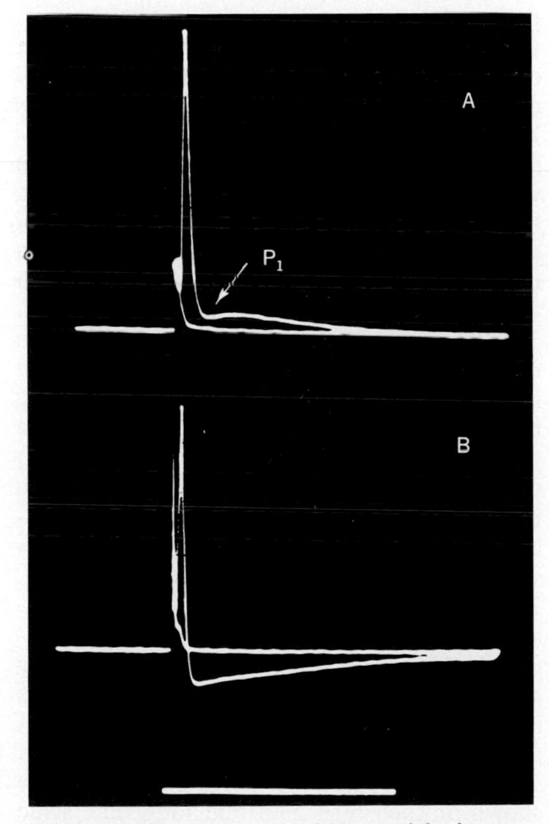

Fig. 2-32. Two types of afterpotentials in mammalian C fibers. **A,** Hypogastric nerve of cat (sC fiber type). **B,** Vagus nerve of rabbit (drC fiber type). Horizontal bar represents 250 msec. Temperature 25° C. Afterpotential shown in **A** is largely negative; a slight postspike positivity is labeled P_1. In **B,** pronounced postspike positivity merges smoothly with positive afterpotential, obscuring any negative afterpotential. (From Armett and Ritchie.[6])

and it is followed by a hyperpolarization of the fiber called the positive afterpotential. The relative magnitude and duration of the afterpotentials shown in Fig. 2-31 are those commonly observed. The negative afterpotential is generally of somewhat larger amplitude and shorter duration than the positive afterpotential. However, both are subject to wide variation, not only from one fiber type to another but especially in response to previous activity of the fiber and with respect to its metabolic condition.

There is in some fibers a third type of afterpotential not illustrated in Fig. 2-31. This deflection, occurring immediately after the falling phase of the spike, is a very transient hyperpolarization variously referred to as the positive phase, the positive underswing, postspike positivity, etc. Such a positive phase is seen in squid axons (Fig. 2-16). Two examples in mammalian C fibers are shown in Fig. 2-32. In the record in Fig. 2-32, *A* (sC fibers), the postspike positivity (P_1) is evident only as a slight dip in the negative afterpotential. The positivity is insufficient to actually hyperpolarize the membrane but only interrupts the smooth contour of the negative afterpotential. By contrast, in another type of C fiber (drC fiber), shown in the lower record (Fig. 2-32, *B*) the postspike positivity is well developed and merges smoothly with the positive afterpotential, obscuring the negative afterpotential.

Certain properties of the afterpotential are listed in Table 2-1. Durations and magnitudes are given only to indicate orders of magnitude; they vary widely among different nerve fibers. Although each afterpotential has been assigned a single mechanism, it will become

Table 2-1. Some properties of afterpotentials

Afterpotential	Typical duration	Typical maximum amplitude	Probable mechanism
Postspike positivity	5 msec	5 mv	Transient increase in potassium permeability
Negative afterpotential	30 msec	5 mv	Potassium accumulation outside axolemma
Positive afterpotential	200 msec	2 mv	Stimulation of sodium pump

clear from the discussion of each that the mechanisms invoked as explanations for the afterpotentials act not sequentially but continuously and simultaneously following an action potential. The net effect on membrane potential of the processes acting in concert, and hence the type of afterpotential produced, will depend on their relative intensities.

Postspike positivity. The postspike positivity occurs after the falling phase of the spike when the relative potassium conductance of the membrane briefly exceeds the resting value. In consequence, the membrane becomes relatively more potassium permeable than it was before the spike, and the membrane potential approaches the potassium electrode potential more closely than it does at rest, producing a transient hyperpolarization relative to the resting level.

Since the transient changes in sodium and potassium conductances are controlled by the voltage-dependent parameters given by the Hodgkin-Huxley analysis, the time course of the postspike positivity is accounted for by the same type of analysis, which describes the action potential and is in fact seen in calculated as well as experimental records (Fig. 2-16). Qualitatively one can see that the magnitude of the postspike positivity will reflect the difference between the resting potential and the potassium equilibrium potential. Postspike positivity is therefore most obvious in cells in which V_m is relatively low compared to E_K. In some cells the phenomenon is probably an artifact resulting from the slight depolarization of resting potential caused by the experimental trauma of dissection and micropuncture.

Negative afterpotential. The negative afterpotential has been widely ascribed to a transient accumulation of the potassium ions released during an action potential in a region immediately external to the axolemma. Calculations based on the known amount of potassium released during an action potential show that if these ions are confined to a sufficiently thin layer around the axon, the immediately external potassium concentration will be increased some 20 to 30% above resting levels following a single impulse, an increment sufficient to depolarize the axon by a few millivolts. The initial magnitude of the negative afterpotential would therefore be determined by the amount of potassium rapidly dumped outside the axolemma during an action potential and by the dimensions of the extra-axonal space. The subsequent time course is determined by the rate at which potassium is removed from the space by either inward active transport into some cell or diffusion away from the axolemma.

In the case of the squid axon there is evidence that the space between the Schwann cell layer and the axolemma has the requisite dimensions and permeability properties to actually retain a small excess potassium concentration for the length of time required to explain the negative afterpotential. In other tissues, e.g., mammalian C fibers, the Schwann cell cleft appears to be too large to allow the requisite concentration increase, and it is necessary to postulate some sort of a diffusion barrier around the axolemma.

The effect of a surrounding cell layer on the afterpotentials can be seen clearly in an interesting experiment performed upon the central nervous system of the leech. The experiment also serves to expose the somewhat tenuous logic used to relate the negative afterpotential to the excess potassium concentration, since the object is to determine the effect of potassium accumulation in reducing the amplitude of the *postspike positivity* (which is known to depend upon E_K) and from this result to infer that potassium accumulation would also depolarize the membrane during the period of the *negative afterpotential,* assuming that external potassium acts upon the membrane potential in accordance with the constant field equation in both situations.

In the leech preparation it is possible to remove the glial cell layer investing certain neurons in isolated ganglia without damaging the neurons. This procedure, which of course removes any restraining barrier for potassium diffusion, also abolishes the slight reduction in amplitude of the postspike positivity produced by the potassium accumulated in response to a volley of rapid shocks. In Fig. 2-33, *A,* single action potentials recorded intracellularly from the neuron are shown before and after destruction of the surrounding glia. The close similarity of the two responses indicates that the neuron membrane is not seriously damaged by the dissection. In Fig. 2-33, *B,* the postspike positivity phases are shown at such high gain and long time scale that the spike of the action potential is not visible on the trace. In the case of the intact neuron the effect of a volley is to reduce the amplitude of each succeeding positive phase by a small amount, presumably because potassium accumulates faster than it can diffuse away. In the case of the "naked" neuron the

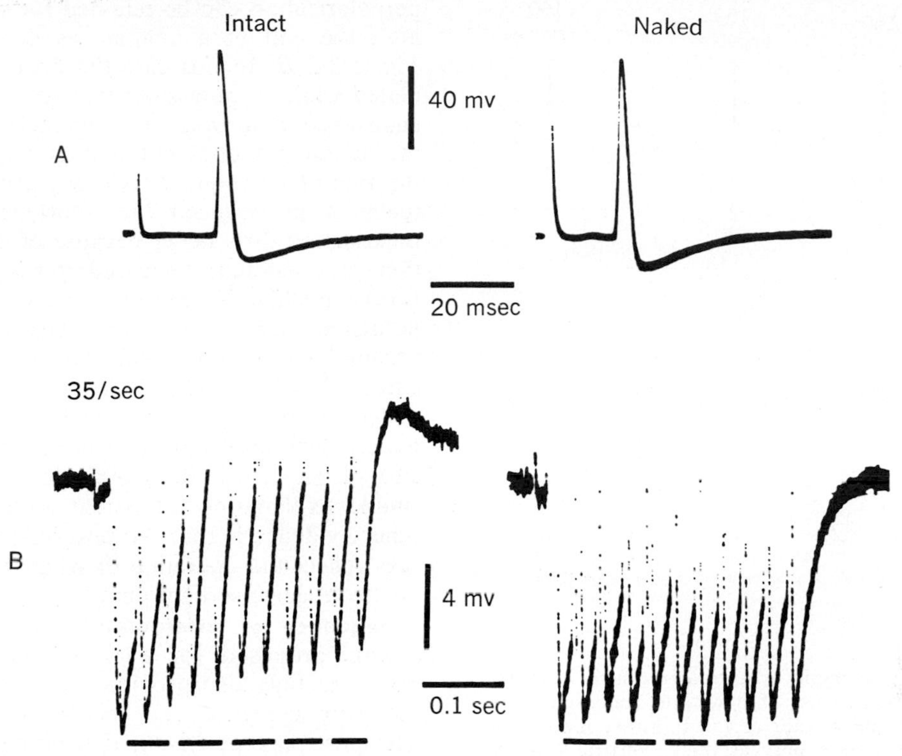

Fig. 2-33. Effect of repetitive activity on postspike positivity in single ganglion cells of leech. Note scale changes between top and bottom records. **A,** Single intracellular action potentials from intact neurons and from neurons lacking glial cell matrix normally investing them. Spikes appear similar, indicating that process of removing surrounding tissue did not damage cells. **B,** Repetitive stimulation at 35/sec of neurons illustrated in **A.** Because of high gain, spike proper does not appear on record; only postspike positive phases are shown. Because of slow time scale compressing wave form, these positive phases appear as downwardly directed spikes. In intact cell, positive phases are reduced in amplitude during repetitive stimulation, presumably as result of potassium accumulation in spaces of glial matrix. In contrast, in cells lacking surrounding glia, attenuation of positive phase is not seen. (From Baylor and Nicholls.[7])

glial permeability barrier has been destroyed, the diffusion no longer exists, and the positive phases are of nearly constant amplitude during a volley.

Although this is an adequate explanation of the negative afterpotential in nerve cells, it does not suffice for other tissues. In some muscle cells, for example, a slow negative afterpotential occurs although there is no histologically identifiable restraining layer. In frog skeletal muscle the explanation appears to be related to the existence of the tubular system, which may provide space for potassium accumulation or be the site of long-lasting conductance changes. Potassium may also accumulate in the glycocalyx, which closely invests some muscle fibers.

Negative afterpotentials in nerve and muscle can also be produced by external application of the veratrum alkaloids. Although these substances do promote potassium leakage from cells, the explanation for the afterpotential appears to be related more to long-lasting conductance changes in the membrane produced by the drug than to potassium accumulation outside it.

Positive afterpotential. The positive afterpotential following a single nerve impulse is too small to study conveniently; consequently, most of the work relating to the positive afterpotential has actually dealt with the pronounced "posttetanic hyperpolarization" that develops following the last spike in a train of impulses, thought to be due to the same mechanism producing the positive afterpotential.

Since the positive afterpotential and the posttetanic hyperpolarization are increased by procedures that stimulate sodium transport and are reduced by treatments that inhibit the sodium pump, it is natural to sup-

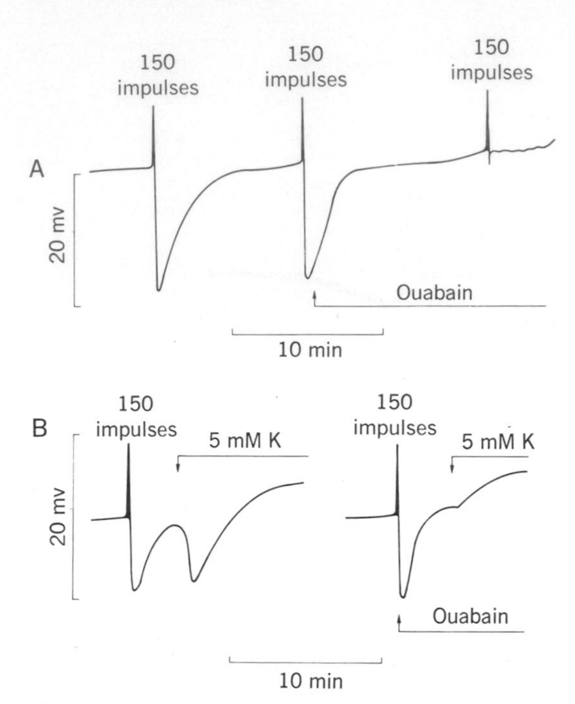

Fig. 2-34. Posttetanic hyperpolarization in mammalian C fibers. Because of slow time scale, individual spikes in stimulus train of 150 impulses are not resolved and appear only as thick, vertical line. **A,** Fibers bathed in normal Ringer's solution. Ouabain was added to bath, producing some curtailment of afterpotential and completely blocking its reappearance during subsequent stimulation. **B,** Fibers bathed initially in potassium-free solution. Stimulation produces attenuated posttetanic hyperpolarization, but further hyperpolarization is induced by adding potassium to bath. Potassium-induced hyperpolarization is blocked by ouabain. (Redrawn from Rang and Ritchie.[22])

pose that the operation of a sodium pump is involved in the production of these postspike hyperpolarizations. In some cases the membrane may actually be hyperpolarized beyond E_K, suggesting that the effect may be due to an electrogenic sodium pump. An example of the evidence supporting this mechanism for posttetanic hyperpolarization is given in Fig. 2-34. The time scale has been compressed so that the individual spikes resulting from the train of stimuli are not seen separately but appear only as a thick line in the figure. The posttetanic hyperpolarization appears as a slow wave developing after the spike and waning with a time constant of 1 to 2 min (Fig. 2-34, *A*). The sodium transport inhibitor ouabain not only curtails the hyperpolarization as soon as it is applied but completely prevents its development following a subsequent series of shocks.

The capacity to produce a posttetanic hyperpolarization can be retained for some time after the end of a tetanus, as illustrated in Fig. 2-34, *B*. In this case the fiber was stimulated while in potassium-free solution, a circumstance that does not significantly affect the action potential but that greatly reduces the rate of subsequent sodium extrusion. Actually a pronounced but shortened hyperpolarization does occur because of accumulation of potassium (released during stimulation) immediately outside the axolemma. This potassium permits a partial extrusion of the accumulated sodium load. The remainder is extruded when potassium is subsequently added to the bath and produces a typical hyperpolarization that, as shown on the right side of Fig. 2-34, *B,* is ouabain sensitive. A simple explanation consistent with this and other evidence is that the positive afterpotentials reflect the operation of what is probably an electrogenic sodium pump.

Sequence of afterpotentials. The mechanisms proposed for each of the separate afterpotentials also provide some explanation for their sequence and relative magnitudes. The postspike positivity, depending on transient conductance changes, might be expected to persist only a brief time after the spike. When the conductances return to approximately the resting level, the membrane potential will be determined by the ionic concentrations inside and outside the axon and the relative resting permeabilities. An accumulation of external potassium would therefore lead to a transient depolarization observable after the positive phase, the negative afterpotential. However, the sodium pump would be turned on immediately after the spike in response to the sodium load imposed by the action potential, and if electrogenic, this pump would contribute a hyperpolarizing component to the membrane potential. The net effect on the postspike membrane potential at any moment will of course depend on the relative magnitudes of the depolarizing effect of potassium accumulation versus the hyperpolarizing effect of an electrogenic pump. One might expect the hyperpolarizing effect to outlast the depolarizing effect and thus lead to a late hyperpolarization, simply because diffusion through the Schwann cell channels is faster than active transport across the axolemma. In accordance with this notion is the observation that most large invertebrate axons that do not have strong electrogenic pumps do not exhibit positive afterpotentials, although they do have negative

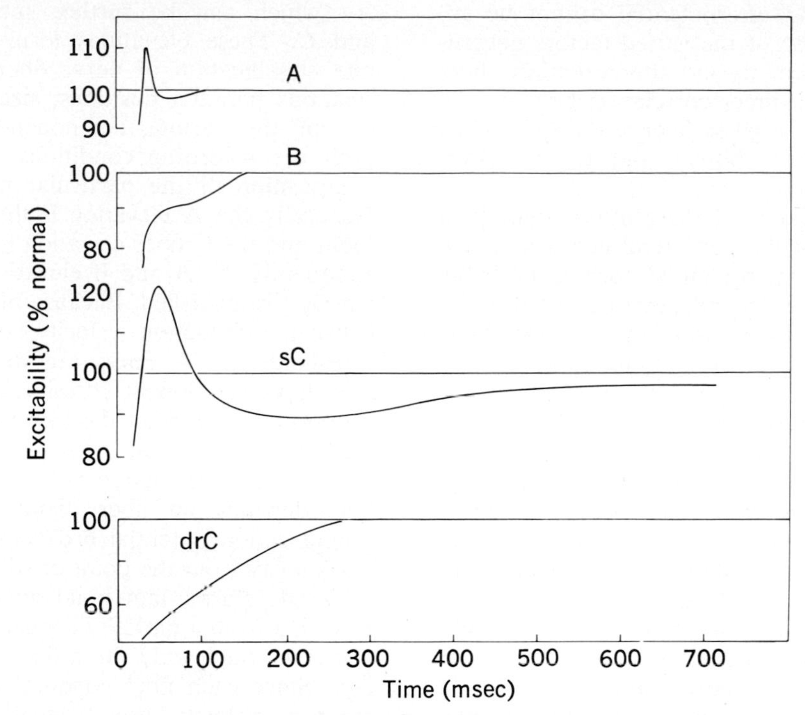

Fig. 2-35. Excitability curves of various fiber classes in myelinated nerve trunks of cat (temperature 37° C). Excitability, as measured by strength of extracellular test shocks, is plotted as percent of ratio (control shock/test shock). (Graphs for A, B, and sC fibers from Erlanger and Gasser[2]; graph for drC fibers from Grundfest and Gasser.[11])

Table 2-2. Some properties of mammalian nerve fibers

Property	Fiber type			
	A	*B*	*sC*	*drC*
Fiber diameter (μ)	1-22	< 3	0.3-1.3	0.4-1.2
Conduction speed (m/sec)	5-120	3-15	0.7-2.3	0.6-2.0
Spike duration (msec)	0.4-0.5	1.2	2.0	2.0
Absolute refractory period (msec)	0.4-1.0	1.2	2.0	2.0
Negative afterpotential (% of spike amplitude)	3-5	None	3-5	None
Duration (msec)	12-20	—	50-80	—
Positive afterpotential (% of spike amplitude)	0.2	1.5-4.0	1.5	10-30*
Duration (msec)	40-60	100-300	300-1,000	75-100*
Order of susceptibility of asphyxia	2	1	3	3

*Refers to postspike positivity.

afterpotentials. Since afterpotentials depend not only on membrane permeabilities but also on factors both internal and external to the axolemma, it is not surprising that they should be much more variable than the spike, which depends largely on potential dependent membrane permeability changes.

Significance of afterpotentials. Since the afterpotentials represent deviations of the resting potential either toward or away from threshold, one might expect them to be associated with postspike excitability changes. The excitability curves of mammalian A, B, and C fibers are given in Fig. 2-35. There are several points of correspondence between the excitability curves of these fiber types and the characteristics of their afterpotentials (Table 2-2). For example, in A and sC fibers the excitability sequence supranormal-subnormal correlates roughly with the duration of the negative and positive afterpotentials. By contrast, B fibers, which have no negative afterpotential and an intermediate-duration positive afterpotential, exhibit only an intermediate-duration period of subnormality. However, a precise correspondence between excitability and afterpotential cannot be expected. Because of the varied factors governing action potentials and afterpotentials, there need not be a direct correlation between the contribution of a given fiber to the spike (i.e., determining excitability) and to the afterpotentials.

The significance of these afterpotentials in the function of the peripheral nervous system is uncertain. Afterpotentials per se probably cannot block a second nervous impulse following closely in the wake of its predecessor, since the density of local circuit currents flowing in front of the spike are more than sufficient to depolarize the membrane to threshold. However, since the time required to depolarize to threshold is in part a function of the absolute level of membrane potential, afterpotentials may have a slight effect on the conduction velocity of a succeeding impulse.

Afterpotentials may have greater significance when they occur in those portions of the neuron that receive inputs resulting in subthreshold local responses, e.g., the synaptic regions of the neuron soma or dendritic tree. Here, an afterpotential, depending on its polarity, could sum either positively or negatively with an incoming postsynaptic potential and thus affect neuron excitability. This type of interaction has been tentatively identified in the ganglia of the leech central nervous system, but its general occurrence and significance remain conjectural.

PROPERTIES OF MULTIFIBER NERVE TRUNKS

The compound action potential

Because of the practical difficulty of isolating single mammalian fibers, most of the electrical data on such fibers derive from studies of the extracellularly recorded compound action potential resulting from simultaneous activation of all or some of the population of fibers comprising the nerve tract. Since the compound action potential is merely the algebraic sum of individual fiber action potentials, most of the properties of the trunk can be inferred from a knowledge of single-fiber electrophysiology and the distribution of fiber sizes in the trunk. Close inspection of compound action potentials recorded from peripheral nerves indicates a number of components. For reference a typical compound action potential (recorded from frog sciatic nerve trunk) is shown in Fig. 2-36. The record shows three elevations: *A* (which can be further subdivided), *B*, and *C*. These elevations form the basis of one classification of nerve fibers. The exact relations between positions, size, and threshold of the various components depend on both the recording conditions and the fiber composition of the particular nerve studied. Generally the A elevation is the most prominent and the C fiber elevation is the smallest. Frequently the A and B elevations cannot be clearly distinguished. Because of the relations between conduction velocity, diameter, and threshold, the A fibers are the largest and also have the lowest threshholds. The fiber properties are further discussed in the section on fiber classification, p. 74.

Since the conduction velocity of single fibers depends on fiber diameter, the compound action potential broadens as it propagates away from the point of stimulation. Fig. 2-37, *A*, shows individual action potentials recorded from a partially dissected frog sciatic nerve in which only three fibers were left intact. Since each fiber conducts impulses at a constant velocity proportional to diameter, the temporal dispersion of the spikes increases at progressively more distant recording sites. Fig. 2-37, *B*, shows the situation in an intact frog sciatic nerve with several hundred fibers. In this case one cannot resolve action potentials from individual fibers because the limited length of the trunk does not permit adequate dispersion of the spikes, but one can clearly observe the separation of the compound action potential into two peaks, indicating a bimodal distribution of fiber sizes. The total area under the action potential remains constant, reflecting the fact that all fibers contribute to the compound action potential regardless of the location of the recording site relative to the stimulating site.

A compound action potential is probably never recorded from a peripheral nerve under physiologic conditions because there is normally both centripedal and centrifugal conduction of impulses from individual fibers in most nerve trunks. Moreover, physiologic stimuli never result in such nearly complete synchronization of activation as that produced by strong electrical stimulation. Physiologic stimulation is usually sufficiently asynchronous that individual spikes can be resolved on the oscilloscope trace. An example of discharge recorded from a peripheral nerve in response to physiologic stimulation during respiration is shown in Fig. 61-3.

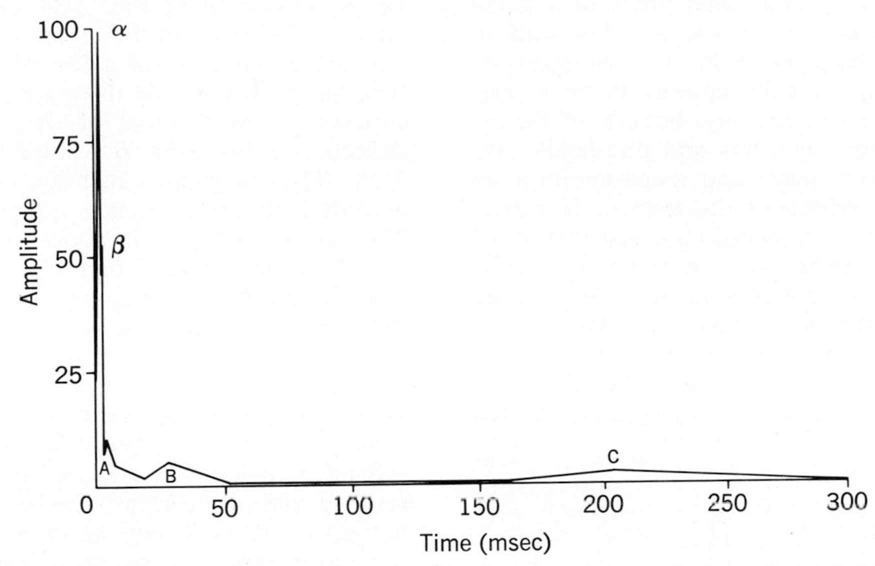

Fig. 2-36. Complete compound action potential recorded from sciatic nerve of bullfrog. Several components of A group of fibers are not clearly resolved because time scale of record was chosen to demonstrate B and C peaks. Amplitude of action potential is plotted relative to A alpha peak. Room temperature. (From Erlanger and Gasser.[2])

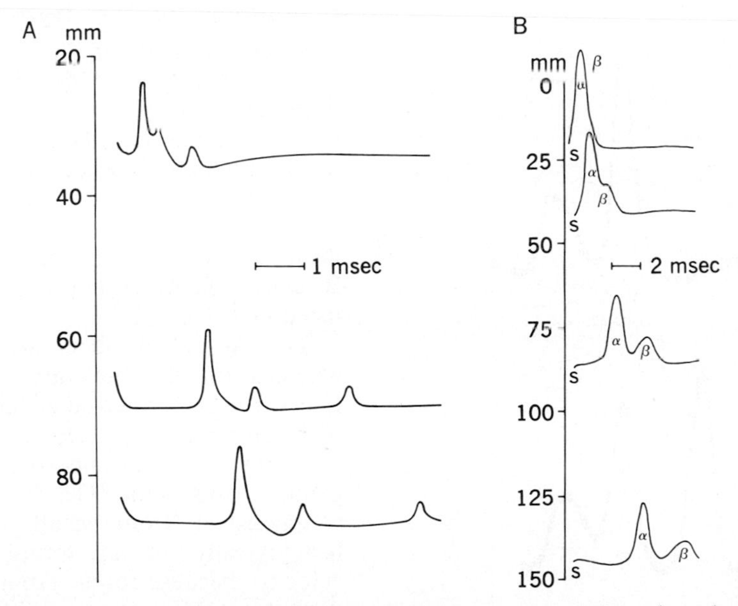

Fig. 2-37. Temporal dispersion of action potentials during propagation along length of frog sciatic nerve. Vertical scale indicates distance of recording site from point of stimulation. Note difference in time scale between **A** and **B. A,** Partially dissected nerve bundle containing only three excitable fibers. Conduction velocity of each fiber is constant. Because conduction velocities are different, separation between spikes increases at recording sites more distant from stimulating electrodes. Initial deflection at start of each trace is shock artifact. **B,** Compound action potential produced by A fibers in intact frog sciatic nerve with many hundreds of fibers. Action potentials cannot be resolved but add together to produce a smooth envelope. In top trace, close to the stimulating site, only a single envelope can be seen. However, at progressively more distant recording sites, separation of compound action potential into two peaks (A alpha and A beta) occurs because of bimodal distribution of fiber sizes. (**A,** redrawn from Tasaki[1]; **B** from Erlanger and Gasser.[2])

DISTRIBUTION OF THRESHOLDS IN MYELINATED FIBERS

Although the individual fibers in a nerve trunk each obey the all-or-none law with an extremely sharp threshold, the compound action potential usually appears to be a continuously graded response because of the individual fiber diameters and thresholds vary by small increments and therefore form an essentially continuous distribution. However, when a clear-cut multimodal distribution of fiber sizes occurs, it is possible by careful adjustment of stimulus intensity to activate the lower threshold (larger and faster) distribution of fibers independently of the higher threshold (smaller and slower) fibers. The intensity of stimulus S_1 in Fig. 2-38, *A,* has been adjusted to activate all of the A-alpha fibers but essentially none of the A-beta fibers. (Close inspection of Fig. 2-38, *A,* shows a minimal deflection at the arrow, corresponding to the activation of a few of the largest beta fibers. This barely discernible deflection corresponds to the foot of the larger beta deflection in Fig. 2-38, *B.*) Stimulus S_2 (Fig. 2-38, *B*) is of greater intensity, sufficient to activate both alpha and beta groups of fibers. The alpha and beta elevations of the compound action potential represent conduction in entirely independent groups of fibers with different thresholds. This is demonstrated by close sequential application of both S_1 and S_2 (Fig. 2-38, *C*). The second alpha elevation (α_2) is somewhat reduced as expected, since some of the alpha fibers will still be absolutely refractory after S_1 and be unable to respond to S_2. By contrast, the beta fibers are not activated by S_1 and hence are not in a refractory state. These fibers are able to respond normally to S_2 even though S_1 has preceded, leading to the conclusion that the A-alpha and A-beta fibers have essentially nonoverlapping distributions of thresholds.

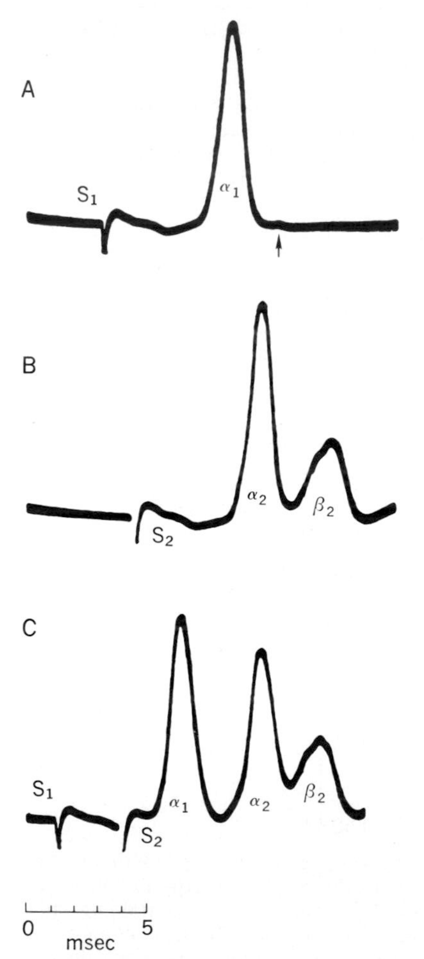

Fig. 2-38. Records showing that A-alpha and A-beta components of compound action potential represent conduction in different groups of fibers with different thresholds. Frog sciatic nerve. S_1 and S_2 represent shock artifacts. Arrow in **A** indicates small elevation in trace produced by stimulation of a few beta fibers. (Redrawn from Erlanger and Gasser.[2])

Classification of fiber types

Several systems are used in the classification of fiber types. One of the earliest was based on electrophysiologic properties, mainly the conduction velocity as revealed by peaks in the compound action potential and differences in afterpotential as revealed by excitability curves. This electrophysiologic classification divides fibers into three groups: A, B, and C. Some properties of these fibers are listed in Table 2-2.

The A and B fibers are all myelinated, whereas the C fibers are unmyelinated. A fibers can be further subdivided, on the basis of mean conduction velocity, and hence fiber size, into several subgroups: alpha, beta, gamma, and delta. The B fibers cannot be distinguished from small A fibers either histologically or in terms of conduction velocity because both groups have similar diameters. However, a distinction can be made on the basis of differences in afterpotentials. A fibers have a short, pronounced negative afterpotential and minimal positive afterpotential. By contrast, B fibers show no negative afterpotential but have a large positive afterpotential. As might be expected, these differences in afterpotentials result in quite different excitability curves (Fig. 2-35). The main difference is that A fibers show a a postspike hyperexcitability, whereas B fibers show only a relatively prolonged period of hypoexcitability.

C fibers can be readily distinguished from either A or B fibers because they are all unmyelinated and of small diameter and hence of very slow conduction velocity. C fibers have been further subdivided on the basis of postspike excitability changes into two classes: sC fibers (postganglionic efferent *sympathetic* C fibers) and drC fibers (afferent *dorsal root* C fibers). Sympathetic C fibers evidence a postspike sequence of hyper- and hypoexcitability, whereas the dorsal root C fibers show only a postspike hypoexcitability. As can be seen from Fig. 2-35, the postspike alternations of excitability are much more prolonged for C fibers than for either A or B fibers, in keeping with the more protracted afterpotentials.

Unfortunately it has been difficult to apply the A fiber classification to all types of nerves. In part the problem is technical; the original gamma elevation in the compound action potential appears to have been a recording artifact. More importantly, however, since the first elevation in the A part of the compound action potential was called A-alpha and represented the fastest conducting fibers, this designation could refer to groups of fibers of different conduction velocities in different nerves, depending on the distribution of fiber sizes. Thus the fastest identifiable group of fibers in afferents from the soleus muscle of the cat have diameters of about 17μ, whereas in the saphenous nerve the largest fibers are only about 14μ. Although these two groups of fibers are clearly distinguishable on the basis of conduction velocity, in the electrophysiologic classification both would be designated as A-alpha. Despite the unsatisfactory nature of the A fiber group terminology, some vestiges remain and have to be learned. The small-diameter motor fibers to muscle spindles are called gamma fibers. The large motor fibers to extrafusal phasic muscle fibers are called alpha fibers. The fastest fibers in cutaneous nerves are sometimes called alpha cutaneous fibers.

A second classification, applied by sensory physiologists to afferent fibers, is based on a division of fibers into four groups, principally on the basis of fiber size, but also on the basis of fiber origin. Table 2-3 lists the four groups and the histologic structure at the afferent terminal of the fiber of each subgroup as well as the approximate corresponding A fiber subgroup. The functional significance of the various terminations is discussed in Chapter 23.

Table 2-3. Classification of afferent fibers

Fiber type	Diameter (μ)	Electrophysiologic grouping	Origin
I-A	12-22	A-α	Annulospiral Golgi tendon organ
II (muscle, skin)	5-12	A-β	Flower spray, touch, pressure, vibratory receptors
III	2-5	A-δ	Free nerve endings, pain-temperature
IV	0.1-1.3	C	Pain, temperature, mechanoreceptors

A similar size-function classification is possible for efferent fibers. Generally the largest motor fibers (12 to 20μ) innervate the extrafusal muscle fibers, whereas the smaller efferent fibers (2 to 8μ) innervate intrafusal fibers within the fusiform spindles.

REFERENCES

General reviews

1. Cole, K. S.: Membranes, ions and impulses, Berkeley, 1968, University of California Press.
2. Erlanger, J., and Gasser, H. S.: Electrical signs and nervous activity, Philadelphia, 1938, University of Pennsylvania Press.
3. Hodgkin, A. L.: The conduction of the nervous impulse, Springfield, Ill., 1964, Charles C Thomas, Publisher.
4. Tasaki, I.: Nervous transmission, Springfield, Ill., 1953, Charles C Thomas, Publisher.
5. Woodbury, J. W.: Action potential. Properties of excitable membranes. In Ruch, T. C., and Patton, H. D., editors: Physiology and biophysics, Philadelphia, 1966, W. B. Saunders Co.

Original papers

6. Armett, C. J., and Ritchie, J. M.: On the permeability of mammalian non-myelinated fibres to sodium and to lithium ions, J. Physiol. **165:**130, 1963.
7. Baylor, D. A., and Nicholls, J. G.: Changes in extracellular potassium concentration produced by neuronal activity in the central nervous system of the leech, J. Physiol. **203:**555, 1969.
8. Frankenhaeuser, B., and Huxley, A. F.: The action potential in the myelinated nerve fiber of Xenopus laevis as computed on the basis of voltage clamp data, J. Physiol. **171:**302, 1964.
9. Frankenhaeuser, B., and Vallbo, A. B.: Accommodation in myelinated nerve fibers of Xenopus laevis as computed on the basis of voltage clamp data, Acta Physiol. Scand. **63:**1, 1965.

10. Gasser, H. S., and Grundfest, H.: Axon diameters in relation to spike dimensions and conduction velocity in mammalian fibers, Am. J. Physiol. **127:**393, 1939.
11. Grundfest, H., and Gasser, H. S.: Properties of mammalian nerve fibers of slowest conduction, Am. J. Physiol. **123:**307, 1938.
12. Hille, B.: Ionic channels in nerve membranes, Progr. Biophys. Mol. Biol. **21:**1, 1970.
13. Hodgkins, A. L., and Huxley, A. F.: Currents carried by sodium and potassium ions through the membrane of the giant axon of Loligo, J. Physiol. **116:**449, 1952.
14. Hodgkin, A. L., and Huxley, A. F.: The components of membrane conductance in the giant axon of Loligo, J. Physiol. **116:**473, 1952.
15. Hodgkin, A. L., and Huxley, A. F.: The dual effect of membrane potential on sodium conductance in the giant axon of Loligo, J. Physiol. **116:**497, 1952.
16. Hodgkin, A. L., and Huxley, A. F.: A quantitative description of membrane current and its application to conduction and excitation in nerve, J. Physiol. **117:**500, 1952.
17. Hodgkin, A. L., and Katz, B.: The effect of sodium ions on electrical activity of the giant axon of the squid, J. Physiol. **108:**37, 1949.
18. Huxley, A. F., and Stämpfli, R.: Saltatory transmission of the nervous impulse, Arch. Sci. Physiol. **3:**435, 1949.
19. Huxley, A. F., and Stämpfli, R.: Evidence for saltatory conduction in peripheral myelinated nerve-fibers, J. Physiol. **108:**315, 1949.
20. Huxley, A. F., and Stämpfli, R.: Direct determination of membrane resting potential and action potential in single myelinated nerve fibers, J. Physiol. **112:**476, 1951.
21. Michalov, J., Zachar, J., and Kostolanský, E.: Automatic computation of changes in membrane potential and underlying processes in the Hodgkin-Huxley model, Physiol. Bohemoslov. **15:**307, 1966.
22. Rang, H. P., and Ritchie, J. M.: On the electrogenic sodium pump in mammalian nonmyelinated nerve fibers and its activation by various external cations, J. Physiol. **196:**183, 1968.
23. Tasaki, I., and Takeuchi, T.: Weitere Studien über der markhaltigen Nervenfaser und über die elektrosaltatorische Übertragung des Nervenimpulses, Pfluegers Arch. **245:**764, 1942.

3

KENNETH L. ZIERLER

Mechanism of muscle contraction and its energetics

It is the function of muscle to contract and, in so doing, to perform work. The work of muscle may be either to produce angular motion about a joint, as in the case of skeletal muscles; to produce linear motions or complex maneuvers that can be resolved to sets of linear maneuvers, as by the tongue or by the byssus retractor by which Mollusca pull themselves to their rocky bed; or to alter pressure and volume so as to propel contents of hollow organs, as in the cases of the air movement effected by the diaphragm, the pumping action of the myocardium, peristalsis by intestinal muscles, emptying of the uterus, and locomotion by jet propulsion, as in the sea anemone. In each case, work is quantifiable from measurements by which it is defined as exertion of force through a changing distance ($W = P \cdot dx$) or of pressure throughout a changing volume ($W = P_r \cdot dV$).

The greatest number of experiments have been performed on cross-striated muscle. Most cross-striated muscle is skeletal muscle; i.e., it causes angular motion of two bones about a joint. In man about 40% of body weight is cross-striated muscle, which makes it the largest tissue in the body. The most abundant protein in man is collagen, and the next most abundant are probably the contractile proteins of muscle.

There is now a large mass of data concerning structure, mechanics, and energetics of muscle that one would hope to assemble to form a clear picture of muscular contraction. Such a synthesis is in many respects premature. Nevertheless, it is possible to describe some plausible notions of how muscle may work, to indicate discrepant data, and to suggest areas in which critical information is needed. In the rest of this chapter, unless specified otherwise, cross-striated muscle will be referred to simply as muscle.

STRUCTURE OF SKELETAL MUSCLE

The great mass of mammalian, avian, amphibian, and reptilian skeletal muscle is made up of bundles of fibers; each fiber is a single cell.

Skeletal muscle fibers may be arranged in a pennate fashion, as in Fig. 3-1, *A*, or in parallel, as in Fig. 3-1, *C*, or in various combinations of these arrangements. Because the force exerted by a muscle fiber is a remarkably constant function of its cross-sectional area (usually between 1 and 2 kg/cm^2) and not of its length, and because more fibers can be accommodated in the same volume when the arrangement is pennate, muscles with this arrangement exert greater force per gram of muscle than do those with long parallel fibers. However, since the direction of shortening of individual fibers is different from that of the whole muscle, the distance by which pennate muscles shorten is less than that by which those with parallel fibers shorten (Fig. 3-1, *B* and *D*). Pennate muscles are designed for small powerful movements. Muscle with parallel fibers can shorten further, but this is seldom important in situ. What is important, as will be shown later, is that this property also permits them to move more rapidly over small distances.

It will be important in considering contraction to recall that in addition to muscle fibers, muscle contains fibrous and elastic tissue, including tendon, that contributes to the complicated strain response of muscle to applied stress. It is also important, of course, in the subsequent consideration of muscle metabolism to recall that in life muscle includes its blood and nerve supply and that adipose tissue is interspersed between muscle bundles.

The single muscle fiber in adult man may have a diameter of about 60μ and a length from a few millimeters to tens of centimeters.

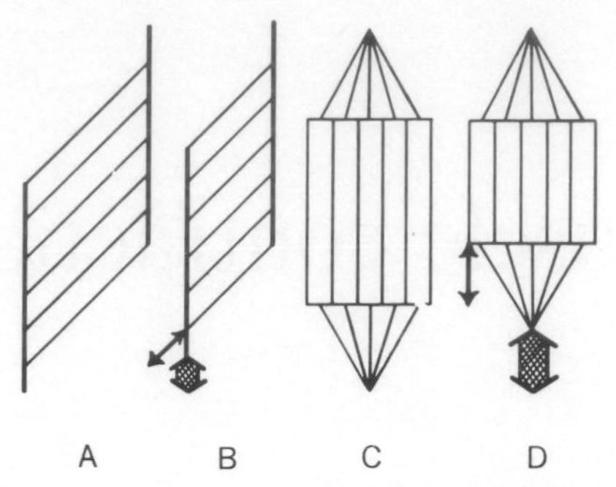

Fig. 3-1. Two arrangements of muscle fibers, both at rest and at contracted length. **A,** Pennate fiber arrangement at rest. Tendons are lines extending from two sides of parallelogram. **B,** Same muscle at maximum shortening. Fibers have shortened by one-third rest length along axis indicated by narrow double arrow, whereas muscle as whole shortens only by amount indicated by length of and in direction of broad double arrow. **C,** Parallel fibers at rest. Lines radiating from rectangles to point at each end represent tendons. **D,** Same muscle at maximum shortening. Fibers have shortened by one-third rest length (direction and magnitude indicated by narrow double arrow) and muscle shortens by same amount (indicated by broad double arrow). Thus, though fibers in **A** and **C** are same length, shortening of muscle **A** is less than shortening of muscle **C.**

It is roughly cylindric. Its ends usually fuse with poorly compliant tissue that often forms tendons or fibrous sheets. Every fiber is wrapped in a compound membrane, the sarcolemma, that separates it from interstitial fluid. In twitch muscles (to be defined later) there is usually only one nerve ending per muscle fiber. The terminal endings of the axon are partially encased by specialized invaginations of sarcolemma, the muscle end plate.

Every fiber contains a bundle of subunits, the fibrils. Fibrils are cylinders on the order of 1μ in diameter and their length is that of the fiber. There may be 1,000 fibrils per fiber. Fibrils have no membrane sheaths. The spaces between them are filled with cytoplasm, called myoplasm or sarcoplasm, which has mitochondria and through which passes a fine tubular network, the sarcoplasmic reticulum, which will be discussed later.

Fibers are composed of serially repeated units called sarcomeres. A sarcomere is a cylinder. Its length, depending on the degree of stretch or of shortening of the fiber, is from about 1.5 to 3.5μ. Conventionally a sarcomere is bounded on each end by a disc, the Z disc or Z line. Fibrils within a given fiber are aligned so that their Z discs are in register. Viewed by light microscopy, fibers appear to have longitudinal striations, which in reality are the fibrils, and transverse striations. The transverse striations, which give cross-striated muscle its name, are alternations of isotropic (I) bands with anisotropic (A) bands. Each I band is bisected transversely by a Z disc. A bands often have a relatively isotropic midzone, the H zone. The A band is about 1.5μ long. It has been known for years that when muscle shortens, the sarcomere shortens and the distance between Z discs decreases. This is accomplished largely if not entirely at the expense of I bands. A bands change length little if any.

Electron microscopy has revolutionized the concept of muscle structure and has been responsible for the dominant current notions of muscle function.

Our understanding of the internal structure of A and I bands emerges from the work of Hanson and Huxley[67] and of Huxley[81] on glycerol-extracted rabbit muscle (Fig. 3-2). Viewed under the electron microscope, fibrils are evidently composed of linear filaments. There are two major families of filaments. The I band contains only one kind of filament, called the thin or I band filament. It is about 50 to 60 Å thick and about 1μ long. It starts at the Z disc and runs longitudinally through the I band, penetrating the A band. The A band contains not only portions of thin filaments but also wider filaments approximately 110 Å in diameter. These are called thick or A band filaments because they are coextensive with the A band; i.e., they are about 1.5μ long. The H zone is the region in the center of the A band in which there are no thin filaments. Isotropicity results from the presence of only one class of filaments; anisotropicity from the presence of both thick and thin filaments.

There may be a third and narrower family of filaments, sometimes called S fibrils, running through H zones to link opposite ends of thin filaments in the A band.

The only visible connection between thick and thin filaments are transverse processes in the A band called bridges. Bridges are more or less regularly spaced at intervals of about 435 Å. They are about 40 Å thick and extend for the 180 Å between the surfaces of thick and thin filaments.

Cross sections reveal the magnificent organization of filaments. Except in the immediate vicinity of the Z disc, thin filaments are arranged in a regular hexagonal array.

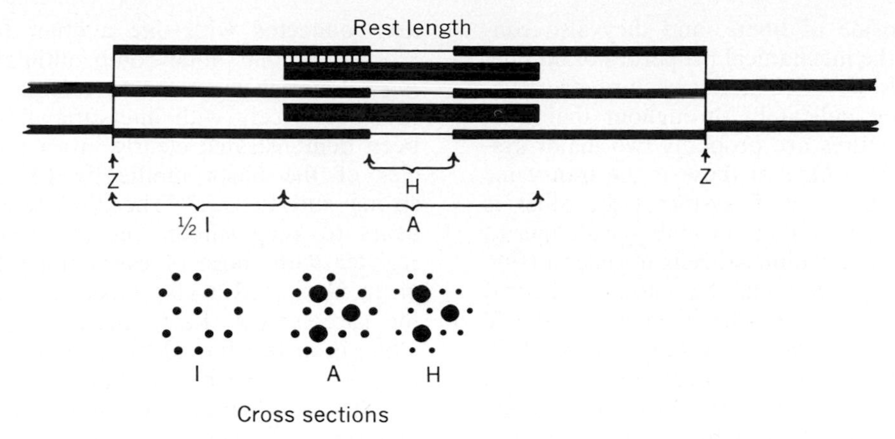

Fig. 3-2. Scheme of organization of myofibrils. Sarcomere is unit bounded by Z lines, which bisect I bands. I bands contain only thin filaments. A bands contain both thick and thin filaments. H zone contains no thin filaments, but there may be very fine threads, illustrated here, connecting pairs of thin filaments. Perpendicular bridges link thick and thin filaments. For simplicity, only one set of bridges is shown. Hexagonal array of thin filaments is illustrated in cross section. In the vicinity of the Z line, hexagonal array is transformed to square array.

The length of the sides of the hexagon is about 260 Å from center to center of thin filaments. The hexagonal array continues through the I band and A band for the entire length of thin filaments. In the A band a thick filament occupies the center of every hexagon of thin filaments. The distance from the center of the thin to the center of the nearest thick filament is about 260 Å, about 180 Å from surface to surface. Three thick filaments are equidistant from a given thin filament.

Regardless of whether the muscle is fixed after it has been permitted to shorten or after passive stretch, the length of thick filaments is invariant. There is less certainty about the independence of the length of the thin filaments, although for sarcomere lengths from about 2μ (the length of two thin filaments) to more than 3.5μ, the length of thin filaments is apparently constant. There is dispute about the length and arrangement of filaments in muscle shortened maximally. In some preparations, thick filaments from one sarcomere have been said to penetrate the Z disc and enter the adjacent sarcomere. In other preparations, thin filaments are said to buckle, i.e., shorten, at the center of the A band. In others, thin filaments are said to double back upon themselves. The matter requires clarification.

The structure of the Z disc may be important in the mechanism of contraction. As thin filaments approach their origin in the Z disc, they are said to rearrange from a hexagonal to a square array. At the Z disc every thin filament breaks up into or is attached to four filaments called Z filaments. Knappeis and Carlsen[84] and Franzini-Armstrong and Porter[58] have shown that the Z filaments form tetrahedral pyramids on square bases. Each I filament on each face of the Z disc is the apex of a pyramid formed by four filaments that completely traverse the Z disc. In turn, each of these four Z filaments links with three other Z filaments on the opposite face of the Z disc to connect with an I filament in the adjacent sarcomere. The angle at the apex of the pyramid is about 15 degrees. Z filaments are about 50 Å thick, and the Z disc is 350 to 800 Å wide. Although it is stated generally that the dimensions of the Z disc are unaffected by muscle contraction, this may not be true; critical measurements are lacking.

Electron microscopy has also pictured the muscle membrane and the space between muscle fibers as complex structures. Each muscle cell, or fiber, is enveloped by a plasma membrane, the bilamellar leaflet, 100 Å thick, that seems to be universal to all cells. The plasma membrane is sheathed in turn by an amorphous structure that stains as a mucoprotein or polysaccharide and is nearly 500 Å thick. This in turn is contained within a mesh of collagen fibers that are woven tightly about the fiber for another 500 Å and more sparsely thereafter, so that the collagen around a fiber becomes indistinguishable from collagen between fibers. These structures play at least two roles. They form the barriers through which material must pass if it is to exchange between interstitial space and the inside of fibers, and they also contribute to the mechanical properties of muscle.

There is a reticulated structure or structures lacing delicately throughout the fiber. These structures are probably two major systems of tubes. One of these is the transverse tubular system (or T system); the other is sarcoplasmic reticulum, probably analogous to endoplasmic reticulum of cells in general (Fig. 3-3). The T system may be unique to skeletal and cardiac muscle. The elements of the T system are, first, a ring around every fibril, located either at every Z disc, in which case there is one annulus per fibril, or at every junction of the A and I bands, in which case there are two annuli per fibril. The location of the annulus in skeletal muscle is species specific. The perifibrillar annuli of the T system

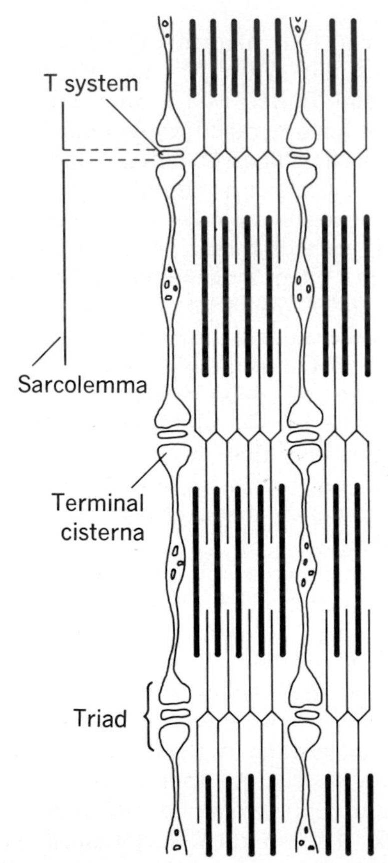

Fig. 3-3. Diagrammatic representation of relation of T system and of SR to each other and to myofibrils. At plane of Z line or at both A-I junctions in some species is the triad, including section through T tubule and terminal sac or cisterna from both adjacent SR tubes. In the middle of every SR tube is a fenestrated or cribriform dilatation. T tubules radiate through fiber to open directly to interstitial space.

are connected with one another to form a kind of tubular honeycomb, ultimately sending tubules through the sarcolemma to communicate freely with interstitial fluid, as has been demonstrated clearly in segmental muscles of the black mollie by Franzini-Armstrong and Porter.[59] The T system thus appears to keep sarcomeres of every fibril in register with those of every other fibril in a given fiber, and it also links every fibril with the sarcolemma. Large molecules, e.g., ^{131}I-albumin or ferritin, added to solutions bathing muscle, are visible within the T system, which is considered to be simply an extension of interstitial space within muscle fibers. Presumably the T system carries interstitial fluid directly to every sarcomere.

The sarcoplasmic reticulum (SR) is composed of tubules that run in a more or less longitudinal direction, parallel to fibrils, though in a tortuous path in the space between fibrils, between two annuli of the T system on the same fibril.[95] Thus in those species in which the T system is localized to the Z disc, a single tubule of SR extends over the whole length of one sarcomere.

The tubule of the SR is composed of three portions. Both ends of the tubule broaden into relatively large sacs that flatten at their ends to make broad contact with an annulus of the T system. As a result, a very large portion of the T system is in close relation to terminal sacs, or terminal cisternae, of the SR. The terminal cisternae taper to form long narrow tubules, a few hundred angstroms in diameter, that travel nearly to the midpoint of the sarcomere at the H zone. There the tubules broaden to form a common central structure that is apparently pierced by small tubules and is therefore cribriform. The total volume of the SR, estimated on the basis of electron microscopic studies of amphibian muscle, may be about 13% of the volume of the muscle fiber.[91] Estimates based on extrapolations of sodium flux studies in mammalian muscle yield figures of about 12% of the wet weight of the muscle or about 14% of fiber volume.

Longitudinal section through a muscle fiber reveals a fragment of the T system as a short segment of tube with an apparently large vacuole on either side of it. In fact, the apparent vacuoles are the terminal cisternae of the SR. The cluster of three structures, T tubule plus two terminal cisternae, is called the muscle triad. There has been much discussion about the morphology of the junction between the terminal cisternae and the T tubule. It has been suggested that it is a tight junction. This is an important suggestion because it carries with it the implication that the junction has relatively great permeability to monovalent ions, said to be a property of tight junctions. However, in electron micrographs of great clarity Franzini-Armstrong[57] has found that the double-layered plasma membrane of the T tubule is separated from the double-layered plasma membrane of the terminal cisternae by 120 Å; hence their association is not a tight junction.

The tubular systems, T and SR, play a major role in excitation of muscle and in the coupling of excitation to contraction. The interdigitating arrangement of the thick (A band) and thin (I band) filaments is the basis of muscle shortening, and as we shall see, links between thick and thin filaments over regions of overlap make it possible for muscle to exert tension. In the following sections we will consider the events leading to muscle contraction.

EXCITATION OF SKELETAL MUSCLE

Each efferent nerve fiber to skeletal muscle branches terminally, and each terminal bud is in close proximity to a specialized segment of a muscle fiber called the motor end plate. (See Couteaux[41] and Tiegs[10] for reviews.) The nerve bud and the motor end plate together constitute the myoneural junction (Chapter 5). Usually one efferent nerve fiber supplies endings to a number of muscle fibers, perhaps as many as 1,000 fibers in man, according to Feinstein et al.[53] In mammals each muscle fiber has one or only a few end plates or groups of end plates.[76] The complex, a single efferent nerve fiber with its multiple terminals and associated muscle fibers, is referred to as a motor unit. The muscle fibers composing a single motor unit are not necessarily adjacent. Indeed, Buchthal et al.[34] find that in human muscle they are in packets of no more than thirty, intermingled with other motor units. Furthermore, since the length of the terminal nerve branches varies within a single motor unit, the impulse in a single efferent nerve fiber will not arrive at all of its end plates simultaneously.

When a motor nerve fiber is stimulated, the impulse arriving at the nerve endings produces a release of acetylcholine (Chapter 5). There is reason to hold that the mitochondria-rich nerve endings synthesize acetylcholine steadily and vigorously and that acetylcholine may be stored temporarily in packets in the nerve endings. Acetylcholine is probably discharged from many points over the broad nerve ending into a wide, shallow, hemispheric trough that separates the ending from the motor end plate. Diffusing only a short distance and arriving at many points on the motor end plate, acetylcholine increases permeability of the end plate, and as a result the end plate becomes electrically less negative with respect to extracellular fluid. The consequent end-plate potential spreads only by electrotonus, depolarizing only the immediately adjacent surfaces of the muscle fiber. If the initial depolarization is sufficiently large, an impulse is propagated along the muscle fiber by a mechanism that is at least qualitatively like that responsible for the propagation of a nerve impulse (Chapter 2).

In amphibians but not in mammals there is a second class of muscle fibers that has been studied in detail by Kuffler.[85] These muscle fibers are covered liberally with end plates that, when depolarized, do not lead to a propagated impulse. Because the impulse is not conducted, it is possible to find contraction of one part of a muscle fiber but not of another. Alternating contraction and relaxation of segments of a single fiber permit the fiber as a whole to exert tension for a considerable period of time. Such fibers, therefore, are said to contract tonically. In this respect they occupy a position midway between smooth muscle and skeletal muscle.

The presence of the T system, and possibly its relation to SR, endows skeletal muscle with electrical properties different from those of nerve. Initial events, however, are similar.

When skeletal muscle is at rest there is an electric potential difference between the inside of the fiber and extracellular fluid, the inside being relatively negative. In excised amphibian muscle, resting membrane potential is about –90 mv, close to that of the potassium equilibrium potential difference, E_K. In mammalian muscle its magnitude is somewhat less; values as low as –70 mv have been reported. This may be due to relatively greater permeability to Na^+ in mammalian muscle than in amphibian muscle, perhaps 5% of K^+ permeability in the former compared to 1% or less in the latter.[24] Since the equilibrium potential difference for Na^+, $E_{Na} = -(RT/F)\ln[(Na^+)_i/(Na^+)_o]$, is of a sign opposite that of both E_K and the resting potential difference, increased Na^+ permeability, which implies increased Na^+ conductance, reduces the absolute size of the resting potential difference.

The presence of the T system gives conducting electrolytes two routes between sarcoplasm and interstitial fluid. One route is directly across sarcolemma. The other route has three steps: between sarcoplasm and SR, between SR and T, and between T and interstitial fluid, although for many purposes movement between T and interstitial fluid is so rapid compared to movement across the other barriers that contents of T and interstitial fluid can be considered to form a single component.

It is the presence of these surfaces in addition to the usual cell membrane, or sarcolemma, that gives skeletal muscle an apparent electrical capacitance greatly in excess of that of nerve, 1 $\mu F/cm^2$. If one neglects these additional surfaces and expresses capacitance in microfarads per square centimeter of cylindric surface, one usually obtains values of from 5 to 9 $\mu F/cm^2$ in amphibian muscle. It has not been possible to determine the total surface area of the T system accurately, but estimates based on electron microscopic studies indicate that it is from about 3 to 7 times that of sarcolemma.[91] Values as high as 40 $\mu F/cm^2$ have been reported in crustacean muscle where sarcolemmal invaginations also occur.

One of the most meticulous studies of resistance and capacitance in a skeletal muscle was by Falk and Fatt.[52] For frog sartorius muscle, their data fitted a model in which a resistance (R_m) of a surface membrane (sarcolemma), estimated at 3,100 ohm-cm^2, shunting a membrane capacitance (C_m) of 2.6 $\mu F/cm^2$, is in series with another path (T system?) with a resistance (R_T) of 330 ohm-cm^2 and a capacitance (C_T) of 4.1 $\mu F/cm^2$, all referred to unit area of fiber surface. Since the reference area is probably not correct, the absolute figures are wrong; but the fact remains that at least 60% of the membrane capacitance of frog sartorius muscle lies in some component other than the surface membrane.

More direct evidence of the quantitative role of the T system was made possible by the discovery that T tubules could be more or less specifically ruptured by soaking the muscle in hypertonic glycerol solution and returning it to isotonic Ringer's solution (Howell[75]). Such fibers could have normal resting potentials and could propagate action potentials, but they did not twitch. Using this technique, Gage and Eisenberg[61, 62] and Eisenberg and Gage[46] were able to calculate membrane resistance (R_m) as 3,700 ohm-cm^2, which is essentially normal; but remember that Falk and Fatt estimated that 90% of the resistance was in sarcolemma. Membrane capacitance (C_m) was only 2.2 $\mu F/cm^2$, compared to 6.1 $\mu F/cm^2$ in control muscles, agreeing well with the earlier calculation of Falk and Fatt.

Skeletal muscle, then, has two kinds of membranes in series, sarcolemma and T tubules, with different resistances and capacitances. It is at least conceivable that the two types of membranes might differ also in their conductance to specific ions.

The first clue came from observations by Hutter and Noble[77] in 1960 that when the external solution was free of Cl^-, membrane conductance of frog sartorius muscle was reduced by 68%; i.e., about two thirds of membrane conductance was to Cl^-. In the same year Hodgkin and Horowicz[74] observed that changes in external Cl^- concentration produced appropriate changes in the membrane potential of single muscle fibers, as predicted by the constant field equation (Chapter 1). These changes were rapid in that they were half complete in less than 0.3 sec. However, when external K^+ concentration was decreased from an abnormally high level, the expected repolarization took 10 times as long. This suggested that K^+ was retained temporarily in what Hodgkin and Horowicz called some special region. They suggested that Cl^- conductance lay mainly through some readily accessible surface but that K^+ conductance was across the T tubules.

Adrian and Freygang[26] in 1962 attacked the question of membrane conductance by means of an elegant technique requiring insertion of three micropipet electrodes in a single muscle fiber. Their data were consistent with the interpretation that the surface membrane was permeable to both Cl^- and K^+ but that the T tubule was permeable only to K^+.

The best evidence comes from experiments conducted by Eisenberg and Gage[46] on muscles in which T tubules were ruptured by glycerol treatment. About 80 to 90% of the conductance through the surface membrane was to Cl^-, and K^+ conductance through the nonsurface membrane (T tubules?) was about twice as great as through the surface membrane. However, if this nonsurface membrane is that of T tubules, since the total surface area of T tubules is about 4 times that of sarcolemma, specific conductance per square centimeter of surface is only half as great through T tubular walls as through sarcolemma.

With this evidence that there is membrane resistance and capacitance in both sarcolemma and T tubules and that specific ion conductances in resting muscle differ through the two surfaces, we are ready to consider propagation of an action potential.

The first suggestion that the function of the T system is to propagate the action potential into the center of the fiber so that core myofibrils can contract more rapidly was made on anatomic grounds by Porter and Palade[95] in 1957. Huxley and Taylor[80] showed in 1958 that subthreshold depolarizations applied very locally to the surface of frog muscle fibers did not produce a twitch unless they were applied at the plane of the T tubules. Small depolarizations caused visible twitches only of surface myofibrils. Larger ones caused all myofibrils to twitch. To this evidence we add the later experiments in which T tubules were ruptured by being soaked first in hypertonic glycerol solution and then in isotonic Ringer's solution. Recall that in this case there were action potentials but no contraction. It is concluded that the normal path of excitation in skeletal muscle is as follows: The motor end plate is depolarized in response to the neurotransmitter, acetylcholine. If there is threshold depolarization, there is a propagated, self-regenerative action potential, a wave of depolarization and repolarization traveling along the sarcolemma. This depolarization is somehow propagated down the walls of T tubules into the core of the fiber.

The question is whether excitation spreads throughout the walls of T tubules by means of electrotonus or by means of a self-regenerative propagated action potential.

We have already referred to the experiments of Huxley and Taylor[80] in which small local depolarizations produced twitches only of outer fibrils, whereas larger ones produced twitches of all fibrils. This lack of all-or-none character suggests that depolarization spreads down the T system passively by electrotonus. Criticism of this interpretation is that the small local depolarizations imposed by Huxley and Taylor were less potent than a normal action potential. Indeed, when larger areas of the surface of fibers were depolarized (Sugi and Ochi[99]), contractions spread completely around or across fibers.

Adrian et al.[28] studied the problem further by blocking action potentials with tetrodotoxin and controlling membrane potentials with a voltage clamp. From application of square steps of depolarization of increasing magnitude, with observation of contraction of superficial and axial fibrils, it was possible to estimate a space constant (λ) for electrotonic spread through the transverse tubular network. The estimate depends on a number of assumptions: diameter and geometric arrangement of tubules, no summation effects of neighboring depolarizations, and the usual assumption in the limiting case of the core conductor equations that wall thickness is negligible by comparison to radius. This last assumption permits one to obtain a tractable expression and is acceptable for large-diameter fibers, but it cannot be exact for the case of T tubules in which wall thickness is more than 100 Å while radius is only about 250 Å. On the basis of what may be tenuous assumptions, Adrian et al. calculated that the passive electrical characteristics of the T system would just allow an action potential to activate the axial myofibrils. However, since electrotonus by these calculations provides no safety factor, an active potential change is necessary if the magnitude of depolarization at the core is to be well above threshold to produce a twitch.

By quite a different technique (involving making myofibrils wavy by setting the muscle fiber in gelatin and compressing it laterally so that active shortening of myofibrils is recognized by straightening out of the waves viewed by high-speed cinemicrography), González-Serratos[65] measured the velocity of the inward spread of activation. This velocity was compatible with that expected for propagation of an action potential, and it was more sensitive to temperature than expected for electrotonus; Q_{10} was more than 2, compatible with temperature dependence of an action potential.

The evidence, then, is suggestive but not conclusive that depolarization spreads down the T system as a propagated action potential rather than by passive electrotonus.

We now consider the question of whether or not depolarization continues from the T system across the triad junction to the terminal cisternae of the SR. We do not know the concentrations of Na^+ and K^+ in SR because there is as yet no technique for their measurement. However, there are strong suggestions that the concentration of these ions is nearly the same in fluid in SR as in interstitial fluid. For example, the rapidly exchanging Na^+ space and the rapidly filled Li^+ space occupy a volume in excess of interstitial volume, and this volume is about 12% of fiber weight, compatible with estimates of SR volume based on electron microscopic findings. The excess of sucrose space over inulin space suggests that small uncharged molecules such as sucrose can enter SR but that large molecules such as inulin cannot. Furthermore, when muscles are soaked in solutions made hypotonic or hypertonic by varying NaCl concentrations, the changes in SR volume are directionally opposite to those in sarcoplasmic volume and are in agreement with predictions based on relatively free exchange of Na^+ between SR and interstitial fluid or bathing solution.

If it is correct that Na^+ and K^+ concentrations in SR are about the same as those in interstitial fluid, then if the wall of SR is permeable to K^+, there must be an electric potential gradient across the wall of SR. Studies of Na^+ flux from extensor digitorum longus muscle of rats lead to the interpretation that Na^+ movement between interstitial fluid and the fiber is about 1,000 times as great across the SR-T system junction as across sarcolemma and furthermore that the flux data are incompatible with the hypothesis that the wall of SR is impermeable to Na^+. Therefore we conclude that there is probably an electric potential difference across the wall of SR. However, there is no evidence that an action potential is propagated along the SR. Indeed the measurements of electrical capacitance to which we referred earlier seem to be incompatible with the idea that depolarization continues from T tubules to SR. The total surface of SR is at least 50 times that of sarcolemma,[91] and if depolarization continued from T tubules to SR, one would expect to detect far greater capacitance than is found.

Since the wall of a terminal sac of SR is separated by about 120 Å from the wall of a T tubule, there is not likely to be continuous current directly through both sets of membranes, such as might possibly occur for the case of a tight junction. Furthermore, as Franzini-Armstrong[57] has pointed out, if capacitance per unit surface of SR is the same as for T tubules, since the total surface of SR is so much greater, there is a mismatch such that if the total membrane charge of the T system were suddenly transferred to SR, it would not depolarize it sufficiently to initiate an action potential.

Despite these arguments against the participation of SR in the spread of excitation, there are reasons for keeping an open mind on the subject. In the first place, as we shall see, excitation must be transferred from the depolarized T tubule to SR in order to explain the activation of processes leading to muscle contraction. In the second place, studies of the afterpotential point to a role played by SR.

In skeletal muscle a train of action potentials is followed by a late afterpotential that is less negative than the resting potential. When T tubules are destroyed by glycerol treatment, there is no late afterpotential (Gage and Eisenberg[62]). This eliminates sarcolemma as the location of the source of the phenomenon. The suggestion that the source is in SR arises from observations by Adrian et al.[26] in studies in which membrane potentials were clamped. Results may be interpreted as follows: During the action potential there is a large net outward K^+ current. This represents K^+ movement out of sarcoplasm into some space in which it accumulates temporarily, raising the local K^+ concentration of this "extracellular" region. Therefore over this region the equilibrium potential for K^+ is less negative than expected from calculations on the basis of K^+ concentration in the bathing solution. From equilibrium potential measurements during voltage clamp experiments, Adrian et al. calculated that the extra K^+ accumulated in a space equal to one sixth to one third of the fiber volume. This is at least an order of magnitude greater than the volume of the T system. The lower estimate of this volume is similar to that for the volume of SR.

Therefore, although we do not know how depolarization spreads from the T system to SR, it is difficult to deny the possibility that there is a resting potential across the SR wall, and that ionic currents flow across that wall during the inscription of what we recognize as an action potential.

The continuation of the journey from depolarization of the end plate along the paths taken by spread of excitation to the link between excitation and contraction, called excitation-contraction coupling, may seem to be a reasonable design. However, current understanding of this latter process depends on views of the molecular organization of myofilaments; therefore consideration of excitation-contraction coupling will be deferred until the chemistry of contraction and molecular morphology has been discussed.

There must be a relation between the chemistry of contraction and the molecular morphology of myofilaments on the one hand and the phenomena observed grossly as mechanical properties of and heat production by whole muscle on the other hand. A desire to understand the latter has led to study of the former. Therefore a description of the behavior of whole muscle is appropriate prior to an examination of the elementary processes.

PHENOMENOLOGIC DESCRIPTIONS OF CONTRACTION

This discussion will consider the mechanical events in muscle contraction, i.e., the relation between the velocity with which a muscle shortens, the distance it shortens, and the force it exerts at any length. Only those muscles called *twitch* muscles, as opposed to *tonic* muscles, will be considered. No tonic muscle fibers in mammals, except for portions of extraocular muscles, have been identified.

When a muscle is stimulated, usually electrically either by stimulating the nerve supplying the muscle or by stimulating the muscle itself, the muscle contracts. If the ends of the muscle or its tendons are fixed so that the muscle cannot shorten, it will develop tension. This sort of contraction is called *isometric*. Measurement of the tension developed during isometric contraction in fact requires that the muscle be permitted to shorten a small but finite distance so that the force it exerts can be recorded. The distance shortened, however, is so small that it is usually neglected, and during isometric contraction a muscle is said to do no work; the force that it exerts moves through no distance. This is not to say that no work is done by subunits of a fiber during isometric contraction. Several lines of evidence—measurement of heat production simultaneously from several regions of muscle by Hill and Howarth[72] and measurement of sarcomere length under polarizing microscopy by Huxley and Peachey[79]—indicate that during isometric contraction not all sarcomeres are the same length and that some sarcomeres shorten, thereby stretching other sarcomeres. Work done by sarcomeres equals work done on sarcomeres, and there is no net work done by the fiber. Furthermore, although the tethered ends of the muscle do not move, contractile elements do shorten at the expense of stretching elastic elements in series with them. In general these series elastic elements are nonlinear in their stress-strain pattern, reaching limiting compliance with relatively small stretch, but this is enough to permit finite shortening of contractile elements. Therefore, even though there is no overall shortening of muscle and hence no measurable external work, contractile elements have shortened under a load and hence have worked. This is often referred to as *internal* work.

If only one of its ends is fixed, a muscle shortens when stimulated. When the free end of the muscle is made to lift a weight or to oppose a resistance held constant during contraction, the contraction is *isotonic* and the work done is the product of distance shortened (Δx) and either the weight lifted or the force exerted by the muscle in overcoming external resistance (P).

If the weight is carried by the muscle throughout contraction, the muscle is free-loaded. If the muscle is permitted to contract when it is partially unloaded or lightly loaded and then the remainder of the contraction is loaded, the muscle is said to be afterloaded. The purpose of afterloading is to permit the muscle to take up the slack in the complex fibroelastic system before it is made to work in order that the performance of the contractile components can be observed in purer form.

If the stimulus is sufficiently brief so that it produces only a single action potential, tension is developed during isometric contraction rather rapidly to a maximum, from which it decays more slowly and curvilinearly. This single response, or *twitch,* is qualitatively the same whether a single fiber or an entire muscle is stimulated with maximum intensity (so that all fibers contract) (Fig. 3-4). When the muscle has returned to resting tension, a second stimulus evokes an identical response. As the interval between successive stimuli is shortened, the muscle does not return to its resting tension between responses. The tension developed in response to each stimulus is additive, and finally, when the stimuli become sufficiently frequent, the tension attains a maximum and the myogram shows no irregularities related to individual stimuli. This is a *fused* or *complete tetanus* in which maximal tension is maintained until the muscle fatigues. Frequencies of stimulation between

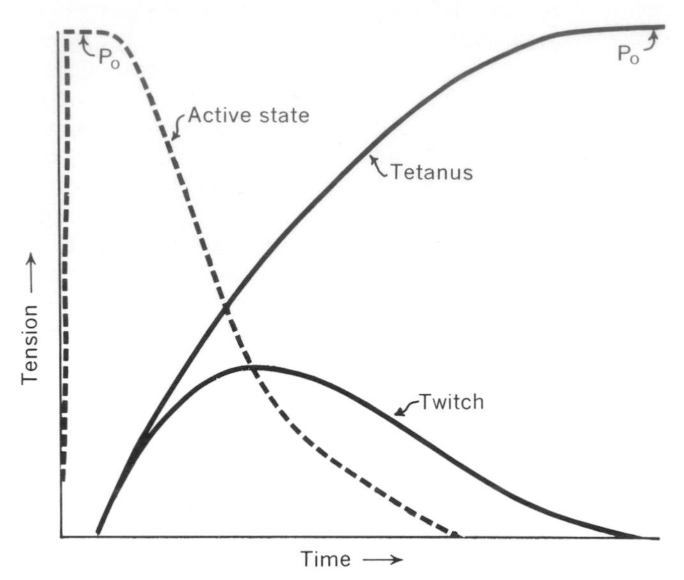

Fig. 3-4. Relation between tension and time for twitch, tetanus, and active state. P_o is maximum tension of which fiber is capable. Stimulus begins at zero time.

those that give a series of twitches and one that elicits a fused tetanus produce *unfused* or *incomplete tetanus*. The frequency of stimulation that will just produce a fused tetanus is to some extent an arbitrary artifact resulting from the sensitivity of the recording apparatus.

Muscles are divided into two categories, fast and slow, according to the least frequency of stimulation required for fusion. For mammalian muscles at 37° C, slow muscles may fuse when stimulated at a rate of only 16/sec. Fast muscles may not fuse until stimulated at a rate of 60/sec or even more. When muscles are stimulated at frequencies well over 100/sec, fatigue occurs early. Such high frequencies probably greatly exceed those to which the muscle is exposed in response to nervous impulses arriving during excitation of voluntary contraction in the intact mammal.

Analogous to the time course of tension development in isometric contraction is the time course of shortening in isotonic contraction. A brief stimulus produces a twitch characterized by rapid shortening and a relatively slow return to rest length. A series of stimuli evokes a series of twitches, an unfused tetanus, or a fused tetanus in that order as the frequency of stimulation is increased.

The rate at which a muscle shortens and the final distance shortened during maximum tetanus depends on the load. An unloaded muscle shortens most rapidly. As the load increases, the velocity of shortening decreases. An unloaded or a lightly loaded muscle shortens to about the same extent. With increasingly heavy loads, the maximum distance shortened decreases.

Latent period and latency relaxation

When a twitch is examined in detail, it is expected that there must be a delay between imposition of the stimulus and development of tension. This delay is the latent period. A surprising observation was made by Rauh in 1922 and later studied in detail by Sandow.[98] When excised frog sartorius muscle is stimulated directly and massively by condenser discharge, no mechanical event is recorded for the first 1.4 msec. During the next 1.8 msec the muscle relaxes; there is a true decrease in tension called latency relaxation. The time course of latency relaxation is sigmoid, followed by abrupt development of tension. Latency relaxation is clearly the initial mechanical event in muscle contraction but it remains unexplained. Some unknown element slackens briefly.

Length-tension diagram

When a muscle or a single muscle fiber at rest is stretched passively, it opposes stretch by a force that increases slowly at first and more rapidly with increased stretch; i.e., it becomes less compliant when it is elongated (Fig. 3-5). This purely elastic property re-

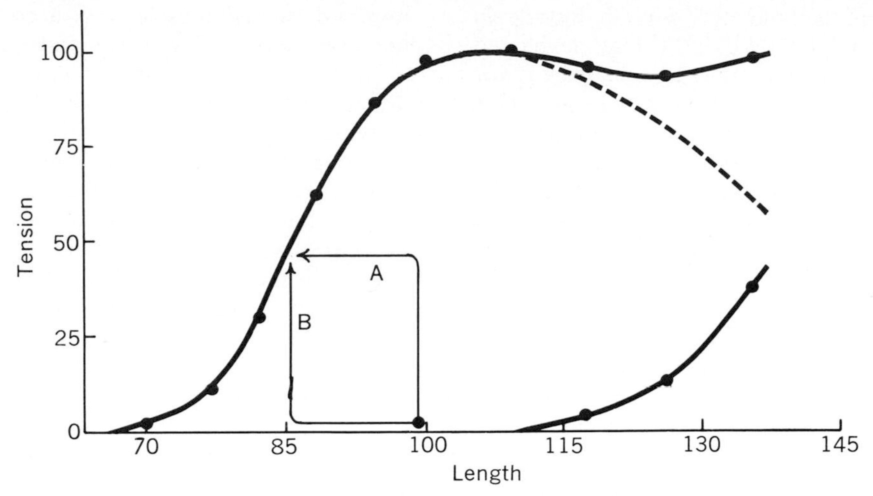

Fig. 3-5. Relation between length and tension. 100 is rest length and maximum tension. Monotonic increasing curve at lower right represents effect of passive stretch on tension. Upper solid curve is obtained from system as a whole. Broken line represents behavior of contractile elements when passive stretch curve is subtracted from upper curve. Set of thin-lined arrows beginning at zero tension and rest length indicates that, apart from effects of plasticity and whether muscle contracts isotonically, *A*, or isometrically, *B*, final relation between tension and length should be the same.

sides largely, if not entirely, outside the contractile elements in elastic tissue, including the sarcolemma. When the muscle is stretched by 30% or less of its rest length, the length-tension curve is usually reversible. Excessive stretch produces irreversible changes. For stretch that exceeds rest length by more than approximately 50%, the limiting compliance is that of the sarcolemma. At about twice rest length the sarcolemma tears. The tearing force required for sarcolemma is about twice that required for cast iron.

If the muscle is fixed at rest length (arbitrarily defined in a number of ways but approximately the length it assumes at rest in the body) and stimulated to contract maximally and tetanically, it develops its greatest tension. As the length is increased or decreased beyond rest length, the force exerted by the muscle during tetanic isometric contraction decreases. (The total force increases with large stretch but this is due to the large contribution of the passive components—the rest length–tension relation—which must be subtracted from the observed tension to obtain that part of the tension due to contraction.) Skeletal muscle, except under unique experimental conditions, shortens only to about 60% of rest length when stimulated to contract isotonically. At this length in isometric contraction it exerts no tension (Fig. 3-5).

The relation between the force exerted by a muscle and the load the muscle lifts is as follows: When a muscle is stimulated under isotonic conditions, it rapidly develops tension or force. The muscle cannot shorten until the force it develops is greater than the load it must lift. When muscle force exceeds the load, the muscle begins to shorten. As it shortens from rest length, its tension drops. When muscle force falls to that of the load, the muscle can no longer lift the load; shortening stops. Since the muscle can no longer change its length, what was originally an isotonic preparation is now isometric. The force exerted at the final length equals the load on the muscle. This force is the maximum force of which the muscle is capable at the final length, given by the isometric length-tension diagram. Because shortening occurs rapidly, the force exerted by the muscle throughout the period of shortening is never greatly in excess of the load; i.e., it is nearly isotonic.

Therefore one might expect that the isotonic length-tension diagram would be superimposable on the isometric diagram. When a muscle is loaded and permitted to shorten from various lengths, it should shorten to that length which corresponds to the tension developed in the isometric length-tension diagram. This is not always quite true experimentally. At a given contracted length the tension developed isotonically may be less than (Buchthal et al.[2] for frog muscle in vitro; Rosenblueth et al.[97] for cat muscle in situ) or equal to (Wilkie[21]) that developed isometrically at the same length, except for shortening to 70% or less of rest length. This inconsistent difference between isometric and isotonic contractions presumably is an expression of plasticity of the system and may be due to differences in the stretch on elastic tissue rather than to any difference in the contractile proteins.

Force-velocity diagram

When isotonic force is plotted against the initial (or maximal) velocity of shortening, a curve like that in Fig. 3-6 is obtained. An unloaded muscle shortens most rapidly. Velocity decreases with force until, of course, the maximum force is reached; at this point there is no shortening, and velocity is therefore zero.

The relation between force and velocity has been analyzed in several ways. In one method the experimental curve has been fitted empirically, yielding an equation not based on a conceptual model but accepted simply as a description of the real system. Hill[70] in 1938 produced such an empirical equation, one form of which is:

$$(P + a)\ (v + b) = \text{Constant} = (P_o + a)b \tag{1}$$

or, in explicit form for velocity:

$$v = (P_o - P)b/(P + a) \tag{2}$$

where $v = dx/dt =$ velocity, P_o is the isometric tension, P is the experimental force imposed on the muscle, b is a constant with the dimensions of velocity and magnitude $b = (av_o)/P_o$, and a is a constant with the dimensions of force. The ratio $a/P_o = b/v_o$, in which v_o is maximum velocity at no load, is about one-fourth and is even less variable from muscle to muscle than P_o. The equation states that the velocity of shortening is proportional to the difference between the applied force and isometric tension, $P_o - P$, and inversely proportional to the applied force. That is, the relation between force and velocity is hyperbolic.

The force-velocity relation has been approached in other ways, both empirical and theoretical. On empirical grounds Fenn and Marsh[56] and later Aubert[1] expressed the relation as exponential, with either a linear or a constant displacement, $P = P_o e^{-k_1 v} - k_2 v$, or $P = P_o{}^{-k_1 v} - k_2$.

On theoretical but different grounds Ramsey[96] and Pollisar[94] developed exponential equations, and Caplan[35] developed an equation identical in form to Hill's empirical equation.

All the equations require three constants. All fit the data about equally well. Distinctions among them must rest on an examination of certain consequences of these various expressions and of the assumptions underlying them.

Length-velocity diagram

Hill's classic force-velocity curve (Fig. 3-6) is obtained as follows: The muscle is fixed at a given length so that it begins to shorten from

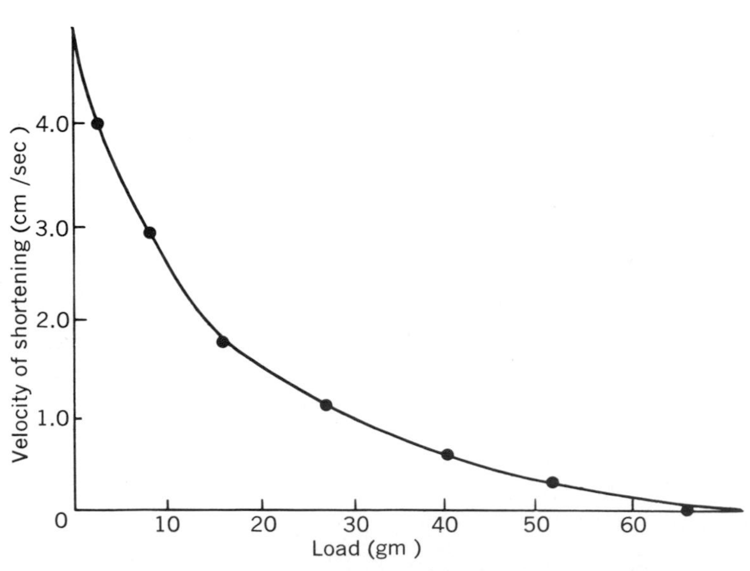

Fig. 3-6. Maximum velocity of shortening as function of load or force. (Redrawn from Hill.[70])

the same initial length against varying loads. In Hill's experiments the initial length was approximately rest length (L_o), the length at which $P = P_o$ (Fig. 3-5). The velocity recorded in the classic force-velocity curve is the initial velocity, which, with the muscle starting from L_o, is also the maximum velocity.

If the experiment is performed at any initial length different from L_o, the muscle cannot lift as heavy a load as it can at L_o, and zero velocity always occurs at loads less than that lifted at L_o, i.e., at loads less than P_o. The length-tension diagram gives the maximum force that can be exerted by a muscle at any length. This maximum force, as a function of length, will be called $P_o(L)$.

Abbott and Wilkie[25] extended Hill's observations to lengths less than L_o and found that the same constants (a and b) in Hill's equation 2 applied to a family of curves, each with its own $P_o(L)$. The equation is simply a modification of Hill's equation 1:

$$(P + a)\ (v + b) = [P_o(L) + a]b \qquad (3)$$

Velocity is therefore a function of length as well as of tension.

Families of force-velocity curves can be obtained over the whole range of initial lengths, from those greater to those less than L_o. All of the curves have the same general shape (Fig. 3-7). At any given force the velocity is always greatest when the initial length is L_o.

As just seen, there is a family of force-velocity diagrams, depending on the initial length from which the muscle shortens. Carlson[36, 37] and Bahler et al.[30] have examined the explicit relation between velocity and length.

A display of a function against its derivative is called a phase-plane diagram. A length-velocity diagram is therefore a phase-plane diagram.

When the muscle is allowed to shorten from rest length (L_o), maximum velocity is attained almost immediately (Fig. 3-8). Maximum velocity is maintained briefly, as shown by the plateau; the muscle then de-

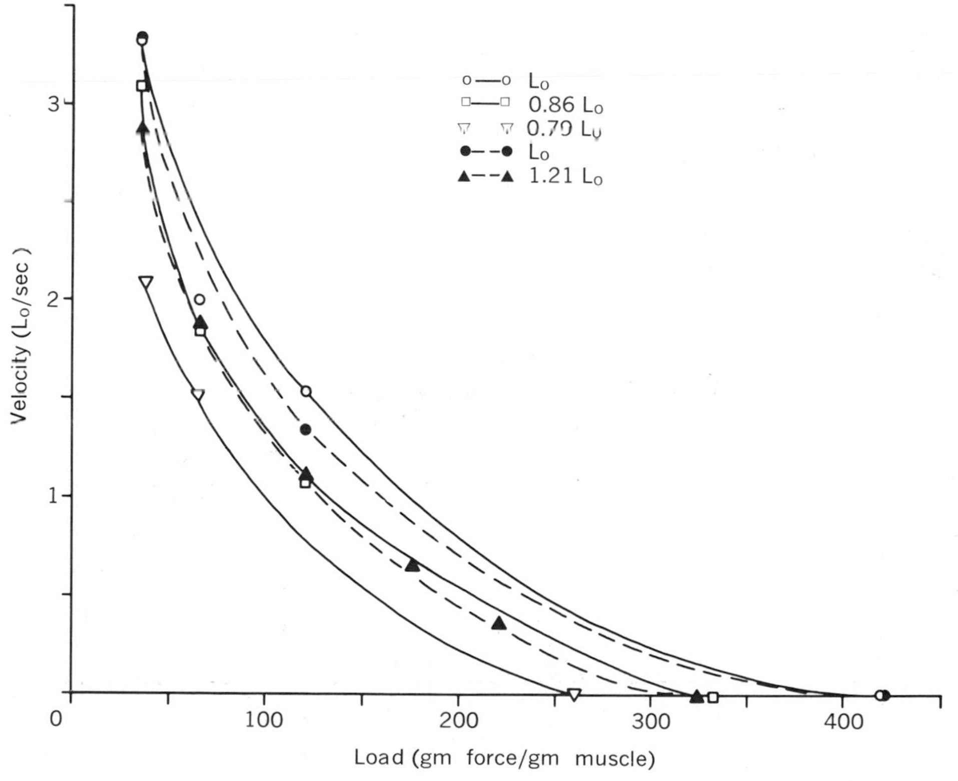

Fig. 3-7. Family of force-velocity curves obtained from gracilis muscle of rat. Temperature, 16.4° C. Muscle weight, 83 mg. Velocity given in units of L_o/sec. L_o, 2.9 cm. Solid lines and open figures are one series of experiments. Broken lines and filled figures are second series of experiments on same muscle. L_o defined as length at maximum isometric tension. Note that any load velocity is greatest at L_o, even at lightest load, 2.9 gm, or 36 gm/gm of muscle. (Based on unpublished data by A. S. Bahler.)

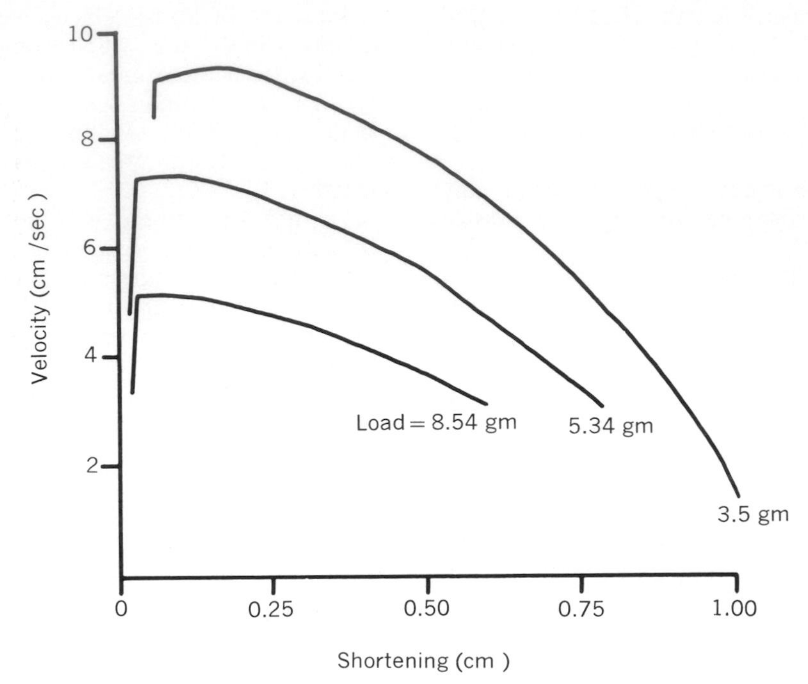

Fig. 3-8. Family of phase-plane or length-velocity curves obtained from gracilis muscle of rat showing effect of different loads at same length. Shortening in every case was from rest length. Maximum velocity was reached in about 5 msec and maintained briefly, followed by increasing deceleration. Three curves, from above downward, were performed on same muscle against increasing loads. Temperature, 16.8° to 17° C. Muscle weight, 54 mg; L_o, 2.55 cm; P_o, 28.8 gm, or 534 gm/gm of muscle. (Traced from unpublished photographic records obtained by A. S. Bahler.)

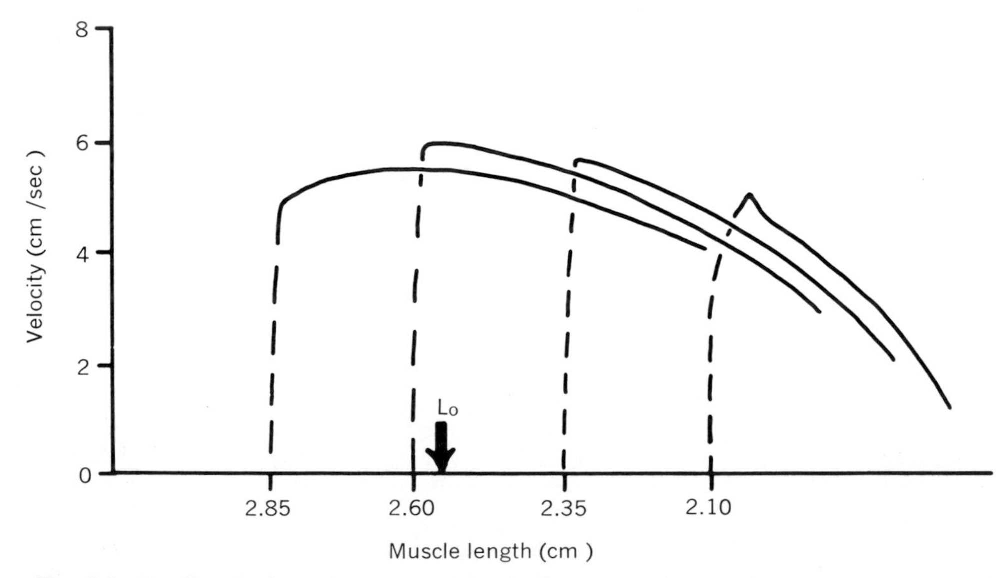

Fig. 3-9. Family of phase-plane or length-velocity curves obtained from gracilis muscle of rat showing effect of different initial lengths at same load. Initial length indicated on ordinate scale. L_o, shown by arrow, was 2.55 cm. Initial lengths corresponded to approximately 1.12, 1.02, 0.92, and 0.82 × L_o. Load was 6.94 gm, or 328 gm/cm² of muscle cross section. Same muscle as in Fig. 3-8. (Traced from unpublished photographic records obtained by A. S. Bahler.)

celerates with increasing rapidity, finally plummeting to zero velocity. The greater the load, the less the velocity at any given displacement.

When the muscle is allowed to shorten from any length less than rest length, it still attains its maximum velocity almost immediately. The velocity reached and the subsequent course of the curve are nearly identical at any length with those described by the muscle starting to shorten from L_0.

When the muscle is stretched beyond L_0 and then allowed to shorten, maximum velocity is not reached immediately. The muscle still accelerates very rapidly but reaches only submaximum velocity. From that point acceleration is relatively small until its length shortens to L_0, at which point velocity is maximum and about the same as that of a muscle beginning to shorten from L_0. From that length on, velocity is nearly the same at any length as that for a muscle shortening from L_0.

Thus velocity is a function of the absolute length of the muscle, independent of the initial length from which shortening started and independent of the distance shortened. Fig. 3-9 displays a family of length-velocity diagrams, each curve obtained at a different initial length.

Close examination of the family of curves shows consistent though usually minor departures. Velocity is slightly higher at any given length as the initial length decreases. The difference in velocity at any length is obvious when the muscle is loaded more heavily. Muscles starting to shorten from longer lengths have been stimulated for a longer time when they reach any given short length than have those muscles that were already shortened before stimulation began. This means that longer muscles are stimulated for a longer time than short muscles, and that there is more dissipation of chemical energy. Because muscles starting from a longer length cannot quite maintain velocities as great as

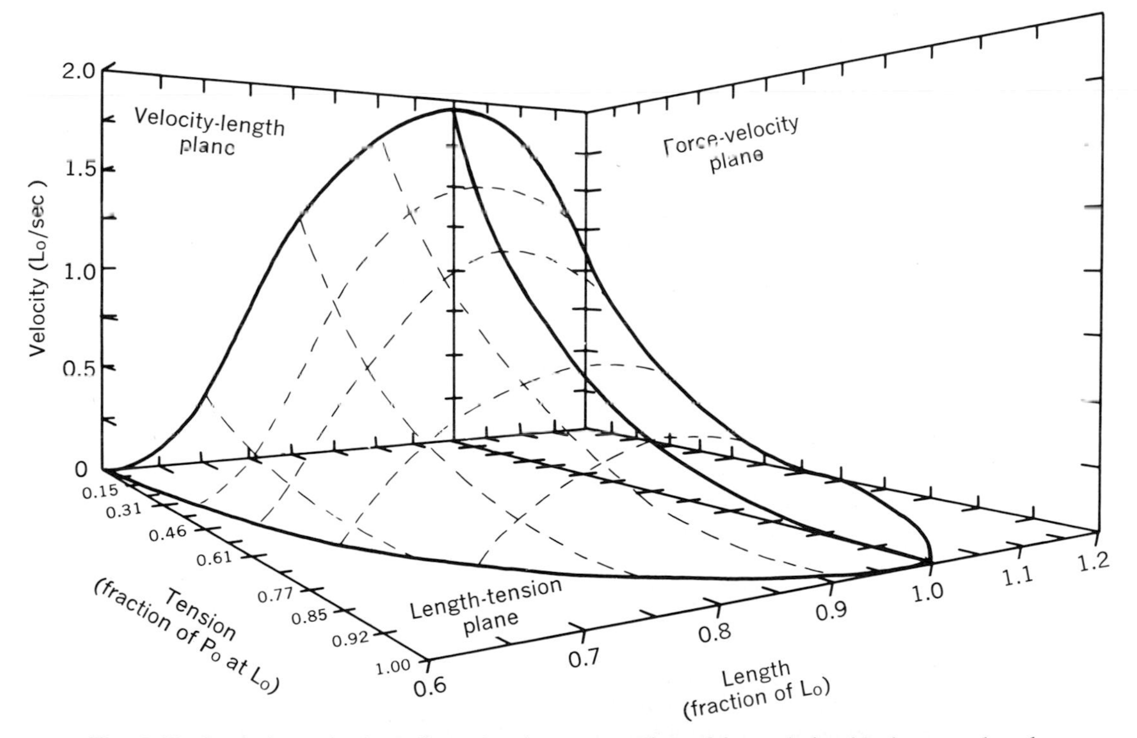

Fig. 3-10. Semischematic three-dimensional representation of interrelationship between length, tension, and velocity. Based on data from gracilis muscle of rat at approximately 16° C. Isometric length-tension diagram forms the base of the figure in the length-tension plane. Force-velocity curve at initial length L_0 is given as heavy continuous line with its own velocity-length axes. The foot on velocity-length curve at short lengths and small loads is extrapolated beyond experimental data in order to meet isometric length-tension diagram. Other force-velocity and velocity-length relations are shown by broken lines to suggest curvature of surface. There is a family of such three-dimensional figures, depending on available chemical energy for contraction.

those starting from shorter lengths, it is likely that velocity of shortening is linked to a chemical process.

The velocity with which a muscle shortens is therefore a function of three things. It is a function of the load on the muscle, of absolute length of the muscle, and of the duration of stimulation. The latter function may indicate dependence of velocity on a chemical energy source. Fig. 3-10 illustrates the relations between velocity, length, and tension, ignoring the time-dependent effect just described. It should be studied with the aid of the separate length-tension, force-velocity, and length-velocity diagrams.

Active state

When a muscle is stimulated, the magnitude of tension developed by the contractile machinery and the time course over which this tension is exerted are obscured by the heterogeneity of the tissue. If the system is slack to begin with, it must first take up the slack before any change in tension appears. Then, as the contractile machinery shortens, it stretches elastic tissue with which it is in series, and stressed elastic tissue contributes to the observed force. It is obviously of great theoretical importance to observe the responses of the contractile machinery itself.

If the muscle is stretched suddenly during the course of a twitch at various times after it has been stimulated, or if it is stretched before stimulation and then released suddenly at various times during the course of a twitch, the tension it develops after a brief delay is presumably the maximum isometric tension of which the contractile machinery is capable at that moment. A plot of this tension against time (Fig. 3-4) expresses the duration and magnitude of what is called the active state. When a muscle is stimulated, the change produced in the contractile machinery is such that, for a given length, the tension that it exerts is greater than that it exerts at rest or, for a given tension, the equilibrium length it assumes is less than that at rest. The rapid change from equilibrium length-tension conditions at rest to different equilibrium length-tension conditions during activity is what is meant by the term "active state."

The active state leads very rapidly to maximum tension development, certainly within a few milliseconds, which, within the limit of resolving power of the experiments, is immediate. Maximum tension is maintained at a constant value for a number of milliseconds and then decays slowly to rest tension. The duration of the plateau of peak tension is arbitrarily called the duration of the active state. At 20° C the active state lasts for only about 10 msec and at 0° C, for 35 to 40 msec. Its duration may be modified by alterations of the ionic environment of the muscle and by certain pharmacologic agents.

Relation between twitch, tetanus, and active state

During a twitch so much time is required to stretch the series elastic components that the active state has begun to decay by the time the muscle as a whole exhibits its peak tension. In fact, the peak tension of a twitch is only about one third of the maximum tension developed during the active state. When a muscle is stimulated tetanically, the active state is never permitted to decay fully. If it is not permitted to decay at all (i.e., if the active state lasts for 10 msec and the frequency of stimulation is 100/sec), muscle

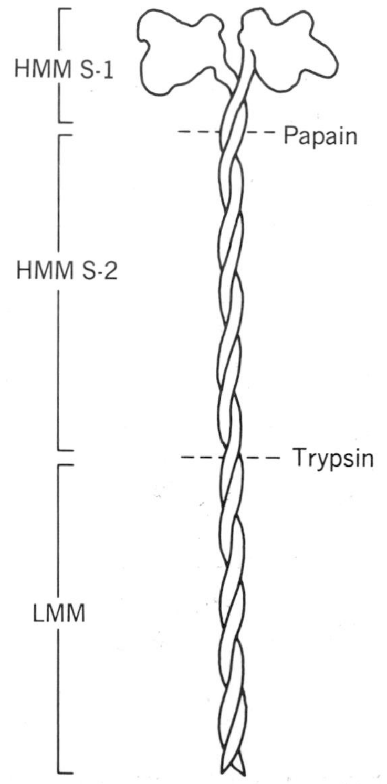

Fig. 3-11. Structure of fundamental myosin unit. Pair of fibrous proteins, terminating in globular portion, form supercoil that is richly α-helical. Trypsin can split fibrous portion into fibrous light meromyosin, *LMM,* about 40% of entire length, and heavy meromyosin, *HMM,* containing globular units. HMM can be split by papain into subfragments *1* and *2.*

tension will rise to a maximum that will be maintained and will equal the plateau tension characteristic of the active state. Fused tetani occur at lower frequencies, i.e., at those in which the active-state plateau is maintained only intermittently, as long as the time constants of the whole muscle are sufficiently large to obliterate minor fluctuations representing slight decay of the active state.

A twitch is therefore not simply a subunit of a tetanus. It differs importantly from a tetanus in that it never fully expresses the active state.

CONTRACTILE PROTEINS, THE CHEMISTRY OF CONTRACTION, EXCITATION-CONTRACTION COUPLING, AND RELAXATION

Contractile proteins

There are two major kinds of myofilaments, the *thick* filaments of the A bands and the *thin* filaments of the I bands. There may be others. There is some question as to whether there is a fine filament connecting the central ends of two thick filaments to each other through the H zone. There are filaments in the Z disc, Z filaments, that are arranged to form pyramids on a square base and connect with the thin filaments of the I bands. But by the term "contractile proteins" we refer only to the proteins that make up the filaments of the A and I bands. (See Driezen et al.,[3] Ebashi and Endo,[4] Ebashi et al.,[5] Szent-Györgyi,[18] and Young[23] for reviews of this subject.)

Thick filaments are probably composed entirely of one protein, myosin (Fig. 3-11). Thin filaments (Fig. 3-12) contain three proteins: actin, tropomyosin, and troponin. Let us first describe these four contractile proteins and then consider their polymeric or aggregated forms and finally their interactions.

Myosin. Myosin is a large protein with a molecular weight approximating 500,000; except for an enlarged end, it is a rod about 1,500 Å long and 15 to 20 Å in diameter. Myosin is capable of splitting ATP. Under certain conditions trypsin splits myosin into two fragments, one about twice the weight of the other. The lighter fragment, light meromyosin (LMM) is a straight rod about 600 Å long and with a molecular weight of about 150,000. LMM is responsible for the self-aggregation properties of myosin. The heavier unit is called heavy meromyosin (HMM). HMM, but not LMM, retains the ATPase activity of myosin. Myosin and HMM, but not LMM, aggregate with actin, a reaction that may be fundamental to contraction. HMM can be split into two subunits, HMM S-1 and HMM S-2. S-2 is linear with a very high α-helical content and a molecular weight of about 60,000. One end of S-2 is attached to LMM, the other to S-1. S-1 consists of a short chain, attached to S-2, bifurcating to terminate in globular units, two globular units per S-1 and hence two globular units per myosin. Each globular unit has a diameter of 70 Å, and each is an ATPase. It is S-1 that attaches to actin. Although there is some suggestion that the two globular units may not be identical, by and large they are about the same.

Myosin polymerizes by forming side-to-side aggregates, probably attaching in the region of LMM. Each additional myosin molecule attaches 143 Å further toward the globular end of the original monomer and at a point rotated 120 degrees from the preceding molecule. In this way the polymer forms a helix with a repeat distance of $3 \times 143 = 429$ Å. Aggregation continues until the macromolecule is about 100 Å thick and about 1.5 to 2 μ long, i.e., about the size of a thick filament occurring naturally in muscle.

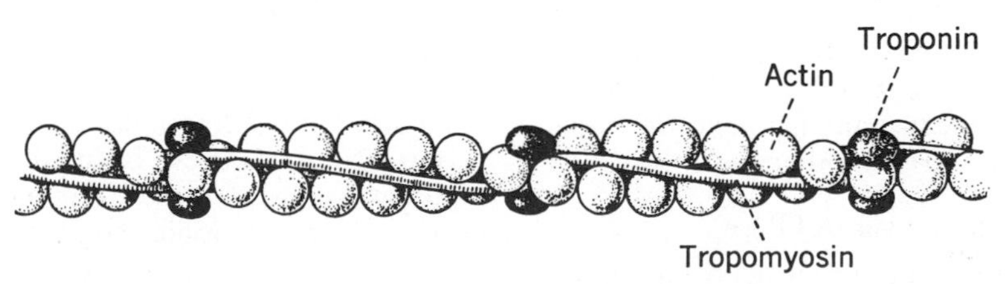

Fig. 3-12. Proposed structure of thin filaments. Units of globular (G) actin polymerize to form fibrous (F) actin. Two strands of F actin coil, with a period of about 13 G actins. In longitudinal notch of actin coil is tropomyosin, a fibrous protein, with period slightly longer than that of actin coil. Therefore there are two tropomyosin per actin coil. One globular troponin is associated with each tropomyosin. (From Ebashi et al.[5])

Actin. The basic monomer of actin, called G actin, is globular, about 55 Å in diameter, has a molecular weight of about 60,000, and contains bound ATP. In the presence of salts the dimer of G actin polymerizes to a large fibrous form, F actin, in the process of which the bound ATP is dephosphorylated to ADP. However, actin is not an ATPase; it will not dephosphorylate added ATP.

Purified F actin is made of two strands twisted about each other so that there are thirteen globular subunits per complete turn of the helix, with a period of 357 Å. There are differences between purified F actin and thin filaments in that the periodicity of thin filaments is about 407 Å. The explanation for this difference lies in the relation between tropomyosin, troponin, and actin.

Tropomyosin. Tropomyosin is a filamentous protein about 400 Å long and with a molecular weight of about 70,000. Hanson and Lowy[68] suggest that two strands of tropomyosin are twisted around the double coil of F actin in such a way as to lie in the hollows of the twisted actin.

Troponin. Troponin is a globular protein that has a molecular weight of about 50,000. It binds to tropomyosin but not to actin. Furthermore, it binds only to a specific region of tropomyosin, to give one troponin globule per approximately 400 Å of tropomyosin filament. This spacing of troponin probably accounts for the periodicity of the natural thin filament. The important function of troponin is thought to be based on its enormous avidity for Ca^{++}, a property that we will consider when we discuss activation of the contractile process.

There are other proteins in the contractile apparatus, probably for the purpose of organizing the myofilaments and possibly also for the purpose of transmitting the force of contraction throughout sarcomeres and from sarcomere to sarcomere. Thus these proteins may be important parts of the series elastic components. They are the M protein, or M substance, which may align thick filaments at the M line, and α-actinin, which is in the Z band and connects thin filaments of adjacent sarcomeres to one another.

Myosin and actomyosin ATPase

Solutions of myosin and of actin are highly viscous and exhibit birefringence of flow. When mixed, particularly in the naturally occurring proportion of about 3 parts of myosin to 1 part of actin, the increase in viscosity is greater than the sum of their separate viscosities. The mixture behaves like a new protein, and in many ways it resembles an extract of muscle prepared by simultaneous solution of myosin and actin. Both the mixture and the extract are said to be actomyosin.

Engelhardt and Ljubimowa[48] opened a new approach to muscle physiology with the discovery in 1939 that both myosin and actomyosin are enzymes capable of splitting the terminal phosphate from ATP. When ATP was added to solutions of either, there was a transient and repeatable decrease in viscosity and in flow birefringence. Actomyosin can be prepared as a hydrous gel. In the presence of small concentrations of ATP the gel shrinks and its water content is markedly reduced, a phenomenon called superprecipitation by Szent-Györgyi,[18] its discoverer. Superprecipitation is probably a manifestation, in the absence of natural structural constraints, of the fundamental contractile process, the association of actin and myosin to form an actomyosin of a certain configuration.

Huxley[82] has examined by electron microscopy complexes of actin and myosin and of actin and HMM, which retains the ATPase activity of myosin. Actin-HMM complexes are filaments with a diameter of 200 to 300 Å and with a strong axial period of about 370 Å. There is remarkable structural polarity. Arrowheadlike objects all point in the same direction along the whole length of the filament. The same appearance occurs when HMM is complexed with I band filaments. Complexes of actin and myosin and complexes of I band filaments and myosin all give the same appearance.

It is likely that the cross bridges between thick and thin filaments, visible by electron microscopy, are the globular heads of heavy meromyosin, HMM S-1, and possibly some portion also of HMM S-2, and that the association between actin and myosin occurs at the ATPase site of one of the globular heads of myosin and a receptor on actin. In this association, ATP is split.

There are differences among the various contractile ATPases. It has been difficult to unravel some of these differences because so many different ionic conditions have been used by various investigators.

It is an old observation that the ATPase activity of what has been called purified myosin is inhibited by Mg^{++} and stimulated by Ca^{++}. When actin is added to myosin, there are changes in properties of the ATPase. Actin is said to activate myosin ATPase. It is not always clear that the ATPase activity of actomyosin exceeds that of Ca^{++}-activated myosin ATPase, but it often does (Bárány[33]). It is definite, however, that ionic requirements change. Actomyosin ATPase requires Mg^{++}. In large part it may not be Mg^{++} but the Mg-ATP complex that is required, but some Mg^{++} is necessary for activation of the interaction between myosin and actin.

There are two preparations called actomyosin. One has been called natural actomyosin because it is simply an extract of muscle and historically was the first actomyosin preparation. The other is synthetic actomyosin, made from purified myosin and purified actin. Synthetic actomyosin does not require Ca^{++} for its ATPase activity but natural actomyosin does. Investigation of this difference in responsiveness to Ca^{++} led to the discovery, first, that natural actomyosin contained in the actin fraction another protein, which we now recognize as tropomyosin, and, second, that the early preparations of tropomyosin were themselves contaminated with a protein we now call troponin. The fact that mixtures of purified actin and myosin do not require Ca^{++} for activation of ATPase, but mixtures of myosin, actin, tropomyosin, and troponin do, suggested to Ebashi et al.[5] that tropomyosin and troponin inhibited actomyosin ATPase activity and that Ca^{++} removed this inhibition.

Activation of contraction and excitation-contraction coupling

Contraction is activated by Ca^{++} because it is the concentration of Ca^{++} that regulates actomyosin ATPase activity, and contraction is a physical manifestation of the union between actin and myosin in conjunction with the splitting of ATP. According to Ebashi et al.,[5] troponin, as it occurs naturally in muscle, binds 87% of the calcium bound in myofibrils (even though it constitutes only 3% of the weight of myofibrils) under controlled ionic conditions in the presence of a low Ca^{++} concentration (3×10^{-6}M). Most of the rest of the calcium is bound by myosin, which constitutes 60% of myofibril weight. Tropomyosin binds virtually no calcium.

It is assumed, with some evidence,[5, 60] that the binding of Ca^{++} to troponin causes an alteration in troponin. Presumably this alteration is a configurational change. Troponin does not bind to actin but tropomyosin does, and tropomyosin extends over the whole length of F actin. It is postulated that the configurational change induced in troponin by calcium binding produces a configurational change in tropomyosin, resulting in the removal of tropomyosin's inhibition of actomyosin ATPase or in the activation of actomyosin ATPase.

Although the correlation is not perfect, Bárány[33] has shown that variations from muscle to muscle in actomyosin ATPase activity tend to be accompanied by variations in velocity of contraction, and others have correlated, again imperfectly, increases in ATPase activity with increases in viscosity of actomyosin solutions, indicating a correlation with the association of actin and myosin.

Further consideration of proposals for mechanisms by which the chemical and mechanical events of contraction are related will be deferred. For the present let us concern ourselves only with the fact that Ca^{++} activates the actomyosin ATPase, initiates the association between actin and myosin,[101] and triggers the mechanical events we call contraction.

How is Ca^{++} concentration regulated? A nice preparation for studying quantitative mechanical responsiveness to Ca^{++} is the skinned muscle fiber, first described by Natori[90] and exploited by Endo et al.[47] and Podolsky[15] and his colleagues. The preparation is essentially a segment of a single muscle fiber, usually frog muscle, from which sarcolemma, with a certain number of underlying myofibrils, has been dissected. With the preparation in mineral oil (or in a controlled aqueous electrolyte solution), known quantities of known concentrations of Ca^{++} can be added, and the tension exerted by the fiber under isometric conditions can be measured. There are differences in reported absolute values, but in general, contraction begins at a Ca^{++} concentration of about 10^{-6}M and is maximum at about 10^{-5}M. Contractions can be caused also by longitudinal depolarizing electrical stimulation of the skinned fiber, which presumably causes the muscle to release its own Ca^{++} from some Ca^{++} stores, and by the addition of isotonic Cl^-, which presumably depolarizes the walls of the SR, since the lumen of the SR is probably filled with a fluid resembling interstitial fluid with respect to its Na^+, K^+, and Cl^- concentrations, while the external environment of the SR of the skinned fiber in mineral oil is simply the normal aqueous solution of sarcoplasm.

Observations of this sort suggest that excitation-contraction coupling is effected simply by the release of Ca^{++} from SR and the subsequent translocation of Ca^{++} to troponin (which has a great avidity for it), with consequent activation of actomyosin ATPase and association of actin and myosin when the Ca^{++} concentration has been raised sufficiently.

The questions are the following: (1) How does the signal for release of Ca^{++} progress from depolarization of the T tubule to the SR? (2) Is Ca^{++} localized within the SR? (3) How is Ca^{++} translocated from the SR to troponin?

We have already touched on the question of whether or not depolarization of the T tubule is propagated to depolarize the membrane of the SR or at least of the terminal cisternae. Arguments against the likelihood that transmission from the T tubule to the SR is electrical are as follows: (1) The space between the outer layer of the T tubule and the outer layer of the terminal cisterna is 120 Å, greater than seen with tight junctions that are readily permeable to monovalent ions. (2) The surface area of the SR is so much greater than that of the T tubule that the SR could not be depolarized sufficiently to propagate an action potential by transfer of the total charge from the T tubule to the SR. (3) Measurements of muscle membrane capacity give values almost compatible with the assumption that total capacity is due to sarcolemma and the walls of the T tubule, on the assumption that capacitance per unit surface is about the same for all surfaces. Certainly the total membrane capacity is far less than it should be if SR contributed to that capacity, on the assumption that capacitance per unit surface of SR is not substantially less than that of other membranes. Arguments in favor of the possibility that the normal signal for release of Ca^{++} from SR is depolarization of SR are as follows: (1) There is no positive evidence for any other mechanism. (2) Probable depolarization of SR, either electrical or chemical, noted in Natori skinned fibers, can release Ca^{++}. (3) The Na^{+}, K^{+}, and Cl^{-} concentration gradients across the walls of SR are probably similar to those across sarcolemma. There is evidence that the walls of SR are permeable to these monovalent ions, and there must therefore be a resting potential difference across the walls of SR with the possibility of generation of a propagated action potential. (4) Electrical stimulation has produced responses propagated as far as 200μ along a skinned fiber segment, although rarely.[40] (5) Skinned fibers prepared from muscle presoaked in solutions containing cardiac glycosides are not responsive to electrical stimulation; it is as though excitability depends on actively established Na^{+} and K^{+} gradients across the walls of the internal membrane system, and this effect is reversible.[40]

The conclusion is that the means by which depolarization of the wall of the T tubule signals release of Ca^{++} from SR is not known. However, it is known that in autoradiography using radioisotopic Ca^{++}, the Ca^{++} in well-rested muscle is localized strongly to the region of terminal cisternae, probably to the walls of these structures (Winegrad[103]). It is not known how Ca^{++} is translocated from the walls of terminal cisternae to troponin. It is postulated that translocation occurs simply by diffusion,[103] but diffusion seems to be too slow and to offer too much dispersion to account for the rapid rise of active-state tension known to occur.

Relaxation

Historically, current concepts about the role of Ca^{++} in activation followed an understanding of its role in relaxation. Marsh[88] first described an extract of muscle that caused relaxation, which became known as relaxing factor. As the result of contributions from many laboratories, it is now known that the relaxing factor is simply granules of SR and that its action is to bind Ca^{++}. Podolsky and Constantin[93] have found that if a sufficiently large quantity of Ca^{++} is added to the Natori skinned fiber, the ability of SR to bind Ca^{++} from solution is saturated and relaxation does not occur.

From a series of experiments based primarily on autoradiography of muscles fixed at various times before, during, and after tetanic contractions, Winegrad[103] proposes the following model for the flow of Ca^{++}. Somehow depolarization of the T tubule provides a signal that causes liberation of Ca^{++} from terminal cisternae. Ca^{++} diffuses to myofibrils and penetrates to myofilaments, where, according to Ebashi and Endo,[4] troponin constitutes a temporary sink, binding Ca^{++}. The binding leads to activation of actomyosin ATPase, to association between actin and myosin, and to the mechanical event we call contraction. When the depolarizing stimulus from T tubules stops, Ca^{++} ceases to flow from terminal cisternae to troponin. Somehow the longitudinal elements of SR become the sink for Ca^{++}, and Ca^{++} diffuses from troponin to some calcium-binding component in the wall of longitudinal SR. There is a Ca^{++} pump that is related to an ATPase in SR. It is postulated that the function of this pump is to translocate Ca^{++} from its binding sites on the wall of longitudinal SR into the lumen of longitudinal SR, from whence it diffuses back to its resting sites in the terminal cisternae.

This Ca^{++} odyssey is speculative. The actual autoradiographic observations demonstrate that at rest Ca^{++} is in or on terminal cisternae, during tetanic contraction it is concentrated in the region of myofilaments and decreased in terminal cisternae, shortly after contraction it is in or on longitudinal SR, and not until remotely after exercise, minutes later, is it again concentrated in terminal cisternae. There are weaknesses in attributing its flow solely to diffusion. It is almost implausible that binding affinities should be static, for how can the release of Ca^{++} from terminal cisternae that is not immediately taken up by the walls of longitudinal SR be explained? Could there be time-dependent changes in affinities for Ca^{++} during the contraction cycle?

We might think of the active state, described previously in the language of the time course of tension exerted by the contractile elements, in terms of the time course of the binding of Ca^{++} to troponin, for if these hypotheses are correct, it is this that regulates the activity of contractile elements. If this is the case, then we must account for the shape of the active-state time curve, which has a small delay, a very rapid rise to, or nearly to, maximum tension, an appreciable plateau or near plateau, and a relatively slow decay (Fig. 3-4). The general shape of the active-state curve is consistent with the diffusion process and paths outlined if the threshold Ca^{++} concentration is substantially less than one-half the Ca^{++} concentration at which maximum active-state tension occurs.

Molecular basis of contraction

Current notions of contraction spring from interference microscopy observations by A. F. Huxley and Niedergerke,[78] and electron microscopy observations by H. E. Huxley and Hanson,[83] who pointed out simultaneously that the lengths of the A band and of the broad myofilaments in the A band are constant and independent of stretch or contraction. Later measurements taken from electron micrographs showed that within an uncertainty of perhaps 5 to 10% there is no shortening of either thin filaments or thick filaments. These observations rendered obsolete all earlier hypotheses of contraction that had depended on shortening of contractile proteins and led to the present concept that contraction is caused by the interdigitation or sliding of myofilaments.

In a sarcomere the sum of the lengths of thick and thin filaments is about 3.5μ. When sarcomere length exceeds 3.5μ, there is no overlap of thick and thin filaments and the muscle does not contract in response to an electrical or chemical stimulus. The muscle can contract from any length less than 3.5μ. When contraction occurs, thick and thin filaments slide past one another, the zone of overlap of filaments increases, and the H zone decreases.

This orderly interdigitation of thick and thin filaments can continue only until the sarcomere has shortened to the length at which opposing thin filaments abut in the same sarcomere, i.e., to about 2μ. Organization of myofilaments with further shortening has not been defined unequivocally. Huxley has some evidence that opposing thin filaments overlap in the center of the A band with further shortening. There is a possibility that the thin filaments may shorten or that the Z line may widen transversely with extreme shortening of the sarcomere. If the Z line widens, thin filaments near it would separate further from thick filaments, perhaps exceeding the 180 Å distance spanned by cross bridges. These possibilities are illustrated in Fig. 3-13.

The existence of cross bridges between thick and thin filaments suggested that these were the means by which thick and thin filaments were joined physically, through which tension was exerted, and by which shortening occurred. The fact that sarcomeres stretched beyond 3.5μ (the length beyond which there is no overlap of thick and thin filaments) cannot shorten and cannot exert extra tension on stimulation suggests that cross bridges are necessary for shortening and for tension of contraction.

A. F. Huxley[9] proposed that thick filaments walked along thin filaments by means of the cross bridges. The process is held to resemble the action of a man climping a rope. The cross bridges are his arms and legs. He climbs by reaching with his arms, gripping, pulling, breaking contact, reaching, gripping, and pulling. Huxley suggests that there might be spatial organization of chemically active sites along myosin and along actin, and that the myosin and actin sites might interact as the filaments pass through certain critical distances relative to one another. In a crude way this model seems intuitively to account for the force-velocity relation. If force is proportional to the number of interactions, and if the probability of interaction between a pair of sites is great when the filaments move slowly past one another and small when they move rapidly, then there will be less force at high velocity.

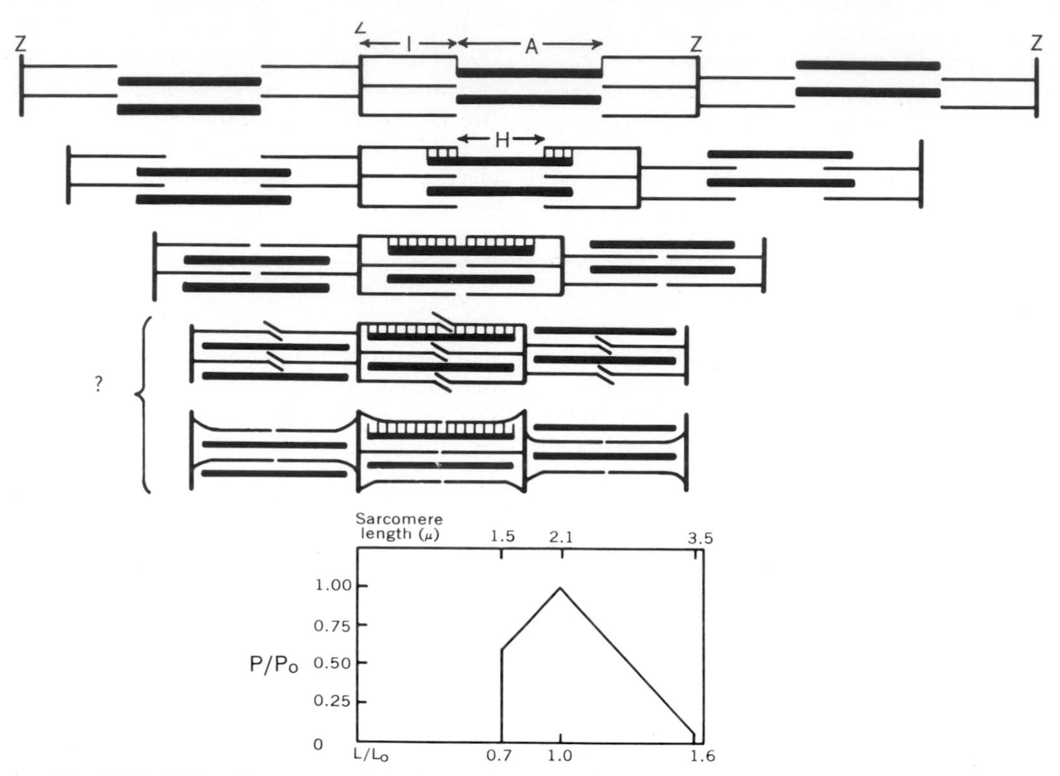

Fig. 3-13. Sliding-filament model of muscle contraction. Three sarcomeres are illustrated in various stages of contraction. Lengths of thick and thin filaments are drawn to scale: length of thick filament is 1.5μ; of thin filament, 1μ. In top fibril, sarcomere length is 3.5μ; thick and thin filaments just meet so that one cross bridge can be formed between an adjacent thick and thin filament. (Cross bridges illustrated only in middle sarcomere.) In second fibril from top, each sarcomere has shortened by 0.6μ; there is 0.3μ overlap of thin filaments on both ends of thick filament. Length of thick and thin filaments remains constant. Number of cross bridges increases linearly with overlap. There is now an H zone, the central region of A band in which there is no overlap. Shortening continues until thin filaments just touch their opposite numbers, shown in third fibril from top. At this point maximum number of cross bridges has formed and, hypothetically, tension should be maximum. The length-tension diagram for a sarcomere is illustrated at bottom. As one proceeds from stretched to shortened muscle (from right to left), contractile machinery exerts no tension until there is just overlap, at 3.5μ. Tension rises linearly with shortening until thin filaments abut. By definition, this is L_o, about 2.1μ. For whole muscle, it is known that tension falls with shortening from rest length. If cross-bridge model is correct, there must be fewer cross bridges as muscle shortens. Two possibilities are illustrated by last two fibrils. In upper one, with further shortening, thin filaments overlap those from opposite end of same sarcomere in center of A band, mutually interfering with cross bridging. In lower one, Z band widens (by distortion of angle between Z filaments), increasing distance between thick and thin filaments at Z line. If this distance exceeds 180 Å, bridges cannot be completed. Shortening continues until the sarcomere is reduced to length of thick filament, 1.5μ, or 70% of L_o. At this length there should remain a substantial number of cross bridges (half or more of maximum), illustrated in hypothetical length-tension diagram. This is unlike real length-tension diagram for whole muscle or fiber in which tension is zero at maximum shortening. Length-tension diagram shown here, drawn on theoretical grounds, agrees very closely with that reported by Gordon et al.[66] for central sarcomere of single fiber of frog semitendinosus muscle.

Weber[20] assigned a rather definite chemical linkage to Huxley's model by proposing that the active sites on actin were a series of SH and OH groups, forming and breaking alternately —S— and —O— bridges between actin and myosin. This suggestion is consonant with those observations indicating that shortening is proton dependent.

Podolsky[14] has carried Huxley's model further by linking the extent of shortening to the extent of the driving chemical reaction, where extent of reaction is a measure of its departure from equilibrium.

There seems to be a generally held assumption that tension is proportional to the number of cross bridges. In Fig. 3-13, formation of cross bridges is indicated as shortening proceeds. As the sarcomere shortens from its initial length (3.5μ in Fig. 3-13, or 1.6 L_0), the number of cross bridges should increase linearly with shortening. If tension is a function only of the number of cross bridges, then tension should also increase linearly with shortening. However, since there are a finite number of HMM projections on myosin, there are a finite number of potential cross bridges. Once all cross bridges have been formed, maximum tension is reached if the model is correct. The length at which maximum tension is reached is defined arbitrarily as L_0, or 2.1μ in Fig. 3-13. Thus this model predicts a linear increase in tension with shortening of a sarcomere from a stretched position to L_0.

At this point some other process must be brought to bear. We know that in the real length-tension diagram, tension falls as the muscle shortens from L_0. In the model, if nothing should intervene, once all bridges are formed their number would remain constant with further shortening, and tension would be constant, independent of shortening at lengths below L_0. If tension is indeed a linear function of the number of cross bridges, then there must be fewer cross bridges as the sarcomere shortens below L_0. Two possibilities for decreasing the number of cross bridges are illustrated in Fig. 3-13. In one case opposing thin filaments overlap in the A band, occluding sites at which actin might combine with HMM. In the other case the Z line widens, separating adjacent actin and myosin molecules by more than 180 Å, so that cross bridges cannot span the full distance. Whether either of these models is true is uncertain, and if true, whether the number of cross bridges decreases linearly with shortening is arbitrary. In either case, however, when the sarcomere has shortened to its limit, defined by the length of the thick filament (1.5μ, or 0.7 L_0), both models predict a residue of a substantial proportion of cross bridges; i.e., both models predict that there will be about 50% or more of P_0 at minimum length or maximum shortening. But in the real length-tension diagram, tension is in fact zero at minimum length. The length-tension diagram predicted by cross-bridge models is illustrated in Fig. 3-13. It is not really a good fit to experimental length-tension diagrams obtained for whole muscles or whole fibers. It is conceivable that the length-tension diagram of a single sarcomere might better resemble the model and that the experimental length-tension diagram is the sum of distributed behavior of sarcomeres. Since the hypothetical sarcomere length-tension diagram illustrated in Fig. 3-13 is skewed, the distribution of sarcomere length-tension diagrams must be skewed in the reverse direction if the model is correct; i.e., at any given overall muscle length there must be more stretched sarcomeres than shortened ones. Whether this is in fact the case is not known.

Objections were raised to the cross-bridge model and alternatives were proposed, but as more is learned about the structure of the contractile proteins, the objections are being met. That is not to say that our understanding is total. In fact, we do not yet know at the molecular level how a muscle contracts. Therefore there are many hypotheses, but now they all begin with the assumption that at least one of the two globular heads of myosin, HMM S-1, combines with a receptor site on a G actin monomer of F actin to form a poorly compliant body, stiffer than resting muscle.

It is further proposed by several investigators that the highly helical HMM S-2 fragment serves as a double hinge that permits HMM to angulate away from LMM and at its other end permits the globular head of myosin to angulate away from HMM S-2. Problems arise with the details. How are we to account quantitatively for the tension developed and for the speed of shortening? It is known from our experience with whole muscles that there is a reciprocal relation between tension exerted by a muscle and the speed of shortening, as though the muscle at a given initial length has a certain power that can be factored between speed of shortening or development of tension, although muscle is neither a constant power machine nor does it have constant enthalpy, as shall be seen.

When contraction is isotonic so that gross shortening occurs, thick and thin filaments must slide past one another. This can only occur if the actomyosin links are made and broken rapidly and synchronously throughout every half sarcomere. If the ends of the fiber are tethered so that the shortening excursion of the contractile elements is limited, it is evident that myofilaments will slide past one another initially, as in isotonic contraction, until the elastic limits of series elastic components are reached, whereupon sliding ceases. This does not mean that links between actin and myosin are no longer made and broken. Presumably the making and breaking of these links depends on the highly localized Ca^{++} concentration, the configuration of troponin, and the configuration of tropomyosin. As pulses of depolarization descend the T tubules, waves of Ca^{++} are released from terminal cisternae. Similarly, there will be phasic uptake of Ca^{++} by longitudinal SR. Consequently, there will be phasic changes in local sarcoplasmic Ca^{++} concentration, and phasic changes in activation of actomyosin ATPase and of mechanical linkage between actin and myosin. The tension exerted must depend on the average number of links between actin and myosin at any instant. Energy must be spent to maintain tension, even in the absence of overall shortening in the absence of external work.

It is much less clear how the velocity of shortening ought to be related to the number of links between actin and myosin. In accounting for shortening, we must explain not only why myofilaments slide past one another but also why they slide in the direction of shortening. We must also account for the known velocity-length and force-velocity relations of muscle as a whole.

There have been a number of proposals designed to account for these events, beginning with one by A. F. Huxley,[9] to whom many others are indebted. These hypotheses were necessarily vague in certain critical respects, but they were useful in leading to further experiments that have clarified important molecular and structural details; as a result, more satisfying hypotheses have been advanced. It is largely due to studies by Pepe[92] of specific antibody-combining elements of myofibrils that we have a better idea that the cross bridges are HMM and that probably the flexible portion is the helical HMM S-2 and not the attachment of the globular HMM S-1 to actin. The crux of the hypotheses is that it is a change in structure of the bridge, probably in HMM S-2, that leads to both mechanical force and shortening.

Pringle,[16] for example, proposed that cyclic activity at the bridges may proceed stepwise as follows: (1) ATP is bound to the active enzyme site of myosin HMM S-1. (2) HMM S-1 attaches to actin.* (3) Actomyosin ATPase is activated. The terminal phosphate of ATP is hydrolyzed, leading to a change in the angle of attachment to actin. As a result, the HMM S-2 portion of the bridge bends, exerting tension between the two ends of the sarcomere that is proportional to the number of bridges. (4) If shortening can occur, this mechanical potential energy is released as work. The direction of sliding, toward shortening, is thus predetermined by the fact that myosin monomers are all oriented so that each HMM is toward the Z line, and the force of bending HMM S-2 drives the myosin toward the Z line, as though myosin were reaching back to actin and pushing itself. (5) Exchange between bound ADP of actomyosin and free nucleotides in sarcoplasm occurs. If ATP replaces ADP, the bridge detaches and the cycle is repeated.

Gergely et al.[63] have shown that in muscles with different contraction speeds (in general, red versus white muscles) the actins are indistinguishable but the myosins are different, suggesting that structural changes determine contraction velocity. Among many other intriguing hypotheses,[43] particularly attractive is the one by Harrington.[69] Only one of the two heads of HMM S-1 is attached to actin at any moment. Harrington proposed that the unattached head of one myosin monomer lies close to the LMM-HMM S-2 junction of its succeeding myosin monomer on the same thick filament. Hydrolysis of ATP by the unattached head produces a rapid elevation in the highly localized concentration of electrons, leading to a cooperative helix-to-coil transition at the LMM-HMM S-2 hinge site and possibly further along HMM S-2. This transition causes HMM to flip out from the core of the thick filament to form a bridge to actin. In the presence of a low Ca^{++} concentration and high Mg-ATP concentration, as in resting muscle, the binding constant of myosin to actin is depressed, and the result is a flick-

*Bárány,[32] by blocking myosin ATPase with SH inhibitors, has shown that the ATPase site and the actin attachment sites are different.

ering oscillation of cross bridges without tension generation. However, if Ca^{++} is injected from the terminal cisternae to remove the inhibition of actomyosin ATPase, then this ATPase is activated. The steady-state rate of ATP hydrolysis by actomyosin is 200 times greater than that by myosin (Eisenberg and Moss[45]), probably because of the increased dissociation rate of the enzyme-product complex.[100] This reduces the depression of the cross-bridge binding constant due to Mg-ATP. Binding of Mg-ATP to cross bridges may be rate-limited by the dissociation rate of the split products ADP and P_i from myosin. When this dissociation rate is increased by linking actin to myosin, the rate of Mg-ATP binding is also increased, so that the cycle of making and breaking cross-bridge connections is more rapid.

Summary of events in contraction and relaxation

Excitation: action potential propagated along sarcolemma

↓

Depolarization of T tubules (active or electrotonic?)

↓

Unknown signal to terminal cisternae of SR

↓

Ca^{++} released into sarcoplasm

↓

Ca^{++} bound by troponin

↓

Deformation of troponin

↓

Cooperative configurational change in tropomyosin

↓

Release of inhibition of actomyosin ATPase and link between actin and myosin

↓

Configurational change in heavy meromyosin

↙ ↘

Tension exerted — Shortening by sliding filaments

Ca^{++} bound by longitudinal SR

\+ ↘

Mg-ATP bound by actomyosin? — Ca^{++} pumped into SR lumen

↓ ↓

Cross bridges disconnected — Diffuses to terminal cisternae

↓

Actomyosin ATPase inhibited

↓

Series elastic elements restore rest length; active tension disappears

Energetics

Muscle expends energy even when at rest. When it contracts, it spends even more energy. In order to restore itself structurally and with respect to its total chemical composition, energy in excess of that spent at rest must continue to be spent for some time after contraction ceases. Several questions arise. What are the magnitudes and time courses of these energy expenditures? For what processes are these energies spent? What are the chemical reactions underlying each phase of energy expenditure?

Answers to these questions can be sought in a number of ways. Like all animals, humans are thermodynamically open systems. All our chemical free energy comes from oxygen breathed and food eaten. Energy is spent when chemical free energy is converted to heat or work. Therefore the total net loss of chemical free energy can be measured by determining the sum of heat dissipated and work performed by muscle. But if one wants to know the details of the chemical reactions and the sources of energy, individual reactions must be studied. It should be recognized at the outset that all the reactions are not known, so that a total chemical balance sheet does not exist; but a great many reactions are known, and it is therefore possible to account for most of the energy cost of contraction.

Heat

Because nearly all chemical processes occurring in muscle at rest are exergonic, resting muscle releases heat to its environment. When a muscle contracts isometrically, heat is liberated in excess of that produced at rest. When a muscle is stimulated isotonically, it produces even more heat.

The time course of heat production can be measured and the quantitative relation of the time course of heat production and of the total heat produced to load, to distance shortened, and to work have been matters of major concern. If one measures work done by a muscle and its heat production, one knows the total energy dissipated by the muscle. Any theory of the nature of muscle contraction and of its chemical basis must satisfy the facts of muscle heat production.

Most knowledge of muscle heat production stems from the instrumentation, observation, and interpretation of experiments by Hill[7, 8] and his many colleagues, nearly all based on study of excised amphibian muscle.

Hill measured heat production by means of a thermopile. The thermopile does not detect all the heat released by muscle, and substantial corrections must be made for its inefficiency. For this reason there is room for uncertainty about the magnitude of heat production, and several investigators are now in the process of measuring muscle heat production by other means. These include the integrating thermopile of Wilkie[102] and the gradient layer calorimeter used by Fales et al.[49-51] as well as further measurements with the thermopile with improved calibration.[64, 73] It is premature to draw some of the important conclusions, certain of which were drawn earlier, that are needed to build a model of muscle contraction. Some facts, however, are clear.

If a muscle is stimulated to contract either isometrically or isotonically, the heat it liberates is much greater than that produced at rest. In an isometric twitch, heat production begins promptly following a stimulus and decreases from its maximum rate more or less simultaneously with decay of the active state. The total heat produced during a single isometric twitch is called *activation* heat, implying that it is the heat dissipated by those processes that got contraction under way. In an isometric tetanus, heat production is maintained approximately at a constant rate and total heat production increases rather linearly, finally decreasing as tetanic tension fails. The total heat produced during an isometric tetanus is called *maintenance* heat and is generally regarded simply as a summation of activation heats produced by iterative activation of muscle with each stimulus.

Fenn[55] discovered in 1923 that when a muscle was permitted to shorten so that it lifted a weight and therefore did work, the muscle liberated an amount of heat that was greater than that released during isometric contraction. The phenomenon, energy transferred from muscle to its environment as work and heat in excess of the energy transfer in an isometric contraction, is known as the Fenn effect. This additional heat is called *shortening* heat and has been investigated extensively by Hill for 30 years. Until 1963, on the basis of observations illustrated in Fig. 3-14, Hill held that shortening heat was proportional only to the distance shortened (Fig. 3-14, *A*) and not to the load (Fig. 3-14, *B*). Hill therefore spoke of this constant of proportion-

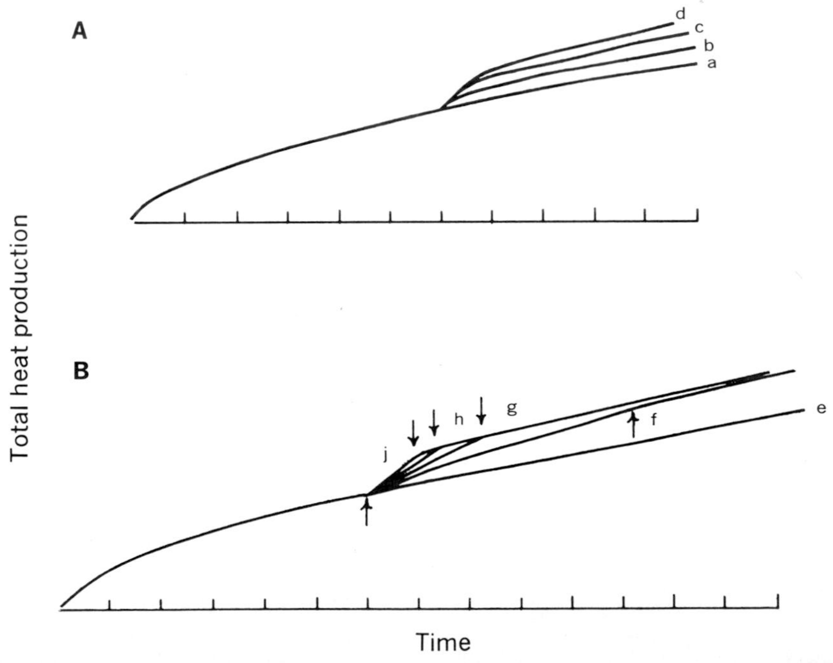

Fig. 3-14. Total heat production by muscle as a function of time, according to Hill. Frog muscle stimulated tetanically at 0° C. Time marks show intervals of 0.2 sec. In **A,** a 29.5 mm long muscle was released 1.2 sec after onset of stimulation and allowed to shorten different distances under constant load. Curves *a, b, c,* and *d* are, respectively, for 0, 1.9, 3.6, and 5.2 mm shortening. During shortening, rate of heat production increases. Final displacements of curves *b, c,* and *d* from *a* are proportional to amount of shortening. In **B,** same muscle shortens a constant distance under varying loads, increasing loads from *j* to *f; e* is isometric.. Muscle released at 1.2 sec. End of shortening indicated by arrows. Because final displacement of *f, g, h,* and *j* is identical, total heat in excess of isometric heat, *e,* is independent of load and velocity and dependent only on distance shortened. (From Hill.[70])

ality as the constant of the heat of shortening. Total heat production during an isotonic tetanus, in excess of resting heat and providing the load was not permitted to stretch the muscle back out again, was:

$$Q = A + ax \qquad (4)$$

where A is maintenance heat, a is the coefficient of shortening heat (heat per unit length shortened), and x is the distance shortened.

However, in more recent studies with improved techniques Hill[71] has found not only that the absolute value of a in equation 4 was falsely high in his earlier work but also that a is not constant and does increase with increasing load. He found an approximately

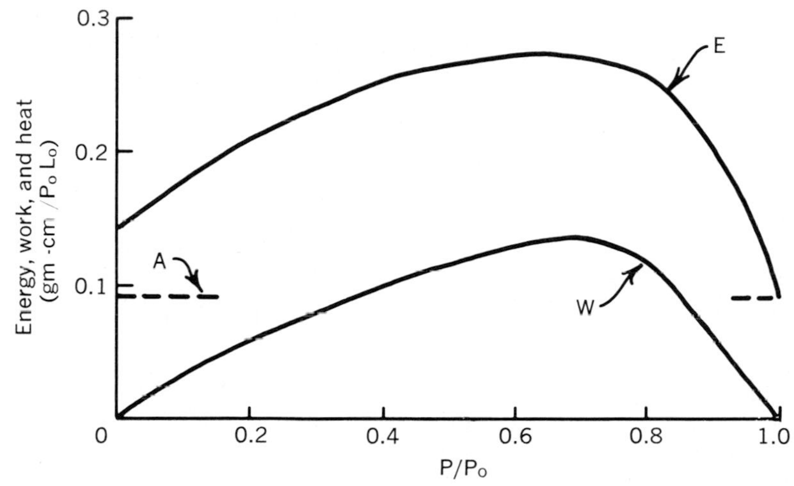

Fig. 3-15. Work, heat, and total energy as a function of load applied to muscle. Work, *W*, is the product of force, *P*, and distance shortened; i.e., it can be obtained by inspection of length-tension diagram. Activation heat, *A*, is independent of load, indicated by dashed line. Total energy, *E*, is sum of W + A + shortening heat (illustrated in Fig. 3-14). Load has been normalized in terms of maximum tension, P_0. Energy components have been expressed per P_0L_0, maximum tension multiplied by length at which maximum tension is exerted. Data are typical for frog sartorius muscle at 0° C, where P_0L_0 = 266 gm-cm/muscle, or about 1,600 gm-cm/gm of muscle.

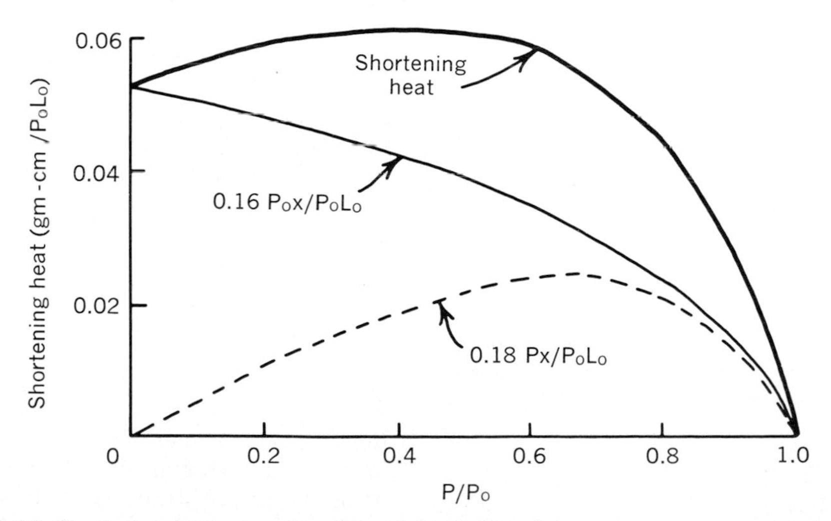

Fig. 3-16. Shortening heat as a function of load. Two lower curves are separate terms of Hill's revised equation for shortening heat; x is the distance shortened. Middle curve is simply left-hand part of length-tension diagram (Fig. 3-5) rotated 90 degrees about L_0 and multiplied by scale factor. Bottom curve is simply the curve for work (Fig. 3-15) multiplied by scale factor. Sum of two lower curves is upper curve, shortening heat, which is nearly independent of load for light and moderate loads, below 0.6 P_0. Data are for same typical muscle as in Fig. 3-15.

linear relation between the coefficient of shortening, which he now calls α, and load, which empirically is as follows:

$$\alpha/P_o = 0.16 + 0.18\ P/P_o \qquad (5)$$

From equation 5, shortening heat, or αx, is:

$$0.16\ P_o x + 0.18\ Px$$

But Px is the external work performed by the muscle, so that, according to Hill:

$$\text{Shortening heat} = 0.16\ P_o x + 0.18\ (\text{work}) \qquad (6)$$

The total energy loss measured during contraction, shown as a function of load in Fig. 3-15, is the sum of external work, found as the product of the ordinate and abscissa on the length-tension diagram (Fig. 3-5), and heat released. Shortening heat, from equation 6, is illustrated in Fig. 3-16.

For many years Hill had been struck by the remarkable coincidence that the old coefficient of shortening heat (a) in equation 4 had the same magnitude as the constant (a) in his force-velocity equation 1, i.e., both equaled 0.25. From equation 5, α/P_o can vary only from 0.16 to 0.34, and α/P_o = 0.25 only when P/P_o = 0.5.

Woledge[22] has a nice way of showing the relation between the force-velocity and heat equations to demonstrate that the coefficient of shortening (a) can be the same as the force constant in the force-velocity equation.

In isometric tetanus the rate of heat production can be expressed simply as some function of muscle length (a) and of the duration (t) of tetanus:

$$\left(\frac{\partial Q}{\partial t}\right)_x = f(x,t) \qquad (7)$$

where the subscript x indicates that the derivative is at constant x or under isometric conditions.

If the muscle is stimulated tetanically and permitted to shorten, under a fixed load, it produces heat at a greater rate then it does isometrically. The increment in rate is empirically proportional to the velocity of shortening:

$$\left(\frac{\partial Q}{\partial t}\right)_P = f(x,t) + av \qquad (8)$$

where the subscript P indicates that the derivative is at constant P, or under isotonic conditions, a is the constant of proportionality to the velocity of shortening (v), where v = dx/dt.

Hill[70] found that an empirical expression for the total rate of change in heat plus work in an isotonic tetanus could be expressed by the term of f(x,t) plus a second term proportional to the load on the muscle:

$$\frac{dW}{dt} + \left(\frac{\partial Q}{\partial t}\right)_P = f(x,t) + b(P_o - P) \qquad (9)$$

where b is a constant, and f(x,t) is the isometric heat function, as in equation 7.

Increase in work per unit time (dW/dt) is power. Since external work done by muscle is P dx:

$$\frac{dW}{dt} = P\frac{dx}{dt} = Pv \qquad (10)$$

Combination of equations 8, 9, and 10 leads to Hill's familiar parabolic force-velocity curve:

$$(P + a)\ (v + b) = (P_o + a)b \qquad (11)$$

Thus the mechanical constant (a) appears from the constant of proportionality for shortening heat.

The notion that there is any shortening heat has been challenged. Aubert[1] suggested that the chemical mechanism that produces shortening heat must be the same phenomenon that maintains activation; therefore separation of initial heat into shortening heat and maintenance (or activation) heat is purely arbitrary. Aubert and Lebacq[29] did, however, clearly demonstrate a net thermal effect of shortening, most of which persists through relaxation, and they calculated an average coefficient of shortening heat (α/P_o) of 0.22. Carlson et al.[39] failed to find evidence of shortening heat, but it is now claimed that the conditions of their experiments were such that they missed it, in that it was a small part of a large total heat. If a muscle, following contraction, is stretched back to its initial resting position by a load or weight hanging on it, then the weight performs work on the muscle during relaxation that is exactly equal to the work of opposite sign done by the muscle in lifting the weight during contraction. The work done by the weight during relaxation appears as heat. The total heat generated in such an experiment is therefore the sum of maintenance heat, shortening heat, and work and can be so large that the contribution of shortening heat may fall within experimental variation. If the load is removed from the muscle during relaxation, and if heat is measured early so that the contribution of maintenance heat is reduced, then shortening heat is observed. Whether or not this heat is precisely proportional to shortening can be debated, but it does seem to exist. Therefore we define *initial* heat as the sum of activation or maintenance heat and shortening heat. Neither the time course nor the magnitude of initial heat is affected by oxygen lack.

When the stimulus ceases and a muscle relaxes, heat is produced in excess of resting heat if and only if the load on the muscle is allowed to stretch the muscle. This is *relaxation* heat. It equals the work done by the load in stretching the muscle and can be prevented if the load is removed before the muscle relaxes.

Not to be confused with relaxation heat is a long and slow rate of heat production that follows a return to the relaxed state. It is assumed that this heat is a manifestation of processes of all sorts that are restoring the muscle to the state in which it was before contraction. These processes include ion pumping, regeneration of ATP, restoration of substrate concentrations, etc. The exact time course of these processes is not known. There is reason to suspect that at least some of them begin almost immediately after the onset of contraction and that recovery is accomplished by a continuing family of processes. In any event, this heat is called *recovery* heat or sometimes *delayed* heat; it is defined operationally as the heat that does not appear when muscles are made to contract in the presence of the glycolytic poison iodoacetic acid and in the absence of oxygen. Most, if not all, of the recovery heat is due to oxidative processes and is absent in muscles contracting in a nitrogen atmosphere. There are reports of a small recovery heat, amounting to only 5 to 10% of total recovery heat, in muscles contracting in nitrogen but not poisoned by iodoacetic acid. How real this quantity is is uncertain. However, some have used these observations to separate recovery heat into two portions, aerobic and anaerobic, with nearly all of it being aerobic. It happens, perhaps only by chance, that recovery heat under favorable conditions is approximately equal to the initial heat of isometric contractions. With repeated contractions, in vitro, recovery heat decreases, presumably as the dying muscle is less successful in maintaining and restoring its normal internal chemical environment.

Resting heat provides the baseline upon which all other heat is measured. It is the summation of all chemical reactions in resting muscle, and these are predominantly exothermic. Although most reactions of intermediary metabolism are reversible, the final aerobic step in which oxygen is consumed is not. When muscle is deprived of oxygen, a condition it can endure temporarily, its resting heat production decreases. It is therefore probable that the energy change during anaerobic metabolism is less than that during aerobic metabolism at rest. Resting heat production is not invariant for any given muscle even when it is amply supplied with oxygen. Under certain circumstances, for example, it may increase with stretch[54] and with an increase in external potassium concentration.

Thermodynamics as related to muscle: energy, work, heat, enthalpy, entropy, and efficiency

The literature on this subject can be confusing. There are different notations for the same thing. Some writers have not rigorously used the words and ideas of thermodynamics. Some writers have made measurements of quantities important in thermodynamic analysis of muscle under one set of conditions, and then used the data in equations that applied to another set of conditions. Some writers have made usually implicit assumptions that have little basis in the reality of muscle contraction in the body under normal circumstances. It is the purpose of this section to provide a brief background that will introduce the subject and may resolve or prevent some confusion.*

The approach will be to examine a simple, controllable thermodynamic system, to derive some elementary relationships, and to determine the extent to which these can be applied to muscle.

A thermodynamic system can be classified according to its relation to its environment. A thermodynamic system has walls or boundaries. If no energy can be exchanged across the walls between the system and its surroundings, the system (and the processes within the system) is adiabatic, or isolated. If matter cannot be exchanged but energy can, the energy so exchanged is called heat and the system is diathermic, or closed. If both matter and heat can be exchanged, the system is open. (In all three of these cases there can be another kind of change in energy in the system produced by a change in the size of the

*For further study, the work by Katchalsky and Curran[12] is recommended.

system. Such a change is called mechanical work, which is defined more completely later.)

A thermodynamic system is at any moment in a definite state or condition that must be specified in detail in order to describe it fully. The state is determined by two classes of parameters. Extensive parameters (e.g., volume, length, mass, number of moles of a chemical substance, quantity of electricity) describe the size of the whole system. Intensive parameters (e.g., pressure, force, concentration, electric potential) can be different at different arbitrarily small regions within the system, do not depend on the size of the system, and really describe its internal structure and organization.

Work is done only when at least one extensive parameter of the system is changed; i.e., there is no work unless some measure of the size of the system is altered. A change in only an intensive parameter may alter the system's capability of (or potential for) work, but it does not create work. The change in an intensive parameter is necessarily a change in the internal structure of the system. A change in any parameter, whether extensive or intensive, is a change in the state of the system, such that the total energy content of the system is altered. This alteration appears as work if the change is in the size of the system, or it appears as heat or as a change in entropy if the change is only in the internal structure of the system. These statements constitute the first law of thermodynamics. A thermodynamic system is considered to be at equilibrium in a specific state. We produce a small change in one or more parameters. The change in energy content of the system is:

$$dE = dQ - dW \tag{12}$$

where Q is heat and W is work. Work done by the system and gained by the outside world is positive.

If we sum a number of such arbitrarily small steps as the system changes from state 1 to state 2, we have a total energy change:

$$\int_{E_1}^{E_2} dE = E_2 - E_1 = \int_1^2 dQ - \int_1^2 dW = Q_{12} - W_{12} \tag{13}$$

A shorthand notation for equation 13 is the specification that in the new steady state 2:

$$E = Q - W$$

This shorthand notation, which seems to substitute absolute values for differences, can lead to both conceptual and practical problems and should be avoided. There is no work (W) in an equilibrium system. There can only be increments of work as something happens to change extensive parameters of the system. Furthermore, the notion of an absolute value of Q leads to difficulties that are avoided if we think only of dQ, the transfer of energy, but not matter, between the system and its surroundings.

Work. Let us define work quantitatively. First, in the most general terms, we designate the *i*th extensive parameter of the system by a_i. We cause it to change by some arbitrary amount (da_i). The force we, the external world, have to apply to the system to produce this change is X_i. The element of work we perform on the system is the product of the force and the change in size, or:

$$dW_i = X_i da_i$$

If there are m extensive parameters, then the quality of work performed is the sum of m such contributions, or:

$$dW = \sum_{i=1}^{m} dW_i = \sum_{i=1}^{m} X_i da_i \tag{14}$$

We now specify the forces (or potentials) and the extensive parameters. The list is as long as we can make it.

$$dw = Pr \cdot dV - P \cdot dx - \psi de - \sum_{i=1}^{m} \mu_i dn_i + \ldots \tag{15}$$

where Pr is pressure, V is volume, P is force, x is distance or length, ψ is electric potential, e is quantity of electricity, μ_i is chemical potential of the *i*th kind of chemical substance, and n_i is the total number of moles of the *i*th kind of chemical substance in the system.

We speak of pressure-volume work, meaning $Pr \cdot dV$, or of force-distance work, meaning Pdx, or of electrical work or of chemical work. We lump pressure-volume and force-distance work into the term "mechanical work." Gibbs called $(dW - Pr \cdot dV) = dW'$ the useful work of the system, possibly meaning that for a steam engine, it was the work obtained by harnessing the pressure-induced volume changes to mechanical devices.

We get into trouble if we think of work as, for example, pressure-volume work, and write:

$$W = PrV$$

and then differentiate totally to obtain:

$$dW = VdPr + PrdV$$

because the VdPr term is not a work term. No work is performed if the size of the system does not change. This does not mean that VdPr cannot exist. It can. But it is a change in the internal structure of the system and appears as a change in heat or, as we shall see, as a change in entropy.

Entropy. The second law of thermodynamics defines the entropy for reversible processes as:

$$dS = dQ/T \tag{16}$$

An element of entropy, dS or increase in entropy, is the heat absorbed per degree of temperature. If no work is done, the first law of thermodynamics states:

$$dE = dQ, \text{ for } dW = 0$$

and the second law then tells us:

$$dE = TdS, \text{ for } dW = 0$$

or the increase in total internal energy of the system, if no work is done, is the temperature multiplied by the increase in entropy.

Entropy is an extensive parameter like total energy and includes in dS all changes in intensive parameters. The equation dQ = TdS is true only for reversible processes. For irreversible processes, and these include those of muscle contraction, the second law requires that $dS > dQ/T$, which we will now show. We will say more about entropy, in the course of which there will be other operational definitions of it.

Irreversible processes. Start with a thermodynamic system in equilibrium with its environment and in a steady state with respect to disposition of internal forces; i.e., the energy and entropy of the system are constant. Call this state 1. We now produce or permit a displacement of the system. There follows a time-dependent change to a new state, state 2, in which the energy and entropy of the system are constant, independent of time. The details of the change between states constitute a thermodynamic process.

If we can produce a small increase in heat, for example, such that the system performs a small amount of work in order to obtain a small change in energy ($dE = dQ - dW$), the process is reversible if either we perform the same small amount of work (but of opposite sign) on the system (–dW) or we permit the system to lose the same amount of heat (–dQ). In either case the resulting energy change is equal and opposite (–dE) and restores the system entirely back to state 1, including, of course, the original entropy. If we cannot reverse the displacement back to the energy and entropy of state 1 by reversing either dQ or dW, the process is irreversible. In order to restore a system displaced by an irreversible process, we have to perform more work on the system than the system can perform on its environment.

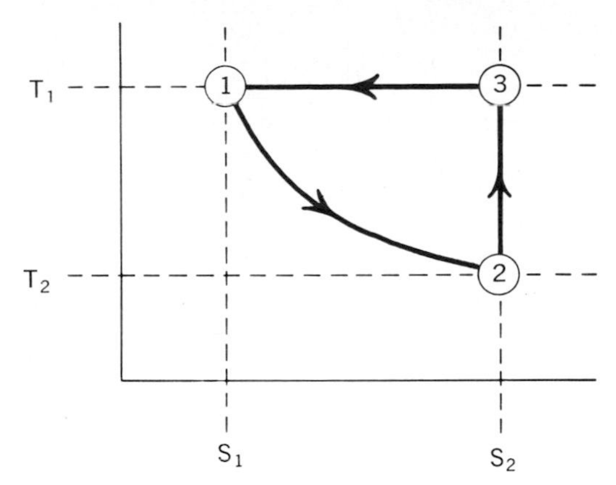

Fig. 3-17. Cyclic process containing irreversible step. See text for description of process. Horizontal axis is entropy, *S*. Vertical axis is temperature, *T*. Thermodynamic system is defined by its state. It passes from initial state, *1*, irreversibly to state *2*. It then passes reversibly from state *2* to *3*, and from state *3* back to initial state.

A helpful example is given by Kirkwood and Oppenheim, adapted by Katchalsky and Curran.[12]

Fig. 3-17 is a temperature-entropy diagram. The initial state (state 1) of the system is characterized by temperature (T_1) and entropy (S_1). If we carried out a reversible adiabatic change, the entropy remains constant (because dQ = 0 since no heat is exchanged), but temperature changes along a perpendicular through the point S_1, T_1. Work done on such a system raises the temperature; work done by the system lowers the temperature. But we do not carry out a reversible adiabatic change. We let the system change irreversibly but adiabatically, and let it do work to reach state 2 at temperature T_2 and entropy S_2. Step 1, then, is:

$$E_2 - E_1 = \int_{E_1}^{E_2} dE = Q_{12} - W_{12} = -W_{12}$$

where:

$$W_{12} > 0$$

and

$$Q_{12} = 0$$

In step 2 we perform work on the system to bring it reversibly at constant entropy back to its original temperature. The system is now in state 3.

$$E_3 - E_2 = \int_{E_2}^{E_3} dE = Q_{23} - W_{23} = -W_{23}$$

where:

$$W_{23} < 0$$

and

$$Q_{23} = 0$$

In step 3 we allow the system to take up heat from a reservoir at constant temperature, and so to bring the system back to initial state 1. In this final step:

$$E_1 - E_3 = \int_{E_3}^{E_1} dE = Q_{31} - W_{31} = Q_{31}$$

where:

$$Q_{31} < 0$$

and

$$W_{31} = 0$$

Now the cycle is complete and we sum the contributions of the steps.

$$\oint dE = \int_{E_1}^{E_2} dE + \int_{E_2}^{E_3} dE + \int_{E_3}^{E_1} dE =$$

$$E_2 - E_1 + E_3 - E_2 + E_1 - E_3 = 0$$

$$\oint Q = \int_3^1 dQ = Q_{31}$$

$$\oint W = \int_1^2 dW + \int_2^3 dW = W_{12} + W_{23}$$

or

$$Q_{31} = W_{12} + W_{23}$$

The step from state 3 to state 1 was a reversible step so that the definition $dQ = TdS$ applies, and since it was isothermal:

$$Q_{31} = T_1 \int_{S_3}^{S_1} dS = T_1(S_1 - S_3)$$

But since $S_2 = S_3$:

$$Q_{31} = T_1(S_1 - S_2)$$

The main question we want answered from this analysis is whether we can determine a relation between S_2 and S_1. In a change of state by an irreversible process, is there necessarily a change in entropy and, if so, is the nature of that change determined?

To answer this question we turn to a formulation of the second law of thermodynamics. It is impossible to devise an engine operating in a cycle that produces positive work as a result of cooling a single reservoir. In a reversible process in which the system works on its environment at constant temperature:

$$\int_1^2 dW_T = W_{12}$$

which exactly equals the work required to bring the system isothermally back to its starting point:

$$\int_2^1 dW_T = W_{21} = -W_{12}, \text{ reversible}$$

Therefore:

$$\oint dW_T = 0, \text{ reversible}$$

But for an irreversible process:

$$\left|\int_2^1 dW_T\right| = \left|\int_2^n dW_T + \int_n^1 dW_T\right| > \left|W_{12}\right|$$

and

$$\int_2^1 dW_T < 0$$

Therefore:

$$\oint dW_T < 0, \text{ irreversible}$$

Thus we have to put more work into a system than we obtain from it if its contribution is by an irreversible process.

The applicability of this statement of the second law to our model is that in the cycle shown in Fig. 3-17 there was only one nonadiabatic step, and in that step the energy of the system was enriched at the expense of the single isothermal reservoir. Therefore:

$$\oint dW_T \leq 0$$

Hence:

$$Q_{31} = \oint dQ \leq 0$$

Since $T_1 > 0$:

$$S_1 - S_2 = Q_{31}/T_1 \leq 0$$

or

$$0 < S_1 \leq S_2$$

The equality holds if $\oint dW_T = 0$, which it does only for a reversible process. Since we deal with an irreversible process, this analysis shows that in a change from state 1 to state 2, with change in entropy from S_1 to S_2 by an irreversible process, there is always a gain in entropy.

We have dealt with state functions only at the completion of transition steps E_1, E_2, S_1, S_2, etc. We now measure entropy at any stage of an irreversible process on the assumption that entropy is determined only by the set of local parameters that fully characterize the details of the system at any moment. Let a_k be an extensive parameter and ξ_k an intensive parameter. We can reach any given distribution of a's and ξ's by a number of routes, both reversible and irreversible. It does not matter how we got there; the entropy of the system at that moment is the same. But the energy expenditure in reaching that state will be different.

In a nonequilibrium system there may be, at a moment in time, a certain entropy (S) defined by a's and ξ's, with total internal energy E. If we wished to transfer this nonequilibrium system to an equilibrium system, we would have to do something to keep the particular momentary nonequilibrium distribution of intensive parameters (hence, entropy) from changing. This act of ours must constitute setting up forces with potential energy $-\sum_{k=1}^{m} \xi_k X_k$. The total energy of this equilibrium system is then:

$$E^* = E - \sum_{k=1}^{m} \xi_k X_k$$

To maintain this distribution of intensive parameters there must be work, $\sum_{k=1}^{m} \xi_k dX_k$. Therefore the work element required to reach reversibly the given configuration (dW^*) exceeds that required to reach the same configuration irreversibly (dW):

$$dW^* = dW + \sum_{k=1}^{m} \xi_k dX_k$$

For the reversible process the combined first and second laws of thermodynamics yield:

$$TdS = dE^* + dW^*$$

Differentiation of the equation for E* yields:

$$dE^* = dE - \sum \xi_k dX_k - \sum X_k d\xi_k$$

From equations for TdS, dE*, and dW*:

$$TdS = dE + dW - \sum_{k=1}^{m} X_k d\xi_k$$

According to the first law of thermodynamics, $dQ = dE + dW$. Therefore:

$$TdS = dQ - \sum_{k=1}^{n} X_k d\xi_k$$

The change in entropy during an irreversible process differs from the heat transfer by a function of the intensive and extensive parameters that can be specified. In an adiabatic process, $dQ = 0$, and there is a change in entropy despite the absence of heat transfer.

With respect to muscle contraction, we deal with isothermic processes. It is helpful in such processes to consider that the change in entropy can be expressed as the sum of entropy created in the system (d_iS) and entropy exchanged with the environment (d_eS), or:

$$dS = d_iS + d_eS \qquad (17)$$

where, by definition:

$$d_eS = dQ/T \qquad (18)$$

the entropy of reversible processes. For irreversible processes we have seen the following:

$$dS > dQ/T$$

Therefore:

$$d_iS \geq 0 \qquad (19)$$

or the production of internal entropy by the system is positive for an irreversible process and zero for one that is reversible. This creation of internal entropy can occur, for example, because of heat flow from a region of higher to a region of lower temperature within the system, because of the flow of a molecular species from a region of higher to a region of lower chemical potential within the system, etc .

Consider an isothermal system with diathermal walls. Let the system change irreversibly from state 1 to state 2 in contact with a constant-temperature reservoir. During this step there is transfer of a quantity of heat (ΔQ_{irr}), the heat exchange of the irreversible step. The system returns to state 1 from state 2 in a manner that is reversible and still isothermal. For this cycle:

$$\oint dE = 0 = \oint dQ - \oint dW$$

According to the second law, since work is the result of the cooling of only a single body:

$$\oint dW = W \leq 0$$

Therefore:

$$\oint dQ = \int_1^2 dQ + \int_2^1 dQ \leq 0$$

$\int_1^2 dQ$ is the heat absorbed from the reservoir in the irreversible process ΔQ_{irr}. But the reservoir operates reversibly, so that from its state:

$$dS = dQ/T$$

for the reservoir. Or

$$\Delta_e S = \Delta Q_{irr}/T$$

where $\Delta_e S$ is the entropy exchanged between the system and the reservoir. For the reversible process, state 2 to state 1:

$$\int_2^1 dQ = T\int_2^1 dS = T(S_1 - S_2) = -T\Delta S$$

Therefore:

$$\oint dQ = T(\Delta_e S - \Delta S) \leq 0$$

If both steps had been reversible, $\Delta S = \Delta_e S$. But because of the presence of the irreversible process, $\Delta S > \Delta_e S$. We introduce the quantity:

$$\Delta_i S = \Delta S - \Delta_e S \geq 0 \tag{20}$$

where the positive value applies to an irreversible process. This is the entropy created by reactions occurring inside the system, and it is the presence of this positive value that characterizes a process as irreversible.

Enthalpy. Enthalpy, an extensive parameter, means heat content (denoted by H), and is defined as the internal energy of the system plus pressure-volume work, or

$$H = E + PrV$$

Since $E = Q - W$:

$$H = Q - (W - PrV) = Q - W'$$

where W' is the useful work of a steam engine. Since muscles shorten without changing volume, they do no pressure-volume work, and enthalpy and internal energy are the same. It is not clear why muscle physiologists should prefer the term "enthalpy" to energy, except that enthalpy is useful in quantitative considerations of the energetics of certain chemical reactions that may be involved in muscle contraction. I prefer to avoid the word "enthalpy" and to speak of heat exchange (dQ) between a muscle and its environment and of the entropy of the system rather than its heat to avoid confusion between Q and H. The heat (Q) and the heat content or enthalpy (H) are the same only when there is no work of any kind.

Free energy. Free energy is an extensive parameter. Helmholtz free energy is defined as:

$$F = E - TS$$

and Gibbs free energy is defined as:

$$G = E - TS + PrV$$

Again, since muscle does not change in volume, there is no pressure-volume work. At constant temperature and volume, therefore:

$$dF = dE - TdS \tag{21}$$

$$dG = dE - TdS + VdPr \tag{22}$$

Substitution of $dE = dQ - dW$ yields:

$$dG = (dQ - TdS) - (dW - VdPr)$$

If we deal not with muscle but with a reversible system and remove the restriction that $dV = 0$, then $dQ = TdS$, and

$$dG = -dW + d(PrV)$$

Substitute for dW all possible forms of work:

$$dG = VdPr + Pdx + \sum_{i=1}^{m} \mu_i dn_i$$

At constant pressure and length and at constant chemical content of all species except one, we add one unit (dn_i) of the *i*th chemical species. The internal energy of the system is changed accordingly and the Gibbs free energy change is:

$$\partial G = \mu_i \partial n_i$$

or

$$\mu_i = \left(\frac{\partial G}{\partial n_i}\right)_{P,T,x,n_j} \tag{23}$$

which defines the chemical potential. Gibbs

free energy is useful in that it states the maximum work that can be produced.

Gibbs free energy is related to enthalpy in an isothermic reaction by the following:

$$dG = dH - TdS$$

Applications to muscle. These considerations are very important in muscle physiology. Activation of muscle for contraction is an irreversible process in which entropy is created, e.g., by Ca^{++} diffusion. There is an increase in heat, which is transferred to the environment. If the muscle is permitted to shorten, mechanical work is done. There is exchange of material with the environment; chemical work is done. Muscle volume is constant during contraction ($dV = 0$); no pressure-volume work is done. There is electrical work, but it is negligible if one can use the results of studies of nerve in which large numbers of action potentials are required to produce detectable heat flow. Therefore, for muscle:

$$dE = dQ - dW$$

$$= dQ + Pdx + \sum_{i=1}^{m} \mu_i dn_i \qquad (24)$$

or the change in energy is expressed entirely as the sum of heat transfer, the force-distance (or mechanical) work, and chemical work.

Chemical work (including gas exchange; uptake of substrates, e.g., glucose and fatty acids; output of metabolic products; translocation of electrolytes; and all other exchanges of matter between muscle and its environment) proceeds during the resting state. If muscle is in a quasi-steady state at rest, then over some period of time:

$$\oint dE = 0 \text{ at rest}$$

and

$$\oint dQ = -\sum \oint \mu_i dn_i \text{ at rest} \qquad (25)$$

or resting heat transfer can be entirely accounted for by chemical work.

When a muscle contracts, there is an increase in heat transfer to the outside. There must also be an increase in chemical work, perhaps more of the same type of work as well as additional types of chemical work. But beyond the heat transfer, which represents external exchange of entropy, there is entropy creation. For the change from rest to contraction:

$$\int_1^2 dE = E_2 - E_1 = T\Delta_e S + P \cdot \Delta x + \sum \int_1^2 \mu_i dn_i \qquad (26)$$

If contraction is isotonic at constant load, then P is constant, and if $\Delta x < 0$, the muscle performs work. If the load remains on the muscle during relaxation, the load stretches the muscle back to its initial length, and the load performs work on the muscle that is exactly equal, and opposite in sign, to that performed by the muscle, or

$$\oint Pdx = 0$$

If the original state is restored, then $\oint dE = 0$, but so also does $\oint dS = 0$, for restoration of the original configuration implies exactly that. Therefore:

$$\oint dQ = -\sum \oint \mu_i dn_i \qquad (27)$$

over the whole contraction cycle, just as at rest.

Let us now consider a contraction cycle, step by step (Fig. 3-18). Consider a cycle of

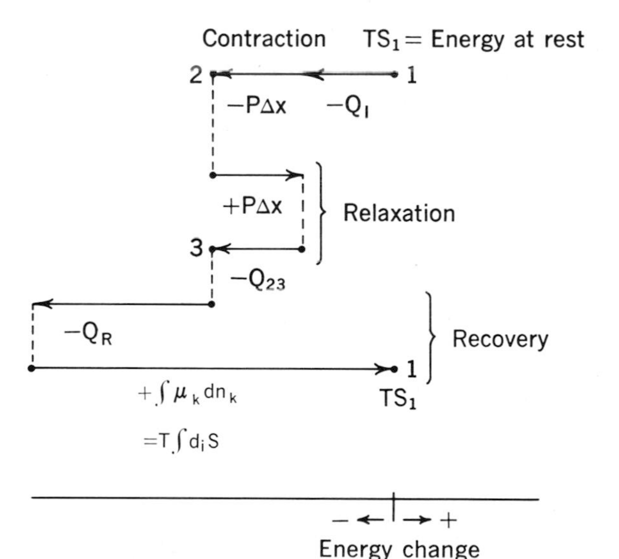

Fig. 3-18. Energy changes during complete cycle of muscle contraction, relaxation, and recovery. Horizontal axis is energy change from initial state. Arrows are drawn to scale. Transition from state *1* to *2* is contraction, from *2* to *3* is relaxation, and from *3* back to *1* is recovery, completing the cycle. There is no vertical scale; vertical displacement is only for convenience in demonstrating sequential and continuous changes in energy. See text for notation and further explanation.

rest → isotonic contraction → rest. A weight (P) is hung on a muscle in a constant-temperature environment. Rest is state 1, with energy defined by:

$$E_1 = TS_1$$

The muscle is stimulated to contract to thermodynamic state 2. The cumulative change in energy in the muscle in transition from state 1 to state 2 is, from the first law of thermodynamics:

$$E_{12} = \int_1^2 dE = Q_{12} - W_{12} \qquad (28)$$

where Q_{12} is the initial heat (Q_I) and W_{12} is the algebraic sum of mechanical work and chemical work. We know from experiment that $Q_I < 0$. The mechanical work is $-P \cdot \Delta x$, where $\Delta x < 0$ because the muscle shortens. Therefore the mechanical work $-P \cdot \Delta x > 0$, because it increases the energy of the surroundings. Chemical work is $-\sum \int_1^2 \mu_i dn_i$. To date, it is not known whether this integral is positive or negative.

Entropy changes in the transition from rest to contraction. First, there is an internal change in entropy ($\Delta_i S_I$) due to altered internal configuration of muscle. We know that it is a necessary characteristic of irreversible processes that $\Delta_i S > 0$. Second, there is an external change in entropy ($\Delta_e S_I$) equal to $(1/T)$ times the transfer of heat (Q_I) to the outside. Therefore $\Delta_e S_I < 0$. The overall change in entropy in the transition to contraction is:

$$S_{12} = \int_1^2 dS = \Delta_i S_I + \Delta_e S_I = \Delta_i S_I + (Q_I/T)$$

One term is positive and one is negative. We cannot tell from these considerations alone whether or not $S_{12} \geq 0$ or $S_{12} < 0$.

A transition from state 2 to state 3 is defined as relaxation. During that transition the weight stretches the muscle back to its rest length, and heat is dissipated from muscle. In general this relaxation heat is nearly the same as the heat equivalent of the work done, but this equality is not essential to the present discussion. The change in energy over the relaxation portion of the cycle is:

$$E_{23} = Q_{23} - P \cdot \Delta x \qquad (29)$$

where we retain the convention that $\Delta x < 0$, so that $-P \cdot \Delta x$ is a gain in energy by muscle. $Q_{23} < 0$ is relaxation heat. The entropy change in the transition from state 2 to state 3 is:

$$S_{23} = \Delta_i S_2 + (Q_{23}/T)$$

Recovery is the transition from state 3 back to rest state 1, with restoration of energy and internal configuration and hence restoration of entropy:

$$E_{31} = Q_{31} - W_{31} \qquad (30)$$

where:

$$Q_{31} = Q_R \text{ (recovery heat)}$$

$$W_{31} = -\sum \int_3^1 \mu_i dn_i$$

the net chemical work performed on muscle by its surroundings during recovery.

We sum all the energy changes, $E_{12} + E_{23} + E_{31} = \oint dE$, to obtain $E_2 - E_1 + E_3 - E_2 + E_1 - E_3 = 0$. From equations 28, 29, and 30:

$$\oint dE = Q_1 + Q_{23} + Q_R + \sum \oint \mu_i dn_i = 0 \qquad (31)$$

Therefore the total chemical work performed by the environment on muscle over a contraction cycle equals the sum of all heat produced by muscle in excess of resting values.

Entropy changes also in the transition from contraction back to rest. As before, this entropy change can be divided into internal and external changes, where the external change equals the heat transferred:

$$S_{31} = \int_3^1 dS = \Delta_i S_R + \Delta_e S_R = \Delta_i S_R + (Q_R/T)$$

The sum of all entropy changes is:

$$\oint dS = S_{12} + S_{23} + S_{31} = \frac{1}{T} \oint dQ + \oint d_i S = 0 \qquad (32)$$

where $\oint d_i S$ is the sum of all internal changes in entropy.

Equations 31 and 32 confirm that the chemical work performed on muscle around the contraction cycle equals the total internal change in entropy.

It is a common observation that, under what appears to be the most favorable of the conditions in which heat production by excised muscle has been measured, recovery heat approximately equals in magnitude the sum of initial heat and mechanical work.

This empirical observation is not implicit in the thermodynamic model we have been considering. If the observation is true, it is a consequence of detailed mechanisms and not of the broad, general relations given here.

It is also interesting to make explicit the relation of the quantity of mechanical work performed by muscle to other energy changes. From equation 28, the mechanical work is:

$$
\begin{aligned}
P \cdot \Delta x &= - E_{12} + Q_I + \sum \int_1^2 \mu_i dn_i \\
&= - T(S_2 - S_1) + Q_I + \sum \int_1^2 \mu_i dn_i \\
&= - T\Delta_i S_I + \sum \int_1^2 \mu_i dn_i \qquad (33)
\end{aligned}
$$

Mechanical work performed by muscle is the difference between the internal change in entropy during activation and contraction and the chemical work done on the muscle during this period. The latter may be relatively small. Nearly all the additional chemical work of the contraction cycle may be performed during recovery. If so, then mechanical work and internal change in entropy are matched. Since mechanical work is a function of muscle length and applied force, the tension the muscle must develop to lift that force is obtained from the internal change in entropy, determined by the preset length and applied force.

Now consider a cycle in which a muscle moves from rest to isometric contraction and back to rest, with total restoration of the initial thermodynamic state. One important difference is that instead of dissipating its energy by performing work on the environment as in isotonic contraction, the muscle changes its potential energy in steps of xdP and dissipates this energy as heat. Restoration of energy to muscle is accomplished entirely by chemical work by the environment.

A second important difference has to do with initial heat. It is a matter of experience that initial heat in preloaded isotonic contraction exceeds that of isometric contraction at the same final tension and of the same duration by a term proportional to the distance shortened:

$$(Q_I)_x = A$$

$$(Q_I)_P = A + ax$$

where A is activation or maintenance heat, depending on whether a twitch or a tetanus is being dealt with. The gross thermodynamic model does not predict this difference in initial heat. The contribution of shortening heat needs explanation in the details of the system.

There has also been interest in the flow of energy or the rate of energy dissipation:

$$dE/dt = \dot{E} = \dot{Q} - \dot{W} \qquad (34)$$

$\dot{W}$, or dw/dt, is the rate at which work is performed. For mechanical work, this rate is called power. At a constant force mechanical work is:

$$dW = Pdx$$

and power is:

$$dW/dt = Pdx/dt = Pv \qquad (35)$$

where v is velocity of shortening. $\dot{W}$, then, is the product of the ordinate and abscissa on the force-velocity diagram. Fig. 3-19 is a plot of the rate of mechanical work against a load normalized to P_0. Work rate is normalized for individual muscles by dividing the observed Pv by bP_0, where b is the velocity constant in Hill's equation 1, or normalized power is $(P/P_0) \cdot (v/b)$. Also in Fig. 3-19 a value $\Delta\dot{E}$ is plotted against a normalized load. $\Delta\dot{E}$ is the energy flow associated with mechanical work in excess of that in an isometric contraction. From equation 6, this is approximately:

$$\Delta\dot{E} = 0.16\ P_0 v + 1.18\ Pv$$

Efficiency. It is understandable that muscle physiologists who deal with mechanical work should have been concerned with the efficiency with which mechanical work is performed. In engineering one builds a machine in which an energy source is coupled to a mechanical device that performs work. The designer and user are interested in the cost of the work obtained from the machine, so that cost is expressed in terms of mechanical work per unit energy input into the machine, and energy input is equated with the heat developed by the boiler. This cost per unit energy expenditure is called the efficiency. If the efficiency strikes the designer as being too small, he redesigns the machine to make it more efficient or less costly per unit mechanical work.

By analogy with such definitions of efficiency, muscle physiologists calculated the cost of mechanical work by muscle per unit energy expenditure. It seemed plausible, on the basis of what was known at the time, to imagine that the total energy dissipated by

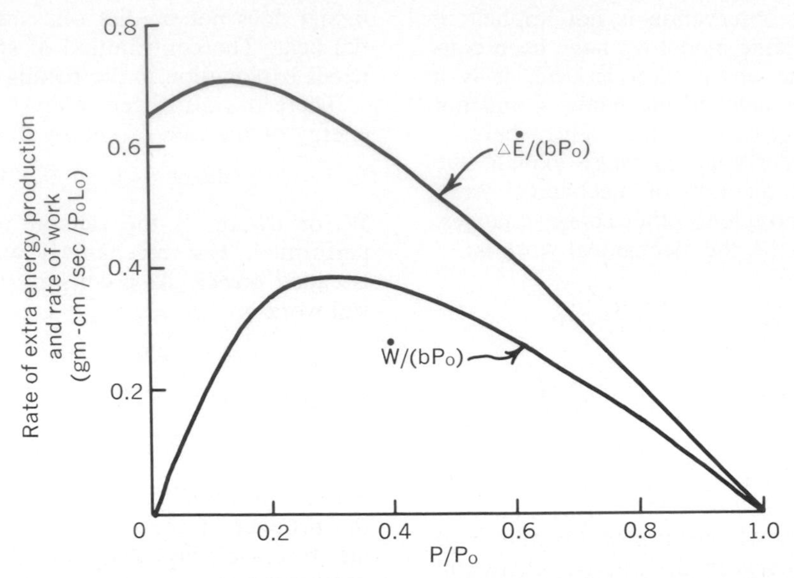

Fig. 3-19. Rate of work and rate of total energy produced in isotonic contraction in excess of that produced in isometric contraction. Rate of extra energy production, $\Delta\dot{E}$, is sum of rate of work, $\dot{W}$, and rate of production of shortening heat plotted against relative load, P/P_o. Rates are given per bP_o, where b is velocity constant in Hill's force-velocity equation. For frog sartorius at 0° C, b is approximately 0.3 L_o/sec, L_o is 3.8 cm, and P_o is 70 gm. Muscle weighs 0.165 gm. Maximum $\Delta\dot{E}$ is 56.6 gm-cm/sec or 343 gm-cm/sec/gm of muscle.

the contracting muscle machine was given by the sum of initial heat and mechanical work and that the energy input necessary to restore the muscle machine to its resting state of readiness was exactly equal to the energy dissipated by the machine as initial heat plus mechanical work. Therefore an early definition of efficiency of muscle work was:

$$\text{Efficiency} = P \cdot \Delta x/(Q_I + P \cdot \Delta x) = P \cdot \Delta x/\Delta H \quad (36)$$

where ΔH was the change in enthalpy.

However, we now have reason to believe that $Q_I + P \cdot \Delta x$ does not express the total energy cost of the mechanical work. Muscle is a chemical machine; there is mechanochemical coupling. The cost of restoring the original state is given by the sum of chemical work performed on muscle, or $\sum_k \oint \mu_k dn_k$. We have seen that the contribution of the *k*th molecular species to a change in Gibbs free energy of the system $(\partial G)_{nj,T,x,V}$ is $\mu_k \partial n_k$. Therefore the total chemical work can be expressed as the sum of all such contributions:

$$dG = \sum_k \mu_k dn_k$$

and we define:

$$\Delta G = \sum_k \oint \mu_k dn_k$$

The efficiency of the mechano-chemical muscle machine, at constant T and V, can then be expressed by:

$$\text{Efficiency} = P \cdot \Delta x/\Delta G \quad (37)$$

But the sum of the chemical work done on the muscle over a contraction cycle is equal, and opposite in sign, to the sum of all heat dissipated by muscle over the contraction cycle, initial heat plus relaxation heat plus recovery heat. Thus even though all the individual reactions and all the changes in chemical potential of individual molecular species are not known, if the contraction cycle is truly successful in restoration of muscle energy and entropy, then the energy cost of mechanical work by muscle can be calculated as the sum of all heat dissipated throughout a contraction cycle, or:

$$\text{Efficiency} = P \cdot \Delta x/\oint dQ \quad (38)$$

This calculation of efficiency gives a lower value than that calculated from the change in enthalpy because $(Q_I + P \cdot \Delta x) < \oint dQ$, not necessarily in theory but in fact. If we equate relaxation heat to mechanical work, $\oint dQ = Q_I + P \cdot \Delta x + Q_R$, and the difference between the calculations of cost is given by Q_R.

Chemical source of initial heat and work

Opinions regarding the chemical reactions thought to provide the energy source for contraction have gone in and out of favor. We have already seen that myosin ATPase is activated by association with actin and that ATP is probably split during some portion of the contraction cycle.

There is now no doubt that ATP is split during some phase of contraction. The fact had been hard to prove because under normal circumstances the ATP that is split is regenerated rapidly from phosphocreatine. The following reactions occur:

1. $ATP \rightarrow ADP + P_i$
2. $PC + ADP \rightarrow ATC + C$

where P_i is inorganic phosphate, C is creatine, and PC is phosphocreatine. The net reaction is:

3. $PC \rightarrow C + P_i$

So long as there is a large reservoir of phosphocreatine, ATP is regenerated. Under in vitro conditions, in which muscle is not able to regenerate PC at a normal rate, prolonged series of contractions lead ultimately to a measurable decrease in PC, with an approximately stoichiometric increase in C and P_i (Carlson and Siger,[38] Mommaerts et al.,[89] Davies et al.,[44] and Kushmerick et al.[86, 87]). Convincing support comes from experiments by Davies et al.[44] in which creatine phosphoryltransferase was inhibited by 2,4-dinitrofluorobenzene so that reaction 2 above could not occur. Under these circumstances frog rectus abdominis muscle could contract only 3 or 4 times rather than 30 or 40 times at 0° C. There was a significant decrease in ATP content without a change in PC content in these poisoned muscles.

The amount of ATP split in a single twitch is too small to measure accurately and has therefore escaped detection. This fact makes it impossible with present techniques to discover the phase of the contraction cycle in which ATP is split. However, since it is known that ATP is split and that in model systems ATP splitting is closely related to the association of actin and myosin, i.e., to the earliest phase of contraction, it is difficult to avoid the suggestion that at least one portion of ATP is split during activation. Therefore this ATP splitting may be the source of activation heat. Of course, if the models are correct and cross bridges are made and broken cyclically and ATP is split repeatedly during this phase, the same ATPase and ATP splitting are the source of maintenance heat and shortening heat. But we have seen that muscle is not a constant-power machine and that consequently an additional energy source is needed to account for all the shortening heat, or for the Fenn effect. This may have been confirmed by measurement of decrease in ATP.

Measurements of changes in ATP and PC content and their correlation with heat measurements are difficult. There are uncertainties in the absolute value of heat measurements, partly because of difficulties in calibration but also because of the great variance among animals. There are also uncertainties in the chemical determinations, which involve not only analytic error but also the uncertainty that the process of stopping the reaction and extracting the ATP and CP has not led to their further and accidental hydrolysis. Finally, measurements of heat production and chemical changes are seldom done on the same muscle and are often performed under different conditions. To these experimental difficulties add the uncertainty that the heats produced by reactions 1 and 2 in vivo are not known with certainty because the heat produced depends on other reactions and concentrations of reactants associated with hydrolysis of ATP or PC, e.g., H^+ changes and buffer reactions.

If one divides the heat produced by the apparent number of moles of PC split using ordinary freezing techniques that require tens of seconds, one obtains a figure of from 8,000 to 12,000 calories/mole (Carlson et al.[39]; Wilkie[102]). The range of these values is large enough to accommodate probable values of heat of splitting of PC, but the uncertainty is so great that the results are not decisive. Gilbert et al.[64] used an ultrarapid freezing technique so that measurements of ATP plus PC splitting during the first half second of contractions could be made. Under these conditions, during the first 10 sec, if ΔH for phosphate splitting from ATP and PC were −11,000 calories/mole, only about one half to two thirds of the observed heat could be accounted for. If these results are technically correct, then there must be some other source of heat production.

Chemical sources of recovery heat

Splitting of some ATP must also occur during recovery. For example, pumping of Ca^{++} across walls of longitudinal SR into the lumen of SR occurs during recovery and is linked to an ATPase in the SR.

However, nearly all the recovery energy must be spent in resynthesis of the phosphorylated energy sources ATP and PC. In some circumstances, regeneration of these phosphorylated compounds is linked to anaeobic glycolysis. Some of the anaerobic recovery heat may be due to formation of hexose phosphate, but the overwhelming majority of resynthesis is linked to oxidative phosphorylation in intact muscle in situ.

There is a large body of data on chemical changes in excised frog sartorius muscle. If our interest is in how muscle functions in the body of man, we must be aware of two important differences between the behavior of excised frog sartorius muscle and mammalian muscle in situ.

In the first place, excised muscle has as its substrates only those stores that happen to be in the muscle plus whatever the investigator adds to the bathing solution. In the latter case, substrate supply is available to deep fibers only by diffusion from the bathing solution. In the body, however, muscle is perfused by blood, which normally can provide an almost unlimited supply of a variety of substrates plus many substances, some probably still unrecognized, that assist in regulating the translocation of substrates into muscle or in their metabolism. Furthermore, material is distributed by blood through diffusion from capillaries supplying every muscle fiber and across distances measured in microns rather than millimeters. As a result of these differences and undoubtedly of others not yet understood, excised muscle is not able to continue to regain the total chemical organization of the resting state with succeeding series of tetani, and it deteriorates.

The first, or first few, isometric tetani in response to trains of stimuli produce much higher tension than ever recorded subsequently. A plateau of responding tension is then maintained for a varying period, up to perhaps 30 tetani, followed by a progressive decrease in tension. At the same time the initial heat, which is entirely activation and maintenance heat, decreases approximately in proportion to the decrease in tension, so that the ratio of initial heat to force generated remains approximately constant. Recovery heat is at first essentially equal to initial heat, but as the muscle fails to generate the same tension, recovery heat becomes a decreasingly smaller proportion of total heat. One suspects that the decrease in recovery heat is due to the inability of the muscle to keep up the pace of resynthesis of PC and ATP, and that it is the decrease in amounts of these substances that leads to decreased production of force during contractions because decreased numbers of cross bridges are formed between actin and myosin.

All of this is very different from conditions in living muscle in the body, which lasts a lifetime.

The second important difference between frog sartorius muscle and the bulk of human skeletal muscle lies in structure and metabolism. Frog sartorius muscle is fast twitch, made of large, pale fibers with a relatively scant sprinkling of mitochondria, and hence is relatively less active in oxidative phosphorylation, laden with glycogen, and has a rather frugal blood capillary network. Mammalian skeletal muscles have both fast and slow twitch fibers. Slow twitch fibers are smaller, rich in myoglobin, denser in mitochondria, and hence are more active in oxidative phosphorylation, relatively poorer in glycogen, active in metabolizing fatty acids, and emmeshed in a luxuriant blood capillary network. Evidently frog sartorius muscle is designed to tolerate anaerobic metabolism more effectively than slow twitch muscles, which are designed primarily for aerobic metabolism.

Metabolism in muscles of the forearm of man in situ has been studied by measurement of brachial arterial blood flow and arterial and venous concentration differences.[31] The product of the blood flow and arterial concentration gives the rate at which a substance is delivered to the forearm. The product of blood flow and venous concentration gives the rate at which it leaves the forearm. The difference between input and output in the steady state is the rate at which the substance is consumed by the forearm (e.g., O_2 consumption = blood flow × arterial − venous O_2 content), or the rate at which the substance is produced by the forearm, if venous concentration exceeds arterial concentration (e.g., CO_2 production = blood flow × venous − arterial CO_2 content). Based on such studies,[104] several statements can be made.

Although skeletal muscle is about 40% of body weight, its O_2 consumption at rest accounts for only about 15% of total body O_2 consumption. With exercise, muscle O_2 consumption increases, until muscle easily accounts for 70% or more of the total body O_2 uptake.

The respiratory quotient (RQ) of muscle is about 0.74 at rest, during mild to moderate exercise, and during recovery from exercise. Since the RQ of carbohydrate is 1 and the RQ of long-chain fatty acids is about 0.7, an RQ of 0.74 implies that only about 13% of the O_2 consumption is spent in oxidation of carbohydrate, while about 87% is spent in oxidation of long chain fatty acids.

With the arm (and the man) at rest in the basal state, nearly half the glucose removed from arterial blood can be accounted for by production of lactic acid. If all the rest of the glucose taken up from arterial blood were oxidized, this would account for only about 15% of the observed forearm O_2 uptake, a result that is in agreement with the prediction based on measurements of the RQ. The remaining 85% of O_2 uptake can be accounted for by oxidation of fatty acids abstracted from arterial blood. These fatty acids are a fraction of blood lipids transported in blood in association with albumin. Although the affinity of albumin for these fatty acids is very great, with association constants of 10^{11} or more, these fatty acids readily cross capillary walls. When a fatty acid labeled with ^{14}C is bound to albumin and injected into the bloodstream, half the label leaves the bloodstream in less than 2 min. This is a turnover time comparable to that of blood gases, and by an order of magnitude more rapid than that of glucose. This rapidly turning over albumin-bound fraction of serum lipids is called free fatty acids (FFA) and is the major fuel oxidized by skeletal (and cardiac) muscle in man at rest and during recovery from exercise. There are many kinds of fatty acids circulating, but only five are present in bulk. The commonest FFA is oleic acid, which is about 45% of arterial FFA and accounts for more than 45% of FFA uptake. If oxidized completely, oleic acid accounts for more than 3 times as much of the O_2 uptake as does glucose. The second most prominent FFA is palmitic acid, slightly less than 30% of arterial FFA and forearm FFA uptake. Its oxidation accounts for about twice as much O_2 uptake as does that of glucose. Far less abundant are stearic, linoleic, and palmitoleic acids. About 95% of arterial FFA has 16 or 18 carbons.

During exercise, muscle cannot immediately remove from the blood all the substrate it oxidizes. In large part this is because during contraction tissue pressure exceeds pressure within small, thin-walled vessels, and even within arteries on occasion, temporarily occluding them. Therefore as tissue pressure waxes and wanes with cyclic contraction and relaxation, local blood flow spurts and ceases. During these brief periods of no local blood flow, lasting for tens to hundreds of milliseconds, muscle O_2 tension can be maintained by myoglobin.* There is therefore O_2 to be consumed and the possibility of continuing oxidative phosphorylation. During these periods the substrates could be supplied in theory by dissimilation of stores in muscle, either of glycogen to hexose phosphates, to pyruvate, to oxalate, and through the Krebs cycle and electron transport chain, or by dissimilation of stored lipid. We have known for decades that muscle glycogen can be broken down during exercise. Indeed, this observation and the measurement of lactic acid evolved from glycogen breakdown led to the old proposal that anaerobic glycolysis was the energy source for muscle contraction. It is true that glycogen is broken down in mammalian muscle also, but it is not the major source of substrate. There is now evidence that there are lipid storage forms, probably triglyceride, in skeletal muscle fibers. Hydrolysis of triglyceride to fatty acids (and glycerol) is an important source of substrate, at least during continued contraction and relaxation cycles of exercise and possibly also at rest. The evidence is that even at rest the FFA removed from arterial blood is not transported immediately to oxidation sites but enters the lipid (triglyceride?) pool in exchange for a molecule of fatty acid produced from the pool by hydrolysis. It is this latter molecule that is oxidized.[42]

Fatty acids and glucose or other carbohydrates are not the only substrates that

*There is enough myoglobin in red muscles to last for about 30 sec of supramaximal tetanic stimulation, a condition not imitated naturally. Myoglobin is nearly saturated at an O_2 tension of 40 mm Hg and is desaturated precipitously at tensions below 20 mm Hg.

muscle can oxidize. Under normal conditions there is so little circulating ketoacid that it cannot provide much substrate material. However, with prolonged fasting, beyond 3 days or so, arterial concentrations of β-OH butyric acid and acetoacetic acid rise by more than an order of magnitude, and oxidation of these substances can then account for substantial portions of muscle O_2 uptake.

In summary, muscle has the capacity to oxidize various substrates, ultimately to link oxidation with phosphorylation. It can also link glycolysis and phosphorylation. Muscles are not structurally or metabolically homogeneous. What they oxidize and whether at any moment they are oxidizing are factors determined largely by blood flow. In myoglobin-poor, capillary-poor muscles that are also mitochondria-poor (e.g., frog sartorius), anaerobic glycolysis is prominent. In the myoglobin-rich, capillary-rich, mitochondria-rich muscles that constitute a large proportion of the muscle in man, recovery is associated with oxidative phosphorylation for which lipid is by far the major substrate. Among the lipids, serum FFA, headed by oleic acid, provides more oxidizable substrate than any other substance, even during mild to moderate exercise, although oxidation is not direct; i.e., most of the serum FFA appears to have to pass through a lipid pool in the muscle fiber before it is oxidized.

REFERENCES

General reviews

1. Aubert, X.: Le couplage énergétique de la contraction musculaire, Brussels, 1956, Éditions Arscia.
2. Buchthal, F., Kaiser, E., and Rosenfalck, P.: The rheology of the cross striated muscle fibre with particular reference to isotonic conditions, Dan. Biol. Medd. **21:**1, 1951.
3. Dreizen, P., Gershman, L. C., Trotta, P. P., and Stracher, A.: Myosin. Subunits and their interactions, J. Gen. Physiol. **50**(suppl.)**:**85, 1967.
4. Ebashi, S., and Endo, M.: Calcium ion and muscle contraction, Progr. Biophys. Mol. Biol. **18:**123, 1968.
5. Ebashi, S., Endo, M., and Ohtsuki, I.: Control of muscle contraction, Q. Rev. Biophys. **2:**351, 1969.
6. Gergely, J., editor: Biochemistry of muscle contraction, Boston, 1964, Little, Brown & Co.
7. Hill, A. V.: Trails and trials in physiology, Baltimore, 1965, The Williams & Wilkins Co.
8. Hill, A. V.: First and last experiments in muscle mechanics, London, 1970, Cambridge University Press.
9. Huxley, A. F.: Muscle structure and theories of contraction, Progr. Biophys. Biophys. Chem. **7:**255, 1957.
10. Huxley, H. E.: The mechanism of muscular contraction, Science **164:**1356, 1969.
11. Jöbsis, F. F.: Energy utilization and oxidative recovery metabolism in skeletal muscle. In Sanadi, D. R., editor: Current topics in bioenergetics, New York, 1969, Academic Press, Inc., vol. 3.
12. Katchalsky, A., and Curran, P. F.: Nonequilibrium thermodynamics in biophysics, Cambridge, Mass., 1965, Harvard University Press.
13. Mommaerts, W. F. H. M.: Energetics of muscular contraction, Physiol. Rev. **49:**427, 1969.
14. Podolsky, R. J.: Thermodynamics of muscle. In Bourne, G. H., editor: Structure and function of muscle, New York, 1960, Academic Press, Inc., vol. 2.
15. Podolsky, R. J.: Membrane systems in muscle cells, Symp. Soc. Exp. Biol. **22:**87, 1968.
16. Pringle, J. W. S.: Mechano-chemical transformation in striated muscle, Symp. Soc. Exp. Biol. **22:**67, 1968.
17. Sandow, A.: Skeletal muscle, Ann. Rev. Physiol. **32:**87, 1970.
18. Szent-Györgyi, A.: Chemical physiology of contraction in body and heart muscle, New York, 1953, Academic Press, Inc.
19. Tiegs, O. W.: Innervation of voluntary muscle, Physiol. Rev. **33:**90, 1953.
20. Weber, H. H.: The motility of muscle and cells, Cambridge, Mass., 1958, Harvard University Press.
21. Wilkie, D. R.: The mechanical properties of muscle, Br. Med. Bull. **12:**177, 1956.
22. Woledge, R. C.: Heat production and chemical change in muscle, Progr. Biophys. Mol. Biol. **22:**37, 1971.
23. Young, M.: The molecular basis of muscle contraction, Ann. Rev. Biochem. **38:**913, 1969.
24. Zierler, K. L.: Some aspects of biophysics of muscle. In Bourne, G. H., editor: Structure and function of muscle, New York, 1973, Academic Press, Inc., vol. 3.

Original papers

25. Abbott, B. C., and Wilkie, D. R.: The relation between velocity of shortening and the tension-length curve of skeletal muscle, J. Physiol. **120:**214, 1953.
26. Adrian, R. H., and Freygang, W. H.: The potassium and chloride conductance of frog muscle membrane, J. Physiol. **163:**61, 1962.
27. Adrian, R. H., Chandler, W. K., and Hodgkin, A. L.: Voltage clamp experiments in striated muscle fibres, J. Physiol. **208:**607, 1970.
28. Adrian, R. H., Costantin, L. L., and Peachey, L. D.: Radial spread of contraction in frog muscle fibres, J. Physiol. **204:**231, 1969.
29. Aubert, X., and Lebacq, J.: The heat of shortening during the plateau of tetanic contraction and at the end of relaxation, J. Physiol. **216:**181, 1971.
30. Bahler, A. S., Fales, J. T., and Zierler, K. L.:

The dynamic properties of mammalian skeletal muscle, J. Gen. Physiol. **51:**369, 1968.
31. Baltzan, M. A., Andres, R., Cader, G., and Zierler, K. L.: Heterogeneity of forearm metabolism with special reference to free fatty acids, J. Clin. Invest. **41:**116, 1962.
32. Bárány, M.: Studies on the functional sulfhydryl groups of myosin and actin. In Benesch, R., et al., editors: Sulfur in proteins, New York, 1959, Academic Press, Inc.
33. Bárány, M.: ATPase activity of myosin correlated with speed of muscle shortening, J. Gen. Physiol. **50:**197, 1967.
34. Buchthal, F., Guld, C., and Rosenfalck, P.: Multielectrode study of the territory of a motor unit, Acta Physiol. Scand. **39:**83, 1957.
35. Caplan, S. R.: A characteristic of self-regulated linear energy converters. The Hill force-velocity relation for muscle, J. Theor. Biol. **11:**63, 1966.
36. Carlson, F. D.: Kinematic studies in mechanical properties of muscle. In Remington, J. W., editor: Tissue elasticity, Washington, D. C., 1957, American Physiological Society.
37. Carlson, F. D.: The kinematics of retraction in contracting striated muscle, J. Cell. Comp. Physiol. **49** (suppl. 1):291, 1957.
38. Carlson, F. D., and Siger, A.: The creatine phosphoryltransfer reaction in iodoacetate-poisoned muscle, J. Gen. Physiol. **43:**301, 1959.
39. Carlson, F. D., Hardy, D. J., and Wilkie, D. R.: Total energy production and phosphocreatine hydrolysis in the isotonic twitch, J. Gen. Physiol, **46:**851, 1963
40. Costantin, L. L., and Podolsky, R. J.: Depolarization of the internal membrane system in the activation of frog skeletal muscle, J. Gen. Physiol. **50:**1101, 1967.
41. Couteaux, R.: Innervation du muscle strié et organisation du sarcoplasme au nive au des terminaisons motrices. In Schapira, G., editor: Le muscle: étude de biologie et de pathologie, Paris, 1950, L'Expansion Scientifique Française.
42. Dagenais, G. R., Tancredi, R. G., and Zierler, K. L.: Evidence for an intramuscular lipid pool in the human forearm, J. Clin. Invest. **50:**23a, 1971.
43. Davies, R. E.: A molecular theory of muscle contraction: calcium-dependent contractions with hydrogen bond formation plus ATP-dependent extensions of part of the myosin-actin cross-bridges, Nature **199:**1068, 1963.
44. Davies, R. E., et al.: Changes in creatine, phosphocreatine, inorganic phosphate and adenosine triphosphate during single contractions of isolated muscles. In Gergely, J., editor: Biochemistry of muscle contraction, Boston, 1964, Little, Brown & Co.
45. Eisenberg, E., and Moos, C.: The adenosine triphosphatase activity of acto-heavy meromyosin. A kinetic analysis of actin activation, Biochemistry **7:**1486, 1968.
46. Eisenberg, R. S., and Gage, P. W.: Ionic conductances of the surface and transverse tubular membranes of frog sartorius fibers, J. Gen. Physiol. **53:**279, 1969.
47. Endo, M., Tanaka, M., and Ogawa, Y.: Calcium induced release of calcium from the sarcoplasmic reticulum of skinned skeletal muscle fibres, Nature **228:**34, 1970.
48. Engelhardt, W. A., and Ljubimowa, M. N.: Myosine and adenosine-triphosphatase, Nature **144:**668, 1939.
49. Fales, J. T.: Muscle heat production and work: effect of varying isotonic load, Am. J. Physiol. **216:**1184, 1969.
50. Fales, J. T., and Zierler, K. L.: Relation between length, tension, and heat: frog sartorius muscle, brief tetani, Am. J. Physiol. **216:**70, 1969.
51. Fales, J. T., Crawford, W. J., and Zierler, K. L.: Gradient-layer calorimetry of muscle: relation between length, twitch tension, and heat, Am. J. Physiol. **213:**1427, 1967.
52. Falk, G., and Fatt, P.: Linear electrical properties of striated muscle fibres observed with intracellular electrodes, Proc. R. Soc. Lond. (Biol.) **160:**69, 1964.
53. Feinstein, B., Lindegaard, B., Nyman, E., and Wohlfart, G.: Morphologic studies of motor units in normal human muscles, Acta Anat. **23:**127, 1955.
54. Feng, T. P.: The effect of length on the resting metabolism of muscle, J. Physiol. **74:**441, 1932.
55. Fenn, W. O.: A quantitative comparison between the energy liberated and the work performed by the isolated sartorius muscle of the frog, J. Physiol. **58:**175, 1923.
56. Fenn, W. O., and Marsh, B. S.: Muscular force at different speeds of shortening, J. Physiol. **85:**277, 1935.
57. Franzini-Armstrong, C.: Studies of the triad. I. Structure of the junction in frog twitch fibres, J. Cell Biol. **47:**488, 1970.
58. Franzini-Armstrong, C., and Porter, K. R.: The Z disc of skeletal muscle fibrils, Z. Zellforsch. Mikrosk. Anat. **61:**661, 1964.
59. Franzini-Armstrong, C., and Porter, K. R.: Sarcolemmal invaginations and the T-system in fish skeletal muscle, Nature **202:**355, 1964.
60. Fuchs, F.: The effect of Ca^{2+} on the sulfhydryl reactivity of troponin: evidence for a Ca^{2+}-induced conformational change, Biochim. Biophys. Acta **226:**453, 1971.
61. Gage, P. W., and Eisenberg, R. S.: Capacitance of the surface and transverse tubular membrane of frog sartorius muscle fibers, J. Gen. Physiol. **53:**265, 1969.
62. Gage, P. W., and Eisenberg, R. S.: Action potentials, afterpotentials, and excitation-contraction coupling in frog sartorius fibers without transverse tubules, J. Gen. Physiol. **53:**298, 1969.
63. Gergely, J., et al.: Comparative studies on white and red muscle. In Ebashi, S., Oosawa, F., Sekine, T., and Tonomura, Y., editors: Molecular biology of muscular contraction, Amsterdam, 1965, Elsevier Publishing Co.
64. Gilbert, C., Kretzschmar, M., Wilkie, D. R., and Woledge, R. C.: Energy balance during muscular contraction, J. Physiol. **207:**15P, 1970.
65. González-Serratos, H.: Inward spread of activation in vertebrate muscle fibres, J. Physiol. **212:**777, 1971.

66. Gordon, A. M., Huxley, A. F., and Julian, F. J.: The variation in isometric tension with sarcomere length in vertebrate muscle fibres, J. Physiol. **184:**170, 1966.
67. Hanson, J., and Huxley, H. E.: Structural basis of the cross-striations in muscle, Nature **172:**530, 1953.
68. Hanson, J., and Lowy, J.: The structure of F-actin and actin filaments isolated from muscle, J. Mol. Biol. **6:**46, 1963.
69. Harrington, W. F.: A mechanochemical mechanism for muscle contraction, Proc. Natl. Acad. Sci. U. S. A. **68:**685, 1971.
70. Hill, A. V.: The heat of shortening and the dynamic constants of muscle, Proc. R. Soc. Lond. (Biol.) **126:**136, 1938.
71. Hill, A. V.: The effect of load on the heat of shortening of muscle, Proc. R. Soc. Lond. (Biol.) **159:**297, 1964.
72. Hill, A. V., and Howarth, J. V.: Alternating relaxation heat in muscle twitches, J. Physiol. **139:**466, 1957.
73. Hill, A. V., and Woledge, R. C.: An examination of absolute values in myothermic measurements, J. Physiol. **162:**311, 1962.
74. Hodgkin, A. L., and Horowicz, P.: The effect of sudden changes in ionic concentrations on the membrane potential of single muscle fibres, J. Physiol. **153:**370, 1960.
75. Howell, J. N.: A lesion of the transverse tubules of skeletal muscle, J. Physiol. **201:** 515, 1969.
76. Hunt, C. C., and Kuffler, S. W.: Motor innervation of skeletal muscle: multiple innervation of individual muscle fibers and motor unit function, J. Physiol. **126:**293, 1954.
77. Hutter, O. F., and Noble, D.: The chloride conductance of frog skeletal muscle, J. Physiol. **151:**89, 1960.
78. Huxley, A. F., and Niedergerke, R.: Interference microscopy of living muscle fibres, Nature **173:**971, 1954.
79. Huxley, A. F., and Peachey, L. D.: The maximum length for contraction in striated muscle, J. Physiol. **146:**55P, 1959.
80. Huxley, A. F., and Taylor, R. E., Local activation of striated muscle fibres, J. Physiol. **144:** 426, 1958.
81. Huxley, H. E.: The double array of filaments in cross-striated muscle, J. Biophys. Biochem. Cytol. **3:**631, 1957.
82. Huxley, H. E.: Electron microscopic studies on the structure of natural and synthetic protein filaments from striated muscle, J. Mol. Biol. **7:**281, 1963.
83. Huxley, H. E., and Hanson, J.: Changes in the cross-striations of muscle during contraction and stretch and their structural interpretation, Nature **173:**973, 1954.
84. Knappeis, G. G., and Carlsen, F.: The ultrastructure of the Z disc in skeletal muscle, J. Cell Biol. **13:**323, 1962.
85. Kuffler, S. W.: The two skeletal nerve-muscle systems in frog, Arch. Exp. Path. Pharmakol. **220:**116, 1953.
86. Kushmerick, M. J., and Davies, R. E.: The chemical energetics of muscle contraction. II. The chemistry, efficiency and power of maximally working sartorius muscles, Proc. R. Soc. Lond. (Biol.) **174:**315, 1969.
87. Kushmerick, M. J., Larson, R. E., and Davies, R. E.: The chemical energetics of muscle contraction. I. Activation heat, heat of shortening and ATP utilization for activation-relaxation processes, Proc. R. Soc. Lond. (Biol.) **174:** 293, 1969.
88. Marsh, B. B.: A factor modifying muscle fibre synaeresis, Nature **167:**1065, 1951.
89. Mommaerts, W. F. H. M., Seraydarian, K., and Marechal, G.: Work and chemical change in isotonic muscular contractions, Biochim. Biophys. Acta **57:**1, 1962.
90. Natori, R.: The property and contraction process of isolated myofibrils, Jikei Med. J. **1:** 119, 1954.
91. Peachey, L. D.: The sarcoplasmic reticulum and transverse tubules of the frog's sartorius, J. Cell Biol. **25:**209, 1965.
92. Pepe, F. A.: The myosin filament. II. Interaction between myosin and actin filaments observed using antibody staining in fluorescent and electron microscopy, J. Mol. Biol. **27:** 227, 1967.
93. Podolsky, R. J., and Costantin, L. L.: Regulation by calcium of the contraction and relaxation of muscle fibers, Fed. Proc. **23:**933, 1964.
94. Polissar, M. J.: Physical chemistry of contractile process in muscle, Am. J. Physiol. **168:**766, 1952.
95. Porter, K. R., and Palade, G. E.: Studies on the endoplasmic reticulum. III. Its form and distribution in striated muscle cells, J. Biophys. Biochem. Cytol. **3:**269, 1957.
96. Ramsey, R. W.: Analysis of contraction of skeletal muscle, Am. J. Physiol. **181:**688, 1955.
97. Rosenblueth, A., Alanis, J., and Rubio, R.: A comparative study of the isometric and isotonic contractions of striated muscles, Arch. Int. Physiol. Biochim. **66:**330, 1958.
98. Sandow, A.: Latency relaxation and a theory of muscular mechano-chemical coupling, Ann. N. Y. Acad. Sci. **47:**895, 1947.
99. Sugi, H., and Ochi, R.: The mode of transverse spread of contraction initiated by local activation in single frog muscle fibres, J. Gen. Physiol. **50:**2167, 1967.
100. Taylor, E. W., Lymn, R. W., and Moll, G.: Myosin-product complex and its effect on the steady-state rate of nucleoside triphosphate hydrolysis, Biochemistry **9:**2984, 1970.
101. Weber, A., and Winicur, S.: The role of calcium in the superprecipitation of actomyosin, J. Biol. Chem. **236:**3198, 1961.
102. Wilkie, D. R.: Heat work and phosphorylcreatine break-down in muscle, J. Physiol. **195:**157, 1968.
103. Winegrad, S.: Intracellular calcium movements of frog skeletal muscle during recovery from tetanus, J. Gen. Physiol. **51:**65, 1968.
104. Zierler, K. L., Maseri, A., Klassen, G., Rabinowitz, D., and Burgess, J.: Muscle metabolism during exercise in man, Trans. Assoc. Am. Physicans **81:**266, 1968.

4 JEAN M. MARSHALL

Vertebrate smooth muscle

Smooth muscles, in contrast to striated muscles, are a heterogeneous group that show great diversity both in their morphologic arrangement and in their physiologic properties. In the walls of the gastrointestinal tract, blood and lymph vessels, uterus, vas deferens, and nictitating membrane, smooth muscle is arranged in sheets or layers of contiguous cells. Single smooth muscle cells, however, are found in the capsule and trabeculae of the spleen, and isolated contractile units called myoepithelial cells occur in some glands. Occasionally smooth muscle cells are arranged in small groups analogous to skeletal muscles; small, cylindric pilomotor muscles are inserted onto the hairs in mammalian skin. The smooth muscles of the hair follicles, most blood vessels, vas deferens, and nictitating membrane are activated only by their motor nerves, whereas visceral smooth muscles exhibit spontaneous activity. The response of various smooth muscles to neurohumoral transmitter substances is not uniform; norepinephrine relaxes intestinal muscle and contracts vascular smooth muscle in most mammals. There is also diversity in the behavior of functionally comparable muscles of different species; norepinephrine causes the uterus of the rabbit to contract but inhibits the contractions of the rat uterus.

Because of these diverse morphologic and physiologic properties, it is difficult to make generalizations about the fundamental properties of smooth muscles. Experiments with smooth muscle are not easy to perform, and the results of investigations are often confusing and ambiguous. There are, nevertheless, three physiologic characteristics of smooth muscles that are of general enough occurrence to be listed here. (1) Smooth muscles are capable of slow, sustained contractions that can be maintained with a minimum expenditure of energy. (2) Their motor innervation is exclusively autonomic. (3) They all exhibit a certain degree of intrinsic "tone," i.e., basal resting tension on which contractions are superimposed.

STRUCTURE AND CHEMISTRY OF SMOOTH MUSCLE CELLS

When examined with the light microscope, smooth muscle cells appear as long, spindle-shaped fibers, 2 to 5μ in diameter and 50 to 100μ in length, with a single nucleus situated in the middle, widest portion of the cell body. Each cell is surrounded by its plasma membrane, the sarcolemma, but there is no protoplasmic continuity between one cell and its neighbors.[33, 54, 63] The intercellular space is small and contains blood vessels, nerve fibers, extracellular matrix, and reticular fibers. Connective tissue cells are rarely found in the interstices of the uniform sheets of visceral muscles. None of the smooth muscles has the distinctive tissue "capsule" or perimysium so characteristic of skeletal muscles.

The cytoplasm of smooth muscle cells appears to be homogeneous when viewed under the light microscope. The contractile proteins, or myofilaments, of smooth muscle cells are not arranged in distinct sarcomeres, and the alternating dark and light bands characteristic of striated muscles are lacking; hence the name *"smooth* muscle." Electron microscopic studies reveal that the myofilaments are 50 to 80 Å wide and about 1μ long. They may occur in tracts or bundles (Fig. 4-1) but are not arranged in any obvious pattern that suggests how they function during contraction of the muscle. Scattered throughout the cytoplasm of smooth muscle cells are streaks of electron-dense material, the so-called fusiform densities.[33] Tracts of myofilaments appear to meet in these "dark bodies," and some observers have compared them to the intercalated discs of heart muscle and the Z lines of skeletal muscle.

The myofilaments are similar in appearance to the actin filaments of striated muscle

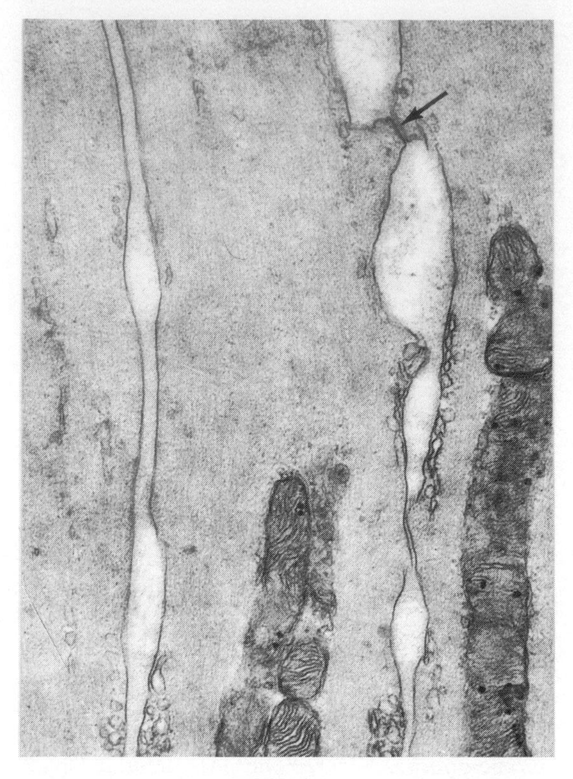

Fig. 4-1. Electron micrograph of portions of three smooth muscle cells in longitudinal section of small intestine of mouse. Myofilaments are visible as thin tracts of uniform density running within cytoplasm. Elements of smooth-surfaced endoplasmic reticulum appear just beneath plasma membrane in certain areas of each of three cells. Arrow indicates region of close contact between lateral surfaces of two muscle cells. (From Rhodin.[63])

cells and, indeed, they seem to contain F actin.[45] X-ray diffraction studies fail to reveal myosin filaments in smooth muscle,[35] and there is no morphologic evidence for the existence of thick filaments comparable to the myosin filaments of striated muscles. However, a protein with chemical properties similar to myosin has been isolated from smooth muscle cells.[60] This myosin protein will react with actin from skeletal muscle to produce actomyosin. Moreover, actin from smooth muscle will react with myosin from skeletal muscle.[60] Hanson and Lowy[45] have suggested that myosin filaments may exist in vivo in smooth muscle, but they may be as thin as actin filaments and easily destroyed by experimental manipulation. Nevertheless, these workers[46] were able to extract myosin filaments from homogenates of intestinal and uterine muscle under special conditions (low temperature and pH and aging of the preparation). It is conceivable that normally in smooth muscle the aggregation of myosin into thick filaments comparable to those of skeletal muscle rarely occurs and can be demonstrated only under special conditions. Further investigations are needed to elucidate the fine structure of the contractile proteins and to clarify the mechanism of contraction in smooth muscle. Recent work suggests that the sliding filament hypothesis generally believed to be the basis of contraction in striated muscle may also apply to smooth muscle.[67a]

The concentration of actomyosin is lower in smooth muscles than in skeletal muscles. Rabbit uterine muscle contains 6 to 10 mg/gm wet weight as compared to 70 mg/gm wet weight in rabbit skeletal muscle.[60] Furthermore, the concentration of high-energy phosphate compounds (ATP, phosphocreatine) is much lower in smooth than in skeletal muscles. Purified actomyosin from uterine smooth muscle shows only about one-tenth as much ATPase activity as actomyosin from skeletal muscle.[60] Smooth muscle actomyosin can be extracted in soluble form in solutions of low ionic strength in the presence of a low concentration of ATP.[60] This soluble actomyosin is called "tono-actomyosin" because of its possible implication in the maintenance of tone in smooth muscle.[39]

Studies with the electron microscope have revealed that the plasma membrane of smooth muscle cells is surrounded by an extraneous glycoprotein coat similar to the basement membrane of epithelial cells. At certain points along the surface, however, the sarcolemma is divested of extraneous coats and the plasma membrane of one cell comes into very close approximation with that of an adjacent cell (Fig. 4-1). High-resolution electron micrographs of these areas of close contact suggest that they are of at least two types, "tight" junctions and "gap" junctions. In the tight junctions the plasma membranes of adjacent cells appear to fuse, completely excluding the intercellular space between the membranes.[33, 54] In the gap junctions there is no fusion of membranes but rather a 20 to 30 Å separation of opposing membranes. Both types of junctions are found along the lateral surfaces as well as at the ends of muscle cells. Although no systematic investigation of the type, distribution, and number of junctions has been made, there is considerable variation in the type, distribution, and number of junctions in different smooth muscles.[25] For example, tight junctions are a prominent feature in intestinal muscle but have not been found in the nictitating membrane.[25] It has been suggested that sites of membrane fusion may be involved in the spread of the excitatory impulse from one smooth muscle cell to another.[3, 13, 25]

The plasma membrane of smooth muscles exhibits an unusual number of pinocytotic vesicles, spherical invaginations about 600 Å in diameter. The vesicles serve to increase the overall surface area of the cell membrane.[38, 63] Although their function remains to be elucidated, it is possible that the vesicles are moving portions of the plasma membrane that are constantly forming and pinching off to incorporate materials into the cell.

In contrast to the well-developed sarcoplasmic reticulum of skeletal muscle, the reticulum of smooth muscle cells is poorly developed (Fig. 4-1). In skeletal muscle cells the sarcoplasmic reticulum consists of an elaborate network of tubules and vesicles surrounding the myofibrils, which is believed to be involved in the spread of excitation from the cell surface to the myofibrils. In such a large cell (50 to 190μ wide) the reticulum may provide a method for rapid activation of the contractile elements. In the smaller smooth muscle cell (2 to 5μ wide) the contractile elements could be activated directly by events occurring across the surface membrane of the cell. The activation of contractile elements need not be as rapid as those of striated muscle since the contractions of smooth muscle are much more slowly developed.

PHYSIOLOGIC CLASSIFICATION OF SMOOTH MUSCLES

Some years ago Bozler[16] proposed that smooth muscles be classified into two groups according to their physiologic properties—unitary muscles and multi-unit muscles. The unitary muscles are characterized by spontaneous activity initiated in pacemaker areas within the tissue that spreads throughout the whole muscle as if the muscle were a single unit. Multi-unit muscles do not contract spontaneously and normally are activated in more than one region by multiple motor nerves.

The smooth muscles of the gastrointestinal tract, uterus, and ureter are good examples of unitary muscles. The contractions of these muscles are not initiated by nerve impulses, but they may be coordinated and regulated by nervous mediation. In this respect the unitary or visceral muscles resemble cardiac muscle. Another property of unitary muscle is its ability to respond to stretch by developing active tension.

The multi-unit muscles include the pilomotors, the muscles of the nictitating membrane, ciliary muscles, iris, and larger blood vessels. These muscles usually respond to a single volley of nerve impulses with a contraction. Multi-unit muscles are organized roughly into "motor units"; hence their name. These "units," however, are diffuse and show considerable overlap. In general, multi-unit muscles do not react to stretch by developing tension.

This classification should not be regarded as rigid, since some smooth muscles do not fall clearly into one or the other of these two groups but combine properties of both. For example, the muscles of the vas deferens are usually activated by stimulation of their extrinsic motor nerves, but when placed in an isolated organ bath, they may become spontaneously active. Smooth muscles in certain arterioles, venules, and veins are autorhythmic; they show spontaneous activity, yet they also respond to extrinsic nerve stimulation. A single volley of nerve impulses can initiate contraction of the urinary bladder, but this muscle also can be made to contract by stretching. Although these and other exceptions exist, the classification of smooth muscles into two general categories will be followed here because it provides a convenient framework for describing physiologic properties of smooth muscle. Visceral muscle will be considered first, followed by a discussion of multi-unit muscle. The chapter ends with a consideration of the transmission of excitation from autonomic nerves to smooth muscle cells.

VISCERAL OR UNITARY SMOOTH MUSCLE

The foundations of visceral smooth muscle physiology were established by the classic experiments of Emil Bozler from 1938 to 1948.[16, 17] He recorded, by means of extracellular electrodes, electrical activity in visceral smooth muscle and correlated it with the contractions of the muscle both in vivo and in vitro. From these studies he concluded that the spontaneous rhythmic contractions of the muscle were initiated and maintained by periodic electric discharges (action potentials) from the muscle cells. These action potentials originated from pacemaker areas within the muscle and were conducted over the muscle in a manner similar to that observed in cardiac muscle. The spontaneous, rhythmic discharge of action potentials persisted in the presence of ganglion and nerve blocking agents. Hence they were myogenic; i.e., they originated within the muscle cells themselves.

Recently these observations have been confirmed and extended with the aid of improved instrumentation and techniques, especially those recording electrical activity with intracellular microelectrodes or with the sucrose gap method. Both of these methods are described in Chapters 1 and 2. The most extensive in vitro studies utilizing these techniques have been performed on two types of muscle, the longitudinal muscles from the cecum of the guinea pig (the taenia coli)[21, 22] and from the mammalian uterus.[52, 56] Thin segments of the external muscle of either organ can be dissected free from the underlying muscle coats and ganglion cells. When placed in an isolated organ bath and bathed in oxygenated physiologic salt solution at 37° C, the strips of muscle will contract spontaneously for many hours. It is possible to penetrate the individual muscle cells with very fine glass capillary microelectrodes (tip diameter 0.5μ) and obtain records of the transmembrane potentials from individual cells while the contractile tension developed by the entire strip is measured simultaneously with a muscle lever or force transducer. At present most of our information about visceral smooth muscle comes from such studies.

When a microelectrode is inserted into a quiescent visceral smooth muscle cell, a potential difference of between 55 and 60 mv is recorded across the cell membrane, with the inside of the cell negative relative to the outside. Spontaneously active unitary smooth muscles are never in a true resting state comparable to that of an inactive skeletal muscle; thus the term *"resting* membrane potential" cannot really be applied to these muscles. Nevertheless, the term is used in describing autorhythmic smooth muscles to refer to the maximal level of membrane polarization attained between periods of activity.[6]* The resting membrane potential of a quiescent visceral smooth muscle cell differs from that of a skeletal muscle cell at rest in two ways: it has a lower value (Table 4-1) and it is more labile. The latter property enables visceral smooth muscle cells to generate action potentials spontaneously.

*In this chapter the membrane potential refers to the potential, measured by an *intracellular* microelectrode, relative to the extracellular fluid. The resting membrane potential will therefore have a negative sign, e.g., –55, –60 mv. Changes in resting potential, e.g., from –60 to –40 mv, will be noted as "depolarization" or "decreased negativity," shifts in the opposite direction, e.g., –60 to –80 mv, will be designated "hyperpolarization" or "increased negativity."

Initiation of spontaneous activity

The spontaneous generation of action potentials in visceral smooth muscle results from at least two types of variations in membrane potentials. The first type resembles that seen in the nodal tissue of the heart and consists of a slow, localized depolarization of the cell membrane culminating in the production of an action potential. An example of this type of activity in a uterine smooth muscle cell appears in Fig. 4-2, where it is compared with the pacemaker discharge from a sinus node fiber in the mammalian heart. Because of its similarity to the cardiac pacemaker, this spontaneous activity in smooth muscle has been termed "pacemaker" type.[6] Unlike the cardiac pacemaker, which is localized in the nodal tissue, pacemaker activity in visceral muscle arises at random and moves about within the tissue because all smooth muscle cells are capable of becoming pacemakers.

The second type of spontaneous activity seen in visceral muscles consists of slow, sinusoidal oscillations of the membrane potential. This form of spontaneous variation in membrane potential differs from the pacemaker type in that action potentials may appear on either the rising or the falling phase of the slow oscillation. Also, the action potentials do not "wipe out" or in any way alter the frequency or amplitude of the slow waves.

Examples of these spontaneous fluctuations in membrane potential appear in Fig. 4-3. An individual smooth muscle cell may exhibit both types of spontaneous activity and in addition show conducted action potentials, i.e., action potentials generated in a pacemaker area some distance from the cell under observation (Fig. 4-3, *d*).

Conduction of excitation

Action potentials produced within the pacemaker areas are conducted throughout the entire muscle in a manner not yet completely understood. Since smooth muscle cells do not have intracellular protoplasmic continuity, how are the action potentials transmitted from cell to cell? It seems most likely

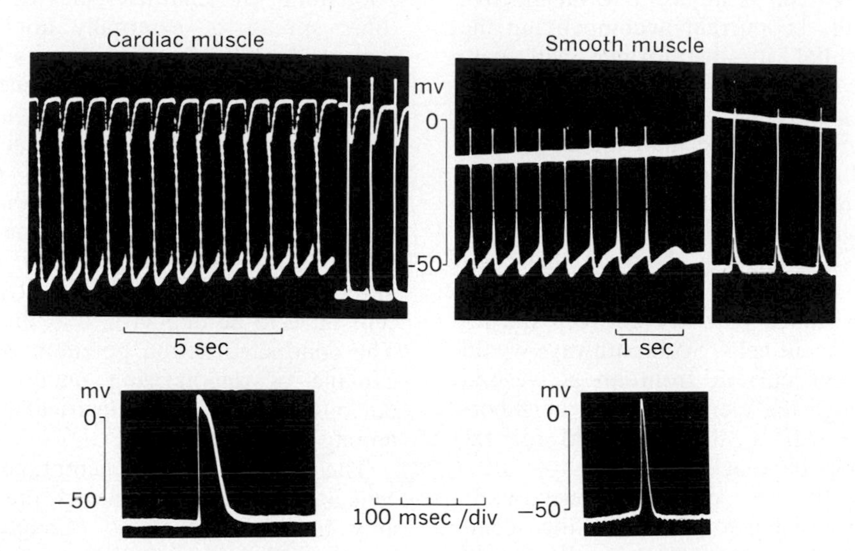

Fig. 4-2. Comparison of pacemaker activity in cardiac and smooth muscle. Top left portion of figure shows transmembrane action potentials (lower trace) and contractions (upper trace) recorded in isolated right atrium of rabbit. Top right, similar records from isolated segment of rat myometrium. Potentials from sinus node cell of atrium (left part of frame) and in uterine pacemaker cell show slow membrane depolarization (prepotential or pacemaker potential), which culminates in production of action potential. Conducted action potentials in both atrial and smooth muscle (right part of both frames) show no prepotentials. Bottom portion of figure illustrates conducted action potentials recorded on a faster time base of oscilloscope beam. Note that duration of action potential (time from onset of depolarization to completion of repolarization) is shorter in smooth than in atrial muscle. Also, rates of depolarization and repolarization are about the same in smooth muscle, while depolarization rate is much faster than repolarization in atrial muscle. (From Marshall.[56])

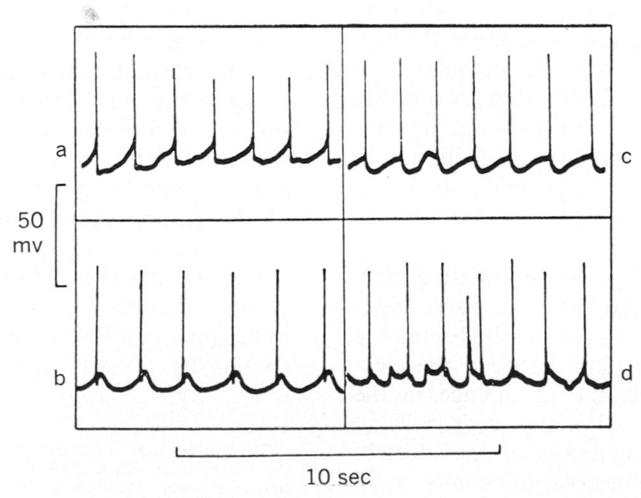

Fig. 4-3. Patterns of spontaneous electrical activity in four different intestinal smooth muscle cells from taenia coli of guinea pig. **a,** Pacemaker type, action potentials continuous with rising phase of prepotential. **b,** Sinusoidal oscillations of membrane potential with action potentials occurring on rising phase of slow oscillations. **c,** Sinusoidal oscillations with action potentials occurring on falling phase. **d,** Mixture of pacemaker, oscillatory, conducted, and local (abortive) potentials. (From Bülbring.[21])

that transmission is electrical and is made possible by the presence of low-resistance pathways between the interiors of adjacent muscle cells.

If an active cell is adjacent to an inactive one, some of the current accompanying the action potential in the former will pass through the membrane of the latter. However, if the membrane of the inactive cell has a much higher resistance than the extracellular fluid, the fraction of current traversing the membrane of the inactive, adjacent cell will be insufficient to reach threshold. In order to devise a feasible electrical theory of transmission, one must include the postulate of a low-resistance pathway between the interiors of adjacent cells. Such pathways would allow sufficient current from an active cell to pass through the membrane of a neighboring, inactive cell so that threshold for excitation would be reached.

If the cells are connected electrically through low-resistance pathways, the membrane potentials of neighboring cells should be affected when current is injected into one cell by means of a microelectrode. The results of such an experiment in intestinal muscle indicate that stimulation of one cell through an intracellular microelectrode produces localized membrane depolarization in some of the nearby cells.[3, 49, 59, 70] The spatial decay of this electrotonic potential is sharp and its time course is rapid (space constant 0.25 mm, time constant 31 msec), suggesting that the current density in the vicinity of the stimulating electrode dissipates quickly. In other words, a significant fraction of the current injected into the cell is shunted through low-resistance paths into the cytoplasm of neighboring cells. This shunting of current through a complex three-dimensional network of cells probably explains why it is difficult to elicit a conducted action potential in many visceral muscles by localized, intracellular stimulation.[70]

On the other hand, if the stimulating electrodes are extracellular and are large relative to the dimensions of a single muscle cell, then the current flow between the electrodes causes a simultaneous change in the membrane potential of many cells.[49, 70] In this situation the magnitude of the electrotonic potential declines exponentially relative to distance from the stimulating electrodes, suggesting that the electrical properties of the tissue resemble those of a uniform core conductor. The externally applied current spreads longitudinally in one dimension, with space and time constants of 1.9 mm and 100 msec, respectively (intestinal muscle).[70] Since a single cell is only 5μ wide and about 200μ long, the characteristics of the electrotonic spread of externally applied current indicate that the functional units for conduction are bundles of interconnecting cells rather than individual cells. This concept is strengthened by the observation that spike generation becomes graded in longitudinal strips of muscle less than 100μ wide and when the diameter of the stimulating electrodes is less than 50μ.[59] Thus, in order to initiate a conducted action potential, many cells have to be depolarized at the same time. The conducted action potential is apparently capable of depolarizing many cells simultaneously,[70] indicating electrical homogeneity among groups of cells.

The regions of cell membrane where the cells are electrically connected are most probably the tight junctions (Chapter 6). The number of such junctions per cell is not known, nor is their electrical resistance. Furthermore, there is no direct evidence indicating that electrical spread of excitation actually does occur at these sites.

Ionic basis of electrical activity in visceral muscle

The ionic basis of electrical activity in visceral muscle is qualitatively similar to that already described for nerve and striated muscle fibers (Chapters 1 and 2), the membrane potentials being related to the distribution of ions across the cell membrane and to the ionic permeability of the membrane.

The resting potential of excitable cells arises from the concentration gradient for potassium ions across the membrane (potassium being more concentrated inside than outside) and from the relative impermeability of the resting membrane to ions other than potassium. If the resting potential is exclusively a potassium diffusion potential, then (1) its magnitude should be equal to the potassium equilibrium potential* and (2) it should be linearly related to the logarithm

*The equilibrium potential for any ion is defined as the potential at which there is no *net* movement of that ion across the cell membrane, i.e., when the electric gradient just balances the concentration gradient. If the concentration of the ion on the two sides of the membrane is known, then the equilibrium potential can be calculated using the Nernst equation:

$$E_{ion} = \frac{RT}{nF} \ln \frac{[ion]outside}{[ion]inside}$$

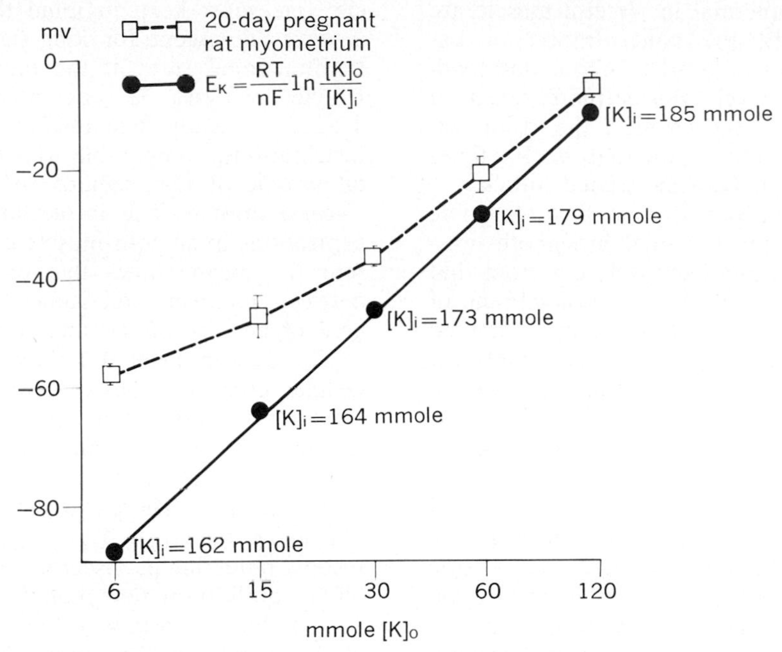

Fig. 4-4. Relation between resting membrane potential, calculated potassium equilibrium potential (E_K), and external potassium concentration ($[K]_o$) in uterine smooth muscle. Squares are measured resting potentials (mean ±SD). Circles are equilibrium potentials calculated from Nernst equation. Note that as $[K]_o$ is raised, $[K]_i$ also increases. (From Casteels and Kuriyama.[29])

of the potassium concentration gradient across the membrane. In skeletal muscle the resting potential and the potassium equilibrium potential are almost identical, but the resting potential in smooth muscle is considerably less than the potassium equilibrium potential (Table 4-1). This suggests that in visceral muscle other ions in addition to potassium contribute to the resting potential. The most likely candidates are sodium and chloride. In intestinal muscle over 95% of the intracellular sodium exchanges with the extracellular sodium with a half-time of less than 1 min.[43] This exchange rate is many times greater than in skeletal muscle at rest. The ratio of sodium to potassium permeability has been estimated to be 7 times higher in intestinal muscle than in resting skeletal muscle.[52] Chloride ions also exchange rapidly in visceral muscle, at least 85% of the tissue chloride exchanging with a half-time of 8 min.[44]

Despite these appreciable sodium and chloride permeabilities, the resting potential of visceral muscle is nevertheless *predominantly* a potassium potential. The magnitude of the resting potential is inversely related to the concentration of potassium ions in the extracellular medium (Fig. 4-4). However, this relationship is linear only at concentrations of potassium above 30 mM. The deviation from linearity at concentrations below 30 mM is probably the result of the contributions of sodium and chloride permeabilities to the resting potential.

According to the ionic theory, the action potential is produced by movements of sodium and potassium through the membrane, successively and respectively depolarizing and repolarizing the membrane (Chapter 2). During the rising phase of the action potential, sodium ions move inward across the membrane, down their concentration gradient, and toward their equilibrium potential. Consequently, the membrane depolarizes, and as the sodium equilibrium potential is approached, the membrane reverses its polarity, producing the "overshoot" of the action potential.

The action potential in visceral muscle, as in striated muscle, is sodium dependent and eventually disappears in a sodium-free medium. In visceral muscle the action potential also "overshoots." Unlike striated muscle, however, where the amplitude and rate of rise of the action potential are directly related to the logarithm of the extracellular sodium ion concentration, these parameters of the action potential in visceral muscle are unchanged until the concentration of external sodium is reduced to about one tenth of its normal level. Furthermore, even in normal sodium concentration the rising velocity of the action potential is 10 times slower in smooth than in striated muscle.[3]

These peculiarities suggest that the ionic basis of the action potential in smooth muscle may be quantitatively different from that in striated muscle. Perhaps the membrane of the smooth muscle cell has only a limited number of sites (or pathways) available for the inward movement of sodium during the rising phase of the action potential. For muscle cells in an environment of normal sodium concentration, these sites may already be partially inactivated because of the low level of the resting potential. Hence the available sites may remain fully saturated over a wide range of extracellular sodium concentration (i.e., from 100 to 50% of normal), so that a very drastic reduction of external sodium is needed before the rising velocity and amplitude of the action potential are altered. Alternatively, other ions in addition to sodium may contribute to the inward current during the rising phase of the action potential. There is convincing evidence that calcium as well as sodium is involved in spike generation in many smooth muscles.[3, 49, 70]

The repolarization phase of the action potential in smooth muscle is presumably associated with an increase in potassium permeability.[3]

Ionic distribution in visceral muscles

When evaluating the experimental findings for the distribution of electrolytes (intracellular vs. extracellular) in visceral muscle, one must keep in mind that this tissue is never quiescent for long periods of time. Its functional state at the time of the electrolyte analyses is not precisely known. Therefore steady-state resting levels of ionic distributions, comparable with those in skeletal muscle at rest, cannot be determined in visceral muscle. The intracellular ionic concentrations in smooth muscle probably represent the mean values that exist somewhere between complete quiescence and some degree of activity of the muscle.

The data in Table 4-1 show that the intracellular concentrations of sodium and potassium in visceral muscle are similar to those in striated muscle, while the internal concentration of chloride is considerably higher in the former. Therefore the chloride equilibrium potential is less negative than the resting potential in visceral muscle. The sodium equilibrium potential, as would be expected, has a positive value. The instability of the membrane, i.e., its tendency for spontaneous depolarization, in visceral muscle may be related to the fact that the distribution of both sodium and chloride is not in equilibrium with the resting potential. A slight increase in sodium or chloride permeability (or decrease in potassium permeability) would drive the potential toward the equilibrium potentials for sodium and chloride. The membrane would then depolarize, and when its potential reached threshold, action potentials would appear.

Despite the appreciable sodium permeability, the concentration of sodium ions within the smooth muscle cell remains constant and at a much lower level than in the extracellular medium (Table 4-1). Hence sodium ions must be actively ejected as fast

Table 4-1. Comparison of electrolyte distribution, ionic gradients, equilibrium potentials, and

	Extracellular electrolytes* (mEq/L)			Intracellular electrolytes (mEq/kg cell H_2O)		
Tissue	$[Na]_o$	$[K]_o$	$[Cl]_o$	$[Na]_i$	$[K]_i$	$[Cl]_i$
Myometrium[29] (20-day pregnant rat)	137	5.9	134	22	162	40
Heart[66, 71] (cat ventricle)	159	4.8	127	7	151	4§
Skeletal muscle[30, 55] (rat)	150	6.4	119	16	150	5

*Values for myometrium are concentrations in physiologic saline solution bathing tissue; others are plasma concentrations.

†Equilibrium potentials at 37° C, calculated as $E_{ion} = \frac{RT}{nF} \ln \frac{[ion]_o}{[ion]_i}$.

‡RMF = resting membrane potential; OS = overshoot of action potential.

§Calculated from membrane potential, assuming that Cl distribution is passive and in equilibrium with resting potential.

as they enter the cell; i.e., smooth muscle must have a very effective sodium pump. The presence of a metabolically dependent sodium pump is suggested by the fact that low temperatures or metabolic inhibitors cause visceral muscles to gain sodium and lose potassium. Rewarming the tissue or washing out the inhibitor results in an extrusion of sodium and a reaccumulation of potassium.[32]

Relation between action potentials and contractions

Contractions of visceral muscles are preceded by a depolarization of the cell membrane and the production of action potentials. The means by which the action potential triggers the contraction of the muscle are unknown. It has been suggested that the action potential liberates a "substance" from the cell surface that diffuses inward to activate the contractile elements.[62] It will be recalled that the sarcoplasmic reticulum, which is thought to be involved in the intracellular spread of excitation in striated muscle, is poorly developed in smooth muscle. However, since smooth muscles develop tension more slowly than striated muscles, the need for a very rapid activation of the contractile elements in the former is not as great as in the latter. It will also be recalled that the smooth muscle cell has a small diameter and a large surface-to-volume ratio. Diffusion time in these cells has been calculated and found to lie well within the limit of the observed latency between membrane excitation and contraction.[62]

An absolute measure of this latency cannot be obtained in smooth muscle because it is not possible to isolate a single smooth muscle cell. However, a reasonable estimate of the "excitation-contraction" latency can be made by simultaneously recording the action potential and the contraction from a very small segment of muscle where the propagation of excitation is rapid and uniform and the conduction distance is short. The segment is stimulated electrically at multiple points along its surface in order to eliminate as far as possible the effect of impulse propagation. In this situation, conduction time is presumably only a small fraction of the excitation-contraction latency. The stimulus intensity is adjusted so that each shock evokes only one action potential. The results of such an experiment are shown in Fig. 4-5, which also includes, for comparison, the excitation-contraction interval in a single fiber of frog skeletal muscle and in a short strip of mammalian ventricle. The excitation-contraction latency is very short (about 10 msec) in skeletal and cardiac muscle, but quite long (about 200 msec) in smooth muscle. The tracings in Fig. 4-5 also illustrate the relatively slow rate of tension development in the smooth muscle.

The nature of the process linking excitation with contraction is unknown, although there is convincing evidence in striated muscle[14, 61] that both sodium and calcium ions are involved. A similar situation may exist in smooth muscle, where action potentials will not evoke a contraction unless both calcium and sodium are present in the external medium.[6] Moreover, calcium influx into smooth muscle cells increases during contraction.[2]

The magnitude of the contraction of visceral muscle is directly related to (1) the frequency of action potential discharge from the individual muscle cells and (2) the total

membrane potentials in smooth, cardiac, and skeletal muscle in vitro

Ionic gradients			Equilibrium potentials†			Membrane potentials‡	
$[K]_o$ / $[K]_i$	$[Na]_o$ / $[Na]_i$	$[Cl]_o$ / $[Cl]_i$	E_K (mv)	E_{Na} (mv)	E_{Cl} (mv)	RMP (mv)	OS (mv)
1/26	4.7	2.8	-86	+41	-27	-58	+15
1/31	22.7	29.6	-91	+79	-90	-90	+30
1/24	9.4	23.8	-84	+58	-82	-80	+25

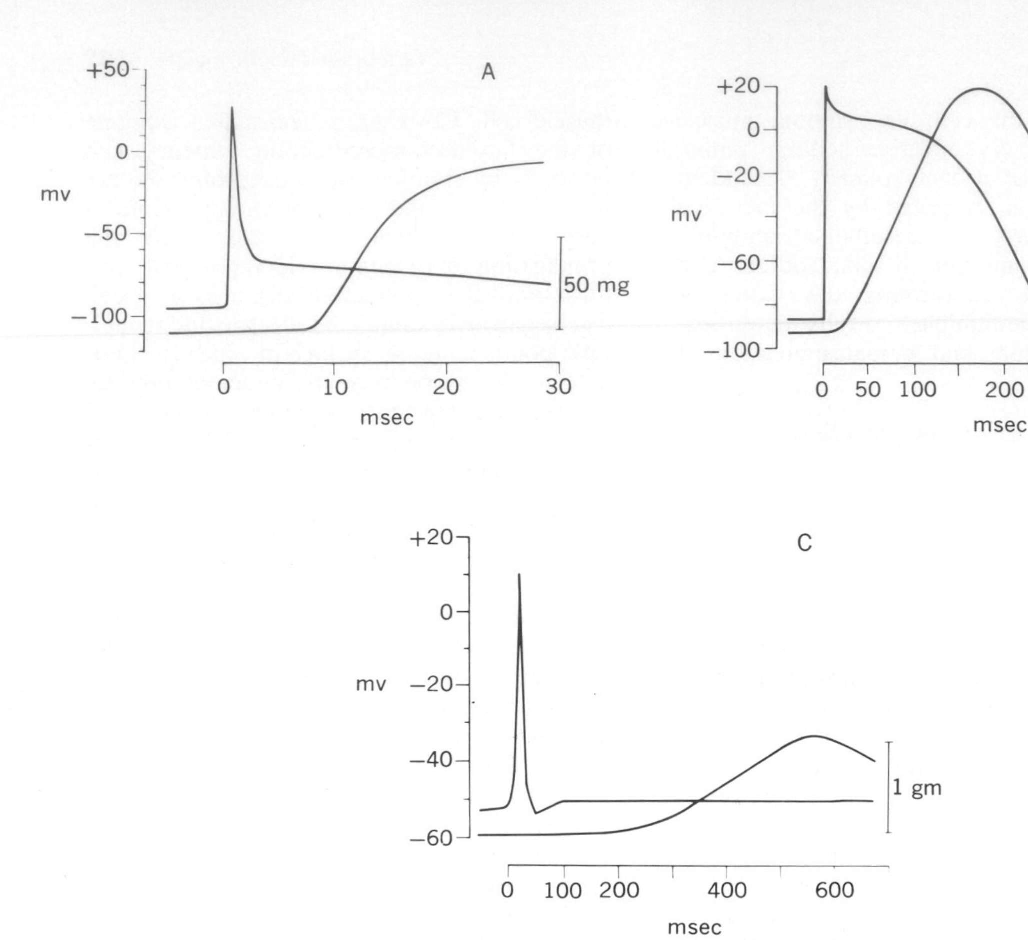

Fig. 4-5. Diagrams showing relation between transmembrane action potential and isometric tension curve. **A,** Single muscle fiber from frog semitendinosus muscle. Temperature 20° C. Excitation-contraction latency (time from onset of depolarization to initial development of tension) about 8 msec. **B,** Small segment of ventricular muscle of cat. Temperature 37° C. Excitation-contraction latency about 10 to 15 msec. **C,** Small segment of uterine muscle of rat. Temperature 37° C. Excitation-contraction latency about 200 msec. (**A** from Hodgkin and Horowicz[48]; **B** modified from Brooks et al.[18])

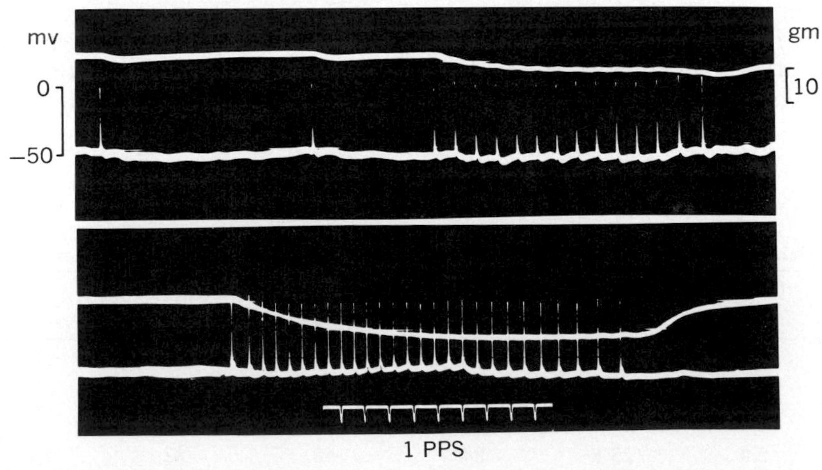

Fig. 4-6. Relation between rate of action potential discharge and force of contraction in isolated segment of estrogen-treated rat myometrium. Isometric tension record, top trace (downward deflection indicates increase in tension); transmembrane potentials, lower trace. Continuous record with microelectrode remaining within same smooth muscle cell during several successive contractions of muscle. Magnitude and duration of contractions are regulated by rate and duration of action potential discharge. (From Marshall.[56])

number of cells synchronously active. The relationship between action potential frequency and force of contraction is best illustrated in very small strips of muscle where conduction distance is short and the propagation of the action potentials is uniform; i.e., electrical activity of one cell is representative of all the cells in the strip. Simultaneous electrical and mechanical records from one such segment of uterine smooth muscle are shown in Fig. 4-6. A single action potential produces a small, discrete increase in muscle tension, multiple action potentials at low frequency produce partially fused contractions, and a further increase in action potential frequency results in a smooth tetanic contraction. The duration of the contraction is controlled by the duration of the action potential train. This direct correlation between action potential frequency and contraction exists, of course, only if all of the individual muscle cells are active in synchrony. If the activity is asynchronous or if some cells do not participate in every contraction, then the electrical record from an individual cell is not necessarily representative of all cells.

Factors influencing performance of visceral smooth muscle

Stretch. Visceral smooth muscles are sensitive to stretch and respond with a depolarization of the cell membrane. In quiescent muscles, if this depolarization reaches threshold, action potentials are generated and the muscle contracts.[6] If the muscle is rhythmically active when it is stretched, the resulting membrane depolarization converts the cyclic discharge of action potentials into a continuous train, producing a sustained contraction of the muscle. The magnitude of the membrane depolarization is related to the degree of the stretch. When a segment of intestinal muscle is progressively stretched in a stepwise manner, there is a graded depolarization of the cell membrane as well as a graded increase in spike frequency and in tension.[21]

The stretch sensitivity is a property of the smooth muscle cell itself and does not depend on the nervous elements within the tissue. It is present in nerve- and ganglion-free preparations and is not affected by ganglion or nerve blocking substances.[6] Smooth muscle thus differs from the stretch-insensitive skeletal muscle fiber and actually resembles the stretch receptor of skeletal muscle, the muscle spindle.

The sensitivity to stretch is of obvious functional importance in visceral muscles. An increase in the intraluminal pressure within an organ such as the ureter, bladder, uterus, or intestine can elicit a contraction of the smooth muscle elements within its walls. The contraction would expel or propel the contents of the lumen.

Humoral factors

Acetylcholine. Acetylcholine stimulates most visceral muscles and increases the force of their contractions.[5, 7] These actions are characterized by depolarization of the cell membrane, bringing the potential to threshold and initiating action potentials, which in turn elicit a contraction of the muscle (Fig. 4-7). If the muscle is active when acetylcholine is administered, there is an increase in the frequency of discharge and in the duration of the action potential train. Conduction of action potentials through the tissue is faster, causing a more synchronous activation of the individual muscle cells. This greater degree of synchrony increases the force of contraction. The stimulatory effect of acetylcholine is prevented by atropine but not by ganglion and nerve blocking agents and is present in nerve-free segments of smooth muscle in the chick amnion. These findings suggest that acetylcholine acts directly on the muscle cell and not via neural elements within the tissue. The cholinergic receptors in visceral muscle are apparently distributed over the entire surface of the muscle rather than localized at one specific site, as at the motor end plate in skeletal muscle.[3]

The mechanism of the stimulatory action of acetylcholine on smooth muscle is probably similar to that at the motor end plate and is associated with a generalized increase in cationic permeability of the cell membrane.[53] Studies with radioactive tracers show that acetylcholine increases the influx and efflux of both potassium and chloride ions in intestinal muscle.[34] The effects of acetylcholine are abolished when the calcium concentration in the external medium is reduced.[53]

Although under normal circumstances acetylcholine acts initially on the excitable membrane of the smooth muscle cell, the recent experiments of Schild and colleagues[36] have shown that even when the membrane is completely depolarized, e.g., by exposure to isotonic potassium solutions, and all action potentials are thereby abolished, acetylcholine can still evoke a contraction. In this

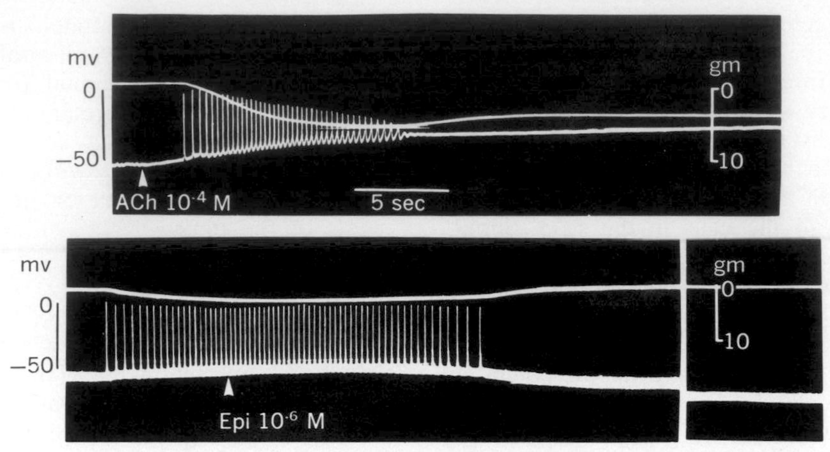

Fig. 4-7. Effect of acetylcholine, *ACh,* and epinephrine, *Epi,* on contractions (upper trace) and transmembrane potentials (lower trace) in uterine muscle of rat. In experiment shown in top frame, high concentration of acetylcholine produced sustained depolarization of cell membrane and prolonged contraction of muscle. In experiment illustrated in lower frame, epinephrine was added during spontaneous contraction of muscle. Within about 15 sec, action potentials disappeared and muscle relaxed. Records in right-hand part of frame were taken 5 min after application of epinephrine, and at this time membrane potential had increased to around -65 mv.

situation the action of acetylcholine depends on the presence of calcium in the environment and is accompanied by an increase in the movement of potassium and chloride ions across the cell membrane.[34] It should be kept in mind, however, that under physiologic conditions the cell is not bathed in isotonic potassium solutions and acetylcholine acts initially by depolarizing the excitable membrane of the smooth muscle cell.

Epinephrine and norepinephrine. The actions of the adrenergic amines epinephrine and norepinephrine can be either excitatory or inhibitory, depending on the type of visceral muscle and on the species of animal from which it comes. For example, in the rabbit and in the human the amines inhibit intestinal muscle and stimulate the uterus, while in the rat they inhibit both the intestine and uterus. These actions are mediated by a combination of the amines with specific receptor sites on or in the muscle cell, the so-called adrenoceptors.[12] The classification of adrenoceptors into alpha and beta groups is now generally accepted. In smooth muscle, activation of the *alpha* recoptors usually results in excitation, while *beta* receptors mediate inhibition, with the exception of the intestine where inhibition is mediated by both alpha and beta receptors.

The stimulatory actions of epinephrine and norepinephrine are indistinguishable from those just described for acetylcholine. It seems reasonable to assume that these actions also result from a generalized increase in the permeability of the membrane to sodium, potassium, calcium, and chloride ions. Thus the membrane will depolarize to a level somewhere between the equilibrium potentials for these ions and in so doing will reach threshold for action potential discharge. The excitatory effects are dependent on the presence of calcium in the external medium.[3, 5, 6]

The inhibitory actions of the adrenergic amines are characterized by a cessation of action potential discharge and a hyperpolarization of the cell membrane; i.e., the membrane potential becomes more negative (Fig. 4-7). The amount of hyperpolarization depends on the magnitude of the membrane potential when the amines are applied. If a strip of intestinal muscle is stretched so that the membrane potential is reduced to about −45 mv, epinephrine increases the potential to about −60 mv. In the absence of stretch or when the muscle is quiescent, the membrane potential may already be around −60 mv, and in this situation the hyperpolarizing effects are minimal.[22]

The mechanisms underlying these inhibitory effects have been studied most extensively in intestinal muscle (taenia coli of the guinea pig), where both alpha and beta receptors mediate the inhibition. The membrane hyperpolarization is an *alpha* effect (since it is abolished in the presence of phentolamine, an alpha blocking agent) and is mainly associated with a selective increase in membrane permeability to potassium and chloride ions.[23] The *beta* effect is primarily associated with a suppression of the pacemaker activity (since this suppression of pacemaker activity is prevented by a beta blocking agent, propranolol).[24] The ionic basis of the pacemaker activity is unknown, and thus the ionic mechanisms underlying the beta adrenergic effects in the intestine are obscure.

There is a considerable amount of evidence suggesting that the beta adrenergic effects are associated with metabolic changes within the muscle cells.[3, 53] Epinephrine stimulates cellular metabolism and the inhibitory effects of this amine on the intestine are accompanied by an increase in the formation of energy-rich compounds within the muscle cells.[20] It has been postulated that a portion of this increased energy supply may be used to accentuate the processes associated with the partition of ions across the cell membrane, i.e., the "ion pumps."[22] In the spontaneously active intestine with its high sodium permeability the membrane potential is normally around –50 mv and near threshold for spontaneous discharge of action potentials. The high sodium permeability implies a continual leakage of sodium into the cell and hence the need for outward transport of sodium against its concentration gradient in order to maintain the membrane potential in its normal range. If a portion of the increased energy supply evoked by epinephrine is used to stimulate ion pumping, the rate of sodium extrusion as well as the rate of potassium uptake will be increased. As a result, the membrane potential would become more negative; i.e., the potential would move closer to the potassium equilibrium potential.

The inhibitory actions of norepinephrine on the intestine are qualitatively similar to those for epinephrine, but a higher concentration of norepinephrine is usually needed to produce an action equivalent to that of a given concentration of epinephrine.[12]

Hormonal factors. Although the behavior of all smooth muscles is probably influenced by their hormonal environment, the only visceral smooth muscle that has been extensively studied with respect to the hormonal regulation of its activity is the myometrium (uterine muscle). This discussion will therefore concern itself only with this muscle, which is uniquely sensitive to the ovarian hormones estrogen and progesterone and to oxytocin, a hormone of the neurohypophysis.

The uterus from an animal in diestrus shows little spontaneous activity and is virtually insensitive to both electrical and chemical stimulation and to stretch. During estrus the uterus becomes increasingly spontaneously active and sensitive to external stimuli and to stretch.[17] A similar evolution of uterine activity can be induced in ovariectomized or immature animals by the administration of estrogen. In addition to its effects on the excitability of the myometrium, estrogen also stimulates the synthesis of actomyosin and high-energy phosphate compounds within the myometrial cell.[31] After priming with estrogen, further treatment with progesterone results in a reduction of uterine motility and of sensitivity to external stimuli and to stretch. Progesterone has no significant effects on the contractile components of the muscle cell.[60]

One of the means by which estrogen and progesterone modify myometrial activity is through alterations in the level of the resting membrane potential and in the propagation of action potentials through the myometrium.[56] The resting membrane potential of the myometrial cells in an immature or ovariectomized animal is only about –35 mv. At this potential the membrane is relatively inexcitable, supposedly because the membrane is in a state of "accommodation,"* and the muscle is quiescent. Estrogen treatment brings the potential into the range (about –50 mv) in which spontaneous discharge of action potentials occurs and the muscle becomes rhythmically active. Estrogen also enhances the conduction of excitation throughout the muscle, producing a well-synchronized activation of many cell groups. Thus the strong contractions typical of the estrogen-dominated uterus result from the actions of estrogen on the excitable membrane as well as on the contractile elements of the myometrial cell.

Progesterone either stabilizes the membrane potential near the resting level or, in some species, increases it to about –65 mv. These effects diminish the excitability to the muscle and produce areas of conduction block within the tissue.[3, 56] Consequently, the propagation of action potentials from cell to cell is impeded. Not all regions of the mus-

*A prolonged, steady depolarization of its membrane makes a cell intrinsically less excitable, presumably because of the high degree of inactivation of sodium conductance. Therefore a stronger stimulus is required to increase sodium conductance sufficiently to produce a regenerative response, i.e., an action potential. This decrease in excitability (increase in threshold) accompanying a sustained membrane depolarization is termed accommodation (Chapter 2).

cle are active during each contraction, and the force of contraction is reduced. The weak, uncoordinated contractions typical of the progesterone-dominated uterus result from the effects of progesterone on the excitable membrane of the myometrial cell and not from an alteration in the contractile machinery.

During the course of pregnancy, changes occur in the myometrium that parallel those just described for estrogen and progesterone. In the rat and rabbit myometrium the membrane potential of the muscle cells gradually becomes more negative, reaching a maximum (about –60 mv) around midpregnancy and remaining at this level until the end of term. During this period the uterus may show some spontaneous contractions, but these are localized, irregular, and weak. About 24 hr before parturition the membrane begins to depolarize and the uterus becomes progressively more active. At parturition the membrane potential is about –50 mv and uterine motility is maximal, with action potentials and waves of contraction spreading uniformly through the muscle.[29] The concentration of high-energy phosphate compounds and actomyosin within the myometrial cells also increases throughout pregnancy, reaching a maximum several days before parturition.[31]

The increased negativity (e.g., about –60 mv) of the membrane potential at midpregnancy comes at a time when the concentration of progesterone in the maternal blood is relatively higher than that of estrogen. The decline in negativity of the membrane potential, with the subsequent increase in uterine conductivity and motility at the terminal stage of pregnancy, may be the result of the rising level of estrogen in the blood that occurs during the latter part of gestation.

These changes in membrane potential and propagation of action potentials may reflect hormonally induced modifications in the transmembrane ionic concentration gradients and in the ionic permeability of the cell membrane. To explore these possibilities, Casteels and Kuriyama[29] measured the ionic distributions (intracellular vs. extracellular) and permeabilities in the myometrium of the rat at various stages of pregnancy. They found no significant differences in the sodium, potassium, or chloride concentration gradients during pregnancy. On the other hand, the potassium permeability gradually increased until midpregnancy and then remained high until near the end of gestation. Sodium permeability was relatively low throughout pregnancy but increased several days before delivery. The electrophysiologic changes in the myometrium that accompany pregnancy might therefore be explained on the basis of these permeability variations.

Oxytocin, a hormone from the neurohypophysis, is one of the most selective of all visceral smooth muscle stimulants, acting primarily on the myometrium and on the myoepithelial cells of the mammary glands. The stimulatory effects of oxytocin on the myometrium are characterized by a depolarization of the cell membrane and an increase in the frequency of action potentials.[57] If the muscle is quiescent, oxytocin initiates rhythmic bursts of action potentials that produce strong contractions. If the muscle is already rhythmically active, oxytocin increases both the force and the frequency of the contractions.

A prerequisite for the effect of oxytocin is that the membrane potential of the myometrial cell has to be close to threshold for the generation of action potentials. Consequently, the myometrium is most sensitive to oxytocin when the uterus is predominantly under the influence of estrogen.

Although there is no direct experimental evidence for the permeability changes accompanying the action of oxytocin, presumably they are similar to those already suggested for acetylcholine.

MULTI-UNIT SMOOTH MUSCLE

Nictitating membrane and pilomotor muscles

Of the two classic examples of multi-unit muscles, the nictitating membrane and the pilomotor fibers, the former has long been a favorite of the investigator interested in the effects of drugs on autonomic neuroeffector systems. In the cat, where the nictitating membrane is more highly developed than in most species, this organ consists of two thin sheets of smooth muscle inserted into adjacent sides of a T-shaped cartilage on the nasal portion of the orbit. The nictitating membrane is innervated exclusively by postganglionic adrenergic nerve fibers arising from cells in the superior cervical ganglion, and it can be easily exposed with both pre- and postganglionic nerve fibers intact. Normally the membrane shows no spontaneous activity but can be made to contract by stimulation of the pre- or postganglionic nerves. Several attempts have been made to record with extracellular electrodes the electrical activity of the smooth muscle cells during nerve stimulation,[67] but the records are a mixture of complex wave forms and are difficult to interpret. The structure of the nictitating membrane has been studied recently with the electron microscope, and it is interesting that the smooth muscle cells in this organ do not show the tight junctions so characteristic of visceral smooth muscle.[25] Each smooth muscle cell in the nictitating membrane is completely surrounded by a basement membrane that separates the cells by gaps of at least 600 Å. Furthermore, many small nerve fibers run between the muscle cells and apparently each individual cell is reached by at least one or more axons. These morphologic characteristics substantiate the physiologic findings that the activation of the smooth muscle cells in the nictitating membrane is neurogenic and that excitation by myogenic conduction from cell to cell, so typical of visceral smooth muscle, does not occur.

Although the nictitating membrane is normally activated only through its motor nerves, it can contract in response to a variety of smooth muscle stimulants, e.g., epinephrine, norepinephrine, histamine, and acetylcholine, when these substances are either topically applied or injected into the blood supplying the organ. The finding that acetylcholine stimulates the muscle is interesting because it shows that the cells are sensitive to the cholinergic transmitter even though the muscle itself does not possess cholinergic innervation. As would be expected for a multi-unit muscle, the nictitating membrane is insensitive to stretch.[6]

The pilomotor muscles are discrete microscopic bundles of smooth muscle cells attached at one end to hair follicles and at the other end to the inner surface of the basal layer of the epidermis. Their motor innervation comes exclusively from sympathetic adrenergic fibers. When the muscles contract in response to nerve stimulation, they throw the tissue into folds and erect the hairs. Recently an in vitro method for studying the activity of the pilomotor fibers in the skin of the cat's tail has been devised.[47] Although individual muscle bundles cannot be isolated, tubes of skin containing many such bundles can be removed from the tail. When these tubes are everted and suspended in an isolated organ bath and the nerves within the tissue are stimulated, the pilomotor muscles contract and the skin tube shortens. The shortening can be registered with a suitable muscle lever or transducer. The effects of nerve stimulation are mimicked by norepinephrine and epinephrine, but unlike the nictitating membrane, the pilomotor muscles are insensitive to acetylcholine. Nothing is known about the mechanism of the effects of nerve stimulation or of exogenously applied neurohumoral agents on the individual pilomotor muscle cells.

Vascular smooth muscle

It is difficult to study vascular smooth muscle in vivo since many factors, both intramural and extramural, influence its performance. In the intact animal as well as in isolated vascular beds the influences of neurogenic control, humoral vasoactive substances, tissue metabolites, and vessel wall thickness cannot be divorced from those of the experimental test substances.

Because of these in vivo complications, much of the recent work on vascular smooth muscle has been done in vitro, where the environment can be more effectively controlled and where the activity of the muscle can be measured directly. Since the smooth muscle in the walls of arteries is oriented in a close spiral or circular manner, helical strips of muscle can be dissected from the wall, so that the long axis of the strip is parallel to the long axes of the individual muscle fibers.[15] Such strips, when cut from large-conduit vessels (carotid artery, aorta, pulmonary artery) or from small-diameter (200 to 300μ) resistance vessels and placed in an isolated organ bath, remain viable for many hours. A prominent longitudinal layer of smooth muscle is found in certain veins (portal, inferior vena cava, renal, adrenal) and segments of these muscles can be easily isolated and arranged for experimental use. The results of experiments with these two types of preparations have considerably advanced our understanding of the physiology of vascular muscle over the past decade.[1, 8-11, 68]

The functional characteristics of vascular muscle are reasonably uniform from species to species but vary considerably within different regions of the vascular bed in any one animal. The behavior of smooth muscle from the large-conduit arteries (aorta, carotid) resembles that of typical multi-unit muscles. The large arteries are not spontaneously active and do not respond to stretch by developing active tension. Their muscle cells are separated from one another by connective tissue elements and few, if any, regions of cell-to-cell contact are present.[10] Although the electrolyte distribution in the smooth muscle of the large arteries is similar to that in visceral muscle, nothing is known about the ionic permeability of the cell membrane in vascular muscle. The resting potential of the muscle cells in large arteries is about –50 mv and within the range reported for visceral muscle.[10, 68] The potential in arterial muscle is quite stable and spontaneous action potentials are rarely, if ever, observed. Stimulants, including epinephrine, norepinephrine, 5-hydroxytryptamine, angiotensin, and acetylcholine, produce a graded depolarization of the cell membrane resulting in a slow, sustained contraction of these muscles in vitro.[1, 68] Even in the stimulated muscles, action potentials are not seen; hence graded depolarization rather than spike generation appears to be the normal electrical trigger for the contractile elements.[10, 11]

The behavior of small-resistance vessels and veins resembles that of visceral muscle in many respects.[10, 68] For example, smooth muscle from these vessels often displays spontaneous contractions that are initiated by action potentials arising from pacemaker areas within the muscle. The muscles are also stretch sensitive and their cells lie quite close together. Cell-to-cell junctions are a prominent morphologic feature of these muscles, suggesting a high degree of electrical continuity between cells.[72] The usual response of these muscles to stimulatory agents, e.g., histamine, vasopressin, 5-hydroxytryptamine, and angiotensin, is depolarization, initiation of action potentials in quiescent muscles, or an increase in the frequency of action potential discharge in contracting muscles. The response of the small arteries, arterioles, and veins to epinephrine and norepinephrine varies from one given vascular bed to another and is determined by the relative sensitivity of the vascular smooth muscle in these different areas to *alpha* (excitatory) and *beta* (inhibitory) adrenergic activation.[1, 68, 74] Excitation and vasoconstriction are the usual responses of most vascular smooth muscles to the adrenergic amines. These excitatory effects are prevented and often reversed by alpha adrenergic blocking agents. In the coronary circulation, however, the inhibitory action of epinephrine and norepinephrine predominates, resulting in a pronounced dilatation of the coronary bed when these agents are present.[74] This inhibitory action is prevented by appropriate beta adrenergic blocking agents. Venous smooth muscle is also frequently dilated by epinephrine and norepinephrine, this action being mediated by activation of beta adrenoceptors within the muscle.[11, 12]

The predominant effect of acetylcholine on small-resistance vessels and veins is usually inhibitory, producing a marked vasodilatation.[1, 68] This action is prevented by atropine, suggesting that acetylcholine is acting directly on the smooth muscle cells rather than on neural elements in the tissue. Acetylcholine has been reported to cause contraction of the carotid and main pulmonary arteries in vitro when applied in relatively high concentrations.[68] Since the contractions are prevented by ganglionic and adrenergic blocking agents, it is likely that the acetylcholine is acting indirectly via activation of adrenergic nerves within the muscle.[1, 68]

The mechanisms underlying the vasodilator and vasoconstrictor actions of these various agents are not known. It is tempting to speculate that the excitatory effects are the result of a generalized increase in ionic permeability of the cell membrane causing membrane depolarization and an increase in spike discharge. The inhibitory actions could result from a specific increase in permeability to potassium and/or chloride ions or a decrease in permeability to sodium. These speculations have yet to be proved experimentally.

AUTONOMIC NERVE: SMOOTH MUSCLE TRANSMISSION

Electrophysiologic aspects

The basis of our knowledge about the mechanism of transmission of excitation by neurohumoral agents from motor nerves to effector cells comes from the classic electrophysiologic studies on neuromuscular transmission in skeletal muscle by Katz.[51] Recent experiments show that many of the principles that apply to skeletal neuromuscular transmission are also relevant for smooth muscle.[4, 50] The subsequent discussion attempts to summarize the pertinent aspects of neuroeffector transmission in smooth muscle.

The two smooth muscle–autonomic nerve preparations most extensively studied are the vas deferens–hypogastric nerve from the guinea pig and the colon–pelvic and colon–colonic nerves from the rabbit. The colon can be isolated with both the parasympathetic (pelvic) and sympathetic (lumbar colonic) nerves attached, so that the motor as well as the inhibitory effects of nerve stimulation can be observed. The spontaneous activity of the colon sometimes complicates the response of the muscle to nerve stimulation. The vas deferens, on the other hand, is not spontaneously active, but it can only be isolated with its sympathetic (hypogastric) nerve supply intact.

When a microelectrode is inserted into a smooth muscle cell in the vas deferens, the resting membrane potential is found to be stable and lies between -60 and -70 mv. If the hypogastric nerve is stimulated repeatedly with pulses of submaximal intensity and about 1 msec in duration, each stimulus produces a small localized depolarization of the cell membrane, the so-called excitatory junction potential (EJP).[26] If the nerve is stimulated with sufficient rapidity, the individual potentials sum, reducing the membrane potential

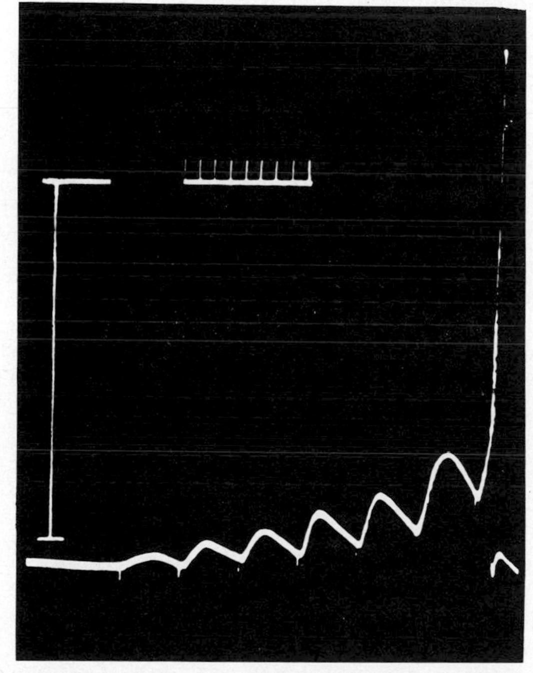

Fig. 4-8. Effect of repeated stimulation of hypogastric nerve on transmembrane potential of single smooth muscle cell in vas deferens of guinea pig. Each stimulus (indicated by small vertical "pip" on trace) is followed by slow depolarization of cell membrane, the so-called excitatory junction potential (EJP). Individual EJPs sum, bringing membrane potential to threshold, around -35 mv, where action potential is initiated.

Vertical calibration, -50 mv, with horizontal top portion of calibration equaling 0 mv. Time mark = 100 msec intervals. (From Burnstock and Holman.[26])

to about -35 mv, where an action potential is initiated and the muscle contracts (Fig. 4-8). Junction potentials are recorded in practically every cell impaled by a microelectrode, and they resemble in many respects the end-plate potentials of skeletal muscle. Adrenergic blocking agents (bretylium and phentolamine) or pretreatment of the animal with reserpine (a substance that depletes the adrenergic nerve terminals of their transmitter —norepinephrine) abolishes the EJPs, and the muscle no longer responds to nerve stimulation. These findings indicate that the EJPs are produced by the release of the adrenergic transmitter, norepinephrine, during stimulation of the hypogastric nerve.

When the nerves are stimulated at frequencies of less than 1 pulse/sec, the individual EJPs are easily seen since they do not sum and can be studied without the complications of action potential discharge and muscle contraction. Under these circumstances the amplitude of the EJPs is remarkably constant from one cell to another, but their latency (time from nerve stimulation to appearance of junction potential) and rate of development vary considerably. Latencies from 20 to 70 msec have been recorded from cells only 2 mm apart. Assuming that the conduction time for the nerve impulse is about 10 msec, then a minimal delay of 10 msec is attributable to the transmission processes, which include the diffusion time for the transmitter from the nerve to the effector site.[26]

Since there are more muscle cells than nerve fibers, the presence of EJPs in every cell suggests that each axon of the hypogastric nerve is able to influence the membrane potentials of many different muscle cells. In order to accomplish this, the individual axons might branch extensively, so that each muscle cell receives a discrete nerve ending, or one branch of an axon might liberate its transmitter substance directly on some cells and in the vicinity of many others. In the latter case the cells would be activated by a general diffusion of transmitter. Activation by generalized diffusion would also account for the marked variations in latency and rising velocity of the individual EJPs. The concept of a generalized diffusion of transmitter was first suggested some years ago by Rosenblueth[67] to explain his finding that in many autonomic neuroeffector systems the effects of variations in strength and frequency of nerve stimulation were "quantitatively interchangeable." Similarly, the amplitude and the rate of depolarization of the EJPs in the vas deferens are increased either by raising the strength of nerve stimulation, i.e., by increasing the number of active fibers (at constant frequencies above 1 pulse/sec), or by increasing the frequency (at constant strength). These findings for the vas deferens also support the idea of a generalized diffusion of transmitter during nerve stimulation.

In addition to the EJPs, which occur only when the nerve is stimulated, another form of electrical activity has been recorded from the smooth muscle cells in the vas deferens. The second type is seen in the absence of nerve stimulation and consists of a random discharge of small, subthreshold potentials that resemble the "miniature" potentials arising from the postjunctional membrane of the nerve-muscle junction in skeletal muscle (Fig. 4-9).[26, 27] These spontaneous potentials do not elicit a contraction of the muscle and are recorded from every cell penetrated with the microelectrode. The amplitude and frequency of the "miniatures" are unaffected by atropine but are reduced by 90% in preparations from reserpine-treated guinea pigs. They presumably reflect the response of the smooth muscle cell to the spontaneous release of norepinephrine from the adrenergic nerve fibers within the tissue.

In contrast to the miniature end-plate potentials recorded from the motor end plate in skeletal muscle, the amplitude and time course of the spontaneous miniature potentials from the vas deferens vary over a wide range (Fig. 4-9). The largest potentials with a fast time course might be produced by the release of transmitter from nerves close to the point of impalement of the microelectrode on the effector cell membrane; the smaller, slower potentials might be caused by transmitter liberated at greater distances from the electrode. The shape of the amplitude distribution curve for the spontaneous potentials is continuous (Fig. 4-10) and similar to that obtained from other muscle cells having multiple innervation, e.g., crustacean muscle and "slow" fibers from frog and chick skeletal muscle.[27] In these cells the microelectrode simultaneously records the activity of many different nerve endings at various distances from the point of impalement. The amplitude distribution curves from muscles that have localized nerve endings, e.g., frog and mammalian "twitch" fibers, show a gaussian or bell-shaped distribution. In such muscles the microelectrode records from only one discrete motor end plate.

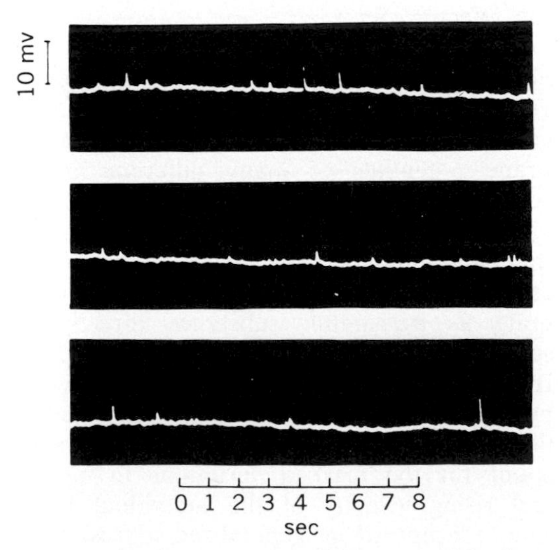

Fig. 4-9. Continuous intracellular recording from smooth muscle cell in vas deferens of guinea pig. Note spontaneous discharge of miniature junction potentials that appear in quiescent muscle cell in absence of hypogastric nerve stimulation. (From Burnstock and Holman.[26])

Katz[51] has shown that the miniature potentials at the motor end plate in frog skeletal muscle are due to the spontaneous release of quanta of acetylcholine and that the neurally evoked end-plate potential arises from the synchronous release of many quanta of transmitter. At present, techniques are not sufficiently refined to permit recording from a single neuroffector junction in the smooth muscle of the vas deferens. Consequently, it is impossible to determine whether the spontaneous potentials in this tissue arise in a quantal fashion, i.e., whether they are produced by the release of equal amounts of transmitter at different distances from the locus of the microelectrode. Furthermore, it is not possible to say whether or not the junction potentials in smooth muscle are composed solely of miniature potentials. Indirect evidence, however, indicates that they are. Pretreatment of the guinea pig with reserpine or partial denervation of the vas deferens reduces the *amplitude* of the neurally evoked potentials and the *rate* of discharge of the spontaneous miniature potentials. Furthermore, when the strength of nerve stimulation is initially low and is gradually increased, the amplitude of EJP increases in a stepwise fashion.[3, 50]

EJPs have also been recorded in smooth

muscle cells from the longitudinal layer of the rabbit colon in response to stimulation of the pelvic (parasympathetic) nerves.[41] Although in this situation the transmitter is acetylcholine, the EJPs have many features in common with those just described for the vas deferens. One advantage of the rabbit colon preparation is that both the right and left pelvic nerves can be isolated and stimulated. In this manner the interactions of the two nerves can be observed on the effector cell membrane. EJPs are recorded from every smooth muscle cell when either the right or left pelvic nerve is stimulated with single or

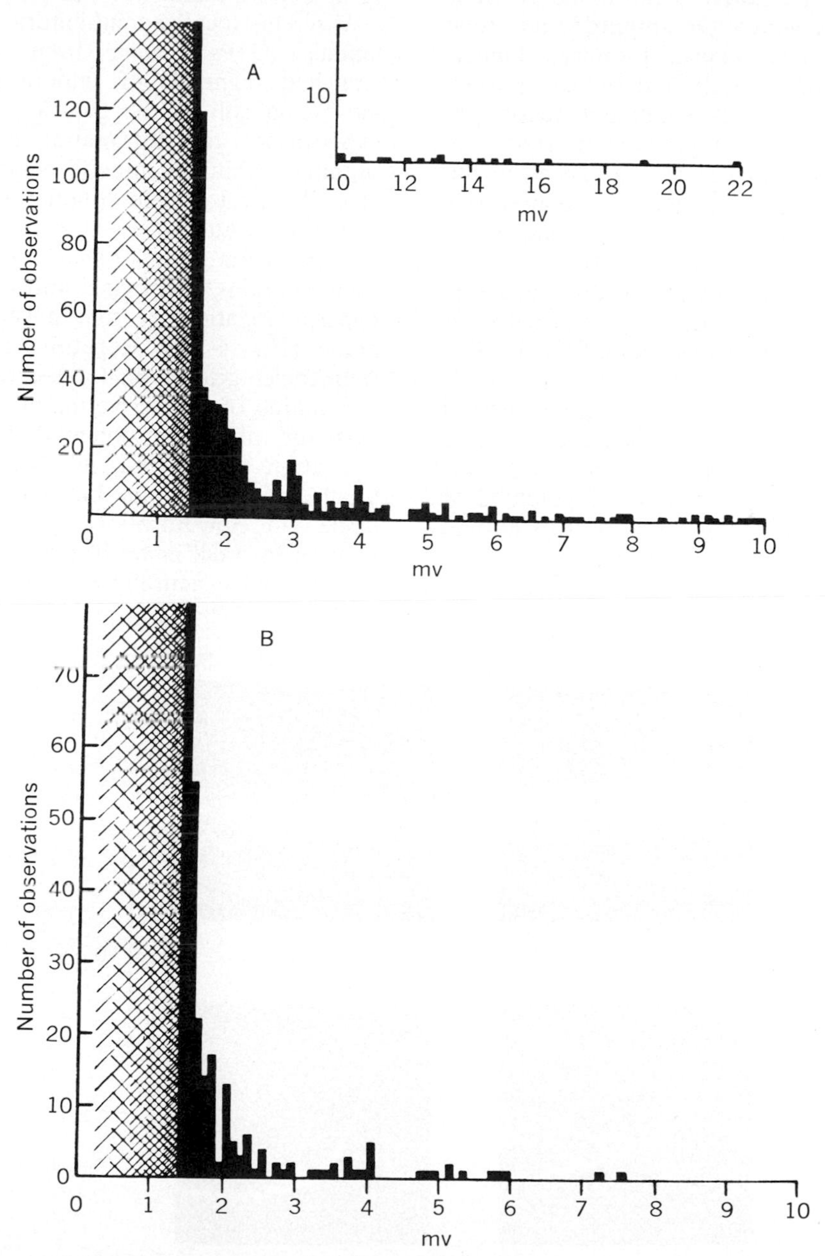

Fig. 4-10. Frequency distribution of amplitudes of spontaneous miniature junction potentials in two typical smooth muscle cells from different preparations of guinea pig vas deferens. Hatched area indicates width of baseline on oscilloscope trace (i.e., "noise" level of recording instrument). Width of baseline determines amplitude of smallest potential resolvable in any one experiment, and therefore calculation of mean amplitude is somewhat arbitrary. Note that shape of histogram is continuous, in contrast to bell-shaped gaussian amplitude distribution of spontaneous miniature potentials at motor end plate in frog skeletal muscle. (From Burnstock and Holman.[27])

repetitive shocks. If both nerves are stimulated simultaneously, there is no increase in the rate of development or in the amplitude of the individual EJPs. These findings show that each pelvic nerve has access to every smooth muscle cell in the colon, and that the effects of simultaneous stimulation of both nerves are not additive.[42] Since the colon is spontaneously active, the amplitude and time course of the EJPs elicited by nerve stimulation are usually complicated by the appearance of spontaneously generated action potentials, and it is not possible to study the EJPs as thoroughly in this muscle as in the vas deferens. The high degree of spontaneous motility also makes it technically difficult to keep the microelectrode securely inside one muscle cell for long periods of time, and it is impossible to demonstrate the existence of "miniature" spontaneous potentials in this tissue.

Stimulation of the lumbar colonic nerves abolishes the spontaneous discharge of action potentials and the muscle becomes quiescent. During inhibition the membrane potential of the smooth muscle cells remains at its normal resting level. If the colon is stretched so that the membrane potential of the muscle cells is reduced and the spike discharge becomes continuous, the inhibitory effects of nerve stimulation are accompanied by a generalized hyperpolarization of the cell membrane. The hyperpolarization is abolished by adrenergic blocking agents, suggesting that the inhibitory transmitter is norepinephrine or a mixture of epinephrine and norepinephrine.[42]

Discrete, localized inhibitory junction potentials (IJPs) have been recorded in stretched segments of smooth muscle from the taenia coli of the guinea pig.[28] In these experiments the intramural nerves (those running within the muscle coats) were activated by electric field stimulation with pulses of short duration (less than 1 msec) at frequencies from 1 to 30 shocks/sec. Following each stimulus there was transient localized hyperpolarization of the muscle cell membrane (Fig. 4-11). Repetitive stimulation at frequencies greater than 1/sec resulted in the summation of successive inhibitory potentials, cessation of action potential discharge, and relaxation of the muscle. These IJPs were unaffected by atropine but abolished by procaine (in concentrations that blocked conduction in small nerve fibers), suggesting that the IJPs were neurally mediated but that the

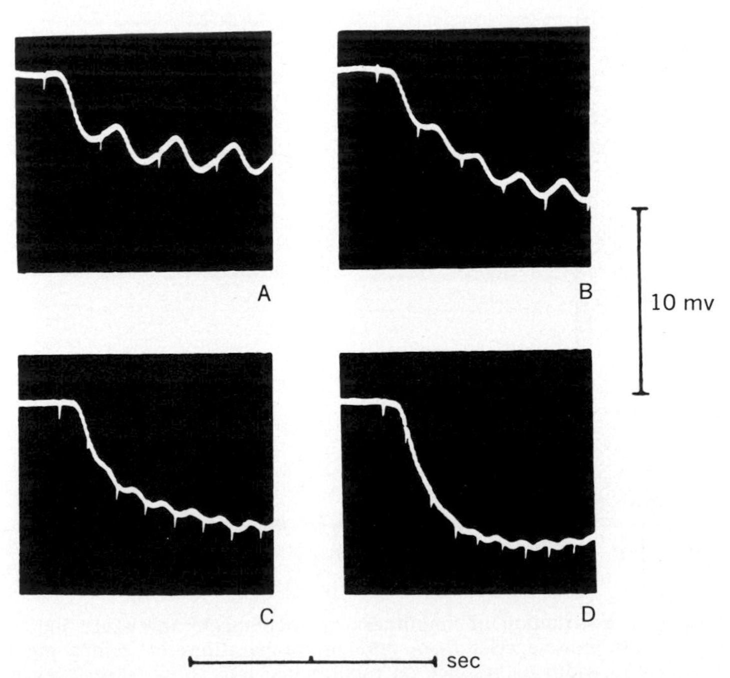

Fig. 4-11. Inhibitory junction potentials (IJPs) recorded from smooth muscle cells in taenia coli of guinea pig in response to stimulation of intrinsic nerves. Electrical records obtained with sucrose gap method. Downward deflection of trace indicates hyperpolarization of membrane. Frequencies of stimulation; **A,** 2/sec; **B,** 3/sec; **C,** 4/sec; **D,** 6/sec. (From Burnstock et al.[28])

neurotransmitter was not acetylcholine. The inhibitory potentials were not prevented by adrenergic blocking agents, suggesting that the transmitter was not an adrenergic amine. The identity of the inhibitory transmitter liberated during stimulation of these intrinsic nerve fibers is not known.[50]

Morphology of neuroeffector junctions—correlation with electrophysiology

The term *"nerve ending"* as applied to an autonomic neuroeffector junction is best defined in the functional sense to mean that part or parts of the neuron from which the release of transmitter substance occurs during stimulation.[25, 65] Studies with the electron microscope have greatly increased knowledge of the neuroeffector junction. Of the various smooth muscles examined, the vas deferens will be discussed here. The fine structure of the innervation of this tissue is essentially similar to that of other smooth muscles (uterus, urinary bladder, intestine, gallbladder, colon, iris),[25, 33, 58] and the electrophysiology of neuroeffector transmission has been more extensively studied in the vas deferens than in these other tissues.

As the hypogastric nerves enter the vas deferens, they divide into numerous branches of nonmyelinated axons supported by a network of Schwann cells, the so-called autonomic ground plexus. Fine strands of the plexus, containing small bundles of two to eight axons, ramify within the muscular tissue, and occasionally single axons, partially or completely free of their Schwann cell sheath, come to lie within 200 to 300 Å of the smooth muscle cell membrane.[58] Sometimes these unsheathed axons cause indentations in the surface of the muscle cell and when viewed in longitudinal section appear as "beadlike" enlargements of the axon lying in a cleft of the muscle cell membrane.[64] In these instances the space separating the nerve and muscle cell membranes may be only 150 Å wide. Since serial sections of this material are not available, it is not possible to state with certainty whether these nerve-muscle contacts are actually nerve terminals or whether the axon emerges from the groove on one muscle cell and proceeds to form close contacts with other muscle fibers, constituting a series of "synapses *en passage*." It has been suggested that the nerve action potential may release transmitter from many points along the axon—from the regions of close contact, the "synapses *en passage*," as well as other points in the autonomic ground plexus more distant from the muscle cells.[25, 58] In this manner one axon may influence many different smooth muscle cells. This arrangement could be the morphologic basis for the electrophysiologic findings that EJPs are recorded from every smooth muscle cell in the vas deferens. The variations in latency and time course of the individual junction potentials may be caused by the release of transmitter, directly on some cells (at the "synapses *en passage*") and at greater distances from other cells (in the autonomic ground plexus).

Until serial sections and three-dimensional reconstructions of the tissue are available, one cannot tell whether each muscle cell lies in close approximation to one or more axons. In the vas deferens of the rat the innervation is so dense that every cell may be intimately associated with one or more axons.[64] In the vas deferens of the guinea pig, however, Merrillees et al.[58] could find only one close axon-muscle contact per 100 muscle cells. These observations led Burnstock and Holman[4] to postulate that in the guinea pig some cells were activated directly by the release of transmitter, while others were activated indirectly, either by generalized diffusion of transmitter from axons in the autonomic ground plexus or by passive electrotonic spread of current from the directly excited cells.

Unlike the motor end plate of skeletal muscle, the nerve and smooth muscle cell membranes are not specialized at their points of contact. The axons lying in close apposition to the smooth muscle cell membrane as well as those running within the autonomic ground plexus, however, are filled with vesicles and mitochondria.[58, 64, 65] It will be recalled that vesicles are also found within the nerve terminals at the neuromuscular junction in skeletal muscle. It has been suggested, but not proved, that acetylcholine might be stored in these vesicles, each vesicle representing a "packet" of transmitters, and that the release of such packets produces the miniature end-plate potentials. The finding of vesicles in the autonomic axons within the vas deferens and the presence of spontaneous miniature potentials in the muscle cells suggest that the transmission processes in the vas deferens and at the skeletal muscle neuromuscular junction are essentially similar.[4, 50]

Since many smooth muscles, including the vas deferens, are innervated by both adre-

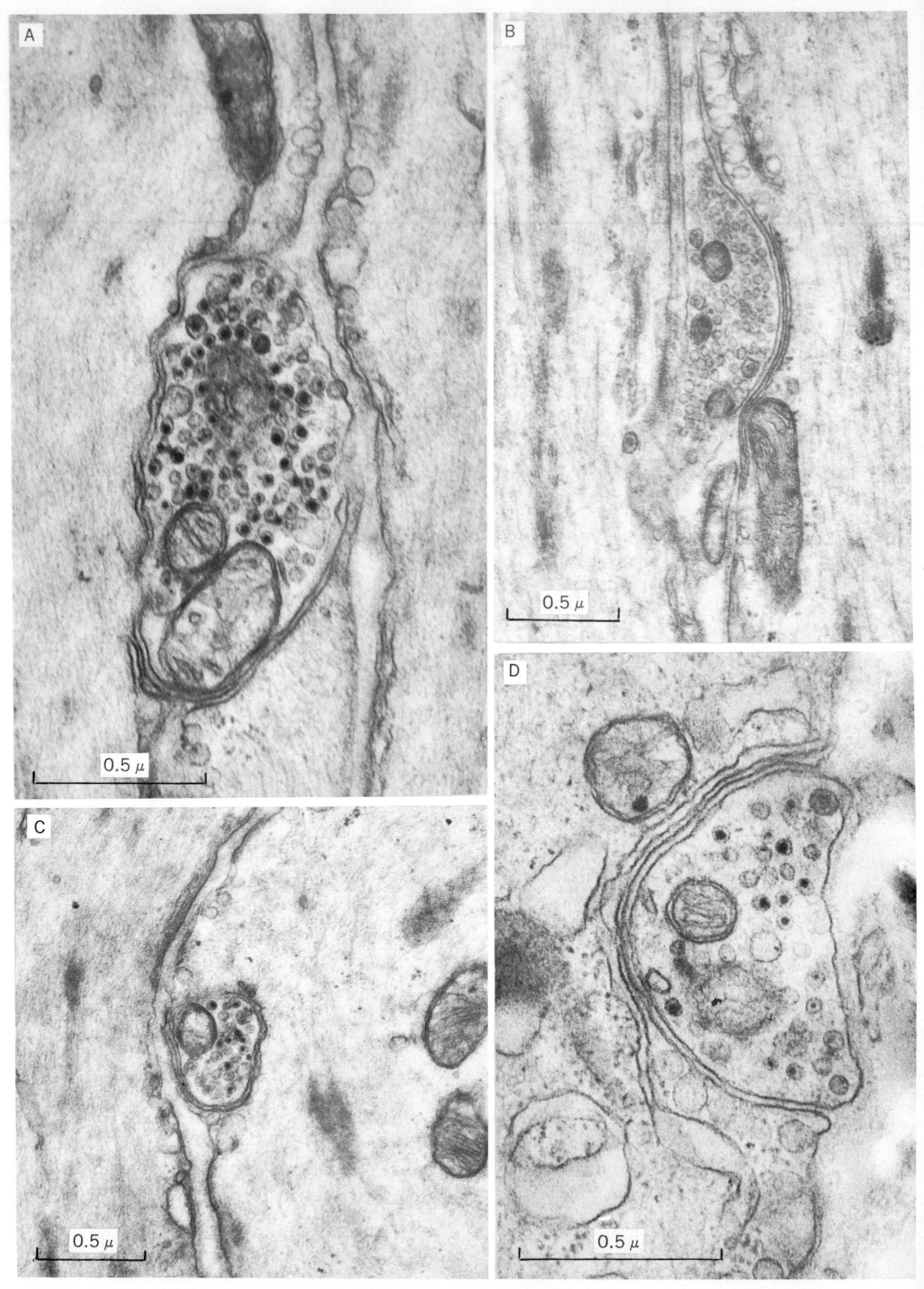

Fig. 4-12. For legend see opposite page.

nergic and cholinergic nerves, the transmitter liberated at the neuroeffector sites can be either norepinephrine or acetylcholine. A crucial problem, therefore, is to distinguish between cholinergic and adrenergic neuroeffector junctions. One experimental approach is to examine the relationship between the axoplasmic vesicles and the transmitter substances and/or their precursors. Two types of vesicles, granular and agranular, have been seen in the axons of both the rat and guinea pig vas deferens.[25, 58, 64] The agranular vesicles are of uniform size (400 to 600 Å) and resemble those found in the nerve terminals at the skeletal neuromuscular junction. The granular vesicles range in diameter from 300 to 900 Å and contain a core of electron-dense material. Richardson[64] suggests that the granular vesicles contain norepinephrine and that the axons containing such granules might be adrenergic. Support for this idea comes from the finding that tritiated norepinephrine, injected intravenously, can be localized by electron microscopic autoradiography within the granular vesicles in sympathetic nerves in the rat heart and pineal gland.[73] At present there is no certain evidence regarding the significance or contents of the agranular vesicles, and Richardson cautions "it would be premature to state that axons containing agranular vesicles are cholinergic in function."[64] The agranular vesicles may contain dopamine, a precursor of norepinephrine that comprises about 50% of the total amine content of sympathetic nerves, the remainder being norepinephrine. Since many axons contain both granular and agranular vesicles, while others have predominantly one or the other (Fig. 4-12), it is conceivable that the two forms of vesicles do actually represent stages in the synthesis of catecholamines.[64] Nevertheless, electron microscopic studies on nerve axons in the sphincter muscle from the rabbit iris, which is believed to be innervated predominantly by cholinergic nerves, show that the axoplasmic vesicles are of the agranular type. The axons in the dilator muscle, thought to be adrenergically innervated, contain mostly granular vesicles.[65] Information is just beginning to accumulate on the effects of procedures that decrease the storage and release of transmitter substances (e.g., reserpine treatment, increased extracellular potassium ion concentration, prolonged nerve stimulation) on the fine structure of autonomic nerves.[25]

The physiologist is particularly interested in identifying vesicles and other intra-axonal structures associated with cholinergic or adrenergic transmission in smooth muscle. In this regard, histochemical techniques have been particularly useful. Catecholamines develop an intense fluorescence when treated with formaldehyde, and this reaction has been exploited to visualize adrenergic neurons in smooth muscle.[37] When freeze-dried muscles are exposed to formaldehyde and then sectioned and examined under the fluorescence microscope, areas of intense fluorescence appear as beadlike varicosities within the cell body and along the entire lengths of the terminal axons of the adrenergic nerves (Fig. 4-13, *A*). Chemical analyses of the catecholamine content of the muscles correlate positively with the size and distribution of these varicosities. Combined studies using fluorescence histochemistry and electron microscopy show that the varicosities are especially prominent where the axons containing granular vesicles come in close contact with smooth muscle cells, providing additional evidence that the adrenergic neurotransmitter is localized and probably released at these sites.[3, 25]

Histochemical identification of cholinergic axons presumes that functionally cholinergic axons contain higher levels of acetylcholinesterase (AChE) than noncholinergic axons. The histochemical procedure for the localization of AChE involves incubation of the

Fig. 4-12. Electron micrographs showing relationship of autonomic nerve axons to smooth muscle cells in vas deferens of rat. **A,** Axon, free of Schwann cell sheath, lying in deep groove in surface of smooth muscle cell. Axon contains numerous granular and agranular vesicles and mitochondria. **B,** "Beadlike" enlargement of nerve axon, packed with agranular vesicles, lying in groove on surface of smooth muscle cell. Space between nerve and muscle membranes is about 150 Å. Obliquely sectioned lower portion of axon reenters muscular cleft and appears to be turning to left to form second "bead" on adjacent muscle cell. **C,** Tiny axon almost submerged beneath surface of muscle cell on right. Axon contains granular and agranular vesicles and one mitochondrion. **D,** Axon containing fewer vesicles and two mitochondria. In this instance, separation between nerve and muscle membrane is about 250 Å. Row of pinocytotic vesicles associated with muscle cell membrane is located within lower part of neuroeffector junction. (From Richardson.[64])

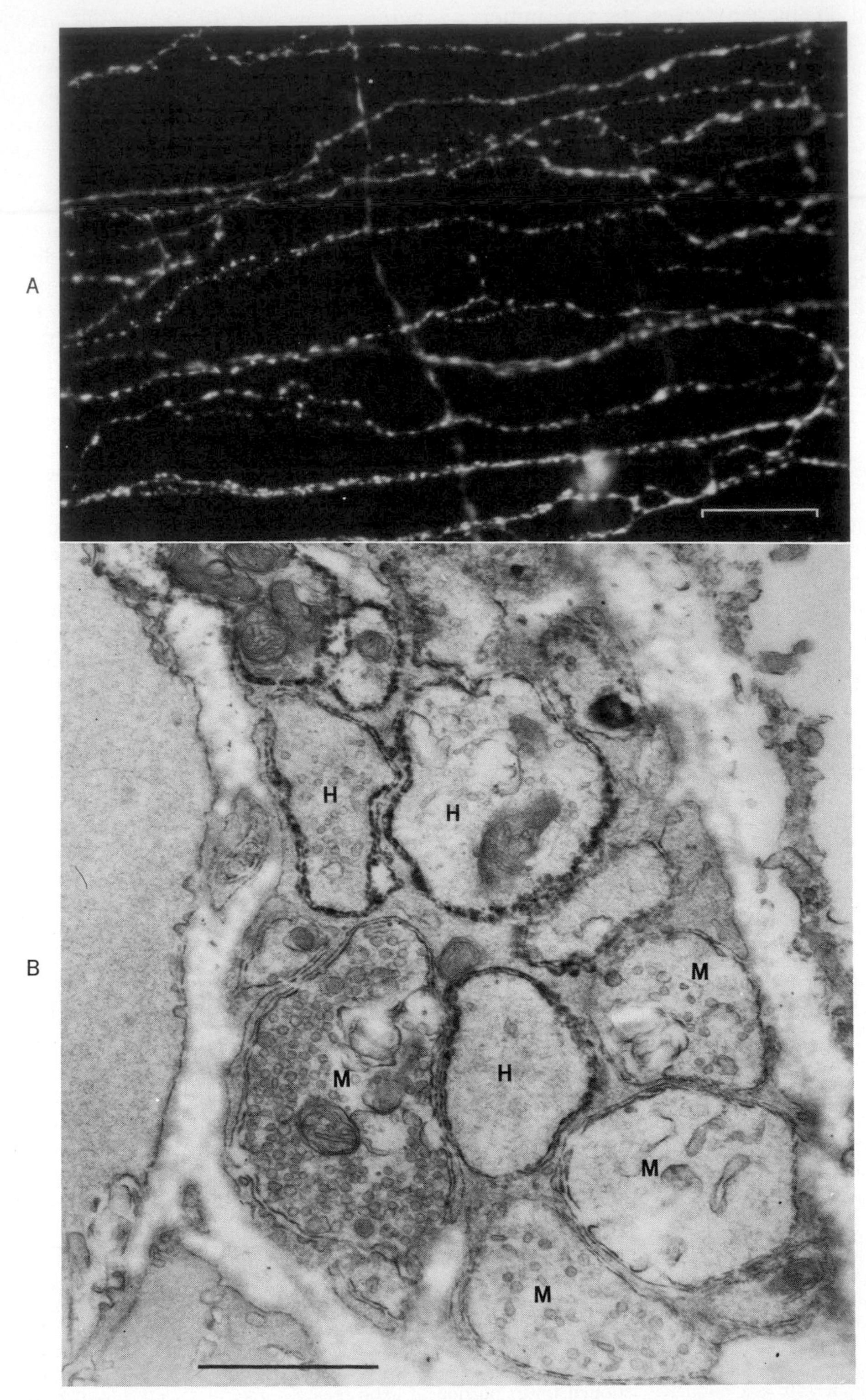

Fig. 4-13. A, Whole mount of mesenteric vein from sheep; freeze-dried preparation incubated with formaldehyde for 1 hr. Bundles of nerve fibers containing noradrenaline are identified by beadlike regions of intense fluorescence along axons. Bar represents 50μ. **B,** Electron micrograph of axon bundle in vas deferens of guinea pig. Acetylcholinesterase (AChE) incubation. Heavy deposits of AChE appear in plasma membranes of some axons, *H*, and lighter deposits in others, *M*. Bar represents 1μ (From Burnstock.[25])

tissue with acetylthiocholine and $CuSO_4$. Tissue AChE hydrolyzes the acetylthiocholine to thiocholine and acetic acid. Thiocholine then precipitates as copper thiocholine, which in the presence of H_2S forms dark-brown deposits at the sites of AChE activity within the tissue. These sites are visible with the light microscope[25] and presumably indicate the areas of acetylcholine release and/or storage. This histochemical method has recently been modified for use in electron microscopy. The innervation of the bladder, vas deferens, and nictitating membrane has been examined with this technique.[25] Heavy deposits of AChE appear in the plasma membranes of some of the axons in these tissues, identifying them as cholinergic (Fig. 4-13, *B*). Many cholinergic axons are found in the bladder but relatively few in the vas deferens and nictitating membrane. AChE activity and norepinephrine (as indicated by autoradiography) do not occur together within the same axon (i.e., a single axon does not liberate both transmitters), but separate cholinergic and adrenergic axons sometimes run together in the same nerve bundle in many autonomically innervated muscles.[3, 25]

Nerve-muscle relationships in vascular muscle

Although the characteristics of autonomic innervation of vascular muscle are similar to those of visceral muscle, the neuroeffector organization in vascular muscle is characterized by a relatively long distance between terminal axons and smooth muscle cells.[10, 69] In most blood vessels the autonomic ground plexus lies within the adventitia and the terminal axons rarely penetrate beyond the outer surface of the media. This anatomic distribution, especially well documented for adrenergic nerves, varies considerably not only among different blood vessels but also within different regions of the same vessel.[10, 40] In large elastic arteries the adrenergic nerves are generally limited to the outer portion of the vessel; in muscular arteries they penetrate for a short distance into the media; some thick-walled cutaneous veins contain many adrenergic fibers in their media, while veins in skeletal muscle have few nerve fibers in their smooth muscle layers. In the proximal part of the main pulmonary artery of the rabbit, adrenergic nerves terminate on the outer surface of the media, while in the distal portion of the same vessel the nerves extend well into the media.[10] The nerve-muscle relationships in most blood vessels become more intimate as vessel size diminishes. In terminal arterioles and precapillary sphincters and venules the ratio of axons to muscle cells is greater than in large arteries. In the smaller vessels the regions of close contact between axons and muscle cell are also more numerous.[1, 10, 68, 69] The cholinergic innervation, although less well studied than adrenergic innervation, is also restricted to the medioadventitial junction.[10, 68]

The long distance between terminal axons and smooth muscle cells (from 5,000 to 10,000 Å in large arteries and 700 to 1,000 Å in arterioles and venules) has functional implications. Muscle cells other than those at the medioadventitial border must be activated either by diffusion of transmitter over relatively long distances or by some form of electrical or mechanical coupling between individual muscle cells. It has been suggested[40] that the outer layer of cells is controlled by nerves and therefore is of the multi-unit type, while the inner layer has myogenic activity and is of the unitary type. If this is the case, then one would expect to find numerous gap or tight junctions between cells of the inner layer and relatively few junctions in the outer layer. Although this possibility has never been systematically studied, in the few instances where arterial muscle has been examined with the electron microscope, the muscle cells are separated by connective tissue elements throughout the entire muscular layer and cell-to-cell junctions are rare.[68] However, in arterioles, precapillary sphincters, and veins the smooth muscle cells lie close together and gap junctions are frequently seen. These vessels manifest spontaneous vasomotion independent of neural activity.[10, 68]

EJPs have been recorded in the outermost cells of the mesenteric artery in the guinea pig during stimulation of the splanchnic nerve.[68] The time course and latency of these potentials are similar to those in the vas deferens during stimulation of the hypogastric nerve. In the one reported electrophysiologic study of a blood vessel having cholinergic innervation (the uterine artery of the guinea pig), stimulation of the cholinergic nerves or the administration of acetylcholine had no effect on the muscle cell membrane potential, although a pronounced vasodilatation occurred.[1] The dilatation might have been mediated by an effect on propagation of electrical activity rather than by a change in membrane potential. Extension of these studies to other blood vessels is difficult because many vascular beds are inaccessible for electrophysiologic techniques.

SUMMARY

Smooth muscles can be divided generally into two groups according to their physiologic properties: (1) unitary and (2) multi-unit. Examples of unitary muscles are those of the viscera, e.g., intestine, uterus, and ureter. These muscles are characterized by their spontaneous motility and by their sensitivity to stretch. The basis of the spontaneous rhythmicity is a periodic fluctuation in the membrane potential in some muscle cells or groups of cells, which is analogous to the activity of the cardiac pacemaker. The fluctuations bring the membrane potential to threshold, and action potentials are generated, which in turn initiate a contraction of the muscle. Unlike the cardiac pacemaker, the pacemaker areas in unitary smooth muscles are not fixed but migrate about within the tissue. Action potentials produced in such areas spread uniformly over the entire muscle as though the tissue were a single unit. Although the spontaneous activity may be independent of nerve stimulation, it is usually modified and coordinated by the autonomic nervous system. The excitability of the muscle is also influenced by stretch since unitary smooth muscle cells behave as stretch receptors, their membranes being depolarized by stretching. The action of muscle stimulants, e.g., acetylcholine, is also characterized by a depolarization of the cell membrane and by an increase in frequency of action potential discharge. Inhibitors of muscle activity, e.g., epinephrine in the intestine, suppress action potential discharge and "stabilize" the membrane at its normal quiescent level or at a slightly more negative (hyperpolarized) level.

Examples of multi-unit muscles are the pilomotor fibers in the skin, the ciliary muscles and the muscles of the nictitating membrane of the eye, and some vascular smooth muscles. All these muscles are normally quiescent and are activated only through their autonomic nerves, each muscle being composed of multiple "motor units." Multi-unit muscles are generally insensitive to stretch. Some of them, e.g., those of the nictitating membrane and of various blood vessels, can be stimulated by the direct application of certain humoral substances, e.g., norepinephrine. The electrophysiologic aspects of such activation are not well established.

Neuroeffector transmission processes associated with neural activation in smooth muscle are qualitatively similar to those occurring at the motor end plate in skeletal muscle. Discrete nerve endings have never been seen in smooth muscle, however, and it seems likely that during nerve stimulation the transmitter is released at various points along the nerve axon. Some of these regions lie in close proximity to the muscle cell membrane, while others are located farther away. Thus some smooth muscle cells are activated directly and others by a generalized diffusion of transmitter. Unlike the motor end plate in skeletal muscle, there is no specialization of the nerve and smooth muscle cell membranes at their points of contact. However, the axons do contain vesicles, a number of which are granular and are believed to be storage sites for norepinephrine. In other axons the vesicles are predominantly agranular, and these may contain acetylcholine or possibly precursors of norepinephrine.

REFERENCES

General reviews

1. Bevan, J. A., Furchgott, R. F., Maxwell, R. A., and Somlyo, A. P., editors: Physiology and pharmacology of vascular neuroeffector systems, Basel, 1971, S. Karger.
2. Bohr, D. F.: Electrolytes and smooth muscle contraction, Pharmacol. Rev. **16:**85, 1964.
3. Bülbring, E., Brading, A. F., Jones, A. W., and Tomita, T., editors: Smooth muscle, Baltimore, 1970, The Wililams & Wilkins Co.
4. Burnstock, G., and Holman, M. E.: Smooth muscle: autonomic nerve transmission, Ann. Rev. Physiol. **25:**61, 1963.
5. Burnstock, G., and Holman, M. E.: Effects of drugs on smooth muscle, Ann. Rev. Pharmacol. **6:**129, 1966.
6. Burnstock, G., Holman, M. E., and Prosser, C. L.: Electrophysiology of smooth muscle, Physiol. Rev. **43:**482, 1963.
7. Daniel, E. E.: Effect of drugs on contractions of vertebrate smooth muscle, Ann. Rev. Pharmacol. **4:**189, 1964.
8. Eichna, L. W., editor: Vascular smooth muscle, Physiol. Rev. **42** (suppl. 5)**:**1, 1962.
9. Furchgott, R. F.: The pharmacology of vascular smooth muscle, Pharmacol. Rev. **7:**183, 1955.
10. Somlyo, A. P., and Somlyo, A. V.: Vascular smooth muscle. I. Normal structure, pathology, biochemistry, and biophysics, Pharmacol. Rev. **20:**197, 1968.
11. Somlyo, A. P., and Somlyo, A. V.: Vascular smooth muscle. II. Pharmacology of normal and hypertensive vessels, Pharmacol. Rev. **22:**249, 1970.

Original papers

12. Ahlquist, R. P.: A study of the adrenotropic receptors, Am. J. Physiol. **153:**586, 1948.
13. Barr, L., and Dewey, M. M.: Electrotonus and electrical transmission in smooth muscle. In Code, C. F., editor, Alimentary canal section: Handbook of physiology, Washington, D. C., 1968, American Physiological Society, vol. 4.
14. Bianchi, C. P.: Cell calcium, Washington, D. C., 1968, Butterworth & Co.
15. Bohr, D. F., Goulet, P. L., and Taquini, A. C.: Direct tension recording from smooth muscle of resistance vessels from various organs, Angiology **12:**478, 1961.
16. Bozler, E.: Action potentials and conduction of excitation in muscle, Biol. Symp. **3:**95, 1941.
17. Bozler, E.: Conduction, automaticity and tonus of visceral smooth muscles, Experientia **4:**213, 1948.
18. Brooks, C. M., Hoffman, B. F., Suckling, E. E., and Orias, O.: Excitability of the heart, New York, 1955, Grune & Stratton, Inc.
19. Bueding, E., and Bülbring, E.: The inhibitory action of adrenaline. Biochemical and biophysical observations. In Bülbring, E., editor: Pharmacology of smooth muscle; proceedings of the second international pharmacological meeting, Oxford, 1964, Pergamon Press, Ltd., vol. 6.
20. Bueding, E., et al.: The effect of adrenaline on the adenosine triphosphate and creatine phosphate content of intestinal smooth muscle, J. Physiol. **193:**187, 1967.
21. Bülbring, E.: Physiology and pharmacology of intestinal smooth muscle. In Lectures on the scientific basis of medicine (British Postgraduate Medical Federation), London, 1957, Athlone Press, vol. 7.
22. Bülbring, E.: Electrical activity in intestinal smooth muscle, Physiol. Rev. **42**(suppl. 5):160, 1962.
23. Bülbring, E., and Tomita, T.: Increase in membrane conductance by adrenaline in the smooth muscle of the guinea-pig taenia coli, Proc. R. Soc. Lond. (Biol.) **172:**89, 1969.
24. Bülbring, E., and Tomita, T.: Suppression of spike generation by catecholamines in the smooth muscle of the guinea-pig taenia coli, Proc. R. Soc. Lond. (Biol.) **172:**103, 1969.
25. Burnstock, G.: Structure of smooth muscle and its innervation. In Bülbring, E., Brading, A. F., Jones, A. W., and Tomita, T.: Smooth muscle, Baltimore, 1970, The Williams & Wilkins Co.
26. Burnstock, G., and Holman, M. E.: The transmission of excitation from autonomic nerve to smooth muscle, J. Physiol. **155:**115, 1961.
27. Burnstock, G., and Holman, M. E.: Spontaneous potentials at sympathetic nerve endings in smooth muscle, J. Physiol. **160:**446, 1962.
28. Burnstock, G., Campbell, G., Bennett, M., and Holman, M. E.: Inhibition of the smooth muscle of the taenia coli, Nature **200:**581, 1963.
29. Casteels, R., and Kuriyama, H.: Membrane potential and ionic content in pregnant and non-pregnant rat myometrium, J. Physiol. **177:** 263, 1965.
30. Conway, E. J., and Hingerty, D.: The influence of adrenalectomy on muscle constituents, Biochem. J. **40:**561, 1946.
31. Csapo, A.: The four direct regulatory factors of myometrial function. In Progesterone: its regulatory effect on the myometrium, Ciba Foundation Study Group No. 34, London, 1969, J. & A. Churchill, Ltd.
32. Daniel, E. E., et al.: The sodium pump in smooth muscle. In Bevan, J., Furchgott, R. F., Maxwell, R. A., and Somlyo, A. P., editors: Physiology and pharmacology of vascular neuroeffector systems, Basel, 1971, S. Karger.
33. Dewey, M. M., and Barr, L.: Structure of vertebrate smooth muscle. In Code, C. F., editor, Alimentary canal section: Handbook of physiology, Washington, D. C., 1968, American Physiological Society, vol. 4.
34. Durbin, R. P., and Jenkinson, D. H.: The effect of carbachol on the permeability of depolarized smooth muscle to inorganic ions, J. Physiol. **157:**74, 1961.
35. Elliott, G. F.: X-ray diffraction studies on striated and smooth muscle, Proc. R. Soc. Lond. (Biol.) **160:**469, 1964.
36. Evans, D. H. L., Schild, H. O., and Thesleff, S.: Effects of drugs on depolarized plain muscle, J. Physiol. **143:**474, 1958.
37. Falck, B.: Observations on the possibilities of the cellular localization of monamines by a fluorescence method, Acta Physiol. Scand. **56** (suppl. 197):1, 1962.
38. Fawcett, D. W.: The fine structure of capillaries, arterioles and small arteries. In Reynolds, S. R. M., and Zweifach, B. W., editors: The microcirculation, Urbana, 1959, University of Illinois Press.
39. Filo, R. S., Rüegg, J. C., and Bohr, D. F.: Actomyosin-like protein of the arterial wall, Am. J. Physiol. **205:**1247, 1963.
40. Folkow, B.: Autoregulation in muscle and skin, Circ. Res. **15** (suppl. 1)19, 1964.
41. Gillespie, J. S.: The electrical and mechanical responses of intestinal smooth muscle cells to stimulation of their extrinsic parasympathetic nerves, J. Physiol. **162:**76, 1962.
42. Gillespie, J. S.: Electrical activity of the colon. In Code, C. F., editor, Alimentary canal section: Handbook of physiology, Washington, D. C., 1968, American Physiological Society, vol. 4.
43. Goodford, P. J.: The sodium content of the smooth muscle of the guinea-pig taenia coli, J. Physiol. **163:**411, 1962.
44. Goodford, P. J.: Distribution and exchange of electrolytes in intestinal smooth muscle. In Code, C. F., editor, Alimentary canal section: Handbook of physiology, Washington, D. C., 1968, American Physiological Society, vol. 4.
45. Hanson, J., and Lowy, J.: Comparative studies on the structure of contractile systems, Circ. Res. **15** (suppl. 2):4, 1964.
46. Hanson, J., and Lowy, J.: The structure of actin filaments and the origin of the axial periodicity in the I-substance of vertebrate striated muscle, Proc. R. Soc. Lond. (Biol.) **155:**449, 1964.
47. Hellman, K.: The isolated pilomotor muscles as an in vitro preparation, J. Physiol. **169:**603, 1963.
48. Hodgkin, A. L., and Horowicz, P. C.: The differential action of hypertonic solutions on the twitch and action potential of a muscle fibre, J. Physiol. **136:**17P, 1957.

49. Holman, M. E.: Introduction to electrophysiology of visceral smooth muscle. In Code, C. F., editor, Alimentary canal section: Handbook of physiology, Washington, D. C., 1968, American Physiological Society, vol. 4.
50. Holman, M. E.: Junction potentials in smooth muscle. In Bülbring, E., Brading, A. F., Jones, A. W., and Tomita, T., editors: Smooth muscle, Baltimore, 1970, The Williams & Wilkins Co.
51. Katz, B.: The transmission of impulses from nerve to muscle, and the subcellular unit of synaptic action, Proc. R. Soc. Lond. (Biol.) **155:**455, 1962.
52. Kuriyama, H.: Ionic basis of smooth muscle action potentials. In Code, C. F., editor, Alimentary canal section: Handbook of physiology, Washington, D. C., 1968, American Physiological Society, vol. 4.
53. Kuriyama, H.: Effects of ions and drugs on the electrical activity of smooth muscle. In Bülbring, E., Brading, A. F., Jones, A. W., and Tomita, T., editors: Smooth muscle, Baltimore, 1970, The Williams & Wilkins Co.
54. Lane, B. P., and Rhodin, J. A. G.: Cellular interrelationships and electrical activity in two types of smooth muscle, J. Ultrastruct. Res. **10:**470, 1964.
55. Li, C. L., Shy, G. M., and Wells, J.: Some properties of mammalian skeletal muscle fibres with particular reference to fibrillation potentials, J. Physiol. **135:**522, 1957.
56. Marshall, J. M.: Regulation of activity in uterine smooth muscle, Physiol. Rev. **42** (suppl. 5):213, 1962.
57. Marshall, J. M.: The action of oxytocin on uterine smooth muscle. In Bülbring, E., editor: Pharmacology of smooth muscle; proceedings of the second international pharmacological meeting, Oxford, 1964, Pergamon Press, Ltd., vol. 6.
58. Merrillees, N. C. R., Burnstock, G., and Holman, M. E.: Correlation of fine structure and physiology of the innervation of smooth muscle in the guinea-pig vas deferens, J. Cell Biol. **19:**529, 1963.
59. Nagai, T., and Prosser, C. L.: Electrical parameters of smooth muscle cells, Am. J. Physiol. **204:**915, 1963.
60. Needham, D. M., and Shoenberg, C. F.: Proteins of the contractile mechanism in vertebrate smooth muscle. In Code, C. F., editor, Alimentary canal section: Handbook of physiology, Washington, D. C., 1968, American Physiological Society, vol. 4.
61. Niedergerke, R.: Movements of Ca in frog heart ventricles at rest and during contractions, J. Physiol. **167:**515, 1963.
62. Peachy, L. D., and Porter, K. R.: Intracellular impulse conduction in muscle cells, Science **129:**721, 1959.
63. Rhodin, J. A. G.: Fine structure of vascular walls in mammals with special reference to the smooth muscle component, Physiol. Rev. **42** (suppl. 5):48, 1962.
64. Richardson, K. C.: The fine structure of the nerve endings in smooth muscle of the rat vas deferens, J. Anat. **96:**427, 1962.
65. Richardson, K. C.: The fine structure of the albino rat iris with special reference to the identification of adrenergic and cholinergic nerves and nerve endings in its intrinsic muscles, Am. J. Anat. **114:**173, 1964.
66. Robertson, W. van B., and Dunihue, F. W.: Water and electrolyte distribution in cardiac muscle, Am. J. Physiol. **177:**292, 1954.
67. Rosenbluetth, A.: The transmission of nerve impulses at neuroeffector junctions and peripheral synapses, New York, 1950, John Wiley & Sons, Inc.
67a. Somlyo, A. P., Devine, C. E., Somlyo, A. V., and Rice, R. V.: Filament organization in vertebrate smooth muscle, Proc. R. Soc. Lond. (Biol.) **101:**225, 1973.
68. Speden, R. N.: Excitation of vascular smooth muscle. In Bülbring, E., Brading, A. F., Jones, A. W., and Tomita, T., editors: Smooth muscle, Baltimore, 1970, The Williams & Wilkins Co.
69. Su, C., and Bevan, J. A.: Adrenergic transmitter release and distribution in blood vessels. In Bevan, J. A., Furchgott, R. F., Maxwell, R. A., and Somlyo, A. P., editors: Physiology and pharmacology of vascular neuroeffector systems, Basel, 1971, S. Karger.
70. Tomita, T.: Electrical properties of mammalian smooth muscle. In Bülbring, E., Brading, A. F., Jones, A. W., and Tomita, T., editors: Smooth muscle, Baltimore, 1970, The Williams & Wilkins Co.
71. Trautwein, W., and Zink, K.: Über Membrane und Aktionspotentiale einzelner Myokardkasern des Kalt- und Warmbluterherzens, Pfluegers Arch. Physiol. **256:**68, 1952.
72. Verity, M. A.: Morphologic studies of the vascular neuroeffector apparatus. In Bevan, J. A., Furchgott, R. F., Maxwell, R. A., and Somlyo, A. P., editors: Physiology and pharmacology of vascular neuroeffector systems, Basel, 1971, S. Karger.
73. Wolfe, D. E., Potter, L. T., Richardson, K. C., and Axelrod, J.: Localizing tritiated norepinephrine in sympathetic axons by electron microscopic autoradiography, Science **138:**440, 1962.
74. Zuberbuhler, R. C., and Bohr, D. F.: Responses of coronary smooth muscle to catecholamines, Circ. Res. **16:**431, 1965.

II

INTERACTIONS BETWEEN EXCITABLE TISSUES

5 Neuromuscular transmission

WILLIAM L. NASTUK

In the body, information in the form of nerve impulses flows from one point to another over elements of the nervous system. The transmission pathways are generally made up of several neurons in a chain. Thus nerve impulses not only travel along the surface membranes of individual neurons but must also pass from cell to cell. In this chapter the mechanisms involved in cell-to-cell transmission will be discussed.

The region at which two neurons make close contact with each other is called a synapse. At their peripheral terminations, motor neurons also make close contact with muscle cells and this region of contact is known as the neuromuscular junction. Synapses and neuromuscular junctions are characterized by unique morphologic and physiologic features. Details of these specializations will be given in this and subsequent chapters.

It is important to recognize that each neuron entering into synaptic contact with another maintains its surface membranes intact. At such synaptic regions the unbroken surface membranes of the individual neurons are separated by a gap called the synaptic cleft, which is about 200 Å wide. This cleft is continuous with the extracellular space and therefore it is presumed to be filled with a solution resembling extracellular fluid. At the neuromuscular junction the motor neuron and muscle fiber are also separated by a cleft but, as we will see, it is larger and more irregularly shaped than the synaptic cleft.

It might be supposed that the transmission of impulses from one excitable cell to another is accomplished by the same mechanisms as those involved in axonal transmission. This idea, which has been described as the "electrical theory" of synaptic or neuromuscular transmission, was much espoused in earlier years. The theory was believed to be generally applicable to all synapses but, as powerful evidence accumulated, it became more and more certain that cell-to-cell transmission is most generally carried out by chemical intermediaries. Thus synaptic and neuromuscular transmission is now widely described in terms of a "chemical theory." Nonetheless, in recent years it has been demonstrated using modern experimental techniques that at some cell-to-cell junctions electrical coupling is operative (p. 188). In fact, there are cases in which the coupling involves both chemical and electrical mechanisms operating in parallel, with one bearing a heavier duty than the other.[115, 116] One should no longer be surprised to find both chemical and electrical cell-to-cell coupling mechanisms being employed in the same organism.

In Chapter 2 the details of the processes underlying excitation and conduction of nerve impulses along axons were presented. In an axon, adjacent regions of the surface membrane are electrically coupled to each other since they are shunted by an electrolyte-containing fluid present in the axoplasm and outside the cell surface. For this reason a change in membrane potential in one region, produced, let us say, as a result of a Na^+ ion inrush, causes changes in membrane potential of the electrically coupled adjacent regions of the axon membrane. When two adjacent regions of axon membrane have unequal potential differences across them, electric currents (eddy currents) flow between these regions. Such interactions occur during axonal conduction and thereby resting regions of the axon that lie ahead of the oncoming nerve impulse become depolarized and liminally excited.

In order for synaptic or neuromuscular transmission to occur via an electrical mechanism, several conditions must be met. (1) The electrical coupling between the cells must be relatively tight so that the eddy currents generated by an active presynaptic cell are forced to cross the membrane of the neighboring quiescent postsynaptic cell. Flow of this electric current causes the membrane potential of the quiescent postsynaptic cell to shift from its resting level. Close electrical coupling between cells can be achieved by interdigitating their membranes, and one example of such a morphologic arrangement is found at the intercalated discs of the ventricular myocardium. (2) The presynaptic membrane, the postsynaptic membrane, and the membrane contiguous with the postsynaptic membranes must all be electrically excitable. In nature there are junctions at which the latter requirement is not met.[17, 67] For example, in the frog the membranes of slow (tonus) fibers are not electrically excitable and neuromuscular transmission is effected by a chemical mechanism.[25] The suppression of electrical excitability is neurally controlled because action potentials can be electrically initiated in denervated slow fibers.[126] (3) The presynaptic elements must be capable of withdrawing a sufficient electric charge from the postsynaptic membranes so that the potential difference across these membranes falls at least to the critical level at which an action potential is initiated. This condition may not be met for several reasons. For example, the presynaptic elements are sometimes physically small, and because of their small membrane area, they have a limited capacity to transfer electric charge. On the other hand, the area of the postsynaptic chemoreceptive membrane can be relatively large and the *total* electric charge carried by it is correspondingly large. Hence the presynaptic elements have to dissipate a large postsynaptic electric charge in order to change the potential difference across the postsynaptic membranes.

Neuromuscular transmission in "twitch" fibers of vertebrate skeletal muscle involves the steps by which the motor nerve impulse leads to the initiation of a propagated action potential in the muscle fiber. This propagated muscle action potential in turn leads to changes that activate the contractile elements of the fiber (excitation-contraction coupling). A neuromuscular junction is classed as excitatory if its activation leads to a reduction of the potential difference across the postjunctional membrane. This localized depolarization is known as an end-plate potential (EPP) because the change in membrane potential is nonpropagated and can be demonstrated to be maximal at the postjunctional region. Because the muscle fiber has distributed electrical properties such as those of an electrical cable, the EPP spreads with decrement along the muscle fiber membrane that is contiguous with the postjunctional membrane. If and when the EPP reaches a critical magnitude, a propagated muscle action potential is initiated. Normally conduction takes place from neuron to muscle fiber (dromic conduction), but under certain circumstances the eddy currents accompanying the muscle action potential can electrically excite the nerve terminals and thereby antidromic conduction occurs in the motor neuron.

Although only excitatory neuromuscular junctions have been found in vertebrate skeletal muscle, it has been shown that the neuromuscular junctiors of muscle fibers in invetebrate animals are not all excitatory; another type of junction classed as inhibitory is also present. Activation of an inhibitory neuromuscular junction can block the initiation of action potentials and lead to muscular relaxation. The reason for this is that arrival of a nerve impulse at an inhibitory junction ultimately causes specific increases in the ionic conductance of the postjunctional membrane that result in either an increase in the potential difference across the postjunctional membrane (hyperpolarization) or a more effective maintenance of the membrane potential near its resting value. These changes, as pointed out in Chapter 2, lead to a diminution in excitability.

Peripheral inhibition has not been observed at the neuromuscular junctions of vertebrate skeletal muscle fibers. In vertebrate animals, inhibitory synaptic activity is seen in the CNS. The details of central inhibition are presented in Chapter 6.

For the skeletal muscle of vertebrate animals, evidence strongly indicates that neuromuscular transmission is accomplished via a chemical intermediary, acetylcholine. This agent, which is released from the motor nerve terminal subsequent to the arrival of a nerve action potential, diffuses across the synaptic cleft and combines with "receptive sites" in the postjunctional membrane. Thereupon the permeability of the postjunctional membrane changes and transmembrane ion movements occur that generate the EPP. Thus the neuromuscular transmission process involves neurosecretion on one hand and chemoreception on the other.

Because of its accessibility, much experimental work on cell-to-cell transmission has been carried out on the neuromuscular junction, and as a result more is known about the details of the transmission process for this junction than of chemically mediated synaptic transmission processes that occur at other regions of the nervous system. However, knowledge of the neuromuscular junction provides a basis for interpreting synaptic transmission at central excitatory synapses, autonomic ganglia, and other neuroeffector junctions. Details of the transmission processes at the autonomic ganglia and that for the smooth muscle or gland cells innervated by autonomic neurons are described in Chapters 4, 6, and 30.

MORPHOLOGIC FEATURES OF NEUROMUSCULAR JUNCTION[11, 15, 42, 150]

The general arrangement of the motor innervation of striated muscle is described in Chapter 20. Each motor nerve fiber and the extrafusal muscle fibers it innervates represent a divergent system. Each branch of the mye-

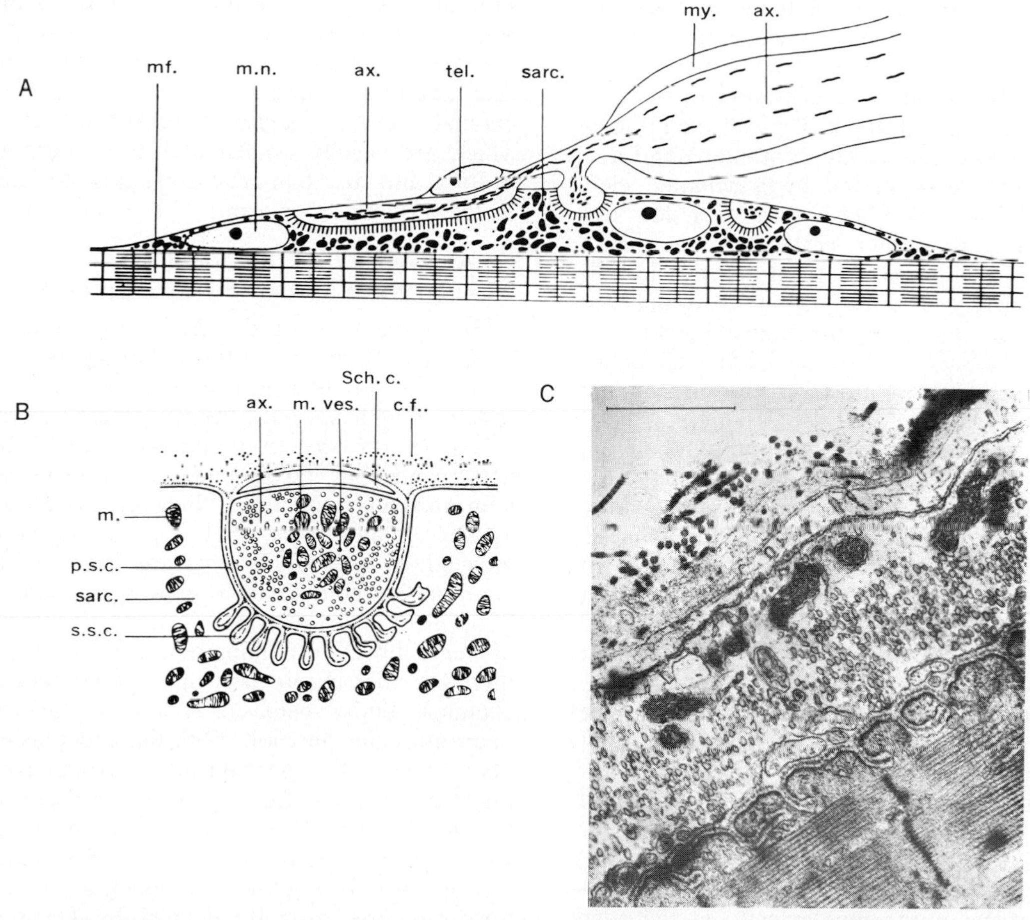

Fig. 5-1. A, Schematic drawing of typical neuromuscular junction. Terminal nerve branches lie in synaptic troughs. At axoplasmic-sarcoplasmic interface, transversely cut subneural lamellae are seen as rodlets about 1μ in length. *ax.,* Axoplasm of motor fiber; *mf.,* myofibrils; *m.n.,* muscle cell nuclei; *my.,* myelin sheath; *sarc.,* muscle cell sarcoplasm and its mitochondria; *tel.,* thin Schwann cell that completely covers nerve endings—only nucleus is indicated. **B,** Schematic drawing of synaptic trough seen in cross section. *ax.,* Axoplasm; *c.f.,* collagen fibrils; *m.,* mitochondria; *p.s.c.,* primary synaptic cleft; *sarc.,* sarcoplasm; *Sch.c.,* Schwann cell; *s.s.c.,* secondary synaptic cleft; *ves.,* synaptic vesicles. **C,** Electron micrograph of neuromuscular junction of frog. From upper left to lower right: connective tissue with collagen fibrils, Schwann cell, motor nerve terminal filled with synaptic vesicles and mitochondria, primary synaptic cleft containing electron-dense ground substance, secondary synaptic clefts, muscle cell sarcoplasm, and myofibrils. Globular-shaped elements of irregular size located in primary synaptic cleft are fingerlike projections of Schwann cell. Calibration mark, 1 μm. (**A** and **B** modified from Robertson[150], from Couteaux[41]; **C** from Birks et al.[15])

linated motor axon approaches the muscle fiber and further divides to form a divergent array of unmyelinated terminal filaments that spread along the muscle fiber in both directions, often occupying several thousand square microns of its surface. The form of this terminal tree as well as its extent varies greatly from one vertebrate species to another and may even differ in different muscles in the same animal. The ultrastructure of the pre- and postjunctional elements, however, appears much the same for all those vertebrate forms in which it has been studied with the electron microscope.

The general features of these elements are illustrated in Fig. 5-1. Each unmyelinated axon terminal lies embedded—often nearly completely so—in an indentation in the surface of the muscle fiber. The cell membranes of nerve and muscle are continuous and distinct and are separated by a synaptic cleft some 200 to 500 Å across. Fig. 5-1 does not show that the outer surface of the nerve terminal, the side away from the muscle cell, is completely covered with a Schwann cell sheath, as indeed is the unmyelinated preterminal axon, so that the nerve cell membrane is nowhere exposed to the surrounding connective tissue.

The neuromuscular junction differs from other synapses by virtue of the remarkable morphologic specialization of the postsynaptic cell membrane. In the region of the synaptic gutter the muscle cell membrane is thrown into folds (secondary synaptic clefts) that open into the primary synaptic cleft. In some forms these folds occur in a fairly regular arrangement, as shown for the frog in Fig. 5-1, but in others the secondary clefts are less regularly distributed. The synaptic cleft is differentiated extracellular space, for within it is a layer of electron-dense material called ground substance. This dense layer follows faithfully the contour of each postjunctional fold, and indeed it covers the entire surface of the muscle cell. At the edge of the synaptic gutter it fuses with a similar layer covering the outside surface of the covering Schwann cells. From the physiologic evidence it must be concluded that the layer of ground substance within the cleft is not a diffusion barrier.

The terminal endings of the motoneurons show the specialized features of nerve terminals (Fig. 5-1). Each terminal ending is packed with mitochondria and with many small globular bodies having structureless interiors. These bodies, called synaptic vesicles, may not be distributed randomly within the terminal. In frog muscle the motor nerve terminals along their course exhibit a succession of nodulelike enlargements that are densely packed with vesicles, while the intervening regions show much lower vesicle density.[117] Vesicles are frequently seen to be congregated within the axoplasm just opposite each postjunctional fold of the muscle cell membrane, and it is at this position that the axon membrane shows local regions of thickening and increased electron density.

The synaptic troughs usually occur in the summit of small, hillocklike elevations of the muscle cell membrane, elevations caused by the accumulation of sarcoplasm, mitochondria, and many muscle cell nuclei. This region may also contain some vesicular inclusions, which are usually smaller and more variable in form and size than are the synaptic vesicles of the terminal axoplasm.

GENERAL FEATURES OF NEUROMUSCULAR TRANSMISSION

From the time of the classic experiments of Claude Bernard, performed more than a century ago, it had been inferred that a process having a special chemical nature is responsible for the excitation of muscle by nerve and that this process is localized at neuromuscular junctions. Bernard observed, as have many since, that after paralysis is induced by the alkaloid curare, the muscle continues to contract in response to a directly applied electrical stimulus, and conduction in nerve fibers is unimpaired. He concluded therefore that curare acts upon some special chemical entity believed to operate at the neuromuscular junction. With time, increasing knowledge of the mechanisms of axonal conduction generated fresh questions concerning the mechanism of neuromuscular transmission. The long-enduring problem of whether transmission is electrical or chemical in nature was first formally defined by DuBois-Reymond in 1877. To explain, electrical transmission means that excitation of the muscle cell is produced by the flow of ionic currents across its membrane, and these ionic currents are generated by arrival of an action potential in the nerve terminals. Chemical transmission means that depolarization releases from the nerve terminal a specific substance that reacts with the muscle cell membrane in such a way as to produce its excitation. It will be seen that the evidence now available supports the second of these, indicates with equal certainty that acetylcholine is the transmitter agent, and provides reciprocally cogent reasons for rejecting the electrical hypothesis.

The development of knowledge of chemical transmission at skeletal neuromuscular junctions is intimately linked with transmission in autonomic ganglia and transmission between postganglionic autonomic fibers and the smooth muscle or gland cells they innervate. The latter subject is discussed in Chapters 4 and 30. Insight into the mechanism of transmission at each of these junctions was derived from the study of drugs or naturally occurring substances that, upon injection or local application, mimic the effects produced by nerve impulses. The muscle postjunctional membrane has been shown to be especially sensitive to many drugs, which lends further support to the idea that it has a chemoreceptor function. It was Langley,[100] for example, who showed that the stimulating action of the drug nicotine is confined to the end-plate regions of muscles, and the stimulating effect is unimpaired following the degeneration of the nerve terminals produced by nerve section. This stimulating action of nicotine is blocked by curare to a degree determined by the relative concentrations of the two drugs. Langley's quantitative studies of the competitive nature of the actions of nicotine and curare led him to generalize that such compounds compete for the same "receptor substance" in the postsynaptic membrane and that the loose union of the drug with this hypothetical substance leads, in the case of curare, to a block of neuromuscular transmission and, in the case of nicotine, to excitation of the muscle cell. The idea that the combination of the receptor substance with one drug inhibited its union with another was used to explain the competitive nature of their actions. During the years following Langley's work the concept of postjunctional membrane receptors was widely used by physiologists and pharmacologists in explaining the junctional activities of various quaternary ammonium compounds and other drugs. More recently, many investigators have attempted to isolate and characterize membrane receptor molecules, but technical and theoretical complications have cast doubts on the meaning of much of this literature (see Ehrenpreis[2]). An interesting recent development in the field concerns the blocking of acetylcholine receptors of eel electroplax by α-bungarotoxin isolated from snake venom.[36, 37, 125] The receptor characterized in these studies appears to be a membrane-bound protein tetramer whose single units have a molecular weight of 80,000. This receptor is distinct from acetylcholinesterase.

Sequence of events[137]

The train of events in neuromuscular transmission is as follows: Acetylcholine (ACh) is synthesized by motoneurons and is stored

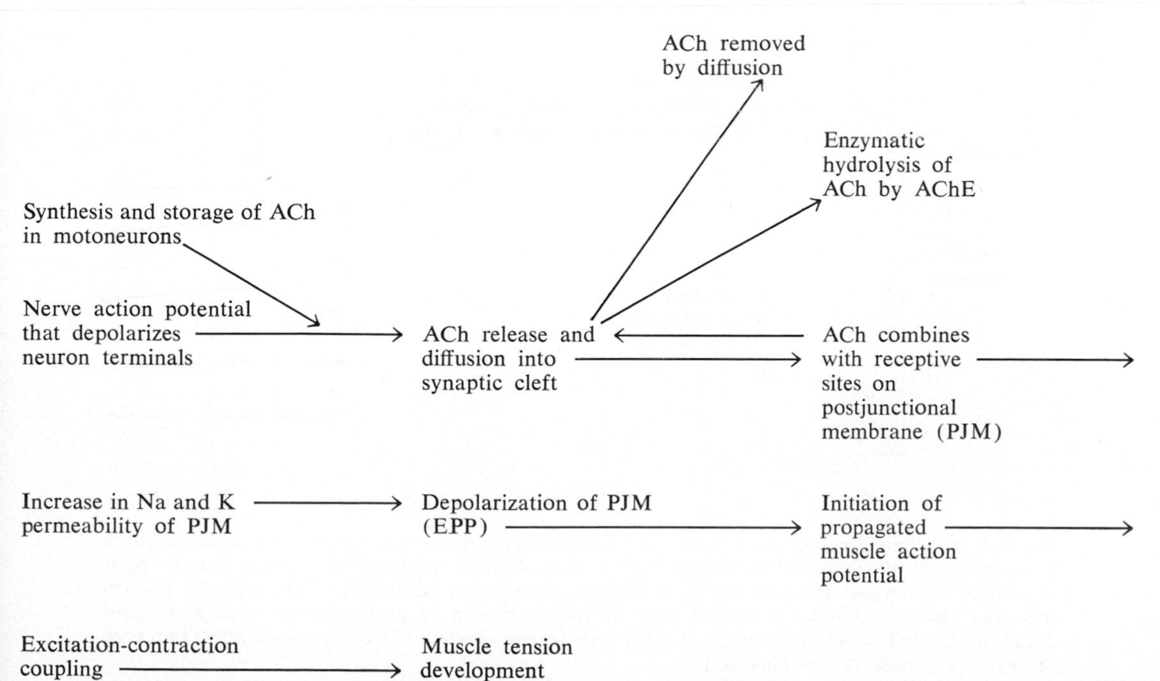

in a sequestered form in the neuron terminals. When liberated from this store by a nerve impulse, ACh diffuses across the synaptic cleft and reacts at special receptive sites of the muscle cell membrane. The formation of this transmitter-receptor complex brings about a change in permeability of that cell which leads to an interposed local reduction in membrane potential called the EPP. When the EPP reaches a critical level, a conducted action potential is initiated in the adjacent muscle fiber membrane and this action potential causes activation of the contractile elements. The ACh is either quickly destroyed by the hydrolytic action of the enzyme acetylcholinesterase (AChE) or it is removed by diffusion, as seen in the schematic diagram at the bottom of p. 155.

Acetylcholine as transmitter agent

The evidence provided by research in recent years, particularly data obtained using biophysical methods, allows a precise description in quantitative terms of several of the sequential steps just described. It is appropriate to pause at this point, however, to indicate the background of knowledge from which these advances have been made. It is largely from the early investigations of Dale, Brown, Feldberg, and their colleagues that we have obtained direct evidence that ACh is the transmitter agent at the neuromuscular junction.[20, 23, 43, 44] That evidence is summarized as follows:

1. By section and cross union of nerves and by observation of the capacity for reinnervation after regeneration, it has been shown that motor nerves to skeletal muscle can replace and be replaced by other cholinergic nerves. It is well known that such reinnervation fails when cross unions are made between nerves that operate with different transmitter agents.

2. Stimulation of motor nerves to perfused voluntary muscle causes the liberation of ACh into the perfusate, provided that enzymatic destruction of ACh is prevented by inhibiting AChE with an appropriate drug such as physostigmine. ACh release by nerve stimulation is essentially unaffected when muscle contraction is blocked by curare. Results from several sources cited in Hubbard's review[4] indicate that the quantity of ACh released per impulse at a single end-plate region is of the order of 10^7 molecules.

3. Close arterial injection of ACh causes a quick twitchlike contraction in both normal and denervated skeletal muscle. This takes the form of a brief asynchronous tetanus, and electrical recording shows that the muscle impulses originate at the region of the end plates.[21, 31, 97] A powerful and now widely used iontophoretic technique for very rapid localized application of ACh and other ionized

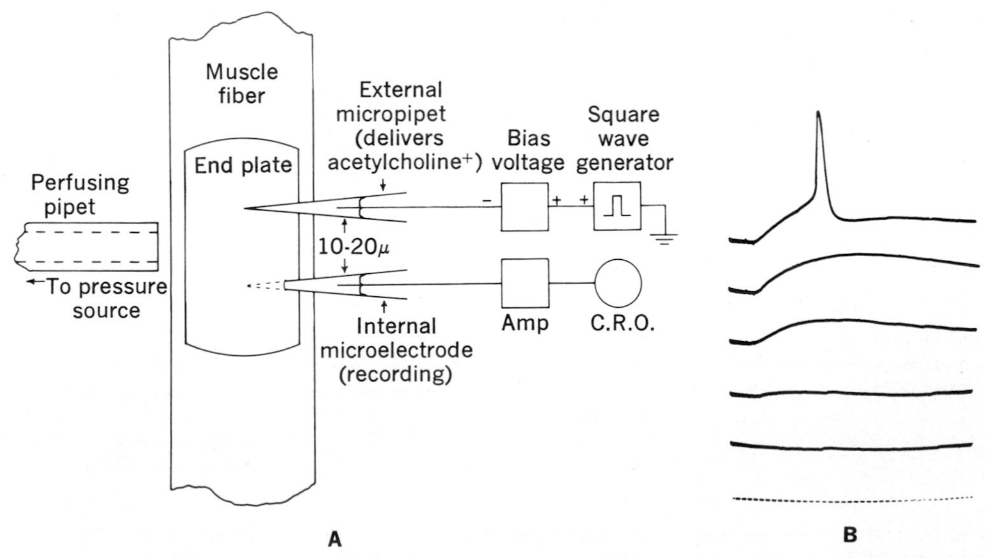

Fig. 5-2. A, Schematic diagram showing experimental arrangement used when recording changes in transmembrane potential produced during iontophoretic application of ACh ions to postjunctional membrane. Pipet to left is used when changing composition of extracellular fluid in this zone. **B,** Intracellular recording showing depolarization of postjunctional membrane produced by iontophoretic application of ACh ions in successively increasing amounts. Time base markers = 1 msec. (From Nastuk.[135])

drugs, as originated by Nastuk,[134] is shown schematically in Fig. 5-2, *A*. By such iontophoresis, a minute quantity of ACh can be rapidly ejected from a loaded micropipet placed at the external surface of the postjunctional membrane. Fig. 5-2, *B,* shows that a stepwise increase in the amounts of ACh applied causes *graded* increases in depolarization of the postjunctional membrane. In the final upper trace the depolarization produced became large enough to initiate a propagated action potential. Another very significant point brought out by using the iontophoretic technique is that when ACh is delivered to the interior of the muscle cell, neither local depolarization nor a conducted action potential results.[31] This indicates that the ACh receptor sites are present on the outside but not on the inside of the postjunctional membrane. This view is further supported by the fact that when *d*-tubocurarine, a drug that competes with ACh for receptor sites, is injected into the muscle cell beneath the postjunctional membrane, it has no effect on the EPP.[35]

Although ACh-sensitive receptors are abundant on the outer surface of the postjunctional membrane, they are more sparsely distributed at extrajunctional regions.[120, 121] The ACh sensitivity of these zones is comparatively low (approximately 1,000 times less) and the time course of the depolarization produced is relatively slow. Recently evidence indicates that junctional and extrajunctional receptors represent two different groups.[60, 61] The sensitivity of the extrajunctional regions to ACh increases over a period of many days following denervation of the muscle fiber. This denervation sensitization has been studied by many investigators over the past years. Some important aspects of this phenomenon were investigated more recently by Miledi,[119] who utilized the advantages of the iontophoretic technique.

4. Normally a single motor nerve impulse initiates a single muscle action potential and this in turn leads to the production of a brief contraction of the muscle fiber ("twitch"). These relationships are changed by the application of drugs that inhibit AChE (physostigmine, neostigmine, edrophonium, etc.). After such treatment a single nerve impulse causes the initiation of a train of muscle action potentials and the muscle contraction becomes "tetanic" in character.[139] When measures are taken to prevent the initiation of muscle action potentials, as was done in the experiment illustrated in Fig. 5-3, it is seen that cholinesterase inhibition causes a considerable prolongation of the EPP produced in response to a single nerve impulse. (See also Fatt and Katz[56] and Nastuk and Alexander.[139])

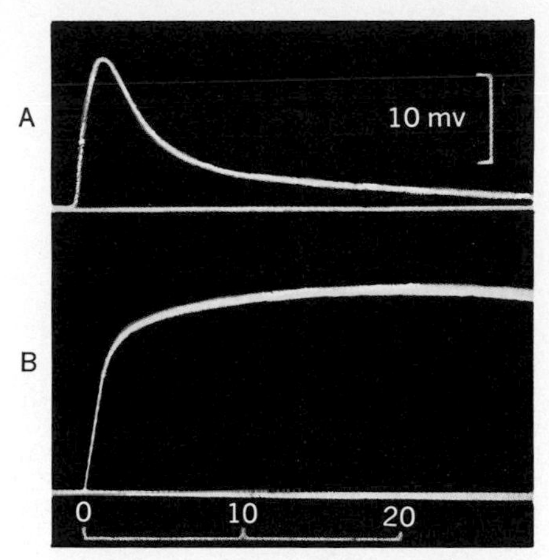

Fig. 5-3. Results of experiment illustrating effect of neostigmine, an inhibitor of AChE, on EPP, which is recorded via intracellular electrode placed in muscle cell just beneath end plate. **A,** EPP evoked by nerve impulse. Successful evocation of action potential in muscle cell is prevented by lowering Na^+ of external fluid (four fifths of Na^+ replaced by sucrose), thus revealing full time course of EPP. **B,** Muscle has now been treated with neostigmine bromide, 10^{-6}M; inhibition of AChE greatly prolongs EPP but alters its amplitude only slightly. Scale: time in msec. (From Fatt and Katz.[56])

End-plate potential[19, 56, 133]

The discovery by Göpfert and Schaefer[66] in 1938 and by Eccles and O'Connor[51] that a third electrical event is interposed between nerve impulse and muscle action potential provided a means for more detailed analyses of the mechanism of neuromuscular transmission. This third event, known as the EPP, is a transient but intense local depolarization of the postjunctional region of the muscle cell. Following the arrival of the impulse in the terminals of the nerve fiber there is a synaptic delay of 0.5 to 1 msec, after which the EPP begins its appearance.[83] Normally the EPP increases in amplitude rapidly and excites the muscle cell, and therefore much of its true time course is masked by the superposition of the initiated muscle action potential (Fig. 5-4). The EPP and the action potential may be separated by utilizing *d*-tubocurarine, which inhibits and reduces the postsynaptic depolarizing action of ACh (*d*-tubocurarine is a pure alkaloid found in crude curare). The differences between the EPP

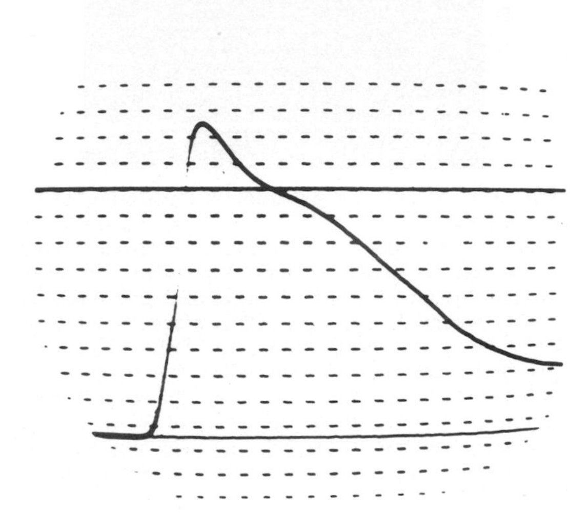

Fig. 5-4. Intracellular recording from muscle fiber at neuromuscular junction showing response to motor nerve stimulation. Calibration grid: ordinate, 10 mv steps; abscissa, 0.2 msec intervals. Upper horizontal trace is zero potential line. Negative potential is downward. (From Nastuk.[135])

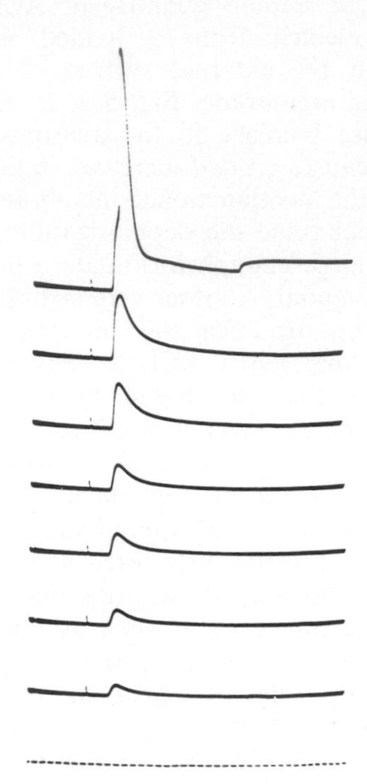

Fig. 5-5. Intracellular recording from muscle fiber showing serial group of EPPs recorded during washout of *d*-tubocurarine from junctional region. Motor nerve was stimulated every 3 sec and washout began with bottom record. Time base markers = 1 msec. (From Nastuk.[135])

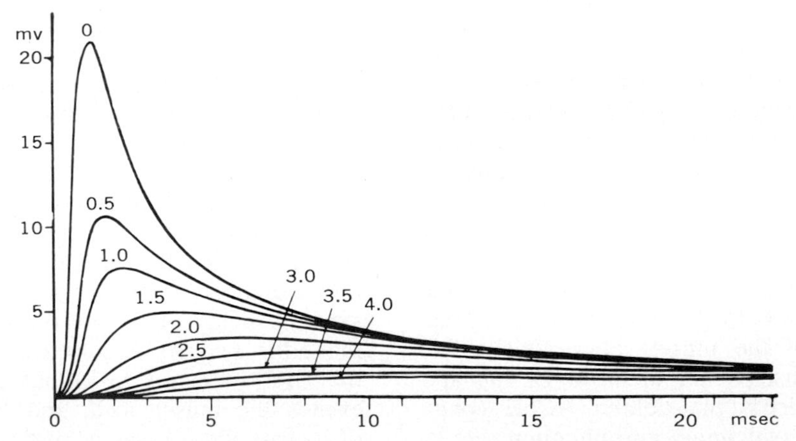

Fig. 5-6. Records taken via intracellular microelectrode at different distances from end-plate focus of curarized frog muscle in response to nerve volleys. Records have been superimposed on single time scale, taking instant of nerve stimulation as common point. Numbers indicate approximate distances in millimeters from end plate at which each record was taken. Displacement of electric charge from muscle cell membrane at end plate reaches maximum about 2 msec after start of EPP. Restoration of charge follows exponential time course. Active phase of neuromuscular transmission is a brief, impulsive event; prolonged time course and spatial spread of change in membrane potential are determined by resistance and capacitance of resting muscle cell membrane. (From Fatt and Katz.[56])

and the muscle action potential are made clear by intracellular records made at a curare-blocked neuromuscular junction during progressive washout of the curare. As the drug is removed from the postjunctional membrane, the EPP elicited by nerve stimulation increases in amplitude until it reaches the critical level at which a muscle action potential is initiated (Fig. 5-5). The EPP is a local depolarization of the postjunctional membrane caused by a transient increase in the ionic permeability of this membrane. As we shall see, the ACh-induced portion of the EPP lasts only about 2 msec and thus much of the falling phase of the EPP merely represents the passive return of the membrane to its resting potential level along a time course determined by the distributed resistance (a function of extrajunctional ionic permeability) and capacitance of the muscle cell membrane. To better appreciate this the reader should refer to the discussion of local response of nerve and the cablelike properties of cell membrane in Chapter 2. The membrane cable properties determine also the extent to which the EPP spreads, with decrement, into the adjacent nonjunctional regions of the muscle fiber. The spatial extent of the EPP is indicated by the records in Fig. 5-6, obtained at a curarized junction. It is important to recognize that when the postjunctional membrane has normal chemosensitivity (no blocking agents such as curare are present) the amount of ACh released by a nerve impulse is more than sufficient to drive the postjunctional membrane potential to the critical level at which an action potential is initiated. Thus there is a three- to fourfold safety factor in the neuromuscular transmission process. It is this safety factor that accounts for the remarkable reliability of transmission at the neuromuscular synapse, and it explains the nearly 1:1 relation between nerve and muscle impulses that prevails under normal conditions.

The EPP possesses other properties characteristic of local responses. The first of these is that it exhibits no refractory period for, if one nerve impulse follows another by the proper interval in a curarized muscle, the depolarizations produced by them summate. If curarization is not too heavy, this summed depolarization may reach the threshold level of the muscle cell, producing in it a conducted action potential.[52] The phenomenon is an example of *temporal summation,* a property seen in synaptic action in the CNS.

Further experimental studies that reveal the processes responsible for the production of the EPP will be described in the following sections. However, at this stage it will be instructive to highlight some of the experimental evidence and arguments that caused the electrical theory of neuromuscular transmission to be discarded.[98]

1. The total electric charge transferred across the muscle cell membrane during a single EPP is larger than the estimated *total* ionic content of the nerve terminals. Thus the EPP cannot result solely from ionic eddy currents associated with an action potential in the neuronal terminals.[56]

2. Recording of the nerve action potential in single telodendrites by carefully placed external microelectrodes shows that this action potential is separated in time from the onset

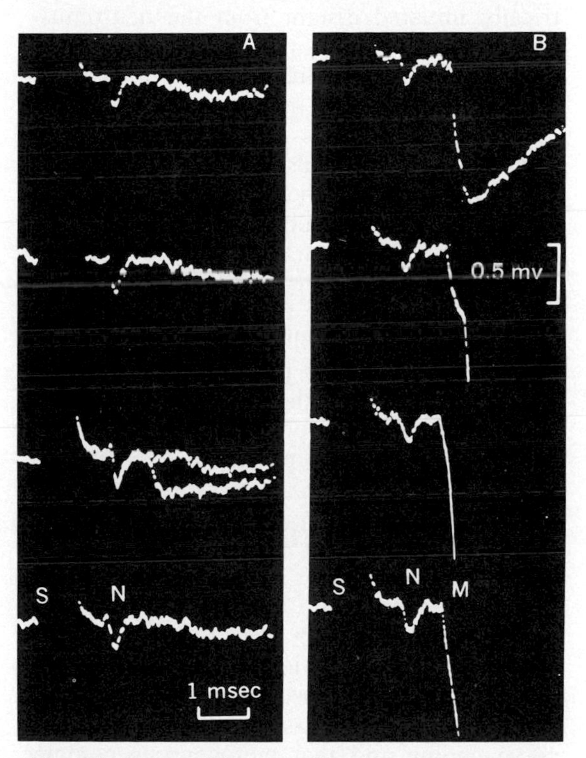

Fig. 5-7. Extracellular recording from frog sartorius muscle obtained using microelectrode containing 0.5M $CaCl_2$ placed at point of contact between terminal filament of motor neuron and its adjacent muscle fiber. Muscle was bathed in low Ca, high Mg solution to block neuromuscular transmission. Negative potential is downward. Stimulus, *S,* to motor nerve evoked action potential, *N,* in nerve terminal followed by variable postsynaptic response, *M.* In **A,** efflux of Ca from electrode is stopped. In **B,** Ca efflux is permitted to occur and thereafter postsynaptic potential increases in amplitude. (From Katz and Miledi.[82]) (See also Katz and Miledi.[83, 85, 86])

of the EPP.[83, 84] This synaptic delay, which has been estimated to average 0.75 msec in the frog, is, by the way, only partly accounted for by the estimated diffusion time of ACh across the synaptic cleft (Fig. 5-7) and other factors related to the probability of ACh release enter in.

3. The flow of subthreshold currents across the nerve terminal membrane produces no change in the membrane potential of the muscle cell. The same result is obtained whether the local currents are produced by local electrical stimulation of the nerve terminals or by an approaching action potential that is blocked just proximal to the nerve terminal arborization. In the second case the terminal membrane supplies the eddy currents that are associated with the early rising phase of the action potential.

4. A muscle action potential can be electrically initiated distant from the neuromuscular junction and allowed to propagate along the muscle fiber into the junctional region. By proper timing, at the moment when the muscle action potential is making the junctional transit, a nerve impulse can be arranged to arrive at the motor nerve terminals. At this stage the muscle membrane is occupied with a muscle action potential and hence it should be absolutely refractory and beyond any significant influence by the eddy currents associated with the "colliding" nerve impulse. However, despite this prediction, experimental results show that the postjunctional membrane can undergo a powerful change in permeability on arrival of the nerve impulse (Fig. 5-12).

ENZYMATIC SYNTHESIS AND HYDROLYSIS OF ACETYLCHOLINE[69, 70, 148]

Synthesis, distribution, and storage

For ACh to function as the transmitter agent at the neuromuscular synapse requires that this ester be released suddenly from the nerve ending and that motoneurons contain a mechanism for its synthesis, for it is clear that cholinergic neurons, when subjected to long-term repetitive stimulation, will release more ACh from their endings than the amount stored in the resting neuron terminals.[14, 22, 146, 147] A second requirement is that released ACh be removed from its site of action at the postsynaptic membrane within 1 to 2 msec. These stipulations have been met by the discoveries that cholinergic neurons contain an enzyme (choline acetyltransferase, earlier called choline acetylase) that catalyzes the acetylation of choline, and that at the postsynaptic regions there is present a second enzyme, AChE, in quantities apparently sufficient to hydrolyze the released ester rapidly. These relationships are of general significance, for ACh has been proved to be the transmitter agent at the synapses of autonomic ganglia, and substantial evidence indicates that ACh serves as a transmitter at certain synapses in the CNS as well.

ACh synthesis in vivo was first demonstrated in 1937 by Brown and Feldberg.[22] Shortly thereafter Mann et al.[113] found that respiring slices of mammalian brain perform this synthesis in vivo and the Stedmans[154] found that the synthetic mechanism survived the cell rupture produced by grinding. An important contribution was made by Nachmansohn and Machado,[131] who found in 1943 that the synthesis of ACh by brain is greatly accelerated by the addition of adenosine triphosphate (ATP) and choline, and that the synthesizing enzyme contained —SH groups. Thereafter the continuing investigations of Feldberg[58] and Nachmansohn[127] and their colleagues established that (1) the synthesizing enzyme system is present in aqueous extracts of acetone-dried brain; (2) the system yields maximal activity on the addition of ATP, choline, K^+, and (if incubated anaerobically) an —SH-containing compound such as cysteine; and (3) dialysis reduces activity, which is restored when the dialysate is returned. This "activator," as described by Feldberg, has been identified as coenzyme A.

It is now apparent that acetylation occurs in two steps, and that choline acetyltransferase catalyzes only the second. The first is the combination of acetate with coenzyme A to form "active acetate," i.e., acetyl coenzyme A:

I. ATP + Co-A + Acetate $\rightleftharpoons$ Acetyl Sco-A + AMP + Pyrophosphate

II. Acetyl Sco-A + Choline $\xrightleftharpoons{\text{Choline acetyltransferase}}$ ACh + HS–CO-A

The overall reaction indicated by step I may occur via a series of intermediate stages. It should be emphasized that this is not the only possible source of acetyl Co-A, for it is likely that there is a common pool to which many acetyl donor systems contribute and from which various acceptor systems draw. There is no evidence that any one donor system is specifically channeled into step II.

Distribution of choline acetyltransferase[59, 69]

In higher vertebrates, choline acetyltransferase has been detected in those parts of the peripheral and autonomic nervous systems that contain cholinergic neurons. Thus in the central and peripheral regions of the nervous system of higher vertebrates the distribution of choline acetyltransferase parallels that of ACh itself; i.e., both are constituents of cholinergic neurons. The quantities of choline acetyltransferase present in any given region appear to depend on the proportion of individual cholinergic neurons present in the neuronal population and not on wide variation in the amount of this enzyme in individual neurons. In the central and peripheral nervous systems the distribution of cholinergic neurons is not uniform, and for this reason certain structures such as the caudate nucleus contain large amounts of the enzyme, whereas others such as the cerebellar cortex contain practically none. Choline acetyltransferase is abundant in ventral spinal roots, but dorsal roots contain little of this enzyme and that which is found is probably confined to those sensory neurons that are cholinergic and that also contain AChE.[63]

The intracellular distribution of various elements of the ACh system has been intensively investigated by Hebb,[69, 70] Whittaker,[162, 164] De Robertis,[47] and many other workers. For practical reasons much of the work is done on subcellular fractions obtained from mammalian cerebral cortex (e.g., see Marchbanks[114] and Whittaker[164]) or other tissues rich in cholinergic neurons. Although results of earlier studies indicated that choline acetyltransferase is a constituent of synaptic vesicles (see below), it now appears that the enzyme is largely found in the cytoplasm. However, some of the enzyme may be bound to negatively charged sites on intracellular membrane surfaces. ACh synthesized in the cytoplasm is taken up into synaptic vesicles, where it is stored, protected from hydrolysis by AChE. Studies of the microphysiology of neuromuscular transmission detailed in a later section indicate that this bound packet of ACh finds its way to the nerve terminal membrane, from which it is released by the depolarization that is associated with the nerve impulse.

Enzymatic hydrolysis of acetylcholine[129, 165, 166]

Esterases of various types are widely distributed in animal tissues and, of these, at least two show some degree of specificity for choline esters. One is found in large quantities in mammalian blood serum, and the second is present in red blood cells and in many of the conducting tissues of the nervous system. The latter, called "true cholinesterase," or AChE, shows a high but not exclusive affinity for ACh, and the rate of substrate hydrolysis decreases with an increase in the length of the acyl chain: acetyl > proprionyl > butyrylcholine. There is a well-defined optimal substrate concentration; the rate decreases for weaker or stronger solutions, a fact that can be depicted by a bell-shaped curve that relates hydrolytic activity to substrate concentration. The serum esterase (called "pseudo cholinesterase"), on the other hand, has a much weaker affinity for ACh, and the rate of hydrolysis increases both with substrate concentration and with increasing length of the acyl chain.[132]

It is now apparent that there is a reasonably good correlation between the nervous system distributions of AChE, ACh, and choline acetyltransferase.[24] The concentration of AChE is generally highest in those regions in which a high concentration of synapses is joined with a high content of choline acetyltransferase, a conjunction to be expected when synaptic transmission is mediated by ACh. The ventral roots, for example, contain a large amount of choline acetyltransferase, and there is a high concentration of AChE at the neuromuscular end plates. Histochemical studies, particularly those by Couteaux,[11, 12] Davis and Koelle,[45] and Koelle,[91-93] have shown that most of the AChE is concentrated on the muscle side of the junction and is associated closely with the folds of the postjunctional membrane. It is most likely that reactive sites of the enzyme are exposed on the membrane outer surface. AChE is more plentiful in the end-plate region than elsewhere along the muscle and this distribution is at least partially explained by the existence of convolutions of the postjunctional membrane, which increase its surface area at that zone. When the nerve terminals degenerate following nerve section, the concentration of AChE is reduced but appreciable amounts continue to remain in this region. Thus it may reasonably be assumed that an appreciable amount of the postsynaptic AChE is of presynaptic origin, and Koelle[93] has suggested on the basis of recent work that this relationship applies to cholinergic synapses of autonomic ganglia. Although biosynthesis of some neuronal constituents such as ACh takes place in neuronal terminals, one should also keep in mind the growing lines of evidence showing that many cellular constituents and organelles found in the terminals of various types of neurons may be synthesized elsewhere in the cell and transported to the axonal terminals by axonal flow. For the reader who wishes to familiarize himself with various ideas and details of such cellular dynamics, there are a number of helpful publications.[9, 10, 13]

As shown in the schematic diagram on p. 155, ACh released into the synaptic cleft is rapidly removed by diffusion or enzymatic hydrolysis. ACh removal by diffusion is more effective than is commonly realized,[50, 144] and even if AChE is inhibited, released ACh is removed in a few milliseconds. However, enzymatically catalyzed hydrolysis of released ACh is an important mechanism for terminating the postsynaptic action of this transmitter. The velocity of the hydrolysis of ACh by AChE is dependent on many factors and past estimates are complicated by many uncertainties. Recently Wilson and Harrison[167] estimated the turnover number at 25° C, pH 7.0, ACh 2.5×10^{-3}M to be 1.2×10^{4}M/sec; they concluded that a biologically reasonable concentration of enzyme would be required to hydrolyze in 2 msec the ACh released during neuromuscular transmission. The turnover time for AChE has been estimated by Lawler[101] at less than 100 μsec.

The process of hydrolytic enzyme action takes place in two stages.[127] The first is the formation of an enzyme-substrate complex. The apparent dissociation constant (Michaelis' constant) of the AChE-ACh complex (1×10^{-4}) indicates that this complex is stable and reversible. A simplified schematic representation of a way in which this binding might occur is shown in Fig. 5-8. The positively charged N^+ of ACh is bound by coulombic forces to an anionic site on the enzyme. The existence of such a negative site has been demonstrated with the aid of competitive inhibitors and appropriate substrates. An additional binding is attributed to the nonspecific van der Waals forces. It is also postulated that a region on the active surface of the enzyme, close to the anionic site, contains a basic group that forms a covalent bond with the acyl carbon of the ester. The further rearrangements leading to the hydrolysis of the bound ester are thought to proceed through a first step involving acetyla-

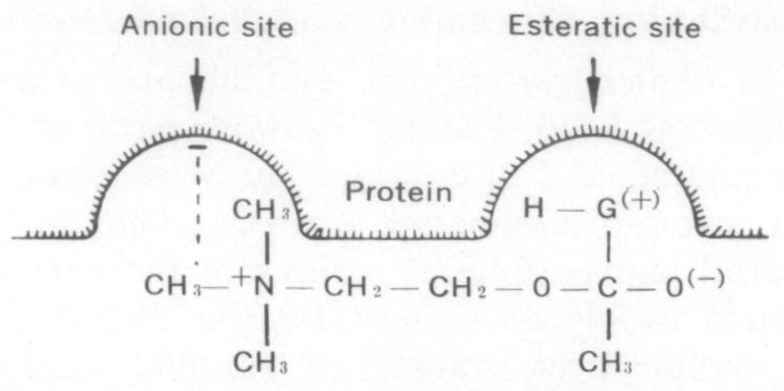

Fig. 5-8. Schematic representation of complex formed between AChE and ACh. Complex is stabilized by coulombic and van der Waals' forces at anionic site and by covalent bond formation between carbonyl carbon and basic group at esteratic site. This basic group is symbolized by *G*, and *H* represents dissociable hydrogen atom not involved in binding. Hydrolysis then follows by a process thought to occur in two consecutive steps, the first involving acetylation of enzyme with elimination of choline and the second, regeneration of enzyme and acetic acid. (From Nachmansohn.[127])

tion of the enzyme and the elimination of choline and a second step involving an acid enzyme complex that leads to regenerated enzyme and acetate ion. Detailed investigations of these hydrolytic mechanisms have been made by Wilson.[165, 166]

Although the functional role of AChE at the neuromuscular junction is well established, the fact that this enzyme is closely associated with extrajunctional axonal and muscle fiber membranes raises questions as to its functional significance at such sites. For many years Nachmansohn[127, 128, 130] has maintained that the release and electrogenic action of ACh are essential first steps that lead to sequentially timed increases in Na^+ and K^+ conductances, which underlie production of action potentials. This hypothesis and its corollaries have been subject to much experimental testing and scientific argument and many investigators find it unacceptable. It is not possible to review here these arguments, many of which hinge on technical details. The interested reader should explore the original literature to form his own opinion. A few examples of the points of discussion are the following: (1) choline acetyltransferase is not universally distributed in the nervous system and some neurons do not contain ACh[24, 69, 70]; (2) AChE is present in the membranes of some nerve cells that are not cholinergic[92, 93]; (3) tetrodotoxin, a powerful poison obtained from the puffer fish, blocks the sodium conductance mechanism of axons and muscle fibers but it has no effect on the response of muscle postjunctional receptors to ACh (see references in the book

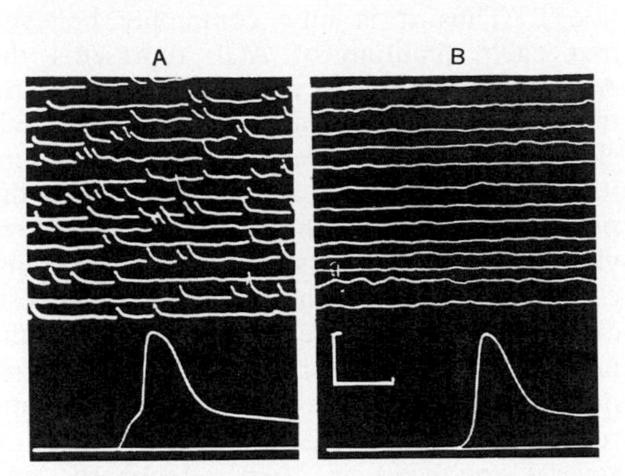

Fig. 5-9. Records illustrating fact that spontaneous miniature end-plate potentials (MEPPs) occur only at end-plate regions of muscle fiber. For **A,** microelectrode was placed inside frog muscle fiber at nerve-muscle junction. Successive lines from above downward were recorded consecutively in time. Note random occurrence of spontaneous potential changes of about the same amplitude, which occasionally superimpose and sum. Record below, obtained at much lower amplification and on a faster time base, is EPP–muscle action potential complex recorded at this end-plate region. For **B,** microelectrode was placed 2 mm away from junction, in same muscle fiber. Records above show amplifier noise but no MEPPs. Record below shows conducted action potential in this fiber set up by nerve impulse, with little sign of EPP. Calibrations for lower records: 50 mv and 2 msec. (From Fatt and Katz.[57])

by Hubbard et al.[5]). For additional facts and viewpoints consult the reviews by Hebb,[69] Castillo and Katz,[33] and Katz.[80]

MICROPHYSIOLOGY OF NEUROMUSCULAR JUNCTION[5-7, 34, 55, 80]

Before proceeding with this section the reader should recall some of the morphologic details given in Fig. 5-1. The presynaptic terminals contain large numbers of globular bodies (synaptic vesicles), which are about 500 Å in diameter and which have a structureless interior. Similar appearing bodies are occasionally seen in the muscle fiber cytoplasm underlying the postsynaptic membrane. A primary synaptic cleft 200 Å wide separates the pre- and postsynaptic membranes except for regions where the latter is infolded (secondary synaptic clefts).

Spontaneous and activated release of acetylcholine from motor nerve terminals[18, 57, 68, 102]

In 1952 Fatt and Katz discovered that the motor nerve endings are not at rest even in the absence of nerve impulses. The records in Fig. 5-9 show that small spontaneous depolarizations of the muscle cell membrane, known as miniature end-plate potentials (MEPPs), occur at the region of the end plate and that these are rapidly attenuated

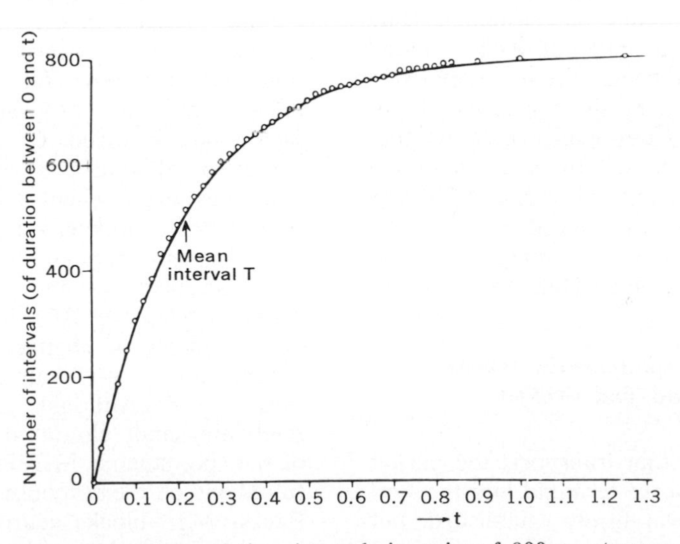

Fig. 5-10. Statistical distribution of time intervals in series of 800 spontaneous MEPPs. That sequence is random is shown by the fact that distribution of intervals is asymmetric and follows simple exponential law. Probability that MEPP will occur in a given interval increases with that time interval along simple exponential curve of type $p = 1 - e^{-t/T}$, in which t is interval chosen for observation and T the mean interval between successive events. (From Fatt and Katz.[57])

as the intracellular recording site is moved away from the junctional region. Extracellular recording, which allows a more precise localization, showed that MEPPs occur at many very discrete loci about a nerve ending.[34] The combined discharge of the ending is disordered in time, for the probability that a MEPP will occur at any given instant fits that predicted for a completely random sequence (Fig. 5-10). The time course of such a MEPP is roughly similar to that of a neurally evoked one, though its amplitude is only 0.5 to 1.0 mv. This amplitude is changed only by factors that alter the responsiveness of the postsynaptic membrane; e.g., it is increased in size and prolonged in time course by such anti-AChE drugs as neostigmine or eserine. The frequency of discharge, on the other hand, is changed by factors that alter the state of the nerve terminal membrane, particularly the membrane potential, for the discharge frequency is powerfully increased by depolarization and decreased by hyperpolarization of the terminal.[29, 80, 104] These facts led Katz and his colleagues to the conclusion that the MEPPs represent the response of the muscle postjunctional membrane to spontaneously released packets of ACh (quanta). Although the amplitude of the MEPP can vary, these variations do not necessarily represent changes in quantal size but they reflect variations in postjunctional chemosensitivity or in the electrophysiologic characteristics of the muscle fiber. From this and various other lines of reasoning, Katz proposed that the quantity of ACh released during each spontaneous event is relatively constant, at least over short periods of observation. The available evidence shows that the MEPP is not caused by reaction of the postsynaptic membrane with one ACh ion; rather, each MEPP is produced by the action of a quantum of ACh amounting to 10,000 to 40,000 ACh ions (see Hubbard,[4] pp. 76-77).

Relation between spontaneous release of acetylcholine and that evoked by nerve impulse[27, 80, 103]

The mechanism that transports the packet of ACh from the nerve terminal into the synaptic cleft is not yet firmly established, but in the last 15 years this subject has been under intensive investigation and many theoretical ideas have appeared in the literature.[4, 71] For one thing, synaptic vesicles have been isolated by differential cell fractionation techniques and they have been shown to contain ACh (see, for example, Whittaker[163, 164] for recent work and other literature citations). Thus it is now commonly believed that each quantum of ACh represents the ACh contained in one presynaptic vesicle and that a MEPP is generated when such a vesicle ruptures and discharges its contents into the synaptic cleft to reach and react with the postsynaptic membrane. Hubbard[4] has reviewed various suggested mechanisms concerning the interaction of the synaptic vesicles with sites on the inner surface of the neuronal terminal membrane. Direct electron micrographic evidence showing presynaptic rupture of synaptic vesicles has appeared in several publications (e.g., Hubbard and Kwanbunbumpen[72]). Further evidence establishing the links between synaptic vesicles and ACh release during neuromuscular transmission also comes from other sources.[39, 107] For example, application of black widow spider venom to the neuromuscular junction causes a marked rise and then a fall in MEPP frequency. After such treatment, electron micrographs show that presynaptic terminals contain very few synaptic vesicles.

The rate of quantal release is accelerated by an increase in the extracellular concentration of K^+, and this increased release has been analyzed by Katz[80] in terms of the depolarizing action of increased extracellular K^+ on the nerve terminals. In his analysis a decrease in membrane potential caused a logarithmic increase in the rate of ACh quantal release. Thus the level of the membrane potential has powerful influence on ACh release but, as we shall see, additional factors are involved. One experimental indication of this fact is that quantal release continues even when nerve terminal membranes are completely depolarized.[32]

A deeper understanding of the relation between spontaneous and the action potential–actuated release of ACh has come from study of the effects of changes in Ca^{++} and Mg^{++} concentrations on neuromuscular transmission.[4, 5, 26, 76] Although such changes cause relatively small alteration in the frequencies of the spontaneous MEPPs, they have powerful effects on the response to a nerve impulse. Excess Mg^{++} blocks neuromuscular transmission by interfering with the presynaptic release of transmitter agent. Ca^{++} opposes the effect of Mg^{++} and relieves the block. Reduction in Ca^{++} concentration, on the other hand, produces neuromuscular block in a way quite

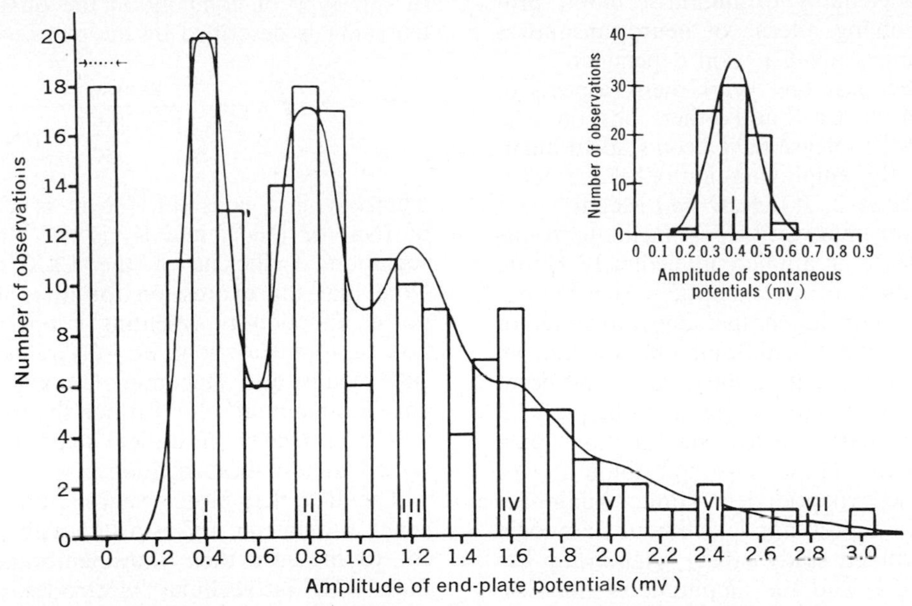

Fig. 5-11. Histograms of neurally evoked EPPs and spontaneous MEPP (inset) amplitude distributions in fiber of mammalian skeletal muscle in which neuromuscular transmission was blocked by increasing magnesium concentration of bathing Krebs' solution to 12.5 mM. Peaks of neurally evoked EEP amplitude distribution occur at 1, 2, 3, etc. times the mean amplitude of spontaneous potentials. Gaussian curve is fitted to spontaneous potential distribution and used to calculate theoretical distribution of neurally evoked EPP amplitude (continuous curve), which fits well with amplitude distribution of those experimentally observed (bar graph). Bar placed at zero indicates number of failures observed in series of trials; arrows and dotted line, number of failures expected theoretically, from Poisson's law. (From Boyd and Martin.[19])

similar to that caused by excess Mg^{++}. As the block produced by Mg^{++} progresses, electrical recording of the events produced by testing nerve volleys shows that the amplitude of the evoked EPP falls in a series of steps and that just before complete block occurs the postsynaptic response is similar in amplitude and duration to the spontaneous MEPPs. Before this stage, it is seen that the amplitudes of a series of neurally evoked EPPs vary in size by a quantal factor that is also equal to the amplitude of the MEPPs. When the average quantum content of the EPP is small, these amplitude fluctuations occur in a manner predictable by Poisson's law. A statistical analysis of this variation is given in Fig. 5-11. This leads to the conclusion that the MEPP is the basic unit of action at the neuromuscular junction. The depolarization of the nerve terminal accompanying the nerve impulse increases enormously the probability of quantal releases. Thus the normal EPP is generated by the synchronous release of quanta of transmitter, the individual spontaneous liberations of which give rise to the miniature potentials.

Under normal conditions the arrival of a nerve action potential at the motor neuron terminals liberates approximately 200 to 300 ACh quanta (quantal content). The quantal content can be estimated by various experimental techniques, some of which involve statistical analysis.[5, 7] Additional experimental methods outlined by Hubbard et al.[5] allow one to estimate that roughly 300,000 quanta are stored in the neuronal terminals at a single neuromuscular junction. One can see therefore that the stores of preformed ACh can be depleted during high-frequency repetitive neuronal discharge and that replenishment of the stores and intracellular deployment of ACh quanta are important factors in maintaining neuromuscular transmission during sustained activity.

Various hypothetical models have been proposed[4, 5] to explain how synaptic vesicles move to the inner surface of the presynaptic terminals and discharge their contents into the synaptic cleft. One of the firmly established facts that must be incorporated in such models is that the ACh release process is dependent on the extracellular [CA^{2+}] ion activity. In fact, many excitation-secretion processes involving release of neurotransmitters and hormones are Ca^{2+} ion dependent.[152]

Over the past few years many aspects of the action of Ca^{2+} and other ions on presynaptic ACh release have been studied quantitatively. By employing iontophoretic techniques (Fig. 5-2, *A*), Katz and Miledi[82, 85, 86] have shown that Ca^{2+} ions facilitate transmitter release if applied immediately before depolarization of presynaptic terminals occurs. They concluded that depolarization of the axon terminal membrane opens a channel for Ca^{2+}, allowing it to move to the inside of the axon membrane where it participates in a reaction that increases the rate of transmitter release. These steps contribute a large part of the synaptic delay time. Additional aspects of the presynaptic action of Ca^{2+} come from a detailed study of the relationship between $[Ca^{2+}]_o$ and the amplitude of the EPP (see Dodge and Rahamimoff[48]). These investigators concluded from their work that the cooperative action of four Ca^{2+} ions is required for neural release of one quantum of transmitter.

It is now well established that the rate of ACh release from motor neuron terminals increases rapidly as the membrane potential of the terminals is decreased. Such membrane depolarization occurs during normal neuromuscular transmission when an action potential traverses the motor neuron terminals. Because the amplitude of the action potential depends on the extracellular $[Na^+]_o$ concentration, one would expect reduction in $[Na^+]_o$ to decrease the quantal content and thus to cause the EPP amplitude to diminish; this result was obtained by Colomo and Rahamimoff[40] when the extracellular $[Ca^{2+}]_o$ was normal. However, with low $[Ca^{2+}]_o$, reduction of $[Na^+]_o$ increased the quantal content and EPP amplitude. This result is interpreted on the basis that Na^+ and Ca^{2+} compete for sites on the terminal membrane. Such Ca^{2+}-Na^+ antagonism is not unique; it has been observed in other systems such as the frog heart.[108] To summarize the work of Dodge, Rahamimoff, and Colomo, three reactions are assumed to occur:

$$\begin{aligned} Ca + X &\rightleftharpoons CaX \\ Mg + X &\rightleftharpoons MgX \\ nNa + X &\rightleftharpoons Na_nX \end{aligned}$$

where X = membrane sites. CaX is effective in ACh release but MgX and Na_nX are not. In this type of competition the quantal content (m) is described by the expression:

$$m = K\left(\frac{W\,[Ca]}{1 + \frac{[Ca]}{K_1} + \frac{[Mg]}{K_2} + f(Na)}\right)^4$$

where W is a constant, f(Na) is a function of [Na] or $[Na^2]$, and K_1 is the "true" dissociation coefficient of the CaX complex. Note that the expression for m is a fourth-power function representing cooperative action of four Ca^{2+} ions in ACh release. Consult Dodge and Rahamimoff[48] and Colomo and Rahamimoff[40] for further theoretical details and results of critical experiments designed to test these relationships.

The fact that motor neuron terminals are small in diameter makes it impossible at present to measure their transmembrane potentials with intracellular electrodes. Because the values of these membrane potentials must be indirectly estimated, our understanding of the physiology and pharmacology of motor neuron terminals is limited. In the stellate ganglion of the squid a favorable synapse between preganglionic fibers and the giant axon of the last stellar nerve allows simultaneous intracellular recording from both pre- and postsynaptic elements. Input-output characteristics, ionic requirements, and other fundamental characteristics of synaptic transmission have been studied and reported in the literature.[87, 89, 122, 123] Details cannot be given here, but the interested reader should consult these references and also Chapter 6.

Postsynaptic response of the muscle cell[5, 33, 64]

From its point of release, ACh must move across the 100 to 200 Å synaptic cleft to reach its point of action on the postjunctional membrane. Because of the short distance involved, this transit can occur by simple diffusion in times well within the synaptic delay. Katz and Miledi[83] estimate the time from the moment of ACh release to the onset of the MEPP at less than 50 μsec. In this estimation they made the reasonable supposition that the rise time of the MEPP represents a rise in ACh concentration at the receptor site. The release of ACh into the perfusing fluid of a stimulated eserinized muscle and the quick excitatory action of ACh when applied locally near the postjunctional membrane indicate that this ion is not bound to macromolecules within the synaptic cleft but diffuses as it would in free solution.

As mentioned earlier in this chapter, the ACh released from neuronal terminals reaches the postjunctional membrane where it reacts with receptor sites and thereby causes an increase in the ionic permeability of this structure. Thereby the postjunctional membrane potential decreases (depolarization), and if the reduction is sufficient, a muscle action potential is initiated in the surrounding conductile membrane of the muscle fiber. The electrogenic action of ACh on the postjunctional membrane is one example of a large number of chemoelectrical transductions that are operative at the chemosensitive membranes found at many synapses, on smooth and cardiac muscle cells, on the endings of some sensory neurons, on many nonconductile cells, etc. At present much experimental work is directed to the isolation and characterization of receptor molecules and to an understanding of the mechanisms whereby membrane ionic permeability is controlled.

Direct evidence has shown that cholinergic receptor sites are found on the external but not on the internal surface of the muscle fiber membrane.[31] These sites are abundantly distributed on the postjunctional membrane, but they are also found on the extrajunctional membranes.[81, 120, 121] Many investigators (e.g., Koelle[93]) who have utilized electron microscopic histochemical localization techniques have shown that AChE is present at both pre- and postjunctional membranes. Thus ACh liberated into the synaptic cleft can reach the postjunctional membrane either to react with receptor sites or with AChE located there, or it can be lost by diffusing out of the synaptic cleft. Mathematical formulation of the kinetics of these competitive processes is incomplete at present.

The hypothesis that ACh exerts its action on the muscle cell by union with a specific receptor substance of the cell membrane derives from the original proposition of Langley, to which reference has been made. Langley's ideas were greatly elaborated by Clark. Further evidence for the existence of a cholinergic receptor has come from study of the action of various drugs on neuromuscular transmission. It is well known that many quaternary ammonium compounds, including ACh, can depolarize the postjunctional membrane when they are applied to this region by iontophoresis or by microperfusion techniques.[138] Such drugs, which are known as receptor activators, cause increases in the ionic permeability of the postjunctional membrane, presumably by producing a conformational change in the receptor molecule. It is worth noting that if such activators are applied for relatively long periods of time (seconds or more), the initially produced receptor activation is lost and postjunctional membrane permeability as well as membrane potential return toward control values.[138, 140] This and some other pharmacologic aspects of neuromuscular transmission will be discussed in more detail in a following section.

In addition to the receptor activators, there exists a large group of drugs that are receptor inhibitors; i.e., when they are combined with receptor sites, no permeability change is produced but this occupation prevents receptor sites from reacting readily with neurally released ACh. *d*-Tubocurarine, a bisquaternary compound, is a good example of such a drug; tetraethylammonium is another. For more information in this very large field, consult various textbooks of pharmacology and appropriate reviews and monographs.

In recent years many attempts have been made to isolate the molecules located in postjunctional membranes that bear ACh receptive sites. The problem is difficult for several reasons. One is that receptor molecules represent a small fraction of the molecular constituents of synaptic membranes and closely associated structures. (For example, Waser[161] used autoradiographic techniques with ^{14}C-labeled curarine to map end-plate receptor sites in the mouse diaphragm. He estimated that 4×10^6 molecules of curarine are bound when a lethal dose of this drug is administered, this amount covering about 1% of the chemosensitive membrane area.) An additional complication in isolation of receptors is that many of the molecular entities separated by fractionation procedures exhibit nonspecific binding for ACh and related compounds. This fact makes it more difficult to distinguish "true" receptor molecules. A final problem in this field is that once isolated from the membranes in which they function, membrane receptors can no longer demonstrate their capacity to control membrane permeability, which is vital in establishing their identity.

A readable account of the history and progress in receptor isolation has been given by De Robertis,[47] who also has reviewed recent work in his own laboratory concerning isolation of a proteolipid both from eel electroplax tissue and cerebral cortex. De Robertis cautiously suggested that this proteolipid may function as a receptor and provided a hypothetical model for it.

As mentioned earlier in this chapter, Changeux et al.[37] and Miledi et al.[125] have used α-bungarotoxin (a snake venom) to isolate and characterize a cholinergic receptor protein from eel electric tissue. The receptor appears to be a tetramer composed of protein units with a molecular weight of about 80,000. Both groups of investigators distinguish this protein from AChE.

Another approach to the characterization of the active site in a receptor molecule is to test its behavior in situ after reaction with various chemical reagents. By this means it has been found that the ACh receptor contains a disulfide bond at the active site and receptor models were proposed.[38, 77] The possibility that phosphate groups are also involved in the receptor molecule was indicated by Liu and Nastuk,[106] who showed that UO_2^{2+} ions, which bind strongly to phosphate groups, inhibit the receptor response to carbamylcholine (a stable analog of ACh).

We may now turn to the electrophysiologic phenomena that occur at the muscle postjunctional membrane during neuromuscular transmission. Ginsborg's review[64] of this subject is very clear and instructive.

When ACh reacts with receptors of the postjunctional membrane, the ionic conductance of this structure is greatly increased and, as a result, rapid ionic movements occur across it. These movements reduce the electric charge on the postjunctional membrane and thus the potential difference across it falls. This membrane depolarization is commonly known as the EPP. In the ACh-activated postjunctional membrane the channels through which ions move are open for a very brief period (about 2 msec) because ACh is quickly dissociated from the receptor site. The reason for this is that the [ACh] concentration in the synaptic cleft is rapidly reduced both by enzymatic hydrolysis and by diffusion. If, for various reasons, the EPP does not reach the critical level (about –50 mv), a muscle action potential is not initiated and the postjunctional membrane potential is restored to its normal resting value by a passive outward diffusion of K^+ ions.

The action potential neurally initiated at the neuromuscular junction has a form different from that seen in action potentials recorded at nonjunctional regions. A typical recording made at the neuromuscular junction is seen in Fig. 5-4. The first phase of the record represents the EPP, which in this case rises to the critical level (about –50 mv) at which a propagated action potential is initiated. Thereafter the record becomes complicated since it represents the combined activity of the postjunctional membrane and the adjacent electrically excitable nonjunctional membrane. Further, the action potential record shows a reduced amplitude and

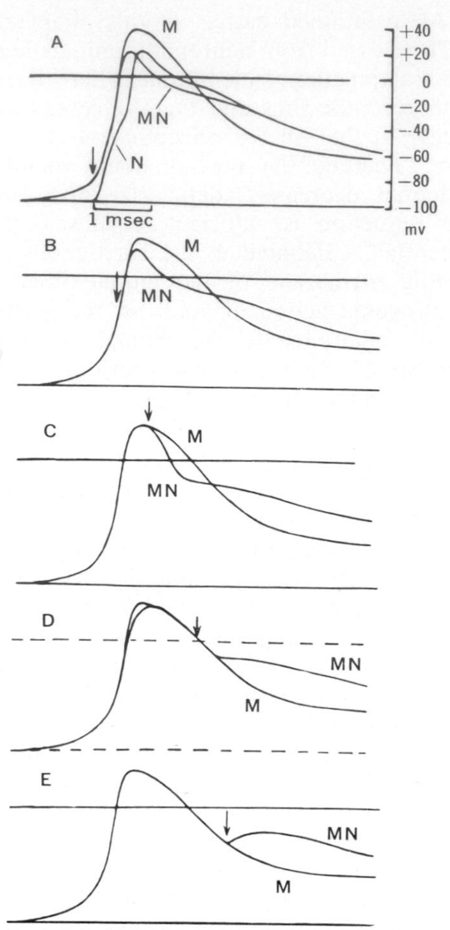

Fig. 5-12. Superimposed tracings of records from single end plate showing interaction, at various phases, between nerve-muscle transmitter, released by nerve impulse, and directly excited muscle action potential passing site of intracellular recording at end plate, *M*. Arrows indicate starts of responses to interjected nerve impulses, *N*. When ACh release is timed to occur during peak of muscle action potential, as in **C,** it tends to drive membrane potential toward a level of about −15 mv, and thus record *MN* is produced. When ACh is released after membrane begins to repolarize, as in **E,** it tends to drive membrane potential back toward −15 mv level. This −15 mv level represents a "reversal potential"; i.e., it is potential toward which ACh activated postjunctional membrane is driven. (From Castillo and Katz.[30])

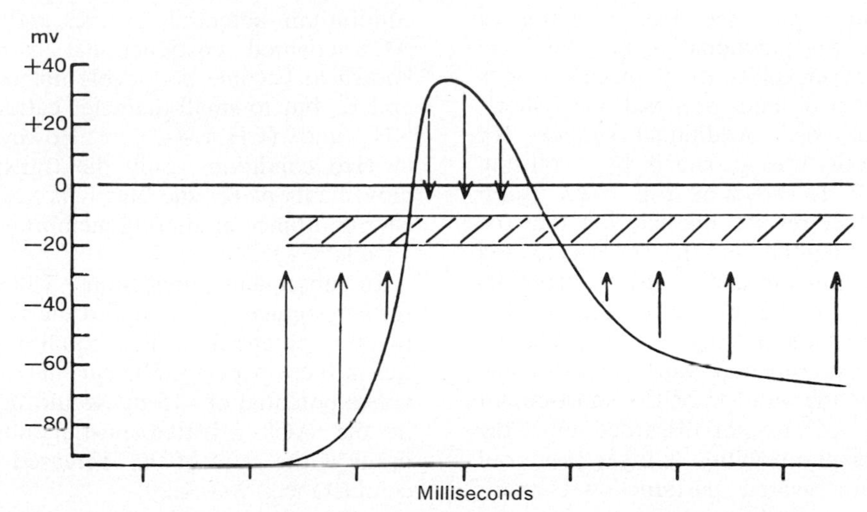

Fig. 5-13. Diagram summarizing results obtained by means of null-point experiment, illustrated in Fig. 5-12, which indicates direction of EPP produced by ACh applied at various values of membrane potential (represented by different phases of muscle action potential). Arrows indicate direction and relative magnitude of potential change due to release of ACh. Hatched area indicates approximate level at which EPP reverses sign, the "reversal potential." (From Castillo and Katz.[31])

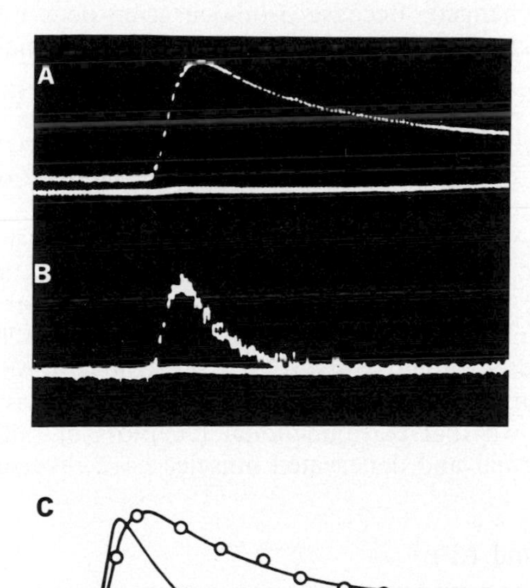

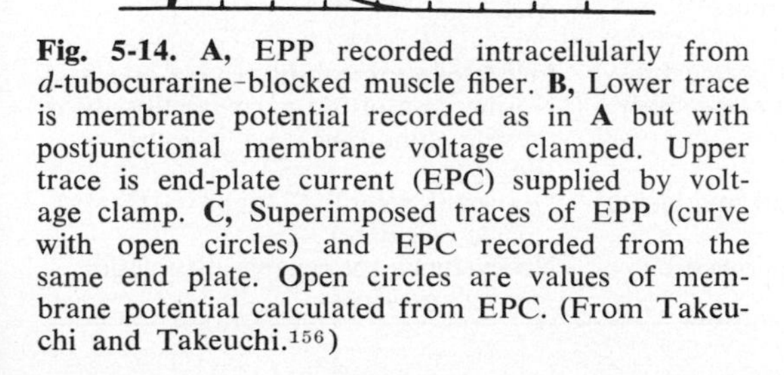

Fig. 5-14. A, EPP recorded intracellularly from *d*-tubocurarine-blocked muscle fiber. **B,** Lower trace is membrane potential recorded as in **A** but with postjunctional membrane voltage clamped. Upper trace is end-plate current (EPC) supplied by voltage clamp. **C,** Superimposed traces of EPP (curve with open circles) and EPC recorded from the same end plate. Open circles are values of membrane potential calculated from EPC. (From Takeuchi and Takeuchi.[156])

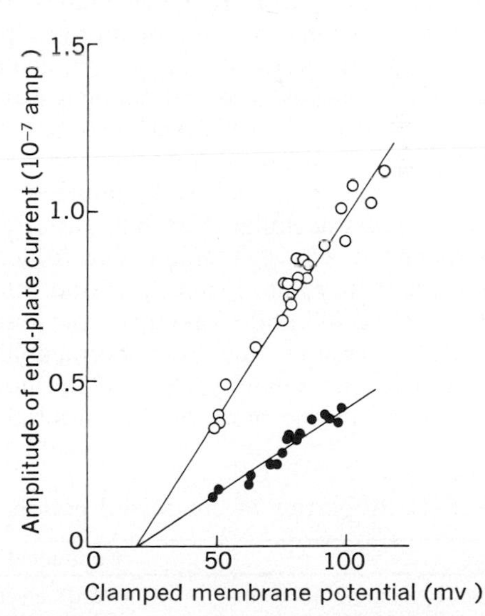

Fig. 5-15. Relationship between EPC measured at various values of postjunctional membrane potential maintained by voltage-clamping *d*-tubocurarine-blocked preparation. Open circles, results with *d*-tubocurarine 3×10^{-6} gm/ml. Closed circles, results with *d*-tubocurarine 4×10^{-6} gm/ml. (From Takeuchi and Takeuchi.[157])

longer falling phase than is seen with action potentials recorded at sites distant from the end plate. In seeking the explanation for this reduced "overshoot" of the action potential, Fatt and Katz[56] proposed that the action of ACh on the postjunctional membrane is to increase its permeability to all species of ions and they termed their proposal the "short-circuit hypothesis." Additional support for this hypothesis was provided by "collision experiments," as shown in Fig. 5-12. From these and other results, one can see that the membrane potential of the ACh-activated postjunctional membrane is driven from its resting value of –92 mv to a value in the –10 to –20 mv range (Fig. 5-13). Although subsequent experimental work raised some doubts about the validity of the short-circuit hypothesis,[156] it was not discarded until the specific ionic permeability changes produced at the ACh-activated postjunctional membrane were clarified by the satisfying and well-conducted experiments of Takeuchi and Takeuchi.[156] Using a curare-blocked nerve-muscle preparation, these investigators applied a voltage clamp to the postjunctional membrane and measured the EPP and the end-plate current (EPC) after nerve stimulation (Fig. 5-14). They then showed (Fig. 5-15) that over a wide range of values, the EPC is proportional to the membrane potential.[157] Extrapolation of the plot indicated that the EPC reaches zero and changes direction at the –10 to –20 mv level (reversal potential). They went on to show that the value of the reversal potential was changed by variation in extracellular Na^+ but variations of extracellular Cl^- (glutamate substituted) did not affect its value. From these and other results, Takeuchi and Takeuchi proposed that ACh increases the Na^+ permeability (P_{Na}) and K^+ permeability (P_K) of the postjunctional membrane more or less simultaneously, and for this reason the potential of the postjunctional membrane moves to a new value lying in the range of –10 to –20 mv. This region is roughly midway between the equilibrium potentials for K^+ and Na^+. The ACh-activated postjunctional membrane is known to become permeable not only to Na^+ and K^+ but to small-diameter cations such as NH_4^+ and $(CH_3)_4N^+$.[62, 136] However, under in vivo conditions, only the transmembrane movements of K^+ and Na^+ ions are of practical importance in altering membrane potential (Table 5-1).

In subsequent publications, Takeuchi[158, 159] further studied and analyzed the basis for the reversal potential (called equilibrium potential in these papers). She calculated that a reversal potential of –15 mv would be produced in the ACh-activated postjunctional membrane if the ratio of the increased Na and K conductances was 1.29.

$$\frac{\Delta G_{Na}}{\Delta G_K} = 1.29$$

She found that raising the extracellular $[Ca^{2+}]_o$ concentration from 1.8 to 18 mM made the reversal potential more negative because increased $[Ca^{2+}]_o$ reduced ΔG_{Na}, leaving ΔG_K unchanged. Because *d*-tubocurarine does not affect the value of the reversal potential, the ratio $\frac{\Delta G_{Na}}{\Delta G_K}$ is unaffected, although the individual conductances are greatly reduced. Thus *d*-tubocurarine reduces the number of channels opened by released ACh.

A number of factors in addition to the $[Ca^{2+}]_o$ concentration can cause variations in the reversal potential. Among the drugs, procaine can shift it in the negative direction,[109] and atropine shifts it in the positive direction.[110] Further, Feltz and Mallart[61] have shown that extrajunctional receptors of both normal and denervated muscles have reversal

Table 5-1. Properties of conducted action potential and EPP*

	Conducted action potential	EPP
Initiated by changes of membrane conductance	(Outward) electric current	ACh
During rising phase	Specific increase of Na permeability	Increased permeability to Na and K
During falling phase	Specific increase of K permeability	No increase of ion permeability above resting condition (i.e., "passive" decay)
Equilibrium potential of active membrane	Na potential (approx. 50 mv, inside positive)	"Reversal potential" (approx. 15 mv, inside negative)
Other distinguishing features	Regenerative ascent; followed by refractory period	No evidence for regenerative action or refractoriness

*Modified from Castillo and Katz.[33]

potentials of –42 mv and this was also true for receptors at "postjunctional areas" of denervated muscles. These results have been interpreted on the basis of variations in ΔG_{Na}, ΔG_K, or both. Maeno[109] supposed that separate membrane channels were available for Na^+ and K^+ ions, but this view is not universally accepted[61, 64, 95] and other explanations have been suggested that do not necessarily require synthesis of different receptor molecules.[61]

SOME FURTHER ASPECTS OF NEUROMUSCULAR TRANSMISSION

Neuromuscular fatigue and facilitation

Under normal conditions, conduction in the motor nerve and transmission at the neuromuscular junction are relatively indefatigable at levels of activity at which the contractile mechanism fails, and it is at this last stage of events that fatigue first makes its appearance—during a sustained maximal effort. This is shown, for example, by Merton's study[118] of the functional capacity of the *adductor pollicis* in normal, waking human subjects. The following conclusions can be drawn from his observations. (1) A maximal voluntary effort of a muscle produces a tension equal to that caused by a high-frequency stimulation of all the motor nerve fibers to the muscle. (2) During intense voluntary effort a single electrical stimulation of the motor nerve evokes a normal muscle action potential, but there is no increase in tension. Such a stimulus produces synchronous impulses in every motor nerve fiber not at that instant refractory because of a passing impulse of natural origin. This is independent evidence that the voluntary effort can activate the contractile mechanism to the full. (3) When the strength of the maintained maximal voluntary effort fails, tension cannot be restored by electrical stimulation of the motor nerve, even though such a stimulus evokes a maximal nerve volley, neuromuscular transmission occurs, and a maximal action potential in the muscle results. This is true even when extreme fatigue of the contractile mechanism has reduced the tension of the voluntary effort to zero.

Fatigue of neuromuscular transmission can occur, however, under certain conditions. For example, if the nerve of an isolated nerve-muscle preparation, which receives no circulation, is stimulated at a low frequency for a period of several minutes, there is a slow fall to zero of the muscle tension produced by each stimulus, and the muscle may relax incompletely. At this time the motor nerve continues to conduct impulses, and direct electrical stimulation of the muscle elicits a response of maximum tension. This neuromuscular fatigue is due to a slow, steady decline in the amount of ACh released by each nerve impulse, so that gradually the EPP produced by the released ACh falls in amplitude and fails to initiate action potentials in the muscle fibers. It has been shown by Castillo and Katz[28] that the decreased output of ACh is due to a reduction in the number of ACh quanta released and not to a drop in the amount of ACh per packet. This is true, at least, over the short term of several minutes. Furthermore, the progressive decline in the number of quanta released per impulse is associated with an increase in the spontaneous release rate, as shown by the increase in MEPP frequency. It might be supposed that the repeated discharge of ACh on the postjunctional membrane might depress the responsiveness of the latter since it is known that such "desensitization" can occur during the sustained application of depolarizing quaternary ammonium compounds such as ACh and its analogs. However, it has been shown that postjunctional desensitization is not an important factor in explaining the reduction of the EPP during a train of nerve impulses.[145]

Under ordinary physiologic conditions, during repetitive stimulation of the motor neuron, depression of neuromuscular transmission develops. This depression is attributed to a reduction in the store of ACh packets readily available for neuronal release and hence the quantal output of ACh per nerve impulse (quantal content) falls. If the $[Ca^{2+}]_o$ concentration in the extracellular fluid is reduced and the $[Mg^{2+}]_o$ concentration is increased, the quantal content is small and thus each neuronal action potential places a very small demand on the store of readily releasable ACh packets. Under these conditions, repetitive neuronal discharge is not associated with depression of neuromuscular transmission. On the contrary, if amphibian[30] or mammalian[18, 104] nerve-muscle preparations are bathed in low Ca^{++}, high Mg^{++} solutions, it is found that the EPP amplitude is *increased* during repetitive nerve stimulation. Statistical analysis indicates that this increase occurs during repetitive discharge ("tetanus") because more and more quanta of ACh are released with each successive nerve impulse.

The physiologic characteristics of neuromuscular facilitation and the mechanisms responsible for its production have been studied by many investigators and the recent publications cited here can guide the interested reader to more details. Mallart and Martin,[111, 112] using both Mg-blocked (low quantal content) and curare-blocked (high quantal content) preparations, showed that potentiation occurs in two phases. The first phase appeared within 5 msec of the conditioning impulse and decayed exponentially with a time constant of about 35 msec. The second phase became evident about 60 to 80 msec after the conditioning impulse, rose to a peak in approximately 120 msec, and decayed with a time constant of about 250 msec. These investigators concluded that in both phases of facilitation an increased probability of ACh release occurs and that the two components can sum linearly.

Rahamimoff[149] and Katz and Miledi[88] studied the effect of variation of $[Ca^{2+}]_o$ concentration on neuromuscular facilitation in order to better understand the mechanisms responsible for it. They systematically varied the extracellular $[Ca^{2+}]_o$ and $[Mg^{2+}]_o$ concentrations in the bathing fluid or relied on iontophoretic techniques to produce local changes in $[Ca^{2+}]_o$ concentration at appropriate times. In both of these papers it is concluded that $[Ca^{2+}]_o$ concentration is an important factor in determining the facilitation produced by a nerve impulse. The models presented involve entry of Ca into the axon terminal during propagation of an impulse. The Ca that enters is thought to combine with specific sites on the inner surface of the membrane, thereby raising the probability of ACh release. Thus neuromuscular facilitation (which is associated with an increased probability of ACh release) would be exhibited during the time that the membrane-Ca complex persists. These workers came to grips with the problem of explaining the relatively long time course of neuromuscular facilitation. Part of the problem may be resolved by assuming that four Ca ions cooperate in release and that the dissociation of the Ca-membrane complex follows a nonlinear rate equation. However, it is recognized by these workers and others that the mechanisms that determine the extent and time course of neuromuscular facilitation are manifold and as yet far from understood.[4, 124] The problem may be better attacked by study of synapses such as exist in the squid stellate ganglion, where it is possible to use intracellular techniques to measure and manipulate membrane potentials and to apply ions and drugs to both pre- and postsynaptic elements.*

The preceding discussion deals with the neuromuscular facilitation that appears early in the wake of a single nerve impulse and during repetitive discharge of the motor neuron. However, after such tetanic discharge is completed, neuromuscular transmission can show immediate depression, followed by a period of facilitation that can last up to several minutes. This slowly developing long-enduring facilitation is known as posttetanic potentiation. It can be demonstrated by applying a train of stimuli to the motor nerve, following which appropriately timed test stimuli are delivered and the EPP or the contractile response is recorded. An example taken from Hutter[73] is shown in Fig. 5-16. In this case the nerve-muscle preparation was partially curarized (a large fraction of the junctions were blocked) as shown by the fall in muscle tension output (Fig. 5-16, *A* and *B*). Tetanic stimulation caused a brief neuromuscular facilitation followed by depression that resulted in a complete neuromuscular block. However, during the posttetanic period the tension output in response to test stimuli rose nearly to the normal level. The plots in Fig. 5-16, *C* and *D,* illustrate that the degree and duration of posttetanic potentiation depend on the frequency and duration of the conditioning tetanus. Posttetanic potentiation was studied more directly by Liley and North[105] and Liley,[102] who showed that during such potentiation the quantal content is increased and the frequency of spontaneously released quanta (MEPPs) is raised. More recently a number of variables affecting posttetanic potentiation were studied by Rosenthal[151] and by Miledi and Thies.[124] They found posttetanic potentiation was prolonged or intensified by raising the extracellular $[Ca^{2+}]_o$ concentration. From this and other evidence they suggested that the phenomenon may be related to accumulation of Ca in the neuronal terminals.

There are two other forms of neuromuscular facilitation that deserve mention. The first is that produced by stretch of the muscle.[75] If in a partly curarized nerve-muscle preparation the muscle is stretched, there is an increase in the number of muscle fibers

*For further information, consult references 87, 89, 90, 122, and 123.

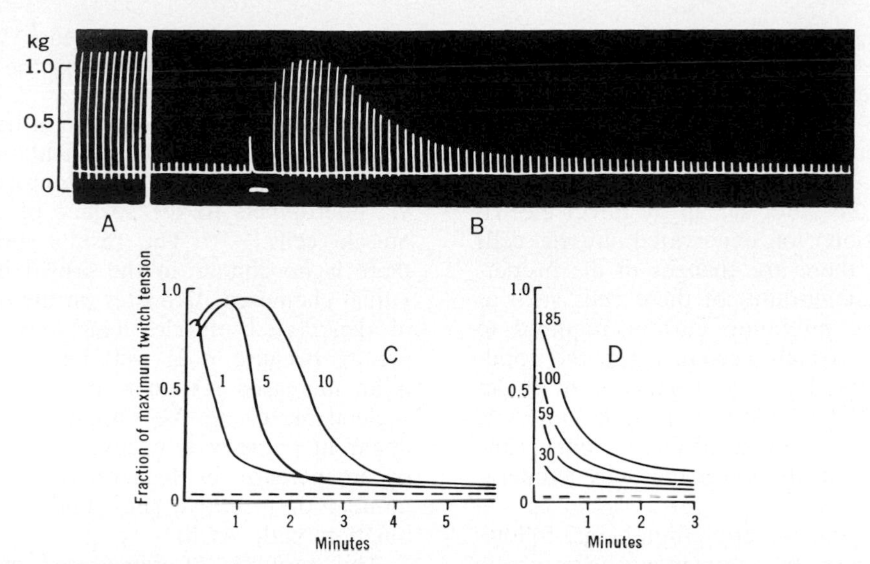

Fig. 5-16. Illustration of phenomenon of posttetanic facilitation in partially curarized muscle.

A and **B,** Contractions of tibialis muscle of anesthetized cat produced by supramaximal single stimuli to sciatic nerve every 10 sec, recorded mechanically. **A,** Untreated preparation. **B,** After partial curarization, contractions greatly reduced in amplitude—many end plates blocked. At horizontal signal, nerve was stimulated for 20 sec at 80/sec, and thereafter stimulation at rate of once in 10 sec resumed. Curare block was completely overcome (decurarization) for a period of about 1 min.

C and **D,** Graphs illustrating time course of posttetanic potentiation in partially curarized tibialis muscle of cat. Experimental arrangement as above. Twitch tension, expressed as fraction of the maximum twitch tension in normal preparation, plotted against time in minutes after end of tetanization of sciatic nerve, which contains efferent motor fibers innervating this muscle. **C,** Duration of potentiation depends on duration of conditioning tetanus; results plotted here following tetani of 1, 5, and 10 sec at 250 stimuli/sec. **D,** Time course of potentiation after 1 sec repetitive stimulation at different frequencies.

(From Hutter.[73])

that respond to a nerve volley, i.e., recruitment of fibers that had previously failed to contract. This facilitation results from an increase in quantal content; that is to say, mechanical deformation in some way causes a larger number of ACh packets to be released by each nerve impulse and thus increases the amplitude of the EPP. Like all forms of facilitation mentioned so far, that produced by stretch is associated with an increased spontaneous release rate.

The last form of facilitation to be mentioned was first described by Orbeli many years ago. He observed that when neuromuscular transmission has been fatigued by prolonged, low-frequency stimulation of the motor nerve, stimulation of the sympathetic innervation of the muscle restores neuromuscular transmission and results in an increase in twitch tension. The effect is due to the recruitment of muscle fibers which, in the fatigued state, had failed to respond. It is produced also by epinephrine and by norepinephrine, and Hutter and Loewenstein[74] have shown that these agents (or sympathetic nerve stimulation) increase the amplitude of the EPP produced by the nerve impulse. They concluded that the facilitation is postjunctional because these agents were found to increase the sensitivity of the postjunctional receptors to ACh. However, in a recent paper, Kuba and Tomita[96] showed that noradrenaline augments the probability of ACh release.

Changes in neuromuscular transmission following motor nerve section[16]

Section of the efferent nerve to a skeletal muscle results in its paralysis, a complete loss of voluntary and reflex contractions. Stimulation of the distal end of the cut nerve will continue to elicit contractions of the muscle for many hours. The first change to occur is a failure of neuromuscular transmission, and there is evidence that for any one junction this breakdown occurs rather suddenly. There is little doubt that the change is entirely due to a failure of transmitter release from presynaptic terminals, for at this stage there is no change in the sensitivity of the postjunctional membranes of the muscle cells to ACh. The mechanism of this failure is unknown. The main axons of the motor fibers will continue to conduct action potentials in response to electrical stimuli for several additional hours. It is of some importance to note that although the resting and action potentials (the latter set up by direct electrical stimulation) of denervated muscle cells are normal, there are changes in the properties of the membranes of these cells such as to lower the minimum current required to excite them, which accounts for the moderately increased electrical excitability of denervated skeletal muscle.[143] In poikilothermic animals the time course of these events is the more prolonged, the lower the body temperature.

In both mammals and amphibians, following denervation, the spontaneously occurring MEPPs cease abruptly when transmission fails.[102, 119] It has been shown for the frog that MEPPs are not, however, permanently abolished, for during the second week after denervation they reappear, though at a frequency some 100 times lower than normal. They occur at the places on the muscle cells where the remnants of the myelinated axons can be seen to terminate. These local potentials appear some weeks before the muscle cell develops hypersensitivity to ACh (see below), and at this time the pharmacologic reactions are similar to those of normal muscle. It seems likely that the slowly recurring MEPPs are produced by the quantal release of ACh from outside the muscle cell, and it has been suggested that the Schwann cell, which survives degeneration of the axon, is responsible for this slow production and release of ACh packets.[16] Upon reinnervation of the muscle cells by regenerating motor fibers, the spontaneous MEPPs once again occur at normal frequencies. Their occurrence during the period of denervation has not been observed in mammalian skeletal muscle.[102]

It has been known for a long time that denervated mammalian skeletal muscle develops a great increase in its sensitivity to ACh. This can be shown by injection of ACh into the arteries supplying a chronically denervated muscle. The increase in the sensitivity of a denervated effector cell is a general phenomenon (Chapter 30) exhibited by muscle cells, nerve cells, and gland cells. Denervated effector cells become increasingly sensitive to the transmitter agent secreted by the nerve fiber that innervates them. In skeletal muscle a 100-fold increase in chemosensitivity may develop during the weeks after denervation.

The sensitization phenomenon has been investigated by the local application of small quantities of ACh, delivered iontophoretically via micropipets to the surface of denervated muscle cells.[12, 119] The results indicate that there is no change in the sensitivity of individual chemosensitive sites on the membranes of denervated muscle. This may seem surprising because it is well known that there is an increased responsiveness of denervated skeletal muscle to ACh applied in bulk. The apparent paradox is believed to be explained by an increase, in denervated muscle, in the number of receptive sites available for combination with ACh.

The paralysis of denervated muscle cannot itself be a main factor in increasing its chemosensitivity because Miledi[119] has shown that for muscle fibers having two or more end plates, section of one motor fiber leads to an increased ACh sensitivity around the denervated neuromuscular junction but this does not occur at the remaining innervated junction. In normal muscle, chemosensitive receptor sites are abundantly distributed on postjunctional membranes but, although less numerous, they can also be detected at extrajunctional regions.[81] Recent studies show that junctional and extrajunctional receptors do not behave in an identical manner.[61] Following denervation, ACh-sensitive sites of the muscle cell can be detected with greater ease in extrajunctional regions of the muscle fiber and, in effect, the region sensitive to ACh spreads from the old junctional area until the entire surface of the fiber can be shown to be responsive to ACh. Anticholinesterases do not potentiate the effects of ACh applied to these non-end-plate regions, which means that AChE is either absent or is less effective at extrajunctional sites. (See Katz and Miledi[81] and additional references therein.) The extension of the ACh-sensitive region to the whole surface of the muscle fiber allows the drug to depolarize the entire length of the fiber and thereby to produce the prolonged ACh contracture observed in denervated mammalian muscle.

These findings have a general significance in that the motor nerve fiber exerts important influence on the chemosensitivity of the muscle cell it innervates. This is only one example of the so-called trophic influence of nerve. An abundant literature (e.g., Drachman[49]) shows that the trophic influence is exerted on many of the physiologic mechanisms of effector cells. The increase in the chemosensitive area of a muscle cell following denervation might be explained as due (1) to an actual spread of ACh receptor molecules outward from the end-plate region or (2) to the loss of some covering or protective substance that is normally present everywhere save at the end-plate region itself (assuming that receptive sites are normally present over the entire muscle membrane).

In mammals, denervated muscle displays very fine, diffuse, spontaneous movements. Usually the tremor is so slight in degree that it is well seen only when the bared surface of the muscle is observed in reflected light. The surface of the muscle is then observed to be "rippled by a restless agitation without either apparent rhythm or obvious center of activity."[46] This confused medley of small twitches constitutes true *fibrillation,* a condition produced by the randomly occurring twitches of individual muscle fibers. The cause of this spontaneous contraction is uncertain. It may be the result of excitation of the hypersensitive muscle cells by small quantities of circulating or locally released ACh. It is quite different from the randomly occurring contraction of entire motor units, called *fasciculation,* that is commonly seen in certain diseases of the CNS, e.g., amyotrophic lateral sclerosis.

PATHOLOGIC DISTURBANCES OF NEUROMUSCULAR TRANSMISSION

It can now be appreciated that neuromuscular transmission is a very complicated process whose many steps are potentially vulnerable to pathologic derangement and interference. For aid in considering the possibilities, the steps in neuromuscular transmission and a few of the factors known to influence them are given below.[137]

- Presynaptic elements
  - ACh
    - Synthesis
    - Storage
    - Mobilization
    - Release
      - Membrane depolarization
        - (via action potential or electrotonically)
      - Ca, Mg, botulinum toxin, etc.
  - Extent of terminal arborization
- Synaptic cleft
  - Diffusion of ACh
- Postsynaptic elements
  - Extent of postjunctional membrane area
  - Reaction of postjunctional membrane receptor sites to ACh
    - Desensitization mechanism
  - Resting potential of muscle fiber
    - Ionic composition of extra- and intracellular fluids
  - Excitability of conductile system
  - Hydrolysis of ACh
- Contractile elements
  - Excitation-contraction coupling
  - Performance of actomyosin system

Myasthenia gravis[3]

One well-known example of a pathologic disturbance in neuromuscular transmission is that which appears in a relatively rare disease known as myasthenia gravis. If a patient with this malady is asked to carry out a muscular task such as the repeated clenching of the fist, he is found to be able to produce strong contractions at the start, but very rapidly thereafter, muscular weakness appears. This weakness may be exhibited in many of the skeletal muscles, including those involved in eye movements, limb movements, deglutition, etc. Some regions may be more greatly affected than others. There is a vast literature dealing with both clinical and basic research in the field of myasthenia gravis. Many other papers have been devoted to problems concerning the treatment of patients with this disease. It is impossible here to give a complete review of the numerous past developments in this field but the general reviews at the end of the chapter should guide the reader to further details.

In myasthenic patients, recording of muscle action potentials (electromyography) from affected regions shows that during repetitive stimulation of the motor nerve, neuromuscular transmission fails. This finding is further supported by the long-known fact that the muscular strength of the myasthenic patient is appreciably improved after administration of anticholinesterase drugs such as neostigmine. Such treatment has no curative value, but many patients afflicted with the disease are able to perform reasonable daily activities if anticholinesterase medication is continuously administered via the oral route.

There have been many attempts to pinpoint the exact nature of the neuromuscular transmission defect in myasthenic patients. Past speculation has included the suggestions that there is a curare-like agent in the circulation, that there is an excessive accumulation of AChE at the neuromuscular junction, that there is inadequate synthesis or release of

ACh from the motor nerve terminals, etc. There is no compelling evidence to support the first two hypotheses, but experimental evidence indicates that a presynaptic defect is involved. Thesleff[160] presented evidence showing that in patients with myasthenia gravis the MEPPs are much reduced in amplitude. Postjunctional chemosensitivity was found to be unchanged and the increase in MEPP frequency from presynaptic depolarization was normal. On the basis of this evidence it was concluded that in this disease ACh quanta are released in the usual number by nerve impulses but that the *size* of each quantum is small. The interpretation of the reduction in MEPP amplitude is still under consideration and a partial explanation may be provided by Engel and Santa's findings.[54] Their electron micrographs show that in myasthenia gravis the primary and secondary synaptic clefts are increased in size and the secondary synaptic clefts are less numerous than usual. Under such circumstances the ACh discharged into the clefts by each quantum undergoes greater dilution, diminishing the ACh concentration at the postsynaptic receptor sites. It is evident that single-fiber techniques applied in vitro to biopsy specimens of affected skeletal muscle provide a critical means for investigation of the pre- and postsynaptic defects in myasthenia gravis.

Although it is well established that neuromuscular transmission is faulty in myasthenic patients, relatively little is known about the mechanisms whereby the fault is produced. The possibility that the disease might have an etiology based on immunologic derangements was first proposed in two independent papers by Nastuk et al.[142] and by Simpson.[153] The former group based their view on experimental work concerning the detection of cytolytic activity and altered serum complement activity in the blood of myasthenic patients and on further powerful evidence provided by Strauss et al.,[155] who demonstrated that the serum of some myasthenic patients contains an antibody against skeletal muscle. In formulating their hypothesis, Nastuk et al. made an effort to take into account and to interpret many aspects of myasthenia gravis, including the following: (1) existence of immunologic abnormalities; (2) presence of enlarged thymus glands in many myasthenic patients (connected with item 1 by virtue of the demonstrations by others that the thymus gland has an important immunologic function); (3) recognition that significant histopathologic changes occur in the muscles of myasthenic patients; and (4) consideration of the possibility that the neuromuscular defects seen in myasthenia gravis might be, to some extent, irreversible or at least that some functional defects might persist long after the etiologic agents that produced them had appeared and vanished.

It is evident that our understanding of the etiology of myasthenia gravis would be advanced if this disease could be produced experimentally in animals. Many investigators have been trying to do this and the status of these efforts has been recently reviewed by Goldstein and Manganaro.[65] The reader who wishes further details should consult the general reviews, including the critique of the autoimmune theory, and the appraisal of past research in this field presented by Nastuk and Plescia.[141] Many aspects of myasthenia gravis remain to be clarified and it appears likely that the combined efforts of immunologists and electrophysiologists will be required in future investigations.

Neuromuscular transmission defects also appear in other clinical situations. A myasthenic syndrome sometimes associated with bronchogenic carcinoma and different from that seen in myasthenia gravis has been studied by Lambert and his colleagues. In this disease the rested patient shows muscular weakness when voluntary movements are begun, but with sustained activity, muscle tension increases progressively. From electrophysiologic evidence obtained on single fibers, Lambert and his co-workers were able to show that the principal defect in this disease is presynaptic in origin in that it is caused by reduced quantal content. Repetitive activity produces marked presynaptic facilitation. An excellent review and discussion of this syndrome is given by Lambert and Elmqvist.[99]

REFERENCES

General reviews

1. Eccles, J. C.: The physiology of synapses, New York, 1964, Academic Press, Inc.
2. Ehrenpreis, S., editor: Cholinergic mechanisms, Ann. N. Y. Acad. Sci. **144:**383, 1967.
3. Fields, W. S., editor: Myasthenia gravis, Ann. N. Y. Acad. Sci. **183:**3, 1971.
4. Hubbard, J. I.: Mechanism of transmitter release, Progr. Biophys. Mol. Biol. **21:**33, 1970.
5. Hubbard, J. I., Llinas, R., and Quastel, D. M. J.: Electrophysiological analysis of synaptic transmission, Baltimore, 1969, The Williams & Wilkins Co.

6. Katz, B.: Nerve, muscle and synapse, New York, 1966, McGraw-Hill Book Co.
7. Martin, A. R.: Quantal nature of synaptic transmission, Physiol. Rev. **46:**51, 1965.
8. Nastuk, W. L.: Fundamental aspects of neuromuscular transmission, Ann. N. Y. Acad. Sci. **135:**110, 1966.
9. Quarton, G. C., Melnechuk, T., and Schmitt, F. O., editors: The neurosciences. A study program, New York, 1967, The Rockefeller University Press.
10. Schmitt, F. O., editor: The neurosciences. Second study program, New York, 1970, The Rockefeller University Press.

Original papers

11. Anderson-Cedergren, E.: Ultrastructure of motor end-plate and sarcoplasmic components of mouse skeletal muscle fiber, J. Ultrastruct. Res. suppl. **1:**1, 1959.
12. Axelson, J., and Thesleff, S.: A study of supersensitivity in denervated mammalian muscle, J. Physiol. **147:**178, 1959.
13. Barondes, S. H., editor: Cellular dynamics of the neuron, New York, 1969, Academic Press, Inc.
14. Birks, R., and MacIntosh, F. C.: Acetylcholine metabolism in a sympathetic ganglion, Can. J. Biochem. Physiol. **39:**787, 1961.
15. Birks, R., Huxley, H. E., and Katz, B.: The fine structure of the neuromuscular junction of the frog, J. Physiol. **150:**134, 1960.
16. Birks, R., Katz, B., and Miledi, R.: Physiological and structural changes at the amphibian myoneural junction, in the course of nerve degeneration, J. Physiol. **150:**145, 1960.
17. Bishop, G. H.: Natural history of the nerve impulse, Physiol. Rev. **36:**376, 1956.
18. Boyd, I. A., and Martin, A. R.: Spontaneous subthreshold activity at mammalian neuromuscular junctions, J. Physiol. **132:**61, 1956.
19. Boyd, I. A., and Martin, A. R.: The end-plate potential in mammalian muscle, J. Physiol. **132:**74, 1956.
20. Brown, G. L.: Transmission at nerve endings by acetylcholine, Physiol. Rev. **17:**485, 1937.
21. Brown, G. L.: Action potentials of normal mammalian muscle. Effects of acetylcholine and eserine, J. Physiol. **89:**220, 1937.
22. Brown, G. L., and Feldberg, W.: The acetylcholine metabolism of a sympathetic ganglion, J. Physiol. **88:**265, 1937.
23. Brown, G. L., Dale, H. H., and Feldberg, W.: Reactions of the normal mammalian muscle to acetylcholine and to eserine, J. Physiol. **87:**394, 1936.
24. Burgen, A. S. V., and Chipman, L. M.: Cholinesterase and succinic dehydrogenase in the central nervous system of the dog, J. Physiol. **114:**296, 1951.
25. Burke, W., and Ginsborg, B. L.: The action of the neuromuscular transmitter on the slow fibre membrane, J. Physiol. **132:**599, 1956.
26. Castillo, J. del, and Katz, B.: The effect of magnesium on the activity of motor nerve terminals, J. Physiol. **124:**553, 1954.
27. Castillo, J. del, and Katz, B.: Quantal components of the end-plate potential, J. Physiol. **124:**560, 1954.
28. Castillo, J. del, and Katz, B.: Statistical factors involved in neuromuscular facilitation and depression, J. Physiol. **124:**574, 1954.
29. Castillo, J. del, and Katz, B.: Changes in end-plate activity produced by presynaptic polarization, J. Physiol. **124:**586, 1954.
30. Castillo, J. del, and Katz, B.: The membrane change produced by the neuromuscular transmitter, J. Physiol. **125:**546, 1954.
31. Castillo, J. del, and Katz, B.: On the localization of acetylcholine receptors, J. Physiol. **128:**157, 1955.
32. Castillo, J. del, and Katz, B.: Local activity at a depolarized nerve-muscle junction, J. Physiol. **128:**396, 1955.
33. Castillo, J. del, and Katz, B.: Biophysical aspects of neuromuscular transmission, Progr. Biophys. Biophysical Chem. **6:**121, 1956.
34. Castillo, J. del, and Katz, B.: Localization of active spots within the neuromuscular junction of the frog, J. Physiol. **132:**630, 1956.
35. Castillo, J. Del, and Katz, B.: A study of curare action with an electrical micromethod, Proc. R. Soc. Lond. (Biol.) **146:**339, 1957.
36. Chang, C. C, and Lee, C.-Y.: Electrophysiological study of neuromuscular blocking action of cobra neurotoxin, Br. J. Pharmacol. **28:**172, 1966.
37. Changeux, J.-P., Kasai, M., and Lee, C.-Y.: Use of a snake venom toxin to characterize the cholinergic receptor protein, Proc. Natl. Acad. Sci. U. S. A. **67:**1241, 1970.
38. Changeux, J.-P., Podleski, T., and Meunier, J.-C.: On some structural analogies between acetylcholinesterase and the macromolecular receptor of acetylcholine, J. Gen. Physiol. **54:**225s, 1969.
39. Clark, A. W., Mauro, A., Longenecker, H. E., and Hurlbut, W. P.: Effects of black widow spider venom on the frog neuromuscular junction: effects on the fine structure of the frog neuromuscular junction, Nature **225:**703, 1970.
40. Colomo, F., and Rahamimoff, R.: Interaction between sodium and calcium ions in the process of transmitter release at the neuromuscular junction, J. Physiol. **198:**203, 1968.
41. Couteaux, R.: Localization of cholinesterases at neuromuscular junctions, Int. Rev. Cytol. **4:**335, 1955.
42. Couteaux, R.: Morphological and cytochemical observations on the postsynaptic membrane at motor end-plates and ganglionic synapses, Exp. Cell Res. suppl. **5:**294, 1958.
43. Dale, H.: Transmission of nervous effects by acetylcholine, Harvey Lect. **32:**229, 1937.
44. Dale, H., Feldberg, W., and Vogt, M.: Release of acetylcholine at voluntary motor nerve endings, J. Physiol. **86:**353, 1936.
45. Davis, R., and Koelle, G. B.: Electron microscopic localization of acetylcholinesterase and nonspecific cholinesterase at the neuromuscular junction by the gold-thiocholine and gold-thiolacetic acid methods, J. Cell Biol. **34:**157, 1967.
46. Denny-Brown, D. E., and Pennybacker, J. B.: Fibrillation and fasciculation in voluntary muscle, Brain **61:**311, 1938.

47. De Robertis, E.: Molecular biology of synaptic receptors, Science **171**:963, 1971.
48. Dodge, F. A., Jr., and Rahamimoff, R.: Cooperative action of calcium ions in transmitter release at the neuromuscular junction, J. Physiol. **193**:419, 1967.
49. Drachman, D. B.: Neuromuscular transmission of trophic effects, Ann. N. Y. Acad. Sci. **183**:158, 1971.
50. Eccles, J. C., and Jaeger, J. C.: The relationship between the mode of operation and the dimensions of the junctional regions at synapses and motor end-organs, Proc. R. Soc. Lond. (Biol.) **148**:38, 1957.
51. Eccles, J. C., and O'Connor, W. J.: The responses which nerve impulses evoke in mammalian striated muscle, J. Physiol. **97**:440, 1939.
52. Eccles, J. C., Katz, B., and Kuffler, S. W.: Nature of the "end-plate" potential in curarized muscle, J. Neurophysiol. **5**:362, 1941.
53. Ehrenpreis, S.: Acetylcholine and nerve activity, Nature **201**:887, 1964.
54. Engel, A. G., and Santa, T.: Histometric analysis of the ultrastructure of the neuromuscular junction in myasthenia gravis and in the myasthenic syndrome, Ann. N. Y. Acad. Sci. **183**: 46, 1971.
55. Fatt, P.: Skeletal neuromuscular transmission. In Magoun, H. W., editor, Neurophysiology section: Handbook of physiology, Baltimore, 1959, The William & Wilkins Co., vol. 1.
56. Fatt, P., and Katz, B.: An analysis of the end-plate potential recorded with an intracellular electrode, J. Physiol. **115**: 320, 1951.
57. Fatt, P., and Katz, B.: Spontaneous subthreshold activity at motor nerve endings, J. Physiol. **117**:109, 1952.
58. Feldberg, W.: Present views on the mode of action of acetylcholine in the central nervous system, Physiol. Rev. **25**:596, 1945.
59. Feldberg, W., and Mann, T.: Properties and distribution of the enzyme system which synthesizes acetylcholine in nervous tissue, J. Physiol. **104**: 411, 1946.
60. Feltz, A., and Mallart, A.: An analysis of acetylcholine responses of junctional and extrajunctional receptors of frog muscle fibers, J. Physiol. **218**:85, 1971.
61. Feltz, A., and Mallart, A.: Ionic permeability changes induced by some cholinergic agonists on normal and denervated frog muscles, J. Physiol. **218**:101, 1971.
62. Furukawa, T., Furukawa, A., and Takagi, T.: Fibrillation of muscle fibers produced by ammonium ions and its relation to the spontaneous activity at the neuromuscular junction, Jap. J. Physiol. **7**:252, 1957.
63. Giacobini, E.: Histochemical demonstration of AChE activity in isolated nerve cells, Acta Physiol. Scand. **36**:276, 1956.
64. Ginsborg, B. L.: Ion movements in junctional transmission, Pharmacol. Rev. **19**:289, 1967.
65. Goldstein, G., and Manganaro, A.: Thymin: a thymic polypeptide causing the neuromuscular block of myasthenia gravis, Ann. N. Y. Acad. Sci. **183**:230, 1971.
66. Göpfert, H., and Schaefer, H.: Uber den direkt und indirekt erregten Aktionsstrom und die Funktion der Motorischen Endplatte, Arch. Physiol. **239**:597, 1938.
67. Grundfest, H.: Synaptic and ephaptic transmission. In Bourne, G. H., editor: The structure and function of nervous tissue, New York, 1969, Academic Press, Inc., vol. 2.
68. Grundfest, H.: Synaptic and ephaptic transmission. In Magoun, H. W., editor, Neurophysiology section: Handbook of physiology, Baltimore, 1959, The Williams & Wilkins Co., vol. 1.
69. Hebb, C. O.: Biochemical evidence for neural function of acetylcholine, Physiol. Rev. **37**: 196, 1957.
70. Hebb, C. O.: Formation, storage, and liberation of acetylcholine. In Koelle, G. B., subeditor: Cholinesterases and anticholinesterase agents. Handbuch der experimentellen pharmakologie, Berlin, 1963, Springer Verlag, vol. 15.
71. Hubbard, J. I.: Mechanism of transmitter release from nerve terminals, Ann. N. Y. Acad. Sci. **183**:131, 1971.
72. Hubbard, J. I., and Kwanbunbumpen, S.: Evidence for the vesicle hypothesis, J. Physiol. **194**:407, 1968.
73. Hutter, O. F.: Post-tetanic restoration of neuromuscular transmission blocked by d-tubocurarine, J. Physiol. **118**:216, 1952.
74. Hutter, O. F., and Loewenstein, W. R.: Nature of neuromuscular facilitation by sympathetic stimulation in the frog, J. Physiol. **130**:559, 1955.
75. Hutter, O. F., and Trautwein, W.: Neuromuscular facilitation by stretch of motor nerve-endings, J. Physiol. **133**:610, 1956.
76. Jenkinson, D. H.: The nature of the antagonism between calcium and magnesium ions at the neuromuscular junction, J. Physiol. **138**: 434, 1957.
77. Karlin, A.: Chemical modification of the active site of the acetylcholine receptor, J. Gen. Physiol. **54**:245s, 1969.
78. Katz, B.: Microphysiology of the neuromuscular junction. A physiological "quantum of action" at the myoneural junction, Bull. Johns Hopkins Hosp. **102**:275, 1958.
79. Katz, B.: Microphysiology of the neuromuscular junction. The chemoreceptor function of the motor end-plate, Bull. Johns Hopkins Hosp. **102**:295, 1958.
80. Katz, B.: The transmission of impulses from nerve to muscle, and the subcellular unit of synaptic action, Proc. R. Soc. Lond. (Biol.) **155**:455, 1962.
81. Katz, B., and Miledi, R.: Further observations on the distribution of acetylcholine-reactive sites in skeletal muscle, J. Physiol. **170**:379, 1964.
82. Katz, B., and Miledi, R.: Localization of calcium action at the nerve-muscle junction, J. Physiol. **171**:10P, 1964.
83. Katz, B., and Miledi, R.: The measurement of synaptic delay, and the time course of acetylcholine release at the neuromuscular junction, Proc. R. Soc. Lond. (Biol.) **161**:483, 1965.
84. Katz, B., and Miledi, R.: Propagation of electrical activity in motor nerve terminals, Proc. R. Soc. Lond. (Biol.) **161**:453, 1965.

85. Katz, B., and Miledi, R.: The effect of calcium on acetylcholine release from motor nerve terminals, Proc. R. Soc. Lond. (Biol.) **161:** 496, 1965.
86. Katz, B., and Miledi, R.: The timing of calcium action during neuromuscular transmission, J. Physiol. **189:**535, 1967.
87. Katz, B., and Miledi, R.: A study of synaptic transmission in the absence of nerve impulses, J. Physiol. **192:**407, 1967.
88. Katz, B., and Miledi, R.: The role of calcium in neuromuscular facilitation, J. Physiol. **195:** 481, 1968.
89. Katz, B., and Miledi, R.: Tetrodotoxin-resistant electric activity in presynaptic terminals, J. Physiol. **203:**459, 1969.
90. Katz, B., and Miledi, R.: The effect of prolonged depolarization on synaptic transfer in the stellate ganglion of the squid, J. Physiol. **216:**503, 1971.
91. Koelle, G. B.: The elimination of enzymatic diffusion artifacts in the histochemical localization of cholinesterases and a survey of their cellular distributions, J. Pharmacol. Exp. Ther. **103:**153, 1951.
92. Koelle, G. B., subeditor: Cholinesterases and anticholinesterase agents. Handbuch der experimentellen Pharmakologie, Berlin, 1963, Springer Verlag, vol. 15.
93. Koelle, G. B.: Current concepts of synaptic structure and function, Ann. N. Y. Acad. Sci. **183:**5, 1971.
94. Koelle, G. B., Davis, R., and Devlin, M.: Acetyl disulfide $(CH_3COS)_2$ and bis (thioacetoxy) aurate I complex Au $(CH_3COS)_2$ histochemical substrates of unusual properties with acetylcholinesterase, J. Histochem. Cytochem. **16:**754, 1968.
95. Kordas, M.: The effect of procaine on neuromuscular transmission, J. Physiol. **209:**689, 1970.
96. Kuba, K., and Tomita, T.: Noradrenaline action on nerve terminal in the rat diaphragm, J. Physiol. **217:**19, 1971.
97. Kuffler, S. W.: Specific excitability of the endplate region in normal and denervated muscle, J. Neurophysiol. **6:**99, 1943.
98. Kuffler, S. W.: Symposium on physiology of neuro-muscular junctions: electrical aspects, Fed. Proc. **7:**437, 1948.
99. Lambert, E. H., and Elmqvist, D.: Quantal components of end-plate potentials in the myasthenic syndrome, Ann. N. Y. Acad. Sci. **183:**183, 1971.
100. Langley, J. N.: On the contraction of muscle, chiefly in relation to the presence of "receptive" substances. IV. The effect of curari and some other substances on the nicotine response of the sartorius and gastrocnemius muscle of the frog, J. Physiol. **39:**235, 1909-1910.
101. Lawler, H. C.: Turnover time of acetylcholinesterase, J. Biol. Chem. **236:**2296, 1961.
102. Liley, A. W.: An investigation of spontaneous activity at the neuromuscular junction of the rat, J. Physiol. **132:**650, 1956.
103. Liley, A. W.: The quantal components of the mammalian end-plate potential, J. Physiol. **133:**571, 1956.
104. Liley, A. W.: The effects of presynaptic polarization on the spontaneous activity at the mammalian neuromuscular junction, J. Physiol. **134:**427, 1956.
105. Liley, A. W., and North, K. A. K.: An electrical investigation of effects of repetitive stimulation on mammalian neuromuscular junction, J. Neurophysiol. **16:**509, 1953.
106. Liu, J. H., and Nastuk, W. L.: The effect of UO_2^{2+} ions on neuromuscular transmission and membrane conduction, Fed. Proc. **25:**570, 1966.
107. Longenecker, H. E., Hurlbut, W. P., Mauro, A., and Clark, A. W.: Effects of black widow spider venom on the frog neuromuscular junction: effects on end-plate potential, miniature end-plate potential, and nerve terminal spike, Nature **225:**701, 1970.
108. Lüttgau, H. C., and Niedergerke, R.: The antagonism between Ca and Na ions on the frog's heart, J. Physiol. **143:**486, 1958.
109. Maeno, T.: Analysis of sodium and potassium conductances in the procaine end-plate potential, J. Physiol. **183:**592, 1966.
110. Magazanik, L. G., and Vyskocil, F.: Different action of atropine and some analogues on the end-plate potentials and induced acetylcholine potentials, Experientia **25:**618, 1969.
111. Mallart, A., and Martin, A. R.: An analysis of facilitation and transmitter release at the neuromuscular junction of the frog, J. Physiol. **193:**679, 1967.
112. Mallart, A., and Martin, A. R.: The relation between quantum content and facilitation at the neuromuscular junction of the frog, J. Physiol. **196:**593, 1968.
113. Mann, P. J. G., Tennenbaum, M., and Quastel, J. H.: Acetylcholine metabolism in the central nervous system. The effects of potassium and other cations on acetylcholine liberation, Biochem. J. **33:**822, 1939.
114. Marchbanks, R. M.: Biochemical organization of cholinergic nerve terminals in the cerebral cortex. In Barondes, S. H., editor: Cellular dynamics of the neuron, New York, 1969, Academic Press, Inc.
115. Martin, A. R., and Pilar, G.: The dual mode of synaptic transmission in the avian ciliary ganglion, J. Physiol. **168:**443, 1963.
116. Martin, A. R., and Pilar, G.: Transmission through the ciliary ganglion of the chick, J. Physiol. **168:**464, 1963.
117. McMahan, U. J., Spitzer, N. C., and Peper, K.: Visual identification of nerve terminals in living isolated skeletal muscles, Proc. R. Soc. Lond. (Biol.) **181:**421, 1972.
118. Merton, P. A.: Voluntary strength and fatigue, J. Physiol. **123:**553, 1954.
119. Miledi, R.: The acetylcholine sensitivity of frog muscle fibres after complete or partial denervation, J. Physiol. **151:**1, 1960.
120. Miledi, R.: Junctional and extra-junctional acetylcholine receptors in skeletal muscle fibers, J. Physiol. **151:**24, 1960.
121. Miledi, R.: Induction of receptors. In Mongar, J. L., and de Reuck, A .V. S., editors: Boston, 1962, Little, Brown & Co.
122. Miledi, R.: Spontaneous synaptic potentials and quantal release of transmitter in the

stellate ganglion of the squid, J. Physiol. **192:** 379, 1967.
123. Miledi, R., and Slater, C. R.: The action of calcium on neuronal synapses in the squid, J. Physiol. **184:**473, 1966.
124. Miledi, R., and Thies, R.: Tetanic and post-tetanic rise in frequency of miniature end-plate potentials in low-calcium solutions, J. Physiol. **212:**245, 1971.
125. Miledi, R., Molinoff, P., and Potter, L. T.: Isolation of the cholinergic receptor protein of Torpedo electric tissue, Nature **229:**554, 1971.
126. Miledi, R., Stefani, E., and Steinbach, A. B.: Induction of the action potential mechanism in slow muscle fibers of the frog, J. Physiol. **217:**737, 1971.
127. Nachmansohn, D.: Chemical and molecular basis of nerve activity, New York, 1960, Academic Press, Inc.
128. Nachmansohn, D.: Role of acetylcholine in neuromuscular transmission, Ann. N. Y. Acad. Sci. **135:**136, 1966.
129. Nachmansohn, D.: Chemical control of the permeability cycle in excitable membranes during electrical activity, Ann. N. Y. Acad. Sci. **137:**877, 1966.
130. Nachmansohn, D.: Proteins in excitable membranes, Science **168:**1059, 1970.
131. Nachmansohn, D., and Machado, A. L.: The formation of acetylcholine. A new enzyme: "choline acetylase," J. Neurophysiol. **6:**397, 1943.
132. Nachmansohn, D., and Rothenberg, M. A.: Studies on cholinesterase. I. On the specificity of the enzyme in nerve tissue, J. Biol. Chem. **158:**653, 1945.
133. Nastuk, W. L.: The electrical activity of the muscle cell membrane at the neuromuscular junction, J. Cell. Comp. Physiol. **42:**249, 1953.
134. Nastuk, W. L.: Membrane potential changes at a single muscle end-plate produced by transitory application of acetylcholine with an electrically controlled microjet, Fed. Proc. **12:**102, 1953.
135. Nastuk, W. L.: Neuromuscular transmission: fundamental aspects of the normal process, Am. J. Med. **19:**663, 1955.
136. Nastuk, W. L.: Some ionic factors that influence the action of acetylcholine at the muscle endplate membrane, Ann. N. Y. Acad. Sci. **81:**317, 1959.
137. Nastuk, W. L.: Fundamental aspects of neuromuscular transmission, Ann. N. Y. Acad. Sci. **135:**110, 1966.
138. Nastuk, W. L.: Activation and inactivation of muscle postjunctional receptors, Fed. Proc. **26:**1639, 1967.
139. Nastuk, W. L., and Alexander, J. T.: The action of 3-hydroxy-phenyldimethylethyl-ammonium (Tensilon) on neuromuscular transmission in the frog, J. Pharmacol. Exp. Ther. **111:**302, 1954.
140. Nastuk, W. L., and Gissen, A. J.: Actions of acetylcholine and other quaternary ammonium compounds at the muscle postjunctional membrane, In Paul, W. M., Daniel, E. E., Kay, C. M., and Monckton, G., editors: Muscle, London, 1965, Pergamon Press, Ltd.
141. Nastuk, W. L., and Plescia, O. J.: Current status of research on myasthenia gravis, Ann. N. Y. Acad. Sci. **135:**664, 1966.
142. Nastuk, W. L., Plescia, O. J., and Osserman, K. E.: Changes in serum complement activity in patients with myasthenia gravis, Proc. Soc. Exp. Biol. Med. **105:**177, 1960.
143. Nicholls, J. G.: The electrical properties of denervated skeletal muscle, J. Physiol. **131:**1, 1956.
144. Ogston, A. G.: Removal of acetylcholine from a limited volume by diffusion, J. Physiol. **128:**222, 1955.
145. Otsuka, M., Endo, M., and Nonomura, Y.: Presynaptic nature of neuromuscular depression, Jap. J. Physiol. **12:**573, 1962.
146. Perry, W. L. M.: Acetylcholine release in the cat's superior cervical ganglion, J. Physiol. **119:**439, 1953.
147. Potter, L. T.: Synthesis, storage, and release of [^{14}C] acetylcholine in isolated rat diaphragm muscles, J. Physiol. **206:**145, 1970.
148. Potter, L. T.: Acetylcholine metabolism at vertebrate neuromuscular junctions. In Costa, E., and Giacobini, E., editors: New York, 1970, Raven Press.
149. Rahamimoff, R.: A dual effect of calcium ions in neuromuscular facilitation, J. Physiol. **195:** 471, 1968.
150. Robertson, J. D.: The ultrastructure of a reptilian myoneural junction, J. Biophys. Biochem. Cytol. **2:**381, 1956.
151. Rosenthal, J.: Post-tetanic potentiation at the neuromuscular junction of the frog, J. Physiol. **203:**121, 1969.
152. Rubin, R. P.: The role of calcium in the release of neurotransmitters and hormones, Pharmacol. Rev. **22:**389, 1970.
153. Simpson, J. A.: Myasthenia gravis: a new hypothesis, Scott. Med. J. **5:**419, 1960.
154. Stedman, E., and Stedman, E.: The mechanism of the biological synthesis of acetylcholine, Biochem. J. **33:**811, 1939.
155. Strauss, A. J. L., et al.: Immunofluorescence demonstration of a muscle binding complement-fixing serum globulin fraction in myasthenia gravis, Proc. Soc. Exp. Biol. Med. **105:** 184, 1960.
156. Takeuchi, A., and Takeuchi, N.: Active phase of frog's end-plate potential, J. Neurophysiol. **22:**395, 1959.
157. Takeuchi, A., and Takeuchi, N.: On the permeability of end-plate membrane during the action of transmitter, J. Physiol. **154:**52, 1960.
158. Takeuchi, N.: Some properties of conductance changes at the end-plate membrane during the action of acetylcholine, J. Physiol. **167:** 128, 1963.
159. Takeuchi, N.: Effects of calcium on the conductance change of the end-plate during action of the transmitter, J. Physiol. **167:** 141, 1963.
160. Thesleff, S.: Acetylcholine utilization in myasthenia gravis, Ann. N. Y. Acad. Sci. **135:** 195, 1966.
161. Waser, P. G.: Receptor localization by autoradiographic techniques, Ann. N. Y. Acad. Sci. **144:**737, 1967.
162. Whittaker, V. P.: The isolation and character-

ization of acetylcholine-containing particles from brain, Biochem. J. **72:**694, 1959.
163. Whittaker, V. P.: Origin and function of synaptic vesicles, Ann. N. Y. Acad. Sci. **183:** 21, 1971.
164. Whittaker, V. P.: Subcellular localization of neurotransmitters. In Clementi, F., and Ceccarelli, B., editors: Advances in cytopharmacology, New York, 1971, Raven Press.
165. Wilson, I. B.: The mechanism of enzyme hydrolysis studied with acetylcholinesterase. In McElroy, W., and Glass, B., editors: A symposium of the McCollum Pratt Institute, Baltimore, 1954, The Johns Hopkins Press.
166. Wilson, I. B.: The inhibition and reactivation of acetylcholinesterase, Ann. N. Y. Acad. Sci. **135:**177, 1966.
167. Wilson, I. B., and Harrison, M. A.: Turnover number of acetylcholinesterase, J. Biol. Chem. **236:**2292, 1961.

6

VERNON B. MOUNTCASTLE and ROSS J. BALDESSARINI

Synaptic transmission

A synapse is a region of close contact between neurons or between a neuron and a muscle or gland cell that is specialized to allow the excitation or inhibition of one by the other. Synapses are, moreover, the intercellular contacts across which one cell exerts a trophic influence upon another. Simple cellular appositions without the specialized features to be described are not synapses.

In Chapter 5, transsynaptic excitation at the skeletal neuromuscular junction was considered at the cellular and microphysiologic levels. The neuromyal synapse is the prototype of what will be called *chemical* synapses, those at which an effect on a postsynaptic element is produced by the neurosecretion from presynaptic endings of specific chemical substances that produce changes in the permeabilities of the postsynaptic membrane. Synapses possessing the basic features of the neuromuscular junction, i.e., chemical synapses, predominate in vertebrates, especially mammals. At many synapses in invertebrates and at some in vertebrates the effect of a presynaptic impulse on a postsynaptic neuron is achieved by the flow of action currents from the former across the plasma cell membrane of the latter. These are called *electrical* synapses, and they possess characteristic morphologic features by which they differ from chemical synapses. Electrical synapses occur rarely in the mammalian CNS, and for this reason greatest attention will be given to chemically operated synapses.

In spite of a variety of patterns of synaptic connectivity in different nervous systems, the physiologic properties of synapses within each of these two major classes are remarkably uniform, so that certain general principles may be applied to synapses in different locations. They have emerged from three lines of research: study of the ultrastructure of synapses by electron microscopy, observations of the electrical events in nerve cells by means of intracellular recording, and identification of transmitter agents by biochemical and histochemical methods.

In nervous systems even simple synaptic actions involve large numbers of neurons, and knowledge of the connections of those elements is requisite to an understanding of function. The action of a presynaptic impulse on a postsynaptic element is almost always *either* excitatory or inhibitory; only in one special case has a sequentially diphasic effect been observed.[139] A central neuron receives a mixture of inputs of opposite action and expresses by its rate of discharge the net—the integrated—influence of all. Moreover, central synaptic transfers are in many places complicated by the action of interneurons, cells intercalated between the incoming and the outgoing elements of a nuclear region; such internuncial cells may modify and control synaptic transfer through the region.

The neuron doctrine

The essence of the neuron doctrine is that each nerve cell is a structural entity bounded by its plasma cell membrane and that a functional discontinuity parallels that of structure. This generalization was established by the anatomist Ramón y Cajal. It is supported by many experimental observations, including those that follow:

1. Electron microscopic studies show a cytoplasmic discontinuity between nerve cells; any central neuron is everywhere separated from the processes of other neurons or glial cells by an intercellular cleft 150 to 200 Å wide.

2. All parts of a neuron depend metabolically on its cell body. When an axon is severed, the ensuing wallerian degeneration extends to its terminals, but usually no change occurs in the cells they contact. In certain cases, if all or nearly all of the presynaptic axons ending upon a cell are cut, that cell may shrink in size, an isolation atrophy occurring, for example, in the cells of the lateral

geniculate body when the optic nerves are cut.

3. When the axon of a neuron is severed, there is a transient change in the distribution of the granular endoplasmic reticulum of the cell body (it may later disappear), and the nucleus moves to an eccentric position. This chromatolysis recedes after several weeks if the axon regenerates, as do those of motoneurons but not those of neurons wholly central. During chromatolysis no change is seen in the synaptic boutons ending upon the altered cell. Impulses in presynaptic fibers are conducted into their terminals at a time when the chromatolysed neurons upon which they impinge either cannot respond or respond abnormally.[95]

4. When neurons of the mammalian CNS are studied by means of intracellular recording techniques, their interiors are observed to be separated from the extracellular surround and from the insides of adjacent neurons and glial cells by a phase possessing a high electrical resistance and capacitance compared to cytoplasm or the extracellular fluid. That phase is the cell membrane.

5. The neuroblasts of the mantle layer of the embryonic neural tube remain distinct cellular entities throughout their development. The growth cones of those destined to become spinal motoneurons, for example, push out through the enveloping mesoderm in complete separation from each other and from the mesoblasts.

Thus each neuron is a separate cellular unit. Once initiated, the conduction of impulses in nerve axons is invariant. Yet central nervous activity displays a variability not found in nerve fibers. It was the major contribution of Sherrington[129] to provide the functional counterpart of the neuron theory and to show that intercellular synaptic linkages confer on the nervous system its basic properties: the ability of a given group of neurons to vary their response to presynaptic impulses, a fractionation of activity; and facilitation and inhibition—in a word, integration. The mechanism of this integrative action of the nervous system, from the level of the cell to that of CNS control of behavior, remains the central problem of experimental neurology.

MICROSCOPIC ANATOMY OF SYNAPTIC LINKAGES

Nerve cells are related to one another at intercellular junctions in a wide variety of ways, but the presynaptic to postsynaptic relation is commonly either axodendritic, axosomatic, or axoaxonic. Synapses may also be classified in terms of the specialized nature of their membrane interfaces and intracellular organelles. A classification made by Bodian and using these criteria is given in the following outline.

Classification of junctions between neurons in vertebrates, based on topography and vesicular types*

- I. Vesicular synapses, chemically operated
  - A. Conventional topography
    1. Axodendritic, predominantly type I with spherical vesicles
    2. Axosomatic, predominantly type II with flattened vesicles
    3. Axoaxonic, terminating upon:
       - a. Axon hillock with predominantly flattened vesicles
       - b. Axonal shaft (e.g., at node of Ranvier); rare; vesicle type unknown
       - c. Axonal terminals with spherical vesicles
  - B. Unconventional topography
    1. Telotelodendritic, reciprocal, e.g., the retina
    2. Dendrodendritic, reciprocal, e.g., the olfactory bulb
    3. Axodendrodendritic, e.g., in sympathetic ganglia
- II. Nonvesicular synapses, electrically operated
  - A. Conventional topography; gap junctions
    1. Axodendritic
    2. Axosomatic
  - B. Unconventional topography, gap or desmosome-like junctions
    1. Dendrodendritic
    2. Dendrosomatic
    3. Somasomatic
- III. Mixed junctions, chemically and electrically operated

*From Bodian.[25]

These varieties of synapses are widely distributed in both vertebrates and invertebrates, and thus far electron microscopic studies have not revealed specific types of synapses unique to complex brains. Barring such a discovery, it appears that the remarkable development of the nervous system in mammals is due to the great increase in the numbers of neurons and of their interconnections.

Nerve fibers commonly branch repeatedly to terminate upon a number of postsynaptic cells; a single neuron may thus have synaptic effects on many others, for its impulses are usually conducted into each branch and terminal. Reciprocally, a given neuron may receive presynaptic fibers from many neurons that may be of quite different types and reside in different locales. These two facts lead to the principles of divergence and convergence, which, it will be seen, are of considerable functional significance. Superimposed divergent and convergent arrangements occur to degrees varying from the rare 1:1 relation to those in which the pre-post and post-pre ratios number several hundred. These ratios determine the synaptic security within a neural system and thus its capacity for rapid action in the temporal domain, the degree of "local sign" or spatial specificity, etc.

The degree to which neuronal connections are fixed at the end of mammalian embryonic life varies among species and among different systems within a single CNS. For example, systems in which synaptic connectivity is highly organized at birth, e.g., the visual system, depend for their maintenance on continued activation by patterned peripheral stimuli. An important experimental problem is to determine the nature and extent of the structural plasticity thought to exist even in the adult mammalian brain.

Synaptic ultrastructure[11a, 25, 61, 62]

The general features of synaptic ultrastructure are illustrated in Fig. 6-1. At chemically operated synapses the cell membranes are separated by a uniform cleft of 200 to 300 Å that is continuous at its edges with the somewhat narrower intercellular cleft system. At the synaptic interface both cellular membranes show regions of increased electron density that may be asymmetric; at some synapses an electron-dense material is seen within the synaptic cleft. Interdigitations between the two neurons occur in some locations, but the separation of the two cell membranes is preserved. Synaptic boutons commonly contain more mitochondria than a comparable volume of cytoplasm elsewhere, a fact that suggests a high and perhaps special metabolic activity. A regular feature of chemical synapses is the presence in the presynaptic, but not the postsynaptic, location of large numbers of intracellular vesicles 200 to 800 Å in diameter. They are called synaptic vesicles because of this location and because chemical analysis of subcellular centrifugation fractions of neural tissue shows that they contain high concentrations of transmitter agents (p. 206). In peripheral and central adrenergic neurons these vesicles may have an electron-dense core, while those of peripheral cholinergic and nonadrenergic central synapses have structureless, electron-translucent interiors. There is good evidence that the synaptic vesicles of peripheral adrenergic neurons are constructed within the neuronal cell body and moved by axonal transport to the synaptic boutons, a transport known to depend on the integrity of the microtubules (p. 215). Those of peripheral and central cholinergic neurons are presumably formed and transported similarly.

The regular association of certain of these ultrastructural characteristics at some synapses, and of other sets at others, has led to classifications of chemically operated synapses. The two major classes (types I and II) originally proposed by Gray include more than 80% of central synaptic boutons.[34, 61] *Type I synapses are commonly axodendritic and excitatory.* They display a slightly wider synaptic cleft and greater assymetry than do type II synapses: there is a continuous and especially dense thickening of the presynaptic membrane, from which dense projections extend at intervals into the cytoplasm; when seen face-on they form a hexagonal array on centers of about 900 Å.[16] A layer of electron-dense material lies within the synaptic cleft. The synaptic vesicles are spheroid in shape and strongly resist form distortion in any fixative. Accumulations are frequently seen at the dense projections. *Type II synapses are commonly axosomatic and inhibitory.* Here the membrane thickenings are restricted to small spots in the area of synaptic contact. Synaptic clefts are about 200 Å wide and contain no electron-dense material. The synaptic vesicles of type II synapses are oblong and flattened in aldehyde-fixed material,[24] a property that clearly distinguishes them from the synaptic vesicles of type I.

Special postsynaptic organelles occur at some chemical synapses. In primate motoneurons, ribosomes and endoplasmic reticulum are aggregated postsynaptically to the large synaptic boutons of monosynaptically related dorsal root fibers[23a] but are separated from the cell membrane by a subsynaptic "cistern." The apical dendrites of cortical pyramidal cells are covered with small cellular protrusions called spines; each spine receives one or several synaptic boutons and contains within it a number of membranous sacs or plates called the "spine apparatus," which also contain ribosomes. The function of such postsynaptic organelles is unknown, but their ribosomal content suggests a special postjunctional protein synthesis. There is evidence that additional morphologic types of chemically operated synapses are present in some centers in the vertebrate CNS, but data

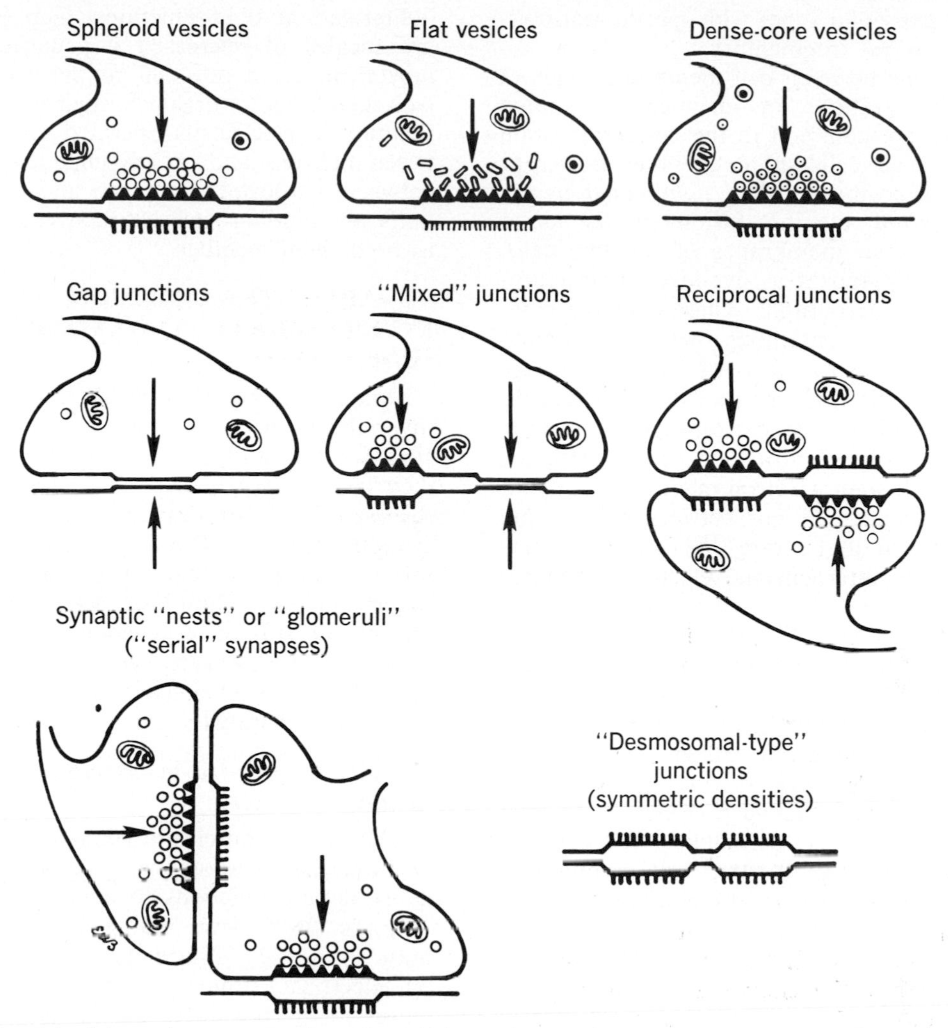

Fig. 6-1. Schematic representation of variations of synaptic structure and topography, presumably associated with functional diversity.

Upper row: "Asymmetric" synaptic contacts characteristic of most central presynaptic bulbs. Rigid cleft of about 160 Å, usually delimited by variable amounts of pre- and postsynaptic electron-dense material after osmium or phosphotungstic acid treatment, is characteristic. Dense material (not shown here) is also present in the cleft. In addition to mitochondria, presynaptic bulbs contain one of several varieties of microvesicles of about 200 to 400 Å, of which three major types are shown. *Spheroid vesicles* that resist deformation in fixation process are associated with certain excitatory synapses. Vesicles that are readily flattened after aldehyde fixation have been related to a number of known inhibitory synapses *(flat vesicles)*. Vesicle populations of intermediate susceptibility to flattening are characteristic of still other types of presynaptic bulbs, including peripheral cholinergic synapses. Microvesicles with electron-dense cores are characteristic of peripheral adrenergic axon terminals, whereas central "adrenergic" endings are thought to be associated with clusters of larger *dense-core vesicles* of about 600 to 800 Å.

Middle row: Narrow clefts of about 20 Å (gap junctions) are characteristic of synaptic contacts at which electrotonic transmission has been demonstrated. These are devoid of junctional densities or clusters of microvesicles. Same synaptic interface may also exhibit separate sites with morphologic characteristics of either "cleft" (chemical) junctions or "gap" (electrotonic) junctions. Reciprocally asymmetric junctions have also been described in same interface in olfactory bulb and retina.

Lower row: Specialized interneuronal junctions that are not known to be sites of synaptic transmission are present in specific sites, e.g., sympathetic ganglia, where dendrites of two cell types are often coupled in this way. "Desmosomal type" of junction is similar to that found in other epithelia and shows characteristic symmetry of interface densities, delimiting rigid cleft of about 200 Å.

(From Bodian.[25])

linking particular types with specific transmitters are as yet fragmentary.

Junctions between cell membranes without the specializations of chemically operated synapses are common in the vertebrate brain and between cells in many other tissues as well.[27] Appositions called *tight junctions* are characterized by the fusion of the outer leaflets of the membranes of the two cells, so that they appear as five-layered structures in electron micrographs. These junctions commonly form continuous belts or zonules around the perimeters of certain surface epithelial cells and are thought to hold cells together and impede the passage of material between them. Contrary to results obtained earlier with some electron micrographic methods, true tight junctions between neurons are now thought to be rare. Whether they may function as interneuronal electrical synapses is uncertain.

Another type of approximation between neurons and occasionally between glial cells presents a seven-layered appearance. At such junctions the intercellular cleft is narrowed from 200 Å to 20 to 30 Å; there are no asymmetric densities of the opposing cell membranes, and the narrow cleft is thought to be crossed by perpendicularly oriented structures linking the two cell membranes. These *gap junctions* are discontinuous plaques and differ from true tight junctions in that labeled protein molecules may pass from adjacent intercellular clefts into the narrow gap, there outlining but not penetrating the transgap structures already mentioned. Both gap and tight junctions resist separation when the intercellular cleft is caused to distend with fluid in response to an induced osmotic change. Gap junctions have now been identified at several electrically operated interneuronal synapses. At them, some vesicles are sometimes seen on both sides but not clustered at the presynaptic membrane as in chemically operated synapses. Gap junctions between astrocytes are thought to account for the low-impedance link between them but do not function here as electrical synapses, for glial cells, unlike neurons, do not possess the capacity for impulse generation (p. 215).

In summary, the neuronal and glial elements of the brain are separated by an intercellular cleft system about 200 Å wide. This cleft is slightly widened and is associated with a variety of membrane and organelle specializations at chemically operated synapses. It is narrowed to a slit at gap junctions and obliterated at tight junctions. Gap junctions are locales of decreased impedance to ion movement from one cell to the other; they lack membrane or organelle specialization and function as electrically operated synapses between neurons and as low-impedance links between astrocytes. Both gap and tight junctions are binding sites between cells, serving to hold them together.

SYNAPTIC TRANSMISSION AT INVERTEBRATE AXOAXONIC JUNCTIONS[29, 84]

Synapses between large fibers and cells in invertebrate nervous systems are favorable sites for the study of synaptic transmission; they can be isolated by microdissection and several microelectrodes can be placed in each synaptic element. The large volumes permit microchemical analysis of the axoplasm of single elements. They have been used in several classic studies of synaptic function. Two excitatory invertebrate synapses will be described; one is chemically and the other is electrically operated.

Chemically operated invertebrate excitatory synapse

The mantle muscles of cephalopods are innervated by giant axons synaptically linked at axoaxonic junctions to equally large presynaptic fibers with diameters as great as 500μ (Fig. 6-2). Presynaptic impulses set up by electrical stimuli invariably evoke impulses in the postfiber, and for short periods of time the synapse can be driven in a 1:1 manner at rates as high as 400/sec. It then exhibits no signs of integrative action. If stimuli are delivered to the postfiber, impulses propagate backward—antidromically—to the synaptic region, but no event is observed in the prefiber. The synapse is thus directionally oriented and provides an example, as does the neuromuscular junction, of the *principle of forward conduction,* a property of all chemically operated synapses.

The three pairs of pre- and postfiber intra-axonal records in Fig. 6-2, *A-C,* were taken just as a high-frequency train of impulses ended in transmission failure. They illustrate several of the general properties of chemically operated synapses, which are listed in Table 6-1. First, there is an irreducible period of time after arrival of the presynaptic impulse (solid line) during which no change occurs in the postfiber membrane potential (dotted line). This period is the *synaptic delay,* which

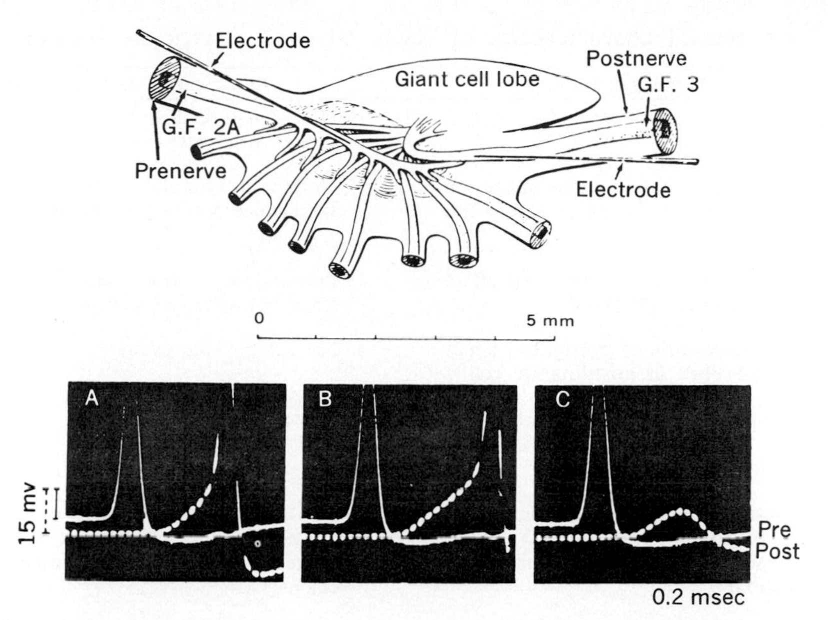

Fig. 6-2. Top: Diagrammatic representation of axoaxonic synapse between second- and third-order giant axons (*G.F. 2A* and *G.F. 3*) of the squid. Microelectrodes are shown in pre- and postfibers at their approximate angles of approach. Pairs of stimulating electrodes are also applied to two nerves during experiment but are not shown in drawing. (From Bullock and Hagiwara.[28])

A to **C,** Simultaneous records taken via intracellular microelectrodes in prefiber (continuous line) and postfiber (broken lines) of squid giant axon synapse shown above. Both electrodes are close to region of synaptic contact. Potential changes occurring at tip of each electrode are amplified and displayed upon the sweeps of a double-beam cathode-ray oscilloscope. Three sets of records were taken just as transmission began to fail, during prolonged high-frequency presynaptic stimulation. Successive frames show prolongation of postsynaptic response time and, in the third, local postsynaptic response remaining after conduction of postsynaptic spike has failed. (From Hagiwara and Tasaki.[68])

for chemically operated neuroneural synapses has values of 0.5 to 1.0 msec. Second, the initial depolarizing postsynaptic event has a slower time course than the action potential, to which it leads at an inflection point. This time is the postsynaptic response time, which varies with conditions of excitability and facilitation in contrast to the nearly constant synaptic delay. As the high-frequency train continues, the initial event decreases in amplitude, then fails to depolarize the postfiber to the critical level for impulse initiation and is revealed alone (Fig. 6-2, *C*). This depolarizing event is the *excitatory postsynaptic potential, or EPSP.* It has properties in common with the local event at the neuromuscular junction: it is not conducted but spreads over the postfiber membrane for a distance determined by the resistance and capacity of that membrane, and its amplitude can be graded, as it is in this case reduced by rapid repetition.

The transmission failure illustrated in Fig. 6-2 occurs at the synaptic region itself, for after failure both pre- and postfibers still conduct impulses at high frequencies. Thus one of the characteristics of synapses is a susceptibility to transmission failure, in contrast to the virtually indefatigable nature of axonal conduction. Study of this synapse after failure reveals that if two presynaptic impulses arrive at a short interval, the EPSP evoked by the second sums with that of the first; their summed depolarization may reach threshold and elicit a postfiber impulse. Under these conditions the synapse is not obligatory and displays a *temporal summation of synaptic excitation, one of the characteristics of integrative action.*

The giant synapse of the squid has been studied using multiple microelectrode recording techniques and the voltage clamp method. The results reveal that the (synaptically delayed) effect of a presynaptic action potential is to drive the inside potential of the postsynaptic fiber close to the potential level of the extracellular cleft.[68] A voltage clamp of the postfiber beyond this point

Table 6-1. Some general characteristics of chemically and electrically operated synapses

Chemically operated synapses	Electrically operated synapses
1. Interface ultrastructure and organelle content usually asymmetric	1. Ultrastructure usually symmetric, without interface or organelle specialization
2. Low-impedance shunt to intercellular space via widened synaptic cleft; no change in cell-to-cell impedance	2. Narrowed synaptic cleft, reducing shunt; low-impedance pathway between cells
3. Presynaptic action currents have minimal effect on postsynaptic membrane potential	3. Presynaptic action currents are immediate agent for synaptic transmission
4. Spontaneous quantal release of transmitter agent produces miniature PSPs in postsynaptic cell	4. No comparable event
5. Presynaptic action potential causes synchronous release of large number of quantal units of transmitter requiring Ca^{++}	5. No such event
6. Event 5 leads to local, nonconducted, gradable, summable response of postsynaptic cell (PSP), which reverses sign with shift of postsynaptic membrane potential	6. Similar response produced by flow of presynaptic action currents across postsynaptic membrane; not reversed by membrane potential changes
7. If of sufficient amplitude, local EPSP leads to postsynaptic action potential; response of opposite sign, IPSP occurs at inhibitory synapses	7. Similar sequence from EPSP to impulse
8. Postsynaptic membrane receptor molecules combine with transmitter, leading to permeability changes and PSPs	8. No such changes
9. Pre- or postsynaptic membrane may contain hydrolysing enzyme for transmitter or other mechanism for transmitter inactivation that occurs pre- or postsynaptically	9. No such mechanism
10. Pre- and postsynaptic events are modifiable by chemical agents	10. No comparable susceptibility
11. Transsynaptic conduction is unidirectional	11. Conduction either uni- or bidirectional, commonly the latter
12. Local PSP allows integrative action by temporal and/or spatial summation	12. Similar, but many are 1:1 with minimal integrative properties
13. Transsynaptic action sensitive to temperature changes	13. Relatively insensitive to temperature changes

causes a reversal of the sign of the EPSP, strong evidence that the synapse is chemically operated. This conclusion is further supported by the observation of miniature EPSPs occurring spontaneously in the quiescent postfiber.[107] The nearly zero equilibrium potential for the EPSP indicates that the permeability change produced in the postfiber by transmitter action is nonspecific in the sense that it is not restricted to Na^+ ions. The permeability change allows a charge movement from outside to inside (the EPSP); if large enough, this movement leads to the conducted action potential. There is some evidence that the glutamate ion is the transmitter agent at this synapse,[107] but this identification is still uncertain.

Electrically operated excitatory synapse[55]

Electric or electrotonic synapses are distinguished from those that are chemically operated by the differing properties listed in Table 6-1. Prominent among these is a specialized low-impedance pathway that provides an effective electric coupling between pre- and postsynaptic cells. A further requirement for an electrically operated synapse is a proper electric match between the two cells, for the presynaptic element must be large enough to supply ionic current adequate to depolarize

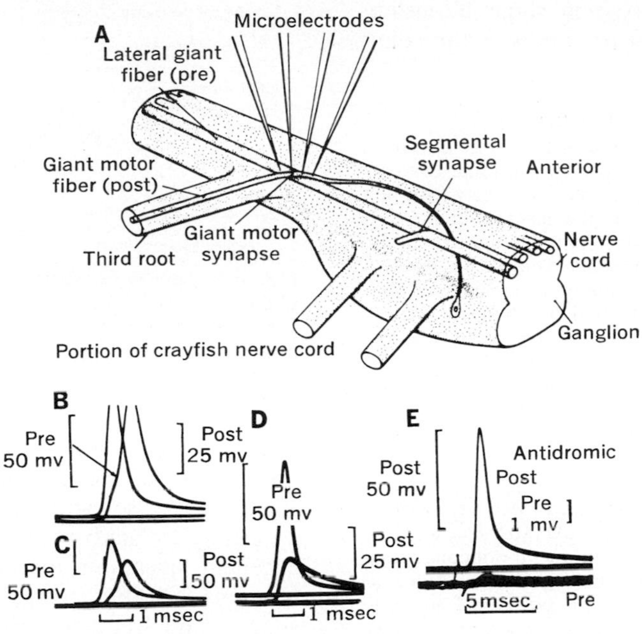

Fig. 6-3. A, Drawing of one ganglion in crayfish abdominal nerve cord showing courses of one giant motor fiber from cell body to exit. Its junction with lateral giant prefiber is shown; other synapses are not. **B** to **D,** Orthodromic nerve impulse transmission recorded, as in **A,** with intra-axonal recording from prefiber (upper trace) and postfiber (lower trace). **B** and **C** at different amplifications. In **B,** arrow indicates inflection from EPSP to nerve impulse. In **D,** EPSP did not reach threshold of postfiber. Postfiber action potential shown in **E,** elicited antidromically, produced only negligible depolarization of prefiber. Minimal synaptic delay is characteristic of electric synapses. This one is unidirectional; others conduct in either direction. (From Furshpan and Potter.[55])

the postsynaptic one. There are now a large number of synapses in invertebrates and submammalian vertebrates proved to be electrically operated and at which narrow gap junctions exist (p. 186). They apparently isolate the synaptic interface from the low-impedance shunt of the intercellular cleft system and provide in some unknown way just such a shunt linking the two synaptic cells.

The electrically operated synapse between the lateral giant fiber and the giant motor fiber of the crayfish is diagrammed in Fig. 6-3. The records displayed were obtained from microelectrodes in pre- and postfibers. They show that conduction occurs only from the pre- to the postfiber; an antidromic impulse in the postfiber produces little change in the prefiber (Fig. 6-3, *E*). Dromic conduction occurs with virtually no synaptic delay. The initial postsynaptic response is a typical EPSP (Fig. 6-3, *D*), which, if it is of sufficient amplitude, leads to an action potential in the postfiber. Its onset after no delay indicates a direct electric spread of depolarizing current across the synaptic gap. Another feature that clearly distinguishes the two types is this: EPSPs at chemically but not at electrically operated synapses may be reversed in sign by depolarization of the postsynaptic membrane to a level beyond the equilibrium potential for transmitter action.

At this crayfish synapse (Fig. 6-3, *A*), currents may spread in the dromic direction only, a rectifying property of the synapse not yet explicable in terms of its ultrastructure. At many electrical synapses, e.g., the septal synapses of the giant axons of annelids or the somatosomatic synapses in certain nuclei of the fish brain, conduction occurs with ease in either direction. The minimal synaptic delay, large safety factor for synaptic action, and reciprocity underlie an important function of electrotonic synapses: cells linked by them tend to discharge in synchrony.[22] The commonly observed reciprocal conduction at electrical synapses indicates that the law of forward conduction can be applied with certainty only to chemically operated synapses. Although electrotonic coupling between cells has been shown to exist in some brainstem motor nuclei,[18] it must be rare in mammalian nervous systems.

Ephaptic interactions between nerve cells

In the intact organism a nerve cell lies in a conducting medium, and the extrinsic ionic currents associated with its action potentials flow through the volume conductor of the extracellular cleft system. An adjacent neuron will be penetrated by those currents to a degree determined by its own membrane resistance relative to that of the cleft system, and its excitability will be changed thereby in a direction determined by the direction and density of the transmembrane current. In the nervous system the packing density of neurons and their processes is very high, and the question arises as to whether ephaptic interactions, as these nonsynaptic influences are called, are of physiologic consequence.

Such ephaptic influences occur between experimentally isolated nerve fibers juxtaposed at a crossing but elsewhere surrounded by a nonconducting medium. They occur also at regions of injury to nerve trunks, and such cross excitation is frequently evoked to explain some of the aberrant sensory phenomena that may follow nerve injury. It is important to emphasize that such effects have only been observed at sites of injury and that under normal conditions they must be very small indeed. The *principle of isolated conduction remains valid;* impulses traveling along normal nerve fibers of a group do not spread laterally to travel along adjacent fibers. The compact intermingling of fibers and neurons of diverse functions in the CNS makes this almost a priori for orderly function.

Although these effects are too small to produce nonsynaptic cross excitation between axons, it is possible that they influence neuronal activity in a more subtle fashion. Terzuolo and Bullock[136] have shown, for example, that extracellular currents of a value comparable to those under discussion, although insufficient to excite adjacent neurons, may affect the frequency of discharge of those already active by influencing the excitability of the spike-generating zone in the cell soma. These ephaptic influences may play a role in causing closely adjacent neurons to discharge at the same frequency. Such a synchronization of neural activity is a common event in large masses of neurons in the CNS, e.g., in the cerebral cortex. While in the cortex the synchronization is due mainly to the driving effects of thalamocortical systems (Chapter 7), the large and interlocking fields of cortical cells provide an optimal geometric arrangement for ephaptic effects. Direct evidence that this is true, however, is lacking.

CONDUCTION THROUGH AUTONOMIC GANGLIA

The synaptic relation in sympathetic ganglia is markedly divergent, for each preganglionic fiber sends terminal branches to each of a large number of cells, e.g., as many as 200 in the human stellate ganglion,[45] while each ganglion cell appears to receive synaptic terminals from only a few preganglionic fibers.

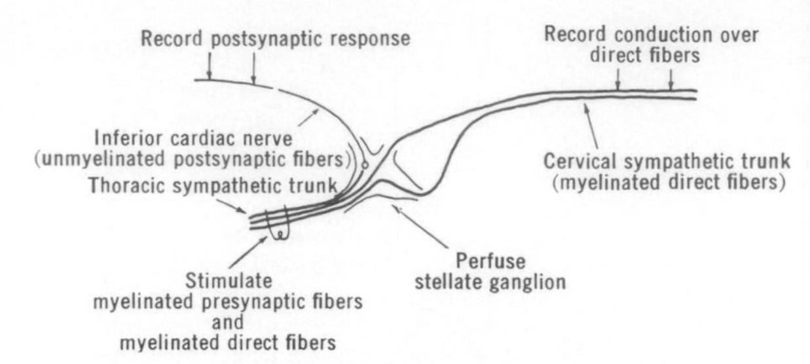

Fig. 6-4. Schematic representation of stellate ganglion of cat indicating some experimental approaches to study of ganglionic synaptic transmission described in text. (From Larrabee and Bronk.[97a])

This divergence accounts in part for the diffuse action of the sympathetic nervous system. The ease with which they may be surgically isolated makes sympathetic ganglia excellent structures for the study of synaptic transmission (Fig. 6-4), for (1) the number, source, and frequency of presynaptic volleys can be controlled and varied, (2) synaptically relayed and directly conducted impulses can be recorded simultaneously, (3) the circulation of the ganglion can be isolated and perfused and the effluent analyzed for substances released or taken up during activity, (4) ganglion cells can be studied by intracellular recording, or from single postganglionic fibers isolated by microdissection, and (5) the nerves and ganglia are easily identified and removed for histochemical or electron micrographic study.

Local synaptic events[47, 100]

The records shown in Fig. 6-5, *II,* were obtained by intracellular recording from a cell of the rabbit superior cervical ganglion. They reveal that the first postsynaptic event is a depolarization, the EPSP, which if it is of sufficient amplitude leads to a postsynaptic impulse. These events are similar in many ways to those occurring at the chemically operated neuromuscular junction and squid giant synapse. (1) There is an irreducible and constant synaptic delay followed by a variable postsynaptic response time. (2) Transsynaptic conduction occurs in the dromic direction only. (3) During the EPSP there is an increased ionic conductance; the EPSP equilibrium potential is close to zero, indicating that the permeability change evoked by the transmitter is not specific, although the charge transfer of the EPSP must be carried largely by Na^+. (4) Transmitter action is susceptible to competitive block by drugs

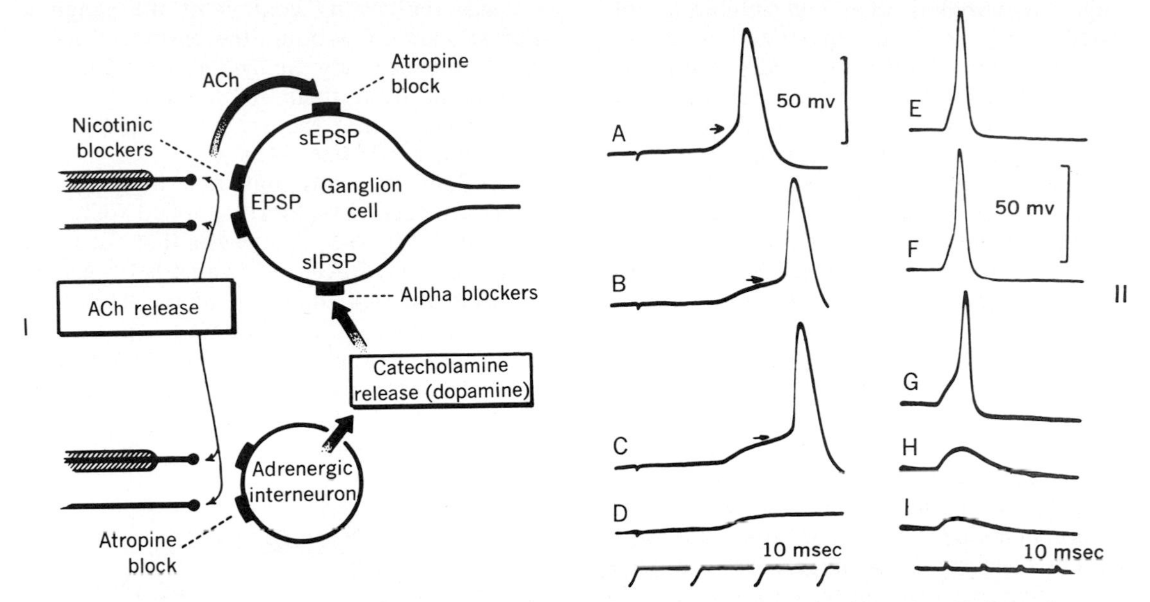

Fig. 6-5. I, Schematic outline of synaptic mechanisms at peripheral sympathetic ganglion of autonomic nervous system. All preganglionic fibers are cholinergic; released ACh acts at two receptor sites, one nicotinic, leading to rapid EPSP, the other muscarinic, leading to prolonged sEPSP. Some presynaptic fibers, operating via interneuron utilizing dopamine as transmitter, exert prolonged inhibitory synaptic action on postganglionic cells. (From Libet.[100])

II, Intracellular records from ganglion cells of isolated superior cervical ganglion of rabbit; potential changes are amplified and displayed on cathode-ray oscilloscope; records shown are photographs of its sweep. Depolarization of cell is indicated by upward movement of record. *A* is action potential of ganglion cell evoked by single maximal preganglionic volley; *B* to *D* are responses when preganglionic volley contains successively fewer fibers. There is then prolongation of postsynaptic response time; finally, in *D*, EPSP evoked by weakest volley fails to reach threshold that, as indicated by arrows, was steady at about 14.5 mv depolarization. Records *E* to *I* show progressive series of responses evoked in another ganglion cell by maximal preganglionic volley before *E*, and at intervals *F* to *I* (4, 5, 7, and 9 min, respectively), after addition of blocking dose of curare-like drug to bathing solution. Resting membrane potential remained at -66 mv throughout series. (From Eccles.[47])

that occupy postsynaptic receptors, e.g., tubocurarine (records *E* to *I*, Fig. 6-5, *II*).

Miniature EPSPs (mEPSPs) occur spontaneously and randomly in autonomic ganglion cells, and their frequency is increased by agents that affect the presynaptic terminals; i.e., their frequency is accelerated by terminal depolarization. The amplitude of the mEPSP is affected by agents influencing the postsynaptic cell; they are reduced by drugs that act as competitive blocking agents, e.g., tubocurarine. The available evidence supports the view that here, as at the skeletal neuromuscular junction, the mEPSPs are responses of postsynaptic cells to spontaneously released multimolecular packets of transmitter agent, thought to be packaged in synaptic vesicles. The release of such packets depends on the presence of external Ca^{++}. The EPSP produced by a presynaptic impulse is the response to a number of such quanta released nearly simultaneously by the rapid depolarization of the synaptic knob during the action potential. The synaptic vesicles of preganglionic, like those of postganglionic, autonomic neurons (p. 206) are thought to be constructed in cell bodies and transported rapidly down the axon to its terminals.

The short-latency EPSP just described is followed by two others of much longer latency and prolonged time course, one hyperpolarizing and one depolarizing.[100] As shown schematically in Fig. 6-5, *I*, the long-latency EPSP (sEPSP) is thought to be elicited by transmitter interaction with a second set of membrane receptors, a union blocked by atropine. The long-latency hyperpolarization (sIPSP) is disynaptically evoked via small interneurons, which are thought to be chromaffin cells within the ganglion that utilize dopamine as a transmitter agent. These two successive slow events produce a prolonged hyperpolarization and inhibition followed by a prolonged depolarization and facilitation. They differ markedly from the EPSPs, described at this and several other synapses for they occur without change in postsynaptic membrane conductance, and their signs are not reversed by changes in membrane potential. They are thought to depend on the selective activation or inhibition of electrogenic pumps in the postsynaptic cell,[141] the nature of which is not yet clear.

Transmitter agent at ganglionic synapses

The possibility that a chemical agent is involved in ganglionic synaptic transmission was suggested by the early observations of Langley and of Dale on the stimulating action of acetylcholine (ACh) on the autonomic nervous system. It is largely to the Dale school that we owe the evidence that ACh is released from presynaptic terminals in autonomic ganglia and produces the local synaptic events previously described. The evidence that ACh is the transmitter agent at this synapse[42] is summarized as follows:

1. ACh and its synthesizing enzyme choline acetyltransferase ("acetylase") (p. 208) are present in the ganglion. They disappear when the presynaptic fibers degenerate after section of the preganglionic nerves.

2. The acetylcholinesterase (AChE) of preganglionic nerves is distributed intracellularly with the endoplasmic reticulum. It is also present external to the cell membrane of the autonomic neurons, and this external AChE is thought to destroy the ACh released from presynaptic fibers. Transmitter destruction is essential for orderly synaptic action; otherwise the postsynaptic cell remains depolarized and thus unresponsive to subsequent synaptic impulses. Just such a depolarization block occurs within a ganglion when the AChE is inhibited. In sympathetic ganglia, AChE is limited to the preganglionic fibers; postganglionic fibers are adrenergic. In parasympathetic ganglia it is found external to both pre- and postsynaptic elements; both are cholinergic.[86]

3. ACh is released spontaneously from the resting ganglion, and the rate of release is greatly accelerated by preganglionic volleys (in a perfused ganglion with inhibition of AChE). Its release is not increased by antidromic activation of the ganglion cells. During long, continued, repetitive preganglionic stimulation the amount of ACh released far exceeds the resting content of the ganglion. For at least a few hours the peripheral axons retain the capacity to synthesize ACh after separation from their cell bodies, a finding that suggests that the synaptic vesicles both store and synthesize ACh.

4. Ganglion cells but not preganglionic fibers are excited by locally applied ACh. The scheme in Fig. 6-5, *I*, suggests that the transmitter action of ACh in sympathetic ganglia is more complex than that at the neuromuscular junction. ACh is thought to have a dual action in the ganglion; the two different postsynaptic effects are believed to result from the effects of ACh at two quite different receptor molecules in the postsynaptic

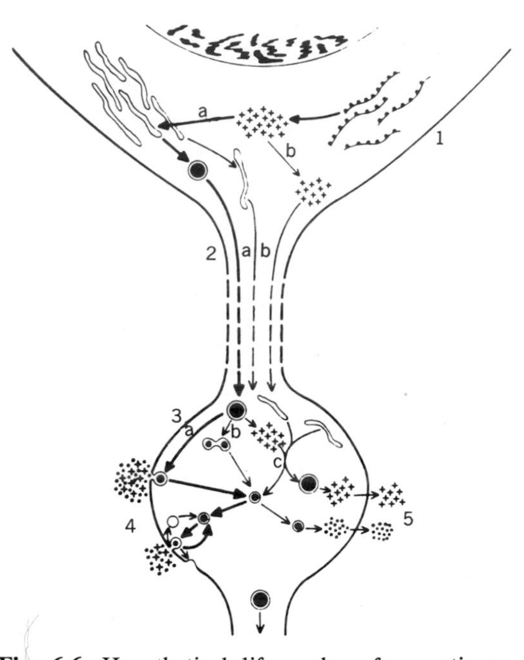

Fig. 6-6. Hypothetical life cycles of synaptic vesicles, shown here for sympathetic postganglionic neuron but applicable in general to chemically operated neurons. The more likely hypothesis, *a*, is traced by heavy arrows and possible alternatives (*b* and *c*) by lighter ones. Dotted lines indicate axon from cell body to varicose region near axon terminal. Crosses represent soluble protein constituents of vesicles; dots represent transmitter, in this case norepinephrine. The successive stages are thought to be as follows: *1,* synthesis of vesicle components in soma of neuron; *2,* axoplasmic transport of vesicles to distal ends of neuron, a transport requiring neurotubules; *3,* transformation of large granular vesicles into smaller variety in axon terminal, perhaps by partial exocytosis, with release of vesicular contents; *4,* vesicles recycle during transmission until they become exhausted; *5,* alternative but unlikely possibility is that vesicles are formed locally and liberate soluble proteins and norepinephrine by diffusion. (From Geffen and Livett.[6])

membrane. At the first receptor molecule the combination leads to the EPSP of short duration, resembling in this action that at the neuromuscular junction. Such actions are called "nicotinic" for they are mimicked by the local application of the alkaloid nicotine; they are blocked competitively by curarelike drugs. Even after such a block, ACh still elicits the long-latency, long-duration sEPSP, presumably by union with receptor molecules not blocked by curare. Such actions are called "muscarinic" for they are mimicked by the quaternary amine muscarine; they are competitively blocked by atropine. There is now evidence that the chromaffin cell inhibitory interneuron shown in Fig. 6-5, *I*, is activated by a muscarinic action of ACh and that it releases dopamine as a transmitter agent that elicits the sIPSP and a prolonged postsynaptic inhibition of the ganglion cell (p. 191). Thus there are three quite different transmitter actions operative in autonomic ganglia.

It is now firmly established that postganglionic neurons are chemically operated, that norepinephrine is the transmitter agent at many sympathetic neuroeffector synapses, and that ACh is the agent at those that are parasympathetic. These subjects are reviewed in some detail in Chapters 28 and 29. Postganglionic autonomic neurons are the most readily accessible of all nerve cells for many types of investigation, and it is in results of such studies that the hypothesis that transmitters are stored in presynaptic vesicles has received its strongest support.[6] A diagram illustrating the postulated life cycle of synaptic vesicles in postganglionic neurons is shown in Fig. 6-6.

Conduction through parasympathetic ganglia

Synaptic transmission in parasympathetic ganglia resembles that in sympathetic ganglia and fits the general paradigm of chemically operated synapses elsewhere. It is well established that ACh is the transmitter agent for *both* pre- and postganglionic parasympathetic neurons. Knowledge of transmission in these ganglia has been greatly increased recently by Kuffler et al. in a study of the postganglionic parasympathetic neurons of the frog heart. These cells are located in the interauricular septum, a thin structure that can be isolated with its innervation intact and maintained for days in tissue culture. The neurons and the presynaptic boutons upon them can be visualized and studied directly.[106] The synapse has these characteristics: (1) miniature (unitary) EPSPs occur spontaneously, (2) neurally evoked EPSPs of multiunit composition lead to action potentials at critical levels of membrane depolarization, (3) neurally evoked EPSPs reverse sign at about −12 mv membrane potential, indicating the nonspecific permeability change produced by ACh, (4) all these events can be mimicked by ACh applied electrophoretically through a micropipet placed close to a synaptic area,[43] and (5) these synaptic areas, where boutons are seen, are much more sensitive to ACh than areas of the cell surface that are free of boutons.[69]

This last observation allowed Kuffler et al.[91] to study the changes in ACh sensitivity following preganglionic nerve section. It has been known for a long time that denervated structures, whether nerve or muscle cells, become more sensitive to the transmitter substances that normally play on them. This *law of denervation* of Cannon is discussed in detail in Chapter 28. Kuffler et al. found that within a few days after denervation the region of ACh sensitivity spread to include the entire surface of the postganglionic neuron but there was no absolute increase in sensitivity at any locus. It is as if the sensitivity of receptor molecules to ACh either spread over the cell or, being there continuously, became manifest following denervation.

CENTRAL SYNAPTIC CONDUCTION

The simplest activities of CNS synapses are reflexes, in which some sensory stimulus evokes from the nervous system a discharge to muscles or glands, producing a response appropriate to the stimulus. The spinal cord can be isolated from the remainder of the CNS by section, thus reducing the complexity of events in distal segments and making anesthesia unnecessary. Each spinal segment possesses reflex arcs of two (monosynaptic) and three or more (polysynaptic) neurons that can be studied simultaneously. In experiments the sensory stimulus is frequently replaced by a nerve volley elicited by an electrical stimulus, which allows precise timing. The fiber groups conducting the afferent volley and their peripheral origin can be chosen selectively. Postsynaptic discharges in motor nerves can be studied by either direct recording or measuring the muscle contractions they produce. Moreover, the full panorama of methods centered around intracellular recording has been applied with brilliant success to the study of synaptic events in motoneurons and provides the base for a modern understanding of synaptic transmission in the CNS.[5]

General properties of central synapses

The relay of impulses through central synapses, like the relay at chemically operated synapses elsewhere, is strictly unidirectional. Indeed, it has been known since at least the time of Bell (1811) and Magendie (1822) that the dorsal and ventral roots consist, respectively, of afferent sensory fibers entering the cord and motor fibers leaving it, and that conduction through spinal reflex arcs is strictly one way, from the former to the latter. Transmission through even the most direct of these synaptic arcs occurs with an *irreducible synaptic delay* of about 0.5 msec at the local region of synaptic contact. An illustrative case is given by the experiment shown in Fig. 6-7.[35] A microelectrode was placed inside a gastrocnemius motoneuron of a cat spinal cord, and afferent volleys to it were generated repetitively in monosynaptically related afferents. Each volley produced a local postsynaptic response after a synaptic delay of about 0.5 msec, which was identical for each of the 30 volleys. These responses are superimposed to compose the record of Fig. 6-7. That record shows also that 10 of the 30 volleys elicited EPSPs just large enough to lead to conducted impulses, which are indicated by lines going off the record. Each was initiated after a variable interval, the *postsynaptic response time*.

Normally, pools of central neurons receive trains of impulses occurring asynchronously in afferent fibers. Postsynaptic discharge may then begin after a period of time called the *nuclear delay,* which allows for a buildup through spatial and temporal summation of the excitatory action of these impulses to the threshold levels of the most excitable cells in the pool. Nuclear delay is not related to any synaptic delay, and its value will vary with the number of active fibers, the state of excitability within the postsynaptic ensemble, etc. Only under the special conditions of synchrony of afferent input and maximal excitability will it approach the limiting value of synaptic delay.

Fractionation of a neuron pool and the functional results of convergence

A single incoming volley in an afferent pathway projecting on a pool of central neurons, e.g., a motor nucleus, may evoke the discharge of some but not all of the postsynaptic elements. They are thus divided into those of the *discharge zone* and of the *subliminal fringe;* neurons of the latter receive subthreshold synaptic excitation. This fractionation depends on some mixture of two

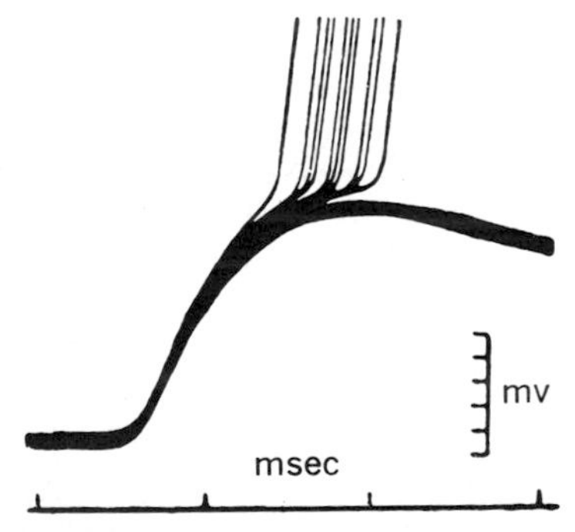

Fig. 6-7. Responses of cat gastrocnemius motoneuron evoked by repetitive stimulation of large afferents from that muscle monosynaptically linked to its motoneurons. Potential changes at tip of intracellular microelectrode displayed on superimposed successive sweeps of cathode-ray oscilloscope. Resting membrane potential of cell was −74 mv. Strength of afferent volley just critical for evoking conducted action potential, which some volleys do (action potentials are strokes off the record), while others evoke only EPSPs which, superimposed, form heavy smooth curve. (From Coombs et al.[35])

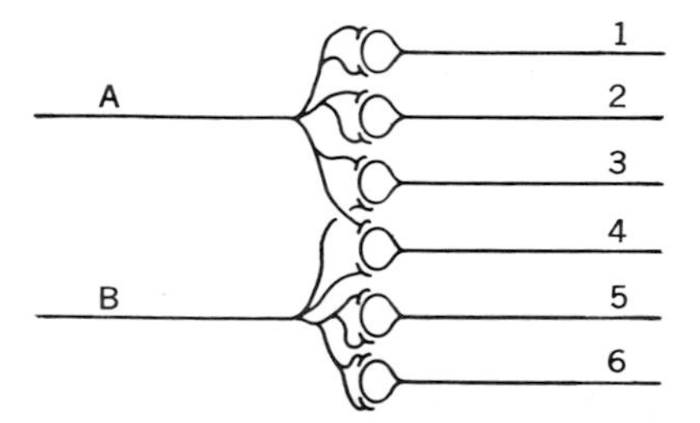

Fig. 6-8. Schematic illustration of functional results of convergence: occlusion and spatial summation. If activation of two synaptic endings is required to secure postsynaptic discharge, impulse in *A* or *B* alone will discharge two postsynaptic cells. However, simultaneous impulses in *A* and *B* will discharge all six cells. Cells *3* and *4* are discharged as result of spatial summation of presynaptic excitatory actions of impulses in *A* and *B*. However, if only one active ending is adequate to secure synaptic transmission, impulse in either *A* or *B* will each discharge four cells. Their simultaneous action can discharge only six cells. There is a discharge deficit of two cells, a phenomenon called occlusion.

variables: (1) neurons of equal excitability may receive different numbers of presynaptic boutons from the active fibers or (2) neurons with equal presynaptic excitation may have different discharge thresholds; e.g., smaller neurons are more readily discharged than large ones.

With the rare exception of 1:1 convergence numbers, each neuron of a central nucleus receives presynaptic boutons from more than one presynaptic fiber and frequently from a large number; these presynaptic fibers may arise from different afferent sources. The functional results of this converging overlap are of great importance for understanding neural function. They are revealed by the simple experiment illustrated schematically in Fig. 6-8. Arrangements are made to record the postsynaptic discharge from a central nucleus, e.g., to record the synaptically evoked discharge from a motor nucleus in a ventral root or motor nerve and to stimulate two sets of dorsal root afferents that each contain fibers projecting directly upon the cells of the motor nucleus. It will then be observed that the postsynaptic discharge evoked by simultaneous volleys in the two sets of converging fibers is less than the sum of the two discharges evoked by each separately. This deficit of discharge is called *occlusion.*

If now the number of fibers composing each volley is reduced, the postsynaptic discharge evoked by simultaneous volleys may be much greater than the sum of those evoked by separate volleys. This *spatial summation* results from the reciprocal overlap of the subliminal fringes, for the effects produced in these fringe cells by the two sets of synaptic endings have summed to bring them to threshold. In a similar way, if two weak volleys in a single group of afferent fibers, each ineffectual when alone, are delivered at a brief interval, the subliminal effects of the second sum with the persisting affects of the first to bring the postsynaptic cells to threshold depolarizations. This is an example of *temporal summation,* quite comparable to that previously described for the squid giant synapse. Spatial and temporal summation in monosynaptically related systems follow time courses identical with those of the local excitatory postsynaptic responses to be described later. Where interneurons participate in overall transmission through a central nucleus, spatial and temporal summation may be prolonged (p. 203).

SYNAPTIC REACTIONS OF NERVE CELLS[5, 8]

The description of synaptic actions on central neurons will point out that both synaptic excitation and postsynaptic inhibition fit the paradigm of chemically operated synaptic actions outlined in Table 6-1. In each case the postsynaptic membrane potential changes (EPSPs and IPSPs) result from changes in ionic conductances of the postsynaptic membrane caused by chemical transmitters.

Synaptic excitation[1]

The local EPSP of a cat motoneuron, recorded intracellularly, is shown in Fig. 6-9. Such a motoneuron receives monosynaptically the excitatory endings of many afferent fibers originating from stretch receptors of the muscle it innervates. A single volley in these afferents generates a transient depolarization, the EPSP, which is increased in am-

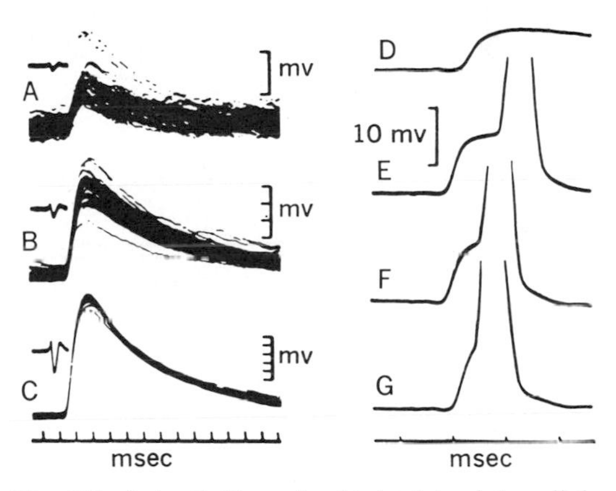

Fig. 6-9. A to **C,** Records obtained by intracellular microelectrode of EPSPs evoked in cat spinal motoneuron by volleys in monosynaptically related afferents. Potential changes occurring at its tip were amplified and displayed on cathode-ray oscilloscope. For each record, 20 to 30 responses were superimposed. Upward deflection indicates depolarization. Size of afferent volley increased from **A** to **C,** as indicated by inset records obtained by recording on adjacent dorsal root with another electrode. Increase in size of EPSP, indicated by decreasing amplification from **A** through **C** (note calibrations), is example of spatial summation. Impulse discharge was prevented by deep anesthesia. (From Coombs et al.[35])

D to **G,** Recordings of potentials in spinal motoneuron generated by volleys in monosynaptically related afferents, obtained as described above. Volleys increased progressively in size from **D** through **G.** Weakest volley evokes EPSP only, while larger volleys evoke EPSP of sufficient size to attain discharge threshold and produce the impulses written partially off records. Note progressive decrease in postsynaptic response time but invariant synaptic delay. (From Eccles.[45a])

plitude but unchanged in time course as the number of fibers conducting impulses is increased (Fig. 6-9, *A* to *C*). Thus active synapses generate local postsynaptic responses of similar time course, and the EPSPs shown in Fig. 6-9 result from the summation of such individual synaptic potentials. This simple experiment provides an example of the classic concept of spatial summation put forward by Sherrington. In this experiment, precautions were taken to prevent impulse discharge, thus revealing the time course of the EPSPs. Under other conditions, as increasingly larger volleys produce larger EPSPs, a point of depolarization is reached at which sodium conductance across the membrane becomes regenerative, and a conducted impulse is generated (Fig. 6-9, *D* to *G*). This point, the *threshold,* is in general a constant property of a given central neuron but differs among neurons. Further analyses have shown that the local EPSPs of central neurons possess the general properties of the local responses of chemically operated synapses previously considered at neuromuscular, axoaxonic, and ganglionic junctions (Table 6-1).

The inset in Fig. 6-10, *A,* shows that the ionic current that produces the EPSP flows inwardly beneath the active excitatory endings and thus outwardly (the depolarizing direction) through surrounding areas of cell membrane. The time course of the inwardly directed current can be calculated (if the time

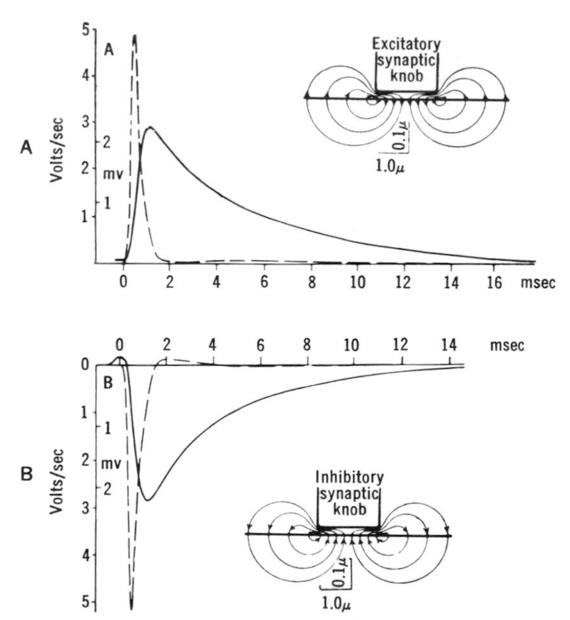

Fig. 6-10. Solid lines are mean values of EPSP and IPSP, **A** and **B**, respectively, evoked by just maximal volleys in muscle afferents respectively excitatory or inhibitory for motoneuron under study by intracellular recording. These potential changes are analyzed as described in text, after time constant of cell was determined experimentally. Excitatory and inhibitory synaptic currents so derived are plotted as broken lines on ordinate scale of volts/sec. Insets show directions of ionic current at excitatory and inhibitory synapses. (From Curtis and Eccles.[37a])

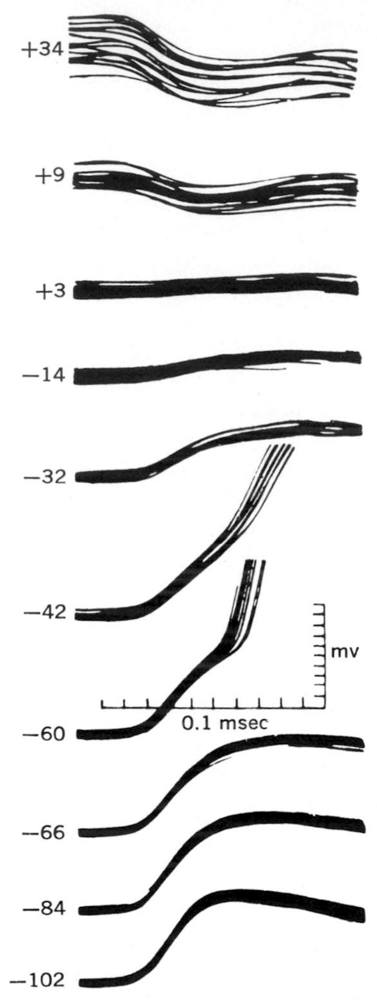

Fig. 6-11. Records of EPSPs in cat spinal motoneuron evoked by volleys in monosynaptically related afferents, obtained by intracellular recording through one barrel of double-barreled electrode. Steady membrane potentials are maintained at series of levels (indicated in mv to left of each series of records) by passing extrinsic current through second barrel. Potentials indicated are those of interior of cell with respect to exterior. Depolarization is upward; resting membrane potential, -66 mv. At steady membrane potentials of -42 and -60 mv, EPSPs led to conducted action potentials; at other levels they did not. EPSPs reversed polarity at about zero transmembrane potential difference. (From Coombs et al.[35])

constant of the membrane is first determined) from the relation dv/dt = I/C – V/γm, in which dv/dt is the slope of the recorded potential curve at any instant (t), C and I are, respectively, the capacity of the membrane and the current flowing through it, and γm is the electric time constant of the membrane. The results of calculations for a series of times is plotted in Fig. 6-10, *A*. The active current flow is a brief event that rapidly depolarizes the membrane—the rising phase of the EPSP. The repolarization phase is prolonged from a purely passive course by a small, persisting subsynaptic current.

Further important facts are revealed by an application of the voltage clamp technique described in Chapter 2. It can be carried out by inserting a double-barreled microelectrode into a motoneuron. Steady current is passed through one barrel, maintaining the membrane potential at any desired level, and the steady membrane potential and changes in it are recorded through the second barrel. The results (Fig. 6-11) show that the equilibrium potential toward which the membrane potential is driven during synaptic action is close to zero. This suggests that the excitatory transmitter substance causes a large nonspecific change in membrane permeability, a change comparable to that produced by ACh at the neuromuscular junction (Chapter 5).

In summary, excitatory synaptic transmission is accomplished in the CNS by means of chemical operations: substances are released from presynaptic terminals by the arrival of nerve impulses, and in each case the substance released appears to change the ionic conductance of the postsynaptic membrane, although, as shown for ganglionic synapses, there are cases in which the postsynaptic action is to activate an electrogenic pump without a directly induced conductance change. The transmitter substance released at the motoneuronal synapses just considered is unknown.

Initiation of impulses in central neurons[5, 54]

In the classic view, synaptic excitation from axon terminals to dendrites leads to an impulse that sweeps over the cell body and down the axon. It is now clear that this is not the case, and that neurons integrate synaptic action by summing local events at a critical site for impulse initiation. Intracellular recording reveals that the motoneuron impulse has three components (Fig. 6-12); the first is the EPSP, and the last two are all-or-nothing events. The EPSP leads, at a critical level of about 10 mv depolarization, to a conducted impulse that arises first in the initial segment of the cell and axon, the IS spike. It is only when depolarization reaches 30 mv that an impulse is set up in the cell body itself; it then sweeps backward over the cell body and proximal dendrites. This seemingly reverse sequence of events is due to the low threshold of the initial segment. Thus synaptically evoked transmembrane ionic currents set up in the cell body will depolarize the initial segment to its threshold before that of the soma is reached, even though the former is spatially more distant.

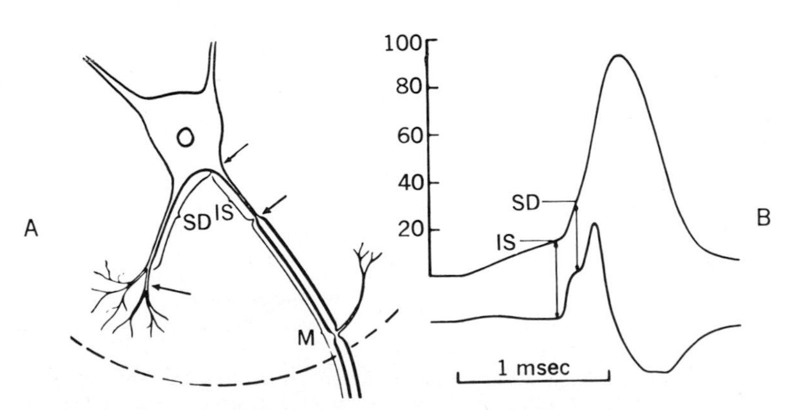

Fig. 6-12. A, Schematic drawing of motoneuron of spinal cord. Recurrent axon collateral leaves axon at point *M*. Portions of cell referred to as initial segment, *IS,* and somadendritic part, *SD,* are indicated by arrows and brackets. (From Eccles.[45b])

B, Intracellular recording of response of motoneuron to synchronous monosynaptically related afferent volley. Record below is taken simultaneously and electrically differentiated, which accentuates change in slope of depolarization from EPSP to *IS* spike and from latter *SD* spike. (From Coombs et al.[34a])

Further direct evidence for this sequence has been obtained by simultaneously recording intracellularly from a motoneuron and from its axon in a small ventral root filament.[5] The conducted axon spike was initiated by the initial segment and under some conditions of high excitability even reached the ventral root *before* the spike was initiated within the cell body itself. The origin of impulses in the initial segment, or under some conditions even at the first node of Ranvier for cells with myelinated axons, seems to be a general property of nerve cells. It has been observed in such widely different locations as mammalian motoneurons, the crustacean stretch receptor,[123] and the giant neuron of *Aplysia.*

Local synaptic events[82, 120]

Intracellular studies of mammalian spinal motoneurons, under conditions that eliminate afferent impulses, reveal spontaneous miniature PSPs comparable to the spontaneous quantal events at the neuromuscular junction[11, 23, 30] (Chapter 5). The number of quanta released from a single synaptic bouton when it is invaded by an impulse is small by comparison with the number released at the neuromuscular junction; such a "unitary" PSP in a motoneuron is commonly composed of one to five quanta. The ways in which such unitary PSPs initiated at different locations on the cell may affect its excitability and may interact with each other have been the subjects of elegant quantitative studies by a number of investigators, and a highly successful quantitative model of the motoneuron has been developed by Rall.[119] Interaction between concept, model, and experiment has provided a solution for the formerly perplexing problem of how synaptic impingements occurring on the distal dendritic branches of a neuron could affect the initial segment and the soma, in which impulses are generated. Taking into account the cable properties of neurons, it is now clear that even in cells that possess very long and widely branching dendritic trees, e.g., motoneurons, the overall electrotonic length of the cell is no more than twice the membrane space constant. This means that synaptic action, even on the most distant dendrites, can cause an appreciable flow of ionic current across the membrane of the initial segment.

The records in Fig. 6-9 indicate that, under the conditions of that experiment, EPSPs generated by volleys of impulses in an increasing number of converging presynaptic fibers were of about the same time course. This identity exists only if the synaptic impingements are of about the same (averaged) electrotonic distance from the cell body. Obviously the closer a synaptic action to the soma, the more powerful its effect on the excitability of the cell. The more distant such a synapse, the less is its effect due to the decrement in amplitude

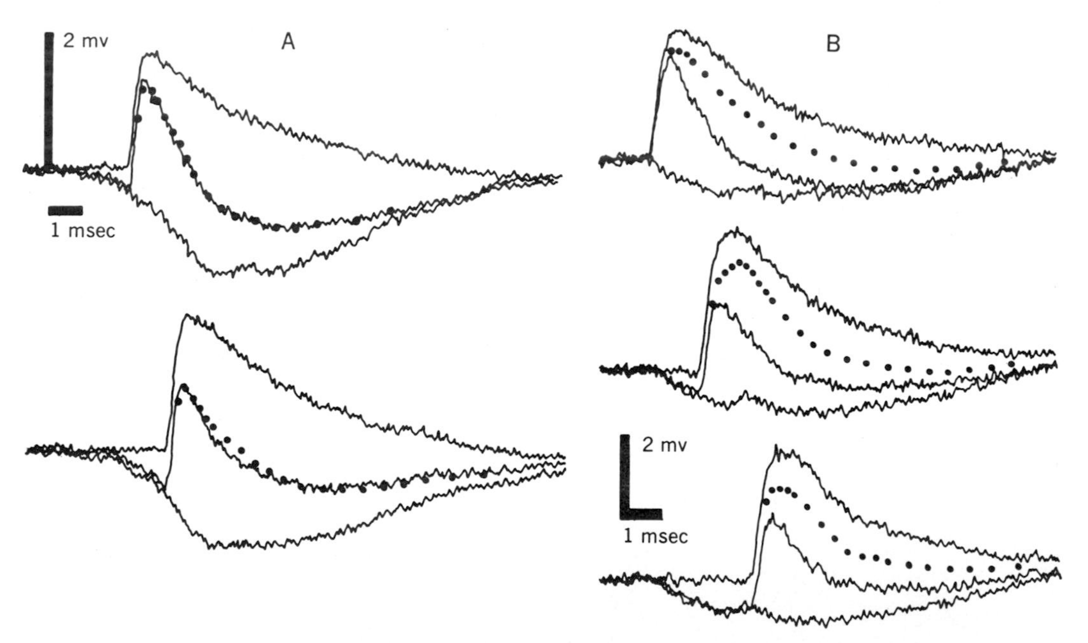

Fig. 6-13. Records of averaged PSPs recorded in spinal motoneuron of cat. Column **A** illustrates linear summation of EPSP generated by lateral gastrocnemius nerve volley (upper set in each trace), with IPSP generated by volley in nerve to antagonistic muscle (lowest trace in each set). Middle records were generated when two volleys were generated so that their synaptic actions would reach the motoneuron simultaneously; they are neatly superimposed on dots that represent algebraic summation of upper and lower records, i.e., linear summation. Opposite case is shown in column **B,** obtained from another gastrocnemius motoneuron, where the summation is obviously markedly nonlinear. Most motoneurons studied in this way showed intermediate degrees of nonlinearity. (From Rall et al.[120])

and slowing of time course caused by electrotonic conduction.

PSPs will sum linearly if they are generated at membrane locations so far apart that they do not interact with each other. When closer, the change in the membrane potential, e.g., at the region of synapse A, caused by synaptic action at synapse B, will change the driving potential at A and hence reduce the amplitude of the PSP caused by synaptic action at A. Examples of the two cases, linear and nonlinear summation, are shown in Fig. 6-13.

Synaptic inhibition[5, 13]

Crustacean muscle cells, unlike the striated muscle of vertebrates, receive a dual innervation, one excitatory and the other inhibitory. The two types of motor nerve fibers end in adjacent loci on the surfaces of muscle cells. Inhibitory action is exerted in two ways.[44] First, by increasing the conductance of the muscle cell membrane to Cl^- and K^+ but not to Na^+, the action of the inhibitory transmitter stabilizes membrane potential at a level between E_{Cl^-} and E_{K^+} or drives it to that level by a hyperpolarizing PSP, the IPSP. The postsynaptic membrane is then less readily depolarized by the action of the transmitter released from excitatory presynaptic terminals: this is *postsynaptic* inhibition. Second, the inhibitory axons also terminate, in axoaxonic synapses, upon the terminals of the excitatory axon, and the action of the transmitter there is to depolarize the axon terminals. Impulses in the excitatory axon invading these partially depolarized terminals release less transmitter per impulse, for the number of quanta released is a function of the total change in membrane potential and its rate of change. The net effect is a reduction of the depolarizing action on the postsynaptic cell of impulses in excitatory presynaptic axons. This is called *presynaptic* inhibition, even though the inhibition results from an excitatory depolarization of the presynaptic terminals and occurs without any conductance or membrane potential change in the postsynaptic cell.

In vertebrates, such an antagonistic innervation of peripheral effectors occurs only in the autonomic nervous system. Yet the arrest of an active contraction or a slackening of existing tension is a common response of skeletal muscle to an afferent input. *Since vertebrate skeletal muscle receives no efferent inhibitory fibers, a reflexly evoked cessation of muscle contraction must result from a decrease in the discharge of excitatory motoneurons. It is a central event.* Inhibition is important for orderly function even in simple reflex actions: active contraction of one muscle is commonly accompanied by a decreased contraction of its antagonist, an example of reciprocal innervation that will be considered from the reflex point of view in Chapter 24. Here it is pertinent to consider the mechanisms of synaptic inhibition, which occurs in the mammalian CNS in both its postsynaptic and presynaptic forms.

Postsynaptic inhibition

The records in Fig. 6-14 show that an afferent volley inhibitory for a motoneuron elicits in it a hyperpolarizing local response, provided the cell is at its resting membrane potential of −70 mv. This inhibitory postsynaptic potential (the IPSP) is, like the EPSP, local and not conducted; it extends electrotonically and can be graded in amplitude by varying the size of the afferent in-

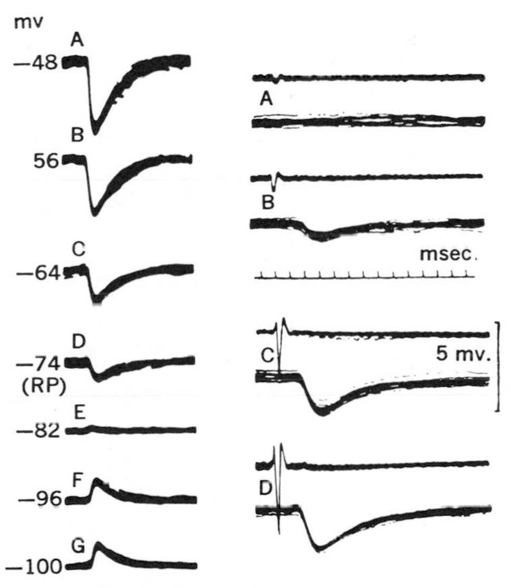

Fig. 6-14. Left: Potentials recorded intracellularly from cat spinal motoneuron via double-barreled microelectrode. Records were formed by superimposition of about 40 faint traces of cathode-ray oscilloscope; they show IPSPs evoked by volleys in nerve fibers inhibitory for cell under study. By means of steady current through one barrel of electrode, membrane potential was preset at voltage indicated on each record. *A* to *G* indicate that IPSP reversed its polarity at about −80 mv transmembrane potential difference.

Right: Lower records of each pair give intracellular recorded responses of cat motoneuron to an inhibitory afferent volley, which was increased in size from *A* to *D* and is indicated by sharp downward deflections in upper record of each pair, obtained via another electrode placed on adjacent dorsal root. Records illustrate gradation of IPSP and spatial summation of inhibitory action.

(From Coombs et al.[35])

hibitory volley evoking it (Fig. 6-14, right, records *A* to *D*). The ionic current producing the IPSP is thought to flow outward across the membrane beneath an active inhibitory synaptic bouton and hence inward, the hyperpolarizing direction, across surrounding areas of cell membrane. The IPSP relaxes passively from its peak toward the resting membrane potential along a time course determined by the membrane time constant (Fig. 6-10, *B*).

When the membrane potential is increased past –80 mv, the IPSP is reversed to a depolarizing potential, as shown by the records in Fig. 6-14. The IPSP "equilibrium potential" is midway between the E_{K^+} of –90 mv and the E_{Cl^-} of –70 mv, leading to the conclusion that the inhibitory transmitter increases membrane conductance to smaller hydrated ions such as K^+ and Cl^-, but has little effect on Na^+ permeability. Although the transmitter has no effect upon a resting membrane potential of –80 mv, an inhibitory action is still exerted, for greater excitatory input is then required to drive the membrane toward its threshold of about –60 mv.

Both afferent fibers and those of long CNS tracts are excitatory in nature; yet activity in them commonly leads to inhibition in some sets of neurons near their terminations. This inhibition is thought to be exerted via small, short-axoned, inhibitory interneurons that release inhibitory transmitter. The linkages of this sort are thought to account for reciprocal inhibition, e.g., in the myotatic reflex pathways of the spinal cord; they are diagrammed in Fig. 6-15, *A* and *B*. Each neuron in such a circuit produces only a single transmitter agent, a general principle known as *Dale's law*.

Direct postsynaptic inhibition is a widespread phenomenon in the CNS and has been studied with detailed microphysiologic methods in many locations. Synaptic inhibition possesses many of the general properties of

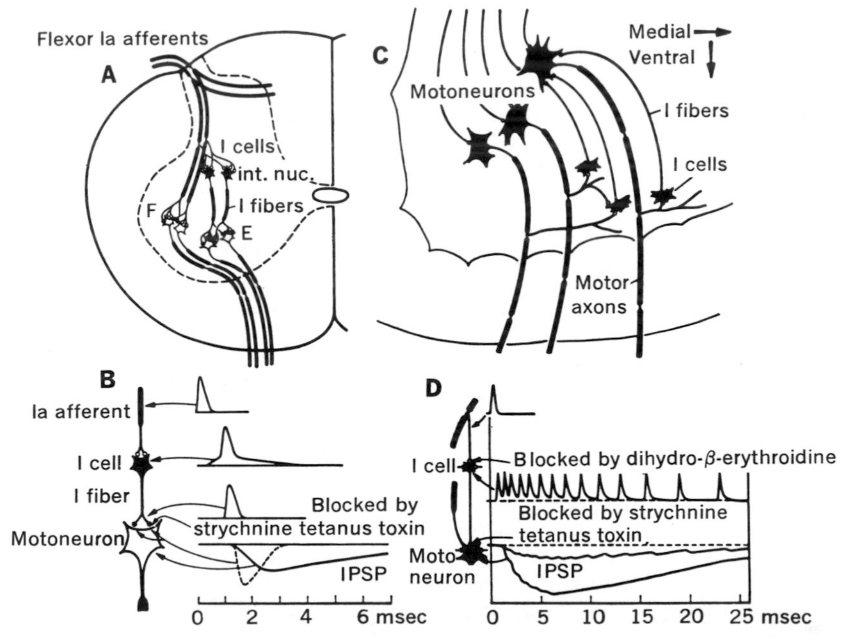

Fig. 6-15. A represents pathway for reciprocal action within spinal cord. Flexor Ia afferents are shown terminating directly upon flexor motoneurons—the monosynaptic excitatory pathway of the stretch reflex. Collaterals of these first-order afferents terminate upon inhibitory interneurons *(I cells)* that project upon extensor motoneurons—the disynaptic inhibitory component of the stretch reflex. Events in this pathway are shown schematically in **B.** Final effect on inhibited motoneuron is shown by inhibitory synaptic current (dotted line) and resulting IPSP. **C,** Recurrent inhibitory pathway. Collaterals of motor axons project upon Renshaw interneurons *(I cells),* which in turn project back upon cells of same (and perhaps other) neuron pool. **D,** Postulated events in this pathway are shown diagrammatically. Impulse in axon collateral evokes prolonged repetitive discharge in Renshaw interneuron, and prolonged IPSPs are evoked in motoneuron upon which it impinges. Heavier rippled line shows greatly increased IPSP that results from convergence upon motoneuron of many Renshaw interneurons. (From Eccles.[5])

synaptic excitation. It exhibits peripheral local sign and central localization, it may be summed temporally and spatially, and like all synaptic actions, it shows in comparison with axonal conduction a greater susceptibility to fatigue, anoxia, and the action of drugs.

Presynaptic inhibition[5, 13]

A volley of impulses in dorsal root afferents leaves in its wake an enduring depolarization of the intraspinal segments of both the active and the adjacent fibers. The depolarization extends electrotonically along the dorsal root fibers and can be recorded from them as the "dorsal root potential." It is not conducted, can be summed spatially and temporally, and displays a latency that allows time for several intraspinal synaptic delays; it thus possesses the properties of a local PSP occurring in axons. This primary afferent depolarization, or PAD, has a striking effect upon the synaptic excitatory capacity of impulses invading the partially depolarized synaptic boutons, and it is thought to operate by the same mechanism of "presynaptic inhibition" as that described at the crustacean neuromuscular junction[44] (p. 199). The terminals of some branches of afferent fibers are linked by one or more excitatory interneurons to the terminals of adjacent fibers via axoaxonic synapses. Activation of these chemically operated excitatory synapses leads to the prolonged depolarization of the axon terminals, which when occurring in dorsal root afferents is recorded as the PAD. When thus depolarized, such terminals release less transmitter when they are themselves invaded by impulses and a degree of "inhibition" results. The occurrence of such a depolarization has been shown by direct intraterminal recording in some cases and in many others by the fact that during presynaptic inhibition the terminal segments of fibers are more readily excited by local, direct electrical stimulation.[140]

Powerful and reciprocal presynaptic inhibitory linkages exist between primary afferent fibers and between the terminals of second- and third-order fibers in subcortical regions of sensory systems, e.g., the dorsal horn of the spinal cord, between fibers terminating in the dorsal column nuclei, and between optic tract fibers in the lateral geniculate, but not at the level of the cerebral cortex. In sensory systems, presynaptic inhibition appears to be distributed in the pattern of surround inhibition. For example, a *local* skin stimulus elicits a train of impulses in fibers innervating the stimulated spot, impulses that, via the interneuronal arcs described, "inhibit" the terminals of fibers whose peripheral branches terminate in adjacent areas of skin.[128] The net effect is to limit and sharpen the profile of central neural activity set up by the stimulus. This mechanism operates as a powerful negative feedback control upon a variety of primary afferent fibers and serves to limit and regulate the further central transmission of intense activity in first-order elements. How and to what degree presynaptic inhibition operates in spinal reflexes is uncertain.

The identity of the transmitter agent at these axoaxonic synapses is not known, and thus far it has not been possible to specify the conductance changes that are caused in the recipient presynaptic endings. Strychnine is known to depress postsynaptic but not presynaptic inhibition, while picrotoxin, another convulsive drug, has the reverse effect.[37] Anesthetics such as barbiturates, and especially chloralose, increase and prolong presynaptic inhibition. The specific but differing actions of these drugs suggest that some operate by competitive occupation of receptor sites and others, by enzyme inhibition.

There are in the nervous system many instances of a decrease or cessation of neural activity that result from mechanisms other than pre- or postsynaptic inhibition. Neurons are, of course, inexcitable immediately after impulse discharge; they are then refractory, and their excitability remains decreased during the period of afterhyperpolarization of the cell membrane, during which the membrane potential is driven nearly to the E_{K^+} of −90 mv. In voltage clamp experiments this afterhyperpolarization reverses at the E_{K^+}, suggesting a temporarily persisting increase in K^+ permeability. The afterhyperpolarization is deeper and longer in tonic motoneurons, which in the main innervate peripheral muscle fibers with slow twitch times and are more common in red than in pale muscles. This process, together with recurrent inhibition, is thought to play an important role in setting the low and rather regular rates of discharge characteristic of these neurons.[46]

Recurrent inhibition[5]

Renshaw discovered that antidromic impulses in motor axons of a ventral root inhibited neighboring motoneurons, and that such a volley evoked a high-frequency discharge in interneurons of the ventral horn, which have since been called Renshaw cells. This recurrent inhibitory pathway is diagrammed in Fig. 6-15, *C;* the inhibition is postsynaptic and chemically operated. The excitatory transmitter at the junction of the motor axon collateral and the Renshaw interneuron is ACh, as it is at the peripheral neuromuscular terminations of the same motoneurons. Thus Renshaw cells are depressed by curare-like drugs, their excitation is prolonged by anticholinesterases, and they are stimulated by locally applied ACh or nicotine.[39] The inhibitory transmitter released by the Renshaw interneuron itself is unknown; its effect is antagonized by strychnine but not by picrotoxin.

The recurrent inhibitory pathway is important in the organization and control of spinal reflex action[74]; it will be discussed in Chapter 24. Recurrent inhibitory pathways are ubiquitous in the CNS, where they stabilize and limit the spread of neuronal activity.

SYNAPTIC MECHANISMS IN INTEGRATIVE ACTION

In the preceding discussions of synaptic transmission, experimental evidence obtained by studies of single cells has been used to elucidate synaptic microphysiology. It is unlikely, however, that the activity of a single neuron is of critical significance in the overall operation of the nervous system. Central actions involve large populations of neurons, and knowledge of the spatial and temporal distributions of activity in neuron pools is required for an understanding of brain function. Central neurons are bombarded by trains of impulses that vary in frequency in any given fiber and that are out of phase in adjacent fibers. It is important to learn how these patterns of activity are transformed at synaptic junctions into postsynaptic patterns of discharge. The natural event is poorly mimicked by experimentally induced repetitive volleys, but studies of postsynaptic responses under these circumstances have yielded facts of importance.

Integrative action at the cellular level

It was from his studies of spinal cord reflexes that Sherrington derived his perceptive generalization concerning the integrative action of the nervous system.[36, 129] He conceived that an afferent volley excitatory for a motoneuron created in the synaptic region an enduring change, a central excitatory state, which outlasted in time the afferent input, was not itself conducted, and could sum in both space and time with other such events. Afferent volleys inhibitory for a motoneuron created the central inhibitory state with similar properties but of opposite sign.

It is possible now to explain how an indi-

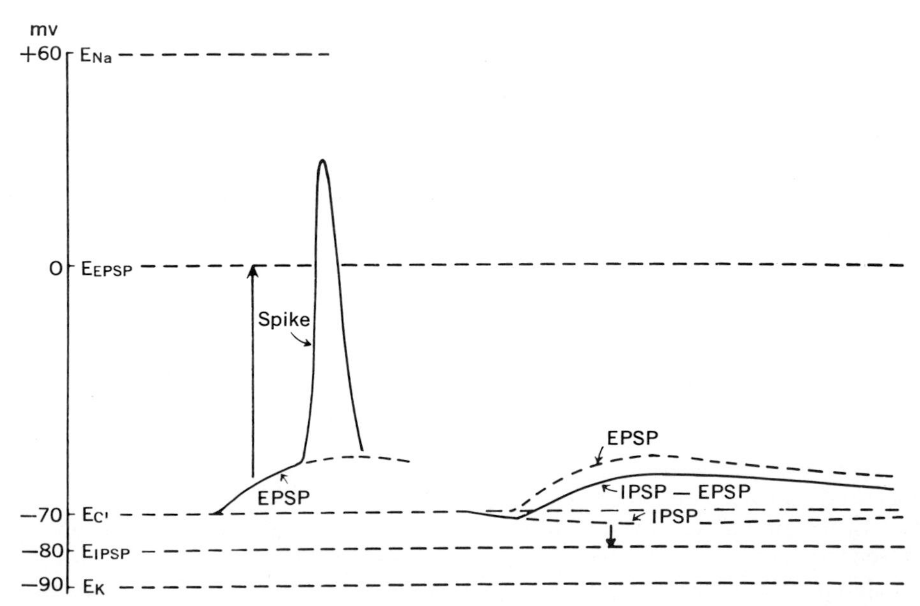

Fig. 6-16. Diagram summarizes mode of action of postsynaptic excitation and inhibition at chemically operated synapses in CNS in terms of ionic hypothesis. Equilibrium potentials for Na^+, K^+, and Cl^- and for EPSP and IPSP given as dotted lines. At left, EPSP is seen driving membrane potential in depolarizing direction, and at threshold eliciting an action potential in the cell. To right, IPSP and EPSP are shown alone (dotted lines) and when they interact (net effect, continuous line). EPSP is now so depressed by simultaneous inhibitory effect that it does not reach cell threshold. Interaction of synaptic influences of opposite signs is essence of integrative action of single neurons. (From Eccles.[45c])

vidual motoneuron—or any nerve cell—integrates its total constellation of excitatory and inhibitory synaptic inputs, as diagrammed in Fig. 6-16. The key discoveries were that the impulse is generated in the initial segment of the cell, and that such an impulse need not traverse the cell body itself and is probably seldom conducted into the dendrites. Thus the membrane potential of the initial segment continuously integrates the excitatory and inhibitory currents flowing across it, currents set in motion by transsynaptic actions occurring at many and variously located sites on the soma and dendrites. The level of depolarization of the initial segment determines the rate of discharge of the cell.

Transsynaptic input-output functions

It has already been noted that an afferent volley of sufficient size will discharge some cells of a neuron pool, activate others subliminally, and at least in some systems leave still other cells unaffected. The quantitative relation between the number of active presynaptic fibers and these fractions of a neuron pool is important, for the extent of the subliminal fringe determines the degree of summation that is possible. Recent evidence suggests that in the stretch reflex arc of the spinal cord all the neurons of a motor nucleus receive synaptic input (of varying degrees) from each monosynaptically related stretch afferent fiber (Chapter 24). Motoneurons are graded, i.e., they enter the discharge zone, in terms of their size, as thresholds are lower for smaller cells. Synaptic actions within motoneuron nuclei thus include a very large subliminal fringe that accounts for the spatial and temporal summation between converging presynaptic inputs, e.g., from two muscle synergists.

In other synaptic systems the gradation between discharge zone and subliminal fringe may be determined by the degree of convergence if cell size is the same throughout the neuron pool. Moreover, the relations between these fractions of a neuron pool vary widely. In the major sensory systems, e.g., at the dorsal column nuclei or the lateral geniculate body, the synaptic link is secure and little subliminal fringe exists; i.e., almost all cells receiving presynaptic impulses will discharge. Undoubtedly all variations of these relations exist in the nervous system, and for any given synaptic region they must also vary with the general level of excitability of the neuron pool, a level frequently controlled by presynaptic inputs other than the one considered.

Central facilitation and depression

It is a characteristic feature of central synaptic action that the postsynaptic discharge evoked by a given afferent input is modified by repetition. The temporal and spatial summations of local responses contribute to an overall facilitation of the responses to repetitive input; inhibitory postsynaptic responses are similarly summed. Repetition produces no known change in the capacity of the postsynaptic cell to respond with EPSPs or IPSPs. Repetition does, however, cause marked increases or decreases in the synaptic stimulating efficacy of the presynaptic elements, presumably due to a change in the amount of transmitter released per terminal per impulse; the size and direction of such a change depend on the frequency of afferent impulses. Once again the monosynaptic link between large muscle afferents and spinal motoneurons has provided a suitable system for study. Intracellular recordings show that the second of two volleys in these afferents produces a larger EPSP in the related motoneuron than does the first, a true facilitation that reaches 120% of the control value, and declines gradually over 15 to 20 msec, during which time the two responses are temporally summed.[38] Both summation and facilitation thus contribute to a greater depolarization of the cell and to a greater likelihood of impulse generation. When, on the other hand, the interval between stimuli is lengthened, there is a decrease in the second response lasting for many seconds. The evidence is that these changes also have a presynaptic origin and leads to the conclusion that at short intervals the active terminals release greater amounts of transmitter by virtue of their prior activation; at longer intervals they release less. Such frequency-bound alterations also occur during repetitive stimulation. At low frequencies there is an enduring depression; at intermediate frequencies (60 to 100/sec) there is a relative facilitation. At still higher rates of synaptic drive (250/sec) a steady state of maintained depolarization is reached, suggesting that the maximal rate of overall transmitter output has been achieved.

In summary, the size of the discharge zone and the frequency of activity of its neurons for any synaptic system expresses the net effect of several factors acting in concert. Among these are (1) the general state of ex-

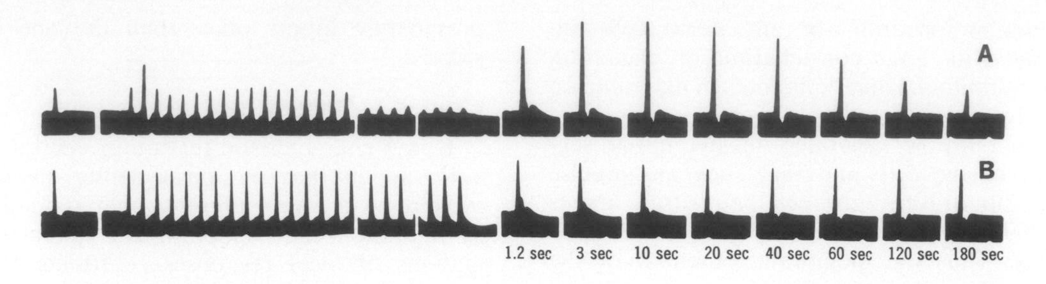

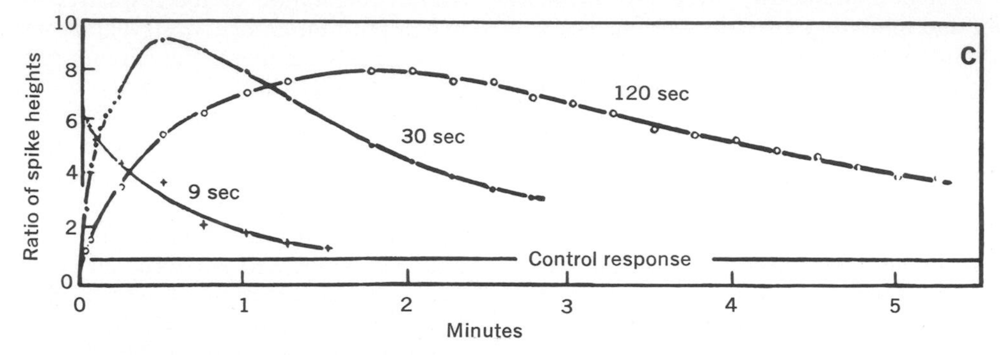

Fig. 6-17. Prolonged facilitation following repetitive synaptic excitation. Postganglionic action potentials before, during, and at intervals after tetanic conditioning stimulation of preganglionic fibers, **A,** at 15/sec for 10 sec. **B,** Control experiment—conditioning tetanus is delivered directly to postganglionic nerve. Absence of facilitation indicates that which is seen in **A** is produced by synaptic phenomenon. **C,** Plots of prolonged facilitation after tetanic preganglionic stimuli for various times as indicated. Period of facilitation is extended by prolonged conditioning. Amplitudes of compound action potentials indicate number of active postganglionic fibers. (From Larrabee and Bronk.[97])

citability in the neuron pool, a level conditioned by the "background" activity in all presynaptic pathways, (2) the proportion of the fibers of the presynaptic pathway that are inhibitory rather than excitatory, (3) the degree of spatial and temporal summation of the PSPs of the neurons, (4) changes during repetitive activity in the rates of production or release of transmitter at the presynaptic terminals, (5) the presence of recurrent inhibition or facilitation produced by impulses in the axon collaterals of the active neurons of this or other neuron pools, and (6) the subnormality of the neurons that discharge, a factor limiting impulse frequency.

Prolonged facilitation or posttetanic potentiation.[73, 97, 102] An account has been given (Chapter 5) of the fact that after a conditioning train of impulses in motor nerve fibers, neuromuscular transmission is thereafter more effective for a considerable period of time. This prolonged facilitation has been shown to be a general property of neuroneural synapses as well. The records and graphs in Fig. 6-17 show that after a short, repetitive preganglionic stimulation there is a succeeding prolonged phase of facilitation. Recording from single postganglionic fibers during a tetanic stimulation of the presynaptic input shows that some ganglionic cells cease to respond, while some in the subliminal fringe are recruited to action. During repetitive activity the two opposing processes of depression and recruitment go on simultaneously, one tending to decrease and the other to increase the number of active ganglion cells. In the posttetanic period the facilitatory influence is greatly increased after the phases of super- and subnormality have run their courses. The degree of this facilitation is limited by the available subliminal fringe; its duration depends on the number and frequency of the presynaptic volleys.

Prolonged facilitation is not due to changes in postsynaptic cells. Since posttetanic testing in converging presynaptic pathways other than the one tetanized reveals no change in the responses evoked by them, the facilitation is not due to increased excitability of the postsynaptic cells. Instead, several converging lines of evidence indicate that at neuroneuronal synapses, just as at the neuromuscular junction, the facilitation is due to increased release of transmitter from each tetanized presynaptic terminal during subsequent activation.

The same phenomenon has been studied by Lloyd[102] and others in the monosynaptic pathway through the spinal cord. In this system the effect can be demonstrated only after repetitive activity, and Lloyd called the effect "posttetanic potentiation." The phenomenon appears to be of general occurrence at synaptic junctions in the CNS.

The action of interneurons[4, 7, 103]

First-order afferents and motoneurons are highly specialized cells that by virtue of their accessibility are by far the best known central neural elements. The monosynaptic link between large muscle afferents and spinal motoneurons has been the locus of the most important studies of central synaptic transmission. The two-neuron arc is a special case, however, for most of the synaptic relay centers of the brain and spinal cord are complicated by the presence of interneuronal cells. In this restricted definition, interneurons are cells whose axons do not enter white matter and which, by virtue of recurrent feedback looping or by feed-forward connections, play important roles in controlling nuclear transmission. For example, note the interneurons of the spinal reflex pathways diagrammed in Fig. 6-15. In a more general sense, however, very large regions of the brain may be thought of as giant chains of interneurons relating to reflex arcs or to long projecting systems and influencing activity in them.

There is a general belief that those aspects of the integrative action of the nervous system that cannot be explained on the basis of the integrative mechanisms of single cells depend on the complex interactions within and between *populations* of neurons. The overall action of a neural population thus reveals emergent properties that are generated by the population action per se and that are not obvious in the action of any single element. It has not as yet been possible to set down general rules concerning these actions that will apply to all or even to many interneuronal populations. Their study, case by case, makes up a large part of central neurophysiology, which is discussed in many chapters of this book. Here it is pertinent to indicate how synaptic transmission at interneurons differs from that at the motoneuron, the prototype cell just discussed.

Interneurons form indirect, multisynaptic, divergent and reentrant, convergent loops in nuclear regions, superimposed upon more direct afferent and efferent pathways.* Thus afferent activity, which may reach efferent neurons directly, will also evoke activity that, delayed and amplified through interneuronal chains, will continue to reach efferent neurons during a considerable period of time. It appears that interneurons play an important role in prolonging and sustaining activity in the CNS. The simplest form of this sustaining action is known as *afterdischarge*. For example, a single volley of impulses in a cutaneous nerve elicits flexor muscle contractions that may continue for 0.1 sec or more due to a repetitive, reflexly evoked discharge of flexor motoneurons that is sustained by the repetitive and prolonged activity in the interneuronal chains linking input to output in the spinal reflex pathway.

Direct study of interneurons has been made most extensively in the spinal cord.[103] Even the interneurons of a single spinal segment are heterogeneous; some link afferent fibers to motoneurons, others are inhibitory neurons in recurrent pathways, others link long descending systems to motoneurons, and still others are intercalated between afferent fibers and the cells of origin of long ascending systems. Moreover, there are powerful interactions between these groups, a subject only recently brought under study.[103] Background or "spontaneous" activity, occurring in the absence of obvious stimulation, is a prominent feature of interneuronal activity, as it is of a very large proportion of all CNS neurons. A second prominent aspect of interneuronal activity alluded to is the tendency to discharge long trains of impulses, even in response to a single excitatory volley. Intracellular recording from interneurons reveals that EPSPs provoked by such a volley are prolonged and irregular.[75] Their action potentials, in contrast to those of motoneurons, are not followed by a period of hyperpolarization; there is no prolonged period of subnormality after impulse discharge. Thus one may suppose that the lowest threshold portion of the cell, the initial segment, will respond with impulses again and again, so long as the ionic current is adequate to drive the EPSP to threshold. The combination of such events in the several elements of divergent-convergent and circular reentrant linkages of

*"Afferent" and "efferent" are used to indicate the major inflow to and outflow from a synaptic center in the nervous system as well as their more traditional use to indicate the dorsal and ventral root fibers of the spinal cord.

neurons will contribute to prolonged interneuronal and efferent neuronal discharge. Moreover, all those aspects of monosynaptic transmission making for facilitation (temporal and spatial summation, temporal facilitation, posttetanic potentiation) will be operative at interneuronal synapses also, all in concert tending to increase and prolong the output of the circuit, be it excitatory or inhibitory. Undoubtedly the actions of interneurons are responsible for one constant feature of central neurons, indeed of the brain as a whole: a steady, slightly oscillating state of high excitability, *a readiness to respond.*

SYNAPTIC TRANSMITTER AGENTS IN THE MAMMALIAN CNS

Identifications of central transmitters are based on techniques and concepts developed in the study of neuromuscular and ganglionic synapses, with added steps to circumvent the blood-brain diffusion barrier and to deal with the anatomic complexity of the CNS. Synaptic delays, polarized conduction, a characteristic ultrastructure, and reversible PSPs all indicate that central synapses are chemically operated. Moreover, central pharmacologic effects are frequently best understood as the action of drugs upon transmitter synthesis and release or upon postsynaptic receptor molecules. Although transmission at some central synapses in mammals may be electrotonic in nature, there is to date no compelling evidence that this is the case.

ACh and norepinephrine are substances commonly considered to be central transmitters, as they are at peripheral junctions, but it is likely that a number of other substances function in a similar way. The positive identification of other candidate transmitters awaits methods with which to locate, isolate, and assay them in low, natural concentrations. Leading candidate central transmitters are indicated in Table 6-2.

There are certain criteria to be met for establishing transmitter identity: (1) The substance and the enzymatic system for its synthesis are present in the CNS neurons, (2) the substance is released from presynaptic neurons by impulses in them, (3) mechanisms exist for the rapid inactivation of the candidate transmitter, (4) local application of the substance elicits postsynaptic permeability changes similar to those produced by synaptic activity, and (5) drug effects upon the actions of naturally or experimentally delivered transmitter are similar. Based on these criteria, it has been established that ACh and norepinephrine are transmitters at peripheral synapses.[5, 138] All of these criteria have been met for the Renshaw synapse (p. 200), while evidence for the identity of transmitters elsewhere in the CNS is still somewhat indirect.

Central cholinergic transmission

It is now well established that ACh is the transmitter agent at skeletal neuromuscular synapses and in autonomic ganglia. Its presence within the CNS has been known since the 1930s, and its nonhomogeneous distribution there suggests a specific function in some regions, not in others. The dorsal roots, optic nerves, and cerebellum contain virtually no ACh, while the ventral roots, spinal cord, caudate nucleus, and retina contain large amounts (as much as 7 μg/gm tissue). The distribution of choline acetyltransferase ("acetylase") generally parallels that of ACh; it is present in activities up to 3,000 units (μg ACh synthesized/gm tissue/hr).[50] The distribution of "true" AChE is also roughly parallel to that of ACh, although it is distributed somewhat more widely than the synthesizing enzyme. Its cellular localization appears to be less restricted in central neurons than at the myoneural junction, for it exists both at the subsynaptic membrane and within the presynaptic axon, where it is bound to small intracellular particles. It may therefore hydrolyze not only the synaptically released ACh but any intra-axonal ACh not protected in vesicular storage as well. AChE-specific histochemical methods have been used in attempts to map certain putatively cholinergic central tracts, and their combination with electron microscopy now appears feasible.[86]

Chemical studies of brain tissue after homogenization and differential density-gradient centrifugation, particularly in combination with electron microscopic studies of the resulting fractions, have proved valuable in correlating chemical composition with subcellular localization.[144] In such a brain homogenate, 70 to 75% of the total ACh is contained within a fraction made up almost wholly of nerve terminals. When these "synaptosomes" are themselves disrupted, the ACh is found within a fraction containing mainly synaptic vesicles. Choline acetyltransferase, a small protein with a molecular weight of 67,000, is also present in nerve endings, but its precise intraneuronal localization is still unsettled. At least a part of this

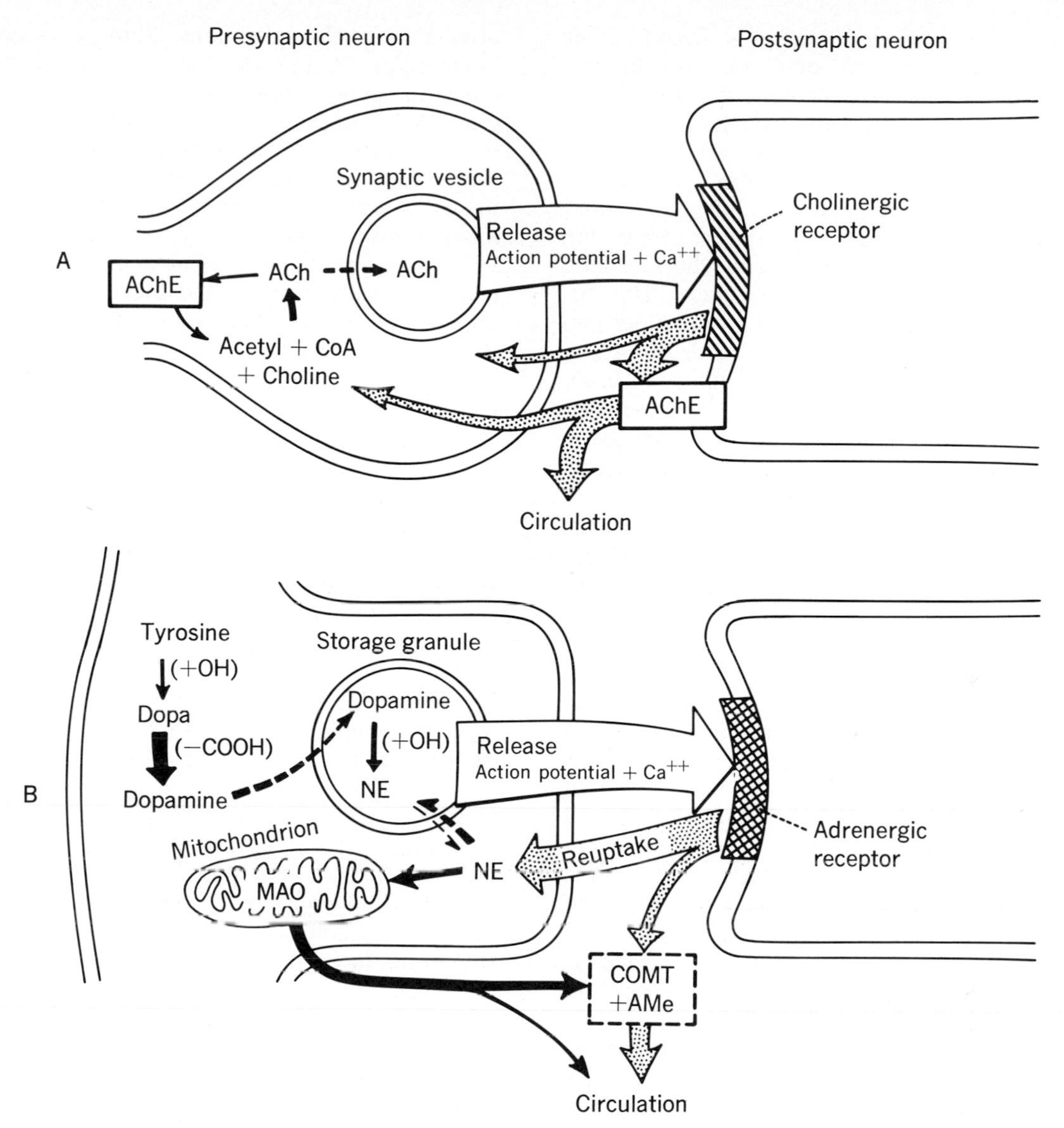

Fig. 6-18. A, Cholinergic synapse. Diagrammatic representation of major events at typical cholinergic synapse, including synthesis of ACh by choline acetyltransferase (probably in cytoplasm), storage in and release from synaptic vesicles, postsynaptic interaction, and subsequent inactivation by AChE, which probably also acts presynaptically to remove excess ACh. Mechanism of release by nerve impulses (white arrow) is not certain but depends on Ca^{++} and perhaps Na^{+} and probably does not involve intermediate discharge into free presynaptic cytoplasm. Black arrow represents metabolic pathway of ACh catabolism at postsynaptic neuron, with choline release into circulation and feedback of choline and perhaps ACh into presynaptic synthetic sites. Nature of "receptor" is poorly understood.

B, Adrenergic synapse. Synthesis of NE occurs from phenylalanine via tyrosine and directly from endogenous tyrosine. First step, hydroxylation of tyrosine to form dihydroxyphenylalanine (dopa), appears to be rate-limiting (small arrow). This step and subsequent decarboxylation to form dopamine occur in cytoplasm. Dopamine is then taken into storage vesicles for final β-hydroxylation to form NE. NE that leaks into the cytoplasm is destroyed by mitochondrial monoamine oxidase (MAO). Release by nerve impulses requires Ca^{++}, and whether there is intermediate discharge from storage granules into presynaptic cytoplasm is unknown. Nature of postsynaptic receptor in CNS is not readily describable in terms of classic alpha or beta adrenergic receptors. After interaction with receptor sites, NE is removed mainly by mechanism of active reuptake but partly also by enzymatic deactivation by catechol-*O*-methyltransferase (COMT), utilizing methyl donor *S*-adenosylmethionine (AMe). This latter enzymatic step produces *O*-methylated metabolites that are detectable in blood, CSF, or urine. Unmetabolized catecholamines are not able to escape rapidly from brain because of blood-brain diffusion barrier.

acetylating enzyme activity is found either free in the cytoplasm or in microsomal particles throughout the neuron; it may be synthesized within the cell body and moved to the nerve endings by axoplasmic transport. The ACh within the synaptic vesicles is thought to be in a storage form, protected from hydrolysis until its release from the nerve endings, and replaced by the synthetic action of choline acetyltransferase. The activity of the released ACh is presumably terminated by hydrolysis with AChE, which is found bound to membrane fragments in brain homogenates. However, there is some evidence that a reuptake mechanism may exist for ACh in brain tissue, and that choline is conserved by vigorous reuptake.

Although perfusion experiments such as those using peripheral autonomic ganglia (p. 190) are not possible within the CNS, there is good evidence that ACh is released from central neural tissue during activity, e.g., from the surface of the cerebral cortex into the cerebrospinal fluid (CSF), and that the rate of release is roughly proportional to the level of activity.[105] Furthermore, in some studies the caudate nucleus has been perfused with small volumes of fluid via double-barreled cannulas (1 mm tip),[105, 108] and concentric micropipets have made it possible to record the activity of single neurons through one barrel while removing small quantities of fluid from the other. The results obtained are preliminary, but they confirm that ACh is released in active neural tissue of the forebrain, presumably from presynaptic terminals in the immediate locale of the micropipet tip.

In a similar way, multiple micropipets allow one to apply substances through one barrel to a very local region by iontophoresis, while recording through the other. The chemoreceptive properties of the synaptic membrane of Renshaw interneurons, described on p. 202, were established in this way. This method of electrical recording combined with electrophoresis has been used very widely in studies of central neurons,[5, 12, 126] with results that are sometimes contradictory and difficult to interpret. They do include evidence for the "nicotinic" properties of some CNS synapses.[126] In many locations, however, the excitatory responses of neurons to ACh resemble more closely the "muscarinic" action of ACh: they are blocked by atropine but not by curarization, they are slow in onset, and they often persist for many seconds after even brief applications. Many cells give intermediate responses that are not clearly either nicotinic or muscarinic. At some locations ACh suppresses the discharge of cells, but the nature of this effect is still uncertain.[126] In the cerebral cortex and caudate nucleus, truly inhibitory cholinergic transmission may occur.

It should be emphasized that in these electrophoretic experiments the local concentrations of ACh and the sites at which it acts are unknown; the actions observed may be pharmacologic. Nevertheless, the evidence suggests strongly that ACh is a synaptic transmitter agent in many brain regions, including the cerebral and cerebellar cortices, thalamus, geniculates, and caudate nucleus.[12] What is certain is that the ACh system (Fig. 6-18, *A*) is present in the brain and is localized to synaptic boutons, and that some central synapses are cholinergic. It is equally certain that others are not.

Central adrenergic transmission

That norepinephrine (NE) is the major transmitter agent of the postganglionic sympathetic neurons is established with some certainty.[9, 138] Its location in these neurons can be demonstrated by histofluorescent and radioautographic methods. The synthetic steps from phenylalanine and tyrosine to NE and the catabolism of NE by oxidative deamination have been described in neurons. NE disappears rapidly from peripheral tissues after section of their sympathetic nerves. The nerve terminals take up exogenous or released catecholamine (CA) by an avid process important in the removal of released NE. The endogenous NE is stored mainly in granules or vesicles of the nerve terminals in a metabolically inactive, ATP-bound form. The stores are replenished by the enzymatic conversion of dopamine (DA) to NE in the granules, while the preceding synthetic steps probably occur in the cytoplasm. A portion of the NE is loosely bound, rapidly metabolized, and readily available for release by nerve impulses. Some of this fraction may exist in the free cytoplasm as well as in vesicles. The level of free NE in the cytoplasm is controlled by mitochondrial monoamine oxidase (MAO). Nerve impulses release NE from perfused organs, for it has been identified in perfusates by specific biochemical assays. The processes of synthesis and reuptake allow a sustained output of NE without exhausting the supply. The mechanism of release is not well understood, but requires Ca^{++}. The postsynaptic effects of impulses in adrenergic nerve fibers are mimicked by administered NE. These effects are generally excitatory (alpha adrenergic), particularly at smooth muscle receptor sites. Released NE is either removed by the circulation and acted upon by catechol-*O*-methyl transferase (COMT) and extraneuronal MAO or taken up again into the presynaptic nerve terminals. These processes are diagrammed in Fig. 6-18, *B*.

The presence of an adrenergic substance in extracts of mammalian brain was discovered in 1946,[58, 138] and regional distribution of NE in the brain was described in 1959. Whole-brain concentrations are about 0.1 to 0.5 μg/gm, and specific regions rank as follows: midbrain, pons, and medulla, followed by a group of structures with low concentrations, including the cerebral cortex, hippocampus, cerebellum, and spinal cord. Concentrations of NE are low in the striatum and particularly so in the caudate nucleus where DA, the immediate precursor of NE, is found in high concentration and is itself thought to function as a transmitter.

The fluorescence of the CAs and 5-hydroxytryptamine, or 5-HT (serotonin), in

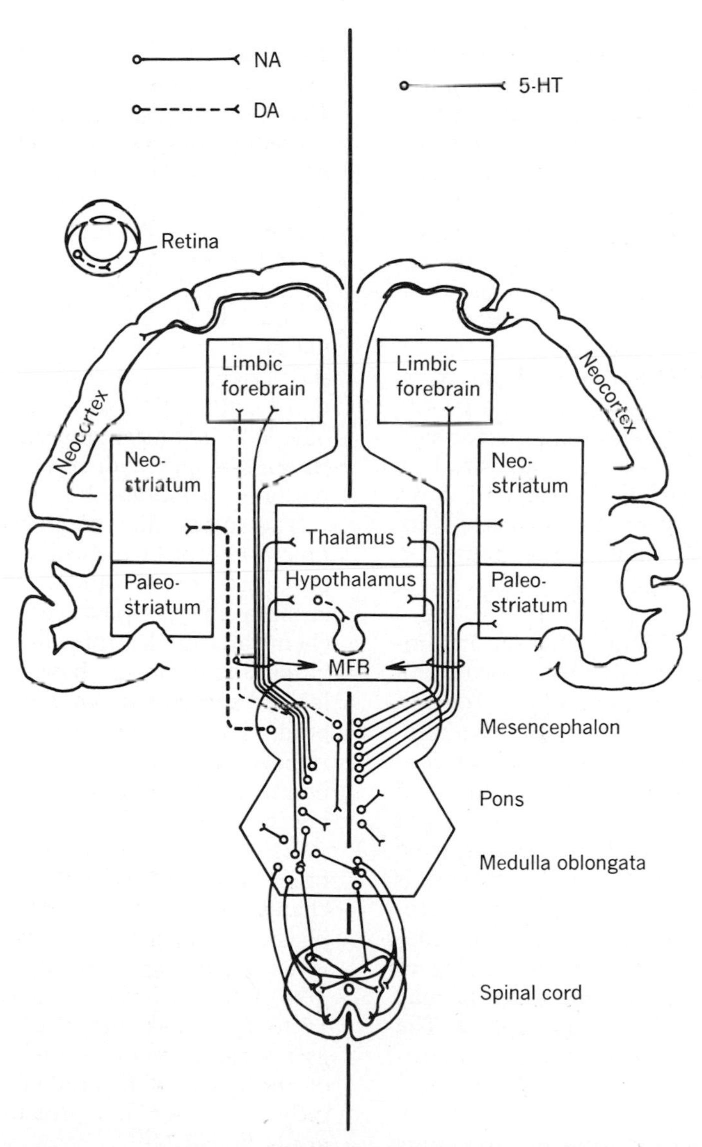

Fig. 6-19. Schematic representation of distribution of neurons containing norepinephrine (NE), dopamine (DA), or serotonin (5-hydroxytryptamine, 5-HT) in mammalian CNS as determined by histofluorescence techniques. (From Anden et al.[17])

brain slices exposed to formaldehyde vapor, particularly if MAO has been inhibited, allows the histologic localization of these biogenic amines.[17, 56] NE and 5-HT can be differentiated by color with this method, but NE and DA can be differentiated only by more complex pharmacologic methods involving the inhibition of CA synthesis. In the peripheral sympathetic system, large amounts of the amines are located in the terminal "varicosities" of small fibers, and these are generally considered to be adrenergic presynaptic terminals. In the CNS (Fig. 6-19), similar NE-containing fibers are most numerous in the hypothalamus, olfactory bulb, retina, median eminence, limbic system, and cranial nerve nuclei. In the spinal cord there is a system of descending adrenergic fibers that terminates about neurons of the intermediolateral horn. Since the latter neurons are cells of origin of the preganglionic fibers and are known to be cholinergic, there is an alternating adrenergic-cholinergic system of synaptically linked neurons. Central "adrenergic" systems have also been mapped by combining the fluorescence technique with brain lesions.[56] After an axon is sectioned, the amines concentrate on the proximal side of the section and may disappear distal to it. The precursors of the transmitters or enzymes and other protein molecules required for synthesis and storage[3] may be synthesized in the cell body and then be transmitted down the axon, although all of the synthetic steps involved in synthesis of CAs are known to take place in the terminals also (Fig. 6-18, *B*).

The intracellular location of CAs and indole amines has been studied by a correlation of differential centrifugation and electron microscopic observations. The same synaptosomal fraction described on p. 206 that contains ACh also contains large amounts of NE and 5-HT. These amines are also present in the isolated vesicles that are prepared from disrupted nerve endings.[144] DA, however, is thought to be localized partly in the cell cytoplasm. Even some of the NE may not be confined to the vesicles, although the finding of amines free in the cytoplasmic fraction may be partly an artifact of tissue disruption. The vesicle-bound CA probably represents a storage form that affords protection from enzymatic destruction.

The barrier to the diffusion of circulating CAs into the brain makes the study of their metabolism difficult and also precludes accumulation of CA in the brain from peripheral sites of synthesis. The enzymes and precursors required for synthesis of NE are present in the brain. The major steps are as follows:

$$\text{Tyrosine} \xrightarrow{(1)} \text{Dopa} \xrightarrow{(2)} \text{DA} \xrightarrow{(3)} \text{NE}$$

The formation of epinephrine from NE may be possible in the brain, but ordinarily this does not appear to be an important step. Tyrosine is a normal constituent of brain and can be formed there by the hydroxylation of phenylalanine. Since dihydroxyphenylalanine (dopa) is readily decarboxylated to form DA, it does not accumulate in the brain. The enzymes required for the synthetic steps just given are (1) tyrosine hydroxylase, (2) aromatic L-amino acid decarboxylase, and (3) DA-β-hydroxylase. The first step is probably normally rate-limiting, and catechol compounds may regulate their own synthesis by exerting negative feedback at this step.[132] There is evidence that the enzymes for steps 1 and 3 are localized in the synaptosomal fraction of a centrifuged brain homogenate. Further localization of step 1 to the cytoplasm of the nerve endings and of step 3 to the vesicles as in Fig. 6-18, *B*, is not yet certain for the CNS and is based largely on analogy with findings in peripheral tissues.[72, 144] Brain tyrosine hydroxylase may be at least partially membrane bound.[49]

The availability of labeled CAs and their precursors, of high specific radioactivities, has permitted further studies of CA metabolism in both the periphery and the CNS.[58] Biochemical and electron microscopic radioautographic techniques have shown that these labeled substances, when injected directly into brain tissue or into the ventricular CSF, are rapidly taken up at nerve terminals and bound to the synaptic vesicles, thus mimicking the distribution of endogenous amines.

Estimates of the turnover of NE in the brain from measurements of the rate of change of specific radioactivity reveal a biphasic disappearance curve in which there is an initial rapid phase with a half-life of 3 to 4 hr, followed by a slow phase with a 17 to 18 hr half-life. These observations suggest a rapid synthesis and utilization of part of the store, with another portion metabolically less active and probably represented by firmly bound NE.

The disappearance of CA (and 5-HT) from central neurons has been observed under a variety of conditions. The prolonged electrical stimulation of various areas of the brain leads to CA decrease, which suggests its synaptic release.[134] More directly, the double-barreled cannula has been used to demonstrate spontaneous release of DA from the caudate nucleus, which is increased by thalamic stimulation.[105] Further, endogenously synthesized labeled NE is released from superfused brain slices by brief electrical stimuli of low intensity.[19] In vivo, spontaneous CA release or release evoked by electrical stimulation or drugs has been demonstrated with the use of push-pull cannulas[57, 105] or the ventricular perfusion technique.[116, 118]

The interpretation of the results of injections of CAs into the CSF or of iontophoretic application of CAs to the brain is difficult for they are frequently contradictory. They indicate that the CAs have widespread effects that are inhibitory or excitatory. The CAs (and indole amines) have been implicated in a wide variety of centrally mediated physiologic functions, including sleep and arousal, affect, memory, the release of pituitary hormones, and most of the essential homeostatic mechanisms, e.g., regulation of temperature, blood pressure, and food and fluid intake.[2]

Mechanisms for NE inactivation and removal also exist in the CNS and are generally similar to those in peripheral adrenergically innervated tissues. Efficient enzymatic mechanisms exist for the inactivation of brain CAs. These include mitochondrial MAO as well as COMT, together with the appropriate methyl donor (*S*-adenosylmethionine). The location of the methylating enzyme is uncertain. The relative importance of the two catabolic processes is unknown, but it is generally held that an avid and stereospecific process of reuptake at the neuronal membrane is itself adequate to account for the removal of NE released by presynaptic impulses from the site of action.[9] One unique feature of CNS metabolism of NE is that the major metabolites (deaminated and 3-*O*-methylated), include large portions of 3-methoxy-4-hydroxyphenylglycol (MHPG) in addition to vanillylmandelate (VMA) and normetanephrine (NMN).[104]

Other proposed transmitters

5-Hydroxytryptamine (serotonin). The idole amine 5-hydroxytryptamine (5-HT) has been proposed frequently as a CNS neurotransmitter. Many aspects of its metabolism parallel that of brain CAs. 5-HT is present in many invertebrate tissues and in all vertebrate brains examined to date. In the brain[137] it is distributed regionally, with concentrations generally below 0.5 μg/gm, but concentrations may be as high as 1 to 2 μg/gm in the midbrain and even 50 to 70 μg/gm in the pineal gland. Areas rich in 5-HT include the hypothalamus, the midbrain raphe, the basal ganglia, and spinal cord. There are systems of ascending and descending fibers, with cell bodies in the midbrain raphe nuclei and fibers to the diencephalon, telencephalon, and cord and with nerve terminals that produce intense 5-HT histofluorescence.[17, 56] 5-HT is retained in nerve ending fractions of brain homogenates and may, like CAs, be stored in granular synaptic vesicles, although some 5-HT may also exist in the nerve ending cytoplasm (Fig. 6-19).[115]

The required synthetic (tryptophan hydroxylase and aromatic L-amino acid decarboxylase) and catabolic (MAO) enzyme activities are present in CNS tissues. The hydroxylation of tryptophan, a relatively minor metabolic pathway for this essential amino acid, is probably the rate-limiting step of 5-HT synthesis, and the enzyme has been at least partially localized to nerve endings.[80] The availability of tryptophan may play an important regulatory role in 5-HT synthesis in the brain. The deaminated metabolite 5-hydroxyindolacetate (5-HIAA) is the major product of 5-HT catabolism. In the pineal gland, 5-HT can also be enzymatically *N*-acetylated and then 5-*O*-methylated by a specific *S*-adenosylmethionine-hydroxyindole-*O*-methyl transferase to form melatonin, a product of uncertain function. The turnover of 5-HT in the brain is relatively rapid; the nominal half-life is less than 1 hr,[101] although it is probable that 5-HT, like other intraneuronally stored substances, exists in multiple pools with both fast and slow turnover rates. Tritiated 5-HT has been used to label central nerve endings, although some of the exogenous 5-HT is taken up from the CSF nonspecifically.[14]

The use of double-barreled cannulas to study the release of labeled 5-HT during neural stimulation has been unsuccessful.[48] Some spontaneous release of endogenous 5-HT into perfused cerebral ventricles[118] as well as into cortical cups[48] has been demonstrated. The spontaneous cortical release of 5-HT occurred at a rate of about 40 pg/min/cm^2 of surface area, and this rate trebled following stimulation of the brainstem raphe;

the spontaneous release of 5-HIAA was nearly 400 pg/min/cm^2. Midbrain stimulation also reduced forebrain 5-HT and increased 5-HIAA concentrations.[15] Release of ^{3}H-5-HT from brain slices by depolarization in vitro has also been demonstrated.[32] Iontophoretic application of 5-HT to central neurons has produced a variety of equivocal excitatory and inhibitory effects,[12] and the possible functions of 5-HT in the brain are still poorly understood.[2]

Histamine.[63, 109] This imidazole amine occurs in mammalian brain tissue and, like NE and 5-HT, is found in highest concentrations in the hypothalamus. In other locations, e.g., the pituitary or pineal glands and the choroid plexus, much of the histamine is contained within mast cells. Some histamine has been located in the synaptosomal fraction of centrifuged brain homogenates and other particulate fractions that may also contain small nerve endings. Enzymes for histamine synthesis from histidine (aromatic L-amino acid decarboxylase or another specific histidine decarboxylase) and inactivation (imidazole-*N*-methyltransferase with its methyl donor, a poorly characterized enzyme that oxidatively deaminates histamine in some species, and MAO, for which *N*-methylhistamine is a substrate) occur in the CNS. The major urinary metabolites of histamine, like those of NE and 5-HT, are both deaminated and ring-methylated. Locally applied histamine has a variety of excitatory and inhibitory actions. An obstacle to further understanding of the role of histamine in brain function has been the lack of convenient tissue assays of sufficient sensitivity and specificity to measure brain histamine in small amounts, on the order of 0.1 μg/gm,[131] although such an assay is now available.[131, 135a]

Amino acids.[12, 40] Several aliphatic amino acids are known from iontophoretic experiments to have profound effects on the state of polarization of many central neurons. Among the compounds generally found to be excitatory, the dicarboxylic amino acids L-glutamate and L-aspartate are the best known. They produce rapid and striking transient depolarization when applied to neurons but not when injected intracellularly. Many natural and synthetic compounds with a terminal carboxyl or sulfonyl moiety separated by one or two carbon atoms from an α-carbon bearing an amino and a carboxyl group possess such activity. The natural L-enantiomers of glutamate and aspartate are widely distributed in the cytoplasm of mammalian central neurons, particularly in isolated nerve endings.[93, 125] They can be released from the cerebral cortex during thalamic and brainstem stimulation at rates many times greater than their spontaneous efflux[83]; they as well as γ-aminobutyrate (GABA) are found to be released rather specifically, compared with other amino acids, from electrically stimulated CNS slices[71] or even cortical synaptosomes in vitro.[26] It is not yet clear whether natural decarboxylation, transamination, cellular reuptake, and diffusion are adequate to serve as physiologic inactivation mechanisms for these putative transmitter substances.

Other amino acids generally found to be inhibitory when applied to central neurons include GABA and other monocarboxylic ω-amino acids such as glycine, taurine, and guanidino-acetate, or synthetic analogs of these natural compounds with terminal amino and acidic groups separated by a chain of one to five carbon atoms. Certain characteristics of physiologic inhibitory transmission, including true hyperpolarization and increased membrane conductance to Cl^- or K^+, that are antagonized by strychnine and related compounds have been produced by glycine when applied to ventral horn motoneurons, where glycine may be an inhibitory transmitter.[71] GABA appears to have a similar action at the lateral vestibular nucleus and cerebral and cerebellar cortices, except that its action is resistant to strychnine. However, the specificity of strychnine is questionable.

At this time more is known about GABA than other proposed mammalian central inhibitory transmitters. In the crustacean peripheral inhibitory motor fibers, GABA is already convincingly identified as the inhibitory transmitter.[44, 87, 114] In mammalian brains, GABA is found by chemical assay in highest concentrations in the colliculi, the diencephalon, and the occipital lobes; much lower levels occur in the pons, medulla, and most of the cerebral cortex.[40, 124] GABA is a unique substance in that its synthesis is almost completely restricted to the CNS, where it is made from the excitatory amino acid glutamate by a specific decarboxylase controlled by end-product inhibition.[87, 124] GABA is further restricted to the CNS by a relatively efficient diffusion barrier. GABA can be inactivated by oxidative metabolism, mainly by an intermediate conversion by a transaminase to succinic semialdehyde, which is then con-

Table 6-2. Proposed central neurotransmitters; summary of candidate substances, their structures, and metabolism

Compound	Structure	Synthesis	Inactivation
Acetylcholine (ACh)	$H_3C\overset{\overset{O}{\|}}{C}—O—(CH_2)_2\overset{+}{N}(CH_3)_3$	Choline acetylation	Hydrolysis by AChE
Dopamine (DA)	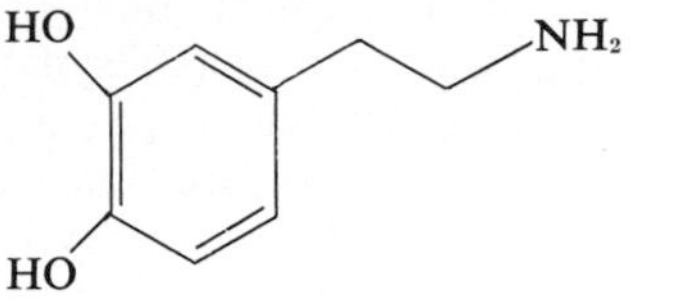 HO, HO, NH_2	Tyrosine hydroxylation, dopa decarboxylation	Reuptake, MAO, COMT
Norepinephrine (NE)	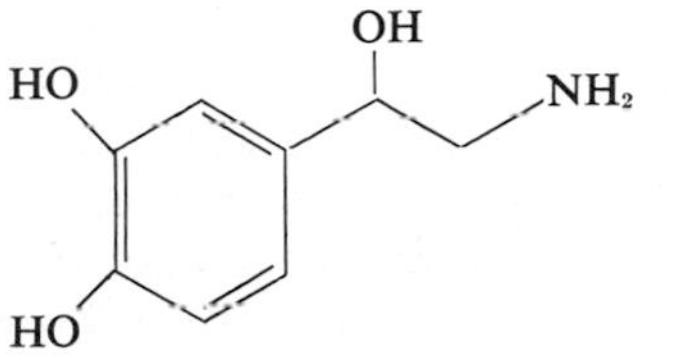 OH, HO, NH_2, HO	Dopamine-β-hydroxylation	Reuptake, oxidative deamination by MAO, 3-*O*-methylation by COMT
5-Hydroxytryptamine (5-HT), or serotonin	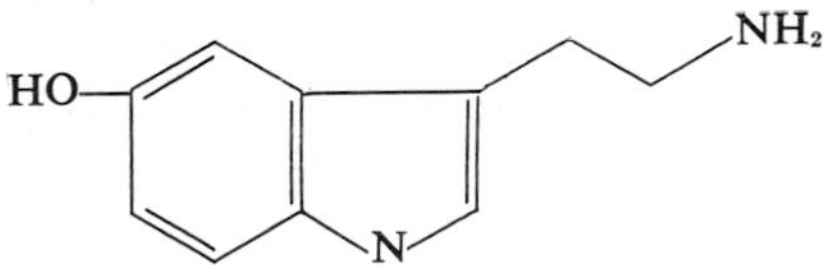 NH_2, HO—, N	Tryptophan hydroxylation, 5—OH—tryptophan decarboxylation	Reuptake, MAO
Histamine	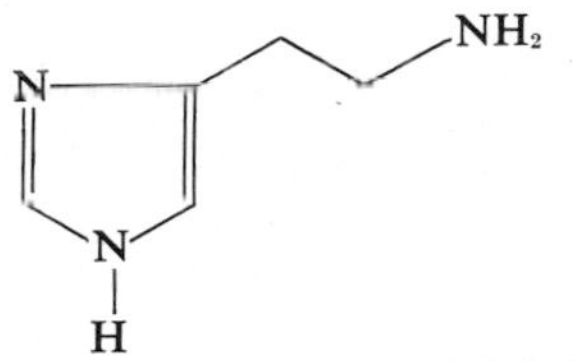 NH_2, N, N, H	Histidine decarboxylation	*N*-methylation by histamine-*N*-methyltransferase; oxidative deamination (MAO ?)
Excitatory amino acids (e.g., glutamate, aspartate)	$CH_2\ (CH_2)_n\ CH\ NH_2$ COOH COOH n = 0-1	—	Reuptake, decarboxylation, NH_3 fixation
Inhibitory amino acids (e.g., GABA, glycine)	$CH_2\ (CH_2)_n\ NH_2$ COOH n = 0-4	GABA by glutamate decarboxylation	GABA—reuptake transamination and oxidation to succinate
Prostaglandins (PG)	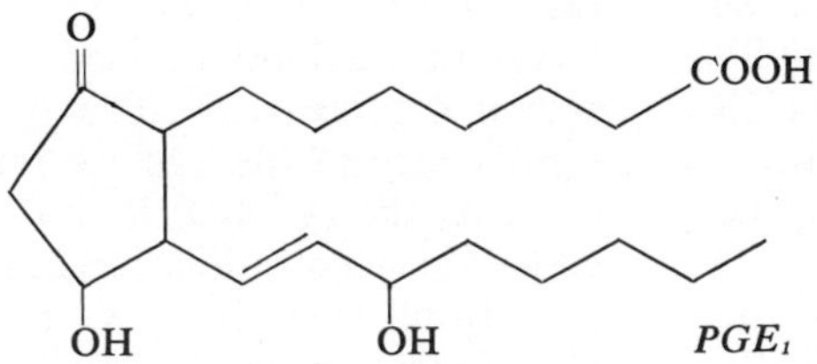 O, COOH, OH, OH, *PGE_1*	—	—
Substance P	Polypeptide, mol wt ∼ 1,600	—	—

verted to succinate by a dehydrogenase. Transaminase activity can be visualized histochemically and can be selectively inhibited by drugs.[124] The distribution of the GABA-synthesizing enzyme appears to correlate with the tissue distribution of GABA, and it has been identified in isolated nerve endings[51] and even in presynaptic vesicles.[96] Labeled GABA is avidly accumulated by brain tissues and is localized in the nerve ending fraction.[110] The degree of uptake correlates well with the regional distribution of endogenous GABA, and this has been confirmed by radioautographic techniques.[70] Reuptake may be an additional mechanism for inactivating released GABA. Efflux of GABA from the cerebral cortical surface has been found to correlate with neural activity,[83] and release of GABA from electrically or ionically stimulated brain slices[133] and synaptosomes[26] has been demonstrated.

Other substances. It has been proposed that two other substances discovered by von Euler in the 1930s also play a role in central synaptic transmission. They are substance P,[99] a basic polypeptide with a molecular weight of about 1,600 and composed of 13 different amino acids, and the prostaglandins.[127] The latter are a group of more than a dozen slightly different long-chain fatty acids, composed of 20 carbon atoms and including a cyclopentane ring, that were originally isolated from prostatic tissue. Both substance P and the prostaglandins have intense action on smooth muscle. The prostaglandins exist in the mammalian CNS. Their tissue distribution has been studied and they have been found in isolated nerve endings.[85] The enzymes that metabolize them are known but not yet described in the brain. The release of prostaglandin-like substances from many brain and spinal cord regions has been described.[122] Exogenous prostaglandins have a variety of central effects, and their iontophoretic application to central neurons produces both excitatory and inhibitory changes.[12] Since their actions tend to be prolonged, it has been suggested that they or other substances may act as "modulators" of neural activity and membrane polarization, in contrast to the more typical phasic effects of "transmitters." It is even thought that they may be released from the postsynaptic cell.

Knowledge of substance P is limited to the facts of its localization to phylogenetically older parts of the CNS and its association with isolated synaptic vesicles.[125] Its central physiologic and bioelectrical effects are not well defined, and study has been limited by the lack of ready availability of pure preparations of substance P. The proposed transmitters are described in Table 6-2.

Postsynaptic receptors

The events that occur between the release of a chemical neurotransmitter and an electrical sign of postsynaptic response are largely unknown. It has long been postulated that specific surface interactions must take place between the transmitter molecule and the subsynaptic plasma membrane. Speculations include analogies with the interaction between a substrate or cofactor and a protein-enzyme. The net result must include a physical change in the membrane and altered permeability to small ions. The descriptive pharmacology of the interactions of transmitter substances and their inhibitors with sensitive cells has been studied intensively, and some conclusions about the likely structural requirements for such interactions have been reached.[81] A new approach has resulted from the description by Sutherland and Rall in the late 1950s of adenosine 3′,5′-monophosphate, or cylic AMP (cAMP), as the receptor cell mediator of the hepatic glycogenolytic effects of epinephrine and glucagon.[64, 135] This discovery has led to the hypothesis that cAMP or similar substances may act at nearly all cells sensitive to the effects of hormones and even neurochemical transmitters.

It is known that brain tissues are particularly rich in cAMP and the enzymes important for its production and breakdown. Brain tissues, particularly the hypothalamus, have high activities of adenyl cyclase, the enzyme that catalyzes the production of cAMP from ATP. It is present in high activity at nerve endings and appears to be associated with membrane components of cells, leading to the theory that it may be a component of the cell membrane and thus closely associated with macromolecules involved in receptor function. The activity of the cyclase in unbroken cells is markedly stimulated by the CAs, prostaglandins, and many other hormonal substances.[135] The possible regulatory significance of the cyclase reaction is emphasized by the fact that it can raise intracellular cAMP concentrations severalfold in about 1 sec. Brain also has the highest activity of phosphodiesterase, the enzyme that catalyzes the breakdown of cAMP to 5′-AMP, which is the only known physiologic mechanism for terminating the activity of cAMP. The overall significance of the many known effects of cAMP on cellular metabolism is not yet well defined. There is some indication that its ability to stimulate the phosphorylation of various proteins and enzymes, thus altering their structure and activity, may be of some general significance.

Working models for a possible function of cAMP in transmitter-receptor function usually include the localization of adenyl cylase to cell membranes, its stimulation by transmitter substances, and subsequent rapid increases in the rate of production of cAMP. The cAMP is often called a "second messenger," as it acts within the receptive cell to mediate the action of a transmitter (or hormone) by modifying important metabolic processes.

AXOPLASMIC TRANSPORT AND NEURONAL DYNAMICS[3, 21, 41, 59]

The great length of neuronal processes, particularly of axons, poses a special problem for the transport of materials from one part of the cell to another. The early experiments of Weiss and Hiscoe[143] showed that following constriction of an axon there was an accumulation of axoplasmic material proximal to and a diminution of axonal diameter and contents distal to the constriction. When the constriction was removed, the accumulated material moved down the nerve fiber at a rate of 1 mm/day, an event interpreted as occurring continuously in normal axons. This discovery of proximodistal axonal flow has been confirmed and extended in recent experiments using other methods, particularly the radioactive labeling of material, usually protein, within the cell body and tracing it radioautographically by light and electon microscopy. A number of molecular species and intracellular organelles are now known to be synthesized or constructed within the cell body and then moved down the axon to its terminals. This axoplasmic transport is thought to function in the maintenance of the neuron, supplying even its most distant parts with materials synthesized in the cell body; it may, moreover, supply substances necessary for the morphologic, biochemical, and functional specification of postsynaptic cells, contributing to the "trophic" influence that neurons exert upon the cells they innervate.[67] Grafstein[60] has recently observed the apparent transsynaptic movement of labeled amino acid, injected into the ocular bulb (of mice), into the visual but not other areas of the cerebral cortex.

Of the protein that is transported slowly (1 mm/day) down the axon, 60% is particulate and the remainder is soluble, together a process thought to provide replacement of major constituents of the axon (including neurofilaments and neurotubules), for the transport rate equals the rate of regeneration of severed fibers. There is a smaller but much more rapidly transported (100 mm/day) protein component consisting almost entirely of synaptic vesicles and mitochondria. This is consonant with other evidence that the enzymes required for synthesis of synaptic transmitters are also synthesized within the cell body and then moved to the nerve endings.

The mechanism of axoplasmic transport is unknown; it is certainly not a movement of liquid or gel propelled by intracellular hydrostatic forces. Indeed, axonal transport from proximal to distal portions of axons continues for a time after inhibition of protein synthesis within the cell body. By using time-lapse cinephotography of living axons, Weiss found that periodically recurring peristaltic waves sweep down axons and appear to propel axonal contents before them. The structure producing these waves is unknown, but it may be the axolemma itself. How vesicles and mitochondria are moved so rapidly remains a mystery, but it is perhaps significant that substances, e.g., colchicine, that bind to and alter the molecular configuration of the neurotubules also selectively block the axonal transport of transmitter-containing vesicles.[20, 130] What controls these transport processes is unknown, but only in rare locations are they affected by the conduction of impulses.

Thus it seems likely that the rapid transport provides components of transmitter systems and perhaps other substances specifically required at nerve endings. The slow transport—in fact a bulk movement of the axon itself—raises the possibility that axons and their endings are continuously growing or being replaced, that synaptic connections are therefore dynamic, and that those changes may be modified by the activity of the neuron.

NEUROGLIA[10, 98]

Neuroglial cells of the brain greatly outnumber its neurons and occupy half its volume. While they have elaborate processes, they do not possess axons and make no

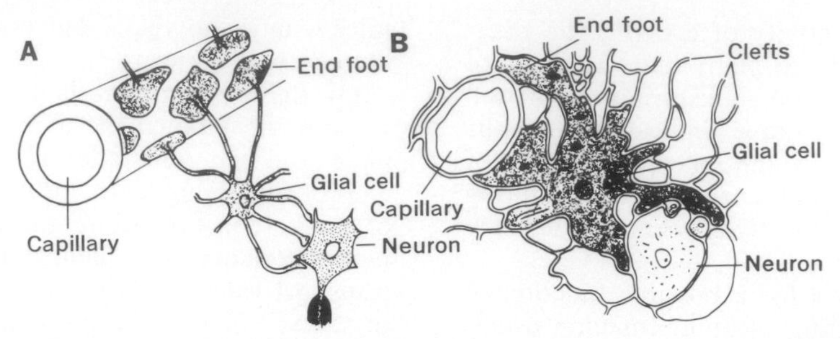

Fig. 6-20. A, Schematic drawing shows neuron-glia-capillary relation as seen with light microscope. Some "end feet" of astrocyte make contact with capillary wall and others with neuron. This relation is basis for original idea that glial cells might serve as channels for passage of substances between blood and neurons, and vice versa. **B,** Sketch shows same relation as seen with electron microscope. All cells, axons, dendrites, and astrocytes are tightly packed but separated by intercellular cleft system that is usually of 150 to 200 Å width. (From Kuffler.[10])

synaptic contacts. Fused membrane contacts, so-called gap junctions, occur between adjacent glial cells; they are regions of high ionic conductance. In contrast the neuron-neuroglial appositions seen in some tissues[33] are not regions of decreased impedance to ionic flow, as are the interglial junctions.[90]

The neuroglia possess the potential to continue to divide throughout life, a property lost by neurons early in development. In classic neurology the glia have been regarded as the supporting or interstitial tissue of the nervous system, while a much more active role is now proposed for glial cells. Electron microscopic studies show that the space between neurons, other than that at synaptic contacts, is occupied by processes of glial cells, leaving clefts of 150 to 200 Å separating cellular elements. These intercellular clefts make up about 5% of total brain volume and constitute the true extracellular space. Glial cells possess the elements of actively metabolizing cells: mitochondria, endoplasmic reticulum, ribosomes, glycogen, fat, ATP, etc. A special transport function was for a long time attributed to glial cells, based upon their apparent interposition between blood vessels and nerve cells (Fig. 6-20). Glial cells congregate in regions of the brain where neurons are destroyed by injury or disease.[98, 121]

There are three types of glial cells. Microglia are of mesodermal origin, and hence not truly neuroglia; they enter the CNS from the bloodstream and function there as macrophages. The *oligodendroglia* and the *astrocytes* are true neuroglia and, like neurons, are of ectodermal origin. The oligodendroglia form myelin sheaths, as do the Schwann cells of peripheral nerves. The function of the astrocytes is less certain. It has been suggested that they play some role in impulse conduction and synaptic transmission, serve as transport channels between vessels and neurons or as reservoirs of metabolites for neurons, or have an active symbiotic role in neuronal metabolism. Direct evidence concerning astrocyte function has been obtained by the application of microphysiologic and microchemical methods to invertebrates, in which glial cells can be identified and observed directly. The important contributions of Kuffler and his colleagues to an understanding of glial function were made in such simple brains.[10, 89]

Experimental preparations.[90, 111, 112] Direct observations of glial cells were made in the CNS of the leech and in the optic nerves of amphibia,[92, 113] which have large glial cells accessible in situ. The neuron-glial relation in a central ganglion of the leech is shown in Fig. 6-21. Substances that pass between its neurons and the surrounding blood sinus must diffuse through or between glial cells or be moved by active or pinocytotic transport across their membranes. Within each packet, 50 to 60 neurons are embedded in invaginations of a single large glial cell, but all cell boundaries are distinct and nearly everywhere separated by 150 Å clefts, except at occasional tight junctions that link adjacent glial cells. The cleft system opens freely to the region of the outer capsule and the capillary endothelium. Both glial cells and neurons can be studied with intracellular micropipets, and the movement of substances from the outside into the ganglion can be measured by their effects, if any, upon the membrane potentials of these cells.

The optic nerve fibers of *Necturus* are similarly encased within invaginations of large glial cells.[92, 113] The blood vessels from which nutrient substances are derived run on the surface of the nerve, to which the intercellular cleft system opens freely. Glial cell membrane potentials can be observed during changes in the composition of the extracellular fluid and when the nerve fibers are excited either electrically or by light flashes to the retina.

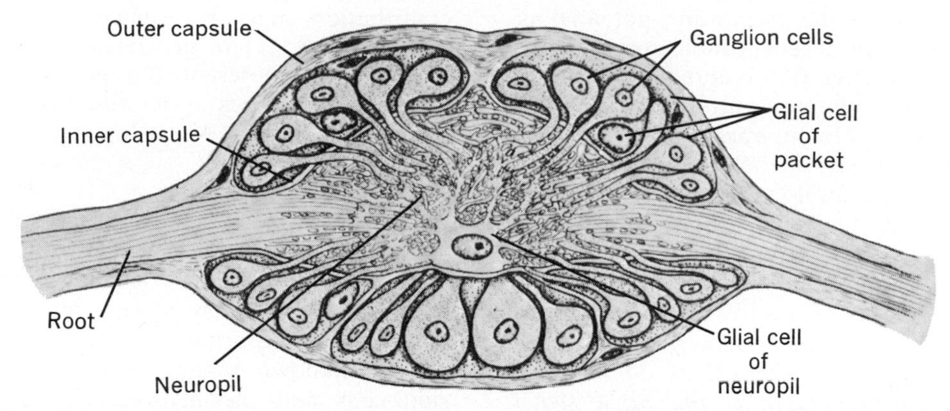

Fig. 6-21. Semidiagrammatic representation of transverse section through ganglion of leech CNS. Each peripheral compartment or packet contains numerous nerve cells, all invested by single large glial cell. Neurons send their processes through inner capsule into neuropil, which occupies central compartment and is where synaptic contacts are made. Neuropil itself is completely invaginated within embrace of other glial cells, one of which is shown on ventral aspect of central compartment. (From Coggleshall and Fawcett.[33])

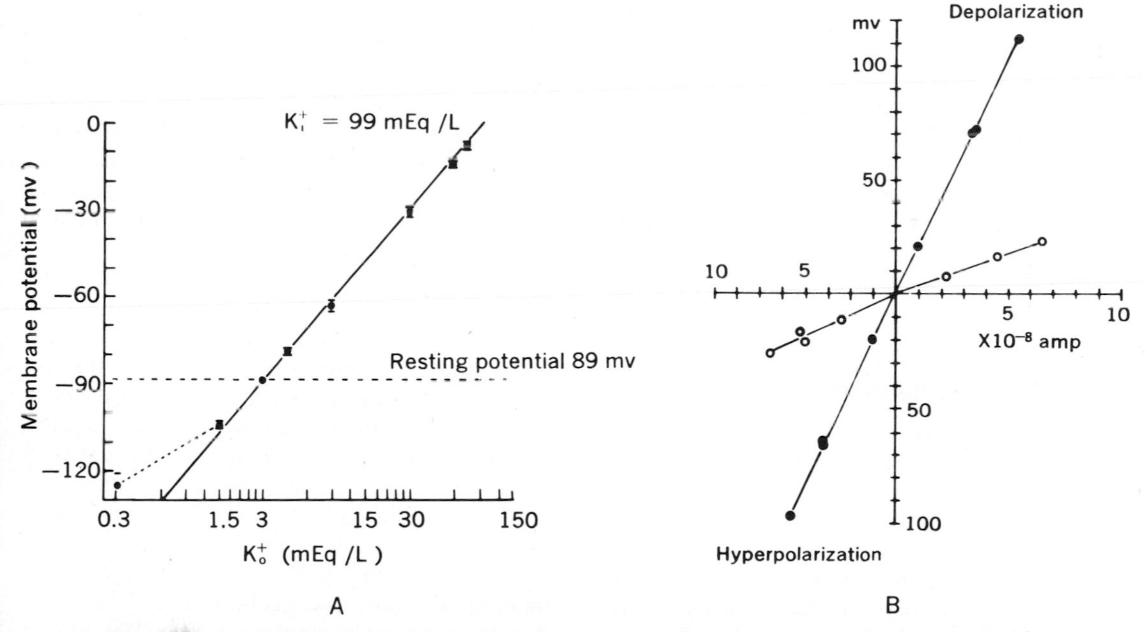

Fig. 6-22. A, Relation between glial membrane potential and K^+ concentration in bathing fluid (K_o^+) in optic nerves of *Necturus*. Forty-two measurements were made with K^+ concentrations below or above normal. Mean resting potential in Ringer's solution was 89 mv. Horizontal bars = ±SD of mean. Solid line has slope of 59 mv/10-fold change in K^+ concentration. It fits observed points except at lowest concentration tested. Membrane potential when $K_o^+ = K_i^+$; K_i^+ is therefore about 99 mEq/L.

B, Current-voltage relation for two glial cells. Square pulses of current were passed through one intracellular electrode; membrane potential changes that resulted were recorded through second electrode in same cell. Resistance of each cell remained constant even though in the series represented by steeper curve, membrane potential was displaced over total range of 200 mv. No active membrane responses were observed, even with rapid depolarizations. Glial cells do not give regenerative responses similar to those seen in neurons.

(From Kuffler et al.[92])

Physiologic properties of glial cells

The membrane potential of a glial cell varies, as predicted by the Nernst equation (p. 28), when external K^+ is changed (Fig. 6-22, *A*). Thus the membrane potential of a glial cell is an excellent indicator of extracellular K^+. Internal $[K^+]$ is about 100 mEq/L, for at that external concentration, membrane potential is zero. In contrast to the membrane potentials of neurons, those of glial cells can be displaced passively over a wide range by transmembrane current without evoking active membrane responses (Fig. 6-22, *B*). The important conclusion is that *glial cells do not possess the capacity for regenerative membrane conductance changes and generation of conducted impulses, as do neurons, and do not participate in the rapid signal transmission function of the nervous system.* After destruction of their investing glial cells, neurons of the leech ganglion still generate and conduct action potentials, use oxygen, store glycogen, and metabolize glucose. Whatever "metabolic support" role glia may play in relation to neurons, it must operate over a longer time scale than the hours during which these denuded neurons were observed.[90, 111] Further, in the intact ganglion, both direct observation and theoretical calculation show (1) that the cleft system is an adequate avenue for diffusion of ions between the external medium and the vicinity of neurons, (2) that it provides an equally open pathway for the ionic currents of action potentials, and (3) that no significant portion of these currents flows through glial cells.

Glial cells of the mammalian CNS

Recently a class of cells presumed to be glial cells has been studied within the mammalian cerebral cortex by means of intracellular recording. Such cells never show postsynaptic responses or action potentials, but their membrane potentials may slowly depolarize in conjunction with activity in adjacent neurons, e.g., during spindle activity (p. 240) or when those neurons are activated by repetitive volleys of thalamocortical impulses.[31, 65, 66] Moreover, these idle cells have now been more positively identified as glia by intracellular injection of marking dyes and later study of histologic sections. Their slow depolarization is thought to be produced by the increase in extracellular K^+ ion during neuronal activity. This depolarization may contribute to the changes in steady potential that are recorded across the cortex when it is active. Whether glial depolarization also contributes to the electroencephalogram (EEG) (Chapter 8) is uncertain; if so, that contribution must be limited to the lower end (up to 6 Hz) of the range of frequencies (1 to 40 Hz) represented in the normal EEG, for the time courses of de- and repolarization of the glial cells are very slow.[66]

Neuron-glia relations

On the basis of the foregoing, the conclusion can be drawn that glial cells are not directly involved in the signaling function of the CNS. Nevertheless, there is reason to suppose that an important functional relation of a still unknown nature does exist between glial cells and the neurons they surround. Only very rarely are central neurons or their processes observed to be free of a glial envelopment, and neurons do not survive long in tissue culture if separated from glial cells. Glia adjacent to a motoneuron vary in number with the length of its axon[52, 53] and increase in number when motoneuron activity increases.[94] It is not known whether the K^+-mediated depolarization described is a signal to glia from neuron that evokes some glial response. Glial cells in tissue culture display rhythmic movements that occur cyclically over seconds or minutes,[117] and it has been suggested that these movements may promote the axoplasmic flow of intracellular material from soma to terminals.[142]

Even though neurons may function normally for many hours after removal of their investing glia,[90] the possibility remains that the latter may provide some metabolic support to neurons over a longer time period. Both neurons and glia contain respiratory enzymes, fat, and glycogen, and their concentrations in both types of cell vary with the functional and the nutritional state of the organism.[145] Direct evidence for an exchange of metabolites between the two is, however, still lacking. Possibly the problem could be clarified by selectively labeling metabolites in one cell type and following their redistribution during activity.

Hyden et al.[76, 77] have presented experimental results that they interpret as indicating that just such a biochemical relation does exist between glia and neurons, and on that basis they have suggested that such a neuron-glial interaction plays a role in a postulated macromolecular mechanism underlying learning and memory.

Neurons and glia maintain very high concentrations of both RNA and proteins. Based upon his microchemical analyses of portions of cells dissected from the brain in a number of behavioral conditions, Hyden has proposed that RNA and proteins are synthesized and respiratory enzyme activities are increased with increased neural activity and that the amount of RNA in a neuron is a measure of its immediate past level of activity. He reported that in states such as sleep and learning there are increases not only in total RNA in certain brain areas but also in the ratio of purine to pyrimidine bases in the neuronal RNA.[78, 79] The accumulation of RNA with highly specific base ratios in neurons is said to parallel a reciprocal decrease in RNA of the same base ratio in adjacent glial cells. In this view, learning is thought to involve the transfer of specific nucleotides from glia to neurons. The experimental observations on which these hypotheses were based have been severely criticized[89] and they stand unproved. Moreover, they have not yet been followed by testable hypotheses about the mechanisms for memory storage and recall.

REFERENCES

General reviews

1. Andersen, P., and Jansen, J. K. S., editors: Excitatory synaptic mechanisms, Oslo, 1970, Universitetsforlaget.
2. Baldessarini, R. J.: Biogenic amines and behavior, Ann. Rev. Med. **23**:343, 1972.
3. Barondes, S. H.: Axoplasmic transport, Neurosci. Res. Progm. Bull. **5**:entire issue, 1967.
4. Brazier, M. A. B., editor: The interneuron, Berkeley, 1969, University of California Press.
5. Eccles, J. C.: The physiology of synapses, Berlin, 1964, Springer Verlag.
6. Geffen, L. B., and Livett, B. G.: Synaptic vesicles in sympathetic neurons, Physiol. Rev. **51**:98, 1971.
7. Horridge, G. A.: Interneurons, their origin, action, specificity, growth, and plasticity, San Francisco, 1968, W. H. Freeman & Co., Publishers.
8. Hubbard, J. I., Llinas, R., and Quastel, D. M. J.: Electrophysiological analysis of synaptic transmission, Baltimore, 1969, The Williams & Wilkins Co.
9. Iversen, L. L.: The uptake and storage of noradrenaline in sympathetic nerves, London, 1967, Cambridge University Press.
10. Kuffler, S. W.: Neuroglial cells: physiological properties and a potassium mediated effect of neuronal activity on the glial membrane potential (the Ferrier Lecture), Proc. R. Soc. Lond. (Biol.) **168**:1, 1968.
11. Kuno, M.: Quantum aspects of central and ganglionic synaptic transmission, Physiol. Rev. **51**:647, 1971.

11a. Pappas, G. D., and Purpura, D. P., editors: Structure and function of synapses, New York, 1972, Raven Press.

12. Phillis, J. W.: The pharmacology of synapses, Oxford, 1970, Pergamon Press, Ltd.
13. von Euler, C., Skoglund, S., and Soderberg, U., editors: Structure and function of inhibitory synaptic mechanisms, New York, 1968, Pergamon Press, Inc.

Original papers

14. Aghajanian, G., and Bloom, F.: Localization of tritiated serotonin in rat brain by electron-microscopic autoradiography, J. Pharmacol. Exp. Ther. **156**:23, 1967.
15. Aghajanian, G., Rosecrans, J., and Sheard, M.: Serotonin: release in the forebrain by stimulation of midbrain raphe, Science **156**: 402, 1967.
16. Akert, K., Moor, H., Pfenninger, K., and Sandri, C.: Contributions of new impregnation methods and freeze etching to the problems of synaptic fine structure, Progr. Brain Res. **31**:223, 1969.
17. Anden, N. E., et al.: Ascending monoamine neurons to the telencephalon and diencephalon, Acta Physiol. Scand. **67**:313, 1966.
18. Baker, R., and Llinas, R.: Electrotonic coupling between neurones in the rat mesencephalic nucleus, J. Physiol. **212**:45, 1971.
19. Baldessarini, R. J., and Kopin, I.: The electrical release of norepinephrine-H^3 from rat brain slices, J. Pharm. Exp. Ther. **156**:31, 1967.
20. Banks, P., Mayor, D., Mitchell, M., and Tomlinson, D.: Studies on the translocation of noradrenaline-containing vesicles in postganglionic sympathetic neurones in vitro. Inhibition of movement by colchicine and vinblastine and evidence for the involvement of axonal microtubules, J. Physiol. **216**:625, 1971.
21. Barondes, S. H.: Further studies of the transport of protein to nerve endings, J. Neurochem. **15**:599, 1968.
22. Bennett, M. V. L.: Similarities between chemically and electrically mediated transmission. In Carlson, F. D., editor: Physiological and biochemical aspects of nervous integration, Englewood Cliffs, N. J., 1968, Prentice-Hall, Inc.
23. Blankenship, J. E., and Kuno, M.: Analysis of spontaneous subthreshold activity in spinal motoneurons of the cat, J. Neurophysiol. **31**: 195, 1968.

23a. Bodian, D.: A suggestive relationship of nerve cell RNA with specific synaptic sites, Proc. Natl. Acad. Sci. U. S. A. **53**:418, 1965.

24. Bodian, D.: An electron microscopic characterization of classes of synaptic vesicles by means of controlled aldehyde fixation, J. Cell Biol. **44**:115, 1970.
25. Bodian, D.: Neuron junctions: a revolutionary decade, Anat. Rec. **174**:73, 1972.
26. Bradford, H. F.: Metabolic response of synaptosomes to electrical stimulation: release of amino acids, Brain Res. **19**:239, 1970.
27. Brightman, M. W., and Reese, T. S.: Junctions between intimately apposed cell membranes in the vertebrate brain, J. Cell Biol. **40**:648, 1969.
28. Bullock, T. H., and Hagiwara, S.: Intracellular recording from the giant synapse of the squid, J. Gen. Physiol. **40**:565, 1957.
29. Bullock, T. H., and Horridge, G. A.: Structure and function in the nervous systems of invertebrates, San Francisco, 1965, W. H. Freeman & Co., Publishers.

30. Burke, R. E.: Composite nature of the monosynaptic excitatory postsynaptic potential, J. Neurophysiol. **30**:1114, 1967.
31. Castelluci, V. F., and Goldring, S.: Contribution to steady potential shifts of slow depolarization in cells presumed to be glia, Electroencephalogr. Clin. Neurophysiol. **28**: 109, 1970.
32. Chase, T., Katz, R., and Kopin, I.: Release of (^{3}H)serotonin from brain slices, J. Neurochem. **16**:607, 1969.
33. Coggleshall, R. E., and Fawcett, D. W.: The fine structure of the central nervous system of the leech, Hirudo medicinalis, J. Neurophysiol. **27**:229, 1964.
34. Colonnier, M.: Synaptic patterns on different cell types in the different laminae of the cat visual cortex, an electron microscope study, Brain Res. **9**:268, 1968.
34a. Coombs, J. S., Curtis, D. R., and Eccles, J. C.: The generation of impulses in motoneurones, J. Physiol. **139**:232, 1957.
35. Coombs, J. S., Eccles, J. C., and Fatt, P.: Excitatory synaptic action in motoneurones, J. Physiol. **130**:374, 1955.
36. Creed, R. S., et al.: Reflex activity of the spinal cord, Oxford, 1932, Clarendon Press.
37. Curtis, D. R.: The pharmacology of central and peripheral inhibition, Pharmacol. Rev. **15**: 333, 1963.
37a. Curtis, D. R., and Eccles, J. C.: The time courses of excitatory and inhibitory synaptic actions, J. Physiol. **145**:529, 1959.
38. Curtis, D. R., and Eccles, J. C.: Synaptic action during and after repetitive stimulation, J. Physiol. **150**:374, 1960.
39. Curtis, D. R., and Ryall, R. W.: The acetylcholine receptors of Renshaw cells, Exp. Brain Res. **2**:66, 1966.
40. Curtis, D. R., and Watkins, J.: The pharmacology of amino acids related to gamma-aminobutyric acid, Pharmacol. Rev. **17**:347, 1965.
41. Dahlstrom, A.: Axoplasmic transport (with particular respect to adrenergic neurons), Trans. Philos. R. Soc. Lond. (Biol. Sci.) **261**: 325, 1971.
42. Dale, H. H.: Transmission of nervous effects by acetylcholine, Harvey Lect. **32**:229, 1937.
43. Dennis, M. J., Harris, A. J., and Kuffler, S. W.: Synaptic transmission and its duplication by locally applied acetylcholine in parasympathetic neurons in the heart of the frog, Proc. R. Soc. Lond. (Biol.) **177**:509, 1971.
44. Dudel, J., and Kuffler, S. W.: Presynaptic inhibition at the crayfish neuromuscular junction, J. Physiol. **155**:543, 1961.
45. Ebbesson, S. O. E.: Quantitative studies of superior cervical sympathetic ganglia in a variety of primates including man. I. The ratio of preganglionic fibers to ganglionic neurons, J. Morphol. **124**:117, 1968.
45a. Eccles, J. C.: Excitatory and inhibitory synaptic action, Harvey Lect **51**:1, 1955.
45b. Eccles, J. C.: The central action of antidromic impulses in motor nerve fibres, Arch. Gesamte Physiol. **260**:385, 1955.
45c. Eccles, J. C.: Modes of communications between nerve cells, Australian Academy of Science year book, Sydney, 1963, Waite & Bull.
46. Eccles, J. C., Eccles, R. M., and Lundberg, A.: The action potentials of the alpha motoneurons supplying fast and slow muscles, J. Physiol. **142**:275, 1958.
47. Eccles, R. M.: Intracellular potentials recorded from a mammalian sympathetic ganglion, J. Physiol. **130**:572, 1955.
48. Eccleston, D., Randic, M., Roberts, H., and Stranghan, D.: Release of amines and amine metabolites from brain by neural stimulation. In Hooper, G., editor: Metabolism of amines in brain, New York, 1969, The Macmillan Co.
49. Fahn, S., Rodman, J., and Cote, L.: Association of tyrosine hydroxylase with synaptic vesicles in bovine caudate nucleus, J. Neurochem. **16**:1293, 1969.
50. Feldberg, W., and Vogt, M.: Acetylcholine synthesis in different regions of the central nervous system, J. Physiol. **107**:372, 1948.
51. Fonnum, F., Storm, A., Mathisen, J., and Walberg, F.: Glutamate decarboxylase in inhibitory neurons, Brain Res. **20**:259, 1970.
52. Friede, R. L.: Relationship of body size, nerve cell size, axon length and glial density in the cerebellum, Proc. Natl. Acad. Sci. U. S. A. **49**:187, 1963.
53. Friede, R. L, and van Houten, W. H.: Neuronal extension and glial supply: functional significance of glia, Proc. Natl. Acad. Sci. U. S. A. **48**:817, 1962.
54. Fuortes, M. G. F., Frank, K., and Becker, M. E.: Steps in the production of motoneuron spikes, J. Gen. Physiol. **40**:735, 1957.
55. Furshpan, E. J., and Potter, D. D.: Transmission at the giant synapses of the crayfish, J. Physiol. **145**:289, 1959.
56. Fuxe, K., Hokfelt, T., Jonsson, G., and Ungerstedt, U.: Fluorescence microscopy in neuroanatomy. In Nauta, W. J., and Ebbesson, S., editors: Contemporary research methods in neuroanatomy, New York, 1970, Springer-Verlag, New York, Inc.
57. Glowinski, J.: Release of monoamines in the central nervous system. In Shumann, H. J., and Kroneberg, G., editors: New aspects of storage and release mechanisms of cathecholamines, Bayer symposium II, Berlin, 1970, Springer Verlag.
58. Glowinski, J., and Baldessarini, R. J.: The metabolism of norepinephrine in the central nervous system, Pharmacol. Rev. **18**:1201, 1966.
59. Grafstein, B.: Axonal transport: communication between soma and synapse. In Costa, E., and Greengard, P., editors: Advances in biochemical psychopharmacology, New York, 1969, Raven Press, vol. 1.
60. Grafstein, B.: Transneuronal transfer of radioactivity in the central nervous system, Science **172**:177, 1971.
61. Gray, E. G.: Electron microscopy of excitatory and inhibitory synapses: a brief review, Prog. Brain Res. **31**:141, 1969.
62. Gray, E. G., and Guillery, R. W.: Synaptic morphology in the normal and degenerating nervous system, Int. Rev. Cytol. **19**:111, 1966.
63. Green, J. P.: Histamine. In Lajtha, A., editor: Handbook of neurochemistry, New York, 1970, Plenum Publishing Corp., vol. 4.
64. Greengard, P., and Costa, E., editors: Role

of cyclic AMP in cell function, Adv. Biochem. Psychopharmacol. **3:**entire issue, 1970.
65. Grossman, R. G., and Hampton, T.: Relationships of cortical glial cell depolarizations to electrocortical surface wave activity, Electroencephalogr. Clin. Neurophysiol. **28:**90, 1970.
66. Grossman, R. G., Whiteside, L., and Hampton, T. L.: The time course of evoked depolarization of cortical glial cells, Brain Res. **14:**401, 1969.
67. Guth, L.: "Trophic" influences of nerve on muscle, Physiol. Rev. **48:**645, 1968.
68. Hagiwara, S., and Tasaki, I.: A study on the mechanism of impulse transmission across the giant synapse of the squid, J. Physiol. **143:**114, 1958.
69. Harris, A. J., Kuffler, S. W., and Dennis, M. J.: Differential chemosensitivity of synaptic and extrasynaptic areas on the neuronal surface membrane in parasympathetic neurons of the frog, tested by microapplication of acetycholine, Proc. R. Soc. Lond. (Biol.) **177:**541, 1971.
70. Hokfelt, T., and Ljungdahl, A.: Cellular localization of labelled gamma-aminobutyric acid (^{3}H-GABA) in rat cerebellar cortex: an autoradiographic study, Brain Res. **22:**391, 1970.
71. Hopkin, J., and Neal, M. J.: Effect of electrical stimulation and high potassium concentrations on the efflux of (^{14}C) glycine from slices of spinal cord, Br. J. Pharmacol. **42:**215, 1971.
72. Hortnagl, H., Hortnagl, H., and Winkler, H.: Bovine splenic nerve: characterization of noradrenaline-containing vesicles and other cell organelles by density gradient centrifugation, J. Physiol. **205:**103, 1969.
73. Hughes, J. R.: Post-tetanic potentiation, Physiol. Rev. **38:**91, 1958.
74. Hultborn, J., Jankowska, E., and Lindstrom, S.: Relative contribution from different nerves to recurrent depression of Ia IPSP's in motoneurons, J. Physiol. **215:**637, 1971.
75. Hunt, C. C., and Kuno, M.: Properties of spinal interneurones, J. Physiol. **147:**346, 1959.
76. Hyden, H.: A molecular basis of neuron-glia interaction. In Schmitt, F. O., editor: Macromolecules and biological memory, Cambridge, Mass., 1962, The M. I. T. Press.
77. Hyden, H.: Biochemical and functional interplay between neuron and glia. In Wortis, J., editor: Recent advances in biological psychiatry, New York, 1964, Plenum Publishing Corp., vol. 6.
78. Hyden, H.: Changes in RNA content and base composition in cortical neurons of rats in a learning experiment involving transfer of handedness, Proc. Natl. Acad. Sci. U. S. A. **52:**1030, 1965.
79. Hyden, H., and Lange, P. W.: Rhythmic enzyme changes in neurons and glia during sleep, Science **149:**654, 1965.
80. Ichyama, A., Nakamura, S., Nishizuka, Y., and Hayaishi, O.: Enzymatic studies on the biosynthesis of serotonin in mammalian brain, J. Biol. Chem. **245:**1699, 1970.
81. Innes, I. R., and Nickerson, M.: Drugs acting on postganglionic adrenergic nerve endings and structures innervated by them (sympathomimetic drugs). In Goodman, L. S., and Gilman, A., editors: The pharmacological basis of therapeutics, New York, 1970, The Macmillan Co.
82. Jack, J. J. B., Miller, S., Porter, R., and Redman, S. J.: The time course of minimal excitatory post-synaptic potentials evoked in spinal motoneurones by group Ia fibres, J. Physiol. **215:**353, 1971.
83. Jasper, H., and Koyama, I.: Amino acids released from the cortical surface in cats following stimulation of the mesial thalamus and mid-brain reticular formation, Electroencephalogr. Clin. Neurophysiol. **24:**292, 1968.
84. Kandel, E. R., and Kupfermann, I.: The functional organization of invertebrate ganglia, Ann. Rev. Physiol. **32:**193, 1970.
85. Kataoka, K., Ramwell, P., and Jessup, S.: Prostaglandins: localization in subcellular particles of rat cerebral cortex, Science **157:**1187, 1967.
86. Koelle, G., and Foroglou-Kerameos, C.: Electron microscopic localization of cholinesterases in a sympathetic ganglion by a goldthiolactic acid method, Life Sci. **4:**417, 1965.
87. Kravitz, E. A., Molinoff, P. B., and Hall, Z. W.: A comparison of the enzymes and substrates of gamma-amino-butyric acid metabolism in lobster excitatory and inhibitory neurons, Proc. Natl. Acad. Sci. U. S. A. **54:**778, 1965.
88. Krnjevic, K., and Schwartz, S.: Some properties of unresponsive cells in the cerebral cortex, Exp. Brain Res. **3:**306, 1967.
89. Kuffler, S. W., and Nicholls, J. G.: The physiology of neuroglial cells, Ergeb. Physiol. **57:**1, 1966.
90. Kuffler, S. W., and Potter, D. D.: Glia in the leech central nervous system. Physiological properties and the neuron-glia relationship, J. Neurophysiol. **27:**290, 1964.
91. Kuffler, S. W., Dennis, M. J., and Harris, A. J.: The development of chemosensitivity in extrasynaptic areas of the neuronal surface after denervation of parasympathetic ganglion cells in the heart of the frog, Proc. R. Soc. Lond. (Biol.) **177:**555, 1971.
92. Kuffler, S. W., Nicholls, J. G., and Orkand, R.: Physiological properties of glial cells in the central nervous system of amphibia, J. Neurophysiol. **29:**768, 1966.
93. Kuhar, M., and Snyder, S.: The subcellular distribution of free ^{3}H-glutamic acid in rat cerebral cortical slices, J. Pharmacol. Exp. Ther. **171:**141, 1970.
94. Kuhlenkampf, H.: Verhalten der Neuroglia in den Vorderhorern des Ruckenmarkes der weissen Maus unter dem Reiz physiolopischen Tatigkeit, Z. Anat. Entwicklungsgesch. **116:**304, 1952.
95. Kuno, M., and Llinas, R.: Alterations of synaptic action in chromatolysed motoneurones of the cat, J. Physiol. **210:**823, 1970.
96. Kuriyama, K., Roberts, E., and Kakefuda, T.: Association of the gamma-amino-butyric acid system with a synaptic vesicle fraction from mouse brain, Brain Res. **8:**132, 1968.
97. Larrabee, M. G., and Bronk, D. W.: Prolonged facilitation of synaptic excitation in

sympathetic ganglia, J. Neurophysiol. **10:**139, 1947.

97a. Larrabee, M. G., and Bronk, D. W.: Metabolic requirements of sympathetic neurons, Symp. Quant. Biol. **17:**245, 1952.

98. Lasansky, A.: Nervous function at the cellular level: glia, Ann. Rev. Physiol. **33:**241, 1971.

99. Lemback, F., and Zetler, G.: Substance P: a polypeptide of possible physiological significance, especially within the central nervous system, Int. Rev. Neurobiol. **4:**159, 1962.

100. Libet, B.: Generation of slow inhibitory and excitatory postsynaptic potentials, Fed. Proc. **29:**1945, 1970.

101. Lin, R., et al.: In vivo measurement of 5-hydroxytryptamine turnover rate in the rat brain from the conversion of C^{14}-tryptophan to C^{14}-5-hydroxytryptamine, J. Pharmacol. Exp. Ther. **170:**232, 1969.

102. Lloyd, D. P. C.: Post-tetanic potentiation of response in monosynaptic pathways of the spinal cord, J. Gen. Physiol. **33:**147, 1949.

103. Lundberg, A.: Convergence of excitatory and inhibitory action on interneurons in the spinal cord. In Brazier, M. A. B., editor: The interneuron, Berkeley, 1969, University of California Press.

104. Maas, J., and Landis, D.: In vivo studies of the metabolism of norepinephrine in the central nervous system, J. Pharmacol. Exp. Ther. **163:**147, 1968.

105. McLennan, H.: The release of acetycholine and 3-hydroxytryptamine from the caudate nucleus, J. Pnysiol. **174:**152, 1964.

106. McMahan, U. J., and Kuffler, S. W.: Visual identification of synaptic boutons on living ganglion cells and of varicosities in postganglionic axons in the heart of the frog, Proc. R. Soc. Lond. (Biol.) **177:**485, 1971.

107. Miledi, R.: Spontaneous synaptic potentials and quantal release of transmitter in the stellate ganglion of the squid, J. Physiol. **192:** 379, 1967.

108. Mitchell, J. F.: Acetylcholine release from the brain. In von Euler, U. S., Rosell, S., and Uvnas, B., editors: Mechanisms of release of biogenic amines, Oxford, 1966, Pergamon Press, Ltd.

109. Monnier, M., Sauer, R., and Hatt, A.: The activating effect of histamine on the central nervous system. Int. Rev. Neurobiol. **12:**265, 1970.

110. Neal, M., and Iversen, L.: Subcellular distribution of endogenous and ^{3}H-gamma-aminobutyric acid in rat cerebral cortex, J. Neurochem. **16:**1245, 1969.

111. Nicholls, J. G., and Kuffler, S. W.: Extracellular space as a pathway for exchange between blood and neurons in central nervous system of leech: the ionic composition of glial cells and neurons, J. Neurophysiol. **27:**645, 1964.

112. Nicholls, J. G., and Kuffler, S. W.: Na and K content of glial cells and neurons determined by flame photometry in the central nervous system of the leech, J. Neurophysiol. **28:**519, 1965.

113. Orkand, R. K., Nicholls, J. G., and Kuffler, S. W.: The effect of nerve impulses on the membrane potential of glial cells in the central nervous system of amphibia, J. Neurophysiol. **29:**788, 1966.

114. Otsuka, M., Iversen, L. L., Hall, Z. W., and Kravitz, E. A.: Release of gamma-aminobutyric acid from inhibitory nerves of lobster, Proc. Natl. Acad. Sci. U. S. A. **56:**1110, 1966.

115. Pellegrino de Iraldi, A., Zieher, L. M., and Etcheverry, G. J.: Neuronal compartmentation of 5-hydroxytryptamine stores, Adv. Pharmacol. **6:**257, 1968.

116. Phillipu, A., Heyd, G., and Burger, A.: Release of noradrenaline from the hypothalamus in vivo, Eur. J. Pharmacol. **9:**52, 1970.

117. Pomerat, C. M.: Functional concepts based on tissue culture studies of neuroglia. In Windle, W. F., editor: Biology of neuroglia, Springfield, Ill., 1958, Charles C Thomas, Publisher.

118. Portig, P. J., and Vogt, M.: Release into the cerebral ventricles of substances with possible transmitter functions in the caudate nucleus, J. Physiol. **204:**687, 1969.

119. Rall, W.: Distinguishing theoretical synaptic potentials computed for different soma-dendritic distributions of synaptic input, J. Neurophysiol. **30:**1138, 1967.

120. Rall, W., et al.: Dendritic location of synapses and possible mechanisms for the monosynaptic EPSP in motoneurons, J. Neurophysiol. **30:** 1169, 1967.

121. Ramón y Cajal, S.: Neuron theory or reticular theory? Objective evidence of the anatomical unity of nerve cells (translated by M. U. Purkiss, and C. A., Fox), Madrid, 1954, Consejo Superior de Investigaciones Cientifices.

122. Ramwell, P., and Shaw, J.: Spontaneous and evoked release of prostaglandins from the cerebral cortex of anaesthetized cats, Am. J. Physiol. **211:**125, 1966.

123. Ringham, G. L.: Origin of nerve impulse in slowly adapting stretch receptor in crayfish, J. Neurophysiol. **34:**773, 1971.

124. Roberts, E., and Kuriyama, K.: Biochemical-physiological correlations in studies of the gamma-aminobutyric acid system, Brain Res. **8:**1, 1968.

125. Ryall, R. W.: The subcellular distributions of acetylcholine, substance P., 5-hydroxytryptamine, gamma-aminobutyric acid and glutamic acid in brain homogenates, J. Neurochem. **11:** 131, 1964.

126. Salmoiraghi, G. C., Costa, E., and Bloom, F. E.: Pharmacology of central synapses, Ann. Rev. Pharmacol. **5:**213, 1965.

127. Samuelson, B.: The prostaglandins, Angew. Chem. **4:**410, 1965.

128. Schmidt, R. F., Senges, J., and Zimmerman, M.: Determination of the peripheral receptive field and excitability measurements of the central terminals of single mechanoreceptive afferents, Exp. Brain Res. **3:**220, 1967.

129. Sherrington, C. S.: The integrative action of the nervous system, New Haven, 1906, Yale University Press.

130. Smith, D. S.: On the significance of cross-bridges between microtubules and synaptic vesicles, Philos. Trans. R. Soc. Lond. (Biol. Sci.) **261:**395, 1971.

131. Snyder, S., Baldessarini, R. J., and Axelrod,

A.: A sensitive and specific enzymatic isotopic assay for tissue histamine, J. Pharmacol. Exp. Ther. **153:**544, 1966.

132. Spector, S., Gordon, R., Sjoerdsma, A., and Udenfriend, S.: End-product inhibition of tyrosine hydroxylase as a possible mechanism for regulation of norepinephrine synthesis, Mol. Pharmacol. **3:**549, 1967.
133. Srinivasan, V., Neal, M. J., and Mitchell, J. F.: The effect of electrical stimulation and high potassium concentrations on the efflux of (^{3}H)γ-aminobutyric acid from brain slices, J. Neurochem. **16:**1235, 1969.
134. Stein, L., and Wise, C. D.: Release of norepinephrine from hypothalamus and amygdala by rewarding medial forebrain bundle stimulation and amphetamine, J. Comp. Physiol. Psychol. **67:**189, 1969.
135. Sutherland, E., and Rall, T.: The relation of adenosine-3′,5′-phosphate and phosphorylase to the actions of catecholamines and other hormones, Pharmacol. Rev. **12:**265, 1960.

135a. Taylor, K. M., and Snyder, S. H.: Isotopic microassay of histamine, histidine, histidine decarboxylase, and histamine methyl transferase in brain tissue, J. Neurochem. **19:**1343, 1972.

136. Terzuolo, C., and Bullock, T. H.: Measurement of voltage gradient across a neuron adequate to modulate its firing, Proc. Natl. Acad. Sci. U. S. A. **42:**687, 1956.
137. van Pragg, H. M.: Indoleamines in the central nervous system, Psychiatr. Neurol. Neurochir. **73:**9, 1970.
138. von Euler, U. S.: Noradrenaline, Springfield, Ill., 1956, Charles C Thomas, Publisher.
139. Wachtel, H., and Kandel, E. R.: Conversion of synaptic excitation to inhibition at a dual chemical synapse, J. Neurophysiol. **24:**56, 1971.
140. Wall, P. D.: Excitability changes in afferent fibre terminations and their relation to slow potentials, J. Physiol. **142:**1, 1958.
141. Weight, F. F., and Votava, J.: Slow synaptic excitation in sympathetic ganglion cells: evidence for synaptic inactivation of potassium conductance, Science **170:**755, 1970.
142. Weiss, P. A.: The concept of perpetual neuronal growth and proximo-distal substance convection. In Kety, S. S., and Elkes, J., editors: Regional neurochemistry, Oxford, 1961, Pergamon Press, Ltd.
143. Weiss, P. A., and Hiscoe, H. B.: Experiments on the mechanism of nerve growth, J. Exp. Zool. **107:**315, 1948.
144. Whittaker, V. P.: The subcellular fractionation of nervous tissue. In Bourne, G. H., editor: The structure and function of the nervous system, New York, 1961, Academic Press, Inc., vol. 3.
145. Wigglesworth, V. B.: The nutrition of the central nervous system in the cockroach Periplaneta americana. I. The role of the perineurium and glial cells in the mobilization of reserves, J. Exp. Biol. **37:**500, 1960.

III

GENERAL PHYSIOLOGY OF THE FOREBRAIN

7

VERNON B. MOUNTCASTLE and GIAN F. POGGIO

Structural organization and general physiology of thalamotelencephalic systems

In previous chapters the mechanisms of axonal conduction and synaptic transmission have been considered in some detail. The ground was thus laid for study of the CNS, which is introduced in this and the following chapter with a discussion of the general aspects of thalamocortical structure and function, and of those systems of the forebrain concerned with controlling levels of excitability, and thus awareness. A nervous system is a collection of cells specialized to convey signals in rapid tempo and with great fidelity. Its forte is the transmission in neural code of information concerning the state of the body's internal measuring devices, of the immediate external environment, and of more distant objects and events in the external world; and the dispatch of command signals to effectors for corrective action and—at will, independent of input—for the initiation of action on the environment. Between these input and output channels there intervenes the working brain, an information-processing, data-storing, decision-making machine: the executant organ of behavior. The complexity of that behavior sets animals endowed with nervous systems apart from other forms of life, and in animal phylogeny increasing complexity of behavior is in general paralleled by an increasing size and complexity of brains. The functions of brains fall into two classes—homeostatic regulation and the initiation of action—though many functions show properties of both.

In the domain of homeostatic regulation, for example, are mechanisms that control body temperature, maintain appropriate levels of blood pressure and pulmonary ventilation, regulate steady and cyclic endocrine function, and preserve or restore an erect body position in the face of the collapsing force of gravity. They are in general invariant and automatic in that they occur without conscious volition: they are reflexes, stereotyped items of behavior destined for execution upon appropriate signal unless consciously withheld, a restraint possible only for those involving somatic musculature. In the domain of initiation of action lie those items of exploratory and appetitive behavior that apparently begin independently of any external evoking stimulus. Although these may serve regulatory ends, at certain levels of complexity they appear to be truly spontaneous and mark that independence and freedom from environmental contingencies characteristic of animals with large brains. Along this spectrum they merge with those activities that are free of both stimulus from and action upon the environment—learning, thinking, remembering—actions known only by introspective evaluation and, at choice, public description.

Study of brains from small to large reveals no striking phylogenetic innovations in cellular morphology or synaptic mechanisms.[2] What is noteworthy along this scale is the increasing number of neuronal units, the extensive elaboration of the number and complexity of their interconnections,[28] and the flowering of interconnected systems and subsystems. Thus the emerging intricacies of brain function will find explanation, it is believed, in the emergent functional capacities of large populations of neural elements, properties not immediately and completely deducible from those of single cells, so far as they are presently known. Thus attention is first directed to the general properties of the largest accumulation of cells, the forebrain, and to the systems and subsystems that exert general, not specific, control of its function.

ANATOMIC ORGANIZATION OF THALAMOTELENCEPHALIC SYSTEMS

The concept of the "sensory system" is abstract but useful for indicating those populations of neurons concerned with sensing certain classes of stimuli and transmitting information about them. Each sensory system originates in a peripheral sheet of receptors and ascends through the nervous system to reach the cerebral cortex; it is composed of a series of neural populations that are synaptically connected in the direction from periphery to center. At each level such a population is arranged in an orderly topographic manner and constitutes a replica in neural space of the spatial density of the receptive sheet itself. With the exception of the olfactory, all these systems project respectively upon one or another region of the dorsal thalamus. These are the so-called thalamic "relay" nuclei, the axons of whose cells project in turn upon the "sensory" areas of the cerebral cortex. *The thalamus is the afferent gateway to the cerebral cortex.* Other nuclear regions of the thalamus are interconnected with neural structures, both cortical and subcortical, that are known to play important roles in somatic motor and autonomic functions. Thus the thalamus and cerebral cortex, with their complex interconnections, constitute an anatomic substratum essential for our perception of the external world and our actions upon it.

By contrast, other thalamocortical systems serve a more general function, for by their modes of distribution and termination they are suitably disposed to control the level of excitability of cortical neurons and, indeed, those of subcortical regions as well. The multiple convergence upon these systems precludes their role as sensory systems in the manner just described. Thus an important generalization is obvious—*a duality exists among thalamocortical systems, many being organized precisely as information-signaling systems and others serving energizing and regulating functions.*

Development of the thalamus

At early stages of development the paired thalamic primordium lies in the diencephalic wall of the third ventricle, separated from the more ventral hypothalamus by the sulcus

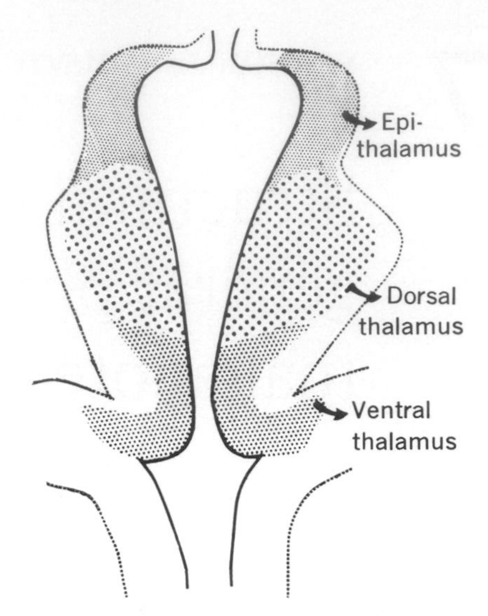

Fig. 7-1. Diagrammatic outline of oral end of thalamic plate from an 18 mm rabbit embryo. The three subdivisions are shown. (Modified from Rose and Woolsey[57]; from Ajmone Marsan.[1])

hypothalamicus. Within the thalamic plate, three areas differentiate: the dorsal area is termed the *epithalamus;* the middle and larger, the *dorsal thalamus;* and the most ventral, the *ventral thalamus* (Fig. 7-1). Within each of these areas clusters of cells differentiate at various times during ontogenesis, the dorsal thalamus being still largely undifferentiated when the nuclei of the epithalamus and ventral thalamus have already developed completely.

General classification of thalamic nuclei and their telencephalic projections

In the adult the thalamic mass may be subdivided on the basis of topographic relations, by differences in the cyto- and myeloarchitectural characteristics of different parts, as well as by the fiber connections between thalamic parts and the remainder of the CNS. Such a division is given in Table 7-1, a classification derived from the studies of Crouch,[30] Walker,[68] and Olszewski[44] on the monkey thalamus and from the studies of Rose and Woolsey[58] on a representative series of mammals. More elaborate classifications are available for the human being (e.g., Hassler[6]), based mainly on a finer nuclear parcellation. The thalamotelencephalic organization in subhuman primates is not thought to differ significantly from that of man, and the description that follows deals in the main with that organization in the monkey. It is thought to be typical of the primate pattern.

Table 7-1. Thalamic subdivisions and nuclear classification

Region	Subdivision	Nuclei	System
Epithalamus		N. habenularis, medialis and lateralis (Hb) N. paraventricularis, anterior and posterior (Pv) Regio pretectalis (PT)	
Dorsal thalamus	Anterior	N. anteromedialis (AM) N. anteroventralis (AV) N. anterodorsalis (AD)	Specific system
	Lateral—Ventral	N. ventralis anterior (VA) N. ventralis lateralis (VL) N. ventralis posterior N. ventralis posterolateralis (VPL) } (VB) N. ventralis posteromedialis (VPM) } (VB) N. ventralis posteromedialis parvicellularis (VPMpc) N. ventralis posteroinferior (VPI)	Specific system
	Lateral—Dorsolateral	N. lateralis dorsalis (LD) N. lateralis posterior (LP) Pulvinar (Pul)	Specific system
	Posterior	N. geniculatus medialis (GM) N. geniculatus lateralis dorsalis (GLd)	Specific system
	Medial	N. medialis dorsalis (MD)	Specific system
		Nuclei of the midline N. reuniens (Re) N. paratenialis (Pt) Massa intermedia (MI)	Generalized system
	Intralaminar	N. centralis medialis (Cm) N. paracentralis (Pc) N. centralis lateralis (Cl) N. centrum medianum (CM) N. parafascicularis (Pf) N. suprageniculatus (Sg) N. limitans (Li)	Generalized system
Ventral thalamus		N. reticularis (R) N. geniculatus lateralis ventralis (GLv)	

Differences exist between the three primordial thalamic areas in their connections with the telencephalon. Epithalamic nuclei do not degenerate after telencephalic removal: the axons of their cells terminate elsewhere. Dorsal thalamic nuclei, on the other hand, do degenerate after such removal, for the axons of their cells are then severed. The cells of the majority of these nuclei send their axons to the cerebral cortex itself and make up the thalamocortical projections. A smaller number of nuclei project upon the putamen and caudate. The relations of the ventral thalamic nuclei and the telencephalon appear to differ from those of the dorsal thalamus.

Epithalamus. Nuclei of epithalamic origin are the habenular complex, the paraventricular complex, and the pretectal group of nuclei, no one of which sends axons to the endbrain. The habenular complex is thought to be related to the olfactory system, for its main afferent connections are through the stria medullaris from the septal nuclei, from portions of the amygdala, and perhaps from the hypothalamus as well. The largest of its efferent pathways terminates in the n. interpeduncularis of the midbrain, which in turn projects upon brainstem structures. The connections of the paraventricular nucleus are not known.

Ventral thalamus. The relations between the two nuclei of ventral thalamic origin and the telencephalon are incompletely understood. The ventral lateral geniculate nucleus remains intact after removal of the endbrain, and its precise projections are unknown. The reticular nucleus is a sheath of cells that surrounds the anterior and lateral aspects of the dorsal thalamus, intercalated between the external medullary lamina and the internal capsule. It is traversed by all thalamocortical and corticothalamic fibers, which produce its reticulated appearance. The prevailing orientation of the den-

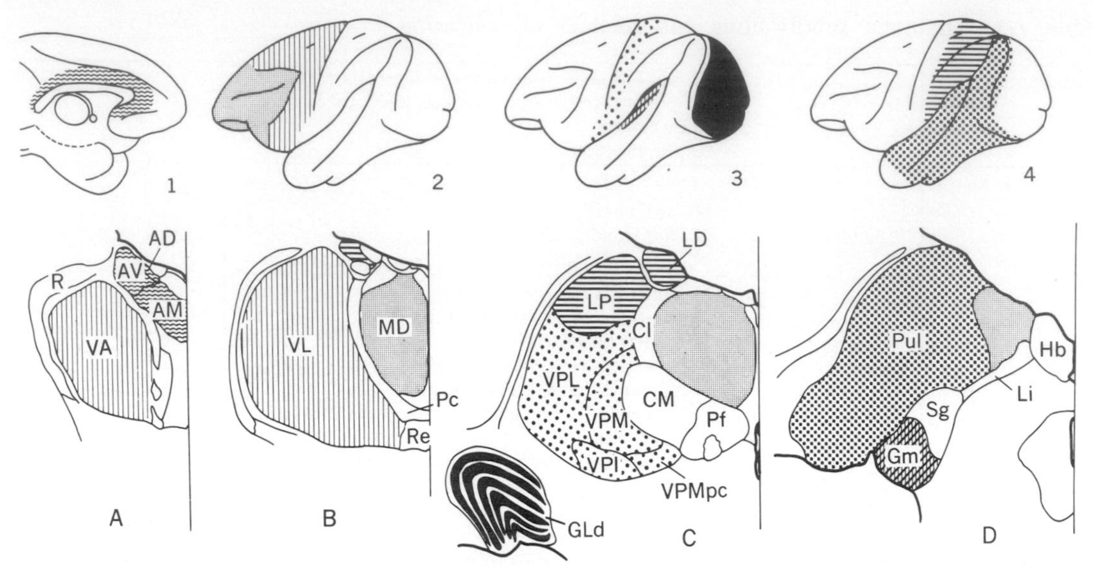

Fig. 7-2. Schematic diagram of anatomic organization of specific thalamocortical systems. Drawings at bottom outline nuclear configuration of monkey thalamus, and those at top, the surface of cerebral hemisphere. Areas similarly marked indicate thalamic nuclear groups and cortical areas to which they project. Sketches of thalamus are of coronal sections 2.1 mm apart in a frontocaudal sequence from **A** to **D** and are derived from plates of the stereotaxic atlas of macaque thalamus by Olszewski.[44] Habenular nuclei, n. centrum medianum, and nuclei of generalized thalamocortical system are not marked with symbols. Abbreviations used are listed in Table 7-1. **1** and **2** show cortical regions that receive connections from thalamic nuclei, which are part of systems concerned with efferent control mechanisms. In **3,** cortical projection areas of sensory nuclei are indicated, and in **4,** those of thalamic structures concerned with high-order integrative mechanisms (see text).

dritic fields of reticular cells is perpendicular to the course of the traversing fibers, from which they receive extensive innervation via axon collaterals. Reticular cell axons project into the dorsal thalamus, where they branch very widely and synapse with neurons of both specific and generalized (intralaminar) thalamic nuclei. Thus the reticular nucleus is optimally placed to regulate the activity of thalamic cells by feeding back upon them some integral of their own activity, and that of the corticothalamic systems projecting upon them.[24, 40, 61]

Dorsal thalamus. The dorsal thalamus is the largest portion of the thalamic mass, and the greatest number of nuclei originate from it (Table 7-1 and Fig. 7-2). The internal medullary lamina separates the dorsal thalamus into medial and lateral nuclear regions and bifurcates dorsally to enclose the anterior group of nuclei. Within the lateral and larger portion, ventral and dorsolateral groups of nuclei are recognized. A group of smaller nuclei differentiates in or about the internal medullary lamina; these are thought to have similar functional properties and together make up the intralaminar nuclei. Two major structures, the medial and lateral geniculate bodies, develop early in fetal life in the caudal thalamus. The external medullary lamina bounds the lateral aspect of the dorsal thalamus and separates it from n. reticularis.

All major sensory afferent systems except the olfactory system terminate upon dorsal thalamic nuclei, as do important elements of the motor system, the cerebellar and pallidal afferents to the thalamus. These nuclei also receive input from other subcortical structures and from the cerebral cortex. It has already been indicated that their efferent axons are telencephalic in destination, terminating in the neocortex (largely), in the striatum, and in other subcortical structures. Corticothalamic connections in general parallel reciprocally the thalamocortical ones; i.e., they project from the cortical area to the dorsal thalamic nucleus that projects upon that area, although exceptionally they may have a wider thalamic distribution.

Essential and sustaining thalamocortical projections

Two major types of thalamocortical projections have been defined by Rose and Woolsey.[59] The first, an *essential* projection, is said to exist from a thalamic nucleus to a cortical area if a removal restricted to that area causes retrograde atrophy of the cells of the nucleus. A *sustaining* projection, on the other hand, is said to exist if two cortical areas are considered, and destruction of either one of them does not lead to atrophic changes in the thalamic nucleus, but simultaneous removal of both causes degeneration within it. Either cortical area will "sustain" the cells of the nucleus and prevent their atrophy.

The observed phenomena can be explained on the assumption that a nerve cell will degenerate when its axonic terminations are destroyed or are in contact only with atrophic cells. An essential projection would then exist if the axons of the cells of a thalamic nucleus terminated wholly within a discrete cortical field—a field often but not always of uniform cytoarchitecture (p. 235). A sustaining projection could exist, and the thalamic nucleus would thus not degenerate when one of the two or more cortical fields concerned is removed, on the assumption that collaterals of the thalamocortical axons terminate outside the cortical area destroyed. It is thought that within a given thalamocortical projection system essential and sustaining projections may coexist.

Generalized and specific thalamocortical systems[9]

Dorsal thalamic nuclei may be grouped in two classes on the basis of internal neuropil organization, afferent and efferent connections, and functional role. The first group is phylogenetically the older; it includes the midline and intralaminar nuclei and the medial portion of n. ventralis anterior[25, 62] (Table 7-1 and Fig. 8-2). These nuclei are heavily interconnected among themselves, reciprocally so with specific thalamic nuclei, and a frontally directed bundle of axons of intralaminar and n. ventralis anterior origin projects upon the cerebral cortex (particularly the frontal lobe)[63] and also upon the striatum.[59] This system receives ascending input from the mesencephalic reticular formation, the paleospinothalamic component of the ascending anterolateral system of the spinal cord, and a descending projection from the cerebral cortex. It is thought to play an essential role in regulating the general excitability of neurons in thalamus and cortex,[52] and for this reason it is called the *generalized thalamocortical system;* it is termed by some the nonspecific thalamocortical system.

The centre médian is located within the intralaminar region, but it is not a part of the generalized thalamocortical system. It grows rapidly in phylogeny, paralleling the rapid enlargement of the putamen upon which it projects; the centre médian receives a major descending input from the motor cortex.[38] It is thus the thalamic component of a major forebrain loop involved in the control of posture and movement.

The nuclei of the *specific thalamocortical system* develop in parallel with the neocortex, and they may be divided on the basis of their cortical connection, as follows.[1]

Sensory relay nuclei. These nuclei of the major afferent systems comprise the medial and lateral geniculates and the n. ventralis posterior (the ventrobasal complex) for the auditory, visual, and somesthetic systems, respectively. They are organized with a precise neuropil relation between afferent and efferent elements that preserves a high degree of spatial and modality specificity and ensures a strong synaptic security across the relay while receiving also the modulating influence of other systems, e.g., the corticothalamic systems.

Nuclei concerned with efferent control mechanisms. These nuclei receive input from extrathalamic sources other than sensory pathways. The lateral parts of n. ventralis anterior and n. ventralis lateralis are part of the motor system, receiving projections from the basal ganglia and the cerebellum, respectively, and projecting upon the motor cortical areas.[14, 17] The anterior nuclear group receives relayed input from the hippocampus via the mammillary body and projects upon the limbic cortex.[57, 72] The mediodorsal nucleus receives projections from the hypothalamus and the amygdala[42] and projects upon the frontal lobe. Thus the mediodorsal and anterior nuclei are parts of neural systems controlling visceral efferent and endocrine mechanisms and are thought to play roles in the CNS mechanisms controlling emotional behavior as well.

Nuclei concerned with higher order integrative mechanisms. These are nuclei of the dorsolateral group: the n. lateralis dorsalis and lateralis and the pulvinar complex. All show a progressive increase in size in phylogeny; the pulvinar, for example, is a rudimentary structure in rodents but one of the largest of the thalamic nuclei in man. The dorsolateral nuclei project upon the areas of the parietal, temporal, and occipital lobes outside the primary sensory receiving areas—cortical areas known to be involved in higher order integrations.

ANATOMIC ORGANIZATION OF CEREBRAL CORTEX[3, 4, 37, 54]

The cerebral cortex is a convoluted and laminated sheet of neurons that develops from the primitive pallial outpouching of the telencephalon, and evolves most extensively in phylogeny, reaching its greatest development in primates. In man it covers some 2,000 cm^2 of surface, is from 2.5 to 4.0 mm thick, occupies about 600 cm^3 volume, contains several billion neurons,[28, 45] and has a large but unknown number of glial cells. Areas of cortex receive afferent fibers from subcortical structures, in particular from the dorsal thalamus, from cortical areas of the same hemisphere, and from usually homologous areas of the opposite hemisphere. Cortical neurons may project their axons intracortically locally, to cortical areas of the some hemisphere through the white matter (association fibers), via the great commissures to the opposite hemisphere (commissural fiber), or to subcortical cellular masses even so far away as the spinal cord (projection fibers). In concert with subcortical nuclear regions and afferent and efferent systems the cerebral cortex receives and analyzes sensory information, stores a record of experience in memory, programs and governs the execution of movements, regulates homeostatic processes, and is the essential but not exclusive neural substratum of those complex aspects of brain function indicated by such words as thinking, remembering, calculating, planning, judging, etc. While no understanding of these events is possible now at the level of mechanism, the available evidence supports the view that the cerebral cortex is their prime seat of action. Yet in these functions the cortex interacts with virtually all other regions of the CNS. *Total behavior is the result of the total function of the nervous system.*

Cortical cell types

Cortical neurons are not all alike, and many classifications have been made based upon the size and form of cell bodies, the length and distribution of their dendritic trees, and the destinations and degree of branching of their axons.[33, 54] There are three general cell classes, and within the first two many subtypes have been identified.[4, 36]

Pyramidal cells. Pyramidal cell bodies are commonly triangular or trapezoidal in silhouette, with base downward and apex directed toward the cortical surface. Their complex dendritic trees usually consist of (1) a basilar dendritic arborization that ramifies in the immediate locale of the cell body, largely horizontally, and (2) an apical dendrite up to 2 mm in length that ascends from the cell body through overlying cellular layers, frequently reaching and branching terminally within the outermost layer. Both the apical trunk and its numerous branches may be covered with the specialized postsynaptic protrusions called spines (Fig. 6-1).[64] Pyramidal cells are frequently classified in terms of axonal destination. Many emerge from the cortex as association, commissural, or projection fibers, frequently sending recurrent collateral branches back upon the cellular regions from which they sprang. Axons of some pyramidal cells (the cells of Martinotti) turn back toward the cortical surface, never leaving the gray matter, to end via their many branches upon the dendrites of other cells. Many subtypes have been described; some of them occur only in restricted cortical areas. Pyramidal cell bodies vary greatly in size, from axial dimensions of $15 \times 10\mu$ up to $120 \times 90\mu$ or more for the giant pyramids of the motor cortex, which are called Betz cells after their discoverer.

Stellate or granule cells. These cells differ remarkably from pyramidal cells. Their cell bodies are small, and dendrites spring from them in all directions to ramify in the immediate vicinity of the cell of origin. The axon may arise from a large dendrite and commonly divides repeatedly to terminate upon the cell bodies and dendrites of immediately adjacent cells. The axons of other granular cells turn upward to end in superficial layers or, uncommonly, may leave the cortex.

Fusiform cells. Fusiform cells have spindle-shaped cell bodies, and branching dendritic trees may arise from both ends of the spindle. The axons usually project from the cortex after emitting recurrent collaterals; less commonly they turn toward the cortical surface to terminate in more superficial cortical layers.

Layered distribution of cortical cells

Cortical cells are not randomly distributed along the axis normal to the cortical surface; this orderliness affects both the distribution of cell types and their packing density. Relative segregation by depth produces a stratification; each stratum is called a cortical layer. Over the greater portion of the cerebral hemispheres the cortex displays six layers. This stratification is apparent in the sixth month of fetal life in the human being and is well developed at birth. Cortex that possesses six layers in adult life or that possessed six layers in ontogeny is termed isocortex, or more commonly neocortex. Neocortex occupies some 90 to 92% of the cortical surface. The cortex lying medial to the rhinal sulcus never possesses six layers, and the degree of its stratification varies from place to place. It is termed allocortex and is characterized in places by an external layer of myelinated fibers; it includes cortex of the hippocampus and the rhinencephalon.

The cortical layers are usually named and defined as follows, from the pial surface inward (Fig. 7-3):

Molecular or plexiform layer (I). This layer contains the terminal branches of the apical dendrites of pyramidal cells of layers II, III, V, and VI and some of the axonal terminals of cortical cells with ascending axons (Martinotti cells) and of unspecific cortical afferents. It contains a few neurons, the horizontal cells of Cajal.

External granular layer (II). It contains a large number of tightly packed, small pyramidal cells. Their basilar dendrites ramify within layer II, their rather short apical ones within the overlying molecular

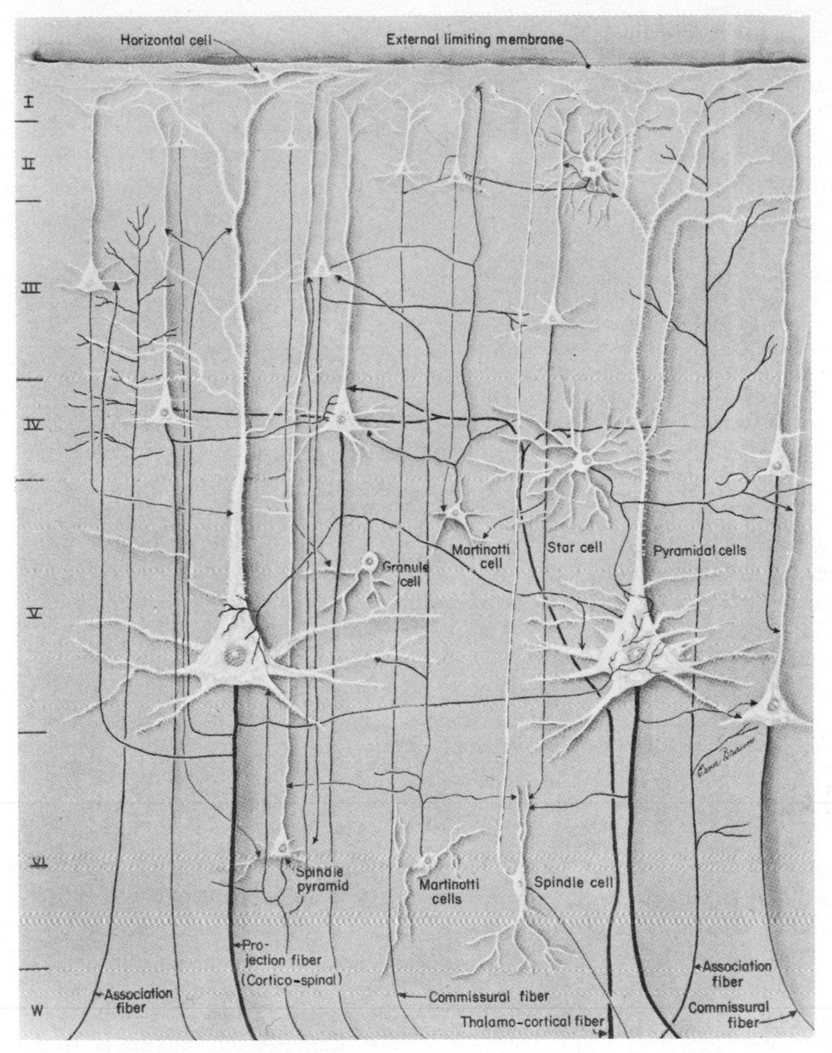

Fig. 7-3. Schematic diagram of different types of nerve cells in cerebral cortex, their connections, and their topographic distribution. Roman numerals indicate various cortical layers; *W*, the subcortical white matter. (From Elias and Pauly.[4a])

layer. Only a few of their axons project from the cortex, the large majority branch repeatedly and terminate about the cells of layers V and VI. Afferent axons to this layer are those of Martinotti cells, of granular cells of layer IV, of association fibers, and of recurrent collaterals of axons of the pyramidal cells of deeper layers.

Pyramidal cell layer (III). The pyramidal cell layer is a continuation of layer II without a sharp line of demarcation but possessing larger pyramidal cells. Their apical dendrites ramify upward, yielding branches in layer II and terminating in layer I; their basilar dendritic field is expanded horizontally in layer III. The cells of the lower edge of layer III receive specific thalamic afferents, and synapses are made throughout the layer from association axons, axons of granular cells of layer IV, and axons of Martinotti cells of layers V and VI. Efferents from layer III end mainly in layers V and VI, but some leave the cortex as projection or association fibers.

Granular cell or internal granular layer (IV). This layer is packed with small stellate cells that give the name "granular" to those regions of cortex in which this layer is best developed, e.g., primary sensory receiving areas, which are also called koniocortex, and the granular cortex of some regions of the frontal lobe. The dendrites of the granule cells arborize locally within layer IV and receive a very heavy synaptic impingement from the axon terminals of the specific thalamocortical afferents. The granular cell axons rarely leave the cortex; some ascend to terminate in layers I to III, while the majority descend to end upon cells of layers V and VI. The motor cortex of the precentral gyrus is called agranular because of the absence of this layer. Layer IV also contains a number of pyramids with apical dendrites that reach to layer I.

Ganglion cell or giant pyramidal layer (V). It is sharply demarcated from layer IV. The large pyramids have apical dendritic shafts that reach layer I, where they end in complicated brushes; the basilar dendrites and the collaterals of the ascending shafts are distributed exclusively in layer V. The efferent axons are usually projection, commissural, or association fibers and with rare exceptions have branches and recurrent collaterals that ascend to terminate in layers II and III, and even I. Cells with ascending axons are found in this as in all other cellular layers. The pyramidal cells of layer V vary greatly in size in different areas of the cortex.

Fusiform or multiform cell layer (VI). The fusiform layer contains many spindle cells. Dendrites may arborize from either or both ends of the cells, and larger ones may ascend to layer I without branching. Smaller spindles have dendritic fields confined to layers V and VI. The spindle cell axons usually project out of the cortex and give off branches and recurrent collaterals before leaving.

Summary of cortical afferents

1. Thalamocortical projection fibers, the so-called specific thalamic afferents (e.g., lateral geniculate to visual cortex), ascend still myelinated through layers V and VI and divide repeatedly to form extensive terminal plexi in layers IV and lower III. Synapses are made in great numbers upon the dendrites of the granular cells of layer IV and also upon the apical dendrites of pyramids of layers V and VI as they pass through layers III and IV.

2. A second type of thalamocortical afferent was identified by Lorente de Nó[37] as projecting to more than one cortical area, giving off collaterals to all layers in passing through the cortex, as high as layer I. Their source is thalamic, but their exact thalamic origin is unknown. They may arise from the generalized thalamocortical system.

3. Association and commissural fibers may give collaterals to cells of layer VI upon entering the cortex, but their main field of termination is in layers I to IV, and especially in II and III.

Summary of cortical efferents

1. Axons of small- and medium-sized pyramidal and granule cells of layers II, III, and IV may be distributed mainly within the cortex itself, although some of the medium-sized pyramids of lower III may emit projection or association axons, and these then emit recurrent collaterals, especially to layer V.

2. The axons of the larger pyramids of layer V and of spindle cells of layer VI compose the large majority of the projection, association, and commissural fibers. They, too, before leaving the cortex, emit recurrent collaterals that terminate mainly in layers II and III.

3. Neurons with short axons that ramify within the cortex are numerous in all layers and increase in relative number with increasing development of the brain in phylogeny. They are especially numerous in the brain of man. They may be the Golgi type II cells (the granule cells), the Martinotti cells with ascending axons, or the horizontal cells of Cajal in layer I.

Synaptic ultrastructure is discussed in Chapter 6.

FUNCTIONAL IMPLICATIONS OF THE CELLULAR CONNECTIVITY OF THE CORTEX: COLUMNAR ORGANIZATION

It should now be clear that the weight of synaptic relations in the cerebral cortex is in a vertical direction, normal to the pial surface, and that pathways and synapses that might allow horizontal intragriseal spread of activity are few in number and limited to the molecular layer. From the functional point of view the cortex must be regarded as made up of complex chains of interneurons that do not differ in principle of organization from such chains elsewhere. These chains are interposed between input and output over a vertical column of cells extending across all cellular layers. Initial input to such a chain, e.g., over the specific thalamic afferents, gains a powerful and rapid access via some granular cells of layer IV downward to the efferent neural elements of layers V and VI and via others upward to the more superficial pyramids of layers II and III. Access to the efferent elements of the cortex may in the first place occur fairly directly, over one or a few synapses, if excitability and convergence are high. Second, a quick translation of activity to the outer cellular layers of the cortex will occur, activating arcs of neurons superimposed upon the initially activated oligosynaptic neurons and delivering repetitive trains of impulses by recurrent and reentrant activation to the efferent cortical neurons. These in turn will influence the activity of those superimposed chains via recurrent collaterals, either excitatory or inhibitory, that are distributed in a patterned, not a random, spatial array. It is likely, for example, that the *efferents from each column, via recurrent inhibition, will depress the activity of neurons in adjacent columns,* thus contributing further to the vertical segregation of cortical neuronal activity. The neurons of the superimposed reentrant chains, lying mainly in layers II, III, and IV, will also receive input from other cortical areas via association and commissural fibers and from the generalized thalamocortical system (p. 231). What is usually referred to as "an integration of neural activity" will result: the influences reaching these groups of neurons will be titrated in such a way that the resulting output is some net result of all. The meaning of this titration in terms of "what cortex does" to its input to produce its output is unknown. *It is the central problem of cortical physiology.*

Physiologic evidence that a vertical column of cells such as that just referred to indeed serves as an elementary functional unit in the primary sensory cortices will be given in Chapters 10 and 16.

VARIATIONS IN THE LAYERED DISTRIBUTION OF CORTICAL CELLS: CYTOARCHITECTURE

Even a casual inspection of cortical sections reveals that (1) the layers vary in packing density and in thickness from place to place, (2) in some areas certain layers are greatly reduced or even absent, e.g., the near absence of layer IV from the precentral gyrus, which is thus termed the "agranular motor cortex," (3) special cell types appear in some areas and are not present in others, and (4) the numbers of afferent and efferent fibers and their segregation at some cortical depths (e.g., the line of Gennari of the visual cortex) are not the same from one place to another. These observations led to the idea that the cerebral cortex might be divided into a number of fields, within each of which these morphologic characters are uniform, and each differing significantly from its neighbors.

Study of serial sections of the cortex reveals to every observer that in some places the juncture between fields is obvious, sharp, and can be marked on a photomicrograph with a line. This is true, for example, of the juncture between the striate (area 17) and the prestriate cortex (area 18). At other conjunctions the change is not so abrupt but occurs gradually over a distance of 1 to 3 mm, a region that is labeled a transition zone; e.g., area 3a in the depths of the central sulcus is an area of transition from the granular cortex of the postcentral gyrus to the agranular motor cortex of the precentral gyrus. All observers agree on facts such as these. When parcellations are attempted in such areas as the granular frontal cortex (areas 9 to 12) or the temporal lobe, however, observers disagree. Their disagreement varies from the statement that in such a large area as the granular frontal cortex no separate fields exist at all, to the position that a very large number of separate fields can be identified. For the student concerned with the complexities of the cortical maps that have resulted from cyto- and myeloarchitectonic studies, a working definition of what may constitute a cortical field may be useful.

Definition of a cortical field

1. A cortical field is a region within which the cell types and packing densities of the cortical layers may be uniform. Within many fields, however, variants of morphology occur. Rose[56] put forward the idea that these variants should be grouped together, a natural series constituting a field. Thus a cortical field as defined by other criteria may display continuing change in structural detail. Changes that vary systematically are designated the gradients of a field; the line dividing two fields is placed at that point at which these gradients, which may be occurring in all layers as one moves from point to point across the field, are changing most rapidly. The important conclusion is that two adjacent fields that have definite overall differences may yet have no sharp border between them.

2. A cortical field may receive an essential thalamocortical projection from a particular thalamic nucleus. When the extent of that projection coincides with the cytoarchitectonic definition of a field determined independently, that definition is greatly strengthened, e.g., the projection of the lateral geniculate body upon area 17 and the projection of each of the anterior thalamic nuclei upon precisely defined architectonic fields of the cingular gyrus. In other areas the thalamocortical projection field is not so precisely defined, nor is the cytoarchitectural definition: in general the more precise the one, the sharper the other.

3. The extent of a field defined by physiologic means is frequently found to coincide with that determined by the two methods previously described. This is particularly true of the primary cortical areas to which the great afferent systems project.

When a cortical area that is a candidate for identification as a separate field meets all three criteria, identification cannot be doubted.

Cortical maps and functional localization

Fig. 7-4 displays maps of the human cerebral cortex that were published by Brodmann.[22] Studies of the human and other primate brains had been made earlier by Brodmann,[21] by Campbell,[23] and by others and have been repeated since.[31] The Brodmann maps are presented here because his system of numeration is the most widely used in both clinical and experimental neurology.

One possible implication of the cytoarchitectonic differences is that cortical areas function differently and deal with separate functions *because* of these cellular differences. This seems altogether unlikely: there is nothing peculiarly *motor* about the cellular arrangements in the *motor* cortex. An alternative hypothesis is that cortical areas function differently and deal with particular functions because of differences in their afferent and efferent connections. Indeed it may be that cytoarchitectural differences are themselves due to variations in the number and destinations of afferent fibers growing into the uniform fetal neocortex and in the numbers and

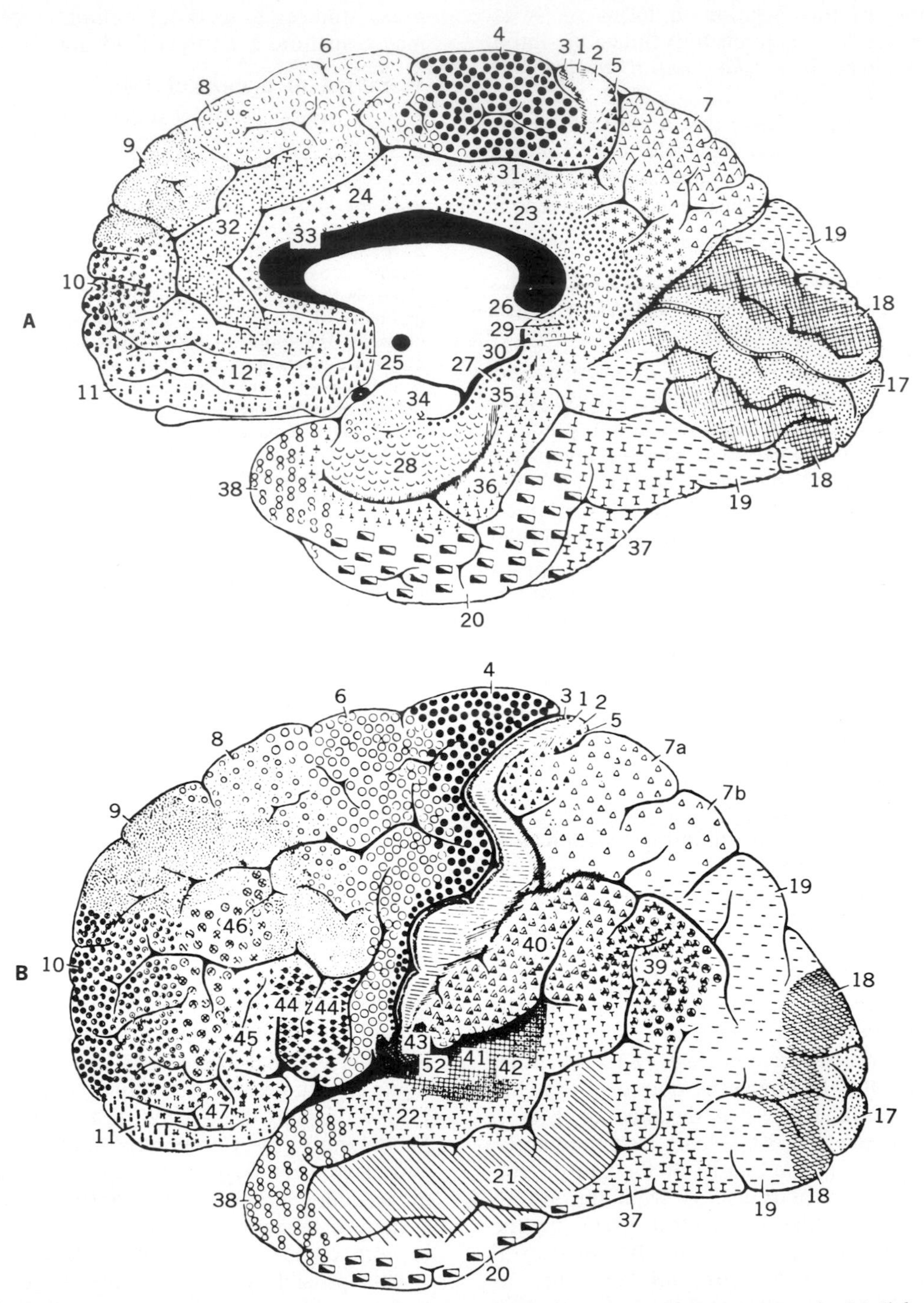

Fig. 7-4. Brodmann's map of cytoarchitectural fields of human cerebral cortex. **A,** Medial surface of cerebral hemisphere. **B,** Lateral surface of cerebral hemisphere. (From Brodmann.[22])

lengths of the efferent axons growing out. A corollary of this hypothesis that function depends on connection is that what cortex does with its input to produce its output may be, at least in principle, identical from one neocortical field to another.

Thus the problem is posed of what is meant by "localization of function" within the cerebral cortex. It does not mean that the neocortex is made up of a number of distinct organs, each dealing with a particular function or "quality" and all fitted together in a complex mosiac. What it does mean is that *some particular areas, by virtue of their connections, are involved with one particular aspect of function.* For example, stimulation and ablation experiments as well as clinical observations support the statement that the precentral motor cortex is intimately involved in the control of movement and the regulation of posture. What is *not* meant is that motor function resides there and there alone, and the partial recoveries of motor function that follow its removal indicate clearly the part played in this function by other regions of the brain. On the other hand, the prepotent executive role of the motor cortex in movement is obvious.

GENERAL PHYSIOLOGY OF THE CEREBRAL CORTEX

Cellular properties and synaptic transmission

The subject of synaptic transmission in the CNS was considered in Chapter 6, largely on the basis of the facts obtained in studies of spinal motoneurons. The electrical phenomena that occur in different parts of nerve cells were used to support a hypothesis of their integrative action. The techniques of intracellular recording have been applied to both neocortical cells[30, 48, 65] and the pyramidal cells of the hippocampus.[36] Results indicate that the general concepts elucidated for spinal motoneurons apply with some modifications to neurons located in more rostral portions of the nervous system.

Cortical cells, like other neurons, are capable of three types of electrical activity: (1) the regenerative change in Na^+ conductance that occurs at a critical level of depolarization and leads to impulse conduction, (2) depolarizing (excitatory) local postsynaptic responses, and (3) polarizing (inhibitory) local postsyaptic responses. There is some evidence that, as for spinal motoneurons, all parts of cortical neurons are not equally capable of local and regenerative processes. The low threshold of the initial segment, or axon hillock region for regenerative action, coupled with the propensity of soma and dendrites to support under some conditions only local PSPs, affords an explanation for the integrative action of cortical cells. The frequency of impulses discharged from the initial segment and down the axon is a running and probably linear function of the net ionic current flowing across its membrane, the intensity and direction of which is determined by the number, sign, and location of the local PSPs occurring elsewhere on the cell membrane. It is important to emphasize certain properties of PSPs, particularly with reference to subjects discussed later. They are local, nonconducted changes in membrane potential produced by local changes in ionic permeability; thus they affect other regions of the cell (e.g., the initial segment) only by electrotonic extension. They may be algebraically summed both in time and space over the cell surface.

When an afferent volley to the cerebral cortex is either purely excitatory or (after interneuronal relay) purely inhibitory for a cell under study, that cell responds with PSPs that are depolarizing or (hyper) polarizing. Such a purity of synaptic input rarely occurs naturally and is difficult to achieve experimentally, so that the PSPs observed are commonly diphasic, a combination of sequentially timed excitatory and inhibitory synaptic responses (Fig. 7-5). The PSPs of cortical cells differ from those of spinal motoneurons in time course, which is longer than can be explained by the membrane time constant. The prolongation of PSPs of cortical cells beyond that predicted by their membrane time constants may be due to a persistence of the transmitter agents released by single presynaptic impulses or, what is more likely, to the repetitive activity of cortical cells, particularly of the excitatory and inhibitory interneurons (i.e., the granular cells) that relay afferent input to the large pyramidal cells.

The membrane potentials of cortical neurons are not constant but oscillate more or less continuously near the threshold level for spike generation. This continuous variation and the accompanying impulse discharges are frequently referred to as *spontaneous,* a word which, when used in central neurophysiology or electroencephalography, designates neural activity in the absence of any overt or intentionally provoked afferent input. Spontaneous activity is not thought to be generated by, or not wholly by, an intrinsic cellular pacemaker but by a continuous and ongoing synaptic input.* The result is that large numbers of cells in the cerebrum are active at rates of discharge of 10 to 30/sec. Thus neuronal transactions take on a second degree of freedom: either an increase or a decrease in rate of discharge may serve as a positive signal.

Cortical neurons tend to discharge repetitive trains of impulses even when the excitatory presynaptic volley consists of but a single impulse per presynaptic fiber. This lends further credence to the idea that the action potential is initiated at a site separate and distal in the cell from the PSP loci and that the action potential does not invade those loci. Upon recovery from the first spike discharge, the spike-generator membrane is again influenced by the persisting currents set up by the persisting synaptic effect and so on until those currents fall below threshold.

Observations made on cortical cells are consistent with the assumption that synaptic transmission in the cerebral cortex is "chemical" in nature (Chapter 6); i.e., the depolarization of synaptic endings by nerve impulses causes them to release substances that diffuse across the synaptic cleft to combine with specific receptor substances in the subsynaptic membrane. That combination leads to changes in the ionic permeability of the postsynaptic cell and the resulting PSPs. Candidate synaptic transmitters have been described in Chapter 6.

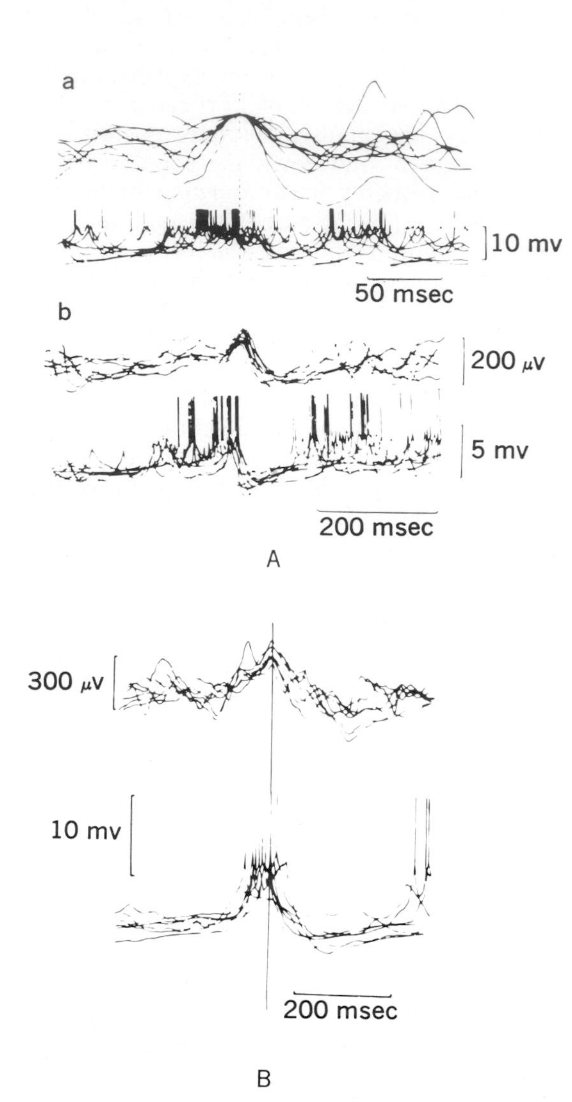

Fig. 7-5. In all records shown, upper trace is electrical activity recorded at the surface of cat cortex with "gross electrode," and lower traces are recorded from micropipet tip within a cortical cell.

A, Drawings of EEG waves of two different types are superimposed and oriented in time by summits of negative waves and correlated with simultaneous intracellular events. In *a,* mainly surface-negative waves are shown to correlate with EPSPs and associated impulse discharges of cortical cells. In *b,* negative-positive waves recorded at surface are shown to correlate well with EPSPs and impulse discharge and an IPSP following in sequence. **B,** Records of mainly negative waves of surface-recorded EEG (upper traces) just at onset of increased electrical activity produced by the drug Metrazol, which later led to convulsions. Lower traces show correlation with these surface-negative waves of EPSPs and synaptic events plus impulse discharges of a subjacent cortical cell. (From Creutzfeldt et al.[29])

Cortical pyramidal cells and the origin of surface-recorded electrical activity[51]

When records of electric potential differences are made between one electrode resting on the cortical surface and a second placed a considerable distance away, the array represents a special case of recording at the boundary of a large conductile medium containing active elements, explicable only in part on the basis of the classic volume conductor theory. It is likely that conducted action potentials in axons contribute little to cortical surface records, for insofar as they occur asynchronously in time in large numbers of axons that run in many directions

*Many cellular neurophysiologists prefer to restrict the descriptive term "spontaneous" to autochthonus forms of activity and use "maintained" or "background" as adjectives descriptive of ongoing CNS activity.

relative to the surface, their net influence upon an electrode at that surface will be zero. An exception is the special case in which large numbers of thalamocortical axons are activated simultaneously by electrical stimulation of thalamic nuclei or their afferent pathways. It will be shown later that surface records obtained under other circumstances signal principally the net effect of local PSPs of cortical cells. These may be of either sign and may occur immediately beneath the electrode or at some distance from it; a potential change recorded at the surface is the measure of the net IR between the surface site and the distant electrode, produced by the extracellular current flows associated with local PSPs. It is obvious, however, that if all the cell bodies and dendrites of cortical cells were randomly arranged in the cortical matrix, the net influence of their synaptic currents would be zero. *Any* electrical change recorded at the surface must be due to the orderly and symmetric arrangement of some class of cells within the cortex.

The cortical pyramidal cells seem the most likely candidates. Their long apical dendrites are arranged parallel to one another and normal to the cortical surface. Potential changes in one part of such a cell relative to other parts create "open" fields of current flow that can be detected at the surface of the cortex or indeed at the surface of the head. The granular cells, on the other hand, are unlikely to contribute substantially to surface records. Their spatially restricted dendritic trees are radially arranged around their cell bodies, so that charge differences between dendrites and somata will produce "closed" fields of current flow that will add to zero when viewed from the relatively great distance of the cortical surface.

The influence on the surface record of a PSP depends on its sign, orientation, and location. Each may be regarded as creating a radially oriented dipole. Thus continuing synaptic input creates a series of potential dipoles and resulting current flows that are staggered but overlapped in space and time. Surface potentials of any form can be generated by one population of presynaptic fibers and the cells upon which they terminate, depending on the proportion that are inhibitory or excitatory, whether predominantly axodendritic or axosomatic, the level of postsynaptic cells in the cortex etc.

A general and empirical description of the oscillatory changes in electric potential that can be recorded from the surface of the brain or from the scalp is given in the following section. The question of the relation of these waves to cellular events, introduced previously with general statements, has recently come under direct study. Comparisons have been made between surface-recorded electrical activity and cellular events recorded via intracellular microelectrodes.[29, 32, 48, 65] Convincing evidence has been adduced that the frequency, signs, and amplitudes of the surface-recorded slow waves reflect the net effect of the PSPs of subjacent cortical cells. The records shown in Fig. 7-5, taken from the work of Creutzfeldt et al.,[29] illustrate some of these correlations. The results obtained may be summarized as follows.

1. Excitatory (depolarizing) PSPs in neural elements located close to the cortical surface generally induce negativity in the surface record; when deep within the cortex, they generally induce positivity in the surface record. Inhibitory (polarizing) PSPs generally induce opposite effects in the surface record.
2. The correlation between the cortical surface wave and the PSP sequence in any single cortical element cannot always be predicted. The result depends on simultaneous events, which may be of either sign or of the two in either sequence, in many neurons located at various depths.
3. Some correlations are very strong, however, especially under certain well-defined circumstances:
   a. The primary evoked potential, the positive-negative wave recorded on the cortical surface in response to sensory, afferent nerve, or thalamic relay nuclear stimulation, is correlated with initially depolarizing synaptic electrogenesis in the cortical depths, with succeeding polarizing PSPs in these same elements, and with slower depolarizing PSPs in more superficial cortical elements. Delayed or prolonged PSPs lead to more complicated "secondary" waves in the evoked potential.
   b. The surface-negative recruiting response (Chapter 8) recorded on the cortical surface in response to stimulation of the generalized thalamocortical system is associated with summated depolarizing PSPs in cortical cells with the same time course. The waxing and waning of the evoked cortical waves during slow iterative stimulation (5 to 7/sec), which

is characteristic of the recruiting responses, is paralleled by a waxing and waning of the cellular EPSPs as well as the size of the thalamocortical volley (p. 256).

c. During the spontaneous waves of the surface record, whether occurring in the anesthetized or the sleeping animal, or when synchronized by a convulsive agent, there is a close positive correlation between cortical surface negativity and cellular depolarization, during which a high-frequency discharge of impulses may occur in the cortical cell. The implication is that the depolarizing PSPs predominate throughout all the layers of the cortex.

To date a detailed analysis of this sort has not been made in the waking and alerted state, which is characterized by higher frequency, asynchronous, lower amplitude oscillations in the surface record. The asynchrony of the cortical waves in the waking state is produced by asynchronous synaptic events in the subjacent cortical cells.

Synchronous activity in large numbers of neural elements temporarily raises the K^+ ion concentration in the extracellular clefts within the brain, producing purely passive changes in the resting membrane potential of glial cells. The close electrical coupling between glial cells (Chapter 6) indicates that these electrical changes may have widespread effects and may contribute to slow potential changes recorded by electrodes placed on the surface of the head or on the scalp.

ONGOING ELECTRICAL ACTIVITY OF THE BRAIN: THE ELECTROENCEPHALOGRAM[5, 69]

Oscillations in electric potential occur almost continuously between any two electrodes placed on the surface of the head or on the cerebral cortex itself, and the records are termed, respectively, the electroencephalogram (EEG) and the electrocorticogram. These oscillations differ in frequency and amplitude from place to place and in different states of awareness. They persist in altered form during excitement, drowsiness, sleep, coma, anesthesia, during epileptic attacks, and through severe changes in blood gas or cerebral metabolite concentrations. They never cease short of massive cerebral catastrophe or impending or actual death. In theory these wavelike potential changes might serve as direct and measurable indices of brain activity. Although this objective is not yet realized, the association between certain brain-wave patterns and some states of altered brain function is so constant and so readily recognized that the study of brain waves has become an important diagnostic tool in clinical medicine.

The spontaneous electrical activity of the brain was discovered by Caton in 1875, though the phenomenon was observed independently at about the same time by others.[20] However, it was Hans Berger who, between 1929 and 1938, showed that electrical activity could be recorded from the scalp surface of human beings, developed methods for making such recordings, and described several varieties of the activity that commonly occur.[18] Adrian et al.[12, 13] confirmed and extended these observations. Berger also studied alterations in the EEG produced in certain disease states, and he is thus rightly regarded as the founder of the medical specialty of electroencephalography.

EEG recordings are ordinarily made from a large number of electrodes positioned on the surface of the head. Three types of electrode connections are used: (1) between each of a pair, (2) between each as a monopolar lead against a "distant" electrode, commonly placed on the ear, and (3) between each as a monopolar lead and the average of all. While the same electrical events are recorded in each of the three ways, they will appear in different format in each. The potential changes that occur are amplified by high-gain, differential, capacity-coupled amplifiers. The output signals are usually displayed by ink-writing oscillographs (writing on moving paper), a method that limits the frequency response to the range between 0.5/sec and 80 to 100/sec. Alternatively the output signals may be displayed by an inertialess system for preservation of high-frequency response, the cathode-ray oscilloscope, or recorded on magnetic tape for later analysis. In ordinary practice, records are analyzed by direct inspection and measurement. More sophisticated analyses are measurement of the frequency-power spectrum and of auto- and cross-correlation functions[10, 35, 55] and by display of potential-space-time three-dimensional plots. Recordings are made directly from the cortical surface in human beings during neurosurgical procedures or from surface and depth electrodes that may be implanted during such procedures and left on or in the brain even for extended periods of time. The development of these techniques has opened a wide and hopefully fruitful field of investigation in which correlations between behavior and the slow-wave electrical events occurring in the brain are attempted.

The potential waves recorded may vary in frequency from 1 to over 50/sec and when led from the head surface are 50 to 200 μv in amplitude. In a normal individual at rest with eyes closed in a quiet room, the dominant rhythm varies from 8 to 13/sec and is seen at greatest amplitude in the parietal and occipital regions; the amplitude may

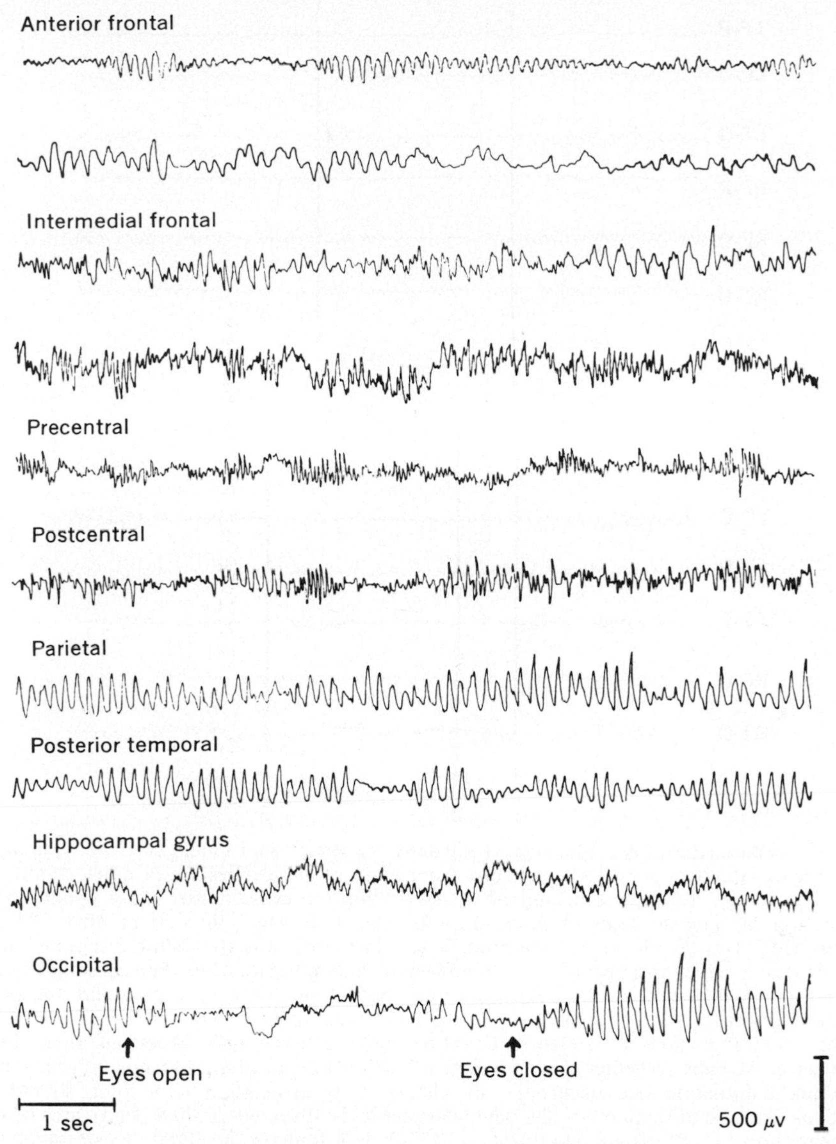

Fig. 7-6. Spontaneous electrical activity, or "resting rhythms," from different cortical areas in man. Sample tracings were taken directly from exposed cortex with silver-chloride, cotton-wick bipolar electrodes. Alpha rhythms are most prominent from entire parietal, posterior temporal, and occipital regions, with exception of the postcentral gyrus itself. More rapid activity is present in anterior regions, with a relatively pure beta rhythm in the precentral gyrus. Note blocking of alpha rhythm in occipital region when eyes are open. (From Penfield and Jasper.[8a])

slowly wax and wane. This is the *alpha* or Berger rhythm, which is illustrated by the records in Fig. 7-6. The normal alpha rhythm varies in amplitude and spatial distribution from one individual to another, and an occasional individual thought to possess a normal and a normally functioning brain may never show an alpha rhythm. In any one person, however, the dominant alpha frequency is remarkably constant from time to time, scarcely varying by as much as 1 Hz. The alpha rhythm is thought to originate from an alert but relatively "unoccupied" brain. Upon sensory stimulation, especially with light, or a conscious effort for vision or purposeful mental activity, it is quickly replaced by a higher frequency (13 to 25 Hz), lower voltage pattern (Fig. 7-6). This *beta* rhythm is commonly referred to as the "activated" or desynchronized pattern, although

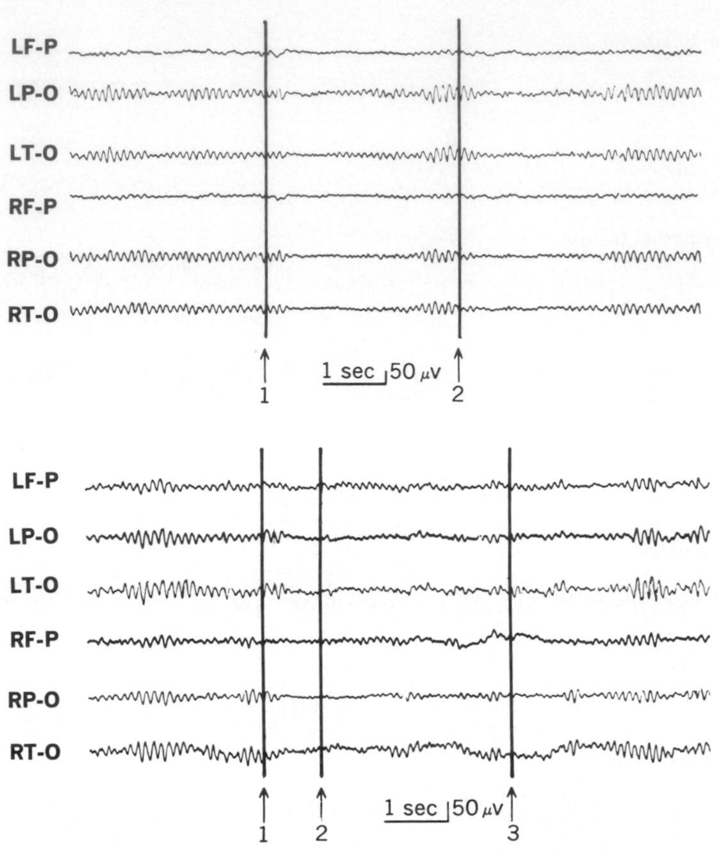

Fig. 7-7. Contingent alpha blocking, frequently called "conditioned cerebral response," in human being. Records are EEGs recorded from scalp of subject resting in quiet, semidarkened room. Upper set illustrates blocking of alpha rhythm, recorded under these conditions, with presentation of light stimulus of 3 sec duration (on at arrow *1* and off at arrow *2*). Sound stimulus (500 Hz, 50 db above threshold, 4 sec duration) was then slowly repeated until *its* alpha blocking effect disappeared, a phenomenon called habituation. Paired sound and light stimuli were then presented, the light appearing about 1 sec after onset of sound, the two then continuing together for 3 sec. Lower records, obtained after several presentations of paired stimuli, show that tone to which habituation had occurred now produces alpha blocking before onset of light. Whether this "contingent" acquisition of alpha blocking capacity by tone (to which habituation had occurred), by virtue of its association with light, should be regarded as true conditioning in the pavlovian sense is disputed. Paired leads are *LF-P,* left frontoparietal; *LP-O,* left parieto-occipital; *LT-O,* left temporo-occipital; *RF-P,* right frontoparietal, etc. (From Wells.[70])

it may also appear in other states[11] (Chapter 8). Slower waves than the alpha occur rarely in normal individuals, and when they do occur in waking subjects other than newborn infants, they usually indicate disease or injury to the brain. These are the *theta* (3 to 7 Hz) and the *delta* (0.5 to 3.5 Hz) rhythms. The changes in the EEG associated with drowsiness and sleep will be considered in Chapter 8. The desynchronization of the alpha rhythm ("alpha blocking") by light[34] can be made contingent upon stimuli delivered over other sensory pathways (Fig. 7-7).[70] If light is preceded by a noise and the pair then associated for a number of trials, the alpha blocking will be produced by the noise delivered alone, even though the noise when delivered previously in isolation had been presented so many times as to lose its own capacity for alpha blocking, a phenomenon termed habituation. Whether this contingent alpha blocking fits the paradigm of classic pavlovian conditioning is unsettled,[39] but it can be shown to be similar to it on a number of counts.

Considerable attention has been given to the ontogenetic development of the EEG.[41] Records of newborn infants are characterized by continuous, irregular, asymmetric waves from all areas; no regular rhythms appear.

Faster activity appears intermittently during the first few weeks and months of life, becoming persistent after 1 year of age. The further development to the adult pattern occurs gradually over the early years to adolescence.

It will be clear from the preceding section that slow waves recorded from the brain, such as those of the EEG, are the compounded result of the local postsynaptic responses of cortical cells.[32] A major conclusion is that the synchronization of cortical potential waves, their rhythmicity, and their spatial progression over the cortical surface results from the postsynaptic effects of ordered patterns of impulses in presynaptic fibers rather than from intrinsic rhythmic properties of cortical cells. The question of whether the spontaneous waves themselves exert some intercellular influence remains open. It has been shown that rhythmically firing neurons are extraordinarily sensitive to voltage gradients in their extracellular surround,[67] but whether the extracellular voltage gradients associated with a local PSP of one cell are sufficiently intense to affect the excitability of adjacent cells is unknown. Such a mechanism would provide an alternative, but not mutually exclusive, explanation (other than that of synchronous presynaptic input) for the fact that large masses of neurons tend to beat in unison, particularly when idling. There is no evidence that slow waves per se constitute a separate signaling mechanism within the brain.

METHOD OF SINGLE-UNIT ANALYSIS AND ITS APPLICATION TO THE STUDY OF BRAIN FUNCTION[15]

It is only by empirical correlation, not by elucidation of mechanism, that the study of brain waves has proved useful in clinical medicine. In a later section and in later chapters it will emerge that one slow-wave event, the potential change evoked by sensory stimulation, has proved a valuable tool for mapping the central topography of sensory systems. The complexities of slow waves and the uncertain knowledge of their relation to cellular events has in the past limited their use in the study of mechanism, for only in special cases can the patterns of activity in a population of neurons be deduced with certainty from the amplitudes, forms, durations, or signs of the slow potential changes recorded from them. Yet it is just those neuron firing patterns that must be defined before the function of large aggregates of nerve cells is understood. In the first place, it must be known how a single neuron, as a representative of a class of neurons in a given locale, performs during the execution of various functions by the neural populations of which it is a member. Equally important are the relations between the activity of such a neuron and that of its neighbors. Experimental objectives such as these can now be reached by use of the method of single-unit analysis and its corollary, the derived reconstruction of population events. This is possible because central nerve cells, which cannot be isolated anatomically, may sometimes be isolated electrically.

The method is applied in various forms in neurophysiologic experiments. The way in which certain aspects of sensory stimuli are encoded in terms of afferent impulses in peripheral nerve fibers is described in Chapter 9, largely on the basis of the study of nerve fibers isolated by microdissection. Observation of cellular events by intracellular recording has led to the present understanding of neuromuscular and synaptic transmission, and this method has been applied successfully at the level of the cerebral cortex. When the experimental objectives are those just given, the method of recording via microelectrodes whose tips lie in an extracellular position is especially appropriate, for single neurons can then be observed for long periods of time and a considerable number of cells of a given population can be studied *seriatim* under standard, well-controlled conditions.

It is now clear that using these methods the neural replication of sensory events can be determined in great detail (Chapters 10, 13, and 16). Most importantly, the study of many cells of a given locale allows a post hoc reconstruction of the total profile of activity in all the neurons of a population that are influenced by a peripheral stimulus. Findings are then correlated closely with histologic identification of recording sites. Thus attention of investigators is directed toward the dynamic and time-dependent aspects of the activity of central sensory neurons, i.e., how they signal such things as changes in stimulus positions or intensity, stimulus shape, quality (e.g., color), temporal cadence, etc. Comparison of the response properties of populations of neurons at different levels of a sensory system, e.g., of the retinal ganglion cells, the cells of the lateral geniculate, and those of the visual cortex, establishes the transformations and abstractions of neural activity that occur between periphery and center. Significant signals are thought to derive from either increases or decreases in the ongoing spontaneous rate of activity present in many but not all central sensory neurons. There is another possibility—that at one and the same rate of discharge two different signals might be made by virtue of differences in the temporal order with which impulses occur. The possibility that pulse-interval modulation constitutes a significant signaling mechanism has attracted intensive study in recent years. It has so far been shown to be a signaling device only in certain special cases. Indeed, it is unlikely on a priori grounds that the precise sequential timing of impulses is important in those neural systems in which a considerable convergence of presynaptic fibers upon postsynaptic neurons occurs at each level of the system. The unique temporal sequence of impulses in any one presynaptic fiber will, after the transformation imposed by the PSPs of the postsynaptic cell, scarcely be identifiable in the temporal pattern of the train of impulses emitted by that cell.

In other systems, notably the great afferent sensory pathways in which convergence is restricted and the security of transmission is very high, the rhythmic pattern in a synchronously active group of first-order sensory fibers is transmitted with great fidelity, at least through the first stage of cortical activation. Therefore there is the possibility that in these systems the temporal order in which impulses occur is of critical importance for information transmission, over and above the general level of activity. In either case the study of this "internal structure of the neural message" is an elegant and frequently a most revealing method of data analysis.[49]

Statistical nature of neural activity[27, 41, 46, 47]

The study of central neurons has confirmed what is intuitively predictable from inspection of brain-wave recordings—that neural activity is a variable, a statistical affair that for quantitative evaluation must be treated from the probabilistic point of view. The spontaneous discharges of central neurons occur irregularly in time. Impulses do not appear randomly, however, for the time spacing is subject to certain constraints such as refractoriness and afferent or recurrent inhibition. This uncertainty exists even at the level of first-order fibers, though their discharge sequences under a steady sensory stimulus are much less variable than are those of central sensory neurons. It is apparent that the asynchronous convergence of trains of impulses in first-order fibers, even though quasiperiodic, will evoke from the postsynaptic element upon which they impinge trains of impulses that show a much wider dispersion of impulse intervals, and so on at successive relays. In some locales these discharge sequences are also influenced by cyclic variations in the likelihood of discharge, an influence undoubtedly exerted by other converging systems.[49]

A certain variability pertains also to the responses of the first- (and the *n*th-) order neurons activated by even brief sensory stimuli, as shown by the variations in the responses of mechanoreceptive afferents from the skin of the monkey's hand to mechanical stimuli (Fig. 10-19). Inspection of this graph reveals that threshold of response is itself a statistical matter, as is the value of the response to a stimulus of a given strength. Thus a central detecting and interpretive mechanism, itself composed of neurons, must identify a neural signal as such against a background of ongoing neural activity and discriminate between two signals, i.e., between two trains of impulses, as being different, when each will from one trial to the next show considerable fluctuation, even though each population of responses is evoked by a repeated stimulus of constant strength. The neurophysiologist uses a variety of methods for averaging responses to estimate true means and for distinguishing signals from background, but it should be emphasized that the central detecting apparatus must frequently decide whether a stimulus has occurred at all or whether one stimulus is stronger than another on the basis of a single brief period of altered neural activity. How this matter may be treated from the standpoint of statistical decision theory is considered in some detail in Chapter 18. Whether brains function in this way is unknown, but some such mechanism seems a prerequisite for orderly function.

PRIMARY EVOKED POTENTIAL: CORRELATION WITH SYNAPTIC ORGANIZATION AND CELLULAR ACTIVITY[16, 51]

The term "evoked potential" identifies the electrical change that may be recorded in some part of the brain in response to the deliberate stimulation of sense organs, the afferent fibers of peripheral nerves, or some point on the sensory pathway leading from periphery to center. It refers to the evoked slow-wave events rather than the cellular discharges that may be associated with it. Although most frequently observed in sensory systems, the electrical changes evoked in one part of the brain by stimulation of a distant part with which it is linked fall into the same category. This provides one method for identifying intracerebral connections, e.g., those that link homologous areas of the two cerebral hemispheres via the corpus callosum. Evoked potentials differ from spontaneous brain waves in several ways. They usually appear in definite time relations to the stimuli that evoke them. They commonly are observed in only a local part of the brain, e.g., in a sensory area upon stimulation of the appropriate receptors, though this restricted distribution depends on the level of excitability of the brain and the particular component of the wave form to which reference is made. Under certain experimental conditions, sensory evoked potentials may be used to map the central representation of sensory systems, for they are more or less predictable and reproducible both from time to time and from one animal to another.

The hypothesis now seems well established that the evoked slow waves recorded within or upon the surface of the cortex or within subcortical nuclei are the net result of the

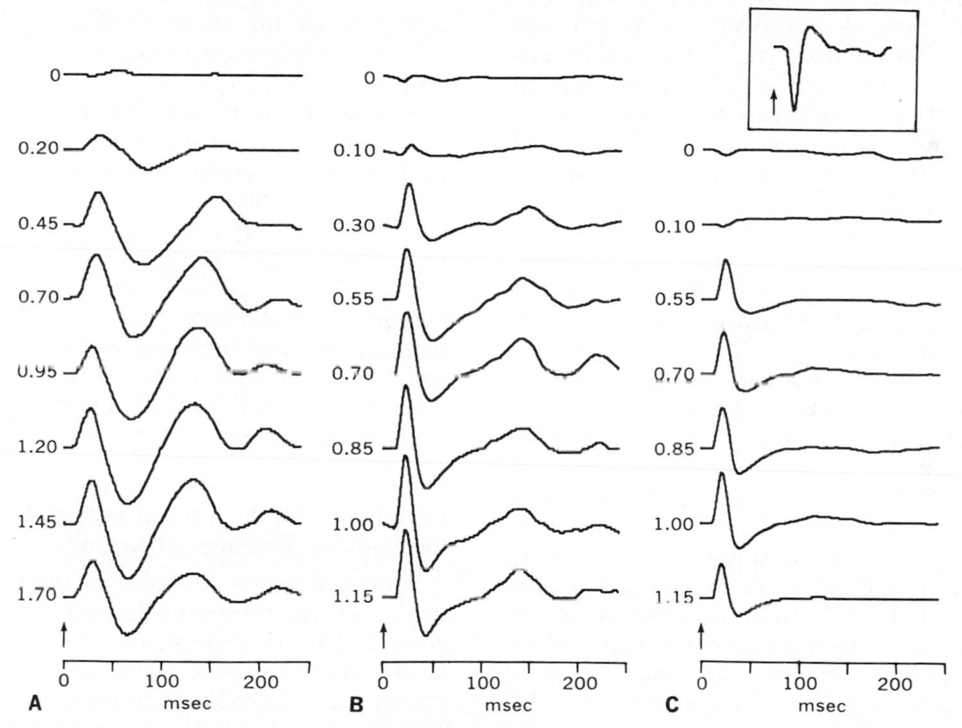

Fig. 7-8. Cortical potentials from postcentral gyrus of monkey evoked by brief mechanical stimulus to skin; positivity at electrode tip downward.

In inset at upper right, a typical evoked potential recorded on cortical surface with 2 mm, ball-tipped platinum wire electrode, a "gross electrode," is shown. Three columns of superimposed tracings (**A** to **C**) show evoked potentials recorded with microelectrode passed through cortex in direction normal to its surface. Numbers to left of each series indicate depth (in mm below cortical surface) at which responses were obtained. Each tracing represents computed average of 16 successive responses. Vertical arrows mark onset time of 10 msec mechanical stimulus to skin. Stimulus was delivered repetitively at rate of 1/2 sec; skin indentation was about 1 mm. In all instances stimulus was applied at peripheral locus from which maximal cortical activation was obtained, located on first digit of contralateral hand.

In **A** are plotted results of microelectrode penetration into cortex of postcentral gyrus of unanesthetized monkey. Subsequently, 10 mg/kg of sodium pentobarbital was administered intravenously to animal, and another cortical penetration was made close to the first. Recorded evoked potentials are shown in **B.** With electrode deep within cortex (1.15 mm), second and equal dose of anesthetic was given, and evoked responses were observed at various depths as microelectrode was withdrawn. These later results are shown in **C.** Sensitivity of recording amplifier was 4 times greater in **A** than in **B** and **C.**

extracellular current flows generated by the local postsynaptic responses of neurons. Action potentials in axons are discernible only in records made from the cortical surface when large numbers of thalamocortical fibers are activated synchronously, as by electrical stimulation of thalamic relay nuclei.

Typical evoked potentials recorded from the postcentral gyrus of a monkey, evoked by brief mechanical stimuli to the skin, are shown in Fig. 7-8. The positive-negative wave form recorded on the surface (inset) could result from one or a combination of the following: (1) depolarization of cell somata deep within the cortex, followed by or overlapped with depolarization of membranes lying immediately beneath the electrode, i.e., the apical dendrites; (2) polarization of dendrites followed by their depolarization; (3) depolarization of deep-lying somata followed by their polarization; etc. Some further evidence, not completely conclusive, is obtained by observing the changes in the evoked potential as a recording microelectrode is passed through the cortex in a direction normal to its surface. As the records in Fig. 7-8 show, the initial positive component dwindles quickly after penetration, while the negative component grows remarkably and remains of great amplitude throughout the traverse of the cellular layers.

Under these conditions the afferent barrage of impulses has produced, either mono- or oligosynaptically, a set of standing dipoles distributed throughout the cellular layers of the cortex, dipoles oriented with positive ends toward the cortical surface. The vertical orientation of synaptic relations within the cortex fits this suggestion. In this particular case the site of microelectrode penetration was at that cortical locus most intensely activated by the peripheral stimulus. The initial and predominantly negative configuration of the slow-wave evoked responses recorded within the cortex is taken to indicate the predominance of an early excitatory inflow to the cells in that region. The presence of PSPs, here recorded extracellularly as slow waves, indicates that the anesthetic agent has not blocked synaptic transmission within the cortex. It has, however, exerted a powerful effect upon the spike-generator mechanism of the cortical neurons, for under deep anesthesia only a few cells, relative to the total number present, discharge action potentials.

When the same experiment is repeated in the absence of anesthesia (Fig. 7-8, *A*), the slow-wave evoked responses are more complex, and averaging methods are commonly used to bring out the signal—the evoked potential—from the ongoing EEG. The sequential reversal of the sign of the initial phase of the evoked potential is identical to that observed in anesthetized animals, and the initial wave is frequently followed by "late" or secondary waves. The secondary waves, even when evoked by sensory stimuli, are commonly not confined to the relevant cortical sensory area and may be very widely distributed over the surface of the hemisphere. When recorded outside the primary receiving area, they may be initially and wholly negative in sign, an event interpreted to indicate an intense depolarization of the apical dendrites. This fact, plus their wide distribution, suggests that they are the sign of collateral activation by the afferent inflow of the generalized thalamocortical system, which is described in Chapter 8.

The technique of deriving the evoked potential from the ongoing EEG waves by the method of averaging makes it possible to study evoked potentials in human beings, using recordings from the surface of the scalp with appropriate precautions to eliminate artifact due to evoked reflex activation of muscle.[19] Investigations have now been undertaken to correlate these electrical changes with the behavioral events of sensation and perception as well as their alterations in patients with disease or injury of the nervous system.[71]

Correlation of the evoked potential with the impulse discharge of neurons

Thus far efforts to correlate the events in single cortical cells with the spontaneous ongoing EEG recorded on the surface have not been very successful, and more statistical studies are required to determine the relation. This is so because the waves of the EEG are the complex result of all the membrane conductance changes in neurons within the "view" of the recording electrode (Fig. 7-5). More precise correlations have been observed between the surface-recorded evoked potential and the EPSP-IPSP sequence of membrane potentials produced in cortical cells by stimulation of thalamic relay nuclei.[29] Further, a very precise correlation can sometimes be shown to exist between the evoked potential recorded with a microelectrode within the cortex and close to the cell for which correlation is sought. However, such a parallelism is not always observed, for some cells may be inhibited when the form of the local intracortical potential predicts excitation.

What is certain is that from the moment a recording microelectrode penetrates layer II of a sensory cortex, down across all the cellular layers of the cortex, an appropriately placed sensory stimulus evokes at the microelectrode an initially negative wave. This is indicative of intense postsynaptic depolarization, and associated with it is the impulse activity of cells in all layers of the cortex.

Volume conductor theory

F. J. BRINLEY, Jr.

The most direct method of recording an action potential from an excitable tissue is by placing a microelectrode inside the cell. However, in some cases, e.g., clinical electroencephalography and electrocardiography, intracellular microelectrode recording of transmembrane potentials is not feasible, and it is necessary to record electric potentials with electrodes placed some distance from the actual site at which the signal is generated. Although the potentials observed in the two recording situations appear dissimilar, they are generated by the same events and are related mathematically. The analysis that describes this relation is called the volume conductor theory.

As an illustration of the sort of reasoning involved, consider the relation between electrical signals recorded by electrodes located in two different positions as an action potential propagates along a nerve. The two recording arrangements to be compared are (1) an intracellular micropipet and (2) an external electrode located some distance from the axon surface. For mathematical simplicity the fiber is assumed to be unmyelinated, of uniform diameter, infinitely long, and contained in an infinite, electrically homogeneous conducting volume.

Since the action potential occupies a finite period of time and propagates along the nerve with a finite velocity, at any given instant the action potential occupies a discrete length of nerve. The length of fiber (L) so involved is given by L = Vt, where V is the velocity of propagation (assumed constant) and t is the time required for a complete action potential. The assumption of a constant velocity of propagation implies that the temporal variation of potential at a fixed point is equivalent to a spatial variation at a fixed time. Experimentally it may be easier to use a fixed electrode and measure the time course of an action potential as it passes by, but conceptually it is simpler to consider the action potential as "frozen in time" and study the spatial variation in potential as an exploring electrode is moved along the nerve.

The theory can be considered conveniently in three steps. (1) The cable theory relates the transverse membrane potential at a particular point to a transverse current density at the same point. (2) Ohm's law relates a flow of current (i.e., current density) in a region to the strength of the electric field in that region. (3) Coulomb's law relates the potential observed at a point in space to a charge distribution.

1. It can be shown from the cable theory (p. 40) that the transmembrane current flow (amps/cm²) is proportional to the second derivative of the transmembrane potential (V_m) at that point (Fig. 7-9).

$$i_m(x = x_1) = \left.\frac{\partial^2 V_m}{\partial x^2}\right|_{x = x_1} \cdot \frac{1}{R_o + R_i} \qquad (1)$$

V_m can represent either the total observed potential across the membrane or only the electrotonic potential (i.e., the numeric value of the deviation of the actual potential from the resting potential, which is assumed to be temporally and spatially invariant). Since the transverse membrane current is proportional to the second derivative of the potential, a constant resting potential will not contribute to i_m. R_o and R_i are the external (tissue fluid) and internal (axoplasmic) electrical resistances, respectively (ohm/cm); x represents distance along the nerve (in centimeters).

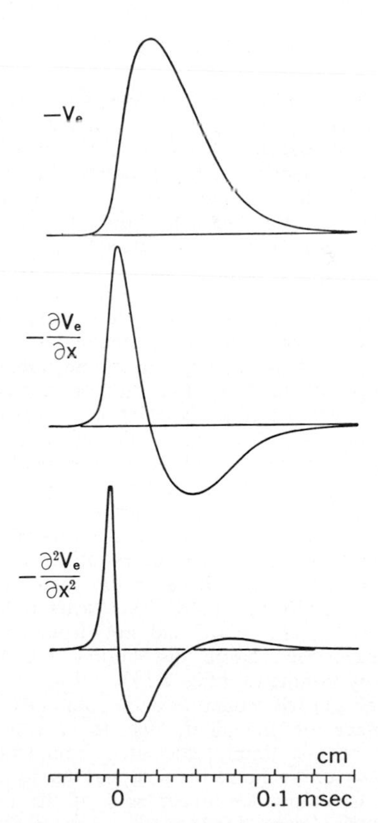

Fig. 7-9. Graphic differentiation of transmembrane action potential, $-V_e$, upper curve, to give membrane current, lower curve. Time and distance scales are given at bottom of figure. (From Lorente de Nó.[36a])

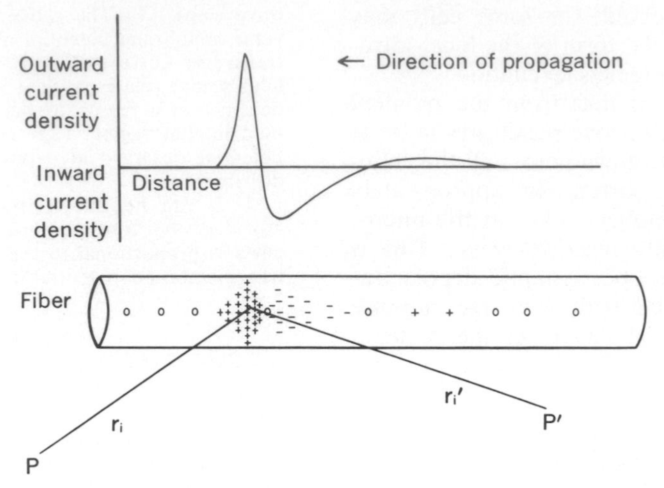

Fig. 7-10. Schematic diagram illustrating relation between membrane current density and conceptual (virtual) charge density on surface of nerve fiber used to calculate external field by Coulomb's law. Density of charge distributed on surface of fiber is indicated qualitatively by number of + or − symbols. The small *o* indicates no net surface charge.

2. The magnitude of the current density at a particular point on the surface of the nerve will be proportional to the electric field at that point (Ohm's law).

$$i_m(x = x_1) \propto E(x = x_1) \qquad (2)$$

This electric field can be considered to result from the presence of electric charges on the surface of the fiber. Therefore the spatial variation of transmembrane potential along a nerve fiber may be regarded as determining a charge distribution on the outside of the fiber. It should be understood that the assumption of net surface electric charge made here is a mathematical artifice to aid the derivation. Such charges do not exist physically and should not be confused with the separated ion charges on either side of the membrane that are responsible for the resting membrane potential. A distribution of fictitious ion charges such as that described here is frequently called a "virtual charge" distribution. The magnitude of this assumed charge, at a specified point, will be proportional to the second derivative of the transmembrane potential at that point. Since the transmembrane current (which is also proportional to the second derivative) can be either positive or negative, the charge can have either sign. Because the membrane potential is radially symmetric, i.e., varies only along the length of the nerve and not around the circumference, the charge distribution will likewise be radially symmetric (Fig. 7-10).

3. The electric potential at a point (P) not on the surface of the fiber, due to this array of charges, can be simply calculated from Coulomb's law ($V_p = q/r_p$). Since the potential due to multiple charges is simply the linear sum of the potential due to each charge, one has:

$$V_p = \sum_n \frac{q(n)}{r_p(n)} \qquad (3)$$

where V_p is the potential at point P and r_p is the distance of the *n*th charge q to point P. The summation extends over all charges.

In the foregoing statements it has been assumed that the diameter of the fiber is very small compared to the distance to the field point (P). Even if the field point is assumed to be so close to the surface of the fiber that the charge must be considered as distributed on a circular surface rather than concentrated at a point, no new physical principles are involved in the analysis. Coulomb's law can still be applied. However, the mathematical formulation is somewhat more complicated.

The cable theory used to derive equation 1 assumes that there is no radial component of current in either the external or internal medium; i.e., the longitudial currents flow only in the regions immediately adjacent to the nerve membrane and do not circulate to any extent in the medium distant from the membrane. While this is approximately true for an isolated fiber immersed in oil or suspended in a moist chamber where the current in the external medium is constrained to flow in the narrow path of conducting fluid outside of the nerve, it is not generally true for the nerve in situ. Clark and Plonsey[26] have presented a formal mathematical solution for a long unmyelinated nerve fiber in a conducting medium conducting an action potential along its surface. The formal solution contains two terms. The first term, which is proportional to the membrane current, is equivalent to the solution of equation 1. The second term is a complicated integral, which can be considered as a correction term indicating the deviation of the simple cable theory from the true physical situation. Clark and Plonsey have evaluated this integral with the aid of a computer. The calculations indicate that the radial current (and hence radial component of the field) in the external medium may be substantial. According to their analysis, an action potential recorded in the external medium remains triphasic, as in the simple analysis given here, but the amplitude and phase relationship of the potential recorded in the volume may be quite different from that predicted from simple cable theory.

Applications of volume conductor theory

The computational complexities of applying equation 3 to each element of a large population of excitable cells with different spatial orientations conducting asynchronously preclude a quantitative calculation of the configuration of extracellular potentials recorded in a volume conductor, even if the basic assumptions such as homogeneity and infinitely large size of the volume conductor were to be met. For this reason, in practice, electrophysiologists and clinicians alike rely upon semiempirical correlations between function of an excitable tissue and configuration of the observed potentials. Nevertheless, a few useful qualitative generalizations can be made on the basis of the simple analysis just presented and illustrated with three specific examples.

Electrocardiogram (ECG). The general configuration of the ECG as recorded with either limb or unipolar leads is obtained from the second derivative of the transmembrane action potential of the myocardial muscle cells. As can be seen from Fig. 7-10, when a conducting fiber is oriented so that the action potential is propagating toward the sensing electrode (e.g., point P), the observed deflection will be positive because the electrode is relatively nearer the positive than the negative charge. When the action potential is propagating away from the electrode (e.g., point P′), the observed deflection will be negative because the electrode is now closer to negative charge. A practical example can be seen in the precordial leads of the ECG. The small initial positive R wave seen in the right precordial leads represents progressive depolarization of the right ventricle muscle fibers moving in the general direction of the right sternal border. This small R wave is quickly superceded by the larger negative S wave that represents activation of the much larger left ventricular mass. Propagation in the left ventricle is directed toward the left axilla and away from the sternum; hence the deflection is negative. The potential configuration is approximately reversed in the left precordial leads, and a large initial R wave representing left ventricular depolarization is recorded.

Extracellular potentials recorded within substance of the brain by small microelectrodes. Diagrams similar to Fig. 7-10 are sometimes used to locate regions of "active" or "passive" membrane along a length of nerve or on the surface of an excitable cell. Negative potentials are presumed to represent sinks of current toward which current flows in the extracellular field. Since the current flow across the membrane in this sink region is inward, the membrane is considered to be actively depolarized, i.e., undergoing regenerative sodium and potassium conductance changes. Positive potentials are presumed to indicate sources of current flow, implying that the membrane nearest the sensing electrode is passively supplying current to a more remote, active region where the ionic current is inward.[60] Although this interpretation is approximately true for a very simple case, e.g., an isolated single nerve fiber, it is not valid generally for the complicated arrangements of conducting elements found in the CNS.

The theoretical extracellular fields of some relatively simple arrangements of dendrites and somata have been calculated by Rall.[53] One such orientation of dendrites is illustrated in Fig. 7-11. Fig. 7-12 shows the extracellular field resulting from this particular orientation during the peak of an action potential generated in the soma; i.e., the soma is considered to be active and all of the dendrites passive. The extracellular field along the dendrites, away from their immediate surface, is negative out to a distance of about eight soma radii, although these elements are not undergoing a regenerative depolarization. The potential field shown in Fig. 7-12 has been calculated assuming an infinite, homogeneous, isotropic extracellular space. Since practically all of the extraneuronal volume in mammalian brain is occupied by glial cells, the true extracellular space is probably rather small. This circumstance could cause considerable quantitative modification of the shape of the potential surfaces illustrated in Fig. 7-12. However, Nelson and Frank[43] have confirmed experimentally the theoretical conclusions for the spinal motoneuron; the extracellular field actually is predominantly negative and is recordable (i.e., is greater than a few microvolts) to at least 500 to 700μ from the soma.

Fig. 7-13 shows the theoretical relations between the entire time course of an intracellular action potential arising from the soma of a hypothetical cell with passive, symmetrically oriented dendrites and extracellular potentials recorded at varying distances from the soma. In actual experiments, Terzuolo and Araki[66] used two parallel microelectrodes joined together and recorded intra- and extracellular potentials simultaneously from spinal motoneurons. Since the tip separations were about 3 to 20μ, the pipet recording the extracellular potential must necessarily have been reasonably close to the cell soma. In this recording circumstance the extracellular potential sequences following antidromic activation was usually positive-negative-positive instead of simply negative-positive, as shown in Fig. 7-13. The first additional positive wave in these triphasic responses represents current flow from the initially inactive soma-dendritic complex into the depolarized initial segment.

In some cases, relatively large, positive, extracellular action potentials can be recorded from neurons in both brain and spinal cord.[42] It is thought that these positive potentials result when the electrode tip is pushed against the cell membrane, causing sufficient local damage to render the spot under the electrode, but not the adjacent areas, inexcitable. The amplitude of the positive extracellular potentials, up to several millivolts, is considerably larger than that of the usual negative extracellular spikes, which are no more than a few hundred microvolts. In the case of the positive spikes the difference in amplitude presumably results because the tip of the microelectrode is pressed against the surface of the cell or perhaps partially plugged with torn glial cell membranes, creating a higher resistance for current flow. The resultant IR drop is therefore greater than in the case of a microtip well out in the extracellular space.

Electric potentials recorded from surface of the brain. A variety of multiphasic electric potentials can be recorded from the cortical surface in response to

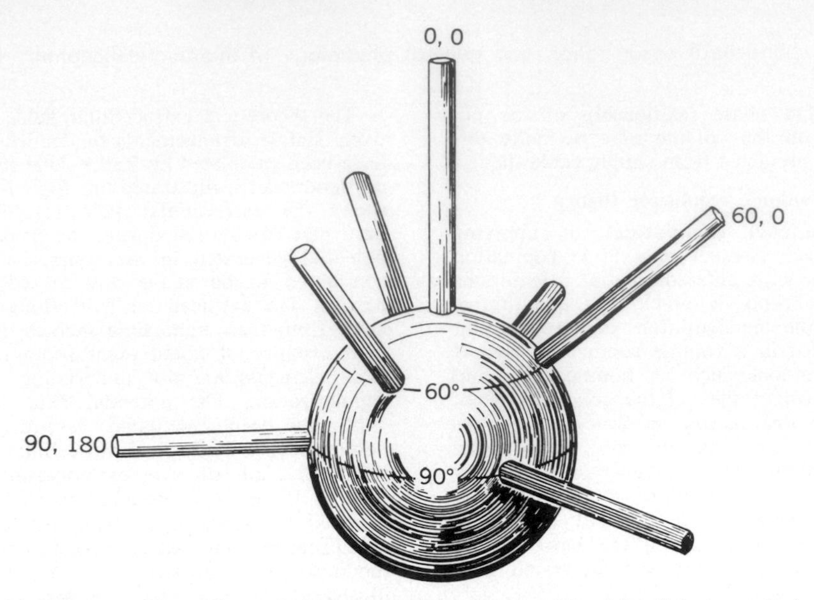

Fig. 7-11. Diagram illustrating relative orientation of seven "dendrites," but not their lengths, around spherical "soma" used to calculate potential field shown in Fig. 7-12. Numbers indicate angular coordinates of the three dendrites shown in Fig. 7-12. (From Rall.[53])

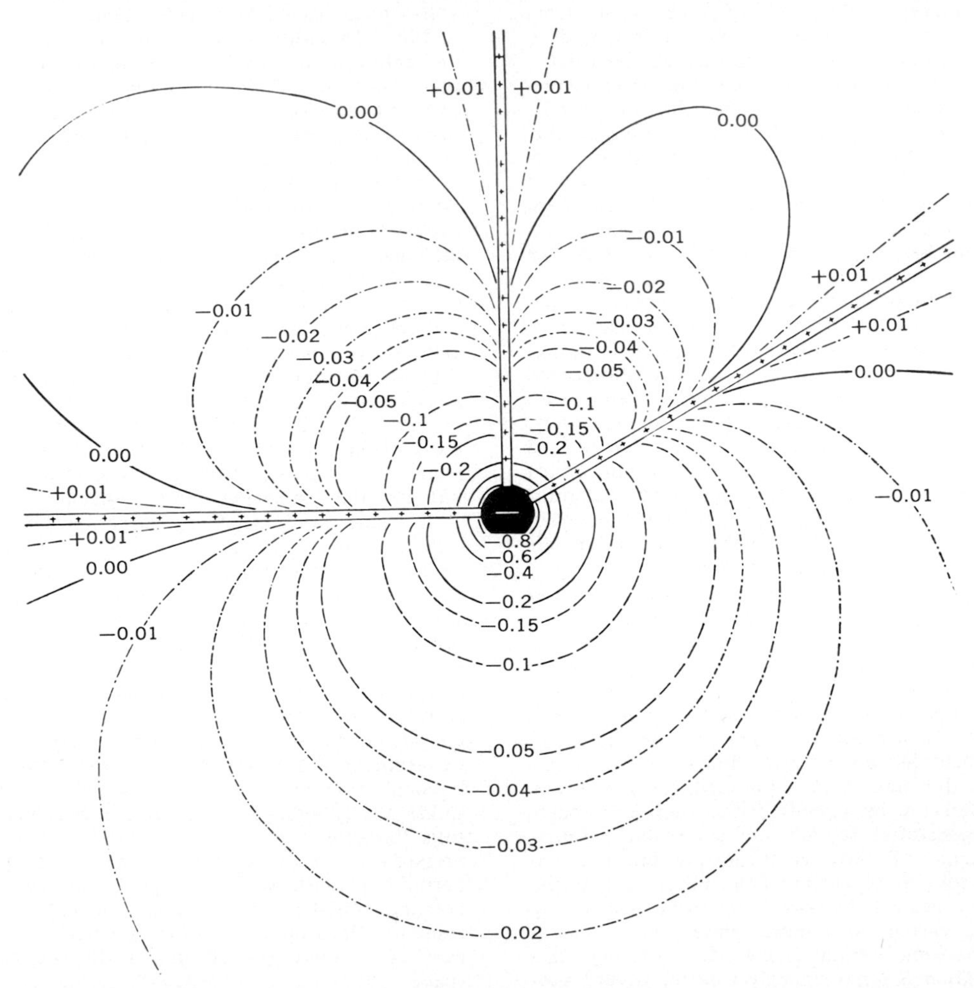

Fig. 7-12. Isopotential contours for dendrite-soma model illustrated in Fig. 7-11. Plane of figure is that defined by three labeled dendrites in Fig. 7-11. Zero potential contour is shown as solid line. Field corresponds to that which occurs at peak of action potential generated in neuron with active soma and passive dendrites. Figures are in millivolts. (From Rall.[53])

diverse stimuli to the nervous system, peripheral as well as central. These electrical responses result from current flow in the extracellular space caused by action potentials or synaptic potentials at the surface of many individual neurons. The observed surface potentials bear only a general relationship to the membrane currents generated by a single neuron or part of a neuron for three reasons: (1) the relatively large surface electrodes commonly used that permit recording from many units, (2) the complex orientation of cortical elements, and (3) asynchronous variation in membrane potential of the units. However, the principles presented in the initial paragraphs of this section can be used to correlate the sign of the observed surface potential with the extracellular currents flowing within the brain substance.

As an example of the application of volume conductor theory to the evoked activity recorded from the cortical surface, consider a large electrode resting on the surface of the exposed brain. A second electrode is placed some distance away, on an inactive portion of the cortex or skull. This second or "indifferent" electrode is placed so far away from the site of electrical activity that its potential does not vary significantly with time. The potential at this point can therefore be taken as defining a "zero of potential," against which fluctuations of potential at the first recording electrode can be compared.

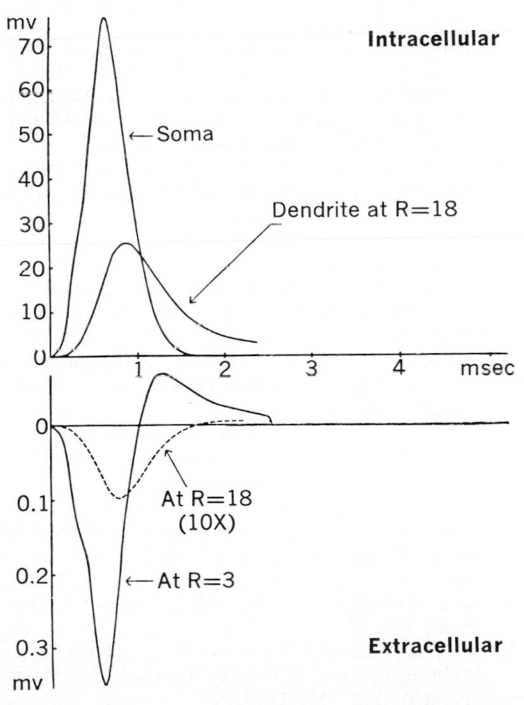

Fig. 7-13. Time course of calculated intra- and extracellular potentials, assuming that action potentials arise in soma (as by intracellular stimulation) and that electrically passive dendrites are symmetrically arranged around soma. Curve labeled "dendrite" in upper record shows delay and attenuation of soma action potential as it spreads electronically along dendrites. (From Rall.[53])

If cells or cell elements in the immediate vicinity of the recording electrode are acting as sources of membrane current, i.e., current is flowing out of the region of the recording electrode, then the potential will go positive with respect to an indifferent electrode.* If the units are acting as electric sinks, the current will flow into the region and the brain substance under the recording electrode will become negative. This analysis identifies only the general direction and intensity of net current flow in the region near the recording electrode. It provides no information concerning the distribution, number, or identity of the cellular elements acting as sources or sinks within the region, nor does it identify the regions elsewhere that are supplying current to or receiving current from the region under the electrode. If the electrical activity generated within the brain by the original stimulus is complex, the evoked surface potentials may also reflect this complexity and consist of multiphasic waves whose component deflections have varying durations and/or polarities.

Since the magnitude of the potential is equal to the IR drop between the tip of the recording electrode and the indifferent electrode, the actual size of the potential will depend to some extent on the amount of fluid shunt upon the surface of the brain and the firmness with which the electrode rests upon the cortical surface as well as on the location and intensity of the intracortical sources and sinks. The size of the potential will not, however, depend on the location of the indifferent electrode as long as it is a great distance away from the recording site.

In certain evoked potentials the sources and sinks are relatively discrete and oriented more or less normal to the cortical surface. In such cases the potential surfaces corresponds roughly to those of a dipole, with one charge (e.g., a negative one) located at or very near the surface and a second charge of opposite sign (in this example, a positive one) located at the appropriate position in the cortical substance. These charges are simply lumped values indicating the intensity and general location of a source-sink pair at a particular instant during the course of an evoked potential. They should not be considered as representing the charge distribution on the surface of any single unit.

The representation of a source-sink pair as a dipole implies that the negative (in this example) amplitude of the surface wave should diminish as the recording electrode passes down into the cortical substance, reach zero at some isopotential surface between the source and sink, and then increase in positive amplitude as the source is approached.

For further details on these and other problems in identifying the cellular elements involved in producing various evoked potentials, refer to other sections in this text.

*These statements are only approximately true, in view of Rall's demonstration that negative potentials (i.e., with respect to an indifferent electrode) can be observed in the vicinity of a passive membrane acting as a source (Fig. 7-13). However, the cell bodies of the majority of cortical neurons lie 100μ or more beneath the cortical surface. The nerve elements in immediate proximity to an electrode on the arachnoid surface are therefore axons or dendrites each relatively distant from its own cell body. In such cases the conventional notions of source or sink apply fairly well.

REFERENCES

General reviews

1. Ajmone Marsan, C.: The thalamus. Data on its functional anatomy and on some aspects of thalamo-cortical integration, Arch. Ital. Biol. **103:**847, 1965.
2. Bullock, T. H., and Horridge, G. A.: Structure and function of the nervous systems of invertebrates, San Francisco, 1965, W. H. Freeman & Co., Publishers.
3. Chow, K. L., and Leiman, A. L.: The structural and functional organization of the neocortex, Neurosci. Res. Symp. Summ. **5:**149, 1971.
4. Colonnier, M.: The structural design of the neocortex. In Eccles, J. C., editor: The brain and conscious experience, Berlin, 1966, Springer Verlag.

4a. Elias, H., and Pauly, J. E.: Human microanatomy, Chicago, 1960, Da Vinci.

5. Glaser, G. H.: The normal electroencephalogram and its reactivity. In Glaser, G. H., editor: EEG and behavior, New York, 1963, Basic Books, Inc., Publishers.
6. Hassler, R.: Anatomy of the thalamus. In Schaltenbrand, G., and Bailey, P., editors: Introduction to stereotaxis with an atlas of the human brain, New York, 1959, Grune & Stratton, Inc., vol. 1.
7. Krupp, P., and Monnier, M.: The unspecific intralaminary modulating system of the thalamus, Int. Rev. Neurobiol. **9:**45, 1966.
8. MacKay, D. M.: Evoked brain potentials as indicators of sensory information processing, Neurosci. Res. Symp. Summ. **4:**397, 1970.

8a. Penfield, W., and Jasper, H. H.: Epilepsy and the functional anatomy of the human brain, Boston, 1954, Little, Brown & Co.

9. Scheibel, M. E., and Scheibel, A. B.: Elementary processes in selected thalamic and cortical subsystems—the structural substrates. In Schmitt, F. O., editor: The neurosciences. Second study program, New York, 1970, The Rockefeller University Press.
10. Walter, D. O., and Brazier, M. A. B.: Advances in EEG analysis, Electroencephalogr. Clin. Neurophysiol. suppl. **27:**1, 1969.

Original papers

11. Adey, R. W.: Spectral analysis of EEG data from animals and man during alerting, orienting and discriminative responses. In Neurophysiology of attention, London, 1968, Butterworth & Co., Ltd.
12. Adrian, E. D., and Matthews, B. H. C.: The Berger rhythm: potential changes from the occipital lobes in man, Brain **57:**355, 1934.
13. Adrian, E. D., and Yamagiwa, K.: The origin of the Berger rhythm, Brain **58:**323, 1935.
14. Akert, K.: Comparative anatomy of frontal cortex and thalamofrontal connections. In Warren, J. M., and Akert, K., editors: The frontal granular cortex and behavior, New York, 1964, McGraw-Hill Book Co.
15. Amassian, V. E.: Microelectrode studies of the cerebral cortex, Int. Rev. Neurobiol. **3:** 67, 1961.
16. Amassian, V. E., Waller, H. J., and Macy, J., Jr.: Neural mechanism of the primary somatosensory evoked potential, Ann. N. Y. Acad. Sci. **112:**5, 1964.
17. Angevine, J. B., Locke, S., and Yakovlev, P. I.: Thalamocortical projection of the ventral anterior nucleus in man, Arch. Neurol. **7:**518, 1962.
18. Berger, H.: Das Elektrenkephalogramm des Menschen, Acta Nova Leopold. **6:**173, 1938.
19. Bickford, R. G., Jacobsen, J. L., and Cody, D. T.: Nature of average evoked potentials to sound and other stimuli in man, Ann. N. Y. Acad. Sci. **112:**204, 1964.
20. Brazier, M. A. B.: A history of the electrical activity of the brain. The first half-century, New York, 1961, The Macmillan Co.
21. Brodmann, K.: Vergleichende Lokalisationslehre der Grosshirnrinde in ihren prinzipien dargestellt auf Grund des Zellenbaues, Leipzig, 1909, Johann Ambrosius Barth.
22. Brodmann, K.: Feinere Anatomie des Grosshirns. In Handbuch der Neurologie: allgemeine Neurologie, Berlin, 1910, Springer Verlag, vol. 1.
23. Campbell, A. W.: Histological studies on the localization of cerebral function, Cambridge, England, 1905, University Press.
24. Carman, J. B., Cowan, W. M., and Powell, T. P. S.: Cortical connexions of the thalamic reticular nucleus, J. Anat. **98:**587, 1964.
25. Carmel, P. W.: Efferent projections of the ventral anterior nucleus of the thalamus in the monkey, Am. J. Anat. **128:**159, 1970.
26. Clark, J., and Plonsey, R.: A mathematical evaluation of the core conductor model, Biophys. J. **6:**95, 1966.
27. Cowan, J. D.: Statistical mechanics of neural networks. In Gerstenhaber, M., editor: Mathematical questions in biology; proceedings of the second annual symposium, Providence, 1969, American Mathematical Society.
28. Cragg, B. G.: The density of synapses and neurones in the motor and visual areas of the cerebral cortex, J. Anat. **101:**639, 1967.
29. Creutzfeldt, O. D., Watanabe, S., and Lux, H. D.: Relations between EEG phenomena and potentials of single cortical cells (parts I and II), Electroencephalogr. Clin. Neurophysiol. **20:**1, 1966.
30. Crouch, R. L.: The nuclear configuration of the thalamus of macacus rhesus, J. Comp. Neurol. **59:**451, 1934.
31. Economo, C. von: The cytoarchitectonics of the human cerebral cortex (translated by S. Parker), Oxford, 1929, Humphrey-Milford.
32. Elul, R.: Brain waves: intracellular recording and statistical analysis help clarify their physiological significance. In Rochester conference on data acquisition and processing in biology and medicine, Oxford, 1966, Pergamon Press, Ltd., vol. 5.
33. Globus, A., and Scheibel, A. B.: Pattern and field in cortical structure: the rabbit, J. Comp. Neurol. **131:**155, 1967.
34. Goldstein, S.: Phase coherence of the alpha rhythm during photic blocking, Electroencephalogr. Clin. Neurophysiol. **29:**127, 1970.
35. Joseph, J. P., Remond, A., Rieger, H., and Lesevre, N.: The alpha average. II. Quantitative model and the proposition of a theoretical model, Electroencephalogr. Clin. Neurophysiol. **26:**350, 1969.

36. Kandel, E. R., Spencer, W. A., and Brinley, F. J., Jr.: Electrophysiology of hippocampal neurons (parts I to IV), J. Neurophysiol. **24:** 225, 1961.
36a. Lorente de Nó, R.: A study of nerve physiology, Stud. Rockefeller Institute Med. Res. **132:** 1, 1947.
37. Lorento de Nó, R.: Cerebral cortex: architecture, intracortical connections, motor projections. In Fulton, J. F., editor: Physiology of the nervous system, London, 1949, Oxford University Press.
38. Mehler, W. R.: Further notes on the center median nucleus of Luys. In Purpura, D. P., and Yahr, M. D., editors: The thalamus, New York, 1966, Columbia University Press.
39. Milstein, V.: Contingent alpha blocking: conditioning or sensitization? Electroencephalogr. Clin. Neurophysiol. **18:**272, 1965.
40. Minderhoud, J. M.: An anatomical study of the efferent connections of the thalamic reticular nucleus, Exp. Brain Res. **12:**435, 1971.
41. Moore, G. P., Perkel, D. H., and Segundo, J. P.: Statistical analysis and functional interpretation of neuronal spike data, Ann. Rev. Physiol. **28:**493, 1966.
42. Nauta, W. J. H.: Neural associations of the amygdaloid complex of the monkey, Brain **85:** 505, 1962.
43. Nelson, P. G., and Frank, D.: Extracellular potential fields of single motoneurons, J. Neurophysiol. **27:**914, 1964.
44. Olszewski, J.: The thalamus of the Macaca mulatta. An atlas for use with the stereotaxic instrument, Basel, 1952, S. Karger.
45. Pakkenberg, H.: The number of nerve cells in the cerebral cortex of man, J. Comp. Neurol. **128:**17, 1966.
46. Perkel, D. H., Gerstein, G. L., and Moore, G. P.: Neuronal spike trains and stochastic point processes. I. The single spike train, Biophys. J. **7:**419, 1967.
47. Perkel, D. H., Gerstein, G. L., and Moore, G. P.: Neuronal spike trains and stochastic point processes. II. Simultaneous spike trains, Biophys. J. **7:**440, 1967.
48. Phillips, C. G.: Intracellular records from Betz cells of cat, Q. J. Exp. Physiol. **41:**58, 1956.
49. Poggio, G. F., and Viernstein, L. J.: Time series analysis of the discharge sequences of thalamic somatic sensory neurons, J. Neurophysiol. **27:** 517, 1964.
50. Powell, T. P. S., and Cowan, W. M.: The interpretation of the degenerative changes in the intralaminar nuclei of the thalamus, J. Neurol. Neurosurg. Psychiatry **30:**140, 1967.
51. Purpura, D. P.: Nature of electrocortical potentials and synaptic organization in cerebral and cerebellar cortex, Int. Rev. Neurobiol. **1:** 48, 1959.
52. Purpura, D. P.: Operations and processes in thalamic and synaptically related neural subsystems. In Schmitt, F. O., editor: The neurosciences. Second study program, New York, 1970, The Rockefeller University Press.
53. Rall, W.: Electrophysiology of a dendritic neuron model, Biophys. J. **2:**145, 1962.
54. Ramón y Cajal, S.: Histologie du systeme nerveus, Paris, 1909-1911, Maloine, vol. 2.
55. Remond, A., et al.: The alpha average. I. Methodology and description, Electroencephalogr. Clin. Neurophysiol. **26:**245, 1969.
56. Rose, J. E.: The cellular structure of the auditory region of the cat, J. Comp. Neurol. **91:** 409, 1949.
57. Rose, J. E., and Woolsey, C. N.: Structure and relations of limbic cortex and anterior thalamic nuclei in rabbit and cat, J. Comp. Neurol. **89:**279, 1948.
58. Rose, J. E., and Woolsey, C. N.: Organization of the mammalian thalamus and its relationships to the cerebral cortex, Electroencephalogr. Clin. Neurophysiol. **1:**391, 1949.
59. Rose, J. E., and Woolsey, C. N.: Cortical connections and functional organization of the thalamic auditory system of the cat. In Harlow, H. F., and Woolsey, C. N., editors: Biological and biochemical bases of behavior, Madison, 1958, University of Wisconsin Press.
60. Rosenthal, F.: Relationships between positive-negative extracellular potentials and intracellular potentials in pyramidal tract neurons, Electroencephalogr. Clin. Neurophysiol. **30:**38, 1971.
61. Scheibel, M. E., and Scheibel, A. B.: The organization of the nucleus reticularis thalami: a Golgi study, Brain Res. **1:**43, 1966.
62. Scheibel, M. E., and Scheibel, A. B.: The organization of the ventral anterior nucleus of the thalamus: a Golgi study, Brain Res. **1:** 250, 1966.
63. Scheibel, M. E., and Scheibel, A. B.: Structural organization of nonspecific thalamic nuclei and their projection toward the cortex, Brain Res. **6:**60, 1967.
64. Scheibel, M. E., and Scheibel, A. B.: On the nature of dendritic spines—report of a workshop, Comm. Behav. Biol. **1:**231, 1968.
65. Stefanis, C., and Jasper, H.: Intracellular microelectrode studies of antidromic responses in cortical pyramidal tract neurons, J. Neurophysiol. **27:**828, 1964.
66. Tersuolo, C. A., and Araki, T.: An analysis of intra- versus extracellular potential changes associated with activity of single spinal motoneurons, Ann. N. Y. Acad. Sci. **94:**547, 1961.
67. Terzuolo, V. A., and Bullock, T. H.: Measurement of imposed voltage gradient adequate to modulate neuronal firing, Proc. Natl. Acad. Sci. U. S. A. **42:**687, 1956.
68. Walker, A. E.: The primate thalamus, Chicago, 1938, University of Chicago Press.
69. Walter, W. G.: Intrinsic rhythms of the brain. In Magoun, H. W., editor, Neurophysiology section: Handbook of physiology, Baltimore, 1959, The Williams & Wilkins Co.
70. Wells, C. E.: Alpha wave responsiveness to light in man. In Glaser, G. H., editor: EEG and behavior, New York, 1963, Basic Books, Inc., Publishers.
71. Whipple, H. E., and Katzman, R., editors: Sensory evoked response in man, Ann. N. Y. Acad. Sci. **112:**entire issue, 1964.
72. Yakovlev, P. I., Locke, S., and Angevine, J. B., Jr.: The limbus of the cerebral hemisphere, limbic nuclei of the thalamus, and the cingulum bundle. In Purpura, D. R., and Yahr, M. D., editors: The thalamus, New York, 1966, Columbia University Press.

8

VERNON B. MOUNTCASTLE

Sleep, wakefulness, and the conscious state: intrinsic regulatory mechanisms of the brain

In treating the problem of consciousness, neurobiologists assume that they deal with a certain aspect of the functional organization of brains that will eventually be defined in terms of neural mechanisms. Consciousness is an aspect of brain function for which, even from the behavioral point of view, there is no ready, brief definition, but several of its observable attributes are known with reasonable certainty.

1. *Consciousness is a neural phenomenon.* Organisms without nervous systems never display the publicly observable attributes of consciousness.

2. *Consciousness exists in other animals as well as in man.*[14] Evidence exists that consciousness appears in animals *pari passu* with the development of a complex nervous system as the master regulatory mechanism of the animal. But a nervous system per se is not enough. The attributes of consciousness do not appear when nervous systems containing relatively few neurons control behavior by linking afferent input via "releasers" to neural mechanisms governing innate behavioral patterns. It is difficult to draw any line separating those animals that are conscious from those that are not, and it is perhaps better to regard the different species as distributed along a continuum of an increasing degree and complexity of consciousness. The presence of the conscious control of action may, as an operational definition, be assumed when an organism displays the capacity for choice of action, the ability to set one goal aside in favor of another, the power to withhold action or reaction. Certainly a high order of consciousness is involved in anticipatory planning for action, in modifying action once initiated in terms of then-current events, and in the preparation of alternative stratagems to deal with abstract conceptualizations of events that may be encountered. All these latter compose a property of brains called intelligence.

What are some of the publicly observable aspects of behavior in man and in animals that suggest the presence of consciousness?

a. The *act of attention* and the capacity to shift attention selectively.
b. *Manipulation of abstract ideas,* preeminently a conscious process characteristic of human beings—the representation of abstract or general ideas by words or other symbols. No animal approaches this level, but evidence for elaborate means of communication between individual animals of many species suggests that this ability is present in higher animals in a rudimentary form.
c. *Capacity for expectancy* is further indicated by the use of tools by animals in the wild, by the organization of troops for hunting or food gathering, by the posting of sentinels, etc.
d. *Self-awareness and the recognition of other selves,* evidenced by the social and familial behavior of animals, by their organized play and elaborately imitative behavior, and by their ability to copy novel acts or utterances. This involves also the ability of some animals to profit by their fellows' experience.
e. *Esthetic and ethical values,* while rarely

observed, certainly exist in animals. Thus the rescue operations of dolphins to aid their comrades in respiratory distress indicate concern for the welfare of others.

3. *Consciousness is a variable quantity in a given individual;* i.e., human beings are from time to time more or less conscious. The state of introspective clarity and awareness of the external world varies in the same person over a range from excitement through normal alertness, to drowsiness, to light sleep, and then to deep sleep. Consciousness has the property of lability, and movement along this scale is relatively free: the cyclic variation from sleep to wakefulness is common experience. Perhaps the temporarily or permanently irreversible loss of consciousness called coma should be set apart from this continuum, as should the anesthetic state, the loss of consciousness associated with some epileptic attacks, and the alteration of consciousness produced by certain drugs. These all give further evidence that the level of consciousness depends on the state of brain activity.

4. *Consciousness presupposes perception,* a continual updating of the central neural reflection of the state of events in the external world. Leaving aside such special states as hypnosis, the available evidence indicates that a reduction or a lack of variety in afferent input leads to a reduction in the level of consciousness. Whether total reduction of input leads to total loss of consciousness is unsettled, perhaps because complex nervous systems may replicate afferent input by imagery, thus maintaining the conscious state.

5. *Consciousness presupposes memory,* a continuous storage of information about internal and external events, a record against which anticipatory plans for action are laid.

The neural mechanisms involved in these actions that give evidence for consciousness are poorly understood. Considerable information is available concerning systems of the brain that control and maintain levels of excitability, however, and some reason attaches to the proposition that the *overall level of consciousness depends on the level of excitability of the brain.* For this reason attention will now be given to those intrinsic systems that regulate the excitability—the readiness for action—of central neuronal populations, particularly those of the cerebral cortex.

INTRINSIC REGULATORY MECHANISMS OF THE FOREBRAIN: THE GENERALIZED THALAMOCORTICAL SYSTEM

Fundamental to orderly function of the brain is the presence within it of systems of neurons that by virtue of their wide distribution and terminal synaptic organization, serve to regulate neuronal excitability, exerting this influence from the level of the cerebral cortex to that of spinal motoneurons. This system occupies part of the tegmentum of the medulla, pons, and mesencephalon, projecting upward upon telencephalic distributors, one of which is the generalized thalamocortical system. A descending projection, the reticulospinal system, serves to funnel excitatory and inhibitory influences from the forebrain upon the spinal segmental apparatus. The *ascending reticular activating system* lies parallel with and receives collateral input from the great afferent systems, and driven thus indiscriminately, tunes appropriately the ongoing levels of excitability of cells of the cerebral cortex, the basal ganglia, and other large gray structures of the forebrain: it sets the stage for action. It is their indiscriminate activation by a variety of afferents, a property of their cells both individually and collectively considered, and the wide distribution of their effects that are indicated by the term "generalized." By contrast, the so-called *specific* afferent systems that they parallel are precisely organized for the transfer of sensory information, while the ascending reticular systems are engaged more generally in arousal and alerting functions. Both are essential for normal function; neither may be categorized as higher or lower than the other, for "without the former the animal is blind, deaf, or anesthetic; without the latter it cannot be aroused from a sleep-like state."[39] Almost all areas of the cortex receive projection fibers of the two types. The term "specific," in this context, indicates not only the thalamocortical stages of the great afferent systems but all others that project upon the cortex in a more or less precise, point-to-point manner, e.g., the projection of the ventralis lateralis upon the motor cortex or of the medialis dorsalis upon the frontal granular cortex. While any given cortical area, with rare exception, receives overlapping thalamic projections from a specific and generalized system, the two engage intracortical synaptic organizations in different ways and thus exert different effects on cortical neurons.

Generalized thalamocortical system and the recruiting phenomenon[1, 13, 70, 124]

Morison and Dempsey discovered in 1942 that electrical stimulation within the nuclei of the generalized thalamocortical system evokes changes in the electrical activity of large areas of the cerebral hemispheres. Further analysis of this phenomenon, particularly by Jasper, suggests that it receives activating influences from the brainstem reticular formation, in turn exerting control over neuronal excitability and thus the ongoing electrical activity of the cortex.

The electrical phenomena are illustrated by the records in Fig. 8-1. In an anesthetized animal, electrical stimulation of nuclei of the generalized thalamocortical system evokes slow waves over a large part of the cerebral cortex, of greater amplitude in frontal areas than elsewhere. The slow waves are usually and predominantly surface negative in sign. They show several remarkable characteristics. First, the amplitude of the evoked response grows from the first response, which may be very small or absent, to the second, to the third, etc. This property is called recruiting. Second, the amplitude of the response tends to cycle through recurrently waxing and waning phases. Third, recruiting responses are of long latency, frequently greater than 25

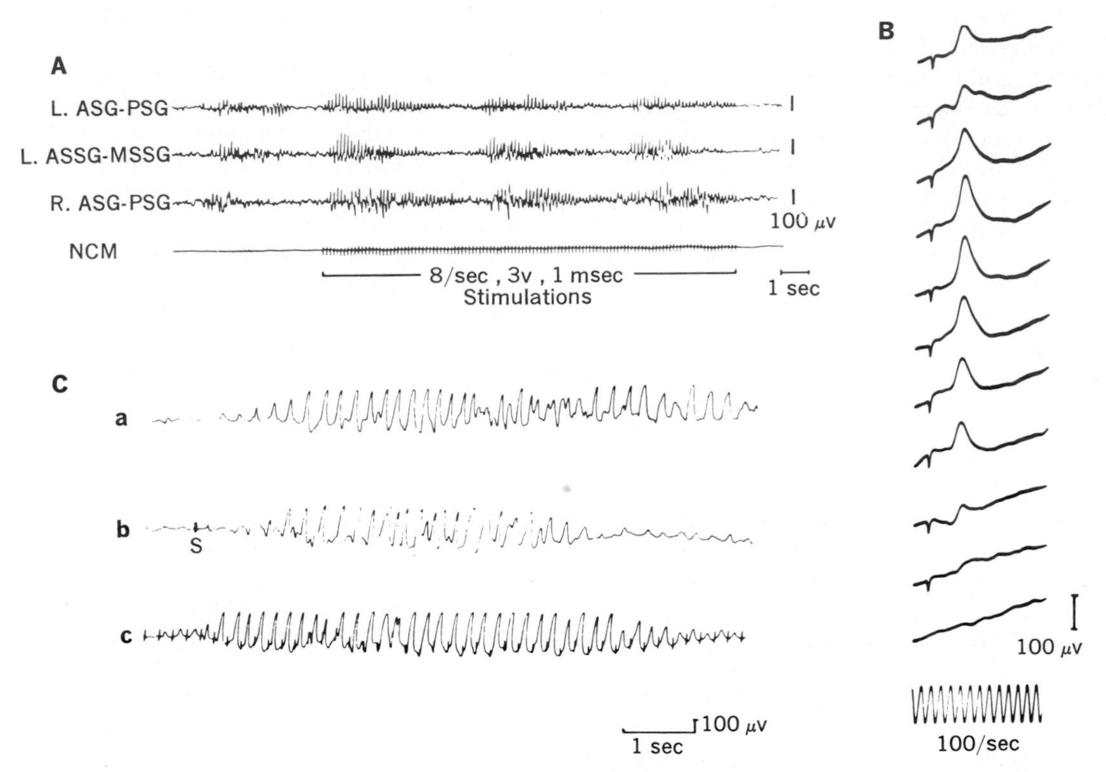

Fig. 8-1. A, Recruiting responses recorded from cerebral cortex of cat anesthetized with sodium pentobarbital, evoked by stimulation of n. centralis medialis at rate of 8/sec. Stimulation of central commissural system evokes responses bilaterally when stimulated locus is near midline of thalamus, as here. Note waxing and waning of recruiting responses during train of stimuli. Spontaneous spindle burst ("barbiturate" spindle) preceded period of stimulation. Bipolar recordings from *L. ASG-PSG,* left anterior sigmoid gyrus versus left posterior sigmoid gyrus; *L. ASSG-MSSG,* left anterior suprasylvian gyrus versus middle suprasylvian gyrus; and *R. ASG-PSG,* right anterior sigmoid gyrus versus right posterior sigmoid gyrus. *NCM,* trace indicating stimuli.

B, Recruiting responses recorded via implanted recording electrodes from anterior sigmoid gyrus of waking cat in response to 8/sec stimulation of n. centralis lateralis via implanted stimulating electrodes. Monopolar tracings, negativity upward. Records read from below upward, one trace for each stimulus. Note waxing and waning of response similar to recruiting responses obtained in anesthetized animals.

C, Records obtained from cortex of cat under pentobarbital anesthesia. *a,* Spontaneous or "barbiturate" spindle; *b,* similar spindle tripped by single stimulus, *S,* delivered to n. centralis medialis; *c,* typical recruiting sequences evoked at some locus by stimulation at 5/sec.

(**A** and **B** from Yamaguchi et al.[135]; **C** from Jasper.[70])

msec, as compared with the latency of 1 to 2 msec for cortical responses to stimulation of specific thalamic nuclei. Fourth, the recruiting waves can be evoked only by stimulation within a frequency range of 6 to 12/sec; indeed, they are best seen when stimuli are delivered at frequencies close to or slightly higher than the dominant frequency of the normal alpha rhythm. In waking animals with implanted recording and stimulating electrodes, as in anesthetized animals, recruiting responses are evoked by stimulation of the nuclei of the central commissural system. They are most readily produced in quiet, relaxed animals and may then, with continued stimulation, lead to the onset of sleep. Recruiting waves are reduced in amplitude in alert animals with desynchronized EEGs, e.g., in the arousal produced by sensory or brainstem reticular stimulation.[130, 131, 135] In animals lightly asleep or under barbiturate narcosis, the slow waves of the EEG are frequently interrupted by spontaneous runs of waves at or slightly above the alpha frequency ("sleep spindles"; Fig. 8-5), and in this state a single stimulus delivered to a nucleus of the generalized thalamocortical system may trip such a spindle (Fig. 8-1). These facts, taken in concert, first suggested the role of this system in regulating, i.e., in pacesetting, the rhythmic electrical activity of the cortex, particularly the alpha rhythm. It is important to note that high-frequency stimulation of the system produces not cortical synchronization or recruiting waves or sleep, but cortical desynchronization resembling that of arousal and the alert state. When so elicited in a drowsy animal, this thalamically induced arousal differs from that evoked by natural sensory or electrical reticular stimulation in that it may not outlast the period of stimulation and, depending on the thalamic nucleus

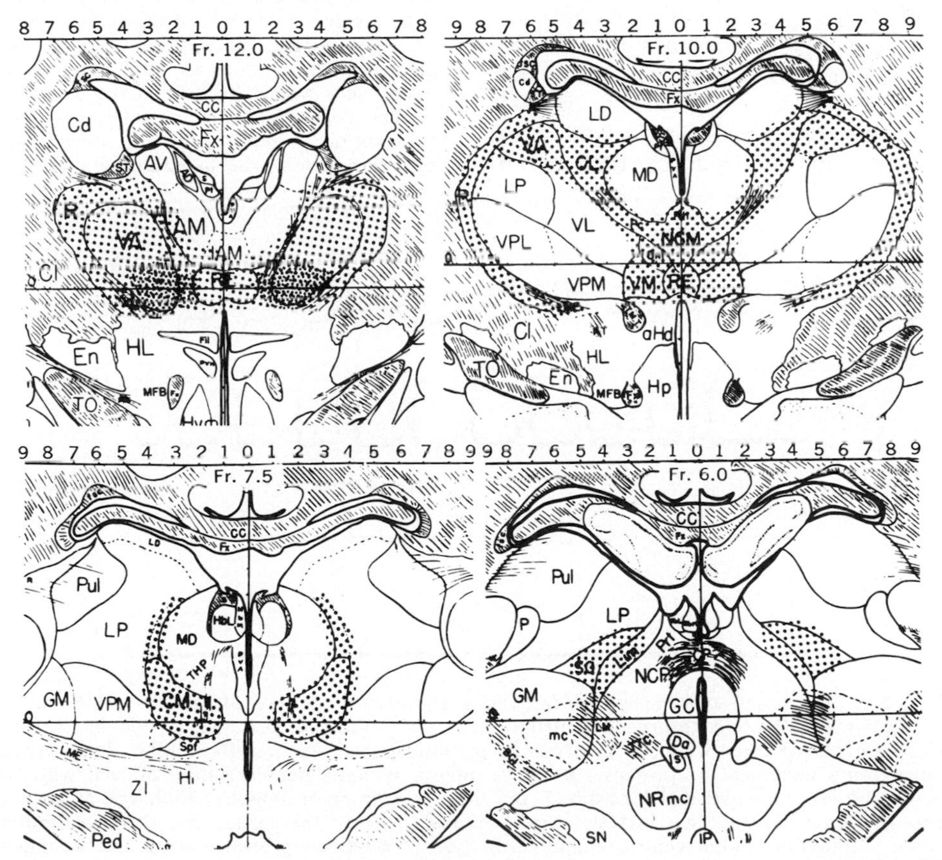

Fig. 8-2. Drawings of cross sections of cat thalamus; four frontal planes from anterior (Fr. 12.0) to posterior (Fr. 6.0). Nuclei of origin of generalized thalamocortical system are indicated by stippling. Areas identified are those within which electrical stimuli at 6 to 12/sec will evoke recruiting responses in cerebral cortex. System has similar distribution in primate brain but is of relatively smaller size there in comparison to greatly enlarged specific relay and associational nuclei. For nuclear designations, see p. 229. (From Jasper.[70])

stimulated, it may be restricted to one hemisphere.

The areas of the cat thalamus from which cortical recruiting responses can be evoked are shown in Fig. 8-2, and the system has a similar distribution in the primate thalamus. Stimulation at one point in an intralaminar nucleus produces recruiting responses that spread throughout the system, perhaps over multisynaptic chains of neurons collaterally activated, forward through its major outflow funnel, the n. ventralis anterior and its adjacent cap of n. reticularis. This is a region from which the most powerful cortical recruiting responses can be elicited by electrical stimulation at shortest latency. In general, strong stimulation at any one locus in the system tends to elicit recruiting activity widely distributed in both thalamus and cortex, and stimulation of midline nuclei does so bilaterally. Nevertheless, there is a pattern of topographic organization within it, for threshold stimulation of the dorsolateral portions evokes responses preferentially in parieto-occipital regions of the cerebral cortex, while medioventral portions activate primarily the cortex of the frontal pole and the medial wall of the hemisphere. There is both anatomic and physiologic evidence that nuclei of the generalized system project intrathalamically,[55] perhaps via axonal collaterals, upon nuclei that are related to "association" cortex as well as upon cells of the specific sensory nuclei. These latter are not, however, essential links in the thalamocortical pathways of the recruiting system, for intralaminar stimulation continues to evoke widespread cortical recruiting responses after their destruction, at which time the responses are especially prominent in the primary receiving areas of the cortex.[66]

The nuclei of the central commissural system project upon the neocortex in a loose anteroposterior topographic arrangement. The fact that local, slowly repeated stimulation at any one point in the system produces recruiting responses that spread throughout the system and produces powerfully synchronizing effects upon other thalamic nuclei as well can be explained by the hypothesis that the system is connected by short intrathalamic axons or by axon collaterals of thalamocortical axons with nearly all thalamic nuclei. In addition, or alternatively, later evidence suggests that the n. reticularis may serve as this distributor of reentrant activity. Its cells receive collaterals of the thalamocortical axons of both specific and generalized nuclei (and of corticothalamic axons as well) and distribute their axons very widely back upon the thalamus itself.[117]

The generalized thalamocortical system receives, in addition to input from the ascending brainstem reticular system, the paleospinothalamic tract. Thus

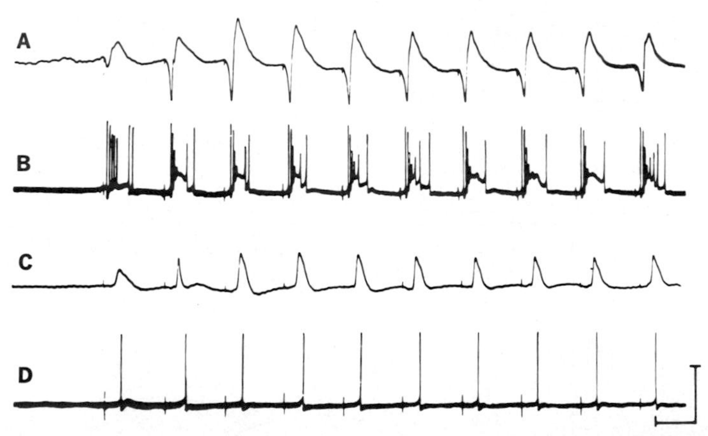

Fig. 8-3. Responses evoked by low-frequency stimulation, 7/sec, of specific relay nucleus of thalamus (VL) and recorded, **A,** from surface of motor cortex and, **B,** via intracellular electrode placed in cortical cell of origin of pyramidal tract axon, a PT cell. **A** shows typical augmenting increment of response. **B** shows intense synaptic depolarization of cell, with "inactivation by excessive depolarization," for there is failure of impulse discharge at peak of depolarization. Lower two records were obtained directly thereafter, but this time during 7/sec stimulation within central commissural system. **C** shows typical waxing and waning of amplitude of almost purely negative wave, recruiting phenomenon recorded at surface of cortex. **D** shows that transsynaptic relay upon cortical cell is much less powerful than for input from specific relay nucleus, for it evokes only slow and weak depolarization that leads to single impulse per response. All records negative up. Calibration: horizontal bar 0.1 sec for all records; vertical bar 50 mv for records **B** and **D.** Augmenting and recruiting responses in **A** and **C** are 0.5 to 1.5 mv in amplitude. (From Purpura et al.[109])

cells of this system receive afferent input both from the cerebral cortex and from all major afferent systems, either directly via the paleospinothalamic system or indirectly via the brainstem reticular system. This global convergence affords a suitable mechanism for titration of input and regulation of the level of excitability of cortical neurons. Such a convergence is achieved with the loss of specific information about driving stimuli and thus should not be regarded as a higher order integrative action, a conclusion that fits the ancient phylogenetic history of the system.

Augmenting phenomenon. Dempsey and Morison[54] first observed that the cortical response evoked by electrical stimulation of the thalamic relay nucleus for a sensory system differs from that evoked by stimulation of the appropriate receptors. When such a thalamic stimulus is repeated at rates of 6 to 12/sec, the cortical response, particularly its second (the negative) component, grows rapidly in amplitude over the first few responses of the train (Fig. 8-3). The augmenting phenomenon is not evoked by repetitive stimulation of a sensory system at any point peripheral to the thalamus. The augmented negative component is distributed over a somewhat wider area than the limited cortical focus in which the primary evoked potential appears. The suggestion is that the electrical stimulation of the thalamic nucleus engages a second set of neural elements that impinges upon cell bodies and dendrites of the more superficial layers of the cortex. Since the generalized thalamic nuclei apparently project collaterals of their axons upon the cells of the specific relay nuclei, the phenomenon of augmentation is probably due to the combined excitation of specific and generalized thalamocortical systems. Certain differences between the two are apparent, however, especially in their relation to spindling waves in the EEG and pyramidal tract excitability.[122, 123]

Synaptic engagement of cortical neurons by thalamocortical systems

Specific and generalized thalamocortical systems engage the neuronal chains of the cortex in different ways. Volleys in generalized afferents produce an intense depolarization of the dendrites and cell bodies in the outer two or three cortical layers, which accounts for the fact that recruiting responses are negative in sign when recorded by a surface electrode and positive when recorded by a microelectrode deep in the cortex.[122, 123] Specific volleys set up by sensory stimuli or by electrical stimulation of a thalamic relay nucleus, on the other hand, evoke an initially positive potential change when recorded at the cortical surface (p. 245). This fits with the notion that these afferents initially produce an intense depolarization of neurons of the middle layers of the cortex, though rapid and powerful synaptic relays lead to EPSPs in cells of all layers. These differences are further revealed by studying the efferent discharge from the cortex produced by volleys of the two types, discharges observed by recording either from axons of the pyramidal tracts[36, 105] or intracellularly from the neurons from which they arise (called PT cells).[109] Synchronous volleys in specific afferents lead to a depolarization of PT cells so intense as to inactivate spike generation in the midst of the high-frequency burst of impulses (Fig. 8-3). Volleys in generalized afferents, on the other hand, produced much weaker EPSPs in PT cells and elicited but a single impulse, as shown in Fig. 8-3. The two systems converge upon the final common path of the pyramidal tract neurons over interneuronal chains that differ, a fact predictable from their different intracortical terminal synaptic distribution.

The sequence of local PSPs of thalamic and cortical neurons evoked by stimulation of a generalized thalamic nucleus gives some indication of how this system may phase cortical activity and the waves of the EEG.[104, 106, 108] Such stimuli evoke short EPSPs followed by large and prolonged IPSPs in cells of other generalized as well as specific thalamic nuclei. For the latter this phasic inhibition is so powerful as to block completely afferent thalamocortical transmission. Thus slow rhythmic discharge in the central commissural system will bring into synchronous oscillation widely distributed populations of thalamic neurons (and thus of the cortical neurons upon which they project) between brief excitation and a powerful and prolonged inhibition. It is this mechanism that is referred to as the "thalamic pacemaker," thought to possess this rhythmic tendency inherently, a tendency displayed especially when the system is free of ascending excitatory drive. This accounts for the spindling cortical waves characteristic of light sleep or barbiturate anesthesia and perhaps for the alpha rhythm of relaxed, quiet wakefulness. The results of studies using intracellular recording now allow a tentative understanding of this shift from synchronization to desynchronization of cortical activity as the generalized system shifts from idling to activation.[107] The prolonged inhibitory PSPs produced by generalized volleys are replaced by sustained depolarization. This is particularly obvious when the system comes under the drive of the ascending reticular activating system, which controls and indeed initiates this change during the shift from sleep to wakefulness. It is thus a major subject for discussion in the section that follows.

SLEEP AND WAKEFULNESS

The nature of sleep[6, 19, 48]

The term "sleep" describes two behavioral states that differ from alert wakefulness by a readily reversible loss of reactivity to environmental events, a reversibility that differentiates sleep from other altered states of consciousness, e.g., anesthesia or coma. The two sleep states are each characterized by profound but quite different alterations in the functional state of the brain. In the first, called slow-wave sleep (SWS), the brain waves are slow and synchronous as compared to waking; cardiovascular, respiratory, and autonomic system levels are somewhat reduced but steady; and subjects awakened from SWS seldom report dreaming. In the second sleep state, variously called activated, paradoxical, or desynchronized sleep (DS), the brain waves are of low voltage and variable frequency, resembling those in the waking state, but the threshold for arousal is much higher than in SWS; there are marked irregularities in the blood pressure, heart rate, respiration, and autonomic activity; outbursts of rapid eye movements (REMs) occur; and subjects awakened from DS commonly report dreaming.

Sleep in mammals is a special example of a more general phenomenon: in their activity all plants and animals to some degree show a periodicity, cyclic variations timed by internal clocks of uncertain nature. In simple animals, rudimentary periodicities may survive destruction of the CNS, but *in mammals, sleep is a neural phenomenon.*

Biologic clocks and the 24 hr sleep-wakefulness cycle[2, 12]

The rhythmic alterations referred to are ubiquitous both in the kinds of organisms and the kind of functions considered.[126, 127] When the period of oscillation approximates the period of the earth's rotation, they are called *circadian rhythms*. The available evidence supports the proposition that the "clocks" governing these rhythms are endogenous and innate and that they are usually self-sustaining and undamped. Whether such oscillations in activity appear at levels of organization simpler than that of single cells is unknown. When free-running, circadian rhythms are remarkably precise, for the observed standard errors may be no more than ±2 min per nearly 24 hr cycle. Over a very wide range, circadian rhythms are nearly independent of temperature and are remarkably insensitive to chemical perturbations. They are, however, sensitive to the intensity of light, and the nearly 24 hr circadian rhythms of many types may be *entrained* to an exact 24 hr rhythm by 12:12 hr light-dark alternations. A free-running cycle ending before the superimposed dawn or dusk resets by stepwise delay, one ending afterward by stepwise advance. The phase of some free-running or nonentrained circadian rhythms can be shifted by a single brief alteration or stimulus, particularly in the light regimen.

Nocturnal mammals such as the rat show cyclic variations between rest and activity, with a period of 60 to 90 min associated with increased motor activity, feeding, grooming, gastric and intestinal activity, etc. Under a 12:12 hr light-darkness regimen or in the natural habitat these periods of activity begin promptly at dusk and cease at dawn. If, however, the animal is never exposed to light or if the eyes are enucleated or the optic nerves cut, this cyclic activity drifts gradually with reference to the 24 hr day, following a period slightly less or greater than 24 hr. This indicates that under normal circumstances the circadian rhythm is entrained to an exact 24 hr period by a light-darkness cycle of that duration. Once set free by light deprivation, this cycle is astonishingly free of atmospheric or climatic conditions, runs independently in a number of similarly deprived animals housed under identical conditions, and is unaffected by a host of experimental manipulations.[107] Removal of the endocrine glands, including the gonads and pituitary, pineal, thyroid, and adrenal glands in different animals, produces no change. The cycle survives periods of anoxia or of epileptic convulsions and is impervious to periods of deep anesthesia, taking up at the appropriate point in the cycle once consciousness is regained. Alcohol intoxication, the administration of tranquilizing drugs, and decreases or increases in activity of the autonomic nervous system (produced by drugs) are all equally without influence. Of lesions of the forebrain, only those that injure the hypothalamus produce an effect, but this effect is profound: the precisely timed cycle dissolves into a random time series of brief periods of rest and activity. While it is clear that this internal clock depends on the hypothalamus for its existence, the nature of the clock and the underlying mechanisms of timing are unknown. Examples of such cyclic changes occur also in the severity of some diseases and

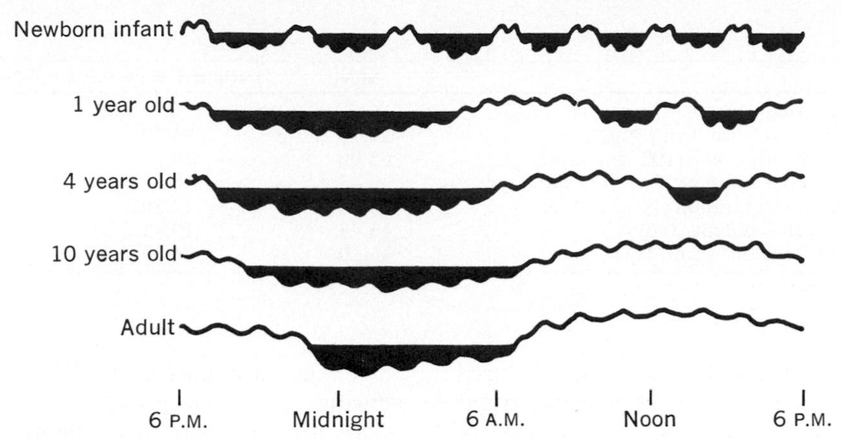

Fig. 8-4. Schematic representations of gradual transition from polycyclic alternations between sleep and wakefulness of newborn infant to monocyclic pattern in adult. Black areas indicate sleep. Secondary undulations indicate rest-activity periodicity, which gradually lengthens from 50 to 60 min in infants to 80 to 90 min in adults. (From Kleitman.[8])

in the presence or absence of symptoms, and these often show periods greater than 1 day. The role of these periodicities in the pathogenesis of disease is still uncertain.[110]

Like the rat, the newborn infant shows rest-activity cycles of about a 90 min duration (Fig. 8-4). When sleeping, the troughs of these cycles are occupied by SWS, and the peaks by DS. During the first year of life these cycles become entrained to the ordinary daily rhythm of human existence, although the 90 min undulations persist, associated during sleep with the periodic appearance of episodes of DS (Fig. 8-12). This entrainment to a 24 hr cycle may be due in part to familial and social pressures; it is associated with a gradual decrease in the need to sleep and increased activity during periods of wakefulness. However, it is of interest to note that a similar 24 hr cycle is inherent and innate in man and other mammals.[111] Efforts to entrain human activities to periods of considerably shorter or longer lengths of time have so far been relatively unsuccessful, though more extensive study is required before the matter can be regarded as settled.

Physiologic variables and the 24 hr sleep-wakefulness cycle[8]

Together with the gradual entrainment of the sleep-wakefulness cycle in the human infant, usually established by the end of the second year of life, cyclic variations appear in other functions. During SWS there is a mild reduction in alveolar ventilation accompanied by a slight rise in mixed venous and alveolar CO_2. The heart slows and there is a moderate drop in blood pressure.[28, 48] The minimal changes in the concentrations of many blood constituents are accounted for by the equally slight dilution that follows the movement of extravascular water into the blood on assumption of the recumbent position. A diurnal cyclic variation in hypothalamic-pituitary function is evidenced by the activity of one of its target organs, the adrenal cortex: there is a decreased blood level and urinary excretion of ketosteroids and an increase in the eosinophil count in the blood. The morning eosinopenia is less marked in blind than in normal individuals. There is a cyclic variation in the rate of release of growth hormone, that rate being greatest during SWS.[98, 128] Water excretion by the kidney is lowered during the night. Perhaps the most constant cyclic variation is in the body temperature, which drops by as much as 2° F during the night. It has been surmised at one time or another that each of these changes stood in a causal relation to sleep, in particular that the slowing of the heart and mild drop in blood pressure would result in decreased blood flow through the brain, i.e., that sleep is produced by a cerebral anemia. That this is not the case has been shown by direct measurement (Table 8-1).[87]

Changes in central nervous activity during sleep. The remarkable changes in CNS activity during sleep are not limited to those regulating the level of consciousness. With sleep, there is an elevation of thresholds for many reflexes, and in very deep sleep (SWS), certain pathologic reflexes may appear, even in healthy individuals. There is simultaneous-

Table 8-1*

	Sleep	Fatigued wakefulness	Normal rested
Number of cases	6	6	11
Mean arterial pressure (mm Hg)	89.70	94.20	86.50
Cerebral blood flow (ml/100 gm/min)	65.00	59.20	54.80
Cerebral O_2 consumption (ml/100 gm/min)	3.42	3.52	3.34
Hemoglobin concentration (gm/100 ml)	14.36	14.30	14.57
Arterial oxygen content (vol%)	19.43	19.42	19.44
Arterial CO_2 tension (mm Hg)	46.30	46.00	41.30

*Data from Mangold et al.[87]

ly a reduction in muscle tone, and the limbs may be completely flaccid. Movements occur during any lighter sleep, and the degree of motility varies cyclically with a period of 60 to 90 min, although motility patterns differ from one individual to another. Striking changes occur in the musculature of the eye: the pupil is narrowly constricted by an increase in frequency of discharge of neurons of the Edinger-Westphal nucleus. Though the pupils are narrow, the light reflex is present. The eyes most frequently assume a position of upward divergence. A decrease in peripheral sympathetic activity is evidenced by increased electrical resistance of the palmar skin; there is, however, a tonic contraction of the vesical and rectal sphincters.

Changes in the EEG characteristic of different stages of sleep have proved to be a useful tool for experimental studies of sleep. Various schemes of classifying EEGs in man and other primates[32, 60, 61] and correlating them with sleep have been proposed, among which is the commonly used one of Dement and Kleitman,[53] given in the following outline and Fig. 8-5. Automatic methods are now commonly used for analyzing and classifying EEG records obtained during sleep.[77, 132]

Stage 1: Record in which EEG alpha waves are slowed very slightly; seen in quiet relaxed wakefulness or as drowsiness supervenes.

Stage 2: Record with spontaneous "sleep spindles," runs of a few seconds' duration consisting of regular 14 to 15/sec waves superimposed upon a low-voltage background with admixture of 3 to 6/sec waves; a state from which the sleeper is readily aroused.

Stage 3: Record with some sleep spindles but now on a background of slower 1 to 2/sec delta waves; associated with sleep of intermediate depth.

Stage 4: Record consisting entirely of high-voltage slow delta waves without spindling, associated with a phase of sleep in which the threshold for awakening is greatly elevated.

Cyclic variations in the EEG pattern occur during normal sleep, with a period of 80 to 120 min, the pattern changing on the peak of each cycle to a modified or "emergent" stage 1. This phase is associated with the presence of REMs, an acceleration of the heart and respiratory rates, and occasionally by gross body movements occurring against a background of decreased muscle tone (Fig. 8-12). This emergent stage 1 differs from the descent through stage 1 on going to sleep in that the threshold for arousal by an auditory stimulus is elevated 5 to 10 times. By awakening sleepers during various phases of the cycle, Dement and Kleitman[53] discovered that the *emergent stage 1 with REMs is commonly associated with dreaming.* This finding has permitted study of dream content, of the ontogenetic development of dreaming, of the time scale in dreaming, of the effect of presleep events and conditions upon dream content, etc.[3, 53] This stage of sleep in man is probably not completely identical to the special state of deep sleep with EEG desynchronization in animals, which will be described in a later section.

Activation or arousal of the EEG in human beings is said to occur when the high-voltage slow waves of sleep are replaced by the low-voltage fast activity usually seen in wakefulness. This transition is commonly abrupt and direct, but in more gradual awakenings the transitional intervening stages of sleep spindles, modified alpha rhythm, etc., may be seen in reverse sequence from that of the transition from waking to sleep. In experimental work in animals (most commonly cat or monkey), three states are usually specified, although intervening states do occur: (1) low-voltage fast activity with widened pupils of wakefulness, (2) high-voltage slow activity with pupillary constriction of moderately deep sleep, and (3) the special state of deep sleep with EEG desynchronization referred to previously as

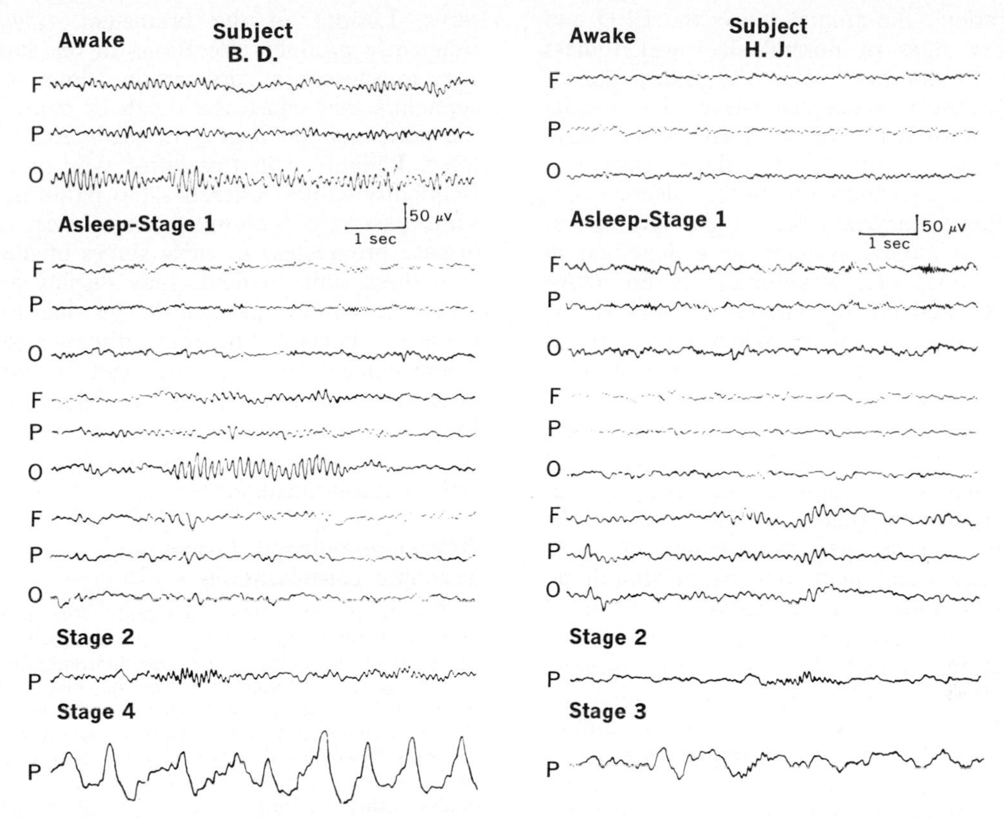

Fig. 8-5. EEG records of two subjects during various stages of sleep to illustrate classification given on p. 262. *F,* Frontal, *P,* parietal, and *O,* occipital locations of leads. (From Dement and Kleitman.[53])

paradoxical or desynchronized sleep (DS). Activation of the EEG is frequently equated with the transition from sleep to wakefulness when the behavioral correlates cannot be evaluated, although the two are occasionally dissociated in otherwise normal individuals, may be dissociated by the action of certain drugs such as atropine, and are characteristically dissociated in DS in animals.

CENTRAL NEURAL MECHANISMS REGULATING SLEEP AND WAKEFULNESS

Two important discoveries generated modern concepts concerning the neural mechanisms involved in sleep and wakefulness. The first of these was made by Hess,[20, 67] who discovered that slow, rhythmic stimulation (8/sec), via implanted electrodes, of a diencephalic region encompassing the thalamic nuclei of the generalized thalamocortical system in waking cats produced, after several minutes of stimulation, all the behavioral signs of normal sleep, i.e., the sequence of postures, etc. that are normally seen in a cat going to sleep and in actual sleep itself. Hess also discovered that stimulation of a more ventral and posterior region, including the posterior hypothalamus and the gray matter at the mesodiencephalic junction, produced all the behavioral signs of awakening. These observations led to the concept of sleep as an active process; i.e., activation by some means of what is now termed the generalized thalamocortical system (the "recruiting" system) influences the excitability and the rhythmic patterns of activity of cerebral cells in such a way that sleep ensues. This hypothesis has been modified in the light of subsequent findings, however, but as will be apparent in the following discussion, the synchronizing mechanisms do play an important part in a unified concept of the cerebral mechanisms in sleep.

In a novel and important experiment, Bremer discovered in 1935 that when the neuraxis of a cat is transected at the level of the first cervical segment, with artificial respiration and precautions for maintenance of blood pressure (Bremer's *encéphale isolé* preparation) the animal shows the EEG and pupillary signs of normal sleep-wakefulness cycles. In contrast, when the transection is made at the mesencephalic level, just caudal to the motor nuclei of the third cranial nerve (Bremer's *cerveau isolé*), there ensued a permanent condition resembling sleep.[34]

Bremer's discovery led to the concept of sleep as a passive process, as a deactivation phenomenon; i.e., wakefulness is an active state maintained by afferent input to the brain, and sleep ensues when that input is removed, as in the *cerveau isolé* cat, or falls below a certain critical level, as in normal sleeping. In the *cerveau isolé* preparation, olfactory input to the brain remains, but strong olfactory stimuli produce only a transient activation that does not outlast the stimulus. Visual pathways from retina to cortex are also intact, but visual stimuli do not evoke widespread activation of the EEG in the *cerveau isolé* with high mesencephalic transection, as they do in the intact animal. While Bremer tentatively concluded that deafferentation per se is sufficient to induce sleep, this last observation concerning visual stimuli indicates that some neural mechanism in addition to the direct sensory pathways is required for the maintenance of wakefulness. Indeed, all the investigations of the last two decades have given results that support the point of view that this additional mechanism involves collateral activation of neural populations occupying the reticular formation of the brainstem, an important concept that derives largely from the work of Moruzzi[92] and Moruzzi and Magoun.[93] This neural system, by virtue of its upward projection upon the cerebrum, is thought to control the level of excitability of the forebrain and thus the state of consciousness. It is of some importance, before presenting the experimental findings concerning the physiology of the reticular formation and the reticular deactivation concept of sleep regulation, to consider the anatomic structure and afferent and efferent connections of the brainstem reticular formation.

A number of clinical observations support the idea that brainstem mechanisms play an essential role in regulating the state of consciousness. Large areas can be removed from any part of one cerebral hemisphere, symmetric parts can be removed from both, or indeed one entire hemisphere can be removed without loss of consciousness or disruption of normal sleep-wakefulness patterns. Lesions of the brainstem, however, frequently produce alterations in the state of consciousness,[41] as do certain forms of encephalitis that cause the death of neurons in the periacqueductal gray of the mesencephalon.[56] Patients with this latter type of lesion frequently exhibit altered sleep patterns, and some pass into prolonged somnolence as the disease progresses. In early stages of the disease these same patients may display a persistent insomnia, thought to be due to the irritative effects of the acute disease process. These clinical observations, together with the discoveries of Bremer and of Moruzzi and Magoun, direct attention to the detailed anatomy and physiology of the reticular formation of the brainstem.

Brainstem reticular formation: anatomic considerations

The term "brainstem" designates the medulla, pons, and mesencephalon, a region extending from the level of the obex to the mesodiencephalic junction. The term "reticular formation (RF) of the brainstem" identifies those areas of the brainstem that are characterized by aggregations of cells of different types and sizes, interspersed with a wealth of nerve fibers traveling in many directions, and excludes more definitely circumscribed groups of cells, e.g., the sensory and motor nuclei of the cranial nerves, the red nucleus, the substantia nigra, and the inferior olive. The RF is frequently referred to as being diffuse in its organization. This generalization is not without some ground, although almost every anatomic investigation of recent years has shown that the RF is not only complexly built but that regions of it differ considerably from one another with regard to their cytoarchitecture and their afferent and efferent connections with other regions of the brain.[35, 97, 113] (For a recent detailed description, see Berman's book.[31]) The RF consists of a variety of cellular fields that grade almost imperceptibly from one to another but within which a large number of nuclei are differentiated. In many physiologic investigations, precise anatomic identification of regions stimulated or ablated or from which recordings of electrical activity were made was not attempted. For this reason it will be necessary in the discussion that follows to use such terms as the *mesencephalic* or the *pontile* RF. The large majority of studies have been carried out in the cat. However, it should be emphasized that there are important changes in this region across the evolutionary series from rabbit to man. For example, the giant neurons that are so conspicuous in the RF of rabbit and cat are far less numerous in that of the primate brainstem. The region emits systems of long axons projecting both caudally and rostrally from the brainstem, but a prominent characteristic is the profuse interconnections of its parts, both by the collaterals of projecting axons and by those which ramify wholly within the brainstem. A large part of what is known of the function of this region has resulted from experiments in which electrical stimuli are delivered to the RF and observations made of the bodily changes that result. It is not surprising, then, in view of the profuse anatomic interconnections of regions of the RF, that the simultaneous engagement by such electrical stimuli of cells in the locale and of fibers of passage gave results that led to the idea that the region is only "diffusely" organized, for with strong stimuli it is possible to produce such diverse physiologic phenomena as changes in the EEG, in reflex excitability of the spinal cord, in respiratory rhythm, and in peripheral blood pressure. How to untangle these complex and interconnecting control systems is a problem of major importance for which no ready solution is apparent.

Reticular regions related to the cerebellum. The most obvious division of reticular structures is between those that are related to the cerebellum and those that are not. The lateral reticular nucleus receives afferents from the lateral funiculus of the spinal cord in a somatotopic arrangement and projects via the restiform body upon the ipsilateral vermis and paramedian lobule. The n. reticularis tegmenti pontis receives afferents mainly from the cerebral cortex, but also from the spinal cord, and it projects the axons of its cells upon the entire cerebellum, bilaterally to the vermis, and only ipsilaterally to the cerebellar hemisphere. The n. reticularis paramedianus projects upon the vermis and paramedian lobule and receives its afferents from the cerebellar cortex. On removal of the cerebellum all cells of these nuclei degenerate, an indication that they contain no interneurons. The functional implications of reticulocerebellar systems will be considered further in Chapter 28.

Major efferent systems of the noncerebellar reticular formation

Reticulospinal connections. Cells of origin of the reticulospinal system are concentrated in, although not exclusive to, two areas of the medial or "effector" regions of the RF. The first is the entire n. gignatocellularis of the medulla, together with adjacent portions of n. reticularis ventralis and n. reticularis lateralis. In the pons the reticulospinal neurons lie in the n. reticularis pontis oralis and part of n.r.p. caudalis. Reticulospinal fibers course caudally in the ventrolateral funiculus of the spinal cord; those from the pons course ipsilaterally; and those from the medulla are bilaterally distributed. Few reticulospinal fibers descend below the thoracic level, and the system is thought to influence more distal spinal segments via the propriospinal system.

Ascending connections.[94] Long ascending connections originate only from the medial two thirds of the RF, from a medullary and pontile region, which in each case are shifted caudally from areas of origin of the reticulospinal systems (Fig. 8-6). The bulk of these fibers pass through the mesencephalon in or near the central tegmental tract. The system divides at the caudal border of the diencephalon. A major division projects upon the generalized thalamocortical system, i.e., upon nuclei of the central commissural system and perhaps also upon the reticular nucleus of the thalamus. It should be pointed out also that elements of the paleospinothalamic tract pass in the same region and make the same terminations (Fig. 11-10). A second component projects upon the subthalamic

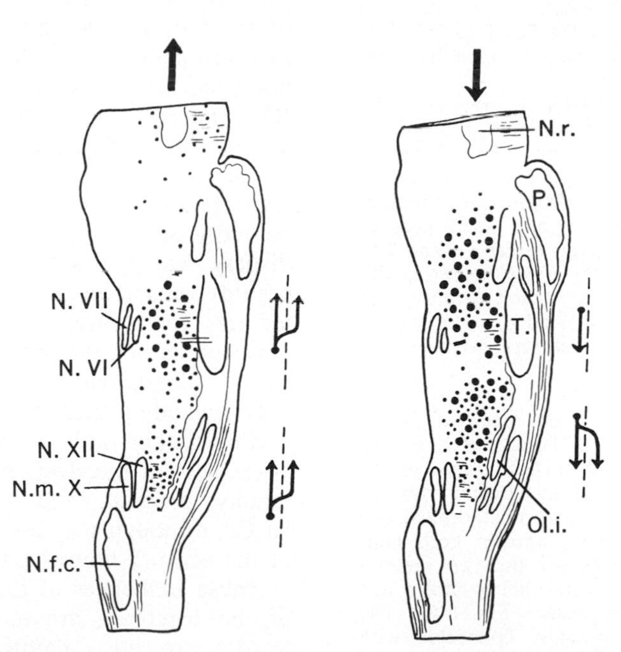

Fig. 8-6. Distribution of cells of RF that send long axons to spinal cord are shown on right, projected upon parasagittal section; large dots indicate large cells. Medullary RF projects bilaterally, pontine RF homolaterally. Similarly, on left, distribution of cells having long axons that extend above mesencephalon. Both medullary pontile areas project bilaterally. Distributions for two projections are not identical, and in both medulla and pons heaviest concentrations of reticulospinal neurons lie oral to those for cells with ascending axons. What is not illustrated is that some cells emit axons that divide and send one branch to spinal cord and one orally, above mesencephalon. (From Brodal.[35])

body of Luys, the zona incerta, and the entopeduncular nucleus and contributes fibers to the fields of Forel. Another major system originates wholly from the mesencephalic RF and projects very widely upon the mammillary nuclei, the periventricular and lateral hypothalamic areas, the septal nuclei, the preoptic region, and the basal ganglia.

In the regions of origin of the long projecting systems, more than one half of the neurons emit such long axons; indeed, many emit axons that divide into an ascending branch that reaches the cerebrum and a descending branch that reaches the spinal cord, branches that emit numerous collaterals to reticular cells as they pass.[116]

In addition to these efferent systems projecting outside the brainstem, there are profuse projections from reticular regions upon the motor and sensory nuclei of the cranial nerves as well as upon cells of the superior colliculus.

Major afferent systems to the noncerebellar reticular formation

Fibers of cortical origin course through the corticobulbar system to terminate in the brainstem,[76] some in the pontile and medullary RF, and in particular upon the nuclei that emit long reticulospinal fibers. These descending elements originate from the frontal cortex, particularly from the motor cortex and less certainly from the cortex of the parietal and temporal lobes and from the medial wall of the hemisphere. The total number of corticofugal fibers reaching the RF is not large, and fibers from widely separate cortical areas project upon the same RF locales. Other fibers of cerebral origin project from the basal ganglia, the subthalamus, and the hypothalamus upon the mesencephalic RF.

Fibers of spinal origin reach the reticular formation in several ways. Collaterals of ascending fibers of the anterolateral system project upon the small-celled, more lateral component of the medial two thirds of the RF, which also receives input of vestibular and auditory origin and itself projects upon the more medial effector region of the RF. However, large numbers of direct spinoreticular fibers reach this same effector region directly, and their terminations are concentrated in those regions that emit long ascending reticulocerebral fibers.

• • •

In summary, the RF can be regarded as encompassing several subsystems. Some portions relay input of spinal and cerebral cortical origin to the cerebellum. The more medial or effector portions of the RF subserve both a cerebroreticulospinal and a spinoreticulocerebral system, and in both pons and medulla the relay zones for the two are not coincident, those of the former being orally displaced from those for the latter (Fig. 8-6). The mesencephalic RF is reciprocally connected with the hypothalamus, septum, basal ganglia, and preoptic region.

It is therefore obvious that the RF is not organized in a wholly "diffuse" fashion. It is just as important to emphasize, however, the extensive overlap between the areas subserving these systems and the profuse synaptic interconnections between them. Thus whether the RF is to be considered as "diffusely" or "specifically" organized is a redundant question: it is both.

Sleep as a passive process: reticular deactivation[91, 113]

In 1949 Moruzzi and Magoun[93] discovered that rapid stimulation (50 to 200/sec) of the brainstem produced activation of the EEG, an effect evoked by stimulation of the central core of the brainstem in a region extending upward from the bulbar RF to the mesodiencephalic junction, the dorsal hypothalamus, and the ventral thalamus, thus including the dynamogenic area of Hess.[67] This phenomenon is illustrated by the records in Fig. 8-7, and the relevant anatomic information is summarized in Fig. 8-8. In many features the activation produced by RF stimulation resembles the arousal produced by natural stimulation. The discovery led to a host of succeeding investigations, the results of which support the idea that *withdrawal of the energizing action of the reticular activating system plays an important role in the genesis of sleep.*

When the RF is stimulated via implanted electrodes in sleeping animals, behavioral awakening and EEG desynchronization result. This is also true in animals after section of the long ascending sensory systems in the mesencephalon (i.e., the lemnisci) but does not occur after lesions of the mesencephalic RF. Indeed, after extensive lesions of the mesencephalic RF, animals may be comatose for many days and unresponsive to any stimuli.[62] If they survive, they may show good recovery of sensory and motor functions but display various and sometimes prolonged periods of somnolence, with marked refractoriness for arousal, which, when evokable, may not outlast the arousing stimuli. By contrast, animals surviving transection of the long ascending and descending tracts of the midbrain, but with no RF lesion, show no alterations of the sleep-wakefulness cycle, are readily aroused, and then show activated EEGs, although they are profoundly deficient in the sensory-motor spheres.

These influences of the RF on the activity of the forebrain are mediated via two upwardly projecting pathways. The first pathway projects upon the central commissural nuclei, thus influencing excitability of the entire cortex by entraining the elements of the generalized thalamocortical system. That a second pathway exists is suggested by the finding that arousal can still be produced by

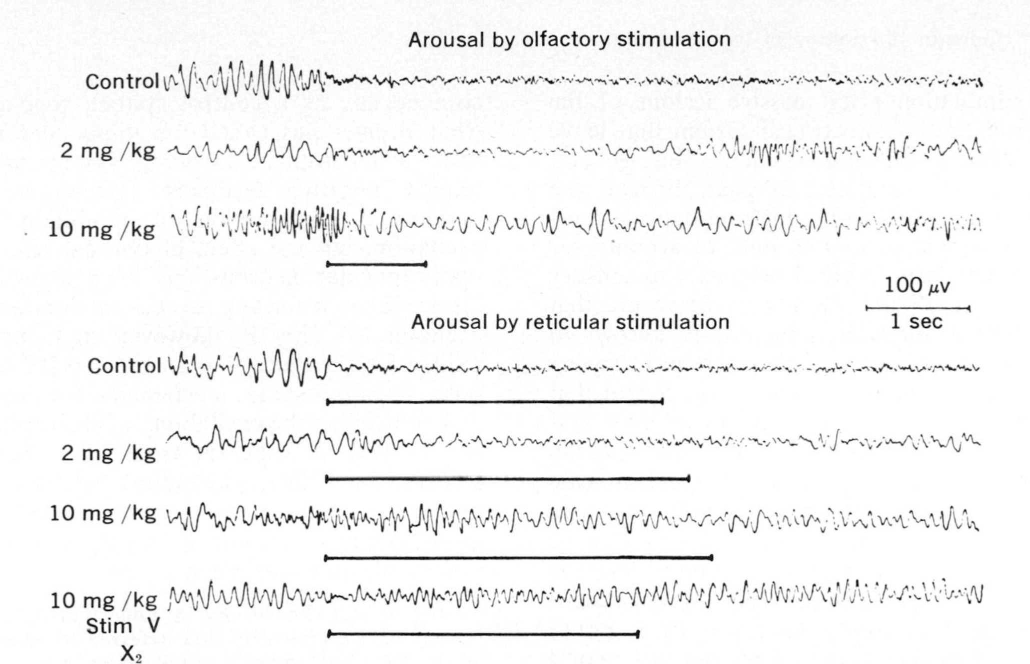

Fig. 8-7. EEG of rabbit showing, for two traces labeled "control," arousal or activation by sensory stimulus, above, and by high-frequency electrical stimulation of reticular activating system, below. In each case, effect is abolished with increasing doses of anesthetic (sodium pentobarbital). (From Arduini and Arduini.[21a])

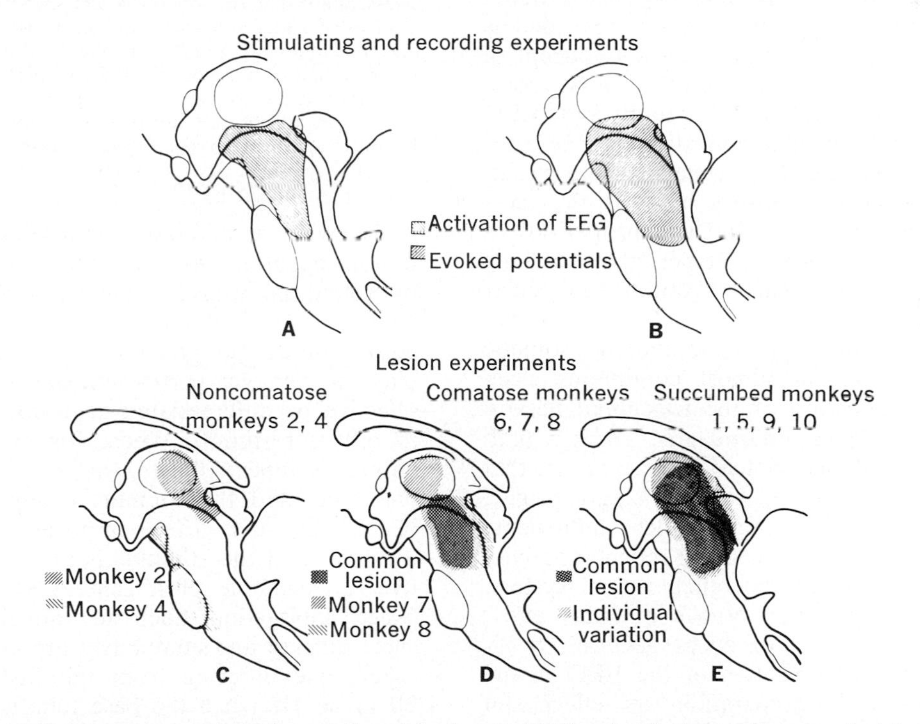

Fig. 8-8. Diagrams summarizing studies of brainstem RF by electrical recording, by electrical stimulation, and by observation of animals after reticular lesions. Reticular regions involved are shown projected upon midsagittal drawings of monkey brainstem. **A,** Area within which high-frequency electrical stimulation produced EEG activation and arousal of sleeping animals. **B,** Area of brainstem from which evoked potentials were recorded in response to a variety of peripheral stimuli, with considerable convergence. Areas shown in **A** and **B** are nearly coincident. Contrast of **C** and **D** indicates that, when lesions involve central core of pontile and mesencephalic RF, permanent loss of consciousness may ensue, certainly when lesions are complete, but that more peripheral lesions that may sever long ascending and descending tracts will not affect consciousness or disturb sleep-waking cycles. **D** indicates that many large reticular lesions are incompatible with life. (From French and Magoun.[62])

RF stimulation after massive lesions of the generalized thalamocortical system that leave the ventral thalamus intact. This second pathway is postulated to pass through the ventral thalamus to the internal capsule, perhaps identical to that thought to account for the "secondary" cortical responses to sensory stimuli. It should be noted, however, that no ventral thalamic structure is known to project directly upon the cerebral cortex.

Electrophysiologic studies have shown that the RF receives input from all afferent systems, including the olfactory and visceral. Studies of single RF cells, however, indicate that under some experimental conditions more than half are insensitive to any sensory stimuli. Of those that are sensitive the convergence is wide but rarely global. Where a wide convergence does exist, the range of response of the reticular neuron provides no neural signal that is unique for each source, nor for the spatial, intensive, or temporal aspects of each sensory input. The important point is that, whereas the ascending system is driven by sensory input, it should not be regarded as a sensory system, fitting the usual concept of that term no better than do the spinal motoneurons.[118, 119] While behavioral and EEG arousal and reticular activation can be produced by any sufficiently intense sensory input, the somatic sensory input is the most potent in this regard. Indeed, the periods of activation that are interspersed with long periods of EEG synchronization and ocular signs of sleep in the *encéphale isolé* cat are largely sustained by the remaining somatic sensory input, for almost continuous sleep follows transection of the trigeminal nerves.

Corticoreticular interactions. There is both anatomic and electrophysiologic evidence that neuronal systems of cortical origin project directly upon the brainstem RF, influencing the transmission of ascending reticular activity as well as that of the reticulospinal system. Electrical stimulation of one of many effective cortical areas produces generalized bilateral desynchronization of the EEG in the *encéphale isolé* preparation, an effect still present after section of the corpus callosum. Behavioral arousal and EEG desynchronization are produced by stimulation via implanted electrodes in otherwise normal, sleeping animals, but cortical activation by such cortical stimulation is no longer evoked after a high mesencephalic transection. These observations indicate that the corticoreticulocortical systems may function as reentrant circuits, i.e., as a control system, regulating what Bremer has called the *tonus cérébrale.* Such a reentrant controlling system would require negative feedback (inhibition) or perhaps a combination of inhibition and excitation, but the effect of cortical efferents upon reticular neurons has been shown by intracellular recording to be predominantly excitatory.[83] This is, however, not incompatible with the idea that cortex and RF compose a homeostatic mechanism for setting and controlling the excitability of the former,[50] for, as will be apparent from the following section (p. 269), ascending reticular influences are not only desynchronizing and arousing but contribute positively to cortical synchronization and sleep as well.

It has been known for a long time that the regions of the brainstem now referred to generally as the RF exert controlling influences over a wide range of bodily functions, particularly in what might be termed the vegetative sphere. Its controlling function in respiration and circulation is described in some detail in Chapters 39 and 63, and it appears likely that this region is also concerned in the control of many autonomic and endocrine functions, e.g., the hypothalamopituitary mechanisms. How all these functions are coordinated by this system in a way that leads to constancy and not chaos is an important subject for future study. Some concepts in this regard have been put forward by Dell,[50] particularly from the standpoints of homeostasis and cybernetic theory.

Bremer's observation[34] that EEG activation in the *encéphale isolé* preparation evoked by sound no longer occurred following bilateral removal of the auditory areas of the cortex raises the possibility that the cortex plays a role via corticoreticular systems in selective or differentiated arousing responses to different stimuli. When, for example, an animal is repeatedly aroused by a sound of one tone, and the stimulus is repeated, the behavioral and EEG arousals gradually dwindle until the stimulus is no longer effective, an example of a general phenomenon called habituation. Such an animal is, however, quickly and completely aroused by another tone differing from the first by only 30 to 40 Hz when the base tone is 200/sec, although there is some spread of the habituation to tones closer to the basic habituating one.[21] Behavioral correlates of this phenomenon are obvious: the awakening of a sleeper when his name is softly called, though he is unaffected by other names called loudly; the awakening of a mother to the cry of her child, though much louder noises have no effect, etc. The integrity of auditory cortex is deemed essential for such differentiated arousals.

Activity of cortical neurons during sleep.[57, 68] Studies of single cortical neurons via implanted electrodes in unanesthetized animals have shown that during the transition from wakefulness to drowsiness to SWS, the large majority of cortical neurons slow their rate of discharge,[96] while such large cells as those from which pyramidal tract fibers originate may accelerate.[57] Although a slow fall in frequency does occur in the net activity of the cortical neuronal population, there does not appear to be a universal inhibition of cortical neurons in SWS. In the transition from SWS to DS, almost all cortical neurons increase their discharge rates to levels even higher than those of quiet wakefulness[57, 78] although still below those of alert activity. What is more striking than rate changes between these various states is the change in the temporal pattern of discharge.[57, 95] During wakefulness, cortical neurons discharge at intervals that compose a time series neither random nor perfectly periodic. As sleep ensues, discharges appear in short, high-frequency trains with long silent periods intervening. During the spindling phase of sleep these clustered discharges are likely to occur in synchrony with the spindle waves, and closely adjacent neurons tend to discharge together. The reverse transition, from sleep to wakefulness, is marked by disappearance of the clusters and differential changes in the discharge frequencies of different neurons but only a moderate increase in overall activity. During brisk arousal and during purposeful activity that overall rate increases very greatly. Cells of origin of the pyramidal tract are no exception to this pattern.

This change in the temporal patterns is thought to reflect similar changes in that of neurons of the generalized thalamocortical system, the thalamic pacemaker. In wakefulness these cells discharge rapidly and asynchronously, producing a smoothly maintained depolarizing pressure on cortical neurons, and thus on their quasi-periodic discharge sequences. With decrease in ascending activating influences at the onset of sleep and a simultaneous increase in ascending synchronizing influences (to be described later), the cells of this system tend to discharge in unison at the rate of 3 to 5/sec, characteristic of cortical slow waves in sleep. Each such thalamocortical volley evokes EPSP-IPSP sequences, which are the basis of the slow waves observed, and entrain cortical neurons to discharge high-frequency bursts in response to each thalamocortical volley.

Sleep as an active process: synchronizing mechanisms of the lower brainstem[9]

Largely due to the recent work of the Italian school of physiology led by Moruzzi, it is now clear that the brainstem activating mechanisms just described are paralleled by others that exert a reciprocal effect, tending to synchronize the EEG and produce behavioral sleep by an *active* process. The initial discovery was made by Batini et al.,[25-27] who found that after complete transection of the brainstem rostral to the medulla—their midpontine, pretrigeminal preparation—animals display a relative insomnia characterized by activated EEGs and the ocular signs of behavioral wakefulness. For example, when observed continuously (up to 9 days), the EEGs of such animals were activated for 78 ± 10% of the time, as compared with 37 ± 12% for normal cats, when both groups were observed under the same conditions and without any intentional stimulation. That a state of true wakefulness existed in these animals was further evidenced by the finding that orienting reflexes were present and true conditional responses could be established.[17, 18] The fact that prolonged periods of somnolence appear in the *encéphale isolé* preparation suggested that the synchronizing structures separated from the forebrain by the midpontine transection must lie in the lower brainstem and that they must be tonically active. That this mechanism is localized to the region of the rostral solitary tract and the adjacent n. reticularis ventralis is shown by the facts that (1) low-frequency (10/sec) stimulation of this region produces EEG synchronization, while high frequencies produce strong arousal[82]; (2) single neurons of the region, observed in free-moving, unanesthetized animals, begin to discharge 1 to 2 min before the onset of sleep and are relatively inactive during wakefulness[45]; (3) following small bilateral lesions in this region, animals show increased spontaneous activation and that produced by stimulation of the reticular activating system is prolonged[33]; (4) local cooling of the bulbar RF in sleeping cats produces arousal[30]; and (5) differential anesthetization of bulbar or mesencephalic regions produces differential EEG effects, an experiment illustrated in Fig. 8-9.

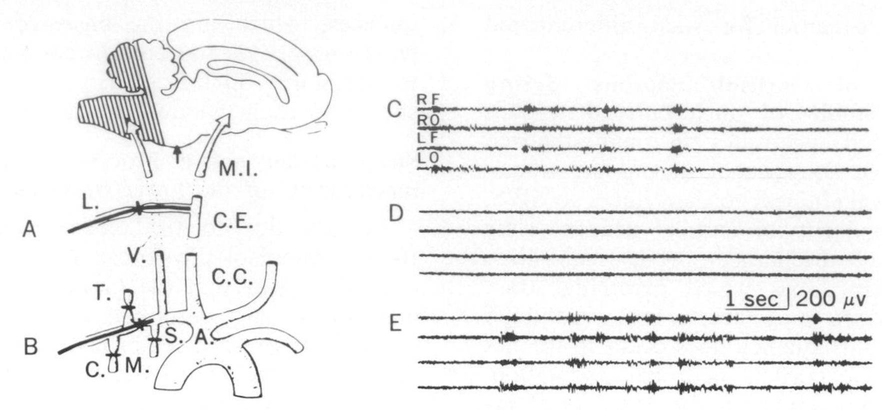

Fig. 8-9. A and **B,** Procedure for separate perfusion of medulla and caudal pons via vertebral circulation, *A,* or all brain rostral to midpontine region via carotid, *B.* Clip placed on basilar artery at arrow.

C to **E,** EEG records of cat prepared as indicated in diagram but also with transection of spinal cord at C_1, an *encéphale isolé* preparation. Four traces are shown: *RF,* right frontal; *RO,* right occipital; *LF,* left frontal; *LO,* left occipital. Five-second intervals between **C** and **D** and between **D** and **E.** At beginning, EEG shows typical spindling activity of drowsing or lightly sleeping *encéphale isolé* preparation. During black signal in **C,** 0.3 µg sodium pentobarbital was injected into vertebral artery. *Anesthetization of medulla produces activation of forebrain,* presumably by anesthetization of medullary synchronizing mechanisms. (From Magni et al.[84])

The proximity of the bulbar synchronizing region to that known to receive afferents influenced by cardiovascular events raises the question of whether the two are identical. Indeed, it has been known for a long time that high pressure within a vascularly isolated but neurally innervated carotid sinus will produce behavioral sleep. The weight of evidence, however, suggests that the synchronizing reticular neurons and those played upon by cardiovascular afferents are not the same, and that the soporific effect of the latter depends on inhibition of the reticular activating system. How pressoreceptors from the abdominal cavity (a full stomach!) exert their sleep-inducing effect is unknown, but it is probably caused by excitation of the bulbar synchronizing region and/or inhibition of the reticular activating system.

The capacity of somatic sensory afferents to induce sleep when activated by low-frequency stimulation has been clarified by ingenious experiments of Pompeiano and Swett.[100-102] These investigators used stimulating and recording electrodes implanted on peripheral nerves and studied the effect of stimulation in unanesthetized, freely moving animals. They found that low-frequency stimulation of cutaneous mechanoreceptive afferents (group II) at rates of 10/sec or lower elicited the onset of natural sleep with EEG synchronization (Fig. 8-10). More rapid excitation of these same afferents or excitation of nociceptive afferents (group III) at any frequency produced prompt arousal in sleeping cats. They also showed that group II afferents preferentially activate bulbar reticular neurons and that group III afferents activate those of the pons and mesencephalon, all via pathways in the anterolateral columns of the spinal cord. Thus an explanation is provided for the everyday observation that sleep may be induced by slowly repeated, monotonous stimuli and for the habituation of the orienting reaction and the induction of sleep described by Pavlov.

The hypothesis that besides the reticular activating system there is an antagonistic group of synchronizing and therefore sleep-inducing structures in the lower brainstem is supported by evidence that is partly indirect but nevertheless strong. How this system may exert its effects and how it interacts with the reticular activating system are important questions yet unanswered. An attractively simple explanation is that each acts in an antagonistic way upon the thalamic pacemaker, the final outcome—sleep or wakefulness—depending on the relative degree of activity in each. However, there is evidence that the synchronizing region may operate by inhibition of the reticular activating system, to which it is reciprocally linked, and both are

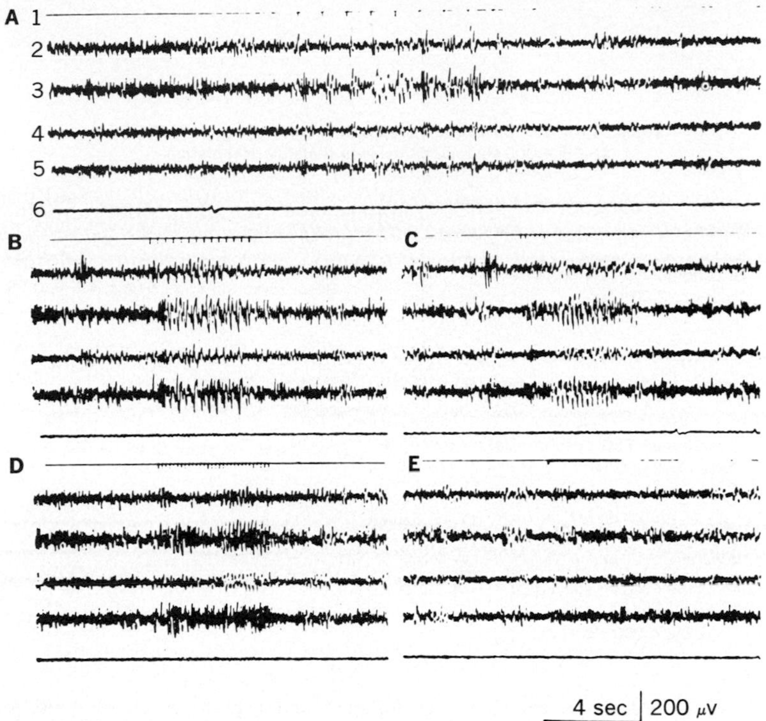

Fig. 8-10. Synchronization of EEG produced by stimulation of right superficial radial nerve at group II strength in intact unanesthetized cat. Cerebral recording and nerve-stimulating electrodes implanted 17 days earlier. Traces: *1*, stimulus marker; *2*, left parieto-occipital lead; *3*, left temporoparietal lead; *4*, right parieto-occipital lead; *5*, right temporo-occipital lead; *6*, left neck electromyogram. **A,** Spindles tripped by each stimulus at 1/sec. **B** to **D,** Stimulation at 3, 5, and 6/sec, respectively, produced synchronization of the EEG waves, more or less at stimulus frequency. **E,** Stimulation at 16/sec produced no clear synchronization. Stimulation at higher frequencies produced clear-cut activation patterns and arousal. (From Pompeiano and Swett.[100])

influenced by corticoreticular systems. Thus the final state of awareness must certainly depend on the result of these more complex interactions, which remain to be defined more precisely.

Basal forebrain system

It has been known for a long time that lesions of the neural structures just anterior to and above the optic chiasm or of the neocortex of the orbital surface of the frontal lobe lead to hyperactivity and insomnia.[81] In an extended series of experiments, Clemente and Sterman and their colleagues[46, 47] have now shown that electrical stimulation of these basal forebrain regions via chronically implanted electrodes will elicit all of the behavioral, electroencephalographic, and reflex signs of synchronized sleep. The effect is best produced by stimuli of low frequency but is not changed by rapid stimuli (up to 150/sec). In acute experiments with electrodes implanted in both the basal forebrain region and the reticular activating system of the brainstem, these investigators were able to demonstrate an interaction of the opposite behavioral and electroencephalographic effects produced by stimulation of these two regions. Moreover, it was possible to establish classic pavlovian conditioning and differential conditioning by pairing electrical stimulation of the basal forebrain region with previously neutral sounds of different frequencies. The results of such an experiment are illustrated in Fig. 8-11; they provide further evidence for a functional relation between the basal forebrain and the cerebral cortex. It is thought that the orbital cortex and the subcortical, preoptic areas give origin to a multisynaptic system of short-axoned cells that exerts an inhibitory influence over the locus of origin of the ascending reticular activating system in the brainstem. However, the role this system plays in regulating the

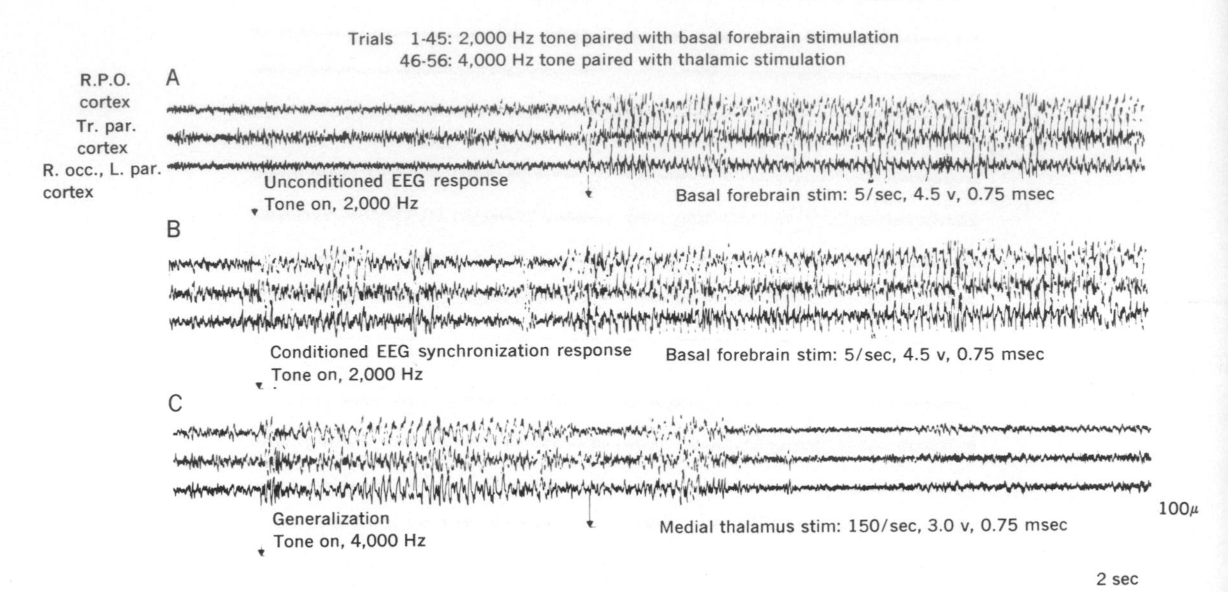

Fig. 8-11. EEG records from cat show development and generalization of conditioned EEG synchronization. **A,** Synchronization elicited by slowly repetitive stimulation of basal forebrain system. Early in series of conditioning trials, presentation of 2,000 Hz tone produced no change in ongoing EEG. **B,** After additional pairings, tone evokes shift to slow-wave, high-voltage EEG activity. **C,** Later presentation of 4,000 Hz tone evokes similar change in EEG records, an example of generalization of conditioned response. **C** also shows that when electrical stimuli are delivered to basal forebrain system, there is (slightly delayed) *desynchronization* of EEG, an example of the fact that at many locations within the brain at which low-frequency stimulation produces synchronization, high-frequency stimulation has opposite effect. (From Clemente et al.[47a])

periodic cycles of sleep and wakefulness is not yet fully understood.

SLEEP WITH DESYNCHRONIZED EEG AND DREAMING[6]

The *synchronized* or *slow-wave sleep* (SWS) of mammals is interrupted periodically by transitions to a different state that is characterized by desynchronization of the EEG and a powerful, active, descending inhibition of the segmental motor apparatus. Episodes of *desynchronized sleep* (DS) are regularly preceded by periods of SWS, reappear at intervals of about 90 min (the length of the basic mammalian rest-activity cycle[125]; Fig. 8-4), and last from 5 to 30 min or more, tending to lengthen as the night's sleep progresses, as shown in Fig. 8-12. DS is dual in nature; its tonic phase, characterized by a low-voltage, variable-frequency EEG and muscle atonia, is frequently broken by phasic outbursts of the rapid conjugate movements of the eyes called REMs and by a phasic deepening of the muscle atonia and superimposed muscle twitches and sometimes general and even violent movements. The REMs are preceded and accompanied by sharply phasic neural activity originating in the pontile RF, which is projected via polysynaptic pathways to the lateral geniculate bodies and from thence to the visual cortex. This activity is seen in the cat electrocorticogram as series of high-voltage waves called ponto-geniculo-occipital (PGO) spikes. They have a likely counterpart in the saw-toothed wave forms sometimes seen in the human EEG preceding and during the REM episodes of DS. It is important to emphasize that during this phase of sleep there is a dissociation between the EEG pattern and the behavioral state, for the desynchronized EEG of DS resembles, at least superficially, that of the waking state, while the threshold for behavioral arousal is greatly elevated over that of SWS.

The discovery by Aserinsky and Kleitman[22]

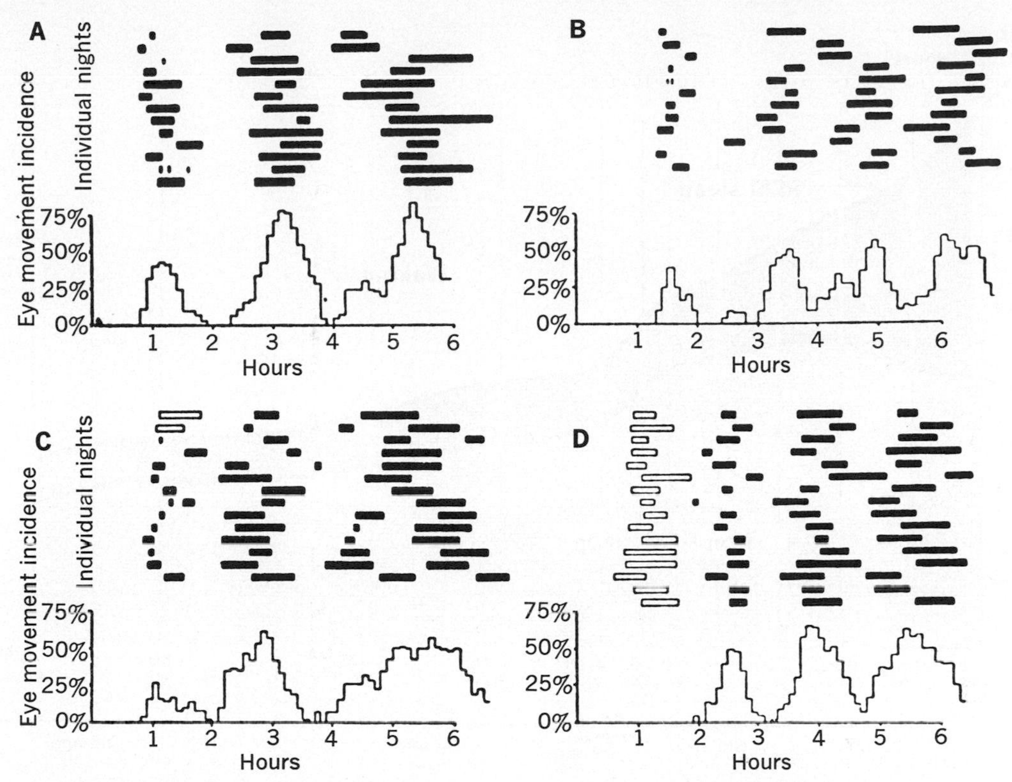

Fig. 8-12. Rhythmic occurrence of desynchronized sleep with REMs, an emergent stage 1 EEG, and dreaming for four subjects studied for a number of nights. Each bar represents single period of eye movements; each row of bars a single night's sleep. Composite histograms of incidence of REMs for several nights' sleep of each subject are placed under the series of bars. Subjects **A** and **D** cycled somewhat more regularly than **B** and **C.** Open bars indicate "cycles" when expected REMs did not appear, and during these times a persisting stage 2 was seen in EEG. (From Dement and Kleitman.[53])

and Dement and Kleitman[53] that human beings commonly report dreaming if awakened during DS but rarely do so if awakened during SWS led to greatly increased study of sleep as a biologic phenomenon. In particular, these studies concerned the cyclic nature of sleep and the periodic relation between SWS and DS, the need for sleep and the effects of sleep deprivation, dreaming, and such pathologic sleeplike states as narcolepsy.[114] The periodicity of DS is illustrated by the study of four human subjects summarized in Fig. 8-12. It is now well documented by a number of studies[80, 99, 112] that DS is most prevalent at birth and decreases both in actual amount and as a percent of total sleep during ontogeny (Fig. 8-13). This has suggested a hypothesis concerning the functional meaning of DS—that the widespread activation of the forebrain that accompanies DS is important for the maturation of the developing nervous system and in the maintenance of synaptic connectivity in the adult, although there is thus far no direct evidence to support this idea.

Physiologic changes during desynchronized sleep

During SWS there is a progressive decline in average heart and respiratory rates and an early though slight fall in blood pressure. During DS there are slight increases in the average levels of these measures but, more characteristically, a marked increase in their short-term variability.[28, 85, 121] Thus the phasic periods of DS may be accompanied by sharp rises or falls in blood pressure and by wide variations in heart rate and respiratory pattern. These changes are thought to precipitate the catastrophic nocturnal events common in patients with cardiovascular and respiratory disease.[6] Other episodic changes occur during DS: there is an increased blood flow through and oxygen consumption by the brain, penile erection,[59] and both renal and endocrinologic signs of CNS activation

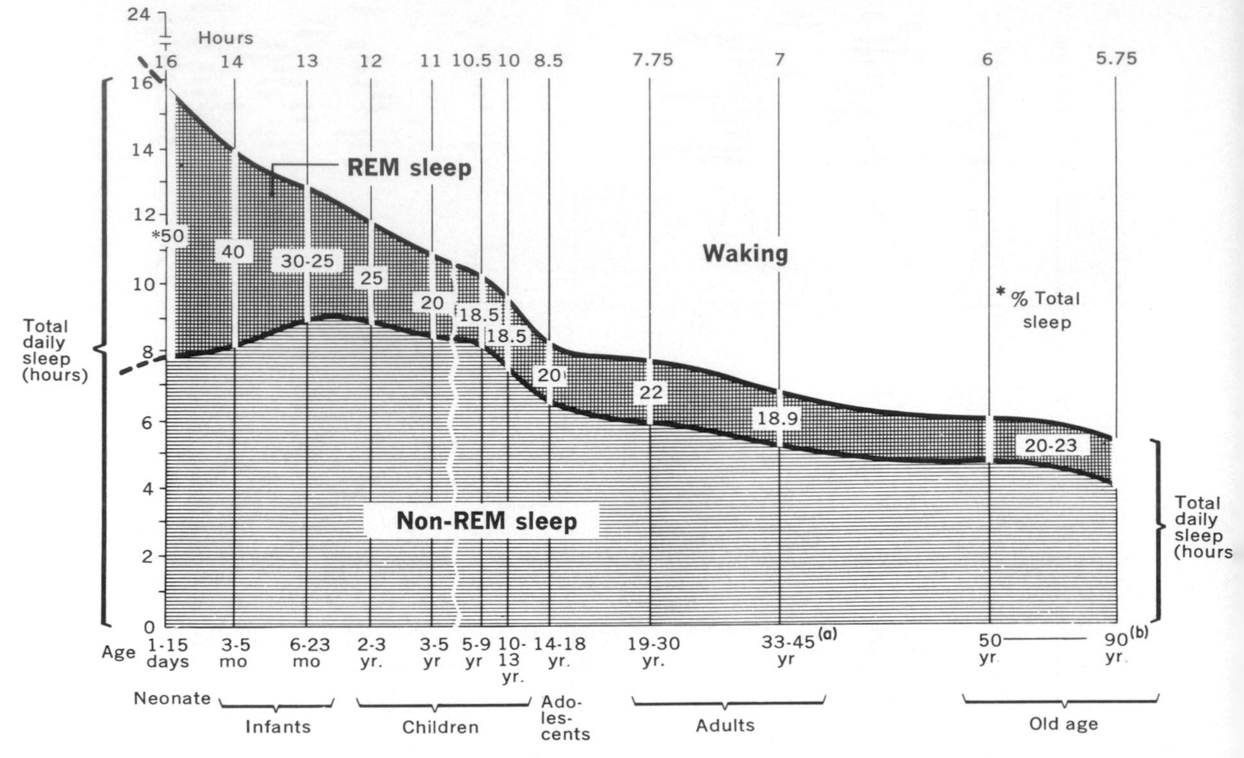

Fig. 8-13. Graph showing changes with age in total amount of daily sleep and percentage of REM (activated) sleep. There is a sharp drop in amount of REM sleep after early years, falling from 8 hr at birth to less than 1 hr in old age. Change in amount of non-REM sleep is much less marked, falling from 8 hr to about 5 hr over life-span. (From Roffwarg et al.[112])

of both the anterior and posterior lobes of the pituitary gland.[86] All of these events occur commonly but irregularly and thus are thought to be results and not causes of the episodes of DS.

Neural mechanisms in desynchronized sleep[4]

Experiments have shown that cats no longer display periods of DS following bilateral lesions of a particular part of the pontile RF and the n. reticularis pontis, oralis, and caudalis.[42] Cats with such lesions continue to show SWS in normal periodicity. It was Jouvet[73] who also showed that all aspects of DS that are mediated through effector systems of the brainstem and spinal cord occur regularly in animals that survive into the chronic state after total transection of the mesencephalon.* The desynchronization of the EEG, observed at cortical, mesencephalic, and diencephalic levels, is accompanied by a moderate increase in the discharge rate of cortical neurons.[78, 96] Each period of DS is preceded by the appearance of large, slow electrical waves in the pontile RF,[37] and McCarley and Hobson[79] found that at this time the neurons of the reticular nuclei just referred to *increase their rates of discharge by factors of 5 to 10.* Simultaneously, large, slow electrical waves appear in the hippocampus, but the role of this and other limbic structures in DS is still uncertain.[23, 40] During the tonic phase of DS, spinal motoneurons are subjected to a powerful postsynaptic inhibition, which Pompeiano et al.[11] have shown to be exerted by descending activity in reticulospinal systems of the ventral quadrants of the spinal cord.

In an important series of experiments, Pompeiano and his colleagues have shown that many of the events that characterized the *phasic* component of DS are initiated by activity that begins in the vestibular nuclei.[11] During outbursts of REMs these vestibular

*Such animals, usually cats and dogs, can be maintained for many months following high mesencephalic transection if an island of hypothalamus is left intact over the pituitary gland with an intact blood supply. This assures normal regulation of water balance. The animals are poikilothermic and must be kept in a temperature-regulated environment, with careful attention to food intake, etc.[24]

neurons discharge high-frequency bursts of impulses. Bilateral destruction of the medial and descending vestibular nuclei in the cat has the following effects on the phasic component of DS: (1) it eliminates REMs and the phasic increases in pupil diameter and heart rate that accompany them, leaving the desynchronized EEG unaffected, (2) abolishes the phasic increment in motor inhibition, shown to be due to a presynaptic inhibition of the terminals of segmental afferents, and (3) eliminates the excitatory pyramidal tract discharges that, breaking through even the phasically increased segmental inhibition, produce the muscular twitches and jerks that accompany the REMs.[90]

The powerful motor suppression exerted during DS has its counterpart on the sensory side, for there is a parallel suppression of afferent transmission, eg., in the dorsal column–medial lemniscal component of the somatic afferent system.[43, 44, 65]

Thus the state of deep sleep with desynchronized EEG is characterized by special and somewhat different sets of neural actions during its tonic and phasic periods. The tonic phase is initiated by an outburst of activity that begins in the pontile RF and seemingly engages the nervous system from the level of the cerebral cortex to that of the segmental efferents. The timing mechanism that controls its appearance is entirely unknown and the further events that trigger the superimposed phasic outbursts of REMs and associated phenomena are not understood. Among the several hypotheses that have been put forward concerning these control mechanisms, that postulating special biochemical mechanisms has received the most attention and is discussed in a following section.

The need for sleep

Although its physiologic role is unknown, there is a clear biologic need for sleep and for each of its two distinct states, SWS and DS.[52] If, for example, a healthy human being is selectively deprived of DS by systematic awakening when it appears or by action of a drug that selectively affects DS, a rebound in DS occurs to levels higher than normal when undisturbed sleep is once again permitted. This increase in DS occurs without affecting SWS; moreover, selective deprivation of SWS produces just the reciprocal rebound effects during the recovery period.[51] If subjects are deprived of all sleep, the recovery period is marked at first by an excess of SWS, and the rebound in DS is delayed until the second or third recovery night.[29, 134]

Thus sleep is an active state of the nervous system that appears in two qualitatively quite different forms that depend on two sequentially linked but distinct patterns of neural activity. It is as if the circadian appearance of sleep and the periodic alternation between SWS and DS are controlled by some clocklike timer that reaches a set point and then, given compatible environmental circumstances and drive, initiates first SWS and then periodically thereafter episodes of DS. The long time course of these cycles and their slow recovery when disrupted by deprivation suggest that these periods are controlled by regulating humoral factors that operate on a longer time scale than do synaptic transmitter actions and other short-term neural effects.

BIOCHEMICAL REGULATORY MECHANISMS IN SLEEP[5, 15, 72, 129]

Recent studies of the candidate biochemical mechanisms regulating sleep suggest that the induction and maintenance of the stages of sleep and waking are controlled, at least in part, by biogenic amines, specifically by the relative concentrations of serotonin (5-hydroxytryptamine, or 5-HT) and norepinephrine (NE), *or their related catabolic products,* at certain pertinent sites of action within the brain. The importance of these amines has been emphasized by the recent discovery, using the histochemical fluorescence technique (p. 211), that neurons of the nuclei of the median raphe of the brainstem contain large amounts of 5-HT and that they project their small axons upward very widely to the diencephalon and cerebral cortex.[49, 63, 64] Similarly, it has been shown that neurons of the brainstem containing large amounts of NE are heavily concentrated in the lateral pontile RF, particularly in the locus coeruleus (Fig. 6-18, p. 207). Lesions of either of these areas lead to a marked reduction in the brain content of the relevant amine.

The evidence that the 5-HT system plays a role in sleep regulation is as follows. (1) Depletion of 5-HT in the brain or blockage of its synthesis by appropriate drugs (e.g., *p*-chlorophenylalanine) produces a sudden decrease in both SWS and DS that lasts 1 to 2 days and is followed by a slow recovery.[103, 133] (2) Lesions of the raphe nuclei in cats lead to effective 5-HT depletion and to a severe and longer lasting insomnia[71] that is quickly relieved by administration of 5-hydroxytryptophan, a precursor of 5-HT which, unlike that amine, crosses the blood-brain barrier. (3) Close arterial injection of 5-HT into the brainstem circulation leads to a brief arousal followed by a prolonged hypersynchrony of the EEG.[115] (4) Surgical destruction of the area postrema of the medulla prevents this synchronizing effect of intra-arterial 5-HT, while local application of 5-HT to the area postrema on the medullary surface leads to immediate hypersynchrony.[74] It is postulated that 5-HT receptors of the area postrema activate the dendrites of neurons that lie in the n. tractus solitarius or its immediate vicinity,[89] part of the medullary synchronizing center (p. 269). (5) Administration of the drug reserpine depletes the brain of both 5-HT and NE and produces insomnia in both cats and men.[69] If reserpine is followed by 5-hydroxytryptophan, the precursor of 5-HT, SWS is immediately restored while DS is not. In the cat, if reserpine is followed by a precursor of NE, dihydroxyphenylalanine (dopa), DS but not SWS is restored. Such a clear-cut effect is not seen in man. (6) Complete bilateral destruction of the locus

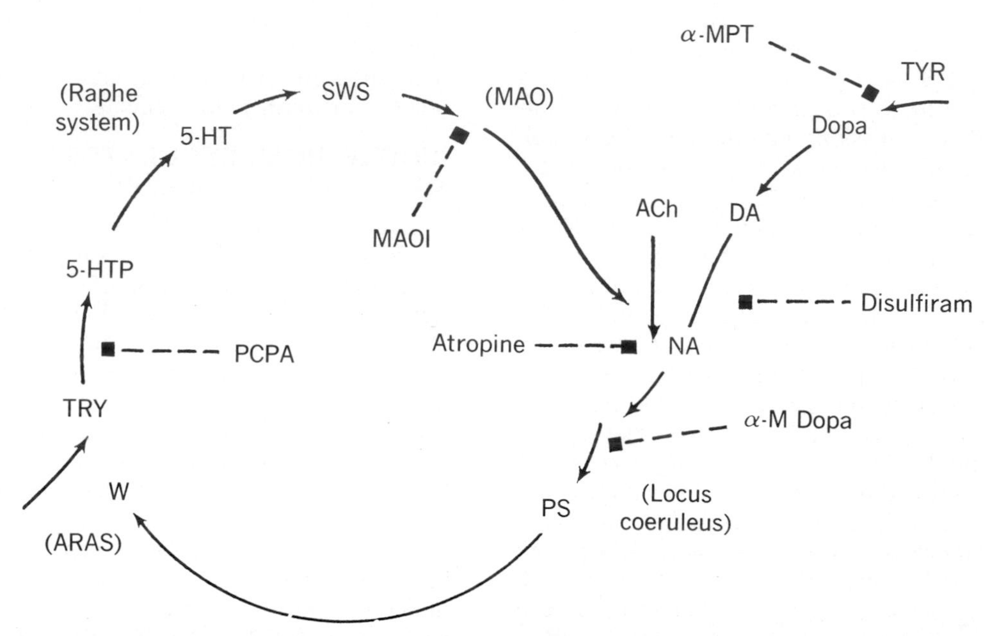

Fig. 8-14. Schema representing monoaminergic mechanisms that may be involved in regulation of sleep states, based largely on work in cat. It is postulated that there are cyclic biochemical changes within brain associated with cyclic change from waking, *W,* to *SWS.* Waking depends on ascending reticular activating system, *ARAS,* of which the chemical modulator or transmitter agents are still unknown. SWS is thought to depend in some way on 5-HT-containing neurons of raphe system, but how 5-HT levels control onset of SWS is uncertain. Paradoxical sleep, *PS,* or desynchronized sleep (referred to as DS in text) is thought to depend on noradrenaline (NA, i.e., norepinephrine)-containing neurons of pontile RF, particularly in locus coeruleus. In normal animal the first two steps are reversible ($W \leftrightarrows SW \leftrightarrows PS$), while final step from *PS* to *W* is never reversed except in pathologic or drug-induced states.

Actions of several drugs that may block certain steps of this cycle are indicated. *p*-Chlorophenylalanine (PCPA), which inhibits the enzyme tryptophan hydroxylase (TRY), decreases brain 5-HT by impairing its synthesis and thereby leads to total insomnia that may be reversed by subsequent injection of 5-HTP, the immediate precursor of 5-HT.

Monoamine oxidase inhibitors (MAOI), which prevent catabolism of 5-HT, lead to increase in SWS and total suppression of desynchronized sleep. This finding suggests that deaminated metabolite of 5-HT may trigger desynchronized sleep mechanisms. Action of atropine, which suppresses final steps in desynchronized sleep, indicates that a cholinergic step may be involved. Noradrenergic mechanisms of desynchronized sleep may be altered by drugs that impair synthesis of dopa and NA at level of tyrosine hydroxylase, such as α-methyl-*p*-tyrosine (α-MPT); by disulfiram, which inhibits the enzyme dopamine-β-hydroxylase and thus prevents the synthesis of NA from dopa; and by α-methyldihydroxylalanine (α-M dopa), which may act as false transmitter when it is converted to α-methylnoradrenaline. (From Jouvet.[5])

coeruleus in cats causes a complete suppression of DS and has no effect on SWS.

There appears to be a functional link between SWS and DS, for after destruction of the raphe system or depletion of brain 5-HT by a drug blocking its synthesis, the amounts of both SWS and DS are related to the remaining levels of brain 5-HT. DS does not reappear until returning SWS reaches about 15% of normal. This leads to the idea that the 5-HT mechanism of SWS somehow "acts as a priming mechanism for triggering paradoxical sleep" (i.e., DS).[71] Nevertheless, there is evidence that the NE-containing neurons of the pontile RF, and particularly those of the locus coeruleus, play an important role in the regulation of DS. (1) Bilateral selective destruction of the locus coeruleus in the cat produces a selective suppression of DS and a fall of NE levels in rostral parts of the brain. (2) Drugs that selectively inhibit NE synthesis suppress DS.[38] However, Williams[15] has shown that humans maintained on a synthetic diet low in phenylalanine (NE precursors) display a reversible syndrome that includes a disruption of sleep patterns, a marked decline in DS, but impaired memory and reduced physical activity as well.

It is obvious from the foregoing that a combination of biochemical and behavioral research on sleep has uncovered a host of new facts of great potential significance for understanding the mechanisms within the brain that regulate its own excitability and program its cyclic oscillation between sleep and wakefulness. Jouvet[52] has attempted a tentative systematization of the role of the biogenic amine systems in this regulation, as shown in Fig. 8-14. It should be regarded as a framework, and many of the steps indicated require further study. Jouvet,[5] Williams,[15] and others suggest that it is the rate of turnover of 5-HT, or of one of its catabolic products, that is monitored by a clocklike mechanism within the brain and regulates the sleep cycles.

Is there a more general hypnogenic factor? The idea that a sleep-inducing substance might be produced in other tissues, or indeed within the brain itself, long antedates the biochemical investigations just described and continues to receive some experimental support. There are reports, for example, that a dialysate of venous blood draining the brain of a sleeping donor animal will induce sleep in an alert animal, while such a dialysate from an alert donor has no such effect.[88] The hypothetical sleep-inducing substance has not been identified. Other investigators report that in a cross-circulation experiment in dogs, the induction of sleep in a donor animal by electrical stimulation of the medial thalamus (in the manner of Hess, p. 263) induced EEG hypersynchrony within the recipient dog within 20 to 30 sec.[75] The candidate hypnogenic substance has not been identified.

Pappenheimer and his colleagues, taking a novel approach, have developed a method for collecting the cerebrospinal fluid of animals (they used goats) in the chronic condition.[58] They found that the intraventricular injection of small amounts of CSF from sleep-deprived animals induced EEG and behavioral signs of sleep in well-rested recipient animals. Identification studies of this substance are not yet complete. The sleep-promoting factor has a molecular weight of less than 500. Its actions are not duplicated by the intraventricular injection of 5-HT or several other candidate hypnogenic substances.

REFERENCES

General reviews

1. Andersen, P., and Andersson, S. A.: Physiological basis of the alpha rhythm, New York, 1968, Appleton-Century-Crofts.
2. Cold Spring Harbor symposia on quantitative biology. Biological clocks, Baltimore, 1960, The Waverly Press, Inc.
3. Hartmann, E., editor: Sleep and dreaming, Int. Psychiatry Clin. **7**:entire issue, 1970.
4. Jouvet, M.: Neurophysiology of the states of sleep, Physiol. Rev. **47**:117, 1967.
5. Jouvet, M.: Biogenic amines and the states of sleep, Science **163**:32, 1969.
6. Kales, A.: Sleep: physiology and pathology, Philadelphia, 1969, J. B. Lippincott Co.
7. Kety, S. S., Evarts, E. V., and Williams, H. L., editors: Sleep and altered states of consciousness, Baltimore, 1967, The Williams & Wilkins Co.
8. Kleitman, N.: Sleep and wakefulness, ed. 2, Chicago, 1963, University of Chicago Press.
9. Moruzzi, G.: Active processes in the brain stem during sleep, Harvey Lect. **58**:233, 1962-1963.
10. Moruzzi, G.: Sleep and instinctive behavior, Arch. Ital. Biol. **107**:175, 1969.
11. Pompeiano, O.: The neurophysiological mechanisms of the postural and motor events during desynchronized sleep, Res. Publ. Assoc. Res. Nerv. Ment. Dis. **45**:351, 1967.
12. Richter, C. P.: Biological clocks in medicine and psychiatry, Springfield, Ill., 1965, Charles C Thomas, Publisher.
13. Scheibel, M., and Scheibel, A.: Structural organization of nonspecific thalamic nuclei and their projection toward cortex, Brain Res. **6**:60, 1968.
14. Thorpe, W. H.: Ethology and consciousness. In Eccles, J. C., editor: Brain and conscious experience, Berlin, 1966, Springer Verlag.
15. Williams, H. L.: The new biology of sleep, J. Psychiat. Res. **8**:445, 1971.
16. Witkin, H., and Lewis, H.: Experimental studies of dreaming, New York, 1967, Random House, Inc.

Original papers

17. Affani, A., Marchiafava, P. L., and Zernicki, B.: Orientation reactions in the midpontine

pretrigeminal cat, Arch. Ital. Biol. **100:**297, 1962.
18. Affani, A., Marchiafava, P. L., and Zernicki, B.: Conditioning in the midpontine pretrigeminal cat, Arch. Ital. Biol. **100:**305, 1962.
19. Akert, K., Bally, C., and Schadé, J. P., editors: Sleep mechanisms, Progr. Brain Res. **18:** entire issue, 1965.
20. Akert, K., Koella, W. P., and Hess, R., Jr.: Sleep produced by electrical stimulation of the thalamus, Am. J. Physiol. **168:**268, 1952.
21. Apelbaum, J., Silva, E. E., Fruck, P., and Segundo, J. P.: Specificity and biasing of arousal reaction habituation, Electroencephalogr. Clin. Neurophysiol. **12:**829, 1960.
21a. Arduini, A., and Arduini, M. G.: Effect of drugs and metabolic alterations on brain stem arousal mechanism, J. Pharmacol. Exp. Ther. **110:**76, 1954.
22. Aserinsky, E., and Kleitman, N.: Regularly occurring periods of eye motility, and concomitant phenomena, during sleep, Science **118:**273, 1953.
23. Bancaud, J., et al.: Les accès épileptiques au cours du sommeil de nuit, Rev. Neurol. Psychiat. **110:**314, 1964.
24. Bard, P., Woods, J. W., and Bleier, R.: The effects of cooling, heating, and pyrogen on chronically decerebrate cats, Comm. Behav. Biol. **5(A):**31, 1970.
25. Batini, C., Palestini, M., Rossi, G. F., and Zanchetti, A.: EEG activation patterns in the midpontine pretrigeminal cat following sensory deafferentation, Arch. Ital. Biol. **97:** 26, 1959.
26. Batini, C., et al.: Neural mechanisms underlying the enduring EEG and behavioral activation in the midpontine pretrigeminal cat, Arch. Ital. Biol. **97:**13, 1959.
27. Batini, C., et al.: Effects of complete pontine transections on the sleep-wakefulness rhythm, the midpontine pretrigeminal preparation, Arch. Ital. Biol. **97:**1, 1959.
28. Berger, R. J.: Physiological characteristics of sleep. In Kales, A., editor: Sleep: physiology and pathology, Philadelphia, 1969, J. B. Lippincott Co.
29. Berger, R. J., and Oswald, I.: Effects of sleep deprivation on behavior, subsequent sleep and dreaming, J. Ment. Sci. **108:**457, 1962.
30. Berlucchi, G., Maffei, L., Moruzzi, G., and Strata, P.: EEG and behavioral effects elicited by cooling of medulla and pons, Arch. Ital. Biol. **102:**372, 1964.
31. Berman, A. L.: The brain stem of the cat. A cytoarchitectonic atlas with stereotaxic coordinates, Madison, 1968, University of Wisconsin Press.
32. Bert, J., et al.: A comparative sleep study of two cercopithecinae, Electroencephalogr. Clin. Neurophysiol. **28:**32, 1970.
33. Bonvallet, M., and Allen, M. B., Jr.: Prolonged spontaneous and evoked reticular activation following discrete bulbar lesions, Electroencephalogr. Clin. Neurophysiol. **15:**969, 1963.
34. Bremer, F.: The neurophysiological problem of sleep. In Adrian, E. A., Jasper, H. H., and Bremer, F., editors: Brain mechanisms and consciousness, Springfield, Ill., 1954, Charles C Thomas, Publisher.
35. Brodal, A.: The reticular formation of the brain stem. Anatomical aspects and functional correlations, Edinburgh, 1957, Oliver & Boyd, Ltd.
36. Brookhart, J. M., and Zanchetti, A.: The relation between electrocortical waves and responsiveness of the cortico-spinal system, Electroencephalogr. Clin. Neurophysiol. **8:** 427, 1956.
37. Brooks, D. C., and Bizzi, E.: Brain stem electrical activity during deep sleep, Arch. Ital. Biol. **101:**648, 1963.
38. Buguet, A., Petitjean, F., and Jouvet, M.: Suppression des pointes ponto-geniculo-occipitales du sommeil par lesion ou injection in situ de 6-hydroxy-dopamine au niveau de tegmentum pontique, Compt. Rend. Soc. Biol. **164:**2293, 1970.
39. Bullock, T. H., and Horridge, G. A.: Structure and function in the nervous systems of invertebrates, San Francisco, 1965, W. H. Freeman & Co., Publishers, chaps. 1 and 5.
40. Cadillac, J., Passouant-Fontaine, T., and Passouant, P.: Modifications de l'activite de l'hippocampe suivant les divers stades du sommeil spontane chez le chat, Rev. Neurol. Psychiat. **105:**171, 1961.
41. Cairns, H.: Disturbances of consciousness with lesions of the brain stem and diencephalon, Brain **75:**109, 1952.
42. Carli, G., and Zanchetti, A.: A study of pontine lesions suppressing deep sleep in the cat, Arch. Ital. Biol. **103:**751, 1965.
43. Carli, G., Diete-Spiff, K., and Pompeiano, O.: Transmission of sensory information through the lemniscal pathway during sleep, Arch. Ital. Biol. **105:**31, 1967.
44. Carli, G., Diete-Spiff, K., and Pompeiano, O.: Presynaptic and postsynaptic inhibition of transmission of somatic afferent volleys through the cuneate nucleus during sleep, Arch. Ital. Biol. **105:**52, 1967.
45. Caspers, H.: Die Veränderungen der corticalen Gleichspannung und ihre Beziehungen zur senso-motorischen Aktivitat (Verhalten) bei weckreizungen am freibeweglichen Tier, Proceedings of the twenty-second congress, Leiden, 1962, vol. I, part 1, p. 443.
46. Clemente, C. D.: Forebrain mechanisms related to internal inhibition and sleep, Cond. Reflex **3:**145, 1968.
47. Clemente, C. D., and Sterman, M. B.: Basal forebrain mechanisms for internal inhibition and sleep, Res. Publ. Assoc. Res. Nerv. Ment. Dis. **45:**127, 1967.
47a. Clemente, C. D., Sterman, M. B., and Wyrwicka, W.: Forebrain inhibitory mechanisms: conditioning of basal forebrain induced synchronization and sleep, Exp. Neurol. **7:**404, 1963.
48. Coccagna, G., et al.: Arterial pressure changes during spontaneous sleep in man, Electroencephalogr. Clin. Neurophysiol. **31:**277, 1971.
49. Dahlstrom, A., and Fuxe, K.: Evidence for the existence of monoamine-containing neurons in the central nervous system. I. Demonstration of monoamines in the cell bodies of

brainstem neurons, Acta Physiol. Scand. **62**(suppl. 232):1, 1964.
50. Dell, P.: Reticular homeostasis and critical reactivity. In Moruzzi, G., Fessard, A., and Jasper, H. H., editors: Brain mechanisms, Amsterdam, 1963, Elsevier Publishing Co.
51. Dement, W.: Effect of dream deprivation, Science **131:**1705, 1960.
52. Dement, W.: The biological role of REM sleep (circa 1968). In Kales, A., editor: Sleep: physiology and pathology, Philadelphia, 1969, J. B. Lippincott Co.
53. Dement, W., and Kleitman, N.: Cyclic variations of EEG during sleep and their relations to eye movements, body motility and dreaming, Electroencephalogr. Clin. Neurophysiol. **9:**673, 1957.
54. Dempsey, E. W., and Morison, R. S.: The electrical activity of thalamocortical relay systems, Am. J. Physiol. **138:**283, 1942.
55. Desiraju, T., and Purpura, D. P.: Organization of specific-nonspecific thalamic internuclear synaptic pathways, Progr. Brain Res. **21:**169, 1970.
56. Economo, C. von: Sleep as a problem of localization, J. Nerv. Ment. Dis. **71:**249, 1930.
57. Evarts, E. V.: Unit activity in sleep and wakefulness. In Quarton, G. C., Melnechuk, T., and Schmitt, F. O., editors: The neurosciences. A study program, New York, 1967, The Rockefeller University Press.
58. Fencl, V., Koski, G., and Pappenheimer, J. R.: Factors in cerebrospinal fluid from goats that affect sleep and activity in rats, J. Physiol. **216:**565, 1971.
59. Fisher, C., Gross, J., and Zuch, J.: Cycle of penile erection synchronous with dreaming (REM) sleep, Arch. Gen. Psychiatry **12:**29, 1965.
60. Freemon, F. R., McNew, J. J., and Adey, W. R.: Sleep of unrestrained chimpanzee. Cortical and subcortical recordings, Exp. Neurol. **25:**129, 1969.
61. Freemon, F. R., McNew, J. J., and Adey, W. R.: Chimpanzee sleep stages, Electroencephalogr. Clin. Neurophysiol. **31:**485, 1971.
62. French, J. D., and Magoun, H. W.: Effects of chronic lesions in central cephalic brain stem of monkeys, Arch. Neurol. Psychiatry **68:** 591, 1952.
63. Fuxe, K.: Evidence for the existence of monoamine neurons in the central nervous system. IV. Distribution of monoamine nerve terminals in the central nervous system, Acta Physiol. Scand. **64**(suppl. 247):37, 1965.
64. Fuxe, K., Hokfelt, T., and Ungerstedt, V.: Localization of indolealkylamines in CNS, Adv. Pharmacol. **6A:**235, 1968.
65. Ghelarducci, B., Pisa, M., and Pompeiano, O.: Transformation of somatic afferent volleys across the prethalamic and thalamic components of the lemniscal system during the rapid eye movements of sleep, Electroencephalogr. Clin. Neurophysiol. **29:**348, 1970.
66. Hanbury, J., and Jasper, H. H.: Independence of diffuse thalamocortical projection system shown by specific nuclear destruction, J. Neurophysiol. **16:**252, 1953.
67. Hess, W. R.: The functional organization of the diencephalon, New York, 1957, Grune & Stratton, Inc.
68. Hobson, J. A., and McCarley, R. W.: Cortical unit activity in sleep and waking, Electroencephalogr. Clin. Neurophysiol. **30:**97, 1971.
69. Hoffman, J. S., and Domino, E. F.: Comparative effects of reserpine on the sleep cycle of man and cat, J. Pharmacol. Exp. Ther. **170:** 190, 1969.
70. Jasper, H. H.: Unspecific thalamocortical relations. In Magoun, H. W., editor, Neurophysiology section: Handbook of physiology, Baltimore, 1960, The Williams & Wilkins Co., vol. 2.
71. Jouvet, M.: Insomnia and decrease of cerebral 5-HT after destruction of the raphe system in the cat, Adv. Pharmacol. **6B:**265, 1968.
72. Jouvet, M.: Serotonin and sleep. In Blum, J. J., editor: Biogenic amines as physiological regulators, Englewood Cliffs, N. J., 1969, Prentice-Hall, Inc.
73. Jouvet, M., and Jouvet, D.: A study of the neurophysiological mechanisms of dreaming, Electroencephalogr. Clin. Neurophysiol. suppl. **24:**133, 1963.
74. Koella, W. P.: Serotinin and sleep, Exp. Med. Surg. **27:**157, 1969.
75. Kornmüller, A. E., Lux, H. D., Winkle, K., and Klee, M.: Neurohumoral ausfeloste Schlafzustande an Tieren mit gekreuztem Kreislauf unter der Kontrolle von EEG-Ableitungen, Naturwissenschaften **48:**503, 1961.
76. Kuypers, H. G. J. M.: Some projections from the peri-central cortex to the pons and lower brain stem in monkey and chimpanzee, J. Comp. Neurol. **110:**221, 1958.
77. Larsen, L. E., and Walter, D. O.: On automatic methods of sleep staging by EEG spectra, Electroencephalogr. Clin. Neurophysiol. **28:**459, 1970.
78. McCarley, R. W., and Hobson, J. A.: Cortical unit activity in desynchronized sleep, Science **167:**901, 1970.
79. McCarley, R. W., and Hobson, J. A.: Single neuron activity in cat gigantocellular tegmental field: selectivity of discharge in desynchronized sleep, Science **174:**1250, 1971.
80. McGinty, D. J.: Encephalization and the neural control of sleep. In Sterman, M. B., McGinty, D. J., and Adinolfi, A. M., editors: Brain development and behavior, New York, 1971, Academic Press, Inc.
81. McGinty, D. J., and Sterman, M. B.: Sleep suppression after basal forebrain lesions in the cat, Science **160:**1253, 1968.
82. Magnes, J., Moruzzi, G., and Pompeiano, O.: Synchronization of the EEG produced by low-frequency electrical stimulation of the region of the solitary tract, Arch. Ital. Biol. **99:**33, 1961.
83. Magni, F., and Willis, W. D.: Cortical control of brain stem reticular neurons, Arch. Ital. Biol. **102:**418, 1964.
84. Magni, F., Moruzzi, G., Rossi, G. F., and Zanchetti, A.: EEG arousal following inactivation of the lower brain stem by selective in-

jection of barbiturate into the vertebral circulation, Arch. Ital. Biol. **97**:33, 1959.
85. Mancia, G., Baccelli, G., Adams, D. B., and Zanchetti, A.: Vasomotor regulation during sleep in the cat, Am. J. Physiol. **220**:1086, 1971.
86. Mandell, A. J., and Mandell, M. P.: Biochemical aspects of rapid eye movement sleep, Am. J. Psychiatry **122**:391, 1965.
87. Mangold, R., et al.: The effects of sleep and lack of sleep on cerebral circulation and metabolism of normal young men, J. Clin. Invest. **34**:1092, 1955.
88. Monnier, M., and Hosli, L.: Humoral transmission of sleep and wakefulness. II. Hemodialysis of a sleep inducing humor during stimulation of the thalamic somnogenic area, Pflügers Arch. Physiol. **282**:60, 1965.
89. Morest, D. K.: Experimental study of the projections of the nucleus of the tractus solitarius and the area postrema in the cat, J. Comp. Neurol. **130**:277, 1967.
90. Morrison, A. R., and Pompeiano, O.: Vestibular influences during sleep, Arch. Ital. Biol. **108**:154, 1970.
91. Moruzzi, G.: The physiological properties of the brain stem reticular formation. In Adrian, E. D., Jasper, H. H., and Bremer, F., editors: Brain mechanisms and consciousness, Springfield, Ill., 1954, Charles C Thomas, Publisher.
92. Moruzzi, G.: Reticular influences on the EEG, Electroencephalogr. Clin. Neurophysiol. **16**:1, 1964.
93. Moruzzi, G., and Magoun, H. W.: Brain stem reticular formation and activation of the EEG, Electroencephalogr. Clin. Neurophysiol. **1**:455, 1949.
94. Nauta, W. H. J., and Kuypers, H. G. J. M.: Some ascending pathways in brain stem reticular formation. In Jasper, H. H., et al., editors: Reticular formation of the brain, Boston, 1958, Little, Brown & Co.
95. Noda, H., and Adey, W. R.: Firing variability in cat association cortex during sleep and wakefulness, Brain Res. **18**:513, 1970.
96. Noda, H., and Adey, W. R.: Changes in neuronal activity in association cortex of the cat in relation to sleep and wakefulness, Brain Res. **20**:263, 1970.
97. Olszewski, J., and Baxter, D.: Cytoarchitecture of the human brain stem, Philadelphia, 1954, J. B. Lippincott, Co.
98. Parker, D. C., Sassin, J. F., and Mace, J. W.: Human growth hormone release during sleep; electroencephalographic correlation, J. Clin. Endocrinol. Metab. **29**:871, 1969.
99. Parmelee, A. H., et al.: Maturation of EEG activity during sleep in premature infants, Electroencephalogr. Clin. Neurophysiol. **24**:319, 1968.
100. Pompeiano, O., and Swett, J. E.: EEG and behavioral manifestations of sleep induced by cutaneous nerve stimulation in normal cats, Arch. Ital. Biol. **100**:311, 1962.
101. Pompeiano, O., and Swett, J. E.: Identification of cutaneous and muscular afferent fibers producing EEG synchronization or arousal in normal cats, Arch. Ital. Biol. **100**:343, 1962.
102. Pompeiano, O., and Swett, J. E.: Action of graded cutaneous and muscular afferent volleys on brain stem units in the decerebrate, cerebellectomized cat, Arch. Ital. Biol. **101**:552, 1963.
103. Pujol, J. F., et al.: The central metabolism of serotonin in the cat during insomnia. A neurophysiological and biochemical study after administration of p-chlorophenylalanine or destruction of the raphe system, Brain Res. **29**:195, 1971.
104. Purpura, D. P., and Cohen, B.: Intracellular recording from thalamic neurons during recruiting responses, J. Neurophysiol. **25**:621, 1962.
105. Purpura, D. P., and Houspian, E. M.: Alterations in corticospinal neuron activity associated with thalamo-cortical recruiting responses, Electroencephalogr. Clin. Neurophysiol. **13**:365, 1961.
106. Purpura, D. P., and Shofer, R. J.: Intracellular recording from thalamic neurons during reticulocortical activation, J. Neurophysiol. **26**:494, 1963.
107. Purpura, D. P., McMurtry, J. G., and Maekawa, K.: Synaptic events in ventrolateral thalamic neurons during suppression of recruiting responses by brain stem reticular formation, Brain Res. **1**:63, 1966.
108. Purpura, D. P., Scarff, T., and McMurtry, J. G.: Intracellular study of internuclear inhibition in ventrolateral thalamic neurons, J. Neurophysiol. **28**:487, 1965.
109. Purpura, D. P., Shofer, R. J., and Musgrave, F. S.: Cortical intracellular potentials during augmenting and recruiting responses. II. Patterns of synaptic activities in pyramidal and nonpyramidal tract neurons, J. Neurophysiol. **27**:133, 1964.
110. Richter, C. P.: Sleep and activity: their relation to the 24-hour clock, Res. Publ. Assoc. Nerv. Ment. Dis. **45**:8, 1967.
111. Richter, C. P.: Inborn nature of the rat's 24-hour clock, J. Comp. Physiol. Psychol. **75**:1, 1971.
112. Roffwarg, H. P., Muzio, J. N., and Dement, W. C.: Ontogenetic development of the human sleep-dream cycle, Science **152**:604, 1966.
113. Rossi, G. F., and Zanchetti, A.: The brain stem reticular formation. Anatomy and physiology, Arch. Ital. Biol. **95**:199, 1957.
114. Roth, B., Bruhova, S., and Lehovsky, M.: REM sleep and NREM sleep in narcolepsy and hypersomnia, Electroencephalogr. Clin. Neurophysiol. **26**:176, 1969.
115. Roth, G. I., Walton, P. L., and Yamamoto, W. S.: Area postrema: abrupt EEG synchronization following close intra-arterial perfusion with serotonin, Brain Res. **23**:223, 1970.
116. Scheibel, M. E., and Scheibel, A. B.: Structural substrates for integrative patterns in the brain stem reticular core. In Jasper, H. H., et al., editors: Reticular formation of the brain, Boston, 1958, Little, Brown & Co.
117. Scheibel, M. E., and Scheibel, A. B.: The organization of the nucleus reticularis thalami: a Golgi study, Brain Res. **1**:43, 1966.
118. Segundo, J. P., Takenaka, T., and Encabo, H.: Electrophysiology of bulbar reticular neurons, J. Neurophysiol. **30**:1194, 1967.

119. Segundo, J. P., Takenaka, T., and Encabo, H.: Somatic sensory properties of bulbar reticular neurons, J. Neurophysiol. **30**:1221, 1967.
120. Snyder, F.: Autonomic nervous system manifestations during sleep and dreaming, Res. Publ. Assoc. Res. Nerv. Ment. Dis. **45**:469, 1967.
121. Snyder, F., Hobson, J. A., Morrison, D. F., and Goldfrank, F.: Changes in respiration, heart rate, and systolic blood pressure in human sleep, J. Appl. Physiol. **19**:417, 1964.
122. Spencer, W. A., and Brookhart, J. M.: Electrical patterns of augmenting and recruiting waves in depths of sensorimotor cortex of cat, J. Neurophysiol. **24**:26, 1961.
123. Spencer, W. A., and Brookhart, J. M.: A study of spontaneous spindle waves in sensorimotor cortex of cat, J. Neurophysiol. **24**:50, 1961.
124. Steriade, M.: Ascending control of thalamic and cortical responsiveness, Int. Rev. Neurobiol. **12**:87, 1970.
125. Sterman, M. B., and Hoppensbrouwers, T.: The development of sleep-waking and rest-activity patterns from fetus to adult in man. In Sterman, M. B., McGinty, D. J., and Adinolfi, A. M., editors: Brain development and behavior, New York, 1971, Academic Press, Inc.
126. Strumwasser, F.: Neurophysiological aspects of rhythms. In Quarton, G. C., Melnechuk, T., and Schmitt, F. O., editors: The neurosciences. A study program, New York, 1967, The Rockefeller University Press.
127. Strumwasser, F.: The cellular basis of behavior in Aplysia, J. Psychiatr. Res. **8**:237, 1971.
128. Takahashi, Y., Kipnis, D. M., and Daughaday, W. H.: Growth hormone secretion during sleep, J. Clin. Invest. **47**:2079, 1968.
129. Torda, C.: Biochemical and bioelectric processes related to sleep, paradoxical sleep, and arousal, Psychol. Rep. **24**:807, 1969.
130. Velasco, M., and Lindsley, D. B.: Effect of thalamocortical activation on recruiting responses, Acta Neurol. Lat. Am. **14**:188, 1968.
131. Velasco, M., Weinberger, N. M., and Lindsley, D. B.: Effect of thalamocortical activation on recruiting responses. I. Reticular stimulation, Acta Neurol. Lat. Am. **14**:99, 1968.
132. Walter, D. O., and Brazier, M. A. B.: Advances in EEG analysis, Electroencephalogr. Clin. Neurophysiol. suppl. **27**:1, 1969.
133. Weitzman, E. D., Rapport, M. M., McGregor, P., and Jocoby, J.: Sleep patterns of the monkey and brain serotonin concentration: effect of p-chlorophenylalanine, Science **160**: 1361, 1968.
134. Williams, H. L., et al.: Responses to auditory stimulation, sleep loss and the EEG stages of sleep, Electroencephalogr. Clin. Neurophysiol. **16**:269, 1964.
135. Yamaguchi, N., Ling, G. M., and Marczynski, T. J.: Recruiting responses observed during wakefulness and sleep unanesthetized chronic cats, Electroencephalogr. Clin. Neurophysiol. **17**:246, 1965.

IV

CENTRAL NERVOUS MECHANISMS IN SENSATION

9

VERNON B. MOUNTCASTLE

Sensory receptors and neural encoding: introduction to sensory processes

Previously the subjects of excitation and conduction within nerve cells and the transmission of excitation from cell to cell have been discussed in some detail. The general physiology of the forebrain and the generalized thalamocortical systems that control its levels of activity have also been explored. Further study of the CNS is begun here with a consideration of the neural mechanisms in sensation, dealing initially with what is called peripheral or sensory transduction, i.e., how the terminal ends of peripheral nerve fibers, often acting in concert with nonneural cells that surround them, convert impinging energy into local excitation at the nerve ending, and how this in turn elicits trains of afferent nerve impulses. This mechanism of peripheral encoding[5, 6, 18, 55] determines what information the nervous system receives about the quality, locus, intensity, and temporal patterns of stimuli that elicit sensations. Ultimately the goal is to learn how this initial neural replication of the external world is further elaborated and transformed in the CNS, and how it leads to perception, discrimination, storage in memory, etc., and, at choice, motor response or public description.

GENERAL SENSORY CAPACITY

It is everyday human experience that different stimuli elicit differing sensory experiences that can be classified and named. Electromagnetic waves in the range of 400 to 760 nm elicit experiences called visual because they are seen. Those visual experiences elicited by lights of different wavelengths are described differently, for they are seen as different colors. Objects that are tactually examined are described as hard or soft, warm or cool, and as having certain spatial contours or temporal patterns. Substances are readily differentiated by the way they taste or smell. Indeed, it seems to be given directly to conscious experience that stimuli of a given kind elicit a sensory experience that, within certain limits of discrimination and resolution, can be defined with some precision and recognized whenever it is encountered. Each of these readily distinguishable classes is termed a *sensory modality,* which may be defined as a class of sensations connected along a qualitative continuum. Sensations of color compose a single modality, as do those of tones. Sensations grouped under the term "general somatic sensibility" do not, for mechanoreception (touch-pressure), warmth, cool, pain, and the sense of the position and movement of the joints can each be identified separately; they differ in quality. The central neural mechanisms responsible for the identification of different qualities are incompletely understood. This major problem in the neurophysiology of sensation will influence treatment of the subject, from the level of first-order fibers to that of total behavior.

Human observers accurately locate the spatial position of stimuli; i.e., the position of a tactile stimulus to the tip of a finger can be reproduced with an error of only 1 to 2 mm. In the central region of the visual field the position of a light can be replicated to 1 degree of arc or less. Paradoxically, the comparable property in audition is not the position of sound in space, but its pitch (Chapter 13). From the standpoint of mechanism, understanding how a stimulus is located in physical space reduces to the question of the meaning of local peaks of activity within neural fields, i.e., in the primary sensory cortices, and this remains a major unsolved problem in experimental neurology.

Beyond recognizing place and quality, humans can differentiate between stimuli of identical character and position but different intensity. We possess a poorly developed capacity to rate such stimuli along physical scales (pressure, heat, luminous flux, sound pressure, etc.) but discriminate nicely between stimuli that differ only in intensity when only two are presented (Chapter 18.) Finally, humans possess a refined ability to identify or discriminate between stimuli that differ in spatial configuration or in temporal pattern: men excel in sensing transients, not steady states.

In summary, a nervous system faces not the external world directly, but the afferent input reaching it over first-order sensory nerve fibers. It derives from that input an ongoing, constantly changing, and nearly up-to-date picture of the external environment and of stimuli that impinge upon the organism. Our perceptual image of the external world is always to some degree an abstraction of physical reality, an abstraction determined in degree and in kind by the initial transformations that occur at the level of first-order nerve fibers, and by those subsequently interposed between stages in the long chain of neural events leading to perception. It is the aim of this and subsequent chapters to explain, so far as is presently possible, the neural mechanisms of these series of transformations. Initially we shall elucidate the extent to which the sensory capacities are determined (are limited) by the transducer action of peripheral nerve endings and, as is sometimes the case, of the specialized nonneural receptor cells associated with them. Those serving what is termed general somatic sensibility will be treated in this chapter.* The stretch receptors of muscle, which play a vital role in the regulation of movement and muscle tone, are not properly regarded as *sensory* receptors and are discussed in Chapter 22, in the section dealing with motor mechanisms.

DEFINITIONS AND PRINCIPLES

1. The terms *"sensory receptor," "sensory ending,"* and *"sensory organ"* are used loosely in sensory physiology. More precisely, sensory endings are the peripheral terminals of the afferent nerve fibers. In some cases these endings terminate in relation to specialized nonneural cells that play an important role in the transducer process, e.g., the hair cells of the cochlea or the specialized cells of the taste buds.

2. Sensory endings possess *thresholds;* i.e., there are physically definable classes of stimuli; some of these elicit afferent nerve impulses and others do not. This threshold is a variable quantity and is therefore usually defined by some statistical estimator, e.g., the strength that will excite on one half the trials. The threshold for sensation will, in the limiting case, approach that of first-order nerve fibers when the observer attends closely to the stimuli under study and excludes others.

3. A *sensory unit* is a single primary afferent nerve fiber, including all its peripheral branches and central terminals. In a more extended sense, such a unit includes associated nonneural transducer cells as well.

4. A *peripheral receptive field* is that spatial area within which a stimulus of sufficient intensity and proper quality will evoke a discharge of impulses in a sensory unit. It may be measured, for example, as the area of skin within which a mechanical stimulus will excite a cutaneous mechanoreceptive unit, or as the area of the visual field within which light will excite an optic nerve fiber—a third-order sensory unit, etc. The threshold varies with the position of a stimulus within the receptive field, usually being lowest near its center, a fact determined, it is thought, by the differential density of terminals within the field.

5. The peripheral branches of adjacent sensory units are intertwined to a considerable degree, and this overlap shifts gradually across a sensory surface such as the skin or retina: it is unlikely that any stimulus ever engages a single afferent nerve fiber alone. This principle of *partially shifted overlap* is important for theories of CNS mechanisms in spatial discrimination. The peripheral receptive fields of afferent fibers terminating in a sensory field may vary greatly in size, usually inversely with the numbers of innervating nerve fibers per unit area of the sensory sheet. The volume of central nervous tissue devoted to the representation of a sensory field varies directly with this *peripheral innervation density,* as indeed it does with sensory acuity. Thus the fingers are innervated densely by mechanoreceptive afferents whose receptive fields are small, the central representation of the fingers in the postcentral gyrus of the cerebral cortex is large, and tactile acuity is highly developed on the fingers. Exactly converse statements hold

*The highly developed receptors of the ear are discussed in Chapters 12 and 13; those of the eye, in Chapters 14 to 16; and those for taste and smell, in Chapter 17.

for the proximal regions of the limbs and the trunk.

6. Johannes Müller suggested that impulses in different afferent nerve fibers elicit different sensations by virtue of "specific energies." This does not mean, of course, that specificity resides in some qualitatively unique property of one or another nerve fiber. To put Müller's doctrine in modern terms, *different sets of nerve fibers, when active, elicit different sensations by virtue of their different central connections,* and a given set of nerve fibers elicits an identical sensation no matter how excited, whether by adequate natural stimuli to the endings or artificially, e.g., by electrical stimulation. An important problem in the study of somatic sensibility is how a given set of nerve fibers, which terminates with other sets in a common spatial distribution within a receptive surface such as the skin, is activated by a stimulus of one quality whereas a second set of nerve fibers is activated by another. The extent to which this may be explained by a differential sensitivity, or "specificity," of the nerve endings and their associated structures will be apparent in later discussions.

7. Sensory fibers differ from one another in the manner in which they respond to continuing stimulation, i.e., in the rate of *adaptation.* Some are detectors of transients only, for they discharge a few impulses upon the application of a steady stimulus and none during its steady continuation; they may discharge again when the stimulus is removed. Such afferents provide little information about the intensity of a stimulus but are activated at low stimulus intensities by temporal or spatial transients. Other afferent fibers respond to stimulus onset with a high-frequency discharge whose time course and peak frequency are functions of the rate of stimulus application and its final intensity. These afferents also provide information about stimulus intensity in the steady state, for they continue to discharge during stimulus application at a frequency determined by that intensity.

Table 9-1. Classification of some first-order afferents

Incident stimulus	Intermediate mechanism	Examples of receptor types and function served
Mechanical force	Unknown; possibilities are: 1. Change in static properties of nerve ending, e.g., in capacitance, resistance, etc. 2. Intermediate release of specific chemical agent and chemoreception at nerve ending	Mechanoreceptors, serving: 1. Touch-pressure in skin and subcutaneous tissues; both organized and free nerve endings 2. Position sense and kinesthesia: mechanoreceptors of joints and vestibular receptors of inner ear 3. Mechanoreceptors of cochlea, serving hearing 4. Stretch receptors of muscle and tendon, which probably do not serve conscious sensation 5. Visceral pressure receptors: carotid and right atrium
Light	Photochemical transduction, leading to excitation of nerve endings (by synaptic mechanism?)	Photoreceptors of eye, serving vision
Heat	Unknown (by regulation of chemical reaction that influences state of nerve ending)	Thermoreceptors, separately for: 1. Warmth 2. Cold
Substances in solution	Uncertain; probably excitation of receptor cell or nerve ending by specific chemical combination, leading to change in permeability	Chemoreceptors, separately for: 1. Taste 2. Smell Osmoreceptors Carotid body receptors
Extremes of mechanical force, heat or its absence, presence of certain chemicals	Incipient or actual destruction of tissue cells (release of substance exciting nerve ending)	Nociceptors, serving pain

CLASSIFICATION OF FIRST-ORDER AFFERENTS

A useful system for classifying first-order afferents is in terms of the form of energy to which they respond at lowest stimulus intensity, i.e., to which they are differentially sensitive. Such a classification is given in Table 9-1, which, though not exhaustive, includes most of those types known to exist in mammals.

The intermediate mechanisms that lead to generator processes in the first-order terminals are poorly understood, and the possibility should be kept in mind that in many cases this intermediate mechanism may eventually be shown to be of a special biochemical nature, involving the formation and release from nonneural cells of substances that produce permeability changes in the peripheral nerve endings.

THE TRANSDUCER PROCESS[2, 3, 50]

Transducer mechanisms in detectors of transients[4]

In general, excitation at afferent nerve endings follows this sequence:

Stimulus → Local change in permeability →
Generator current (charge transfer) →
Local depolarization (generator potential) →
Conducted action potential

This scheme is descriptive also of neuromuscular transmission and of the events that take place in retina or cochlea when they are stimulated appropriately; study of a variety of receptors indicates its generality. Mammalian mechanoreceptive nerve endings are commonly so embedded in others tissues that it is difficult to apply microphysiologic methods to their study. The quickly adapting pacinian corpuscle is one exception, for it is found in

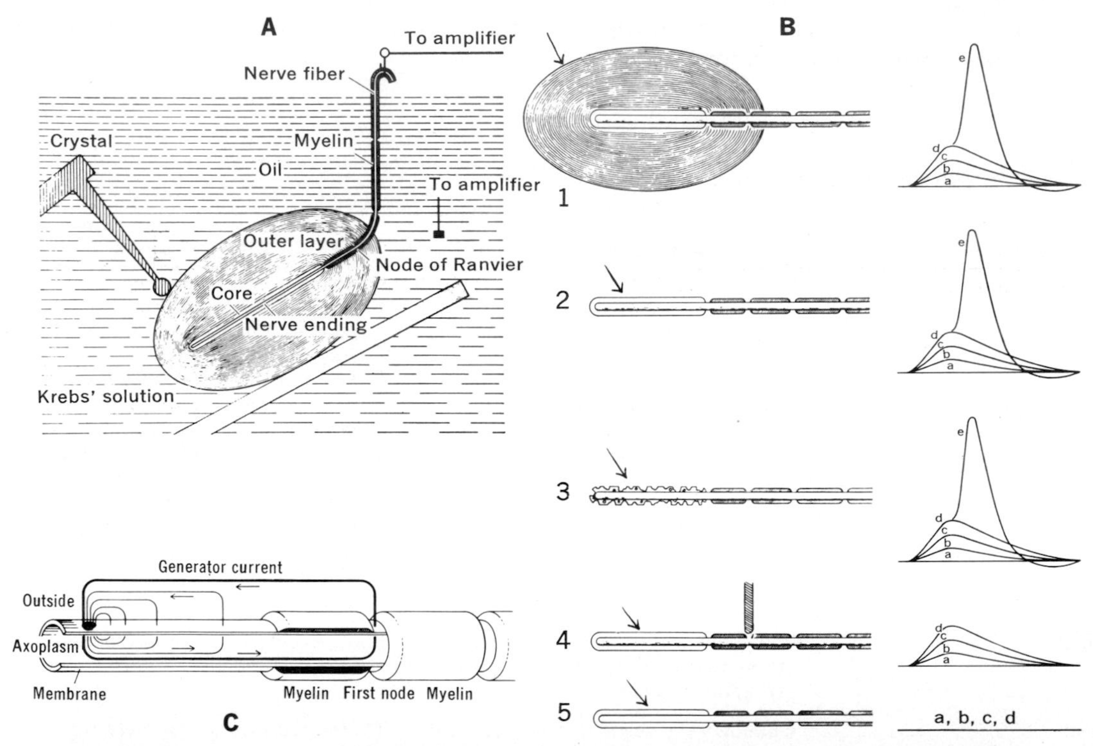

Fig. 9-1. A illustrates method of stimulation of pacinian corpuscle after it and its axon have been dissected free of cat mesentery. Recordings of generator potential and action potentials were made with one electrode on axon and the other in volume conductor surrounding corpuscle. Recording electrode is occasionally placed closer to corpuscle for better observation of generator potentials. Glass stylus attached to Rochelle salt crystal moves when crystal is activated by short electric pulses. **B,** *1,* Records *a* to *d* show increasing generator potential produced by successively stronger stimuli that for *e* produced generator potential that reached firing level for axon, resulting in conducted action potential. Similar sequence for *2* is unchanged after removal of all outer lamellae of corpuscle or, *3,* after bits of inner core have been removed. In *4,* pressure at node of Ranvier blocks production of action potential without affecting generator process. Records in *5* illustrate that mechanical stimulation to decapsulated core produces no response when axon has been caused to degenerate by section several days earlier. **C** indicates concept of local nature of change in membrane permeability and resulting flow of generator current reaching first node of Ranvier. (Modified from Loewenstein.[47])

the mesentery from which individual corpuscles and/or their axons can be isolated for study, either in situ with normal or perfused circulation or after complete removal. This organized receptor is also present in the dermis, in subcutaneous and intramuscular connective tissue, and in periosteum.

The pacinian corpuscle is an ellipsoidal body made up of a number of concentric lamellae (Fig. 9-1). The myelinated nerve fiber enters at one end, a final node of Ranvier occurs within the corpuscle, the myelin and Schwann sheaths are then lost, and the long, bare, nearly straight nerve terminal occupies the center of the inner core of the corpuscle. Inner-core lamellae differ from those of the outer core in that they are hemiconcentric; they are the thin and flattened protoplasmic extensions of cells whose nuclei lie in the intermediate zone of the corpuscle. The intralaminar spaces are filled with extracellular fluid and contain collagenous fibers. The corpuscle is turgid and scarcely compressible, so that a mechanical stimulus to its surface can only affect the nerve ending by a differential displacement of corpuscular elements.[50]

Such a corpuscle and its innervating axon can be isolated for study in the manner shown in Fig. 9-1. The upper records in Fig. 9-1, *B*, show that weak mechanical stimuli elicit a local change in membrane potential which, if of sufficient amplitude, leads to regenerative depolarization and a conducted nerve impulse in the stem axon. This is called the generator potential, a local change in membrane potential caused by local transmembrane flow of ionic current. This process has the following properties. (1) It is generated in the nerve terminal, not in corpuscular elements. (2) It is local and is not conducted. (3) It may be summed both spatially and temporally; i.e., the generator responses set up by two weak stimuli delivered sequentially at one spot or to two different spots on the nerve terminal may add to depolarize the nerve terminal membrane to "firing" level. The generator event has many properties in common with the end-plate potential (EPP) of muscle cells and the EPSP of nerve cells.

Studies of the generator event during polarization of the nerve terminal indicate that its equilibrium potential is close to zero membrane potential.[51] The permeability change produced by mechanical stimulation is therefore nonspecific, but since Na^+ is the most prevalent mobile univalent cation in the extracellular space, the charge transfer is likely to be carried by Na.[60, 69]

Close mapping with microelectrodes shows that the generator response evoked in the nerve terminal by a very small stimulating probe (20μ) declines exponentially with distance.[49] It seems likely that the nerve terminal can support both this local nonconducted event and the regenerative change in Na^+ conductance leading to a conducted action potential, for dromic impulses can originate within the terminal and antidromic ones can invade it.[34, 63] Whether these two membrane processes occur in the same or adjacent patches of membrane is unknown. They can, however, be separated by use of tetrodotoxin, a substance that at low concentrations blocks impulse initiation by poisoning regenerative Na^+ conductance, leaving the local generator response only mildly affected.[53, 62] Under these conditions it is possible to show that over a considerable range of stimulus intensity the relation between stimulus strength and the amplitude of the generator response (i.e., the local membrane depolarization) is linear (Fig. 9-2). However, it is likely that in intact corpuscles, impulses are initiated at the first node of Ranvier. This would be the case if the firing threshold for conducted action potentials is lower there than in the terminal membrane; the first node might thus function much as does the initial segment of the motoneuron (Chapter 6).

There is no evidence that a special chemical event intervenes between the mechanical stimulus and the evoked change in permeabil-

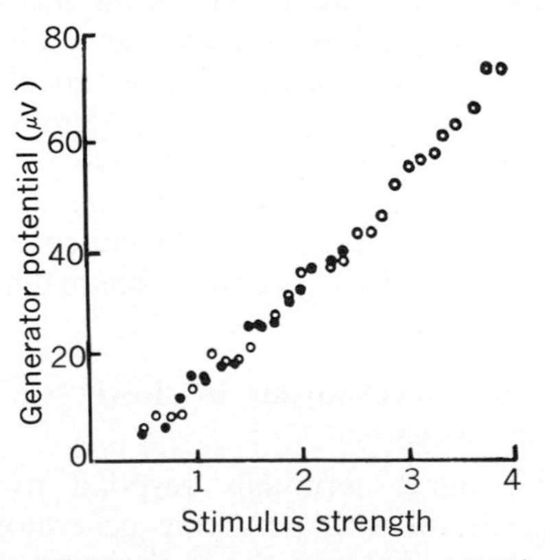

Fig. 9-2. Relation between generator potentials and strengths of mechanical stimuli evoking them for pacinian corpuscle before (filled circles) and after (open circles) block of action potential generation by application of tetrodotoxin, a toxic substance obtained from puffer fish. Relation is approximately linear over considerable range of stimulus strengths. (From Loewenstein et al.[53])

ity of the terminal nerve membrane. An alternate explanation is that the permeability change results from an evoked change in the purely physical properties of the membrane, i.e., a conformational change in large membrane molecules.* It should be emphasized that the membrane of the stem axon is by comparison very insensitive to mechanical stress; stretch of the axon blocks conduction without prior excitation.[31] The capacity of the nerve terminal membrane to respond to minute mechanical deformations with generator responses is an example of the specialization of function of a nerve ending.

The level of membrane depolarization required to initiate an action potential and the amplitude of action potentials in stem axons are both nearly constant over a temperature range of 12° to 40° C.[40] Charge transfer through the receptor membrane of the pacinian nerve terminal is, on the other hand, markedly affected by temperature; the amplitude and rate of rise of the generator potential increase with temperature with Q_{10}'s of 2.0 to 2.5. Nevertheless, Loewenstein[48] showed that no generator changes and no conducted action potentials were produced in a pacinian corpuscle by even large and rapid changes in temperature. Thus, although the charge transfer process is sensitively temperature dependent, the ending is a highly specific mechanoreceptor.

These facts suggest that the generator process occurs in local patches of nerve terminal membrane that are not electrically excitable, i.e., cannot support a regenerative change in Na^+ conductance, and that these local areas are interspersed with membrane that can. In the limit the two separable processes may take place within identical small areas of membrane. The nerve terminal and its stem axon illustrate the two types of activity of which nerve cells are capable: one is all or none and explosively conductile, the other is local, gradable, summable, and not conducted.

Transducer mechanisms in slowly adapting receptors

Other nerve terminals respond to long-lasting stimuli with persisting generator processes that elicit repetitive discharge of impulses in their stem axons. Such a fiber may discharge initially at a high frequency that is sensitively determined by the rate of stimulus application; following this *onset transient,* the discharge declines to a more or less steady rate that is determined by stimulus amplitude. Obviously the generator processes at the terminals of quickly and slowly adapting afferents differ in some fundamental way.

Many of these slowly adapting afferents that have been identified are clearly sensory as well as afferent, e.g., those that terminate in the palmar skin of the hand and play a role in the sense of touch-pressure. Their peripheral terminals are, however, so intimately embedded in other tissues that the full array of biophysical methods of study has not been applied successfully to them. Important information has come from the study of certain slowly adapting afferents that although not proved to be sensory, can be isolated for study by microdissection. Katz[41] discovered that stretch of the nerve terminals of the frog muscle spindle produced in them a local depolarization. This process affects the stem axon by electrotonic extension, shows no refractory period, and can be summed; if of sufficient strength, it elicits the discharge of action potentials in the axon. Both the generator potential and the afferent discharge evoked by it persist for as long as the spindle is stretched (Fig. 9-6).

The stretch receptor found in the tail muscle of Crustacea, schematically shown in Fig. 9-3, *A,* consists of specialized muscle cells to whose surface the dendrites of the sensory cells are closely applied.[58] The muscle fiber receives ordinary efferent motor fibers; a smaller efferent fiber, not illustrated, terminates synaptically on the sensory cell dendrites, exerting an inhibitory action. The entire organ can be isolated by dissection, its muscle bundles stretched, intracellular recording micropipets placed in the cell body, and the ionic environment controlled.[42] Results obtained by intracellular recording from the "slow cell" during stretch are illustrated in Fig. 9-3, *B* and *C.* In *B,* minimal stretch evoked a generator process alone, which persisted as long as the stretch; slightly greater stretch produced a slightly greater cell depolarization, and the latter produced a train of action potentials. These impulses are initiated in the axon itself at some distance from the cell body, at a trigger zone depolarized by electrotonic extension to it of the generator process originating in the dendrites.[23] The records in Fig. 9-3, *C,* show that greater increments of stretch elicit gen-

*For some theoretical considerations on the problem of mechanoelectrical transduction, see Teorell's review.[73]

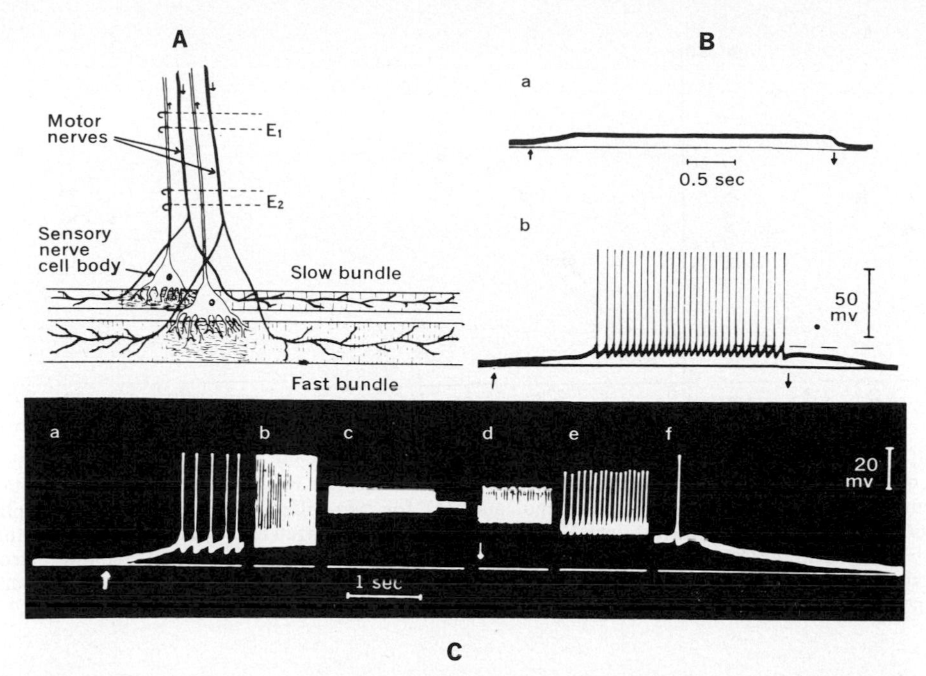

Fig. 9-3. A shows crustacean stretch receptor neurons and their relation to cells of peripheral muscle. Two fine muscle strands, each composed of several individual muscle units, are separated in midportion by segment of connective tissue in which are embedded dendrite terminals of receptor cells. Only two of several motor efferents are shown. Smaller efferent axons that are inhibitory for receptor cells and terminate upon their dendrites are not shown. **B,** Records made via intracellular electrode placed in cell body of neuron whose dendrites end in "slow bundle." *a,* Very weak stretch of bundle beginning at first arrow and ending at second elicits small but steadily maintained generator potential that is subthreshold for discharge of impulses; *b,* greater stretch elicits greater generator depolarization that reaches critical potential level (broken line) at which impulses recur as long as stretch is maintained. **C,** Generator response and impulse discharges recorded from neuron innervating slow bundle during continuous stretch that progressed from just above threshold in *a*. Only segments of total record are shown. With greater stretch and increased generator depolarization in *c,* impulse amplitudes declined to about 10 mv before complete block occurred, when membrane depolarized to about 35 mv level. Blocking stretch was maintained for about 2 min. *d* to *f* show events as stretch was slowly reduced. (**A** from Kuffler[41a]; **B** from Kuffler[42]; **C** from Eyzaguirre and Kuffler.[24a])

erator depolarization of greater amplitude and higher frequencies of axonal discharge. The relations between these three variables are given in Fig. 9-4, *A* and *B;* they are linear over the range of stretch below that, causing inactivation by excessive depolarization.[7, 76] Terzuolo et al. have used the systems analysis approach to study the dynamic behavior of this receptor cell and have shown that linearity holds at frequencies up to 10 Hz for small stretch amplitudes.[32]

Polarization of the membrane of the crustacean receptor cell by passing currents of different values through an intracellular micropipet, while equal muscle stretches were applied intermittently, indicated that the equilibrium level for the generator process is close to zero membrane potential[76] (Fig. 9-4, *C*). This suggests that the permeability change produced by the stretch is not ion specific; however, it is likely that under normal conditions the generator current is carried largely by Na^+ ions,[24] as is that of the muscle spindle.[64]

Sensory receptor adaptation and nerve accommodation

Two rate-sensitive processes determine the frequency range of stimuli to which different mechanoreceptors are differentially sensitive. These are the mechanical properties of tissue about the nerve ending and the accommodative property of the axonal membrane to generator current. Hubbard[33] and Loewenstein[49] have shown that the lamellated outer portion of the pacinian corpuscle acts as a mechanical high-pass filter, thought to be composed of both viscous and elastic ele-

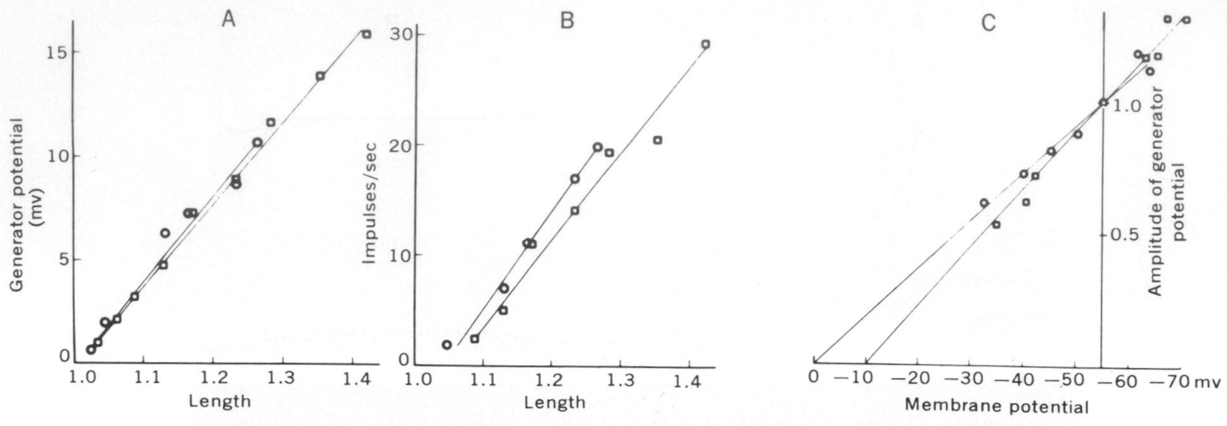

Fig. 9-4. Graphs to left show linear relations between generator potentials, **A,** and impulse frequency, **B,** and relative degree of stretch of slow muscle bundle innervated by receptor neuron of crayfish. Different symbols indicate data for two different neurons. **C** plots amplitude of generator potential produced by constant amount of stretch as function of membrane potential when latter is displaced by passing polarizing current through impaling microelectrode. By extrapolation, equilibrium potential for generator process is estimated at 5 to 10 mv depolarization. Different symbols for two different receptors. (From Terzuolo and Washizu.[76])

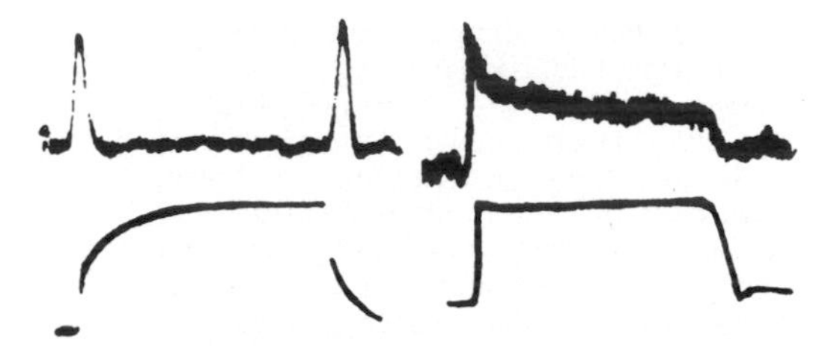

Fig. 9-5. Records to left illustrate generator potential (upper trace) evoked in nerve ending of intact pacinian corpuscle at application and release of mechanical compression pulse approximately 50 msec in duration, indicated by lower trace. Records to right were obtained similarly, but after removal of outer lamellae of corpuscle. Generator on-response in second case declines slowly, but is maintained throughout duration of stimulus. (From Loewenstein and Mendelsohn.[52])

ments (Fig. 9-1). Rapid compressional forces are transmitted directly to the central core, and the time course of this dynamic component is paralleled by that of the receptor potential. The steady component of an applied force is stored in stretched elastic elements and transmitted to the nerve terminal with a marked attenuation (100:1). Upon impulsive release of steady compression this stored force is transmitted from stretched elastic to viscous elements, producing a second dynamic event and the "off" generator response shown in Fig. 9-5. After removal of this mechanical filter by microdissection, steady compression of the decapsulated central core elicits a maintained generator potential (Fig. 9-5).[52] Nevertheless, even a quasi-steady generator potential of 5 times threshold for discharge of the nodal membrane evokes only one or two impulses at its transient onset. This quickly accommodating property of the nodal membrane plays an important role in the frequency selectivity of the pacinian afferent, for it prevents repetitive discharge in response to the steady component of generator potential that occurs by temporal summation in response to high-frequency stimuli. The axonal membrane discharges only at the transient peaks of that oscillating generator process. These two properties thus account for the fact that pacinian afferents are selectively sensitive to sinusoidal mechanical stimuli of high frequency, with a best frequency at about 250 Hz. At this level, stimuli of 0.1 to 0.2μ peak-to-peak amplitude elicit one impulse in phase with each sine-wave cycle.[68, 72]

In slowly adapting receptors, e.g., the muscle spindle, no mechanical high-pass filter exists at the nerve ending, for maintained generator potentials are produced by steady stretch (Fig. 9-6). The decline from the onset transient response of such a spindle, both of impulse discharge rate and receptor potential, is due in part to the fact that the tension within the spindle itself declines along such a time course during the early period following application of a quick stretch.[35, 36, 64] The remaining component of the early adaptation may be due to changes in the

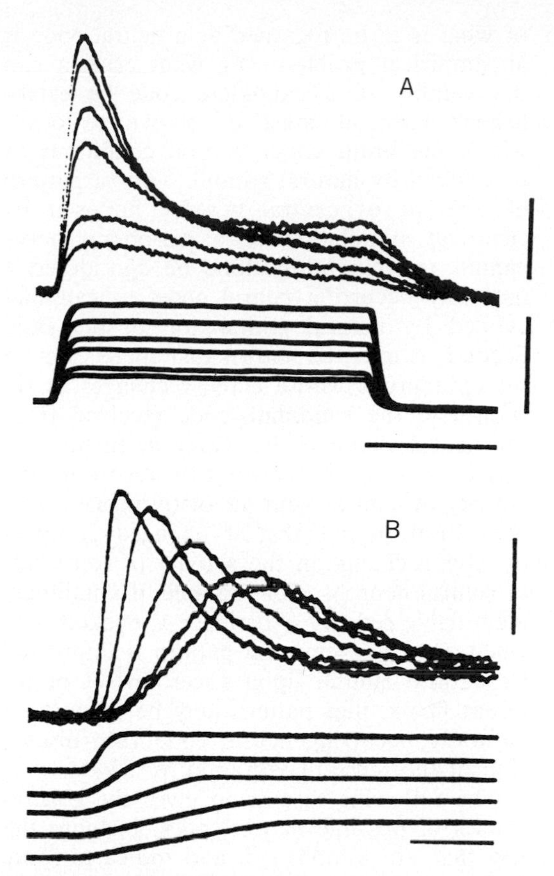

Fig. 9-6. A, Early adaptation of muscle stretch with increasing amounts of steplike stretch of isolated spindle from m. ext. dig. long. IV of frog. Impulse generation blocked by weak concentration of local anesthetic. Upper traces, receptor potential; lower traces, stretch monitor. Vertical bars: upper, 0.5 mv; lower, 300 μm. Resting length of spindle, 950 μm. **B,** Early adaptation of muscle spindle with changes in velocity of stretch. Upper records show receptor potentials produced by five velocities of linearly rising stretches (lower traces) to same final total extension of 250 μm. Resting spindle length, 950 μm. (From Husmark and Ottoson.[36])

ionic mechanisms underlying the receptor potential. The decline in discharge rate is not due to an accommodative characteristic of the nodal membrane of the stem axon, for passage of a steady current through the node produces a nearly instantaneous rise to a regular discharge rate that is then maintained.[46] Other mechanoreceptors, however, differ in the relative importance of mechanical and membrane properties in determining their rates of adaptation. It has been shown, for example, that in the crustacean stretch receptor cells (Fig. 9-3) the marked difference in rate of adaptation between the two cells is due largely to differences in properties of their impulse-generating membranes rather than to differences in generator processes.[59]

Wherever mechanoreceptors are found they fall into these two classes, quickly and slowly adapting. Chemo-, thermo-, and photoreceptors are usually slowly adapting, although wide variations exist. Quickly adapting mechanoreceptors can provide little information about the intensity of a mechanical stimulus. They are properly regarded as detectors of transients, although by virtue of the signals in a population innervating a sensory sheet they may also signal the direction of a spatial translation. Slowly adapting mechanoreceptors play a part in signaling both rates of change and steady states.

In summary, it has been shown for a large number of receptors that a local event intervenes between the application of a stimulus to a sensory receptor and the discharge of nerve impulses in its stem axon. The local event is a transfer of charge across the nerve terminal membrane that produces a depolarization called the generator potential. This generator event is not conducted but invades adjacent regions of the parent axon by electrotonic extension. It can be summed both temporally and spatially, shows no refractory period, and when the depolarization in spike-generating membrane reaches threshold, conducted action potentials result. The differences in the rates of adaptation of receptors to steadily applied stimuli, seen most notably in mechanoreceptors, are due to both the special properties of the nonneural elements of the receptors, which may in some cases act as high-pass filters, and also, in varying proportions, to the fact that the spike-generating membranes of stem axons of different receptors vary in the rates at which they accommodate to generator current.

Chemosensitivity of mechanoreceptors

It is clear from foregoing sections that the immediate source of energy used in the generator process is stored in the form of electrochemical gradients of ions across the terminal nerve membrane. There are several reasons for testing the idea that an intermediate step leading to excitation is the release and action of a special chemical transmitter upon the nerve membrane, although there is as yet no direct evidence for this effect. First, in many cases the relation between the nerve ending and the related nonneural cells has some characteristics of a synapse. Second, the generator process closely resembles the EPSPs of neurons and the EPPs of muscle cells, local events known to occur in response to transmitter release. Third, the generator process is markedly more susceptible to the application of local anesthetic agents than is spike generation in the stem axon.[61] Fourth, many mechanoreceptors are sensitive to naturally occurring substances that might be considered candidate transmitters at sensory endings. Acetylcholine (ACh), for example, lowers the threshold of and may cause spontaneous discharges in carotid pressure receptors,[21] and this is also the case in mechanoreceptors of the tongue and skin.[25] The effects of ACh and natural stimuli may be summed. Epinephrine may influence the generator processes of the pacinian afferent. Succinylcholine excites the nerve endings of stretch receptors of the mammalian muscle spindle.[29] Slowly adapting mechanoreceptors of the cat's skin are excited by ACh, histamine, and 5-hydroxytryptamine. However, when the action of ACh or other candidate transmitters is blocked in some way, the responses of these mechanoreceptors to natural mechanical stimuli are not changed.[30]

Thus while mechanoreceptors appear to be sensitive to many naturally occurring substances, no one of them has yet been shown to be essential for natural activation. The question of whether there is an intermediate chemical step remains unanswered.

COMMUNICATION AND CODING IN THE NERVOUS SYSTEM[5, 6]

The cellular elements of the nervous system are interconnected in divergent-convergent arrays of varying degrees of complexity. The ratio numbers for those relations determine to a considerable extent the functional properties of any neural subsystem within the brain. The business of this network is communication; it receives, may transform, transmits, stores, retrieves, generates, and distributes information—a word used here in its general, not Shannon, sense. This is achieved by successive processes of encoding, transmission, readout, and reencoding. It is important to emphasize that beyond transmission something akin to computation occurs in each nuclear region, where the often diverse influences arriving over a few or many channels combine with the intrinsic properties of the recipient neurons and those of the local neural network to determine the nature of the output: integrative action is achieved. Thus the problem of neural encoding includes both the coding of sensory events into impulse patterns in first-order fibers and the transmission and modification of these coded signals between different populations of cells within the nervous system.

The term "code" is used in its everyday sense to describe the manner in which information is represented in neural activity. The very great variety of those activities allows for many candidate codes, and it is now clear that there is no single code that is universal: the number of codes is finite, probably not large, but greater than one. The identification of what is to be regarded as a neural code is an empirical problem: by what criteria can the validity of a candidate code be established? First, it must be shown to occur within the brain under natural conditions or be evoked by natural stimuli. Thus a pattern of afferent or central activity imposed by electrical stimulation of a peripheral nerve cannot on that basis alone be considered a naturally occurring neural code, as can that evoked by natural stimulation of the skin. Second, it must be shown that there exists a set of neurons whose activity changes in response to the candidate code received from the neural channels that carry it. In practice this is frequently measured by recording the activity of central neurons or by measuring a behavioral output that if changed is taken to infer a change in the activity of some sets of central neurons. For example, if a uniquely identifiable sensory experience is evoked only when a certain temporal pattern is impressed by natural stimuli upon a certain set of afferent fibers, that pattern may be taken as a naturally occurring neural code (see discussion of the sense of vibration, p. 334).

The following discussion gives the general classes of possible neural codes, includes the few that are established, and indicates some of the many suggested codes that have not yet been established with certainty.

Code of labeled lines. The information in a message depends on the fiber or set of fibers that is active; once the identity of an active line is known, the information stored within the system tells the meaning of the activity. This is at once the most pervasive and well established of neural codes, encompassing the modality specificity of sensory afferents but generalized to apply to any neuron, sensory or not. It includes also the proposition that the locus of activity within a neural field as well as the place of the field itself within the brain carries information. For example, increases in neural activity within the visual cortex lead solely to the perception of light, and to no other perception; the intra-areal position of that activity determines the sector of the visual field to which the sensation of light is referred.

It is important to emphasize that the information carried by a given neural line may be embedded in more than one code. Activity of any sort within a labeled set of mode-specific afferents always evokes a qualitatively specific sensory experience; different frequencies of that neural activity may specify different intensities of that sensation.

Rate or frequency codes. These include, among a number of candidates, the common frequency code referred to previously, i.e., the coding of the intensity of a sensory stimulus or the level of presynaptic input by the frequency of the output neuronal discharge, averaged over some short period of time that is compatible with the integrating time constants of target neurons. Thus this code may apply to slowly but not to quickly adapting first-order afferents, and similarly to the sets of central neurons upon which the two classes of fibers project. Frequency modulation of a quasi-steady rate of discharge is one variant of the basic rate code.[74] In multi-stable systems the rate code may provide an all-or-none signal, with frequency above or below a certain level setting one or the other system state.

Sequential order of impulses in single neurons or in a small number of neurons with common properties and connections.[65] The time occupied by a nerve impulse is usually a small fraction of the time that elapses between impulses; the impulse train may be regarded as a point process, and all the information transmitted may thus reside in the interval sequence and the appropriate statistical descriptors. Obviously a powerful addition is made to the carrying capacity of a line if the system can encode and read out different micropatterns at each mean rate. This is the aspect of neural coding studied most intensively in recent years; examples of the results are given in Chapters 10 and 18. Thus far no code of this general microstructure class has been established with certainty, as have the limiting cases. These vary from the simple "go–no go" command signal of a motoneuron to the periodic activity entrained in an afferent fiber by oscillating mechanical stimulation of the skin (see Fig. 10-21, p. 332), in which the period signals vibration frequency.

Distributed or ensemble codes.[28] In these, information is coded in terms of the profile of activity within a neural field, i.e., the distribution both of levels of activity and the specific times of occurrence of impulses in members of a neural population. It is likely that the spatial contours of sensory stimuli are coded in this way. The class includes a number of variants, e.g., coincidence gating, by which is meant that postsynaptic discharge occurs only if impulses in two specified afferent lines reach a target neuron within some short period of time.

Nonimpulse codes. Prominent among these is the transmission of signals between cells by virtue of local depolarizing and hyperpolarizing events, which by passive extension over very short "axons" control synaptic transmission to target neurons. This certainly occurs in the retina[79] and is perhaps found in short axon cells of the CNS. Electrotonic coupling between cells (Chapter 6) provides for rapid information transmission between them, often in the positive feedback mode. On slower time scales it is possible that the transmission of molecules other than transmitter agents from the interior of one neuron to the interior of another provides a mechanism for information transmission.

Slow electrical events. Both the PSPs and action potentials of central neurons generate extracellular current flows. These flows can be recorded as potential changes locally within neural tissue or from the surface of the brain or head. A record made in the latter manner is called an electroencephalogram (EEG) (Chapter 7). The component of these currents, often small, that penetrates the cell membranes of other neurons may condition their levels of excitability and hence to some degree their activity: such extracellular currents may function as carriers of information. Evidence that such slow events are active informational agents of major consequence for brain function is not yet convincing.

• • •

In summary, certain neural codes that have been studied meet the criteria that have been stated. Others under study appear as likely candidate codes, while still others that can be impressed artificially on neural elements or extracted from neural activity by statistical methods are unlikely to be of importance naturally. Little is known about means of information transmission by other than impulse codes, but transmission by transfer of large molecules between neurons is a particularly interesting possibility. Virtually nothing is known about how impulse-coded messages, and particularly those coded in the activity of ensembles of neurons, are read by the neurons receiving them.

ANALYSIS OF RECEPTOR FUNCTION IN TERMS OF AFFERENT NERVE IMPULSES AND THEIR RELEVANCE FOR SENSORY EXPERIENCE

Five distinct qualities of sensation can be evoked by stimulation of body tissues; touch-pressure, warmth, cold, pain, and the sense of position. The first four are best developed in the skin, the fifth depends on joint afferents. More complex sensory experiences are synthesized by the CNS from combinations of activity in sets of afferent fibers, each of which, acting alone, evokes a primary sense quality. Varieties of sensory experience depend in some cases on varying quantitative aspects of input in the same homogeneous set of fibers and in other cases on afferent discharges in sets serving different elementary modalities. Thus the perception of the size and shape of an object grasped in one's hand results from the combination of input from cutaneous mechanoreceptive afferents with that from position detectors of the finger joints. The examples are many, and in general it can be said that human nervous systems may identify the elementary sense qualities on those rare occasions when appropriately simple stimuli occur and may compound a wide range of sensory experiences by a synthesis derived from the spatial and temporal patterns of input in different sets of first-order fibers.

Can the specific sensitivity of afferent nerve fibers be correlated with other characteristics, e.g., the axon diameters or the nature of peripheral endings?[13, 14] Fibers of different sizes do not occur in equal numbers in peripheral nerves; the distribution of sizes present in one human sensory nerve is shown in Fig. 9-7, together with the compound action potential to be anticipated in such a nerve, after conduction for several centimeters, if all its myelinated fibers were to be excited by a brief electrical stimulus.[26, 44] What is not revealed by the spectrum of Fig. 9-7 is that cutaneous nerves contain large numbers of unmyelinated fibers, of which 85 to 90% are dorsal root afferents, the remainder being postganglionic sympathetic efferents. The ratio of C fibers to A fibers ranges from 5:1 in nerves innervating the trunk and proximal limbs to 1:1 or less in nerves that innervate the face and hands.[67] Thus those regions of the body in which innervation is most dense and sensation is most acute and that are represented in the CNS in disproportionately large volumes of tissue receive the highest proportion of myelinated sensory fibers.

Relevant evidence comes from experiments in humans following differential block of fibers of different sizes. Local anesthetics block C fibers first, and then A fibers are blocked in an ascending order of size, as can be proved by electrical recording. The first sensations lost as such a block progresses are the burning pain called second pain and itching, followed by pricking or first pain (Chapter 11). Senses of cool and warm are lost almost *pari passu* with pricking pain; these occur as the delta fibers succumb to

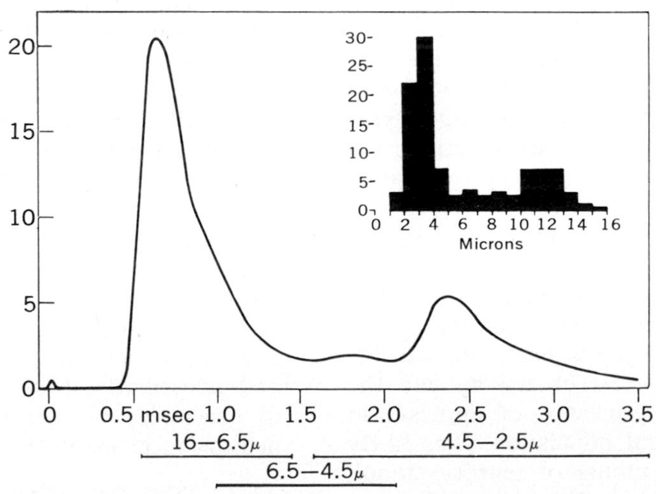

Fig. 9-7. Form of compound action potential in human sensory nerve (medial cutaneous) calculated from distributions of fiber diameters shown in inset histogram, which plots numbers of fibers for each diameter.[56] Data were recorded as fiber diameters but converted to axon diameters for calculation of reconstruction. Ordinate: relative amplitude of expected compound action potential. Abscissa: time scale; latency is that expected after 4 cm of conduction; micron scale for fiber, not axon, diameters. Construction made on basis of conversion factor of velocity in meters/second equals 9.2 times axon diameter in microns. Largest afferent fibers of peripheral nerves, group I, are not present in cutaneous nerves. (From Gasser.[26])

the block. Touch-pressure and position sensibilities survive until the very largest fibers in the nerve fail to conduct.[43, 45, 70, 71]

Block of a nerve by asphyxia, as with a compressive cuff around the arm, blocks A fibers first, and then after a delay of 30 to 45 min occlusion of blood flow, C fibers as well. The sequence of block within the A fiber band is not as clear as with anesthetic block, but a majority of observers agree that position sensibility and some components of touch-pressure are the first to go; a remaining component of contact sensibility, pricking pain, and low-threshold temperature sensibility is lost as the A fiber block progresses to include the delta fibers.[26, 43] At this time the C fibers appear to conduct normally, and the remaining sensibility is confined to second pain and to itch,[27] and perhaps some remnants of temperature sense, at least to the extremes of heat and cold.

In summary, there is a correlation between the sensations that are mediated and the diameters of the first-order afferent fibers mediating them. The beta fibers serve a number of sensations that can be grouped together as mechanoreceptive: touch-pressure, flutter-vibration, and position sensibility; some cruder forms of contact sensibility are served by fibers of delta size. Pricking pain is mediated by fibers in the delta range, as are cold and warmth. These statements deal with central tendencies only. The distributions are sufficiently broad that it is not incompatible to say that the quality of sensation served by any single fiber of the A band cannot be unequivocally determined by knowledge of its axon diameter.

Certain C fibers mediate second pain as well as the low-frequency sensory experience that we term "itch." When only the C fiber group is conducting, the tissues so innervated are completely insensitive to noninjurious mechanical stimuli. Yet a large proportion of C fibers, at least of those innervating hairy skin, are sensitive to light mechanical stimulation of their endings; they apparently do not evoke a tactile experience when acting alone. If these observations are confirmed for human skin, they mark an instance of a group of afferent fibers that provoke no sensory experience when active. Thus they stand in common with many afferents from the cardiovascular and respiratory systems that play important roles in reflex regulatory mechanisms but none in sensation.

Punctate nature of cutaneous sensibility

It was Blix[15] who discovered the existence of regions of differential sensitivity to the four elementary qualities of cutaneous sensibility, so that one can speak of touch, warm, cold, and pain spots. Histologic studies of identified spots show that each receives terminal branches from several afferent fibers and that any single nerve fiber may innervate several sensory spots. Study of the sensations produced by stimulation of cutaneous spots provides evidence for the modality specificity of sensory nerve fibers. No matter how a particular spot is excited, as long as the excitation is local to it only one elementary sensory experience is elicited, although its derived variations are evoked by temporal or intensive changes in the stimulus, e.g., touch to flutter by increasing frequency; itch to pain by increasing intensity.[11, 12] The temporal or intensive variation of a purely *local* stimulus does not produce a change in sensation, e.g., from touch to warmth, cold to touch, etc. This is consonant with the fact that electrical stimulation of a human peripheral nerve that engages only the A beta fibers elicits mechanical sensations of some kind, no matter what its frequency or the details of its temporal pattern.

The phenomenon of "paradoxical cold" provides another example of specificity of action. When the tip of a cool rod is applied locally to a cold spot, the experience of cold is elicited, as it is if the spot is stimulated mechanically or electrically. As the temperature of the stimulating rod slowly rises, the sensation of cold disappears, and no sensation is evoked by neutral temperatures. The sensation of cold may reappear, however, if the temperature of the rod is raised to about 45° C! A warm stimulus is in this case reported as cold because it is confined to a cold spot and thus to the terminals of mode-specific cold fibers. If the endings of some cold fibers are heated to 45° C, they are indeed caused to discharge; 45° C is a temperature capable of producing tissue damage, and it is an injury discharge that elicits the paradoxical perception of cold because it is conducted in cold fibers.

Morphology of nerve endings and associated structures and its relation to sensation

The modality specificity of myelinated afferent nerve fibers raises the question of whether this depends on the special properties of the multicellular noneural structures that in many places encapsulate their peripheral terminals. The discovery of sensory spots in the skin and the results of differential blocking experiments (cited previously) led morphologists to seek a structurally unique end-

Table 9-2. Classification of peripheral nerve endings by structure

	Glabrous skin	
Epidermis and associated dermal papillae		
Free endings (in epidermis and in dermal papillae)	Endings with expanded tips (Merkel's discs)	Encapsulated endings (Meissner's corpuscles)
Subpapillary dermis		
Free endings	Endings with expanded tips (Ruffini's endings)	Encapsulated endings (Krause's end bulbs and pacinian corpuscles)
	Hairy skin	
Free endings (in epidermis and dermis)	Endings with expanded tips (Merkel's discs, endings of palisade fibers, and Ruffini-like endings)	Encapsulated endings (endings with expanded tips when enveloped by connective tissue sheath produce structure similar to Meissner's, Krause's, or pacinian corpuscles)
	Deep fibrous structures	
Free endings	Endings with expanded tips (Ruffini-type endings)	Encapsulated endings (small pacinian corpuscles or Golgi-Mazzoni endings)

ing to account for each of the primary qualities of sensation—to account for the differential sensitivity of different nerve fibers. A large number of supposedly different end organs were identified, each thought to serve in this way some single sensory quality. It is now clear that the plethora of such identifications can be accounted for by the vagaries and artifacts of the methods used.[78] Indeed, it has repeatedly been confirmed that in spots of skin subjected to careful sensory testing, then marked and excised, histologic study revealed no organized endings—only free nerve endings—beneath some loci at which the pre-excision stimulation had evoked a single sensory experience. Evidence accumulates that the well-recognized organized endings, e.g., the hair follicle apparatus, Meissner's corpuscles of the glabrous skin, and Rufini's endings of deep connective tissue and joint capsules, are all linked to mechanoreceptive afferents. It seems likely that they determine the dynamic response characteristics of these fibers rather than their mode specificity. The pacinian corpuscles, for example, act as high-frequency mechanical filters, thus contributing in part to the quickly adapting nature of the myelinated afferents they adorn.

Table 9-2 summarizes results of studies revealing that a basically similar triad of endings is found in the skin and in subcutaneous and deep connective tissue, with variations from place to place in the detailed morphology of the encapsulated organs.[54] In the epidermis and papillary dermis of the human palm, for example, the three sets of nerve endings terminate in specific relations to the skin structure (Fig. 9-8). Free nerve terminals end among the connective tissue cells of the papillary dermis as well as between dermal cells; indeed, they penetrate almost to the stratum corneum. They are densely distributed throughout the skin. On the other hand, the expanded tip endings (Merkel's discs) are closely apposed to the lowest cells of the stratum germinativum of the intermediate ridges and their laterally extending septa.[57] The dermal papillae contain Meissner's corpuscles, and in young people nearly every cross section of a papillary ridge shows a pair. Each receives the terminals of two to five large myelinated fibers, and each fiber branches to send terminals to corpuscles distributed over a receptive field of 2 to 3 mm^2. The dermis contains encapsulated groups of endings with expanded tips (Ruffini's endings) and, on occasion, pacinian corpuscles. These latter are found in large numbers in the deeper connective tissue of the hand.

The form, number, and distribution of the myelinated fibers with encapsulated endings are not rigidly set, for they vary with age, region of the body surface, and even occupation. Meissner's corpuscles of the glabrous skin, for example, are rarer at birth than in childhood, more plentiful on hands than feet, and decrease in numbers with advancing years. Many atypical and transitional forms occur.

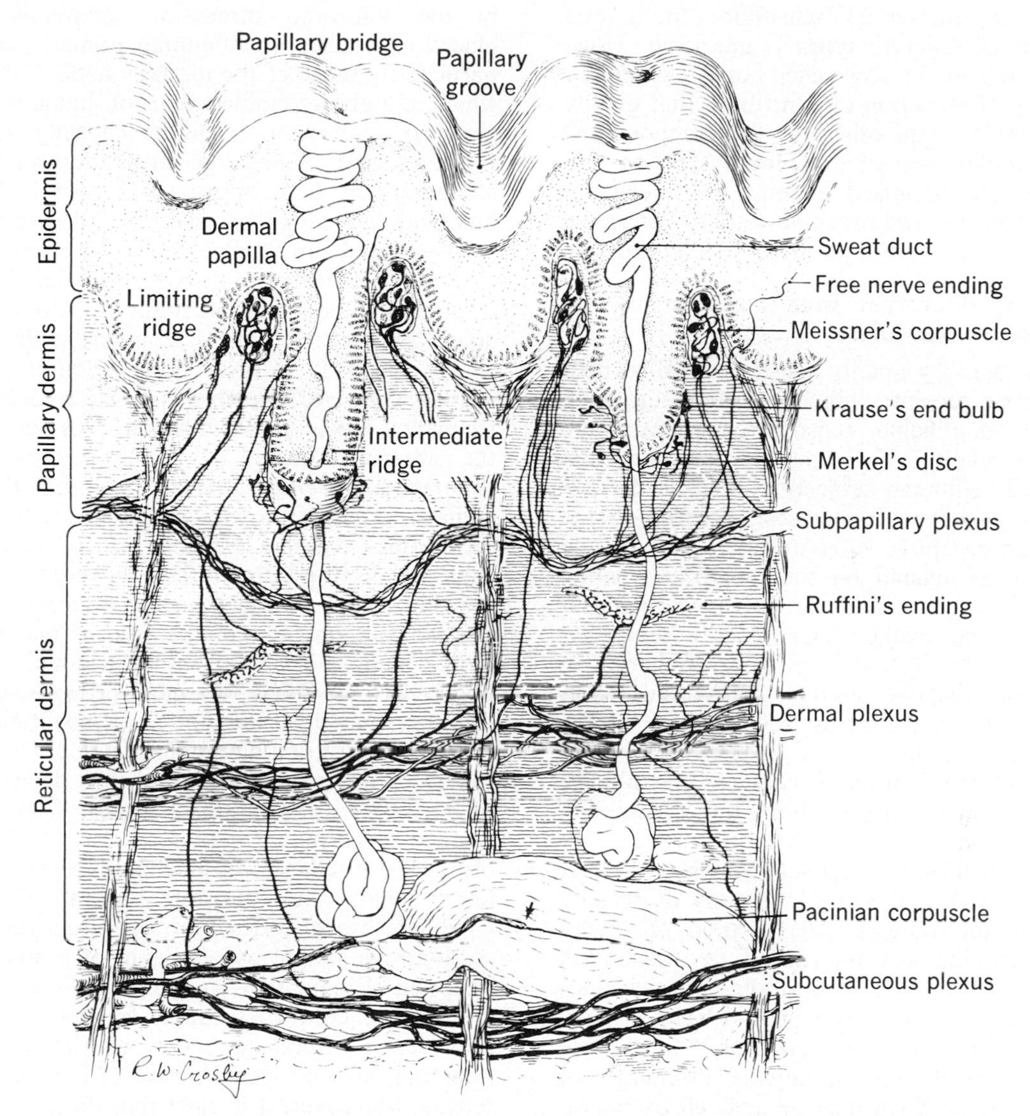

Fig. 9-8. Section of glabrous skin cut transversely across papillary ridges. Glabrous skin can be divided into three layers: epidermis, papillary dermis, and reticular dermis. Epidermis forms characteristic papillary ridges that contain row of sweat orifices in their center. Area deeper between adjacent papillary ridges is papillary groove. There are three major networks of nerves, corresponding generally to vascular plexuses: subcutaneous, dermal, and subpapillary plexuses. Deeper plexuses contain more and larger myelinated fibers. Axon may travel successively in each plexus or go directly from any one to more superficial dermis and epidermis. Limiting ridge, located beneath papillary groove, is tightly adherent to dermal collagen, fixed to underlying bone, and associated with relatively few nerve endings. Intermediate ridge, corresponding to overlying papillary ridge, tends to lie free in papillary dermis, is mobile, and its surface is rich in nerve endings with tapered or expanding tips (Merkel's discs). Pacinian corpuscles lie in deep dermis or subcutaneous tissue. Meissner's corpuscles occupy uppermost reaches of the papillary dermis. Location of other encapsulated or expanded tip endings (e.g., Ruffini or Krause's) is less orderly, and they may appear at any place in dermis. Tapered-tip ("free") endings occur throughout all layers of dermis and many cross basement membrane to terminate in lower reaches of epidermis. (Courtesy Dr. M. E. Jabaley.)

In summary, the peripheral endings of myelinated sensory fibers are differentially sensitive to one or another form of impinging energy, regardless of whether their terminals are encapsulated by sensory end organs. They are mode-specific labeled lines; the mechanism of their differential sensitivity to natural stimuli of different types is unknown. However activated, they elicit one elementary quality of sensation or contribute that quality to combine with others in the compounding of complex sensations. Fibers that do terminate in organized endings are mechanoreceptors; the end organ appears to determine the dynamic sensitivity of the nerve ending.

Afferent discharges from mechanoreceptors and tactile sensibility

The sensory quality sometimes called light touch or pressure will be treated here as a single identifiable sensory experience, the tactile quality. Variations of this sensation evoked in human subjects as well as the different patterns of impulses observed in the mechanoreceptive afferent fibers of animals appear to depend on the temporal, spatial, and intensive variations of the mechanical stimuli that evoke them. Tactile sensibility may also be elicited by afferent discharges in mechanoreceptive afferents that terminate in subcutaneous tissues as well as those of the skin, for the threshold of each is so low that any but the faintest mechanical disturbance of the body surface will bring both to action.

This and the following chapters rely heavily on the results of studies of single peripheral nerve fibers and central neurons of sensory and motor systems. The method of single-unit analysis was introduced by Adrian[1] in 1926, in the first such studies of mechanoreceptive afferents. The method has now been used directly in human subjects.[77] In practice the electrical signs of impulse discharges of single nerve fibers may be isolated by microdissection of the fiber bundles until only one remains in contact with recording electrodes; those of either fibers or nerve cells may be observed by placing recording microelectrodes either close to or actually within the neuron. Once a single element is brought under observation, precise control of natural peripheral stimuli allows the investigator to measure the sensitivity of its endings to stimuli of various kinds, to outline the receptive field of the neuron, and to measure the relation between stimulus intensity and neuronal discharge. Successful use of the method requires that a large population of neural elements be studied *seriatim,* for only then can the functional properties of different sets of neural elements be characterized.

The peripheral nerves of a very large number of species have been studied in this way; in the following discussion, emphasis is placed on studies in subhuman primates, and particularly those of the monkey hand,[27, 56, 72] which closely resembles that of humans in structure, innervation, and its capacity as a sensing agent. The large mechanoreceptive afferents (the beta or group II fibers) that innervate the palmar or "glabrous" skin of the monkey hand can be divided into two classes. Fibers in the first class discharge a brief burst of impulses upon step indentation of the skin, adapt quickly to the steady pressure, and discharge again in a brief burst upon its removal. They are selectively sensitive to low-frequency mechanical sinusoids, thus serving the submode of flutter (p. 331): *they are detectors of movement.* It is likely that their endings terminate in Meissner's corpuscles of the dermal papillae.[19] Those of the second class respond to a similar step indentation of the skin with an onset transient discharge, its frequency a function of the rate of indentation, which then declines along a double exponential time course to a more or less steady rate determined by the degree of indentation, i.e., by the stimulus intensity (Fig. 9-9). Fibers of this second class are thus detectors of both transient and steady states. They are thought to terminate peripherally as Merkel's discs. The discharges of mechanoreceptive afferent fibers thus illustrate several of the candidate neural codes described in an earlier section: the simple on-off signal of a quickly adapting element that an event has occurred, its periodic code of stimulus frequency, and the rate or frequency code of the slowly adapting afferent that signals stimulus intensity. Moreover, it is clear that the code for stimulus movement, direction, speed, and contour must reside in the ensemble of activity.

This *concept of population* is of great importance for understanding CNS action. Consider, for example, the events that occur when even a local stimulus is delivered to the skin of the finger. It will fall at different positions in the overlapping receptive fields of a number of different fibers. Thus the degree of activity will grade from maximum for those engaged at their field centers to minimum for those engaged only marginally. The stimulus

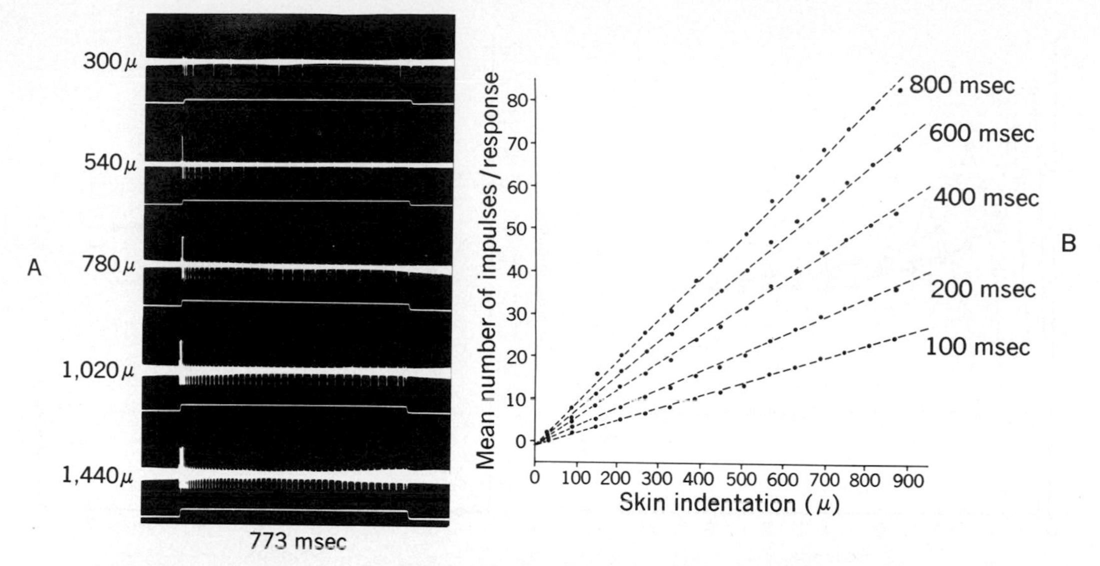

Fig. 9-9. A, Records of electrical signs of impulses in nerve fiber of median nerve of monkey, isolated for study by microdissection. Fiber terminated in small receptive field in glabrous skin of hand. Stimulus (lower trace of each pair) was mechanical indentation of skin by a 3 mm diameter probe tip that covered receptive field of fiber completely. On assumption of *frequency code,* intensity of stimulus is signaled by mean frequency of discharge. **B,** Graph of results obtained in study of similar fiber shows linear relation between skin indentations and mean discharge rates. Moreover, linear relation holds for successively additive 100 msec epochs of 800 msec stimulus duration.

will activate a population of fibers and that activity will be graded across this "neural space" in a manner that reflects the size and shape of the stimulus. Obviously the localization of a stimulus, identification of two points on the skin as two and not one, and the recognition of more complex spatial contours of stimuli cannot be explained in terms of the activity of insulated neural lines, as indeed the reciprocally arranged divergence and convergence at central relays makes certain. A central detecting apparatus must read profiles of neural activity.

Afferent discharges from mechanoreceptors and sense of position and movement of limbs (kinesthesia)

Human beings are able to appreciate the movements and steady positions of the joints in a remarkably accurate way. The thresholds for the perception of passive motion at proximal joints such as the shoulder and the hip are as low as 1 degree of movement at rates as low as 1 degree/sec. These thresholds are successively higher at successively more distal joints, rising to 5 to 10 times these figures at the interphalangeal joints of the hand and even higher for those of the toes.[17, 66] This gradient of sensory acuity with respect to the passive sense of body position is thus the reciprocal of that for tactile sensibility. The steady position of the limbs is sensed with an accuracy that allows the reproduction of position in space, e.g., of the shoulder joint to within 2 degrees.[10] There are thus three components of the senses of position and movement: rate, direction, and steady position. A set of afferent fibers qualifying for this function must provide information about all three. It has been known for a long time that it is receptors in and about the joints, and not those within the muscles themselves, that are responsible for these forms of sensibility.

The connective tissues surrounding joint capsules receive a rich innervation. Slowly adapting afferent fibers termininating in organized endings of the Ruffini type are found in the capsular tissues, others terminate in Golgi-type endings on the ligaments. Those terminating in pacinian corpuscles occur rarely. All these afferents are in the beta (group II) size range; they travel centrally in both muscle and cutaneous nerves—a purely "cutaneous" or a purely "muscle" nerve does not exist. The response properties of joint afferents have been studied in detail.[8, 16] Only a small percentage are of the quickly adapting type, which can provide little information other

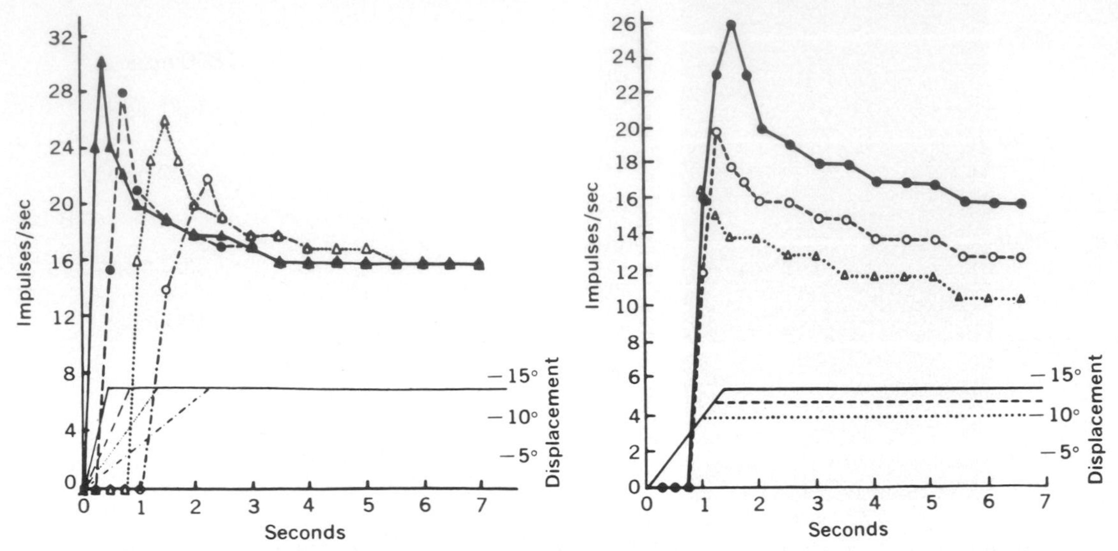

Fig. 9-10. Graphs of frequencies of discharge of single fiber innervating knee joint of cat. **A,** Response to flexion of leg through angle of 14 degrees at four different rates: closed triangles, 35 degrees/sec; closed circles, 17 degrees/sec; open triangles, 10 degrees/sec; open circles, 6 degrees/sec. Displacements are indicated by lines below. Steady impulse frequency with leg in final position is same in each case, but onset transient is function of velocity of movment. **B,** Frequencies of discharge of such a fiber during flexion of leg at knee at rate of 10 degrees/sec through three different angles, as indicated. (From Boyd and Roberts.[16])

than that some mechanical event has occurred. The slowly adapting afferents, on the other hand, respond to movement of the joint into their "excitatory angles" with high-frequency discharges (Fig. 9-10), the rate of change of which is a linear function of the rate of change of joint angle: they are *absolute detectors of the direction and speed of movement.* If the movement comes to an end within the excitatory angle of the fiber, the frequency of discharge declines over a period of seconds to a lower and more or less steady frequency that may persist undiminished for several hours. This continuing discharge is a variable influenced only by the joint angle (Fig. 9-10), not by the rate or direction of the movement that brought the joint to that angle: *these fibers function also as absolute detectors of position.* Movements within the excitatory angle from a more to a less excitatory position produce a reciprocal sequence, a transient decrease in the rate of discharge with recovery to a lower steady rate. Since the excitatory angles of individual joint afferents are placed differently along the range of movement of the joint, it is obvious that the entire population of afferents supplying a given joint codes with great accuracy both its steady position and the direction, rate, and extent of movements.

A correlation between the receptive angles of these first-order joint afferents and the central neurons upon which they play is given in Chapter 10, together with a discussion of the integrative action and code transformations involved in the central signaling of joint position.

Both the peripheral and central neural mechanisms serving the third major mechanoreceptive modality, that of flutter-vibration, will be considered in Chapter 10.

Unmyelinated fibers and multiplicity of function in peripheral nervous system

The number of C fibers in sensory nerves is equal to or greater than the number of those that are myelinated; particularly large numbers occur in regions of low sensory acuity.[26] Yet blocking experiments show that the qualities of sensory experience elicited by stimulation of tissues innervated only by C fibers are second pain and pain produced by the extremes of heat or cold—the tissues are sensitive only to noxious stimuli. Recent studies of C fibers by the method of single-unit analysis[9, 10, 37-39] have revealed that at least in the hairy skin of furred mammals there are unmyelinated afferents that as a group cover as wide a range in their qualitative sensitivity as do the myelinated fibers of the A fiber band. Many C fibers innervating cat or monkey hairy skin, for example, are excited by weak mechanical stimuli that are not noxious in nature. They innervate circumscribed receptive fields (4 to 25 mm^2), as do A fibers, and some may continue to discharge steadily during steady stimuli of a few seconds' duration and display a brief afterdischarge as well. They are particularly insensitive to temporally changing aspects of stimuli. Such fibers could by their functional properties serve some aspects of mechanical contact sensibility. Other C fibers respond when the skin is cooled by fractions of a degree centigrade[32]; only a few that have been observed respond to slight warming of the skin. Still other C fibers are only brought to action by very intense mechanical stimuli or by extremes of heat or cold. These latter fibers are properly identified as nociceptive afferents, and their presence in the C fiber group fits with the resulting of blocking experiments. *If the sensitivity of C fibers innervating human skin resembles those described in animals,* a question arises regarding the function served by the low-threshold C fiber mechano- and thermoreceptors. First, they could be concerned only with the powerful reflex effects that C fiber activity commonly evokes—somatic, cardiovascular, and respiratory. Second, and less parsimoniously, they might contribute to mechanoreceptive sensibility only when mechanoreceptive A fibers are simultaneously active. There is no evidence for this idea and one fact against it; there are no clear losses of mechanoreceptive capacity when only C fibers are blocked by local anesthetics.

It is clear that when considered as *afferent* fibers, whether sensory or not, the full range of qualitative sensitivities is represented in the delta group of A fibers and a second time in the C fiber group. The mechanoreceptive modes are represented a third time among the larger sensory fibers of the beta group innervating peripheral tissues. It is these that are tuned so finely to the spatial, intensive, and temporal aspects of peripheral stimuli. Indeed, it is these larger fibers that gain access to the medial lemniscal system, the rapid-tempo highway to the somesthetic cortex, while the smaller myelinated fibers of the delta group and unmyelinated fibers relay via the dorsal horns into pathways of the anterolateral columns of the spinal cord.[11] Thus the multiplicity of functional representation in the peripheral nervous system has its central counterpart, a subject for consideration in Chapters 10 and 11.

Specificity of action of first-order fibers

The evidence has steadily mounted that the specific quality of the different elementary sensations that can be evoked from body tissue is due to the action of groups of nerve fibers that differ in their central connections and in their sensitivity to stimuli of different kinds. These groups form mutually distributed, overlaid mosaics of nerve endings in the skin and function as differential filters, each preferentially signaling a particular quality of peripheral events, either in isolation (a rare occurrence) or in combination with others. These views deserve critical examination.

No doubt exists concerning the differential sensitivity of such quickly adapting mechanoreceptive afferents as those terminating in the hair apparatus or in pacinian corpuscles. The mechanoreceptors of the joints are not only highly specific stretch receptors, but their orientation to the forces produced in the pericapsular tissues is such that they signal the direction and velocity of joint movement as well as steady position. The true thermoreceptors are so differentially sensitive that they show changes in discharge rate in response to temperature changes of a few hundredths of a degree centigrade. One might assume that nociceptive afferents are "nonspecific," for they respond to stimuli of any form that is sufficiently intense to destroy cells. This is, however, a highly specific sensitivity in itself, and the painful experience uniformly evoked by their activation is unmistakenly identified by all humans save those rare individuals possessing a congenital indifference to pain.

Doubts might be raised by the following facts. Those cutaneous mechanoreceptive afferents that adapt slowly to steady stimuli are occasionally observed to discharge at low rates in the absence of any overt peripheral stimulus, a form of "spontaneous" activity. This activity may be changed by varying the temperature of the skin. For the majority of fibers in which there is any effect this change is irregular and small, but for some it is monotonic over a considerable temperature range, with slopes in the range of 0.35 to 0.65 impulses/sec/degree C in the steady state. These same fibers respond to mechanical stimuli with the remarkable sensitivity of 1 impulse/6 to 8μ of skin indentation in the steady state. Now, temperature change and the force producing skin indentation are not readily equated on a common scale. It is in comparison to the sensitivity of the true thermoreceptors that the inadequacy of these mechanoreceptors to function as thermoreceptors becomes obvious: they differ in sensitivity by factors of 10 to 50 times. Such a differential sensitivity is all that is required to establish that these fibers are specific mechanoreceptors. The changes in spontaneous activity and in thresholds produced by changes in temperature may provide an explanation for certain common mechanoreceptive illusions, e.g., the fact that weights feel heavier when cold than when warm, the Weber illusion.

The mechanism of action of temperature changes on mechanoreceptive afferents is unknown. Explanation may lie in the fact that the intermediate step leading to the generator process, or that process itself, has a high Q_{10} (p. 290). Changes in discharge rate with changes in temperature would be predicted for a nerve ending that oscillates around the point of regenerative change in Na^+ conductance, the presumed origin of spontaneous activity, and possesses such a high Q_{10}. There is no evidence that these afferents play a role in thermal sensibility.

Confusion concerning the problem of quality specificity has arisen in the past from the idea that each morphologically identifiable receptor organ should be matched with a specific sensitivity of the fiber terminating within it: the number of such organs very quickly exceeded the number of identifiable primary sensory qualities. It is now clear that the mechanism of differential sensitivity of different fibers to mechanical deformation, warmth, cold, or the products of cell injury, while still unknown, is likely to be a property of the nerve terminal itself, for all these classes are represented among the unencapsulated delta fibers whose nerve endings are bare. The highly organized encapsulated endings apparently surround the terminals of myelinated mechanoreceptive afferents, conditioning the quantitative nature of their response, not determining their mode specificity.

REFERENCES

General reviews

1. Adrian, E. D.: The physical background of perception, Oxford, 1947, Clarendon Press.
2. Catton, W. T.: Mechanoreceptor function, Physiol. Rev. **50:**297, 1970.
3. Cold Spring Harbor Symposia on Quantitative Biology, Sensory receptors, Cold Spring Harbor, New York, 1965, vol. 30.
4. Loewenstein, W. R.: Mechano-electric transduction in the Pacinian corpuscle. Initiation of sensory impulses in mechanoreceptors. In Loewenstein, W. R., editor: Handbook of sensory physiology: principles of receptor physiology, New York, 1971, Springer-Verlag New York, Inc., vol. 1.
5. Perkel, D. H., and Bullock, T. H.: Neural coding, Neurosci. Res. Symp. Summ. **3:**405, 1969.
6. Segundo, J. P.: Communication and coding by nerve cells. In Schmitt, F. O., editor: The neurosciences. Second study program, New York, 1970, The Rockefeller University Press.

Original papers

7. Albuqerque, E. X., and Grampp, W.: Effects of tetrodotoxin on the slowly-adapting stretch receptor neurone of lobster, J. Physiol. **195:** 141, 1968.
8. Andrew, B. L., and Dodt, E.: The deployment of sensory nerve endings at the knee joint of the cat, Acta Physiol. Scand. **28:**287, 1953.
9. Bessou, P., and Perl, E. R.: Response of cutaneous sensory units with unmyelinated fibers to noxious stimuli, J. Neurophysiol. **32:**1025, 1969.
10. Bessou, P., Burgess, P. R., Perl, E. R., and Taylor, C. B.: Dynamic properties of mechanoreceptors with unmyelinated (C) fibers, J. Neurophysiol. **34:**116, 1971.
11. Bishop, G. H.: Responses to electrical stimulation of sensory units of skin, J. Neurophysiol. **6:**361, 1943.
12. Bishop, G. H.: Neural mechanisms of cutaneous sense, Physiol. Rev. **26:**77, 1946.
13. Bishop, G. H.: The relation between nerve fiber size and sensory modality: phylogenetic implications of the afferent innervation of cortex, J. Nerv. Ment. Dis. **128:**89, 1959.
14. Bishop, G. H.: The relation of nerve fiber size to modality of sensation. In Montagna, W., editor: Cutaneous innervation, Oxford, 1960, Pergamon Press, Ltd.
15. Blix, M.: Experimentelle beiträge zur Lösung der Frage über die specifische Energie der Hautnerven, Z. Biol. **20:**141, 1884.
16. Boyd, E. A., and Roberts, T. D. M.: Proprioceptive discharges from stretch-receptors in the knee joint of the cat, J. Physiol. **122:**38, 1953.
17. Browne, K., Lee, J., and Ring, P. A.: The sensation of passive movement at the metatarso-phalangeal joint of the great toe in man, J. Physiol. **126:**448, 1954.
18. Bullock, T. H.: Signals and neuronal coding. In Quarton, G. C., Melnechuk, T., and Schmitt, F. O., editors: The neurosciences. A study program, New York, 1967, The Rockefeller University Press.
19. Cauna, N., and Ross, L. L.: The fine structure of Meissner's touch corpuscles of human fingers, J. Biophys. Biochem. Cytol. **8:**467, 1960.
20. Cohen, L. A.: Analysis of position sense in human shoulder, J. Neurophysiol. **21:**550, 1958.
21. Diamond, J.: Observations on the excitation by acetylcholine and by pressure of sensory receptor in the cat's carotid sinus, J. Physiol. **130:** 513, 1955.
22. Dykes, R. W.: Mechanisms of thermal sensibility in the primate. Thermoreceptive afferent fibers in the glabrous skin of the monkey, Ph. D. dissertation, 1970, The Johns Hopkins University.
23. Edwards, E., and Ottoson, D.: The site of impulse initiation in a nerve cell of crustacean stretch receptor, J. Physiol. **143:**138, 1958.
24. Edwards, C., Terzuolo, C. A., and Washizu, Y.: The effect of changes of the ionic environment upon an isolated crustacean sensory neuron, J. Neurophysiol. **26:**948, 1963.

24a. Eyzaguirre, C., and Kuffler, S. W.: Processes of excitation in the dendrites and in the soma of single isolated sensory nerve cells of the lobster and crayfish, J. Gen. Physiol. **39:**87, 1955.

25. Fjalibrant, N., and Iggo, A.: The effect of histamine, 5-hydroxytryptamine and acetylcholine on cutaneous afferent fibers, J. Physiol. **156:** 578, 1961.
26. Gasser, H. S.: Conduction in nerves in relation to fiber types, A. Res. Nerv. Ment. Dis., Proc. **15:**35, 1935.
27. Gasser, H. S.: Pain producing impulses in peripheral nerves, A. Res. Nerv. Ment. Dis., Proc. **23:**44, 1943.
28. Gerstein, G. L.: Functional association of neurons: detection and interpretation. In Schmitt, F. O., editor: The neurosciences. Second study program, New York, 1970, The Rockefeller University Press.
29. Granit, R., Skoglund, S., and Thesleff, S.: Activation of muscle spindles by succinylcholine and decamethonium. The effects of curare, Acta Physiol. Scand. **28:**134, 1953.
30. Gray, J. A. B., and Diamond, J.: Pharmacological properties of sensory receptors and their relation to those of the autonomic nervous system, Br. Med. Bull. **13:**185, 1957.
31. Gray, J. A. B., and Ritchie, J. N.: Effects of stretch on single myelinated nerve fibres, J. Physiol. **124:**84, 1954.
32. Hensel, H., Iggo, A., and Witt, I.: A quantitative study of sensitive cutaneous thermoreceptors with C afferent fibres, J. Physiol. **153:**113, 1960.
33. Hubbard, S. J.: A study of rapid mechanical events in a mechanoreceptor, J. Physiol. **141:** 198, 1958.
34. Hunt, C. C., and Takeuchi, A.: Responses of the nerve terminal of the pacinian corpuscle, J. Physiol. **160:**1, 1962.
35. Husmark, I., and Ottoson, D.: Relation between tension and sensory response of the isolated frog muscle spindle during stretch, Acta Physiol. Scand. **79:**321, 1970.
36. Husmark, I., and Ottoson, D.: The contribution of mechanical factors to the early adaptation of the spindle response, J. Physiol. **212:** 577, 1971.
37. Iggo, A.: Cutaneous mechanoreceptors with afferent C fibers, J. Physiol. **152:**337, 1960.
38. Iggo, A.: Non-myelinated visceral, muscular and cutaneous afferent fibres and pain. In Keele, C. A., and Smith, R., editors: The Universities Federation for Animal Welfare symposium on assessment of pain in man and animals, London, 1962, E. & S. Livingstone.
39. Iriuchijima, J., and Zotterman, Y.: The specificity of afferent cutaneous C fibers in mammals, Acta Physiol. Scand. **49:**267, 1960.
40. Ishiko, N., and Loewenstein, W. R.: Effects of temperature on the generator and action potentials of a sense organ, J. Gen. Physiol. **45:**105, 1961.
41. Katz, B.: Depolarization of sensory terminals and the initiation of impulses in the muscle spindle, J. Physiol. **111:**261, 1950.

41a. Kuffler, S. W.: Mechanisms of activation and motor control of stretch receptors in lobster and crayfish, J. Neurophysiol. **17:**558, 1954.

42. Kuffler, S. W.: Excitation and inhibition in single nerve cells, Harvey Lect. **54:**176, 1960.
43. Landau, W., and Bishop, G. H.: Pain from dermal, periosteal, and fascial endings and from inflammation, Arch. Neurol. Psychiatry **69:**490, 1958.
44. Landau, W., Clare, M. H, and Bishop, G. H.: Reconstruction of myelinated nerve tract action potentials: an arithmetic method, Exp. Neurol. **22:**480, 1968.
45. Lewis, T., Pickering, G. W., and Rothschild, P.: Centripetal paralysis arising out of arrested bloodflow to the limbs, Heart **16:**1, 1931.
46. Lippold, O. C. J., Nicholls, J. G., and Redfearn, J. W. T.: Electrical and mechanical factors in the adaptation of a mammalian muscle spindle, J. Physiol. **153:**209, 1960.
47. Loewenstein, W. R.: Biological transducers, Sci. Am. **203:**98, 1960.
48. Loewenstein, W. R.: On the "specificity" of a sensory receptor, J. Neurophysiol. **24:**150, 1961.
49. Loewenstein, W. R.: Excitation processes in a receptor membrane, Acta Neuroveg. **24:**184, 1962.
50. Loewenstein, W. R.: Rate sensitivity of a biological transducer, Ann. N. Y. Acad. Sci. **156:**892, 1969.
51. Loewenstein, W. R., and Ishiko, N.: Effects of polarization of the receptor membrane and of the first Ranvier node in a sense organ, J. Gen. Physiol. **43:**981, 1960.
52. Loewenstein, W. R., and Mendelsohn, M.: Components of receptor adaptation in a pacinian corpuscle, J. Physiol. **177:**377, 1965.
53. Loewenstein, W. R., Terzuolo, C. A., and Washizu, Y.: Separation of transducer and impulse-generating processes in sensory receptors, Science **142:**1180, 1963.
54. Miller, M. R., Ralston, H. J., and Kasahara, M.: The pattern of cutaneous innervation of the human hand, foot and breast. In Montagna, W., editor: Cutaneous innervation, Oxford, 1960, Pergamon Press, Ltd.
55. Mountcastle, V. B.: The problem of sensing and the neural coding of sensory events. In Quarton, G. C., Melnechuk, T., and Schmitt, F. O., editors: The neurosciences. A study program, New York, 1967, The Rockefeller University Press.
56. Mountcastle, V. B., Talbot, W. H., and Kornhuber, H. H.: The neural transformation of mechanical stimuli delivered to the monkey's hand. In de Reuck, A. V. S., and Knight, J., editors: Ciba Foundation symposium on touch, heat and pain, London, 1966, J. & A. Churchill, Ltd.
57. Munger, Bryce L.: Patterns of organization of peripheral sensory receptors. In Loewenstein, W. R., editor: Handbook of sensory physiology. Principles of receptor physiology, New York, 1971, Springer-Verlag New York, Inc.
58. Nadol, J. B., Jr., and de Lorenzo, A. J. D.: Observations on the organization of the dendritic processes and receptor terminations in the abdominal muscle receptor organ of Homarus, J. Comp. Neurol. **137:**19, 1969.
59. Nakajima, S., and Onodera, K.: Adaptation of the generator potential in the crayfish stretch receptors under constant length and constant tension, J. Physiol. **200:**187, 1969.

60. Nishi, K.: Modification of the mechanical threshold of the pacinian corpuscle after its perfusion with solutions of varying cation content, Jap. J. Physiol. **18**:216, 1968.
61. Nishi, K.: Abolition of impulse initiation at the pacinian corpuscle by local anesthetics, Jap. J. Physiol. **18**:536, 1968.
62. Nishi, K., and Sato, M.: Blocking of the impulses and depression of the receptor potential by tetrodotoxin in non-myelinated nerve terminals in pacinian corpuscles, J. Physiol. **184**:376, 1966.
63. Oseki, M., and Sato, M.: Initiation of impulses at the non-myelinated nerve terminal in pacinian corpuscles, J. Physiol. **170**:167, 1964.
64. Ottoson, D., and Shepherd, G. M.: Transducer properties and integrative mechanisms of the frog's muscle spindle. In Loewenstein, W. R., editor: Handbook of sensory physiology. Principles of receptor physiology, New York, 1971, Springer-Verlag New York, Inc.
65. Perkel, D. H.: Spike trains as carriers of information. In Schmitt, F. O., editor: The neurosciences. Second study program, New York, 1970, The Rockefeller University Press.
66. Provins, K. A.: The effect of peripheral nerve block on the appreciation and execution of finger movements, J. Physiol. **143**:55, 1958.
67. Ranson, S. W., Droegemueller, W. H., Davenport, H. K., and Fisher, C.: Number, size, and myelination of the sensory fibers in the cerebrospinal nerves, A. Res. Ment. Nerv. Dis., Proc. **15**:3, 1935.
68. Sato, M.: Response of pacinian corpuscles to sinusoidal vibration, J. Physiol. **159**:391, 1961.
69. Sato, M., Ozeki, M., and Nichi, K.: Changes produced by sodium-free condition in the receptor potential of the non-myelinated terminal in pacinian corpuscles, Jap. J. Physiol. **18**:232, 1968.
70. Sinclair, D. C., and Hinshaw, J. R.: A comparison of the sensory dissociation produced by procaine and by limb compression, Brain **73**:480, 1956.
71. Sinclair, D. C., and Stokes, B. A. R.: The production and characteristics of "second" pain, Brain **87**:609, 1964.
72. Talbot, W. H., Darian-Smith, I., Kornhuber, H. H., and Mountcastle, V. B.: The sense of flutter-vibration: comparison of the human capacity with response patterns of mechanoreceptive afferents from the monkey hand, J. Neurophysiol. **31**:301, 1968.
73. Toerell, T.: A biophysical analysis of mechanoelectrical transduction. In Loewenstein, W. R., editor: Handbook of sensory physiology. Principles of receptor physiology, New York, 1971, Springer-Verlag New York, Inc.
74. Terzuolo, C. A.: Data transmission by spike trains. In Schmitt, F. O., editor: The neurosciences. Second study program, New York, 1970, The Rockefeller University Press.
75. Terzuolo, C. A., and Knox, C. K.: Static and dynamic behavior of the stretch receptor organ of crustacea. In Loewenstein, W. R., editor: Handbook of sensory physiology. Principles of receptor physiology, New York, 1971, Springer-Verlag New York, Inc.
76. Terzuolo, C. A., and Washizu, Y.: Relation between stimulus strength, generator potential and impulse frequency in stretch receptor of Crustacea, J. Neurophysiol. **25**:56, 1962.
77. Vallbo, A. B., and Hagbarth, K.-E.: Activity from skin mechanoreceptors recorded percutaneously in awake human subjects, Exp. Neurol. **21**:270, 1968.
78. Weddell, G., Palmer, E., and Paillie, W.: Nerve endings in mammalian skin, Biol. Rev. **30**:159, 1955.
79. Werblin, F. S., and Dowling, J. E.: Organization of the retina of the mudpuppy, Necturus maculosus. II. Intracellular recording, J. Neurophysiol. **32**:339, 1969.

10

VERNON B. MOUNTCASTLE

Neural mechanisms in somesthesia

General somatic sensibility designates sensations aroused from body tissues other than those of sight, hearing, taste, and smell. Normal observers identify the five primary qualities of touch-pressure, pain, warmth, cold, and joint position. The usual sensory experiences are amalgams of these; this blend, yielding readily recognized and easily remembered experiences, is a property of the somatic afferent system and indeed of the brain more generally considered.

A sensory system is composed of those afferent pathways, subcortical nuclei, and cortical areas that receive their major afferent input via one sensory portal and that can be shown to play an important role in the sensation considered. Such a segregation might be regarded as artificial, for given the divergent relations between neural structures, the reciprocal interplay between thalamus and cortex, and the activation of certain generalized systems by all sensory inputs, an impulse in any given afferent fiber can, in the limit, affect the excitability of any neuron anywhere. Yet order reigns: lights are not felt, pains are not heard, and lesions of the visual cortex produce visual, not auditory, deficiences. Such functional segregation suggests that at least through the initial CNS processing levels there is a parallel degree of anatomic segregation.

The somatic afferent system is dual in nature. One major part is composed of the large myelinated fibers ($<$ 4 to 6μ) of the dorsal roots that traverse the ipsilateral dorsal column to terminate in its nuclei. These latter project via the medial lemniscus upon the contralateral ventrobasal thalamic complex, whose cells send their axons to the postcentral gyrus. Within this system there is a precise replication of the body form, a preservation of the modality specificity of the myelinated first-order fibers, and a degree of synaptic security well suited for neural action in rapid tempo. It is precisely organized to serve the discriminative aspects of mechanoreceptive sensibility, and the term "lemniscal" will be used to designate this system and the functional properties of its neurons.

At the level of dorsal root entry there is a massive offshoot composed of smaller myelinated fibers, collaterals of some larger myelinated fibers, and unmyelinated fibers, as well. These project upon interneurons of the dorsal horn to serve local and multisegmental reflexes. A second major part of the somatic system arises from the interneuronal system of the dorsal horn and runs cranially in the anterolateral columns. This major part has three subsystems: the spinobulbar, paleospinothalamic, and neospinothalamic systems. This *anterolateral* system serves a general form of mechanoreception and the senses of pain and temperature. The replication of the body form is less precise here than in the lemniscal system; there is cross-modality convergence upon some of its elements and a less secure synaptic drive of its neurons by any particular input.

These two major divisions of the somatic system, divergent through prethalamic pathways, probably converge again to a certain extent at thalamocortical levels of the brain. How they work synergistically in signaling sensory events is a major problem in understanding the neural mechanisms in somesthesia. A similar duality obtains in the trigeminal components of the system.

HUMAN CAPACITIES IN THE SPHERE OF SOMESTHESIS

It is important to quantify the human capacity for sensing, to establish man's detection threshold and his ability to scale and discriminate between stimuli that evoke sensations of the same quality but that differ in place, amplitude, or temporal pattern.[4] Such measures indicate the parametric ranges over which the relevant neural mechanics operate; when developed for experimental animals (p. 339), they allow simultaneous measurement of sensory behavior and those neural events.

Touch-pressure

The threshold for pressure sensibility is determined by counting the number of times a human subject detects mechanical stimuli of different intensities delivered to the skin. The threshold strength is a matter of the probability of its recognition; it is customary to designate the 50% point on such a function as the threshold; it is lowest in body regions with the greatest innervation density, e.g., the hands and face (Fig. 10-1). It is probable that regional differences reflect differences in the number of mechanoreceptive afferents engaged by a stimulus probe of a given size applied to different body parts, i.e., that threshold is a function of the degree of spatial summation. However, it is not known whether all touch spots have similar thresholds, and the regional differences may be due in part to differences in the thresholds of small groups of fibers terminating in different single spots. Thresholds are influenced by temperature, being lowest at about 37° C, some 4° to 5° higher than the normal palmar skin temperature. Thresholds are lower when the observer is allowed to move the receptor sheet (the exploring finger) in an oscillating fashion over the test object. For example, using the fingernail tip or a sharp tool held in the hand to make rapid to-and-fro movements, a human can detect a 1μ trough cut in a smooth metal surface.[65] Thresholds are virtually identical on the two sides of the body, in both right- and left-handed individuals.[38]

Intensity discrimination and rating. For brief mechanical stimuli delivered to the fingertip, the human observer's estimate of stimulus intensity is a nearly linear function of the extent of skin indentation (Fig. 10-19, p. 330), or of applied force. The just detectable stimulus increment is surprisingly high for touch-pressure: it is 10 to 15% of the base comparison over the midrange of stimulus intensities, and this Weber fraction rises for both weak and intense base stimuli (Fig. 18-7, p. 557).

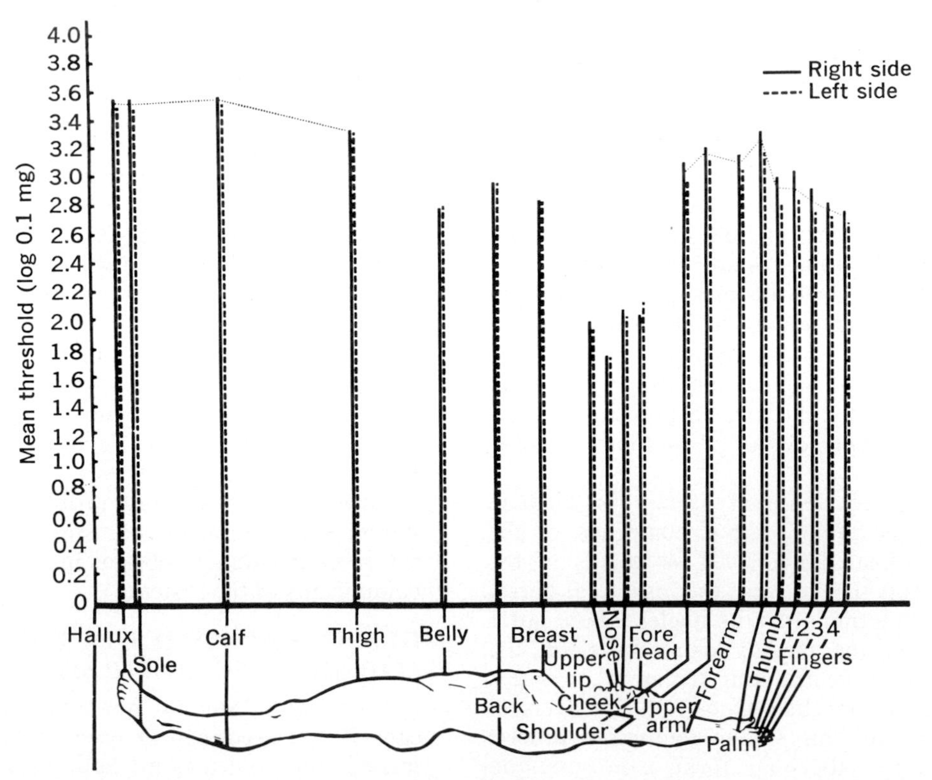

Fig. 10-1. Pressure sensitivity thresholds for different regions of male body surface. (From Weinstein.[125])

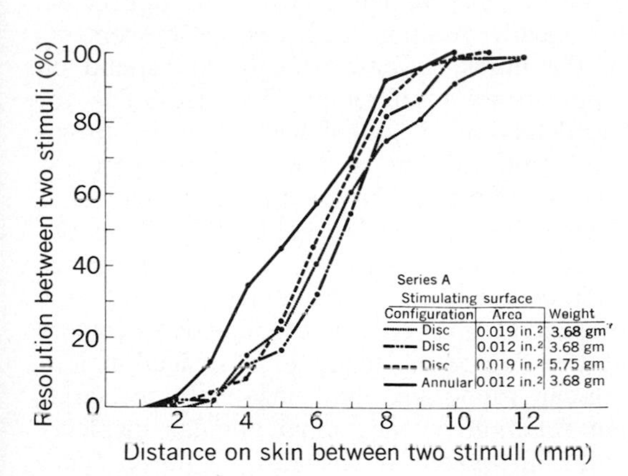

Fig. 10-2. Percentage correct resolutions of two points pressed on palmar skin as two and not one as a function of point separation. Averages for six normal human subjects; method of constant stimuli. Psychometric functions differ slightly for differing stimulus parameters, specified in inset. (From Carmon.[24])

Spatial and temporal discrimination.[1] An important feature of human somesthesia is the capacity to recognize and identify both static and dynamically changing spatial arrays of stimuli delivered to the skin surface. At the level of mechanism the question is how a central neural apparatus specifies a locus of increased activity in a neural field and discriminates between different and often rapidly changing contours of activity within it. Two simple tests of this capacity are accuracy in reproducing the localization of the point stimulated and identification of two points stimulated as two and not one. A frequency of sensing curve for the latter is given in Fig. 10-2 and the regional variations in two-point threshold in Fig. 10-3. Threshold is lowest for mechanical stimuli of intermediate intensity and may be further lowered if the two points are successively stimulated, suggesting that in the simultaneous case the overlap of the central zones of activity tends to limit acuity. The potential overlap of these two zones as the two peripheral stimuli are brought closer to one another is much restricted, it is thought, by afferent inhibitory mechanisms (p. 328).

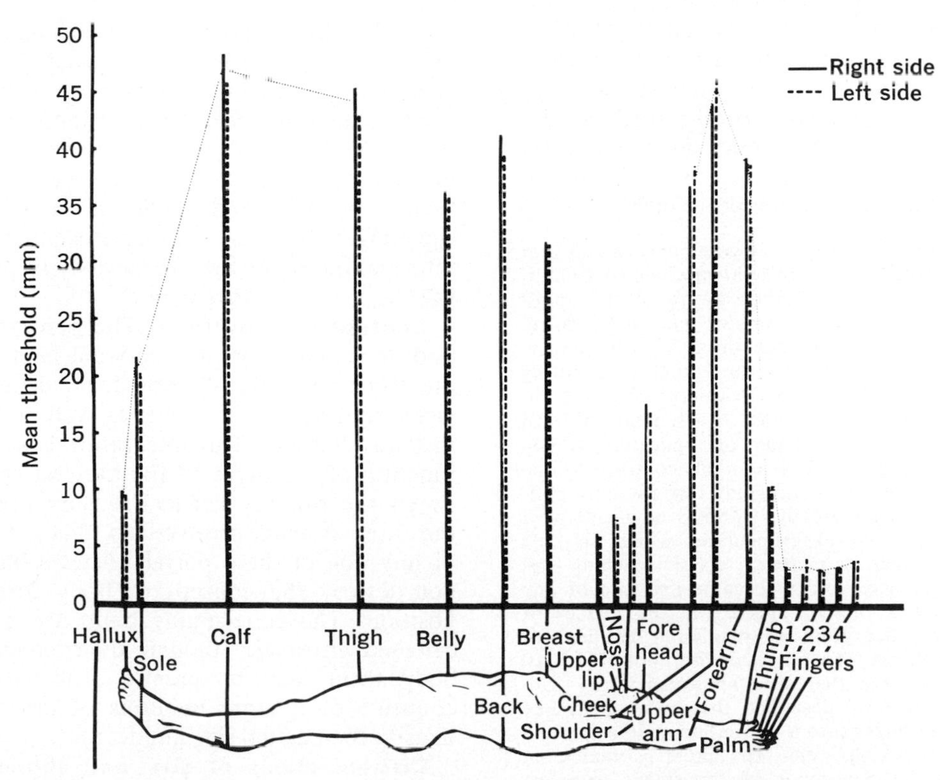

Fig. 10-3. Two-point discrimination thresholds for males for different parts of body surface. (From Weinstein.[125])

Humans recognize and discriminate between more complex two-dimensional arrays such as differing grades of sandpaper or weaves of cloth. Discrimination is made more acute by moving the receptor sheets. Examples of the importance of serial order in somesthesia is the ability to recognize letters or numerals written on the skin or the performance of the blind in reading braille type.[29] The blind have no greater capacity for touch than the sighted, but with intensive training their tactile discriminatory capacity reaches high levels. In braille an alphanumeric system is coded by combinations of as many as six embossed dots that represent each letter or number. The hemispheric dots stand 1 mm above the page surface and are arranged in two vertical groups of three, each separated from its neighbors by 2.5 mm—just above the two-point limen for the fingertips. There are thus 64 alternative spatial arrays. The practiced braillist, using one preferred index finger, frequently the left even in right-handed individuals, scans the line rapidly; more detailed attention to any difficulty is reflected by rapid oscillatory movements of the finger over the letter. Highly skilled persons read about 100 words (600 letters)/min, approximately the rate at which the sighted read aloud. The very high information flow suggested (60 bits/sec) is greatly reduced by the serial dependencies and redundancies of language. Braille reading, like other complex somesthetic discriminations, depends on the detection of a blend of spatial, intensive, and temporal cues.

Problem of adaptation. It is commonly stated that humans "adapt rapidly" to light mechanical stimuli; certainly one is more readily aware of the application and removal of a stimulus than of its steady presence. This fits with the brief, high-frequency, on-and-off discharges of the quickly adapting mechanoreceptive afferents. However, after a high-frequency onset discharge that is a function of both stimulus intensity and its rate of application, slowly adapting mechanoreceptive afferents continue to discharge at a more or less regular rate for long periods of continuing steady stimulus application. I have repeated the experiment in which subjects immerse a finger in mercury; classic teaching had it that the subject quickly became unaware of any stimulus save a ring of pressure at the surface of the mercury, the point of maximum gradient of skin deformation. To my surprise, not 1 of 10 healthy adults reported such an isolated ring feeling. They variously described the sensation as like "pressing my finger into a toy balloon," like "pulling on a silk stocking," etc. The point is that even though the large number of movement detectors among the population of mechanoreceptive afferents innervating the skin do adapt rapidly to a mechanical stimulus, others do not. Information is provided concerning steadily maintained deformation that the subject may attend or disregard at will.

Position sense and kinesthesis

In Chapter 9 a general description was given of the human capacity to appreciate the steady position and passive movements of the limbs, a sense thought to depend on input in joint afferents. The thresholds for the detection of joint deflection at the proximal joints may be as low as 0.2 degree for passive adduction of the hip at a speed of 2 degrees/sec; these thresholds depend on the speed of joint rotation, as the discharge characteristics of the first-order joint afferents suggest (Fig. 9-10). The subjective estimate of the degree of steady joint deflection is a regular, monotonic, and negatively accelerating function of the actual angular displacement, adequately described by a power function with an exponent of 0.7 to 0.8[49] (p. 334).

The sense of the rate and direction of active movement called kinesthesis depends in part on signals in joint afferents. It may persist, however, in their absence. Merton[75] has shown that the angular deflection of a finger may be reproduced accurately by a subject without the aid of vision after all afferent input from the moving joint has been blocked. Yet it is clear that an afferent feedback from the contracting muscles themselves cannot provide information of the rate or degree of joint movement,[43] and Merton suggests that this "sense of effort" depends on a reentrant signaling within the nervous system itself of information about the neural activity producing the external movement.

Concept of body form. The perception of body position in the gravitational field and of the three-dimensional body form depend on visual and vestibular inputs as well as signals in joint afferents. This integration is a special function of the areas of the parietal lobe between the primary cortical sensory areas for the three systems involved (p. 341). Closure of any one of these portals distorts but does not destroy the concept of body form and position. The congenitally blind are not disoriented in space, though their concept of body form may be greatly distorted—they conceive of it more in terms of innervation density than actual size and form.

Discriminations of size and shape. The ready manual recognition of the shape and size of objects is thought to depend on a

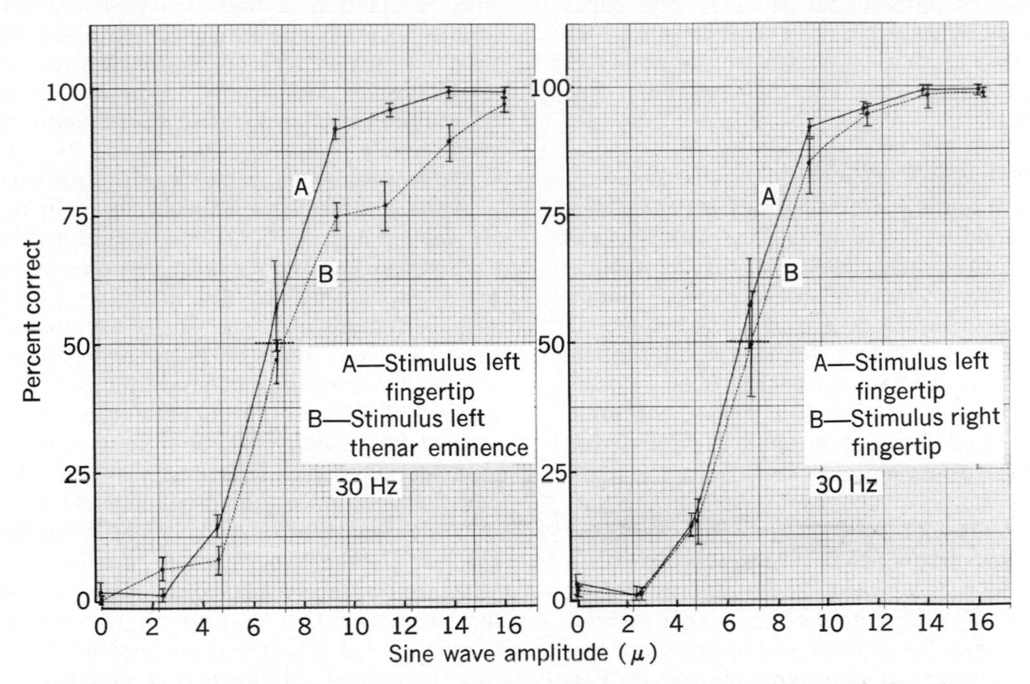

Fig. 10-4. Psychometric functions for experienced human observer; each graph plots the percent correct detection of 30 Hz mechanical oscillation, delivered to the hand, as a function of sine wave amplitude. Each curve is average of 6 separate runs, each of 256 trials distributed over 8 amplitude classes and delivered in random order; vertical lines plot ±1 SE of means. Graph to left shows intramanual transfer of task; that to right, interhemispheric transfer, which was immediate in both cases.

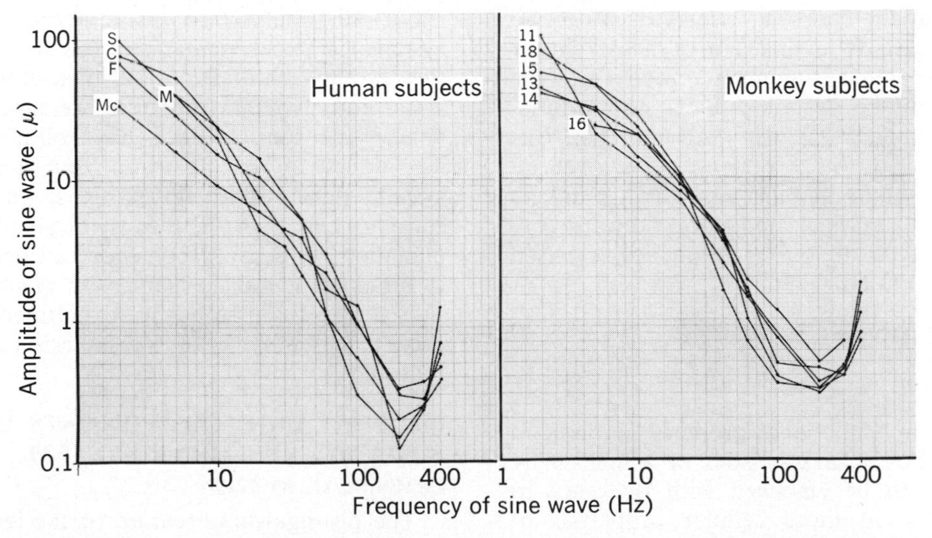

Fig. 10-5. Left: Frequency threshold functions for five human subjects. Thresholds at each frequency determined as 50% point on psychometric functions, e.g., those illustrated in Fig. 10-4. Right: Similar functions for six monkey subjects. Results show that monkey and human subjects detect vibrating mechanical stimuli delivered to their palms at about same stimulus amplitudes and over about same frequency range. Not illustrated is fact that monkey and human subjects respond to such stimuli in reaction times that are virtually identical. (From Mountcastle et al.[86])

combination of tactile and joint afferent input. Adults differentiate between solid cubes that vary by only 1 mm in edge lengths of 40 mm. This capacity is shown by the ability to recognize coins that differ by only minor embossment.

Sense of flutter-vibration

A sinusoidally vibrating mechanical stimulus applied to the body surface evokes a sensation that humans readily identify as qualitatively different from other mechanical stimuli. At frequencies from 5 to 40 Hz the feeling is one of a well-localized cutaneous flutter; higher frequencies elicit the deep, poorly localized hum of vibration. Flutter-vibration is a derived form of mechanoreception that depends for its unique quality on the temporal patterns of activity evoked in two sets of afferents, one tuned to low and the other to higher frequencies.[118] The results of a threshold measurement at 30 Hz are shown in Fig. 10-4. Repetition of this experiment at a number of frequencies yields threshold-frequency curves of the type shown in Fig. 10-5. Thresholds are lower at high than at low frequencies; the best frequency is about 250 Hz. Above 500 to 600 Hz, vibrations are commonly felt as stationary. Humans discriminate between frequencies whose cycle lengths differ by about 20%, a fraction nearly constant over the range of frequencies perceived.[90] Vibratory stimuli of different amplitudes in the frequency range of 40 to 80 Hz are rated by humans in a nearly linear fashion to actual stimulus amplitude; i.e., the relation can be described by a power function with an exponent of 1. Some investigators find that this exponent holds over all frequencies that humans sense, while others describe a gradual reduction toward exponents of 0.6 to 0.7 as frequency is raised to 200 to 300 Hz.[114, 115, 123]

DEFINITION: FUNCTIONAL PROPERTIES OF CENTRAL SENSORY NEURONS

The term "functional property" is used to indicate those characteristics of a neuron by which it can be classified with reference to others, beyond those cellular properties of membrane potential, energy metabolism, synaptic mechanism, etc., which are in one variant or another common to all central nerve cells. It is thought that for central cells of a sensory system these properties determine the immediate aspects of sensation, i.e., whether information is transmitted that permits recognition of the quality of a stimulus and its location, the rating of and discrimination between stimuli ordered along intensive or extensive continuums, and identification of stimulus rhythms. These properties are of two classes.

Static properties. This class of properties comprises those set by the synaptic linkages between first-order fibers and central neurons; they are invariant over wide changes in forebrain excitability or the influence of converging systems or of anesthetic agents. Static properties are those of place and modality. The term "modality" commonly refers to the quality of a sensation experienced and described by human observers. It is used here in a special way to indicate the nature of the peripheral stimulus that will activate a sensory neuron. Obviously it must be defined carefully in the experimental situation.

Dynamic properties. This class includes the properties that depend on the quantitative aspects of synaptic transmission; they are thus influenced by the action of converging and/or reentrant systems that impinge on sensory synaptic relays and are most sensitive to the action of anesthetics and other drugs and to changes in brain excitability, e.g., between sleep and wakefulness. Obviously certain properties of central neurons are both static and dynamic, e.g., afferent inhibition.

The successful application of the method of single-unit analysis to the study of sensory systems in experimental animals depends on the careful identification and measurement of static and dynamic properties of large numbers of neurons at each level of a system. The course of events throughout a sensory system during and following a peripheral stimulus can then be reconstructed, and explanations of sensory behavior can be sought in terms of neural mechanics. In much of the subsequent discussion this experimental approach to the study of somesthesis has been followed.

PRETHALAMIC COMPONENTS OF SOMATIC AFFERENT SYSTEM

LEMNISCAL SYSTEM[81, 107]

The distinguishing feature of the lemniscal system is that information concerning the location, spatial form, quality, and temporal sequence of stimuli that impinge on the body is transmitted with great fidelity at each synaptic station. A precise neural transform of certain stimulus attributes, signaled by spatially and temporally distributed patterns of neural activity, is preserved at least through the first stage of cortical activation. Certain of the spatial—particularly the topographic—temporal, and intensive relations of events within the system are invariant across nuclear relays. This invariance is due to the fact that the peripheral sheet of receptors is represented centrally in a precise and detailed pattern preserved through successive stages of the system. Moreover, the system encompasses within this single topographic pattern the several submodalities of mechanoreception. These properties define a *lemniscal neuron,* a cell of the lemniscal system that subtends a restricted peripheral receptive field, is activated by a single type of first-order mechanoreceptive afferent, and possesses through its synaptic relations powerful transmitting capacities in the temporal domain. Interneurons within the system may possess these properties, to varying degrees or not at all. In this section these properties of lemniscal neurons at prethalamic levels of the system will be discussed.

Dorsal ascending systems of spinal cord and dorsal column nuclei

The dorsal ascending systems that project via synaptic zones of the lower brainstem into the medial lemniscus appear in rudimentary form in submammalian vertebrates; they enlarge disproportionately to other ascending systems in mammals, particularly in primates. Largest of these is the dorsal column (DC) itself, composed of the ascending stem axons of the myelinated fibers (larger than 4 to 6μ) of the dorsal roots. Just cephalad to each dorsal root entry the DC contains a replicate of the large-fiber composition of that root, including the types given in Table 10-1. The fibers from each root are arranged in a narrow band applied laterally to similar bands from more caudal segments (Fig. 10-6). Of those fibers from each root that enter the DC, some 25% terminate within the dorsal column nuclei (DCN). At least that is true in the cat,[45] and the proportion is probably higher in primates. Study of the receptive fields and modality types of DC fibers just at entry into the DCN reveals that in their course upward from the level of dorsal root entry, there is a reshuffling of fibers as regards both topography and modality.[127, 131, 133]

The topographic sorting is such that first-order fibers with contiguous peripheral receptive fields, which enter the same or adjacent dorsal roots, are themselves contiguous at the highest level of the DC.[131] In contrast, fibers that enter distantly separate dorsal roots are not contiguous, even though their peripheral receptive fields may be. For example, detailed mapping of the pattern at the upper DC level

Table 10-1. First-order fibers feeding lemniscal system via dorsal columns

Source	Type	Mechanoreceptive submode
Hairy skin	Quickly adapting fibers sensitive to hair movement (movement detectors)	Touch-pressure (and flutter component of flutter-vibration)
	Slowly adapting fibers innervating tactile organs (Iggo) (detectors of transients and steady states)	Touch-pressure
	Slowly adapting fibers innervating distributed fields in skin surface; few in number (ending unknown)	Touch-pressure
Glabrous skin	Quickly adapting fibers innervating dermal ridges, probably Meissner's corpuscles (movement detectors)	Touch-pressure (and flutter component of flutter-vibration)
	Slowly adapting fibers innervating dermal ridges, probably Merkel's discs of intermediate ridges (detectors of both movements and steady states)	Touch-pressure
	Slowly adapting fibers innervating distributed receptive fields; few in number (ending unknown)	Touch-pressure
Dermis and deep tissues	Fibers ending in pacinian corpuscles (detectors of high-frequency transients)	Vibratory sensibility
	Slowly adapting fibers ending in fascia and periosteum (Ruffini-like)	Touch-pressure
	Fibers ending in joint capsules and joint ligaments (Ruffini-like)	Position sense and kinesthesia

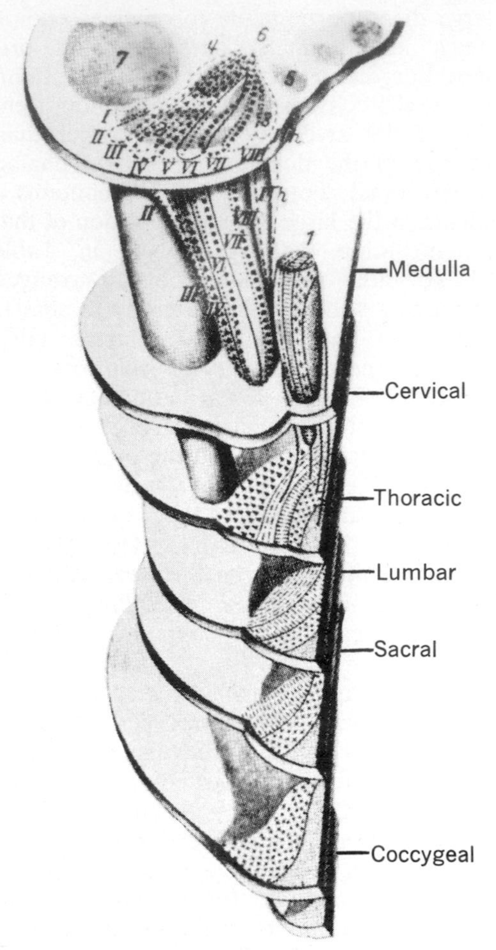

Fig. 10-6. Topical organization of fibers in posterior column and in posterior column nuclei. Upper two cross sections refer to medulla, lower five to coccygeal, sacral, lumbar, thoracic, and cervical levels of spinal cord. Relative positions of fibers are indicated by dots for coccygeal fibers and by crosses, dashes, dots and dashes, and triangles for fibers of successively higher segments. *1*, Nucleus gracilis; *2* to *4*, complex of nucleus cuneatus; *7*, descending root of trigeminal nerve. (From Glees et al.[46])

shows that the fibers originating from either the foot or the hand are clustered together, separating the preaxial and postaxial sides of arm and leg, respectively. The general result is that the map represents dermatomic patterning that is distorted by differences in innervation density and is not completely intact topologically. The map does preserve, however, a maximum number of trajectories on the surface of the body that are mapped continuously within the central pattern. Moreover, the representation of the body form produced by this topographic reshuffle is maintained at successive central replications in the ventrobasal complex of the thalamus and the cortical somatic sensory area in the postcentral gyrus (Figs. 10-8 and 10-11).[6, 130]

Nearly all DC fibers that reach the upper cervical level are quickly adapting cutaneous mechanoreceptive afferents.[6] Fibers originating peripherally in deep tissues, as well as the slowly adapting cutaneous afferents, leave the DC after only short ascents to synapse upon second-order cells whose axons are thought to project upward in the dorsolateral columns. In some species, some of these are collected into the spinocervical system (p. 315), but this is apparently not the case in man.

Fig. 10-6 shows that the reshuffled lamellar arrangement of the fibers of the DC is impressed in toto upon the DCN, each dermatomal lamella upon a long, thin, dorsoventrally oriented sheath of cells; these projections together compose the quasi-topologic transform of the body form. Study of DCN cells, however, has revealed that the modality spectrum of the dorsal root has to a certain extent been reconstituted. From caudal to rostral within each segmental lamina of DCN cells, there is a shift from a predominance of cutaneous mechanoreceptive elements posterodorsally to those activated by mechanical stimulation of deep tissue or by joint rotation anteroventrally.[134] This shift parallels a change in cytoarchitecture from the closely packed cell nests posteriorly to the "reticular" arrangement anteriorly. Obviously those first-order afferents projecting to "deep" cells of the anterior region do not reach the DCN via the DC; it is surmised they do so via the dorsolateral column. A striking feature of synaptic relations within the DCN, as at higher levels of the lemniscal system, is that modality specificity is as precise for central as for first-order elements; indeed, the facts in Table 10-1 apply with equal validity to each. The implication is that each DCN cell receives its presynaptic input only from first-order fibers of a single modality type. Within this mode-specific pattern, spatial convergence and divergence occur within each class, for the receptive fields of DCN cells are on the average 10 times as large as those of the first-order fibers. The problem is how to understand the spatial discrimination of which humans are capable, given this apparent decrease in spatial resolution. One mechanism thought to contribute to spatial resolution is afferent inhibition (p. 328).

Efferent axons of the lemniscal cells of the DCN cross in the decussation of the medial lemniscus and project upon the ventrobasal complex of the contralateral thalamus.[26] In crossing there is an inversion of the body form so that dorsal parts come to be represented dorsally, tips of limbs ventrally, and sacral regions laterally; successive medial application of secondary ascending trigeminal elements completes the natural replica of the body form. An important feature of the medial lemniscus is that its axons, after leaving the DCN, emit no other collaterals in their passage through the brainstem: its sole projection is upon the diencephalon.

Synaptic organization within the dorsal column nuclei. The DCN are complex, for within them interactions occur between primary afferent inputs as well as between DCN themselves and descending systems arising in the forebrain (p. 337). Neurons of the dorsal and caudal area of the DCN possess the bushy, concentrically arranged dendritic trees typical of sensory relay neurons[67] and are arranged in small clusters, each receiving its input from one or at most a few DC fibers via axodendritic synapses. They are activated monosynaptically, subtend small receptive fields, follow rapid peripheral stimuli, receive only a sparse direct input via corticobulbar systems, and project their axons into the medial lemniscus.

Ventrally and more cephalad in the DCN these cluster cells are gradually replaced by neurons with wide dendritic fields of the "reticular" type. They subtend large peripheral receptive fields, respond to input from these fields with delays suggesting polysynaptic activation, receive a heavy corticobulbar projection, and while many do, others do not project their axons lemniscally. The latter are thought to be interneurons,[14] cells interposed between the ascending afferent and descending systems and the DCN cells with lemniscal fibers.

Lateral cervical system.[78, 120] This system is one of the components of the ascending dorsolateral columns activated by cutaneous stimuli. It projects upon the forebrain and, in some species, plays a role in somesthesis. It has been identified in carnivores, ungulates, cetaceans, and some primates, but if present at all is rudimentary in man.[122] In the cat its fibers originate from large neurons of layer IV of the dorsal horn, ascend in the most medial corner of the dorsolateral column, and terminate in the lateral cervical nucleus, a group of cells located lateral to the dorsal horn of the first and second cervical segments. Axons of these third-order cells cross the cord to join the contralateral medial lemniscus, terminate in a thin lamina of cells surrounding the ventrobasal complex, and transmit neural activity that reaches the somatic areas of the cerebral cortex.[10] Its cells of origin receive segmental collaterals of DC fibers less than 10μ but not delta fibers; i.e., they are activated by mechanical stimulation of small cutaneous receptive fields.[95] In those species possessing it, the system may account for the persistence of certain aspects of mechanoreceptive sensibility following DC lesions.

ANTEROLATERAL SYSTEM[26, 73, 74, 107]

The anterolateral system is phylogenetically older than the DC system, is less precisely organized topographically, and contains many delta and C fibers. These originate from cells of the dorsal horn; the majority cross via the anterior commissure and turn cephalad, while a smaller contingent projects ipsilaterally. Fibers from each segment are applied medioventrally to those from more caudal regions. All of the fibers turning cephalad in the lateral columns do not reach supraspinal levels: those that do not are part of the propriospinal system, those that do compose the anterolateral system. The latter may be grouped in terms of their central destinations.

1. The *spinobulbar component* is a major afferent pathway to the mesencephalic reticular formation, from which activity is relayed upward via the reticulothalamic systems to terminate in the intralaminar nuclei of the thalamus.

2. The *paleospinothalamic component* projects directly from the cord to the n. centralis lateralis of the intralaminar group, as defined by Mehler.[74] It contains delta and C fibers and converges with the relayed spinobulbar projections upon common thalamic targets.

3. The *neospinothalamic component* projects upon the posterior nuclear group and the ventrobasal complex of the thalamus, reaching the latter in a topographic pattern generally merged with that of the medial lemniscus, although its terminations are concentrated in the posterior parts of the ventrobasal complex. The neospinothalamic system grows rapidly in phylogeny. It is scarcely discernible in rodents and carnivores but increases in size at an accelerating rate in primates. In man the tract contains 1,500 to 2,000 myelinated fibers 4 to 6 μm in size; whether it also contains C fibers is unknown. It is the only topographically organized component of the anterolateral system.

In Chapter 8 evidence is given that the anterolateral system provides a major driving input for the ascending reticular formation (RF) and thus plays an important role in controlling levels of excitability in the forebrain. Its role in pain and temperature sensibility is considered in Chapter 11, together with the problems of synaptic processing in the dorsal horn.

TRIGEMINAL SYSTEM[31]

The system serving facial sensation recapitulates the dual system just described, though it differs in detail. In mammals other than primates, facial sensation is important in environmental exploration, a fact reflected in the dominance of the head and face in central replications of the body form (Figs. 10-8 and 10-11). While in man this tactile exploratory function is greatly reduced, sensory input from intraoral structures plays an important role in deglutition and is thought to be an important regulatory factor in speech production. Indeed, for many tests of somesthetic capacity the tongue is the most sensitive area of the body; e.g., two-point discrimination on its tip is 3 times better than on the tips of the fingers.[106]

First-order fibers

Nerve fibers innervating the skin of the face are of the same modality types as those of the body: mechano-, thermo-, and nociceptive. Their cell bodies are distributed in a topographically orderly way in the semilunar ganglion. The centrally directed axonal branches are as a group partially rotated in the trigeminal root, so that at entry to the pons there is an inverted dorsoventral representation of the face, a pattern retained throughout their central course and in the second-order trigeminal cellular regions of the medulla upon which they project. Upon entering, most but not all of the myelinated fibers divide into short ascending and long descending branches, the former terminating within the main sensory nucleus, the latter making up the bulk of the trigeminal spinal tract. The level of termination of these descending fibers is not related to their peripheral origin, for the topographic representation of the face is complete to the full caudal extension of the tract and the spinal nucleus of the trigeminal. In addition, some cutaneous afferents from the head enter the brainstem via the seventh, ninth, and tenth cranial nerves, join the trigeminal tract, and descend to the n. caudalis.

The trigeminal tract contains a lower proportion of unmyelinated fibers than do comparable spinal nerves; many of those that are present arise from regions of the facial and buccal surfaces that are highly sensitive to thermal and noxious stimuli. These C fibers pass without bifurcation into the spinal tract and terminate almost wholly within n. caudalis. Not all C fibers of the trigeminal tract are of external origin, however, for many arise from neurons of n. caudalis and constitute a trigeminal linkage system that is perhaps comparable to Lissauer's tract of the spinal cord.[116]

Organization of brainstem trigeminal nuclei

The main sensory and spinal trigeminal nuclei are structurally distinct and are related to the ascending and descending branches of the entering trigeminal tract fibers, respectively. The spinal nucleus is further separable into three cytoarchitecturally distinct nuclei: oralis, interpolaris, and caudalis. The main sensory nucleus and the n. oralis appear functionally analogous to the DCN and the lemniscal system.[92] The pattern of representation within them is inverted, with perioral and intraoral areas represented most medially (Fig. 10-7). Most of the cells of these nuclei send their axons to the contralateral ventrobasal complex of the thalamus, ascending closely applied to the dorsomedial aspect of the medial lemniscus. Some cells of the main sensory nucleus project to the ipsilateral thalamus, constituting the dorsal trigeminothalamic tract, and thus account for the small ipsilateral face representation in lemniscal patterns at thalamic and cortical levels.

The n. caudalis, in contrast, cytoarchitecturally resembles the dorsal horn of the spinal cord, and its cells project their axons, some contra- and some ipsilaterally, to the thalamic targets of the anterolateral system, the intralaminar and posterior nuclear groups.[112] Its similarity to the anterolateral system in function is further strengthened by the fact that section of the trigeminal spinal tract at the level of the obex and thus cephalad to the n. oralis produces in both man and animals loss of facial sensations of pain and temperature but results in only trivial changes in mechanoreceptive sensibility.

Static properties of trigeminal neurons

Neurons of the main sensory nucleus and of n. oralis possess static properties quite similar to those of the cells of the DCN. They respond to only one form of mechanical stimulation and subtend small and continuous receptive fields that together compose the topologically intact pattern shown in Fig. 10-7.[31] Studies of n. caudalis by the method of single-unit analysis have so far yielded puzzling results. Here, too, cells are found to be modality specific, related to small continuous receptive fields. Yet the thalamic projection of n. caudalis resembles that of the anterolateral system, and section of the tri-

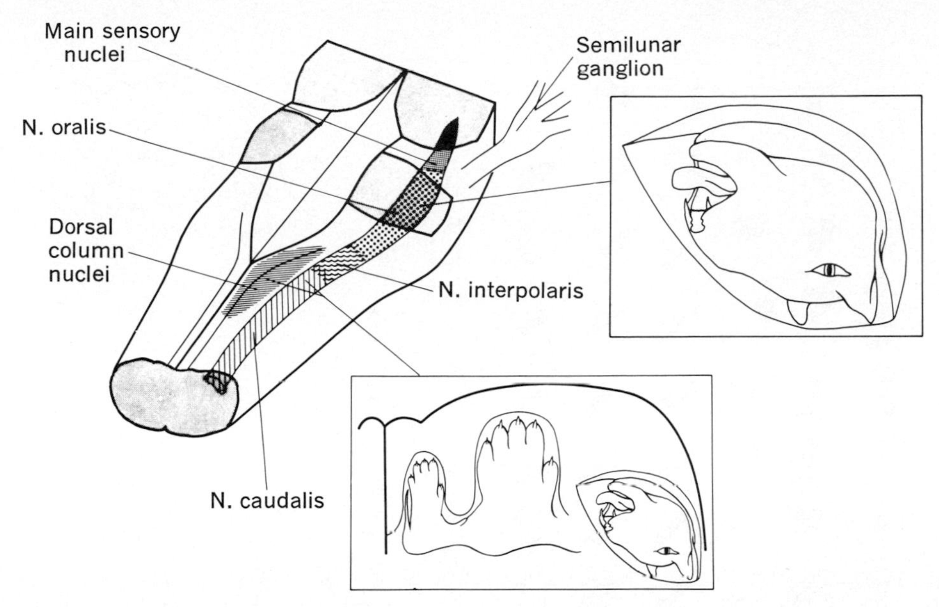

Fig. 10-7. Diagram illustrating different subdivisions of trigeminal nuclear complex in cat and pattern of somatotopic projection within these components. Each area of ipsilateral face is represented centrally at all rostrocaudal levels within complex, except for n. interpolaris, which lacks somatotopic organization. Pattern of projection is shown at two levels. Projections of ipsilateral limbs and trunk to DCN and that of face to n. caudalis are juxaposed but not continuous.

geminal tract results in facial insensitivity to noxious and thermal stimuli. Further studies are needed, particularly because those thus far completed were made in deeply anesthetized animals, a state in which other converging connections to these second-order cells may have been inoperative.

The n. interpolaris cytoarchitecturally resembles the external cuneate nucleus and, like the latter, many but not all of its cells project upon the cerebellar cortex; neither is thought to contribute directly to somesthesia.

In summary, the medial lemniscal analog within the trigeminal system includes the main sensory nucleus and n. oralis, though the latter differs from the medial lemniscus system in sending collateral projections upon the RF. N. caudalis, on the other hand, resembles the anterolateral system in its cytoarchitecture and its cephalad connections. The ipsilateral face is represented in toto in both components of the trigeminal system. Like that in the DCN, this central pattern is a somewhat distorted image of the peripheral body form, reflecting the heavy peripheral innervation of peri- and intraoral areas. Any single locus on the ipsilateral face or inside the mouth is represented centrally by a column of cells extending the whole length of the trigeminal complex, with a differential distribution for modality within the column.

THALAMOCORTICAL COMPONENTS OF SOMATIC AFFERENT SYSTEM

The thalamic somatic sensory areas of the two major divisions of the system are described separately, even though the two afferent pathways overlap in their thalamic termination, and neural activity in the two is thought to interact at both thalamic and cortical levels.

Lemniscal thalamic relay and the static properties of ventrobasal neurons

The ventral thalamic mass receives in an anteroposterior sequence three ascending systems, the dentatothalamic projection from the cerebellum, the medial lemniscus, and the neospinothalamic tract. At the mesodiencephalic junction the medial lemniscus contains the latter two systems in a single topographic pattern, together with the trigeminothalamic projections, and terminates in a region of distinctive architecture and functional properties, the *ventrobasal (VB) complex.*[82] There are three salient features of this region: a detailed topographic representation of the contralateral body form, the highly

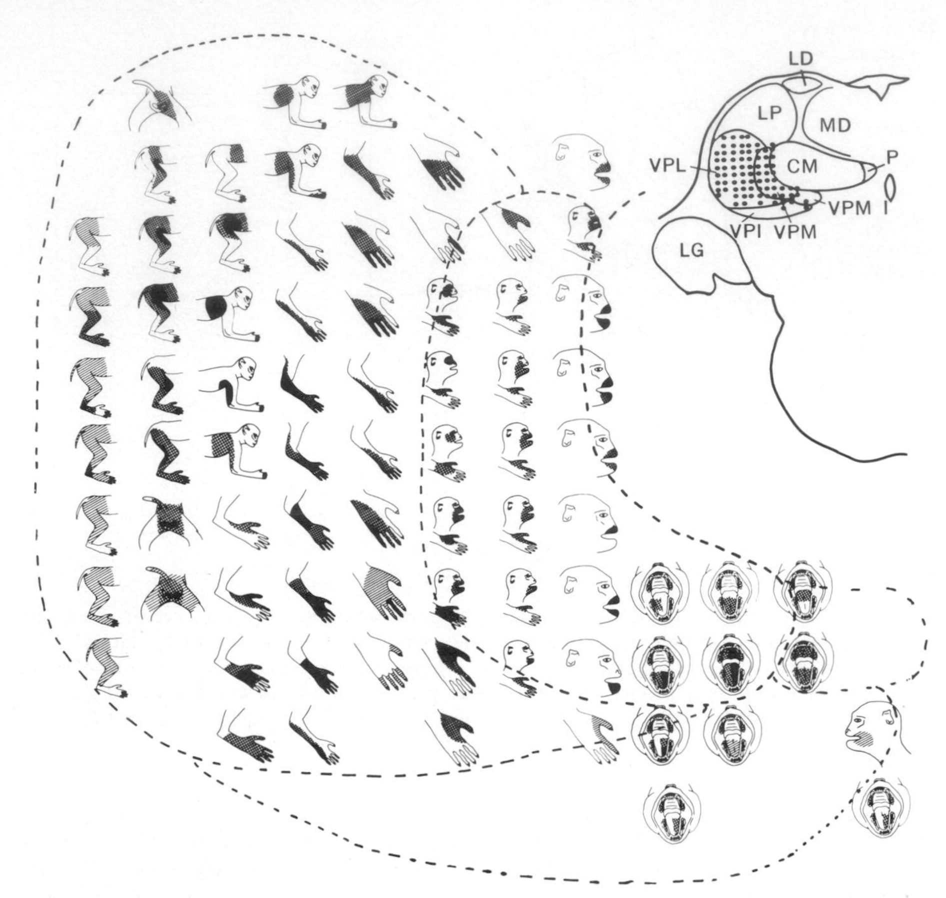

Fig. 10-8. Representation of cutaneous tactile sensibility in one frontal plane of thalamus of monkey, as determined by evoked potential technique. Inset drawing prepared from frontal section of brain in plane of electrode penetrations; dots indicate positive points, and each figurine drawing is arranged accordingly. Tactile stimulation of skin of areas marked on figurines evoked responses at points indicated. Gradation in intensity of projection, from most to least, indicated by solid shading and cross-hatching. With exception of ipsilateral intraoral and perioral regions, all responses were obtained only from stimulation of contralateral side of body and head. *VPL, VPM,* and *VPI* indicate divisions of ventral posterior nuclei, also called the ventrobasal complex. *CM,* Centre médian; *P,* parafascicularis; *MD,* medialis dorsalis; *LP,* lateralis posterior; *LD,* lateralis dorsalis; and *LG,* lateral geniculate body. (From Mountcastle and Henneman.[82])

specific static properties of its neurons as regards place and modality, and the great synaptic security between incoming elements of the medial lemniscus and the thalamocortical cells.[103] Moreover, it receives a heavy corticothalamic projection (p. 337).

The topographic details in the monkey are shown in the figurine map of Fig. 10-8: a similar pattern has been observed directly by recording in human patients. The distorted pattern is set at entry to the DCN by the place sorting of the DC and it also reflects the fact that the central replication is related to peripheral innervation density rather than to body geometry. From locus to locus across the thalamic pattern there is a gradually shifting overlap of the peripheral zones represented.[72] The afferents from any single spinal segment project upon a thin curving lamella of thalamic tissue, concave medially, with the proximal part of the dermatome dorsally and the distal portion, ventrally. The general patterns for the rabbit, cat, and monkey (Fig. 10-9) indicate the dominance of those body parts where sensation is highly developed.

VB neurons are highly specific for place and modality. They are activated by stimuli delivered within discrete, contralateral, continuous receptive fields and are preferentially responsive to but one stimulus quality. The classification of first-order fibers given in Table 10-1 applies with equal certainty to

Fig. 10-9. Schematic outline drawings of representation of body surface in ventrobasal thalamic complex in rabbit, cat, and monkey. Drawings were made from figurine maps such as that in Fig. 10-8 and aim to emphasize dominance of trigeminal representation in rabbit and relatively balanced one in monkey. In all, pattern reflects peripheral innervation density, not body geometry. (From Rose and Mountcastle.[107])

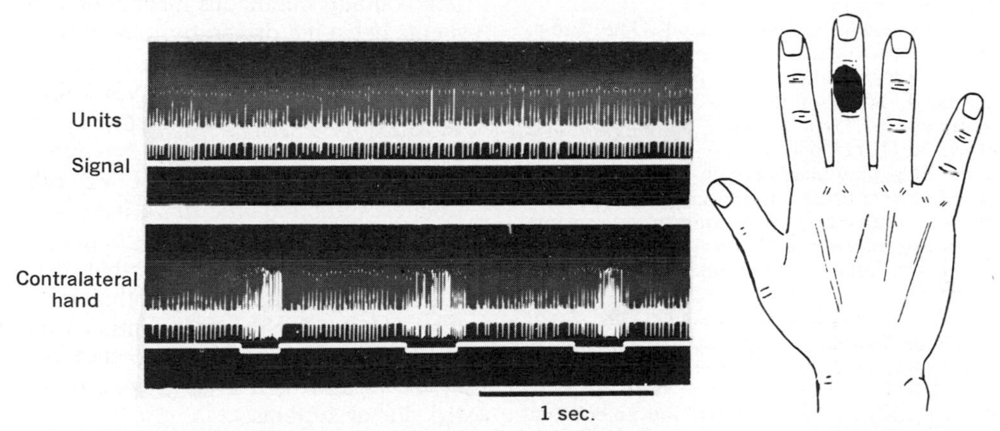

Fig. 10-10. Records, obtained during microelectrode exploration of thalamus in waking human being undergoing stereotactic thalamotomy for parkinsonism, of impulse discharges of single cell located in posterior portion of ventral thalamic group, homolog of ventrobasal complex in monkey. Cell was activated by light tactile stimulation of area of skin on dorsum of contralateral middle finger, shown in black. Upper records: spontaneous activity. Lower records: acceleration of spontaneous activity produced by three light strokes across receptive field, indicated by signal line. (From Jasper and Bertrand.[58])

DCN, VB, and postcentral neurons, implying that DCN neurons receive presynaptic terminals from fibers of only one of the classes of Table 10-1, and that in turn each VB neuron is linked to but one class of DCN neurons, etc. It is not known what factors are operative in ontogenesis to produce such a precise ordering of relations in the specific sensory systems.

Within the topographic pattern there is a differential distribution of the neurons of different modality classes. Those related to cutaneous afferents are concentrated posteriorly, while those activated from deep fascia, periosteum, and joint afferents are shifted anteriorly. Still further anteriorly, clearly in front of the VB complex and thus within the ventralis lateralis, is a zone thought to receive relayed activity originating in low-threshold fibers of muscle nerves (p. 336). This differential distribution becomes more marked in the transition from cat to monkey to man.[18]

Studies of human thalamus[17, 54, 58, 119]

Observations on the somatic sensory nuclei of the thalamus have been made in human beings, the opportunity being afforded by the discovery that local destruction of a small area of the ventrolateral thalamic mass, in front of the VB complex, reduces both the tremor and the increased muscle tone of some patients with Parkinson's disease. Thalamic explorations are carried out under local anesthesia, and recording-stimulating electrodes are guided into the thalamic target by a stereotaxic instrument that ensures an exact orientation of the electrode traverse with reference to identifiable landmarks. In humans the ventrolateral thalamic mass is even more differentiated than in the monkey, so that in its ventral portions four nuclei can be defined.[53]

N. ventralis caudalis (V.c.). The V.c. receives medial lemniscal and neospinothalamic input and projects upon the postcentral gyrus. Electrical stimulation elicits mechanoreceptive paresthesias referred to local regions of the contralateral body surface, and neurons of the region are activated from small contralateral cutaneous receptive fields (Fig. 10-10). These may be surrounded by inhibitory zones in the pattern of afferent inhibition described later. Taken together, the receptive fields of these cells compose a precise topographic representation of the contralateral body surface.

N. ventralis intermedius (V.im.). The V.im. apparently receives only selections of the lemniscal and no spinothalamic input, for its cells are activated by the passive rotation of the contralateral limbs at their joints and not by skin stimulation. These cells project upon the postcentral gyrus. The region is homologous to the anterodorsal parts of VB in the monkey.[103] A puzzling report is that electrical stimulation here produces no sensory experience in waking humans, but more observations are needed on this point.

N. ventralis oralis posterior (V.o.p.). The V.o.p. receives the dentatothalamic system from the cerebellum and projects upon the precentral motor cortex, area 4. Some of its cells are activated by muscle stretch (p. 336) but none by skin stimulation; others discharge only during voluntary movement, and they may discharge in synchrony with the 4 to 5/sec tremor of parkinsonians. Stimulation produces an acceleration of voluntary movements but no sensory experience. Lesions here are sometimes effective in relieving the tremor of Parkinson's disease.

N. ventralis oralis anterior (V.o.a.). The V.o.a. receives the pallidothalamic system, a major output from the basal ganglia, and projects upon Brodmann's area 6. Electrical stimulation here increases muscle tone and slows voluntary movement but elicits no sensory experience, and it is here that small lesions may relieve the hypertonia of Parkinson's disease.

Neither electrical stimulation nor destruction of the loci in which lesions modify the signs of Parkinson's disease produce alterations in somatic sensation. The implication is that the thalamic relay nuclei for the cerebellothalamocortical systems play important roles in the forebrain regulation of posture, tone, and movement but no essential role in somesthesia.

Thalamic areas activated by anterolateral system and functional properties of anterolateral neurons[73, 74]

The spinobulbar and paleospinothalamic systems funnel somatic afferent inflow to the ascending reticular and generalized thalamocortical systems and thus play an important role both in regulating levels of excitability in the forebrain (Chapter 8) and in pain and temperature sensibilities (Chapter 11). Elements of these systems possess static and dynamic properties quite different from those characteristic of lemniscal neurons.

In carnivores the neospinothalamic component of the anterolateral system contains only a small number of fibers; these project upon the posterior nuclear group and perhaps upon the ventralis lateralis, but not, it is thought, upon the VB complex.[21] In primates and especially in man this newer anterolateral tract increases greatly in the number and the size of its fibers, and a major projection upon the VB complex appears. After section of the dorsal ascending systems of the spinal cord in monkeys, stimulation of cutaneous nerves elicits the response of some cells of the VB complex and of the postcentral gyrus in a predominantly contralateral projection. Little is known, however, of the details of the neospinothalamic-VB projection in primates save that it is in topographic register with that of the medial lemniscus and contains cutaneous mechanoreceptive elements but none driven by joint rotation.[100, 129] It is not known whether the two sets of fibers preempt exclusive sets of VB cells or converge commonly upon them, or what the dynamic properties of the synaptic linkages are; the static properties of VB cells under isolated spinothalamic drive have so far been studied only in deeply anesthetized animals. Those observations suggest that the phyletically burgeoning neospinothalamic system may account for the rudimentary capacity for mechanoreceptive sensibility that remains in monkeys and men after lesions or disease of the dorsal systems.

The neospinothalamic system also projects upon the posterior nuclear group (PO), an area of thalamus intercalated between the VB complex anteriorly, the geniculate bodies laterally and posteriorly, and the pulvinar above.[102] Although this projection exists also in primates, the PO has thus far been studied by the method of single-unit analysis only in cats. In them, its cells may be activated from large receptive fields that may be contralateral or ipsilateral, or both may be discontinuous and may in the limiting case cover nearly the entire body surface. There is no regular topographic representation of the body form in this region, in contrast to the detailed representation in the VB complex. Unlike the mode-specific cells of the lemniscal system, the large majority of PO cells are polyvalent with respect to the adequate stimuli that activate them, some being responsive to light mechanical stimuli in one part of their receptive fields, only to noxious stimuli delivered to other parts, and to sound and light as well. The majority of PO cells, whether polyvalent or not, are activated by stimuli destructive of tissue.[102]

In summary, the lemniscal and anterolateral projections upon the thalamus differ strikingly. The former is organized as a large-fibered, rapid-tempo system, possessing those functional properties required to serve the discriminative aspects of somatic sensibility: identification of place, contour, and quality of mechanical stimuli, the sensing of the position and movement of the limbs, the resolution of serial order, etc. The anterolateral system serves a variety of more general functions. Its phyletically ancient projection, either directly or via the ascending RF of the brainstem, serves an energizing, not an elaborative, function by driving the generalized thalamocortical system. Its most recent development, the neospinothalamic projections upon the VB complex, reveals rudimentary lemniscal properties. Its projection upon the PO in combination with anterolateral projections upon the generalized thalamocortical system are essential neural substrates in pain sensation, a subject discussed in Chapter 11.

SOMATIC SENSORY AREAS OF CEREBRAL CORTEX

Somatic sensory area I: postcentral gyrus

It has been known for a long time that lesions of somatic sensory area I (SI), the postcentral gyrus, produce defects in somatic sensibility in human beings[30] and that electrical stimulation in this region elicits somatic sensory experiences referred by a conscious patient to a local part of the contralateral body. Indeed, electrical stimulation of the paracentral cortex, combined with observation of movements induced or sensory experiences described by such a patient (under local anesthesia) is of great localizing value in neurosurgical procedures; its use by Penfield[99] and other surgeons has provided a general map of the body representation in the postcentral gyrus (Fig. 10-11). The postcentral gyrus is but one of a number of cortical areas within which electrical stimulation will evoke such sensory experiences, but it is here that they can be provoked most consistently at lowest current thresholds, be referred so consistently to local contralateral parts and when the reports are ordered from point to point, reveal a regular topographic pattern. On these grounds alone it is possible to conclude that the postcentral gyrus stands apart from those other areas from which somatic sensations can occasionally be elicited, in a hierarchy ordered in terms of the importance of the areas for the primary attributes of somesthesia. The general representation pattern originally revealed in the human in this way has now been confirmed and detailed by

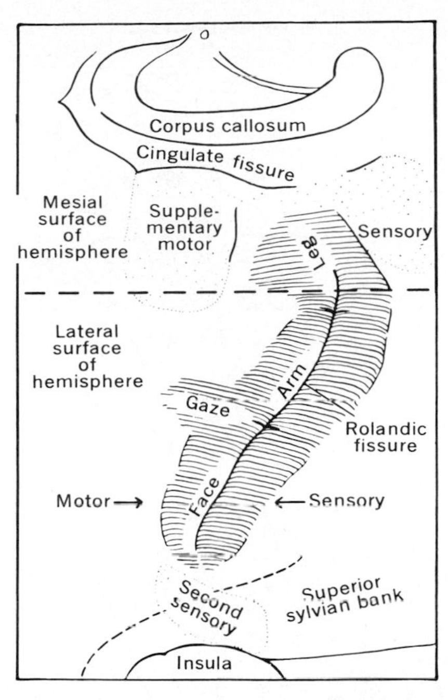

Fig. 10-11. Map of somatic motor and sensory areas in cerebral cortex of man determined by method of electrical stimulation of unanesthetized human beings, observing movements produced, and recording patient's verbal description of any sensory experience evoked. (From Penfield and Jasper.[99])

the application in man of the evoked potential method described in Chapter 7.

It is largely to the evoked potential studies of Woolsey[136] that we owe a detailed knowledge of the somatic sensory cortical representation in animals. That for the monkey is shown in Fig. 10-12. Inspection suggests that the pattern is produced by a projection in toto of the three-dimensional representation of the body form in the VB complex (partially illustrated in Fig. 10-8) on the two dimensions of the cortical surface. There is the same distortion to provide disproportionately expanded areas of cortex for the face, hand, and foot as compared with the area given to proximal body parts; the same point-to-area and area-to-point linkage between periphery and center; the same partially shifted overlap; and the same disruption of the representation of hand and foot. The maps for four mammalian species studied by Woolsey (Fig. 10-13) display general diagrams of the motor and sensory representations in the cerebral cortices of the rat, rabbit, cat, and monkey, as determined by the methods of electrical stimulation and evoked potentials, respectively. Such diagrams do not reveal the continuously shifted overlap that exists within

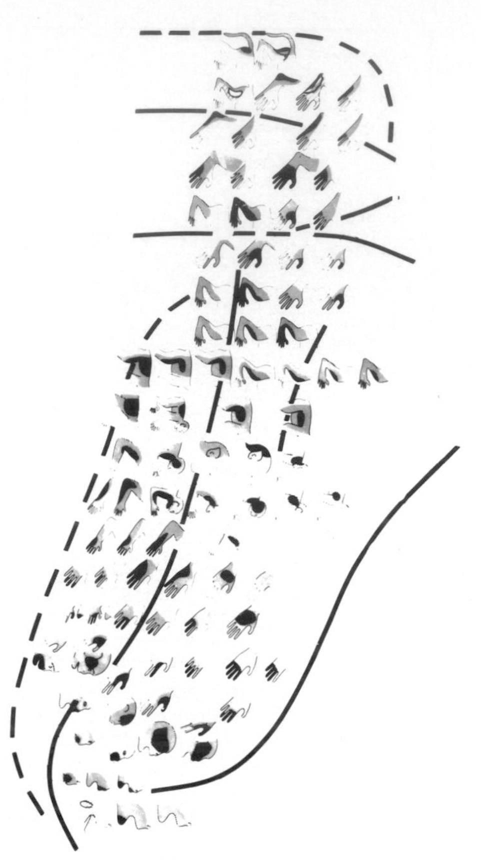

Fig. 10-12. Figurine map of postcentral tactile area of monkey *(Macaca mulatta)*. Each drawing designates area of body in which light mechanical stimulation evoked slow-wave potentials recorded via electrode resting on cortical surface, at position of drawing. Black and shading indicate degrees of intensity of projection. Taken together, drawings compose pattern of representation of body form that appears to be determined by peripheral innervation density rather than body geometry. Vertical dotted line shows depths of central sulcus, so that posterior bank of central sulcus is represented in same plane as external surface of postcentral gyrus. Similarly, medial wall of hemisphere and superior bank of cingular fissure are swung upward into same plane. Upper dotted line represents depth of cingular fissure. (From Woolsey.[136])

the patterns. Comparison of the rat and rabbit patterns with that of the monkey emphasizes the expansion of cortical areas for those parts best developed as manipulative and somesthetic organs. The drawings also indicate the relative expansion in primates of cortical regions not *primarily* either motor or sensory, i.e., the association areas that occupy such a large proportion of the cerebral cortex of man.

Recently Werner and Whitsel have applied the method of single-unit analysis to the study of postcentral topography. Their results lead to the general conclusion that fibers of the dorsal roots are arranged in dermatomal subassemblies that then function as the elementary units in the mapping process.[5] The somatotopic pattern resulting from the orderly combination of these units is already evident in the DCN (p. 317), composing there a pattern that is replicated at thalamic and cortical levels. The result is that the receptive fields of a linear mediolateral array of postcentral neurons compose a continuously connected trajectory on the body surface in the manner shown for the monkey in Fig. 10-14.

Functional organization of postcentral gyrus

Lemniscal neurons from DCs to postcentral gyrus are each preferentially activated by a particular type of mechanical stimulation of peripheral tissues. The modality classes of postcentral cells replicate those of the myelinated fibers listed in Table 10-1. This specificity, for mode as well as for place, has been documented by the study of several thousands cells of the VB complex and of the postcentral gyrus in anesthetized and in waking, behaving monkeys. This has suggested a functional correlate for the anteroposterior cytoarchitectural gradient across the three differentiable areas of the postcentral region, Brodmann's areas 3, 1, and 2. Almost all neurons of the most posterior area 2 are activated by the rotation of joints or mechanical stimulation of periosteum or fascia. As the locale considered shifts anteriorly through area 1 into area 3, the proportion of cells activated by cutaneous stimulation gradually increases, so that area 3 is almost exclusively a cutaneous projection zone. Indeed, a further separation exists,[98] for the majority of the cutaneous neurons of area 3 are linked to slowly adapting first-order fibers, those of area 1 to quickly adapting fibers. The neurons of area 3a,[101] transitional between postcentral sensory and precentral motor cortices, may be activated by stretch afferents from muscle (p. 336).

There are two candidate explanations for this differential distribution for modality. First, it has been suggested that this distribution reflects the body form and its innervation pattern; i.e., in each cortical "dermatomal" strip the more proximal body parts innervated by a given dorsal root are represented posteriorly, the distal anteriorly, and distal body parts receive by comparison a heavy

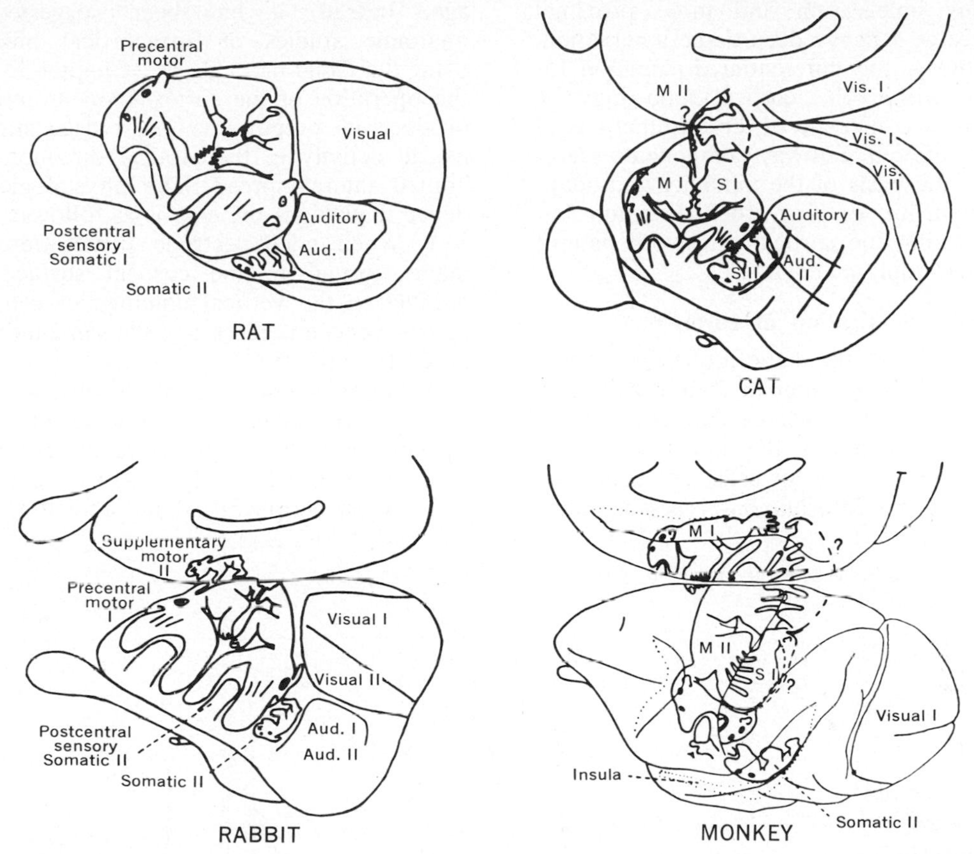

Fig. 10-13. Diagrams of cortices of rat, rabbit, cat, and monkey *(Macaca mulatta)* showing locations and general plans of organization of precentral motor *(MI)*, supplementary motor *(MII)*, postcentral sensory *(SI)*, and second somatic sensory *(SII)* areas. Relations to auditory and visual cortices are shown, except for monkey auditory area, which lies hidden on lower bank of sylvian fissure. For rabbit, cat, and monkey, medial walls of hemispheres are swung upward in drawings to occupy same planes as lateral surfaces. (From Woolsey.[136])

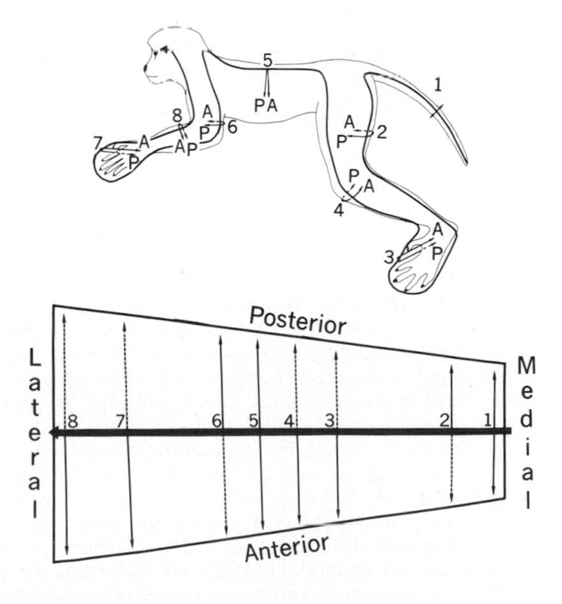

Fig. 10-14. Diagram below illustrates and is key for local neighborhood relations of cortical representation of body surface. Medial, lateral, anterior, and posterior designate position on right postcentral gyrus of macaque. Heavy uninterrupted arrow corresponds to continuous mediolateral array of cell columns that extends across entirety of SI and maps sequence of body regions encountered along heavy arrow drawn on figurine shown above. Narrow arrows orthogonal to heavy arrow in lower diagram correspond to arrays of neurons aligned in anteroposterior dimension of SI. Interrupted portions of narrow arrows correspond to parts of body surface hidden from view in figurine above and indicate cortical cell columns that represent those same regions. *A* and *P* at extremes of peripheral paths indicated above designate body regions encountered as one moves in anteroposterior dimension of SI map: *A* = anterior in SI; *P* = posterior in SI. (From Whitsel et al.[130])

cutaneous innervation and more proximal parts receive a heavy deep-tissue innervation. Alternatively, the differential distribution for modality within the cortical map may be viewed as a composite of the terminations of different afferent pathway components segregated on the basis of the modality and adaptive properties of first-order afferents but aligned within the cortical projection pattern in a somatotopic manner.

Columnar organization of cortex[80, 105]

An analysis of the static properties of cortical neurons in relation to their distribution in depth provides evidence that the elementary functional unit of the cortex is a vertically oriented column of cells that composes an input-output information-processing linkage. Indeed, it has been suggested by anatomic studies of intracortical linkages, using the Golgi method[27, 70] (Chapter 7), that the operation of the cortex upon its input to produce its output involves a translation of neural activity in the vertical direction with limited lateral spread. The physiologic evidence that this is the case is as follows:

1. When microelectrode penetrations are made normal to the cortical surface and parallel to the vertical columns of cells, all neurons encountered are of the same modality type (Fig. 10-15).

2. Neurons encountered in vertical penetrations are related not only to the same modality but also to nearly identical peripheral receptive fields (Fig. 10-15, inset).

3. When peripheral stimuli activating each

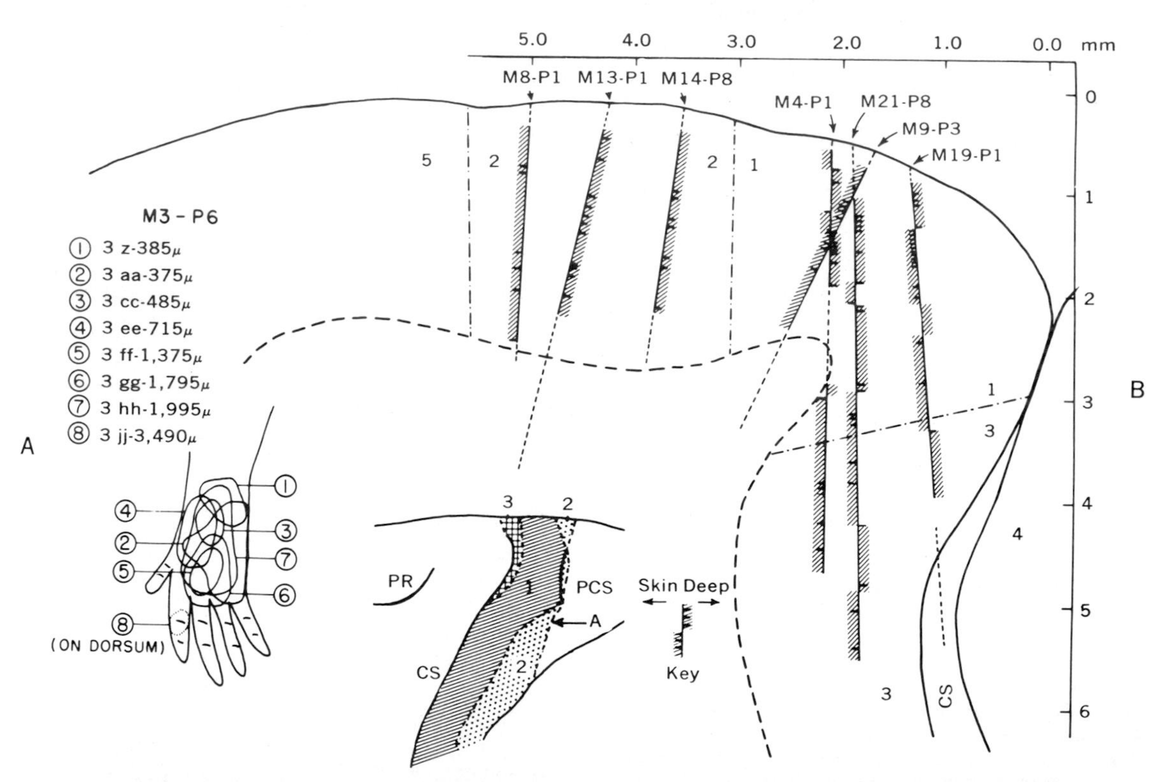

Fig. 10-15. Reconstruction from serial sections of seven microelectrode penetrations made in postcentral gyrus of macaque at level marked *A* on inset drawing of postcentral gyrus. Reconstructed tracts shown as if all were occurring on outline drawing of gyrus in cross section, with central sulcus, *CS,* now placed to right. Along each penetration, cross-hatching and horizontal lines to left indicate, respectively, a multi-unit record observed and an "isolated" neuron studied that was activated by stimulation of skin. Similar markings to right for those activated by stimulation of joint, periosteum, and fascial receptors. When penetrations are made normal to cortical surface and pass through cortex parallel to vertical columns of cells, all those encountered are of same modality type, are activated from superimposed peripheral fields (as shown for penetration *M3-P6* on inset, upper left), and respond with minimal differences in latency. Where penetrations pass at angles across columns, en bloc reversal of modality types are observed, for electrode tip is then passing successively from one cell column to another. (From Powell and Mountcastle.[105])

cutaneous cortical neuron are carefully positioned at the most sensitive loci in their receptive fields, the latencies of responses of cells in the different cortical layers fall within a narrow range, within 2 to 4 msec of each other. Cells of layers III and IV are activated earliest, those of the deeper layers later. Thus within the time allowing for only a few synaptic relays, activity is translated from the cells initially activated to those whose axons project out of the cortex. It is emphasized that this course of events holds only for the initial cortical activity evoked by a brief peripheral stimulus; continuing activity will be conditioned by the superimposed and reentrant chains of cortical neurons described in Chapter 7.

The cortical columns activated by different modes of stimulation are sharply demarcated from one another, and there is evidence that activation of one produces inhibition of neurons in the immediately surrounding columns.[84] It seems likely that such a vertical column of cells is the basic unit of the functional organization of the somatic sensory cortex, and this form of organization has now been found to exist in the second somatic,[132] the auditory,[7] the visual,[56, 57] and the motor cortices as well. The interdigitated mosaic of columns of different natures provides a means for plotting more than two variables onto a two-dimensional surface.

Somatic sensory area II

Adrian[8] and Woolsey[135] discovered that the body surface is represented in a second cortical area now known to exist in a number of animals including man. This second somatic area (SII) occupies the parietal cortex of the superior bank of the sylvian fissure in primates. The body regions are represented in SII, as in SI, in the sequence of the dermatomes, although dermatomal overlap is greater in SII. A distinctive feature of SII is that *both the contralateral and the ipsilateral body halves are mapped to SII* in a single, superimposed image of the body surface, the receptive field of each neuron consisting of pairs of matched and symmetric body areas.[131] The two parts of a field are disjoint if they occupy the apices of the limbs, but for more proximal body parts the ipsilateral and contralateral receptive fields may be continuous across the midline.

Tabulation of the static properties of neurons of SII with regard to cortical location and depths reveals a columnar organization as precise as that in SI. On other counts, however, there are marked differences. First, no neurons of SII are activated by joint rotation. Second, SII neurons are less securely linked to peripheral input than those of SI, for with slowly repeated stimuli their response gradually diminishes, a form of habituation. On the other hand, neurons of SII appear particularly sensitive to the direction of movement of a mechanical stimulus, and for a given neuron the preferred direction may be reciprocal in the two (the contra- and ipsilateral) parts of the receptive field. For this reason Werner and Whitsel[128] suggest that SII neurons could "compute logical functions over the stimulus categories of direction of stimulus motion and sidedness on the body."

The neurons of the transitional cortex linking SII with the auditory areas resemble in their properties those of the PO of the thalamus: they show a polysensory convergence, many may be activated only by noxious mechanical stimulation, and they are commonly related to very large and often asymmetric receptive fields. In the cat, neurons of this type are distributed in columns that overlap the posterior border of SII,[25] but in the monkey they appear segregated in a well-circumscribed and homogeneous area.[132] Its importance in pain sensibility will be discussed in Chapter 11.

Thalamo- and corticocortical connectivity in somatic afferent system

Much of the foregoing description of the somatic sensory thalamic and cortical regions is synthesized from the results of the large number of electrophysiologic studies of the last two decades. More recently, important advances have been made in specifying the detailed neuronal connections between the somatic sensory areas of thalamus and cortex by identifying the antegrade degeneration of terminal boutons, after local destruction of neuronal cell bodies. These studies have now been extended to include the intra- and interhemispheric connections between somatic sensory and other cortical areas, connections that are thought to compose the anatomic substratum for a successive elaboration and integration of the neural activity that is set in motion by somatic sensory stimuli and that eventually leads to somesthetic perceptions and, at will, motor response.

It is now clear that the VB complex projects upon both SI and SII,[62, 79] thus accounting for the lemniscal properties of their neurons that, at least in the monkey, are as clear for SII neurons as for those of SI, with the exceptions noted above.[50] It remains uncertain whether this dual projection derives from separate VB cell populations or whether a given VB neuron may project, for example, upon SI and deliver a collateral branch to SII, in the style of a sustaining projection (p. 230); some evidence indicates that both cases exist.[71]

In the cat, PO projects in the main upon SII but also upon SI, providing the anterolateral properties just described for the cortical cells upon which it projects. In all species, neurons with such anterolateral neurons are rare indeed in SI.[80, 105] In the cat, such a population exists in columns interdigitated with those of a lemniscal nature in the posterior third of SII, but in the monkey these two subsections of SII appear to have been completely differentiated and are identifiable separately.

Precisely reciprocal to these thalamocortical projections are corticothalamic projections from SI strongly to VB and less so to PO, and from SII strongly to PO and less so to VB. The details of these synaptic terminations are of great importance in terms of the intrinsic functioning of thalamic nuclei and cortical areas; the interested student should consult Jones and Powell.[61, 63]

The first somatic postcentral cortex sends interhemispheric fibers that terminate within somatotopically homologous areas of both SI and SII of the opposite side, while SII projects only to its contralateral fellow SII.[60, 97] The SI projection, however, derives only from the body, head, and limb girdle areas of the cortical representation pattern and projects in a point-to-point fashion to the opposite pattern. The apices of the limbs, those areas most highly developed for somesthetic function (in the monkey, *both* hand and foot), are free of such connections, as indeed are the primary cortical receiving areas of the visual and auditory systems.

As concerns intrahemispheric corticocortical connections, the precentral motor (area 4) and the postcentral sensory (areas 3 ,1, 2) cortices are heavily and reciprocally linked.[59, 96] In addition, there is a "stepwise outward progression from the main sensory area within (i.e., into) both the parietotemporal *and* frontal lobes, with an interlocking of each new parietotemporal and frontal step."[64] Briefly, the postcentral gyrus projects backward upon area 5, and area 5 is reciprocally connected with area 6 of the frontal lobe. Area 5 in turn projects farther backward upon area 7, and area 7 is reciprocally connected with area 46 of the frontal lobe; areas 46 and 6 are reciprocally linked. In these successive projections, general somatotopy is certainly preserved in the first step, from the postcentral gyrus to area 5, but thereafter there is increasing intrasystem convergence. The next projection areas in turn project upon other areas of the frontal, orbitofrontal, and temporal cortices, and in that step converge with the similarly elaborated projections of the visual and auditory systems. With this intersystem convergence, there appears a projection from the convergent areas upon those areas of the brain thought to be concerned with memory (the entorhinal cortex and the hippocampus) and with affective behavior (the limbic cortex). There can be little doubt that these important anatomic concepts have laid the groundwork for a fruitful combination of behavioral and electrophysiologic experiments aimed at elucidating cortical mechanisms in perception.

Recapitulation: summary and comparison of lemniscal and anterolateral systems

The meaning for somesthetic function of the duplicate representation of the somatic afferent system in the cerebral cortex and of the multiple prethalamic systems projecting upon the forebrain is still to a considerable degree uncertain. The lemniscal system, by virtue of the properties of its neurons, appears designed to serve the discriminative forms of somesthesis, for it can present at the cortical level neural transforms signaling the senses of touch-pressure, kinesthesia, and flutter-vibration. This conclusion is consonant with clinical and experimental observations, for lesions of the system produce severe deficiencies in these quantitative aspects of somatic sensibility.

The anterolateral system seems to serve more general aspects of sensation and to transmit information about certain qualitatively distinct sensations, namely those of a thermal and painful nature (Chapter 11). It provides much less precise information concerning the place, spatial pattern, and temporal cadence of stimuli. Its wide projection via the spinobulbar and spinoreticular systems upon the ascending reticular and the generalized thalamocortical systems indicates its potent role in arousal and in the control of forebrain excitability.

At the cortical level it is not yet possible to define precisely the contributions of the first and second somatic areas to function. SI and SII are not exclusively representatives of the lemniscal and anterolateral systems, respectively, in terms of functional properties. To what extent each contributes and how they must work synergistically in the cortical elaboration of afferent activity leading to the perception of somatic sensory events in all the shades and degrees of which human beings are capable is an important problem for future study. Important contributions toward its solution have come from the study of humans and animals after lesions of afferent pathways or of the cerebral cortex, a body of information detailed in a following section.

SOME FURTHER ANALYSES OF SOMESTHESIA: LEMNISCAL MECHANISMS IN TACTILE, VIBRATORY, AND KINESTHETIC SENSATIONS

The static and dynamic functional attributes of the lemniscal system will be used to describe, so far as is presently possible, the central neural processing mechanisms in the three forms of mechanoreceptive sensibility that humans can identify and scale quantitatively: touch-pressure, flutter-vibration, and kinesthesia. The system transmits a precise neural transform of peripheral events for initial cortical processing. Little is known of how this information— sharpened, preserved, and relayed within the system—is further processed centrally. A current assumption is that there are intracortical neuronal mechanisms that by some further integrative action lead to recognition, discrimination, and indeed, to perception itself. Little is known of this neuronal machinery, though it may involve the corticocortical systems previously described. It is hoped that study of the initial cortical transformation of afferent signals (the system's input) and overt human reactions to sensory stimuli as well as subjective descriptions of sensory experiences (two of its outputs) will lead to testable hypotheses concerning its nature.

Neural mechanisms in tactile sensibility[84]

The tactile sense depends on information about the place, intensity, and form of mechanical stimuli delivered to the skin; identification of the last requires information about the spatial distributions of the intensity of a stimulus. Detection of the direction and speed of stimulus movement requires an analysis of both the sequential spatial translation of neural activity across a neural population and temporal change as well. More subtle aspects of the tactile sense are likely to depend on combinations of these three: place, intensity, and serial order.

The lemniscal system is organized to preserve at successive neural levels the local sign of a peripheral stimulus, although we do not know how an increase in activity at one locale in a neural field, and not in another, identifies peripheral location. Even within this specific system the divergence between periphery and center is such that the capacity for localization or two-point discrimination, for example, (Fig. 10-3), can hardly be explained in terms of isolated, parallel neural lines evoking two completely separated central locales of neural activity. This capacity is to be understood in terms of the profiles of activity evoked at successively more central neural levels by the two stimuli, and especially

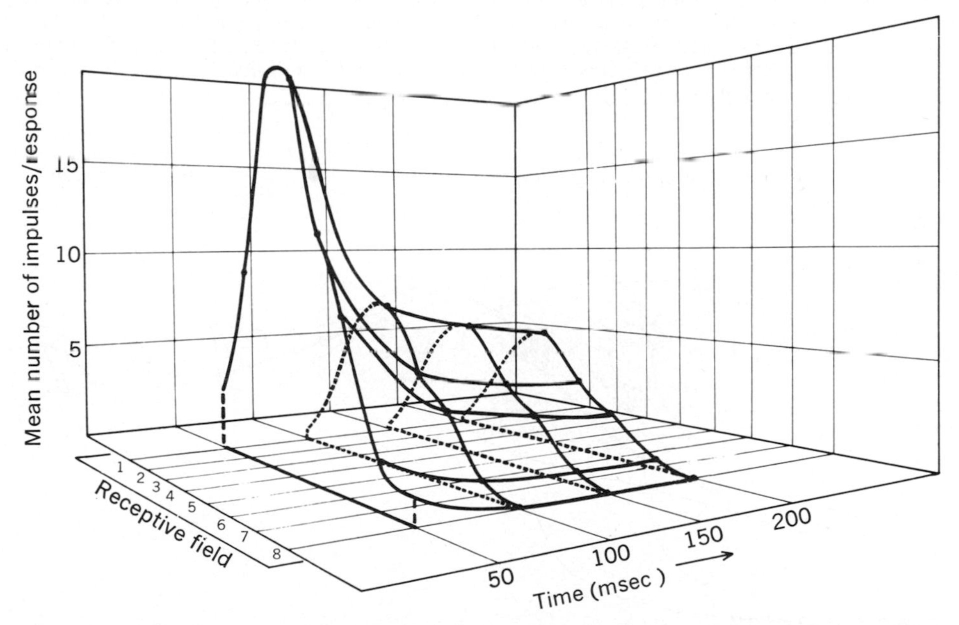

Fig. 10-16. Results of receptive field study in monkey of mechanoreceptive myelinated afferent of median nerve (slowly adapting type) that innervated peripheral receptive field extending across seven to eight dermal ridges of distal pad of thumb. A line bisecting receptive field forms left horizontal axis. Numbered dermal ridges occur at about 300μ intervals; full field measured about 2.4 mm in diameter. Right horizontal axis is time, and vertical axis is mean number of impulses, occurring in each successive 50 msec periods, evoked by abrupt indentation of skin of supramaximal strength. Stimulating probe tip was 0.5 mm in diameter, machined to a one-third spherical surface. Contoured lines can be thought to represent distribution of activity in entire population of first-order afferents activated by stimulus, on the assumption that as stimulus is moved from place to place across receptive field from edge to center to edge again, fiber under observation occupies similar series of positions in populations of fibers activated at each stimulus position. Contours show high-frequency onset transient and decline to quasi-steady rate of discharge. (From Mountcastle et al.[85])

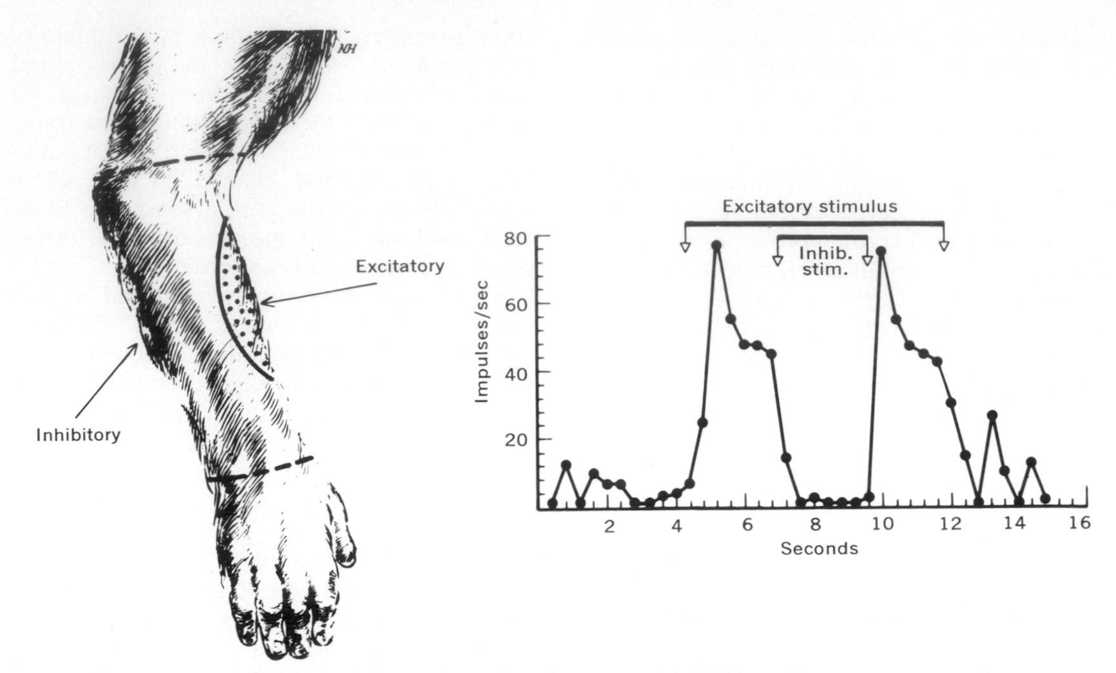

Fig. 10-17. Illustration of interaction of excitatory and inhibitory effects on neuron of post-central gyrus of monkey that was produced by stimuli delivered to its peripheral receptive field on contralateral forearm. Cell was excited by stimuli delivered to field on preaxial side of arm and inhibited by stimuli delivered anywhere within large surrounding area (only dorsal half is shown). Graph plots impulse frequency versus time during excitatory-inhibitory interactions. Application of excitatory stimulus evoked high-frequency onset transient discharge that declined toward steady plateau until interrupted by application of inhibitory stimulus. On removal of latter, sequence was repeated in response to continuing excitatory stimulus. (From Mountcastle and Powell.[105])

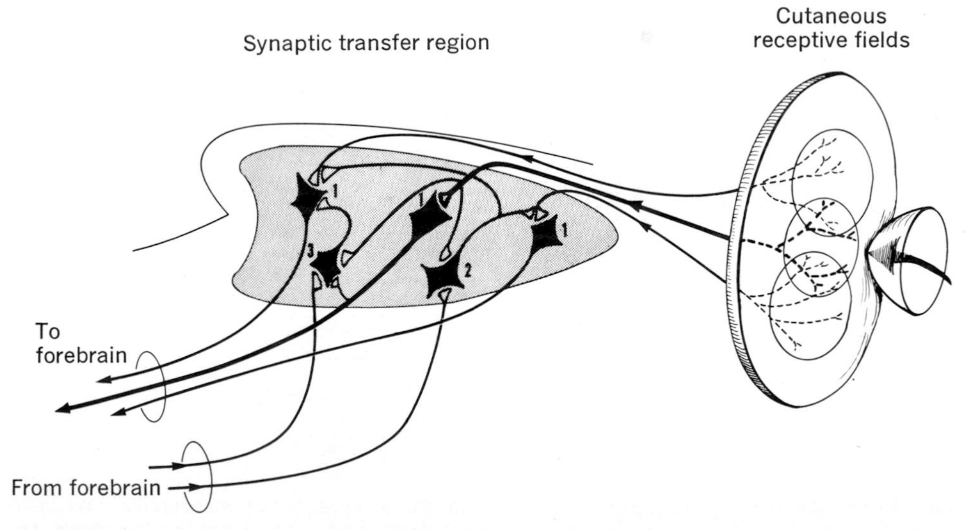

Fig. 10-18. Types of synaptic linkage subserving afferent inhibition in synaptic transfer region within lemniscal system. Input axons may terminate on primary relay neurons, *1*, on interneurons, *2*, with axoaxonal synaptic contact on neighboring entering fiber terminals, or on interneurons, *3*, which themselves possess polarizing synaptic terminals on adjacent relay neurons. Type 2 cells subserve presynaptic inhibitory mechanism and type 3 cells are interneurons in postsynaptic inhibitory pathway. Both presynaptic and postsynaptic inhibitory interneurons may be discharged via recurrent collaterals of axons of relay cells and via descending pathways from forebrain. Not all such interneurons, however, will have excitatory inputs from each of these sources. Relative inputs from these different sources will vary with different interneurons in nucleus and in different nuclei.

in terms of (1) the mechanisms that exist for preserving and sharpening those activity profiles, i.e., for preventing a smearing overlap between the two, and (2) the mechanism that might read such a profile.

A suggestion of the shape of such a neural profile is gained by studies of the responses of a single somatic neuron to stimuli delivered serially to a number of positions arranged in a line across its receptive field, as shown in Fig. 10-16. Use is made of the reciprocal interpretation, i.e., the supposition that the single element has occupied in each step of the serial experiment a series of different positions in an active central population, positions occupied by other neurons when the stimulus is delivered to one spot. The contour of such a graph can then be interpreted to represent the spatial and temporal "contours" of the neural activity in the central population. *It indicates the transform in neural space of the intensity, contour, and location of the peripheral stimulus.* The cascaded divergence of the lemniscal system means that the population of neurons activated by such a stimulus has a considerable spatial extent in, for example, the postcentral gyrus, a fact that limits the discriminations possible on the basis of spatial relations alone. Something more is needed to limit lateral divergence and sharpen profile edges; that something is afferent inhibition.

This phenomenon is illustrated in Fig. 10-17. This postcentral neuron of a monkey responded to light pressure applied to its excitatory receptive field, and both its spontaneous and evoked activities were inhibited by light pressure delivered to a region of skin surrounding the excitatory receptive fields. The graph plots the interaction of the two opposite influences on the response of the neuron. This form of spatially ordered afferent inhibition does not occur between mammalian first-order fibers, as is the case in the *Limulus* eye: it is a central neural event and has been observed at each level of the lemniscal system. Afferent inhibition, by limiting and shaping neural profiles, may contribute to two-point discrimination and hence to all spatial discriminations. The phenomenon is a general one in mammalian sensory systems; its role in the CNS mechanisms in hearing and vision is described in Chapters 13 and 16.

Recent studies of synaptic stations of the lemniscal system have shown that the mechanisms of afferent inhibition are the presynaptic and postsynaptic mechanisms already familiar in segmental reflex pathways. The presynaptic form is prevalent at the DCN[16]; the postsynaptic, in thalamus and cortex. A scheme of these relations applicable to any level of the system is given in Fig. 10-18. It is a general principle that inhibition is exerted by local, intranuclear mechanisms; i.e., no primary afferent neuron and no neurons with long axons linking DCN-thalamus-cortex in either direction exert inhibition by direct monosynaptic action. The shaping effect of inhibition might occur in either or all of three ways:

1. *Feed-forward mechanisms,* via branches of axons entering a nuclear region and terminating (a) upon excitatory interneurons that in turn end upon the terminals of neighboring entering axons, thus exerting presynaptic inhibition (p. 201), or (b) upon inhibitory interneurons that terminate upon the cell bodies of adjacent relay neurons, exerting postsynaptic inhibition (p. 199).
2. *Local feedback mechanisms,* via recurrent collaterals of axons leaving a relay nucleus, inhibiting the relay of activity in the belt of surrounding neurons, by either of the mechanisms listed under 1.
3. *Reflected feedback mechanisms,* via descending elements from more cephalad regions of the forebrain that, impinging upon more inferior relay regions from above, may shape the spatial extent and form of the relayed activity, again by either of the mechanisms given under 1. There is evidence that such reflected feedbacks may also be positive in nature. The possibility of positive feedback to the center of an active nuclear relay zone, combined with negative feedback to the surround, would at the same time limit the spatial spread of that discharge zone and facilitate the relay of impulses through its center. The function of these descending systems is described in more detail in a following section.

Intensity functions and linearity of transmission in lemniscal system[2]

The neural scaling of stimulus intensity is particularly important for tactile sensibility, for which the degrees of freedom are only place, intensity, and temporal pattern. The slowly adapting, myelinated, mechanoreceptive afferents of large caliber that innervate the glabrous skin of the hand are, unlike quickly adapting afferents, sensitive to both the rate of a stimulus application and to its intensity (Chapter 9). Together with the central lemniscal neurons to which they are linked, these fibers can account for the intensity discriminations that humans make with their hands. A stimulus-response relation for such an afferent is shown in Fig. 10-19, *A*.[89] It is linear, and the degree of variability between responses is small and independent of response level. Thus for this afferent the initial transformation across the skin is linear, on the assumption that the operative neural code is that of frequency (p. 294). Experiments of this same type carried out for neu-

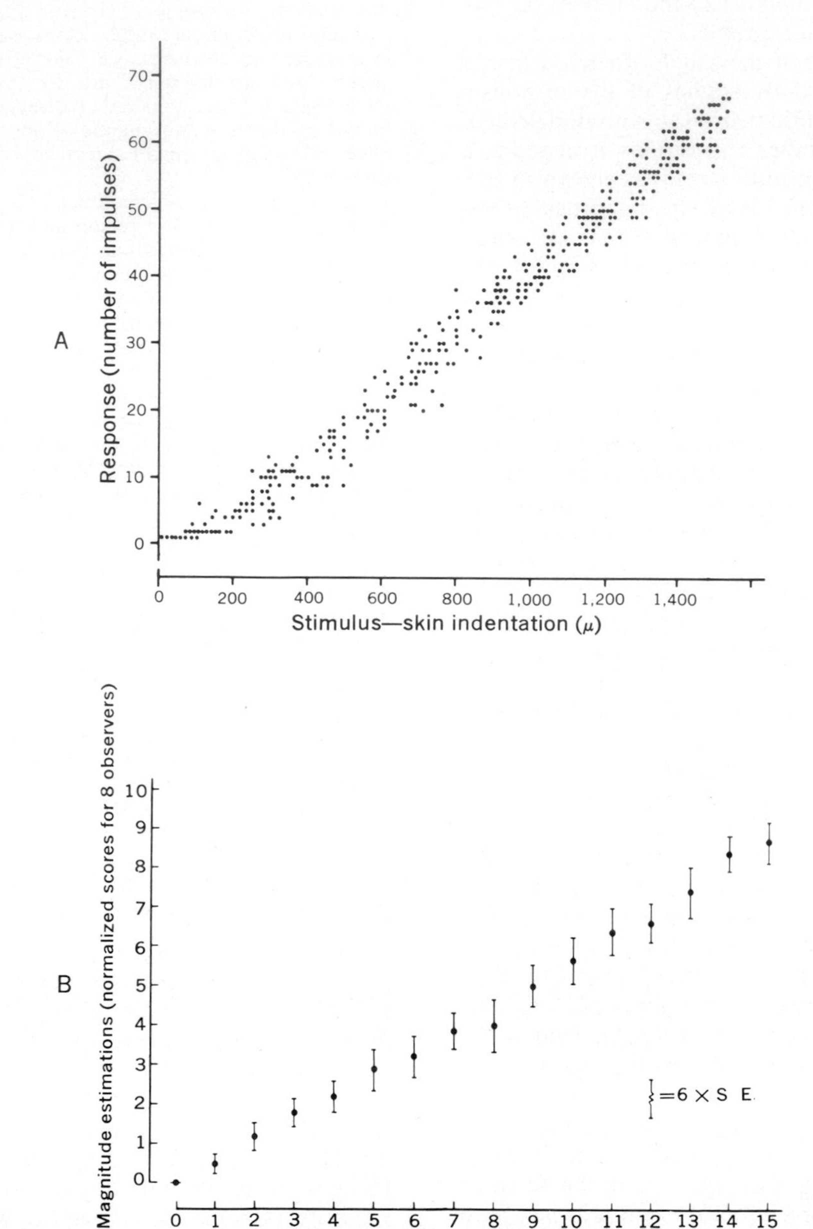

Fig. 10-19. A, Stimulus-response relation for slowly adapting myelinated fiber innervating glabrous skin of monkey hand. Stimuli were step indentations of skin of 600 msec duration, delivered in random order of intensity at rate of 12/min. Stimulator tip 2 mm in diameter, machined to one-third spherical surface. Number of impulses evoked by each stimulus is plotted as function of measured intensity of that stimulus. **B,** Result of psychophysical experiment in humans in which subjects are asked to rate stimuli along intensive continuum by assigning numbers to them. Stimulator used and experimental arrangements otherwise similar to experiment on monkey hand. Points are means for numbers called by eight healthy young adults after normalization of individual subjective scales. Relation obviously linear. Implication is that all those neural transformations that intervene between neural input and final verbal description of magnitude estimation are *in sum* linear. Second important implication is that human relation to external world as regards intensity of sensory stimuli is set by properties of peripheral receptors. (**A** From Mountcastle et al.[85])

rons of the VB complex of the thalamus and the postcentral gyrus in the monkey have yielded similar results. The implication is that transmission across the intervening nuclear relays imposes no further transformation upon the first-order input, as regards intensity. Of course, this does not imply a one impulse in–one impulse out relation at those relays, for the frequencies of discharge of neurons at different levels of the system may differ. The quantitative relation between the peripheral stimulus measured on some physical scale and the neural activity scaled in frequency is identical at all levels.

Between the postcentral gyrus and a behavioral output evoked by such a mechanical stimulus, the perceptive and interpretive mechanisms of the brain intervene; although little is known of those mechanisms, the output can be measured. The result of such an experiment in a human subject is given in Fig. 10-19, *B;* in this experiment the series of stimuli delivered was exactly similar to the series used in the study of monkey first-order fibers. The response variable was, however, the human subject's estimate of the magnitude of each stimulus. The result indicates that whatever the intervening transformations may be, they are linear in sum. Findings for other sets of afferents in this and other systems are similar and lead to the general conclusion that *the relation of the human to the external environment, as regards the variable of stimulus intensity, is set by the transfer properties of the peripheral terminals of first-order fibers.*

Harrington and Merzenich[52] have recently tested this generalization by study of the hairy skin of monkey and man. They found that if the stimulus probe tip is moved only a few centimeters from the palmar to the hairy skin of the adjacent forearm, the subjective magnitude estimation function for human observers changes from a linear to a negatively accelerating one, best described by a power function wth an exponent of about 0.5.[126] They then confirmed the fact that the stimulus-response function for large, myelinated, slowly adapting mechanoreceptive afferents innervating the hairy skin of the monkey's arm is also negatively accelerating and also best described by a power function with an exponent of about 0.5. Such a relation holds as well in the visual system, in which the initial transduction is nonlinear, and both central neuronal activity and behavioral performance are linearly related to the first-order signal.

Neural mechanisms in flutter-vibration

The nature of vibratory sensibility has been described in an earlier section. Fig. 10-5 shows the detection thresholds over the frequency range of mechanical oscillations that humans and monkeys detect with their hands. This derived form of mechanoreception is of general interest because the periodic nature of the stimulus and of the neural activity it evokes allow study of neural coding in the temporal domain as well as the dynamic frequency range of the system of lemniscal neurons linking periphery and cortex. The explanatory power of the results of such experiments has been strengthened by the development of methods for measuring sensory performance in subhuman primates. The results of one such experiment are illustrated in Fig. 10-20, which shows psychometric functions constructed after several measures of the likelihood with which the monkey detected the presence of 30 Hz mechanical sinusoids of different amplitude that were delivered to his hand. Repetition of such measurements at a number of frequencies yields frequency-threshold functions

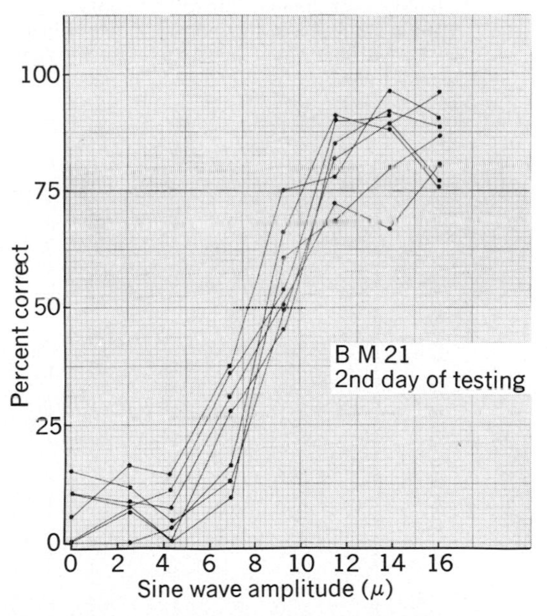

Fig. 10-20. Psychometric functions for monkey subject trained to indicate when he detected presence of mechanical oscillation at 30 Hz delivered to his palm. Each curve plots percent correct detection as function of sine wave amplitude and represents results of run of 256 trials distributed equally between 8 amplitude classes. Stimuli of different amplitudes were delivered in random sequential order. Dotted line indicates point of 50% detection, which is taken as threshold. Congruence of curves for seven separate runs in 1 day's working session indicates accuracy with which threshold can be measured in well-trained and highly motivated monkey. Correct detections were rewarded; no punishment for errors.

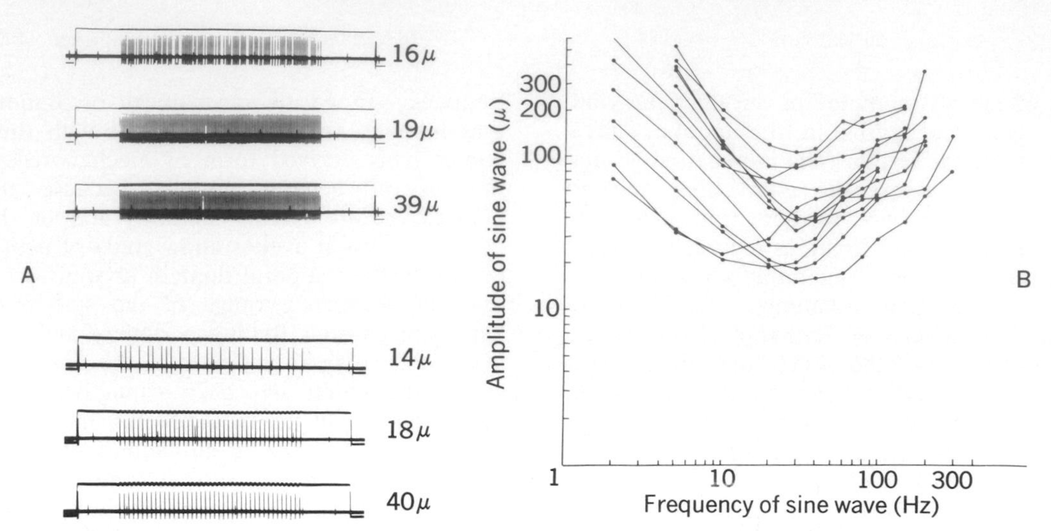

Fig. 10-21. A, In each of six pairs of records, upper trace indicates movement of 3 mm probe tip against skin of palm of anesthetized monkey. Upward deflection indicates movement into skin; note superimposed sine wave oscillations. Lower trace in each pair displays electrical signs of impulses in single fibers isolated from median nerves of monkeys by microdissection. Top three pairs were obtained while pacinian afferent was under study; frequency of superimposed sine wave is 150 Hz. Note that with increase of only 3μ, discharge "locked in" to stimulating sine wave. This amplitude is called "tuning point" of fiber. Further increases in amplitude of sine wave produced no further change in discharge. Lower three sets of records were obtained in similar manner, while quickly adapting fibers innervating glabrous skin were under study; stimulating sine wave is 40 Hz. **B,** Tuning curves for a number of quickly adapting mechanoreceptive afferents terminating peripherally in glabrous skin of hand. Tuning point of each fiber was determined at a number of frequencies in manner illustrated in **A.** All fibers have approximately the same "best frequency," about 30 Hz, but they differ in sensitivity.

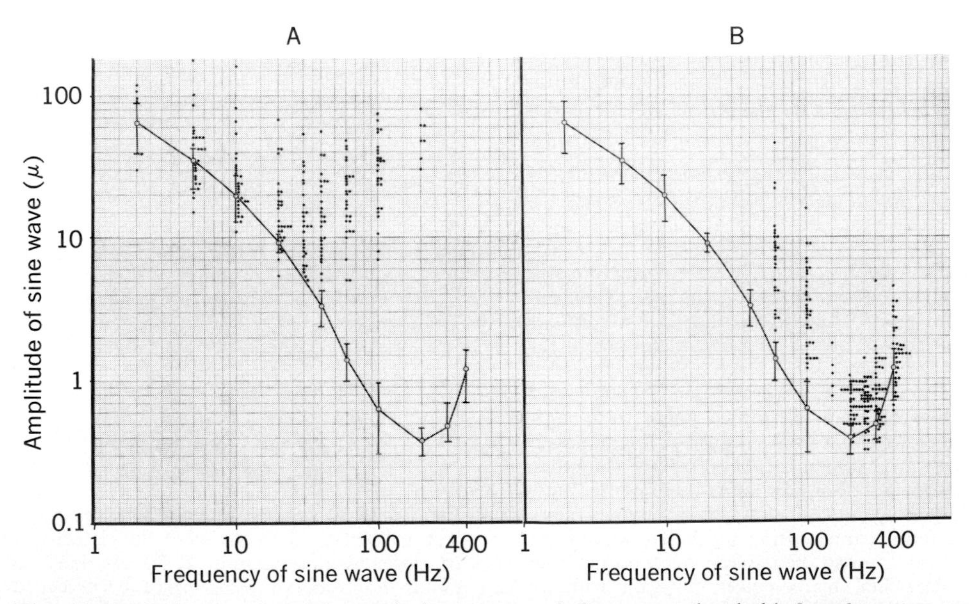

Fig. 10-22. Solid line in each graph is average of frequency threshold functions for six monkey subjects whose individual curves are shown in Fig. 10-5. Vertical lines at each tested frequency indicate ±1 SE of means. **A,** Each dot indicates tuning point for quickly adapting, large, myelinated mechanoreceptive afferent fiber innervating glabrous skin of monkey hand. Fibers were isolated for study by microdissection of median nerves; response of each was determined for a number of sine wave amplitudes at each of several frequencies. Thresholds of monkeys for perception of vibration in low-frequency range from 5 to 40 Hz could be accounted for by hypothesis that what is required for determination that mechanical stimulus is oscillating and not steady is appearance of tuned discharge in some small number of fibers of this class. **B,** Similar results for a number of pacinian afferent fibers terminating in hand. Similar statement can be made concerning appearance of tuned discharges in pacinian fibers in frequency range of 60 to 400 Hz and concerning sensation of vibration.

such as those shown for a number of animals in Fig. 10-5. The similarity of the results obtained in the two primates is obvious and supports the conclusion that monkeys and men detect mechanical sinusoids at about the same amplitudes, over the same range of frequencies, and (what is not illustrated) respond to such stimuli with similar reaction times.[86]

Local anesthesia of the skin of the hand differentially elevates thresholds in the low-frequency range, leaving those for 80 to 400 Hz unchanged.[118] This result implies the duality of the sense of flutter-vibration and its dependency on afferent signals in two sets of primary afferent fibers, one tuned to low frequencies and terminating in the glabrous skin, a second terminating beneath the epidermis and differentially sensitive to high-frequency stimuli. Examination of all the candidate myelinated fibers innervating the monkey hand revealed that the quickly adapting fibers ending in the dermal ridges and the pacinian afferents fit the requirements. Records of the impulses evoked in a fiber of each type by a mechanical sinusoid of the appropriate frequency are shown in Fig. 10-21, *A*. Studies of the type illustrated reveal that as the amplitude of the sinusoid is increased by small increments, from one stimulus to the next in a series, there is a very sharp point at which the response "locks in" with 1 impulse/cycle, the "tuning point" of the fiber. Repetition of this experiment at a number of frequencies allows construction of tuning curves such as those shown in Fig. 10-21, *B*. These findings lead to the hypothesis that it is the appearance of such periodic discharges in some small number of first-order fibers that is a necessary signal for the detection of a vibratory stimulus, and that the propagation of that neuronal periodicity in the somatic afferent system provides signals used for discriminations such as those between different frequencies.

The charts in Fig. 10-22 show, respectively, measures of the tuning points for populations of glabrous skin, quickly adapting afferents

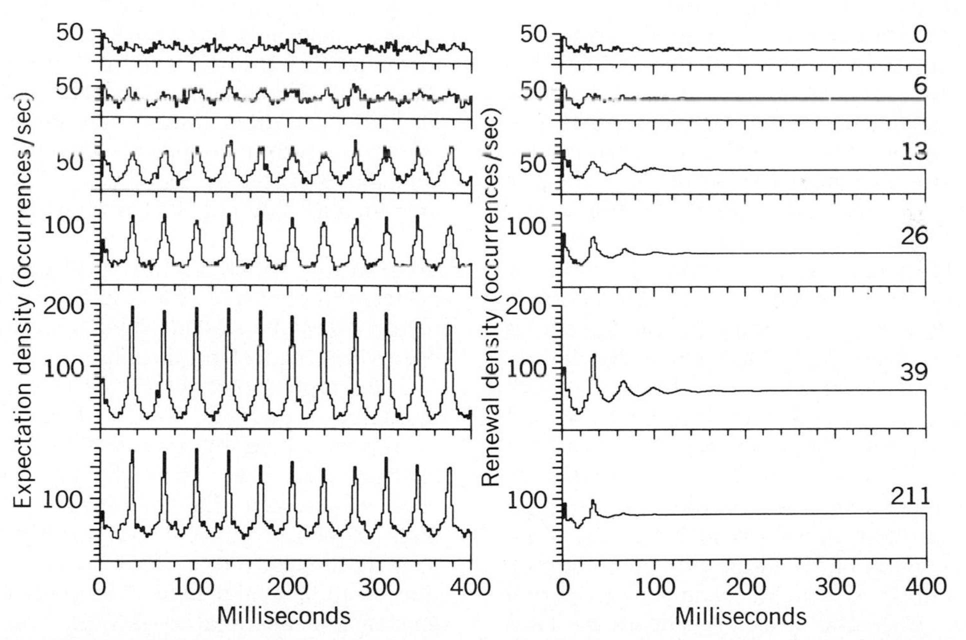

Fig. 10-23. Expectation density and renewal density histograms for responses of neuron of postcentral gyrus of unanesthetized monkey linked to quickly adapting mechanoreceptive afferents innervating glabrous skin of contralateral hand. Driving peripheral stimulus at 30 Hz, delivered at a number of different amplitudes, indicated in microns to far right for each pair of histograms. Expectation density analysis resembles autocorrelation function and likelihood of occurrence of impulse in each small sequential period of time. Histograms show strong replication of driving stimulus in rhythmic discharge of cortical neuron, suggested at stimulus amplitude of 6μ and clear and strong at 12μ. Monkey detection threshold at 30 Hz is 6 to 8μ. Renewal density analysis is essentially repetition of expectation density after random shuffling of impulse sequence. Results show importance of *temporal order in which impulses occur* for this particular neuronal code.

(A) and pacinian afferent fibers *(B)* innervating the monkey hand, together with the average threshold function for monkeys. The correlation shown is compatible with the periodicity hypothesis and is another example of the generalization that the sensory capacity and performance are in the limit set by the functional properties of first-order fibers.

The idea that periodicity is the peripheral and central neural code for flutter-vibration has been tested further in studies of the responses of postcentral neurons to mechanical sinusoids delivered to the hands of monkeys trained to indicate when they detect them.[90] The analysis of the records of such a neuron, shown in Fig. 10-23, reveals the periodic nature of the discharge of a cortical neuron linked to glabrous skin, quickly adapting fibers, a periodicity that appears at about the average monkey detection threshold and increases gradually with increases in stimulus amplitude. Recent experiments in which sensory detection and cortical neuronal activity were observed simultaneously[85] reveal striking parallel increases in periodicity and the likelihood of stimulus detection, but no causal relation has yet been established.

Recent measurements have shown that both monkeys and men can discriminate between very small differences in the frequency of mechanical sinusoids, e.g., between 30 and 35 Hz, at the 75% level of confidence. Moreover, if the discriminable increment is treated as cycle length rather than as frequency, the discriminable increment or decrement in cycle length is a constant fraction of the base cycle length over a considerable range. It is obvious from the tuning curves shown in Fig. 10-21 that such a discrimination cannot be made on the basis of which fibers are active. It is unlikely, moreover, that such discriminations can be made on the basis of differences in the total number of impulses per stimulus, for the cycle-to-cycle "jitter" in the responses is such that the overall frequencies of discharge evoked by stimuli of 30 and 35 Hz might, for any given period of time, e.g., 200 msec, be identical. Thus all the available facts suggest that it is the dominant interval in the periodic impulse interval trains that is the relevant signal and that the differential recognition of the frequency of an oscillating mechanical stimulus requires a CNS mechanism for discriminating between different dominant periods.

Thus the lemniscal system can transmit a single impulse or a brief volley of impulses from periphery to cortex within a short time after transmitting such a signal, e.g., from cycle to cycle at 40 Hz, and thus code the frequency of the stimulus in the temporal domain. This transmission capacity requires a rapid recovery of excitability after each transmission, and measurements of the recovery capacity at the input stage to the postcentral gyrus, in unanesthetized monkeys, show that the system has recovered to at least 70% of resting excitability within 5 msec after transmitting a volley.[103] This rapid recovery of excitability is one of the dynamic properties of the lemniscal system that is most profoundly affected by anesthetic agents.

Neural mechanisms in position sense and kinesthesis[83, 87, 88]

Afferent information in the somatic sphere plays an important role in the perception of body position and attitude, together with information of visual and vestibular origin. The relevant sensory inflow arises mainly in the joints, for position sense is severely affected in its absence. The ligaments and capsules of the joints are innervated by myelinated afferents that provide precise signals of the steady angle and the speed and direction of movement of the joints (Fig. 9-10, p. 302). The joint afferents enter the DC and exit to an undefined dorsolateral column ascending system to terminate upon cells of the more cephalad portion of the DCN.[6] These latter are linked, free of cross-modality convergence, with thalamic neurons of similar properties,[87] which in turn project upon the postcentral gyrus.[83] This linked system of neural elements presents signals of kinesthetic events for cortical processing.

The graph shown in Fig. 10-24 plots the discharge frequency of a thalamic joint neuron activated by step movements of the contralateral knee through the excitatory angle of the cell. Each movement elicited a high-frequency transient that quickly subsided to a quasi-steady rate whose value is a function of joint angle; here it is minimal in extension and maximal in flexion. This neuron subtended an "excitatory angle" of about 75 degrees. The steady discharge rate of such a neuron is determined by the steady angular position, not by the speed or direction of movement to the position. The relation between joint angle and discharge rate is negatively accelerating and best described by a power function with an exponent of 0.5 to 0.6, a figure similar to that of the function

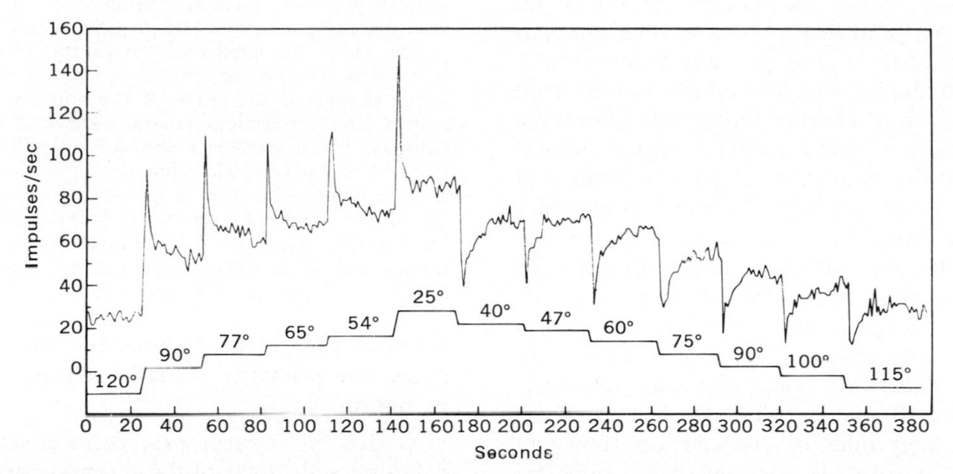

Fig. 10-24. Results of excitatory angle study of ventrobasal thalamic neuron activation by flexion of contralateral knee in unanesthetized monkey. Knee was rotated in short steps from position outside excitatory angle (from 120-degree joint angle) to full flexion (25-degree joint angle) and back again. Each movement in flexor direction produced transient discharge that declined over a few seconds to more or less steady discharge rate determined by joint angle. Movements in extensor direction produced, reciprocally, off-transients and recovery to plateau determined, once again, by joint position, but with obvious hysteresis. Plateau of discharge obtained on first limb of study used to plot excitatory angle functions such as those shown in Fig. 10-25. (From Mountcastle et al.[88])

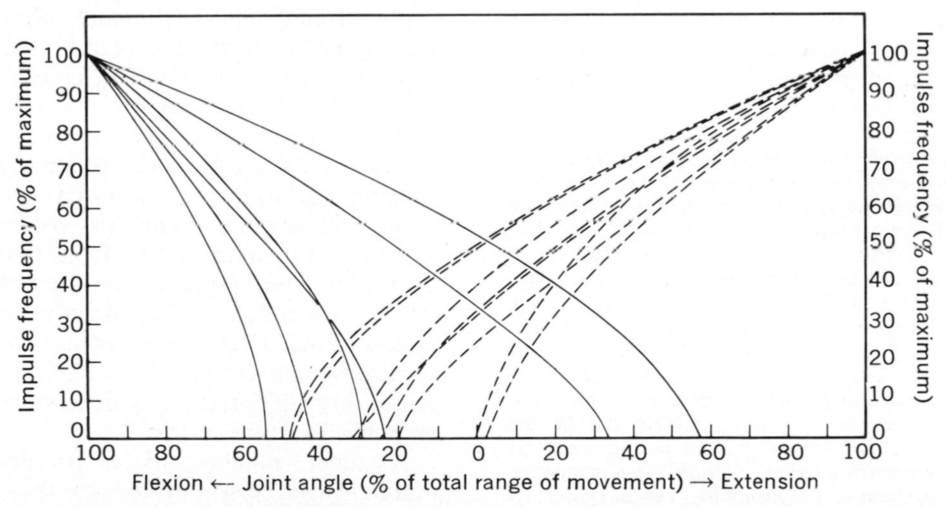

Fig. 10-25. Plots of excitatory angle functions for 14 ventrobasal thalamic neurons related to hinge joints of contralateral limbs studied in unanesthetized monkeys in manner of experiment illustrated in Fig. 10-24. Abscissal scale is normalized for different joints by changing each to percent of maximal movement possible in either direction from midposition of joint. Ordinal scale is normalized for each neuron by converting maximal discharge rate to 100. Majority of neurons subtend excitatory angles greater than one half the total range of movement of joint to which they are related; maximally excitatory positions are always at either full flexion or full extension, never both: curves are monotonic. Considered as a population, neurons will discharge reciprocally as joint moves back and forth through middle range of its movement, and such a reciprocal action of cortical and thalamic joint neurons is regularly observed. (From Mountcastle et al.[88])

characteristic of the human subject when estimating joint position.[49]

Fitted functions for the responses of a number of thalamic joint neurons are shown in Fig. 10-25, where the data are presented on normalized scales to allow a synthetic reconstruction of the events in the relevant population of thalamic cells as a joint is moved from one position to another through its full range. A majority of cells subtend angles greater than half the total range of joint rotation, and all the curves are "single-ended." First-order afferents from the knee joint of the monkey are related to excitatory angles of only 10 to 30 degrees, and some are located at different positions in the range of movement and are "double-ended," although others are located at the full extent of flexion or extension. Thus thalamic and cortical neurons, by varying their rates of discharge as shown in Fig. 10-25, provide running spatial integrals of the activity of a considerable number of first-order fibers that must, across the intervening synaptic relays, converge upon them. In a sense a combined spatial and intensive pattern of representation of joint position within the first-order neural population is converted by the third- and fourth-order levels to a pattern varying only intensively in the domain of discharge frequency. At these higher levels the datum of the *position* of a given neuron within the neural field provides information concerning the joint that is moved; the degree of joint displacement is signaled by impulse frequency.

Can stretch afferents from muscle provide information concerning the position or movement of the limbs? It is an old idea in neurology that kinesthetic sensation is part of a more general "muscle sense," including the sense of effort described on p. 310. Neither of the stretch afferents from muscle, however, discharges in any uniform relation to muscle length, the datum relevant for signaling joint angle. The discharge rate of tendon organs is determined by tension, not length; the total tendon organ discharge varies independently of joint angle. The spindle organs are excited by muscle stretch and cease to discharge when muscle is shortened by alpha motoneuron action. Spindle organs are influenced by action of fusimotor neurons that innervate the intrafusal muscle fibers. At any given length, spindle organs may be silent or discharge at any frequency, up to very high ones, depending on the level of activity in fusimotor neurons. It is therefore unlikely that spindle receptors can provide reliable signals of muscle length.

Central projection of muscle afferents. The major central projection of the fibers from these two muscle stretch afferents is over the spinocerebellar systems[94]; they form a massive feedback mechanism that is operative, after cerebellar processing, on all executant levels of the motor system (Chapter 28). A number of investigations have shown that they also project via the somatic afferent system, it is thought via the most ventral portion of the DCN[108, 110] and the most anterior portion of the ventral posterior thalamic group,[48, 109] to project upon Brodmann's area 3a, a transitional cortical region lining the depths of the central fissure and interposed between the sensory and motor cortex.[101] When compared to others of the somatic afferent system this projection possesses several unusual features, for it has been shown that volleys confined to group I muscle afferents (1) evoke no sensory experiences in waking human subjects, (2) do not elicit behavioral responses in animals except for local reflex action, and (3) cannot serve as conditioning stimuli in waking animals.[104, 117] The functional significance of this projection is not known.

Somatic afferent projections to cortical areas not primarily somatic sensory in nature[23]

Studies of intracerebral connectivity indicate that wide areas of the cerebral cortex and subcortical areas may, after a number of relays, receive neural activity originally set in motion by sensory stimuli. There are three major pathways for relay of sensory activity to "nonsensory" cortex. The first is via primary sensory areas and corticocortical connections; e.g., the postcentral gyrus is linked to a succession of more posterior parietal areas, each of which in turn is reciprocally connected with successively more rostral areas of the frontal lobes. Ultimately these interlocking chains, from each of the visual, auditory, and somesthetic systems, converge upon the frontal pole, the orbitofrontal cortex, and parts of the temporal lobe. Second, association areas receive projections from intrinsic thalamic nuclei, the lateral posterior and pulvinar groups, which themselves receive intrathalamically relayed signals from sensory relay nuclei. Third, each major afferent system projects collaterally either directly or via the ascending RF upon the generalized thalamocortical system (Chapter 7); afferent activity can, in principle, reach any neocortical region over this divergent system.

A great number of electrophysiologic studies have been devoted to these "nonprimary sensory" projections. It has been shown that the areas of cortex just described (and also discussed on p. 341) that receive convergent corticocortical linkages from the three major primary areas may be activated by either visual, auditory, or somatic stimuli.[19, 20] Moreover, this activation may occur via the other two major channels that have been described. While the role of these areas in somesthesis is still uncertain and is presently under study (p. 341), there is little doubt that these anatomic and electrophysiologic results have laid the groundwork for an understanding of the function of association areas.

It is useful to define a cortical "sensory area." Based on the previous discussion, almost any might qualify. However, the selective processing of mode-specific sensory input and the intersensory integration displayed by humans suggests a selective isolation of specific sensory signals up to some vaguely defined level of multisensory convergence. Such a definition has two parts: (1) *A sensory projection area receives a major input from an ascending sensory system such that its own cells preserve, in their functional properties, some signs of the quality and local sign characteristic of that sensory system* and (2) *lesions of such an area result in functional deficiencies recognized as belonging to one sensory sphere or another.* On these grounds the postcentral gyrus and area 5 of the parietal lobe are somatic sensory cortical areas. The motor cortex, which in some species (particularly subprimates) receives somatic afferent input, is not a somatic sensory area, nor are those more distantly linked areas of the limbic or temporal lobes. The neurons of these areas do not preserve the quality and local sign of somesthetic neurons, and/or lesions in them do not result in functional losses that are specifically somesthetic in nature.

Precentral motor cortex. This area of the primate CNS is one major source of the descending outflow via the pyramidal and extrapyramidal systems that terminate in the basal ganglia, RF, and spinal cord, systems discussed in chapters dealing with the central control of movement. The region receives projections from the cerebellothalamocortical and the paloidothalamocortical systems, avenues allowing modulations of cortical efferent output by the cerebellum and basal ganglia, respectively—the great reentrant avenues of the motor system. The pre- and postcentral gyri are interconnected by corticocortical fibers that are thought to provide for an integration of sensory and motor activities at the cortical level.

There have been no more vexed questions in CNS physiology than whether the "motor cortex" receives a direct somatic sensory input from the thalamus, and what the nature of such an input might be. This confusion is due in part to the rapid evolutionary changes in the motor and sensory cortices themselves. In marsupials the two are mutually coextensive.[68] In mammals there is an increasing structural and functional separation of the two, culminating in the central fissure in primates. In carnivores, cells of the motor cortex may be activated by stimulation of skin and deep tissues of the limb, particularly by stimuli that lead to movement.[22] Experiments employing microstimulation reveal a columnar arrangement within the cat motor cortex with a tight coupling between input and output; i.e., peripheral stimulation excites cortical columns whose stimulation via intracortical microelectrodes elicits movement of the limb in the direction of the stimulus delivered.[15] In monkeys, however, there appears to be no direct projection of cutaneous afferents to the motor cortex; its cells may be activated by joint afferents, providing reafferent signals of joint position to the highest level of motor control.[9] This projection is thought to arise from the transition zone between n. ventralis lateralis and the VB complex of the thalamus. Muscle stretch afferents project to area 3a (p. 336), which in turn projects upon both the motor and sensory cortices.[59] Studies in both waking and anesthetized human subjects using the method of evoked potentials, however, give equivocal results regarding this question; it appears that an afferent input to the motor cortex is present in some individuals and not in others.[47]

DESCENDING SYSTEMS INFLUENCING AFFERENT TRANSMISSION IN SOMESTHETIC PATHWAYS[51]

All the major sensory systems contain descending as well as ascending elements, corticofugal neurons that terminate either directly on or after RF relay upon subcortical relay zones of the sensory pathways. It is possible that this descending system plays a role in that selective attention to one sensory portal to the exclusion of others that is so characteristic of human behavior. In this sorting process, humans may consciously attend to only one of many inputs and remain unaware of others. It seems likely, however, that the major part of this action occurs in the cerebral cortex, the primary representation of peripheral events being continually present at the first stage of cortical processing and available for "inspection" at will. Descending systems may contribute to the shaping and funneling of the ascending neural activity evoked by sensory stimuli, perhaps by reinforcing or modifying the afferent inhibition known to improve resolution in the neural replication of stimulus contours or adding contrast to the spatial display. It will be seen that the corticofugal elements of the somatic system are disposed to contribute by reflected feedback to this sharpening effect.

Anatomic studies.[66, 91] It has been shown by anatomic methods that there are heavy descending projections to the spinal and bulbar relay zones of the somatic system and to the trigeminal component as well. In primates those terminating in the DCN arise mainly in the postcentral gyrus, with less dense projections from the motor cortex and the posterior parietal areas. The projection is topographically organized to complete a feedback loop, from hind-leg cortical area to hind-leg portion of the DCN pattern, etc. The descending projections from one hemisphere are distributed bilaterally but largely contralaterally. The major projection is by direct fibers from cortex to DCN; a less dense component relays in the brainstem RF.

The cortical projection to the first synaptic relay zone of the anterolateral system, the dorsal horn of the spinal gray, arises from a more restricted area of the cerebral cortex. It is topographically organized, i.e., from hind-leg area to the dorsal horns of the lumbar cord, etc. The majority of these corticofugal fibers reach their spinal terminations via the pyramidal tract. The latter is therefore a complex system, for it contains (1) fibers of precentral origin projecting to bulbar and spinal motoneurons or their associated interneurons—the efferent motor components of the pyramidal system; (2) fibers of precentral origin that project upon the RF; and (3) the fibers mainly of postcentral origin but also some from other cortical areas that project upon the synaptic transfer regions of the somatic afferent system—the reflected feedback system now under discussion.

The VB complex of the thalamus receives corticothalamic fibers from both SI and SII, so that cortical and thalamic patterns are linked in a reciprocal somatotopy.

Study of these recurrent systems has thus far been limited, for they can only be activated experimentally by electrical stimulation of their origins. This permits identification of the direction of their effects and of topographic relation, but not of the manner in which normal input to the cortex drives the feedback loops. The dominant effect of corticofugal activation upon sensory synaptic transmission is inhibition.[11, 13, 14, 51, 69, 121] The regions of cortex from which this inhibition of the DCN can be elicited are nearly coincident with the origins of the corticofugal systems that have been described previously. The inhibition is largely but not completely abolished after transection of the pyramidal tract.[69, 121] Studies of DCN neurons[13, 14, 69, 134] have shown this inhibition to occur in the absence of postsynaptic IPSPs. The terminals of myelinated cutaneous afferents in the DCN and in the dorsal horn of the cord as well are depolarized by electrical stimulation of the appropriate cortical area[36] or descending pathway.[41, 124] The inhibition is therefore at least partially presynaptic in nature: the corticofugal volleys activate interneurons within the DCN or the dorsal horn that in turn synaptically depolarize the primary afferent terminals, making afferent impulses in them less effective synaptic stimulating agents (Chapter 6). The net effect is inhibition. Many of these same interneurons may also be excited by cutaneous stimulation, and they appear to be common to the presynaptic inhibitory pathways serving the cortical and the afferent presynaptic inhibition observed in these nuclei. The fact that many sensory relay cells of the DCN are excited rather than inhibited by cortical stimulation indicates that the reflected feedback influence may be positive as well as negative.[134] A similar inhibitory action of the corticothalamic system upon the neurons of the VB complex has been described.[35, 113]

In summary, efferent systems take origin from the sensory and motor areas of the cortex and project upon the subcortical sensory relay nuclei of both the lemniscal and anterolateral systems; the projection is mainly direct via pyramidal tract fibers but also indirect via the RF. So far the most intensively studied of the actions of this system is that upon the DCN, within which a descending corticofugal volley exerts a prolonged presynaptic inhibitory effect upon the terminals of DC fibers. Other lemniscal neurons are excited by such a volley. While it is possible that this system acts as an on-off valve serving the faculty of selective attention, that process is thought to operate largely at a cortical level. The reflected cortical feedback to sensory relay nuclei may contribute to funneling and shaping of the discharge zones set up in them by sensory stimuli, thus contributing powerfully to spatial contrast.

DEFECTS IN SOMESTHESIA PRODUCED BY NERVOUS SYSTEM LESIONS IN ANIMALS AND MAN

Lesions of the lemniscal system produce defects in the discriminative aspects of somesthesia. What is lost is not the simple appreciation that a stimulus has occurred; for this, the remaining anterolateral input may suffice. What is lost is information about stimuli that in the normal individual permits exact localization; determination of quality, relative and absolute intensities, and temporal patterns; recognition of spatial extent and form; and appreciation of bodily movement and position. The effects will also be considered of lesions at those higher levels of neural organization that although concerned with somesthesia, can no longer be specified as lemniscal or anterolateral or even as exclusively somesthetic in nature. These effects are complex disturbances of appreciation and awareness of body form, spatial orientation, and the relation of the body to extrapersonal space—functions that obviously depend on integration of information in several sensory spheres.

Lesions of the lemniscal system, in certain locations in certain patients, produce not only defects but abnormalities of sensation, pathologic events that are attributed to the released and unbridled action of the anterolateral system upon the forebrain. These latter, together with the effects produced by lesions of the anterolateral system per se, will be considered in Chapter 11 on pain mechanisms.

The study of sensory defects in humans has proved to be of the greatest importance for an understanding of sensation. A verbal description of a sensory experience has a unique value in itself and allows the use of objective measures of the quantitative aspects of sensory experiences.

In animals with lesions deliberately placed in sensory pathways, measures of sensory capacity and of discrimination are made frequently in training experiments such as that illustrated in Fig. 10-26. It is important to emphasize three aspects of these experimental methods that occasionally make their results difficult to interpret.

1. In both man and animal a great deal more is tested than the sensory experience itself. The human report depends on introspection, immediate recall, and verbal expression;

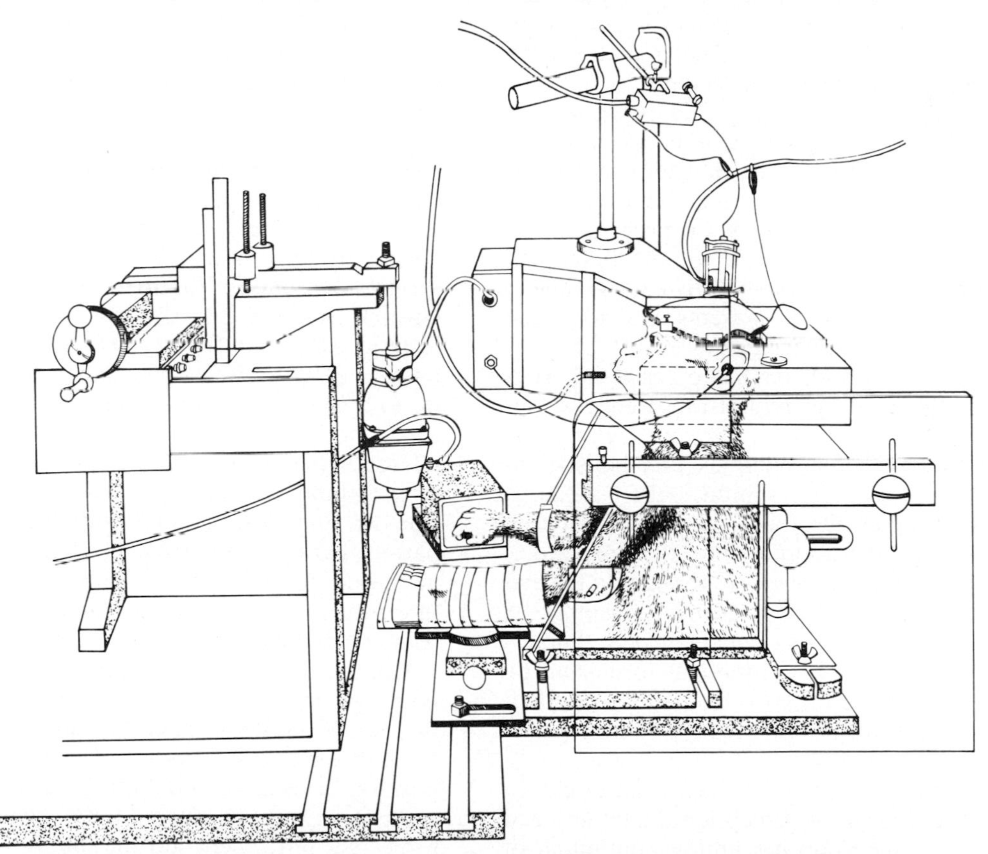

Fig. 10-26. Waking monkey in experimental apparatus for combined psychophysical and neurophysiologic experiments. Mechanical stimulator carried on rack and pinion devices at left delivers stimuli of variable intensity, duration, and temporal pattern to exposed glabrous skin of immobilized left hand. Animal manipulates signal key with his right hand to initiate trials, indicate detection, etc. Head is fixed in a "halo" device attached to skull for periods of study which are repeated from day to day. Tube delivers liquid reward to monkey's mouth. Drawings above indicate microelectrode chamber implanted over right postcentral gyrus and its attached microdriver, input preamplifier, etc. In this way, simultaneous studies can be made of animal's response to mechanical stimuli delivered to left hand and responses of cells of contralateral sensory cortex that are activated by same stimuli.

the animal report depends on the capacities to learn and to respond with an overt, directed motor act.

2. In such experiments *one observes the remaining capacity for sensory function; one infers from any deficiency the function of the lesioned part.* The function of a part (e.g., a cortical area) is not confined to its intrinsic neural mechanisms but includes all its complex relations with other parts. It is not likely that the defects in the remaining capacity will be simple negative reflections of the function of the part removed.

3. In both animal and man, particularly in young individuals, the lesioned nervous system possesses a remarkable power to *compensate* for structural loss by an unknown process. With increasing time there is frequently a considerable recovery of functional capacity, and the long-term residual deficiency following a CNS lesion is commonly much less than that manifested in the days immediately following the lesion.

In traditional neurology and psychology a clear distinction was drawn between sensation and perception. The first term described the "pure" and "simple" experiencing of the primary sensory qualities, a process to be explained directly in terms of the functional properties of receptors, afferent nerve fibers, and central sensory systems up to some poorly defined "intermediate" level of complexity. Perception, on the other hand, subsumed the total behavioral experience of apprehending an object, more particularly the experiencing of the complex and holistic aspects of sensory stimuli, the patterns, selectivity, contrasts, and similarities between stimuli, and the apprehension of serial order. Somesthetic *sensations* were thought to depend on the primary sensory pathways and to be elaborated in the postcentral gyrus. Somesthetic *perceptions* were thought to depend on these and, in addition, a further neural elaboration and integration in the posterior parietal association areas.

Studies of recent years have cast doubt on this strict sensation-perception dichotomy and indicate a sensation-perception continuum of increasing complexity. Indeed, the presumed "simple" sensory experiences may themselves be complex and may not be governed by isomorphic relations between stimulus qualities and those experiences. It appears likely that the progressively more elaborate processing of sensory input, its comparison with information stored in immediate and long-term memory, its evolution into efferent motor activity or into that undefined central neural reflection subject to introspective analysis and verbal description, and its recombination into neural activities that lead to total sensory experiences all involve not only the postcentral gyrus but other cortical areas as well, in addition to their interactions with subcortical referrants. This process of progressively more elaborate action is undoubtedly a continuum, and lesions at one level or another will affect function differentially at one or more points along it. Examples of these variations follow.

DEFECTS IN DISCRIMINATIVE ASPECTS OF SOMESTHESIA PRODUCED BY SUBCORTICAL LEMNISCAL SYSTEM LESIONS

Lesions of the myelinated afferents entering the lemniscal system may occur in peripheral nerves or along their central trajectories in the dorsal systems of the spinal cord. These afferents are destroyed in certain dietary deficiencies or may be reversibly blocked by ischemic pressure exerted on peripheral nerves for experimental purposes. Precise lesions of the DC or dorsolateral column systems occur rarely in humans, but they are frequently interrupted along with other spinal systems by tumors, injuries, or vascular occlusion. For example, the constellation of defects termed the Brown-Séquard syndrome, produced by hemisection of the cord, results in (1) loss of the discriminative aspects of somesthesia on the ipsilateral side, a dorsal afferent system defect; (2) loss of pain and temperature sensibility on the contralateral side, an anterolateral system defect; and (3) partial paralysis of movement and changes in muscle tone on the ipsilateral side, due to interruption of descending motor control systems, including the pyramidal tract.

Conduction defects in lemniscal first-order fibers or their ascending central trajectories in the DC systems result in specific alterations in somatic sensation.[39] There is a loss of position sense, of kinesthesis, and an accompanying sharp drop in the accuracy of projected movement of the limbs in space (see subsequent discussion) in reaching for and grasping targets. The capacity for two-point discrimination is markedly affected, as are vibratory sensibility and the ability to judge lifted weights. The subject no longer recognizes the shape of objects explored tactually, a defect called stereoanesthesia. What remains in the mechanoreceptive sphere after large fiber or *dorsal system* lesions is the capacity to recognize that a mechanical stimulus has occurred, though it is no longer possible to specify its location, intensity, or shape. The senses of pain and temperature are unaffected by dorsal system lesions, although certain abnormalities in the pain sense may appear (Chapter 11).

Some confusion concerning the effect of DC lesions made at different segmental levels has now been cleared by the discoveries of the dorsal system trajectories of lemniscal afferents, described on p. 313. A transection of the DC at upper cervical levels in the monkey, for example, produces the full syndrome just described in the arm without serious sensory defects in the leg. Thoracic DC lesions produce in the leg elevations of thresholds for the sense of position, for two-point discrimination, and vibratory sensibility, but no absolute losses. If, however, a lesion of the dorsolateral columns is added, all are lost permanently.[111] The relevant dorsolateral system is apparently not the spinocervical system, for its elements do not possess the requisite functional characteristics. These ascending dorsolateral systems from the lumbar region project upon the DCN in conformity with the topographic and modality patterns described on p. 314.

Cervical DC lesions produce, in addition to the sensory losses described, a permanent loss of the tactile placing and hopping reactions (Chapter 29). A marked disability appears in the motor sphere, particularly in the arm. Animals are reluctant to move the denervated limb. When used at all it makes poorly directed, flailing movements; does not participate normally in righting, feeding, or ambulation; and may be left for long periods of time in awkward positions of which the animals seem unaware. There is "a loss of projected movement into space."[44] These defects in voluntary movement are thought to result from the removal of afferent feedback from the limbs, providing information concerning position and movement, that plays upon the executant motor apparatus.

Isolated lesions of the DCN or the medial lemniscus rarely occur in humans and are difficult to produce experimentally. Local DCN lesions in monkeys produce losses similar to those just described, but no quantitative study of them has been made.[40]

Similarly, lesions rarely destroy the VB complex entirely without other damage. The destruction of this complex (e.g., by vascular occlusion), together with adjacent areas, may produce in addition to somesthetic defects that constellation of abnormal pain sensations termed the thalamic syndrome (Chapter 11). In stereotaxic explorations of the human thalamus, aimed at making lesions to alleviate intractable pain, smaller lesions within the VB complex have been made. They produce a profound contralateral mechanoreceptive sensory loss but little or no change in pain sensibility.

Defects in discriminative aspects of somesthesia produced by cerebral cortex lesions[23, 37, 93]

Lesions of postcentral sensory area. Removal of SI in monkeys produces sensory defects contralaterally. The defect in position sensibility is obvious as a lack of awareness of movement of the contralateral limbs and the frequent arrest of movements, always reluctantly begun, in awkward positions maintained for long periods. The animal responds to contralateral tactile, thermal, and noxious stimuli, though thresholds may be high and reaction to the last excessive. There is a profound loss of the discriminative aspects of somesthesia: gross impairment or loss of tactile shape, size, and roughness discrimination, which is never relearned. The contralateral placing reactions are permanently lost.

Cortical lesions limited to SI are rare in man. In patients in whom partial removals have been made to excise epileptogenic foci, the resulting sensory defects resemble those in the monkey.[44] Appreciation of contact, heat, cold, and noxious stimuli is preserved, although at elevated thresholds.[28] There are severe losses in the sense of position and movement of the contralateral limbs, tactile shape, and two-point discrimination. Large errors are made in point localization. These deficiencies duplicate in many ways those that follow lesions of the DCN or the VB complex of the thalamus. Removals of SII produce no detectable sensory abnormalities in man or monkey, either when its is removed alone or when its removal is added to removal of the postcentral gyrus.

Lesions of posterior parietal cortex. There is now good evidence that the regions of the parietal lobe and of the adjacent temporal area, which lie between the primary somesthetic, visual, and auditory cortices, are intimately concerned with higher order integrative aspects of sensation and perception. Indeed, their anatomic connections predict just such functions. Lesions confined to these areas may produce complex abnormalities of somatic sensation and of the image of the body form in man without major defects in the primary aspects of somesthesia, provided these latter are measured in simple tests. Circumscribed lesions of area 5 or area 7 in the monkey produce impairments in tactile size,

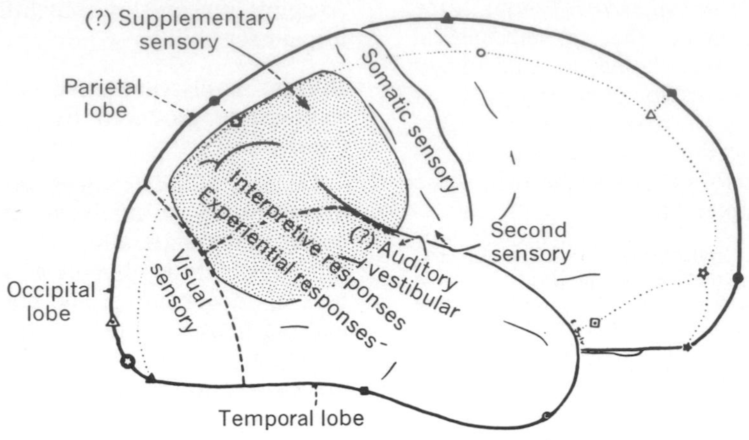

Fig. 10-27. Drawing of lateral surface of right cerebral hemisphere of man. Shaded area includes those posterior parietal and superior temporal cortical areas within which lesions produce disturbance in spatial orientation for contralateral half of personal and extrapersonal space. This disorder is termed *amorphosynthesis*. Normal perception of body form and its relation to extrapersonal space depends on integration in these areas of information arriving over several sensory channels: somatic, visual, vestibular, and auditory. When lesions of shaded area occur in dominant hemisphere, e.g., left in right-handed individuals, they will produce, in addition to amorphosynthesis, defects in recognition of symbols and use of symbolic concepts called agnosia and defects in use of language. Supplementary sensory area is indicated to lie on medial wall of hemisphere, auditory and vestibular cortices on inferior bank of sylvian fissure, and second somatic on its superior bank. (From Hecaen et al.[55])

shape, or roughness *discriminations,* but not in simple detection.[76, 176] The degree of sensory impairment in animals with lesions of the posterior parietal areas is a direct function of the complexity of the somesthetic task.

Such circumscribed lesions occur only rarely in humans, so that commonly disease or injury of the parietal lobe produces a number of defects. The typical parietal lobe syndrome is seen in man in its greatest purity after lesions of the nondominant hemisphere. This syndrome of *amorphosynthesis* includes a defective perception of the spatial aspects of sensations arising in the contralateral side of the body.[32, 34, 55] A patient with such a lesion may fail to dress or care for the contralateral side of his body; e.g., he may shave half his face, comb only half his hair, etc. He may even deny the presence of the contralateral half of his body altogether, or deny that illness exists when in fact it does. He may fail to complete the contralateral side of drawings or to construct correctly that side of three-dimensional forms he is asked to copy, e.g., with building blocks. There is in addition a perceptual rivalry, shown by the phenomena of extinction and proximal dominance.[16] Extinction exists when the two sides of the body are stimulated at homologous points; the patient may recognize only the stimulus opposite the normal hemisphere. Dominance refers to the fact that when both stimuli are delivered to the side opposite the diseased hemisphere the patient may recognize only the more proximal stimulus.

Such a patient is not disoriented in space and can recognize and name the body parts. The best studied cases are those in which precisely known lesions in the parietal cortex were made to remove epileptogenic foci (Fig. 10-27). The more dorsal the lesion, the more predominantly somesthetic the disorder, while more ventral lesions that include the marginal gyrus add disorders in the visual sphere as well. Lesions produced by disease frequently also include the postcentral gyrus, and such patients display mixtures of primary sensory loss and amorphosynthesis. Behavioral defects resembling amorphosynthesis are produced in monkeys by posterior parietal lesions.[176]

The purely contralateral disturbances classed together as amorphosynthesis rarely occur alone in lesions of the dominant hemisphere (i.e., the left in right-handed individuals). In patients with such lesions these signs are more commonly mixed with disorders in the formulation and use of concepts and confusion in naming objects, recognizing numbers, or following topographic schemes or plans. These together are called the agnosias, which are described in greater detail in Chapter 19.

In summary, it is evident that deficiencies in somesthesia and in behavior elaborated upon somesthetic input, which follow cortical lesions, occur at three successively more complex levels of neural organization: (1) lesions of the postcentral gyrus produce primary defects in the discriminative aspects of somesthesia on the opposite side of the body, (2) lesions of either parietal lobe may produce a strictly contralateral disorder of spatial orientation coupled with loss of discriminative capacities and a disorder of the concept of the body form, and (3) lesions of the parietal lobe of the dominant hemisphere in man may, in addition, produce agnosic syndromes.

Interhemispheric integration in somesthesia[3, 42]

Considerable attention has been given in recent years to the role of the great interhemispheric commissures in higher order integrative function, including perception. Animals have been studied following section of the commissures and, if vision is under study, the section of the optic chiasm in the sagittal plane. A number of humans in whom the commissures were sectioned to control the interhemispheric propagation of epileptic discharges have also been studied intensively. If monkeys or chimpanzees are trained to solve a series of somesthetic tasks of increasing complexity with one hand, the learning is almost totally and immediately transferred if the animal is then tested through the opposite, untrained hand. If the cerebral commissures are first sectioned and then training delivered to one hand, transfer to the other fails completely. If the commissural section is made after training one hand, there is partial retention of performance in the postoperative testing of the previously untrained hand. These observations are taken to mean (1) that in the normal animal the "memory trace" is either laid down bilaterally upon its formation or is readily available from the trained to the untrained hemisphere when the latter is tested and (2) that any residuum in the untrained hemisphere is insufficient to support criterion performance when, in the last experiment cited, testing is carried out through the untrained hand. A last important conclusion for somesthesis is that whatever the projection from hand or foot reaching the ipsilateral hemisphere via the anterolateral system, it does not provide sufficient information for even the simplest sensory discrimination.

REFERENCES

General reviews

1. Boring, E. G.: Sensation and perception in the history of experimental psychology, New York, 1942, Appleton-Century-Crofts.
2. Mountcastle, V. B.: The problem of sensing and the neural coding of sensory events. In Quarton, G. C., Melnechuk, T., and Schmitt, F. O., editors: The neurosciences. A study program, New York, 1967, The Rockefeller University Press.
3. Sperry, R. W.: Mental unity following surgical disconnection of the cerebral hemispheres, Harvey Lect. **62:**293, 1968.
4. Stevens, S. S.: Neural events and the psychophysical law, Science **170:**1043, 1970.
5. Werner, G.: The topology of the body representation in the somatic afferent pathway. In Schmitt, F. O., editor: The neurosciences. Second study program, New York, 1970, The Rockefeller University Press.
6. Werner, G., and Whitsel, B. L.: The somatic sensory cortex: functional organization. Handbook of sensory physiology. The somatosensory system, New York, 1972, Springer-Verlag New York, Inc.

Original papers

7. Abeles, M., and Goldstein, M. H.: Functional architecture in cat primary auditory cortex: columnar organization and organization according to depth, J. Neurophysiol. **33:**172, 1970.
8. Adrian, E. D.: Afferent discharges to the cerebral cortex from peripheral sense organs, J. Physiol. **100:**159, 1941.
9. Albe-Fessard, D., and Liebeskind, J.: Origines des messages somato-sensitifs activant les cellules du cortex moteur chez le singe, Exp. Brain Res. **1:**127, 1966.
10. Andersen, P., Andersson, S. A., and Landgren, S.: Some properties of the thalamic relay cells in the spino-cervico-lemniscal path, Acta Physiol. Scand. **68:**72, 1966.
11. Andersen, P., Eccles, J. C., and Sears, T. A.: Cortically evoked depolarization of primary afferent fibers in the spinal cord, J. Neurophysiol. **27:**63, 1964.
12. Andersen, P., Etholm, B., and Gordon, G.: Presynaptic and postsynaptic inhibition elicited in the cat's dorsal column nuclei by mechanical stimulation of skin, J. Physiol. **210:**433, 1970.
13. Andersen, P., Eccles, J. C., Schmidt, R. F., and Yokota, T.: Depolarization of presynaptic fibers in the cuneate nucleus, J. Neurophysiol. **27:**92, 1964.
14. Andersen, P., Eccles, J. C., Schmidt, R. F., and Yokota, T.: Identification of relay cells and interneurones in the cuneate nuclei, J. Neurophysiol. **27:**1080, 1964.
15. Asanuma, H., Stoney, S. D., Jr., and Abzug, C.: Relationship between afferent input and motor outflow in cat motorsensory cortex, J. Neurophysiol. **31:**670, 1968.

16. Bender, M. B.: Disorders in perception, Springfield, Ill., 1952, Charles C Thomas, Publisher.
17. Bertrand, G., Jasper, H., and Wong, A.: Microelectrode study of the human thalamus: functional organization of the ventrobasal complex, Confin. Neurol. suppl. **29:**81, 1967.
18. Bertrand, G., Jasper, H., Wong, A., and Mathews, G.: Microelectrode recording during stereotactic surgery, Clin. Neurosurg. **16:** 328, 1969.
19. Bignall, K. E.: Auditory input to frontal polysensory cortex of the squirrel monkey: possible pathways, Brain Res. **19:**77, 1970.
20. Bignall, K. E., and Imbert, M.: Polysensory and cortico-cortical connections to frontal lobe of squirrel and rhesus monkey, Electroencephalogr. Clin. Neurophysiol. **26:**206, 1969.
21. Boivie, J.: The termination of the spinothalamic tract in the cat. An experimental study with silver impregnation methods, Exp. Brain Res. **12:**331, 1971.
22. Brooks, V. B., and Stoney, S. D., Jr.: Motor mechanisms: the role of the pyramidal system in motor control, Ann. Rev. Physiol. **33:**337, 1971.
23. Buser, P., and Bignall, K. E.: Nonprimary sensory projections on the cat neocortex, Int. Rev. Neurobiol. **10:**111, 1967.
24. Carmon, A.: Stimulus contrast in tactile resolution, Percept. Psychophys. **3:**241, 1968.
25. Carreras, M., and Andersson, S. A.: Functional properties of neurons of the anterior ectosylvian gyrus of the cat, J. Neurophysiol. **26:**100, 1963.
26. Clark, W. E. L.: The termination of the ascending tracts in the thalamus of the macaque monkey, J. Anat. **71:**7, 1936.
27. Colonnier, M. L.: Structural design of the neocortex. In Eccles, J. C., editor: Brain and conscious experience, Berlin, 1966, Springer Verlag.
28. Corkin, S., Milner, B., and Rasmussen, T.: Effects of different cortical excisions on sensory thresholds in man, Trans. Am. Neur. Assoc. **89:**112, 1964.
29. Critchley, M.: Tactual thought with special reference to the blind, Brain **76:**19, 1953.
30. Critchley, M.: The parietal lobes, Baltimore, 1953, The Williams & Wilkins Co.
31. Darian-Smith, I.: The trigeminal system. In Iggo, A., editor: Handbook of sensory physiology. The somatic afferent system, New York, 1972, Springer-Verlag New York, Inc.
32. Denny-Brown, D., and Banker, B.: Amorphosynthesis from left parietal lesion. Arch. Neurol. Psychiatry **71:**302, 1954.
33. Denny-Brown, D., and Chambers, R. A.: The parietal lobes and behavior, Res. Publ. Assoc. Res. Nerv. Ment. Dis. **36:**35, 1958.
34. Denny-Brown, D., Meyer, J. S., and Horenstein, S.: The significance of perceptual rivalry resulting from parietal lesion, Brain **75:**433, 1952.
35. Dormont, J. F., and Massion, J.: Duality of cortical control of ventrolateral thalamic activity, Exp. Brain Res. **10:**205, 1970.
36. Eccles, J. C., Schmidt, R. F., and Willis, W. D.: Depolarization of the central terminals of cutaneous afferent fibers, J. Neurophysiol. **26:**646, 1963.
37. Ettlinger, G., and Kalsbeck, J. E.: Changes in tactile discrimination and in visual reaching after successive and simultaneous bilateral posterior parietal ablations in the monkey, J. Neurol. Neurosurg. Psychiatry **25:**256, 1962.
38. Fennell, E., Satz, P., and Wise, R.: Laterality differences in the perception of pressure, J. Neurol. Neurosurg. Psychiatry **30:**337, 1967.
39. Ferraro, A., and Barrera, S. E.: Effects of experimental lesions on the posterior columns in macacus rhesus monkeys, Brain **57:**307, 1934.
40. Ferraro, A., and Barrera, S. E.: The effect of lesions of the dorsal column nuclei in the macacus rhesus, Brain **59:**76, 1936.
41. Fetz, E. E.: Pyramidal tract effects on interneurons in the cat lumbar dorsal horn, J. Neurophysiol. **31:**69, 1968.
42. Gazzaniga, M. S.: The bisected brain, New York, 1970, Appleton-Century-Crofts.
43. Gelfan, S., and Carter, S.: Muscle sense in man, Exp. Neurol. **18:**469, 1967.
44. Gilman, S., and Denny-Brown, D.: Disorders of movement and behaviour following dorsal column lesions, Brain **89:**397, 1966.
45. Glees, P., and Soler, J.: Fibre content of the posterior column and synaptic connection of nucleus gracilis, Z. Zellforsch. Mikrosk. Anat. **36:**381, 1951.
46. Glees, O., Livingston, R. B., and Soler, J.: Der intraspinal Verlauf und die Endigungen der sensorischen Wurzeln in den Nucleus gracilis und cuneatus, Arch. Psychiatr. Nervenkr. **187:**190, 1951.
47. Goldring, S., Aras, E., and Weber, P. C.: Comparative study of sensory input to motor cortex in animals and man, Electroencephalogr. Clin. Neurophysiol. **29:**537, 1970.
48. Goto, A., et al.: Thalamic potentials from muscle afferents in the human, Arch. Neurol. **19:**302, 1968.
49. Grigg, P., Finerman, G. A., and Riley, L. H.: Joint position sense after total hip replacement, 1972, Orthopedic Research Society.
50. Guillery, R. W., Adrian, H. O., Woolsey, C. N., and Rose, J. E.: Activation of somatosensory areas I and II of cat's cerebral cortex by focal stimulation of the ventrobasal complex. In Purpura, D. P., and Yahr, M. D., editors: The thalamus, New York, 1966, Columbia University Press.
51. Hagbarth, K. E.: Centrifugal mechanisms of sensory control, Ergeb. Biol. **22:**47, 1960.
52. Harrington, T., and Merzenich, M. M.: Neural coding in the sense of touch: human sensations of skin indentation compared with the responses of slowly adapting mechanoreceptive afferents innervating the hairy skin of monkeys, Exp. Brain Res. **10:**251, 1970.
53. Hassler, R., Anatomy of the thalamus. In Shaltenbrand, G., and Bailey, P., editors: Introduction to sterotaxis with an atlas of the human brain, Stuttgart, 1959, Georg Thieme, vol. 1.
54. Hassler, R.: Thalamic regulation of muscle tone and speed of movement. In Purpura,

D. P., and Yahr, M. D., editors: The thalamus, New York, 1966, Columbia University Press.
55. Hecaen, H., Penfield, W., Bertrand, C., and Malmo, R.: The syndrome of apractognosia due to lesions of the minor cerebral hemisphere, Arch. Neurol. Psychiatry **75**:400, 1956.
56. Hubel, D. H., and Wiesel, T. N.: Receptive fields and functional architecture of monkey striate cortex, J. Physiol. **195**:215, 1968.
57. Hubel, D. H., and Wiesel, T. N.: Anatomical demonstration of columns in the monkey striate cortex, Nature **221**:747, 1969.
58. Jasper, H. H., and Bertrand, G.: Thalamic units involved in somatic sensation and voluntary and involuntary movements in man. In Purpura, D. P., and Yahr, M. D., editors: The thalamus, New York, 1966, Columbia University Press.
59. Jones, E. G., and Powell, T. P. S.: Connexions of the somatic sensory cortex of the rhesus monkey. I. Ipsilateral connections, Brain **92**:477, 1969.
60. Jones, E. G., and Powell, T. P. S.: Connexions of the somatic sensory cortex of the rhesus monkey. II. Contralateral connections, Brain **92**:717, 1969.
61. Jones, E. G., and Powell, T. P. S.: An electron microscopic study of the mode of termination of cortico-thalamic fibres within the sensory relay nuclei of the thalamus, Proc. R. Soc. Lond. (Biol.) **172**:173, 1969.
62. Jones, E. G., and Powell, T. P. S.: Connexions of the somatic sensory cortex of the rhesus monkey. III. Thalamic connexions, Brain **93**:37, 1970.
63. Jones, E. G., and Powell, T. P. S.: An electron microscopic study of the laminar pattern and mode of termination of afferent fibre pathways in the second somatic cortex of the cat, Philos. Trans. R. Soc. Lond. (Biol. Sci.) **257**: 45, 1970.
64. Jones, E. G., and Powell, T. P. S.: An anatomical study of converging sensory pathways within the cerebral cortex of the monkey, Brain **93**:793, 1970.
65. Kesten, W.: Die Grenzen der haptischen Leitungsfahigkeit, Z. Biol. **109**:24, 1956.
66. Kuypers, H. G. J. M.: The descending pathways to the spinal cord, their anatomy and function, Progr. Brain Res. **11**:178, 1964.
67. Kuypers, H. G. J. M., and Tuerk, J. D.: The distribution of cortical fibres within the nuclei cuneatus and gracilis in the cat, J. Anat. **98**: 143, 1964.
68. Lende, R. A.: A comparative approach to the neocortex: localization in monotremes, marsupials, and insectivores, Ann. N. Y. Acad. Sci. **167**:262, 1969.
69. Levitt, M., Carrerras, M., Liu, C., and Chambers, W.: Pyramidal and extrapyramidal modulation of somato-sensory activity in gracile and cuneate nuclei, Arch. Ital. Biol. **102**:197, 1964.
70. Lorente de Nó, R.: Cerebral cortex: architecture, intracortical connections, motor projections. In Fulton, J. F., editor: Physiology of the nervous system, London, 1949, Oxford University Press, ed. 3.
71. Manson, J.: The somatosensory cortical projection of single nerve cells in the thalamus of the cat, Brain Res. **12**:489, 1969.
72. Marshall, W. H., and Talbot, S. A.: Recent evidence for neural mechanisms in vision leading to a general theory of sensory acuity, Biol. Symp. **7**:117, 1942.
73. Mehler, W. R.: Some observations on secondary ascending afferent systems in the central nervous system. In Knighton, R. S., and Dumke, P. R., editors: Pain, Boston, 1966, Little, Brown & Co.
74. Mehler, W. R.: Some neurological species differences—a posteriori, Ann. N. Y. Acad. Sci. **167**:424, 1969.
75. Merton, P. A.: Human position sense and the sense of effort, Symp. Soc. Exp. Biol. **18**: 387, 1964.
76. Moffett, A., and Ettlinger, G.: Tactile discrimination performance in the monkey: the effect of unilateral posterior parietal ablations, Cortex **6**:47, 1970.
77. Moffett, A., Ettlinger, G., Morton, H. B., and Percy, M. F.: Tactile discrimination performance in the monkey: the effect of ablation of various subdivisions of the posterior parietal cortex, Cortex **3**:59, 1967.
78. Morin, F.: A new spinal pathway for cutaneous impulses, Am. J. Physiol. **183**:245, 1955.
79. Morrison, A. R., Hand, P. J., and O'Donoghue, J.: Contrasting projections from the posterior and ventrobasal thalamic nuclear complexes to the anterior ectosylvian gyrus of the cat, Brain Res. **21**:115, 1970.
80. Mountcastle, V. B.: Modality and topographic properties of single neurons of cat's somatic sensory cortex, J. Neurophysiol. **20**:408, 1957.
81. Mountcastle, V. B.: Some functional properties of the somatic afferent system. In Rosenblith, W. A., editor: Sensory communication, Cambridge, Mass., 1961, The M. I. T. Press.
82. Mountcastle, V. B., and Henneman, E: The representation of tactile sensibility in the thalamus of the monkey, J. Comp. Neurol. **97**:409, 1952.
83. Mountcastle, V. B., and Powell, T. P. S.: Central nervous mechanisms subserving position sense and kinesthesis, Johns Hopkins Med. J. **105**:173, 1959.
84. Mountcastle, V. B., and Powell, T. P. S.: Neural mechanisms subserving cutaneous sensibility, with special reference to the role of afferent inhibition in sensory perception and discrimination, Bull. Johns Hopkins Hosp. **105**:201, 1959.
85. Mountcastle, V. B., Carli, G., and LaMotte, R. H.: Central neural mechanisms in the detection of peripheral events: simultaneous measurement of behavioral thresholds in monkeys and activity evoked in postcentral neurons by the stimuli detected, Unpublished paper.
86. Mountcastle, V. B., LaMotte, R. H., and Carli, G.: Detection thresholds for vibratory stimuli in humans and monkeys; comparison with threshold events in mechanoreceptive first order afferent nerve fibers innervating monkey hands, J. Neurophysiol. **35**:122, 1972.
87. Mountcastle, V. B., Poggio, G. F., and Wer-

ner, G.: The relation of thalamic cell response to peripheral stimuli varied over an intensive continuum, J. Neurophysiol. **26:**807, 1963.
88. Mountcastle, V. B., Poggio, G. F., and Werner, G.: The neural transformation of the sensory stimulus at the cortical input level of the somatic afferent system. In Gerard, R., editor: Information Processing in the Nervous System, Amsterdam, 1964, Excerpta Medica Foundation.
89. Mountcastle, V. B., Talbot, W. H., and Kornhuber, H. H.: The neural transformation of mechanical stimuli delivered to the monkey's hand. In de Reuck, A. V. S., and Knight, J., editors: Ciba Foundation symposium on touch, heat and pain, London, 1966, J. & A. Churchill, Ltd.
90. Mountcastle, V. B., Talbot, W. H., Sakata, H., and Hyvarinen, J.: Cortical neuronal mechanisms in flutter-vibration studied in unanesthetized monkeys. Neuronal periodicity and frequency discrimination, J. Neurophysiol. **32:**452, 1969.
91. Nyberg-Hansen, R.: Functional organization of descending supraspinal fibre systems to the spinal cord. Anatomical observations and physiological correlations, Ergeb. Anat. Entwicklungsgesch. **39:**6, 1966.
92. Olszewski, J.: On the anatomical and function organization of the spinal trigeminal nucleus, J. Comp. Neurol. **92:**401, 1950.
93. Orbach, J., and Chow, K. L.: Differential effects of resection of somatic areas I and II in monkeys, J. Neurophysiol. **22:**195, 1959.
94. Oscarsson, O.: Functional organization of the spino- and cuneocerebellar tracts, Physiol. Rev. **45:**495, 1965.
95. Oswaldo-Cruz, E., and Kidd, C.: Functional properties of neurons in the lateral cervical nucleus of the cat, J. Neurophysiol. **27:**1, 1964.
96. Pandya, D. C., and Kuypers, H. G. J. M.: Cortico-cortical connections in the rhesus monkey, Brain Res. **13:**13, 1969.
97. Pandya, D. N., and Vignolo, L. A.: Interhemispheric neocortical projections of somatosensory areas I and II in the rhesus monkey, Brain Res. **7:**300, 1968.
98. Paul, R. L., Merzenich, M., and Goodman, H.: Representation of slowly and rapidly adapting cutaneous mechanoreceptors of the hand in Brodmann's areas 3 and 1 of Macaca mulatta, Brain Res. **36:**229, 1972.
99. Penfield, W., and Jasper, H.: Epilepsy and the functional anatomy of the human brain, Boston, 1954, Little, Brown & Co.
100. Perl, E., and Whitlock, D. G.: Somatic stimuli exciting spinothalamic projections to thalamic neurons in cat and monkey, Exp. Neurol. **3:**256, 1961.
101. Phillips, C. G., Powell, T. P. S., and Wiesendanger, M.: Projection from low-threshold muscle afferents of hand and forearm to area 3a of baboon's cortex, J. Physiol. **217:**419, 1971.
102. Poggio, G. F., and Mountcastle, V. B.: A study of the functional contributions of the lemniscal and spinothalamic systems to somatic sensibility. Central nervous mechanisms in pain, Bull. Johns Hopkins Hosp. **106:**266, 1960.
103. Poggio, G. F., and Mountcastle, V. B.: The functional properties of ventrobasal thalamic neurons studied in unanesthetized monkeys, J. Neurophysiol. **26:**775, 1963.
104. Pompeiano, O., and Swett, J. E.: Identification of cutaneous and muscular afferent fibers producing EEG synchronization or arousal in normal cats, Arch. Ital. Biol. **100:**343, 1962.
105. Powell, T. P. S., and Mountcastle, V. B.: Some aspects of the functional organization of the cortex of the postcentral gyrus of the monkey: a correlation of findings obtained in a single unit analysis with cytoarchitecture, Bull. Johns Hopkins Hosp. **105:**133, 1959.
106. Ringel, R. L.: Oral region two-point discrimination in normal and myopathic subjects. In Bosma, J. F., editor: Oral sensation and perception, Springfield, Ill., 1970, Charles C Thomas, Publisher.
107. Rose, J. E., and Mountcastle, V. B.: Touch and kinesthesis. In Magoun, H. W., editor, Neurophysiology section: Handbook of physiology, Baltimore, 1959, The Williams & Wilkins Co., vol. 1.
108. Rosen, I.: Afferent connexions to group I activated cells in the main cuneate nucleus of the cat, J. Physiol. **205:**209, 1969.
109. Rosen, I.: Excitation of group I activated thalamocortical relay neurones in the cat, J. Physiol. **205:**237, 1969.
110. Rosen, I.: Localization in caudal brain stem and cervical spinal cord of neurons activated from forelimb group I afferents in the cat, Brain Res. **16:**55, 1969.
111. Schwartzman, R. J., and Bogdonoff, M. D.: Proprioception and vibration sensibility discrimination in the absence of the posterior columns, Arch. Neurol. **20:**349, 1969.
112. Shende, M. C., Stewart, D. H., Jr., and King, R. B.: Projections from the trigeminal nucleus caudalis in the squirrel monkey, Exp. Neurol. **20:**655, 1968.
113. Shimazu, H., Yanagisawa, N., and Garoutte, B.: Cortico-pyramidal influences on thalamic somatosensory transmission in the cat, J. Physiol. **15:**101, 1965.
114. Stevens, S. S.: Tactile vibration: change of exponent with frequency, Percept. Psychophys. **3:**223, 1968.
115. Stevens, S. S.: On predicting exponents for cross-modality matches, Percept. Psychophys. **6:**251, 1969.
116. Stewart, W. A., and King, R. N.: Fiber projections from the nucleus caudalis of the trigeminal nucleus, J. Comp. Neurol. **121:**271, 1963.
117. Swett, J. E., and Bourassa, C. M.: Comparison of sensory discrimination thresholds with muscle and cutaneous nerve volleys in the cat, J. Neurophysiol. **30:**530, 1967.
118. Talbot, W. H., Darian-Smith, I., Kornhuber, H. H., and Mountcastle, V. B.: The sense of flutter-vibration: comparison of the human capacity with response patterns of mechanoreceptive afferents from the monkey hand, J. Neurophysiol. **31:**301, 1968.
119. Tasker, R. B.: Thalamotomy for pain: lesion

localization by detailed thalamic mapping, Can. J. Surg. **12**:62, 1969.

120. Taub, A., and Bishop, P. O.: The spinocervical tract: dorsal column linkage, conduction velocity, primary afferent spectrum, Exp. Neurol. **13**:1, 1965.
121. Towe, A. L., and Jabbur, S. J.: Cortical inhibition of neurons in dorsal column nuclei of cat, J. Neurophysiol. **24**:488, 1961.
122. Truex, R. C., Taylor, M. J., Smythe, M. Q., and Gildenberg, P. L.: The lateral cervical nucleus of cat, dog, and man, J. Comp. Neurol. **139**:93, 1970.
123. Verrillo, R. T., Fraioli, A. J., and Smith, R. L.: Sensation magnitude of vibrotactile stimuli, Percept. Psychophys. **6**:366, 1969.
124. Wall, P. D.: The laminar organization of dorsal horn and effects of descending impulses, J. Physiol. **188**:403, 1967.
125. Weinstein, S.: Intensive and extensive aspects of tactile sensitivity as a function of body part, sex and laterality. In Kenshalo, D. R., editor: The skin senses, Springfield, Ill., 1968, Charles C Thomas, Publisher.
126. Werner, G., and Mountcastle, V. B.: Neural activity in mechanoreceptive afferents: stimulus-response relations, Weber functions, and information transmission, J. Neurophysiol. **28**: 359, 1965.
127. Werner, G., and Whitsel, B. L.: The topology of dermatomal projection in the medial lemniscal system, J. Physiol. **192**:123, 1967.
128. Werner, G., Whitsel, B. L., and Petrucelli, M.: Data structure and algorhythms in the primate somatosensory cortex. In Karczmer, A. G., and Eccles, J. C., editors: Brain and human behavior, Berlin, 1972, Springer Verlag.
129. Whitlock, D. G., and Perl, E.: Thalamic projections of spinothalamic pathways in monkey, Exp. Neurol. **3**:240, 1961.
130. Whitsel, B. L., Dreyer, D. A., and Roppolo, J. R.: Determinants of the body representation in the postcentral gyrus of macaques, J. Neurophysiol. **34**:1018, 1971.
131. Whitsel, B. L., Petrucelli, L. M., and Sapiro, G.: Modality representation in the lumbar and cervical fasciculus gracilis of squirrel monkeys, Brain Res. **15**:67, 1969.
132. Whitsel, B. L., Petrucelli, L. M., and Werner, G.: Symmetry and connectivity in the map of the body surface in somatosensory area II of primates, J. Neurophysiol. **32**:170, 1969.
133. Whitsel, B. L., Petrucelli, L. M., Ha, H., and Dreyer, D.: Modality representation and fibre sorting in the fasciculus gracilis of squirrel monkeys, Exp. Neurol. **29**:227, 1970.
134. Winter, D. L.: N. gracilis of cat. Functional organization and corticofugal effects, J. Neurophysiol. **28**:48, 1965.
135. Woolsey, C. N.: Patterns of sensory representation in the cerebral cortex, Fed. Proc. **6**: 437, 1947.
136. Woolsey, C. N.: Organization of somatic sensory and motor areas of the cerebral cortex. In Harlow, H. F., and Woolsey, C. N., editors: Biological and biochemical bases of behavior, Madison, 1958, University of Wisconsin Press.

11

VERNON B. MOUNTCASTLE

Pain and temperature sensibilities

Pain is that sensory experience evoked by stimuli that injure or threaten to destroy tissue, defined introspectively by every man as that which hurts. Pain is nevertheless such a universal experience of everyday life that verbal descriptions of it provide recognizable signals from one person to another concerning its presence, nature, intensity, duration, location, reference, and temporal course. Indeed, descriptions of particular pains are telltale signs to the inquiring physician of the nature and extent of the pathologic processes that arouse them.

Pain is composed, first, of a separate and distinct sensation and, second, of the individual's *reaction to pain,* with accompanying emotional overtones, widespread reflected activity in both somatic and autonomic effectors, and volitional efforts of avoidance or escape. This reaction, akin to suffering and related to the sometimes life-threatening nature of the experience, differs widely among persons, being influenced by age, sex, and race, by the nature, duration, and intensity of the pain, and above all by the personality of the sufferer. Pain, moreover, is commonly produced by stimuli that at weaker intensities evoke other somatic sensations, e.g., warmth, cold, or mechanical contacts. Experimentally, pain can rarely be produced in isolation.

All this has led investigators to devise methods for standardizing stimuli reported to be painful and for quantifying the reports of the subjective experiences evoked by them. Human subjects can concentrate upon the sensation itself and describe it accurately, ignoring accompanying nonpainful sensations and controlling their feeling states, i.e., their reactions to pain.[2] Results obtained this way, however, cannot be transferred directly to the case of a patient suffering pain produced by injury or disease. Attempts to quantify this latter experience have frequently been cast in terms of the amounts of narcotics or analgesics required to reduce the patient's suffering, as gauged by his verbal reports and by observation of his discomfort.[13] While measures of therapeutic effectiveness, these are not direct measures of pain.

Until recently, most information concerning pain has come from the study of human subjects. Neurophysiologic studies in animals, however, give promise of important results based on the assumption that stimuli that evoke pain in human beings also evoke in the animal a subjective sensory experience of the same qualitative nature, i.e., avoidance, escape, signs of anger, attack, and the correlated circulatory and respiratory reflex changes characteristic of pain in man. To reject this assumption renders the subject of pain forever insulated from experimental techniques applicable in their full power only in animals.

The weight of evidence indicates that pain is a separate and identifiable sensation with its own first-order nerve fibers. Uncertainty as to the central nervous pathways activated by impulses in nociceptive afferents is not regarded as a sign that pain involves any unique or mystical operations of the nervous system unknown in the neural mechanisms of other sensory phenomena.

HUMAN PERCEPTION OF PAIN

Pains are described by a variety of words, and many problems and disagreements about pain are semantic in nature.[84] As regards the qualities of pains, however, there are at least three on which most observers agree. The first is the bright, pricking pain so readily evoked by a brisk needle jab of the skin. It is accurately localized, subsides quickly, and is so much less potent in evoking that emotional overtone common with other pains that, of all, it is the most readily studied in a quantitative manner. It is first or "fast" pain, and evidence will be presented to show that it depends on activation of a certain set of delta fibers in peripheral nerves. The second is burning pain, a sensation of slower onset, greater persistence, and less certainly localized, which may continue as an "afterimage" for many seconds after removal of the provoking stimulus. It is this pain that is so difficult to endure and that readily evokes the cardiovascular and respiratory reflexes characteristic of pain. This second or "slow" pain is produced by activation of a certain set of unmyelinated fibers (C fibers) in peripheral nerves. Thus pain from the skin is dual in nature; the two qualities are subjectively distinguishable and quite independent of the duration of the noxious stimulus.[74]

Visceral and somatic deep structures, when noxiously stimulated, give rise to pain most commonly described as aching, sometimes with an additional quality of burning. This aching pain is of great importance for the physician, for it commonly signals life-threatening disturbances of vital organs. It is just this pain that is most difficult to define precisely, for the location of the structures concerned makes experimental observation difficult, and naturally occurring pains of visceral origin are frequently referred to sites distant from the actual location of this disturbing process (p. 368).

Each of the qualities of cutaneous pain has been brought under experimental observation, and to a lesser extent so has aching pain of visceral origin, but with varying degrees of success. It is proper next to consider the experimental results and the methods used to obtain them.

Adequate stimuli for pain and the human threshold for perception of pain[50]

In contrast with other sensations, the application of a pain-provoking stimulus commonly produces local tissue changes such as erythema of the skin, and continued noxious stimulation alters the tissue and its function. This leads to the suggestion that *the adequate stimulus for pain is the rate of destruction of a tissue innervated by pain fibers,* including changes short of cell rupture. The importance of rate of change in reaching the pain threshold is shown by the strength-duration curve in Fig. 11-3. When this fact is taken into account, no mystery surrounds the observation that even large and mutilating wounds may, in some instances and at some time after their infliction, be nearly painless.[12] The actual transection of some pain fibers innervating the injured area, coupled with a drop in the rate of tissue destruction, may decrease pain input to tolerable levels.

Pain can be evoked by energy in any one of four forms: electrical, mechanical, chemical, or thermal. Electrical stimulation of nerves is useful in determining which peripheral nerve fibers will elicit pain when active and in determining certain temporal aspects of pain. Pain is readily elicited by pressure upon the skin. Mechanical stimuli have been used successfully in the study of visceral pain, e.g., by distention of a hollow viscus with inflatable balloons. Chemical agents have been used in attempts to identify neurohumoral agents that may excite pain endings (p. 357). Of all the methods used experimentally to produce pain, that of thermal radiation has been of the greatest value, particularly in the elegant studies of Hardy et al.,[2] who have brought it to a high state of precision.

The dolorimeter designed by Hardy consists of a heat source (a lamp) and a lens condensing system that produce an even distribution of heat at the surface of the subject's skin. Shutter and aperture diaphragm allow control of the duration of the stimulus and the area of the skin exposed

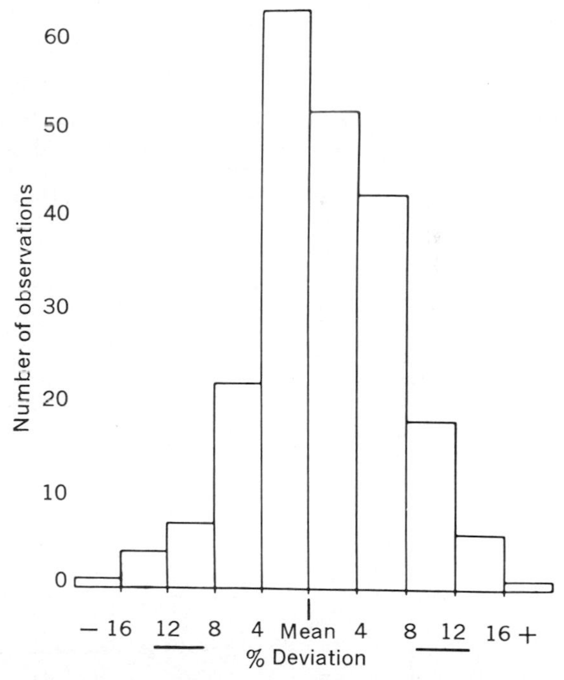

Fig. 11-1. Distribution of pain thresholds measured by thermal radiation method in large population of individuals. Threshold is taken as that intensity of thermal radiation delivered to blackened skin of forehead that produces trace of pricking pain at end of 3 sec period of exposure. It is expressed as rate of heat transfer: mcal/sec/cm²; for this population the mean is 206 ± SD = 21 mcal/sec/cm² and corresponds to skin temperature at end of 3 sec period of about 45° C. (From Hardy et al.[2])

to thermal radiation. The intensity is varied by varying the voltage on the lamp; radiometric calibrations of the instrument allow designation of the stimuli in terms of heat transfer per unit area of skin: milligram-calories per second per square centimeter (mcal/sec/cm²). Blackening the skin surface raises heat transfer to better than 95%, makes it independent of the degree of natural pigmentation, and prevents penetration of the rays to deeper layers of the skin. Skin temperature is measured radiometrically or with a small bead thermister. Successful use of the method developed by Hardy et al. requires an understanding on the part of the subject of the pain end point sought, which must be distinguished from accompanying sensations of warmth and heat. For the majority of their studies these investigators chose as end point a small but distinct and sharp stab of pricking pain occurring just at the end of exposure to thermal radiation for 3 sec.

The histogram in Fig. 11-1 shows the distribution of pricking pain thresholds measured on the skin of the forehead in a large number

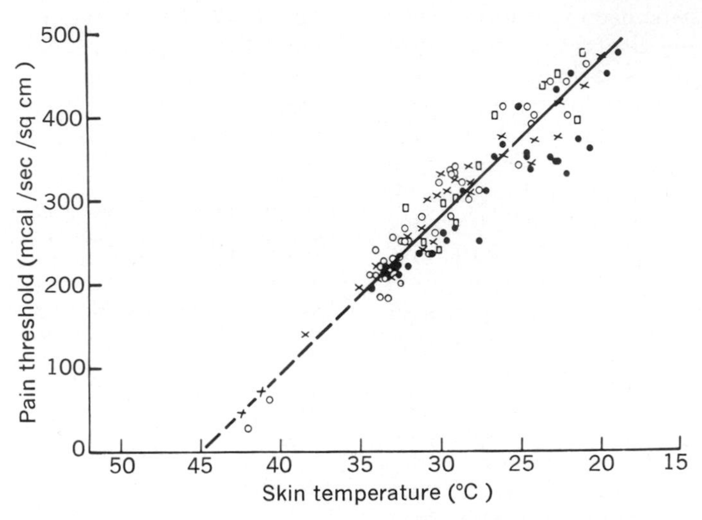

Fig. 11-2. Relation of initial temperature of skin of forehead (site of stimulation) to rate of heat transfer required to elicit threshold pricking pain, studied in three normal subjects. Linear relation intercepts abscissa at about 45° C, a skin temperature that, if continued, will produce irreversible skin damage. (From Hardy et al.[2])

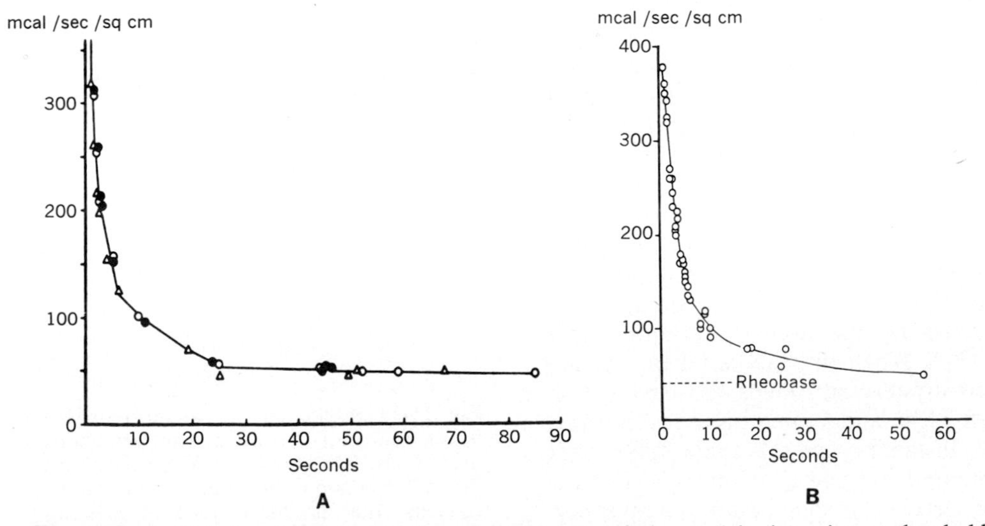

Fig. 11-3. A, Relation between duration of thermal irradiation and its intensity, at threshold for pricking pain, in three normal subjects. **B,** Relation between duration of thermal irradiation and intensity required to elicit minimal flexion reflex in human being with complete transection of spinal cord. Stimulus was delivered to skin of foot innervated by segments of spinal cord below transection. (**A** from Hardy et al.[2]; **B** from Hardy.[49])

of instructed but untrained individuals who varied in age (14 to 70 years),[106] sex, race, and occupation. The mean value is 206 ± SD = 21 mcal/sec/cm². With training this variability can be reduced by one-half, but when uninstructed subjects are used, as in the series studied by Chapman and Jones,[28] it may double. Presumably this is due to the choice by some of burning or aching pain and by others of the reaction to pain as the end point for report. The skin of the forehead was chosen for stimulation in these experiments because in ordinary room environments its temperature is remarkably constant at about 34° C. The graph in Fig. 11-2 shows that skin temperature is an important variable in such measurements; the threshold for thermally evoked pricking pain is the rate of heat transfer to the skin that exceeds the rate of heat loss by an amount just sufficient to drive the skin temperature to about 45° C within any given period of time. Indeed, pain occurs spontaneously at skin temperatures of about 44° to 45° C, regardless of previous thermal history. The precision of this end point is shown by the strength-duration curve in Fig. 11-3, *A*. The pricking pain threshold measured by thermal radiation varies from place to place on the body surface, but Hardy et al.[2] found that for every locus tested the final skin temperature at threshold fell within a degree or so of 45° C.

A skin temperature of about 45° C is therefore the one invariant requirement for eliciting pricking pain by thermal irradiation; if maintained for a number of hours, it will lead to epidermal necrosis.[87] The difference in time scale is important, for pain and the associated avoidance reactions are elicited quickly at 45° C, so that irreversible tissue damage does not occur.

The constancy of the pain threshold in terms of skin temperature raises the question of whether the threshold for the reflex avoidance reactions to pain is similar. Hardy[49] studied a patient who had survived for many years a complete transection of the spinal cord in the thoracic region. The pricking pain threshold on the normally innervated skin of the dorsum of the hand was found to be 204 ± 20 mcal/ sec/cm², equivalent to a final skin temperature at threshold of 43.6° ± 1.1° C, well within the normal range. The stimulus was then shifted to the skin of the dorsum of the foot, a region linked only with the isolated segments of the spinal cord. A strength-duration curve for stimuli just sufficiently intense to evoke the first sign of flexion withdrawal is shown in Fig. 11-3, *B*. The skin temperature reached at the thresholds for stimuli of different durations was 44.1° ± 0.75° C, a value not distinguishable from the pain threshold in the same individual. The suggestion is that the two thresholds may be identical, although that for reflex withdrawal, normally under descending control, may vary from time to time in the normal individual.

Burning pain is a cutaneous sensation qualitatively different from pricking pain, and evidence adduced from blocking experiments (Chapter 9) indicates that it is served by unmyelinated or C fibers, a proposition for which further evidence will be given. It is this burning pain that is so prominent a feature of pain produced by disease or injury arising from the skin and skin deprived of its myelinated fiber innervation. The threshold for burning pain, when evoked by the method of thermal radiation, is some 20 to 40 mcal/ sec/cm² *below* that for pricking pain, and the variation among individuals is somewhat greater.

Aching pain arises from deeper tissues, in particular the viscera, the periosteum, and the joints and the tissues that surround them. It can be evoked by thermal radiation sufficiently intense to raise the temperature of deeper tissues; its threshold is in the range of 300 to 320 mcal/sec/cm². When evoked by mechanical pressure upon the forehead, delivered via a blunt probe, the threshold was found to be 550 ± SD = 130 gm.[2] Pains that arise from the viscera and from the body cavities and the special quality of pain called headache will be discussed in later sections.

Factors influencing the pain threshold

Under the conditions just described the threshold for pain induced by thermal radiation is not affected by age, sex, fatigue, or the minor alterations of mood that all regard as normal. It is affected by the local conditions of the area stimulated, particularly by those that affect the transfer of heat to or its dissipation from the skin: its absorbent capacity, resting temperature, wetness or dryness, etc. The threshold of the skin is also influenced by its immediate past history: the number, intensities, and lengths of previous stimuli and the interval of time since the last. These latter factors are related to the appearance of primary hyperalgesia (p. 357), which is accompanied by a lowering of threshold

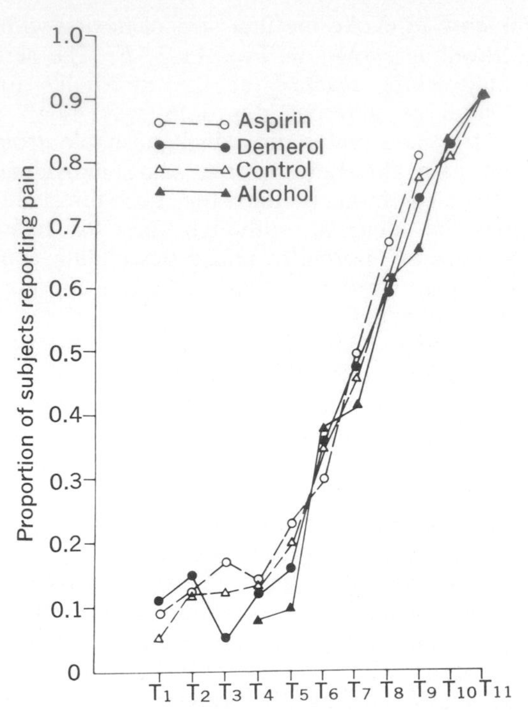

Fig. 11-4. Distribution functions of pain thresholds for a number of subjects after administration of no drug, after analgesic aspirin, after narcotic Demerol, and after ethyl alcohol. Functions plot percentage of positive identifications of pain when stimuli of each intensity were delivered for a number of times. Stimuli were thermal irradiations and are plotted here as temperature of irradiated skin reached at end of 3 sec period of exposure. $T_1 = 39.1°$ C, $T_2 = 40.0°$ C, $T_3 = 40.7°$ C, $T_4 = 41.4°$ C, $T_5 = 42.2°$ C, $T_6 = 43.0°$ C, $T_7 = 43.7°$ C, $T_8 = 44.5°$ C, $T_9 = 45.3°$ C, $T_{10} = 46.0°$ C, $T_{11} = 46.8°$ C. (From Chapman et al.[25])

for pain. Even in trained subjects the response bias may be influenced by distraction or suggestion, e.g., by the administration of a drug with the information that an analgesic has been given when in fact it has not (the placebo effect).[30] Pain elsewhere in the body regularly elevates threshold at a test site.[14] Most impressive concerning the complexity of central mechanisms is the fact that some individuals may be rendered completely analgesic by suggestion given during hypnotic trance.[71] When all these factors are controlled or avoided, however, the threshold for pain determined by the method of thermal radiation is a stable physiologic quantity. It is nevertheless a statistical affair, and stimuli that on the average do not may sometimes evoke pain and vice versa, a property pain shares with all other sensations. This phenomenon is illustrated in Fig. 11-4, which shows that the skin temperature producing reports of pain in 50% of the trials was about 44° C in the particular experiments represented. The figure also illustrates the finding that analgesics and narcotics, which exert a profound effect upon the total pain experience, may at the same time produce no change in the threshold for the perception of pain.[25]

Is there adaptation to painful stimuli?

It is a common experience that the pain produced by a brief noxious stimulus subsides quickly, an event attributable to removal of the stimulus and disappearance of afferent input in first-order pain fibers (or a drop in frequency below a liminal level). It is equally common to observe that pain produced by a continuing noxious stimulus continues so long as one attends it. One may be distracted from or ignore such a pain and one may control the reaction to it, but when attended to, the pain is present. This suggests that adaptation to pain does not occur in either of the two senses of the word—that a discharge in first-order fibers may cease despite continuing stimulation, or that a continuing first-order discharge may gradually lose its power to evoke sensation. An experiment bearing on this point is illustrated in Fig. 11-5. Five trained observers were asked to maintain by their own control of an unseen dial an intensity of thermal radiation that just continued to evoke pricking pain.[48] Note that the intensities of radiation that they chose are similar in strength to the rheobase of the strength-duration curves in Fig. 11-3. Rather than adaptation, the curves show that there was a gradual lowering of the skin temperature required for the pain threshold, a change attributed to the onset of *primary hyperalgesia.*

On the other hand, Lele et al.,[70] report that sudden immersion of the skin of human subjects in water baths at 36° to 41° C elicited responses of a short, transient stab of pain during the first few seconds of immersion. This suggests adaptation and, further, that thermal pain may not depend in every case on such clear destruction of tissue as that beginning with skin temperatures of 44° to 45° C. Hardy and Stolwijk[51] have confirmed this observation and attribute it to a phasic activation of pain fibers, an onset transient discharge that at low temperatures quickly subsides to subthreshold rates, but that at temperatures of 44° to 45° C persists into a steady rate of discharge—the threshold for persisting pain. It is then an adaptation in the first sense given previously. Following a particularly thorough analysis of the temperature gradients in the skin during such a stimulus, Hardy

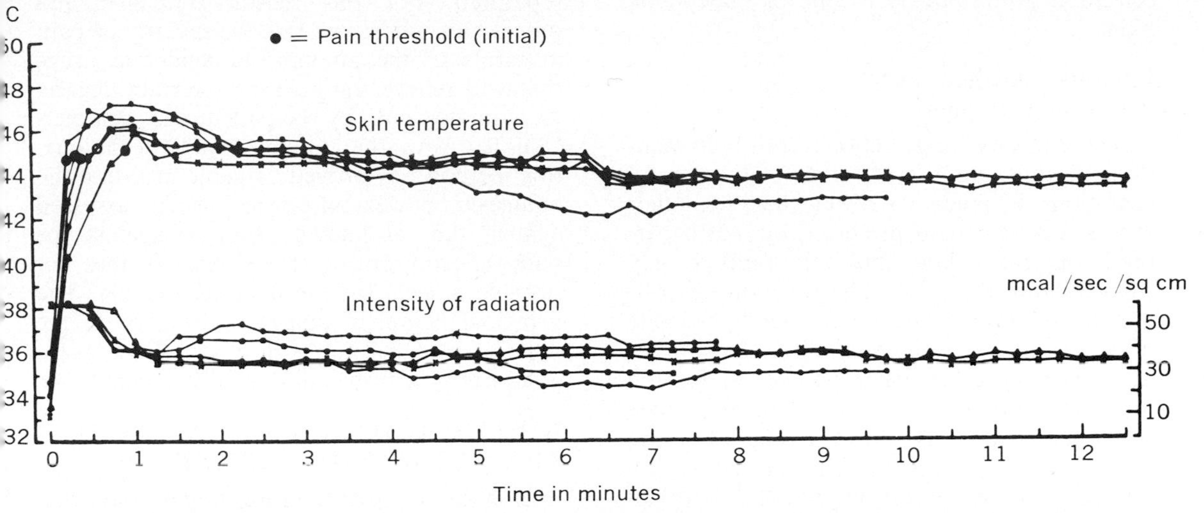

Fig. 11-5. Results of experiment aimed at testing degree of adaptation to painful stimuli, using thermal radiation method, in five subjects. Each controlled for himself intensity of thermal radiation just required to produce continued pricking pain. Intensities they chose are plotted below, skin temperatures above. Rather than adaptation, skin temperature required just to maintain a sensation of pain acually dropped about 1° C during exposure, a change thought to be due to onset of primary hyperalgesia. (From Hardy.[50])

and Stolwijk were unable to support the hypothesis put forward by Lele et al. that activation of temperature fibers and the thermal activation of pain fibers is due to such gradients. They support the view that the elevated temperature initiates processes leading to protein breakdown in tissue cells, releasing substances that excite pain endings.[52]

Is there spatial or temporal summation of pain?[53, 57]

The term "spatial summation" as applied to sensations refers to two related ideas. The first is the possibility that a stimulus that is subthreshold for sensation when applied to a small area will evoke sensations when applied over a larger area of the receptor surface. Presumably this is due to the additive central effects of the activation of an increasing number of primary sensory fibers. The second is that if the stimulus is initially supraliminal when applied to a small area it will, when applied to larger areas, elicit more intense sensations. Spatial summation is a classic attribute of vision (Ricco's law) and occurs for the cutaneous senses of touch, warmth, and cold. It occurs to only a limited extent for pricking, burning, or aching pain. For example, comparison of the senses of warmth and pain, elicited in the two cases by thermal radiation of different intensities, revealed that for warmth there was more than a 200-fold increase in threshold as the area of skin stimulated was reduced from 200 to 0.2 cm^2; the pricking pain threshold remained constant over the range of skin areas from 0.5 to 28 cm^2, the largest area tested.[2]

A sharp distinction should be made between the lack of spatial summation in the threshold and in the intensity of pain sensation and its importance in certain reactions to pain. Spatial summation is a functional property of the segmental reflex mechanisms of the spinal cord, for example, and most powerfully so for the flexion reflexes evoked by noxious stimuli. Moreover, the total experience of pain, in contrast to the sensation of pain, is undoubtedly influenced by the spatial extent of pain-provoking injury or disease: the life-threatening effect of an extensive as contrasted to a local burn of the body surface, for example, may certainly influence in some subjects the total pain experience associated with the injury.

Temporal summation, on the other hand, plays an important role in pain sensation; indeed, it is required for sensing certain pains. Collins et al.[35] studied the effect in conscious human subjects of afferent volleys in C fibers evoked by electrical stimulation of peripheral nerves after block of myelinated fibers. A single C fiber volley, proved to have entered the nervous system, evoked no sensation whatsoever! At frequencies of 3/sec the subjects always experienced pain, and with continued stimulation at this or higher rates the sensation grew quickly to one of unbearable pain.

Intensity functions and the scaling of pain

The intensity range between threshold stimuli and those that evoke maximally painful sensations is much narrower than for other senses; e.g., for pain produced by thermal irradiation, the ceiling/threshold ratio is only slightly more than 2:1. The function describing the relation between a subject's estimate of a pain and the physical value of the stimulus evoking it has been established for the first two thirds of the ceiling/threshold range by Adair et al.[10] As shown in Fig. 11-6, the relation is linear, for the left-hand curve is a straight line in log-log coordinates with an exponent of 1.0, after subtraction of the constant that represents absolute threshold. In these experiments, threshold was measured at 180 mcal/sec/cm^2, a value somewhat lower than that of the subject population studied by Hardy et al.[54] (Fig. 11-1). The threshold value has been measured in seven other investigations; the results range between 190 to 220 mcal/sec/cm^2. A considerable part of that variation is undoubtedly due to differences in initial skin temperature (Fig. 11-2).

The pain scales established by these methods have not proved useful in estimating the severity of pain suffered by patients with injury or disease, largely, it is believed, because in this setting the sensation of pain and the total experience of it cannot be subjectively separated. For this reason Beecher[13] and others have chosen to rate the severity of pain in terms of the amounts of analgesic drugs that will relieve the pain by a certain degree, as estimated both by the patient and observer. Coupled with the "double-blind" technique, this method has proved valuable in rating the analgesic power of drugs when assessed against that of known drugs or against the pain-relieving power of placebos. It does not provide a scale for the intensity of pain. The principal action of analgesic drugs is on the reaction to pain (not its threshold, Fig. 11-4), a subject now to be considered in some detail.

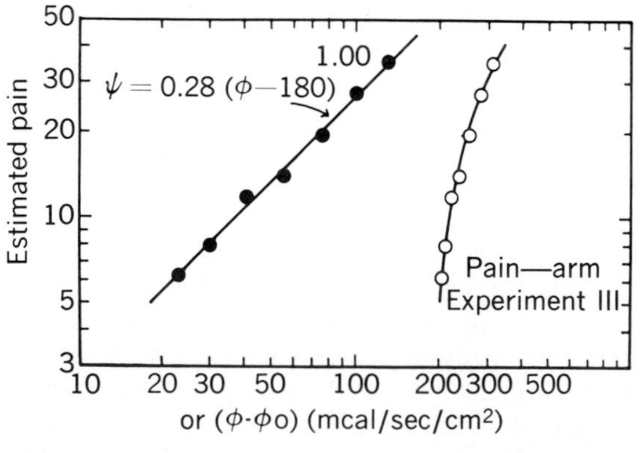

Fig. 11-6. Results of subjective magnitude estimation of intensity of painful sensations produced by radiant heat stimuli. Unfilled circles are geometric means of estimates made by 10 subjects of stimuli of 7 different intensities. Filled circles are same means plotted as function of irradiance minus calculated threshold of 180 mcal/sec/cm^2. (From Adair et al.[10])

HUMAN REACTION TO PAIN AND THE TOTAL PAIN EXPERIENCE

It is the thesis of the foregoing account that human observers can identify the sensation of pain, examine it introspectively, and describe it publicly in spite of accompanying reactions to pain. These reactions are of three types: local, reflex, and behavioral. Local reactions will be discussed in a following section. Reflex actions are provoked by noxious stimuli and may occur whether or not those stimuli also evoke the sensation of pain, although they commonly do. The behavioral changes that may accompany the sensation of pain involve the function of the nervous system at higher integrative levels. Many of them take on the attribute of suffering.

Somatic reflex reactions. The reflex contraction of flexor muscles evoked by noxious stimulation is part of the withdrawal response to pain. It shows spatial summation, in contrast to pain sensation, and in experimental animals or in paraplegic man its threshold may equal that for human pain sensation (Fig. 11-3). In normal man, flexion withdrawal may be suppressed completely or it may occur with the anticipation or threat of pain. Its threshold is thus variable and indefinable, in contrast to that of pain sensation.

Continuing noxious stimulation, particularly that arising in deep somatic or visceral tissues, commonly elicits a steadily maintained reflex contraction, usually of adjacent but sometimes of distant muscles. Splinting of the muscles of the abdominal wall, for example, may be the first sign of intra-abdominal disease. It may appear at a stage when the frequency of discharge in pain fibers innervating the area of tissue injury is too low to evoke the sensation of pain. With continued contraction, the muscles may themselves become painful, a process greatly intensified and accelerated if they are deprived of blood supply or their venous drainage is blocked. It is thought that this muscle tenderness or "soreness" is due to the release of pain-producing substances from the continually contracting and thus partially ischemic muscle cells. This pain of muscle origin frequently becomes more severe than the pain that initiates it, and it may dominate the clinical picture. It should not be confused with referred pain (p. 368), although referred pain may accompany the initiating process.

Autonomic reflex reactions. Reflex reactions engaging autonomic effectors commonly accompany the somatic motor responses to noxious stimulation, particularly when the pain provoked is severe and of sudden and unexpected onset. They are parts of the more generalized mobilization for defensive or aggressive reaction to attack, or to the threat of it, which depend for integration on executant neural mechanisms of the forebrain, particularly of the hypothalamus (Chapter 31). They may include cardiac acceleration and peripheral vasoconstriction, with a resulting rise in blood pressure, dilatation of the pupils, and secretion from the sweat glands and the adrenal medulla—all signs of intense activity in sympathetic efferent nerve fibers. All pains do not evoke similar reflex patterns, however, for visceral pain of sudden onset (testicular crushing, pain of chemical peritonitis, etc.) may provoke vasodilatation and a drastic fall in blood pressure, together with a decrease in somatic motor tone. In the intact mammal these autonomic effects are welded into the total behavioral pattern, even though some are basically spinal reflexes. In chronic spinal man, for example, with a high thoracic transection, distention of the bladder may elicit intense vasoconstriction, sweating, and piloerection in the parts of the body innervated by the isolated spinal segments.

The phasic sympathetic discharge to the sweat glands evoked by a painful stimulus is accompanied by a change in the electrical resistance of the skin. While this change is not wholly due to the excretion of sweat, it is a ready measure of the integrity of the sympathetic efferent pathways: it is called the galvanic skin reflex (GSR). How completely such a segmental reflex may be dominated by descending influences of forebrain origin is shown by the fact that the GSR adapts or 'habituates" to repeated stimuli, but more especially by the fact that its presence and amplitude vary greatly from one individual to another and from time to time in a single person. It appears in part to reflect the affective state of the subject, his attitude toward the situation in which he receives the painful stimulus, and the meaning he attaches to such stimuli. As a component of the "alarm" reaction it has found some use in "lie detector" tests. Thus many of the reflex reactions to pain cannot be separated from the overall emotional attitude of the subject toward pain. It is the meld of both with the sensation itself that constitutes the total pain experience.

PERIPHERAL MECHANISMS IN PAIN

Peripheral nerve fibers and pain[19]

Several lines of evidence converge to support the statement that *pain is served by two specific sets of peripheral nerve fibers that are differentially nociceptive in function, and that the dual quality of cutaneous pain depends in the first instance on the differential sensitivity of these fibers to destructive stimuli: certain fibers of delta size subserve pricking pain; certain C fibers, burning pain.* The evidence is as follows:

1. The duality referred to has recently been reconfirmed.[20] The short-latency pricking pain evoked by a noxious stimulus is followed by a second long-latency pain of a burning and less bearable quality.

2. When conduction in peripheral nerves is blocked by pressure, the double nature of pain persists so long as all fibers of delta size and smaller continue to conduct. At this stage the more discriminative aspects of the mechanoreceptive modes are lost, leaving warmth, cold, and an imprecise form of mechanoreception, together with both qualities of pain. When conduction in the small myelinated fibers fails, pricking pain disappears; the second long-latency burning pain persists when only C fibers are conducting. It is, in fact, often greatly exaggerated, a release phenomenon to be discussed. When only C fibers are conducting, subjects are insensitive to any nonnoxious mechanical stimuli delivered to the skin innervated by the blocked nerve. A reverse sequence occurs when a cutaneous nerve is blocked by a local anesthetic agent.[21] A similar sensory dissociation—elevation of threshold for or loss of pricking pain, with continued presence and exaggeration of second pain—is produced by peripheral neuropathies that principally affect myelinated fibers.[18, 123]

3. Collins et al.[34, 35] recorded the descriptions by human subjects of their sensory experiences when their peripheral cutaneous nerves, exposed under local anesthesia, were stimulated electrically at different strengths. When beta fibers were activated at various frequencies and temporal patterns, all experi-

ences were appreciated by the subjects as nonpainful. When, however, a stronger stimulus added the delta fibers to the entering volley, a sharp change in the sensation evoked occurred after even a single volley. With frequencies as low as 3/sec, the sensation was sharp and prickingly painful. Then, in these same experiments on human subjects, conduction was blocked in the myelinated fibers, and a much stronger stimulus was used to elicit volleys of impulses in C fibers. A single volley evoked no sensation. At frequencies of 3/sec the sensation of pain was always evoked, and the experience was so unbearable that the subjects would not allow repetition. The difference between the emotional overtone and the degree of suffering produced by volleys in C fibers as compared to those in delta fibers was a striking aspect of these experiments and fully confirmed earlier but less direct observations.

4. The thesis put forward is strongly supported by studies of peripheral nerve fibers. Perl[91] and Burgess and Perl[21] have used the method of single-unit analysis to specify the properties of large numbers of delta-sized myelinated fibers that innervate the hairy and the glabrous skins of both monkeys and cats. In each group, approximately 20% of fibers required for their excitation strong mechanical stimuli that were destructive of tissue and that evoked pain when applied to the skin of man. The receptive fields of these *nociceptive afferents* were usually no larger than 3 × 2 cm, and within such a field the sensitive areas were distributed in a spotlike manner. These delta fibers were markedly differentially sensitive to noxious stimuli, for they were unaffected by large changes in skin temperature, and their sensitivity to mechanical stimuli was several orders of magnitude less than that of the true mechanoreceptive afferents of delta size.

Similar studies of unmyelinated C fibers have been made by Iggo[61] and by Bessou and Perl.[15] Of those innervating the skin, fully one third were classed as nociceptive afferents, being sensitive only to stimuli of an order of magnitude more intense than other fibers classed as mechano- or thermoreceptive afferents. Moreover, C fibers susceptible to noxious stimuli differ; some are properly designated as high-temperature, some as low-temperature, and still others as strong-pressure nociceptive afferents.

In summary, according to observations of normal humans and humans subjected to differential nerve block, it can be predicted that a certain set of delta fibers subserve the short-latency, pricking form of cutaneous pain, and a certain set of C fibers subserve the long-lasting, burning pain, with its unbearable quality of suffering and prolonged afterimage. These predictions have been fully borne out by nerve stimulation and recording experiments in waking human subjects, in which correlations with the evoked sensory experiences were direct, and by electrophysiologic experiments in animals. These latter have revealed beyond any persisting doubt that certain sets of delta fibers and C fibers are nociceptive in nature. All delta and all C fibers are not nociceptive; indeed, their range of sensitivities provides a puzzle of considerable proportions, for among C fibers only the nociceptive and perhaps the high-threshold temperature fibers appear to function in a sensory as well as an afferent manner. Whether the mechanoreceptive fibers subserve itch is discussed later.

These conclusions do not imply that other fibers, when discharging in one or another temporal pattern or temporospatial conjunction with nociceptive afferents, may not contribute to the total perceptual experience of which pain may only be a part. What they do state is that afferent impulses in nociceptive delta or C fibers are both necessary and sufficient, as regards peripheral input, to evoke painful sensations in conscious man under the experimental conditions described.

Peripheral transducer mechanisms and nociceptive afferents[1]

Given that two sets of peripheral nerve fibers are differentially sensitive to noxious stimuli and evoke the sensation of pain when active, attention is directed to the peripheral transducer mechanism involved. Whether the distal terminals of nociceptive afferents can be excited *directly* by noxious stimuli is uncertain, but evidence accumulates to indicate that injury to tissue cells releases from them chemical substances that are potent activators of the nociceptive fiber endings; i.e., these latter are chemoreceptors.

Shortly after injury to the skin there appears in the injured area intense vasodilatation that leads to local edema and formation of a wheal a few millimeters in diameter, soon surrounded by a less intense but much wider (several centimeters) vasodilatory flushing of the skin. Local reddening, wheal formation, and surrounding flare make up the triple response of Lewis.[73] Within the injured area and in part of the surrounding region of flare, a pinprick produces more intense pain than before injury and the pain threshold is lowered; stimuli not previously painful become so, and these changes may persist for days. This heightened sensibility is termed *primary hyperalgesia*. The hypersensitivity to pinpricks that also develops in a wide region outside the flare may persist for only a few hours; within this region the threshold for pain is not lowered. This is termed the region of *secondary hyperalgesia*. The two hyperalgesias are thought to differ in causal mechanism.[1]

After acute section of a peripheral nerve the triple response persists as long as the peripheral axons. After their degeneration (5 to 6 days), only local reddening and wheal formation follow local injury. The full triple response is produced by cutaneous injury after section of the dorsal roots central to the ganglion, for then the peripheral axons do not degenerate. The vasodilatation seen as flare is therefore an *axon reflex;* impulses traveling centrally from the region of injury are conducted antidromically over other branches of the nociceptive afferents, invade their endings, and elicit the vasomotor changes. Indeed some branches of dorsal root C fibers end in close proximity to small peripheral blood vessels, thought not directly innervating them. Autonomic efferents are not involved, for the triple response survives complete sympathectomy.

This phenomenon is closely related to the vasodilatation produced by electrically induced antidromic volleys in dorsal root C fibers, discussed in some detail in the following section. The mechanism is thought to be identical for the two cases of cutaneous vasodilatation and to be important for understanding the chemoreceptive nature of nociceptive afferents. Foerster,[44] for example, showed that in man, if an adjacent root on either side of a transected dorsal root were intact, stimulation of the peripheral portion of the transected root resulted in cutaneous vasodilation in its peripheral field of distribution and in sensations of pain. The adjacent roots, of course, have overlapping peripheral fields. The proposition that antidromic invasion of C fiber terminals either releases or causes the formation in the intercellular fluid of a substance having the properties of relaxing arteriolar smooth muscle *and* activating the endings of nociceptive afferents (thus accounting for Foerster's observation) leads to the idea that the normal excitation of nociceptive afferents after cellular injury is similar in nature.

Neurohumoral mediators and pain.[4, 5] The identity of the pain-producing neurohumor activated by cellular injury has proved elusive during nearly 40 years of intensive research, though recent findings narrow the field of candidates considerably. Histamine pricked into the skin elicits intense itching and pain and the triple response; therefore Lewis[73] postulated that a histamine-like substance is released from damaged cells to produce pain and serves as the vasodilator mediator between antidromically activated C fibers and blood vessels. The antidromic vasodilatation, however, is not blocked by antihistaminic agents; thus histamine is not a mediator at the end of the axon reflex, and its role as the exciting agent at its origin in injured skin is now doubted. Potassium ions, 5-hydroxytryptamine (5-HT), acetylcholine (ACh), adenosine triphosphate (ATP), and a number of other naturally occurring substances that produce pain on intradermal injection have been considered the pain-producing neurohumor on the basis of one or another line of evidence. Recent findings, however, support the thesis that cell injury releases into the intracellular fluid one or more proteolytic enzymes that then act upon gamma globulins to produce a series of polypeptides possessing extraordinary potencies as vasodilators and as excitants of nociceptive afferents.

The experiments of Chapman et al.[1] have illuminated the subject considerably. Using a method for perfusing the subcutaneous spaces in man, they have examined the perfusate collected for pain-producing neurohumors under a variety of circumstances. When the skin above the perfused area was untouched, the perfusate was no more potent in producing pain when injected intradermally than is sodium chloride solution in physiologic concentration. When, however, the triple response was induced by injury to the skin, the perfusate contained a proteolytic enzyme and a polypeptide that produced vasodilatation and pain, even when greatly diluted (Fig. 11-7). Upon bioassay this substance paralleled in its properties many but not all those of the nonapeptide bradykinin. It was also released into the perfusate during stimulation of the distal end of a transected dorsal root innervating the region perfused. It did not appear when chronically denervated skin was

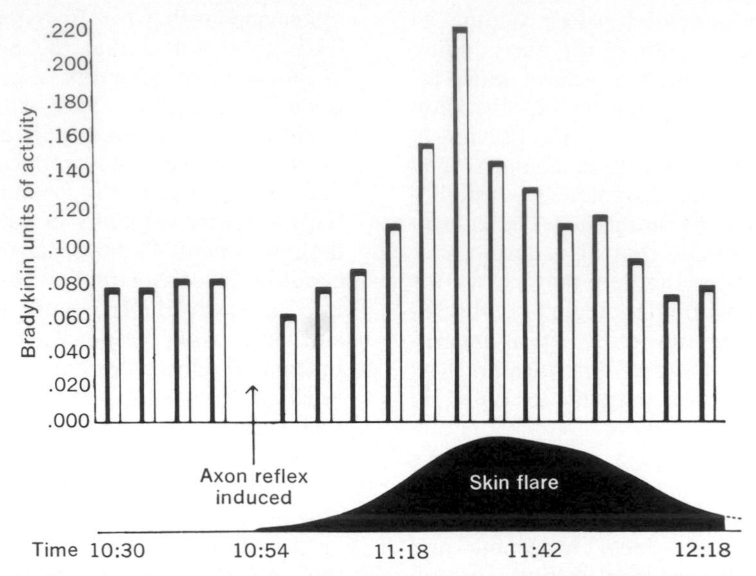

Fig. 11-7. In this experiment subcutaneous space was perfused with fluid via inflow and outflow needles. Lewis' triple response was induced by intracutaneous injection of histamine in region of skin overlying perfused area. Polypeptide with bradykinin-like properties appeared in perfusate in quantities measured in "equivalent bradykinin" units. Fluid collected during period of skin flare also produced pain upon intracutaneous injection in a remote region of skin; that collected during control period did not. (From Chapman et al.[26])

locally injured. When severe thermal injury was delivered to the skin, a polypeptide not differentiable by any bioassay from bradykinin appeared in the perfusing fluid. Bradykinin has been shown by Keele and Armstrong[65] to be present in blister fluid of human skin and in other painful tissue exudates and to be among the most potent pain-producing substances yet examined. It was not released or formed as a result of vasodilatation per se, for it did not appear in the perfusate during the vasodilatation produced by other causes, e.g., reactive hyperemia.

Whether bradykinin or some other polypeptide (or the proteolytic enzyme itself) is actually released from an intracellular store upon antidromic invasion of nociceptive afferents or whether those impulses in some way elicit its release from adjacent tissue cells is unknown. If the former, the possibility is raised that it is also released at the central terminals of those fibers when orthodromic impulses reach them under natural circumstances. Indeed, a proteolytic enzyme is found in small quantities in the cerebrospinal fluid (CSF) of normal individuals and in increased quantities after intense CNS activity (convulsions) and in certain diseases of the brain.[24] Its slow rate of action probably precludes its function as a synaptic transmitter in the usual sense, but the possibility exists that it plays a role in local vasomotor control within the CNS and influences neuronal excitability.

To return to the mechanisms of cutaneous hyperalgesia, the sensory changes at the local site of injury and in the flaring surround are classed together as primary hyperalgesia, for in either case the presence of pain-producing substances lowers threshold and causes a stimulus of a certain strength to evoke more discharges in more afferent fibers than it normally would—the hyperalgesic phenomenon. In the region of local injury the substance is released from the injured cells; in the region of flare it is released by the action of antidromic impulses in C fibers. However, no such peripheral mechanism can account for the secondary hyperalgesia of the nonerythematous surround, for local changes do not occur there and the area is too wide to be reached by axon reflexes. It has been suggested that secondary hyperalgesia results from a heightened level of excitability in interneuron pools receiving afferent input from both the injured and the secondarily hyperalgesic regions.[2] This would account for hyperalgesia without change in the pain threshold. No direct evidence relating to this idea is yet available. If, however, afferent volleys in C fibers release centrally the same bradykinin they cause to be formed peripherally, its spread through an interneuron pool could produce the heightened excitability required by the hypothesis.

In summary, noxious stimuli of any type possess the common property of evoking pain by virtue of a capacity to injure tissue cells. Injury releases from these cells a proteolytic

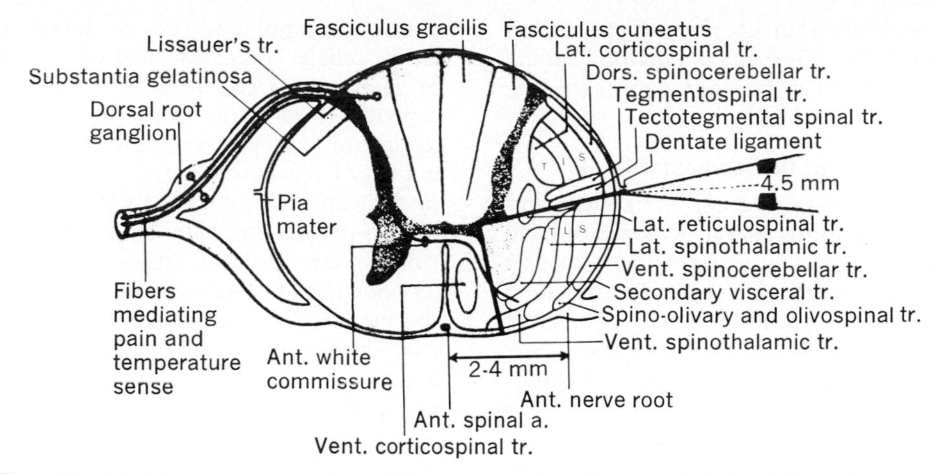

Fig. 11-8. Drawing of cross section of human spinal cord at level of third thoracic segment to illustrate some cord pathways and extent of section required for adequate anterolateral cordotomy. *T, L,* and *S* refer to thoracic, lumbar, and sacral, respectively, and illustrate relatively crude topographic organization in anterolateral system. Visceral afferents probably do not run in a separate tract, but lie most medial because in large part they enter cord over thoracic dorsal roots. (From Taren and Kahn.[116a])

enzyme that forms from the globulins of the intercellular fluid a series of polypeptides, potent excitors of the endings of nociceptive afferents. There is much that is speculative about these statements, and it is not known whether they are plausible for all nociceptive afferents or only the C fibers. It is possible that the endings of the delta fibers are activated directly by noxious stimuli, without an intervening step such as that described. It is important to note, however, that the unmyelinated terminals of the delta fibers differ from those of the C fibers only in location—they seem to end somewhat more superficially in the skin (in the epidermis)—otherwise the two sets are structurally identical.

CENTRAL NEURAL MECHANISMS IN PAIN

Of all the major sensory systems, those for olfaction and pain are undoubtedly the least well understood. Present ideas concerning the CNS mechanisms in pain are largely inferences from anatomic facts and from observations of altered pain sensibility in patients with disease or surgical lesions of one or another part of the nervous system.

Two facts serve as starting points. First, pain is evoked by impulses in both small myelinated and C fibers in peripheral nerves (p. 355), and these sets of fibers feed via the synaptic mechanisms of the dorsal horn into the anterolateral system.[76, 77] They do not activate the dorsal column (DC) system. Second, a *full* surgical transection of the anterolateral column of the human spinal cord, extending from the base of the corticospinal tract to within 1 mm or so of the median fissure, results in pronounced alterations in pain (and temperature) sensibility over the contralateral body surface, amounting in the case of perfect surgical execution to hemianesthesia, beginning a few segments below the level of transection (Fig. 11-8).[8, 122] Injury or section of the DCs does not lead to loss of pain sensibility. The anterolateral system, like the two sets of peripheral nerve fibers, is a necessary condition for the evocation of painful sensations by nociceptive stimuli. It is important to remember, however, that large numbers of both delta and C fibers are activated by light mechanical stimuli, and that a form of cutaneous mechanoreceptive sensibility survives transection of the DCs. Therefore, while the anterolateral column contains the neural elements essential for pain sensation, it is by no means exclusively a "pain tract" or a "pain and temperature" tract. Furthermore, while transection of the DC does not produce analgesia, lesions of it or indeed of the lemniscal system at any level may lead to profound alterations in pain sensibility, most commonly to the hyperpathias described later. It is therefore correct to say that while the anterolateral component of the somatic system is essential for the perception of pain, activity conducted over the lemniscal system conditions that sensory experience, as do, undoubtedly, efferent systems descending from the forebrain to impinge upon subcortical relays.

Pain mechanisms at the segmental level

Nociceptive afferents enter the spinal cord largely over the more lateral divisions of each dorsal root. Many turn into the extrinsic division of Lissauer's tract before terminating in the substantia gelatinosa (SG—equivalent to Rexed's lamina II). Several physiologic observations support this anatomic finding: (1) Multiple shallow incisions of Lissauer's tract produce a zone of analgesia in the related dermatomes in both animals and humans.[60, 95, 96] (2) After section of its lateral division, stimulation of a dorsal root no longer elicits reflected actions characteristic of pain or evoked responses in the mesencephalic tegmentum, a major forebrain projection of the spinoreticulodiencephalic component of the anterolateral system. Responses in the ventral posterior nucleus of the thalamus, a projection of the lemniscal system, are still evoked by stimulation of that partially sectioned root. Converse results obtain when the medial, large-fibered division of a dorsal root is sectioned.[110] This is substantial evidence that impulses in the large myelinated fibers of a dorsal root *do not directly activate* ascending pain pathways. It will be seen, however, that via their segmental collaterals these large fibers may modulate the transmis-

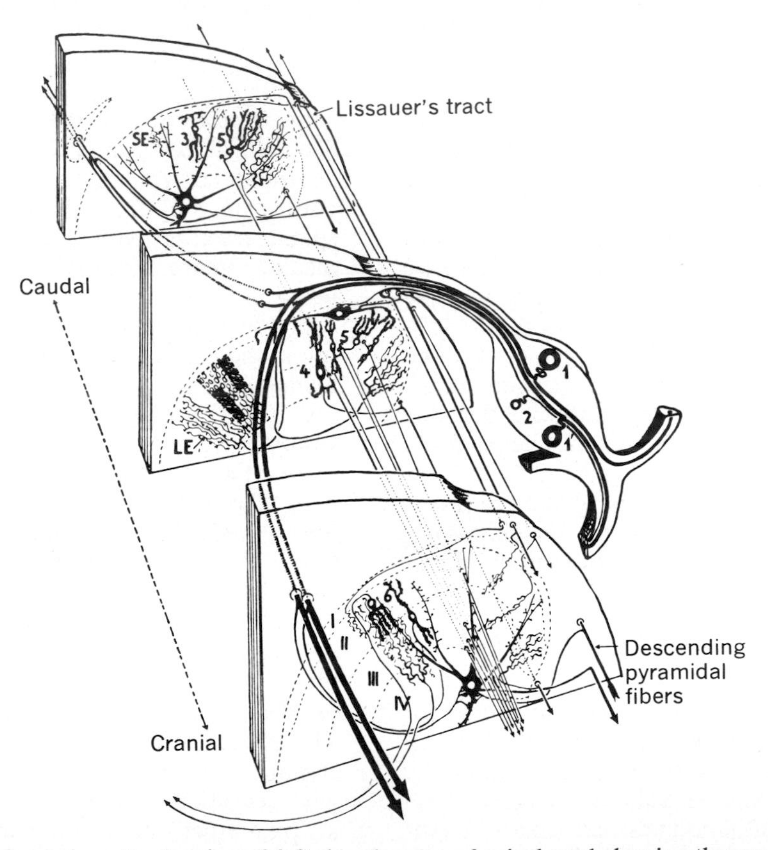

Fig. 11-9. Semischematic drawing of left dorsal sector of spinal cord showing three successive levels less than one segment apart to illustrate neuronal arrangement in substantia gelatinosa. Cytoarchitectural layers according to Rexed[99] indicated by Roman numerals. *1* and *1'*, Large-caliber cutaneous afferents making central terminations via branches in SG and on large neurons of lamina IV; their stem axons project into dorsal columns. *2*, Small-caliber cutaneous afferent with axon turning into medial part of Lissauer's tract and giving small endings, *SE*, to SG as well as axosomatic knobs to large marginal neurons of lamina I (middle section). *3*, SG neurons proper that send their axons into lateral division of Lissauer's tract. *4*, Larger SG neurons sending their axons into lateral fasciculus proprius. *5*, SG neurons that emit axons that synapse with dendrites of large lamina IV neurons. *6*, SG neurons that project their axons to contralateral SG. (From Szentágothai.[115])

sion of activity in smaller myelinated and unmyelinated fibers through the synaptic relays of the dorsal horn into the anterolateral system. That latter projection is largely contralateral, with a minimal ipsilateral projection. Moreover, the SGs of the two sides are heavily interconnected via the posterior white commissure. The dominance of the contralateral projection, however, is emphasized by the fact that transection of the ventral commissure of the spinal cord in man produces a profound analgesia in the related dermatomes.

Synaptic mechanisms of the dorsal horn.[93, 94, 104] Anatomic studies have shown that the dorsal horn may be divided into six horizontal layers, Rexed's laminae I-VI[99]; each differs from the other in size, distribution, and orientation of its constituent neurons. Supraspinally directed anterolateral fibers are axons of dorsal horn cells that are activated by impulses in both thinly myelinated and unmyelinated primary afferent fibers.[33, 76, 77] It is not known precisely which cells of the dorsal horn give rise to ascending anterolateral fibers, although they appear to be concentrated in its deeper layers.[41, 117] One interpretation of synaptic connectivity in the dorsal horn is shown in Fig. 11-9. Myelinated fibers of the medial division of the dorsal roots terminate upon neurons of lamina IV; other branches of these same fibers deploy in dorsally directed radiating sprays of axodendritic synaptic endings upon the small SG neurons. While many of these fibers of the medial division are segmental collaterals of afferents that enter the DCs, others may be the stem axons of fibers that do not. The thinly myelinated and unmyelinated fibers of the lateral divisions of the dorsal roots enter Lissauer's tract but travel in it for no more than a segment before penetrating the SG in a radial direction to terminate upon the large marginal cells and the small SG neurons. Fibers of extrinsic origin account for only 25% of the fibers of Lissauer's tract; the remainder are axons of SG cells that run within it only a short way before reentering the SG to terminate upon other SG neurons. The axons of some SG neurons run longitudinally through SG, synapsing *en passage* upon the radially directed dendrites of the large neurons of lamina IV. A prominent feature of the neuropil of laminae II and III is the large number of axoaxonic synapses within it.[93, 98]

The SG proper appears to be a closed system, for the axons of its cells either remain within or return to it after short longitudinal courses or project via the dorsal spinal commissure to the SG of the opposite side. An output might be channeled via its synaptic projection upon the dendrites of neurons of layer IV. The latter are not thought to be cells of origin of ascending anterolateral fibers, however, although their axons may make connections upon cells of deeper layers that are. Should that be the case, the SG mechanism is in a strategic place to exert a controlling action upon the transsegmental pathways for pain and temperature. Quite obviously the synaptic connectivity of this important region deserves further intensive study.

Functional properties of neurons of the dorsal horn. Electrophysiologic studies of the dorsal horn are among the most difficult in the nervous system: many of its cells are small, and its synaptic relays are sensitive to anesthetic agents, decreases in vascular perfusion, mechanical injury by microelectrodes, and particularly to changes in the levels of activity in descending controlling systems. For these reasons an adequate description of the functional properties of dorsal horn cells cannot be given, but certain important observations have been made.

First, the large majority of dorsal horn cells are activated by impulses in delta fibers and in C fibers, and frequently the two converge upon single central neurons.[59] Second, Christensen and Perl,[25] using the method of single-unit analysis, have been able to characterize the functional properties of one class of dorsal horn cells, the marginal cells of lamina I, in cats. These cells are activated only by impulses in delta and/or C fibers. They vary in properties according to their input from selected sets of peripheral fibers. One group responds only to frankly noxious mechanical stimulation of the skin, a second to either noxious heat or intense mechanical stimulation, and a third to mild temperature changes as well as the two classes of noxious stimuli. Thus the marginal cells could, according to these findings, relay afferent signals concerning pain and temperature sensibility.

Thus far, such studies of laminae II and III appear not to have been successful, undoubtedly because of the small sizes of their neurons. The cells of lamina IV are activated by gentle mechanical stimulation delivered to small receptive fields in the skin,[59, 119, 121] while those of lamina V appear to receive converging input from many cells of lamina IV. Those of lamina V also display excitatory fields that are bordered or surrounded by inhibitory ones. The cells of lamina VI are activated by gentle movements of the limbs and thus presumably by impulses in afferents from muscles and joints.

Obviously this meager knowledge allows no understanding of the complex synaptic circuitry of the dorsal horn. What can be said is that many of its cells possess the functional properties as concerns place, modality, and diameter of related primary afferents suitable for relays in a pain and temperature pathway.

Control mechanisms of the substantia gelatinosa. A volley of impulses in cutaneous afferents of a dorsal root leaves in its wake a complex series of potential changes. Most prominent is a prolonged depolarization of the primary afferent terminals, which may be recorded in dorsal root fibers after electrotonic extension.[75] It is accompanied by a lengthy depression of the segmental reflex, evoked by testing volleys delivered in the afferents with depolarized terminals. Depolarizations occur in the nerve fibers conducting the volley and in passive fibers of the two or three adjacent roots as well as in contralateral roots.[62] There is some preference for fibers of one modality type to exert the strongest depolarizations upon fibers of the same type. This primary afferent depolarization is produced by an internuncially relayed synaptic excitatory action upon the axonal terminals. Inhibition of the synaptic action of testing volleys in the affected fibers may occur (1) by a reduction in the transmitter-releasing capacity of impulses invading the partially depolarized terminals or (2) by a depolarization block of invasion of preterminal branches. It thus fits the paradigm of *presynaptic inhibition* described in Chapter 6. Thus the function of the closed polysynaptic mechanisms of the SG might be to control the excitability of primary afferent terminals and thus to control the synaptic relay of afferent activity. The mechanism is thought to operate in both a tonic and a phasic manner and to be arranged in the "surround" manner, so that it may function as a negative feedback upon the primary afferent input, serving to limit the spatial spread of activity and to shrink the subliminal fringe in the postsynaptic population of neurons.[63]

It has been suggested that this mechanism of presynaptic inhibition might account for the well-known facts that (1) activity in large myelinated cutanous afferent fibers suppresses pain arising in the region innervated by them and (2) spontaneous pain may arise in such a region if it is denervated of its large-fibered innervation. However, it appears that volleys in large cutaneous fibers do not suppress the central postsynaptic effects of volleys in delta or C fibers entering over the same root by this mechanism. Discharge into the anterolateral system by C fiber volleys sums with that elicited by volleys in large fibers.[76]

The existence of just such a local control in the dorsal horn of the synaptic efficacy of impulses in small fibers by large fibers was a basic assumption in the so-called "gate-control" theory of pain put forward by Melzack and Wall.[85] The theory was based also upon the observation of Mendell and Wall[86] that volleys in C fibers hyperpolarized the terminals of large myelinated fibers of the same dorsal root. It was thus suggested that the levels of activity in these two sets of afferent fibers would have mutually antagonistic effects upon the frequency of discharge of a set of postulated "T cells," which were thought to feed into the anterolateral system, and that central monitors of that level determined the sensation of "pain" or its absence. The theory sought to dispense with any differential sensitivity of peripheral fibers for noxious stimuli. The postulated presynaptic facilitation of the terminals of large fibers by small fibers has not been observed by the several sets of investigators who tested it.[46, 118, 124] This, combined with the clear demonstration of the specific sensitivity of certain groups of both C fibers and delta fibers to noxious stimuli,[15, 21, 61, 91] makes the gate-control theory untenable, at least in its present form. Nevertheless, the idea that the SG is a site of control and interaction between various types of afferents and descending systems is very likely and is suggested by the sources of input to the SG and its synaptic connectivity. The problem of elucidating its physiologic mechanisms remains.

Forebrain projections of the anterolateral system

The anesthesia produced in humans by complete anterolateral cordotomy establishes that ascending spinal systems in this region are essential for the appreciation of pain. Further, local electrical stimulation of the anterolateral quadrant via penetrating electrodes elicits sensation of pain in conscious humans.[114] In a series of 200 such stimulations, sensations of pain were evoked at 54% of the loci stimulated, feelings of heat at

37%, and feelings of cold at the remaining 9%. The pain sensation was referred to the contralateral side of the body in 82%, to the ipsilateral side only in 12%, and to both sides in 6%. The locations of peripheral reference agreed roughly with the pattern shown in Fig. 11-9, with considerable overlap between segments. So far as the peripheral and spinal ascending components of the nervous system are concerned, therefore, *it is neural action in certain sets of peripheral nerve fibers and ascending central axons that provoke pain,* and no special "spatiotemporal" pattern of impulses in fibers otherwise unspecific for pain is required. The pattern imposed by electrical stimulation at either locus does not occur under any physiologic circumstances; yet pain results from that stimulation. Variation of the electrically imposed pattern may change the intensity and the temporal cadence of the sensory experience elicited, but not its qualitative nature. At the same time it is emphasized that the patterns of impulses in a variety of afferents, including those other than nociceptive, will contribute to the total sensory experience—to its temporal pattern, its spatial location and extent, its combination with other sensory qualities—but the pain per se is independent of that pattern and of those attributes.

The projection upon the forebrain of the various components of the anterolateral system has been studied intensively in recent years in both man and other animals.[6, 80, 83]

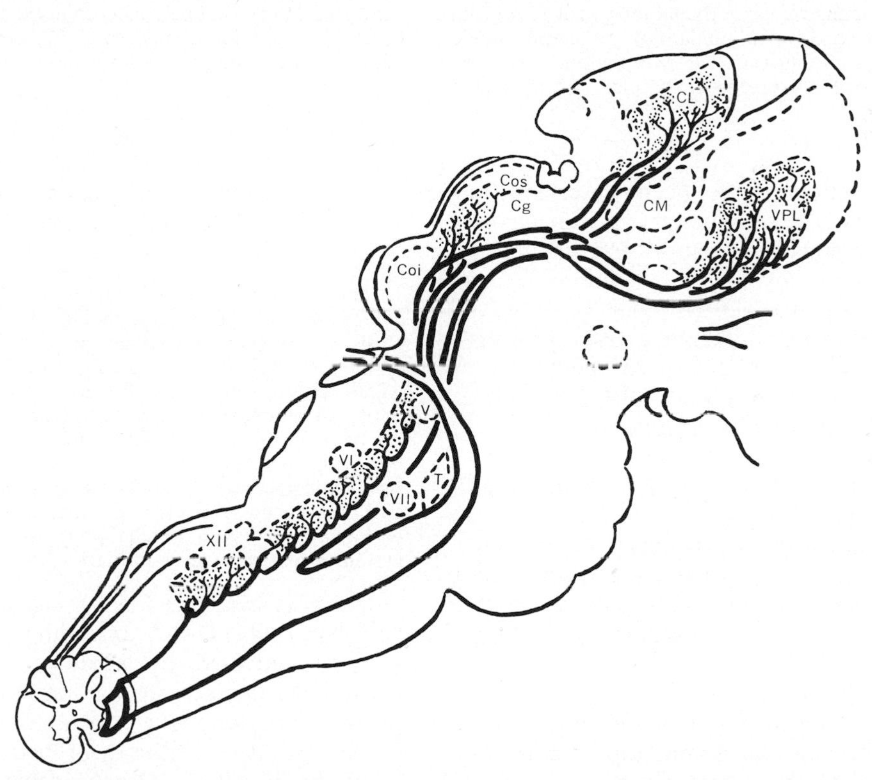

Fig. 11-10. Illustrating ascending fiber projections of anterolateral system in man. Fiber systems of anterolateral fasciculus are shown by heavy outlines; their projection regions are stippled. These include various medullary and pontine reticular nuclei; regions of periaqueductal gray matter, *Cg,* and tectum of superior colliculus, *Cos;* and ultimately the nucleus ventralis posterolateralis, *VPL,* a part of ventrobasal nuclear complex. This portion is termed neospinothalamic system in text. System also projects upon nucleus centralis lateralis, *CL,* member of "intralaminar" group of nuclei, which are part of what is called generalized thalamocortical system in text. This portion is called paleospinothalamic system. At level of facial nucleus, *VII,* and trapezoid body, *T,* some spinothalamic fibers course superficially with recurrent trajectory of ventral spinocerebellar tract, but at level of inferior colliculus, *Coi,* they rejoin deeper coursing fibers ascending to midbrain and thalamus. Less than one third of ascending fibers of anterolateral system actually reach thalamus. Relation of anterolateral system to central nervous mechanisms in pain is discussed in detail in text. (Courtesy Dr. William R. Mehler.)

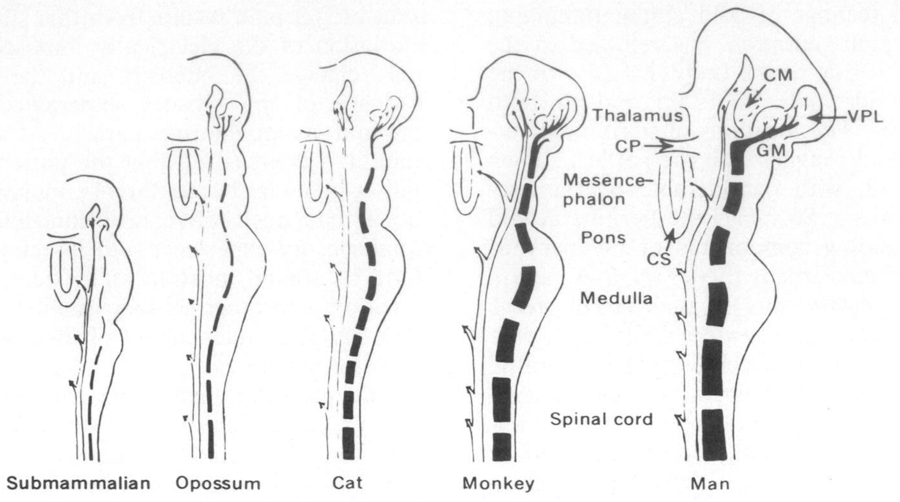

Fig. 11-11. Schematic representations of course and distributions of ascending spinal fiber projections in such submammalian forms as frog or pigeon, and in opossum, cat, monkey, and man. Neospinothalamic system is indicated by interrupted black lines, paleospinothalamic (and spinoreticulothalamic) by uninterrupted outline. *CP,* Posterior commissure; *Cs,* central gray substance of mesencephalon; *CM,* centre médian; *VPL,* ventral posterolateral nuclei, also called ventrobasal complex. (From Mehler.[6])

These projections have been described in Chapter 10. To recapitulate, a convenient division with some implications for function is as follows (Fig. 11-10):

1. *Lateral components*
   a. The neospinothalamic tract projects directly upon the more caudal parts of the venterobasal (VB) complex of the dorsal thalamus in a detailed somatotopic pattern.[23, 83] Its relative increase in phylogeny[58] is indicated in Fig. 11-11.
   b. The neospinothalamic component projects also upon the posterior nuclear complex, with only a vague preservation of somatotopic representation.[81]
2. *Medial components*
   a. The paleospinothalamic tract projects without somatotopy upon nuclei of the intralaminar group, principally upon the centralis lateralis. It does not project upon the centre médian.[82]
   b. The spinoreticulodiencephalic component projects upon elements of the reticular formation (RF) at medullary, pontile, and mesencephalic levels, and thence via the upwardly projecting reticulodiencephalic systems, without somatotopy, upon the intralaminar nuclei of the dorsal thalamus and upon the ventral thalamus and the hypothalamus as well.

The spinoreticular projections are the cephalad extensions of the intersegmental systems of the cord commonly referred to as the propriospinal system, or Bechterew's "ground bundle" of fibers, which closely surround the gray columns of the cord. In animals such as carnivores, activity set up by nociceptive stimulation may reach the forebrain via this multi-chain, polysynaptic system. The results of accurate anterolateral cordotomy indicate that this is not the case in man, emphasizing the continuing phylogenetic change in the anterolateral system in the series of mammals.

There is evidence for a loose matching of peripheral fibers and ascending prethalamic systems activated by noxious stimuli. The neospinothalamic system is preferentially activated by the delta fibers of peripheral nerves and less certainly or less powerfully by C fiber volleys. On the other hand, the "medial components" described previously are powerfully activated by repetitive C fiber volleys, but may also receive delta fiber input.[31-33]

Thalamocortical mechanisms in pain: facts and hypotheses

It has evolved from the preceding discussion that the sensation of pain is dual in nature. To recapitulate, two types of pain are differentiated by the human observer: one of short latency, sharp and pricking in quality, the other slow in onset, burning and penetrating in character, and persisting for seconds after removal of the stimulus. The first is accurately localized, although not quite so well as is touch; the second, when evoked without the first, is diffuse and difficult to locate precisely. The first type can be observed subjectively as a sensation, evaluated quantitatively, and accurately scaled; the second is characterized by its capacity to evoke suffering, with appropriate emotional reactions and widespread reflected events in both somatic and autonomic spheres; it is difficult to evaluate quantitatively. First pain is served by the delta fibers of peripheral nerves, and second pain is served by C fibers; both are specific nociceptive afferents. Impulses in both sets relay via synaptic mechanisms of the dorsal horn into the anterolateral system, within which at spinal levels they are not geographically separate: an accurate and complete anterolateral cordotomy eliminates both, for then analgesia results. In its cephalad distribution this system once again displays duality in the form of the lateral and medial components just described, and the two can there once again be dissociated by surgical lesions or disease process.

Much less is known concerning thalamocortical mechanisms in pain. The dualities described, encompassing pain as a behavioral phenomenon and the relevant anatomic substratum, suggest a duality of forebrain mechanisms. The following hypotheses derive from this idea and appear to fit the facts presently available; their hypothetical nature is emphasized.

1. Pricking pain is served by the lateral components of the anterolateral system that project via the VB complex and the PO group of the dorsal thalamus upon the somatic sensory areas of the cerebral cortex.

2. Second or burning pain is served by the medial, phylogenetically older components of the system, projecting upon the generalized thalamocortical system, the ventral thalamus, and the hypothalamus. These projections account for the arousal potency of intense noxious stimuli and the affective and autonomic reactions produced by them. It is possible that the so-called pathologic pains of disease states are relayed over this system, and it is thus of paramount importance in the clinical control of pain—indeed, it is this system that is preferentially influenced by narcotic agents that relieve the suffering of pathologic pain, while leaving the perception of pricking pain and its threshold unchanged (Fig. 11-4).[25]

3. Activity relayed over either system, after destruction of the other, is capable of evoking the conscious perception of pain—of the kind characteristic for the remaining system.

4. Powerful interactions occur between the two subsystems, both by the reflected action of descending systems (Chapter 10) and by feed-forward interaction at subcortical levels. Indeed, lesions of the lateral component at thalamic or cortical levels result in release phenomena, for they are commonly followed by the severe hyperpathia of central origin described in later sections. It appears, moreover, that afferent activity in the lemniscal system also affects transmission in the anterolateral system, for lemniscal lesions at any level may on occasion result in hyperpathia. The mechanisms of these release phenomena are unknown.

These statements are based primarily on the results of anatomic studies in experimental animals and in man.[80, 81, 83] What direct evidence there is has been obtained by observing the effects of brain stimulation and of brain lesions upon pain sensibility in human beings. These have been greatly facilitated in recent years by the development of human stereotaxic instruments, devices that allow the passage of electrodes to many selected brain regions with considerable accuracy and with minimal damage to overlying structures. Such interventions are attempts to reduce or eliminate intractable pains of many types; e.g., invasive cancers commonly cause pain so severe it cannot be controlled by analgesics or narcotics.

Thalamocortical mechanisms in pain: effects of brain stimulation and lesions in man[8]

Brainstem. Transection of the spinothalamic systems low in the medulla produces an effective contralateral analgesia[36] because the site chosen for section, at the level of the inferior olive, is caudal to the major egress from the system of the medial components projecting upon the central reticular core. Transection of the descending tract of the trigeminal system is an effective method for pain denervation of the face (p. 373). Lesions have not been made elsewhere in the medulla and pons, for here neural systems concerned with pain are closely packed with others upon which life depends.

Mesodiencephalic junction. Stimulation of the spinothalamic and trigeminothalamic systems at this level elicits severe contralateral pain in waking human beings, as does stimulation of the medially adjacent region of the mesencephalic tegmentum—the region of ascent of the reticulothalamic systems.[7, 88] Stimulation of the tectum also evokes painful experiences.[109] Thus excitation of either the medial or the lateral components of the anterolateral system is effective in eliciting pain. To be effective in relieving pain, lesions at this level must include both components, though the desired extension of the lesion medially into the tegmentum is limited by unwanted effects.[108] Transection of the lateral components alone produces an initial contralateral hypalgesia for pricking pain but is soon followed by an accentuation of the pain for which the procedure is performed. It is a hyperpathia of central origin.[42]

Dorsal thalamic relay nuclei.[8, 55, 56] Repetitive stimulation in the most caudal portion of the human homologue of the VB nuclear complex (n. ventralis caudalis of Hassler) produces severe pain in waking human subjects that is referred to the contralateral side of the body in a somatotopic pattern. The more posterior and ventral the electrode tip, the more severe the pain evoked and the more widespread the peripheral reference, especially when stimuli are delivered to what Hassler terms the n. ventralis caudalis portae, or limitans, which is probably homologous to part of the PO group in other animals. Stimulation of other dorsal thalamic relay nuclei does not evoke pain.

Electrolytic lesions of the more caudal parts of the VB complex produce initially a contralateral analgesia to pinprick and relief from intractable pain. This relief may be short-lived, however, as it is commonly followed in days or weeks by a reappearance of the pain for which surgical therapy was initiated and may proceed to a state of excruciating hyperpathia.[16]

Intralaminar (generalized) thalamic nuclei. Stimulation in this region elicits peculiar feelings of anxiety in waking humans, feelings that are termed unpleasant but not overtly painful and that are occasionally associated with changes in levels of awareness. Lesions in the intralaminar region are reported to be effective in relieving severe chronic pain without producing significant neurologic changes. Such lesions should destroy n. centralis lateralis, the major thalamic target of the paleospinothalamic tract, as well as some of the zones of termination of the more medial spinoreticulodiencephalic small-fibered system. It should therefore relieve chronic pain that is caused largely by afferent input in C fibers without a resulting central hyperpathia.

Sensory areas of cerebral cortex. Considerable confusion still surrounds the question of the cortical mechanisms in pain sensibility. Available information has come almost entirely from observations of human subjects. Electrical stimulation of the postcentral gyrus under some circumstances elicits reports of pain in humans, although rarely.[90] In some patients, removal of the postcentral gyrus produces an immediate elevation of the pain threshold on the contralateral body surface, but this is soon followed, in many cases, by an exaggeration of any preexisting pain and by an accentuation of the pain elicited by noxious stimuli sufficiently strong to exceed the elevated threshold; a hyperpathic state ensues. Of the reasons for differing observations, the following appear to be of special significance. (1) A scanty but functionally significant ipsilateral projection exists in the anterolateral system, and interconnections between the two sides occur at segmental and thalamic levels. (2) The cortical representation of the contralateral skin surface is buried deep in the posterior bank of the central sulcus, which receives the heaviest cortical projection from the most caudal part of the VB complex, which in turn receives the ascending projection of the neospinothalamic system. This buried cortex may be inaccessible to stimulation of the cortical surface and may be spared by lesions or removals of the exposed surface of the postcentral gyrus. (3) Many of the clinical reports referred to were made before discovery of the second somatic sensory area (SII) or without consideration of the possibility that this area may play a role in cortical pain mechanisms. (4) The rapid onset of hyperpathia may obscure changes in threshold. For these reasons, in the following discussion the greatest weight will be given to positive observations.

Stimulation of the *postcentral gyrus* elicits contralaterally referred sensations ("paresthesias") that are occasionally painful in nature,[7, 45, 61, 113] more commonly so when stimuli are delivered to its buried portion.[43, 72] Lesions or surgical removal of one or another topographic area of the postcentral gyrus result in an immediate hypalgesia in the related body area, accompanied by loss in some of the discriminative aspects of somatic sensibility on the contralateral side.[43, 72, 79, 103] In some cases the change in pain sensibility is permanent, and relief from the pain for which removal was made persists. In the majority of cases, however, the relief of pain and the hypalgesia are soon followed by hyperpathia. It appears, therefore, that while pain, particularly pricking pain, is represented in a somatotopic pattern in the postcentral gyrus the frequent appearance of hyperpathia following postcentral removals indicates that this representation is but one part of the cortical mechanism in pain. This region is probably the locus of origin of descending systems controlling transmission of activity evoking pain, as evidenced by the release phenomena that follow its removal.

It was Biemond[17] who first called attention to the fact that a lesion of the cortex and white matter of the parietal operculum, which contains SII in man, may result in disturbances in pain sensation without appreciable loss of discriminative components of somatic sensibility. Talairach et al.[116] have shown that electrical stimulation of the white matter just beneath this region elicits responses that are often painful, and that destruction of these thalamocortical (and corticothalamic) fibers alleviated a severe pain of pathologic origin on the contralateral side of the body in each of eight patients, without the development of hyperpathia or any other sensory defect. These clinical observations are consonant with the results of studies in animals that show that many neurons of the PO complex of the thalamus and of cortical area SII are activated only by noxious stimuli,[22, 92] but the role of this system in pain sensibility is uncertain.

Other areas of the cerebral cortex. Disturbances in pain sensibility may occur as one of the many alterations in function produced by lesions of the *posterior parietal areas* described in Chapter 10. Lesions of the dominant hemisphere, particularly when they include the marginal gyrus, produce what Schilder termed an asymbolia for pain.[8, 102] A patient with such a lesion may retain a normal threshold to pain but no longer appreciate its destructive significance; e.g., he may no longer withdraw from threatening gestures. These patients also show the disturbances of body schema termed amorphosynthesis by Denny-Brown, associated with agnosias of various degrees and types. The observations indicate that there is a further higher order processing in the parietal lobe of activity evoked by noxious stimuli, and that pain plays a role in sensory and perceptive phenomena at the highest level.

When the severe and intractable pain of advancing disease is not relieved by attempted interruption of afferent systems concerned with pain, more drastic procedures are occasionally undertaken in attempts to alleviate suffering during the terminal months or years of life. Among these, lesions of the mediodorsal nucleus of the thalamus or transection of the fibers linking it to areas 9 to 12 of the frontal lobe upon which it projects ("frontal leukotomy") are often undertaken. Bilateral lesions of the mediodorsal nucleus or bilateral frontal leukotomy may diminish the anguish of constant pain, but such lesions also produce drastic changes in personality and intellectual capacities. The reactions of these patients to individual noxious stimuli may even be exaggerated, but they state that while their pain persists as before, it is no longer so disturbing and they may require little or no pain-relieving medication. With time there is usually some regression of the disorders in personality and intellectual functions caused by these lesions, but with it a recurrence of severe suffering from pain. Such interventions are considered by neurosurgeons as a last resort in the sequence of procedures aimed at the relief of intractable pain. The results do not indicate a projection of a pain system upon the mediodorsal nucleus and thence upon the frontal lobe.

PAIN ARISING FROM VISCERAL AND DEEP SOMATIC STRUCTURES[2]

The pain that arises from the viscera is of great importance to the physician, for knowledge of its mechanisms and particularly of the phenomenon of reference is essential for the diagnosis and location of disease processes in the thoracic and abdominal cavities. These depend on visceral afferent nerve fibers whose cell bodies lie in dorsal root ganglia or in the ganglia of certain cranial nerves. Despite the fact that their axons course peripherally through autonomic nerves, they are in every way analogous to afferent fibers of somatic nerves.

Visceral pain

In considering the nociceptive innervation of the viscera it is important to differentiate those fibers and pathways that are both afferent and sensory from those that evoke regulatory reflexes of various sorts when active, but when stimulated elicit no sensory experience. Aside from a sensory innervation of the upper trachea and esophagus, the vagus nerve, for example, contains no afferent *sensory* fibers. Stimulation of the vagus at any point below the recurrent laryngeal evokes no sensation in conscious human subjects and no reflexes characteristic of pain in animals. Yet about 80% of vagal fibers are afferent and an even larger percent are unmyelinated. They are mechano- and chemoreceptors apparently concerned with reflex regulation of gastric motility and secretion. Nociceptive afferents from the lower urinary and reproductive tracts reach the CNS via the pelvic parasympathetic nerves. Otherwise those innervating the thoracic and abdominal viscera run in sympathetic pathways, the cardiac and splanchnic nerves. These contain large numbers of unmyelinated and delta-sized myelinated afferent fibers. The splanchnic nerves contain in addition a few hundred larger myelinated fibers that end in the pacinian corpuscles of the mesentery. The function of this mechanoreceptive innervation is uncertain, but it is known that these receptors are exquisitely sensitive to vibratory stimuli. The parietal pleura and peritoneum and the outer borders of the diaphragm are innervated by branches of the thoracic and lumbar spinal nerves, and the center of the diaphragm is innervated by afferents of the phrenic nerve. Thus both visceral and somatic afferents may signal noxious processes that involve both the viscera and the body wall.

The parietal pleura and peritoneum are exquisitely sensitive, and light mechanical stimuli to them evoke pain in conscious humans. Under normal conditions, excessive distention of the gut or bladder evokes pain, as does strong contraction of their muscular walls, especially when working against obstructions to movement of their contents. Indeed, increased tension in the gut wall appears to be the only *normal* adequate stimulus evoking pain. The exposed normal intestinal mucosa is insensitive to mildly noxious stimuli. However, if the mucosa is hyperemic or inflamed, even light mechanical stimuli or dilute solutions of acid or alkali suffice to evoke the deep aching pain so characteristic of intra-abdominal disease. This is frequently attributed to a sparse innervation, but it seems more likely that the inflammatory process changes the threshold of nociceptive afferents, perhaps by the polypeptide mechanisms described on p. 356.

Experiments carried out by Lim[5] and associates have given further support to this idea. They studied a number of substances (bradykinin, substance P, 5-HT, histamine, ACh) known to produce pain in man when injected into the skin, when injected intra-arterially, or when applied to the exposed base of a cutaneous blister.[4] Of these, they found the nonapeptide bradykinin to be by far the most potent in eliciting action potentials in visceral afferents of the splanchnic nerve when injected directly into the splenic artery. Of course, the excitatory action of the bradykinin in this experiment may have been upon the nociceptive, the mechanoreceptive, or the chemoreceptive splanchnic afferents, or upon all three. Lim therefore devised procedures allowing perfusion of the spleen of a recipient from the circulatory system of a donor, leaving intact the normal innervation of the cross-perfused spleen. Injection of bradykinin, even in minute amounts, via cannula into the cross-perfused organ produced behavioral signs of pain and its characteristic reflex respiratory and circulatory actions. The experimental arrangement also allowed study of the site of action of analgesics and narcotics, whether peripheral or central, by testing their pain-suppressing action when injected into the donor dog and thus reaching only the peripheral ending of the visceral afferents, or into the recipient dog and thus reaching the CNS, but not the relevant peripheral nerve endings. Results indicate that the nonnarcotic analgesics such as aspirin act mainly peripherally, while the narcotics such as morphinelike synthetic drugs act centrally.

The release of proteolytic enzymes from damaged cells during hyperemia and inflammation of the gut wall and the resulting formation of pain-provoking polypeptides of the bradykinin type could thus account for the spontaneous pain that arises under these conditions and for the greatly lowered threshold of the inflamed mucosa, as compared with its normal insensitivity.

Referred pain

Pain that arises in the viscera of the thoracic and abdominal cavities may be felt at the site of primary stimulation, though poorly localized. Frequently such pain may be perceived as if occurring at a distant site on the skin surface innervated from the same spinal segment as the visceral locus of origin. Referred pain may appear either simultaneously with or indeed without pain appreciated as arising at the site of noxious stimulation. To a lesser extent this is also true of the deep somatic pain arising from muscles, joints, and periosteum. Hyperalgesia of the second type may occur in the region of reference (p. 357). A knowledge of the reference patterns produced by visceral disturbances is necessary for the accurate diagnosis and localization of disease processes; they are detailed in clinical treatises.

It has been observed that when the intensity of the disease process initiating a visceral pain is low or is just beginning, the pain may disappear when the site of reference or the afferent nerves innervating it are locally anesthetized. When, however, the pain-provoking process increases in severity, this may not be the case, and the pain may be referred to an area of skin even when the latter is completely anesthetic. The first observation suggests that the low-frequency sustained afferent discharge in some cutaneous nociceptive afferents, not normally sufficient to maintain a suprathreshold frequency in ascending anterolateral elements, sums with convergent input from visceral nociceptive afferents to do so. The convergence and summation could of course occur at a higher level of the anterolateral system. Why the sensation of pain then evoked is interpreted as arising from the skin and not the viscera is unknown. It is frequently stated that this false reference is due to the fact that any given segment receives many times more somatic than visceral afferents, and that during life experience reference is "learned" for the dominant input. There is no experimental evidence for such a statement.

The second observation, that local anesthesia of the area of reference may not affect that reference when visceral nociceptive input is intense, suggests that this latter may then alone drive anterolateral elements at suprathreshold frequencies. The problem of the mechanism of reference is the same.

In some cases the mechanisms of referred pain may be even more complex. If, as seems likely, nociceptive afferents from deep and superficial tissues are related to one another in the presynaptic inhibitory mode described in Chapter 6, then afferent impulses in one may at a certain frequency elicit antidromic impulses in the other. This antidromic activity in C fibers innervating the skin would be expected to produce there a vasodilatation and primary hyperalgesia (p. 357); indeed, this is exactly what is seen in some loci of referred pain. The possibility is open that referred pain, like hyperalgesia, may be produced by two quite different mechanisms, one peripheral and the other central.

Deep somatic pain

Pain of a particularly severe and aching quality is evoked by injury to deep structures of the body other than the viscera—from the muscles, fascia, joints, periosteum, and tendons. There is reason to believe that these deep structures are innervated by both delta and C fiber sizes of nociceptive afferents. Pressure, cutting, and heat—any stimulus destructive of tissue cells—are adequate stimuli for evoking pain from deep structures. The most potent stimulus for evoking pain from muscle is sustained or repetitive contraction; the pain is especially severe if the working muscle is deprived of its blood supply. This is thought to be due to the release of algesic substances from the anoxic contracting cells, probably one or more of the bradykinin-like polypeptides.

Visceral and deep somatic pains commonly evoke powerful reflex contraction of skeletal muscle, e.g., the continuous splinting contraction of the abdominal musculature provoked by intra-abdominal pain or the cocontraction of all muscles acting to fixate an injured joint. Muscles steadily contracting over long periods of time become themselves sources of painful afferent input, probably due to the release of proteolytic enzymes from the continually contracting muscle cells and the pain-provoking action of the resulting polypeptides. This muscle "soreness" may persist for hours or days after the original source of pain has disappeared.

The central nervous projections of visceral and deep somatic nociceptive afferents are thought to be identical with those previously described for the pain input from the skin.

HEADACHE MECHANISMS[9]

The most common deep somatic pain is headache, perhaps the most frequent complaint made by patients to their physicians. Regardless of their source, headaches are remarkably similar; the pain is of a deep, aching, diffuse nature, quick to arouse reflex contractions of extracranial muscles of the head, which commonly then become additional sources of nociceptive input themselves. Pain is frequently referred to a region of the head distant from the site of its initiation. Headache as a pain is similar in quality to that evoked from deep somatic tissues elsewhere in the body or, indeed, that arising from extracranial deep tissues of the head itself, e.g., the teeth or the orbital tissues.

Pain from within the cranium does not arise from the brain itself but from its supporting tissues, except when pain pathways are stimulated directly or when an increase in their activity occurs as a release phenomenon following brain lesion, the hyperpathia of central origin. Neurons and glia are not themselves innervated by nociceptive afferents. The intracranial structures that *are* so innervated have been identified by direct electrical or mechanical stimulation during neurosurgical procedures performed upon conscious patients after local anesthesia of extracranial tissues.[97] They are (1) the great venous sinuses and their large venous tributaries on the surface of the brain; (2) parts of the dura, particularly at the base of the skull; and (3) the cerebral arteries, particularly the middle meningeal and the great cerebral vessels close to their origin. The cranium itself, the brain parenchyma, most of the dura and the pia-arachnoid, the ependyma, and the choroid plexuses are insensitive. Thus pain may be evoked from within the head by traction on large veins or the venous sinuses, traction or dilatation of the middle meningeal artery, traction on the large arteries at the base, pulsatile distention of those arteries, or inflammatory processes about any pain-sensitive structure. Pain may also result from an irritating pressure on an afferent nerve, e.g., the root of the trigeminal, in its intracranial course. Headaches of intracranial origin are explained in terms of these mechanisms. Pain originating on the upper surface of the tentorium or anywhere in the cranial vault above it is referred anterior to a frontal intra-aural plane and is mediated via nerve V. Pain originating in the the posterior fossa is referred behind this plane and is mediated via IX and X and the upper two or three cervical nerves.

Headache associated with changes in intracranial pressure. In the adjustments made to changes in posture the CSF follows closely the venous pressure measured at the same level of the intracraniospinal system. In the erect position, both reach negative values, while pressure in the lumbar sac rises. The difference in specific gravity between the brain and the fluid in which it is suspended leaves a net weight of about 40 gm, which is carried by the intracranial supporting structures, principally the large veins and venous sinuses, and the tentorium. When CSF is removed, this net suspended weight evokes pain by stretch or distortion of the suspending tissues. It is of course greatly increased when CSF is replaced by air. The pain evoked by drainage is at first referred to the vertex and front of the head but may then increase in severity and spread widely. It is relieved by restoration of the CSF and at least partially by assumption of a recumbent position. The headache that commonly follows a needle puncture of the lumbar sac, a frequently em-

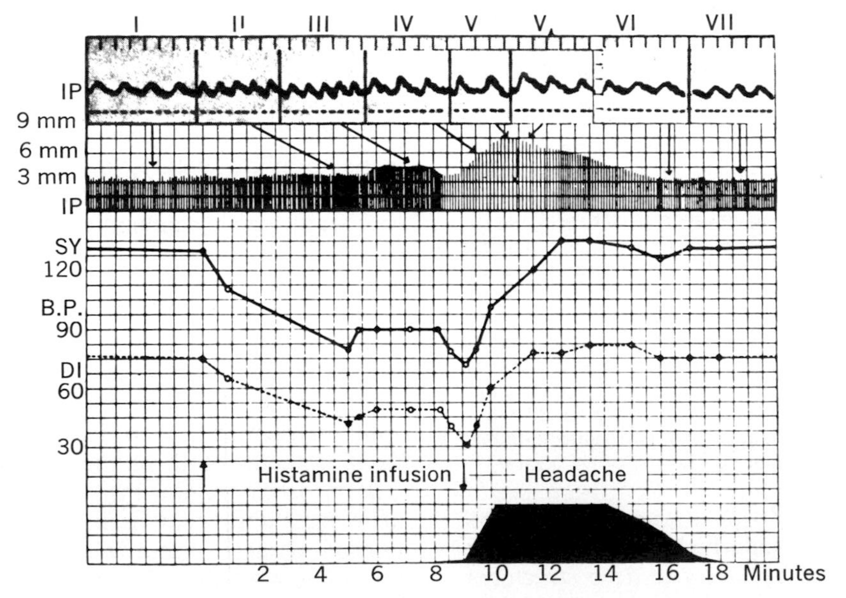

Fig. 11-12. Representation of several physiologic variables in human subject when potent vasodilator, histamine phosphate, was infused continuously for 9 min at rate of 0.1 mg/min. *SY* and *DI* = systolic and diastolic blood pressures. *IP* = record of intracranial pulsations, parts of which are displayed on expanded time scale in top record. Systemic blood vessels recover tone more rapidly than do intracranial ones; resulting increased intracranial pulsations produced severe headache. Both increased pulsations and headache can be prevented by *increasing* intracranial pressure. (From Wolff.[9])

ployed diagnostic procedure, is thought to be due to a continued leak of CSF through the dural hole.

Increased intracranial pressure per se does not produce headache. Brain tumors cause increased intracranial pressure when they block the free flow of CSF through the ventricular system of the brain; they produce pain by direct mechanical traction upon or distortion of pain-sensitive structures within the head.

Headache produced by distention of cerebral arteries. Severe headache is produced by direct mechanical distention of cerebral vessels resulting from excessive pulsation. Such an event follows the intravenous injection of histamine. As shown in Fig. 11-12, histamine injection first produces a drop in systemic blood pressure due to a generalized vasodilatation. Recovery proceeds more rapidly in extracranial than in intracranial vessels, so that the recovering blood pressure then distends the passive cerebral vessels with each systole of the heart, producing a severe throbbing headache that subsides with the recovery of intracranial vascular tone and decreased intracranial pulsation. The extracranial and meningeal arteries are thought to play only a minor role in the headache produced by histamine. This experimentally induced pulsation headache can be abolished by increases in the intracranial pressure. Headaches associated with fever and those produced by the inhalation of vasodilator substances are thought to be of this pulsation type.

Headache arising from extracranial tissues of the head. Of all headaches that occur, those just described are relatively rare. Much more common are those that arise from extracranial tissues. Diseases of the nose, the paranasal sinuses, and the orbital contents, for example, frequently announce their presence by producing headache, and it is these that elicit powerful and pain-provoking contraction of the temporalis muscle and the muscles of the neck. Knowledge of their origin and particularly their patterns of reference is important for accurate diagnosis; they are detailed in monographs[7, 9] and in clinical texts. Many headache syndromes of particular interest arise from the extracranial vessels, particularly from the temporal arteries, which are exquisitely sensitive to manipulation or distention. Intense headache is produced by inflammation of the walls of these vessels. More common are those severe headaches grouped under the generic term "migraine," a periodically recurring headache that is frequently unilateral in onset but that may become generalized and is sometimes associated with a more generalized vascular disorder. The pain is produced by excessive pulsation of the temporal arteries, and an attack can frequently be aborted by a vasoconstrictor agent (Fig. 11-13). If a migraine attack is allowed to proceed, edema fluid collects in the extracranial perivascular tissues. This fluid contains proteolytic enzymes and pain-provoking polypeptides of the neurokinin-bradykinin type (p. 358).[27, 89] Thus the mechanism of headache production in migraine is dual in nature: an initial vasodilation and increased pulsatile distention of the arteries, followed by the extravasation of pain-producing polypeptides.

Headache is frequently associated with the continued elevation of blood pressure called "essential" hypertension. The headache, like

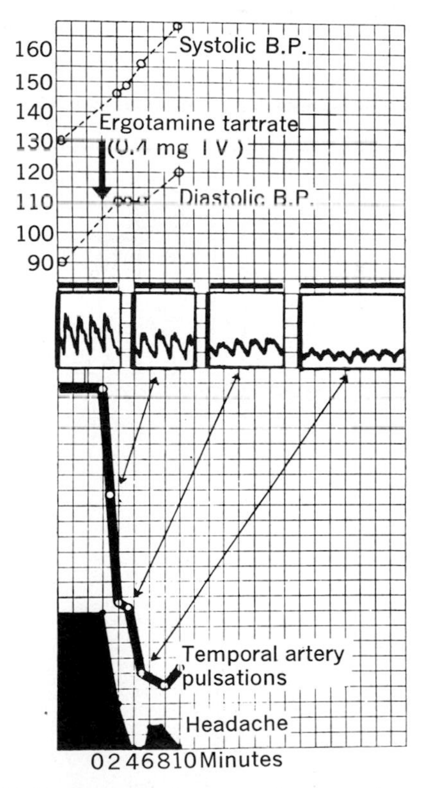

Fig. 11-13. Records of course of events when headache of migraine type is interrupted by administration of potent vasoconstrictor. Decreased amplitude of pulsations of temporal artery (middle line of records and plot below) is associated with remarkable decrease in intensity of headache, although systemic blood pressure rises. (From Wolff.[9])

that of migraine, is due to excessive pulsation of the extracranial arteries. It occurs most commonly when smooth muscle tone is low and pressure is high, but it may occur if the former is low enough when the latter is normal.

DISORDERS OF THE PERCEPTION OF PAIN AND PAIN OF CENTRAL ORIGIN

Many disorders of the peripheral nerves and of the CNS lead to a condition of spontaneous pain or to pain that is produced by trivial, normally innocuous stimuli. These pains are among the most severe known to man, may come to dominate completely the behavior of the sufferer, and present extraordinarily difficult therapeutic problems. Of these, only a few that cast light upon normal pain mechanisms will be discussed.

Painful states produced by disorders of peripheral nerves

Nerve section. Head made a major contribution to sensory physiology by describing in detail the changes in sensation that follow transection of a cutaneous nerve innervating the distal parts of an extremity.[3, 101] The speculative elaborations he made, based upon his initial observations, have stimulated controversy; nevertheless, his ideas concerning somesthesia have a modern ring when transferred from a peripheral to a central setting. The facts are these. When a peripheral nerve innervating the arm and hand is sectioned, as Head did in his own arm, the changes in sensation produced are not uniform throughout the area affected (Fig. 11-14). Between the zone of normal sensation proximally and the small area of complete anesthesia distally, there are three zones of altered sensibility, characterized by the fact that normally neutral stimuli produced unpleasant sensations when applied to them. In the first, bordering skin with normal sensation, an extremely unpleasant, poorly localized sensation is produced by a moving cotton wisp, by light pinprick, or by cold. An extremely light touch or warmth is not felt. Distal to this, in the second area, pinpricks, heat, cold, and pressure are all identified as painful. Last, next to the zone of total anesthesia, the skin is insensitive, but deep pressure evokes a deep throbbing pain that radiates up the wrist and arm. It is important to emphasize, as Denny-Brown[40] has done, that these changes appear *immediately* after nerve transection and do not depend upon differential rates of nerve regeneration as Head supposed, although the picture may be altered during regeneration. The present interpretation placed on these findings is that the overlap in the fields of distribution of peripheral nerves differs for different fiber groups, being least for those myelinated afferents above approximately 6μ in size that project into the DC–medial lemniscal system and greatest for the small myelinated delta fibers and the C fibers innervating the skin and deep somatic tissues that project into the anterolateral system. Section of a nerve therefore exposes a zone

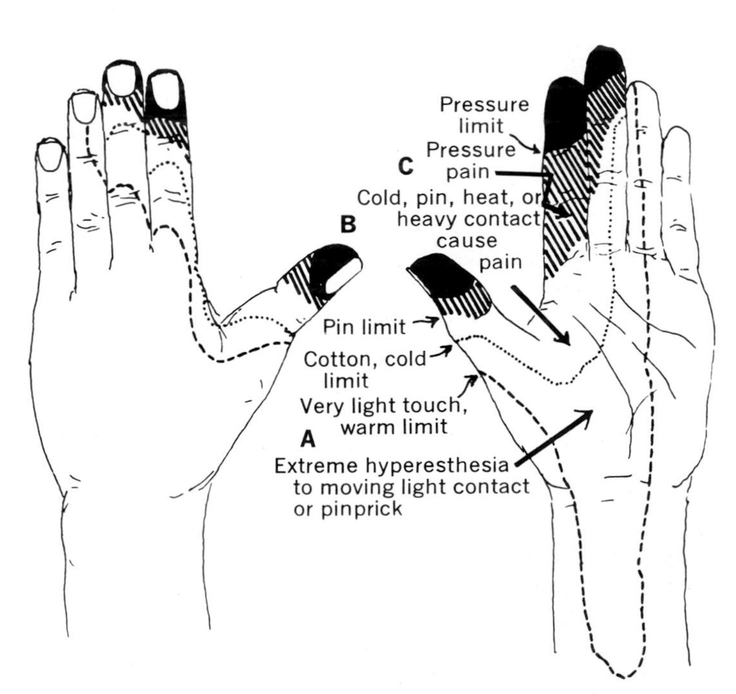

Fig. 11-14. Representation of sensory findings in right hand of 39-year-old patient following transection of medial nerve, with no regeneration. Areas *A* and *B* exhibit hyperesthesia, it is thought, because these regions are now represented centrally only in anterolateral system and not in lemniscal system: symptoms are release phenomena. Excessive pain on deep pressure in zone *C* is thought to be similar in origin. (From Denny-Brown.[40])

of peripheral tissue, both skin and subcutaneous, which is represented centrally only in the anterolateral system. The lack of the discriminative capacities in these zones, the diffuse and radiating character of sensations evoked from them, and above all the hyperpathia that appears immediately after nerve section are all consistent with the view that under normal circumstances the transmission of activity into and up the anterolateral system is controlled by simultaneous activity in the lemniscal system. It is apparent that this control may be exerted at every level of the somatic system from dorsal horn to cerebral cortex, for the release phenomena described may follow lesions at any level. Head labeled these two aspects of somatic sensibility epicritic and protopathic, and he postulated that they are served by two completely different sets of peripheral nerve fibers. This latter is certainly not literally correct, for peripheral nerve fibers are distributed along a continuum as regards size and other properties, but a division placing the slower delta and the dorsal root C fibers in the latter category and all other myelinated cutaneous afferents in the former is not far from the mark. Head's concept of duality of the somatic afferent system, although he applied it only at the periphery, fits well what is known of the lemniscal and anterolateral systems described in Chapter 10. It is still uncertain, but there is evidence to suggest that the neospinothalamic system—or at least that part that projects upon the VB complex and that develops rapidly in the series of primates—possesses in man properties more lemniscal than anterolateral in nature.

Causalgia.[8] The classic syndrome of causalgia was described in men surviving gunshot wounds of the extremities by Mitchell in 1864; it occurs in less than 5% of such wounds that damage the nerves, most commonly when the median and sciatic nerves are partially injured but not transected; injury to arteries is not a necessary concomitant. A severe, burning hyperpathia appears soon after injury in the distribution of the nerve. The hand or foot is first warm and dry and is later cool and sweats profusely. When severe, the pain is so great that the patient cannot bear the slightest touch, even of clothing or puffs of air. The pain is greatly accentuated during emotional disturbances or with any increased peripheral sympathetic activity. Stimulation of sympathetic efferents to the extremity greatly accentuates the pain,[120] and sympathectomy is a specific and almost uniformly successful treatment. It is therefore not a hyperpathia of the type that follows complete section of a nerve as described earlier. It is frequently hypothesized that injury to the nerve damages both sympathetic efferents and nociceptive afferents in such a way that the extrinsic currents accompanying impulses in the former elicit afferent impulses in the latter at the site of injury, a theory that fits well the exacerbation of the pain during increased sympathetic activity and its relief by sympathectomy. There is, however, no direct evidence that this is so in causalgia, and it is possible that a zone of nerve anoxia is produced by vascular injury within the nerve, so that afferent impulses in C fibers normally of subthreshold frequency now elicit at the site of injury prolonged and repetitive trains of impulses that enter the CNS, a well-known property of anoxic nerve. According to this theory, sympathectomy would be effective by eliminating vasomotor efferents to the vasa nervorum, thus improving local blood supply.

Trigeminal neuralgia.[8] One of the most severe of the neuralgias occurs in the distribution of the fifth cranial nerve. It is characterized by brief, often serially repetitive attacks of severe, lightning-like pains, limited unilaterally to the distribution of the trigeminal nerve, interspersed with intervals free of pain. When severe, the attacks occur rapidly and without any obvious provoking stimulus; more commonly they are elicited by any light mechanical stimulus delivered to what may be very small trigger zones on the upper lip or the nasolabial folds.[68] Cold, warmth, or noxious stimuli do not evoke attacks; it is activity in the smallest myelinated and C fiber mechanoreceptive afferents that is most potent in doing so. Between attacks no abnormalities of facial sensation can be demonstrated; indeed, only rarely is any neurologic deficit apparent, and hyperesthesia of the face does not occur. The condition is relieved by transection or block of peripheral branches of the trigeminal or, more permanently, by section of its root between the gasserian ganglion and the brainstem. Presumably the resulting anesthesia is effective by eliminating the background inflow of impulses on which, it is inferred, the spontaneous attacks depend and by preventing provocation by trigger stimuli. Sjöqvist[107] has shown that transection of the descending trigeminal tract at the level of the obex effectively relieves the condition and produces a dissociation of facial sensation so that pain and temperature sensibilities disappear, with only a slight reduction in tactile acuity. This is strong evidence that the n. caudalis of the trigeminal complex is essential for the transmission of pain from the face. The pathophysiology of trigeminal neuralgia is not known with certainty. Frequently it appears to be produced by an irritative compression of the gasserian ganglion or the root by vessels, dural strictures, etc.; at least temporary relief is afforded in many cases by a "decompression" of the ganglion and root. According to this theory, damage to the fibers of the root or ganglion allows a "cross-talk" between mechanoreceptive and nociceptive fibers, similar to one explanation of the causalgia-like pain of the extremities that may follow nerve injury. Indeed, demyelination of trigeminal fibers in the root and ganglion has been described. Other investigators believe that the condition is produced by a disorder of synaptic mechanisms within the n. caudalis itself. The evidence supporting these conflicting points of view has been summarized.[66, 67]

Pain of central origin: the thalamic syndrome[113]

Pain in the absence of overt noxious stimulation may be produced by acute "irritative" disease processes involving the central pain pathways or may occur when central neurons concerned with pain discharge at high frequencies during epileptic attacks. There is, however, another form produced by stable, nonirritative lesions of the CNS, called central pain, defined as spontaneous pain with a characteristic overreaction to external stimuli. The prototype is the thalamic syndrome originally described by Dejerine and Roussy,[39] which may follow vascular lesions that destroy parts of the ventral or lateral nuclear complexes of the dorsal thalamus. It is characterized by a contralateral, severe, and persistent pain, frequently so intolerable as to be uncontrolled by narcotic agents. The threshold for pain sensation may be normal or elevated, but in either case once threshold is reached by a stimulus there is an explosive onset of pain that may spread to regions far removed from the offending stimulus. In severe cases such explosive episodes are produced by normally neutral stimuli. When the lesion involves the VB complex, there is some loss of discriminative cutaneous sensibility, which may be absolute, and position sense, and thus a variable degree of astereognosis. These general features of the central pain syndrome may appear with lesions at any level of the somatic afferent system. They may occur following the degeneration of the myelinated fibers of peripheral nerves caused by vitamin deficiencies or tabes dorsalis and are produced commonly by lesions at the thalamic level, only rarely by lesions of the DCs,[44] and occasionally by lesions of the parietal cortex itself. The appearance of spontaneous pain of central origin caused by destructive but nonirritative lesions is interpreted as a release phenomenon similar in principle to that occurring after section of a peripheral nerve in the zone between normal and totally anesthetic tissue. Such lesions interrupt afferent inflow over myelinated fibers that produce in their central projections a tonic level of inhibition at synaptic junctions concerned with the transmission of nociceptive input. It may occur also by driving those descending control mechanisms of the somatic system described in Chapter 10. What is uncertain is whether the control is exerted and evoked mainly by activity in the lemniscal system or in the neospinothalamic component of the anterolateral system, or in both.

Control of pain by stimulation of suppressing mechanisms.[8] The release phenomena just described have led to the idea that electrical stimulation of the larger myelinated fibers of cutaneous nerves or of the central systems through which they project might be an effective therapeutic method for controlling pain. It has been shown that stimulation of the beta-sized mechanoreceptive afferents, either by electrical stimulation of a nerve trunk or by natural means (e.g., vibratory stimulation of the skin), will suppress pain arising from the peripheral innervation field of the nerve stimulated. This suppression may outlast the period of stimulation, in some cases up to 1 hr. A similar suppression is produced by stimulation of the DCs via chronically implanted electrodes,[105] and there are scattered reports of the suppression of pain during electrical stimulation of one or another central locale. Certainly these observations are tentative, but they do suggest that further research may demonstrate the feasibility of controlling otherwise unbearable pain by stimulation of portions of the brain via permanently implanted electrodes and stimulating devices.

ITCH AND TICKLE[65]

Itching is that sensation, recognized by all persons with normal innervation of the skin, that evokes a desire to scratch. It may vary from a barely perceptible annoyance to a sensation so intense that it dominates behavior as totally as does severe pain. Tickle is the itching component of the sensation evoked by a light moving mechanical stimulus. Both itch and tickle are followed by a strong sensory "afterimage" that may last for seconds after removal of the stimulus. Examination of the skin with a fine mechanical stimulus or by electrical stimuli delivered via an intracutaneous microelectrode reveals a punctate distribution of itch spots in the skin; they appear to be identical with the pain spots. Itch differs from pain in that it can be evoked only from the skin, the palpebral conjunctiva, and portions of the skin and mucous membranes of the nose, but not from any deep tissue of the body.

Several lines of evidence indicate that impulses evoking itch are conducted in C fibers: (1) the reaction time to itch evoked by electrical stimulation of itch spots varies from 1 to 3 sec with body location; (2) itch disappears *pari passu* with burning pain with the onset of the block of cutaneous nerves produced by local anesthetic when the discriminative aspects of somesthesia are still intact; and (3) itch persists along with burning pain when nerve trunks are blocked by pressure at a time when only C fibers are still conducting. In common with second pain, itch is exaggerated, diffuse, and poorly localized in skin deprived of A fiber innervation. A more highly localized type of "pricking itch" may depend upon delta fibers in parallel with pricking pain, but this is uncertain.

Itch spots identified by stimulation have, like pain spots, been shown to be regions of increased density of bare nerve terminals of unmyelinated and the thinnest myelinated fibers. Those evoking itch appear to terminate more superficially in the skin than do those of nociceptive afferents, for itch is not evoked by stimuli directed beneath the epidermodermal border.

Local mechanical stimuli or the intradermal injection or natural release of chemicals are adequate itching stimuli (Fig. 11-15). Of these, histamine is a powerful pruritogenic agent. It evokes itch with a long latency (5 to 10 sec) compared to that of mechanical stimuli. When pruritic, it invariably produces wheal and flare in the skin. It can be diluted to a point at which it no longer evokes itching, but it then still produces wheal and flare. Spicules of cowhage (the common itching powder) produce itching by virtue of their content of a proteolytic enzyme, mucanain, and indeed it has now been shown that pro-

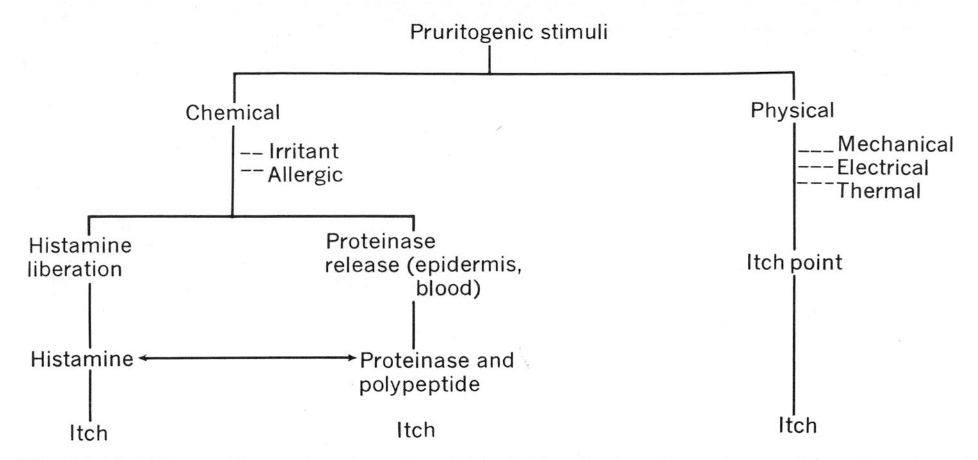

Fig. 11-15. Scheme illustrating ways in which itching is thought to be produced in skin: (1) by direct physical excitation of nerve endings or (2) by release of histamine and/or proteolytic enzymes from damaged cells. (From Arthur and Shelley.[11])

teinases from a wide variety of sources produce itching upon intradermal injection, even in a very low concentration.[11] Whether they act directly or by formation of polypeptides is uncertain; the latter seems likely, for polypeptides of the bradykinin type are themselves pruritogenic in very low concentrations. Proteinases or the polypeptides they produce may act independently of histamine release, for they still produce itching in (1) histamine-insensitive subjects and (2) in regions of skin previously depleted of histamine. Some proteolytic enzymes, like trypsin, are not pruritogenic under these conditions and presumably do act by histamine release.

A region of local itch is surrounded by an area of itchy skin in the hyperalgesic paradigm, presumably produced by the axon reflex mechanism. Painful stimuli, on the other hand, inhibit itching sensation: scratching must usually be painful to be effective.

The central neural mechanisms responsible for itching are on every count similar to those described for pain, and only very rarely have reports of their dissociation by central lesions been made. However, some central lesions, e.g., tumors of the spinal gray or of the pons, may early in their course produce intolerable itching, presumably by an excitatory irritation of central neurons concerned with itching sensation.

From much of the previous discussion it might be concluded that itch is but another form of pain, served by the same peripheral C fibers and central neural mechanisms. Indeed, this hypothesis has frequently been put forward, with the statement that itch is produced by low-frequency discharges in pain fibers. There are several facts that make this unlikely:

1. Low-frequency discharge in nociceptive C fibers, where brought under direct control, evoke severe pain with frequencies as low as 3/sec.
2. Increasing frequency of electrical stimulation of an itch spot, without increasing intensity, may increase the intensity of the itch without evoking pain.
3. The reflex effects evoked by noxious and by itching stimuli are totally different; that of the first is flexion withdrawal, that of the second, the temporally and spatially organized patterns of the scratch reflex.
4. The two can be dissociated peripherally. Immersion of the skin in water at 41° C quickly abolishes itch but intensifies pain. Skin stripped of its epidermis is insensitive to any itch-provoking stimulus but exquisitely sensitive to noxious ones.

A more likely hypothesis is that itch is served peripherally by a set of highly sensitive mechanoreceptive C fibers.

THERMAL SENSIBILITY

Temperature sensibility is composed of two separate and distinct qualities, warmth and cool, which are served by two separate and differentially sensitive sets of first-order afferent fibers. Sensitivity to temperature change is not uniform over the skin and mucous membranes, but distributed in a spot-like fashion; sensitivity is greatest where spots are most dense. In general, cold spots are more numerous than warm spots by ratios of 4:1 to 10:1, and both are more common on face and hands than elsewhere on the body.

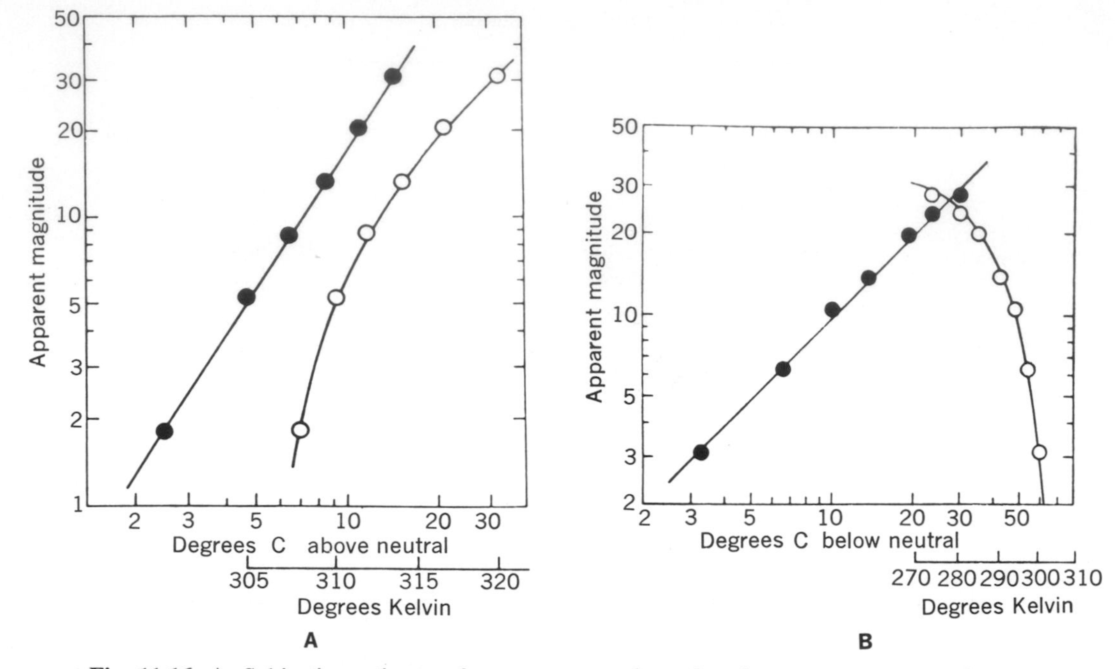

Fig. 11-16. A, Subjective estimate of apparent warmth; each point represents geometric mean of 36 estimates made by 12 observers of warmth of metallic cylinders, at various temperatures, applied to skin of forearm. Lower abscissa, for unfilled points, is log scale of absolute temperature (in degrees Kelvin). Upper abscissa, for filled points, is log scale of difference in temperature (degrees centigrade) between stimulus and assumed "physiologic zero" of 32.5° C. Line connecting filled points has a slope of 1.6. **B,** Results of similar experiment using cold stimuli and an assumed "physiologic zero" of 31.0° C. Slope of line connecting filled points is about 1.0. (From Stevens and Stevens.[112])

The ratio and hence relative sensitivity vary markedly; e.g., the forehead is very sensitive to cold but relatively insensitive to warmth, and some regions are completely insensitive to the latter. The sense of temperature differs markedly from pain in two ways: over a certain range of temperature, adaptation to a state of sensory neutrality occurs, and spatial summation is a salient feature of warmth and cool. The important role of the peripheral thermoreceptive afferents in the regulation of body temperature is described in Chapter 56.

Using the method of subjective magnitude estimations, Stevens and Stevens[112] have shown that the relations between the human estimates of the intensities of thermal stimuli and their temperatures are adequately described by power functions (Chapter 18), as shown in Fig. 11-16. This is true, however, only when the skin temperature at the level of thermal neutrality is entered as a subtractive constant, thus creating a ratio scale for the independent variable in terms of physiologic meaning as well as physical rating. A major factor in the detection of temperature changes and the scaling of temperature sensations is the area of skin stimulated, for there is a continuing spatial summation of the input as the exposed area is increased, especially for warmth. Stevens and Marks[111] have recently shown that for any areal extent stimulated, the degree of apparent warmth does indeed grow as a power function of the intensity level, but the smaller the area, the greater the exponent of that function. The exponent increases from about 1.0 for large areas to 1.6 for small ones, as in the experiment illustrated in Fig. 11-16. These families of power functions for warmth extrapolate to a point of convergence at about the level of heat intensity that leads to tissue damage and pain. Intensity and area trade one for the other to evoke the same sensations of warmth over a considerable range. When the entire body is exposed in a climate chamber, the region of thermal indifference is greatly narrowed, and even at rates of increase in skin temperature as low as 0.001° C/sec, the warmth threshold is reached at a skin temperature slightly less than 35° C.[78]

The subjective magnitude estimation function for the sense of cool is nearly linear (Fig. 11-16). This sense shows another property of interest. Removal of a cool stimulus is followed by a persisting, gradually fading feeling of cool lasting for many seconds. This is due to the fact that the cool afferents continue to discharge as long as the temperature at their terminals is below a certain level, regardless of the direction of the thermal gradient, which is reversed when the stimulus is removed from the skin.

Relation between temperature change in skin and thermal sensations

A constant thermal sensation persists indefinitely when the temperature of the skin is below 20° or above 40° C. Adaptation occurs in the range between, so that the thermal sensation produced by an imposed temperature step gradually fades to one of thermal neutrality. This adaptation takes longer when the deviation of the imposed step from the initial skin temperature is greater. The period of persisting sensation long outlasts the time required for the intracutaneous temperature to reach a constant value. It is clear from this fact as well as from the persistence of a constant sensation when skin temperature is below 20° or above 40° C that the adequate stimulus is the absolute skin temperature, not a transcutaneous temperature gradient. Thermoreceptors are likely to be exquisitely sensitive to the temporal gradients of thermal stimuli, but to signal the instantaneous temperature, *not* its spatial gradient, e.g., between the endings and the stem axons.

Within the range of 20° to 40° C, for uniform rates of change of different slopes, adaptation produces a shift in threshold that depends on the initial skin temperature and the slope of the imposed change. Very rapid rates of temperature change in either direction evoke a thermal sensation when the skin temperature has shifted by only 1° to 2° C. Very gradual rates of change, however, produce very large shifts in skin temperature before any report of thermal sensation is made. Outside the range of 20° to 40° C, the rate of change required to evoke sensation is zero: these are the zones of constant thermal sensation.

Peripheral thermoreceptive nerve fibers

Nerve blocking experiments using pressure or local anesthetic agents show that the two thermal senses disappear together, simultaneously with pricking pain. When all myelinated fibers are blocked, leaving only C fibers conducting, human subjects are sensitive only to those extremes of heat and cold that are known to excite hot and cold nociceptive afferents. This suggests that both warm and cold thermoreceptive afferents in man are members of the thinly myelinated delta group of fibers. These terminate in the skin as bare nerve endings, scarcely discriminable by light microscopy from the similar endings of C fibers.

Knowledge of the functional properties of cold and warm fibers has been greatly advanced by a recent series of studies by Darian-Smith and his colleagues,[37, 38, 64] particularly because their studies were made on fibers innervating the glabrous skin of the hands of monkeys. The palmar skin and its innervation appear identical in man and monkey. This investigation takes on added value by virtue of the close quantitative comparisons made between the electrophysiologic studies of nerve fibers innervating the monkey hand and psychophysical measures of human subjects' capacity to discriminate between thermal stimuli delivered to their hands. The results are summarized as follows:

1. Cold fibers are unresponsive to mechanical deformation of the skin. They constitute about one third of the A-delta fibers innervating the monkey hand and have a mean conduction velocity of about 14 m/sec; each terminates in a spotlike receptive field (< 1 mm diameter) in the glabrous skin. These afferents respond with a high-frequency onset transient to the onset of a cooling pulse that declines to a more or less steady rate of discharge for pulses up to a few seconds in duration. There is a remarkably small variability in their responses to slowly repeated cooling pulses of the same intensity. The stimulus-response function is linear over a range of 8° to 10° C downward from neutral skin temperature. It is only this input in cold fibers that can account for the human capacity to differentiate the intensity of cooling pulses. Although many large-fibered, slowly adapting mechanoreceptive afferents innervating the glabrous skin are excited by cooling pulses, they are more than an order of magnitude less sensitive to cold than are the cold afferents themselves. *Such a differential sensitivity is the essence of modality specificity.*

2. Warm fibers are unresponsive to mechanical deformation of the skin but are exquisitely sensitive to warming pulses. They are among the smallest of delta fibers (mean

conduction velocity 2.5 m/sec) and terminate in the glabrous skin in single small spots. Their response differs from that of cold fibers, for there is no onset transient discharge at the onset of an abrupt warming pulse; the frequency of discharge gradually rises over the first second after pulse onset to a steady level set by the intensity of that pulse. The variability of responses to trains of slowly repeated pulses of equal intensity is low; the stimulus-response function is linear. It is only these warm fibers that can account for the human subject's capacity to discriminate between warming pulses of slightly different intensities.

3. The human discriminable increment between pairs of warming or pairs of cooling pulses is fully accounted for by the differences in the neural discharge evoked by members of the pairs. This increment is identical regardless of the overall level of activity during the two pulses, which may of course be quite different depending on the general intensity level of the two pulses. The form of the Weber function, which relates the fractional increment in cool or warmth that can be discriminated to the intensity level at which the discrimination is made, is set by the characteristics of the populations of thermoreceptive afferents and not by subsequent central neural events.

REFERENCES

General reviews

1. Chapman, L. F., Ramos, A. O., Goodell, H., and Wolff, H. G.: Neurohumoral features of afferent fibers in man. Their role in vasodilatation, inflammation, and pain, Arch. Neurol. **4:**617, 1961.
2. Hardy, J. D., Wolff, H. G., and Goodell, H.: Pain sensations and reactions, Baltimore, 1952, The Williams & Wilkins Co.
3. Henson, R. A.: Henry Head's work on sensation, Brain **84:**535, 1961.
4. Keele, C. A.: Chemical causes of pain and itch, Ann. Rev. Med. **21:**67, 1970.
5. Lim, R. J. S.: Pain, Ann. Rev. Physiol. **32:** 269, 1970.
6. Mehler, W. R.: Some observations on secondary ascending afferent systems in the central nervous system. In Knighton, R. S., and Dumke, P. R., editors: Pain, Boston, 1966, Little, Brown & Co.
7. Pain, Res. Publ. Assoc. Res. Nerv. Ment. Dis. **23:**entire issue, 1943.
8. White, J. C., and Sweet, W. H.: Pain and the neurosurgeon: a forty year experience, Springfield, Ill., 1969, Charles C Thomas, Publisher.
9. Wolff, H. G.: Headache and other pain, New York, 1963, Oxford University Press, Inc.

Original papers

10. Adair, E. R., Stevens, J. C., and Marks, L. E.: Thermally induced pain, the dol scale and the psychophysical power law, Am. J. Psychol. **81:**147, 1968.
11. Arthur, R. P., and Shelley, W. B.: The peripheral mechanism of itch in man. In Wolstenholme, G. E. W., and O'Connor, M., editors: Pain and itch; nervous mechanisms, Boston, 1959, Little, Brown & Co.
12. Beecher, H. K.: Pain in men wounded in battle, Ann. Surg. **123:**96, 1946.
13. Beecher, H. K.: The measurement of pain, Pharmacol. Rev. **9:**59, 1957.
14. Berlin, L., Goodell, H., and Wolff, H. G.: Studies on pain: relation of pain perception and central inhibitory effect of noxious stimulation to phenomena of extinction of pain, Arch. Neurol. Psychiatry **80:**533, 1958.
15. Bessou, P., and Perl, E. P.: Response of cutaneous sensory units with unmyelinated fibers to noxious stimuli, J. Neurophysiol. **32:**1025, 1969.
16. Bettag, W., and Yoshida, T.: Stereotaktische Eingriffe in verschiedenen thalamischen Kerngebieten zur Behandlung unbeeinflussbarer Schmerzzustande, Arzt. Forsch. **14:**527, 1960.
17. Biemond, A.: The conduction of pain above the level of the thalamus opticus, Arch. Neurol. Psychiatry **75:**231, 1956.
18. Biglow, N., Harrison, I., Goodell, H., and Wolff, H. G.: Studies on pain: quantitative measurements of two pain sensations of the skin with reference to the nature of the "hyperpathia" of peripheral neuritis, J. Clin. Invest. **24:**503, 1945.
19. Bishop, G. H.: The relation between nerve fiber size and sensory modality: phylogenetic implications of the afferent innervation of cortex, J. Nerv. Ment. Dis. **128:**89, 1959.
20. Bishop, G. H., and Landau, W.: Evidence for a double peripheral pathway for pain, Science **128:**712, 1958.
21. Burgess, P. R., and Perl, E. R.: Myelinated afferent fibres responding specifically to noxious stimulation of the skin, J. Physiol. **190:** 541, 1967.
22. Carreras, M., and Andersson, S. A.: Functional properties of the anterior ectosylvian gyrus of the cat, J. Neurophysiol. **26:**100, 1963.
23. Chang, H.-T., and Ruch, T. C.: Topographical distribution of spinothalamic fibres in the thalamus of the spider monkey, J. Anat. **81:** 150, 1947.
24. Chapman, L. F., and Wolff, H. G.: Studies of proteolytic enzymes in cerebrospinal fluid, Arch. Intern. Med. **103:**86, 1959.
25. Chapman, L. F., Dingman, H. F., and Ginzberg, S. P.: Failure of analgesic agents to alter the absolute sensory threshold for the simple detection of pain, Brain **88:**1011, 1965.
26. Chapman, L. F., Goodell, H., and Wolff, H. G.: Augmentation of the inflammatory reaction by activity of the central nervous system, Arch. Neurol. **1:**557, 1959.
27. Chapman, L. F., et al.: A humoral agent implicated in vascular headache of the migraine type, Arch. Neurol. **3:**223, 1960.

28. Chapman, W. P., and Jones, C. M.: Variations in cutaneous and visceral pain in normal subjects, J. Clin. Invest. **23**:81, 1944.
29. Christensen, B. N., and Perl, E. R.: Spinal neurons specifically excited by noxious or thermal stimuli: marginal zone of the dorsal horn, J. Neurophysiol. **33**:293, 1970.
30. Clark, W. C.: Sensory-decision theory analysis of the placebo effect on the criterion for pain and thermal sensitivity (d'), J. Abnorm. Psychol. **74**:363, 1969.
31. Collins, W. F., and O'Leary, J. L.: Study of a somatic evoked response of midbrain reticular substance, Electroencephalogr. Clin. Neurophysiol. **6**:619, 1954.
32. Collins, W. F., and Randt, C. T.: Evoked central nervous system activity relating to peripheral unmyelinated or C fibers in cat, J. Neurophysiol. **21**:345, 1958.
33. Collins, W. R., and Randt, C. T.: Midbrain evoked responses relating to peripheral unmyelinated or C fibers in cat, J. Neurophysiol. **23**:47, 1960.
34. Collins, W. F., Nulsen, F. E., and Randt, C. T.: Relation of peripheral nerve fiber size and sensation in man, Arch. Neurol. **3**:381, 1960.
35. Collins, W. F., Nulsen, F. E., and Shealy, C. N.: Electrophysiological studies of peripheral and central pathways conducting pain. In Knighton, R. S. and Dumke, P. R., editors: Pain, Boston, 1966, Little, Brown & Co.
36. Crawford, A. S., and Knighton, R. S.: Further observations on medullary spinothalamic tractotomy, J. Neurosurg. **10**:113, 1953.
37. Darian-Smith, I., Johnson, K. O., and Dykes, R. D.: "Cold" fiber population innervating palmar and digital skin of the monkey: responses to cooling pulses, J. Neurophysiol. **36**:325, 1972.
38. Darian-Smith, I., Johnson, K. O., and LaMotte, C.: Warming and cooling the skin: peripheral neural determinants of perceived changes in skin temperature. In Kornhuber, H. H., editor: The somatic sensory system,
39. Dejerine, J., and Roussy, G.: The thalamic syndrome, Arch. Neurol. **20**:560, 1969.
40. Denny-Brown, D.: The release of deep pain by nerve injury, Brain **88**:725, 1965.
41. Dilly, P. N., Wall, P. D., and Webster, K. E.: Cells of origin of the spinothalamic tract in the cat and rat, Exp. Neurol. **21**:550, 1968.
42. Drake, C. G., and MacKenzie, K. G.: Mesencephalic tractotomy for pain, J. Neurosurg. **10**:457, 1953.
43. Erickson, T. C., Bleckwenn, W. J., and Woolsey, C. N.: Observations on the postcentral gyrus in relation to pain, Trans. Am. Neurol. Assoc. **77**:57, 1952.
44. Foerster, O.: Die Leitungsbahnen des Schmerzegefuhls und die chirurgische Behandlung der Schmerzzustande, Berlin, 1927, Urban und Schwarzenberg.
45. Foerster, O.: In Bumke, O., and Foerster, O., editors: Handbuch der Neurologie, Berlin, 1936, Springer Verlag, vol. 6.
46. Franz, D. N., and Iggo, A.: Dorsal root potentials and ventral root reflexes evoked by non-myelinated fibers, Science **162**:1140, 1968.
47. Grant, F. C., and Wood, F. A.: Experiences with cordotomy, Clin. Neurosurg. **5**:38, 1958.
48. Greene, L. C., and Hardy, J. D.: Adaptation of thermal pain in the skin, J. Appl. Physiol. **17**:693, 1962.
49. Hardy, J. D.: Threshold of pain and reflex contraction as related to noxious stimuli, J. Appl. Physiol. **5**:725, 1953.
50. Hardy, J. D.: The pain threshold and the nature of pain sensation. In Keele, C. A., and Smith, R., editors: International symposium on assessment of pain in man and animals, London, 1962, E. & S. Livingstone.
51. Hardy, J. D., and Stolwijk, J. A. J.: Tissue temperature and thermal pain. In de Reuck, A. V. S., and Knight, J., editors: Ciba Foundation symposium on touch, heat and pain, London, 1966, J. & A. Churchill, Ltd.
52. Hardy, J. D., Stolwijk, J. A. J., and Hoffman, D.: Pain following step increase in skin temperature. In Kenshalo, D. R., editor: The skin senses, Springfield, Ill., 1968, Charles C Thomas, Publisher.
53. Hardy, J. D., Wolff, H. G., and Goodell, H.: Studies on pain. A new method for measuring pain threshold: observations on the spatial summation of pain, J. Clin. Invest. **19**:649, 1940.
54. Hardy, J. D., Wolff, H. G., and Goodell, H.: Studies on pain. An investigation of some quantitative aspects of the dol scale of pain intensity, J. Clin. Invest. **27**:380, 1948.
55. Hassler, R.: Die zentralen systeme des Schmerzes, Acta Neurochir. **8**:353, 1960.
56. Hassler, R., and Reichert, T.: Klinishe und anatomische Befunde bei stereotaktischen Schmerzoperationem im thalamus, Arch. Psychiatr. Nervenkr. **200**:93, 1959.
57. Hazouri, L. A., and Mueller, A. D.: Pain threshold studies on paraplegic patients, Arch. Neurol. Psychiatry **64**:607, 1950.
58. Herrick, C. J., and Bishop, G. H.: A comparative survey of the spinal lemniscal systems. In Jasper, H. H., et al., editors: Reticular formation of the brain, Boston, 1958, Little, Brown & Co.
59. Hillman, P., and Wall, P. D.: Inhibitory and excitatory factors influencing the receptive fields of lamina 5 spinal cord cells, Exp. Brain Res. **9**:284, 1969.
60. Hyndman, O. R.: Lissauer's tract section, J. Int. Col. Surg. **5**:394, 1942.
61. Iggo, A.: Non-myelinated visceral, muscular and cutaneous afferent fibres and pain. In Keele, C. A., and Smith, R., editors: International symposium on assessment of pain in man and animals, Edinburgh, 1962, E. & S. Livingstone, Ltd.
62. Janig, W., and Zimmermann, M.: Presynaptic depolarization of myelinated afferent fibres evoked by stimulation of cutaneous C fibers, J. Physiol. **214**:29, 1971.
63. Janig, W., Schmidt, F. R., and Zimmermann, M.: Two specific feedback pathways to the central afferent terminals of phasic and tonic mechanoreceptors, Exp. Brain Res. **6**:116, 1968.
64. Johnson, K. O., Darian-Smith, I., and LaMotte, C.: Peripheral neural determinants of

temperature discrimination in man: a correlative study of responses to cooling skin, J. Neurophysiol. **36**:347, 1973.
65. Keele, C. A., and Armstrong, D.: Substances producing pain and itch, London, 1964, Edward Arnold, Ltd.
66. Kerr, F. W. L.: Evidence for a peripheral etiology of trigeminal neuralgia, J. Neurosurg. **26**:168, 1967.
67. King, R. B.: Evidence for a central etiology of tic douloureux, J. Neurosurg. **26**:175, 1967.
68. Kugelberg, E., and Lindbloom, U.: The mechanism of pain in trigeminal neuralgia, J. Neurol. Neurosurg. Psychiatry **22**:36, 1959.
69. Landau, W., and Bishop, G. H.: Pain from dermal, periosteal, and fascial endings and from inflammation, Arch. Neurol. Psychiatry **69**:490, 1953.
70. Lele, P. P., Weddell, G., and Williams, C. M.: The relationship between heat transfer, skin temperature and cutaneous sensibility, J. Physiol. **126**:206, 1954.
71. Lenox, J. R.: Effect of hypnotic analgesia on verbal report and cardiovascular responses to ischemic pain, J. Abnorm. Psychol. **75**:199, 1970.
72. Lewin, W., and Phillips, C. G.: Observations on partial removal of the postcentral gyrus for pain, J. Neurol. Neurosurg. Psychiatry **15**: 143, 1952.
73. Lewis, T.: Pain, New York, 1942, The Macmillan Co.
74. Lewis, T., and Pochin, E. E.: The double pain response of the human skin to a single stimulus, Clin. Sci. **3**:67, 1937.
75. Lloyd, D. P. C., and McIntyre, A. K.: On the origins of dorsal root potentials, J. Gen. Physiol. **32**:409, 1949.
76. Manfredi, M.: Modulation of sensory projections in anterolateral column of cat spinal cord by peripheral afferents of different size, Arch. Ital. Biol. **108**:72, 1970.
77. Manfredi, M., and Castellucci, V.: C-fiber responses in the ventrolateral column of the cat spinal cord, Science **165**:1020, 1969.
78. Marechaux, E. W., and Shafer, K. E.: Ueber Temperaturempfindungen bei Einwirkung von Temperaturreizen verschiedener Steilheit auf den ganzen Körper, Arch. Ges. Physiol. **251**: 765, 1949.
79. Marshall, J.: Sensory disturbances in cortical wounds with special reference to pain, J. Neurol. Neurosurg. Psychiatry **14**:187, 1951.
80. Mehler, W. R.: The anatomy of the so-called "pain tract" in man: an analysis of the course and distribution of the ascending fibers of the fasciculus anterolateralis. In French, J. D., and Porter, R. W., editors: Basic research in paraplegia, Springfield, Ill., 1962, Charles C Thomas, Publisher.
81. Mehler, W. R.: The posterior thalamic region in man, Confin. Neurol. **27**:18, 1966.
82. Mehler, W. R.: Further notes on the centre médian nucleus of Luys. In Purpura, D. P., and Yahr, M. D., editors: The thalamus, New York, 1966, Columbia University Press.
83. Mehler, W. R., Feferman, M. E., and Nauta, W. H. J.: Ascending axon degeneration following anterolateral cordotomy, Brain **83**:718, 1960.
84. Melzack, R., and Torgerson, W. S.: On the language of pain, Anesthesiology **34**:50, 1971.
85. Melzack, R., and Wall, P. D.: Pain mechanisms: a new theory, Science **150**:971, 1965.
86. Mendell, L. M., and Wall, P. D.: Presynaptic hyperpolarization: a role for fine afferent fibers, J. Physiol. **172**:274, 1964.
87. Moritz, A. R., and Henriques, F. C., Jr.: Studies of thermal injury. II. The relative importance of time and surface temperature in the causation of cutaneous burns, Am. J. Pathol. **23**:695, 1947.
88. Nashold, B. S., Jr., Wilson, W. P., and Slaughter, D. G.: Sensations evoked by stimulation in the midbrain of man, J. Neurosurg. **30**:14, 1969.
89. Ostfeld, A. M., Chapman, L. F., Goodell, H., and Wolff, H. G.: Studies in headache. Summary of evidence concerning a noxious agent active locally during migraine headache, Psychosom. Med. **19**:199, 1957.
90. Penfield, W., and Boldrey, E.: Somatic motor and sensory representation in the cerebral cortex of man as studied by electrical stimulation, Brain **60**:389, 1937.
91. Perl, E. R.: Myelinated afferent fibres innervating the primate skin and their response to noxious stimuli, J. Physiol. **197**:593, 1968.
92. Poggio, G. F., and Mountcastle, V. B.: A study of the functional contributions of the lemniscal and spinothalamic systems to somatic sensibility. Central nervous mechanisms in pain, Bull. Johns Hopkins Hosp. **106**:266, 1960.
93. Ralston, H. J., III: The fine structure of neurons in the dorsal horn of the cat spinal cord, J. Comp. Neurol. **132**:275, 1968.
94. Ralston, H. J., III: Dorsal root projections to dorsal horn neurons in the cat spinal cord, J. Comp. Neurol. **132**:303, 1968.
95. Rand, R. W.: Further observations on Lissauer tractolysis, Neurochirugia **3**:151, 1960.
96. Ranson, S. W.: Cutaneous and sensory fibers and sensory conduction, Arch. Neurol. Psychiatry **26**:1122, 1931.
97. Ray, B. S., and Wolff, H. G.: Experimental studies on headache. Pain sensitive structures of the head and their significance in headache, Arch. Surg. **41**:813, 1940.
98. Rethelyi, M., and Szentágothai, J.: The large synaptic complexes of the substantia gelatinosa, Exp. Brain Res. **7**:258, 1969.
99. Rexed, B.: The cytoarchitectonic atlas of the spinal cord of the cat, J. Comp. Neurol. **100**: 297, 1954.
100. Riddoch, G.: The clinical features of central pain, Lancet **1**:1093, 1150, 1205, 1938.
101. Rivers, W. H. R., and Head, H.: A human experiment in nerve division, Brain **31**:323, 1908.
102. Rubins, J. L., and Friedman, E. D.: Asymbolia for pain, Arch. Neurol. Psychiatry **60**:554, 1948.
103. Russell, W. R.: Transient disturbances following gunshot wounds of the head, Brain **68**: 79, 1945.
104. Scheibel, M. E., and Scheibel, A. B.: Terminal axon patterns in cat spinal cord. II. The dorsal horn, Brain Res. **9**:32, 1968.
105. Sheally, C. N., Motimer, J. T., and Hagfors,

N. R.: Dorsal column electroanalgesia, J. Neurosurg. **32:**560, 1970.
106. Shumacher, G. A., Goodell, H., Hardy, J. C., and Wolff, H. G.: Uniformity of the pain threshold in man, Science **92:**110, 1940.
107. Sjöqvist, O.: Studies on pain conduction in the trigeminal nerve. A contribution to the surgical treatment of facial pain, Acta Psychiatr. Scand. suppl. **27:**1, 1938.
108. Spiegel, E. A., and Wycis, H. T.: Present status of stereoencephalotomies for pain relief, Confin. Neurol. **27:**7, 1966.
109. Spiegel, E. A., Kletzkin, M., and Szekely, E. G.: Pain reactions upon stimulation of the tectum mesencephali, J. Neuropathol. Exp. Neurol. **13:**212, 1954.
110. Spivy, D. G., and Metcalf, J. S.: Differential effect of medial and lateral dorsal root sections upon subcortical evoked potentials, J. Neurophysiol. **22:**367, 1959.
111. Stevens, J. C., and Marks, L. E.: Spatial summation and the dynamics of warmth sensation, Percept. Psychophys. **9:**391, 1971.
112. Stevens, J. C., and Stevens, S. S.: Warmth and cold: dynamics of sensory intensity, J. Exp. Psychol. **60:**183, 1960.
113. Sweet, W. G.: Pain. In Magoun, H. W., editor, Neurophysiology section: Handbook of physiology, Baltimore, 1959, The Williams & Wilkins Co., vol. 1.
114. Sweet, W. H., White, J. C., Selverstone, B., and Nilges, R. G.: Sensory responses from anterior roots and from surface and interior of spinal cord in man, Trans. Am. Neurol. Assoc. **75:**165, 1950.
115. Szentágothai, J.: Neuronal and synaptic arrangements in the substantia gelatinosa Rolandi, J. Comp. Neurol. **122:**219, 1964.
116. Talairach, J., Tournax, P., and Bancaud, J.: Chirugie parietale de la douleur, Acta Neurochir. **8:**153, 1960.
116a. Taren, J. A., and Kahn, E. A.: Thoracic anterolateral cordotomy. In Knighton, R. S., and Dumke, P. R., editors: Pain, Boston, 1966, Little, Brown & Co.
117. Trevino, D. L., Willis, W. D., and Maunz, R. A.: Location of cells of origin of the spinothalamic tract in the cat. In Proceedings of the twenty-fifth international congress of physiological sciences, Berlin, 1972, Springer Verlag.
118. Vyklicky, L., Rudomin, P., Zajac, F. E., III, and Burke, R. E.: Primary afferent depolarization evoked by a painful stimulus, Science **165:**184, 1969.
119. Wagman, I. H., and Price, D. B.: Responses of dorsal horn cells of M. mulatta to cutaneous and sural nerve A and C fiber stimuli, J. Neurophysiol. **32:**803, 1969.
120. Walker, A. E., and Nulsen, F.: Electrical stimulation of the upper thoracic portion of the sympathetic chain in man, Arch. Neurol. Psychiatry **59:**559, 1948.
121. Wall, P. D.: The laminar organization of dorsal horn and effects of descending impulses, J. Physiol. **188:**403, 1967.
122. White, J. C., and Sweet, W. H.: Pain. Its mechanisms and neurosurgical control, Springfield, Ill., 1955, Charles C Thomas, Publisher.
123. Wortis, H., Stein, N. H., and Joliffe, N.: Fiber dissociation in peripheral neuropathy, Arch. Intern. Med. **69:**222, 1942.
124. Zimmerman, M.: Dorsal root potentials after C-fiber stimulation, Science **160:**896, 1968.

12 The auditory periphery

MOÏSE H. GOLDSTEIN, Jr.

The vestibular and auditory structures of the inner ear and the lateral line organ found in fish and some amphibia derive from the same mesectodermal anlage and can be grouped together on morphologic grounds. Although differences in the various sensory structures of the inner ear and lateral line exist, they all have the following common features. The sensory cells (hair cells) have cilia that project into a gelatinous covering structure, the sensory epithelium contains supporting cells as well as hair cells, and the hair cells are innervated by neurons that synapse in nuclei of the medulla. The gross anatomy of the inner ear will be considered first, with indications of the functions of the different organs. Then recent evidence that these receptors are similar in structure and function will be cited.

Gross anatomy of the labyrinth

The orientation of the bony labyrinth in the compact petrous part of the temporal bone and its relationship to the middle ear

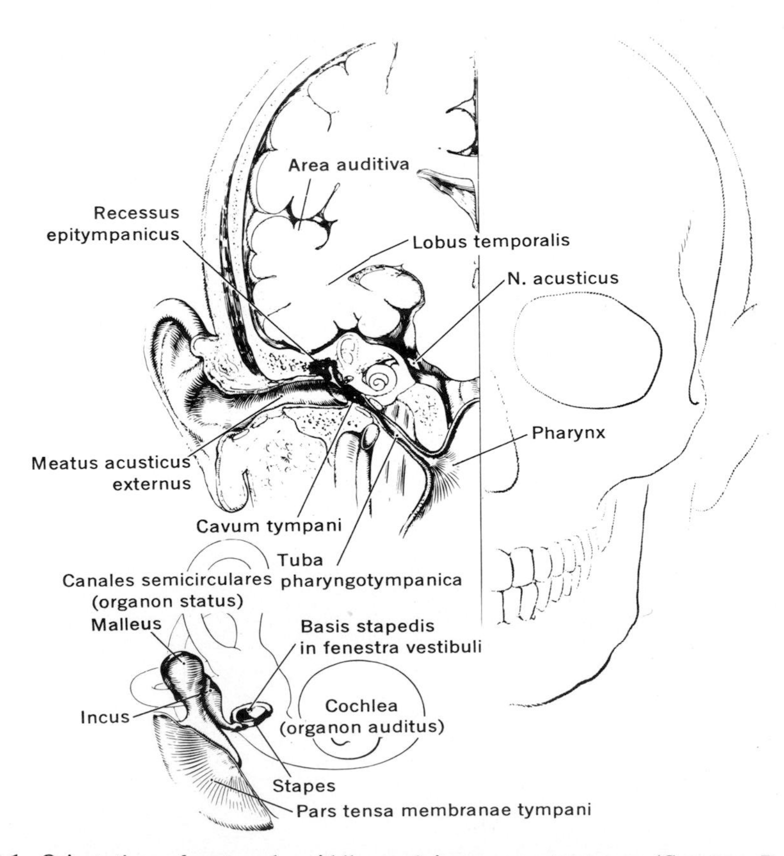

Fig. 12-1. Orientation of external, middle, and inner ear structures. (Courtesy B. Melloni.)

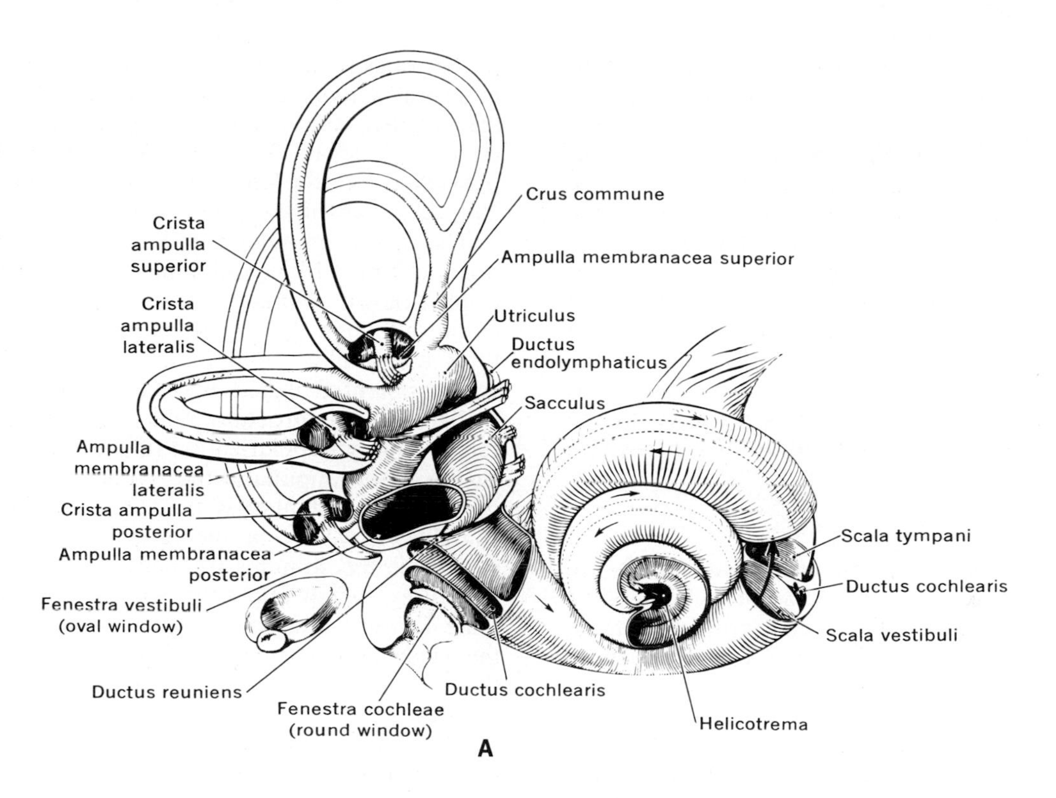

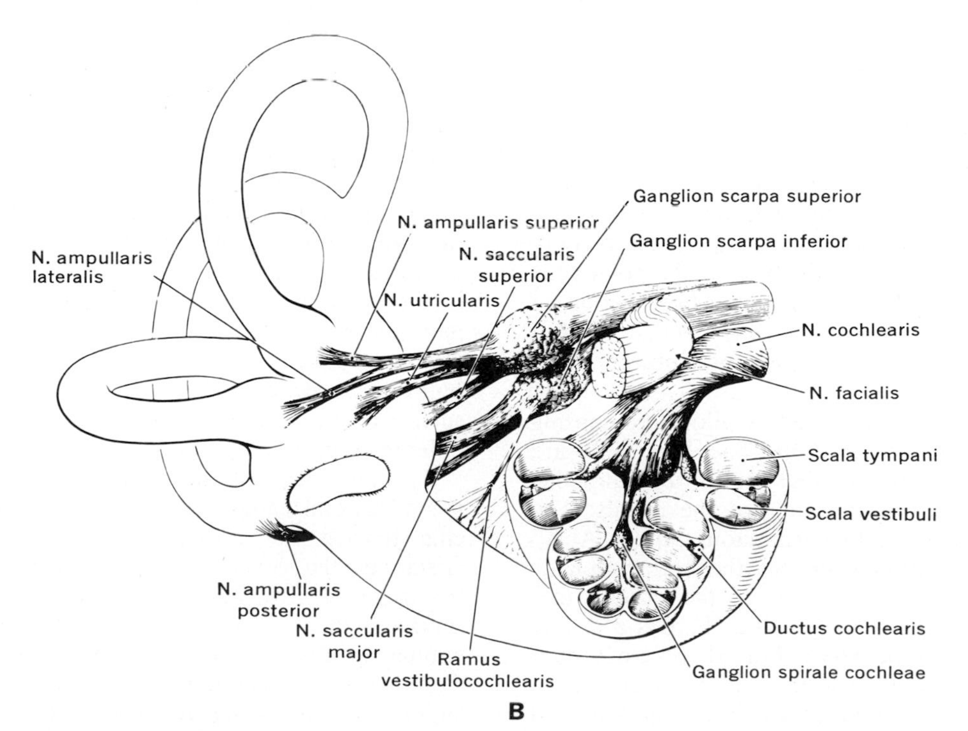

Fig. 12-2. A, Inner ear, showing location of end organs. **B,** Innervation of inner ear end organs, also showing cross section of cochlea. (Courtesy B. Melloni.)

bones, the tympanic membrane, the eustachian tube (tuba pharyngotympanica), and other structures are shown in Fig. 12-1. The membranous labyrinth consists of thin-walled sacs and ducts filled with a clear fluid, endolymph. These structures and the perilymphatic fluid that surrounds them are illustrated in Fig. 12-2, *A,* which also shows the location of the sensory receptors. Fig. 12-2, *B,* illustrates the innervation of the inner ear structures by ramifications of the two branches of the eighth nerve. The vestibular organs of the inner ear are surrounded by endolymph, and for these organs the perilymph serves as a cushion between the membranous labyrinth and the walls of the bony labyrinth. On the other hand, the cochlea is situated so that it is sensitive to the acoustic energy imparted to the perilymph by the stapes.

Function of the receptors[3]

The hair cells of the utricle and saccule are covered by a gelatinous pad in which otoliths are embedded. The specific gravity of these calcium carbonate granules is several times greater than that of the surrounding endolymph. When the head is in its normal position, the macula (receptor organ—literally, "spot") of the utricle is in an approximately horizontal position, with the otoliths lying on the hair cells. Tilting of the head and linear acceleration are adequate stimuli for the utricle. Presumably these motions acting on the otoliths cause the transmission of a shearing or bending force to the hair cells. The saccular maculae are situated obliquely so that when the head is in an erect position, the otoliths of the saccule are placed laterally to the hair cells, embedded in the gelatinous substance that covers them. The function of the saccule is not completely understood, but it appears to be a receptor for vibratory stimuli.

The semicircular canals allow the sensing of angular acceleration. The hair cells are located on cristae that sit in the ampullae (Fig. 12-2) at the ends of the canals. The crista is covered by a gelatinous cupula. Angular acceleration displaces the endolymphatic fluid, causing a motion of the cupula that stimulates the sensory cells. There is a resting discharge in the nerve in the absence of stimulation. In the posterior and vertical canals, utriculofugal movement of the cupula excites the nerve and utriculopetal movement suppresses its discharge; in the horizontal canal the situation is reversed. This difference is explained in the following discussion.

Morphologic polarization of hair cells[34, 95]

It seems likely that mechanisms involved in the transduction of mechanical stimulation to neural activity are similar in hair cells of the cochlea, the vestibular organs, and the lateral line organ. More is known about the physiology of the lateral line and vestibular hair cells than of the cochlear hair cells.

The crista ampullaris and the lateral line organ exhibit striking differences in their electrophysiologic responses to sinusoidal stimulation, although their gross morphologic features are similar. Both organs exhibit a receptor potential called the microphonic potential, which is probably related to depolarization and hyperpolarization of the hair cells. In the crista ampullaris the frequency of the microphonic potential is that of a sinusoidal stimulus; however, in the lateral line organ it is twice the stimulus frequency. These two structures, which appear grossly to have the same features, are found when examined by the electron microscope to differ in a way that explains the different microphonic responses.

Electron micrographs reveal that the end of the hair cell toward the cupula contains a number of stereocilia and one kinocilium, as shown in Fig. 12-3. The kinocilium is always at one side of the hair cell. This and other features give the hair cells a morphologic polarization.

A difference in the orientation of the hair cells in the receptors explains the difference in the microphonic potentials.[34] In Fig. 12-4 the orientation of the sensory hair bundles is indicated by the arrows on each hair cell: in the crista ampullaris, all of the hair cells have the same orientation; in the lateral line organ, adjacent hair cells are polarized in opposite directions. Fig. 12-5 shows how such an arrangement could account for the difference in microphonic potentials.

The hair cells of the crista ampullaris of the horizontal canals are oriented with kinocilia toward the utricle; the hair cells of cristae of the other canals are polarized in the opposite direction. This explains the observation that utriculopetal movement of the cupula results in increased firing in nerve fibers of the horizontal canal organs but in suppression of firing for fibers innervating the other cristae. Recent intracellular recordings obtained from hair cells of the lateral

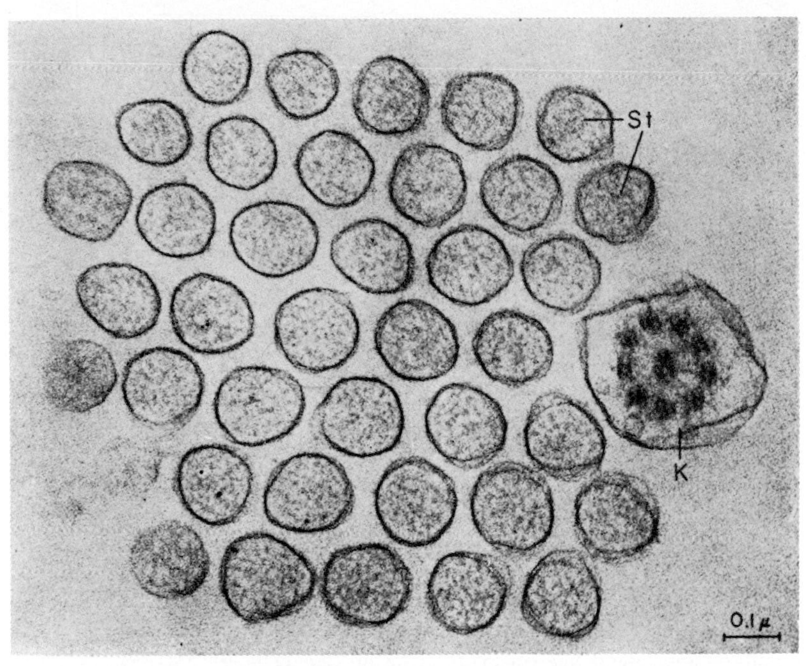

Fig. 12-3. Cross section through sensory hair bundle of hair cell of lateral line organ of teleost fish *Lota volgaris.* Kinocilium, *K,* is located in V-shaped indentation formed by advanced peripheral rows of stereocilia, *St.* Osmium tetroxide fixation. (From Flock and Wersäll.[34])

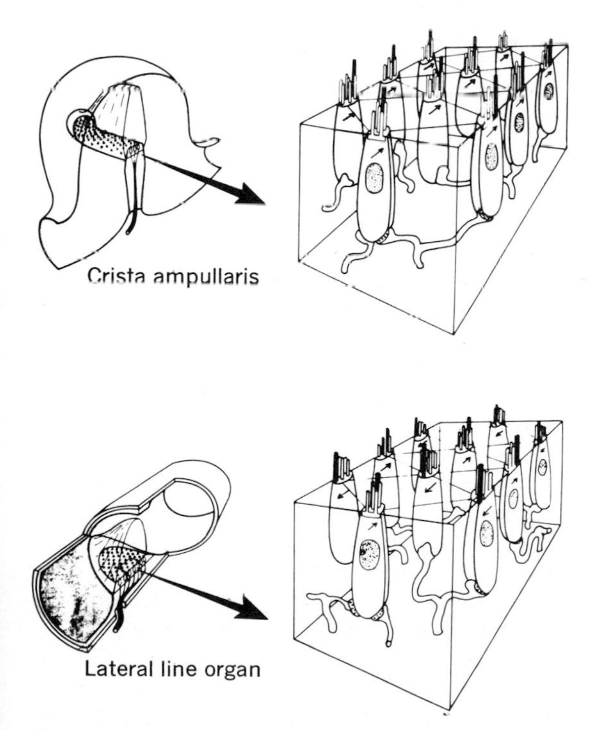

Fig. 12-4. Diagram of crista ampullaris and lateral line organ with enlarged areas of sensory epithelia. Kinocilium is painted black. Cupula overlying epithelium is omitted in drawings at higher magnification. (From Flock and Wersäll.[34])

line[11] provide direct support for the concept of morphologic polarization. Even for strong stimulation the receptor potential peak-to-peak amplitude is less than 1 mv, far smaller than receptor potentials recorded in visual receptor cells, but on the order of magnitude of the cochlear microphonic potential (p. 396).

We will now turn to the auditory periphery. It will become apparent that the mechanical events that lead to the vibration pattern of the cochlear partition are rather well understood. Further, considerable information has recently become available concerning the coding of acoustic stimuli in patterns of neural discharges in the auditory nerve. The details of the process by which the mechanical stimulation of the hair cell is converted to nerve impulses is not so well understood and provides an excellent area for further studies. The work just discussed, which indicates the possibility that all of the end organs of the acousticolateralis system (auditory, vestibular, lateral line) share a similar mechanism for transducing a stress on the hair cell to a neural signal, may serve as an important guide in continuing investigation.[6, 33-34]

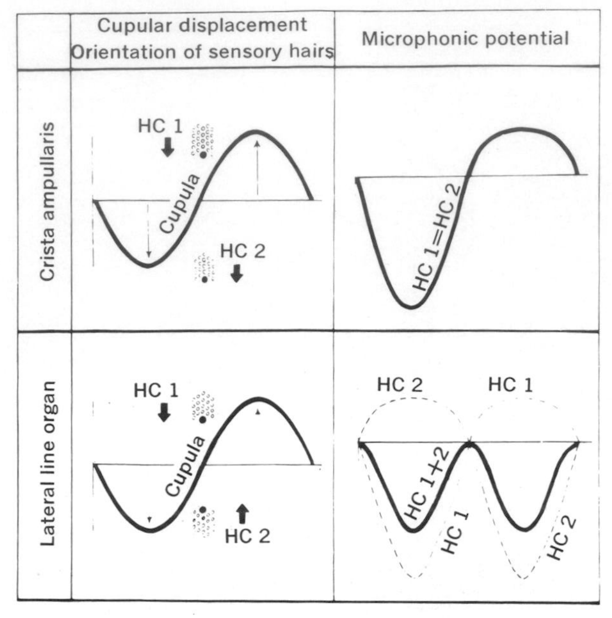

Fig. 12-5. Illustration of hypothesis accounting for difference in microphonic potentials from crista ampullaris and lateral line organ. In crista all hair cells are polarized in same direction, as indicated by *HC 1* and *HC 2*. Cupular displacement in this direction is excitatory for both cells and is followed by decrease in potential (depolarization), while displacement in other direction is accompanied by increase in potential (hyperpolarization), as is shown by right curve representing sum of responses from *HC 1* and *HC 2*. In lateral line organ, *HC 1* and *HC 2* are polarized in opposite directions. Potential changes induced by cupular displacement contributed by *HC 1* will follow course indicated by dotted curve marked *HC 1* in right figure, while potential changes contributed by *HC 2* will follow course indicated by curve *HC 2*. Recorded microphonic potential represents sum of these two partially canceling wave forms, curve *HC 1* + *HC 2*, and will consequently show a frequency double that of cupular displacement. (From Flock and Wersäll.[34])

MECHANICAL PROCESSES IN AUDITION

If the spatial and temporal pattern of evoked neural events in the auditory nerve could be identified for a specified sound wave, a great deal could be understood concerning the neural coding of acoustic signals by the auditory periphery. The modification of the acoustic stimulus by the outer and middle ears and the spatiotemporal pattern of motion of the cochlear partition are the principal topics of this section. Neural coding, with emphasis on the current knowledge of coding in the fibers of the auditory nerve, will be discussed later.

Physics of sound[10]

The movements of the cone of a loudspeaker produce compression and rarefaction of the air particles in front of it. These variations of pressure and displacements of particles are propagated by reason of the elastic nature of the medium. The situation for a sinusoidal plane wave is illustrated in Fig. 12-6. The distribution at an instant in time of air particles in a sound wave traveling to the right is shown at the top of Fig. 12-6. The two variables, pressure and particle velocity, are respectively the differential pressure due to the acoustic disturbance and the average particle velocity that excludes the random brownian motion of gas molecules. Note the region indicated by the vertical line in the top portion of Fig. 12-6. Particles just to the left are moving toward this region, and particles just to the right are also moving toward this region, as indicated by the arrows. A compression of air particles and an increase of pressure result, so that a short time later the particles are distributed as shown in the bottom portion of Fig. 12-6, and the pressure and particle velocity are as indicated below. By continuing this sort of reasoning it can be seen that the sound wave will travel, although the individual air particles simply move back and forth.*

*For a more detailed treatment see Zwislocki.[105]

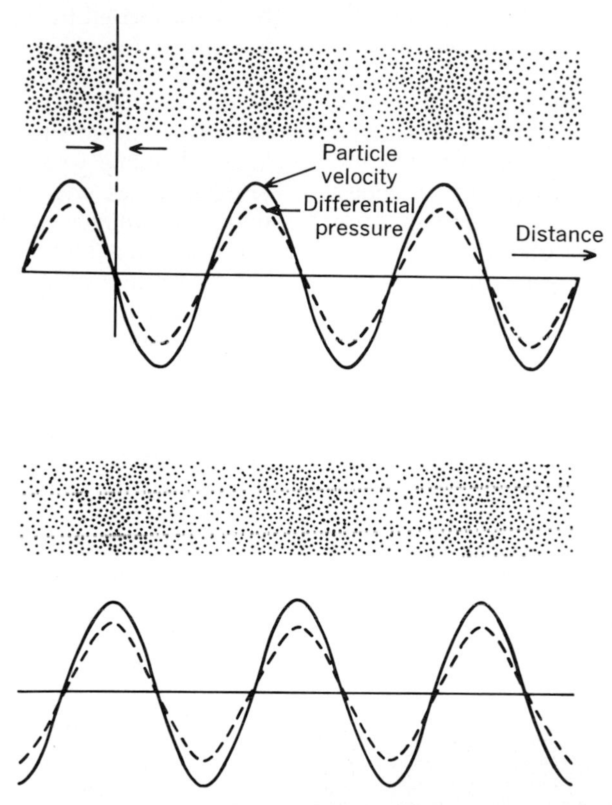

Fig. 12-6. Fluctuation of particle velocity and differential pressure as a function of distance for traveling wave at two different instants. (Modified from Pierce and David.[62])

The source of sound in speech is the air expelled from the lungs through the larynx. During the articulation of a vowel the velocity of air entering the vocal tract is made to vary more or less periodically by the opening and closing of the vocal folds. The configuration of the tract leaves its characteristic imprint on the resulting sound wave, so that the propagated acoustic signal carries information about both the source and the configuration of the vocal tract. Near a sound generator the propagating waves may have a curved front, but as they move away from the source, they approximate plane waves. Mathematical analysis of plane waves yields the wave equation that indicates that they will travel undistorted at a constant velocity, $c = \sqrt{\beta/\rho}$. β is the bulk modulus of elasticity and ρ is the density of the medium (see Beranek[10] for the derivation).* For sinusoidal traveling waves, e.g., those illustrated in Fig. 12-6, the distance between peaks of pressure or particle velocity is called the wavelength. The wavelength (λ), velocity of propagation (c), and frequency in Hertz (f) are related by $c = \lambda f$. Sound waves will travel in fluids and solids as well as in gases; however, unlike electromagnetic waves, they will not propagate in a vacuum.

*The wave equation neglects dissipation that causes the wave to diminish as it travels. These effects limit the transmission of sound over long distances.

Propagation of sound in the middle ear[9, 43, 55, 97, 105]

The auditory system operates over a remarkable range. In discussing the energy or intensity levels of acoustic stimuli the common measure is a relative one, the decibel (db). The decibel measure is equal to 10 $\log_{10} \frac{E}{E_R}$, or to 10 times the logarithm of the ratio of energy of the signal to some reference energy. If, as is usual, we measure pressure rather than energy, the corresponding measure is 20 $\log_{10} \frac{P}{P_R}$ (since energy varies as the square of pressure). The reference pressure most often used in audition is 0.0002 dynes/cm^2, which is close to human threshold at 1,000 Hz. Intensity measures relative to this reference are given in decibels as the sound pressure level (SPL). For stimuli in the region of maximum auditory sensitivity (2,000 to 4,000 Hz) the range of intensities between threshold and discomfort is about 120 db, or an energy range of 1 million million and a pressure range of 1 million.

There are a number of ways to specify the acoustic stimulus. In experimental studies for which accurate specification is important, it is preferable to know the pressure signal at the tympanic membrane. Other measures such as pressure in a free field or at the entrance to the ear canal require that effects of sound diffraction caused by the listener's head and the effect of the ear canal be taken into account. The former are important at high frequencies. The main effect of the ear canal is to introduce a broadly tuned resonance at around 3,000 Hz.[92]

The physical properties of a medium that are relevant to the propagation of sound can be characterized by its impedance. Impedance is defined for sinusoidal variations of pressure and particle velocity; it equals the ratio of the amplitude of the pressure variation to the amplitude of the variation of volume velocity. Volume velocity is defined as particle velocity times the area through which the sound is propagated. In general, impedance has both magnitude and phase and is thus a complex quantity. A sound wave is not completely transmitted through a boundary between media of different impedances. Instead, part of the incident wave is reflected. The amount of reflection depends on the mismatch of impedances. When one is swimming with ears submerged, it is hard to hear airborne sounds. The impedance mismatch between air and water is such that only 0.1% of the incident energy is transmitted through the boundary—the remainder is reflected.

There is an impedance mismatch problem in transmitting the energy of airborne sound

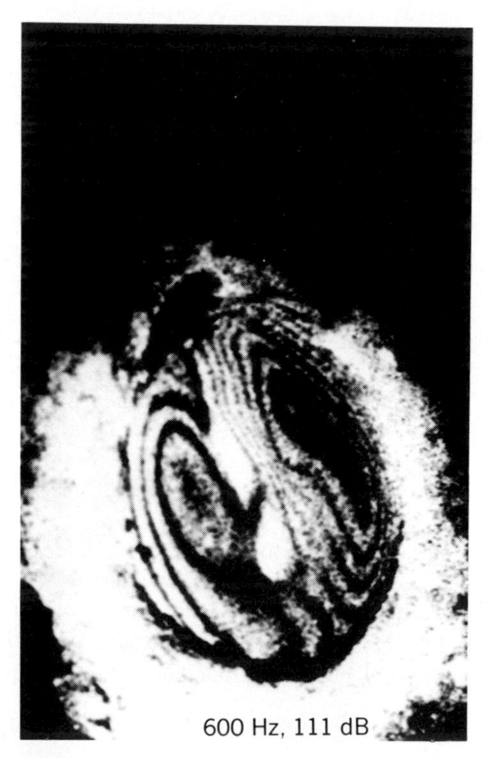

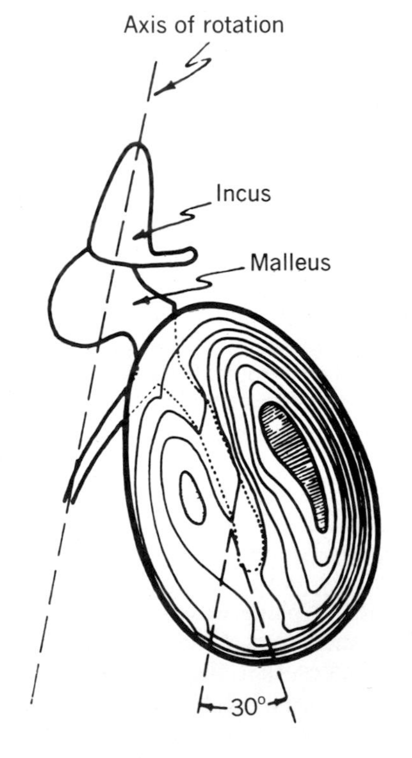

Fig. 12-7. A, Photograph of time-averaged hologram of left tympanic membrane of cat (frequency, 600 Hz; pressure, 111 db SPL). **B,** Schematic drawing of hologram. Dark and bright fringes in photograph indicate isoamplitude deflection. Vibration of posterior (right) peak is 14.6×10^{-5} cm; vibration of anterior peak is 7.52×10^{-5} cm. Method detects motions as small as 1.2×10^{-5} cm. Note that dark fringes visible along manubrium run parallel to ossicular axis, indicating that manubrium rotates about that axis. (From Tonndorf and Khanna.[90])

to the cochlea, which has a much higher impedance. When both impedances are resistive, it can be shown that the ratio of transmitted energy (E_t) to incident energy (E_i) is:

$$\frac{E_t}{E_i} = \frac{4r}{(r + 1)^2}$$

in which r is the ratio of impedances. The ratio for the cat was experimentally determined to be 10^4.[91] From this equation it is clear that the impedance mismatch would cause a loss of 34 db.

The middle ear structures are illustrated in Figs. 12-1 and 12-8. One of the functions of the tympanic membrane and the delicate middle ear bones that provide the mechanical linking between the tympanic membrane and the stapes is to effect an impedance match. There are three factors that cause pressure at the tympanic membrane to be lower than at the footplate of the stapes and volume velocity to be higher.[90] First, the major factor is the ratio of the area of the tympanic membrane to the stapes footplate. The second factor is the lever system that consists of the malleus and incus. The third factor is the leverage that results from the curvature of the tympanic membrane and the manner in which it vibrates. The curved membrane mechanism was first formulated by Helmholtz[4] more than a century ago. Until recently, Helmholtz's scheme was rejected in favor of the idea that at low frequencies the tympanic membrane moves as a stiff, hinged plate.[1] However, using the technique of time-averaged laser holography, Tonndorf and Khanna[90] have demonstrated the correctness of Helmholtz's formulation. The pattern of vibration they obtained is shown in Fig. 12-7. Their calculations indicate an almost complete impedance match in cats, so that the 34 db that would be lost due to mismatch of impedances is recovered.

Such an analysis of middle ear action is valid only for frequencies below about 3,000 Hz, at which the motion of the tympanic membrane is frequency independent[90] and the ossicles move as a rigid body.[43] The analysis also assumes that the stapes footplate motion is pistonlike.[43] Even in this frequency range the analysis gives an overly simplified picture, since the mechanical properties of the middle ear structures and cavities and of the cochlear windows affect the transmission of acoustic energy to the cochlea.[104]

The middle ear apparatus of man exhibits two resonances, a principal one at about 1,200 Hz and a second at about 800 Hz. The former appears related to elements between the eardrum and incudostapedial joint, the latter to elements beyond the incudostapedial joint.[51] At frequencies above approximately 3,000 Hz the vibratory pattern of the eardrum breaks up into quasi-independent subpatterns.[90] Also at higher frequencies the elasticities in the joints of the middle ear chain must be taken into account.[43]

The middle ear apparatus has two muscles—the tensor tympani muscle, which is inserted on the manubrium of the malleus, and the stapedius muscle, which is inserted on the neck of the stapes. The former is innervated by a branch of the trigeminal cranial nerve, the latter by the facial nerve. The situation of the middle ear muscles is shown in Fig. 12-8. Excitation of the tensor tympani muscle pulls the manubrium of the malleus inward, resulting in a movement of the drum. Excitation of stapedius muscle moves the stapes but does not result in a significant movement of the drum. When the mobility and transmission properties of the middle ear mechanism are considered, the two muscles are seen to act as synergists, with the effect of the separate contraction of each of the muscles of the same order of magnitude. The action of the muscles causes an upward shift in the principal resonance of the middle ear mechanism, resulting in a loss in transmission for the lower frequencies.[55, 98] The muscles provide a protective mechanism, producing an attenuation of cochlear excitation for frequencies below 1,000 Hz of as much as 40 db.[102] This can make the difference between a permanent hearing loss and a temporary shift of threshold when one is exposed to an intense sound field.

The middle ear muscles are reflexly activated by intense sound. In man the threshold of activation is more than 70 db above the threshold of hearing,[15, 53] and contraction increases with increasing stimulus intensity over a range of about an additional 30 db. The reflex in one ear can be excited by stimulation of that ear, of the opposing ear, or of both. Sensitivity of the reflex is greatest for bilateral, least for contralateral, and intermediate for ipsilateral stimulation.[52] The muscles may contract in the absence of intense sound,[14] a phenomenon that has been demonstrated in a cat tense and expectantly observing a mouse.[84] In humans they contract during periods of REM sleep.[7] The reflex affords no protection for very brief and intense

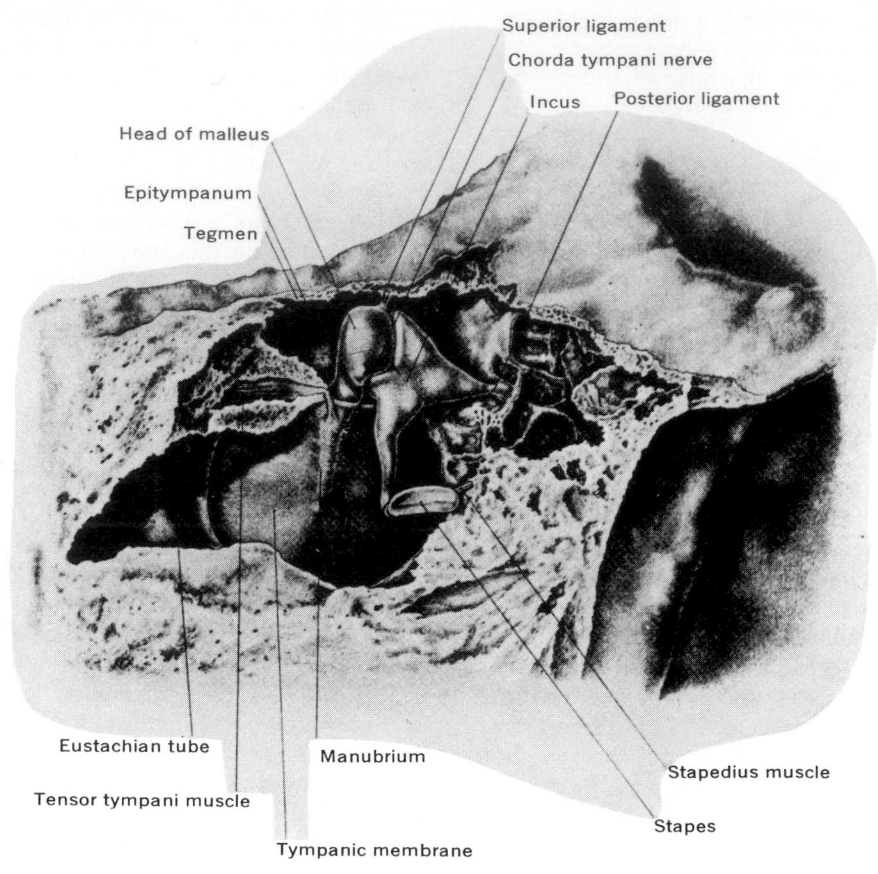

Fig. 12-8. Middle ear on right side, view from within. For view from front, see Fig. 12-1. (From Deaver: Surgical anatomy of the human body, Philadelphia, 1926, The Blakiston Co.)

sounds such as pistol shots, for its latency is 40 to 160 msec. Békésy[1] has shown that at very high intensities the mode of vibration of the ossicular chain changes to one that reduces the volume displacement of the cochlear fluid. This is a very fast-acting mechanism that supplements the middle ear reflex in protecting the cochlea.

The eustachian tube, which connects the tympanic cavity with the pharynx, is normally closed but opens during swallowing, yawning, chewing, and sneezing. It functions to equalize the static pressure on the two sides of the eardrum. An imbalance of static pressure reduces the transmission of the middle ear structures, an effect easily observed during the landing of an aircraft as the cabin pressure increases above the level maintained during flight.

Otosclerosis is a disease in which the footplate of the stapes is immobilized in the oval window by an abnormal bone growth. Modern practice is to replace the stapes. An older method, the fenestration operation, consists of creating a new window in the ampullated end of the horizontal semicircular canal. Then the transmission of sound to the cochlea is without the aforementioned impedance transformation. The dynamic range of audition, normally more than 120 db, is such that an acceptable level of hearing is achieved although with reduced sensitivity.

Summary

Transmission of sound from the ear to the fluid of the cochlea may be characterized by three resonances—one at approximately 3,000 Hz for the ear canal, and two at approximately 1,200 and 800 Hz that can be assigned to middle ear elements. The middle ear muscles act to stiffen the middle ear transmission system, moving the principal resonance at 1,200 Hz to a higher frequency. This decreases transmission for frequencies below 1,000 Hz. The amplitude of motion of the cochlear partition is greatest for these lower frequencies, so that the middle ear reflex is an effective protective mechanism.[73, 74] The

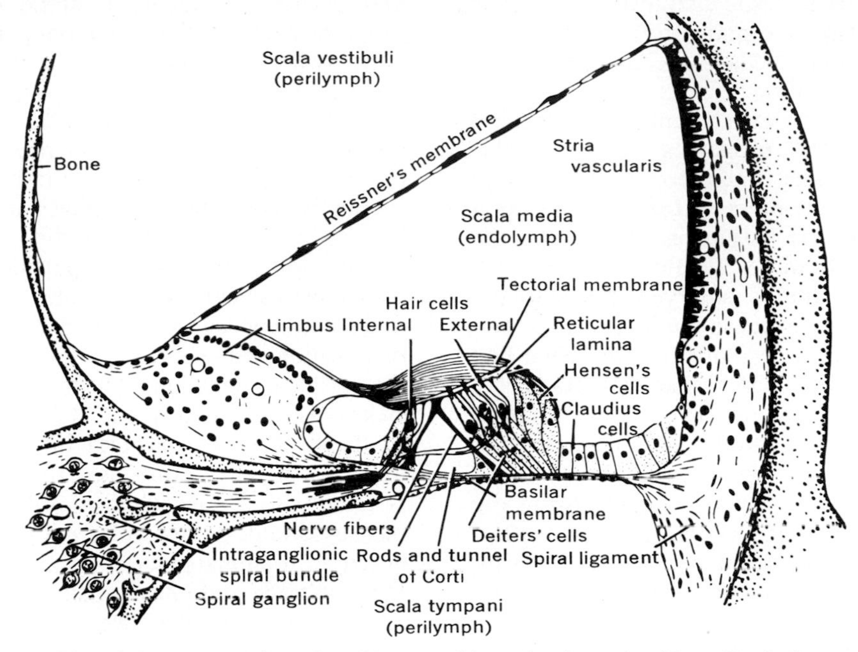

Fig. 12-9. Cross section of cochlear partition of guinea pig. (From Davis.[2])

technique of deriving a network model of the middle ear from measurements of the input impedance at the eardrum has played an important part in obtaining and substantiating these data.[104] In the cat the inverse input impedance at the eardrum and the potential measured near the round window for constant pressure at the eardrum are proportional in the frequency range from 400 to 4,000 Hz.[51] Moreover, the round window potential is proportional to the vibration velocity of the cochlear fluid in this frequency range, thus giving direct justification for the derivation of transmission properties from measurements made at the eardrum.

The resonances of the outer and middle ear are broadly tuned, and although they have a part in determining that the human auditory system is most sensitive for frequencies between 1,000 and 4,000 Hz, they do not play a major role in the frequency (or spectral) analysis that is so important in audition. That analysis is performed by the mechanical vibration of the cochlear partition and then sharpened in the neural "readout" of these vibrations. Descriptions of these functions are topics considered in following sections.

COCHLEAR VIBRATION[1]

Fig. 12-9 shows the cochlea in cross section. The human cochlea is a spiral canal of about two and a half turns, 35 mm in length. The cochlear duct or partition runs the length of the cochlea. It separates the scala vestibuli and the scala tympani, which communicate at the apical end of the cochlea by a small opening called the helicotrema. The oval window is the seat of the footplate of the stapes and thus the point of excitation of the normal cochlea. The round window is a flexible membrane at the basal end of the scala tympani. The relationship of the structures to the middle ear and external meatus is shown in Figs. 12-1 and 12-2. By direct observation of the motion and mechanical properties of cochlear structures in the temporal bones of fresh cadavers and in animal preparations, Békésy indicated how cochlear partition vibration provides a basis for frequency analysis of the acoustic stimulus. The reader is urged to consult the collection of his papers[1] for a detailed description of the experiments.

The cochlear partition contains the hair cells, the receptor cells of audition. Mechanical excitation of these cells leads to excitation of primary fibers of the cochlear nerve. The detailed structure of the end organ will be considered later; at present the cochlear partition can be considered as a structure with mechanical properties that change continuously from the basal to the apical end of the cochlea, providing the basis for a spectral analysis of the acoustic stimulus. The basilar membrane plays the major role in producing these changes in mechanical characteristics of the cochlear partition.

In one series of experiments, Békésy measured the elastic properties of the basilar membrane. Under microscopic observation, glass threads were pressed at right angles upon the surface of the basilar membrane and the resulting pattern of deformation was observed. The circular patterns observed indicated the same elastic properties in the longitudinal and transverse directions. Furthermore, when fine cuts were made in the basilar membrane, the cut surface did not draw apart. Thus the basilar membrane is not under tension; it can best be likened to a gelatinous sheet covered by a thin, homogeneous layer of fibers.[1] Since the hairs would bend at a certain maximum pressure, this method could be used to measure the compliance of the structures of the cochlear partition. Only the basilar membrane exhibited elastic properties that change along the length of the cochlea. Reissner's membrane, the organ of Corti, and the basilar membrane showed circular patterns of depression, but the tectorial membrane exhibited a much elongated hollow. The tectorial membrane is considerably stiffer in the longitudinal than in the transverse direction.

The compliance of the basilar membrane was also measured by attaching a fluid-filled tube to the pierced round window and blocking the helicotrema with a mixture of agar and gelatin, producing a static differential pressure between the scala tympani and the scala vestibuli.[1] Under these conditions the displacement of the cochlear partition is proportional to its compliance. The results indicated an increase in compliance by a factor of more than 100 from near the oval window to the helicotrema. This physical characteristic to a large extent underlies the pattern of vibration of the cochlear partition.

A most difficult task and one crucial to an understanding of cochlear dynamics is the observation of the actual pattern of motion of the cochlear partition. The method used by Békésy is illustrated in Fig. 12-10. The dome of the cochlea was ground off under water, exposing the apical turn, and the opening was covered with a glass window. Vibrations were measured under a binocular microscope. Stroboscopic illumination was used to determine the amplitude and phase of sinusoidal motion. Small crystals of silver were scattered on the cochlear partition to make it more clearly visible and to allow accurate measurement. Excitation was produced by an electromechanical driving unit coupled by fluid to the small tube, *R*, that was fastened into the round window after removal of the membrane. An artificial stapes was fitted over the oval window and attached to an electromagnetic device, allowing accurate measurement of the volume displacement of the cochlear fluid. Because the uppermost turn of the spiral is only 8 mm long, observation of the

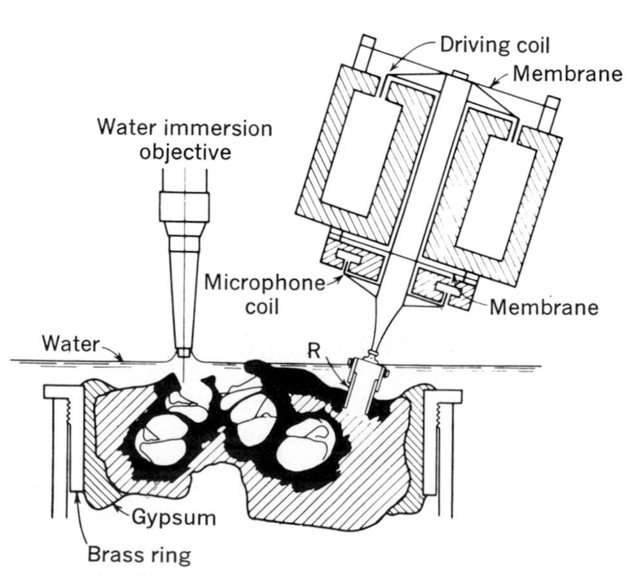

Fig. 12-10. Method of measuring amplitude of vibration of cochlear partition in response to volume displacements of stapes. (From Békésy.[1])

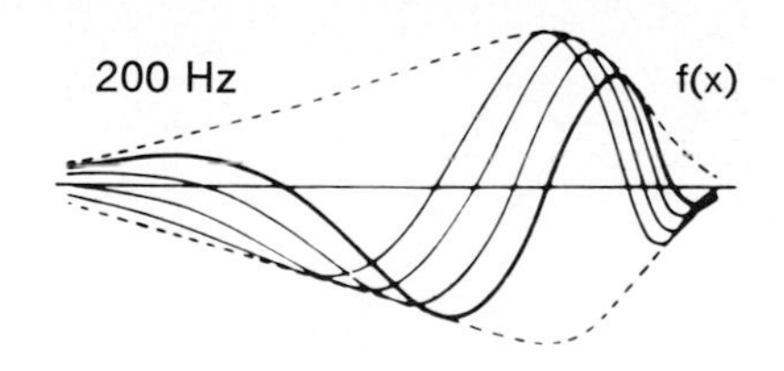

Fig. 12-11. Traveling wave along basilar membrane produced by 200 Hz stimulus. Displacements are shown at a number of sequential instants within a cycle, the dark curve occurring last. (From Békésy.[1])

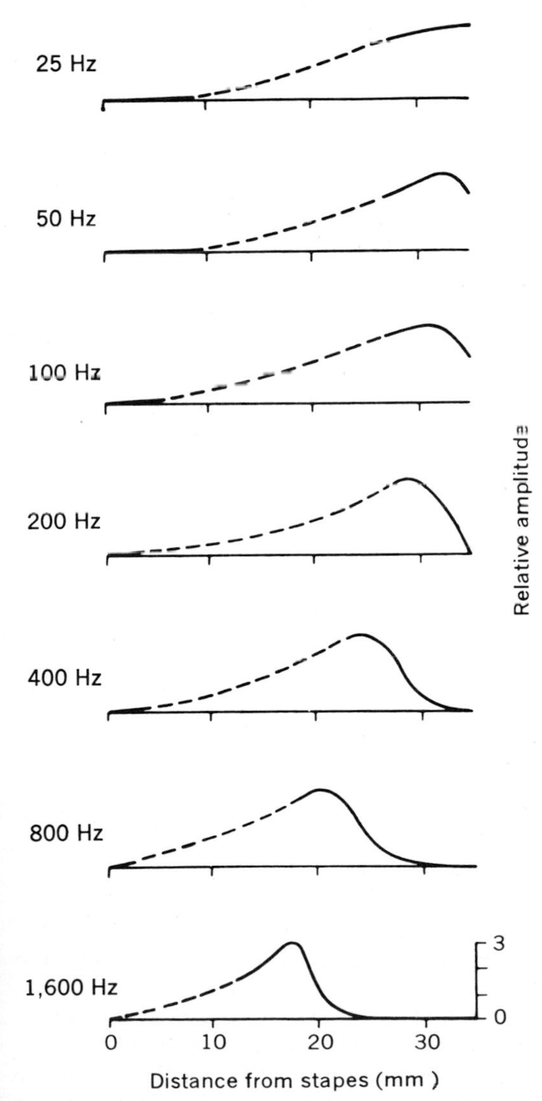

Fig. 12-12. Envelope of patterns of vibration of cochlear partition of cadaver specimen for various frequencies. (From Békésy.[1])

apical turn permitted study of only that section of the partition 27 to 35 mm from the stapes. In order to study the vibrations closer to the stapes, it was necessary to grind away more of the upper end of the cochlea. To observe the vibration near the stapes a method was devised for opening the cochlea at the base.

With these measurements, Békésy reconstructed the pattern of vibration for sinusoidal stimulation of a given frequency. The results indicated a traveling wave moving from base to apex, shown in Fig. 12-11 for a stimulus of 200 Hz. The actual form of displacement is shown at a number of sequential instants of time within a cycle, the dark line occurring last. The vertical scale is greatly magnified, since even for the high pressures Békésy used (about 140 db SPL), the maximum amplitude of vibration of the cochlear partition is only about 3×10^{-3} mm.

The movement at any given position is sinusoidal at the stimulus frequency, a characteristic of a linear system. The *envelope* of the pattern of vibration (indicated by the dotted lines in Fig. 12-11) shows a rather broad maximum at about 28 mm. Fig. 12-12 shows the envelope of vibration for a sequence of increasing frequencies. (The dotted lines are extrapolations later verified by direct observation.) The point of maximum vibration moves toward the base with increasing frequency, and for any given frequency the pattern of vibration drops off more sharply toward the apex than toward the base.

The resonance characteristics of the cochlea can be displayed in another way by plotting the amplitude of vibration of a given point as a function of the frequency of the sinusoidal stimulus. Fig. 12-13 shows such curves for six positions along the cochlea. The positions are from approximately 30 to 15 mm from the stapes. The curve on the far left corresponds to the most apical position. Fig. 12-14 shows both amplitude and phase* characteristics for one of the positions.

Békésy was not able to study vibration of the basal turn on the cochlea. Recently resonance curves for basal positions have been measured using the Mössbauer technique.[46, 64] The curves are similar to Békésy's, but steeper on the high-frequency side. There is some indication in the curves of Fig. 12-13 that this

*Flanagan[32] and Siebert[71] have pointed out from physical considerations that as frequency approaches zero the phase should approach $\pi/2$ radians leading rather than zero.

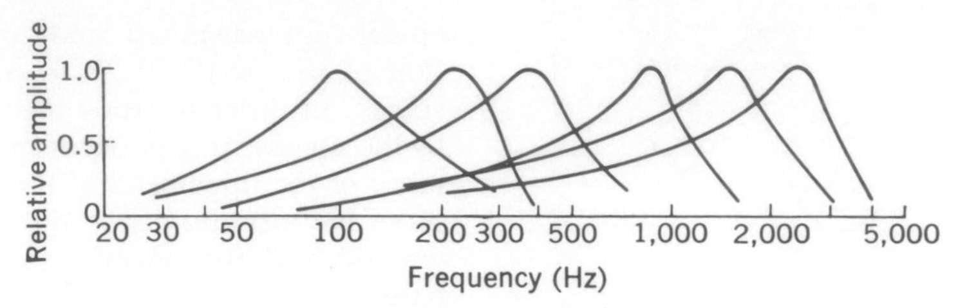

Fig. 12-13. Forms of resonance curves for six positions along cochlear partition. (From Békésy.[1])

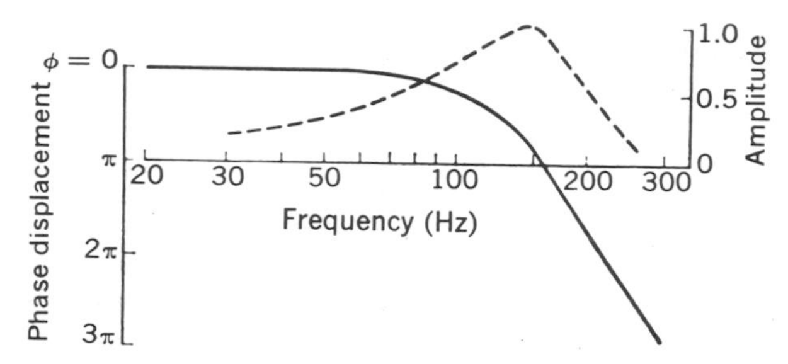

Fig. 12-14. Phase displacement (solid line) and resonance curve (broken line) for point on cochlear partition 30 mm from stapes. Phase angle is phase lag relative to stapedial motion. (From Békésy.[1])

slope increases as the point of measurement moves basalward.

The application of new techniques for measuring small vibrations, e.g., laser interferometry,[89] laser holography,[90] and the Mössbauer technique,[46, 64] is an important new development in study of the mechanics of the ear. Conventional measuring technique has shown that the growth of middle ear and cochlear vibration with increasing stimulus intensity is quite linear for sinusoidal stimulation at sound levels up to about 130 db SPL.[1, 43] The resolution of the method is such that measurement for levels below approximately 110 db is not possible. Thus there has not been a thorough investigation of the linearity of the mechanics of the ear. With the more sensitive new methods, the question of linearity can be explored further at levels nearer threshold and with more complex stimuli as well as single sinusoids. Rhode's recent work[64] indicates nonlinear mechanical vibration of the basilar membrane for frequencies that produce the largest deflections at the place of measurement. Studies of hydrodynamic models of the cochlea have indicated nonlinearities[1, 88] that have yet to be measured directly.

The sort of traveling wave illustrated in Fig. 12-11 is not an ordinary traveling wave as in an acoustic delay line. The acoustic wave travels the length of the cochlea in about 25 μsec, or two orders of magnitude smaller than the propagation time of the traveling waves of cochlear partition displacement.[93] In Fig. 12-13 it is shown that the basal positions for which the basilar membrane is relatively stiff give maximum response for high-frequency sinusoidal stimulation, and the apical positions for which the basilar membrane is relatively compliant respond maximally to low frequencies. Similarly, the displacement response at the basal end of the cochlea is much faster than the response at the apical end, primarily due to compliance change. Since the physical parameters of the cochlear partition and thus the lag time vary continuously, the displacement wave travels from base to apex. Békésy demonstrated the paradoxical nature of the traveling wave by stimulating the apical end of the cochlea and showing that the wave still traveled from base to apex, toward the driving source! It is the almost instantaneous travel of the acoustic wave along the cochlea relative to the slow response times of cochlear partition displacements that gives rise to the phenomenon.

The first step in transduction of the mechanical vibration of the cochlear partition to a pattern of neural impulses in the eighth nerve must involve some deformation or stressing of the hair cells, a fact that led

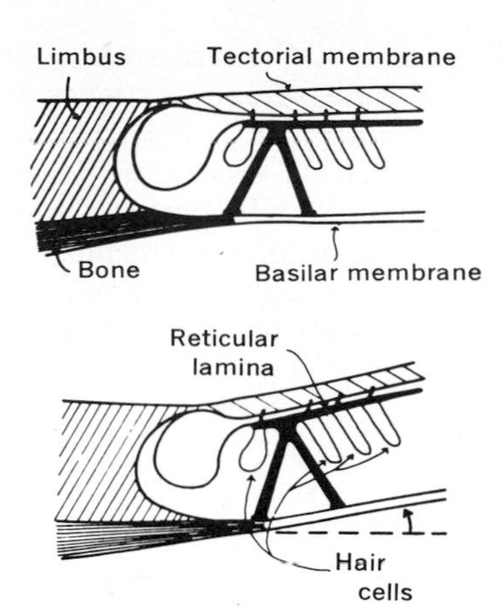

Fig. 12-15. Movement of organ of Corti and tectorial membrane, based on descriptions by Békésy.[1] Shearing action between two stiff structures (tectorial membrane and reticular lamina) bends hairs of hair cells. (From Davis.[2])

Békésy to examine the movement of the different tissues relative to the basilar membrane. The movement of the ends of the hair cells under the tectorial membrane and of the cells of Hensen is in a radial direction on the stapes side of the point of maximum vibration. Near the position of maximum vibration the inner and outer hair cells showed only up-and-down movement. Farther in the direction of the heliocotrema, the direction of the movement is longitudinal.

The ends of Deiters' cells that face the scala media form a stiff open network, the reticular lamina. The hair-bearing ends of the sensory cells are firmly held in the openings of this lamina. The opposite ends of the hair cells rest in cuplike supports that are also formed by Deiters' cells and receive the nerve endings of the afferent and efferent auditory fibers. The rods of Corti form a stiff triangular supporting structure for the reticular lamina. The flask-shaped inner hair cells are in a single row along the inner edge of the reticular lamina; in humans, they number about 3,500. The external hair cells are cylindric in shape and are arranged in three rows on the opposite (outer) side of the tunnel of Corti. They number about 20,000. These relationships are shown schematically in Fig. 12-15.

The tectorial membrane is attached at one edge to the spiral limbus and overlies the hair cells, probably serving as a point of attachment for the ciliary processes. A combination of small elastic and large frictional forces makes the tectorial membrane stiff to vibrations but pliable for slow static displacements. The means by which a displacement of the basilar membrane during vibration in response to an acoustic stimulus could cause bending of the hairs of the hair cells is shown in Fig. 12-15. The illustration applies to the region of the cochlear partition toward the stapes from the position of maximum vibration (i.e., where Békésy observed radial movement of the ends of the hair cells under the tectorial membrane).

Exactly how these forces on the hair cells lead to excitation of the fibers of the eighth nerve is not known. One approach to a better understanding of the mechanical-neural transduction has been through studies of the electric potentials of the cochlea.

COCHLEAR POTENTIALS

A number of potentials may be recorded from the cochlea by gross electrodes placed inside the cochlea through small holes drilled in the bony wall or by an external electrode placed near the round window. The potentials that can be recorded with gross electrodes from the auditory end organ will be described here, and their sources will be specified.

Resting potentials[1, 22, 50, 86]

The DC or steady potentials of the unstimulated cochlea (the resting potentials) have been studied most commonly in the guinea pig, an animal in which the four and a half turns of the cochlea are readily explored. The scala tympani and scala vestibuli are filled with perilymph, which has the chemical composition of cerebrospinal fluid. The scala media is filled with endolymph, which differs markedly in chemical composition, having high potassium and low sodium concentrations. Endolymph probably is secreted by the stria vascularis and is reabsorbed in the endolymphatic sac. The intercellular spaces between the reticular lamina and the basilar membrane, including the tunnel of Corti, are also fluid filled, but the exact composition of this fluid is unknown.

As an electrode is advanced from the scala tympani through the basilar membrane, the potential drops 80 to 90 mv when the electrode enters the organ of Corti. In this region the potential fluctuates considerably, and it seems likely that the large negative potentials are intracellular potentials from cells in the organ of Corti, including hair cells, although no systematic study of them has been made.[16] As the electrode passes into the scala media, the potential rises to a value that is 80 to 90 mv positive with respect to the scala tympani. This is called the endocochlear potential (EP). Further penetration through Reissner's membrane into the scala vestibuli returns the potential close to that of the scala tympani. The source of the EP is the stria vascularis.[20, 50, 86]

Cochlear potentials in response to sound[18, 20]

A number of cochlear potentials are evoked by acoustic stimulation. Differential recording from the scala vestibuli and scala tympani yields the cochlear microphonic (CM) potential, which is related to the instantaneous pattern of vibration of the cochlear partition in the vicinity of the position of the electrodes, and two summating potentials (SP+ and SP–), which follow the envelope of the acoustic stimulus.[18]

There is a linear relationship between CM amplitude and a sinusoidal acoustic stimulus at low sound levels that breaks down at moderate and high-stimulus intensities. At moderate intensities, distortion components of CM potentials are localized to the region of the cochlear partition of significant excitation by the primary stimuli. At high intensities the locus of distortion components is determined by their own frequency and the spectral filtering properties of the cochlear partition.[17] One aspect of nonlinearity in the CM potential is illustrated by the input-output curve in Fig. 12-16. For high-frequency tones the wave form of CM potentials is almost undistorted for intensities that far exceed the linear region of the input-output relationship (insets, Fig. 12-16). It is important to remember that the CM potential is a weighted summation of hair cell potentials from a region of the cochlea

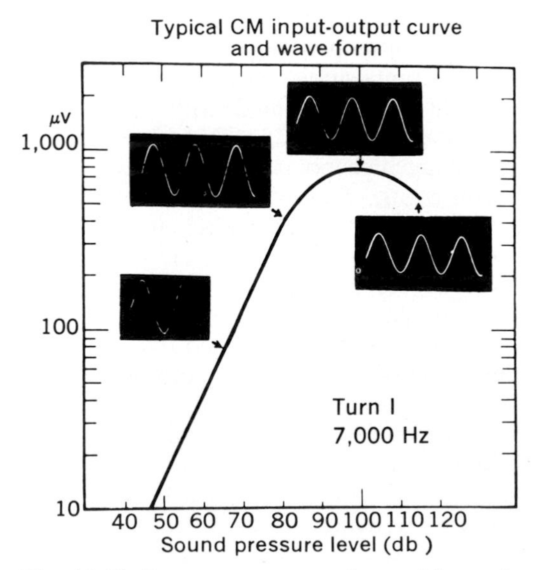

Fig. 12-16. Input-output curve for cochlear microphonic (CM) response of first turn of guinea pig cochlea to 7,000 Hz tone bursts. Insets show wave form of differentially recorded CM. Note absence of peak limiting, even at highest sound intensity. (From Davis.[2])

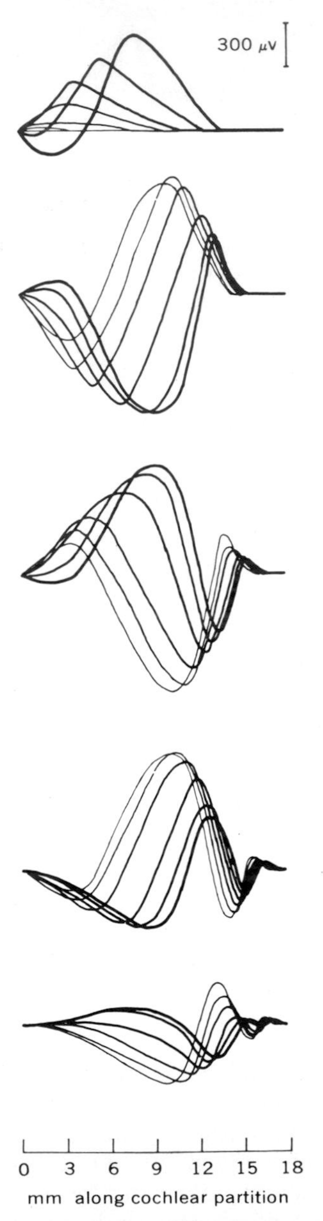

Fig. 12-17. Instantaneous CM voltage as a function of distance along cochlear partition at intervals of 0.1 msec. Thicker lines represent more recent times. Progression from top to bottom set is a continuous sequence. Between sets go from thickest line in set above to thinnest line in set below. Acoustic signal is diphasic transient with about 1 msec between peaks. (From Teas et al.[87])

and is not equivalent to hair cell potentials from a single longitudinal position. The summation of hair cell responses from a region can partially cancel harmonic distortion products.[99] For resonance curves of the amplitude of CM potentials as a function of frequency,[49] the same effect would attenuate high-frequency responses more than responses to low frequencies. In both situations there is a greater change in phase over the region of the cochlear partition contributing to the recording for the higher frequencies, and thus a relatively great cancellation effect.

By recording CM potentials at a number of positions along the cochlea it is possible to reconstruct the "traveling wave" of this potential.[87] The traveling wave observed in the guinea pig cochlea (Fig. 12-17) is similar to that of the human cochlear partition produced by a sudden displacement of the stapes. Békésy found that it takes somewhat over 3 msec for the wave in response to a brief transient to traverse the length of the human cochlea; in the guinea pig this time is nearer 2 msec.

Schema of receptor mechanisms*

A microphonic potential can be recorded[1] when the tectorial membrane is vibrated in a direction parallel to its surface. These poten-

*This discussion is quite speculative. Although I have drawn heavily from the work of several scientists, who have been cited in the references, there is no intention to unnecessarily burden them with the responsibility for this schema.

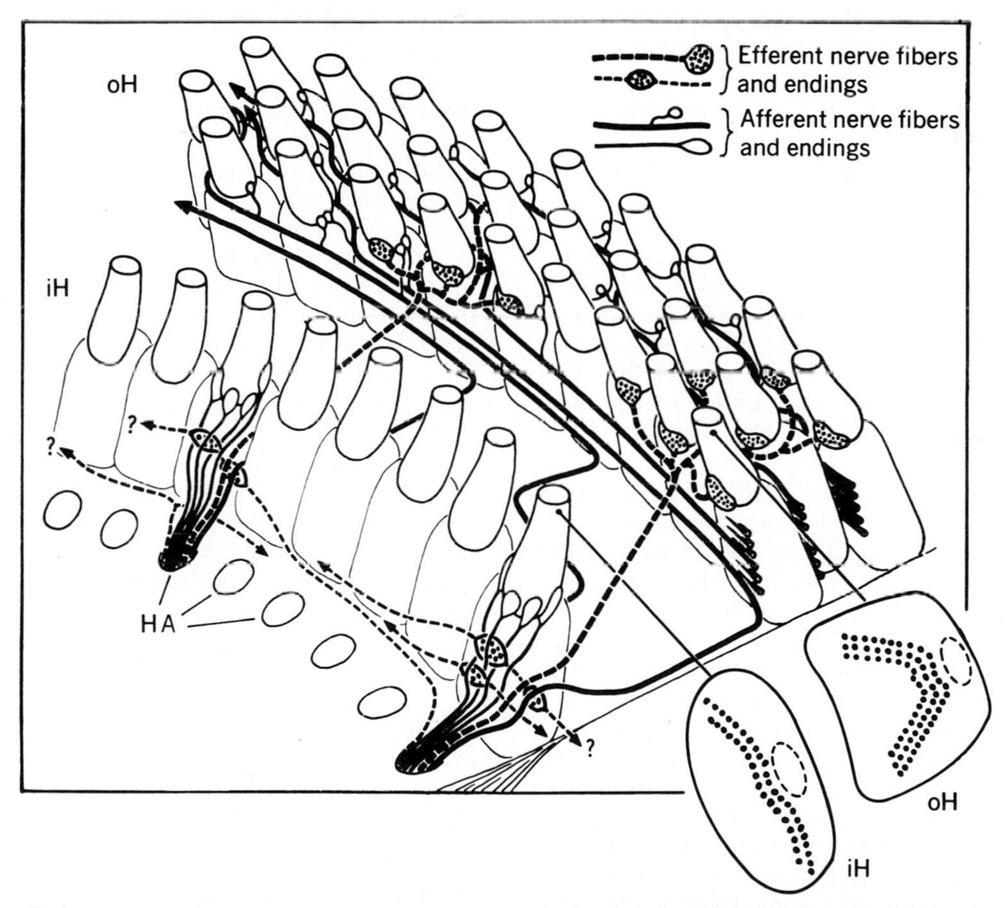

Fig. 12-18. Schema of innervation pattern of organ of Corti in cat. At lower right of illustration is top view of one inner hair cell, *iH*, and one outer hair cell, *oH*, showing stereocilia (dots) and place where cuticle is missing (dashed oval). Only representative examples of nerve fibers arriving to the organ of Corti through two habenular openings, *HA*, are shown. Afferent fibers are shown as continuous lines; efferent fibers are dashed. (See text for quantitative information.) Full spiral basalward extension of afferent fibers from outer hair cell (outer spiral fibers) cannot be shown because of limited space. Indicated nerve fibers and endings do not correspond to their actual numbers. (Schema from Spoendlin[80]; tops of hair cells after Flock et al.,[35] Engström et al.,[28] and Spoendlin.[81])

tials depend on the direction of vibration; for the outer portion of the tectorial membrane a radial deflection produces maximum microphonic potential, while for the inner edge a longitudinal deflection produces maximum microphonic potential. Results from diverse experiments support a theory for the mechanism by which vibration of the cochlear partition, giving rise to shearing forces on the hair cells, leads to excitation of the auditory fibers.

The relevant data include (1) Békésy's studies of cochlear vibration; (2) Békésy's studies of the motion of the tectorial membrane relative to the basilar membrane; (3) Békésy's studies of microphonic potentials due to displacement of the tectorial membrane; (4) the observation that the positive EP between endolymph and perilymph is increased during a deflection of the cochlear partition toward the scala tympani and decreased for a deflection in the opposite direction[2]; (5) the observation that neural excitation occurs when the cochlear partition is deflected toward the scala vestibuli[21, 59]; and (6) a morphologic basis for the directional sensitivity of the hair cells.[35]

Fig. 12-18, *A*, presents a surface view of the tops of the inner and outer hair cells, showing the orientation of the hairs, which is constant throughout the cochlea.[28, 35] Outer hair cells of the cat have about 100 stereocilia; inner hair cells have about 50.[81] The morphologic polarization of both outer and inner hair cells is in a radial direction, with the stereocilia increasing in length toward the outer edge of each hair cell, and with a spot in which the cuticle is missing near the outer edge. Some species[35] (but not the cat[81]) typically have a basal body near the outer edge. Cochlear hair cells of all adult mammals lack kinocilia.

Hypothetically, when the cochlear partition is deflected toward the scala vestibuli, a shearing force in the radial direction results (Fig. 12-15), accompanied by a depolarization of the hair cells.[35] It is assumed that reduction of the positive EP and excitation of the afferent neurons accompany and are related to these events. If attention is focused on the outer hair cells, the compatibility of the various experimental results is good. It is not clear whether the hypothesis of Flock et al.[35] can be applied to the inner hair cells. At their location, Békésy[1] observed a maximum microphonic potential for vibration in the longitudinal direction, yet the morphologic polarization is the same as for the outer hair cells. More detailed discussions of possible generator mechanisms can be found in other sources.[18, 20, 59]

Compound action potential

None of the potentials discussed thus far is a direct sign of nerve impulses in the auditory fibers. If, instead of recording differentially across the cochlear partition, as in the study of CM potentials, the potentials from the scala vestibuli and scala tympani are summed and referred to a reference electrode on the neck, the compound action potential of nerve evoked by sound may be obtained almost free of CM potentials and SP.

Action potentials may equally well be obtained by recording between a point near the round window and a distance reference electrode. In this case a microphonic component is also obtained. Fig. 12-19 illustrates responses to rarefaction and to condensation clicks. The microphonic potential reverses polarity with the stimulus, but the neural responses do not. The microphonic potential recorded in this manner is a spatial average of CM potentials, weighted so that the basal portion is predominant.

The neural potential depends on a synchronous discharge of auditory fibers.[41] Since it is the fibers innervating the basal turn that fire synchronously to a brief transient stimulus,[87] the neural component shown in Fig. 12-19 also disproportionately weights electrical activity from the basal portion of the cochlea. The two negative peaks (N_1 and N_2) of the action potential have been studied in detail. The latency between the microphonic and neural responses decreases with increasing stimulus intensity, and the neural component grows in a characteristic way, since it is a

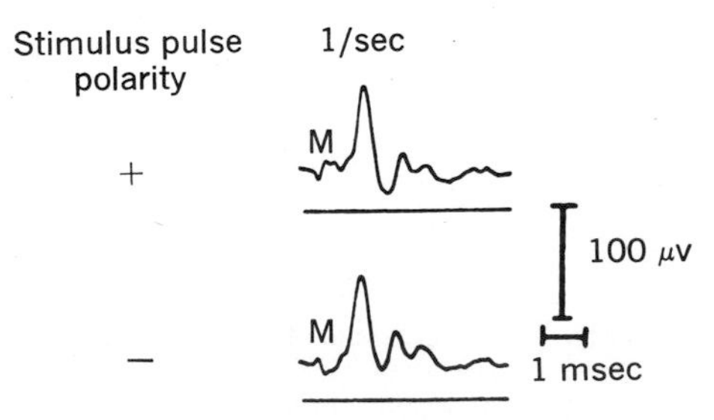

Fig. 12-19. Responses to rarefaction (+) and condensation (−) clicks at repetition rate of 1/sec. Only microphonic *M* reverses with reversal of stimulus polarity. Upward deflection of electric signal indicates negative potential at electrode located near round window. (From Peake et al.[60])

summation of many all-or-none events having different thresholds.[59] It follows repetitive stimuli to rates of approximately 3,000/sec.[60]

The most direct method of investigating the patterns of neural activity in the eighth nerve is to record impulses from individual fibers, an experimental method now commonly used. The results obtained are the subject of the next section.

CODING OF ACOUSTIC STIMULI IN FIBERS OF THE EIGHTH NERVE[5, 57, 67, 85]

There are 30,000 afferent neurons in the human auditory nerve. The cell bodies are arranged in a spiral ganglion that parallels the organ of Corti but is within the bony modiolus. The portion of such a neuron peripheral to the cell body is myelinated until it passes from the habenula perforata into the organ of Corti. A schema of the innervation pattern of the hair cells in the cat is shown in Fig. 12-18.[80] This animal has about 50,000 afferent neurons,[81] approximately nine tenths of which innervate inner hair cells in the radial fashion shown. The other afferent fibers cross the tunnel of Corti at the bottom, are frequently embedded in the cytoplasm of the pillar cells, spiral basalward in regular rows between Deiters' cells, and finally terminate on the outer hair cells.[80,81] The afferent neurons are bipolar, and the myelinated portion functions as the axon. The centrally directed portions of the fibers pass through the hollow core of the cochlea to form the cochlear portion of the eighth nerve. They enter the cranial cavity through the internal auditory meatus and end on the cells of the cochlear nucleus of the medulla.

Each radial afferent fiber seems to make a single synapse with one inner hair cell. Each inner hair cell has terminations from 10 to 20 radial afferent fibers. On the other hand, each spiral afferent fiber innervates about 10 outer hair cells, and outer hair cells have about 4 afferent terminals. The spiral afferent fibers run along the cochlea for a distance of 0.7 to 1 mm, sending out terminal branches to the outer hair cells from approximately the final 0.2 mm of their length.[80, 81] It is not yet clear to what extent the pattern of innervation in other mammals corresponds to that seen in the cat.

It is important to know how the temporal-spatial pattern of vibration of the cochlea is encoded as a temporal-spatial pattern of nerve impulses in cochlear nerve fibers, which are the initial portion of the intricately interconnected structures of the auditory pathway. For levels of acoustic stimulation within the linear range of vibration of the middle and inner ear, it is possible to entirely reconstruct the time course of the acoustic signal from the pattern of vibration of the cochlear partition. But when the nonlinear processes that lead to and include the excitation of neural impulses in the auditory fibers are encountered, this reversibility is lost. Thus coding in the primary fibers should indicate the properties of the acoustic signal that are lost and the properties that are not in the translation into nerve impulses and thus indicate those properties that can possibly be sensed by the organism.

Ongoing activity[5, 47]

When advancing a microelectrode into the auditory nerve of an experimental animal, it is usual to present an acoustic stimulus. Trains of clicks or noise bursts are commonly used, since they excite the cochlea broadly. Study proceeds after the nerve impulses of a single fiber are isolated. The first question is whether a fiber exhibits "spontaneous" activity, i.e., whether impulses are present in the absence of acoustic stimulation. Most (if not all) primary auditory fibers of mammals exhibit spontaneous activity. The average rate of this discharge may be above 100/sec, although it is usually considerably lower. The impulses occur in a random time sequence closely modeled by a modified Poisson process. In this model the occurrence of an impulse is equally likely at any time and does not depend on previous events; it is modified only to account for a brief refractory period after the occurrence of an impulse.

Tuning curves

Cochlear nerve fibers are differentially sensitive to sounds of different frequency. A set of tuning curves displaying this property is shown in Fig. 12-20. It is of interest to compare these neural tuning curves with the tuning properties of the vibration of the cochlear partition observed by Békésy. If curves such as those in Fig. 12-13 are used to plot the intensity needed to give a specified position along the cochlear partition a certain vibration amplitude as a function of frequency, we obtain a type of mechanical tuning curve. Fig. 12-21 illustrates several neural tuning curves together with vibration tuning curves. It is apparent that a considerable sharpening of the mechanical tuning is present in the neural readout.

Temporal coding

The responses of primary auditory fibers to continuous stimuli are of a tonic nature. At the onset of a noise or tonal stimulus that excites the unit the discharge rate is relatively high; subsequently it adapts to a more or less constant value.[5, 103] For wide-band noise or high-frequency tone stimulation the cadence of nerve impulses is not regular but has a random pattern, similar to the spontaneous activity but with a higher average rate.

For sinusoidal stimuli of low and middle frequencies, fibers of the auditory nerve are

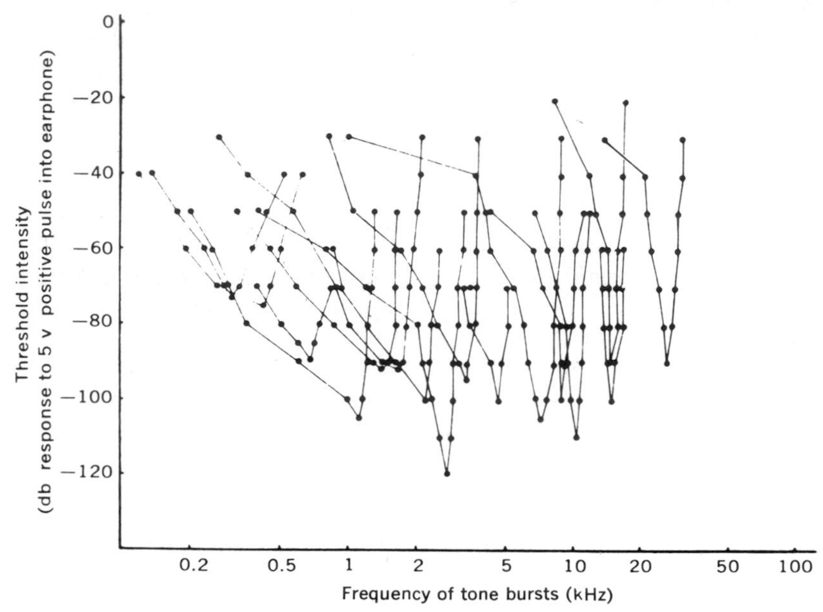

Fig. 12-20. Representative "response areas," or "tuning curves," for 16 different auditory nerve fibers in 9 cats. Each curve is obtained by setting intensity of tone bursts and measuring frequency range for which spike responses are obtained. Limits of this frequency range for a number of intensities are represented by points that have been joined to produce a "curve." Abscissa of lowest point of each curve is defined as characteristic frequency (C.F.) of that unit. Tone-burst stimuli had rise-fall times of 2.5 msec and duration of 50 msec. (From Kiang et al.[47])

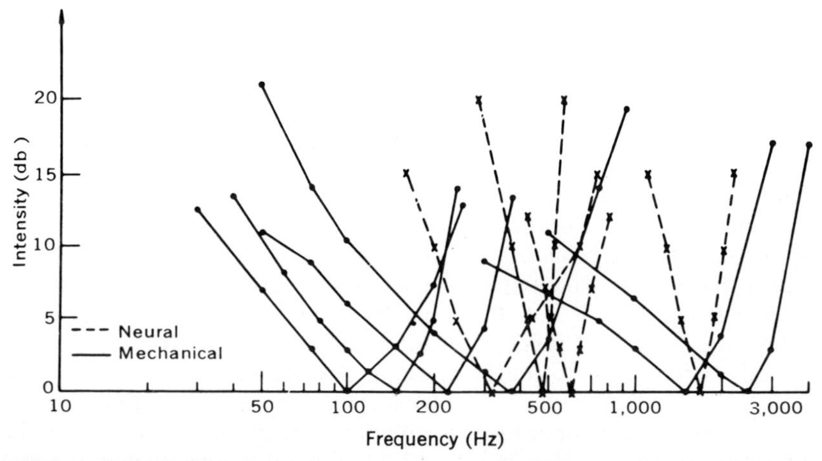

Fig. 12-21. Mechanical and neural tuning curves. Mechanical tuning curves are based on data obtained by Békésy from human cadavers. Neural tuning curves were obtained from data on eighth nerve fibers of cats by Kiang et al.[47] (From Weiss.[93])

more likely to discharge in one part of the stimulus cycle than in others. This phase-locking appears in recordings from monkeys for frequencies as high as 4,000 to 5,000 Hz.[65] The interspike interval histogram in Fig. 12-22 illustrates the stochastic nature of the responses. Some intervals are clustered about the period of the pure tone, some about twice this value, etc. Such a distribution is observed even with low frequencies, for which the period is considerably longer than the neural refractory time. This is not a simple frequency division in which the cell discharges on every cycle and then at

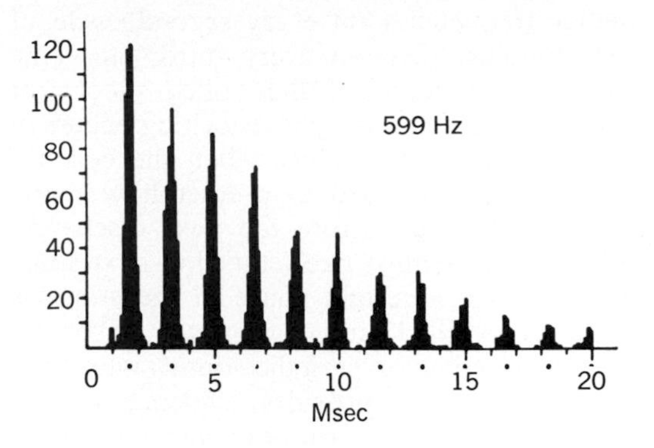

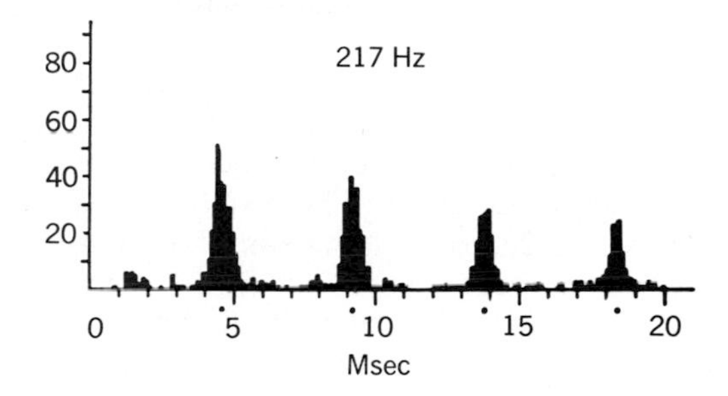

Fig. 12-22. Interval histogram for responses to pure tones. Abscissa: interval in msec. Ordinate: number of intervals in discharges to a 100 db SPL tone of 20 sec duration. Tone frequencies: 599 Hz, near best frequency of unit, and 217 Hz. Intervals cluster around integral multiples of period of tone indicated by dots below ordinate scale. (From Rose et al.[65])

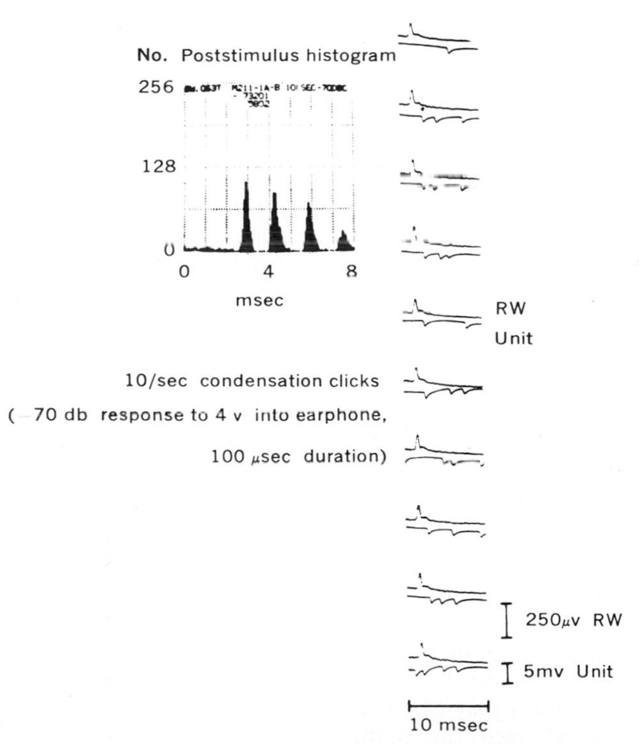

Fig. 12-23. Time pattern of responses of single auditory nerve fiber to clicks, as shown by poststimulus time histogram. Column of traces at right shows individual responses to 10 successive click presentations. Start of each trace is synchronized with click presentation. Upper trace of each pair shows gross potential changes as recorded by large electrode placed near round window; large upward deflection is N_1. Visual detection level (VDL) for N_1 potential was −90 db. Lower trace of each pair shows spike discharges recorded simultaneously with microelectrode in auditory nerve. Each downward deflection represents spike discharge by auditory nerve fiber. Poststimulus time histogram at left shows distribution of latencies of these spikes during 1 min sample of recording. Vertical axis represents number of spikes having a particular latency. Horizontal axis represents time after click presentation. (From Kiang et al.[47])

higher frequencies on every second cycle of the stimulus, then on every third, etc. (the simplest statement of the volley theory of Wever[96]). Instead, the phase-locking occurs in a probabilistic way. Thus, when the cell has discharged, it is hard to predict how many cycles will occur before the next discharge. However, it is most likely that the next spike will occur in a certain phase of the stimulus cycle. A detailed study of primary fiber responses to complex periodic sounds with two-tone components provides evidence of the coding of the wave form of cochlear partition vibration by the temporal pattern of discharges.[12]

Kiang et al.[5, 47] discovered a remarkable ability of primary fibers to code the temporal patterns of vibration of the cochlear partition in response to clicks. The result is illustrated in Fig. 12-23. Many hundreds of stimulus presentations were used to obtain the poststimulus time histogram of unit discharges. This statistical display gives the distribution of latencies of the unit impulses relative to the onset of the click. Note that the poststimulus time histogram has several evenly spaced peaks. Examination of the individual records to the right shows that in any given stimulus presentation the fiber does not discharge at the time of each and all of the peaks. These are, however, the times when the probability of discharge is greatest. The multipeaked pattern was observed for units with characteristic frequencies up to about 5,000 Hz. Furthermore, the space between peaks was found to be almost exactly equal to the inverse of the characteristic frequency of the unit under study.

Evidence has been cited to indicate that neural excitation occurs only on the movement of the cochlear partition toward the scala vestibuli. Further evidence that neurons are excited for only one direction of movement of the cochlear partition comes from an extension of studies such as that illustrated in Fig. 12-23. If a fiber is studied with opposite polarity clicks (rarefaction clicks and condensation clicks), one observes that when the polarity of the click is reversed the multipeaked pattern remains; however, a shift occurs so that peaks appear where valleys for the opposite polarity were found.

Place and time in peripheral coding

As fibers leave the cochlea, they maintain their relative positions in the spiraling bundle,[70] so that one finds an orderly progression of characteristic frequencies as an electrode is advanced through the cochlear nerve.[5] In the previous section, evidence of the temporal coding of acoustic stimuli by the pattern of impulses in the primary fiber population was reviewed. Thus the afferent fibers innervating the hair cells exhibit the capability of coding both temporal and spectral patterns of acoustic signals. The spectral coding is in terms of those fibers in the auditory nerve that are active and is called a place code.

An important question is the extent to which animals use temporal coding in discriminating pitch of sinusoidal stimuli. For high frequencies a place code must be used, but for low and middle ranges both place and temporal (phase-locking) patterns code tone frequency. Siebert[72] has applied a model from decision theory to the question, and it seems that human psychophysical judgments are far poorer than one would predict if the phase-locking code were used efficiently. It is likely that for frequencies as low as about 200 Hz, pitch discriminations of pure tones are based on different average rates of firing in different fibers (the place code). This does not mean the phase-locking temporal code is not used at all. It is clearly implicated in allowing judgments of interaural phase of sinusoidal stimuli below 4,000 Hz, judgments that are important in localization of a sound source. Furthermore, pitch discrimination of nonsinusoidal stimuli, e.g., chopped noise or pulse trains with the fundamental frequency removed, may well employ a phase-locking type of temporal code.

Most experimental results and the models pertaining to them attribute a homogeneity to the coding of acoustic stimuli by the primary afferent fibers, which is surprising in view of the separate innervation of inner and outer hair cells. Perhaps differences in coding by the two groups of primary fibers are sufficiently subtle, and one group (fibers from the inner hair cells[81]) is numerically so predominant that an extensive study is required to provide definite evidence of dual functional populations.

Models and nonlinearities

For some years now, scientists have been seeking appropriate mathematical models for the coding of information by primary auditory fibers. The ability of investigators to control the acoustic stimulus and to obtain reliable data, albeit usually best described in probabilistic terms, motivates the search for useful models. One may ask the purpose of such models. First, a model can be a very compact form of reporting experimental findings. But more important, the extent to which experimental findings can be modeled is often a reflection of the understanding of the mechanisms underlying observed phenomena.

One type of model for auditory fiber activity could be described as a transformation model. Given the acoustic stimulus, the model predicts the pattern of discharges in a primary unit. The nerve fiber is characterized by one parameter, usually its characteristic frequency. One such model, which includes computer simulation, successfully fits data for a restricted set of stimuli, i.e., clicks[94] and low-frequency tones.[38] However, experimental data for more complex stimuli, e.g., two tones or two clicks, are not well fit by this model.

When two-tone stimuli are presented, a suppression of activity in response to one (excitatory) tone can result when the second (inhibitory) tone is presented.[68, 69] This phenomenon is illustrated schematically in Fig. 12-24. The dashed, V-shaped curve is the tuning curve of a unit. A pure tone within this curve increases firing rate by 20% or more above the spontaneous rate. When an excitatory stimulus of frequency and intensity indicated by the triangle has added to it a second tone with frequency-intensity parameters in the shaded regions, the rate drops from that in response to the excitatory tone alone by 20% or more. All primary units show two-tone inhibition, which has been described by a mathematical model specifically intended to fit data obtained with these stimuli.[68]

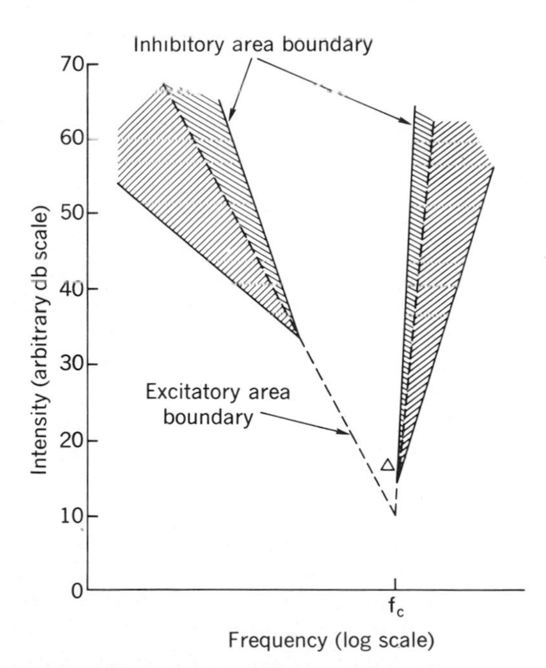

Fig. 12-24. Idealized inhibitory and response areas for typical auditory nerve fiber. Tuning curve as in dashed lines. Second tone in shaded areas suppresses response to excitatory tone (triangle). Regions for which the second tone presented alone would result in excitation are shaded differently than regions for which the second tone alone would not excite the fiber. (From Sachs.[68])

Helmholtz[4] observed that when two tones of frequencies (f_1 and f_2) are presented, a subject can detect combination tones including one of frequency $2f_1 - f_2$. Recent evidence shows that coding by first-order fibers parallels the psychophysical results.[40]

These findings for two-tone stimulation and results obtained when paired click stimuli were used[39] clearly indicate the limitations of the early mathematical model[94] for the transformation from acoustic stimulus to primary fiber code. It appears that a general model will have to be more fundamentally nonlinear.

Often the term "linear" is used to suggest that the amplitude of a response grows proportionally with stimulus intensity. However, the definition of a linear system is broader than this. The salient feature of a linear system is that superposition holds; i.e., if one stimulus gives a certain response and a second stimulus gives another response, the response to the addition of the two stimuli is the addition of the two responses.

Weiss' model[94] had nonlinear elements, but the tuning properties of the fiber were expressed by a linear transformation. A tentative first step toward a more general model has been made by Pfeiffer.[61] Although each year brings new advances, we are still a long way from a general transformation model for the auditory periphery or a real understanding of the underlying mechanisms in the transformation from sound signal to neural code.

In addition to the afferent fibers considered in this section, there are other neural systems innervating the cochlea. The autonomic nervous system sends nerve endings to the labyrinthine artery and its larger branches, and also forms a dense terminal plexus in the area of the habenula perforata.[83] Another more extensively studied efferent system innervating the cochlea and intimately connected with the afferent fiber system is the olivocochlear afferent system. Anatomic and physiologic studies

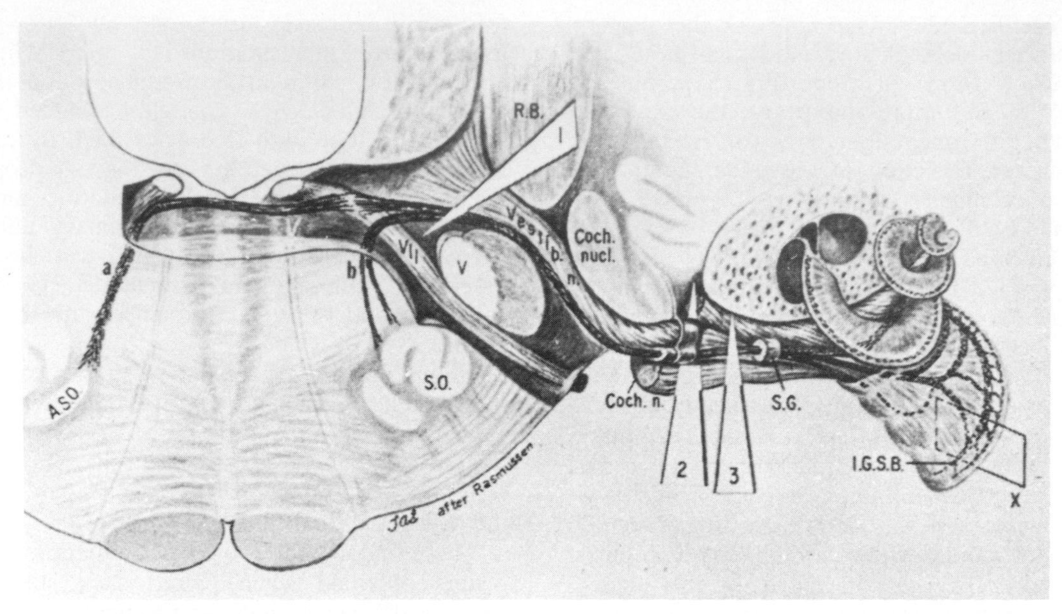

Fig. 12-25. Olivocochlear bundle (in black) and associated structures in cat. Three types of lesions indicated by *1, 2,* and *3. a,* Contralateral limb of olivocochlear bundle; *b,* homolateral limb of olivocochlear bundle; *A.S.O.,* accessory superior olive; *S.O.,* superior olive; *R.B.,* restiform body; *V,* descending trigeminal root; *VII,* facial root; *coch. nucl.,* cochlear nucleus; *I.G.S.B.,* intraganglionic spiral bundle; *S.G.,* saccular ganglion; *X,* plane of section through modiolus taken to verify lesion. (From Spoendlin and Gacek.[82])

of this system are the topics of the next section.

OLIVOCOCHLEAR EFFERENT SYSTEM[23, 27, 30, 36]

The auditory pathway includes descending as well as ascending tracts throughout its course. Two of these descending conduction systems project into the cochlea; they are the crossed and uncrossed olivocochlear fibers. This system[63] is illustrated in Fig. 12-25. In the cat there are about 600 efferent fibers in the bundle, with the uncrossed component contributing about one fifth of this number. There is also, at least in rodents, a reticulocochlear pathway of efferent fibers ending in the cochlea.[66]

Morphologic considerations

By making selective lesions (Fig. 12-25) and observing patterns of terminal degeneration, anatomists have determined the manner in which the efferent fibers innervate the cochlea.[45, 48, 76, 82] Results of findings in the cat are summarized in Fig. 12-18. Note that efferent innervation of the outer hair cells is radial and innervation of the inner hair cells is spiral. Also, the inner hair cell efferent synapses appear to be postsynaptic on the afferent dendrites, and the outer hair cell efferent synapses are presynaptic on the hair cells.

A number of electron microscopic studies of the nerve endings on the hair cells have been made. One type of synaptic pattern found for guinea pig outer hair cells is shown in Fig. 12-26. The pattern for cat outer hair cells is similar. The first row of outer hair cells has the most abundant supply of efferent (type 2) synapses found in the cochlea. In the upper (more apical) turns, they gradually disappear from the second and third row of hair cells.[81] The number of efferent endings on the outer hair cells has been estimated at 40,000.[79] Obviously there is a great deal of branching of the fibers of the olivocochlear bundle.[58]

The innervation of a typical internal hair cell is shown schematically in Fig. 12-27. The elongated terminals belong to the radial afferent fibers that terminate on the internal hair cells. The small, heavily vesiculated nerve endings rarely terminate on the hair cell, usually making synaptic contact with the afferent dendrites by means of vesiculated enlargements. These efferent endings belong to the small nerve fibers of the inner spiral bundle.

Serial reconstructions of electron micro-

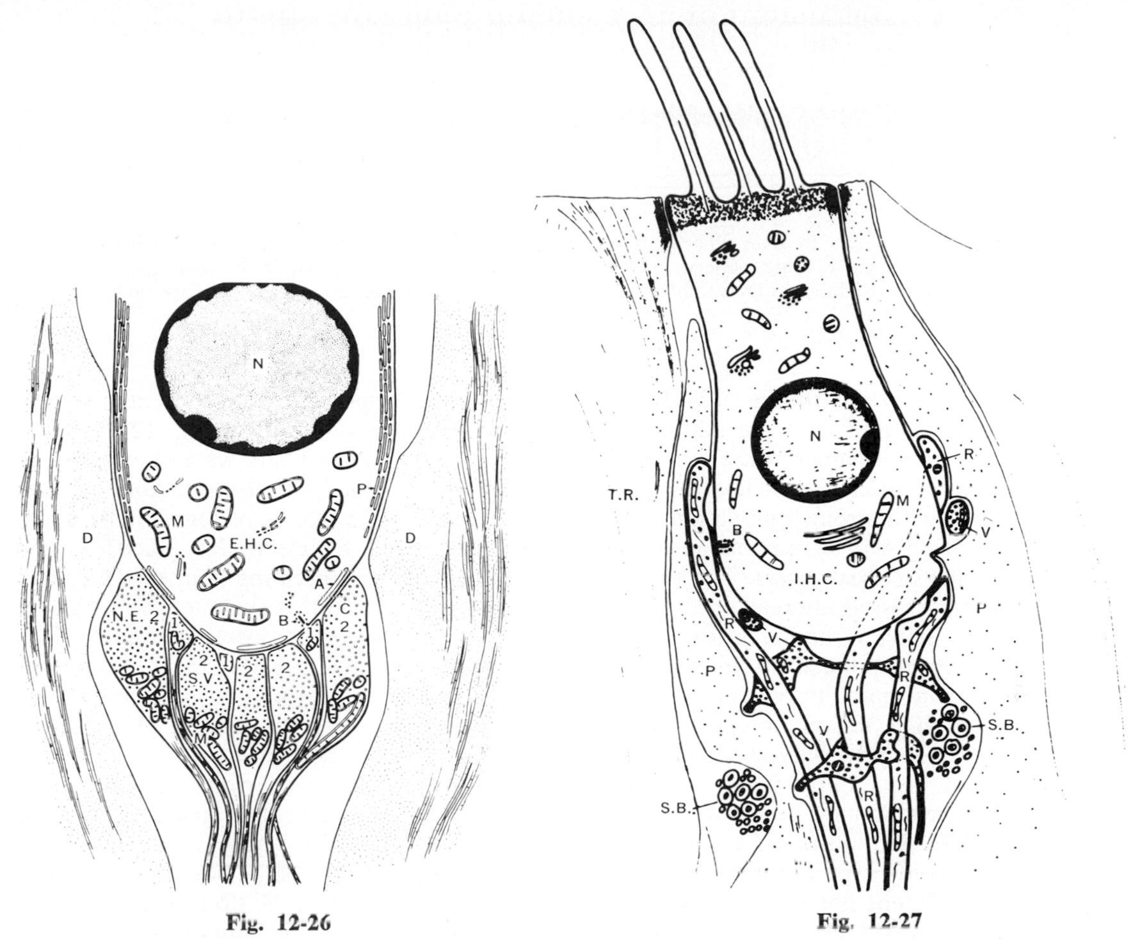

Fig. 12-26

Fig. 12-27

Fig. 12-26. Schematic drawing of outer hair cell from third cochlear turn of guinea pig. This pattern of synapses is typical for hair cells in inner row (all turns) and more basal portions of second and third rows. *A,* Accessory membrane; *B,* synaptic bar; *C,* region of vesicle concentration; *D,* Deiters' cell; *E.H.C.,* external hair cell; *M,* mitochondrion; *N,* nucleus; *1,* type 1 nerve ending; *N.E. 2,* type 2 nerve ending; *P,* vesicular peripheral membranes; *S.V.,* synaptic vesicles. (From Smith and Sjöstrand.[77])

Fig. 12-27. Schematic drawing showing innervation of internal hair cell of guinea pig cochlea. *B,* Synaptic bar; *I.H.C.,* internal hair cell; *M,* mitochondrion; *N,* nucleus; *P,* phalangeal cell; *R,* radial nerve fiber; *S.B.,* spiral nerve bundle; *T.R.,* tunnel rod; *V,* vesiculated nerve with enlargement. (From Smith.[75])

graphs indicate that for essentially all afferent synapses on inner hair cells there is a synaptic bar in the hair cell near the region of synaptic contact. In cats the synaptic bars are not consistently found near the afferent terminals on the outer hair cells.[79]

Electrophysiologic studies

This discussion of the electrophysiology of the olivocochlear pathways will have two main parts. First, the nerve discharge activity evoked by sound in the olivocochlear fibers will be described. Second, the effect of electrical stimulation of the olivocochlear bundle on activity of the auditory nerve fibers will be considered.

Fex[30] was able to record from individual fibers of the crossed olivocochlear bundle of cats at the vestibular-cochlear anastomosis. In order to place the electrode in the olivocochlear bundle under visual control, it was necessary to destroy the cochlea on that side. Thus acoustic stimulation of only the contralateral ear was possible, although the afferent fibers of the ipsilateral ear could be stimulated electrically.

Fig. 12-28 illustrates the pattern of response of an efferent fiber to a brief sinusoidal

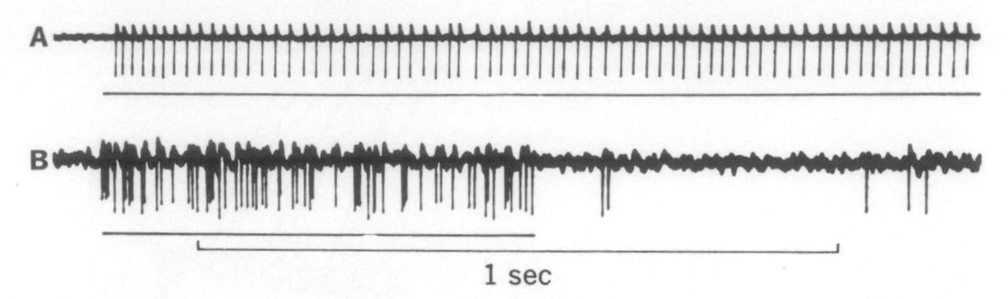

Fig. 12-28. Responses to tone pips from single efferent fiber, **A,** and from primary auditory fiber, **B.** Note regular firing pattern in **A** in contrast to irregular pattern in **B.** Records are from different experiments. Duration of sound stimuli are indicated by horizontal bars under records. (From Fex.[30])

signal. The firing pattern in Fig. 12-28, *A,* is quite regular in contrast to the pattern of response of an afferent fiber, shown in Fig. 12-28, *B*. Rates of discharge of less than 50 spikes/sec, even with strong stimulation, and an absence of firing with no stimulation were also characteristic of these fibers. The tuning properties of the crossed efferent fibers indicated that axons going to the basal turn of the cochlea generally responded to higher frequencies than axons going to the apical turns. Thus excitation of a region of one cochlea would result in efferent activity at roughly the same region of the contralateral cochlea. Thresholds of excitation by sound were usually above 40 db SPL, although a few fibers with lower thresholds were observed.

The crossed efferent fibers were also activated by electrical stimulation of the ipsilateral auditory nerve. When sound was presented to the contralateral ear during this electrical stimulation, responsiveness was increased, indicating a convergence of inputs from the two ears on the cells of origin of the crossed efferent pathway. The properties of the uncrossed efferent fibers are similar to those of the crossed fibers, with the following exceptions[31]: (1) many showed resting activity that with a few exceptions could be inhibited by a tonal stimulus presented to the contralateral ear; (2) inhibition of response to one tonal signal is produced by another tone of a different frequency; and (3) when activated by sound, the uncrossed efferents in the basal fascicle of the anastomosis generally respond to higher tone frequencies than do fibers in the apical fascicle, suggesting that afferents and uncrossed efferents from homotopic cochlear points in opposite cochleae are connected.

In another series of experiments, Fex studied the effects of electrical stimulation of the crossed olivocochlear bundle on single-unit activity of primary afferent fibers. The inhibition of the response of an afferent fiber excited by a tone is shown in Fig. 12-29. Here the crossed efferent fibers were stimulated by a tetanus of repetitive electric shocks (rates from 250 to 425/sec were found most effective) delivered by electrodes placed in the fiber tract as it crosses the floor of the fourth ventricle. The effect of such activation of the efferent pathway was always inhibitory.

Recently these results have been extended.[100, 101] Wiederhold has demonstrated, in nearly all fibers in the cat, that the effect of shocks to the crossed efferent bundle on the intensity function of a primary afferent fiber is to shift the curve to the right. The result is shown in Fig. 12-30. The average amount of shift in the region of rapid growth of the intensity function was 10 db in this case. The effect of efferent bundle stimulation and thus the shift varied from fiber to fiber, and values ranging from 1 to 25 db were obtained. Shifts were greatest for units with characteristic frequencies in the range of 6 to 10 kHz.

Earlier, Galambos[36] found that the neural potential recorded from near the round window (action potential) may be reduced or abolished by tetanic stimulation of the crossed efferent fibers. Close observation reveals not only that the neural potential is abolished by stimulation of the efferent fibers but also that the microphonic component is increased in magnitude.[29] This effect, although small, is quite consistent and also can be demonstrated with tonal acoustic stimulation. The parameters of the tetanus most effective for reducing the neural potential are also most effective for increasing the size of the microphonic potential. Furthermore, the time course of the two effects is similar, as is their dissipation after cessation of the tetanus.[23] It has been demonstrated that the electrical stimulation of the uncrossed olivocochlear bundle will also reduce the size of the click-evoked action potential.[24] The effect is smaller

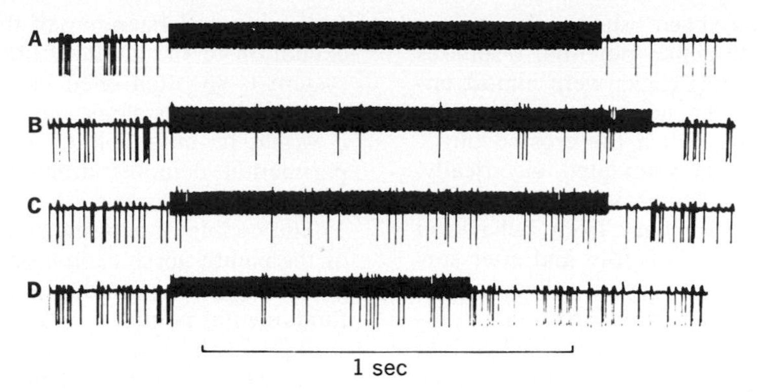

Fig. 12-29. Effect of tetanic electrical stimulation of crossed efferents on tone-evoked activity in primary auditory neuron. Primary afferent is stimulated with a tone of its best frequency, 950 Hz. In **A,** sound pressure was kept at constant level between 5 and 10 db relative to threshold of fiber. Between each record sound pressure was increased 5 db. Tetanic bursts were applied to efferents. Note total inhibition in **A** and some poststimulatory inhibition after cessation of stimulation. Note that in **B, C,** and **D** there was dissipation of inhibition from efferent stimulation after about a quarter of a second. (From Fex.[30])

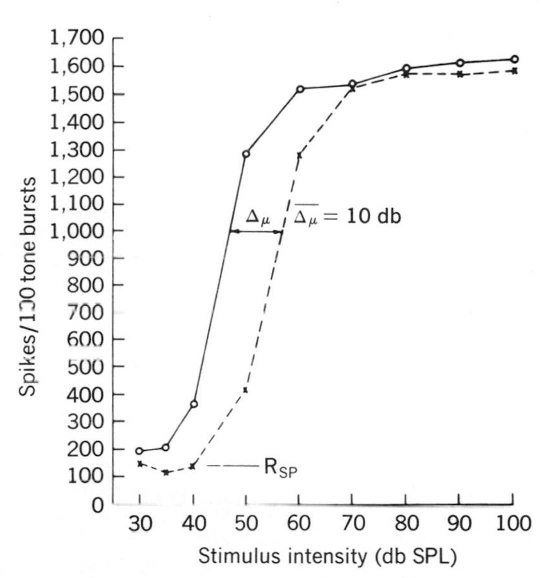

Fig. 12-30. Intensity functions for primary afferent unit with and without efferent stimulation. Solid curve for 40 msec tone bursts at unit's characteristic frequency. Dashed curves for same acoustic stimulus preceded by 32 shocks presented to crossed olivocochlear bundle at 400/sec. Tone burst starts at 10 msec after last shock. Characteristic frequency, 16.5 kHz. Intensity scale gives peak-to-peak sound pressure level. R_{SP} = rate of spontaneous activity; Δ_U = shift of intensity function by efferent stimulation. (From Wiederhold.[100])

than that observed with stimulation of the crossed bundle,[24, 78] and there is no increase in the CM potential.[78]

All of the phenomena just described that result from electrical stimulation of the olivocochlear fibers are diminished or disappear following intravenous injection of strychnine sulfate or the related alkaloid brucine.[25] A similar block of inhibition following administration of these drugs is seen for postsynaptic inhibition in the cat's spinal cord,[11] a phenomenon that suggests that the mechanism of inhibition in the two systems is similar (i.e., through release of a potassium and/or chloride permeability–increasing transmitter that has the effect of "clamping" the nerve membrane at a subthreshold potential).

Function of efferent fibers in audition

Note that in Fig. 12-30 electrical stimulation of the crossed olivocochlear bundle has very little effect on the response of the primary afferent fiber for intensities greater than 70 db SPL. The result is typical and in keeping with the finding that the intensity function is shifted to the right, since intensity functions of most afferent fibers saturate at levels approximately 30 db above threshold.[5] Similarly, with click stimulation there is a greater reduction of action potentials by olivocochlear bundle stimulation for low-intensity clicks than for high-intensity clicks.[23, 78]

Noting such results, Dewson[26] hypothesized that the function of the crossed olivocochlear bundle is to raise the threshold of auditory nerve fibers, making them insensitive to masking noise but still responsive to signals of moderate or high intensity. He demonstrated, in cats, that the attenuation of the click-evoked action potential by noise masking or by electrical stimulation of the olivocochlear bundle did not summate when the conditions were delivered simultaneously.[26] A more dramatic result has been shown in guinea pigs.[56] In this case action potential responses to moderately intense clicks were almost entirely masked by the addition of high-rate, low-intensity clicks. When the crossed olivocochlear bundle was stimulated electrically, the responses were "unmasked." Support for Dewson's hypothesis came from behavioral tests of rhesus monkeys before and after surgical section of the crossed olivocochlear bundles.[27] Discrimination of moderately intense vowel sounds presented in background noise was poorer following section of the bundles.

There has also been a suggestion, based on behavioral findings in squirrel monkeys, that the olivocochlear efferent system has a role in frequency discrimination.[13] Unfortunately in these studies the control discriminations were poor, although they did become even worse after transection of the olivocochlear bundle.

It is interesting to note that although the action potential is reduced by electrical stimulation of the olivocochlear bundles, attempts to demonstrate changes that could be attributed to the olivocochlear pathways in action potential recordings from chronically studied cats have been without positive results. When the cats were asleep, awake, placid, or interestedly observing a mouse, no changes assignable to the efferent pathways could be observed in the characteristic N_1, N_2 neural evoked potential.[8, 37] Perhaps in view of the hypothesis that the efferent system acts to reduce masking by background noise, future experiments of this nature should employ masking and masked stimuli.

CONCLUSION

In this chapter four aspects of the periphery of the auditory system have been considered: (1) the mechanical events by which an acoustic signal is transformed to a spatiotemporal pattern of vibration of the cochlear partition; (2) the mechanisms by which this vibration leads to excitation of the fibers of the cochlear branch of the eighth nerve; (3) the spatiotemporal patterns of nerve fiber discharges in the eighth nerve (that have been called coding), and (4) the olivocochlear efferent system.

In all of these areas there are interesting questions yet to be answered. Certainly the studies of morphologic polarization of the hair cells should stimulate biochemical and electrophysiologic investigations aimed at furthering understanding of the mechanism of excitation of the auditory fibers. The efferent system is so often cited in speculations concerning the physiologic basis of behavior that it would be most helpful to have further experimental demonstrations of its functional significance.

The coding of acoustic stimuli by the fibers of the eighth nerve cannot yet be described in general terms for a wide class of stimuli. Some fundamental nonlinearities have been brought to light in the last few years, and the search for a good mathematical model is just beginning. However, a good deal is now known about the temporal and spatial patterns of response to certain stimuli. Using models from detection theory, it is even possible to ask meaningful questions about those aspects of the neural code that are used by an animal in making psychophysical judgments.

The eighth nerve serves as the gateway to more central structures that are interconnected and anatomically organized in a most intricate manner. Within the central structures the patterns of activity of the eighth nerve fibers are processed and reprocessed in ways that are only beginning to be indicated by experimental investigation. The representation of acoustic signals in the CNS is one of the topics of the next chapter.

REFERENCES

General reviews

1. Békésy, G. von: Experiments in hearing (Research articles from 1928 to 1958), New York, 1960, McGraw-Hill Book Co.
2. Davis, H.: Excitation of auditory receptors. In Magoun, H. W., editor, Neurophysiology section: Handbook of physiology, Baltimore, 1959, The Williams & Wilkins Co., vol. 1.
3. Gernandt, B. E.: Vestibular mechanisms. In Magoun, H. W., editor, Neurophysiology section: Handbook of physiology, Baltimore, 1959, The Williams & Wilkins Co., vol. 1.
4. Helmholtz, H. L. F. von: Die Lehre von den Tonempfindungen als physiologische Grundlage für die Theorie der Musik, ed. 1, Brunswick, Germany, 1863, Vieweg-Verlag. (Translated and adapted by Ellis, A. J.: Sensations of tone, New York, 1954, Dover Publications, Inc.)
5. Kiang, N. Y-S., Watanabe, T., Thomas, E. C., and Clark, L. F.: Discharge patterns of single fibers in the cat's auditory nerve, Cambridge, Mass., 1965, The M.I.T. Press.
6. Wersäll, J., Flock, Å., and Lundquist, P. G.: Structural basis for directional sensitivity in cochlear and vestibular sensory receptors, Symp. Quant. Biol. **33:**115, 1965.

Original papers

7. Baust, W., Berlucchi, G., and Moruzzi, G.: Changes in the auditory input in wakefulness

and during synchronized and desynchronized stages of sleep, Arch. Ital. Biol. **102:**657, 1964.
8. Baust, W., Berlucchi, G., and Moruzzi, G.: Changes in auditory input during arousal in cats wth tenotomized middle ear muscles, Arch. Ital. Biol. **102:**675, 1964.
9. Békésy, G. von, and Rosenblith, W. A.: The mechanical properties of the ear. In Stevens, S. S., editor: Handbook of experimental psychology, New York, 1951, John Wiley & Sons, Inc.
10. Beranek, L. L.: Acoustics, New York, 1954, McGraw-Hill Book Co.
11. Bradley, K., Easton, D. M., and Eccles, J. C.: An investigation of primary or direct inhibition, J. Physiol. **122:**474, 1953.
12. Brugge, J. F., Anderson, D. J., Hind, J. E., and Rose, J. E.: Time structure of discharges in single auditory nerve fibers of the squirrel monkey in response to complex periodic sounds, J. Neurophysiol. **32:**386, 1969.
13. Capps, M. J., and Ades, H. W.: Auditory frequency discrimination after transection of the olivocochlear bundle in squirrel monkeys, Exp. Neurol. **21:**147, 1968.
14. Carmel, P. W., and Starr, A.: Acoustic and nonacoustic factors modifying middle-ear muscle activity in waking cats, J. Neurophysiol. **26:**598, 1963.
15. Dallos, P.: Dynamics of the acoustic reflex: phenomenological aspects, J. Acoust. Soc. Am. **36:**2175, 1964.
16. Dallos, P.: On the negative potential within the organ of Corti, J. Acoust. Soc. Am. **44:** 818, 1968.
17. Dallos, P., Shoeny, Z. G., Worthington, D. W., and Cheatham, M. A.: Cochlear distortion: effect of direct-current polarization, Science **164:**449, 1969.
18. Davis, H.: Biophysics and physiology of the inner ear, Physiol. Rev. **37:**1, 1957.
19. Davis, H.: Mechanism of excitation of auditory nerve impulses. In Rasmussen, G. L., and Windle, W. F., editors: Neural mechanisms of the auditory and vestibular system, Springfield, Ill., 1960, Charles C Thomas, Publisher.
20. Davis, H.: Some principles of sensory receptor action, Physiol. Rev. **41:**391, 1961.
21. Davis, H., Fernandez, C., and McAuliffe, D. R.: The excitatory process in the cochlea, Proc. Natl. Acad. Sci. U.S.A. **36:**580, 1950.
22. Davis, H., et al.: Modification of the cochlear potentials produced by streptomycin poisoning and by extensive venous obstruction, Laryngoscope **68:**596, 1958.
23. Desmedt, J. E.: Auditory-evoked potentials from cochlea to cortex as influenced by activation of the efferent olivo-cochlear bundle, J. Acoust. Soc. Am. **34:**1478, 1962.
24. Desmedt, J. E., and LaGrutta, V.: Function of the uncrossed olivo-cochlear fibers in the cat, Nature **200:**472, 1963.
25. Desmedt, J. E., and Monaco, P.: Suppression par la strychnine de l'effect inhibiteur contrifuge exerce par le faisceau olivo-cochleaire, Arch. Int. Pharmacodyn. Ther. **129:**244, 1960.
26. Dewson, J. H., III: Efferent olivocochlear bundle: some relationships to noise masking and to stimulus attenuation, J. Neurophysiol. **30:**817, 1967.
27. Dewson, J. H., III: Efferent olivocochlear bundle: some relationships to stimulus discrimination in noise, J. Neurophysiol. **31:**122, 1968.
28. Engström, H., Ades, H. W., and Hawkins, J. E.: II. Structure and functions of the sensory hairs of the inner ear, J. Acoust. Soc. Am. **34:**1356, 1962.
29. Fex, J.: Augmentation of the cochlear microphonics by stimulation of efferent fibres to cochlea, Acta Otolaryngol. **50:**540, 1959.
30. Fex, J.: Auditory activity in centrifugal and centripetal cochlear fibers in cat. A study of a feedback system, Acta Physiol. Scand. suppl. **189:**1, 1962.
31. Fex, J.: Auditory activity in uncrossed centrifugal cochlear fibers in cat, Acta Physiol. Scand. **64:**43, 1965.
32. Flanagan, J. L.: Models for approximating basilar membrane displacement, Bell Syst. Tech. J. **39:**1163, 1960.
33. Flock, Å.: Transducing mechanisms in the lateral line canal organ receptors, Symp. Quant. Biol. **30:**133, 1965.
34. Flock, Å., and Wersäll, J.: A study of the orientation of the sensory hairs of the receptor cells in the lateral line organ of fish, with special references to the function of the receptors, J. Cell Biol. **15:**19, 1962.
35. Flock, Å., Kimura, R., Lundquist, P. G., and Wersäll, J.: Morphological basis of directional sensitivity of the outer hair cells of the organ of Corti, J. Acoust. Soc. Am. **34:** 1351, 1962.
36. Galambos, R.: Suppression of auditory nerve activity by stimulation of efferent fibers to cochlea, J. Neurophysiol. **19:**424, 1956.
37. Galambos, R.: Studies of the auditory system with the implanted electrodes. In Rasmussen, G. L., and Windle, W. F., editors: Neural mechanisms of the auditory and vestibular system, Springfield, Ill., 1960, Charles C Thomas, Publisher.
38. Geisler, D. C.: A model of the peripheral auditory system responding to low-frequency tones, Biophys. J. **8:**1, 1968.
39. Goblick, T. J., and Pfeiffer, R. R.: Time domain measurements of cochlear nonlinearities using combination click stimuli, J. Acoust. Soc. Am. **46:**924, 1969.
40. Goldstein, J. L., and Kiang, N. Y-S.: Neural correlates of the aural combination tone $2f_1 - f_2$, Proc. IEEE **56:**981, 1968.
41. Goldstein, M. H., and Kiang, N. Y-S.: Synchrony of neural activity in electric responses evoked by transient acoustic stimuli, J. Acoust. Soc. Am. **30:**107, 1958.
42. Guinan, J. J., Jr., and Peake, W. T.: Motion of the middle ear bones, Quarterly Progress Report No. 74, Cambridge, Mass., 1964, The M.I.T. Press.
43. Guinan, J. J., Jr., and Peake, W. T.: Middle-ear characteristics of anesthetized cats, J. Acoust. Soc. Am. **41:**1237, 1967.
44. Harris, G. G., Frishkopf, L. S., and Flock, Å.: Receptor potentials from hair cells of the lateral line, Science **167:**76, 1970.

45. Iurato, S.: Efferent fibers to the sensory cells of Corti's organ, Exp. Cell Res. **27**:162, 1962.
46. Johnstone, B. M., Taylor, K. J., and Boyle, A. J.: Mechanics of the guinea pig cochlea, J. Acoust. Soc. Am. **47**:504, 1970.
47. Kiang, N. Y-S., Watanabe, T., Thomas, E. D., and Clark, L. F.: Stimulus coding in the cat's auditory nerve, Ann. Otol. Rhinol. Laryngol. **71**:1009, 1962.
48. Kimura, R., and Wersäll, J.: Termination of the olivocochlear bundle in relation to the outer hair cells of the organ of Corti in guinea pig, Acta Otolaryngol. **55**:11, 1962.
49. Laszlo, C. A., Gannon, R. P., and Milsum, J. H.: Measurement of the cochlear potentials of the guinea pig at constant sound-pressure level at the eardrum. I. Cochlear-microphonic amplitude and phase, J. Acoust. Soc. Am. **47**:1063, 1970.
50. Misrahy, G. A., De Jonge, B. R., Shinberger, E. W., and Arnold, J. E.: Effects of localized hypoxia on the electrophysiological activity of cochlea of the guinea pig, J. Acoust. Soc. Am. **30**:705, 1958.
51. Møller, A. R.: Network model of the middle ear, J. Acoust. Soc. Am. **33**:168, 1961.
52. Møller, A. R.: Acoustic reflex in man, J. Acoust. Soc. Am. **34**:1524, 1962.
53. Møller, A. R.: The sensitivity of contraction of the tympanic muscles in man, Ann. Otol. Rhinol. Laryngol. **71**:86, 1962.
54. Møller, A. R.: Transfer function of the middle ear, J. Acoust. Soc. Am. **35**:1526, 1963.
55. Møller, A. R.: An experimental study of the acoustic impedance of the middle ear and its transmission properties, Acta Otolaryngol. **60**: 129, 1965.
56. Nieder, P. C., and Nieder, I.: Crossed olivo-cochlear bundle: electric stimulation enhances masked neural responses to loud clicks, Brain Res. **21**:135, 1970.
57. Nomoto, M., Suga, N., and Katsuki, Y.: Discharge pattern and inhibition of primary auditory nerve fibers in the monkey, J. Neurophysiol. **27**:768, 1964.
58. Nomura, Y., and Schuknecht, H. F.: The efferent fibers in the cochlea, Ann. Otol. Rhinol. Laryngol. **74**:289, 1965.
59. Peake, W. T., and Kiang, N. Y-S.: Cochlear responses to condensation and rarefaction clicks, Biophys. J. **2**:23, 1962.
60. Peake, W. T., Goldstein, M. H., and Kiang, N. Y-S.: Responses of the auditory nerve to repetitive acoustic stimuli, J. Acoust. Soc. Am. **34**:562, 1962.
61. Pfeiffer, R. R.: A model for two-tone inhibition of single cochlear-nerve fibers, J. Acoust. Soc. Am. **43**:1373, 1970.
62. Pierce, J. R., and David, E. E.: Man's world of sound, New York, 1958, Doubleday & Co., Inc.
63. Rasmussen, G. L.: The olivary peduncle and other fiber projections of the superior olivary complex, J. Comp. Neurol. **84**:141, 1946.
64. Rhode, W. S.: Observations of the vibration of the basilar membrane in squirrel monkeys using the Mössbauer technique, J. Acoust. Soc. Am. **49**:1218, 1971.
65. Rose, J. E., Brugge, J. F., Anderson, D. J., and Hind, J. E.: Phase-locked response to low frequency tones in single auditory nerve fibers of the squirrel monkey, J. Neurophysiol. **30**: 771, 1967.
66. Rossi, G., and Cortesina, G.: Research on the efferent innervation of the inner ear, J. Laryngol. Otol. **77**:202, 1963.
67. Rupert, A., Moushegian, G., and Galambos, R.: Unit responses to sound from the auditory nerve of the cat, J. Neurophysiol. **26**:449, 1963.
68. Sachs, M. B.: Stimulus-response relation for auditory nerve fibers: two-tone stimuli, J. Acoust. Soc. Am. **45**:1025, 1969.
69. Sachs, M. B., and Kiang, N. Y-S.: Two-tone inhibition in auditory-nerve fibers, J. Acoust. Soc. Am. **43**:1120, 1968.
70. Sando, I.: The anatomical interrelationships of the cochlear nerve fibers, Acta Otolaryngol. **59**:417, 1965.
71. Siebert, W. M.: Models for the dynamic behavior of the cochlear partition, Quarterly Progress Report, No. 64, Cambridge, Mass., 1962, The M.I.T. Press.
72. Siebert, W. M.: Frequency discrimination in the auditory system: place or periodicity mechanism? Proc. IEEE **58**:723, 1970.
73. Simmons, F. B.: Middle ear muscle protection from the acoustic trauma of loud continuous sound, Ann. Otol. Rhinol. Laryngol. **69**:1063, 1960.
74. Simmons, F. B.: Perceptual theories of middle ear muscle function, Ann. Otol. Rhinol. Laryngol. **73**:724, 1964.
75. Smith, C. A.: The innervation pattern of the cochlea: the internal hair cell, Ann. Otol. Rhinol. Laryngol. **70**:504, 1961.
76. Smith, C. A., and Rasmussen, G. L.: Recent observations on the olivo-cochlear bundle, Ann. Otol. Rhinol. Laryngol. **72**:489, 1963.
77. Smith, C. A., and Sjöstrand, F. S.: Structure of the nerve endings in the external hair cells of the guinea pig cochlea as studied by serial sections, J. Ultrastruct. Res. **5**:523, 1961.
78. Sohmer, H.: A comparison of the efferent effects of the homolateral and contralateral olivo-cochlear bundles, Acta Otolaryngol. **62**: 74, 1966.
79. Spoendlin, H.: The organization of the cochlear receptor, Basel, 1966, S. Karger.
80. Spoendlin, H.: The innervation pattern of the organ of Corti, J. Laryngol. Otol. **81**:717, 1967.
81. Spoendlin, H.: Structural basis of peripheral frequency analysis. In Plomb, B., and Smörenburg, G. F., editors: Frequency analysis and periodicity detection in hearing, Leiden, The Netherlands, 1970, Sijthoff.
82. Spoendlin, H. H., and Gacek, R. R.: Electromicroscopic study of the efferent and afferent innervation of the organ of Corti in the cat, Ann. Otol. Rhinol. Laryngol. **72**:660, 1963.
83. Spoendlin, H., and Lichtensteiger, W.: The adrenergic innervation of the labyrinth, Acta Otolaryngol. **61**:423, 1966.
84. Starr, A.: Influence of motor activity in click-evoked responses in the auditory pathway of waking cats, Exp. Neurol. **10**:191, 1964.
85. Tasaki, I.: Nerve impulses in individual audi-

tory nerve fibers of guinea pig, J. Neurophysiol. **17:**97, 1954.

86. Tasaki, I., and Spyropolous, C. S.: Stria vascularis as a source of endocochlear potential, J. Neurophysiol. **22:**149, 1959.
87. Teas, D. C., Eldredge, D. H., and Davis, H.: Cochlear responses to acoustic transients: an interpretation of whole-nerve action potentials, J. Acoust. Soc. Am. **34:**1438, 1962.
88. Tonndorf, J.: Nonlinearities in cochlear hydrodynamics, J. Acoust. Soc. Am. **47:**579, 1970.
89. Tonndorf, J., and Khanna, S. M.: Submicroscopic displacement amplitudes of the tympanic membrane (cat) measured by a laser interferometer, J. Acoust. Soc. Am. **44:**1546, 1968.
90. Tonndorf, J., and Khanna, S.: The role of the tympanic membrane in middle ear transmission, Ann. Otol. Rhinol. Laryngol. **79:**743, 1970.
91. Tonndorf, J., Khanna, S. M., and Fingerhood, B. J.: The input impedance of the inner ear in cats, Ann. Otol. Rhinol. Laryngol. **75:** 752, 1966.
92. Weiner, F. M., and Ross, D.: The pressure distribution in the auditory canal in the progressive sound field, J. Acoust. Soc. Am. **18:** 401, 1946.
93. Weiss, T. F.: A model for firing patterns at auditory nerve fibers, Technical Report No. 418, Cambridge, Mass., 1964, The M.I.T. Press.
94. Weiss, T. F.: A model of the peripheral auditory system, Kybernetik **4:**153, 1966.
95. Wersäll, J., and Flock, Å.: Physiological aspects of the structure of vestibular end organs, Acta Otolaryngol. suppl. **192:**85, 1964.
96. Wever, E. G.: Theory of hearing, New York, 1949, John Wiley & Sons, Inc.
97. Wever, E. G., and Lawrence, M.: Physiological acoustics, Princeton, 1954, Princeton University Press.
98. Wever, E. G., and Vernon, J. A.: The effect of the tympanic muscle reflexes upon sound transmission, Acta Otolaryngol. **45:**433, 1955.
99. Whitfield, I. C., and Ross, H. C.: Cochlear microphonics and summating potentials and the output of individual hair cell generators, J. Acoust. Soc. Am. **38:**126, 1965.
100. Wiederhold, M. L.: Variations in the effects of electric stimulation of the crossed olivocochlear bundle on cat single auditory-nerve fiber responses to tone bursts, J. Acoust. Soc. Am. **48:**966, 1970.
101. Wiederhold, M. L., and Kiang, N. Y-S.: Effects of electric stimulation of the crossed olivocochlear bundle on single auditory-nerve fibers in the cat, J. Acoust. Soc. Am. **48:**950, 1970.
102. Wiggers, H. C.: The functions of the intra-aural muscles, Am. J. Physiol. **120:**771, 1937.
103. Young, E. D., and Sachs, M. B.: Recovery of single auditory-nerve fibers from sound exposure, J. Acoust. Soc. Am. **50:**94, 1971.
104. Zwislocki, J.: Analysis of the middle-ear function. I. Input impedance, J. Acoust. Soc. Am. **34:**1514, 1962.
105. Zwislocki, J.: Analysis of some auditory characteristics. In Luce, R. D., Bush, R. R., and Galanter, E., editors: Handbook of mathematical psychology, New York, 1965, John Wiley & Sons, Inc., vol. 3.

13

VERNON B. MOUNTCASTLE

Central neural mechanisms in hearing

The preceding account of the mechanical transducer action of the ear and the excitation of auditory nerve fibers that results provides considerable evidence supporting the place theory of function of the peripheral auditory apparatus, originally proposed by Helmholtz, which now takes on some of the certainty of established fact. A pure tone elicits a differential movement of the cochlear partition that is maximal at a given locus and is displaced monotonically along the cochlear partition as the frequency of the stimulating tone is changed. The degree of local deformation and therefore the degree of nerve excitation are functions of sound intensity. The frequency of a pure tone determines the profile of activity in the distributed array of eighth nerve fibers; other stimulus properties determine the overall frequencies, durations, temporal modulation, and internal timing of events in the trains of impulses in those active fibers. From this time-dependent distribution, humans with normal hearing make accurate discriminations for pitch and loudness (the perceptive derivatives from frequency and intensity), locate precisely the position of a sound source in space, and identify the temporal modulations of sound as well. The purpose of this chapter is to describe those central neural structures and mechanims that play a role in hearing. That the peripheral analysis made in the cochlea is not itself sufficient is obvious from the fact that the extent of the active region on the cochlear partition is considerable even for a pure tone and that this region enlarges with increases in stimulus intensity. There cannot be uniquely determined locales of activity in the neural field of eighth nerve fibers for each pure tone discriminated. In this as in other neural systems, spatial displacements of the stimulus produce a partially shifted and overlapping series of active regions as a stimulus is shifted across the spatial sheet of receptors projecting into the system; in this case the change in frequency of a sound is equivalent to a spatial shift, e.g., of a mechanical stimulus across the skin or light across the visual field. It is thought that such a partially shifted neural distribution will, when compared to that from which it is shifted, provide a sufficient cue for a central mechanism capable of reading its profile to a fine degree. While cogent descriptions can be given of the central neural transforms of sensory stimuli, upon which a central detecting apparatus might operate, little is known at present of the nature of that neuronal detecting, measuring, and interpretive device. The gap between the study of afferent neural representations of sensory stimuli and the mechanisms of the introspective sensory experience is very great; it is a major objective for present and future research.

To define the problem a brief description is first given of the human capacity to hear. A review of the anatomy and topography of the central auditory system then follows, in which evidence from both purely anatomic and electrophysiologic experiments is blended to elucidate the central representation of the cochlear partition, and thus of frequency, at various levels of the system. Studies of the remaining capacity of animals and man to hear after lesions of the central auditory pathways are used to indicate the differing functional complexities of which those various levels are capable. Certain questions will then be asked and answers sought in modern electrophysiologic studies, incomplete as they may be for the present: what are the central neural signals upon which discrimination for pitch, loudness, lateralization, etc. might be based? Some discussion will be devoted to the subject of attention to a sound stimulus, with particular reference to the efferent components of the auditory system, as a prototype of the more general problem of how one may attend to one sensory input to the relative exclusion of others.

THE HUMAN CAPACITY TO HEAR[3, 6-8]

The measurement of the sensory attributes of auditory stimuli and of their relation to the physical characteristics of sounds is the business of auditory psychophysics. It sets a measure of the overall sensory performance of an observer, with which the results of neurophysiologic experiments may be balanced. The comparison, it is hoped, will stimulate new generalizations concerning the neural mechanisms in sensation and perception and new experiments to test them. A major problem in psychophysics is that of scaling: along what scales should subjective sensory experiences be measured? Surely the physical scale of the stimulus is not appropriate, for the total sensory experience is determined by the physical properties of a stimulus, its transduction by peripheral receptors, and the CNS operations upon the resulting neural input. Considerable attention is given in Chapter 18 to this problem. Here it is intended to indicate briefly the human capacity to hear, relative to several physical dimensions of heard sounds, and thus to establish the range of what hopefully will be understood eventually in terms of neural mechanisms.

Sensitivity

Fig. 13-1 presents a series of audibility curves. It is obvious from them that a major factor determining the threshold for hearing is the frequency of the testing sound, for the sensitivity at best frequencies (1,000 to 3,000 Hz) is some 100 to 10,000 times better than at the extreme upper and lower ranges of audible frequencies, respectively. The threshold for hearing in the best frequency range is generally found to be about 0.0002 dyne/cm^2 sound pressure for young adults free of ear or CNS disease. The two ordinates in Fig. 13-1 indicate the two ways in which sound stimuli are commonly scaled for intensity: that on the right indicates sound pressures in decibels above a level that is near threshold at 1,000 Hz; that on the left, pressures as decibels below or above an arbitrary standard of 1 dyne/cm^2. A reversed scale that is often used in clinical audiometry expresses

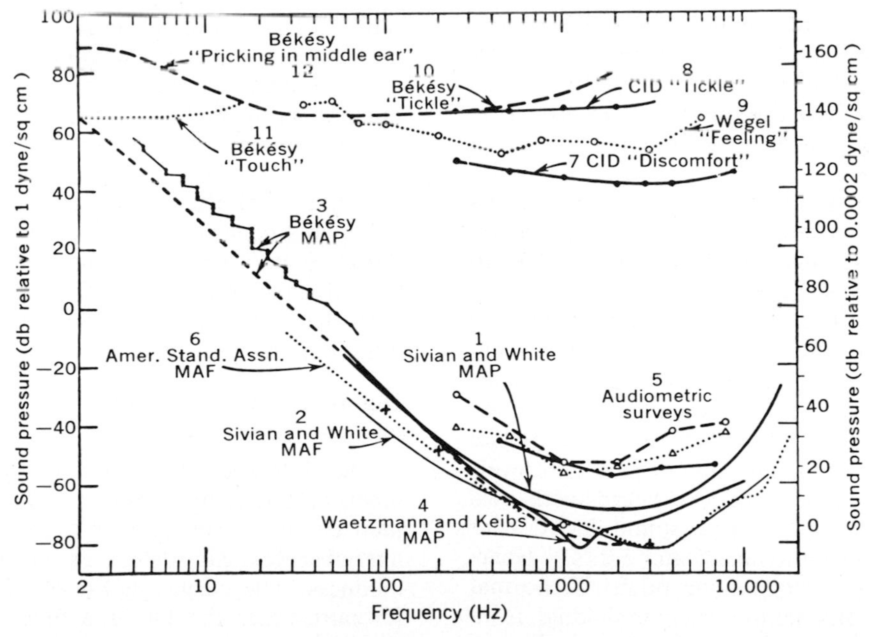

Fig. 13-1. Determinations of threshold of hearing at different frequencies and threshold for nonauditory sensations produced by sound: definition of auditory area. Curves *1* to *6* represent various measurements of threshold audibility curve; audiometric surveys may include individuals with impaired hearing. *MAP* = minimum audible pressure at eardrum; *MAF* = minimum audible pressure in free sound field measured in space occupied by observer's head during test. Curves *7* to *12* represent different measurements of threshold of nonauditory sensations produced by sounds and which are uncomfortable. (Curves collected from various sources by Licklider.[3])

hearing loss as the difference in decibels between threshold for normal individuals and that of the patient tested at a series of frequencies. As sounds become very intense, a level is reached at which discomfort and/or mechanoreceptive sensations are evoked; these ranges are indicated by the upper set of curves in Fig. 13-1. This level of discomfort is virtually independent of sound frequency. The area between these two sets of curves measures the dynamic range of human hearing and is sometimes referred to as the auditory area.

Auditory sensations can also be produced when vibrations are delivered directly to the bones of the head, and this is the basis for certain types of hearing aids. It is thought to be due to direct activation of the cochlear partition by the conducted vibrations. The thresholds for hearing via bone conduction are about 50 db above that for hearing via air conduction, and they rise very rapidly at the upper end of the frequency scale.

The absolute threshold for hearing is influenced by a variety of factors. When testing sounds are delivered via an earphone and the minimal audible pressure (MAP) is measured at the subject's eardrum, thresholds are some 4 to 6 db higher than when sounds are delivered in a free sound field and pressures are later measured in the center of the region previously occupied by the subject's head (MAF). This apparent paradox is at least partially explained by the fact that a closely applied earphone generates a certain amount of physiologic "noise" (the seashell effect) and because there is some enhancement of pressure at the eardrum due to standing waves in the ear canal.

Over a considerable frequency range the threshold for hearing with two ears is significantly lower than for monaural listening: neural inputs from the two ears sum for the property of loudness as well as for subliminal stimuli. At moderate sound pressure levels the ratio of the loudness heard with two ears to that of the same sound heard with one ear varies between 1.5 and 2.0.[52, 88] Hearing thresholds vary among otherwise normal individuals and in the same individual from one testing session to the next; these variations are of the order of 5 to 10 db. There is, moreover, a consistent loss of hearing acuity with advancing age, particularly for frequencies above 3,000 to 4,000 Hz.

The threshold for hearing is limited by properties of the ear and CNS, not by the physical characteristics of the air. A considerable range exists between the minimum audible pressure of even the best observer and that level of sensitivity at which random thermal agitation of gas molecules in the air would become audible. For example, if the region of greatest *sensitivity* for human hearing (1,000 to 6,000 Hz) is considered, the summed pressure level of thermal fluctuation over this range is 10 db below the human threshold at 1,000 Hz.[96] It is unlikely that thermal noise affects hearing even in those individuals with the most sensitive ears.

A continuing question of great theoretical interest is how the physiologic process set up by a sound that is just audible may differ from that evoked by one just weak enough to be inaudible. In Chapter 18 this matter is treated from the standpoint that what is threshold is a continuous variable over a certain range, and it is described in the context of the signal detection theory.[44, 45] Some investigators, however, have presented evidence that the difference between the two sets of neural activity—the one just evoking the perception of sound and the other just not—is a single "neural quantum," an elementary increase not further defined but regarded as a step change in neural activity that raises the total to a supraliminal level.[79] It seems obvious that these two are compatible: neural activity must certainly increase by some quantal jump, in the limit by one additional impulse in one or a group of neurons. The decision theory model might well describe the changing level along the neural quantal scale at which a central detecting apparatus accepts the decision—yes, a stimulus has been delivered versus no, it has not. It is this decision level that is likely to be influenced by factors other than stimulus intensity.

Difference thresholds and loudness functions

The difference threshold, or the difference limen (DL), is the measure of how small an increase in the intensity of a sound an average observer can detect as an increase in loudness. The curves in Fig. 13-2 show that for pure tones the DL is a function of both the frequency and the intensity of the sound, when its loudness is compared with that of the test stimulus. For tones well above threshold and in the best frequency range for hearing, normal observers can detect changes as small as 1 db in sound intensity. As the intensity of the comparsion stimulus is lowered, how-

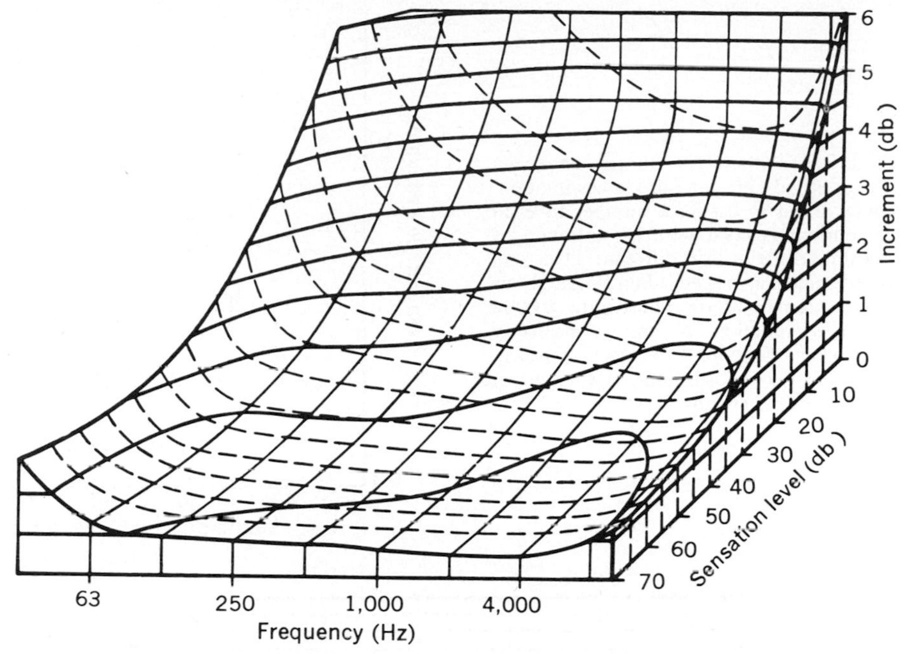

Fig. 13-2. Three-dimensional graph showing differential intensity threshold as function of frequency and intensity of standard tone. Threshold is given as difference in decibels between standard intensity and standard plus increment. (Based on data by Riesz, 1928; from Licklider.[3])

ever, the detectable increment rises rapidly. Thus the DL for loudness is not constant over the intensive continuum, nor is it a constant fraction of the comparison intensity, as Weber believed. This fact leads to the nonlinear Weber function for sound intensity shown in Fig. 18-7, p. 557.

Loudness is the intensive attribute of an auditory sensation, in terms of which sounds may be ordered on a scale extending from soft to loud. A scale for the sensory attribute called loudness cannot be constructed by summing the number of DLs over the intensity range of audible sounds. A suitable scale can be constructed by asking observers to assign numbers to different sounds (magnitude estimation) or to adjust the intensity of a sound to produce a loudness that matches a series of numbers (magnitude production). When this is done, the results show that a loudness ratio of 2:1 is produced by a pair of stimuli that differ by 10 db, i.e., by a ratio of 3:1, and that this relation holds over a wide range of audible intensities.[54, 101, 103] It follows that the loudness of a sound (L) can be approximated by a power function of the intensity of the sound (I) as follows:

$$L = kI^{0.3}$$

where I is measured in units of energy and k is a constant of proportionality. The exponent is therefore 0.6 when sound is scaled in units of pressure, as was done in Fig. 13-3. Some alinearity in the loudness functions (in log-log relation) is usually observed for soft sounds in the intensity region up to 40 db above threshold. Fig. 13-4 indicates this deviation in the region of soft tones and that a proper loudness function must take into account the loudness required of a tone that is just threshold. The curved section of the loudness function at low levels is thought to be due to masking by physiologic or other "noise." Thus a loudness function most accurate for all intensities up to at least 100 db is:

$$L = k(I^n - I_o^n)$$

where I_o is the threshold intensity. For the data illustrated the exponent n was determined to be 0.27, with I scaled in energy units.[51, 64]

The accepted scale for loudness commonly used in acoustics is the *sone* scale; a sone is arbitrarily defined as the loudness produced by a 1,000 Hz tone 40 db above 0.0002 dyne/cm^2.

Fig. 13-3. Loudness function for human observers determined by method of magnitude estimation. Each subject was asked to assign a number to each of a series of tones in accord with his estimate of its intensity. Number scales for different observers were normalized and averaged data plotted as shown. For tones above 40 db relative to 0.0002 dyne/cm^2, loudness function is almost perfectly described by power function of form shown, plotted here in log-log coordinates. Exponent n = 0.3 is in energy units. (From Stevens.[100])

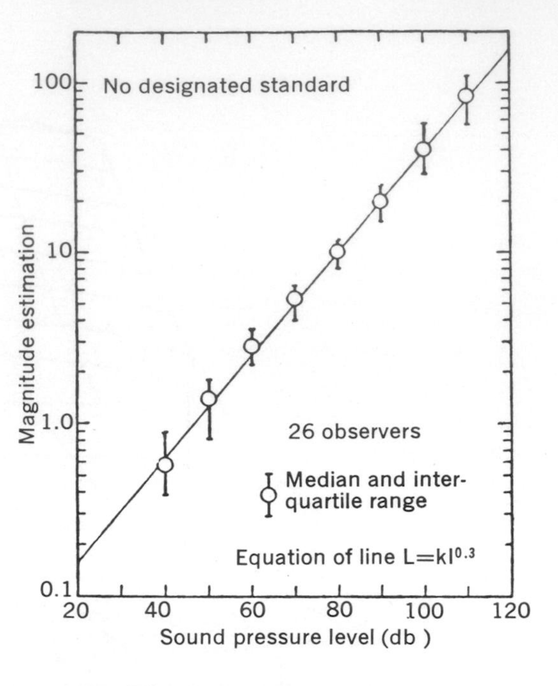

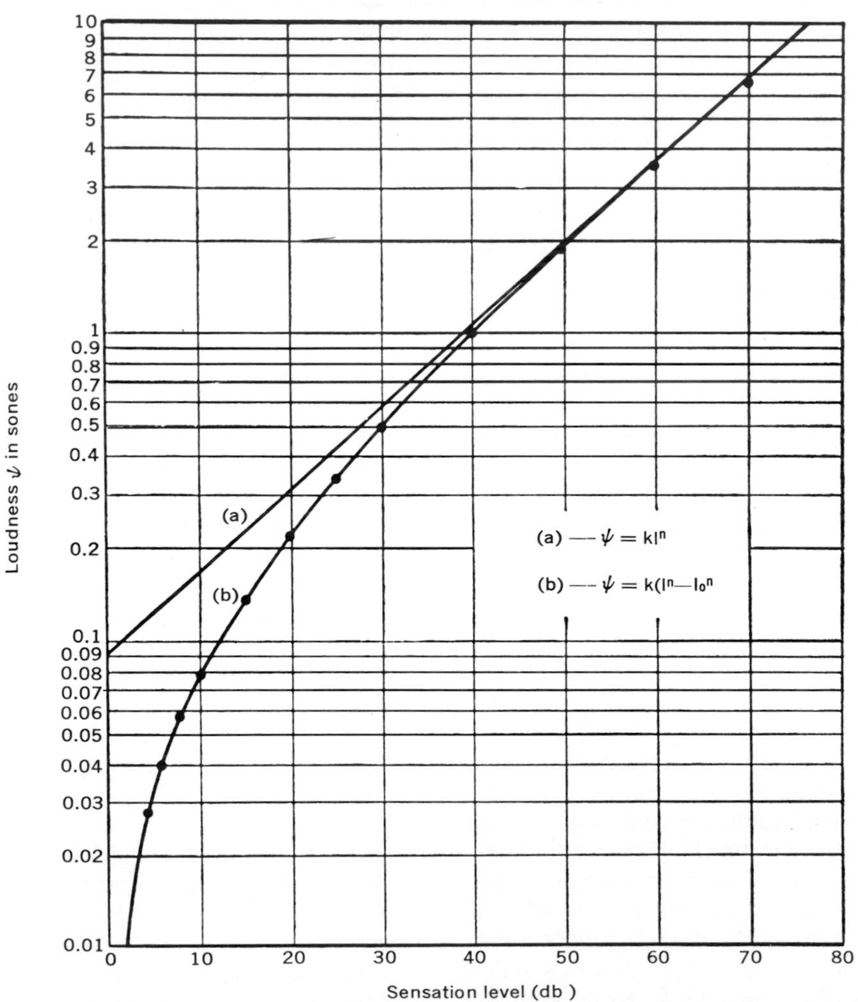

Fig. 13-4. Loudness function extended into range of very weak sounds and complementary to that in Fig. 13-3. Equation *b* provides almost perfect fit to observed magnitude estimations. I_o^n is factor introduced to represent masking of testing noise by "physiologic noise," i.e., threshold. (Based on data from Hellman and Zwislocki[51, 52]; from Lochner and Burger.[64])

Pitch and pitch discrimination

Pitch is the subjective attribute of hearing that is largely but not completely determined by the frequency of the sound heard. The pitch experience is evoked in human beings by sounds of sufficient intensity over a frequency range from about 15 to about 20,000 Hz. Sounds below 15 Hz do evoke auditory experiences, but these rough rumblings differ markedly in quality from the musical note produced by sounds of higher frequency. Very weak tones of any intensity evoke no subjective experience of pitch; this atonal area covers the first 3 to 5 db above threshold on the intensity scale. For more intense stimuli, further changes in intensity usually produce changes in the subjective estimation of pitch of no more than ±2% of stimulus frequency.[25] A precise identification of pitch requires tonal stimuli of 15 to 20 msec duration, a time invariant with frequency. Thus the lower audible tones can be identified after only one or a few cycles are heard, very high ones only after many.

Loudness and pitch differ in a fundamental way.[6, 102] Loudness is an attribute scaled along a prothetic continuum: an increase in loudness can be presumed to produce increases in the levels of activity, i.e., rates of discharge, in a centered and slowly expanding population of neurons. Pitch, on the other hand, is an attribute scaled along a metathetic or extensive continuum: changes in the frequency of sound can be presumed to result in a translation in neural space of the activated population of neurons, within the accessible neuronal fields. The populations activated by any two frequencies may overlap, but it is further presumed they must be separated to some unknown degree before a discrimination between the two pitches can be made. That is, the cochlear partition is mapped onto the sensory projection areas of the auditory system in such a way as to preserve ordinal relations, which must also be preserved in whatever other neural elaborations subserve perception.

The validity of the identification of loudness and pitch as different *types* of sensation is indicated by the fact that while the DL for the former is a nonlinear function of intensity, that of the latter is uniform over a considerable range of the auditory area. The DL for frequency is 2 to 3 Hz over an intensity range from 20 to 70 db and for frequencies up to about 3,000 Hz. The DL rises very rapidly for weak sounds or for higher frequencies. Obviously these values will result in a nonlinear Weber function for pitch.

It is of some value to construct a scale for the subjective sensory experience of pitch and to determine the pitch function for man. The unit of the scale is the *mel,*[104, 105] and a value of 1,000 mels is assigned to the pitch of a 1,000 Hz tone sounded at 40 db above threshold for normal observers. When such observers are asked to scale frequencies along the continuum of pitch, it appears that they are asked to make a subjective estimate of a sense interval or distance rather than a sense magnitude, as in the case of loudness. The magnitude estimation of pitch is a difficult judgment, particularly if the observer is previously trained in the musical scales, but for the most part the pitch judgment in mels varies as a power function of sound frequency, with some departures close to the low-frequency end of the audible range.[102]

Masking and the critical bands[112]

The threshold for the perception of a pure tone may be raised by the simultaneous presence of a second auditory stimulus, a phenomenon called masking. The degree of masking depends on the frequency components of the masking stimulus, for as its spectrum is widened around the frequency of the testing tone, a width is reached at which there is no further increase in masking of the test tone: this width is known as the critical band. Masking may be of various kinds. The first curve to the left in Fig. 13-5 shows that when a masking noise with its spectrum centered around the test tone frequency is increased in intensity, with both sounds delivered to the same ear, the degree of masking increases linearly: a 10 db increase in the masker produces a 10 db elevation of threshold for the testing tone. A complex masking function is produced when the same masking sound is delivered to the ear contralateral to the ear tested: it is thought to be due to a combination of sound escape around the head and central nervous interaction. When masking noise segments are raised some 60 to 80 db above threshold, they acquire the capacity to mask testing tones remotely placed on the frequency continuum, a result thought to be due to the combination of nonlinear distortion within the cochlea itself and the attenuation for the test tone caused by the evoked contraction of the middle ear muscles. The contralateral remote masking by the higher frequency noise segment (curve

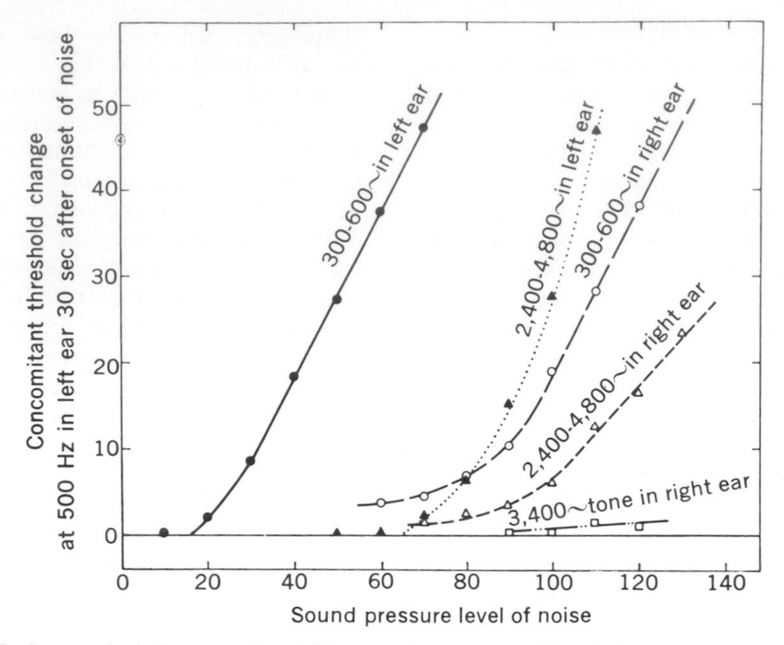

Fig. 13-5. Curves depicting growth of direct and remote masking in human beings. Threshold at 500 Hz in left ear was measured in presence of low-frequency (300 to 600 Hz) or high-frequency (2,400 to 4,800 Hz) noise or a 3,400 Hz tone delivered to left ear or right ear. (From Ward.[112])

marked 2,400 to 4,800 in right ear in Fig. 13-5) must be due to reflex attenuation and/or to central neural interaction, for the transcranial escape of sound at this frequency is negligible.

An experiment designed to measure the critical bandwidth, carried out by Greenwood,[46, 47] has shown that the degree of masking increases in proportion to total power in the masking noise. At a certain critical width of the masker the degree of masking increases no more, but extends farther over the frequency range as the masking bandwidth broadens. A number of experiments of this sort have established the fact that the critical bandwidth is an exponential function of distance along the basilar membrane, beginning with the helicotrema; i.e., bandwidth is a nearly linear function of frequency. Comparison of this function with that for frequency localization in the cochlea indicates that a bandwidth is equivalent to about 1 mm along the basilar membrane, and this may be regarded as the area over which the cochlea integrates the power of signals. The masked threshold is that intensity of a testing tone required to produce a discriminable change in the cochlear disturbance and resulting neural input produced by the critical band of the masking stimulus, when the two are summed. Obviously masking phenomena are important for everyday hearing, which for the most part requires the discrimination of signals against background noises of widely varying intensity.

The use of the masked audiogram also enabled Greenwood to measure the resolving power for hearing of two tones as two distinct pitches when sounded simultaneously. In this case a narrow band of noise was used as the test stimulus, this being masked by two tones that were varied in their frequency difference. The critical band measured in this manner turned out to be nearly equal to the ΔF required for two tones to be heard distinctly as two different pitches. It is important to note that this resolving power, about 440 Hz at the tested frequency of about 3,000 Hz, is some 100 times the DL for frequency discrimination in this range for two tones sounded sequentially (Fig. 13-2). Apparently for audition the sensory resolving power for simultaneous contrast discrimination is not nearly as acute as for sequential discriminations.

Auditory fatigue[112]

Auditory fatigue designates a temporary decrease in threshold sensitivity following exposure to an auditory stimulus; it occurs *after* exposure and thus differs from masking. The temporary threshold shift is a time-linked process, usually declining in degree as an exponential function of time. In some pathologic states the temporary shift may be greatly prolonged, even for many hours. The degree of the shift is determined by the intensity and, up to moderate intensity levels, by the frequency of the fatiguing stimulus. With higher intensities, the fatigue effect and the associated temporary shifts in threshold are also associated with moderate shifts in pitch perception in the fatigued ear. At least one factor contributing to the temporary threshold shift is a change in the sensitivity of auditory nerve fibers.[115] It is likely that changes in CNS function contribute also,[61] but nothing is known of the mechanism involved.

Localization of sound in space[16, 26]

The ability of human beings to locate the position in space of a pure tone varies with the frequency of the tone. For low tones the error for the azimuth is about 5 to 10 degrees; it rises in the range of 2,000 to 4,000 Hz to 20 to 25 degrees and falls again for higher frequencies. Obviously there are at least two possible cues, for both the amplitude and the phase of a plane sound wave emanating from an eccentrically placed sound source will differ at the two ears. For low frequencies, significant phase differences are possible, but only negligible ones in amplitude; while for higher frequencies the sound shield of the head is much more effective, thus increasing the interaural difference in amplitude, while the possible time differences due to phase shift are quite small. The two possible cues—time and intensity—are present to different degrees for different tones, and neither can be regarded as the sole basis for localization of a sound.[55] The localization of complex sounds is much better than is that for pure tones, but for either the sensing of inclination is poor; in normal behavior it depends largely on kinesthetic cues generated by tilting the head to direct an ear toward the source.

The time and intensity cues can be separated by delivering two sounds, tones or short clicks, to the ears via earphones, thus gaining independent control of the amplitude of each and of the time interval between the two. When the sounds delivered are identical, the sound is localized to the center of the head. If the sound in one ear is more intense than that in the other or if it leads in time, the sound is localized toward the side of the first ear. There is evidence that lateralization of sounds in this manner is essentially the same task as localizing external sources. A lag in time can be compensated by an increase in amplitude of the lagging signal and vice versa; i.e., a time-intensity trading relation exists. The ratio has been variously estimated, but a figure of 20 μsec/db fits many observations.

The mechanisms of sound localization and the time-intensity trade are neural, not cochlear; they will be described on p. 433.

• • •

This brief review has indicated only the simpler aspects of hearing in human beings, for nothing has been said of their performance when listening to more complex sounds or of the mechanisms of binaural interaction and fusion.[3, 16, 26] What has been said, however, is already much more than can be explained in terms of central neural mechanism. It is now appropriate to examine in some detail the central neural apparatus played upon by the cochlear afferents, in an attempt to elucidate those mechanisms. It will soon be evident that at the present time explanations can be given—at least at the first level of analysis—for those aspects of hearing that depend on the topographic or tonotopic mapping of the cochlea within the auditory system and of those properties of its neurons that are uniquely determined by neural connections. The most active field of research in auditory neurophysiology at the present time, however, is concerned with the more dynamic and time-dependent aspects of the activity of auditory neurons and with how they relate to the parameters of auditory stimuli, a field of enquiry and body of knowledge to which some attention will be given.

ANATOMY AND TOPOGRAPHY OF THE AUDITORY SYSTEM

General statement

A general problem for this as for all sensory systems is to state with some precision which afferent pathways, nuclear complexes, and cortical fields compose it, a problem considered also in relation to the somesthetic and visual systems. This specification is in a certain sense an abstraction, for the processing of sensory input, its complexion with the ongoing "current status" reports in other sensory systems, its storage and comparison with other items in memory, and the composition of appropriate responsive action may certainly involve very large parts of the CNS. What then earns the label "auditory" and what does not? It seems reasonable to say that some regions are more directly concerned with the sensory mechanisms of audition than are others, particularly when the concept *sensory* is separated from what is merely *afferent,* and it is both practically and conceptually advantageous to regard such regions as composing a system. To do so does not imply that they alone are concerned with hearing and that other mechanisms are not involved but rather that hearing is their principal concern, which allows an orderly study of parts, hopefully leading to a synthesis.

Thus those regions that (1) can be shown to receive a major share of their input over the eighth nerves or via subsequent central relays of eighth nerve activity and (2) can be shown on other grounds to play an important role in sensory auditory mechanisms, e.g., by behavioral testing of auditory capacity after lesion, will be classified as auditory structures. Such a definition establishes a hierarchy of auditory regions, from the purity of the eighth nerve itself to the auditory areas of the cerebral cortex, which of course receive input from the generalized thalamocortical system, from other cortical areas, and from the cerebellum, in addition to their more specific thalamocortical inflow generated by auditory stimuli—one that, as will be shown, varies in density and security from one area to another, all within areas that can yet be classified as auditory. Such a definition excludes from the auditory system regions of the brain that receive auditory as one of many inputs, even though from the overall behavioral point of view such a region may play an important role in phenomena evoked by auditory stimuli.

The auditory system has many subcortical nuclei, each concerned in the afferent relay of auditory input and many with the integration of afferent and other input as well. A striking characteristic of the system is that it is disposed in a combined series-parallel arrangement (Fig. 13-6), in which some efferents of a nucleus may end in the next and others may bypass it to end in the next one, etc. The result is that one cannot, at central levels of the system, designate any given elements as "second order" or "third order," etc., even though on a statistical basis one may classify an entire population as largely of one order or another.

A further characteristic is that the cochlear partition is represented more than once at each "level" of the system. Throughout the system there is a central core of a specific and highly differentiated representation of the cochlear end organ and thus of frequencies. Surrounding this central core region are one, two, or even more additional representations of the cochlea, although these may differ in characteristics from the central one.

Two points should be emphasized. First, not only are successive stages of the system organized in a series-parallel arrangement in the afferent sense but also in the efferent sense as well. Large numbers of efferent fibers descend from one level to the next, frequently with bypass, providing a neural substratum for an efferent control of afferent input at every level.[27] One final stage in this multiply reentrant controlling mechanism, that of the

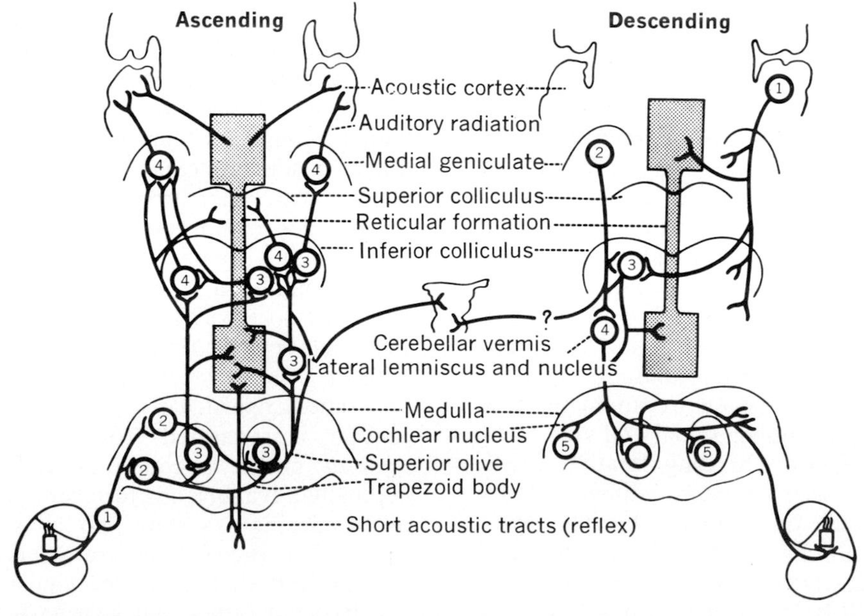

Fig. 13-6. Schematic outline of auditory system. Ascending afferent pathways and nuclei are shown to left and descending efferent components to right. Numerals indicate approximate order of neurons, counting from origin of each system. Crosshatched areas indicate central core of brainstem reticular formation and diencephalic nuclei of generalized thalamocortical system origin. (From Galambos.[34])

efferent olivocochlear bundle of Rasmussen, has already been described in some detail in Chapter 12. Second, there is good evidence that a heavy offshoot of elements of the auditory system, both directly and by means of axon collaterals, projects upon the reticular formation of the brainstem and upon pontile, medullary, and spinal motor systems, providing in the first case avenues for influencing forebrain excitability and in the second for low-order, more or less automatic auditory reflexes.

Cochlear nuclei

All afferent fibers from the cochlea terminate centrally within the cochlear nuclei, a complex located on the lateral aspect of the brainstem at the level of eighth nerve entry, where it covers the dorsal and posterior aspect of the restiform body.[81, 82] The complex can be divided into three nuclei: the dorsal, the anteroventral, and the posteroventral cochlear nuclei (Dc, Av, Pv).[94] The disposition of these nuclei is shown in Fig. 13-7, which also illustrates the fact that each of the cochlear nerve fibers divides to send terminal branches to each nucleus, and that they do so in such a way as to compose a complete and orderly representation of the cochlear partition in each of the three cochlear nuclei.

This triplication of cochlear projection upon the first central station of the auditory system, suggested by anatomic studies,[24] has been confirmed by electrophysiologic experiments.[91] An orderly sequence of best frequencies exists in each of the three nuclei, and in each these range from highs most dorsally to lows most ventrally; i.e., the cochlear partition is completely represented in each, with the basilar region oriented in a dorsal and the apex in a ventral direction.

There is a ratio of about 1:3 between cochlear nerve fibers and cells of the cochlear nuclear complex.[24] Incoming fibers and the neurons upon which they impinge are arranged in a reciprocally convergent-divergent relation: each fiber sends terminal branches to a number of cells, and each cell receives terminals from a number of fibers. An important question then is whether there is a marked increase in the range of frequencies to which a cochlear nuclear cell is responsive, as compared to the impinging first-order fibers. The question is settled by comparing the "tuning curves" of first-order fibers of the eighth nerve with those of neurons of central stations of the auditory system. Such a

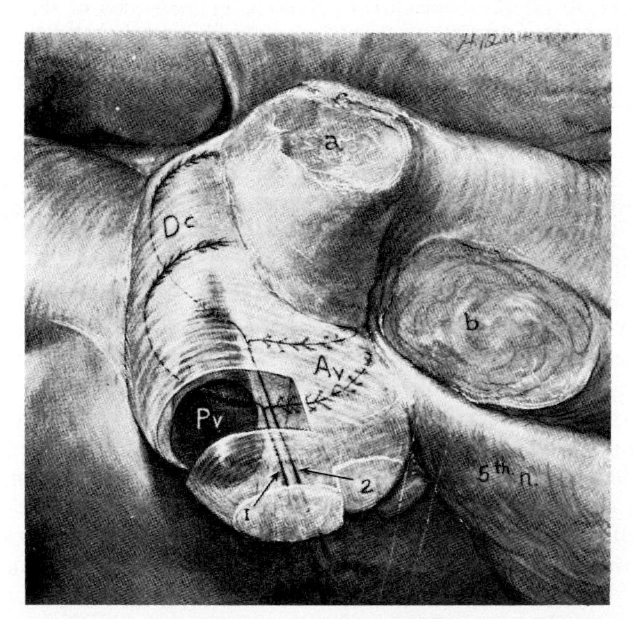

Fig. 13-7. Schematic representation of distribution of fibers of auditory nerve within cochlear nuclear complex of cat. An orderly array of distributions of fibers with best frequencies from high to low is thought to exist in a dorsoventral direction in all three divisions. Fiber *1* derives from middle turn of cochlea and fiber *2* from basilar end. Each fiber distributes terminals within each of three divisions of complex. *a,* Cut edge of restiform body and of superior cerebellar peduncle; *b,* cut edge of brachium pontis; *Av,* anteroventral nucleus; *Pv,* posteroventral nucleus; *Dc,* dorsal cochlear nucleus. Brainstem seen from right side. (From Rose et al.[91])

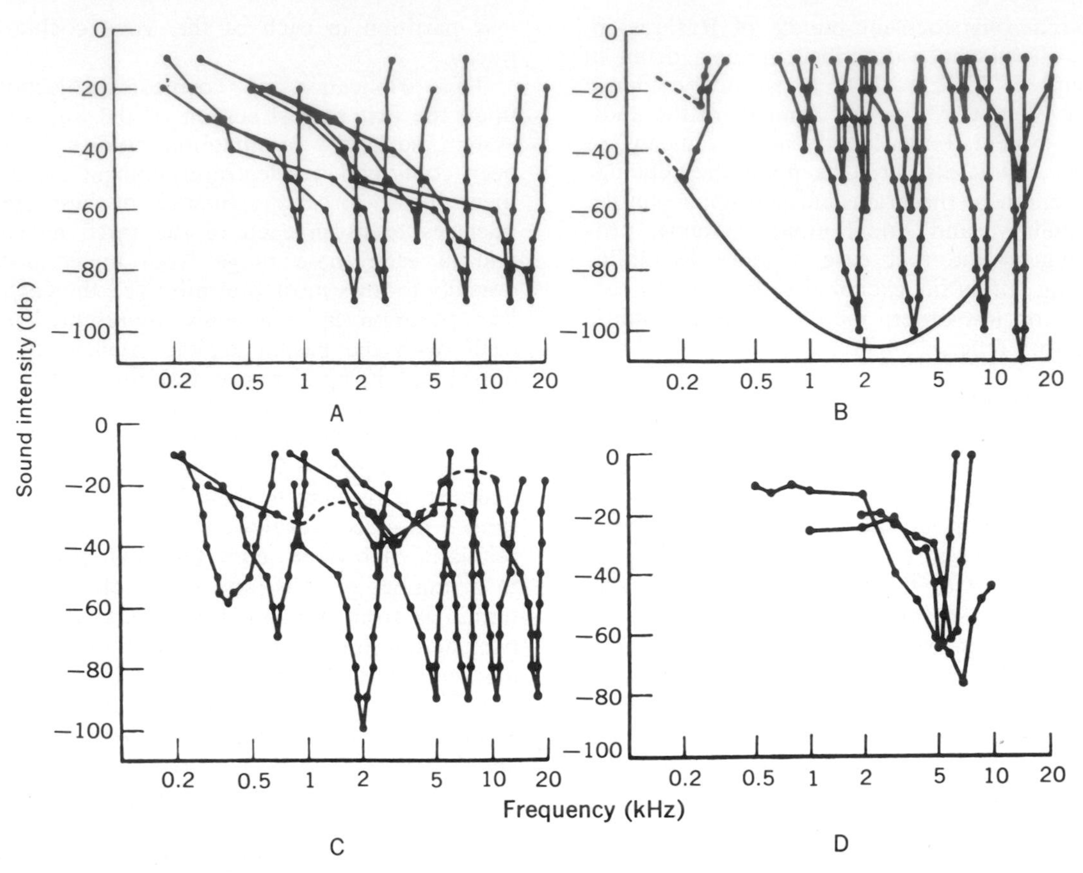

Fig. 13-8. Tuning curves for neurons located at various levels of auditory system of cat. **A,** Cochlear nerve fibers. **B,** Inferior colliculus. **C,** Neurons of the region of the trapezoid body. **D,** Medial geniculate body. Curve for each neuron plots, in decibels below a reference level, intensity of tones of different frequencies required to just excite it. Later, more extensive studies have shown that tuning curves of medial geniculate cells are no broader than are those of neurons at lower levels of auditory system. (From Katusi.[59])

tuning curve (Fig. 13-8) is constructed by determining, for each frequency, the intensity of sound just sufficient to evoke a response. Results of this sort are shown for four regions of the system in Fig. 13-8. Several facts are evident from these and similar studies. (1) The range of sound frequencies to which a single neuron is responsive is a function of sound intensity; (2) for many neurons the curves are slightly asymmetric, although this appearance is exaggerated by the logarithmic method of plotting; and (3) the tuning curves for cochlear nuclei cells are not obviously broader than are those of the first-order fibers of the cochlear nerve.[89]

A powerful factor that tends to restrict the spread of the tuning curves of cochlear neurons is *afferent inhibition*. In the auditory as in the visual and somesthetic systems, afferent inhibition is topographically disposed in such a way that delivery of a stimulus that excites a certain number of cells in a neural field will suppress the activity of those that surround it, thus tending to sharpen and isolate the island of activity. The phenomenon is illustrated in Fig. 13-9, which shows that the steady discharge of a cell of the cochlear nucleus under drive by a steady tone at the characteristic frequency for the cell is completely inhibited by an interpolated tone of a slightly lower frequency.[91] The degree and duration of the inhibition are determined by the relative intensities of the two tones. Whether the inhibition is of the pre- or postsynaptic variety[32, 37, 97] or a combination of both is uncertain. For many central neurons the afferent inhibition is "single-sided," but others may be inhibited by tones on either side of the best frequency, as illustrated by the plots in Fig. 13-10.[48]

To recapitulate, on the basis of what may be termed the static properties of single neu-

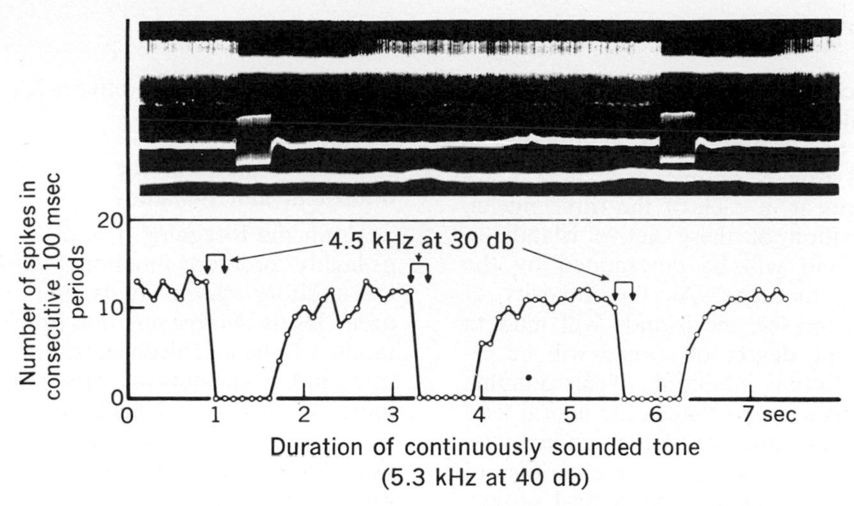

Fig. 13-9. Top trace displays electrical signs of nerve impulses of cell of cochlear nucleus of cat, driven by continuously sounded tone at best frequency of cell (5,300 Hz) delivered at 40 db below a reference level and about 60 db above threshold. Signals on second trace indicate when short bursts of second tone at 4,500 Hz were delivered at 30 db below reference level. Complete inhibition during sounding of second tone and its persistence for part of a second after its end are shown by records and graph given below. (From Rose et al.[91])

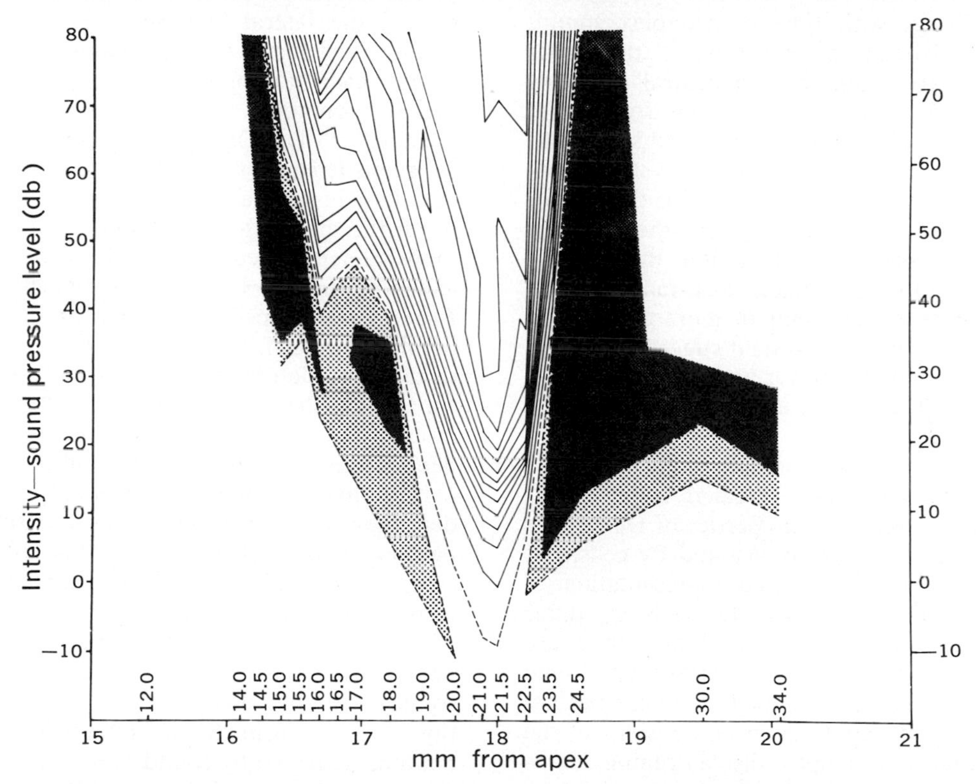

Fig. 13-10. Contour representation of excitatory and inhibitory response areas of neuron of cochlear nuclear complex of cat. Each solid contour line connects all frequency-intensity combinations that evoked same number of nerve impulses. Dashed line connects points at which number of impulses evoked exceeded mean of spontaneous rate of discharge by 1 SD; first solid line above dashed line is contour for total count of 200, and subsequent lines above represent counts that increase by counts of 100 impulses. Crosshatched areas indicate regions in which stimuli inhibited spontaneous activity by at least 3 SD. By making a reciprocal interpretation, this map for single neuron studied at series of different frequencies can be thought to represent distribution of activity in cochlear neural field produced by intense tone at single frequency: a center of very intense activity surrounded by inhibition. Abscissae: upper, frequency of testing tones in kilocycles; lower, estimated positions along basilar membrane for best response to given frequency. Ordinate: intensity of testing tones in decibels relative to 0.0002 dyne/cm^2. (From Greenwood and Maruyama.[48])

rons, one can reconstruct events produced in the cochlear nuclei when a pure tone is sounded in the ipsilateral ear. If the tone is weak, small islands of increased neural activity will appear in each of the three nuclei, and the position of these active islands in the neural field will be determined by the frequency of the tone. As the intensity of the sound increases, the islands will tend to spread, but the degree of spread will be restricted by afferent inhibition of surrounding neurons. Thus a sound shapes the neural field in terms of increased or decreased activity, and it is the preservation of these profiles of neural activity throughout the central projection of the system that is thought to form the basis for the discrimination of pitch. How complex the distribution of activity within the central neural fields must be when complex sounds are heard and how rapidly they must change with time, as complex sounds do, are obvious and indicative of the power and subtlety required of a central detecting mechanism. Although, for low frequencies of sound, cochlear nerve fibers and some cells of the cochlear nuclei may follow the sound waves beat for beat (Chapter 12), there is little other evidence to support the "volley" theory of pitch discrimination that is frequently elaborated upon this fact. Indeed, study of neural elements of more central stations of the auditory system suggests that this frequency following over a part of the audible sound spectrum is not important for pitch discrimination.

So far electrophysiologic studies of the cochelar nuclei have revealed some differences in the response properties of Dc cells as compared with those of Av and Pv cells, but the meaning of the triple representation of the cochlea is unknown. Dc cells do differ from the others in the fact that they are inhibited when certain sounds that excite them at weak and intermediate intensities are made very strong. Cells in all divisions of the central nuclear complex display tuning curves that are virtually identical in shape and sharpness.[60] The mean frequencies of the spontaneous activity of Dc cells are lower than are those of Av and Pv cells, and the histograms of their interspike interval distributions are more symmetric.[62, 84] The fact that Dc cells receive a much heavier efferent or descending innervation than do Av and Pv cells suggests that reflexly evoked descending inhibition may produce these properties, but there is no direct evidence that this is so and the meaning of these differences for hearing is unknown.

Auditory regions of the brainstem and thalamus

From the foregoing it is clear that there is a highly ordered functional topography in the auditory sytsem at its first- and second-order levels. Moreover, this precise representation of the cochlear partition—and hence of sound frequency—is preserved in one or more components of each central nucleus of the auditory system that plays a role in audition. Such frequency specificity may not be preserved in those pathways that lead, via brainstem and spinal mechanisms, to reflex action or, via the ascending reticular formation, to the generalized systems of the forebrain.

The superior olivary complex and the nuclei of the lateral lemniscus are the major auditory regions of the medulla and pons.[86, 87] Within each grouping there is one or more precise tonotopic representations for frequency.[13, 38] The Dc cells project via the dorsal acoustic striae upon the n. lateral lemniscus, largely contralaterally; few of its axons reach the midbrain or thalamus directly. The more numerous ventral acoustic striae contain axons of the Av and Pv cells, which project (1) upon the ipsilateral superior olive and preolivary nuclei, (2) via the trapezoid body to the contralateral nucleus of that body and (sparsely) the contralateral n. lateral lemniscus, and (3) upon the laterally directed dendrites of cells of the ipsilateral medial superior olive and upon the medially directed dendrites of the neurons of its contralateral partner. Thus each neuron of this latter nucleus receives input from each ear and projects its axon upward via the lateral lemniscus; what role cells of this type may play in a neural mechanism for the lateralization of sound will be discussed in a later section.[19, 39, 113] From this medullary level some neurons activated by sound (those of the preolivary nuclei and the immediate surround) make local connections with cells of the reticular formation, with the cranial motor nerve nuclei, and by descending connections relayed in the ventral columns, with the spinal segmental apparatus, connections thought to serve reflexes evoked by sound and that orient the head and ears toward a sound source.

The large majority of the neurons of the superior olivary-trapezoid group, however, project upward via the lateral lemniscus to terminate upon the cells of its nucleus; more do so contralaterally than ipsilaterally. Few of these axons reach the inferior colliculus directly, and probably none reach the medial geniculate body.[87] The cells of the n. lateral lemniscus project in turn upon those of the inferior colliculus, although some make direct and many make collateral connections with neurons of the adjacent reticular formation.

The inferior colliculus must thus be considered a major relay station of the afferent (and efferent) auditory pathway. The lamination of the central nucleus of the inferior colliculus and the patterning of the ascending fibers as they enter it suggest a precise topographic representation.[73] This is fully borne out by the results of electrophysiologic studies, which show a precise replication of the cochlear partition in both the external and central nuclei of the inferior colliculus.[93] The tuning curves of cells of the central nucleus are as narrow as those of cochlear nerve fibers (Fig. 13-8), a restriction that depends on collateral afferent inhibition. Many collicular neurons are extraordinarily sensitive to the timing of afferent volleys reaching them from the two ears and thus, like the cells of the medial superior olive, are appropriately disposed to play a role in the lateralization of sound.[94]

The medial geniculate body is an extrinsic nucleus of the dorsal thalamus. It receives auditory fibers from the inferior colliculus and descending ones from the auditory cortex, upon which it projects.[57, 58] It has three subdivisions.[72, 74] The ventral division is activated by sound, contains a precise tonotopic representation,[1] and its neurons have very narrow tuning curves. The dorsal division is also activated by sound, apparently via a fine-fibered lateral tegmental system of unknown origin.[71] The dorsal and ventral divisions make up the *pars principalis* of some nomenclatures. The medial or magnocellular division is a region of nonuniform cytoarchitecture that receives converging input from both auditory and somatic sensory systems. The central core, i.e., the ventral division, emits an essential projection destined for the auditory cortex, while the surrounding areas project upon a number of cortical fields in a sustaining fashion.[90]

Auditory areas of the cerebral cortex

The peripheral auditory apparatus effects a transform in such a way that the frequency of a given sound determines the position and distribution of the active elements within the total band of eighth nerve fibers. This selection of locus by frequency is preserved through a central core of the subcortical components of the auditory system. If the perception of pitch depends on the identification of the locus of activity within a neural field, then precise representation of the peripheral sensory sheet must be preserved through some early stage of cortical processing. There is evidence that the region concerned is the first auditory area of the cerebral cortex. This "primary" sensory cortex is surrounded by a number of fields, within which the preservation of place may be less precise and for some of which the auditory is but one of several inputs.

The auditory area has been characterized by its cytoarchitecture, in terms of thalamocortical relations, and by the evoked potential method. There is now general agreement that the central core of the system, within which there obtains a precise tonotopic representation, is the first auditory area. In man the auditory cortex (not further subdivided) occupies the anterior transverse temporal gyrus of Heschl, Brodmann's areas 41 and 42, on the superior bank of the first temporal convolution. Stimulation of this region in waking humans elicits a variety of auditory illusions.[83] The auditory cortex has been identified in a similar location in the brain of the monkey by means of the evoked potential technique. (Fig. 13-11).[110] The most extensive studies of the auditory cortex have been made in the brains of carnivores, mainly cats, for in these forms the areas of cortex activated by sound are exposed on the lateral surface of the hemisphere. These areas for the cat cortex are displayed by the map in Fig. 13-12, which was made by observing the electrical responses of the cortex evoked by stimulating one or another small group of eighth nerve fibers along the bony spiral lamina of the opened cochlea. This drawing summarizes the extensive study of this projection by Woolsey.[9] Fig. 13-12 illustrates that the central region, auditory I, is surrounded by three other regions, within which there is at least a differential representation of the apical and basilar turns of the cochlea. Reference will be made to these and to the regions indicated as auditory association areas in a later section.

It should be emphasized that the projection to the cortex is bilateral: the two cochleae are represented in registered overlay. The thresholds for cortical excitation by tones de-

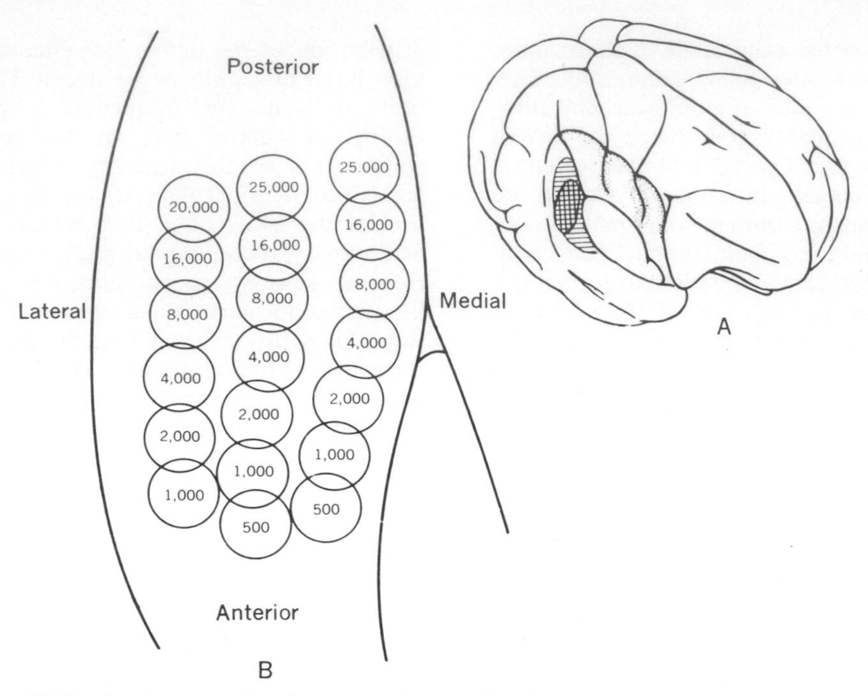

Fig. 13-11. A, Cutaway drawing of monkey brain exposes auditory cortical areas on supratemporal plane. Horizontally shaded area indicates region in which click stimuli evoke slow-wave potentials in deeply anesthetized animals. Crosshatched area is region within which lesions produce retrograde degeneration in medial geniculate. Exact homologies with auditory areas of cat cortex are uncertain, but crosshatched area must contain auditory I. **B,** Enlarged drawing corresponding to crosshatched area in **A** and showing patterns of representation of frequencies within this region determined by evoked potential method. (**A** from Ades[11]; **B** based on data from Kennedy and reported by Neff.[75])

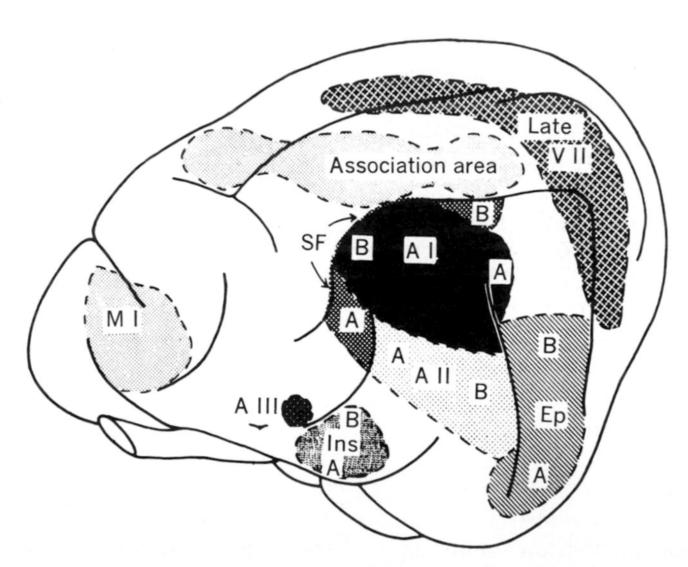

Fig. 13-12. Summary diagram showing areas of cat cerebral cortex that can be activated by auditory input under variety of experimental conditions. In each area, *A* = apex and *B* = base of cochlea, i.e., low and high frequencies, respectively. Central core is auditory I, *A I,* partially surrounded by auditory II, *A II,* ectosylvian posterior area, *Ep,* and suprasylvian sulcus, *SF.* Both insular, *Ins,* and infratemporal fields lateral to A II, not indicated in this drawing, receive patterned input from cochlea. Small area marked *A III* also activated by sound, but whether in detailed pattern is unknown. Association and motor cortices, *M I,* respond to sound under special conditions and in all-areas or no-areas fashion, as described in text: they are thought to be activated via generalized thalamocortical system. *Late* indicates visual area II, *V II,* of cortex, within which responses to brief auditory stimuli occur with 100 msec delay under chloralose anesthesia. (From Woolsey.[114])

livered to the ipsilateral ear are several decibels higher than for the contralateral ear.

Studies of the auditory cortex in cats and monkeys[18, 53, 68] have confirmed the tonotopic pattern of representation previously revealed by evoked potential experiments. The tuning curves for many cortical neurons are as narrow as those of subcortical auditory neurons, while those of others spread rapidly as stimulus intensity increases. This diversity, together with variations in the temporal pattern of response, is to be expected from a group of neurons that constitute a complicated input-output chain, not a uniform population. The auditory cortex, like the visual and somesthetic cortices, is organized on the basis of columns of cells oriented normally to the cortical surface, cells whose best frequencies range no more than an octave from lowest to highest. The best frequencies of binaural cells are nearly identical from the two ears. A powerful factor in determining the maximum excitation of a cortical cell is the time-intensity relation between stimuli reaching the two ears; this is consonant with the essential role of auditory cortex in sound localization (p. 433).

Auditory fields surrounding auditory I. The map in Fig. 13-12 indicates that auditory I (A I) is nearly surrounded by a cortical belt composed of three fields, within each of which there is a spatial representation of the cochlea. Whether these tonotopic representations are as detailed as is that of A I is unknown. They may, however, sustain behavior that depends on complex discriminations between auditory inputs in the absence of the central core area, as will become evident in a later section. The removal of A I results in retrograde degeneration of the ventral division of the pars principalis of the medial geniculate. Removal of any one or all of the belt areas produces only insignificant degeneration in the medial geniculate, while extirpation of all these fields, together with A I, the inferior temporal fields, and the second somatic area (Chapter 10), results in a profound degeneration of all subdivisions of the medial geniculate. Thus the first auditory area can be regarded as receiving an essential thalamocortical projection from the ventral division, while the other auditory cortical fields receive sustaining projections, within the definitions of these projection types given in Chapter 7.[90]

Cortical association fields receiving auditory input. The map in Fig. 13-12 shows that in addition to the fields just described there exist four regions that may be activated by sound, but that are located well outside the regions that may qualify as the cortical components of the auditory system.[17, 109, 110] They differ markedly from the auditory fields: (1) they are indiscriminately activated without tonotopic patterning; (2) they may be activated also by visual or somesthetic stimuli in an equally unspecific manner; (3) they may still be activated after chronic ablation of all auditory cortical fields; and (4) responses in each of the four are almost perfectly correlated with each other as regards presence and amplitude, but show no correlation with simultaneously recorded evoked potentials in A I. These facts led to the hypothesis that a peripheral stimulus delivered to any portion of the auditory, visual, or somesthetic receptive fields activates one and the same central "association" system in an undifferentiated way, and evidence has been put forward that this central system is some part of the generalized thalamocortical system. This idea is supported by the finding that complete transection of the brachium of the inferior colliculus permanently abolishes, in both anesthetized and unanesthetized animals, evoked responses in the auditory cortical fields, but has little or no effect upon responses in the auditory association fields.[78] Little is known about what sort of associations between diverse sensory inputs might be accomplished by a system conveying no information other than that *some* stimulus has been presented *somewhere*. Further behavioral studies of the function of these cortical areas are needed.

REMAINING CAPACITY FOR HEARING AFTER BRAIN LESIONS

The bilaterality of the auditory system at thalamic and cortical levels accounts for the fact that only symmetric lesions on the two sides will produce very obvious defects in hearing. Although such lesions occur rarely in man, a number have been described and it is clear from them that if the auditory cortical fields or the auditory radiations are destroyed bilaterally a near-total deafness ensues. This is further evidence for the progressive phyletic encephalization of the auditory system, for such a complete deafness does not follow auditory cortex removals in cats or monkeys. Patients with unilateral lesions show an elevated threshold for hearing with the ear contralateral to the lesion[35, 114] and a disorganization of the contralateral spatial field with a marked decrease in the capacity to localize sounds on that side.[95]

A major field of research in audition is the study of the hearing capacities of animals before and after destruction of a part of the auditory system. Techniques have been developed for measuring the auditory capacity of subhuman primates using operant conditioning and a psychophysical method in combination.[70, 98] While many variations are used, the experimental protocol is in general as follows:

Monkeys are trained to sit quietly in training chairs with their heads loosely restrained and earphones applied. They learn to initiate a trial by depressing one key. After a variable delay a pure tone is delivered; a response on a second key during the presentation of the tone is rewarded with food. The intensity of the tone is varied, and a psychometric function is constructed by plotting the percent of correct detections as a function of tone intensity. Threshold is taken as the point at which 50% of the tones are detected. The same method is used to determine differential sensitivity to frequency. In this case the standard pure tone is always present as a pulsed signal; initiation of a trial produces alternations between the standard and a comparison tone of a different frequency. A response on the second key during this alternation is rewarded. The DL for intensity can be determined in a similar way. For tests of the capacity to localize sounds, an animal may be required to select a particular position of a sound source to be rewarded with food for the correct choice.

The results of studies of this sort made by Stebbins[5] in subhuman primates are shown in Fig. 13-13. They show that the macaque monkey's range of hearing extends about one octave above that of man, to nearly 45 kHz. At frequencies below 8 kHz the monkey's absolute sensitivity is comparable to man's, although the monkey is slightly less sensitive than is man to changes in either frequency or intensity. Measurements of this sort are made both before and after brain lesion, in the hope that correlation between changes in sensory performance and the location and extent of the brain lesion will indicate the role played in audition by different components of the auditory system. Postoperatively the animals are tested for (1) retention of the learned act and (2) the capacity for relearning, if lost. The method is a geographic one and is unlikely to disclose underlying mechanisms. Some findings obtained with its use are summarized in the following paragraphs.

Stimulus detection and intensity discrimination

After removal of all auditory cortical areas, cats[80] and monkeys[32] retain or can relearn a conditional response to the appearance of a sound at threshold or to differences in the intensity levels of two sounds at thresholds that are little changed from normal. When a lesion of the inferior colliculus is added to one of the cortex, or following section of the brachium of the inferior colliculus, the DLs may be increased by 7 to 10 db. Thus the tonotopically organized thalamocortical component of the auditory system ensures in the

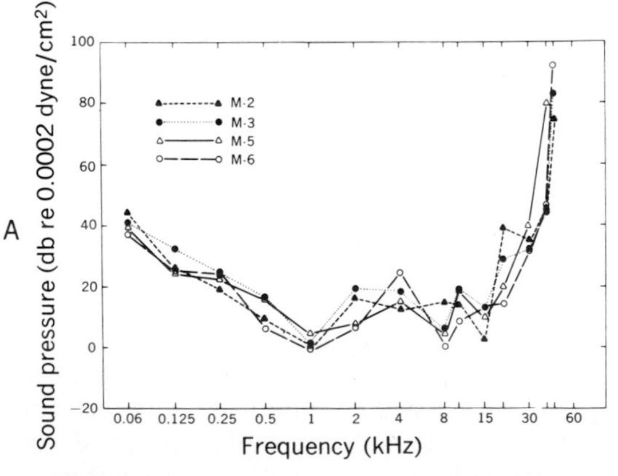

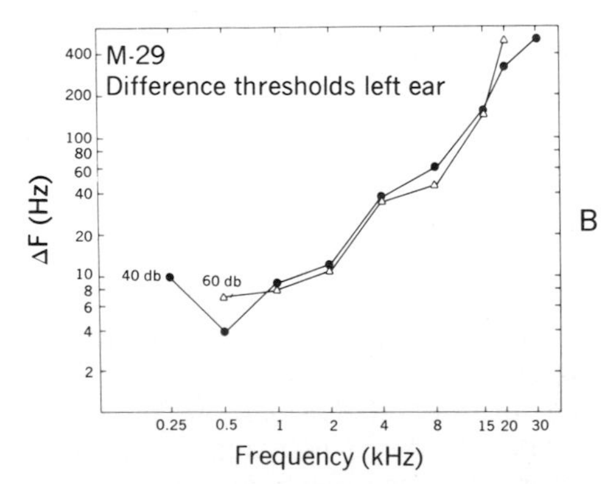

Fig. 13-13. A, Normal auditory thresholds for four macaques, animals able to hear sound frequencies as high as 45 kHz, at least one octave above upper limit for man. Sensitivity functions are otherwise quite similar to those for man, except for slight elevation at about 4 kHz. Compare with human sensitivity functions shown in Fig. 13-1. **B,** Frequency difference thresholds ($\triangle$F) for one macaque, shown as function of base frequency, at intensity levels 40 and 60 db above absolute threshold. Macaque's capacity to discriminate between sounds of different frequencies is only slightly less than that of man himself. (**A** from Stebbins et al.,[98a] copyright by American Association for the Advancement of Science; **B** from Stebbins.[98])

normal animal a low DL for intensity (we cannot speak of loudness for an experimental animal). However, these results indicate that activation of the forebrain by sounds, via other avenues such as the central reticular core and the generalized thalamocortical system (Chapter 7), suffices to call attention to novel stimuli or to intensity differences between stimuli.

Simple frequency discrimination

Removal of all tonotopically organized auditory areas in the cat eliminates a learned avoidance reaction to tones of different frequency. A visual discrimination learned simultaneously is unaffected. The capacity for relearning depends on how the task is set. If a single short tone of one frequency repeated regularly, e.g., at 3/sec, is the negative signal and an interpolated regular alternation between it and a second tone of another frequency is the positive signal (i.e., AAAABABAAA), avoidance conditioning is reestablished after about the same training as is required in normal animals. If, however, several seconds elapse between tones, the positive signal appearing de novo out of silence and the animal required not only to identify that it is different but asked to indicate whether it is of higher or lower frequency than the negative stimulus, the capacity for frequency discrimination is lost and cannot be relearned.[42, 69, 108] Bilateral section of the brachium of the inferior colliculus in cats is followed by complete failure to discriminate frequency.[41]

Following removal of the auditory cortical areas, monkeys immediately lose a learned response to tones of different frequency. The deficit is severe and permanent, thresholds for even the simplest frequency discrimination remaining 15 to 20 times above preoperative levels.[65] It seems likely that in man an even more severe deficit exists, for patients with bilateral lesions of cortical auditory areas are described as deaf. However, such patients occur rarely, and they have not yet been studied in a quantitative manner.

Discrimination of tonal pattern

Cats can be trained to discriminate between sequences of tones that appear in different order, e.g., ABA versus BAB. This capacity is immediately lost after bilateral removal of all auditory cortical areas and cannot be relearned.[31] Such a task requires identification of the temporal order of events and their immediate recall. Studies of animals with more restricted lesions suggest that the central core region, A I, is essential for this performance but is not itself sufficient, for severe deficiencies in pattern discrimination follow cortical removals in the cat that are confined to the insular and temporal fields lateral to A I and A II.

A similar loss in the capacity to recognize auditory patterns occurs in monkeys after removal of homologous auditory association areas of the superior temporal gyrus,[99] and the loss is particularly severe if the task requires discrimination between complex sequences or qualities, e.g., between different speech sounds.[29, 30] These results suggest a functional differentiation among auditory cortical fields. The central core region suffices for simple identification of frequencies and discrimination between them. Those auditory tasks of greater complexity that require the processing of information about temporal pattern and complex qualities depend, in addition, on the auditory association fields of the superior temporal gyrus.

Localization and lateralization of sound in space[76]

It has long been known that localizing a sound in space is a different task when using two ears than when using one ear alone. The human or animal subject makes scanning head movements to elicit differences in quality and intensity of the sound when hearing with one ear. Even after the task is learned, fixation of the head causes a severe deterioration of performance. The subject with normal hearing in two ears uses differences in time of onset, phase, and intensity of sound at the two ears to derive information about the spatial location of a sound source. This suggests that the CNS locales essential for sound localization are those that receive convergent input from the two ears, and studies of animals with brain lesions confirm that this is the case.

Transection of the trapezoid body, the lowest commissural pathway of the auditory system, produces a severe deficit in localizing capacity, as do lesions of the superior olivary complex or the lateral lemniscus. These findings are consonant with the implication of anatomic facts as well as electrophysiologic observations (p. 433) of a convergent excitatory-inhibitory projection from the two ears upon single neurons of the superior olivary complex.[67] Lesions of the inferior colliculus or of its brachium produce a similar deficit.[107] It has been known for a long time that bilateral removal of all auditory cortical areas results in as severe a deficit as do those of afferent projection pathways,[66, 77] and it is now clear that unilateral lesions in both men and animals produce defects in the location of sound in the contralateral half-space.[76, 106]

In summary, animals and men function at many levels of complexity as regards audition, and when the sensory system is reduced by partial ablation, the remainder may suffice to process information about sound stimuli of simple configuration and thus provide cues for differential action. That this processing of sensory input goes on in the absence of cortex does not mean that cortex does not participate in this function in the intact animal. Cortical auditory areas are essential, moreover, for any tasks in the sphere of audition that require for execution any of the following: (1) a short-term memory storage and immediate recall of information about auditory stimuli; (2) *identification* of what is new, a step beyond the simple *detection* that something new has occurred—indeed, it seems likely that the ascending core of the reticular formation and the generalized thalamocortical system upon which it projects may suffice for signaling that something novel has occurred, but not for identification; (3) when the *temporal order* of events in the auditory stimuli must be identified, stored, and recalled; and (4) when the location of the sound in space must be identified.

For carnivores, auditory stimuli are obviously potent guides for behavior and gain priority in attention, in contrast to the visual domination of the total sensory inflow so characteristic of the life of primates. It is nevertheless likely that the progressive encephalization and corticalization of the specific sensory systems in the latter means that in them the capacity for function of the subcortical components of the auditory system are much reduced over that which obtains in carnivores. Certainly this is suggested by reports that near-total deafness follows bilateral lesions of the cortical auditory areas in human beings, but more intensive and particularly more quantitative studies of hearing in human beings with brain lesions are required to eleborate this point.

DYNAMIC PROPERTIES OF AUDITORY NEURONS

The static properties of the auditory system are those determined by its synaptic connections alone, properties that persist unchanged with variations in the temporal and spatial pattern of activity within it. Of these, the "best frequency" of an auditory neuron seems the most obvious case, for it is invariant over a wide range of excitability of the cell, with changes in the temporal patterning of the stimulating tone, etc., even when its tuning curve is modified by afferent inhibition. Dynamic properties are, on the other hand, those very aspects of the activity of a cell that are sensitive to temporal and intensive variations in impinging activity. Study of these latter properties has intensified in recent years: what is sought is a further understanding of the dynamic and time-dependent operations of the nervous system, ones that are nevertheless bounded within certain limits by its gross geography and its synaptic microstructure. Such a study encompasses much more besides, and a principal aim is to discover in what way the auditory performance of an organism can be explained in terms of neural mechanism.

The method used is that of single-unit analysis, with which the activity of auditory neurons at any level can be observed for long periods of time in anesthetized or in waking, freely moving animals and under a host of stimulus conditions. It is important to emphasize again that the aim of this method is not to study a single neuron per se but to study large numbers in such a detailed way that a post hoc reconstruction of events in a large population of neurons can be synthesized. Given records of the impulse trains of auditory neurons, it is not at all obvious what scaling procedures should be used for their analysis, a vexing problem common to psychophysics and neurophysiology. The similarity of the intermittent, all-or-nothing delivery of impulses along axons to certain interval-modulated, pulse-operated informational systems has led to an intensive application of what may be termed the method of interval analysis. A major question is this: given two instances in which the average frequencies of impulse transmission in a neural element are identical, is it possible that the information transmitted in the two cases may differ by virtue of differences in the interspike interval sequences in the two trains of impulses? There is no certain answer, although preliminary evidence indicates, that such a principle applies, if at all, only in certain special circumstances. Nevertheless, the study of interspike interval sequences is an elegant method of data analysis, and its use has revealed certain aspects of the dynamic activity of auditory neurons hitherto unknown. Instances will be cited in the discussion that follows. While this line of investigation has so far yielded little information that allows useful correlation between the behavior of auditory neurons and the auditory performance of animals and man, it is worth examining to what extent this is possible and to pose some general problems for future study.

Spontaneous activity

Auditory neurons from the level of the eighth nerve to the cerebral cortex are spontaneously active in the absence of auditory stimuli of any sort. The origin of the activity in the peripheral end of the eighth nerve fibers is unknown; for central neurons it is undoubtedly the result of both the spontaneous end-organ input and the influence of other systems upon the auditory neurons. In particular, the central reticular–generalized thalamocortical system influences the degree of activity at all levels, from cochlear nuclei to cerebral cortex. How central and peripheral drives interact is unknown, but it is known that the cells of the cochlear nucleus are still spontaneously active after destruction of the ipsilateral cochlea.[62] A third major influence that must condition spontaneous activity is that of the descending components of the auditory system, about which more will be said in a later section.

Given spontaneous activity that will vary with the general level of brain excitability, an important question is how spontaneous activity interacts with that evoked by auditory stimuli: are the two simply additive or is interaction multiplicative? For neurons of the inferior colliculus, Bures and Buresova[21] observed both types of interaction: for some neurons there was simple addition of the two; for others there was a constant increment-background ratio. What form these interactions may take is important for hypotheses concerning how a cortical discriminative mechanism may operate, and it deserves further intensive study,

The interval sequences of the spontaneous activity of auditory neurons appear in different forms. At lower levels, interval histograms are commonly poissonian in form, with a dead-time to allow for neuronal refractoriness: the sequences approach randomness.[84] For other neurons the intervals cluster in quasi-gaussian distributions around the mean. At higher levels of the system the interval distributions observed are more variable, change with the state of alertness of the animal, and in sleep or depressed states such as anesthesia are frequently bimodal in form. The meaning for function of the ubiquitous spontaneous activity observed in this as in other central neuronal systems remains obscure.

Intensity functions of auditory neurons

The rate of discharge of auditory neurons commonly increases rapidly along an S-shaped curve when plotted against a log scale of stimulus intensity (Fig. 13-14), functions described only by complex equations.[43] Characteristically the peak rates reached are much higher for cells at lower levels of the system than for those of the cerebral cortex. For many auditory neurons further increases in stimulus strength produce not higher but *lower* rates of discharge: the intensity-frequency functions are nonmonotonic.[40, 43] This is thought to be due to the fact that stronger

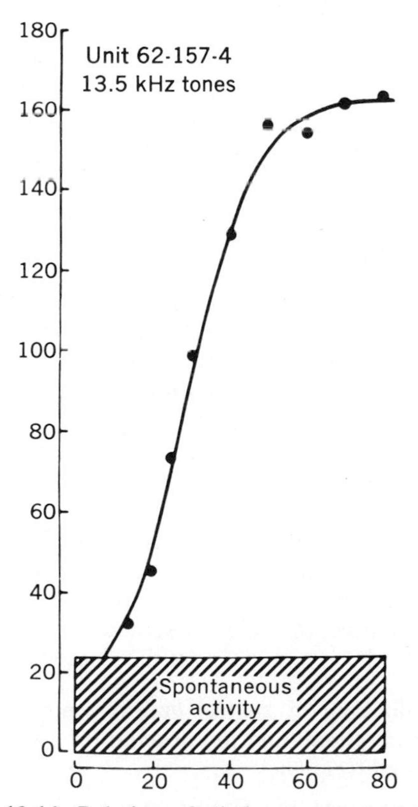

Fig. 13-14. Relation of discharge rate to intensity of tone at best frequency for neuron of superior olivary complex of cat. Ordinate plots mean rate in impulses/sec during seventh to tenth second after onset of tone at best frequency of 13,500 Hz delivered to contralateral ear. (From Goldberg et al.[43])

stimuli spread the active zone along the cochlear partition, thus recruiting to the afferent inflow fibers of adjacent best frequencies that are inhibitory for the neuron under observation.

In an earlier section, evidence was presented that on the basis of magnitude judgments the loudness function for man approximates a power function with an exponent of 0.6, with stimulus in pressure units. No correlation has yet been found in the intensity functions of auditory neurons. Increasingly intense auditory stimuli not only produce increases in frequency of discharge for neurons tuned to the stimulus frequency but cause a spread of activity to adjacent neurons. It is thought that a central discriminating mechanism must generate the overall input-output power function descriptive of human auditory behavior by some combination of intensive and extensive changes in the active neural field, but how this might be accomplished is a mystery.

Adaptive properties of auditory neurons

Neurons at lower levels of the auditory system respond to a steadily maintained tone of best frequency with very high-frequency onset transient discharges, after which the discharge rate declines to a much lower level but may then persist for very long periods of time.[43] A smaller number respond only to tonal onset with a burst discharge, remain silent although the stimulus continues, and may discharge a second burst when the stimulus is turned off. The sequence is thought to be due to a temporally mixed input complexed of both excitatory and inhibitory influences evoked by the stimulus. Preliminary observations indicate that at successively higher levels of the system a higher proportion of cells respond to transients and a lower proportion respond steadily to maintained stimuli. Like other sensory systems, the auditory seems organized to detect and accentuate preferentially *changes* in the environment, the steady nature of which is signaled by only a portion of the available neural elements.

Phase-locking of central auditory neurons to sound stimuli and the problem of coding for sound frequency

In the preceding discussion it has been emphasized that the most likely way in which the nervous system codes for sound frequency is by place, i.e., by the differential in the loci of activity within auditory neural fields set up by different sounds, a differential mapping shown to be carried out at every level from the cochlear nerve to the auditory cortex. Indeed, it is obvious that a volley theory for pitch discrimination is untenable if it demands a neural discharge of central auditory neurons for each cycle of a sound wave, for the overall rates of discharge of those neurons scarcely ever exceeds 100 to 200/sec, even under intense peripheral drive. Close study of the temporal sequences of auditory neuronal discharges, however, has revealed a second candidate, a period-time code. It has been shown for auditory nerve fibers[92] and for the central neurons upon which they project,[63, 94] with best frequencies below 2,000 to 3,000 Hz, that neuronal discharges occur in a preferred relation to the phase of the stimulating sound wave. The periods of time between successive impulses in single elements are always integral—multiples of the wavelength of the stimulating sound wave, even though the overall neuronal discharge rate is some small fraction of sound frequency. Given that adjacent elements responding to a tone respond to different successive cycles of the sound wave, it would be possible for a central detecting network to decode sound frequency by intregrating the information carried in a number of converging elements. How such a detecting network might achieve that integration is unknown, for the time constants of the membranes of central neurons are long compared to the sound wavelengths. For sounds of frequencies above 2,000 to 3,000 Hz this phase-locking does not occur, and for all higher audible frequencies, place remains the best possible candidate code for sound frequency.

Neural mechanisms for sound localization

Cats, after unilateral ablation of auditory cortical fields, and human beings with unilateral disease of the temporal lobes show disturbances of sound localization in the contralateral half-field that may increase DLs from about 3 degrees to as much as 15 degrees. Human beings with total loss of hearing in one ear show a most profound loss of the capacity to localize sounds when the head is fixed in space. These observations suggest that both binaural input and central interactions of that input are required for localization. Evidence has already been presented that of the two possible binaural cues of localizing importance the one due to unequal phase might operate only for low-frequency sounds, while the one for intensity will be important only for those sounds (above 3,000 Hz) for which the head acts as an effective sound shadow. What neural mechanisms are sensitive to these differences that might therefore indicate the direction of a sound source?

The first locale of the auditory system at which binaural interaction occurs is in the medial superior olive. Each of its cells receives a contralateral input upon its medially directed dendrite and an ipsilateral input upon its lateral dendrite. For most of these cells the former is excitatory and the latter is inhibitory. The discharge rate of neurons with low best frequencies (3,000 Hz) is strongly affected by variations in *interaural phase delay,* being maximal at a "characteristic delay" that differs for different neurons.[35, 39, 113] The factors of phase delay and stimulus intensity can be traded and an imbalance produced by one partially compensated by increase in the other.[50] High-frequency neurons excited by input from one ear and inhibited by that from the other are sensitive to *interaural intensity differences* and relatively insensitive to overall intensity level.[39] Thus the interaural phase delay for low frequencies and intensity differences for high frequencies will produce a net increase in the level of activity in the contralateral nucleus and a net decrease in the ipsilateral one. The result is a weighting of ascending activity toward the contralateral side of the cerebral cortex for contralateral stimuli (providing a signal for lateralization) and a spatial selection of that imbalance (providing a cue for localization). Thus the superior olive is a major integrating center for sound localization (Fig. 13-15).[16]

These differential cues in the output of the superior olive are reflected in the response properties of neurons upon which they project, in the nucleus of the lateral lemniscus,[41] the inferior colliculi,[15, 36, 94] and auditory cortex.[14, 18, 20, 68] For example, as the phase relation between a pure tone delivered to the two ears at equal intensity is changed, neurons of the inferior colliculus (with best frequencies below 3,000 Hz) inscribe periodic alterations in discharge rate, there being a certain "characteristic delay" (of the ipsilateral stimulus behind the contralateral) at which the net excitatory effect of the two stimlui upon the cell is maximal (Fig. 13-16).[94] This characteristic delay is invariant over the frequency range to which a given neuron is sensitive. If it is assumed that maximal discharge rate is an important signal, then each cell can be regarded as subtending a hyperboloid in the contralateral sound field, as illustrated in Fig. 13-16. For neurons with very short characteristic delays the arms will diverge and at simultaneity will transcribe a line in the sagittal plane. With longer delays, the hyperboloid becomes more acute and in the limit the arms fuse to form a line perpendicular to the sagittal plane, along which

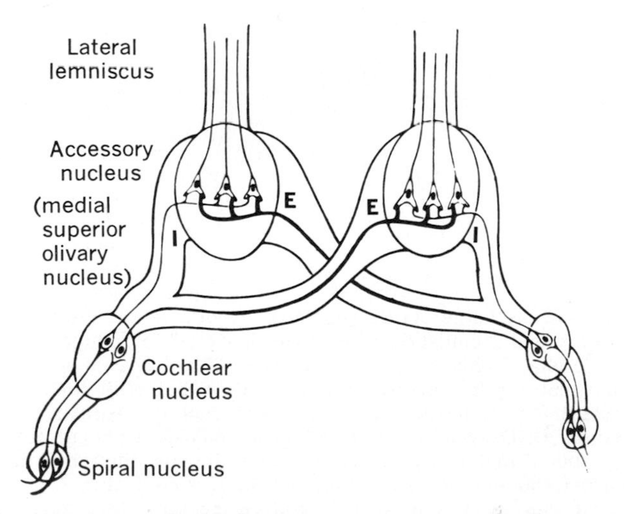

Fig. 13-15. Outline of features of Békésy–van Bergeijk model of role neurons of medial superior olive may play in localization of sound in space. Neurons of cochlear nuclei predominantly inhibit cells of ipsilateral accessory nucleus and excite those of contralateral one. Thus if neural input from one ear is more intense than that from the other or leads it in time, ascending neural activity will be weighted toward contralateral side of auditory system. (From van Bergeijk.[16])

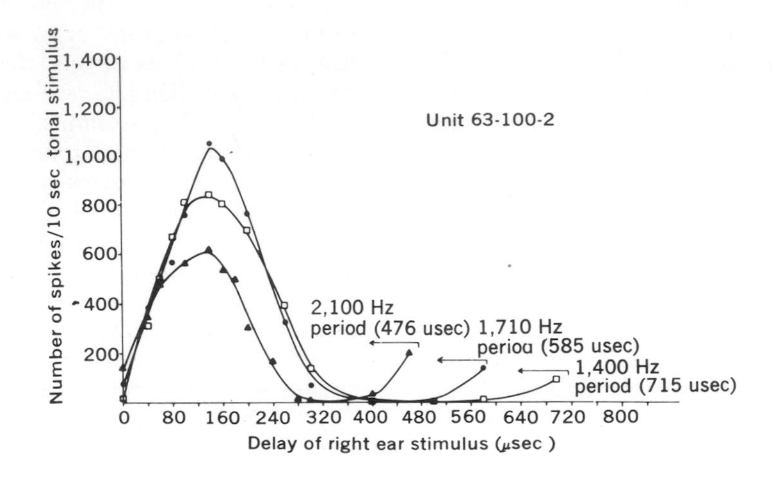

A

B

Fig. 13-16. A, Periodic discharge curves generated by neuron of inferior colliculus of cat at three frequencies of sound stimulation when tone delivered to right ear was successively delayed in time (i.e., in phase) relative to that in left ear. Characteristic delay at which excitatory effects sum to maximal for this neuron is about 140 msec. With further delay, response is completely suppressed. Characteristic delay is independent of frequency of stimulating tone for any given neuron. **B,** Drawing of hyperboloid to indicate schematically possible locations in space of sound source that would result in interaural time differences matching characteristic delay for a particular neuron of contralateral inferior colliculus. *L* and *R* represent locations of ears. Sound sources at *A* or *B* will produce the same time delay at *L* relative to *R* ($a - b = d - e$), which describes a hyperbola locating all sound sources in one plane which produce the same time delay. Rotation of hyperbola around *L-R* axis locates all possible sound sources giving this same interaural time difference, neglecting sound shadows, etc. For neurons with different characteristic delays a nest of such hyperboloids would indicate all possible sound sources. (From Rose et al.[94])

sound localization should be most acute. For other neurons, confusion between the locus in the anterior and posterior quadrants might arise for sounds placed at equal angles of incidence to the sagittal plane, and indeed this is just what is found for human beings. Similar observations have been described for cells of the auditory cortex with best frequencies below 3,000 Hz.[20] The hypothesis under consideration is that such auditory neurons subtend different *receptive fields* within the contralateral space, and that the spatial field for sound is mapped onto a spatially distributed population of neurons in much the same way that spatial position is mapped in other sensory systems. A group of neurons with differing characteristic delays can be regarded as subtending a nested series of hyperboloids such as that in Fig. 13-16. It is an interesting example of a time-space transform in the CNS.

Auditory neurons with best frequencies above 3,000 Hz that are sensitive to binaural intensity differences also provide cues for spatial position. This mechanism is exquisitely developed in bats, in whom the inferior collicular neurons may display differential sensitivities as high as several decibels per degree shift in space: thus some maximal rate of discharge will signal location at 90 degrees to the sagittal plane, some intermediate rate a sound in the saggittal plane, and some minimal rate a sound located 90 degrees in the opposite direction.[49] This appears to be a system highly developed for the echo-location of prey while on the wing.

DESCENDING COMPONENTS OF THE AUDITORY SYSTEM AND THE PROBLEM OF ATTENTION TO AUDITORY STIMULI[10]

In everyday life states of selective attentiveness are easily observed aspects of human behavior: we may attend to a particular sensory channel to the relative exclusion of others.* This selectivity in the adaptive behavior of organisms appears to operate at all psychologic levels, whether observed introspectively by man or in behavioral studies of men and animals, and particularly for sensation. If, for example, a subject wearing earphones is required to repeat words delivered to one ear, he may ignore a different set of words delivered to the other until items of special significance for him (e.g., his name) are inserted in the unattended set.[22, 23] This suggests at once that exclusion from recognition of unattended stimuli is a phenomenon occurring at a high functional level of the CNS, and that neural signals concerning the unattended input are immediately available when *attention* is directed toward them.[85] Moreover, experiment has shown that information concerning unattended stimuli is stored in the memory for short periods of time and can be recalled, with decreasing accuracy over the first 10 sec or so after delivery over the unattended channel. Indeed, if attending to one channel meant excluding input from the environment over all other channels, directed attention would be even more dangerous for survival than it is. Even when attention is narrowly focused, it seems necessary to conclude that information concerning other peripheral stimuli is admitted to the CNS and evaluated in terms of their significance for survival.

The presentation of a novel stimulus evokes in animals a reaction of orientation, interest, and investigation. If the stimulus is slowly repeated and if it possesses for the animal no other interest value than that it has occurred, attention quickly wanes and the orienting reaction disappears after about 10 to 15 stimuli have been presented: this phenomenon is termed habituation. The continued slow repetition of an uninteresting stimulus has a powerful soporific effect. What neural mechanisms are involved in attention and habituation? Interest in this question directs attention toward the descending components of sensory systems.

The efferent or descending components of the auditory system are schematized in Fig. 13-6. They take origin from the auditory cortex, particularly from the infratemporal fields, and course with relays through the more medial components of the medial geniculate complex and the nucleus of the lateral lemniscus to terminate in the cochlear nuclei. It is not known whether the descending system impinges also upon the olivary origin of the olivocochlear bundle, which is known to exert a suppressing or inhibiting effect upon the cochlear hair cells (p. 404). Desmedt[27] has shown that electrical stimulation of the descending system causes a reduction in the

*"The mere outward sense, being passive in responding to the impression of the objects that come in its way and strike upon it, perhaps cannot help entertaining and taking notice to everything that addresses it, be it what it will, useful or unuseful: but, in the exercise of his mental perception, every man, if he chooses, has a natural power to turn himself upon all occasions, and to change and shift with the greatest ease to that which he shall judge desirable." (Attributed to the emperor Marcus Aurelius.)

slow-wave response evoked in the cochlear nucleus by sounds, without necessarily diminishing eighth nerve input. Thus the descending system can exert an influence upon afferent transmission at the level of first synaptic relay and independently of olivocochlear action. The latter, however, does exert a powerful control over the cochlear (and hence the CNS) response to sound. In Desmedt's experiments he was able to show that repetitive stimulation of the olivocochlear bundle reduces eighth nerve response to a click by an amount equivalent to an 18 to 25 db reduction in sound intensity. An interesting observation made also by Desmedt[28] is that reduction in the eighth nerve volley produced a linearly proportional decrease in the evoked response at each central level of the auditory system. The implication is that, whatever transforms occur in the stimulus-response relation of the auditory system, they are largely cochlear in location, and that thereafter transmission is linear, at least as measured by the slow-wave evoked potentials. The neurons of the olivocochlear bundle may be activated by sound, possess tuning curves not noticeably different from those of auditory neurons elsewhere, and when active discharge impulses at very regular rates. Little is known about what central neural mechanisms control this system.

The middle ear muscles provide a third mechanism for controlling auditory input, as detailed in the previous chapter. Their reflex contractions decrease sensitivity of the ear to sounds of the lower audible frequencies, and thus they may function as a protective device. The direct activation of the muscles during vocalization, however, suggests a more important function: to minimize sensitivity to sounds that we ourselves produce.

What function can then be attributed to the descending component of the auditory system? No evidence presently exists, but an hypothesis worthy of intensive study is that, far from "blocking" or "valving" sensory input in a nonselective manner, this system plays an important role in shaping sensory input and thus in discrimination.

REFERENCES

General reviews

1. Erulkar, S. D., Nelson, P. G., and Bryan, J. S.: Experimental and theoretical approaches to neural processing in the central auditory pathway. In Neff, W. D., editor: Contributions to sensory physiology, New York, 1968, Academic Press, Inc., vol. 3.
2. Green, D. M., and Henning, G. B.: Audition, Ann. Rev. Psychol. **20:**105, 1969.
3. Licklider, J. C. R.: Basic correlates of the auditory stimulus. In Stevens, S. S., editor: Handbook of experimental psychology, New York, 1951, John Wiley & Sons, Inc.
4. Sachs, M. B., editor: The physiology of the auditory system, proceedings of workshop, Baltimore, 1971, National Educational Consultants.
5. Stebbins, W. C.: Hearing. In Schrier, A., et al., editors: Behavior of nonhuman primates, New York, 1970, Academic Press, Inc., vol. 3.
6. Stevens, S. S.: The psychophysics of sensory function, Am. Sci. **48:**226, 1960.
7. Stevens, S. S., and Davis, H.: Hearing, New York, 1938, John Wiley & Sons, Inc., chaps. 2 to 8.
8. Wever, E. G., and Lawrence, M.: Physiological acoustics, Princeton, 1954, Princeton University Press.
9. Woolsey, C. N.: Organization of cortical auditory system. In Rasmussen, G. L., and Windle, W. F., editors: Neural mechanisms of the auditory and vestibular systems, Springfield, Ill., 1960, Charles C Thomas, Publisher.
10. Worden, F. G.: Attention and auditory electrophysiology. In Stellar, E., and Sprague, J., editors: Progress in physiological psychology, New York, 1966, Academic Press, Inc., vol. 1.

Original papers

11. Ades, H.: Central auditory mechanisms. In Magoun, H. W., editor, Neurophysiology section: Handbook of physiology, Baltimore, 1959, The Williams & Wilkins Co.
12. Aitkin, L. M., and Webster, W. R.: Tonotopic organization in the medial geniculate body of the cat, Brain Res. **26:**402, 1971.
13. Aitkin, L. M., Anderson, D. J., and Brugge, J. F.: Tonotopic organization and discharge characteristics of single neurons in nuclei of the lateral lemniscus of the cat, J. Neurophysiol. **33:**421, 1970.
14. Barrett, T. W.: The response of auditory cortex neurons in cat to various parameters of acoustical stimulation, Brain Res. **28:**579, 1971.
15. Benevento, L. A., and Coleman, P. D.: Responses of single cells in cat inferior colliculus to binaural click stimuli: combinations of intensity levels. Time differences and intensity differences, Brain Res. **17:**387, 1970.
16. Bergeijk, van, W. A.: Variation on a theme of Békésy: a model of binaural interaction, J. Acoust. Soc. Am. **34:**1431, 1962.
17. Bignall, K. E.: Auditory input to frontal polysensory cortex of the squirrel monkey: possible pathways, Brain Res. **19:**77, 1970.
18. Brugge, J. F., and Merzenich, M. M.: Representation of the cochlea in auditory cortex in the macaque. (In press.)
19. Brugge, J. F., Anderson, D. J., and Aitkin, L. M.: Responses of neurons in the dorsal nucleus of the lateral lemniscus of cat to binaural tonal stimulation, J. Neurophysiol. **33:**441, 1970.
20. Brugge, J. F., Dubrovsky, N. A., Aitkin, L. M., and Anderson, D. J.: Sensitivity of single neurons in auditory cortex of cat to

binaural tonal stimulation: effects of varying interaural time and intensity, J. Neurophysiol. **32:**1005, 1969.
21. Bures, J., and Buresova, O.: Relationship between spontaneous and evoked unit activity in the inferior colliculus of rats, J. Neurophysiol. **28:**641, 1965.
22. Cherry, E. C.: Some experiments upon the recognition of speech, with one and with two ears, J. Acoust. Soc. Am. **25:**975, 1953.
23. Cherry, E. C.: Two ears—but one world. In Rosenblith, W. A., editor: Sensory communication, Cambridge, Mass., 1961, The M.I.T. Press.
24. Chow, K. L.: Numerical estimates of the auditory central nervous system of the rhesus monkey, J. Comp. Neurol. **95:**159, 1951.
25. Cohen, A.: Further investigation of the effects of intensity upon the pitch of pure tones, J. Acoust. Soc. Am. **33:**1363, 1961.
26. Deatherage, B. H.: Examination of binaural interaction, J. Acoust. Soc. Am. **39:**232, 1966.
27. Desmedt, J. E.: Neurophysiological mechanisms controlling acoustic input. In Rasmussen, G., editor: Neural mechanisms of the auditory and vestibular systems, Springfield, Ill., 1960, Charles C Thomas, Publisher.
28. Desmedt, J. E.: Auditory-evoked potentials from cochlea to cortex as influenced by activation of the efferent olivo-cochlear bundle, J. Acoust. Soc. Am. **34:**1478, 1962.
29. Dewson, J. H., III, Cowey, A., and Weiskrantz, L.: Disruptions of auditory sequence discrimination by unilateral and bilateral cortical ablations of superior temporal gyrus in the monkey, Exp. Neurol. **28:**529, 1970.
30. Dewson, J. H., III, Pribram, K. H., and Lynch, J. C.: Effects of ablations of temporal cortex upon speech sound discrimination in the monkey, Exp. Neurol. **24:**579, 1969.
31. Diamond, I. T., Goldberg, J. M., and Neff, W. D.: Tonal discrimination after ablation of auditory cortex, J. Neurophysiol. **25:**223, 1962.
32. Erulkar, S. D., Butler, R. A., and Gerstein, G. L.: Excitation and inhibition in cochlear nucleus. II. Frequency-modulated tones, J. Neurophysiol. **31:**537, 1968.
33. Faglioni, P., Spinnler, H., and Vignolo, L. A.: Contrasting behavior of right and left hemisphere-damaged patients on a discriminative and a semantic task of auditory recognition, Cortex **5:**366, 1969.
34. Galambos, R.: Neural mechanisms in audition, Laryngoscope **68:**388, 1958.
35. Galambos, R., Schwartzkopff, J., and Rupert, A.: Microelectrode study of superior olivary nuclei, Am. J. Physiol. **197:**527, 1959.
36. Geisler, C. D., Rhode, W. S., and Hazelton, D. W.: Responses of inferior colliculus neurons in the cat to binaural acoustic stimuli having wide-band spectra, J. Neurophysiol. **32:**960, 1969.
37. Gerstein, G. L., Butler, G. L., Butler, R. A., and Erulkar, S. D.: Excitation and inhibition in cochlear nucleus. I. Tone-burst stimulation, J. Neurophysiol. **31:**526, 1968.
38. Goldberg, J. M., and Brown, P. B.: Functional organization of the dog superior olivary complex: an anatomical and electrophysiological study, J. Neurophysiol. **31:**639, 1968.
39. Goldberg, J. M., and Brown, P. B.: Response of binaural neurons of dog superior olivary complex to dichotic tonal stimuli: some physiological mechanisms of sound localization, J. Neurophysiol. **32:**613, 1969.
40. Goldberg, J., and Greenwood, D. D.: Response of neurons of the dorsal and posteroventral cochlear nuclei of the cat to acoustic stimuli of long duration, J. Neurophysiol. **29:**72, 1966.
41. Goldberg, J. M., and Neff, W. D.: Frequency discrimination after bilateral section of the brachium of the inferior colliculus, J. Comp. Neurol. **116:**265, 1961.
42. Goldberg, J. M., and Neff, W. D.: Frequency discrimination after bilateral ablation of cortical auditory areas, J. Neurophysiol. **24:**119, 1961.
43. Goldberg, J. M., Adrian, H. O., and Smith, F. D.: Responses of neurons of the superior olivary complex of the cat to acoustic stimuli of long duration, J. Neurophysiol. **27:**706, 1964.
44. Green, D. M.: Psychoacoustics and detection theory, J. Acoust. Soc. Am. **32:**1189, 1960.
45. Green, D. M.: Detection of auditory sinusoids of uncertain frequency, J. Acoust. Soc. Am. **33:**897, 1961.
46. Greenwood, D. D.: Auditory masking and the critical band, J. Acoust. Soc. Am. **33:**484, 1961.
47. Greenwood, D. D.: Critical bandwidths and the frequency coordinates of the basilar membrane, J. Acoust. Soc. Am. **33:**1344, 1961.
48. Greenwood, D. D., and Maruyama, M.: Excitatory and inhibitory response areas of auditory neurons in the cochlear nucleus, J. Neurophysiol. **28:**863, 1965.
49. Grinnell, A. D.: The neurophysiology of audition in bats: directional localization and binaural interaction, J. Physiol. **167:**97, 1963.
50. Hall, J. L., II: Binaural interaction in the accessory superior olivary nucleus of the cat, J. Acoust. Soc. Am. **37:**814, 1965.
51. Hellman, R. P., and Zwislocki, J.: Some factors affecting the estimation of loudness, J. Acoust. Soc. Am. **33:**687, 1961.
52. Hellman, R. P., and Zwislocki, J.: Monaural loudness function at 1000 cps and interaural summation, J. Acoust. Soc. Am. **35:**856, 1963.
53. Hind, J. E.: Unit activity in the auditory cortex. In Rasmussen, G. L., and Windle, W. F., editors: Neural mechanisms of the auditory and vestibular systems, Springfield, Ill., 1960, Charles C Thomas, Publisher.
54. Irwin, R. J., and Corballis, M. C.: On the general form of Stevens' law for loudness and softness, Percept. Psychophys. **3:**137, 1968.
55. Jeffress, L. A., and McFadden, D.: Differences of interaural phase and level in detection and lateralization, J. Acoust. Soc. Am. **49:**1169, 1971.
56. Jones, E. G., and Powell, T. P. S.: Electron microscopy of synaptic glomeruli in the thalamic relay nuclei of the cat, Proc. R. Soc. Lond. (Biol.) **172:**153, 1969.
57. Jones, E. G., and Rockel, A. J.: The synaptic organization in the medial geniculate body of

afferent fibres ascending from the inferior colliculus, Z. Zellforsch. Mikrosk. Anat. **113:** 44, 1971.
58. Karp, E., Belmont, I., and Birch, H. G.: Unilateral hearing loss in hemiplegic patients, J. Nerv. Ment. Dis. **148:**83, 1969.
59. Katusi, Y.: Neural mechanisms of auditory sensations in cats. In Rosenblith, W. A., editor: Sensory communication, Cambridge, Mass,. 1961, The M.I.T. Press.
60. Kiang, N. Y-S., Pfeiffer, R. R., Warr, W. B., and Bakus, A. S. N.: Stimulus coding in the cochlear nucleus, Ann. Otol. Rhinol. Laryngol. **74:**463, 1965.
61. Kitzes, M., and Buchwald, J.: Progressive alterations in cochlear nucleus, inferior colliculus and medial geniculate responses during acoustic habituation, Exp. Neurol. **25:**85, 1969.
62. Koerber, K. D., Pfeiffer, R. R., Warr, W. B., and Kiang, N. Y-S.: Spontaneous spike discharges from single units in the cochlear nucleus after destruction of the cochlea, Exp. Neurol. **16:**119, 1966.
63. Lavine, R. A.: Phase-locking in response of single neurons in cochlear nuclear complex of the cat to low-frequency tonal stimuli, J. Neurophysiol. **34:**467, 1971.
64. Lochner, J. P. A., and Burger, J. F.: Pure-tone loudness relations, J. Acoust. Soc. Am. **34:**576, 1962.
65. Massopust, L. C., Jr., Wolin, L. R., and Frost, V.: Increases in auditory middle frequency discrimination thresholds after cortical ablations, Exp. Neurol. **28:**299, 1970.
66. Masterson, R. B., and Diamond, I. T.: Effects of auditory cortex ablation on discrimination of small binaural time differences, J. Neurophysiol. **27:**15, 1964.
67. Masterson, R. B., Jane, J. A., and Diamond, I. T.: Role of brainstem auditory structures in sound localization. I. Trapezoid body, superior olive, and lateral lemniscus, J. Neurophysiol. **30:**341, 1967.
68. Merzenich, M. M., and Brugge, J.: Representation of the cochlea in auditory cortex in the macaque. (In press.)
69. Meyer, D. R., and Woolsey, C. N.: Effects of localized cortical destruction upon auditory discriminative conditioning in the cat, J. Neurophysiol. **15:**149, 1952.
70. Mitchell, C., Gillette, R., Vernon, J., and Herman, P.: Pure-tone auditory behavioral thresholds in three species of lemurs, J. Acoust. Soc. Am. **48:**531, 1970.
71. Morest, D. K.: An ascending extra-brachial pathway to the medial geniculate body of the cat, Anat. Rec. **145:**262, 1963.
72. Morest, D. K.: The neuronal architecture of the medial geniculate body of the cat, J. Anat. **98:**611, 1964.
73. Morest, D. K.: The laminar structure of the inferior colliculus of the cat, Anat. Rec. **148:** 314, 1964.
74. Morest, D. K.: The laminar structure of the medial geniculate body of the cat, J. Anat. **99:**143, 1965.
75. Neff, W. D.: Neural mechanisms of auditory discrimination. In Rosenblith, W. A., editor: Sensory communication, Cambridge, Mass., 1961, The M.I.T. Press.
76. Neff, W. D.: Localization and lateralization of sound in space. In de Reuck, A. V. S., and Knight, J., editors: Hearing mechanisms in vertebrates, a Ciba Foundation symposium, Boston, 1968, Little, Brown & Co.
77. Neff, W. D., Fisher, J. F., Diamond, I. T., and Yela, M.: Role of auditory cortex in discrimination requiring localization of sound in space, J. Neurophysiol. **19:**500, 1956.
78. Nieder, P. C., and Strominger, N. L.: Evoked potentials in auditory cortex after bilateral transection of the brachium of the inferior colliculus in the cat, J. Neurophysiol. **28:** 1185, 1965.
79. Norman, D. A.: Sensory thresholds and response bias, J. Acoust. Soc. Am. **35:**1432, 1963.
80. Oesterreich, R. E., Strominger, N. L., and Neff, W. D.: Neural structures mediating differential sound intensity discrimination in the cat, Brain Res. **27:**251, 1971.
81. Osen, K. K.: Cytoarchitecture of the cochlear nuclei in the cat, J. Comp. Neurol. **136:**453, 1969.
82. Osen, K. K.: Course and termination of the primary afferents in the cochlear nuclei of the cat, Arch. Ital. Biol. **108:**21, 1970.
83. Penfield, W., and Rasmussen, T.: The cerebral cortex of man. A clinical study of localization of function, New York, 1950, The Macmillan Co.
84. Pfeiffer, R. R., and Kiang, N. Y-S.: Spike discharge patterns of spontaneous and continuously stimulated activity in the cochlear nucleus of anesthetized cats, Biophys. J. **5:** 302, 1965.
85. Picton, T. W., Hillyard, S. A., Galambos, R., and Shiff, M.: Human auditory attention: a central or peripheral process? Science **173:** 351, 1971.
86. Powell, T. P. S., and Erulkar, S. D.: Transneuronal cell degeneration in the auditory relay nuclei of the cat, J. Anat. **96:**249, 1962.
87. Rasmussen, G. L.: The olivary peduncle and other fibre projections of the superior olivary complex, J. Comp. Neurol. **84:**141, 1946.
88. Reynolds, G. S., and Stevens, S. S.: Binaural summation of loudness, J. Acoust, Soc. Am. **32:**1337, 1960.
89. Rose, J. E.: Organization of frequency sensitive neurons in the cochlear nuclear complex of the cat. In Rasmussen, G. L., and Windle, W. F., editors: Neural mechanisms of the auditory and vestibular systems, Springfield, Ill., 1960, Charles C Thomas, Publisher.
90. Rose, J. E., and Woolsey, C. N.: Cortical connections and functional organization of the thalamic auditory system of the cat. In Harlow, H. F., and Woolsey, C. N., editors: Biological and biochemical bases of behavior, Madison, 1958, University of Wisconsin Press.
91. Rose, J. E., Galambos, R., and Hughes, J. R.: Microelectrode studies of the cochlear nuclei of the cat, Johns Hopkins Hosp. Bull. **104:** 211, 1959.
92. Rose, J. E., Brugge, J. F., Anderson, D. J., and Hind, J. E.: Phase-locked response to

low-frequency tones in single auditory nerve fibers of the squirrel monkey, J. Neurophysiol. **30:**769, 1967.

93. Rose, J. E., Greenwood, D. D., Goldberg, J. M., and Hind, J. E.: Some discharge characteristics of single neurons in the inferior colliculus of the cat. I. Tonotopical organization, relation of spike-counts to tone intensity, and firing patterns of single elements, J. Neurophysiol. **26:**294, 1963.
94. Rose, J. E., Gross, N. B., Geisler, C. D., and Hind, J. E.: Some neural mechanisms in the inferior colliculus of the cat which may be relevant to localization of a sound source, J. Neurophysiol. **29:**288, 1966.
95. Sanchez-Longo, L. P., and Forster, F. M.: Clinical significance of impairment of sound localization, Neurology **8:**119, 1958.
96. Sivian, L. J., and White, S. D.: On minimum audible sound fields, J. Acoust. Soc. Am. **4:** 288, 1933.
97. Starr, A., and Britt, R.: Intracellular recordings from cat cochlear nucleus during tone stimulation, J. Neurophysiol. **33:**137, 1970.
98. Stebbins, W. C., editor: Animal psychophysics; the design and conduct of sensory experiments, New York, 1970, Appleton-Century-Crofts.

98a. Stebbins, W. C., Green, S., and Miller, F. L.: Auditory sensitivity of the monkey, Science **153:**1646, 1966.

99. Stepien, L. S., Cordeau, J. P., and Rasmussen, T.: The effect of temporal lobe and hippocampal lesions on auditory and visual recent memory in monkeys, Brain **83:**470, 1960.
100. Stevens, S. S.: The measurement of loudness, J. Acoust. Soc. Am. **27:**815, 1955.
101. Stevens, S. S.: On the validity of the loudness scale, J. Acoust. Soc. Am. **31:**995, 1959.
102. Stevens, S. S., and Galanter, E. H.: Ratio scales and category scales for a dozen perceptual continua, J. Exp. Psychol. **54:**377, 1957.
103. Stevens, S. S., and Guirao, M.: Loudness, reciprocality, and partition scales, J. Acoust. Soc. Am. **34:**1466, 1962.
104. Stevens, S. S., and Volkman, J.: The relation of pitch to frequency: a revised scale, Am. J. Psychol. **53:**329, 1940.
105. Stevens, S. S., Volkmann, J., and Newman, E. B.: A scale for the measurement of the psychological magnitude pitch, J. Acoust. Soc. Am. **8:**185, 1937.
106. Strominger, N. L.: Localization of sound in space after unilateral and bilateral ablation of auditory cortex, Exp. Neurol. **25:**521, 1969.
107. Strominger, N. L., and Oesterreich, R. E.: Localization of sound after section of the brachium of the inferior colliculus, J. Comp. Neurol. **138:**1, 1970.
108. Thompson, R. F.: Function of cerebral cortex in frequency discrimination, J. Neurophysiol. **23:**321, 1960.
109. Thompson, R. F., Johnson, R. H., and Hoopes, J. J.: Organization of auditory, somatic sensory, and visual projection to association fields of cerebral cortex in cat, J. Neurophysiol. **26:**243, 1963.
110. Thompson, R. F., Smith, H. E., and Bliss, D.: Auditory, somatic sensory, and visual response interactions and interrelations in association and primary cortical fields of the cat, J. Neurophysiol. **26:**365, 1963.
111. Walzl, E. M.: Representation of the cochlea in the cerebral cortex, Laryngoscope **57:**778, 1947.
112. Ward, W. D.: Auditory fatigue and masking. In Jerger, J., editor: Modern developments in audiology, New York, 1963, Academic Press, Inc.
113. Watanabe, T., Liao, T., and Katsuki, Y.: Neuronal response patterns in the superior olivary complex of the cat to sound stimulation, Jap. J. Physiol **18:**267, 1968.
114. Woolsey, C. N.: Organization of cortical auditory system. In Rosenblith, W. A., editor: Sensory communication, Cambridge, Mass., 1961, The M.I.T. Press.
115. Young, E. D., and Sachs, M. B.: Recovery of single auditory-nerve fibers from sound exposure, J. Acoust. Soc. Am. **50:**94(A), 1971.

14 The eye

GERALD WESTHEIMER

The human eyeball (Fig. 14-1) approximates in shape a sphere with a diameter of just less than an inch. The function of the organ is to produce an optical image of the world on light-sensitive cells. The method used for this in the vertebrate eye (but not, for example, in the insect eye) is the one we are familiar with in cameras: a transparent lens system producing an inverted image on a screen, the retina. The light-transducing properties and neural characteristics of the retina are discussed in detail in Chapter 15 and the central mechanisms of the visual process in Chapter 16. Here some of the physiologic aspects of the eye's protective mechanism and of the processes involved in retaining the eye's shape and transparency will be discussed. We then go on to consider the three major muscular systems of the human eye: the pupil, accommodation, and eye movement. The chapter will be concluded with a consideration of the eye's primary function of optical image formation, including some of its simpler anomalies and methods of their correction.

PROTECTIVE MECHANISMS

The eye, unlike the ear, has lids to close it off from the environment. The process of closing the lids, mediated by a relaxation of the levator palpabrae muscle, supplied by nerve III, coupled with a contraction of the orbicularis oculi muscle supplied by nerve VII, occurs in the following circumstances:

1. *As a protective reflex* whose afferent arc is the sensory system of the external eye, chiefly of the cornea, and nerve V. Protective lid closure also occurs with certain visual stimuli such as sudden bright lights (dazzle reflex) or rapidly approaching objects (menace reflex).
2. *During sleep,* when there is a tonic contraction of the orbicularis.
3. *During spontaneous blinking,* which takes place throughout waking life once every few seconds, the rate depending on many factors. A blink is bilateral, lasts about a quarter of a second, and the eyeballs move up during a blink, as indeed they do also during sleep. Blinking serves to keep the corneal surface clear of mucus and the lacrimal fluid spread evenly over it.
4. *In voluntary lid closure,* where it can be utilized to effectually disrupt the visual process. About 1% of the incident light, evenly distributed over the whole visual field, passes through the closed eyelid with average skin pigmentation.

On the margins of the eyelids there are *eyelashes,* or cilia. Totaling about 200 in each eye, they each have an average life of a few months. The follicle of each cilium is innervated by mechanoreceptive nerve endings that discharge impulses when the cilium is bent. Activation of these receptors evokes a blink reflex when a foreign body touches the lid aperture.

The upper outer portion of each orbit contains the *lacrimal glands.* They are innervated by the midbrain autonomic outflow traveling first along nerve VII, synapsing in the sphenopalatine ganglion, and reaching the glands via the mandibular division of nerve V and the zygomatic nerve. Secretion of tears, as well as being of psychic origin, is produced by activation of receptors of the lids and conjunctivae.

The normal secretion of tears amounts to less than 1 ml/day. The tear fluid has a pH of about 7.4 and osmotic pressure equivalent to about 0.9% NaCl. Reports of higher tonicity (up to 1.4% NaCl equivalent) can probably be accounted for by evaporation. An important constitutent of tears is *lysozyme,* a mucolytic (hydrolyzing mucopolysaccharides) enzyme with bactericidal action. The tear fluid serves (1) to provide lubrication for lid movements, (2) to provide a good

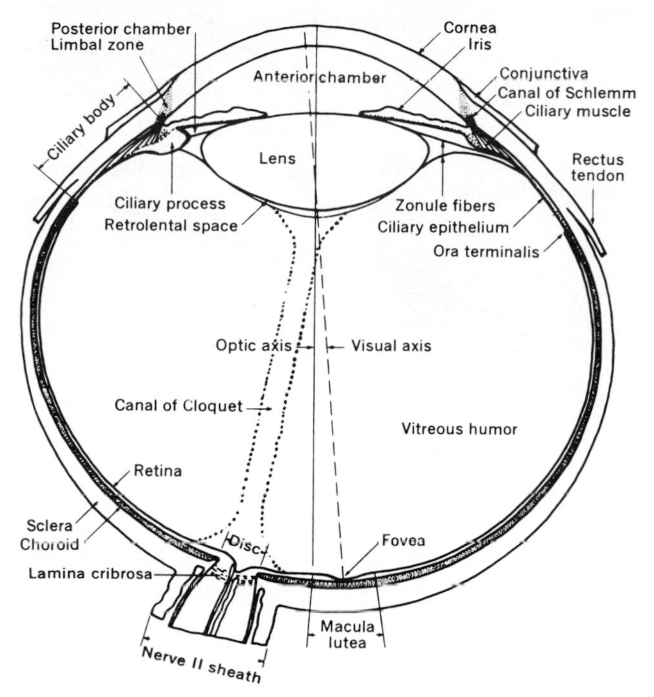

Fig. 14-1. Horizontal section of right human eyeball. (From Walls.[16])

optical surface for the cornea by forming a thin film over it, and (3) as an emergency mechanism to wash away noxious agents.

Most of the tear fluid normally formed is lost by evaporation. The normal drainage channels are the lacrimal canaliculi, sac, and duct leading into the inferior nasal meatus.

OPTIC MEDIA

In its passage through the eye, light successively traverses the cornea, the aqueous humor, the crystalline lens, and the vitreous humor. The cornea is anatomically continuous with and histologically similar to the sclera. Together they form the outer shell of the eyeball, which contains the lens, and the gel-like vitreous humor, whose internal pressure and hence shape are maintained by the appropriate balance between the rates of formation and elimination of aqueous humor.

Cornea

The cornea is about 11 mm in diameter and has a thickness of 1 mm near its junction with the sclera and 0.5 mm near its center. Its outer surface, the epithelium, is continuous with the epithelial layers of the conjunctiva. Most of the cornea consists of the so-called stroma, a lamellar structure of submicroscopic collagen fibrils in an organized arrangement that makes it transparent to light. There are normally no blood vessels in the cornea; as a consequence, it derives its oxygen supply by diffusion from the air and from surrounding structures. This is also the pathway of CO_2 loss. The supply of glucose and transfer of lactic acid seem to involve diffusion from the aqueous and other sites.

The transparency of the cornea depends on the maintenance of normal intraocular pressure and many other factors as well. Severe elevation of the intraocular pressure, lack of proper osmotic equilibrium with the surrounding fluids, changes in temperature, and improperly fitted contact lenses are some of the factors that may cause edema of the cornea, with a resulting increase in the light scattered by this structure. The earliest signs of this are halos seen around bright lights, but later visual acuity is lowered. The effect is usually reversible. A seriously injured cornea is repaired by scar tissue, which is usually not transparent. Because the cornea is avascular, corneal grafts may be uncomplicated by immunologic reactions.

The cornea is supplied by sensory fibers from nerve V. They lose their myelination shortly after entering the cornea and terminate as free nerve endings in the epithelium and the stroma. Weak mechanical stimuli to the cornea evoke tactile sensations. Stronger

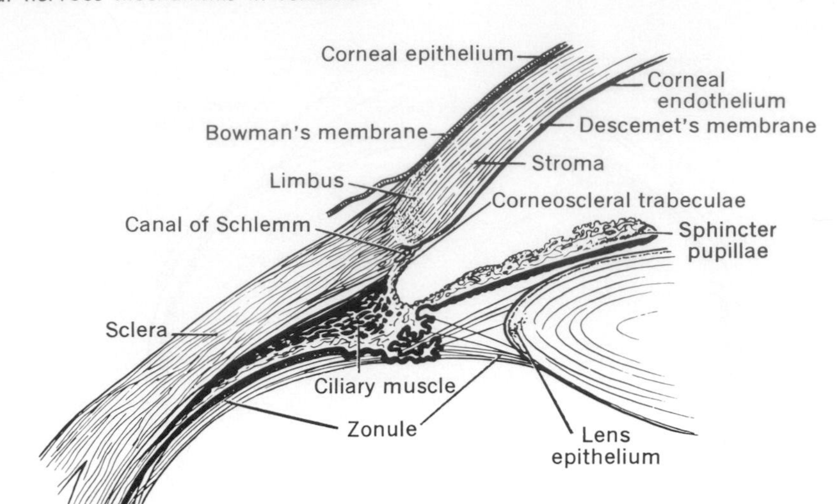

Fig. 14-2. Section showing details of anterior segment of human eye. (From Davson.[6])

ones are painful. Sensory localization is poor on the cornea, and there is some doubt whether heat or cold can be sensed there. If the corneal nerves are cut, e.g., by an incision for removal of cataract, they regenerate within a few weeks.

Lens

The crystalline lens also is an avascular structure. It is suspended by fibers, the zonule of Zinn (Fig. 14-2). The lens consists of an elastic membrane, the capsule, which encloses a single layer of epithelium and a complicated system of transparent fibers derived from it. New fibers are continually laid down, but the volume of the lens changes little throughout life. In youth the lens substance is malleable, which allows the changes in shape called accommodation to occur, but with age it becomes harder, has a higher solid content, and usually loses some of its transparency. An eye with an opaque lens is said to have a cataract, but there are many degrees of loss of transparency, all resulting in some loss of definition in the retinal image. Cataracts may be caused by physical trauma, radiation, or chemical factors such as the elevated glucose concentration of the aqueous in a patient with uncontrolled diabetes.

The lens uses oxygen and glucose and produces lactic acid. Ascorbic acid seems to be an important agent in lens metabolism. All these substances are carried in the aqueous humor and diffuse across the lens capsule, inside which a potential of about –70 mv is maintained, a result of the differential concentration of sodium and potassium ions actively maintained between the inside of the lens fibers and the anterior chamber.

Vitreous

Little is known about the distinctive features of the vitreous. Its gel-like physical consistency is thought to be the result of a network of submicroscopic fibers containing within its spaces large molecules of hyaluronic acid; the enzyme hyaluronidase produces an increased fluidity of the vitreous.

Aqueous

The composition of the aqueous humor closely approximates that of protein-free plasma (Table 14-1). The mechanism of formation and drainage of the aqueous is important not only because the aqueous is the principal carrier of metabolites for the lens and the cornea but also because it regulates the intraocular pressure, on which the maintenance of the eye's shape and transparency depends. The aqueous fluid is formed in the processes of the ciliary body. Comparison of the constituents of aqueous and blood plasma shows that it is not formed by simple filtration; e.g., bicarbonate ions are in considerably higher concentration in the aqueous than in plasma. The mechanism of secretion of aqueous is not fully understood, but it is known that carbonic anhydrase inhibitors such as acetazolamide (Diamox) reduce the rate of aqueous formation. This drug is of considerable assistance in controlling glaucoma, a condition in which the intraocular

Table 14-1. Values of two distribution ratios*

	Concentration in aqueous / Concentration in plasma	Concentration in dialysate / Concentration in plasma
Na	0.96	0.945
K	0.955	0.96
Mg	0.78	0.80
Ca	0.58	0.65
Cl	1.015	1.04
HCO_3	1.26	1.04
Urea	0.87	1.00

*From Davson.[4]

pressure is elevated because of interference with normal aqueous drainage. Since the aqueous bathes the lens, the iris, and the posterior surface of the cornea and is in contact with the vitreous, diffusion at the surface of these structures will affect its composition.

After the aqueous is formed, it flows forward between the lens and the iris and in this way into the anterior chamber. The temperature gradient in the anterior chamber (the cornea is nearer the external environment and hence cooler) causes a thermal circulation of the aqueous, the layers nearer the posterior surface of the cornea rising and those nearer the lens falling. Drainage channels for the aqueous are located near the junction between the sclera and the cornea (Fig. 14-2). A trabecular meshwork leads into the canal of Schlemm, which drains into the venous system of the eye.

Intraocular pressure

Normally the intraocular pressure is about 20 mm Hg above atmospheric pressure. It varies with respiration and with the pulse, and there is also a diurnal cycle. These changes are usually no more than a few millimeters of mercury. Intravenous injection of adrenaline in the intact animal causes a sudden rise in the intraocular pressure. External pressure such as that caused by contraction of the lid and the extraocular musculature may cause a marked increase. Prolonged contraction of the sphincter of the iris opens the drainage channels and may therefore lower intraocular pressure in eyes in which these are obstructed; the lowering due to reduction in the rate of formation of aqueous produced by acetazolamide has already been referred to.

A good estimate of the intraocular pressure in an intact eye may be obtained by an instrument called the tonometer, which measures the resistance that is encountered in producing a small indentation of the cornea by a plunger. Since this depends to some extent on the rigidity of the cornea, these instruments must be calibrated on cadavers.

OCULAR MUSCULATURE

There are three muscular systems associated with the eye. The first controls the state of focus and permits a clear retinal image to be formed for a wide range of object distances. The second controls the aperture of the eye. Both of these are situated within the eye and are called intraocular. They are made up of smooth muscles innervated by the autonomic system. The third system functions to control the direction in which the eye is pointing. The muscles are attached to the outside of the eye and hence this system is called the extraocular motor system.

Ciliary muscle and accommodation

The ciliary muscle is in its action a sphincter muscle and is innervated by the parasympathetic bulbar outflow via nerve III and the short ciliary nerves.[22] The crystalline lens is suspended by the fibers of the zonule of Zinn, which run from the ciliary body to the lens. In the unaccommodated state the ciliary muscle is relaxed and the zonular fibers taut. The result is that the anterior surface of the lens is flatter than the posterior surface. Contraction of the ciliary muscle makes the ciliary processes bulge inward toward the anteroposterior axis of the eye. The effect is a slackening of the tension in the zonular fibers, now leaving unopposed the elastic forces in the lens capsule that tend to mold the malleable lens substance into a different shape: the middle of the anterior surface is more highly curved and bulges into the anterior chamber (Fig. 14-3). This process is called accommodation and depends on (1) the contraction of the ciliary muscle and (2) the capacity of the lens substance to assume a more biconvex shape when the tension on the zonule is released. Failure of accommodation to occur when neural impulses arrive at the ciliary muscle may be due to muscular factors or lenticular factors. The ciliary muscle is a smooth muscle innervated by the parasympathetic system, and it may be paralyzed by atropine, homatropine, or similar drugs that block the myoneural junction to acetylcholine. (It may also be stimulated by drugs such as eserine that inactivate cholinesterase.) While such an effect is usually

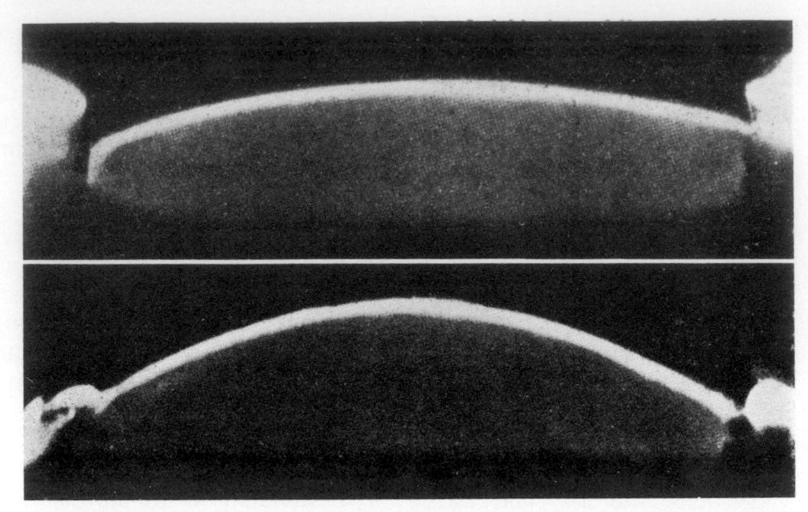

Fig. 14-3. Slit-lamp section showing anterior surface of normal human crystalline lens in unaccommodated state (upper) and accommodated state (lower). (From Fincham.[9])

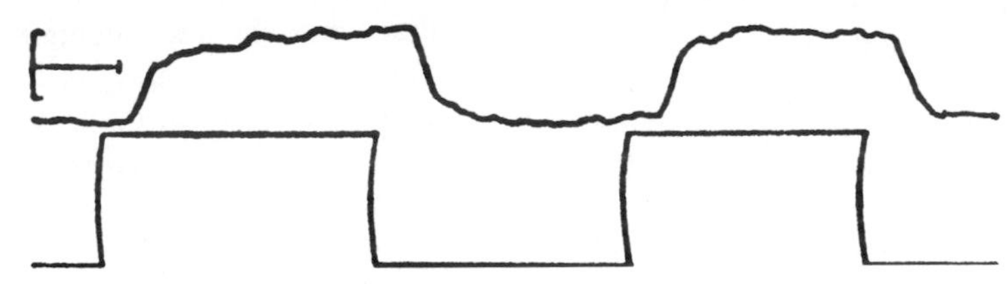

Fig. 14-4. Accommodation responses to step focus changes in normal human eye. Top line: Accommodation (length of horizontal line, 1 sec; height of arc, 1 D). Upward movement represents far-to-near accommodation. Bottom line: Stimulus signal (same scale). (From Campbell and Westheimer.[3])

produced by local application of drugs, occasionally it may be the result of systemic administration.

A normal contraction of the ciliary muscle may also fail to evoke a change in lens shape if the lens substance is unable to yield to the physical molding force of the lens capsule. We have seen that the lens substance gets harder with age. This process is a progressive one, but it becomes noticeable at an age of 40 years and more when only a small response in the lens shape results even on maximal contraction of the ciliary muscle. In the fifties a stage is usually reached when the lens cannot change shape at all, even when completely freed from the zonule of Zinn.

Accommodation serves as a focusing device in the eye, a response evoked by the presence of a blurred image on the retina. The reaction time of this response is about ⅓ sec, and its execution occupies another ½ sec (Fig. 14-4). The two eyes respond equally and simultaneously.

During steady viewing of a close-up target there is an unsteadiness of the ciliary muscle response. The resulting oscillations of accommodation are also binocular and have a frequency of about 2 Hz. They have been interpreted as the inevitable consequence of the accommodation system acting as a servomechanism. When a person is confronted with an empty visual space, as in complete darkness or when looking at a vast expanse of sky, one might expect the ciliary muscle to be relaxed. In fact there is a slight contraction, focusing the eye not at infinity but at about 1 m. This is called night or empty field myopia.

The neural pathway subserving accommodation is coupled with that for convergence of the two eyes, for the effort to produce accommodation usually results in some degree of convergence, even when one eye is covered—the so-called accommodation-convergence synkinesis. It exists even in people who never had binocular vision, e.g., when there is uniocular congenital cataract. A simi-

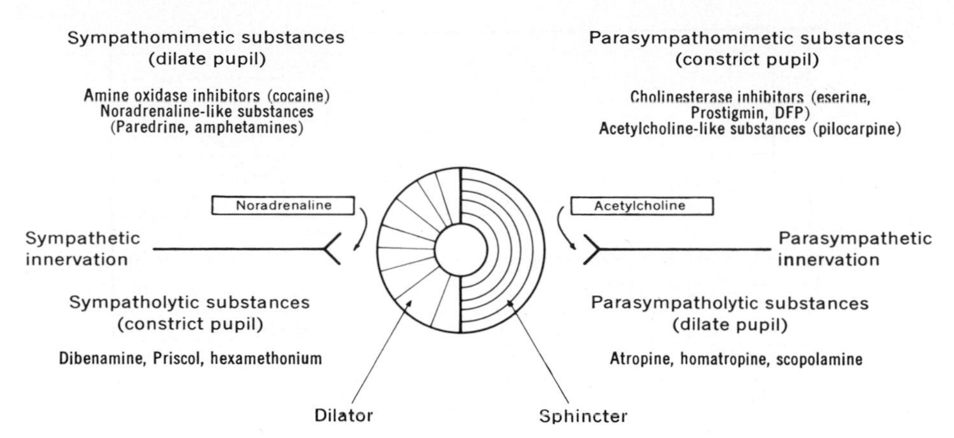

Fig. 14-5. Pharmacology of iris musculature.

lar synkinesis exists between accommodation and pupil constriction.

Iris musculature

The aperture of the iris, the pupil, is controlled by the state of contraction of the two muscles in the iris, the sphincter and dilator. The sphincter is a parasympathetically innervated muscle; its innervation and drug susceptibility closely resemble those of the ciliary muscle, but unlike the latter it synapses on the ciliary ganglion.[21] The dilator's innervation comes from the thoracic sympathetic outflow.

Contraction of the sphincter alone or in association with relaxation of the dilator causes constriction of the pupil, or miosis. Contraction of the dilator and relaxation of the sphincter (the sphincter muscle is stronger than the dilator) causes dilation of the pupil, or mydriasis. Drugs producing miosis (Fig. 14-5), so-called miotics, are either those interfering with the normal mode of operation of cholinesterase (e.g., eserine) or those competing with acetylcholine for the receptor sites of the sphincter muscle (e.g., pilocarpine). There are two major kinds of drugs producing mydriasis, so-called mydriatics: the parasympatholytic drugs such as atropine and its derivatives and the sympathomimetic drugs (various catecholamines, such as phenylephrine, and the amphetamines).

The size of the pupil at any time reflects the relative state of contraction of its two muscles. There is some doubt whether the two major effector substances have a reciprocal pharmacologic action on the two muscles or whether one muscle, the sphincter, is the more powerful and the pupil is predominately affected by its influence. States such as fright that are characterized by high sympathetic activity produce mydriasis, as does pain also. Morphine causes a pronounced constriction of the pupil, an effect probably produced by the central nervous action of this drug.

Constriction of the pupil, effected largely by the parasympathetic activity via nerve III and the sphincter muscle, is a reflex for which the stimulus is light. The pathway for this pupillary light reflex includes afferents from the retina to the pretectal area in the midbrain, internuncial fibers to the Edinger-Westphal portion of nerve III nucleus, preganglionic efferents synapsing in the ciliary ganglion, and postganglionic fibers that reach the sphincter of the iris through the short ciliary fibers. Illumination of one eye causes pupillary constriction of the same eye (direct reflex) and the other eye (consensual reflex). The pupillary light reflex has a reaction time of about 0.2 sec (Fig. 14-6). The pupil constricts also when a person regards a close-up object (near reflex), a response associated with accommodation. An interesting neurologic sign is the Argyll Robertson pupil, commonly encountered in tertiary syphilis, where the pupil response to light is lost but that to near objects is retained.

The iris musculature exerts only a moderate control over the quantity of light admitted to the eye. The pupil diameter of the normal human eye ranges from 8 mm in total darkness to 2 mm under very bright conditions. The maximal possible change in area is thus 16 times, which is far too small to maintain a constant incident light flux in the total range of brightness over which the eye operates.

The pupils undergo a characteristic series of changes with the depth of anesthesia. They

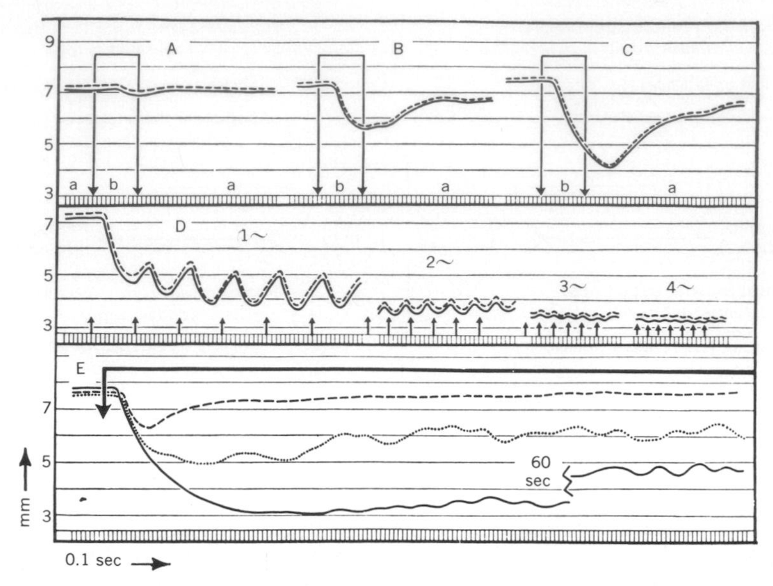

Fig. 14-6. Pupillary reactions to light in normal man. Pupillary diameter (in millimeters) against time (in 0.1 sec units). Solid lines represent right pupil's reactions; broken lines, left pupil's simultaneous responses (except in **E**). First row (**A** to **C**): At *a,* subject's eyes were in darkness. During 1 sec periods, *b* (outlined by double arrows), white light flashes were presented. Light intensity was varied by neutral gray filters. In **A,** light was 1, in **B** it was 4, and in **C** it was 8 $\log_{10}$ units brighter than subject's absolute visual threshold. With increasing stimulus intensity, pupillary reflexes increased in extent and speed and latent period shortened. Reactions were equal on both sides, although only right eye was stimulated while left eye remained in darkness. Second row (**D**): At moments marked by small arrows, short bright light flashes (5 msec duration, same brightness as in **C**) were presented to subject's right eye at rates of 1 to 4/sec. With increasing rate of stimulation, pupillary oscillations become smaller and mean pupillary diameter decreases. Third row (**E**): Pupillary movements elicited by prolonged light stimulation (duration framed by arrow). Only right pupil's movements are shown. Intensities were 8 (solid line), 5 (dotted line), and 2 (broken line) $\log_{10}$ units above subject's visual threshold. After initial contraction, pupil dilated partially, more so when light was dim than when it was bright. Pupillary oscillations appeared that were faster in bright than in dim light. (From Lowenstein and Loewenfeld.[11])

are dilated in the beginning stages and also in the deepest stage. In intermediate stages of surgical anesthesia as well as during sleep the pupils are constricted. Curiously, accommodation follows a similar pattern.[22]

Under normal conditions the pupil exhibits small oscillations called physiologic hippus. These changes occur synchronously in the two eyes and may reach an amplitude of 1 mm or more.

Extraocular muscle system

The muscular apparatus that moves the eye in the orbit serves to stabilize the eye with respect to the surrounding environment, compensating for changes in head position; these positional changes depend in large part on vestibular input. In primates and especially in man there has evolved a specialized region of the retina called the fovea within which spatial resolution is markedly better than elsewhere on the retina. The extraocular muscles function to "point" the eye toward a desired target so that its image is located upon the fovea.

The eye moves within the orbit like a ball in a socket joint; no lateral movement occurs; i.e., one point within the eye remains fixed in the orbit during movement. The center of rotation of the eye is about 13 mm behind the apex of the cornea. There is one pair of muscles corresponding to each degree of freedom for movement of such an apparatus, six muscles in all (Fig. 14-7). They originate in the wall of the orbit and are wrapped around the eye for some distance before inserting thereon. Horizontal movements are produced by reciprocal contraction and relaxation of the medial and lateral recti. Vertical and rotational (about the anteroposterior axis of the eye) movements are produced by similar reciprocal contractions of

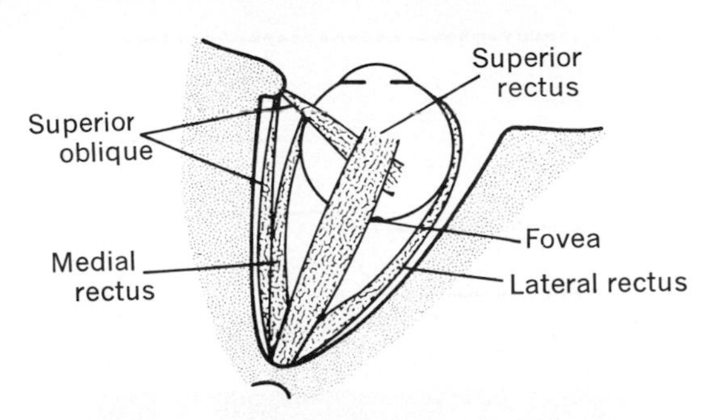

Fig. 14-7. Extraocular muscles of right eyeball as viewed from above. Inferior oblique runs parallel to final portion of superior oblique but is situated below eyeball. Inferior rectus runs parallel to superior rectus but is situated below eyeball. Eyeball is tightly packed into orbit with orbital fat and content of orbit is held back by lids. As a consequence, when muscles contract, they will rotate eyeball around axis through center of rotation normal to plane containing points of origin and insertion of muscles and center of rotation.

the superior and inferior recti coordinated with actions of the two oblique muscles. The muscles are striated, relatively powerful for their load, have fast contraction times, and their motor units have a low ratio of muscle-to-nerve fibers. There are stretch-sensitive receptors within the extraocular muscles, but their central reflex or sensory functions are unknown: they do not evoke the stretch reflexes common to other muscles. The levator palpebrae elevates the upper lid and frequently operates in conjunction with the superior rectus. The neuromuscular junctions in these muscles are particularly susceptible to the action of curare-like agents, and the earliest signs of neuromuscular disorders such as myasthenia gravis (Chapter 5) may appear in them as drooping of the eyelid (ptosis) or divergence or convergence of the eyes themselves. In some species the orbit contains smooth muscle innervated by autonomic nerves, which upon contraction protrudes or retracts the eye relative to the orbit, but if present at all, the muscles are rudimentary in man.

Although the superior oblique is innervated by the fourth cranial nerve, the lateral rectus by the sixth, and all other extraocular muscles by the third, their actions are so well coordinated that it is possible to regard the 12 muscles of the two orbits as a single functional unit. This remarkable coordination is a central neural event and depends in the first instance on heavy reciprocal interconnections between the motor nerve nuclei of the brainstem by nerve fibers that run in the medial longitudinal fasciculus. This interconnected system controlling the extraocular system receives a heavy projection from the vestibular nuclei of the medulla.

During deep sleep or anesthesia no impulses reach the ocular muscles along their nerves and the eye assumes its anatomic position of rest. During the waking state or during activated sleep (Chapter 8) all the muscles are in a graded, reciprocal state of contraction and the eye position is determined by the net balance of all the forces acting on the eyeball. Eye movements may be initiated in several places in the CNS; they are produced by a simultaneous change in rate of firing of the motoneurons supplying the various extraocular muscles. The result is a simultaneous change in contraction level of the relevant muscles and the eye assumes a new position as governed by the new balances of forces. In this way, exquisite control is maintained with no waste of energy. Such an arrangement, of course, is aided by the relatively small and constant load involved.

Vestibulo-ocular reflexes

Some eye movements are simple reflexes, the vestibular system providing the afferent arc. For example, when the head is tilted sideways, the eyes rotate around their anteroposterior axes, tending to keep their vertical axes aligned with the direction of gravity. This response can compensate for greater transients than steady head deviations. The afferent input evoking this reflex response originates in the sacculus of the vestibule. A very prominent response is observed when a person undergoes angular acceleration. Suppose a subject is rapidly spun toward his left around a vertical axis, say, while he is sitting on a piano stool. He is soon spinning at a high angular velocity. Suppose he is now suddenly stopped. This is equivalent to giving him a high angular acceleration to his right. The cristae in his semicircular canals will signal "angular acceleration to the right," since this is what happens when a constant-velocity rotation to the left is suddenly stopped. A simple reflex arc is activated, operating via the vestibular nuclei, the medial longitudinal fasciculus, and the oculomotor nuclei, the result of which is a smooth movement of both eyes toward the left. In intact animals the eyes do not remain fixedly looking toward the left, but when their excursion has come near its anatomically limited maximum, the eyes will suddenly return toward the right and then repeat the maneuver many times over, until the vestibular stimulus has subsided. The whole pattern of eye movements, i.e., the smooth binocular movement in the direction opposite that of the angular acceleration and the quick jerky binocular return movement, is called *nystagmus,* and since in this case the stimulus is evoked by vestibular afferents, it is called vestibular nystagmus. The same kind of pattern can be induced in different directions, depending on the direction of the vestibular stimulus. While the adequate stimulus for vestibular nystagmus is angular acceleration of the head, the same response will occur when the vestibular receptors are activated in some other way, e.g., by electrical stimulation or by thermal stimulation through irrigation of the ear with hot or cold water.

In contrast to the relatively simple brainstem reflex pathways responsible for vestibular nystagmus, the majority of eye movements are more complex, and they are influenced by neural action in many locales of the nervous system. The predominant afferent influence upon eye position and movement is that of the retina itself, an activity relayed centrally via the optic nerve and tract, and the geniculocortical system.

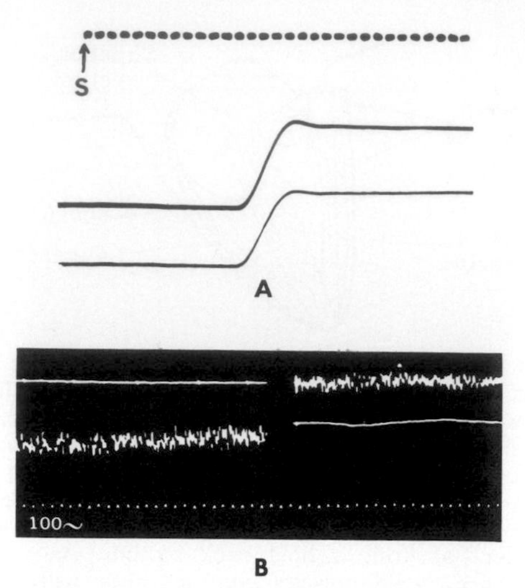

Fig. 14-8. A, Time course of 20-degree human saccadic eye movement. Time trace interrupted every 0.01 sec. Stimulus presented at *S.* **B,** Electromyogram of human extraocular muscles. Upper trace, right lateral rectus; lower trace, left lateral rectus. Left half, extreme left gaze; right half, extreme right gaze. There is reciprocal innervation of eye muscles, and eye position is determined by balance of forces applied to eyeball. Eye movements are produced by changes in neural impulses to, and hence tension in, eye muscles. Sudden changes lead to saccadic movements, **A;** gradual changes to smooth movements (Fig. 14-9). (**A** from Westheimer[17]; **B** from Breinin and Moldaver.[2])

Conjugate eye movements

The two main types of movement of the human eye are saccadic and smooth pursuit movements. In each the two eyes move in a parallel fashion, and they are therefore called conjugate movements. A saccadic eye movement is a sudden displacement of both eyes (Fig. 14-8). Such a rapid movement is executed by a simultaneous change in the rates of motoneuron discharge and hence in muscle tension of the relevant extraocular muscles. The duration of a saccadic movement is 20 to 50 msec, depending on the extent of the displacement. If the movement is initiated as a consequence of target displacement, it follows the latter with a reaction time that varies between ⅛ and ¼ sec. Saccadic movements do not, however, occur only in response to target displacement; they occur continually during environmental scanning. During reading or looking at a picture the eye movements are only of the saccadic type, following each other repetitively at intervals of about ¼ sec. Smooth pursuit movements, on the other hand, occur only when there is a moving target in the visual field. The eyes then move smoothly with good binocular coordination, and the velocity of the tracking movement matches that of the target as long as the latter does not exceed about 30 to 40 degrees/sec. The two kinds of movements differ markedly in the temporal and spatial organization of cranial motoneuron discharge rates required for their execution; they appear to be due to entirely different control mechanisms that utilize the same final common path (Fig. 14-9).

The two types of eye movements are occasionally employed together, e.g., in opticokinetic nystagmus. If a subject watches a continuously moving target such as a train, the eyes fix a point on the target and move with the angular velocity required to keep the image of that point stationary on the fovea. When they reach the limit of the field of eye movements, a quick saccadic movement occurs in the opposite direction, bringing the eyes to a new point on the moving target, and the maneuver is repeated.

In the pauses between larger saccadic movements, even when there are no pursuit movements, the eyes are not perfectly still.

Fig. 14-9. Photograph of human eye movement in response to target moving with constant angular velocity of 5 degrees/sec. Heavy trace, stimulus; light trace, eye; vertical lines, time in 0.01 sec. There is a reaction time, and then eye makes up accumulated steady-state error with saccadic movement and tracks target with constant-velocity "smooth" movement. All features of these types of eye movements are conjugate, i.e., parallel in the two eyes. (From Westheimer.[18])

Small, spontaneous saccades and other movements occur when the eyes may otherwise appear at rest. These small oscillations are of interest because some theories of visual acuity are based on the supposition that the small scanning oscillations convert a steady light stimulus into a time-varying one to which the sensory receptive mechanism of the retina is much more sensitive than it is to steady illumination. It is now thought unlikely that these oscillations aid in resolution, but it has been shown that if the visual image is stabilized on the retina by special optical techniques, a curious "washing out" of contrast occurs, even though resolution per se may not be diminished.

Disjunctive eye movements

In addition to the binocularly parallel or conjugate movements so far described, animals with binocular vision have a further mechanism that allows the two foveas to be directed to a close-up target—the disjunctive movements. They are slower than saccadic movements but have a similar latency. Not only can the two eyes converge but they can also diverge somewhat, and there is a facility for small vertical and cyclorotational vergences as well. Exact registration of the images of a given target on corresponding points in the retina of each of the two eyes is a necessary prerequisite to good binocular vision with its process of accurate spatial localization, stereoscopy; hence the need for a mechanism to allow this to take place when the target is closer than infinity and to make small adjustments when, owing to anatomic or other factors, the two eyes are not quite accurately aligned.

When the images of a single target on the two retinas give rise to a single coordinated sensory impression, they are said to be fused. When fusion is prevented by covering one eye, it is possible to measure the natural alignment of the two eyes. A slight misalignment, but one that can be overcome by the normal process of convergence (or vertical vergence if the misalignment is vertical), is called a *heterophoria.* A patient in whom normal vergence movements are not adequate to maintain fusion has strabismus, or *squint.* The relationship between accommodation and convergence has already been referred to. It depends on a neural linkage between the midbrain centers subserving the two functions.

OPTICAL IMAGE FORMATION IN THE EYE

To produce an image of the outside world on the retina the eye has a refracting apparatus consisting of the cornea and the crystalline lens. The optical properties of refracting surfaces are determined by their radius of curvature and the index of refraction of the media that they separate, and the characteristics of the optical apparatus as a whole depend on those of the surfaces and their separation.

The details for each of the individual surfaces in a typical normal eye are given in Table 14-2. Included in the table is the specification of the refracting power of each surface. The unit in which refracting power is

Table 14-2. Optical constants of a typical normal eye

Surface	Radius of curvature (mm)	Refractive index		Distance from anterior surface of cornea (mm)	Refractive power (D)
		Anterior	Posterior		
Anterior cornea	7.8	1.000 (air)	1.376	0	+48.2
Posterior cornea	6.8	1.376	1.336 (aqueous)	0.5	- 5.9
Anterior lens	10.0*	1.336	1.386†	3.6*	+ 5.0†
Posterior lens	-6.0	1.386†	1.336 (vitreous)	7.2	+ 8.3†
Retina				24.0	

*During maximum accommodation the anterior surface of the lens has a radius of curvature of 5 mm and its anterior surface is moved forward to be nearly 3 mm behind the anterior surface of the cornea. Partial accommodation will produce values between these values and those given in the table.

†The index of refraction of the lens varies from 1.386 near each surface to 1.406 in the center. The indicated refractive power is for the lens surfaces only. The gradient of refractive index within the lens produces additional refractive power.

measured is the diopter, the reciprocal of the focal distance in meters. It is seen from the table that the anterior surface of the cornea is the major refracting surface of the eye—hence any irregularities in it are of major significance in image formation in the eye.

The total refracting power of the eye cannot be obtained by simple addition of the values in the last column of Table 14-2 for two reasons. First, the separation of the surfaces affects their cumulative refractive effect, and second, the crystalline lens behaves in quite an unusual way. The refractive index of the lens is not uniform but varies between 1.386 near its surfaces to 1.406 at its center. A structure with such a refractive index gradient produces an optical effect corresponding to a much higher refractive index than is actually found. It has been estimated that if the human lens were filled with a medium of uniform refractive index, this would have to be 1.416 to produce the same refractive power as that of the actual lens. This phenomenon is even more marked in the fish eye, for here the cornea, being immersed in water, has little influence on the light rays, and most of the refraction necessary to bring an image onto the retina is provided by the lens. Some fish lenses have a gradient of refractive index reaching values as high as 1.5 in the center, and their power is equivalent to homogeneous lenses filled with medium that has a refractive index of 1.7.

While human eyes show considerable individual differences, it is useful to devise an optical model for the typical eye, with the aid of which a number of important calculations can be done very simply and with surprisingly good general validity. These optical models are called schematic eyes, and by far the least complicated of these is the

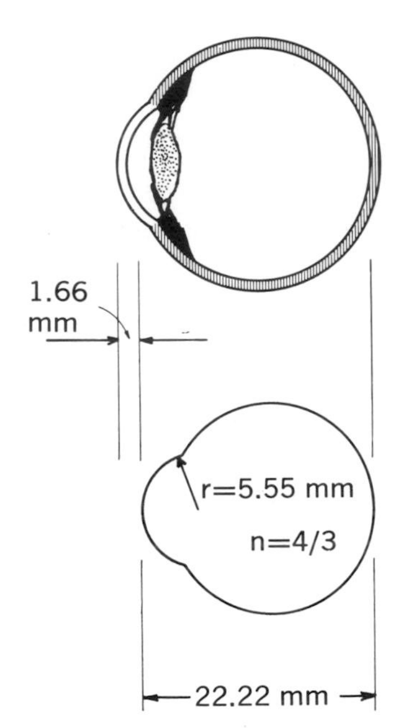

Fig. 14-10. Schematic diagram of optic media of eye (upper) and of "reduced" eye (lower). Reduced eye consists of single surface of 5.55 mm radius of curvature separating air from water, and its image-forming properties are similar to those of typical human eye.

reduced eye. In it, a single surface has been substituted for a cornea and lens, just as it is helpful at times to substitute a single thin lens for a complicated many-component optical system. In the reduced eye the single surface, which separates the vitreous from air, obviously has to have a different curvature from that of the real cornea, since it alone must focus rays on the retina. The position of this single surface is also not quite the same as that of the cornea; rather it occupies the place at which a single lens or surface would have the same effect as that of the whole optical system of the eye—the principal plane. A more detailed com-

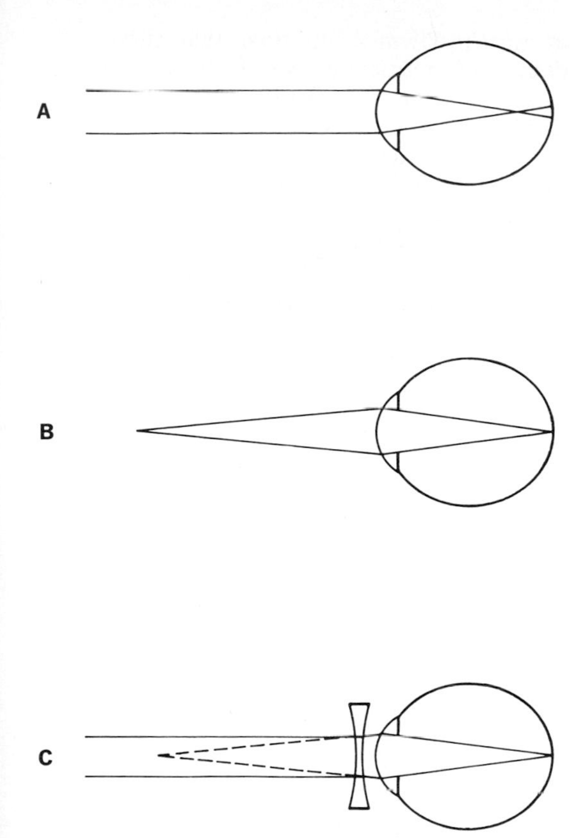

Fig. 14-11. Myopic eye. **A,** Rays from point object at infinity form an image in vitreous and are intercepted as a blur patch on retina. Size of blur patch for given state of out-of-focusness depends on pupil size. **B,** In the same eye, close-up target is imaged on retina. Reciprocal of object distance in meters gives amount of myopia in diopters. **C,** Myopia may be corrected by a lens of such diverging power that it makes rays from target at infinity appear to come from far point of myopic eye.

parison between the reduced eye and a typical real eye may be made from Fig. 14-10.

The optical constants of the reduced eye have been chosen so that in its most salient features it operates as an image-forming mechanism most like the real eye. There are some important differences—the reduced eye cannot accommodate in the way the real eye does—but the optical properties are sufficiently similar and so very much simpler that it is an ideal model on which to illustrate the major optical effects and defects occurring in a normal human eye.

Refractive errors

A normal eye with accommodation relaxed will produce an image of an infinitely far object in its focal plane, which is nearly 24 mm behind the corneal vertex and 22.22 mm behind the vertex of the reduced eye. If the retina is situated in this plane, a sharp image will be formed on the receptors and the optical prerequisite for clear vision is met. This matching of focal length of the optical part of the eye and axial length is the condition of *emmetropia.* It must be remembered that this merely satisfies the prerequisite for clear retinal imagery; whether, in fact, the patient will *see* clearly depends on the integrity of the photochemical or neurophysiologic stages of the visual process, the subject of later chapters.

What is the effect of a mismatch between focal length of the optics of the eye and the eye's axial length? A typical example of this is myopia—a condition in which the image of a distant object is formed not on the retina but in front of it (Fig. 14-11). This may occur because the particular eye has an optical refracting system more powerful (shorter focal length) than usual but has a normal axial length or, more commonly, because the focal length is normal but the eye is too long. In any case the rays from a point object at infinity come together in a point focus in the vitreous and will then diverge again. When the rays reach the retina, the cross section of the bundle will form not a point but a patch, the so-called blur patch, whose size depends on the extent of out-of-focusness and the diameter of the pupillary aperture. Objects at a great distance, as a consequence, will appear blurred. The same laws of optical image formation, according to which the image of a distant object will be formed in the focal point and hence in the myopic eye in front of the retina, allow us to determine the position of a target so that its image in such an eye *is* formed on the retina. This point in object space is called the *far point.* The reciprocal of the distance, in meters, between the eye and the far point is the amount of myopia in diopters. For example, a reduced eye whose focal point is at a distance of 22.22 mm behind the surface but whose retina is 23.4 mm behind the surface will have its far point ⅓ m in front of the eye and will therefore be 3 D myopic. For an axial length of 25.6 the figures will be ⅛ m and 8 D, respectively.

The only way to produce a clear retinal image in such an eye when the object is at a long distance is to place a lens in front of the eye, which changes the vergence of the bundle of rays emanating from this object

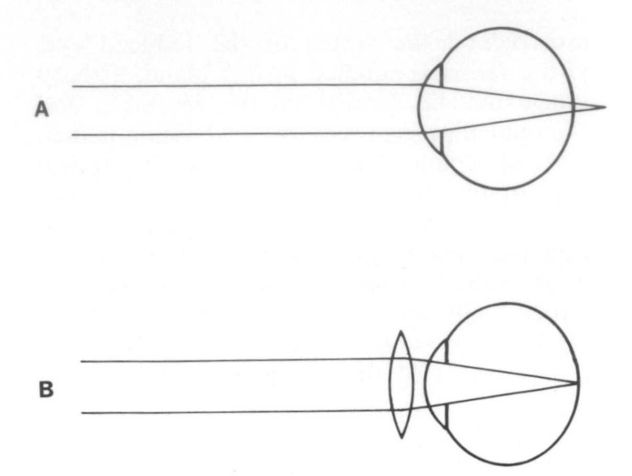

Fig. 14-12. Hyperopia. **A,** In uncorrected hyperopic eye, rays from a point target at infinity are intercepted by retina before they form an image. **B,** Positive lens placed in front of eye will converge rays and then refracting power of eye can produce clear image on retina.

and in such a way that when the rays emerge from the lens they appear to be coming from the patient's far point. There is thus a simple relationship between the far point distance and the focal length of the correcting lens. In myopia the latter has to be a negative lens, since a parallel incident bundle of rays must emerge from it as a diverging bundle to impinge on the eye as if it were coming from the far point. In turn this bundle is then imaged on the retina by the optical system of the eye. There is thus optical conjugacy between the distant object and the patient's far point with respect to the correcting lens, and the far point and the retina with respect to the eye's optics. This concept of optical conjugacy permits ready handling of such a problem as finding the lens power necessary to correct an eye when the lens is placed farther away than usual—the lens now has to have the power necessary to make the object point conjugate to the far point with respect to the new lens position.

In any case a myopic eye associated with the lens that brings a distant object conjugate to its far point is equivalent to an emmetropic eye: in each case the retina receives a clear image when the target is far away.

When the axial length of an eye is too short relative to its focal length, the retina will intercept the bundle of rays from a distant object before it comes to a focus. An eye with this kind of defect is said to be *hyperopic* (Fig. 14-12). A good way of looking at this problem is to note that there is a deficit of refracting power in the optical system of the eye. Bringing the target in from infinity does not help as it does in myopia—it merely moves the image even farther behind the retina. What this eye needs is additional refracting power. This can be provided by a positive lens placed in front of the eye, which gives the bundle of rays from a distant object some convergence and this, added to the refracting power of the eye, will produce a clear image on the retina.

The eye is usually not perfectly symmetric, but as a first approximation one may postulate the existence of an anteroposterior axis of symmetry, which is also the optical axis of the eye, i.e., the line on which the centers of curvature of the various surfaces lie. A plane section of the eye containing this anteroposterior axis is called a meridian of the eye. Thus the horizontal (0 to 180 degrees) meridian of the eye bisects it into an upper and a lower hemisphere, and the vertical (90 to 270 degrees) meridian bisects it into the nasal and temporal hemispheres.

An important optical anomaly is one in which the curvature of an optical surface, usually that of the cornea, is not the same in all meridians. For example, a cornea may have a higher curvature in the vertical than in the horizontal meridian. As a result, the sheet of rays lying in the vertical meridian will be refracted more than that in the horizontal meridian, and the point object will give rise in image space to a complicated bundle (Fig. 14-13) whose cross section is never a point, but either an ellipse or in one special situation a vertical line and, in another, a horizontal line. This condition is called *astigmatism.* The uncorrected astigmatic eye, depending on the position of the retina, may have generally blurred vision or a particular kind of imagery in which line targets in just one meridian appear sharp. The correction of an astigmatic eye is achieved by means of a lens that also has different curvatures in its meridians, so that its astigmatism is exactly complementary to that of the eye. When this lens has no power in one principal meridian, it is called a cylindric lens; when it has power, but differing in the two principal meridians, it is a spherocylindric or toric lens.

Myopia, hyperopia, or astigmatism is called a *refractive error,* or *ametropia.* It occurs when the optical refracting apparatus of the

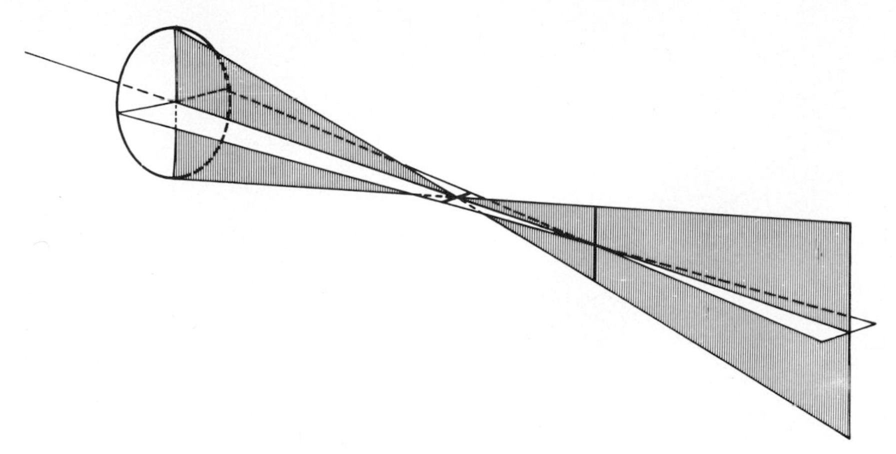

Fig. 14-13. Astigmatic imagery. When optical system has unequal refracting power in two meridians at right angles to each other, bundle of rays from object at infinity will undergo refraction to produce configuration illustrated in this figure. Cross section of bundle in image space is never a point, but an ellipse in general and a line in two special positions.

eye is incapable of producing a sharp image of a distant object or when it does so but the retina is not situated in the correct position. Such conditions are quite common. There is a strong genetic influence and the evidence that environmental factors contribute materially to its origin is uncertain. The newborn eye is often not far from being emmetropic. Myopia, when it occurs, usually develops during the teens and is irreversible. There is little change in the refractive error during adult life, but when lenticular changes commence with age, there may be associated anomalies of refraction.

For most purposes an ametropic eye with the appropriate spectacle correction functions like an emmetropic eye. The lens may be worn in frames in front of the eyes, or in suitable patients they may be made in the form of thin shells worn directly on the cornea—contact lenses. There is always a thin layer of tear fluid or specially prepared buffer solution between the cornea and the contact lens, a fact that makes the optical properties of the total contact lens correction somewhat complicated. Contact lenses have the advantage not only of being less conspicous than spectacles but also of effectively eliminating the anterior surface of the cornea as a refracting surface, an invaluable aid to clear retinal imagery when this surface is irregular or deformed by scarring or disease.

Accommodation

The diopter, the reciprocal measure of distance in meters, which is used to scale the degree of ametropia and the correcting lens power of an eye, is also of value in expressing the amount of accommodation exerted by an eye. The following discussion applies alike to emmetropic eyes and eyes made artificially emmetropic by spectacles or contact lenses. Such eyes have a target at optical infinity imaged on the retina when the ciliary muscle is relaxed. In order to image a close-up target on the retina it is necessary to increase the optical refracting power of the eye. This occurs when the crystalline lens is allowed to assume a more biconvex (i.e., more highly refracting) shape as a result of contraction of the ciliary muscle. If the distance of the target that has its image formed on the retina when the eye is in a given state of accommodation is measured in meters, its reciprocal is the amount of accommodation in diopters. Thus an emmetropic eye that has changed its refractive power to bring a target at ¼ m into focus is exerting 4 D accommodation. Due to the progressive sclerosis of the lens substance the amplitude of accommodation decreases with age; i.e., the nearest point to which an eye can accommodate recedes. This is illustrated in Fig. 14-14 both in terms of this near point of accommodation (in meters) and also of its reciprocal, the accommodative amplitude (in diopters). When accommodation is no longer adequate to achieve or maintain clear focus on the desired close-up target, e.g., reading material, it may be supplemented by positive lenses that can supply the necessary additional refractive power. Such a patient

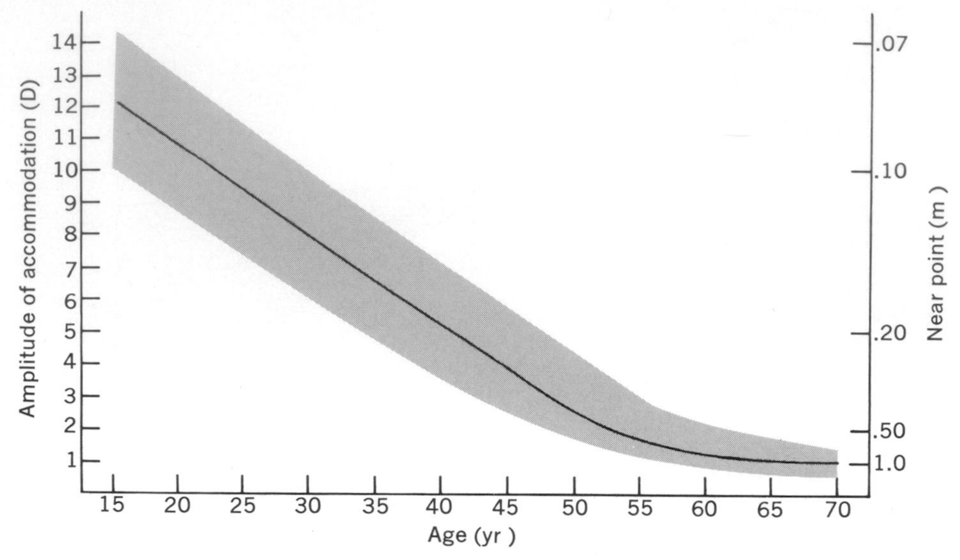

Fig. 14-14. Changes in accommodation with age. Graph shows typical value and range of accommodative amplitude accepted as normal for a given age. Ordinates at left, diopters, and at right, nearest point of clear vision for an emmetrope.

is said to be *presbyopic* and can be helped by bifocal or trifocal lenses that provide the refractive correction for distant objects, if necessary, and also the additional power ("the add") for one or two specified near viewing distances.

Insofar as accommodation adds refractive power to the eye, it can also be of help to the uncorrected hyperopic eye: clear vision of objects at optical infinity can be achieved by an effort of accommodation in diopters equal to its amount of hyperopia, also in diopters.

THE RETINAL IMAGE AND VISUAL ACUITY

The image, i.e., the replica of the outside environment, formed by the optical system of the eye on the retina is two dimensional. The geometric relationship between the outside world and the retinal image is based on a projection system, using the center of the eye's entrance pupil as the center of projection. An object sends out light energy in many directions, but only a small proportion—that which passes through the pupil—contributes to the formation of the retinal image. The most representative light "ray" therefore is the one passing through the center of the eye's entrance pupil—the chief ray. Since each light ray in object space has as its conjugate a light ray in the image space, the most representative ray for the purposes of defining the retinal position of the image (clear or blurred) is the image-sided chief ray. It passes from the center of the eye's exit pupil to the geometric image or, if a blur patch is present on the retina rather than a clear image, to the center of the blur patch. If it is desired to state where on the retina a target in the outside world is being imaged, one consequently draws a straight line from the object to the center of the eye's entrance pupil and then finds the intersection of its image-sided correlate with the retina. In any given eye there is a single and unique relationship between these two, and it suffices therefore to state object position and dimensions in terms of angles subtended at the center of the entrance pupil.

If the detailed calculations are carried out in a typical emmetropic human eye, it is found that an object subtending an angle of 1 degree at the eye will form an image on the retina that is 0.3 mm in size. One can thus look at the process of image formation in the eye as a mapping of directions in object space (determined by the line between the object point and the center of the eye's entrance pupil) and position occupied by the image of the object point. Along each direction only one radial distance will be in focus, that for which the eye is accommodated. This explains the common practice of interchanging retinal distances (in millimeters) and angular object subtense (in degrees or radians); e.g., one describes the width of the optic nerve head as 5 degrees rather than 1.5 mm.

Field of fixation and visual field

The most important target direction for any eye is the one that corresponds to the center of the fovea, for here the anatomic and functional organization of the retina is most favorable for good resolution. Normally the two eyes will be positioned to image the object of regard in the centers of the foveas. This process of fixation is the major function of the whole oculomotor system, a fact attested to by the relative absence of eye movements in species possessing no foveas. The object fixated is called the fixation point, and the line joining the fixation point and the center of the eye's entrance pupil is called the primary line of sight.

If the head is held still, the subject is asked to fixate a small object, and this object is moved about, it is possible to map out the *field of fixation*. It is a measure of the capacity of the extraocular muscles to move the eyes and it usually extends about 45 degrees in all directions, often more in a downward direction. When there is paralysis of an extraocular muscle or a mechanical obstruction in the orbit, the field of fixation may be restricted.

Consider, on the other hand, the situation when a stationary target is presented to an eye and the eye steadily fixates it. This target now represents the center of the fovea, and all other retinal locations may be specified by other points on a sphere, with the center of the entrance pupil as its center. On this sphere one may map out the extent of all those regions from which a light sensation may be elicited. This is called the *visual field*—it measures the regions of peripheral retina of the stationary eye that respond to light, as distinct from the field of fixation, which measures the range of target positions that can be foveally fixated with the moving eye.

Visual acuity

The analysis of this capacity to discriminate small differences in spatial configuration of targets involves a study of the quality of the replica available at the retinal level, the anatomic and functional characteristics of the retina, and the CNS. Only the first one of these factors will be considered here; the others are dealt with in subsequent chapters.

If the retinal image is a point-by-point replica of the target, resolution would in theory be unlimited. However, even a perfect optical instrument does not form a point image, but spreads the light coming from a point object into a diffraction pattern whose size is inversely related to the pupil aperture. When the pupil of the eye is small, the spread of the focused image of a point, the so-called point-spread function, is entirely that given by diffraction theory. As the pupil increases in diameter beyond about 2 mm, the point-spread function no longer conforms to the diffraction limitation (which would demand its continued narrowing) but stays near the shape shown in Fig. 14-15, only to widen again when the pupil diameter exceeds 5 mm. The failure of the eye to perform as an ideal optical instrument when its pupil is larger than 2 mm is due to the aberrations inherent in its optical system. The eye has some correction for spherical aberration in its cornea, which is aspherical, but none for chromatic aberrations. These aberrations become progressively more important as the pupil is enlarged.

The best optical performance of the eye, then, is found when the pupil is in the range of 2 to 5 mm, but this is only true when the eye is in good focus. We have seen that out-of-focusness results in the imaging of a point as a blur patch whose size is directly proportional to the pupil diameter, so that out-of-focus imaging is better the smaller the pupil, often down to very narrow pinholes.

Fig. 14-15 shows that a point object such as a star will even under the best possible conditions spread its light over an appreciable retinal region. When two point sources such as two stars are close together, each will produce its own light spread, but the two light distributions will summate: what a retinal receptor responds to is the sum of all the light reaching it from all sources and over its whole acceptance area. The shape of the summed light distribution from two closely adjacent point sources differs insignificantly from that of a single, brighter target, and it is therefore impossible to make the distinction between a single and a double light source. If the two point sources are now separated, the intensity configuration on the retina will show two distinct humps with a trough between them. The condition for resolution has now been reached provided, however, that the receptor mosaic has a "grain" such that the trough and its flanking humps each fall

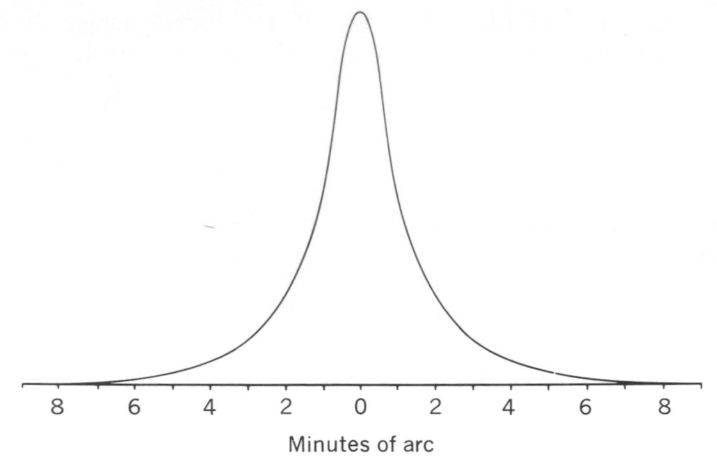

Fig. 14-15. Light spread in human eye. Optics of eye, even in best state of focus, spread light from line source over finite distance. This is in part due to diffraction and in part due to aberrations. Spread function, illustrated in this diagram, provides physical limitation to resolution in eye. (From Westheimer.[19])

on individual mosaic units, and that the physiologic pathway from each of these units is capable of transmitting independent signals. This example illustrates the factors involved in the resolution of spatial details or, as it is called, visual acuity: optical (point-spread function), anatomic (receptor mosaic), and functional (independent pathways for transmitting signals). In an eye in good focus, in the center of the fovea, when the light stimulus lasts sufficiently long and is sufficiently intense, these three factors operate optimally and have evolved to about the same limit of performance. It follows that the total resolution performance suffers when there is an abnormality in any of its components.

Tests of visual acuity involve the discrimination between spatial configurations that lead to differing sensations only if details of a given angular subtense are resolved. Examples are the Landolt C's, which are seen as C's of the correct orientation and not O's if the subject can resolve the gap. The scale for these letters is gap size in minutes of arc, usually the gap is one-fifth the overall size of the letter. Snellen letters are based on the same principle; the letter is 5 times the width of a limb.

It has been accepted for at least a century that normal visual acuity is 1′ of arc, i.e., separation of two stars by 1′ of arc leads to the sensation of two stars, and by less than 1′ of arc to the sensation of one star. On the same basis, if the gap in a Landolt C is 1′ of arc or larger, the C can be distinguished from an O. This, together with the convention of carrying out these tests with objects at 6 m or 20 ft, is the origin for designating a resolving capacity of 1′ as 20/20 visual acuity; i.e., a letter with a feature subtending 1′ of arc at 20 ft can then be resolved at 20 ft. If an eye requires a 2′ of arc gap to recognize a letter, this corresponds to a letter whose gap would subtend 1′ of arc at 40 ft; hence the designation of such a visual acuity as 20/40.

Under good conditions most normal eyes have a resolving capacity somewhat better than 20/20; 20/15 is not unusual. And with attention to all factors and a repetitive target like a grating, good observers can resolve down to nearly ½′ of arc—which for a 2 mm pupil represents just about the limit of resolution for even an ideal optical system like the eye's.

• • •

This completes our survey of the physical basis of image formation in the eye. How this information about the environment is encoded in the light-sensitive cells in the retina and then transmitted to and processed by the higher visual centers in the CNS is the concern of the following two chapters.

REFERENCES

1. Alpern, M.: Movements of the eyes; also Accommodation. In Davson, H., editor: The eye, ed. 2, New York, 1969, Academic Press, Inc., vol. 3.
2. Breinin, G. M., and Moldaver, J.: Electromyography of human extraocular muscles; nor-

mal kinesiology; divergence mechanism, Arch. Ophthalmol. **54:**200, 1955.

3. Campbell, F. W., and Westheimer, G.: Dynamics of accommodation responses of the human eye, J. Physiol. **151:**285, 1960.
4. Davson, H.: Physiology of the ocular and cerebrospinal fluids, London, 1956, J. & A. Churchill, Ltd.
5. Davson, H.: The physiology of the eye, ed. 3, Boston, 1970, Little, Brown & Co.
6. Davson, H.: The intra-ocular fluids; also The intraocular pressure, In Davson, H., editor: The eye, ed. 2, New York, 1969, Academic Press, Inc., vol. 1.
7. Duke-Elder, W. S.: Textbook of ophthalmology, St. Louis, 1942, The C. V. Mosby Co., vol. 1.
8. Emsley, H. H.: Visual optics, ed. 5, London, 1955, Hatton Press, Ltd., vol. 1.
9. Fincham, E. F.: Mechanism of accommodation, Br. J. Ophthalmol. suppl. **8:**5, 1937.
10. Helmholtz, H.: Treatise on physiological optics, New York, 1962, Dover Publications, Inc., vols. 1 and 2.
11. Lowenstein, O., and Loewenfeld, I. E.: The pupil. In Davson, H., editor: The eye, ed. 2, New York, 1969, Academic Press, Inc., vol. 3.
12. Maurice, D. M.: The cornea and sclera. In Davson, H., editor: The eye, ed. 2, New York, 1969, Academic Press, Inc., vol. 1.
13. Moses, R. A.: Adler's physiology of the eye, clinical application, ed. 5, St. Louis, 1970, The C. V. Mosby Co.
14. Robinson, D. A.: Eye movement control in primates, Science **161:**1219, 1968.
15. Sorsby, A., et al.: Emmetropia and its aberrations, Medical Research Council Special Report No. 293, London, 1957, Her Majesty's Stationery Office.
16. Walls, G. L.: The vertebrate eye, Bloomfield Hills, Mich., 1942, Cranbrook Institute of Science.
17. Westheimer, G.: Mechanism of saccadic eye movements, Arch. Ophthalmol. **52:**710, 1954.
18. Westheimer, G.: Eye movement responses to horizontally moving-visual stimulus, Arch. Ophthalmol. **52:**932, 1954.
19. Westheimer, G.: Optical properties of vertebrate eyes. In Fuortes, M. G. F., editor, section 4: Handbook of sensory physiology, Berlin, 1972, Springer Verlag, vol. 7.
20. Westheimer, G.: Visual acuity and spatial modulation thresholds. In Jameson, D., and Hurwich, L. M., editors, section 2: Handbook of sensory physiology, Berlin, 1972, Springer Verlag, vol. 4.
21. Westheimer, G., and Blair, S. M.: The parasympathetic pathways to internal eye muscles, Invest. Ophthalmol. **12:**193, 1973.
22. Westheimer, G., and Blair, S. M.: Accommodation of the eye during sleep and anesthesia, Vision Res. **13:**1035, 1973.

15 Physiology of the retina

KENNETH T. BROWN

In Chapter 14 it has been shown that the major optical function of the eye is to gather light rays from the primary stimulus, which is the external world, and focus them into a secondary stimulus, which is an image upon the retina. This image is a two-dimensional representation of the external world, reduced in size to fit the retina, and it is the stimulus that leads to spatially patterned excitation of the photoreceptors. The ultimate end products of the visual system are sensations that represent the external world with considerable accuracy and that can therefore guide adaptive behavior in relation to the external environment. Since these sensations are part of conscious experience, they fall within the subject matter of psychology. Thus one of the overall tasks of visual physiology is to bridge the gap between physics and psychology by explaining how light, as a phenomenon of physics, is translated into light as a conscious experience. Our visual experiences contain much information and attain considerable complexity, so the problems of visual physiology are correspondingly complex. In this chapter only retinal aspects of visual physiology will be considered; more central levels of the visual system will be taken up in Chapter 16. Emphasis will be placed on evidence that applies most clearly to man, since the human retina is of primary interest in medicine.

FUNCTIONAL ANATOMY

It is helpful to look first at some major features of retinal structure, because retinal histology tells much about how the retina functions. Most knowledge of retinal structure, which may be applied with confidence to the human eye, has been obtained from macaque monkeys. These monkey retinas are very similar to those of humans but are more readily obtained under good conditions for histologic study.

Cells and their synaptic relations in the peripheral retina

Synaptic relations between retinal cells have been revealed largely by Golgi preparations,[1, 10, 11] and Fig. 15-1 summarizes Polyak's findings in the macaque monkey. The retina possesses two main classes of receptors —the *rods* and the *cones*. Each type of receptor possesses both outer and inner segments, which are joined by a connecting cilium. In retinal anatomy the terms "outer" and "inner" designate relative distances of structures from the center of the eye. Thus in Fig. 15-1 the outer segments of the receptors are in layer *2a,* while the inner segments are in layer *2b*. In the peripheral retina both the outer and inner segments of rods are slender and roughly cylindric; hence their appearance is rodlike. In peripheral cones the outer segments are rather conical in form, and the inner segments are thicker than those of rods. The outer segments of both cones and rods contain visual photopigment that absorbs the light leading to excitation. These receptors are best regarded as neurons with outer and inner segments that are highly specialized. In cones the cell soma is close to the inner segment, but in rods a slender fiber joins the inner segment to the soma. In both rods and cones an axon arises from the soma and proceeds to the axon terminal. The axon terminals of the receptors exhibit expansions that show typical differences between rods and cones. The expansion is roughly spherical in rods and is termed a rod spherule; in cones it spreads more widely and is called a cone pedicle.

The second-order neurons, in the direct line of transmission, are bipolar cells. These have been classified into several types, as illustrated in Fig. 15-1, based on detailed differences. Most bipolar cells make dendritic synapses with receptors, while their axons make synaptic contact with ganglion cells. The ganglion

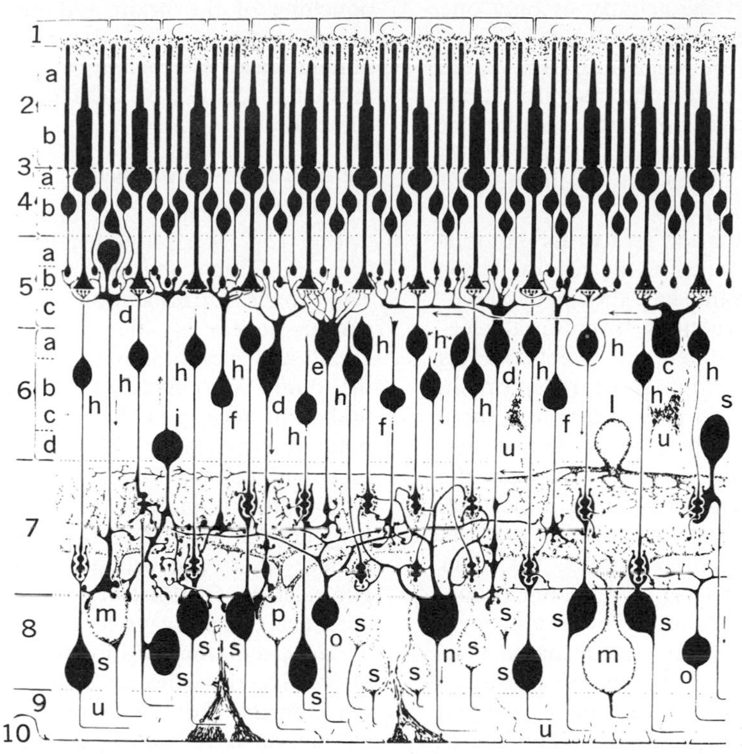

Fig. 15-1. Reconstruction of primate retina showing major types of neurons and their synaptic relationships, as revealed by Golgi impregnations. Layers are as follows: *1,* pigment epithelium; *2a,* outer segments of rods and cones; *2b,* inner segments of rods and cones; *3,* external limiting membrane; *4,* outer nuclear layer containing cell bodies of cones, *4a,* and rods, *4b; 5,* outer plexiform layer; *6,* inner nuclear layer; *7,* inner plexiform layer; *8,* ganglion cell bodies; *9,* optic nerve fibers; *10,* internal limiting membrane. Cell types are labeled as follows: *c,* horizontal cells; *d, e, f, h,* various types of bipolar cells; *i, l,* amacrine cells; *m, n, o, p, s,* various types of ganglion cells; *u,* radial glial cells of Müller. (From Polyak.[10])

cells are therefore third-order neurons; their axons lie close to the retinal surface in the optic fiber layer. These axons proceed directly to the optic disc, where they leave the eye and course through the optic nerve and tract to terminate upon fourth-order neurons of the lateral geniculate body. The axons of these fourth-order cells complete the line of transmission to the visual cortex.

The fine structure of retinal synapses has been studied by electron microscopy[7] but will not be considered in this chapter because its functional significance remains to be clarified by further work in retinal neurophysiology. It is evident from Fig. 15-1 that the most direct line of information flow through the retina includes a minimum of three cells. It is also evident that all information in the retinal image that enters into visual sensations must be transmitted to the brain by coded impulse discharges in the ganglion cell axons. If only a single aspect of light stimuli, e.g., brightness, were perceived, the problem of coding in the optic nerve fibers would be relatively simple. But there are many aspects of visual stimuli that can be discriminated; the more important of these are brightness, color, size and shape, and location in a three-dimensional space. Thus a major problem is to understand what the code is and how the code is established for each type of information.

Retinal histology has revealed a number of points that seem fundamental to an eventual solution of this problem. First, most photoreceptors of the peripheral retina do not possess an isolated line of transmission to the ganglion cell. As shown in Fig. 15-1, most bipolar cells receive activity from a number of receptors, including both rods and cones; also a single ganglion cell receives activity from a number of bipolar cells. Hence there is considerable convergence of pathways from receptors to ganglion cells. If a large number of receptors affected the activity of only one ganglion cell, visual acuity would be very poor. Thus it is not surprising that there is much overlap between the receptor areas that affect the discharge patterns of adjacent ganglion cells. In other words, the excitation of a given receptor affects the discharge patterns of many ganglion cells.

The anatomic situation is further complicated by lateral connections at several levels of the retina. Sjöstrand has shown by three-dimensional reconstructions from serial electron micrographs that there are direct contacts between the axon terminals of receptors in the guinea pig retina.[99] This means that there are interreceptor synapses prior to the first synapse in the direct pathway through the retina. These findings seem quite important, but their functional significance is not yet clear. In addition to bipolar cells, the inner nuclear layer contains cells with long, transversely oriented processes. There are horizontal cells that have rather widespread contacts with the axon terminals of rods and cones; also amacrine cells make contact with the dendritic processes of a number of ganglion cells. These lateral connections confer an anatomic complexity on the retina rivaling that of certain parts of the brain. This is probably related to the fact that the retina is formed embryologically as an outpouching of the brain and may be regarded as a sample of brain tissue. The retina is a unique case of receptors and brain tissue intimately connected at a peripheral level of the nervous system.

Retinal layers and their circulatory supply

Fig. 15-2 is a photomicrograph that shows some additional features of retinal structure. The precise stratification of the retina into various layers is illustrated, and the major layers and membranes are labeled. Note that the outer nuclear layer, containing the cell bodies of receptors, has more cells than the inner nuclear layer. This occurs despite the fact that the inner nuclear layer contains cell types other than bipolars, illustrating a convergence of pathways from receptors to bipolars. There are still fewer cells in the ganglion cell layer, illustrating an additional convergence of pathways to the ganglion cells.

The circulatory supply of the retina has two main divisions, the so-called retinal circulation and the choroidal circulation. The retinal circulation is formed from the central retinal artery, which enters the eye at the optic disc. This artery branches into smaller arteries that course over the retinal surface; the collecting veins return over the retinal surface, and all join to form the central retinal vein

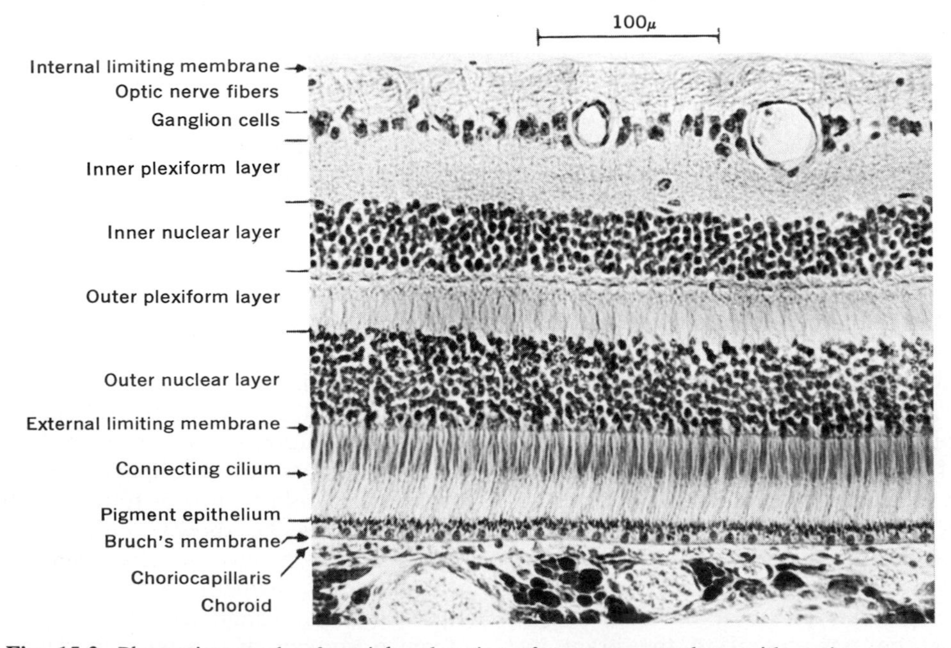

Fig. 15-2. Photomicrograph of peripheral retina of macaque monkey, with major structures and layers labeled. Histologic preparation as described by Brown and Tasaki,[31] stained with hematoxylin-eosin. (From Brown.[2])

that leaves the eye at the optic disc. The major blood vessels of the retinal circulation tend to run in artery-vein pairs; one such pair is shown in Fig. 15-2. The capillaries of the retinal circulation occur in strata at several retinal levels, and the deepest capillary plexus occurs with consistent precision at the outer margin of the inner nuclear layer.[11] Major vessels of the choroidal circulation penetrate the sclera to one side of the optic disc and form a rich system of large vessels in the choroid[11]; some of these may be seen in Figs. 15-2 and 15-3. An exceedingly rich capillary plexus, called the choriocapillaris, is derived from the choroidal circulation. As shown in Fig. 15-2, this lies immediately against the basement membrane of the pigment epithelium, which is called Bruch's membrane. Note that the entire deeper half of the retina, from Bruch's membrane to the synapses between photoreceptors and second-order neurons, is entirely devoid of blood vessels. Thus there are no capillaries in direct contact with any portion of the photoreceptors.

Because of the dual circulatory supply of the retina, it long seemed likely that cells of the ganglionic and inner nuclear layers are served primarily by the retinal circulation, whereas the receptors are supported partly from the retinal and partly from the choroidal circulation. This has been confirmed by studies in which needles were inserted into the monkey eye and used as channels for recording microelectrodes and other devices.[3, 32] A steel rod was introduced through one needle and its rounded end was pressed on the optic disc to occlude the retinal circulation without occluding the choroidal circulation. Within a few minutes after this procedure the electrical responses of ganglion cells and most cells of the inner nuclear layer were abolished, but receptor potentials remained for many hours. Thus the photoreceptors derive an important portion of their metabolic requirements from the choriocapillaris.

Retinal supporting tissue

The retina is mechanically delicate and bound together mainly by a type of glial cell called the Müller cell. Portions of these cells are shown in Fig. 15-1; they are perpendicular to the retinal surface and send processes in both directions from their somata in the inner nuclear layer. The processes directed toward the vitreous humor form expansions when they reach the inner surface of the retina; these fuse together to form the internal limiting membrane. The processes passing the other direction terminate at the bases of receptor inner segments. There they form the external limiting membrane, which is not a true membrane at all but an aggregation of the distal tips of Müller cells. The distal processes of Müller cells consistently fill the intercellular spaces between receptors[42, 99] and may serve to insulate receptors from each other from the base of the inner segment to the axon terminal. The Müller cells serve more clearly as connective tissue that binds the retina together over the greater part of its thickness. There is no true connection between receptors and pigment cells, but the pigment cells have long microvilli on their inner surface that interdigitate with the outer segments of receptors, thus forming a connection of sorts. Normally the retina is held in close contact with the pigment epithelium by intraocular pressure. In cases of retinal detachment, caused by severe blows or other pathologic effects, the detachment usually occurs between the pigment epithelium and the outer segments of the photoreceptors. The severe visual consequences, in the portion of the visual field served by the detached retinal area, are readily understood because of the metabolic dependence of the photoreceptors upon the choriocapillaris.

Optical effects of retinal tissue

Light must pass through the entire retina before it is absorbed by photopigment in the receptor outer segments. This probably limits visual acuity in the peripheral retina. Such a thin sheet of nervous tissue is almost transparent, but its optical effects cannot be neglected, and the stimulating light must also pass through the retinal blood vessels.

Light not absorbed in receptor outer segments passes on to the pigment epithelium or choroid. If this light were reflected back, it would degrade the retinal image. However, this light is largely absorbed by melanin, a black pigment contained in granules in the pigment epithelium and choroid, as shown in Figs. 15-2 and 15-3. By greatly reducing reflected light, the optical quality of the image at the photoreceptors is thus improved.

Anatomic specialization of the fovea

Fig. 15-3 shows a histologic section through the center of the fovea, a small and highly specialized area of the retina. The fovea is

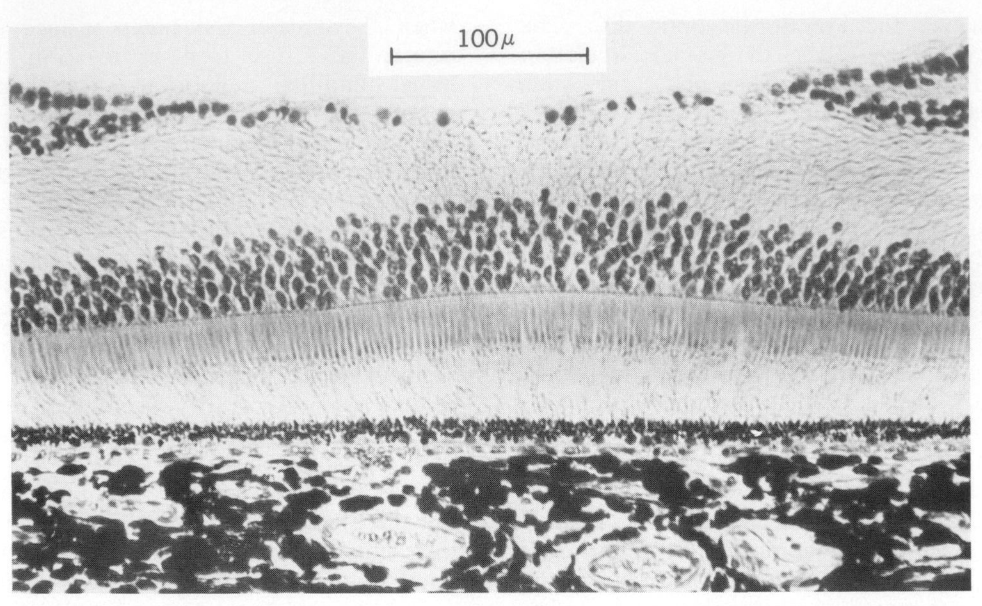

Fig. 15-3. Photomicrograph of central fovea of macaque monkey. Histologic methods similar to those used for Fig. 15-2. Very thin outer segments of foveal cones are poorly preserved. (From Brown et al.[3])

the part of the retina that is used whenever we look directly at an object; it occupies the subjective center of the visual field. This is the retinal area of highest visual acuity, and the fovea has a number of anatomic specializations that favor visual acuity. Note that the pigmentation of the pigment epithelium is stronger in the fovea than in the peripheral retina; reflected light is thus especially well suppressed in the fovea. The central fovea contains only cones, being entirely free of rods. The outer and inner segments of foveal cones are rather elongated and cylindric by comparison with peripheral cones. Hence foveal receptors look superficially like rods but may be recognized as cones by the shape of the axon terminal, the cone pedicle. The foveal cones are very slender; this increases the fineness of the "grain" of the foveal receptor surface, thus enhancing visual acuity. In the very center of the fovea the receptors are especially slender, having a diameter in humans of 1 to 1.5μ.[10] The increase of receptor density in the central fovea results in a thickening of the outer nuclear layer, as shown in Fig. 15-3. The length of the inner segment does not increase in the central fovea, but in humans the length of the outer segment approximately doubles.[10] Since the outer segment contains the photopigment, this increased length probably increases the photopigment content of these receptors. Thus the slenderness of receptors in the central fovea favors visual acuity, and the increased length of the outer segment reduces the loss of sensitivity that would otherwise occur.

In the central fovea the ganglionic and inner nuclear layers are almost completely absent, and there are no blood vessels of the retinal circulation. Thus retinal impediments in the path of light to the photopigment are greatly reduced in this area. The axons of foveal photoreceptors turn laterally and make synaptic contact with bipolar cells in the parafoveal region. In Fig. 15-3 the inner nuclear and ganglionic layers may be seen forming on either side of the foveal center. Thus the second- and third-order neurons that serve foveal receptors are swept outward to the parafoveal region, where strong development of these layers creates a parafoveal ridge that encircles the fovea.

Convergent pathways from receptor to ganglion cell, as found in the peripheral retina, probably limit visual acuity. In the case of foveal cones, Polyak[10, 11] describes an essentially "private line" pathway from each receptor to its ganglion cell. The interneuron in this pathway is the midget bipolar, which is the typical type of bipolar in the parafovea. In this region each midget bipolar contacts a single cone pedicle and carries activity to only one midget ganglion cell. This is the typical type of ganglion cell in the parafovea, and it makes synaptic contact with only the one midget bipolar cell. The axon of each midget ganglion cell therefore appears to carry impulse messages that represent the activity of a single cone. This is probably an important specialization of the fovea for attainment of high visual acuity.

Duplicity theory

In 1866 the anatomist Max Schultze[98] first advanced the theory that rods and cones serve different retinal functions. His view has become known as the duplicity theory, which states that in the human retina the cones function at the high intensities of daylight vision, conferring advantages of high visual acuity and color vision, while the rods have the greater sensitivity required for night vision but do not mediate color and cannot resolve fine details. His theory was developed largely by comparing receptor structure with known visual abilities in many vertebrate species. Thus he noted that the retinas of diurnal birds (e.g., the falcon) are rich in cones, while the retinas of nocturnal species (e.g., the owl and bat) are rich in rods. He also compared anatomic and functional differences between the fovea and periphery of the human retina. The duplicity theory is now so well supported and so universally accepted that it is more doctrine than theory, and it is probably the most useful single principle in visual physiology. In view of contemporary evidence, however, the principle applies better to the *rod system* and the *cone system,* rather than to the rods and cones alone. This is because some of the functional differences between rod and cone systems are not due entirely to differences between rods and cones themselves but result partly from synaptic connections in the rod and cone pathways. The division of the retina into rod and cone receptor systems is of such profound importance that it affects all major retinal functions, as shown in the following section.

ABILITIES OF THE VISUAL SYSTEM

Resolution of stimuli in space: visual acuity

Detection in the visual field of a border, or an abrupt change in the spatial distribution of stimulus intensity, is a fundamental ability of the eye. Under certain conditions the eye can discriminate very fine details in the stimulus pattern. This ability is referred to as visual acuity, which has already been discussed from certain points of view and which

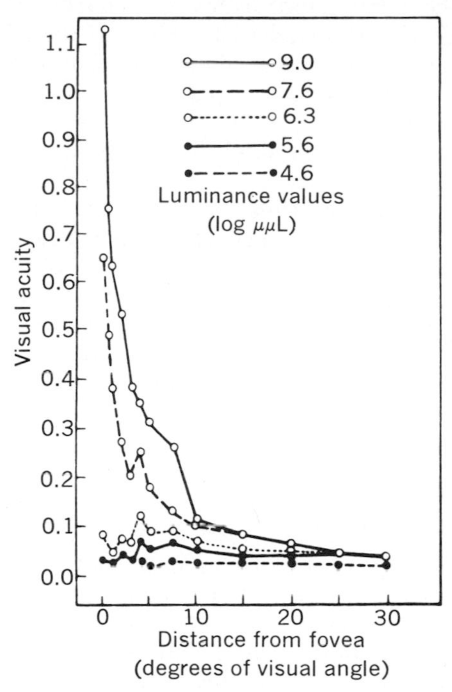

Fig. 15-4. Visual acuity as a function of distance from fovea at five different luminance levels of test stimulus. Test stimulus was the gap in a ring known as a Landolt C, and visual acuity is expressed in arbitrary units on a relative scale. Measurements were made along horizontal meridian of retina on temporal side. For definition of the unit of luminance, the lambert (L), see *The Science of Color.*[4, p. 231] (From Mandelbaum and Sloan.[78])

may be measured by determining the fineness of stimulus detail that can be seen. Thus measurements of visual acuity indicate the resolving power of the eye in two-dimensional space. Stereoscopic acuity, by comparison, is a measure of ability to discriminate differences in the distance of stimuli from the eye. Since depth perception depends partly on integration of signals from the two eyes, it will not be discussed in this chapter.

The manner in which visual acuity varies with brightness of the stimulus and its distance from the fovea is shown in Fig. 15-4. At the lowest brightness level the stimulus could not be seen in the fovea. It was first detected when presented at about 4 degrees from the fovea, and there was no distinct change of visual acuity as the stimulus was moved further toward the periphery. This illustrates foveal blindness at low levels of illumination; astronomers have long known that dim stars, which cannot be seen in direct vision, can often be seen by looking slightly to one side. At the highest brightness level, on the other hand, acuity was maximum in the fovea and decreased rapidly as the stimulus was displaced to about 10 degrees from the fovea; beyond that point, the decrease of acuity was more gradual. Thus Fig. 15-4 illustrates how the duplicity theory applies to visual acuity.

Note in Fig. 15-4 that although stimulus intensity has a marked effect upon acuity in the pure-cone fovea, it has very little effect in the predominantly rod area at the extreme periphery of the retina. The improvement of visual acuity with stimulus intensity, which occurs most strongly in the fovea, is probably due to a finer "grain" in the *cone response pattern* that results from increased intensity. Let us assume for simplicity that there is no background illumination. For any given receptor in an illuminated stimulus area, the probability of absorbing light and reacting will increase with stimulus intensity. Thus an increased stimulus intensity will increase the proportion of receptors that respond within the stimulus area. As this occurs, the "grain" of the receptor response pattern becomes finer and more closely represents the details of the stimulus itself. Of course, this increased detail in the cone response pattern could not be effective unless each cone possessed a pathway to the brain that approximated to a private line. Such a pathway also seems necessary to account for the fact that under light-adapted conditions, acuity increases as the fovea is approached, in a manner that corresponds roughly with increased cone density.[10] Convergent pathways from the rods, on the other hand, are probably responsible for the acuity of rod vision being relatively fixed, at a lower level than that of cones, with little improvement resulting from either increased stimulus intensity or increased rod density. Although the rod density increases toward the periphery and reaches a maximum at about 20 degrees from the human fovea,[78] the lowest curve of Fig. 15-4 shows that there are no corresponding changes of visual acuity. Although detailed pathways from receptor to brain are not yet worked out, it seems established that there is much less convergence for peripheral cones than for rods and that convergence is minimal in the case of foveal cones. Cell counts have shown that in the predominantly rod retina of the guinea pig there are 220,000 rods, 50,000 cones, and 7,000 ganglion cells; in a pure-cone squirrel retina there are 200,000 cones, no rods, and 90,000 ganglion cells.[110] Thus the ratio of rods to ganglion cells in the guinea pig has a minimum value of about 31:1, but the ratio of cones to ganglion cells in the squirrel retina is only slightly greater than 2:1. It appears that the functional duplicity of the human retina, as it relates to visual acuity, results largely from this basis. This illustrates why the duplicity theory is best applied to rod and cone systems rather than to rods and cones alone.

Resolution of stimuli in time: critical fusion frequency

The eye can also resolve stimuli that are separated in time. The limit of this ability is indicated by measuring the minimum frequency at which repetitive stimuli appear to fuse together into a continuous stimulus. This minimum frequency of subjective fusion is called the critical fusion frequency (CFF). As visual acuity is a measure of the spatial-resolving power of the eye, CFF is an analogous measure of the time-resolving power of the eye. The resolution of stimuli in time is limited because the response to a given stimulus does not cease exactly when the stimulus ceases but persists for a time thereafter. The phenomenon of subjective fusion has useful applications; e.g., it provides the physiologic basis for motion pictures. The visual continuity of "movies" is attained by flashing successive frames of the film at a frequency in excess of CFF; motion pictures were called "flickers" during the period when the rate at which the frames were flashed was not sufficiently high.

Fig. 15-5 shows CFF as a function of both retinal illumination (intensity of the flickering stimulus) and retinal location. In the

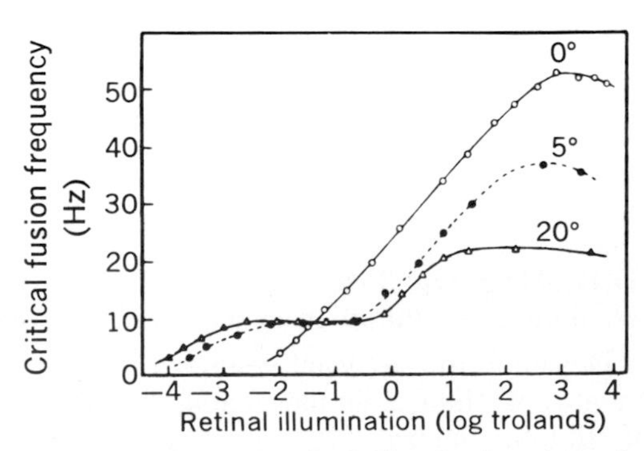

Fig. 15-5. Effect of retinal illumination (stimulus intensity) on critical fusion frequency in fovea (0 degree) and at 5 and 20 degrees above fovea. For definition of the unit of retinal illumination, the troland, see *The Science of Color*.[4, p. 232] (From Hecht and Verrijp.[61])

fovea, CFF rises rapidly with stimulus intensity and then falls slightly after attaining a maximum value. The foveal CFF can exceed 50 Hz, so the time-resolving ability of the eye is well developed. In the curves at both 5 and 20 degrees from the fovea (Fig. 15-5), CFF first increases with stimulus intensity and reaches a plateau, which is followed by a second and more rapid rise of the curve to a higher level. In the fovea the first limb of these curves is not found, but the second limb is present and maximally developed. This indicates that in the peripheral retina the first limb of each curve represents rod function, while the second and higher limb represents cone function. This interpretation is supported by many lines of evidence. Hence the rod and cone receptor systems also affect the time-resolving power of the visual system. The fact that cone CFF can go much higher than that of rods appears to be explained by the different time courses of cone and rod receptor potentials, as shown later in this chapter.

Spectral sensitivity

The sensitivity of the eye to a flash of light varies with the wavelength of the stimulus flash. If the wavelength of the light is varied, and the relative energy required to produce a sensation of constant brightness is determined, the reciprocals of these energies will give a curve called a *luminosity function.* Such a function shows the relative effectiveness (luminosity) of various wavelengths of light, and hence the spectral sensitivity of the eye. The scotopic and photopic luminosity functions are obtained under dark-adapted and light-adapted conditions, respectively. Such functions are shown in Fig. 15-6. Note that these are relative functions in which the maximum luminosity of each curve has been set equal to one; this permits convenient comparison of the two curves. If expressed on an absolute basis, luminosity is greater in the scotopic case.

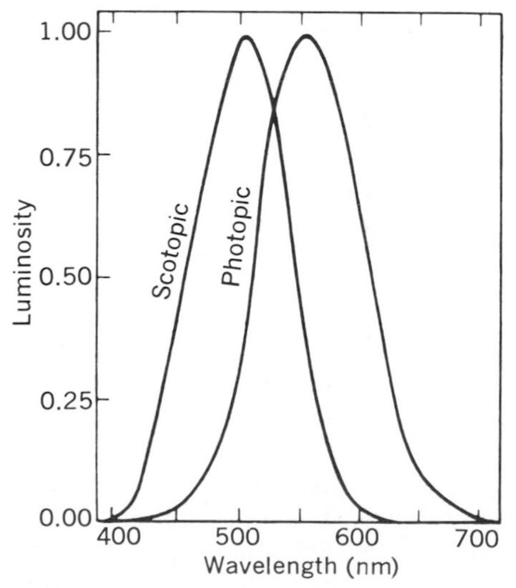

Fig. 15-6. Relative scotopic and photopic luminosity functions of normal human eyes. These functions were adopted as standards by International Commission on Illumination in 1951 (scotopic) and 1924 (photopic). (From *The Science of Color.*[4])

Fig. 15-6 shows that the eye responds best to a limited band of wavelengths called the *visible spectrum,* extending from slightly below 400 nm to somewhat above 700 nm. This places its sensitivity in the wavelength range between ultraviolet and infrared radiation. Wavelengths shorter than 400 nm are normally absorbed by the lens of the eye.[111] In the lensless *(aphakic)* eye the visible spectrum extends downward to at least 315 nm, and the ultraviolet light in this extended spectrum is not injurious to the retina.[111] Wavelengths shorter than 315 nm tend to be injurious to tissue; they are absorbed by the cornea and do not reach the retina.[111] The upper limit of the visible spectrum is determined by the sensitivity of the retina itself, and very low levels of sensitivity to infrared radiation extend to approximately 1,000 to 1,050 nm.[54]

Fig. 15-6 shows that the photopic luminosity function is shifted toward longer wavelengths than the scotopic function. This shift of spectral sensitivity is called the *Purkinje shift* because it was first recognized by Johannes Purkinje in 1825. Maximum sensitivity of the scotopic function is at 507 nm, while maximum sensitivity of the photopic function is at 555 nm. The scotopic curve has been shown to represent rod function, while the photopic curve depends on the cones, so the Purkinje shift is another aspect of the duplicity theory.

Rod photopigment. The form of the scotopic luminosity function has been explained by studying the visual photopigment in the outer segment of human rods. This has been done best by macerating a human retina, a process that fragments the outer segments. Particles of the rod outer segments were separated from the retina by centrifugation. The absorption spectrum of the suspension of rod particles was then determined and plotted as one of the curves in Fig. 15-7. For comparison, scotopic sensitivity was determined

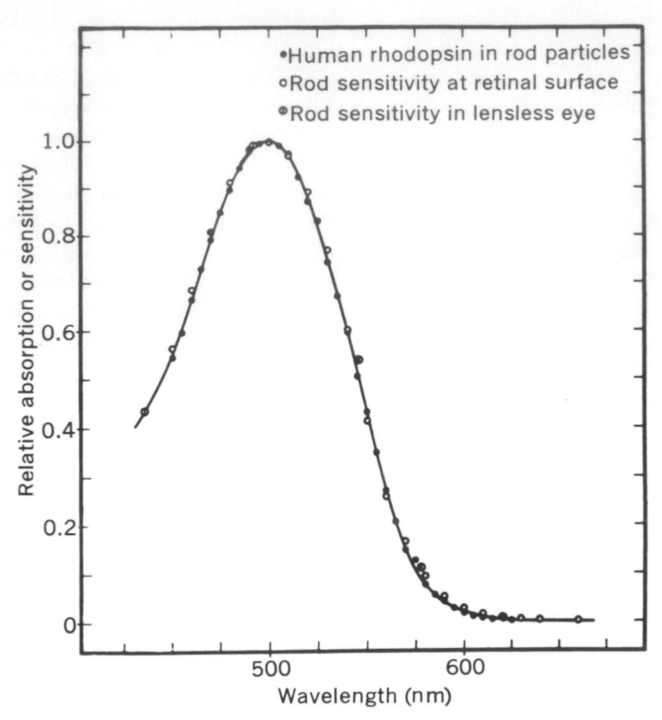

Fig. 15-7. Relative absorption spectrum of human rhodopsin, as measured in rod particles, compared with human rod sensitivity determined by two methods. See text for explanation. (From Wald and Brown.[113])

by two separate methods. In one method the scotopic luminosity function was determined and corrected for light absorption in the ocular media in order to represent sensitivity to light arriving at the retina. In the other method the scotopic luminosity function was determined in an aphakic eye; the principal colored structure of the ocular media, the yellow lens, had been removed in a cataract operation. Because of these corrections for light absorption in the ocular media, maximum scotopic sensitivity in Fig. 15-7 is at 500 nm instead of 507 nm. Furthermore, rod sensitivity is expressed as reciprocals of the number of light quanta at each wavelength, rather than as reciprocals of light energy; this is required for accurate comparison of sensitivity data with absorption data. A single curve fits all three sets of measurements in Fig. 15-7, thus establishing that the scotopic luminosity curve results from the absorption spectrum of the photopigment in rod outer segments.

If rod outer segments are treated with digitonin, the photopigment may be extracted and studied in solution. The photopigment from human rods is called *rhodopsin;* it is similar to the rod pigment of other mammals. The results given in Fig. 15-7 show that equal numbers of quanta absorbed by rhodopsin are equally effective throughout the visible spectrum in initiating excitation. When passing from maximum sensitivity at 500 nm to longer or shorter wavelengths, the sensitivity falls to half at a wavelength at which twice the number of quanta are required to be incident upon the retina. However, Fig. 15-7 shows that relative absorption of these quanta by the rhodopsin correspondingly falls to half, so that the number of quanta absorbed is the same as at maximum sensitivity. Thus equal numbers of absorbed quanta produce a constant brightness, since constant brightness is the criterion in determining a luminosity curve. This means that a quantum of light *absorbed* by rhodopsin is equally effective at any wavelength in the visible spectrum, although the energy content of a single quantum is inversely proportional to wavelength. Fig. 15-7 also shows that all quanta are not absorbed with equal ease, the relative sensitivity of the dark-adapted retina to different wavelengths of light being governed by the relative effectiveness of rhodopsin in trapping quanta at different wavelengths.

Cone photopigments and color vision. The study of cone photopigments has always presented greater difficulties than the study

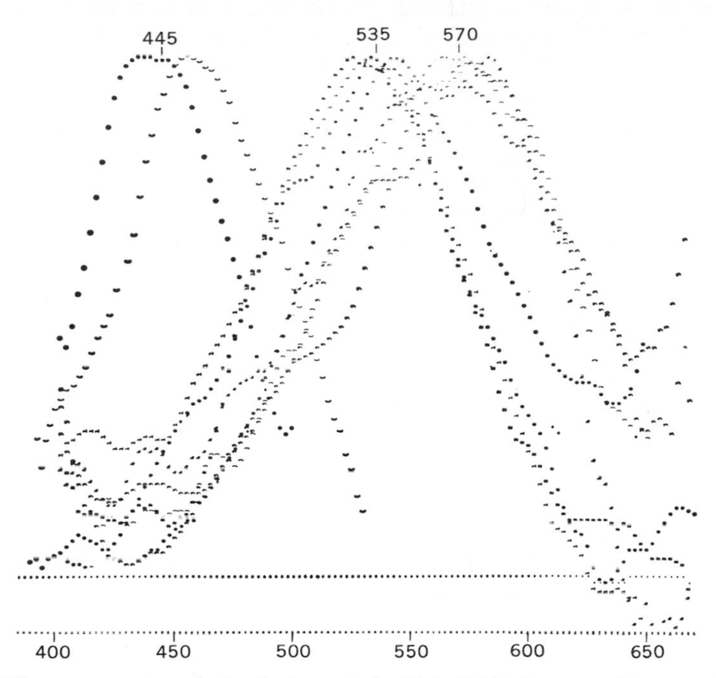

Fig. 15-8. Difference spectra of visual pigments in 10 individual cones of macaque and human retinas. Curves recorded from macaque monkey cones are represented by numbers, while those from human cones are shown by open parentheses. Maximum absorption of all curves has been adjusted to same level. (From Marks et al.[80])

of rods, partly because cone pigments are difficult to extract, and partly because color vision requires the wavelength of the stimulating light to be coded in the receptor response. The Young-Helmholtz theory of color vision assumed three different photopigments with absorption maxima in the blue, green, and red portions of the spectrum, respectively. In subsequent work it was generally assumed that each of these photopigments is present in only one of three major classes of cones. This theory was long supported by indirect evidence, but final confirmation required the determination of absorption spectra on the pigments contained in single cones.

The necessary technique of microspectrophotometry was first developed in Japan in 1957 by Hanaoka and Fujimoto.[58] Intact outer segments were obtained from frog or carp retinas, placed on a microscope slide, and spectrophotometry was performed only upon the beam of light passing through a single outer segment. The effects of pigments other than photopigments were eliminated by taking difference spectra. In this procedure an absorption spectrum is obtained, a strong light is then used to partially or wholly bleach the photopigment, and another absorption spectrum is obtained. The difference between the two absorption spectra is the difference spectrum. It excludes the effects of any photostable pigments in the measuring beam and therefore represents the absorption characteristics only of the photopigment bleached by light. Outer segments of frog rods were found to have difference spectra resembling that of rhodopsin, with maximum absorption at around 500 nm. Cone outer segments from the carp retina gave difference spectra that fell into several groups, each with its characteristic absorption maximum. Studies by Marks,[79] who examined 113 cone outer segments of the goldfish retina, showed that the difference spectra fell into three separate groups with absorption maxima that were respectively in the blue, green, and red portions of the spectrum. This confirms the Young-Helmholtz theory for the goldfish retina and gives background data from a species in which many cones may be examined. Similar experiments were then reported on human and monkey retinas.[14, 39, 80] In these studies the retinas were removed and placed on a slide with the receptors facing upward. Light was passed axially through the receptors from below, in the normal direction, and spectrophotometry was performed on the light passing through single cones of the parafovea where the cones are especially large.

Although the number of receptors studied was small because of technical difficulties, the results indicate that the distribution of photopigments in single cones of human and monkey retinas is similar to that in goldfish. Fig. 15-8 shows the results of one of these studies in which the difference spectra of the pigments in single cones fell into three distinct groups. The peak absorptions of the three groups were at approximately 445, 535, and 570 nm. The first two of these are called blue and green pigments, which are the colors of the wavelengths maximally absorbed. The third pigment absorbs best in the yellow part of the spectrum, but it is called the red pigment because of historical precedent and the fact that it absorbs red light better than any of the other pigments. Note that there is considerable overlap between the difference spectra of the three cone classes. Thus a given wavelength will stimulate at least two, and in some cases all three, types of cones. This work reveals a response code for color at the receptor level, since each wavelength of light will elicit a unique ratio of responses among the three fundamental cone classes. By contrast, single rods of the human parafovea all appear to contain the same visual pigment, rhodopsin.[39]

Since the scotopic luminosity function is determined by the absorption spectrum of rhodopsin in rods, the photopic luminosity function may be expected to result from cone photopigments. There are at least three different cone pigments, and the simplest possibility is that the summed absorption spectrum of all the cone pigments will fit the photopic luminosity function. The relative amounts of the three cone pigments are critical in determining the summed absorption spectrum, so results are best obtained on a sample of retina containing the normal proportions. This has been done by applying the technique of microspectrophotometry to the human retina, using a light beam 0.2 mm in diameter that is confined to the pure-cone portion of the fovea.[38] The difference spectrum of a representative sample of foveal cone pigments was obtained in this way. The foveal photopic luminosity function was then corrected for absorption by the ocular media and yellow pigment of the macula in order to represent light sensitivity at the cone outer segments, and expressed on a quantum basis. The resulting pair of curves matched satisfactorily. Thus the photopic luminosity curve represents cone function, and its shape appears determined by the relative absorption of different wavelengths of light by the combined effects of all the cone pigments.

The Young-Helmholtz theory has long provided the most accepted basis for explaining many of the facts of color mixing. The dominant *hues* (color sensations) resulting from stimulation by narrow-wavelength bands of the visible spectrum are four in number: blue, green, yellow, and red. But it is well established that any color sensation that can be experienced may be elicited by an appropriate mixture of three pure primary colors from the blue, green, and red portions of the spectrum. Thus the sensation of yellow may be produced by mixing pure red and green stimuli; and the sensation of purple, which is not elicited by any pure wavelength, may be produced by mixing red and blue. These empirical rules of color mixing hold only when the colors are additive, as when mixing pure spectral bands of light. By comparison, a paint gains color because its pigment absorbs all light except that which is reflected; hence the mixing of paints is referred to as subtractive color mixture. The empirical *rules* of color mixing are different in the two cases, but the *principles* are the same if one considers only the light that reaches the observer's eye. If all three spectral primaries are mixed in appropriate proportions, cancellation of color can be achieved. Variation of stimulus intensity will then produce the entire achromatic gray scale. These are some of the major rules of color mixing; they may be explained by assuming that all color sensations (plus achromatic sensations) result from an appropriate ratio of stimulation of three kinds of receptors, each responding strongly to one of the primary colors. Although the receptor code for color now seems established, the photoreceptors are only the first stage in our understanding of color vision. There is evidence that the response code for color changes markedly at later stages of the visual system and that the later type of response code influences color sensations. Hence these later stages are likewise important for understanding color vision and will be taken up in connection with single cell responses.

Color blindness. The three cone pigments also provide a basis for understanding most aspects of color blindness.[14] Normal color vision is said to be *trichromatic,* since all of the hues seen by a normal individual may be produced by using three primary colors.

Trichromatic color vision also includes *protanomalous* and *deuteranomalous* types of color defects. If red and green primary colors are mixed to match a spectral yellow, the protanomalous subject will require more red than normal, while the deuteranomalous subject will require more green than normal. Thus defects in the cone pigment systems are not all or none but can occur in varying degrees. Dichromatic vision represents a more severe color defect, in which all of the hues that can be seen may be produced by mixing only two primary colors. This classification includes three subcategories, *protanopia, deuteranopia,* and *tritanopia,* depending on whether the nonfunctional pigment system is the first (red), second (green), or third (blue). The *monochromatic* subject may be assumed to be totally color blind, seeing the world as it appears in black and white photographs, since the sensation produced by any wavelength of light may be matched by varying the intensity of only one primary color. The typical totally color blind subject is a *rod monochromat,* who possesses little or no cone function.

As shown in Fig. 15-8, either red or green spectral bands will stimulate both the red and green pigment systems. Thus the sensations of both red and green may be expected to depend on a particular ratio of response in the two types of cones, and loss of either pigment system should cause color deficits in both red and green portions of the spectrum. Correspondingly, both the protanope and deuteranope confuse red and green and are said to be *red-green* blind. The color confusions of color-blind subjects are the basis for all of the common tests of color vision.

Although color blindness may result from acquired retinal disorders, including vitamin A deficiency,[114] most cases are inherited and cannot be corrected. Red-green blindness is inherited as a simple sex-linked recessive trait. It is passed from a man to all of his daughters, who are color-normal carriers if the man's wife is not also a carrier, and it reappears in half of his daughters' sons. Some degree of this type of defect is estimated to be present in about 8% of men and less than 0.5% of women.[15] Tritanopia is much more rare, and total color blindness is likewise very rare.

Absolute sensitivity: dark adaptation

The sensitivity of the eye to a flash of light cannot be represented by any fixed value. It changes dramatically with light and dark adaptation, the effects of which are known from common experience. When passing from bright light into a darkened place, e.g., a movie theater, it is sometimes difficult at first even to see which seats are empty, but after a few minutes the surroundings can be seen quite well. On leaving the theater one is dazzled by the bright light, but the eyes quickly become less sensitive and the discomfort disappears. The increased sensitivity of the eye in the dark is called dark adaptation, and its converse is light adaptation.

Fig. 15-9 shows dark-adaptation functions, which represent absolute threshold as a function of time in the dark. The *absolute visual threshold* is the minimum intensity of light that evokes a light sensation in the absence of background illumination. It is absolute only in the sense of being the threshold in complete darkness, and it falls during dark adaptation. It may be contrasted with the *incremental threshold,* the stimulus intensity that must be added to a background illumination in order to be seen against that background, a subject dealt with later in this chapter. In Fig. 15-9 the results in the normal

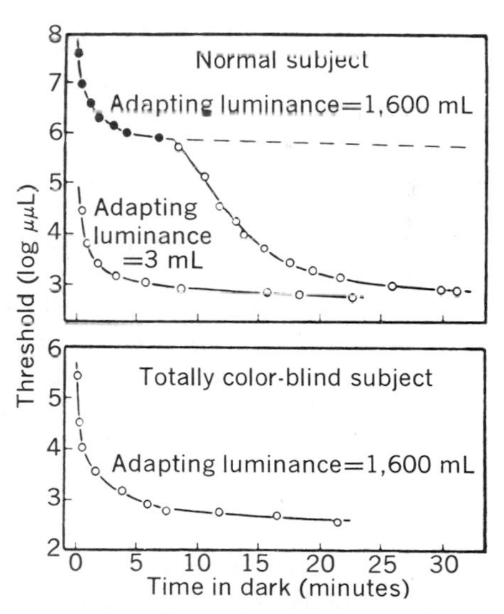

Fig. 15-9. Absolute thresholds as a function of time in the dark. Results are shown from normal subject after both high and low levels of light adaptation, and from totally color-blind subject following high level of light adaptation. Thresholds were measured with circular test field subtending 3 degrees of visual angle, presented 7 degrees from fovea on temporal side of retina. In results from normal subject, dashed line was not part of original experiment; it is added here to show approximate time course of later part of dark adaptation when only cones are stimulated. (From Hecht et al.[63])

subject, following strong light adaptation, illustrate the classic dark-adaptation function for the peripheral retina. This curve has two separate limbs. The first limb shows a rapid drop in threshold to an approximate plateau. The onset of the second limb occurs after about 8 min in the dark and may be seen as a second rapid decrease in threshold. Dark adaptation is almost complete after 30 min, but the threshold continues to fall slowly for more than 1 hr.

It has been demonstrated by several methods that the first limb of this classic dark-adaptation curve represents cone function, whereas the second limb represents rod function The test stimulus in the experiment of Fig. 15-9 was violet in color. The solid black dots represent thresholds at which this violet color was seen, and this occurred only during the first limb of the curve The open circles indicate thresholds at which the stimulus was colorless, and this was the case during the second limb of the curve. Pure-cone dark-adaptation curves have been obtained primarily by two methods. In one case the dark-adaptation function is measured by a small test stimulus confined to the fovea. This test stimulus may also consist only of red light, which is shown in Fig. 15-6 to activate cones and have minimal effects upon rods. Dark-adaptation functions under these conditions show an initial portion that coincides with the cone limb of the curve illustrated; a second limb of the curve does not occur, and the later part of the foveal curve follows the approximate time course shown by the dashed line in Fig. 15-9.[60] The second method employs subjects with severe congenital cases of night blindness *(nyctalopia),* in whom rod function appears entirely absent. When a dark-adaptation function is obtained in any portion of the retinas of these subjects, the results are similar to those obtained from the foveas of normal subjects.[60]

The second limb of the classic dark-adaptation curve has also been isolated by two major methods. When the light-adapting luminance (Fig. 15-9) was decreased from 1,600 to 3 millilamberts (mL), the dark-adaptation curve of the normal subject was quite different from that following the high adapting luminance. In this case the color of the test stimulus was not seen at any time during dark adaptation The threshold was already relatively low at the beginning of dark adaptation, and it dropped much more rapidly than following the high adapting luminance. Thus if light adaptation is not sufficiently strong to make cone function predominant, only the rod limb of the dark-adaptation curve is found in the periphery of the normal retina. The other method is to use subjects with rare total color blindness (rod monochromats) who show little or no evidence of cone function. The lower part of Fig. 15-9 shows results from the peripheral retina of such a subject following the same high level of light adaptation that was used for the normal subject. The dark-adaptation curve of the rod monochromat does not exhibit a cone limb, and sensitivity falls quickly to about the same level that was attained only after a considerably longer time in the normal subject.[63, 95] Thus dark-adaptation studies reveal another aspect of the duplicity theory. Whereas the cones are capable of a small amount of dark adaptation, the rods can dark adapt to a much greater extent.

Note in Fig. 15-9 that when exposure to strong light was followed by 30 min of dark adaptation, the sensitivity of the normal peripheral retina increased by about 5 log units. Sensitivity can increase by 1 additional log unit during the first 0.5 sec of dark adaptation, prior to the first measurement of Fig. 15-9.[18] Thus dark adaptation can increase retinal sensitivity by at least 6 log units (1 million times).

The combined results of Figs. 15-6 and 15-9 explain the fact that red light permits one to perform tasks involving high visual acuity while maintaining readiness for rapid dark adaptation. The red light stimulates primarily cones, which can engage in photopic tasks such as reading, and has little light-adapting effect on the rods. On entering the darkness, thresholds are already relatively low and the eye rapidly dark adapts. This is why red goggles are used in situations requiring readiness for dark-adapted requirements, e.g., by military night pilots on defensive alert.

The ability of the retina to alternate between cone and rod dominance of visual functions is at the heart of the duplicity theory. One may now ask why the cones are dominant in the light-adapted state and early dark adaptation, while the rods become dominant in the more fully dark-adapted state. It is also important to understand the processes that underlie the increased sensitivity of both cones and rods during dark adaptation.

A technique of reflection densitometry has been developed for measuring photopigment densities in the living human eye.[40] This technique has been used to study both cone and rod pigments, including bleaching of the pigments by a strong light and the time course of regeneration in the dark.[40, 96] The cone pigment in the fovea of the protanope regenerates more rapidly than rhodopsin.[96] Hence one reason why cone thresholds are lower than rod thresholds during the early part of dark adaptation is the more rapid regeneration of cone pigments. Also the cone system may exert a suppressive effect on the rod system at photopic intensities and during early dark adaptation. The totally color-blind subject is typically *photophobic,* or abnormally sensitive to bright light, and Fig. 15-9 shows that the early part of the dark-adaptation curve is much lower in a rod monochromat than in a normal subject.

As light-adapting intensity is increased, the break between the cone and rod limbs of the dark-adaptation curve occurs at a progressively later time during dark adaptation. But regardless of when the break occurs, it appears when rhodopsin is about 90% regenerated.[94] Hence rod thresholds become lower than cone thresholds when rhodopsin has regenerated to about 90% of its fully dark-adapted concentration. This suggests that rhodopsin concentration is critical in the changeover from cone to rod dominance of visual sensitivity during dark adaptation.

We may next ask why the fully dark-adapted sensitivity of the rod system is greater than that of the cone system. An important fact is that a single human rod can be excited by the absorption of a single quantum of light. This was first shown by Hecht et al.[62] from analysis of human threshold data, and this conclusion is now well supported. But activation of a single rod by an absorbed quantum of light does not result in a threshold sensation. This requires the activation of at least two, and perhaps as many as five to eight receptors, during the period of time within which temporal summation can occur.[19] The activation of a rod by absorption of a single quantum of light means that the mechanism of exciting a single rod has achieved the theoretical ultimate in sensitivity, being limited only by the physical unit of energy in which light is delivered.

Most dark-adaptation functions, such as those shown in Fig. 15-9, are obtained with relatively large stimulus spots. The threshold to a large stimulus spot will be determined not only by the threshold of each individual receptor, but by the extent to which area summation occurs. There is little or no convergence of cone pathways, so increasing the size of the stimulus spot will have little effect upon the cone threshold. But if the activities of many receptors converge upon a common site, as indicated by the anatomy of rod pathways, increased size of the stimulus spot will be quite effective in decreasing the rod threshold. If dark-adaptation functions are obtained with very small stimulus spots, the difference between final thresholds in the fovea and periphery is much reduced.[17] Thus when the test stimulus is fairly large, as in Fig. 15-9, another reason for the final rod threshold being lower than that of cones is that the rods can take greater advantage of the large stimulus area. Further information on mechanisms of dark adaptation will be given later in this chapter, in conjunction with techniques required to study this subject at various retinal levels.

PHOTORECEPTOR STRUCTURE

Inner and outer segments

The fine structure of the connecting zone between inner and outer segments is similar in cones and rods and is shown schematically in Fig. 15-10. The connection itself is a true cilium. In cross section it shows an outer membrane and nine pairs of peripheral fibrils; this is the usual structure of nonmotile cilia throughout the animal kingdom. The intracellular fibrils of the connecting cilium arise from a basal body (centriole) located at the outer end of the inner segment. These fibrils penetrate into the outer segment and traverse a portion of its length. De Robertis has shown that the entire outer segment is derived from a cilium. In morphogenetic studies of mice sacrificed at various times after birth, sequential stages were demonstrated in differentiation of the outer segment from a primitive cilium.[46] Hence the vertebrate photoreceptor is a ciliated cell in which the main portion of the cilium is highly differentiated and specialized, with the primitive ciliary characteristics maintained only at the connection between inner and outer segments.

Fig. 15-10 also shows that in longitudinal section the rod outer segment contains a tightly packed stack of discs. Each disc appears to be a closed membranous saccule with a flattened form similar to that of a red blood corpuscle. These saccules are formed in both rods and cones by infolding of the outer segment membrane, as shown schemati-

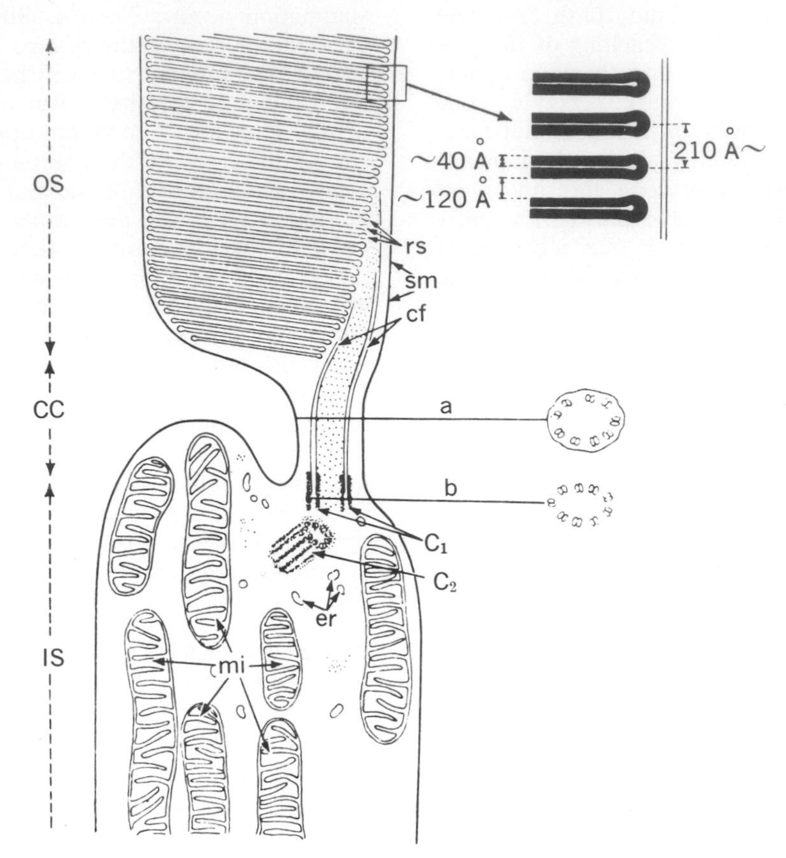

Fig. 15-10. Diagram of connecting zone between outer and inner segments of mammalian rod, based on electron microscopy. Main structures shown include base of outer segment, *OS;* connecting cilium, *CC;* and outer portion of inner segment, *IS.* C_1 is major centriole, or basal body, of connecting cilium; C_2 is secondary centriole. Other structures are labeled as follows: *cf,* ciliary filaments; *rs,* rod saccules; *sm,* surface membrane; *mi,* mitochondria; *er,* endoplasmic reticulum. At right are cross sections through, *a,* connecting cilium and, *b,* basal body. Portion of outer segment is also shown at greater magnification. (From De Robertis.[46])

cally in Fig. 15-11. At the base of the rod outer segment, saccules are shown to be formed by infolding along an incision passing around the outer segment for a portion of its circumference. Proceeding distally from the base of the rod outer segment, this incision becomes shorter and quickly disappears. Thus saccules in the greater part of rod outer segments have no direct connection with the outer membrane, as depicted also in Fig. 15-10. By comparison, the outer segment of a frog cone typically tapers, is much shorter than the rod outer segment, and its saccules are continuous with the outer membrane by a zone of infolding that extends the entire length of the outer segment. Thus rod and cone outer segments follow a similar structural plan but show distinct variations from that plan. Recently the outer segments of living frog eyes were infiltrated with procion yellow, a highly diffusible fluorescent dye, which clearly confirmed that cone saccules have patent openings to the extracellular medium throughout the outer segment, while rod saccules have such patent openings only at the base of the outer segment.[75] Studies of lanthanum infiltration into glutaraldehyde-fixed retinas indicate that in monkeys also it is only the cone saccules that have openings to the outside along the entire length of the outer segment.[43]

Electron microscopy has produced evidence that the visual photopigment is contained in the saccule membranes. After extraction of rhodopsin by digitonin an electron-dense component of the saccule membranes in rod outer segments is absent, and it is more difficult to demonstrate after light adaptation.[50]

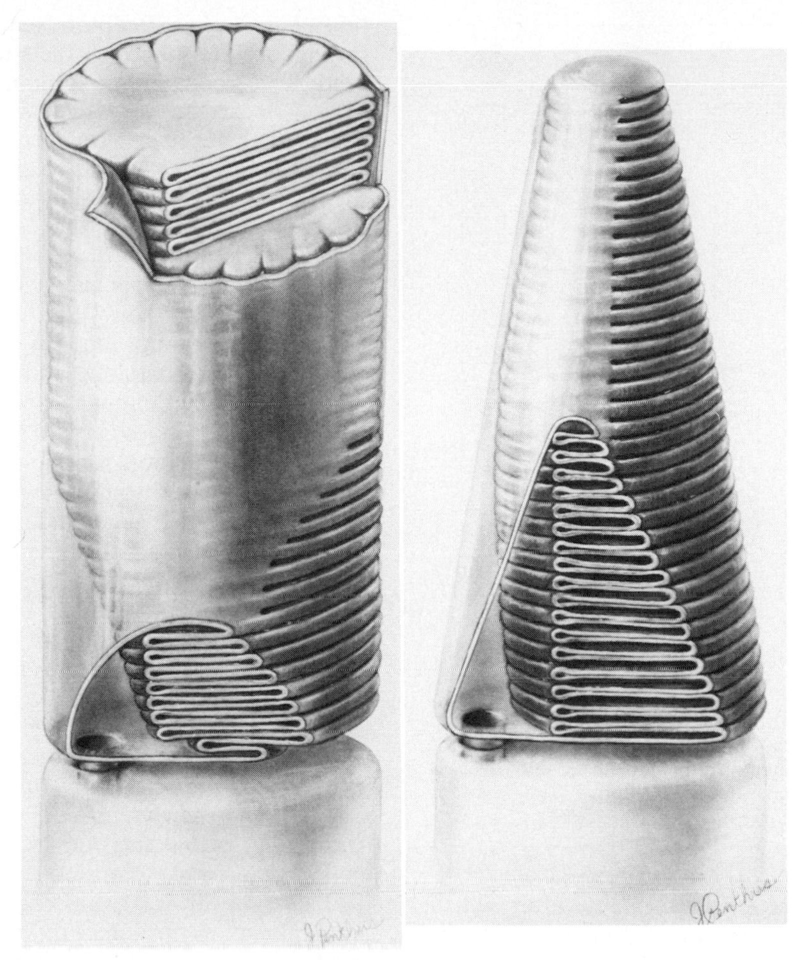

Fig. 15-11. Schematic structure of outer segments of frog rod (left) and cone (right), based on electron microscopy. Drawings show entire cone outer segment, but only base of rod outer segment because of its greater length with no significant change of structure beyond point shown. (Courtesy Dr. R. W. Young.)

Rhodopsin has a molecular weight of 27,000 to 28,000 in several mammals,[64] and biochemical evidence also indicates that it is a major component of saccule membranes. Thus infolding of the outer membrane to produce the saccules seems to have the functional significance of greatly increasing the amount of membrane available to contain photopigment. This increases the photopigment content of the outer segment, improves its light-trapping efficiency, and hence increases the light sensitivity of both rods and cones.

Renewal of outer segments

Young[119] has shown recently that rod outer segments have a remarkable renewal mechanism. When radioactive protein precursor was injected into rats, mice, and frogs, it quickly appeared in rod outer segments in the sequence of stages shown in Fig. 15-12. In stage 1, shown at left, the black dots in the inner segment represent radioactive protein revealed by a high-resolution autoradiograph. This radioactive protein migrated through the connecting cilium (stage 2) and formed a dense narrow band at the base of the outer segment (stage 3). This band then migrated along the axis of the outer segment (stage 4) and reached the distal end of the outer segment (stage 5). Finally, the distal tip of the outer segment that contained the radioactive band was phagocytized by the pigment epithelium cell, within which it appeared as a cytoplasmic inclusion called a phagosome (stage 6). Since the protein is incorporated into the outer segment in a narrow band, saccules appear to be forming constantly at

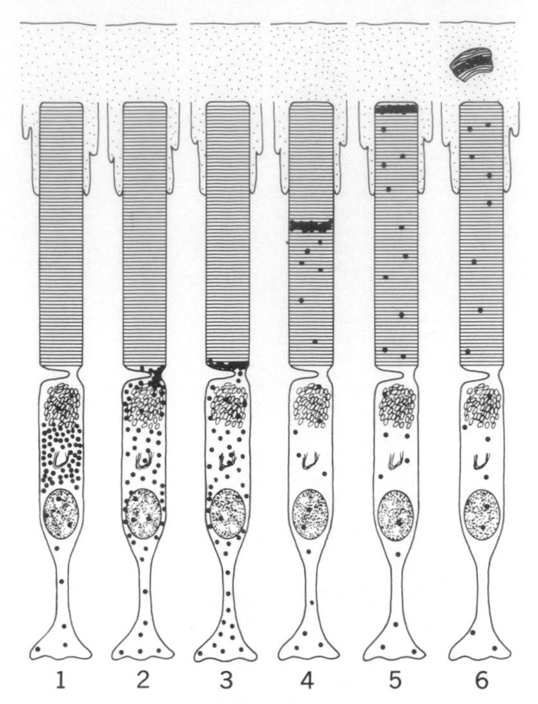

Fig. 15-12. Continuous renewal of base of rod outer segment and removal of distal tip of outer segment by pigment epithelium cells, as revealed by high-resolution autoradiography. See text for explanation. (Courtesy Dr. R. W. Young.)

the base of the outer segment, after which they move along the outer segment, with the oldest material at the distal tips being phagocytized and digested by the pigment epithelium cells. In rats and mice (the mammals studied), it takes only about 10 days for the radioactive band to traverse the outer segment, so the turnover is quite rapid. Furthermore, extraction of visual pigment while the radioactive band was traversing the outer segment recovered 80 to 85% of the total radioactivity.[57] Thus the main protein synthesized from radioactive precursor in this work appears to be the visual photopigment itself.

By comparison, a single injection of radioactive protein precursor appears in cones as a diffuse radioactive labeling at scattered sites throughout the outer segment. Hence the cone outer segment is not continuously renewed from its base. Rather, the sites at which protein molecules appear to be renewed at any given time are widely dispersed in the outer segment. Thus rods and cones also differ in the manner of renewal of their outer segments.

These results answer the important question of how photoreceptors remain functional throughout life, although their high sensitivity renders them vulnerable to damage by normal intensities of the very light stimuli they are designed to receive.[83] Being nerve cells, they cannot renew themselves by division. Instead, they have evolved mechanisms for renewing the specific portion of the cell that absorbs light and is thus most subject to damage. Since the rods are especially sensitive to light, it is also noteworthy that the rods have evolved an elaborate renewal mechanism that appears more complete and may be more effective than that of cones.

This line of investigation has suggested a plausible explanation for the different shapes of cone and rod outer segments in most species.[120] In the salamander, for example, Young[120] has found that the outer segments of both rods and cones are conical during early development. This is because the first saccules formed are small, but saccules formed later at the base of the outer segment become larger. In the cones, saccule development is arrested at this stage. In the rods the small early saccules are shed as saccule formation continues at the base of the outer segment, the size of the newly formed saccules finally stabilizing so that the outer segment becomes cylindric in form.

This type of work has likewise suggested an explanation of human retinitis pigmentosa, a blinding disease in which the postmortem histology shows a marked accumulation of cellular debris between the receptors and pigment epithelium.[66] These results would be expected if phagocytic activity of the pigment epithelium cells was markedly defective or absent. As rod outer segments continued to be formed, debris would then accumulate and eventually block the normal exchange of materials between the receptors and choriocapillaris. This suggestion is greatly strengthened by the study of a hereditary retinal degeneration in rats in which the postmortem histology is similar, and for which detailed studies have confirmed the suggested etiology of the condition.[23, 65, 66]

PHOTOCHEMISTRY OF VISUAL PIGMENTS

Effects of light on rhodopsin

The photochemistry of rhodopsin has been determined primarily by extracting it from the outer segments of rods and studying it in isolation.[9, 67] In common with all visual pigments studied to date, rhodopsin is a *protein*

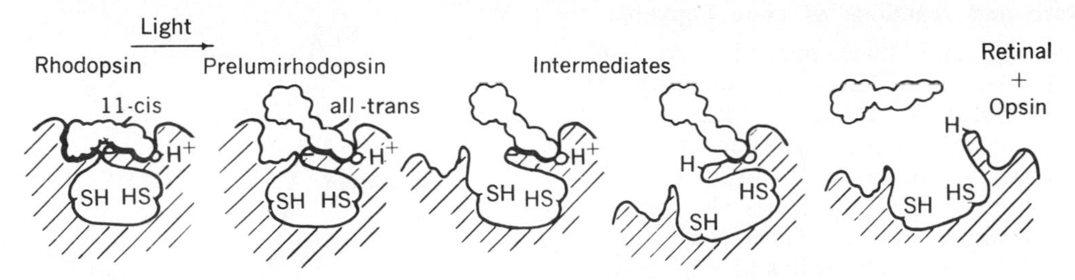

Fig. 15-13. Action of light on rhodopsin. See text for explanation. (Modified from Wald and Brown.[14])

bearing a light-absorbing *chromophore,* to which it owes its color and sensitivity to light. Its general structure and the major effects of light absorption are shown in Fig. 15-13. The protein part of the molecule is called *opsin.* The chromophore is called *retinal$_1$,* an abbreviation of retinaldehyde$_1$, a name resulting from the chromophore being an aldehyde of vitamin A_1. In the resting state this chromophore is in the specific 11-*cis* form, which fits closely into the opsin portion of the molecule. The absorption of light isomerizes the chromophore, which straightens into the all-*trans* configuration, and this stage is called prelumirhodopsin. The reaction may be stopped at this stage if the temperature is below about –140° C. But prelumirhodopsin is highly unstable, presumably because the chromophore no longer fits closely to the opsin. With progressive warming, the opsin opens up in several stages, which are designated intermediates in Fig. 15-13. Above about –140° C the first intermediate, called lumirhodopsin, is formed. Above about –40° C, another stage, metarhodopsin I, occurs, and metarhodopsin II is formed when the temperature exceeds about –15° C. Finally, at temperatures above 0° C and in the presence of water, the molecule is hydrolyzed into its two main fractions of all-*trans* retinal and opsin. Low temperatures are convenient for studying this sequence of events, which can thus be stopped at any stage. The same sequence of events occurs under physiologic conditions,[9] but the sequence occurs very rapidly, and the unstable stages between rhodopsin and its hydrolysis are more difficult to detect because they are short lived. Note that in this sequence of events the only action of light is to isomerize retinal from the 11-*cis* to the all-*trans* form.[112] All of the succeeding reactions are energy yielding and proceed spontaneously under normal physiologic conditions. Even the isomerization of retinal is an energy-yielding reaction,[112] so the only energy required of light is that which triggers an energy-yielding reaction. This is a typical characteristic of highly sensitive cellular reactions in which the stimulus is only required to trigger the release of a preloaded source of potential energy, so that there is an energy amplification at the initial stage of the reaction.

Resynthesis of rhodopsin

Following its hydrolysis from the opsin, retinal is reduced by enzymatic action to vitamin A, with which the retinal comes into equilibrium. Some of this vitamin A appears to be stored in the pigment epithelium, at least if the light is sufficiently strong and prolonged for vitamin A to be formed from retinal in large quantities.[47] Since the photopigment molecule is broken down after light absorption, a process often referred to as bleaching, it must regenerate in the dark to prepare for another response to light. Following its hydrolysis from the opsin, retinal is in the all-*trans* form; the first step required for resynthesis of rhodopsin is conversion of this retinal back to the 11-*cis* form.[112] This can take place in the dark through the action of an enzyme, and this reaction consumes energy. Combination of the 11-*cis* retinal with opsin to form rhodopsin then occurs spontaneously. This reaction promotes the re-formation of retinal from vitamin A in two ways.[112] First, this spontaneous reaction yields energy that is used for the oxidation of vitamin A to retinal. Second, by removing retinal from one side of the reaction the equilibrium is altered to favor conversion of vitamin A to retinal. When all of the opsin present has recombined with retinal, there can be no further energy-yielding reactions to support further accumulation of retinal. Thus the action is limited by the amount of available opsin.[112]

Structure and reactions of cone pigments

Cone pigments have proved more difficult to extract in sufficient quantity for study in solution. However, one pigment called iodopsin has been extracted from the predominantly cone retina of the chicken. The chromophore of this pigment is also $retinal_1$, but the opsin is different from that of rhodopsin.[9, 112] The breakdown and synthesis of iodopsin appear to occur in basically the same manner as in rhodopsin.[9]

Spectrophotometry of the human fovea has detected the presence of pigments that absorb strongly in the red and green portions of the spectrum, respectively,[38] and may be called red and green pigments for convenience. Iodopsin appears to correspond with the red pigment of the human fovea.[38] When foveal pigments were bleached by a strong light, addition of 11-*cis* $retinal_1$ resulted in resynthesis of both pigments.[38] Thus both the red and green pigments of the human retina contain the same chromophore as rhodopsin. This makes it probable that all human photopigments possess the same chromophore, $retinal_1$, and differ among themselves only in the opsin component. The fact that retinal is formed from vitamin A explains why an experimental vitamin A deficiency in humans results in higher dark-adapted thresholds for both cones and rods, and why normal cone and rod thresholds are recovered following vitamin A administration.[114]

Early receptor potential

An intense brief flash of light elicits from the retina an extremely rapid electrical response. This was first found by microelectrode studies in the macaque monkey retina, and the response was shown to have maximum amplitude when the electrode was near the distal tips of the photoreceptors.[27] It was designated the *early receptor potential* (ERP) because of its receptor origin and the absence of any detectable latency.[27] It is typically biphasic in form,[28] and the two phases have been designated R_1 and R_2. As shown in Fig. 15-14, this response may also be recorded between wick electrodes on opposite sides of an isolated retina.

The ERP shows little or no sensitivity to ionic composition of the bathing medium and hence cannot be generated by a transmembrane ion flux.[84] It is, however, dependent on the integrity of the visual photopigment[27] and on normal orientation of the photopigment molecules in the saccule membranes.[45]

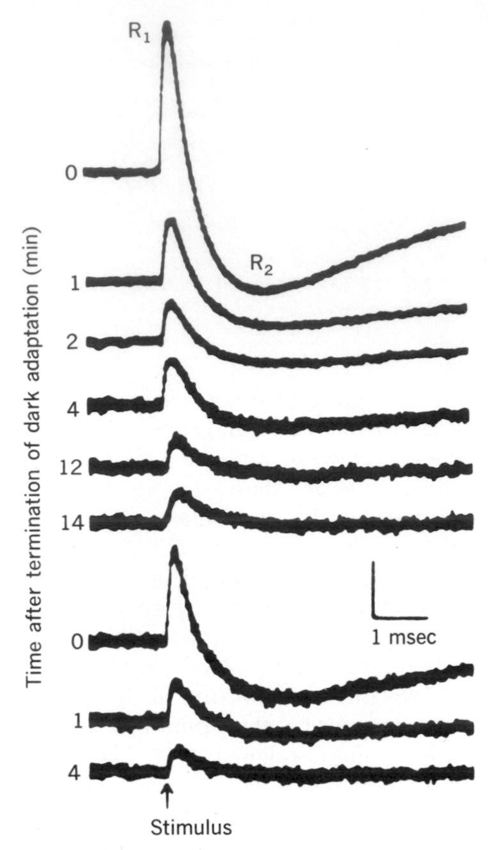

Fig. 15-14. Influence of adaptation on first phase, R_1, and second phase, R_2, of early receptor potential of isolated toad retina. Response recorded between wick electrodes on opposite sides of retina; upward deflection indicates positivity of vitreous side of retina, as in ERG recording. Stimulus was intense 0.7 msec flash covering retina, which was initially dark adapted. Retina was then light adapted by 15 stimulus flashes at 1 min intervals, dark adapted again for 15 min, and finally light adapted by another 5 flashes at 1 min intervals. Voltage calibration indicates 100 μv for top three records and 50 μv for all other records. (From Brown.[2])

In a pure-rod eye the spectral sensitivities of both R_1 and R_2 match well to the absorption spectrum of rhodopsin.[86] A variety of methods have now shown that the response is a direct electrical consequence of light-induced changes in the form of the photopigment molecule.[85] Apparently the altered molecular form shifts the position of electric charges on the molecule, thus generating a transient electrical response. The potential developed by a single molecule must be extremely small, but the intense flash required to elicit the ERP causes synchronous events in a great many molecules, yielding a measurable response. With lowered temperature, R_2 may be abolished selectively, thus isolating R_1, which has been recorded at temperatures as low as −35° C.[87] It appears likely that R_1 results from a molecular step in the conversion of prelumirhodopsin to metarhodopsin I, while R_2 results from a later step that is more sensitive to lowered temperature.[87] These results indicate that the ERP is a convenient method for in vivo detection of rapid light-induced changes of molecular form in the photopigment, and it is being applied in this type of investigation.

As expected from the manner of its generation, the amplitude of the ERP is linearly proportional to the number of photopigment molecules that are activated by absorbing quanta from the light flash.[44] With the flash intensity constant, the number of quanta absorbed will increase linearly with concentration of the photopigment. Under these conditions the amplitude of the ERP is linear with photopigment concentration, thus giving a long-needed method for the rapid and convenient measurement of photopigment concentration. In Fig. 15-14 this method is applied to show the bleaching of photopigment by light and its regeneration in the dark. Beginning with an isolated and dark-adapted toad retina, a series of 15 flashes at 1 min intervals quickly light adapted the retina and reduced the photopigment concentration, as shown by the reductions in both R_1 and R_2. The small stable response finally remaining at the end of this series of flashes is probably generated by the heating effect of light absorption, a separate phenomenon that has also been demonstrated.[55] After 15 min in darkness the ERP was increased, and it was abolished again by a series of flashes at 1 min intervals. Hence toad photopigment regenerates in darkness, even when the retina is isolated from the pigment epithelium. Photopigment regeneration is more efficient, however, when the toad retina is in contact with the pigment epithélium.[26]

An ERP may be recorded from the intact human eye by a gross corneal electrode and a remote reference electrode that is effectively behind the eye. In this application it may be used diagnostically, as it is a convenient method for detecting pathologies involving the photopigment.[22]

An ERP is generated by both rods and cones, but the signal generated by a single cone is larger than that generated by a single rod.[3] This effect is so great that although cones contain only about 10% of the total photopigment in human and macaque monkey retinas, the ERP from both species is contributed mainly by cones.[51, 52] This is probably a consequence of all the cone saccules being open to the extracellular medium, so that electrical events in the cone photopigments are more readily recorded by extracellular electrodes.

Functions of photochemical cycle

The breakdown of visual pigment by light and its regeneration in the dark is one mechanism by which the retina adapts its sensitivity to the prevailing illumination. During light adaptation the reduced photopigment concentration decreases the light-trapping efficiency of the outer segment, thus lowering sensitivity; converse effects occur during dark adaptation. This increases the range of light intensities to which the retina can respond, which is perhaps the main functional advantage underlying the evolution of such an elaborate photochemical cycle. Under physiologic conditions, rhodopsin in the squid retina isomerizes, but it does not bleach or hydrolyze into its two major fractions following light absorption.[68] Thus the more elaborate photochemical cycle of vertebrates may represent a higher evolutionary development in the service of light and dark adaptation.

Absorption of light by the photopigment also triggers the process of excitation. Initial changes in the molecular form of rhodopsin are exceedingly rapid, but hydrolysis appears to be too slow to account for the rapidity of light sensations.[13] It is generally assumed that one of the early molecular events shown in Fig. 15-13 plays a role in excitation by serving as a direct causal step in initiating electrical activity that is transmitted along the photoreceptor. This is a strong possibility, but there is no evidence to this effect, and the critical excitatory step in the photopigment molecule could well be some molecular event that has not yet been revealed.

Another approach to the excitation problem is to record the electrical activity of the photoreceptor and determine the characteristics that molecular events must have in order to initiate the electrical activity. This approach seems required to select among the candidates for excitatory events, and a beginning has now been made in this direction.

MAMMALIAN ELECTRORETINOGRAMS

The electroretinogram (ERG) is a relatively slow and complex electrical response

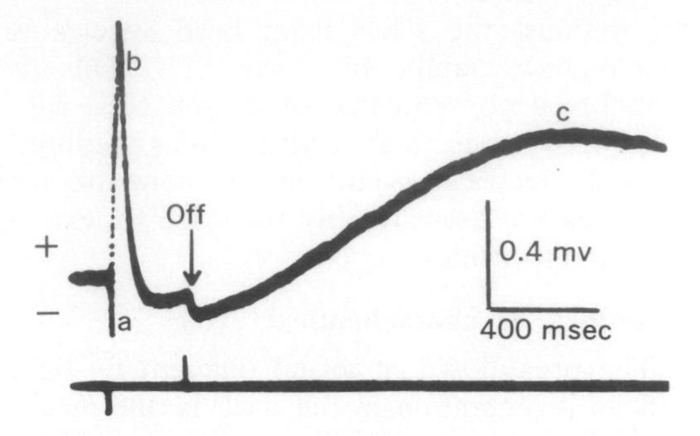

Fig. 15-15. ERG of cat, with major deflections conventionally labeled. Eye was light adapted and retina was stimulated by a large and intense light spot. Onset and termination of stimulus shown in lower record. See text for further explanation. (From Brown.[2])

of the retina to illumination; it was first recorded by Holmgren in 1865. The most conventional recording method is to place an electrode in contact with the cornea, while the reference electrode is at a remote location that is effectively behind the eye. Recent analytic work in mammals has been done primarily by an electrode introduced through a needle into the vitreous humor of an experimental animal. If a microelectrode is used, it may be introduced into the retina itself to record the ERG from various retinal levels.[35-37]

Fig. 15-15 is an ERG recorded from the cat retina by an active electrode in the vitreous humor and a reference electrode in the orbit behind the eye. Positivity of the active electrode is shown as an upward deflection, which is the convention for ERG work and also for intracellular recording. That convention will be followed for all electrical responses illustrated in this chapter. The cat ERG is typical of that elicited from a predominantly rod retina. Immediately after onset of the stimulus there is a negative a-wave, followed by a larger positive b-wave. The positive c-wave is also initiated by onset of the stimulus, but the c-wave rises so slowly that with short stimuli its peak occurs long after the stimulus. On termination of the stimulus, the off-response is a simple negative deflection.

Electroretinogram components and their origins

It is obvious from Fig. 15-15 that the ERG is not a unitary response but consists of several components. Since the ERG may be recorded by gross electrodes, each component must consist of the summed activity of many cells of a given class that are responding in synchrony to the light stimulus. Hence the ERG offers the unique advantages of a readily recorded response that reveals the electrical activity of certain classes of retinal cells. The basic problems have always been to analyze this complex response into its components and to identify the type of cell generating each component. Solution of these problems is required so that the ERG may be used to maximum advantage in studying retinal physiology and diagnosing retinal disorders.

A combination of several methods has shown that the mammalian ERG consists of four major components. These are (1) the *c-wave,* (2) an ionically generated *receptor potential,* (3) *the b-wave,* and (4) a *dc component,* so called because its time course resembles that of a direct current pulse. The c-wave is generated by the pigment epithelium, as shown most clearly by intracellular recording of the c-wave from single pigment epithelial cells of the cat retina.[103] The c-wave results from light absorption by outer segments of the photoreceptors,[103] which apparently causes some chemical change in the extracellular fluid, to which the pigment epithelium cells respond. The physiologic significance of the c-wave is not clear, but it cannot be in the direct chain of nervous activity that leads to visual sensations. Thus Fig. 15-16 shows the general form of the ERG in the absence of the c-wave, and how this response can be accounted for by summation of the three major neural components that have been identified. Such an analysis is shown for the ERG from a predominantly rod retina and also for the ERG of a pure-cone eye. The macaque monkey and human are intermediate cases, since their retinas contain both rods and cones.

The neural components of the ERG, and their origins, have been revealed most clearly by studies with penetrating microelectrodes.[2, 36, 37] An intraretinal electrode can record a local ERG that is only from a small retinal area surrounding the electrode.[36] Since the recorded response is only from the area penetrated, the amplitude of each component should be maximum at the retinal level where it is generated. Electrode depth is determined by physiologic methods and confirmed by electrode marking.[31, 37] Such studies in the cat retina showed that the a-wave attains maximum amplitude near the distal tips of the photoreceptors, indicating that the a-wave

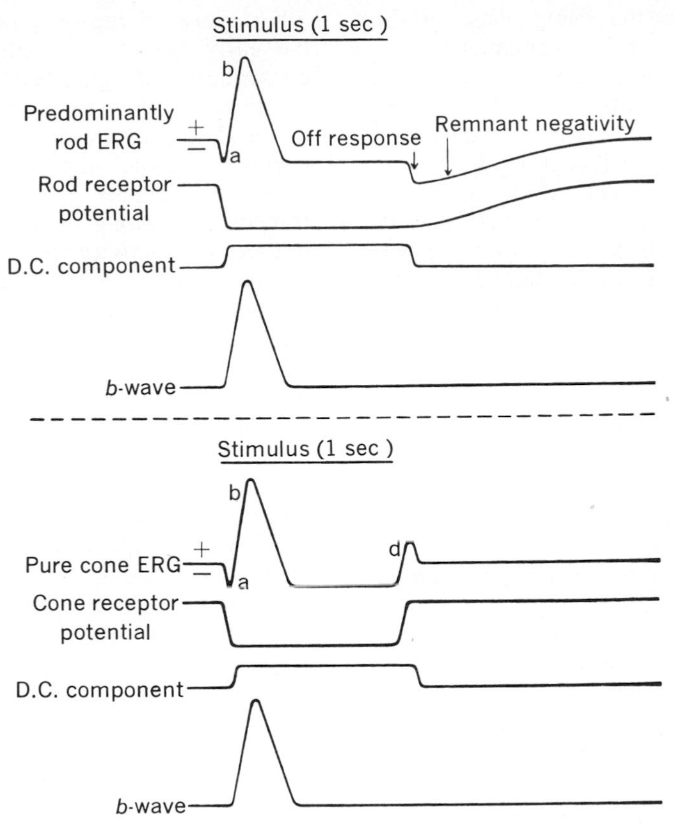

Fig. 15-16. Schematic component analyses of mammalian ERGs from predominantly rod and from pure-cone eyes. In both rod and cone analyses, c-wave from pigment epithelium is omitted, and remaining ERG is shown to be accounted for by algebraic summation of the three major neural components that have been identified. (From Brown.[2])

results from the rising phase of a receptor potential.[37] By contrast, maximum amplitudes of both the b-wave and dc component were in the inner nuclear layer.[37] The b-wave and dc component were then separated by two methods. Lidocaine (Xylocaine), a local anesthetic, abolishes the b-wave without affecting the dc component.[37] Reduction of stimulus intensity can isolate the dc component, thus permitting its shape to be determined.[36]

The type of cell that generates the dc component has not been identified. Intracellular recording in Müller cells of *Necturus* has indicated, however, that the b-wave is generated across the membranes of these glial cells.[81] Analytic work on glial cells indicates that they respond only to ionic composition of the intercellular fluid, as when the potassium concentration is increased by the activity of nearby nerve cells.[74] Hence generation of the b-wave by the Müller cells is probably secondary to the activity of nerve cells of the inner nuclear layer. In any event the b-wave is closely comparable to ganglion cell activity as an indicator of retinal sensitivity,[19] and the b-wave is especially useful for this purpose because it is so readily recorded.

Foveal recording in the macaque monkey shows a great reduction of both the dc component and b-wave, corresponding to almost complete absence of the inner nuclear layer, while the a-wave from the fovea is larger than from the peripheral retina.[2] When the retinal circulation of the macaque monkey is mechanically clamped at the optic disc, the dc component and b-wave are quickly abolished, but the receptor potential survives for many hours, being well supported by the choroidal circulation.[2, 32] The c-wave also survives but may be abolished in the macaque monkey by light adaptation, thus isolating the receptor potential. This method has shown that the time course of the rod receptor potential is much slower than that of the cone receptor potential (Figs. 15-17 and 15-20).

Analysis of the rod ERG (Fig. 15-16) shows that the initial rise of the rod receptor potential contributes the rising phase of the a-wave, which is abruptly terminated by onset of the large b-wave and dc component, both of which are of opposite polarity from the receptor potential. These components have longer latencies than the receptor potential, as they are generated by cells of the inner nuclear layer. After the b-wave has run its course, there is usually a steady level of negative potential during the remainder of the stimulus, apparently because the negative receptor potential is larger than the positive dc component. Immediately after the stimulus, rapid decay of the dc component gives the negative deflection called the off-response. This is followed by a slowly decaying phase, usually referred to as "remnant negativity,"[8] which results from the very slow decay of the rod receptor potential.

The form of the cone ERG is very similar to that of the rod ERG until the stimulus terminates. At that time the cone ERG typically shows a d-wave, consisting of a positive rising phase and a negative falling phase, following which there is no remnant negativity. The cone ERG appears to consist of the same neural components as the rod ERG, but the cone receptor potential decays rapidly when the stimulus terminates.[2, 32] In the cone ERG, termination of the stimulus is followed first by decay of the cone receptor potential, giving the positive rising phase of the d-wave. The falling phase of the d-wave then results from decay of the dc component, which is opposite in polarity and which decays later because it is generated in the inner nuclear layer.[2] Since the cone receptor potential has decayed by the time the d-wave is completed, remnant negativity is not found in the cone ERG. Thus the main differences in form between rod and cone ERGs result from the different decay rates of rod and cone receptor potentials.

ROD AND CONE RECEPTOR POTENTIALS

The early receptor potential does not usually appear in the ERG, as it requires higher stimulus intensities than are used in most ERG work. When the early receptor potential does appear, the peaks of both R_1 and R_2 precede the onset of the a-wave.[27, 28] Thus the slower cone and rod receptor potentials of the ERG are often referred to as late receptor potentials but here they are called simply receptor potentials; they are generated by ionic fluxes in common with other types of receptor potentials. Although the early receptor potential is an inevitable consequence of light absorption by the photoreceptor, there is strong evidence that it is not a causal step in the direct chain of excitatory events.[85] On the other hand, the slower cone and rod receptor potentials appear to be transmitted along photoreceptors and to initiate the activity of second-order cells.[29] Hence these receptor potentials are especially important to the understanding of visual functions.

Decay rates

Many lines of evidence have shown that decay of the rod receptor potential is consistently much slower than decay of the cone receptor potential,[2, 3, 108, 109] so that this is a third major aspect of the duplicity theory. Cones and rods differ not only in their anatomy, and in their contained photopigments, but also in the time courses of their electrical responses. This is shown by Fig. 15-17 in the responses from monkey retinas. After clamping the retinal circulation to the predominantly rod retina of the night monkey, the negative receptor potential decays very slowly. Following the same procedure in the macaque monkey, the receptor potential shows a rapid decay phase followed by a phase of much slower decay. In this case the receptor potential is recorded from an entire retina that contains large populations of both rods and cones. Hence the receptor potential is the summation of a rapidly decaying cone component superimposed upon a slowly decaying rod component. This is emphasized in Fig. 15-17 by the dashed line showing the approximate time course of the rod receptor potential during the stimulus, as determined in other experiments. Note that the rise of the rod receptor is also slower than that of the cones. The results shown in Fig. 15-17 were obtained at high stimulus intensities at which decay of the rod receptor potential is exceptionally slow. With lowered stimulus intensity, decay of the rod receptor potential becomes more rapid, but it appears never to become as rapid as decay of the cone response.

It is highly probable that the different time courses of their receptor potentials determine some of the major functional differences between cones and rods, especially those involving the dimension of time; e.g., the rapid recovery of the cone response prepares the cones quickly for response to a second stimulus. Rapid recovery is a requirement for re-

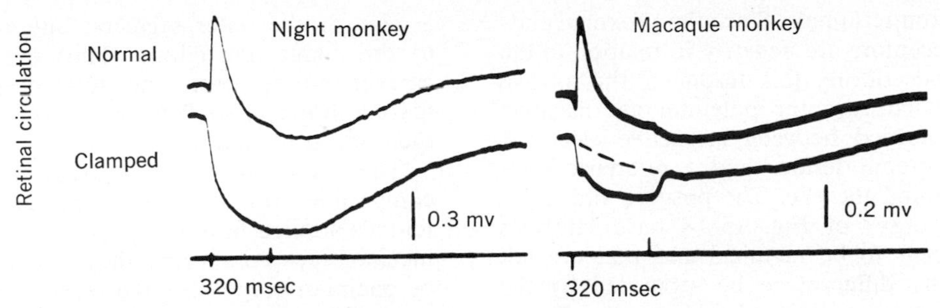

Fig. 15-17. Receptor potentials from predominantly rod retina of night monkey and from mixed rod-cone retina of macaque monkey. In both cases the normal ERG was recorded first; retinal circulation was then clamped to isolate receptor potential. Note slow rise and fall of rod receptor potential of night monkey. In macaque monkey a dashed line indicates initial time course of rod receptor potential, on which cone receptor potential that rises and decays much more rapidly is superimposed. Responses recorded between active electrode in vitreous humor and reference electrode behind eye. Stimulus was large intense flash delivered every 30 sec for night monkey and every 10 sec for macaque monkey. For pure-cone receptor potential of macaque fovea, see Fig. 15-20. (Compiled from Brown and Watanabe[33] and Brown et al.[3])

sponses to flickering stimuli at high frequency and is probably the main physiologic basis for the CFF of cones being much higher than that of rods.

Depth distribution of cone and rod receptor potentials

After clamping the retinal circulation to isolate the receptor potential of a macaque monkey, the distribution of this potential along the length of the receptor may be determined by a penetrating microelectrode. Fig. 15-18 shows results of such an experiment in the peripheral retina. This problem is best studied with intense brief stimuli that give large responses while minimizing light adaptation by the stimulus. At very high flash intensities the decay of the cone receptor potential occurs only after a delay, that is a function of stimulus intensity, and that in Fig. 15-18 was about 150 msec.[29] This delay permits both the onset and decay of the cone receptor potential to be seen in the response to a brief flash. Decay of the cone receptor potential is followed by the slower decay of the rod receptor potential, which is not completed within the time course of these records. At the retinal side of the pigment epithelium (100% total retinal depth), polarity of the receptor potential was positive. As the electrode was withdrawn, this positive response increased and reached maximum amplitude near the connection between outer and inner segments. With further withdrawal the positive response decreased, then response polarity reversed, and maximum amplitude of the negative response was at about the level

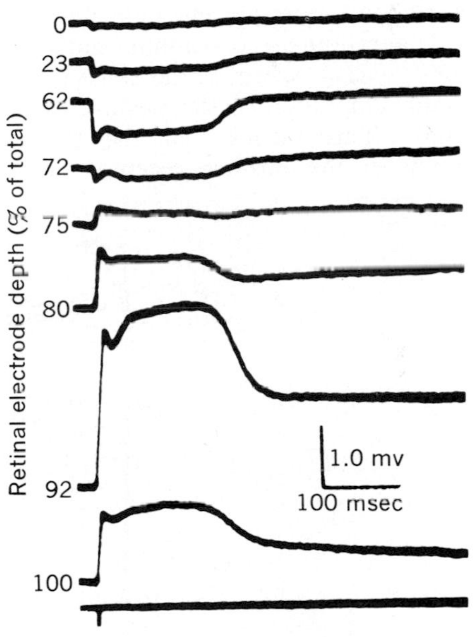

Fig. 15-18. Polarity reversal of cone receptor potential of macaque monkey as microelectrode traverses long axis of photoreceptors. Responses recorded from peripheral retina, with reference electrode in vitreous humor. Electrode depth shown as a percentage of total distance along electrode track from internal limiting membrane to pigment epithelium. Stimulus was intense 20 μsec flash of light; time of its delivery is shown in lowest record. Stimulus repeated every 10 sec to maintain constant light adaptation. For further explanation, see text. (From Brown et al.[3])

of the axon terminals. Since the proximal ends of the receptors are negative in relation to the distal ends during this response, the a-wave and isolated receptor potential are negative when recorded between an active electrode in the vitreous humor and a reference electrode behind the eye. The positive and negative responses of Fig. 15-18 have latencies that appear to be identical and that are not sufficiently different to be separated by the synaptic delay of a chemical synapse.[3] Hence they must represent the distribution of extracellular potential along a single type of cell, resulting from extracellular current flow along that cell. By convention, current is said to flow from a region of maximum positivity (the source) to a region of maximum negativity (the sink). Fig. 15-18 shows this source-sink relation for cones, and similar investigative methods have shown that with small stimulus spots the rod receptor potential gives corresponding results in the night monkey.[3, 33] The early receptor potential is generated in the outer segment, and during any given penetration the maximum amplitude of the ERP always occurs at the same depth as maximum amplitude of the positive receptor potential.[3] This means that the amplitude maximum of the positive receptor potential is also at the outer segment, although close to the connecting cilium. Thus extracellular current during cone and rod receptor potentials appears to flow from the outer segment to the axon terminal.

The direction of this current flow is incompatible with the actions of receptor potentials studied in other types of receptors. In mechanoreceptors (e.g., the stretch receptor or pacinian corpuscle) the receptor potential results from an active depolarization of the distal part of the cell; this draws extracellular current from more proximal regions, thus depolarizing a proximal portion of the membrane to its firing level and initiating a conducted impulse. The electrical activity of the squid photoreceptor appears to follow this same principle.[56] But Fig. 15-18 shows that the extracellular current flow associated with the receptor potential of the mammalian retina is in the *opposite direction* from that found to date in other types of receptors. Thus the mammalian photoreceptor is a special case with respect to other receptors whose slow potentials have yielded to analysis.

Intracellular receptor potentials

Intracellular recording from photoreceptors was first achieved by Bortoff with *Necturus,* a

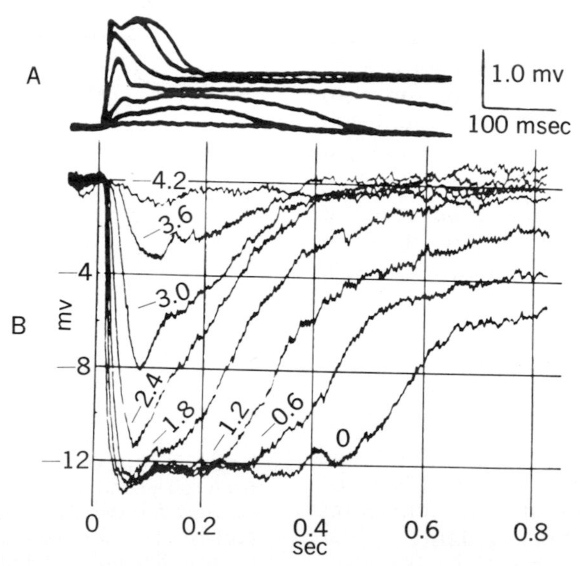

Fig. 15-19. A, Extracellular receptor potentials from peripheral retina of macaque monkey, recorded by microelectrode near distal tips of receptors. Responses evoked by 20 μsec flash covering large area around microelectrode. Superimposed responses are shown to maximum flash intensity and to flash intensities that decreased by steps of 0.6 log unit. (From Brown and Murakami.[29])

B, Intracellular receptor potentials from single turtle cone, evoked by 20 μsec flashes covering area of 160μ diameter around impaled cone. Here also, superimposed responses are shown to maximum flash intensity (0), and to flash intensities reduced by steps of 0.6 log unit. (From Baylor and Fuortes.[21])

salamander in which all of the cells are exceptionally large.[24] The recording site was identified by electrophoretic deposition from the intracellular electrode of a dye substance, which was then found by histologic study to be in the inner segment. Similar work has been done now in gecko, frog, turtle, and certain fish,[21, 82, 107-109] and also in gecko outer segments.[82] This work has shown that during the receptor potential the plasma membranes of both outer and inner segments become hyperpolarized. In agreement with other evidence,[3] it also shows that the vertebrate photoreceptor does not generate a nerve impulse.

In Fig. 15-19 the effects of stimulus intensity on the amplitude and time course of the receptor potential are shown by extracellular and intracellular recording. The records in Fig. 15-19, *A*, were obtained from the peripheral retina of a macaque monkey after clamping the retinal circulation, using an extracellular microelectrode near the distal tips of the photoreceptors. The records in Fig. 15-19, *B*, were recorded intracellularly from the inner segment of a turtle cone. In both experiments the responses were evoked by a 20 μsec flash covering the area around the electrode, and superimposed responses are shown to flashes in which intensity was reduced below the maximum available by steps of 0.6 log unit. The extracellular responses from the monkey contain both rod and cone contributions. The lowest stimulus intensities elicited only a rod potential, which rose and decayed very slowly. With increased intensity the amplitude of the rod potential increased, and in the middle of the seven records a clear cone potential with rapid rise and decay has appeared. This cone potential then increased to its maximum amplitude; beyond this point the effect of higher intensity was to increase response duration. The intracellular responses show only the cone potential and likewise illustrate that when the cone response reaches maximum amplitude, further increase of intensity increases the response duration. Note that response polarity inverts between extracellular and intracellular recording and that the intracellular polarity shows hyperpolarization of the cell membrane. Note also that with comparable stimulus conditions the effects of stimulus intensity on the amplitude and time course of the cone potential are very similar for the two species and the different recording conditions. The effects of stimulus intensity upon intracellularly recorded rod potentials of the frog retina have also been reported recently[109] and agree well with extracellular results obtained in the monkey. The monkey records in Fig. 15-19 illustrate that low stimulus intensities affect only the rod potential, while the upper range of the intensity scale alters only the cone receptor potential. In fact the rod and cone receptor potentials respond to different but somewhat overlapping ranges of stimulus intensity,[3, 29] indicating that this aspect of the duplicity theory is determined largely at the receptor level.

Intensity discrimination by foveal cones

After clamping the retinal circulation, a pure-cone receptor potential may be recorded by a microelectrode near the distal tips of photoreceptors in the monkey fovea.[32] Fig. 15-20 shows the typical time course of this cone receptor potential to a high-intensity stimulus of long duration. Under natural conditions the task of cone vision is usually to detect an increment or decrement of stimulation against a prevailing steady background illumination, the intensity of which can vary by many log units. Thus Fig. 15-20 shows the amplitude of the foveal cone receptor potential as a function of stimulus intensity in darkness and also when the stimuli were presented against a large range of background illuminations. The results fit well to the theoretical curves drawn. The basic response curve in darkness is well fit by the following expression[25]:

$$R_t = \frac{I_t^n}{I_t^n + K_r^n}$$

In this equation, R_t is response amplitude to the test flash, and I_t is intensity of the test flash in trolands, while the constant K_r^n is the number of trolands required to elicit a cone receptor potential of half-maximum amplitude. This equation is a power law at low to moderate intensities, in agreement with the effects of stimulus intensity upon sensation magnitudes as determined psychophysically in a variety of sensory modalities.[104] In the case of monkey cones the exponent of this power law has a value of about 0.70.[25] As background intensity increases, the response curves in Fig. 15-20 move to the right, indicating that higher stimulus intensities are required to elicit an incremental receptor potential of constant amplitude.

In Fig. 15-21 the intensity discrimination of the human fovea is compared directly with

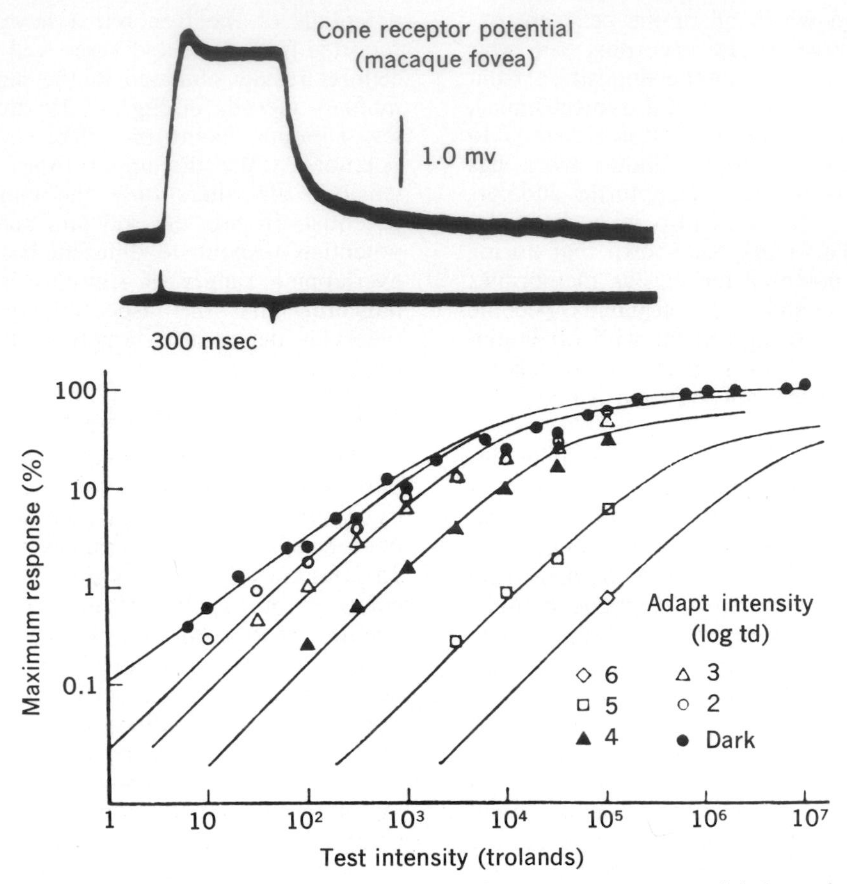

Fig. 15-20. Above: Record shows time course of cone receptor potential from fovea of macaque monkey, as recorded by extracellular microelectrode near distal tips of foveal receptors after clamping of retinal circulation. Stimulus was intense spot with retinal diameter of 0.25 mm, carefully focused and centered on fovea and repeated every 10 sec. (From Brown and Watanabe.[32])

Below: Amplitude of foveal cone receptor potential as function of stimulus intensity. Recording conditions same as for top record, and pure-cone origin of receptor potential confirmed by spectral response functions. Response amplitudes were measured at termination of stimulus. Stimuli were presented in complete darkness and against variety of adapting background intensities. Both background and test flashes had a retinal diameter of 1.1 mm and were centered on fovea. Stimulus was 150 msec flash of light at 580 nm. Data are mean values from six animals, and curves through data points are theoretical. (From Boynton and Whitten.[25])

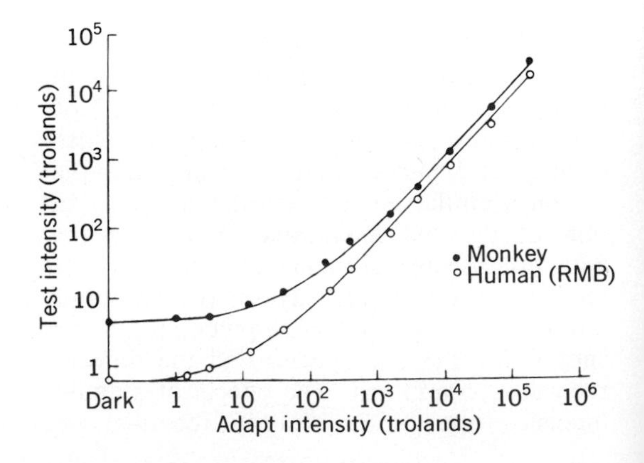

Fig. 15-21. Incremental thresholds as function of steady adapting intensity for human observer and for receptor potential of macaque monkey. Both curves obtained with same optical stimulator, using yellow adapting field 1.1 mm in diameter and centered on fovea. Test stimulus was 150 msec flash of yellow light. For human, test stimulus was confined to fovea, and thresholds were obtained by adjusting test stimulus to minimum detectable intensity. In monkey, pure-cone foveal receptor potential was recorded as in lower half of Fig. 15-20 and evoked by 1.1 mm test flash centered on fovea. Responses were averaged for detection of small responses, and 10 μv amplitude of receptor potential was taken as threshold. (From Boynton and Whitten.[25])

detection of incremental intensities by cone receptor potentials of the monkey fovea. The psychophysical function in humans shows the effect of steady adapting intensity upon the incremental test intensity required to elicit a threshold sensation. This function rises at first slowly and then more rapidly. The monkey results show the effect of steady adapting intensity upon the incremental test intensity required to elicit a very small cone receptor potential of constant amplitude. The similarity of the two curves indicates that the effect of background intensity upon incremental sensitivity is determined largely at the receptors, and the reasons for this have been analyzed.[25]

At greater than approximately 10^3 trolands of adapting intensity the two curves of Fig. 15-21 have similar slopes, indicating that in this range the psychophysical effects of adapting intensity are strongly determined by the receptors. In this high range of intensities it might be expected that a background intensity would be reached that would elicit a cone receptor potential of maximum amplitude, causing the incremental threshold function to rise vertically as the threshold became infinitely high. Fig. 15-21 shows that this does not happen; instead, cones are able to give incremental brightness sensations against indefinitely high background illuminations. The basis for this important aspect of cone function now seems clear.[25] It is only above approximately 10^3 trolands that the bleaching of cone photopigment by a steady background illumination becomes significant. Above this level the cone receptor potential elicited by a steady background illumination is only about half the amplitude elicited by a brief flash of equal intensity. This is because the long duration of the steady background causes photopigment bleaching, thus reducing receptor sensitivity and the amplitude of cone receptor potential elicited by a steady background. In the intensity range in which this effect occurs, each doubling of background intensity will roughly halve the photopigment concentration, so that the number of quanta absorbed per unit time remains constant, and the cone receptor potential elicited by the background illumination is unchanged. Thus photopigment bleaching serves to hold the cone receptor potential elicited by intense backgrounds to only about half the maximum cone response, keeping the cone response in its dynamic range and permitting an additional cone response to the incremental stimulus. Of course, the incremental threshold rises in this range of background intensity, as shown in Fig. 15-21, because progressively less photopigment is available to absorb the incremental light flash. Thus bleaching of photopigment reduces sensitivity to incremental light flashes, but serves to extend greatly the upper limit of background illuminations against which cones are capable of intensity discrimination. This mechanism is not available to rods, apparently because rod receptor potentials reach maximum amplitude at background intensities that do not cause significant bleaching of photopigment.[25] Thus with increased background intensity the incremental threshold of rods rises sharply to infinity at the upper end of the rod response range.[16] This does not impair intensity discrimination because at these background intensities the cones, endowed with a remarkable mechanism of intensity discrimination up to extremely high levels of background intensity, have already taken over.

RECEPTOR EXCITATION

The outlines of this subject are now defined and some of the critical steps are known. When the time of decay of the cone receptor potential is altered by stimulus intensity, as shown in Fig. 15-19, the timing of off-responses at later levels of retinal activity is correspondingly affected,[29, 100] a strong indication that the receptor potential, or events associated with this potential, play a causal role in activating second-order cells. Current underlying the receptor potential appears to flow from the outer segment to the axon terminal, where it presumably initiates release of a synaptic transmitter to activate bipolar and horizontal cells. Hence this current probably carries the entire burden of transmission through the photoreceptor, which does not generate nerve impulses.

Steady currents flow along the photoreceptor in darkness, which are thus called *dark currents,* and it is now believed that light-induced transient modulations of these dark currents give rise to the receptor potentials. Penn and Hagins[88] first showed that in the dark a steady extracellular current flows from the inner to the outer segment; the type of dark current they described appears to result mainly from a higher Na^+ permeability of the outer segment membrane.[121] This dark current maintains the outer and inner segments in a relatively depolarized state, as indicated by their low resting membrane potentials.[21, 82, 108] Light causes an increased electrical resistance of the outer segment membrane,[108] which results largely or entirely from a decreased membrane Na^+ conductance.[72] This reduces the depolarizing dark current,[88] so that the outer and inner segments become hyperpolarized with respect to their membrane potentials in the dark. Note that light does not initiate an extracellular current flow from outer to inner segment but reduces a steady current flowing the opposite direction. Relative to the inactivated state, this would make the outer segment more positive than the inner segment, as shown in Fig. 15-18.

Zuckerman has demonstrated a second type of dark current, generated by an electrogenic pump,[121] that has been localized to the inner segment,[122] a site of intense metabolic activity as indicated by the high concentration of mitochondria (Fig. 15-10). This electrogenic pump extrudes Na^+ more rapidly than K^+ enters, thus ejecting current from the inner segment. The ejected current then divides and reenters the cell, partly at the outer segment and partly at the axon terminal.[122] There is evidence that the electrogenic pump acts as a virtually constant source of current during the rising phase of the receptor potential.[122] It follows that when a light flash increases the electrical resistance of the outer segment, it would decrease pump current flowing into the outer segment, with a corresponding *increase* of pump current flowing into the axon terminal. These findings indicate why current underlying the receptor potential appears to flow all the way from the outer segment to the axon terminal. These results also explain how light-induced current is transferred efficiently to the axon terminal, where synaptic transmission must be initiated.

The transduction problem now involves the means by which light increases electrical resistance of the outer segment membrane by reducing its Na^+ conductance. The critical photochemical events have not been identified, but many investigators are now thinking along a common line. The general hypothesis is that the action of light upon the photopigment in the saccule membrane opens channels to release from inside the saccule a chemical transmitter that diffuses to the outer segment membrane and decreases its Na^+ conductance by preempting the Na^+ channels. This hypothesis is especially attractive with respect to the distal tips of rod outer segments, where the saccules have no connection with the outer membrane and where a chemical transmitter seems to be required. The measured latency from stimulus to initial rise of the macaque monkey's receptor potential decreases markedly with stimulus intensity in the low range of intensities; it then decreases more slowly and appears to have a minimum value of about 1.5 msec.[3] This minimum latency of the receptor potential must be accounted for; it may well be required for the diffusion of a chemical transmitter from saccule to plasma membrane of the outer segment. If this is the case, the receptor potential should have much in common with postsynaptic potentials.

SINGLE-CELL ACTIVITY OF INNER NUCLEAR LAYER

Horizontal cells

It has long been possible to record intracellular responses from single cells near the outer margin of the inner nuclear layer. These responses are called S potentials after their discoverer, Svaetichin, who first found them in fish retinas in 1953.[105] Intracellular marking has now identified them as horizontal cell responses in both lower vertebrates[71, 77, 115] and mammals.[102] In common with photoreceptors, horizontal cells of the cat retina yield only hyperpolarizing slow potentials.[30, 35] Furthermore, cat S potentials follow closely the time courses of both rod and cone receptor potentials.[30, 101] Thus mammalian hori-

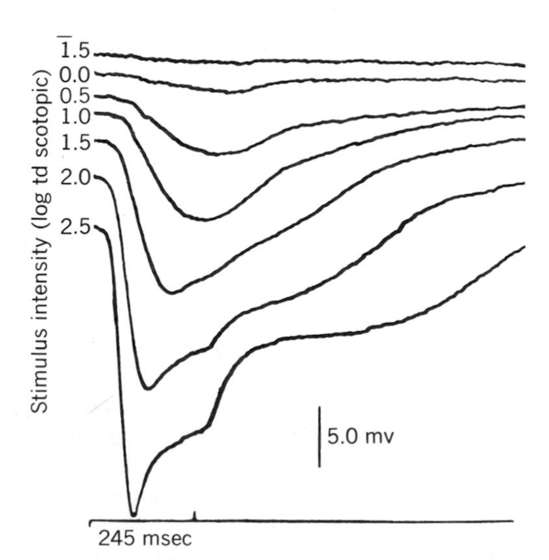

Fig. 15-22. Intracellular potentials from horizontal cell of cat retina as a function of stimulus intensity. Stimulus spot of 2.0 mm diameter was centered on electrode. For further explanation, see text. (From Steinberg.[101])

zontal cells are activated directly by the receptors and have inputs from both rods and cones. Fig. 15-22 shows intracellular responses from a cat horizontal cell as a function of stimulus intensity. At low stimulus intensities there is only a slowly rising and decaying rod response, which increases in amplitude and decays more slowly as stimulus intensity increases. Then a rapidly decaying cone response appears and increases in amplitude, as seen in responses to the two highest intensities, and at still higher intensities the cone response becomes much larger than is shown. The respective rod and cone origins of these responses were confirmed by spectral response curves.[101] Thus horizontal cells of the cat retina integrate rod and cone responses.

Effects of varying the stimulating wavelength on S potentials have been studied most intensively in fish retinas.[77, 106] The responses of some cells are hyperpolarizing and vary only in amplitude, as in the cat. Such cells yield spectral sensitivities that correspond well with luminosity functions, so they appear to register the luminosity of the stimulus and are called L-type cells.[106] Other cells exhibit changes in both amplitude and *polarity* of the response. Since they register chromaticity of the stimulus in terms of response polarity, they are called C-type cells.[106] Fig. 15-23 shows responses from two C-type horizontal cells when stimulated by brief flashes containing equal energy but varying in wavelength. In one cell the maximum response of positive polarity occurred in the red part of the spectrum, while the maximum negative response was in the green, so it is labeled an R-G cell. Another cell gave maximum positive response in the yellow and maximum negative response in the blue and is labeled a Y-B cell. These results suggest a mechanism of color discrimination first proposed by Hering and now referred to as the opponent-response theory of color vision.[70] Hering's theory stated that there is one class of cell giving opposite reactions to red and green stimuli, while another class of cell gives opposite responses to yellow and blue. This theory appears to hold for C-type horizontal cells and helps to account for certain aspects of color vision that are difficult to understand by the Young-Helmholtz mechanism alone. Although all color sensations may be elicited by an appropriate mixture of three primary colors, a purely subjective analysis of the visible spectrum shows four primary hues, including red, yellow, green, and blue. Since the hue of yellow is as distinctive as the other three, one may expect that the sensation of yellow is associated with a physiologic response as distinctive as that underlying the other three. Thus the findings shown in Fig. 15-23 can give an account of the four *subjective* primary colors in the spectrum, which result from opposing responses in two classes of cells. It is well known that red and green are complementary colors, while yellow and blue are also complementary colors; this means that each pair can be mixed to give cancellation of color sensations. This aspect of color mixing is also difficult to explain based solely on the Young-Helmholtz mechanism but is readily accounted for by the findings presented in Fig. 15-23; e.g., when the R-G cell is stimulated simultaneously by red and green wavelengths in an appropriate intensity ratio, the depolarizing and hyperpolarizing responses exactly cancel each other.[106] Hence it appears likely that the response code for color is converted from a three-receptor code to an opponent-response code at the horizontal cells. This could be accomplished by combining the effects of receptor activity on horizontal cells in specific ways. For example, the R-G cell appears to receive inputs from cones containing red and green photopigments, being depolarized by the red cones and hyperpolarized by the green cones. The

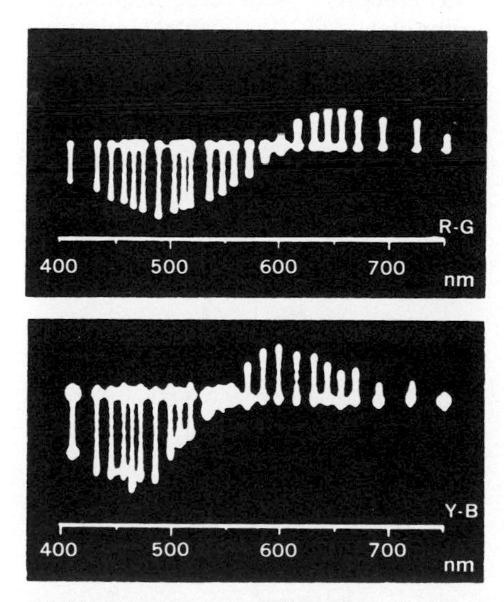

Fig. 15-23. Responses as a function of stimulating wavelength from two different horizontal cells of fish retina. See text for explanation. (From Svaetichin and MacNichol.[106])

Y-B cell is maximally depolarized by the yellow region of the spectrum, which strongly stimulates both red and green cones. Hence the Y-B cell appears to receive depolarizing inputs from both the red and green cones and a hyperpolarizing input from blue cones. Once established at the level of horizontal cells, there is evidence that the basic opponent-response type of code is maintained for color responses at higher levels of the visual system.

Since the retina is unique in combining receptors with brain tissue, compactly joined at a peripheral level of the system, a question arises regarding the functional advantage that may have endowed this arrangement with survival value during its evolution. The horizontal cells suggest an answer by mixing receptor signals on the basis of slow potentials alone. It is theoretically efficient for the early stages of signal mixing to be accomplished without impulse activity, thus avoiding the necessity for changing an amplitude-modulated receptor potential to an impulse frequency code for purposes of transmission, then changing back to an amplitude-modulated signal code in the postsynaptic potential of the first synapse. Since transmission over any appreciable distance requires impulses, their elimination requires neurons to be quite short. The retina has achieved this arrangement by evolving an outpouching of the brain to meet the receptors. There is a definite advantage in the brain tissue going out to meet the receptors, rather than vice versa; the retina can thus move independently of the brain for scanning the visual scene.

Bipolar and amacrine cells

In *Necturus* and goldfish, intracellular recordings have been made from all three major classes of neurons in the inner nuclear layer, each cell type being identified by intracellular dye marking.[71, 115] Results on the two species agree well, and results on horizontal cells agree with previous work in other species. Responses from bipolar and amacrine cells of the goldfish are shown in Fig. 15-24. The bipolar cell likewise gives only slow potentials in response to stimulation, showing no sign of nerve impulses. In this cell the polarity of the response depends on location of the stimulus in relation to the cell. To obtain the upper record in Fig. 15-24 the stimulus spot was centered upon the electrode, thus activating a small receptor area concentric with the impaled bipolar cell and eliciting a hyperpolarization of the bipolar cell. The stimulus for the lower record was an annulus, which elicited a depolarization. The *receptive field* of any given visual cell may be defined as the retinal receptor area within which a light stimulus affects the response of that cell. The receptive field of a goldfish bipolar cell is organized into central and peripheral portions that elicit opposing responses and are mutually antagonistic when both regions are stimulated simultaneously. Thus far all goldfish bipolar cells have been found to have this general type of receptive field organization, but in some bipolar cells the central stimulus de-

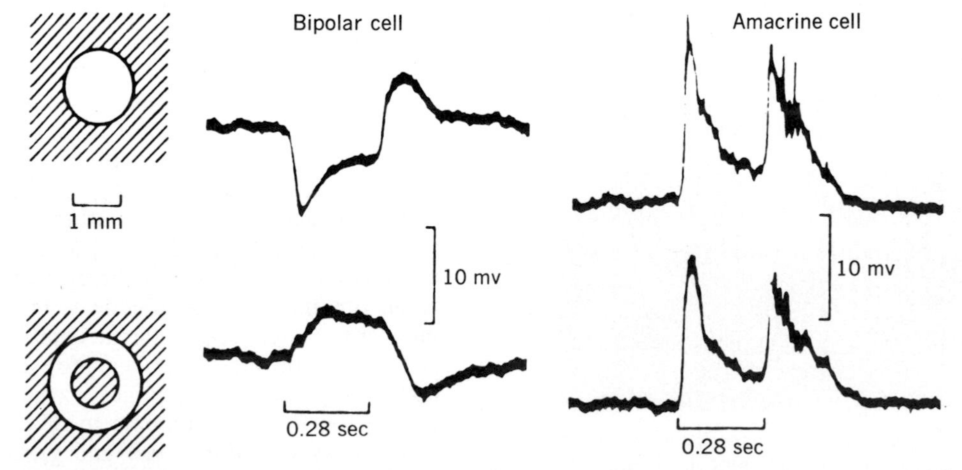

Fig. 15-24. Functional organization of receptive fields of bipolar and amacrine cells of goldfish retina, as revealed by intracellular recording. For explanation, see text. (From Kaneko.[71])

polarizes while the peripheral portion of the receptive field hyperpolarizes.

By contrast, Fig. 15-24 shows that an amacrine cell gave transient depolarizations when the stimulus was turned either on or off, and the same type of response was elicited by stimuli to either the center or periphery of the cell's receptive field. All amacrine cells of the goldfish have thus far shown these response characteristics. The amacrines appear to be the first cells in the direct neural pathway to discharge impulses, typically during the transient depolarizations, and several such impulses may be seen in the top record in Fig. 15-24. The detailed functions of amacrine cells remain to be clarified, but they appear primarily to register *changes* in illumination. In this they differ from the responses of all other retinal nerve cells, which seem well designed to register not only changes in illumination but also the intensities of steady levels of illumination.

GANGLION CELLS

The retinal cells that have been studied in most detail are the ganglion cells. Hartline's early work in cold-blooded vertebrates was

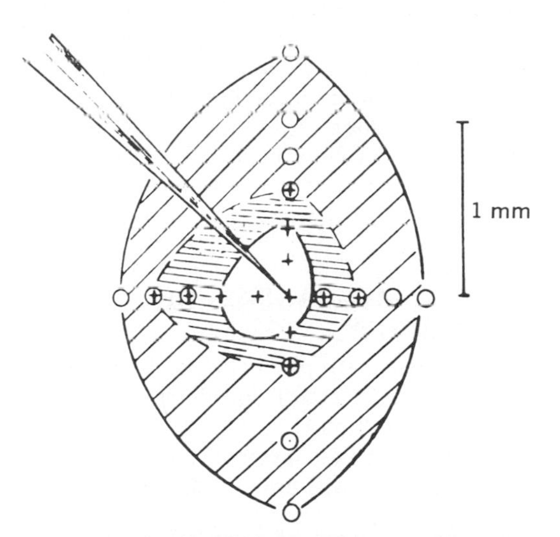

Fig. 15-25. Distribution of discharge patterns within receptive field of cat ganglion cell in light-adapted state. Recording electrode is shown, at tip of which ganglion cell soma was located, and retina was light adapted by steady background illumination. Exploring stimulus spot was 0.2 mm in diameter and of intensity about 2 log units above threshold for evoking response in center of field. In central part of receptive field (crosses) only on-discharges were found, while in diagonally hatched periphery of field only off-discharges occurred (circles). In intermediate zone (horizontally hatched), discharges were on-off. (From Kuffler.[73])

done by dissecting out single fibers of the optic fiber layer.[59] Early mammalian studies were accomplished by Granit's method of removing the cornea and lens and placing upon the retinal surface a glass-insulated wire electrode, which in favorable cases can detect the activity of a single ganglion cell.[8] Both techniques showed that ganglion cells fire impulses, as expected from the length of their axons. These studies also classified responses to light in three major categories: on-discharges, off-discharges, and on-off–discharges, depending on whether the cell responded with a burst of impulses at the beginning or at the termination of the stimulus, or at both.

When Kuffler[73] developed the technique of inserting a needle through the temporal side of the cat eye and introducing a Granit-type electrode through this needle, the normal optics of the eye were maintained and it became possible to stimulate the mammalian retina with well-focused exploring spots of light. Kuffler discovered that the response of a single ganglion cell varies with the position of the stimulus spot. The schematic organization of responses within the receptive field of a ganglion cell in the *light-adapted state* is shown in Fig. 15-25. Stimulation of the center of the receptive field (immediately beneath the soma of the cell) yielded on-discharges. Stimulation of the periphery of the field gave off-discharges, and an intermediate zone yielded on-off–discharges. If stimuli were delivered simultaneously in on- and off-zones, each stimulus inhibited the response evoked by the other. Thus the receptive field of the cell may be divided roughly into an on-center and an off-periphery, with mutual inhibition occurring between the two areas. Responses from a single ganglion cell that illustrate these principles are shown in Fig. 15-26. In record *A* a small stimulus in the center of the receptive field elicited an intense on-discharge. The response frequency was so high that the individual impulses were not resolved in the photograph. Also, because each impulse was following in the relative refractory period of the preceding impulse, when the membrane had not fully repolarized, impulse amplitude decreased markedly during the early part of the on-discharge. It is typical that on-discharges are followed by inhibition on termination of the stimulus, as shown in this record. In record *B* an annular stimulus was used to stimulate primarily the peripheral portion of the receptive field. The onset of this stimulus gave inhibition, which

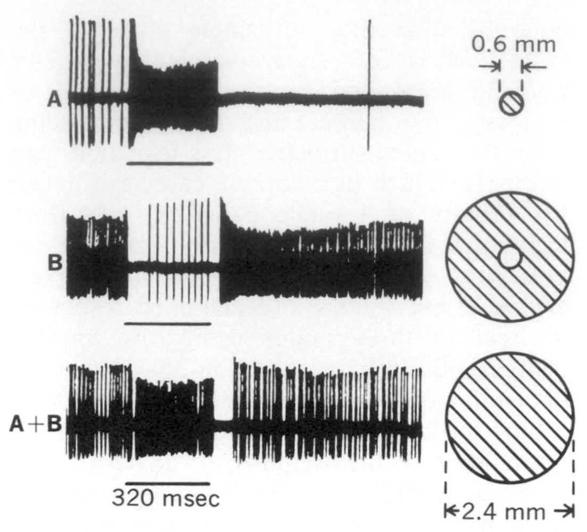

Fig. 15-26. Functional organization of light-adapted receptive field of cat ganglion cell. Retina light adapted by steady background illumination on which stimuli were superimposed. Stimulus pattern centered on electrode, and stimulus area evoking each response shown at right. Stimulus repeated every 10 sec, which was insufficient for complete recovery from previous off-discharge. Thus in record *B*, prestimulus discharge rate was higher than in other records, to better illustrate inhibitory effects of light stimulation. Principles illustrated still hold with constant prestimulus discharge rate. For further explanation, see text.

decreased during the stimulus, as shown by a gradual increase of impulse frequency. When the stimulus terminated, the off-discharge that occurred was of such high frequency that impulse amplitude again decreased. When the stimuli of *A* and *B* were combined into a single large stimulus, shown as *A* + *B*, the on-discharge was weaker than in record *A*, and none of the off-discharge of record *B* may be seen. Thus each stimulus area inhibited the discharge produced by the other.

The only important difference found thus far among ganglion cells of the cat, with respect to qualitative receptive field organization, is that some cells have an on-center and off-periphery while others have an off-center and on-periphery.[35, 73] The two types seem present in approximately equal numbers, and the same principles of receptive field organization hold for most ganglion cells of the spider monkey.[69] Intracellular recording from cat ganglion cells has shown that on-discharges result from depolarization of the ganglion cell membrane, while light stimuli that hyperpolarize the ganglion cell give rise to off-discharges when the cell is released from hyperpolarization at the end of the stimulus.[35, 116] The recent results on bipolar cells show that their receptive fields are organized similarly to those of ganglion cells, although the bipolar cells do not fire impulses. This makes it likely that some of the basic characteristics of ganglion cell receptive fields are already organized at the bipolar cell and passed on rather directly to the ganglion cell. For example, consider a bipolar cell in which the receptive field has a depolarizing center and a hyperpolarizing periphery. If these signals were passed on directly to a ganglion cell without inverting response polarity, the receptive field of that ganglion cell would have an on-center and an off-periphery. Similarly, the bipolar cell that has a receptive field with hyperpolarizing center and depolarizing periphery could account for the ganglion cell whose receptive field has an off-center and an on-periphery. If response polarity inverts between bipolar and ganglion cells, so that depolarization of the bipolar causes hyperpolarization of the ganglion cell, and vice versa, this would only alter which type of bipolar cell accounts for which type of ganglion cell.

Work in the spider monkey indicates that there are also a few specialized ganglion cells that give on-responses to short wavelengths and off-responses to long wavelengths, with mutual inhibition occurring between the responses to short and long wavelengths.[69] These ganglion cells may carry the color response, coded in terms of on- and off-discharges, and further analysis of their activity is required.

NEURAL MECHANISMS OF ADAPTATION

In addition to photochemical adaptation, there are also neural mechanisms that influence adaptation.[20, 34, 48, 76, 97] The methods for studying these mechanisms having been presented, they will now be discussed.

Early neural stage of adaptation

Fig. 15-27 shows an experiment that reveals and localizes a neural stage of adaptation in the monkey retina at an early level of retinal signal processing. Following a period of dark adaptation, the retina was light adapted by repeated stimuli. The effects of this light adaptation were determined by measuring amplitude of the b-wave, and again by measuring amplitude of the receptor potential after its isolation by clamping the retinal cir-

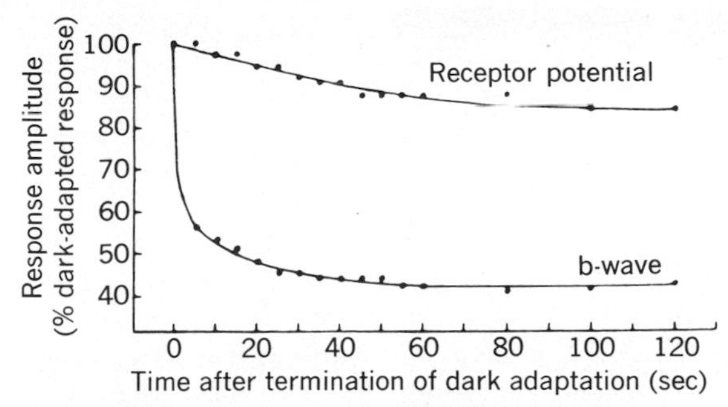

Fig. 15-27. Changes in amplitude of receptor potential and b-wave of macaque monkey during light adaptation by successive stimuli. All responses recorded from same area of peripheral retina by intraretinal electrode near pigment epithelium. Results were first obtained for b-wave by recording normal ERGs; then retinal circulation was clamped to isolate receptor potential and experiment was repeated. Prior to each series of responses, retina light adapted to steady level, followed by 15 min of dark adaptation. Then retina light adapted by repetitive stimulation with large stimulus spot centered on electrode. The 320 msec flash was repeated every 5 sec, and response to each stimulus recorded. Thus the same stimuli that light adapted the retina evoked responses showing the effects of light adaptation. (From Brown and Murakami.[30])

culation. The receptor potential showed only a small and gradual decline, while the b-wave decreased more rapidly and to a much greater extent. Since the effects of light adaptation were much greater for the b-wave, there is a neural stage of adaptation between the receptor potential and generation of the b-wave by cells in the inner nuclear layer.

Neural adaptation at ganglion cell level

Further neural mechanisms have been demonstrated by determining the effects of adaptation on the organization of the receptive field of a single ganglion cell. When the retina is dark adapted, the discharge produced by the peripheral part of the receptive field changes to match that of the center.[20] Thus an on-center cell, which gives off-discharges from the periphery in the light-adapted state, gives on-discharges from the periphery after dark adaptation. The same principle applies to off-center cells. Hence the central and peripheral portions of the receptive field have summative effects in the dark-adapted state. This change of receptive field organization shows that dark adaptation changes the functional pathways from the receptors to the ganglion cell. This must occur because stimulation of the same receptors in the peripheral portion of the receptive field of a ganglion cell can elicit an off-discharge in one case but an on-discharge in another case. The mechanism whereby the functional pathways are altered between light-adapted and dark-adapted states is not known, but it is of special interest since it has important consequences for vision.

The receptive field organization of ganglion cells in the light-adapted state has a theoretical advantage for the detection of a border in the retinal stimulus. Although a detailed explanation will not be developed here, it may be assumed that detection of a border requires different discharge rates for ganglion cells that are close together but on opposite sides of the border. It appears that the effect of mutual inhibition between the central and peripheral parts of the receptive field is to *enhance* this difference between discharge rates of ganglion cells on opposite sides of the border, thus improving the ability of the retina to detect borders. After dark adaptation the ganglion cell can integrate the effects of stimuli throughout the receptive field, thereby improving its ability to detect weak light stimuli. In summary, the light-adapted receptive field seems to be organized for the detection of borders and fine details of the stimulus, at the expense of sensitivity, whereas the dark-adapted receptive field seems to be organized for improved light detection at the expense of detailed vision. This indicates that although the anatomy of the retina is exceedingly complex, it is highly organized in the service of visual requirements. There is even a flexibility of the functional pathways through the retina, which can change to meet the demands of different conditions.

EFFECTS OF LIGHT DEPRIVATION

The effects of severely reduced visual stimulation are of interest because of the information provided about the normal role that stimulation plays in the maturation and functioning of the visual system. The eye is unique for this type of study because light stimulation may be prevented by raising animals in darkness or fitting opaque covers over the eyes. In other sense organs the total absence of stimulation is difficult or impossible to ensure without destruction of the sense organ, and the ensuing tissue degeneration complicates interpretation of the results.

Systematic experimental work in this field was first conducted by Riesen and co-workers[41, 90-93] in chimpanzees. Animals were placed in total darkness at various times after birth and for varying periods. Visual abilities were determined by training animals to respond differentially to paired stimuli that tested color discrimination, visual acuity, and pattern vision. Visual reflexes were also tested, and certain animals were sacrificed at appropriate times for histologic study. Immediately after light deprivation, visual abilities were markedly affected. These effects included difficulty in fixating objects and in following moving stimuli. Visual acuity was also restricted. But the most marked effects were in discrimination of patterns. This was shown most dramatically by obvious behavior; e.g., although normal animals quickly recognize a feeding bottle by sight, as shown by vocalizations and other anticipatory responses, the light-deprived animals did not respond until actually touched by the feeding bottle. Effects were more severe when the duration of light deprivation was increased or when it was introduced earlier in infancy. Complete recovery occurred following total darkness from birth to 7 months of age. But when kept in total darkness from birth to 16 months or more, the effects were severe and irreversible. Aside from pupillary reactions, startle responses to turning on a light, and primitive pursuit movements of the eyes to follow a gross visual stimulus, such animals were blind. Histologic study of their retinas showed almost complete degeneration of the ganglion cell layer.[41] As might be expected, extensive degeneration also occurred in the lateral geniculate bodies of such animals,[93] since transneuronal degeneration occurs in cells of the lateral geniculate body after damage to optic nerve fibers. An important question is whether light deprivation prevents normal development of the visual system, or whether it damages a visual system after it is fully developed. One animal was raised normally for 8 months, and at that age was comparable to other normal animals of similar age on all tests. Following 16 months of subsequent light deprivation, severe deficits of visual abilities were found that were typical of those resulting from light deprivation, and from which there was very little recovery. When sacrificed, this animal also displayed an almost complete degeneration of the ganglion cell layer. The parafovea of this retina, compared with that of a histologically normal retina, is shown in Fig. 15-28. These findings show that light is necessary not only for the development of a normal retina and visual functions, but for the maintenance of normal function and even the anatomic integrity of retinal ganglion cells. In lower mammals such as the cat and rat the effects of light deprivation are less severe.[89] In the cat retina there is a significant thinning of the inner plexiform layer, and histochemical techniques have shown a reduced RNA content of the ganglion cells.[89] However, there is no disappearance of ganglion cells, even in cats raised in darkness from birth to 40 months.[89] In the cat a more marked type of degeneration occurs only more centrally, in the lateral geniculate body, where the individual cells are reduced in size.[117] Thus there are important species differences in the histologic effects of light deprivation.

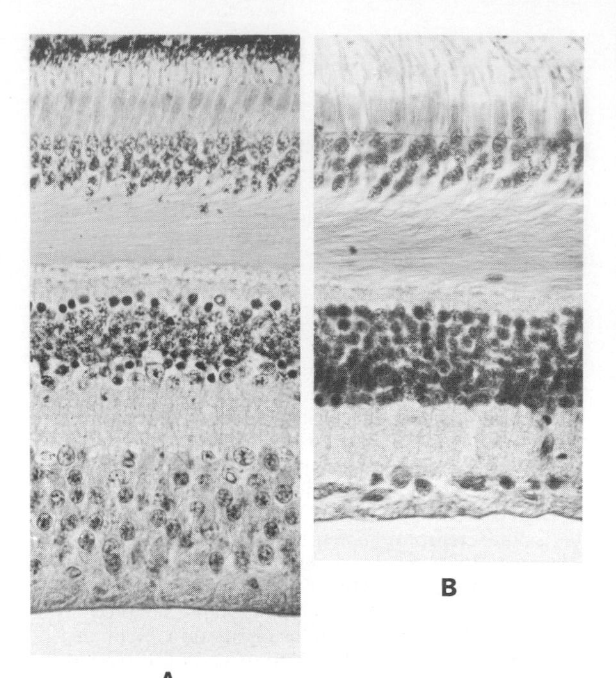

Fig. 15-28. Effect of prolonged light deprivation on ganglion cell layer of chimpanzee retina. **A,** Histologically normal control parafovea with typical thick ganglion cell layer. **B,** Parafovea from animal reared normally for first 8 months, then in total darkness for 16 months, and sacrificed at about 8 years of age. Although inner nuclear layer appears normal, ganglion cell layer is severely reduced. Thickness of experimental section was about twice that of control; thus it has generally darker appearance, especially in inner nuclear layer. Otherwise sections were similarly prepared and these photographs were at same magnification. (From Chow et al.[41])

If each day chimpanzees are given 90 min of diffuse light that is supplied through a diffusing plastic dome to the immobilized animal, retinal degeneration is prevented but the animals are still highly deficient in pattern vision.[91, 92] A control animal given 90 min of normal light experience each day was scarcely distinguishable from animals raised under completely normal conditions. Thus specific experience with forms seems necessary to develop normal pattern vision. If either chimpanzees or kittens are given only diffuse light to one eye but normal patterned stimulation to the other eye, monocular form blindness develops.[91, 118] Furthermore, when such a chimpanzee or kitten learns a pattern discrimination with one eye, it must be learned independently by the other eye.[91] Thus monocular form deprivation can disrupt the normal equivalence of the two visual pathways with respect to pattern discrimination. In more recent years the effects of light and form deprivation have also been studied by recording the impulse activity of cells at various levels of the cat's visual system.[117, 118] These investigations have confirmed many of the qualitative principles revealed by earlier methods and have shown that monocular form deprivation is correlated with an inability of the form-deprived eye to elicit activity in cortical cells.

The effects of form deprivation are also seen in congenital cataract patients after the cataracts are removed and corrective lenses are supplied. Triangles and squares are initially distinguished only by *counting* the number of corners, and a long learning period is required before recognition of even such simple forms is immediate. If the cataracts are removed by the end of the first year, reasonably good vision can be attained, but if they are removed too late, the effects of form deprivation are serious and cannot be overcome. Thus the relatively complex ability to discriminate patterns, which is accomplished primarily at central levels of the visual system, requires some type of learning process that must be started early in life. The role of prior experience in appreciating even more subtle visual qualities is well expressed by Thoreau's statement that "beauty is in the eye of the beholder." To those who study vision, his words are doubly meaningful. For the eye itself is an object of infinitely ordered beauty, which even dimly viewed is scarcely forgotten. Like all true beauty it grows with deeper understanding, for which the following references are recommended.

REFERENCES

General reviews

1. Boycott, B. B., and Dowling, J. E.: Organization of the primate retina: light microscopy, Phil. Trans. R. Soc. Lond. **255B:**109, 1969.
2. Brown, K. T.: The electroretinogram: its components and their origins, Vision Res. **8:** 633, 1968.
3. Brown, K. T., Watanabe, K., and Murakami, M.: The early and late receptor potentials of monkey cones and rods, Symp. Quant. Biol. **30:**457, 1965.
4. Committee on Colorimetry, Optical Society of America: The science of color, New York, 1953, Thomas Y. Crowell Co.
5. Davson, H., editor: The eye, New York, 1962, Academic Press, Inc., vol. 2.
6. Davson, H.: The physiology of the eye, ed. 2, London, 1963, J. & A. Churchill, Ltd.
7. Dowling, J. E., and Boycott, B. B.: Organization of the primate retina: electron microscopy, Proc. R. Soc. Lond. (Biol.) **166B:**80, 1966.
8. Granit, R.: Sensory mechanisms of the retina, New York, 1963, Hafner Publishing Co.
9. Hubbard, R., Bownds, D., and Yoshizawa, T.: The chemistry of visual photoreception, Symp. Quant. Biol. **30:**301, 1965.
10. Polyak, S.: The retina, Chicago, 1941, University of Chicago Press.
11. Polyak, S.: The vertebrate visual system, Chicago, 1957, University of Chicago Press.
12. Tomita, T.: Electrical activity of vertebrate photoreceptors, Q. Rev. Biophys. **3:**179, 1970.
13. Wald, G.: Molecular basis of visual excitation, Science **162:**230, 1968.
14. Wald, G., and Brown, P. K.: Human color vision and color blindness, Symp. Quant. Biol. **30:**345, 1965.
15. Wright, W. D.: Researches on normal and defective colour vision, London, 1946, Henry Kimpton.

Original papers

16. Aguilar, M., and Stiles, W. S.: Saturation of the rod mechanism of the retina at high levels of stimulation, Opt. Acta **1:**59, 1954.
17. Arden, G. B., and Weale, R. A.: Nervous mechanisms and dark-adaptation, J. Physiol. **125:**417, 1954.
18. Baker, H. D.: Initial stages of dark and light adaptation, J. Opt. Soc. Am. **53:**98, 1963.
19. Barlow, H. B.: The physical limits of visual discrimination. In Giese, A. C., editor: Photophysiology, New York, 1964, Academic Press, Inc., vol. 2.
20. Barlow, H. B., Fitzhugh, R., and Kuffler, S. W.: Change of organization in the receptive fields of the cat's retina during dark adaptation, J. Physiol. **137:**338, 1957.
21. Baylor, D. A., and Fuortes, M. G. F.: Electrical responses of single cones in the retina of the turtle, J. Physiol. **207:**77, 1970.
22. Berson, E. L., and Goldstein, E. B.: The early receptor potential in sex-linked retinitis pigmentosa, Invest. Ophthalmol. **9:**58, 1970.
23. Bok, D., and Hall, M. O.: The etiology of retinal dystrophy in RCS rats, Invest. Ophthalmol. **8:**649, 1969.
24. Bortoff, A.: Localization of slow potential responses in the Necturus retina, Vision Res. **4:**627, 1964.
25. Boynton, R. M., and Whitten, D. N.: Visual adaptation in monkey cones: recordings of late receptor potentials, Science **170:**1423, 1970.

26. Brown, K. T.: An early potential evoked by light from the pigment epithelium-choroid complex of the eye of the toad, Nature **207:** 1249, 1965.
27. Brown, K. T., and Murakami, M.: A new receptor potential of the monkey retina with no detectable latency, Nature **201:**626, 1964.
28. Brown, K. T., and Murakami, M.: The biphasic form of the early receptor potential of the monkey retina, Nature **204:**739, 1964.
29. Brown, K. T., and Murakami, M.: Delayed decay of the late receptor potential of monkey cones as a function of stimulus intensity, Vision Res. **7:**179, 1967.
30. Brown, K. T., and Murakami, M.: Rapid effects of light and dark adaptation upon the receptive field organization of S-potentials and late-receptor potentials, Vision Res. **8:**1145, 1968.
31. Brown, K. T., and Tasaki, K.: Localization of electrical activity in the cat retina by an electrode marking method, J. Physiol. **158:** 281, 1961.
32. Brown, K. T., and Watanabe, K.: Isolation and identification of a receptor potential from the pure cone fovea of the monkey retina, Nature **193:**958, 1962.
33. Brown, K. T., and Watanabe, K.: Rod receptor potential from the retina of the night monkey, Nature **196:**547, 1962.
34. Brown, K. T., and Watanabe, K.: Neural stage of adaptation between the receptors and inner nuclear layer of monkey retina, Science **148:** 1113, 1965.
35. Brown, K. T., and Wiesel, T. N.: Intraretinal recording with micropipette electrodes in the intact cat eye, J. Physiol. **149:**537, 1959.
36. Brown, K. T., and Wiesel, T. N.: Analysis of the intraretinal electroretinogram in the intact cat eye, J. Physiol. **158:**229, 1961.
37. Brown, K. T., and Wiesel, T. N.: Localization of origins of electroretinogram components by intraretinal recording in the intact cat eye, J. Physiol. **158:**257, 1961.
38. Brown, P. K., and Wald, G.: Visual pigments in human and monkey retinas, Nature **200:** 37, 1963.
39. Brown, P. K., and Wald, G.: Visual pigments in single rods and cones of the human retina, Science **144:**45, 1964.
40. Campbell, F. W., and Rushton, W. A. H.: Measurement of the scotopic pigment in the living human eye, J. Physiol. **130:**131, 1955.
41. Chow, K. L., Riesen, A. H., and Newell, F. W.: Degeneration of retinal ganglion cells in infant chimpanzees reared in darkness, J. Comp. Neurol. **107:**27, 1957.
42. Cohen, A. I.: The fine structure of the extrafoveal receptors of the rhesus monkey, Exp. Eye Res. **1:**128, 1961.
43. Cohen, A. I.: Further studies on the question of the patency of saccules in outer segments of vertebrate photoreceptors, Vision Res. **10:**445, 1970.
44. Cone, R. A.: The early receptor potential of the vertebrate eye, Symp. Quant. Biol. **30:** 483, 1965.
45. Cone, R. A., and Brown, P. K.: Dependence of the early receptor potential on the orientation of rhodopsin, Science **156:**536, 1967.
46. De Robertis, E.: Some observations on the ultrastructure and morphogenesis of photoreceptors, J. Gen. Physiol. **43**(suppl.)**:**1, 1960.
47. Dowling, J. E.: Chemistry of visual adaptation in the rat, Nature **188:**114, 1960.
48. Dowling, J. E.: Neural and photochemical mechanisms of visual adaptation in the rat, J. Gen. Physiol. **46:**1287, 1963.
49. Dowling, J. E., and Ripps, H.: Visual adaptation in the retina of the skate, J. Gen. Physiol. **56:**491, 1970.
50. Fernandez-Moran, H.: The fine structure of vertebrate and invertebrate photoreceptors as revealed by low-temperature electron microscopy. In Smelser, G. K., editor: The structure of the eye, New York, 1961, Academic Press, Inc.
51. Goldstein, E. B.: Contribution of cones to the early receptor potential in the rhesus monkey, Nature **222:**1273, 1969.
52. Goldstein, E. B., and Berson, E. L.: Cone dominance of the human early receptor potential, Nature **222:**1272, 1969.
53. Granit, R.: Neural activity in the retina. In Magoun, H. W., editor, Neurophysiology section: Handbook of physiology, Baltimore, 1959, The Williams & Wilkins Co., vol. 1.
54. Griffin, D. R., Hubbard, R., and Wald, G.: The sensitivity of the human eye to infra-red radiation, J. Opt. Soc. Am. **37:**546, 1947.
55. Hagins, W. A., and McGaughy, R. E.: Molecular and thermal origins of fast photoelectric effects in the squid retina, Science **157:**813, 1967.
56. Hagins, W. A., Zonana, H. V., and Adams, R. G.: Local membrane current in the outer segments of squid photoreceptors, Nature **194:** 844, 1962.
57. Hall, M. O., Bok, D., and Bacharach, A. D. E.: Biosynthesis and assembly of the rod outer segment membrane system. Formation and fate of visual pigment in the frog retina, J. Mol. Biol. **45:**397, 1969.
58. Hanaoka, T., and Fujimoto, K.: Absorption spectrum of a single cone in carp retina, Jap. J. Physiol. **7:**276, 1957.
59. Hartline, H. K.: The response of single optic nerve fibers of the vertebrate eye to illumination of the retina, Am. J. Physiol. **121:**400, 1938.
60. Hecht, S.: Rods, cones, and the chemical basis of vision, Physiol. Rev. **17:**239, 1937.
61. Hecht, S., and Verrijp, C. D.: Intermittent stimulation by light, J. Gen. Physiol. **17:**251, 1933-34.
62. Hecht, S., Shlaer, S., and Pirenne, M. H.: Energy, quanta, and vision, J. Gen. Physiol. **25:**819, 1942.
63. Hecht, S., et al.: The visual functions of the complete colorblind, J. Gen. Physiol. **31:**459, 1948.
64. Heller, J.: Comparative study of a membrane protein. Characterization of bovine, rat, and frog visual pigments$_{500}$, Biochemistry **8:**675, 1969.
65. Herron, W. L., Riegel, B. W., and Rubin, M. L.: Outer segment production and removal in the degenerating retina of the dystrophic rat, Invest. Ophthalmol. **10:**54, 1971.
66. Herron, W. L., Riegel, B. W., Myers, O. E.,

and Rubin, M. L.: Retinal dystrophy in the rat—a pigment epithelial disease, Invest. Ophthalmol. **8**:595, 1969.

67. Hubbard, R., and Kropf, A.: Molecular aspects of visual excitation, Ann. N. Y. Acad. Sci. **81**:388, 1959.
68. Hubbard, R., and St. George, R. C. C.: The rhodopsin system of the squid, J. Gen. Physiol. **41**:501, 1957.
69. Hubel, D. H., and Wiesel, T. N.: Receptive fields of optic nerve fibres in the spider monkey, J. Physiol. **154**:572, 1960.
70. Hurvich, L. M., and Jameson, D.: Perceived color, induction effects, and opponent-response mechanism, J. Gen. Physiol. **43**(suppl.):63, 1960.
71. Kaneko, A.: Physiological and morphological identification of horizontal, bipolar and amacrine cells in goldfish retina, J. Physiol. **207**: 623, 1970.
72. Korenbrot, J. K., and Cone, R. A.: Ionic permeabilities of rod outer segments determined by osmotic behavior, Biophys. Soc. Abstr. **11**:45a, 1971.
73. Kuffler, S. W.: Discharge patterns and functional organization of mammalian retina, J. Neurophysiol. **16**:37, 1953.
74. Kuffler, S. W., and Potter, D. D.: Glia in the leech central nervous system: physiological properties and neuron-glia relationship, J. Neurophysiol. **27**:290, 1964.
75. Laties, A. M., and Liebman, P. A.: Cones of living amphibian eye: selective staining, Science **168**:1475, 1970.
76. Lipetz, L. E.: A mechanism of light adaptation, Science **133**:639, 1961.
77. MacNichol, E. F., Jr., and Svaetichin, G.: Electric responses from the isolated retinas of fishes, Am. J. Ophthalmol. **46**:26, 1958.
78. Mandelbaum, J., and Sloan, L. L.: Peripheral visual acuity, Am. J. Ophthalmol. **30**:581, 1947.
79. Marks, W. B.: Visual pigments of single goldfish cones, J. Physiol. **178**:14, 1965.
80. Marks, W. B., Dobelle, W. H., and MacNichol, E. F., Jr.: Visual pigments of single primate cones, Science **143**:1181, 1964.
81. Miller, R. F., and Dowling, J. E.: Intracellular responses of the Müller (glial) cells of mudpuppy retina: their relation to b-wave of the electroretinogram, J. Neurophysiol. **33**: 323, 1970.
82. Murakami, M., and Pak, W. L.: Intracellularly recorded early receptor potential of the vertebrate photoreceptors, Vision Res. **10**: 965, 1970.
83. Noell, W. K., and Albrecht, R.: Irreversible effects of visible light on the retina: role of vitamin A, Science **172**:76, 1971.
84. Pak, W. L.: Some properties of the early electrical response in the vertebrate retina, Symp. Quant. Biol. **30**:493, 1965.
85. Pak, W. L.: Rapid photoresponses in the retina and their relevance to vision research, Photochem. Photobiol. **8**:495, 1968.
86. Pak, W. L., and Cone, R. A.: Isolation and identification of the initial peak of the early receptor potential, Nature **204**:836, 1964.
87. Pak, W. L., and Ebrey, T. G.: Visual receptor potential observed at sub-zero temperatures, Nature **205**:484, 1965.
88. Penn, R. D., and Hagins, W. A.: Signal transmission along retinal rods and the origin of the electroretinographic a-wave, Nature **223**: 201, 1969.
89. Rasch, E., Swift, H., Riesen, A. H., and Chow, K. L.: Altered structure and composition of retinal cells in dark-reared mammals, Exp. Cell Res. **25**:348, 1961.
90. Riesen, A. H.: Arrested vision, Sci. Am. **183**: 16, 1950.
91. Riesen, A. H.: Plasticity of behavior: psychological aspects. In Harlow, H. F., and Woolsey, C. N., editors: Biological and biochemical bases of behavior, Madison, Wisc., 1958, University of Wisconsin Press.
92. Riesen, A. H.: Effects of stimulus deprivation on the development and atrophy of the visual sensory system, Am. J. Orthopsychiatry **30**:23, 1960.
93. Riesen, A. H.: Effects of early deprivation of photic stimulation. In Osler, S. F., and Cooke, R. E., editors: The biological basis of mental retardation, Baltimore, 1965, The Johns Hopkins University Press.
94. Rushton, W. A. H.: Dark-adaptation and the regeneration of rhodopsin, J. Physiol. **156**: 166, 1961.
95. Rushton, W. A. H.: Rhodopsin measurement and dark-adaptation in a subject deficient in cone vision, J. Physiol. **156**:193, 1961.
96. Rushton, W. A. H.: Cone pigment kinetics in the protanope, J. Physiol. **168**:374, 1963.
97. Rushton, W. A. H., and Westheimer, G.: The effect upon the rod threshold of bleaching neighboring rods, J. Physiol. **164**:318, 1962.
98. Schultze, M.: Zur anatomie und physiologie der retina, Arch. Mikr. Anat. **2**:175, 1866.
99. Sjöstrand, F. S.: Ultrastructure of retinal rod synapses of the guinea pig eye as revealed by three-dimensional reconstructions from serial sections, J. Ultrastruct. Res. **2**:122, 1958.
100. Steinberg, R. H.: High-intensity effects on slow potentials and ganglion cell activity in the area centralis of cat retina, Vision Res. **9**:333, 1969.
101. Steinberg, R. H.: Rod-cone interaction in S-potentials from the cat retina, Vision Res. **9**:1331, 1969.
102. Steinberg, R. H., and Schmidt, R.: Identification of horizontal cells as S-potential generators in the cat retina by intracellular dye injection, Vision Res. **10**:817, 1970.
103. Steinberg, R. H., Schmidt, R., and Brown, K. T.: Intracellular responses to light from cat pigment epithelium: origin of the electroretinogram c-wave, Nature **227**:728, 1970.
104. Stevens, S. S.: On the psychophysical law, Psychol. Rev. **64**:153, 1957.
105. Svaetichin, G.: The cone action potential, Acta Physiol. Scand. **29**(suppl.):106, 565, 1953.
106. Svaetichin, G., and MacNichol, E. F., Jr.: Retinal mechanisms for chromatic and achromatic vision, Ann. N. Y. Acad. Sci. **74**: 385, 1958.
107. Tomita, T.: Electrophysiological study of the mechanisms subserving color coding in the fish retina, Symp. Quant. Biol. **30**:559, 1965.
108. Toyoda, J., Nosaki, H., and Tomita, T.: Light-

induced resistance changes in single photoreceptors of Necturus and Gekko, Vision Res. **9:**453, 1969.
109. Toyoda, J., Hashimoto, H., Anno, H., and Tomita, T.: The rod response in the frog as studied by intracellular recording, Vision Res. **10:**1093, 1970.
110. Vilter, V.: Histologie et activité electrique de la rétine d'un Mammifère strictement diurene, le Spermophile (Citellus citellus), Compt. Rend. Soc. Biol. **148:**1768, 1954.
111. Wald, G.: Alleged effects of the near ultraviolet on human vision, J. Opt. Soc. Am. **42:**171, 1952.
112. Wald, G.: The photoreceptor process in vision. In Magoun, H. W., editor, Neurophysiology section: Handbook of physiology, Baltimore, 1959, The Williams & Wilkins Co., vol. 1.
113. Wald, G., and Brown, P. K.: Human rhodopsin, Science **127:**222, 1958.
114. Wald, G., Jeghers, H., and Arminio, J.: An experiment in human dietary night-blindness, Am. J. Physiol. **123:**732, 1938.
115. Werblin, F. S., and Dowling, J. E.: Organization of the retina of the mudpuppy, Necturus Maculosus. II. Intracellular recording, J. Neurophysiol. **32:**315, 1969.
116. Wiesel, T. N.: Recording inhibition and excitation in the cat's retinal ganglion cells with intracellular electrodes, Nature **183:**264, 1959.
117. Wiesel, T. N., and Hubel, D. H.: Effects of visual deprivation on morphology and physiology of cells in the cat's lateral geniculate body, J. Neurophysiol. **26:**978, 1963.
118. Wiesel, T. N., and Hubel, D. H.: Single-cell responses in striate cortex of kittens deprived of vision in one eye, J. Neurophysiol. **26:** 1003, 1963.
119. Young, R. W.: Visual cells, Sci. Am. **223:**81, 1970.
120. Young, R. W.: An hypothesis to account for a basic distinction between rods and cones, Vision Res. **11:**1, 1971.
121. Zuckerman, R.: Mechanisms of photoreceptor current generation in light and darkness, Nature New Biol. **234:**29, 1971.
122. Zuckerman, R.: Photoreceptor dark current: role for an electrogenic sodium pump, Biophys. Soc. Abstr. **12:**101a, 1972.

16

GIAN F. POGGIO

Central neural mechanisms in vision

One form of experience of the physical world of objects is furnished by the capacity man and other creatures possess to receive and interpret information carried by a particular type of radiant energy called light. This capacity for *vision* is provided by a specialized biologic system that accomplishes physicophysiologic transformations and neural decision processes leading to an integration of the information received into perceptual experience.

In previous chapters the optical properties of the receiving apparatus, the eye, and the mechanisms for the formation of images on the retinal surface have been discussed. Reference has also been made to the motor component of the system by which the position of the eye can be controlled either voluntarily or automatically and directed in space. The transducer function of the retinal receptors, the rods and cones, has been described as well as the anatomic and functional relation of these receptors to the retinal ganglion cells, the neural elements in which the neurally coded visual message originates.

This chapter deals with the portion of the visual system that extends from the retinal ganglion cells to the primary receiving area of the cerebral cortex, and beyond it, with those central systems that appear to be implicated in the elaboration of the visual experience. The evidence presented is obtained, whenever possible, from experimental investigations in primates and from studies in man. Findings from other animals, the cat in particular, are described when they are the only available ones and it is thought they may contribute to a better understanding of the central mechanisms in vision.

ORGANIZATION OF THE VISUAL AFFERENT SYSTEM

The axons of the retinal ganglion cells, the third-order neurons of the visual pathway, form the innermost layer of the retina (Chapter 15) and converge at the optic papilla into a bundle of fibers named the optic nerve, the chiasma, and the optic tract in different portions along its centripetal course. In primates the great majority of these axons terminate on cells of the dorsal lateral geniculate nucleus. A contingent of these fibers, or collaterals of them, pass to regions of the midbrain, particularly to the superior colliculus and pretectal region. The axons of geniculate neurons form the visual radiation and project wholly upon the striate cortex of the occipital lobe. Beyond the area striata the visual pathway progresses to other cortical regions, both adjacent and distant, to thalamic nuclei other than the dorsal lateral geniculate, and to other subcortical structures, including the superior colliculus–pretectum complex.

Like other major afferent systems, the visual system is topographically organized: within most of its components, the retinal surface is spatially represented in an orderly manner, and patterns of retinal activity correspond topologically to patterns of activity in central nervous structures. Details of the anatomic connections between structures subserving visual functions have come from the study of neural degenerations resulting from lesions that occur pathologically in man or are produced experimentally in animals. In humans, correlation of the functional defect in the field of vision with the anatomic location of the lesion has made it possible to determine in detail the central representation of the retinal surface. In the experimental animal, accurate retinotopic maps have been constructed at the superior colliculus, lateral geniculate nucleus, and especially at the cerebral cortex by means of the evoked potential method. Recent studies with microelectrodes have confirmed the nature of these maps and increased our understanding of the organization of the visual system.

Topographically the retina may be subdivided with respect to its specialized functional center, the fovea, into four quadrants. A vertical straight line, the vertical meridian, passing through the center of the fovea separates the retinal surface into a temporal and a nasal half. Because of the eccentric position of the fovea, these two halves are unequal in extent, the nasal half being larger than the temporal half (Fig. 14-1). A horizontal straight line, the horizontal meridian, divides each half into dorsal and ventral quadrants. Any number of retinal sectors may be defined by straight lines radiating from the center of the fovea.

Along the centroperipheral radial direction, two concentric principal retinal regions or zones are commonly recognized: a *central* and a *peripheral* region. Further zonal subdivisions may be made by any number of circles concentric to the fovea. Anatomically the division between central and peripheral retina rests upon the different accumulation of ganglion cells in the two regions. The central retina contains small cells that are densely packed in six to seven rows at the edge of the foveal depression and reduce to one row at the outer margin of the area. In the peripheral retina the ganglion cells form a continuous, occasionally broken, single layer.[8] The central area of the retina measures 5 to 6 mm in diameter, which corresponds to about 15 to 20 degrees of visual angle. In clinical literature the term "central area" generally designates the more restricted *macular* region occupied by the fovea and its vicinity, which is characterized by yellow pigmentation (macula lutea) and measures about 3 mm in diameter (10 degrees).

A similar topographic subdivision can be made of the visual field, i.e., of that portion of space which the retina subtends when the eye is held stationary (Fig. 16-6). Because of the transparent lens system, there exists, of course, an inverse spatial relation between the position of any point in the field and its image on the retinal surface.

THE PRIMARY VISUAL PROJECTION

The retinogeniculostriate pathway is the core of the visual afferent system and is termed the primary visual projection. In primates this pathway is elaborately organized topographically and forms a *unilateral* and *forward* system with no fibers passing over to the contralateral hemisphere and no "return" connections between striate cortex, geniculate nucleus, and retina. In the following discussion the anatomic organization of its components is described in some detail.

Retinogeniculate projection

In their intraretinal portion the axons of ganglion cells of the peripheral retina run toward the papilla of the optic nerve in an approximately radial course, while axons from the central retina, a large number of which are of small caliber, follow a straight or slightly arched horizontal path. At the papilla the macular fibers are located laterally and are flanked above and below by the fibers from the dorsal and ventral temporal retina, respectively, while the fibers from the nasal retina are located medially. On leaving the eyeball the fibers acquire their myelin sheathes and run backward in the optic nerve.

At the level of the chiasm the fibers from the nasal half of the retina cross the midline and enter the contralateral optic tract, while the fibers from the temporal half pass into the ipsilateral tract. This is true for macular fibers as well as for fibers from the peripheral retina, and the vertical meridian separates the portion of retina with a crossed projection from that with an uncrossed projection. Accordingly, the suprachiasmatic portion of the afferent visual system on each side of the brain carries a representation from each of the two retinas and subserves the contralateral hemifield of vision. Topographically the fibers from the temporal retina maintain their relative position on the outside of the chiasm, while the crossing nasal fibers occupy the central portion. Crossing macular fibers are situated posteriorly and dorsally and spread over a large extent of the central chiasm.[9, 66]

On leaving the chiasm, crossed and uncrossed fibers from the two homonymous dorsal retinal quadrants move to occupy the medial dorsal portion of the optic tract and those from the ventral quadrants its lateral ventral aspect. The macular fibers come to lie dorsocentrally in the tract and mix diffusely in medial and lateral areas containing the peripheral projections. Within the limits of this gross topographic arrangement, the fibers from the two eyes are fairly evenly distributed and intermingled in an apparently random fashion.[32, 66]

Thus there exists in all portions of the retinogeniculate projection a topographic organization of the optic fibers with respect to their retinal origin, but this organization is not a precise one. On the other hand, as the optic fibers reach the lateral geniculate nucleus, they are sorted in accord with their origin and terminate within this structure in a highly detailed topographic order.

Dorsal lateral geniculate nucleus (LGN)

The cells of this thalamic nucleus are arranged in layers, or laminae, separated by interlaminar bundles of fibers. In primates, six major cellular layers may be recognized, numbered 1 to 6 beginning at the hilus and continuing toward the dorsal aspect of the nucleus. The two most ventral layers, 1 and 2, contain cells larger and more uniform in size and shape than do the four dorsal layers. Two neuron types are distinguished in the LGN of the monkey: a large cell, thought to be a geniculostriate projection cell, and a small cell forming about 10% of the total neuron population, considered to be an interneuron.[83] Synaptic junctions are almost as densely distributed between cell laminae as within them. The large retinogeniculate axon terminals are confined to the cellular layers, while two other types of axons of unknown origin (intra- and/or extrageniculate) form synaptic contacts in all parts of the nucleus.[59, 83]

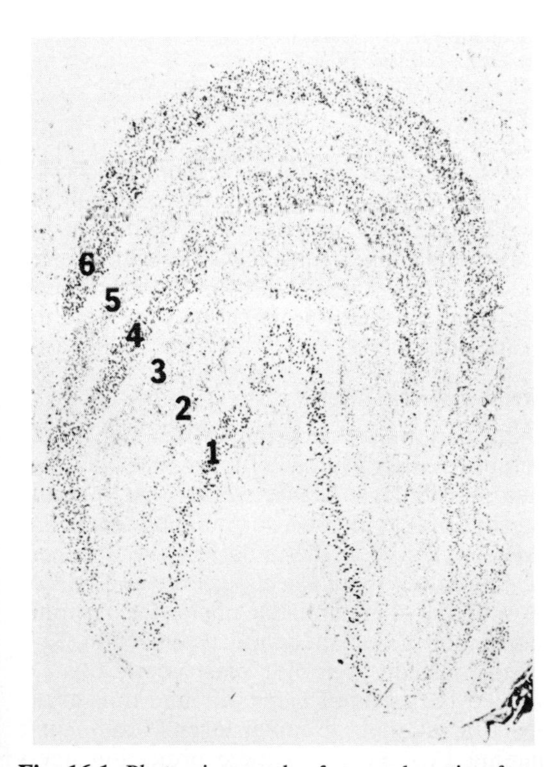

Fig. 16-1. Photomicrograph of coronal section from posterior half of left lateral geniculate nucleus of macaque 6 days after enucleation of ipsilateral eye. There is distinct transneuronal atrophy of neurons of uncrossed layers *(2, 3,* and *5),* while cells of crossed ones are normal *(1, 4,* and *6).* Section 25μ thick, stained with thionine. (×16.5.) (From Matthews et al.[85])

The laminar differentiation of the LGN is accompanied by a characteristic sorting of the incoming fibers from the two eyes: crossed and uncrossed projections terminate in a segregated fashion upon different cell layers. The fibers from the contralateral nasal hemiretina reach layers 1, 4, and 6, while the ipsilateral temporal fibers distribute to layers 2, 3, and 5 (Fig. 16-1). The termination of these fibers is sharply circumscribed, and minute lesions anywhere in the retina are followed by transneuronal degeneration of discrete local clusters of cells in the appropriate layers.[33] It is important to recognize that afferent fibers from the central area of the retina terminate mainly in the small-celled dorsal layers of the nucleus, whereas the ventral magnocellular layers receive fibers from the peripheral retina almost exclusively.

The topographic representation of the retinal surface within the LGN is highly organized and precisely defined. The replication of the horizontal meridian coincides with the median dorsoventral axis of the nucleus along its entire anteroposterior extent. Anatomically corresponding areas of the two hemiretinas project—each to the respective layers —onto successive radial sectors on either side of the median axis, the dorsal quadrants medially and the vertical ones laterally. The projection locales in the six laminae lie in exact register so that any small area in the contralateral field of vision can be shown as a straight dorsoventral column through the nucleus. The fibers from the central retina have an extensive projection to the middle portion of the caudal two thirds of the LGN; fibers from progressively more peripheral zones terminate on either side of the wedge-shaped central projection region and anterior to it. A schematic outline of the retinogeniculate projection is shown in Fig. 16-2.

Geniculostriate projection

The axons of the principal geniculate cells form the visual radiation and terminate wholly in the striate cortex of the ipsilateral occipital lobe, the primary visual cortex. On leaving the confines of the LGN the visual radiation fibers pass through the posterior part of the internal capsule between the somatic and auditory radiations, spread out, and sweep backward to form an orderly stratified lamina along the lateral wall and the posterior horn of the lateral ventricle. The fibers arising from the median posterior portion of the geniculate nucleus (the site of termination of the projection from the central retina) follow

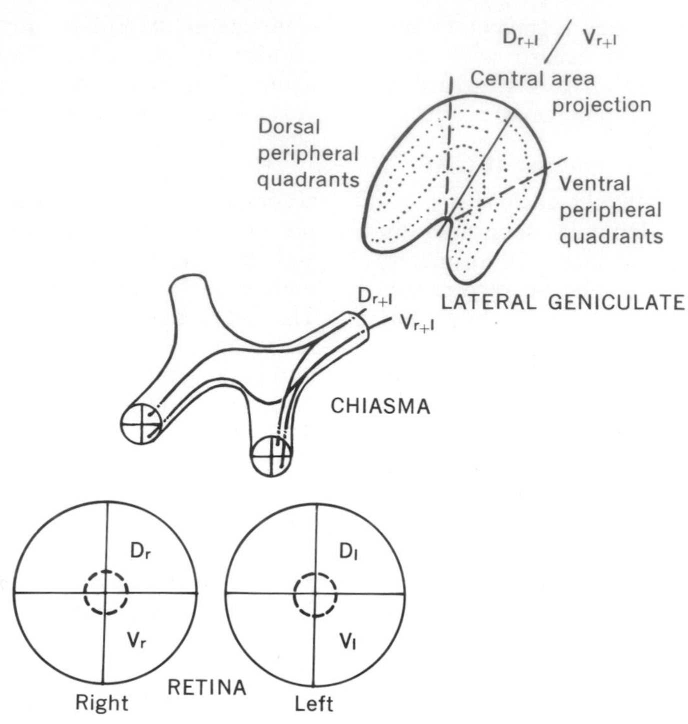

Fig. 16-2. Schematic diagram of topographic arrangement of retinogeniculate projection in monkey. Optic fibers from dorsal, *D,* and ventral, *V,* quadrants of nasal half of right, *r,* retina cross in chiasm and join uncrossed projection from temporal half of left, *l,* retina. In optic tract and at lateral geniculate nucleus, homonymous dorsal quadrants, D_{r+l}, are represented mediodorsally; ventral quadrants, V_{r+l}, ventrolaterally.

a fairly straight anteroposterior course and are flanked dorsally by the fibers that originate from the medial part of the nucleus (homolateral dorsal retinal quadrants). The fibers that arise from the lateral part of the LGN (homolateral ventral retinal quadrants) loop downward and forward into the temporal lobe to sweep around the lateral horn of the ventricle (temporal loop of Meyer) before turning backward to join the other projection fibers and form the ventral portion of the visual radiation. If a small lesion is made in the area striata of a monkey, the ensuing retrograde cellular degeneration in the LGN is localized to a sharply defined region. Circumscribed lesions in the optic radiation, on the other hand, lead to much less clear-cut foci of degeneration within the geniculate nucleus. Thus the orderly topographic representation of the retina at the LGN becomes somewhat smeared in the optic radiation, and it is reconstituted again in an exquisitely precise manner at the primary visual cortex.[32]

The retrograde degeneration in the geniculate affects all six cell laminae along a given narrow sector of the nucleus, even when the cortical lesion is only 1 mm^2.[9] This indicates that cell groups in adjoining laminae, receiving afferents from corresponding points in the two hemiretinas, project close together in the area striata, an arrangement thought to provide the initial condition for the interaction of visual messages from the two eyes at the visual cortex.

Striate cortex

In man and other primates this area of the cerebral cortex (area 17) possesses distinctive morphologic characteristics that separate it sharply from the surrounding area 18. Approximately at the middle of the thickness of this cortex there is a zone composed mainly of a band of medullated fibers, the terminal portion of the geniculocortical projection. This band, usually visible macroscopically, is named the "stria of Gennari," and runs at the level of the inner granular layer (Brodmann's lamina IV). Geniculostriate projection fibers terminate mainly in the layer of densely packed small neurons (layer IVc) immediately below the stria and to a lesser extent within the stria itself (IVb) and above it (IVa). A very few fibers ascend to layer I.[55, 72] The large majority of their terminals

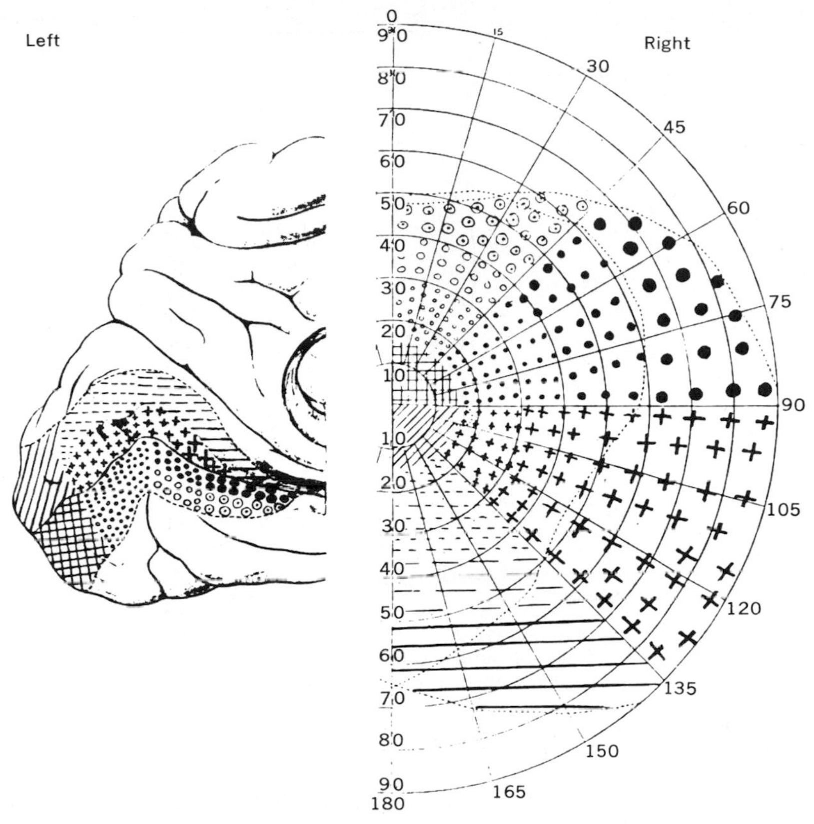

Fig. 16-3. Cortical representation of contralateral half of field of vision according to Holmes.[65] Striate area of left hemisphere of brain is shown with calcarine fissure widely opened. Areas of right half of visual field are marked corresponding to their areas of cortical representation. Macular region of retina is represented posteriorly in brain and is relatively large; peripheral retina is represented anteriorly and is relatively small. Horizontal meridian is projected along depth of calcarine fissure and vertical meridian along upper and lower margins of striate cortex. (From Duke-Elder.[51])

make synaptic contacts with dendritic spines of pyramidal cells and likely also of stellate cells. Other terminals end directly on dendritic shafts and a very few on stellate cell bodies.[55]

Few commissural fibers reach the striate cortex; sparse callosal connections exist only in a very narrow zone bordering on area 18.[92] Efferent projections from the striate cortex are to the ipsilateral prestriate cortex (p. 507) and, subcortically, to the pulvinar of the dorsal thalamus, to the superior colliculus and pretectal region, and to other midbrain structures.[27]

The retinotopic geography of the striate cortex of man has been determined largely by clinical studies of war wounds and of surgical ablations of the occipital lobe.[9, 10, 60, 65, 108, 109] These observations have outlined the spatial representation of the retina on the cortical surface but have provided only limited information on its detailed organization. More accurate maps of the visual cortex have been obtained experimentally in animals with electrophysiologic techniques.[34, 43, 113, 125]

The topographic organization of the primary visual cortex appears to be similar in nonhuman primates and in man. In man the area striata is almost totally confined to the superior and inferior lips of the calcarine fissure on the medial aspect of the occipital lobe, and its posterior boundary is located at about the occipital pole (Fig. 16-3). In the monkey, on the other hand, a large portion of the striate cortex extends over the lateral surface of the hemisphere (Fig. 16-4). The cortical replication of the horizontal retinal meridian runs longitudinally along the depth of the calcarine fissure and around the occipital pole onto the lateral surface of the brain. It divides the primary visual cortex into upper and lower portions, where dorsal and ventral retinal quadrants are represented, re-

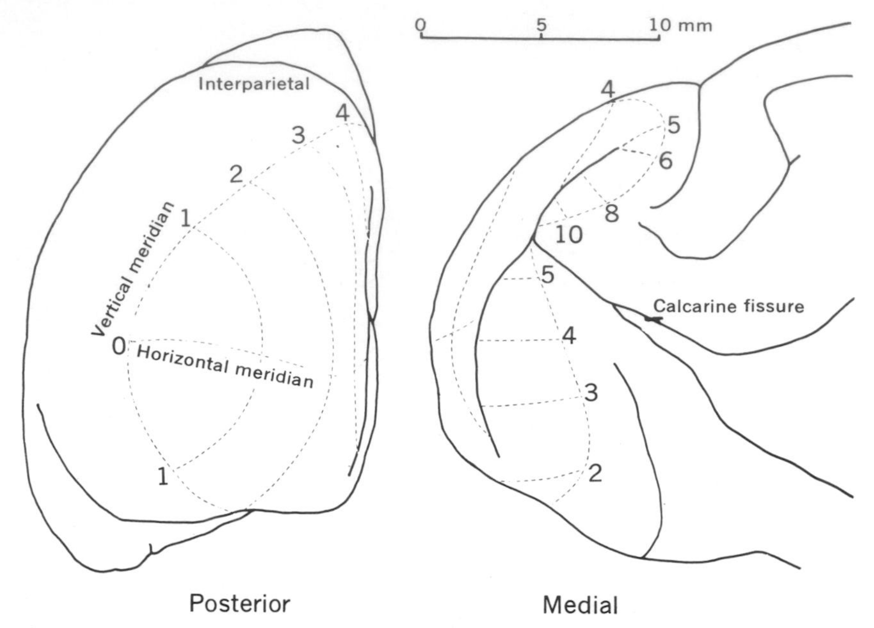

Fig. 16-4. Diagrammatic representation of projection of central area of retina (10 degrees) to lateral and medial striate cortex of squirrel monkey. Numerals refer to radii of visual field. Map was constructed from data obtained using method of evoked potentials. (From Cowey.[34])

spectively. The replication of the vertical retinal meridian coincides with the upper and lower borders of the striate area. The central regions of the homolateral hemiretinas project posteriorly in the hemisphere, with the representation of the fovea at about the middle of the posterolateral boundary of the primary area. Successively more peripheral zones are arranged in an orderly sequence, the most peripheral ones being represented at the anterior end of the calcarine fissure (peripheral nasal retina of the opposite side) (Figs. 16-3 and 16-4).

FUNCTIONAL ASPECTS OF TOPOGRAPHIC ORGANIZATION

The primary visual system, like other primary sensory systems, is organized to replicate in neural space the *spatial relationships* of the physical world of objects that we perceive. The precise topographic arrangement of the retinogeniculocortical pathway represents the more general aspect of the organization subserving the maintenance of spatial relations. Thus contiguous areas of the retinal surface are represented contiguously in the LGN and in the cerebral cortex: the topologic relations of the external world are maintained within the visual system.

As in other primary afferent systems, the number of central neural elements assigned to various parts of the sensory periphery is not proportional to the physical size of that part, but it is an expression of the degree of *innervation density* of the receptive surface, i.e., of the number of sensory elements per unit area. The sensory visual receptors, the rods and cones, are connected through a complicated intraretinal network to the ganglion cells (Chapter 15). Each ganglion cell thus defines a retinal receptive unit, i.e., the population of receptors distributed over an area of retinal surface that is connected to the brain by a single fiber. In considering the functional relation between retina and central structure, the ganglion cell may then be taken as the representative member of the peripheral surface.

When one compares the topographic anatomy of the ganglion cell layer in the retina and that of the striate cortex, an important difference is apparent. While the retinal ganglion cells are distributed along a centroperipheral density gradient,[9, 116] the cell density of various parts of the striate cortex appears remarkably uniform.[30, 34] As an approximate generalization, the area relation between retina and striate cortex may be visualized by imagining the population of ganglion cells spread out in a uniform layer and this layer superimposed upon the receiving cortical surface. The representation of

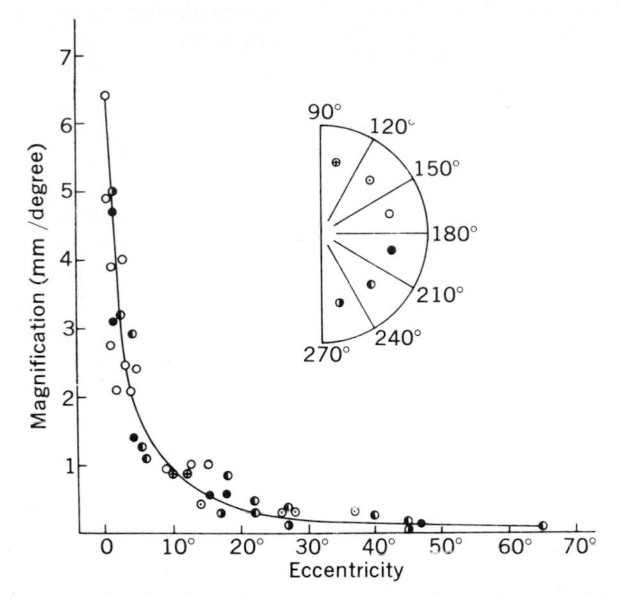

Fig. 16-5. Plot of "magnification factor" versus degrees of retinal eccentricity in macaque (see text). Data for various radii explored have been grouped in six sectors; decrease in magnification from center to periphery appears to be equal in all directions. (From Daniel and Whitteridge.[43])

the central retina, where the density of ganglion cells is higher, occupies an area of cortex larger than the area covered by the peripheral retina. A numerical estimate of this relation may be obtained by measuring the extent of cortical surface from which a response can be evoked by stimulation of a discrete retinal area with a small spot of light.[43, 113] The results of such an experiment performed in the macaque are shown in Fig. 16-5, where the "magnification factor," i.e., the linear millimeters of cortex related to 1 degree of visual field, is plotted with respect to retinal eccentricity.

The following facts suggest that the magnification factor represents an important anatomic basis for the visual capacity for two-dimensional resolution.[36, 43] In the macaque the diminution in magnification factor from the foveal projection area to the peripheral projection area resembles the reduction in visual acuity in man from the fovea to the periphery (Fig. 15-4). Macaque and squirrel monkeys possess very similar visual acuity and similar magnification factors (6 mm) for the central 1 degree of visual field.[35] Finally, in the squirrel monkey the cortical representation of the extrafoveal retina is much more compressed than in the macaque, and indeed, removal of the foveal projection area in the striate cortex leads to greater impairment of visual acuity in the squirrel monkey than in the macaque.[36]

In the macaque, approximately 1 million optic fibers terminate in the LGN. The total number of cells in this structure is estimated to be of the order of 2 million. In the macaque, therefore, the ratio between optic fibers and geniculate cells is about 1:2, while in man the relation appears close to 1:1. In the monkey, approximately 1,300 geniculate cells project to each square millimeter of striate cortex, whose total area is of the order of 1,400 mm^2.[30, 31, 34] The cell density in the visual cortex of the monkey is approximately 150,000 cells/mm^2. Thus assuming the thickness of the cortex to be only 1 mm (1.4 mm at the region of the foveal projection and 1 mm at the anterior calcarine fissure),[30, 34] the total number of cells in the area striata is of the order of 200 million. Since the spatial distribution of the projection fibers in the radiation and of the cells in the receiving cortical visual area is remarkably uniform, we may conclude that in the macaque there are on the average at least 100 cortical neurons available to process information transmitted from a single geniculate cell. This, of course, does not imply that the retinogeniculostriate system is simply organized on the basis of a 1:2:200 neuronal multiplication, i.e., that one optic fiber terminates upon two geniculate cells, each of which in turn makes synaptic contact with hundredfold elements in the cortex. While precise numbers are unknown, what is certain is that each single incoming fiber synapses with more than one cell (divergence) and that each cell receives afferent terminals from more than one fiber (convergence). The first arrangement would then be responsible for multiplication of pathway from periphery to centers; the second would bring about reciprocal overlap between the populations of neurons relating to contiguous peripheral units. Moreover, anatomic connections between central neurons transmit the afferent signals over a population of cells larger than the original receiving population, the final functional effect depending also on the intrinsic nature of the connections, i.e., on their excitatory or inhibitory nature. It is important to note, however, that in the striate cortex, as in other primary sensory areas, the extent of divergence-convergence is limited, and various parts of the central representation of the sensory periphery remain to a large extent functionally independent. Convergence and interaction within the afferent visual system occur chiefly beyond the striate cortex.

EFFECTS OF STRIATE CORTEX LESIONS IN PRIMATES

In man and other primates, injury to the occipital cortex produces a permanent loss of vision in the relevant portion of the visual field. In the monkey, however, the visual deficit caused by lesions of the striate area appears to be less severe than that observed in man. Klüver[5] has shown that the visual behavior of monkeys after bilateral removal of the primary visual cortex depends on the ability they retain to use differences in the total amount of light entering the eye as a factor for discrimination. The destriated monkey, which possesses an absolute visual threshold similar to that of a normal animal, behaves visually like a photocell that merely registers and integrates light energy. No form or color vision is left. Recent experimental evidence indicates that after ablation of the striate cortex the monkey is able to make certain primitive visual discriminations, for it can be trained to detect, orient itself toward, and reach out for localized events of certain kinds, especially moving events. It seems highly unlikely, however, "that the animal can discriminate many of the events from each other except for their spatial discreteness."[119]

After partial damage of the striate cortex the loss of function is limited to a portion of the visual field, the size, shape, and position of the defect being dictated by the topographic arrangement of the primary projection to the cortex. Because of their specificity of place and modality, visual field defects are of great clinical significance for the topical diagnosis of lesions involving any portion of the visual pathway; they will be summarized in the next section. In the monkey, partial lesions of the area striata are followed by partial defects in the visual field, with the expected topographic characteristics. In the defective portion of the field, however, the animal still possesses a certain degree of visual capacity. Although the recognition of objects that lie within the area of the defect is lost, the animal can respond to flashes of light within it: the deficit is a greatly reduced sensitivity rather than total blindness.[38, 119] In addition to the visual field defect, partial striate lesions that include the foveal area impair visual acuity and, transiently, visually guided reaching. Such lesions have little or no effect on postoperative learning or retention of visual discrimination habits, whereas this capacity is severely impaired by lesions of cortical areas beyond the striate cortex (p. 507).[39, 124] The lack of a discrimination deficit after foveal striate lesions may be attributed to the ability of man and monkey to use extrafoveal parts of the retina by fixating eccentrically.[10, 38] The orderly topographic representation of the retinal surface in the striate cortex and the absence of any extensive spatial convergence within it make it possible for visual information received at the extrafoveal region of the retina to be processed—largely unaffected by the lesion—by extrafoveal cortex and transmitted forward to other cortical areas for further elaboration.

There appear to be no basic differences in the organization of the primary visual system in man and monkey. In both, removal of the striate cortex leads to complete degeneration of the dorsal LGN. The residual visual capacity of the destriated monkey is not surprising, for we know that neural activity evoked by light is transmitted from the retina not only to the geniculostriate system but also to regions of the midbrain, to the superior colliculus and pretectum in particular (p. 510), which may provide the basis for a response to light. The functional significance of the retinomesencephalic projection diminishes in phylogeny, and its role in vision seems to be inversely related to the extent to which the geniculostriate system has developed. Moreover, there is some evidence that in man some light sensitivity also remains after geniculostriate damage.[101] Thus the differences observed between monkey and man are relative ones, the monkey being able to use as cues the secondary effects of light stimulus better than man.[38] The neural processing of visual information is likely to be organized in the same manner in the two species.

THE VISUAL FIELD AND ITS ABNORMALITIES

The visual field is defined as the bounded portion of space that is visible at any one time while the

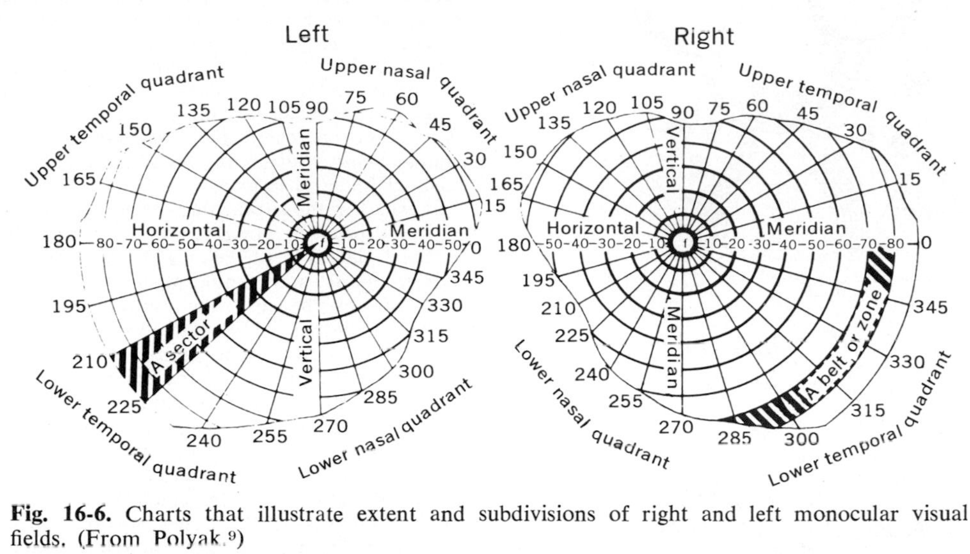

Fig. 16-6. Charts that illustrate extent and subdivisions of right and left monocular visual fields. (From Polyak.[9])

eye is fixating a stationary point in it. The image of any point in the field falls spatially inverted on the retina. Accordingly, the visual field may be topographically subdivided, like the retinal surface, into four quadrants by vertical and horizontal straight lines crossing at the center of the fixation point as well as into concentric central and peripheral regions.

Accurate mapping of the visual field in humans may be done by means of a simple instrument, the perimeter, which allows the presentation of a visual stimulus at any point in the subject's field of view. In this way, functional defects may be outlined and charted. Perimetric mapping has also furnished important information on several other aspects of the functional organization of the visual system such as acuity, sensitivity, extent of the color field, and response to movement.

Because of the normally recessed position of the eye in the orbit, the portion of space subtended by the retina of the eye fixed upon a point straight ahead is limited in its extreme extent by the nose, eyebrows, and cheek bones. The *monocular visual fields* (Fig. 16-6) are slight irregular ovals that extend from the fixation point, approximately 60 degrees nasally and above, to about 70 degrees below and 90 degrees temporally. The papilla of the optic nerve produces the physiologic *blind spot* in the temporal half of the field of vision, a vertical oval area, located about 12 degrees from the point of fixation and 1.5 degrees below it. The *binocular visual field* is the combination of the partially overlapping right and left monocular fields. Its central part represents the portion of space visible simultaneously with the two eyes directed toward a common fixation point; it is roughly circular in shape and has a diameter of about 120 degrees. On either side of this paired portion there extends the nonoverlapping part of each monocular field, the so-called temporal crescent, which corresponds to the most peripheral nasal retina.

It was previously mentioned that the study of the visual field defects has helped us to understand the anatomic organization of the visual afferent system in man, in particular its topographic arrangement. Clearly the characteristics of the visual field defect can provide most valuable information for the clinical diagnosis of the site, extent, and in some cases also the nature of a pathologic process involving the visual projection system.

Lesions at any level of the visual pathway may produce functional defects in the area of the visual field represented in the damaged part of the system. The intensity of the defect may range from a slight reduction of visual acuity to total loss of function within the affected area. The size, shape, and position of the defect depends on the location and extent of the lesion within the visual pathway (Fig. 16-7).

Visual field defects are fundamentally different when the interrupting lesion is located in front of or behind the optic chiasm:

1. Unilateral damage to the portion of the visual pathway peripheral to the decussation of the optic fibers gives rise to a functional defect limited to the field of vision of the eye on the affected side. Complete severance of the optic nerve is, of course, followed by total loss of function in the corresponding eye. Partial lesions determine a variety of defects that correspond to the area of the field represented by the fibers that are interrupted. This is particularly evident for damage to the optic fibers in their intraretinal course, where, because of their convergent arrangement toward the optic papilla, lesions similar in all respects produce an extensive field defect when situated near the optic nerve head and a small one when located in the retinal periphery. When two unilateral lesions exist, two visual defects are produced, one in each monocular field of view. In general the two defects are different in size, shape, and other characteristics, a condition termed "incongruity," which may also be produced by unilateral lesions of more proximal portions of the afferent pathway.

2. Interruption of the visual pathway anywhere from the chiasm to and including the area striata results in field defects in which the essential feature is some degree of *hemianopsia,* i.e., loss of function

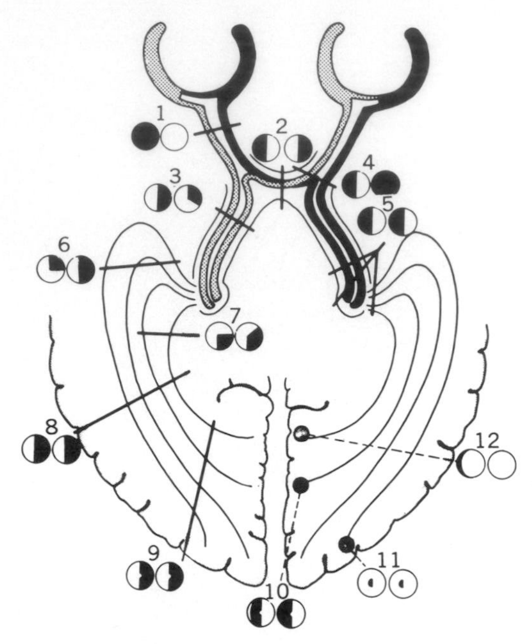

Fig. 16-7. Visual field defects produced by interruption of visual pathways at various sites. *1,* Optic nerve: blindness in eye on side of lesion, with normal vision in contralateral eye. *2,* Sagittal lesion of chiasm: bitemporal hemianopsia. *3,* Optic tract: contralateral incongruous homonymous hemianopsia. *4,* Optic nerve–chiasmal junction: blindness on side of lesion with contralateral temporal hemianopsia. *5,* Posterior optic tract, lateral geniculate nucleus, and posterior limb of internal capsule: contralateral homonymous hemianopsia. *6,* Optic radiation, temporal loop: contralateral incongruous homonymous hemianopsia or superior quadrantic hemianopsia. *7,* Optic radiation, medial fibers: contralateral incongruous homonymous inferior quadrantic hemianopsia. *8,* Optic radiation in parietal lobe: contralateral homonymous hemianopsia. *9,* Optic radiation in posterior parietal lobe and occipital lobe: contralateral homonymous hemianopsia with macular sparing. *10,* Midportion of calcarine cortex: contralateral homonymous hemianopsia with macular sparing and sparing of contralateral temporal crescent. *11,* Pole of occipital lobe: contralateral homonymous hemianoptic scotomas. *12,* Anterior tip of calcarine fissure: contralateral loss of temporal crescent with otherwise normal visual fields. (From Harrington.[61])

that extends over one half of the visual field. Hemianopsic defects are termed homonymous or heteronymous, depending on whether the corresponding or the noncorresponding halves of the two respective monocular visual fields are affected. As determined by the anatomic arrangement of the visual pathway, heteronymous hemianopsia is characteristic of lesions of the chiasm and homonymous hemianopsia of unilateral damage to the suprachiasmal, intrahemispheric portion of the system, i.e., optic tract, geniculate nucleus, visual radiation, and striate cortex.

Pathologic processes at the level of the chiasm usually affect the intermediate, bridgelike portion of this structure. Consequently, they interrupt the two sets of decussating fibers, i.e., the fibers originating from the nasal half of each retina, and give rise to the characteristic visual defect in the temporal half of the two monocular fields of view—bitemporal hemianopsia. Very infrequent is the occurrence of the reciprocal condition, i.e., the interruption of the uncrossed fibers and consequent binasal hemianopsia, which can only be explained on the basis of bilateral lesions.

Lesions in the suprachiasmal portion of the visual pathway produce homonymous hemianopsia. Each cerebral hemisphere relates to the contralateral halves, nasal and temporal, of both monocular fields of view, hence to the contralateral half of the binocular visual field. Thus right homonymous hemianopsia results from lesions of the optic tract, geniculate nucleus, visual radiation, and striate cortex of the left side; and left homonymous hemianopsia results from damage to the same structures in the right hemisphere. Total hemianopsia implies, of course, total interruption of the visual pathway behind the chiasm on one side. Hemispheric lesions most often involve the upper portion of the visual system, the visual radiation, and the visual cortex. The large spread of the optic pathway in the posterior portion of the hemisphere offers a possibility for partial lesions and consequently limited defects, always hemianopsic in character, in the contralateral field of view. These partial defects may take a great variety of forms and involve only a small area (scotoma), a segment of a quadrant, an entire quadrant, or more. Unilateral field defects from lesions to the suprachiasmal visual pathway may occur as the result of the interruption of the unpaired central projection of the most peripheral nasal fibers of the ventrolateral hemiretina (temporal crescentic defects).

A common occurrence in cases of homonymous hemianopsia is the preservation of visual function in a portion of the field of view of each eye around the fixation point. Perimetric mapping shows that in these cases the border between functioning and nonfunctioning halves of the visual field does not pass through the point of fixation but curves around it. In most cases the preserved central vision subtends 2 to 5 degrees, but it may vary from 1 degree or less from fixation to a large portion of the affected half-field. This remarkable phenomenon, termed central or "macular" sparing, may occur with lesions of any portion of the suprachiasmal visual pathway, but more commonly with lesions of the posterior portion.

Central sparing has been taken at times as evidence that either the macula is bilaterally represented (i.e., each homonymous half of the macula projects to both striate areas) or alternatively that the unilateral macular projection is much larger than usually considered and perhaps diffuse to the entire primary visual area. There is no anatomic support for either of these interpretations and much against them.

Some forms of central sparing may be accounted for on the basis of a lesser vulnerability of central fields as compared to peripheral fields.[10] The representation of the macula is very great and occupies a large proportion of the visual pathway. The occipital pole, where the macula is represented, receives a double vascular supply, from both the middle and the posterior cerebral arteries.[9] These anatomic features make it possible for even large hemispheric lesions, particularly those of a vascular nature, to leave intact the macular projection and thus to spare an island of central vision.

In a number of cases it can be demonstrated that the sparing of the macula is apparent rather than real. In these cases, central vision is extended to the blind half of the field by employing in fixation not the fovea but an area of retina on the functioning side (pseudofovea). The development of eccentric fixation may be facilitated by the physiologic movement of the eye scanning the image of the object on which attention is directed. If a more complete picture is obtained from an area outside the fovea, then this area becomes the center of attention and an automatic fixation is shifted to it.[10, 65]

VISUAL PATHWAYS BEYOND THE STRIATE CORTEX

The devastating functional consequences of the destruction of the geniculostriate pathway clearly indicate that in primates the primary projection system plays a fundamental role in vision, possibly an essential one in man. Moreover, it has long been recognized that visual disturbances may result from lesions of cortical areas outside the area striata. These disturbances reflect, in general, an inability of the affected individual to utilize correctly the visual information provided by a healthy primary projection system. In humans they include difficulties in the recognition and identification of objects, patterns, or color and visual disorientation in space and are often associated with other disorders, particularly of speech, reading, and writing.

These clinical observations, strongly supported by experimental investigations in monkeys, have led to the notion that from the striate cortex visual information is brought to converge and interact within other cortical areas in which a further elaboration of the visual message occurs, an elaboration that is essential for the total perceptual experience.

The cortical progression of the visual pathway beyond the striate cortex is an orderly one and its organization in the monkey has been outlined in recent years both anatomically[40, 76, 81, 95, 127] and functionally.[34, 37, 88] In brief, the visual projection extends forward from the *striate cortex* (area 17) to the *prestriate cortex* (areas 18 and 19), and from the latter to the inferior convexity of the temporal lobe, the *inferotemporal cortex* (areas 20 and 21). All these regions send fibers to the *frontal cortex:* the striate and prestriate areas project to area 8 (the frontal eye field); the inferotemporal cortex also projects to area 8 as well as to other premotor and prefrontal areas (Fig. 16-8). Homologous regions of the two sides, with the notable exception of the area striata, are connected through commissural fibers: the frontal and prestriate cortex via the corpus callosum and the inferotemporal cortex chiefly via the anterior commissure. In addition, cortical visual areas are part of complex corticosubcortical circuits whose functional significance is little understood. These connections will be mentioned in the following paragraphs.

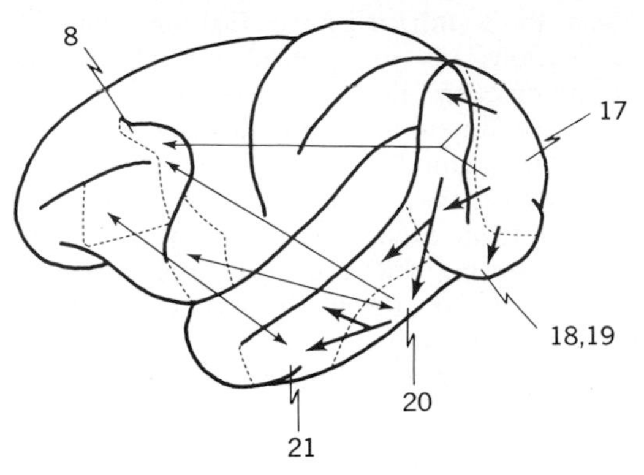

Fig. 16-8. Schematic diagram of progression of visual pathway beyond striate cortex. Forward connections exist from primary visual cortex (area 17) to prestriate cortex (areas 18 and 19) and from the latter to inferotemporal cortex (areas 20 and 21). Areas 17, 18, 19, and 20 also project to area 8 on lateral aspect of frontal lobe. Reciprocal connections exist between inferotemporal cortex and other premotor and prefrontal regions. (Refer to Jones and Powell[76] for details.)

Prestriate cortex

In man and other primates the primary visual area is surrounded by the prestriate cortex, which Brodmann subdivided in two concentric areas: area 18, adjoining the striate cortex, and area 19, rostral to the former and extending into the parietal and temporal lobes. Morphologically the transition between these two areas is difficult to determine, and in recent years the terms "area 18" and "area 19" have been used with different meanings by different investigators, often without reference to their different cytoarchitectures. In the cat also, similar areas have been recognized, areas 17, 18, and 19 being located side by side in a mediolateral sequence.[70]

In 1942 Talbot[112] described in the cat a topographically organized representation of the contralateral field of vision in a cortical region (visual area II) situated laterally to the primary projection area (visual area I). The sensory periphery is represented in these two areas in a mirror-image fashion, with the replications of the vertical retinal meridian adjoining each other.[70, 125] Hubel and Wiesel[70] have confirmed and extended these findings with the method of single-unit analysis and have shown that there exists an efferent connection from area 17 to both areas 18 and 19, suggesting that the visual message is relayed from striate to prestriate cortex for further processing.

A similar organization exists in the monkey. The major striate-prestriate projection is arranged topographically with the representation of the vertical retinal meridian at the 17-18 border. The upper portion of striate cortex (dorsal retinal quadrants' representation) projects to dorsolateral prestriate cortex, while the lower striate cortex (ventral retinal quadrants) projects to ventrolateral prestriate cortex. The representation of the fovea occupies a large extent of prestriate cortex, at a position in which the representations of the upper and lower visual field juxtapose. Largely buried in sulci, the foveal prestriate cortex extends over the preoccipital gyrus and toward the temporal lobe.[37, 40, 127] The striate projection does not occupy the entire circumstriate belt, and it is largely confined to Brodmann's area 18.[40] Additional nontopographic projections from striate cortex have been described recently, e.g., to the middle of the caudal bank of the superior temporal sulcus, and intracortical connections within the prestriate cortex are known to exist.[40, 76, 81]

At the prestriate cortex the visual pathways of the two sides are interconnected through commissural fibers. Callosal projections of the prestriate cortex are complexly organized and exist chiefly, if not exclusively, between areas of representation of the vertical retinal meridian and of the fovea.[40, 92, 128] At the striate cortex, commissural fibers are few and sparse and only in the vicinity of the 17-18 border. It has been suggested that this interhemispheric link of the visual pathways provides the basis for the functional union of the right and left halves of the field of vision.[28, 70, 91]

Subcortically, the prestriate cortex is reciprocally connected with the pulvinar of the dorsal thalamus, and it sends an efferent projection to the superior colliculus, pretectum, and other midbrain structures.[27]

The specific role of the circumstriate cortex in visual functions is not fully understood. As previously described, a monkey who has had the striate cortex removed bilaterally retains some elementary form of visually induced behavior. Under this condition the animal is still able to perform efficiently in a familiar surrounding; visual recognition of still objects is lost but moving objects attract gross movements of the eyes and hand and may evoke prehensile reactions; visual placing reactions may still be observed.[5, 49, 119] Additional removal of the prestriate cortex, area 18 in particular, results in a loss of all visual reactions, leaving only pupillary reflexes and blink to bright flashes. Bilateral removal of prestriate cortex alone is followed by disturbances in spatial judgment and confusion of moving objects, while prehensile and placing reactions remain intact.[49, 50] Moreover, in the total absence of this region the monkey is not able to learn, or retain without deficit, pattern discrimination tasks (see Zeki[126] for discussion). With subtotal ablations, this deficit is particularly severe only when the lesion involves the prestriate area receiving a projection from that part of the striate cortex linked to the central retina.[37] Indeed, foveal prestriate cortex is crucial for pattern discrimination whereas, as already indicated, foveal striate cortex is not. Electrical stimulation of prestriate cortex produces an orderly pattern of conjugate eye movements related to the visual field representation in this area.[117]

These findings indicate that the prestriate cortex is the fundamental link over which information regarding visual stimuli progresses from striate cortex to other cortical regions of the same and opposite hemisphere. The prestriate cortex plays a major role in regulating visuospatial adjustments and it may serve visual functions in the absence of input from the primary visual projection. It may be then assumed that the contribution of areas 18 and 19 to the organization of visual behavior depends *also* on the connections these areas entertain with subcortical structures, which in turn must receive visual information directly or indirectly via retinal afferents that do not synapse in the LGN. The cortical pathway, however, is essential for higher visual functions, for bilateral destruction of the pulvinar does not affect pattern discrimination learning,[29] whereas cortical lesions do.

Inferotemporal cortex

In 1938 Klüver and Bucy described a complex syndrome of behavioral alterations in monkeys that follows bilateral removal of the temporal lobe. They observed that "while the monkey showed no gross defect in the ability to discriminate visually, she seemed to have lost entirely the ability to recognize and detect the meaning of objects on the basis of optic criteria alone."[78] Many neuropsychologic investigations have since shown that the neocortex of the inferior convexity of the temporal lobe serves visual functions exclusively and that it is of crucial importance in visual discrimination learning.[3, 37, 39, 88, 124]

The inferotemporal cortex includes the middle and inferior temporal gyri and corresponds closely to areas 20 and 21 of Brodmann. Visual information reaches the temporal lobe chiefly via the striate-prestriate cortical pathway. The first evidence of the importance of this projection was provided by the experiments of Mishkin,[88] in which both occipital and temporal cortices were removed *unilaterally* in monkeys trained to perform a visual discrimination task. When the two lesions were on the same side of the brain, the animal relearned the problem almost immediately. When, on the other hand, the occipital cortex was removed in one hemisphere and the temporal cortex in the other, the monkey required a large number of trials to relearn the discrimination. Finally, when the corpus callosum was also sectioned in animals with "crossed" lesions, relearning was very difficult if possible at all. Interruption of callosal fibers eliminated the essential link between the intact occipital lobe on one side with the temporal lobe on the other. This link, as already indicated, is almost exclusively between the prestriate cortices of the two sides.

Recent anatomic investigations have detailed the steps in the progression of the visual pathway to the inferotemporal cortex (Fig. 16-8). The circumstriate belt projects chiefly to the inferior temporal gyrus (area 20) and possibly also to the posterior part of the middle temporal gyrus. Area 20 in turn projects to the middle temporal gyrus (area 21).[76, 81, 95] According to Jones and Powell,[76] no backward connections exist from any of these extrastriate areas to preceding ones: like the retinogeniculostriate projection, the cortical visual pathway beyond striate cortex appears to be essentially a "forward" pathway. From the inferotemporal cortex, fibers pass over to cortical areas of the frontal lobe, to the tip of the temporal lobe, and to a part of the cortex buried in the superior temporal sulcus.[76] Subcortically, the inferotemporal cortex is connected with a number of structures, among them the amygdala, pulvinar of the dorsal thalamus, superior colliculus, and pretectum.[76, 120] There is no detailed topographic replication of the field of vision in the temporal lobe. Subtotal lesions of the prestriate cortex are followed by degeneration that spreads over large areas of the inferotemporal cortex, and microelectrode analysis of the latter region has shown that single inferotemporal neurons possess large unilateral or bilateral receptive fields, greater than 10×10 degrees in size, and almost always including the fovea.[58]

When the inferotemporal cortex is removed bilaterally, the behavior of the monkey appears normal, with no detectable visual field defects, impairment of visual acuity,[39] or impairment of threshold for brightness discrimination.[53] On the other hand, certain learning capacities are severely disturbed: the animal is unable to perform a variety of visual choice tasks even though it had learned them preoperatively. The deficit produced by inferotemporal ablations is exclusively visual: no somesthetic or auditory defects are present. Its nature is not clearly understood; as Diamond and Chow[3] point out, it cannot be attributed to "sensory" changes, to the animal's inability to comprehend the testing situation, or to the complete disappearance of specific memory traces corresponding to specific visual habits. What is known is that the inferotemporal impairment in discrimination learning does not depend critically on the dimension in which the visual cues differ (e.g., size, brightness, hue, or pattern), but rather on the nature of the discrimination problem. A task that requires a large number of trials for a normal animal to learn is usually impossible for a monkey with inferotemporal lesions, while a simple task may be learned correctly.[89, 90, 100] The impairment is most pronounced for visual discrimination involving alternatives,[53] particularly for concurrent discrimination, i.e., parallel learning of several simple discriminations.[37] There is some evidence that certain visual defects that occur in man after temporal lobe damage are similar to those observed in monkeys following inferotemporal lesions.[87]

The foregoing neurologic observations on the functional significance of cortical regions beyond the primary receiving area provide evidence that these regions are essential for certain complex dimensions of vision. On the other hand, the notion that the area striata is a "receptive" area serving relatively simple functions and the "associative" cortical areas beyond it privileged integrative centers and sites of higher functions is not tenable in its extreme form. It is suggested that the total replica of visual events projected in neural code from the eye to the brain undergoes a series of successive transformations, each essential to that following, along the central visual system. The perceptual process develops out of the continuum of these transformations. "Associative" cortical regions are, with regard to function, part of the whole system, and they assume their particular significance not because they are structurally or operationally more complex but because of the nature of the visual information they receive and of their interconnections with other cortical and subcortical regions.

Frontal cortex

It has long been known that in man and monkey electrical stimulation of an area on the lateral aspect of the frontal lobe (the frontal eye field), with lowest threshold at Brodmann's area 8, elicits horizontal and oblique deviations of the eyes to the contralateral side, the precise direction of which depends on the locus of stimulation. These movements are often accompanied by turning of the head toward the opposite side. Adversive eye deviations can also be produced by stimulation of several other brain regions, most readily from prestriate and striate cortex, again in a topographically organized pattern.[42, 101, 117]

In humans, lesions involving the lateral aspect of the frontal lobe interfere with conjugate eye movements, producing paralysis of gaze or forced contralateral deviation of eyes and head. These disturbances are usually transitory. In monkeys the two main effects observed after unilateral frontal eye-field lesions are ipsilateral deviation of eyes and head and a neglect of contralateral visual stimuli, the result of a contralateral relative visual field defect, a hemiamblyopia.[82] Complete recovery from these impairments is commonly observed within a few weeks.

Anatomically the frontal eye fields, area 8 in particular, receive cortical projections from the occipital and temporal visual areas (Fig. 16-8). Subcortically, there exist reciprocal connections with the dorsomedial nucleus of the thalamus and efferent projections to other thalamic nuclei, basal ganglia, and the superior colliculus–pretectal complex. There is no evidence of a direct projection from area 8 to the oculomotor nuclei.[12]

Recent electrophysiologic studies in the alert monkey have shown that neurons in the frontal eye fields discharge with movements of the eyes and head.[22, 23] Some neurons discharge only during quick jumps of the eyes (saccades; Chapter 14), others when the eyes are slowly drifting or held stationary at a specific position, and still others when the head is turned in a particular direction. Most important, in nearly all instances the discharge of such neurons occurs after the beginning of the eye movements.

These findings suggest that frontal "visual" neurons do not participate in the initiation of oculomotor action, but rather monitor oculomotor activity, e.g., eye shifts and relative positions in the orbit. On the other hand, neither evoked visual or vestibular activity nor impulses from the neck or eye muscle proprioceptors appear to be a source of afferent input to these neurons. Thus the frontal eye fields do not possess specific "motor" or "sensory" functions and may be regarded as an interactive site of neural events operating in the control of eye movements and more generally in the complex coordination of motor behavior in the visual sphere.

THE RETINOMESENCEPHALIC PROJECTION AND ROLE OF THE MIDBRAIN IN VISION

Visual information is forwarded from the retina to central structures along two separate pathways: the retinogeniculostriate projection described previously and the retinomesencephalic projection. Fibers of retinal origin terminate bilaterally in the midbrain, chiefly in the superior colliculus and pretectal region. From these structures, descending connections exist to the brainstem and cervical spinal cord and ascending ones to the posterior thalamus (pulvinar), which is in turn reciprocally connected with the prestriate and inferotemporal cortex. Moreover, the striate cortex and all cortical visual areas beyond it send efferent projections to the superior colliculus and pretectum, which also receive fibers ascending from lower centers.

The primary visual projection and its forward extension, on the one hand, and the extrageniculate visual pathways, on the other, constitute two interconnected systems that are thought to subserve essentially different visual functions. The former provides the neural mechanisms for the capacity to discriminate and recognize objects on the basis of their shapes and motions; the latter subserves the ability to visually orient to these objects, to relate movements of the body to their location in space.[62, 106, 115] The extent of functional dissociation between the two systems is not clearly defined. While there is little doubt that in higher mammals the midbrain structures play a fundamental role in visual orientation and the visual cortex in form discrimination there is also evidence that the extrageniculate system may operate in visual discrimination, at least in some species. For example, Sprague et al.[110] have found that extensive midbrain lesions including the superior colliculus and the pretectum severely impair form discrimination learning in the cat.

Anatomic and electrophysiologic findings have shown that in the tectum of all vertebrates there exists an orderly topographic representation of the visual field. In the monkey, as in the cat, homonymous halves of the field of view of the two eyes are represented superimposed in the contralateral colliculus. Upper and lower field quadrants are represented in medial and lateral parts of the tectum, respectively, and are separated by the replication of the horizontal meridian. The central region of the field is represented rostrally and the periphery caudally. As in the geniculostriate projection, the map of the visual field is distorted in that the central 20 degrees of field occupy about two thirds of the colliculus, the remaining 70 degrees being crammed in the posterior third. As previously indicated, the superior colliculus also receives a projection from the ipsilateral striate cortex: retinotectal and corticotectal projections are topographically corresponding.

There is evidence that in the macaque the tectal map of field representation is not wholly determined by the retinomesencephalic projection. Anatomically, no fiber degeneration can be found in the superior colliculus after lesions within the central 7 degrees of retina, while the corticotectal projection is complete and replicates the entire contralateral hemifield.[123]

These findings indicate that in the macaque visual information from the peripheral retina reaches the tectum directly as well as via the cerebral cortex, whereas information from the central retina arrives at the colliculus only through the cortical route. The possibility exists that anatomic techniques are not sufficiently sensitive to reveal degeneration of fibers of small calibers that could form the projection from the central retina. An electrophysiologic analysis of collicular neuron properties in the monkey after striate cortex ablation might resolve this question. It is interesting to note in this regard that residual vision in the destriated monkey is, as Trevarthen points out, "vision of poor acuity but highly sensitive to motion and brightness, like vision normally associated with the peripheral field, or scotopic illumination."[115]

The importance of the superior colliculus in the control of eye and head movements and more generally in visuomotor mechanisms has been shown by results of stimulation experiments, by observations of the effects of lesions on the animal's behavior, and by the analysis of the functional properties of collicular neurons.

Stimulation of the surface of the colliculus produces a pattern of eye movements that relates to the topographic map of the visual field in this region. In unrestrained animals, collicular stimulation evokes orienting movements of the eyes, head, and body.[17] Anatomic connections from the superior colliculus to the pretectal region, midbrain reticular formation, pontine nuclei, and motoneurons of the upper cervical segments of the spinal cord are known to exist. Direct connections from the tectum to extrinsic oculomotor nuclei have not been demonstrated, and whichever influence the superior colliculus has on motoneurons of the eye muscles is likely to be exerted through intermediate centers.

After unilateral removal of the superior colliculus, the monkey can look up and down or to either side and reach out to a moving object with either hand, but it appears unconcerned with events in the visual field contralateral to the lesion. In that field the animal cannot fixate but can only look approximately in the direction of an object. Bilateral destruction of the colliculus results in profound disturbances in visual and general behavior. The monkey stares fixedly into space, shows no orientation to visual events, and spends most of its time silently in a corner of the cage.[48]

The functional properties of neurons in the superior colliculus of the alert monkey have been described recently by Schiller and Koerner.[105] There are two major types: cells that respond to visual stimuli only and cells that discharge in relation to eye movements and generally to visual input as well. Neurons responding exclusively to retinal stimulation are located in the superficial layers of the colliculus. Usually activated equally well from both eyes, these neurons respond best to small, luminous moving patterns and display little selectivity for shape, orientation, and direction of motion of the stimulus. These cells appear sensitive to nonspecific "events" appearing in the field of vision.

Cells in the deeper collicular layers discharge in association with saccadic eye movements of specific directions and sizes. Cells in the right colliculus are associated with saccades toward the left and neurons in the left colliculus with movements to the right. In all instances the neural discharge *precedes* the eye movement by 70 to 500 msec and thus these cells are likely to operate in the initiation of oculomotor action. Moreover, the majority of these neurons are also influenced by visual stimuli, and the receptive field of each neuron is typically located in that area of the visual field to which the eyes move as the result of the associated saccade. Neurons with receptive fields within 5 degrees of the center of the fovea respond also during smooth pursuit movements of fixation.

The experimental findings just described provide evidence of the importance of the midbrain in visuomotor activity. The superior colliculus receives a topographically organized input from the contralateral hemifield of vision and its neurons send signals to the motoneurons of the eye, head, and trunk muscles that activate a sequence of orienting movements. The functional organization of collicular neurons supports the view that the superior colliculus is part of the oculomotor mechanisms by which an animal is capable of detecting the "where" in the visual world a stimulus of relevance has appeared and to quickly move its eyes to fixate on that stimulus and to maintain that fixation. It is likely that similar visual orienting reactions of the head and body also are mediated by collicular neurons.

In man and other primates, fixation is largely reflex in nature and depends on the occipital cortex. Striate and prestriate cortex project to the midbrain in an orderly topographic fashion, and stimulation of these regions evokes eye and head movements similar to those elicited by tectal stimulation. These projections are likely to operate as part of the fixation mechanisms just described and to be essential for the reflex shift of the eye toward an object attracting the attention of the animal, and also in voluntary visuospatial adjustments.

The role of the frontal eye field in visuomotor control is still unclear. It is important to remember that eye movement–related neurons in the frontal cortex discharge after the beginning of the movement, whereas those in the colliculus discharge before the movement. Frontal neurons do not participate in the initial oculomotor activity; tectal neurons most likely do. Finally, midbrain visual centers are part of the subcortical mechanisms mediating pupillary constriction on illumination of the eye (the light reflex). There is good evidence that the pretectal region is the important structure for this reflex.

FUNCTIONAL PROPERTIES OF NEURONS OF THE VISUAL SYSTEM

It may be assumed that in its course from the retina to the cerebral cortex the visual message, translated in neural code, undergoes a series of transformations that are reflected in the activity of the neurons of the system. These transformations depend on the anatomic connections within and between the neuronal aggregates; on their mode of action, which can be either excitatory or inhibitory; and finally on the qualitative and quantitative nature of the interaction between them. A single visual neuron, as representative of the population of cells of which it is a member, reflects to a certain degree the ability of the visual system to perform those discriminating and abstracting functions on which the recognition and interpretation of the visual stimulus is based.

Visual neurons at all levels of the system discharge impulses in the absence of any specific retinal stimulation. This *spontaneous activity,* in light and dark, is generally irregular and changeable; it exhibits a great variety of patterns and occasionally rhythmic fluctuations.[21, 80, 102] Like all types of neural activity, it is greatly influenced by anesthetic agents, particularly barbiturates, and—at least at the geniculate and cortical levels—by the state of altertness of the animal. Variations in the rate and pattern of discharge occur spontaneously and concomitantly with the changes in the EEG and behavioral signs that are characteristic of sleep and wakefulness.[111] Functional elimination of the visual input by temporary retinal ischemia, produced by applying high intraocular pressure, modifies but does not abolish the spontaneous activity of central visual neurons.[11]

Against this background of maintained activity, changes in illumination of the retina with either white or monochromatic flashes of light modify the frequency and the pattern of firing of visual neurons, whose impulse activity may increase or decrease or be completely suppressed for the duration of such a stimulus. These responses and, by extension, the neural elements that manifest them may be described as excitatory and inhibitory, with no implication as to how they are determined in the neural networks. These transient changes are also termed *on, off,* and *on-off,* depending on whether the neuron yields a burst of impulses when the stimulus is turned on or off or whether it yields impulses to both. If the light is not turned off, the neural activity tends, with time, to return to the frequency and pattern it exhibited in darkness. This maintained activity under conditions of light adaptation is almost independent of the intensity of illumination.

Visual neurons respond in a variety of qualitative and quantitative ways to various physical characteristics of the stimulus. Variations in the intensity of the stimulating light are reflected in changes of the firing rate of the neuron, and the degree of change is related to the luminance of the stimulus as well as to the level of adaptation of the retina upon which it is shone. Some neurons respond uniformly to all wavelengths of the spectrum; others respond differently to different spectral bands. The duration of the stimulus and the rate at which it recurs also modify the pattern of the neural response.

Visual neurons, like other sensory neurons, do not relate anatomically and functionally to the entire peripheral receptive surface, but only to a discrete region of it. The *receptive field* of a neuron of the visual system may be defined as that area in the visual field—or corresponding area of retinal surface—within which an adequate stimulus influences the activity of that neuron. If one explores the retina with a small spot of light, it becomes apparent that the receptive fields of visual neurons are seldom functionally homogeneous. In their simpler form they are organized in two spatially separated but contiguous zones arranged either concentrically (retinal ganglion cells, geniculate neurons) or side to side (cortical neurons). Neural impulse activity may be increased or suppressed, depending on where within the receptive field the light falls and thus on the spatial configuration of the stimulus. More complex forms of receptive field organization exist at all levels of the system, particularly for neurons of the cerebral cortex. Moreover, there is evidence that under certain conditions the excitability of visual neurons may be modified by stimulation of retinal areas outside the receptive field as previously defined.[86]

In summary, the activity of visual neurons in response to a photic stimulus depends on the intensity, the spectral characteristics, and the spatial and temporal configuration of the pattern of illumination falling on their receptive fields. The central replication of the visual message is determined not only by the physical dimensions of the stimulus and by the intrinsic properties and state of excitability of the neural network implicated but also by the spatial relationships of the excitatory and inhibitory components within the receptive field of each element of the population of neurons involved.

RETINAL GANGLION CELLS AND GENICULATE NEURONS

Receptive field organization

In the light-adapted state the receptive field of retinal ganglion cells (Fig. 15-26) as well as those of neurons of the LGN are grossly round in shape and are composed of a central region, either excitatory (on-center neurons) or inhibitory (off-center neurons), surrounded by an annular region giving the opposite response. Illumination of part or all of the excitatory region increases the firing rate of the neuron, whereas illumination of the inhibitory region reduces or suppresses it and generally evokes a discharge at *off.* This form of center-surround organization of the receptive field, circular in shape or with various degrees of radial symmetry, obtains for the large majority of retinal and geniculate neurons of the cat and the monkey.[67, 79, 103, 107, 121]

Measurements in the cat indicate that the *size* of the center region of retinal ganglion cell receptive fields is inversely related to the ganglion cell density.[103] Fields located near the fovea have smaller centers than fields in the retinal periphery. In the cat, field centers range from 0.5 to 8 degrees of arc in diameter, although larger ones have been found occasionally. In the monkey, centers as small as 2′ of arc have been reported at 1 or 2 degrees from the fovea.[122] The total extent of the receptive field is more difficult to determine. It appears, however, that it is more uniform than the center region alone and that there is a tendency, at least at the retinal level, for fields with small centers to have large peripheral zones and vice versa.[121] These findings support the notion that the centro-peripheral gradient of visual acuity in man "may well be related to variations in receptive field center size similar to those found in the monkey and cat."[67]

Within each region of the receptive field

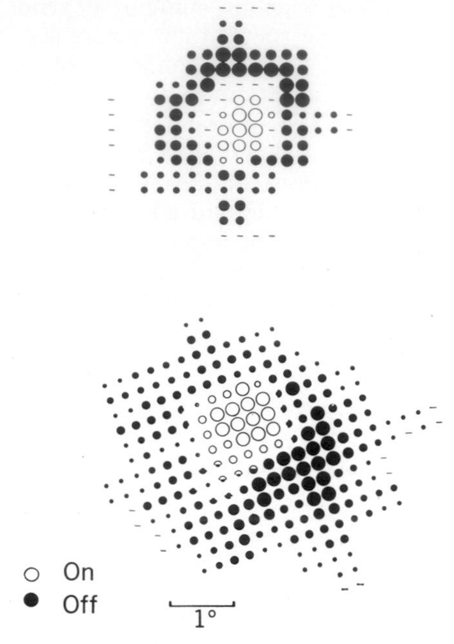

Fig. 16-9. Receptive fields of two on-center retinal ganglion cells of cat. Stimulus was a light spot 2′ to 4′ in diameter, flashed 1 sec on and 1 sec off. Open circles indicate an *on*-response; filled circles an *off*-response; and half-filled circles, an *on-off*-response. Size of circle is measure of magnitude (strength) of response at that point. Short line indicates no detectable response. Note asymmetric off-surround of receptive field plotted in lower half of figure. (From Rodieck and Stone.[103])

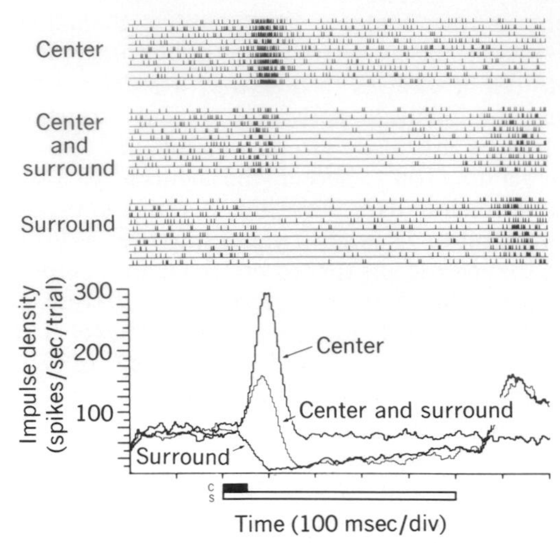

Fig. 16-10. Response of on-center cat geniculate neuron to discrete flash stimulation of center and surround regions of its receptive field with spot and annulus, respectively. Upper part of figure shows replicas of sequences of neural impulses before, during, and after illumination of the two regions separately and together. In lower part of figure, corresponding time histogram of impulses, constructed over 50 stimulus repetitions, are plotted superimposed. Times of presentation of stimuli are shown at bottom of figure: Center spot, *C,* 0.6 degrees in diameter, 50 msec flash duration; surround annulus, *S,* 1 to 2.7 degrees inner-outer diameters, 500 msec flash duration. Stimulation of center region evokes on-response; stimulation of surround, suppression of activity and off-response following cessation of flash. Simultaneous stimulation of center and surround elicits reduced excitatory response that reflects antagonistic interaction between the two regions; on termination of brief center stimulus, response continues with characteristics similar to those of response evoked by illumination of surround alone. (From Poggio et al.[99])

there exists a direct relation between the area of the stimulating spot and the magnitude of the neural response (spatial summation). However, each region is not equally sensitive over its entire extent. When a spot of light of constant size and intensity is flashed at various positions within the field, the center region generally gives the strongest response when stimulated near its geometric center, the surround near its internal boundary (Fig. 16-9).

Between the two regions of the receptive field there exists mutually antagonistic interaction, so that upon simultaneous stimulation of center and surround the individual effects tend to cancel; in the limit no response at all will be evoked. Usually, the net final result on the activity of the neuron will be an on-response, an off-response, or an on-off–response, depending on the relative effectiveness of the two antagonistic components; in general the response will reflect the characteristics of the center region (Fig. 16-10). There is evidence that excitatory and inhibitory mechanisms do not occupy spatially separated regions, but overlap over the entire extent of the receptive field. One mechanism, however, is more sensitive over the center region of the field and the opponent mechanism over the surround.[13, 20, 75, 99, 103] The antagonistic effect of the surround on the center response varies with the location of the receptive field on the retinal surface, being more marked for fields near the fovea (small center, relatively large surround) than for those in the retinal periphery. Moreover, the surround action is more effective at the geniculate than at the retinal level, and geniculate neurons respond less to diffuse light than retinal ganglion cells do.[68] This phenomenon of reduced sensitivity to diffuse light stimulation is even more evident and more often observed for neurons of the cerebral cortex.

The functional organization of the recep-

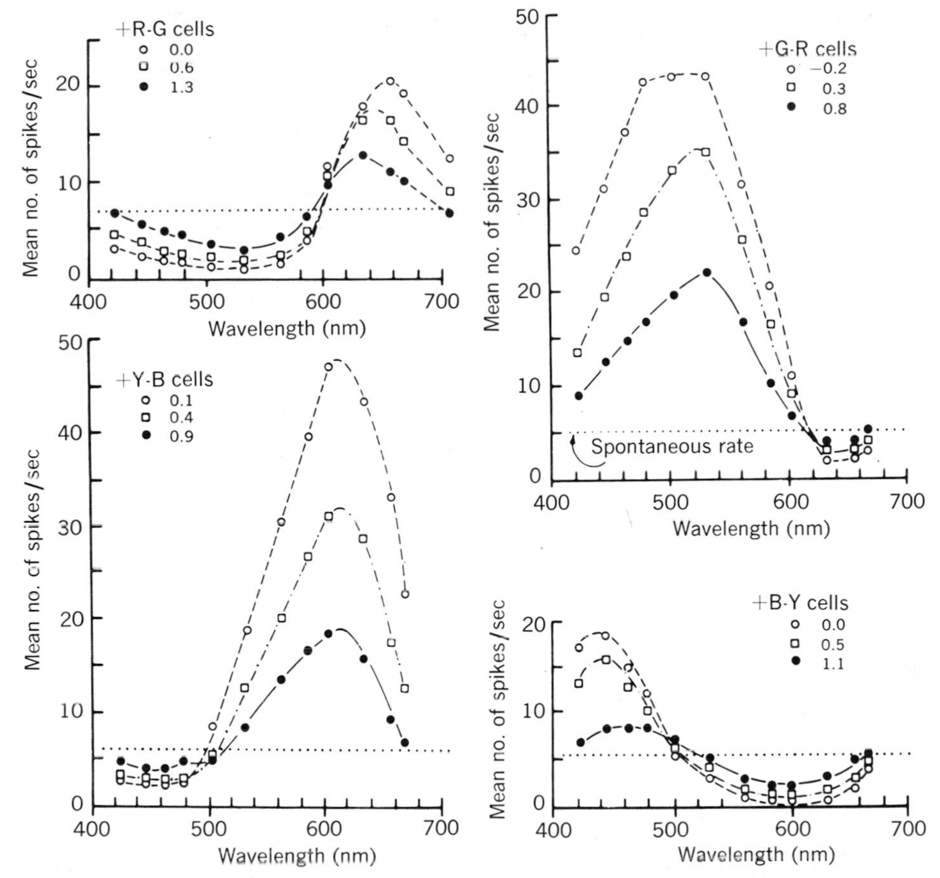

Fig. 16-11. Average firing rate of samples of the four classes of spectrally opponent cells described by De Valois et al.[40] in macaque geniculate nucleus. Diffuse flash stimulation with monochromatic lights of equal energy. *+R-G* = red excitatory, green inhibitory; *+G-R* = green excitatory, red inhibitory; *+Y-B* = yellow excitatory, blue inhibitory; *+B-Y* = blue excitatory, yellow inhibitory. For each class of cells, spectral response curves to three levels of light intensity are shown; numbers next to each curve represent log attenuation relative to maximum available. Open symbols and vertical lines at each point enclose 1 SE of mean. Dotted horizontal line indicates mean firing rate in absence of stimulation for each type. (From De Valois.[15])

tive fields of visual neurons is profoundly influenced by the state of *adaptation* of the eye, which is determined by the level of ambient illumination.[15, 79, 103, 104, 121] The surround region appears to be more sensitive in this regard, and adaptational effects always occur first at its outer border. Under conditions of dark adaptation the influence of the surround diminishes and may actually disappear, the size of the center region increases, and the response to illumination of the center becomes stronger. Upon light adaptation the antagonistic surround becomes more and more effective and the receptive field acquires the duplex character described. With a very bright background, surround responses are difficult to elicit in isolation, but the strength of the center response decreases, the field shrinks, and again it may appear to be composed only of a small area with the functional properties of the center.

Neural response to lights of different wavelengths

The chromatic properties of neurons of the LGN have been investigated in some detail in the macaque, an animal that behavioral tests have shown to possess color vision essentially identical to that of man. An important series of studies by De Valois[2, 44, 45] and colleagues[46] and the detailed analysis by Wiesel and Hubel[122] have furnished experimental evidence that geniculate neurons may be subdivided in two major categories: spectrally opponent cells, which respond with an increase in firing rate to some parts of the visible spectrum and with a decrease to other parts, and broad-band cells, which are uniformly excited or inhibited by light of all wavelengths and by white light. Cells of the two categories are thought to subserve different visual functions: the former carry information about the color of the light, the latter about its brightness.

Spectrally opponent cells. In the macaque more than two thirds of the cells in the four dorsal layers of the geniculate nucleus display a differential spectral sensitivity. On the basis of their response to diffuse flash stimulation of the retina with monochromatic light, De Valois et al.[46] have recognized four classes of opponent cells (Fig. 16-11). Some cells are excited at long wavelengths and inhibited at short wavelengths: (1) red excitatory–green inhibitory and (2) yellow excitatory–blue inhibitory; and other cells have the opposite chromatic organization: (3) green excitatory–red inhibitory and (4) blue excitatory–yellow inhibitory. R-G cells are found more frequently than Y-B cells.

The color names used to define these neurons do not refer to retinal photopigments, but rather to the color that we see at those wavelengths that in any one cell produce excitation and inhibition, and that correspond to the four subjective primary colors in the spectrum. As the graphs in Fig. 16-11 show, neither the shape of the spectral curves nor the wavelength of maximal effects that obtain for the four classes of geniculate opponent neurons are similar to the absorption spectra of any of the photopigments found in cone receptors of the retina of primates (Fig. 15-8). On the other hand, cone receptors subserve color vision, and geniculate cells are known to receive an input from them.[122]

The component inputs to geniculate neurons have been determined by De Valois[44, 45] by means of chromatic adaptation experiments such as that illustrated in Fig. 16-12. In this example the neural response to diffuse flashes is plotted as a function of the stimulus wavelengths for three different conditions of eye adaptation. Under dark adaptation (no bleach) the neuron responds to wavelengths in the middle spectral range with an increase of its impulse frequency over the spontaneous rate (the zero point on the ordinate scale) and to longer wavelengths with a decrease in frequency (+G–R cell). This response may be separated into excitatory and inhibitory components by selective bleaching of retinal pigments with intense

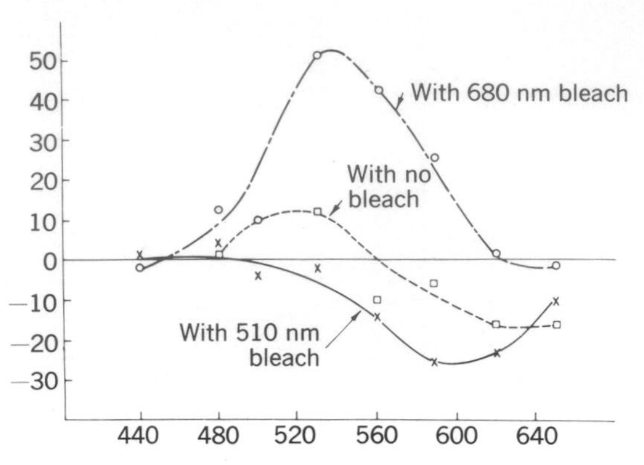

Fig. 16-12. Spectrally opponent cell of lateral geniculate nucleus of macaque (green excitatory; red inhibitory). Plot of response of neuron to diffuse retinal stimulation with various monochromatic lights during dark adaptation and during chromatic adaptation from two different spectral regions. Abscissa: wavelength in nanometers. Ordinate: number of impulses/sec during light stimulus with respect to spontaneous rate taken as zero for this plot. (From De Valois.[2])

adapting lights (chromatic adaptation). When the system of retinal receptors with red-absorbing pigments is selectively adapted out with a red light (680 nm bleach), only excitatory responses can be evoked that reflect the excitatory input to the cell from the middle-wavelength sensitive system. Similarly, an isolated inhibitory input from the long-wavelength system is revealed under conditions of green adaptation (510 nm bleach). The spectral response curves for the two antagonistic components are broad and overlap over a large extent; the points of maximal effect have shifted from their positions under dark adaptation and correspond closely to the points of maximum absorption of the middle- and long-wavelength photopigments (Fig. 15-8).

These chromatic adaptation studies indicate that the response of opponent geniculate neurons is the result of neural interaction (algebraic summation[45]) between pairs of functionally antagonistic inputs from cone receptor systems. According to De Valois, any one cell receives one input—either excitatory or inhibitory—from the cone receptors sensitive to long wavelengths (maximum absorption at 570 nm) and an opponent input from either the middle- or short-wavelength receptors (535 and 445 nm). Long- and middle-wavelength combinations produce the two reciprocal classes of R-G opponent neurons and long- and short-wavelength combinations the Y-B cells.

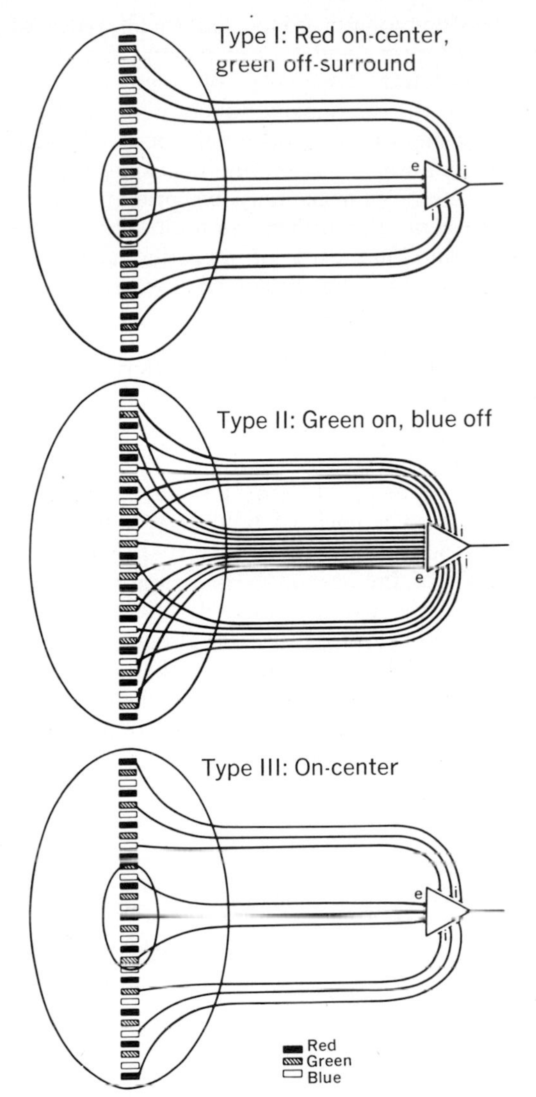

Fig. 16-13. Schematic diagrams to illustrate contribution of cones to type I, II, and III geniculate neurons, as proposed by Wiesel and Hubel. Three types of cones are indicated with different symbols, and for simplicity, receptors are shown only along one line through receptive field. Cones project to geniculate cells via intervening synapses, not indicated in drawings, and activation of receptors leads to excitation, *e,* of geniculate neuron or to inhibition, *i.* Example of type I cell: excitatory input from red-sensitive cones in field center and inhibitory input from green-sensitive cones in periphery. Three cone types from center are arbitrarily shown as being present in ratio of 1:1:1, and their ratio is the same in periphery. Example of type II cell: excitatory input from green-sensitive cones and inhibitory input from blue-sensitive cones. Relative contribution of two cone types is same in all parts of receptive field. Example of type III cell: input from all three types of cones, excitatory in nature from center and inhibitory from periphery. Proportion of three types of receptors is the same for two antagonistic regions of receptive field. (Redrawn from Wiesel and Hubel.[122])

The analysis of Wiesel and Hubel[122] of the spatial and chromatic properties of geniculate neurons has confirmed and extended these observations and has shown that spectrally opponent cells possess two types of receptive fields (Fig. 16-13). Type I cells have fields with a typical center-surround organization to white light. Monochromatic light stimulation reveals that within these receptive fields the two spectrally antagonistic systems are spatially separated and concentrically arranged. From these neurons an excitatory or inhibitory response is obtained by stimulation of the field's center with lights in a given spectral range and the opposite response from the surround with another band of wavelengths. For type II neurons the spatial distribution of the two spectral components is identical, and no center-surround organization exists within the field. These cells respond in one manner, excitatory or inhibitory, to stimulation over the entire receptive field at some wavelengths and in the opponent fashion at other wavelengths. White-light stimulation is relatively ineffective. Type I neurons are much more common than are those of type II, and both types are found in all dorsal layers of the LGN, but not in the ventral layers.

These experiments have confirmed that the peak sensitivity of the component elements of the opponent systems at the geniculate levels agrees closely with the peak absorption of the three cone photopigments in the retina of primates. Some type I neurons also appear to receive an input from the rod receptor system. Nearly all combinations of green and red and green and blue have been observed, the former more often than the latter. No R-B cells have been found.

To sum up, geniculate neurons of the macaque that are differentially sensitive to wavelengths receive functionally opponent inputs from paired sets of the three cone receptor systems maximally sensitive in different regions of the spectrum. In most instances these dichromatic mechanisms are spatially segregated over the receptive field of these neurons. Spectrally opponent cells are part of a neural system carrying information about the chromatic as well as the spatial characteristics of the photic stimulus. There is evidence that intraretinal mechanisms may account for the sorting of the excitatory and inhibitory components and that the opponent type of neural code originates at the ganglion cell level and is essentially maintained at higher levels of the system[56] (Chapter 15).

Broad-band cells. Another group of geniculate neurons is not differentially sensitive to monochromatic light. The concentrically arranged antagonistic regions of the receptive fields of these neurons (type III of Wiesel and Hubel, Fig. 16-13) have identical spectral sensitivity, and from each of them the same uniform response, either excitatory or inhibitory, is obtained at all wavelengths and with white light. Broad-band neurons appear to be functionally connected to all types of cone receptors and to rod receptors as well. The spectral sensitivity of these neurons matches the luminosity function of the animal, as measured by behavioral tests of visual discrimination.[2] These findings suggest that broad-band neurons may be primarily concerned with the transmission of information about the luminosity of the stimulus. Their properties will be discussed further in the following section.

Neural response to lights of different intensities

All other parameters of stimulation being constant, the change in the pattern and frequency of firing of central visual neurons in response to a flash of light is a function of the intensity of the flash. Fig. 16-14 shows an analysis of the responses of a neuron of the LGN of the cat to lights of different intensities. The stimulus was a circular spot of light, spatially restricted to the excitatory center of the neuron's receptive field, flashed for 50 msec against a large adapting background of moderate luminance. The neural

log I=log I₀+0.0
0.2
0.4
0.6
0.8
1.0
1.2
Average number of impulses/trial (0.4/div)
Time (100 msec /div)
2.2° I₀=2.1 x 10⁻³ lm/sq m
Background I=3.0 x 10⁻³ lm/sq m
A

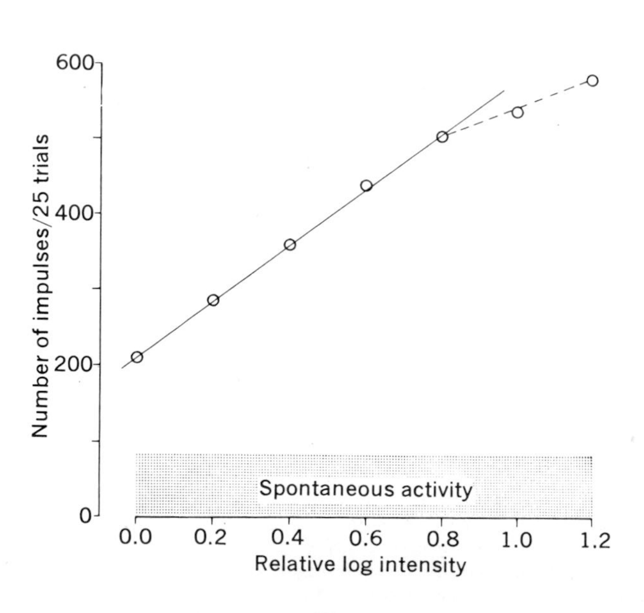

Fig. 16-14. Response of neuron of lateral geniculate nucleus to lights of different intensities. Unanesthetized cat; on-center neuron. **A,** Time histograms of neural impulses occurring before, during, and after stimulation (4 msec histogram classes). Intensity of illumination is indicated. Intensity increases from top to bottom histograms in 0.2 log unit steps. **B,** Plot of neural response versus intensity of illumination light stimulus. Magnitude of response is measured at each intensity by total number of impulses occurring in 25 trials during period of 100 msec. Beginning of counting period was different for different intensities and chosen on basis of latency measurements. (From unpublished experiments by Poggio and Baker, 1966.)

activity, averaged over 25 samples, is represented in the form of a histogram of the impulses occurring before, during, and after the presentation of the stimulus (Fig. 16-14, *A*). Each histogram refers to a different intensity of stimulation (0.2 log unit steps). As the intensity of the flash increases, the neural response, as measured by the number of impulses, also increases and over a large part of the range investigated stands in an approximately linear relation to the logarithm of the stimulus intensity (Fig. 16-14, *B*).

The general features of the neural replication of the intensive dimension of the visual stimulus are indicated by studies of Jacobs[74] of the LGN of the squirrel monkey in which the behavior of broad-band neurons, characteristically sensitive to changes in stimulus intensity, was analyzed. In this experiment, while the impulse activity of a single geniculate neuron was continuously recorded, the eye of the animal was first adapted to a luminous field subtending 15 degrees of visual angle. The intensity of the field was then shifted to a higher or lower luminance for a period of 1 sec and returned to the adapting level. A number of such shifts was made for each adaptation test condition investigated. The plots in Fig. 16-15 indicate the results obtained at four different adapting luminances for a series of stepped increments and decrements around each of them, over a total range of ±0.6 logarithmic unit. The neural response is measured in terms of the increase or decrease of mean rate of impulses during the intensity shift from the mean rate of discharge at the relative level of adaptation.

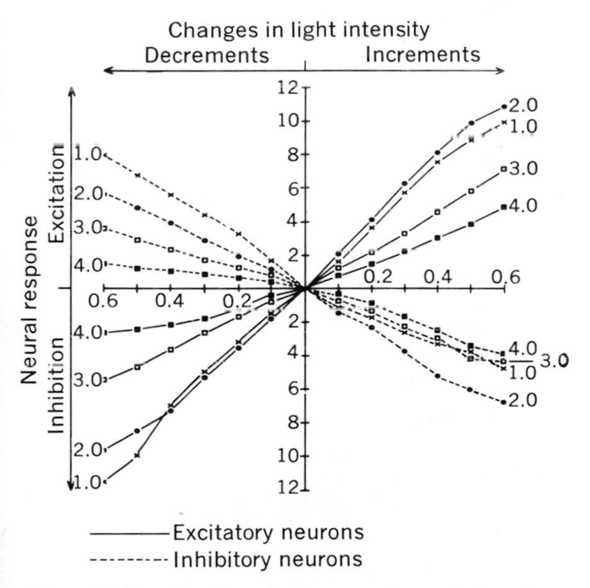

Fig. 16-15. Graphs plotting number of impulses discharged by excitatory and inhibitory broad-band squirrel monkey geniculate neurons in response to increments and decrements in light intensity about different levels of adaptation (see text). Four adaptation luminances from which stimulus shifts were made are indicated at both ends of each curve as log attenuations. Abscissa: size of stimulus shift in relative log units. Ordinate: mean change in frequency of firing (impulses/sec) above and below mean spontaneous discharge rate. (Replotted from data by Jacobs.[74])

When the retina is diffusely illuminated with flashes of light whose luminance is greater than the luminance level at which the eye is adapted, broad-band geniculate neurons respond with either an increase of their firing frequency (excitatory neurons) or a decrease (inhibitory neurons). Conversely, when the intensity of the test stimulus is less than that of the adapting level, a decrease of the impulse rate of excitatory neurons and increased activity of inhibitory neurons are produced. At any given level of adaptation of the eye the total range of luminances over which a graded response, either excitatory or inhibitory, is obtained is a limited one, of approximately ±1 log unit of intensity about the relative adaptation luminance. Beyond this range no further increase in the magnitude of the response is observed.

These observations allow certain generalizations with regard to the central neural coding of information about the intensity of the visual stimulus. Under common natural conditions the eye is adapted at some moderate level of illumination. From time to time the luminance of one part of the field of vision changes; either it increases or decreases. The population of visual neurons anatomically related to that part of the field will undergo changes in its total pattern of activity: the discharge rate of some neurons will increase, that of others decrease, and the magnitude of change will be a function of the relative difference, or contrast, between the intensity of the modified part of the field and the luminance of the adapting background. The range of luminances over which these neurons operate efficiently, i.e., the range over which they might contribute to the visual discrimination of intensity, is a limited one and its magnitude is relative to the adaptation state of the retina. As shown by the slopes of the curves in Fig. 16-15, the effective range of change in neural activity is greater at higher than at lower adaptative luminances.

Neural response to intermittent light stimulation

While there exists considerable information on the psychophysics of the time-dependent effects of photic stimuli,[6, 98] neurophysiologic investigations on their neuronal correlates and in general on the temporal properties of the activity of visual neurons have not been numerous. Experimental data of interest are, however, available on the capacity of retinal ganglion cells of the cat to follow repetitive light stimulation,[52, 94] and some results from recent studies of this kind will be described.

As already described in Chapter 15, a measure of the visual capacity to resolve photic stimuli in time may be obtained in human beings by determining the minimal frequency at which regularly alternating periods of light and darkness are subjectively perceived as a continuous stimulus of constant luminance (critical fusion frequency, or CFF).

The behavior of a single visual neuron under conditions of intermittent light stimulation is shown in Fig. 16-16, with an example from an off-center ganglion cell of the cat's retina. At low frequencies of stimulation and up to 32 Hz in Fig. 16-16, *A*, the neuron responds

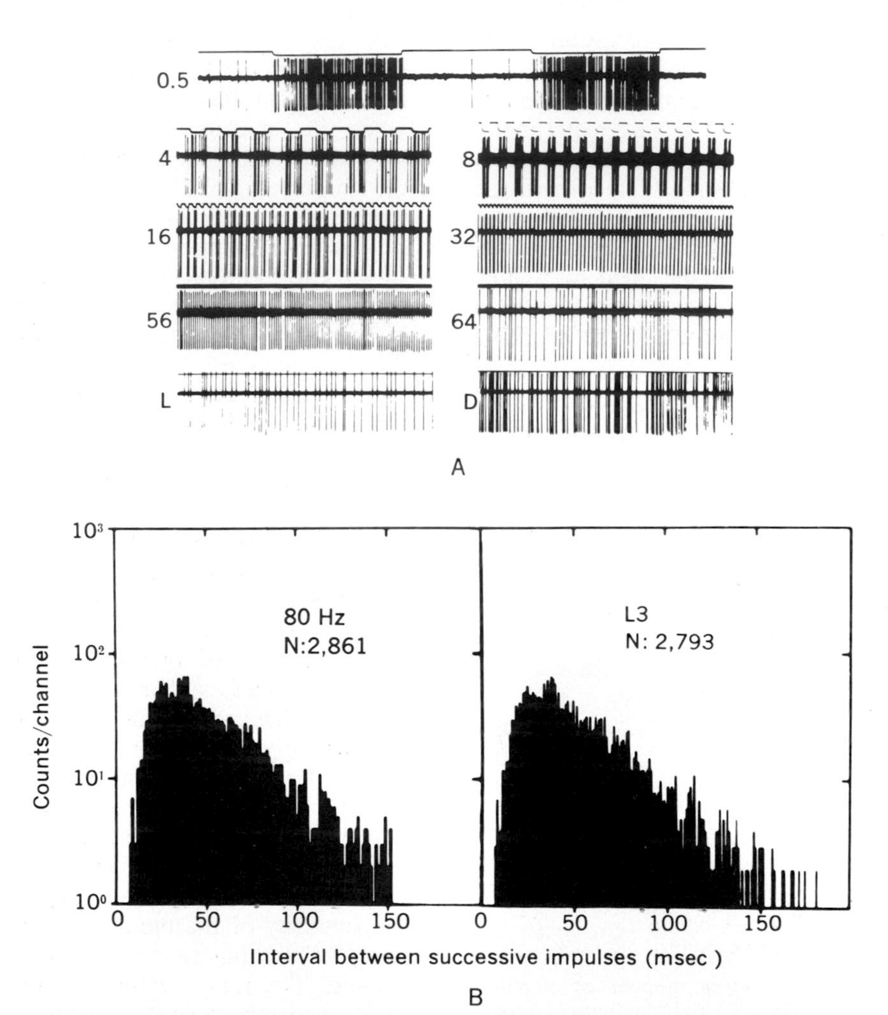

Fig. 16-16. Response of off-center retinal ganglion cell of cat to intermittent light stimulation. **A,** Neural activity in response to light spot flashed at frequency (Hz) indicated by numbers to left of each record. Light spot was 17.5 degrees of arc in diameter; intensity was 8.4 lm/m^2. No background illumination. For each sample, phases of flash cycle are indicated by upper trace; downward, light off. *L,* Response to 8.4 lm/m^2 steady retinal illumination. *D,* Discharge in darkness after dark adaptation. **B,** Interval histograms of impulse activity during light spot stimulation at 80 Hz and during steady illumination (intensity 8.4 $1m/m^2$). Recording time for each histogram = 150 sec. Histogram bin = 1 msec. Counts per channel indicate number of interspike intervals in each histogram bin. Two histograms suggest that at a frequency of 80/sec there is no discernible neural signal of stimulus frequency. (From Ogawa et al.[94])

to *every* dark phase of the cycle; the rate and pattern of firing are related to the duration of the stimulus. This ability of the neuron to respond to each flash (complete following) does not obtain at higher frequencies. At 56 Hz the response, consisting now almost entirely of a single impulse per cycle, begins to fail. As the frequency of stimulation increases, the number of failures also increases; and at 64 Hz there is no obvious sign of response following, although finer analysis reveals that the time of occurrence of the impulses is still influenced by the frequency of the intermittent light. At a frequency of 80 Hz the neuron's activity equilibrates; i.e., the sequence of impulses is not statistically different from that recorded when the retina is illuminated with a continuous light (Fig. 16-16, *B*). The frequency at which equilibration occurs may then be defined as the CFF for the neuron under study.[94]

The relation between the ability of the neuron to follow intermittent stimulation and the intensity of the stimulating light is plotted in Fig. 16-17. The duality of the retina is clearly seen, the two limbs of the curve reflecting the effects of the rods and cones, respectively. The CFF/intensity relation for these neurons is, in its general features, similar to that which obtains psychophysically for the extrafoveal region of the retina (Fig. 15-5).

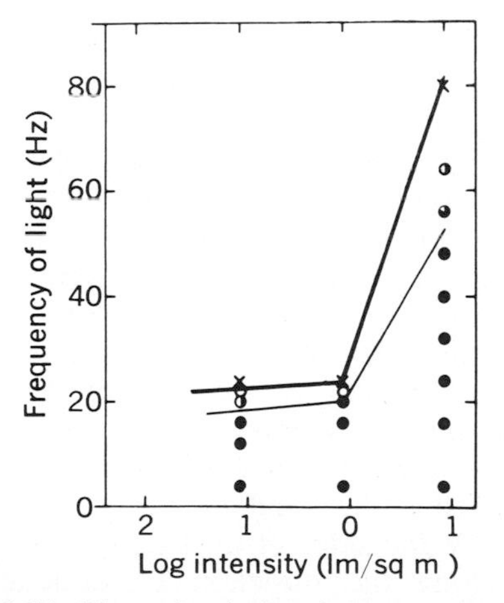

Fig. 16-17. Plot of relation between capacity of ganglion cell of cat retina to follow repetitive stimulation and intensity of stimulating light. Same neuron as in Fig. 16-14. Same parameters of stimulation. Filled circles, complete frequency following; half-filled circles, partial following; open circles, very weak following; crosses, no following. Thick line, CFF/intensity relation. Thin line, maximum frequency for complete following. (From Ogawa et al.[94])

These findings indicate that the subjective phenomenon of flicker fusion may be determined at the level of the retinal ganglion cells. That this may indeed be the case for other forms of simple visual discrimination as well is apparent from experimental observations described in preceding paragraphs. It must be emphasized, however, that there is clear evidence that even these elementary aspects of vision depend on neural transformations and integration at more central levels of the visual pathways. For example, the visual ability to discriminate stimuli in time, as measured by the level of CFF, is reduced over the entire visual field by lesions to the geniculostriate system that produce only localized homonymous field defects.[114]

NEURONS OF THE VISUAL CORTEX

Receptive field organization

Investigations of the functional properties of cortical visual neurons have shown that in the cerebral cortex there occur important changes in the organization of the receptive fields of single neurons, changes that are thought to reflect neural mechanisms subserving the analysis of spatial features of the visual stimulus and leading to the perception of complex forms.

Cortical visual neurons respond more readily to straight border patterns than to stationary flashes. Narrow rectangles of light (slits), borders between areas of different luminous intensities (edges), and dark bars against a light background are among the most effective stimulus configurations, and for any one neuron, certain patterns are often more effective than others. Most characteristically, the orientation of the line stimulus within the receptive field is critical and specific for the cortical cell. The direction of stimulus movement and also its speed are in many neurons important parameters for optimal response.[16, 47, 63, 69-71, 96] In the following discussion an account will be given of the spatial properties of cortical visual neurons, using the classification of Hubel and Wiesel.[69, 70]

A number of cortical neurons have receptive fields with intrinsic properties similar to those of cells at lower levels of the system. These neurons are found only in the primary visual cortex and, within it, most frequently at about cortical layer IV. Some of these neurons have functionally uniform receptive fields; others display properties not essentially different from those of geniculate cells: they possess fields roughly circular in shape and

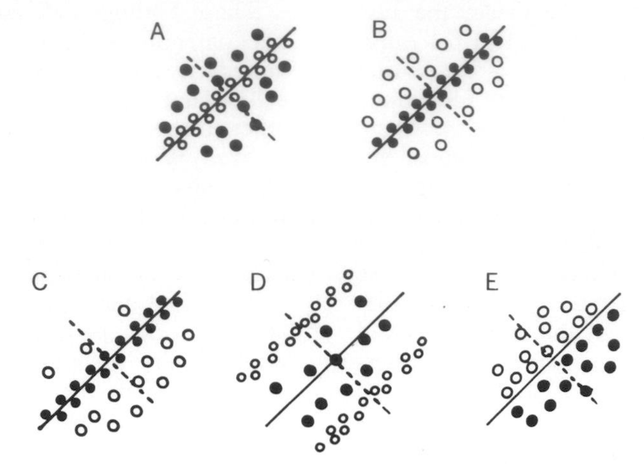

Fig. 16-18. Various arrangements of receptive fields of simple cortical neurons. Open circles, excitatory responses; filled circles, inhibitory responses. Receptive field axes are shown by continuous lines through field centers; here they are all oblique, but among population of cortical neurons each arrangement occurs in all orientations. (Redrawn from Hubel and Wiesel.[69])

respond equally well to stimuli at all orientations *(nonoriented neurons).* More typically, cortical receptive fields have excitatory and inhibitory regions arranged side by side and separated by straight boundaries rather than circular ones *(simple neurons).* Some fields are composed of an elongated center flanked by symmetric or asymmetric peripheral regions; in other fields, excitatory and inhibitory areas are juxtaposed. Characteristically the orientation of the borders between antagonistic regions is specific for each cortical cell: it may be vertical, horizontal, or oblique (Fig. 16-18). For the cell to respond well, the stimulus must be positioned accurately within the field so as not to activate opponent regions simultaneously, and it must be specifically oriented. Stimulation with diffuse light evokes little or no response in many cells, indicating that the effects from the antagonistic regions are nearly balanced. Moving stimuli evoke a brisker and stronger response than do stationary ones, most likely because of the sequential activation of the two components of the receptive field whose individual effects summate temporally and act in a synergistic way.

The majority of neurons in the striate cortex, and all of them in extrastriate cortex, possess receptive fields that are more elaborately organized than those of simple cells. *Complex neurons* are found in all cortical layers; they respond optimally to a properly oriented slit, edge, or dark bar, and for any one neuron certain spatial patterns are more effective than others. Their receptive field, however, is functionally uniform and cannot be subdivided in antagonistic regions typical of fields with simpler organization. Moreover, the specificity for stimulus position, so critical for simple neurons, is no longer present, and with a properly oriented pattern, responses may be obtained from anywhere in the field (Fig. 16-19). For many complex neurons, movements of the stimulus in an opposite direction perpendicular to optimal orientation evoke asymmetric responses, a directional preference (Fig. 16-20). The next level of receptive field complexity defines the *hypercomplex neurons.* These cells, which are more frequently encountered in superficial and deep cortical layers, also are sensitive to stimulus orientation, and in addition they show specificity toward another parameter of the stimulating pattern—its *length* along the axis of the field—which for optimal response has to be limited in one or both directions. Within these receptive fields, functionally opponent regions can be defined, but unlike simple fields the inhibitory regions are arranged on the sides of an "activating" center area, along the direction of optimal field orientation. No effects on the impulse discharge of the cell are obtained by stimulation of the inhibitory regions alone (Fig. 16-21).

The various degrees of receptive field complexity just described are thought to reflect progressively increasing degrees of specificity of cortical neurons toward the spatial parameters of the stimulus. Hubel and Wiesel[69-71]

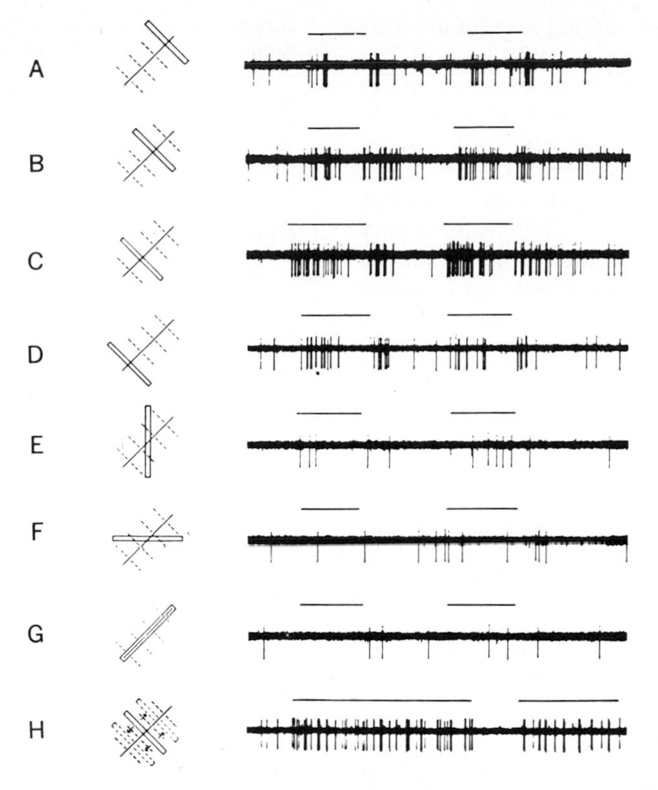

Fig. 16-19. Complex cortical neuron (area 17 of cat). Stimulation of contralateral eye with slit ⅛ × 2½ degrees. Receptive field in area centralis, 2 × 3 degrees in size. **A** to **D,** Neural response to slit oriented parallel to axis of receptive field. **E** to **G,** Slit oriented at 45 and 90 degrees to receptive field axis. **H,** Slit parallel to axis and moved rapidly from side to side. Time, 1 sec. (From Hubel and Wiesel.[69])

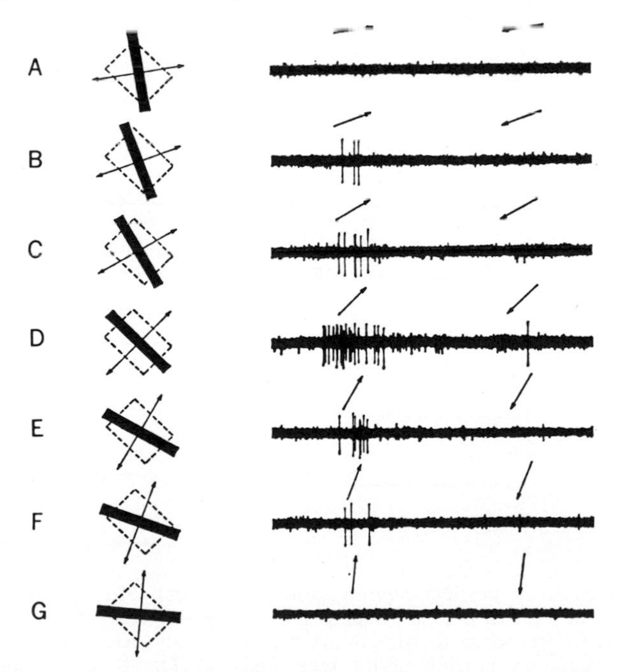

Fig. 16-20. Responses of a complex cell in right striate cortex (layer IV A) of macaque to various orientations of moving black bar. Receptive field in left eye indicated by interrupted rectangles, approximately ⅜ × ⅜ degrees in size. Duration of each record, 2 sec. Arrows indicate direction of stimulus motion. (From Hubel and Wiesel.[71])

have made the suggestion that any small cortical region analyzes some small part of the visual field in terms of the orientation, direction of motion, and type of light-dark contours and, at the hypercomplex level, also of a detection of change in direction (curvature) of the contour. In this small area of cortex only a portion of cells is influenced at any one time since inappropriate stimuli or diffuse light have little or no effect on cortical neurons.

On the basis of receptive field characteristics and on considerations of the anatomic location within the cortex of neurons with different spatial properties, Hubel and Wiesel[69, 70] propose that the organization of the receptive fields of cortical neurons results from successive convergence of a set of neurons with similar properties on other cortical cells thought to subserve a higher order integrative function than the former. The first transformation would occur at those neurons in the striate cortex upon which the axons of geniculate cells terminate. The following stages are reflected by increasing receptive field complexity: simple cells converge onto complex cells, and complex cells onto hypercomplex cells, both within the primary visual area and beyond it in extrastriate cortical regions. There is some evidence that this hierarchical paradigm may be an oversimplication. Recent studies of conduction velocities in the retinocortical pathway of the cat indicate that the characteristics of both simple and complex fields may be determined by geniculate input, and thus that cortical neurons with different spatial properties may process visual information in parallel rather than in sequence.[64] Moreover, cells with simple field organization respond preferentially to slower stimulus speed than do cells with complex fields.[96] Also, experimental observations suggest that inhibition of cortical neural activity is mediated chiefly if not exclusively by interneurons within the cortex.[41, 118] These findings, taken together, indicate that the spatial properties of cortical visual neurons, as reflected by their receptive field organization, may result from more complicated neural mechanisms than a simple hierarchical convergence from neuron to neuron in an "ascending" direction.

Chromatic properties

As in the geniculate nucleus, there are cells in the striate cortex of the macaque that can be excited at all wavelengths of the visible spectrum and whose spectral sensitivity is similar to the overall spectral sensitivity of the animal, and there are cells that possess chromatically selective properties. In the cortex, color-coded cells appear to be less numerous and have more complex chromatic organization than cells in the geniculate nucleus, at least in the extrafoveal cortical projection area.[71]

The majority of chromatically selective cortical neurons have nonoriented or simple receptive fields. In general, these cells display opponent color properties, light exciting them at some wavelengths and inhibiting them at others. Other cortical neurons, particularly neurons with more complex spatial properties, respond preferentially or exclusively over a restricted band of wavelengths, usually toward one or the other end of the spectrum.[71, 84]

It is likely that dichromatic mechanisms, similar to those De Valois described at the geniculate level (p. 516), are the basis of

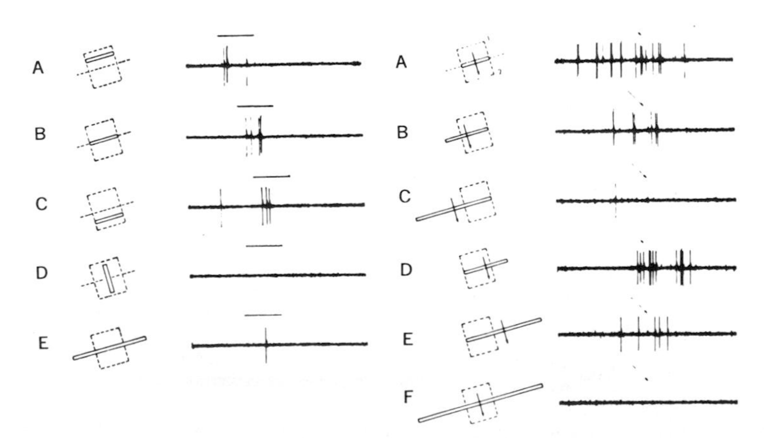

Fig. 16-21. Hypercomplex cortical neuron (area 19 of cat). Stimulation of ipsilateral eye. Excitatory portion of receptive field roughly 2 × 2 degrees, represented by interrupted rectangle. Left: **A** to **C,** Response to stimulation with optimally oriented slit, ⅛ × 2 degrees in various parts of excitatory portion of the receptive field. **D,** Slit at 90 degrees to optimum orientation. **E,** Slit 5½ degrees long, extending beyond excitatory regions on both sides. Right: **A** to **F,** Response to slits ⅛ degree wide and of various lengths moved downward across receptive field at about 3 degrees/sec. Note that antagonistic flank on right of excitatory region is weaker than that on left. Sweep duration, 1 sec. (From Hubel and Wiesel.[70])

the functional organization of many color-selective cortical cells. On the other hand, there is evidence that all three cone mechanisms can intersect on a single striate neuron, not only on neurons with broad-band spectral characteristics but also on neurons with evident color-selective properties.[57] The organization of the visual cortex with respect to color and its operation for the development of color vision are as yet unknown.

Binocular interaction

The majority of neurons in the visual cortex are influenced by stimulation of the two eyes. In the striate cortex of the cat some 84% of cells are binocularly activated; in area 17 of the monkey, binocular neurons are less common (62%), but their number increases considerably in area 18. For all these neurons a receptive field can be found in either eye, and the two fields have similar size, shape, and functional organization. Accordingly, the most effective stimulus in one eye—in form, orientation, direction of motion, etc.—is also the most effective in the other eye. With binocular stimulation, mutual facilitation or summation of the neural response is obtained from functionally corresponding regions of the two receptive fields and suppression or inhibition from functionally opponent regions.[14, 63, 69-71, 73, 97]

In two important respects, however, the binocular input on single cortical neurons is not entirely symmetric. First, the amplitude of the response evoked from each eye is not necessarily identical. For some cells the two monocular responses are about equal, but in general there is a relative ocular dominance, the cell being driven more effectively from one eye than from the other. At the limit are those neurons that respond to stimulation of one eye only. Second, for any given cell the location of the receptive fields in the two eyes is not always exactly the same. At a given

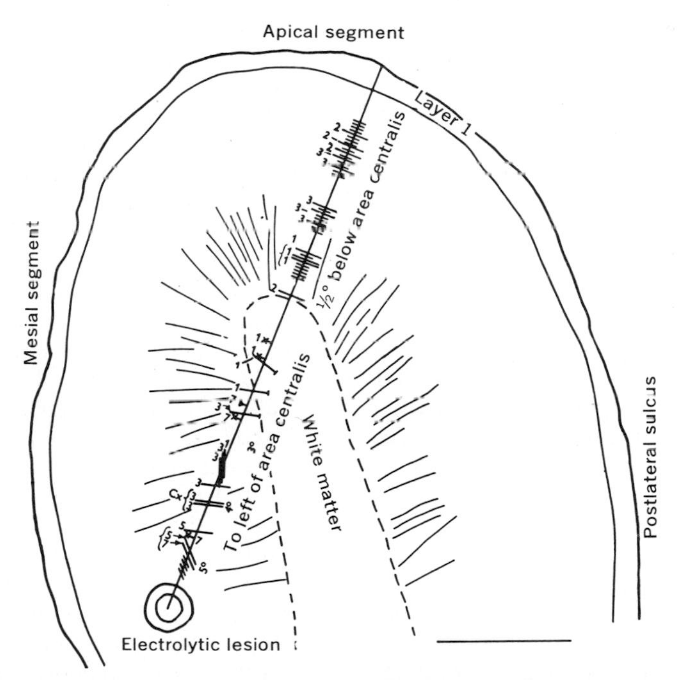

Fig. 16-22. Reconstruction of microelectrode penetration through lateral gyrus of cat. Electrode entered normal to cortical surface and remained parallel to deep fiber bundles (indicated by radial lines) until it reached white matter; in cortex of mesial segment, electrode course was oblique. Field axis orientation is shown by direction of each line; lines perpendicular to track represent vertical orientation. Braces indicate pairs or groups of neurons observed simultaneously. Complex receptive fields are indicated by *Cx*. Afferent fibers from lateral geniculate nucleus are indicated by *x* for on-center and ▲ for off-center. Approximate positions of receptive fields on retina are shown to right of penetration. Shorter lines show regions in which unresolved background activity was observed. Numbers left of penetration refer to ocular dominance. Neurons are classified *1* to *7*, from marked dominance of contralateral eye, *1*, through no obvious differences between the two eyes, *4*, to exclusively ipsilateral driving, *7*. (From Hubel and Wiesel.[69])

cortical locale there are neurons whose receptive fields occupy topographically corresponding regions in the right and left retinas and other neurons whose fields have slightly different positions in the two eyes. This phenomenon of "receptive field disparity" is thought to represent an important aspect of the neural mechanisms operating in binocular vision and will be discussed in a later section.

Functional organization of the visual cortex

As described in other chapters, the somatosensory cortex and the auditory cortex may be subdivided into discrete vertical columns of neurons, oriented perpendicularly to the cortical surface and extending to the white matter, columns that are functionally defined by certain identical properties of their constituent elements. A similar organization is present in the visual cortex.

In the somatic cortex the functional identity of the column is provided by common modality properties and location of the receptive fields of the cells, and in the auditory cortex by the specificity to the frequency of the stimulating sound and by the characteristic delay of the neurons, i.e., their sensitivity to the location of a sound in space. In the visual cortex all neurons in a column, whether simple, complex, or hypercomplex in type, have the same receptive field orientation and binocularly the same eye preference (Fig. 16-22). On the basis of these common properties, two overlapping but completely independent columnar systems have been recognized: one system of narrow columns, identified by receptive field orientation, and another system of coarser columns, often including several orientation columns, in which cells are aggregated according to ocular dominance.[69, 71]

Thus all sensory areas of the cerebral cortex appear to be organized in terms of vertical systems, of which the column is the elementary unit. Such an organization coexists with the two-dimensional representation of the peripheral receptive surface. This representation is continuous over the cortical surface and upon it the column-systems are engrafted.[71] The "horizontal" topographic organization reflects the relation between the "where" in the periphery to the "where" in the brain. The vertical organization, on the other hand, provides the superimposed chains of interneuronal mechanisms placed between cortical input and cortical output by which the cortex operates upon the former to produce the latter. The vertical systems subserve parameters of the stimulus other than topography.

BINOCULAR VISION AND DISCRIMINATION OF DEPTH

Psychophysics

During the course of evolution there occurs a shift in the position of the two eyes from a symmetric location on opposite sides of the head to a frontal binocular position. This allows the development in primates of exquisite manual skills executed under visual guidance. Most important, in binocular vision there emerges a new and vivid perception of depth not possible in monocular vision.*

Binocular visual space comprises a solid angle determined by the overlapping part common to the two monocular fields. Within this binocular field, the two eyes are stimulated from a single point in space at the same time, but not necessarily at the same position on each retina. Because of the horizontal separation between the two eyes, the two-dimensional images of a tridimensional object occupy slightly different horizontal positions on right and left retinas (binocular parallax). This horizontal shift between the two images, or *retinal image disparity,* is the essential cue for binocular depth discrimination or stereopsis.

The main features of binocular vision are summarized in Fig. 16-23. When the two eyes are directed simultaneously to a given fixation point, the images of that point fall on the center of left and right retinas (Fig. 16-23, *A*—point *F* and its retinal projections f_1 and f_2). The two images are combined, or "fused," by central neural mechanisms and the point is seen as single and localized "in one and the same visual direction, no matter whether the stimulus reaches the retinal elements in one eye alone, its corresponding partner in the other eye alone, or both simultaneously."[26] By definition, retinal elements that give rise to a localization in the same subjective visual direction when stimulated are said to be *corresponding retinal elements* in that one may be suppressed without affecting the result.

*When we see the world with one eye, the sense of the third dimension can be added to the visual image by indirect cues such as movements of the head to produce successive images of the same object on different retinal elements (monocular parallax), knowledge of the angular subtense of the object, and shadow effects on distant objects by nearer ones. The process by which depth is perceived in monocular vision is quite different from that operating in binocular depth perception. Its nature is not as yet fully understood.

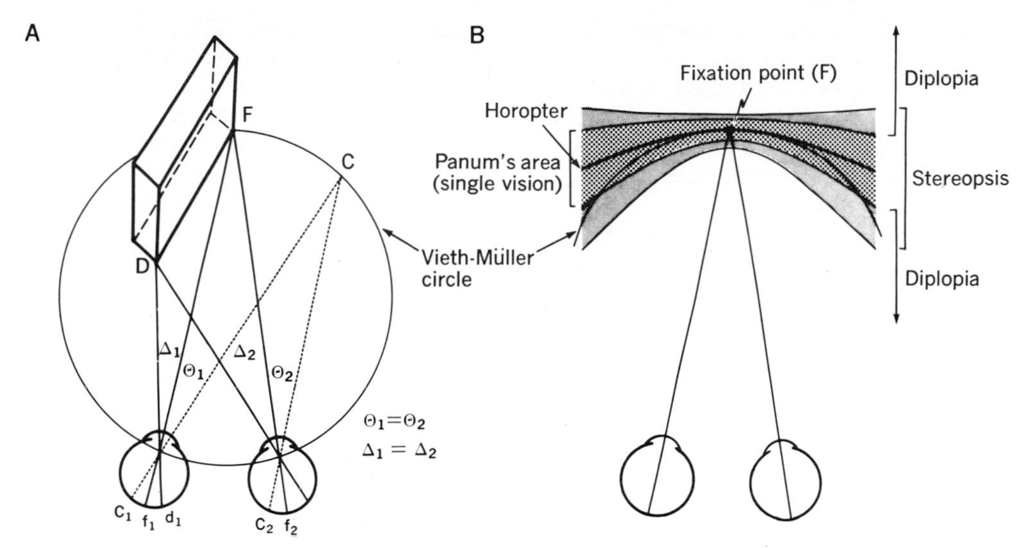

Fig. 16-23. Binocular vision. **A,** Diagram showing geometric construct of corresponding and disparate retinal images and Vieth-Müller horopter circle. **B,** Schematic representation of relative positions of Vieth-Müller circle and empirical horopter and of overlapping regions of binocular single vision and of stereopsis.

For a given position of the eyes, there exists in space a locus of points, about the fixation point, whose images fall on corresponding retinal elements and thus are seen as single and localized in the same, common visual direction. This locus of points is called the *horopter*. Its simplest geometric construct is a circle that passes through the fixation point and the optical centers of the two eyes: the Vieth-Müller circle (Fig. 16-23, *A*—point *C* and its retinal projections c_1 and c_2: $\Theta_1 = \Theta_2$). Experimental determinations, however, have shown that the horopter does not coincide with the circle (see Ogle[7]). For a fixation distance of approximately 1 m the empirical horopter is concave toward the observer and lies between the Vieth-Müller circle and the frontoparallel plane through the fixation point (Fig. 16-23, *B*). With increasing viewing distances, the shape of the horopter changes from concave to convex, but there is evidence that the set of corresponding retinal elements remains the same.[54, 77]

Points in the visual world farther or nearer to the observer than the special position just described project optically to different horizontal positions in the two eyes (Fig. 16-23, *A*—point *D* and its retinal projections d_1 and d_2: $\Delta_1 \neq \Delta_2$). The retinal elements stimulated are noncorresponding or disparate retinal elements, and the images of that point, being localized in separate directions, may be seen as double. Horizontal retinal disparity is the unique cue for binocular depth discrimination. Vertical disparities may also occur, determined by such factors as slight differences in the elevation of the eyes or magnification differences between the two eyes for close objects. Vertical disparity does not contribute to the stereoscopic depth experience.

When we fixate on some point in space, we are not ordinarily aware that most of the objects in the visual world can be seen as double. In fact, there exists only a restricted spatial region in front of and behind the horopter plane within which binocular single vision is obligatory. Clearly, optical correspondence of retinal images is not a prerequisite for singleness of vision, and Panum made the suggestion that each point on the retina should be regarded as corresponding not with a point but with an area in the retina of the other eye (see Boring[24]). As long as the retinal image disparities are within the limits of this area, the images in the two eyes are combined in the brain and seen as a fused, single image in depth. This retinal area, and by extension the corresponding region in space, is called *Panum's fusional area:* it is smallest at the center of the fovea and it increases toward the retinal periphery. Outside the limits of Panum's area, vision is double (Fig. 16-23, *B*).

The experience of depth in binocular vision is a vivid and independent sensation (much as the sensation of color and brightness) by which the tridimensional quality of the visual world emerges in a striking manner. The stereoscopic experience is one of relative depth only: one object is perceived as farther or nearer than another, in an ordinal scale of depth. Stereopsis alone does not provide for the sensation of absolute distance of objects from the observer.

With small retinal disparities, e.g., those occurring within Panum's area as well as in a limited region in front of and behind it (Fig. 16-23, *B*), the experience of stereoscopic depth is evident and compelling, and within this range the subjective depth is correlated with the magnitude of the disparity.[7] Clearly, fusion of disparate images (single vision) is not a prerequisite for binocular depth perception, and it is still possible to localize objects correctly even though their images have retinal disparities so large that the objects may appear double (diplopia). However, for objects beyond this region of patent or fine stereopsis, depth may still be experienced but only as a less specific and purely qualitative coarse sensation of "nearer" and "farther."[7, 19] With increasingly large image disparities, the sense of depth gradually fades and eventually disappears.

Stereoscopic acuity, i.e., the ability to discriminate differences in depth, is comparable in sensitivity to monocular visual acuity; i.e., the ability to recognize two lines as separate when situated on the same plane in space (Chapter 15). Considering that binocular depth perception involves the alignment of the two eyes, this fact is indeed proof of the remarkable capacities of the visual apparatus. Stereoscopic acuity is maximal in the vicinity of the horopter and declines rapidly with departure from it. At the limit of stereoscopic acuity it is possible to discriminate a difference in depth as small as 0.15 mm between two objects placed a distance of approximately 1 m from the eyes.

The extensive investigations by Julesz[4] have provided conclusive evidence that retinal image disparity is the unique visual determinant of binocular depth perception. Julesz has used computer-generated random-dot patterns as stereo pairs, an example of which is given in Fig. 16-24. The right and left images are identical random-dot patterns, except for certain areas that, while identical, are shifted relative to each other in the horizontal direction by an integral number of dots. When viewed monocularly, the two figures appear as homogeneously random without recognizable features. When binocularly combined (e.g., by crossing the eyes), there emerges a vivid depth impression of a pattern suspended above the background (a central diamond in Fig. 16-24).

These results indicate that stereoscopic depth may be experienced without any monocular or binocular cues to depth except horizontal retinal image disparity, and that the neural integration of dissimilarities in the images seen by the two eyes is established without monocular recognition of any pattern or form. Moreover, these findings suggest that the neural mechanisms operate on the basis

Fig. 16-24. Random-dot stereo pair. When viewed monocularly, two fields appear as random sets of dots without recognizable features. However, when stereoscopically combined (e.g., by crossing eyes), there is a vivid depth impression of a diamond suspended over the background. (From Julesz.[4])

of a point-to-point analysis since line contours of any appreciable length are not required for the stereoscopic experience.[18]

Neural mechanisms

Our understanding of the central neural mechanisms operating in binocular vision rests on considerations of the anatomic organization of the visual afferent system and of the functional properties of its constituent neurons. The decussation of the optic fibers at the chiasm brings together in the same hemisphere the central projections of the two homolateral hemiretinas, and within this pathway, the representation of topographically corresponding retinal areas are kept closely associated with one another in a precisely organized retinotopic pattern. The inputs from the two eyes pass through the LGN with minimal or no interaction, and come together for the first time at the striate cortex. During the past several years the central mechanisms of binocular vision have been investigated by the method of single-unit analysis, particularly at the level of simple cells in the striate cortex of the cat and to a lesser extent in the visual cortex of the monkey as well.

As previously indicated, the majority of cortical visual neurons can be activated from both eyes independently, and each of these neurons has two receptive fields, one in each eye, with closely similar functional organization. The two receptive fields, however, do not always occupy anatomically identical regions in the two retinas. At a given locus in the retinotopic cortical map there are cells whose fields have exactly corresponding positions in right and left retinas and cells whose fields have slightly different positions in the two eyes, both in the horizontal and vertical directions. This phenomenon of "receptive field disparity" forms the basis of a neurophysiologic theory of binocular vision suggested by Barlow et al.[14] and Nikara et al.[93] and further developed by Bishop and coworkers.[18, 19, 77]

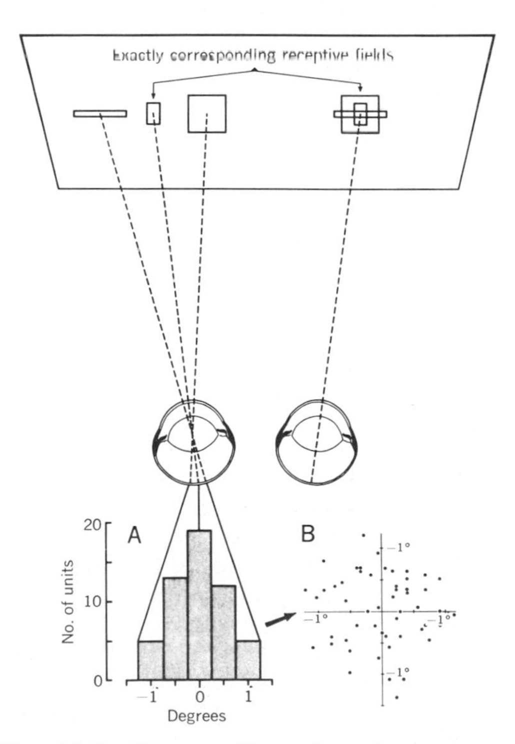

Fig. 16-25. Diagram illustrating phenomenon of receptive field disparity (see text for explanation). (From Bishop.[18])

Fig. 16-25 illustrates receptive field disparity.[18] The upper part of the figure shows the eyes of the cat looking at a tangent screen onto which luminous patterns can be projected for visual stimulation. On that screen the receptive field pairs of three representative cortical neurons are drawn as rectangles of different shapes. In these experiments the extraocular muscles of the animal are paralyzed; the eyes assume fixed, divergent positions and the receptive fields in right and left retinas can be defined independently. To simplify the description, neurons are chosen whose receptive fields in the right eye have the same visual direction and thus fall on each other on the tangent screen. The members of the receptive field pairs in the left eye, on the other hand, are not in register and are scattered at different horizontal positions. Clearly, there exist different horizontal separations between binocular receptive fields of different cortical neurons; i.e., the fields occupy different positions in the two eyes. The distribution of these differences in separation describes the binocular receptive field disparity in the horizontal direction. Fig. 16-25, *A*, shows such a distribution for a population of 54 cells in the striate cortex of the cat with receptive fields within 5 degrees of the visual axis. In the foregoing description it has been assumed that there are no differences in receptive field separation in the vertical direction. The scattergram of Fig. 16-25, *B*, illustrates the receptive field disparities for the same group of cortical neurons when the vertical dimension is taken into account. In both the horizontal and vertical directions the distribution is approximately normal with a standard deviation of 0.5 degree.[93] Fields in the left eye whose position is at the mean of the disparity distribution can be taken as exactly corresponding to fields in the right eye (zero disparity). At all locations in the retinotopic cortical map, i.e., over the entire visual field, there are populations of neurons whose receptive fields in the two eyes are spatially distributed in the two directions, much as illustrated in Fig. 16-25, and each distribution overlaps extensively with the distribution of surrounding populations.

The importance of horizontal receptive field disparity in the neural mechanisms underlying binocular depth discrimination may be appreciated from Fig. 16-26, which is developed from Fig. 16-25 by converging the eyes of the cat, as under natural conditions, to fixate a point *F* on the tangent screen. Again, all receptive fields in the right eye have the same visual direction. Corresponding retinal elements for central vision in the two eyes will coincide at *F,* and the pairs of receptive fields with zero disparity (corresponding receptive fields) will superimpose precisely in the plane of the tangent screen. Disparate receptive field pairs, on the other hand, will not superimpose on that plane but will do so on planes nearer or farther than the plane of fixation, depending on the direction of the horizontal disparity.

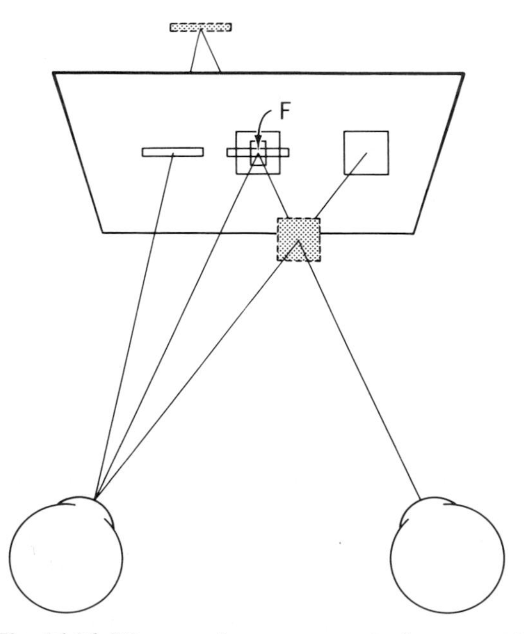

Fig. 16-26. Diagram of rearrangement of upper part of Fig. 16-25 to illustrate how horizontal receptive field disparity might form basis of neural mechanism for binocular depth discrimination (see text). (From Bishop.[18])

Cortical visual neurons are particularly sensitive to the spatial characteristics of visual stimuli and respond optimally to patterns of specific shape, orientation, size, motion, etc. Binocular cortical neurons have receptive fields in the two eyes that are closely similar in organization, and for these neurons the same specific stimulus that is most effective when presented in one eye is also the most effective when presented in the other eye. Moreover, optimal activation of binocular neurons occurs when the same specific stimulus is applied simultaneously to functionally identical parts of the receptive fields in the two eyes. Thus in normal binocular vision, conditions of optimal stimulation obtain when the two receptive fields superimpose in the plane of the stimulus. At planes other than that of field superposition the cell will be stimulated at the same time by two separate stimulus features or by the same feature at functionally different parts of the receptive fields in the two eyes, and its response will be reduced or abolished.

Taken together, these experimental findings suggest that under conditions of normal binocular vision, a tridimensional object in space, imaged two dimensionally in slightly different positions onto the two retinal surfaces and translated in neural code, is "reconstructed" in the visual cortex by the profile of activity in the population of neurons topographically related to that part of the visual field in which the object is located. Because of the horizontal scatter in the separation between receptive field pairs in the two eyes, different neurons in the population will be optimally excited by objects at different distances. The majority of cells have exactly corresponding receptive fields and will be maximally activated by the two identical images of single specific features in object space that lie on the plane containing the point of fixation, i.e., on the horopter plane. Cells with horizontally disparate receptive fields, on the other hand, will be optimally excited by features both in front of and behind that plane, and with departure from it, the number of cells binocularly influenced will decline in accord with the distribution of receptive field disparities. Accurate feature-by-feature pairing takes place because the two receptive fields of each cell have the same highly specific stimulus requirement.

Moreover, when the specific feature and the two receptive fields of the binocular neuron coincide in space, that feature will be treated by the brain as single, whether its retinal images are corresponding or disparate, either in the horizontal or vertical direction. At planes other than that of field superposition, visual stimulation of binocular neurons will be inappropriate, their activity reduced or suppressed, and double vision prevented. Beyond the spread of receptive field disparities, double vision occurs because the two receptive fields simultaneously stimulated by a single object feature are not binocular pairs of a single neuron but fields of different cortical neurons.

Thus the phenomenon of receptive field disparity may provide the neural basis of singleness of vision, and Panum's fusional area may represent, at least to a first approximation, "the extent to which receptive fields in one eye, all with the same visual direction, are linked to fellow members in the other eye over a range of receptive field disparities."[93] Supporting this view is the fact that in humans Panum's dimensions are in the vicinity of the fixation point and are about the same in both the vertical and horizontal directions, much as the dimensions of receptive field disparities in the cat[77] (Fig. 16-25, *B*).

As previously indicated, the experimental findings just described were obtained by recording from cells in the striate cortex of the cat. Recently, Hubel and Wiesel[73] have investigated binocular mechanisms in the visual cortex of the monkey. They found that in the striate area (area 17) of this animal, cells activated from both eyes are fewer than in the cat, and that those that are do not possess receptive field disparity. On the other hand, in area 18 of the monkey, the next level in the cortical progression of the visual input, practically all cells are binocularly activated, and about half of them have specialized properties that are thought to be of particular importance in binocular depth discrimination mechanisms. These cells, termed *binocular depth cells,* do not usually respond to stimulation of either eye separately, but appropriate simultaneous stimulation of the two eyes together evokes a brisk response. The position of the receptive fields of these cells in the two eyes may be corresponding or disparate. In addition, these cells have special properties not described for the cat. First, the displacement of the field in one eye, relative to the field in the other, is usually at right angles to the receptive field orientation. Second, cells with vertically oriented fields, and thus with possible horizontal receptive field disparity, are the more common and cells with oblique field orientation, for which there is a vertical component to the disparity as well, also respond briskly to purely horizontally displaced stimuli. It is then apparent that the majority of the "depth cells" in area 18 of the monkey are sensitive to horizontal image disparity and thus possess the essential requirement for a neural mechanism of steropsis. Recently Bough,[25] using random-dot stereoscopic patterns, has provided convincing behavioral evidence for stereoscopic vision in the macaque.

REFERENCES

General reviews

1. Brindley, G. S.: Physiology of the retina and the visual pathway, ed. 2, London, 1970, Edward Arnold, Ltd.
2. De Valois, R. L.: Behavioral and electrophysiological studies of primate vision, Contrib. Sens. Physiol. **1:**137, 1965.
3. Diamond, I. T., and Chow, K. L.: Biological psychology. In Koch, S., editor: Psychology: a study of a science, New York, 1962, McGraw-Hill Book Co.
4. Julesz, B.: Foundations of cyclopean perception, Chicago, 1971, University of Chicago Press, pp. xiv and 406.
5. Klüver, H.: Functional significance of the geniculo-striate system, Biol. Symp. **7:**253, 1942.
6. Le Grand, Y.: Light, colour and vision, London, 1957, Chapman & Hall, Ltd.
7. Ogle, K. N.: The optical space sense. In Davson, H., editor: The eye, New York, 1969, Academic Press, Inc., vol. 4.
8. Polyak, S. L.: The retina, Chicago, 1941, University of Chicago Press.
9. Polyak, S. L.: The vertebrate visual system, Chicago, 1957, University of Chicago Press.
10. Teuber, H. L., Battersby, W. S., and Bender, M. B.: Visual field defects after penetrating missile wounds of the brain, Cambridge, Mass., 1960, Harvard University Press.

Original papers

11. Arduini, A.: Influence of visual deafferentation and of continuous retinal illumination on the excitability of geniculate neurons. In Jung, R., and Kornhuber, H., editors: The visual system: neurophysiology and psychophysics, Berlin, 1961, Springer Verlag.
12. Astruc, J.: Corticofugal connections of area 8 (frontal eye field) in Macaca mulatta, Brain Res. **33:**241, 1971.
13. Baker, F. H., Riva Sanseverino, E., Lamarre, Y., and Poggio, G. F.: Excitatory responses of geniculate neurons of the cat, J. Neurophysiol. **32:**916, 1969.
14. Barlow, H. B. D., Blakemore, C., and Pettigrew, J. D.: The neural mechanisms of binocular depth discrimination, J. Physiol. **193:**327, 1967.
15. Barlow, H. B., Fitzhugh, R., and Kuffler,

S. W.: Change of organization in the receptive fields of the cat's retina during dark adaptation, J. Physiol. **137**:338, 1957.

16. Baumgartner, G., Brown, J. L., and Schulz, A.: Responses of single units of the cat visual system to rectangular stimulus patterns, J. Neurophysiol. **28**:1, 1965.
17. Bender, M. B., and Shanzer, S.: Oculomotor pathways defined by electric stimulation and lesions in the brainstem of monkey. In Bender, M. B., editor: The oculomotor system, New York, 1964, Harper & Row, Hoeber Medical Division.
18. Bishop, P. O.: Beginning of form vision and binocular depth discrimination in cortex. In Schmitt, F. O., editor: The neurosciences. Second study program, New York, 1970, The Rockefeller University Press.
19. Bishop, P. O., and Henry, G. H.: Spatial vision, Ann. Rev. Psychol. **22**:119, 1971.
20. Bishop, P. O., and Rodieck, R. W.: Discharge patterns of cat retinal ganglion cells. In Proceedings of symposium on information processing in sight sensory systems, Pasadena, 1965, California Institute of Technology.
21. Bishop, P. O., Levick, W. R., and Williams, W. O.: Statistical analysis of the dark discharge of lateral geniculate neurones, J. Physiol. **170**:598, 1964.
22. Bizzi, E.: Discharge of frontal eye field neurons during saccadic and following eye movements in unanesthetized monkeys, Exp. Brain Res. **6**:69, 1968.
23. Bizzi, E., and Schiller, P. H.: Single unit activity in the frontal eye fields of unanesthetized monkeys during eye and head movements, Exp. Brain Res. **10**:151, 1970.
24. Boring, E. G.: Sensation and perception in the history of experimental psychology, New York, 1942, Appleton-Century-Crofts.
25. Bough, E. W.: Stereoscopic vision in the macaque monkey: a behavioral demonstration, Nature **225**:42, 1970.
26. Burian, H.: Sensorial retinal relationship in concomitant strabismus, Trans. Am. Ophthalmol. Soc. **43**:373, 1945.
27. Campos-Ortega, J. A., Hayhow, W. R., and Clüver, P. F. de V.: The descending projections from the cortical visual fields of Macaca mulatta with particular reference to the question of a cortico-lateral geniculate-pathway, Brain Behav. Evol. **3**:368, 1970.
28. Choudhury, B. P., Whitteridge, D., and Wilson, M. E.: The function of the callosal connections of the visual cortex, Q. J. Exp. Physiol. **50**:214, 1965.
29. Chow, K. L.: Anatomical and electrographical analysis of temporal neocortex in relation to visual discrimination learning in monkeys. In Delafresnaye, F. J., editor: Brain mechanisms and learning, Oxford, 1961, Blackwell Scientific Publications.
30. Chow, K. L., Blum, J. S., and Blum, R. A.: Cell ratios in the thalamocortical visual system of Macaca mulatta, J. Comp. Neurol. **92**:227, 1950.
31. Clark, W. E. L.: The laminar organization and cell content of the lateral geniculate body in the monkey, J. Anat. **75**:419, 1941.
32. Clark, W. E. L.: The sorting principle in sensory analysis as illustrated by the visual pathways, Ann. R. Coll. Surg. Engl. **30**:299, 1962.
33. Clark, W. E. L., and Penman, G. G.: The projection of the retina in the lateral geniculate body, Proc. R. Soc. Lond. (Biol.) **114**: 291, 1934.
34. Cowey, A.: Projection of the retina on the striate and prestriate cortex in the squirrel monkey, Saimiri sciureus, J. Neurophysiol. **27**:366, 1964.
35. Cowey, A., and Ellis, C. M.: Visual acuity of rhesus and squirrel monkeys, J. Comp. Physiol. Psychol. **64**:80, 1967.
36. Cowey, A., and Ellis, C. M.: The cortical representation of the retina in squirrel and rhesus monkeys and its relation to visual acuity, Exp. Neurol. **24**:374, 1969.
37. Cowey, A., and Gross, C. G.: Effects of foveal prestriate and inferotemporal lesions on visual discrimination by rhesus monkeys, Exp. Brain Res. **11**:128, 1970.
38. Cowey, A., and Weiskrantz, L.: A perimetric study of visual field defects in monkeys, Q. J. Exp. Psychol. **15**:91, 1963.
39. Cowey, A., and Weiskrantz, L.: A comparison of the effects of inferotemporal and striate cortex lesions on the visual behaviour of rhesus monkeys, Q. J. Exp. Psychol. **19**: 246, 1967.
40. Cragg, B. G.: The topography of the afferent projections in the circumstriate visual cortex of the monkey studied by the Nauta method, Vision Res. **9**:733, 1969.
41. Creutzfeldt, O., and Ito, M.: Functional synaptic organization of primary visual cortex neurones in the cat, Exp. Brain Res. **6**:324, 1968.
42. Crosby, E. C., Yoss, R. E., and Henderson, J. W.: The mammalian midbrain and isthmus region. II. The fiber connections. D. The pattern for eye movements on the frontal eye field and the discharge of the specific portions of this field, to and through midbrain levels, J. Comp. Neurol. **97**:357, 1952.
43. Daniel, P. M., and Whitteridge, D.: The representation of the visual field on the cerebral cortex in monkeys, J. Physiol. **159**:203, 1961.
44. De Valois, R. L.: Analysis and coding of color vision in the primate visual system, Symp. Quant. Biol. **30**:567, 1965.
45. De Valois, R. L.: Physiological basis of color vision. In Tagungsbericht Internationale Farbtagung, Color 69, Stockholm, 1969, pp. 29-47.
46. De Valois, R. L., Abramov, I., and Jacobs, G. H.: Analysis of response patterns of LGN cells, J. Opt. Soc. Am. **56**:966, 1966.
47. Denney, D., Baumgartner, G., and Adorjani, C.: Responses of cortical neurones to stimulation of the visual afferent radiations, Exp. Brain Res. **6**:265, 1968.
48. Denny-Brown, D.: The midbrain and motor integration, Proc. R. Soc. Med. **55**:527, 1962.
49. Denny-Brown, D., and Chambers, R. A.: Visual orientation on the macaque monkey, Trans. Am. Neurol. Assoc. **83**:37, 1958.
50. Denny-Brown, D., and Chambers, R. A.: The

parietal lobe and behavior, A. Res. Nerv. Ment. Dis., Proc. **36**:35, 1958.
51. Duke-Elder, S.: Text-book of ophthalmology, St. Louis, 1952, The C. V. Mosby Co., vol. 1.
52. Enroth, C.: The mechanism of flicker and fusion studied on single retinal elements in the dark-adapted eye of the cat, Acta Physiol. Scand. **27**(suppl. 100):1, 1952.
53. Ettlinger, G.: Visual discrimination with a single manipulandum following temporal ablations in the monkey, Q. J. Exp. Psychol. **3**: 164, 1959.
54. Flom, M. C., and Eskridge, J. B.: Changes in retinal correspondence with viewing distance, J. Am. Optom. Assoc. **39**:1094, 1968.
55. Garey, L. J., and Powell, T. P. S.: An experimental study of the termination of the lateral geniculo-cortical pathway in the cat and monkey, Proc. R. Soc. Lond. (Biol.) **179**:41, 1971.
56. Gouras, P.: Identification of cone mechanisms in monkey ganglion cells, J. Physiol. **199**: 533, 1968.
57. Gouras, P.: Trichromatic mechanisms in single cortical neurons, Science **168**:489, 1970.
58. Gross, C. G., Bender, D. B., and Rocha-Miranda, C. E.: Visual receptive fields of neurons in inferotemporal cortex of the monkey, Science **166**:1303, 1969.
59. Guillery, R. W., and Colonnier, M.: Synaptic patterns in the dorsal lateral geniculate nucleus of the monkey, Z. Zellforsch. Mikrosk. Anat. **103**:90, 1970.
60. Halstead, W. C., Walker, A. E., and Bucy, P. C.: Sparing and nonsparing of "macular" vision associated with occipital lobectomy in man, Arch. Ophthalmol. **24**:948, 1940.
61. Harrington, D. O.: The visual fields: a textbook and atlas of clinical perimetry, ed. 3, St. Louis, 1971, The C. V. Mosby Co.
62. Held, R.: Dissociation of visual functions by deprivation and rearrangement, Psychol. Forsch. **31**:338, 1968.
63. Henry, G. H., and Bishop, P. O.: Simple cells of the striate cortex, Contrib. Sens. Physiol. **5**:1, 1971.
64. Hoffmann, K. P., and Stone, J.: Conduction velocity of afferents to cat visual cortex: a correlation with cortical receptive field properties, Brain Res. **32**:460, 1971.
65. Holmes, G.: The organization of the visual cortex in man, Proc. R. Soc. Lond. (Biol.) **132**:349, 1945.
66. Hoyt, W. F., and Luis, O.: The primate chiasm: details of visual fiber organization studied by silver impregnation techniques, Arch. Ophthalmol. **70**:69, 1963.
67. Hubel, D. H., and Wiesel, T. N.: Receptive fields of optic nerve fibers in the spider monkey, J. Physiol. **154**:572, 1960.
68. Hubel, D. H., and Wiesel, T. N.: Integrative action in the cat's lateral geniculate body, J. Physiol. **155**:385, 1961.
69. Hubel, D. H., and Wiesel, T. N.: Receptive fields, binocular interaction and functional architecture in the cat's visual cortex, J. Physiol. **160**:106, 1962.
70. Hubel, D. H., and Wiesel, T. N.: Receptive fields and functional architecture in two nonstriate visual areas (18 and 19) of the cat, J. Neurophysiol. **28**:229, 1965.
71. Hubel, D. H., and Wiesel, T. N.: Receptive fields and functional architecture of monkey striate cortex, J. Physiol. **195**:215, 1968.
72. Hubel, D. H., and Wiesel, T. N.: Laminar and columnar distribution of geniculocortical fibers in the macaque monkey, J. Comp. Neurol. **146**:421, 1972.
73. Hubel, D. N., and Wiesel, T. N.: Cells sensitive to binocular depth in area 18 of the macaque monkey cortex, Nature **225**:41, 1970.
74. Jacobs, G. H.: Effects of adaptation on the lateral geniculate response to light increment and decrement, J. Opt. Soc. Am. **55**:1535, 1965.
75. Jacobs, G. H., and Yolton, R. L.: Distribution of excitation and inhibition in receptive fields of lateral geniculate neurones, Nature **217**: 187, 1968.
76. Jones, E. G., and Powell, T. P. S.: An anatomical study of converging sensory pathways within the cerebral cortex of the monkey, Brain **93**:793, 1970.
77. Joshua, D. E., and Bishop, P. O.: Binocular single vision and depth discrimination. Receptive field disparities for central and peripheral vision and binocular interaction on pheripheral single units in cat striate cortex, Exp. Brain Res. **10**:389, 1970.
78. Klüver, H., and Bucy, P. C.: An analysis of certain effects of bilateral temporal lobectomy in the rhesus monkey, with special reference to "psychic blindness," J. Psychol. **5**:33, 1938.
79. Kuffler, S. W.: Discharge patterns and functional organization of mammalian retina, J. Neurophysiol. **16**:37, 1953.
80. Kuffler, S. W., Fitzhugh, R., and Barlow, H. B.: Maintained activity in the cat's retina in light and darkness, J. Gen. Physiol. **40**:683, 1957.
81. Kuypers, H. G. J. M., Szwarcbart, M. D., Mishkin, M., and Rosvold, H. E.: Occipitotemporal corticocortical connections in the rhesus monkey, Exp. Neurol. **11**:245, 1965.
82. Latto, R., and Cowey, A.: Visual field defects after frontal eye-field lesions in monkeys, Brain Res. **30**:1, 1971.
83. Le Vay, S.: On the neurons and synapses of the lateral geniculate nucleus of the monkey, and the effects of eye enucleation, Z. Zellforsch. Mikrosk. Anat. **113**:396, 1971.
84. Lennox-Buchthal, M. A.: Spectral sensitivity of single units in the cortical area corresponding to central vision in the monkey, Acta Physiol. Scand. **65**:101, 1965.
85. Matthews, M. R., Cowan, W. M., and Powell, T. P. S.: Transneuronal cell degeneration in the lateral geniculate nucleus of the macaque monkey, J. Anat. **94**:145, 1960.
86. McIlwain, J. T.: Receptive fields of optic tract axons and lateral geniculate cells: peripheral extent and barbiturate sensitivity, J. Neurophysiol. **27**:154, 1964.
87. Milner, B.: Visual recognition and recall after right temporal-lobe excision in man, Neuropsychologia **6**:191, 1968.
88. Mishkin, M.: Visual mechanisms beyond the striate cortex. In Russell, R. W., editor: Fron-

tiers in physiological psychology, New York, 1966, Academic Press, Inc.
89. Mishkin, M., and Hall, M.: Discrimination along a size continuum following ablation of the inferior temporal convexity in monkeys, J. Comp. Physiol. Psychol. **48**:97, 1955.
90. Mishkin, M., and Pribram, K. H.: Visual discrimination performance following partial ablations of the temporal lobe. I. Ventral vs. lateral, J. Comp. Physiol. Psychol. **47**:14, 1954.
91. Mitchell, D. E., and Blakemore, C.: Binocular depth perception and the corpus callosum, Vision Res. **10**:49, 1970.
92. Myers, R. E.: Organization of visual pathways. In Ettlinger, E. G., editor: Functions of the corpus callosum, Ciba Foundation Study Group No. 20, London, 1965, J. & A. Churchill, Ltd.
93. Nikara, T., Bishop, P. O., and Pettigrew, J. D.: Analysis of retinal correspondence by studying receptive fields of binocular single units in cat striate cortex, Exp. Brain Res. **6**: 353, 1968.
94. Ogawa, T., Bishop, P. O., and Levick, W. R.: Temporal characteristics of responses to photic stimulation by single ganglion cells in the unopened eye of the cat, J. Neurophysiol. **29**: 1, 1966.
95. Pandya, D. N., and Kuypers, H. G. J. M.: Cortico-cortical connections in the rhesus monkey, Brain Res. **13**:13, 1969.
96. Pettigrew, J. D., Nikara, T., and Bishop, P. O.: Responses to moving slits by single units in cat striate cortex, Exp. Brain Res. **6**:373, 1968.
97. Pettigrew, J. D., Nikara, T., and Bishop, P. O.: Binocular interaction on single units in cat striate cortex: simultaneous stimulation by single moving slit with receptive field in correspondence, Exp. Brain Res. **6**:391, 1968.
98. Piéron, H.: Vision in intermittent light, Contrib. Sens. Physiol. **1**:179, 1965.
99. Poggio, G. F., Baker, F. H., Lamarre, Y., and Riva Sanseverino, E.: Afferent inhibition at input to visual cortex in the cat, J. Neurophysiol. **32**:892, 1969.
100. Pribram, K. H., and Mishkin, M.: Simultaneous and successive visual discrimination by monkeys with inferotemporal lesions, J. Comp. Physiol. Psychol. **48**:198, 1955.
101. Robinson, D. A., and Fuchs, A. F.: Eye movements evoked by stimulation of frontal eye fields, J. Neurophysiol. **32**:637, 1969.
102. Rodieck, R. W.: Maintained activity of cat retinal ganglion cells, J. Neurophysiol. **30**: 1043, 1967.
103. Rodieck, R. W., and Stone, J.: Analysis of receptive fields of cat retinal ganglion cells, J. Neurophysiol. **28**:833, 1965.
104. Sakmann, B., and Creutzfeldt, O. D.: Scotopic and mesopic light adaptation in the cat's retina, Pfluegers Arch. **313**:168, 1969.
105. Schiller, P. H., and Koerner, F.: Discharge characteristics of single units in superior colliculus of the alert rhesus monkey, J. Neurophysiol. **34**:920, 1971.
106. Schneider, G. E.: Two visual systems, Science **163**:895, 1969.
107. Singer, W., and Creutzfeldt, O. D.: Reciprocal lateral inhibition of on- and off-center neurones in the lateral geniculate body of the cat, Exp. Brain Res. **10**:311, 1970.
108. Spalding, J. M. K.: Wounds of the visual pathway. I. The visual radiation, J. Neurol. Neurosurg. Psychiatry **15**:99, 1952.
109. Spalding, J. M. K.: Wounds of the visual pathway. II. The striate cortex, J. Neurol. Neurosurg. Psychiatry **15**:169, 1952.
110. Sprague, J. M., Berlucchi, G., and Di Berardino, A.: The superior colliculus and pretectum in visually guided behavior and visual discrimination in the cat, Brain Behav. Evol. **3**: 285, 1970.
111. Taira, N., and Okuda, J.: Sensory transmission in visual pathway in various arousal states of cat, Tohoku J. Exp. Med. **78**:76, 1962.
112. Talbot, S. A.: A lateral localization in the cat's visual cortex, Fed. Proc. **1**:84, 1942 (abstract).
113. Talbot, S. A., and Marshall, W. H.: Physiological studies on neural mechanisms of visual localization and discrimination, Am. J. Ophthalmol. **24**:1255, 1941.
114. Teuber, H. L.: Physiological psychology, Ann. Rev. Psychol. **6**:267, 1955.
115. Trevarthen, C. B.: Two mechanisms of vision in primates, Psychol. Forsch. **31**:299, 1968.
116. Van Buren, J. M.: The retinal ganglion cell layer, Springfield, Ill., 1963, Charles C Thomas, Publisher.
117. Wagman, I. H.: Eye movements induced by electric stimulation of cerebrum in monkeys and their relationship to bodily movements. In Bender, M. B., editor: The oculomotor system, New York, 1964, Paul B. Hoeber, Inc.
118. Watanabe, S., Konishi, M., and Creutzfeldt, O. D.: Postsynaptic potentials in the cat's visual cortex following electrical stimulation of afferent pathways, Exp. Brain Res. **1**:272, 1966.
119. Weiskrantz, L., and Cowey, A.: Filling in the scotoma: a study of residual vision after striate cortex lesions in monkeys. In Stellar, E., and Sprague, J. M., editors: Progress in physiological psychology, New York, 1970, Academic Press, Inc., vol. 3.
120. Whitlock, D. G., and Nauta, W. J. H.: Subcortical projections from the temporal neocortex in Macaca mulatta, J. Comp. Neurol. **106**:183, 1956.
121. Wiesel, T. N.: Receptive fields of ganglion cells in the cat retina, J. Physiol. **153**:583, 1960.
122. Wiesel, T. N., and Hubel, D. H.: Spatial and chromatic interactions in the lateral geniculate body of the rhesus monkey, J. Neurophysiol. **29**:1115, 1966.
123. Wilson, M. E., and Toyne, M. J.: Retinotectal and cortico-tectal projections in Macaca mulatta, Brain Res. **24**:395, 1970.
124. Wilson, W. A., and Mishkin, M.: Comparison of the effects of inferotemporal and lateral occipital lesions on visually guided behavior in monkeys, J. Comp. Physiol. Psychol. **52**: 10, 1959.
125. Woolsey, C. N.: Comparative studies on cor-

tical representation of vision, Vision Res. suppl. **3:**365, 1971.

126. Zeki, S. M.: The secondary visual areas of the monkey, Brain Res. **13:**197, 1969.
127. Zeki, S. M.: Representation of central visual fields in prestriate cortex of monkey, Brain Res. **14:**271, 1969.
128. Zeki, S. M.: Interhemispheric connections of prestriate cortex of monkey, Brain Res. **19:** 63, 1970.

17

LLOYD M. BEIDLER

The chemical senses: gustation and olfaction

Animals live in a chemical world. Many rely on chemical clues to serve two of their most basic needs—nutrition and reproduction. In addition, some animals use a chemical means of communication. Humans usually associate the senses of taste and olfaction with the pleasures of foods and perfumes. However, the gustatory and olfactory organs serve as vital monitors of all food ingested and all air breathed. This is reflected, for example, in man's ability to self-regulate food intake under conditions of dietary and endocrine deficiencies. An adrenal deficiency results in a lowering of the threshold of sodium chloride preference and a resultant increased intake.[69] The importance of odors in human reproduction is still not certain, but other mammals, e.g., mice, react strongly to sex odors. The odor of a strange male may result in abortion in a pregnant female mouse![26] The importance of knowledge concerning the role of tastes and odors in the behavior of mammals is just being appreciated.

The interaction of chemicals with living systems is of prime importance to anyone interested in the physiology of man. The study of chemoreceptors may aid in understanding the molecular basis of other functions, particularly drug-receptor interactions.

GUSTATION

Morphology

Taste is associated with specialized organs, taste buds. These were first discovered in fish, where they are found not only in the mouth but also over the entire head and in some species over much of the body surface. In man they are confined to the oral cavity, principally the tongue and palate, although at birth they are found on the lips, pharynx, and epiglottis. Human taste buds appear in the fetus at 3 to 4 months, and the fetus begins to swallow at about the same time.[23] Saccharin injected into human amniotic fluid increases the rate of fetal swallowing.

The taste buds of the tongue are always associated with specialized papillae. In man, each fungiform papilla on the front of the tongue contains three to five taste buds on its dorsal surface. This number varies in other mammals. These papillae can easily be identified as red spots since the rich blood supply beneath the taste bud is seen through the thin epithelium above. Numerous taste buds are also found in the folds of the foliate papillae, which are located on the lateral border of the posterior of the tongue. The larger circumvallate papillae on the posterior tongue surface contain thousands of taste buds located on the walls of the surrounding troughs (Fig. 17-1). Since the taste stimuli must enter the trough before stimulating the taste buds, the latency of taste sensation is greater than that associated with the exposed taste buds of the fungiform papilla. Stimulus removal is enhanced by tongue movement, which forces the liquid in and out of the trough, and also by von Ebner's glands at the bottom of the trough, which produce a secretion to flush the stimulus away from the taste buds. The difficulty of stimulus removal is expressed as a lingering taste sensation. Since the circumvallate papilla taste buds have a higher sensitivity to bitter tastes than those of the fungiform papilla, bitter aftertastes are common at the back of the tongue. The sensitivity to sweet is greatest at the tip of the tongue; sensitivity to salty and sour is greatest at the side of the tongue. Although most of the taste organs are located on the tongue, many that are also present on the palate contribute disproportionately to the total taste sensation. Their function is usually not appreciated until they are inactivated by dentures.

The taste bud is a bulblike structure about

Fig. 17-1. Scanning electron photomicrograph of dog circumvallate papilla showing encircled trough in which taste buds are located.

50μ in diameter that contains 40 to 50 cells (Fig. 17-2). Electron microscopists classify the cells into three or four types based on the anatomic details and electron density.[60] It is now known that these different cell types are associated with different physiologic functions. Beidler and Smallman[18] studied the life-span of rat taste cells by injecting a pulse of tritiated thymidine, a DNA precursor, into the intraperitoneal fluid to label all newly dividing cells. Rats were then sacrificed at different intervals for autoradiographic examination of the fungiform taste buds. It was thus determined that epithelial cells surrounding the taste bud undergo mitotic division and that some of the daughter cells may enter the taste bud, differentiate, function as sensory cells, and finally degenerate. The average life-span of these cells is about 10 days. Continuous administration of tritiated thymidine for 42 days labels all cells of the taste bud. Thus all cells of the taste bud are continually renewed. It is quite likely that the various cell types described by the electron microscopists represent different stages of development of the same cell.

The taste bud communicates with the surface of the tongue by means of a narrow channel, the taste pore. The cells contain microvilli at their apical end that project into the taste pore and come in direct contact with the saliva (Fig. 17-3). These microvilli are about 2μ long and 0.2μ wide and are thought to be the structures with which the chemical stimuli interact. Near the base of the taste cell are synaptic connections with taste nerves.

The taste cells are innervated by many fine nerve endings of taste fibers that branch profusely beneath the taste bud. Thus one taste nerve fiber may innervate several taste buds. In addition, a single taste fiber may branch and innervate several neighboring taste cells. Miller[55] recorded from a single taste fiber of the rat chorda tympani nerve and stimulated

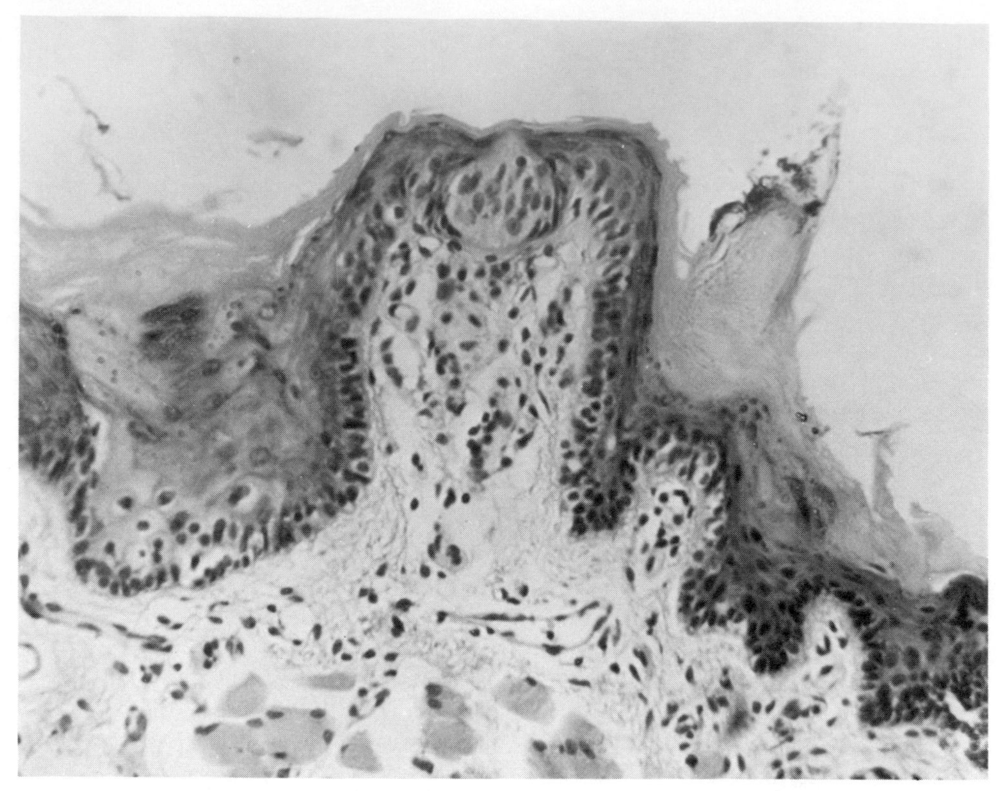

Fig. 17-2. Fungiform papilla of rat showing taste bud on its dorsal surface.

individual fungiform papillae, each containing but one taste bud. As many as four taste buds were found to be innervated by one taste fiber. The response of a single fiber to 0.3M NaCl stimulation of one papilla increased as additional neighboring papillae were also stimulated, as shown in Fig. 17-4. This complex innervation suggests that information processing may already occur at the level of the taste buds.

The integrity of the taste bud depends on its innervation.[43] If the taste nerve is cut, the buds degenerate and disappear within 1 week; On nerve regeneration, the taste buds reappear. Thus the taste nerve may have a trophic influence on the cells of the taste bud. The chemical basis of the tropism has not yet been discovered, although it is known that protein molecules manufactured in the cell bodies of the taste nerves of the nodose ganglion will travel quickly down the axon and concentrate specifically in the taste buds of the epiglottis.

Taste pathways

The facial, glossopharyngeal, and vagus nerves innervate a majority if not all of the primate taste buds. The cell bodies of their taste fibers are located in the geniculate, superior petrosal, and nodose ganglia, respectively (Fig. 17-5).

The afferent taste fibers travel to the medulla and terminate in the ipsilateral nucleus of the tractus solitarious (NTS). Neural responses have been recorded in the NTS of mammals when taste substances were applied to the tongue, although responses to thermal and mechanical stimuli were also often recorded at the same site.[44, 54, 67]

The thalamic taste relay has been identified in the monkey, cat, and rat as being in the ventromedial division of the ventral posterior thalamus. Single-unit recording has shown that most taste neurons respond also to thermal but not mechanical stimuli.[20, 33, 50]

The taste fibers leave the thalamic relay and pass via the internal capsule to terminate either on the surface of the parietal cortex or in the anterior opercular-insular cortex overlying the claustrum.[21, 22] These two taste nerve projections are close to those of the somatic sensory system.

Neural coding

It is commonly assumed that there are four distinct taste qualities: sour, bitter, salty, and

Fig. 17-3. Electron photomicrograph of apical end of taste bud (rabbit foliate) showing numerous taste microvilli extending into taste pore. (From Murray and Murray.[60])

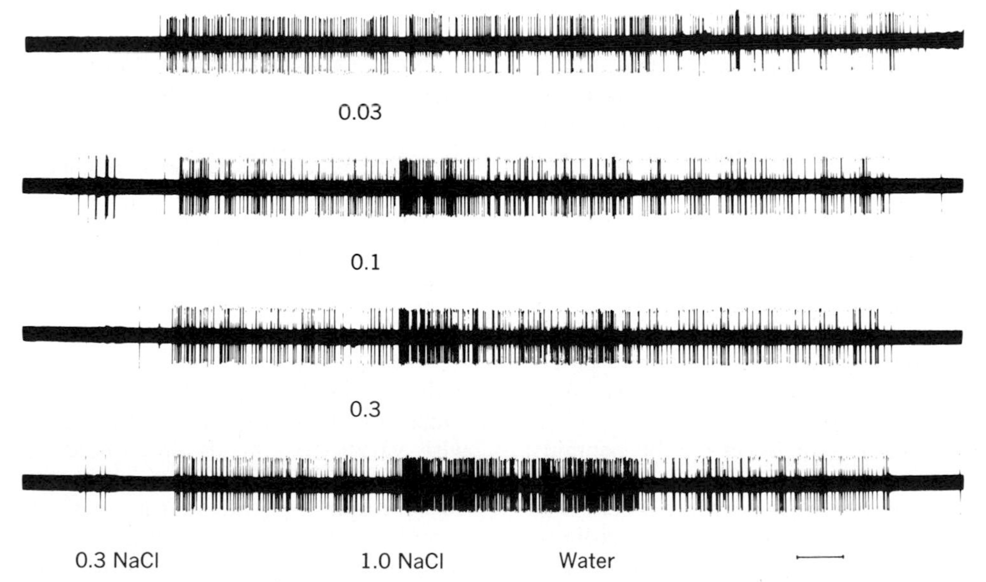

Fig. 17-4. Single rat fungiform papilla was stimulated with 0.3M NaCl for 30 sec and response was recorded from single taste fiber. After 10 sec of stimulation, surrounding papillae were also stimulated for 10 sec by NaCl concentrations as indicated. Time mark = 2 sec. (From Miller.[55])

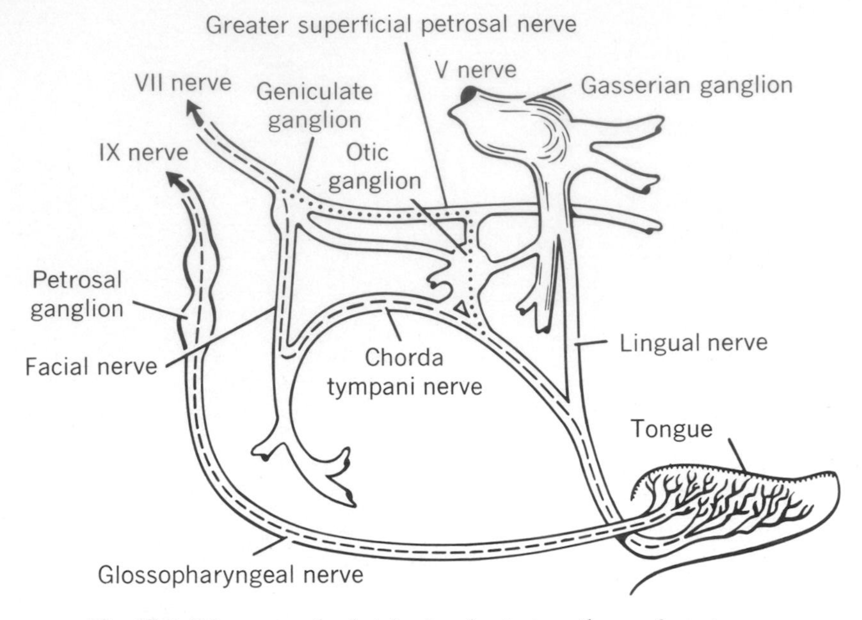

Fig. 17-5. Diagrammatic sketch of major taste pathways from tongue.

sweet. This concept has had a great influence on the study of taste. Earlier authors ascribed taste specificity to taste buds and taste fibers that respond to stimuli representing but one of the four basic taste qualities. However, the first systematic study of the response of single taste fibers by Pfaffmann[65] showed that a single taste fiber of the cat can respond to a large variety of chemical stimuli and that great heterogeneity exists in the response characteristics of the population of the fibers studied. He concluded that sensory quality is not coded by excitation of a highly specific group of taste fibers but rather by relative excitation of a large population of taste fibers of varying specificity and sensitivity. Other investigators have since studied single taste fiber responses in a variety of different mammals and have also observed the lack of high specificity.[34, 35, 81] Ogaawa et al.[61] recently recorded from 50 single taste fibers (chorda tympani) of the rat in response to four chemical stimuli representing the sour, bitter, salty, and sweet qualities of taste. He found that no two of the fibers had identical response characteristics (Fig. 17-6).

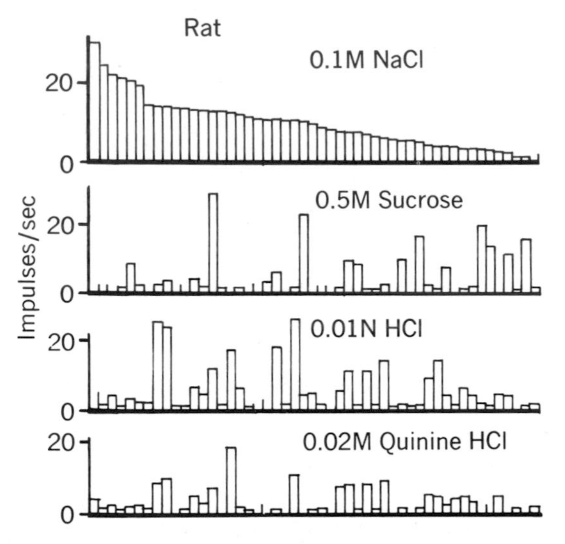

Fig. 17-6. Responses of 50 single rat taste fibers (chorda tympani) to stimuli corresponding to salty, sweet, sour, and bitter qualities. (Redrawn from Ogaawa et al.[61])

A single human fungiform papilla contains three to five taste buds; each is innervated by a small number of taste nerve fibers that branch to form hundreds of nerve endings. These in turn are in close association with 150 to 250 cells of the taste buds. What amount of specificity of taste sensation can be ascribed to this complex neural structure within a single papilla? Békésy[19] stimulated single fungiform papillae of the human tongue with various chemicals and found that most papillae are specific to but one taste quality. These results indicate that the sensory-neural complex associated with three to five human taste buds is sufficient to discriminate among the sour, bitter, salty, and sweet tastes, even though the individual taste fibers respond to a variety of stimuli.

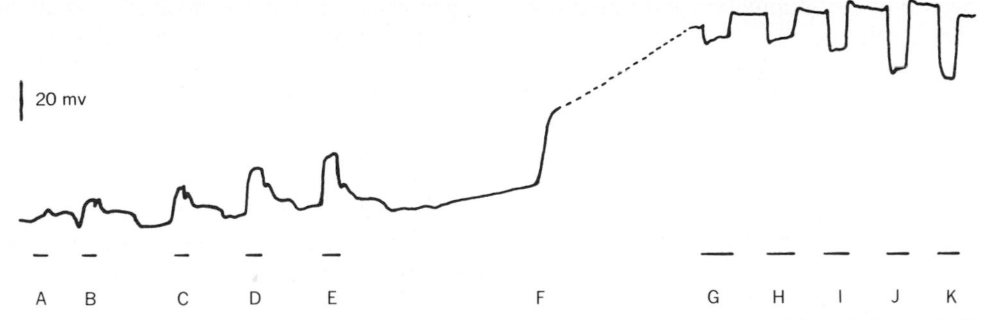

Fig. 17-7. Response of single taste cell of rat to various concentrations of NaCl (*A* and *G*, 0.05M; *B* and *H*, 0.1M; *C* and *I*, 0.25M; *D* and *J*, 0.5M; *E* and *K*, 1.0M) before and after application of cocaine at *F*. (From Tateda and Beidler.[77])

There are many shades of taste within one taste quality. The taste of fructose is not identical to that of sucrose. The relative magnitude of responses to a series of sugars has been found to differ from one single taste fiber to another, as does the rank order.[56, 66] Moreover, each rat single taste fiber may respond differently to a series of salts; e.g., some taste fibers may respond better to sodium than to equimolar concentrations of potassium, while others do just the reverse. The percent of the fiber population that may respond better to Na^+ than to K^- also varies from one species to another.[13]

Transduction

The tongue is richly innervated; in addition to the nerves associated with the taste buds, there are also numerous free nerve endings that may also respond to the chemical stimuli. However, the epithelium covering the tongue surface is a barrier to the penetration of most chemicals applied to the oral cavity. Permeability measurements with radioactive chemicals indicate that the rat tongue epithelium is almost as impermeable as the skin of the belly.[57] Ethyl alcohol is exceptional in that it quickly penetrates the epithelium and stimulates the nerve endings. However, most chemical stimuli interact with the taste buds whose receptors are in direct contact with the stimulus. They may respond within 15 to 50 msec after NaCl is applied to the tongue surface.[14]

Microelectrodes have been inserted into the single taste cells of frogs,[70] rats,[46, 64, 77] and hamsters[14] to record the receptor potentials in response to chemical stimuli. It has been found that most taste cells respond to a variety of stimuli that are not restricted to those representing but one taste quality.

If the resting potential of the taste cell is made positive by prolonged application of cocaine or $FeCl_3$, the receptor potential elicited by stimulation with NaCl reverses in sign,[77] as shown in Fig. 17-7. This observation is in agreement with the concept that membrane permeability increases when a stimulus such as NaCl is applied. This permeability change does not occur at the site of stimulation on the microvilli, but rather at more distant membrane areas near the site of synaptic innervation.[17] Thus information concerning stimulation is transmitted in some manner to other areas of the taste cell membrane.

Quantitative analysis of taste cell and taste nerve responses to a variety of chemical stimuli has been undertaken in an attempt to determine the transduction mechanism. Beidler[12] has demonstrated that taste receptors respond in a manner consistent with the concept that chemical stimuli are adsorbed to the microvilli membranes and that conformational changes in their molecular architecture result in permeability changes and nerve excitation. If it is assumed that the taste cell membrane has a finite number of equivalent but independent sites for stimulus adsorption, the simple mass action law can be applied to obtain the equation:

$$R = \frac{CKR_s}{1 + CK}$$

where R is the magnitude of response to a given stimulus, C is the stimulus concentration, R_s is the maximum response at high concentration, and K is the binding constant. Responses to chemicals that evoke pure taste qualities, e.g., NaCl and sucrose, can be described by this equation. More complex stimuli that evoke mixed tastes require the additional assumption that several different receptor sites are involved.[14] The cation is most important for stimulation, but the anion also plays a role.

If the chemical stimulus is adsorbed to a particular macromolecule of the taste cell membrane, then it should be possible to isolate the macromolecule and study its adsorption properties. Dastoli and Price[27] isolated from the tongue epithelium a protein that binds sugars and other sweet substances in a manner predicted by the previous taste equation. Since this protein has not yet been solely identified with taste cells, it is not known whether it is a taste protein specifically involved in chemical transduction.[47]

Structural determinants of taste quality

Sweet. The human chorda tympani taste nerve passes through the middle ear and is accessible for electrical recording. Diamant et al.[30] recorded the responses of the human chorda tympani to various sugars applied to the tongue (Fig. 17-8). The structures of those compounds eliciting the same taste quality can be examined to see if they possess any common physicochemical properties. The weak forces bonding nonionic taste stimuli to the receptor membrane suggested that hydrogen bonding may be involved.[14] Shallenberger and Acree[71] postulated that all sweet stimuli must have an electronegative atom (A) to which a hydrogen atom is attached by a single covalent bond and another electronegative center or atom (B) within a 3 Å distance. An assorted group of molecules having these properties is shown in Fig. 17-9. The sweet-tasting compound can then bind by two hydrogen bonds to two similar atoms or centers of the taste bud receptor site.

Taste bud receptor site	—A—H- - - - - - - -B— —B- - - - - - - -H—A—	Sweet-tasting compound

The ε-amino group of lysine in the taste receptor protein could provide the A—H structure, and the carbonyl atom of the peptide bond could provide the B structure. The A—H, B unit is a necessary but insufficient requirement for predicting the sweet taste of a molecule. The molecular conformation, including stearic factors, is also important.

Bitter. Many sweet compounds can be made bitter by small changes in molecular structure. However, no single structural component can be related to all bitter substances. Kubota and Kubo[48] have shown that certain bitter diterpenes also contain an A—H, B system in which the two units are separated by about 1.5 Å.

Different individuals show a marked difference in threshold to a group of bitter molecules containing the N—C=S group. This difference depends on a single autosomal allelic gene pair that presumably controls specific bitter sites of the receptor membrane. The so-called tasters represent two genotypes with the heterozygous Tt or the homozygous TT genetic constitution, whereas the nontasters have the recessive homozygous tt.

Sour. The presence of the hydrogen ion is a necessary requirement for a sour-tasting substance, although other factors play a role in the determination of the intensity of sourness at a given pH. Not all acids are equally sour at equi-pH and equinormality, as shown in Table 17-1. Two factors appear to be of particular importance; these are the ionic strength of the solution and the electric charge of the receptor membrane. The latter becomes more positive as an increasing number of hydrogen ions are adsorbed, which in turn partially repels the addition of more hydrogen ions. This can be overcome by simultaneous adsorption of the negative anions of the acid. Thus the ability of a particular anion of an acid to bind to the membrane

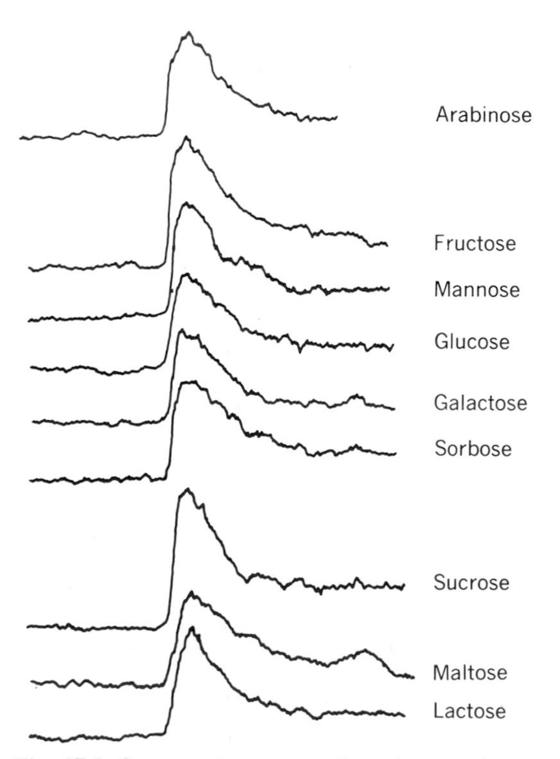

Fig. 17-8. Summated responses from human chorda tympani in response to various 0.5M sugar solutions flowed over tongue. (From Diamant et al.[30])

Table 17-1. Equal sour response*

Acid†	Acid (mM)	$[H^+]$ (mM)	pH	$[H_3A]$ (mM)	$[H_2A]$ (mM)	[HA] (mM)
Sulfuric	2.2	4.20	2.38	0	0	0.20
Oxalic	3.3	3.50	2.46	0	0.20	2.70
Hydrochloric	5.0	5.00	2.30	0	0	0
Citric	5.5	1.97	2.71	3.71	1.61	1.16
Tartaric	5.9	2.38	2.62	0	3.85	1.69
Nitric	5.9	5.90	2.23	0	0	0
Maleic	6.4	4.82	2.32	0	1.59	4.79
Dichloroacetic	9.0	7.70	2.11	0	0	1.30
Succinic	10.0	0.824	3.08	0	9.20	0.74
Malic	10.0	1.91	2.72	0	8.20	1.68
Monochloroacetic	10.4	3.13	2.50	0	0	7.27
Glutaric	11.0	0.73	3.14	0	10.30	0.67
Formic	11.6	1.32	2.88	0	0	10.30
Adipic	14.0	0.74	3.13	0	13.30	0.68
Glycolic	15.0	1.45	2.84	0	0	13.50
Lactic	15.6	1.41	2.85	0	0	14.20
Mandelic	25.0	3.11	2.51	0	0	21.80
Acetic	64.0	1.06	3.00	0	0	63.00
Propionic	130.0	1.32	2.88	0	0	129.00
Butyric	150.0	1.44	2.84	0	0	149.00

*From Beidler.[15]

†These acid concentrations produce a summated electrophysiologic response equal in magnitude to that produced by 5 mM HCl applied to the tongue of the rat.

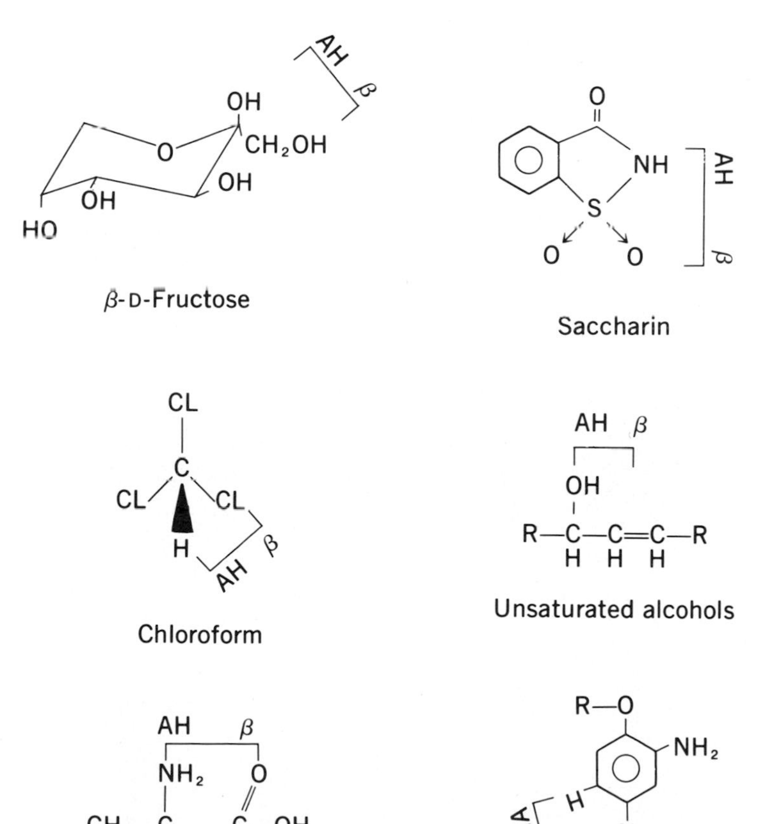

Fig. 17-9. Molecules that possess AH—B structural relationship that Shallenberger and Acree[71] predict necessary for sweet taste.

helps to determine the number of the hydrogen ions bound at a given pH and the resultant degree of sourness.[15]

An increase in the ionic strength of an acid solution results in an increase in the amount of acid bound to wool protein.[4] In a like manner, addition of a salt to an acid solution may increase the sourness, even though the pH may be higher. This has been shown to be true in taste buds of such diverse species as man,[79] rat,[15] and fish.[45]

Salty. The taste of NaCl is pure salty except near threshold and at very high concentration. Most other salts produce mixed tastes. The stimulating efficiency of salts depends primarily on the cation, although the anion is also important. The ability of a taste receptor cell to bind a given cation varies from cell to cell, and thus the response varies from one taste fiber to the next. The overall response of a population of taste fibers also varies from one species to another. For example, rodents respond better to Na^+ than to K^+, whereas the opposite is true for carnivores.[13] These observations have led to the concept that cation binding depends on the relative size of the hydration shell surrounding the cation, which is partially determined by the binding strength of the receptor site.

OLFACTION

Morphology

The olfactory receptors are found in the posterior of the nasal cavity in a mucosal area of about 5 cm^2.[5] This mucosa is readily distinguished from the pink respiratory area by its dark yellow pigmentation, the absence of rhythmically beating cilia, and the presence of Bowman's glands. The nature of the yellow olfactory pigment is not yet fully understood, although Kurihara[49] has found insoluble chromoproteins in cattle, and vitamin A and β-carotene have been implicated in other species.[24] Most of the pigment is contained in Bowman's glands and in the supporting cells, although Gerebtzoff and Shkapenko[37] have also found some pigment in the receptor cells. The olfactory mucosa of certain fish and the vomeronasal organs of mammals, which are also chemosensitive, do not contain the yellow pigments.

Olfactory sense cells, supporting cells, and basal cells comprise the olfactory mucosa (Fig. 17-10). A 10 to 40μ layer of mucus

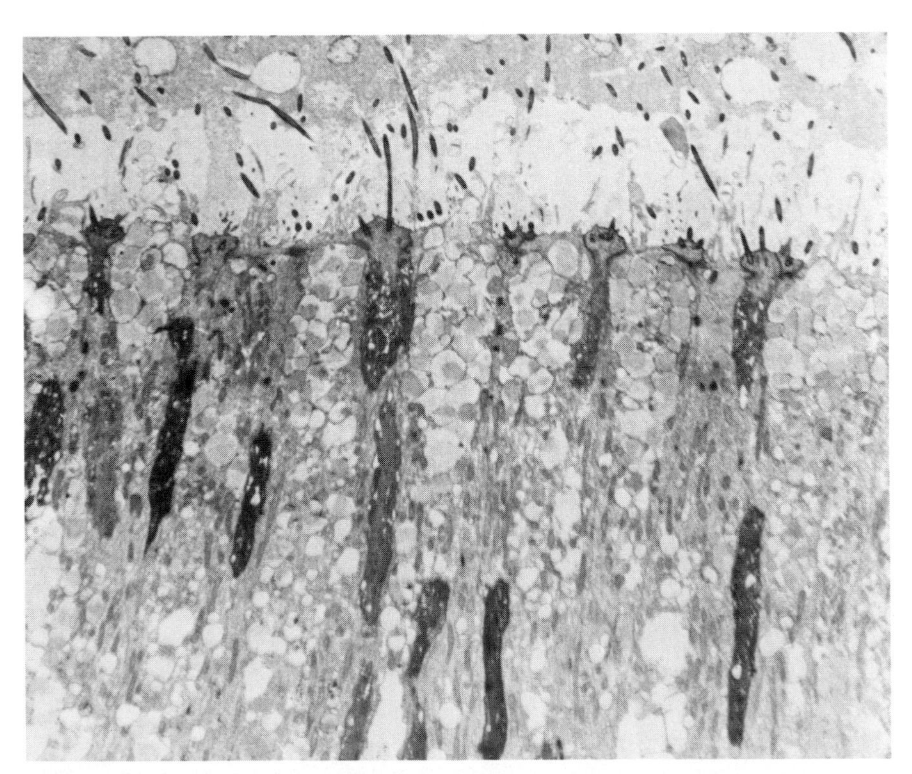

Fig. 17-10. Electron photomicrograph of section through turtle olfactory epithelium showing dark olfactory cells with cilia and light supporting cells indicating secretion. (Courtesy Dr. P. P. C. Graziadei.)

covers this tissue. The olfactory receptor is a bipolar sensory neuron. Its apical region is slightly enlarged to form the olfactory vesicle (2 to 3μ diameter), from which numerous cilia extend into the mucus for 50 to 150μ (Fig. 17-11).[38] The olfactory receptors of some birds contain microvilli as well as cilia,[25] whereas some fish olfactory receptors do not possess cilia.[9] Thus the necessity of cilia for olfaction is not certain. Tucker[79] used detergents to remove the cilia from the olfactory receptors of the turtle, and their neural activity in response to odors was not diminished.

The basal end of each olfactory receptor constricts to form an unmyelinated axon about 0.2μ in diameter. Hundreds of axons group together and are enclosed by a Schwann sheath.[36] It has been estimated that there are 100 million olfactory receptors in the rabbit and thus 100 million axons in the primary olfactory nerve, which courses through the cribriform plate to terminate in the olfactory bulb. In man the length of these axons may be a fraction of a centimeter, whereas their length in the garfish may be greater than 30 cm.

The olfactory receptors are partially separated from each other by supporting epithelial cells. They extend to the surface of the olfactory epithelium, where they project numerous microvilli into the mucus. Their endoplasmic reticulum is very prominent and many granules are found in their distal end. This suggests that they have a secretory as well as a supporting function.[41]

Prismatic basal cells are seen between the basal ends of the supporting cells. Mitotic figures are often observed in these cells, and they may be the source of newly regenerating olfactory receptors.[42]

Central pathways

The nerve fibers of the olfactory receptors enter the olfactory bulb, where they converge and synapse with the dendrites of the mitral cells.[5] Reciprocal inhibition of mitral cell activity is effected by horizontal interneurons that synapse with dendrites of neighboring mitral cells. Centrifugal fibers also synapse in this region of the olfactory bulb. Spontaneous but irregular firing is observed when microelectrodes record from single mitral cells. Odor stimulation can

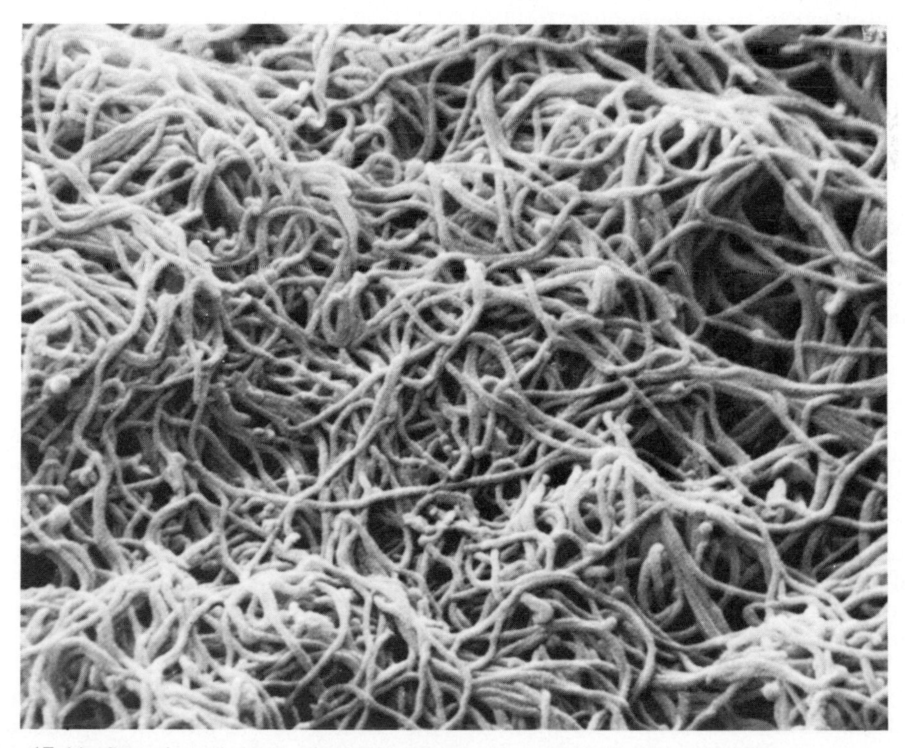

Fig. 17-11. Scanning electron photomicrograph of surface of olfactory epithelium of turtle showing numerous cilia. (Courtesy Dr. D. Tucker.)

either excite or inhibit this electrical activity.[31]

The axons of the mitral cells, the number of which is about a thousandth of the number of primary olfactory nerves, enter the olfactory tract and travel to the olfactory tubercle, the prepiriform cortex, and the periamygdaloid area.[53, 80] Olfactory messages are then relayed by projections to the thalamus and hypothalamus,[68] where they play an important role in the control of feeding and reproductive behavior. Centrifugal fibers may also enter the olfactory tract and exert a predominantly inhibitory influence on the olfactory bulb.

Transduction

There are over 100 million receptor cells in the olfactory epithelium of the rabbit.[51] Their cilia are thought to contain the sites of transduction, but it is not certain that they are the only sites; the olfactory vesicles may also be involved. The odorous molecules are believed to be adsorbed to the cilia membrane and to cause conformational changes within the membrane structure. Thus the transduction mechanism may be similar to that of taste. In order to reach the receptors the odorous molecules must have a reasonable vapor pressure. Aqueous solubility enables the molecules to penetrate the mucus covering the olfactory epithelium and lipid solubility accelerates the interaction with the receptor membrane. Odors are not the only molecules that can stimulate the olfactory receptors; amino acids and salt solutions also stimulate when applied to the epithelium.[40, 74] It has been suggested that only one molecule of certain odors may be sufficient to stimulate a single receptor cell.[73] In this respect the transduction mechanism resembles that of visual rather than that of taste receptors.

Davies[28] and Davies and Taylor[29] have developed the most complete model of olfactory receptor transduction. They proposed that one or more odorant molecules are adsorbed at specific sites, penetrate the receptor membrane, and leave behind a temporary hole in the lipid bilayer through which inorganic ions can pass to depolarize the membrane. Healing of the lipid bilayer occurs in 10 to 100 msec. Takagi et al.[76] showed that it is the permeability to Cl^- and K^+ ions that changes with odor stimulation. The calculations of Davies indicate that each receptor site is about 64 $Å^2$ and that 44,000 of these receptor sites are contained on each olfactory receptor cell. Calculated thresholds are in close agreement with those observed (Fig. 17-12). Odor quality is determined by the cross-sectional area of the odorant molecule and the ease of desorption of the molecule from the lipid-water interface into air.

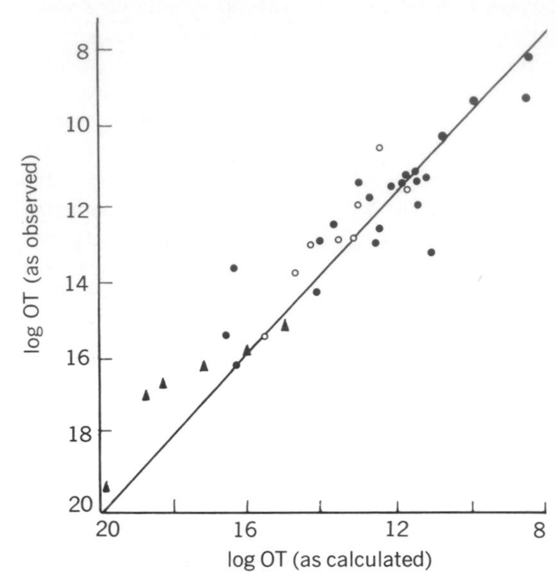

Fig. 17-12. Relationship between values of observed olfactory thresholds, *OT*, and values calculated according to theory of Davies. (From Davies.[28])

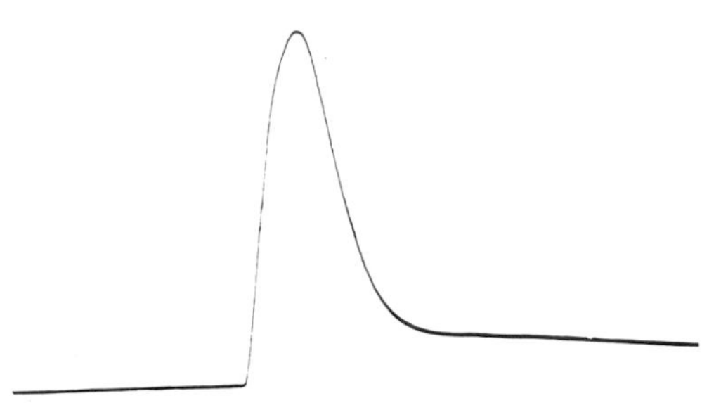

Fig. 17-13. Electro-olfactogram recorded from surface of olfactory epithelium of box turtle in response to puff of geraniol odor. Peak amplitude, 8 mv. (Courtesy Dr. D. Tucker.)

Ash[8] reasoned that the odorant molecules interact with specific membrane proteins. He prepared a water-soluble fraction of proteins from scrapings of the olfactory epithelium surface. Odorants mixed with this fraction produced a change in absorbance at 267 nm, a result that he attributed to a protein conformational change.

A monophasic negative potential (electro-olfactogram, or EOG) can be recorded from the surface of the olfactory epithelium when odor is applied (Fig. 17-13).[62] It has a latency of 100 msec or less, rises to a height

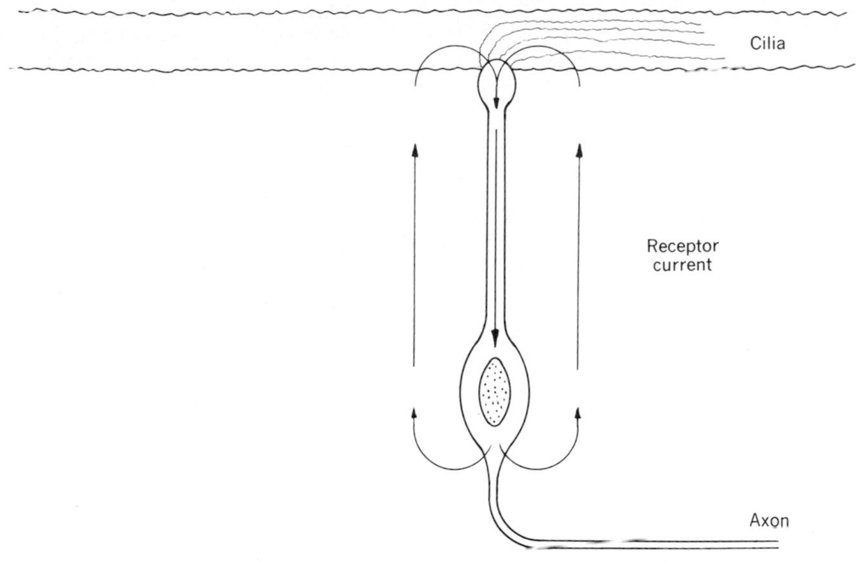

Fig. 17-14. Increase in membrane permeability of cilia or olfactory vesicle results in current flow into and through receptor and out near axon. Receptor current may generate action potentials that propagate down axon.

that is odor-concentration dependent, and declines exponentially after the odor is removed. Degeneration experiments imply that the presence of olfactory receptors is a necessary condition for the recording of the EOG.[75] The EOG may reflect the generator potentials of a large group of receptors, but to date a systematic study of intracellularly recorded generator potentials has not been undertaken.

The amplitude of the EOG is maximum at the surface of the epithelium. Ottoson[63] has suggested that adsorption of the odor to the cilia leads to a change in their permeability. This creates an increased current flow into the cilia, through the receptor, and out below the body near the axon hillock (Fig. 17-14). This latter flow would then elicit olfactory nerve activity.

Neural basis of odor quality

Man can distinguish odors of a vast number of different molecules. It has been suggested that there are specific olfactory receptors that respond but to one group of molecules associated with one odor quality. Gesteland et al.[38] recorded multispike activity with metal microelectrodes inserted into the frog olfactory epithelium and then used electronic devices to select spikes from single receptors. Stimulation with a number of different odors showed that highly specific receptors do not exist; rather each receptor responds to a variety of odors. Grouping of receptors according to odor sensitivity proved difficult. Shibuya and Tucker[72] later recorded unit activity with micropipet electrodes in the vulture and obtained similar results. Thus olfactory receptors resemble taste receptors in their lack of stimulus specificity and their response to a wide spectrum of stimuli.

Perhaps greater specificity will be observed in the higher nervous centers. Doving[31, 32] recorded from a large number of single mitral cells in the olfactory bulb in response to a variety of odors. Excitatory, inhibitory, and zero responses were recorded when odors were applied to the olfactory epithelium. Statistical methods were used to evaluate the degree of similarity in responses to pairs of odors. Pairs of musk odors showed great response similarities, whereas floral odors, e.g., geranial and salicylic aldehyde, showed little similarity. The receptor sites that interact with musks may be very homogenous, whereas the receptor sites associated with floral odors may be heterogenous. No simple arrangement of the odors studied could be made to support the suggestion that there are a small number of primary odors.

Mozell[58, 59] has questioned the assumption that odor quality discrimination is based only on the existence of receptors of highly selective sensitivity. He studied the response of various areas of the olfactory epithelium to given odors as well as differences in response latency. He successfully ranked odors

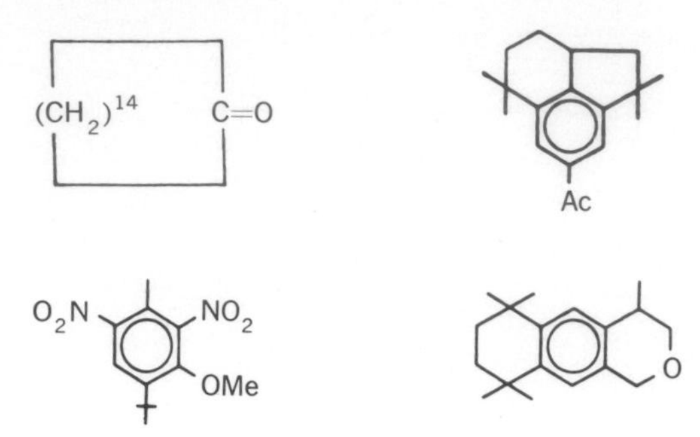

Fig. 17-15. Wide variety of molecular structures associated with given odor quality of musk. (From Beets.[11])

according to differences in their spatial-temporal patterns. It was noted that spatial-temporal patterning is the basis of chromatographic separation of molecules and that a similar process may be utilized in odor discrimination by the olfactory epithelium.

Structural determinants of odor quality

Many authors have tried to classify all odors into a finite number of groups according to their odor quality. For example, Amoore[6] used seven classifications: camphoraceous, musky, floral, pepperminty, etheral, pungent, and putrid. It was thought that these might be primary odors and that specific receptors might exist for each group. Single-fiber analysis proved this to be wrong, and Amoore later abandoned his idea of only seven primary odors. It was also thought that there must be structural similarities in the molecules identified with any one class of odors. Amoore[7] studied the structures of a large number of odorant molecules and concluded that their gross profiles (size and shape) could be related to the odor quality that they evoked. He postulated specific shapes of receptor sites into which suitable odorant molecules could be fitted. For example, an oval pan 9 Å wide, 11.5 Å long, and 4 Å deep was described as the receptor site for musky odorants. The musks have been thoroughly studied by Beets.[11] He agreed that molecular profiles are important but he also stressed other physicochemical features, particularly the characteristics of functional groups. The variety of molecular structures that are associated with the musk odor is shown in Fig. 17-15.

REFERENCES

General reviews

1. Beidler, L. M., editor: Handbook of sensory physiology. I. Chemical senses—olfaction. II. Chemical senses—taste, New York, 1971, Springer-Verlag New York, Inc., vol. 4.
2. Moncrieff, R. W.: The chemical senses, ed. 3, Cleveland, Ohio, 1967, Chemical Rubber Company Press.
3. Pfaffmann, C., editor: Olfaction and taste, proceedings of the third international symposium, New York, 1969, The Rockefeller University Press.

Original papers

4. Alexander, P., Hudson, R. F., and Earland, C.: Wool—its chemistry and physics, Palisades, N. J., 1963, Franklin Publishing Co., Inc.
5. Allison, A. C., and Warwick, P. T. T.: Quantitative observations on the olfactory system of the rabbit, Brain **72:**186, 1949.
6. Amoore, J. E.: The stereochemical theory of olfaction. I. Identification of seven primary odors, Proc. Sci. Soc. Toilet Goods Assoc. **37** (suppl.):1, 1962.
7. Amoore, J. E.: Stereochemical theory of olfaction. In Schultz, H. W., Day, E. A., and Libbey, L. M., editors: Chemistry and physiology of flavors, Westport, Conn., 1967, Avi Publishing Co.
8. Ash, K. O.: Chemical sensing: an approach to biological molecular mechanisms using difference spectroscopy, Science **162:**452, 1968.
9. Bannister, L. H.: The fine structure of the olfactory surface of teleostean fishes, Q. J. Microscop. Sci. **106:**333, 1965.
10. Beets, M. G. J.: Quelques aspects du probleme de l'odeur, Parf. Cosm. Sav. **5:**167, 1962.
11. Beets, M. G. J.: A molecular approach to olfaction. In Ariens, E. J., editor: Molecular pharmacology, New York, 1964, Academic Press, Inc., vol. 2.
12. Beidler, L. M.: A theory of taste stimulation, J. Gen. Physiol. **38:**133, 1954.
13. Beidler, L. M.: Species differences in taste responses, Am. J. Physiol. **181:**235, 1955.
14. Beidler, L. M.: Taste receptor stimulation, Progr. Biophys. Biophys. Chem. **12:**109, 1962.
15. Beidler, L. M.: Anion influences on taste receptor response. In Hayashi, T., editor: Olfaction and taste II, Oxford, 1967, Pergamon Press, Ltd.
16. Beidler, L. M.: Innervation of rat fungiform papilla. In Pfaffmann, C., editor: Olfaction and taste III, New York, 1969, The Rockefeller University Press.
17. Beidler, L. M., and Reichardt, W. E.: Sensory transduction neurosciences, Res. Program Bull. **8:**461, 1970.
18. Beidler, L. M., and Smallman, R.: Renewal of cells within taste buds, J. Cell Biol. **27:**263, 1965.
19. Békésy, G. von: Taste theories and the chemical stimulation of single papillae, J. Appl. Physiol. **21:**1, 1966.
20. Benjamin, R. M.: Some thalamic and cortical mechanisms of taste. In Zotterman, Y., editor: Olfaction and taste I, Oxford, 1963, Pergamon Press, Ltd.
21. Benjamin, R. M., and Burton, H.: Projection of taste nerve afferents to anterior opercular insular cortex in squirrel monkey (Saimiri sciureus), Brain Res. **7:**221, 1968.

22. Benjamin, R. M., Emmers, R., and Blomquist, A. J.: Projection of tongue nerve afferents to somatic sensory area I in squirrel monkey (Saimiri sciureus), Brain Res. **7**:208, 1968.
23. Bradley, R. M., and Stern, I. B.: The development of the human taste bud during the foetal period, J. Anat. Physiol. **101**:743, 1967.
24. Briggs, M. H., and Duncan, R. B.: Pigment and the olfactory mechanism, Nature **195:** 1313, 1962.
25. Brown, H. E., and Beidler, L. M.: The fine structure of the olfactory tissue in the black vulture, Fed. Proc. **25**:329, 1966.
26. Bruce, H. M.: Role of olfactory sense in pregnancy block by strange males, Science **131:** 1526, 1960.
27. Dastoli, F. R., and Price, S.: Sweet sensitive protein from bovine taste buds: isolation and assay, Science **154**:905, 1966.
28. Davies, J. T.: The penetrating and puncturing theory of odor, J. Colloid Interface Sci. **29:** 296, 1969.
29. Davies, J. T., and Taylor, F. H.: Molecular shape, size and adsorption in olfaction, Proc. Int. Congr. Surface Activity **4**:329, 1957.
30. Diamant, H., Funakoshi, M., Strom, L., and Zotterman, Y.: Electrophysiological studies on human taste nerves. In Zotterman, Y., editor: Olfaction and taste I, Oxford, 1963, Pergamon Press, Ltd.
31. Doving, K. B.: An electrophysiological study of odour similarities of homologous substances, J. Physiol. **186**:97, 1966.
32. Doving, K. B.: Analysis of odour similarities from electrophysiological data, Acta Physiol. Scand. **68**:404, 1966.
33. Emmers, R.: Separate relays of tactile pressure, thermal and gustatory modalities in the cat thalamus, Proc. Soc. Exp. Biol. Med. **121**:527, 1966.
34. Erickson, R. P., Doetsch, G. S., and Marshal, D. A.: The gustatory neural response function, J. Gen. Physiol. **49**:247, 1965.
35. Fishman, I. Y.: Single fiber gustatory impulses in rat and hamster, J. Cell. Comp. Physiol. **49:** 319, 1957.
36. Gasser, H.: Olfactory nerve fibers, J. Gen. Physiol. **39**:473, 1956.
37. Gerebtzoff, M. A., and Shkapenko, G.: Recherches sur le pigment de la muquense olfactif, Compt. Rend. Assoc. Anat. **68**:511, 1952.
38. Gesteland, R. C., Lettvin, J. Y., and Pitts, W. H.: Chemical transmission in the nose of the frog, J. Physiol. **181**:525, 1965.
39. Gesteland, R. C., Lettvin, J. Y., Pitts, W. H., and Rojas, A.: Odor specificities of the frog's olfactory receptors. In Zotterman, Y., editor: Olfaction and taste I, Oxford, 1963, Pergamon Press, Ltd.
40. Getchell, T. V.: The interaction of the peripheral olfactory system with nonodorous stimuli. In Pfaffmann, C., editor: Olfaction and taste III, New York, 1969, The Rockefeller University Press.
41. Graziadei, P. P. C.: Personal communication, 1973.
42. Graziadei, P. P. C., and Metcalf, J. F.: Autoradiographic and ultrastructural observations of the frog's olfactory mucosa, Z. Zellforsch. Mikrosk. Anat. **116**:305, 1971.
43. Guth, L.: Taste buds on the cat's circumvallate papillae after reinnervation by glossopharyngeal, vagus, and hypoglossal nerves, Anat. Rec. **130**:25, 1958.
44. Halpern, B. P., and Nelson, L. M.: Bulbar gustatory response to anterior and to posterior tongue stimulation in the rat, Am. J. Physiol. **209**:105, 1965.
45. Hidaka, I.: Effects of salts and pH on fish chemoreceptor responses, Nature **228**:1102, 1970.
46. Kimura, K., and Beidler, L. M.: Microelectrode study of taste receptors of rat and hamster, J. Cell. Comp. Physiol. **58**:131, 1961.
47. Koyama, N., and Kurihara, K.: Do unique proteins exist in taste buds? J. Gen. Physiol. **57:** 297, 1971.
48. Kubota, I., and Kubo, I.: Bitterness and chemical structure, Nature **223**:97, 1956.
49. Kurihara, K.: Isolation of chromoproteins from bovine olfactory tissues, Biochim. Biophys. Acta **148**:328, 1967.
50. Landgren, S.: Thalamic neurons responding to tactile stimulation of cat's tongue, Acta Physiol. Scand. **48**:238, 1960.
51. Le Gros Clark, W. E.: Inquiries into the anatomical basis of olfactory discrimination, Proc. R. Soc. Lond. (Biol.) **146**:299, 1957.
52. Liljestrand, G.: Uber den Schwellenwert des sauren Geschmacks, Arch. Neerl. Physiol. **7:** 523, 1922.
53. Lohman, A. H. M., and Lammers, H. G.: On the structure and fibre connections on the olfactory centres in mammals. In Zotterman, Y., editor: Sensory mechanisms, Amsterdam, 1967, Elsevier Publishing Co.
54. Makous, W., Nord, S., Oakley, B., and Pfaffmann, C.: The gustatory relay in the medulla. In Zotterman, Y., editor: Olfaction and taste I, Oxford, 1963, Pergamon Press, Ltd.
55. Miller, I. J.: Peripheral interactions among single papilla inputs to gustatory nerve fibers, J. Gen. Physiol. **57**:1, 1971.
56. Mistretta, C. M.: A study of rat chorda tympani fiber discharge patterns in response to lingual stimulation with a variety of chemicals, D.Sc. dissertation, Florida State University, 1970.
57. Mistretta, C. M.: Permeability of tongue epithelium and its relation to taste, Am. J. Physiol. **220**:1162, 1971.
58. Mozell, M. M.: The spatiotemporal analysis of odorants at the level of the olfactory receptor sheet, J. Gen. Physiol. **50**:25, 1966.
59. Mozell, M. M.: Evidence for a chromatographic model of olfaction, J. Gen. Physiol. **56**:46, 1970.
60. Murray, R. G., and Murray, A.: Fine structure of taste buds of rabbit foliate papillae, J. Ultrastruct. Res. **19**:327, 1967.
61. Ogaawa, H., Sato, M., and Yamashita, S.: Multiple sensitivity of chorda tympani fibers of the rat and hamster to gustatory and thermal stimuli, J. Physiol. **199**:223, 1968.
62. Ottoson, D.: Analysis of the electrical activity of the olfactory epithelium, Acta Physiol. Scand. **35**(suppl. 122):1, 1956.
63. Ottoson, D.: Electrical signs of olfactory transducer action. In Wolstenholme, G. E. W.,

and Knight, J., editors: Ciba symposium on taste and smell in vertebrates, London, 1970, J. & A. Churchill, Ltd.

64. Ozeki, M.: Hetero-electrogenesis of the gustatory cell membrane, Nature **228:**868, 1970.
65. Pfaffman, C.: Gustatory afferent impulses, J. Cell. Comp. Physiol. **17:**243, 1941.
66. Pfaffmann, C.: Taste preference and reinforcement. In Tapp, J., editor: Reinforcement and behavior, New York, 1969, Academic Press, Inc.
67. Pfaffmann, C., Erickson, R. P., Frommer, G. P., and Halpern, B. P.: Gustatory discharges in the rat medulla and thalamus. In Rosenblith, W. B., editor: Sensory communication, Cambridge, Mass., 1961, The M.I.T. Press.
68. Powell, T. P. S., Cowan, W. M., and Raisman, G.: The central olfactory connexions, J. Anat. **99:**791, 1965.
69. Richter, G. P.: Salt taste thresholds of normal and adrenalectomized rats, Endocrinology **24:**367, 1939.
70. Sato, T.: The response of frog taste cells (Rana nigromaculata and Rana catesbeana), Experientia **25:**709, 1969.
71. Shallenberger, R. S., and Acree, T. E.: Molecular theory of sweet taste, Nature **216:**480, 1967.
72. Shibuya, T., and Tucker, D.: Single unit response of the olfactory receptors in vultures. In Hayashi, I., editor: Olfaction and taste II, Oxford, 1967, Pergamon Press, Ltd.
73. Stuiver, M.: The biophysics of smell, Ph.D. dissertation, Groningen University, 1958.
74. Suzuki, N., and Tucker, D.: Olfactory receptor response to skin extracts in the catfish, Fed. Proc. **30:**552, 1971.
75. Takagi, S. F., and Yajima, T.: Electrical activity and histological change in degenerating olfactory epithelium, J. Gen. Physiol. **48:**559, 1965.
76. Takagi, S. F., Wyse, F. A., Kitamura, H., and Ito, K.: The roles of sodium and potassium ions in the generation of the electro-olfactogram, J. Gen. Physiol. **51:**552, 1968.
77. Tateda, H., and Beidler, L. M.: The receptor potential of the taste cell of the rat, J. Gen. Physiol. **47:**479, 1964.
78. Tucker, D.: Personal communication, 1973.
79. Tucker, D.: Olfactory cilia are not required for receptor function, Fed. Proc. **26:**544, 1967.
80. White, L. E.: Olfactory bulb projections of the rat, Anat. Rec. **152:**465, 1965.
81. Yamashita, S., Ogaawa, H., and Sato, M.: Analysis of response of hamster taste units to gustatory and thermal stimuli, Kumamoto Med. J. **20:**159, 1967.

18

GERHARD WERNER

The study of sensation in physiology: psychophysical and neurophysiologic correlations

Research in psychology and physiology closely interrelate in the study of sensation. Sensory processes can be explored at various levels of analysis—at the introspective level, with the subject providing verbal reports of his sensory experience; at the level of externally observable behavior, such as adjusting a certain stimulus under the subject's control to a criterion value; and finally, in terms of neural processes engendered by sensory stimuli. Each of these levels of analysis will be illustrated with examples intended to demonstrate how sensation is studied and in what form sensory performance in psychophysical and behavioral tasks can be correlated with spatial and temporal patterns of neural activity. The underlying notion is that events in the physical environment of a given organism find a representation in its nervous system; the goals are to discover the relation between sensory performance and spatiotemporal patterns of neural activity and to account for the former in terms of the latter.

Representation and coding in the nervous system

In general the representation of physical events involves the transformation of input signals into output messages by means of transducers. In this process of transformation any physical resemblance between the input signal and the output may be discarded, as is the case when chemical or physical stimuli impinging on the body's receptors generate trains of nerve impulses in sensory pathways. Another illustration is the representation of verbal or numerical messages as magnetized points on a metallic tape, as in computer technology. Thus a representation may be quite abstract, reflecting only some of the qualitative properties or quantitative relations in the input signal and excluding others. Yet the transformation of the input signal to its representation retains some correspondence between the two, to the extent that certain events (or combinations of events) of the repertoire of input signals are unambiguously matched by certain features of the output.

The nature of the correspondence between signals and their representation may be of different forms. In the simplest case there is a well-defined quantitative relation between input into and output from the transducer, e.g., between sound pressure acting on a microphone and the electric current thus generated. In such cases the rules of correspondence between input and output are of the form of a continuous function, expressed in mathematical terms. The study of stimulus-response relations in psychophysics and in some areas of quantitative neurophysiology follow this paradigm.

In other cases, it is more appropriate to view the representation of a signal as a discrete mapping from the domain of input signals into the domain of output symbols: this is a form of coding.

To illustrate the concept of coding, consider the set of events that can occur at the input as consisting of the elements $V = \{v_1, v_2, v_3, v_4\}$; let $A = \{0, 1\}$ be the elements of which the encoded representation of events of V can be composed. An example of a code could be the following mappings (M) from V to A:

$$M(v_1) = 1$$
$$M(v_2) = 011$$
$$M(v_3) = 010$$
$$M(v_4) = 00$$

A code such as this can be thought of as a simple kind of automaton, i.e., a device that accepts symbols from the set V and produces symbols composed of the elements of A according to predetermined rules.[1]

Clearly, there can be a sequence of consecutive encoding schemes interposed between the stimulus domain and an interpretive device, such that the output of one encoder serves as input to an encoder of a higher level of abstraction. This is a distinct possibility in the CNS wherever stimulus-evoked afferent neural activity is transmitted through a succession of sensory relay stations and cortical receiving areas, the neurons of which tend to signal stimulus configurations of progressively increasing complexity.

A simple illustration of this possibility is the interpretation by Hubel and Wiesel[21] of some of their own experimental findings with the receptive field organization of neurons in the lateral geniculate body and of the neurons with "simple" receptive fields in cortical area 17 (Chapter 16). For these cortical neurons, specifically oriented lines tend to replace circular light spots as the optimal stimuli. Fig. 18-1 illustrates a possible scheme that is capable of generating a "code" of a higher order abstraction (consisting of line segments as elements of the code alphabet) from a lower order representation of light stimuli in the form of neurons with circular receptive fields.

These introductory remarks on the concepts of coding and representation are intended to set the stage for an appreciation of the experimental strategy employed in the investigations of the nature of correspondences between neural and sensory events.

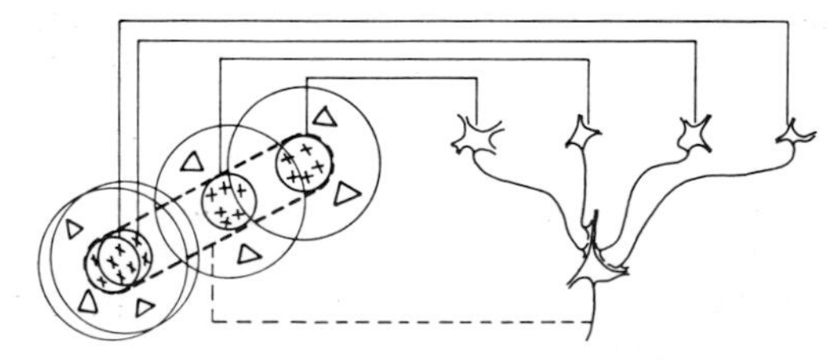

Fig. 18-1. Possible scheme for explaining organization of simple receptive fields. Large number of lateral geniculate cells (four are shown at right) have receptive fields with "on" centers arranged along straight line on retina. All project on single cortical cell, and synapses are supposedly excitatory. Receptive field of cortical cell will then have elongated "on" center (indicated by interrupted lines in receptive field diagram to left). (From Hubel and Wiesel.[21])

In the attempts to recognize in neural activity the features that represent the properties of stimuli, several experimental approaches in neurophysiology have proved to be informative. On the one hand, electrical responses in populations of neurons, recorded as "evoked potentials" with relatively gross electrodes, can be taken as indicative of the location and magnitude of a neural response to a stimulus. On the other hand, the stimulus response of individual neurons can be observed with microelectrodes. In this case, it is possible to analyze the neural response in terms of a mean discharge rate and also in terms of the sequential ordering of the neuron discharge, in the expectation that the latter might bear some relation to certain stimulus attributes.

Applications of these approaches will be illustrated later in this chapter, after an introductory discussion of the principles that underlie the evaluation of sensory functions in psychophysics.

PSYCHOPHYSICS: INTROSPECTIVE AND BEHAVIORAL METHODS USED TO MEASURE SENSATION

G. T. Fechner,[2] who is generally considered to be the founder of psychophysics as it is currently understood, developed the idea that sensations, differing in degree, could be assigned numerical values, and that their relative magnitudes would stand in a certain mathematical relation to the magnitudes of the corresponding physical stimuli.

Fechner set out to determine how "sensation intensity" varies with physical stimulus magnitude. The method was essentially that used earlier by Weber: he would set a variable stimulus (S_v) equal to a standard stimulus (S_s) and then change S_v by small steps until the observer reported that he felt a just noticeable difference (jnd). The method was therefore based on introspection, a procedure that uses one and the same person as subject and as instrument of measurement as well. With the early empiricism, the belief grew that introspection as a form of immediate experience is incorrigible and, by definition, not subject to error; it provides privileged access to private experience that can be shared with others through the medium of language. However, in the absence of objective referents, the words of the language of introspection can never be unambiguously defined, except perhaps in the restricted context of a well-defined psychophysical test situation consisting of observation of a single physical event in isolation.

Criticisms such as this and the lack of reliability of the introspective method were some of the forces that contributed to the rise of the school of behaviorists in the early 1910s. J. B. Watson,[12] as did others, found that animals can learn to respond differentially to different stimuli. In the language of the then current sensationalistic psychology this implied that they could discriminate, say, one tonal frequency from another, i.e., that their behavior could be viewed as indicating differences in sensory experiences. Consequently, there would be no longer any need for considering subjective sensory experience as an independent experimental variable in psychological studies; rather, psychophysics could be brought into harmony with the principle of the scientific method by examining correlations between two sets of objectively observable phenomena, physical stimuli and overt behavior. Actually, Watson showed that it was possible to describe all psychological facts established at that time in terms of behavioral manifestations only, with no need to bring consciousness or sensation into the picture. By 1930 the mainstream of behaviorism was fused with and largely superseded by the dominant philosophic movements of that period, namely, logical positivism and operationism. The principal repercussions of this were twofold. In the first place, Stevens[37] could rightly point out the possibility of measuring subjective sensation magnitude by requiring no more than that the subject make some observable, behavioral response: this was not to deny the existence of a subjective response but to replace it, for the purpose of measurement, by a behavioral expression. Accordingly, sensation is not judged from introspective reports; instead, its measurement is based on overt behavioral acts. This concept of operationist psychophysics makes it possible to study sensation in animal experiments by essentially the same procedures as are applied in man, except that differential reinforcement techniques for control of animal behavior take the place of verbal instruction to human subjects. The common feature is the type of stimulus control technique that was developed by Békésy for human audiometry studies.[13]

The principle of this procedure is that the intensity of a certain stimulus increases continuously as long as a signal button is pressed and decreases automatically when this button is released. Through control of the button the subject is able to let the stimulus intensity fluctuate between just above and just below threshold. By recording these intensity fluctuations, a difference limen for intensity and an absolute intensity threshold can be determined without any verbal (introspective) report from the subject.

In an analogous situation, experimental animals can be made to confront two response keys (A and B) and a small lighted stimulus patch. In a training period (which is, in effect, the counterpart to the verbal instruction given a human subject) the light patch changes in a random sequence between illumination and darkness: pressing key A blanks out the stimulus patch, and pressing key B is followed by food reward if the light patch is dark. In this manner, the animal acquires the behavior of pressing key A when light is visible and pressing key B when the light patch is dark. In addition, appropriate switching circuitry reduces the luminance of the stimulus if key A is pressed, while pressing key B increases luminance. As a result, the stimulus is kept oscillating about threshold. This procedure has been successfully employed to determine, for instance, the scotopic spectral sensitivity in various species.[14]

The extent to which behavioral and neural data can jointly be brought to bear on studies of sensation is illustrated with an example of recent work on vision. As an important attribute of human vision, stereopsis has been studied for many years as the sensation of relative visual depth that results from the dissimilarities of the images seen by the two eyes. To avoid interference by various familiarity cues and other uncontrollable circumstances, Julesz[25] devised a novel technique for random-dot stereo images. An example of such a stereo pair is shown in Fig. 18-2. When viewed monocularly, both fields of Fig. 18-2 give a homogeneous random impression without any apparent features, but when viewed stereoscopically, a center square is vividly perceived in front of its surround.

Fig. 18-3 illustrates how Fig. 18-2 and similar randon-dot stereo images are generated. The description of the procedure is here reproduced in Julesz's own words:

Fig. 18-2. Basic random stereo pair. When two fields are viewed stereoscopically, center square appears in front of background. (From Julesz.[25])

1	0	1	0	1	0	0	1	0
1	0	X	A	A	B	B	0	0
0	0	Y	B	A	B	A	1	1
0	1	0	0	1	1	1	0	1
1	1	A	B	A	B	A	0	0
0	0	B	A	B	A	B	1	0
1	1	0	1	0	1	1	0	0
1	0	A	A	B	A	X	0	1
1	1	B	B	A	B	X	1	0
0	1	0	0	0	1	1	1	1

1	0	1	0	1	0	0	1	0
1	0	A	A	B	B	Y	0	0
0	0	B	A	B	A	X	1	1
0	1	0	0	1	1	1	0	1
1	1	B	A	B	A	B	0	0
0	0	A	B	A	B	A	1	0
1	1	0	1	0	1	1	0	0
1	0	Y	A	A	B	A	0	1
1	1	Y	B	B	A	B	1	0
0	1	0	0	0	1	1	1	1

Fig. 18-3. Illustration of method by which stereo pair in Fig. 18-2 was generated. (From Julesz.[25])

The equally probable randomly selected black and white picture elements which are contained in corresponding areas in the left and right fields are labeled in three categories: (1) Those contained in corresponding areas with 0 disparity (which when viewed stereoscopically are perceived as the surround) are labeled 0 or 1. (2) Those contained in corresponding areas with non 0 disparity (which when viewed stereoscopically are perceived in front of or behind the surround) are labeled A or B. (3) Those contained in areas which have no corresponding areas in the other field (that is, project on only one retina and thus have no disparity) are labeled X and Y. The 0 and 1 picture elements are identical in corresponding positions on the two fields. The positions of the A and B picture elements belonging to corresponding areas in the fields are also identical, but are shifted horizontally as if they were a solid sheet. Because of this shift some of the picture elements of the surround are uncovered and must be assigned new brightness values (X and Y). Since these areas lack disparity, they can be regarded as undetermined in depth. Fig. 18-3 contains three rectangles in the left and right fields, composed of A and B picture elements. Each field contains an upper, middle and lower rectangle which can be regarded as corresponding left and right "projections" of a rectangular planar surface located in depth when viewed from different angles. The projections of the upper rectangle (that is the corresponding upper rectangles in the left and right fields) are horizontally shifted relative to each other in the nasal direction by one picture element, the corresponding lower rectangles are shifted in the temporal direction to the same extent, while the corresponding middle rectangles have a one-picture-element periodicity and may be regarded as being shifted in either direction. The low density of picture elements and the large disparities would prevent stereopsis in a pattern corresponding to Fig. 18-2. In order to achieve stereopsis the number of picture elements would have to be decreased considerably. For this reason a computer is used.*

By combining the random-dot pattern technique with standard operant conditioning procedures, Bough[15] succeeded in determining that, like man, macaques have stereoscopic vision. The animals were required to distinguish random-dot patterns that, if viewed binocularly, would seem to have an inner square displaced behind the plane of the surround, from random-dot patterns in which no inner square was distinguishable during binocular vision. In 200 trials the two animals in this study reached correct performance levels of 90 and 98%, respectively.

In neurophysiologic studies with macaques, Hubel and Wiesel[23] ascertained that some 40% of the neurons in area 18 function as "binocular depth cells." These neurons usually respond most vigorously if both eyes are stimulated simultaneously. In some instances the responses are elicited by stimulation of exactly corresponding retinal regions; in other instances there is a disparity in the positions of the two receptive fields. This suggests that neurons in area 18 play a role in the elaboration of the stereoscopic depth mechanism that was shown to exist in this species in the behavioral tests of Bough.[15]

This example has been described in some detail, for it illustrates with particular clarity how a *qualitative* comparison between neurophysiologic and behavioral observations can contribute to the recognition of the neural processes that play a part in a particular sensory function. In the following sections, similar comparisons between sensory and neural activity will be discussed, with emphasis on *quantitative relations.*

Quantitative correlations between sensation and neural activity

The neurophysiologic line of inquiry essentially parallels the traditional approach of the psychophysicist, except that the latter tried to correlate stimulus values with sensation magnitudes, while the neurophysiologist attempts to correlate stimulus values with differences in the neural responses. A distinct stimulus attribute with well-defined physical properties is selected, and the magnitude of this attribute is varied over a range of values; e.g., in the case of tactile stimuli restricted to a point of the skin, the displacement of the skin (measured in units of length), the force acting on the skin, or the impulse of a rapidly applied stimulus of the cutaneous surface (measured as the product of mass times impact velocity of the stimulating agent) could constitute suitable scales for the quantification of the stimulus. Similarly, the pitch and loudness of tones are stimulus attributes that can be directly measured on appropriate scales.

The next step involves measuring the stimulus-evoked neural activity and seeking a

*Julesz, B.: Binocular depth perception without familiarity cues, Science **145**:356, 1964.

correspondence between the stimuli of different values on their measurement scale and the neural responses they evoke.

This raises some difficult questions. What is an appropriate way to "measure" neural activity? And what is a suitable rule for assigning numerical values to neural activity, such that differences in magnitude of neural activity can be represented as relations of greater and smaller and, under more exacting conditions, as numerical differences on a continuous and monotonic scale?

These questions are not unique to the measurement of neural activity: they underlie all measurement. But most conventional measurement scales (e.g., weight or length) are embedded in some conceptual framework from which they derive their justification. In employing these conventional scales of measurement we do not ordinarily reflect on what Wiener referred to: "Things do not, in general, run around with their measures stamped on them like the capacity of a freight car: it requires a certain amount of investigation to discover what their measures are."[48]

In the absence of a suitable conceptual framework to guide the selection of an appropriate scale for measuring neural activity, the procedure is purely experimental and pragmatic: it is necessary to vary the stimulus, to "measure" the neural activity in various ways, and to discover the scale of neural activity measurement that makes the functional relation between stimulus and magnitude of the neural response tally with that between the stimulus and a psychophysical measurement of sensation. It is then possible to test further whether these relationships also tally under a variety of conditions, e.g., variations of the attentive set, or under the influence of pharmacologic agents with known effects on behavior and subjective experience in man.

The ultimate goal is to find a particular means of characterizing neural responses in quantitative terms (i.e., a scale to measure neural responses), such that the functional relation between stimulus and neural response is of the same form as the relation of stimulus to "sensation," as measured in an independent psychophysical or behavioral experiment. It may then be said that this particular functional relation between stimulus and neural response qualifies as the "neural code" of the stimulus attribute under study.

The selection of distinct stimulus attributes for study reflects the idea that sensation can be viewed as being composed of individual and mutually independent attributes, of which classic psychophysics defined four—namely, quality, intensity, duration, and extension. The independence of these attributes is fictitious, for there cannot be an intensity in sensation except as an intensity *of* a quality, and this argument can be applied *mutatis mutandis* to the relation of the other attributes as well. Nevertheless, for experimental purposes the concept of isolated attributes (notably the stimulus attribute of intensity) can be useful and informative, as long as one remains mindful of its restricted validity.

INTENSITY ATTRIBUTE IN PSYCHOPHYSICS AND NEUROPHYSIOLOGY

There are four types of questions that can be asked concerning the magnitude of a sensation.[3] The question *"Is anything there?"* leads to the *detection problem;* the *problem of recognition* includes the systematic attempts to answer the question *"What is it?";* we cannot meaningfully and consistently answer the question *"How much of it is there?"* unless we give a systematic account of the *scaling problem;* and finally the question that, historically, led to the foundation of quantitative psychophysics and its original preoccupation with the *discrimination problem: "Is this different from that?"*

Each of these questions has its counterpart in quantitative neurophysiology. In observing neural activity, we wish to know whether the response to stimulus A differs from that to stimulus B and, if so, in what quantitative manner. This reflects the close interrelation of discrimination and scaling. When we wish to detect from ongoing neural activity whether a stimulus has been applied and, if so, which one of a given set of alternative possibilities it was, we seek to determine the way in which the CNS could conceivably perform the task of stimulus detection and recognition.

Discrimination problem

Investigations of sensory discrimination were begun early in the 19th century by the physiologist E. H. Weber. Weber's procedure was based on what is now known as "comparative unidimensional judgments": one particular stimulus of a certain intensity (the standard stimulus) is applied in alternation with one of a number of other stimuli (the comparison stimulus) that are of the same type but that differ in physical stimulus magnitude. A set of standard and comparison stimuli could, for instance, be a series of tones of equal pitch and different loudness or tactile stimuli differing in the force applied to the skin. The subject is instructed to note whether the second stimulus in each pair is stronger or weaker than the standard stimulus, which is applied as the first stimulus in each pair. When this is done a large number of times for each stimulus pair, the probability with which the comparison stimulus in each pair is judged stronger or weaker than the standard can be estimated. Fig. 18-4 presents experimental data on loudness discrimination, and Fig. 18-5 displays schematically the typical properties of these data. Note that at the point with the ordinate $p = 0.5$, comparison and standard are judged to be of equal intensity; this point of subjective intensity frequently differs from objective equality. The difference generally depends on the time interval between the presentation of comparison and standard stimuli and tends to decrease as this interval becomes larger.

The interquartile range of this discrimination function is that interval on the abscissa that is delineated by the probabilities 0.25 and 0.75, respectively, of judging the comparison stimulus stronger than the standard (Fig. 18-6). By convention, this difference in stimulus intensity at the 0.25 and 0.75 point is divided by two and then measures the so-called just noticeable difference (jnd): in the example in Fig. 10-6 the jnd is 0.9 stimulus units. Obviously good discrimination capacity means a small jnd and vice versa.

The discrimination function depicted in Fig. 18-6 is designated the "psychometric function." It is informative with respect to the nature of stimulus discrimination in the neighborhood of the standard stimulus chosen for this particular series of measurements. One can now ask how this function varies with changes in the magnitude of the standard stimulus? This Weber determined, and he came to the empirical conclusion that the jnd varies with S in the form $\Delta S = K \cdot S$, where $K > 0$ is called the Weber fraction. This relation is known as Weber's law. It has become customary to plot such data as $\Delta S/S$ versus S, in which case Weber's law describes a line parallel to the abscissa.

Later investigators extended such discrimination measurements over a wider range of stimulus intensities than were applied by Weber. Their results revealed a considerable departure from the linear relation of Weber's law for most sense modalities tested, notably for low intensities of the standard stimulus (Fig. 18-7). Following a recommendation of Luce and Edwards,[6] it is now general usage to designate the empirical relation $\Delta S/S =$ f(S) as the Weber function, of which Weber's law would be the special case of linearity over the entire range of S.

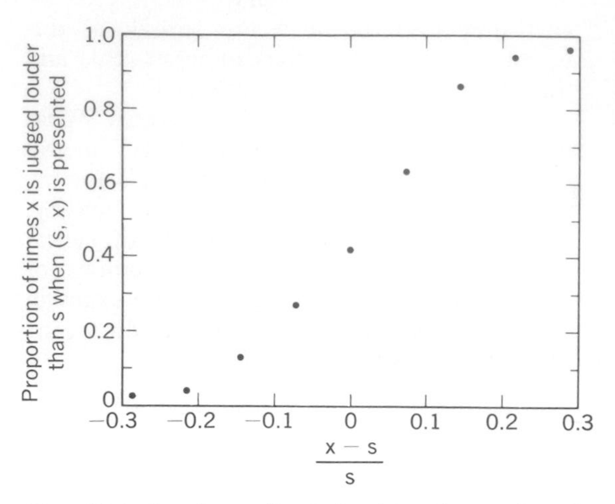

Fig. 18-4. Loudness discrimination when pairs of stimuli, x, s, are presented: s = standard stimulus (50 db or 0.0002 dynes/cm^2); x = comparison stimulus. Each data point is based on 105 observations collected over three sessions. Note that abscissa is plotted on linear dimension scale with $x - s/s$. (Unpublished data by Galanter; from Luce et al.[6a])

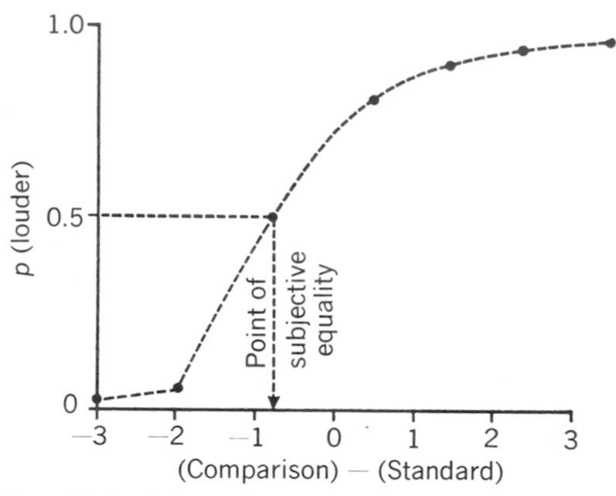

Fig. 18-5. Schematic representation of series of measurements of stimulus discrimination obtained with method of comparative unidirectional judgment. Ordinate plots proportion of times comparison stimulus is judged more intense as probability; abscissa plots difference between standard and comparison stimulus intensity. (From Galanter.[3])

Weber functions for tactile discrimination depart in monkeys as they do in man from the linearity of Weber's law.[35] This may now be used for comparison with observations in neurophysiologic studies. However, while the criterion for the stimulus discrimination in psychophysical measurements is the just noticeable sensation difference, what might constitute a comparable, liminal increment in

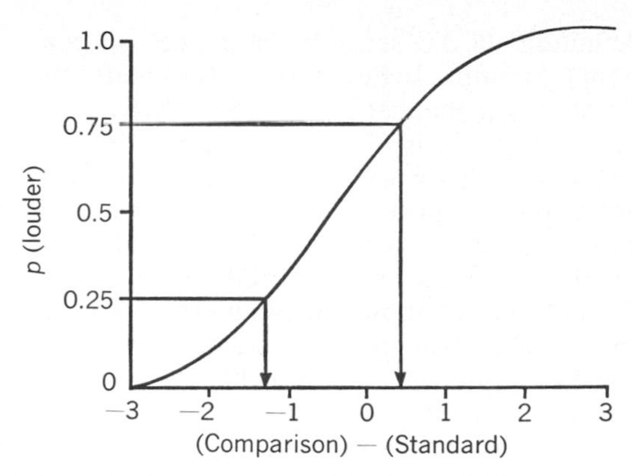

Fig. 18-6. Illustration of definition of "just noticeable difference" (jnd). See text for discussion. (From Galanter.[3])

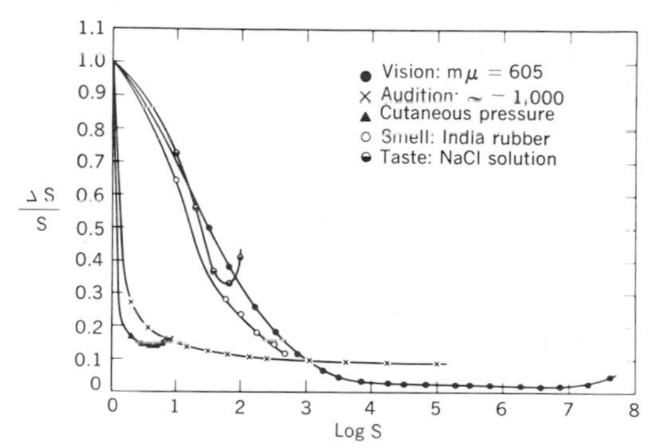

Fig. 18-7. Plots of Weber functions obtained for five different sense modalities: on ordinate, Weber fractions; on abscissa, intensity of base stimulus in logarithmic units. Weber fractions are normalized to unity at threshold. Note that if Weber's law were valid, individual graphs would assume form of straight lines parallel to abscissa. (Based on data by Holway and Pratt; from Luce et al.[6a])

neural activity is not at all certain. Therefore the objective of such comparisons between psychophysical and neurophysiologic discrimination studies is to find, essentially by trial and error, units and a scale for measuring neural activity that would bring the Weber function of neural responses in a certain modality into correspondence with that of its psychophysical Weber function. This approach was applied to neural activity evoked by tactile stimuli in first-order cutaneous afferents[45]: it was found that correspondence between neural and psychophysical Weber functions could be attained if the critical neural response increment (ΔR) for "recognition" of a stimulus increment was assumed to be a constant number of impulses, equal at any response level for any standard stimulus. This assumption about the nature

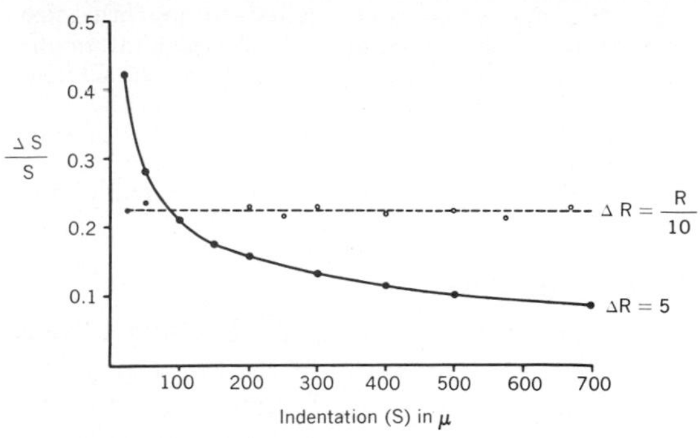

Fig. 18-8. Graphs of two Weber functions, ΔS/S = f(S), computed from stimulus-response curve of first-order cutaneous afferent nerve fiber responsive to mechanoreceptor stimulation. Stimulus-response curve for skin indentations of 1 sec duration was $R = 7.004 \times S^{.444}$ (R = number of nerve discharges in 1 sec; S = skin indentation in microns). Solid points connected by curve were obtained by assuming that least discriminable response increment, ΔR, is 5 impulses/response at any intensity of standard stimulus, S. Open circles connected by dashed line were obtained by assuming that ΔR is constant fraction (i.e., 1/10) at response at any intensity of standard stimulus. Note that only curve ΔR = 5 is of the same form as psychophysically determined Weber functions. (From Werner and Mountcastle.[45])

of the discriminable ΔR is parsimonious, and it is conceivable that other, more complex assumptions concerning ΔR could lead to the same correspondence. On the other hand, the assumption that a constant fractional increment ($\Delta R/R = C$) in neural activity is discernible over the entire stimulus range does *not* lead to a correspondence of neural and psychophysical Weber functions (Fig. 18-8). This supports the suggestion that the primate nervous system may measure equal increments of neural activity, rather than equal fractional increases, in order to generate the type of Weber functions established in psychophysical experiments.

Although concerned with an extensive rather than an intensive stimulus property, size and distance are aspects of the discrimination problem. Since Weber's work on cutaneous sensibility, it has been customary to use the minimal distance at which two points touching the skin can be recognized as separate stimuli as a measure of sensitivity for spatial stimulus attributes. This sensitivity differs in a characteristic manner at different body sites: e.g., it is maximal at the apices of the extremities and diminishes with a proximodistal gradient toward the trunk.[43] The comparison of two-point discrimination thresholds with the capability to recognize size differences of disks gently placed on the skin revealed a far superior performance in the latter test. Thus the cutaneous sense appears much more suited to perform size than distance discriminations.[41] The general implication is that the choice of the test to characterize the capability of a sensory system is critical: in general, the task should match the role this sensory system plays in the life of the species under natural conditions, for these conditions shaped the design of the nervous system in evolution.

Scaling problem

As outlined on p. 552, Fechner's idea was this: for many of the primary physical stimulus attributes, e.g., loudness, brightness, weight, etc., there would exist corresponding dimensions of sensory experience; changes in the magnitude of the one would be related to changes in the magnitude of the other, in the form of a single-valued, monotonic, and everywhere differentiable (smooth) function. This requires a distinction between two kinds of jnd's: one that measures the just discriminable increment of the physical stimulus magnitude as defined earlier by Weber and a second, distinctly different, that is measured in the units of the appropriate sensation continuum. Fechner defined the latter jnd (the sensation jnd) as the unit on the sensation scale and assumed all sensation jnd's to be subjectively equal over the entire sensation continuum. The implication of this is shown in Fig. 18-9: two stimuli separated by one (stimulus) jnd at the low end of the intensity scale give rise to the same subjective difference in sensation intensity as two stimuli at the upper end of the intensity scale that are also one stimulus jnd apart. Accepting the validity of Weber's law ($\Delta S/S = K$), this

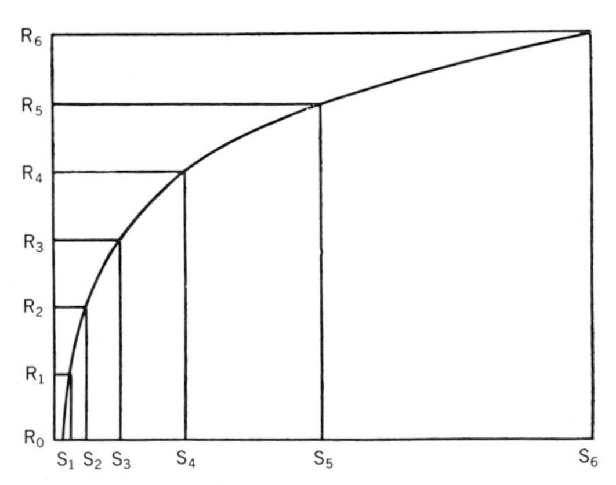

Fig. 18-9. Schematic representation of psychophysical scale proposed by Fechner (see text). Equidistant segments on ordinate are sensation jnd's; abscissa plots stimulus magnitude and indicates stimulus increments (stimulus jnd's) that correspond to different sensation jnd's at different points along sensation continuum.

definition of the sensation jnd (ΔR = constant) permitted Fechner to formulate the simple mathematical relationship $\Delta S/S = A \cdot \Delta R$, where A is, by the definition of the magnitudes involved, a constant. Next Fechner applied a principle often used in mathematical physics: what is a valid relation for small differences remains valid in the limit as the differences approach infinitesimally small values. Accordingly, the previous equation can be written in the form $dS/S = A \cdot dR$, which on integration yields Fechner's law in the form $R = A \cdot \log S/S_0$; S_0 is a constant of integration, to be interpreted as the threshold stimulus strength below which there would be no corresponding sensation elicted. This seemed to accomplish Fechner's objective of bringing physical stimulus intensity and sensation magnitude into a unique mathematical relationship of the same general form for all sensation continuums.

For more than 50 years, Fechner's law remained almost entirely unrivaled and his assumptions unchallenged. Finally, criticisms arose. First, evidence accumulated to indicate that Weber's law is not in all cases entirely valid (Fig. 18-7). This leads to an internal logical inconsistency in the derivation of Fechner's law, for Luce and Edwards[6] proved that, except for the special case of Weber's law, the assumption of the equality of the sensation jnd's and the transition, in the limit, to the differential equation are mutally contradictory. A second source of dissatisfaction relates to Fechner's "indirect" method of scaling, in which the unit of measurement is derived from the discriminatory capacity and its statistical fluctuations (p. 563).

To circumvent this latter objection, two alternative methods for psychophysical scaling were proposed and are presently in use. These are, first, the scaling procedures that require the observer to partition a given segment of a certain sensation continuum into a predetermined number of subjectively equal intervals and, second, those procedures that require the observer to make a direct judgment of apparent sensation magnitudes. In the former case the general protocol of the experiment is as follows: the subject is told the designation of a set of categories, e.g., the first 11 integers; he is then exposed to the weakest and the strongest stimuli that delineate the segment of the stimulus range that will be used in the final test, and he is told to place the weakest stimulus in category 1 and the strongest in category 11. The task is now to distribute the subsequently presented stimuli among the 11 categories in such a manner that the sensation intervals between categories are perceived as subjectively equal. The simplest analysis of the final data consists in calculating the mean category assignment for the different stimulus intensities used in the test; these mean values are the sensation scale values of the respective stimulus intensities. In Fig. 18-10 the ordinate is

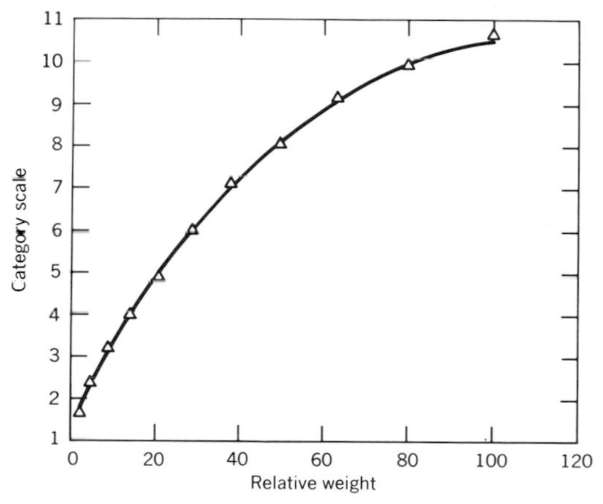

Fig. 18-10. Category scale for apparent weight. Lifted weights cover range from 43 to 292 gm. Weights were logarithmically placed within that range. (Modified from data by Stevens and Galanter.[38])

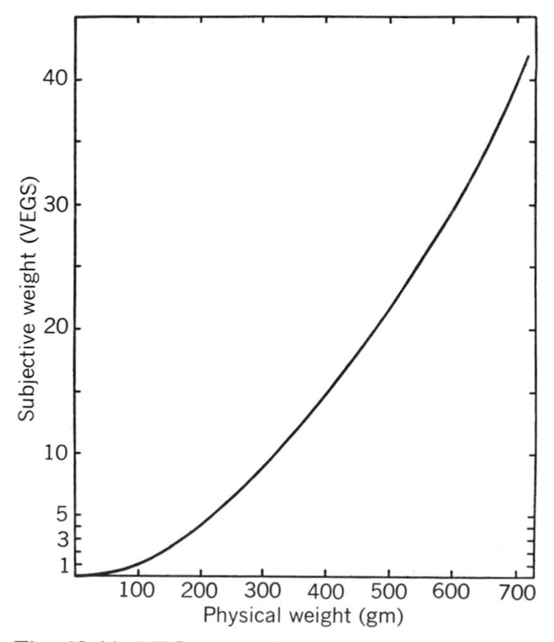

Fig. 18-11. VEG scale of subjective weight. Curve shows how subjective weight varies as function of physical weight. One VEG is weight experienced by lifting 100 gm. (From Harper and Stevens.[19])

marked off in 11 equal intervals that correspond to the subjectively equidistant sensation categories. The abscissa plots the relative intensity of the stimuli used (in this case, weights). The category scale for apparent weight results directly from this.

The second group of psychophysical scaling methods, aimed at a direct estimation of the apparent sensation magnitude, is based on the following general protocol: a certain stimulus is presented, and the subject is asked to adjust a variable stimulus to a value that is subjectively either half or twice as large. Using weight as an example, we define as a unit of subjective weight the apparent weight perceived when 100 gm is lifted; Harper and Stevens[19] have designated this unit of subjective weight as 1 VEG. The subject now determines in a series of trials that a weight of, say, 72 gm is subjectively half as heavy as the standard of 100 gm. Therefore a subjective scale value of 0.5 is assigned to correspond to 72 gm physical weight. Next, we may find that 140 gm is judged twice as heavy as 100 gm: accordingly, a subjective scale value of 2 VEG is assigned to correspond to the physical weight of 140 gm. The systematic continuation of this procedure leads to the scale shown in Fig. 18-11. In this case, fractionation of sensation magnitude elicited by a standard stimulus were the operations that led to the scale construction.

Of course, subjects do not invariably assign the same number (or category) to a particular stimulus. The variation in the values assigned to the same stimulus in repeated trials may be considerable: the standard deviation of the estimated magnitude may be as much as 20 to 40% of the mean value. Most of the data published in the psychophysical literature are based on "averages" obtained in trials with different subjects and are intended to represent a "typical" scale.

The comparison of the psychophysical scales obtained with the three different methods discussed so far (i.e., the fechnerian method, category estimation, and magnitude estimation) discloses some important general relations. First, the resulting scales often do not correspond with one another. This is shown in Fig. 18-12; the departure of the curves for magnitude and category scales from the psychometric function depicting the fechnerian scale is indicative of the failure of the former two scales to fulfill Fechner's assumption of the equality of the sensation jnds. On the other hand, there are certain sensation continua for which scales obtained by the three different scaling procedures are indeed

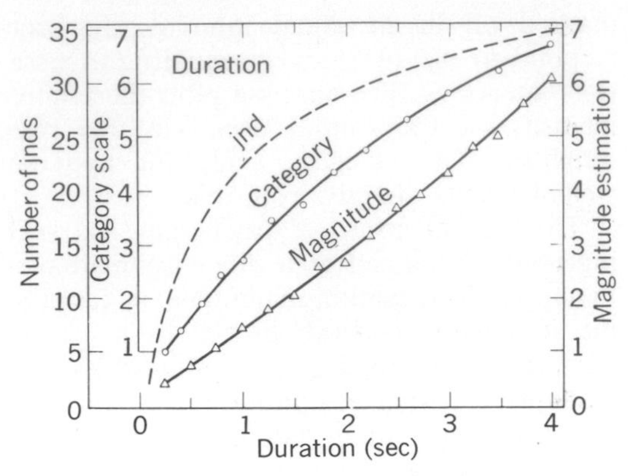

Fig. 18-12. Three kinds of psychological measures of apparent duration. Triangles represent mean magnitude estimations by 12 observers who judged apparent durations of white noises. Circles represent mean category judgments by 16 observers on scale from 1 to 7. Two end stimuli (0.25 and 4 sec) were presented at outset to indicate range, and each observer twice judged each duration on 7-point scale. Dashed curve is discriminability scale obtained by counting off jnd's. (Modified from Stevens.[10])

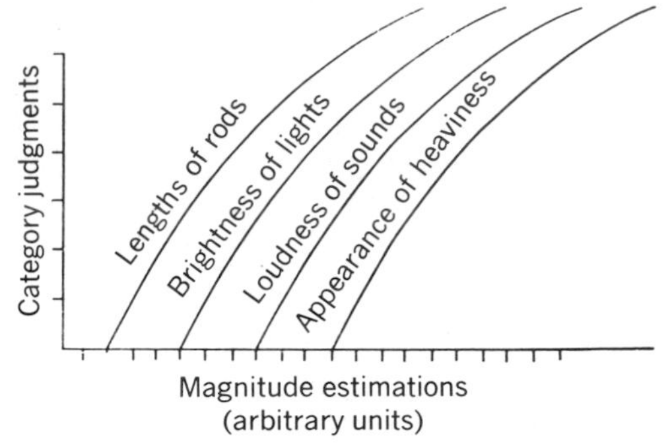

Fig. 18-13. Relation between magnitude and category judgments. (From Galanter.[3])

linearly related. Stevens and Galanter[38] have shown that sense continua can be divided into two classes, depending on whether or not the relation between magnitude and category scales is linear. They have designated as class I (or prothetic) continua those for which the relation of the two scales always deviates from linearity in a manner illustrated in Fig. 18-13. These examples, depicted schematically, indicate (and other instances corroborate) that this relation applies to the intensity dimension in sensation. It has been suggested that it is characteristic for those sensation attributes for which increase in stimulus intensity is accompanied by some neural additive process of excitation. Sense continua of the second type (class II, or metathetic continua), for which magnitude

Table 18-1. Representative exponents of power functions relating psychophysical magnitude to stimulus magnitude on prothetic continua*

Continuum	Exponent	Stimulus conditions
Loudness	0.60	Binaural
Loudness	0.54	Monaural
Brightness	0.33	5° target—dark-adapted eye
Brightness	0.50	Point source—dark-adapted eye
Lightness	1.20	Reflectance of gray papers
Smell	0.55	Coffee odor
Smell	0.60	Heptane
Taste	0.80	Saccharine
Taste	1.30	Sucrose
Taste	1.30	Salt
Temperature	1.00	Cold—on arm
Temperature	1.60	Warmth—on arm
Vibration	0.95	60 Hz—on finger
Vibration	0.60	250 Hz—on finger
Duration	1.10	White-noise stimulus
Repetition rate	1.00	Light, sound, touch, and shocks
Finger span	1.30	Thickness of wood blocks
Pressure on palm	1.10	Static force on skin
Heaviness	1.45	Lifted weights
Force of hand-grip	1.70	Precision hand dynamometer
Autophonic level	1.10	Sound pressure of vocalization
Electric shock	3.50	60 Hz, through fingers

*From Stevens.[37a]

and category scales may be linearly related, include pitch, position, and others for which there is reason to believe that the stimulus-evoked neural activity shifts from one locale to another, as the stimulus parameter in question varies over a series of values.

The second generally valid conclusion that has emerged from the studies on psychophysical scaling concerns a surprising uniformity of the scales obtained with the magnitude estimation procedure for a large number of different sense continua. In the cases listed in Table 18-1, and in others as well, Stevens[10] and co-workers have shown that the sensation magnitude relates to the physical stimulus magnitude in the form of a power function $R = K \cdot S^n$, with the exponent n being a numerical parameter, characteristic for the individual sense continua.

It is important to note that the general validity of this "power law" depends on the appropriate selection of the physical stimulus scale. It was pointed out earlier that magnitude estimation techniques involve essentially a fractionation of sense

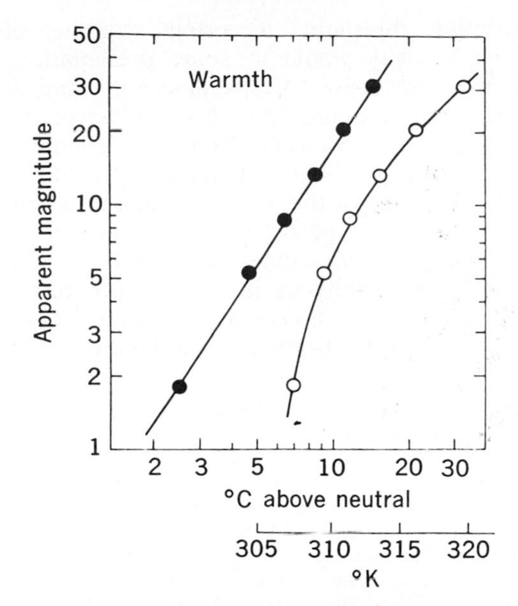

Fig. 18-14. Magnitude estimation of apparent warmth. Each point is geometric mean of 36 estimates (12 observers). Upper abscissa, for filled points, is log scale of difference in temperature (Celsius) between stimulus and "physiologic zero." Lower abscissa, for unfilled points, is log scale of absolute temperature (Kelvin). (From Stevens and Stevens.[36])

magnitudes; therefore these procedures lead to a numerical scale for sensation that is a "ratio scale," i.e., a scale that reflects quantitative relations of the measured property faithfully only if no algebraic operations other than multiplication or division are performed on the measured values. The addition or subtraction of a constant leads to loss of information. Therefore the origin of a ratio scale (its zero value) must be an invariably fixed reference point, like a constant of nature. In psychophysical measurements the threshold of a stimulus subserves this role of a fixed reference point; it is then possible to measure stimuli in terms of a ratio scale of *distance* from threshold. This becomes particularly important at low values of sensation magnitude. As shown in Fig. 18-14, when apparent temperature is scaled by magnitude estimation and the results are plotted against the absolute temperature scale (with both scales in logarithmic units), the data fall on a concave curve, but when plotted in terms of degrees above threshold intensity for warmth sensation, the data plot on a straight line. Thus the data fit a power function of the general form:

$$R = k(S - S_o)^n$$

where S_o is the threshold stimulus intensity.[36]

The third important consideration pertains to the influence that the selection of the stimulus values and the temporal sequence in which the stimuli of different intensities are applied has on the sensation estimate of the subjects. While category scales of one and the same subject vary with these parameters, it is a notable property of magnitude scales that they are invariant with respect to changes in the stimulus ensemble. As a result, it has been suggested that magnitude scales reflect the events on which sensory judgments are based more directly than psychophysical measurements involving category scales.

Scaling of stimulus and response in quantitative neurophysiologic studies. Mountcastle et al.[28] have conducted an investigation in which the discharges of individual neurons responding to steady position and movement of joints were recorded in the ventrobasal thalamic nuclei of macaques. The experimental records permitted evaluation of the quantitative relation between the angular position of a joint and the mean discharge rate of a thalamic neuron that was responsive to differences in position of that same joint.

For an analysis of the precise functional relation between these variables, the principle of converting measurements to ratio scales was applied; i.e., the numerical values of the variables were referenced to their natural zero values as scale origin. This required expressing the angular position as increment over the joint angle that is the threshold position for the neuron under study (Θ_T) and plotting as response the increment of mean discharge rate over the spontaneously ongoing activity (c) that persists while the joint is outside the excitatory angle of the neuron under study. The straight-line relation between the variables displayed in Fig. 18-15 indicates that the relation between joint position and mean neuronal discharge rate, after appropriate scaling of the variables, can be described by a power function.

The essential point of this study is the demonstration of a lawful, quantitative relation between the appropriately scaled stimulus intensity (provided, in this case, by the angular position of joints) and the mean rate of thalamic cell discharges; a stimulus-response relation that is of the same general form as that which describes the functional dependence of the subjective sensation magnitude in man on the intensity of stimuli in a large number of different sensation continua.

In this study the neural activity was measured in the third-order sensory relay station of the medial lemniscal pathway, i.e., at a level of the somatosensory system at which extensive convergence of neural activity from peripheral receptors of the joint under study has occurred. The neural activity elicited in first-order cutaneous nerve fibers by graded indentation of slowly adapting cutaneous touch corpuscles was the subject of a similar

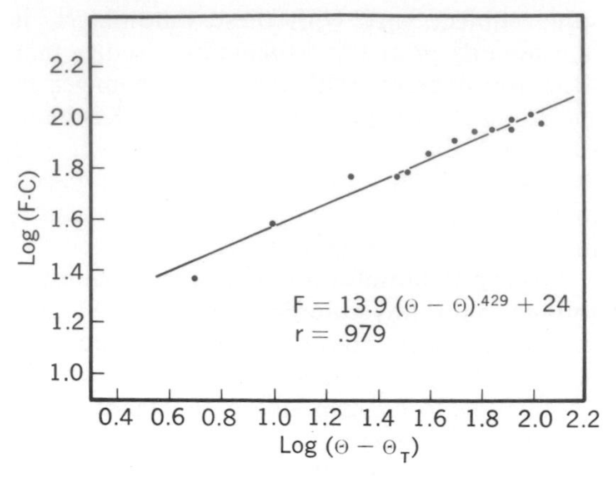

Fig. 18-15. Plot of mean rate of discharge (impulses per second) versus angle for ventrobasal thalamic neuron, driven by extension of contralateral knee joint. Abscissa marks in logarithmic units departure of angular position of joint, Θ, from threshold position for excitation of neuron, Θ_T. Ordinate plots on logarithmic scale the increment of discharge rate above spontaneous activity, *C*. Thus both dependent and independent variables are expressed in reference to absolute zero point (Θ and *C*, respectively) and are ratio scales. Equation of functional relation between dependent and independent variables is written as inset (value of 24 represents spontaneous activity in impulses per second; *r* = Pearson). (From Mountcastle et al.[28])

experimental study[45]: in this case, too, the stimulus-response relation is best described by a power function, as in the human observer's estimate of stimulus magnitudes.[24]

One implication becomes immediately apparent: the serially superposed neural transforms leading from the site of neural activity recorded in these studies to the final behavioral (verbal) response must also be power functions, at least as regards the intensity attribute.

Other more qualitative arguments can be adduced to strengthen this contention. One is that the sensation of passive movements of joints in man depends only on the function of the receptors in the joint capsule; muscle and tendon receptors are not involved.[32] The other argument is provided by the finding of Hensel and Bowman[20] that a single impulse in a single or in each of a very few cutaneous afferent fibers is sufficient to evoke a conscious sensation in man.

Detection problem

Historically the detection problem developed out of the question of the absolute threshold in sensation. Fechner's concepts and methods incorporated the idea of a lower limit of sensitivity, thought to be characteristic of sensory systems. He and his successors were aware of the inherent instability of this sensation threshold: a certain stimulus of fixed intensity would at some presentations elicit the response "Yes, I detect it," and at others the response "No, I don't detect it." It was also recognized from the beginning that the observer's attitude affected this threshold estimate. In this form the idea of a sensory threshold persisted as an essentially fictitious and sometimes controversial concept in psychophysics until the more recent availability of methods that permitted the isolation and quantification of the factors determining the limits of stimulus detection at any one time and their dependence on certain experimental variables.

One important step in this development was a change in the experimental design: instead of determining the threshold by varying the stimulus, the stimulus was kept at a fixed value and the question became how often the presentation of this stimulus elicited the correct response of detection, and how often the stimulus was reported to have occurred although it was actually not applied ("false alarm"). Accordingly, two kinds of errors can be made in this situation: failures to detect a signal when it is there, and reports of the perception of a signal when it is not there. Detection is concerned with the factors that influence the relation between these two types of errors, and their implications for sensory mechanisms, in general. With this, the emphasis shifted to the problem of specifying and controlling the criteria the observer may use in making perceptual judgments.

The typical experimental situation is this[11]: sensory events occurring in a fixed interval of time are observed by the subject, who must then decide whether the observation interval contained only background interferences or a signal (stimulus) as well. The background interference is thought to be random; it may either be introduced by the experimenter or be inherent in the sensory process. It is commonly designated as noise (N). In the alternative case the sensory event is judged to consist of noise on which a signal is superimposed (SN); e.g., the signal may be a flash of light in a known location of a uniformly illuminated background or a brief sound. The trials are repeated a certain number of times, under identical conditions, with observation periods *with* and *without* signals randomly intermingled. The outcome of a series of trials is expressed as the probabilities of "false alarm" and "correct detection."

One of the experimental variables found to affect the subject's performance under such conditions was prior information regarding the probability of signal presentation in the observation periods, as shown schematically

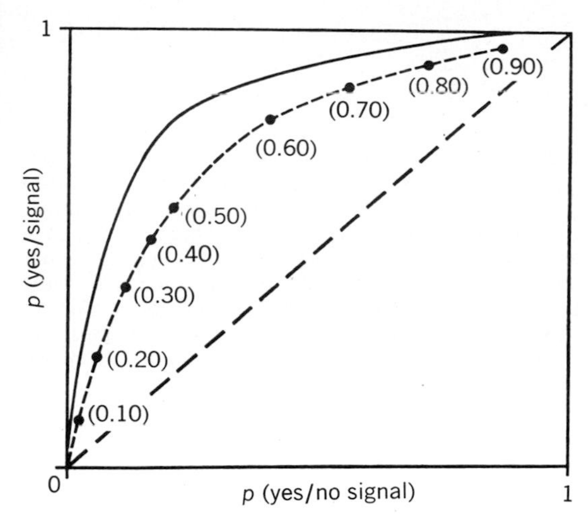

Fig. 18-16. Likelihood of correct signal detection —p(yes/signal)—and of false alarm—p(yes/no signal)—in relation to probability of signal presentation and signal intensity. Dashed curve with dots represents this relation for signal of medium intensity; numbers in parentheses are respective probabilities of signal presentation. Solid curve depicts this relation for signal of higher intensity. (From Galanter.[3])

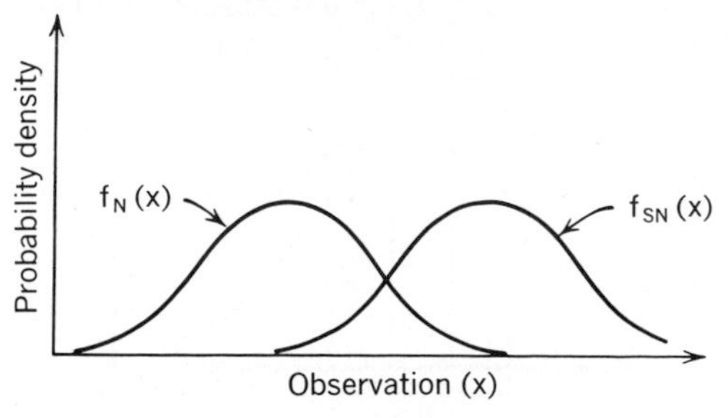

Fig. 18-17. Probability density functions of signal and signal plus noise (see text). (From Swets et al.[11])

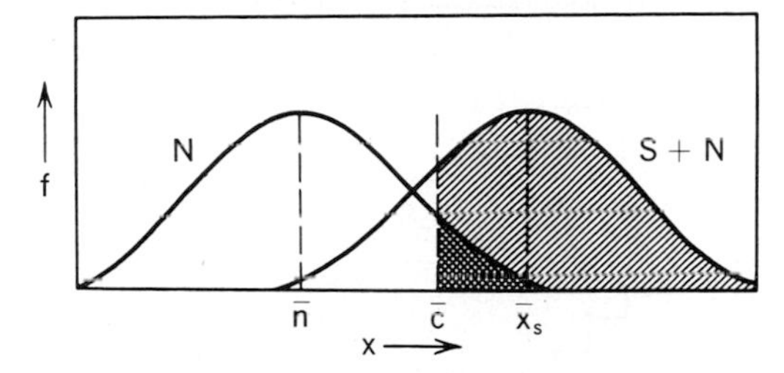

Fig. 18-18. Probability density of neural activity: x, for noise alone and for noise plus signal; c is criterion. (From Eijkman and Vendrik.[16a])

in Fig. 18-16. When the subject is told that the signal will be present in a large percentage of the observation periods, the false alarm rate is very high, approaching almost 100%. On the other hand, with prior information as to a low relative rate of signal presentation, the subject often fails to detect the signal. Similarly, it is possible to modify the subject's performance in accord with rewards offered for correct recognition and costs incurred with false alarms.

Although it is not necessary for the quantitative evaluation of the psychophysical data, it may be helpful to think of the sensory events that form the basis for the perceptual judgment as some measure of neural activity, perhaps the number of discharges in a specific period of time in a certain locale in the CNS. This enables one to visualize in hypothetical form the probability that a certain sensory event of magnitude (x) will occur when noise alone is present, $f_N(x)$, and the probability of occurrence of the same sensory event when a signal is also present, $f_{SN}(x)$ (Fig. 18-17).

With this general model in mind, one may formulate the factors governing signal detection in terms of a decision between two alternative hypotheses: the subject must decide whether a certain observation (x) is more likely to be a member of one of two partially overlapping probability density functions. This can be done by setting a criterion cutoff value for x (=c) such that for any $x > c$ the observation will be placed in the category SN, and for any $x < c$, in the category N (Fig. 18-18). Depending on the position of c along the x axis, the ratio of errors in correct detection and false alarms will vary: a moment's reflection will make it clear that the area to the left of c under the distribution curve SN represents the likelihood of stimulus presentations that the observer labels "noise," while the area to the right of c under N (crosshatched in Fig. 18-18) represents the likelihood of false alarm. Accordingly, the shaded area under SN is the likelihood of correct decision. Thus the fundamental quantities in the evaluation of the performance of the subject are the two areas under the distribution curves in Fig. 18-18 that are marked by shading and cross-hatching. These areas under the respective probability density curves of the hypothetical model correspond to the experimentally measured correct decision and false alarm probabilities. The hypothetical criterion value enables one to visualize one means by which the actual perceptual judgment could be under the control of a variety of factors, e.g., anticipation and motivation. In this form the hypothesis of signal detection is accessible to experimental neurophysiologic tests.

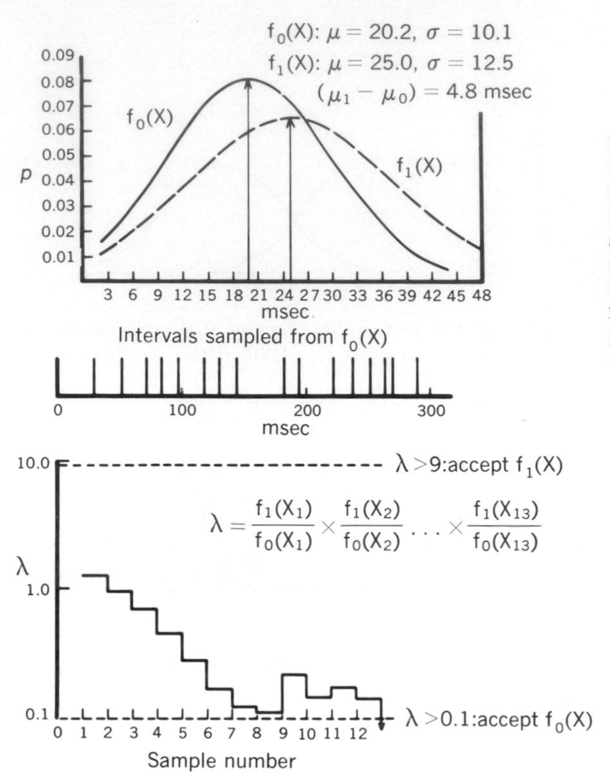

Fig. 18-19. Illustration of use of sequential probability ratio test (see text). Above are two probability density functions between which discrimination is sought, with actual mean value, μ, for each as observed in study of thalamic neuron under slightly different intensities of sensory drive. Next is sample of sequence of intervals belonging to $f_0(X)$. Graph below plots successive changes in probability ratio, λ, determined as shown by inset. (From Mountcastle et al.[21])

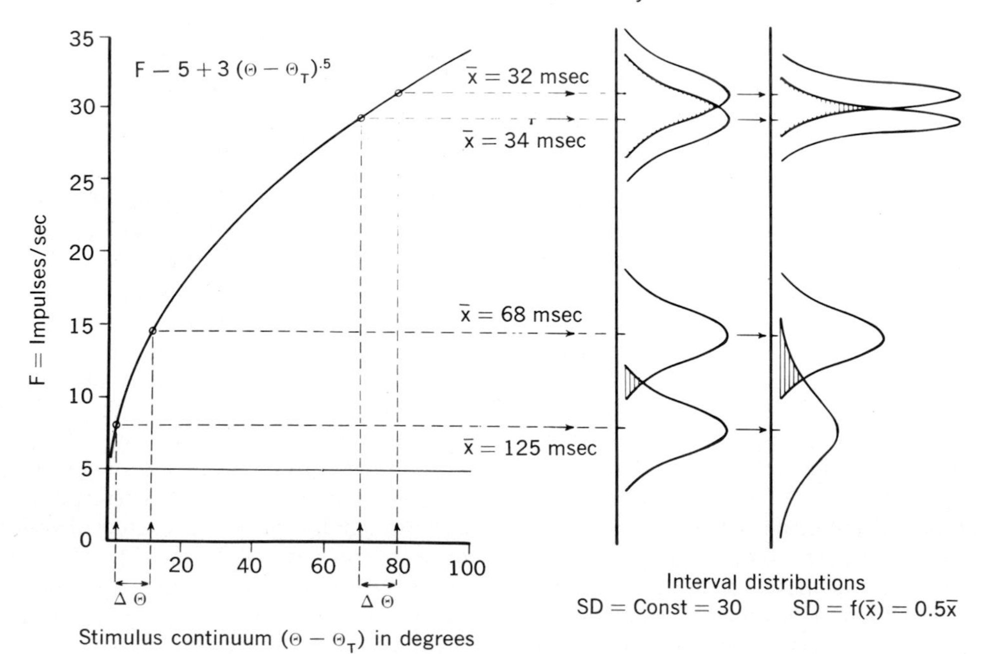

Fig. 18-20. Schematic presentation of manner in which constancy of relative variability (co efficient of variation) of discharge intervals during steady-state activity of individual neurons could affect statistical decision between different intensities of sensory drive (see text). (From Werner and Mountcastle.[44])

Mountcastle et al.[27] and Werner and Mountcastle[44] designed a statistical estimator for the discrimination between two trains of neuron discharges evoked, for example, in thalamic joint neurons by two slightly different positions of the relevant joint within the excitatory angle for that neuron. The statistical method chosen was the sequential probability ratio test; its application to the specific problem is illustrated in Fig. 18-19. Beginning with the first discharge interval of the impulse train to be tested, a ratio is established between the probabilities with which that interval occurs in each of the parent populations between which a decision is to be attempted. The ratios for successive intervals are similarly established, and these ratios are successively multiplied. A decision is reached when the consecutive product thus formed reaches either of two levels that are preset by the degree of statistical validity desired; one level indicates that the impulse train examined should be considered as belonging to the interval distribution $f_0(x)$ (and the respective joint position) and vice versa. The difference in joint positions that could be detected from comparisons of the respective neuronal impulse trains with this test were of the same order as can be discriminated in psychophysical experiments in man.

The potential relevance of statistical tests of this kind for sensory functions of the CNS is perhaps in part related to the startling phenomenon that the relative variability of discharge intervals in impulse trains of central sensory neurons is practically constant over a large range of mean discharge rates. The possible implication of this for the discrimination of neural activity evoked by sensory stimuli of different intensities is schematically illustrated in Fig. 18-20; on the left, a curve plots the relation between the mean discharge rate of a thalamic joint neuron and the intensity of the exciting stimulus, the joint position (this is the power law relation described earlier). From this, we read off the mean discharge interval for the neural activity at two pairs of positions along the continuum: the members of each pair are separated by equal increments in stimulus intensity, a $\Delta\Theta$ of 10 degrees. In the first case, illustrated by the next column to the right, we calculate the interval distributions about each of the four means, assuming normality and in this case a constant value of the standard deviation (= 30 msec), for each of the four distributions. The overlap of the distributions reflects pictorially what was also confirmed by the application of the sequential probability ratio test, that position discrimination would, under these circumstances, become increasingly difficult as the test increment is moved to the right along the stimulus continuum. This is due to the negatively accelerating nature of the power law relation between mean response frequency and joint angle. The task of position discrimination is quite different, however, if the constant relation between mean and standard deviation, observed experimentally, is taken into account. This situation is illustrated to the right in Fig. 18-20 for the case of a constant coefficient of variation (CV) of 50%. The relatively much smaller overlap of the pair of distributions at the upper end of the stimulus response curve is apparent and preserves discriminability. Thus the diminution of the spread of the interval distribution with increasing mean frequency offsets, at least to some extent, the negatively accelerating nature of the stimulus response relation.[44]

Recognition problem

When the task is one of identifying a particular stimulus with a single name or number, perceptual judgments are probably made by comparing the presented stimulus with some subjective standard. To set this clearly aside from discriminatory judgments of whether one stimulus is greater than, less than, or the same as another stimulus, the former type of judgment is often designated as "stimulus rating" or "absolute judgment." Stevens has estimated that the young human listener can distinguish 350,000 different tones when they are presented in pairs for *discrimination.* But if tones of the same frequency range (100 to 8,000 Hz) are presented for *individual* identification, there are no more than five to seven different pitches that the untrained individual never confuses.[31] Garner[4] and Miller[7] have shown that the precise number of stimulus categories that can be distinguished by absolute judgment can be calculated from experimental data by applying concepts and algorithms of information theory; thus the recognition problem relates directly to some quantitative considerations of information transmission in sensory systems.

The relation between stimuli and the subject attempting identification of these same stimuli is thought to be analogous to that between a source emitting messages and a receiver of these messages. Information exists in a message only if there is a priori some uncertainty about what the message will be; the *amount* of information transmitted is equal to the amount by which the uncertainty in the receiver is reduced by the message received. Uncertainty, like information, is measured in *bits,* where 1 bit is the uncertainty involved in an event with two possible and a priori equally likely outcomes. The measure that satisfies the condition that uncertainties of independent events are additive is logarithmic, and it has become common practice to use logarithms to the base 2 for this measure.

For the actual design and evaluation of experiments, consider a certain number of stimuli chosen from a set of different, discrete stimulus intensities. Depending on the number of different intensities in the set and on the relative frequency, p(S), with which stimuli of each intensity occur in the group, there exists some uncertainty that any one given stimulus of the group is of a certain specific magnitude. This stimulus uncertainty can be computed as:

$$H(S) = -\sum p(S) \cdot \log_2 p(S)$$

The absolute value of these expressions is plotted on the abscissae of the graphs presented later in this section. For a given set of stimuli, defined by its un-

Table 18-2. Matrix of probabilities of joint occurrences of fixed value of R (R_j) and fixed value of S (S_i)*†

Stimulus categories (S_i)	Response categories (R_j) 1	2	3	4	Marginal sums
1	0.08	0.02	0	0	0.10
2	0.17	0.20	0.03	0	0.40
3	0	0.03	0.18	0.09	0.30
4	0	0	0.04	0.16	0.20
Marginal sums	0.25	0.25	0.25	0.25	

*From Garner.[4]
†Internal entries are joint probabilities p(SiRj). Marginal sums give values of p(R) and p(S), respectively.

certainty value, the experimental data can be displayed in the form of a matrix with the joint occurrences of a certain stimulus and a certain response as entries. Table 18-2 depicts the absolute number of paired occurrences expressed as probabilities.

The matrix of Table 18-2 represents a bivariate distribution of the relative number of instances (normalized to probabilities) at which the stimulus S_i was associated with the response R_j. A "joint uncertainty" may be calculated for this bivariate distribution in a manner analogous to that used for the computation of the stimulus uncertainty:

$$H(S,R) = -\sum P(S,R) \cdot \log P(S,R)$$

This result can then be compared with the uncertainty in a reference matrix containing as entries those values that would result if stimuli and responses were not lawfully associated with one another but occurred in a strictly random association only. The entries to this latter reference matrix are calculated as the products of the respective marginal values of R and S, and its uncertainty is maximal (H_{max}). The difference by which the uncertainty value of the experimentally determined matrix is decreased from the theoretical maximal value determines the actual amount of information transmitted in the psychophysical experiment. This value is plotted on the ordinate of the graphs in Figs. 18-21 and 18-23.

The information transmitted in experiments involving absolute judgments varies in a characteristic fashion with the number of different stimulus categories used (and, accordingly, with the stimulus uncertainty). An example is shown in Fig. 18-21. In this particular case, information transmission equals the stimulus uncertainty up to a value of the latter of about 2.3 bits; at higher values of stimulus uncertainty (i.e., when a larger number of different stimulus categories are presented), information transmission reaches a plateau value of about 2.5 bits. The interpretation of this is that, regardless of the number of cate-

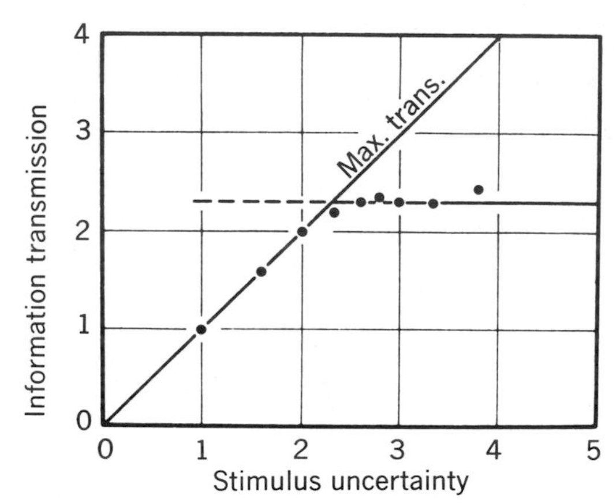

Fig. 18-21. Information transmission for absolute judgments of auditory pitch. (Based on data from Pollack[31]; from Garner.[4])

gories into which a given stimulus range is divided, the observer can single out no more than about five or six (i.e., the antilogarithm of 2.3 with the base 2) values for the stimulus that he will never confuse with one other.

Fig. 18-21 reflects the characteristic feature of any imperfect information transmitting system, namely, an upper limit of information transmission that is a characteristic magnitude of the system and code under study. This is sometimes also defined as channel capacity.

The startling result of psychophysical studies of this kind is that the maximum amount of information transmission in a large number of sense continua, notably intensive ones, is on the order of about 2.5. This is the basis for the statement that the number of stimuli that can be identified by absolute judgment is far below the number of discriminable stimulus pairs.

The procedure outlined for the informa-

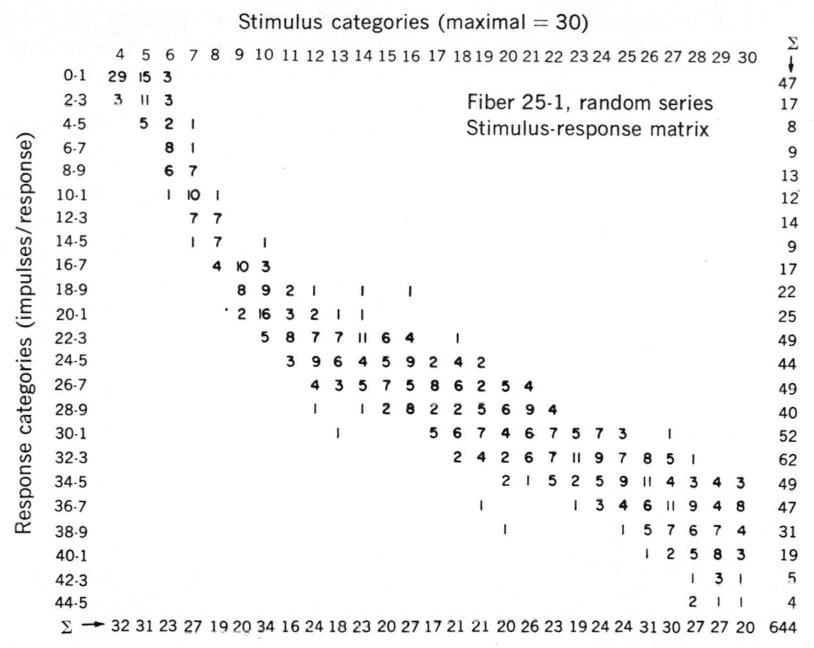

Stimulus categories (maximal = 30)

Fiber 25-1, random series
Stimulus-response matrix

Response categories (impulses/response)	4	5	6	7	8	9	10	11	12	13	14	15	16	17	18	19	20	21	22	23	24	25	26	27	28	29	30	Σ ↓
0-1	29	15	3																									47
2-3	3	11	3																									17
4-5		5	2	1																								8
6-7			8	1																								9
8-9			6	7																								13
10-1			1	10	1																							12
12-3				7	7																							14
14-5				1	7		1																					9
16-7					4	10	3																					17
18-9						8	9	2	1		1		1															22
20-1						2	16	3	2	1	1																	25
22-3							5	8	7	7	11	6	4		1													49
24-5								3	9	6	4	5	9	2	4	2												44
26-7									4	3	5	7	5	8	6	2	5	4										49
28-9									1		1	2	8	2	2	5	6	9	4									40
30-1										1				5	6	7	4	6	7	5	7	3		1				52
32-3															2	4	2	6	7	11	9	7	8	5	1			62
34-5																	2	1	5	2	5	9	11	4	3	4	3	49
36-7																1				1	3	4	6	11	9	4	8	47
38-9																	1					1	5	7	6	7	4	31
40-1																							1	2	5	8	3	19
42-3																									1	3	1	5
44-5																									2	1	1	4
Σ →	32	31	23	27	19	20	34	16	24	18	23	20	27	17	21	21	20	26	23	19	24	24	31	30	27	27	20	644

Fig. 18-22. Stimulus-response matrix for mechanoreceptive fiber activated by light mechanical stimuli delivered to its corpuscular ending. Afferent fiber was isolated for study by dissection of saphenous nerve in thigh. Maximal movement of stimulus probe was 690μ. This was divided into 30 equal steps. Threshold lay between steps 4 and 5 throughout study; stimulus duration was 500 msec. Stimuli were delivered once every 3 sec. Stimuli of different intensities were delivered in random order until 644 were given. Numbers of stimuli of different intensities are indicated by lower horizontal row of numbers. Responses to each stimulus are categorized by number of impulses in steps of two impulses. (From Werner and Mountcastle.[45])

tion evaluation of absolute psychophysical judgments can also be applied to a quantitative evaluation of stimulus-evoked neural activity. In this case, a measure of the stimulus-evoked neural response (e.g., the number of discharges evoked) forms the entries in the (S-R) matrix. Fig. 18-22 gives such a display, constructed from the activity recorded from a mechanoreceptive afferent nerve fiber activated by graded mechanical stimulation of its cutaneous receptive field. The numerical evaluation of data such as those in Fig. 18-22 indicated that the maximal information transmission as regards skin indentation is, on the average, 2.5 bits for slowly adapting touch corpuscles of the hairy skin and 3.0 bits for receptors of the glabrous skin (Fig. 18-23). These results correspond to those cited earlier for information transmission in psychophysical judgments. The implication is that the limiting link in information transmission across the skin, insofar as intensity is concerned, is at the level of the first-order fiber, and that certain central mechanisms preserve all, or nearly all, of the information they receive.[46]

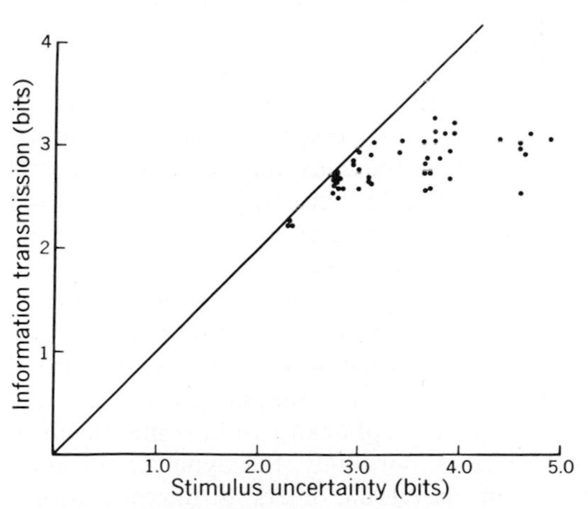

Fig. 18-23. Information transmission calculated from a series of stimulus-response data obtained for each of 17 mechanoreceptive fibers ending in palm of hand (macaque). Stimuli were applied at rate of 6 or 12/min; observation times were 500 msec or longer but not exceeding 1 sec. (From Werner and Mountcastle.[46])

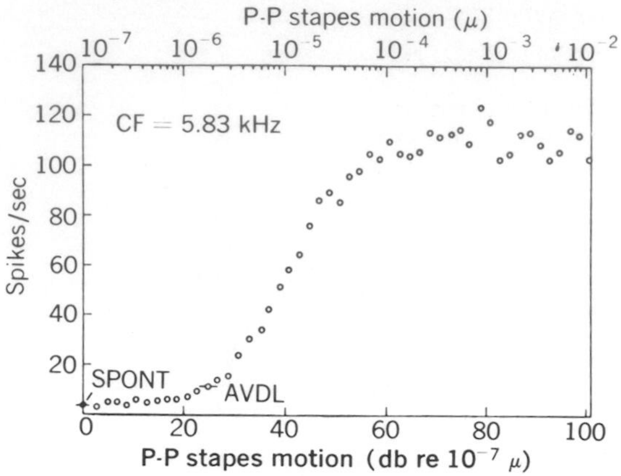

Fig. 18-24. Discharge rate of single auditory-nerve fiber as function of stimulus level. Stimulus was continuous tone at characteristic frequency of fiber (5.83 kHz). Horizontal scale is expressed in terms of stapes displacement. Each point on graph represents 1 min run. There were no rest periods between runs, which were made for a series of ascending levels in 2 db steps. Filled circle, *SPONT*, represents level of spontaneous discharge; point *AVDL* represents discharge rate at "threshold," defined as audiovisual detection level. Stimulus at *AVDL* corresponds to sound-pressure level of 2 db re 0.0002 dyne/cm². (From Kiang.[5])

Code for stimulus intensity

The examples chosen to illustrate instances of the neural representation of the intensity attribute of stimuli did not require more than measuring the mean rate of discharges or the average number of responses of individual neural elements in order to obtain the correspondence between neural and psychophysical stimulus-response relationships. Thus, depending on the test situation, mean discharge rate and average response numbers can be considered in these examples to be a neural code for stimulus intensity, at least at the levels of the afferent pathways, at which these measurements were taken.

A more complex situation appears to exist with the representation of loudness by discharges of auditory nerve fibers, as illustrated in Fig. 18-24. The discharge rate in these fibers reaches a plateau within some 40 db of threshold. Accordingly, the dynamic response range of individual fibers is narrow. Moreover, thresholds of different afferent fibers do not differ markedly. Yet psychophysical loudness estimates vary monotonically over a stimulus range of 100 db or more without attaining a saturation level. This discrepancy suggests that the neural code for loudness is complex and possibly related to the temporal spacing of discharges in a population of afferent fibers or affected, in some not yet understood manner, by the activity in the olivocochlear efferent system.[5]

TIME STRUCTURE OF DISCHARGES AS A NEURAL CODE

In Chapter 12, evidence was presented to indicate that the discharges of auditory nerve fibers and of neurons of the cochlear nuclear complex and the inferior colliculus may be time-locked to a segment of the cycle of a sinusoidal auditory stimulus. Accordingly, these neural elements reflect the frequency of a persisting tone by producing discharges at intervals grouped around the value of the tone period and its integral multiples, at least at tone frequencies up to the order of 2,000 to 5,000 Hz, depending on the animal species. At these low frequencies, phase-locking of discharges in auditory nerve fibers reflects the tone frequency unambiguously, for it is not at all affected by stimulus duration or strengths.[33]

This leads to the idea that the auditory system represents tone frequency by a period-time code, at least at low frequencies and at its input level from the periphery. In pursuit of this notion, Rose et al.[34] obtained some further evidence from the use of complex periodic sound stimuli.

Fig. 18-25 illustrates, in addition to period histograms and the appropriate fitted wave forms, distributions of interspike intervals for four stimulus conditions, as indicated. In Fig. 18-25, *A*, approximately 98% of all intervals are aa and bb intervals grouped around integral multiples of the period of the lower frequency tone. This indicates that the fiber transmitted information pertaining almost exclusively to the lower primary tone of 600 Hz. Increase of the loudness of tone 2 in Fig. 18-25, *B*, leads to the appearance of some ab and ba intervals, although the lower tone is still by far dominant. In data shown in Fig. 18-25, *C*, all three possible interval values appear in substantial numbers. Approximately 90% are ab intervals with a mean duration of 563 μsec; ba intervals and ab intervals occur with equal frequency and their main duration is 1,103 μsec; the remaining 62% are aa and bb intervals, with the mean duration of 1,666 μsec, or an integral multiple of this value. The pattern of interspike intervals in Fig. 18-25, *C*, consists of triads that will repeat themselves every 1,666 μsec. The members of the first four triads are identified as modes 1, 2, and 3 in the histogram.

From these experimental findings stems the evidence that discharges can recur periodically at frequencies for which there is no spectral component in the acoustic stimulus. Instead, these latter discharges reflect the peaks of unidirectional elevations of the wave

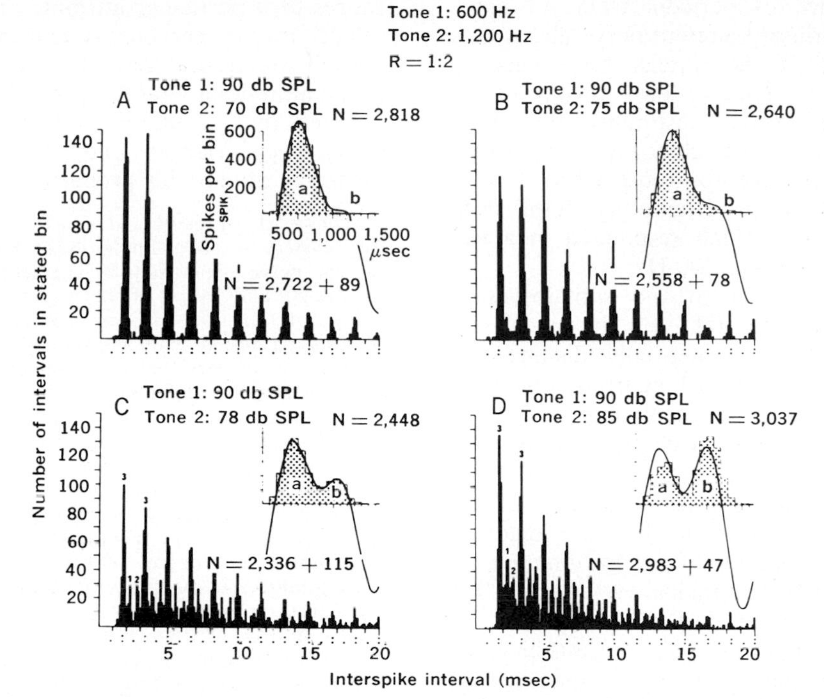

Fig. 18-25. Relation between interspike intervals and distribution of spikes in period histograms. Tone of 600 Hz was locked to tone of 1,200 Hz in ratio of 1:2. **A** to **D,** Interspike interval and period histograms obtained when lower frequency tone was at 90 db SPL, while strength of higher frequency tone varied as indicated. Each sample based on responses to two tonal presentations. Duration of each presentation: 10 sec. Interspike-interval histograms: abscissa, time in milliseconds; each bin = 100 μsec. Dots below abscissa indicate integral multiples of periods of both primary tones. Upper row of dots indicates integral multiples of period of 600 Hz tone; second row of dots indicates integral multiples of period of 1,200 Hz tone. Ordinate: number of interspike intervals in each bin. Number of intervals, *N*, is given by two numbers; the first indicates number of plotted intervals; the second is number of intervals that exceeded value of abscissa. Period histograms: abscissa, period of complex wave in microseconds; each bin = 100 μsec. Ordinate: number of spikes in each bin. Period of complex sound: 1,666 μsec. *N* = number of spikes in sample. Each period histogram is fitted by curve that is the sum of two sine functions. Different subpopulations in period histograms are identified in alphabetical order from left to right. (From Rose et al.[34])

form of the stimulus. In man the same auditory stimuli elicit the sensation of combination tones; i.e., tones that are heard but for which there is no spectral component in the acoustic stimulus. Thus it appears that combination tones can be a consequence of the way in which tone frequency is encoded in the ear. Moreover, the occurrence of periodicities in the auditory nerve discharge pattern and the sensation of tones, both unrelated to spectral components in the auditory input, strengthens the contention that the temporal pattern of neural discharges in auditory nerve fibers is indeed a significant determinant of the perceived tone, at least at low tone frequencies.

The situation is unresolved as regards tone frequencies in excess of 5,000 Hz when refractoriness of the auditory nerve fibers prevents the occurrence of time intervals between consecutive discharges shorter than some 700 to 800 μsec. In this case the discharges accumulate at integral multiples of the period, which are markedly longer than the refractory period. Some question also remains as to the mechanism in the central auditory pathways that is capable of interpreting the time-period code of the afferent nerve fibers, and as to the manner in which the time-period code may be transformed into a different central representation. A possible candidate for an alternative form of tone frequency coding is the tonotopic organization of the subcortical structures of the auditory system (Chapter 13).

This complex of questions is, to some extent, similar to those raised by the investigations of Mountcastle and co-workers[29, 39] regarding the cutaneous sense of flutter vibration (Chapter 10). The conclusion of these studies was that the perception of regular oscillatory movements of the skin depends on the appearance of periodic trains of nerve impulses in primary afferent nerve fibers, with the periodicity in the impulse train reflecting the stimulus frequency. Two classes of afferent fibers were distinguished: those originating from the glabrous skin, which were most sensitive to frequencies of 5 to 40 Hz, and those originating from deep cutaneous tissue, which responded in a frequency range of 60 to 300 Hz.

In the projection to the somatosensory area I of the postcentral gyrus the two classes of afferents remain essentially distinct and activate different cortical neurons; one class of quickly adapting cortical neurons readily follows the sinusoidal mechanical stimulus of the skin over a frequency range of 5 to 80 Hz, with a strong periodic recurrence of impulses at intervals close to the cycle length of the vibratory stimulus. Thus information on stimulus frequency remains preserved in the form of the temporal discharge pattern of these cortical neurons. This suggests that the capability of discriminating frequencies of vibratory stimuli in behavioral and psychophysical tests may be attributable to a central neural mechanism that can detect differences between the period lengths in impulse trains of this class of neurons.

The situation is different in another class of cortical neurons, which receives its afferent input from the pacinian corpuscles located in the deeper cutaneous layers. Vibratory stimuli with a frequency of 80 to 400 Hz, which entrain periodic discharges in the afferents originating from pacinian corpuscles, are reflected only by an increase of discharge rates in these cortical neurons. But there is no relation between the magnitude of the increase in firing rate and the stimulus frequency, nor is there any periodicity in the discharges that would reflect the stimulus frequency. Yet the human observer is capable of discriminating between different frequencies of vibratory stimuli, irrespective of whether the stimuli engage, according to their frequency, the quickly adapting system of afferents from the glabrous skin or the pacinian elements.

The general implication is that a certain stimulus attribute, which by virtue of the choice of the selected physical measure (e.g., frequency, as in the case under consideration) can be represented on a continuous and monotonic scale, need not be processed by the nervous system in an equally continuous and homogeneous fashion. Instead, different ranges of a particular attribute may be represented in the nervous system by means of entirely different types of codes. In the extreme a mode of neural representation may be adopted that is entirely unlike the scales of sensory experiences and the conventional measures of stimulus properties.

The work of Gesteland et al.[17] with the olfactory system is an example. Neural activity in olfactory nerve fibers may be augmented or reduced by odors, and every olfactory cell is affected by many different odors, some excitatory, some inhibitory. If tested with a group of different odors *seriatim,* it appears that every olfactory nerve fiber ranks these odors differently according to the degree of excitation (or inhibition) each of them elicits. The chances of encountering two nerve fibers with characteristically different responses to one pair of odors selected at random from the group would be small, but with many fibers engaged in the response, each of them ranking the odors differently according to the degree of excitation or inhibition they produce, it would be possible for the entire population as a whole to distinguish many different odors. In this way an entire population of neural elements generates a code, the effectiveness of which cannot be appreciated when the responses of individual elements are examined in isolation.

SENSE MODALITY AND TEMPOROSPATIAL PATTERNS OF NEURAL DISCHARGES

Since the time of Helmholtz it has been customary to categorize sensations according to modalities, i.e., classes of sensations that form qualitative continua. Thus tone perception is a single modality, since hearing encompasses a continuous series of tones with no qualitative gaps. Unlike hearing or color vision, the cutaneous sense consists of several classes of sensations with different introspective qualities (Chapter 9).

Two contrasting points of view have been proposed to explain the multiplicity of modalities of the cutaneous sense in neuroanatomic and neurophysiologic terms. One viewpoint, which originated with Johannes Müller, states in current terms that experienced sense quality is determined by the central connections of the activated nerve fibers, irrespective of the physical nature of the stimulus that elicited this activity. Some of the ramifications of this view are that afferent nerve fibers with different diameters are essentially specific for different sense modalities, and that the different modalities of cutaneous sensation have their own peripheral receptors that are preferentially sensitive to a particular form of stimulus energy (e.g., heat, mechanical deformation, etc.).

The second viewpoint originated with Nafe[30] and found active proponents in Weddell[42] and Sinclair.[9] Its principal claim is that the complex spatially and temporally dispersed pattern of neural activity is the determining factor in the experience of sensory quality. Thus activity in a given group of fibers or neurons could at one point in time contribute to the experience of touch, and at another point, contribute to the experience of pain, cold, or warmth.[9]

More recently, it has been proposed that these contrasting viewpoints (commonly designated as the "specific modality" and "pattern" theories, respectively) are not necessarily mutually exclusive. Recognition of receptor and neural pathway specialization for the transduction and transmission of particular types of cutaneous stimuli does not preclude that differences in temporal impulse spacings in individual afferents and differences in the relative distribution of afferent activity in a spectrum of nerve fibers could also be an important aspect of the central neural representation of a stimulus. The implication is, for instance, that a discharge pattern that consists of a rapid rise followed by a slow decline of discharge rate could be transmitted and perceived differently than a discharge pattern in the same afferent that rises slowly and declines rapidly.[41] Whether there are neurons in the CNS that are specifically triggered by one and not another temporal pattern of afferent input is still uncertain. The possibility exists, however, in view of several studies in invertebrates, since certain postsynaptic neurons responded differently when stimulated with impulse trains of different temporal spacing, although the mean stimulus rate was held constant.

STIMULUS FEATURE DETECTION BY NEURONS

Our central theme thus far in this chapter has been the proposition that mean rate and periodicity of neural discharges represent stimulus intensity and frequency in afferent neural pathways. The argument revolved around the demonstrations of correspondence between these parameters of neural discharges and measures of sensation, the latter obtained largely in man. In the conversion of physical stimulus attributes to neural activity, each magnitude value, which is measured on a continuous scale of the one, is made to correspond to an appropriate magnitude value, which is measured on a continuous scale of the other. The relation between the physical and the neural scale may in some instances be a linear and in others a nonlinear function. This general concept was shown to satisfy certain kinds of experimental data, but it also became apparent that this principle of stimulus representation does not exhaust the scope of stimulus encoding by the CNS.

There are now also a number of known instances that attest to a different principle, namely, that neurons, notably in cortical sensory receiving areas, respond quite specifically to more complex "stimulus features." A characteristic example is provided by a class of neurons in the auditory cortex that respond most effectively to an appropriate phase difference between binaurally applied tone stimuli, with that phase difference itself being a function of the frequency of the tone employed, while another class of neurons responds preferentially to the intensity difference of binaural tones.[16] Other examples are

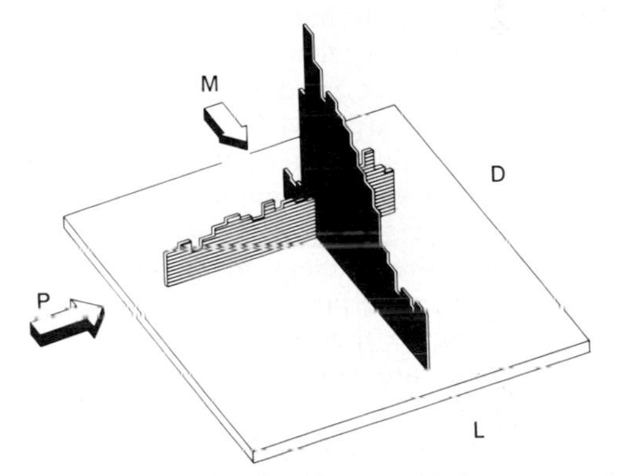

Fig. 18-26. Responses of neuron in somatosensory area I (postcentral gyrus) of unanesthetized macaque to movements of fine brush across cutaneous receptive field of neuron, which was located on dorsum of contralateral foot. Two directions of stimulus movement are shown: from medial to lateral (black; $M \rightarrow L$), and from proximal to distal (striped, $P \rightarrow D$). Height of bars composing vertical displays signifies number of discharges in each of consecutive 50 msec bins of duration of stimulus motion across receptive field, averaged over 25 consecutive and identical stimuli and expressed as discharges per second. Stimuli moved at constant velocity of 30 mm/sec. Spontaneous activity of neuron was 19 (± 2.0) impulses/sec, exactly the height of striped plane at beginning of stimulus motion near border of receptive field. Intersection of vertical planes reflecting responses to stimulus movement in two orthogonal directions shows that an area in receptive field that responds most vigorously when stimulus moves across it from medial to lateral does not respond at all when stimulus moves across it from proximal to distal. (Based on unpublished experiments of Whitsel, Petrucelli, and Werner.)

the neurons with "simple," "complex," and "hypercomplex" receptive fields in the visual cortex,[22] and the relative selectivity with which neurons in somatic sensory area I respond to cutaneous stimuli moving in a particular direction across the receptive field of the skin (Fig. 18-26).[47]

A recent observation on single neural responses in the inferotemporal cortex of macaques underscores the nature of the problem the investigator faces when searching for the "optimal" stimulus for a certain neuron. In some instances, neurons responded most strongly when the animal was presented with displays of relatively complex geometric figures. This may indicate that factors other than the stimulus configuration per se, e.g., the significance of the stimulus for the animal's behavioral repertoire, entered into determining the magnitude of the neural response.[18]

The common denominator of these experimental findings and others as well,[26] is that individual neurons can be shown to be "triggered" to discharge by a relatively specific spatial or temporospatial stimulus configuration; there is some gradation of the density of discharge with variation of the stimulus properties, but there is generally a clearly defined stimulus context (e.g., shape or motion of a stimulus in the neuron's receptive field) that elicits a maximal response. Thus neurons in a population can be classified according to their "best" stimuli. These "best" stimuli are contexts of spatial simultaneity (in the case of shape) or of temporal succession (in the case of stimulus motion) of excitation in the neuron's peripheral receptive field. A less than maximal response indicates only departure from the optimal stimulus, but not the direction of departure or which component of the context was altered. The occurrence of the maximal discharge in a particular neuron can be thought of as the "code" of the particular stimulus context that this neuron represents.

In the visual and the somatosensory cortical receiving areas in which this form of stimulus-feature encoding has been investigated in some detail,[22, 47] it has been found to be associated with a functional architecture that places neurons with common response properties in juxtaposition. In the vertical dimension of the cortex, neurons with near-identical receptive fields are assembled to cell columns; and in the horizontal dimension, neurons with similar optimal stimuli are assembled to cortical laminae, each lamina differing from the other in the complexity of stimulus features its neurons represent. Thus in a first approximation, it appears that a sensing surface (retina or cutaneous body surfaces) is mapped many times over; each map represents a particular stimulus feature, and all maps are in register with one another as regards place on the peripheral receptor sheet.

One of the most intriguing problem areas in the study of sensation in neurophysiology is understanding the mechanisms and the sequence of events that intervene between the piecemeal fragmentation of stimulus attributes and features as "seen" by individual cortical neurons and the perception of objects with permanent identities that move in a perceptual space along continuous trajectories.

There is, perhaps, some analogy to the approaches used in automated pattern-recognition technology: the recognition of handwritten characters by computers is an example. A relatively crude and inefficient procedure would be to store in the computer memory descriptions of templates for each letter; the input pattern with unknown classification would then be compared with the stored templates, and "recognition" would be based on finding the template that matches the input test. An alternative and more powerful approach consists in performing the classification on a set of selected measurements taken on the input. The items selected for measurement are "features" that are relatively insensitive to variation and contain minimal redundancies. The "feature extractor" supplies the input to a classifier that has stored and executes the algorithms that allow a decision as to whether a certain set of input features can be recognized as a particular pattern.

The afferent systems, up to at least the primary sensory cortical areas, function in this analogy as the feature extractors. It was suggested that these physiologic feature extractors themselves might be under some "program" control from associated cortical regions.[8]

REFERENCES

General reviews

1. Chomsky, N., and Miller, G. A.: Introduction to the formal analysis of languages. In Luce, R. D., Bush, R. R., and Galanter, E., editors: Handbook of mathematical psychology, New York, 1963, John Wiley & Sons, Inc., vol 2.
2. Fechner, G. T.: Elements of psychophysics (translated by H. E. Adler; edited by D. H. Howes and E. G. Boring), New York, 1966, Holt, Rinehart & Winston, Inc.
3. Galanter, E.: Contemporary psychophysics. In Brown, R., Galanter, E., Hess, E. H., and Mandler, G., editors: New directions in psychology, New York, 1962, Holt, Rinehart & Winston, Inc.
4. Garner, W. R.: Uncertainty and structure as psychological concepts, New York, 1962, John Wiley & Sons, Inc.

5. Kiang, N. Y. S.: A survey of recent developments in the study of auditory physiology, Ann. Otol. Rhinol. Laryngol. **77**:656, 1968.
6. Luce, R. D., and Edwards, W.: The derivation of subjective scales from just noticeable differences, Psychol. Rev. **65**:222, 1958.
6a. Luce, R. D., Bush, R. R., and Galanter, E.: Handbook of mathematical psychology, New York, 1963, John Wiley & Sons, Inc., vol. 1.
7. Miller, G. A.: The magical number seven, plus or minus two; some limits on our capacity for processing information, Psychol. Rev. **63**:81, 1956.
8. Pribram, K. H.: The amnesic syndromes: disturbances in coding? In Talland, G. A., and Waugh, N. C., editors: The pathology of memory, New York, 1969, Academic Press, Inc.
9. Sinclair, D.: Cutaneous sensation, London, 1967, Oxford University Press.
10. Stevens, S. S.: On the psychophysical law, Psychol. Rev. **64**:153, 1957.
11. Swets, J. A., Tanner, W. P., and Birdsall, T. G.: Decision processes in perception, Psychol. Rev. **68**:301, 1961.
12. Watson, J. B.: Psychology from the standpoint of a behaviorist, Philadelphia, 1919, J. B. Lippincott Co.

Original papers

13. Békésy, G., von: A new audiometer, Acta Otolaryngol. **35**:411, 1947.
14. Blough, D. S., and Schrier, A. M.: Scotopic spectral sensitivity in the monkey, Science **139**: 493, 1963.
15. Bough, E. W.: Stereoscopic vision in the macaque monkey: a behavioral demonstration, Science **225**:42, 1970.
16. Brugge, J. F., Dubrovsky, N. A., Aitkin, L. M., and Anderson, D. J.: Sensitivity of single neurons in auditory cortex of cat to binaural tonal stimulation—effects of varying interaural time and intensity, J. Neurophysiol. **35**:1005, 1969.
16a. Eijkman, E., and Vendrick, A. J.: Detection theory applied to the absolute sensitivity of sensory systems, Biophys. J. **3**:65, 1963.
17. Gesteland, R. C., Lettvin, J. Y., and Pitts, W. H.: Chemical transmission in the nose of the frog, J. Physiol. **181**:525, 1965.
18. Gross, C. G., Bender, D. B., and Rocha-Miranda, C. E.: Visual receptive fields of neurons in inferotemporal cortex of the monkey, Science **166**:1303, 1969.
19. Harper, R. S., and Stevens, S. S.: A psychological scale of weight and a formula for its derivation, Am. J. Psychol, **61**:343, 1948.
20. Hensel, H., and Bowman, K. K. A.: Afferent impulses in cutaneous sensory nerves in human subjects, J. Neurophysiol. **23**:564, 1960.
21. Hubel, D. H., and Wiesel, T. N.: Receptive fields, binocular interaction and functional architecture in the cat's visual cortex, J. Physiol. **160**:106, 1962.
22. Hubel, D. H., and Wiesel, T. N.: Receptive fields and functional architecture of monkey striate cortex, J. Physiol. **195**:215, 1968.
23. Hubel, D. H., and Wiesel, T. N.: Cells sensitive to binocular depth in area 18 of the macaque monkey cortex, Science **225**:41, 1970.
24. Jones, F. N.: Some subjective magnitude functions for touch. In Hawkes, G. R., editor: Symposium on cutaneous sensibility, Report No. 424, Fort Knox, 1960, U. S. Army Medical Research Laboratory.
25. Julesz, B.: Binocular depth perception without familiarity cues, Science **145**:356, 1964.
26. Maturana, H. R., Lettvin, J. Y., McCulloch, W. S., and Pitts, W. H.: Anatomy and physiology of vision in the frog, J. Gen. Physiol. **43**:129, 1960.
27. Mountcastle, V. B., Poggio, G. F., and Werner, G.: The neural transformation of the sensory stimulus at the cortical input level of the somatic afferent system. In Gerard, R. W., and Duyff, J. W., editors: Information processing in nervous system, Proceedings of the International Union of Physiological Sciences, New York, 1962, vol. 3.
28. Mountcastle, V. B., Poggio, G. F., and Werner, G.: The relation of thalamic cell response to peripheral stimuli varied over an intensive continuum, J. Neurophysiol. **26**:807, 1963.
29. Mountcastle, V. B., Talbot, W. H., Sakata, H., and Hyvarinen, J.: Cortical neuronal mechanisms in flutter vibration studied in unanesthetized monkeys. Neuronal periodicity and frequency discrimination, J. Neurophysiol. **32**: 452, 1969.
30. Nafe, J. P.: The psychology of felt experience, Am. J. Psychol. **39**:213, 1957.
31. Pollack, I.: The information of elementary auditory displays, J. Acoust. Soc. Am. **24**:745, 1952.
32. Provins, K. A.: The effect of peripheral nerve block on the appreciation and execution of finger movements, J. Physiol. **143**:55, 1958.
33. Rose, J. E., Brugge, J. F., Anderson, D. J., and Hind, J. E.: Phase-locked response to low frequency tones in single auditory nerve fibers of the squirrel monkey, J. Neurophysiol. **30**:769, 1967.
34. Rose, J. E., Brugge, J. F., Anderson, D. J., and Hind, J. E.: Some possible neural correlates of combination tones, J. Neurophysiol. **32**:386, 1969.
35. Ruch, T. C., Fulton, J. F., and German, W. J.: Sensory discrimination in monkey, chimpanzee and man after lesions of the parietal lobe, Arch. Neurol. Psychiatry **39**:919, 1938.
36. Stevens, J. C., and Stevens, S. S.: Warmth and cold–dynamics of sensory intensity, J. Exp. Psychol. **60**:183, 1960.
37. Stevens, S. S.: The operational basis of psychology, Am. J. Psychol. **47**:323, 1935.
37a. Stevens, S. S.: The psychophysics of sensory function. In Rosenblith, W. A., editor: Sensory communication, Cambridge, Mass., 1961, The M.I.T. Press.
38. Stevens, S. S., and Galanter, E. H.: Ratio scales and category scales for a dozen perceptual continua, J. Exp. Psychol. **54**:377, 1957.
39. Talbot, W. H., Darian-Smith, I., Kornhuber, H. H., and Mountcastle, V. B.: The sense of flutter vibration—comparison of the human capacity with response patterns of mechanoreceptive afferents from the monkey hands, J. Neurophysiol. **31**:301, 1968.
40. Vierck, C. J., and Jones, M. B.: Size discrimination on the skin, Science **158**:488, 1969.

41. Wall, P. D., and Cronly-Dillon, J. R.: Pain, itch and vibration, Arch. Neurol. **2:**365, 1960.
42. Weddell, G.: The anatomy of cutaneous sensibility, Br. Med. Bull. **3:**167, 195, 1945.
43. Weinstein, S.: Intensive and extensive aspects of tactile sensitivity as a function of body parts, sex and laterality. In Kenshalo, D. R., editor: The skin senses, Springfield, Ill., 1968, Charles C Thomas, Publisher.
44. Werner, G., and Mountcastle, V. B.: The variability of central neural activity in a sensory system, and its implications for the central reflection of sensory events, J. Neurophysiol. **26:**958, 1963.
45. Werner, G., and Mountcastle, V. B.: Neural activity in mechanoreceptive cutaneous afferents—stimulus response relations, Weber functions and information transfer, J. Neurophysiol. **28:** 359, 1965.
46. Werner, G., and Mountcastle, V. B.: Quantitative relations between mechanical stimuli to the skin and neural responses evoked by them. In Kenshalo, D. R., editor: The skin senses, Springfield, Ill., 1968, Charles C Thomas, Publisher.
47. Whitsel, B. L., Roppolo, T. R., and Werner, G.: Cortical information processing of stimulus motion on primate skin, J. Neurophysiol. **35:** 691, 1973.
48. Wiener, N.: A new theory of measurement—a study in the logic of mathematics, Proc. Lond. Math. Soc. **19:**181, 1920.

19

GERHARD WERNER

Higher functions of the nervous system

INTRODUCTION: THE ASSOCIATION CORTEX

The sensory deficits after partial or total ablation of primary cortical receiving areas are in large measure predictable on the basis of the modality specificity and the topologic organization of the projections they receive from the periphery *via* the specific thalamic relay nuclei (Chapters 10 and 16). Rose and Woolsey[114] set these "extrinsic" neural systems apart from the "intrinsic" systems that, by contrast, consist of the cortical projections from thalamic nuclei without any major extrathalamic or extratelencephalic input. Traditionally these latter cortical areas are known as "association cortex"; a designation that reflects the idea of 19th century psychophysics that perception is based on "associating" the elementary units of sensation to more complex ideas, patterns, and relations. In this tradition the "association cortex" was thought to be the neural structure in which the modality-, place-, and intensity-specific inputs from primary cortical receiving areas would interact and eventually initiate an integrated motor output by way of transcortical action.

To the extent to which this dichotomy between sensation and perception in psychology was discredited under the influence of Gestalt psychology, phenomenology, and certain trends in operational behaviorism (Chapter 18), the considerable evidence that has accumulated during the past two decades from neurobehavioral studies and clinical neurology has demanded a different conceptualization of the function of the association cortex. These current ideas are no longer compatible with a role of "associating" diverse sensory inputs and serving as a link between sensory and motor cortex. Instead, the range of functions in which the "intrinsic" cortical areas play a role is now thought to encompass the spatial structure of perception, associative learning, and behavior and involving transactions at a symbolic level in which "meaning" of stimulus context and "intentions" of behavioral acts appear to play some role.

The primate cortical areas in question are located within the parietopreoccipital convexity (the classic sensory association areas) and include, in addition, the prefrontal area occupying the anterior pole of the frontal lobe (the classic frontal association area). These intrinsic cortical areas are phylogenetically more recent and become myelinated later in development than the sensory and motor areas. There is some regional specificity for different sense modalities and functions: the parietal-occipital portion is mainly concerned with the somesthetic sphere of perception and behavior[106, 132]; the anterior temporal cortex is related to taste[54]; the middle temporal region is involved in auditory discrimination[131]; and the inferior temporal convolution is important for vision.[29] There is also evidence that some functions transcend any single modality and involve the capability of dealing with the spatial arrangement of the objects of perception and with the temporal sequencing of behavior.

POSTERIOR PARIETAL AREA

Parietal lobe syndrome

A patient with a gross lesion of the parietal lobe presents obvious and striking abnormalities of behavior that are not attributable to defects in the elementary aspects of sensation. The manifestations differ according to the hemisphere involved. Hemispheric specialization was first recognized in connection with language function, which is attributed to the left hemisphere in right-handed subjects; hence the notion of hemispheric dominance. However, lesions of the parietal lobe have also produced evidence of hemispheric specialization for functions other than language.

Damage of the parietal lobe in the right (minor) hemisphere of right-handed subjects leads to difficulties in recognizing spatial relationships in the contralateral half of body and extracorporeal space: the patient may act as though the limbs on the side opposite to the lesion did not exist; he may not use the affected hand, yet deny abnormality in it; or he may claim that a contralateral limb belongs to another person. He may disregard events taking place in the side of the environment that is contralateral to the lesion, and he will tend to give naive and irrational explanations of these aberrations. He may deny the existence of the contralateral body half altogether or deny illness present in it. The patient may fail to complete that portion of drawings or to correctly construct (e.g., with building blocks) that side of a scene that is contralateral to his lesion. The latter symptom is known as constructional apraxia.[78]

Denny-Brown[63] characterized this class of disorders as amorphosynthesis and interpreted it as a physiologic deficit similar to the behavioral disorders observed in monkeys after parietal cortex ablation. One of the characteristic behavior abnormalities in parietal lobe–ablated macaques is the gross exaggeration of the tactile avoiding reaction, which consists in rapid withdrawal of the hand after light contact with an object; on the other hand, the instinctive tactile grasp reaction is largely abolished. This latter response consists of an orienting movement of the hand that is elicited by contact with any of its parts to bring the palm of the hand in full contact with the stimulus, followed by grasp.

The common denominator of these abnormalities and the clinical signs of amorphosynthesis are thought to consist in a deficit of stereotaxic exploratory behavior and orientation in space; both factors lead to a fragmentation of the perceptual process. At least to some extent this deficiency may be the result of a more fundamental process: a frequent manifestation of parietal lobe damage consists in the lack of appreciation of a stimulus on the affected side of the body when an equivalent stimulus is simultaneously presented to the unaffected side. This phenomenon is known as *extinction:* the patient may recognize only the stimulus delivered to the side opposite to the normal hemisphere; or when the two stimuli are simultaneously delivered to the side opposite the diseased hemisphere, the patient may recognize only the proximal and disregard the distal stimulus, particularly if the former has been applied to the face (*proximal dominance*).[4] Extinction may also occur between stimuli of different modalities: e.g., a visual stimulus opposite the normal hemisphere can obliterate perception of a somatic stimulus. Thus there arises some *perceptual rivalry* between simultaneously applied stimuli that normal subjects can differentiate and interpret in terms of their relation to one another without difficulty.[62, 64] Accordingly, some of the behavioral defects associated with right parietal lobe lesions may not entirely be due to neglect of the left but may also be caused by excessive preoccupation with the right side of body and extracorporeal space: constructional apraxia from right-sided lesions or confusion of garments or in putting them on (apraxia of dressing) is in this sense an excessive distraction by small, irrelevant, and mostly right-sided parts. The same can be found in the types of dyscalculia and dysgraphia resulting from right-sided lesions, in which preoccupation with small parts of numerals and letters distracts and prevents normal perception.

Disorders caused by parietal lobe lesions in the left (dominant) hemisphere differ from those attributable to the minor hemisphere by characteristically involving *both* sides of body and extracorporeal space.[78] These disorders, which involve disturbances related to classification and naming and to manipulating the symbol of a particular class of percepts, are traditionally subsumed under the term *"agnosias."*

The following example, quoted from Denny-Brown, illustrates this aspect of the defect and its difference from the minor hemisphere lesion in not being related to the synthesis of complex exploratory movements: "In such disorders, the patient cannot name a pencil or suggest a use for it when it is presented to him as an object; yet, a moment later, given some paper and asked to write his name, picks up the pencil from among other objects and begins to write."[62]

The conventional notion of agnosia is that of failure to recognize objects by some particular sensory channel, although the patient appreciates the appropriate elementary sense data and is able to recognize the same object by means of some other sense. This conception of agnosia implies the 19th century two-stage interpretation of sensation versus perception and recognition, alluded to earlier. Accordingly, an agnosiac would be a person who lives in a state of denuded stimulus reception, unable to combine the primary sense data of sensation to percepts.

In the presently prevailing unitary approach to perception, this view of agnosia is under renewed scrutiny. For instance, object recognition by palpation must meet rather special requirements to be successful[5]: exposure to the stimulus is ordinarily sequential, and successive impressions must be matched against an anticipatory schema of what the palpated object might be. Certain critical details must be singled out, retained in short-term memory, and rechecked before a decision can be made about the object being felt. Rapid fading or abnormal interaction of successive impressions would therefore be especially disruptive for object recognition by palpation, as would persistent aftersensations. Partial or total deficiency in any one of these participating processes and others (to be discussed later) as well could conceivably lead to tactual object agnosia.[28] Patients with focal parietal lesions have in fact been reported to overestimate the second of two weights put successively on their palm.[130]

The distinction between left and right parietal lobe function does not imply that the dominant hemisphere lacks the capability for morphosynthesis; rather, it appears that this function is overlaid, and its defects obscured, by a process of higher abstraction that encompasses the perception of the entire body and extracorporeal space in a unitary, indivisible manner that is one important aspect of cerebral dominance: purely contralateral disturbances of morphosynthesis are therefore rarely seen after parietal lesions in the dominant hemisphere. Denny-Brown[62] suggests that the condition of asymbolia for pain is an instance of bilateral amorphosynthesis and not an agnosia, for such patients can feel pain and discuss it, although they do not exhibit the normal behavioral response to painful stimuli.

Body image. A particularly dramatic demonstration of the important, though not exclusive, role of the parietal lobes in the spatial organization of perception and behavior is associated with pathologic disturbances of the mental image that a person possesses about his own body and its physical attributes. Before the turn of the century, Bonnier called attention to some striking distortions shown by some of his patients in the attitude toward their own body: he described individuals who actually felt that their entire body had vanished (aschematia). Slightly later, A. Pick designated as autotopagnosia the disturbances in orientation on one's own body surface (e.g., the inability to distinguish left and right). Pick suggested that each individual develops a spatial image of his body from information supplied by the sensory systems. The idea of this body image may be so persistent that a person may generate a phantom sensation after losing a leg or an arm. Phantom limbs are to be expected in the majority of amputees, unless the limb is lost very early in life.[60]

Henry Head developed the concept of body image into a central theme of his neurologic thinking[100]: he concluded that each person constructs in the course of development a model of himself that becomes the frame of reference against which all body movements and postures are judged. This spatial scheme is not confined to the anatomic limits of the body; it may also incorporate instruments held in the hand or objects otherwise attached to the body.

Although frequently associated with parietal lobe damage, body-image pathology may also be associated with damage to other brain areas.[6] Some of the best examples of spatial alterations of body image are to be found in states of drug intoxication. Moreover, schizophrenic patients can show almost the same range of abnormal body-image phenomena.[39]

Haptic sensitivity: tactile apprehension of object quality and shape

Identification of objects through the sense of touch presupposes the combined and coordinated contribution of two submodalities—cutaneous touch and kinesthesis. Together, these two sources contain the information needed to specify the layout of the surfaces of an object being manipulated. J. J. Gibson[11] argued that the perception of the layout of surfaces *is* the perception of space, and that the concurrent sensory influx from skin and articular angle and motion conveys stimulus information in its own right, namely that of haptic touch. This implies a functional unity between sensory information originating within the body space (i.e., sensory events signaling joint position) and sensory information related to physical stimuli impinging on the body from without (i.e., cutaneous sensory events).

The sequence of events that Gibson noted when an observer is required to discriminate the shapes illustrated in Fig. 19-1 is typically the following:

> (1) He curves his fingers around its face, using all fingers and fitting them into the cavities; (2) he moves his fingers in a way that can only be called exploratory, since the movements do not seem to become stereotyped, or to occur in any fixed sequence, or even to be clearly repeated; (3) he uses oppositions of thumb and finger, but different fingers; he rubs with one or more fingers and occasionally he seems to trace a curvature with a single finger. The activity seems to be aimed principally at obtaining a set of touch-postures, the movement as such being incidental to this aim. Introspection bears out the hypothesis that the phenomenal shape of the subject does emerge from such a series of covariant transformations. No subject ever tried to run his finger over the whole array of

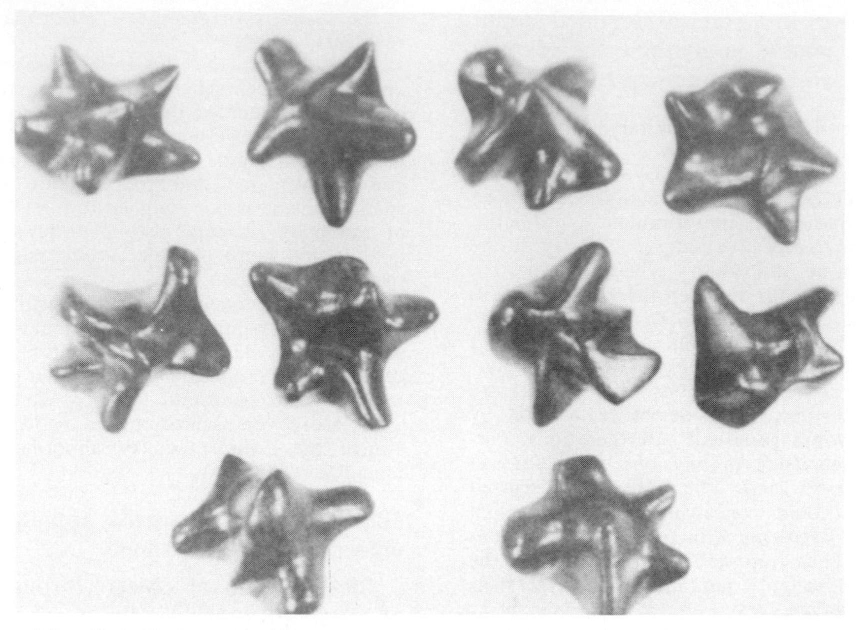

Fig. 19-1. Sculptured objects for studying haptic perception. (From Gibson.[11])

curves in a systematic manner such as that of the scanning beam of a television tube.*

Indeed, the eye and the hand sample different properties of the object. In a haptic task, left-right reversals and up-down changes of the position of test figures are felt to be very dissimilar; visually, the same changes produce little sense of differences. Conversely, changes in curvature of contours elicit a sharp sense of difference visually and little sense of difference haptically.[74]

Psychological studies of the blind confirm that the haptic sense contributes to the subject's idea or notion of surrounding space. Révész[36] considers the blind person a "pure haptic." The blind are conscious of the space generated by the surfaces that are accessible to haptic exploration. The particular role of the hand in the generation of this haptic space was recognized some 80 years ago by Féré,[69a] who wrote, "The hand merits the notice of physiologists and psychologists who have up to now rather neglected it: for the hand is both an agent and an interpreter in the growth of the spirit."

Supramodal mechanism in stereognosis. When the inability to recognize the form of objects by touch, known as astereognosis, occurs in the absence of any overt signs of sensory deficits as measured by the conventional tests of somatic sensation (i.e., two-point discrimination, joint position sense, etc.), the concept of tactual agnosia has traditionally been invoked and associated with damage to the posterior parietal area. While this aspect of the posterior parietal lobe function is well established, there is now also evidence that brain lesions that spare this and primary sensory and motor areas can also produce some impairment of shape discrimination.[40]

In the chronic state after parietal injury, when the dramatic manifestations of apraxias and agnosias for the body and for surrounding space recede, more subtle tests of spatial orientation are required to bring deficiencies into prominence. One such task, employed by Semmes et al.,[40] consisted of a series of maps, reproduced in Fig. 19-2, that are presented to the subjects to test their capacity for route finding.

The following description is taken from Teuber's account of the test.

The person to be tested is led into a large room where nine dots are marked on the floor, all equidistant from each other in a square array. These dots are represented on each of the maps with a line connecting them, indicating where in the room the person tested has to start and which path he is to follow by active locomotion, from dot to dot, until he reaches the goal symbolized by the arrowhead. All maps are carried by the patient, one map at a time; he is not permitted to reorient the map as he looks about. However, one edge of the map, as well as the appropriate wall of the room, have been marked to indicate north. Five of the maps [top row, Fig. 19-2] are presented visually, being line drawings on a square of cardboard. The remaining maps are made of raised tacks with cords running between them and these are carried on a tray attached to the patient's chest who palpates them under a black cloth.*

*Gibson, J. J.: The senses considered as perceptual systems, Boston, 1966, Houghton Mifflin Co.

*Teuber, H. L.: Alterations of perception after brain injury. In Eccles, J. C., editor: Brain and conscious experience, New York, 1966, Springer-Verlag New York, Inc.

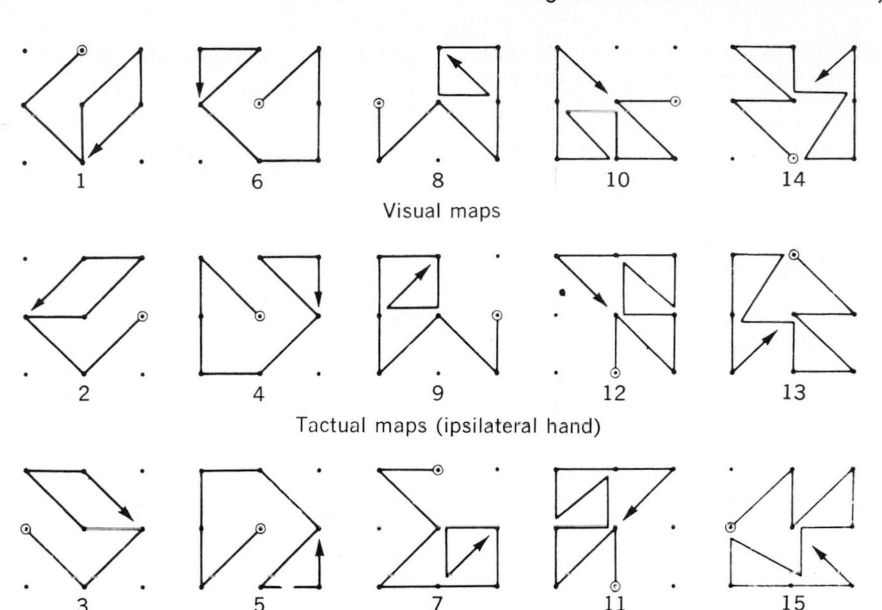

Fig. 19-2. Maps used in tests of route finding. (From Teuber.[126])

The patients who have difficulties with this task show equally inferior performance for the visual and the tactual-kinesthetic (or haptic) modes of presentation. Yet this general difficulty in route finding by means of maps is specifically dependent on the parietal lobes, since only the patients with parietal lesions show a significant deficit. This deficiency cannot be classified as an *agnosia* in the usual sense, as it implies disorders limited to "higher functions" *within* a given sensory modality. Instead, these observations suggest that one component of the parietal lobe syndrome consists in an inability to evaluate sensory information in terms of its spatial organization, and that the contribution of the parietal lobe to the structuring of sensory information into a spatial framework is not limited to the somesthetic senses. The integrity of some structure in this region may be necessary for the utilization of maps and schema as a device to represent space internally.[115, 124]

Although not demonstrated by the route-finding test, there is other evidence to indicate that defects in the apprehension of spatial data or topographic concepts are preferentially associated with lesions in the right parietal lobe.[78]

Hemispheric asymmetry in sensory-motor function

Specialization of hemispheric function does not appear limited to those manifestations commonly implied by the concept of cerebral dominance. This concept may even be misleading, for it suggests superiority of the dominant over the minor hemisphere, whereas it is now understood that the hemispheres differ with respect to the particular function for which they are specialized, each performing its assigned role in the sense of a reciprocal specialization rather than a hierarchy of importance.[127] Systematic studies of sensory-motor capacities of the hands of human subjects with brain injury led to some generalizations regarding the functional organization of each hemisphere. The nature of these differences is such that the hemispheric specialization for language and spatial tasks would then merely be consequences of the more fundamental difference in the basic modes of operation of each hemisphere.[116]

In tests of four of the classic aspects of discrimination sensibility (sense of passive movement, touch-pressure thresholds, two-point discrimination, and point localization), maximal deficits in the right hand occurred when the lesion was in the left sensorimotor cortex, and there was little if any deficit caused by lesions in cortex surrounding the sensorimotor region. Surprisingly, this is not the case in the right hemisphere: there is little difference in the deficit, irrespective of whether the lesion encompasses the sensorimotor region or not. Thus there is a suggestion for two differing modes of neural organization: a focal, restricted representation of elementary sensorimotor function in the left hemisphere and a diffuse representation in the right hemisphere. In a generalization from these findings, Semmes[116] suggested that the discrete organization of the left hemisphere favors the integration of *similar* functional units to more complex behavioral performances, thus predisposing for performances that demand exquisite sensorimotor control, e.g., manual skills and speech. On the other hand, the diffuse organization of the right hemisphere may facilitate the integration of *dissimilar* elementary functions required for multimodal coordination, such as occurs in the various spatial abilities.

Associative learning of tasks in somesthesis

There is firm evidence that the posterior parietal lobe lesions in macaques produce well-delineated deficiencies in the acquisition and performance of tasks involving discrimination of tactual stimuli, deficiencies of a nature that excludes the involvement of the ataxic component of the parietal lobe syndrome described earlier.

The general method used in these studies involves one or the other variant of the so-called Wisconsin General Test Apparatus (WGTA), described below. As a rule the subjects are first given trials on a large number of problems similar to that which will later constitute the test selected for the definitive study; this is to develop the monkey's ability to "learn to learn," based on the learning-set procedure of Harlow.[15]

There are two different aspects to the role of the posterior parietal lobe in associative learning of somesthetic stimulus discriminations. In the first place, in tests involving the discrimination of roughness of textures (sandpapers of different grades), the deficit induced by posterior parietal ablations consists in a decrement of the capacity to resolve fine differences between stimuli, while the performance remains unimpaired when the difference between the stimuli is great.[135] Thus the posterior parietal lobe determines, as it were, a "set point" for the degree of resolution in stimulus discrimination.

The second aspect concerns the selection of the strategy for handling multiple-object, problem-solving tasks. The test procedure in the WGTA is as follows:

The animals are initially confronted with two objects placed over two holes, on a board containing 12 holes in all (with a peanut under one of the objects). An opaque screen is lowered between the monkey and the objects as soon as the monkey has displaced one of the objects from its hole (a trial). When the screen is lowered, separating the monkey from the 12 hole board, the objects are moved (according to a random number table) to two different holes on the board. The screen is then raised and the animal again confronted with the problem. The peanut remains under the same object until the animal finds the peanut five consecutive times (criterion). After the monkey reaches criterion performance, the peanut is shifted to the second object and testing continues (discrimination reversal). After an animal again reaches criterion performance a third object is added. Each of the three objects in turn becomes the positive cue; testing proceeds as before until the animal reaches criterion performance with each of the objects positive, in turn. Then a fourth object is added and the entire procedure repeated. As the animal progresses, the number of objects is increased serially through a total of 12 (Fig. 19-3).*

The experimenter is interested in analyzing the monkey's performance in terms of recur-

*Pribram, K. H.: On the neurology of thinking, Behavioral Sci. **4**:245, 1959.

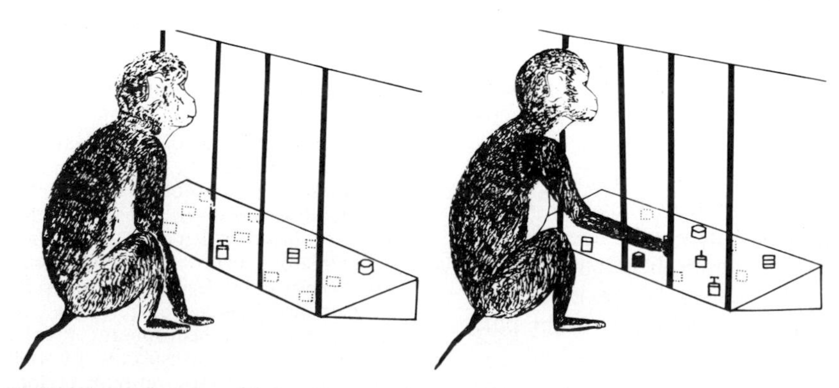

Fig. 19-3. Diagrams of multiple-object, problem-solving examples of three- and seven-object situations. Food wells are indicated by dashed circles, each of which is assigned a number. Placement of each object over food well was shifted from trial to trial according to random-number table. Record was kept of object moved by monkey on each trial, only one move being allowed per trial. Trials were separated by lowering an opaque screen to hide objects from monkey as they were repositioned. (From Pribram.[34])

rent and persistent trends. If a consistent pattern of choices can be seen, one tends to interpret this as a reflection of a "hypothesis" the monkey may have adopted regarding the consistencies in the environment he faces in order to guide his behavior toward optimization of reward. Some typical questions the experimenter wishes to answer are: Given a particular stimulus situation on consecutive trials, what pattern of responses does a monkey show? Does he choose the object on the right side consistently? Or does he alternate?

The problem just described is particularly interesting because for optimal performance it requires two different strategies that should be followed in alternation: (1) During search, move on successive trials each of the objects until the peanut is found. (2) After search, select the object under which the peanut had been found on the preceding trial (i.e., lose-shift, win-stay). These strategies should be appropriately alternated to obtain the maximal number of rewards.

The result of this experiment by Pribram[107] was that animals with posterior *intrinsic sector* ablations show a striking reduction in searching; i.e., they will sample fewer objects for the reward. This deficit appears specific for the parietal lesions; the deficit found after frontal intrinsic sector ablations is of an entirely different nature (p. 584).

Based on these and similar observations, Pribram[107] formulated a generalization that has significant implications for an understanding of the operational principles of the posterior intrinsic cortex. The essential nature of the posterior parietal lobe deficit becomes apparent when one compares the failure in the multiple-choice discriminative task with a behavioral situation in which normal animals and those with parietal lobe lesions perform indistinguishably: a monkey with drastic resection of the posterior parietal lobe is able to catch a flying gnat in midair and perform similar isolated acts with undiminished skill and precision; his reaction is entirely normal in simple "go–no go" situations that merely require the decision of whether or not to execute a certain behavior. In contrast, when presented with a series of stimuli for selection and required to choose different responses according to preceding outcomes of the test, the animal with a parietal lobe lesion fails.

Accordingly, it appears that the posterior intrinsic cortical mechanism enables the subject to deal with more complex units of behavior requiring a selection from "contextual alternatives" into which the sensory information can be partitioned. It is as if a set of environmental circumstances were divided into different mutually exclusive classes, as if one of these classes were selected as the relevant environmental state (its "image"); and as if a plan were adopted that would control an appropriate sequence of actions to be performed.[26]

M. Wilson[133] suggested that the failure of the monkey with a posterior parietal lesion to select a certain consistent strategy of action or plan of behavior is equivalent to his failure in assigning "meaning" to a stimulus array in accord with previously structured experience. The agnosias of patients with damaged parietal lobes may be attributable to the same fundamental deficiency.[125]

INFEROTEMPORAL CORTEX

The recognition of the role of the temporal lobe in vision originated with an observation of Klüver and Bucy[88]: one of the most startling defects following bilateral temporal lobectomy in macaques was their inability to recognize objects by vision alone ("psychic blindness"). Discrimination of two-dimensional patterns is the most severely impaired visual ability, but there is also some deficiency in the discrimination of object quality in terms of color, size, and brightness. While widespread within vision, the defect does not extend beyond it. Subsequent studies led to a progressively more precise delineation of a circumscribed region in the temporal lobe, ablation of which abolished the monkey's ability to discriminate between pairs of visual patterns that he was successful in discriminating preoperatively. The area in question is the inferior temporal convexity, comprising the middle and inferior temporal convolutions.[29, 82]

On gross observation the monkey with an inferotemporal lesion is indistinguishable from a normal animal; the visual discrimination impairment becomes evident only in more formal training situations and becomes more pronounced as the difficulty of the discrimination tasks is increased. The defect consists primarily in a retardation of the rate at which a particular visual discrimination can be acquired. This raises the question of whether the trained monkey with an inferotemporal lesion responds to the same "distinctive stimulus features" as the normal monkey, or whether he perhaps utilizes cues provided by the stimulus in a different manner.

Experimental evidence supports this latter

idea[58]: if complex stimuli are presented (e.g., panels depicting a series of parallel bars that are alternately black and colored, with both angular orientation of the bars and color hue as variable stimulus attributes) the inferotemporal monkey performs, at least at a certain stage of training, as though he had only learned at most to respond to "colored bars" on the panel and not to the more narrowly defined stimulus of "blue vertical bar," as would the normal monkey. These and similar findings suggest that the nature of the deficit of the inferotemporal monkey is not merely attributable to a deficit in perception, learning, or retention.

On theoretical grounds, one would predict that the discrimination performance of normal subjects would improve with stimulus patterns of increasing redundancy; concomitantly, retention would diminish. Normally, there is a trade off between attending to redundant stimulus features to optimize discrimination, on the one hand, and focusing on a particular aspect of the stimulus to optimize retention, on the other.

In a study of this type, Wilson[134] employed the three dimensions of color, form, and ground texture. If a one-dimensional problem was to be presented, only one of these stimulus properties was varied while the others remained constant; in this case, there are no redundant cues. On two-dimensional problems, two cues were paired, e.g., color and form or texture and form would be varied concomitantly, thus providing one redundant cue, while the third stimulus dimension would be held constant in each case.

Characteristically, the monkeys with inferotemporal lesions perform with stimuli of different redundancy as if they were attempting to process *all* the available stimulus information: it seems that they try to attend to all of the choices, correct or incorrect, and attempt to remember also the redundant features of the stimuli.[134] These results fit with the idea that the inferotemporal cortex exerts some efferent control over the input from the primary visual system, with the effect of reducing redundancy in visual input.

There is some experimental support for this idea. In the first place, continued electrical stimulation of the inferotemporal cortex slows the recovery of responses evoked in occipital cortex by pairs of light flashes, an effect that would keep some neurons in this system "occupied" for a longer than normal period. Thus iterated visual input would find different neurons of the population in a responsive condition, which would lead to a form of multiplexing in the visual system.[120] Furthermore, resection of the inferotemporal cortex drastically alters the electrical activity in the visual cortex of monkeys that, in normal controls, is characteristically related to stimulus, response, and reinforcement variables of a discrimination task.[121]

Inferotemporal neurons have a high degree of specificity and individuality in their responses to visual stimuli: length, width, and shape of the stimulus, as well as contrast with background, color, and direction and speed of movement affect the response. The receptive fields are relatively large, may be unilateral or bilateral, and invariably include the fovea. Characteristically, the responses diminish with repeated stimulus presentation.[75, 76]

The anatomic connections of inferotemporal cortex are complex and place it in a strategic position for visual information processing. On the one hand, inferotemporal cortex has two-way connections to the optic tectum, which accommodates the primary central visual mechanisms in lower vertebrates; in the primate, this pathway involves also the pulvinar. On the other hand, there are reciprocal connections to the phylogenetically more recent geniculostriate visual system *via* the circumstriate belt of cortical tissue. Consequently, the inferotemporal cortex receives information related to visual orienting processes (via the superior colliculus and pulvinar) and also the information extracted from the visual scene by the "feature-detecting" neurons of visual cortex. There is evidence for a defect in visuomotor integration after inferotemporal lesion: the patterns of eye movements during visual discrimination are altered. This may reflect a primary defect of scanning and information gathering in the visual scene as the cause of the visual deficit, or it may be the result of some perceptual or mnemonic defect that interferes with search or selection of cues.[14, 55]

Visual deficits after temporal lobe damage in man. In contrast to the language disturbances caused by temporal lobe damage in the dominant hemisphere, a constellation of deficits in nonverbal task performance arises after cortical excision in the right temporal lobe; some of these deficits appear closely related to the visual deficits of monkeys with inferotemporal lesions.[95] Among these visual deficits are disturbances in tachistoscopic recognition and dot numerosity,[87] in classification of pictures of faces in which normal contour lines have been eliminated,[89] and in the detection of small differences in complex patterns.[92] The recall of geometric drawings, but not of stories or word pairs, is also impaired.[97]

MIDDLE TEMPORAL AREA: AUDITORY DISCRIMINATION

The evidence for modality-specific auditory discrimination deficits after temporal lobe lesions is less definite than for tactile and visual deficits with regard to their respective intrinsic cortical areas. To some extent this is related to the fact that monkeys, the preferred species for studies of this general type, have great difficulty in acquiring behavior guided by auditory stimuli. However, Penfield and Kristiansen's observations[33] indicate the existence of a circumscribed area in the temporal lobe of man that, if stimulated electrically during brain surgery, elicits the patient's recall of past auditory experiences, e.g., the specific rendition of a piece of music he may have heard years ago. As during temporal lobe epileptic discharges, the patient feels "distant" and removed from the reenacted experience.

The experiments of Diamond and Neff,[65] carried out in cats, demonstrated that bilateral removal of a cortical area corresponding to primate temporal neocortex produces a specific deficit that is not sensory. Ablation of the superior temporal convolution is followed by a lasting deficit in the discrimination of tonal patterns in the absence of any obvious frequency or intensity discrimination deficit. In addition, the cat's ability to localize the source of the sound is disrupted: it is as though the sound image produced by a pair of clicks had lost the attribute of locus in an extracorporeal space. Thus location of a sound source in space is no longer a component of the animal's perception of the external environment.[7, 91]

Auditory perception deficiencies in man. In its most dramatic form the auditory deficit entails a failure to comprehend spoken language (receptive aphasia). Most patients with such a condition also have severe expression disorders; they produce a rush of speechlike but incomprehensible sounds, often without appreciating their own disability. An important and not yet entirely resolved question is whether there is in the human brain a regional specialization for processes underlying perception and comprehension of speech patterns and other nonspeech sound patterns.

The evidence suggestive of lateralization of specific auditory functions is derived from the responses to dichotic tasks involving the separate presentation of two different strings of signals (e.g., the numbers 234 and 357 or other verbal material), one to each ear. The normal listener will first deal with the input to one ear (i.e., 234) while holding the other ear's input in short-term memory and then reproduce the input to the other ear (i.e., 357). When the stimuli are presented in strict synchrony, there is a superiority in performance on the side opposite the hemisphere that mediates speech (normally, the left); this asymmetry is established early in life.[52] A preference for the stimulus delivered to the left ear exists for nonverbal material, e.g., brief melodies. As one would expect from this asymmetry, patients with long-standing right temporal lobe lesions perform in the dichotic number test as normal subjects do, while left temporal lobe lesions lower the performance score.[28] The situation is reversed for dichotic presentation of melodies.

Information processing in perception

The preceding sections were presented in order to demonstrate that behavior controlled by perceptual processes appears to involve internal models of reality that permit the structuring of sense experiences into spatial order and temporal sequence; but current sensory information never mediates behavior directly.[13] This conception is at variance with the more traditional stimulus-response reflex arc concept that ties current action rather closely to current sensory input.

The term "model" is here used in the sense proposed by Minsky: "To an observer B, an object A* is a *model* of an object A to the extent that B can use A* to answer questions about A."[98] This implies that the model subserves a dual function and consists, as it were, of two compartments: one that contains the required knowledge that the organism has about itself and its world (like a data file on a computer tape or disk) and one that contains the machinery for addressing questions and decoding answers. Although bipartite, this constitutes one information-processing system.

Information-processing theories of perceptual processes in problem solving, including those employing computer simulation as their means of formalization and analysis, have in recent years generated valuable insights. It is now generally agreed that problem solving can be represented as searching through a large set of possibilities until a satisfactory solution is found. For instance, a characteristic way in which people respond to a set of symbol sequences is to adopt some hypotheses as to the regularities in the sequence; look for evidence to see whether the hypothesis holds true; if needed, search for additional evidence; and eventually adopt the hypothesis or test another alternative. In this sense, patterns in sets of elements are "conceived," and in some sense generated in the perceiver's mind by some iterated process of devising a plan, testing its consequences, and either accepting or rejecting it.[26, 117]

In solving problems that require actions as outcome this process of "pattern conception" becomes an integral part of the problem-solving process itself. Evidence for the character of the initial perceptual activity comes largely from situations where problems are presented to subjects in visual form. The evidence takes two forms: (1) a record of the subject's eye movements during the first few seconds after problems (e.g., a chess position) are presented to them, showing the succession of fixations during this time, and (2) the subject's ability to retain information about complex visual displays after brief exposure.

As regards the second point, the important feature is that configurations in which the subject recognizes some "meaning" can be retained in short-term memory as one "chunk"[25]: a chess master can reproduce a position on the board almost without error, placing all or almost all of the pieces correctly, but with random boards his performance drops to that of an amateur. As regards the "perception" of the chess position while selecting the next move, the observations indicate that the succession of fixations involves not only the positions of the figures actually on the board but also the connections between present and future positions after a hypothetical move. In this sense the eye movements reflect thought processes.[50] Accordingly, there are two processes in operation that focus on the "map" provided by the chess board: (1) the recoding of information into "meaningful" chunks, where meaning is related to the subject's past knowledge and familiarity with the situation, and (2) a goal-directed exploration of alternative changes to be brought about in the position on the board with these alternative changes actually carried out, one by one, by the appropriate eye movement. Both processes are ingredients of the computer programs that successfully mimic the chess player's performance.[42]

This last example has been discussed in some detail because it offers an analogy or perhaps a prototype for the kind of information processing the association cortex is part of: one could conceive of the primary cortical receiving areas, which are topologic maps of their respective receptor sheets, as fulfilling the role of the chess board in the preceding example. The figures on the chess board would in this view represent the fractionated and fragmented stimulus-feature representations in the specific sensory systems (discussed in Chapter 18). The role of the modality-specific association cortex would be to provide the programs and strategies that operate on the "data structure" in the primary cortical receiving areas, to partition the fractionated stimulus feature to meaningful "chunks" in accord with past experience, to test on these same maps various ways of combining "chunks" to objects of perception with "meaning" in the light of past experience, and to test the outcomes of plans for future action.

FRONTAL INTRINSIC SECTOR

One of the most striking difficulties of the macaque with a defective frontal lobe is manifested in behavioral situations involving delays between presentation of a stimulus and the opportunity to respond to it. The tasks involving such delays were originally devised to demonstrate that animals were capable of symbolic processes in the sense that an "idea" would be required to solve a problem for which the clues to a solution are given prior to, and not at the time of, opportunity for solution.[81]

The delayed reaction test is conducted in the following form or some variant of it: a monkey is allowed to view through bars a piece of food being deposited beneath one of two or more cups on a sliding tray. An opaque door is then lowered in front of the animal for a chosen interval. At the end of this interval the tray is pushed forward to the cage, the door is raised, and the animal is permitted to reach the cups. The animal is allowed to select one cup, the reward being obtained if the proper cup is selected. With training, a normal monkey makes successful choices after delays as long as 90 sec between seeing the food and making the selection.

The delayed alternation test is similar. In this case the animal is required to make alternate right (R) and left (L) turns and to remember which turn comes next in a sequence; there is a delay imposed between completion of one turn and the initiation of the next in the sequence. A special case is that of double alternation, in which the correct choices are RRLLRRLL.

In the studies of Jacobsen[82a] the delayed response task proved to be a selective index of primate frontal lobe injury: after bilateral orbital-frontal lesions, delays as short as 5 sec between seeing the reward and selecting its location reduces success in the choice to a matter of chance; the animal is simply at a loss in choosing the container that conceals the food.

The interpretation of this phenomenon has been the subject of much discussion: Jacobsen[82a] favored the idea that an impairment of recall, manifested as a defect in an "immediate-memory" process, could account for the deficit. Others emphasized increased distractibility. It was also suggested that locomotor hyperactivity occurring after frontal surgery in the monkey might interfere with the utilization of mnemonic devices (e.g., spatial cues) adopted to help bridge the interval between stimulus exposure and response opportunity. None of these suggestions appeared to adequately explain the behavior deficit.

Additional and more refined experiments by Pribram,[108] and Pribram et al.[112] provided evidence for a more precise characterization of the consequences of frontal intrinsic sector ablation. First, there are several factors that can be eliminated as elements contributing to the behavior deficit. The deficit is not due to a deficiency of function in one particular sense modality; it is independent of whether visually discriminable or visually indistinguishable cues are to be relied on for problem solving, and impairment of proprioception is not responsible.

Second, there is some more postive evidence as to what constitutes the basic defect: in the multiple-object discrimination problem (p. 580), which requires alternation at appropriate times between a systematic search strategy and a strategy to persevere once a successful solution is obtained, the monkey with a defective frontal lobe fails to take the cue that indicates the point at which to switch strategies.

When a monkey chooses the positive cue five times in a row, he attains criterion (p. 580). From this moment on, his best strategy would be to return to the original search (i.e., to move on successive trials each of the objects until the peanut is found again). The time for changing from the previous strategy (win-stay) to search is signaled to him by the fact that a response to the previously rewarded object is no longer rewarded. The animal with a defective frontal lobe does not take this cue; instead, it perseveres with the object that was previously but is no longer rewarded.

It appears as though the deficit in an animal with a frontal lobe lesion implies an inability to cope with situations in which an element of remoteness and ambiguity is introduced into the relation between stimulus and success of response. A particularly clear-cut experimental situation devised by Pribram[112] affirmed this idea. The experimental protocol in question was as follows.

Two objects, a small tobacco tin and a flat ash tray, served as cues. All subjects were given 30 trials a day and were initially rewarded only when they chose the tobacco tin. When the tobacco tin had been chosen for ten consecutive responses, the reward (a peanut) was placed under the ash tray until ten consecutive correct responses were again obtained. Another reversal was then instituted. Reversals were continued to the "ten correct" criterion until 500 trials were accomplished. The procedure was then changed so that reversals were given after an animal had reached a criterion of only five consecutive correct responses; the reversals to the five correct criterion were continued until another 500 trials were completed. After this, the monkeys were run to criteria "four correct," "three correct," and "two correct," in that order.*

The prominent feature of this situation is that the number of consecutive successful discriminations required to meet criterion keeps changing; the animal with a defective frontal lobe is unable to cope with this problem.

Behavior guided by stimulus contingencies that are no longer present when the response is to take place implies some temporal extension of the reaction to the stimulus presentations that can bridge the period to the response. What is the nature of this process? As stated before, monkeys with frontal ablations routinely fail the delayed alternation task, but this result changed radically when Pribram[110] used the following stimulus schedule: between each pair of R-L presentations, a 15 sec delay was interposed, so that the temporal pattern of the task had this appearance:

R-L——R-L——R-L——R-L, etc.

This change in stimulus program enabled the animal with a defective frontal lobe to perform the delayed alternation task as well as normal animals. It was critical that the 15 sec alternation occurred between pairs of R-L sequences and not between each R-L alternation. Thus external "pacing" provided by the environment can compensate for the failure of the mechanism that enables the normal animal to overcome the temporal separation between stimulus presentation and response opportunity. The conjecture from this observation is that the role of the frontal intrinsic sector consists in imposing some temporal organization on the stream of stimuli impinging on the organism, and that the proper division or "chunking" of the stimulus stream is an important factor in guiding the organism's behavior when the relation between stimulus and response is fraught with ambiguous expectations.

A suggested analogy is the parsing of sentences and its relation to the meaning a string of words may take on. The well-known example by Chomsky illustrates this point for the word string "they are flying planes," which according to the segmentation in components, may have entirely different meanings: "(they) (are flying) (planes)" or "(they) (are) (flying planes)."

The point of view derived from these observations is that the "short-term memory," which intervenes between the presentation of stimuli and the opportunity for response, involves active working processes of input coding and programming and not merely the deposition of a memory trace of some sort, which fades according to some law of decay.[111]

SENSORY-MOTOR COORDINATION

The example of the relation between eye movements, perception, and planning an activity that was discussed earlier emphasizes the intimate interrelations between sensory and motor processes in the information transactions of the CNS. Another example is that active movements of the eyes leave the perceived world stable, while passive movements (produced by pushing against one's eyeball) make the visual scene jump. Similarly, a patient experiences a recently acquired paralysis of extrinsic ocular muscles as a subjective jumping of the visual scene every time he in-

*Pribram, K. H., et al.: A progress report on the neurological processes disturbed by frontal lesions in primates. In Warren, J. M., and Akert, K., editors: The frontal granular cortex and behavior, New York, 1964, McGraw-Hill Book Co.

tends to move his eyes and is unsuccessful in doing so.

These and similar phenomenon have been interpreted as indicating a mechanism whereby a self-produced movement involves not only the classic efferent discharge to the musculature of the effector organ but also a concomitant (corollary) discharge from motor to sensory areas within the CNS. This corollary discharge is assumed to be present in all voluntary movements but absent in purely passive ones; it presets the sensory systems for the changes in the input that are anticipated consequences of the intended movement.[46, 118a, 129]

The significance of the interrelations between sensory and motor systems has become apparent from experiments on restricted rearing (with motor rather than sensory deprivation), in which a normal adult was fitted with distorting (prismatic) spectacles. The prisms imposed tilts and curvatures upon contours and displaced the visual scene. Held and Freedman[79] have shown that far-reaching adaptation to the optically induced distortions and displacements can be achieved, provided that the subject actively moves about while viewing his visual environment through the distorting prisms. In contrast, when the subject is moved about in a wheelchair while wearing the distorting spectacles, he fails to produce any adaptation. This proves that self-initiated movement is a necessary prerequisite for perceptual adaptation under conditions of deranged sensory input. It is thought that the hypothetical corollary discharges associated with the self-produced movement permit the development of this perceptual adaptation.

Held and Hein[80] performed an additional experiment that seems to indicate that movement plays a similar role in the initial acquisition of sensory-motor coordination during early development, as it does in adaptation to rearranged sensory input. A kitten whose visual exposure is limited from birth to the inside of the drum shown in Fig. 19-4 will exhibit essentially normal form and depth perception when tested after several months, provided that he moves himself about actively (Kitten *A* in Fig. 19-4). On the other hand, the same exposure under passive conditions, e.g., within a gondola (kitten *P* in Fig. 19-4), in which a "passive" kitten is carried about by an "active" littermate, precludes the normal development of visual motor coordination. The interpretation is that the passively moved kitten lacks the opportunity to correlate its self-produced motor output with the corresponding changes of visual input.

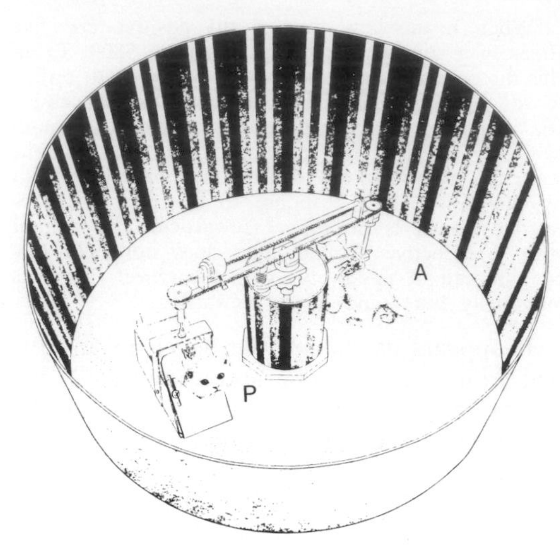

Fig. 19-4. Active and passive movements of kittens were compared using this apparatus. Active kitten walked about more or less freely; its gross movements were transmitted to passive kitten by chain and bar. Passive kitten, carried in gondola, received essentially same visual stimulation as active kitten because of unvarying pattern on wall and on center post. Active kittens developed normal sensory-motor coordination; passive kittens failed to do so until after being freed for several days. (From Held.[18])

There is evidence that profound and lasting disorders of the postulated corollary discharge mechanism are related to lesions of the frontal lobe or the basal ganglia in man.[47] A more direct demonstration exists for the monkey; Bossom[57] has shown severe deficiencies in adaptation to distorting prisms in monkeys with bifrontal lobectomies or bilateral lesions in the head of the caudate nucleus.

INTERHEMISPHERIC INTEGRATIONS

This topic directs attention immediately to the large forebrain commissures, of which the corpus callosum is the most prominent structure. Evidence that has been forthcoming since 1950 dealt at first with the role of the corpus callosum in the interhemispheric transfer of visual discrimination learning[31] and later with more general aspects of interhemispheric relations.[43]

Sectioning the optic chiasm in a mammal with crossed optic fibers leaves the major part of the visual field intact, but each eye then projects only to its homolateral hemisphere (Fig. 19-5). Cats with sectioned chiasms, trained to discriminate visual patterns through

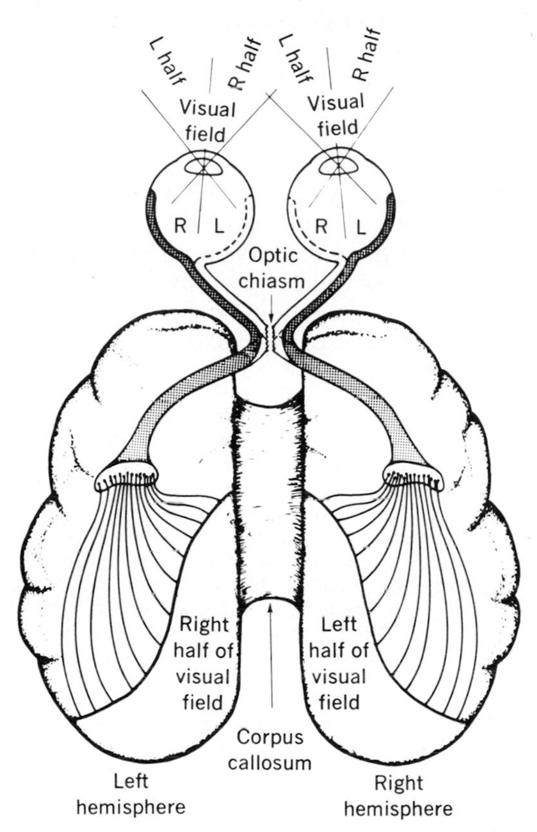

Fig. 19-5. Sketch indicating effects of midsagittal division of optic chiasm. (From Sperry.[44])

one eye while the second eye is covered with a mask, can also execute the same task through the other, previously covered eye. Thus there is intercommunication between the two brain halves. However, this cross-availability of information between the two hemispheres can be abolished by sectioning the corpus callosum. Accordingly, with both optic chiasm and corpus callosum sectioned, the cat can learn a different, even opposite, discrimination task with each eye without mutual interference. This functional independence of the surgically separated hemispheres with respect to discrimination learning has since been amply substantiated.

There are some exceptions to this general rule. For instance, easy brightness discriminations learned with one eye do transfer to the other hemisphere in split-brain preparations.[93] Moreover, split-brain cats and monkeys can discriminate brightness between one stimulus presented to one eye and another stimulus presented to the second eye.[113] Thus certain cross comparisons are still achieved between the two divided hemispheres: it appears that discriminations involving visual identification of objects are a highly corticalized process that requires the corpus callosum for interhemispheric transfer; on the other hand, discriminations between spatial distributions of otherwise identical stimulus components (e.g., five-pointed star *versus* six-pointed star) transfer from one side to the other even in the absence of the corpus callosum, presumably *via* brainstem pathways.[128]

Corpus callosum section has been carried out in humans in an attempt to cope with otherwise unmanageable epileptic seizures. These patients have undergone extensive tests of their perceptual and cognitive abilities, and many interesting findings were uncovered, notably with regard to the subtleties of the differentiation between the major and the minor (subordinate) hemisphere. Right-handed patients whose left hemisphere is equipped with the expressive mechanism for speech and writing and with the main centers for comprehension and organization of language can communicate verbally about the visual experiences of the right half of the visual field and about the somesthetic experiences of the right body half. The right hemisphere in aphasic and agraphic patients who are unable to express themselves verbally can be shown to perform inter- and cross-modal generalizations of perceptual or mnemonic material. For example, after a picture of some object, e.g., a cigarette, has been presented to the minor hemisphere through the left visual field, the subject can retrieve the item pictured from a collection of objects using blind touch with the left hand, which is mediated through the right hemisphere. Unlike the normal person, however, the commissurotomy patient is obliged to use the corresponding hand (the left hand, in this case) for retrieval and fails when he is required to search with the right hand. Using the right hand, the subject recognizes and calls off the names of each object, but the right hand, or its hemisphere, does not know what it is looking for. The minor hemisphere can also perform certain generalizations and abstractions, e.g., directing the sorting of objects into groups by touch on the basis of shape, size, or texture. Moreover, subjects give correct answers with manual responses of the left hand in tests requiring the determination of sums or products of numerals perceived stereographically, but they are unable to report the same answers verbally. In some cases the minor hemisphere is found to be superior to the major hemisphere, notably in tasks that involve drawing spatial relationships and performing block design tests.[44]

Interesting examples of "cross-cuing" from

one hemisphere to the other have been observed in some subjects. Gazzaniga describes the following incident:

> At first, after either a red or a green light was flashed to the right hemisphere, the patient would guess the color at a chance level, as might be expected if the speech mechanism is solely represented in the left hemisphere. After a few trials, however, the score improved whenever the examiner allowed a second guess. We soon caught on to the strategy the patient used. If a red light was flashed and the patient by chance guessed red, he would stick with that answer. If the flashed light was red and the patient by chance guessed green, he would frown, shake his head and then say, "Oh no, I meant red." What was happening was that the right hemisphere saw the red light and heard the left hemisphere make the guess "green." Knowing that the answer was wrong, the right hemisphere precipitated a frown and a shake of the head, which in turn cued in the left hemisphere to the fact that the answer was wrong and that it had better correct itself!*

The general implication is that patients with neurologic lesions have various strategies at their command to compensate for their defects. Therefore it is very difficult to obtain a conclusive neurologic description of a patient with brain damage.

Another factor is the difference in symptoms caused by a certain lesion, depending on the stage of neural maturation at which the defect occurred. A case of agenesis of the corpus callosum, recently studied by Sperry,[119] illustrates the high degree of functional reorganization and compensation that is possible with congenital defects in the CNS. This patient was able to rapidly read words that fell partly in one visual half-field and partly in the other, a task at which patients with surgical separation of the hemispheres in later life fail entirely. Interestingly, persons with congenital corpus callosum defects are very often ambidextrous, suggesting that speech is represented bilaterally. Compensation is not equally effective for all tasks and in all sense modalities; information of tactile, positional, and kinesthetic patterns acquired by one hand cannot be utilized as fully to guide motor action of the opposite hand as it is in normal subjects. This deficit is particularly obvious in manual maze tests, in which each hemisphere has to acquire, essentially independently, the skill to solve the problem.

MECHANISMS OF LEARNING AND MEMORY

The process of learning is concerned with establishing a relation between a present and a past event; furthermore, when successive stimuli can be compared and a response that is contingent on that comparison can be made, we speak of memory. The implication is that the original event must leave some change (a trace or engram)[41] in the nervous system that permits the present event to be related to it.

Research on memory has focused on the properties and nature of this hypothetical trace. However, with the exception of some measurements of human recall, it is for all practical purposes impossible to study memory in isolation without also involving associative learning, notably in animal experiments applying procedures to elucidate the character of this trace.

Weiskrantz has analyzed this complex situation lucidly: "In order to know whether an event can be stored and retrieved (more strictly, can produce a change within the organism that permits a subsequent repetition of the event to be related to the original), we must have a response that we can measure. We obtain such a response by associating our event with another that is known to control behavior in a predictable manner, i.e., a reward or punishment."[49] Learning, then, is bound up with the study of memory as a matter of practical necessity.

Natural history of the memory trace. In the typical experimental situation the subject initially undergoes one of a series of *exposure trials* for learning; e.g., an exposure trial may consist of the application of a visual or an auditory stimulus, followed by an aversive stimulus that the subject can avoid by a specific act or instrumental performance (like pressing a bar). On the subsequent *retention trials* the subject is expected to "remember" the association between the neutral and the aversive stimulus. The overt evidence for the establishment of this association is provided by the subject's act to prevent the aversive stimulus from occurring.

When an electroconvulsive shock (ECS) is administered immediately after the exposure trial, the performance at the retention trial is usually impaired; however, if the ECS is administered after some longer interval (e.g., some 6 hours after the exposure trials), the impairment, if any, is unappreciable.[59] Various other interferences such as brain concussion[38] and anesthetics[102] have the same retrograde amnesic effect. These and similar findings have led to the idea that a gradual change occurs in the trace initially set up by the exposure trial, in the sense of a *consolidation.*

The original theory of consolidation, put forward by Müller and Pilzecker,[99] was that information is initially stored as some form of persistent, reverberating neural activity that

*Gazzaniga, M. S.: The split brain in man, Sci. Am. **217**:24, 1967.

is then gradually converted into structural changes at synapses. In more general terms the implication is that there are at least two kinds of traces: one that is short-lasting and susceptible to disrupting treatments and one that is more stable and long lasting. There is a suggestive parallel between the distinction between these types of traces and the concepts of a more immediate short-term and more indirect long-term memory, based on observations in man.

This latter distinction has a long and colorful history in human psychology. William James[22] distinguished between immediate knowledge of the past and what he called "properly recollected objects." The latter required knowledge that what we are recovering differs from what we are presently experiencing. Along similar lines, William Exner wrote in 1880:

> Impressions to which we are inattentive leave so brief an image in the memory that it is usually overlooked. When deeply absorbed, we do not hear the clock strike. But our attention may awake after the striking has ceased, and we may then count off the strokes. Such examples are often found in daily life. We can also prove the existence of this primary memory-image, as it may be called, in another person, even when his attention is completely absorbed elsewhere. Ask someone, e.g., to count the lines of a printed page as fast as he can, and whilst this is going on walk a few steps about the room. Then, when the person has done counting, ask him where you stood. He will always reply quite definitely that you have walked. Analogous experiments may be done with vision. This primary memory-image is, whether attention has been turned to the impression or not, an extremely lively one, but is subjectively quite distinct from every sort of after-image or hallucination. . . . It vanishes, if not caught by attention, in the course of a few seconds. Even when the original impression is attended to, the liveliness of its image in memory fades fast.*

Storage versus accessibility. The retrograde amnesia that occurs after ECS does not necessarily imply that this treatment interferes with consolidation. One may argue that the disturbance affects the subsequent availability of the trace for retrieval. The question is: How permanent is the apparent retention impairment? Experimental as well as clinical evidence[38, 136] indicates improvement of retention with time after the disruptive treatment; e.g., subjects who had no retention 24 hr after ECS may have some residual when tested 48 hr later. This suggests that the interfering trauma does not cause total trace destruction but that the retrieval of traces may be temporarily interfered with.

Neurophysiologic mechanisms related to learning. A first approach to the study of neural mechanisms that may be involved in the formation of the memory trace is to study the electroencephalographic (EEG) changes during acquisition of new behavior.

In the nonattentive, relaxed state the EEG exhibits the characteristic 8 to 13 Hz oscillations of the alpha rhythm. Any novel and unexpected stimulus results in an arrest or blocking of this alpha activity and the electric record assumes the form of low-voltage, fast activity. There are concomitant somatic and autonomic signs: eye movements toward the stimulus source, respiratory irregularities, change in electric skin resistance, and an increased muscle tone. The sum total of these changes is the *orienting* response, of which the blockade of the alpha rhythm is the EEG sign. If the stimulus has no behavioral significance to the animal and is repeated a number of times, the orienting response quickly subsides: *habituation* has occurred. This habituation is quite specific for the repeatedly applied stimulus. The degree of selectivity with which habituation occurs can be seen in Fig. 19-6. Certain kinds of stimuli lead more readily to habituation than others. For instance, visual stimuli may require hundreds of presentations before the alpha blockade over the occipital area fails to take place, whereas the cortical arousal with auditory stimuli fades after some 20 to 30 stimulus presentations.[30]

An important new aspect was introduced in 1935 when Durup and Fessard, in experiments with habituation to visual stimuli, accidentally discovered that the click of the shutter of a camera used for intermittent photography of the oscilloscope record also elicited alpha block after some time, even when the light stimulus was withheld. The click had not previously produced a change of electrocortical activity and had apparently acquired the ability of altering the occipital alpha rhythm as a result of its temporal association with the visual stimulus. When the click was subsequently presented without association to the light stimulus, it soon lost the acquired capacity of altering the occipital rhythm.

These experiments, and others performed later in different animal species and man,[84] clearly demonstrated that the occipital alpha

*Exner, W.: Physiologie der Grosshirnrinde. In Hermann, L., editor: Handbuch der Physiologie, Leipzig, 1880, F. C. W. Vogel, vol. 2.

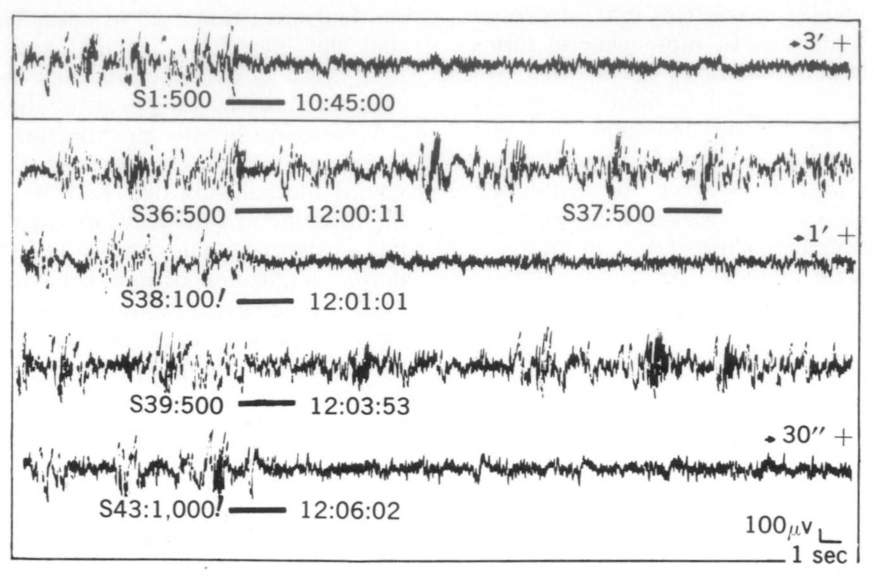

Fig. 19-6. Cortical EEGs from suprasylvian gyrus of normal cat showing habituation of arousal reaction to 500 Hz tone after about 30 stimulus presentations. Top tracing shows response to first presentation of 500 Hz tone (S1:500). Solid bar shows duration of stimulus, followed by time in hours, minutes, and seconds (10:45:00). Second tracing shows 36th and 37th trials (S36 and S37). Then novel tone of 100 cycles is presented in 38th trial (S38:100!), followed by repetition of habituated tone (S39:500) and then another novel tone (S43:1,000). Figures at right above EEG traces indicate duration of EEG activation in each trial. (From Sharpless and Jasper.[116a])

rhythm of the EEG can be depressed or blocked by sound, a previously ineffective stimulus, if it has been repeatedly paired with light as an unconditioned stimulus. Prior to this finding it was generally accepted that responses that could be conditioned in the classic pavlovian sense were of three kinds: (1) glandular, smooth-muscle, and blood pressure responses, (2) involuntary responses in striate muscle (e.g., eye movements), and (3) potential voluntary responses (e.g., withdrawal movements or locomotion). To these a fourth group must now be added—electrical activity of cortical centers themselves.

Among the cortical electrical events that can be brought under the control of conditioning stimuli is the *cortical steady* potential, i.e., a maintained, nonrhythmic difference of potential between the surface of the cortex and the immediately underlying white matter. As an example of this, Rowland[37] demonstrated that the changes in cortical steady potential that normally occur in hungry cats during feeding can be conditioned to flash or click stimuli.

The work of Adey[1] and co-investigators introduced a new perspective to this line of investigation: the emphasis in these studies shifted to attempts to discern, in the simultaneously recorded electrical activity of various cortical and subcortical structures, consistent mutual relations that would characterize the participation of and interaction between different neuronal populations in the learning process. Before learning started, the electrical waves in the hippocampus were found to lead those in the entorhinal cortex; but when the cat learned to reach the food reward in the T maze, this phase relation was reversed.[51] One possible interpretation of this phenomenon is the assumption that the information acquired during the learning process was stored in the entorhinal cortex and served subsequently as a "standard" against which sensory input was compared to decide which action should be taken. Observations such as these may be taken to support Penfield's suggestion[103] that the hippocampus functions in consolidation and recall of memory in conjunction with other neuronal systems, rather than solely as an independent and isolated respository of memory traces.

The idea of the participation of a multiplicity of neural structures in the information transactions that underlie learning is particularly well illustrated by an experiment of John and Killam[85]; this experiment also directs attention to differences in the involvement of different neural structures as learning progresses. The conditioning stimulus was a light flashing 10 times/sec, called a tracer-conditioned stimulus because it allowed the investigators to trace the signal in the brain wherever it would appear in the form of evoked potentials at 10 Hz or multiples thereof. The cat had to learn to jump from one compartment of a double-grid box to the other to avoid shock whenever the conditioned stimulus was presented. Later a differential stimulus consisting of a light flashing 7 times/sec was introduced. No shock followed the presentation of this stimulus.

The conditioning stimulus initially gave rise to photic driving at many points in the brain, which faded out with repeated presentation as evidence for habituation. As soon as the conditioning started (when shock was presented after the conditioning stimulus), the tracer potentials reappeared at all recording sites, including the visual and auditory cortices, the reticular formation, nucleus ventralis anterior of the thalamus, the fornix, the septum, the hippocampus, and the amygdala. As the cat learned to avoid shocks, the tracer potentials disappeared again, except from the visual cortex, hippocampus, and midbrain reticular formation. Later still, when the cat had completely learned the avoidance response, the tracer potentials also disappeared from the hippocampus but reappeared in nucleus ventralis anterior of the thalamus.

During differential training, it was found that the 7/sec flicker first generated a 10/sec evoked response in the visual cortex and that the cat would make its usual avoidance response. As the cat began to discriminate behaviorally, however, the cortical responses also became differentiated, and the 7/sec flashes evoked 7 Hz waves. During the intermediate phase the cat usually made an error when the frequency of the evoked potential was not that of the presented stimulus. In other words, the behavior corresponded to the cortical activity rather than to the objective conditioning stimulus.

Experiments such as those just described highlight the complexity of the neural changes that accompany the acquisition of new (learned) behavior. Various experimental approaches have also been pursued to determine whether prior stimulation can induce temporary changes in the functional characteristics of individual neurons of the CNS. For this purpose, Morrell[30] studied a population of neurons in the visual cortex that was responding to a particular "preferred" stimulus (e.g., an illuminated line oriented at a certain angle of the visual field) and also to some other stimulus (e.g., a click). As a first step in the experimental procedure the temporal discharge pattern of the neuron in response to the presentation of the visual and the auditory stimulus in isolation was ascertained as being quite different and characteristic for each stimulus mode. Next the two stimuli were presented simultaneously for some 40 trials; the temporal discharge pattern of the neuron was then different from that obtained by separate presentation of either of the two stimuli. The interesting finding was that presentation of the light stimulus alone, after this period of "training," elicited a temporal sequence of discharges that resembled closely the sequence obtained with the simultaneous visual and auditory stimulation. These and similar observations demonstrate that temporal discharge patterns of certain neurons can temporarily be altered by prior exposure to combination stimuli.

The analysis of the synaptic events that may be related to this and similar observations of "plasticity" in complex neural systems has been attempted in studies of simple neuronal aggregates: the underlying idea is that one may be able to characterize changes of synaptic function that reflect prior stimulus exposure and interpret them under better-defined conditions than are attainable in complex neuronal systems.[23]

The marine gastropod mollusc *Aplysia* was found to be a suitable object of study: its behavioral repertoire includes as part of a defensive response a quill-withdrawal reflex that can be elicited by potentially harmful mechanical stimuli. This reflex is supported by a single reflex arc involving only a few synaptic relays. The reflex "habituates" rapidly if the same mechanical stimulus is applied a number of times in succession. Pinsker et al.[105] succeeded in characterizing the neural event underlying this habituation as a reduction of effectiveness of the afferent excitatory elements to the motorneuron that supports the reflex response.

Molecular basis of memory trace. The clarification of the mechanisms for coding genetic information by the nucleotide sequence of chromosomal DNA, and for its decoding during morphogenesis, led to the question of whether an analogous biochemical transcription mechanism might also be operative for behavioral information in the nervous system. Over the past decade, Hyden[20] and co-workers developed refined microtechniques to determine the amount of nuclear RNA and the relative proportions of the four bases that it contains. The analysis of small groups of neurons of glia cells revealed a characteristic change of the nuclear RNA composition during acquisition of new behavior: this change consisted in an alteration of the relative distribution of the bases in the RNA molecule, reflecting possibly an increase in a messenger-type RNA. The possible exchange of proteins and RNA between glia and neurons in the course of learning has also been suggested.[19] In other studies it was found that the incorporation of uridine, one of the four bases, increases by about 50% in animals under training, in comparison to control animals.[12]

The possible role of protein synthesis in the

formation of the memory trace has also attracted attention. The experimental basis for these studies lies in the effect of inhibitors of protein synthesis (such as puromycin or acetoxycycloheximide) on learning.

Barondes and Cohen[56] report that acetoxycycloheximide will disrupt memory formation, but it must be present in the brain at the time of the presentation of the conditioning stimulus or given immediately (up to 5 min) afterward. In this case the rats learned at first as well as saline-injected controls, remembered the conditioned response normally for 3 hr, but had almost entirely forgotten it after 6 hr. This amnesia persisted when the animals were tested again 7 days later. Accordingly, the inhibition of protein synthesis appears to affect only permanent memory, but it must be present at the time of conditioning or very shortly thereafter.

The precise relation of these and related experimental observations to information processing at the level of neural and behavioral events poses considerable conceptual difficulties that are thus far unresolved.[48] One of the questions is whether there are biochemical mechanisms in neurons or glia that would transcribe different firing patterns of neurons, carrying different sensory information (Chapter 18) into different associations of molecules. It is perhaps conceivable that the biochemical changes that accompany changes in environmental influences are early manifestations and, in some sense, forerunners of changes in growth and connectivity of short-axoned neuronal elements whose extensive postnatal differentiation is known to be subject to environmental influences.[2]

These and related questions cannot be answered at present, but the experimental methods available today provide the tools for a meaningful approach.

Human memory processes: a conceptual scheme. Contemporary memory research in normal individuals has characterized several stages in human memory processes (Table 19-1).[8, 32, 118] In the first place, there is evidence for a *sensory memory* of limited capacity that holds raw data of sensory events for a period of a few hundred milliseconds for attention, scanning, and further processing. This first stage in the processing of sensory information prior to further storage imposes in some sense a discontinuous, quantal character on the continuous stream of incoming sense impressions, much like a sample-and-hold device does in electronic data-gathering systems: the continuous flow of sensory signals becomes "packaged" in discrete units, each unit being either selected for transfer into another storage medium or discarded. Information is removed from sensory memory either by spontaneous decay or by a process of "erasure" in response to incoming new information.

The successful transfer from this short-lived store requires that it be attended to and recoded into a more stable form. This may occur in two different forms[32]: a verbal label may be attached to the content of the sensory memory, which is then thought to be transferred to another limited-capacity storage system, the *primary memory*. Although not accurately determined, there is a suggestion that its information capacity matches the 7 ± 2 bits of information content that are also the upper

Table 19-1. Human memory processes*

	Storage system			
	Sensory memory	*Primary memory*	*Secondary memory*	*Tertiary memory*
Capacity	Limited by amount transmitted by receptor (?)	The 7 ± 2 of memory span	Very large (no adequate estimate)	Very large (no adequate estimate)
Duration	Fractions of a second	Several seconds	Several minutes to several years	May be permanent
Entry into storage	Automatic with perception	Verbal recoding	Rehearsal	Overlearning
Organization	Reflects physical stimulus	Temporal sequence	Semantic and relational	?
Accessibility of traces	Limited only by speed of readout	Very rapid access	Relatively slow	Very rapid access
Types of information	Sensory	Verbal (at least)	All	All
Types of forgetting	Decay and erasure	New information replaces old	Interference: retroactive and proactive inhibition	May be none

*From Ervin and Anders.[8]

limit of information capacity in sensory systems (p. 565). Forgetting occurs in primary memory as new items enter and displace old ones. In the absence of such interference, the information storage extends over several seconds.

An alternate route, if verbal recoding does not take place, consists in transfer from sensory to *secondary memory,* a large and more permanent storage system. Forgetting in secondary memory appears to require "unlearning" and interference by either previously (proactive inhibition) or subsequently (retroactive inhibition) learned material. The secondary memory can also serve as recipient for information displaced from primary memory. The transfer mechanisms in this case are associated with rehearsal, which consists in recycling of material through primary memory. The probability of successful transfer appears to increase with the amount of time an item spends in primary memory: thus the success of transfer increases with the number of rehearsals associated with transfer cycles.

Present evidence suggests that "meaning" is a significant factor in the organization of secondary memory, in contrast to the operating mode of the primary memory. This difference is illustrated by the kinds of errors made on recall from the one as opposed to the other store[53]: in primary memory most errors result from phonetic confusion; items that sound alike (e.g., B's and V's) are substituted for one another. Errors in secondary memory, on the other hand, involve items with similar meaning. Thus part of the recoding in secondary memory is semantic, but other types of relationships, e.g., temporal and spatial contiguity, also play a role.

Finally, there are memory traces of extreme durability, ready access, and high resistance to brain damage and disease. This is the store for which Erwin and Anders[8] proposed the designation of *tertiary memory*: it is thought to be the repository of such highly overlearned material as one's own name or how to read and write, i.e., material and skills outlasting retrograde amnesias that may affect other, less stable components of memory.

Pathologic memory: anterograde amnesia. A failure to retain newly acquired information is commonly described as anterograde amnesia: when a patient with this disorder is presented with new sensory information, he is capable of responding appropriately for as long as he remains exposed to the test material. However, if the stimulus material is removed, the patient loses his ability to produce the correct response within a few seconds.

A clinical syndrome that resembles this description of anterograde amnesia is named after Korsakoff; this syndrome is commonly associated with diffuse degenerative changes in the brain, notably in chronic alcoholics, and consists in the inability to remember recent events although memory for remote events remains intact. Frequently the patients attempt to fill the gap between the present, which they do not remember, and the remembered past with confabulations.

Bilateral ablation or lesion of the mesial aspect of the temporal lobes (sparing the lateral neocortex) produces a generalized (global) memory impairment that cuts across the distinction between verbal and nonverbal material and different sense modalities.[27] Sensory and primary memory seem to function normally in these patients, as demonstrated by their normal ability to reproduce short sequences of digits. The performance deficit becomes apparent as soon as the task approaches the capacity for the primary memory or when recall is delayed. Apparently there is some failure in the efficient use of rehearsal: interrupting rehearsal, even after the patient may have engaged in it for several minutes, is often sufficient to erase any recall of the rehearsed material. Thus it appears that the mnemonic capability of these patients depends entirely on the function of the primary memory.[8] This implicates a failure of secondary memory as a principal cause of the disorder: the difficulty may be one of encoding for transfer from primary to secondary memory, of storage in secondary memory, or of retrieval from it. The weight of current opinion favors the first alternative, i.e., that the amnesiac's basic deficiency is one of encoding and categorizing experience in a form suitable for subsequent storage.

Material-specific memory loss. In contrast to the generalized memory defect discussed in the preceding section, one also encounters memory changes that are limited to one particular kind of stimulus material and that need not be restricted to a single sense modality.

One of the most notable examples is the defect in verbal learning and recall after left anterior temporal lobectomy, sparing the speech area. Such patients have difficulty in remembering what they read or hear, but they retain nonverbal material normally. Conversely, right temporal lobectomy impairs memory for both visual and auditory patterned stimuli that are nonverbal.[96] Thus specific memory defects appear as a sequel to unilateral temporal lobectomy, the specific nature depending on the hemisphere involved.

Remembering or reconstructing. The attempts to distinguish between the processes of remembering and reconstructing are a current theme in memory research and are based on techniques of inferring the properties of memory from the types of errors made by subjects. Bartlett[3] developed a theory of remembering that was based on the observation that when subjects tried to recall stories he had asked them to learn, their version was shorter, the phraseology updated, and the entire tale more coherent and consequential. Moreover, the subjects were often unaware that they were substituting rather than remembering, and often the very part that was created anew was the part the subject was most certain about. Thus there was some invention or, in Bartlett's term, *"constructive rendering."*

Bartlett was led to propose that recall from past experience is seldom merely reduplicative; instead, we reconstruct and schematize the remembered material in accord with past experiences, circumstances, and expectations.[3] Thus memory and recall have more nearly the features of dynamic encoding and decoding processes than the properties one would expect from an immutable trace or engram.[109]

Current trends in thought on human information processing suggest that in order for material to be stored within secondary memory, it must be assimilated into an existing organization of rules and schemata, and that retrieval from memory is a generative process in which these same rules and schemata reenact, as it were, the event we attempt to remember.[32]

NEURAL MECHANISMS OF VERBAL BEHAVIOR

Present knowledge of the organization of language in the brain is based almost exclusively on observations on adults in whom certain areas of the brain have been damaged, mostly by occlusion of blood vessels. However, there are suggestions of a certain degree of linguistic capability in nonhuman primates, notably the chimpanzee.[70] Furthermore, it has been argued that the striking development of the angular gyrus region in man that appears to be related to human language[10] is foreshadowed in the macaque by the extensive corticocortical connections from all sensory projection fields to the premotor area.[101]

During the second half of the 19th century, neurologists became increasingly impressed by observations suggesting that disorders of language resulting from brain damage *(aphasias)* can occur in the absence of generalized motor and sensory deficits or intellectual impairment; e.g., speech can be disturbed by incoordination of the muscles of articulation, although these same muscles function normally in nonlinguistic activities; similarly, a deficiency in comprehension may occur for spoken or written words in the absence of auditory or visual deficits for nonverbal material.

Of the various classifications of aphasic disorders proposed since that time, the theory of Wernicke, published in 1874, continues to serve as a lucid and valid guide for the analysis of the neural mechanisms of normal and pathologic language behavior.[10] Wernicke's investigations followed by some 10 years the discovery by Broca that a circumscribed lesion in the dominant hemisphere, slightly anterior to the cortical motor representation of tongue, lips, and vocal cords, caused the characteristic deficit now known as Broca's aphasia: the patient typically produces little speech with

Table 19-2. Aphasia classification by syndromes*

Type of aphasia	Site of lesion	Spontaneous speech	Comprehension	Repetition	Naming
Broca's aphasia	Posterior, inferior frontal	Nonfluent	Intact	Limited	Limited
Wernicke's aphasia	Posterior, superior, temporal	Fluent	Impaired	Impaired	Impaired
Conduction aphasia	Arcuate fasciculus	Fluent	Intact	Impaired	Impaired
Isolation syndrome	Association cortex	Fluent, echolalic	Impaired	Intact	Impaired
Anomic aphasia	Angular gyrus	Fluent	Intact	Intact	Impaired

*From Green.[74a]

great effort and poor articulation, even if it merely involves repeating a sentence just heard; yet comprehension of spoken and written language is normal, as is the capacity to reproduce a melody. Because of the close proximity of Broca's area to the motor cortex, these patients frequently also suffer from some paralysis on the right side of the body.

Wernicke's observations delineated a form of aphasia that differs strikingly from Broca's type (Table 19-2): speech is produced rapidly and without effort; grammatical structure and normal speech rhythm are preserved, but there are frequent errors in word usage (paraphasias) and circumlocutions. Most prominent is the inability to understand spoken or written language. The localization of the cortical area identified with this form of aphasia is an important point in Wernicke's theory: Wernicke's area lies next to the cortical representation of hearing, a location that suggests that this area plays a role in the recognition of the patterns of spoken words. A natural assumption was that Wernicke's and Broca's areas must be connected: the act of speaking would then consist in retrieving in some manner the auditory form of words; this information would be transmitted to Broca's area, where it would be transformed into the complex programming of the speeech organs (Fig. 19-7).

This basic model is easily extended to account for a remarkable variety of aphasias. For instance, it accounts for the aphasic symptoms due to lesions in the lower parietal lobe that sever the arcuate fasciculus, which connects Wernicke's and Broca's areas: in such cases, there is fluent paraphasic speech

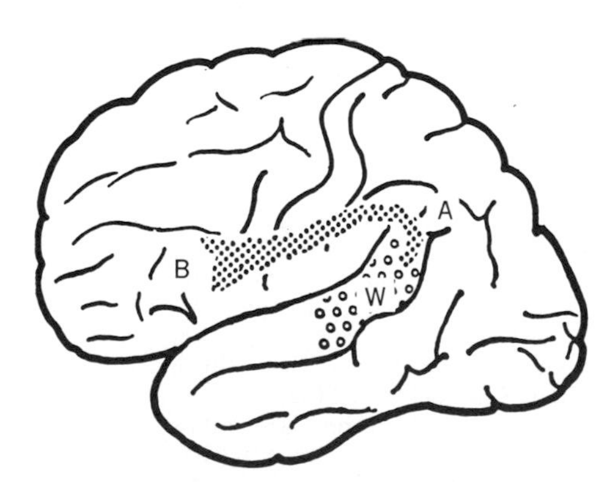

Fig. 19-7. Lateral surface of left hemisphere of human brain. *B*, Broca's area, which lies anterior to lower end of motor cortex; *W* (open circles), Wernicke's area; *A* (filled circles), arcuate fasciculus, which connects Wernicke's area to Broca's area. (From Geschwind.[10])

and writing, while comprehension of the spoken and written word remains intact.[71] Another aphasic syndrome, known as *isolation syndrome,* can also readily be explained (Table 19-2). Language comprehension is completely lacking, and the patient is totally unable to spontaneously produce propositional speech; however, the patient can repeat sentences perfectly, with normal articulation, and he can complete sentences after having been told the beginning: there is a certain degree of verbal learning. Postmortem examination of one patient with this syndrome showed that the connections between Broca's and Wernicke's areas were intact, but in the regions surrounding the speech area, either the cortex or underlying white matter was destroyed.[73] Consequently, the speech region was totally isolated from the sensory cortical projection fields.

More restricted language defects can arise as a consequence of partial disconnection of the speech area from the cortical representation of one particular sense modality. In these instances the patient may find it impossible to produce the name of an object if confronted with it in one sense modality but not in another (*anomic aphasia,* Table 19-2), although spontaneous speech is fluent.

Geschwind and Kaplan described a particularly illustrative case of a patient whose lesion in the corpus callosum was subsequently confirmed by postmortem examination:

> When writing with the right hand the patient produced linguistically correct words and sentences and carried out calculations correctly. When writing with the left hand he produced incorrect words (for example, "run" for "go") and performed calculations incorrectly. The theory outlined above implies that, for writing to be carried out correctly with the left hand, the information must be transmitted from the speech areas across the corpus callosum, whose interruption in our patient explained his failures. Similarly, the patient could correctly name objects (concealed from vision) which he palpated with the right hand. On the other hand he would misname objects palpated with the left hand, although it could be shown by nonverbal means that his right hemisphere recognized the object. Thus, if a pencil was placed in his left hand the patient could draw the object previously held in that hand. Again, the Wernicke theory implies that, for an individual to correctly name an object held in the left hand, the information must be transmitted from the sensory regions in the right hemisphere to the speech regions via the corpus callosum, which had been destroyed in this patient.*

*Geschwind, N., and Kaplan, E.: A human cerebral deconnection syndrome, Neurology **12:**675, 1962.

Psycholinguistic considerations in the study of aphasias

In addition to the classic neurologic approach to the study of aphasias, which is firmly rooted in neuroanatomic considerations and exemplified by Wernicke's theory, attempts have also been made to characterize aphasic impairments in terms of linguistic theory. H. Head,[16] as early as 1926, and more recently Luria[90] drew attention to the condition of *semantic aphasia,* in which there is neither distortion of articulation nor of comprehension of word meaning, but in which patients are unable to perceive the complex relations between separate concepts presented in the logicogrammatical system of language: grammatical constructions requiring a system of subordination of words (e.g., "the brother's father") or constructions with double negations (e.g., "I am not used not to submit to rules") are most susceptible to disruption. Typically, this form of language disturbance is coupled with some disorientation in space, constructive apraxia, and acalculia.

Luria[90] introduced the term "*dynamic aphasia*" to designate a particular type of breakdown in contextual speech that leaves the fundamental code of language essentially preserved. This is a condition attributed to damage of the left frontal lobe. The principal deficit concerns the predicative structure of speech: the use of verbs is considerably more impaired than that of nouns, and the difficulty becomes most apparent when the patient changes from rehearsed and habituated speech to independent, extemporaneous expression.

The addition of these two forms of language disorders broadened the scope of syndromes considered in the preceding section and became part of an attempt to categorize aphasias in terms of functional criteria. The starting point for these considerations is the view that normal language is based on two distinct processes that combine to form an integral unit[21]: the selection of appropriate constituents and their combination to grammatically valid and meaningful expressions. The production of linguistic expressions starts with the selection of constituents that are to be combined and integrated into a context: this process may be thought of as *encoding,* with the selection process being the antecedent and the building of the context the consequent. In the reception of language the sequence of these processes, which may be thought of as *decoding,* is reversed: the listener is faced with a context and must retrieve from it the constituents. The thesis is that in aphasic disorders, the second (consequent) step is impaired[83]: thus there are *aphasias of encoding,* with the selection process remaining intact and the context formation being subject to deterioration. This situation is characterized by the loss of the so-called little tools of language, i.e., connectives, articles, and pronouns; ultimately speech is reduced to independent words in isolation, i.e., nouns and nominal forms of verbs, with breakdown of contextual speech. Conversely, *aphasias of decoding* consist in primary impairment of the word selection and identification. Thus the overriding symptoms are the improverishment of the variety of means available, a tendency to supplant them by generalized substitutes, and an inability to furnish synonyms and antonyms.

In the extreme, aphasias of the decoding type approach the character of Wernicke's aphasia, while Broca's aphasia has the features of the disturbance of encoding.

REFERENCES

General reviews

1. Adey, W. R.: Spontaneous electrical brain rhythm accompanying learned responses. In Schmidt, F. O., editor: The neurosciences. Second study program, New York, 1970, The Rockefeller University Press.
2. Altman, O.: Postnatal growth and differentiation of the mammalian brain, with implications for a neurophological theory of memory. In Quarton, G. C., Melnechuk, T., and Schmitt, F. O., editors: The neurosciences. A study program, New York, 1967, The Rockefeller University Press.
3. Bartlett, F. C.: Remembering, London, 1932, Cambridge University Press.
4. Bender, M. B.: Disorders in perception, Springfield, Ill., 1952, Charles C Thomas, Publisher.
5. Birkmeyer, W.: Hirnverletzungen, Vienna, 1951, Springer Verlag.
6. Critchley, M.: The parietal lobe, Baltimore, 1953, The Williams & Wilkins Co.
7. Diamond, I. T.: The sensory neocortex, Contrib. Sens. Physiol. **2:**51, 1967.
8. Ervin, F. R., and Anders, T. R.: Neural and pathological memory, data and conceptual scheme. In Schmitt, F. O., editor: The neurosciences. Second study program, New York, 1967, The Rockefeller University Press.
9. Gazzaniga, M. S.: The split brain in man. Sci. Am. **217:**24, 1967.
10. Geschwind, N.: The organization of languages and the brain, Science **170:**940, 1970.
11. Gibson, J. J.: The senses considered as perceptual systems, Boston, 1966, Houghton Mifflin Co.
12. Glassman, E.: The biochemistry of learning: an evaluation of the role of RNA and protein, Ann. Rev. Biochem. **38:**605, 1969.
13. Gregory, R. L.: On how so little information

controls so much behaviour. In Waddington, C. H., editor: Towards a theoretical biology, Chicago, 1969, Aldine-Atherton, Inc., vol. 2.
14. Gross, C. G.: Visual functions of inferotemporal cortex. In Jung, R., editor: Handbook of sensory physiology, Heidelberg, 1973, Springer Verlag, vol. 7.
15. Harlow, H. F.: Learning set and error factory theory. In Koch, S., editor: Psychology—a study of a science, New York, 1959, McGraw-Hill Book Co., vol. 2.
16. Head, H.: Aphasia and kindred disorders of speech, London, 1926, Cambridge University Press.
17. Hebb, D. O.: The organization of behavior, New York, 1949, John Wiley & Sons, Inc.
18. Held, R.: Plasticity in sensory-motor systems, Sci. Am. **213**:84, 1965.
19. Hyden, H.: Trends in brain research on learning and memory. In Bogoch, S., editor: The future of brain science, New York, 1967, Plenum Press, Inc.
20. Hyden, H.: Learning and memory. In Pribram, K. H., editor: On the biology of learning, New York, 1969, Harcourt, Brace, Jovanovich, Inc.
21. Jakobson, R., and Halle, M.: Fundamentals of language, The Hague, 1956, Mouton Publishers.
22. James, W.: Principles of psychology, New York, 1890, Henry Holt & Co.
23. Kandel, E. R., and Spencer, W. A.: Cellular neurophysiological approaches in the study of learning, Physiol. Rev. **48**:65, 1968.
24. Levine, M.: Hypothesis behavior. In Schrier, A. M., et al., editors: Behavior of nonhuman primates, New York, 1965, Academic Press, Inc., vol. 1.
25. Miller, G. A.: The magical number seven, plus or minus two; some limits on our capacity for processing information, Psychol. Rev. **63**:81, 1956.
26. Miller, G. A., Galanter, E., and Pribram, K. H.: Plans and the structure of behavior, New York, 1960, Holt, Rinehart & Winston, Inc.
27. Milner, B.: Amnesia following operation of the temporal lobe. In Whitty, C. W. M., and Zangwill, O. L., editors: Amnesia, London, 1966, Butterworth & Co.
28. Milner, B., and Teuber, H. L.: Alterations of perception and memory in man: reflections on method. In Weiskrantz, L., editor: Analysis of behavioral change, New York, 1968, Harper & Row, Publishers.
29. Mishkin, M.: Visual mechanisms beyond the striate cortex. In Russell, R. W., editor: Frontiers in physiological psychology, New York, 1966, Academic Press, Inc.
30. Morrell, F.: Electrical signs of sensory coding. In Quarton, G. C., Melnechuk, T., and Schmitt, F. O., editors: The neurosciences. A study program, New York, 1967, The Rockefeller University Press.
31. Myers, R. E.: Corpus callosum and visual gnosis. In Delafresnaye, J. F., editor: Brain mechanisms and learning, Springfield, Ill., 1961, Charles C Thomas, Publisher.
32. Norman, D. A.: Memory and attention; an introduction to human information processing, New York, 1969, John Wiley & Sons, Inc.
33. Penfield, W., and Kristiansen, K.: Epileptic seizure patterns, Springfield, Ill., 1951, Charles C Thomas, Publisher.
34. Pribram, K. H.: The intrinsic systems of the forebrain. In Field, J., editor, Neurophysiology section: Handbook of Physiology, Baltimore, 1960, The Williams & Wilkins Co., vol. 2.
35. Pribram, K. H.: The physiology of remembering. In Bogoch, S., editor: The future of brain science, New York, 1969, Plenum Press.
36. Révész, G.: Psychology and art of the blind, New York, 1950, Longmans, Green, & Co., Inc.
37. Rowland, V.: Cortical steady potential, Progr. Physiol. Psychol. **2:1, 1968.**
38. Russell, W. R.: Brain, memory and learning, London, 1959, Oxford University Press.
39. Schilder, P.: The image and appearance of the human body, New York, 1964, John Wiley & Sons, Inc.
40. Semmes, J., Weinstein, S., Ghent, L., and Teuber, H. L.: Somatosensory changes after penetrating brain wounds in man, Cambridge, Mass., 1960, Harvard University Press.
41. Semon, R.: The mneme, London, 1904, Allen & Unwin, Ltd.
42. Simon, H., and Barenfeld, M.: Information-processing analysis of perceptual processes in problem solving, Psychol. Rev. **76**:473, 1969.
43. Sperry, R. W.: Some general aspects of interhemispheric integration. In Mountcastle, V. B., editor: Interhemispheric relations and cerebral dominance, Baltimore, 1962, The John Hopkins University Press.
44. Sperry, R. W.: Hemisphere deconnection and unity in conscious awareness, Am. Psychol. **23**:723, 1968.
45. Talland, G. A.: Deranged memory, a psychonomic study of the amnesic syndrome, New York, 1965, Academic Press, Inc.
46. Teuber, H. L.: Perception. In Field, J., Magoun, H. W., and Hall, V. E., editors, Neurophysiology section: Handbook of physiology, Baltimore, 1960, The Williams & Wilkins Co., vol. 3.
47. Teuber, H. L.: The riddle of frontal lobe function in man. In Warren, J. M., and Akert, K., editors: The frontal granular cortex and behavior, New York, 1964, McGraw-Hill Book Co.
48. Ungar, G.: Molecular mechanisms in information processing, Int. Rev. Neurobiol. **13**:223, 1970.
49. Weiskrantz, L.: Memory. In Weiskrantz, L., editor: Behavioral change, New York, 1968, Harper & Row Publishers.
50. Yarbus, A. L.: Eye movements and vision, New York, 1967, Plenum Press.

Original papers

51. Adey, R.: Studies of hippocampal electrical activity in approach learning. In Delafresnaye, J. F., editor: Brain mechanisms and learning, Springfield, Ill., 1961, Charles C Thomas, Publisher.
52. Averbach, E., and Coriell, A. S.: Short term memory in vision, Bell Syst. Tech. J. **40:** 309, 1961.
53. Baddeley, A. D., and Dale, M. C. A.: The

effects of semantic similarity on retroactive interference in long and short term memory, J. Verb. Learn. Behav. **5**:417, 1966.
54. Bagshaw, H., and Pribram, K. H.: Cortical organization in gustation (Macaca mulatta), J. Neurophysiol. **16**:499, 1953.
55. Bagshaw, H., Mackworth, N. H., and Pribram, K. H.: The effect of inferotemporal cortex ablation on eye movements of monkeys during discrimination training, Int. J. Neurosci. **1**:153, 1970.
56. Barondes, S. H., and Cohen, M. D.: Memory impairment after subcutaneous injection of acetoxy cycloheximide, Science **160**:556, 1968.
57. Bossom, J.: The effect of brain lesions on prism-adaptation in monkey, Psychosom. Sci. **2**:45, 1965.
58. Butter, C. M., Mishkin, M., and Rosvold, H. E.: Stimulus generalization in monkeys with inferotemporal and lateral occipital lesions. In Mostofsky, D. I., editor: Stimulus generalization, Stanford, 1965, Stanford University Press.
59. Chorover, S. L., and Schiller, P. H.: Short term retrograde amnesia in rats, J. Comp. Physiol. Psychol. **59**:73, 1965.
60. Critchley, M.: The body-image in neurology, Lancet **1**:335, 1950.
61. Critchley, M.: Tactile thought, with special reference to the blind, Brain **76**:19, 1953.
62. Denny-Brown, D.: Discussion, In Mountcastle, V. B., editor: Interhemispheric relations and cerebral dominance, Baltimore, 1962, The Johns Hopkins University Press.
63. Denny-Brown, D., and Chambers, R. A.: The parietal lobe and behavior, Res. Pub. Assoc. Res. Nerv. Ment. Dis. **36**:35, 1958.
64. Denny-Brown, D., Meyer, S., and Horenstein, S.: The significance of perceptual rivalry resulting from parietal lesion, Brain **75**:29, 1952.
65. Diamond, I. T., and Neff, W. D.: Ablation of temporal cortex and discrimination of auditory cortex, J. Neurophysiol. **20**:300, 1957.
66. Ettlinger, G., and Kalsbeck, J. E.: Changes in tactile discrimination and in visual reaching after successive and simultaneous bilateral posterior parietal ablations in the monkey, J. Neurol. Neurosurg. Psychiatry **25**:256, 1962.
67. Ettlinger, G.: Defective identification of fingers, Neuropsychologia **1**:39, 1963.
68. Evans, S. H.: Redundancy as a variable in pattern perception, Psychol. Bull. **67**:104, 1967.
69. Exner, W.: Physiologie der Grosshirnrinde. In Hermann, L., editor: Handbuch der Physiologie, Leipzig, 1880, F. C. W. Vogel, vol. 2.
69a. Féré, C.: Le main, la préhension et le toucher, Rev. Phil. **41**:621, 1896.
70. Gardner, R. A., and Gardner, B. T.: Teaching sign language to a chimpanzee, Science **165**:669, 1969.
71. Geschwind, N.: Disconnection syndrome in animals and man, Brain **88**:237, 585, 1965.
72. Geschwind, N., and Kaplan, E.: A human cerebral deconnection syndrome, Neurology **12**:675, 1962.
73. Geschwind, N., Quadfasel, F. A., and Segarra, J. M.: Isolation of the speech area, Neuropsychologia **6**:327, 1968.
74. Goodnow, J.: Eye and hand: differential sampling of form and orientation properties, Neuropsychologia **7**:365, 1969.
74a. Green, E.: On the contribution of studies in aphasia to psycholinguistics, Cortex **6**:216, 1970.
75. Gross, C. G., Bender, D. B., and Rocha-Miranda, C. E.: Visual receptive fields of neurons in inferotemporal cortex of monkey, Science **166**:1303, 1969.
76. Gross, C. G., Schiller, P. H., Wells, C., and Gerstein, G. L.: Single-unit activity in temporal association cortex of the monkey, J. Neurophysiol. **30**:833, 1967.
77. Hecaén, H.: Brain mechanisms suggested by studies of parietal lobes. In Darley, F. L., editor: Brain mechanisms underlying speech and language, New York, 1967, Grune & Stratton, Inc.
78. Hecaén, H., Penfield, W., Bertrand, C., and Malmo, R.: The syndrome of apractognesia due to lesions of the minor cerebral hemisphere, Arch. Neurol. Psychiatry **75**:400, 1956.
79. Held, R., and Freedman, J.: Plasticity in human sensorimotor control, Science **142**:455, 1963.
80. Held, R., and Hein, A.: Movement produced by stimulation in the development of visually guided behavior, J. Comp. Physiol. Psychol. **56**:872, 1963.
81. Hunter, W. S.: The delayed reaction in animals and children, Behav. Monogr. **2**:1, 1913.
82. Iwai, E., and Mishkin, M.: Further evidence on the locus of the visual area in the temporal lobe of the monkey, Exp. Neurol. **25**:585, 1969.
82a. Jacobsen, C. F.: Studies of cerebral function in primates. I. The function of the frontal association areas in monkeys, Comp. Psychol. Monog. **13**:3, 1936.
83. Jakobson, R.: Towards a linguistic typology of aphasic impairments. In de Reuck, A. V. S., and O'Connor, M., editors: Ciba Foundation symposium on disorders of language, Boston, 1964, Little, Brown & Co.
84. Jasper, H. H., and Shagass, C.: Conditioning of the occipital alpha rhythm in man, J. Exp. Psychol. **28**:373, 1941.
85. John, E. R., and Killam, K. F.: Electrophysiological correlates of avoidance conditioning in the cat, J. Pharmacol. Exp. Ther. **125**:252, 1959.
86. Kimura, D.: Speech lateralization in young children as determined by an auditory test, J. Comp. Physiol. Psychol. **56**:899, 1963.
87. Kimura, D.: Right temporal lobe damage, Arch. Neurol. **8**:264, 1963.
88. Klüver, H., and Bucy, P. C.: An analysis of certain effects of bilateral temporal lobectomy in the rhesus monkey with special reference to "psychic blindness," J. Psychol. **5**:33, 1938.
89. Lansdell, H. C.: Effect of extent of temporal lobe ablation on two lateralized deficits, Physiol. Behav. **3**:271, 1968.
90. Luria, A. R.: Factors and forms of aphasia. In de Reuck, A. V. S., and O'Connor, M.,

editors: Ciba Foundation symposium on disorders of language, Boston, 1964, Little, Brown & Co.
91. Masterton, R. B., and Diamond, I. T.: Effects of auditory cortex ablation on discrimination of small binaural time differences, J. Neurophysiol. **27**:15, 1964.
92. Meier, M. J., and French, L. O.: Lateralized deficits in complex visual discrimination and bilateral transfer of reminiscence following unilateral temporal lobectomy, Neuropsychologie **3**:261, 1965.
93. Meikle, T. H., Jr., and Sechzer, J. A.: Interocular transfer of brightness discrimination in "split-brain" cats, Science **132**:734, 1960.
94. Miller, G. A., Heise, G. A., and Lichten, W.: The intelligibility of speech as a function of the context of the test materials, J. Exp. Psychol. **41**:329, 1951.
95. Milner, B.: Intellectual function of the temporal lobes, Psychol. Bull. **51**:42, 1954.
96. Milner, B.: Laterality effects in audition. In Mountcastle, V. B., editor: Interhemispheric relations and cerebral dominance, Baltimore, 1962, The Johns Hopkins University Press.
97. Milner, B.: Visual recognition and recall after right temporal lobe excision in man, Neuropsychologia **6**:191, 1968.
98. Minsky, M. L.: Matter, mind and models. In Minsky, M., editor: Semantic information processing, Cambridge, Mass., 1968, The M. I. T. Press.
99. Müller, G. E., and Pilzecker, A.: Experimentelle beitrage zer Lehre vom Gedachtnis, Z. Psychol. Physiol. Sinnesorg. Erg. Bd. **1**:1, 1900.
100. Oldfield, R. C., and Zangwill, O. L.: Head's concept of the schema and its application in contemporary British psychology, Br. J. Psychol. **32**:18, 1942.
101. Pandya, D. N., and Kuypers, H. G. J. M.: Cortico-cortical connections in the rhesus monkey, Brain Res. **13**:13, 1969.
102. Pearlman, C. A., Sharpless, S. K., and Jarvik, M. E.: Retrograde amnesia produced by anesthetic and convulsant agents, J. Comp. Physiol. Psychol. **54**:109, 1961.
103. Penfield, W.: Functional localization in temporal and deep sylvian areas, Res. Publ. Assoc. Res. Nerv. Ment. Dis. **36**:210, 1958.
104. Peterson, L. R., and Peterson, M. J.: Short term retention of individual verbal items, J. Exp. Psychol. **58**:193, 1959.
105. Pinsker, H., Kandel, E. R., Castellucci, V., and Kupfermann, J.: An analysis of habituation and dishabituation in Aplysia, Adv. Biochem. Psychopharmacol. **2**:351, 1970.
106. Pribram, B., and Barry, J.: Further behavioral analysis of parietotemporo-preoccipital cortex, J. Neurophysiol. **19**:99, 1956.
107. Pribram, K. H.: On the neurology of thinking, Behav. Sci. **4**:245, 1959.
108. Pribram, K. H.: A further experimental analysis of the behavioral deficit that follows injury to the primate frontal cortex, Exp. Neurol. **3**:432, 1961.
109. Pribram, K. H.: The amnestic syndromes: disturbances in coding? In Talland, G. A., and Waugh, N. C., editors: The pathology of memory, New York, 1969, Academic Press, Inc.
110. Pribram, K. H.: The primate frontal cortex, Neuropsychologia **7**:259, 1969.
111. Pribram, K. H., and Tubbs, W. E.: Short-term memory, parsing and the primate frontal cortex, Science **156**:1765, 1967.
112. Pribram, K. H., Ahumada, A., Hartog, J., and Ross, L.: A progress report on the neurological processes disturbed by frontal lesions in primates. In Warren, J. M., and Akert, K., editors: The frontal granular cortex and behavior, New York, 1964, McGraw-Hill Book Co.
113. Robinson, J. S., and Voneida, T. J.: Interocular perceptual integration in cats with optic chiasma and corpus callosum sectioned, Am. Psychol. **16**:447, 1961.
114. Rose, J. E., and Woolsey, C. N.: Organization of the mammalian thalamus and its relationships to the cerebral cortex, Electroenceph-alogr. Clin. Neurophysiol. **1**:391, 1949.
115. Semmes, J.: A non-tactual factor in astereognosis, Neuropsychologia **3**:295, 1965.
116. Semmes, J.: Hemispheric specialization, a possible clue to mechanism, Neuropsychologia **6**:11, 1968.
116a. Sharpless, S., and Jasper, H.: Habituation of the arousal reaction, Brain **79**:655, 1956.
117. Shipstone, E. I.: Some variables affecting pattern conception, Psychol. Monogr. **74**:1, 1960.
118. Sperling, G.: Successive approximation to a model for short term memory. In Sanders, A. F., editor: Attention and performance, Amsterdam, 1967, North Holland Publishing Co.
118a. Sperry, R. W.: Neural basis of the spontaneous optokinetic response produced by visual inversion, J. Comp. Physiol. Psychol. **43**:482, 1950.
119. Sperry, R. W.: Cerebral dominance in perception. In Young, F. A., and Lindsley, D. B., editors: Early experience and visual information processing in perceptual and reading disorders, Washington, D. C., 1970, National Academy of Sciences.
120. Spinelli, D. N., and Pribram, K. H.: Changes in visual recovery functions produced by temporal lobe stimulation in monkeys, Electroencephalogr. Clin. Neurophysiol. **20**:44, 1966.
121. Spinelli, D. N., and Pribram, K. H.: Neural correlates of stimulus response and reinforcement, Brain Res. **17**:377, 1970.
122. Stengel, E.: Loss of spatial orientation, constructional apraxia and Gerstmann's syndrome, J. Ment. Sci. **90**:753, 1944.
123. Talland, G. O.: Amnesia—a world without continuity, Psychol. Today **1**:43, 1967.
124. Teuber, H. L.: Space perception and its disturbances after brain injury in man, Neuropsychologia **1**:47, 1963.
125. Teuber, H. L.: Somatosensory disorders due to cortical lesions; preface: disorders of higher tactile and visual functions, Neuropsychologia **3**:287, 1965.
126. Teuber, H. L.: Alterations of perception after brain injury. In Eccles, J. C., editor: Brain and conscious experience, New York, 1966, Springer-Verlag New York, Inc.

127. Teuber, H. L.: Lacunae. In Milligan, C. H., and Darley, F. L., editors: Brain mechanisms underlying speech and language, New York, 1967, Grune & Stratton, Inc.
128. Trevarthen, C. B.: Two mechanisms of vision in primates, Psychol. Forsch. **31:**299, 1968.
129. von Holst, E., and Mittelstaedt, H.: Das Reafferenzprinzip (Wechselwirkungen zwischen Zentralnervensystem und Peripherie), Naturwissenschaften **37:**464, 1950.
130. Weinstein, S.: Time error in weight judgment after brain injury, J. Comp. Physiol. Psychol. **48:**203, 1955.
131. Weiskrantz, L., and Mishkin, M.: Effect of temporal and frontal cortical lesions on auditory discrimination in monkeys, Brain **81:**406, 1968.
132. Wilson, M.: Effect of circumscribed cortical lesions upon somesthetic discriminations in the monkey, J. Comp. Physiol. Psychol. **50:**630, 1957.
133. Wilson, M.: Tactual discrimination learning in monkeys, Neuropsychologia **3:**353, 1965.
134. Wilson, M., and Kaufman, H. M.: Effect of inferotemporal lesions upon processing of visual information in monkeys, J. Comp. Physiol. Psychol. **69:**44, 1969.
135. Wilson, M., Stamm, J. S., and Pribram, K. H.: Deficits in roughness discrimination after posterior parietal lesions in monkeys, J. Comp. Physiol. Psychol. **53:**535, 1960.
136. Zinkin, S., and Miller, A. J.: Recovery of memory after amnesia induced by electroconvulsive shock, Science **155:**102, 1967.

V

NEURAL CONTROL OF MOVEMENT AND POSTURE

20

ELWOOD HENNEMAN

Organization of the motor systems—a preview

The motor systems of the brain exist to translate thought, sensation, and emotion into movement. At present the initial steps in this process lie beyond analysis. We do not know how voluntary movements are engendered, nor where the "orders" come from. Most of the information that is available concerns the circuits that execute these shadowy commands.

Movement is the end product of a number of control systems that interact extensively. Their complexity demands that we proceed logically by (1) defining the nature of movement in terms of muscles and joints, (2) presenting an outline of the motor systems so that the relation of the parts to the whole is apparent from the outset, and (3) explaining how "control" is achieved. This introductory chapter is designed to meet the first two of these needs. The next chapter deals with the general principles of control in motor systems and provides further orientation for more detailed treatment of individual topics in subsequent chapters.

THE NATURE OF MOVEMENT: MUSCLES AND JOINTS

In most forms of skeletal movement the motion occurs at joints, where two or more bones come together to form a nearly frictionless pivot. Muscles are arranged so that their ends are attached on opposite sides of the joint, which may act as a fulcrum (Fig. 20-1). Since individual muscles can pull but not push, at least two muscles on opposite aspects of the joint are generally required to provide a full range of movement in both directions. At some joints there are several pairs of antagonistic muscles disposed so as to produce abduction, adduction, and rotation as well as flexion and extension.

There is a great deal of interaction between the various muscles in a limb. One reason for this is that muscular forces are exerted as much on the point of origin as on the insertion. Movement of one member of a joint therefore requires fixation of the other member by muscles at the next joint. In movements of the forearm, for example, the muscles of the shoulder generally contract to fix the humerus. If powerful forces are involved, fixation may be required at joints some distance away. A series of interconnections has developed in the spinal cord to link the motor neurons of muscles that commonly work together. The nature of the connections depends on the kind of mechanical interactions involved. The most closely related muscles are the direct antagonists such as biceps and triceps (Fig. 20-1), for when either one of the pair shortens, the opposing muscle is necessarily lengthened. Direct antagonists are of great importance because precise control of movements frequently involves application of braking action. Each muscle acts as a brake for its antagonist, serving to bring rapid movements to a quick, smooth stop. Recordings of the electrical activity of the human biceps and triceps muscles, for example, reveal that during a contraction of the biceps there is usually some activity in the triceps and that toward the end of a rapid contraction this activity increases sharply.

In general, muscles can be divided into two groups, depending on their action in the body. One group plays a special role in upright posture, standing, and locomotion. These are the muscles that normally oppose the force of gravity. Special reflexes and postural reactions involving them are highly developed in vertebrates. As a matter of tradition, physiologists refer to all antigravity muscles as "extensors" and to their antagonists as "flexors," regardless of their particular actions at joints. The flexor digitorum longus in the leg, for example, is regarded as an extensor muscle

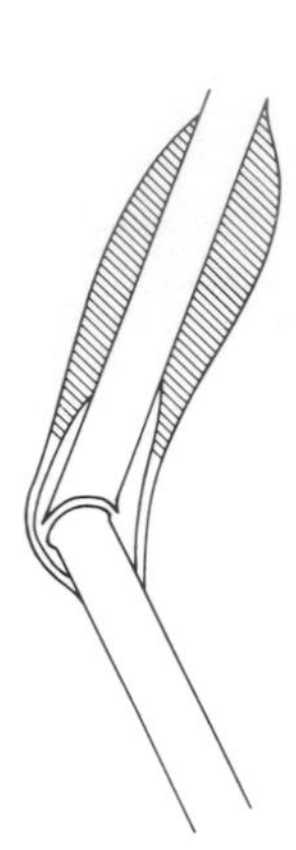

Fig. 20-1. Schematic representation of typical joint showing pair of antagonistic muscles that flex and extend it.

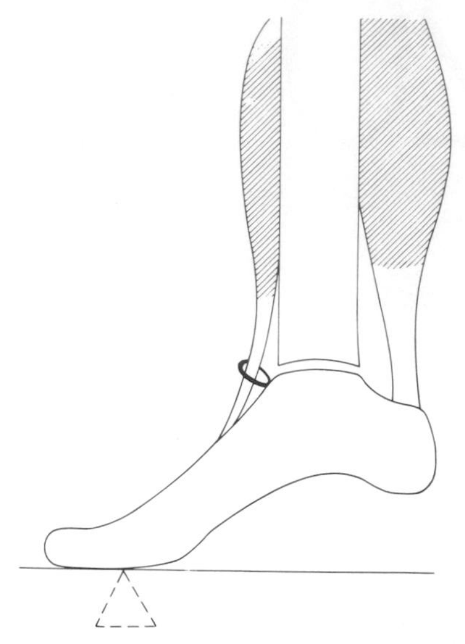

Fig. 20-2. Antigravity action of gastrocnemius muscle that raises the heel and the weight of the body transmitted through the tibia.

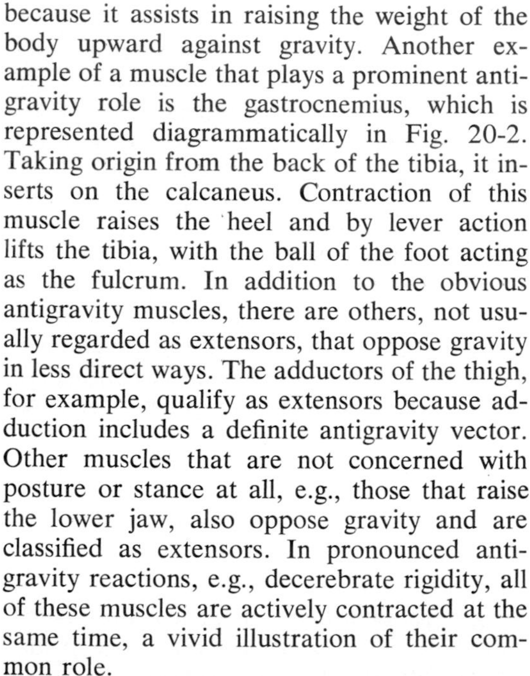

because it assists in raising the weight of the body upward against gravity. Another example of a muscle that plays a prominent antigravity role is the gastrocnemius, which is represented diagrammatically in Fig. 20-2. Taking origin from the back of the tibia, it inserts on the calcaneus. Contraction of this muscle raises the heel and by lever action lifts the tibia, with the ball of the foot acting as the fulcrum. In addition to the obvious antigravity muscles, there are others, not usually regarded as extensors, that oppose gravity in less direct ways. The adductors of the thigh, for example, qualify as extensors because adduction includes a definite antigravity vector. Other muscles that are not concerned with posture or stance at all, e.g., those that raise the lower jaw, also oppose gravity and are classified as extensors. In pronounced antigravity reactions, e.g., decerebrate rigidity, all of these muscles are actively contracted at the same time, a vivid illustration of their common role.

The other group of muscles, which is antagonistic to the extensors, has in common the function of withdrawing the body reflexly from painful stimuli. Any muscle that takes part in flexor withdrawal reflexes is classified as a flexor regardless of its joint action. The tibialis anterior, which opposes the gastrocnemius, is a flexor at the ankle; the extensor digitorum longus, despite its anatomic name, is a physiologic flexor of the toes and participates in most withdrawal reactions of the foot.

Finally, it should be emphasized that even simple movements are deceptively complex in their underlying mechanics. The number of muscles involved and the interactions between them greatly complicate the problem of neural control. In contrast to most engineering control systems, there is not one process or action to be regulated, but many simultaneously. As we shall see, the task of coordinating them is so formidable that it requires an elaborate center containing, according to recent estimates, a total of 10^{11} neurons!

OUTLINE OF THE MOTOR SYSTEMS

In this section an attempt has been made to isolate the parts of the nervous system that are primarily motor in function and to describe their roles in motor activity as concisely as possible. The aim is to provide a functional outline that will permit the reader in subsequent chapters to fit each new circuit into its proper relation with the whole system.

Fig. 20-3 is a block diagram of the motor systems, greatly simplified for the sake of initial orientation. Each block represents a major subdivision of the nervous system. The flow of signals between blocks is indicated by arrows (unshaded for sensory connections, shaded for motor and nonsensory connec-

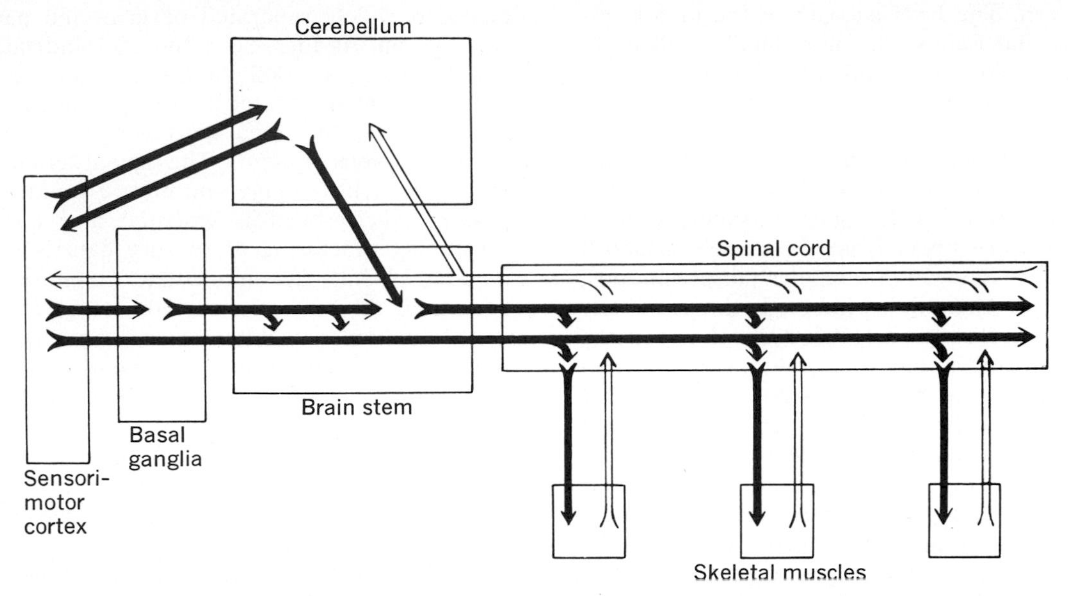

Fig. 20-3. Block diagram of motor systems. Each block represents a major subdivision of the nervous system. Unshaded arrows indicate sensory connections; shaded arrows represent motor and nonsensory connections.

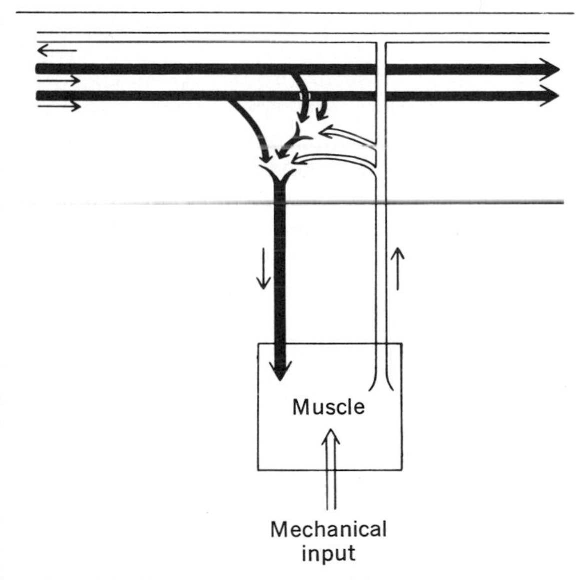

Fig. 20-4. Schematic representation of closed loop that links each muscle with spinal cord. Shaded arrow to muscle represents motor neuron. Sensory and motor connections to it are indicated by unshaded and shaded arrows, respectively.

tions). To begin with, it should be emphasized that all of the various motor pathways shown in this schematic nervous system ultimately converge on a series of simple circuits that link each muscle with the spinal cord. As Fig. 20-4 indicates, the sensory neurons carrying impulses from a given muscle are connected with the motor neurons, which transmit impulses back to it. A closed loop that regulates the activity of each muscle individually is thus formed. This is the basic control mechanism of the motor systems, for no movement, reflex or voluntary, can occur except through the agency of this circuit. Although these circuits may function autonomously in simple reflexes, their activity is largely controlled by centers at higher levels. From these centers, descending tracts run the entire length of the spinal cord, giving off fibers at all levels. One such tract, the corticospinal, originates in the cerebral cortex and transmits signals from the sensorimotor cortex directly to the spinal level. All of the other tracts that descend into the spinal cord arise in the brainstem. The brainstem is a prespinal integrating center of great complexity that receives signals from all higher centers and processes them for transmission to the spinal cord. Superimposed on the upper end of the brainstem are the basal ganglia, a heterogeneous group of nuclei that receive signals from the sensorimotor cortex and discharge into the brainstem. At the highest level is the sensorimotor cortex, which presides over the entire motor system. Interconnected with all levels and functioning as an overall coordinator of motor activities is the cerebellum, a highly organized center with an extensive cortex of its own. The brief account of the motor systems that follows this introduction will begin at the spinal level and will take up the other subdivisions of the nervous system in ascending order.

The closed loop that links each muscle with the spinal cord is shown schematically in Fig. 20-4. As the figure indicates, some of the sensory fibers from the muscle establish a direct connection with the motor neurons, whereas others do so indirectly through internuncial cells. In the latter type of circuit the loop includes the following parts: the cell bodies of the motor neurons in the spinal cord, their axons running out in the ventral roots, the neuromuscular junctions, the muscle itself, the sensory receptors with their fibers entering the spinal cord through the dorsal roots and their terminations in the spinal cord, and finally one or more internuncial cells sending axons to the motor neurons. These segmental circuits automatically regulate the length and tension of the muscle in accordance with various requirements. They are responsive peripherally to mechanical input, e.g., stretching of the muscle, and centrally to neural input, e.g., signals from higher centers. Mechanical input stimulates the stretch receptors, provoking circuit action that elicits a reflex response of the muscle; neural input initiates activity in the central parts of the loop that also travels around the entire circuit. Signals from higher centers impinge chiefly on the internuncial cells and, to a lesser extent, directly on the motor neurons themselves.

It has been emphasized that muscles interact extensively. Much of the interaction is controlled automatically by spinal circuits. The sensory feeedback from muscles is not restricted to their own motor neurons; rather, it spreads out through collaterals of primary sensory neurons and through internuncial circuits to reach the motor neurons of all closely related muscles and, to a decreasing extent, those of more distant muscles. A stretch or a contraction of one muscle affects its own motor neurons most strongly, those of its direct antagonist somewhat less, and those of other synergists and antagonists around it still less. Thus every primary loop is a part of a larger feedback network serving a group of muscles.

It is a general rule of the nervous system that sensory impulses arising in a primary sensory neuron are distributed by way of collaterals, projection tracts, and secondary relays to widely separated parts of the nervous system. As Figs. 20-3 and 20-4 indicate, signals arising in skeletal muscles not only serve in local segmental circuits but are also transmitted up the spinal cord to higher levels of the nervous system. The spinal circuits involve very little delay and ensure rapid responses when speed is essential and time-consuming processing of sensory data is not required. While this immediate response is occurring, the same signals are being transmitted to higher centers for more elaborate analysis of their information content and for combination with signals from other types of receptors. After variable delays, signals from the higher centers are relayed back to segmental levels. Thus "long loops" involving higher centers and more delay help to regulate the activity of the spinal circuit.

Superimposed on the orderly, segmental organization of the spinal cord is the more complex brainstem. Its structure provides no easily read clues to its functions. Its central role in motor activities and in immediate control over the spinal cord is obvious, however, for *all of the descending tracts to the spinal cord except one, the corticospinal tract, arise in the brainstem.* As indicated in Fig. 20-3, these descending tracts terminate at all levels of the spinal cord. They can thus control or regulate activity at many spinal levels simultaneously.

It is not yet possible to specify the function of the brainstem centers and their exact relations to the segmental circuits, but it is clear that they extend motor capacity far beyond the stereotyped regulatory behavior of the spinal animal. Animals with a brainstem (but without higher centers) are capable of integrative activities such as standing, walking, and postural adjustments. These activities require the participation of various righting reflexes and antigravity mechanisms, they demand precise control of equilibrium, and they involve coordination of muscles throughout the body.

Running the length of the brainstem is a core of gray matter called the reticular formation (RF), which is centrally involved in motor activities. It receives connections from the spinal cord, the cerebral cortex and basal ganglia, and the cerebellum. It functions as an integrating center, adding, subtracting, and combining the influences of all these centers. Two subsystems, which have not been adequately defined anatomically, have been identified within the RF. A reticular inhibitory center in the medulla gives origin to a reticulospinal tract that transmits impulses that are predominantly inhibitory to extensor motor neurons. A reticular facilitatory system located in the higher portions of the brainstem transmits impulses that are predominantly facilitatory to extensor motor neurons. The functions of these two systems are far from clear, but it appears that they serve as common efferent pathways through which the entire extrapyramidal system exerts its influence on stretch reflex circuits.

Closely related to the reticular systems in their postural activities are the elaborate sensing devices in the vestibule of the ear and their ramifications in the nervous system. Through its connections with the RF and through the vestibulospinal tract, this system regulates muscle in response to changes in the position of the head in space and in response to all accelerations that the body experiences.

At the next higher level of the motor system are the basal ganglia, a group of nuclei that receive descending connections from the sensorimotor cortex and project to the RF of the brainstem. The fact that they occupy a very large volume of the brain suggests that their motor functions are of major importance. The details of their structure and connections, however, provide few clues as to their precise roles. Certain pathologic lesions confined to the basal ganglia result in characteristic motor disturbances without sensory deficits or mental impairment. Although these syndromes are often dramatic and very disabling, they offer surprisingly little insight into the normal function of these systems. The basal ganglia, with the descending fibers they receive from the cerebral cortex, and with the RF and the reticulospinal tracts, comprise the bulk of the "extrapyramidal" motor system.

At the highest level of the nervous system there are several cortical areas that are partially or wholly devoted to motor functions. They include the area just anterior to the central sulcus, generally known as the "motor cortex," and the area just posterior to the central sulcus, which is a "receiving area" of the somatic sensory system. Together these areas are often referred to as the sensorimotor cortex, a term that properly emphasizes the important role of sensory systems in motor control. In each of the subdivisions of the sensorimotor cortex the parts of the body are represented in orderly fashion. Electrical stimulation within any of these areas results in movements of the particular muscle groups represented under the electrodes. The effects of stimulation are due to activation of two different descending systems, which are indicated in Fig. 20-3. One of them consists of corticifugal fibers from all subdivisions of the sensorimotor cortex. These fibers come together and form the pyramidal or corticospinal tract, which runs without interruption from the cerebral cortex to all levels of the spinal cord. Most of these long fibers end on internuncial cells that form part of the feedback loop from muscle to motor neurons. The organization of the pyramidal tract suggests that it is designed for precise control of individual muscle groups. The other corticifugal system is the extrapyramidal system, which sends fibers to some of the basal ganglia and to the RF of the brainstem. The pyramidal and extrapyramidal systems, though dissimilar in their organization, work together harmoniously.

Acting as a coordinating center for motor systems as a whole is a large, highly organized outgrowth of the brainstem known as the cerebellum. It receives signals relayed from receptors in muscles, tendons, joints, and skin as well as from visual, auditory, and vestibular end organs. This great volume of sensory input serves purely motor functions, chiefly regulatory in effect. Signals from the cerebral cortex and from other motor regions also reach the cerebellum in great numbers. All of this enormous influx is analyzed and integrated in an elaborate but orderly cortical network. From the cerebellum, efferent fibers pass to the thalamus, red nucleus, vestibular nuclei, and RF. Through these connections the cerebellum influences motor centers from the cerebral cortex down to the spinal motor neurons, coordinating the activity of motor circuits at all levels of the CNS.

21

JAMES HOUK and ELWOOD HENNEMAN

Feedback control of muscle: introductory concepts

Before embarking on a detailed physiologic description of the motor system it is helpful to consider some of the general principles of feedback control and show how they are embodied in the nervous system. An understanding of these principles should make it easier to appreciate the significance of the anatomic and physiologic facts presented in subsequent chapters.

INTRODUCTION TO FEEDBACK CONTROL

The essential feature of all feedback control systems is a provision for a more or less continuous flow of information from the element controlled to the device that controls it. This flow of information, which is called "feedback," ensures a continuous adjustment of the control system in accordance with the changing conditions of the control task. Controlling a system without feedback is like driving an automobile with closed eyes. If the driver has a vivid recollection of the curves in the road and if no other objects are in his way, he may be able to steer his vehicle several hundred yards before he makes an error sufficiently large to throw him off the road. However, if he opens his eyes, he may observe his errors before he leaves the road, correcting for them as they occur. Adjustment of the automobile's course in accordance with the amount of error is, in fact, an example of feedback control.

All feedback control systems have certain features in common. A block diagram with these features clearly identified provides the best guide to a new system. Fig. 21-1 illustrates a typical, generalized control system. The *controlled system* has an input, represented by an arrow, that may be varied by the *controller.* This input interacts with external influences called *disturbances* and produces changes in the *actual output,* which is the variable to be controlled. In the example just given the controlled system is the automobile. Its input is the position of the steering wheel and its output is the position of the automobile on the road. An example of a disturbance is a gust of wind, since it may cause an undesired shift in the output, the position of the automobile.

The *transducer* is a sensitive device that is placed where it can measure the actual output of the controlled system. It converts this measurement into a *feedback signal,* which is transmitted to the *error detector.* Here it is compared with the *control signal,* which is the primary input to the entire feedback system. Since the control signal dictates a desired output and the feedback signal measures the actual output, the difference computed at the error detector is called the *error signal.* Applied to the controller, the error signal alters the input to the controlled system so as to reduce the amount of error. The actual output is thus brought closer to the output specified by the control signal. In the example of driving an automobile the transducer, the error detector, and the controller are all within the human operator. Through his visual system he measures the position of his automobile and compares this with the position of the winding road. If an error exists, he uses his muscle system, the controller, to turn the steering wheel of his automobile in the direction that will bring his car back to the proper place on the road.

As illustrated in this example, feedback automatically performs two important functions. (1) It compensates for disturbances such as gusts of wind that could otherwise cause the actual output to deviate from the desired output. In performing this function the feedback system is called a *regulator.* (2) It follows the changes in the control signal (the winding course of the road) quickly and

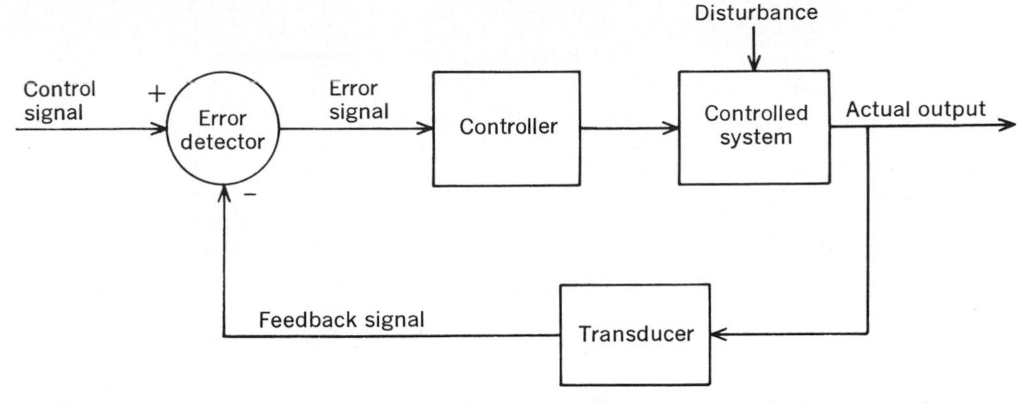

Fig. 21-1. Block diagram of generalized feedback control system. Any feedback control system can be cast into this form. The operation performed by each component is disclosed by its name. Arrows indicate unidirectional flow of information.

accurately. In performing this function the feedback system is called a *servomechanism.* Both of these functions of feedback are utilized in the control of movement and posture.

FEEDBACK CONTROL OF MUSCLE LENGTH

As Liddell and Sherrington[3] demonstrated, stretching a muscle results in a prompt contraction of that muscle, which tends to restore it to its former length. They proved the reflex nature of the reaction by showing that it depended on afferent signals from the muscle to the spinal cord that elicited efferent responses returning to the muscle. They named the reaction the "myotatic" (muscle-stretching) reflex, also called the stretch reflex. It has since been shown that one group of afferent fibers, those that innervate primary spindle receptors in the muscle, make direct connections with the motor neurons of the same muscle, forming a monosynaptic reflex circuit. The following is an interpretation of this reflex arc as a feedback system that controls muscle length.

The anatomic connections between the components of the stretch reflex are illustrated in a schematic manner in Fig. 21-2, *A*. If one identifies the role that each of the various components serves in carrying out a stretch reflex, a block diagram of the system resembling that in Fig. 21-1 can be constructed (as in Fig. 21-2, *B*).

We may begin by recognizing a disturbing force as the cause of an initial stretch to a muscle. This disturbance may be the force of gravity acting upon the mass to which the muscle is attached. Moreover, the mass may be identified as the controlled system whose position the muscle is attempting to regulate. For convenience, we may choose muscle length, which is one measure of the position of the mass, as the actual output of the controlled system.

Muscle spindles, described in more detail in the following chapter, are located within the belly of the muscle (Fig. 21-2, *A*). Because they are attached in parallel with the other muscle fibers, they are stretched whenever the muscle is stretched. A primary receptor in each spindle responds to an increase of muscle length by generating action potentials at an increased rate. These receptors are therefore physiologic transducers that measure muscle length (Fig. 21-2, *B*). Their signals are conducted to the spinal cord in type Ia nerve fibers, which involves a small but significant afferent delay.

In addition to monosynaptic excitatory connections formed by Ia fibers, motor neurons also receive connections from various other fibers (Fig.21-2, *A*). In Fig. 21-2, *B,* the excitatory nature of the input from spindles is shown with a positive sign. All the remaining excitatory and inhibitory inputs have been combined into a single input—the control signal. Among the many sources of control signals are the centers where voluntary actions originate. The effects of feedback signals and control signals are combined synaptically in the motor neuron, where an efferent signal that is analogous to the error signal in Fig. 21-1 is generated.

Thus increases in muscle length that are produced by disturbing forces are sensed by spindle receptors that generate a larger feedback signal. As a result the motor neurons send a larger efferent nerve signal to the mus-

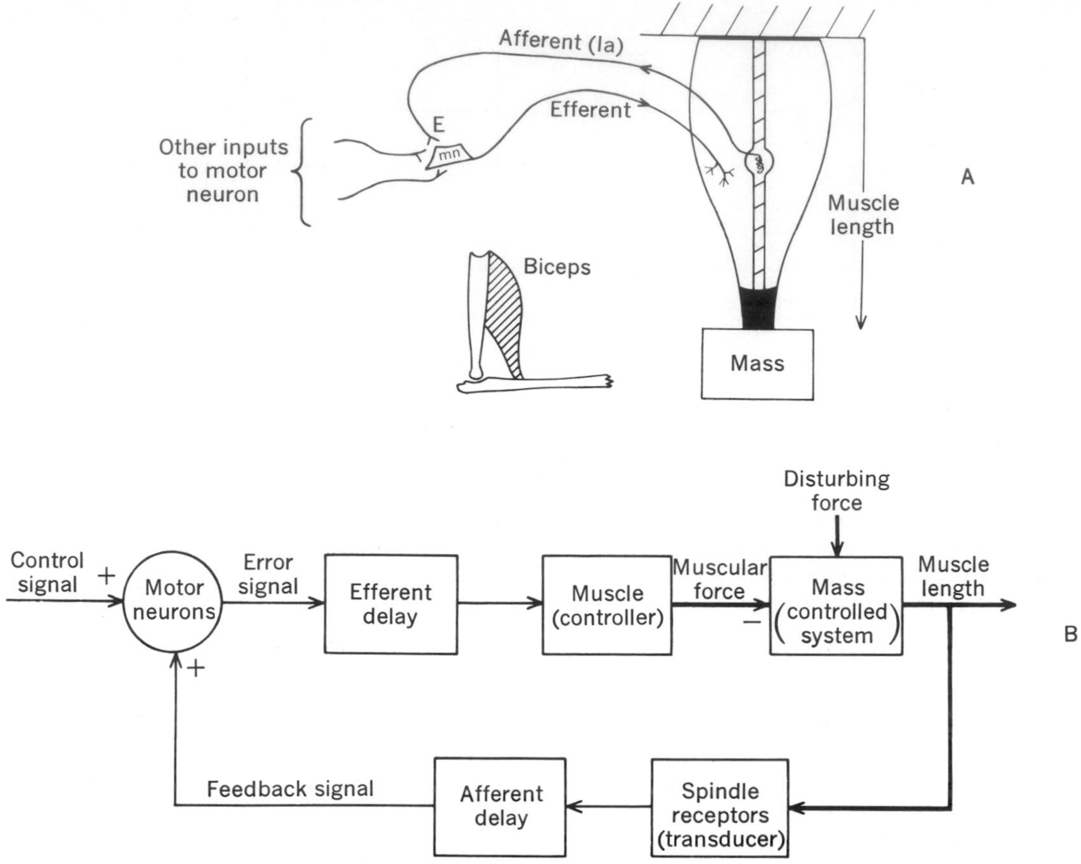

Fig. 21-2. Length control system. **A,** Schema showing anatomic connections between physiologic components that participate in stretch reflex. For simplicity, only a single spindle receptor (in the belly of the muscle), a single motor neuron, *mn,* and single afferent and efferent nerve fibers are shown. Inset shows similarity between human biceps muscle supporting the forearm and isolated muscle of stretch reflex circuit. (*E* stands for excitatory synaptic connection.) **B,** Block diagram of stretch reflex, a feedback system that controls muscle length. Neural connections are shown with thin lines, whereas mechanical interactions are shown with heavy lines.

cle (the controller), where the force of contraction is enhanced. The larger muscular force will counteract the disturbing force, tending to move the mass back toward its former position and the muscle toward its former length.

Now let us review Figs. 21-1 and 21-2, relating them to each other. There are two points of dissimilarity between Figs. 21-1 and 21-2, *B*. The first is the appearance of time delays in the feedback loop of Fig. 21-2, *B*. The effect of these delays on transient responses of the system will become clear from the examples to be discussed subsequently. The second dissimilarity is in the postion of negative signs. The generalized diagram (Fig. 21-1) showed a negative sign where the feedback signal enters the error detector, whereas the corresponding feedback signal in Fig. 21-2, *B,* has a positive sign. Although the impulses fed back to the motor neurons are excitatory and thus positive in sign, the feedback *around the whole loop* is actually negative. To illustrate this a negative sign is shown beneath the arrow representing muscular input to the controlled system to indicate that an *increase* in muscular force causes a *decrease* in muscle length and therefore a *decrease* in the signal fed back from the spindle. There is thus one negative sign in each feedback loop. Its location is of minor importance. What is essential is that any increase in error signal must eventually lead to a decrease in the error signal by way of the feedback pathway.

A detailed example is required to illustrate

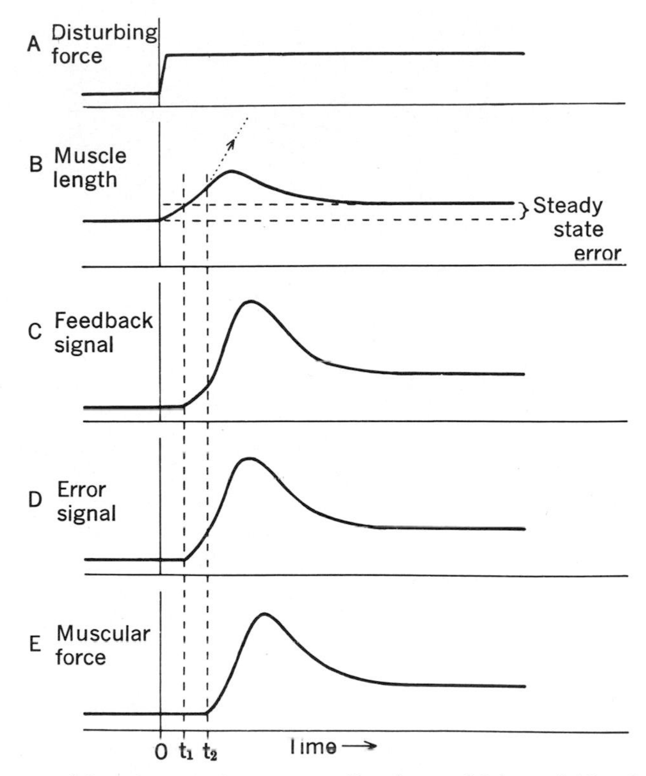

Fig. 21-3. Response of length control system to disturbance. Main variables that may be measured at various places in length control system (Fig. 21-2, *B*) are plotted as functions of time. Afferent time delay is t_1; afferent plus efferent delay is t_2. In the plot of muscle length, effects of feedback are first apparent at time t_2. Dotted line indicates the response that might be observed in absence of feedback. Even with feedback there is a small steady-state error in length due to disturbance.

the interactions occurring among the several physiologic components during normal operation of the stretch reflex. Imagine that the muscle shown in Fig. 21-2, *A*, is the biceps muscle of a human arm (see inset) and that it is supporting the forearm against gravity. Before our example begins, the length control system for the biceps muscle is in a *steady state* with the arm motionless. This steady state is illustrated in Fig. 21-3 by the constant values of all the signals in the length control system prior to time zero (to the left of the origin of the abscissa).

At time zero an additional weight is placed in the man's hand, producing a sudden rise in the disturbance force (Fig. 21-3, *A*). This disturbance immediately begins to lengthen the biceps muscle (Fig. 21-3, *B*). (It will of course have other consequences; however, we are focusing our attention on the biceps muscle alone.) The lengthening is measured by the spindles in the muscle, which relay this information to the spinal cord. Hence after a brief afferent delay (t_1) the feedback signal increases as shown in Fig. 21-3, *C*. The biceps motor neurons almost immediately respond by producing a larger error signal (Fig. 21-3, *D*). (The control input to the neurons remains constant throughout this example.) The new error signal is transmitted in efferent nerve fibers to the muscle, where after an efferent delay ($t_2 - t_1$) an increased muscular force is produced (Fig. 21-3, *E*). Thus at time t_2, muscular forces begin to build up in opposition to the disturbance caused by the added weight. Had feedback not been

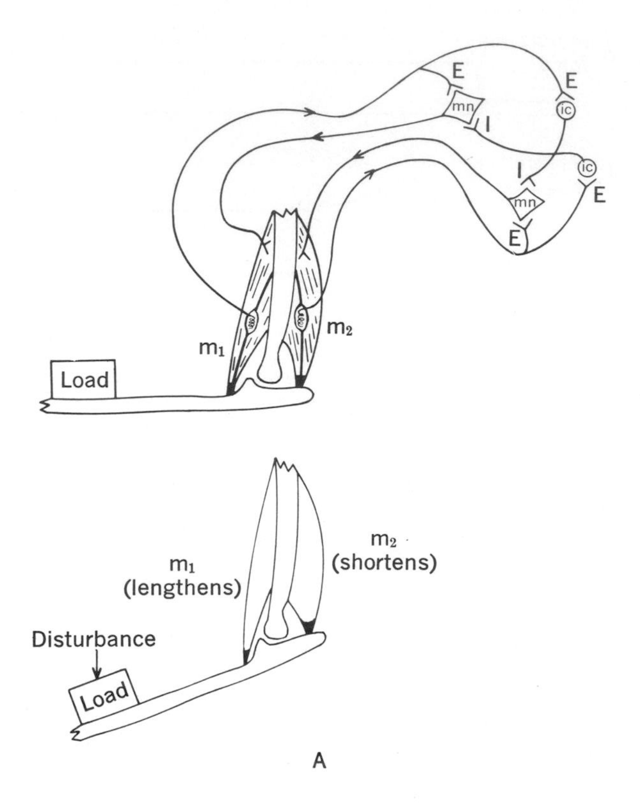

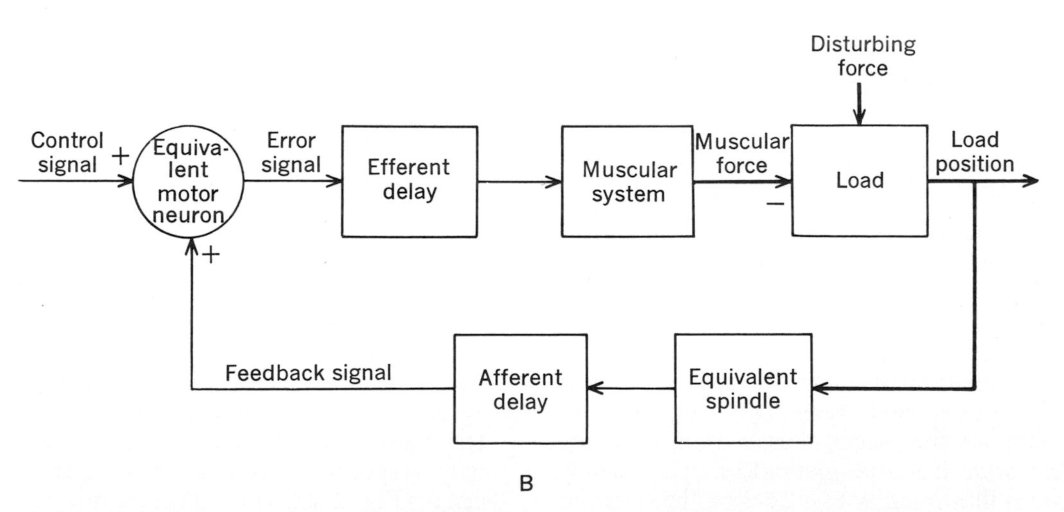

Fig. 21-4. Representation of two antagonistic muscles and their spinal connections as an "equivalent" length control system. **A,** Reciprocal connections between antagonistic muscles are illustrated schematically. In addition to excitatory monosynaptic connections, Ia afferents from spindles also make disynaptic inhibitory, *I,* connections via internuncial cells, *ic,* with antagonistic motor neurons. This reciprocal innervation allows antagonists to work together closely and leads to the formulation of the block diagram shown in **B.** Inset shows changes in lengths of the muscles resulting from an increased disturbance. **B,** Block diagram of equivalent system in which the two antagonistic systems shown above are combined into one. Although this system closely resembles that for a single muscle (Fig. 21-2, *B*), the blocks are now replaced by "equivalent" components whose signals can take on positive or negative values.

present the muscle would have continued to lengthen, as shown by the dotted trace in Fig. 21-3, *B*. The muscular force rapidly rises to a value greater than the disturbing force, causing the mass to cease its downward motion and begin moving back toward its former position. As the muscle shortens, the spindles inform the motor neurons of this change by decreasing the feedback signal. This results in a decreased error signal and consequently a decreased muscular force. The reaction continues until a new steady state is reached at which the muscle is producing a force equal to the new disturbing force and there is no further movement.

Note in Fig. 21-3, *B,* that the length control system does not return the muscle exactly to its former length; i.e., it does not compensate completely for the disturbance. A small *steady-state error* in length remains, causing an increased output from the spindles, which in turn produces the additional force that is required to counteract the disturbing force. A larger disturbing force will cause a larger steady-state error, since a larger change in the length of the muscle is required to elicit reflexly a muscular force sufficient to counter the larger disturbing force. This feature of the length control system resembles the behavior of a spring; increasing forces applied to a spring will stretch it by increasing amounts. We will return to this analogy later in the chapter.

This example demonstrates how the length control system functions as an automatic regulator that quickly compensates for disturbances. Another manifestation of the system is its ability to execute any changes in length directed by the control signal. Before we discuss this aspect of length control, a more physiologic system consisting of two muscles, an agonist and an antagonist working in opposition at a joint, will be introduced.

CONTROL OF LENGTH IN ANTAGONISTIC MUSCLES

In Fig. 21-4, *A,* a pair of antagonistic muscles, the biceps and triceps, are represented diagrammatically with some of their reflex connections. The direct or monosynaptic connections that the afferent fibers from muscle spindles make with their own motor neurons are indicated as they were in Fig. 21-2, *A*. These are excitatory *(E)* pathways. In addition, an inhibitory *(I)* connection from the same fibers to the motor neurons of the antagonist muscle is shown. The latter pathway probably includes an internuncial cell. Connections such as these link all pairs of direct antagonists. This reciprocal innervation tends to prevent opposing muscles from working against each other and, in fact, binds them into a single working unit.

In Fig. 21-4, *B,* the two muscles shown in Fig. 21-4, *A,* and their reflex connections are combined into one so-called equivalent system. The two antagonistic muscles are represented as one "muscular system," which can move the joint (load) in either direction. The spindles in these two opposing muscles are represented by one block labeled "equivalent spindle," which can signal "positively" or "negatively," depending on which way the load is moved, i.e., which muscle is stretched. The feedback itself is always negative in effect, regardless of the particular sign of the signal. The motor neurons of the two muscles are represented by one "equivalent motor neuron," which can send "positive" or "negative" signals to the "muscular system," i.e., to either of the two opposing muscles.

The operation of this antagonistic pair of length control systems as an equivalent system having a single feedback path is best appreciated by following an example (Fig. 21-4, *A*). Assume that the forearm is initially motionless, in which case the net muscular force—the force produced by m_1, the biceps, minus the force produced by m_2, the triceps—is just sufficient to support the limb and its load against the gravitational force. If an additional external disturbance causes a downward movement of the load (forearm), m_1 lengthens and m_2 is allowed to shorten (see inset in Fig. 21-4, *A*). Four reflex reactions follow. The spindles in m_1 fire more rapidly, leading to increased excitation of m_1 and, by reciprocal innervation, increased inhibition of m_2. Simultaneously the spindles in m_2 fire less rapidly, resulting in decreased excitation of m_2 and, by reciprocal innervation, decreased inhibition of m_2. These four separate reflex actions all tend to increase the net muscular force. Because both of these length control systems act together in such a well-knit reciprocal manner, we are justified in representing the antagonistic pair of systems by the single equivalent length control system, which is shown in Fig. 21-4, *B*.

RESPONSES TO CONTROL SIGNALS

A second input to the length control system is the control signal, shown in Figs. 21-2,

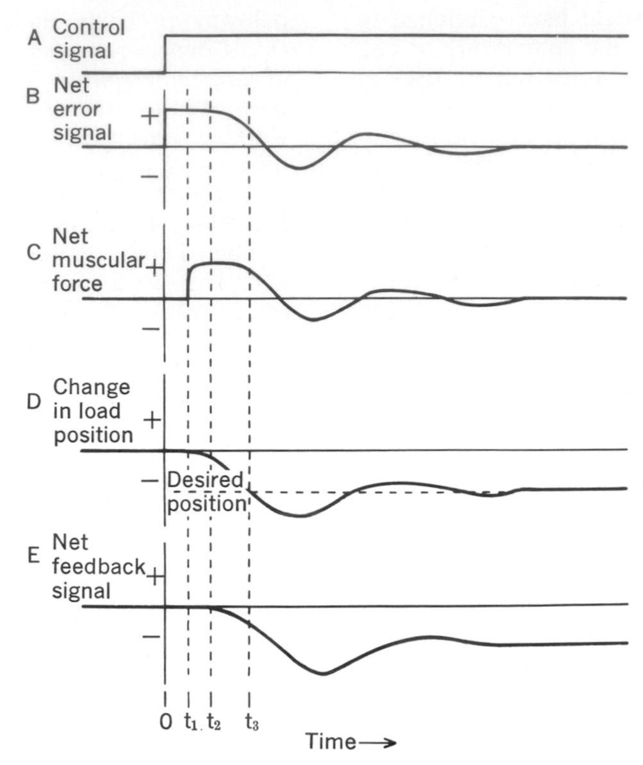

Fig. 21-5. Sequence of events in execution of desired change in position by length control system. This figure is similar to Fig. 21-3 except that input to system is a control signal rather than a disturbance. Several variables that would appear in the equivalent system shown in Fig. 21-4, *B*, are plotted as functions of time. Complete description is given in text.

B, and 21-4, *B*. It represents a neural command to change the position of the load. Let us use the antagonistic system to see how a change in position is automatically executed. The arguments should be followed with Figs. 21-4 and 21-5.

For simplicity we will consider the case in which disturbing forces are absent; consequently, all the variables shown in Fig. 21-5 are initially zero. A command in the form of a sudden increase in the control signal comes to the system at time zero (Fig. 21-5, *A*). Since the feedback signal is initially zero, a postive error signal is produced by summation at the equivalent motor neuron (Fig. 21-5, *B*). After a brief delay in the efferent fibers (t_1), the error signal excites the muscular system, which produces a large positive muscle force (Fig. 21-5, *C*). This force begins to move the load toward the desired position (Fig. 21-5, *D*). In this example the desired position is negative, corresponding to a decrease in the length of m_1 (Fig. 21-4, *A*). The equivalent spindle measures this negative change in position, which after an afferent delay appears in the feedback signal (Fig. 21-5, *E*). The negatively moving feedback signal informs the motor neuron that the load is being moved in the appropriate direction. In response the motor neuron (at time t_2) begins to decrease its output, the error signal.

Eventually the load reaches the desired position (at time t_3). However, because of the afferent delay the motor neuron is not informed of this until a short time later. At that point the load has moved a bit too far; i.e., it has overshot the desired position. The overshoot is subsequently corrected by the generation of a force in the opposite direction (muscle force becomes negative), which brings the load to a stop and begins to move it in the opposite direction. The movement again goes too far, and another reversal of muscle force becomes necessary. Normally these overshoots become smaller and smaller until the load finally remains fixed at the desired position. This ordinarily takes only a fraction of a second. Under these conditions the system is said to be stable. The neurologic syndrome "clonus" is a less stable state of the system in which the oscilla-

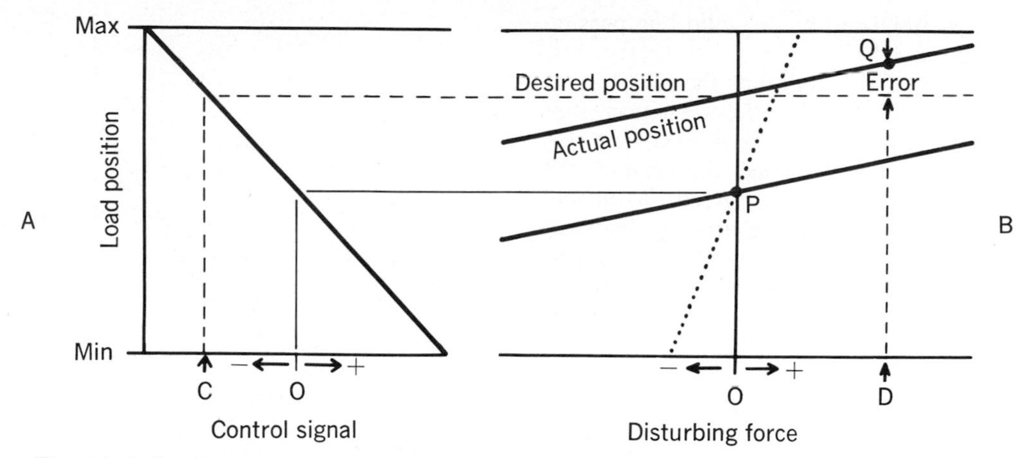

Fig. 21-6. Steady-state responses of length control system. **A,** In the absence of disturbing forces, position of load is determined only by magnitude of control signal. **B,** Disturbing forces cause actual position to deviate from desired position designated by control signal.

tions (back and forth movements) may continue for some time before eventually dying out.

The output of the spindle is not only proportional to the position of the load but also has a component that is proportional to the rate of change (first derivative) of load position. The effect of this second component is to predict where the load would be at some future time if it continued to move at the same velocity. This prediction overcomes some of the undesirable consequences of the time delays in the nerve fiber and also compensates for some of the sluggishness caused by an inertial load. Thus the muscular system may begin to brake the movement of the load before it reaches the desired position and there may be no overshoot in the response.

If one examines the complex changes that the error signal to the muscles must go through in order to bring the load to a new position (Fig. 21-5, *B*), the power of feedback control is appreciated more fully. Higher centers in the nervous system have only to initiate a new control signal; they may then go on to other tasks, knowing that the length control system will automatically move the arm (the load) to this position. Furthermore, if any disturbances should occur, the control system would also automatically regulate against them.

GRAPHIC REPRESENTATION OF STEADY-STATE RESPONSES

In analyzing the responses of the length control system to its two inputs, control signals and disturbances, we have assumed that the one is absent and have then studied the response of the other. In animals, however, both will be present and both may be changing simultaneously. A simple graphic analysis will be very helpful for appreciating the overall result of length feedback. It will be possible here to neglect the transient phases in the responses shown in Figs. 21-3 and 21-5. We shall focus instead on the steady-state responses.

Following the arrival of a control signal to the motor neurons (Fig. 21-5), the position of the load (or the length of the muscle) goes through a transient phase and then settles down to a new position that is designated as "desired position" in Fig. 21-5, *D*. It was stated earlier that in response to a larger control signal a larger change in load position would automatically be executed. The dependence of the steady-state position on the magnitude of the control signal is shown for a continuous range of signals in Fig. 21-6, *A*. The load position faithfully assumes any value directed by the control signal within maximal and minimal values that are determined by restraining ligaments.

The application of a disturbing force also results in a change in the position of the load, which was shown as the steady-state error in muscle length in Fig. 21-3, *B*. It is called an error because it represents a deviation in the position of the load from the desired position designated by the control signal. Let point *P* in Fig. 21-6, *B,* represent the desired position when the control signal is zero. In the absence of disturbing forces the load will be maintained in this position; but disturbing forces in either direction will cause the load position to deviate, and the extent of deviation will depend on the magnitude of the disturbance as indicated by the solid line passing through point P.

One index of the ability of the length control system to minimize the deviation that results from a disturbance is the slope of this curve. The smaller the slope, the smaller will be the deviation for any given disturbing force. Earlier we made the analogy between the responses of the length control system to disturbing forces and the behavior of a spring. If this analogy is followed in Fig. 21-6, *B*, the line passing through P defines the changes in the length of the spring that result when various forces are applied to it. Positive forces cause extension and negative forces cause compression. The shallow slope of the line signifies that the spring is stiff. The dotted line through P represents the deviations that would result if the spring were not so stiff. The *stiffness* of the spring is thus defined as the inverse of the slope of these lines. Stiffness is a term that is also applied to the length control system. In this case, however, the amount of stiffness depends on the strength of the regulation provided by the stretch reflex.

Responses of the length control system to various combinations of control signals and disturbing forces can be easily understood from Fig. 21-6. For example, let us assume that the control signal has a value C (Fig. 21-6, *A*) and the disturbing force has a value D (Fig. 21-6, *B*). The dashed construction shows the corresponding desired position that would be attained in the absence of disturbing forces. However, the actual position attained will also depend on the disturbance and, in our example, will assume the value designated by point Q in Fig. 21-6, *B*. The deviation of the actual position from the desired position is again the steady-state error.

The mechanical analog in Fig. 21-7 serves to summarize the two basic operations performed by the length control system. The antagonistic muscular system and its reciprocal stretch reflexes shown in Fig. 21-4, *A*, have been replaced by a mechanical analog of the length control system. (This is not meant to be an analog of muscle but rather of the whole control system.) The spring accounts for the behavior of the system in response to disturb-

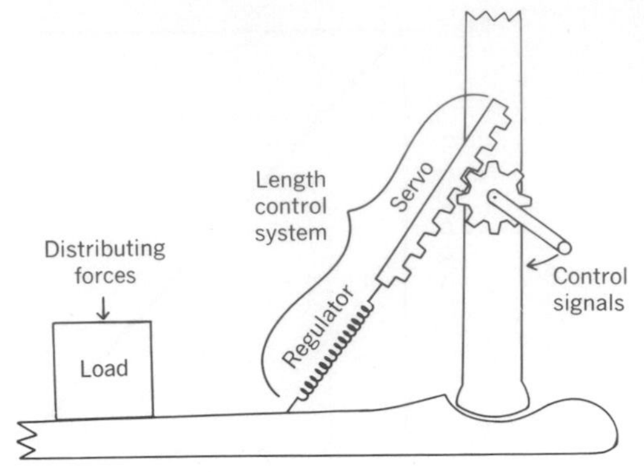

Fig. 21-7. Mechanical analog of length control system. Disturbing forces stretch spring, which is used to represent regulator function of length control system. Control signals crank rack and pinion, used to represent servo function of length control.

ing forces, the regulator function of length feedback. Disturbing forces stretch the spring, causing the actual position of the load to deviate from its desired (or equilibrium) position. The rack and pinion account for the behavior of the system in response to control signals, the servo function of length feedback. Control signals can be imagined to crank the rack and pinion to new desired positions. The change in position of the limb may be slowed by the inertia of the load, but it will be executed automatically.

The account of muscular control given here has been greatly simplified in order to illustrate some fundamental properties of feedback systems. Chapters 22 to 24 will describe the properties of muscle, its receptors, and the spinal reflexes that influence its contraction. In Chapter 25 we will then return to a more complete analysis of the feedback systems that regulate the contractile responses of muscles.

REFERENCES

1. Dorf, R. C.: Modern control systems, Reading, Mass., 1967, Addison-Wesley Publishing Co., Inc.
2. Houk, J., and Henneman, E.: Feedback control of skeletal muscles, Brain Res. **5**:433, 1967.
3. Liddell, E. G. T., and Sherrington, C. S.: Reflexes in response to stretch (myotatic reflexes), Proc. R. Soc. Lond. (Biol.) **96**:212, 1924.
4. Milhorn, H. T., Jr.: The application of control theory to physiological systems, Philadelphia, 1966, W. B. Saunders Co.

22

ELWOOD HENNEMAN

Peripheral mechanisms involved in the control of muscle

In the previous chapter, muscle was treated as a component of a feedback control system, capable of producing desired tensions and of shortening to desired lengths. In this chapter we shall examine the various mechanisms that make muscle highly responsive to control by the nervous system. We shall see that it is not a simple motor, but a complex one requiring a continuous and heavy stream of control signals between it and the spinal cord. During active use of the triceps surae of the cat, for example, there may be as many as 50,000 impulses/sec coming from it into the spinal cord. During a powerful contraction there may be approximately the same number of impulses passing from it to the spinal cord. Even during complete relaxation the gamma system, which never rests, transmits a few thousand impulses per second to the muscle spindles of the triceps.

The first part of this chapter deals with the efferent aspects of muscle control, with special reference to the functional basis of activity—the motor unit. In the second part the specialized receptors in which feedback signals originate will be described and the nature of the information they send to the spinal cord will be discussed.

EFFERENT CONTROL OF MUSCLE

What requirements must muscles meet?

Before considering *how* a muscle is controlled it is important to appreciate fully what it may be called on to do in various circumstances. Hill pointed out that "every muscle or group of muscles will show qualitatively or quantitatively the sort of properties that a very intelligent engineer, knowing all the facts, would have designed for them in order to meet, within wide limits, the requirements of their owner. These properties must provide a compromise between various needs. For example, the possibility of speed of movement is often a desirable quality, but so is economy of energy in maintaining force or posture."[14] In addition, one could add that a muscle must be capable of developing tensions varying from the most delicate contractions to maximal efforts of the greatest power. It must shorten sufficiently to provide a full range of movement at its joint. It must often function without fatiguing. Above all, it must be susceptible to fine control over a wide range of lengths, tensions, speeds, and loads. The problem of combining all of the necessary contractile properties in a homogeneous muscle was probably insoluble. As we shall see, three kinds of muscle fibers with different properties have evolved to meet these requirements. The majority of skeletal muscles consist of a mixture of these three types. A few muscles consist of just one type of fiber.

Types of muscle fibers

The three types of fibers can be distinguished best in muscles that are specially treated to demonstrate mitochondrial adenosine triphosphatase (ATPase). Their appearance and distribution under low magnification are illustrated in Fig. 22-1, which is a photomicrograph of fibers in the cat's gastrocnemius muscle.[12] Large, pale (type A) fibers, containing relatively few mitochondria, predominate in all parts of the muscle. Scattered among them are small, dark (type C) fibers, which are loaded with ATPase. A third type of fiber (B), intermediate in size and in the intensity of its ATPase reaction, has mitochondria situated at the periph-

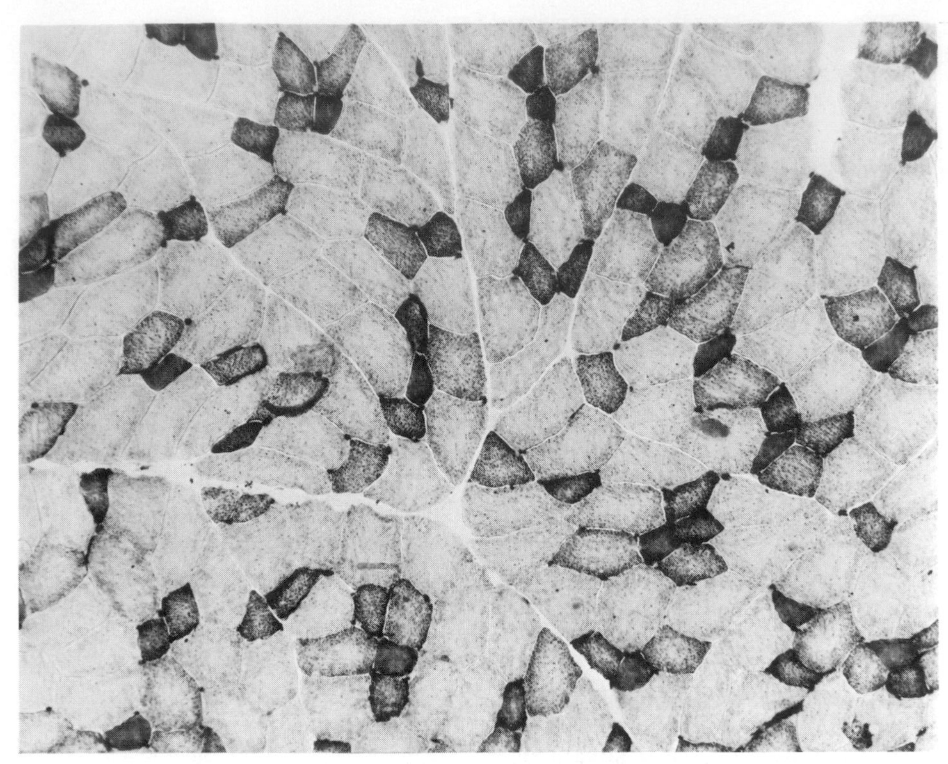

Fig. 22-1. Cross section of m. gastrocnemius muscle of cat showing mitochondrial ATPase activity of three types of fibers. Large, pale fibers are type A; small, dark fibers are type C; and fibers of intermediate size and density are type B. (From Henneman and Olson.[12])

Fig. 22-2. Characteristic features of A, B, and C fibers in m. gastrocnemius at high power: type A, large, with few mitochondria; type B, intermediate in size and enzymatic activity, with subsarcolemmal distribution of large mitochondria; type C, small, with marked background activity and numerous small mitochondria. ATPase section. (From Henneman and Olson.[12])

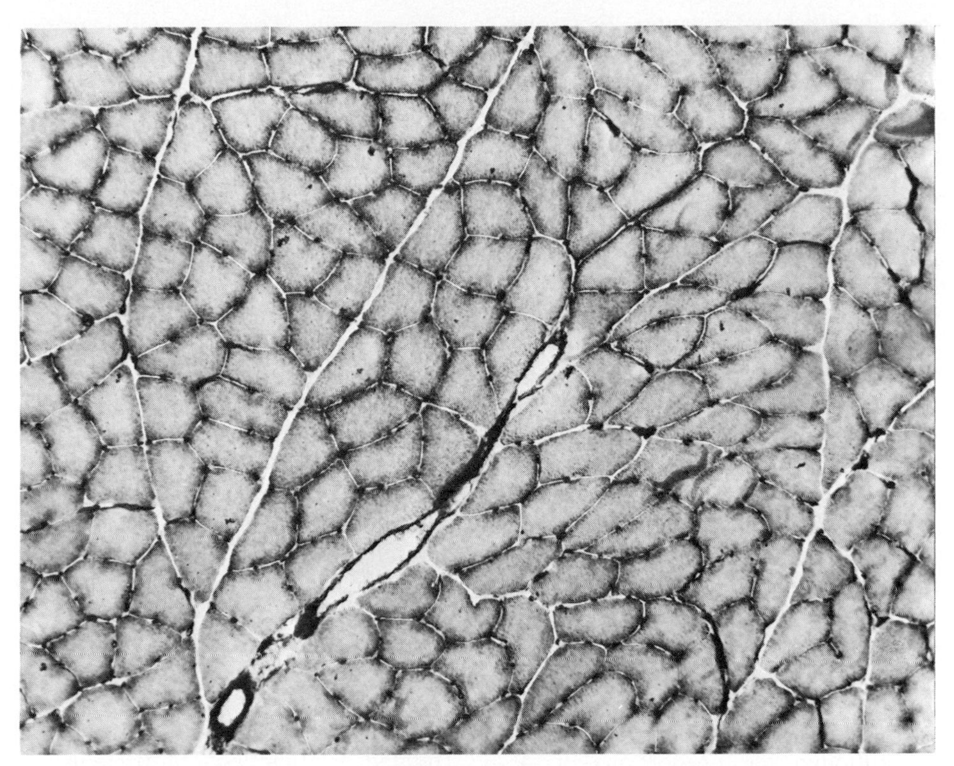

Fig. 22-3. Cross section of soleus muscle of cat showing uniformity of ATPase activity and fiber size. Note intense ATPase activity of capillaries around each fiber. (From Henneman and Olson.[12])

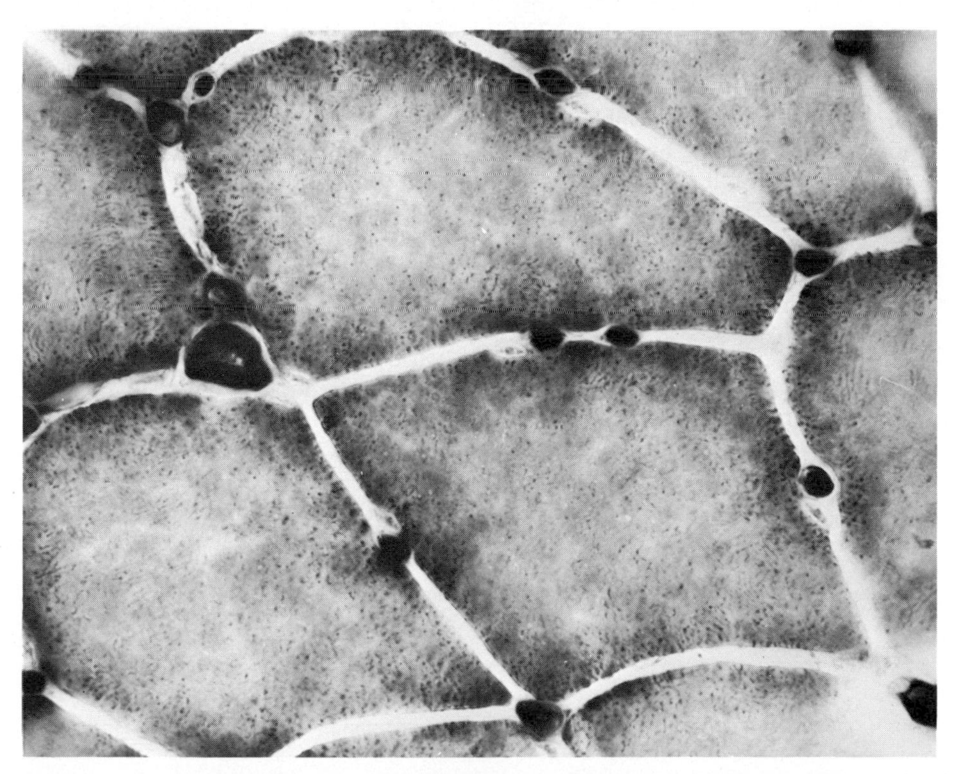

Fig. 22-4. Characteristic features of soleus fibers at high power. Mitochondrial density is greatest at periphery of fibers and especially near the numerous capillaries. ATPase section. (From Henneman and Olson.[12])

ery of the cell with a paler center. Examples of the three types of fibers are shown at higher magnification in Fig. 22-2. A close examination of Figs. 22-1 and 22-2 will reveal the presence of darkly stained capillaries located at the interstitial angles between fibers. The density of capillaries is greatest around the small, dark C fibers, slightly less around B fibers, and least around the pale A fibers. Unless they are adjacent to B or C fibers, A fibers may have no associated capillaries at all. The differences in size, enzymatic activity, and vascularity of these three types of fibers are directly related to their specific roles in contractile activity. Of great importance is the fact that different types of muscle fibers are subject to widely varying degrees of use due to the nature of their innervation. The most active are the smallest fibers; hence they have the richest blood supply, the most ATPase, and because of their small diameter, the shortest distance across which oxygen and other substances must diffuse.

The majority of muscles consist of a mixture of the three types of fibers discussed previously, although the proportions of A, B, and C fibers vary considerably in different muscles. All such heterogeneous muscles are known as *pale* muscles because their gross appearance is, in general, lighter than that of another type, known as *red* muscles. The latter consist of type B fibers throughout and therefore look relatively homogeneous under the microscope. Their appearance is illustrated in Figs. 22-3 and 22-4, which are photomicrographs of fibers in the cat's soleus muscle.[12] These photographs clearly show the numerous small mitochondria, more densely grouped at the periphery of the cell near each capillary.

In subsequent paragraphs we shall see how these histologic and histochemical features are reflected in the properties of the motor units. There is good reason to believe that a single motor neuron innervates a group of muscle fibers that are all of the same type. Hence we may expect to find three types of motor units, corresponding to the three types of muscle fibers.

Variability in contractile properties of different muscles

Since the studies of Ranvier in the last century, it has been known that muscles differ widely in their speed of contraction, fatigability, and response to different rates of stimulation and that the differences are correlated to some extent with their gross paleness or redness. Fig. 22-5 illustrates the great range of contraction speeds found in three different muscles of the cat. The extraocular muscles, which rotate the eyeball, require only 7.5 msec to reach their maximal tension in a twitch. The soleus, a very slow antigravity muscle in the hind limb, requires about 100 msec. The contraction speeds of most pale muscles fall somewhere between these extremes. Speed of contraction is clearly related to function. Movements of the eye require muscles of great speed; movements of the limbs require muscles of less speed but greater endurance. Each muscle in the body is adapted for its special role. It is often forgotten, however, that even very specialized muscles such as those that are designed primarily for great speed must also be capable of some degree of maintained activity without undue fatigue. It is therefore not surprising that such muscles usually do not consist exclusively of rapidly contracting fibers, but a mixture of three types.

As we shall see, observations on whole muscles yield a very inadequate view of their varied capacities. The characteristics of the individual motor units in a muscle differ greatly, forming a broad spectrum of con-

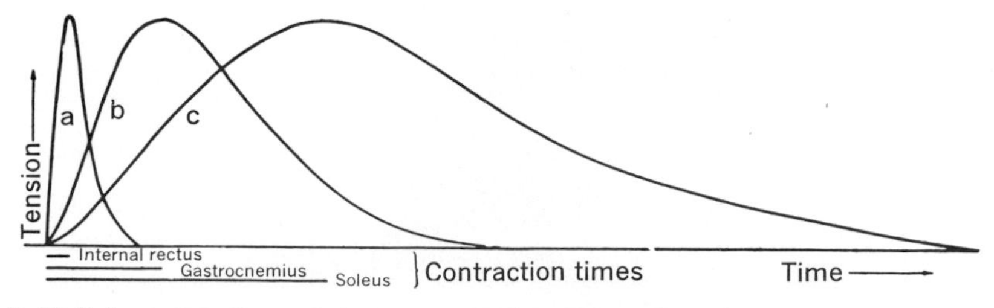

Fig. 22-5. Twitch contractions of three cat muscles, arranged to show the great differences in their speeds. Curve *a* represents internal rectus, curve *b* represents gastrocnemius, and curve *c* represents soleus. (From Cooper and Eccles.[7])

tractile properties even in a homogeneous muscle. The distribution of these properties confers on the muscle functional characteristics that it would not otherwise possess. These characteristics must be described in some detail in order to explain how control is achieved.

Properties of motor units in pale and red muscles

A recently developed technique has made it possible to activate a single motor unit in an otherwise quiescent muscle and study its properties in isolation. A single unit is separated from its fellows by repeatedly dividing and subdividing ventral root filaments under a dissecting microscope until a thin strand is found that contains only one axon supplying the muscle under study.[19] From such a filament it must be possible to record a single, all-or-none antidromic action potential in response to electrical stimulation of the muscle nerve; stimulation of the same filament must produce all-or-none twitches of the muscle accompanied by all-or-none muscle action potentials. When a filament fulfilling these criteria is found, observations can be made on the axonal conduction velocity of the unit, on the twitch tension and contraction speed of its muscle fibers, on the maximal tetanic tension they can produce, and on the fatigability of the unit. These properties are all closely related and each has a bearing on control of muscle. In order to obtain a fair sample of the population of units making up a muscle, more than 100 units from the medial gastrocnemius, a pale muscle, were studied[23] and compared with a similar group from the soleus, a red muscle.[19] Both of these muscles insert on the same tendon and both extend the ankle, yet their properties differ considerably.

One of the first differences between gastrocnemius and soleus motor units is the size of the nerve fibers innervating them. As Fig. 22-6 indicates, the conduction velocities of soleus axons fall chiefly between 50 and 80 m/sec, with a peak distribution between 60 and 70 m/sec. Gastrocnemius axons fall chiefly between 60 and 110 m/sec, with a peak between 80 and 90 m/sec. The greater conduction velocities of gastrocnemius axons are due to their larger diameter. Since the diameter of an axon is directly related to the size of its cell body,[21] it may be concluded that the neurons supplying gastrocnemius are larger than those innervating soleus. The presence of large, rapidly conducting fibers in the nerve to gastrocnemius is presumably correlated with the presence of the large, pale muscle fibers found in this muscle, but not in soleus. As Fig. 22-6 indicates, all muscle nerves also contain a group of smaller, more slowly conducting (10 to 40 m/sec) *gamma* fibers, which are the motor innervation of the tiny muscle fibers inside muscle spindles.

When individual motor units are stimulated at various rates and the resulting tensions are recorded (as in Fig. 22-7), significant differences between units are noted. The most striking of these is the variation in maximal tetanic

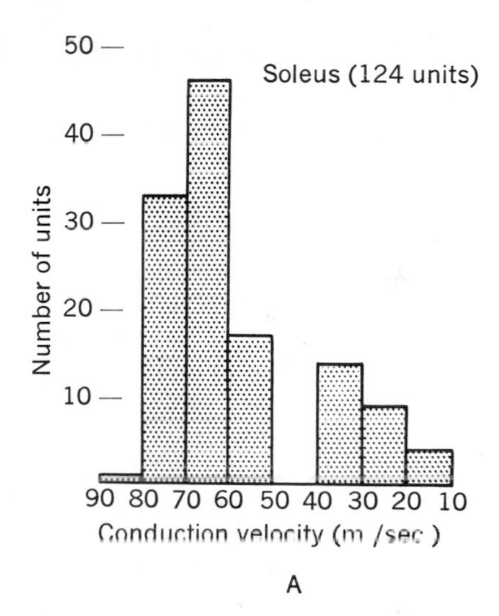

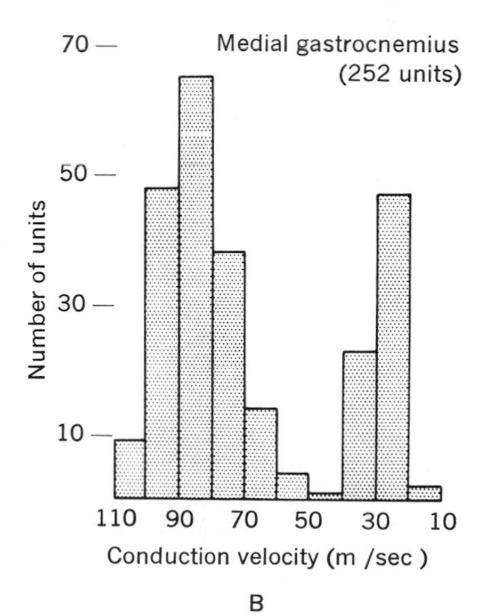

Fig. 22-6. A, Conduction velocities of 124 soleus axons, grouped in decades. **B,** Conduction velocities of 252 axons of m. gastrocnemius. (**A** from McPhedran et al.[19]; **B** from Wuerker et al.[23])

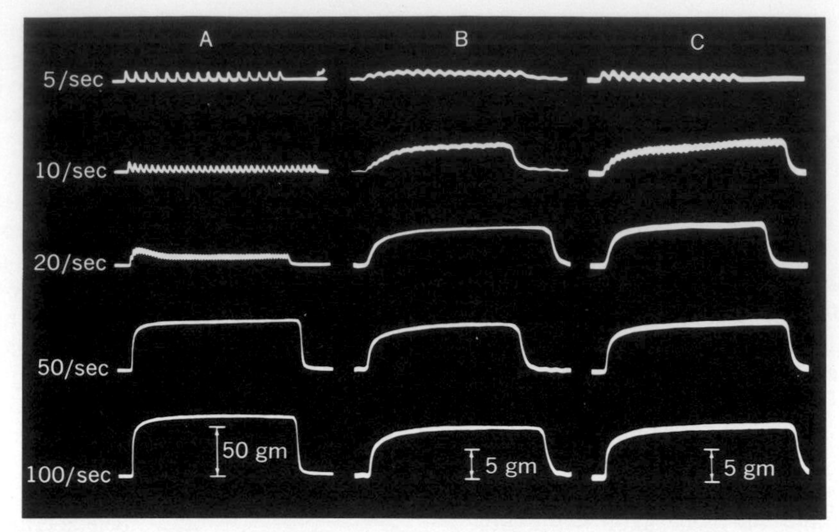

Fig. 22-7. Tetanic tensions developed by three units (**A**, **B**, and **C**) stimulated at 5, 10, 20, 50, and 100/sec. Units **A** and **B** from m. gastrocnemius; unit **C** from soleus. (From Wuerker et al.[23])

tensions. In the case of the soleus muscle this ranges from about 3 to 40 gm; in gastrocnemius some single units can develop no more than 0.5 gm, whereas others may develop as much as 120 gm. The distributions of maximal tetanic tensions for soleus and gastrocnemius are plotted in Fig. 22-8. In both of these histograms the largest number of units is found in the lowest decade (0 to 10 gm), and in general there is a progressive decrease in the number of units in each higher decade. The functional significance of this distribution pattern will be explained in detail later. For the present it is sufficient to state that the abundance of small units makes possible precise, finely graded control of muscle tension in the lower ranges, whereas the presence of a smaller number of very large units makes possible the addition of large increments of tension during maximal efforts.

The maximal tetanic tensions of motor units in the soleus permit one to estimate the total number of muscle fibers innervated by each parent axon. By counting all the muscle fibers in the soleus and dividing this by the total number of motor fibers in the nerve to the soleus, Clark[6] calculated that each motor neuron innervated 120 muscle fibers. Cor-

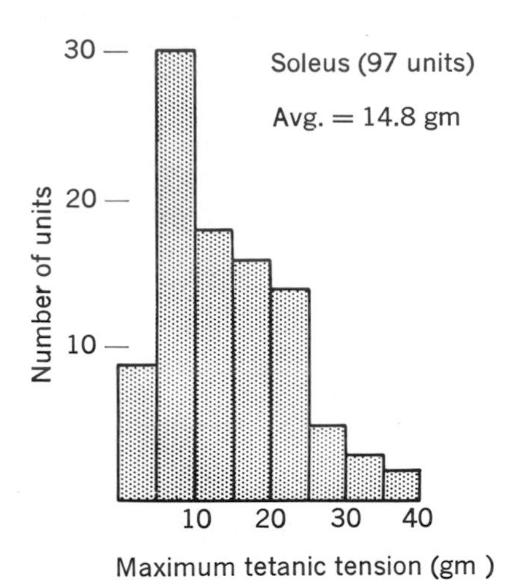

Medial gastrocnemius (103 units)
Number of units
30
20
10
20 40 60 80 100 120
Maximum tetanic tension (gm)
B

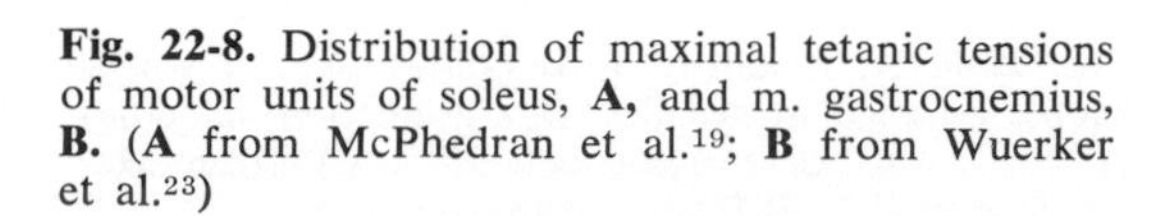

Fig. 22-8. Distribution of maximal tetanic tensions of motor units of soleus, **A**, and m. gastrocnemius, **B**. (**A** from McPhedran et al.[19]; **B** from Wuerker et al.[23])

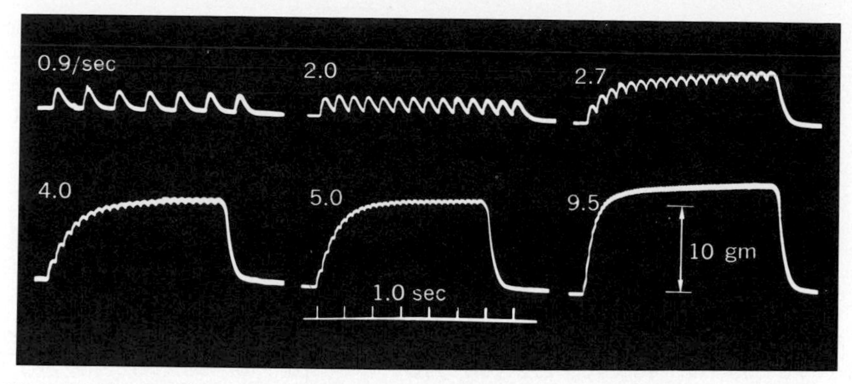

Fig. 22-9. Oscillographic records of tensions developed by a soleus motor unit with a very long contraction time (193 msec) when stimulated at rates of 0.9 to 9.5/sec. (From McPhedran et al.[19])

recting this for 33% gamma fibers, we arrive at an innervation ratio of 180:1. This would apply to the average motor unit that develops 14.8 gm of tension. By extrapolation the smallest unit (3.2 gm) and the largest unit (40.4 gm) would consist of 39 and 491 muscle fibers, respectively. The muscle fibers of the medial gastrocnemius have not been counted in the cat; hence no estimate can be made of the innervation ratios for this muscle. By making several reasonable assumptions, however, we can make a speculative estimate of 1,000 to 2,000 muscle fibers for the largest gastrocnemius units. Feinstein et al.[9] counted the muscle fibers in the human medial gastrocnemius and the large nerve fibers in its nerve, and by subtracting the estimated number of afferent fibers in the nerve, they arrived at an innervation ratio of 1,730:1. These estimates are of general interest because they supply the only available data indicating the number of terminals a single nerve fiber can give off. They also serve to emphasize the magnitude of the trophic influence of nerve on muscle, for they show that one motor neuron can trophically maintain as many as 1,000 muscle fibers.

In Fig. 22-7 it may be noted that the large motor unit whose tension records appear in column *A* developed relatively little tension at rates of stimulation between 5 and 20/sec, whereas the units in columns *B* and *C* developed a considerable proportion of their maximal tension at these rates. The frequency of contraction at which consecutive individual twitches overlap one another and undergo summation depends on the total time required for each twitch contraction and relaxation. This varies widely from unit to unit. Fig. 22-9 illustrates the process of summation in a soleus unit with an extremely slow contraction. As a means of comparing the speed of contraction in all motor units, the times from onset of tension to peak of twitch were measured and plotted as in Fig. 22-10. These "contraction times" vary widely (58 to 193 msec for soleus and 18 to 129 msec for gastrocnemius). The majority of the gastrocnemius units contracted more rapidly than the fastest soleus units, a finding that is related to the presence of large, pale fibers in the gastrocnemius and their absence in the soleus. It was found that large motor units always contract rapidly, whereas small units, with some exceptions, tend to be slow. Slow units are invariably small.

The fact that individual motor units differ so greatly in contraction speed indicates that each unit is homogeneous, with a complement of muscle fibers that resemble each other very closely. The correlation between the size of a unit and its speed of contraction reinforces this conclusion. By conservative estimate the large motor units of gastrocnemius consist of at least 1,000 muscle fibers. A sample of that size would surely include a considerable number of slow fibers if the distribution of fiber types were random. Large units, however, are uniformly fast. Neither the time course, their twitch, nor their fusion characteristics suggest that such units contain a mixture of slow fibers. It also seems clear that very slow units do not include any fast fibers, for they develop their maximal tetanic tensions at low frequencies of stimulation. This would not occur if they contained fast fibers.

The homogeneous character of a motor unit is the natural result of the uniform neural control exerted by the motor neuron

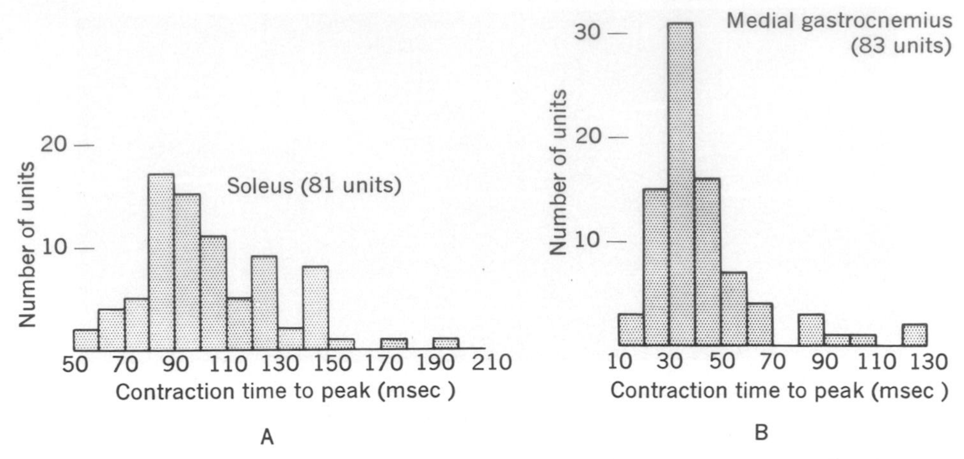

Fig. 22-10. Distribution of contraction times of soleus, **A,** and m. gastrocnemius, **B,** motor units. (**A** from McPhedran et al.[19]; **B** from Wuerker et al.[23])

over its muscle fibers during their period of differentiation. This control may be trophic in nature or it may be due to neural regulation of the intensity of muscular activity. In recent experiments on fast and slow muscles whose nerves had been transposed at an early age, Buller et al.[4] concluded that nerves exert a controlling influence over muscles at the time of their differentiation into fast and slow types. Whatever the nature of this influence may be, it is evidently powerful and precise, for it enables each motor neuron to specify the contractile properties of its muscle fibers within narrow limits. As a consequence, the functional properties of each motor neuron are closely matched with those of all its muscle fibers.

If the maximum tetanic tensions developed by individual motor units are plotted against the conduction velocities of the nerve fibers innervating them, a linear relationship is apparent. The three plots in Fig. 22-11 illustrate the relationship observed in three different soleus muscles. The larger the nerve fiber, the more tension its unit can develop. Presumably a large axon is required to give off a large number of terminals and a large cell body is required to maintain them trophically. In the gastrocnemius muscle a similar relationship between axonal conduction velocity and unit size has been demonstrated for the slower fibers, but not for the faster fibers.

Finally, there is a relationship between the size of a motor unit and its vulnerability to fatigue. If the tension and the electrical activity developed by a motor unit are recorded simultaneously during repetitive stimulation of its axon at 100/sec, a decline in the amplitude of both may occur, denoting progressive failure of transmission at the neuromuscular junction (Fig. 22-12). As a general rule, the larger a motor unit, the more vulnerable it is to such fatigue. Small units can maintain their maximum tension for long periods without fatigue. If neuromuscular fatigue occurs to the same extent under normal conditions, it must seriously limit the intensive use of large motor units.

With these facts it is possible to put together a fairly detailed picture of the functional organization of muscle. In view of the apparent homogeneity of individual motor units it is reasonable to assume that there are three kinds of motor units in pale muscles, corresponding to the A, B, and C types of muscle fibers, that can be distinguished histochemically. It is sometimes difficult, however, to identify the three types by physiologic criteria, for the properties of each type are distributed over a wide range, and overlapping may occur. Type A includes all the rapidly contracting motor units and all of the large units, for they are not found in red muscles. Although pale fibers predominate in muscles such as the gastrocnemius, the actual number of large, fast units is small because each includes so many muscle fibers. The tendency for large motor units to show rapid fatigue is probably correlated with the scanty blood supply and comparative lack of mitochondria of their pale muscle fibers. Type B units range widely in size and speed, as the data on soleus illustrate, but none of them contracts very rapidly and none of them is very large. They are relatively insusceptible to fatigue if they are driven at

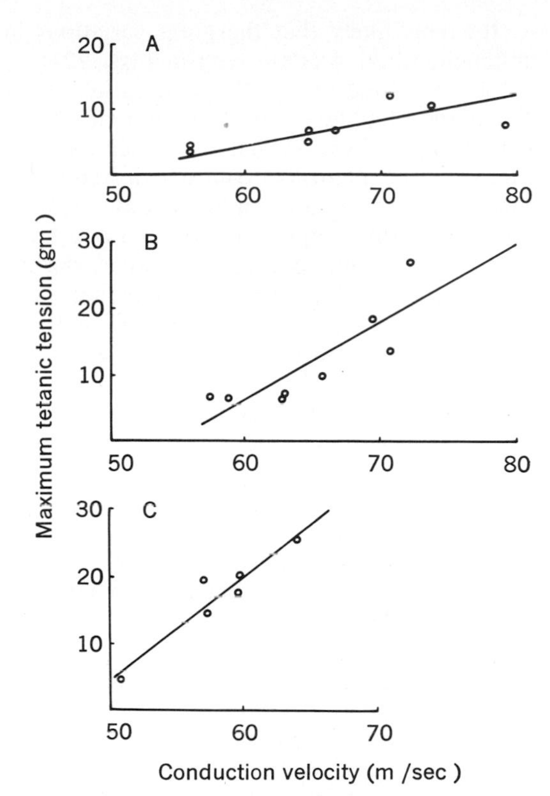

Fig. 22-11. Plots of the relation between maximum tetanic tension of individual motor units and conduction velocity of their axons in three different experiments. (From McPhedran et al.[19])

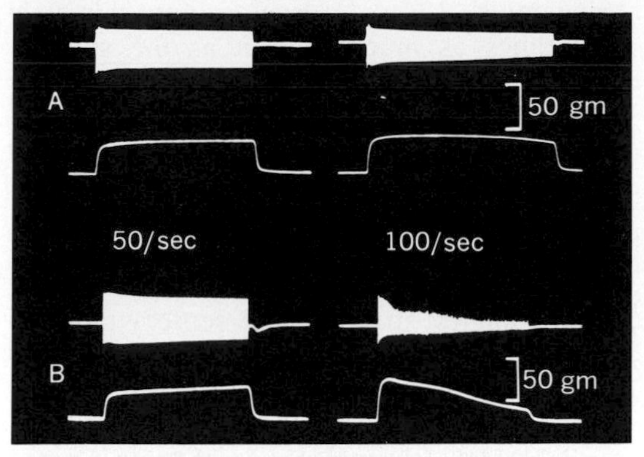

Fig. 22-12. Effect of repetitive stimulation at 50 and 100/sec upon the electromyogram and tension of two motor units of m. gastrocnemius. (From Wuerker et al.[23])

rates within their normal limits, a phenomenon that may be correlated with their ample supply of mitochondria and capillaries. Although there are fewer B and C fibers than A fibers in the gastrocnemius, there are a great many small motor units. Each unit, however, consists of fewer muscle fibers than a single large fast unit. The facts do not allow us to state whether there are any small units consisting of A fibers. Although the properties of A and B fibers are reasonably clear from comparisons of the gastrocnemius and soleus muscles, it is not yet possible to characterize C fibers satisfactorily.

Functional significance of properties of motor units

It might be supposed that the CNS could select out of the available supply any group of motor units it required for use according to the needs of the moment. This is not the case. As the following chapter will explain in detail, the excitabilities of motor neurons depend on their dimensions, which extend over a very wide range. As a consequence, the motor units of a muscle can be fired in only one particular order, as determined by the sizes of their motor neurons. The smaller a motor neuron is the more easily it can be fired; the larger it is the greater is the amount of excitatory input required to discharge it. Since the size of a motor neuron and the size of a motor unit are directly related, it follows that the participation of a motor unit in graded motor activity is dictated by its size. A necessary corollary of this rule is that the total activity of a unit per day decreases as the size of the unit increases.

These simple principles governing the activity of motor neurons and motor units and the inferences they permit are sufficient to explain many of the physiologic and histochemical findings in a consistent manner. Since small motor units are innervated by small motor neurons, which are the most excitable of the alpha cells, they are always the first units in a muscle to become active. As Fig. 22-8 indicates, there are a great many small units in a muscle; each has a slightly different susceptibility to discharge because of the size of its motor neuron. Small tensions are produced and precisely controlled by selective mobilization of varying numbers of these small units. The order of recruitment is the same as the left-to-right order of unit sizes in Fig. 22-8. When the total output of the muscle is increased, larger motor units, providing larger increments of tension, are activated. The evidence indicates that larger units never become active without the participation of all smaller units because the order of activity is fixed centrally. As total tension rises, fewer and fewer additional motor units are required to provide the increases. The largest unit in medial gastrocnemius (120 gm) develops approximately 240 times as much tension as the smallest unit. By comparison the dynamic range of soleus is only 13 to 1. It may be noted that the grading of output in a muscle is analogous to the grading of sensation. The smallest increment that can be added to the force exerted by a muscle becomes greater as the total force of contraction itself increases, just as the smallest increment of sensory stimulation required to produce a "just noticeable difference" in sensation becomes larger as the sensation itself increases in intensity. With the available data it can be shown that if all increments of tension are viewed as percentages of the *preexisting total tension,* there is no loss of fine control as the total tension approaches its maximum.[13]

Since, as was pointed out previously, small motor neurons are more readily discharged than large cells, they are, in the course of normal activity, fired much more often than large ones. In consequence the small motor units with which they are connected are heavily "used" in comparison with large units. *The "usage" of any motor unit, in fact, is probably in inverse ratio to its size.* Since units of small size are often used intensively and for prolonged periods, they must of necessity consist of muscle fibers that function economically and are not subject to fatigue. They must always be ready to respond despite a preceding period of prolonged activity. Rapidly contracting pale fibers cannot meet these requirements, but slowly contracting red fibers can. Hence the smallness of a unit, which implies heavy usage, necessarily specifies red fibers and these, in turn, are fibers that contract slowly.[12]

It seems likely that the great variations in mithochondrial ATPase seen in Figs. 22-1 to 22-4 are related to equally great variations in the usage of different types of muscle fibers. Large, pale fibers, which are innervated by large motor neurons, contract infrequently and therefore require little ATPase. Correspondingly they require less oxygen and have a very scanty blood supply. Smaller, darker cells, which are innervated by smaller neurons, contract much more frequently and require ample supplies of ATPase as well as a rich supply of blood.

The picture of muscular activity that emerges from these studies is surprising and somewhat unexpected. We are naturally inclined to think that all parts of a muscle contribute an equal share of its work. We find, however, that a small fraction of the fibers in a pale muscle actually do most of the work. The larger, pale fibers, particularly those in large motor units, may remain inactive for long periods. Only when a maximal effort is required do they participate. Even then, their activity is apt to be relatively brief, for maximal effort is very fatiguing. We may speculate that athletic "conditioning" improves the performance of the rarely used large, pale muscle fibers much more than that of the continually active red fibers.

FEEDBACK SIGNALS FROM MUSCLE

In the preceding chapter we showed that the length and tension of muscles are controlled through the operation of closed feedback systems connecting each muscle with the spinal cord. There are three types of stretch receptors in skeletal muscle that serve

Table 22-1. Number and density of spindle capsules*

Muscle	Mean weight (gm)	Spindle capsule (content)		No. of spindle capsules/gm
		Range	*Mean and SD*	
Lateral gastrocnemius	7.61	21 (25 to 45)	35 ± 7	5
Mesial gastrocnemius	7.34	35 (46 to 80)	62 ± 9	9
Rectus femoris†	8.36	56 (77 to 132)	104 ± 14	12
Tibialis anterior†	4.57	38 (52 to 89)	71 ± 9	15
Semitendinosus	6.41	62 (80 to 141)	114 ± 14	18
Soleus	2.49	31 (40 to 70)	56 ± 7	23
Flexor digitorum longus lateral	3.25	34 (58 to 91)	75 ± 8	23
Tibialis posterior	0.78	19 (21 to 39)	31 ± 4	39
Flexor digitorum longus mesial	1.06	24 (36 to 59)	48 ± 6	45
Vth interossei (foot)‡	0.33	12 (22 to 33)	29	88
Vth interossei (hand)†	0.21	11 (21 to 31)	25 ± 2	119

*From Chin et al.[5]
†Data from Baker and Chin.[2a]
‡Four muscles counted; data from Ip.[15a]

as the transducers in these control systems. It is their function to measure muscle length and tension and to translate these measurements into nerve signals. This section will describe the receptive apparatus in detail and examine the signals that are sent to the spinal cord. It will also describe the specialized efferent system through which the nervous system influences the signals sent back to it.

Structure and innervation of muscle spindles and tendon organs[1-3]

Two of the three types of stretch receptors are located within a highly specialized receptive organ called a muscle spindle. Spindles are macroscopic in size, fusiform in shape, and widely scattered throughout the fleshy parts of the muscle. They are usually attached at both ends to the ordinary or "extrafusal" muscle fibers. As indicated in Table 22-1, their number per gram of muscle tissue varies widely. Small distal muscles such as the interossei of the forepaw that require delicate control have a far greater density of spindles (119/gm) than large, powerful muscles, e.g., the lateral gastrocnemius (5/gm). Examination of Table 22-1 suggests that spindle density may be inversely related to mean innervation ratio, which is reasonable if both are related to fineness of muscular control.

Each spindle consists of a connective tissue sheath containing from 2 to 12 thin muscle fibers known as *intrafusal* fibers. These tiny, striated muscle fibers are innervated by axons called *gamma* fibers because of their diameter (1 to 8μ). At the center of each spindle there is an expanded, fluid-filled region; inside, the

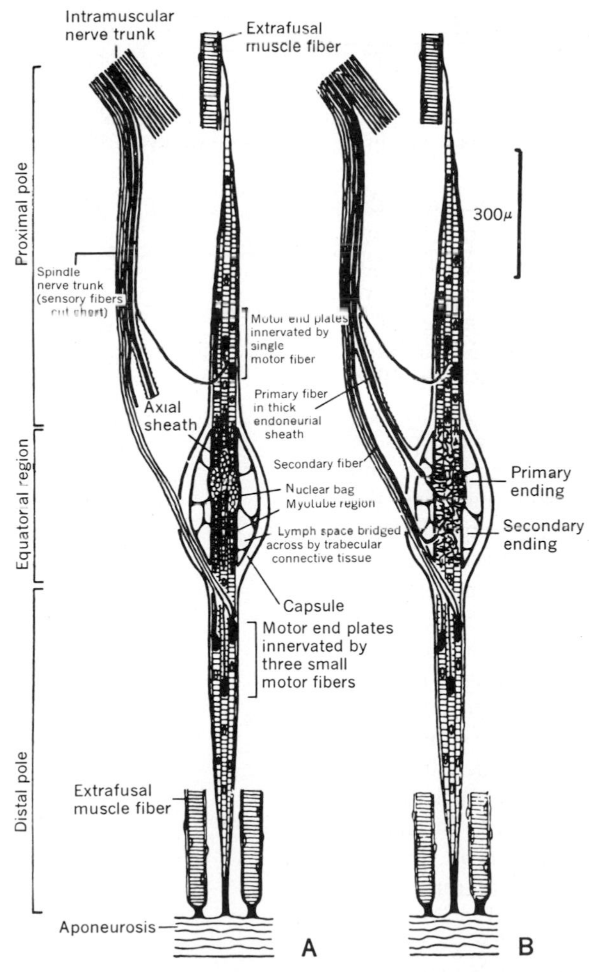

Fig. 22-13. Schematic drawing of structure and innervation of typical spindle organ from mammalian muscle. In **A** only motor innervation is shown. In **B** both sensory and motor innervation are given. (From Barker.[1])

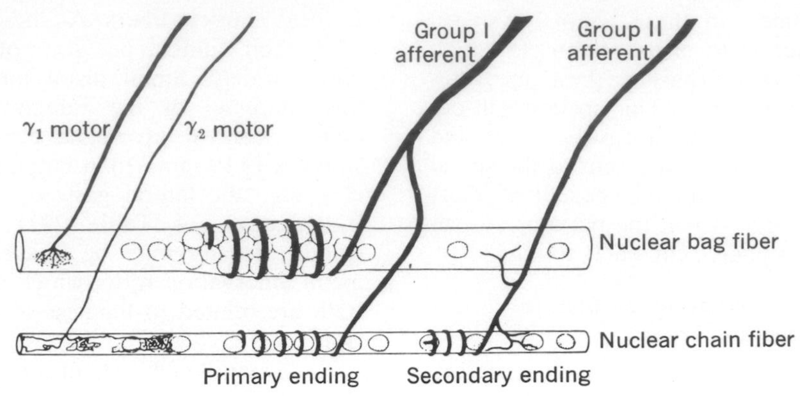

Fig. 22-14. Types of intrafusal fibers in mammalian spindles. (Modified from Boyd; from Matthews.[18])

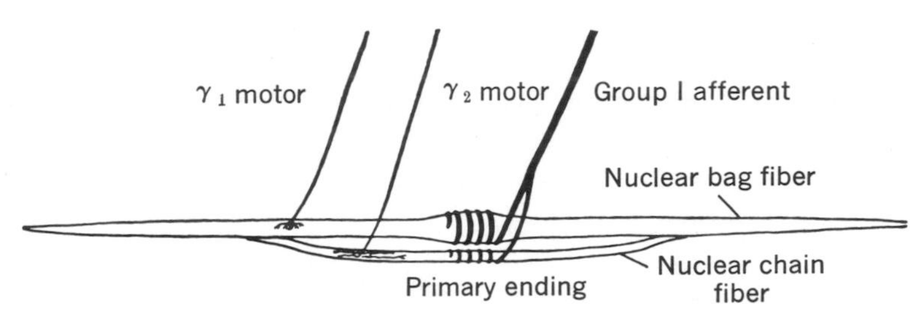

Fig. 22-15. Diagram showing innervation of two types of intrafusal fibers and attachment of nuclear chain fiber to nuclear bag fiber. (From Crowe and Matthews.[8])

terminals of sensory fibers are attached to the intrafusal fibers. There are two types of sensory endings in most spindles—primary endings derived from group Ia (12 to 20μ) nerve fibers and secondary endings coming from group II (6 to 12μ) fibers. Fig. 22-13 illustrates these features in a highly simplified way.

Mammalian spindles contain two types of intrafusal fibers (Fig. 22-14). The longer and larger of these, which are less than half the diameter of extrafusal fibers, contain numerous large nuclei closely packed in a central bag; hence they are called "nuclear bag fibers." The shorter and thinner fibers, which are about half the length and diameter of nuclear bag fibers, contain a single row of central nuclei resembling a chain and are known as "nuclear chain fibers." As indicated in Fig. 22-15, the ends of the nuclear chain fibers are attached to the longer nuclear bag fibers. There are usually two of the larger fibers and four of the smaller type in each spindle, but the numbers of both may vary. The efferent or motor innervation of these fibers comes chiefly from 7 to 25 gamma or fusimotor fibers. These fibers are of two types. One type terminates in discrete motor end plates, which lie chiefly at each pole of the spindle. The other type terminates as a fine

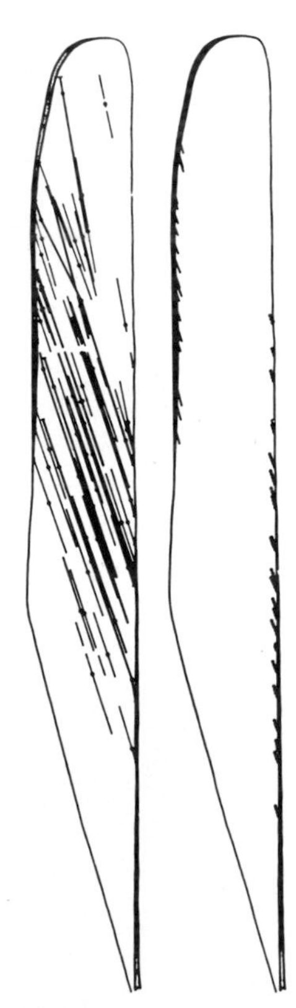

Fig. 22-16. Distribution of spindles (left) and tendon organs (right) in m. gastrocnemius of cat, as if projected onto imaginary midsagittal plane. (From Swett and Eldred.[22])

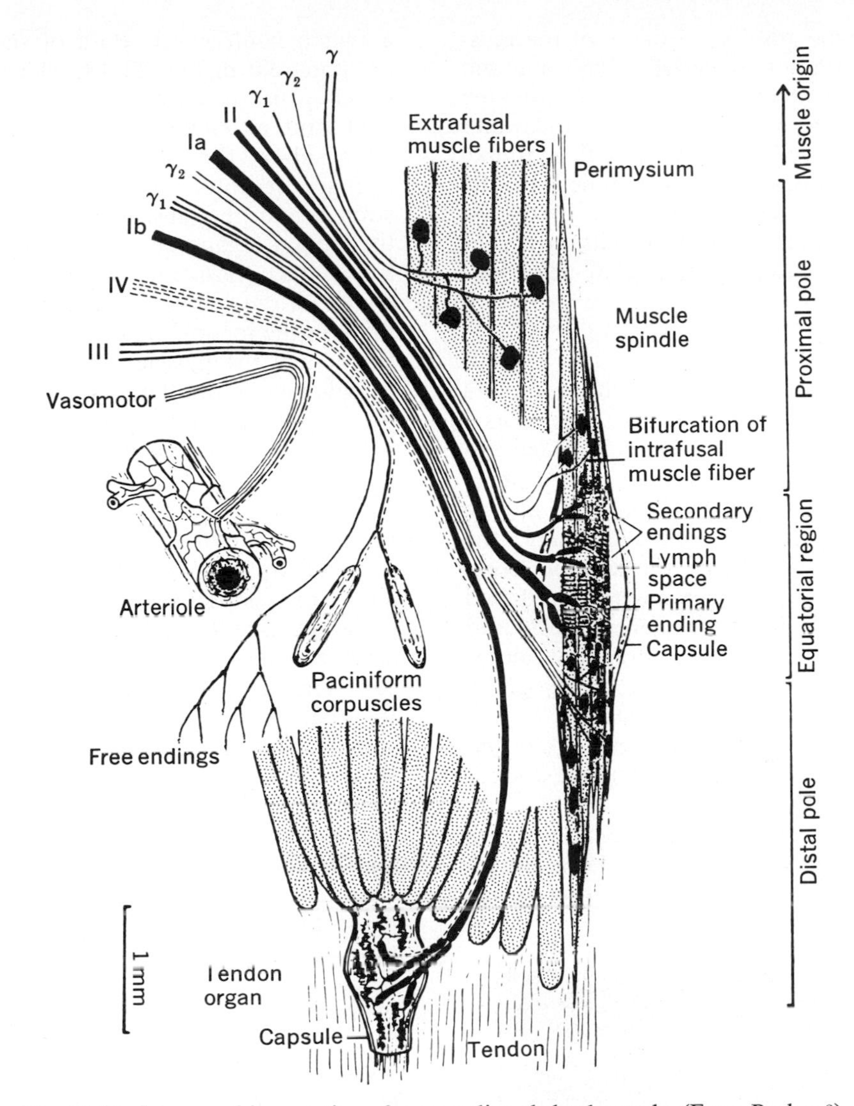

Fig. 22-17. Schema of innervation of mammalian skeletal muscle. (From Barker.[2])

network or "trail" ending, which may extend from the central portion of the spindle all the way to the poles. Authorities do not agree on the distribution of these two types of endings. According to one view, plate endings are found chiefly on nuclear bag fibers, and trail endings are found chiefly on nuclear chain fibers. Another opinion is that each of the two types of intrafusal muscle fibers usually receives both plate and trail endings. Some spindles may also receive plate endings from type A-beta fibers, which also innervate extrafusal muscle. Each spindle has one primary sensory ending supplied by a group Ia fiber and 0 to 5 secondary endings, each supplied by a group II fiber. The primary endings lie on both nuclear bag and nuclear chain fibers; the secondary endings lie predominantly on nuclear chain fibers.

The third type of stretch receptor found in mammalian muscle is the so-called tendon organ of Golgi. In contrast with spindles that lie "in parallel" with the muscle fibers, tendon organs are located at the musculotendinous junctions and hence are "in series" with the contractile elements. Fig. 22-16 shows the contrasting distribution of spindles and tendon organs in the medial gastrocnemius of the cat. In general, tendon organs are less numerous than spindles; the ratio varies from about 1:1 to 1:3, tendon organs being relatively more numerous in slowly contracting muscles.

A tendon organ is formed around a bundle of small tendon fascicles, immediately adjacent to the musculotendinous junction (Fig. 22-17). The receptor, which is about 500μ long, is enclosed in a delicate capsule that blends with the connective tissue of the muscle. Its innervation is derived from a group Ib nerve fiber that divides into myelinated branches inside the capsule and then into sprays of unmyelinated fibers with clasplike granular swellings that are applied to the surfaces of the tendon fascicles. Distortion of these terminals probably results in a generator potential that gives rise to nerve impulses.

Spindle receptors[2, 18]

The activity of individual stretch receptors can be recorded most readily from dorsal root filaments that have been subdivided until they contain the afferent fiber from a single active receptor. By stimulating the muscle nerve distally and recording centrally, the conduction velocity of the afferent fiber coming from this ending can be measured and its diameter calculated. The tension produced by passive stretch or by contraction can be recorded simultaneously with a myograph attached to the tendon of the muscle.

The responses obtained from fibers innervating primary (annulospiral) and secondary (flower-spray) receptors in spindles can be distinguished from those of tendon organs by their greater sensitivity to passive stretch and their characteristic pause in firing during a twitch contraction. Both of these features are apparent in Fig. 22-18, which shows side by side the responses of a primary receptor (*A*) and a tendon organ (*B*). The upper traces show the rhythmic discharge of the primary receptor in response to a 10 gm stretch and the absence of discharge from the tendon organ under 20 gm of tension. The lower traces illustrate the pause in the discharge of the primary ending and the onset of firing in the tendon organ during a twitch contraction of the muscle. This striking contrast in behavior is due largely to the locations of the receptors. Spindles are placed "in parallel" with the contracting muscle fibers, whereas tendon organs lie "in series" with them. When a muscle is stretched, the intrafusal fibers in the spindle are also stretched and the endings attached to them are deformed. Tendon organs are less affected by passive stretch because they are located on inelastic tendon that lengthens very little in comparison with muscle. When the muscle contracts, the shortening that occurs relieves the stretch on the spindle, and its receptors cease firing. At the same time the shortening of the contractile elements lengthens the noncontractile part of the muscle where the tendon organ is located, and its endings are stimulated to fire.

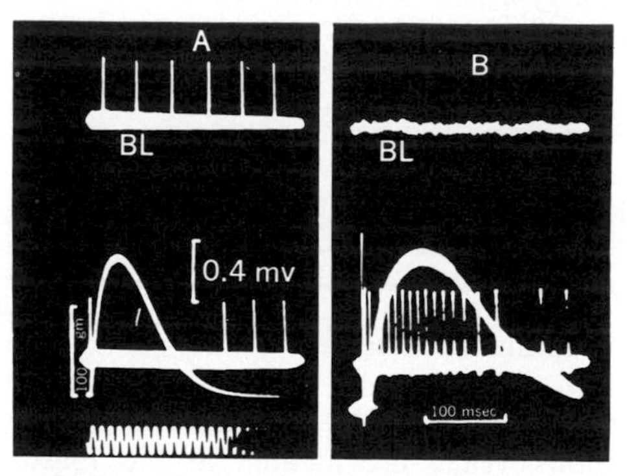

Fig. 22-18. A, Electrical signs of impulses in single afferent fiber from soleus muscle of cat; nerve fiber was isolated by microdissection of dorsal root. It innervated A-type spindle receptor. Initial tension of 10 gm causes baseline discharge, *BL*. For lower records, stimulus maximal for large-nerve motor fibers delivered to soleus nerve produces twitch contraction, shown by curving line of strain-gauge recording. During twitch contraction, afferent discharge ceases completely. Time line, 50 Hz. **B,** Another single afferent from cat soleus that innervated receptor with B-type properties, a tendon organ. Muscle tension of 20 gm caused no baseline discharge. Twitch contraction of muscle caused outburst of nerve impulses, shown in lower records. (From Kuffler and Hunt.[16])

In recent years considerable effort has been made to distinguish between the signals sent to the spinal cord by the two types of spindle receptors. It now appears that they respond in much the same way to static stretch, but respond quite differently during the dynamic phase of lengthening or shortening. Fig. 22-19 illustrates the differences very clearly. The upper recording from a primary ending shows a uniform, rapid rate of discharge during the period of actual extension, a slower rate during the period of static stretch, and a cessation of discharge during release of stretch. The lower recording from a secondary ending shows a gradual increase in rate of discharge during the phase of extension to a rate that is maintained during static stretch and a gradual decrease in rate during release of stretch. From comparisons of this type it appears that the rate of discharge of a secondary ending is chiefly a function of the instantaneous length of the muscle and is only slightly related to the rate of change in length. The discharge frequency of primary endings is proportional to both velocity of stretching and length. During actual extension their sensitivity to the dynamic phase predominates, and their rate of discharge is a function chiefly of velocity of stretching; during static stretch their rate of firing is determined by muscle length. Thus *the primary ending measures length plus velocity and the secondary ending mainly length.* Gamma innervation modifies these characteristics, as will be described later.

Efferent control of spindle receptors[17, 18]

Muscle spindles are designed to enable the CNS to modify or control the activity of the receptors they contain. This influence is exerted by way of two types of gamma fibers that innervate the two types of intrafusal fibers in the spindle. The nuclear bag fibers consist of a central noncontractile region situated between two contractile regions (Fig. 22-13), each with its own motor innervation. Shortening of the contractile ends of the fiber stretches the central portion where the primary receptors are located and causes them to discharge. Since extension of the muscle and contraction of intrafusal fibers both result in stretch of the primary receptor, the response of this ending is determined by the additive effect of these two kinds of stimuli. Fig. 22-20 illustrates the separate and combined effects of stretch and gamma activity on a

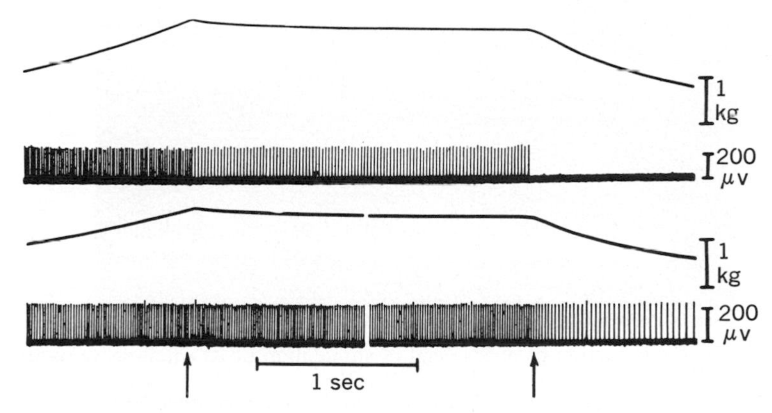

Fig. 22-19. Records obtained during stretch and release of primary ending (above) and secondary ending (below). Rate of stretch is 2.3 mm/sec in both cases. (From Harvey and Matthews.[11])

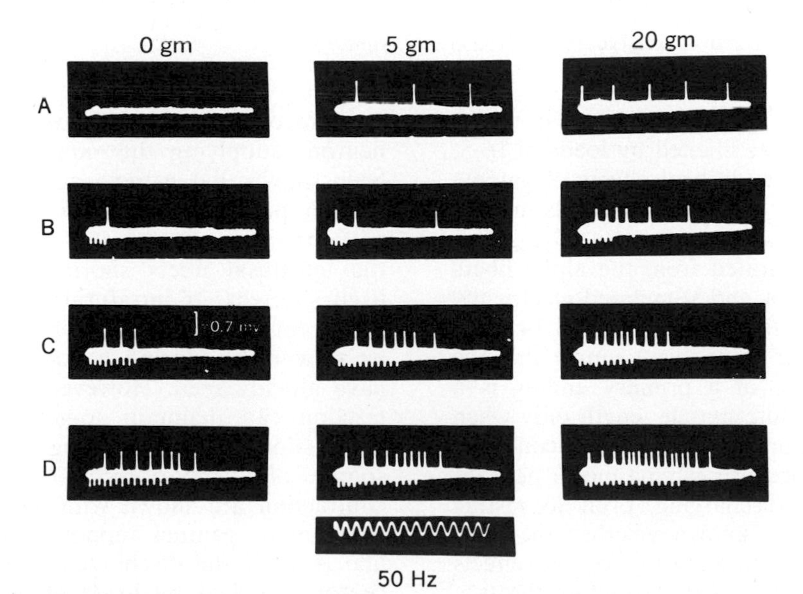

Fig. 22-20. Discharges of primary spindle receptor in soleus muscle recorded from dorsal root filament. Records obtained at 0, 5, and 20 gm of passive stretch. **A,** Baseline discharge. Stimulation of single efferent gamma fiber to soleus is indicated by small downward deflections (4 to 6 stimuli in **B,** 9 to 11 stimuli in **C,** and 14 to 16 stimuli in **D**). (From Kuffler et al.[17])

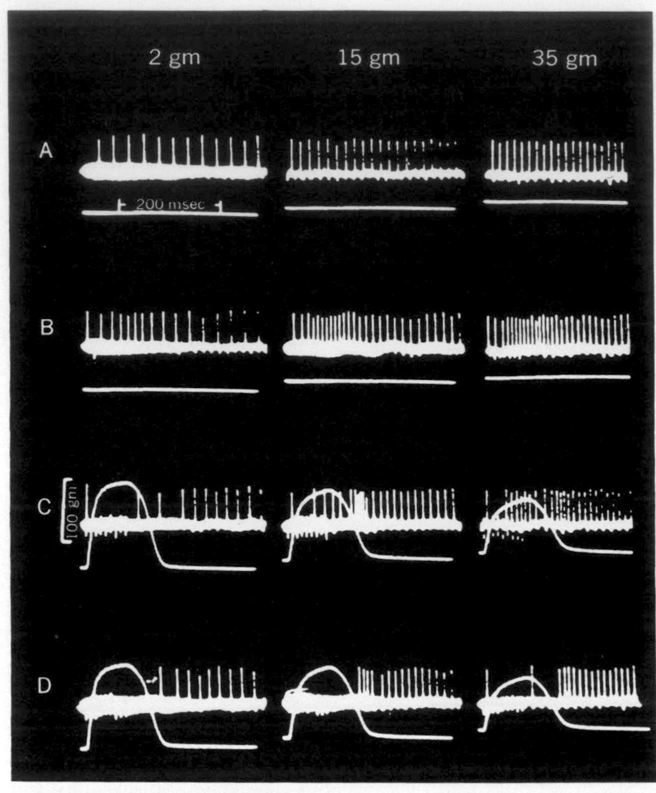

Fig. 22-21. Effect of contraction of muscle and of stimulation of its small-nerve motor innervation on sensory discharge from a spindle receptor. Recording from single afferent nerve fiber from flexor digitorum longus of cat, isolated by microdissection of dorsal root. Second trace of each pair is strain-gauge recording of muscle tension, initially 2, 15, and 35 gm. **A,** Baseline discharge. **B,** Stimulation of small-nerve motor fibers (isolated by microdissection of ventral roots); 9 stimuli at 10 msec intervals delivered at beginning of each record. No muscle contraction was produced although increase in discharge rate occurred. **C,** Stimulation of large- and small-nerve motor fibers produces, at higher tensions, afferent discharges during twitch contraction. **D,** Stimulation of only large-nerve motor fibers causes twitch contraction and cessation of afferent discharge. (From Kuffler and Hunt.[16])

primary receptor. The three tracings in line *A* show the discharges elicited by loads of 0, 5, and 20 gm. The additional effects of gamma stimulation are shown by the records in lines *B, C,* and *D*. Gamma fibers supplying this spindle were separated from the alpha fibers in the ventral root and stimulated electrically (4 to 6 pulses in *B*, 9 to 11 in *C*, and 14 to 16 in *D*). It is clear from this example that the rate of discharge of a primary ending is a measure of absolute muscle length only when there is no gamma activity. This condition seldom occurs because gamma motor neurons are continually discharging, even to resting muscles. It is not known whether the CNS has some means of allowing for the effects of gamma activity and of extracting the true length of the muscle from the combination of various inputs, or whether it is not "interested" in absolute length.

By recording from alpha and gamma motor neurons supplying the same muscle, it has been shown that gamma activity usually occurs in parallel with alpha activity in most patterns of contraction.[15] As a consequence, the intrafusal fibers shorten as the muscle itself shortens. If intrafusal contraction does not occur, the spindle receptors cease firing or slow down during a contraction, as we have already seen. However, intrafusal contraction can maintain spindle firing during contraction, as shown in Fig. 22-21. The responses of a primary receptor during a twitch contraction are shown with and without accompanying gamma support. The records in line *A* show the discharges of a primary receptor stretched by loads of 2, 15, and 35 gm. Line *B* illustrates the effect of stimulating a single gamma fiber to this spindle (9 stimuli at 100/sec). Line *D* shows the pause in firing

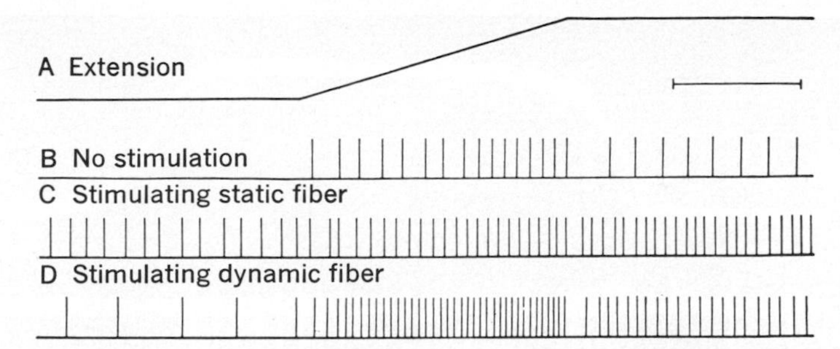

Fig. 22-22. Effects of stimulating single fusimotor fibers on the response of a primary ending to stretch of 6 mm at 30 mm/sec. Throughout **C,** single static fiber was stimulated at 70/sec. Throughout **D,** single dynamic fiber was stimulated at 70/sec. Action potentials drawn from computed data. Time bar: 0.1 sec. (From Crowe and Matthews.[8])

that occurs when the muscle contracts and there is no gamma stimulation. Line *C* illustrates how simultaneous gamma stimulation prevents this cessation of firing. This illustration suggests that one possible function of the gamma system is to adjust the length of intrafusal fibers so that spindle receptors always operate on a sensitive portion of their response scale. During a powerful contraction with considerable shortening, for example, it may be advantageous for the spindle receptors to continue firing in order to reinforce the contraction reflexly.

As noted previously, gamma fibers terminate as plate endings or trail endings. If single gamma fibers are stimulated electrically, the two types of functional effects that can be distinguished presumably correspond to the two types of endings. One type of gamma fiber influences responses to phasic stretch far more than responses to static stretch and hence is called a "dynamic" fiber. The other type influences responses to static stretch more than responses to phasic stretch and is called a "static" fiber. Fig. 22-22 illustrates the effects of these two types of stimulation on a primary receptor. Stimulation of dynamic fibers during phasic stretch results in marked acceleration of primary endings, which are extremely sensitive to phasic stretch, but usually has no effect on secondary endings. At constant length, stimulation of dynamic fibers generally has only a slight effect on primary endings and no effect on secondary endings. Stimulation of static fibers results in marked acceleration of primary and secondary endings if they are at constant length, but causes minor and variable effects during phasic stretch.

The role of the gamma system cannot be defined more precisely until additional facts are available. It is clear that there are two routes leading to muscular contraction; one a direct activation of alpha motor neurons and the other indirect by way of gamma fibers, muscle spindles, and afferent fibers. As indicated in the previous chapter, each of these routes has advantages. In normal functioning they must work together. According to one view,[20] the spindle is essentially a device that measures the difference in the length of extrafusal and intrafusal muscle fibers. Whenever the extrafusal fibers are longer than the intrafusal fibers, the latter are stretched and an afferent discharge occurs that tends to shorten the extrafusal fibers. A gamma input to the spindles causes shortening of intrafusal fibers that may also lead to extrafusal shortening. This mechanism is called a "follow-up servo" because the length of the spindle controls the length of the muscle as a whole.

Functional properties of tendon organs

As mentioned previously, tendon organs are relatively insensitive to passive stretch because they lie in series with a compliant, contractile element that absorbs most of the stretch and prevents elongation of the tendinous region. Changes in muscle length therefore affect tendon organs very little, and these receptors provide no information about length. When a muscle contracts, however, the tendon organs in it discharge in proportion to the tension that is developed. If contraction merely shortens the muscle without developing much tension, the tendon organs are weakly excited. If contraction occurs when the muscle is lengthened and its ends are fixed, the shortening of the contractile part of the muscle necessarily lengthens the noncontractile regions where tendon organs are located and vigorous firing results. A particular tension may be developed at various

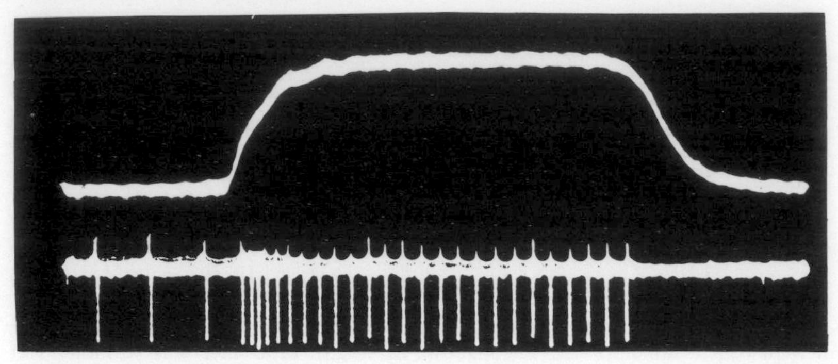

Fig. 22-23. Responses of a single tendon organ in soleus (lower tracing) to tetanic stimulation of ventral root filament containing only one soleus axon. Tension developed by contraction of this single motor unit was 18 gm (upper tracing). (From unpublished records of Houk and Henneman.)

muscle lengths. By virtue of their location, tendon organs measure this tension regardless of length.

The relative insensitivity of these endings to ordinary degrees of passive stretch gave rise to the belief that they functioned primarily to protect muscle from damage by signaling when excessive stretch occurred. Actually the tendon organ is an extremely sensitive receptor if the stimulus is appropriate. This has been demonstrated by recording the activity of single tendon organs while stimulating individual motor units in the same muscle by means of the technique described earlier in this chapter. By systematically testing all the motor units in the soleus muscle it has been found that about 15 different units will each cause firing of a single tendon organ. Fig. 22-23 illustrates the discharge of a tendon organ produced by stimulation of a single motor unit. Histologic studies have shown that 3 to 25 muscle fibers insert on each tendon organ. Hence a motor unit in the soleus that has an average of 180 fibers contributes only one or two of them to a tendon organ. Apparently a tendon organ samples the local tensions in a muscle by monitoring a minute fraction of the muscle fibers of a number of motor units. The response of a tendon organ is determined by the number of active motor units inserting on it and the tension developed by each of the contributing fibers. A single muscle fiber producing 0.1 gm tension is sufficient to cause the receptor to fire at 25/sec. This great sensitivity indicates clearly that tendon organs must be active whenever muscles are contracting and doing work. The fact that tension receptors reflect local conditions within muscle rather than average tension may also have special significance, but this is not yet apparent.

Stretch receptors and position sense

It is now generally recognized that stretch receptors in muscle do not contribute to conscious proprioception; i.e., they do not furnish information regarding the position of a joint or a limb. As Goldscheider[10] originally demonstrated and as has been confirmed many times since, injection of local anesthetics into joints abolishes the sense of position, although muscle receptors are unaffected. Conversely, position sense is retained if muscles are denervated or anesthetized and joints are unaltered.

The main function of the three types of receptors in muscle is in the subconscious control of muscles. Deafferenting a limb by cutting all its dorsal root fibers abolishes all ordinary use of the limb and it hangs as if paralyzed. In the following chapter the profound effect of deafferentation will be explained in terms of the circuits that control motor neurons, and the central effects of signals from muscle will be discussed.

REFERENCES

1. Barker, D.: The innervation of the muscle spindle, Q. J. Microsc. Sci. **89:**143, 1948.
2. Barker, D., editor: Symposium on muscle receptors, Hong Kong, 1962, Hong Kong University Press.

2a. Barker, D., and Chin, N. K.: The number and distribution of muscle-spindles in certain muscles in the cat, J. Anat. **94:**473, 1960.

3. Boyd, I. A.: The structure and innervation of the nuclear bag muscle fibre system and the nuclear chain muscle fibre system in mammalian muscle spindles, Phil. Trans. R. Soc. Lond. (Biol.) **245:**81, 1962.

4. Buller, A. J., Eccles, J. C., and Eccles, R. M.: Differentiation of fast and slow muscles in the cat hind limb, J. Physiol. **150:**399, 1960.
5. Chin, N. K., Cope, M., and Pang, M.: Number and distribution of spindle capsules in seven muscles of the cat. In Barker, D., editor: Symposium on muscle receptors, Hong Kong, 1962, Hong Kong University Press.
6. Clark, D. A.: Muscle counts of motor units: a study in innervation ratios, Am. J. Physiol. **96:**296, 1931.
7. Cooper, S., and Eccles, J. C.: Isometric responses of mammalian muscles, J. Physiol. **69:** 337, 1930.
8. Crowe, A., and Matthews, P. B.: The effects of stimulation of static and dynamic fusiform fibres on the response to stretching of the primary endings of muscle spindles, J. Physiol. **174:**109, 1964.
9. Feinstein, B., Lindegard, B., Nyman, E., and Wohlfart, G.: Morphological studies of motor units in normal human muscles, Acta Anat. **23:**127, 1955.
10. Goldscheider, A.: Untersuchungen über den Muskelsinn. In Gesammelte Abhandlungen von A. Goldscheider, Leipzig, 1898, Barth, vol. 2.
11. Harvey, R. J., and Matthews, P. B.: The response of de-efferented muscle spindle endings in the cat's soleus to slow extension of the muscle, J. Physiol. **157:**370, 1961.
12. Henneman, E., and Olson, C. B.: Relations between structure and function in the design of skeletal muscles, J. Neurophysiol. **28:**581, 1965.
13. Henneman, E., Somjen, G., and Carpenter, D. O.: Functional significance of cell size in spinal motoneurons, J. Neurophysiol. **28:**560, 1965.
14. Hill, A. V.: The design of muscles, Br. Med. Bull. **12:**165, 1956.
15. Hunt, C. C.: Muscle stretch receptors; peripheral mechanisms and reflex function, Symp. Quant. Biol. **17:**113, 1952.
15a. Ip, M. C.: The number and variety of proprioceptors in certain muscles of the cat, M.Sc. thesis, University of Hong Kong, 1962.
16. Kuffler, S. W., and Hunt, C. C.: Mammalian small-nerve fibers: System for efferent nervous regulation of muscle spindle discharge, Res. Pub. Assoc. Res. Nerv. Ment. Dis. **30:**24, 1952.
17. Kuffler, S. W., Hunt, C. C., and Quilliam, J. P.: Function of medullated small-nerve fibers in mammalian ventral roots: efferent muscle spindle innervation, J. Neurophysiol. **14:**29, 1951.
18. Matthews, P. B. C.: Muscle spindles and their motor control, Physiol. Rev. **44:**219, 1964.
19. McPhedran, A. M., Wuerker, R. B., and Henneman, E.: Properties of motor units in a homogeneous red muscle (soleus) of the cat, J. Neurophysiol. **28:**71, 1965.
20. Merton, P. A.: Speculations on the servo-control of movement. In Malcolm, J. L., Gray, J. A. B., and Wolstenholme, G. E. W., editors: The spinal cord, Boston, 1953, Little, Brown & Co.
21. Ramón y Cajal, S.: Histologie du système nerveux de l'homme et des vertébrés, Paris, 1909, Maloine, vol. 1.
22. Swett, J. E., and Eldred, E.: Distribution and numbers of stretch receptors in medial gastrocnemius and soleus muscles of the cat, Anat. Rec. **137:**453, 1960.
23. Wuerker, R. B., McPhedran, A. M., and Henneman, E.: Properties of motor units in a heterogeneous pale muscle (m. gastrocnemius) of the cat, J. Neurophysiol. **28:**85, 1965.

23

ELWOOD HENNEMAN

Organization of the spinal cord

The purpose of this chapter is to provide a brief introduction to the spinal cord that stresses functional organization without undue involvement in specific circuits and reflexes. Attention is therefore directed to a typical segment of the cord and the signals entering and leaving it. In the first portion of the chapter the principles governing distribution of incoming sensory impulses are stated; in the second part the processing of these signals by internuncial cells is discussed; and in the third part the relation of descending motor systems transmitting control signals from higher centers to the spinal mechanism is considered. The last part deals with the individual and collective properties of the motor neurons that translate a complex input into a relatively simple output. The chapter that follows supplies detailed information on specific spinal mechanisms that are of special importance in the control of movement.

The lumbosacral part of the spinal cord and the hind limb structures that it innervates have been used extensively for physiologic experiments. In the seventh lumbar segment of the dog there are, by actual count, about 375,000 cell bodies.[10, 11] Their disposition is indicated in Fig. 23-1, which shows the numbers of large and small cells in the dorsal horn, intermediate region, and ventral horn. Small cells greatly outnumber large cells in all parts of the spinal cord, as they do generally throughout the CNS. In the ventral root itself large fibers predominate. These are the axons of motor neurons that innervate skeletal muscles. There are about 12,000 sensory fibers in each dorsal root and about 6,000 motor fibers in each ventral root of this particular segment. The existence of 375,000 cells between the input and output suggests that the signals coming into this segment via dorsal roots and descending tracts are being subjected to extensive processing before they reach the motor neurons.

SENSORY INPUT

All the sensory input to the spinal cord arrives via the dorsal roots. The entering fibers range in size from about 22μ to about 0.2μ. Unmyelinated C fibers of 1μ or less make up more than half the total. Fibers of large diameter from stretch receptors in muscle constitute a small percentage of the input. For reference purposes a classification of afferent fibers according to fiber diameter is supplied in Table 23-1.

The central end of a dorsal root breaks up into a series of rootlets, each of which separates into a medial and lateral division. The fibers of the medial division, which are relatively large, pass medially over the dorsal horn into the posterior columns. The fine fibers of the lateral division pass directly into the tract of Lissauer at the apex of the dorsal horn. Shortly after entering the cord each fiber divides into an ascending and a descending branch. Some fibers ascend all the way to the gracile and cuneate nuclei. Others run a short distance up or down and end on cells in the posterior columns. Although they take various courses, all entering fibers apparently project to higher centers. In addition, with no known exceptions, all afferent fibers establish connections with the central gray matter of the spinal cord. Thus is appears to be a basic principle that all sensory fibers serve a double function. None project to higher centers without giving off collaterals to spinal centers and none serve a purely spinal function.

The consequences of this arrangement may be observed in a chronic "spinal" cat. In such a preparation, with no interfering responses from higher centers, it is possible to show that every type of sensory stimulus

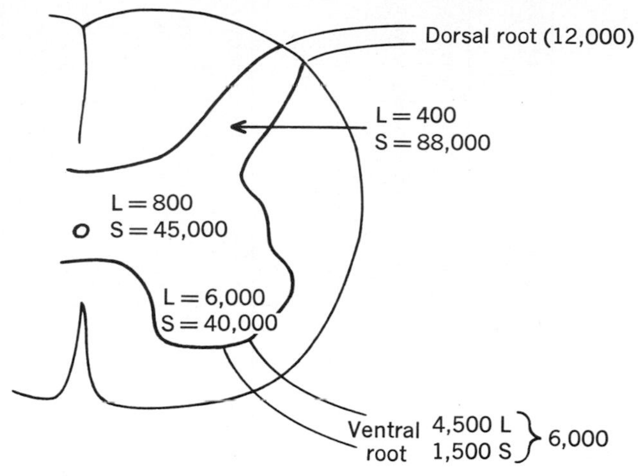

Fig. 23-1. Numbers of large, *L,* and small, *S,* cells and fibers in seventh lumbar segment of dog's spinal cord. Neurons with diameters of less than 34μ (corresponding to surface area of 920μ^2) were classified as small. Classification of fiber size based on position of trough between two peaks in histogram (approximately 8μ). (Based on data from Gelfan and Tarlov.[11])

(touch, pressure, pain, heat, cold, joint movement, muscle stretch) will elicit a reflex response of some kind. This is an important observation because it indicates that all sensory signals have a spinal route to motor neurons.

The obvious inference is that immediate and delayed use is made of incoming sensory signals. At the spinal level, sensory signals serve urgent or pressing needs by eliciting rather stereotyped motor responses with very brief delays. The same signals are also transmitted to higher centers for more eleborate processing of their information content. After variable delays the effects of these signals, blended with other afferent impulses, may be felt again, greatly modified, in the output of integrating centers such as the cerebellum.

As illustrated in Fig. 23-2, the local or segmental terminations of dorsal root fibers are of two types. (1) The majority of them end on internuncial neurons situated in different parts of the central gray matter. Fibers that innervate skin terminate chiefly on cells located in the dorsal horn of the spinal cord (Fig. 23-2, *c*). Fibers from muscle are distributed to various other internuncial nuclei, a prominent example being the intermediate nucleus (Fig. 23-2, *g*). Incoming signals in these fibers affect motor neurons through the agency of one or several intervening cells that may transform the original signals. (2) A small percentage of dorsal root fibers, those coming from annulospiral endings in muscle spindles, send collaterals directly to motor neurons in the ventral horn (Fig. 23-2, *e*).

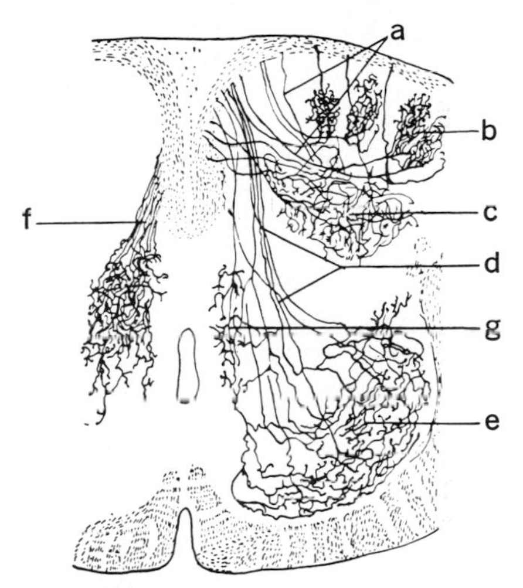

Fig. 23-2. Sketch of cross section of spinal cord indicating various terminations of reflex collaterals of primary dorsal root afferents. Some, *a,* reach neurons of substantia gelatinosa, *b,* whereas others end upon cells of nucleus proprius of dorsal horn, *c.* Direct reflexomotor collaterals, *d,* reach motoneurons at *e* without an intervening synapse. Other collaterals, shown for convenience on second side at *f,* end upon cells of intermediate nucleus of Ramón y Cajal, *g.* (From Ramón y Cajal.[21])

These collaterals complete a two-neuron reflex arc from muscle receptor to motor neuron to muscle fibers, a so-called monosynaptic reflex pathway.

INTERNUNCIAL TRANSACTIONS

Situated between the incoming and the outgoing fibers of the spinal cord are the internuncial cells or interneurons, which trans-

Table 23-1. Classification of dorsal root fibers

Classification of nerve fibers	Range of fiber size and velocity	Peripheral origin	Receptor organs	Effective stimulus	Type of synaptic relay	Central destination	Reflex action	Other properties
Group Ia (A-alpha)	12 to 20μ 70 to 120 m/sec	Muscle	Annulospiral spindle endings (A-2 of Matthews)	Stretch–low threshold	Two-neuron arc "monosynaptic"	1. Motoneurons of muscle of origin	Direct excitation and residual facilitation	Myotatic reflex; receptor discharge may cease during contraction unless small nerve motor fibers to spindles are active
						2. Motoneurons of synergists in myotatic unit	Usually residual facilitation Direct excitation when excitability high	
						3. Motoneurons of antagonists in myotatic unit	Direct inhibition*	Reciprocal component of myotatic reflex action
Group Ib (A-alpha)	12 to 20μ 70 to 120 m/sec	Muscle	Tendon organs of Golgi (B of Matthews)	Active contraction of muscle	Three-neuron arc "disynaptic"	1. Motoneurons of muscle of origin and its synergists in myotatic unit†	Inhibition	Receptor discharge increases during contraction and is not affected by small-nerve motor activity
						2. Motoneurons of antagonists in myotatic unit	Excitation	
Group II (A-beta and gamma)	5 to 12μ 30 to 70 m/sec	1. Extensor muscles	Flower spray of spindle (A-1 of Matthews)	Stretch–low threshold	Multineuron arc	Inhibition of extensors and facilitation of flexors throughout limb		Stretch evoked flexion withdrawal reflex; action is identical, regardless of muscle of origin of afferent discharge
		2. Flexor muscles	Flower spray of spindle (A-1 of Matthews)	Stretch–low threshold	Multineuron arc			
		3. Skin‡	Touch-pressure receptors	Mechanical deformation of skin	Multineuron arc	Excitation of flexors and inhibition of extensors throughout limb		Flexion withdrawal reflex; contralateral component is crossed extensor reflex
Group III (A-delta)	2 to 5μ 12 to 30 m/sec	1. Muscle	Unknown, pain receptors ?	Destructive (?)	Multineuron arc	From either muscle or skin these afferents produce excitation of flexors and inhibition of extensors throughout limb		Flexion withdrawal reflex; contralateral component is crossed extensor reflex
		2. Skin‡	Pain—fast (?), cold, heat	Destructive (?) temperature change	Multineuron arc			
Group IV (C fibers)	0.5 to 1μ 0.5 to 2 m/sec	From muscle and skin	Pain–slow	Destructive	Multineuron arc			

*There is evidence suggesting that an interneuron is interposed in the direct inhibitory pathway (Eccles et al.[7])
†These afferents make some connections with motoneurons of muscles at distant joints as well, but these are yet poorly understood.
‡For an earlier classification of skin nerves see p. 71. At present the terms "a" and "Group I" are used synonymously to designate large afferent fibers from muscle.

mit signals from the dorsal root fibers to the spinal motor neurons. They constitute the great majority of neurons in the spinal cord, outnumbering motor neurons 30:1.[11] They do not merely intervene passively to link up afferent and efferent fibers, but serve to transform incoming signals into new and different patterns. In part the transformation may be due to the characteristics of the interneuron and in part it may result from the way that groups of internuncial cells are connected with each other to form circuits with special properties.

Interneurons are a heterogeneous group of cells that are either collected into several nuclei, as shown in Fig. 23-2, or dispersed without obvious grouping. With few exceptions they are rather small cells, with a mean diameter of only 16μ, as compared with 48μ for motor neurons. The surface area of the average motor neuron accommodates some 5,500 synaptic knobs, whereas the interneuron has only about 640 knobs/cell.[10] Although it is possible to record from interneurons with microelectrodes and to ascertain what afferent signals excite or inhibit them, it is not easy to determine where their axons terminate or how other neurons are affected by them. Information about interneurons is therefore limited mainly to observations of their behavior in response to various inputs.

Properties of internuncial cells[16]

The properties of interneurons do not differ radically from those of other neurons. Excitation is accompanied by membrane depolarization, and inhibition is frequently associated with hyperpolarization. In recording with a microelectrode it is usually easy to distinguish an interneuron from a motor neuron by its tendency to fire "spontaneously," i.e., without applied stimulation, and by its characteristic repetitive response to a single afferent volley (Fig. 23-3). The frequency of its discharge may be as high as 1,500/sec in short trains,[23] whereas that of spinal motor neurons seldom exceeds 100/sec. When excited by intracellular pulses of current through a microelectrode, interneurons tend to fire repetitively to a much greater extent than do motor neurons. This capacity to discharge repetitively is apparently related to a lack of prolonged subnormality following discharge and to minimal accommodation. The cause of prolonged firing is usually prolonged excitation due to the arrival of temporally dispersed presynaptic impulses. There is no evidence for a prolonged duration of excitatory transmitter action.

Functions of internuncial cells

Internuncial cells perform a wide variety of specific functions that we are only begin-

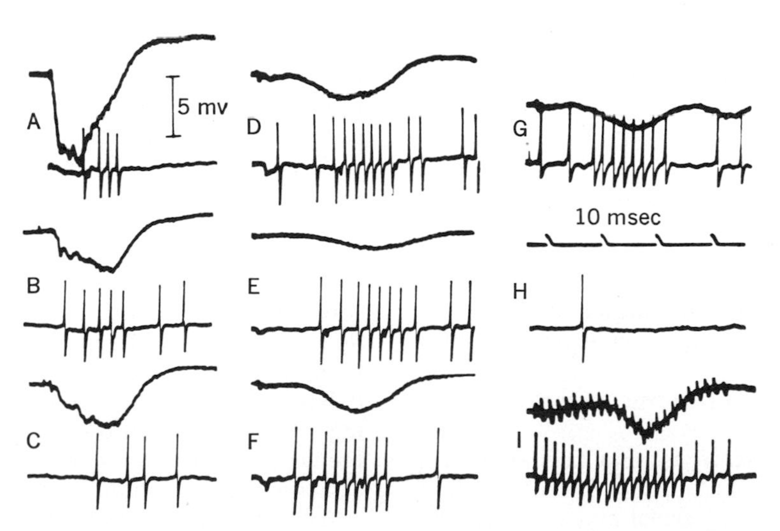

Fig. 23-3. Responses of interneuron in intermediate region of spinal cord to afferent volleys in various nerves. Upper traces are incoming volley recorded from L_7 dorsal root entry zone. Lower traces are responses of interneuron recorded with extracellular microelectrode. Stimulus applied to superficial peroneal, **A;** sural, **B;** saphenous, **C;** flexor digitorum longus, **D;** plantaris, **E;** gastrocnemius, **F;** and biceps-semitendinosus, **G. H** is spontaneous response. **I** is stimulation of biceps-semitendinosus nerve at 680/sec. (From Eccles et al.[6])

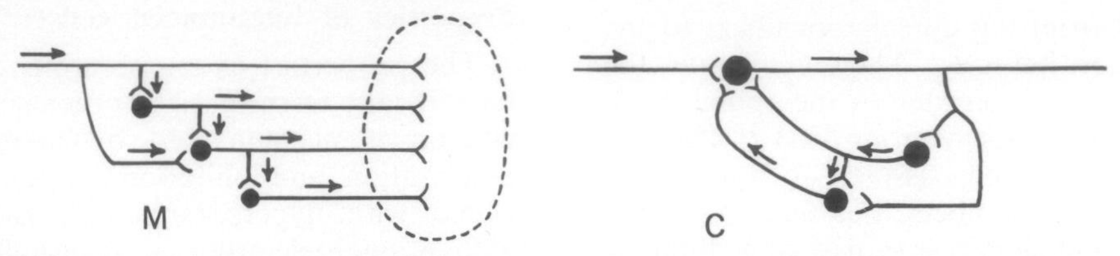

Fig. 23-4. Diagrams of two types of chains formed by internuncial neurons: *M*, multiple chain, and *C*, closed chain, which are self-reexciting. (From Lorente de Nó.[19])

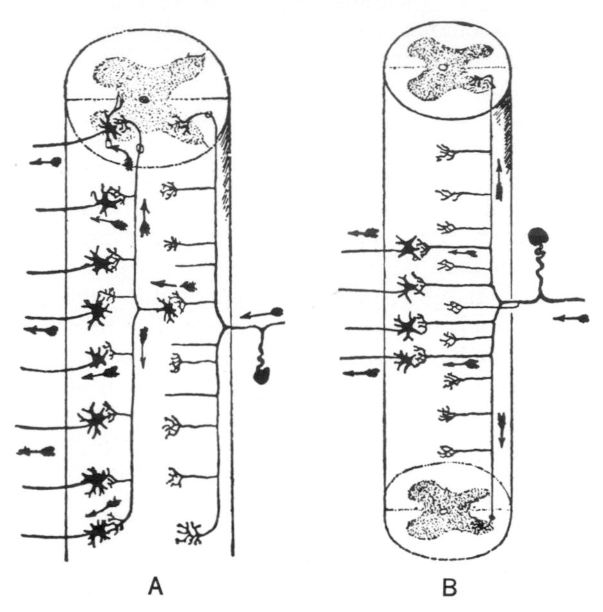

Fig. 23-5. A, Diagrammatic representation of diffuse reflex mechanism of spinal cord showing polysynaptic relay from primary dorsal root afferent, via interneuron, to motoneurons of *several* spinal cord segments. **B,** Diagrammatic representation of circumscribed reflex mechanism showing monosynaptic relay from primary dorsal root afferents to motoneurons of a *few* segments. (From Ramón y Cajal.[21])

ning to appreciate. It is clear that some of these functions cannot be carried out by single cells but require a network or circuit of interconnected cells. To date, such circuits have not been demonstrated in the spinal cord, although they exist elsewhere in the nervous system.

Internuncial cells perform several types of functions:

1. They may serve as amplifiers of three different types. (a) Amplification of the intensity of an incoming signal may be achieved either by *cascading* of cells (Fig. 23-4, *M*) or through an *increase in the rate* of discharge of individual cells. In the first instance the number of active cells may be greatly increased; in the second, the frequency of firing in a primary afferent fiber may be multiplied by a factor of 10 to 100. One internuncial circuit might combine both effects. (b) Amplification in time may be accomplished by means of a closed loop of interneurons (Fig. 23-4, *C*). Each of the cells in such a loop excites others, resulting in a self-perpetuating discharge. Circuits of this type may account for the fact that an incoming volley lasting only 1 msec can elicit a response of motor neurons lasting 1,000 msec. (c) Amplification in space is brought about by interneurons that distribute an incoming volley to widely separated groups of motor neurons. Ramón y Cajal recognized this type of circuit on purely anatomic grounds (Fig. 23-5, *A*).

2. Internuncial cells may serve as valves that pass or prevent transmission of afferent impulses to motor neurons. There is evidence from microelectrode studies that some interneurons are affected not only by incoming volleys from muscle but also by activity in descending systems such as the corticospinal tract. Depending on the control exerted by a higher center, activity arising in tendon organs or secondary spindle receptors may or may not reach the motor neuron. In feedback terminology this function might be referred to as "variable gain control."

3. Internuncial cells may function as signal inverters, changing an incoming excitatory signal into an inhibitory signal. For example, it is well known that impulses in group Ia fibers excite their own motor neurons and inhibit those of direct antagonists. Since release of different transmitters at different endings of the same neuron is unlikely, it has been proposed[4] that the effect on antagonistic motor neurons is mediated not directly but via interneurons that "convert" excitation to inhibition.

4. Internuncial cells probably serve as final common pathways for either excitation or inhibition. Since a number of inputs of different types may each fire a particular interneuron, these cells should be regarded as elements that are common to many afferent

pathways. By taking the sum of all the excitatory and inhibitory effects impinging on it and emitting a signal that is the resultant of all of them, an interneuron may greatly simplify the input to the motor neuron. In addition, by substituting one set of terminals for many, an internuncial cell may serve to promote neuronal economy.

The functions just described are probably the simplest and most easily recognized of a great many types of data processing by internuncial cells. In time, subtler and more complex types will certainly be uncovered.

MOTOR INPUT

In order to understand how higher centers of the brain exert their control over muscles it is necessary to know the exact terminations of the various fiber tracts that transmit these control signals to spinal levels. The required information, however, is not yet available. From both anatomic and physiologic studies it appears that the majority of descending fibers end on internuncial cells, but that some fibers pass directly to motor neurons. In the cat, for example, anatomic and electrophysiologic studies are in agreement that corticospinal fibers end chiefly or exclusively on internuncial cells. Lloyd's diagram (Fig. 23-6) summarizes the results of his investigation of this problem, showing that corticospinal fibers terminate on interneurons in the dorsal horn immediately adjacent to the fiber tract and upon interneurons in the intermediate region of the gray matter. His original studies[18] and several done subsequently[17] indicate that some of these interneurons also receive input from stretch receptors in muscle. It thus appears that in the cat the corticospinal system exerts its effects through the feedback circuits that control the length and tension of muscle. Much of the control exerted via other descending systems is probably of this general type. It is possible that some descending systems function through the length control system and that others act through the tension-regulating system. As indicated in Chapter 21, control signals from any source necessarily bring the "positional control system" into operation. The precise point at which control signals from higher centers converge with local feedback control systems may be of great significance. In monkey and man some corticospinal fibers terminate directly on motor neurons.[1, 15] This suggests that animals with highly developed forebrains require a direct, unconditional means of controlling motor neurons that is not subject to regulation by internuncial valves.

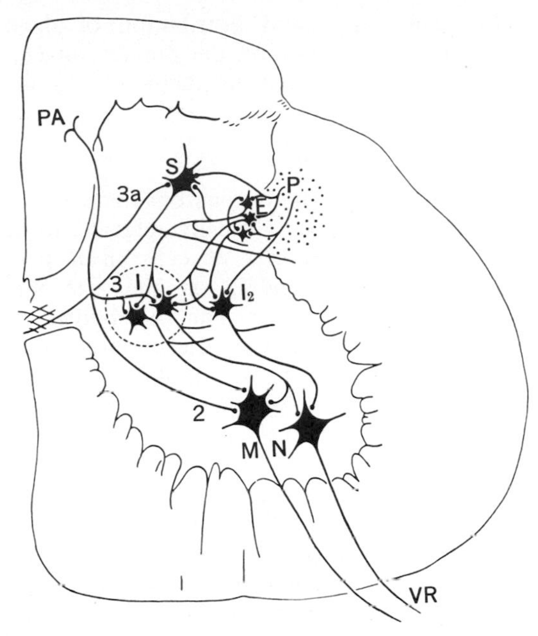

Fig. 23-6. Termination of corticospinal fibers on interneurons of dorsal horn and intermediate region. *P*, Corticospinal tract; *E*, small cells of external basilar region; *I*, intermediate nucleus of Ramón y Cajal; I_2, other neurons of intermediate region; *MN*, motor neurons; *PA*, primary afferent collaterals; *S*, solitary cells of dorsal horn; *VR*, ventral root; *2*, *3*, and *3a*, terminal collaterals of primary afferent system. (From Lloyd.[18])

OUTPUT

The neurons that transmit signals from the spinal cord to muscle have cell bodies located in the ventral horn of the cord. Their axons leave the spinal cord in the ventral roots and run without interruption to skeletal muscles. In the seventh lumbar ventral root of the dog there are about 6,000 fibers. This group of cells receives signals from skin, muscles, joints, cerebellum, vestibular apparatus, brainstem, basal ganglia, and cerebral cortex. These sources of input are mentioned only to stress the great complexity of the total influx to motor neurons. As this complexity evolved, an output system capable of responding with sufficient flexibility and subtlety to the vast input had to develop with it. A homogeneous population of motor neurons with similar properties could not meet the demands. Instead, a highly organized population of cells evolved whose individual characteristics differ widely and systematically. The range and distribution of these characteristics confer on the population *collective* properties that its members do not possess individually.

Motor neuron pool

The group of motor neurons innervating a particular muscle is usually referred to as a "pool." The number of cells in a pool varies widely, depending on the size and function of the muscle. The cell bodies lie in a column that, in the case of large muscles, may extend over two or three segments of the cord. The axons passing out to such muscles emerge from the cord in two or three ventral roots.

If the motor fibers in a muscle nerve are examined histologically (after degeneration of the sensory fibers),[5] it is found that two groups can be distinguished: (1) *alpha* fibers ranging in diameter from about 9μ to about 14 to 20μ, depending on the muscle, and (2) *gamma* fibers ranging from about 1 to 8μ. As Ramón y Cajal[21] first pointed out, the diameter of a nerve fiber is related directly (though not linearly) to the size of its cell body. The largest alpha cells have surface areas of 50,000 to 60,000μ^2 and may have as many as 10,000 synaptic knobs on their soma and dendrites.[10] The smallest alpha cells have surface areas of 10,000 to 15,000μ^2 and correspondingly fewer synaptic knobs. In the previous chapter it was shown that axons of large diameter innervate motor units that develop large tensions and that axons of small diameter supply motor units that develop small tensions. The sizes of the cells in a pool are therefore related directly to the number of muscle fibers innervated. Since a motor neuron is responsible in some mysterious way for trophically maintaining all the muscle fibers it innervates, it is not surprising that those with large trophic responsibilities are bigger than those with few dependents. It should also be recalled that there are three types of fibers in pale mus-

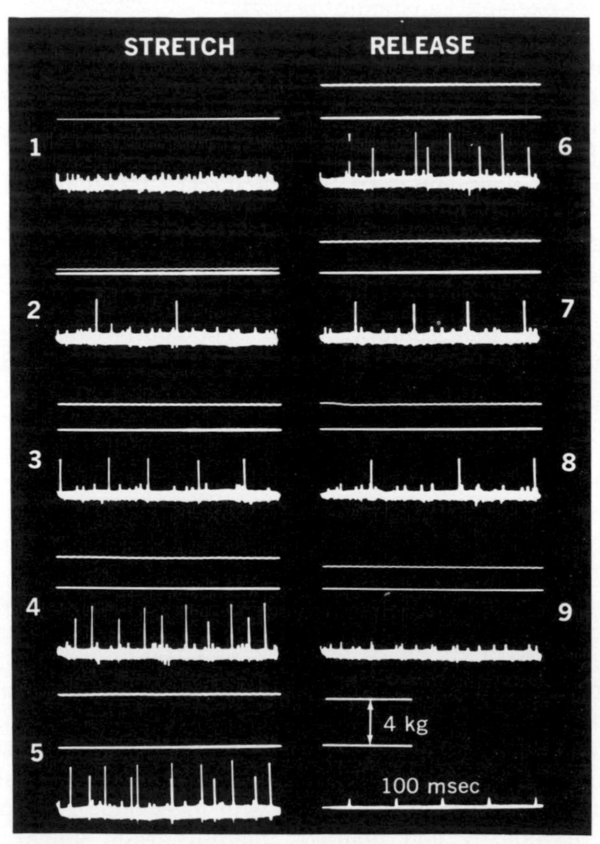

Fig. 23-7. Stretch-evoked responses of two alpha motor neurons, recorded from filament of seventh lumbar ventral root. Amount of tension applied and developed reflexly is indicated by separation of two upper beams in each frame. Several seconds elapsed between successive frames while muscle was stretched, **1** to **5,** and released, **6** to **9.** (From Henneman et al.[13])

cles[24]; hence it is possible that there are three varieties of motor neurons.

The electrical properties of motor neurons and the nature of synaptic transmission in them are described in a previous chapter on synaptic transmission. In this section, attention will be directed chiefly to the functional organization of the motor neuron pool and its relation to muscular activity.

Functional significance of the size of motor neurons

The enormous differences in cell size found in various types of neurons in the CNS intrigued early histologists and provoked many speculations, but the functional significance of cell size did not become apparent until recently. With the introduction of the oscilloscope, it was found that the properties of peripheral nerve fibers differ systematically with size. Of particular significance here is the finding of Gasser[9] that the amplitude of a nerve impulse recorded externally from a peripheral nerve is directly related to the diameter of the fiber transmitting it. Thus when impulses of several different amplitudes are recorded from the same nerve filament, the largest impulse signifies the firing of the largest fiber and the smaller impulses signify the discharges of correspondingly smaller fibers. Identification of single impulses is facilitated by recording from thin filaments of ventral roots, which can be teased apart more easily than peripheral nerves because they contain less connective tissue. In this way samples of the output of a pool of motor neurons can be investigated using a variety of natural or artificial stimuli to elicit responses.

A convenient means of evoking a reflex discharge of motor neurons is to stretch an extensor (antigravity) muscle in the hind limb of a decerebrate cat. In such preparations the stretch reflex is greatly exaggerated and the slightest degree of passive stretch results in the discharge of motor neurons and the development of muscular tension. The reflex is highly specific, the response being limited to the motor neurons of the muscle that is stretched.

Fig. 23-7 illustrates the results of stretching the triceps surae of a decerebrate cat and recording from a filament of the seventh lumbar ventral root. The tension applied to the tendon of the muscle is measured by the separation of the two upper beams in each frame, and the ventral root discharge appears in the lowest tracing. With the muscle completely relaxed (line *1*) the only activity recorded was a steady stream of impulses of very low amplitude. These small impulses were recorded from gamma motor neurons whose axons are of smaller diameter than those of the alpha fibers also present in the filament. With slight stretch (line *2*), two discharges of a much larger nerve fiber were recorded. With further stretch, the discharges of this unit increased in rate and regularity. It was obvious that they represented the action potentials of a single alpha motor neuron responding to the excitatory effects of stretch evoked impulses. With considerably greater stretch, the responses of a second unit generating larger action potentials appeared in the record (line *4*). The firing rate of this unit was accelerated by further stretch, but no additional units were recruited. Upon release of stretch (lines *6* to *9*) the larger unit was the first to cease firing, followed by the smaller unit.

In Fig. 23-8 a similar series of tracings is reproduced to illustrate recruitment in a filament containing the axons of five triceps motor neurons. Again, only gamma activity was recorded prior to stretch. With increasing degrees of stretch, progressively larger impulses (*1* to *5*) appeared, the order of recruitment being 1, 2, 3, 5, 4 and the order of dropout 5, 4, 3, 2, 1.

Fig. 23-9 shows the distribution of thresholds of response for 208 motor neurons. The abscissa indicates the threshold in kilograms of tension required to elicit tonic responses of motor neurons. The range was from less than 0.4 kg to more than 8 kg. The distribution of thresholds in Fig. 23-9 is sufficiently similar to the distribution of maximal tensions of motor units shown in the preceding chapter to indicate that the size of a motor unit and the stretch threshold of its motor neuron are closely interrelated.

These observations[13] on stretch reflexes indicated that orderly recruitment of motor neurons according to their sizes must depend on differences in the excitabilities of the motor neurons themselves or on some systematic difference in the input from stretch receptors resulting in more effective stimulation of small cells. In order to distinguish between these two possibilities a variety of different spinal reflexes were used to discharge flexor or extensor motor neurons. The inputs and spinal circuits involved in these experiments varied widely. In general the smaller of any two responding units in a ventral root filament

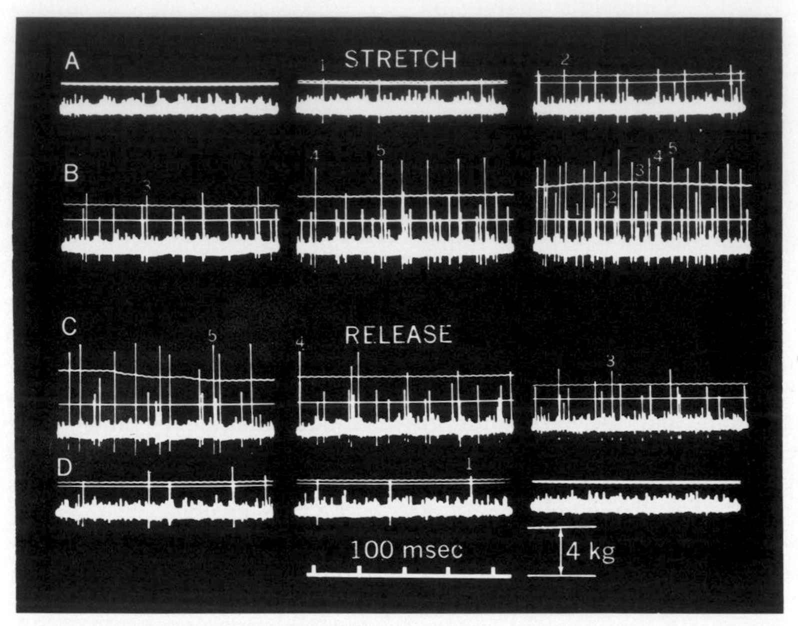

Fig. 23-8. Stretch-evoked responses of five alpha motor neurons recorded from filament of first sacral ventral root during stretch, **A** and **B,** and release, **C** and **D,** of triceps surae muscle. Small numerals above action potentials indicate rank of units according to size. (From Henneman et al.[13])

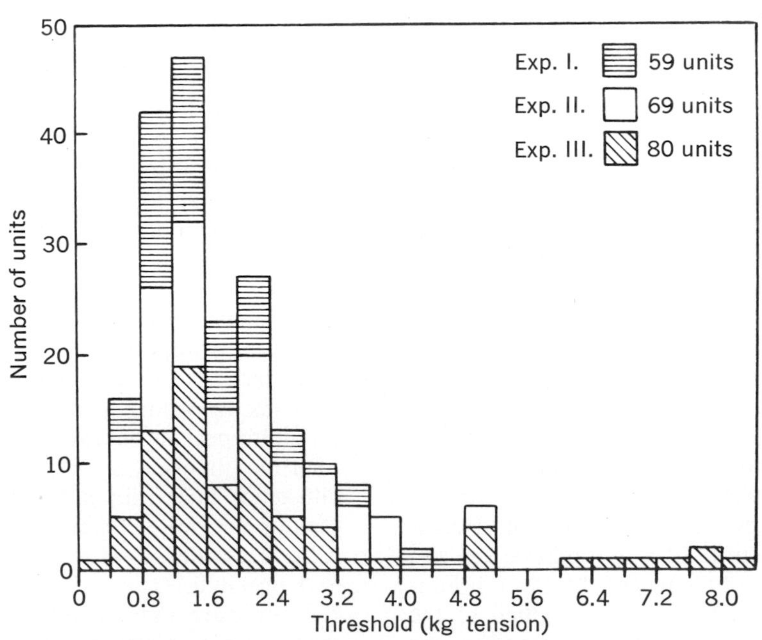

Fig. 23-9. Frequency distribution of thresholds of 208 tonically responding motor neurons. Data obtained in three experiments as indicated. Abscissa: threshold in kilograms of tension applied to deefferented triceps surae muscle. Ordinate: number of units whose thresholds fall between values indicated on abscissa. (From Henneman et al.[13])

was discharged at a lower intensity of stimulation regardless of whether the excitatory stimuli arose ipsilaterally or contralaterally, whether physiologic stimuli or electric "driving" was employed, whether the responses were elicited monosynaptically or polysynaptically, and whether the motor neurons were flexor or extensor. These results left no doubt that *it is the size of a motor neuron that determines its threshold and relative excitability.* In fact, excitability of motor neurons is simply an inverse function of their size.

In view of the findings previously described it is natural to inquire whether susceptibility to inhibition is also a function of cell size. The records reproduced in Fig. 23-10 supply a clear answer to this question. Portion *A* of the figure shows the orderly recruitment of three triceps motor neurons with increasing stretch. After the normal pattern of recruitment had been established, a 4 kg stretch, sufficient to elicit tonic firing of all three motor neurons, was applied by elongating the muscle to a fixed length. This stretch was maintained while the effects of inhibition, shown in part *B,* were recorded. The tracings labeled *1* to *4* on the left side of Fig. 23-10, *B,* show the responses of these neurons before *(1)* any inhibitory stimulation and after weak *(2),* moderate *(3),* and strong *(4)* stimulation. The tracings on the right were obtained during the application of the inhibitory stimuli that lasted throughout the 500 msec duration of the sweep. Weak inhibition silenced the largest of the three units, leaving the two smaller units still firing, though at a lower frequency. Inhibition of moderate intensity suppressed the unit of intermediate size as well, leaving the smallest unit discharging until a still stronger inhibition eliminated all responses to stretch. During each of the brief periods after inhibition there was a partial but

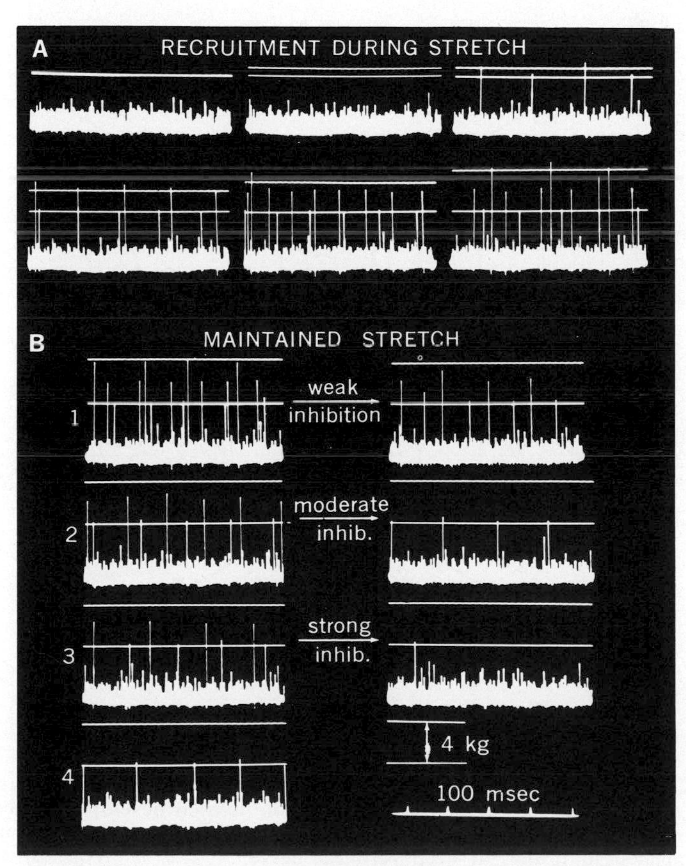

Fig. 23-10. A, Orderly recruitment of three triceps motor neurons of different sizes in response to increasing degrees of stretch. **B,** Orderly inhibition of same three units during constant stretch of 4 kg. Records *1* to *4* on left show control responses to stretch before, between, and after each of three inhibitory stimulations. Largest unit was silenced first (line *1*), intermediate unit next (line *2*), and smallest unit last (line *3*). (From Henneman et al.[13])

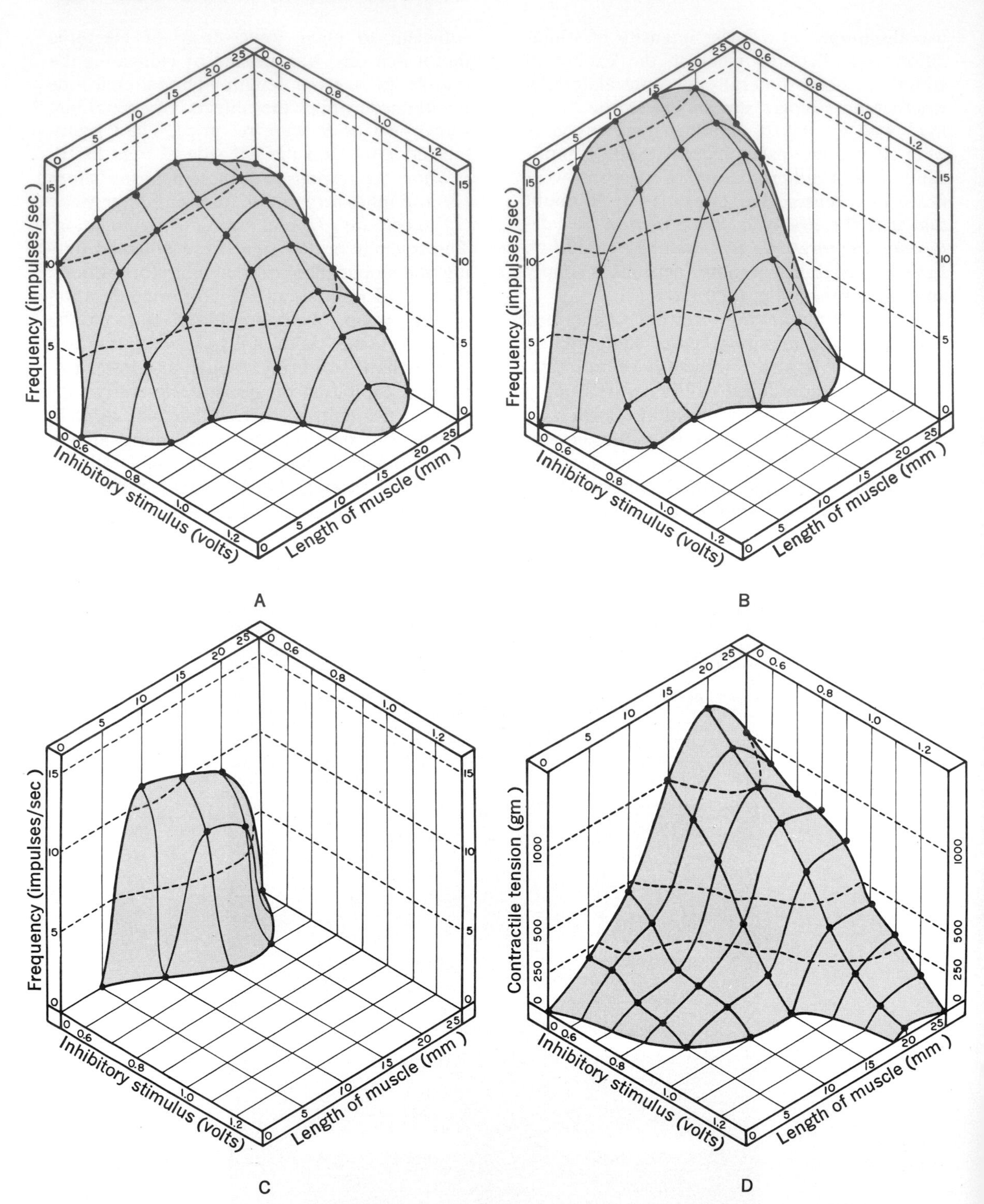

Fig. 23-11. Graphic representation of discharge frequency of three motor neurons of different sizes, **A, B,** and **C,** of triceps surae and of contractile tension developed by the muscle itself, **D,** in response to varying degrees of excitation and inhibition. X axis is intensity of inhibitory stimulation applied to ipsilateral deep peroneal nerve. Y axis is frequency of discharge in **A, B,** and **C** and contractile tension in **D.** Z axis is stretch of triceps surae in millimeters. Data for all four graphs obtained simultaneously. Each plotted point represents mean of two successive determinations. Note that unit **A** (the smallest) fired spontaneously without stretch, while unit **B** began to discharge between 0 and 5 mm of stretch and unit **C** (the largest) between 5 and 10 mm. (From Henneman et al.[13])

incomplete recovery of the original pattern of response. Results of this kind were obtained in a variety of experiments. In general the larger the unit, the more susceptible it was to inhibition. Regardless of the existing level of excitatory drive, regardless of whether the motor neurons were responding rhythmically to stretch or monosynaptically to synchronous volleys, and regardless of whether the inhibition was "direct," internuncially mediated, autogenetic, or recurrent, the inhibitability of each cell was strictly size dependent. It was concluded that, whereas the excitability of motor neurons is an inverse function of cell size, their inhibitability is a direct function of cell size. Thus size is an important factor in determining a cell's response to all synpatic influences.[12]

The foregoing observations indicate that the overt response of a motor neuron varies with the size of the cell, the level of excitation, and the intensity of inhibition. Each motor neuron functions as an integrating device that adds all excitatory effects, subtracts all inhibitory effects, and emits an appropriate output. The net effects of cell size, excitatory input, and inhibition on the responses of a small group of motor neurons can be represented quantitatively by means of three-dimensional graphs as in Fig. 23-11. In parts *A, B,* and *C* the rate of discharge of each of three different motor neurons is shown on the vertical Y axis. The cells were subjected to various mixtures of excitation and inhibition, whose intensities can be read on the X axis (inhibition) and Z axis (excitation). Of the three units whose responses are plotted in Fig. 23-11, the smallest and most excitable was that of part *A,* which was spontaneously active with the muscle completely relaxed. The stretch threshold of the intermediate unit *(B)* was between 0 and 5 mm extension and that of the largest and least excitable unit was between 5 and 10 mm. At all levels of stretch the intensity of inhibitory stimulation required to silence a unit was always the greatest for the most excitable cell *(A).* Fig. 23-11, *D,* reveals how the responses to stretch and inhibition of all the individual motor units in a muscle are combined in the contractile response of the whole muscle. The graph in *D* has the same general features as those in *A, B,* and *C,* although the tension of the whole muscle replaces frequency of firing on the Y axis.

Organization of motor neuron pool

In the course of organizing and explaining his observations on spinal reflexes, Sherrington evolved the concept of the motor neuron "pool." He conceived of the pool as a fairly uniform population of motor neurons supplying the same muscle but differing in their afferent connections. Owing to these differences the pool was "fractionated" by an excitatory input into a "discharge zone" of

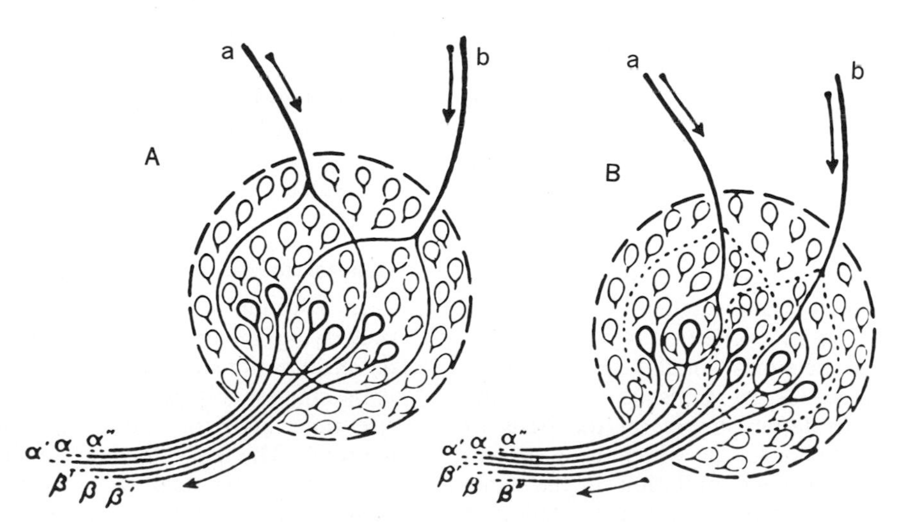

Fig. 23-12. Sherrington's representation of motor neuron "pool" showing "summation" of subliminal effects due to overlap of two subliminal fields, **B,** and "occlusion" of reflex discharge, **A,** due to overlap of two discharge zones. In **B,** fiber *a* and fiber *b* each discharge one unit separately; together they discharge four units. In **A,** fiber *a* and fiber *b* each discharge four units separately; together they discharge not eight, but six units. (From Creed et al.[3])

cells that were fired by the incoming impulses and a "subliminal fringe" of cells that were excited but not sufficiently to be fired.[2] "Facilitation" of transmission resulted when the subliminal fringes associated with different inputs overlapped (Fig. 23-12, *A*) and the combined output was greater than the sum of the separate outputs. "Occlusion" occurred when the discharge zones overlapped and the combined output was less than the sum of the separate outputs (Fig. 23-12, *B*).[3]

In the light of recent results the original concepts of pool organization require modification as follows: Fractionation occurs not because input is distributed to different groups of cells as in Sherrington's diagram but because of differences in the excitability of cells of different sizes. The input is presumably distributed to all members of the pool in proportion to their surface area. Whatever the extent of the discharge zone at any time, it consists of all the alpha motor neurons whose cells are less than a certain size. The larger cells with higher thresholds constitute the subliminal fringe. Recruitment of additional cells as input increases involves the group of neurons that form the shifting boundary between discharge zone and subliminal fringe. Facilitation simply represents an expansion of the discharge zone due to additional excitation. Cells that are well within the discharge zone due to their size discharge tonically during a maintained stretch reflex. Cells that respond phasically, i.e., only during the onset of stretch, are presumably those on the border between discharge zone and subliminal fringe. The remainder of the cells in the pool are those that are too large to be fired at all by a particular input. It appears that motor neurons do not belong to fixed tonic and phasic types, because, depending on the prevailing level of input, they may behave in either mode.

An inhibitory input to a pool of motor neurons produces a net effect that is similar to a reduction in excitatory input. The discharge zone becomes smaller and the subliminal fringe larger. Cells shift from the former to the latter simply on the basis of their size. The order in which cells are silenced by increasing inhibition is the same as the order of dropout when excitation is gradually decreased. The response of a pool to a mixture of excitatory and inhibitory input is therefore no different in principle from its response to excitation alone.

The essence of the motor neuron pool, as redefined, is the spectrum of sizes represented in it. Insofar as can be judged at present, the input to the pool is distributed "equally" to all the alpha cells in the pool in accordance with their relative sizes. This is not an experimentally established fact but an inference based on reasoning that cell size can only determine the firing order in a group of cells if the excitatory influx impinging on each cell is of equal intensity. How this equalization of input is achieved is not known. Like the distribution of cell sizes the distribution of input is anatomically "built in." The whole mechanism functions automatically in the sense that any input elicits an appropriately "sized" output.

An important corollary of the size principle is that the various cells in a pool are fired rarely, moderately, or frequently according to their sizes. As the input to a pool fluctuates during the normal range of motor activities, the smallest and most excitable cells are necessarily discharged most often, the largest and least excitable cells the least often. Although the metabolic consequences of activity are not significant for motor neurons, the cost of activity is far greater for the muscle fibers they innervate. The relation between the size of a motor neuron and the type of muscle fibers it innervates may thus depend in part on the average level of activity of the neuron. The fact that large motor units in pale muscles are composed of pale muscle fibers with few mitochondria and are supplied by few capillaries is clearly correlated with the large size of their motor neurons and the infrequent discharge of these cells. Similarly, the higher density of mitochondria in the red muscle fibers of small motor units and the presence of numerous capillaries around them are correlated with the smaller size of their motor neurons and the higher level of activity.[14]

General implications of neuronal size

In addition to explaining the mechanism underlying the grading of muscular activity, the observations on motor neurons and motor units suggest a general theory regarding cell size that might be tested on other types of neurons. The experimental findings indicate that large cells deal with large amounts of neural energy and small cells with small amounts. The tetanic discharge of a large motor neuron releases 200 times as much contractile energy as that of a small one and perhaps 1,000 times as much as that of a gamma motor neuron. The anatomic basis for this is that large cells have large axons that give off a great number of terminals. A small cell is evidently not capable of supporting many terminal branches, for none that do so have been found. In keeping with these facts is the evidence that a great deal more synaptic energy is required to fire a large cell than a small one. It would be unwise to generalize too widely from observations on one type of neuron. Nevertheless, it may be noted that throughout the nervous system there are enormous variations in the size and surface area of different types of neurons. Certain systems are composed largely or exclusively of small neurons; others consist entirely of large cells. It is tempting to speculate whether the small cells comprising the system that mediates pain, for example, serve principally to maintain a high level of excitability in a vital, protective system, whether they are small because small amounts of energy are involved in this system, or whether other considerations are paramount. Clearly the dimensions of a neuron may determine its utility as a component in a neural circuit.

Gamma motor (fusimotor) neurons[16, 20]

About one third of the fibers in the ventral roots innervate intrafusal fibers in muscle spindles. These fibers range from about 1 to 8μ in diameter. Their cell bodies are scattered among the alpha motor neurons in the ventral horn of the spinal cord.

Gamma motor neurons differ from alpha motor neurons in several respects: they tend to discharge spontaneously, they often discharge at higher frequencies, and their response is more frequently repetitive than that of alpha motor neurons. They apparently lack monosynaptic excitatory connections from primary afferent fibers.

The "spontaneous" activity of gamma motor neurons is a striking phenomenon. As was noted in Figs. 23-7 and 23-8, it goes on in resting or anesthetized animals even when no alpha discharges are occurring and is never normally absent in the nerves supplying ordinary skeletal muscles. It appears to be a natural consequence of the small size of gamma motor neurons and their greater excitability. The afferent inflow from receptors in resting muscles and from other sources is evidently sufficient to maintain activity in a considerable number of gamma neurons at all times.

Various forms of natural or artificial stimulation will elicit reflex discharge in gamma motor neurons. The pattern of discharge often but not invariably parallels that of the alpha fibers innervating the same muscle. Several investigators have reported that gamma motor neurons are discharged by reflex or suprasegmental stimuli that are insufficient to discharge alpha motor neurons.

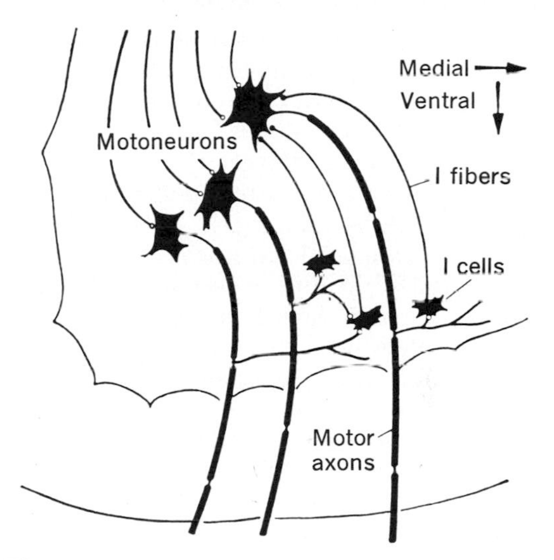

Fig. 23-13. Schematic representation of three motor neurons showing their recurrent collaterals ending on Renshaw internuncial cells, *I*, which in turn send their axons to motor neurons. (From Eccles et al.[8])

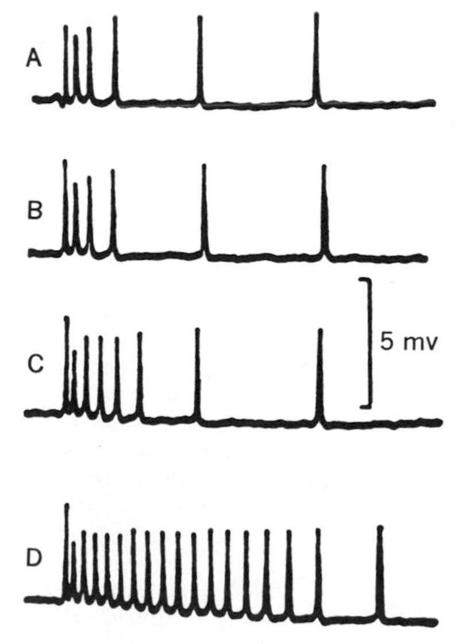

Fig. 23-14. Responses of Renshaw cell to single antidromic volleys in nerves to plantaris, **A;** soleus, **B;** medial gastrocnemius, **C;** and lateral gastrocnemius, **D.** (From Eccles et al.[8])

Again this would appear to be due at least in part to the greater excitability of small neurons. From a control standpoint it is appropriate that the gamma system, which is an essential part of the servomechanism regulating muscle length, should be composed throughout of smaller and therefore more sensitive elements than the alpha system that it serves.

Recurrent collaterals of motor neurons

As Ramón y Cajal first demonstrated, the axons of many spinal motor neurons give off branches that turn back into the ventral horn. These "recurrent collaterals" end on small interneurons known as Renshaw cells (Fig. 23-13). A number of recurrent collaterals converge on each Renshaw cell. When ventral roots are stimulated electrically, the antidromic volley in the recurrent collaterals causes a burst of high-frequency discharges from each Renshaw cell, as illustrated in Fig. 23-14. The axons of Renshaw cells terminate on motor neurons, forming a special type of feedback circuit whose significance is not yet established. Activity of Renshaw cells is usually associated with inhibition of motor neurons in nearby portions of the spinal cord, but under some conditions certain adjacent motor neurons may be facilitated. The exact distribution of these effects is uncertain. Control of the recurrent collateral system is quite complex. Renshaw cells are excited and inhibited by input from skin and muscles and by descending impulses from higher centers. In spite of extensive investigation, the functional role of the recurrent collateral system remains obscure.[22]

REFERENCES

1. Bernhard, C. G., Bohm, E., and Petersen, I.: Investigations on the organization of the corticospinal system in monkeys, Acta Physiol. Scand. **29** (suppl. 106) **:**79, 1953.
2. Cooper, S., Denny-Brown, D. E., and Sherrington, C. S.: Reflex fractionation of a muscle, Proc. R. Soc. Lond. (Biol.) **100:**448, 1926.
3. Creed, R. S., Denny-Brown, D. E., Eccles, J. C., Liddell, E. G. T., and Sherrington, C, S.: Reflex activity of the spinal cord, London, 1932, Oxford University Press, Inc.
4. Eccles, J. C.: The physiology of nerve cells, Baltimore, 1957, The Johns Hopkins Press, p. 153.
5. Eccles, J. C., and Sherrington, C. S.: Numbers and contraction values of individual motor-units examined in some muscles of the limb, Proc. R. Soc. Lond. (Biol.) **106:**326, 1930.
6. Eccles, J. C., Eccles, R. M., and Magni, F.: Monosynaptic excitatory action on motoneurones regenerated to antagonistic muscles, J. Physiol. **154:**68, 1960.
7. Eccles, J. C., Fatt, P., and Landgren, S.: Central pathways for direct inhibitory action of impulses in largest afferent nerve fibres to muscle, J. Neurophysiol. **19:**75, 1956.
8. Eccles, J. C., Eccles, R. M., Iggo, A., and Lundberg, A.: Electrophysiological investigations on Renshaw cells, J. Physiol. **159:**461, 1961.
9. Gasser, H.: The classification of nerve fibers, Ohio J. Sci. **41:**145, 1941.
10. Gelfan, S.: Neurone and synapse populations in the spinal cord: indication of role in total integration, Nature **198:**162, 1963.
11. Gelfan, S., and Tarlov, I. M.: Altered neuron population in L_7 segment of dogs with experimental hind-limb rigidity, Am. J. Physiol. **205:**606, 1963.
12. Henneman, E., Somjen, G., and Carpenter, D. O.: Excitability and inhibitability of motoneurons of different sizes, J. Neurophysiol. **28:**599, 1965.
13. Henneman, E., Somjen, G., and Carpenter, D. O.: Functional significance of cell size in spinal motoneurons, J. Neurophysiol. **28:**560, 1965.
14. Henneman, E., and Olson, C. B.: Relations between structure and function in the design of skeletal muscles, J. Neurophysiol. **28:**581, 1965.
15. Hoff, E. C., and Hoff, H. E.: Spinal terminations of the projection fibers from the motor cortex of primates, Brain **57:**454, 1934.
16. Hunt, C. C., and Perl, E. R.: Spinal reflex mechanisms concerned with skeletal muscles, Physiol. Rev. **40:**538, 1960.
17. Kuno, M., and Perl, E. R.: Alteration of spinal reflexes by interaction with suprasegmental and dorsal root activity, J. Physiol. **151:**103, 1960.
18. Lloyd, D. P. C.: The spinal mechanisms of the pyramidal system in cats, J. Neurophysiol. **4:** 525, 1941.
19. Lorente de Nó, R.: Analysis of chains of internuncial neurons, J. Neurophysiol. **1:**207, 1938.
20. Matthews, P. B. C.: Muscle spindles and their motor control, Physiol. Rev. **44:**219, 1964.
21. Ramón y Cajal, S.: Histologie du système de l'homme et des vertébrés, Paris, 1909, A. Maloine.
22. Renshaw, B.: Central effects of centripetal impulses in axons of spinal ventral roots, J. Neurophysiol. **9:**191, 1946.
23. Woodbury, J. W., and Patton, H. D.: Electrical activity of single spinal cord elements, Symp. Quant. Biol. **17:**185, 1952.
24. Wuerker, R. B., McPhedran, A. M., and Henneman, E.: Properties of motor units in a heterogeneous pale muscle (m. gastrocnemius) of the cat, J. Neurophysiol. **28:**85, 1965.

24

ELWOOD HENNEMAN

Spinal reflexes and the control of movement

REFLEXES OF MUSCULAR ORIGIN

The best introduction to the modern studies of muscle reflexes is an examination of the response of a single muscle to passive stretch. In such an experiment the three types of feedback circuits involved in response to stretch are all intact and capable of contributing and interacting in a normal manner. The central connections and reflex effects of each of these circuits will be described separately in later sections.

Stretch reflexes

The first significant progress in the study of stretch reflexes did not come until 1924 when Liddell and Sherrington published their famous paper.[23] Sherrington, who was professor of physiology at Oxford University, had been interested for some time in the phenomenon of decerebrate rigidity, a condition in which there is a rigid extension of the four limbs of an animal due to a transection of the brainstem. The condition has been called a caricature of standing because the limbs are thrust out stiffly as they are in the standing position. Sherrington was intrigued by the fact that there was considerable resistance to passive movement of the extended limbs and that this resistance was so marked in the muscles used in standing, i.e., the antigravity muscles. He therefore began his study by using the fully isolated quadriceps muscle of a decerebrate cat. All of the nerves to the hind limb were severed except that innervating the quadriceps muscle. The pelvis was securely fixed to an experimental table and the patellar tendon was attached to a rigid myograph.

When the quadriceps muscle was stretched by lowering the table slightly, a relatively large tension was developed in response to a few millimeters of extension, as shown in Fig. 24-1. This was far more tension than the elasticity of the muscle and tendon could possibly develop as a result of passive stretch. Sherrington concluded that the tension was the result of active contraction of the muscle and was largely reflex in origin because it depended on the integrity of the reflex arc. After cutting the nerve to the quadriceps muscle, stretch caused only a slight increase in the measured tension, which was due to the elasticity of the muscle. Similarly, interrupting any other part of the reflex pathway or damaging the spinal cord abolished the reflex tension. Finally, he noted that stimulation of a sensory nerve in the same leg, which normally elicits a potent flexion reflex, also abolished the tension.

This demonstration that stretching an extensor muscle causes it to contract reflexly became at once the most important fact in reflex physiology. It led to a long series of investigations in Sherrington's laboratory at Oxford University. Out of these studies came a number of important principles.

1. All skeletal muscles exhibit stretch reflexes in some degree.
2. Stretch is the adequate stimulus, but the sensitivity of the receptors is so great that jarring or vibration will also discharge them. Ordinarily gravity or the shortening of an opposing muscle causes the stretch. In a sensitive preparation a stretch of less than 1 mm is sufficient to elicit a reflex contraction.
3. The response is specific. The muscle that is stretched is the muscle that contracts. This specificity is so pronounced that stretching one head of a two-headed muscle will ordinarily elicit a contraction only in the stretched portion.
4. The reflex has a very brief latency.
5. The reflex contraction does not outlast the stretch. There is no "afterdischarge" of the motor neurons as there is in some reflexes.

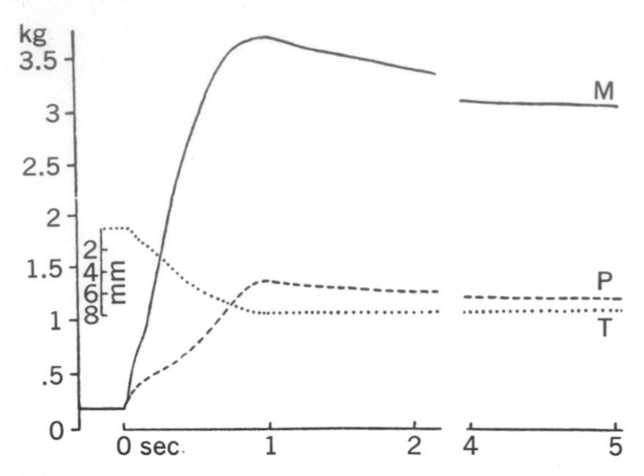

Fig. 24-1. Myotatic reflex of quadriceps muscle of decerebrate cat. *M,* Myographic record of tension change in muscle caused by its stretch, indicated by dotted line, *T. P* records purely elastic tension caused by exactly similar stretch of muscle after section of muscle nerve. Tension difference *(M-P)* is that produced by reflex action. (From Liddell and Sherrington.[23])

6. The reflex is spinal. It remains after the cord is separated from higher centers.

7. The reflex is best developed in extensor muscles. All skeletal muscles respond to stretch with a reflex contraction, but as a rule only certain muscles show a maintained contraction, i.e., one that lasts as long as the stretch continues. These antigravity muscles are called "extensors" by physiologists, regardless of whether they flex or extend a joint. In general, flexor muscles respond to a sudden stretch with a brief phasic contraction, but they do not ordinarily exhibit a maintained contraction.

It should be emphasized that the response to stretch in the decerebrate cat, though useful for demonstrating the existence of an important mechanism, is a greatly exaggerated reflex that is never seen in normal circumstances. It is probably the result of unopposed activity in the length control system, reinforced by powerful facilitation from brainstem centers. Elongation of the muscle results in discharge of spindle receptors, which causes vigorous firing of motor neurons, but the muscle is prevented from shortening and the motor neurons continue to fire as long as the stretch is maintained. An important factor in bringing about the exaggerated response to stretch is a marked increase in the intensity of firing of gamma motor neurons. This causes continual discharge of the spindle receptors in extensor muscles, regardless of whether they are stretched or not. Passive stretch combined with gamma activity causes intense firing of spindle receptors, which results in powerful, well-maintained contractions.

Reciprocal innervation in stretch reflexes

With similar techniques it can also be shown that stretch-evoked impulses from a muscle exert an inhibitory effect on the motor neurons of antagonistic muscles. An example of this reflex inhibition is given in Fig. 24-2. The experiment is a modification of the procedure illustrated in Fig. 24-1. A stretch of the quadriceps muscle elicits a reflex contraction of that muscle which is partially inhibited by stretching the semitendinosus muscles *(S)* and almost completely inhibited by an additional stretch of the biceps femoris *(B)*. Both of these latter muscles are flexors, acting in direct opposition to the quadriceps at the knee joint. The lower records in Fig. 24-2 illustrate the progressive inhibition of quadriceps caused by a progressive stretch of biceps. The basis for this grading of inhibition was described in the preceding chapter, where it was shown that the largest motor neurons were the first to be silenced and the smallest motor neurons were the last. Since muscles are arranged as antagonistic pairs, reciprocal effects occur constantly. The "equivalent" system described in Chapter 21 was simply a convenient way of collapsing all of the various mechanisms into a single mechanism that could be manipulated more easily.

Length control system

Following Sherrington's work the first of the three feedback systems from muscle to be investigated in detail was that involving the primary spindle receptors. It was found by Lloyd that electrical stimulation of the nerve to a muscle results in a reflex discharge that returns to that same nerve but not to other nerves. The reflex response observed in these experiments had the same distribution as the stretch reflex itself, which suggested that it was simply the electric manifestation of the stretch reflex. The very short latency of the reflex indicated that it might be transmitted by the two-neuron arc that Ramón y Cajal's early studies had revealed (Chapter 23). By carefully accounting for all the time elapsing between the sudden stretch of a muscle and the appearance of a reflex discharge in the ventral roots, Lloyd[28] established the existence of a two-neuron or monosynaptic circuit that mediated stretch reflexes (Fig. 24-3). A variety of anatomic and physiologic studies indicated that the receptor involved in this

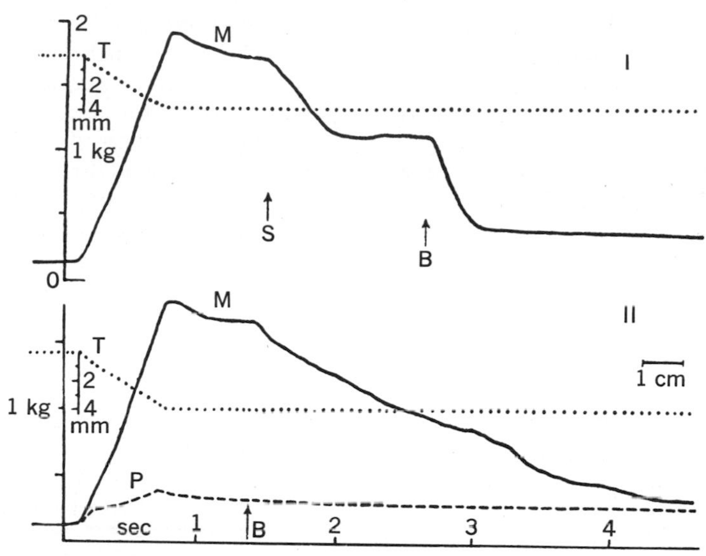

Fig. 24-2. I, Reciprocal inhibition between antagonistic muscles acting at same joint. *M,* Reflex response of quadriceps muscle of decerebrate cat produced by stretch indicated by dotted line, *T,* and recorded myographically. Stretch of semitendinosus muscle beginning at *S* produced partial inhibition of quadriceps reflex tension. Stretch of biceps begun at *B* produced complete inhibition. Inhibitory action on quadriceps motoneurons produced by stretch of two muscles antagonistic to it summed to produce complete effect. **II,** Record of quadriceps stretch reflex obtained as in **I.** Gradually increasing stretch of biceps, beginning at *B,* produced gradually increasing inhibition of quadriceps stretch reflex. Inhibitory effect can be graded by grading intensity of peripheral stimulus producing it. *P,* Passive elastic tension of quadriceps produced by stretching it after section of its nerve. (From Liddell and Sherrington.[24])

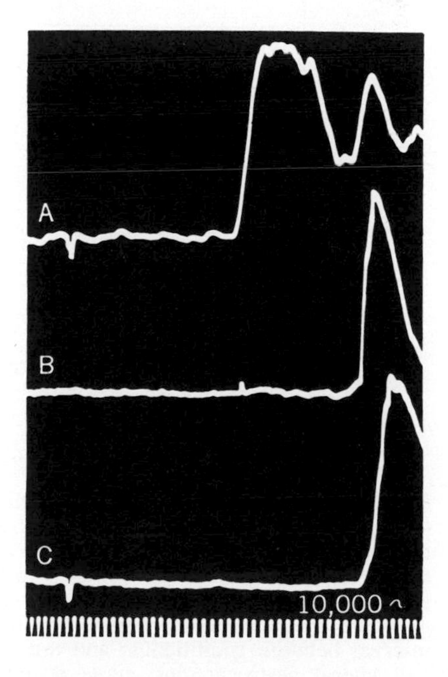

Fig. 24-3. Myotatic reflex may be transmitted through central pathway containing only one synaptic relay; it is monosynaptic. In each record, first small deflection indicates instant of stimulation. **A,** Electric signs of *afferent* discharge evoked by brief stretch of gastrocnemius muscle of acutely spinal cat, recorded from first sacral dorsal root at a given point. **B,** Electric signs of segmental monosynaptic reflex discharge evoked by stimulation at same point on first sacral dorsal root and recorded at given point on first sacral ventral root; initial discharge of segmental reflex is monosynaptically relayed. **C,** Electric signs of reflex discharge into first sacral ventral root evoked by brief stretch of gastrocnemius muscle. Sum of latencies in **A** and **B** approximates that of **C,** which shows that reflex response evoked by stretch was conducted through monosynaptic arcs. Time line identical for all records. (From Lloyd.[28])

reflex was the primary ending in the muscle spindle, which is innervated by a single afferent fiber of large diameter (12 to 20μ) called a Ia fiber. These fibers give off collaterals that run directly to motor neurons in the ventral horn. The efferent limb of the reflex arc consists of the alpha motor neurons innervating the same muscle. The monosynaptic reflex elicited by a single shock to a muscle nerve is the electric analog of the so-called tendon jerk, which is simply the phasic component of the stretch reflex. When the tendon of a muscle is tapped sharply, the whole muscle is stretched slightly and a volley of impulses goes up from the primary endings. These impulses arise simultaneously and are conducted synchronously to the motor neurons. As a consequence, a synchronous discharge of motor neurons occurs and evokes a twitch contraction of the muscle. This contraction is referred to as a knee jerk, ankle jerk, or jaw jerk, depending on the muscles involved. It should be stressed that although a tendon is frequently tapped to elicit the re-

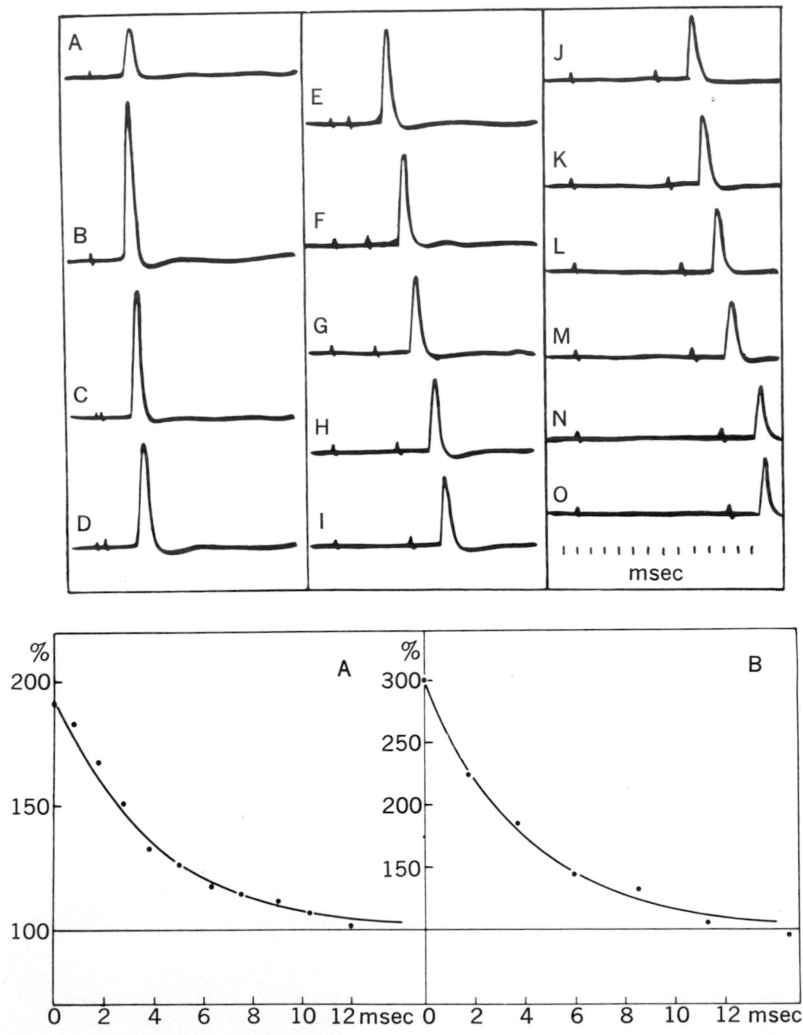

Fig. 24-4. Time course of monosynaptic facilitation. Oscillographic tracings in upper portion of figure are monosynaptic reflexes recorded from ventral root in response to stimulation of two branches of nerve to biceps femoris. Test reflex elicited by stimulation of one branch is shown in frames **A** and **O.** Frames **B** to **N** show facilitated responses obtained when test volley was preceded by a conditioning volley in the other nerve. Curves **A** and **B** in lower portion of figure plot amplitude of conditioned reflex as percentage of test reflex (ordinates) against interval between conditioning and test shocks. Curve **A** obtained from experiment on medial and lateral gastrocnemius; curve **B** obtained from experiment on biceps femoris illustrated above. (Tracings from Lloyd[29]; graphs from Lloyd.[31])

flex the effective afferent volley does not arise in tendon organs but in muscle spindles.

Monosynaptic relations between synergists and antagonists[30]

As a rule there are no monosynaptic excitatory connections between muscles that act upon different joints. Between two muscles or two heads of a single muscle acting synergistically at the same joint, however, there are direct excitatory connections that link their actions closely. Whenever group Ia impulses from a particular muscle are exciting the motor neurons of that muscle, collaterals of the Ia fibers are transmitting similar but less intense effects to the motor neurons of direct synergists. Fig. 24-4 illustrates the time course of this monosynaptic facilitation. In order to appreciate the experiment from which Fig. 24-4 is derived, it should be understood that a volley in Ia fibers arriving at a pool of motor neurons ordinarily discharges only a fraction of the cells. The remainder are excited subliminally but do not fire unless some additional excitatory input is supplied. The amplitude of the monosynaptic reflex evoked by a "test" volley indicates the number of motor neurons fired. Any increase or decrease in the amplitude of this reflex occurring as a result of a previous subliminal "conditioning" volley indicates the reflex effect of that volley in a quantitative fashion. The oscillographic tracings in the upper part of Fig. 24-4 show the facilitation of monosynaptic reflexes of one portion of the biceps femoris muscle when the test volley in the nerve to that part of the muscle is preceded at varying intervals by a conditioning Ia

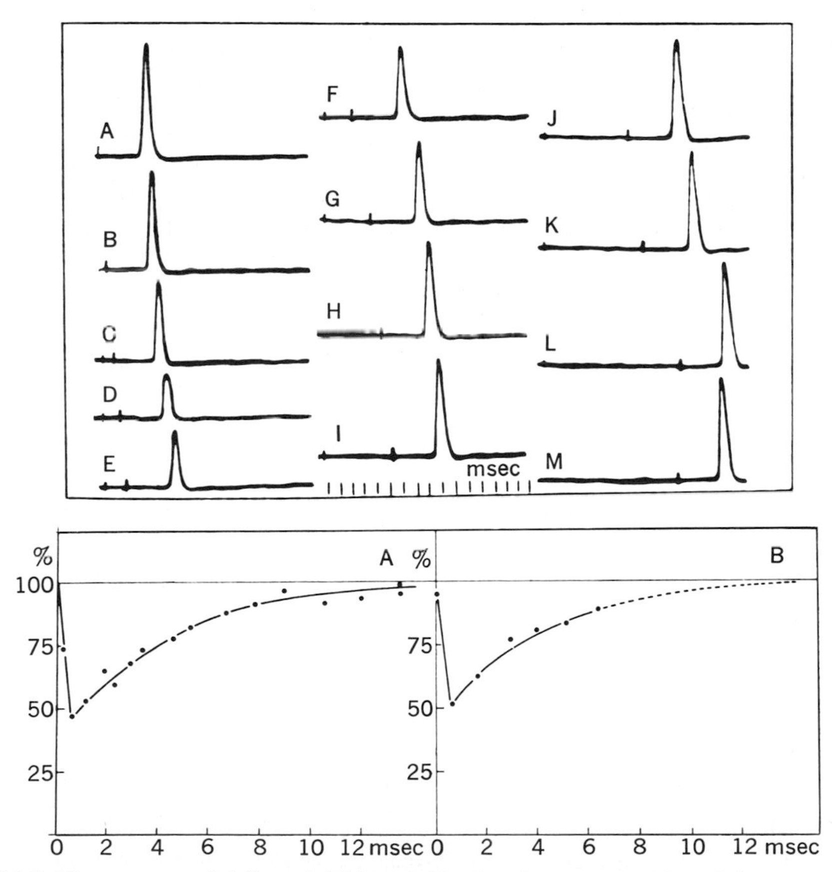

Fig. 24-5. Time course of "direct inhibition." Tracings in upper portion of figure are monosynaptic reflexes recorded in response to stimulation of nerve to gastrocnemius. Unconditioned test reflex shown in frames **A** and **M.** Frames **B** to **L** show inhibition of test reflexes by a conditioning volley in peroneal nerve (a flexor) at increasing intervals. In lower portion of figure curve **A** shows time course of inhibition of tibialis anterior motor neurons by group Ia volleys in nerves to triceps surae. Curve **B** shows time course of inhibition of triceps motor neurons by group Ia volleys in nerves of ankle and knee flexors as illustrated above. (Tracings from Lloyd[29]; graphs from Lloyd.[31])

volley in the nerve to another part of the muscle. Curve *B* in the lower part of Fig. 24-4 indicates the time course of this facilitation. A similar curve *(A)* is obtained when the conditioning and test volleys are set up in the nerves of synergistic extensor muscles. These curves show that the subthreshold Ia excitatory effect of the conditioning volley lasts about 12 msec, decaying exponentially, and that temporal summation of the effects of activity in different fibers plays an important role in synaptic transmission.

Similarly, impulses in group Ia afferent fibers from a given muscle inhibit the motor neurons of direct antagonists. This inhibition is illustrated by the records and curves in Fig. 24-5. The experiments were similar to those that formed the basis of Fig. 24-4. Although the time course of this inhibition appears to be similar to that of monosynaptic excitation, some investigators believe that the "inhibitory" collaterals of Ia fibers end on internuncial cells that in turn cause the inhibition.

The muscles that act together or in opposition at a given joint are called a *myotatic unit*. The members of such a unit are mechanically interdependent, for any change in the length or tension of a particular muscle has a direct effect on its synergists and antagonists. The neural interconnections are the functional expression of this relationship.

Tension control system

The feedback system associated with the tendon organs of Golgi is sometimes called the group Ib system because the afferent fibers belong to group I and are only slightly smaller in average diameter than those from primary spindle receptors. It is well established that these afferent fibers do not end on motor neurons but on internuncial cells that project to motor neurons, completing a disynaptic reflex arc.[21] The evidence for this conclusion comes from experiments similar to those illustrated in Figs. 24-4 and 24-5, in which volleys from two muscle nerves are allowed to interact. If the first or "conditioning" volley is produced by a weak shock to a muscle nerve, it sets up afferent impulses that are limited to group Ia fibers. The effect of such a volley on the motor neurons of a synergist is illustrated by the facilitation curves in Fig. 24-4. If the conditioning shock is strengthened slightly, group Ib fibers are added to the group Ia volley. The effect of this addition is to alter the smooth facilitation curve, as shown in Fig. 24-6. Instead of curve *A*, curve *B* is obtained. Within 0.5 to 0.6 msec after the excitatory effect of the

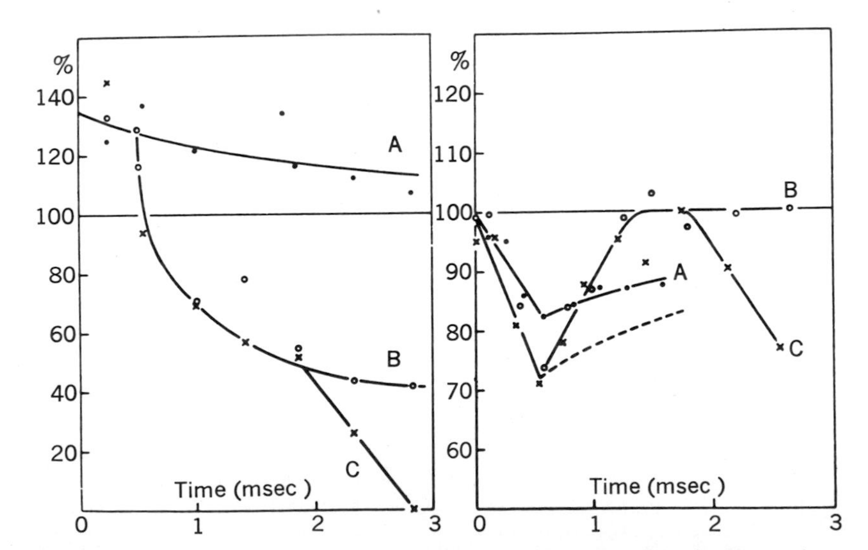

Fig. 24-6. Monosynaptic, disynaptic, and polysynaptic effects of volleys in fibers of groups Ia, Ib, and II. Left: Curve *A*, facilitation of plantaris (extensor) monosynaptic reflexes by group Ia volleys from its synergist, flexor digitorum longus. Curve *B*, inhibition of same reflexes by volleys in fibers of group Ia + Ib. Curve *C*, further inhibition of same reflexes by volleys in fibers of groups Ia + Ib + II. Right: Curve *A*, inhibition of plantaris monosynaptic reflexes by volleys in group Ia fibers of its antagonist, extensor digitorum longus. Curve *B*, excitatory effect of volleys in fibers of group Ib, combined with inhibitory effect of group Ia. Curve *C*, further inhibition of same reflexes by volleys in fibers of groups Ia + Ib + II. (From Laporte and Lloyd.[21])

group Ia volley begins, a sudden decrease in the amplitude of test monosynaptic reflexes occurs, indicating the arrival of an inhibitory volley at the motor neurons. The inhibitory volley is more potent in its effect than the Ia excitatory volley, for their combined effects cause a net inhibition. The delay of 0.5 to 0.6 msec in the onset of inhibition signifies the existence of a single extra synapse in the afferent circuit and indicates the presence of an internuncial cell.

The effect of Ib volleys is inhibitory to the motor neurons of the same muscle and to those of direct synergists, as illustrated in Fig. 24-6, left. Motor neurons of direct antagonists, however, are facilitated by volleys in Ib fibers. As illustrated in Fig. 24-6, right, the Ia effect is inhibitory (curve *A*) and the Ib effect is excitatory (curve *B*). Thus, as in the case of the Ia system, Ib effects are distributed according to the principle of reciprocal innervation.

Following Lloyd's suggestion,[21] it has been customary to refer to the group Ib effect as the "inverse myotatic reflex" because it is opposed to the myotatic reflex. If the Ib system were actually myotatic, as is the Ia system, this terminology would be appropriate. As we have already indicated, however, the adequate stimulus for tendon organs is tension produced by active contraction of the muscle. Changes in length are ineffective. It is not logical to refer to a tension-regulating system as the inverse of a length-regulating system merely because the synaptic effects of the two are opposed; hence the term "inverse myotatic reflex" should be abandoned.

Group II control system

If the conditioning effects of an afferent volley in the group II fibers from the secondary receptors in muscle spindles are examined in the manner previously described for Ia and Ib volleys, a third reflex effect can be distinguished. When the group II volley is added to the Ia and Ib volleys, it produces an effect that is not noticeable in the conditioning curve until 2 msec after the onset of the Ia effect. This delay is due partly to the slower conduction in group II fibers but chiefly to the presence of two or more internuncial cells in the reflex circuit. The effect of group II impulses varies with the type of motor neurons on which they impinge. The motor neurons of physiologic flexors are excited, whereas those of extensors are inhibited. The latter effect is illustrated by the curves labeled *C* in Fig. 24-6. The group II system is sometimes referred to as the "stretch flexor reflex" because it resembles ordinary flexor reflexes in the distribution of excitatory and inhibitory effects. This is not to suggest that it is elicited by pain or excessive stretch. The secondary spindle receptor is extremely sensitive to stretch. Its role would seem to be that of a length-measuring device, but the functional significance of the system as a whole is not apparent. It may in fact be considered the major mystery of this area.

Significance of the three types of circuits

The reflex connections of afferent fibers belonging to groups Ia, Ib, and II are summarized in Fig. 24-7. The presence of internuncial cells in two of these three circuits suggests that afferent activity arising in their receptors is conditional in its effect. Impulses set up in group Ia fibers are transmitted directly and without modification to the motor neurons of the same muscle. Impulses arising in the sensory terminals of group Ib and group II fibers, however, may or may not reach the motor neurons, depending on the other activity that is converging upon the internuncial cells in these circuits.

The full significance of the circuitry found in these three systems is not yet established, but it has recently become apparent that transmission in the circuits containing interneurons is influenced by descending activity from higher centers and by afferent impulses in cutaneous nerves.[20] In the decerebrate preparation, for example, an afferent volley in group II fibers elicits little or no reflex discharge of flexor motor neurons. Impulses descending from the brainstem evidently inhibit the interneurons that transmit excitatory impulses to flexor motor neurons. After spinalization these interneurons are apparently released from descending inhibition, for the previously ineffective afferent volley now elicits a large reflex response. Afferent impulses in certain fibers of cutaneous nerves may cause a similar suppression of group II reflexes. Since depression of reflex transmission can be produced without a corresponding depression in the flexor motor neurons themselves, the internuncial cell is presumably the site at which transmission is regulated in these instances.

As indicated in Chapter 21, control signals that impinge directly on motor neurons necessarily elicit regulatory activity in all three types of feedback control systems. Control

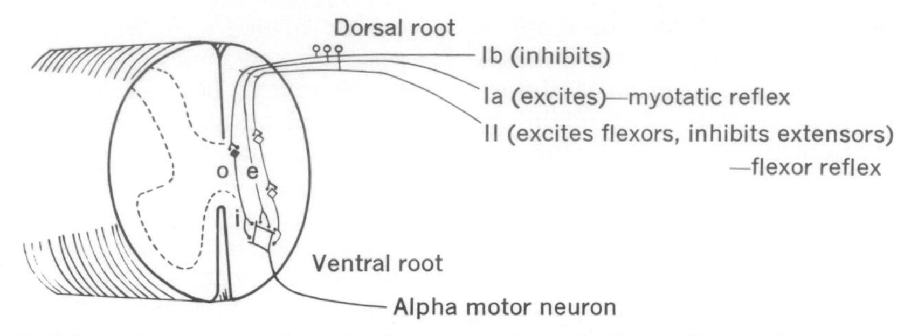

Fig. 24-7. Schematic representation of reflex connections of afferent fibers belonging to groups Ia, Ib, and II of muscle nerves: *e* indicates excitatory effect on motor neurons; *i* indicates inhibitory effect.

signals, which impinge on internuncial cells, however, may reinforce or suppress activity in any one of the three systems selectively. It is not yet understood why the length-regulating system is direct and therefore free from this selective control, whereas the other two systems are subject to it.

It is now apparent that stretch of a muscle produces effects that are not fixed and invariant but that vary according to the conditions of the experiment. The exaggerated reflex response to stretch in decerebrate preparations is thus due in part to the virtual shutdown of the group II control system that is inhibitory to extensor motor neurons. In the "spinal" animal this circuit is apparently "open," for passive stretch may result in net inhibition of extensor motor neurons[14] rather than net excitation. It must be concluded that under normal circumstances descending motor systems actively regulate incoming stretch evoked impulses to produce the most suitable reflex effect under the prevailing conditions.

REFLEXES OF CUTANEOUS ORIGIN

The sensory receptors in skin and subcutaneous tissues respond to touch, pressure, heat, cold, and tissue damage. The signals from all of these receptors exert reflex effects on spinal motor neurons via internuncial cells. They provide feedback that serves to orient the animal to its immediate environment and protect it from injury. The dominant pattern of response to cutaneous stimulation of a limb is ipsilateral flexion and contralateral extension. This suggests that cutaneous activity of all kinds, not only that arising in response to injury, may elicit aversive responses that tend to withdraw the limb from a source of injury. Reflexes of cutaneous origin are not wholly flexor, however, as will be pointed out a little later.

As noted previously, the primary afferent fibers from skin send collaterals to internuncial cells in the dorsal horn. A microelectrode inserted in this region encounters many cells that respond vigorously to various kinds of cutaneous stimulation. Inhibitory as well as excitatory effects are common. The responses of motor neurons are frequently very similar to those of certain interneurons in their time course, which indicates that the patterning of motor responses may be quite a direct consequence of internuncial activity.

Flexor reflexes

The most thoroughly investigated of the cutaneous reflexes is the so-called flexor reflex, consisting of contraction of physiologic flexor muscles and relaxation of physiologic extensors. There are two kinds of flexor reflexes: those resulting from innocuous stimulation of the skin and those resulting from potentially painful and injurious stimuli. The former type consists of a weak contraction of one or more flexor muscles with little actual withdrawal of the limb. The latter type consists of a widespread contraction of flexor muscles throughout the limb that causes an abrupt withdrawal of the injured part from the source of damage.

Stimulation of almost any nerve in a limb will cause a flexor reflex. As indicated in Table 24-1, the reflex tension developed by a particular flexor muscle varies widely with the nerve that is stimulated. It is clear that some mechanism in the spinal cord distributes the afferent impulses to all the flexor muscles of the limb. This distribution is carried out partly by the branching of the primary afferent fibers but largely by internuncial cells. Ramón y Cajal referred to this as the "diffuse reflex mechanism" in contrast to the "circumscribed mechanism" (Fig. 23-5). The distribution of afferent impulses, though

Table 24-1. Maximal reflex tensions developed by tibialis anterior muscle in response to tetanic stimulation of various ipsilateral hind limb nerves (M. tibialis anticus—maximum motor tension 2,160 gm)*

Afferent nerve stimulated	Tension of maximal reflex tetanus (gm)	Reflex tension expressed as percentage of maximal motor tetanus
Internal saphenous	800	32
Superficial obturator	165	6.7
Deep obturator	400	16
Nerve to quadriceps and sartorius	1,190	44
Musculocutaneous branch of peroneal	1,700	69
External plantar	1,240	50
Internal plantar	1,330	54
Small sciatic	680	28
Hamstring	565	23
Nerve of sural triceps	300 (rather low)	12
Total	8,370	

*From Creed et al.[5]

Table 24-2. Patterns of reflex tension resulting from stimulation of three hind limb nerves*

Nerve	Reflex tensions (gm)		
	Hip flexor (tensor fasciae femoris)	*Knee flexor (semitendinosus)*	*Ankle flexor (tibialis anticus)*
Internal saphenous	100	56	87
Popliteal	3 or less	42	100
Peroneal distal to tibialis anticus nerve	14	100	69

*From Creed et al.[5]

diffuse, is not entirely nonspecific. Depending on the nerve that is stimulated, the ipsilateral limb will take up different final positions. Table 24-2 shows how the pattern of reflex flexion varies with the nerve that is stimulated. Hence the reflex exhibits local sign to some extent.

Comparison of the discharges evoked in the ventral root by stimulation of a muscle nerve and a skin nerve brings out the differences in their central transmission very clearly. The response to an afferent volley in group Ia fibers is a synchronous discharge of minimal delay, indicating transmission over a two-neuron pathway (Fig. 24-8, *A*). The response to a volley in a skin nerve (Fig. 24-8, *B*) consists of a more prolonged series of discharges with no monosynaptic component. The earliest deflections are the result of transmission in a reflex arc of four or more neurons; the later deflections are those that have been delayed by passage through additional interneurons. If a peripheral nerve is stimulated with shocks of increasing strength, the reflex discharge recorded from a nerve to a flexor muscle increases progressively in duration as smaller afferent fibers are added to the afferent volley. The response to a single strong shock exciting delta fibers and C fibers may last a full second or more. If the recording electrodes are placed on a very fine muscle nerve so that individual impulses can be distinguished, it can be seen that the prolonged discharge consists of repetitive discharges of the motor neurons of small size. This prolongation of reflex effect serves the purpose of keeping a flexed limb withdrawn from a painful stimulus so that restimulation does not occur.

Crossed extension

As a "spinal" animal withdraws his limb in response to a noxious stimulus, the contralateral limb is simultaneously extended. The function of this reflex is to support the weight of the body when the opposite limb is lifted. Crossed extension is "grafted on" to the flexor reflex at the internuncial level. Collaterals of the interneurons that excite ipsilateral flexor motor neurons cross to the other

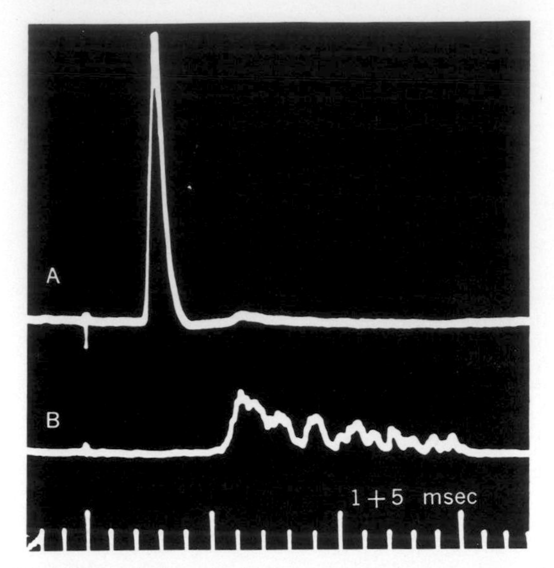

Fig. 24-8. Reflex discharges recorded from ventral root in response to stimulation of, **A,** nerves to gastrocnemius muscle and, **B,** sural nerve (a skin nerve). (From Lloyd.[27])

side of the spinal cord and excite extensor motor neurons. At the same time contralateral flexor muscles are relaxed due to reciprocal inhibition.

Reflexes associated with specific regions

Cutaneous stimulation of certain regions may elicit responses that do not fall into the flexor pattern but serve more specific purposes. Light pressure applied to the pads of the forepaw of a chronic spinal dog, for example, causes reflex extension of the whole limb.[6] This serves to reinforce the stretch reflexes and other supporting reactions brought into play when a foot is placed in contact with the ground. It is important to note that even the slightest pinch of the same toe pads will produce flexor withdrawal rather than extensor thrust. These two very different reactions serve to illustrate that the nature of a reflex action is determined by the quality of the stimulus as well as its locus. Another example of specific reflex action is the bilateral extension of the hind limbs that results from stimulation of the skin of the inner aspects of the thighs. The significance of this reflex in copulation is obvious.

REFLEXES ORIGINATING IN JOINTS

Since the position of a joint specifies the length of the muscles acting across it, the signals from joints should be of great importance in regulating movement at the spinal level, just as they unquestionably are at higher levels where they lead to a conscious sense of position. Relatively little is known, however, about reflexes originating in joints. Several types of sensory receptors are found in and around joints, the most important of which are Ruffini's endings and tendon organs. They are located in all the ligaments and in the fibrous capsule. Every position of a joint involves stretch of some ligaments or capsular bands and relaxation of others. As might be expected, it has been found experimentally[2] that there is no position at which all joint receptors are silent. In general each ending discharges at its maximal steady rate when the joint is at some particular angle and less rapidly when the angle is increased or decreased. Different receptors respond over different portions of the total range of movement. The pattern of firing in all the receptors signals the exact position of the joint to higher centers. These signals unquestionably exert reflex effects as well, for stimulation of articular nerves elicits discharges in ventral roots.[1, 12] The various patterns of signals resulting from each of the possible positions of a joint probably influence the motor neurons of the muscles which move that joint in a systematic manner, although this has not been demonstrated experimentally.

Long spinal reflexes and the propriospinal system

In animals with four weight-bearing limbs the activity of the forelimbs and hind limbs is coordinated by so-called long spinal reflexes.[40] This can readily be observed in preparations with spinal transections in the upper cervical region. A noxious stimulus applied to any extremity results in reflex responses in all four limbs, as illustrated in Fig. 24-9. Forelimb flexion is accompanied by extension of the ipsilateral hind limb. On the other side of the body the pattern is reversed. As a result of these actions the animal tends to move away from the stimulus. A considerable amount of normal locomotor activity in four-footed animals is probably patterned in this way. It is believed that the stimuli responsible for long spinal activity arise in the secondary spindle receptors of muscle and in cutaneous end organs innervated by fibers of groups II, III, and IV.[26, 32] Group I activity from muscles apparently has little effect on distant motor neurons.

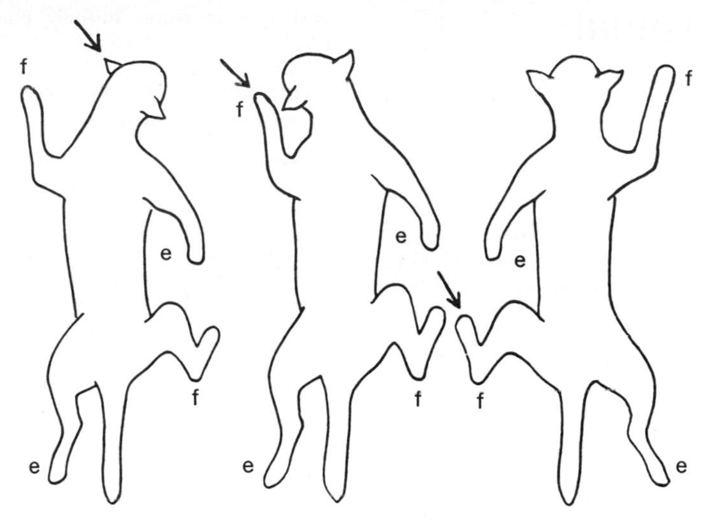

Fig. 24-9. Reflex figures produced in cat with neuraxis sectioned just above first cervical segment (at calamus scriptorius). Noxious stimulation at points marked by arrows; *e*, leg extended in response; *f*, leg flexed in response. (After Sherrington.[38])

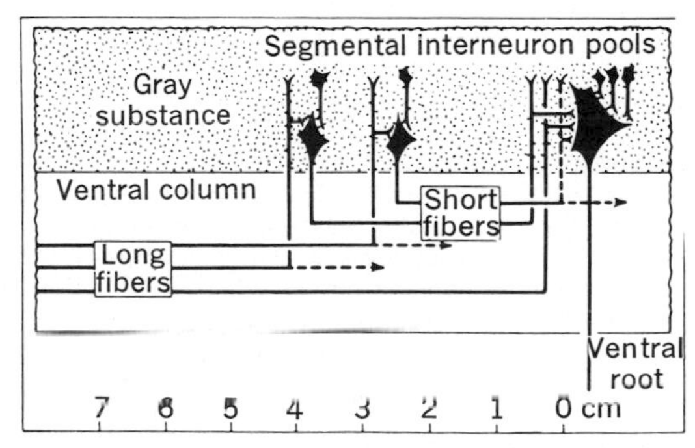

Fig. 24-10. Schema illustrating pathways connecting descending fibers of ventrolateral columns of spinal cord, the bulbospinal and long spinal systems, with motoneurons. They end in greatest concentration about cell bodies of short propriospinal neurons, located in segmental interneuron pools. Activity is relayed through them to motoneurons some 3 to 4 cm farther caudally. A direct pathway exists, but motoneurons are controlled by it only when activity is intense and motorneuron excitability high. (From Lloyd.[25])

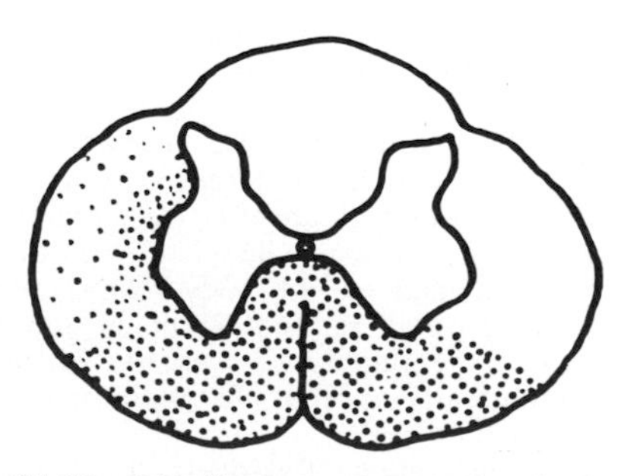

Fig. 24-11. Distribution of propriospinal neurons concerned with long spinal reflexes. Ventral columns are involved bilaterally, lateral columns unilaterally. (From Lloyd.[26])

Long spinal reflexes are mediated by *propriospinal* neurons whose cell bodies lie in the central gray matter and send their axons into the adjacent white matter where they run up or down for variable distances before reentering the gray matter (Fig. 24-10). There is no evidence that nervous activity is ever conducted more than a minimal distance within the gray matter itself. Short propriospinal fibers are found in all columns of the white matter; long fibers are apparently found only in the lateral and ventral columns (Fig. 24-11).

Suggestions for review and further study

As a means of integrating the information that has been presented, the reader is urged to review Chapter 21 and endeavor to translate the block diagrams into suitable anatomic and physiologic terms. In the course of further study, several books will be particularly useful. Ramón y Cajal's classic account[35] of the fine structure of the spinal cord with its numerous illustrations is invaluable. An excellent summary of the studies of the Sherrington school, containing much that has not been included in these chapters, is given by Creed et al.[5] Two monographs by Eccles[8, 9] offer the most complete account of the modern work on synaptic transmission of the spinal cord, and finally, a critical review by Hunt and Perl[15] will serve as a convenient point of departure for a survey of the recent literature on spinal reflexes.

Effects of spinal transection

VERNON B. MOUNTCASTLE

SPINAL SHOCK

When the spinal cord is completely severed, two functional disasters are at once evident: (1) all voluntary motion in body parts innervated by the isolated spinal segments is permanently lost and (2) all sensation from those parts, which depends on the integrity of ascending spinal pathways, is abolished. A third conspicuous sign is an immediate spinal areflexia, a state by usage termed spinal shock.

Spinal shock is a transient condition of decreased synaptic excitability of neurons lying aboral to a transverse section of the spinal cord.[22] It varies in depth and duration with the degree of cerebral dominance of the spinal mechanisms in the species considered, i.e., with the degree of encephalization. Thus in the frog the depression of reflex excitability after high spinal section lasts only a minute or so. In carnivores the duration of reflex depression is measured in hours; in the monkey it may extend over many days or weeks, whereas in anthropoid apes and in man the course of recovery of reflex excitability may extend over many months.

Spinal neurons bear on their cell bodies and dendrites hundreds of terminal end feet, many of them endings of axons descending from supraspinal portions of the nervous system. Under normal circumstances these deliver low-frequency trains of impulses to spinal neurons, so that at any given instant many local postsynaptic responses (subliminal depolarizations) are occurring at widely scattered sites on the postsynaptic cell. Although these may not be dense enough to provoke complete depolarization and firing of the cell, they serve to maintain it in a slightly oscillating state of high excitability, ready to respond to any spatially or temporally more concentrated presynaptic inflow. When that portion of the "background tone" taking origin in more cephalad regions is removed by spinal section, the resting excitability of the spinal cell is, for a time, greatly reduced: this is spinal shock.

Spinal shock cannot be due to any form of irritation at the cut surface of the cord, for a second transection made some days later, below the first, is followed by only a trivial reflex depression in the spinal remnant. Nor is it the result of the drop in blood pressure that follows any but the lowest transection, for oral portions of the nervous system are equally exposed to the decreased blood flow, yet display no decreased excitability, and spinal shock still appears when the hypotension is prevented by previous administration of vasoconstrictor drugs.

The phenomena of spinal shock are confined to regions aboral to the section. With the exception of the slight increase in extensor tonus in the forelegs following thoracic or lumbar section (Schiff-Sherrington phenomenon), no change in function is evident in the cephalad portions of the nervous system. Indeed, a cat or monkey immediately after cord section seems blithely unaware of this catastrophe.

Temporal course of recovery of reflex excitability

Somatic reflexes. In carnivores the period of complete areflexia is so short that it may not be observed if a long-acting anesthetic agent is used for the operation of spinal transection. Within minutes a tightening of quadriceps muscle is palpable on tapping the patellar tendon. Soon visible knee jerks appear, and shortly thereafter feeble flexion of the leg is produced by nociceptive stimulation of the foot. With the passage of hours the flexion withdrawal reflex gains in strength, the threshold of the reflex is lowered, and the peripheral receptive zone from which it can be evoked enlarges. Simultaneously with flexor contraction extensors are inhibited. Cross extension reflexes and long spinal reflexes such as the scratch reflex may not reappear for days, and in general the time of recovery of a reflex action depends on the number and complexity of the synaptic relays in its central pathway. The time course of this recovery varies greatly from one individual to another and may be adversely affected by intercurrent infections, nutritional deficiencies, or other debilitations. How these conditions affect reflex excitability is unknown.

With further passage of time, reflex excitability continues to rise, now to abnormally high levels. Sustained myotatic reflexes appear, and positive supporting reactions are evokable; in such chronic spinal animals these reactions may suffice to support the weight of the animal for 2 or 3 min ("spinal standing") before collapse occurs. When such animals are suspended in air, the hindlegs may execute alternating flexions and extensions ("spinal stepping").

A clear example of the capacity of the isolated cord to combine elementary reflexes into movement patterns having useful purposes is given by the response of the spinal cat to immersion of the feet in water.[33] When the paw of a normal cat is immersed in water, it is withdrawn and shaken vigorously. In the intact animal the response is independent of the temperature of either the water or of the animal, and it is evident that a tactile stimulus evokes the reflex. In the chronic spinal animal no response is evoked by immersion of the foot in water at body temperature. However, when the animal is tested with either cold or hot water, a sharp flexion withdrawal of the foot occurs, followed by vigorous shaking of the paw. This shaking succeeds in ridding the limb of a considerable amount of water. The response to cold water is evoked by stimulation of cold receptors and is related to the temperature differences between body and water. Hot water is an effective stimulus only when sufficiently hot to excite pain receptors, and its action is independent of body temperature. In either case the total sensory input in the intact animal signals the quality of wetness, which produces a functionally appropriate response from the spinal reflex mechanism.

Mass flexion reflex. Relatively mild nocuous stimulation produces widespread contraction of the flexor musculature of the limb and of the abdominal wall; evacuation of the bladder or rectum may occur at the same time but not necessarily as a part of this mass flexion reflex. It should not be supposed that the irradiation of this reflex spreads along neural channels other than those already completely formed, for there is no evidence to suggest that axons may grow to make new synaptic connections in the isolated spinal remnant. On the other hand, it is as though synaptic connections, normally sparse and tenuous, now assume a supraliminal potency.

Some light is thrown on this problem by a consideration of the phenomenon of the hypersensitivity of denervated neurons. Cannon and Rosenblueth[4] had originally shown that viscera innervated by autonomic nerves become, some days after denervation, abnormally sensitive to the synaptic transmitter agent normally active at those nerve endings. This hypersensitivity is so great that the acceleration of the denervated heart serves as a most sensitive measure of minute quantities of epinephrine in the blood. Later Cannon and Haimovici[3] demonstrated that the motoneurons below a cord semisection become hypersensitive to acetylcholine, compared to those on the normal side of the cord, as well as to dorsal root afferent impulses and to excitant drugs. Stavraky[41] has extended these observations on central neurons, but little is known of the basic mechanisms concerned. The phenomenon is of considerable interest relative to altered states of excitability in diseases or injuries of the nervous system.

The course of reflex recovery after spinal transection in monkeys[37] resembles that in carnivores but is displaced on an extended time scale. The depth of spinal shock is profound; the body parts below the transection are flaccid and motionless. The most severe nociceptive stimulation of the skin and direct electrical stimulation of a large afferent nerve are equally ineffective in producing a motor response. In some individuals the knee jerk is the first reflex to appear, occasionally within the first hour, but it is frequently seen only at the end of the first week. In many cases the flexion-adduction of the hallux in response to scratching the sole of the foot (the plantar reflex) is the first response obtainable, and it is only rarely delayed in appearance beyond the fourth day. If adequate care is given to bladder and bowel function, to the avoidance of pressure necrosis of peripheral nerves and skin, and to the nutritional state of the animal, the subsequent course of recovery of the somatic reflexes much resembles that described in carnivores. Insufficient numbers of animals have been studied in the chronic spinal state to know whether complex extensor reflexes such as those responsible for "spinal standing" recover full reflex excitability in monkeys, but it is reasonable to suppose that they do so, for they have been observed in chronically surviving paraplegic humans.

Spinal shock following partial transections. Studies of cats with partial transverse lesions of the spinal cord[11] have shown that it is the reticulospinal and the vestibulospinal tracts that, when severed, produce the phenomena of spinal shock, suggesting that their nuclei of origin are principal sources of descending facilitation playing upon the spinal neurons. In monkeys, however, these seem less important, for isolated section of the lateral columns of the cord (the corticospinal tracts) produces nearly as severe spinal shock as does cord transection. This species difference seems due to the greatly increased direct control of the spinal mechanisms exerted by the pyramidal system in the monkey as compared with the cat. Considerable effort has been made by investigators using electrical recording techniques to determine whether interneurons or motoneurons are the more severely affected by the depression of spinal shock. In the primate the motoneurons are more severely affected than are the interneurons of the dorsal horn, whereas in carnivores the interneurons are involved and the motoneurons suffer only a slight change in excitability.

Visceral reflexes. Spinal transection also results in widespread alterations in visceral functions.

Bladder. Immediately after section of the spinal cord there is complete atony of the smooth muscle of the bladder wall. At the same time there is an increase in constrictor tone in the sphincter muscle, presumably due to loss of inhibitory influence of central origin. As a result, urine accumulates until intravesicular pressure is sufficient to overcome sphincter resistance. Even then only small driblets of urine escape—the so-called overflow incontinence. With recovery of the somatic reflexes, tone returns to the bladder muscles and reflex emptying of the bladder occurs, produced by simultaneous contraction of its smooth muscle walls and, at least to a certain extent, relaxation of tone in the sphincter. Nevertheless, a considerable residuum of urine remains after each reflex emptying. More complete emptying is promoted if this residuum is removed each time by catheterization. Reflex emptying is greatly facilitated or at times even provoked by cutaneous stimuli, either tactile or nociceptive, delivered to abdomen, perineum, or lower extremities. The development of the reflex bladder action differs little among carnivores, monkey, and man except in its time course; in the first two the cord bladder is usually established in the first week; in the latter, only 25 to 30 days after cord section (see also Chapter 79).

Intestine. Little is known of the function of the intestine after cord section. In dogs a loose diarrhea may develop during the first week, but this is transient and it appears that in general the processes of digestion and absorption proceed normally. Greater difficulty is met in the evacuation of waste products from the intestinal canal. Normally the presence of fecal material in the lower bowel and rectum, passively stretching the muscles of its wall, produces their active contraction and peristaltic action, combined with inhibitory relaxation of the sphincter tone, and defecation results. That this mechanism is inherent in the nerve plexus of the bowel wall is shown by the fact that such "reflex" defecation recovers after complete removal of the spinal cord. However, the presence of the superimposed reflex arcs of the cord brings a progressive character and greater fusion to rectal contractions, which results in a more massive and complete reflex response. This mechanism is depressed during the period of spinal shock, when the sphincter ani relaxes only slightly in response to passive distention of the rectum, and retention of rectal contents may occur. With recovery of reflex excitability, reflex defecation occurs; it is greatly facilitated by tactile stimulation of the skin area of the sacral segments or by manual dilatation of the sphincter muscle.

Vasomotor reflexes. Reflex actions on the peripheral vessels and other effectors innervated by the autonomic nervous system that are mediated by spinal arcs are considered in Chapter 30. Here it is pertinent to state that they, in company with reflex arcs debouching upon the striated musculature, are profoundly affected during the period of spinal shock. The loss of background vasoconstrictor tone precipitates a hypotension that persists for some time, and during the early hours after cord section even intense stimulation of an afferent nerve trunk will not produce a rise in blood pressure. With time, however, the tonic discharges return, reflex actions can once again be elicited, and an almost normal level of blood pressure is maintained in the chronic spinal animal. Spinal neurons innervating peripheral effectors concerned in the control of body temperature are, of course, permanently disconnected from descending influences originating in thermoregulatory centers, and animals with transection in the cervical region are almost wholly poikilothermic (Chapter 56). Those with lower lesions show proportionately greater degrees of temperature control, depending on the level of transection.

Sexual reflexes. The postural adjustments and reflex actions operating in the act of copulation are organized at supraspinal levels, but the reflex pathways are inherently spinal. In the spinal male dog, tactile stimulation of the penis produces erection and frequently ejaculation and a flexion of the back and hindquarters that tends to thrust the penis forward. The posture is best seen with the animal prone, belly and thighs contacting the table surface.[17] In the spinal bitch, contractions of uterus and vaginal orifice result from stimulation of the sexual skin. Little is known of the relation of cord function to the endocrine regulation of the organs of reproduction. However, the menstrual cycle of female monkeys is little disturbed by spinal section,[42] and in several instances impregnation, gestation, delivery, and nursing, all in relatively normal fashion, have been recorded in chronic female spinal dogs.

FUNCTIONAL CAPACITY OF THE ISOLATED HUMAN SPINAL CORD

Complete transection of the spinal cord in man due to injury or disease is a rare event in civilian life. Unfortunately such injuries caused by missiles of high velocity are not uncommon in warfare, and some of these paraplegic individuals have survived in the chronic state. Considerable success has been achieved in rehabilitating them to a certain degree for useful and comfortable lives. Descriptions of the ingenious therapeutic methods by which their survival is assured must be sought in clinical texts and papers, particularly those of Munro,[34] Kuhn,[19] Freeman,[10] and others. Here it is more pertinent to describe the reflex activity of which the isolated remnant of spinal cord is capable.

The most informative of the early reports concerning the reflex function of the human cord is that of Theodor Kocher, published in 1896,[16] detailing a comprehensive study of 15 patients suffering a sudden cord transection, the completeness of which was subsequently verified at autopsy. Kocher's patients lived for only short periods. The classic descriptions of spinal man were provided in 1917 by Riddoch[36] and by Head and Riddoch,[13] based on studies of soldiers injured in World War I. Their detailed descriptions should be read by all those pursuing the subject. However, the periods of observation available to these clinical neurologists were frequently cut short by the deaths of patients, and their observations were made difficult to interpret by the presence of severe debilitation and frequently of infection of the urinary tract. Advanced methods of medical care permitted surgeons of World War II to study spinal man for months and years without the complications of intercurrent disease. Particular reference is made to the painstaking studies of R. A. Kuhn,[17, 18] from which much of the following description has been taken. His report gains great validity also from the fact that *in every case the existence of a complete transection and its segmental level were established directly by surgical exploration of the spinal canal.* In all, 29 men were studied, their transections ranging in level from the second to the twelfth thoracic segment of the cord. Many of these men survived transection for several years. It should be emphasized that the time course of recovery of reflex activity is variable, as is the sequence of appearance of various reflexes. The sequence described below is only the usual one, and wide variations from it are to be expected in individual cases.

Stages of spinal shock. Immediately following transection there is complete loss of motor and sensory function below the level of injury, a complete areflexia is observed, and if the transection is high, a degree of hypotension exists. The duration of the complete areflexia varies greatly from one case to another; in a few some reflex activity is elicitable within 24 hr of injury, but more commonly none is observed for a period of 2 to 6 weeks. If complete absence of reflex actions continues for many months, as was the case in 4 of Kuhn's 29 patients, it is likely that the cord itself has been destroyed, either by direct injury or by interference with its blood supply. Paralysis of the bladder, manifested by retention of urine, is invariably complete. If the transection is high, there is also complete paralysis of the mechanism of defecation, with distention of the bowel. There are many suggestions in the reports available that in some cases certain reflexes, particularly those evoked by stimulation of genital regions, may never completely disappear. There is no record, however, of a case studied from the moment of transection; even the earliest accounts begin several hours after the injury had been sustained. In general it can be said that the reappearance of any reflex that has been completely abolished by cord section is a sign of emerging cord activity and recovery from spinal shock.

Period of minimal reflex activity. As the profound reflex depression begins to lift, nocuous stimulation of the soles of the feet elicits reflex responses. Most commonly, the first movements seen are tremulous twitchings of the toes and a brief flexion or extension of the hallux in response to plantar stimulation. At about this time contraction of the sphincter ani appears in response to nocuous plantar stimulation or to tactile stimulation of the perianal skin. The genital reflexes may also appear for the first time. This period of the onset of activity may last from 2 weeks to several months, and in the rare patient it may represent the end stage of recovery. When this occurs, there is reason to believe that the isolated remnant has sustained severe injury.

Development of flexor activity. With further recovery the tremulous toe movements, the first elicitable, change to a typical Babinski pattern (i.e., dorsiflexion of the hallux and fanning of the toes in response to plantar stimulation), flexor withdrawal movements of the foot and ankle are evokable and with further passage of time, flexion of the knee and the hip appear. Such a *mass flexion reflex* could be elicited in 22 of the 25 men in Kuhn's series who survived cord section for long periods of time and who manifested any reflex activity. In some instances the mass flexion was accompanied by the crossed extension reflex, and in many instances extension occurred during the relaxation phase from flexion. The foot, especially its plantar surface, is the reflexogenous zone *par excellence,* and here purely tactile stimuli suffice to elicit widespread flexion, which shows again that afferent fibers from tactile receptors of the skin project upon the reflex mechanisms of the spinal cord as well as into its great ascending tracts. With the passage of months it is possible to elicit mass flexion or components of it from a wider surface of the leg and, at low threshold, from the genital zone.

Development of extensor activity. In a few individuals the stage of mass flexor activity is the final one reached in recovery. However, in 18 of the 22 reflexly active patients in Kuhn's group who survived cord section for 2 or more years, a stage was reached in which the predominant reflex activity was extensor in nature, the so-called extensor spasms.* The movements of extensor muscles become evident as early as 6 months after injury and may continue for many years. The tendon reflexes are hyperactive and clonus occurs; sustained stretch reflexes are at times observed. The inverse myotatic reflex can be elicited (the lengthening reaction, or "clasp-knife" phenomenon) and with it, increased extension of the contralateral leg—Phillippson's reflex. In many individuals strong positive supporting reflexes appear, and a few are able to stand for a short period of time without mechanical support (spinal standing). After development of extensor reflex activity, reflex flexion is still easily elicited by nocuous plantar stimulation, so that the increase in excitability apparently affects all spinal reflex arcs.

It should be noted that the stimulus of greatest potency for strong extensor activity is a sudden, brief stretch of the flexor muscles, particularly the flexors of the thighs. It is difficult to understand this action as other than the inverse myotatic reflex, arising from flexor muscle tendon organs.

Genital reflexes. The genital zone is one from which mass flexion can be elicited by nocuous stimulation at low threshold. In those patients showing any reflex activity, the local genital reactions are invariably present. These are penile erection, contraction of the bulbocavernosus and sphincter ani muscles, and a delayed slow contraction of the scrotal dartos. The optimal stimulus is gentle tactile stimulation of the penile frenulum. Only rarely is penile erection accompanied by ejaculation of seminal fluid. No observations are available concerning genital reflexes in chronic spinal women.

Reflex activity of bladder and bowel. The recovery of reflex emptying of these viscera differs in no significant way from that occurring in spinal animals. However, it should be noted that reflex evacuation of the bladder, which may occasionally follow stimulation of the lower limbs, is not a part of the mass flexion reflex. On the contrary, nocuous

*It is apparent, therefore, that the clinical dictum that the presence of extensor spasms indicates a partial cord transection is valid only if the patient is seen within 6 months after injury.

stimulation most commonly produces an increase in sphincter contraction.

Subjective sensations. Of the 29 men studied by Kuhn, 17 reported some subjective sensation referable to portions of the body below the level of the transection. Most commonly this was described as a dull, burning sensation in the buttocks, perineum, or lower abdomen. It was elicited by long-continued pressure on the ischial tuberosities, by stimulation of the anal or urethral canal, or by distention of the bladder. It seems likely that the afferents concerned travel upward over the splanchnic nerves to enter the cord over the dorsal roots rostral to the level of transection.

Sudomotor activity. Outbursts of sweating commonly occur in spinal men. In favorable cases paroxysmal sweating can be elicited by almost any afferent inflow to the segment of cord remaining—scratching the foot, distention of the bladder, etc. Little is known of the central reflex pathways concerned.

• • •

In summary, it should be emphasized that study of men surviving cord transection for many months and years reveals that the human spinal cord contains all the inherent capacity for reflex action of which the cord of lower mammals is capable. The increasingly prolonged period of reflex recovery after cord section that is observed as one ascends from less to more complex nervous systems is paralleled by the increasing dominance of spinal mechanisms by those resident at suprasegmental levels of the brain.

REFERENCES

1. Beswick, F. B., Blockey, N. J., and Evanson, J. M.: Some effects of the stimulation of articular nerves, J. Physiol. **128:**83P, 1955.
2. Boyd, I. A., and Roberts, T. D. M.: Proprioceptive discharges from stretch-receptors in the knee-joint of the cat, J. Physiol. **122:**38, 1953.
3. Cannon, W. B., and Haimovici, H.: The sensitization of motoneurones by partial "denervation," Am. J. Physiol. **126:**731, 1939.
4. Cannon, W. B., and Rosenblueth, A.: The supersensitivity of denervated structures, New York, 1949, The Macmillan Co.
5. Creed, R. S., et al.: Reflex activity of the spinal cord, London, 1932, Oxford University Press, Inc.
6. Denny-Brown, D. E., and Liddell, E. G. T.: Extensor reflexes in the forelimb, J. Physiol. **65:**305, 1928.
7. Dusser de Barenne, J. G., and Koskoff, Y. D.: Further observations on flexor rigidity in the hindlegs of the spinal cat, Am. J. Physiol. **107:**441, 1934.
8. Eccles, J. C.: The physiology of nerve cells, Baltimore, 1957, The Johns Hopkins Press.
9. Eccles, J. C.: The physiology of nerve cells, New York, 1964, Academic Press, Inc.
10. Freeman, L. W.: Treatment of paraplegia resulting from trauma to spinal cord, J.A.M.A. **140:**949, 1015, 1949; **141:**275, 1949.
11. Fulton, J. F., Liddell, E. G. T., and Rioch, D. M.: The influence of experimental lesions of the spinal cord upon the knee-jerk. I. Acute lesions, Brain **53:**311, 1930.
12. Gardner, E.: Reflex muscular responses to stimulation of articular nerves in the cat, Am. J. Physiol. **161:**133, 1950.
13. Head, H., and Riddoch, G.: The automatic bladder, excessive sweating and some other reflex conditions, in gross injuries of the spinal cord, Brain **40:**188, 1917.
14. Henneman, E.: Excitability changes in monosynaptic reflex pathways of muscles subjected to static stretch, Trans. Am. Neurol. Assoc. **76:** 194, 1951.
15. Hunt, C. C., and Perl, E. R.: Spinal mechanisms concerned with skeletal muscle, Physiol. Rev. **40:**538, 1960.
16. Kocher, T.: Die Verletzungen der Virbelsäule zugleich als Beitrag zur Physiologie des menschlichen Rüchenmarcks, Mit. Grenzgeb. Med. Chir. **1:**415, 1896.
17. Kuhn, R. A.: Functional capacity of the isolated human spinal cord, Brain **73:**1, 1950.
18. Kuhn, R. A., and Macht, M. B.: Some manifestations of reflex activity in spinal man with particular reference to the occurrence of extensor spasm, Bull. Johns Hopkins Hosp. **84:** 43, 1949.
19. Kuhn, W. G., Jr.: The care and rehabilitation of patients with injuries of the spinal cord and cauda equina, J. Neurosurg. **4:**40, 1947.
20. Kuno, M., and Perl, E. R.: Alteration of spinal reflexes by interaction with suprasegmental and dorsal root activity, J. Physiol. **149:**374, 1959.
21. Laporte, Y., and Lloyd, D. P. C.: Nature and significance of the reflex connections established by large afferent fibers of muscular origin, Am. J. Physiol. **169:**609, 1952.
22. Liddell, E. G. T.: Spinal shock and some features in isolation-alteration of the spinal cord in cats, Brain **57:**386, 1934.
23. Liddell, E. G. T., and Sherrington, C. S.: Reflexes in response to stretch (myotatic reflexes), Proc. R. Soc. Lond. (Biol.) **96:**212, 1924.
24. Liddell, E. G. T., and Sherrington, C. S.: Reflexes in response to stretch (myotatic reflexes), Proc. R. Soc. Lond. (Biol.) **97:**267, 1925.
25. Lloyd, D. P. C.: Activity in neurons of bulbospinal correlation systems, J. Neurophysiol. **4:** 115, 1941.
26. Lloyd, D. P. C.: Mediation of descending long spinal reflex activity, J. Neurophysiol. **5:**435, 1942.
27. Lloyd, D. P. C.: Reflex action in relation to pattern and peripheral source of afferent stimulation, J. Neurophysiol. **6:**111, 1943.
28. Lloyd, D. P. C.: Conduction and synaptic transmission in the reflex response to stretch in spinal cats, J. Neurophysiol. **6:**317, 1943.

29. Lloyd, D. P. C.: Facilitation and inhibition of spinal motoneurons, J. Neurophysiol. **9:**421, 1946.
30. Lloyd, D. P. C.: Integrative pattern of excitation and inhibition in two-neuron reflex arcs, J. Neurophysiol. **9:**439, 1946.
31. Lloyd, D. P. C.: On reflex actions of muscular origin, Res. Publ. Res. Assoc. Nerv. Ment. Dis. **30:**48, 1950.
32. Lloyd, D. P. C., and McIntyre, A. K.: Analysis of forelimb-hindlimb reflex activity in acutely decapitate cats, J. Neurophysiol. **11:**455, 1948.
33. Macht, M. B., and Kuhn, R. A.: Responses to thermal stimuli mediated through the isolated spinal cord, Arch. Neurol. Psychiatry **59:**754, 1948.
34. Munro, D.: The rehabilitation of patients totally paralyzed below the waist, with special reference to making them ambulatory and capable of earning their own living. II. Control of urination, New Engl. J. Med. **234:**207, 1946.
35. Ramón y Cajal, S.: Histologie du système nerveux de l'homme et des vertébrés, Paris, 1909, A. Maloine.
36. Riddoch, G.: The reflex functions of the completely divided spinal cord in man compared with those associated with less severe lesions, Brain **40:**264, 1917.
37. Sahs, A. L., and Fulton, J. F.: Somatic and autonomic reflexes in spinal monkeys, J. Neurophysiol. **3:**258, 1940.
38. Sherrington, C. S.: Decerebrate rigidity and reflex coordination of movements, J. Physiol. **22:** 319, 1897-1898.
39. Sherrington, C. S.: The integrative action of the nervous system, New Haven, 1906, Yale University Press.
40. Sherrington, C. S., and Laslett, E. R.: Observations on some spinal reflexes and the interconnections of spinal segments, J. Physiol. **29:**58, 1903.
41. Stavraky, G. W.: The action of adrenaline on spinal neurons sensitized by partial isolation, Am. J. Physiol. **150:**37, 1947.
42. van Wagenen, G.: Uterine bleeding of monkeys in relation to neural and vascular processes. I. Spinal transection and menstruation, Am. J. Physiol. **105:**473, 1933.

25

JAMES HOUK

Feedback control of muscle: a synthesis of the peripheral mechanisms

The control of posture and movement demands the moment-to-moment coordination of the several hundred muscles that make up the motor apparatus of the human body. The CNS must distribute appropriate control signals to each of these muscles, a problem that is discussed in subsequent chapters. The purpose of this chapter is to show how the muscular and neural components that have already been described function together as a *peripheral control system*. This system compensates for disturbances and improves the reliability of the muscular response to control signals.

FACTORS INFLUENCING CONTRACTILE RESPONSES OF MUSCLE

The force generated by a muscle is graded by a progressive recruitment of motor units and by variations in the frequencies at which individual motor units are activated (Chapters 22 and 23). Thus the efferent *error signal* that was referred to frequently in the introduction to feedback control (Chapter 21) is actually a rather complex signal whose magnitude depends on the number and discharge rate of active units. The block diagram in Fig. 25-1, *A*, illustrates the subsequent events leading to movement: an *efferent signal* controlling *muscular force* causes a change in the *length* of a muscle, which determines the *position* of a joint. The diagram includes other influences on contractile force and movement.

There is considerable interaction between a muscle and its load. Normally the load is inertial (a mass), but it may also be elastic (a spring return on a door) or viscous (the resistance to underwater movements). The mass of the forearm is a typical load on the biceps muscle (Fig. 21-2, *A*), whereas the mass of the body loads the extensor muscles of the knee and ankle when a person stands with knees bent (Fig. 25-1, *B*). In both of these examples, muscular forces tend to lift the load in opposition to an *external force* due to gravity.

The changes in the *length* and *velocity* of a muscle, which are coupled with the movement of the load, influence the force of contraction, as indicated in Fig. 25-1, *A*. The dependence of contractile force on the length and velocity of a muscle was illustrated previously by length-tension and force-velocity curves (Figs. 3-5 and 3-6, pp. 87 and 88). Recent studies indicate that the relationships are highly nonlinear and more complex than suggested by these curves, particularly when a muscle is stimulated at the lower frequencies at which motor units normally discharge.[9, 10, 18, 21] Muscular force also depends on other factors, designated as *internal* disturbances in Fig. 25-1, *A*. A common example is the muscular fatigue that accompanies vigorous exercise.

An example will serve to illustrate some of these interactions and also to introduce a graphic analysis that will be used in subsequent figures in this chapter. Curves relating length and force in an extensor muscle (soleus of the cat) when it is activated with three different efferent signals are shown in Fig. 25-2. In curve *1* the efferent signal is one that activates all the fibers in the muscle at a frequency of 35 pulses/sec. The force developed at each length is less when the fibers are stimulated at a frequency of 10 pulses/sec (curve *2*). The latter rate corresponds better to the frequency at which motor units discharge during a tonic stretch reflex (about 8 pulses/sec).[6] Curve *3* shows the forces that would be developed if only half of the muscle

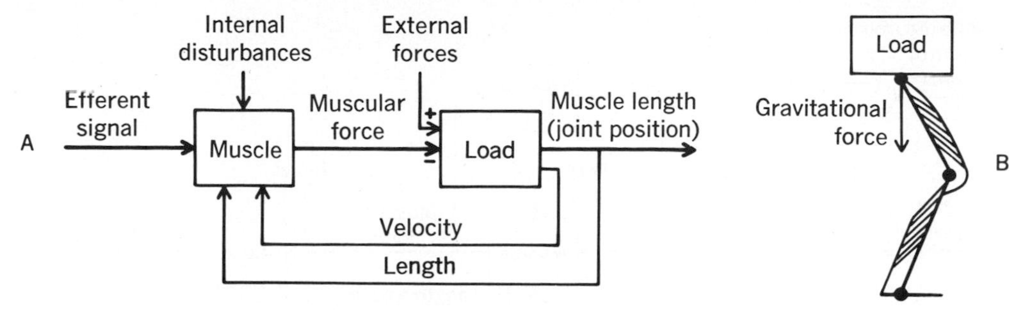

Fig. 25-1. A, Block diagram of factors influencing contractile responses of a muscle. Efferent signal controls muscular force that acts on load to cause changes in length of muscle, thus determining position of joint. Internal disturbances and variations in length and velocity of muscle influence force of contraction; external forces influence movement of load. **B,** Diagram illustrating action of extensor muscles of knee and ankle in man standing with knees bent. Body mass provides load and gravitational force acting on mass creates external force opposed by muscular contraction.

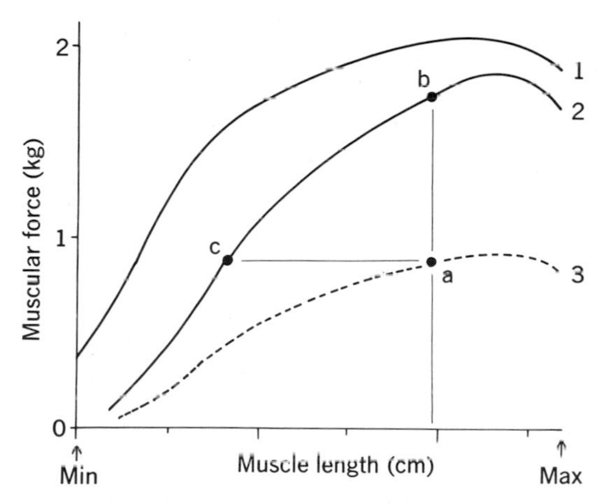

Fig. 25-2. Effect of efferent signals on relation between length and force in muscle. Curves *1* and *2* show forces developed by soleus muscle at various lengths when whole muscle is activated at frequencies of 35 and 10 pulses/sec, respectively. Curve *3* shows smaller forces that are produced if only half of muscle is activated at 10 pulses/sec. Points *b* and *c* illustrate two possible responses to efferent signal that initially activates half of muscle (point *a*) and later activates all of it. Depending on the type of load, such an efferent signal can cause either increase in force (point *b*) or shortening of muscle (point *c*). "Min" and "max" indicate range of muscle length in intact cat. (Curves *1* and *2* redrawn from Rack and Westbury[21]; curve *3* based on data from Houk et al.[8])

fibers were activated at a frequency of 10 pulses/sec. Points *a, b,* and *c* will be used to illustrate the interaction between length and force when a muscle is controlled by an efferent signal.

Begin with the assumption that the muscle is operating at point *a* due to an efferent signal that activates half of the muscle fibers. In response to a larger efferent signal, one that activates all of the fibers, the muscle may generate a larger force (point *b*), shorten (point *c*), or perform some combination of both. The actual response will depend on the load. If the load is unyielding, e.g., a door that is stuck, the efferent signal will cause an increased force; if the load is elastic, it will cause shortening. An efferent signal thus produces a shift of the entire length-tension curve of the muscle rather than the simple change in force that was assumed for the sake of simplicity in Chapter 21.

This brief discussion of the mechanical properties of a muscle interacting with its load serves to illustrate that the control of movement is complex, even if only a single muscle is involved. There are two general methods that the nervous system might use to compensate for this complex behavior. One method depends on past experience. Neural circuits in the brain might compute signals that are "corrected" for anticipated variations in muscular responses. This type of feedback would not compensate for sudden, unexpected variations. The second method of compensation for variability is to correct for it as it occurs. Spinal reflexes provide a continuous regulation of contraction that might be capable of compensating for variations, whether they are anticipated or not. There is little doubt that the nervous system uses both methods in controlling the response of muscles. Although little is known about the neural mechanisms for computing "corrected" signals, the continuous regulation of contraction provided by certain spinal reflexes is understood well enough to permit analysis of the performance of these feedback systems.

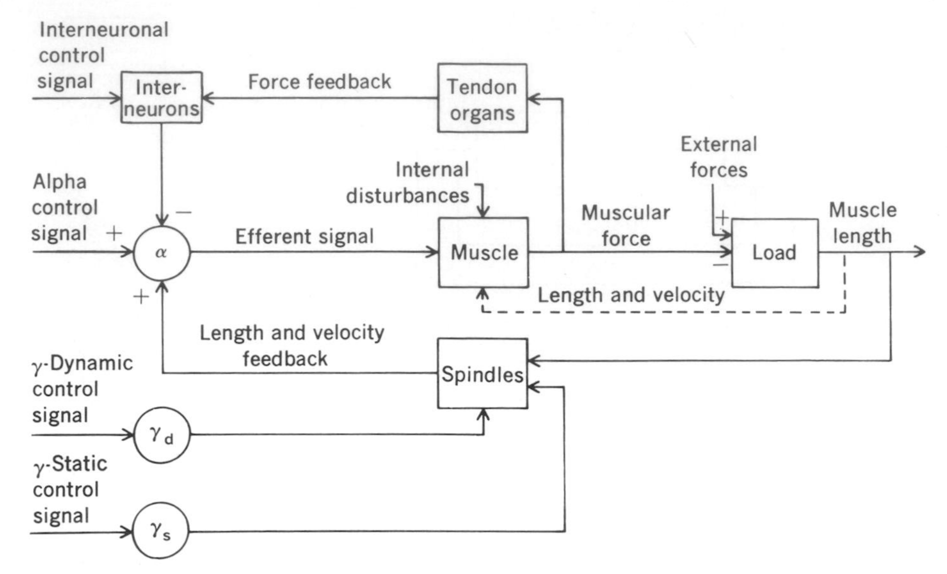

Fig. 25-3. Block diagram of peripheral control system. Muscle and its load are regulated by two feedback pathways, one signaling length and velocity through spindle receptors and the other signaling muscular force through tendon organs. Dashed line represents a nonneural feedback from load to muscle that results from inherent mechanical properties of muscle (Fig. 25-1). CNS initiates movement and modifies feedback by sending control signals to various neurons in the spinal cord (α-, γ_d-, and γ_s-motor neurons and interneurons in tendon organ pathway).

PERIPHERAL CONTROL SYSTEM

In Chapter 21 a very simple length control system involving primary spindle receptors was analyzed in detail in order to illustrate the principles of feedback control. It was found that this system provided a partial compensation for the disturbances caused by external forces such as gravity. The term "stiffness" was introduced as a measure of the capacity of the length control system to resist external forces. Stiffness was defined as the ratio of the disturbing force to the steady-state error in length caused by the disturbing force.

This simplified system must now be modified, since several physiologic features were not included in the length control system. As described in the preceding chapters, primary spindle receptors are sensitive to velocity as well as to length, and their responses are modified by efferent signals from γ-motor neurons. Secondary spindle receptors also function as length transducers and participate reflexly in regulating contraction. Golgi tendon organs monitor the force of muscular contraction, and signals in their afferent fibers cause inhibition of homonymous motor neurons. As indicated in the preceding section, the inherent properties of the muscle also influence contraction. These additional features are summarized in the block diagram of the peripheral control system given in Fig. 25-3. The peripheral control system is more elaborate than the generalized feedback system shown in Fig. 21-1, but the same functional components are present.

The controller, a muscle, and the controlled system, a load, appear in the center of the diagram (compare with Figs. 21-1 and 25-1). The transducers are tendon organs that monitor muscular force and spindle receptors that monitor the length and velocity of the muscle. Responses of spindle receptors are also influenced by signals from gamma-static (γ_s) and gamma-dynamic (γ_d) motor neurons. The various inputs that γ-motor neurons receive are collectively referred to as γ_d- and γ_s-control signals. The efferent signal that triggers muscular contraction represents the net error signal of two feedback pathways, one signaling force and the other signaling length and velocity. The signals conveying force are transmitted by interneurons on their way to α-motor neurons and consequently they are influenced by interneuronal control signals. The net length and velocity feedback originating from both primary and secondary spindle receptors appears as a single pathway in Fig. 25-3. Pathways from secondary receptors are polyneuronal and until recently were

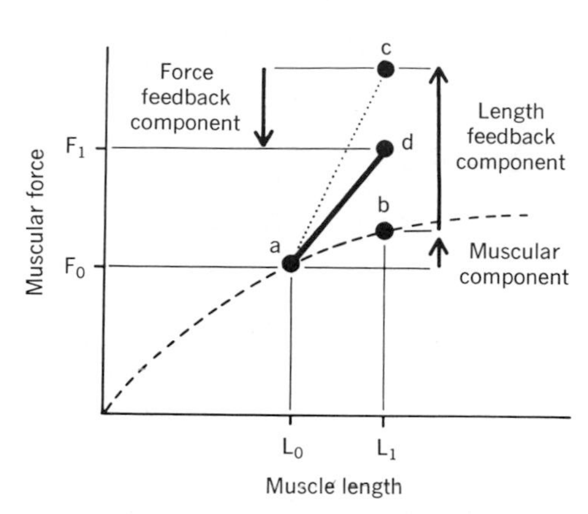

Fig. 25-4. Components of stretch reflex. When muscle is stretched slowly from L_0 to L_1, its force increases from F_0 to F_1. This diagram dissects response into three components: muscular component is due to length-tension properties of muscle, length feedback component increases stiffness of peripheral control system, and force feedback component decreases stiffness. Stiffness is defined by ratio of change in force to change in length.

believed to be inhibitory to extensor motor neurons. Experiments by Matthews[15] suggest instead an excitatory function that contributes quite significantly to the stretch reflex. Pathways from primary receptors are monosynaptic and excitatory.

The systems controlling length, velocity, and force are inextricably linked in nature and function simultaneously. They all act on the same load with forces produced by a single muscle that is innervated by a single group of α-motor axons. Consequently, it is difficult to evaluate the contribution of each feedback pathway to the overall regulation of contraction. For example, the reflex response to stretch must be influenced by length, velocity, and force feedback as well as by the inherent mechanical properties of the muscle. In order to appreciate the combined operation of these systems it is helpful to construct an artificial stretch reflex based on the known characteristics of each influence. For the time being, let us assume that the stretch is applied slowly, so that the contribution of velocity feedback can be postponed for discussion later in this chapter.

In Fig. 25-4, muscular force is plotted as a function of muscle length. Assume that the muscle is initially held at a length L_0. At this length a certain number of motor units are active, causing a contractile force, F_0. These motor units have length-tension properties shown by the dashed curve passing through point *a*. How will muscular force change after the muscle has been slowly stretched from L_0 to L_1?

Because of the length-tension properties of the initially active motor units, the force would increase along the dashed curve to point *b* even if additional motor units were not reflexly recruited. This response of the muscle is analogous to that of a weak spring that offers some resistance to stretching. *Muscular stiffness* thus accounts for one component of the stretch reflex.

Stretch of the muscle would also activate spindle receptors, causing an increased length feedback to α-motor neurons (Fig. 25-3). This would lead to a reflex recruitment of additional motor units and the production of a larger force (point *c*, Fig. 25-4). Thus length feedback accounts for a second component of the stretch reflex, one that augments the stiffness of the overall response, as indicated by the dotted line in Fig. 25-4.

The larger force produced by the muscle would cause an increased discharge of tendon organs (Fig. 25-3). This would lead to a reflex inhibition of α-motor neurons that would reduce the number of additional motor units recruited by the stretch and would reduce the muscular force to point *d* in Fig. 25-4. Thus force feedback accounts for a third component of the stretch reflex, one that reduces the stiffness of the overall response.

It is clear that force feedback interferes with the regulation of muscle length since it causes the overall stiffness of the stretch reflex to decrease. In a later section the consequences of these effects on stiffness will be discussed. A consideration of situations in which there is no length feedback will illustrate the capacity of force feedback to compensate for internal disturbances.

REGULATION OF MUSCULAR FORCE

If the load on a muscle is unyielding, the muscle can develop force but cannot shorten. For example, in writing with a pen or pencil the force perpendicular to the writing surface must be strong enough to cause a uniform trace but not so strong as to break the tip of the writing instrument. The table presents an unyielding surface to the perpendicular component of muscular force. If a muscle in an experimental animal is maintained at a

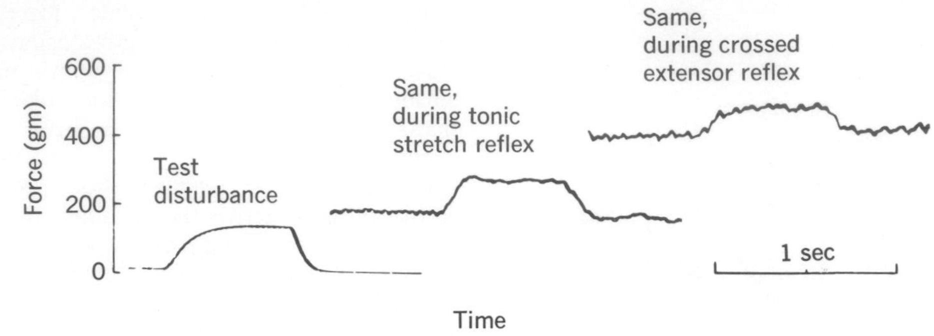

Fig. 25-5. Compensation for internal disturbance. Test disturbance was produced by stimulating ventral root filament. Smaller responses to same stimulus during tonic stretch reflex and during crossed-extensor reflex illustrate extent of compensation for disturbance provided by force feedback. These are responses of an isometric soleus muscle in decerebrate cat. (Modified from Houk et al.[8])

constant length, the situation is similar. In both cases length feedback is effectively eliminated, since changes in length are not allowed. Under these conditions it is possible to study force feedback in isolation.

Since tendon organs monitor the force of contraction, their inhibitory reflex connections (Fig. 25-3) provide a feedback pathway that is capable of compensating for internal disturbances. One such disturbance is muscular fatigue. If the muscular force should decrease because of fatigue, the α-motor neurons would receive less inhibition and motor units would be recruited to compensate for the fatigue.

The effectiveness of force feedback in compensating for internal disturbances has been studied in decerebrate cats by eliciting small contractions in the soleus muscle and directly observing the extent of compensation for these disturbances.[8] Fig. 25-5 shows some myographic records obtained during these experiments. The disturbance was elicited by stimulating repetitively a small filament of ventral root whose connections with the spinal cord had been severed. During a tonic stretch reflex the change in force caused by stimulation of the filament was smaller than the original amplitude of the disturbance (Fig. 25-5), suggesting that the responses of tendon organs to the increase in force reflexly inhibited the tonic stretch reflex.* During a crossed-extensor reflex, elicited by stimulating a nerve in the contralateral limb, the initial reflex force was larger. When the filament of ventral root was again stimulated, the increase in force was only 55% of the amplitude of the disturbance, suggesting that force feedback caused the reflexly generated force to decrease in compensation for the disturbance.

Although opposite in sign, the disturbance elicited by stimulating the filament of ventral root is similar to fatigue. Both cause the net force produced by a muscle to vary independently of the efferent signal sent by motor neurons. The compensation that was observed can be attributed to force feedback since the length of the muscle remained constant. While the extent of compensation demonstrated in these experiments was modest, one can nevertheless conclude that force feedback contributes to the regulation against variations in the properties of a muscle.

As pointed out by Merton,[17] length feedback is also capable of compensating for such variations, provided the muscle is not constrained isometrically. Fatigue of a muscle loaded with a mass leads to a lengthening of the muscle, which is resisted by the recruitment of motor units that reflexly accompanies a stretch of spindles. What, then, is the additional advantage of force feedback?

INTERACTIONS BETWEEN LENGTH AND FORCE FEEDBACK

Length and force feedback are synergistic in regulating against internal disturbances. However, they are antagonistic in regulating against external disturbances such as gravitational forces, since length feedback tends to increase the stiffness of the peripheral control system, whereas force feedback tends to decrease its stiffness (Fig. 25-4). It might seem that this decrease in stiffness would be a dis-

*Spindle receptors probably also contribute to this response and to force feedback, since they may be unloaded slightly as series elastic elements lengthen during the isometric contraction.[8] As force increases, the discharge rate of spindles would decrease, giving rise to feedback signals related to muscular force that act synergistically with signals from tendon organs.

advantage of force feedback, since it would increase the amount of displacement experienced when the external force supported by a muscle changes.

If one considers the higher functions of muscular systems, it becomes apparent that an unyielding system is not always advantageous. For example, one function of muscular systems is to control the position of the head in space. The musculature that moves the limbs of terrestrial animals provides the suspension system for the body and head.[4] This suspension system must absorb the shocks encountered, for example, during locomotion over rough terrain or while standing on an unstable surface. Under such circumstances a yielding suspension system would reduce the displacement of the head caused by mechanical disturbances. Force feedback, by decreasing the stiffness of the peripheral control systems regulating the contraction of individual muscles, would cause the suspension system to yield.

The stiffness of the peripheral control system might also be reduced by decreasing the gain* of length feedback (Fig. 25-4). But in this case the system becomes more sensitive to muscular fatigue and other variations in the properties of the muscle. Decreasing the stiffness by increasing the gain of force feedback has an advantage in that the peripheral control system continues to be well regulated against these internal disturbances. Thus the primary function of length and force feedback in the peripheral control system may be to ensure consistent responses of muscle rather than to keep the length of the muscle constant.

Current evidence suggests that interneuronal control signals (Fig. 25-3) originating from centers in the brainstem are capable of turning transmission on and off in the reflex pathways from tendon organs.[12] By means of this mechanism the CNS may regulate the stiffness of the peripheral control system and hence the properties of the suspension system for the body and head. While this notion is intuitively attractive, it must be pointed out that a change in the stiffness of the stretch reflex produced by control signals to these interneurons has never been demonstrated experimentally.

Spindle receptors and tendon organs subserve functions other than those analyzed here. Their signals influence the activity of synergistic and anatagonistic motor neurons; they are a source of control signals to the peripheral control systems in other limbs; they project to the brainstem, the cerebellum, and the cerebral cortex. At present it is difficult to evaluate the importance of spindles and tendon organs in these integrative functions.

*Gain is defined by the ratio of the output of a system to its input. It provides a measure of the sensitivity of a receptor or the efficacy of a feedback pathway.

CONTROL OF MOVEMENT

In Chapter 21 it was concluded that the length control system could function in two ways—as a *regulator* against disturbances and as a *servomechanism* for executing movements (summarized in Fig. 21-7). The peripheral control system can also perform these two functions. Its operation as a regulator has been discussed in the preceding pages. We must now turn to its operation as a servomechanism.

Control signals to either α- or γ-motor neurons can initiate movement, i.e., changes in the length of the muscle. The α-control signals activate the muscle directly, whereas the γ-control signals must first activate spindle receptors, which then reflexly activate the α-motor neurons and the muscle (Fig. 25-3). Once initiated, the continuation and conclusion of the movement is influenced by feedback from spindle receptors and tendon organs and by the mechanical properties of the muscle. The interactions among force, length, and velocity feedback are best understood by considering initially a more static situation. The effects of control signals on the stretch reflex will be described and these results will then be related to the problem of movement.

Stretch reflex curves are shown in Fig. 25-6. The curve passing through points *a* and *d* illustrates the typical relationship between the length of a soleus muscle and its reflex force in a decerebrate cat. The stretch reflex was elicited by extending the muscle at a slow rate. The curve passing through points *b* and *c* was obtained when a peripheral nerve in the opposite limb was stimulated to elicit a cross-extensor reflex at the same time that the muscle was being stretched. The entire stretch reflex curve was shifted to the left by this excitatory input.

The crossed-extensor reflex in the soleus muscle is one component of a postural adjustment that is initiated by a noxious stimulus in the contralateral hind limb (Chapter 24). An intact animal will withdraw the stimulated limb and shift the weight formerly supported

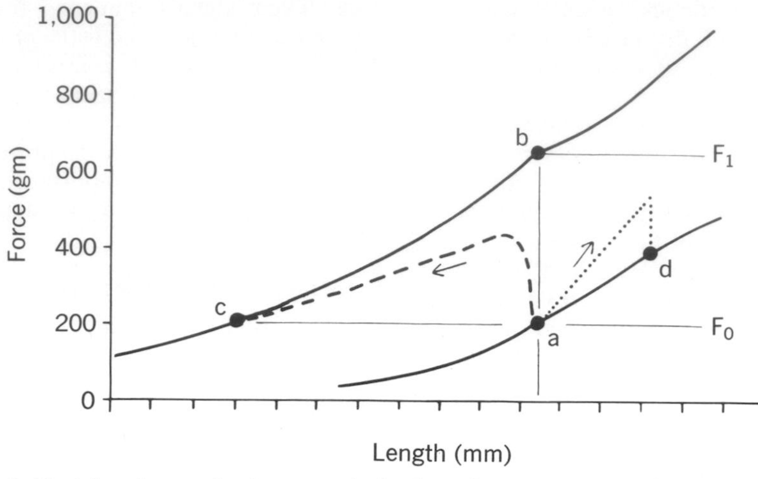

Fig. 25-6. Stretch reflexes of soleus muscle in decerebrate cat. Curve through *a* and *d* shows relation between length and force in unreinforced stretch reflex; curve through *c* and *b* shows same relation during a crossed-extensor reflex. Muscle with constant load would shorten from point *a* to point *c* along dashed trajectory in response to control signal provided by crossed-extensor reflex. Response to rapid stretch would follow dotted trajectory. (Solid curves redrawn from Matthews.[13])

by it to the opposite hind limb. The increased contraction of extensor muscles in the latter, due to a crossed-extensor reflex, prevents the limb that bears the extra weight from sagging. In Fig. 25-6, if we imagine that a force (F_0) is required of the soleus to assist in the initial support of the weight of the animal, a noxious stimulus delivered to the contralateral limb would cause the force to increase to F_1. This prevents the soleus muscle from being stretched, as it is obliged to support the additional weight shifted to it. The excitatory input to soleus motor neurons responsible for this increased output is one example of a control signal. In this example the control signal does not produce movement; rather, it prevents movement. (Whether this input represents an α-control signal or a γ-control signal is of no concern for the moment.)

If the same control signal were to reach the motor neurons at a time when the weight supported by the limb remained constant, the soleus would shorten and the body would be lifted. This possibility can also be followed in Fig. 25-6. Initially a force (F_0) is required to assist in the support of the animal. In response to the postulated control signal, operation would shift from point *a* on the lower stretch reflex curve toward point *b* on the upper curve. The larger force would accelerate the mass of the body upward but also allow the muscle to shorten until it reached point *c* and the muscle again produced the force, F_0, required to support the weight of the animal.

This example illustrates that in the absence of other changes a control signal delivered to the peripheral control system will produce movement. The end point of the movement, point *c* in Fig. 25-6 in this example, is determined by at least three factors in addition to the load: (1) the gain of length feedback, (2) the gain of force feedback, and (3) the length-tension properties of the muscle. It was not necessary to consider these factors individually because their effects are all combined in the two stretch reflex curves that appear in Fig. 25-6 (also review 25-4). Note, however, that the dashed trajectory from points *a* to *c* lies below the upper stretch reflex curve in Fig. 25-6. It has been drawn in this way to illustrate the effects of velocity feedback, which will now be considered in more detail.

DAMPING OF MOVEMENT PROVIDED BY VELOCITY FEEDBACK

Mechanical systems have a tendency to oscillate if they are not *damped* by forces that slow down movement. (More precisely, damping forces oppose movement in proportion to its velocity.) Feedback may add to this instability, particularly if it is delayed as it is in reflexes by conduction in afferent and efferent axons. A tendency of the length control system to oscillate as a result of conduction delay was described in Chapter 21 (Fig. 21-5).

Damping of the peripheral control system is provided by velocity feedback from primary spindle receptors and certain mechanical properties of the muscle (Fig. 25-3). The contribution of these factors to the stretch reflex is evident in Fig. 24-1. The force developed by the muscle rises rapidly as the muscle is stretched and then decays to a lower value while the stretch is maintained. The extra force that accompanies lengthening is related to the velocity at which the muscle is stretched.[25] Part of this force is due to the inherent mechanical properties of the muscle[9, 10] (as revealed by force-velocity curves) and part originates from the sensitivity of primary endings to the velocity of stretch.[14, 17, 25] The velocity component of the stretch reflex is illustrated in Fig. 25-6 by the dotted trajectory from points *a* to *d*. When the muscle is rapidly stretched, its force deviates from the reflex curve that was obtained by stretching the muscle slowly. Note that the deviation is in a direction that opposes the stretch; hence it will tend to damp movement.

The dashed trajectory in Fig. 25-6 was used earlier to describe the response of the peripheral control system to a signal for movement. In the absence of velocity feedback the force produced by the muscle would jump from point *a* to *b* and the trajectory followed during the movement would be along the upper curve toward point *c*. The actual force produced by the muscle as it shortens will be less than this because velocity feedback causes a component of force that opposes the movement. In this example, velocity feedback slows the movement sufficiently to prevent it from "overshooting" and oscillating about point *c* (Fig. 25-6). Rapid voluntary movements of human subjects, in contrast, often slightly overshoot their final position.[24] The overshoot would be greater and would give rise to large oscillations if there were no damping in the peripheral control system.

Another manifestation of instability in the peripheral control system that warrants mention is *clonus*. Even muscular systems of normal human subjects may demonstrate this tendency to oscillate at approximately 8 to 12 Hz, a behavior that is believed to be caused by excessive feedback from spindle receptors.[11] An interesting feature of clonus is that its amplitude is normally limited to small excursions, which suggests that the phenomenon may be related to the high gain of primary spindle receptors when changes in length are small.[16, 20]

CENTRAL CONTROL OF ALPHA AND GAMMA MOTOR NEURONS

Although movement can in theory be initiated by control signals to either α- or γ-motor neurons, each of these routes may have certain advantages. Signals to α-motor neurons avoid the delay caused by conduction to the spindles in the small γ-motor axons and back to the spinal cord in the afferent axons from the receptors (the delays are not shown explicitly in Fig. 25-3). Movements initiated by signals to γ-motor neurons maintain the discharge of the spindle receptors at a high level during the movement; in contrast, muscle shortening produced by a large α-control signal may allow the discharge of spindle receptors to drop below threshold. The spindle endings would then cease to signal changes in length, and there could be no feedback compensation for disturbances during the movement.

Studies from many laboratories indicate that α- and γ-motor neurons are normally coactivated in the initiation of movement,[1, 3, 19, 23] although activation of γ-motor neurons may prevail when the stimulus used to provoke control signals is weak.[1, 2] With α-γ-coactivation, the CNS can use to advantage both the speed of the α-route for the initiation of movement and the continued regulation against disturbances provided by the γ-route. Matthews[14] suggested that if coactivation were appropriately balanced the discharge of spindles might not change during normal movements; there would be no feedback from spindle receptors unless a disturbance, e.g., an unexpected load or fatigue of a muscle, caused the movement to deviate from its "desired" course. If a deviation occurred, signals from spindle receptors would generate muscular forces tending to compensate for the deviation. Current evidence supports α-γ–coactivation but has not provided an adequate test for the more demanding hypothesis offered by Matthews.

The two functional types of γ-motor neurons shown in Fig. 25-3 have quite different effects on the responses of spindle receptors (Chapter 22). Signals in γ_s-axons provoke a marked increase in the discharge frequency of both primary and secondary endings in muscle spindles. Signals in γ_d-axons have no effect on secondary endings but cause a small increase in the discharge of primary endings; their most significant effect is to increase the sensitivity of these endings to stretch (Chapter 22).[7, 14]

The γ_s-motor neurons are more suited to the initiation of movement because of their strong acceleratory effect on spindle discharge at all muscle lengths.[7] Inasmuch as it has been investigated, the normal pattern of coactivation appears to involve α_s- and γ_s-motor neurons.[5] The γ_d-motor neurons are sometimes coactivated and sometimes inhibited by stimuli that activate α-motor neurons.[5] The functional role of γ_d-motor neurons is not yet apparent. Because they control the sensitivity of primary endings to changes in the velocity and length of the muscle, one would expect them to modify the damping and stiffness of the stretch reflex.

SUMMARY

This chapter has focused on regulation of the contraction of individual muscles. The mechanical properties of a muscle, its load, length and velocity feedback from spindle receptors, and force feedback from tendon organs, all interact in this regulation. The system that includes all of these components has been called the peripheral control system.

The stiffness with which the peripheral control system resists changes in muscle length has its origin in the length-tension properties of the muscle and is modulated by feedback from muscle spindle receptors and Golgi tendon organs. Length feedback increases stiffness, whereas force feedback decreases it. Velocity feedback from primary spindle receptors and certain mechanical properties of the muscle serve to damp the system, thus diminishing its tendency to oscillate. Length, velocity, and force feedback function together to compensate in part for the nonlinear and variable behavior of muscular systems.

Control signals sent by the CNS to α- and γ_s-motor neurons cause the peripheral control system to behave like a servomechanism. The muscle exerts forces on the load that lead to changes in the length of the muscle and the position of the load. Length, velocity, and force feedback provide a continuous source of compensation for the various disturbances that may interfere with movement. Control signals sent by the CNS to γ_d-motor neurons or to the interneurons that transmit signals from tendon organs probably serve less to initiate movement than to modulate the gain of the regulatory reflex pathways. They may provide a mechanism for adjusting the stiffness and damping of the peripheral control system.

REFERENCES

1. Diete-Spiff, K., Carli, G., and Pompeiano, O.: Comparison of the effects of stimulation of the VIIIth cranial nerve, the vestibular nuclei or the reticular formation on the gastrocnemius muscle and its spindles, Arch. Ital. Biol. **105:** 243, 1967.
2. Eldred, E., Granit, R., and Merton, P. A.: Supraspinal control of muscle spindles and its significance, J. Physiol. **122:**496, 1953.
3. Granit, R.: Receptors and sensory perception, New Haven, 1955, Yale University Press.
4. Gray, J.: Animal locomotion, New York, 1968, W. W. Norton & Co.
5. Grillner, S.: Supraspinal and segmental control of static and dynamic γ-motorneurons in the cat, Acta Physiol. Scand. suppl. **327:**1, 1969.
6. Grillner, S., and Udo, M.: Motor unit activity and stiffness of the contracting muscle fibers in the tonic stretch reflex, Acta Physiol. Scand. **81:**422, 1971.
7. Henatsch, H. D., and Schafer, S. S.: Fusimotor-sensor and fusimotor-trigger functions: a re-interpretation of the dual control of mammalian muscle spindles, Brain Res. **6:**385, 1967.
8. Houk, J. C., Singer, J. J., and Goldman, M. R.: An evaluation of length and force feedback to soleus muscles of decerebrate cats, J. Neurophysiol. **33:**784, 1970.
9. Joyce, G. C., and Rack, P. H. M.: Isometric lengthening and shortening movements of cat soleus muscles, J. Physiol. **204:**475, 1969.
10. Joyce, G. C., Rack, P. H. M., and Westbury, D. R.: The mechanical properties of cat soleus muscles using controlled lengthening and shortening movements, J. Physiol. **204:**461, 1969.
11. Lippold, O. C. J.: Oscillation in the stretch reflex arc and the origin of the rhythmical, 8-12 c/s component of physiological tremor, J. Physiol. **206:**359, 1970.
12. Lundberg, A.: The supraspinal control of transmission in spinal reflex pathways, Electroencephalogr. Clin. Neurophysiol. Suppl. **25:** 35, 1967.
13. Matthews, P. B. C.: A study of certain factors influencing the stretch reflex of the decerebrate cat, J. Physiol. **147:**547, 1959.
14. Matthews, P. B. C.: Muscle spindles and their motor control, Physiol. Rev. **44:**219, 1964.
15. Matthews, P. B. C.: The origin and functional significance of the stretch reflex. In Andersen, P., and Jansen, J. K. S., editors: Excitatory synaptic mechanisms, Oslo, 1970, Universitetsforlaget.
16. Matthews, P. B. C., and Stein, R. B.: The sensitivity of muscle spindle afferents to small sinusoidal changes in length, J. Physiol. **200:** 723, 1969.
17. Merton, P. A.: Speculations on the servo-control of movement. In Malcolm, J. L., Gray, J. A. B., and Wolstenholme, G. E. W., editors: The spinal cord, Boston, 1953, Little, Brown & Co.

18. Partridge, L. D.: Signal handling characteristics of load-moving muscle, Am. J. Physiol. **210:**1178, 1966.
19. Phillips, C. G.: Motor apparatus of the baboon's hand, Proc. R. Soc. Lond. (Biol.) **173:**141, 1969.
20. Poppele, R. E., and Bowman, R. J.: Quantitative description of linear behavior of mammalian muscle spindles, J. Neurophysiol. **33:** 59, 1970.
21. Rack, P. M. H., and Westbury, D. R.: The effects of length and stimulus rate on tension in the isometric cat soleus muscle, J. Physiol. **204:**443. 1969.
22. Roberts, T. D. M.: Neurophysiology of postural mechanisms, London, 1967, Butterworth & Co., Ltd.
23. Severin, F. V., Orlovskii, G. N., and Shik, M. L.: Work of the muscle receptors during controlled locomotion, Biophysics **12:**575, 1967. (Translation.)
24. Stark, L.: Neurological control systems, New York, 1968, Plenum Press.
25. Takano, K., and Henatsch, H.: The effect of the rate of stretch upon the development of active reflex tension in hind limb muscles of the decerebrate cat, Exp. Brain Res. **12:**422, 1971.

26

ELWOOD HENNEMAN

Motor functions of the brainstem and basal ganglia

The majority of the motor tracts that descend into the spinal cord originate in the brainstem. The areas from which these tracts arise are integrative centers receiving signals from the cerebral cortex, the cerebellum, and the basal ganglia as well as from the spinal cord. In this chapter the organization of the motor areas of the brainstem will be discussed and the nature of their influence on spinal mechanisms will be described. An account of the capacities and abnormalities of animals in the decerebrate and decorticate states will also be given to indicate the contributions of the brainstem to standing, walking, and postural adjustments. The latter part of the chapter will be devoted to the basal ganglia, a group of nuclear masses that play a major role in motor activity and are closely related to the brainstem centers. In this section, recent findings on the role of some of the biogenic amines will be summarized to illustrate the importance of central transmitters in motor function.

Reticular formation of the brainstem

The brainstem contains a central core of nerve cells and fibers running more or less continuously through the mesencephalon, pons, and medulla, as indicated in Fig. 26-1. Its upper end extends into the diencephalon and its lower end extends into the spinal cord. In Weigert-stained sections it shows up as a meshwork or reticulum of interlacing fiber bundles. Lying between and among the fibers are cells of different types and sizes. In some areas the cells are gathered together more densely as nuclei that have names. This loose aggregation of cells and fibers is called the reticular formation (RF). It lies in the central part of the brainstem. Circumscribed cell groups such as the red nucleus, the superior olive, and the cranial nerve nuclei are not ordinarily included within it.

Although the RF is not to be regarded as a simple functional entity, physiologic studies indicate that it exerts powerful facilitatory and inhibitory influences on all types of motor

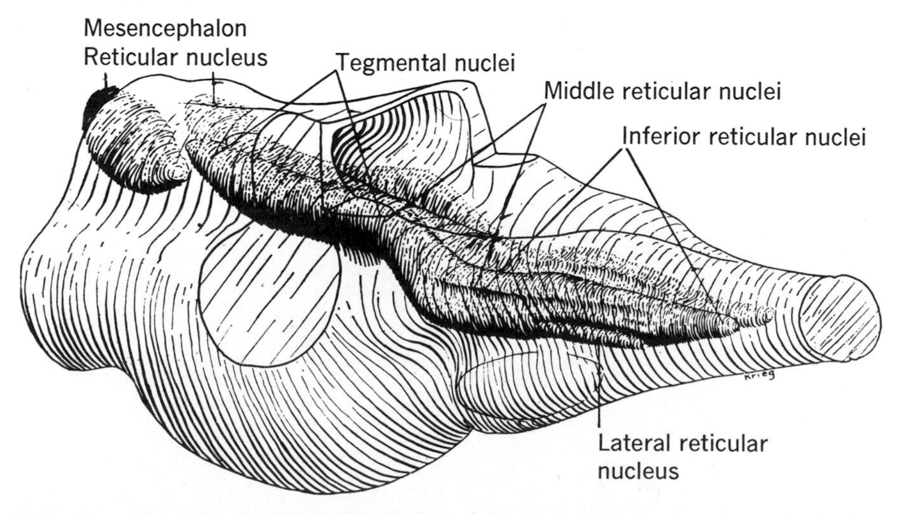

Fig. 26-1. Three-dimensional sketch illustrating position of reticular formation in brainstem. (From Krieg.[22a])

activities at all levels of the nervous system. The medial two-thirds contains many large cells and gives rise to long ascending and descending fibers. The effector functions of the RF are mediated chiefly by this medial two-thirds, whereas the lateral one-third evidently acts as an association area for the medial portion, influencing it through short, medially directed axons. Within the medial portion there is some segregation of cells into groups; some exert their main action on the spinal cord and others act chiefly on more rostral portions of the brain, as illustrated in Fig. 26-2.

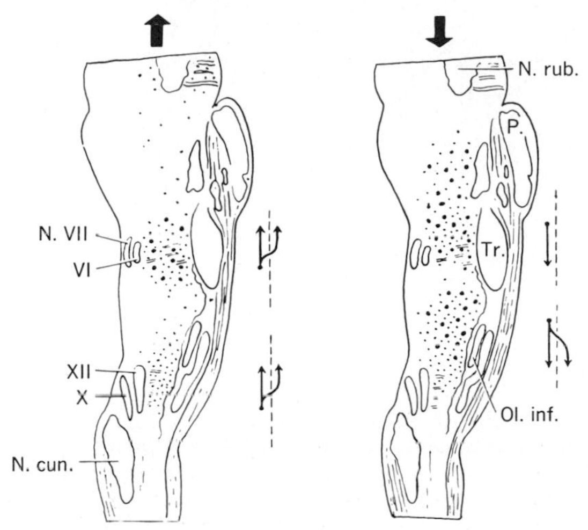

Fig. 26-2. Location of cells in reticular formation (RF) sending long axons to spinal cord (right); location of cells with axons ascending beyond mesencephalon (left). Drawn from parasagittal sections of cat brainstem. Large dots represent giant cells. Note that ascending and descending fibers take origin from almost the entire longitudinal extent of RF, but that they differ somewhat in their regions of maximal origin. Arrows at right of each drawing indicate that pontine reticulospinal fibers descend homolaterally, while all other fibers are crossed as well as uncrossed. (From Brodal.[8a])

As Scheibel and Scheibel[38] have shown with Golgi stains (Fig. 26-3), a considerable number of reticular cells give off an axon that has a long ascending as well as a long descending branch. The latter was often traced down into the spinal cord and the former up into the thalamus. Thus one cell, by means of collaterals at many levels, may influence a vast number of cells above and below it in the neuraxis. All morphologic evidence, in fact, suggests that there must be a significant correlation between the effects exerted by the RF in the rostral and caudal directions.

Golgi-stained sections of the brainstem in the parasagittal plane (Fig. 26-4) reveal that the dendrites of many reticular cells are oriented in a plane perpendicular to the long axis of the brainstem and that the collaterals of fibers that send impulses to them are flattened in the same plane (inset, Fig. 26-4).

These features suggest an arrangement for the collection and distribution of signals from diverse regions of the brain, i.e., an integrat-

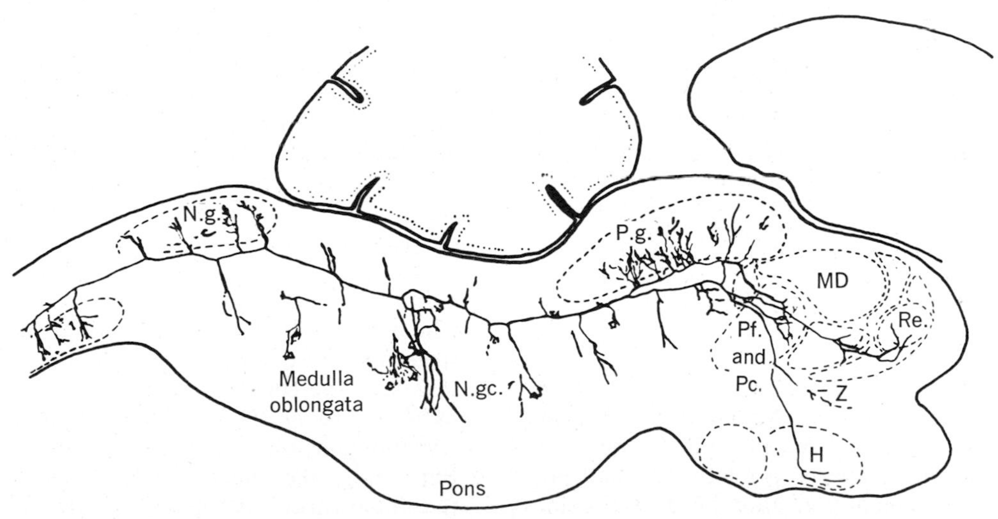

Fig. 26-3. Drawing of Golgi-stained sagittal section of brainstem of 2-day-old rat. Single large cell in reticular formation (RF) is shown. Its axon divides into ascending and descending branches. Latter gives off collaterals to adjacent parts of RF, to nucleus gracilis, *N.gc.*, and to ventral horn in spinal cord. Ascending branch gives off collaterals to RF; periaqueductal gray, *P.g.*, several thalamic nuclei; the hypothalamus, *H*, and zona incerta, *Z*. Other areas identified in drawing are nucleus reuniens, *Re.*, nucleus medialis dorsalis, *MD*, nucleus parafascicularis, *Pf.*, and nucleus paracentralis, *Pc.* (From Scheibel and Scheibel.[38])

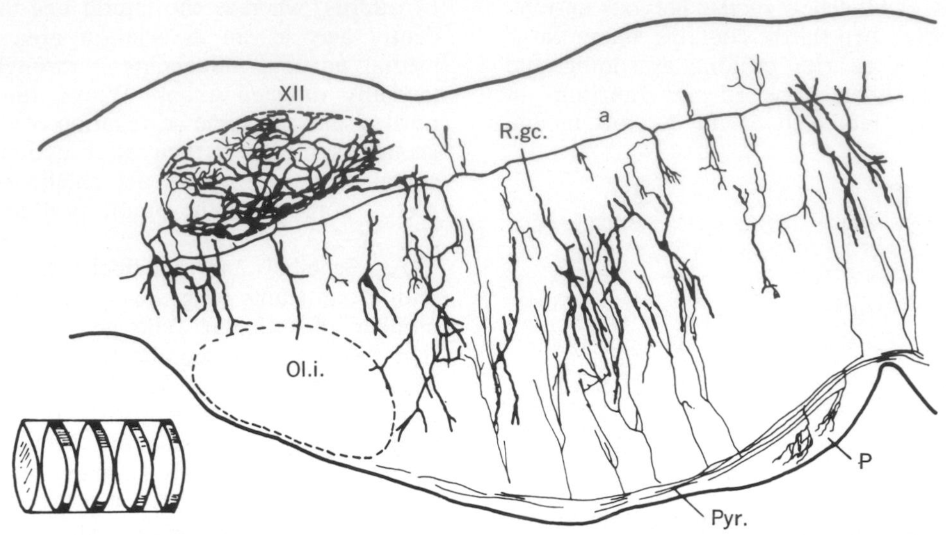

Fig. 26-4. Drawing of Golgi-stained section of brainstem of young rat showing prevailing orientation of dendrites of reticular cells in a plane perpendicular to long axis of brainstem. Collaterals of fibers that send impulses to reticular cells are flattened in same plane. *Pyr.,* Pyramidal tract; *XII,* hypoglossal nucleus; *Ol.i.,* inferior olive; *P,* pons; *R.gc.,* nucleus reticularis gigantocellularis; *a,* long axon of a reticular cell. Inset at left shows how reticular formation may be represented as series of neuropil segments. (From Scheibel and Scheibel.[38])

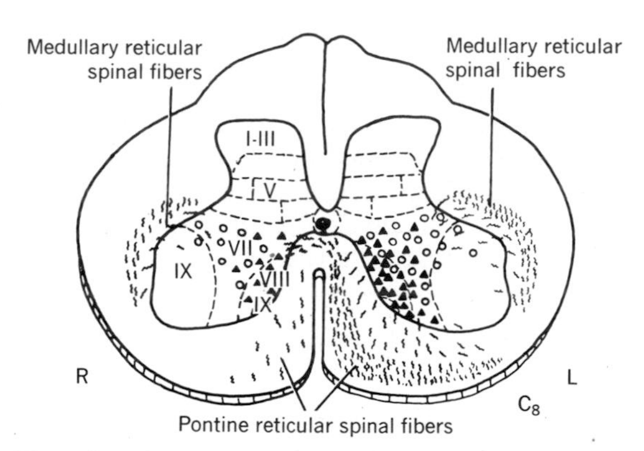

Fig 26-5. Diagram of transverse section of spinal cord of cat at C_8 showing position in cord of reticulospinal fibers from medullary and pontine reticular formation and sites of termination. ▲ = Sites of termination of pontine reticular spinal fibers. ○ = Sites of termination of medullary reticular spinal fibers. (From Nyberg-Hansen.[33a])

ing mechanism. Afferent connections come from many sources. Among the most important are those from the spinal cord. Contrary to early reports, a relatively small portion of the spinal activation of the RF comes from collaterals of secondary sensory fibers. There is a massive influx of direct spinoreticular fibers that ascend in the ventrolateral funiculus; the majority are distributed to the medial two thirds of the RF. A second major afferent influx comes from the cerebellum, particularly the fastigial nucleus. These fibers are distributed chiefly to the medullary portion of the RF and supply both rostrally and caudally projecting areas. In addition there are afferents from the *lateral hypothalamus* and the *pallidum* to the mesencephalic RF, from the *tectum* to the same regions, and from the sensorimotor areas of the cerebral cortex to several levels of the brainstem that give off reticulospinal fibers.

Reticulospinal tracts are derived from small and large cells at all levels of the medullary and pontine RF. Two maximal areas of origin can be distinguished, one from the pons, and the other from the medulla (Fig. 26-2). Most of the medullary fibers come from the *nucleus reticularis gigantocellularis* and are crossed as well as uncrossed. They run bilaterally in the lateral funiculus of the cord. The pontine fibers, which are uncrossed, come from the entire *nucleus reticularis pontis caudalis* and the caudal part of the *nucleus reticularis pontis oralis.* They descend in the ventral funiculus of the spinal cord. As shown in Fig. 26-5, the pontine reticulospinal fibers terminate more ventrally than the medullary fibers. The former end in laminae VII and VIII, whereas the latter end chiefly in lamina VII. The reticulospinal tracts may be regarded as the final common pathways for a variety of extrapyramidal centers at higher levels. The signals from these centers converge in the RF, and the resultant discharge is directed to the spinal cord. Later in this chapter some recent electrophysiologic findings on the synaptic actions of reticulospinal fibers will be described. Before taking up the functional organization of the RF and its spinal projections, however, it will be helpful to describe briefly the motor capacities of bulbospinal (low decerebrate), mesencephalic (high decerebrate), and decorticate animals. These preparations have provided invaluable clues regarding brainstem function and they are frequently used in the experimental analysis of motor systems.

LEVELS OF INTEGRATION IN THE BRAINSTEM*

Bulbospinal (low decerebrate) animal. If the brainstem is transected near the caudal extremity of the mesencephalon or at an upper pontile level so that the medulla oblongata remains connected with the entire spinal cord, the resulting preparation is capable of more complex behavior than is the spinal animal. In contrast to spinal transection, this truncation of the lower brainstem is not followed by any depression of simple spinal reflexes. Immediately after separation of the bulbar region from the suprabulbar structures one may elicit in rabbit, cat, dog, and monkey all those spinal reflexes that have been described in the preceding chapters. Indeed, there is enhancement, as regards excitability and execution, of all extensor reflexes, whereas reciprocally the flexor responses have a higher threshold than in the spinal state. Thus the bulbospinal animal vigorously exhibits the very reflexes that are most affected by spinal shock. In fact, without any external stimulation the extensor muscles remain in a state of steady contraction. Further examination shows that the entire system of musculature that posturally resists gravity in the standing position is in a state of increased tonic activity. This is spoken of as *decerebrate rigidity*.

In the acute bulbospinal preparation the abnormal distribution of tone in the muscles of the body continues regardless of the position of the animal. If placed upright it remains standing, provided it is balanced, but a normal relation to the force of gravity is not at all necessary for the continuance of the rigidity. The extension of the legs and the retraction of the head persist, even becoming stronger, when the preparation is placed on its back.

Further examination of the bulbospinal animal reveals the presence of several capacities for response that are either absent from the acutely spinal preparation or evocable only on strong stimulation. Reflexes can be evoked that activate the neck, trunk, and the four limbs as a whole, thus producing an attitude of the body that is definitely related to the place and quality of the stimulus. In the spinal animal, nociceptive stimulation of a hind foot leads to ipsilateral flexion with crossed extension. Such stimulation, if sufficiently strong, may also evoke extension of the ipsilateral foreleg and flexion of the contralateral foreleg. Such groups of responses are much more readily obtained in the bulbospinal animal. They form *reflex figures* that may be considered to be compensatory, inasmuch as in each the movement of certain parts is adapted to restore the balance disturbed by the movement of other parts, and the result is an orderly change in the position of the body as a whole, which is significant in meeting the exigency that has given rise to the reflex response. The reflex figure shown in Fig. 26-6, *D,* might be brought into play, for example, when a cat steps on some object that hurts its hind foot. The foot is lifted from the ground and the weight of the hindquarters is thus thrown on the contralateral hind leg, which extends to support this weight. At the same time the cat must prepare to move away so that the stimulus will not be encountered again, and for this act the extension of the crossed hind leg and of the ipsilateral forefoot, with their backward thrust, tend to throw the body forward and support it while the flexion of the stimulated hind leg and the crossed foreleg move these limbs forward preparatory to supporting the body at the next step. The reflex figure is thus seen to be an integral mechanism out of which is built the functional act of stepping away from a stimulus that endangers the hind foot.

In spite of an ability to support its weight and to prepare attitudinally for progression, the bulbospinal cat or dog cannot fully right itself, assume a sitting or standing position, run, walk, or jump. That these deficits are due to the fact that the neural mechanisms essential for these more complex acts are situated rostrally to the bulbar region and not to some kind of "shock" has been shown by experiments in which bulbospinal animals

*This section, which was prepared by Philip Bard as part of Chapter 76 in the twelfth edition of this book, is included here with only slight modifications.

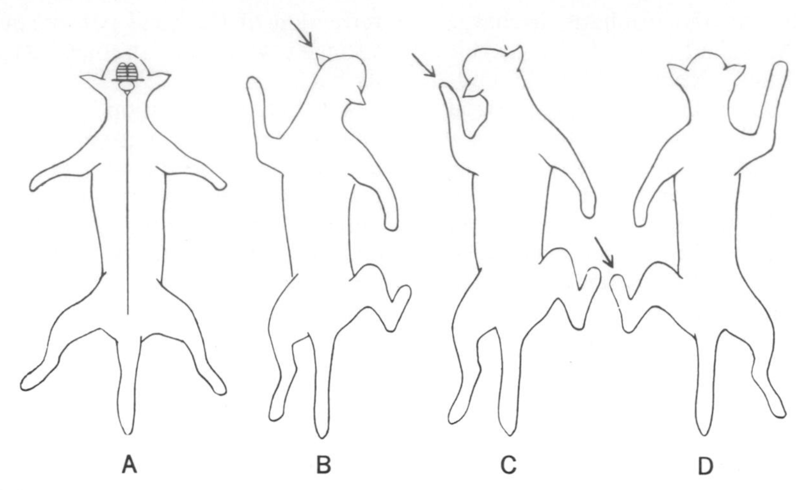

Fig. 26-6. Reflex figures. **A,** Position of cat in decerebrate rigidity. **B, C,** and **D,** respectively, are reflex figures resulting from stimulating left pinna, left forefoot, and left hind foot. (After Sherrington.[40])

have been kept alive for considerable lengths of time.

Bazett and Penfield[7] were the first to maintain decerebrate animals for more than the period of the usual "acute" experiment. Seven of their bulbospinal cats lived more than 1 week, and of these, two survived slightly longer than 2 weeks. In every case decerebrate rigidity persisted, although it was subject to some variations. No effective righting occurred and, of course, locomotion was not possible. Bard and Macht[5] studied a number of bulbospinal ("pontile") cats over somewhat longer survival periods (up to 40 days). None ever spontaneously righted head or body but, when placed on either side, lay with the under limbs rigidly extended and the upper ones either flaccid or slightly flexed. This posture or attitude is the first and most fundamental overt manifestation of the "body righting reaction acting on the body." On strong stimulation (e.g., of tail) the upper forelimb is flexed strongly at the elbow and retracted at the shoulder; the lower forelimb is extended and then quickly flexed so as to bring it under the chest at the same time the head is raised and rotated. Thus the under paw acts as a fulcrum while the upper limb is flexed with claws sunk in the underlying surface. In this way the shoulders and chest are pulled into an upright position. In chronic mesencephalic cats, but never in the bulbospinal preparations, the same series of acts occurs in the hind legs and leads to the assumption of a crouching or standing position. Further, in mesencephalic animals the righting occurs spontaneously.

Mesencephalic (high decerebrate) animal. When the brainstem is transected at a mesencephalic level, the preparation can scarcely be distinguished from the bulbospinal animal unless it is kept alive and in good condition for at least a week to 10 days. As indicated previously, Bard and Macht observed that in the cat, effective righting of the body and typical quadrupedal walking depend on mesencephalic mechanisms. Adequate but not wholly normal standing and locomotion are possible after a truncation that excludes all brain tissue lying rostral to the caudal extremities of the red nuclei. The capacity to right, stand, and walk develops slowly and is not fully gained until at least 10 days after decerebration. In the low mesencephalic animal these acts rarely, if ever, occur spontaneously. Spontaneous standing and walking are exhibited by cats with brainstems truncated just above the level of exit of the third pair of cranial nerves. Animals with transections passing just caudal to the mamillary bodies exhibit locomotor abilities of the same general quality as those exhibited by decorticate cats. They are even able to run and to climb. Such defects in standing and walking as they exhibit are due to the absence of placing reactions and to the high threshold of proprioceptive corrective reactions of the legs (only imperfect hopping and proprioceptive placing reactions are present).

After decerebration the hypertonia of the antigravity muscles (decerebrate rigidity) remains in evidence throughout the survival of the animal, but it is seen only when the chronic preparation is not engaged in some phasic activity or crouching. At any time when the animal is placed on its back (and fails to execute phasic movements of the legs), it is found that the legs are rigidly extended and resist passive flexion as strongly as in the case of the freshly decerebrated animal. High mesencephalic cats in the chronic state rarely lie on their sides, and if not standing or walking, they assume a crouching position on chest and belly with all limbs flexed. If undisturbed in this attitude, they may close the eyes and allow the head to droop; indeed, they appear to sleep, and on being subjected to a gentle tap or a slight sound they show all the overt signs of waking.

Certain postural responses of mesencephalic cats in the chronic state appear to be executed in a manner that is wholly or nearly normal. Tactile stimulation of the anal region results in the assumption of a defecatory posture; i.e., there are dorsiflexion of the tail, flexion and protraction of the hind limbs, extension and retraction of the forelimbs, and ventriflexion of the pelvis. This reflex posture is evocable in bulbospinal as well as in mesencephalic animals, but in the former it is achieved in the lateral position without righting. If the truncation is at or above the level of exit of the third nerves, the cat invariably assumes the normal feline position before defecating. After administration of estrogenic material, high mesencephalic female cats will spontaneously assume the posture characteristic of feline estrus (crouching on chest and forearms with pelvis and tail elevated), and this attitude together with treading movements of the hind legs, typical of estrual behavior, can be induced by genital stimulation. But other more elaborate items of estrual behavior cannot be induced in these animals.

Decorticate animal. When the ablation of cerebral tissue is limited to the most recently acquired part of the cerebrum, its mantle or cortex, the deficiencies are less than those produced by mesencephalic truncation of the brain. In both cases, modes of behavior dependent on past experience are lost; all that has been learned disappears and new types of responses can be induced only by special methods of training. Conditioned reflexes can be established in decorticate cats and dogs, but the process requires special methods. There is no evidence available at present to indicate that wholly decorticate carnivores ever develop learned responses under laboratory conditions in which the normal animal readily acquires them. This combination of a loss of what are ordinarily regarded as signs of higher nervous functions with retention of many complex activities was first established experimentally in 1892 by Goltz, who succeeded in keeping a dog from which he had removed all cerebral cortex alive for many months. Goltz's achievement has since been repeated, and descriptions of a number of decorticate cats and dogs are available (for references see Bard and Rioch[6]). Because it is able to regulate its body temperature in an almost normal fashion, the decorticate mammal is much easier to maintain over long postoperative periods than is the decerebrate preparation.

Following complete decortication a cat or dog can at once right itself, stand, and walk. In view of the fact that the high mesencephalic animal does not engage in these acts until a postoperative period of from 7 to 30 days has elapsed, it seems likely that subcortical influences originating rostral to the midbrain facilitate the mesencephalic and bulbar mechanisms essential for these acts. The chief postural defects of decorticate animals are due to the loss of placing reactions and to defective hopping reactions, responses that are described later in this chapter. When animals without cortex are held suspended, they exhibit, so long as they are quiet, a marked extensor rigidity, and there can be no doubt that this *decorticate rigidity* is closely allied to *decerebrate rigidity*. Indeed, as we shall see, decerebrate rigidity is in part due to the exclusion of influences of cerebral cortical origin. The contraction of antigravity muscles observed in the acutely decerebrate animal is maintained only because the preparation is incapable of the phasic activities that the decorticate animal shows from the beginning and that emerge in the chronically decerebrate animal after some time.

DECEREBRATE RIGIDITY AND SPASTICITY*

The contraction of the antigravity muscles in decerebrate rigidity is reflex in origin, for it can be abolished by section of appropriate nerves or elimination of certain reflex stimuli. In the case of the hind legs the extensor hypertonia may be reduced by section of the dorsal roots that carry the afferent impulses from the extensor muscles themselves. It may also be abolished by eliminating any pull on

*This section, which was prepared by Philip Bard as part of Chapter 76 in the twelfth edition of this book, is included here with only slight modifications.

them. Section of nerves from other muscles in the leg or from the skin does not affect the rigidity at all.[40] These facts indicate that the myotatic reflexes of the extensors constitute an important factor in the production of their rigidity. It is not, however, the only factor, for it has been found that some time after their deafferentation the hind legs of a cat develop typical extensor hypertonia when the animal is decerebrated.[34] Under such circumstances the rigidity has its reflex origin in the neck and labyrinthine proprioceptors (see tonic neck and labyrinthine reflexes, p. 693). The rigidity of the forelegs is not reduced by deafferentation, and it is clear the proprioceptors of ear and neck constitute its essential source.[34] In any case it is definite that after decerebration the myotatic reflexes in extensor muscles have extremely low thresholds.

Spasticity is a term that has had its widest use in the language and literature of clinical neurology. It is employed to describe a combination of symptoms—*muscle hypertonia* (increased resistance to lengthening), *hyperactive tendon reflexes,* and *clonus* (repetitive contractions in response to a suddenly applied but sustained stretch of a muscle). Spasticity is well and classically exemplified by the state of the arm and leg of a patient who has become hemiplegic as the result of severe injury or destruction of a certain portion of the frontal lobe of the cerebral hemisphere opposite the affected extremities or, more commonly, as the result of interruption of the extrapyramidal motor projection of that lobe at the level of the internal capsule. The leg is stiffly extended and resists passive flexion; the arm is held flexed and resists passive extension. Tendon reflexes are hyperactive and clonus is evoked if the stretch is prolonged. Again it can be said that the hypertonia is chiefly encountered in antigravity muscles, for in man the flexors of the arm are the muscles that resist the force of gravity. This state of hypertonia in the arm flexors and leg extensors can be regarded as *decorticate rigidity* in man, for it is induced by a lesion that interrupts descending impulses of cortical origin, but it is to be distinguished from the decorticate rigidity of carnivores, in which the extensor hypertonia is seen in the upper as well as in the lower limbs. Actual decerebration in man, produced, for example, by a massive hemorrhage at a mesencephalic level, is followed by extreme extensor rigidity of all four extremities.

Central nervous mechanisms involved in production of spasticity and decerebrate rigidity

In his original description of decerebrate rigidity, Sherrington stated that it followed transection of the brainstem through the caudal part of the diencephalon or at any mesencephalic level. It is common practice to produce it by sectioning the brainstem at an intercollicular level. The typical rigidity then appears on removal of the anesthetic. It survives successive transections until these reach the level of entry of the eighth pair of cranial nerves. Cutting through the bulb in this region or at any level below it produces the postural and reflex condition characteristic of the spinal state. Sherrington clearly recognized that decerebrate rigidity is due to the interruption of descending pathways that originate from parts of the brain situated rostral to the mesencephalon. The previously noted fact that the rigidity persists for periods of time more than sufficient to assure degeneration of all descending fibers cut by the transection demonstrates conclusively that the hypertonia is not due to any "irritation" set up by the mechanical insult of truncating the brainstem.

Until quite recently it was generally accepted that decerebrate rigidity and the spasticity characteristic of hemiplegia are release phenomena in the jacksonian sense (p. 689), i.e., are conditions that result simply from removal of inhibitory influences of suprabulbar origin. Already sufficient evidence has been given to show that the greater portion of the inhibitory influence has its origin above the mesencephalon. Experimental analyses, particularly those carried out by Magoun and collaborators (see especially Magoun,[28] Magoun and Rhines,[30] and Lindsley[23]) have demonstrated that *decerebrate rigidity and spasticity are due not only to the removal of inhibitory influences acting on the spinal stretch reflex but also to the maintained activity of supraspinal influences that facilitate these reflexes.* These important discoveries have implications that transcend the subject under immediate discussion, for the myotatic reflex is by no means the only spinal response that is subject to these opposing influences.

Vestibulospinal facilitatory influence. It has long been recognized that cutting through the medulla oblongata just below the level of the vestibular nuclei destroys the rigidity of the decerebrate cat and changes the status of the preparation to that of the spinal animal. This and other evidence suggested that in the carnivore, decerebrate rigidity depends on the integrity of one or more of the vestibular nuclei, more especially Deiter's nucleus, and that it is due to impulses descending by way of the uncrossed vestibulospinal tracts. This suggestion received strong support when Fulton et al.[19] showed that if a localized lesion is made in the region of Deiter's nucleus so that degeneration of the corresponding vestibulospinal tract occurs, subsequent decerebration fails to produce rigidity on the side of the injury and degeneration, although hyperextension and increased tendon reflexes appear in the legs of the side with intact vestibulospinal tracts.

The role of the vestibulospinal system has been further elucidated by Bach and Magoun.[1] They confirmed the observation that vestibular nuclear lesions abolish rigidity on the injured side and that this effect is chiefly if not wholly due to destruction of Deiter's nucleus. Significantly they found that the result does not depend on any inadvertent injury of the adjacent caudal portion of the brainstem RF (Fig. 26-7) from which facilitation of spinal reflexes (including the myotatic reflexes) can so readily be obtained.

As indicated previously, an extensor hypertonia, comparable to that of decerebrate rigidity together with overactive myotatic reflexes and clonus (in short, spasticity), is produced by decortication. Ablation of the pericruciate cortex in cat or dog suffices to induce this condition in the contralateral limbs.[3, 51] Destruction of the vestibular nuclei has much less effect on this condition than on decerebrate rigidity; the extensor rigidity is reduced, but the hyperactive tendon reflexes and clonus are essentially unaffected.[1] If in a carnivore pericruciate decortication is combined with removal of the anterior lobe of the cerebellum (Fig. 26-7), there ensues an antigravity posture even more pronounced than that which follows decerebration (cerebellum intact), and stretch reflexes show extremely low thresholds[44]; the animal is rendered immobile by the continued, excessive, and unbalanced activity of the central antigravity mechanism. In such preparations, Schreiner et al.[39] found that lesions limited to Deiter's nucleus were without appreciable effect; but when such lesions were superimposed on a transection of the tegmentum, spasticity was virtually abolished, the stretch reflexes being more depressed than after tegmental injury alone. As indicated before and as will be explained more fully later, there is in addition to the vestibulospinal system another lower brainstem facilitatory system, the reticulospinal. In the low decerebrate animal, destruction of Deiter's nucleus abolishes decerebrate rigidity because, as originally suggested by Bach and Magoun,[1] the truncation of the brainstem has already largely removed the source of the reticulospinal facilitation. On the other hand, the spasticity of the decorticate animal or that of the animal with pericruciate areas and anterior cerebellar lobe

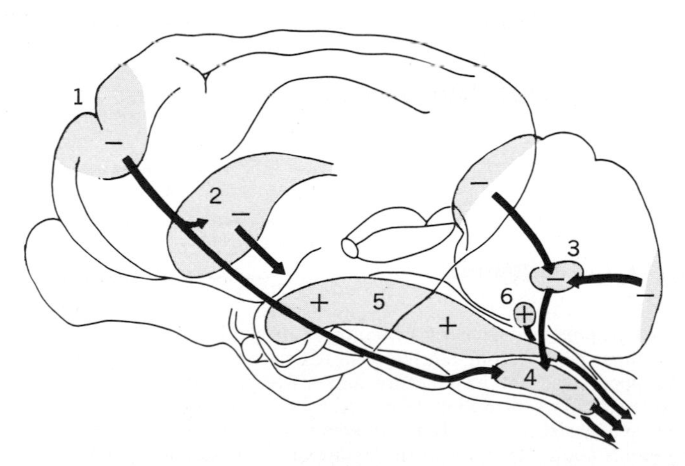

Fig. 26-7. Reconstruction of cat brain showing suppressor or inhibitory and facilitatory systems concerned in spasticity. Suppressor pathways are, *1,* corticobulboreticular, originating in pericruciate cortex; *2,* caudatospinal (probably caudatoreticulospinal); *3,* cerebelloreticular (originating in anterior lobe and paramedian lobules and relaying in nucleus fastigii); *4,* reticulospinal. Facilitatory pathways are, *5,* reticulospinal and, *6,* vestibulospinal. (From Lindsley et al.[24])

removed is but little affected by unilateral or bilateral destruction of Deiter's nucleus, for in these cases the extensive brainstem facilitatory mechanism is still wholly intact and discharges to the cord over reticulospinal paths.

Reticulospinal facilitatory influence. In 1946 Rhines and Magoun[36] reported the very significant finding that, in cat and monkey, stimulation anywhere within a large brainstem area (approximately the same as that marked *5* in Fig. 26-7) augments the response to stimulation of the cerebral motor cortex and also markedly facilitates the knee jerk. It was made clear that these effects are produced by impulses that descend to spinal neurons from the point of stimulation and that their final common spinal path is made up of reticulospinal fibers which course chiefly in the middle portion of the lateral funiculus. The facilitatory area comprises elements of the dorsal diencephalon, hypothalamus, subthalamus, central gray matter and tegmentum of the mesencephalon, pontile tegmentum, and a large part of the bulbar RF.[23, 28, 30] Evidence has been obtained that the facilitatory effects can be evoked by elements at all levels of the area and that they are exerted bilaterally, the crossing taking place in the brainstem as well as in the cord.[33]

The disclosure by Rhines and Magoun that facilitation of a myotatic reflex (the knee jerk) can be elicited by stimulation of the brainstem reticulospinal facilitatory system indicated that this system may be of importance in the production of decerebrate rigidity and spasticity. As already related, Bach and Magoun[1] have shown that after destruction or interruption of the vestibulospinal system, cortically or reflexly induced movements are still markedly facilitated by stimulation of basal diencephalon, brainstem tegmentum, or RF. They also found that stimulation within the rostral portion of this extensive facilitatory area is rendered ineffective by destruction of its caudal extremity. Subsequently, Sprague et al.[45] demonstrated in decerebrate

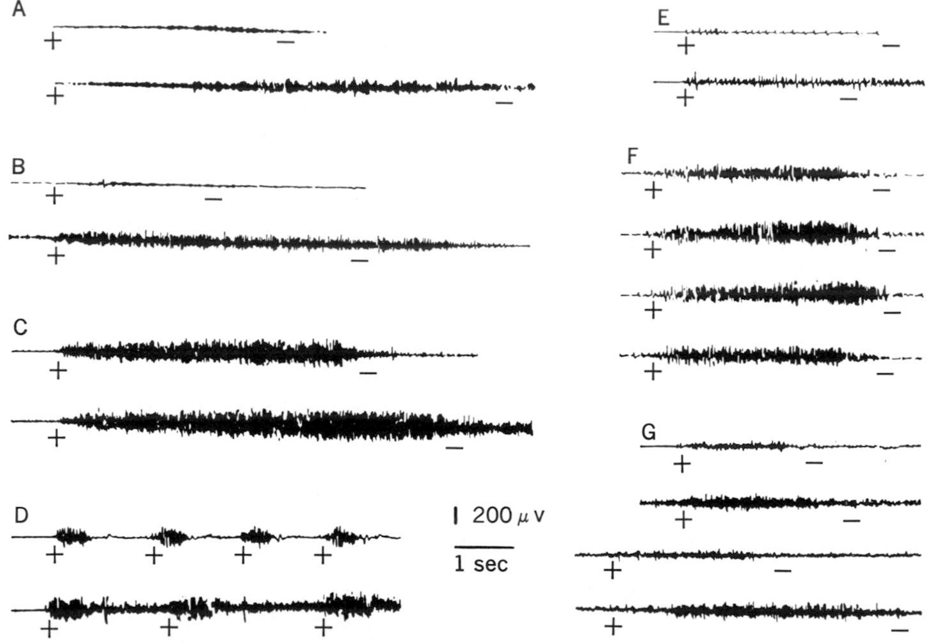

Fig. 26-8. Electromyographic evidence of reticulospinal facilitation of stretch reflexes in decerebrate cat. In such a preparation, stretch reflexes are already exaggerated. In each pair of records, **A** to **E,** upper was obtained before and lower was obtained during stimulation of facilitatory brainstem reticular formation (RF). **A** to **C,** Records of responses of quadriceps **(A),** hamstrings **(B),** and gastrocnemius **(C)** evoked by stretch (indicated between + and −). **D,** Record of gastrocnemius contractions in positive supporting reaction, upward pressure on footpad being exerted at each +. **E,** Record of gastrocnemius responses in crossed extensor reflex, with contralateral toes pinched between + and −. **F** (from above down), Gastrocnemius stretch reflex before, during, 50 sec after, and 100 sec after stimulation of facilitatory RF. **G** (from above down), Contralateral quadriceps stretch reflex before and during stimulation and ipsilateral quadriceps stretch reflex before and during stimulation of the facilitatory RF on one side of brainstem. (From Sprague et al.[45])

cats that stimulation of the portion of the facilitatory RF that remained in the truncated brainstem increased the already exaggerated stretch reflexes both ipsilaterally and contralaterally. This effect is well illustrated in Fig. 26-8, which shows that the experimental excitation of this reticulospinal mechanism augments and prolongs the stretch reflexes of various muscles. Also, what may be termed the intrinsic maintenance of the facilitatory process is prolonged (Fig. 26-8, *F*). The facilitation of the stretch reflex of a flexor muscle shown in Fig. 26-8, *B*, is greater than was usually found in flexors. The general application of this reticulospinal effect is indicated by the facilitation of the crossed extension reflex (Fig. 26-8, *E*). In this context it will be recalled that movements induced by stimulation of the motor cortex or of the pyramidal tract after decortication are similarly facilitated, the effect being produced at a spinal level.[36]

The influence of this brainstem facilitatory system on extensor tone has been especially well demonstrated in experiments on cats rendered chronically spastic by ablation of the pericruciate region of the cortex. The effect has been described by Magoun and Rhines as follows:

> The animal's symptoms of spasticity were markedly reduced by the anesthesia, under which such experiments are performed, and its knee jerk, evoked recurrently, resembled a normal animal's under the same conditions. During stimulation of the brain stem facilitatory mechanism, however, this reflex not only gained in force and amplitude, as occurs in a normal animal, but a background of hypertonus appeared in the quadriceps and increased with each succeeding reflex, and a repetitive component or clonus developed and became so perseverative and extreme that the last reflex could not be detected above it.*

Here the spastic state, which had been eliminated by anesthesia, was not only restored but actually augmented by direct stimulation of the brainstem facilitatory system. Because inhibitory influences normally arising from the pericruciate area had been removed, the effect of the stimulation was greater than can be obtained under the same conditions in the normal animal. The central inhibitory influences that oppose the spastic or decerebrate state are several; these important mechanisms are considered below.

*Magoun, H. W., and Rhines, R.: Spasticity: the stretch reflex and extrapyramidal systems, Springfield, Ill., 1947, Charles C Thomas, Publisher.

The experimental evidence just presented makes it very clear indeed that decerebrate rigidity and spasticity depend on the facilitation of spinal stretch reflexes by impulses that descend from a brainstem facilitatory mechanism. The fact that this continues to discharge powerfully after removal not only of forebrain but also of cerebellum raises the question of how this activity is maintained.

Cerebral, cortical, and cerebellar facilitatory influences. There is abundant evidence (Chapter 28) that facilitation arises from the cerebellum and acts via the reticulospinal facilitatory mechanism. This is doubtless more prominent in the primates than in the carnivores, but in all instances it accompanies an inhibitory cerebellar influence. Another facilitatory influence is transmitted to the spinal cord from the cerebral cortex by way of the pyramidal tracts: the discharge occurs without any specific or special stimulation (Adrian and Moruzzi), is subliminal in the sense that it does not produce any overt movements, and therefore probably plays a role in determining the activity of spinal neurons. Since there is evidence that pyramidal section alone, in the absence of any other central lesion, leads to some degree of hypotonia,[48] it is reasonable to assume that these subliminal pyramidal impulses, detectable only by electrical recording, act to facilitate the stretch reflexes. Obviously they could not contribute to decerebrate rigidity, but it is possible that they may be involved in certain cases of spasticity.

Inhibitory influences of bulbar RF and other supraspinal central mechanisms. As already pointed out, the removal of one or another cerebral or cerebellar area leads to a lasting spastic or rigid state. This can only mean that normally these areas act to hold in check or inhibit central mechanisms essential for the development of spasticity or decerebrate rigidity. The several origins of such inhibitory action and the pathways over which it is exerted are described in some detail in Chapters 28 and 29. Here it is only necessary to keep in mind that this influence originates in a number of places, the most important (Fig. 26-7) being (1) certainly one and probably several of the cerebral motor areas, (2) the cortex of the anterior lobe and of the paramedian lobules of the cerebellum, and (3) the striatum (caudate nucleus and putamen). The cerebral influences are all mediated by way of extrapyramidal pathways;

again it must be emphasized that the pyramidal tracts do not exert any inhibitory action on spinal stretch reflexes and that the spasticity that accompanies paralysis of voluntary movement in the primate is due to an extrapyramidal lesion.[46-48]

A most significant development in our knowledge of the central mechanisms involved in the control of muscle tone was the discovery by Magoun (see Magoun and Rhines[29]) of a powerful inhibitory mechanism in the bulbar RF. The locus of this mechanism is indicated in Fig. 26-7, where it is designated by the number *4;* it occupies the ventromedial part of the bulbar RF and appears to extend as far forward as the level of the trapezoid body. The threshold of its neural elements to direct electrical stimulation is very low, and the inhibitory effects produced are evident against almost any background of motor activity. Thus in both cat and monkey, localized weak stimulation completely abolishes cortically induced movement and all spinal reflexes (multineuronal flexor reflexes as well as monosynaptic stretch reflexes). In the decerebrate animal such stimulation causes complete loss of tone in the rigidly extended legs and renders the stretch reflexes of extensor muscles inelicitable (Fig. 26-9). These effects are bilateral, but the ipsilateral reflexes are affected to a greater extent than the contralateral. Because of its proximity to the vital respiratory and circulatory centers of the bulb, it is difficult if not impossible to determine the effects of destruction of this inhibitory area or region. It should also be understood that lesions placed here are likely to destroy neurons or fibers of the facilitatory system (Fig. 26-7, *5*). The marked inhibitory effects are exerted at spinal levels and are mediated by reticulospinal pathways coursing in the anterolateral white matter of the cord.[33]

It appears that most if not all of the suppressor or inhibitory areas of the cerebrum and the cerebellum act through the bulbar inhibitory mechanism. This is indicated in Fig. 26-7. There is also evidence that the inhibitory influence of the cerebellar cortex (anterior lobe and paramedian lobules) activates this system via a projection from the fastigial nucleus (Chapter 28). It may be supposed that inhibition of spinal activity of striatal origin is also mediated through this reticulospinal channel. The activity of the bulbar inhibitory mechanism depends on these and doubtless other inflows from higher parts of the brain. When these are eliminated, e.g., by high decerebration and decerebellation, bulbospinal inhibitory discharge ceases and can only be evoked by local electrical stimulation of the reticular inhibitory area. Thus

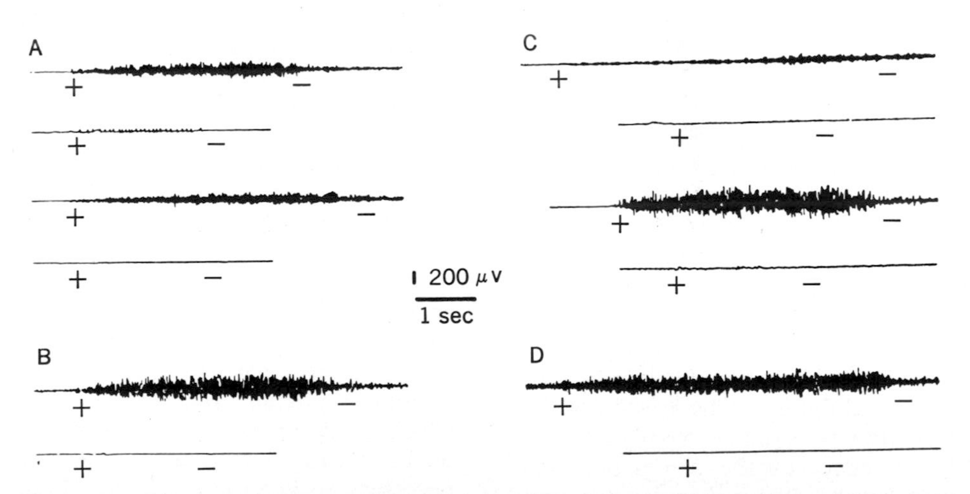

Fig. 26-9. Electromyographic evidence for inhibition of stretch reflexes in decerebrate cat by electrical stimulation of bulbar reticulospinal system. Top tracing in each pair of records was taken without stimulation; bottom tracing during stimulation. Stretch applied between + and −. **A** (from above down), Ipsilateral quadriceps stretch reflex before and during stimulation and contralateral quadriceps stretch reflex before and during stimulation of reticular formation (RF) on one side. **B,** Gastrocnemius stretch reflex before and during stimulation at level of trapezoid body. **C** (from above down), Quadriceps stretch reflex before and during stimulation and gastrocnemius stretch reflex before and during stimulation. **D,** Gastrocnemius stretch reflex before and during stimulation of medial RF overlying inferior olive. (From Sprague et al.[45])

this bulbospinal mechanism for inhibition seems incapable of intrinsic activity.

"Release of function" and "influx" as dual causes of decerebrate rigidity and spasticity. The fact that injury or removal of certain parts of the brain that are known to exert inhibitory effects on stretch reflexes produces spasticity or decerebrate rigidity can properly be interpreted in terms of Hughlings Jackson's concept of "release of function." Thus we may conclude that normally these "higher" portions of the nervous system hold in check the "lower" central mechanisms essential for the development of spasticity or decerebrate rigidity. In view of the evidence just presented it can be said that normally a number of cerebral and cerebellar mechanisms activate the bulbar reticulospinal inhibitory mechanism and thus prevent the spinal stretch reflexes from becoming so prominent that they impose an abnormal posture and limitation of movement on the organism. But this is only part, not more than half, of the story. To finish it one must invoke, as Magoun and colleagues[23-30, 39] have so clearly indicated, another jacksonian concept—that of "influx." We have seen that facilitatory influences are essential for the production of decerebrate rigidity and spasticity. Here again the final common path is reticulospinal (corticospinal facilitation over the pyramidal tracts being of relatively minor significance). Without this "influx" the spinal neurons involved in the stretch reflex would be deprived of a most essential source of facilitation. Thus decerebrate rigidity and spasticity must be regarded as the results of an imbalance between the activities of two systems that exert opposite effects upon spinal motor neurons.

NATURE OF CONTROL EXERTED BY RETICULOSPINAL TRACTS

Although the studies described thus far reveal the existence of powerful facilitatory and inhibitory systems in the brainstem, they do not provide many clues regarding the basic principles of motor control embodied in these systems. A careful analysis of how and where reticulospinal influences are exerted was required to get at these principles. Lundberg and colleagues have carried out a series of studies on the reticulospinal tracts that reveal the extent of the control they exercise over spinal circuits. As illustrated in Fig. 26-10, electrical stimulation of the brainstem inhibitory center as described by Magoun and Rhines elicits large, postsynaptic inhibitory potentials (IPSPs) in both flexor and extensor motoneurons.[21] These IPSPs are produced by stimuli that also cause a collapse of rigidity in decerebrate preparations. They are evoked by activity in fibers that descend in the ventral quadrant of the spinal cord, i.e., in the portion of the spinal cord occupied by the ventral reticulospinal tract described by anatomists (Fig. 26-5). The synaptic delays in this circuit suggest that the reticulospinal

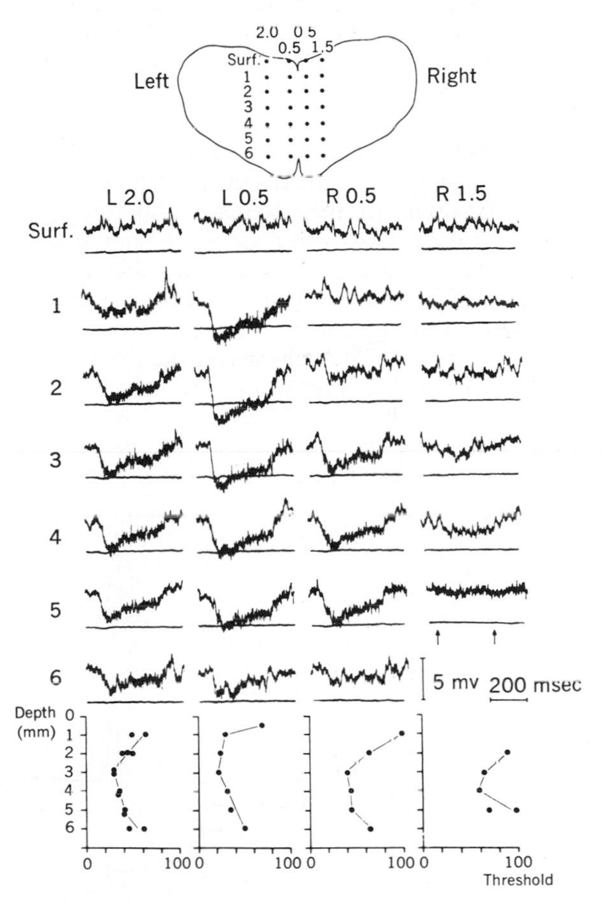

Fig. 26-10. Inhibitory postsynaptic potentials elicited in a posterior biceps or semitendinosus motoneuron by stimulation of brainstem at a series of sites 3 mm rostral to obex. Sites are shown in drawing at top. Distances (in mm) from midline and floor of fourth ventricle are indicated above and to left. Potentials recorded in response to these stimuli are reproduced in central portion of figure; in each pair of records upper traces are intracellular records and lower traces are recorded from dorsal root entry zone. Stimulus intensity was constant throughout, except when measuring the thresholds for IPSPs at different depths. Thresholds for IPSPs are plotted in diagrams below potential records in arbitrary units on abscissa. Arrows mark onset and end of stimulation. (From Jankowska et al.[21])

fibers in question do not end directly on motoneurons but on interneurons that send their axons to the motoneurons. Since the IPSPs are consistently of the same magnitude in flexors and extensors and are evoked by the same strength of stimulation at the same brainstem sites, there can be little doubt about the existence of a reticular center that produces *generalized* inhibition of motoneurons as originally described by Magoun and Rhines. After the original discovery of this inhibitory area, other investigators reported that stimulation elicited mixed excitatory and inhibitory effects on motoneurons instead of pure inhibition. Although mixed effects may be elicited from some areas, Lundberg believes that at least one portion of this brainstem region can evoke pure inhibition of motoneurons when stimulated electrically.

In addition to this indirect inhibition of motoneurons, the ventral reticulospinal tract also inhibits interneuronal transmission from certain sensory fibers to motoneurons. Transmission from cutaneous and high-threshold muscle fibers is definitely suppressed, but it is not clear whether transmission from Ia and Ib afferents is influenced. Many of the descending pathways that influence interneuronal transmission in reflex paths have parallel effects on transmission to motoneurons and to primary afferent terminals.[11] This ventral reticulospinal tract is no exception, for it apparently is responsible for inhibition of transmission from Ia and Ib afferents and from flexor reflex afferents to primary afferent terminals.

Another group of inhibitory effects is mediated by a *dorsal* reticulospinal system originating in approximately the same part of the brainstem.[17] After transection of the ventral quadrant of the spinal cord to eliminate effects mediated by the ventral reticulospinal tract, stimulation of the brainstem inhibitory center no longer elicits IPSPs in motoneurons. It is still possible, however, to suppress the excitatory and inhibitory synaptic actions of flexor reflex afferents and Ib afferents on motoneurons. Fig. 26-11 illustrates this effect very clearly. The monsynaptic reflex of gastrocnemius-soleus motoneurons recorded in *A* is largely inhibited by a preceding volley in high-threshold afferents from the anterior biceps–semimembranosus group. A train of five stimuli in the brainstem (*C* and *D*) effectively removes most of this inhibition *(D)* without affecting the test response *(C)* directly. The specificity of this inhibition is shown by the absence of any effect on postsynaptic po-

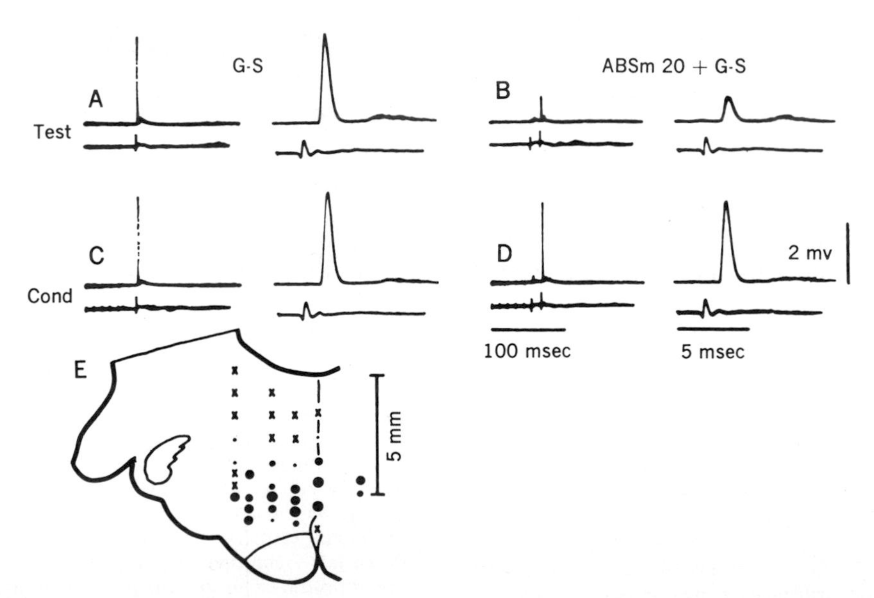

Fig. 26-11. Reticulospinal inhibition of transmission in reflex pathway. **A,** Monosynaptic reflex elicited by stimulating nerve to gastrocnemius and soleus muscles and recorded at two sweep speeds on ventral root (upper trace). **B,** Same reflex inhibited by preceding volley in high-threshold afferents in nerve to anterior biceps and semimembranosus muscles. **C** and **D,** Same as **A** and **B,** respectively, but conditioned by train of five stimuli in brainstem, which removes most of the reflex inhibition. **E,** Diagram of transverse section of brainstem 6 mm above obex. Filled circles of different sizes indicate points of stimulation and magnitude of reticulospinal inhibition. X = points stimulated without effect on reflex transmission. Lower traces in each set of records are recorded from dorsal root entry zone in L_7. (From Engberg et al.[17])

tentials from Ia afferents (compare *A* and *C*) or on IPSPs from Renshaw cells (not illustrated). The inhibitory effect is exerted on interneurons; there is no increase in conductance in the motoneurons themselves. The same dorsal reticulospinal system also inhibits transmission from flexor reflex afferents (but not from Ia and Ib afferents) to primary afferent terminals and to ascending spinal pathways. The inhibitory actions observed in these experiments were transmitted by fibers in the dorsal part of the lateral funiculus of the spinal cord. They cannot be definitely identified with any established descending tract. It is possible that there is no continuous fiber tract in this region and that the effects are relayed in propriospinal fibers; hence this pathway is referred to as the dorsal reticulospinal *system*.

An analysis of the mechanism by which interneuronal transmission is inhibited by this reticulospinal system was carried out by recording intra- and extracellularly from interneurons in the dorsal horn and intermediary regions of the spinal cord.[18] It was found that reticular stimulation evoked IPSPs in a few of these interneurons but not in the majority of them. There was, however, an effective depression of the EPSPs and IPSPs evoked in interneurons by flexor reflex afferents. It is probable that the dorsal reticulospinal system directly inhibits only a few cells early in the chain of interneurons and that later cells in the chain are thus silenced indirectly without having any IPSPs evoked in them. The inhibition may, in fact, be exerted on the first-order interneurons that receive monosynaptic effects from primary afferents. This must be so in the Ib pathways to motoneurons, which are largely disynaptic. Four of the five interneurons (out of 78 investigated) showing IPSPs with brainstem stimulation also had monosynaptic EPSPs from cutaneous afferents. The profound inhibition of reflex transmission exerted by the reticulospinal system may thus be due to selective inhibition of a few first- order interneurons that activate many second-order interneurons. These experiments reinforce the conclusion that the reticulospinal system is primarily inhibitory. Transmission in excitatory and inhibitory paths from primary afferents to interneurons was inhibited to an equal extent, and no trace of an EPSP was evoked in any of the 78 intracellularly recorded interneurons.

A number of diverse inhibitory effects from the RF on the spinal cord have been described. In addition to those already mentioned, it has been demonstrated that reticular stimulation may produce large primary afferent depolarization that results in presynaptic inhibition of transmission from some afferent systems. These different types of inhibition do not necessarily operate in parallel. They must, in fact, be differentially active some of the time. In the decerebrate state the ventral reticulospinal pathways are probably not tonically active, whereas the dorsal reticulospinal system is tonically active, being partly responsible for decerebrate inhibition of reflex transmission.

These brainstem inhibitory centers and their spinal effects have been described in considerable detail to give specific examples of the variety and subtlety of supraspinal control. This control is exerted on sensory input, on motor output, and on the intervening processes in internuncial cells. The initial effect of these descending fibers is apparently directed exclusively to interneurons.

Very little can be said about the pathways that are responsible for reticulospinal facilitatory effects. Their spinal terminations have not yet been established anatomically. It may be assumed that facilitatory actions are transmitted over more than one route and influence several different spinal mechanisms, as in the case of reticulospinal inhibition, but experimental evidence concerning the nature of these effects is lacking.

Other descending systems

The preceding description of the reticulospinal tracts indicates that supraspinal centers exert their influences on the spinal cord in several different ways. Similar studies have been carried out on the rubrospinal and vestibulospinal systems. They will not be described because, as in the case of the reticulospinal tracts, the functional significance of the observations is not yet apparent. It is clear that descending systems control or regulate all the intrinsic motor mechanisms of the spinal cord. Sensory input is regulated by means of presynaptic endings on primary afferent terminals. Transmission between sensory afferents and motor neurons is controlled by excitatory and inhibitory endings on internuncial cells. This control may be exerted on first-order internuncial cells, as in the case of the dorsal reticulospinal system, or on some later stage in transmission. Output from the spinal cord is regulated to some extent by descending fibers that end directly on motoneurons, as in the case of some corticospinal fibers. The majority of descending signals, however, produce their effects on motoneurons indirectly by way of vestibulospinal cells. Renshaw cells, the special internuncial cells that receive input from the recurrent collaterals of motoneurons and project back to motoneurons, are among those that are subject to descending control. It is evident that all the spinal mechanisms involved in movement are under the control of higher centers. The next major step in the analysis of descending control will be to devise methods for identifying the precise aspects of movements controlled by each of these descending systems.

ATTITUDINAL AND POSTURAL REACTIONS*

Attitudinal reflexes. We have seen that the hyperextension of the extremities of a decerebrate animal may be modified by elicitation, through nociceptive stimulation, of one or another reflex figure (Fig. 26-6). Magnus[26, 27] and de Kleijn showed that a similarly harmonious relation of the different parts of the body may be produced by passively changing the position of the head. If the head is forcibly ventroflexed, the postural contraction of the extensor muscles of the forelegs is inhibited and the forequarters sink, while at the same time the postural contraction of the extensors of the hind legs increases, raising the hindquarters. The animal assumes the proper attitude for looking under a shelf or down a hole. On the other hand, if the head is passively dorsiflexed, the forelegs extend further and the hind legs flex. The attitude is now that of a cat looking up at a shelf and ready to spring. If the head is turned to one side, say the right, the displacement of the center of gravity is compensated for by the increased extension of the right legs, while the left are flexed, ready, as it were, to take the initial step. These are the *tonic neck reflexes.* They originate in neck proprioceptors, chiefly if not entirely in the upper joints of the neck,[31] and they last as long as the head retains the new position. *Tonic labyrinthine reflexes,* elicited by change of the position of the head in space, affect all four legs similarly. The influence of the labyrinth in the decerebrate preparation can be studied separately after excluding the neck reflexes by section of the upper cervical dorsal roots or by fixating the head and body in a plaster cast. Extensor tonus of labyrinthine origin is maximal when the animal is supine with the angle of the mouth 45 degrees above the horizontal; it is minimal when the animal is prone with the angle of the mouth 45 degrees below the horizontal. When the head is brought into other positions by rotation of the body around its transverse or longitudinal axis, intermediate degrees of extensor tonus result. The tonic neck and labyrinthine reflexes sum algebraically when both are elicitable. These reactions may be observed in normal animals, in which they may be evoked by active as well as passive movements of the head. They are easily elicited in monkeys after removal of the frontal lobes and in children with severe damage of the higher motor mechanisms of the brain.

These postural reactions are brought about by motor discharges of slow rate and they operate through the same spinal mechanism as that which manages the stretch or myotatic reflexes. It was shown by Denny-Brown[14] that the tonic neck and labyrinthine reflexes influence the stretch reflex just as they influence the posture of the limb. They involve the slow extensor muscles far more than the rapid ones; when the response is small, it is seen only in the former group. It is plain that the stretch reflexes, which as we have seen are essentially spinal in origin, can be modified by impulses originating in the labyrinth or in proprioceptors of the neck. When in a decerebrate animal the afferents from all limb muscles are interrupted by section of the appropriate dorsal roots, these and other sources of excitation may maintain slow rates of motor discharge to the postural muscles. It is for this reason that extensor rigidity reappears in deafferented limbs some time after decerebration.

Normal standing.[35] Decerebrate rigidity has been described as "reflex standing." The standing of the *bulbospinal* decerebrate preparation is, however, a mere caricature of normal standing. Comparison shows that the two conditions differ in several important respects as regards this fundamental postural achievement.[35] An enumeration of these differences will reveal something of the character of normal standing and the extent to which suprabulbar mechanisms contribute to normal postural activity.

1. The distribution of tone in the muscu-

*This section, which was prepared by Philip Bard as part of Chapter 76 in the twelfth edition of his book, is included here with only slight modifications.

lature is quite different in the two cases. In the acutely decerebrate animal the postural contraction of the extensors is accompanied by diminution of the tone of the flexors, whereas in normal standing all the muscles around the joints of the limb are strongly contracted so as to fix the limb in the form of a rigid pillar.

2. If the bulbospinal decerebrate animal is placed on its side or back, it remains in that position, never showing the slightest tendency to right itself. Normal animals succeed in maintaining themselves right side up under the greatest variety of disturbing conditions.

3. The acutely decerebrate animal keeps its legs rigidly extended regardless of its position. It never places its feet in the correct position for standing. In contrast, the normal animal actively places its feet in such a way that a normal stance can be assumed.

4. If a decerebrate animal is placed in a standing position, the slightest shifting of its body to one side causes it to topple over. By executing correcting movements of the legs, however, the normal animal succeeds in maintaining the standing posture during horizontal displacement of the body.

5. When made to stand on an inclined surface, the normal animal adjusts the distribution of tone in its muscles in such a way as to buttress itself against the effect of the consequent shift in its center of gravity. No trace of such a response is seen in a decerebrate preparation subjected to these conditions.

These deficiencies of decerebrate standing are the result of the absence of certain specific postural reflexes. We shall next consider the more important of these reactions.

Supporting reactions. Limbs that are freely movable when they are used in stepping or as instruments for prehension, scratching, fighting, etc. are transformed during standing into rigid pillars that give the impression of being solid columns for the support of the body. Experiments by Magnus[27] and collaborators (Schoen, Pritchard, and Rademaker) have shown that this is accomplished by a series of local static reflexes. The development of the pillarlike state is called the *"positive supporting reaction."* It is brought about in part by an extensor response to the exteroceptive stimulus evoked by the contact of the feet with the ground *(magnet reaction).* It is more especially the result of the proprioceptive stimuli set up in the flexor muscles of the distal joints (digits, wrist, ankle) when these are stretched by the pressure of the foot on the ground. A complex stretch reflex is thereby set up that involves not only the stretched muscles themselves but also the co-contraction of extensors and flexors, abductors and adductors, around each joint, so as to fix the entire limb for standing. In addition, the muscles of the back take part and their participation is seen especially clearly when the whole reaction has been exaggerated by decerebellation. The resolution of this response is not wholly due to cessation of the stimulus, for active processes set up by stretching the extensors of the distal joints take part in loosening the limb. This is the *negative supporting reaction.* In the decerebrate animal these reactions are not present; instead there is only the overactive stretch reflex of the extensors. After removal of the cerebral cortex the supporting reactions are disturbed profoundly but are not entirely absent (Rademaker).

Righting reactions. The ability to stay right side up is a universal property of animal organisms. Magnus[26, 27] and de Kleijn have shown how in the higher animals, including ourselves, this capacity is dependent on a group of specific righting reflexes. These responses can best be studied in decorticate animals in which their reflex nature is quite apparent. If such an animal is held up by the pelvis, the head is kept in its normal position as the body is turned from one side to the other or allowed to hang down. The compensatory movements of the head are *labyrinthine righting reactions,* which are set up by stimulation of the otolithic apparatus. The responding muscles are those of the neck. These righting movements fail to occur in animals without labyrinths, and they are abolished by removing the otoliths from their maculas. So long as a decorticate or blinded animal without labyrinths is held in the air, its head takes a passive position imposed by gravity, but the moment it is placed on its side on the ground, the head is righted into the normal position. The response is due to asymmetric stimulation of the body surface, as can be shown by the fact that placing a weighted board on the uppermost side causes the head to go back into the lateral position. This is a *body righting reflex acting on the head.* When the head has been righted by either of the reflexes acting on it, the resultant twisting of the neck proprioceptively excites a *neck righting reflex* that first causes the thorax, then the lumbar region, and finally the pelvis to follow the head into the normal upright position. *Body righting reflexes acting on the body,* elicited by asymmetric stimulation of the body, may, however, cause the body to right itself even when the head is held in the lateral position. Finally, in the higher mammals (cats, dogs, monkeys) righting reflexes may be initiated through visual stimuli. The *optical righting reflexes* result in the orientation of the head and are capable of bringing this about in the absence of labyrinthine or body stimulation. They are dependent on the cerebral cortex. The centers for the nonvisual righting reflexes lie chiefly in the medulla and mesencephalon, but chronically decorticate cats show a very marked deficiency of the labyrinthine righting reaction.

Placing reactions. A primary requirement for normal standing is that the feet should be placed in the proper position. The normal animal accomplishes this through the agency of a group of placing reactions that have been described by Rademaker[35] and by Bard.[3, 4]

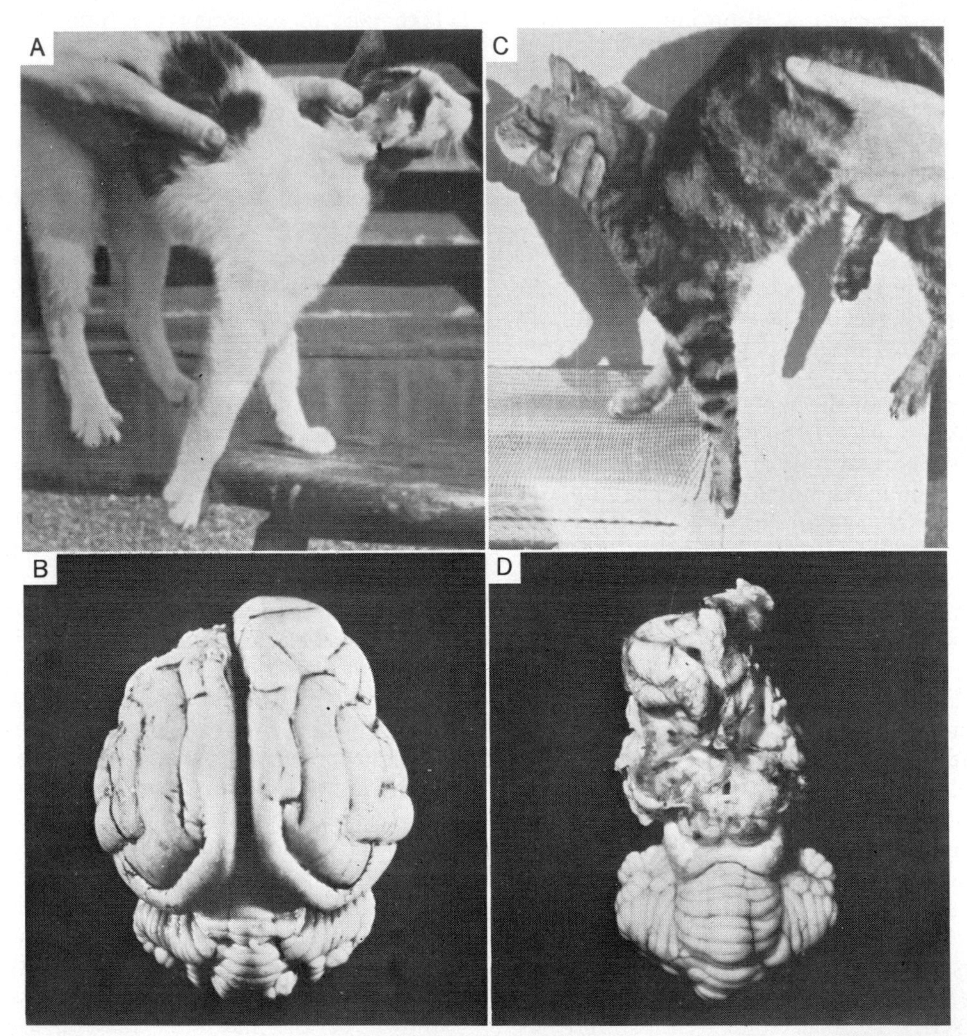

Fig. 26-12. Photographs that indicate nature and localized cortical control of placing reactions of cat. **A,** Picture of animal whose brain is shown below in **B.** Removal of small area of cortex at frontal pole of left hemisphere has permanently abolished placing reaction of right foreleg that occurs when dorsum of foot is lightly touched to edge of supporting surface (top of stool). As can be seen in **A,** ablation has not affected response of ipsilateral foreleg. Defect is as great as that shown by animals from which entire cortex of one side has been removed. **C,** Cat that had been subjected to removal (in two stages) of all cortex except frontal area, which was ablated in other animal. Its brain is shown in **D.** Note that cat has placed the right foot. This and other tests show that small remnant of left cortex was capable of managing in normal fashion the nonvisual placing reactions and the hopping reactions of opposite (right) foreleg. (Adapted from Bard.[3])

When the animal is lowered toward a supporting surface, visual stimuli cause the forelegs to be put down in such a way that without further adjustment they support the body in standing. With vision excluded, various exteroceptive and proprioceptive stimuli resulting from contact, position, or movement elicit other placing reactions that serve the same purpose—to bring the legs from any nonsupporting pose into a standing position.

Five nonvisual placing reactions may be distinguished. A brief description of them will serve to show how rich is the equipment for the attainment of this one postural result.

1. If a cat is held in the air with the legs free and dependent and with the head held up (so that it cannot see its forefeet or any object below and in front), the slightest contact of any aspect of either pair of feet with the edge of a table results in an immediate and accurate placing of the feet, soles down, on the table close to its edge (Fig. 26-12).
2. If the forelegs of a cat suspended in the air are held down and the chin is brought in contact with the edge of a table, both forefeet on being released are instantly raised and placed beside the jaws. Usually this is followed by extension, so that a standing position is quickly assumed. If a blinded animal is used, the forefeet, even though not held down, remain hanging until the chin touches the table.
3. If the forelegs or hind legs of a cat that is standing, sitting, or crouching on a table are thrust over the edge, they are immediately lifted so that the feet quickly regain their original positions on the table.
4. If any leg of a standing cat is passively abducted without being held, it is at once adducted and lowered so as to restore the foot to its normal standing position.
5. Although each of the foregoing reactions may be adequately studied in animals with vision intact, this final one can be evoked in pure form only after blindfolding, enucleating the eyes, or removing the visual cortex. The animal held in the air with the forelegs free is moved toward some solid object. As soon as the tips of the vibrassae of one or both sides touch the object, both forefeet are accurately placed on it. Unless the influence of the eyes is excluded, a visual placing reaction of the forelegs will be evoked under these circumstances.

Hopping reactions. These are essentially corrective movements of the legs that serve to maintain a standing posture under conditions involving displacement of the body in the horizontal plane. They are demonstrated by holding the animal so that it stands on one leg. Then on movement of the body forward, backward, or to either side, the leg hops in the direction of the displacement so that the foot is kept directly under shoulder or hip. Rademaker[35] showed that a disappearance of the supporting tone of the leg is an integral part of each hopping reaction. With the leg in the median standing position, the positive supporting reaction is strong, and the leg is acting effectively as a rigid pillar, but with any displacement of the body that induces a deviation from the median position, the supporting tone diminishes, the foot is raised, transposed in the direction of the displacement, and put down again to give a median support for the body in its new position. It is probable that these reactions are caused by the stretching of one or another group of muscles, i.e., that they are myotatic in origin.

Central control of placing and hopping reactions, an example of strict localization of function.[3, 4] When in a cat the entire cortex of one cerebral hemisphere is removed, the nonvisual placing reactions of the contralateral legs completely disappear and do not return during long survival periods (6 months to 6 years).* The hopping reactions are permanently depressed; some cannot be elicited, and others are retarded, weak, and so slow that the movement is not repeated rapidly enough to keep pace with even slow displacements of the body. Unilateral decortication does not affect the reactions of the legs ipsilateral to the removal. The control is entirely contralateral. After bilateral decortication the loss and deficits are bilaterally equal. An ablation of the frontal pole of one cortex of the extent shown in Fig. 26-12, *B*, produces as great a deficit of the reactions of the opposite legs as does complete unilateral decortication. The cortical area necessary for the elicitation of these responses includes most of the cortical points, stimulation of which causes movements of the contralateral legs. Removal of any other cortical area has no effect on the reactions. Indeed, one may remove all cortex except the frontal pole of one hemisphere (Fig. 26-12, *D*) without disturbing the reactions in the legs contralateral to the remnant. The fact that a remnant of rostrally situated cortex is able to manage the placing and hopping reactions in normal fashion shows conclusively that the cortical control of these responses is strictly localized and functionally independent of all other cortical areas.

This is an instance of the strictest kind of localization of function in the cerebral cortex. In general it may be said that the term *"localiza-*

*A proprioceptive correcting or placing response allied to the hopping reactions remains and may become quite prominent. It consists in a placing of the foot, clumsily and often slowly, when in attempting to elicit placing reaction (1), the leg is retroflexed. This is not to be confused with the tactile, placing reaction.

tion of function" has been used in a rather loose sense. Since evidence indicates that different functions have quite different degrees of dependence on specific cortical areas, localization may be regarded as a relative matter. To apply a rigorous definition, however, it may be said that a cortical function is localized only when a small area contains all the tissue essentially concerned in the cortical control of that function. To demonstrate such localization it is not enough to show that the function is regularly and permanently abolished by extirpation of a restricted area, for remaining areas may act through it. To conclude that a cortical function has its sole residence in a restricted area it is necessary to show that the function is not affected by removal of all cortex except this area. As can be seen, this has been accomplished in the case of the cortical management of the placing and hopping reactions.

It is in accord with our general conception of cortical functions that these cortically managed reactions have to do with the finer adjustments of postures that are developed by subcortical levels of integration.

From the point of view of comparative neurophysiology it is of interest that in the monkey all hopping reactions as well as the placing reactions are completely abolished by a cortical removal (ablation of motor cortex).[4] In the reptile such of these reactions as are present depend on central mechanisms lying below the telencephalon. In the rat and rabbit, reactions quite similar to those of the cat occur, but they are less well developed, and they are definitely less affected by decortication. Thus in the ascending scale of vertebrate quadrupeds the central control of these particular postural responses becomes more and more "corticalized." In the monkey the control can be analyzed in terms of the cortical sensory and motor components.[4]

BASAL GANGLIA

The basal ganglia are a group of nuclei in the upper brainstem and forebrain that have motor functions of great importance, as attested by the variety and severity of the abnormalities that appear when they are involved in naturally occurring diseases. Attempts to produce counterparts of these abnormalities with localized lesions in laboratory animals have been largely unsuccessful. Experimental studies of the basal ganglia, using the techniques of electrical stimulation and stereotaxic lesions as summarized by Jung and Hassler,[22] have in fact produced such a bewildering array of results that few reliable conclusions can be drawn. Under these circumstances it seems appropriate to limit this section to a brief description of the basal ganglia and their principal connections, an account of some recent electrophysiologic studies that offer promise, a description of the major syndromes of the basal ganglia as seen by clinical neurologists, and a summary of recent findings on the biogenic amines that are involved in synaptic transmission in these nuclei.

Anatomic aspects

The basal ganglia are among the most primitive portions of the forebrain. They are relatively large and well developed in birds, which may have almost no cortex, and in reptiles, which have little cortex. In mammals there is considerable reorganization of the basal ganglia; this is presumably related to the appearance of the cerebral cortex, which is extensively interrelated with the basal ganglia.

Authorities differ somewhat regarding the composition of the basal ganglia. They are agreed, however, that the *caudate nucleus* and *putamen* (known together as the *striatum*) and the *globus pallidus* are, as their size suggests, the most important parts of the basal ganglia. Three other structures, the *subthalamic body,* the *red nucleus,* and the *substantia nigra,* all located in the mesencephalon, are usually included because of their close relations with the striatum and globus pallidus.

The globus pallidus may be regarded as the principal efferent mechanism for the striatum (caudate and putamen), from which it receives a great many fibers. The principal motor outflow of the basal ganglia is by way of the pallidal efferents known as the *ansa lenticularis.* This fiber bundle distributes signals to the thalamus, hypothalamus, zona incerta, subthalamic body, red nucleus, substantia nigra, and RF of the brainstem.

There are many fibers passing directly from the cerebral cortex to the basal ganglia. These cortical extrapyramidal fibers are distributed in large numbers to the caudate and putamen as well as the globus pallidus. Many of them were unrecognized for a long time because they are unmyelinated. In addition, many fibers in the internal capsule give off collaterals to the basal ganglia, as first pointed out by Ramón y Cajal.

Activity of neurons in globus pallidus at rest and during movement

Standard neurophysiologic techniques have not provided many clues regarding the functions of the basal ganglia. Electrical stimulation and the production of discrete lesions have yielded confusing and often contradictory results. Recordings of single-unit activity, carried out in anesthetized or paralyzed animals, have offered little insight into the normal functioning of the basal ganglia in the moving animal. Recently, however, DeLong[13] has recorded from single neurons in the globus pallidus of unanesthetized monkeys at rest and during the production of trained, voluntary movements of the forelimb or hind limb. The results of this study supply reliable information on the activity of neurons that constitute an important part of the motor outflow of the basal ganglia. In primates the globus pallidus is a composite structure, consisting of an internal and an external segment that are separated by a fiber bundle called the internal medullary lamina. A microelectrode was inserted obliquely from the surface of the brain so that it passed successively through the cerebral cortex, putamen, globus pallidus externus, globus pallidus internus, and the *substantia innominata* lying ventral to the pallidum. In this study attention was directed chiefly to neurons in the globus pallidus.

With the monkey at rest, the discharge patterns of neurons in the two segments of the globus pallidus were distinctively different. As shown in Fig. 26-13, the external segment contains two types of units. Record *A* shows a pattern of activity characterized by recurrent periods of high-frequency discharge (HFD) separated by intervals of silence lasting up to several seconds. Record *B* shows a pattern consisting of low-frequency discharges interspersed with more rapid bursts of 5 to 20 impulses (LFD-B). These two types of neurons were randomly distributed in the external segment, 85% being high-frequency units and 15% low-frequency units with more rapid bursts. In the internal segment, only one type

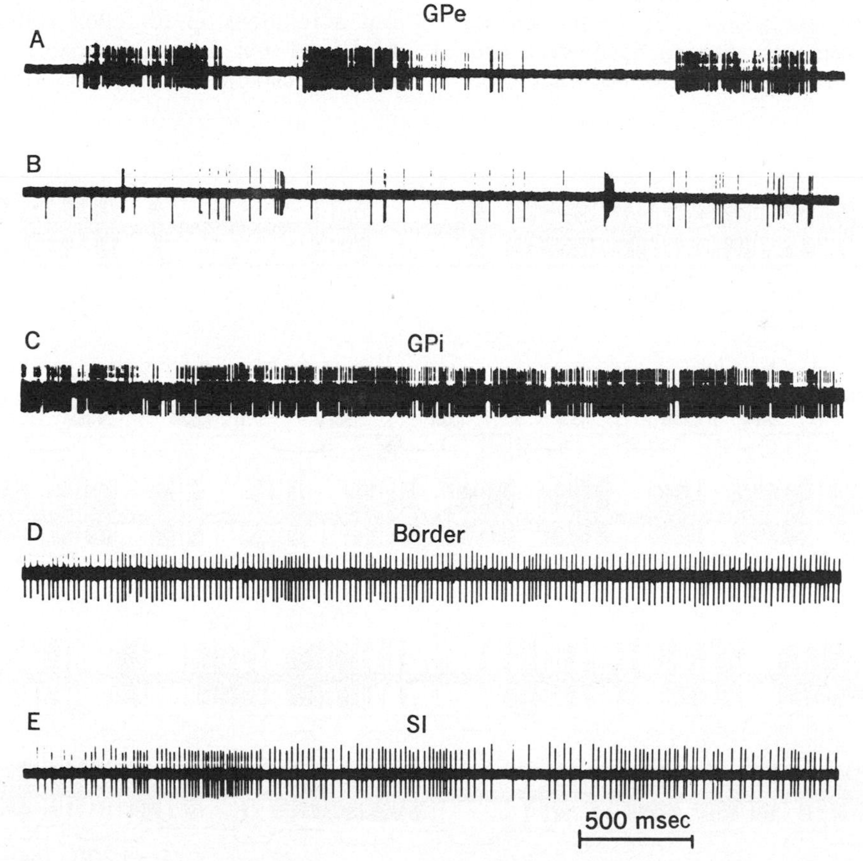

Fig. 26-13. Single-unit discharges recorded from globus pallidus of unanesthetized monkey during rest. **A,** High-frequency discharge with recurrent periods of silence seen in 85% of units in globus pallidus externa, *GPe*. **B,** Low-frequency discharge with bursting seen in 15% of units in *GPe*. **C,** Continuous, high-frequency discharge without long intervals of silence, seen in globus pallidus interna, *GPi*. **D,** Units located along borders of two pallidal segments, exhibited fourth type of discharge pattern, similar to that of neurons in **E,** substantia innominata, *SI*. (From DeLong.[13])

of unit could be distinguished. As illustrated in record *C*, this type discharges continuously at relatively high frequencies, without long intervals of silence but with frequent fluctuations in rate.

In monkeys that had been trained to carry out rapidly alternating movements of one arm, numerous units in both segments of the globus pallidus were found to discharge in consistent temporal relation to the various phases of the movements. A clear-cut correlation was shown by 66 out of 346 units in the external segment and 39 out of 205 units in the internal segment. Of the 66 units in the external segment, 54 were of the *"HFD"* type and 12 were of the *"LFD-B"* type just described. An example of a high-frequency discharge unit in the external segment is reproduced in Fig. 26-14. Record *A* illustrates the resting discharge pattern. During push-pull *(B)* and side-to-side *(C)* movements of the contralateral arm the patterns of discharge bore a consistent relationship to the phases of the movements. During ipsilateral arm movements (*D* and *E*), however, there was no consistent relationship. Contralateral movements were associated with 88% of movement-related units in the external segment and 82% of those in the internal segment of the globus pallidus; the remaining units were correlated almost equally with movements on both sides of the body. Some neurons discharged most intensely during flexion and others during extension. Increases and decreases in frequency were seen in relation to different phases of the movement cycles. In general, activity patterns were related to either arm or leg movements, but not to both.

With a wider range of movements involving more of the muscles in the limbs or perhaps more combinations of muscles, a higher percentage of units related to movement would probably have been found. DeLong's studies indicate a definite relationship between the activity of pallidal neurons and voluntary movements of the limbs, but they do not rule out a relationship to reflex activity. It is to be hoped that the precise nature of the relationship established in these studies can be

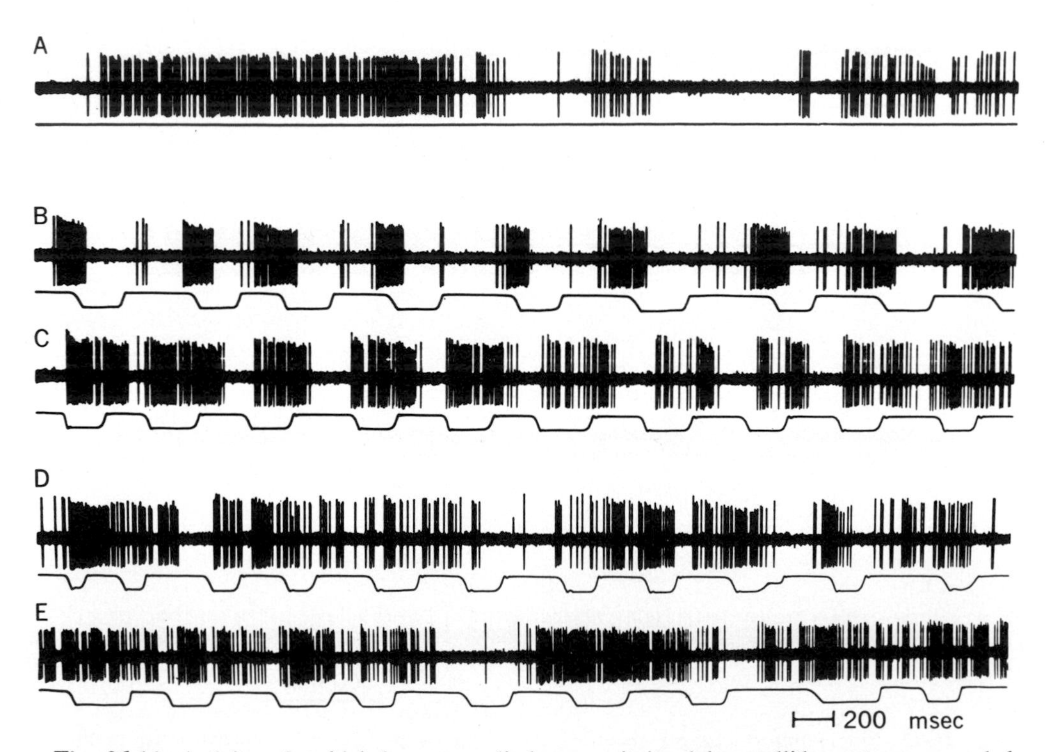

Fig. 26-14. Activity of a high-frequency discharge unit in globus pallidus externa recorded from an unanesthetized monkey. **A,** During rest. **B,** During alternating push-pull. **C,** During side-to-side movements of contralateral arm. **D,** During push-pull. **E,** During side-to-side movements of ipsilateral arm. Line below record of unit activity represents position of rod that monkey is grasping. For push-pull movements, up = pull and down = push; for side-to-side movements, up = extension and down = flexion. (From DeLong.[13])

analyzed further and that a correlation with some highly specific parameter of movement can be established.

Effects of diseases of basal ganglia on movement and posture

The most useful body of information we have regarding the functions of the basal ganglia is that provided by naturally occurring diseases that attack these structures. The signs and symptoms of these diseases indicate that the basal ganglia are concerned primarily, perhaps exclusively, with motor functions. Although sensory loss or mental changes may occur in conjunction with motor abnormalities, as in Huntington's chorea, they are attributable to pathology in the cerebral cortex, thalamus, or internal capsule rather than to lesions of the basal ganglia. The principal motor effects of basal ganglia disease are akinesia, alterations in muscle tone, and several types of involuntary movement that are typical of lesions in particular nuclei.

Akinesia is believed by some neurologists to be the cardinal "deficit" symptom of extrapyramidal disease. It manifests itself as a disinclination to use the affected part of the body in a normal manner. The patient with advanced *paralysis agitans,* for example, exhibits a starched or masklike expression, an unblinking, reptilian stare, a tendency to sit motionless for long intervals, and loss of certain movements normally associated with particular activities. In looking to one side, he moves his eyes but not his head; in walking, he fails to swing his arms and he turns around in a stiff, en bloc fashion; and in rising from a chair, he does not pull his feet back under him and does not use his arms to push himself up. There is a general poverty of movement, which despite the name of the disease is not due to true paralysis, for the patient can activate his immobile muscles. Rigidity is not the cause of the akinesia, for stereotaxic lesions may abolish all rigidity without relieving akinesia. The most profound akinesia is produced by lesions in the substantia nigra; the extent to which lesions elsewhere in the basal ganglia may have a similar effect is not settled.

Alterations of muscle tone characterize most diseases of the basal ganglia. Most frequently seen is an increase in the tone of opposing muscle groups that is called *rigidity.* The muscles in the involved part are firm and tense due to a continuous discharge of their motoneurons, which can easily be verified by electromyographic studies. Resistance is encountered to passive movement in any direction and is accompanied by an increased discharge of motor units. Rigidity may be widespread but seems most pronounced at the larger joints due to the greater mass of muscle there. Small muscles anywhere may be affected, including those of the face, tongue, and larynx. In general, tendon jerks are not enhanced in rigid limbs because contraction of opposing muscles limits the reflex response. Reduction in muscle tone or hypotonia is seen in certain diseases of the basal ganglia, most notably in Huntington's chorea. The limbs are usually slack and the tendon reflexes are often pendular due to lack of restraint from antagonists.

Involuntary movements are an important and characteristic feature of diseases of the basal ganglia. Several of the common types will be described here to illustrate the strong influence that the basal ganglia have on motor activities. Chorea is a dramatic example of involuntary movement over which the patient has little or no control. It is characterized by rapid, somewhat jerky or spasmodic movements involving one limb, one side of the body, or all parts of the body. It is difficult to give an adequate description of them because they are so variable and irregular. Early in the development of the disease they consist of twitches or brief, abrupt contractions of one muscle or a small group of muscles, like fragments of a purposeful act. Later, as these contractions become more frequent, they appear to flow into one another to produce an elaborate sequence of meaningless actions. Normal movements are possible, but they may have choreic features incorporated into or grafted onto them. In Huntington's chorea the cause of the abnormal activity appears to be atrophy of the caudate nucleus and putamen, resulting in "release" of some other center that discharges in a continuous, random fashion.

Athetosis is characterized by slowly spreading contractions of closely related muscles that result in sinuous, writhing movements of the extremities. The muscles of the fingers and wrist are most commonly affected, but in severe cases the forearm, arm, face, tongue, and throat may be involved. The disturbance often appears to start in the distal part of a limb and spread to the proximal parts. The individual contractions are generally slower and of longer duration than those of chorea. As in chorea, they tend to become confluent and lead to bizarre postures. In the fingers, alternating flexion and extension is often seen; in the forearm, extension and pronation tend to alternate with flexion and supination. In general, an athetotic patient cannot prevent his involuntary movements and may have great difficulty in keeping his fingers, toes, or tongue in any fixed position. As in chorea, the pathology underlying athetosis is usually in the striatum.

Torsion spasm is a condition, apparently allied to athetosis, in which there are powerful, tonic contractions of the muscles of the trunk, neck, and proximal segments of the limbs. These spasms result in grotesque postures that are uncontrollable and irresistible. Highly abnormal retractions of the head, extreme lordosis, unnatural rotations of the trunk, and bizarre contortions of an entire limb may be seen. Often the spasms appear on voluntary movement but are absent during rest. It has been pointed out that chorea, athetosis, and torsion spasm all require cortical mechanisms for their expression and that cortical or capsular lesions that cause any degree of paralysis tend to abolish the involuntary movements.

Tremor consists of rhythmic, to-and-fro movements caused by alternating contractions of agonists and antagonists. The rate is about 3 to 6/sec and remains fairly constant, but the range of movement may vary widely. As a rule, tremor is most pronounced in the distal parts of the limbs, but it may involve the trunk, head, face, tongue, or jaw. It is frequently associated with rigidity, as in paralysis agitans. There are many types of tremor; only a few of them are the result of lesions of the basal ganglia. Of these latter, one of the most important is the *static tremor* or *tremor at rest* of paralysis agitans. It usually occurs in the hands but may occasionally involve the tongue or the jaw. Characteristically it is seen when the limb is at rest. It is usually abolished during willed movements, so that it interferes very little with voluntary activity. In paralysis agitans and postencephalitic Parkinson's syndrome the lesions are predominantly in the substantia nigra.

Basal ganglia and biogenic amines

KAREL V. S. TOLL

More rewarding than the results of physiologic studies have been the recent biochemical, pharmacologic, and therapeutic approaches that have resulted in extraordinary advances in knowledge about the biogenic amines and the role played by these compounds and by their inhibitors and potentiators in the normal function of the basal ganglia. There is now little doubt that some monoamines serve as neurotransmitters in the CNS[9] (Chapter 6). The major pathways for the metabolism of the amines involved in this discussion are as follows:

Phenylalanine
↓
L-Tyrosine
↓
L-Dihydroxyphenylalanine (L-dopa) → Melanin
↓
Dopamine (3-Hydroxytyramine) → Homovanillic acid
↓
Noradrenaline
↓
Adrenaline

and:

L-Tryptophan
↓
L-5-Hydroxytryptophan
↓
5-Hydroxyindolacetic acid
↓
5-Hydroxytryptamine (5-HT or serotonin)

In 1953 Twarog and Page[49] found serotonin in the brain, and in 1954 Vogt[50] established the distribution pattern of noradrenaline in various portions of the brain both normally and after drug administration. In the same year lysergic acid diethylamide was found to have a blocking effect on serotonin and in 1955 it was demonstrated that serotonin was released from its tissue store by reserpine.[41] Indeed, the introduction in the early 1950s of the first tranquilizing drugs resulted in major advances when it was shown that the rauwolfia alkaloids (of which reserpine is a prototype) resulted in the release of norepinephrine and other amines and that following large doses of these compounds a state resembling parkinsonism develops in man and in animals. This extrapyramidal symptomatology is due to the depletion of dopamine and can be reversed by early discontinuation of the tranquilizer or by treatment with anti-parkinsonian medication. The parkinsonian syndrome produced by the phenothiazine and butyrophenone derivatives seems to be due not to dopamine depletion per se but rather to blocking of dopamine receptors.[10]

Subsequently,[16] it was found that Parkinson's disease is characterized by a loss of melanin and by a deficiency of dopamine in the substantia nigra–striatal system. There is a decrease in the CSF concentration of homovanillic acid (HVA), the principal breakdown product of dopamine metabolism, lower levels of HVA being correlated with more severe degrees of akinesia. The CSF levels of 5-hydroxyindolacetic acid, a metabolite of serotonin, are also significantly decreased in patients with parkinsonism.[37]

Dopamine is found chiefly in the caudate nucleus, putamen, and substantia nigra. The main dopamine-carrying neuronal pathway arises in the substantia nigra. Axons from this nucleus run in the ventral portion of the crus cerebri and in the internal capsule to the neostriatum where dopamine, its concentration increasing as the fibers approach the target organ, is stored in synaptic vesicles located in the nerve terminals. On neuronal stimulation the transmitter molecules are released into the synapse and combine with receptor sites to affect postsynaptic neurons. Inactivation of transmitters occurs through enzyme action (monoamine oxidase in the case of serotonin; MAO or catechol-O-methyltransferase for the others) or by recapture by the membrane pump. Synthesis of new transmitter occurs within the neuron system as previously indicated.

The noradrenaline-carrying system is more widely distributed than the dopamine system, extending to most parts of the CNS. The serotonin-carrying pathways are also very extensive.

The basic pathology of advanced idiopathic Parkinson's disease, as indicated, consists of degeneration of the melanin-containing cells of the substantia nigra and of the nigrastriatal pathway. Since dopamine itself cannot pass the blood-brain barrier but its metabolic precursor dopa can, this compound was a logical choice for treatment.[2, 8] Although the earlier optimism has had to be modified somewhat because of therapeutic failures and side effects, dopa has been responsible for gratifying clinical results: two thirds of the patients who maintain treatment show an improvement of 50% or more.[32]

The abnormal involuntary movements that are a striking side effect of dopa treatment throw further light on the balance of activity in the basal ganglia. These dyskinesias (choreic, athetotic, or dystonic) resemble the symptoms of other diseases of the basal ganglia. Just as some tranquilizers produce parkinsonism as a side effect, L-dopa produces a picture of choreoathetosis. The mechanism by which L-dopa produces these side effects has not yet been established, although various possibilities have been suggested, including denervation hypersensitivity.

Efforts to enhance the effect of L-dopa by using dopa decarboxylase inhibitors have also proved fruitful. Although the exact mechanisms of their action are not understood, they make adequate treatment with much smaller doses of L-dopa possible.[8] Dopa decarboxylase is dependent for its functioning on the presence of pyridoxine,[25] but when given to L-dopa–treated parkinsonian patients in an effort to enhance the formation of dopamine from L-dopa, it was found to have a detrimental effect.[15] This puzzling result is prevented if the patients receive not only L-dopa but also a dopa decarboxylase inhibitor, a phenomenon that suggests that the reversal of the dopa effect produced by pyridoxine is due to increasing the dopa decarboxylase activity outside the CNS.[22]

A third neurotransmitter of great importance in the basal ganglia is acetylcholine (ACh). This substance, its synthesizing enzyme, choline acetylase, and its inactivating enzyme, acetylcholinesterase (AChE), are present in high concentrations. With new techniques designed to trace cholinergic neurons in the rat brain, Shute and Lewis[42, 43] have demonstrated cholinergic pathways between the substantia nigra, the pallidum, and the caudate-putamen complex, and others have shown that electrical or chemical stimulation of the caudate nucleus causes release of ACh into the CSF. Microinjection of cholinergic drugs into the caudate results in a 20/sec tremor in rats[12] and systemically administered physostigmine produces, in addition to tremor, a rigidity that is electrophysiologically similar to that induced by reserpine. This rigidity is apparently supraspinal in origin, since it is abolished by destruction of the striatum. Physostigmine also exaggerates parkinsonian symptoms in patients not maintained on medication. This exacerbation can be reversed by anticholinergic drugs. These adverse effects of physostigmine do not occur in normal subjects, nor do they occur in patients on an optimal L-dopa dosage.

It should be noted that patients with extrapyramidal symptoms that are due to cortical lesions do not improve after administration of L-dopa, as all known dopaminergic pathways are subcortical; patients also generally do better on combined dosages of L-dopa and anticholinergic medication than on L-dopa alone. This observation lends further support to the belief that both cholinergic and dopaminergic neurons participate in the regulation of activity in the basal ganglia.

From information on the physiology and neurochemistry of acetylcholine and dopamine in the basal ganglia, it is evident that these two "neurohumors" influence the activity of some of the neuronal units within these areas in an antagonistic way. When tested on single caudate cells, acetylcholine is usually excitatory and dopamine inhibitory. It can therefore be assumed that under physiological conditions, a delicate functional equilibrium exists within the striatum between the excitatory cholinergic and the inhibitory dopaminergic mechanisms. The existence of such an equilibrium is postulated to explain the normal functioning of the striatum as a higher control center for extrapyramidal motor activity.*

*Hornykiewicz, O.: Neurochemical pathology of Parkinson's disease. In McDowell, F. H., and Markham, C. H., editors: Recent advances in Parkinson's disease, Philadelphia, 1971, F. A. Davis Co.

REFERENCES

1. Bach, L. M. N., and Magoun, H. W.: The vestibular nuclei as an excitatory mechanism for the cord, J. Neurophysiol. **10:**331, 1947.
2. Barbeau, A.: The pathogenesis of Parkinson's disease: a new hypothesis, Can. Med. Assoc. J. **87:**802, 1962.
3. Bard, P.: Studies on the cerebral cortex. I. Localized control of placing and hopping reactions in the cat and their normal management by small cortical remnants, Arch. Neurol. Psychiatry **30:**40, 1933.
4. Bard, P.: Studies on the cortical representation of somatic sensibility, Harvey Lect. **33:**143, 1938.
5. Bard, P., and Macht, M. B.: The behaviour of chronically decerebrate cats. In Ciba Foundation symposium on the neurological basis of behaviour, Boston, 1958, Little, Brown & Co.
6. Bard, P., and Rioch, D. McK.: A study of four cats deprived of neocortex and additional portions of the forebrain, Bull. Johns Hopkins Hosp. **60:**73, 1937.
7. Bazett, H. C., and Penfield, W. G.: A study of the Sherrington decerebrate animal in the chronic as well as the acute condition, Brain **45:**185, 1922.
8. Birkmayer, W., and Hornykiewicz, O.: Der L-Dioxyphenylalanin effekt bei der Parkinson-akinese, Wien Klin. Wochenschr. **73:**787; 1961.

8a. Brodal, A.: Anatomical aspects of the reticular formation of the pons and medulla oblongata, Progr. Neurobiol. **1:**240, 1956.

9. Carlsson, A.: Basic concepts underlying recent developments in the field of Parkinson's disease. In McDowell, F. H., and Markham, C. H., editors: Recent advances in Parkinson's disease, Philadelphia, 1971, F. A. Davis Co.
10. Carlsson, A., and Lindquist, M.: Effects of chlorpromazine and haloperidol on formation of 3-hydroxytyramine and normetanephrine in the mouse brain, Acta Pharmacol. Toxicol. **20:**140, 1963.
11. Carpenter, D., Engberg, I., and Lindberg, A.: Primary afferent depolarization evoked from the brain stem and the cerebellum, Arch. Ital. Biol. **104:**78, 1966.
12. Connor, J. D., Rossi, G. V., and Baker, W. W.: Analysis of the tremor induced by injection of cholinergic agents into the caudate nucleus, Int. J. Neuropharmacol. **5:**207, 1966.
13. DeLong, M. R.: Activity of pallidal neurons during movement, J. Neurophysiol. **34:**414, 1971.
14. Denny-Brown, D.: On the nature of postural reflexes, Proc. R. Soc. Lond. (Biol.) **104:**252, 1929.
15. Duvoisin, R., Yahr, M., and Cote, L.: Pyridoxine reversal of L-dopa in parkinsonism, Trans. Am. Neurol. Assoc. **94:**81, 1969.
16. Ehringer, H., and Hornykiewicz, O.: Verteilung von noradrenalin und dopamin im gehirn des menschen und ihr verhalten bei erkrankungen des extrapyramidalen systems, Wien Klin. Wochenschr. **38:**1236, 1960.
17. Engberg, I., Lundberg, A., and Ryall, R. W.: Reticulospinal inhibition of transmission in reflex pathways, J. Physiol. **194:**201, 1968.
18. Engberg, I., Lundberg, A., and Ryall, R. W.: Reticulospinal inhibition of interneurons, J. Physiol. **194:**225, 1968.
19. Fulton, J. F., Liddell, E. G. T., and Rioch, D. McK.: The influence of unilateral destruction of the vestibular nuclei upon posture and the knee jerk, Brain **53:**327, 1930.
20. Hornykiewicz, O.: Neurochemical pathology of Parkinson's disease. In McDowell, F. H., and Markham, C. H., editors: Recent advances in Parkinson's disease, Philadelphia, 1971, F. A. Davis Co.
21. Jankowska, E., Lund, S., Lundberg, A., and Pompeiano, O.: Inhibitory effects evoked through ventral reticulospinal pathways, Arch. Ital. Biol. **106:**124, 1968.
22. Jung, R., and Hassler, R.: The extrapyramidal motor system. In Field, J., Magoun, H. W., and Hall, V. E., editors, Neurophysiology section: Handbook of physiology, Baltimore, 1960, The Williams & Wilkins Co., vol. 2.

22a. Krieg, W. J. S.: Functional neuroanatomy, Philadelphia, 1942, The Blakiston Co.

23. Lindsley, D. B.: Brain stem influences on spinal motor activity, Res. Publ. Res. Assoc. Nerv. Ment. Dis. **30:**174, 1952.
24. Lindsley, D. B., Schreiner, L. H., and Magoun, H. W.: An electromyographic study of spasticity, J. Neurophysiol. **12:**197, 1949.

25. Lovenberg, W., Weissbach, H., and Udenfriend, S.: Aromatic L-amino acid decarboxylase, J. Biol. Chem. **237:**89, 1962.
26. Magnus, R.: Korperstellung, Berlin, 1924, Julius Springer.
27. Magnus, R.: Studies in physiology of posture, Lancet **211:**531, 1926.
28. Magoun, H. W.: Caudal and cephalic influences of brain stem reticular formation, Physiol. Rev. **30:**459, 1950.
29. Magoun, H. W., and Rhines, R.: An inhibitory mechanism in the bulbar reticular formation, J. Neurophysiol. **9:**165, 1946.
30. Magoun, H. W., and Rhines, R.: Spasticity: the stretch reflex and extrapyramidal systems, Springfield, Ill., 1947, Charles C Thomas, Publisher.
31. McCouch, G. P., Deering, I. D., and Ling, T. H.: Location of receptors for tonic neck reflexes, J. Neurophysiol. **14:**191, 1951.
32. McDowell, F. H., et al.: The clinical use of levodopa in the treatment of Parkinson's disease. In McDowell, F. H., and Markham, C. H., editors: Recent advances in Parkinson's disease, Philadelphia, 1971, F. A. Davis Co.
33. Niemer, W. T., and Magoun, H. W.: Reticulospinal tracts influencing motor activity, J. Comp. Neurol. **87:**367, 1947.
33a. Nyberg-Hansen, R.: Sites and modes of termination of reticulospinal fibers in the cat. An experimental study with silver impregnation methods, J. Comp. Neurol. **124:**71, 1965.
34. Pollock, L. J., and Davis, L.: Studies in decerebration. VI. The effect of deafferentation upon decerebrate rigidity, Am. J. Physiol. **98:**47, 1931.
35. Rademaker, G. G. J.: Das Stehen, Berlin, 1931, Julius Springer.
36. Rhines, R., and Magoun, H. W.: Brain stem facilitation of cortical motor response, J. Neurophysiol. **9:**219, 1946.
37. Rinne, U. K., and Sonninen, V.: Acid monoamine metabolites in the cerebrospinal fluid of patients with Parkinson's disease, Neurology **22:**62, 1972.
38. Scheibel, M. E., and Scheibel, A. B.: Structural substrates for integrative patterns in the brain stem reticular core. In Jasper, H. H., et al., editors: Henry Ford Hospital symposium on the reticular formation of the brain, Boston, 1958, Little, Brown & Co.
39. Schreiner, L. H., Lindsley, D. B., and Magoun, H. W.: Role of brain stem facilitatory systems in maintenance of spasticity, J. Neurophysiol. **12:**207, 1949.
40. Sherrington, C. S.: The integrative action of the nervous system, New Haven, 1906, Yale University Press.
41. Shore, P. A., Silver, S. L., and Brodie, B. B.: Interaction of reserpine, serotonin and lysergic acid diethylamide in brain, Science **122:**284, 1955.
42. Shute, C. C. D., and Lewis, P. R.: Cholinesterase-containing systems of the brain of the rat, Nature **199:**1160, 1963.
43. Shute, C. C. D., and Lewis, P. R.: The ascending cholinergic reticular system: neocortical, olfactory and subcortical projections, Brain **90:**497, 1967.
44. Snider, R. S., and Woolsey, C. N.: Extensor rigidity in cats produced by simultaneous ablation of the anterior lobe of the cerebellum and the pericruciate areas of the cerebral hemispheres, Am. J. Physiol. **133:**454, 1941.
45. Sprague, J. M., Schreiner, L. H., Lindsley, D. B., and Magoun, H. W.: Reticulo-spinal influences on stretch reflexes, J. Neurophysiol. **11:**501, 1948.
46. Tower, S. S.: The dissociation of cortical excitation from cortical inhibition by pyramid section, and the syndrome of that lesion in the cat, Brain **58:**238, 1935.
47. Tower, S. S.: Extrapyramidal action from the cat's cerebral cortex: motor and inhibitory, Brain **59:**408, 1936.
48. Tower, S. S.: Pyramidal lesion in the monkey, Brain **63:**36, 1940.
49. Twarog, B. M., and Page, I. H.: Serotonin content of some mammalian tissues and urine and a method for its determination, Am. J. Physiol. **175:**157, 1953.
50. Vogt, M.: The concentration of sympathin in different parts of the central nervous system under normal conditions and after the administration of drugs, J. Physiol. **123:**451, 1954.
51. Woolsey, C. N.: Postural relations of the frontal and motor cortex of the dog, Brain **56:**353, 1933.

27

LAURENCE R. YOUNG

Role of the vestibular system in posture and movement

Function of vestibular system

The nonauditory portion of the inner ear consists of a highly specialized set of fluid-filled tubes (semicircular canals) and otoliths, collectively known as the labyrinth or vestibular system. The vestibular apparatus in man serves three major functions: (1) As the primary organ of equilibrium, it plays a dominant role in the subjective sensation of motion and spatial orientation. (2) Vestibular inputs to the postural control system elicit adjustment of muscle activity and body position to prevent falling. (3) Vestibular influences on eye movements tend to stabilize the eyes in space during head movements, thereby reducing the movement of the image of a fixed object on the retina.

The "inputs" to the system are linear and angular motions of the head, as indicated in the functional block diagram in Fig. 27-1. In the classic view of the system the three semicircular canals in each ear respond primarily to angular acceleration of the head, and the sensory afferents indicate angular velocity of the head over the usual physiologic range of movements. The otoliths, responding to gravity and linear acceleration, are also known as statoliths or gravireceptors, terms that emphasize their role in signaling the static orientation of the head with respect to the real or apparent vertical. Labyrinthine signals are combined with visual, exteroceptive, interoceptive, and occasionally auditory sensory signals in the complex closed-loop processes of spatial orientation and maintenance of equilibrium.

The effect of the acceleration stimuli on the individual canals and otoliths depends on their orientation, which in turn is determined by the orientation of the head in space. Nystagmoid eye movements are generated on the basis of vestibular outputs, but these eye movements do not in turn affect the vestibular system directly. However, the vestibular regulation of equilibrium through head and body movements is a closed-loop feedback system, since the labyrinth moves with the head. The resulting head and body motions can be used to null the "closed-loop error," which signals departure from equilibrium or from a planned trajectory.[73]

ANATOMY OF THE VESTIBULAR ORGAN: MECHANICAL SENSORS

The fluid-filled membranous labyrinth is contained within a similarly shaped, membrane-lined bony labyrinth within the temporal bone. The relative sizes of the bony and membranous laybrinths vary with the species; in man the lumen of the semicircular canal is about one fourth that of the bony canal.[32] The space between the membranous labyrinth and bone is filled with perilymph and the supporting tissue; the sacs and canals of the membranous labyrinth itself are filled with endolymph.[68, 77, 79] These fluids, slightly denser and more viscous than water, protect the delicate mechanism from mechanical shock and account for its behavior as a motion transducer. The endolymph conveys information on the mechanical motion of the head to the sensory hair cells and possesses an electrolyte composition suitable for the function of the sensory cells, but does not appear to carry nutrients or oxygen to them.[25]

In man the three fluid-filled canals in each labyrinth lie in nearly orthogonal planes, although the lateral or horizontal canal is pitched up by an angle of 25 to 30 degrees from the horizontal plane of the head. The superior and posterior canals lie roughly at right angles to each other and to the horizontal canal and at 45 to 55 degrees from the sagittal and frontal planes. The two horizontal canals of the two ears are parallel, and the posterior canal of one ear is in nearly the

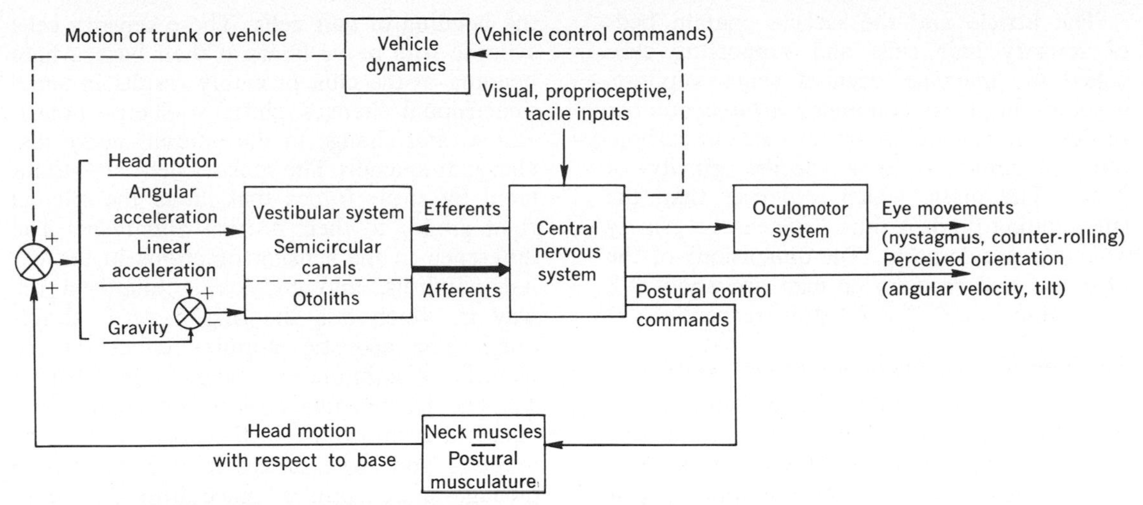

Fig. 27-1. Representation of vestibular role in control of posture, eye movements, and perception of orientation.

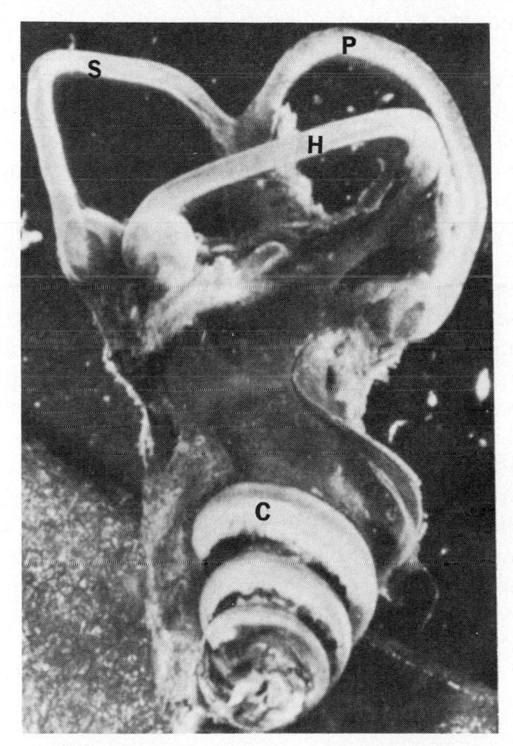

Fig. 27-2. Membranous labyrinth of guinea pig showing superior, *S*, posterior, *P*, and horizontal, *H*, semicircular canals and cochlea, *C*. (From Engstrom et al.[4])

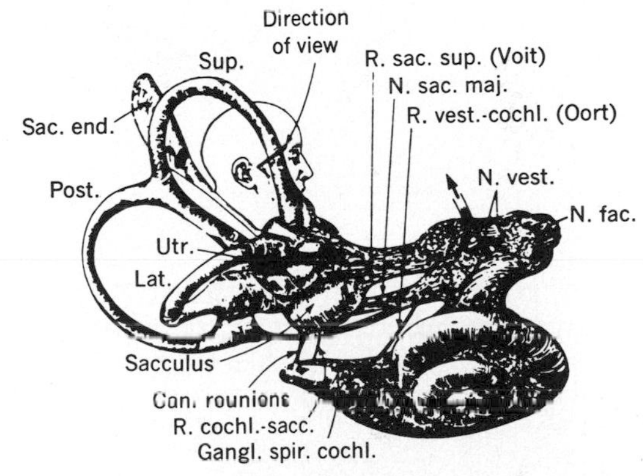

Fig. 27-3. Human vestibular apparatus. (From Hardy.[44])

same plane as the superior canal of the other ear. This arrangement of orthogonal axes and parallel pairs acting synergistically permits the generation of afferent signals indicating rotation in any direction, which evoke corresponding compensatory eye movements.[2]

Each canal originates from and returns to a common sac called the utricle, which also contains an important gravireceptor. The posterior and superior canals share the same duct (common crus) for over 15% of their length, increasing the possibility for fluid coupling between the two canals. Each canal has an enlarged region, the ampulla, where the transduction of mechanical torque or displacement to neural afferent signals by the sensory cells takes place. The ampulla is normally sealed by the cupula, a gelatinous wedge of approximately the same density as endolymph, and coupled to hairs projecting from hair cells in the crista below. As can be seen in Figs. 27-2 and 27-3, the ampullae of the superior and horizontal canals are near each other at the ventral end of the utricular sac, and the posterior canal ampulla is at the other end of the utricular sac, which is a thin-walled tube that is oval in cross section.[4]

The utricle and the saccule contain beds of sensory hair cells and supporting cells called the maculae, each of which supports a gelatinous mass containing a large number of dense hexagonal prisms of calcium carbonate (otoconia) with a specific gravity of 2.94.[3] The mass, which is denser than the surrounding fluid, is thus displaced by gravity or linear acceleration. The dimensions of the utriculus and sacculus in man are about 2.8 × 2.1 mm and 2.2 × 2.1 mm, respectively.[22]

Mechanical transduction—sensory cells

The mechanical to electrical transduction, or "signal generation" function, of all vestibular organs as well as of the organ of Corti in the auditory portion of the inner ear is probably carried out by a similar mechanism, the bending of hair cells. These sensory cells synapse with nerve fibers at their base, where bending of the cilia probably results in small dimensional changes, shifts in electric polarization, and change in the afferent nerve discharge frequency. The sensory cells are stimulated by shear forces that move the cilia at right angles to their normal alignment. The difference in the sensing functions in the canals, otoliths, and cochlea is found in the way in which this shearing force is developed. The acoustic stimulus effect on the tectorial membrane is discussed in Chapter 12. The sensory hair cells in the semicircular canals are embedded in the crista within the ampulla and have processes that extend up through a subcupular space into the gelatinous wedge forming the cupula, as shown schematically in Fig. 27-4. Angular acceleration of the head in the plane of any canal induces an inertial reaction torque in the ring of endolymph, which thus tends to remain fixed in space. However, any small movement of endolymph within the thin duct is opposed by the viscous drag it develops. The endolymph displacement tends to deflect or bend the cupula, which in turn exerts an elastic torque to return the cupula to its rest position. It is this motion of the cupula that moves the hair cells and is finally coded in the afferent neural discharge. The hair cells and supporting structure in the utricular and saccular maculae cover the entire bed and extend upward into the otolith from the base, where they are innervated by myelinated nerve fibers (Fig. 27-5). During linear acceleration or tilting of the head about a horizontal axis, the denser otolith structure moves over the macula, thereby inducing shearing movements between the otoliths and hair cells and thus evoking an afferent signal in vestibular nerve fibers.

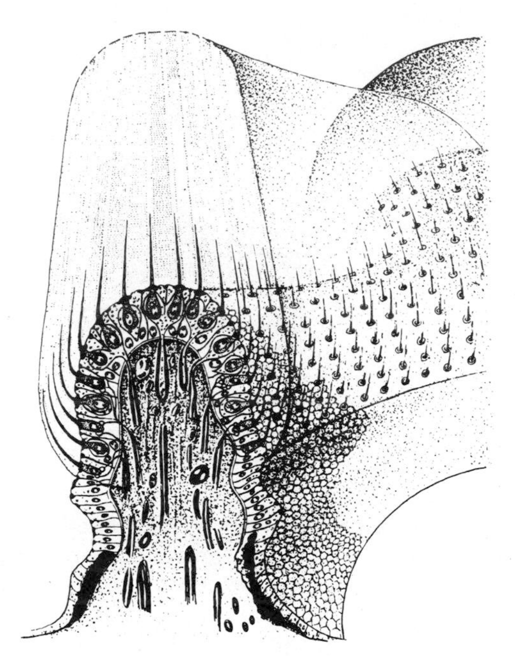

Fig. 27-4. Schematic drawing of crista ampullaris. (From Wersäll.[89])

The vestibular sensory cells in both the

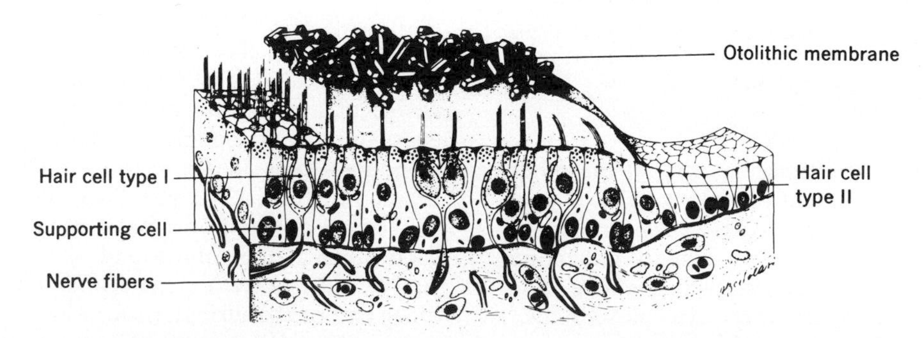

Fig. 27-5. Schematic drawing of macular epithelium. (From Iurato.[46])

cristae and the maculae are of two types,[89] as illustrated in Fig. 27-6. The type I hair cell has a bottle shape, with a narrow neck, rounded bottom, and widened top from which the hair processes protrude. It is surrounded by a nerve chalice that innervates the sensory cell and is in contact with afferent nerve endings. The type II hair cell has an irregular cylindric shape and is in direct contact with afferent and efferent nerve endings at its base. A clear functional difference between the two types of cells has not yet been demonstrated. The density of sensory cells on the cristae is higher on the slopes than at the top, and the density of type I cells relative to type II cells is similarly higher on the slopes.

The morphology and electrophysiology of the hair cells complement each other. Surmounting each sensory cell is a bundle of hairs containing one kinocilium at the apex of a regular hexagonally packed geometric bundle and 60 to 100 shorter clublike stereocilia whose lengths increase regularly with their proximity to the kinocilium.[10] The kinocilium, containing 11 filaments, 9 of which are double, projects from a centriole on one side of the hair bundle. The orientation of the sensory cells is very regular, with the kinocilia always on the same side of the bundle in an area of sensory epithelium. Lowenstein and Wersäll[58] showed that the morphologic polarization of vestibular epithelia corresponded to the direction of motion of the hair cell, which would increase the firing rate of the innervating nerve. Shearing movements that bend the hair processes toward the kinocilium increase the discharge rate in the afferent fiber, and motions away from the kinocilium decrease it.

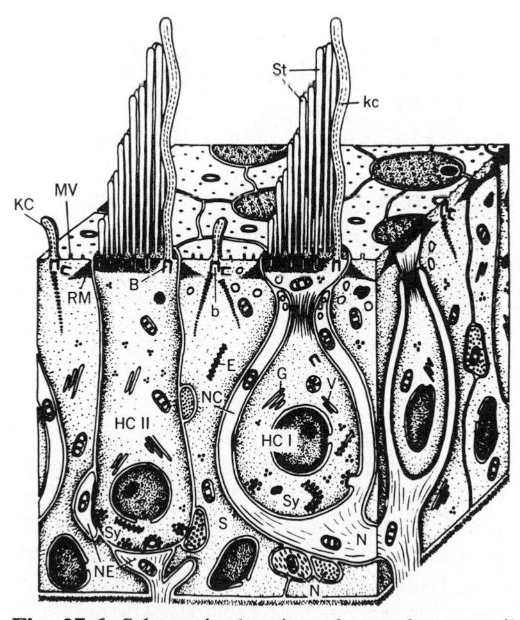

Fig. 27-6. Schematic drawing of area from vestibular epithelium, with hair cells of type I, *HC I,* and type II, *HC II.* Typical arrangement of stereocilia, *St;* kinocilia, *KC;* and modified kinocilia, *kc,* with their basal bodies and roots, *b,* in supporting cells, *S. N,* Nerve fibers; *NC,* nerve chalice; *NE,* nerve endings; *Sy,* Synaptic structures; *G,* Golgi membranes; *RM,* multivesicular reticular membrane; *MV,* microvilli. (From Spoendlin.[10])

In the crista ampullaris all bundles are oriented in the same direction. This direction, for the horizontal canal, is with the kinocilia on the utricular apex of the hair bundle, so that the endolymph movements deflecting the cupula toward the utricle (utriculopetal flow) increase the discharge rate in all afferent fibers. For the vertical canals the situation is reversed, and utriculofugal deviation of the cupula increases the discharge rate. Opposite displacement of the cupula decreases the afferent fiber discharge down to a saturation at zero frequency.

In the maculae the hair cell orientation follows a regular pattern, but the morphologic polarization varies. The orientation of the hair cells at various regions of the utricular and saccular maculae is shown in Fig. 27-7. The arrows point in the direction of the kinociliar pole; hair displacement in that direction increases discharge rate. The utricular maculae are sensitive to displacement in any direction, and the afferent fibers innervating them have been shown to respond to tilt in any direction.[57] The saccular macula, oriented principally in the vertical plane, has a similar structure. Although the function of the saccule in man is still not entirely clear, it appears to respond to low-frequency vibration and to gravity and linear acceleration, especially along the longitudinal axis. Particular areas of the utricular and saccular maculae are stimulated by linear acceleration in a specific direction, and the resulting afferent discharge is thought to determine the direction of compensatory eye movements. For example, forward linear acceleration stimulates the utricular area containing backward oriented sensory cells and causes downward eye movement.[30] The mechanisms by which mechanical energy is transduced into receptor potentials by bending of hair cells and the resulting excitation of nerve endings are still unknown, although a number of theories have been suggested.[29]

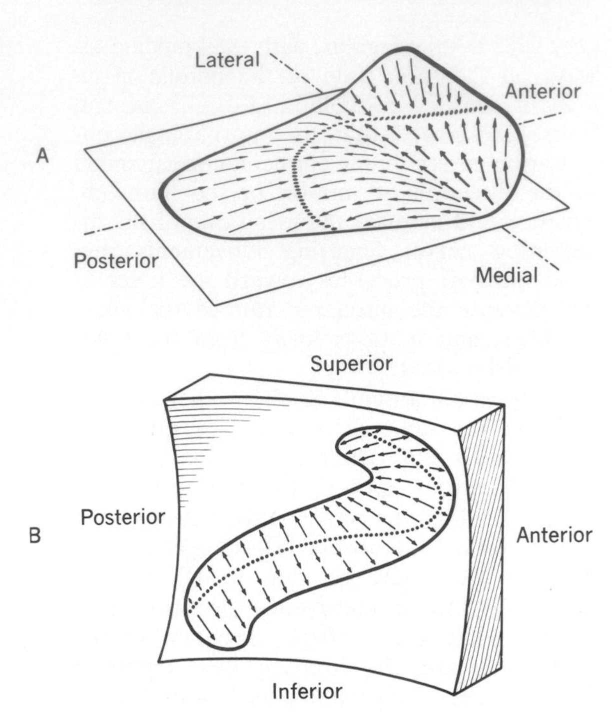

Fig. 27-7. Schematic drawing of directions of polarization of sensory cells in maculae. **A,** Utricle. **B,** Saccule. (From Spoendlin.[10])

Afferent signals

Neural afferent signals from the two maculae and the cristae of the three canals are transmitted along the vestibular branch of the eighth nerve almost exclusively to the vestibular nuclei. In addition to the afferent fibers a smaller number of efferent fibers have been identified in the vestibular nerve. Gacek[5, 31] found efferent nerve fibers coming from the lateral vestibular nucleus and possibly other sources distributed more or less evenly to all of the vestibular end organs. He estimates a total of approximately 200 efferent fibers in the vestibular nerve of the cat, as compared to approximately 12,000 afferent fibers. The function of the vestibular efferents is still unknown.

Nerve fibers innervating hair cells in the cristae may be either spontaneously active or silent.[40, 54, 56, 57] Goldberg and Fernandez,[34] recording from first-order afferents innervating the semicircular canal of the squirrel monkey, found only nerve fibers with a resting discharge. The average rate was 91 impulses/sec but the rate of discharge ranged from only a few to more than 200 impulses/sec. Sensitivity in the monkey ranged, for different fibers, from 0.5 to 4.0 impulses/sec/degree/sec^2. This finding is supported by the observation that the effect of unilateral labyrinthectomy is to decrease but not eliminate the vestibular-ocular reflex in one direction. The effect of this synergistic "push-pull" operation of the two labyrinths is to combine high-output capability (since each side can increase firing rate from its resting discharge up to its saturation in one direction) with a higher degree of linearity than would be achieved from a single sensor.

The CNS receives nearly equal discharge rates from the sensors at zero input. To the extent that there is an imbalance between the two labyrinths and the "zero input" afferent signals from the two ears do not cancel, one would expect a slow drift or directional preponderance that could be inhibited by central habituation or by the use of cues from other sensory modalities.

The relationship between mechanical stimulation (cupula displacement) and impulse frequency is a sigmoid curve, as shown in Fig. 27-8.[82] Changes in the shearing force on the hair cell produce a change in direct-current potential inside the hair cell layer of the crista and an associated change in discharge rate, as shown in Fig. 27-8. The "operating point" of a cell (discharge rate at zero stimulus) can be shifted up or down by polarizing currents to convert bidirectional to unidirectional units.[56]

Operation of semicircular canals as angular rate sensors

Torsion pendulum model. The synergistic pairs of canals in orthogonal planes permit the measurement of angular orientation of the head to be made for rotation about any axis merely by summing the vector components measured along the axes defined by the planes of the canals. The neural signals from the canals, and the resulting subjective orientation and nystagmoid eye movements, suggest that the canals measure *angular velocity* about an axis normal to the canal over the range of head-motion frequencies normally encountered. The ability of the semicircular canals to indicate angular velocity when they are, of course, acted on by the forces and torques resulting from angular acceleration will be explained on the basis of their structure.*

Current theories regarding the operation of the semicircular canals stem from the observation of Steinhausen.[78] He showed that during

*The sensitivity of the semicircular canals to linear acceleration and gravity as well as to angular motion is discussed on p. 717.

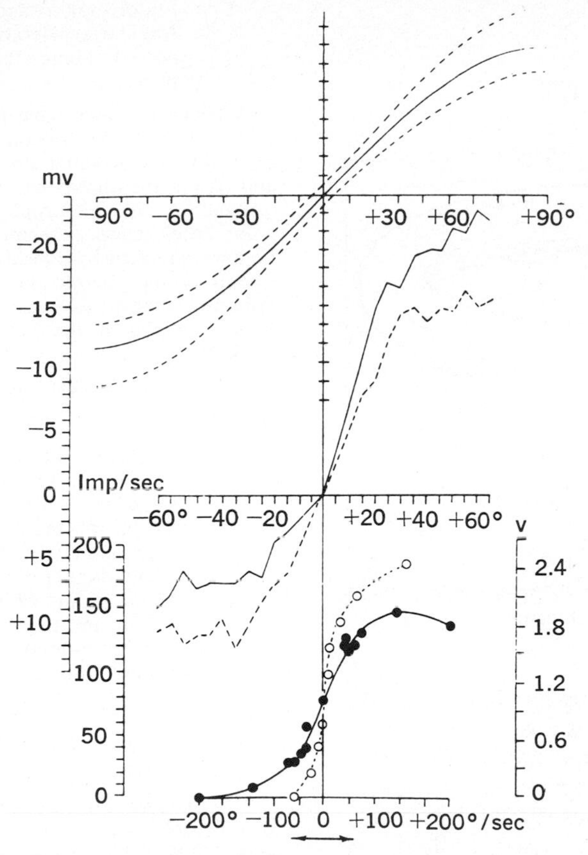

Fig. 27-8. Characteristic curves of semicircular canal receptor system.

Top: Tangential component of shearing force acting on sensory epithelium of crista; continuous line represents middle zone of crista's surface, broken lines are for declining regions; abscissa, degree of cupula deflection (arbitrary units).

Center: Changes in dc potential in millivolts inside hair cell layer of crista of horizontal canal in guinea pig. Ordinate, on left of figure, in millivolts, taking resting potential as zero; abscissa, degree of deflection as in top diagram.

Bottom: Relation between frequency of action potentials of single first-order neurons (left-hand ordinate) and angular velocity in degrees per second (kept constant during prolonged rotation before sudden stop). Right-hand ordinate: amplitude of summed action potentials corresponding to number of synchronously discharging units. Continuous line: single units of ray, after Lowenstein and Sand[57]; broken line: massive discharge from whole ampullar nerve of frog, after Ledoux.[54]

(From Trincker.[82])

angular acceleration endolymph does not flow continuously past the bent cupula, as proposed earlier. Steinhausen[78] and Dohlman,[24] by injecting ink into the semicircular canal on one side of the cupula, showed that the cupula entirely seals the ampulla and that a pressure difference across the cupula causes it to deflect but not allow endolymph to flow from one side to the other. These early observations showed the cupula swinging about its base at the crista through angles of up to ±45 degrees. These measurements are probably very large in terms of expected cupula motion in the normal physiologic range.[71] However, they clearly laid the basis for the torsion pendulum model of the semicircular canal, in which the end organ operates as an integrating angular accelerometer.[11, 36, 39]

The actual motion of the cupula for the physiologic stimulus is still open to question. Dohlman[26] pointed out that the "subcupular space" between the crista and the cupula contains elastic "veils" around each hair cell. Such an arrangement would permit sliding of the cupula over the crista and produce large hair cell shear from small forces and minimal endolymph displacement. It is possible that the large swing of the cupula may only represent a "safety valve" for accelerations beyond the physiologic range. Yet another mode of

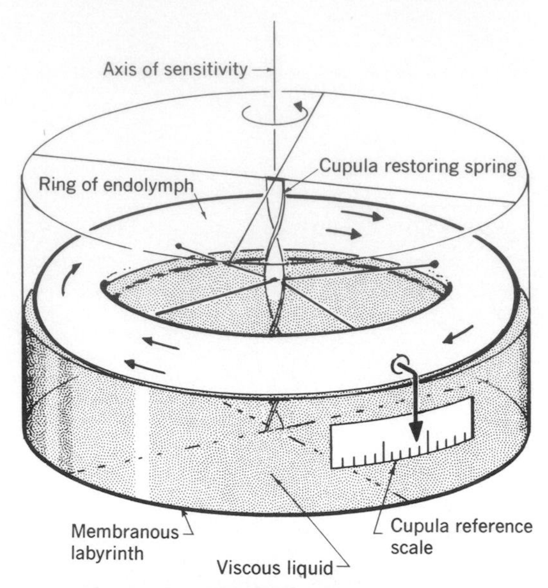

Fig. 27-9. Representation of damped torsion pendulum—analogy to semicircular canal dynamics.

motion, in which the cupula deflects at the center like a diaphragm, is possible.

The torsion pendulum model for canal function draws the analogy to an instrument consisting of a mass suspended by a wire and surrounded by a viscous fluid, as shown schematically in Fig. 27-9. The moment of inertia of the sensor is merely the moment of inertia of the ring of endolymph. The endolymph tends to remain stationary as the head is rotated, resulting in a relative flow of endolymph within the ducts of the membranous labyrinth. However, any flow of endolymph within the canal is immediately opposed by a viscous damping torque proportional to the flow rate. Finally, since endolymph cannot normally flow past the cupula, an elastic restoring force is produced on the endolymph by the displacement of the cupula from its neutral position.

The torsion pendulum mechanism is described by the second-order differential equation:

$$\ddot{\xi} + \frac{\Pi}{\Theta}\dot{\xi} + \frac{\Delta}{\Theta}\dot{\xi} = \alpha$$

in which:

- Θ = Moment of inertia of the endolymph ring, including the fluid in the ampulla
- Π = Viscous drag coefficient of the cupula-endolymph system; torque about the center of the canal per unit angular velocity of flow in the duct
- Δ = Spring constant; torque on the cupula-endolymph system per unit angular deflection of endolymph in the duct
- ξ = Endolymph deflection in the duct
- α = Angular acceleration of the labyrinth (component along the normal to the canal plane)

Various methods have been used to estimate the coefficients of the torsion pendulum model, based on theoretical calculations from labyrinth geometry and from measurements of nystagmus, subjective orientation, and electrophysiologic response to angular acceleration. The moment of inertia (Θ) can be approximated by considering the volume of endolymph in the narrow duct ($2\pi^2a^2R$) and the total volume in the utricle and the ampulla ($\backsim 2\pi B^3$) to be all located at the radius of gyration, taken to be R[71]:

$$\Theta \simeq (2\pi^2a^2R + 2\pi B^3)\rho R^2$$

where:

- a = Radius of the canal cross section (approximately 0.015 cm in human horizontal canal)[45]
- B = Radius of the cupula, modeled as a disc (approximately 0.06 cm in human horizontal canal)
- R = Mean radius of the torus (approximately 0.3 cm in human horizontal canal)
- ρ = Specific gravity of endolymph, approximately 1.00

These figures yield a value of moment of inertia:

$$\Theta \simeq 2.4 \times 10^{-4} \text{ gm-cm}^2$$

The frictional moment can be calculated from Poiseuille's law, in which the torque on the endolymph per unit average angular velocity of the endolymph with respect to the membrane is given by:

$$\Pi = 16\eta\pi^2R^3$$

for a full circle.

η is the viscosity of endolymph, approximately that of water, which is 1.0 centipoise (cp). (Measurements from 0.85 cp[77] to 1.04 cp[74] have been reported.)

The actual canal is, of course, only about three fourths of a circle in man, which would reduce Π accordingly. However, the additional contribution of cupula friction in the ampulla (regardless of its mode of motion) and of endolymph frictional effects would increase this estimate, so the expression given above is probably as good an approximation as any. For man this yields:

$$\Pi \simeq 0.043 \text{ dyne-cm-sec}$$

The ratio of the moment of inertia to the frictional moment (Θ/Π) yields the "short-time constant" of the canal and indicates the time constant involved in the mechanical integration process going from labyrinth angular acceleration to endolymph angular velocity. For man, this would be:

$$\Theta/\Pi \simeq 1/200 - 1/500 \text{ sec}$$

Other calculations showing values of $\Theta/\Pi \simeq 1/10 - 1/35$ sec appear to overestimate the time constant, although they are in closer agreement with some subjective response data. The remaining parameter, the elastic coefficient Δ, has not been estimated on theoretical grounds because of a lack of any direct torque (or pressure difference) versus displacement curves for an intact semicircular canal.

Based on curves for nystagmus and subjective response decline and on animal neural records, the important long-time constant (Π/Δ) is estimated at approximately 20 sec for man's horizontal canal; a somewhat shorter time has been estimated for the vertical canals.[49] The torsion pendulum equation for man's horizontal canal is therefore approximately:

$$\ddot{\xi} + 200\dot{\xi} + 10\xi = \alpha$$

Adaptation

The dynamic response of the torsion pendulum model for the semicircular canal is not sufficient to describe the long-term responses of nystagmus or subjective sensation during rotation or even some of the electrophysiologic records from first-order afferents.[34] The presence of processes that reduce the response to a sustained vestibular stimulus has long been recognized[38] and has recently been described mathematically.[93]

The effect of adaptation on nystagmus or on the subjective sensation of velocity is as though the cupula position, based on the torsion pendulum expression, were passed through a simple first-order adaptation filter. One interpretation is that the reference level is always moved toward the cupula output signal (thereby reducing the response) at a rate proportional to the response. Mathematically this could be expressed by:

$$R(t) = \zeta(t) \quad \frac{1}{T_a} \int_0^t R(t)dt$$

An effect of this adaptation operator is to convert a step of endolymph or cupula displacement to a decaying exponential response with time constant T_a. (Approximate time constants T_a are 30 sec for subjective sensation of velocity and 80 to 120 sec for nystagmus.)

The frictional component in canal dynamics is so great that the system is highly overdamped. For the physiologic frequency range of head movements, the cupula displacement is therefore proportional to head angular velocity (an integrating angular accelerometer); the elastic restoring torque only plays a role at very low frequencies or during prolonged motion.[60] The subjective response to vestibular stimulation, whether by natural means or through caloric or galvanic stimulation (to be described later), appears to be a sensation of angular velocity. Similarly, semicircular canal stimulation results in conjugate movements of the eyes, in the plane of the canal or canals stimulated, in a direction to compensate for the head movement.[2] Vestibular nystagmus occurs for sufficiently strong stimulation; it consists of slow conjugate eye movements that tend to stabilize the eyes in space, alternating with rapid return eye movements in the opposite direction. The slow-phase eye movements tend to compensate for head movements, so that the image of a fixed object tends to move less on the retina. This eye movement compensation accounts for better visual acuity when the head is vibrated than when the visual field is vibrated. The action of this vestibulo-ocular reflex is analogous to the stabilization of a radar platform or a gun mount on a moving ship, except that the vestibulo-ocular reflex alone normally compensates for only about 60% of head motion. (Visual tracking and the neck receptors add to the ocular stabilization.) The strength and duration of vestibular nystagmus are a strong function of the subject's mental set. Vigilance to other mental tasks increases the nystagmoid response.

Responses to rotational stimuli

The simplest test input is the acceleration (or deceleration) impulse, or velocity step. This is most easily achieved by rotating a subject at a constant speed (γ) about a vertical axis for a prolonged period (greater than 40 to 60 sec) and then bringing him suddenly to a stop. The torsion pendulum equation predicts a cupula response:

$$\xi(t) = \gamma \frac{\Theta}{\Pi} [e^{-(\Delta/\Pi)t} - e^{-(\Pi/\Theta)t}]$$

The cupula response shown in Fig. 27-10 is at first a very rapid increase, with time constant θ/Π (~1/200 sec), as the inertia of the endolymph ring carries it past the suddenly stopped labyrinth and is slowed to a stop by the Poiseuille resistance. The sensation and nystagmus correspond to rotation in a direction opposite that of the original spin. There follows a very slow decay of cupula position

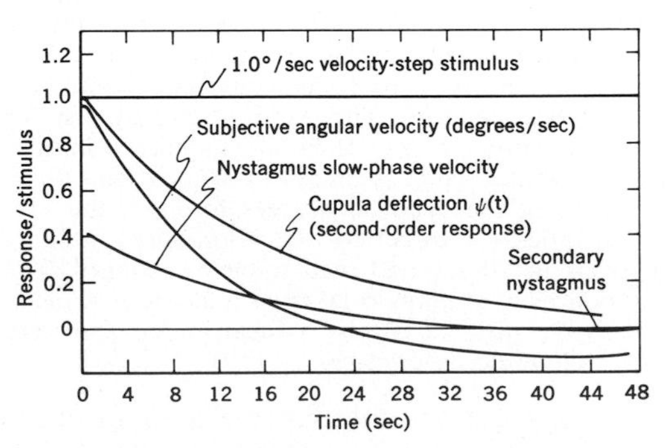

Fig. 27-10. Semicircular canal model response to acceleration impulse (velocity step). (From Young and Oman.[93])

back to normal, with time constant Π/Δ (~20 sec), which represents the weak cupula-restoring torque pushing back against the frictional torque. Adaptation comes into play to yield the subjective angular velocity and slow-phase nystagmus curves in Fig. 27-10, which are seen to decay more rapidly than the cupula and recross the origin. The overshoot in sensation of turning and the secondary phase of nystagmus are frequently seen with impulses on the order of 50 degrees/sec.

Occasional occurrence of "third effects" and later reversals, seen especially in pathologic cases, are not treated by this mathematical description. Among the interesting measures of the impulse response are the decay of slow-phase velocity and the time until nystagmus drops below threshold (usually taken as the last fast beat). Clearly, the cupula decay is an exponential with time constant Π/Δ, and when plotted on semilog paper (log response versus time), it would yield a straight line of slope Π/Δ. Indeed, the decay of a single afferent fiber discharge frequently shows such a course.[39] Subjective velocity and nystagmus also show such a course for 20 to 30 sec, since their decline is roughly exponential, and this discovery led to the clinical technique of *cupulometry* for clinical and research testing. Cupulometry uses a rotation stimulus of a sudden stop from constant velocity and measures either the slope of the decay or the duration of nystagmus.

If duration were assumed to correspond to the period when cupula deflection exceeded some threshold value of endolymph displacement (ξ_{th}), then the duration following an impulsive stop from speed γ would be given by:

$$T = \frac{\Pi}{\Delta} \ln \frac{\Theta\gamma}{\Pi\xi_{th}}$$

The difference in slope between decay of subjective velocity (typically 8 to 10 sec) and decay of nystagmus (typically 15 sec) made it clear that it was not the single mechanical time constant Π/Δ that was being measured, nor could cupula stiffness be correctly deduced. The standard Bárány test for canal function is just such an impulse stimulus. The horizontal canal is placed in a horizontal plane by inclining the head 30 degrees forward; the subject is quickly brought to a rotation speed yielding 10 turns in 20 sec and then suddenly stopped. The duration of nystagmus is taken as a measure of canal sensitivity, with 22 sec as a mean value and very large individual differences.

One cannot assume that the afferent fibers all follow the torsion pendulum model or that no adaptation is reflected at the periphery. Goldberg and Fernandez[34] found evidence that many first-order afferents in the squirrel monkey show just the type of adaptation response shown here but that others exhibit little or no adaptation.

A simple extension of the acceleration impulse is the pattern of brief acceleration followed almost immediately by rapid deceleration to a stop, such as is experienced in a sudden head movement or in the rapid head movements between fixation points executed by a spinning ballet dancer. The adaptation effects are negligible for periods of a few seconds, and the subjective velocity, given by the torsion pendulum model, closely matches the true velocity. The mental integration of angular velocity, yielding a subjective estimate of the angle turned, is a close approximation to the actual angle for rapid movements.

The subjective response and nystagmus associated with constant angular acceleration about a vertical axis are shown in Fig. 27-11. After the first few milliseconds, the inertial reaction torque is balanced by the viscous drag, and, as the cupula endolymph displacement increases, by the elastic restoring torque. Cupula deflection approaches a constant steady-state level proportional to applied acceleration. The time required for the response to exceed a small threshold depends on the magnitude of the acceleration and becomes very long as the acceleration is reduced to a level at which the steady-state response is close to the threshold. The measurement of threshold depends strongly on the indication desired, as discussed in the section on adequate stimulus.

Sinusoidal oscillation of the head and body provides a convenient input for testing the behavior of the vestibular system over the entire frequency range of normal motion. Many nerve fibers in the vestibular nerve respond linearly, with a roughly sinusoidal variation of firing rate during oscillation. Similar re-

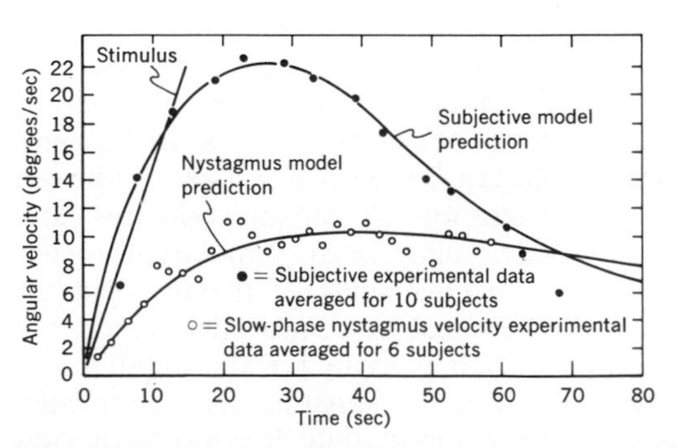

Fig. 27-11. Semicircular canal model response to acceleration step; comparison with data from Guedry and Lauver.[42] (From Young and Oman.[93])

sponses are observed in cells of the vestibular nucleus, although others discharge to only a part of the oscillation.[17]

The same sinusoidal response is apparent in the slow phase of nystagmus. As shown in Fig. 27-12, as the head and trunk are rotated passively from left to right, the eyes undergo a compensatory slow movement from right to left (at about 60% of the velocity required for stabilization), which is periodically interrupted with a fast beat to the left. (By convention this is called nystagmus to the left, or left-beating nystagmus.) All the slow phases taken together compose a sinusoidal curve of cumulative eye position.[65, 81] This curve can be compared with the stimulus sinusoid to yield the "frequency response" of the input-output system, as shown in Fig. 27-13. Both the nerve-firing frequency records and the human behavioral records for nystagmus and sensation of angular velocity show a broad frequency range from approximately 0.02 Hz to greater than 1 Hz, over which the system response is proportional to and nearly in phase with head velocity. At oscillation frequencies much below 0.02 Hz, normally encountered only on a mechanical device or vehicle, the amplitude of the response declines and the phase of the response leads the input velocity, tending at very low frequencies toward a measure of acceleration.

At frequencies above 1 to 2 Hz the situation is less clear. Both subjective sensation and neural discharge rate (at least in the ray) lag the stimulus at frequencies above 0.2 Hz. The vestibulo-ocular reflex shows no such lag, however, but indicates an increase in response amplitude. First-order canal afferents in the squirrel monkey show a similar increase in discharge rate gain (regarding velocity) and in phase lead for sinusoidal rotation at frequencies above 1 Hz.[27]*

Adequate stimulus

The estimation of behavioral thresholds to angular acceleration depends very strongly on the instructions to the subject, his level of attention, and the definition of response.[19] The lowest thresholds appear in tests in which

*The possibility of further velocity-sensitive terms of importance at high frequencies is still open. The relationship of the dimension of the semicircular canal to its sensitivity and frequency range has been explored for a wide variety of species[49] and for growth over a large range of body mass in the pike.[79] The sensitivity of the canal is nearly constant for pike of all sizes and is nearly the same for a large number of birds, mammals, and beasts having the same size cupulae.[79]

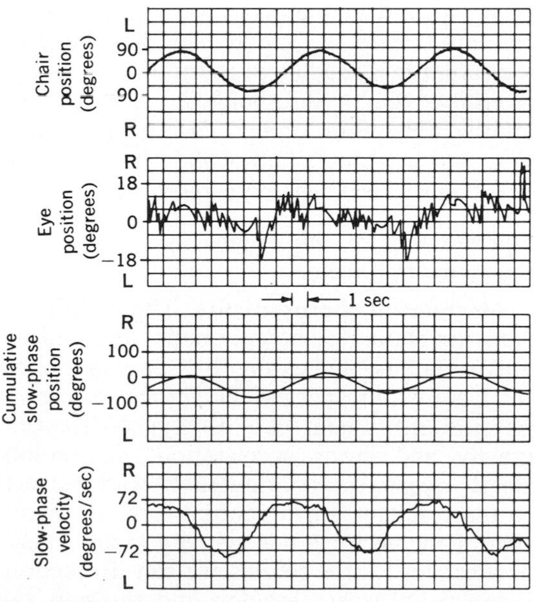

Fig. 27-12. Vestibular nystagmus in response to 0.1 Hz oscillation about a vertical axis. (From Tole and Young.[81])

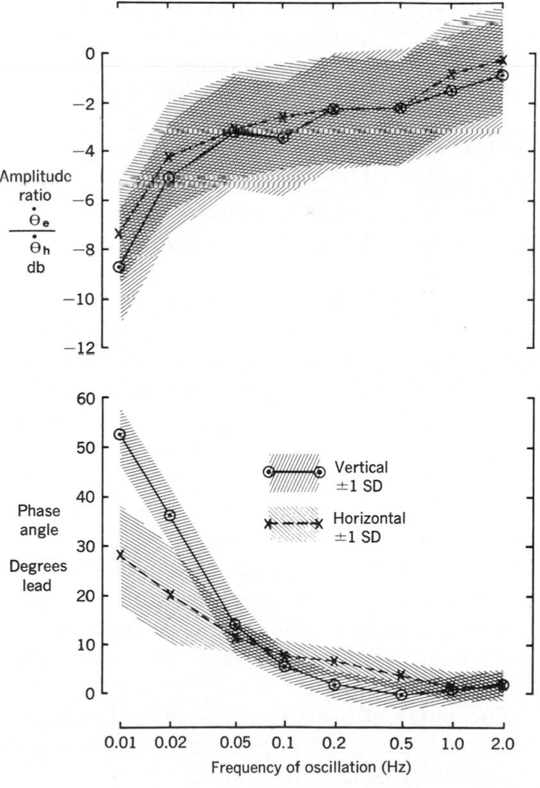

Fig. 27-13. Vestibulo-ocular frequency response evoked by sinusoidal oscillation about longitudinal axis placed in vertical and horizontal axes. Peak angular velocity of stimulus was 30 degrees/sec. Phase lag of 0 degrees indicates eye velocity in opposite direction of stimulus velocity. (Courtesy of Dr. A. J. Benson.)

the oculogyral illusion is used as an indicator. In this illusion a fixation light that is stationary with respect to the subject appears to rotate in the direction consistent with vestibular-induced sensation of movement. Experiments using this method yielded acceleration thresholds as low as 0.036 degree/sec^2, as reported by Oosterveld.[72] Thresholds for sensation of rotation about a vertical axis are approximately 0.1 to 0.2 degree/sec^2 for the horizontal canal plane and possibly up to 0.5 degree/sec^2 for the vertical canal planes.[1, 65] As mentioned previously, a brief acceleration impulse resulting in a velocity change of approximately 2 degree/sec also corresponds to the subjective sensation threshold. Nystagmus thresholds are 5 to 10 times higher than those for sensation.

The pressure drop across the cupula corresponding to these thresholds has been calculated as approximately 10^{-4} dyne/cm^2, or only 10^{-7} cm H_2O—essentially the same as the auditory threshold pressure at the tympanic membrane.[71] Because peak accelerations of the head in the normal environment are usually very brief and thus filtered by the hydrodynamic properties of the canal, pressures across the cupula rarely exceed 10^{-2} dyne/cm^2 (less than 4×10^{-5} cm H_2O).

Caloric stimulation of horizontal semicircular canal

The most common clinical technique for unilateral testing of the semicircular canals is the caloric test.[28, 43] In the Hallpike procedure the head is elevated 30 degrees from the supine position to place the horizontal semicircular canals in the vertical plane. Each ear is irrigated for 40 seconds, in turn, with water 7° C above and below body temperature. Several different measures of caloric response are used, not all of which correlate with one another. These include measurement of nystagmus duration, frequency, and especially peak slow-phase velocity with eyes open or closed.[87] Reduced vestibular response is revealed by a decrease in nystagmus response for both hot and cold stimulation in one ear.

The physical basis for the caloric test rests simply on the density gradient in endolymph produced by heating or cooling.

The canal primarily affected is the horizontal or external canal because of its proximity to the external ear. The water temperature difference from body temperature creates a smaller temperature gradient across the horizontal canal after a thermal lag with a time constant of approximately 25 sec.[77] The warmer endolymph expands with a volumetric temperature coefficient of expansion of 4.4×10^{-4}/degree C. With the canal placed in the vertical plane in a gravity field, the temperature gradient creates a torque that tends to raise the warmer, lighter endolymph and results in a force on the cupula that deflects it toward the colder side. Thus a warm stimulus in the right ear causes the cupula of the right horizontal canal to be deflected toward the utricle. The same deflection could be produced by angular acceleration to the right. Both stimuli produce right-beating nystagmus.

An expression for the torque on the cupula during caloric stimulation[77] is:

$$M \simeq 4 \times 10^{-8} \text{ gm}(\cos \phi)K(\Delta T)$$

where:

g = Gravitational constant (980 cm/sec^2)
ϕ = Angle between the plane of the horizontal canal and the vertical
ΔT = Temperature difference between 37° C and the irrigation water
K = Constant relating temperature difference across the canal to ΔT ($K \simeq 0.1$ to 0.2, according to Dohlman[24] and Cawthorne and Cobb.[18])

The minimum irrigation temperature difference that produces nystagmus is ΔT of 0.2° C to 0.5° C,[63] which is roughly consistent with the torques and pressure drops calculated across the cupula at threshold angular acceleration.[71]

Galvanic stimulation

Horizontal nystagmus and the sensation of rotation can also be produced in man by moderate galvanic stimulation. Direct current passing between each mastoid or between a single ear and a second electrode on the back of the neck produces nystagmus with a fast phase toward the cathode and a sensation of rotation toward the anode. The magnitude of the response is roughly proportional to current intensity, with a threshold for sensation of approximately 600 μamp. The site of excitation is uncertain,[88] although the vestibular nerve must be intact for a response.

Operation of otolith organs as linear accelerometers

Mechanical characteristics. The otolith organs, or statoliths, are normally regarded as gravireceptors, indicating the orientation of the head with respect to the vertical. Since no physical instrument can distinguish between gravity and linear acceleration, the otolith clearly responds to orientation with respect to the resultant gravitoinertial force or direction of the vector $\bar{g} - \bar{a}$ (gravity minus linear acceleration). (Of course, we can distinguish between linear acceleration and tilt with respect to the vertical by use of additional in-

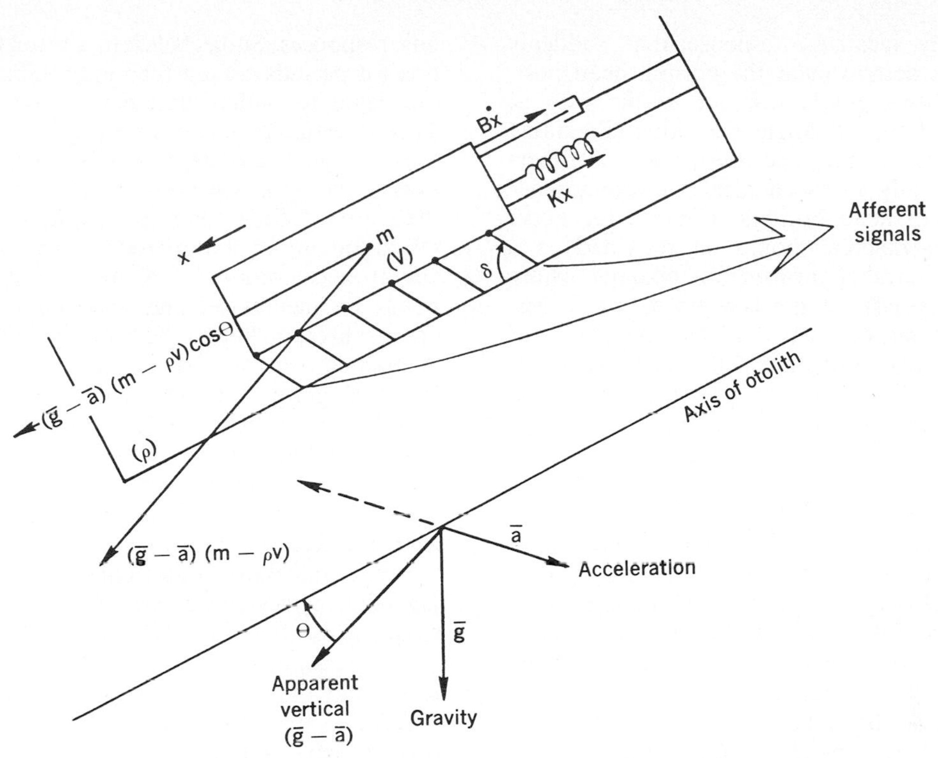

Fig. 27-14. Simplified representation of otolith function.

formation, especially sensation of rotation from the semicircular canals and orientation to the visual vertical.) A simple model for action of the utricular otolith, shown in Fig. 27-14, is a second-order system, a linear accelerometer.

The mass (M) has the following forces acting in the shearing direction of the hair cells: gravitational and linear acceleration reaction force ($m|\bar{g} - \bar{a}|\cos\Theta$); buoyant forces from the endolymph ($\rho V|\bar{g} - \bar{a}|\cos\Theta$); viscous damping forces associated with movement of the otolith through endolymph and friction in the hair cells and supporting structure ($B\dot{x}$); and elastic restoring forces resulting from the bending of hair cells and supporting structure (Kx). The equation describing movement over the macula is thus:

$$\ddot{x} + \frac{B}{m_e}\dot{x} + \frac{K}{m_e}x = \frac{1}{m_e}(m - \rho V)|\bar{g} - \bar{a}|\cos\Theta$$

where:

ρ = Density of endolymph
V = Volume of otolithic gelatinous mass
m_e = "Effective mass" of the moving sac, including the inertia of the endolymph displaced by movement of the otolith

The parameters of this equation have not been satisfactorily determined either theoretically or experimentally. The observation that some saccular nerve fibers seem to follow vibration with a sinusoidal modulation of pulse frequency up to several Hertz and the very short "indication time," on the order of 0.1 to several sec, speak in favor of a rapidly responding system.[6, 23, 84, 86] On the other hand, transient responses from most utricular nerves show an adapting response at sustained peak tilt, reducing firing rate to 40 to 60% of peak several seconds after tilting.[18] This could implicate a higher order "jerk receptor" system that is sensitive to rate of change of tilt or acceleration, as well as acceleration level. The adaptation could be either mechanical, in setting of the gelatinous mass or movement of fluid,[61] or neural.

A more realistic representation of the utricular and saccular otolith emphasizes the nonrigid characteristics of the system, especially of the otolithic membrane. The different areas of the irregular macula are covered with otoconia to varying degrees, and there is dynamic coupling among the different portions. Hair cells in the maculae are polarized in different directions, as discussed earlier, and are responsive to shearing displacement.[30]

Sensory cell responses

Electrophysiologic recordings from utricular units in the vestibular nerve in the ray, as the animal is tilted, indicate that there are several types of receptors. Some respond to the angle of tilt chiefly about one particular horizontal axis. The pattern may be either a continuous variation of firing rate with tilt angle for maximum or minimum response or a sharply localized response that suddenly increases activity near the normal head position.[57] Other fibers respond to the rate of change of the tilt angle (in either direction) as well as to the angle itself, and still others respond only to mechanical vibration, independent of head position. Clearly, the nerve from the macula should be regarded as a complex parallel information channel whose signal depends on the concert of many sensory cell responses to indicate movement of the head and direction of the apparent vertical. It is also clear from the structure of the organs that they must respond to some extent to angular as well as linear acceleration.

Perception of the vertical and of linear motion

Rotation of the head about a horizontal axis produces a tilting sensation. This sensation is closely related to the component of gravitoinertial force parallel to the "major plane" of the utricular otolith. (Although the maculae are irregularly shaped, a "major plane" may be defined; it is pitched up 30 degrees from the horizontal plane of the head.) A change in this parallel or "shear component" of reaction force can be produced on a parallel swing[6] or on a centrifuge. When the resultant gravitoinertial force is displaced from the earth vertical, the subject perceives himself as tilted (the "oculogravic illusion"[20]). The "compressive component" of force on the macula also seems to play a role. The ability to localize orientation in the absence of visual cues is quite good when the head is close to the normal erect position, consistent with the large number of position receptors with high sensitivity to small movements away from the upright. Angular localization shows a mean deviation of only approximately 0.5 degree from lateral tilt near the upright, but the error grows to about 5 degrees for tilts of 90 degrees and increases sharply for an inverted position. Nonstatolith information (visual, semicircular canal, exteroceptive) is given increased weight in determining the subjective vertical as the head is deviated from its normal position. Sudden changes in stimulus to the gravireceptors alone do not lead to an immediate change in perception of orientation, but may take as long as 1 min as the cues from other sensing modalities are "forgotten."[20]

The dynamic situation for sudden shifts in body tilt results in an overshoot in the perceived orientation as well as in most utricular unit responses. Subjects left in a tilted orientation for periods greater than approximately 15 sec begin to underestimate the angle of tilt. This tilt estimate decays with time.

Although gravireceptors indicate the direction of the apparent vertical quite accurately, they are of little help in distinguishing between tilt about a horizontal axis and horizontal acceleration. The interpretation depends on mental set and contributions from other senses, including the semicircular canals. A number of motion illusions result from this ambiguity; e.g., in airplanes a forward acceleration in steady flight, which pushes the passenger back in his seat, may be misinterpreted as a pitch, nose upward. Although the otolith neural response is rapid, the subjective indication of direction of motion in a horizontal plane shows considerable phase lag relative to stimulus velocity at oscillation frequencies above 0.05 Hz and phase lead at lower frequencies.[65, 85, 92] Subjective velocity tests indicate a threshold acceleration of 0.005 *g* in the "plane of the utricle"[92] and for vertical accelerations normal to this plane. Similar thresholds (0.002 to 0.006 *g*) are elicited on the parallel swing, in which subjects are oscillated in approximately horizontal motion.[51]

Eye movements

Stimulation of the otoliths by linear acceleration or tilt produces compensatory eye movements that tend to stabilize the eyes in space around all three axes. Lateral acceleration causes the eyes to deviate in the direction opposite the acceleration, free fall produces ocular deviation upward, etc. Perhaps the most thoroughly studied of these movements is "counterrolling," or conjugate compensatory torsional movement of the eyes about the primary axis during lateral tilt.[66, 67, 90] Although the maximum angle of counterroll in man is small (5 to 6 degrees at 60 to 90 degrees of lateral head tilt) and difficult to measure, it is of interest because it is one of the few measurable signs of reflex otolith function. Counterrolling depends on the presence of gravity or linear acceleration and the functioning of otolith organs but does not require canal function.

Cross coupling between linear and angular sensors

The notion that semicircular canals respond only to angular motion and otoliths respond only to linear motion is clearly an oversimplification. There is fragmentary but increasing experimental evidence that the outputs of the entire vestibular apparatus are integrated for sensation of motion, postural control, and eye movements, and that both canals and otoliths are subject to some influence from both linear and angular acceleration.[8, 54] As for the structural possibilities for this interaction, there are many, none of them proved.

Ter Braak[80] pointed out that only a very small difference in density between cupula and endolymph or between endolymph and perilymph would be sufficient to allow gravity to influence canal responses, since the cupula would tend to "sink" or "float" if not neutrally bouyant. Although no difference between endolymph and cupula density has been found, Money et al.[68] found that pigeon perilymph is 0.11% denser than endolymph and that the membranous ducts are even denser, facts that could account for a deformation of the ampulla under linear acceleration with resulting displacement of the cupula. Steer[76] showed how a small distention of the duct could act as a "roller pump" to displace the cupula under conditions of a rotating linear acceleration vector. Although experiments with plugged semicircular canals tend to indicate that the canal responses to linear acceleration are not responsible for some of the interesting cross-coupling effects, the results are not conclusive. Lowenstein[8] has shown the influence of gravity on semicircular canal function directly by demonstrating how tilt angle of the head modifies the firing rate of canal nerves.

The possibility of angular acceleration affecting the otolith organs is evident from their structure. Angular acceleration of the head would tend to rotate the gelatinous mass with respect to the macula.

Some observations on cross coupling include the influence of linear acceleration on vestibular nystagmus.[53, 62, 70] The low-frequency sensitivity of the slow-phase velocity is 5 to 10 degree/sec/g[91] (g is the unit of gravitational acceleration, 980 cm/sec^2). Linear acceleration on a centrifuge can similarly add to the canal-induced nystagmus,[53] and the orientation of the head seems to inhibit canal-induced nystagmus for the head-upright and facilitate nystagmus for the head-inverted positions. Motions that produce a rotating linear acceleration vector are particularly effective in inducing continuous unidirectional nystagmus with no rotation of the head. Rotation at constant velocity about a horizontal axis produces a sustained compensatory nystagmus (unlike such rotation about a vertical axis) and a periodic component, both of which decay rapidly when the subject is brought to a stop.[15, 21] A similar phenomenon occurs when a subject is counterrotated on the end of a centrifuge so that his head is nonrotating but subject to a rotating linear acceleration stimulus.[48, 76] Benson et al.[16] have shown this same effect of a rotating linear acceleration vector on a horizontal canal–sensitive unit in the brainstem. Both mechanical influence of linear acceleration on the semicircular canal and otolith influence through integration at the vestibular nucleus or elsewhere must be considered as possibilities for this cross coupling.

Habituation

The principal behavioral measures of vestibular function exhibit a high degree of plasticity in the system as vestibular cues are given different weight and interpretation with exposure to a new environment. Perhaps the most elementary habituation of this type takes place in the first few weeks of life. Vestibular nystagmus develops with low threshold and long decay time in the first weeks of life.[37] This gradually changes over the next 2 months, the child then exhibiting the typical adult response to rotation. The ability of people who are regularly subjected to repeated or unusual motions (e.g., ballet dancers, figure skaters, fighter pilots, sailors, etc.) to control their vestibular responses, both in terms of nystagmus and sensation, supports a theory of central regulation of habituation.[38] The particular habituation is reduced when not reinforced with the motion pattern. On an electrophysiologic level, habituation to prolonged oscillation has been observed in recordings from frog ampullar nerve.[33] The habituation was apparently mediated by central efferent activity, since it was present to a certain extent before the animal was oscillated and disappeared entirely in the isolated preparation.

The usefulness of vestibular habituation is evident in exposure to unusual environments, e.g., getting one's "sea legs" or becoming accustomed to the bizarre vestibular responses when moving one's head in a rotating environment.[41] This habituation is customarily retained for several days after prolonged exposure. The process of vestibular habituation takes place more rapidly during active movements than during passive motion.

Vestibular role in posture control

The control of posture in mammals and especially in bipeds involves a complex integration of exteroceptive, vestibular, and visual sensory information. The narrow base for standing and the high position of the center of gravity allow man's rigid body mechanics to be modeled as an inverted pendulum, which must be actively balanced to keep him from falling. The stretch reflex acts rapidly during sway but is normally not stiff enough to prevent falling. Vision seems to play a role in defining the vertical and, to a lesser extent, in signaling rate of fall. The vestibular apparatus plays a role in initiating postural reflexes, damping the movement to prevent oscillation, and defining the vertical for realignment.[69] The vertical semicircular canals signal angular velocity of the head about a horizontal axis and initiate appropriate postural reflexes, stabilizing the head in space except during voluntary head movements. Thus head movements initiate movements of the limb and neck muscles that return the head to its previous position and line up the neck and trunk.[9] Each vertical canal is linked to appropriate antigravity muscles so that an animal's head tilt forward and to the right stimulates the right superior and left posterior canals, resulting in increased antigravity activity in the right forelimb to oppose the fall.

The canals also appear to provide system damping by rate feedback of head position that is necessary to prevent oscillation of the inverted pendulum postural control system. In this role the vestibular system acts in concert with the cerebellum in postural damping. The otoliths are stimulated by both linear acceleration and tilt with respect to the gravity vector and are responsible for labyrinthine reflexes that act to reestablish the "normal position" of the head with respect to the apparent vertical. The labyrinthine reflexes act in consonance with, and opposite in direction to, the neck reflexes, so that the two effects cancel for head movements about a fixed trunk.

When one labyrinth is removed completely, postural reactions are at first disturbed, and the subject maintains a posture with the operated side down. In time the bidirectional signals from the remaining labyrinth are used to provide postural and nystagmus responses that appear almost normal.

A man with no labyrinthine function is normally not handicapped when no great demands are placed on his capacities for orientation and balance. Under normal circumstances and in the absence of rapid head movement, visual, exteroceptive, and tactile cues can substitute for vestibular regulation, provided that there is sufficient light for adequate vision and that there are firm regular surfaces for support.

REFERENCES

General reviews

1. Brodal, A., and Pompeiano, O., editors: Basic aspects of central vestibular mechanisms, Progr. Brain Res. **37:**entire issue, 1972.
2. Cohen, B.: Vestibulo-ocular relations. In Bach-y-Rita, P., Collins, C. C., and Hyde, J. E., editors: The control of eye movements, New York, 1971, Academic Press, Inc.
3. Engstrom, H.: The first order vestibular neurons. In Fourth symposium on the role of the vestibular organs in space exploration, NASA SP-187, Washington, D. C., 1968, National Aeronautics and Space Administration.
4. Engstrom, H., Lindeman, H. H., and Ades, H. A.: Anatomical features of the auricular sensory organs. In Second symposium on the role of the vestibular organs in space exploration, NASA SP-115, Washington, D. C., 1966, National Aeronautics and Space Administration.
5. Gacek, R. R.: Anatomical evidence for an efferent vestibular pathway. In Third symposium on the role of the vestibular organs in space exploration, NASA SP-152, Washington, D. C., 1967, National Aeronautics and Space Administration.
6. Jongkees, L. B. W.: On the otoliths: their function and the way to test them. In Third symposium on the role of the vestibular organs in space exploration, NASA SP-152, Washington, D. C., 1967, National Aeronautics and Space Administration.
7. Lowenstein, O.: The functional significance of the ultrastructure of the vestibular end organs. In Second symposium on the role of the vestibular organs in space exploration, NASA SP-115, Washington, D. C., 1966, National Aeronautics and Space Administration.
8. Lowenstein, O.: Physiology of vestibular receptors. In Pompeiano, E. O., and Brodal, A., editors: Basic aspects of central vestibular mechanisms, Amsterdam, 1972, Elsevier Press.
9. Roberts, T. D. M.: Neurophysiology of postural mechanisms, New York, 1967, Plenum Press.
10. Spoendlin, H. H.: Ultrastructure of the vestibular sense organ. In Wolfson, R. J., edtor: The vestibular system and its diseases, Philadelphia, 1966, University of Pennsylvania Press.
11 Van Egmond, A. J., Groen, J. J., and Jongkees, L. B. W.: The mechanics of the semicircular canal, J. Physiol. **110:**1, 1949.
12. Young, L. R.: The current status of vestibular system models, Automatica **5:**369, 1969.

Original papers

13. Adrian, E. D.: Discharges from vestibular receptors in the cat, J. Physiol. **101:**389, 1943.
14. Benson, A. J.: Interactions between semicircular canals and gravireceptors. In Busby, D. E., editor: Recent advances in aerospace medicine, Dordrecht, Holland, 1969, D. Reidel Publishing Co.
15. Benson, A. J., and Bodin, M. A.: Interaction of linear and angular acceleration on vestibular receptors in man, Aerospace Med. **36:**144, 1966.
16. Benson, A. J., Guedry, F. E., Jr., and Jones, G. M.: Response of lateral semicircular canal units in brain stem to a rotating linear acceleration vector, J. Physiol. **191:**26P, 1967.
17. Benson, A. J., Guedry, F. E., Jr., and Jones, G. M.: Response of semicircular canal-dependent units in vestibular nuclei to rotation of a linear acceleration vector without angular acceleration, J. Physiol. **210:**475, 1970.
18. Cawthorne, T., and Cobb, W. A.: Temperature changes in the perilymph space in response to caloric stimulation in man, Acta Otolaryngol. **44:**580, 1954.
19. Clark, B.: Thresholds for the perception of angular acceleration in man, Aerospace Med. **38:**443, 1967.

20. Clark, B., and Graybiel, A.: Some factors contributing to the delay in the perception of the oculogravic illusion. In The role of the vestibular organs in the exploration of space, NASA SP-77, Washington, D. C., 1965, National Aeronautics and Space Administration.
21. Corriera, M. J., and Guedry, F. E., Jr.: Modification of vestibular responses as a function of rate of rotation about an earth-horizontal axis, NAMI-957, Pensacola, Fla., 1966, Naval Aerospace Medical Institute.
22. Corvera, J., Hallpike, C. S., and Schuster, E. H. J.: A new method for the anatomical reconstruction of the human macular planes, Acta Otolaryngol. **49:**4, 1958.
23. De Vries, H.: The mechanics of the labyrinth otoliths, Acta Otolaryngol. **38:**262, 1950.
24. Dohlman, G. F.: Some practical and theoretical points in labyrinthology, Proc. R. Soc. Med. **50:**779, 1935.
25. Dohlman, G. F.: Secretion and absorption of the endolymph. In Third symposium on the role of the vestibular organs in space exploration, NASA SP-152, Washington, D. C., 1967, National Aeronautics and Space Administration.
26. Dohlman, G. F.: The attachment of the cupulae, otolith and tectorial membranes to the sensory cell areas, Acta Otolaryngol. **71:**89, 1971.
27. Fernandez, C., and Goldberg, J. M.: Physiology of first order afferents innervating the semicircular canals of the squirrel monkey. II. The response to sinusoidal stimulation and the dynamics of the peripheral vestibular system, J. Neurophysiol. **34:**660, 1971.
28. Fitzgerald, G., and Hallpike, C. S.: Studies in human vestibular function. I. Observations on the directional preponderance of caloric nystagmus resulting from cerebral lesions. In Wolfson, R. J., editor: The vestibular system and its diseases, Philadelphia, 1966, University of Pennsylvania Press.
29. Flock, A.: Sensory transduction in hair cells. In Lowenstein, W. R., editor: Handbook of sensory physiology, Berlin, 1971, Springer Verlag, vol. 1.
30. Fluur, E.: The interaction between the utricle and the saccule, Acta Otolaryngol. **69:**17, 1970.
31. Gacek, R. R.: Efferent component of the vestibular nerve. In Rasmussen, G. L., and Windle, W. F., editors: Neural mechanisms of the auditory and vestibular systems, Springfield, Ill., 1960, Charles C Thomas, Publisher.
32. Gernandt, B.: Vestibular mechanisms. In Field, J., Magoun, H. W., and Hall, V. E., editors, Neurophysiology section: Handbook of physiology, Baltimore, 1960, The Williams & Wilkins Co.
33. Goetmakers, R.: Vestibular adaptation in Rana, Adv. Otorhinolaryngol. **17:**107, 1970.
34. Goldberg, J. M., and Fernandez, C.: Physiology of first order afferents innervating the semicircular canals of the squirrel monkeys (parts I, II, and III), J. Neurophysiol. **34:**635, 661, 676, 1971.
35. Groen, J. J.: Cupulometry, Laryngoscope **67:** 894, 1957.
36. Groen, J. J.: Vestibular stimulation and its effects, from the point of view of theoretical physics, Confin. Neurol. **21:**380, 1961.
37. Groen, J. J.: Postnatal changes in vestibular reactions, Acta Otolaryngol. **56:**390, 1963.
38. Groen, J. J.: Central regulation of vestibular function. In Busby, D. E., editor: Recent advances in aerospace medicine, Dordrecht, Holland, 1970, D. Reidel Publishing Co.
39. Groen, J. J., Lowenstein, O., and Vendrik, J. H.: The mechanical analysis of the responses from the end organs of the horizontal semicircular canals in the isolated elasmobranch labyrinth, J. Physiol. **117:**329, 1952.
40. Gualtierotti, T., and Alltucker, D.: The relationship between the unit activity of the utricle-saccule of the frog and information transfer. In Second symposium on the role of the vestibular organs in space exploration, NASA SP-115, Washington, D. C., 1966, National Aeronautics and Space Administration.
41. Guedry, F. E., Jr.: Habituation to complex vestibular stimulation in man: transfer and retention of effects from twelve days of rotation at 10 rpm, Percept. Mot. Skills **21:**494, 1965.
42. Guedry, F. E., Jr., and Lauver, L. S.: The oculomotor and subjective aspect of the vestibular reaction during prolonged constant angular acceleration, Report No. 438, Fort Knox, Ky., 1960, U. S. Army Medical Research Laboratory.
43. Hallpike, C. S.: The caloric test: a review of its principles and practice with especial reference to the phenomenon of directional preponderance. In Wolfson, R., editor: The vestibular system and its diseases, Philadelphia, 1966, University of Pennsylvania Press.
44. Hardy, M.: Observations on the innervation of the maculi saculi in man, Anat. Rec. **59:** 403, 1934.
45. Igareshi, M.: Dimensional study of the vestibular and organ apparatus. In Second symposium on the role of the vestibular organs in space exploration, NASA SP-115, Washington, D. C., 1966, National Aeronautics and Space Administration.
46. Iurato, S.: Submicroscopic structure of the inner ear, Oxford, 1967, Pergamon Press, Ltd.
47. Johnson, W. H.: The importance of otoliths in disorientation, Aerospace Med. **35:**9, 1964.
48. Jones, G. M.: Transfer function of labyrinthine volleys through the vestibular nuclei. In Pompeiano, O., and Brodal, A., editors: Basic aspects of vestibular control mechanisms, Amsterdam, 1971, Elsevier Press.
49. Jones, G. M., and Spells, K. E.: A theoretical and comparative study of the functional dependence of the semicircular canal upon its physical dimensions, Proc. R. Soc. Lond. (Biol.) **157:**403, 1963.
50. Jones, G. M., Barry, W., and Kowalsky, N.: Dynamics of the semicircular canals compared in yaw, pitch and roll, Aerospace Med. **35:** 984, 1964.
51. Jongkees, L. B. W., and Groen, J. J.: The nature of the vestibular stimulus, J. Laryngol. **61:**529, 1946.
52. Jongkees, L. B. W., and Philipszoon, A. J.: Nystagmus provoked by linear acceleration, Acta Physiol. Pharmacol. Neerl. **10:**239, 1962.

53. Lansberg, M. P., Guedry, F. E., Jr., and Graybiel, A.: Effect of changing resultant linear acceleration relative to the subject on nystagmus generated by angular acceleration, Aerospace Med. **36:**456, 1965.
54. Ledoux, A.: Activate electrique des nerfs des canaux semicirculaire de saccule et de l'utricle chez la grenauille, Acta Otorhinolaryngol. **3:** 335, 1949.
55. Lowenstein, O.: The effect of galvanic polarization on the impulse discharge from sense endings in the isolated labyrinth of the thornback ray (Raja clavata), J. Physiol. **127:**104, 1955.
56. Lowenstein, O., and Roberts, T. D. M.: Oscillographic analysis of the responses of the otolith organs of the thornback ray (Raja clavata), J. Physiol. **110:**392, 1949.
57. Lowenstein, O., and Sand, A.: The mechanism of the semicircular canal. A study of responses of single fibre preparations to angular accelerations and to rotation of constant speed, Proc. R. Soc. Lond. (Biol.) **129:**256, 1940.
58. Lowenstein, O., and Wersäll, J.: A functional interpretation of the electron microscopic structure of the sensory hair cells in the cristae of the elasmobranch Raja clavata in terms of directional sensitivity, Nature **184:** 1807, 1959.
59. Mach, E.: Grundlinien der Lehre von den Bewegungsempfindungen, Leipzig, 1875, Engelman.
60. Mayne, R.: The dynamic characteristics of the semicircular canals, J. Comp. Physiol. Psychol. **43:**309, 1950.
61. Mayne, R.: The functions and operating principles of the otolith organs. II. The mechanics of the otolith organs, Arizona Div. Report GERA-1112, Litchfield Park, Ariz., 1966, Goodyear Aerospace Corp.
62. McCabe, B. F.: Nystagmus response of the otolith organs, Laryngoscope **74:**372, 1964.
63. McLeod, M. E., and Meek, J. C.: A threshold caloric test: results in normal subjects, NASA R-47, Washington, D. C., 1962, U. S. Naval School of Aviation Medicine.
64. McNally, W. J., and Tait, J.: Some results of section of particular nerve branches to the ampullae of the four vertical semicircular canals of the frog, Q. J. Exp. Physiol. **23:** 147, 1933.
65. Meiry, J. L.: Vestibular and proprioceptive stabilization of eye movements. In Bach-y-Rita, P., Collins, C. C., and Hyde, J. E., editors: The control of eye movements, New York, 1971, Academic Press, Inc.
66. Miller, E. F., II: Counterrolling of the human eyes produced by head tilt with respect to gravity, Acta Otolaryngol. **54:**479, 1961.
67. Miller, E. F., II, and Graybiel, A.: Otolith function as measured by ocular counterrolling. In The role of the vestibular organs in the exploration of space, NASA SP-77, Washington, D. C., 1965, National Aeronautics and Space Administration.
68. Money, K. E., et al.: Physical properties of fluids and structures of vestibular apparatus of the pigeon, Am. J. Physiol. **220:**140, 1971.
69. Nashner, L. M.: A model describing vestibular detection of body sway motion, Acta Otolaryngol. **72:**429, 1971.
70. Niven, J. I., Carroll, H. W., and Correia, M. J.: Elicitation of horizontal nystagmus by periodical linear acceleration, NAMI-953, Pensacola, Fla., 1965, Naval Aerospace Medical Institute.
71. Oman, C. M., and Young, L. R.: The physiological range of pressure difference and cupula deflections in the human semicircular canal, Acta Otolaryngol. **74:**324, 1972.
72. Oosterveld, W. J.: Threshold stimulation of horizontal semicircular canals, Aerospace Med. **41:**386, 1970.
73. Outerbridge, J. S., and Jones, G. M.: Reflex vestibular control of head movement in man, Aerospace Med. **42:**935, 1971.
74. Rauch, S.: La biochimie de l'endolymphe et de la perilymphe. C. R. Soc. France Otorhinolaryngol. **56:**238, 1959.
75. Schöne, H.: On the role of gravity in human spatial orientation, Aerospace Med. 35:764, 1964.
76. Steer, R. W., Jr.: Progress in vestibular modelling. I. Response of semicircular canals to constant relation in a linear acceleration field. In Fourth symposium on the role of the vestibular organs in space exploration, NASA SP-187, Washington, D. C., 1968, National Aeronautics and Space Administration.
77. Steer, R. W., Jr., Li, Y. T., Young, L. R., and Meiry, J. L.: Physical properties of the labyrinthine fluids and quantification of the phenomenon of caloric stimulation. In Third symposium on the role of the vestibular organs in space exploration, NASA SP-152, Washington, D. C., 1967, National Aeronautics and Space Administration.
78. Steinhausen, W.: Über den Nachweis der Bewegung der Cupula in der intakten Bogengangsampulle des Labyrinthes bei der naturlichen rotatorischen und calorischen Reïzung, Pfluegers Arch. **228:**322, 1931.
79. Ten Kate, J. H., van Barneveld, H. H., and Kuiper, J. W.: The dimensions and sensitivities of semicircular canals, J. Exp. Biol. **53:**501, 1970.
80. Ter Braak, J. W.: Kann der Bogengangsapparat durch geradlinige Beschleunigung gereizt werden? Pfluegers Arch. **238:**327, 1936.
81. Tole, J., and Young, L. R.: MITNYS: A hybrid program for on-line analysis of nystagmus, Aerospace Med. **42:**508, 1971.
82. Trincker, D.: The transportation of mechanical stimulus in nervous excitation, Symp. Soc. Exp. Biol. **16:**289, 1962.
83. Udo de Haes, H. A., and Schone, H.: Interaction between semicircular canals and statolith organs in apparent vertical investigations on the effectiveness of the statholith organs, Acta Otolaryngol. **69:**250, 1970.
84. Von Békésy, G.: Pressure and shearing forces as stimuli of labyrinthine epithelium, Arch. Otolaryngol. **84:**122, 1966.
85. Walsh, E. G.: Role of the vestibular apparatus in the perception of motion on a parallel swing, J. Physiol. **155:**506, 1961.
86. Walsh, E. G.: The use of impulsive mechanical stimuli of variable duration in the evaluation

of the indication time of the otolith organs, J. Laryngol. **80:**1218, 1966.

87. Weiss, A. D.: Calorization in relation to otoneurologic diagnosis. In Wolfson, R. J., editor: The vestibular system and its diseases, Philadelphia, 1966, University of Pennsylvania Press.
88. Weiss, A., and Tole, J.: Effect of galvanic vestibular stimulation on rotation testing, Adv. Oto-Rhinol-Laryngol. **19:**311, 1973.
89. Wersäll, J.: Studies on the structure and innervation of the sensory epithelium of the cristae ampullares in the guinea pig, Acta Otolaryng. suppl. 126, 1956.
90. Woellner, R. C., and Graybiel, A.: Counterrolling of the eyes and its dependence on the magnitude of gravitation or inertial force acting laterally on the body, J. Appl. Physiol. **14:**632, 1959.
91. Young, L. R.: Cross-coupling between effects of linear and angular acceleration on vestibular nystagmus. In Dichgens, J., and Bizzi, E., editors: Cerebral control of eye movements and motion perception, Basel, 1972, S. Karger.
92. Young, L. R., and Meiry, J. L.: A revised dynamic otolith model, Aerospace Med. **39:**606, 1965.
93. Young, L. R., and Oman, C. M.: Model for vestibular adaptation to horizontal rotation, Aerospace Med. **40:**1076, 1969.

28

ELWOOD HENNEMAN

The cerebellum

The cerebellum is a highly organized center that exerts a regulatory influence on muscular activity. It receives a continuous stream of impulses relayed from receptors in muscles, joints, tendons, and skin and from visual, auditory, and vestibular end organs. These impulses do not mediate conscious sensations, but they supply the sensory cues essential to the control of movement. Signals from the cerebral cortex and other motor regions also reach the cerebellar nuclei. Some of this influx terminates in the cerebellar nuclei, but most of it is distributed through an elaborate network to the Purkinje cells of the cortex. The axons of these cells, which are the sole efferent fibers of the cortex, terminate in the cerebellar nuclei. The nuclear cells are thus subject to the effects of afferent impulses, which reach them directly, and to the integrated outflow of the cortex. From the nuclei, fibers pass to the thalamus, red nucleus, vestibular nuclei, and reticular formation, (RF) of the brainstem. Through these connections the cerebellum influences motor centers from the cerebral cortex to the spinal motor neurons, modifying the control of muscular action from the level of its inception to that of its execution.

MORPHOLOGIC SUBDIVISIONS

Early in the course of its development the cerebellum is subdivided by transverse fissures into a series of primary lobules. The continuity between the medial and lateral parts of each of these lobules is apparent at this time. Later the appearance of secondary fissures and the disproportionate growth of the lateral portions of some lobules obscure this simple arrangement. In the 175 mm human embryo, for example (Fig. 28-1), the continuity of the pyramis and paramedian lobules is still obvious, whereas that of the uvula and paraflocculi is disappearing. Further growth makes it appear that the medial and lateral parts of the cerebellum are separate and unrelated, except in the anterior lobe and lobulus simplex. Vestiges of the original continuity remain, however, in the form of slender cortical bridges or fibrous cords in the depths of the fissures. They are of physiologic as well as anatomic significance because experimental studies indicate that the medial and lateral portions of each lobule are closely related functionally.

Authorities differ somewhat in their modes of subdividing the cerebellum.* According to the latest comparative studies of Larsell,[40] there are ten "primary" lobules in mammals. These are shown in Fig. 28-2 as they appear in sagittal section and on the surface of the macaque cerebellum. In sequence they are as follows: I, the *lingula,* a small lobule adjacent to the anterior medullary velum; II and III, a pair, called together the *lobulus centralis*; IV and V, a larger pair called the *culmen;* VI, the *lobulus simplex;* VII, the *folium* and *tuber vermis*, which are continuous laterally with the *ansiform lobules;* VIII, the *pyramis,* which is continuous with the *paramedian lobules;* IX, the *uvula,* which is continuous with the *ventral paraflocculi*†; and X, the *nodulus,* adjacent to the posterior medullary velum and continuous with the *flocculi.*

Lobules I to V, the lingula, centralis, and culmen, constitute the *anterior lobe,* which is separated from the rest of the cerebellum by the *primary fissure*. The lateral portion of lobule VII, the ansiform lobule, is divided

*On the basis of embryologic studies, Larsell[38] proposed that there are two fundamental morphologic subdivisions, the corpus cerebelli and the flocculonodular lobe. The posterolateral fissure separating them is the first to appear in the embryo. The corpus cerebelli may in turn be divided into anterior and posterior lobes by the primary fissure, which is the second to appear. The best physiologic evidence, however, makes it unlikely that the three lobes thus defined correspond to functionally homogeneous entities as some have proposed.[22] Subdivision of the cerebellum into archi-, paleo-, and neocerebellar portions on the basis of their phylogeny is justifiable but not very useful.

†The paraflocculus consists of dorsal and ventral portions. According to Larsell[40] and Jansen and Brodal,[35] the dorsal paraflocculus is usually connected with the base of the pyramis, sometimes with the uvula, and occasionally with both.

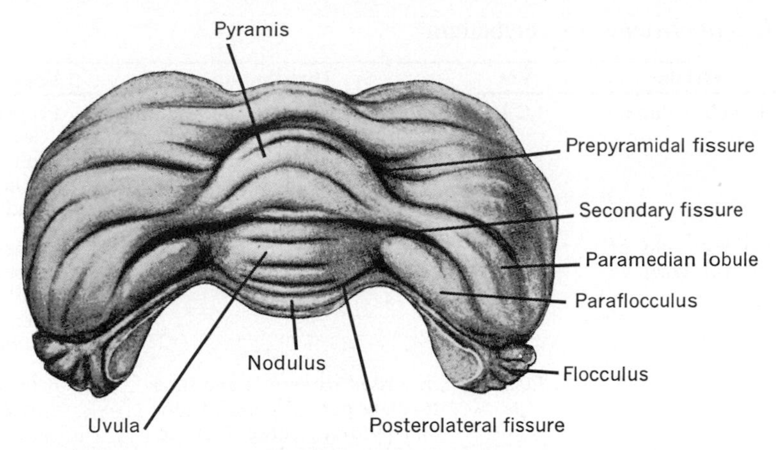

Fig. 28-1. Posterior surface of cerebellum in 175 mm human fetus. (Relabeled from Larsell.[39])

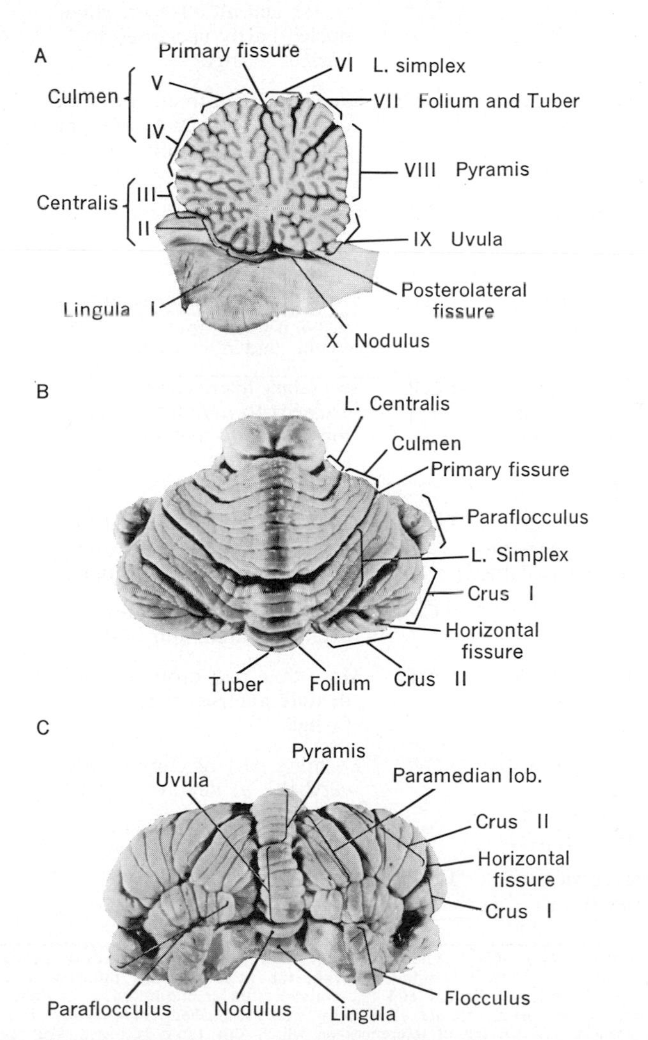

Fig. 28-2. Photographs of three specimens of cerebellum of macaque, prepared to show morphologic subdivisions as described in text. Fissures were opened and some were packed with cotton. **A,** Sagittal section. **B,** Dorsal view. **C,** Ventral view.

Table 28-1. Afferent systems to cerebellum*

Tract	Origin	Via	Distribution	Impulses transmitted
Dorsal spino-cerebellar	Clarke's column (T_1-L_2)	I.C.P.	Chiefly uncrossed to vermis and intermediate part of anterior lobe and pyramis; some fibers to tuber, uvula, and medial part of paramedian lobule	Proprioceptive (muscles and joints) and exteroceptive (skin), from trunk, hind limb, and tail
Ventral spino-cerebellar	"Border" cells of ventral horn	S.C.P.	Crossed and uncrossed to vermis of anterior lobe	Proprioceptive (muscles and joints) and exteroceptive (skin), from all parts of body
Cuneocerebellar	External arcuate nucleus	I.C.P.	Uncrossed to vermis and intermediate part of anterior lobe and pyramis; some fibers to uvula and tuber	Proprioceptive, from upper limb and neck
Olivocerebellar	All parts of inferior olive	S.C.P.	Chiefly crossed to all parts of cortex and all intracerebellar nuclei; partly uncrossed to nucleus fastigii	From all levels of spinal cord; from higher nuclei and from cerebral cortex
Pontocerebellar	All parts of pontine gray	M.C.P.	Chiefly crossed to all cortex except nodulofloccular lobe; partly uncrossed to vermis	From all 4 lobes of cerebrum, spinal cord, and other centers
Reticulo-cerebellar	Lateral reticular nucleus	I.C.P.	Uncrossed to entire cerebellar cortex	From all levels of spinal cord and from higher levels
	Paramedian reticular nucleus	I.C.P.	More than half uncrossed to anterior lobe; some to pyramis, uvula, and nucleus fastigii	From higher levels, including cerebral cortex
Vestibulo-cerebellar	Vestibular nuclei, chiefly medial and descending; some direct vestibular root fibers	I.C.P.	Secondary fibers (crossed and uncrossed) to nodulofloccular lobe, some to uvula and nucleus fastigii; primary fibers to same areas, uncrossed	Vestibular
Perihypo-glosso-cerebellar	Nucleus of Roller Nuclear praepositus Nucleus intercalatus	I.C.P.	More than half uncrossed to anterior lobe; some to pyramis, uvula, and nucleus fastigii	Unknown
Tectocerebellar	Quadrigeminal bodies	S.C.P.	Chiefly crossed, probably to declive, folium, and tuber	Auditory and visual
Rubrocerebellar	Caudal two thirds of red nucleus	S.C.P.	More than half crossed, chiefly to dentate nucleus; some to nucleus fastigii	Cerebral?
Trigemino-cerebellar	Direct sensory fibers; secondary fibers from all parts of trigeminal nucleus	I.C.P.	Forming part of commissura cerebelli; to dentate nucleus	Tactile and proprioceptive, from face and jaw
Lateral cervical cerebellar	Lateral cervical nucleus ($C_{1,2}$)	I.C.P.	Unknown	From all levels of spinal cord

*Condensed, with modifications, from a table supplied by Brodal. I am indebted to Professor Brodal for compiling these data on the afferent connections. In addition to older, well-known work, the table contains information derived from more recent investigations, especially those of Professor Brodal and his collaborators, including some of their unpublished results. It should be emphasized that the connections listed are not all equally well established anatomically. For a detailed description of the afferent systems to the cerebellum and the list of references on which this table is based, the reader may consult Jansen and Brodal.[35]

by the *horizontal* or *intercrural fissure* into *crus I* and *crus II,* which are continuous medially with the folium and tuber vermis, respectively.

It is customary to refer to the unpaired median portion of the cerebellum as the *vermis* and to the lateral masses as the *hemispheres.* In the anterior lobe and lobulus simplex the vermis is not sharply marked off from the hemispheres as it is elsewhere, and the names of the lobules apply to both medial and lateral portions.

CONNECTIONS

The cerebellum is joined to the rest of the nervous system by the superior, middle, and inferior peduncles, each of which contains both afferent and efferent fibers.

Afferent connections. The origin and distribution of the afferent systems are summarized, for reference purposes, in Table 28-1. The wealth of fibers passing to the cerebellum indicates the extent to which the coordination of movement depends on many factors. The impulses carried by these fibers are of two kinds: (1) those arising in sensory receptors and relayed more or less directly to the cerebellum and (2) those from higher levels of the nervous system. Studies using electrical techniques have revealed in considerable detail how the cerebellar cortex is organized into "receiving areas" for these afferent projections.

Efferent connections. Each region of the cortex sends the axons of its Purkinje cells to a definite part of the cerebellar nuclei. This corticonuclear system is organized in an orderly fashion.[34] The vermis projects to the medial or *fastigial nucleus* and is represented there in the same rostrocaudal order as the folia. The most lateral portions of the cortex, except the flocculus, project to the lateral or *dentate nucleus.* Intermediate areas send fibers to the *nucleus interpositus (n. globosus* and *n. emboliformis* of man).

Probably all of the efferent fibers of the cerebellum arise in its nuclei, except for some running directly from the nondulofloccular lobe to the vestibular nuclei. The fastigial nucleus sends fibers to the vestibular nuclei and to the RF of the medulla. The dentate and interposed nuclei project through the superior peduncle to the red nucleus and thalamus. The influence of the cerebellum is therefore exerted on motor neurons through the vestibulospinal, reticulospinal, and rubrospinal pathways and on the precentral motor cortex through the ventrolateral nucleus of the thalamus.

CEREBELLAR CIRCUITS

Cortex

The cortex of the cerebellum represents nature's supreme effort to pack a maximum of cortex into a minimum of space. This is accomplished by extensive infolding of the cortex to form a series of deep folia (Fig. 28-2, *A*). The fine structure of the folia is remarkably uniform throughout the cerebellum; any folium could serve as a model and any section of the cortex could be accepted as representative. The cells and fibers form an orderly array, so precisely arranged and oriented in three-dimensional space that they resemble a regular lattice. The repeating pattern of this lattice indicates that although incoming signals may differ from folium to folium, they are subject to the same neural processing throughout the cerebellum.

The cortex (Fig. 28-3) consists of an outer or *molecular* layer separated from an inner or *granular* layer by a single layer of regularly spaced *Purkinje cells.* The Purkinje axons, which descend directly to the cerebellar nuclei, are the only efferent fibers of the cortex. The dendrites of Purkinje cells ascend through the molecular layer, branching repeatedly until they reach the pial surface. The dendritic tree is virtually two dimensional, being flattened in a plane oriented at right angles to the long axis of the folium. The lower, thicker branches of the tree are smooth; the higher, thinner branches are extremely dense (Fig. 28-4) and are covered with spines, providing an enormous surface area for synaptic contacts.

The rest of the cortical mechanism serves to distribute afferent impulses to the Purkinje cells. Although afferent impulses come from many different parts of the nervous system, there are only two basic types of input to the cortex.

Climbing fibers. The climbing fibers, carrying signals mainly from the inferior olive, emerge from the white matter, cross the granular layer, and climb like ivy up the smooth surface of the primary and secondary branches of the Purkinje dendrites. In electron micrographs it is apparent that some synaptic contacts are also made with dendritic spines. Each Purkinje cell receives a potent, all-or-none excitatory connection from a single climbing fiber. Whenever the climbing fiber discharges, it fires the Purkinje cell.

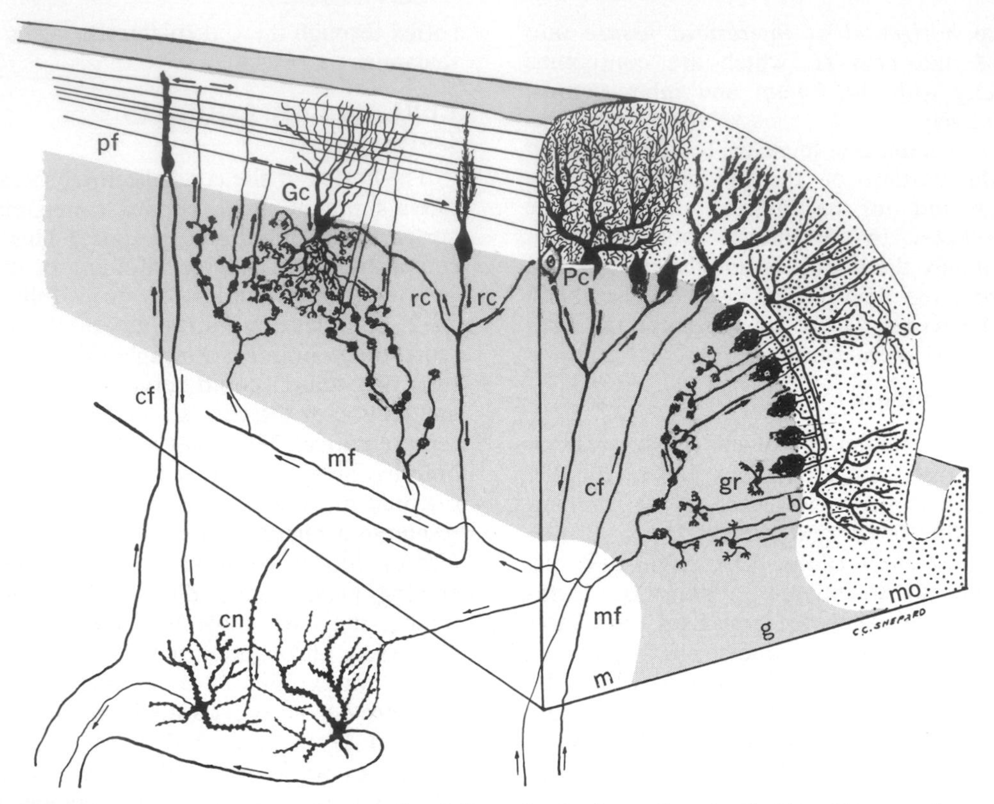

Fig. 28-3. Schematic view of cerebellar folium. *bc,* Basket cells; *cf,* climbing fiber; *cn,* deep cerebellar nuclei; *g,* granular layer; *Gc,* Golgi cell; *gr,* granule cell; *m,* medullary layer (white matter); *mf,* mossy fiber; *mo,* molecular layer; *Pc,* Purkinje cell; *pf,* parallel fiber; *rc,* recurrent collateral; *sc,* stellate cell. (From Fox.[20])

Although the chief target of a climbing fiber is the dendrites of a Purkinje cell, recent studies show that its collaterals also end on stellate cells, basket cells, and Golgi cells.

Mossy fiber–granule cell system. This system of cells provides for the cortical distribution of the great majority of impulses that reach the cerebellum. According to Ramón y Cajal,[45] the large mossy fibers divide in the white matter, often sending branches to several folia and in some instances to widely separated parts of the cerebellum. On entering a given folium, they give off branches into various parts of the granular layer. Each of these subdivides further and the final branches form bulbous expansions called *rosettes.* Within a complex structure called a *glomerulus,* each rosette makes excitatory synaptic

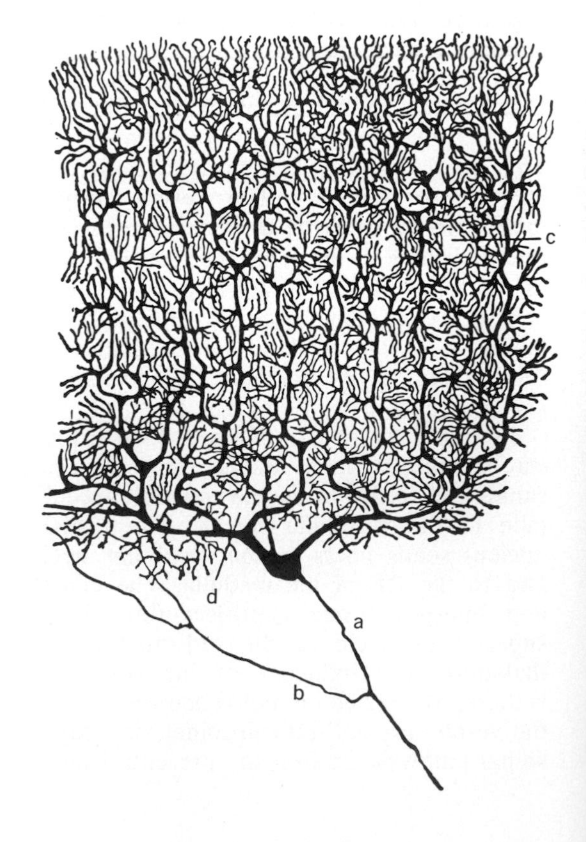

Fig. 28-4. Purkinje cell of adult man showing extensive ramification of dendritic tree. Golgi preparation. *a,* Axon; *b,* recurrent collateral; *c,* space for capillaries; *d,* location of basket cell. (From Ramón y Cajal.[45])

contact with the dendrites of approximately five *granule* cells. The terminals of these dendrites are deeply embedded in the surface of the rosette, an arrangement that probably ensures effective synaptic transmission. Through its numerous branches and multiple synapses within the glomeruli, each mossy fiber excites hundreds of granule cells. The latter are extremely small cells with axons less than 1μ in diameter. According to a recent estimate,[7] there are about 5 million granule cells per cubic millimeter of granular cortex in man. The axons of the granule cells ascend into the molecular layer, where they bifurcate to form a T. The two branches of the T run great distances in the long axis of the folium without crossing or branching and hence are called parallel fibers. They pass through the Purkinje arborization like wires through the cross arms of telegraph poles and make synaptic contacts with the dendritic spines. All of these synapses are apparently excitatory. Since each mossy fiber excites a large number of scattered granule cells whose axons pass through a long series (300 to 500) of Purkinje arborizations, a single mossy fiber impulse exerts a very small excitatory effect on a great many Purkinje cells spread out over a wide area of cortex.

• • •

There are two other types of neurons in the cortex, basket cells and Golgi cells, which are often referred to as inhibitory interneurons.

Basket cells. Basket cells, situated in the lower part of the molecular layer just above the Purkinje cells, and similar *stellate cells* in more superficial locations, distribute a powerful inhibitory input to Purkinje cells. Their dendritic trees, which are much less dense than those in Purkinje cells, are oriented in the same transverse plane and receive an excitatory input from the parallel fibers as well as from collaterals of climbing fibers. The axons of basket cells *(transverse fibers)* run for long distances across the folium (i.e., at right angles to the parallel fibers) just above the bodies of the Purkinje cells. As they pass each cell they give off collaterals that descend to form a dense basketlike network around the cell body and axon hillock. One transverse fiber may send collaterals to 15 to 17 Purkinje cells. Right-angle collaterals from transverse fibers extend a distance of 5 to 6 Purkinje cell bodies on each side of the transverse fibers. Thus each basket cell impulse may inhibit a rectilinear patch of Purkinje cells 10 to 12 cells wide and 10 to 20 cells long. The axons of stellate cells also run across the folium and are believed to form inhibitory synapses with the smooth portions of the Purkinje cell dendrites.

Golgi cells. Golgi cells, usually situated just below the Purkinje cell layer, are the second type of inhibitory interneuron. They are large cells with dendritic trees that extend outward into the molecular layer. Their branches are not as dense as those of Purkinje cells and they spread out in all directions. Excitatory inputs reach Golgi dendrites via parallel fibers and the cell body through collaterals of climbing fibers. An inhibitory input also reaches the cell body via recurrent collaterals of Purkinje axons. The axon of the Golgi cell forms a dense and often extensive arborization in the granular layer beneath the overlying dendritic field. The terminals of the axons end within the glomeruli, where they make inhibitory synaptic contacts with the outer surface of granule cell dendrites, i.e., on the side opposite to their contacts with the mossy rosette. Descending dendrites of the Golgi cell may also enter the glomeruli and make contact with the rosette. Thus within a glomerulus the mossy rosette excites the dendrites of both Golgi cells and granule cells, and the latter are inhibited by impulses in Golgi axons.

Nuclei

According to recent studies of Ito and Yoshida,[31, 32] Purkinje cells are, apparently without exception, inhibitory to the cells on which they synapse. Since the only axons leaving the cerebellar cortex are those of Purkinje cells, it must be concluded that the net effect of all the elaborate cortical processing is the production of a purely inhibitory output. All areas of the cortex, with the possible exception of the flocculus, project to the deep nuclei: the vermis to the fastigial nucleus, the intermediate zone to the nucleus interpositus, and the lateral zone to the dentate nucleus. (The lateral vestibular or Deiters nucleus is sometimes grouped with the cerebellar nuclei because it receives direct projections from Purkinje cells in the anterior lobe and the vestibular parts of the cerebellum.) After reaching its nucleus, each Purkinje axon divides to form a dense, preterminal arborization extending from the upper to the lower surface of the nucleus. These terminals completely envelop the large nuclear cells within their territory, surrounding the cell body and dendritic tree of each with a pericellular nest.

In spite of the uniformly inhibitory input that nuclear cells receive from the cortex, they respond actively to a variety of stimuli, a phenomenon that suggests that they receive excitatory inputs from outside the cerebellum. The recent studies of Ito[30] indicate that both mossy and climbing fibers give off excitatory collaterals to the cerebellar nuclei before they reach the cortex.

The efferent fibers from the nuclei project to the red nucleus and ventrolateral nucleus of the thalamus, the vestibular nuclei, and the RF. These efferent connections are all excitatory, regardless of their precise origins or terminations. There are apparently no exceptions to this general rule.[17]

Function of cerebellar circuits

The orderly arrangement of the cerebellar cortex has made possible a systematic study of its neural circuits with intra- and extracellular microelectrodes. Using a variety of stimulating and recording combinations, Eccles and his colleagues have analyzed the circuit action of each of the different types of cortical neurons. The results of these studies are summarized in Fig. 28-5. The highly selective, excitatory action of climbing fibers on Purkinje cells is shown in Fig. 28-5, *A*. The response of a Purkinje cell to a single impulse in a climbing fiber is illustrated in Fig. 28-6. An intracellular recording (right) shows the initial large spike that results and the series of smaller, irregular oscillations that follow it. A similar response, recorded extracellularly, appears as a short train of impulses, the first being the largest. A burst of impulses at these high frequencies (300 to 500/sec) presumably travels down the axon of the Purkinje cell whenever the climbing fiber

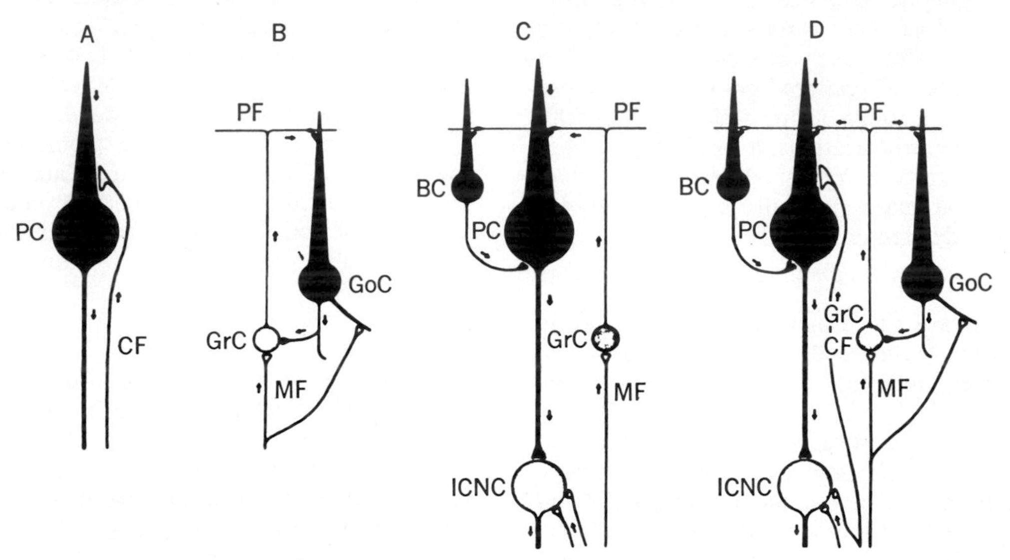

Fig. 28-5. Diagrams of some neural circuits of cerebellum, as described in text. *BC,* Basket cell; *CF,* climbing fiber; *GoC,* Golgi cell; *GrC,* granule cell; *ICNC,* nuclear cell; *MF,* mossy fiber; *PF,* parallel fiber; *PC,* Purkinje cell. All cells in black are inhibitory. (From Eccles et al.[17])

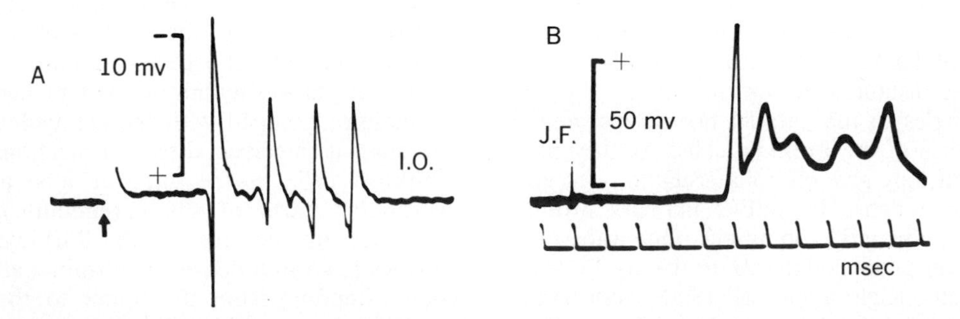

Fig. 28-6. Responses of Purkinje cells to impulses in climbing fibers. Inferior olive stimulation, *I.O.;* juxtafastigial stimulation, *J.F.* **A,** Extracellular recording. **B,** Intracellular recording. (From Eccles et al.[18])

discharges. Climbing fibers have such an intense excitatory action that even at relatively high frequencies each climbing fiber impulse may evoke several discharges from a Purkinje cell. Powerful inhibition may prevent the full repetitive response of the Purkinje cell but rarely abolishes the initial spike. After administration of a light barbiturate there is a sporadic input to the cortex via climbing fibers at a rate of 0.5 to 3.0/sec, a rate that changes in relation to movement.

Fig. 28-5, *B,* illustrates one of the several effects of input via mossy fibers. Afferent activity in mossy fibers excites granule cells, and the impulses set up in them are conducted via parallel fibers to excite Golgi cells. The axons of the Golgi cells end in the glomeruli, where they inhibit the granule cells. Fig. 28-7, *A,* shows the spontaneous activity of several granule cells recorded by one electrode; Fig. 28-7, *B,* shows the inhibition of this activity that results from stimulation of parallel fibers and discharge of Golgi cells. The repetitive discharge of a granule cell firing in response to a single shock to the radial nerve is shown in Fig. 28-7, *E;* in Fig. 28-7, *F* to *H,* this response was inhibited by stimulation of parallel fibers that preceded the radial nerve stimulus at increasing time intervals. Golgi inhibition of granule cells exerts two opposite effects on the Purkinje cell: (1) direct removal of the parallel fiber excitation and (2) removal of the inhibition elicited by parallel fibers through basket and stellate cells.

Fig. 28-5, *C,* is an illustration of the circuit that allows mossy fiber input to exert a powerful inhibitory effect on Purkinje cells by way of granule cells, parallel fibers, and basket cells. A volley of impulses in mossy fibers sets up a similar volley in parallel fibers, which causes a high-frequency burst of discharges from basket cells lasting up to 30 msec. The intensity and duration of these discharges and the concentration of basket cell endings on the axon hillock of the Purkinje cell probably account for the potency of this inhibition.

Fig. 28-5, *D,* combines parts *A, B,* and *C* of the same illustration. Two kinds of excitatory input reach the Purkinje cell, causing it to respond in two different ways, as shown in Fig. 28-8. The constant barrage of impulses in the mossy fiber–granule cell–parallel fiber system causes a maintained discharge of "simple" spikes at rates varying from 0 to 200/sec. At irregular intervals an afferent impulse in a climbing fiber causes a larger and often more prolonged discharge called a "complex spike," which is frequently followed by a brief silent period.

Wherever the axons of Purkinje cells terminate in the cerebellar nuclei or in the ves-

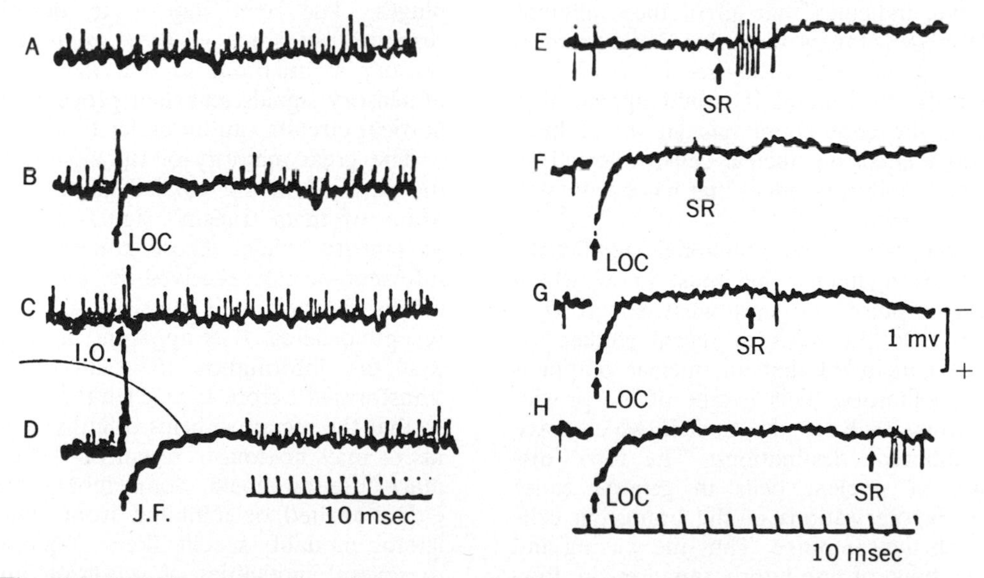

Fig. 28-7. Inhibition of granule cell firing by activation of Golgi cells. **A,** Spontaneous activity of granule cells. **B,** Inhibition by stimulation of parallel fibers on surface of cortex, *LOC.* **C,** Inhibition by stimulation of inferior olive, *I.O.* **D,** Inhibition by stimulation of juxtafastigial region, *J.F.* **E,** Repetitive response of individual granule cell to stimulation of superficial radial nerve, *SR.* **F** to **H,** Inhibition by a preceding stimulus to parallel fibers, *LOC,* at three different intervals. See text for further explanation. (From Eccles et al.[19])

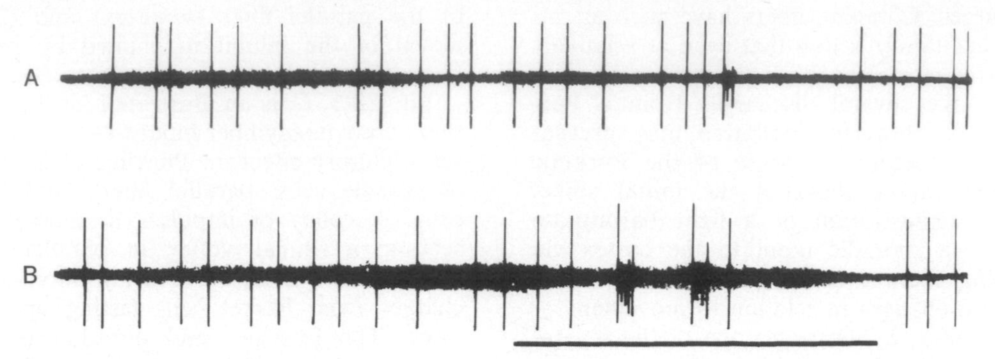

Fig. 28-8. Simple and complex spikes of Purkinje cell. **A,** "Spontaneous" discharges. **B,** Vigorous pinch of skin and deep structures, evoking complex spikes and interrupting discharge of simple spikes. (From Thach.[54])

tibular nuclei, they produce postsynaptic inhibitory responses. This corticonuclear projection is very precise. For example, each neuron in the dentate nucleus can be inhibited monosynaptically by stimulation of a certain small area of cortex in the ipsilateral hemisphere. Microelectrode recordings in the cerebellar nuclei reveal that nuclear cells are discharging continuously, even when no movement is occurring.[55] This activity is presumably due in part to the influx of afferent impulses in collaterals of the same mossy and climbing fibers that carry impulses to the cerebellar cortex. In addition, there may be some afferent fibers that go exclusively to the nuclei. The available evidence, which is not extensive, indicates that all of these afferent inputs are excitatory to nuclear cells firing in spite of the inhibition they receive from Purkinje cells. In general it would appear that input to the cerebellum has an initial brief excitatory effect on nuclear cells, followed by a delayed inhibitory effect that has been cortically processed.

To understand the functional significance of this arrangement, one must know where nuclear impulses go and what effects they produce. On the basis of recent studies[17] it may be concluded that all nuclear output is purely excitatory. This is true of the projections from each of the nuclei, which have quite different destinations. The tonic discharges of nuclear cells in general cause steady depolarizations of the brainstem cells on which they impinge. Thus the waxing and waning flow of inhibitory impulses in Purkinje cells results in less or more excitation of cells in the ventrolateral nucleus of the thalamus, the red nucleus, the vestibular nuclei, and the RF.

The essential features of cerebellar circuitry can be summarized in three statements: (1) All input to the cerebellum is apparently excitatory, both to the cortex and the nuclei. (2) The cortical circuits are similar in all parts of the cerebellum and all cortical output is inhibitory, whether it goes to the deep nuclei or the vestibular complex. (3) All nuclear output is excitatory, regardless of its destination.

Responses of cortical cells to sensory inputs

We have seen how each of the different neural components in the cerebellar circuit is affected by the signals that reach it and how its output affects the cells on which it impinges. The next step is to describe the responses of some of these cells to natural sensory stimulation, so that the distribution of sensory signals and their processing by the cortical circuits can be understood.

The great majority of the signals entering the cerebellum are carried by mossy fibers. Many of them transmit signals that are not of sensory origin. Those conveying sensory information have received the most attention, chiefly because it is possible to study their receptive fields. It is apparent in some cases that the information they carry has been transformed before it reaches the cerebellum, so that the receptive fields of individual mossy fibers may no longer resemble those of primary afferent fibers. Some mossy fibers are either excited or inhibited from small, ipsilateral, modality-specific fields; others respond to several modalities of sensation and have receptive fields, varying widely in size, that are located in ipsilateral, bilateral, or contralateral portions of the body.

Since each granule cell receives afferents from several mossy fibers and inhibitory influences from Golgi cells, its receptive field would presumably differ from that of a single mossy fiber. Thach[54] has used microelectrodes to record from cells in the granular layer. If the units he records are granule cells, as he cautiously concludes, there is apparently little difference between the receptive fields of granule cells and mossy fibers responding during light barbiturate anesthesia. All but 14 of 138 presumed granule cells responded to only one modality of stimulation. Most of them responded specifically to manipulation of muscle (43 units), hair (20 units), skin (10 units), or joints (7 units). Forty-one units required pinching or squeezing of the skin in a small area. "Muscle" units usually had receptive fields limited to one muscle. "Skin" units had fields with distinct borders, whose size ranged from a few hairs to the hair or skin over the entire limb and flank. Stimulation increased the firing rate of most units but decreased the rate of 14 of the 138 units. A few granule cells were apparently excited by one stimulus and inhibited by another of the same or a different modality. There was no evidence of an orderly arrangement of granule cells with respect to modality or location of their receptive fields. Local groups of granule cells usually had receptive fields scattered over a wide area.

According to Thach,[54] the receptive fields of Purkinje cells are large, usually comprising most of an ipsilateral limb, and they overlap extensively with those of nearby cells. The modality-nonspecific field of these cells is apparently the sum of the small, modality-specific fields of the nearby granule cells that give rise to the 200,000 parallel fibers making contact with each Purkinje cell. All of these parallel fibers influence the Purkinje cell, but apparently some types of input are much more effective than others. Recent experiments of Tarnecki and Konorski[53] reveal the predominant role of proprioceptive input to the anterior lobe. In their experiments on unanesthetized decerebrate cats the activity of Purkinje cells was found to be determined chiefly by the static position of one limb. Of the 250 Purkinje cells examined, 92% reacted to passive displacements of the limbs, 35% to squeezing of distal parts of the limbs, and 7% to touching of the skin. Flexion and extension had opposite effects on Purkinje cells, one causing a cell to discharge steadily at rates of 25 to 80/sec (Fig. 28-9), the other causing it to cease firing. These effects lasted as long as the new position was maintained, with no evidence of adaptation. Most Purkinje cells reacted to displacement of only one joint, some to two or more joints in the same limb, and a few to joints in two limbs. Units reacting to the position of a proximal joint were most numerous, but this may have been due to the selection of Purkinje cells from a particular region of cortex. Very few units reacted to displacement of both proximal and distal joints, and units near each other tended to behave similarly. The end organs responsible for these effects have not been definitely identified. It appears that they are located in muscle spindles rather than joints. Since the velocity of a displacement had little effect and its amplitude was all important, the group II endings, or "length receptors," may be primarily involved. Occasionally the discharge of a unit was altered without any change in joint position, under circumstances suggesting that gamma activity was responsible. Under normal conditions, of course, gamma regulation is a major factor in determining the rate of group II discharge. These studies suggest that the so-called spontaneous discharge of simple spikes from Purkinje cells is actually due to a steady excitatory input from one or more muscles. Inactivity presumably results from an absence of excitatory input when a muscle is relaxed. It is noteworthy that a steady input from muscle does not silence granule cells due to inhibitory feedback from the Golgi cells that it excites, and that it does not result in sufficient inhibition via basket cells and stellate cells to prevent a maintained response from Purkinje cells, as one might expect from the configuration shown in Fig. 28-5. It should be emphasized that the geometry of the cortical lattice plays a crucial role in determining exactly how these inhibitory effects are distributed. As several investigators[5, 17] have pointed out, mossy fiber impulses can either fire or inhibit a given Purkinje cell, depending on whether the cell is directly in the path of the active parallel fibers (i.e., "on beam") or lateral to them ("off beam"), where inhibitory effects due to basket cells are unopposed.

The predominant influence of signals from muscle is in accord with anatomic and physiologic studies showing a massive input from muscle via a number of different routes. Although signals from skin, joints, and deep structures reach Purkinje cells by way of mossy fibers and climbing fibers, they apparently do not cause maintained firing. Tar-

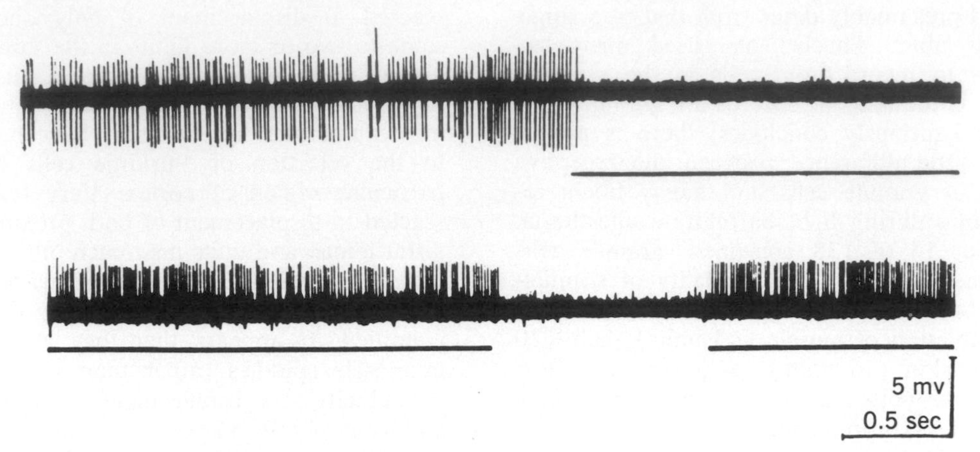

Fig. 28-9. Responses of single Purkinje cell to passive displacements of limbs. Above: Cessation of discharge during flexion of ipsilateral foreleg, with silence during extension. (From Tarnecki and Konorski.[53])

necki and Konorski[53] found that squeezing of the limbs caused brief, phasic bursts of simple spikes in some units or, more frequently, a brief inhibition of discharge. In some cases squeezing or touching caused phasic responses from units that were unresponsive to passive displacements, but most units reacted more strongly to passive movements. Although many granule cells are readily discharged by light touch, very few Purkinje cells were excited sufficiently to discharge.

Activity of cerebellar neurons during simple movements[55-57]

Recent advances in recording techniques have made it possible to relate the normal activity of the principal types of cerebellar cells to movement and posture. The cell type whose activity should be most revealing in this respect is the nuclear cell, since it is the final stage of cerebellar output. In studies by Thach,[56, 57] recordings were therefore taken from individual cells in the dentate and interpositus nuclei of trained monkeys while they moved a lever horizontally by flexion and extension of the wrist in response to light signals. Although these movements were limited to a range of 1 cm, electromyographic recordings showed that at least 12 muscles in the arm, forearm, shoulder, and trunk participated. Since the particular neuron whose activity was recorded during a movement could be related to any of these active muscles, it is not surprising that the discharge patterns of different units varied widely. Most cells in the appropriate parts of these nuclei discharged continuously even when the wrist was not moving and the animal was at rest. The rate of discharge usually increased or decreased during a movement. The pattern of change was generally different for flexion and extension. Some neurons, predominantly those in the nucleus interpositus, altered their frequency *after* the onset of movement. In these instances feedback from active muscles or joints may have played a role. Other neurons, predominantly in the dentate nucleus, altered their frequency well *before* any movement occurred, eliminating the possibility of feedback. It was not clear whether the first changes in the latter cells preceded or followed the onset of activity in the pyramidal tract.

In order to determine the extent to which Purkinje cells control nuclear cells, a similar study was carried out on Purkinje cells. The results of this study were surprisingly similar to those just described. The Purkinje cells and the interpositus cells to which they project apparently have the following characteristics in common: (1) changes in frequency at similar times before and after movement, (2) different patterns of discharge before or after movement in many of them, (3) different frequencies during maintained flexion and extension of the wrist in some of them, and (4) an increase in frequency as the earliest detected change in relation to movement in the majority of them. At first glance these similari-

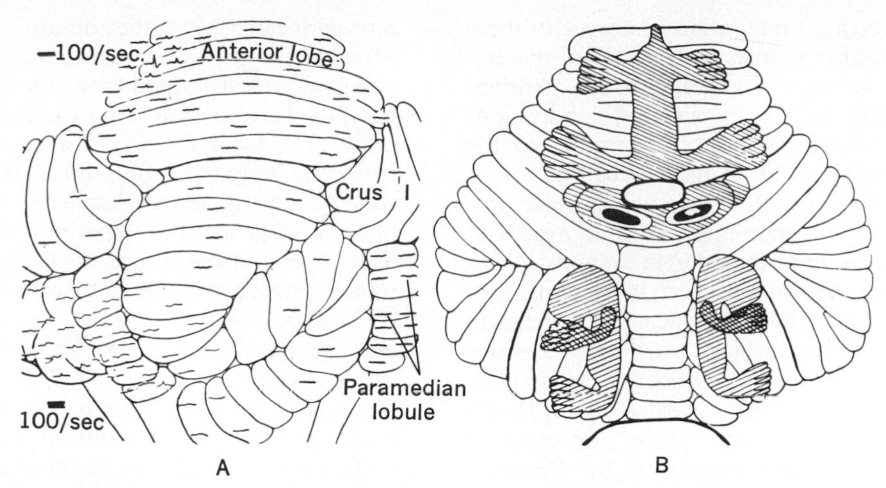

Fig. 28-10. Tactile receiving areas of cerebellum. **A,** Distribution of potentials evoked in cerebellar cortex of cat by movement of hairs around pads of left hind paw (sodium pentobarbital anesthesia). **B,** Pattern of tactile representation in cerebellum of macaque. Representation is ipsilateral in anterior lobe and bilateral in paramedian lobules. (**A** from Snider and Stowell[50]; **B** from Snider.[46])

ties might suggest that Purkinje cells are responsible for the changes in nuclear cells. If this were true, however, the firing rate of nuclear cells should decrease rather than increase in frequency (point 4) whenever Purkinje cells increase their rate of discharge. It seems more likely that the similarities are due to the fact that nuclear cells receive excitatory inputs from many of the same mossy fibers and climbing fibers that directly or indirectly excite Purkinje cells. Although the experiments described in this section do not supply any major insights, the approaches adopted appear to be very promising.*

ANTERIOR LOBE AND LOBULUS SIMPLEX

The anterior lobe and lobulus simplex together constitute a subdivision of the cerebellum that, by available criteria, is functionally homogeneous. The fibers conveying sensory impulses to the anterior lobe are either of spinal origin or arise in brainstem nuclei that relay spinal impulses. Those that terminate in the lobulus simplex transmit signals from the head and neck. The receptors in which this afferent activity arises are located chiefly in muscles, joints, and skin. Impulses from the cerebrum and other higher centers reach the same areas via the corticopontocerebellar system. The anterior lobe and lobulus simplex are thus in receipt of signals from the regions in which voluntary movements originate as well as from the muscles that execute them.

Receiving areas. If physiologic stimuli are applied to the end organs of certain sensory projection systems, it is possible to record the pontentials evoked in the region where they terminate. Utilizing this technique, Snider and Stowell[50] showed that movement of a few hairs of a cat's skin elicited responses in discrete areas of the cerebellar cortex. As the example in Fig. 28-10, *A,* indicates, these responses are sufficiently localized to permit one to "map" the tactile representation of the body in considerable detail on the surface of the cortex. The "pattern" that results is similar for cat and monkey, the tail being represented in lingula, the hind limb in centralis, the forelimb in culmen, and the neck and head in simplex. Each half of the body surface is represented ipsilaterally in the anterior lobe and lobulus simplex, with the trunk medially and the extremities laterally (Fig. 28-10, *B*). A second receiving area is found in the pyramidoparamedian lobule, as will be described later. These experiments provided the first proof of detailed functional localization in the cerebellum.

The somatotopic pattern demonstrated for touch apparently holds also for the projections from muscles and joints. Impulses set up physiologically by pressure on footpads or manipulation of muscles and joints reach the same areas as do those from the overlying

*For those interested in further study of cerebellar circuits, the recent studies by Brodal[9] and Eccles et al.,[17] and the review by Bell and Dow[5] will provide further accounts of the subjects discussed in this section and serve as a useful guide to the literature.

skin.[1] Electrical stimulation of nerves to these structures also evokes responses in the appropriate areas.[41] In addition, as Adrian[1] showed, the face, arm, and leg subdivisions of the cerebral motor cortex project to the corresponding areas in the cerebellum.

The common pattern into which these projections fall is apparently only one aspect of the organization of the afferent systems. Anatomists have shown that certain systems, notably the spinocerebellar tracts, do not terminate in localized fashion. Furthermore, some experiments on sensory projections using electrical techniques have revealed an apparent absence of localization. These discrepancies have recently been reconciled by Combs,[14] who points out that demonstration of localization in the cerebellum depends on anesthesia. In its absence the potentials evoked by tactile or electrical stimuli may be recorded throughout the anterior lobe and lobulus simplex, whereas with barbiturate anesthesia they are restricted to local areas. Whether the widespread responses are due to impulses in collaterals of fibers that terminate locally or to activity in independent systems is not yet known.

Effects of stimulation. Each subdivision of the anterior lobe and lobulus simplex serves as an effector area for the part of the body from which it receives sensory impulses. In the absence of anesthesia, movements can be elicited by stimulating these regions. Such movements involve relaxation as well as contraction of muscles, indicating that both inhibitory and facilitatory processes are set up by excitation. Depending on the existing muscular tone, and perhaps on the type of stimulation, either may predominate. In the decerebrate preparation, inhibition of ipsilateral extensor tone is the more striking result of exciting a medial cortical site, but careful observation will reveal simultaneous contraction of opposing flexor muscles. This occurrence of reciprocal changes in antagonistic muscle groups is characteristic of movements elicited from the cortex of the cerebellum as well as from its nuclei. Excitation of lateral foci in the anterior lobe usually yields ipsilateral extension with inhibition of flexors. Opposite changes often occur as "rebounds" after stimulation, and reciprocal movements are frequently seen in the contralateral limbs as an accompaniment of the direct and rebound effects.[25]

The representation of muscles in the anterior lobe and lobulus simplex fits the pattern delineated by studies on afferent systems. Movements of the head, including the face and jaws, are activated from the lobulus simplex, of the forelimbs from the culmen, of the hind limbs from the centralis, and of the tail from the lingula. Trunk muscles are represented medially and extremities laterally. This pattern, discovered in experiments on decerebrate preparations,[25] has also been found in intact animals through the use of implanted electrodes.[41] Other studies[44] showing that movements elicited from the cerebral cortex can be facilitated or inhibited by stimulation of specific cerebellar foci reveal a similar arrangement of effector points in the cortex.

These motor effects are mediated in part by connections from the cerebellar nuclei to the facilitatory and inhibitory portions of the brainstem RF. Mollica et al.[42] have given a physiologic demonstration of this system in action, showing that the discharges of single reticular cells can be facilitated or inhibited by polarizing the cortex of the anterior lobe with direct currents.

In addition to its effects on large motor neurons that supply skeletal muscle fibers, the anterior lobe also exerts an influence on the small motor neurons that innervate muscle spindles (p. 670). Excitation of its vermal portion inhibits small motor nerve discharge to spindles of extensor muscles, thereby resulting in a decrease in the rate of discharge of these stretch receptors. Stimulation of lateral sites has an opposite effect, producing an increase in the rate of firing from extensor spindles. These important observations of Granit and Kaada[23] provide a further explanation of how the anterior lobe participates in the regulation of muscular tone.

Results of ablation. In subprimate forms, removal of the anterior lobe in its entirety causes a marked increase in extensor tone throughout the body. Stretch reflexes become hyperactive, with well-defined lengthening and shortening reactions; positive supporting reflexes are accentuated. Lesions restricted to one side of the vermis result in extensor rigidity in the ipsilateral legs and marked flexion in the contralateral limbs. Lateral removals cause opposite changes. Effects are confined to the forelimb if the culmen is removed and to the hind limb if the centralis is damaged.[12] Similar though less pronounced alterations in tone and posture occur in monkeys after such removals. The effects of lesions confined to the anterior lobe in man are not known.

Since changes in muscle tone, in its broadest sense, necessarily underlie all movement, reflex or voluntary, lesions that produce such changes should impair any type of muscular performance. Carrea and Mettler[12] found that anterior lobe removals in monkeys cause pronounced ataxia* and tremor during all actions. Errors in rate, range, force, and direction similar to ataxia in man are seen. Vermian lesions result in ataxia of the trunk; lateral removals affect the extremities more. It should be stressed that although there may be great incapacity due to incoordination, no paralysis is present. The motor functions under discussion are *regulative* rather than executive. The observations of Carrea and Mettler appear to be in harmony with the experimental facts already noted. Fulton,[21] however, limits the function of the anterior lobe to regulation of posture, stating that ataxia and tremor during voluntary movements result from "posterior" lobe lesions. The majority of textbook accounts follow his lead in this respect. To the extent that "posterior" lobe lesions involve the pyramidoparamedian lobule, which appears to resemble the anterior lobe functionally, they may indeed produce ataxia. There is no evidence, however, that "voluntary" movements are coordinated in a separate portion of the cerebellum or that this function is centered exclusively in the "posterior" lobe. The view that seems most consistent with experimental facts is that the anterior lobe and lobulus simplex are prominently, though not exclusively, implicated in the regulation of all movement.

**Ataxia* (fr. *a,* not, + *taktos,* ordered) = *incoordination.* Medically it refers to any irregularity of muscular action that results from failure of coordination.

FOLIUM AND TUBER VERMIS AND ANSIFORM LOBULE

On the presumption that the medial and lateral portions of each lobule are functionally related, the folium and tuber vermis will be considered together with their lateral extensions, crus I and crus II of the ansiform lobule. The latter increases in relative size in primates, and in man makes up a large part of the cerebellar hemispheres.

Receiving areas. Using clicks as auditory stimuli, Snider and Stowell[50] recorded evoked potentials in the folium and tuber vermis and the medial part of the lobulus simplex. Within this area there are usually two distinct zones of response, one anteriorly in the lobulus simplex and folium vermis and a second in the tuber vermis (Fig. 28-11). These may correspond to auditory areas I and II of the cerebral cortex. The projection probably reaches the cerebellum via the inferior colliculus and tectocerebellar tract. Destruction of the colliculi abolishes all cerebellar auditory responses, whereas precollicular decerebration does not.

By stimulating the retina with brief flashes of light, these investigators[50] also demonstrated a visual receiving area in the same portion of the cerebellum. In some experiments there were two zones of response, one in the lobulus simplex and folium vermis, the other in the tuber vermis and sometimes in the rostral folia of the pyramus. Although the latency of the responses is long, they do

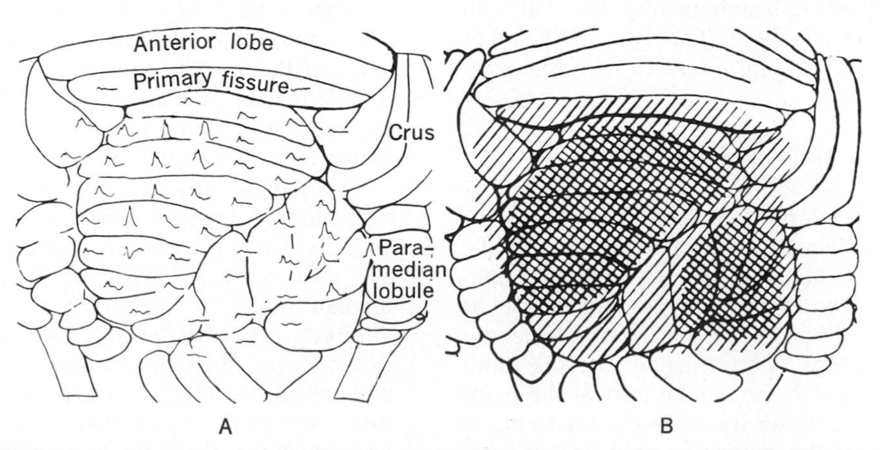

Fig. 28-11. Location and extent of auditory receiving area of cerebellum in cat. **A,** Distribution of potentials evoked by click stimulation in experiment under Chloralosane anesthesia. **B,** Extent of auditory area as determined in a series of experiments. Crosshatched regions represent zones of maximal response. (From Snider and Stowell.[50])

not depend on a relay from the cerebrum, for they are obtainable in cats from which all neocortex has been removed. The visual projection probably involves the superior colliculus.

It has not yet been shown that the ansiform lobule receives afferent impulses from end organs. The suggestion has been advanced that, like the folium and tuber vermis, it is concerned with functions that involve the head.[60]

Results of stimulation. Excitation of the folium and tuber vermis results in turning of the head and eyes to the side of stimulation.[25] No overt movements have been noted from excitation of the ansiform lobule.

Results of ablation. There is little information on the effects of lesions restricted to the areas under consideration. Several investigators stress the absence of obvious disability after removals that include these areas. Keller et al.[36] found that ablations that included the tuber vermis, paramedian lobules, paraflocculus, and ansiform lobules did not elicit noticeable dysfunction in dogs or monkeys. One of their dogs was able to walk upright on his hind legs 3 days after operation. The results obtained by Carrea and Mettler[12] in monkeys are in the same tenor: "Removal of the lobulus ansiformis produces no detectable defect." In a group of monkeys studied by Botterell and Fulton,[6] after hemispheral excisions there were several animals with a defect perhaps attributable to removal of the ansiform lobe. When allowed to run in a corridor, these animals, which were apparently not ataxic in other ways and had keen vision as judged by other tests, persistently crashed headlong into the end wall as if it were not there. This type of behavior is suggestive of some sort of "visuomotor" incoordination.

PYRAMIS AND PARAMEDIAN LOBULE

In the developing cerebellum (Fig. 28-1) the paramedian lobule is continous with the pyramis. In the adult no trace of this continuity may be evident on the surface or, as indicated in Fig. 28-2, there may be a slender bridge of cortex connecting one of the paramedian folia with the caudal part of the pyramis. It is becoming increasingly apparent, as indicated by the evidence cited below, that the pyramis and paramedian lobule are functionally an entity as they are anatomically. As regards its afferent connections, this "pyramidoparamedian" lobule resembles the anterior lobe. It receives impulses of spinal origin via spinocerebellar fibers and brainstem relays and cerebral projections by way of the pons and inferior olive.

Receiving areas. Fig. 28-10, *A*, shows that tactile stimulation of the left hind paw elicits potentials in the left paramedian lobule as well as in the anterior lobe. This "second" tactile receiving area is the cerebellar counterpart of the "second" somatic area in the cerebral cortex (Chapter 10). As in the latter, the cutaneous surface of the body is represented bilaterally and with considerable overlapping. The electrical responses evoked on the side contralateral to the stimulation (not shown in Fig. 28-10, *A*) are smaller and have a longer latency than those evoked ipsilaterally. The somatotopic pattern, as determined for the monkey, is illustrated in Fig. 28-10, *B*. The face is represented in the superior folia, the arm in the middle folia, and the leg in the inferior folia. The location of the trunk is not well established. It may be represented not as shown but in the pyramis.

Lam and Ogura[37] have demonstrated an extremely discrete representation of the larynx in the paramedian lobule. By stimulating the superior laryngeal nerve, which is the principal sensory supply of the larynx, they elicited potentials in a small zone located on the medial third of the adjacent lips of folia III and IV of the paramedian lobule bilaterally. They point out that this is a logical receiving area for impulses from the larynx, which is derived from the fourth and fifth branchial arches, and hence should be represented between the face and upper extremity. A "first" pair of receiving areas, located more anteriorly, was also found. The existence of these areas concerned with laryngeal functions may provide an explanation for the common observation of neurologists that cerebellar lesions result in ataxic speech.

From these observations and from others not cited it appears that the same sensory systems that project to the anterior lobe are also represented in the pyramidoparamedian lobule.

Effects of stimulation. Excitation of the pyramidoparamedian lobule in cats, dogs, and monkeys elicits ipsilateral and sometimes bilateral movements that are usually less localized than those evoked from the anterior lobe. A reproducible motor pattern is found in all three species. It is illustrated in Fig. 28-12, which summarizes the results of stimulating other areas as well. Movements

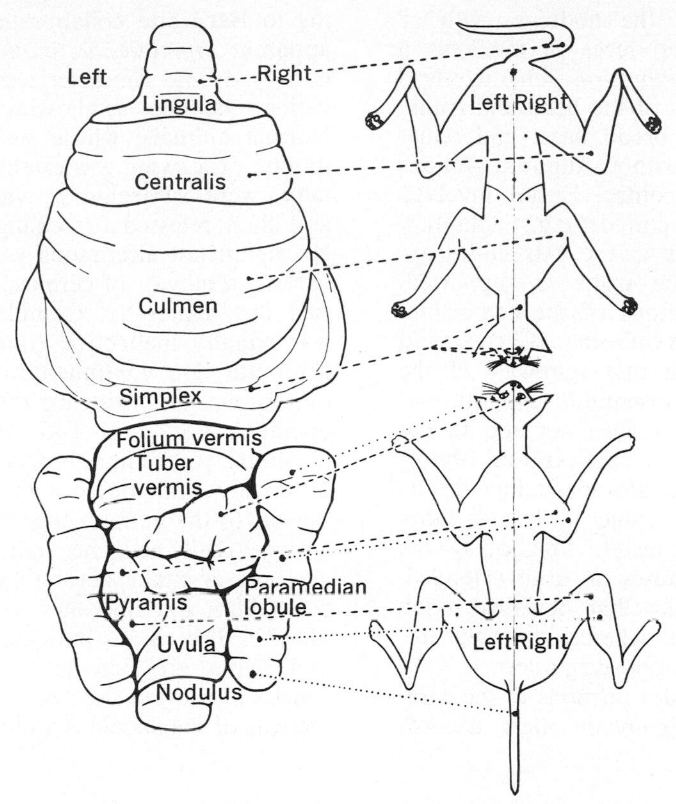

Fig. 28-12. Diagram illustrating somatotopic localization of motor function in cerebellum of cat. Cortical areas are shown, stimulation of which elicits movements of various parts of body in decerebrate animals. Regions not represented in diagram are not necessarily lacking in motor capacities. (From Hampson et al.[25])

of facial muscles are evoked from the upper folia of forelimbs from the middle folia, and of hind limbs and of tail from the lower folia. Active flexions or extensions of the limbs are the most common movements. Rebound contractions and inhibitions are observed less regularly than in the case of the anterior lobe. Movements of the limbs, generally bilateral, and of the tail are evoked by stimulating the pyramis.

Effects of ablation. The results of localized ablations in the pyramis and paramedian lobule await detailed analysis. Unilateral ablation of the pyramis causes effects somewhat similar to those that follow unilateral ablations of the vermis of the anterior lobe.[51]

UVULA AND PARAFLOCCULUS

Little is known about the part of the cerebellum comprising the uvula and paraflocculus. The uvula is a relatively constant structure, but the paraflocculus varies greatly in different species. Its dorsal division is usually continuous with the pyramis and its ventral portion with the uvula. In diving mammals such as the whale and the porpoise the ventral paraflocculus is enormous; in some whales it constitutes almost half the total mass of the cerebellum.[33] Possibly it is concerned with the autonomic regulations involved in diving, such as cardiovascular adjustments in response to changes in water pressure. In this connection it is perhaps significant that the part of the inferior olive that projects to the paraflocculus receives fibers from the periaqueductal gray,[35] a region closely associated with the hypothalamus.

The uvula apparently receives some impulses of spinal origin[15] as well as a contribution from the vestibular system. Its relation to motion sickness is noted later.

NODULOFLOCCULAR LOBE

The noduloflocccular lobe is sometimes called the archicerebellum because it is derived from the most primitive part of the cerebellum. Its afferent and efferent connections are primarily vestibular. The adjacent uvula also receives vestibular fibers.[15]

Effects of ablation. Removal of the flocculus or the nodulus causes functional changes in the vestibular sphere. Dow[16] has shown that the ablation of the nodulus, with or without the uvula, produces in monkeys a *syndrome of disequilibration,* characterized by falling, oscillations of the head and trunk, staggering gait on a broad base, and reluctance to move about without support. Actions in which vestibular control is not involved are apparently unimpaired; e.g., voluntary and reflex movements of the extremities are well performed if the trunk is supported. Lesions in other portions of the cerebellum do not cause the syndrome. Carrea and Mettler[12] have shown that removals of the flocculus also cause disequilibrium but that excision of the lingula does not add to the defect. They emphasize that postural abnormalities resembling the attitudes adopted during elicitation of the tonic neck and labyrinthine reflexes also result. Ablation of the nodulus and uvula causes an arms-extended, legs-flexed posture like that observed when the neck is extended, whereas floccular removals result in the opposite pattern.

Removal of vestibular portions of the cerebellum has another significant effect, according to Bard and collaborators[2, 58]; it confers apparent *immunity to motion sickness.* This is a syndrome characterized in dogs by salivation, licking, swallowing, and vomiting. Normal animals, whose susceptibility to the motion of a swing was established experimentally, were subjected to various operations, and then retested for changes in sensitivity. No significant alterations were detected after various removals of cerebral cortex were carried out bilaterally. Complete decerebration was equally ineffective (one chronically decerebrate dog continued to vomit promptly in response to swinging throughout a postoperative period of 145 days). Ablation of the entire cerebellum, however, rendered dogs outwardly immune to motion sickness. Removals of the nodulus together with the uvula and pyramis had the same effect, whereas excision of other accessible portions of the cerebellum did not alter susceptibility (Fig. 28-13). The results indicate that the nodulus and uvula are definitely implicated in the genesis of motion sickness but leave uncertain the role of the flocculus and lingula.

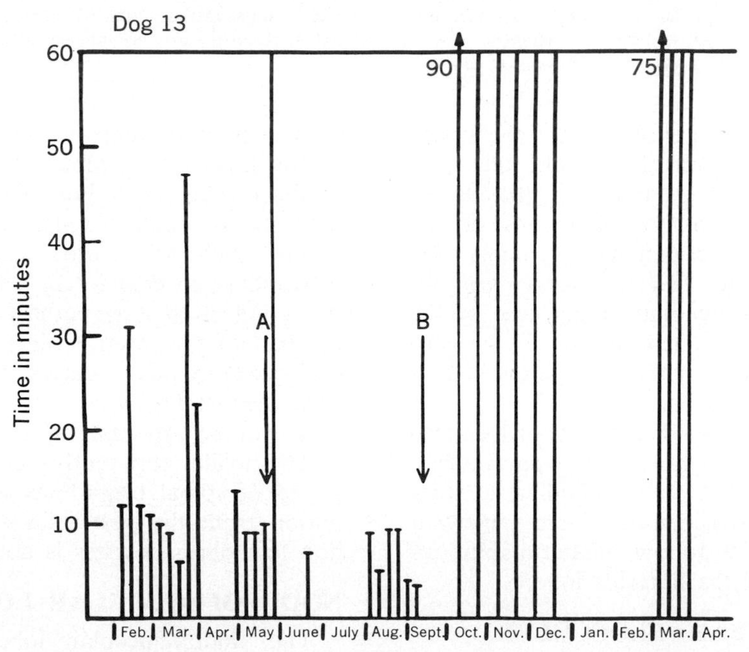

Fig. 28-13. Chart illustrating loss of emetic response to prolonged swinging that follows removal of certain vestibular portions of cerebellum in dog. *A,* Removal of vermal portions of L. simplex, declive, and tuber. *B,* Removal of pyramis, uvula, and nodulus. Height of vertical lines indicates duration of exposure to motion. Occurrence of vomiting is indicated by short horizontal line at top of vertical line. Arrows show when ablation of nonvestibular, *A,* and vestibular, *B,* portions of cerebellar cortex were carried out. Following second operation, *B,* animal failed to vomit or exhibit any other signs of motion sickness. (From unpublished experiments of Bard et al.[3])

CEREBELLAR NUCLEI

Although the nuclei of the cerebellum are the major source of all its efferent connections, relatively little is known about their functional role. Presumably they serve as integrative centers where efferent impulses from the cerebellar cortex converge and interact with impulses from elsewhere. Some fibers of the vestibulocerebellar[15] and olivocerebellar[8] systems and apparently all of the rubrocerebellar[10] projection terminate in the nuclei. Whether they end on the same nuclear cells as the axons of Purkinje cells is not known. It is possible that they constitute part of a nuclear short circuit through the cerebellum, bypassing the cortex. A report[49] describing efferent responses in the brachium conjunctivum approximately 1 msec after stimulation of certain afferent systems could be explained on this basis.

Stimulation of the nuclei results in movements that are similar to, but less localized than, those evoked from the cerebellar cortex.[26] They are generally biphasic: during stimulation there is inhibition of a preexisting posture or assumption of a new one; after stimulation reverse effects, often of long duration, occur. Excitation of the fastigial and globose nuclei usually produces widespread postural effects, often involving the trunk and all four limbs. There is a strong element of reciprocal innervation underlying these movements. While the flexors in a limb contract, the extensors relax; in the contralateral limb the changes may be exactly the opposite. Movements evoked from the emboliform nucleus are generally limited to the ipsilateral forelimb. Under the usual experimental conditions, excitation of the dentate nucleus causes no overt movements.

Severe disabilities result from injury to the nuclei.[12] Experimental lesions of the nucleus fastigius cause profound disturbances of equilibrium, ataxia of trunk muscles, and tremor of the head. Involvement of the dentate and emboliform nuclei produces ipsilateral ataxia and tremor during voluntary and involuntary activity. The defects that occur are ipsilateral but otherwise unlocalized. Recovery is never complete after nuclear lesions (one macaque showed little or no improvement 264 days after operation). For this reason surgeons who are obliged to remove a portion of the cerebellar cortex during certain procedures try not to encroach upon the nuclei.

INTERCORTICAL CONNECTIONS OF CEREBELLUM

Using oscillographic techniques, Barnard and Woolsey[4] set out to determine whether there are functional connections between the various portions of the cerebellar cortex. Investigating first the dual receiving areas for tactile impulses, they found evidence of extensive interconnections between them. Cortical areas that receive impulses from the same part of the body were found to be so linked that excitation of either one elicited responses in the other. It was evident, especially in the larger cerebellum of the chimpanzee, that these linkages are highly specific. Each part of the anterior lobe and lobulus simplex projects to the somatotopically corresponding part of the pyramis or paramedian lobule. Stimulation of lateral foci in the central lobule, for instance, evokes potentials in the lateral portions of the "leg" folia of the paramedian lobule, whereas excitation of medial sites in centralis elicits responses in the pyramis. Projections have also been demonstrated linking the audiovisual area in the folium with its counterpart in the tuber vermis. The most revealing aspect of these studies, however, is the finding[61] that crus I is interconnected with crus II. The intercrual (horizontal) fissure that separates them also passes between the folium and tuber vermis. As a consequence, it takes on special significance as a boundary line between one group of receiving areas anterior to it and a second group of functionally related receiving areas posterior to it.*

The intercortical connections described have not been identified histologically. They are probably intracerebellar, for neither decerebration nor cutting the cerebellar peduncles ipsilaterally abolishes the responses mediated by them. The long latency (6 msec) of the latter makes it unlikely that they are conducted from area to area by association fibers. It is more probable that impulses are relayed through the cerebellar nuclei and reach the cortex again via the climbing fiber system, which, according to Carrea et al.,[13] originates in the nuclei.

INTERRELATIONS BETWEEN CEREBRUM AND CEREBELLUM

Anatomic studies show that the cerebrum and cerebellum are interconnected by two major systems. The frontal, temporal, parietal, and occipital lobes of the cerebrum send fibers to the pontine nuclei, which project through the middle cerebellar peduncle to nearly all parts of the cerebellar cortex.[11] In return the cerebellar nuclei send fibers via the superior cerebellar peduncle to the contralateral thalamus.[35] Most of these fibers end in the ventrolateral nucleus, which projects to the motor cortex.

Cerebrocerebellar projections. By stimulating the cerebral cortex with brief electrical pulses it is possible to set up synchronous volleys in its efferent projections and to record responses in the regions where they terminate. With this technique, Adrian[1] demon-

*The anatomic research of Walberg (Chapter 4 in Jansen and Brodal[35]) also indicates that the intercrual fissure has greater morphologic and functional significance than was previously believed.

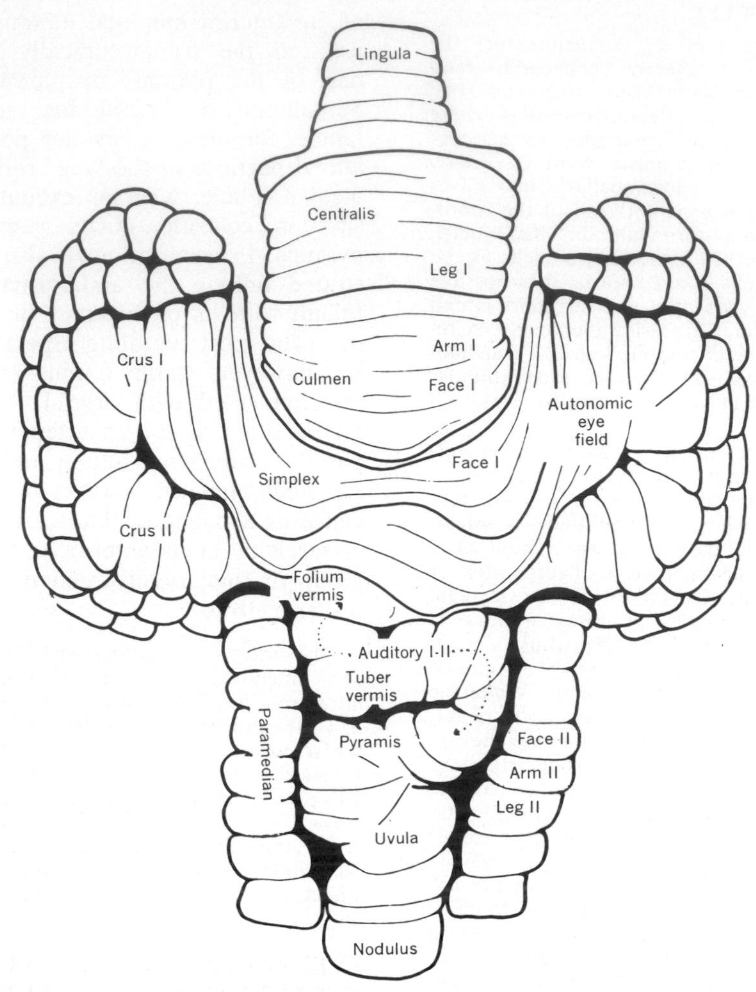

Fig. 28-14. Diagram of cat's cerebellum showing regions to which various functional areas of cat's cerebral cortex project. These relationships were established by stimulating cerebral cortex and recording responses evoked in cerebellar cortex. (From Hampson.[24])

strated that the face, arm, and leg areas of the motor cortex in the monkey project to the simplex, culmen, and centralis, respectively. Subsequent investigations have revealed, in addition, that cortical areas of the cerebrum and cerebellum that receive impulses from the same sense organs are themselves specifically interconnected. This is illustrated in Fig. 28-14, which summarizes the results of Hampson's studies[24] on the cerebrocerebellar projections in the cat. Stimulation of somatic area I (the postcentral homolog) evoked responses in the contralateral anterior lobe and lobulus simplex, with the face area projecting to the simplex and upper culmen, the arm area to the lower culmen, and the leg area to the centralis. Excitation of somatic area II elicited responses in the contralateral paramedian lobule and sometimes in the pyramis. The evidence was suggestive that the leg area projects to caudal folia and the face area to more rostral folia. Auditory areas I and II send impulses to the folium and tuber vermis, the projection from auditory area II being somewhat the more caudal of the two. The autonomic center for the eyes, located on the medial wall of the cerebral cortex, projects to the ansiform lobe and regions adjacent to it. In a similar study on the monkey, Snider and Eldred[48] found a projection from the visual cortex to the folium and tuber vermis. It is noteworthy that the distribution of cerebrocerebellar projections, as determined oscillographically, is in accord with the distribution of pontocerebellar fibers as reported in recent anatomic studies. According to Brodal and Jansen,[11] pontine fibers go to almost all parts of the cerebellar cortex, not exclusively to the "neocerebellum" as was once believed.

Cerebellocerebral projections. The efferent projections of the cerebellum are particularly susceptible to depressing anesthetics, which makes it difficult to investigate them under favorable experimental conditions. In unanesthetized cats (isolated encephalon) stimulation of the anterior lobe and lobulus simplex evokes responses in the contralateral somatic and motor areas I and II.[26] Stimulation of the paramedian lobules elicits responses in the same areas bilaterally. Excitation of the folium and tuber vermis evokes potentials in the auditory and occasionally in the visual areas of the cerebrum. It should be emphasized that the only one of these "projections" that has the full sanction of anatomy is that to the motor cortex. The effects that cerebellar stimulation produces in other cerebral areas do not necessarily indicate the existence of projections in the usual sense. They may depend on the connections of the cerebellum with the RF. Mollica et al.[42] have shown that polarization of the cerebellar cortex may cause a generalized EEG "arousal reaction" associated with increased discharge of reticular cells in the brainstem. Others[47] report that repetitive stimulation of the cerebellum alters the spontaneous activity of the cerebral cortex, even to the extent of suppressing an electric seizure in progress.

FUNCTIONAL ORGANIZATION SUMMARIZED

Most if not all of the sense organs of the body project to the cerebellum. It is probable, moreover, that all of its cortex receives impulses of sensory origin. There are several regions to which no such projection has as yet been demonstrated, but it seems likely that they will prove to be receiving areas for sensory systems not yet investigated and will fall into their proper place in the general plan of organization. The latter is shown schematically in Woolsey's[60] simplified diagram (Fig. 28-15). The main anatomic subdivisions are labeled on the left, and the part of the body represented in each, according to oscillographic studies, is also indicated. The arm and leg are each represented twice, once in the anterior lobe and again in the pyramidoparamedian lobule. Between these two regions is an area of cortex bounded (heavy lines) by the primary fissure anteriorly and the prepyramidal fissure posteriorly. This area is labeled "head" be-

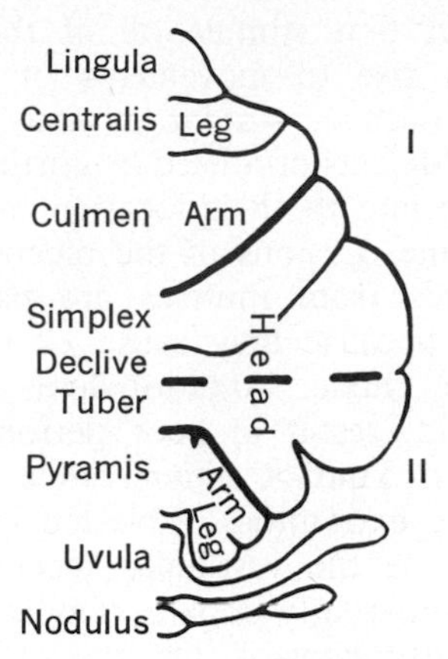

Fig. 28-15. Diagram illustrating dual representation of body in cerebellar cortex. Boundary between areas *I* and *II* is horizontal or intercrural fissure (heavy broken lines). (From Hampson et al.[25])

cause afferent impulses from the eyes, ears, and skin of the head terminate within it. Although little is known about the ansiform lobule, it is reasonable to assume that it too is concerned with sensory and motor systems involving the head.

This somatotopic pattern may be divided into two parts that correspond to the "first" and "second" sensory receiving areas of the cerebrum. The intercrural fissure is the line of separation between them, for analysis of the intracerebellar connections reveals that every cortical site anterior to this sulcus (broken lines) is interconnected with a locus posterior to it. Medially it passes between the folium and tuber vermis, dividing the audiovisual region into two parts that apparently correspond to receiving areas I and II. In this partition, crus I and crus II emerge as the "first" and "second" areas of other unidentified systems. The significance of this double representation is unknown. Its occurrence in both cerebrum (Chapter 10) and cerebellum probably reflects some fundamental duality of the sensory and motor systems.

The place of the uvuloparafloccular lobe in this plan remains unsettled until there is information concerning its sensory connections or motor functions. The nodulofloccular lobe is generally held to be vestibular. Inasmuch as vestibular impulses might be expected to terminate in the "head" region near the auditory area, as they apparently do in the cerebrum, this projection deserves further investigation.

That the cerebellum is organized along somatotopic lines as described is also attested by the fact that stimulation of the various folia gives rise to movements of the parts represented there. As far as can be ascertained, the defects produced by cortical lesions are appropriate to their location in the previous scheme. Lesions of the receiving areas for impulses from muscles are particularly significant because they cause all movement to become ataxic. Anterior lobe removals, for example, result in poor performance of almost every kind of action, reflex or voluntary, of the extremities. Ablation of what is thought to be the vestibular receiving area, however, apparently impairs only acts requiring vestibular control. On the other hand, lesions of the auditory and visual receiving areas should cause defects limited to the spheres in which auditory and visual regulation is vital. Considerations such as this may explain why some lesions produce no apparent disability, the explanation perhaps being that detection of the defects that occur requires very specific tests, which have not yet been applied.

No consideration has been given to the role the cerebellum plays in visceral and autonomic regulations. The interested reader may consult Moruzzi[43] and Jansen and Brodal.[35]

EVIDENCE FROM CLINICAL STUDIES

Cerebellar lesions in man cause disorders of movement resembling those produced experimentally in monkeys. In addition, there is usually some degree of hypotonia, which is seldom prominent in animals. Sensation is not impaired.

The most conspicuous cerebellar "signs" are the errors in rate, range, force, and direction of movements, known collectively as *ataxia.* They appear only when muscles are in use and are most pronounced in precise actions involving several joints. Neurologists employ a number of simple tests designed to elicit these defects and distinguish between them as follows[28]: *intention tremor,* an oscillating tremor most marked at the end of fine movements; *asynergia,* a lack of cooperation between muscles (e.g., failure of the wrist extensors to contract during flexion of the fingers, allowing the wrist to flex also); *decomposition of movement,* the performance of actions in successive parts rather than as a whole (e.g., in touching the nose, first flexing the forearm, then the arm, and lastly adjusting the wrist and finger); *dysmetria,* errors in the range of movement (in touching a point, arresting the action before reaching it, or shooting past it); *deviation from the line of movement* (e.g., carrying food to the ear instead of the mouth); *adiadochokinesia,* inability to perform alternating movements (e.g., tapping) rapidly and smoothly. Although they are useful clinically, such distinctions are not fundamental. The form that ataxia takes depends on the particular muscles involved and how their action is tested. In fixing the gaze on some object, for example, the eyes alternately turn quickly toward it and slowly away from it *(nystagmus)*; in speech there is a tendency to decompose words into separate syllables pronounced in staccato rhythm *(scanning speech);* in walking, the gait is reeling, and associated movements of the arms are lost.

Less conspicuous but equally characteristic of cerebellar disease are several related signs that may be defined as *hypotonia.* These include absence of normal resistance to stretch, easy displacement of a limb from a given posture, hyperextensibility of joints, and undamped or pendular reflexes. These abnormalities apparently indicate a lack of normal response to stretch, in which the cerebellum, through its proprioceptive connections, plays an important role. The opposite condition, *hypertonia,* though common in animals after certain lesions, is rarely seen in man. The reason for this difference is not presently understood.

In man, as in experimental animals, there is considerable recovery of function after cerebellar lesions. Due to this *compensation* a patient may appear to regain his powers of coordination completely. Careful testing, however, will usually reveal motor deficiencies months or years later. Recovery is less complete after nuclear than after cortical lesions.

Patients with cerebellar lesions report no alterations in sensation. They may reel and fall, but they do not feel dizzy; their movements may deviate from the desired mark, but their sense of position in space is unimpaired. Evidently the sensory data coming to the cerebellum do not reach consciousness. It is true, in comparing equal weights in the two hands, that that on the affected side is usually judged the heavier. This should not be considered a test of primary sensation, however, for as in many paretic limbs, weights feel heavier than they actually are because of the greater effort involved in all performance on the affected side.

Localization in human cerebellum

The disabilities found in patients are rarely limited to a single function such as speech or to a small part of the body such as the hand. They usually involve most of the muscles on the side of the lesion. From an analysis of cerebellar injuries in World War I, Holmes[29] concluded that there was little evidence of functional localization in the human cerebellum. Few of his cases came to autopsy, however, and the extent of the damage could not be ascertained. There is unfortunately no well-studied series in which the clinical findings are correlated with the exact sites of the lesions. The apparent lack of localization encountered in human subjects, however, is suggestively similar to that observed in monkeys with nuclear lesions.[12] Whether patients with serious deficiencies always have nuclear involvement is not certain. Surgeons removing lateral portions of the cerebellar cortex in the course of operations on eighth-nerve tumors usually find that only minor and transient disabilities result, whereas injuries to the dentate nucleus leave serious sequelae.

In view of the well-documented evidence of cerebellar localization in experimental animals, its occurrence in man cannot be doubted. Several observations support this contention. Holmes,[29] for example, stated that "the relative prominence of different symptoms such as tremor, slowness, and incoordination of movement, as well as nystagmus, varied with the site of the lesion." The type of disturbance produced by cerebellar medulloblastomas in children also speaks for localization. These tumors develop from cell rests in the nodulus. In their early stages, before damage is extensive, they cause unsteadiness in walking, loss of balance, and falling. There is no incoordination of arms and legs. The syndrome is essentially a disorder of equilibrium similar to that which follows removal of the nodulus in monkeys.

At present, comparative anatomic and physiologic studies are the best guide to the pattern of localization in the human cerebellum. Authorities differ on some details, but there is general agreement that the subdivisions of the human cerebellum may be homologized with those of lower forms. From a comparison of fissures and folia in the opossum, rabbit, pig, sheep, cat, dog, monkey, and man, Woolsey has proposed that the human cerebellum is organized in the manner illustrated in Fig. 28-16. Dorsal and ventral views of a human specimen are shown in Fig. 28-17 for comparison with the diagrams in Fig. 28-16.

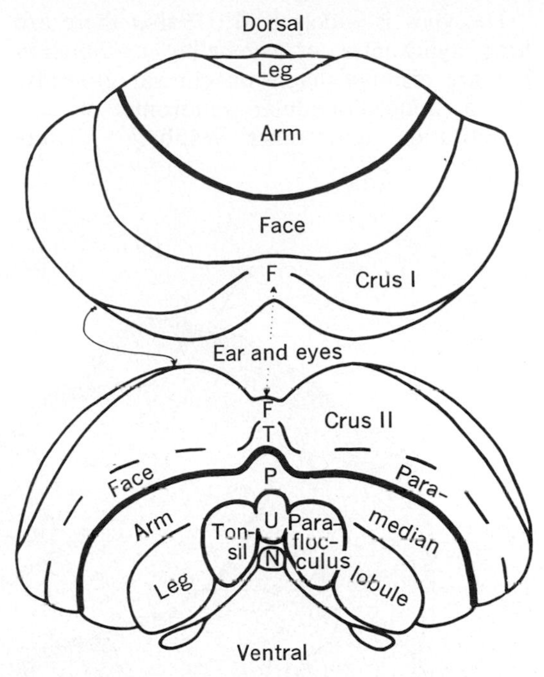

Fig. 28-16. Diagram of human cerebellum with provisional pattern of somatotopic localization as suggested by comparative physiologic and anatomic data. (From Hampson et al.[25])

On the whole, the clinical evidence regarding somatotopic localization in the human cerebellum is vague and contradictory, neither affirming nor denying the results obtained in animal experiments. A recent clinicopathologic study, however, by Victor and et al.[59] confirms the pattern of representation in the anterior lobe very convincingly. A remarkably uniform cerebellar syndrome was noted in a group of 50 patients with a long history of alcoholism. The most striking abnormalities were a severe ataxia of gait and marked incoordination of the legs in voluntary movements. Postmortem examination revealed a degeneration of the cerebellar cortex, confined chiefly to the anterior lobe. The parts most consistently and severely involved were the central lobules, i.e., the homolog of the "leg" area. Although the culmen was frequently affected, its lateral extensions, the "arm" areas, were spared, a finding that correlates with the mild clinical signs in the arms. Most of the region considered representative of the head and neck was intact; correspondingly there were few clinical signs referable to this zone.

The view is widely held[21, 22] that there are three syndromes of cerebellar dysfunction that are distinguishable on clincial grounds: (1) a floucculonodular syndrome of disequilibration and other vestibular disturbances, (2) an anterior lobe syndrome of postural abnormalities, and (3) a neocerebellar syndrome of ataxia of voluntary movements. It seems likely that specific clinical signs are associated with lesions of these three regions,

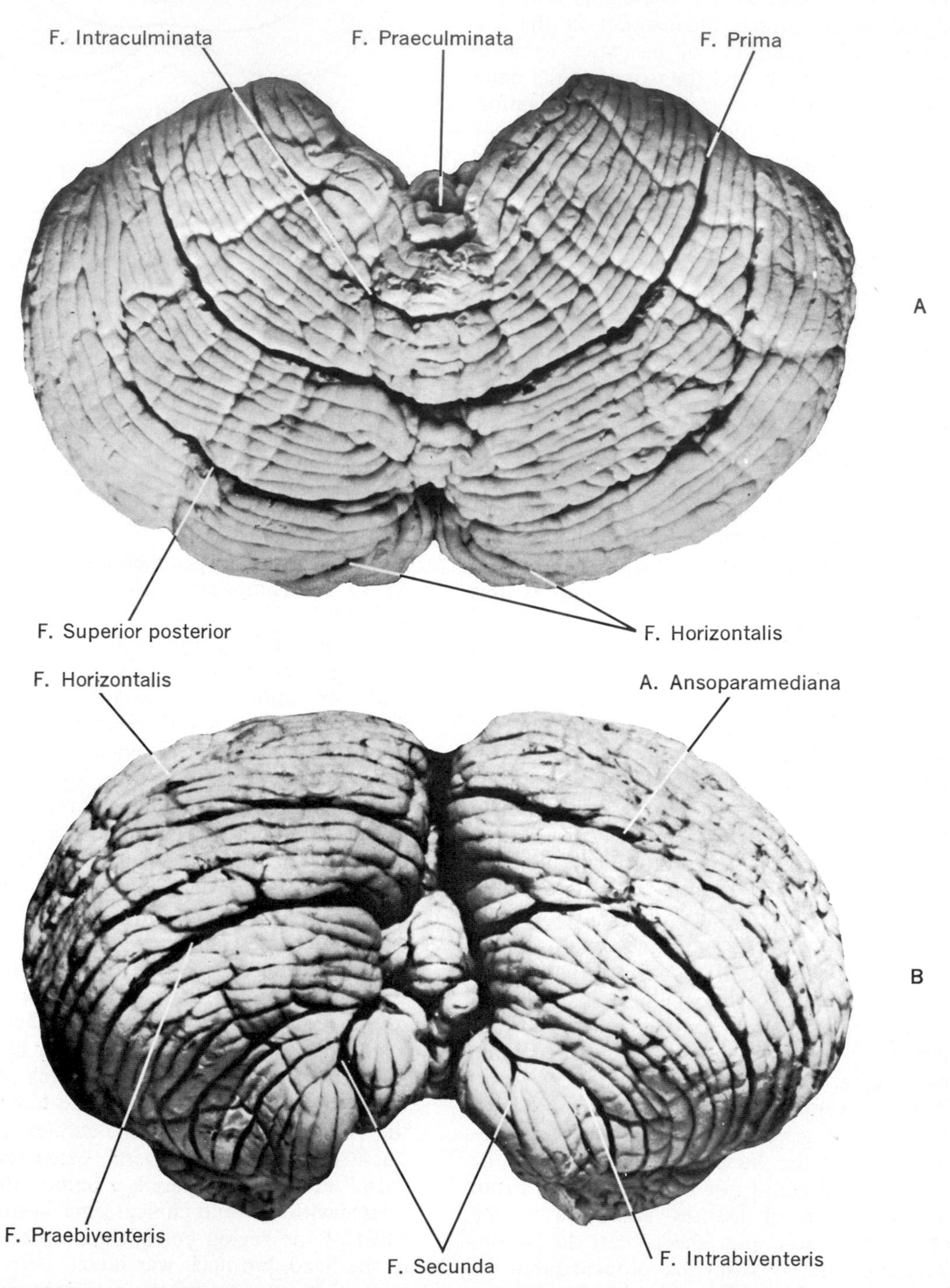

Fig. 28-17. A, Dorsal view and, **B,** ventral view of cerebellum of human adult. Some of the chief fissures were opened and packed with cotton. (From Angevine et al.[2])

but the evidence regarding 2 and 3 is insubstantial. The study just described indicates that lesions of the anterior lobe result in ataxia of voluntary movement, which is similar to that observed following experimental lesions in monkeys. The nature of the deficit produced by neocerebellar lesions in man remains to be established.

REFERENCES

1. Adrian, E. D.: Afferent areas in the cerebellum connected with the limbs, Brain **66:** 289, 1943.
2. Angevine, J. B., Mancall, E. L., and Yakovlev, P. I.: The human cerebellum, an atlas of gross topography in serial sections, Boston, 1961, Little, Brown & Co.
3. Bard, P., et al.: Delimitation of central nervous mechanisms involved in motion sickness, Fed. Proc. **6:**72, 1947.
4. Barnard, J. W., and Woolsey, C. N.: Interconnections between the anterior lobe and the paramedian lobule of the cerebellum, Arch. Neurol. Psychiatry **65:**238, 1951.
5. Bell, C. C., and Dow, R. S.: Cerebellar circuitry, Neurosci. Res. Symp. Sum. **2:**515, 1967.
6. Botterell, E. H., and Fulton, J. F.: Functional localization in the cerebellum of primates; lesions of hemispheres (neocerebellum), J. Comp. Neurol. **69:**63, 1938.
7. Braitenberg, V., and Atwood, R. P.: Morphological observations on the cerebellar cortex, J. Comp. Neurol. **109:**1, 1958.
8. Brodal, A.: Experimentelle Untersuchungen über die olivo-cerebellare Lokalisation, Z. Neurol. Psychiatr. **169:**1, 1940.
9. Brodal, A.: Neurological anatomy in relation to clinical medicine, New York, ed. 2, 1969, Oxford University Press.
10. Brodal, A., and Gogstad, A. C.: Rubro-cerebellar connections. An experimental study in the cat, Anat. Rec. **118:**455, 1954.
11. Brodal, A., and Jansen, J.: The ponto-cerebellar projection in the rabbit and cat; experimental investigations, J. Comp. Neurol. **84:** 31, 1946.
12. Carrea, R. M. E., and Mettler, F. A.: Physiologic consequences following extensive removals of the cerebellar cortex and deep cerebellar nuclei and effect of secondary cerebral ablations in the primate, J. Comp. Neurol. **87:** 169, 1947.
13. Carrea, R. M. E., Reissig, M., and Mettler, F. A.: The climbing fibers of the simian and feline cerebellum, experimental inquiry into their origin by lesions of the inferior olives and deep cerebellar nuclei, J. Comp. Neurol. **87:**321, 1947.
14. Combs, C. M.: Electro-anatomical study of cerebellar localization; stimulation of various afferents, J. Neurophysiol. **17:**123, 1954.
15. Dow, R. S.: The fiber connections of the posterior parts of the cerebellum in the cat and rat, J. Comp. Neurol. **63:**527, 1936.
16. Dow, R. S.: Effect of lesions in the vestibular part of the cerebellum in primates, Arch. Neurol. Psychiatry **40:**500, 1938.
17. Eccles, J. C., Ito, M., and Szentagóthai, J.: The cerebellum as a neuronal machine, New York, 1967, Springer-Verlag New York, Inc.
18. Eccles, J. C., Llinas, R., and Sasaki, K.: The excitatory synaptic action of climbing fibers on the Purkinje cells of the cerebellum, J. Physiol. **182:**268, 1966.
19. Eccles, J. C., Llinas, R., and Sasaki, K.: The mossy fiber granule cell relay of the cerebellum and its inhibitory control by Golgi cells, Exp. Brain Res. **1:**82, 1966.
20. Fox, C. A.: The cerebellum. In Crosby, E. C., Humphrey, T., and Lauer, E. W., editors: Correlative anatomy of the nervous system, New York, 1962, The Macmillan Co.
21. Fulton, J. F.: Cerebrum and cerebellum. In Fulton, J. F., editor: A textbook of physiology, ed. 16, Philadelphia, 1949, W. B. Saunders Co.
22. Fulton, J. F., and Connor, G.: The physiological basis of three major cerebellar syndromes, Trans. Am. Neurol. Assoc. **65:**53, 1939.
23. Granit, R., and Kaada, B. R.: Influence of stimulation of central nervous structures on muscle spindles in cat, Acta Physiol. Scan. **27:**130, 1952.
24. Hampson, J. L.: Relationships between cat cerebral and cerebellar cortices, J. Neurophysiol. **12:**37, 1949.
25. Hampson, J. L., Harrison, C. R., and Woolsey, C. N.: Cerebro-cerebellar projections and the somatotopic localization of motor function in the cerebellum, Res. Pub. Res. Assoc. Nerv. Ment. Dis. **30:**299, 1950.
26. Hare, W. K., Magoun, H. W., and Ransom, S. W.: Localization within the cerebellum of reactions to faradic cerebellar stimulation, J. Comp. Neurol. **67:**145, 1937.
27. Henneman, E., Cooke, P. M., and Snider, R. S.: Cerebellar projections to the cerebral cortex, Res. Pub. Res. Assoc. Nerv. Ment. Dis. **30:**317, 1950.
28. Holmes, G.: The Croonian lectures on the clinical symptoms of cerebellar disease and their interpretation, Lancet **42:**1177, 1231, 1922; **42:**59, 111, 1922.
29. Holmes, G.: The cerebellum of man, Brain **62:**1, 1939.
30. Ito, M.: Cited in Bell, C. C., and Dow, R. S.: Cerebellar circuitry, Neurosci. Res. Symp. Sum. **2:**569, 1967.
31. Ito, M., and Yoshida, M.: The cerebellar-evoked monosynaptic inhibition of Deiters neurones, Experientia **20:**515, 1964e.
32. Ito, M., and Yoshida, M.: The origin of cerebellar-induced inhibition of Deiters neurones. I. Monosynaptic initiation of the inhibitory postsynaptic potentials, Exp. Brain Res. **2:**330, 1966b.
33. Jansen, J.: Studies on the cetacean brain, hvalrådets skrifter (Scientific results of marine biological research), No. 37, Norske Videnskaps-Akad. I, Oslo, 1953.
34. Jansen, J., and Brodal, A.: Experimental studies on the intrinsic fibers of the cerebellum; the cortico-nuclear projection in the rabbit and the monkey, Avhandl. Norske Videnskaps-Akad. I, Mat. Naturv. Kl. No. 3, p. 1, Oslo, 1942.
35. Jansen, J., and Brodal, A.: Aspects of cere-

bellar anatomy, Oslo, 1954, Johan Grundt Tanum Forlag.
36. Keller, A. D., Roy, R. S., and Chase, W. P.: Extirpation of the neocerebellar cortex without eliciting so-called cerebellar signs, Am. J. Physiol. **118**:720, 1937.
37. Lam, R. L., and Ogura, J. H.: An afferent representation of the larynx in the cerebellum, Laryngoscope **62**:486, 1952.
38. Larsell, O.: The cerebellum: a review and interpretation, Arch. Neurol. Psychiatry **38**: 580, 1937.
39. Larsell, O.: The development of the cerebellum in man in relation to its comparative anatomy, J. Comp. Neurol. **87**:85, 1947.
40. Larsell, O.: The cerebellum of the cat and the monkey, J. Comp. Neurol. **99**:135, 1953.
41. McDonald, J. V.: Responses following electrical stimulation of anterior lobe of cerebellum in cat, J. Neurophysiol. **16**:69, 1953.
42. Mollica, A., Moruzzi, G., and Naquet, R.: Décharges réticulaires induites par la polarisation du cervelet: leurs rappaports avec le tonus postural et la réaction d'éveil, Electroencephalogr. Clin. Neurophysiol. **5**:571, 1953.
43. Moruzzi, G.: Problems in cerebellar physiology, Springfield, Ill., 1950, Charles C Thomas, Publisher.
44. Nulsen, F. E., Black, S. P. W., and Drake, C. G.: Inhibition and facilitation of motor activity by the anterior cerebellum, Fed. Proc. **7**:86, 1948.
45. Ramón y Cajal, S.: Histologie du système nerveux de l'homme et des vertébrés, Paris, 1909, A. Maloine, vol. 1.
46. Snider, R. S.: Interrelations of cerebellum and brain stem, Res. Pub. Res. Assoc. Nerv. Ment. Dis. **30**:267, 1950.
47. Snider, R. S, and Cooke, P. M.: Cerebellar activity in relation to the electrocardiogram before, during and after seizures states, Electroencephalogr. Clin. Neurophysiol. supp. **3**:78, 1953.
48. Snider, R. S., and Eldred, E.: Cerebro-cerebellar relationships in the monkey, J. Neurophysiol. **15**:27, 1952.
49. Snider, R. S., and Goldman, M. A.: Monosynaptic and multisynaptic arcs within the cerebellum, Fed. Proc. **11**:150, 1952.
50. Snider, R. S., and Stowell, A.: Receiving areas of the tactile, auditory, and visual systems in the cerebellum, J. Neurophysiol. **7**:331, 1944.
51. Sprague, J. M., and Chambers, W. W.: Regulation of posture in intact and decerebrate cat; cerebellum, reticular formation, and vestibular nuclei, J. Neurophysiol. **16**:451, 1953.
52. Szentagóthai, J.: Ujabb adatok a synapsis funkcionalis anatomiajahoz (New data on the functional anatomy of synapses), Magy. Tud. Akad., Biol. Orv. Tud. Osztal. Kozl. **6**:217, 1963.
53. Tarnecki, R., and Konorski, J.: Patterns responses of Purkinje cells in cats to passive displacements of limbs, squeezing and touching, Acta Neurobiol. Exp. **30**:95, 1970.
54. Thach, W. T.: Somatosensory receptive fields of single units in cat cerebellar cortex, J. Neurophysiol. **30**:675, 1967.
55. Thach, W. T.: Discharge of Purkinje and cerebellar nuclear neurons during rapidly alternating arm movement in the monkey, J. Neurophysiol. **31**:785, 1968.
56. Thach, W. T.: Discharge of cerebellar neurons related to two maintained postures and two prompt movements. I. Nuclear cell output, J. Neurophysiol. **33**:527, 1970.
57. Thach, W. T.: Discharge of cerebellar neurons related to two maintained postures and two prompt movements. II. Purkinje cell output and input, J. Neurophysiol. **33**:537, 1970.
58. Tyler, D. B., and Bard, P.: Motion sickness, Physiol. Rev. **29**:311, 1949.
59. Victor, M., Adams, R. D., and Mancall, E. L.: A restricted form of cerebellar cortical degeneration occurring in alcoholic patients, Arch. Neurol. **1**:579, 1959.
60. Woolsey, C. M.: Summary of the papers on the cerebellum, Res. Publ. Res. Assoc. Nerv. Ment. Dis. **30**:334, 1950.
61. Woolsey, C. M.: Personal communication.

29

ELWOOD HENNEMAN

Motor functions of the cerebral cortex

The existence of an area in the cerebral cortex specifically concerned with motor functions was suggested as early as 1691 by Robert Boyle, who described the case of a knight who had sustained a depressed fracture of the skull and developed what Boyle called a "dead palsy" of the arm and leg on one side of the body. A barber surgeon elevated the depressed bone and promptly relieved the paralysis. The first notion of an orderly "representation" of motor control came in 1864 from the studies of Hughlings Jackson, an English neurologist, who made careful observations on the development of local seizures in his epileptic patients. As these attacks spread from their original focus, the order in which the parts of the body began to jerk suggested that a disturbance of some kind was spreading by local extension over a region of the brain in which movements of the parts of the body were represented in an orderly plan. Jackson's brilliant clinical deduction was confirmed in1870 by Fritsch and Hitzig in Germany, who reported that electrical stimulation of the frontal cortex of cat and dog caused movements of the limbs on the opposite side of the body. In a monograph that appeared in 1874, Hitzig accurately defined the limits of the "motor area" from which these responses could be elicited in the dog and monkey. David Ferrier, who was working at about the same time in England as a physician in the West Riding Lunatic Asylum, made a remarkably accurate motor map of the monkey cortex (Fig. 29-1). Ferrier borrowed the term "localization" from Jackson to describe his findings. In further studies Ferrier first delimited a specific part of the motor area, e.g., the "hand area," by means of electrical stimulation and then removed this area surgically, causing a paralysis of the hand. Ferrier took some of his "hemiplegic" monkeys to the International Medical Congress of 1881 in London. Their resemblance to hemiplegic patients was so striking that the famous French neurologist Charcot, seeing one of the monkeys, declared "It is a patient." The motor cortices of several types of apes were stimulated by Sherrington and his collaborators in the late 19th century, but there was no systematic study of the human motor cortex until the German neurosurgeon Foerster published his report in 1925.[27]

As clinical and experimental studies have progressed, it has become apparent that certain parts of the cerebral cortex such as the precentral region are concerned exclusively with motor functions; other parts such as the somatic sensory areas have both sensory and motor functions; and still other regions, not usually regarded as having any direct role in movement, are doubtless involved in the volitional aspects of motor activity. Closely associated with the cortex in its motor functions are a group of nuclei known as the basal ganglia, which occupy a large volume of brain. They have important connections with the cortex and their function appears to be chiefly motor. The little that is known about their functions has been described in Chapter 26.

Very little can be said about the origin of the neural commands that lead to movement. No precise "command center" has been identified, and no particular cortical or subcortical region seems to be essential for the formulation of the orders to which the motor cortex responds. Most of the available information concerns the systems that execute these commands. There are two main subdivisions of motor control, the pyramidal and extrapyramidal systems. They work together to produce movement, muscle tone, and posture, but they play very different roles—so much

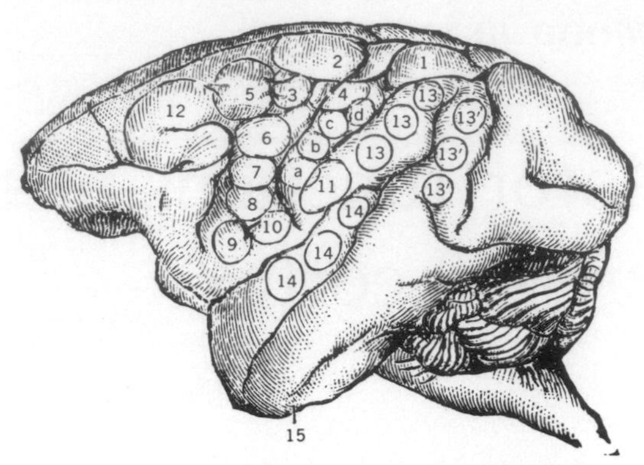

Fig. 29-1. Ferrier's original (1875) motor map of left hemisphere of monkey describing effects of electrical stimulation at series of numbered sites. Ferrier's original legend follows: "The left hemisphere of the monkey. 1, The opposite hind limb is advanced as in walking; 2, flexion with outward rotation of the thigh, rotation inwards of the leg, with flexion of the toes; 3, the tail; 4, the opposite arm is adducted, extended, and retracted, the hand pronated; 5, extension forwards of the opposite arm; *a, b, c, d,* movements of fingers and wrist; 6, flexion and supination of the forearm; 7, retraction and elevation of the angle of the mouth; 8, elevation of the ala of the nose and upper lip; 9 and 10, opening of the mouth, with protrusion (9) and retraction (10) of the tongue; 11, retraction of the angle of the mouth; 12, the eyes open widely, the pupils dilate, and head and eyes turn to the opposite side; 13 and 13′, the eyes move to the opposite side; 14, pricking of the opposite ear, head and eyes turn to the opposite side, pupils dilate widely." (From Ferrier.[22a])

so that the clinical neurologist classifies his patients' motor "signs" as either pyramidal or extrapyramidal.

Pyramidal system

The pyramidal system is much the easier to deal with because it is far simpler to define anatomically. It consists of neurons whose cell bodies lie in the cerebral cortex and whose axons pass through the *pyramid* of the medulla to form the pyramidal or corticospinal tracts of the spinal cord. In man the longest of these fibers runs at least 1 m. There are about 1 million fibers in each human pyramid. About 60% of them arise in precentral cortex and 40% in postcentral cortex.[34] Only about 2% of the fibers come from the large Betz cells. Since 90% of them are small (1 to 4μ) and about half of the total are unmyelinated, the pyramidal tract is, on the whole, a slowly conducting pathway.

Some pyramidal axons end in the motor nuclei of the brainstem; the majority pass through the pyramids and form three corticospinal tracts. The largest of the three in primates is the crossed corticospinal tract, which makes up three fourths of the total. Two uncrossed tracts, somewhat variable in size run in the lateral and ventral parts of the spinal cord and end ipsilaterally. With a special technique for staining terminals of degenerating axons, it has been shown that 80 to 90% of the corticospinal fibers in primates make synaptic contacts with internuncial cells in the spinal cord, whereas 10 to 20% end directly on motoneurons.

Anatomic studies carried out by cutting the pyramidal fibers and studying the retrograde degeneration that follows show that they arise from a very extensive region in front of and behind the central fissure of Rolando. The areas from which they arise include Brodmann's areas 8, 6, and 4 in the precentral region and areas 3, 1, 2, 5, and 7 postcentrally (see Fig. 7-4). Since only about 10% of the pyramidal fibers are large enough ($> 5\mu$) to show up well in Weigert and Marchi preparations, neuroanatomists were uncertain about the origin of the smaller fibers. The neatest way of determining the cortical origin of the pyramidal tract, and certainly one of the quickest, is by an electrophysiologic method devised by Woolsey and Chang.[73] They exposed the pyramid in the medulla and stimulated it electrically with single shocks, taking special precautions to avoid exciting other fiber tracts nearby. An antidromic volley that was set up in the pyramidal fibers was conducted back to the cortex wherever the cells of origin were located. The pattern of antidromic responses recorded in this way precisely defined the origins of the pyramidal tract in the cat and monkey. As shown in Fig. 29-2, responses were de-

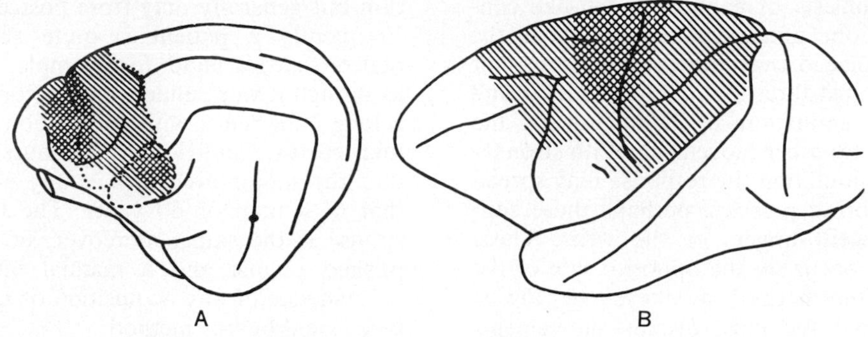

Fig. 29-2. Dorsolateral views of left cerebral cortex of cat, **A,** and monkey, **B,** showing distribution of potentials (hatching and cross-hatching) evoked by antidromic volleys of impulses set up by stimulation of medullary pyramid. (From Woolsey and Chang.[73])

tected over most of the parictal lobc and a great deal of the precentral cortex. This pattern of origin agrees well with anatomic studics but shows somewhat more extended boundaries in areas 8 and 6. It includes all four of the cortical areas now known to have somatic motor functions. A stimulating electrode placed anywhere within this large area will elicit a movement of some part of the body.

The pyramidal tract is late to develop in evolution. Its size and functional importance increase progressively in moving from mammals to monkeys to the higher apes and to man. This suggests that it developed to serve the needs of recently acquired cortical areas for a more direct and autonomous control over motor neurons than that provided by the extrapyramidal system. Differences in the degree of development of the pyramidal system sometimes make comparisons of experimental results in different species difficult.

Extrapyramidal system

The pyramidal tract is a compact bundle of fibers arising in the cortex that is easily recognized as an anatomic entity. In contrast, the extrapyramidal system consists of many separate components arising at different levels of the nervous system, often without obvious relation to each other, and hardly justifying the term "system." The pyramidal tract transmits impulses directly from cortex to spinal cord; the extrapyramidal system transmits signals indirectly from the cortex through one or more relays to the cord. Cortical extrapyramidal fibers take origin from all of the sensory-motor cortex, including the supplementary motor area and the "second" motor areas, and from many other parts of the cerebral cortex as well.[12] It is not known whether they are the axons of separate extrapyramidal cells in the cortex or are collaterals of other corticofugal fibers. Impulses in these cells may take several different routes to the motor neurons. There are two relatively direct routes available: one from the sensorimotor cortex to the reticular formation (RF) in the brainstem and from there to the spinal cord via the reticulospinal tracts; the other from the motor cortex to the red nucleus and from there to the spinal cord via the rubrospinal tract. In addition, a considerable number of cortical fibers descend to the basal ganglia, and from them a series of connections are established to the red nucleus and RF.

MOTOR AREAS OF CEREBRAL CORTEX—RESULTS OF STIMULATION

Motor effects can be elicited by adequate excitation of any site within the cross-hatched region shown in Fig. 29-2. Electrical stimulation is the easiest and most readily controlled type of stimulus to use. The excitable region extends into the depths of the central fissure and other fissures within this extensive area and includes a large portion of the medial surface of the brain as far down as the cingulate sulcus. In man and in animals with highly fissured brains one half to two thirds of the total motor area is buried and rather inaccessible to study unless the buried portions are exposed by removal of one bank of a fissure.

Responses to electrical stimulation

The stimulus most frequently used is a 60 Hz alternating current of a few volts' intensity and 1 to 3 sec duration. If the stimulus is not too strong and the cortex is not too depressed by anesthesia, the typical response generally consists of a brief, twitch-like contracture of one or several muscles. With the electrodes placed over the "thumb" area, for example, a just-threshold stimulus may result in a brief abduction or adduction of the thumb with no other movement. With stronger or longer stimulation the response may spread to include other muscles, perhaps those supplying adjacent fingers or the wrist. These movements occur on the opposite side of the body. In anesthetized monkeys the lowest threshold and the most discrete movements are usually produced by stimulating just in front of the central fissure in area 4. In general, responses elicited from cortex behind the central fissure require a stronger stimulus, a phenomenon that suggests a lower density of the neurons that project to motor centers. Under deep anesthesia, responses are essentially simple contractions of one or more muscles. With anesthesia light enough to permit sustained tonic innervation of muscles, it can be seen that stimulation causes relaxation of some muscles and contraction of others. The pattern of such effects is one of reciprocal innervation, i.e., opposing effects in antagonistic muscles. If the stimulus applied to a cortical site is too intense, the response spreads to involve a whole group of muscles, and there may be alternating jerks due to contractions in opposing muscles. This is really a local seizure, quite comparable to local epilepsy in a patient. If the stimulus is strong enough or the cortex is excitable enough, the spread of activity may continue until the seizure involves the entire body.

Stimulation of human cortex

The movements elicited by stimulating the motor cortex of a conscious patient during surgery are similar to those obtained in animals but they are more discrete, owing to finer control of distal musculature in man. In a conscious, cooperative patient it can also be shown that stimulation produces paralysis of the voluntary control over muscles. For example, a patient gripping a rod tightly with both hands will relax his grip on one side when the contralateral hand area is excited. A patient who is speaking will stop in the middle of a word if the motor area controlling his laryngeal muscles is stimulated. During stimulation a conscious subject may report merely an awareness of the movement as it occurs. Occasionally a desire to move is noted. Sensations of numbness or tingling are occasionally reported from precentral stimulation but generally only from postcentral sites. Frequently a patient is quite surprised to realize that his hand, for example, has moved as though it were under outside control. From a long experience with stimulation of the human cortex, Penfield and Rasmussen[48] report that the motor area of a young child is like that of a man of 60 years. The type of response is the same, moreover, in an accomplished pianist and a manual laborer; i.e., no evidence of any acquisition of motor skills is revealed by this method.

Mechanism of cortically elicited movements

Relatively little is known about the mechanisms set in operation by the cortical stimulus. In most instances the electrode excites both pyramidal and extrapyramidal cortical efferents.The delicacy and discreteness of the responses, however, stamp them as primarily pyramidal. When the pyramids were cut in the medulla, leaving the remainder of the medulla undamaged, Tower[62] reported that somatotopically organized, discrete movements were largely abolished. Stimulation continued to elicit motor responses, but they had lost their fineness and delicacy. Removal of the contralateral motor areas does not abolish these remaining responses; hence they must be mediated by extrapyramidal connections. It is apparent from the electrical studies of Lloyd[39] that a single shock applied to the motor cortex of a cat will evoke descending volleys in at least two systems. There is a widespread activation of the RF in the brainstem 4 msec after the cortical shock. This reticular activity is in turn projected by high-velocity tract fibers to the ventral horn of the spinal cord. A calculation based on known conduction rates and delays involved shows that the volleys relayed through the reticulospinal system could reach the local segmental mechanism at the same time or a little earlier than the more slowly conducted but more direct pyramidal impulses.

Pattern of representation in motor areas

The earliest investigations with electrical stimuli showed clearly that muscles in the different parts of the body are represented in orderly sequence in area 4, just anterior to the central fissure. This sequence is a reflection of the segmental origin of the nervous system. As illustrated to the left of Fig. 29-3, the toes, ankle, and leg are represented on the medial wall of the brain, with the knee, hip, and trunk following in order on the convexity

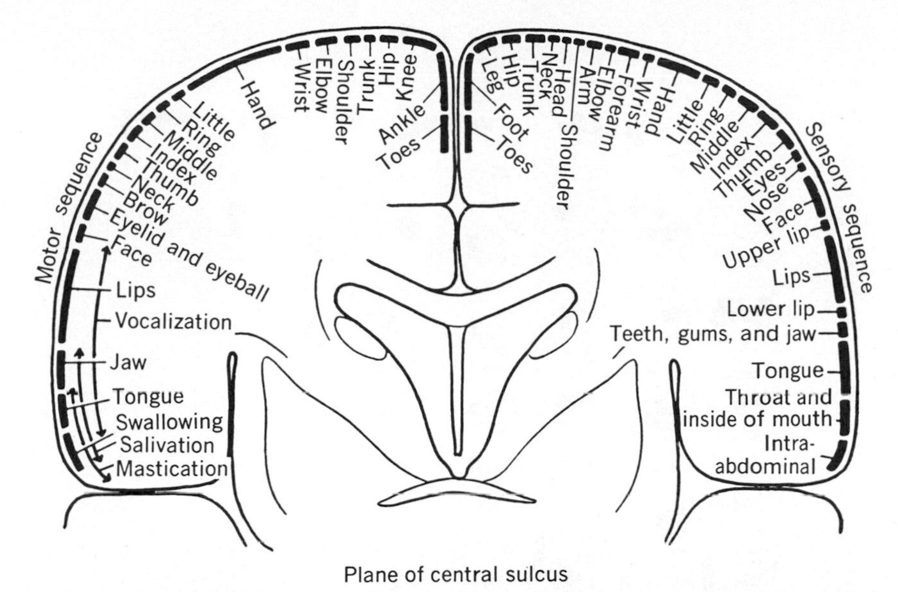

Fig. 29-3. Diagram of sensory and motor sequences in cerebral cortex of man as determined by electrical stimulation. Extent of cortex devoted to each part of body is indicated by length of heavy black lines. (From Rasmussen and Penfield.[51])

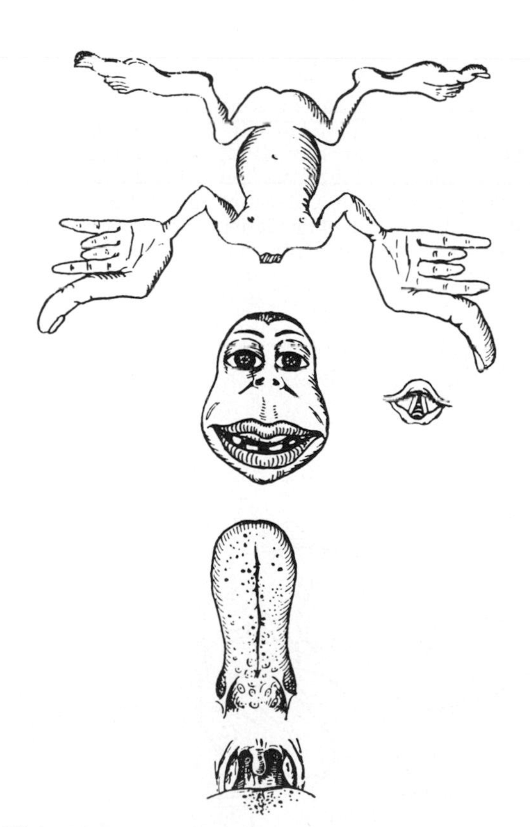

Fig. 29-4. Penfield and Boldrey's homunculus. Parts of body are drawn to illustrate relative amounts of motor cortex devoted to each. (From Penfield and Boldrey.[47])

of the cortex; and then the shoulder, elbow, wrist, and fingers are succeeded by the neck, brow, face, lips, jaw, and tongue most laterally. At the oral and anal ends of this sequence the parts of the body that are "turned inward" are apparently represented in the sylvian fissure and the cingulate fissure.

The amounts of cortex devoted to various parts of the body are very unequal. The thumb, for example, is controlled by an area 10 times as great as that which controls the whole thigh. The parts of the body that are capable of fine or delicate movement have a large cortical area devoted to them; those that perform relatively gross movements, e.g., trunk muscles, have comparatively little cortical representation. The area of cortex devoted to a part is related to the density of the innervation of that part in the periphery. From an analysis of their studies on human subjects, Penfield and Boldrey[47] constructed a small figure of a man called a homunculus (Fig. 29-4). The figure was drawn so that the size of each part corresponded to the amount of cortex given over to it. The result was an extremely distorted figure of a man with a very large thumb, a relatively big face, and an enormous tongue. It is not surprising that focal epilepsy most commonly occurs in these parts of the body, since they comprise a large percentage of the total motor representation.

In spite of thousands of cortical stimulations by a great many investigators, the com-

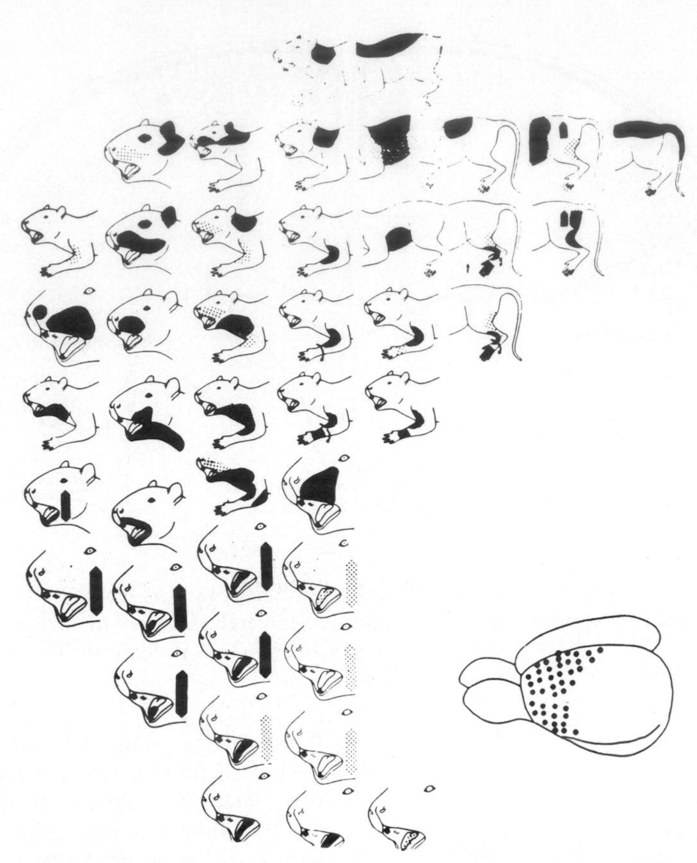

Fig. 29-5. Pattern of representation in left cortical motor area of rat. Figurines correspond to points marked on inset drawing. Intervals between points are approximately 1 mm. Each figurine indicates peripheral location of musculature activated by electrically stimulating cortical point to which figurine corresponds. Regions of maximal responses are indicated in black; those of weaker muscular actions in cross-hatching and stippling. Reacting muscles are in right half of body, contralateral to cortex stimulated, but to maintain desired orientation within pattern, left-sided figurines are used. Motor pattern is essentially a mirror image of cortical pattern of tactile representation in somatic sensory area I of this same species. (After Settlage et al.[54]; from Woolsey et al.[75])

plete and now rather obvious plan of motor representation eluded everyone until 1947. Then, as a result of studies by Mountcastle and Henneman[44] on the localization of tactile responses in the thalamus, it became apparent that within the three dimensions of one nucleus, the entire surface of the contralateral side of the cat was represented in continuity and with no missing parts. This study suggested that a complete representation of skeletal muscles might be demonstrable within the confines of the motor cortex. Woolsey et al.[75] first examined the motor cortex of the rat, which had the advantage of being smooth and unfissured. It soon became evident that all previous maps of the motor cortex were incomplete and partially erroneous. The essense of their study was that the muscles of the opposite side of the body (Fig. 29-5) are

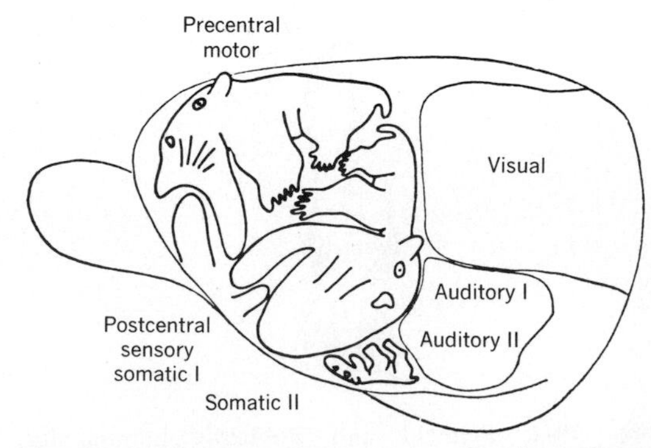

Fig. 29-6. Diagram of rat cortex showing mirror-image representation in precentral motor cortex and somatic sensory area I (S I). (From Woolsey.[72])

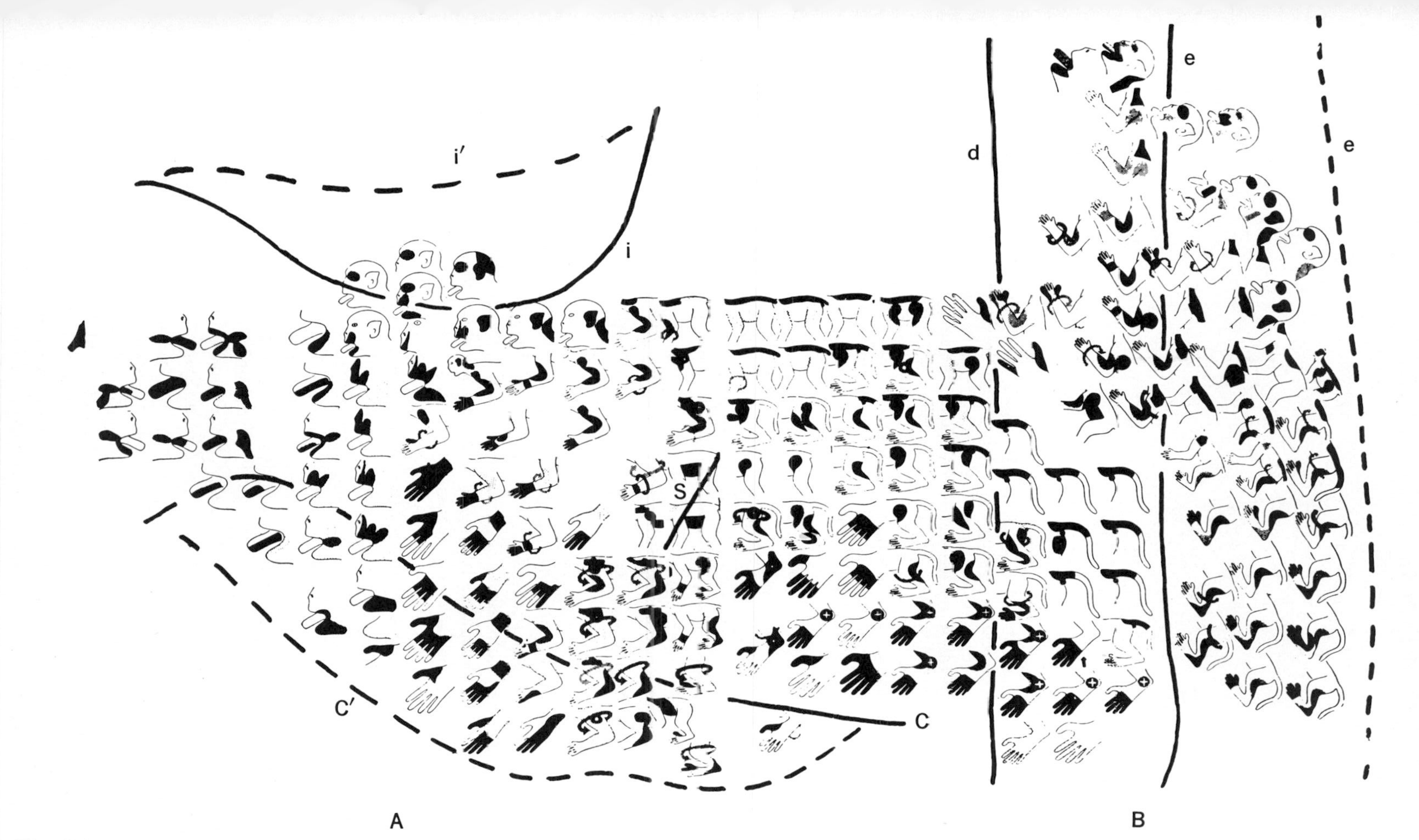

Fig. 29-7. Composite figurine charts of precentral and supplementary motor areas of monkey brain, derived from several experiments in which left cortex was mapped by systematic punctate electrical stimulation. Except for responses from points in ipsilateral motor face area (at extreme left of **A**), muscle reactions are in right half of body, but (as in Fig. 29-5) to maintain desired orientation within total pattern, left-sided figurines are used; to appreciate laterality of movement, reader should imagine that he is looking through each figurine to its opposite side. Strongest and earliest movements are indicated in black; cross-hatching signifies intermediate, and stippling, weakest effects. In many cases, these shadings show nature of movement. Symbols with crosses on ankles indicate eversion of foot; those with open centers on hip and ankle signify adduction and inversion; curved lines with arrows designate rotation. Supplementary motor area, situated almost entirely on medial surface of hemisphere and on upper bank of sulcus cinguli, is quite separate from precentral motor area but adjoins medial and rostromedial limits of the latter. Extents and spatial relationships of the two areas are shown diagrammatically in Fig. 29-8; general arrangement of patterns in Fig. 29-9. Note that both representations are partly on medial surface and that important portions of each are located in one or more deep sulci. Boundaries of these sulci are marked in manner indicated in legend of Fig. 29-8. (From Woolsey et al.[75])

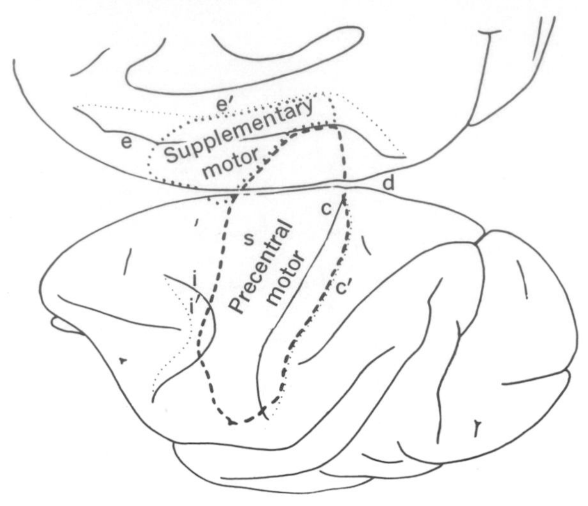

Fig. 29-8. Drawing of dorsolateral (below) and medial (above) views of left hemisphere of macaque showing extents of precentral and supplementary motor areas. (Patterns of localization within each are shown in diagrammatic form in Fig. 29-9.) Caudal bank of central sulcus, anterior bank of inferior precentral sulcus, and lower bank of sulcus cinguli have been cut away to show extension of excitable cortex into opposite banks of these sulci. Different sulci are indicated as follows: *c*, central sulcus; *c'*, bottom of central sulcus; *i*, inferior precentral sulcus; *i'*, bottom of inferior precentral sulcus; *d*, medial edge of hemisphere; *e*, sulcus cinguli (on medial surface of hemisphere between rim and corpus callosum); *e'*, bottom of sulcus cinguli; *s*, superior precentral sulcus. (From Woolsey et al.[75])

represented without separation of face and trunk areas by arm (as previously believed), with the axial muscles arranged in sequence from neck to tail along the rostral and medial portions of the brain and with the limb areas more laterally. As shown in Fig. 29-6, the somatic sensory and motor representations are essentially mirror images.

The results obtained in the simple brain of the rat led to a reexamination of the precentral motor area of the monkey. Experiments were carried out under barbiturate anesthesia. Cortical sites were stimulated systematically at 0.5 to 1.0 mm intervals and several investigators made careful observations of the resulting movements, which were recorded by shading or crosshatching the appropriate part of an outline drawing of the monkey. Fig. 29-7 is a composite chart derived from several experiments showing the cortical location of each stimulation and the part of the body that moved in response. The portion of the cerebral cortex shown in Fig. 29-7 is outlined in Fig. 29-8, in which the fissures and markings are identified. As illustrated in these figures, Woolsey et al.[75] found a complete and continuous pattern of motor representation extending from the bottom (*c'*) of the central fissure posteriorly forward into the inferior precentral sulcus (*i*) anteriorly and from the bottom of the cingulate fissure (*e'*) on the medial wall to the lateral end of the central fissure. Within this region two complete motor representations could be distinguished, as shown in Fig. 29-9. The larger and more detailed of the two patterns, occupying the precentral gyrus (area 4), is the "motor cortex" of classic description; the smaller pattern of the medial wall is a more recent addition called the "supplementary motor area," which will be described later in this chapter. Examination of Fig. 29-7 reveals that the individual responses fit together like the pieces of a jigsaw puzzle, with adjacent muscles located contiguously. The muscles of the axial parts of the body, the back, neck, and head, are represented most anteriorly; those controlling the digits of the hindlimb and forelimb are represented most posteriorly on the anterior bank of the central fissure. The unnatural proximities of certain parts of the body such as the thumb and lower lip or the hallux and little finger result because all of the cortex is used for representation, and there are no unoccupied areas to separate different portions of the body. All of the movements elicited by stimulation within this primary motor area occurred contralaterally, except for those evoked from the most lateral sites where movements of the *ipsilateral* face, lips, and tongue were produced (Fig. 29-9). The existence of this ipsilateral face area is prob-

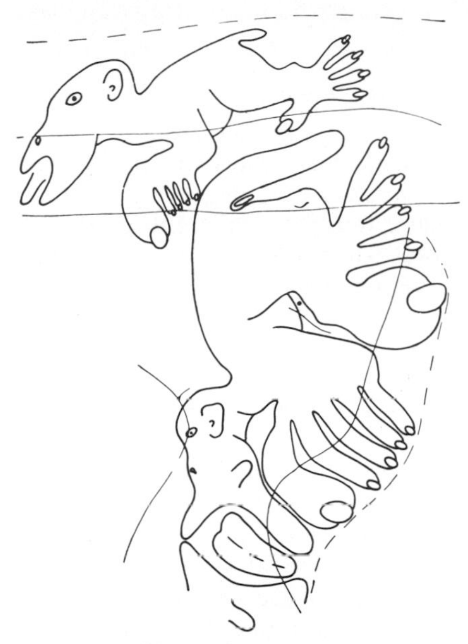

Motor simiusculi

Fig. 29-9. Diagram illustrating general arrangement of patterns of representation of somatic musculature in precentral and supplementary motor areas of monkey. Sulci and medial edge of hemisphere (left) are indicated by lines as in Fig. 29-8. Bodily distortion displayed by each simiusculus (simian equivalent of homunculus) is due to very unequal representation of different parts of musculature in the two areas. Ipsilateral motor face area is indicated at bottom of drawing. (From Woolsey et al.[75])

ably related to the need for bilateral control and coordination of these particular muscles.

Fig. 29-7 gives some indication of the extent to which the representation of one muscle overlaps with that of others nearby. Due to the depth of anesthesia used in these experiments and the uncertain amount of current spread during stimulation, it was not possible to determine whether the true overlap was greater or less than that revealed in Fig. 29-7. More recent and refined observations on this matter will be described later in this chapter.

Supplementary motor area

As already noted, Woolsey et al.[75] confirmed the existence of a second complete motor representation located almost entirely on the medial surface of the hemisphere. It lies on the upper bank of the sulcus cinguli and the medial surface of the brain dorsal to this fissure, with the tip of the thumb area extending up on the convexity. This representation adjoins the hind limb and tail areas of the precentral motor area, but is quite separate from it. By stimulating this region, Penfield and Welch[49] had previously elicited movements of the contralateral extremities in anesthetized monkeys. In patients under local anesthesia they produced vocalization with associated movements of face and jaw, combinations of movements in the extremities, trunk, and head, inhibition of voluntary activity, and certain autonomic effects. They named this region the "supplementary motor area." Woolsey et al.[75] worked out the complete pattern of representation shown in Fig. 29-9. Instead of the rapid, phasic motor responses observed in conjunction with precentral cortical stimulations, they noted that the contractions lasted much longer and that postures assumed as a result of brief stimulation were often maintained for many seconds. Although movements were less discrete, the whole animal was represented in continuity—much of it in the depths of the sulcus cinguli.

Postcentral motor areas

In 1900 Schaefer[53] published a study of the monkey brain showing that movements could be elicited by stimulating not only the frontal areas now identified as the precentral and supplementary motor areas but also the postcentral gyrus. Similar results were reported by the Vogts[67] in animals and by Foerster[26] and Penfield and Rasmussen[40] in man. However, as Woolsey has pointed out, " . . . since the work of Leyton and Sherrington,[37] the preeminence of the precentral gyrus in motor function has dominated teaching and thinking concerning cortical control of the somatic musculature and the motor effects of postcentral stimulation generally have been explained as the result of the spread of excessive stimulating currents to the precentral area, or on the basis of corticocortical connections with this area."[72]

In 1943 Kennard and McCulloch[31] stimulated the postcentral gyrus of infant monkeys after removal of the precentral motor areas. They found that focal movements like those usually evoked from the precentral gyrus were easily produced. Later Woolsey et al.[74] stimulated the postcentral gyrus of adult monkeys months after removal of both precentral and supplementary motor areas. In these experiments they demonstrated a "well-organized postcentral motor outflow after complete degeneration of the motor pathways from both frontal lobes."[74] The detailed pattern of motor representation that was found apparently coincided with the somatic sensory pattern in the same area and appeared to be a mirror

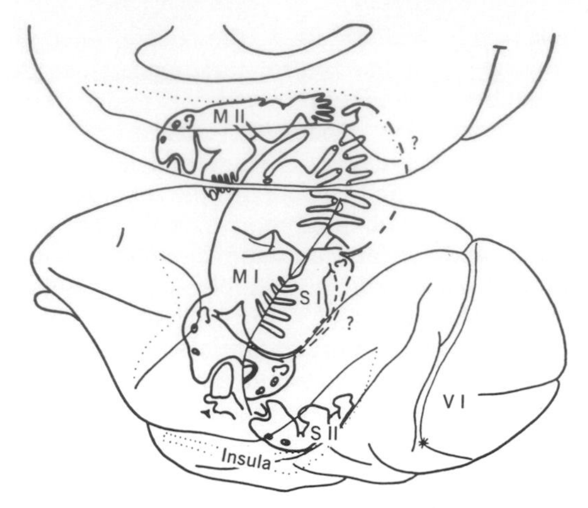

Fig. 29-10. Diagram of monkey cortex showing locations of four principal motor areas, precentral motor, *M I;* supplementary motor, *M II;* somatic sensory I, *S I;* and second sensory, *S II. S II* lies on upper bank of sylvian fissure next to insula and auditory area on lower bank. (From Woolsey.[72])

image of the precentral motor pattern just in front of the central fissure. The chief difference between the pre-and postcentral responses was that the electrical threshold was 2 to 3 times higher for the latter. From these results it appears that there is a well-organized postcentral motor system that ordinarily functions in cooperation with the precentral systems but can function independently of them.

In addition to the sensorimotor area in the postcentral gyrus, there is another parietal motor area coinciding closely with the *second somatic sensory area* first described by Adrian.[1] This is a receiving area for somatic sensory signals from both sides of the body. In monkeys and man it is located on the superior bank of the sylvian fissure (Fig. 29-10), with its face area immediately adjacent to the face area of the postcentral somatic sensory area. Sugar et al.[59] found that strong stimulation in this area elicited movements on both sides of the body, and they named it the *second motor area.* Later it was shown that the sensory and motor patterns in this small area coincide somatotopically.[7, 69]

• • •

In summary, there are four functionally distinguishable motor areas, as shown in Fig. 29-10, two in front of the central fissure and two behind it. Three of them definitely contribute fibers to the pyramidal tract (the supplementary motor area has not been examined in this respect). It is probable that all four areas send connections to extrapyramidal centers as well.

Special motor areas

In addition to the four somatotopically organized cortical areas just described, each of which exercises some degree of control over the skeletal muscles of the entire body, there are several cortical areas concerned with control of the muscles associated with the special senses.

Early in the development of our knowledge of cortical motor areas, it was found that stimulation of a localized region in the frontal lobe of the primate produced movements of the eyes, eyelids, and pupils. According to W. K. Smith,[56] the "frontal eye field" of the monkey is extensive, and on the basis of its responses to electrical stimulation it can be divided into five parts, as shown in Fig. 29-11: (1) an area situated just caudal to and ventral to the angle of the inferior precentral (arcuate) sulcus, from which closure of the eyes is obtained; (2) a field lying just across the sulcus, which yields nystagmus bilaterally (fast component to the opposite side) and opening of the eyes if they are closed; (3) an adjacent area, situated more medially within the upper arm of the arcuate sulcus, which yields conjugate turning of the eyes to the opposite side; (4) a field centered on the upper extremity of the arcuate sulcus, from which an "awakening reaction" (opening of the eyes with slow deviation to the opposite side, blinking, and pupillary dilatation) can be obtained; and (5) an area lying between the medial end of the arcuate sulcus and some point on the medial surface of the hemisphere, excitation of which results in equal dilatation of both pupils. Somewhat similar results have been reported for both chimpanzee and man. The frontal eye fields may represent an elaboration and outgrowth of the precentral motor area controlling the eyes, or it may be a much more specialized motor area functionally quite distinct from it.

Another motor eye field is located in the occipital lobe within or adjacent to the visual receiving areas. Stimulation of this region generally causes conjugate deviation of the eyes to the opposite side. When the electrodes were applied to the cortex of area 17 superior to the calcarine fissure, the movements were lateral and downward. When the stimuli were applied below the calcarine fissure, the deviation was lateral and upward. The frontal and occipital eye fields are interconnected by long

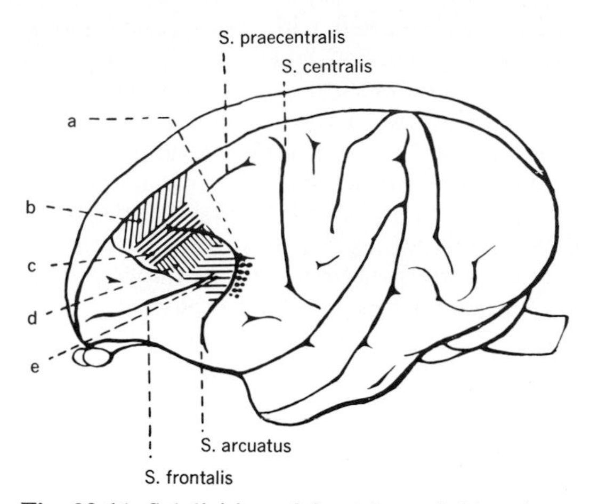

Fig. 29-11. Subdivision of frontal eye field and area yielding closure of eyes in monkey; *a*, closure of eyes; *b*, pupillary dilatation; *c*, "awakening"; *d*, conjugate deviation to opposite side; *e*, nystagmus to opposite side. (From Smith.[56])

corticocortical fibers, but responses to stimulation in either area do not depend on the integrity of the other.

Adjacent to the auditory receiving area in the cat and monkey is a small region whose stimulation causes movements of the ear in these species. In the cat, electrical responses to clicks can be recorded within this "pinna area," suggesting that it responds reflexly to auditory stimuli by turning the ear toward the source of the sound.

From this brief account it can be seen that each of the main sensory receiving areas has a motor area adjacent to it. Whether this rule applies to the receiving areas for taste and olfaction is not yet known.

ORGANIZATION OF MOTOR CORTEX

In order to understand how the motor cortex is organized, it is necessary to know (1) how the efferent cells and fibers are arranged and grouped functionally, (2) what types of signals the cortex receives and how they are distributed to the efferent cells, and (3) how cortical input and output are related. These topics will be considered in order.

Efferent organization

Columnar structure of cortex

Sections of the cortex cut perpendicularly to the surface reveal a columnar arrangement of the cells in the precentral motor cortex similar to that found in other regions. Lorente de Nó's studies with Golgi stains showed a wealth of radial interconnections between the cells in the different layers of these columns and led him to suggest that these "vertical chains" of cells might be considered as elementary units of the cortex.[40] Mountcastle's discovery that the cells in somatic sensory columns all receive signals of the same sensory modality from the same part of the body provided the first physiologic demonstration that these cortical columns are functional entities.[43] Since his initial work, several studies have shown that the columns in the motor areas are the "building blocks" around which the cortex is organized, for input as well as output.[2, 3, 9]

Asanuma and Sakata[2] have developed a technique of intracortical microstimulation that has been very useful in investigating the efferent projections of individual motor columns. Monosynaptic reflexes were used as sensitive indicators of the effects of these stimulations. Electrical thresholds for definite but minimal effects were as low as 1/100 of the threshold required for surface stimulation and therefore entailed far less spread of current. Facilitation and inhibition were generally evoked from distinct, nonoverlapping regions, with facilitation having its lowest thresholds at 1.0 to 1.4 mm from the surface and inhibition 1.0 to 1.5 mm deeper. When the effects of stimulations at various depths in numerous penetrations were correlated with the exact site of each in histologic reconstructions of the electrode tracks, it was found that the lowest thresholds for facilitation of a particular pool of motoneurons were distributed in a radially aligned column estimated to be about 1 mm in diameter. As one would expect with electrical stimuli, some overlapping of the effects obtained from adjacent efferent zones was noted, but not enough to invalidate the clear indication of cylindric arrays of cells with a common spinal projection. Section of the bulbar pyramid abolished these effects, indicating that they were purely pyramidal. Extrapyramidal effects have not yet been identified with this technique.

In further studies with intracortical microstimulation,[3] trains of pulses at 300/sec were used for periods up to several seconds, and the contractions of a pair of antagonistic muscles at the cat's wrist were recorded. With the electrode in layers V and VI it was possible to produce sustained contractions of one of these muscles lasting as long as the stimulation continued. The regions from which sustained contractions could be elicited were very limited in extent. Stimulations in superficial layers produced only phasic contractions, which died out before the stimulation ended. Sustained contractions elicited from one site could be enhanced or reduced by intracortical stimulation of another site. The cortical zone from which contraction of one muscle was elicited was not coextensive with the zone that inhibited its antagonist. Sometimes stimulation at one site caused two antagonistic muscles to contract simultaneously, indicating that their contractions were not mutually exclusive. With two electrodes it was possible to produce well-maintained contractions of antagonists. These observations indicate that there must be discrete, spatially separate zones for the excitation and inhibition of each muscle. Furthermore, the corticofugal path that produces contraction of one muscle is different from the path that inhibits contraction of its antagonist. Although the excitatory and inhibitory areas for a given muscle are spatially separate, they are located close to each other and both of them are also near the excitatory zone of the direct antagonist.

Activity of single cortical neurons during movements

Very little is known about the functional organization of the cells and circuits comprising a motor column, but recordings from individual cortical cells are beginning to reveal how their activity is related to movement. In unanesthetized monkeys, Evarts[17] has examined the discharge patterns of cortical cells during a variety of spontaneous movements of the contralateral arm. Pyramidal cells, identified by backfiring them from the medullary pyramid, were divided into two groups on the basis of their axonal conduction velocities and discharge patterns. The larger cells with rapidly conducting axons were active only during limb movements. Small cells were continuously active even when the arm was at rest, but their rate of firing increased or decreased during a movement. On many occasions it was possible to record simultaneously from these two types of cells with the same microelectrode. Although these pairs of adjacent cells were obviously in the same motor column, their firing patterns were often completely different. During a particular movement of the arm one cell might discharge while the other was silent; during a different movement the two units might discharge in parallel; occasionally their firing patterns were completely uncorrelated. In the great majority of pairs there was no fixed relationship between the two firing patterns, which varied with the type of movement being carried out. This changing or "plastic" relationship is what one would expect if the two cortical cells projected to motoneurons of two different muscles that acted sometimes as synergists, sometimes as antagonists, and at other times behaved independently of each other. The flexors and extensors of the wrist, for example, would discharge in parallel in fixing the wrist during grasping and would discharge reciprocally during simple flexions and extensions of the wrist. It appears, then, that adjacent pyramidal cells usually project to different pools of motoneurons. In view of the fact that there are thousands of cells in a single column, it is surprising that only a few pairs displayed a fixed, positive correlation such as one would expect if both of them projected to the same pool of motoneurons. Although Evarts does not comment on this feature of the results, it suggests that cells projecting to the same motoneurons are systematically separated in the motor columns. It is not surprising that pairs of adjacent cells were never found to have a fixed, reciprocal relationship to each other because such relationships between muscles do not occur, even between direct antagonists.

The precise nature of the relationships between pyramidal cells in the same column did not become apparent until the movements to which the discharges were related were carefully studied. Then it was discovered that when one neuron of a pair was related to movement at a particular joint, the other was almost always related to the same joint. Evarts pointed out that the "common denominator is the joint to whose movements the (pyramidal neurons) are related rather than the way in which they are related to the movement."[18] The same observations were made on the series of cells encountered in a single penetration of the cortex. If the most superficial cell discharged during shoulder movements, the deeper units usually had some relation to the shoulder also.

These observations indicate that the cortical cells controlling a given muscle are not assembled in one compact group, as some have imagined. The cells have different projections even within the narrow limits of a single column. A column, then, is a functional entity responsible for directing a group of muscles acting on a single joint. This conclusion supports the original notion of Hughlings Jackson that movements, not muscles, are represented in the motor cortex. Evarts emphasizes that Jackson "did not hold that muscles were not represented at all, but rather that they were represented again and again and again, in different combinations. . . ."[21]

Recent studies of Phillips and his colleagues give an indication of the extent to which a muscle may be represented in many columns. In one group of experiments,[33] single shocks were applied to the surface of the motor cortex and the excitatory volleys reaching the motoneurons monosynaptically from the pyramidal tract neurons were detected by intracellular recording. With this technique the extent of the cortical field controlling single muscles in a baboon's hand or forearm was determined. The cortical fields were often so extensive that they eliminated the possibility that spread of current to a single "best point" within the area was the cause of a large projection. One motoneuron received projections from a cortical area of 8 × 2.5 mm, and another was found to receive monosynaptic excitation from corticospinal neurons lying 10 mm from the stimulating electrode in all directions! These results indicate that a single pool of motoneurons receives projections from cells in many motor columns and that cortical cells projecting to motoneurons of different muscles are extensively intermingled. Each column is presumably a slightly different combination of these intermingled cells. As Phillips points out, fine control of movements "must require a marvelously subtle routing of activity in the outer cortical layers to pick up, in significant functional grouping, the required corticofugal neurons, which are scattered and intermingled with unwanted ones which may be suppressed."[50] At present there is no evidence to indicate whether entire columns normally contribute as units to the output or whether individual cells within columns are activated selectively. The striking differences in the firing patterns of large and small pyramidal cells have not yet been satisfactorily explained. Evarts suggested that these two types of cells may be the pyramidal counterparts of large and small alpha motoneurons, which tend to discharge phasically and tonically. If this were so, both types of cells should be silent when the arm is completely at rest. Another possibility is that the two types of cortical cells control alpha and gamma motoneurons. The latter are small and have a strong tendency to fire in the absence of any movement or muscle tone just as the small cortical cells do; the former are large and discharge only during active contraction of a muscle, as do the large cortical cells. It has often been suggested that there are independent pathways from the cortex to alpha and gamma motoneurons, but they have not yet been identified.

Inputs to motor areas*

A great many factors control and regulate movement; hence it is not surprising that the primary motor cortex receives inputs from many sources. Somatic sensory impulses from skin, joints, and muscles are relayed to the cortex by way of the ventrobasal nucleus of the thalamus. These signals provide the cortical cells with the sensory cues that are required to guide and direct movements, and they supply the feedback in a long loop that includes the motor cortex and its efferent projections back to the muscles. Cerebellar signals are relayed to the motor cortex via the brachium conjunctivum, red nucleus, and ventrolateral nucleus of the thalamus. This projection may also be regarded as part of a closed loop linking the cerebellum and motor cortex. Input from the "nonspecific" thalamic nuclei reaches the motor cortex, presumably influencing its general level of excitability as it does that of other cortical areas. Signals from the opposite hemisphere, transmitted via the corpus callosum, are important in coordinating motor activity on the two sides of the body. Finally, signals from other, ipsilateral cortical areas reach the motor cortex directly via corticocortical fibers and indirectly via several subcortical centers. The four cortical motor areas are interconnected with each other, the U-shaped fibers linking the pre- and postcentral motor areas being particularly plentiful. Among other inputs are those from the visual cortex, which are of great importance in visually guided movements. The origins of the signals that convey voluntary "commands" to the motor cortex are as yet unidentified.

Afferent signals are distributed to layers I to IV of the motor cortex; efferent projections take origin chiefly from layers V and VI. Fibers from the somatic sensory relay nuclei of the thalamus end as a dense terminal plexus in the fourth layer. These "specific" afferents do not give off collaterals in the deeper layers.

*Further discussion of the problem of somatic sensory input to motor areas will be found on p. 321. The difference between cortical areas that are "sensory" in function and those that merely receive sensory input is carefully defined, and the reasons for some of the discrepancies and confusion in the current literature are explained.

"Cerebellar" input also ends in the fourth layer. Thalamocortical fibers from the "nonspecific" thalamic nuclei pass up to layer I, giving off a few collaterals to other layers on the way. "Association" fibers from other cortical areas end chiefly in the upper four layers. The two deepest layers of the cortex receive little direct sensory input. They are influenced chiefly by the discharges of cells in the four superficial layers and by the numerous internuncial cells of the cortex, whose axons are distributed only within the cortex. In attempting to assess the efficacy of various types of input to cortical cells, as investigators are now doing, it is essential to determine the type of neuron that is responding. An input may "drive" a cell in one of the superficial layers or an internuncial cell without having a comparable effect on a pyramidal tract neuron.

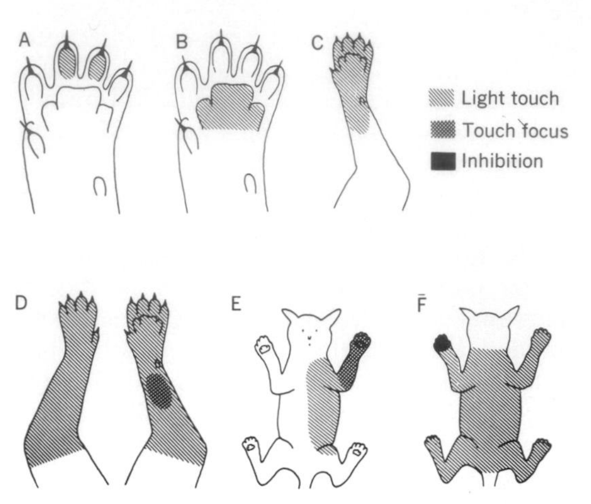

Fig. 29-12. Some typical peripheral "receptive fields" of neurons in cat pericruciate (motor-sensory) cortex. **A** to **C**, Fixed local fields of different types. **D** to **F**, Fixed nonlocal fields (areas greater than 25 cm^2). **D**, Stockinglike field with "focus." **E** and **F**, Fixed wide fields. (From Welt et al.[70])

Distribution of somatic sensory input

Early studies on the distribution of sensory input to the motor cortex, carried out under chloralose anesthesia and often involving electrical rather than natural stimulation, led to the conclusion that there is "widespread convergence of different modalities onto individual cells . . . with no recognizable pattern of spatial localization based either on modality of input or on the production of excitation versus inhibition."[41] Demonstrations that "one single pyramidal tract cell may react to somatosensory stimuli from wide receptive fields of the body surface, to acoustic stimuli and visual input"[71] made it "difficult to imagine that local sensory inflow could have a dominant role in providing sufficiently precise information to accurately initiate or guide limb movements."[9] It now appears that a large proportion of the responses recorded under chloralose, an agent that renders the nervous system abnormally sensitive to sensory stimuli, was mediated by the nonspecific projection system. As its name implies, this system responds to all types of stimuli without regard to modality and evokes cortical activity that may on occasion be confused with that evoked by the specific sensory projections. Responses evoked by the nonspecific projections are not regarded as "sensory."

Recent studies carried out under local anesthesia and with natural stimulation are beginning to resolve the confusion that has arisen. In a series of studies on cats, Brooks et al.[10, 11] and collaborators[70] have extracellulary recorded the discharges of single cortical cells in the motor cortex in response to natural stimulation of skin, joints, and deeper tissues. With this technique, inputs of sufficient intensity to cause frank discharge of cells or inhibition of existing activity were detected, but subthreshold inputs may have been overlooked. In the somatic sensory cortex the cells in a particular column all respond to inputs of the same sensory modality coming from the same part of the body.[43] In a single motor column, however, different cells respond to different types of input. According to Brooks,[8] only about 6% of cells respond to more than one modality of "specific" sensory input. About three fourths of all cells have small (< 25 cm^2), contralateral receptive fields that are sensitive to light touch. Examples of the sizes of these fields are shown in Fig. 29-12, *A* to *C*. The remaining cells have wide (> 25 cm^2) contralateral or bilateral receptive fields, as shown in Fig. 29-12, *D* to *F*, or they respond to joint movement. By careful reconstruction of the electrode tracks in the cortex it was shown that all of the cells within a particular column having local receptive fields received signals from the same part of the body. The wide fields of the other neurons in the same column consistently overlapped the local fields. About half of all wide field units had *focal areas* within the wide fields, in which stimulation produced more intense effects. The location of these focal areas was usually identical with that of the local fields, or it overlapped them. In general, receptive fields were smallest at the periphery of the limb and increased progressively toward the elbow and then decreased somewhat toward the shoulder. Of 59 cells receiving superficial input exclusively from the forepaw, 52 responded only to stimulation of the ventral surface, 3 only to stimulation of the dorsal surface, and 4 to stimulation of both surfaces. This ventral preponderance existed chiefly for the forepaw area.

These findings indicate that the cortical cells in a single motor column receive different kinds of somatic sensory signals that all come from a single local part of the body. Thus far the experimental observations have not revealed significant differences in the receptive fields of the large and small pyramidal cells or in the fields of pyramidal and nonpyramidal cells within a column. The distribution of adequate stimuli is about 60% from superficial receptors (hair bending or light touch), 30% from deep receptors (deep pressure or joint movements), and about 10% from unidentified sources. These data on the cat assign a surprisingly small role to input from joints and muscles. In the monkey, signals from joints and muscles are much more important. Fetz and Baker[24] found that 189 of 233 units in the precentral cortex responded to passive movement of one or more joints of the contralateral leg. Of these, 148 fired only during movement of the joint; 4 fired tonically at rates proportional to the maintained angle of the joint, and 37 discharged both tonically and phasically. Units responding to cutaneous stimulation made up a small percentage of the total in the monkey.

In addition to its excitatory effects, sensory input produces three types of inhibitory effects according to Brooks et al.[11] "Cross-modality" inhibition, a cessation of response to one type of stimulus when another type is applied, was observed frequently. Pressure or joint movement, for example, frequently inhibited cutaneous responses. In one good example a nonpyramidal cell could be driven by touch of the right forepaw only while the right elbow was partly extended. Flexion inhibited both evoked and spontaneous responses of this unit. "Surround" inhibition, a reduction in spontaneous or evoked firing to sensory stimulation when a stimulus of the same type is applied to a peripheral area surrounding or adjacent to that giving rise to excitation, was observed in 13 out of 40 cases in which careful testing was done. "Complementary" inhibition, a form of interaction in which the two receptive areas are located on opposite limbs (right and left forepaws or right and left hind paws), was observed occasionally.

• • •

In **summary**, it appears that motor columns are the basic multicellular units around which sensory input is organized. Some of the cells within a column receive a specific excitatory input from skin or joint muscle in a particular part of the body. The receptive fields of these cells are small and modality specific. Other cells in the column receive convergent inputs from widespread areas of the body and a few may respond to more than one type of sensory stimulation. Inhibitory inputs from the same region, from surrounding areas, and even from contralateral limbs interact with these excitatory inputs. Afferents from the nonspecific thalamic nuclei modulate the excitability of cells in all parts of the column. Intracolumnar circuits permit extensive interaction between cells in different layers. The corticofugal cells, both pyramidal and nonpyramidal, presumably respond to the integrated sum of all these inputs that reach them via the cortical internuncial circuits.

Relation between cortical input and output

It is obvious that there is a close spatial relationship between the peripheral origin of the sensory input to a motor column and the muscles to which the column projects. The important question is how input and output are related.

By using the same microelectrode alternately for recording and stimulating, Asanuma et al.[4] have investigated the relationship between input and output in greater detail. In order to identify minimally effective outputs with maximal precision, they recorded the electromyograms (EMGs) of eight muscles in the cat's forelimb and reduced the stimulus strength until the EMG response was limited to one muscle.To ensure that the stimulating current excited only those pyramidal tract cells whose activity could be recorded by the same electrode (i.e., within a radius of about 90μ), only sites with thresholds of 10 μamp or less were selected for study. In Fig. 29-13 all of the cutaneous receptive fields of cells located within the efferent zone of one muscle are shown projected onto a common map. Although the fields varied considerably in size, they shared a common skin area. Density of overlap is indicated by shading: black for maximum, lined for medium, dotted for minimum. As shown in the map at the upper left, the pyramidal cells that control EDCL (a muscle that dorsiflexes the paw and digits) had receptive fields mainly on the dorsal surface of the paw and digits. The area of maximal overlap is restricted to the dorsolateral paw, where 11 of the 17 fields overlapped. In contrast, the cells controlling PAL (a muscle that ventroflexes the paw and digits) had receptive fields with maximal overlap (20 out of 24 fields) on the ventral surface of the paw. Cells controlling ECU (a muscle that causes dorsiflexion of the paw) received maximal input from the ventrolateral paw. TRB (which extends the elbow) is controlled by

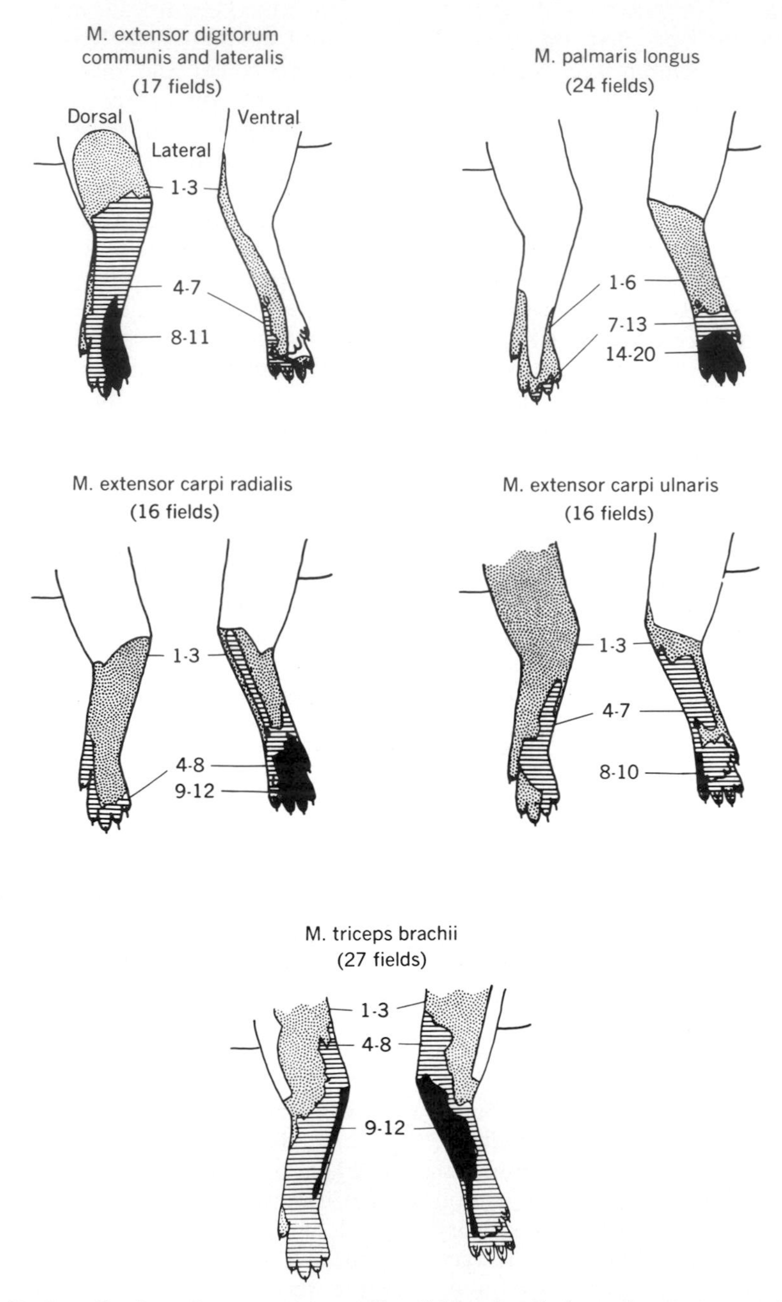

Fig. 29-13. Localization of cutaneous receptive fields of cells in motor-sensory cortex of cat. Receptive fields of all cells responding to light touch or hair bending within efferent zones of each five muscles have been superimposed on individual figurines. Density of overlap is indicated by shading: minimum, dotted; medium, lined; maximum, black. Total number of receptive fields used is shown below each figurine. Numbers associated with dotted, lined, and black areas give number of overlapping fields. (From Asanuma et al.[4])

cells that receive input from a wide area of skin with maximal overlap on the ventrolateral surface of the forearm. *These findings indicate that each efferent zone receives cutaneous inputs predominantly from skin regions that lie in the pathway of limb movement produced by contraction of the muscle to which the zone projects.* Similarly, cortical cells receiving their input from a joint were found to be located in efferent zones controlling muscles that moved that joint.

Using similar techniques, Rosén and Asanuma[52] have recently carried out more detailed studies on the "hand area" of the monkey. In this species, only 45% of the cells studied could be influenced by peripheral stimulation clearly enough to permit description of the receptive field and the adequate stimulus. In this part of the motor cortex the motor columns project to a single distal forelimb muscle or to a few muscles. As in the cat, a single column receives a polymodal input from deep as well as superficial receptors. Fig. 29-14 illustrates the input-output relations of cells that were encountered in a sagittal row of penetrations through the "thumb area." Four adjacent radial columns projecting to different thumb muscles were identified. Cells responding to superficial stimulation were arranged as follows. Two cells in the "thumb flexion" column were both activated from the ventral aspect of the thumb. The cells in the columns projecting to the extensor, adductor, and abductor of the thumb were activated by stimulation of the distal tip, medial aspect, and lateral aspect of the thumb, respectively. The cells in these penetrations responding to stimulation of deeper structures were all activated by passive movements of the thumb or by thenar pressure. The efferent zones in this part of the cortex received tactile input chiefly from that side of the finger or hand that was in the direction of the movement produced by intracortical microstimulation. This finding applies to the zones causing thumb flexion, thumb

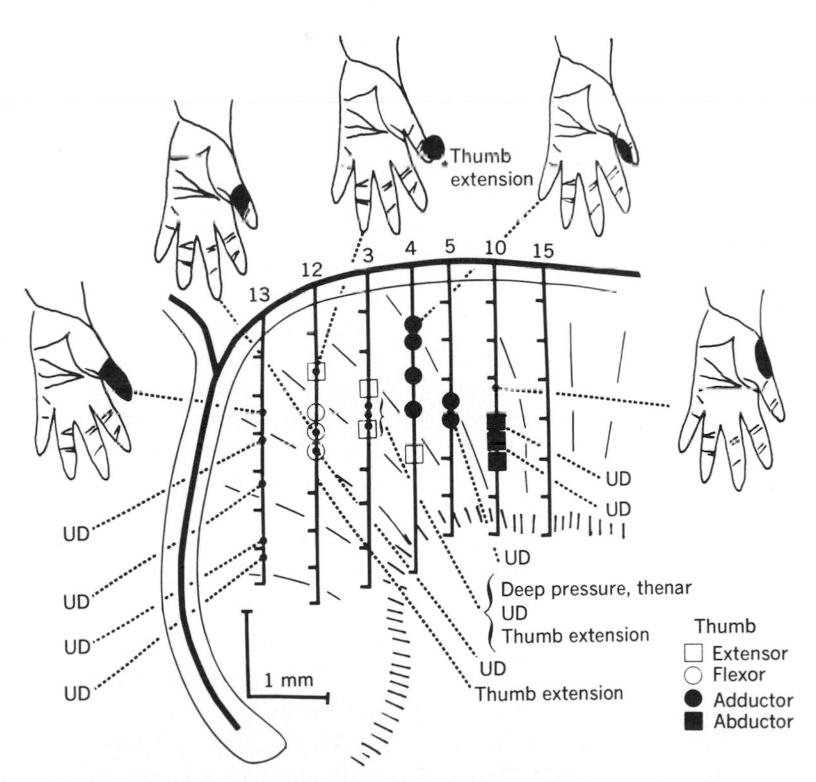

Fig. 29-14. Input-output relations of cells in "thumb area" of monkey. Reconstruction of electrode tracks and cell locations from experiment in which several electrode penetrations (solid lines, identified by numbers) passed through efferent zones projecting to various thumb muscles. Peripheral motor effects produced by intracortical microstimulations are indicated by symbols. Cortical sites stimulated with 5 μamp without evoking motor effects are shown by short horizontal lines perpendicular to electrode tracks. Positions of cells encountered are indicated by dots and are connected to descriptions of receptive fields and adequate stimuli. (From Rosén and Asanuma.[52])

adduction, and finger flexion, all of which would tend to be stimulated during grasping or manipulatory movements of the hand. Zones less likely to be involved in these activities, such as those projecting to extensor muscles of the fingers or to wrist muscles, apparently receive much less input from superficial receptors.

Of 56 cells in this study responding to stimulation of deep structures, 42 were activated by passive movement of joints (38 of them by movement of the joint involved in the motor effect) and the remainder by pressure applied to deep tissues. In almost all cases the direction of joint movement required for activation was the same as that produced by stimulation. Only two cells responded to joint movements in more than one direction. In general it was found that cells driven by movements in the opposite direction were responding to impulses from muscle receptors. These results suggest that cortical cells receive information from joint receptors activated by the movements they produce and from receptors in the muscles contracting during the movement.

Most cells driven from the periphery were located in the superficial and intermediate layers of the cortex, with the largest number in layer IV. There was apparently no difference in the location of cells responding to superficial and deep receptors. Undriven cells were most frequent in layers V and VI, the sites most responsive to microstimulation.

The results of Rosén and Asanuma add considerably to our understanding of the coupling between sensory input and motor output. The distribution of the cells responding to specific afferent inputs (chiefly in layers I to IV) indicates that most of them are intracortical interneurons. Lorente de Nó's studies of cortical architecture and the results described previously indicate that the activity evoked in these superficial layers is directed to deeper layers, where the corticospinal and other efferent neurons are located. Motor effects evoked by intracortical stimulation in superficial layers are presumably relayed synaptically to the corticospinal cells that lie in deeper layers. In support of this view is the recent observation, already cited, that phasic and tonic contractions of muscles result from stimulation of superficial and deep layers of the cortex, respectively.

The functional role of the tight coupling between input and output is not yet fully apparent. Presumably the sensory input is used as feedback to modulate the output of individual cortical columns during the performance of movement. Under certain circumstances it elicits cortical reflexes that are important in normal support and locomotion, as described in Chapter 26. According to Bard, "If a cat is held in the air with the legs free and dependent and with the head held up (so that it cannot see its forefeet or any object below and in front), the slightest contact of any aspect of either pair of feet with the edge of a table results in an immediate and accurate placing of the feet, soles down, on the table close to its edge."[5] Similarly, if a cat is held with one paw in a normal supporting position on a table and then is moved in any direction so that the limb is displaced, a rapid "hopping reaction" will be carried out in the appropriate direction to restore the normal supporting position. Both of these reactions depend on cortical mechanisms. "Placing reactions" in the monkey are abolished by removal of either the pre- or postcentral gyrus on the side opposite the stimulus but are retained if all other cortex is removed. Hopping reactions require only the integrity of the precentral gyrus, which indicates the importance of muscle and joint input to area 4.

Another cortically mediated reaction in which tight coupling between input and output is evidently essential is the "instinctive tactile grasping reaction." According to Denny-Brown, this is an "orientation of the hand or foot in space such as to bring a light contact stimulus into the palm (or sole) when very facile grasping then ensues."[15] This is an example of positive feedback, with the limb moving toward the stimulus. It apparently occurs because skin regions that project most heavily to pyramidal neurons "lie in the pathway of muscle action and therefore, in the course of a manipulatory sequence, they are likely to be excited following muscle contraction."[52]

ROLE OF FEEDBACK IN CONTROL OF MOVEMENT BY HIGHER CENTERS

Whenever a movement occurs, it is accompanied by changes in activity at all levels of the nervous system, in the mechanical state of the muscles and joints that execute the movement, and often in the relation between the organism and its environment. As the movement progresses there are extensive interactions between these different components.

If the control centers are to adapt their output to the changes occurring throughout the control and effector systems, they must be kept informed of them. They must receive a continuous flow of signals from the parts of the body carrying out the movement and from the receptors that provide information about the outside world. This "sensory feedback" has been recognized as essential for control. It has not been as obvious that "internal feedback" from the motor centers themselves is required to monitor their performance. Anatomic evidence suggests that samples of the motor signals are fed back to control centers from many levels of the nervous system and may become an essential part of the information used for control. In this section the evidence concerning these different types of feedback and the role they play will be described. Feedback from sensory receptors will be considered prior to a discussion of feedback from the CNS.

Sensory feedback

One of the most direct ways of assessing the functional significance of a particular type of feedback is to cut or block the input pathway and observe the resulting defect in movement. If the nerves carrying visual, auditory, or vestibular signals into the CNS are interrupted, the resulting loss of motor control is confined to a specific sphere of activity; other types of activity are unaffected. If the sensory fibers from the skin, joints, and muscles of a limb are sectioned, ability to use that limb is impaired for all types of skilled movements. For example, patients with syphilitic lesions of the dorsal root ganglia (tabes dorsalis) supplying the lower limbs have great difficulty in walking and often develop a characteristic gait in which the feet are brought down hard with each step in an effort to increase the number of afferent signals in the remaining nerve fibers. To control their ataxia, they walk with eyes glued on the ground and become so dependent on visual cues that they may fall over if they lift their eyes or close them. When the upper limb is involved, all fine and precise movements suffer. The eyes are no longer able to guide the arm or fingers to their goal, and delicate manipulations are impossible. In 1895 Mott and Sherrington[42] prepared monkeys with unilateral sections of dorsal roots from C_4 to T_4 or from L_2 to S_4, thus eliminating all sensory input from one forelimb or hind limb. From the time of the section and for as long as the animals were kept, almost all movements of the hand or foot were abolished. Grasping movements were completely absent. Movements of the elbow or knee were somewhat less impaired, but neither the hind limb nor the forelimb was used thereafter in locomotion or in climbing. If the intact limbs were restrained, a hungry animal would not make the slightest effort to use the operated limb to obtain food. During struggling, some movements occurred at both proximal and distal joints, but no grasping was observed. Associated movements were impaired less than prime actions. In spite of the profound loss of motility, there was no change in the response to cortical stimulation. This suggests that the pathways from cortex to motor neurons were able to transmit normally, and that the cause of the paralysis was a failure of input to the motor cortex and other centers. If only C_7 and C_8 or T_1 and T_2 dorsal roots were cut, there was obvious clumsiness in the deafferented region but no loss of ability to move. Even section of all except one dorsal root to a limb did not prevent the animal from using that limb. These experiments do not explain the cause of the near-total paralysis (removal of the entire "arm area" does not result in such a complete loss of motility), but they clearly indicate the importance of somatic input for initiation and control of movements. It is possible that complete absence of input results in loss of awareness of a limb, somewhat like the loss of "body image" in lesions of the parietal cortex. Discharge of pyramidal neurons may require convergence of afferent impulses from several sources; elimination of sensory input removes one of the most important of these inputs. However, as will be noted later in this section, monkeys with total sensory denervation can make a variety of simple movements if they had previously been conditioned to do so.

Although many types of motor performance depend on sensory input for control, some movements are too fast to allow time for feedback to influence performance. Stetson and Bouman[58] pointed out that rapid, repetitive movements carried out by expert typists or pianists are essentially ballistic in nature and cannot be modified during their course. According to Brooks and Stoney, "Alternations of opposing agonists and antagonists, which may last about 0.1 sec, are probably beyond control by feedback loops, since minimal feedback times from the periphery to higher centers in man are about 0.1 sec."[9] It is not surprising that errors associated with rapid movements are greater than with slow actions.[46] Although sensory input is continuously available to the CNS, it probably cannot be used continuously by higher centers for control of rapid movements. A number of observations indicate that when movement becomes too rapid for continuous control, it is subject to "intermittent" control. If input is predictable, as in certain repetitive tracking tasks, intermittent control is sufficient to allow movement to follow the target accurately at high tracking rates. With unpredictable inputs, however, control deteriorates.[57]

Parameters of movement controlled by motor cortex

At the conscious level, voluntary actions are formulated in terms of goals, positions, and postures. The motor cortex, with its direct and indirect connections to the spinal cord, must convert these and other "orders" into command signals suitable for the motor neurons. The nature of these control signals has been investigated by recording the activity of single neurons in the motor cortex of monkeys conditioned to move a lever back and forth by movements of the wrist. The forces exerted on the lever and the displacements occurring during these movements have been measured with transducers, so that the correlations between cortical activity and specific parameters of movement could be examined. Although the conditioned movements are relatively simple, usually involving alternate flexion and extension of the wrist, electromyographic recordings have shown that more than a dozen different muscles participate either as prime movers and antagonists or in fixation of adjacent joints. Unfortunately, it has usually been difficult to determine which of the active muscles the particular cortical unit under study was controlling, so that interpretation of experimental data has been handicapped.

In the initial studies carried out by Evarts,[18] monkeys were trained to respond to a light with an extension of the wrist. Many of the pyramidal tract neurons (PTNs) in the hand area of the precentral motor cortex discharged repetitively just before extension began, an indication that the cortical discharge was not due to feedback from active muscles or joints. Other PTNs showed abrupt cessation of firing prior to extension, suggesting inhibition of cells controlling muscles antagonistic to the prime movers. PTNs in the postcentral hand area generally discharged after movement had commenced, suggesting that they were responding to feedback from the active limb.[22] The minimal latency of PTN discharge in the precentral region following the onset of the light was 100 msec, or 70 msec more than the minimum latency required by the anatomic connections between the retina and motor cortex. If the sequence of events occurring during this 70 msec interval could be identified, many clues regarding the origins of movement would be provided.

In a later study, Evarts[19] attempted to determine whether the discharge of PTNs is related to the *force* exerted by the moving part or to the *displacement* resulting from this force. The basic conditioning experiment was modified so that at certain times a load opposed flexion of the wrist; as a result, both flexor and extensor displacements involved activity of the flexor muscles and exertion of flexor force. At other times the load opposed extension, so that both flexor and extensor displacements were associated with extensor force. With this technique, force was dissociated from direction of displacement and it was possible to determine whether PTN activity was related to direction of force or direction of displacement. For the majority of PTNs cortical discharge was related primarily to the force (F) and was only secondarily related to the direction of displacement. For example, when flexor or extensor displacements were associated with flexor force of sufficient magnitude, the unit whose activity is reproduced in the upper records of Fig. 29-15 showed intense activity *regardless of the direction of displacement;* when displacements were associated with extensor force (lower records), the unit was virtually silent *regardless of the direction of displacement.* In many instances it was apparent that the cortical discharge was related both to force and the rate of change of force (dF/dt). A strong relation to dF/dt brings to mind the dynamic response of the Ia afferents of the "nuclear bag" fibers in the muscle spindle. As noted previously, group I muscle afferents have been shown to project to the motor cortex.

It is possible that the parameters of movement to which cortical activity is related may change with the task or the type of motor behavior. In the experiments just described,

PTN activity was related to force under conditions in which accurate joint displacement was rewarded. In another set of experiments, Evarts[20] trained monkeys to maintain a fixed wrist position while supporting a load that sometimes required predominant activity of flexor muscles and at other times required predominant activity of extensors. In this task the monkey tried to avoid wrist movement regardless of the load that opposed maintenance of this position. It was found that PTNs involved in maintaining the required wrist position showed marked variations in discharge frequency for different loads. During a fixed posture, PTN activity was related to the pattern of muscular contraction maintaining the posture rather than to the posture per se. The records in Fig. 29-16 illustrate the variability encountered in experiments of this type that makes interpretation difficult. The microelectrode recorded the activity of two adjacent PTNs. The unit with the smaller spike showed little activity during insertion of the arm through the tube leading to the lever but discharged throughout the performance of the primary task. The unit with the larger spike discharged somewhat more intensely during arm insertion than during the task. The most striking thing about this pair of adjacent units, both of which were related to performance of the task, was that they were related

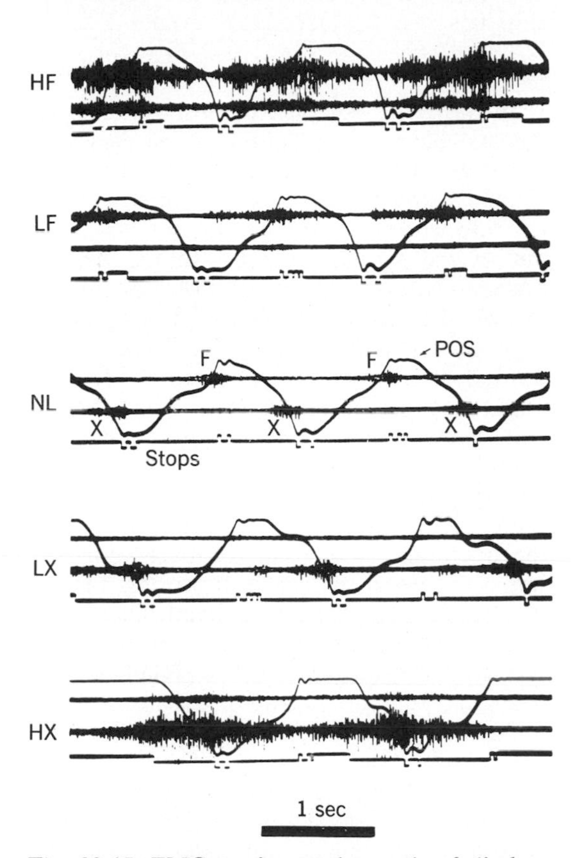

Fig. 29-15. EMG tracings and records of displacement for five of the different loads used in experiment. In middle set of traces, *NL*, the line labeled *POS* is potentiometer output indicating wrist position. Potentiometer output is up for wrist flexion and down for wrist extension. Line labeled *Stops* can assume one of three positions: down for wrist maximally extended, intermediate for wrist in intermediate position (handle not contacting either of stops), and up for wrist maximally flexed. *X* indicates EMG from extensor musculature; *F* indicates the flexor musculature. When heavy load (400 gm) opposed flexion, *HF*, flexor muscles had predominant activity. When heavy weight opposed extension, *HX*, predominant activity was in extensor musculature. When no load, *NL*, opposed movement, there was alternate activity of flexor and extensor musculature. With heavy flexor, *HF*, load there was predominant activity in flexor musculature but also considerable activity in extensor musculature. Sets of traces labeled *LF* (100 gm opposing flexion) and *LX* (100 gm opposing extension) show EMG patterns at intermediate loads. (From Evarts.[19])

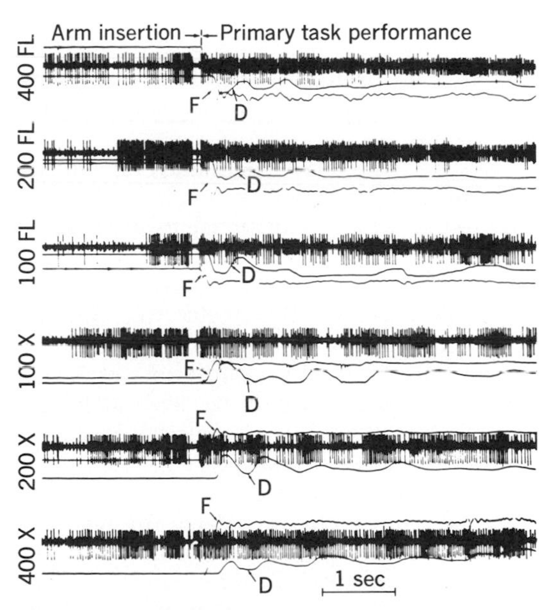

Fig. 29-16. Activity of two pyramidal cells recorded with same microelectrode during insertion of arm through tube leading to lever and during performance of task (maintenance of fixed wrist position). Unit with larger spike was more active during insertion; unit with smaller spike was more active during primary task performance with flexor loads than during insertion. One unit was more active with flexor loads and the other with extensor loads. Numbers at left of records indicate number of grams of flexor load, *FL*, or extensor load, *X*, that wrist was supporting during task. (From Evarts.[20])

in opposite ways: one discharged maximally against a flexor load and the other discharged maximally against an extensor load.

According to Evarts, all of these studies point to one main conclusion, "that the output of precentral motor cortex PTNs is related to the muscular activity of the moving part rather than to the joint displacement or steady-state joint position which the subject may intend to achieve as a result of this muscular activity."[20] This conclusion was not anticipated by Evarts, who expected a correlation between cortical activity and joint position, but it appears almost inevitable when certain basic facts are considered. There is no unique relationship between the muscle forces controlling a joint and the position of that joint. Whenever the load changes, the muscle forces must change accordingly in order to maintain a given position. Since motor neurons control these forces and PTNs control motor neurons, the activity of a single PTN could scarcely correlate with joint position for a wide range of loads. It is not surprising that the activity of PTNs usually correlates with force or rate of change of force. If PTNs do not have the primary responsibility regarding joint position, the centers that do probably lie elsewhere and must use the PTNs to produce the forces required for this task.

Recognizing that thousands of cortical cells must be involved in simple movements, Humphrey et al.[30] attempted to improve the correlation between cortical activity and the parameters of movement by recording from five cells at a time. By suitable weighting and summing of the discharge frequencies of these cells, the time course of certain response measures could be accurately predicted. In spite of this empirical success in prediction, these investigators concluded that their data did not reveal the response variable that was most closely related to and controlled by the activity of cortical neurons. All of these studies were severely handicapped because the investigators did not know whether a cortical cell influences a single muscle, a pair of antagonistic muscles, or a group of muscles. Recently Fetz and Finocchio[25] advised caution in interpreting temporal correlations as evidence for functional relations. They have found that although the activity of a cortical cell may be strongly correlated with activity of one or more muscles in the forelimb, operant conditioning will allow reinforcement of cortical activity along with suppression of all EMG activity in the muscles. Attempts to reinforce muscle activity along with suppression of the cortical unit were only partially successful. They conclude that a "consistent temporal correlation between two events, such as precentral cell activity and some component of the motoneuron response, is necessary but not sufficient evidence for a causal relationship between the correlated events."[25]

Internal feedback

In designing complex control systems, engineers have often found it necessary to monitor the performance of the individual elements and to make use of internal feedback to correct errors before the final output of the system emerges. In the nervous system, where control is exerted through chains of neurons and relatively long delays may occur before sensory feedback can be utilized to correct errors, internal feedback might have great advantages. There are, in fact, many claims that this type of feedback is utilized in various parts of the CNS (for a review of this subject, see Evarts[21]). Although there are a number of studies that suggest the operation of internal feedback, none offers satisfactory evidence regarding the actual neural mechanisms involved. One of the most important series of investigations in this area will be described in detail to illustrate the approaches and problems.

As noted previously, Mott and Sherrington's study[42] on monkeys with deafferented limbs led them to conclude that somatic sensation is essential for the performance of voluntary movement. In later studies by Sherrington, Denny-Brown,[15] Lassek,[35] and Twitchell,[66] similar results were obtained and the same general conclusions were drawn. All of these experiments tended to reinforce the belief, widespread since Sherrington's work, that spinal reflexes were the basic elements out of which movements were synthesized and that all movement was reflex in origin. In recent years, however, a series of investigations by Taub and Berman and their colleagues on conditioned movements of deafferented limbs has provided a new set of experimental data with different implications that have led to reevaluation of the original observations of Mott and Sherrington. In the first experiments, monkeys were trained to avoid a shock to their right arm by flexing their left forearm in response to a buzzer.[32] Both forelimbs were hidden from view. After deafferentation of the executive limb (C_2 to T_3), all animals showed an initial deficit in retention of the conditioned response, but in each case reconditioning back to acquisition criterion was possible. An additional group of monkeys was able to learn the avoidance response without any preoperative training. As previous investigators had reported, none of these animals used their deafferented limbs effectively in the freedom of their cages. If their intact forelimb was immobilized, however, these monkeys learned to push their deafferented limbs through the bars of their cages to obtain food when no other access to it was possible.[32] In the next set of experiments the experimental technique was modified in several ways in order to reduce the amount of feedback from sources outside the deafferented limb. A brief click was substituted for the buzzer without altering the results. Instead of flexing the forearm the animals were trained to squeeze a fluid-filled cylinder taped to the palm of the hand in order to avoid shock. The arm was immobilized so that no response other than flexion of the fingers could exert pressure on the cylinder. With movement limited in this way and consequently with less possibility of feedback from nearby innervated areas, the monkeys still learned to squeeze the manipulandum and exert as much pressure as normal animals.

In another group of experiments, both forelimbs were deafferented.[60] The ability of the monkeys to carry out conditioned responses similar to those already described was not affected, but to the surprise of the investigators, bilateral deafferentation resulted in far less impairment of forelimb movement than the unilateral procedure. During the first week or two after operation the deafferented limbs were virtually useless. During the next 2 months there was considerable recovery of function, so that eventually the animals were able to use the forelimbs rhythmically and in good coordination with the hind limbs during slow and even moderately rapid ambulation. The forelimbs were usually placed palms down on the floor and they bore weight. Grasping was possible, and the animals could climb to the top of an 8 ft bank of cages with reasonable speed. Several monkeys were able to pick up raisins between thumb and forefinger. The investigators emphasized the degree to which the actions of these animals approximated normal patterns of movement. Finally, the animals with bilateral deafferentation of forelimbs were reoperated in two stages to section all remaining dorsal roots in the spinal cord. Total deafferentation was accomplished in three animals. These monkeys showed little or no decrease in the use of their forelimbs. They were not able to use their lower limbs effectively, but none of them survived total spinal deafferentation long enough for recovery such as that observed previously in the forelimbs to take place in the legs.

How can animals with deafferented limbs that they cannot see learn to repeat certain movements until they become conditioned when they do not receive the normal sensory signals that inform them of where the limb is and whether it has moved? Taub and Berman conclude that since the required information was not available from the peripheral nervous system, it must have been provided by purely central mechanisms. They propose, as have others,[21] that signals from motor centers and descending motor pathways provide information about future movements to the CNS before the impulses that will produce these movements have reached the periphery. They suggest that this internal feedback would allow an animal to determine the general position of its limbs in the absence of peripheral sensation.

Anatomic and physiologic evidence of internal feedback

Judging from the connections of the descending motor pathways, there are many examples of internal feedback in the CNS. Some of them may have little to do with the capacities of deafferented limbs that have been described, but others may feed back samples of "command" or "executive" signals and thus provide information regarding intended movements. The pyramidal tract may be taken as a convenient example of a motor pathway that gives off collaterals at many levels of the nervous system. Anatomic and physiologic information regarding these collaterals is summarized in the following paragraphs.

At the cortical level, signals from pyramidal cells are fed back by recurrent collaterals to two different types of stellate cells that exert inhibitory and excitatory effects back on pyramidal cells. These effects may result in alterations in the receptive fields of pyramidal cells. The functional significance of this arrangement has been the subject of considerable speculation.[16] At a subcortical level, fibers from the motor areas of the cerebral cortex send terminals to the caudate nucleus and putamen as well as to other nuclei of the basal ganglia. Some of these fibers may be collaterals of pyramidal neurons. The caudate and putamen project to the pallidum, which sends its efferents to thalamic nuclei that project to the motor cortex. The complexity of this cortico-strio-pallido-thalamo-cortical circuit has defied functional analysis to date. Electrophysiologic experiments indicate that pyramidal fibers send collaterals to the nucleus ventralis lateralis of the thalamus[14] and to the ventrobasal complex.[55] Activity in these collaterals is excitatory to thalamocortical cells and thus might result in rapid feedback to the precentral and postcentral motor areas.

In the brainstem, pyramidal fibers or analogous corticobulbar fibers establish connections with many different cell groups. The most important are the cranial motor nuclei, the trigeminal complex, the mesencephalic and medullary core of the RF, and the lateral reticular nucleus.

Several anatomic studies utilizing different techniques have shown a considerable contribution from the pyramidal tract to the dorsal column nuclei.[13, 38] After transsection of the pyramidal tract, degenerated terminals are seen throughout the gracile and cuneate nuclei, particularly in their rostroventral portions. The degeneration is chiefly contralateral to the lesion, and terminals are found on cell bodies and dendrites. An increasing number of physiologic studies indicates that corticofugal activity arising in the pre- and postcentral cortex exerts excitatory and inhibitory effects on a substantial number of cells in these nuclei. Levitt et al.[36] have shown that the excitatory effect is produced by activity in fibers descending from somatic sensory areas I and II. Lesions of the brainstem that spared the pyramidal tract but interrupted other descending fibers revealed that this excitation was pyramidal in origin. Complete severance of the bulbar pyramids without major damage to other structures indicated that the inhibitory effects originated in the precentral motor cortex and were extrapyramidal. Fig. 29-17 illustrates the excitatory effects observed by Levitt and

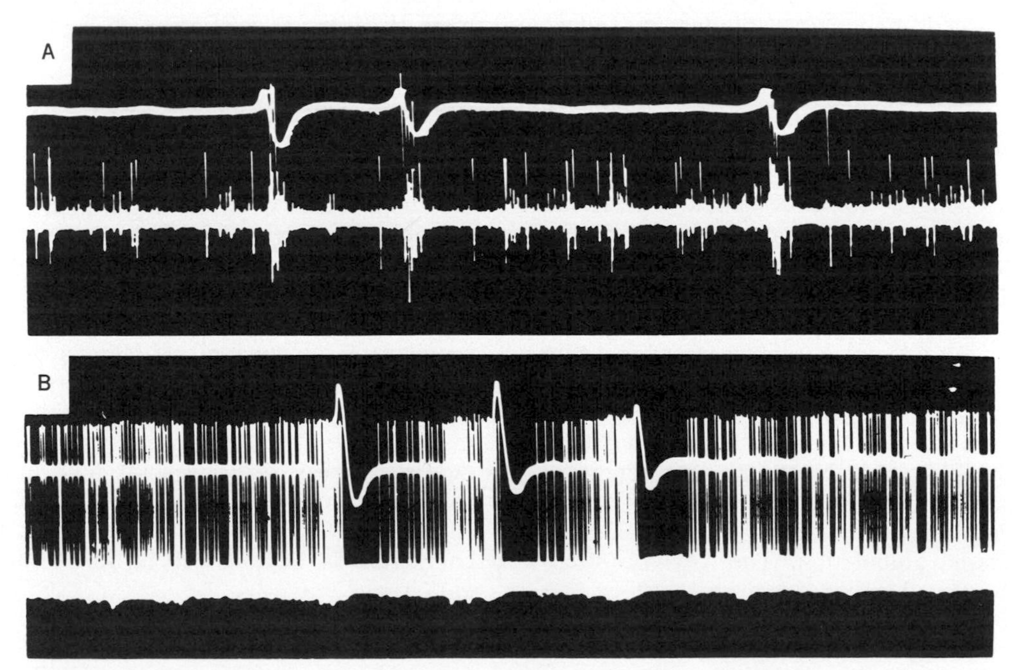

Fig. 29-17. Evocation of unit discharges in gracile nucleus of cat (lower records in **A** and **B**) by synchronous firing of cells in sensorimotor cortex (upper records in **A** and **B**). **A,** Each cortical volley (provoked by application of strychnine) elicits burst of increased firing from units in gracile nucleus that also respond to touch of ipsilateral hind leg. **B,** Similar effect in tactile unit of gracile nucleus that was discharging vigorously, with well-marked post-excitatory depression following it. (From Levitt et al.[36])

the postexcitatory depression that may follow them. In the experimental situation the inhibitory effects may completely block the relay of sensory signals from lower levels, attenuate the spontaneous activity, or increase the signal-to-noise ratio of sensory activity. Under normal conditions the excitatory projection may have a dual function—modulating the sensory feedback from the periphery and supplying samples of the pyramidal outflow for internal feedback.

Although some pyramidal neurons send terminals directly to motor neurons in primates, the majority of them end on cells at the base of the dorsal horn and in the "intermediate zone" of the spinal gray matter on cells, which, like those in the dorsal column nuclei, respond to passive limb movements and cutaneous stimuli. Nyberg-Hansen and Brodal[45] used silver techniques to show that pyramidal fibers from the motor area differed in their terminal distribution from fibers arising in the somatic sensory areas. Fibers from the motor cortex end chiefly in lamina VI of the spinal gray matter and in the dorsal part of lamina VII, as shown in Fig. 29-18. Fibers from the somatic sensory cortex terminate more dorsally, chiefly in laminae IV and V. A corresponding distribution or pre- and postcentral pyramidal fibers has been noted in the monkey. Wall[68] has reported that the receptive fields and thresholds of neurons in this part of the spinal cord are altered during pyramidal stimulation. According to Fetz,[23] the more dorsal cells of laminae IV and V are chiefly inhibited by pyramidal stimuli, whereas the more ventral cells in laminae V and VI are predominantly excited. Pyramidal tract fibers also send terminals to neurons of the spinocerebellar tracts. Hongo et al.[29] report that neurons of the dorsal spinocerebellar tract can be excited or inhibited by pyramidal stimuli, depending on their input. Neurons excited monosynptically by group I afferents are mainly inhibited, whereas neurons responding to volleys in flexor reflex afferents but not to group I are usually excited. In addition to these effects on spinal interneurons, pyramidal fibers are believed to exert a presynaptic inhibitory control over somatosensory input. The details of this circuit are not clear, but pyramidal stimulation results in depolarization of the terminals of primary afferent fibers. This reduces the afferent input to segmental reflex paths, to ascending paths, and to relay nuclei.

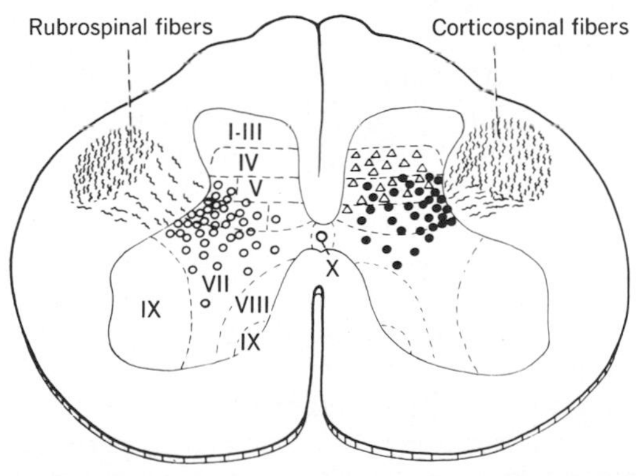

Fig. 29-18. Diagram of cross section of cat spinal cord showing more dorsal terminations of corticospinal fibers from sensory cortex and more ventral terminations of those arising in motor cortex. ○ = Sites of termination of rubrospinal fibers. ● = Sites of termination of corticospinal fibers from "motor" cortex. △ = Sites of termination of corticospinal fibers from "sensory" cortex. (Based on studies by Nyberg-Hansen and Brodal[45]; from Brodal.[7a])

As this brief review indicates, the terminals of pyramidal tract neurons are directed chiefly to cells that form different parts of the somatic sensory pathways from the spinal cord to the cerebral cortex or to cells forming parts of other afferent systems projecting to the cerebral cortex and cerebellum. It would appear that the pyramidal tract is organized primarily to control input to motor centers even at spinal levels and perhaps only secondarily to cause discharge of motor neurons. The well-established observations that the pyramidal tract controls gamma motor neurons that regulate sensory input from muscle spindles lends strong support to this view. Pyramidal influences on presynaptic terminals of primary afferent fibers are also consistent with it. Without additional information it is impossible to say precisely how or why the pyramidal tract influences sensory projections. Modulation of sensory input by impulses arising in the motor cortex suggests that this particular portion of the input is used for "motor" purposes rather than for conscious sensation. The absence of sensory deficits in patients with lesions limited to the pyramidal tract is consistent with this conclusion. The existence of pyramidal collaterals at every level of the nervous system indicates that the same set of signals may be fed back into different portions of the sensory projections with slightly different delays. This arrangement might be of particular advantage when the time relations between ascending sensory signals and descending pyramidal signals are critical.

Although the experimental data regarding the operation of internal feedback circuits are not very revealing, the available evidence indicates that they play important roles. Analysis of these roles will be a fruitful area for future research.

EFFECTS OF LESIONS

It should be obvious from the preceding sections that understanding of the motor systems has not progressed far enough to permit satisfactory explanation of the effects of lesions on movement and posture. Despite extensive clinical and experimental studies there is a surprising amount of disagreement regarding the exact effects of various lesions and their significance. This disagreement is due in part to differences in the species of animals used in experiments, the duration of survival after surgery, the general condition of the animal and his opportunities for maximal recovery of function, and in some cases to failure to distinguish between early and late effects of lesions. Without careful anatomic studies of the extent and severity of lesions, misleading conclusions can easily be drawn. Inability to carry out such studies at all or within a reasonable interval after clinical observations have been made has limited the usefulness of human material. In addition, there is a basic difficulty in the evaluation of lesions that is not always apparent; i.e., a cortical area or a pathway may be *necessary* to a particular function but not *sufficient* for it.

Despite these difficulties and limitations, the study of lesions has contributed significantly to the understanding of motor systems. In this section, emphasis will be placed primarily on the value of cortical lesions in revealing important aspects of cortical motor function and only secondarily on their usefulness in localizing clinical disorders in man.

Contribution of cerebral cortex to movement

As Bard and Rioch[6] and others have pointed out, a decorticate cat or dog is capable of a great many complex motor activities. After a short time he regains the ability to right himself, walk, and run, although defects such as the loss of placing and hopping reactions are permanent. These facts indicate that righting and locomotion can be managed fairly well by subcortical mechanisms in carnivores. A primate is more seriously incapacitated by total decortication, but as Travis and Woolsey[65] have demonstrated, if the surgery is carried out in several steps and sufficient care is taken postoperatively to allow maximal recovery of function after each partial removal of cortex, the decorticate monkey regains the ability to right itself and to move about in an awkward but effective manner. These observations reveal that the cerebral cortex is not essential for many motor performances in primates and they raise the question of what the cortex does contribute to the control of movement.

In discussing this difficult question it is useful to conceive of voluntary actions as a sequence of events occurring in several stages. In the first stage the notion of an act is formed in the mind in response to internal stimuli such as thoughts or emotions or in response to an external stimulus of some kind. In the next stage this "idea" of a movement is presumably translated into patterns of neural signals in a part of the brain in which motor "programs" acquired by learning or practice are stored for use. In the third stage the appropriate "program" is perhaps executed by assemblies of neurons in the principal motor areas of the cortex where the final cortical precision and delicacy are added with the help of various feedback circuits. Pyramidal and extrapyramidal systems working cooperatively transmit these cortical commands to the segmental mechanisms in the spinal cord. This formulation is based on clinical studies of *apraxia* in man, which indicate that the events leading to movement actually occur in this order. A brief description of apraxia is appropriate at this point, because it will help to provide a conceptual framework for a consideration of the types of effects produced by cortical lesions.

Apraxia is the inability to perform an act in the absence of any significant paralysis, sensory loss, or deficit in comprehension. It may take several forms. In one type a patient appears to understand simple commands but has apparently lost his memory of how to perform them, especially if they are called for in unnatural settings. This is called "ideational apraxia." If asked to show "how to wave goodbye" or "how to light his pipe," the patient cannot do so, but if a situation appropriate for these actions arises naturally, he can perform them easily. In this type of disorder the basic defect seems to be in the conception of the act, not in its execution. The patient may make surprising mistakes, such as scratching a cigarette against a match box, without being aware of them. In another type of apraxia a patient may have a clear idea of what he would like to do but be unable to translate this idea into a precise, well-executed act. This "ideomotor apraxia" is characterized by an inability to carry out a simple command, to imitate a gesture, or to perform these actions under natural circumstances. A patient with this defect apparently cannot draw on his store of previously acquired movement patterns. He is aware of the errors he makes, but he cannot correct them and is often irritated with himself. A third and more common form of apraxia is characterized by awkward, inept use of some part of the body for all of its actions. This is "motor" or "kinetic" apraxia. In this form of the disorder the patient knows exactly what he wishes to do, but he cannot carry out the movements properly.

This brief account of apraxia indicates that even relatively simple actions, such as waving goodbye or lighting a pipe, cannot be performed de novo; they utilize acquired patterns of movement or motor memories to a greater extent than we often recognize. There is also considerable evidence that voluntary movements depend on and perhaps even evolve out of basic reflex responses to peripheral stimuli. These reflex elements are most apparent in goal-directed and manipulative activity. Denny-Brown has made an extensive investigation of this aspect of movement in a series of studies on clinical and experimental material. The following account of his findings is drawn largely from his monograph *The Cerebral Control of Movement*.[15]

In patients with lesions of the frontal lobes, automatic prehensile movements of two types could be distinguished, both of which were called "forced grasping."[15] In one type a distally moving tactile stimulus applied to any part of the palm of the hand elicits a simple closing of the hand. After closure begins, traction on the fingers reinforces the reaction, which is termed a "grasp reflex." A similar response occurs in the foot. A second, more complex type of forced grasping can be elicited by contact with a much larger area of skin, including parts of the lateral and dorsal aspects of the hand and wrist. The response consists of extension of the arm followed by flexion, associated with other movements of the forearm that serve to bring further contacts nearer the palm of the hand. When the contact reaches the palm, the hand closes with a rapid grasp reflex. This is an automatic reaction best elicited from patients who are in a deep stupor or have their attention directed elsewhere. This sequence of movements, designed to orient the hand to a stimulus in the environment, is called an "instinctive grasp reaction."

In patients with lesions of the parietal lobes, two essentially opposite reactions to contact, called "tactile avoiding reactions," can be evoked.[15] Contact with any part of the terminal phalanges results in extension of the fingers and wrist, i.e., a simple "avoiding reflex." A more complex withdrawal of the whole hand from the stimulus, i.e., an "instinctive tactile avoiding reaction," can sometimes be evoked from wider areas of skin.

These grasping and avoiding reactions can also be elicited in blindfolded monkeys with frontal or parietal lesions. Each type of response appears to be the converse of the other. If both frontal and parietal cortex are removed, heavy contact on the palm may still evoke a basic grasp reflex, and stimulation of the terminal phalanges may still elicit a coarse avoiding reflex, indications that both responses are, at least in part, subcortical. Denny-Brown believes that there is a natural equilibrium between these two types of cortical reflexes. Damage to areas 8, 6, and 24 destroys tactile avoiding and releases tactile grasping, whereas parietal lesions abolish tactile grasping and release tactile avoiding.

Grasping and avoiding reflexes appear to be the basic elements from which voluntary, goal-directed activity develops. According to Denny-Brown, the ability of the human infant "to reach out with the hand toward a desired object in the field of vision is the outcome of a slow learning process that is preceded by the appearance of automatic reflex grasping and later reflex avoiding and still later by those grasping movements projected into space in response to contact that we have called 'instinctive grasping.' The ability to make a gesture without any visual or tactile stimulus is learned only much later after a long period of random movements."[15] After cerebral lesions, recovery of the ability to make a willed movement is apparently preceded by the appearance of the same movement as a reflex response to a specific stimulus. For example, a patient who was unable for some time to flex his forearm found that he could do so shortly after he recovered a similar type of reflex response, and after a grasp reflex returned voluntary flexion of the fingers reappeared. From observations such as these, Denny-Brown concluded that "purposive movement is indeed the utilization of reflex function as part of the response to a more elaborate stimulus situation."[15] The disuse of deafferented limbs described by Mott and Sherrington may thus be due to the loss of a reflex basis for movement.

Pyramidal tract lesions

By sectioning the pyramidal tract selectively it is possible to study motor function in animals whose movements are carried out almost entirely by the extrapyramidal system. The deficits that result indicate what role the pyramidal tract normally plays, whereas the capacities that remain indicate the extent of the extrapyramidal contribution. The only way of producing a pure pyramidal deficit experimentally is to cut the pyramid in the medulla where there is no admixture of extrapyramidal fibers. This has been done by Tower in experiments on cats[61] and monkeys.[62] The results are of great interest. According to Tower, "The most conspicuous result of unilateral pyramidal lesion in the monkey is diminished general usage and loss of initiative in the opposite extremities. The loss of initiative is grave, but not complete. When both sides are free to act, initiative of almost every sort is delegated to the normal side, but if the normal side is restrained the affected side can, with sufficiently strong excitation, be brought to act."[62] This loss of initiative is somewhat like the effect of deafferentation previously described, though not as dramatic. On the affected side there is a striking loss of fine or discrete control of movement. This *paresis,* as it is called, involves movement in proportion to its delicacy and skill. All fine usage is abolished, especially in the distal muscles of the extremities. The usage that survives, such as postural activity, progression, reaching, and grasping, is stripped of its finer qualities of control, e.g., aim and precision. These remaining performances are still very useful but can hardly be called skilled. Some of them may require intense attention and effort. Obviously, voluntary actions are not the exclusive function of the pyramidal tract. The extrapyramidal systems, contrary to some opinion, can be employed just as voluntarily as the pyramidal.

Section of the pyramid also results in hypotonia on the contralateral side. This is demonstrable as diminished resistance to passive movements and is accompanied by slow, full tendon reflexes. It is sometimes apparent on direct palpation of muscles; after a time there may be loss of muscle bulk due to atrophy.

Superficial reflexes such as local reactions to pinprick as well as the abdominal and cremasteric reflexes are raised in threshold and become slow and full because they are unchecked by antagonistic contraction. Contact and visual placing reactions are abolished in the paretic limbs. Proprioceptive placing and hopping reactions are difficult to elicit. Although stereotyped reaching and grasping remain, the ability to hold onto, grasp, and manipulate objects is greatly impaired. The unopposed action of the extrapyramidal system is apparent in the animal's inability to terminate a grasp while there is tension on the flexor muscles.

In the chimpanzee, pyramidal lesions cause a similar set of deficits, but since the pyramidal tract is relatively larger, the deficits are more severe. Discrete control of the digits is more impaired and use of the affected limb is greatly reduced. The grasp reflex, normally more prominent than in the monkey, becomes so hyperactive that there is considerable difficulty in disengaging the hand from the bars of the cage. In the chimpanzee as in man there is a characteristic response to stimulation of the plantar surface of the foot, which alters radically after a lesion of the pyramid. Stimulation with a stick, a key, or a fingernail drawn along the lateral edge of the sole causes plantar flexion of the toes in a normal subject. After a pyramidal lesion the great toe is dorsiflexed and the other toes fan out. This is called the *sign of Babinski.* Its significance is obscure, but it is very useful clinically as an indication of a pyramidal lesion.

Thus far there is no record of an uncomplicated case of a pure pyramidal lesion in man. A few nearly pure pyramidal lesions have been reported, however, and the parallel between them and the studies on the chimpanzee is striking.

Tower's observations have been confirmed and extended by Denny-Brown[15] and his colleagues, whose studies lead them to conclude that "the pyramidal system is concerned not so much with 'discrete' movements of individual muscles or individual joints as with those spatial adjustments that accurately adapt the movement to the spatial attributes of the stimulus. Thus grasping is adapted to the shape of the thing to be grasped, whether a particle of food, a pen or a surface, only in the presence of the pyramidal tract."[15] This emphasis on the sensory aspects of exploratory or manipulative activity brings to mind the collaterals of pyramidal fibers that establish connections with somatic afferent systems at all levels of the nervous system.

Extrapyramidal lesions

The extrapyramidal system was described in Chapter 26. Lesions of the various portions of this system result in a distinctive group of motor disabilities. These include disorders of muscle tone (spasticity and rigidity), involuntary movements (tremor, chorea, and athetosis), and postures, hyper- and hypokinesis, and paresis. In this chapter only the effects of lesions of the cortical portion of the extrapyramidal system will be considered.

After cutting the pyramids in the medulla, stimulation of the cerebral cortex still results in movements, indicating the existence of cortical connections to extrapyramidal motor centers. The four principal cortical motor areas are probably the most important sources of these fibers, but recent anatomic studies indicate that *all parts of the cerebral cortex give efferent fibers to the basal ganglia.*[12] From this it may be inferred that all cortical lesions result in some deficit in motor function, defined in the broadest terms. Quite often, however, no motor deficit can be identified by the neurologist. Lesions in visual or auditory receiving areas, for example, result in no loss of primary motor function, but they remove an area that supplies important inputs to the motor cortex and they impair a particular sphere of motor control. Deficits resulting from lesions of areas concerned with intellectual function may be extremely subtle and their influence on motor function may be difficult to detect. In the following sections some of the best-known disabilities that result from cortical lesions will be briefly described.

Precentral motor cortex (area 4) lesions

Since the earliest observations on the effects of lesions in the precentral cortex, there has been a long series of confusing and contradictory reports on the amount of paralysis and paresis that results, on the occurrence of flaccidity or spasticity, and on the relation between these effects and the location of the lesion. Until the studies of Woolsey and his associates established the existence of separate precentral and supplementary motor areas and defined their boundaries, experimental ablations often included parts of both areas. Until the patterns of representation in these areas were fully worked out, investigators did not know exactly which part of the body was likely to be affected by a discrete lesion. Furthermore, until the studies of Hines and Tower,[28] showing that section of the pyramidal tracts caused hypotonia and subsequent ablation of area 4 caused spasticity, it was not clear that spasticity was extrapyramidal in origin and due to the release of subcortical centers from cortical inhibition.

As Woolsey et al.[75] have demonstrated, any given part of the skeletal musculature will show some degree of *paralysis and spasticity* when its representation in the precentral motor cortex is removed. With very small lesions in the caudal part of the precentral area where the digits are represented, the spasticity may be quite minimal and difficult to detect in an active, uncooperative monkey. Slightly larger removals of cortex just in front of the central fissure result in mild spasticity of the toes and ankle with a positive Babinski reaction or in mild spasticity of the fingers and wrist with a positive Hoffman sign, according to Denny-Brown.[15] Removal of the entire precentral hand area in the chimpanzee caused persistent spastic flexion of the fingers, initial paralysis of the hand with a return of clumsy grasping after 25 days, and poor orientation of the hand to new surfaces.[15] No specific movements of the hand were lost, but actions were slow and inept and seemed to require visual guidance in the absence of tactile orientation. If the removal was extended forward to include the anterior part of area 4, spasticity occurred in the more proximal joints as well. If the lesion was extended medially or laterally, the distribution of spasticity began to resemble that seen in hemiplegia, with the development of hypertonia in the flexors of the upper limb and the extensors of the lower limb. After removal of all of the cortex of area 4, paralysis and spasticity were maximal. Very localized spasticity and paralysis of an elbow, knee, or hip could not be produced by small lesions, perhaps because muscles are represented in overlapping fashion over wide areas of cortex.[15]

Removal of the anterior half of the precentral cortex also results in the appearance of a reaction wherein stretch of the adductors and retractors of the shoulder leads to increased tone in the flexors of the elbow, wrist, and fingers. This *traction reaction* is enhanced if the flexors of the fingers are stretched at the same time. A similar reflex is often seen in hemiplegic patients, but according to Denny-Brown,[15] it should be distinguished from the previously described "grasp reflex," which is due to lesions of areas 8, 6, and 24.

After a lesion of area 4, ability to flex the limbs in response to a painful stimulus reappears at an early stage of recovery. At first it occurs only in response to pinprick, but later light contact is sufficient. Ability to reach out voluntarily is recovered more slowly. After 3 to 4 weeks a monkey learns to pick up small objects by opposing the thumb to the other fingers, but these digital movements remain clumsy. Denny-Brown concludes that "the precentral gyrus is therefore essential for movement directed into space that accurately orients the hand or foot to the object. It is not essential for retraction from a contact, or a painful or visual stimulus. Exploratory reactions with the hand and fingers, foot or lips to contact stimulus do not recover after area 4 lesions."[15]

Supplementary motor cortex lesions

Relatively little attention has been given to lesions of the supplementary motor cortex compared with those of the precentral motor area. The following brief account will doubtless have to be amended when more extensive studies are carried out. Early reports indicated that lesions confined to this area caused considerable spasticity,[63, 64] but later studies by Denny-Brown revealed only a flexed posture of the arms and legs and a soft resistance to extension. The most notable effect of bilateral removal was the appearance of pronounced grasp reflexes.[15] In these preparations, instinctive tactile avoiding was abolished, and all contacts with the hand or foot led to instinctive grasping. The least contact was sufficient to elicit the reaction. The slight withdrawal that is usually observed after an unexpected contact was completely absent, and even noxious stimuli elicited grasping. These observations led Denny-Brown to conclude that "the supplementary area is the focal point of motor projection of avoiding reactions . . . though it is likely that avoidance behavior requires much more widespread cortical areas for its full elaboration."[15]

Postcentral gyrus (areas 3, 1, and 2) lesions

As noted earlier in this chapter, the somatic sensory receiving area in the postcentral gyrus makes a substantial contribution to the pyramidal tract and, on stimulation, yields discrete movements. Removal of the postcentral gyrus alone does not cause spasticity but greatly intensifies the spasticity that follows precentral lesions. This latter effect indicates that extrapyramidal fibers arise in areas 3, 1, and 2. Most of the motor deficits that follow lesions are attributable to somatic sensory defects. The loss of position sense results in complete lack of awareness of movement in the contralateral limbs, in unnatural positions of the limbs, and in reluctance to use them. Under visual control a wide variety of movements can be carried out, but they are clumsy and awkward. Without visual guidance, goal-directed and manipulative activity is essentially abolished. Denny-Brown points out that there is a severe defect in all exploratory reactions directed into space.[15] Tactile placing reactions cannot be elicited and hopping reactions are depressed. Although movements of the hands and feet are inept, the limbs show well-developed avoiding reactions to contact and visual threats. As part of this picture, monkeys appear to be extremely timid and much of their behavior is dominated by complex visual avoiding reactions. Bilateral ablations of the postcentral gyrus leave animals so dependent on vision that without it they become entirely inactive and unresponsive to any form of tactile stimulation for more than 6 weeks. The instinctive grasp reflex is lost and never returns. Avoiding reactions to pinprick are exaggerated and tend to spread to other parts of the body. It is clear that in the absence of the postcentral gyrus, visually directed movements of the hands and legs and visual avoiding can be managed by the basal ganglia.

Second somatic sensory area (somatic area II) lesions

As noted in Chapter 10, no sensory abnormalities have yet been detected following removal of the second somatic sensory area in man or monkey. Furthermore, such lesions apparently do not accentuate the recognized sensory defects resulting from removal of the postcentral gyrus. To date there is no clear and unequivocal evidence that they impair motor performance or significantly affect motor behavior. Denny-Brown[15] reports some minor effects on instinctive grasping and tactile avoiding, but nothing that points to a distinctive role for this area.

REFERENCES

1. Adrian, E. D.: Double representation of the feet in the sensory cortex of the cat, J. Physiol. **98:**16P, 1940.
2. Asanuma, H., and Sakata, H.: Functional organization of a cortical efferent system examined with focal depth stimulation in cats, J. Neurophysiol. **30:**35, 1967.
3. Asanuma, H., and Ward, J. E.: Patterns of contraction of distal forelimb muscles produced by intracortical stimulation in cats, Brain Res. **27:**97, 1971.
4. Asanuma, H., Stoney, S. D., and Abzug, C.: Relationship between afferent input and motor outflow in cat motorsensory cortex, J. Neurophysiol. **31:**670, 1968.
5. Bard, P.: Studies on the cerebral cortex. I. Localized control of placing and hopping reactions in the cat and their normal management by small cortical remnants, Arch. Neurol. Psychiatry **30:**40, 1933.
6. Bard, P., and Rioch, D. McK.: A study of four cats deprived of neocortex and additional portions of the forebrain, Bull. Johns Hopkins Hosp. **60:**73, 1937.
7. Benjamin, R. M., and Welker, W. I.: Somatic receiving areas of cerebral cortex of squirrel monkey (Saimiri sciureus), J. Neurophysiol. **20:**286, 1957.

7a. Brodal, A.: Neurological anatomy, ed. 2, New York, 1969, Oxford University Press.

8. Brooks, V. B.: Personal communication, 1971.
9. Brooks, V. B., and Stoney, S. D., Jr.: Motor mechanisms: the role of the pyramidal system in motor control, Ann. Rev. Physiol. **33:**337, 1971.
10. Brooks, V. B., Rudomin, P., and Slayman, C. L.: Sensory activation of neurons in the cat's cerebral cortex, J. Neurophysiol. **24:**286, 1961.
11. Brooks, V. B., Rudomin, P., and Slayman, C. L.: Peripheral receptive fields of neurons in the cat's cerebral cortex, J. Neurophysiol. **24:**302, 1961.
12. Carman, J. B., Cowan, W. M., and Powell, T. P. S.: The organization of the cortico-striate connexions in the rabbit, Brain **86:**525, 1963.
13. Chambers, W. W., and Liu, C. N.: Corticospinal tract of the cat. An attempt to correlate the pattern of degeneration with deficits in reflex activity following neocortical lesions, J. Comp. Neurol. **108:**23, 1957.
14. Clare, M. H., Landau, W. M., and Bishop, G. H.: Electrophysiological evidence of a collateral pathway from the pyramidal tract to the thalamus in the cat, Exp. Neurol. **9:**262, 1964.
15. Denny-Brown, D.: The cerebral control of movement, Liverpool, 1966, Liverpool University Press.
16. Eccles, J. C.: Cerebral synaptic mechanisms. In Eccles, J. C., editor, Brain and conscious experience, Berlin, 1966, Springer Verlag.
17. Evarts, E. V.: Relation of discharge frequency to conduction velocity in pyramidal tract neurons, J. Neurophysiol. **28:**216, 1965.
18. Evarts, E. V.: Pyramidal tract activity associated with a conditioned hand movement in the monkey, J. Neurophysiol. **29:**1011, 1966.
19. Evarts, E. V.: Relation of pyramidal tract activity to force exerted during voluntary movement, J. Neurophysiol. **31:**14, 1968.
20. Evarts, E. V.: Activity of pyramidal tract neurons during postural fixation, J. Neurophysiol. **32:**375, 1969.
21. Evarts, E. V.: Feedback and corollary discharge: a merging of concepts, Neurosci. Res. Program Bull. **9:**86, 1971.
22. Evarts, E. V.: Contrasts between activity of pre- and postcentral neurons of cerebral cortex during movement in the monkey. In Buser, P., et al., editors: Neural control of motor performance, Amsterdam, 1971, Elsevier Publishing Co.

22a. Ferrier, D.: The functions of the brain, London, 1876, Smith, Elder & Co.

23. Fetz, E. E.: Pyramidal tract effects on interneurons in the cat lumbar dorsal horn, J. Neurophysiol. **31:**69, 1968.
24. Fetz, E. E., and Baker, M. A.: Response properties of precentral neurons in awake monkeys, Physiologist **12:**223, 1969.
25. Fetz, E. E., and Finocchio, D. V.: Operant conditioning of specific patterns of neural and muscular activity, Science **174:**431, 1971.
26. Foerster, O.: Motorische Felder und Bahnen. In Bumke, O., and Foerster, O., editors: Handbuch der Neurologie, Berlin, 1936, J. Springer, vol. 6.
27. Foerster, O.: The motor cortex in man in the light of Hughlings Jackson's doctrines, Brain **59:**135, 1936.
28. Hines, M.: Control of movements by the cerebral cortex in primates, Biol. Rev. **18:**1, 1943.
29. Hongo, T., Okada, Y., and Sato, M.: Corticofugal influences on transmission to the dorsal spinocerebellar tract from hindlimb primary afferents, Exp. Brain Res. **3:**135, 1967.
30. Humphrey, D. R., Schmidt, E. M., and Thompson, W. D.: Predicting measures of motor performance from multiple cortical spike trains, Science **170:**758, 1970.
31. Kennard, M. A., and McCulloch, W. S.: Motor responses to stimulation of cerebral cortex in absence of areas 4 and 6 (Macaca mulatta), J. Neurophysiol. **6:**181, 1943.
32. Knapp, H. D., Taub, E., and Berman, A. J.: Movements in monkeys with deafferented forelimbs, Exp. Neurol. **7:**305, 1963.
33. Landgren, S., Phillips, C. G., and Porter, R.: Cortical fields of origin of the monosynaptic

pyramidal pathways to some alpha motoneurons of the baboon's hand and forearm, J. Physiol. **161:**112, 1962.

34. Lassek, A. M.: The pyramidal tract: basic considerations of corticospinal neurons, Res. Pub. Res. Assoc. Nerv. Ment. Dis. **27:**106, 1948.
35. Lassek, A. M.: Inactivation of voluntary motor function following rhizotomy, J. Neuropathol. Exp. Neurol. **3:**83, 1953.
36. Levitt, M., Carreras, M., Liu, C. N., and Chambers, W. W.: Pyramidal and extrapyramidal modulation of somatosensory activity in gracile and cuneate nuclei, Arch. Ital. Biol. **102:**197, 1964.
37. Leyton, A. S. F., and Sherrington, C. S.: Observations on the excitable cortex of the chimpanzee, orang-utan and gorilla, Q. J. Exp. Physiol. **11:**135, 1917.
38. Liu, C. N., and Chambers, W. W.: An experimental study of the cortico-spinal system in the monkey (Macaca mulatta). The spinal pathway and preterminal distribution of degenerating fibres following discrete lesions of the pre- and postcentral gyri and bulbar pyramid, J. Comp. Neurol. **123:**257, 1964.
39. Lloyd, D. P. C.: Functional organization of the spinal cord, Physiol. Rev. **24:**1, 1944.
40. Lorente de Nó, R.: Cerebral cortex: architecture, intracortical connections, motor projections. In Fulton, J. F., editor: Physiology of the nervous system, ed. 3, New York, 1949, Oxford University Press.
41. Marchiafava, P. L.: Activities of the central nervous system: motor, Ann. Rev. Physiol. **30:**359, 1968.
42. Mott, F. W., and Sherrington, C. S.: Experiments upon the influence of sensory nerves upon movement and nutrition of the limbs. Preliminary communication, Proc. R. Soc. Lond. (Biol.) **57:**481, 1895.
43. Mountcastle, V. B.: Modality and topographic properties of single neurons of cat's somatic sensory cortex, J. Neurophysiol. **20:**408, 1957.
44. Mountcastle, V. B., and Henneman, E.: The representation of tactile sensibility in the thalamus of the monkey, J. Comp. Neurol. **97:**409, 1952.
45. Nyberg-Hansen, R., and Brodal, A.: Sites of termination of corticospinal fibers in the cat. An experimental study with silver impregnation methods, J. Comp. Neurol. **120:**369, 1963.
46. Partridge, L. D.: Motor control and the myotatic reflex, Am. J. Phys. Med. **40:**96, 1961.
47. Penfield, W. G., and Boldrey, E.: Somatic motor and sensory representation in the cerebral cortex of man as studied by electrical stimulation, Brain **60:**389, 1937.
48. Penfield, W., and Rasmussen, T.: The cerebral cortex of man, ed. 1, New York, 1950, The Macmillan Co.
49. Penfield, W., and Welch, K.: The supplementary motor area of the cerebral cortex, a clinical and experimental study, Arch. Neurol. Psychiatry **66:**289, 1951.
50. Phillips, C. G.: Changing concepts of the precentral motor area. In Eccles, J. C., editor: Brain and conscious experience, New York, 1966, Springer-Verlag New York, Inc.
51. Rasmussen, T., and Penfield, W.: Further studies of sensory and motor cerebral cortex of man, Fed. Proc. **6:**452, 1947.
52. Rosén, I., and Asanuma, H.: Peripheral afferent inputs to the forelimb area of the monkey cortex: input-output relations, Exp. Brain Res. **14:**257, 1972.
53. Schaefer, E. A.: Textbook of physiology, New York, 1900, The Macmillan Co.
54. Settlage, P. H., et al.: The pattern of localization in the motor cortex of the rat, Fed. Proc. **8:**144, 1949.
55. Shimazu, H. N., Yanagisawa, N., and Garoutte, B.: Corticopyramidal influences on thalamic somatosensory transmission in the cat, Jap. J. Physiol. **15:**101, 1965.
56. Smith, W. K.: The frontal eye fields. In Bucy, P. C., editor: The precentral motor cortex, ed. 2, Urbana, 1949, University of Illinois Press.
57. Stark, L.: Neurological control systems—studies in bioengineering, New York, 1968, Plenum Press.
58. Stetson, R. H., and Bouman, H. D.: The coordination of simple skilled movements, Arch. Neerl. Physiol. **20:**179, 1935.
59. Sugar, O., Chusid, J. G., and French, J. D.: A second motor cortex in the monkey (Macaca mulatta), J. Neuropathol. Exp. Neurol. **7:**182, 1948.
60. Taub, E., and Berman, A. J.: Movement and learning in the absence of sensory feedback. In Freedman, S. J., editor: The neuropsychology of spatially oriented behavior, Homewood, Ill., 1968, Dorsey Press.
61. Tower, S. S.: The dissociation of cortical excitation from cortical inhibition by pyramid section, and the syndrome of that lesion in the cat, Brain **58:**238, 1935.
62. Tower, S. S.: Pyramidal lesion in the monkey, Brain **63:**36, 1940.
63. Travis, A. M.: Neurological deficiencies after ablation of the precentral motor area in Macaca mulatta, Brain **78:**155, 1955.
64. Travis, A. M.: Neurological deficiencies following supplementary motor area lesions in Macaca mulatta, Brain **78:**174, 1955.
65. Travis, A. M., and Woolsey, C. N.: Motor performance of monkeys after bilateral partial and total cerebral decortications, Am. J. Phys. Med. **35:**273, 1956.
66. Twitchell, T. E.: Sensory factors in purposive movement, J. Neurophysiol. **17:**239, 1954.
67. Vogt, C., and Vogt, O.: Allegemeinere Ergebnisse unserer Hirnforschung, J. Psychol. Neurol. (Leipzig) **25:**277, 1919.
68. Wall, P. D.: The laminar organization of dorsal horn and effects of descending impulses, J. Physiol. **188:**403, 1967.
69. Welker, W. I., Benjamin, R. M., Miles, R. C., and Woolsey, C. N.: Motor effects of cortical stimulation in squirrel monkey (Saimiri sciureus), J. Neurophysiol. **20:**347, 1957.
70. Welt, C., Aschoff, J. C., Kameda, K., and Brooks, V. B.: Intracortical organization of cat's motorsensory neurons. In Yahr, M. D., and Purpura, D. P., editors: The neurophysiological basis of normal and abnormal motor activities, New York, 1967, Raven Press.
71. Wiesendanger, M.: The pyramidal tract. Recent investigations on its morphology and function, Ergeb. Physiol. **61:**73, 1969.

72. Woolsey, C. N.: Organization of somatic sensory and motor areas of the cerebral cortex. In Harlow, H. F., and Woolsey, C. N., editors: Biological and biochemical bases of behavior, Madison, 1958, University of Wisconsin Press.
73. Woolsey, C. N., and Chang, H. T.: Activation of the cerebral cortex by antidromic volleys in the pyramidal tract, Res. Publ. Res. Assoc. Nerv. Ment. Dis. **27:**146, 1947.
74. Woolsey, C. N., Travis, A. M., Barnard, J. W., and Ostenso, R. S.: Motor representation in the postcentral gyrus after chronic ablation of precentral and supplementary motor areas, Fed. Proc. **12:**160, 1953.
75. Woolsey, C. N., et al.: Patterns of localization in precentral and "supplementary" motor areas and their relation to the concept of a premotor area, Res. Publ. Res. Assoc. Nerv. Ment. Dis. **30:**238, 1950.

VI

THE AUTONOMIC NERVOUS SYSTEM, HYPOTHALAMUS, AND INTEGRATION OF BODY FUNCTIONS

30

KIYOMI KOIZUMI and CHANDLER McC. BROOKS

The autonomic nervous system and its role in controlling visceral activities

The autonomic nervous system is a part of the CNS, not a distinct entity as the term might suggest. It is an efferent outflow, a complex of efferent neurons innervating the visceral organs. Although afferent nerves are included in autonomic nerve trunks, these afferents serve the somatic as well as the autonomic system, except in a few cases to be discussed later.

The autonomic outflow is segmental and parallels to some degree the somatic motor outflow to skeletal muscles of the body. Directives from the brain and reflex responses initiated by general and specific stimuli course out these two motor pathways to affect visceral organs and other body parts as reactions appropriate to the occasion are organized and executed. Neither autonomic nor somatic reactions occur in isolation, and it can be said that the autonomic nervous system organizes the visceral support of somatic behavior. The CNS integrates the activities of the body through these two complexes.

In this field of physiology, as in any other, there are certain fundamental facts and conclusions that must be known. It is important to identify the major involvements of this system in behavior and to emphasize the functional processes that initiate and control its action. The problems and recent advances must also be considered, but the significance of the new cannot be appreciated in the absence of knowledge of basic mechanisms and relationships. For this reason emphasis has not been placed on new advances to the exclusion of well-established principles, but these advances have not been neglected.

The history of our knowledge of the autonomic nervous system is extensive. The earliest accounts of man's physical experiences contain descriptions of autonomically regulated reactions. It was observed that these became most conspicuous under conditions of emotional stress, pain, and injury. Thus the autonomic system has long interested those who have been concerned with psychosomatic and emotional behavior. As a matter of fact, the terminology used today had its origin in these earlier concepts of the control of emotional behavior and visceral reactions. It was observed that activities in and injuries to one part of the body were accompanied by reactions in other organs, as though a sympathetic relationship existed between the two. When the early anatomists located nerves that supplied the heart and other organs, they concluded quite correctly that these nerves mediated the associated reactions. Winslow in 1732 discovered the lateral and collateral ganglia, fibers extending therefrom to visceral organs, and the rami of the thoracolumbar outflow. He called these structures parts of a sympathetic system. Others have spoken of the involuntary nervous system (Gaskell), or the vegetative nervous system (Myer and L. R. Müller), but for the most part we have adopted the terminology of Langley,[67] to whom we owe much of our knowledge of the structure and function of the autonomic nervous system.

During the first quarter of this century the relationships of the adrenal medulla to the autonomic system and its role in the function of this complex were clarified.[17] Shortly thereafter Loewi,[73] Cannon and Bacq,[19] and Dale[25] demonstrated chemical mediation of autonomic nervous system action on peripheral organs. The existence of this humoral transmission of nerve action had been suggested much earlier by Elliot (1904) on the basis of the observation that sympathetic nerve stimulation and injection of epinephrine had much the same effect. Langley, who described many of the functions and effects of autonomic nerves, introduced the concept of receptors. Just what these receptors are and what role they play are still questions of major concern. These same men demonstrated denervation sensitization, the action of blocking agents on transmission, and began many investigations that have resulted in some of the most notable advances of the day.

A number of the major concepts of physiology were developed by those who studied the autonomic nervous system: the importance of tonic activity, the maintenance of homeostasis, the response of the total organism to conditions of stress, and adaptation phenomena are examples of the work of those who pioneered in this field. There are numerous accounts of the development of our knowledge of the anatomy and physiology of the autonomic nervous system.[1a, 58, 63, 74, 77, 87] These are well worth reading because they describe the formation of ideas and present the knowledge that must become familiar to all physiologists.

ANATOMIC ORGANIZATION OF THE AUTONOMIC NERVOUS SYSTEM

Langley spoke of the autonomic system as consisting of two major divisions, the sympathetic, or thoracolumbar outflow, and the

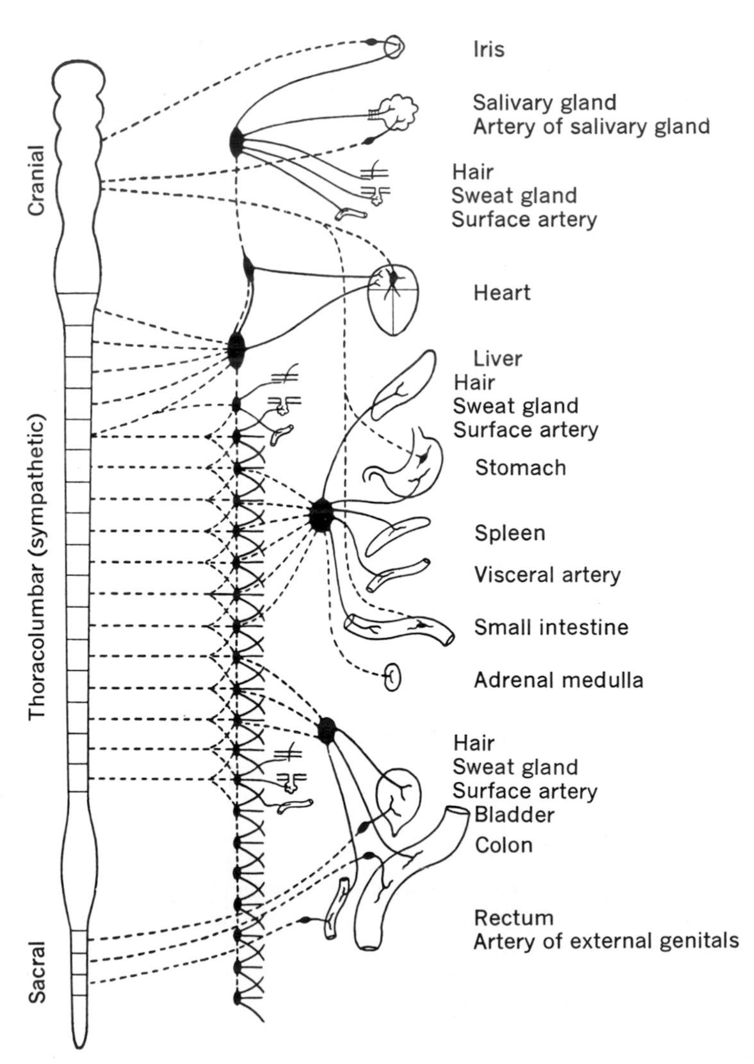

Fig 30-1. Diagram of general arrangement of autonomic nervous system showing one side of bilateral outflow. Brain and spinal cord are represented at left, but nerves of somatic system are not shown. Preganglionic fibers are indicated by broken lines and postganglionic fibers by solid lines. For further description, see text. (After Cannon; from Bard.[1])

parasympathetic, or cranial and sacral outflow. These two divisions stand in somewhat antagonistic relationship and perform unique functions.

Fig. 30-1 shows in diagrammatic form the subdivisions of the system, their anatomic characteristics, the relationships and distribution of fibers within the system, and the innervations of visceral organs.

Sympathetic division (thoracolumbar outflow)

There is a preganglionic fiber outflow from each segment of the spinal cord from the first thoracic to the third lumbar level. The white rami that carry the preganglionic fibers from the spinal cord to chains of lateral ganglia are much shorter than indicated in Fig. 30-1. They are more accurately portrayed in Fig. 30-2, in which preganglionic neurons are shown to correspond to interneurons of the somatic reflex pathway. Their cell bodies lie in the lateral horns and not in the ventral horns, which contain somatic motoneuron pools. The preganglionic fibers are included in the ventral roots and accompany the somatic nerve trunks as they leave the spinal column. Shortly thereafter, these thinly myelinated fibers branch off to form the white rami. Most of these fibers make synaptic connections in the lateral ganglia but some extend to the collateral ganglia before synapsing. The innervation of the adrenal medulla is preganglionic, but this is not exceptional since cells of the adrenal medulla are modified postganglionic neurons.

One of the features shown diagrammatically in Fig. 30-1 is the considerable divergence of preganglionic fibers from each segment. Some of them extend forward in the chain, making synaptic connections at levels three or four segments above the point of exit. The caudal extension is even greater, since the lateral chain extends to the sacral levels despite the fact that the preganglionic outflow terminates in the upper lumbar regions. The phenomenon of convergence is also illustrated here, since a single ganglion receives fibers from several segments. Many preganglionic fibers converge to form with postganglionic neurons the large celiac and inferior mesenteric collateral ganglia.

At the upper end of the lateral chain there is a large stellate ganglion, and the chain ends in a superior cervical ganglion. It appears that a number of ganglia fuse to produce

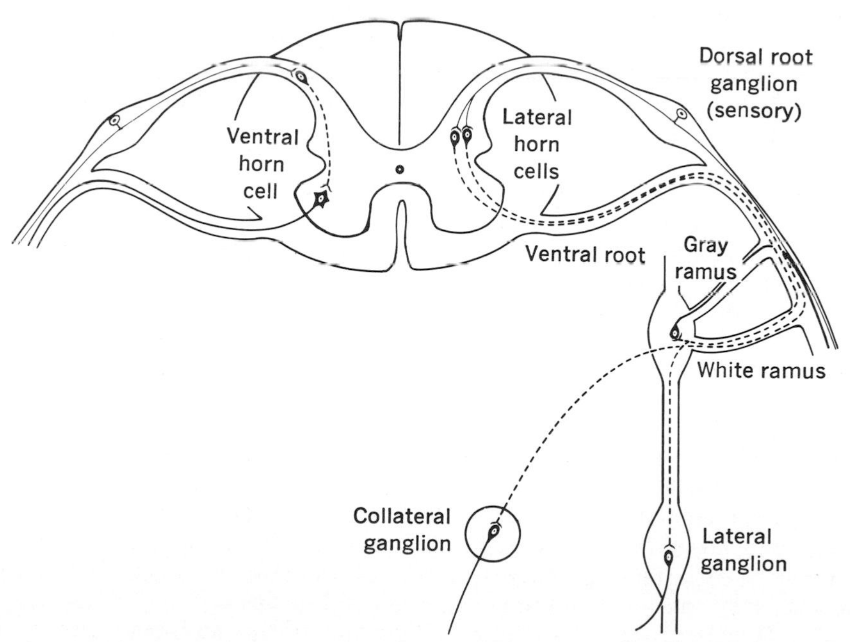

Fig. 30-2. Diagram illustrating different arrangements of neurons in somatic and autonomic nervous systems. Single polysynaptic spinal reflex arc of somatic system is shown at left; that of sympathetic division of autonomic system is shown at right. These segmental arcs are bilateral and superimposed. At right, preganglionic neurons are represented by broken lines and postganglionic neurons by solid lines. Note that preganglionic fibers coming out through white ramus make synaptic connection in more than one lateral chain ganglion or a collateral ganglion.

these larger aggregates of postganglionic neurons. For the most part the preganglionic fibers are short. Postganglionic fibers leave the lateral and collateral ganglia and pass by way of discrete nerve trunks to the visceral organs that they innervate. These are unmyelinated fibers of small diameter, and for the most part they are considerably longer than the preganglionic fibers.

The ratio between post- and preganglionic fibers is large; a single preganglionic fiber innervates many postganglionic neurons. In the superior cervical ganglia of the cat the ratio is 1 preganglionic to 11 to 17 postganglionic fibers. In other ganglia the ratios are somewhat smaller.

Parasympathetic division (cranial and sacral outflows)

In the parasympathetic system, preganglionic fibers end in ganglia close to or within the organ innervated, and very short postganglionic fibers extend to the tissues. There are no interconnections between the components of the cranial and sacral outflows. Furthermore, the parasympathetic system does not innervate all smooth muscle tissues or visceral organs.

In the cranial division, preganglionic neurons originating from the oculomotor nuclei travel out the third cranial nerve to the ciliary ganglia. The postganglionic fibers comprise the short ciliary nerves that innervate the ciliary muscle and the sphincter muscle of the iris. Preganglionic fibers from the superior and inferior salivary nuclei travel through the seventh and ninth cranial nerves to ganglia that supply the innervation of the salivary glands. Submaxillary and sublingual glands are innervated from the superior salivary nuclei through the chorda tympani. The parotid glands receive their innervation from the inferior salivary nuclei.

A major component of the cranial parasympathetic outflow is the vagus nerve. This originates in the vagus nuclei of the medulla and comprises the tenth cranial nerve. It consists of preganglionic fibers that synapse in ganglia within the heart, lung, esophagus, stomach, pancreas, liver, intestine, and upper colon. Vagal fibers synapse in ganglia of the pulmonary plexus and postganglionic neurons therefrom innervate the bronchi and blood vessels of the lung. None of these parasympathetic ganglia are readily visible, and their postganglionic fibers are confined within the tissues of the organs innervated.

The pelvic viscera are innervated from a parasympathetic sacral outflow originating from the second, third, and fourth sacral segments of the cord. Preganglionic axons form the nervi erigentes that end in ganglia adjacent to blood vessels of the erectile tissues. Other components of the outflow go to ganglia and connect with postganglionic fibers that innervate the uterus, intestines, bladder, lower colon, and rectum as well as adjacent tissues. It should be noted that there is very little overlap, convergence, or divergence in the parasympathetic system except in the ganglia, where one preganglionic fiber makes contact with a number of postganglionic fibers (1:2 is the ratio in the ciliary ganglion).

Another anatomic feature of considerable physiologic significance is the fact that the eye, the salivary glands, the heart, the digestive system, and the pelvic viscera receive a dual innervation. Both sympathetic and parasympathetic divisions converge on these organs, although not on the same cells in all instances. In contrast, as shown in Fig. 30-1, the sweat glands, adrenal medulla, piloerectors, and the majority of blood vessels receive only sympathetic innervation.

Autonomic synapses

There are three remaining anatomic considerations that should be reviewed briefly. These are the central, ganglionic, and peripheral synaptic structural relationships. Although these matters have been discussed elsewhere in this text (Chapters 4 and 6), the following review is appropriate. The autonomic neurons located in the lateral horns of the cord form a distinct group. They are considerably smaller than motoneurons of the ventral horns and vary in form, being either spindle shaped or polygonal. Structural details of the synaptic connections between afferent fibers and these neurons are not well known. It is assumed that the same structural arrangements occur as in the case of afferent connections to interneurons and motoneurons of the somatic system.

Much more work has been done on ganglionic synapses,[30, 44, 77, 85] at which preganglionic fibers terminate on postganglionic neurons. A ganglion is composed of many postganglionic cells. These are usually multipolar, possess numerous long dendrites, and are embedded in a fine fibrillary meshwork. The cells range in size from 14 to 55μ, a majority being 20 to 30μ in diameter. Studies of fine structure have shown that synapses are found on the cell body as well as on neuronal processes. The presynaptic endings contain numerous clear vesicles and a few dense-core vesicles. A specialized zone of increased density is found in the postsynaptic membrane where synaptic contacts occur.

At the sites where the peripheral terminals of postganglionic fibers make contact with effector cells there appear to be no specialized end plates, as in the case of somatic neuromuscular junctions; rather, fibers end in close contiguity with the membranes of the cells innervated. In intestinal smooth muscles, in which the fine structures have been carefully studied,[11] autonomic fibers become varicose and form bulbous expansions every 1 to 3μ. The varicosities contain vesicles that are thought to be the storage sites for transmitters. It has been suggested that transmitter is released not only from the varicosity in which the nerve terminates but also from other varicosities that are found along the axons and that make contact with the smooth muscle fibers.

NATURE OF AUTONOMIC FUNCTIONS

Autonomic activity adjusts body states and supports somatic reactions. It is not essential that the system be primarily or immediately involved in the initial phase of a response. Activity within the autonomic nervous system is not necessarily secondary to somatic activities but tends to parallel and support these functions. The autonomic system may even prepare the body for behavioral reactions, and it effects anticipatory adjustments in emotional states. It can increase cardiac output, adjust blood flow, and make sources of energy more readily available, thus preparing the organism for violent activity. Nonetheless, these visceral reactions are basically supportive or anticipatory of body need.

The fibers comprising the autonomic nerves are small myelinated and unmyelinated fibers. The preganglionic fibers for the most part are myelinated, have diameters of less than 3μ, and conduct at a speed of approximately 2 to 14 m/sec. The postganglionic fibers are largely unmyelinated, are approximately 0.3 to 1.3μ in diameter, and conduct at speeds of less than 2 m/sec. The system is therefore not equipped for the speed of response demonstrated by the somatic motor outflow. Examples of the action potentials of these fiber types are shown in Fig. 30-3.

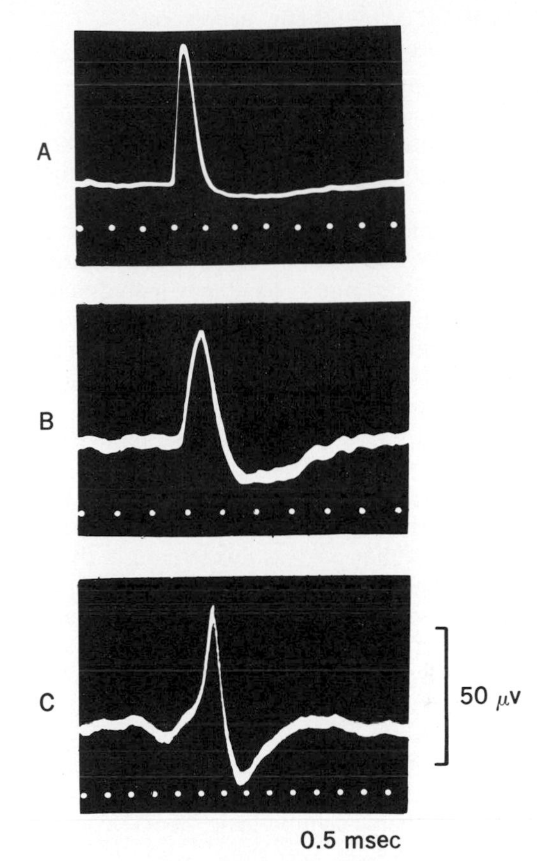

Fig. 30-3. Action potentials from various nerve fibers (cat). **A,** Action potential from anatomically isolated somatic motor fiber, diameter 20μ. **B,** Action potential recorded from single preganglionic fiber of white ramus. **C,** Action potential of single postganglionic sympathetic fiber supplying skeletal muscle. **B** and **C** recorded from spontaneously active single fibers in small nerve filament. Amplification indicator applies to **B** and **C;** amplification for **A** was one-twentieth of that required for **B** and **C** since fiber was isolated and no shunting occurred. (**B** and **C** courtesy Dr. A. Sato.)

Another characteristic of autonomic fibers that is appropriate to the function of this system is their tonic activity. The origin of this tonic discharge will be discussed later, but it is significant in that because of it the visceral organs are held in a state of intermediate activity and can be controlled by either diminution or augmentation of the rate of fiber firing. This tonic action is not entirely an intrinsic property of these neurons but is in part a contribution of central control and peripheral reflex mechanisms that maintain a state of low-level activity in many of the neurons of this system.

Actions of sympathetic nerves on specific effectors

Early investigators stimulated pre- and postganglionic fibers of the autonomic system innervating various organs of the body in order to determine the actions of these fibers.

Stimulation of sympathetic fibers to the eye causes the eyeball to protrude (exophthalmos), the pupil to dilate (mydriasis), and the nictitating membrane to constrict. Vasoconstriction has also been observed. Claude Bernard was among the first to describe the effects of stimulating and cutting the superior cervical sympathetic trunk. Section of these nerves causes pupillary constriction (miosis), relaxation or protrusion of the nictitating membrane in those species such as cats that possess this structure, drooping of the eyelid (ptosis) associated with retraction of the eyeball, local vasodilatation of the skin, and anhidrosis (less sweating) (Horner-Bernard or Horner's syndrome). The effects of nerve section reveal that the dilator muscles of the pupil are normally held in a state of partial contraction by the tonically active postganglionic sympathetic fibers that innervate them. Sympathetic fibers from the superior cervical ganglion also innervate salivary glands. They cause vasoconstriction in these glands and may reduce salivary secretion from its maximum by a reduction of blood supply, but they also act directly on secretory cells, at least in certain glands, causing secretion of saliva. Other sympathetic fibers coursing to the head and neck tissues in the cervical sympathetic trunk cause sweating, piloerection, and vasoconstriction of skin vessels as well as those of the lacrimal glands.

Postganglionic fibers from the stellate (inferior cervical) ganglia and ganglia of the sympathetic chain innervate the heart. They produce acceleration of pacemaker activity in nodal tissues, speed up conduction of impulses, and increase the strength of contraction. Thus the autonomic fibers produce an augmentation of stroke volume and cardiac output. These cardiac fibers are also tonically active. Postganglionic fibers from the upper thoracic ganglia pass to the lungs and innervate muscle fibers of the bronchi, causing them to dilate. They produce a degree of constriction of pulmonary blood vessels. Many preganglionic fibers from the midthoracic region pass through the lateral ganglia to the celiac and other small collateral ganglia. They connect there with postganglionic fibers that innervate abdominal visceral organs. Fibers innervating the liver cause vasoconstriction and inhibit contraction of the gallbladder. These visceral efferents, by a combination of direct and indirect action through the release of epinephrine, produce glycogenolysis and a consequent liberation of glucose. There is also a reduction in blood clotting time consequent to augmented sympathetic activity.

Fibers to the spleen cause it to contract, discharging pooled erythrocytes and other components of blood into the vascular system. The mechanical as well as the secretory activities of the stomach and intestines are inhibited, and the cardiac and pyloric sphincters are constricted by sympathetic activity; digestive functions cease. Sympathetic fibers innervate the kidney, causing vasoconstriction of renal vessels.

Another striking effect of visceral sympathetic fiber action is vasoconstriction, which shunts blood away from the viscera, thus making it available to skeletal muscle and the CNS. Sympathetic nerve action also causes the blood vessels supplying the skin of the trunk, abdomen, and limbs to constrict. Piloerection and sweat gland secretion occur when there is a general discharge of the sympathetic system. All sympathetic vasoconstrictor nerves are tonically active; they increase the rigidity of large blood vessel walls and hold many arterioles and other small blood vessels in a state of partial or total occlusion. These nerves are highly important in the control of peripheral resistance and thus blood pressure. Vasoconstrictors also act on the veins and can effect a constriction therein that reduces the capacity of the venous reservoir (Chapter 41).

Sympathetic fibers to blood vessels of the skeletal musculature contain some cholinergic vasodilator fibers along with adrenergic constrictor fibers. These cholinergic fibers are said to have no tonic action and are not involved in regulation of blood pressure or cardiovascular reflexes. They are said to act during exercise and emotional states and to be controlled by pathways in the CNS that are distinct from those that regulate the vasoconstrictors.[86] The physiologic significance of these fibers, however, has not been fully demonstrated. The sympathetic supply to sweat glands (eccrine glands) is cholinergic. The apocrine sweat glands that are present in some regions, e.g., the axilla, do not participate in thermoregulatory sweating but are thought to secrete in response to mental stress. It appears that they are influenced humorally by epinephrine. It has been claimed that sympathectomy in man does not abolish secretion from the apocrine glands.

Innervation of the adrenal medulla is of great significance, since sympathetic activity causes liberation of norepinephrine and epinephrine (adrenaline). These catecholamines reinforce the action of all postganglionic sympathetic fibers, except those going to sweat glands and those vasodilator fibers that are cholinergic.

The sympathetic supply to pelvic visceral organs tends to have a mixed action on the bladder. It causes contraction of the internal sphincter and relaxation of the bladder wall. There is, however, an increase in the frequency of micturition and a tendency to void at a lower bladder volume as a consequence of emotional disturbance.

Peristalsis in the lower colon and rectum is inhibited by sympathetic nerve action. Vasoconstriction occurs in most of the pelvic organs when the sympathetic fibers are active, but a low level of erection can be effected in the male genitalia and sympathetic fibers can also produce an ejection of semen. Thus the effects of the sympathetic system on pelvic viscera are not all inhibitory.

In concluding this analysis it can be said that all of the effects of sympathetic nerve activity are physiologically compatible, and, if evoked simultaneously, they mimic rage and stress reactions. It appears that the thoracolumbar sympathetic complex or the sympathoadrenal system can and does discharge as a whole. Pupillary dilation, retraction of the nictitating membrane, piloerection, vasoconstriction, sweating, acceleration of the heart, elevation of blood pressure, increase in blood glucose, decrease in blood clotting time, and inhibition of the digestive system are all sympathetically induced phenomena and produce a picture that we have all observed in angry or frightened animals. The anatomic unification of this system would lead one to predict a generalized discharge. This seems to be less of a mystery than the channeling of activity through this system, which permits rather discrete reflex actions and the occurrence of tonic activity in only certain proportions of the sympathetic fibers. The centers that control this system can, to a great degree, determine whether it discharges selectively or totally. Under basal conditions, selective tonic and reflex activities occur, but under stress or in anger the system can and does discharge as a whole.

Actions of parasympathetic nerves on specific effectors

The principal effect of the parasympathetic supply to the eye is pupillary constriction. These fibers have a tonic activity, and section of the third cranial nerve through which they pass produces a degree of pupillary dilation. The lacrimal glands of the eye are innervated by parasympathetic fibers of the seventh cranial nerve, which cause lacrimation. The chorda tympani and other parasympathetic nerves running to the salivary glands produce copious salivation and dilatation of the glands' blood vessels. It is presently thought that vasodilatation is caused by dilator fibers and indirectly by secretory fibers that liberate active dilator substances.[31]

The vagus or tenth cranial nerve, which is a component of the cranial parasympathetic division, has an inhibitory effect on the heart. Its basic action is on the pacemaker; it retards the depolarization process and decreases heart rate as a consequence. It slows conduction and can actually block transmission of impulses; this block occurs in the atrioventricular nodal region. Thus the vagus tends to produce a reduction in cardiac output and blood pressure. Vagus nerves innervating the bronchi and lungs cause constriction of the bronchioles and possibly an increased secretion from the bronchial glands.

Vagal fibers also innervate the stomach, intestine, and upper colon and have an excitatory action on these organs, increasing peristaltic activities, shortening emptying time, and increasing secretion from the gastric and other digestive glands. This parasympathetic fiber action relaxes all sphincters of the digestive tract. It stimulates pancreatic secretion, the secretion of insulin, and the release of bile by causing contraction of the gallbladder.

These vagus fibers to the heart and to the gastrointestinal system are tonically active. However, since smooth muscles of the stomach and gut are intrinsically active, the vagi as well as the sympathetic fibers are merely modulators of gastrointestinal activity.

The sacral parasympathetic system is responsible for bladder contraction, reflex micturition, and defecation. These fibers increase peristaltic activity in the colon and rectum and cause relaxation of sphincter muscles during micturition and defecation. Other fibers of the system have a vasodilator action, producing erection of external genitalia.

It is obvious that there could be no physiologic rationale for the discharge of the parasympathetic system as a whole. Simultaneous dilation of the pupil, salivation, slowing of the heart, increased activity of the gut, defecation, urination, and erection of the penis have no sensible functional association. These phenomena are associated only in abnormal circumstances that produce a mass reflex. The components of the parasympathetic system behave independently, participating in specific reflexes or well-integrated reactions.

Interaction of the two divisions

The somatic musculature is controlled by an on-off mechanism; i.e., impulses go to muscle fibers to produce contractions and, in the absence of such impulses, the muscles relax. Due to reactions through central mechanisms, some somatic motor nerves show tonic activity that can be augmented or minimized depending on the intensity of afferent stimulation. Activity of the antigravity system provides an example of this phenomenon. Control of those visceral tissues that receive a single autonomic innervation is quite similar to the regulating activity of the somatic system.

Sympathetic nerves to piloerectors, sweat glands, visceral and skin blood vessels, and the spleen and liver operate by an on-off action. Vasomotor nerves are tonically active; vasoconstriction can be produced by an increase in tonic activity, and vasodilatation can be produced by a decrease or cessation of tonic activity. The other visceral organs, which receive a dual innervation, are controlled in a much more complex fashion. There are numerous variants of interactions between sympathetic and parasympathetic fibers that innervate common organs. These are best described individually.

In the eye, parasympathetic nerves act on the sphincter muscle of the pupil to produce pupillary constriction and on the ciliary muscle to evoke accommodation for near vision. The sympathetic innervation causes pupillary dilation and distance focusing; this is antagonistic to the parasympathetic action. However, the sympathetic fibers act on radial fibers of ciliary muscles and also on dilator muscles of the pupil, not on a common muscle. Here we have a situation very much like the antagonism seen in flexion and extension of a limb. Also similar is the reciprocal action occurring in reflex responses; when the pupillary constrictor nerve is reflexly excited, the dilator nerve is simultaneously inhibited. Both these fiber outflows are tonically active; thus pupillary constriction and dilation can be affected by increasing or decreasing parasympathetic tonic activity. Similar reactions can be evoked by decreasing or increasing sympathetic tonic activity. Normally, however, both outflows are reciprocally active.

The dual innervation observed in certain salivary glands has a slightly different consequence. Parasympathetic and sympathetic fibers end on different types of secretory cells in some salivary glands, but in other glands they innervate the same cells; however, both produce secretory activity. When both have been stimulated together in laboratory experiments, more saliva is produced than when one component alone is stimulated. This is a synergistic action. There is an antagonistic action between the two autonomic nerves on the blood vessels of the salivary glands; the sympathetic fibers constrict and the parasympathetic fibers dilate these vessels. Blood flow, of course, affects salivary secretion, and under certain conditions the sympathetic fibers can reduce salivation by decreasing blood volume flow through the glands. The innervation that is dominant in controlling salivary secretion depends on the nature of the stimulation. Thus the composition and quantity of saliva produced depend on the balance of the contributions made by each of these two innervations. Tonic activity of these fibers may be involved in controlling the blood flow to the salivary glands, but vasomotor changes appear not to be significant in the normal regulation of salivary secretion.

The heart provides the most beautiful example of control by an antagonistic innervation. Both postganglionic sympathetic and parasympathetic fibers act on the same tissues. Here we have a true excitatory sympathetic action that can accelerate pacemaker firing and conduction of excitatory impulses and increase the strength of contraction (positive inotropic effect). Vagal parasympathetic fibers, on the other hand, exert a true inhibitory action, which is well described in Chapter 4. Vagal nerve fibers, under most circumstances, slow spontaneous depolarization in the pacemaker cells, and they can also retard and block conduction. There is no definite negative inotropic action, and slowing of the heart may not decrease cardiac output because it may increase stroke volume. The parasympathetic and sympathetic fibers have antagonistic actions on the coronary circulation (sympathetic stimulation increases coronary flow) but the physiologic importance of this is not certain (Chapter 43).

As stated earlier, the lung is also innervated by both the parasympathetic and sympathetic divisions of the autonomic nervous system. Parasympathetic impulses constrict and sympathetic impulses dilate the bronchi. Here we have the two systems playing a role that is somewhat the reverse of that played in the heart.

There is also dual innervation of the digestive system. Fibers from the two divisions are tonically active and antagonistic in their effects on the same musculature. In a first approximation the situation may be summarized as follows: The parasympathetic innervation, through the vagus, augments the rhythm and strength of contraction of the muscles in the walls of the esophagus, stomach, intestine, and upper colon; the sphincters are relaxed by parasympathetic action, and this constitutes the major motor control of the gastrointestinal tract. The sympathetic innervations appear to play a major role only during states of stress or emotional disturbance. At such times activity in various segments of the gastrointestinal tract is reduced by active inhibition through catecholamine release. The sphincters tend to close more securely. Sympathetic activity also produces vasoconstriction in these tissues.

With respect to secretory activity in component parts of the digestive system, the parasympathetic has a definitely positive action. The opposite effects produced by sympathetic activity probably are secondary to change in blood flow rather than to a direct secretory cell inhibition.

The interrelationship of sympathetic and parasympathetic actions on the pelvic viscera is somewhat less clear. In the colon and rectum the sacral parasympathetic fibers are activators, while the sympathetic supply is inhibitory. These effects are similar to those occurring in the small intestine and stomach. Sacral parasympathetic fibers exert a powerful contraction-inducing action on the bladder that initiates micturition, while sympathetic nerve effects are mixed; although sympathetic nerves generally cause relaxation of the bladder it has been reported that they can cause contraction. The parasympathetic and sympathetic fibers have a more clearly antagonistic action on the sphincters, the parasympathetic relaxing and the sympathetic contracting these muscles. Not much is known about the significance of autonomic innervation of the uterus and associated tissues. With respect to erection in the male, both parasympathetic and sympathetic nerves contribute to some degree, while ejaculation is under the control of sympathetic fibers; thus cooperation or supplementation is illustrated here.

The parasympathetic and sympathetic fibers cooperate in the control of visceral organ activity. They may have a reciprocal or opposite action, they may have a similar action, or their effects may be supplementary. One cannot say that the sympathetic division is invariably inhibitory; like the parasympathetic division, it has inhibitory effects on some tissues and excitatory effects on others. It must not be forgotten that the CNS plays a major role in adjusting the balance of action of these two outflows.

One of the most difficult problems in this field of study has been to determine the role of autonomic fibers in the control of the peripheral vascular system. In skeletal muscle it appears that the vascular innervation is entirely sympathetic in origin, but these trunks contain both constrictor and dilator components that are controlled separately. Since adrenergic vasoconstrictors are tonically active, vasodilatation can be produced by diminution of tonic discharges. Vasodilatation also can be produced by cholinergic dilator fiber action and by accumulation of metabolites[86] (Chapter 41). There are true parasympathetic cholinergic vasodilators running in the nervi erigentes to blood vessels of the external genitalia; similarly, there are cholinergic dilators to salivary gland and cerebral blood vessels. The literature relative to vasodilator actions in various organs in the body is extensive and controversial but this topic is dealt with in Chapters 41 and 43.

Overaction and underaction of the autonomic system

As stated previously, autonomic fibers are not essential to the functioning of visceral tissues. However, overaction of this innervation has major physiologic consequences and underaction also impairs normal function, at least until tissues recover from their dependency on this innervation.

Low vasomotor tone produces the condition of orthostatic hypotension. If a person with low vasomotor tone is suddenly brought to an upright position on a tilt table, fainting results. Fainting, which occurs because of enforced immobility while erect, is due to insufficient vasomotor compensation and reduction of blood flow through the brain. Surgical removal of the sympathetic system or blocking the system with drugs produces hypotension. A condition described as familial dysautonomia is occasionally reported in the clinical literature. In such cases there is defective lacrimation, vasomotor instability, and defective temperature control.

Overactivity of vasomotor nerves can result in hypertension. One sees frequent reference to neurogenic hypertension and certainly a hyperresponsiveness of vasomotor reactions can develop. A paroxysmal type of hypertension that occurs in individuals with tumors of the adrenal medulla is occasioned by massive discharges of catecholamine from the gland. The hypertension commonly encountered, however, cannot be explained on the basis of overactivity of the sympathetic division of the autonomic system.

In Raynaud's disease there is a spasmodic contraction chiefly of blood vessels supplying the digits, which consequently receive an insufficient blood supply. The hand, for example, becomes cold and pale in color, and gangrene may actually set in. Since sympathetic denervation of the extremities in such cases is often highly beneficial, an overaction may be involved. Preganglionic denervation is preferable, since it prevents the denervation hypersensitivity that is a consequence of postganglionic fiber removal.

Hyperhidrosis (excessive sweating) is another example of overactivity of the sympathetic system. The sweat glands are overactive, and the condition is quite debilitating. Cases are described in the literature.[63, 87]

Abnormal activity of the sympathetic or parasympathetic innervation of the gastrointestinal tract has been considered in relation to the development of ulcers. It has been thought that changes in the blood flow in the mucosa, excessive producion of hydrochloric acid, and reduced production of mucin might be due to abnormal neuronal actions. Gastric ulcers are thought to result from reduction of mucosal defenses, while duodenal ulcers are associated with hypersecretion of acid. Studies of the possible role of the autonomic system in the production of gastric or duodenal ulcers have produced rather contradictory results. It does appear, however, that vagus resection combined with a partial gastrectomy does produce permanent relief, especially of chronic duodenal ulcers.

In the late 1920s, complete sympathectomy was attempted by bilateral removal of the lateral chains of ganglia,[22] a procedure that also disconnected the collateral ganglia from the CNS. It was found that sympathectomized animals had certain inabilities and vulnerabilities, such as hypersensitivity to heat and cold and to hypoglycemic agents such as insulin. Although these animals could live normal lives, the male was rendered infertile but the female was not; there was also an initial hypotension and some susceptibility to fatigue. The totally denervated heart was found to function effectively, and most organs assumed an autonomy of function that enabled them to play an adequate physiologic role after severance of the external innervation. This work merely provided evidence in support of the concept that although the autonomic system plays a very important physiologic role, neither division is completely essential to the maintenance of basic body functions.

In general, it can be said that the parasympathetic system has a conserving, protective function and tends to promote emptying of the hollow viscera. The sympathoadrenal system functions in emergency situations, coming into strong action during stress, under conditions producing fear and rage or when an animal suffers pain. Both systems tend to preserve an essential balance of body states; they react to correct imbalances and generally support behavioral activities. To use the terminology of Cannon, they illustrate the wisdom of the body by reacting to meet emergencies and by maintaining homeostasis.

CHEMICAL TRANSMISSION OF NERVE IMPULSE

One of the most exciting developments of the first third of this century was the discovery and analysis of the chemical transmission of nerve action. It seems advisable to refer students to previously published accounts,[14, 20, 33] which do more justice to this story, rather than attempt a full description here.

The major initial discoveries were those of

Otto Loewi, Walter B. Cannon, Henry H. Dale, and their associates. In 1914 Dale,[24] studying acetylcholine (ACh), a derivative of choline that he had found in ergot extracts, showed that this chemical in minute amounts mimicked very accurately the effects of parasympathetic stimulation. In 1921 Loewi[73] demonstrated that the perfusate from an innervated isolated heart would inhibit a second denervated heart, perfused thereby, when the vagus nerve was stimulated. He called this substance *Vagusstoff* and recognized that it had to be protected by anticholinergic agents such as eserine (physostigmine).

In 1931 Cannon and Bacq[19] initiated a series of classic experiments that they and others carried out at Harvard. They showed that an epinephrine-like substance, which they originally called sympathin, is liberated from most postganglionic sympathetic fibers on stimulation. It took some time to demonstrate that postganglionic fibers of the sympathetic system are largely adrenergic, while parasympathetic fibers are cholinergic. We owe this terminology to Dale,[25] who, with his associates, also showed that somatic motor fibers are cholinergic in their action on the skeletal muscle. His work and that of many others made possible the recognition that ganglionic transmission in both the parasympathetic and sympathetic divisions is cholinergic[26] and that some sympathetic postganglionic fibers, such as those innervating sweat glands, are cholinergic. Fig. 30-4 is a diagram of historic interest, since it gives the original complete analysis of this problem.

It was von Euler[32] who first produced convincing evidence that norepinephrine is the normal adrenergic transmitter of the postganglionic sympathetic neuron. The adrenal gland produces both norepinephrine and epinephrine. Usually more epinephrine than norepinephrine is released, but the proportions may change according to the nature of the physiologic stimulus. Also, the proportions of these catecholamines stored and released from the medulla varies from species to species. Tissues differ in their sensitivity to these two adrenergic compounds released by the autonomic system. Norepinephrine is much more effective in producing peripheral vasoconstriction than is epinephrine, but the two compounds have a quantitatively similar effect on the heart. Epinephrine is the more potent agent in producing effects on some tissues, e.g., the cecum of the fowl.[32]

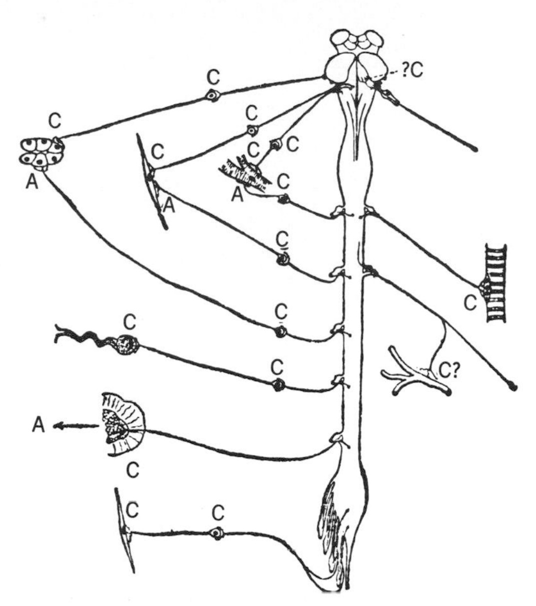

Fig. 30-4. Schematic representation of sites of release and action of adrenergic, *A*, and cholinergic, *C*, mediators as originally presented by Dale. At right is somatic outflow to skeletal muscles and representation of the then current theory of axon reflex producing skin blood vessel dilatation. At left are representative components of outflow to salivary gland, eye, heart, sweat gland, adrenal medulla, and smooth muscle of pelvic organs. Diagram contains certain inaccuracies and should be regarded as an approximation of historic importance. (From Dale.[25])

Recent studies of adrenergic transmitters

There has been a tremendous amount of work in recent years on the analysis, synthesis, storage, metabolism, release, and mechanism of action of adrenergic transmitters.[8, 9, 34, 52]

The individual terminals of adrenergic neurons innervate many smooth muscle fibers. As previously described, these fibers show swellings or varicosities along their course, each of which serves as an ending en passant. Within these swellings are granules or granulated vesicles approximately 300 to 1,000 Å in diameter; each contains an electron-dense core. On isolation by selective centrifugation, these granules are found to contain dopamine-β-oxidase, monoamine oxidase (MAO), and adenosine triphosphate as well as norepinephrine. Unquestionably, norepinephrine is the transmitter agent stored in these granules. These granules are released or release the transmitter stored in the en passant or terminal varicosities. There is a rather continual release of norepinephrine at these sites that is greatly accelerated by

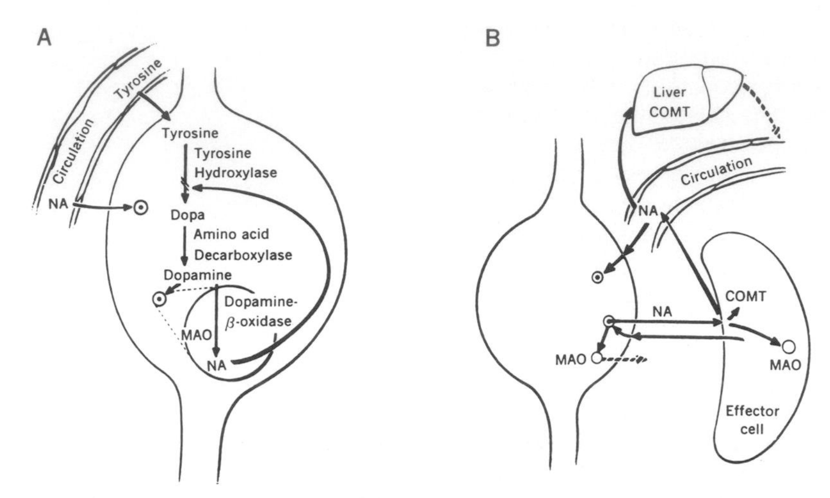

Fig. 30-5. A, Biosynthesis of norepinephrine (noradrenaline) in sympathetic nerve. NA, noradrenalin; MAO, monoamine oxidase. Note storage granules shown as small circle with central dot, enlarged in lower portion of figure to show processes occurring therein. Note feedback action of NA on formation of dopa and transport of NA between blood vessels and neuron. **B,** Fate of noradrenaline released in sympathetic nerves and tissues. COMT, catechol-*O*-methyltransferase. Dotted line represents inactivated NA. Note multiple pathways of movement of NA. (From Axelrod and Kopin.[8])

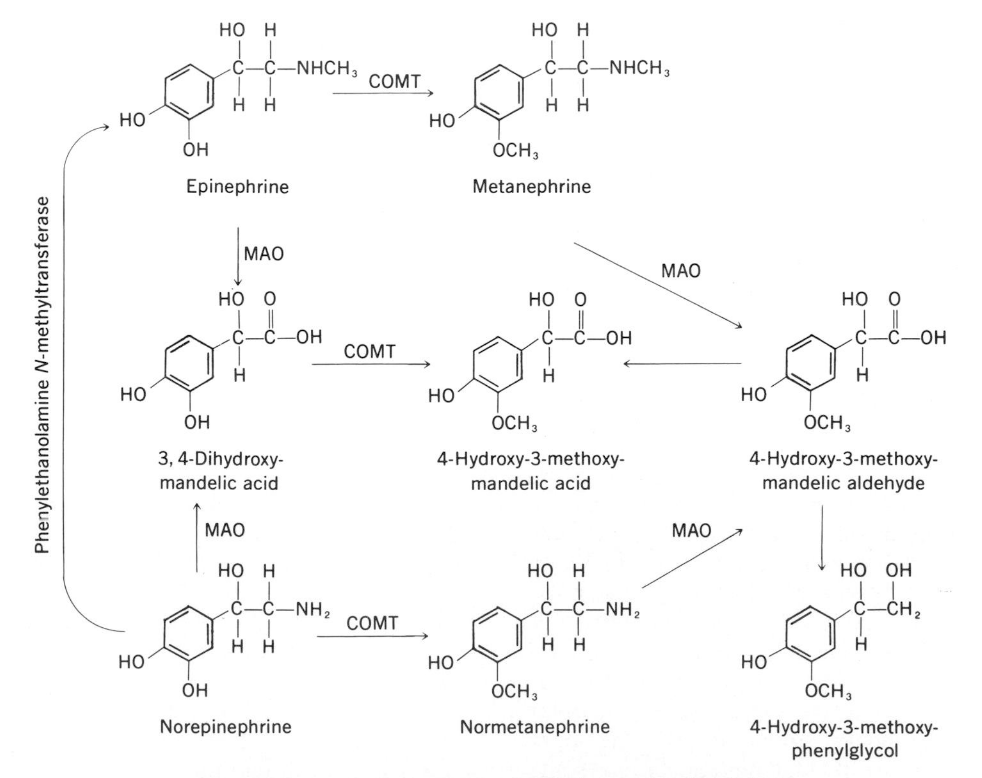

Fig. 30-6. Catabolism of norepinephrine and epinephrine.

the arrival of a nerve impulse. Once released, norepinephrine acts on the receptor of the effector cell and is then removed from synaptic sites in a manner that will be discussed later.

Norepinephrine is synthesized in fiber terminals and the cell body and is moved down to the axon terminals by axonal transport processes. The process of norepinephrine synthesis within the nerve cell is now known. The precursor is considered to be tyrosine, and the steps of synthesis are shown in Fig. 30-5, *A*. This figure indicates that dopamine enters granulated storage vesicles, where it is β-hydroxylated to form norepinephrine. MAO is also present in the granules. A feature shown here, which is of considerable interest, is that the concentration of norepinephrine appears to block new production by a negative feedback process that controls the conversion of tyrosine to dopa. Another process worthy of specific mention is that preformed norepinephrine may enter the cell from the surrounding tissue fluid. Thus nerve terminals that are depleted of stored norepinephrine either by intense activity or by action of a drug can be charged up again by norepinephrine injected into the circulation or introduced into the surrounding medium. This uptake of norepinephrine into the postganglionic sympathetic neuron is accomplished by an active transport system requiring sodium and potassium ions that can occur against concentration gradients as high as 10,000 to 1.

Norepinephrine undergoes a complex fate after it is discharged from the nerve. A fraction is removed by diffusion into the bloodstream, a small amount is metabolized locally by catechol-*O*-methyltransferase (COMT), but the greatest proportion is taken up again by the nerves and is bound again in granules. This uptake maintains what appears to be a dynamic equilibrium between granules and extragranular, intra-axonal stores. Norepinephrine in granules, which constitutes the major proportion of the transmitter present in an axon, is continuously discharged from them into the axoplasm. On release, the major fraction is inactivated within the neuron through deamination by MAO. Norepinephrine escaping into the circulation is mainly metabolized in the liver by COMT. The ultimate fate and the metabolism of norepinephrine are shown in Figs. 30-5, *B*, and 30-6.

Studies of the release of norepinephrine during nerve activity have yielded some interesting figures of physiologic significance. It has been found that a stimulus frequency of 30 impulses/sec causes the maximum release per stimulus from splenic nerves. At a faster rate of stimulus, less is released. Calculation of amount released per stimulus shows that approximately 0.8 ng or 0.01% of the total amount present (8 μg in 5 gm of tissue) at the nerve terminals in the spleen is released. This corresponds to the release of 18% of the total tissue content of norepinephrine per minute if a nerve is stimulated at a rate of 30/sec. In a muscle the rate of release is slightly less. It is obvious that if resynthesis were inhibited, the store of norepinephrine would be depleted by 50% in a few minutes' time.

Cholinergic transmitter

ACh is the transmitter for the preganglionic fibers of the total autonomic system, postganglionic fibers of parasympathetic nerves, and some of the sympathetic postganglionic neurons. It is stored in vesicles in the nerve terminals. These vesicles are 400 to 500 Å in size and occur in clumps or heavy concentrations near specialized areas of the terminal membrane. They appear to move progressively in the cytoplasm toward sites of transmitter release.

ACh is synthesized in the terminals and probably also in the cell body and transported down the axons to the nerve terminals where it can be stored in quantity in vesicles accumulated there. Choline acetylase is present wherever ACh is found and is involved in its synthesis from acetate and choline. There is a constant leakage of ACh as vesicles move down to the terminal membrane and release the transmitter. The rate of this release can be calculated at the motor end plate, for example, by the number of miniature endplate potentials that occur during a given time interval. The arrival of impulses greatly augments this discharge. Quantal release of ACh is of the order of 10^{-4} molecules under basal conditions. The amount released by a single impulse is 100 times greater. Most of our information comes from studies of events at the neuromuscular junctions of skeletal muscle, but available evidence indicates that the processes are the same at autonomic neuroeffector junctions.[62a]

The released ACh diffuses across the synaptic cleft to the subsynaptic membrane, where it combines with a receptor substance to produce permeability changes and a depolarization of the postsynaptic membrane. Thus a response is initiated in the effector cell, where this action is excitatory.

The action of ACh on the heart is inhibitory. There is a hyperpolarization of muscle cell membrane associated with an increase in membrane permeability to potassium in the pacemaker cell. The atrial muscle action potential is also shortened due to the fact that the rate of membrane repolarization is greatly increased (Chapter 34).

ACh is hydrolyzed and inactivated by acetylcholinesterase (AChE), which is present in high concentrations in synaptic regions. This action is extremely rapid and prevents diffusion of liberated ACh away from the site of release. A diffuse action of this compound would, of course, have debilitating physiologic effects. The action of ACh is greatly augmented by anticholinesterases that prolong and augment the action of ACh by retarding its breakdown.

As stated previously, Dale suggested that neurons of the autonomic system might be classified as adrenergic and cholinergic on the basis of their production and release of one or the other compound. It has been suggested from time to time that neurons might produce both these transmitters or, more particularly, that ACh is involved in the release of catecholamines from postganglionic neuron endings, but the evidence for this is indirect and largely pharmacologic.[39] Of course, preganglionic neurons initiate release of catecholamine from the adrenal medulla by the liberation of ACh, which depolarizes the chromaffin cell membrane, thus increasing its permeability to sodium and calcium ions. These ions initiate an unknown process that moves the granules toward the cell surface and releases catecholamines from storage.[27]

Central synaptic and ganglionic transmission

Preganglionic sympathetic neurons have been studied by direct intracellular and extracellular recording methods. The reactions of these lateral horn cells of the cord to antidromic excitation by stimulation of white rami and to orthodromic excitation through the dorsal root or splanchnic nerve have been determined. Intracellular recordings have shown responses to be similar to but slightly longer in duration than those of anterior horn cells.[38] These preganglionic neurons are readily activated from higher centers, but only a few cells of a pool respond to peripheral nerve stimulation. This suggests that very few of these neurons are available to participate in spinal reflex action but that many are activated directly by input from higher centers.

Parasympathetic preganglionic neurons of the sacral spinal cord have been studied in a similar manner. They also resemble anterior horn cells in their responses. An interesting observation is that incoming impulses from somatic afferents that excite the anterior horn cells may cause inhibition of parasympathetic neurons or a mixture of hyperpolarization and depolarization in these lateral horn preganglionic cells.[45] Such excitatory and inhibitory influences on spinal preganglionic neurons by afferent impulses may be involved in normal reflex actions. Obviously some integrative control of function from higher centers must be exerted on these cells to relate their activity to peripheral requirements.

Not much is known about the transmitters operating at central synapses involved in control of the autonomic system. It appears that in the CNS there are some adrenergic, some cholinergic, and still other synapses employing different transmitter substances, e.g., 5-hydroxytryptamine (5-HT) and certain amino acids.[46] Direct iontophoretic applications of ACh or norepinephrine have thus far failed to detectably affect the preganglionic autonomic lateral horn cells.

The events of ganglionic transmission are much better understood, but although this is presently an active field of study, the situation is quite complicated (Chapter 6). Basically preganglionic fibers act on postganglionic neurons by the release of ACh, which produces postganglionic cell discharges. The preganglionic fiber endings are packed with vesicles; the miniature end-plate potentials that have been recorded are similar to those of the neuromuscular junction. AChE is present in high quantity at these terminals. ACh escapes from the ganglia during preganglionic neuron stimulation and can be detected if anticholinesterase is used. ACh applied to the ganglia will evoke postganglionic responses; unquestionably this is the ganglionic transmitter agent.

The realization that other events can occur within the ganglia has come from the observation that in addition to the regular excitatory postsynaptic potential there are also slow excitatory and inhibitory postsynaptic potentials under certain circumstances. Catecholamines also have a depressant action on ganglionic transmission, and it is assumed that the ganglia contain adrenergic terminals as well as cholinergic terminals.[70, 76] Catecholamine-containing cells are found within mammalian ganglia,[53] and there may be adrenergic interneurons or recurrent adrenergic collaterals that contain these compounds and are responsible for the inhibitory reactions evoked by orthodromic and antidromic stimulation of pre- and postganglionic fibers.

It has been known for many years that stimulation of some postganglionic fibers can evoke activities in other fibers. This phenomenon, which was identified as an axon reflex, implies a high branching of a neuron, which permits a stimulus to travel antidromically in one branch to another neuron or out to the periphery orthodromically in the other branch. This may occur, and it now appears that there is much more interaction within ganglia than was previously realized.

The physiologic significance of interaction within the ganglia and of the possible adrenergic inhibition occurring there is not fully understood. For general purposes it suffices to say that the impulses entering the ganglia over preganglionic fibers liberate ACh, which acts on receptor substances and initiates activity in postganglionic neurons. These cholinergic receptors differ somewhat from those found elsewhere because they have a different susceptibility to blocking agents.

The receptor concept

The mode of action of neurotransmitters on target cells has been much discussed, and the concept of the "receptor" has been considerably modified[10] since its introduction by Langley[66] in 1905. It is generally assumed that a neurotransmitter has to unite in some way with the target cell before exerting its action. Often the sites of binding between the cell and the active molecule are referred to as receptors. These postulated receptors, in or on the cell membrane, occupy only a very small portion of the cell surface. Although there is frequent mention of receptor sites, morphologic evidence of specific receptor "patches" on the cell surface is still lacking.

Pharmacologic evidence indicates that *cholinergic receptors* of the various tissues differ markedly. Blocking agents are assumed to compete with the transmitter substances for the binding site in the receptor. Curare has a powerful blocking action on the neuromuscular junction and prevents the action of nerve impulses or injected ACh. Curare, however, is relatively ineffective in the autonomic ganglia and fails to block inhibitory action of the vagus on the heart. Nicotine is an excellent ganglionic blocking agent but is less effective than curare on neuromuscular junctions. Atropine is not a ganglionic blocking agent but tends to prevent the action of ACh and postganglionic parasympathetic fibers on effector organs.

Another type of evidence indicative of differences in receptors is provided by excitatory agents. Muscarine, the alkaloid responsible for the toxicity of certain mushrooms, has little effect on cholinergic stimulation of postganglionic neurons in autonomic ganglia. However, it excites visceral effectors innervated by cholinergic postganglionic neurons. Atropine blocks muscarinic action. These various cholinergic receptors have not yet been thoroughly classified and studied.

It has long been recognized that there are *adrenergic receptors*, which may also differ, since the potencies of sympathomimetic agents and adrenergic transmitters vary widely when tested on different tissues and organs. The effectiveness of blocking agents also varies from tissue to tissue. It has been concluded that there are at least two types of receptors, α- and β-adrenergic receptors.[43] Both α- and β-receptors are responsive to various catecholamines, norepinephrine, and epinephrine, but they have different sensitivities. For example, α-receptors are blocked by phenoxybenzamine (Dibenzyline), phentolamine, and ergot alkaloids. The β-receptors are blocked by dichloroisoproterenol and pronethalol. The α-receptors mediate vasoconstriction, contraction of pupillary dilators, constriction of the gastric, intestinal, and bladder sphincters, and contracture of the spleen. The β-receptors mediate such actions as an increase in cardiac rate and strength of contraction, and they are involved in the inhibition of gastric motility.

Denervation hypersensitivity of autonomic effector organs

When autonomic effector organs are denervated, they become increasingly sensitive to transmitters and to chemical agents.[18, 21] This sensitivity is much greater when postganglionic fibers are destroyed. In the course of degeneration, transected autonomic fibers release stored transmitters and this tends to give a brief paradoxical overaction. After removal of the superior cervical ganglia, for example, a nictitating membrane initially relaxes; then after a few hours it may contract strongly only to relax again. After 7 to 14 days, maximal denervation sensitivity develops, and denervated organs respond to minute quantities of ACh or catecholamine to which they previously would not have been responsive.

The mechanism of denervation hypersensitivity is not fully understood, but at least two contributing factors are known.[31a] In the absence of nerve terminals, transmitters are not reabsorbed and inactivated quickly. Furthermore, following denervation, cholinesterase and MAO, which normally inactivate transmitters, disappear. This is only a partial explanation because denervated tissues become sensitive to many other compounds, both drugs and ions. A development of hypersensitivity has been observed as a result of attempts to effect surgical relief of Raynaud's disease and other autonomic hyperactivities. Laboratory experiments have indicated that preganglionic denervation should be much more beneficial, since peripheral hypersensitivity is minimal and denervated tissues do not overrespond to changes caused by release of catecholamine or cholinergic agents from adjacent tissues. Following removal of postganglionic fibers, sensitivity becomes maximal and, on exposure to cold or excitement, the catecholamines released from the adrenals or adjacent sympathetic fibers cause a very strong overconstriction.

Immunosympathectomy and chemical sympathectomy

Those who are interested in autonomic nervous system physiology should be aware of certain relatively recent studies that have yielded very striking results, the physiologic significance of which is not yet fully understood.

In the late 1940s and early 1950s it was found that certain tumors apparently released substances that served as a nerve growth factor. These factors affected autonomic fiber growth much more than they did somatic neurons. Snake venom was used in attempts to block certain metabolic processes in tumors, and the venom was also found to be rich in autonomic growth factor. Venoms come from modified salivary glands, and it was soon learned that the salivary glands of mice and other mammals contain a protein that is an autonomic nerve growth factor.[68, 69] When this agent is injected into rabbits, it forms an antibody, and serum collected from such animals can be used to produce an immunosympathectomy in newborn animals. It appears that this antibody neutralizes a nerve growth factor essential to the development of sympathetic fibers. Ganglia of the sympathetic chain are most affected. The cell population is found to be only 2 to 10% of normal, and the autonomic system becomes minimally or completely nonfunctional. Organs in immunosympathectomized animals contain much lower amounts of norepinephrine than are normally present. Some ganglia are not as affected as are others, indicating the complexity of the problem.

Chemical sympathectomy can be produced by the use of drugs such as reserpine, which deplete and prevent storage of catecholamine, and by bretylium and guanethidine, which act by preventing norepinephrine release as well as storage. At present the use of most of these agents is still confined to the experimental laboratory, and our knowledge of their action is limited.

TONIC ACTIVITY WITHIN THE AUTONOMIC NERVOUS SYSTEM

It has already been pointed out that many, but not all, of the components of the sympathetic and parasympathetic divisions of the autonomic nervous system are tonically active. In speaking of nerve function the term "tonic activity" is used to imply a continuing discharge in the nerve fiber innervating an organ. This discharge determines the level of activity in the effector organ or tissue. Tonic activity occurs in fibers that have an excitatory action and in fibers that are inhibitory. In the latter case, activity in an organ can be increased by diminution of tone; e.g., a decrease of vagal tone accelerates the heart. In vasoconstriction, which is maintained by sympathetic fiber tonic discharge, a diminution of tone produces vasodilatation. This reduction in vasoconstrictor tone results in a decrease in peripheral resistance which, if sufficiently generalized, causes a fall in blood pressure.

Frequency of discharge in tonically active fibers

A number of methods have been employed in an attempt to determine the levels of tonic activity in various nerves and to assay the frequency of firing that produces certain intensities of peripheral effects. Folkow[40, 41] has stimulated sympathetic trunks innervating blood vessels of the leg and of visceral organs while assaying peripheral resistance by measuring volume flow under a constant perfusion pressure. By varying the frequency of stimulation of sympathetic trunks he obtained different degrees of vasoconstriction that reduced the flow accordingly. Thus he was able to estimate the relationship between the frequency of stimulation and the effector organ response. From such data one can calculate the rates of overall tonic activity that are required to maintain normal peripheral resistance and elevate it to the maximum (see Chapters 36 and 38 for a detailed discussion).

Another method involves recording tonic activity directly from single nerve fibers, fine strands of fibers, or preganglionic cell bodies in the cord. Isolation of single fibers requires special dissection techniques, and the recording of activity from cell bodies of the lateral horn requires the use of microelectrodes. It is not possible, of course, to accurately determine the total outflow of impulses to an effector organ in a unit of time by measuring activity in a single fiber. However, one does obtain a very accurate picture of rates and changes in rates of tonic activity in individual neurons.

The rates of tonic discharge in individual pre- and postganglionic sympathetic fibers have been measured in white rami, in fibers of the cervical sympathetic trunk, and in fibers innervating skeletal muscles, skin, and heart.[12, 54, 56, 58] The rate varies greatly in individual fibers under basic conditions and it ranges between 1 impulse in 10 sec to several impulses in 1 sec. Tonic activities in fibers within the same nerve trunks vary considerably. This is not surprising, since their functions differ. In some the rate is one every 3 or 5 sec, while in others it is 3 to 5/sec. The individual fiber rates tend to remain relatively constant under the same experimental conditions. There are some sympathetic fibers that do not discharge unless they receive a particular stimulus. Rates are affected by internal as well as external influences. Some sympathetic fibers show a very clear respiratory rhythm; i.e., the rate of discharges is augmented during inspiration and reduced during expiration. These same fibers may also show an arterial pulse–related rhythm. Tonic discharges of sympathetic nerve fibers change during reflex action; under conditions of stress such as asphyxia and hemorrhage, the frequency of discharge can increase to 20 or even 30 impulses/sec. Of course, the maximum frequency attained by sympathetic fibers during augmentation varies from one fiber to another.

Parasympathetic tonic activity has been recorded from vagal fibers innervating the heart and stomach and from sacral parasympathetic neurons innervating the bladder.[35, 45, 50, 51, 64] In this system the fibers discharge at a highly variable rhythm and frequency. The rhythms and rates change with pulse pressures and the degree and frequency of smooth muscle contraction in the bladder or stomach. Although a few fibers show a constant rate of discharge, most of them are silent until a highly varied rhythm of action is evoked by some intrinsic or extrinsic stimulus such as a significant increase in blood pressure or bladder pressure. It is therefore difficult to state the exact rates of tonic discharges. In general they are low but approximate those recorded from sympathetic fibers. As in sympathetic nerves, during augmented activity parasympathetic fiber discharges may increase to 30 impulses/sec. Fibers vary in the maximum rates they attain and the range is usually 5 to 35/sec.

Origins of tonic activity in autonomic fibers

In many fibers of the autonomic system, tonic activity is of reflex origin but in other instances it may depend on some instrinsic stimulatory process. The tonic discharge in the vagus nerve supply to the heart depends chiefly on baroreceptor activity, since section of all baroreceptor afferents causes cessation of such discharges. Some vagal fibers show arterial pulse–related bursts of activity. As pulse pressure changes occur, afferent discharges from the baroreceptors excite the vagus nuclei in the medulla. Vagus nerve activity is also related to the respiratory rhythm. Contrary to the respiration-related sympathetic activity, vagal discharges increase during expiration and decrease during inspiration. This fluctuation in vagal tone creates a cardiac arrhythmia of respiratory origin.

Denervation of baroreceptors abolishes the arterial pulse–related surge in sympathetic fibers innervating the cardiovascular system. A basic rhythm remains, indicating involvement of factors other than baroreceptor action.[60]

Tonic activity recorded in peripheral autonomic nerves originates from tonic activity in the preganglionic neurons. The rates recorded from pre- and postganglionic fibers, however, may not be the same because of the events occurring within the ganglia. The tonic activity of the spinal preganglionic neurons is normally under control of higher centers, mainly the medulla, at least in the case of those fibers responsible for cardiac and vasomotor tone. Higher centers may modulate activity in the medulla and the spinal cord, as will be discussed later.

It is difficult to ascertain whether tonic activity initiated in the medulla is of intrinsic origin or whether it is imposed by intrinsic or extrinsic stimuli. Certainly it can be modified by afferent signals and changes in the local cellular environment. There may be a degree of intrinsic rhythmicity due to some internal reverberating circuit or to some type of pacemaker action similar to that occurring in cardiac cells. Although we have no direct evidence of this in mammalian neurons, such a process is seen in the neurons of primitive animals.

Following transection of the spinal cord there is a marked reduction of sympathetic tonic vasomotor discharge and a consequent fall in blood pressure. There is a gradual return of tone, and in species such as the cat this recovery becomes evident within a few hours. It appears that the spinal preganglionic fibers develop their own tonic activity by either assuming an intrinsic rhythm of discharge or a rhythm initiated by unknown local processes.

AUTONOMIC REFLEXES[58]

The components of the autonomic system can be reflexly activated. Even in those reflexes normally considered to be somatic there is an autonomic involvement. For example, painful stimuli that cause a flexion or withdrawal of a limb also produce a rise in blood pressure and modulation of cardiac rate. These responses result from a reduction of parasympathetic and an augmentation of sympathetic accelerator fiber tone. In some reflexes, particularly those to the heart, both parasympathetic and sympathetic outflows are involved; a reciprocal action occurs centrally. The reflex responses observed in sympathetic and parasympathetic fibers are not altogether easy to interpret, however, and they will be discussed in greater detail.

Baroreceptor reflex[47]

Afferents from the carotid sinus that traverse the ninth cranial nerve to the medulla are sensitive to blood pressure changes. The receptors involved are probably stretch receptors, but we speak of pressor receptors since firing is initiated with each pulse pressure wave. Also, an elevation of arterial pressure increases the basic level of afferent discharge. It has been stated previously that these afferent signals act on the vagus nuclei to increase vagal tone, which slows the heart. Simultaneously there is a reciprocal inhibition of the vasomotor and cardiac accelerator "centers" in the medulla and thus a reduction of sympathetic vasoconstrictor and cardiac accelerator nerve discharges. It has been shown that sinus afferents make synaptic connections in or near the nucleus of the tractus solitarius and in the RF of the medulla, but the central organization of this reflex has not been fully analyzed.[75, 84] Since this reflex tends to maintain blood pressure within normal limits, we speak of the reflex as a barostatic reflex.

A rise of pressure in the aorta stimulates receptors in the aortic wall, and impulses that travel upward in a depressor nerve tend to bring blood pressure back to normal again by exerting an inhibitory effect on the heart and causing a diminution of peripheral resistance. This depressor reflex, originally discovered by Cyon and Ludwig in 1866, provided the first known example of a negative feedback mechanism. The carotid and depressor reflexes are mutually supportive. The afferents responsible for these reflexes provide the best examples of "autonomic afferents," since they have little effect on the somatic system except for a possible interaction in the medulla between this input and somatic system reflex pathways. Practically all other afferents evoke both autonomic and somatic reflexes, with the possible exception of afferents from muscle stretch receptors. Stimulation of large afferents from the muscle spindle has little effect on sympathetic neuron discharge, at least under normal conditions.

Cardiac accelerator reflex

It has been known for many years that a stretch of atrial muscle produces acceleration of the heart. It was assumed by Bainbridge that an increase in venous return caused reflex cardiac acceleration acting through vagus afferents to inhibit vagal discharge and augment accelerator fiber tone. Recent studies have shown that a reflex of this nature is present, but the stretch of cardiac fibers, the sinoatrial pacemaker in particular, will accelerate the heart in the absence of extrinsic innervation. Stretched isolated nodal tissues accelerate their action, but the mechanism of this is not understood.[15]

Sympathetic reflexes from somatic afferents[58]

Stimulation of the skin, manipulation of the viscera, and any kind of nociceptive and other stimuli resulting from pressure, stretch, temperature changes, etc. applied to various parts of the body will evoke sympathetic reflex actions. Recent studies of sympathetic reflex responses, using action potential recordings directly from sympathetic fibers, have shown that two pathways are involved in the production of sympathetic reflexes.[60, 82] Supraspinal reflexes that are dependent on structures superior to the spinal cord, probably the medulla, are recognizable because of their long latency (since impulses must travel to the medulla and back) and because they are abolished by disconnection of the medulla from the cord. Spinal reflexes are of short latency (Fig. 30-7).

Supraspinal sympathetic reflexes are bilateral and can be evoked from any region of the body. They affect the entire thoracolumbar outflow in a similar manner and have a characteristic "silent period" or "postexcitatory depression" following the initial phase of augmented discharge. It has been found that practically all myelinated fibers, except large group Ia and Ib fibers from

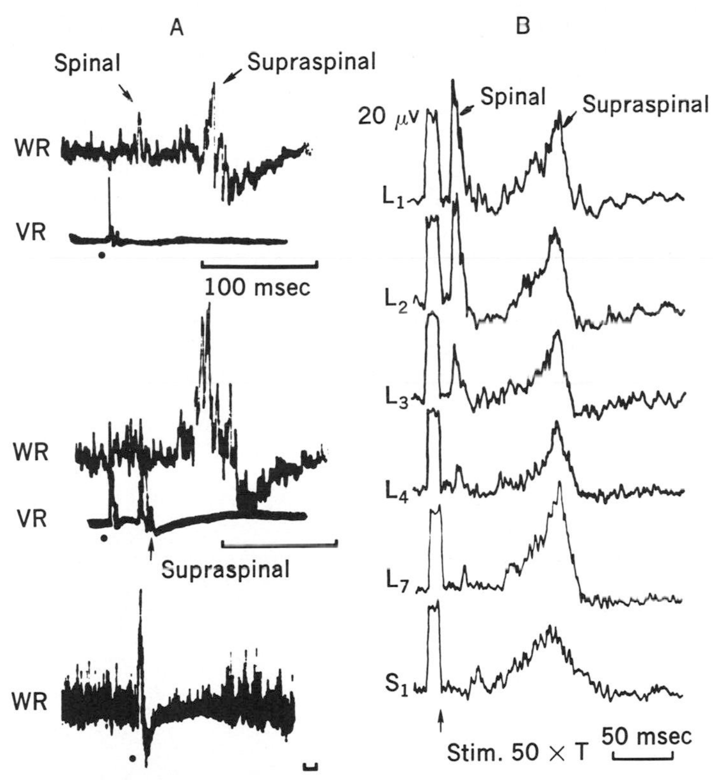

Fig. 30-7. A, Sympathetic reflex responses recorded from lumbar white ramus, *WR,* following single-shock stimulations (marked by dots) of sciatic nerve. Tracings labeled *VR* are somatic reflex responses simultaneously recorded from lumbar seventh ventral root. Sympathetic reflex consists of short-latency spinal and long-latency supraspinal response. Somatic reflex shows mono- and polysynaptic components. In middle tracings, stronger stimulus is shown to evoke greater autonomic reflexes and larger somatic spinal reflex (mono- and polysynaptic reflexes fused in initial response) as well as somatic supraspinal reflex (second large deflection, *VR*). Bottom record was taken at slow-sweep speed from white ramus and shows silent period that follows initial responses. (From Koizumi et al.[60])

B, Sympathetic reflex recorded from L_1 white ramus (L_1WR) but stimuli were applied to spinal nerves at various segmental levels at times indicated by arrow. Each record is average of 10 individual reflexes. Initial rectangular deflection is calibration pulse (20 μv). Early spinal reflex tends to be segmental and is reduced as afferent impulses enter more distant segments, while supraspinal reflex remains constant. (From Sato and Schmidt.[81])

muscle and tendon receptors, can evoke both spinal and supraspinal reflexes. Unmyelinated (group IV, C fiber) afferents, when activated, produce sympathetic reflex discharge with a very long latency.[59, 83] Whether this reflex has a supraspinal or spinal origin is not clearly known.

It has been known for many years that stimulations of somatic afferents at varying intensities and frequencies produce diverse vasomotor changes. It has now been revealed that at low rates of stimulation of myelinated afferents the silent periods tend to fuse in such a fashion that there is a diminution of total discharge during periods of repetitive reflex action. This expresses itself in a depressor response indicating peripheral vasodilatation. More rapid rates of stimulation produce a predominance of active discharge and a pressor response is evoked. Excitation of unmyelinated afferents in addition to myelinated nerve fibers, however, produces a marked pressor response, even at a low rate of stimulation (less than 1/sec). The long-latency reflex discharges caused by activation of the unmyelinated afferents fall during the silent period and reverse its inhibitory influence, which results from the excitation of myelinated afferents.

Spinal sympathetic reflexes are evoked by stimulation of somatic afferents in intact and decerebrate as well as in spinal preparations. This reflex response has a very short latency and initial discharges are not followed by conspicuous silent periods. Reflexes are bilateral but most spinal reflexes are confined to a few segments of the cord, as is the case in spinal somatic reflexes[81] (Fig. 30-7, *B*). These spinal sympathetic reflexes are much more conspicuous in chronic spinal than in normal animals. In conclusion, it should be pointed out (Fig. 30-7, *A*) that stimuli eliciting sympathetic supraspinal and spinal reflex responses also evoke spinal and supraspinal somatic reflexes.

Parasympathetic reflexes

Cranial parasympathetic reflexes do occur and baroreceptor reflexes can be so classified. Excitation of somatic afferents produces changes in cardiac rate along with pressor or depressor responses. Both sympathetic and parasympathetic nerves cooperate in this reflex action. Others are reflex salivation in response to the presence of food in the mouth and the pupillary constriction induced by shining light in the eye. The light reflex certainly involves reciprocal inhibition of sympathetic pupillary dilator fibers. Vagal parasympathetic effects on secretory glands of the stomach and on the motility of the gastrointestinal wall are reflexly evoked by the smell and taste of appetizing food.

The chief reflexes of the sacral parasympathetic outflow are those controlling the bladder and rectum. Basically distention of the bladder and the rectum initiate impulses that reach the sacral portion of the spinal cord; these evoke micturition due to contraction of the bladder wall and relaxation of the urethral sphincter muscle and defecation due to contraction of the distal colon and rectum and the relaxation of anal sphincters. Control of the acts of micturition and defecation are rather complex, involving the somatic system as well as the autonomic system; therefore they are described in some detail.

Micturition

Fig. 30-8 is a diagrammatic representation of the motor innervation of the urinary bladder. The sympathetic innervation of the bladder wall (body, neck, and internal sphincter) originates from the upper lumbar

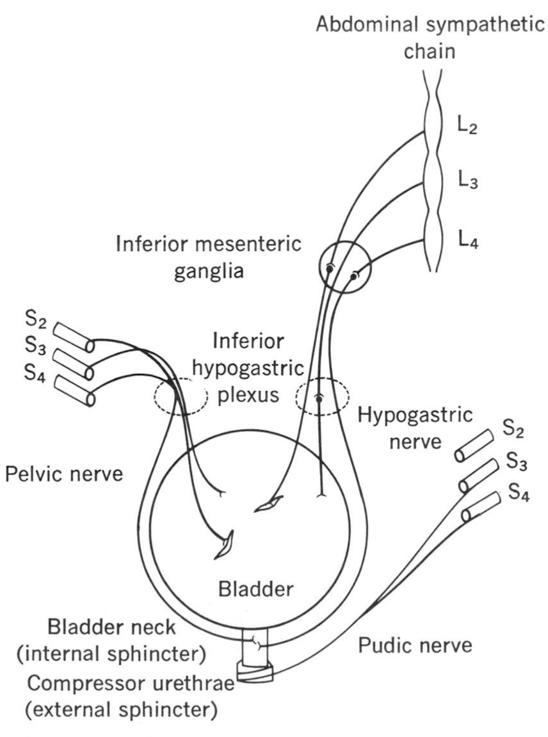

Fig. 30-8. Diagrammatic representation of motor innervation of urinary bladder. Postganglionic parasympathetic neurons are not shown. (After Learmonth.)

segments of the cord. The preganglionic fibers pass down the abdominal sympathetic chains before diverging to form synaptic connections with postganglionic fibers in the inferior mesenteric or inferior hypogastric ganglia. These preganglionic and postganglionic fibers comprise the hypogastric nerves.

The parasympathetic fiber supply comes from the second, third, and fourth sacral segments, and these preganglionic trunks unite on each side to form the pelvic nerves (nervi erigentes). Sympathetic and parasympathetic nerves intermingle in the hypogastric (pelvic) and vesical plexuses. Short postganglionic parasympathetic fibers that innervate the body and neck of the bladder originate in the vesical ganglia.

The powerful external sphincter, or compressor urethrae, is comprised of striated muscle and innervated by the pudendal (pudic) nerves. These sacral somatic nerves originate in the first and second sacral segments of the cord and are under voluntary and reflex control.

The bladder musculature is normally under the influence of low-level tonic discharge from parasympathetic nerves. However, this does not prevent the well-known bladder wall plasticity, which enables this organ to increase greatly in volume without change in intravesical pressure. Normally the bladder maintains an intravesical pressure below 10 cm H_2O. As the volume increases, the bladder wall accommodates with very little increase in pressure until a volume of 400 to 500 ml (in man) is reached. This is considered to be the micturition threshold under normal circumstances. A desire to void may begin to develop above volumes of 150 ml, but micturition normally does not occur much below volumes of 400 to 500 ml. An increase in volume to 700 ml creates pain and often loss of control. Attainment of a threshold volume initiates a sudden rise in pressure and a micturition reflex. As volume increases and the threshold is approached, small rhythmic or periodic fluctuations in pressure tend to appear. It has been shown that the absolute volume of the bladder is not the stimulus for the micturition reflex but the intravesical tension. The cystometrogram in Fig. 30-9 shows an experimental demonstration of filling without immediate increase in pressure and the ultimate initiation of the micturition reflex in man.

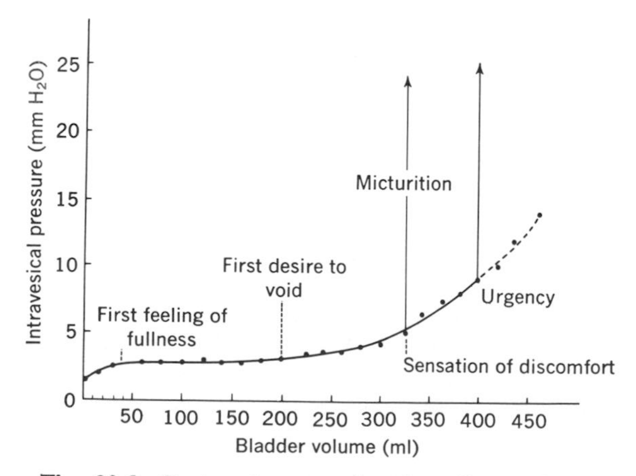

Fig. 30-9. Cystometrogram showing effects of progressive filling of bladder. Note absence of marked change in intravesical pressure until volume of approximately 300 ml is attained. (Modified from Simeone and Lampson.[84a])

It has been suggested that the ability of the bladder to accommodate large volumes of fluid without an increase in pressure is not purely myogenic but is due in part to an inhibitory reflex involving its sympathetic innervation.[36, 37] Section of the hypogastric nerves as well as the lumbar ganglionic chains in experimental animals and sympathectomy in man produces increased frequency of micturition. The bladder becomes hypertonic and contracts at a lower volume.

In emotional states, bladder tone is frequently raised and the micturition threshold lowered; thus greater frequency of voiding occurs at a lower volume. Sex hormones also have a marked effect on micturition threshold and frequency of urination. Following parturition the bladder may reach an enormous volume before voiding occurs. This may be the result of an estrogen-progesterone "deprivation" effect.

The events of the micturition reflex are as follows: When, through distention, intravesical pressure reaches a critical level, stretch receptors in the bladder wall initiate afferent impulses that travel mostly up the pelvic nerves, ultimately reaching the sacral cord. There they evoke reflex contraction of the detrusor muscle and relaxation of the internal sphincter, chiefly through the pelvic nerves. Fluid flowing through the urethra also causes a bladder contraction through parasympathetic nerves, a phenomenon that is important for continuation of micturition. Higher centers, however, are essential to the normal micturition reflex, which involves relaxation of the external sphincter as well as bladder contraction. Facilitatory as well as inhibitory influences on micturition are exerted by structures in the midbrain, pons, and medulla. These "centers" integrate normal association of bladder contraction and internal and external sphincter relaxation. They are also responsible for the very important cooperative action of associated muscles. In voluntary micturition there is normally relaxation of the perineal and levator ani and contraction of abdominal muscles and of the diaphragm.

Sympathetic nerves are not essential to micturition. The force of contraction is largely generated by the parasympathetic supply. Pressures as high as 150 cm H_2O normally develop during evacuation. The powerful abdominal and diaphragmatic muscles that are involved can generate still higher pressure to overcome any resistance to flow. Hypertrophy of bladder musculature can occur in cases of obstruction.

Higher centers have some control over the brainstem-mediated automatic reflex actions. The cortex, which is assumed to be responsible for voluntary directives, can prevent these reflexes from occurring and maintain a closed external sphincter. This voluntary control actually inhibits the parasympathetic reflex and is not exerted solely on the external sphincter. After transection of the cord at levels above the sacral segments, a degree of bladder paralysis results. Low sacral reflexes may eventually be reestablished, but they do not completely empty the bladder. Residual urine remains and bladder infections easily develop. In the "mass reflexes" of chronic spinal animals and man, micturition as well as defecation may occur along with somatic and other autonomic discharges. Destruction of the sacral cord or section of the external innervation to the bladder produces an "automatic bladder." The bladder musculature acquires some intrinsic tone and automatic action. Periodic voiding may occur. More elaborate analyses of the bladder reflex and its central control can be found in various reviews.[7, 65, 79]

Defecation

Fig. 30-10 is a diagram of the innervation of the colon and rectum and shows the pathways involved in the defecation reflex. This external innervation and the intrinsic myenteric plexus of the colon wall are important to the initiation and control of the activity of the colon.

The sympathetic supply delivered through hypogastric nerves and lumbar colonic nerves

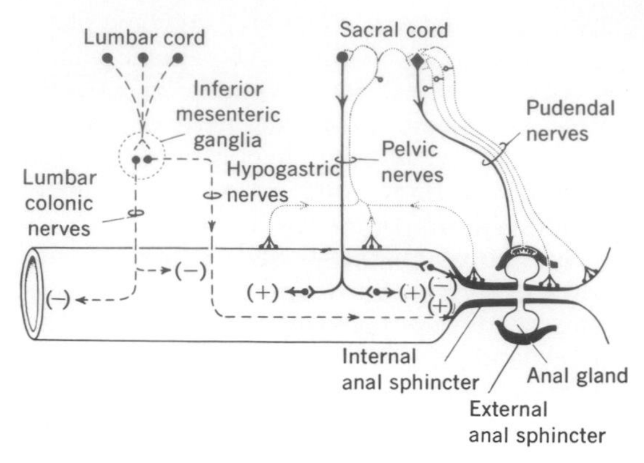

Fig. 30-10. Diagram of innervation of distal colon, rectum, and anal sphincters of cat. (From Garry et al.[43a])

tends to inhibit activity in the colon but causes the internal sphincter to contract. The parasympathetic supply from the sacral cord through the pelvic nerves has an excitatory action on motility of the colon and an inhibitory action on the internal anal sphincter. It was once thought that megacolon occasioned by relaxation and failures of the viscus to empty might be due to overaction of the sympathetic innervation, but it is now thought to be due to degeneration within Auerbach's myenteric plexus.

Activity within the colon is essential to defecation, since peristaltic activity within it forces material into the rectum. The internal and external anal sphincters prevent escape of the accumulated fecal mass. Tonic contraction of the external sphincter is maintained by the pudendal (pudic), which is a somatic nerve; thus this sphincter is under voluntary control.

The sequence of action in defecation is as follows: Distention of the colon initiates, through afferents in the pelvic nerves, a reflex that opens the internal anal sphincter by inhibiting its tone. Simultaneous contraction of the colon distends the rectum and propels the fecal material into the rectum. The desire to defecate is felt when the rectum is considerably distended and feces begin to be propelled through it by mass peristaltic movement. The sympathetic nerves do not contribute to this reflex, but their activity may be inhibited reciprocally as the parasympathetic outflow comes into action. The distention of the rectum and associated stimuli evoke a reflex opening of the anal sphincter and associated contraction of abdominal muscles and diaphragm, much as in the case of the micturition reflex. Although higher centers can inhibit various elements in the defecatory reflex, the act of defecation does not depend on higher centers but is dependent on lower sacral segmental levels. Destruction of the sacral cord does not prevent defecation, and reflexlike contractions of the bowel seem to be initiated by the peripheral neural plexus. However, the sacral cord segments and parasympathetic fibers are essential to a strong and effective defecatory reflex.

Sexual reflexes

The major remaining sacral parasympathetic reflexes are those resulting in erection and vasodilatation in the external genitalia. Under conditions of sexual excitation the discharge of parasympathetic fibers causing erection is unquestionably of cerebral origin. However, erection can occur in the spinal animal and it can be evoked reflexly by genital stimulation.

It appears that the action of parasympathetic erector fibers of the pelvic nerves can be supplemented by sympathetic vasodilator nerves to the penis. The sympathetic fibers are still capable of producing erection in the absence of the parasympathetic component, but of the two supplies the parasympathetic is the more potent in its effect. Complete sympathectomy does not significantly alter the erectile response, although it does prevent emission of semen in the sexual act.

AUTONOMIC CONTROL OF RECEPTORS[58]

Physiologists have long been aware of the fact that reactions occur that affect sensory perception. Reflex adjustments of the pupil that are autonomically mediated affect visual acuity. There are reactions that modify the acuity of hearing by actions on the tensor tympani. Only in recent years, however, have we become fully aware of feedback mechanisms that modify the responsiveness of sensory receptors. The best known example of this is the control of muscle spindles by the fusimotor fiber (gamma efferents) system.

Recent evidence that has begun to accumulate shows that there is an autonomic supply to receptors. These components of the autonomic system appear in some cases to affect sensitivity and response characteristics of the receptors. As in any new field, however, there is a good deal of controversy; evidence for and against various hypotheses cannot be fully presented here, but the best examples of this process will be described.

Chemoreceptors.[78] There is a full description of the carotid body and its role as a chemoreceptor in the control of respiration in Chapter 61. The blood flow through the carotid body and oxygen uptake by cells of the glomus are very high. It is also known that sympathetic fibers innervate the carotid body; these are postganglionic fibers originating in the superior cervical ganglia. When the cervical sympathetic trunk is stimulated, blood vessels in the carotid body and surrounding tissues are constricted and there is a decrease in blood flow to this area. This tends to produce augmentation in chemoreceptor activity as estimated from the rate of afferent discharges in the carotid nerve. This mechanism is not a part of normal respiratory control and section of the cervical sympathetic nerves does not affect respiration nor the rate of chemoreceptor afferent discharge. However, during hemorrhage, asphyxia, or other conditions such as strenuous exercise that greatly augment generalized sympathetic system activity, the increase in chemoreceptor discharges is due in part to sympathetic nerve action on the carotid body. Evidence supporting this conclusion is provided by the observation that in normal animals there is a quick rise in chemoreceptor activity and tidal volume within 1 to 2 sec after exercise is begun. If cervical sympathetic nerves are cut, this rise is much delayed and it takes 15 to 20 sec before a significant change is detectable.

This seeming "sensitization" of the chemoreceptors by sympathetic nerve activity may be due entirely to blood flow changes and a decreased oxygen supply, but the possibility of catecholamine action also cannot be excluded, since norepinephrine and epinephrine applied to the carotid body also cause the rate of discharge to increase.

Mechanoreceptors. The best example of sympathetic influence on receptors has been demonstrated in the mechanoreceptor of the frog skin. Stimulation of sympathetic fibers supplying this receptor in vitro lowers the threshold and slows adaptation to stimuli. Epinephrine directly applied has a similar effect, and these changes have been found to be due to an increase in amplitude and rate of rise of generator potential.[71, 72]

Baroreceptors. Sympathetic fibers innervate the carotid sinus as well as the carotid body. There is some evidence that these innervations may affect the sensitivity and response of baroreceptors. This concept is still controversial, but it does appear that in certain species of mammals, stimulation of the sympathetic fibers to the sinus or reflex augmentation of their activity increases the rate of discharge in baroreceptor afferents.[55, 80] It is an attractive idea that intensified sympathetic activity, causing a rise in blood pressure, might also sensitize those receptors that tend to restore normality.

Muscle spindles.[29, 49] Sympathetic fibers innervate muscle spindles. Repetitive stimulation of these nerves increases the afferent discharge from this receptor. Epinephrine has effects similar to those of sympathetic stimulation. To what degree a vasomotor effect is involved in this change in response has not been settled. Finally, there is some work showing that a sympathetic innervation may influence olfactory and taste receptors.[23]

CENTRAL CONTROL OF AUTONOMIC FUNCTIONS

The autonomic nervous system is comprised of efferent fibers supplying visceral tissues of the body. It parallels the somatic motor outflow to the skeletal musculature. Although there are reactions within the ganglia, the CNS controls the activity of this autonomic complex.

Earlier, when knowledge of the CNS was less extensive than it is now, it was thought that there were autonomic centers or special regions within the CNS that were responsible for the control of autonomic functions. It is true that there are central regions that are more important to the mediation of autonomic reflex action and tonic activity than are others. But these are parts of total complexes that integrate somatic and autonomic reactions to meet behavioral requirements, and it is no longer feasible to think in terms of single centers that are responsible for specific actions.

Autonomic and somatic reactions cooperate to maintain essential body states and to effect necessary adaptations. Emphasis is now placed on the interrelationships of centers and the organization of the nervous system as a whole. There is, however, a hierarchy in the control mechanism, and by using experimental procedures or observing the consequences of injury and spontaneously occurring abnormalities, one can observe the subdivisions of the patterns of control, which parallel those seen in the somatic system. For the purpose of analysis the role played by various divisions of the CNS in the regulation of autonomic system activity will be discussed in the following sections, although separation of somatic and autonomic control systems is arbitrarily made.

Spinal cord

Transection of the spinal cord causes a state of "shock." Spinal shock is characterized by the absence of reflexes, by low blood pressure, and by other evidence of complete or partial paralysis of all motor systems, both somatic and autonomic (Chapter 24). Eventually some degree of activity returns, blood pressure rises to normal, and patterns of somatic and autonomic responses can be evoked. Within a few days after transection of the cord at the sixth cervical level in lower forms, the isolated cord is able to mediate reflex action in response to afferent nerve stimulation. Increments in blood pressure and heart rate, together with the contraction of nictitating membranes and evidence of adrenomedullary secretion (increases in blood sugar level and decreases in clotting time), can be evoked.[13] These changes, however, are not as great as those obtained in anesthetized normal or unanesthetized decerebrate animals when the same stimuli are applied. As stated previously, more refined techniques have shown that spinal reflexes are segmental and tend to be localized unless massive stimuli are used. Generalized responses are very difficult to evoke in the absence of higher centers, particularly the medulla. Tonic discharges recorded from preganglionic sympathetic fibers are greatly diminished but not completely abolished by cord transection. Tonic activity returns to virtually normal levels within a few days in lower forms; this is responsible at least in part for recovery of blood pressure. Destruction of the cord of the spinal animal again reduces the blood pressure to very low levels, an indication that the peripheral resistance is maintained by spinal sympathetic activity.

Deficiency and release phenomena are seen after cord transection. Failure of the bladder to empty completely is due to an ineffective reflex response or a failure in the coordination of the reactions that are normally organized by higher centers. This is a deficiency. The "mass reflex" best seen in spinal man is an example of a release phenomenon.

In man, spinal shock may last as long as 2 months. Autonomic reflexes are even more completely suppressed during shock than are the somatic reactions. During the first month or two the skin is completely dry and sweating is absent, but skin surfaces may be warm and pink because of vasodilatation. Somatic reactions apparently recover first. Flexor reflexes may be elicited within a few days after injury, and by the end of the third or fourth week the withdrawal reflex tends to become more vigorous. The mass reflex that requires several months to develop is elicited by rather weak stimulation or may even occur spontaneously without obvious stimulation. If the plantar surface is scratched, both legs may withdraw violently. The patient may sweat profusely, and both bladder and rectum may contract. Such flexor contractions and their autonomic concomitants are very disturbing and may even interrupt sleep.

A number of explanations have been offered for this hyperreflexia that eventually develops in chronic spinal preparations. Obviously reciprocal action is not present to provide feedback control as in the case of the cardiovascular system. It has also been claimed that new collaterals may sprout from dorsal root fibers to produce new synaptic connections in the absence of trophic influence from higher centers. Finally, these phenomena may be examples of denervation hypersensitivity. It has been reported that in the semitransected cord the ipsilateral reflex has greater responsiveness to afferent impulses, ACh, and other excitant drugs than does the contralateral reflex of the normal side. There is convincing evidence that postganglionic fibers become hypersensitive to transmitters and to drugs when preganglionic neurons are cut. It is reasonable to assume that preganglionic neurons develop a similar sensitivity when they are deprived of their innervation from medullary and other higher centers.

Medulla oblongata

The major reflex actions of the somatic and autonomic system are dependent on the medulla. The function of the medulla in the regulation of heart action first became known about 1845. In 1873 Dittmar demonstrated its role by showing that transection of the brainstem above the midpontine region had little effect on blood pressure. Transection further down in the medulla or below the medulla caused a precipitous fall in arterial pressure. In 1916 Ranson and Billingsley found that both pressor and depressor responses could be elicited by stimulating various areas in the medulla. Subsequent work defined pressor and depressor areas in this structure[6] (see Chapter 42 for detailed discussions). Studies of other reactions involving the autonomic system revealed major dependencies on the medulla such as reflex changes in blood sugar level, reflex excitation of the adrenal medulla, production of salivation, and the initiation of vomiting, swallowing, and micturition.

The baroreceptor reflex illustrates the importance of the medulla to reciprocal control of the cardiovascular system. Respiratory center activity and the respiratory reflexes affect the autonomic system. Control of respiration is mediated largely through the medulla, and the central complex responsible lies in close proximity to the central mechanisms regulating autonomic system functions.

In summary, the medulla oblongata is chiefly responsible for tonic activities of the autonomic system and their normal modulations. It also organizes reciprocal actions between the two divisions of the system and maintains the potential for initiating a massive discharge of the sympathoadrenal system. The bulbospinal preparation has very considerable deficiencies in its responses, and the normality of integrated reactions of somatic and autonomic systems requires the presence of higher centers.

Midbrain

It should be pointed out that the brain functions as a whole and that the subdivisions described are defined by purely arbitrary decision and chiefly on the grounds of anatomic configurations and structural features. The midbrain is a narrow isthmus in which ascending and descending fiber tracts are abundant; it does contain extensions of the RF and discrete nuclear structures. Most of the evidence indicating that it is involved in the regulation of autonomic function is based on stimulation and ablation experiments. It plays an important role in control of the urinary bladder; inhibition and facilitation of micturition can be effected by stimulation of this region. Injury to the midbrain interferes with the normality of micturition. Changes in skin resistance due to sweating or vasomotor reactions, namely the galvanic skin reflex, are also affected by midbrain lesions. Stimulation of the midbrain RF causes modifications of the blood pressure and heart rate with or without obvious modification of somatic responses. The difficulty of estimating the role of this region in regulation of the autonomic nervous system is due to the fact that neurons from the higher centers that regulate the system must traverse this region on their way to the medullary or spinal integrative centers. One of these higher centers that plays a major role in the control of autonomic functions is the hypothalamus.

Hypothalamus

The hypothalamus has been referred to frequently as the center for regulation of the autonomic system. It is certainly of great importance as an integrator of somatic, autonomic, somatoautonomic, and endocrine functions. It confers abilities not possessed by other components of the brain for adjustment of body states. The hypothalamus integrates responses to heat and cold, it mediates those reactions that maintain homeostasis, it plays an essential role in the resistance to stress, and it organizes the visceromotor and other reactions of rage and emotional states. The hypothalamus is the region of the brain that possesses the means to relate humoral and neural control of body processes. Control of the endocrine system is affected by regulation of the secretory activity of the hypophysis. The role of the hypothalamus in integration and control of body functions is discussed in Chapter 31.

Subsequent to removal of those brain structures superior to the hypothalamus, autonomic functions are relatively normal. Refinements of control are lacking, and reactions do not attain normal levels of effectiveness. Temperature regulation, for example, is somewhat deficient; thresholds of body temperature–regulating reactions are high and responses tend to overshoot. Other centers act on the hypothalamus directly or in association with the hypothalamus to influence autonomic system functions.

Limbic system

The limbic system has a close anatomic and functional relationship to the hypothalamus. It is intimately concerned not only with emotional expression but probably also with the genesis of emotions. It does play a role in the control of the autonomic system. Evidence of this has been provided by stimulation as well as ablation experiments.

This system consists of the limbic cortex (orbitofrontal areas, cingulate gyrus, hippocampal gyrus, and pyriform area) and the structures with which it has primary connections, namely the preoptic area, septum, amygdala, and adjacent areas. The hypothalamus is closely connected with this system and is included in it by some investigators.

Stimulation of the amygdala produces a mixture of both parasympathetic and sympathetic reactions as well as hormone secretions (gonadotropins and corticotropins) and respiratory responses. There have been reports of changes in arterial pressure, cardiac rate, gastrointestinal motility and secretion, defecation and micturition, pupillary dilation and constriction, and piloerection evoked by stimulation of this structure. Sexual activity and ragelike reactions have also apparently been evoked through mediation by the hypothalamus. Stimulation of the hippocampus likewise elicits involuntary movements, rage reactions, sexual phenomena, and hyperexcitability. Application of stimuli to the septum affects blood pressure and apparently produces a "sense of reward" that appears to be basic to the well-known drive for "self-stimulation." Such effects, however, are not elicited exclusively from the septum.

Ablation of the amygdala produces placidity and thus is thought by some to be involved in the control of emotions. Injury to the septum likewise reduces susceptibility to fear and anxiety. These structures are said to have the power to evoke the autonomic components of behavioral reactions.

Stimulation of the limbic cortex produces visceral effects, such as vasomotor changes and gastrointestinal activity, as well as somatic and autonomic reactions that typify excitement and rage. Removal causes behavioral changes, a loss of emotional expression, and associative powers.

In summary, it can be said that the limbic system, in addition to playing a role in olfaction and regulation of feeding behavior, is concerned with motivation, the control of sexual behavior, and the expression of rage and fear. It exerts control over the autonomic system, which is superimposed on that organized by the hypothalamus.

Cerebellum[28]

The cerebellum is also involved to some degree in control of the autonomic nervous system. Stimulation of both anterior and posterior lobes produces changes in blood pressure and heart rate. Pupillary constriction and dilation as well as changes in intestinal motility also result from cerebellar stimulation. Chemical stimulation of the posterior lobe by local application of glutamate augments both sympathetic and parasympathetic tonic discharges.[57] Stimulation of the fastigial nuclei as well as the cerebellar cortex produces reactions indicative of augmented sympathetic system action. Injury to or ablation of the cerebellum modifies autonomic reactions. The bladder reflexes are markedly changed and an increase in frequency of micturition is reported, at least in some species. The refinements of cardiovascular regulation may also be impaired.

The significance of this cerebellar autonomic representation is not fully understood, and all that can be said at the present time is that in addition to regulating somatic behavior the cerebellum can also modulate functions within the autonomic nervous system. During motion sickness there is salivation, vomiting, sweating, and other evidence of abnormal autonomic action. It has been shown that removal of the cerebellum abolishes experimentally produced motion sickness (Chapter 28), a result that may also be interpreted to indicate some involvement of the cerebellum in control of autonomic system function.

Cerebral cortex

The cerebral cortex, neocortex to be specific, contributes a refinement of control over autonomic as well as somatic system reactions. The significance of the role of the cerebral cortex varies remarkably within the mammalian kingdom. In man and the primates it has a more major responsibility and its removal abolishes reactions that occur quite adequately in lower species following decortication.

Autonomic reactions are much less affected by cortical lesions in man than are somatic activities. There is evidence, however, of a higher degree of cephalization of control of the autonomic system in man than in lower forms. Certainly the cortex plays a role in elicitation of autonomic involvements in emotional reactions and in the support of voluntarily selected patterns of response. Evidence that the cerebral cortex is involved in the control of autonomic nervous system function has been accumulating for well over 100 years. Much information has been provided by observation of the effects of injury or ablation and by experiments utilizing electrical stimulation techniques.

Ablation experiments have shown that removal of the cerebral cortex produces at least minimal abnormalities in many autonomic reactions. Micturition frequency is changed, and temperature regulation is altered to the extent that compensatory autonomic reactions are delayed in onset and termination. Ablation of the precentral cortex definitely interferes with the normality of blood pressure control, and the occurrence of both hypertension and hypotension has been reported. In the absence of the cerebral cortex there is loss of refined control of autonomic reactions.

A survey of stimulation studies[48] reveals that there are discontinuous loci on the dorsolateral surfaces of both hemispheres that, on stimulation, yield a rise or fall of blood pressure and an increase or decrease in heart rate. Stimulation of other areas can evoke dilation of the pupils, retraction of the nictitating membrane, piloerection, salivation, sweat secretion, and secretory and motor functions of the stomach. For the most part, autonomic localization in the motor and premotor regions corresponds closely with related somatic representations.

A significant role of the cortex in regulation of the autonomic system is that of controlling the distribution of blood to various parts of the body. For example, blood can be shunted from the renal circuit to the limbs by cortical stimulation, and it is also known that the cortex is involved in shunting the blood to and from the skin as an aspect of temperature regulation. Unquestionably the cerebral cortex plays an important role in adjustments of autonomic system function.

CONCLUSION

The various parts of the CNS play specific roles in the control of somatic and autonomic reactions. Students of neurophysiology should seek to identify the exact sites and pathways of control, but they also should bear in mind that every part of the CNS relates to all other parts directly or indirectly. Some regions are primarily responsible for specific functions, but that does not mean that other structures are not involved.

The purpose of any textbook in a field, it seems to us, is to review its significant features and present a conceptual approach. The arguments for and against specific conclusions, the evidence provided by experimentation, and the contents of recent papers cannot be included. It is more important to identify good classic contributions, reviews, and monographs than it is to name those who are working now. However, the reader must be informed of the problems that exist and the apparent current directions of interest. Students need to know how to proceed in case a field becomes one of special interest. There are numerous reviews and monographs that may be consulted.[2-5, 16, 42, 58, 62] The best procedure is to scan the leading journals that publish current work and review articles.

REFERENCES

General reviews

1. Bard, P.: Emotion: I. The neurohumoral basis of emotional reactions. In Murchison, C., editor: Foundations of experimental psychology, Worcester, Mass., 1929, Clark University Press.

1a. Cannon, W. B.: The wisdom of the body, ed. 2, New York, 1939, W. W. Norton & Co., Inc.

2. Eichna, L. W., and McQuarrie, D. G., editors: Central nervous control of circulation, Physiol. Rev. **40**(suppl. 4):entire issue, 1960.

3. Ingram, W. R.: Central autonomic mechanisms. In Field, J., editor: Handbook of physiology, Baltimore, 1960, The Williams & Wilkins Co., vol. 2, section 1.

4. Hillarp, N.: Peripheral autonomic mechanism. In Field, J., editor: Handbook of physiology, Baltimore, 1960, The Williams & Wilkins Co., vol. 2, section 1.

5. Monnier, M.: Functions of the nervous system. General physiology: autonomic functions (neurohumoral regulations), Amsterdam, 1968, Elsevier Publishing Co., vol. 1.

Original papers

6. Alexander, R. S.: Tonic and reflex functions of medullary sympathetic cardiovascular centers, J. Neurophysiol. **9**:205, 1946.

7. Appenzeller, O.: Neurogenic control and disorders of micturition. In The autonomic nervous system, an introduction to basic and clinical concepts, Amsterdam, 1970, Elsevier Publishing Co.

8. Axelrod, J., and Kopin, I. J.: The uptake, storage, release and metabolism of noradrenalin in sympathetic nerves, Progr. Brain Res. **31**: 21, 1969.

9. Axelsson, J.: Catecholamine functions, Ann. Rev. Physiol. **33**:1, 1971.

10. Bealleau, B.: Steric effects in catecholamine interactions with enzymes and receptors, Pharmacol. Rev. **18**:131, 1966.

11. Bennett, M. R., and Burnstock, G.: Electrophysiology of the innervation of intestinal smooth muscle. In Code, C. F., editor, Alimentary canal section: Handbook of physiology, Baltimore, 1968, The Williams & Wilkins Co., vol. 4.

12. Bronk, D. W., Ferguson, L. K., Margaria, R., and Solandt, D. Y.: The activity of the cardiac sympathetic centers, Am. J. Physiol. **117**:237, 1936.

13. Brooks, C. McC.: Reflex activation of the sympathetic system in the spinal cat, Am. J. Physiol. **106**:251, 1933.

14. Brooks, C. McC.: Chemical mediation of the neural control of peripheral organs and the humoral transmission of mediators. In Brooks, C. McC., Gilbert, J. L., Levey, H. A., and Curtis, O. R., editors: Humors, hormones, and neurosecretions, New York, 1962, State University of New York Press.

15. Brooks, C. McC., and Lu, H. H.: Sinoatrial pacemaker of the heart, Springfield, Ill., 1972, Charles C Thomas, Publisher.

16. Burn, J. H.: The autonomic nervous system for students of physiology and pharmacology, ed. 3, Oxford, 1968, Blackwell Scientific Publications, Ltd.

17. Cannon, W. B.: Bodily changes in pain, hunger, fear and rage, ed. 2, New York, 1929, Appleton & Co.

18. Cannon, W. B.: A law of denervation, Am. J. Med. Sci. **198**:737, 1939.

19. Cannon, W. B., and Bacq, Z. M.: Studies on the conditions of activity in endocrine organs. XXVI. A hormone produced by sympathetic action on smooth muscle, Am. J. Physiol. **96**: 392, 1931.

20. Cannon, W. B., and Rosenblueth, A.: Autonomic neuro-effector systems, New York, 1937, The Macmillan Co.

21. Cannon, W. B., and Rosenblueth, A.: The supersensitivity of denervated structures: a law of denervation, New York, 1949, The Macmillan Co.

22. Cannon, W. B., et al.: Some aspects of the physiology of animals surviving complete exclusion of sympathetic impulses, Am. J. Physiol. **89**:84, 1929.

23. Chernetski, K. E.: Sympathetic enhancement of peripheral sensory input in the frog, J. Neurophysiol. **27**:493, 1967.

24. Dale, H. H.: The action of certain esters and ethers of choline, and their relation to muscarine, J. Pharmacol. Exp. Ther. **6**:147, 1914.

25. Dale, H. H.: Chemical transmission of the effects of nerve impulses, Br. Med. J. **1**:834, 1934.

26. Dale, H. H.: Transmission of nervous effects by acetylcholine, Harvey Lect. **32**:229, 1937.

27. Douglas, W. W.: Stimulus-secretion coupling: the concept and clues from chromaffin and other cells, Br. J. Pharmacol. **34**:451, 1968.

28. Dow, R., and Morruzzi, G.: Physiology and pathology of the cerebellum, Minneapolis, 1958, University of Minnesota Press.

29. Eldred, E., Schnitzlein, H., and Buchwald, J.: Response of muscle spindles to stimulation of sympathetic trunk, Exp. Neurol. **2**:13, 1960.

30. Elfvin, L. G.: The ultrastructures of the superior cervical sympathetic ganglion in the cat, J. Ultrastruct. Res. **8**:403, 1963.

31. Emmelin, N.: Nervous control of salivary glands. In Code, C. F., editor, Alimentary canal section: Handbook of physiology, Baltimore, 1967, The Williams & Wilkins Co., vol. 2.

31a. Emmelin, N., and Trendelenburg, U.: Degeneration activity after parasympathetic or sympathetic denervation, Ergeb. Physiol. **66**: 147, 1972.

32. von Euler, U. S.: Noradrenaline: chemistry, physiology, pharmacology and clinical aspects, Springfield, Ill., 1956, Charles C Thomas, Publisher.

33. von Euler, U. S.: Autonomic neuroeffector

transmission. In Field, J., editor: Handbook of physiology, Baltimore, 1959, The Williams & Wilkins Co., vol. 1, section 1.
34. von Euler, U. S.: Adrenergic neuroeffector transmission. In Bourne, G. H., editor: The structure and function of the nervous tissue, New York, 1969, Academic Press, Inc., vol. 2.
35. Evans, J. P.: Observations on the nerves of supply to the bladder and urethra of the cat, with a study of their action potentials, J. Physiol. **86:**396, 1936.
36. Evardsen, P.: Nervous control of urinary bladder in cats. I. The collecting phase, Acta Physiol. Scand. **72:**157, 1968.
37. Evardsen, P.: Nervous control of urinary bladder in cats. II. The expulsion phase, Acta Physiol. Scand. **72:**172, 1968.
38. Fernandez de Molina, A., Kuno, M., and Perl, E. R.: Antidromically evoked responses from sympathetic preganglionic neurons, J. Physiol. **180:**321, 1965.
39. Ferry, C. B.: Cholinergic link hypothesis in adrenergic neuroeffector transmission, Physiol. Rev. **46:**420, 1966.
40. Folkow, B.: Impulse frequency in sympathetic vasomotor fibres correlated to the release and elimination of the transmitter, Acta Physiol. Scand. **25:**49, 1952.
41. Folkow, B.: Nervous control of blood vessels, Physiol. Rev. **35:**629, 1955.
42. Folkow, B., and Neil, E.: Circulation, New York, 1971, Oxford University Press, Inc.
43. Furchgott, R. I.: The receptors for epinephrine and norepinephrine (adrenergic receptors), Pharmacol. Rev. **11:**429, 1959.
43a. Garry, R. C., Bishop, B., Roberts, T. D. M., and Todd, J. K.: Control of external sphincter of the anus in the cat, J. Physiol. **134:**230, 1956.
44. Grillo, M. A.: Electron microscopy of sympathetic tissues, Pharmacol. Rev. **18:**387, 1966.
45. de Groat, W. C., and Ryall, R. W.: Reflexes to sacral parasympathetic neurones concerned with micturition in the cat, J. Physiol. **200:** 87, 1969.
46. Hebb, C.: CNS at the cellular level: identity of transmitter agents, Ann. Rev. Physiol. **32:** 165, 1970.
47. Heymans, C., and Neil, E.: Reflexogenic areas of the cardiovascular system, Boston, 1958, Little, Brown & Co.
48. Hoff, E. C., Kell, J. F., and Carroll, M. N.: Effects of cortical stimulation and lesions on cardiovascular function, Physiol. Rev. **43:**68, 1963.
49. Hunt, C.: Effect of sympathetic stimulation on mammalian muscle spindles, J. Physiol. **151:** 332, 1960.
50. Iggo, A., and Leek, B. F.: An electrophysiological study of single vagal efferent units associated with gastric movements in sheep, J. Physiol. **191:**177, 1967.
51. Iriuchijima, J., and Kumada, M.: Activity of single vagal fibers efferent to the heart, Jap. J. Physiol. **14:**479, 1964.
52. Iversen, L. L.: The uptake and storage of noradrenaline in sympathetic nerves, Cambridge, 1967, Cambridge University Press.
53. Jacobowitz, D.: Catecholamine fluorescence studies of adrenergic neurons and chromaffin cells in sympathetic ganglia, Fed. Proc. **29:** 1929, 1970.
54. Jänig, W., and Schmidt, R.: Single unit responses in the cervical sympathetic trunk upon somatic nerve stimulation, Pfluegers Arch. **314:**199, 1970.
55. Koizumi, K., and Sato, A.: Influence of sympathetic innervation on carotid sinus baroreceptor activity, Am. J. Physiol. **216:**321, 1969.
56. Koizumi, K., and Sato, A.: Reflex activity of single sympathetic fibres to skeletal muscle produced by electrical stimulation of somatic and vasodepressor afferent nerves in the cat, Pflueger's Arch. **332:**283, 1972.
57. Koizumi, K., and Suda, I.: Induced modulation in autonomic efferent neuron activity, Am. J. Physiol. **205:**738, 1963.
58. Koizumi, K., and Brooks, C. McC.: The integration of autonomic system reactions: a discussion of autonomic reflexes, their control and their association with somatic reactions, Ergeb. Physiol. **67:**57, 1972.
59. Koizumi, K., Collins, R., Kaufman, A., and Brooks, C. McC.: Contribution of unmyelinated afferent excitation of sympathetic reflexes, Brain Res. **20:**99, 1970.
60. Koizumi, K., Sato, A., Kaufman, A., and Brooks, C. McC.: Studies of sympathetic neuron discharges modified by central and peripheral stimulation, Brain Res. **11:**212, 1968.
61. Koizumi, K., Seller, H., Kaufman, A., and Brooks, C. McC.: Pattern of sympathetic discharges and their relation to baroreceptor and respiratory activities, Brain Res. **27:**281, 1971.
62. Korner, P. I.: Integrative neural cardiovascular control, Physiol. Rev. **51:**312, 1971.
62a. Kuno, M.: Quantum aspects of central and ganglionic transmission in vertebrates, Physiol. Rev. **51:**647, 1971.
63. Kuntz, A.: The autonomic nervous system, ed. 4, Philadelphia, 1953, Lea & Febiger.
64. Kunze, D. L.: Reflex discharge patterns of cardiac vagal efferent fibers, J. Physiol. **222:**1, 1972.
65. Kuru, M.: Nervous control of micturition, Physiol. Rev. **45:**425, 1965.
66. Langley, J. N.: On the reaction of cells and of nerve endings to certain poisons (nerve endings and receptive substance), J. Physiol. **33:**374, 1905.
67. Langley, J. N.: The autonomic nervous system, Cambridge, 1921, W. Heffer & Sons.
68. Levi-Montalcini, R.: The nerve growth factor: its mode of action on sensory and sympathetic nerve cells, Harvey Lect. **60:**217, 1965.
69. Levi-Montalcini, R., and Angeletti, P. U.: Immunosympathectomy, Pharmacol. Rev. **18:**619, 1966.
70. Libet, B.: Generation of slow inhibitory and excitatory postsynaptic potentials, Fed. Proc. **29:**1945, 1970.
71. Loewenstein, W.: Modulation of cutaneous mechanoreceptors by sympathetic stimulation, J. Physiol. **132:**40, 1956.
72. Loewenstein, W., and Altamirano-Orrego, R.: Enhancement of activity in a pacinian corpuscle by sympathomimetic agents, Nature **178:**1292, 1956.

73. Loewi, A.: Über humorale Übertragbarkeit der Herznervenwirkung. Pfluegers Arch. **189:**239, 1921; **193:**201, 1922.
74. Mitchell, G. A. G.: Anatomy of the autonomic nervous system, Edinburgh, 1953, E. & S. Livingstone.
75. Miura, M., and Reis, D. J.: Termination and secondary projections of carotid sinus nerves in the cat brainstem, Am. J. Physiol. **217:**142, 1969.
76. Nishi, S.: Cholinergic and adrenergic receptors at sympathetic preganglionic nerve terminals, Fed. Proc. **29:**1957, 1970.
77. Pick, J.: The autonomic nervous system, morphological, comparative, clinical and surgical aspects, Philadelphia, 1970, J. B. Lippincott Co.
78. Purves, M. J., and Biscoe, T. J.: Cervical sympathetic activity and the sensitivity of the carotid body chemoreceptors. In Torrance, P. W., editor: Arterial chemoreceptors (proceedings of the Wates Foundation), Oxford, 1968, Blackwell Scientific Publications, Ltd.
79. Ruch, T. C.: Central control of the bladder. In Field, J., editor: Handbook of physiology, Baltimore, 1960, The Williams & Wilkins Co., vol. 2, section 1.
80. Sampson, S. R., and Mills, E.: Effects of sympathetic stimulation on discharge of carotid sinus baroreceptors, Am. J. Physiol. **218:**1650, 1970.
81. Sato, A., and Schmidt, R.: Spinal and supraspinal components of the reflex discharges into lumbar and thoracic white rami, J. Physiol. **212:**839, 1971.
82. Sato, A., Tsushima, N., and Fujimori, B.: Further observation of the reflex potential in the lumbar sympathetic trunk, Jap. J. Physiol. **15:**532, 1965.
83. Schmidt, R. F., and Weller, E.: Reflex activity in the cervical and lumbar sympathetic trunk induced by unmyelinated somatic afferents, Brain Res. **24:**207, 1970.
84. Seller, H., and Illert, M.: The localization of the first synapse in the carotid sinus baroreceptor reflex pathway and its alteration of the afferent input, Pfluegers Arch. **306:**1, 1969.
84a. Simeone, F. A., and Lampson, R. S.: A cystometric study of the function of the urinary bladder, Ann. Surg. **106:**413, 1937.
85. Taxi, J.: Observations on the ultrastructure of the ganglionic neurons and synapses of the frog, Rana esculenta. In Hyden, H., editor: The neuron, Amsterdam, 1967, Elsevier Publishing Co.
86. Uvnäs, B.: Central cardiovascular control. In Field, J., editor: Handbook of physiology, Baltimore, 1960, The Williams & Wilkins Co., vol. 2, section 1.
87. White, J. C., Smithwick, R. H., and Simeone, F. A.: The autonomic nervous system, ed. 3, New York, 1952, The Macmillan Co.

31

CHANDLER McC. BROOKS and KIYOMI KOIZUMI

The hypothalamus and control of integrative processes

It is improper to think of the hypothalamus as an independent entity; neither should it be associated exclusively with the autonomic or endocrine systems because it plays an equally important part in the control of somatic reactions. The hypothalamus is unique in that it confers on animals the ability to maintain essential body states, to react to conditions of stress, and to carry out certain specific components of functions such as reproduction, locomotion, and postural adjustment. It is involved in the process of sleep and awakening, maintenance of body water and weight balance, and regulation of body temperature. The function of virtually no organ system can be adequately discussed without mention of this complex structure identified as the hypothalamus. It is an entity but also a way station that affects those regions of the nervous system lying caudal to it and those that lie above and laterally.

One approach to understanding the importance of this region is to consider the state of the organism in its absence. The spinal cord, medulla, and midbrain are capable merely of mediating reflexes and automatically executed patterns of reactions. In lower forms, integrated behavior and most of the responses characteristic of an intact animal are present so long as the hypothalamus retains its connections with lower structures and remains intact. The cerebral cortex and subcortical regions contribute additional abilities and exert some control over these hypothalamic functions. In man and the primates a higher degree of cephalization has occurred, but the hypothalamus evidently retains many of the same integrative functions it performs in subprimates.

Prior to this century the hypothalamus was known chiefly as an area lying beneath the thalamus. It was not considered to be an independent functional entity, as it is now, and it is quite probable that in order to emphasize its very important role we have tended to isolate it too much in our thinking from other portions of the brain. In 1842 Rokitanski pointed out that lesions involving the base of the brain were commonly associated with gastric hemorrhage and duodenal ulcers. In 1887 a Dr. Story, of Dublin, described a case in which a tumor of the pituitary fossa caused stoutness, irregularity in menstruation, and unusual drowsiness. A year later Bramwell, in a book on *Intracranial Tumours,* reported the association of obesity, presence of sugar in the urine, simple polyuria (diabetes insipidus), and other symptoms now known to be due to hypothalamic injury with the presence of tumors in the pituitary or basal areas of the brain. Mauthner in 1890 associated somnolence with infections involving the base of the brain lying adjacent to the sella turcica. Thus much of our early knowledge of the hypothalamus was obtained by observing the symptoms produced by pituitary tumors or tumors and lesions occurring in this basal region of the brain.

These initial observations have been confirmed both clinically and experimentally. In early attempts to remove hypophyseal tumors or to perform hypophysectomies in efforts to determine the functions of this gland, hypothalamic lesions were inadvertently produced. Between 1901 and 1912, Paulesco in Rumania and Harvey Cushing at the Johns Hopkins Medical School attempted and finally were able to perform hypophysectomies. They observed the effects of pituitary deficiency but also many other symptoms due to injury to the hypothalamus. It was Fröhlich who, in 1901, first described a case of adiposogenital dystrophy. Cushing observed similar symptoms caused by tumors and brain lesions. Aschner (1909) and Erdheim (1904) must be given credit for recognizing that the adiposogenital syndrome is due not to pituitary but to hypothalamic disturbances.

In the course of studies of these phenomena, Camus and Roussy (1920) found that punctures of the brain that left the pituitary intact produced not only the adiposogenital syndrome but also diabetes insipidus. In 1921 Bailey and Bremer were able to publish a monograph on diabetes insipidus that tended to clarify the hypothalamic as well as pituitary role in this disease.

Studies of the role of the hypothalamus in emotional expression and in control of the autonomic system might be said to have begun in 1892 with the work of Goltz on chronic decorticate dogs in which he observed anger or its characteristic phenomena. Woodworth and Sherrington in 1904 described "pseudoaffective reflexes" in animals following high decerebrations. These reflexes involved reactions characteristically seen in rage. In 1909 Karplus and Kreidl, by stimulation of the hypothalamus, produced various visceral and autonomically mediated reactions, some of which gave the picture of rage. Cannon and Britton in 1925 produced the phenomenon of sham rage by removing cerebral cortex and a portion of the thalamus, and in 1928 Bard showed that this release phenomenon was dependent on caudal portions of the diencephalon, the posterior hypothalamus.

One of the greatest aids to the study of hypothalamic function has been the stereotaxic instrument and methodology invented and introduced by Horsley and Clarke in 1908.[40] With this instrument it has been possible to record from, stimulate, and produce lesions in any discrete area or nucleus of the hypothalamus.

In 1940 a symposium was held and a monograph published that very adequately reviewed all anatomic and physiologic studies to date. Much of the work done since then has been an elaboration of findings reported in or suggested by contributors to that symposium. However, progress has been very great, particularly in the field of neuro-

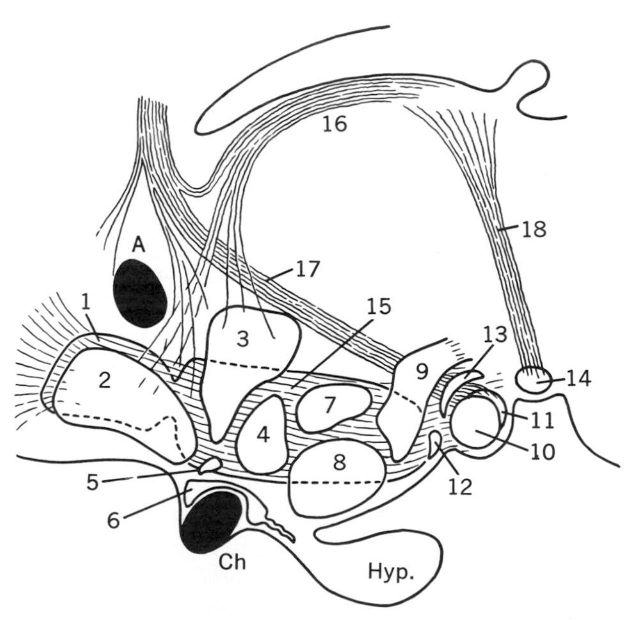

Fig. 31-1. Diagram showing relative positions of hypothalamic nuclei of typical mammalian brain and some interconnecting pathways. *A,* Anterior commissure; *Ch,* optic chiasm; *Hyp,* hypophysis; *1,* lateral preoptic nucleus (permeated by medial forebrain bundle); *2,* medial preoptic nucleus; *3,* paraventricular nucleus; *4,* anterior hypothalamic nucleus; *5,* suprachiasmatic nucleus; *6,* supraoptic nucleus; *7,* dorsomedial hypothalamic nucleus; *8,* ventromedial hypothalamic nucleus; *9,* posterior hypothalamic area; *10,* medial mammillary nucleus; *11,* lateral mammillary nucleus; *12,* premammillary nucleus; *13,* supramammillary nucleus; *14,* interpeduncular nucleus (a mesencephalic element in which habenulopeduncular tract terminates); *15,* lateral hypothalamic nucleus (permeated by medial forebrain bundle). (From Clark et al.[1])

endocrinology, autonomic system physiology, and the study of emotions. This latter area involves the relationship of the hypothalamus to the limbic system. Both the early work and that of more contemporary times is described in recently published monographs.[3-7, 35]

ANATOMIC COMPONENTS OF THE HYPOTHALAMUS

The hypothalamus is a relatively small area of the brain. It is rather uniform in size and structure within the mammalian kingdom. Consequently, in lower forms it occupies a much greater proportion of the total brain mass than it does in man. It lies beneath the thalamic nuclei and extends from the mesencephalon forward to the region of the optic chiasm and the preoptic area.

Subdivisions

Very elaborate anatomic studies have been made of the component structures of the hypothalamus and of their afferent and efferent connections with other portions of the brainstem.[51, 56] For most physiologic purposes it suffices to speak of the anterior, medial, posterior, and lateral regions of the hypothalamus (Fig. 31-1). Only in a few cases can function be ascribed to specific nuclei, and in most instances we merely know that a certain area or a region of the hypothalamus is essential to the performance of certain functions or the production of specific compounds. The anterior region is considered to include the preoptic, supraoptic, suprachiasmatic, and paraventricular nuclei. The medial region, sometimes called the tuberal region, surrounds the third ventricle and is highly cellular; it contains distinct groups such as the ventromedial and dorsomedial nuclei. Ventrally it is comprised of tuberal nuclei and the tuber cinereum, to which is attached the hypophyseal stalk. The lateral hypothalamic region contains cells that are more diffusely arranged and interspersed among fibers of the medial forebrain bundle. This highly complex bundle extends throughout the entire lateral hypothalamus and continues rostrally through the preoptic area to the olfactory regions and caudally to the midbrain. It is composed of many short relays as well as longer fibers, and most of the extrinsic connections to the hypothalamus are made through this medial forebrain bundle. The posterior hypothalamic region includes principally the posterior hypothalamic nuclei and mammillary nuclei.

These hypothalamic nuclei are bilateral. In physiologic studies it appears that one nucleus of the pair suffices to maintain a function. To produce maximum diabetes insipidus, for example, the supraoptic and paraventricular nuclei of both sides must be destroyed. The principal nuclei and subdivisions of the hypothalamus are listed here, and their relative positions are diagrammed in Fig. 31-1.

Anterior hypothalamus
- Lateral preoptic nuclei *(1)*
- Medial preoptic nuclei *(2)*
- Supraoptic nuclei *(6)*
- Paraventricular nuclei *(3)*
- Suprachiasmatic nuclei *(5)*
- Anterior hypothalamic nuclei *(4)*

Medial hypothalamus
- Dorsomedial hypothalamic nuclei *(7)*
- Ventromedial hypothalamic nuclei *(8)*
- Tuberal nuclei—tuber cinereum
- Arcuate nuclei (infundibular nuclei)
- Tuberomammillary nuclei

Lateral hypothalamus
- Lateral hypothalamic nuclear masses *(15)*
- Medial forebrain bundle

Posterior hypothalamus
- Posterior hypothalamic area *(9)*
- Supramammillary nuclei *(13)*
- Premammillary nuclei *(12)*
- Lateral mammillary nuclei *(11)*
- Medial mammillary nuclei *(10)*

There are other nuclear groups not identified in Fig. 31-1. Their functions are relatively unknown or unimportant to this discussion.

Principal connections

The principal connections between the hypothalamus and other regions of the brain are shown diagrammatically in Fig. 31-2. The principal afferents to the hypothalamus arise from the pyriform cortex and hippocampus, two phylogenetically primitive cortical areas. These projections to the hypothalamus are supplemented by secondary connections from corresponding subcortical nuclear masses—the septum in the case of the hippocampus and the amygdala in the case of the pyriform cortex. The septum and amygdala are themselves related to the overlying cortical areas by two-way connections. For each of these afferents to the hypothalamus there are reciprocal connections formed by fibers of hypothalamic origin.

The phylogenetically more recent cortical areas such as the cingulate gyrus affect the

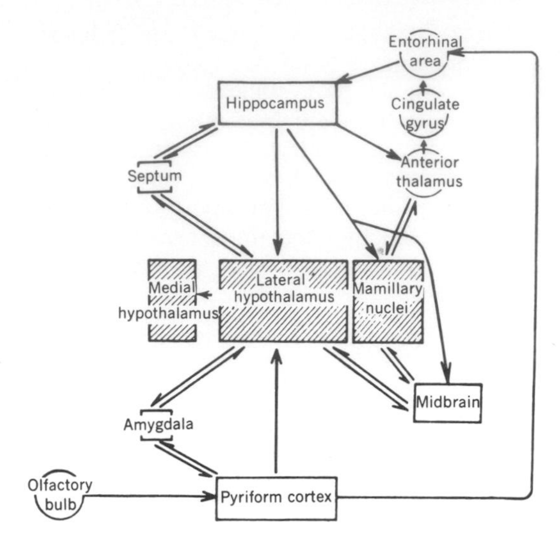

Fig. 31-2. Schematic diagram of principal fiber connections between hypothalamus and other regions of brain. (From Raisman.[56])

hypothalamus through the hippocampus. This region in turn is affected by efferent mammillary fibers relaying through the anterior thalamus. It is not certain that the hypothalamus is linked directly to any other neocortical areas or other thalamic nuclei that relay to the neocortex.

The hypothalamus has a considerable fiber input from the medial parts of the midbrain tegmentum and periaqueductal gray matter. The hypothalamus in turn sends massive descending projections to these areas.

The hypothalamico-hypophyseal tract, which is a major efferent projection of the hypothalamus, originates from cells in the supraoptic and paraventricular nuclei as well as from tuberal and other scattered cells within the medial hypothalamic area. The afferent connections of these cells and their relationships to surrounding areas of the hypothalamus are not known.

Relationships between the medial forebrain bundle and medial hypothalamus are not clearly understood, but the general direction of conduction appears to be from lateral areas to more medial regions.

In conclusion, it may be said that recent work has shown the hypothalamus to be influenced directly or indirectly by almost all parts of the brain, while it in turn can modify virtually all body functions.

Hypothalamico-hypophyseal tract

Ramón y Cajal, an early neurohistologist, described nerve fibers passing from the hypothalamus to the hypophysis in the hypophyseal stalk. This neuroconnection brings the pituitary gland, or at least its posterior lobe, under control of the hypothalamus. According to Ranson and his collaborators,[33] this hypothalamico-hypophyseal tract consists of two parts. The major portion of the tract is comprised of fibers originating in the supraoptic and paraventricular nuclei; it loops over the chiasm and, after coursing medially along the base of the anterior hypothalamus, enters the median eminence, the infundibular stem, and terminates in the posterior lobe. The second component is the tuberohypophyseal tract, which is thought to be derived from the tuberal nuclei of the tuber cinereum. It originates caudolaterally, occupying the caudal or dorsal sector of the stalk. These fibers are all unmyelinated and relatively numerous; there are about 60,000 in the monkey and 100,000 in man. The supraoptic and paraventricular nuclei contain many more cells than there are fibers in the lower part of the stalk, an indication that the axons of some neurons end in the median eminence or at the base of the stalk and do not course to the neural lobe. Additional evidence that all cells of the nuclei do not send fibers into the stalk and hypophysis is provided by the fact that section of the stalk near the posterior lobe causes fewer cells of the nuclei to degenerate than when a higher section involving the median eminence is made. About 90% of the cells degenerate when the entire stalk and median eminence are destroyed.

Most of the fibers comprising the hypophyseal stalk originate in the supraoptic nuclei, but the paraventricular nuclei contribute a significant number. Little is known of the exact origin and role of fibers occupying the dorsal portion of the stalk, those of the tuberohypophyseal tract. They end on capillary loops in the median eminence and proximal part of the stalk and may serve as connectors from the tuberal region to the portal system serving the anterior lobe. There is abundant evidence to show that axons from the supraoptic and paraventricular nuclei control the release of hormones from the neurohypophysis. A few fibers from this bundle do enter the anterior lobe, at least in certain species, but their functional significance, if any, is unknown.

The medial anterior portion of the tuber cinereum is referred to as the median eminence. Portal system capillaries are abundant in this region and nerve terminals containing many granules and vesicles concentrate there. Despite the fact that it has a structural continuity with the tuber, it appears to be part of the neurohypophysis rather than of the hypothalamus.

In the subsequent discussion the following nomenclature will be used:

Adenohypophysis
- Pars infundibularis (pars tuberalis)
- Pars intermedia (intermediate lobe)
- Pars distalis (anterior lobe)

Neurohypophysis
- Infundibulum (median eminence)
- Infundibular stem
- Infundibular process (pars neuralis, neural lobe)

Posterior lobe: Infundibular process (neural lobe) + Pars intermedia

Pituitary stalk: Infundibulum + Pars infundibularis adenohypophysis + Infundibular stem

Blood supply to the hypophysis

A simple diagram of the humoral link between the hypothalamus and hypophysis is shown in Fig. 31-3. The literature on the blood supply to the pituitary gland, especially to the anterior lobe, is fairly extensive. In 1930 Popa and Fielding first described the system of vessels that we now call the hypothalamico-hypophyseal portal system. They found that the vessels coursing along the stalk originated in a capillary bed in the hypothalamus and ended in a capillary bed in the hypophysis. They assumed that blood flowed from the hypophysis into the hypothalamus. In 1936 Houssay, Biasotti, and Sammartino observed that portal flow in the toad was downward toward the anterior lobe from the brain; this type of flow was confirmed in mammals in 1936 by Wislocki and King, who demonstrated conclusively that blood flows from the hypothalamus to the pituitary gland. During the last 30 years it has become clear that the secretory activities of the cells of the anterior lobe are largely controlled by substances formulated in the hypothalamus and transported humorally to the hypophysis via this portal system.[24] The anterior lobe receives virtually all of its blood supply via the hypophyseal portal vessels. The superior hypophyseal arteries are branches of the internal carotid arteries; they supply the upper part of the stalk and form the primary capillary beds that drain into the portal vessels. The capillaries comprising

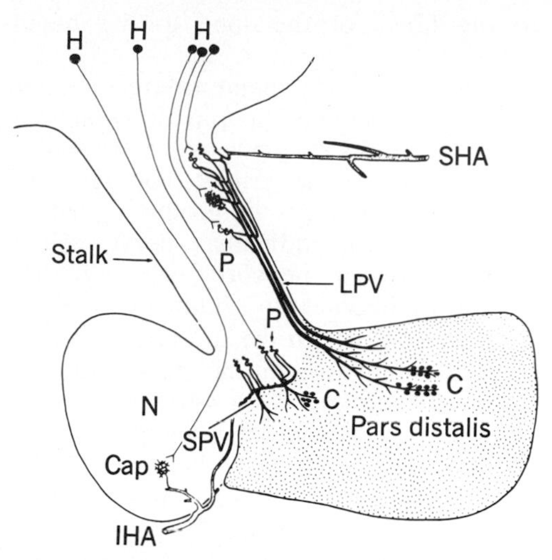

Fig. 31-3. Diagram of midsagittal section of human pituitary gland and stalk showing vessels of portal system. Note complicated coiled capillary loops and nets adjacent to nerve terminals. *H,* Hypothalamic neurons; *N,* posterior lobe; *SHA* and *IHA,* superior and inferior hypothalamic arteries; *P,* upper and lower capillary beds; *LPV* and *SPV,* long and short portal vessels; *C,* representation of secretory epithelial cells adjacent to terminal capillaries; *CAP,* representation of the capillary bed through which hormones enter systemic circulation. (From Adams et al.[8])

these beds are of two types: some are of the simple tubular form seen elsewhere in the nervous system, but others have a most unusual pattern. These capillaries are looped, coiled, and arranged in a highly complicated manner, suggesting that blood flows slowly through them and that they may provide the sites at which neurohumors are transferred from nerve terminals into the bloodstream. The other primary capillary bed is in the lower part of the infundibular stem and derives its blood supply mainly from the inferior hypophyseal arteries. This bed is also comprised of coiled capillaries. The short portal vessels originating from this second capillary bed also carry neurohumors from stalk fibers to certain cells in the anterior lobe. They ensure the survival of a well-defined core of anterior lobe cells, even after pituitary stalk section.

The inferior hypophyseal arteries are the sole source of blood supply to the capillary bed serving the posterior lobe. These capillaries have a simple branching pattern in the young, but in adults the capillaries become markedly convoluted and tend to be arranged in lobular units. This pattern is thought to facilitate the transfer of hormones from the fibers of this lobe to the bloodstream.

Another matter of major interest relative to the hypothalamico-hypophyseal system is the fact that the supraoptic and paraventricular nuclei have the richest capillary bed density of any nucleus or group of cells in the CNS.[32] Each individual nerve cell is surrounded by a network of capillaries; there are approximately 2,600 mm of capillaries/mm^3 of tissue in the supraoptic nucleus and 1,650 mm/mm^3 in the paraventricular nucleus, as compared with 440 mm/mm^3 in the motor cortex. The blood supply to the white and gray matter of other parts of the hypothalamus is not markedly different from that in other areas of the CNS.

FUNCTIONAL ROLE OF THE HYPOTHALAMUS

General nature of function

As stated previously, the hypothalamus integrates many functions and is involved in the flow of reactions essential to the maintenance of homeostasis and to the initiation and control of many behavioral responses. These numerous roles must be described separately. It should be remembered, however, that the reactions do not occur in isolation. For example, reproductive phenomena are not carried out independently of metabolic and water-balance adjustments. The hypothalamus executes its control over body functions through the endocrine, autonomic, hypothalamico-hypophyseal, and somatic neural efferent systems. The functions in which the hypothalamus is primarily involved are the following:

1. Control of the endocrine system through the pituitary gland—neurohypophysis and adenohypophysis (anterior and intermediate lobe)
2. Reproduction
3. Thirst and control of water balance
4. Control of body weight (hunger and satiation)
5. Temperature regulation
6. Reactions to stress
7. Emotional reactions and their control
8. Sleep and arousal
9. Control of somatic reactions

Control of endocrine system—pituitary gland

The hypothalamus is involved in the regulation of endocrine system functions. It has a close relationship with the hypophysis, and this hypothalamico-hypophyseal axis regulates the activity of nearly all endocrine organs. The neural lobe of the hypophysis is, of course, an evagination from the CNS and might be considered a part thereof. Anatomically, this neural lobe is closely related to the adenohypophysis, which originates as an evagination from the buccal mucosa and differentiates to form the intermediate and anterior lobes of the pituitary. Although a direct functional relationship between these parts of the hypophysis has been suggested, it has not been proved. Unquestionably, however, all parts and functions of the hypophysis are under the control of the hypothalamus.

Discussion of the role of the hypothalamus in the control of the body economy through the endocrine system can be subdivided into at least three parts: its role in the control of reproductive functions, water balance, and metabolic processes. The first thing to consider, however, is how the hypothalamus regulates endocrine activities. It acts through neural mechanisms to effect the release of certain hormones from the pituitary gland. Through humoral mechanisms it controls the release of tropic hormones that affect the functions of peripheral endocrine glands. Before proceeding with an analysis of the role of the hypothalamus in specific body functions, the mechanisms whereby the hypothalamus regulates output of hormones from the hypophysis should be considered in some detail.

Control of release of hormones from the neurohypophysis

Antidiuretic hormone or vasopressin. As early as 1924, Abel demonstrated the presence of vasopressin not only in the posterior lobe of the pituitary but also in the tuber cinereum of the hypothalamus. In the 1930s E. Scharrer, Roussy, and Mosinger postulated on the basis of histologic work that the neurons of the supraoptic and paraventricular nuclei have a secretory or endocrine role. It was Bargmann and his associates, however, who in the late 1940s and early 1950s accumulated morphologic evidence that indicated that neurons of the hypothalamus synthesize vasopressin and oxytocin.[13, 14] These compounds are transported down the fibers of the hypophyseal tract and stored in the nerve endings of the neurohypophysis, from which they are secreted into the bloodstream. The idea that there are neurosecretions implies that a neuron can synthesize specific compounds and transport, store, and release them much as can the cells of an endocrine gland.[22] These neurons also possess behavior typical of nerve cells, in that they are electrically and chemically excitable and can conduct action potentials along axons. These action potentials release transmitters or trigger the release of other synthesized materials. Practically all nerve cells are thought to release transmitter agents, but the term "neurosecretion" refers to the processes by which unique materials are produced by specific groups of neurons and eventually released to act on distant target organs, not merely on those with which the neurons make contact.

Bargmann and his associates, using chrome alum hematoxylin (Gomori stain) or other basic dyes, were able to stain secretory materials within the cell bodies and the axons. The concentration of these granules coincides with concentrations of antidiuretic hormone (ADH) in special hypothalamic nuclei and the neurohypophysis. That these neurosecretory granules are accumulated in or released from the supraoptic nucleus has been demonstrated by hydration and dehydration experiments. If the stalk is sectioned, neurosecretory materials that normally would have been transported down the axon accumulate above the site of section, an indication that they are synthesized in the cell body.

Lesion experiments have produced diabetes insipidus of intensities proportional to the degree of posterior lobe and/or supraoptic nucleus destruction. This disease is clearly the result of an absence of production or secretion of ADH from the supraoptic nuclei. Some ADH is also liberated from cells of the paraventricular nuclei, but these nuclei are not totally devoted to its production. Fig. 31-4 shows changes in urine output following stalk section.

In recent years it has been possible to make electrical recordings from cells of the supraoptic nuclei. Since most of these neurosecretory cells send axons to the posterior lobe, electrical stimulation of this lobe evokes antidromic excitation of neurosecretory cells in the hypothalamus. The method has been

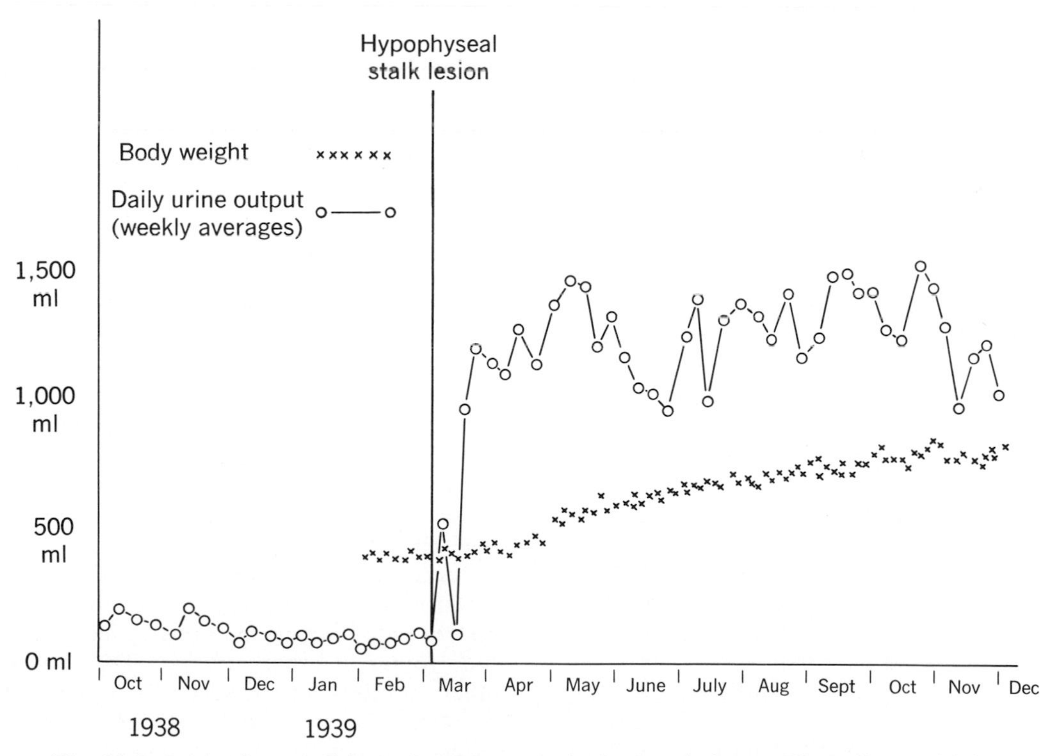

Fig. 31-4. Production of diabetes insipidus and obesity by pituitary stalk section and injury to hypothalamus. Note initial polyuria, thought to be due to transection of "secretory fibers," followed by a brief period of reduced urine output occasioned by release of stored ADH from degenerating cells. Maximum degree of diabetes insipidus developed as cells degenerated and no more ADH was produced. (From Brooks.[17])

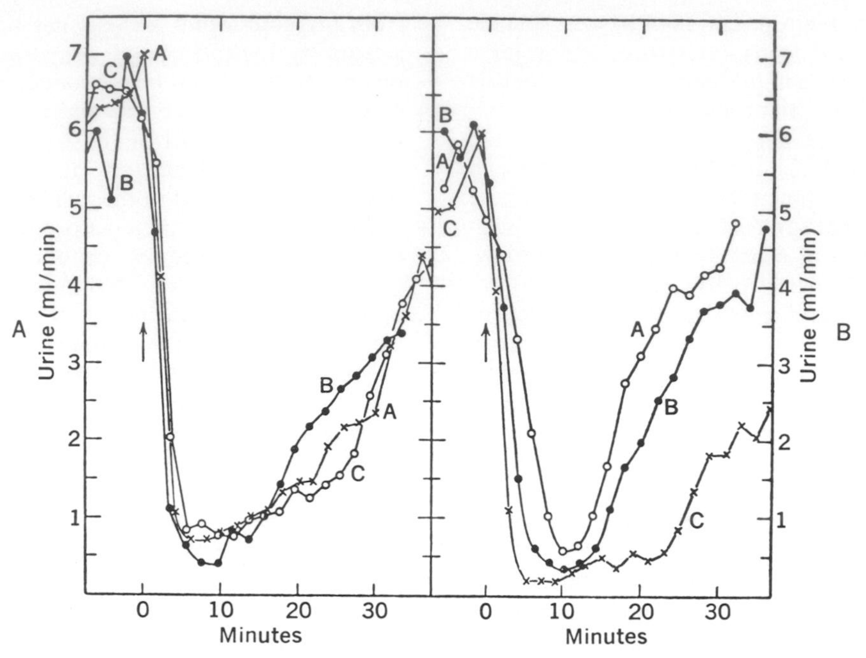

Fig. 31-5. A, Effects on urine output of injections into right carotid artery (↑); *A,* 10 ml 0.428M NaCl in 12 sec; *B,* 10 ml 0.855M dextrose in 11 sec; *C,* 2 ml 1.37M NaCl in 10 sec. Tests done on same conscious dog on different dates. **B,** Assay curves obtained from same dog by injection (↑) into malleolar vein of, *A,* 1 mU; *B,* 2 mU; and *C,* 3 mU of posterior lobe extract. Test dose of 300 to 400 ml of water was given 45 min previously in each case in **A** and **B.** (From Verney[60]; by permission of The Royal Society.)

used for identification of neurosecretory neurons.[44] Many cells of the supraoptic nuclei have a basic rhythm of firing (less than 1 to 5/sec) but this rate varies greatly and at least some cells are periodically silent. Action potentials originating in the soma of these neurosecretory cells are conducted along axons down the stalk to the pituitary gland at a speed of 0.3 to 1.4 m/sec and release hormone at the endings.[28, 41] The mechanism of hormone release from the pituitary gland by nerve impulses has not been clarified. It appears that nerve impulses produce a change in cell membrane permeability and some type of perturbation of hormone-containing granules; this results in exocytosis. Calcium has been found necessary for this process. It is thought that calcium ions are involved in producing an approximation of granules to plasma membranes and a chemical action that releases the stored materials.[25, 26]

Secretion of ADH is influenced by many neural and humoral stimuli that excite or inhibit supraoptic neuron activity. Increase in osmotic pressure of the blood increases ADH secretion. Injections of hypertonic saline solution into a carotid artery produce effective antidiuresis, thus revealing the operation of an osmosensitive system (Fig. 31-5). Such stimuli produce an increase in the firing rate of neurosecretory cells[44, 45] as well as an increase in plasma ADH level as determined by bioassay[27] (Fig. 31-6). Under normal conditions, plasma ADH levels are approximately 1 to 3 μU/ml. Plasma osmolarity is maintained at about 290 mOsm/L, and ADH output is significantly increased when this relationship is changed by as little as 2%. These cells are also responsive to acetylcholine (ACh) and tend to be inhibited by epinephrine. The excitatory and inhibitory stimuli delivered neurally to supraoptic nuclei come from various other parts of the nervous system and not from within the hypothalamus alone.

Intracellular recordings from these neurosecretory cells have shown that they possess resting potentials of the same order of magnitude found in other neurons. They resemble other neurons in electrical behavior.[44] Unquestionably these cells of the supraoptic nuclei are neurons that produce a hormone (ADH) essential to the conservation of body fluids through reabsorption, a process that occurs in the kidney tubules (Chapter 46).

In conclusion, the cells of the supraoptic

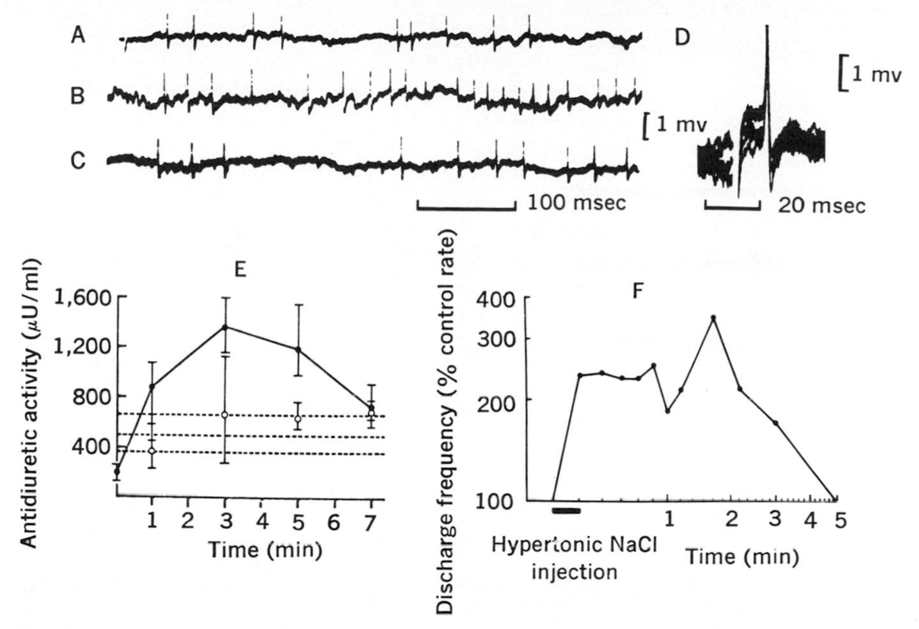

Fig. 31-6. Change in rate of firing of supraoptic nucleus cell in cat and plasma ADH level following osmotic stimulation. **A** to **C**, Records showing rate of firing of neurosecretory cell. This cell identified by antidromic excitation following posterior lobe stimulation as shown in **D**. **A**, Firing rate before and, **B**, rate immediately after injection of hypertonic NaCl into carotid artery; **C**, 5 min later. **F**, Average increase in firing rate in 52 supraoptic nucleus cells of cat following injection of 0.5 ml of 1M NaCl into carotid artery. Range of variations not shown but average values given were statistically valid at a 1% level of probability. **E** shows change in plasma ADH level in rat following intracarotid injection of hypertonic NaCl (filled circles) and isotonic NaCl (open circles) at zero time. Bars indicate 95% confidence limits of estimates. Middle dashed line represents mean concentration; upper and lower dashed lines indicate 95% confidence limits of plasma ADH before NaCl injection. (**E** from Dyball[27]; **F** from Koizumi et al.[45])

nuclei and some in the paraventricular nuclei are sensitive to changes in the surrounding environment, as in long-term hydration and dehydration. They respond to osmotic pressure changes as though they were osmoreceptors or were affected by osmoreceptors located in the same area. Their activity is modified by stimuli arriving from other parts of the brain or distant parts of the body. There is a unanimity of opinion that changes in blood volume and blood pressure modify ADH levels in the blood. Decreases in blood pressure, even those produced by rising to a standing position, result in an ADH increase (1.9 to 3.0 μU/ml), while increases in blood volume reduce ADH and produce a diuresis. Vagus afferents appear to be involved. Distention of the left atrium by artificial means reduces ADH output, probably through action of these afferents. ADH output increases during exercise and emotional states and as a result of hemorrhage. Despite general agreement on this information, quantitative figures found in the literature vary remarkably. In some accounts, changes of only a few microunits per milliliter of blood are reported, while in others, basic levels are stated to be much higher and changes observed are in hundreds of microunits per milliliter. Most measurements of ADH levels are done by bioassay; this and differences in procedures may account for these quantitative discrepancies.

Oxytocin. In 1909 Dale reported that posterior pituitary extracts cause the uterus to contract. This was called oxytocic action, and the existence of an oxytocic hormone was suggested. In 1910 Ott and Scott showed that pituitary extracts increased the output of milk and played a role in the milk "letdown" reaction, or milk ejection response. Separation of the oxytocic from the vasopressor principle was first accomplished by Kamm and co-workers in 1928. Du Vigneaud and his colleagues in 1953 separated the two compounds (oxytocin and ADH), deter-

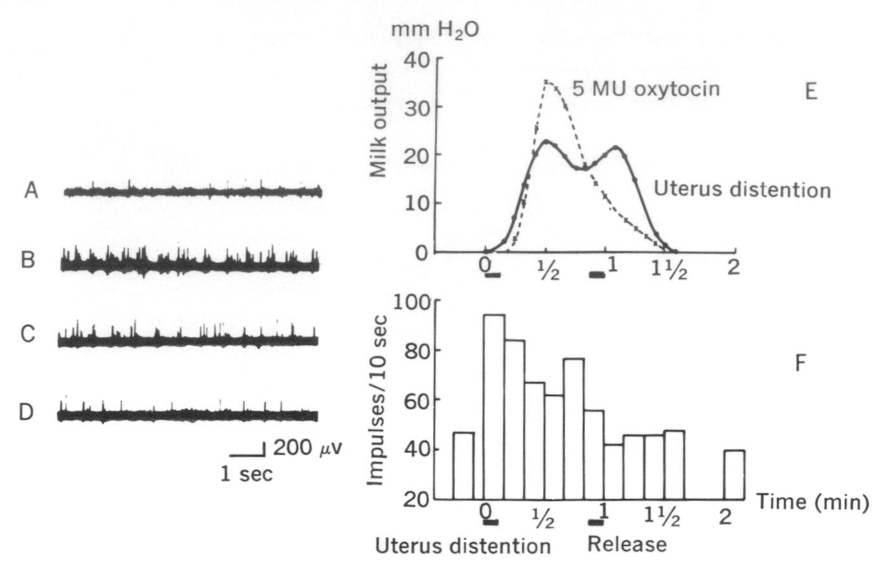

Fig. 31-7. Increase in rate of firing of paraventricular nuclear cells and milk ejection response produced by distention of uterus in postpartum cats. Time between bars in **E** and **F** indicates interval during which balloon inside uterine cavity was inflated by introduction of 7 ml of water. At left, unit discharges of paraventricular nucleus neuron are shown. **A,** Control. **B** and **C,** During distention of uterus. **D,** After distention of uterus. **F,** Rates of this same neuron's discharge in 10 sec intervals before, during, and after distention. **E,** Changes in mammary duct pressures resulting from oxytocin release caused by same stimulus and by injection of 5 MU oxytocin into radial vein during similar interval. (From Brooks et al.[21])

mined their exact chemical structures, and synthesized them. They were found to be octopeptides (Chapter 67).[61]

Oxytocin is primarily produced by cells in the paraventricular nuclei and to a lesser degree by cells in the supraoptic nuclei. Presumably this hormone, like vasopressin, is transported down axons of the pituitary stalk, stored, and released from the neural lobe. It has been shown that stimulation of the hypophyseal tract liberates oxytocin as determined by milk ejection assay.[23] Injury to the stalk or paraventricular nuclei interferes with normal lactation and the milk ejection response.

Oxytocin normally is liberated by a reflex action following stimulation of the female genital tract, distention of the uterus and vagina being quite effective. However, the most remarkable oxytocin release response occurs during suckling. The afferent pathway of this reflex has not been clearly identified. It is known, however, that stimulation of the nipples produces excitation of afferent nerves, and impulses travel through the spinal cord to reach the hypothalamus.

Recent combinations of electrical recording from single units in the paraventricular nuclei with measurements of oxytocin release during physiologic stimulation have illustrated very clearly the quantitative correspondence between paraventricular cell activity and oxytocin output (Fig. 31-7). It has been found that similar stimuli produce an increase in the electrical activity recorded from nerve fibers in the pituitary stalk.[41] Apparently excitation of the hypothalamic neurons generates impulses that course along the hypophyseal tract to the pituitary gland to release stored oxytocin.

Activity of neurons in the paraventricular nuclei, like that of neurons in the supraoptic nuclei, is much affected by excitatory and inhibitory influences from other regions of the nervous system. It is well known that psychic stimuli greatly modify milk ejection. Emotional disturbances affect lactation in women as well as the milk ejection reflexes of many species of animals. There are certainly very strong central inhibitory mechanisms as well as excitatory drives. The osmosensitivity of these neurons appears to be similar to that of supraoptic nuclear cells. Close arterial injection of hypertonic sodium chloride solution or of small amounts of ACh increases activity of these cells and evokes an accompanying milk ejection response.[21]

Oxytocin has a strong effect on uterine motility and its secretion may contribute to normal labor and parturition, but it probably is not indispensable thereto. Sensitivity of the uterus to oxytocin is enhanced by estrogen and inhibited to some degree by progesterone. Late in pregnancy the uterus becomes very sensitive to oxytocin. The administration of physiologic doses of oxytocin is effective in producing labor in women and in animals. Dilation of the cervix and the descent of the fetus down the birth canal probably initiates, through the paraventricular nuclei, a reflex secretion of oxytocin. Certainly it has been shown that distention of the uterus and cervix increases oxytocin output and that oxytocin enhances labor and is increased in concentration in the bloodstream during labor.

It was once thought that oxytocin might act to facilitate the passage of sperm up the female genital tract to the fallopian tubes, but the evidence for this is not convincing. There is a release of oxytocin following coitus and in some species also during uterine contraction. Secretion of oxytocin, like that of ADH, is strongly inhibited by alcohol.

Control of anterior lobe hormone release

Many phenomena indicate a neural control over the release of various hormones from the anterior pituitary. Ovulation associated with coitus, the effect of light on the development of ovaries, reversal of seasonal breeding habits when animals are transferred from one hemisphere to another, and the phenomenon of induced pseudopregnancy all indicate a neural influence over the secretion of gonadotropins. Experimental exposure of animals to cold has given evidence that the hypothalamus is involved in an increased thyroid activity. In the late 1930s hypothalamic stimulation was found to increase the discharge of various anterior pituitary hormones such as luteinizing hormone (LH), adrenocorticotropin (ACTH), and thyrotropin (TSH). Although hypothalamic lesions cause a decrease in the production of certain anterior lobe hormones and a block of their release by usually effective stimuli, they are followed by a increase in the secretion of certain other hormones, such as prolactin.

Initially a search was made for a neural connection between the hypothalamus and anterior lobe that might mediate these actions, and it was shown that transection of the pituitary stalk did abolish some of them. However, no very convincing anatomic evidence was found to support the idea that there is a functionally important innervation of the anterior lobe from the hypothalamus. Attention then turned to the hypothalamico-hypophyseal portal system as a possible pathway for this relationship. It was soon shown that this portal supply is essential to the maintenance of normal anterior lobe function. It was then not much of a step to conclude that special neurosecretions formulated in the hypothalamus are transported humorally to the anterior lobe, where they act to release hormones produced and stored there. These compounds produced in the hypothalamus are known as releasing factors.

There are now at least nine releasing factors, some inhibitory, that are recognized as products of the hypothalamus.[58, 62] They are small polypeptides that are relatively species nonspecific. Unquestionably, active compounds with very special releasing actions have been extracted from the hypothalamus. Ideally, evidence in support of present concepts of the formation and transport of releasing factors requires the demonstration of their presence in portal vessel blood in higher concentration than in the systemic circulation. There must also be a correlation between the concentration in the portal blood and the release of tropic hormone from the anterior lobe. At present only a few releasing factors have been assayed in portal blood.[31, 55, 55a] It is generally accepted, however, that all are transported to the pituitary gland by this portal system.

Presumably, releasing factors are produced by certain neurons of the hypothalamus. These neurons lie in or near the median eminence, or the releasing factors are synthesized elsewhere and carried to the median eminence by axonal transport, since they are found there in highest concentrations. However, no specific cell groups have yet been identified as producers of these particular compounds. How and where they are formulated within the neurons is not known. The median eminence where they appear to be held and from which they are released contains relatively few cells but does contain many nerve fiber endings that lie in close proximity to the capillary loops from which the portal vessels originate. The releasing factors are probably secreted by these endings. It is assumed that these specific compounds accumulate in separate zones within the median eminence, but evidence for this is not as convincing as one might wish.[43] Fig. 31-8 shows the various releasing factors and

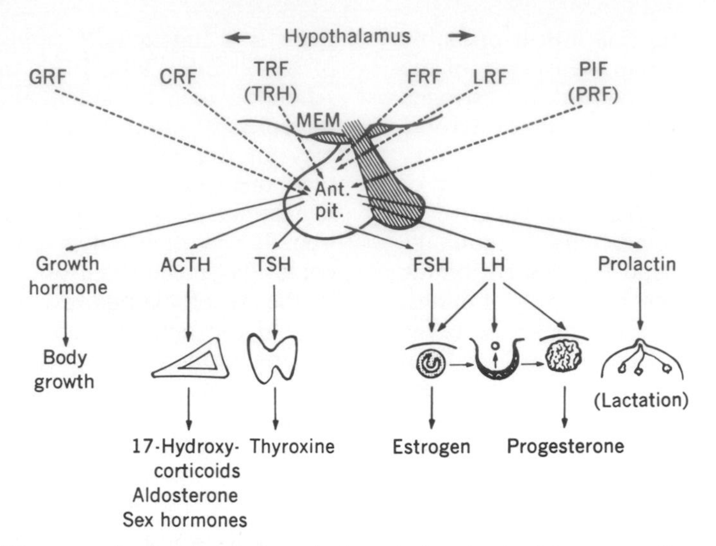

Fig. 31-8. Diagram of releasing factors that control output of hormones from anterior lobe of pituitary gland and ultimate hormone release from peripheral glands. *GRF,* Growth hormone-releasing factor; *CRF,* corticotropin-releasing factor; *TRF (TRH),* thyrotropin-releasing factor or hormone; *FRF,* follicle-stimulating hormone-releasing factor; *LRF,* luteinizing hormone–releasing factor; *PIF (PRF),* prolactin inhibitory or releasing factor; and *MEM,* median eminence. Note that combination of FSH and LH produce maturation of follicle, ovulation, and corpus luteum formation. (Modified from Ganong.[2])

the hormones they cause to be secreted from the anterior lobe.

Corticotropin-releasing factor (CRF or CRH). The first of these compounds to be actively investigated in detail was CRF. In 1955 Guillemin and his collaborators obtained an extract that caused release of ACTH from the pituitary gland. This substance has now been extracted from the pituitary stalk and median eminence. It stimulates not only the release of ACTH but also the synthesis of this tropic hormone in the anterior lobe.

At present there appear to be two variants, α-CRF, which is closely related to α-melanocyte–stimulating hormone (α-MSH), and β-CRF, which is closely related but not identical to vasopressin (ADH). CRF is liberated from the hypothalamus under conditions of stress or by noxious stimuli. There is a diurnal rhythm in its output, which is possibly responsible for the diurnal variation in plasma corticosteroid levels.

Stimulation of the amygdala, septal area, reticular formation, and hypothalamus is known to cause ACTH release, probably by augmenting the output of CRF. The adrenal cortical control system, however, retains its function in the presence of extensive damage to the forebrain and hypothalamus. High levels of corticosteroids in plasma block ACTH release from the anterior lobe by an inhibitory feedback mechanism acting directly on that gland. It is not definitely known, however, whether ACTH or corticosteroids also act on the hypothalamus itself to effect CRF release, although this has been suggested frequently.[58, 62]

Thyrotropin-releasing factor (TRF or TRH). In 1960 Schreiber and associates found that hypothalamic extracts administered to anterior pituitary grafts in the eyes of hypophysectomized rats prevented the involution of the thyroid glands that otherwise occurred. Guillemin and his co-workers in 1962 obtained conclusive evidence that a hypothalamic extract could cause increased liberation of a thyroid-stimulating hormone (TSH) from the anterior lobe. These investigators obtained highly active extracts of sheep hypothalamus that caused a definite release of TSH in the normal animal. This finding was soon confirmed by others, and it is now accepted that there is a thyrotropin- (TSH) releasing factor (TRF) produced in the hypothalamus of the mammal.

By 1966 TRF was isolated and found to contain histidine, glutamic acid, and proline. A compound that was synthetized in 1969 and 1970 had an action identical with that of TRF.[58, 62] This polypeptide, like other releasing factors, has little or no species specificity. It has been stated that the hypothalamic region in which TRF is formulated is anterior to the region that produces CRF. It is liberated under various conditions of stress but exposure to cold is a specific stimulus for TRF release and production. Thyroxine increase is known to inhibit TSH release at the pituitary level, but whether TRF production and release are affected at the hypothalamic level has not as yet been clarified.

Growth hormone–releasing or somatotropin-releasing factor (GRF or SRF). The earliest evidence that the hypothalamus might contain compounds affecting growth hormone release from the anterior pituitary was obtained in 1964 and 1965. It was found that intracarotid injection of an extract of the hypothalamus depleted growth hormone content of the pituitary gland and that such an extract evoked increased secretion of growth hormone from incubated pituitary tissue. Thus evidence was obtained regarding the existence of a factor that could cause release and depletion of growth hormone from the pituitary.

Highly purified GRF has been obtained from the hypothalamic tissues of pigs and sheep. It is most concentrated in the median eminence, as are other releasing factors. It is also a polypeptide containing a high proportion of glutamic acid and alanine. The concentration of growth hormone in plasma is at present measured by radioimmunoassay methods, a technique that has greatly facilitated the study of factors controlling GRF and growth hormone release.

It is interesting that growth hormone, presumably under the drive of GRF, is liberated in remarkably greater quantity during "slow-wave sleep" than in the waking state. The release of the hormone is also related to the concentration of metabolites such as glucose and arginine in the blood. Insulin-induced hypoglycemia and arginine infusions are standard and reliable clinical tests for the growth hormone–releasing function of the brain and the anterior pituitary. GRF release is also influenced by thyroid hormone, glucocorticoid, and estrogen levels in the blood. Catecholamines affect GRF output; norepinephrine or dopamine injected into the lateral ventricles release the hormone. The importance of such actions to normal growth hormone release has not yet been determined. Finally, production of a growth hormone inhibitory factor in the hypothalamus has been suggested. There thus may be a dual control of growth hormone release.[58]

Prolactin inhibitory factor (PIF). A prolactin-releasing factor (PRF) has been found in all birds, and it has recently been demonstrated that such a compound is also produced by the mammalian hypothalamus.[52] In mammals, however, there is strong evidence that the CNS is chiefly involved in the restraint of the secretion of prolactin from the anterior pituitary gland. In 1965 an extract of the mammalian hypothalamus that inhibited prolactin release was obtained. This compound (PIF) is found in high concentration in the median eminence. The isolated anterior lobe secretes prolactin spontaneously, and this secretion is inhibited by PIF. Prolactin release is increased by estrogens and is altered during the normal estrous cycles; this may be due to the inhibition of PIF by estrogens. High prolactin blood levels inhibit further release of that hormone, but whether or not an increase in PIF action or release is involved is unknown. Catecholamines can inhibit and TRF can release prolactin, but there is a PRF. The chemical nature of PIF and PRF remain unknown, but they are probably small polypeptides.[58]

Gonadotropin-releasing factors: follicle-stimulating hormone–releasing factor (FRF) and luteinizing hormone–releasing factor (LRF). In 1964, at approximately the same time that other releasing factors were being identified, crude extracts of the hypothalamus were found that depleted the anterior pituitary of follicle-stimulating hormone (FSH) and initiated ovarian development and estrogen release. When added to pituitary tissue cultures, these extracts increased FSH concentration in the medium, a phenomenon that raised the possibility of a releasing factor produced in the hypothalamus that increases FSH output from the anterior pituitary. It has been called FSH-RF or just FRF.

In 1964 and 1965 minute quantities of another active material were also obtained from the hypothalamus; this substance caused the release of LH from the pituitary glands of animals in estrus. Stepwise purifications of these crude LH-RF or LRF extracts were made and the compound has now been obtained in pure form, isolated, and identified.[29, 58] A recent review claims that the LH- and FSH-releasing factors are parts of one molecule, and certainly their actions are usually supplementary.[58]

FRF is thought to initiate cyclic changes in the ovaries and production of estrogens therefrom by causing a release of FSH from the anterior lobe. Subsequent discharge of LRF liberates LH, which produces ovulation and development of the corpus luteum. Ovulation requires a combination of FSH and LH and thus a combined FRF and LRF action. A cyclic change in the LRF content of the median eminence has been demonstrated; it is highest during proestrus and diestrus and low throughout estrus. LRF concentrations are particularly high in the median eminence at the time of puberty. It is common practice to refer to these synergistic releasing factors, FRF and LRF, as gonadotropin-releasing factors (GTHRF).

Ovarian steroids (estrogens and progesterone) are known to have multiple negative feedback effects that block FSH and LH release. This blocking action on GTHRF release occurs in the basal medial region of the hypothalamus as well as in the pituitary. Recent reviews also mention a stimulatory effect of steroids on FSH and LH release mediated through the preoptic area as well as through the pituitary.[62] FSH and LH can also inhibit their own release through inhibitory feedback effects on the brain. These effects, however, seem not to be an essential feature of their normal functions (Chapter 71).

In concluding the discussion of hypothalamic control of the hypophysis, mention should be made of the release of intermedin or melanocyte-stimulating hormone (MSH) from the pars intermedia.

Visual stimuli cause release of MSH in fish and amphibia, but its importance in mammals is unknown. MSH-releasing factors (MRF) and MSH inhibitory factors (MIF) have been obtained from the hypothalamic structures of amphibia. The pars intermedia receives nerve fibers through the hypophyseal stalk. Their endings appear to contain neurosecretory granules. It is thought that releasing as well as inhibitory factors (MRF and MIF) are transported through the nerve fibers that connect the hypothalamus with the pars intermedia rather than through portal vessels. These hypothalamic hormones have been identified as small peptides.[58]

In considering the control of chromatophores and melanophores, the existence of melatonin, a melanophore-contracting principle found in high concentration in the pineal gland and the basal brain structures, should not be ignored. It has been suggested that melatonin acts on MIF secretion (Chapter 68). In lower vertebrates, blanching as well as melanocyte expansions are seen in emotional and visually induced responses. Very complex reciprocal reactions must occur, involving the hypothalamus as well as other structures of the nervous system, in forming the complex patterns of skin coloration that are visually controlled.[30]

Reproduction

Sexual behavior and reproduction involve the hypothalamus and the hypophysis. Stimuli that initiate seasonal development of the gonads and the resulting sexual expressions depend on the hypothalamico-hypophyseal complex, as do the automatic "lunar" cycles of the primates and the estrous cycles of other mammalian species.

Estrogens and androgens affect the behavior of animals by action on the hypothalamus. Implants of estrogens in the mammillary region produce the full complex of natural estrus. This estrogen-sensitive region extends anteriorly into the preoptic area. Radioactive-labeled estrogens accumulate in highest concentrations in the anterior hypothalamus, but they also accumulate in the amygdaloid complex. Even electrical recordings from the hypothalamus show changes in cellular activity on administration of estrogens, although these changes are not confined to any specific circumscribed area. The estrogens and presumably the androgens affect structures of the nervous system other than the hypothalamus. Estrogens cause some signs of estrus even in decerebrate animals, but the hypothalamus is essential to the full expression of estrous behavior.

In its simplest description, sexual activity and the reproductive cycle in seasonal breeders begins as follows. With a change in season there is a gradual change in the intensity and duration of daylight. This initiates, through the hypothalamus, a discharge of GTHRF, which in turn causes a release of gonadotropins. The gonads, as they develop, release ovarian hormones that heighten secondary sexual characteristics and evoke, again by action on the hypothalamus and other parts of the brain, the sexual display and mating behavior of the species. Many other factors are known to be involved, and species differ markedly in their responsiveness to one or another environmental condition. The precise occurrences that initiate the cycle are not fully known, but the hypothalamus is certainly essential.

In the case of periodic reproductive and sex cycles, the trigger of the initial GTHRF output and the consequent gonadotropin release that produces ovarian development of the menstrual and estrous cycles is also not known. The most that can be said is that an intrinsic rhythm or "clock" has been established in these species by some fundamental natural influence. The medial preoptic area is considered to be responsible for the cyclic nature of the female reproductive system. It evidently acts on the anterior part of the medial-basal hypothalamic region, which is capable of sustaining an acyclic, continuous gonadotropin release that would otherwise produce a continuous estrus. The inhibitory action of estrogens on GTHRF release is thought to occur in this latter region.[62] Lesions in the hypothalamus that destroy these structures essential to gonadotropin release cause genital dystrophy and the abolition of cycles and sexual behavior.

Ovulation and impregnation constitute the next major events of the reproductive cycle. In some species, ovulation is evoked reflexly by special stimuli and/or an emotional reaction. In other species it occurs spontaneously. Electrical stimuli applied to the cervix or the hypothalamus can, at proper times in the cycle, provoke ovulation and persistence of the corpus luteum (pseudopregnancy). It appears that release of the proper admixture of FSH and LH triggers escape of the ovum from the follicle and the formation of a corpus luteum. Since the hypothalamus is involved, release of FRF and LRF must occur.

The pituitary hormones and the hypothalamic "releasing factors" are essential to the nurture and growth of the fetus, but the next dynamic action of the hypothalamus relates to the initiation of parturition and the control of lactation. Evidence of the involvement of oxytocin release from paraventricular nuclei cells during parturition has already been cited. The effects of PIF prevent a continuous liberation of prolactin from the anterior hypophysis. Suckling and the emotional stimuli evoked by nursing are thought to block PIF output; prolactin is released and milk secretion is augmented. The anterior lobe is found to be depleted of prolactin by nursing.

Finally, the suckling stimulus causes a reflex release of oxytocin, which has a "letdown reaction," facilitating the ejection of milk. Simultaneous recording of unit activity within the paraventricular nuclei and measurement of mammary gland ejection pressures during suckling have provided confirmatory evidence of the existence of this hypothalamic oxytocin-releasing reaction.[21]

Thirst and control of water balance

Unquestionably the hypothalamus is of importance to the perception of thirst and the control of water intake. It is also involved in the regulation of body water loss, particularly via the kidney. Lesions in the hypothalamus have long been known to produce the polydipsia and polyuria of diabetes insipidus.

Analysis of the control of water intake presents many complexities. The external and internal sensors undoubtedly contribute to the sensation of thirst; various theories of thirst have been adequately described elsewhere.[34, 59] Dryness of the mouth produced by mouth breathing, ingestion of dry foods, and removal of the salivary glands all create thirst. Dehydration produces the desire to ingest liquids; a loss of body water equivalent to 0.5% body weight in the dog and 0.8% body weight in man appears to be the threshold for a strong water-intake drive. A hyperosmotic state also causes the urge to drink, as does an increase in extracellular fluid osmotic pressure of 1 to 2% in both dog and man. High temperature is also a stimulus to drink, even before there has been a detectable change in body water balance.

The receptors involved have not been identified, but unquestionably the hypothalamus plays an essential role. The evidence for this is much as follows: Lesions in the lateral hypothalamus that impair food intake also create adipsia. Furthermore, such lesions abolish the drinking response to an intraperitoneal injection of hypertonic sodium chloride, a phenomenon that has been most clearly demonstrated in the rat. In the goat and dog, lesions in the dorsal hypothalamus lateral to the paraventricular nuclei seem to be most effective in the production of adipsia.

Stimulation experiments have identified areas in the hypothalamus from which excessive drinking can be evoked. Injection of hypertonic solutions into these same areas has a similar effect. There appear to be some species differences with respect to the exact location of the regions most crucially concerned, but even in a given species the areas from which the effects can be obtained are not very clearly delimited.

The most that can be said at the present time is that stimuli applied to the dorsal hypothalamic area, lateral and somewhat caudal to the paraventricular nuclei, initiate drinking. Also, stimuli applied to a more caudal part of the dorsal hypothalamus, a region lying between the mammillothalamic tracts and descending columns of the fornix, cause polydipsia. It has been reported also that application of ACh to these lateral and/or more dorsal centers causes vigorous and prolonged drinking.

Satiation of thirst can be accomplished by hydration of tissues, gastric distention, and hydration or moistening of the mucous membrane of the mouth and throat. Thus far, however, no area within the hypothalamus that might be called a "thirst satiation center" has been located. Finally, it should be noted that, as in the case of the control of appetite and food intake, the literature is voluminous and confusing. There is a great deal of information that leads to no clarifying explanations. The conclusions given here should suffice, however, as a first approximation relative to the control of water intake. (See Chapters 17 and 46 for further discussion.)

Body water balance is maintained by relating water intake to water loss. Thirst regulates intake; water loss is controlled to some degree by the kidney. Loss in respiration or by perspiration is not controlled by the mechanisms regulating water balance; control operates primarily on the kidney. The role of the supraoptic nuclei in ADH production and release has been discussed previously. Here it suffices to say that cells of these hypothalamic nuclei, or neighboring osmoreceptors that can control their action, monitor the salt concentration or osmotic state of the blood and tissue fluids. These cells respond appropriately; if maximum water conservation is necessary, they release more ADH and the kidney reabsorbs all the water not required for excretion of materials in the urine.

Control of body weight (hunger and satiation)

Early clinical studies resulted in controversy as to whether obesity was the result of pituitary abnormality or hypothalamic injury. We now know that the latter position is correct. Philip E. Smith in 1927 produced obesity by injecting chromic acid into the hypothalamus, supposedly without injury to the hypophysis. Keller and his associates in 1933 made small surgical lesions in the hypothalamus of dogs that produced an "enhanced appetite." The use of the Horsley-Clarke stereotaxic instrument eventually enabled Hetherington and Ranson,[38] Brobeck,[16] Brooks,[18, 19] and others to localize those regions of the hypothalamus that are specifically responsible for hyperphagia and other abnormalities that produce obesity.

As might be expected, obesity is not of uniform magnitude. However, there is some uniformity of occurrence following hypothalamic injury. The development of obesity can be subdivided into a number of phases. There is an early dynamic phase of rapid weight gain; this is followed by a period of slower gain and an eventual establishment of a static phase in which body weight remains relatively constant. When senescence begins there is a gradual loss of body weight accompanied by a considerable decline in food intake.

There is now general agreement that hyperphagia and the failure of body weight–regulating mechanisms are produced by bilateral destruction of the ventromedial hypothalamic nuclei. Obesity apparently will also result from lesions that interrupt fibers passing caudally from these nuclei to the mesencephalic tegmentum. There is still some discussion as to the neural structures that are most crucial to the maintenance of body weight balance, and considerable uncertainty as to the abnormality created by their destruction remains.[39] Behavior of such animals suggests that a release or deficiency phenomena, or a combination of both, may be involved.

Almost immediately after production of an effective lesion, animals begin to eat in an abnormal fashion. The diurnal pattern is lost; meals are larger and more frequent. Certainly the use of the term "hyperphagia" is justified. Other investigators have felt that a loss of the sense of satiety or the mechanism producing that sense, rather than an augmentation of appetite, is responsible for the increase in food intake. There is a doubling or tripling of the amount of food eaten at any one time. In addition to these seemingly primary abnormalities, there are other factors that must contribute to the accumulation of body fat and weight gain.

A reduction in metabolic rate is often, if not invariably, present; following ingestion of food the respiratory quotient is elevated above unity, a phenomenon suggesting a high rate of conversion of carbohydrate to fat; activity is reduced. These changes could explain why a rat rendered potentially obese by a hypothalamic lesion can outgain a litter mate when, by paired feeding, both animals ingest the same amount of food. Such work indicates that obesity is not due simply to hyperphagia and loss of a sense of satiation.

Still other abnormalities are produced by the hypothalamic tumors and lesions that result in obesity. The ventromedial nuclear regions are involved in the control of emotional reactions, and lesions that create obesity may also convert placid animals into rather fierce, intractable individuals. Sexual dystrophy is a frequent accompaniment of obesity, as are disturbances in carbohydrate metabolism and water balance. Large lesions in a small structure interfere with multiple functions; thus it is hard to determine specificities.

Means have been sought to destroy specific cells by using their unique affinities. For example, ventromedial nuclear cells appear to have a specific affinity for glucose. Systemically injected gold thioglucose produces obesity in mice, apparently because this highly toxic compound accumulates in and destroys the ventromedian nuclei. Certainly these nuclei are important to the maintenance of body weight balance, but they are not solely responsible.

It has been demonstrated by Anand and Brobeck[9] that lesions in the lateral hypothalamic regions close to the ventromedial nuclei create hypophagia. The general concept at present is that such lesions destroy bilateral centers essential to hunger and appetite. These are reciprocally related to the ventromedial nuclei, which are responsible for satiation. Those centers of the lateral hypothalamus essential to the food intake drive have not yet been exactly located. It has been demonstrated recently by simultaneous recording from units in the lateral hypothalamic area and the ventromedial nuclei that stimuli which augment activity in one simultaneously diminish firing rates in the other. This reciprocal action has been observed by a number of investigators.[10, 20, 54]

The conditions that activate cells of the ventromedial nuclei and those of the lateral hypothalamus are not fully known at present. One suggestion offered is that these cells are glucose "receptors," which sense the level of glucose within or outside the cell. On the other hand, the rate of glucose utilization may be the determinant of activity of these cells, and this in turn may depend on the availability of glucose and the amount of insulin in the circulating blood. This latter dependency would indicate that these cells, unlike other neural tissues, require insulin for the metabolism of glucose. The special affinity of ventromedial nucleus cells for glucose is indicated by their greater uptake of radioactive glucose and the gold thioglucose experiments mentioned previously. Their dependency on glucose levels is indicated also by the fact that the rate of discharge of ventromedial nucleus neurons increases and that of the lateral hypothalamic cells simultaneously decreases when the level of glucose in the surrounding tissue fluid is raised. Vagus afferent fiber stimulations can affect the firing rate of both groups of neurons, but their role in the control of food intake has not been determined.[20] Finally, iontophoretic application of glucose to cells in the ventromedial nucleus increases their rate of firing, but its effect on cells of the lateral hypothalamus is inconclusive.[53]

It has long been known that obese animals have very low postprandial blood sugar levels. This tendency toward hypoglycemia was thought to be due to the fact that the livers of obese animals are so loaded with fat that the capacity for carbohydrate storage and release is impaired. Modern work has shown, however, that there is an abnormally high blood insulin level in the obese animal; the level remains high and does not follow the normal course of postprandial rise and decline.[42] The level of circulating insulin may determine the level of activity of these hypothalamic neurons, but it should also be remembered that the metabolisms of carbohydrate and lipids are closely interlinked. The level of nonesterified fatty acid (NEFA) may also be a determinant of hunger and satiety. Another idea that has been advanced is that the ventromedial nuclei control the level of fat storage in the body and that injury to these nuclei either abolishes this control or establishes a new level of fat storage.[42] Very fat animals rendered thus by forced feeding do not show as marked a hyperphagia and tendency toward weight gain following hypothalamic lesions as do normal animals with the same lesions. This suggests that there is an upper limit to fat storage and that the body weight limit is established by the fat storage potentiality of an individual.

In a cold environment, animals eat more and eat more continuously. Application of heat or even heating the anterior hypothalamus inhibits food intake. The anterior hypothalamic areas that sense and control body temperature may normally act in conjunction with those centers that control food intake. It is thought by some that the specific dynamic action of food materials produces a sense of body overheating that may contribute to the feeling of satiation. Hypothalamic lesions may impair the sense of satiation by interfering with this temperature-detecting mechanism. Exposure to heat does not depress food intake in animals with obesity-producing lesions to the same degree as in normal animals (Chapter 55).

Growth hormone levels in the blood tend to be low after eating; they then rise progressively. This hormone is known to participate in making fat available as an energy source to the body. Hypothalamic lesions that disturb growth hormone output may also contribute to metabolic abnormalities and changes in feeding behavior that result in obesity.

Attempts have recently been made to determine whether cells in the ventromedial nuclei and lateral hypothalamic areas, which are concerned with regulation of food intake and body weight, are under the control of cholinergic or adrenergic transmitters.[39, 48] Local applications of these chemicals, their blocking agents, and many other drugs normally used in the study of synaptic transmission have yielded a bewildering set of results that the authors are unable to interpret. One can find in the current literature extensive discussions of the role of α- and β-adrenergic receptors and cholinergic receptors in the control of these hypothalamic structures.[39]

Extrahypothalamic centers also cannot be ignored. Unquestionably, psychological factors, conditioned behavior, and motivations are important in the control of food intake. Olfaction, taste, and other gustatory sensations are involved in determining the strengths of food intake drives. Ablations and stimulations of various parts of the brain, particularly the limbic system, have been shown to affect food intake. It may be pointed out again that the functions of the nervous system and the neuroendocrine system are well integrated and that no single isolated part is exclusively involved in the determination of a body state and the maintenance of normality.

Temperature regulation

Chronic spinal and decerebrate animals have no ability to regulate body temperature and are therefore poikilothermic. Removals of the cerebral cortex and structures superior to the hypothalamus somewhat impair the niceties of temperature regulation, but animals remain essentially homeothermic.

Richet in 1885 and Ott in 1887 held that there were thermosensitive centers in the brain involved in detecting and correcting changes in body temperature. It is now generally agreed that the hypothalamus is the region of the brain chiefly responsible for thermosensitivity, maintenance of body temperature homeostasis, and the required reactions to extremes of heat and cold.

Peripheral receptors can initiate generalized reactions to heat and cold. The reflexes evoked by general or localized heating and cooling must act through the hypothalamus to arouse the appropriate body responses. In the absence of the hypothalamus, such stimuli merely provoke localized or ineffective reflex actions.

Changes in the temperature of the arterial blood reaching the brain have been shown to evoke temperature-regulating reactions in the absence of any excitation of peripheral afferents. The first experiments providing evidence of central temperature receptors were performed in dogs by Sherrington in 1924 and in man by Gibbon and Landis in 1932. By cooling or heating the denervated lower extremities of spinal dogs and spinal men, it was found that the upper innervated parts of the body showed temperature-regulating changes within a very few minutes after warmed or cooled blood reached the brain. Heating of the denervated extremities evoked sweating and vasodilation of the innervated parts, while cooling produced vasoconstriction, piloerection, and shivering. It was later shown by other techniques that the hypothalamus was the region responsible for sensing blood temperature variations and for evoking the appropriate somatic and autonomic reactions.

Ablation experiments have provided evidence that the hypothalamus is important to the maintenance of body temperature, and that various regions play specific roles in performing the required adjustments. Destructive lesions in the posterior hypothalamus create deficiencies in responses to cold, while lesions in the anterior hypothalamus impair responses to heat. Lateral hypothalamic regions are essential to the adjustment to both heat and cold. The general conclusions drawn from this work are that the anterior hypothalamus is of major importance in reactions such as panting, sweating, and vasodilatation of skin vessels, and when lesions are made in the anterior hypothalamus, animals become heat sensitive. These lesions, however, do not impair reactions to cold. Vasoconstriction, piloerection, sympathicoadrenal discharge, and shivering depend on the posterior hypothalamus. Lesions within this area, however, do not cause hypothermia in animals kept at normal room temperatures (72° to 74° F), and deficiencies are apparent only on exposure to cold.

Febrile reactions to pyrogens are dependent on the integrity of the posterior hypothalamus, an indication that this region is important to body heat production and conservation. Pyrogens also raise thresholds for sweating and heat loss reactions. Thus lesions in the anterior hypothalamus render animals more susceptible to fever-producing agents. Both of these centers that control peripheral reactions to heat and cold are affected. It is obvious that injury to tracts and connections within the hypothalamus would impair maintenance of normal homeothermia. In man, postoperative hyperthermia is not infrequently observed, especially following surgery in which the hypothalamus is affected. The body temperature changes observed consequent to injury of the hypothalamus caused by tumors or localized hemorrhages illustrate the importance of this structure in the regulation of body temperature in man.

Additonal information concerning temperature-regulating reactions has been obtained by electrical and thermal stimulation of parts of the hypothalamus. In unanesthetized animals, stimulation through electrodes permanently implanted in the rostral hypothalamus causes panting and cutaneous vasodilation, with a resultant drop in body temperature. Cold-induced shivering is stopped by such anterior hypothalamic stimulation. Cooling raises the threshold for induction of responses from the anterior hypothalamus. Localized heating of the anterior hypothalamus evokes responses similar to those caused by electrical stimulation.

Electrical stimulation of the posterior hypothalamus induces a muscular tremor resembling shivering as well as other reactions characteristic of the response to cold. Electrical stimulations of midbrain structures and the septal areas also evoke shivering and a resultant temperature increase. Thus the reactions to heat and cold are not mediated exclusively by the hypothalamus.

The posterior hypothalamus, unlike the anterior region, does not respond to localized heating or cooling; this suggests that no thermoreceptors are present. Investigations of the localization of thermosensitive elements in the hypothalamus have led to the development of some of the most refined techniques employed in this field. The procedures used involve heating or cooling very minute areas and accurately recording the temperature of a neuron as well as its activity. It has been shown that approximately 40% of cells in the anterior hypothalamus respond to local temperature changes. Of these thermosensitive cells, 80% respond by firing more rapidly when the temperature is increased by 1° to 2° C or by firing more slowly when the temperature is dropped 1° to 2° C below the control level. In recent studies numerous heat-sensitive cells with a Q_{10} greater than 2 have been found in the preoptic area.[50] It appears that the highest concentration of such thermosensitive neurons occurs there. Cold-sensitive receptors are less numerous and more diffusely scattered. For example, only 20% of all thermosensitive neurons in the anterior hypothalamus show an increase in firing when the temperature is lowered below control levels. Such cells are not found in the posterior hypothalamus but are present in the midbrain reticular formation, septal, and preoptic regions.

Some thermosensitive cells respond to local temperature changes, but the rate of firing in others is affected only by peripheral warming or change of temperature in distant regions of the brain. Studies of individual units reconfirm the concept that the hypothalamus is involved in peripherally evoked responses to heat and cold and also in the monitoring of blood temperatures relative to the maintenance of normal body temperature homeostasis.

Various attempts have been made recently to determine the significance of transmitter substances to hypothalamic temperature regulation.[49] Direct injections into the anterior hypothalamus of 5-hydroxytryptamine (5-HT), or serotonin, have been found to produce a fever in conscious primates, while similarly applied catecholamines produce hypothermia. Other investigators have felt that the actions of injected amines may be nonspecific since they produce blood flow changes and also directly affect the membranes of many types of neurons. Additional information concerning the role of the hypothalamus in body temperature regulation is given in Chapter 55.

Reactions to stress

Individuals are exposed to a wide variety of stimuli and conditions that tend to drastically alter body homeostasis in a general rather than a specific fashion. Injury, surgical trauma, great extremes of heat and cold, infections, and intense emotional disturbances all create conditions of stress. These injurious influences are termed "stressor" stimuli. A great deal has been written about reaction to stress, and these reactions have been described as producing a "general adaptation syndrome." It has also been pointed out that long-continued reactions to stress eventually have harmful effects; it is important to alleviate the causative situation as quickly as possible. Briefly summarizing, it can be said that these stressor stimuli produce, by way of the hypothalamus, a strong sympathicoadrenal discharge. Stored catecholamines are released and produce their characteristic cardiovascular and metabolic reactions.

The most significant stress reaction, however, is a release of corticosteroids from the adrenal cortex. The stressor stimuli act through afferent neurons or directly on the hypothalamus to cause discharge of CRF and a liberation of ACTH from the pituitary gland. It was once thought that the released epinephrine caused ACTH discharge, but now it is generally held that these stimuli evoke reactions in the hypothalamus through the limbic and reticular systems that organize the resistive responses. Animals have a greatly impaired resistance to infection and injury if these resistive reactions to stress are abolished by hypothalamic lesions or removal of the adrenals.[36]

Situations that create long-lasting tensions or low-level stress tend to produce gastric ulcers. These stimuli presumably act by influencing the hypothalamic control of the digestive tract. It has been repeatedly demonstrated in animals that stimulation of the anterior hypothalamus increases secretory and motor activity in the stomach and intestine. Chronic stimulation produces ulceration.[46] Lesions in the anterior hypothalamus are chiefly responsible for the production of gastric bleeding, but posterior hypothalamic lesions also produce a similar condition. The adrenal gland is thought by some to be involved in the production of a contributory hyperacidity, but the most that can be said at present is that a hypothalamic disorder created by intrinsic or extrinsically imposed disturbances is responsible for this unfortunate consequence of stress. Rather intriguing work conducted at Walter Reed Army Hospital has shown that ulcers occur only in "executive" monkeys as a result of the application of electrical shocks to the body. The executive is the one who, by proper manipulations, can prevent receipt of unpleasant stimuli that he as well as a "nonexecutive" monkey will experience. The nonexecutive monkey cannot do anything to prevent the shocks, so he gives up and gets no ulcer. This work also shows that frequent periodic stress is more debilitating than continuing stress to which adaptation may ultimately be made.[15]

Control of emotional reactions[47]

Studies of animals such as cats and dogs that have long survived the removal of cerebral cortex have shown conclusively that the capacity to display anger or rage reactions depends on subcortical mechanisms. Goltz found that decorticate dogs, in addition to demonstrating rage, revealed a high degree of emotional instability. As stated previously, it was finally shown in studies of both acute and chronic animals that the hypothalamus is necessary for the vigorous expression of rage reactions. After transection of the brain behind the mammillary bodies, only fragmentary elements of emotional response can be initiated.[11, 12]

Subsequent to these early ablation experiments it was found by others that small localized lesions also can produce a release of the rage phenomenon. Ingram (1939) and Wheatley (1944) showed that injury to the ventromedial nuclei in cats produced enduring savage behavior, a result that also occurs in rats. Bilateral lesions in the lateral part of the hypothalamus, however, tend to create placidity in the characteristically rather wild and excitable rhesus monkey. This work has certainly demonstrated that regions in the medial and caudal hypothalamus integrate emotional expression and suggests that a degree of reciprocal action operates in the control of emotional phenomena.

Electrical stimulation has also produced evidence that the hypothalamus plays some role in the arousal and expression of anger. Hess and Brügger (1943) were among the earliest to perfect a technique for brain stimulation in awake, unrestrained animals.[37] They concluded that a center exists in the perifornical region of the anterior hypothalamus that is responsible for anger and defensive behavior, at least in cats. Hunsperger, using both stimulation and ablation procedures, confirmed and extended the work of Hess. He claimed that the essential neural apparatus for integrating an expression of anger is not located solely in the posterior hypothalamus but involves two central zones, one in the perifornical region described by Hess and the other posterior to the hypothalamus in the central gray of the midbrain.

In an attempt to differentiate between centers and pathways to which current might spread during electrical stimulation, various workers have implanted crystalline ACh or injected other drugs into these areas under discussion. Positive reactions were induced from the lateral hypothalamic area and from the dorsal part of the ventromedial nuclei and the periventricular region of the posterior hypothalamus. This technique, however, did not give a more definite localization of the centers primarily involved in emotional expression than did the stimulation and lesion experiments.

Rage reactions are not the only signs of emotional expression that can be evoked from decorticate animals and those with a further reduction of brain components. Behavior characteristic of fear or terror has been evoked by high-pitched sounds or other stimuli in cats and dogs from which all parts of the forebrain except the preoptic and hypothalamic areas have been removed. Even in chronic mesencephalic animals, high-pitched sounds evoke cringing and other signs indicative of fear. Hess also found points in the hypothalamus from which an escape reaction could be elicited by electrical stimulation. These areas are close to those from which rage responses are evoked.

Reactions of pleasure and those associated with sexual functions are generally considered to be emotional in nature. Unquestionably the hypothalamus is involved, but it appears to play a more secondary role than in the expression of fear and anger. Many physiologic studies in both animals and man, using stimulation as well as recordings of evoked activity, have demonstrated that limbic system–hypothalamic cooperation is involved in a wide variety of somatic and autonomic phenomena closely related to the broad range of behavioral activities conventionally associated with emotional expression.

Heath and co-workers (1954) observed that stimulation of the septum frequently yielded pleasurable sensations in patients. Shortly thereafter others observed the phenomenon of self-stimulation in rats. These animals voluntarily pressed levers to cause the delivery of stimuli through electrodes implanted in the septal area, as though they received some intrinsic "reward" therefrom. Positive loci for self-stimulation could be followed along the medial forebrain bundle into the lateral parts of the hypothalamus.

There appears to be an extensive overlap in the control of the various diverse responses dependent on this limbic system. The very considerable mass of clinical and experimental observations, however, does not yield a fully satisfactory picture of the integrative role of the limbic system in the control of emotional manifestations, which most certainly involve the hypothalamus.

Sleep and arousal

The older literature contains rather elaborate discussions of evidence that lesions in the posterior hypothalamus produce sleep, while stimulations of that area arouse a sleeping animal. Stimulations of the dorsal hypothalamus at critical frequencies and for relatively long periods have been reported to produce sleep and sleep-preparation behavior. Following stimulation, cats walk to an "appropriate place," turn around a few times, curl up, and go to sleep as normal animals do. It is now thought, however, that there are no specific "sleep centers" in the hypothalamus, but that sleep-producing stimuli actually act through the general thalamocortical system. Posterior hypothalamic stimuli act on the reticular activating system to cause arousal, while lesions in the posterior hypothalamus interrupt these pathways (Chapter 8).

There is some evidence, however, that the hypothalamus is not completely quiescent during sleep; cardiac and vasomotor surges occur, indicating activity of the autonomic system and its hypothalamic representation. There is also a well-documented release of growth hormone during sleep, and it is thought that GRF secretion is augmented during slow-wave sleep particularly. A diurnal fluctuation in ACTH secretion, indicative of a periodic CRF release, also occurs. In man, ACTH and blood cortisol levels are highest before waking in the morning (6:00 A.M.) and lowest at night, a fluctuation not caused by stress or external circumstance.

The limbic and other portions of the CNS are probably involved in the control of sleep cycles, but brain lesions affecting the biologic clock that is responsible for setting sleep-wakefulness rhythms are those that involve the hypothalamus.[57] It is clear that this internal clock depends on the hypothalamus for its existence, but the nature of the clock and the mechanisms of its control and actions are unknown.

Control of somatic reactions

The hypothalamus plays a role not only in the control of visceral functions and activity of the endocrine system but also in the regulation of somatic behavior. When the hypothalamus is intact there is a better integration of reflex activity than that found in a decerebrate preparation. Activities that one would define as behavioral reactions rather than reflex pattern responses are seen. Walking, running, and righting reactions are merely slightly deficient rather than imperfectly displayed and coordinated, as they are in the absence of the hypothalamus.

Stimulations of the hypothalamus are said to produce directional movements and rather complex patterns of response that involve the total body.[37] Also, as pointed out in an earlier section, hypothalamic stimuli elicit a picture of emotional arousal that involves somatic as well as autonomic and endocrine reactions. On exposure to heat and cold, the hypothalamus organizes compensatory somatic as well as autonomic and endocrine reactions. The hypothalamus functions as a part of the brain, not merely as an adjunct to the autonomic or endocrine systems. It is integrative in function, influencing the entire body and all processes therein. Higher centers contribute refinements of control and special sensitivities, but the basic processes of integration require participation of the hypothalamus.

REFERENCES

General reviews

1. Clark, W. E. L., Beattie, J., Riccoch, G., and Dott, N. M.: The hypothalamus: morphological, functional, clinical and surgical aspects, Edinburgh, 1938, Oliver & Boyd.
2. Ganong, W. F.: Review of medical physiology, ed. 5, Palo Alto, 1971, Lange Medical Publications, Inc.
3. Haymaker, W., Anderson, E., and Nauta, W. J. H., editors: The hypothalamus, Springfield, Ill., 1969, Charles C Thomas, Publisher.
4. McCann, S. M., and Porter, J. C.: Hypothalamic pituitary stimulating and inhibiting hormones, Physiol. Rev. **49:**240, 1969.
5. Martini, L., and Ganong, W. F., editors: Neuroendocrinology, New York, 1966, Academic Press, Inc., vols. 1 and 2.
6. Martini, L., Motta, M., and Fraschini, F., editors: The hypothalamus, New York, 1970, Academic Press, Inc.
7. Recent studies on the hypothalamus, Br. Med. Bull. **22:**195, 1966.

Original papers

8. Adams, J. H., Daniels, P. M., and Pritchard, M. M.: Distribution of hypophysial portal blood in the anterior lobe of the pituitary gland, Endocrinology **75:**120, 1964.
9. Anand, B. K., and Brobeck, J. R.: Hypothalamic control of food intake in rats and cats, Yale J. Biol. Med. **24:**123, 1951.
10. Anand, B. K., et al.: Activity of single neurons in the hypothalamic feeding centers: effect of glucose, Am. J. Physiol. **207:**1146, 1964.
11. Bard, P.: A diencephalic mechanism for the expression of rage with special reference to the sympathetic nervous system, Am. J. Physiol. **84:**490, 1928.
12. Bard, P.: Central nervous mechanisms for emotional behavior patterns in animals, Res. Publ. Res. Assoc. Nerv. Ment. Dis. **19:**190, 1939.
13. Bargmann, W.: Neurosecretion, Int. Rev. Cytol. **19:**183, 1966.
14. Bargmann, W., Hild, W., Ortmann, R., and Schiebler, T. H.: Morphologische und experimentelle Untersuchungen über das hypothamisch-hypophysäre System, Acta Neuroveg. **1:** 16, 1950.
15. Brady, J. V.: Ulcers in "executive" monkeys, Sci. Am. **199:**95, 1958.
16. Brobeck, J. R., Tepperman, J., and Long, C. N. H.: Experimental hypothalamic hyperphagia in the albino rat, Yale J. Biol. Med. **15:** 831, 1943.
17. Brooks, C. McC.: Relation of the hypothalamus to gonadotrophic function of the hypophysis, Res. Publ. Res. Assoc. Nerv. Ment. Dis. **20:** 525, 1940.
18. Brooks, C. McC.: The relative importance of changes in activity in the development of experimentally produced obesity in the rat, Am. J. Physiol. **147:**708, 1946.
19. Brooks, C. McC., and Lambert, E. F.: A study of the effect of limitation of food intake and the method of feeding on the rate of weight gain during hypothalamic obesity in the albino rat, Am. J. Physiol. **147:**695, 1946.
20. Brooks, C. McC., Koizumi, K., and Zeballos, G. A.: A study of factors controlling activity of neurons within the paraventricular, supraoptic and ventromedial nuclei of the hypothalamus, Acta Physiol. Latinoamer. **16:**83, 1966.
21. Brooks, C. McC., Ishikawa, T., Koizumi, K., and Lu, H. H.: Activity of neurons in the paraventricular nucleus of the hypothalamus and its control, J. Physiol. **182:**217, 1966.

22. Burn, H. A., and Knowles, F. G. W.: Neurosecretion. In Martini, L., and Ganong, W. F., editors: Neuroendocrinology, New York, 1966, Academic Press, Inc., vol. 1.
23. Cross, B. A., and Harris, G. W.: The role of the neurohypophysis in the milk-ejection reflex, J. Endocrinol. **8:**148, 1952.
24. Daniel, P. M.: The blood supply of the hypothalamus and pituitary gland, Br. Med. Bull. **22:**202, 1966.
25. Douglas, W. W.: Stimulus-secretion coupling: the concept and clues from chromaffin and other cells, Br. J. Pharmacol. **34:**451, 1968.
26. Douglas, W. W., and Poisner, A. M.: Stimulus-secretion coupling in a neurosecretory organ and the role of calcium in the release of vasopressin from the neurohypophysis, J. Physiol. **172:**1, 1964.
27. Dyball, R. E. J.: Oxytocin and ADH secretion in relation to electrical activity in antidromically identified supraoptic and paraventricular units, J. Physiol. **214:**245, 1971.
28. Dyball, R. E. J., and Koizumi, K.: Electrical activity in the supraoptic and paraventricular nuclei associated with neurohypophysial hormone release, J. Physiol. **201:**711, 1969.
29. Fawcett, C. P.: The present status of the chemistry of PIF, FRF and LRF. In Meites, J., editor: Hypophysiotropic hormones of the hypothalamus: assay and chemistry, Baltimore, 1969, The Williams & Wilkins Co.
30. Fingerman, M.: Comparative physiology: chromatophores, Ann. Rev. Physiol. **32:**345, 1970.
31. Fink, G., Naller, R., and Worthington, W. C., Jr.: The demonstration of luteinizing hormone releasing factor in hypophysial portal blood of pro-oestrous and hypophysectomized rats, J. Physiol. **191:**407, 1967.
32. Finley, K. H.: Angio-architecture of the hypothalamus and its peculiarities, Res. Publ. Res. Assoc. Nerv. Ment. Dis. **20:**286, 1940.
33. Fisher, C., Ingram, W. R., and Ranson, S. W.: Diabetes insipidus and the neuro-hormonal control of water balance: a contribution to the structure and function of the hypothalamico-hypophysial system, Ann Arbor, 1938, Edward Brothers, Inc.
34. Fitzsimons, J. T.: Hypothalamus and drinking, Br. Med. Bull. **22:**232, 1966.
35. Harris, G. W.: Neural control of the pituitary gland, London, 1955, Edward Arnold, Ltd.
36. Harris, G. W., and George, R.: Neurohumoral control of the adenohypophysis and the regulation of the secretion of TSH, ACTH and growth hormone. In Haymaker, W., Anderson, E., and Nauta, W. J. H., editors: The hypothalamus, Springfield, Ill., 1969, Charles C Thomas, Publisher.
37. Hess, W. R.: The functional organization of the diencephalon, New York, 1957, Grune & Stratton, Inc.
38. Hetherington, A. W., and Ranson, S. W.: Hypothalamic lesions and adiposity in the rat, Anat. Rec. **78:**149, 1940.
39. Hoebel, B. G.: Feeding: neural control of intake, Ann. Rev. Physiol. **33:**533, 1971.
40. Horsley, V., and Clarke, R. H.: The structure and functions of the cerebellum examined by a new method, Brain **31:**45, 1908.
41. Ishikawa, T., Koizumi, K., and Brooks, C. McC: Electrical activity recorded from the pituitary stalk of the cat, Am. J. Physiol. **210:**427, 1966.
42. Kennedy, G. C.: Food intake, energy balance and growth, Br. Med. Bull. **22:**216, 1966.
43. Kobayashi, H., and Matsui, T.: Fine structure of the median eminence and its functional significance. In Ganong, W. F., and Martini, L., editors: Frontiers in neuroendocrinology, New York, 1969, Oxford University Press.
44. Koizumi, K., and Yamashita, H.: Studies of antidromically identified neurosecretory cells of the hypothalamus by intracellular and extracellular recordings, J. Physiol. **221:**683, 1972.
45. Koizumi, K., Ishikawa, T., and Brooks, C. McC.: Control of activity of neurons in the supraoptic nucleus, J. Neurophysiol. **27:**878, 1964.
46. Long, D. M., Leonard, A. S., Chou, S. N., and French, L. A.: Hypothalamus and gastric ulceration. I. Gastric effects of hypothalamic lesions. II. Production of gastrointestinal ulceration by chronic hypothalamic stimulation, Arch. Neurol. **7:**167, 176, 1962.
47. MacLean, P. D.: The hypothalamus and emotional behavior. In Haymaker, W., Anderson, E., and Nauta, W. J. H., editors: The hypothalamus, Springfield, Ill., 1969, Charles C Thomas, Publisher.
48. Morgane, P. J., editor: Neural regulation of food and water intake, Ann. N. Y. Acad. Sci. **157:**531, 1969.
49. Myers, R. D.: Temperature regulation: neurochemical systems in the hypothalamus. In Haymaker, W., Anderson, E., and Nauta, W. J. H., editors: The hypothalamus, Springfield, Ill., 1969, Charles C Thomas, Publisher.
50. Nakayama, T.: Single unit activity in the hypothalamus and its relation to temperature regulation, Proc. Int. Union Physiol. Sci. **6:** 287, 1968.
51. Nauta, W. J. H., and Haymaker, W.: Hypothalamic nuclei and fiber connections. In Haymaker, W., Anderson, E., and Nauta, W. J. H., editors: The hypothalamus, Springfield, Ill., 1969, Charles C Thomas, Publisher.
52. Nicoll, C. S., Fiorindo, R. P., McKennee, C. T., and Parsons, J. A.: Assay of hypothalamic factors which regulate prolactin secretion. In Meites, J., editor: Hypophysiotropic hormones of the hypothalamus: assay and chemistry, Baltimore, 1970, The Williams & Wilkins Co.
53. Oomura, Y., Ono, T., Ooyama, H., and Wayner, M. J.: Glucose and osmosensitive neurons of the rat hypothalamus, Nature **222:**282, 1969.
54. Oomura, Y., et al.: Reciprocal activities of the ventromedial and lateral hypothalamic areas of cats, Science **143:**484, 1964.
55. Porter, J. C., Goldman, B. D., and Wilber, J. F.: Hypophysiotropic hormones in portal vessel blood. In Meites, J., editor: Hypophysiotropic hormones of the hypothalamus: assay and chemistry, Baltimore, 1970, The Williams & Wilkins Co.
55a. Porter, J. C., Kamberi, I. A., and Grazia, Y. R.: Pituitary blood flow and portal vessels. In Martini, L., and Ganong, W. F., editors: Frontiers in neuroendocrinology, New York, 1971, Oxford University Press.

56. Raisman, G.: Neural connexions of the hypothalamus, Br. Med. Bull. **22:**197, 1966.
57. Richter, C. P.: Sleep and activity: their relation to the 24-hour clock, Res. Publ. Res. Assoc. Nerv. Ment. Dis. **45:**8, 1967.
58. Schally, A. V., Arimura, A., and Kastin, A. J.: Hypothalamic regulatory hormones, Science **179:**341, 1973.
59. Stevenson, J. A. F.: Neural control of food and water intake. In Haymaker, W., Anderson, E., and Nauta, W. J. H., editors: The hypothalamus, Springfield, Ill., 1969, Charles C Thomas, Publisher.
60. Verney, E. B.: The antidiuretic hormone and the factors which determine its release, Proc. R. Soc. Lond. (Biol.) **135:**25, 1947.
61. Du Vigneaud, V.: Hormones of the posterior pituitary gland: oxytocin and vasopressin, Harvey Lect. **50:**1, 1954.
62. Yates, F. E., Russell, S. M., and Maran, J. W.: Brain-adenohypophysial communication in mammals, Ann. Rev. Physiol. **33:**393, 1971.

INDEX

C

D

E

I

U

W

X

Y

Z